TELFORD
CB015302

Human Anatomy & Physiology

Eighth Edition

Elaine N. Marieb, R.N., Ph.D.
Holyoke Community College

Katja Hoehn, M.D., Ph.D.
Mount Royal College

Benjamin Cummings

San Francisco Boston New York
Cape Town Hong Kong London Madrid Mexico City
Montreal Munich Paris Singapore Sydney Tokyo Toronto

Editor-in-Chief: Serina Beauparlant
Development Managers: Claire Alexander and Barbara Yien
Project Editor: Sabrina Larson
Development Editor: Anne A. Reid
Art and Design Director: Mark Ong
Managing Editors: Debbie Cogan and Wendy Earl
Production Supervisor: Michele Mangelli
Art Development Manager: Laura Southworth
Art Editor: Elisheva Marcus
Media Producer: Erik Fortier
Assistant Editor: Nicole Graziano
Text Designer: Mark Ong
Cover Designer: Riezebos Holzbaur Design Group
Principal Artists: Imagineering STA Media Services Inc. and Electronic Publishing Services Inc.
Art Coordinator: David Novak
Image Rights and Permission Manager: Zina Arabia
Photo Researcher: Kristin Piljay
Copyeditor: Anita Wagner
Production Assistant: Giuseppe Tassone
Compositor: Aptara, Inc.
Senior Manufacturing Buyer: Stacey Weinberger
Marketing Manager: Derek Perrigo

Cover Photograph: Sports Illustrated/Getty Images

Photo and illustration credits follow the Glossary.

Benjamin Cummings
is an imprint of

www.pearsonhighered.com

ISBN 10: 0-321-60261-7; ISBN 13: 978-0-321-60261-9
1 2 3 4 5 6 7 8 9 10— DOW—12 11 10 09 08
Manufactured in the United States of America.

About the Authors

We dedicate this work to our students both present and past, who always inspire us to "push the envelope."

Elaine N. Marieb

For Elaine N. Marieb, taking the student's perspective into account has always been an integral part of her teaching style. Dr. Marieb began her teaching career at Springfield College, where she taught anatomy and physiology to physical education majors. She then joined the faculty of the Biological Science Division of Holyoke Community College in 1969 after receiving her Ph.D. in zoology from the University of Massachusetts at Amherst. While teaching at Holyoke Community College, where many of her students were pursuing nursing degrees, she developed a desire to better understand the relationship between the scientific study of the human body and the clinical aspects of the nursing practice. To that end, while continuing to teach full time, Dr. Marieb pursued her nursing education, which culminated in a Master of Science degree with a clinical specialization in gerontology from the University of Massachusetts. It is this experience, along with stories from the field—including those of former students now in health careers—that has informed the development of the unique perspective and accessibility for which her texts and laboratory manuals are known.

In her ongoing commitment to students and her realization of the challenges they face, Dr. Marieb has given generously to provide opportunities for students to further their education. She contributes to the New Directions, New Careers Program at Holyoke Community College by funding a staffed drop-in center and by providing several full-tuition scholarships each year for women who are returning to college after a hiatus or attending college for the first time and who would be unable to continue their studies without financial support. She funds the E. N. Marieb Science Research Awards at Mount Holyoke College, which promotes research by undergraduate science majors, and has underwritten renovation and updating of one of the biology labs in Clapp Laboratory at that college. Dr. Marieb is also a contributor to the University of Massachusetts at Amherst where she generously provided funding for reconstruction and instrumentation of a cutting-edge cytology research laboratory that bears her name. Recognizing the severe national shortage of nursing faculty, she underwrites the Nursing Scholars of the Future Grant Program at the university.

In 1994, Dr. Marieb received the Benefactor Award from the National Council for Resource Development, American Association of Community Colleges, which recognizes her ongoing sponsorship of student scholarships, faculty teaching awards, and other academic contributions to Holyoke Community College. In May 2000, the science building at Holyoke Community College was named in her honor.

Dr. Marieb is an active member of the Human Anatomy and Physiology Society (HAPS) and the American Association for the Advancement of Science (AAAS). Additionally, while actively engaged as an author, Dr. Marieb serves as a consultant for the Benjamin Cummings *Interactive Physiology*® CD-ROM series. This text—*Human Anatomy & Physiology*, Eighth Edition—is the latest expression of her commitment to the needs of students in their pursuit of the study of A&P.

When not involved in academic pursuits, Dr. Marieb is a world traveler and has vowed to visit every country on this planet. Shorter term, she serves on the board of directors of the famed Marie Selby Botanical Gardens and on the scholarship committee of the Women's Resources Center of Sarasota County. She is an enthusiastic supporter of the local arts and enjoys a competitive match of doubles tennis.

Katja Hoehn

Dr. Katja Hoehn is an instructor in the Department of Chemical and Biological Sciences at Mount Royal College in Calgary, Canada. Dr. Hoehn's first love is teaching. Her teaching excellence has been recognized by several awards during her 14 years at Mount Royal College. These include a PanCanadian Educational Technology Faculty Award (1999), a Teaching Excellence Award from the Students' Association of Mount Royal College (2001), and the Mount Royal College Distinguished Faculty Teaching Award (2004).

Dr. Hoehn received her M.D. (with Distinction) from the University of Saskatchewan, and her Ph.D. in Pharmacology from Dalhousie University. In 1991, the Dalhousie Medical Research Foundation presented her with the Max Forman (Jr.) Prize for excellence in medical research. During her Ph.D. and postdoctoral studies, she also pursued her passion for teaching by presenting guest lectures to first- and second-year medical students at Dalhousie University and at the University of Calgary.

Dr. Hoehn has been a contributor to several books and has written numerous research papers in Neuroscience and Pharmacology. She oversaw the recent revision of the Benjamin Cummings *Interactive Physiology®* CD-ROM series modules, and coauthored the newest module, *The Immune System.*

Dr. Hoehn is also actively involved in the Human Anatomy and Physiology Society (HAPS). When not teaching, she likes to spend time outdoors with her husband and two boys, compete in triathlons, and play Irish flute.

To the Student: How to Use This Book

Introduce yourself to the chapter

Chapter outlines provide a preview of the chapter and let you know where you're going.

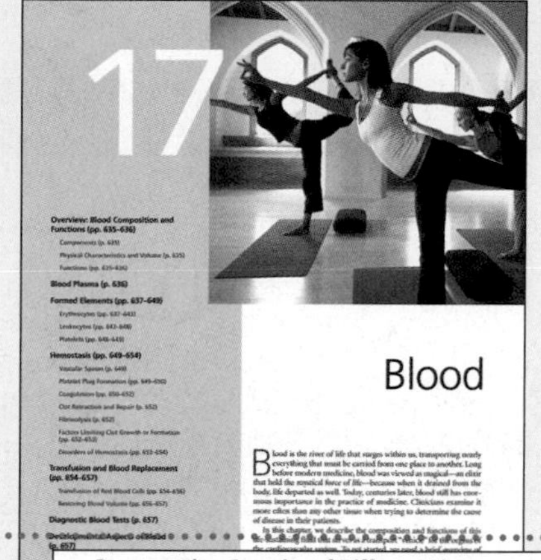

Focus on key concepts

Student objectives have been integrated into the chapter and give you a preview of what content is to come and what you are expected to learn.

NEW! Check Your Understanding questions ask you to stop, think, and to check your understanding of key concepts at the end of major sections.

Homeostatic Imbalance sections are integrated within the text and alert you to the consequences of body systems not functioning optimally. These pathological conditions are integrated with the text to clarify and illuminate normal functioning.

Regenerative Capacity of Different Tissues

The different tissues vary widely in their capacity for regeneration. Epithelial tissues, bone, areolar connective tissue, dense irregular connective tissue, and blood-forming tissue regenerate extremely well. Smooth muscle and dense regular connective tissue have a moderate capacity for regeneration, but skeletal muscle and cartilage have a weak regenerative capacity. Cardiac muscle and the nervous tissue in the brain and spinal cord have virtually no *functional* regenerative capacity, and they are routinely replaced by scar tissue. However, recent studies have shown that some unexpected (and highly selective) cellular division occurs in both these tissues after damage, and efforts are under way to coax them to regenerate better.

In nonregenerating tissues and in exceptionally severe wounds, fibrosis totally replaces the lost tissue. Over a period of months, the fibrous mass shrinks and becomes more and more compact. The resulting scar appears as a pale, often shiny area composed mostly of collagen fibers. Scar tissue is strong, but it lacks the flexibility and elasticity of most normal tissues. Also, it cannot perform the normal functions of the tissue it has replaced.

HOMEOSTATIC IMBALANCE

Scar tissue that forms in the wall of the urinary bladder, heart, or other muscular organ may severely hamper the function of that organ. The normal shrinking of the scar reduces the internal volume of an organ and may hinder or even block movement of substances through a hollow organ. Scar tissue hampers muscle's ability to contract and may interfere with its normal excitation by the nervous system. In the heart, these problems may lead to progressive heart failure. In irritated visceral organs, particularly following abdominal surgery,

adjacent organs together. Such adhesions can prevent the nor mal shifting about (churning) of loops of the intestine, da gerously obstructing the flow of foodstuffs through Adhesions can also restrict heart movements and immobil joints. ■

CHECK YOUR UNDERSTANDING

21. What are the three main steps of tissue repair?
22. Why does a deep injury to the skin result in abundant scar tissue formation?

For answers, see Appendix G.

Developmental Aspects of Tissues

▶ Indicate the embryonic origin of each tissue class.

▶ Briefly describe tissue changes that occur with age.

One of the first events of embryonic development is the form tion of the three **primary germ layers**, which lie one atop next like a three-layered cellular pancake. From superficial deep, these layers are the **ectoderm**, **mesoderm** (mez′o-der and **endoderm**. As shown in Figure 4.13, these primary ge layers then specialize to form the four primary tissues epithelium, nervous tissue, muscle, and connective tissue—t make up all body organs.

By the end of the second month of development, the p mary tissues have appeared, and all major organs are in pla In general, tissue cells remain mitotic and produce the ra growth that occurs before birth. The division of nerve ce however, stops or nearly stops during the fetal period. Af birth, the cells of most other tissues continue to divide u adult body size is achieved. Cellular division then slows grea

Illustrated tables summarize complex information and serve as a "one-stop shopping" study tool.

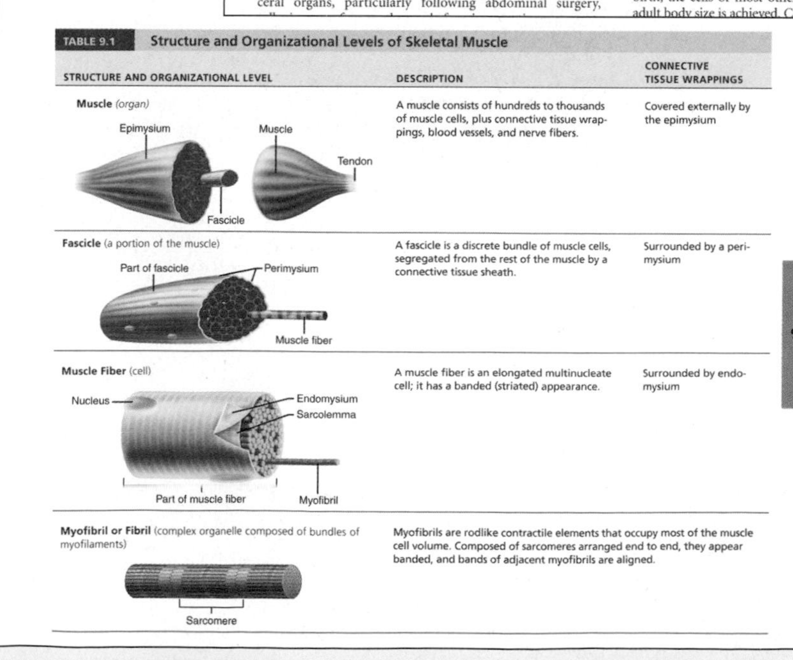

TABLE 9.1 Structure and Organizational Levels of Skeletal Muscle

STRUCTURE AND ORGANIZATIONAL LEVEL	DESCRIPTION	CONNECTIVE TISSUE WRAPPINGS
Muscle *(organ)*	A muscle consists of hundreds to thousands of muscle cells, plus connective tissue wrappings, blood vessels, and nerve fibers.	Covered externally by the epimysium
Fascicle (a portion of the muscle)	A fascicle is a discrete bundle of muscle cells, segregated from the rest of the muscle by a connective tissue sheath.	Surrounded by a perimysium
Muscle Fiber (cell)	A muscle fiber is an elongated multinucleate cell; it has a banded (striated) appearance.	Surrounded by endomysium
Myofibril or Fibril (complex organelle composed of bundles of myofilaments)	Myofibrils are rodlike contractile elements that occupy most of the muscle cell volume. Composed of sarcomeres arranged end to end, they appear banded, and bands of adjacent myofibrils are aligned.	

Follow complex processes step-by-step

NEW! Focus figures help you grasp tough topics in A&P by walking you through carefully developed step-by-step illustrations that use a big-picture layout and dramatic art to provide a context for understanding the process.

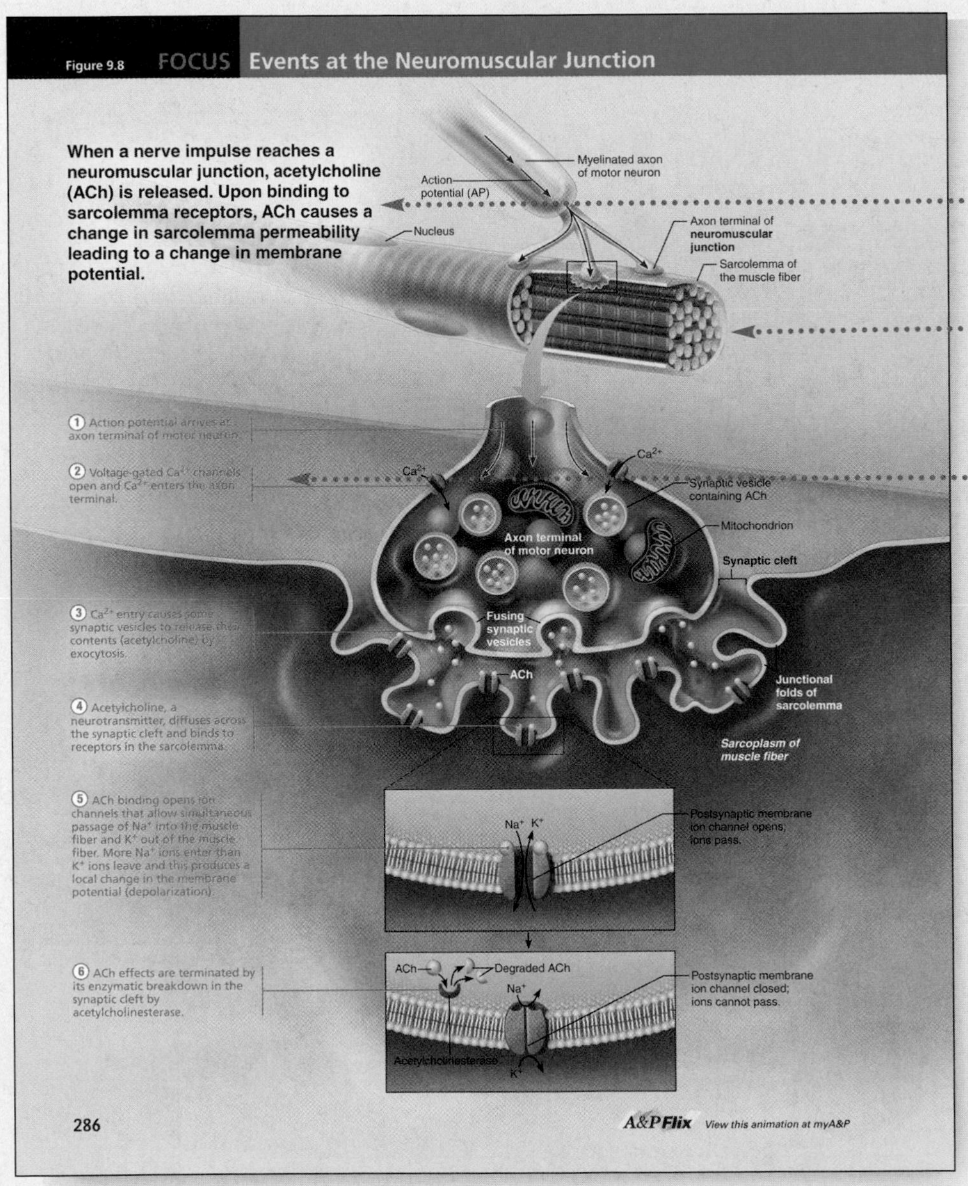

Figure 9.8 FOCUS Events at the Neuromuscular Junction
When a nerve impulse reaches a neuromuscular junction, acetylcholine (ACh) is released. Upon binding to sarcolemma receptors, ACh causes a change in sarcolemma permeability leading to a change in membrane potential.
Myelinated axon of motor neuron
Action potential (AP)
Nucleus
Axon terminal of neuromuscular junction
Sarcolemma of the muscle fiber
① Action potential arrives at axon terminal of motor neuron.
② Voltage-gated Ca^{2+} channels open and Ca^{2+} enters the axon terminal.
Ca^{2+}
Ca^{2+}
Synaptic vesicle containing ACh
Mitochondrion
Axon terminal of motor neuron
Synaptic cleft
③ Ca^{2+} entry causes some synaptic vesicles to release their contents (acetylcholine) by exocytosis.
Fusing synaptic vesicles
ACh
Junctional folds of sarcolemma
④ Acetylcholine, a neurotransmitter, diffuses across the synaptic cleft and binds to receptors in the sarcolemma.
Sarcoplasm of muscle fiber
⑤ ACh binding opens ion channels that allow simultaneous passage of Na^+ into the muscle fiber and K^+ out of the muscle fiber. More Na^+ ions enter than K^+ ions leave and this produces a local change in the membrane potential (depolarization).
Na^+ K^+
Postsynaptic membrane ion channel opens; ions pass.
⑥ ACh effects are terminated by its enzymatic breakdown in the synaptic cleft by acetylcholinesterase.
ACh
Degraded ACh
Na^+
Postsynaptic membrane ion channel closed; ions cannot pass.
Acetylcholinesterase
K^+
286
A&P Flix View this animation at myA&P

Overview provides a quick summary of the key idea of the figure.

Big picture orientation provides you with a concrete starting point for the process.

Step text walks you through the process step-by-step.

Visualize structures

NEW! Stunning 3-D anatomy art is rendered in a dramatically more dynamic, realistic style with vibrant, saturated colors to help you visualize key anatomical structures.

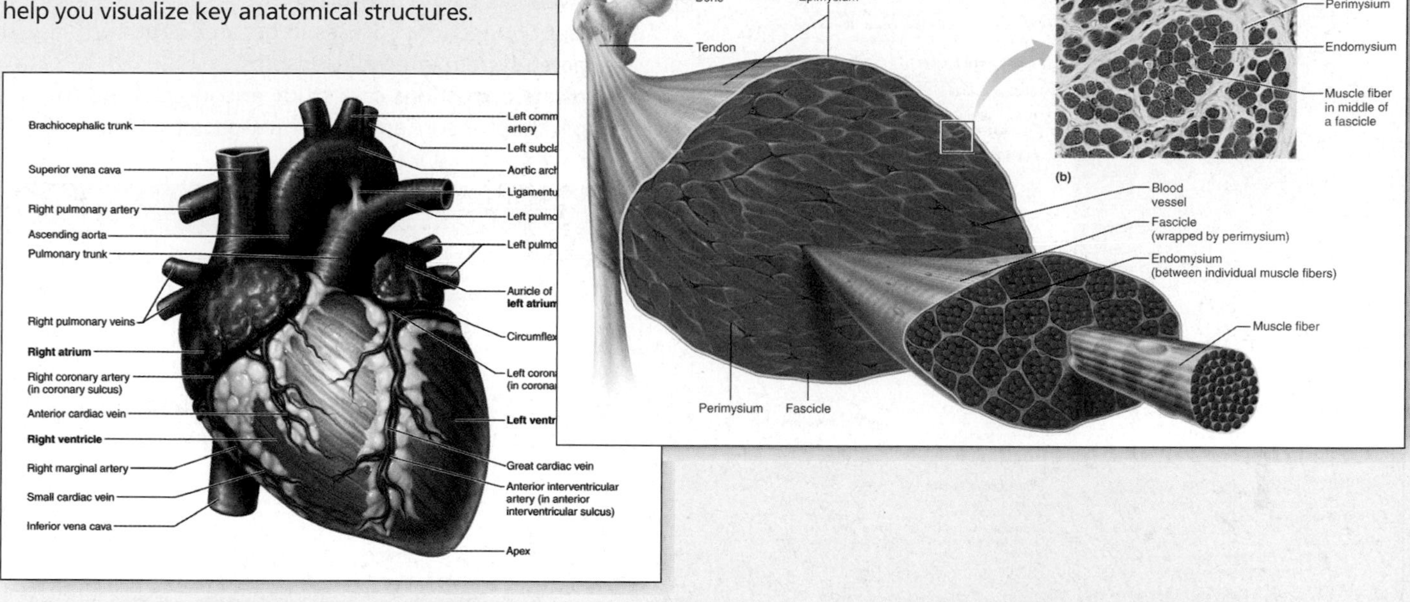

NEW! Bone art features realistic bone color and texture with a consistent style from figure to figure.

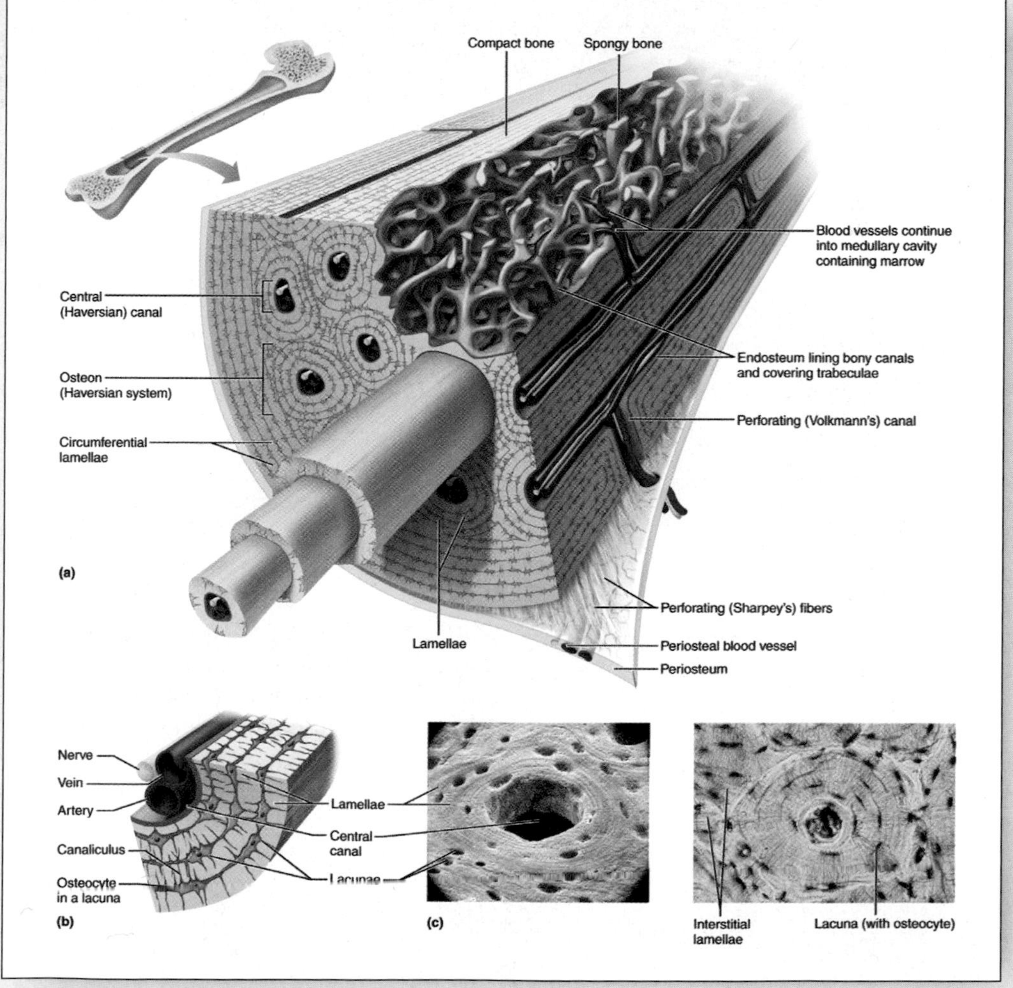

Make connections

MAKING CONNECTIONS

System Connections

Homeostatic Interrelationships Between the Integumentary System and Other Body Systems

5

Endocrine System

- Skin protects endocrine organs; converts some hormones to their active forms; synthesizes a vitamin D precursor
- Androgens produced by the endocrine system activate sebaceous glands and are involved in regulation of hair growth

Cardiovascular System

- Skin protects cardiovascular organs; prevents fluid loss from body; serves as blood reservoir
- Cardiovascular system transports oxygen and nutrients to skin and removes wastes from skin; provides substances needed by skin glands to make their secretions

Lymphatic System/Immunity

- Skin protects lymphatic organs; prevents pathogen invasion; dendritic cells and macrophages help activate the immune system
- Lymphatic system prevents edema by picking up excessive leaked fluid; immune system protects skin cells

Respiratory System

- Skin protects respiratory organs; hairs in nose help filter out dust from inhaled air
- Respiratory system furnishes oxygen to skin cells and removes carbon dioxide via gas exchange with blood

Digestive System

- Skin protects digestive organs; provides vitamin D needed for calcium absorption; performs some of the same chemical conversions as liver cells
- Digestive system provides needed nutrients to the skin

Urinary System

- Skin protects urinary organs; excretes salts and some nitrogenous wastes in sweat
- Urinary system activates vitamin D precursor made by keratinocytes; disposes of nitrogenous wastes of skin metabolism

Reproductive System

- Skin protects reproductive organs; cutaneous receptors respond to erotic stimuli; highly modified sweat glands (mammary glands) produce milk. During pregnancy, skin stretches to accommodate growing fetus; changes in skin pigmentation may occur

Skeletal System

- Skin protects bones; skin synthesizes a vitamin D precursor needed for normal calcium absorption and deposit of bone (calcium) salts, which make bones hard
- Skeletal system provides support for skin

Muscular System

- Skin protects muscles
- Active muscles generate large amounts of heat, which increases blood flow to the skin and may promote activation of sweat glands of skin

Nervous System

- Skin protects nervous system organs; cutaneous sensory receptors for touch, pressure, pain, and temperature located in skin (see Figure 5.1)
- Nervous system regulates diameter of blood vessels in skin; activates sweat glands, contributing to thermoregulation; interprets cutaneous sensation; activates arrector pili muscles

166

Making Connections at the end of each body system helps you understand the relationships between body systems with this three-tiered presentation:

- **System Connections** highlights the interrelationship between all of the body systems.
- **Closer Connections** focuses in greater depth on selected system interrelationships.
- **Clinical Connections** case study encourages you to apply chapter concepts to clinical situations.

Closer Connections

The Integumentary System and Interrelationships with the Nervous, Cardiovascular, and Lymphatic/Immune Systems

First and foremost, our skin is a barrier. Like the skin of a grape, it keeps its contents juicy and whole. The skin is also a master at self (wound) repair, and interacts intimately with other body systems by making vitamin D (necessary for hard bones) and other potent molecules, all the while protecting deeper tissues from damaging external agents. Perhaps the most crucial roles of the skin in terms of overall body homeostasis are those it plays with the nervous, cardiovascular, and lymphatic systems. These interactions are detailed next.

Nervous System

The whole body benefits from the skin's interaction with the nervous system. The skin houses the tiny sensory receptors that provide a great deal of information about our external environment—its temperature, the pressure exerted by objects, and the presence of dangerous substances. What if we stepped on broken glass or hot pavement but did not have neural monitors in our skin? If we did not actually see, hear, taste, or smell such a damaging event, no reports would be sent to the nervous system. Consequently, the nervous system would be left "in the dark," unable to evaluate the need for a response and to order the steps needed to protect us from further damage or to get first aid.

Nervous and Cardiovascular System

Skin provides the site both for sensing external temperature and for responding to temperature changes. Dermal blood vessels (cardiovascular system organs) and sweat glands (controlled by the nervous system) play crucial roles in thermoregulation. So too do blood and the hot and cold receptors in the skin. When we are chilled, our blood loses heat to internal organs and cools. This loss alerts the nervous system to retain heat by constricting dermal blood vessels. When body and blood temperature rises, dermal vessels dilate and sweating begins.

Thermoregulation is vital: When the body overheats, life-threatening changes occur. Chemical reactions speed up, and as the temperature continues to rise vital proteins are destroyed and cells die. Cold has the opposite effect; cellular activity slows and ultimately stops.

Lymphatic System/Immunity

The role of the skin in immunity is complex. Keratinocytes in the skin manufacture interferons (proteins that block viral infection) and other proteins important to the immune response. Epidermal dendritic cells in the skin interact with antigens (foreign substances) that have penetrated the stratum corneum. The dendritic cells then migrate to lymphatic organs, where they present bits of the antigens to cells that will mount the immune response against them. This "messenger" function alerts the immune system early on to the presence of pathogens in the body. 5

Even a mild sunburn disrupts the normal immune response because UV radiation disables the skin's presenter cells. This effect may explain why many people infected by the cold sore virus tend to have a cold sore eruption after sun exposure.

Clinical Connections

Integumentary System

A terrible collision between a trailer truck and a bus has occurred on Route 91. Several of the passengers are rushed to area hospitals for treatment. We will follow a few of these people in clinical case studies that will continue through the book from one organ system to the next.

Case study: Examination of Mrs. DeStephano, a 45-year-old woman, reveals several impairments of homeostasis. Relative to her integumentary system, the following comments are noted on her chart:

- Epidermal abrasions of the right arm and shoulder
- Severe lacerations of the right cheek and temple
- Cyanosis apparent

The lacerated areas are cleaned, sutured, and bandaged by the emergency room (ER) personnel, and Mrs. DeStephano is admitted for further tests.

Relative to her signs:

1. What protective mechanisms are impaired or deficient in the abraded areas?
2. Assuming that bacteria are penetrating the dermis in these areas, what remaining skin defenses might act to prevent further bacterial invasion?
3. What benefit is conferred by suturing the lacerations? (Hint: See Chapter 4, p. 144.)
4. Mrs. DeStephano's cyanotic skin may hint at what additional problem (and impairment of what body systems or functions)?

(Answers in Appendix G)

167

Chapter 9 Muscles and Muscle Tissue 313

CLOSER LOOK

Athletes Looking Good and Doing Better with Anabolic Steroids?

9

Society loves a winner and top athletes reap large social and monetary rewards. It is not surprising that some will grasp at anything that might increase their performance—including "juice," or anabolic steroids. These drugs are variants of the male sex hormone testosterone engineered by pharmaceutical companies. They were introduced in the 1950s to treat anemia and certain muscle-wasting diseases and to prevent muscle atrophy in patients immobilized after surgery. Testosterone is responsible for the increase in muscle and bone mass and other physical changes that occur during puberty and converts boys into men.

Convinced that megadoses of steroids could produce enhanced masculinizing effects in grown men, many athletes and bodybuilders were using them by the early 1960s. Investigations of the so-called Balco Scandal have stunned fans of major league baseball players with revelations in 2004 and since of rampant steroid use by Barry Bonds, of the San Francisco Giants, and many other elite athletes. It is still going on. In October of 2007, Marion Jones, one of the most celebrated of women athletes of all time, admitted that she was using performance-enhancing steroids when she won five gold medals in the 2000 Olympics. Furthermore, steroid use today is not confined to athletes. Indeed, it is estimated that nearly one in every 10 young men has tried them, and the practice is also spreading among young women.

It is difficult to determine the extent of anabolic steroid use because most international competitions ban the use of drugs. Users (and prescribing physicians or drug dealers) are naturally reluctant to talk about it, and users stop doping before the event, aware that evidence of drug use is hard to find a week after its use is stopped. Additionally, "underground" suppliers of performance-enhancing drugs keep producing new versions of designer steroids that evade standard antidoping tests. The Olympic Analytical Laboratory in Los Angeles rocked the sports world in November of 2003 when it revealed that a number of elite athletes tested positive for tetrahydrogestrinone (THG), a designer steroid not previously known or tested for. Nonetheless, there is little question that many professional bodybuilders and athletes competing in events that require muscle strength (e.g., shot put, discus throwing, and weight lifting) are heavy users. Sports figures such as football players have also admitted to using steroids as an adjunct to training, diet, and psychological preparation for games. These athletes claim that anabolic steroids enhance muscle mass and strength, and raise oxygen-carrying capability owing to greater red blood cell volume.

Typically, bodybuilders who use steroids combine high doses (up to 200 mg/day) via injection or transdermal skin patches with heavy resistance training. Intermittent use begins several months before an event, and commonly entails the use of many anabolic steroid supplements (a method called stacking). Doses are increased gradually as the competition nears.

Do the drugs do all that is claimed? Research studies report increased isometric strength and body weight in steroid users. While these are results weight lifters dream about, for runners and others requiring fine muscle coordination and endurance these changes may not translate into improved performance. The "jury is still out" on this question.

Do the alleged advantages of steroids outweigh their risks? Absolutely not. Physicians say they cause bloated faces (Cushingoid sign of steroid excess); acne and hair loss; shriveled testes and infertility; damage to the liver that promotes liver cancer; and changes in blood cholesterol levels that may predispose users to coronary heart disease. In addition, females can develop masculine characteristics such as smaller breasts, enlarged clitoris, excess body hair, and thinning scalp hair. The psychiatric hazards of anabolic steroid use may be equally threatening: Recent studies indicate that one-third of users suffer serious mental problems. Depression, delusions, and manic behavior—in which users undergo Jekyll-and-Hyde personality swings and become extremely violent (termed *'roid rage*)—are all common.

A more recent arrival on the scene, sold over the counter as a "nutritional performance-enhancer," is androstenedione, which is converted to testosterone in the body. Though it is taken orally (and much of it is destroyed by the liver soon after ingestion), the few milligrams that survive temporarily boost testosterone levels. Reports of its use by baseball great Mark McGwire before he retired, and of athletic wanna-bes from the fifth grade up sweeping the supplement off the drugstore shelves, are troubling, particularly since it is not regulated by the U.S. Food and Drug Administration (FDA) and its long-term effects are unpredictable and untested. A study at Massachusetts General Hospital found that males who took androstenedione developed higher levels of the female hormone estrogen as well as testosterone, raising their risk of feminizing effects such as enlarged breasts. Youths with elevated levels of estrogen or testosterone may enter puberty early, stunting bone growth and leading to shorter-than-normal adult height. Some people admit to a willingness to try almost anything to win, short of killing themselves. Are they unwittingly doing this as well?

Closer Look boxes on timely subjects such as medical technology, new discoveries in medical research, and important societal issues broaden your horizons and present scientific information that can be applied to your daily life.

Learn the language

Phonetic spellings are provided for words that may be unfamiliar to you to help you with pronunciation.

Color-coded chapter and unit tabs help you find information quickly and easily.

Word roots listed on the inside back cover will help you learn the special vocabulary of anatomy and physiology.

Review what you've learned

Review questions at the end of each chapter, including multiple choice/matching, short answer, and Critical Thinking and Clinical Application questions, help you evaluate your progress.

Chapter summaries with page references provide excellent study aids.

Answers to Check Your Understanding, Clinical Connections, and end-of-chapter Multiple Choice and Matching Review Questions can be found in Appendix G.

To the Student: How to Use myA&P

Check your readiness

Get Ready for A&P gets you prepared for your A&P course. Take the diagnostic test to see where you need review.

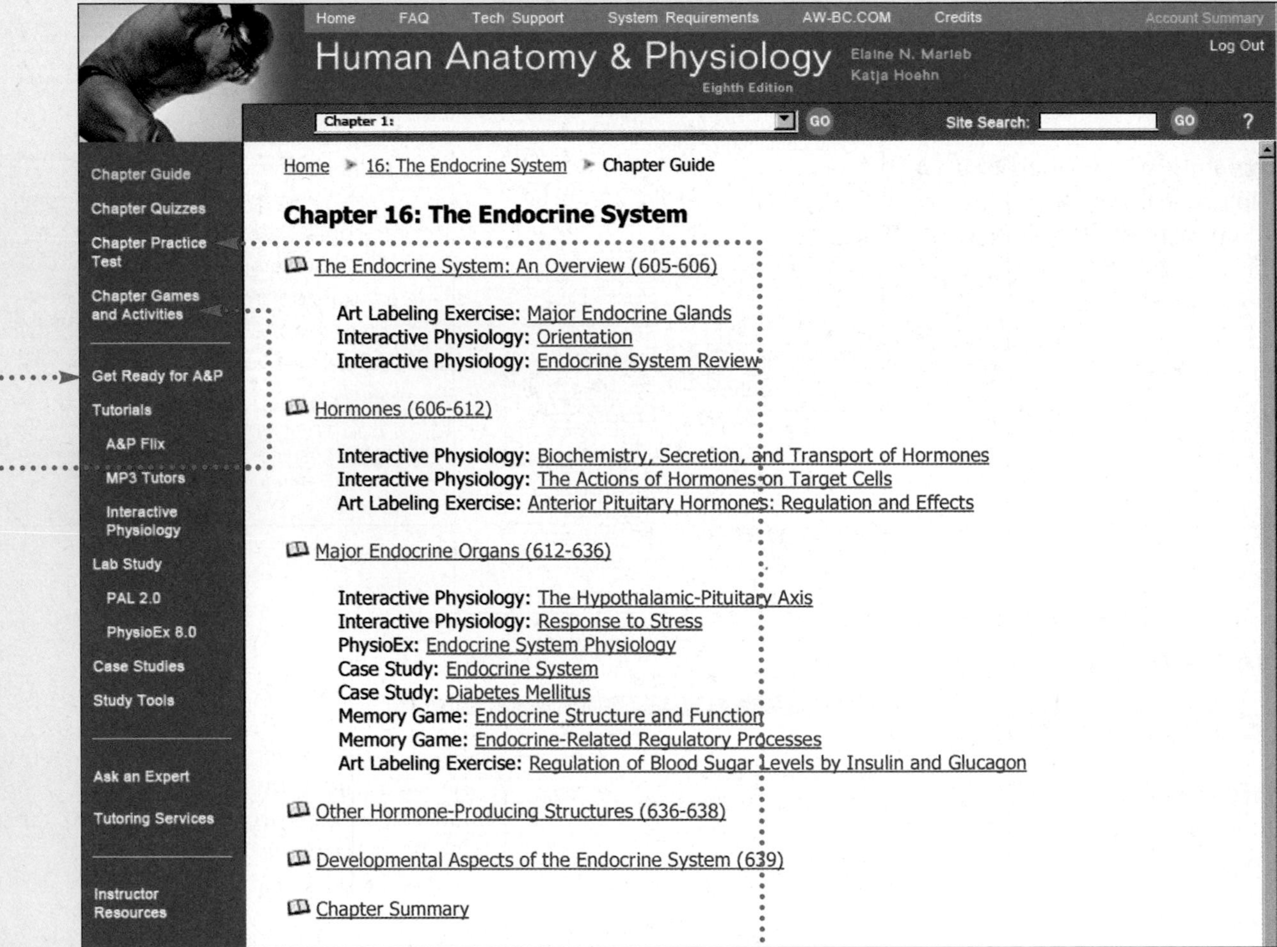

Prepare for exams

Chapter Quizzes and Practice Tests help you assess your understanding of the chapter and prepare for your exams.

Games and activities featuring Art Labeling Exercises, Memory Games, and Crossword Puzzles and Histology, Bone, and Muscle Reviews. Also included are new **MP3 Tutor Sessions** that carefully coach you through the most difficult A&P concepts including calcium regulation, the visual pathway, and gas exchange during respiration.

Chapter Guide

Chapter Quizzes

Chapter Practice Test

Chapter Games and Activities

Get Ready for A&P

Tutorials

A&P Flix

MP3 Tutors

Interactive Physiology

Lab Study

PAL 2.0

PhysioEx 8.0

Case Studies

Study Tools

Ask an Expert

Tutoring Services

Instructor Resources

Master tough concepts in A&P

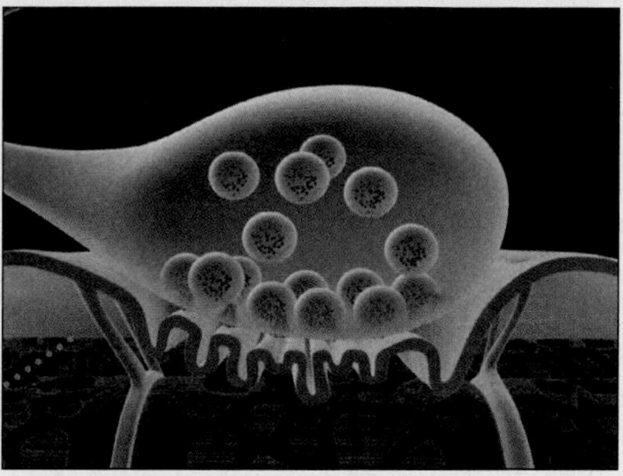

NEW! A&P Flix animations provide carefully developed step-by-step explanations with dramatic 3-D representations of structures that show action and movement of processes, thereby bringing difficult-to-teach A&P concepts to life. Each animation includes gradable quizzes as well as study sheets for practice and assessment.

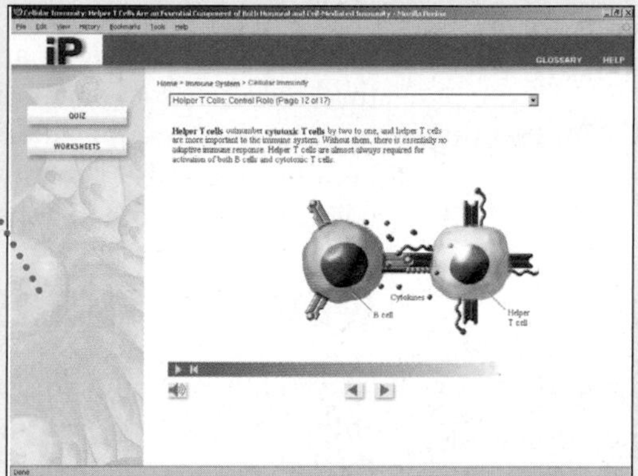

***Interactive Physiology®* 10-System Suite,** an award-winning tutorial program with a new module on the Immune System, tutors you in key physiological concepts and helps you advance beyond memorization to a genuine understanding of complex processes.

Access the A&P lab 24/7

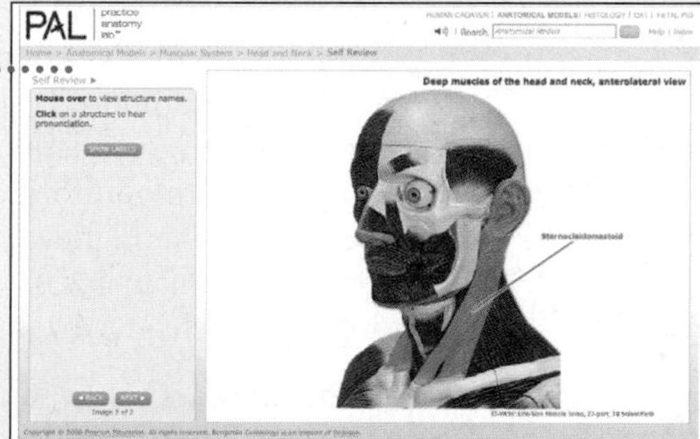

NEW! Practice Anatomy Lab™ 2.0 allows you to view hundreds of images of the human cadaver, anatomical models, histology slides, the cat, and the fetal pig and to take practice quizzes and simulated lab practical exams.

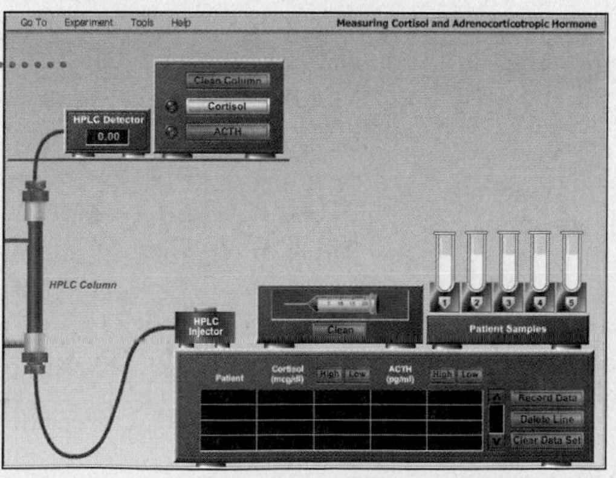

PhysioEx™ laboratory simulations allow you to conduct simulated experiments as part of your A&P lab.

Preface

As educators, clinically trained individuals, and perennial students, we are continually challenged by the learning mind. What works best to get students over conceptual hurdles and to help them apply new information to the world they personally understand? Our clinical backgrounds have served our teaching and writing purposes well. Perhaps even more important, our clinical experience has allowed us to view our presentations through our students' eyes and from the vantage points of their career interests.

For this edition, as for those preceding it, feedback from both student and instructor reviews indicated areas of the text that needed to be revised for clarity, timeliness, and just plain reduction of verbal meatiness. Overall, feedback was positive, verifying that the approach of explaining fundamental principles and unifying themes first as a strong base for all that comes later is still viable. Furthermore, it is clear that backing up these explanations with comfortable analogies and familiar examples enhances the students' understanding of the workings of the human body.

Unifying Themes

Three integrating themes that organized, unified, and set the tone of the first edition of this text continue to be valid and are retained in this edition. These themes are:

Interrelationships of body organ systems. The fact that nearly all regulatory mechanisms require interaction of several organ systems is continually emphasized. For example, Chapter 25, which deals with the structure and function of the urinary system, discusses the vital importance of the kidneys not only in maintaining adequate blood volume to ensure normal blood circulation, but also in continually adjusting the chemical composition of blood so that all body cells remain healthy. The unique Making Connections feature is the culmination of this approach and should help students think of the body as a dynamic community of interdependent parts rather than as a number of isolated structural units.

Homeostasis. The normal and most desirable condition of body functioning is homeostasis. Its loss or destruction always leads to some type of pathology—temporary or permanent. Pathological conditions are integrated with the text to clarify and illuminate normal functioning, not as an end in and of themselves. For example, Chapter 19, which deals with the structure and function of blood vessels, explains how the ability of healthy arteries to expand and recoil ensures continuous blood flow and proper circulation. The chapter goes on to discuss the effects on homeostasis when arteries lose their elasticity: high blood pressure and all of its attendant problems. These homeostatic imbalances are indicated visually by a pink symbol with a fulcrum:

Whenever students see the imbalance symbol in text, the concept of disease as a loss of homeostasis is reinforced.

Complementarity of structure and function. Students are encouraged to understand the structure of an organ, a tissue, or a cell as a prerequisite to comprehending its function. Concepts of physiology are explained and related to structural characteristics that promote or allow the various functions to occur. For example, the lungs can act as a gas exchange site because the walls of their air sacs present an incredibly thin barrier between blood and air.

NEW TO THE EIGHTH EDITION

The Eighth Edition represents a monumental revision with an entirely new art program and text presentation that build upon the hallmark strengths of the previous seven editions. With every edition, our goal is powerful but simple—to make anatomy and physiology as engaging, accurate, and relevant as possible for both you and your students. The changes to the Eighth Edition are all driven by the needs of today's students, as we seek to make the learning of key concepts in A&P as easy as possible for them. Key concepts are important because of the overwhelming amount of material in this course. Mastering this material gives students an anchor and structure for managing this wealth of information. Below are the ways in which we've revised the Eighth Edition to make this book the one where learning happens most effectively, followed by a detailed list of specific chapter-by-chapter content changes.

A whole new art program. The drive for this revision began as a simple list. We sat down together and created a chapter-by-chapter list of the key concepts in A&P where students struggle the most. This list became the basis for our art revision plans. We first boiled it down to some of the toughest topics to get our list of Focus figures. This new Focus feature highlights tough topics in A&P and walks students step-by-step through complex processes that are difficult to teach and visualize. In each case, we scrutinized the process and worked through countless revisions to break it down in the most logical and easy-to-follow way possible for students. We hope you'll be as pleased with the results as we are.

We also revised and reconceptualized many of the process figures in the book to make them easier to follow and to learn from. Where appropriate we have added blue step text that serves as our author voice guiding students step-by-step through complex processes. The blue text clearly separates the process steps from the labels, making the figures easy to navigate.

Flipping through the Eighth Edition, you can see that our new art is dynamic, three-dimensional, and realistic, with dramatic views and perspectives that use vibrant, saturated colors. Using our list of key concepts, we targeted critical figures in anatomy and worked closely with the artistic team on making these figures superior in rendering and in conveying the key pedagogical information and structures that students need to learn from the figure, striking a perfect balance between realism and teaching effectiveness.

Finally, we've also added a wealth of new figures and photos to enhance learning, many of which are listed below.

Improved text presentation. New text features also serve to focus students on key concepts. We have integrated the student objectives to fall within the chapter, giving students a preview in smaller chunks of what they are expected to learn in a given section. We've also added new Check Your Understanding questions that ask students to stop, think, and check their understanding of key concepts at the end of major sections. These changes along with a brand-new design make the book easier than ever to study from and navigate. We have also edited the text throughout with a refined writing style that retains our hallmark analogies and accessible, friendly style while using simpler, more concise language and shorter paragraphs. These changes make the text easier for students to manage as they face the challenging amount of information in this course.

Factual updates and accuracy. As authors we pride ourselves on keeping our book as up-to-date and as accurate as possible in all areas—a monumental task that requires painstaking selectivity. Although information changes even as a textbook goes to press, be assured that our intent and responsibility to update was carried out to the best of our ability. We have incorporated updates from current research in the field as much as possible; many of these updates are included below in the chapter-by-chapter changes. A more complete list, along with references for selected updates, is available from your Pearson sales representative and in the Instructor Guide to Text and Media.

Chapter-by-Chapter Changes

Chapter 1 The Human Body: An Orientation

- New PET scan for A Closer Look on medical imaging

Chapter 2 Chemistry Comes Alive

- Updated information on molecular chaperones

Chapter 3 Cells: The Living Units

- New step art for exocytosis (Figure 3.14)
- Updated discussion of types of endocytosis accompanied by new endocytosis step art (Figure 3.12)
- New Figure 3.13 provides a comparison of three types of endocytosis
- New Figure 3.20 with step text on the signaling mechanism for targeting new proteins to the ER
- New Focus on Primary Active Transport: The Na^+-K^+ Pump (Figure 3.10)
- New Focus on G Proteins (Figure 3.16)
- New Focus on Mitosis (Figure 3.33)
- New diagrams accompany photos in figure showing the effects of varying tonicities on living red blood cells (Figure 3.9)
- New photomicrographs accompany all cell organelle illustrations, including new Figure 3.28 on microvilli
- Revised text and new figures for transcription (Figure 3.35) and translation (Figure 3.7).
- New information on the origin of peroxisomes based on recent research

Chapter 4 Tissue: The Living Fabric

- New Figure 4.1: Overview of four tissue types
- New photomicrographs for pseudostratified ciliated columnar epithelium (Figure 4.3d), goblet cells (Figure 4.4), and elastic connective tissue (Figure 4.8f)
- New Table 4.1 compares four main classes of connective tissue
- Updated A Closer Look on cancer

Chapter 5 The Integumentary System

- New Figure 5.3: Two regions of the dermis, with three new photomicrographs
- New Figure 5.4: Dermal modifications result in characteristic skin markings, with one new photomicrograph
- New photos: partial and full thickness burns (Figure 5.10)

Chapter 6 Bones and Skeletal Tissue

- New Figure 6.4 shows comparative morphology of bone cells
- New Figure 6.14 explains how vigorous exercise can lead to large increases in bone strength
- Updated information on homocysteine as a marker of low bone mass density and bone frailty; additional information on age-related bone changes and treatments

Chapter 7 The Skeleton

- New photo of midsagittal section of the skull (Figure 7.5c)
- New photos for inferior and superior views of the skull (Figures 7.6b, 7.7b)
- New photos of the sphenoid bone, superior and posterior views (Figure 7.9)
- New photo of right lateral view of the maxilla (Figure 7.11)
- New MRI of lumbar region in sagittal section showing herniated disc (Figure 7.17)
- New photo of midsagittal section of the thorax (Figure 7.22)
- New X ray of the foot (Figure 7.34)
- New Figure 7.37: The C-shaped spine of a newborn infant
- New Homeostatic Imbalance: xiphoid process projecting posteriorly

Chapter 8 Joints

- Figure 8.1 expanded to show a comparison of different types of fibrous joints; added gomphosis
- Added new views for knee, shoulder, and mandible joint

Chapter 9 Muscles and Muscle Tissue

- New Focus on Events at the Neuromuscular Junction (Figure 9.8)
- New Focus on Excitation-Contraction Coupling (Figure 9.11)
- New Focus on the Cross Bridge Cycle (Figure 9.12)
- New Figure 9.7: Phases leading to muscle fiber contraction
- New Figure 9.20: Comparison of energy sources used during short-duration and prolonged-duration exercise
- New Figure 9.24: Cross section of the three types of fibers in skeletal muscle
- New Figure 9.30: Formation of a multinucleate skeletal muscle fiber by fusion of myoblasts

Chapter 10 The Muscular System

- New cadaver photo of the anterior and lateral regions of the neck (Figure 10.9c)
- New cadaver photo of superficial muscles of the thorax (Figure 10.13b)
- New cadaver photo of muscles crossing the shoulder and elbow joint (Figure 10.14d)
- New cadaver photo of superficial muscles of the superior gluteal region (Figure 10.20b)

Chapter 11 Fundamentals of the Nervous System and Nervous Tissue

- New Focus on Resting Membrane Potential (Figure 11.8)
- New Focus on Action Potential (Figure 11.11)
- New Focus on Chemical Synapse (Figure 11.17)
- Updated role of satellite cells
- Updated discussion of nitric oxide and carbon dioxide; added paragraph on new class of neurotransmitter endocannabinoids
- Updated the roles of neurotropins in signaling the growth cone during neuronal development
- Updated information in A Closer Look on overcoming cocaine addiction
- Updated information on neurotransmitters (histamine, somatostatin, substance P, CCK) in Table 11.3
- New Figure 11.10: The spread and decay of a graded potential
- New Figure 11.15: Action potential propagation in unmyelinated and myelinated axons
- New photo, a neuronal growth cone (Figure 11.24)

Chapter 12 The Central Nervous System

- Updated location of cortex receiving vestibular input based on new fMRI studies
- New Homeostatic Imbalance on brain tumors in different regions of the brain: the anterior association area and the posterior parietal region
- Updated discussion of regulation of respiratory rhythm in the medulla
- Updated discussion of occurrence of theta waves in adult electroencephalogram
- Updated mechanisms of onset of sleep and wakefulness, the role of orexins (hypocretins) in narcolepsy, and recent finding that orexin antagonists promote sleep in humans
- Updated survival of strokes and stroke treatment
- Updated cause and treatment of Parkinson's disease
- Updated treatments for Alzheimer's disease
- New Figure 12.17 on the cerebellum with side-by-side illustration and photo showing a sagittal view
- New photo of frontal section of the brain (Figure 12.10)
- New photo of inferior view of the brain showing the regions of the brain stem (Figure 12.14)
- New EEG photo (Figure 12.20)

Chapter 13 The Peripheral Nervous System and Reflex Activity

- Updated axon regrowth and treating spinal cord injuries
- Updated Homeostatic Imbalance on cause and treatment of trigeminal neuralgia
- Updated origin and course of the accessory nerves (CN XI)
- New Focus on the Stretch Reflex (Figure 13.17)
- New cadaver photo of the brachial plexus (Figure 13.9)
- New cadaver photo of the sacral plexus (Figure 13.11)
- New Homeostatic Imbalance on hyperalgesia and phantom limb pain

Chapter 14 The Autonomic Nervous System

- New Homeostatic Imbalance on autonomic neuropathy

Chapter 15 The Special Senses

- Updated laser procedures to correct myopia
- Updated the mechanism of light adaptation in rods
- Updated odor signal processing
- Updated taste cell specificity
- Updated the mechanism of transduction for all five taste modalities
- Updated treatment of age-related macular degeneration

Chapter 16 The Endocrine System

- New Figure 16.7 on regulation of thyroid hormone secretion
- Updated hormones released by the thymus and by adipose tissue
- Added new information about incretins and osteocalcin
- Simplified and updated A Closer Look on diabetes mellitus

Chapter 17 Blood

- Updated discussion of erythropoietin—new understanding of how hypoxia induces erythropoiesis
- Updated treatment of sickle-cell anemia—new drug clotrimazole

Chapter 18 The Cardiovascular System: The Heart

- New cadaver photo of frontal section of the heart (Figure 18.4f)
- New photomicrograph of cardiac muscle (Figure 18.11)

Chapter 19 The Cardiovascular System: Blood Vessels

- Updated function of pericytes
- Updated relationship between obesity and hypertension
- Updated development of arteries and veins
- Updated systolic blood pressure as a better predictor of complications of hypertension in those older than 50
- Updated hypertension and its treatment—angiotensin II receptor blockers

Chapter 20 The Lymphatic System and Lymphoid Organs and Tissues

- Updated information on Hassall's corpuscles from current research

Chapter 21 The Immune System: Innate and Adaptive Body Defenses

- Added dermcidin—an important antimicrobial in human sweat
- Updated number of types of human TLRs
- Updated information that dendritic cells can obtain foreign antigens from infected cells through gap junctions
- Updated role of the T_H2 type of helper T cells in immunity
- Updated statistics on HIV/AIDS
- Updated treatments of autoimmune diseases and multiple sclerosis
- Added new type of T_H cell, T_H17
- New Figure 21.2 on phagocytosis
- New SEM of a dendritic cell (Figure 21.10)
- New computer-generated image of an antibody (Figure 21.14)
- New Homeostatic Imbalance on parasitic worms

Chapter 22 The Respiratory System

- Updated role of alveolar type II cells in innate immunity
- Updated mechanism for hypercapnia following administration of oxygen to patients with COPD
- Updated therapy for cystic fibrosis
- New photomicrograph showing a portion of the tracheal wall (Figure 22.6)

Chapter 23 The Digestive System

- New X ray of the mouth of a child showing the permanent incisors forming (Figure 23.10)
- New photomicrograph of small intestine villus (Figure 23.22)
- New photo of a peptic ulcer lesion and SEM of *H. pylori* bacteria (Figure 23.16)
- Updated discussion of the process of HCl formation within the parietal cells
- Expanded section on histology of the small intestine wall; added function of Paneth cells' secretions

Chapter 24 Nutrition, Metabolism, and Body Temperature Regulation

- Vitamin and mineral tables have been simplified for ease of student learning
- New sections and coverage of obesity, short- and long-term regulation of food intake, and additional regulatory factors
- New photo, atomic force microscopy, reveals the structure of energy-converting ATP synthase rotor rings (Figure 24.10)

Chapter 25 The Urinary System

- New photo of the frontal section of kidney (Figure 25.3)
- New photomicrograph of cut nephron tubules in new figure of renal cortical tissue and renal tubules (Figure 25.6)
- New intravenous pyelogram (Figure 25.19)
- Updated structure and possible function of extra-glomerular mesangial cells
- New Homeostatic Imbalance on chronic renal disease and renal failure

Chapter 26 Fluid, Electrolyte, and Acid-Base Balance

- Added clarification of difference between edema and hypotonic hydration
- New paragraph on angiotensin II

Chapter 27 The Reproductive System

- New SEM of sperm (Figure 27.8)
- New photomicrograph of the endometrium and its blood supply (Figure 27.13)
- New photo of mammogram procedure, plus new photos of a normal mammogram compared to one showing a tumor (Figure 27.16)
- New photomicrographs showing stages of follicular development (Figure 27.18)
- New section on erectile dysfunction
- Added new human papillomavirus vaccine
- Expanded discussion of interactions along the hypothalamic-pituitary-ovarian axis with reconceptualized figure
- Updated transmission of herpes virus
- Updated descent of the testes
- Updated hormone replacement therapy for women

Chapter 28 Pregnancy and Human Development

- New photomicrograph of a blastocyst that has just adhered to the uterine endometrium (Figure 28.5)
- New Figure 28.8 showing detailed anatomy of the vascular relationships in the mature decidua basalis
- New Figure 28.13, flowchart showing major derivatives of the embryonic germ layers
- Updated information on the initiation of labor and on contraception

Chapter 29 Heredity

- New photomicrograph of human sex chromosomes (Figure 29.5)
- New Figure 29.8 comparing amniocentesis and chorionic villus sampling
- Updated discussion of stem cells
- Updated discussion of epigenetics and nontraditional methods of gene regulation

Supplements for the Instructor

NEW! Instructor Resource DVD

(0-321-50704-5)

This media tool organizes all instructor media resources by chapter into one convenient package that allows you to easily and quickly pull together a lecture and to show animations, including brand-new A&P Flix, from your PowerPoint® presentations. The IRDVD contains:

- **NEW! A&P Flix**
  Movie-quality A&P Flix animations of key concepts invigorate classroom lectures. These animations provide carefully developed, step-by-step explanations with dramatic 3-D representations of structures that show action and movement of processes, bringing A&P concepts to life. Using the A&P Flix animations, you can help students visualize tough-to-teach A&P concepts such as muscle actions, excitation-contraction coupling, generation of an action potential, and more. These animations can be launched directly from your PowerPoint presentations.

  Note: *these animations are available on the myA&P™ companion website with gradable quizzes as well as printable study sheets for practice and assessment.*

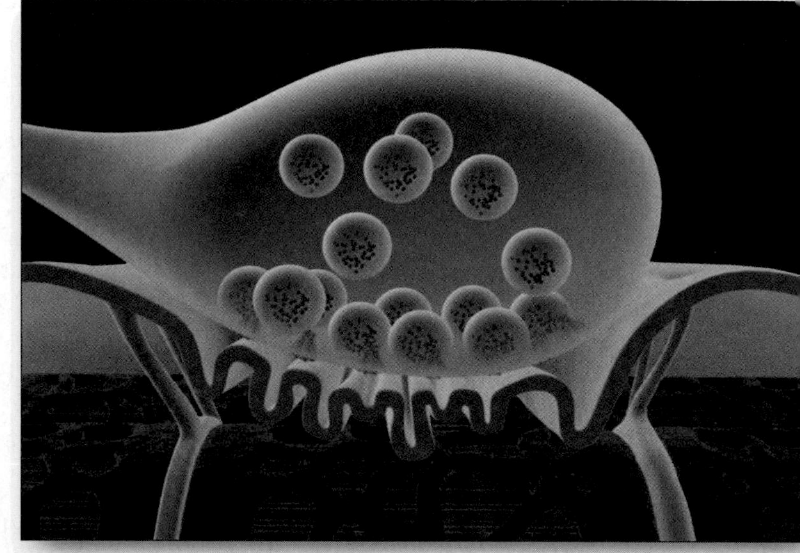

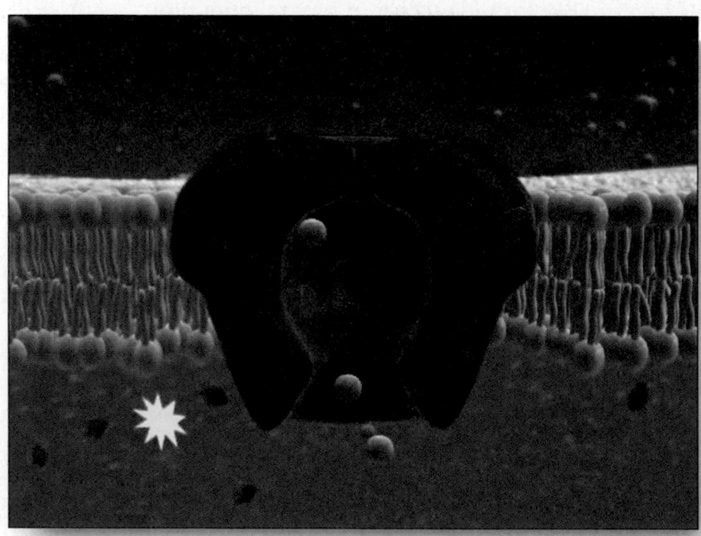

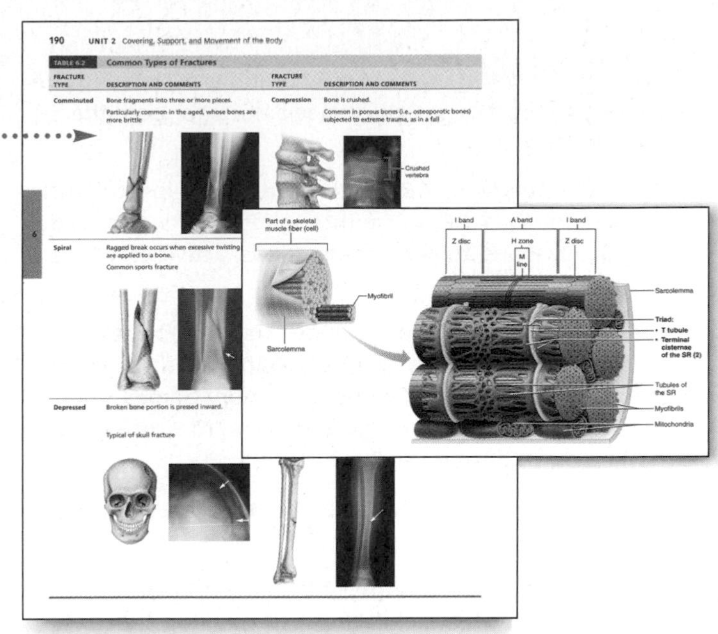

- **All art, photos, and tables** from the book in JPEG and PowerPoint format, as well as all photos from *A Brief Atlas of the Human Body,* Second Edition. Labels have been enlarged in easy-to-read type for optimal viewing in large lecture halls.

- **Instructor Guide to Text and Media**

- **Test Bank**

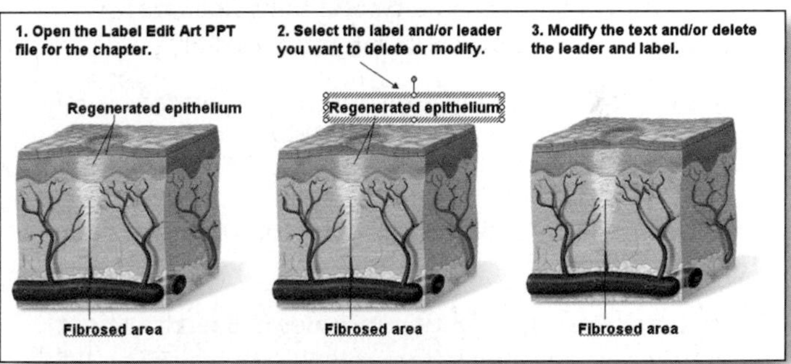

- **Illustrations offered in customizable PowerPoint formats,** including Label-Edit Art with editable leaders and labels and Step-Edit Art that walks through multistep figures step-by-step.

- **Quiz Show Game chapter reviews** that encourage student interaction

Osteoarthritis (OA)

- Most common chronic arthritis; often called "wear-and-tear" arthritis
- Affects women more than men
- 85% of all Americans develop OA
- More prevalent in the aged, and is probably related to the normal agi...

- **Updated, customizable PowerPoint Lecture Outline slides,** available for every chapter, that combine lecture notes, illustrations with editable labels, photos, tables, and animations.

The touch sensors of the epidermis are the ________.

a. keratinocytes
b. tactile cells
c. epidermal dendritic cells
d. melanocytes

- **Active Lecture Questions** (for use with or without clickers) that stimulate effective classroom discussions and check comprehension

BONUS!
IRDVD includes Practice Anatomy Lab 2.0 Instructor Resource DVD

PAL IRDVD includes customizable images from PAL 2.0 in JPEG and PowerPoint format. PowerPoint slides also include embedded links to relevant animations and PRS-enabled active lecture questions for use with or without clickers. Quizzes and lab practical are available in Microsoft Word and Computerized Test Bank formats.

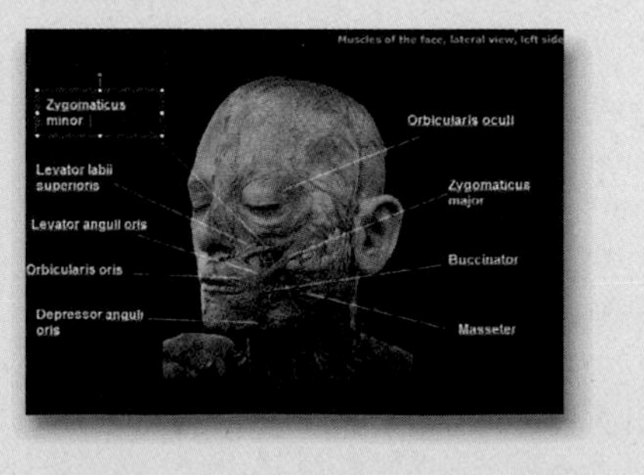

(For a description of PAL, please see page xix.)

myA&P™ Website now includes everything students need to practice, review, and self-assess for both the A&P lecture *and* lab.

- **NEW! *Get Ready for A&P,* Second Edition** helps students prepare for the A&P course through Pre-Tests, Post-Tests with Study Plans, tutorials, animations, activities, and an integrated E-Book.
- Chapter-specific resources include Chapter Quizzes and brand new Chapter Practice Tests; Games and Activities, featuring Art Labeling Exercises, Memory Games, and Crossword Puzzles; Histology, Bone, and Muscle Reviews; Flashcards; a Glossary; and more!
- ***Interactive Physiology*® 10-System Suite** includes a new module on the immune system.
- **NEW! Practice Anatomy Lab™ 2.0** is an indispensable virtual anatomy practice tool that gives students 24/7 access to the most widely used lab specimens (includes self-study quizzes and gradable lab practicals).
- **PhysioEx™ 8.0** supplements traditional wet labs safely and cost-effectively (includes gradable quizzes and printable review sheets).
- **Instructor Gradebook** allows instructors to track student assessment.
- **Instructor Resource Section** includes IP Exercise Sheet Answer Key, and items from the IRDVD, including JPEG images (labeled and unlabeled sets), Label-Edit Art and Step-Edit Art, Active Lecture Questions, and Quiz Show Game Questions.

Course Management

CourseCompass

This nationally hosted, dynamic, interactive online course management system is powered by Blackboard, the leading platform for Internet-based learning tools. This easy-to-use and customizable program enables professors to tailor content to meet individual course needs. Includes all of the content from myA&P™.

WebCT and Blackboard

These open-access cartridges are loaded with rich content, including assessment items, self-study quizzes, case studies, a histology tutorial, reference tools, and hundreds of fun games, animations, and anatomy labeling exercises.

NEW! New assessment items in the course management system of your choice, including CourseCompass, Blackboard, WebCT, and others. In addition to the Gradeable Quizzes from the myA&P Website and the Test Bank, you will now have access to Instructor Test Item assessments for:

- *Get Ready for A&P* (Diagnostic and Cumulative Tests and Chapter Pre- and Post-Tests).
- *Interactive Physiology*®
- *PhysioEx*™ 8.0
- Quizzes and lab practicals from Practice Anatomy Lab™ 2.0, including images and questions not available in the student product. Instructors can modify the questions to reflect the content they want their students to be quizzed and tested on.
- Post-Test versions of the new Chapter Practice Tests on the myA&P Website.

Instructor Guide to Text and Media

(0-321-55876-6)

This fully revised guide includes detailed objectives, lecture outlines, activities, online media resources, answers to end-of-chapter questions, and *Interactive Physiology*® exercise sheets and answer key. All the illustrations from the text are indexed as thumbnails in the Visual Resource Guide so you can easily locate and make the best use of the available media.

Printed Test Bank

(0-321-55884-7)

With more than 3600 test questions, this Test Bank has been updated with new and revised questions that cover all major topics at a range of difficulty levels. All questions in the printed Test Bank are available in Word and TestGen formats on the IRDVD. Both electronic options are cross-platform and allow instructors to easily generate and customize tests.

Transparency Acetates

(0-321-55888-X)

This package includes all illustrations, photos, and tables from the text—approximately 800 images—with labels that have been enlarged for easy viewing in the classroom or lecture hall.

Human Anatomy & Physiology Laboratory Manuals

Elaine N. Marieb's three widely used and acclaimed laboratory manuals complement this textbook and are designed to meet the varying needs of most laboratory courses: *Human Anatomy & Physiology Laboratory Manual: Cat Version,* Ninth Edition Update; *Main Version,* Eighth Edition Update; and *Pig Version*, Ninth Edition Update. Included with each laboratory manual is the PhysioEx™ 8.0 CD-ROM and a registration code for online access. PhysioEx™ 8.0 features 12 experiments and a Histology Tutorial.

Supplements for the Student

NEW! Practice Anatomy Lab™ 2.0 CD-ROM

(0-321-54725-X)

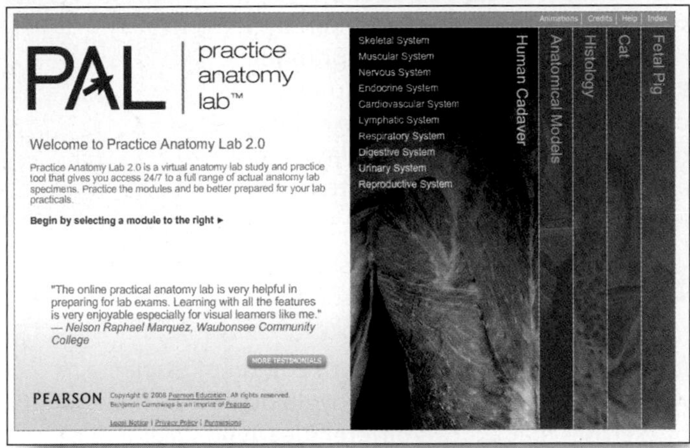

Practice Anatomy Lab™ 2.0 is a virtual anatomy study and practice tool that gives you 24/7 access to a full range of actual lab specimens, including:

- Human cadaver
- Anatomical models
- Histology slides
- Cat dissections
- Fetal pig dissections

Each module includes hundreds of images as well as interactive tools for reviewing the specimens, learning and hearing the names of anatomical structures, seeing animations, and taking multiple choice quizzes and fill-in-the-blank lab practical exams.

PAL 2.0 features include:

- All-new Human Cadaver module
- Fully rotatable human skull and 17 other rotatable skeletal structures
- 3-D animations of origins, insertions, actions, and innervations of over 65 individual muscles
- Greatly expanded Histology module

Interactive Physiology® 10-System Suite

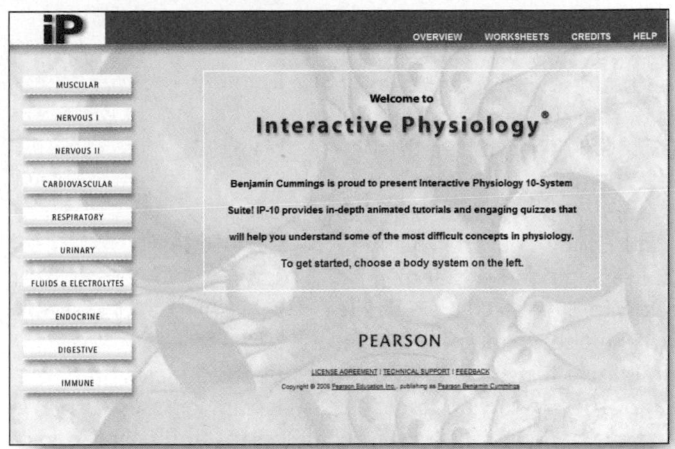

Interactive Physiology® will give you the help you need to grasp some of the most difficult concepts in A&P. This award-winning tutorial system features ten modules containing in-depth, fully narrated, animated tutorials and engaging quizzes covering key physiological processes and concepts. *Interactive Physiology®* is a highly effective program that provides the tools you need to advance beyond simple memorization to a genuine understanding of the most difficult concepts in A&P.

Modules

- Muscular System
- Nervous System I
- Nervous System II
- Cardiovascular System
- Respiratory System
- Urinary System
- Fluids & Electrolytes
- Endocrine System
- Digestive System
- **NEW!** Immune System

myA&P™

Please see How to Use myA&P for a description (p. x).

Get Ready for A&P, Second Edition

(0-321-55695-X)

This book and online component was created to help you be better prepared for your A&P course. This hands-on book helps you get up to speed in your knowledge of basic study skills, math review, basic chemistry, cell biology, anatomical terminology, and the human body. Features include pre-tests, guided explanations followed by interactive quizzes and exercises, and end-of-chapter cumulative tests. The online component includes a gradable diagnostic pre-test and post-test, self-study quizzes with feedback, animations and links, a glossary, and flashcards. It is available via myA&P™.

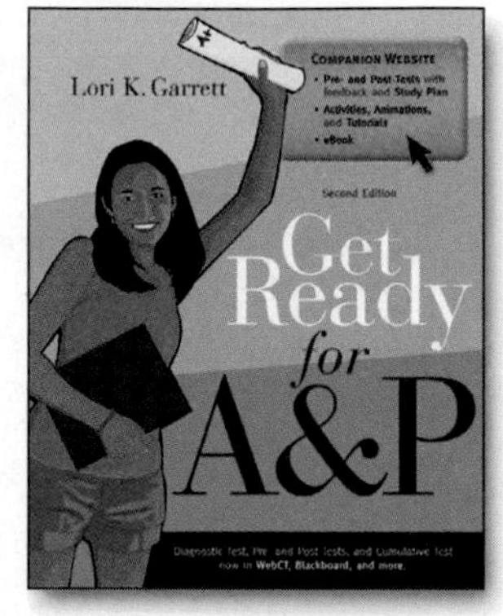

New to the Second Edition

- New topics have been added, including coverage of pH, energy, and meiosis, as well as tips on how to minimize anxiety surrounding tests, and more.
- A more robust Companion Website includes new activities and tutorials on key topics and new myeBook content.
- A new preface for instructors explains how to use the book.
- All assessments are now available in course management platforms, including WebCT, Blackboard, and CourseCompass™. Separate Instructor Test Item versions of the Diagnostic Test, Cumulative Test, and the chapter Pre- and Post-Tests can now be easily imported into these course management systems.

A Brief Atlas of the Human Body, Second Edition

This full-color atlas is bundled with every new copy of the text, and includes 107 bone and 47 soft-tissue photographs with easy-to-read labels. This new edition of the atlas contains a brand-new, comprehensive histology photomicrograph section with more than 50 slides of basic tissue and organ systems. Featuring photos taken by renowned biomedical photographer Ralph Hutchings, this high-quality photographic atlas makes an excellent resource for the classroom and laboratory, and is referenced in appropriate figure legends throughout the text.

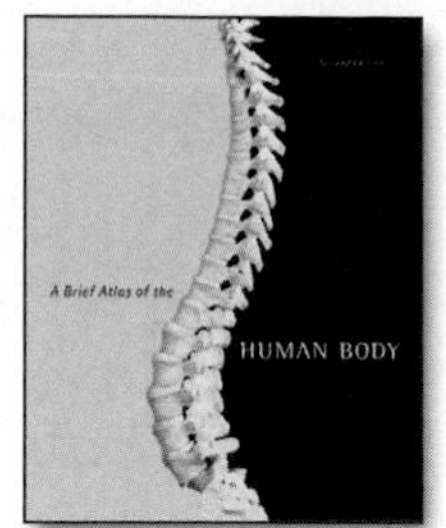

Study Guide

(0-321-55873-1)

Revised to accompany the Eighth Edition of *Human Anatomy & Physiology,* the study guide offers a wide variety of exercises that address different learning styles and call on students to develop their critical-thinking abilities. The three major sections, Building the Framework, Challenging Yourself, and Covering All Your Bases, help students build a base of knowledge using recall, reasoning, and imagination that can be applied to solving problems in both clinical and nonclinical situations.

Additional Supplements Available from Benjamin Cummings

Anatomy & Physiology Coloring Workbook: A Complete Study Guide, Ninth Edition By Elaine N. Marieb

The Physiology Coloring Book, Second Edition By Wynn Kapit, Robert I. Macey, Esmail Meisami

The Anatomy Coloring Book, Third Edition By Wynn Kapit and Lawrence M. Elson

Human Cadaver Dissection Videos By Rose Leigh Vines, et al.

Student Video Series for Human Anatomy & Physiology, Volume 1

Student Video Series for Human Anatomy & Physiology, Volume 2

Acknowledgments

Harmonizing "Oh, happy day," we could hardly express our joy at coming to the end of this Eighth Edition update. This revision, like the ones that preceded it, is a major one. It has consumed most of our day and night hours for nearly two years. Text has been streamlined, rewritten for currency, or reorganized. As usual, it's been difficult to keep up to date in the very broad field of anatomy and physiology, and then a wrenching ordeal to decide how much of it to use in the new edition. We want our students to be informed intellectually, but how rich a diet that requires is a persistent question.

Though text revision is demanding, our task there was nothing compared to the Herculean effort of a whole new art program for the book. Literally every figure in the book has been touched in some way. These modifications ranged from light touches such as color changes or text editing for brevity, to art modifications for enhanced three-dimensionality (particularly in anatomy pieces), revisions for greater clarity, or a whole new conceptualization and rendering for better pedagogy. We are really excited about an entirely new feature—the Focus Feature. These one- to two-page displays showcase major concepts that students traditionally have trouble with, such as generation of an action potential, excitation-contraction coupling, or G protein mechanisms. In most cases the feature uses text that concisely states each step, accompanied by exceptionally instructive and evocative art. We are excited about this feature and hope that you will be too.

The sponsoring editor for this edition, Serina Beauparlant, has supported every aspect of this revision—sometimes too much! Serina is insightful and dedicated to producing the best educational product possible (both text and multimedia). She is a true human dynamo. Frank Ruggirello, publisher for this edition, has backed Serina up in the most effective way possible (with $). Thank you Frank. Project editor for this text is again the conscientious and competent Sabrina Larson. Wrestling with schedule deadlines, picking up manuscript loose ends, and trying to make overwhelmed authors work more quickly have made her life these many months a nightmare, we are sure. But she survived and still speaks to us. Derek Perrigo, our marketing manager, has efficiently kept us abreast of the pulse of the marketplace—keeping in touch with professors and students and providing feedback on what they do or don't like about the text and media products. Stacey Weinberger's expertise as manufacturing buyer has served us well. A resounding dollop of gratitude also goes to those on the editorial team with whom we have had little or no personal contact but who have provided valuable services for the revision: Erik Fortier, media producer, and Nicole Graziano, assistant editor, for the Eighth Edition.

We also want to thank the following reviewers who provided us with their expertise and constructive criticism to improve the art and text presentation. Their input resulted in the continued excellence and accuracy of this text.

Kurt Albertine, *University of Utah Medical School*
C. Thomas G. Appleton, *University of Western Ontario*
Barbie Baker, *Florida Community College–Jacksonville*
Sherry Bowen, *Indian River College*
Virginia Brooks, *Oregon Health and Science University*
Bruce Buttler, *Canadian University College*
Mary Beth Dawson, *Kingsborough Community College*
Trevor Day, *University of Calgary*
Randall Fameree, *Athens Technical College*
Patty Finkenstadt, *Phoenix College*
Lynn Gargan, *Tarrant County Community College*
Cynthia Gill, *Hampshire College*
William Hoover, *Bunker Hill Community College*
Mark Hubley, *Prince George's Community College*
Jody Johnson, *Arapahoe Community College*
John C. Koch, *John Tyler Community College*
John Lepri, *University of North Carolina–Greensboro*
Jane Marone, *University of Illinois at Chicago*
Justin Moore, *American River College*
Karen Payne, *Chattanooga State Technical Community College*
Louise Petroka, *Gateway Community College*
Brandon Poe, *Springfield Tech Community College*
David Quadagno, *Florida State University–Emeritus*
Saeed Rahmanian, *Roane State Community College*
Charles Roselli, *Oregon Health and Science University*
Pamela Siergiej, *Roane State Community College*
Tom Swensen, *Ithaca College*
Carol Veil, *Anne Arundel Community College*
Amy Way, *Lock Haven University*
Kelly Young, *California State University, Long Beach*

Additionally, we want to acknowledge Katja's colleagues at Mount Royal College (Trevor Day, Janice Meeking, Izak Paul, Michael Pollock, and Ruth Pickett-Seltner) for stimulating discussions of the text, which they embraced as their own cause, providing feedback on changes and options. We are also grateful to Yvonne Baptiste-Szymanski, Niagara County Community College, Mark Taylor, Baylor University, and Linda Porter, Midlands Technical College, for conducting student user diaries with their classes and gathering valuable feedback directly from our student users.

Once again Wendy Earl Productions handled the production of this text. As always, Wendy Earl, the managing editor, selects highly qualified people to work with her. Michele Mangelli, production supervisor extraordinaire, guided us through the schedule with an expert hand. Her efficient management style and excellent problem-solving skills headed off many a storm, which the whole team appreciated. Laura Southworth, the art development manager, was charged with being the developmental editor for all the art, ensuring the consistency of the art program to encourage automatic learning, helping to conceptualize and construct the Focus Features and all new pieces. All we had to say to Laura is "I don't like this piece of art; can't we do something to reconfigure it for more interest . . . more clarity . . . ?" Two days later there would be a mailer with a great new conceptualization of that figure. Laura is an enthusiastic, can-do woman and we love her! Recognition for a job well done is due Kristin Piljay, photo researcher, and David Novak, art coordinator. Jim Perkins helped us to reorganize key flowcharts in the book, for which we thank him. The talented artists at Imagineering and at Electronic Publishing Services generated the new art pieces conceived for this edition, and made changes requested on existing art. It is their art that puts the zing in the book. We'd also like to thank Precison Graphics who assisted in establishing the new art style for this edition. The new text design is the handiwork of Mark Ong, and the exciting cover image is a stunning photo of 16-time Olympic medalist, Michael Phelps. As always, a monumental thank-you goes to Anita Wagner, ultra-conscientious copyeditor. She edits with a light hand, all the while double-checking all statistics, drug names, and other technical data. Our text developmental editor, Annie Reid, did a great job helping us to rid the book of redundancies and unclear text. We are also fortunate to have Martha Ghent returning as proofreader for this edition. We thank Izak Paul for meticulously proofreading each chapter as well. In addition, Michael Wiley from the University of Toronto checked many pieces of anatomy art for accuracy. As has been the practice with this text, Aptara assembled the final pages with their customary expertise led by their extremely talented and organized Los Angeles Project Manager, Sandie Sigrist Allaway.

Once again, Dr. Marieb's husband, Harvey Howell, served as a sounding board for some of her ideas, manned the copy machine, and ran the manuscript to the FedEx box daily with nary a complaint during the unbelievably busy days. Thanks also to Katja's husband, Dr. Lawrence W. Haynes, who as a fellow physiologist has provided invaluable assistance to her during the course of the revision. She also thanks her children, Eric and Stefan Haynes, who are an inspiration and a joy.

Well, our tenure on this edition is over, but there will be another edition three years hence. We would really appreciate hearing from you concerning your opinion—suggestions and constructive criticisms—of this text. It is this type of feedback that provides the basis of each revision, and underwrites its improvement.

Elaine N. Marieb

Katja Hoehn

Elaine N. Marieb and Katja Hoehn
Anatomy and Physiology
Benjamin Cummings Science
1301 Sansome Street
San Francisco, CA 94111

Brief Contents

Contents

9 Muscles and Muscle Tissue 275

10 The Muscular System 320

UNIT THREE Regulation and Integration of the Body

11 Fundamentals of the Nervous System and Nervous Tissue 385

1

The Human Body: An Orientation

Welcome to the study of one of the most fascinating subjects possible—your own body. Such a study is not only highly personal, but timely as well. We get news of some medical advance almost daily. To appreciate emerging discoveries in genetic engineering, to understand new techniques for detecting and treating disease, and to make use of published facts on how to stay healthy, you'll find it helpful to learn about the workings of your body. If you are preparing for a career in the health sciences, the study of anatomy and physiology has added rewards because it provides the foundation needed to support your clinical experiences.

In this chapter we define and contrast anatomy and physiology and discuss how the human body is organized. Then we review needs and functional processes common to all living organisms. Three essential concepts—*the complementarity of structure and function, the hierarchy of structural organization,* and *homeostasis*—will unify and form the bedrock for your study of the human body. The final section of the chapter deals with the language of anatomy—terminology that anatomists use to describe the body or its parts.

An Overview of Anatomy and Physiology

- ▶ Define anatomy and physiology and describe their subdivisions.
- ▶ Explain the principle of complementarity.

Two complementary branches of science—anatomy and physiology—provide the concepts that help us to understand the human body. **Anatomy** studies the *structure* of body parts and their relationships to one another. Anatomy has a certain appeal because it is concrete. Body structures can be seen, felt, and examined closely. You don't need to imagine what they look like.

Physiology concerns the *function* of the body, in other words, how the body parts work and carry out their life-sustaining activities. When all is said and done, physiology is explainable only in terms of the underlying anatomy.

To simplify the study of the body, when we refer to body structures and/or physiological values (body temperature, heart rate, and the like), we will assume that we are talking about a healthy young (22-year-old) male weighing about 155 lb (the *reference man*) or a healthy young female weighing about 125 lb (the *reference woman*).

Topics of Anatomy

Anatomy is a broad field with many subdivisions, each providing enough information to be a course in itself. **Gross**, or **macroscopic**, **anatomy** is the study of large body structures visible to the naked eye, such as the heart, lungs, and kidneys. Indeed, the term *anatomy* (derived from the Greek words meaning "to cut apart") relates most closely to gross anatomy because in such studies preserved animals or their organs are dissected (cut up) to be examined.

Gross anatomy can be approached in different ways. In **regional anatomy**, all the structures (muscles, bones, blood vessels, nerves, etc.) in a particular region of the body, such as the abdomen or leg, are examined at the same time.

In **systemic anatomy** (sis-tem′ik),* body structure is studied system by system. For example, when studying the cardiovascular system, you would examine the heart and the blood vessels of the entire body.

*For the pronunciation guide rules, see the Preface to the Student.

Another subdivision of gross anatomy is **surface anatomy**, the study of internal structures as they relate to the overlying skin surface. You use surface anatomy when you identify the bulging muscles beneath a bodybuilder's skin, and clinicians use it to locate appropriate blood vessels in which to feel pulses and draw blood.

Microscopic anatomy deals with structures too small to be seen with the naked eye. For most such studies, exceedingly thin slices of body tissues are stained and mounted on glass slides to be examined under the microscope. Subdivisions of microscopic anatomy include **cytology** (si-tol′o-je), which considers the cells of the body, and **histology** (his-tol′o-je), the study of tissues.

Developmental anatomy traces structural changes that occur in the body throughout the life span. **Embryology** (em″bre-ol′o-je), a subdivision of developmental anatomy, concerns developmental changes that occur before birth.

Some highly specialized branches of anatomy are used primarily for medical diagnosis and scientific research. For example, *pathological anatomy* studies structural changes caused by disease. *Radiographic anatomy* studies internal structures as visualized by X-ray images or specialized scanning procedures.

Subjects of interest to anatomists range from easily seen structures down to the smallest molecule. In *molecular biology*, for example, the structure of biological molecules (chemical substances) is investigated. Molecular biology is actually a separate branch of biology, but it falls under the anatomy umbrella when we push anatomical studies to the subcellular level.

One essential tool for studying anatomy is a mastery of anatomical terminology. Others are observation, manipulation, and, in a living person, *palpation* (feeling organs with your hands) and *auscultation* (listening to organ sounds with a stethoscope). A simple example illustrates how some of these tools work together in an anatomical study.

Let's assume that your topic is freely movable joints of the body. In the laboratory, you will be able to *observe* an animal joint, noting how its parts fit together. You can work the joint (*manipulate* it) to determine its range of motion. Using *anatomical terminology*, you can name its parts and describe how they are related so that other students (and your instructor) will have no trouble understanding you. The list of word roots (at the back of the book) and the glossary will help you with this special vocabulary.

Although you will make most of your observations with the naked eye or with the help of a microscope, medical technology has developed a number of sophisticated tools that can peer into the body without disrupting it. Read about these exciting medical imaging techniques in *A Closer Look* on pp. 18–19.

Topics of Physiology

Like anatomy, physiology has many subdivisions. Most of them consider the operation of specific organ systems. For example, **renal physiology** concerns kidney function and urine production. **Neurophysiology** explains the workings of the nervous system. **Cardiovascular physiology** examines the operation of the heart and blood vessels. While anatomy provides us with a static image of the body's architecture, physiology reveals the body's dynamic and animated workings.

Physiology often focuses on events at the cellular or molecular level. This is because the body's abilities depend on those of its individual cells, and cells' abilities ultimately depend on the chemical reactions that go on within them. Physiology also rests on principles of physics, which help to explain electrical currents, blood pressure, and the way muscles use bones to cause body movements, among other things. We present basic chemical and physical principles in Chapter 2 and throughout the book as needed to explain physiological topics.

Complementarity of Structure and Function

Although it is possible to study anatomy and physiology individually, they are really inseparable because function always reflects structure. That is, what a structure can do depends on its specific form. This key concept is called the **principle of complementarity of structure and function.**

For example, bones can support and protect body organs because they contain hard mineral deposits. Blood flows in one direction through the heart because the heart has valves that prevent backflow. Throughout this book, we accompany a description of a structure's anatomy with an explanation of its function, and we emphasize structural characteristics contributing to that function.

CHECK YOUR UNDERSTANDING

1. In what way does physiology depend on anatomy?
2. Would you be studying anatomy or physiology if you investigated how muscles shorten? If you explored the location of the lungs in the body?

For answers, see Appendix G.

Levels of Structural Organization

- Name the different levels of structural organization that make up the human body, and explain their relationships.
- List the 11 organ systems of the body, identify their components, and briefly explain the major function(s) of each system.

The human body has many levels of structural organization **(Figure 1.1)**. The simplest level of the structural hierarchy is the **chemical level**, which we study in Chapter 2. At this level, *atoms*, tiny building blocks of matter, combine to form *molecules* such as water and proteins. Molecules, in turn, associate in specific ways to form *organelles*, basic components of the microscopic cells. *Cells* are the smallest units of living things. We examine the **cellular level** in Chapter 3. All cells have some common functions, but individual cells vary widely in size and shape, reflecting their unique functions in the body.

The simplest living creatures are single cells, but in complex organisms such as human beings, the hierarchy continues on to the **tissue level**. *Tissues* are groups of similar cells that have a common function. The four basic tissue types in the human body are epithelium, muscle, connective tissue, and nervous tissue.

Each tissue type has a characteristic role in the body, which we explore in Chapter 4. Briefly, epithelium covers the body surface and lines its cavities. Muscle provides movement. Connective tissue supports and protects body organs. Nervous tissue provides a means of rapid internal communication by transmitting electrical impulses.

An *organ* is a discrete structure composed of at least two tissue types (four is more common) that performs a specific function for the body. The liver, the brain, and a blood vessel are very different from the stomach, but each is an organ. You can think of each organ of the body as a specialized functional center responsible for a necessary activity that no other organ can perform.

At the **organ level**, extremely complex functions become possible. Let's take the stomach for an example. Its lining is an epithelium that produces digestive juices. The bulk of its wall is muscle, which churns and mixes stomach contents (food). Its connective tissue reinforces the soft muscular walls. Its nerve fibers increase digestive activity by stimulating the muscle to contract more vigorously and the glands to secrete more digestive juices.

The next level of organization is the **organ system**. Organs that work together to accomplish a common purpose make up an *organ system*. For example, the heart and blood vessels of the cardiovascular system circulate blood continuously to carry oxygen and nutrients to all body cells. Besides the cardiovascular system, the other organ systems of the body are the integumentary, skeletal, muscular, nervous, endocrine, lymphatic, respiratory, digestive, urinary, and reproductive systems. (Note that the immune system is closely associated with the lymphatic system.) Look ahead to Figure 1.3 on pp. 6 and 7 for an overview of the 11 organ systems, which we discuss in the next section and study in more detail in Units 2–5.

The highest level of organization is the *organism*, the living human being. The **organismal level** represents the sum total of all structural levels working together to keep us alive.

CHECK YOUR UNDERSTANDING

3. What level of structural organization is typical of a cytologist's field of study?
4. What is the correct structural order for the following terms: tissue, organism, organ, cell?
5. Which organ system includes the bones and cartilages? Which includes the nasal cavity, lungs, and trachea?

For answers, see Appendix G.

1

1

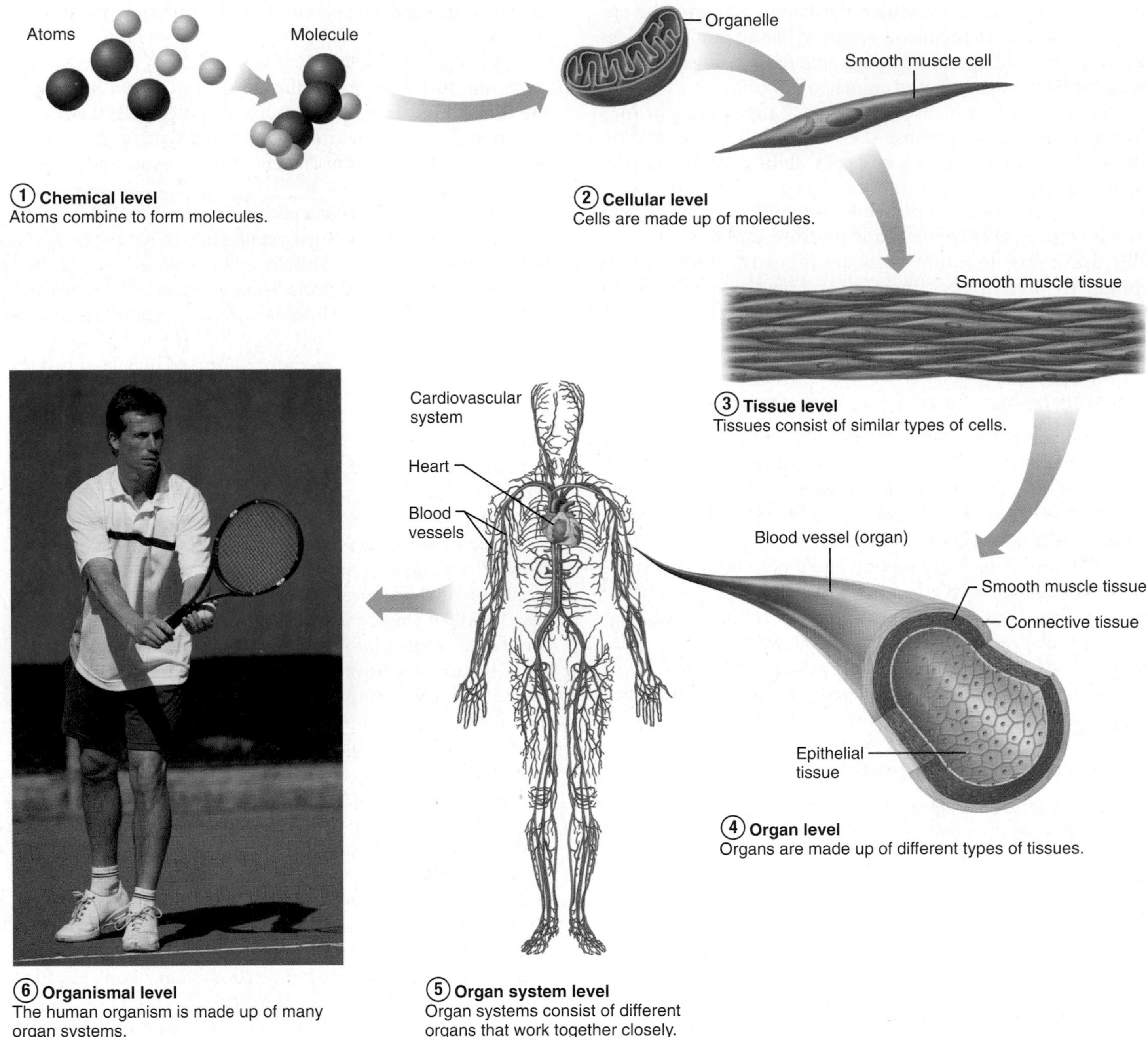
Atoms
Molecule
Organelle
Smooth muscle cell
① Chemical level
Atoms combine to form molecules.
② Cellular level
Cells are made up of molecules.
Smooth muscle tissue
③ Tissue level
Tissues consist of similar types of cells.
Cardiovascular system
Heart
Blood vessels
Blood vessel (organ)
Smooth muscle tissue
Connective tissue
Epithelial tissue
④ Organ level
Organs are made up of different types of tissues.
⑥ Organismal level
The human organism is made up of many organ systems.
⑤ Organ system level
Organ systems consist of different organs that work together closely.

Figure 1.1 Levels of structural organization. Components of the cardiovascular system are used to illustrate the levels of structural organization in a human being.

Maintaining Life

- List the functional characteristics necessary to maintain life in humans.
- List the survival needs of the body.

Necessary Life Functions

Now that you know the structural levels of the human body, the question that naturally follows is: What does this highly organized human body do?

Like all complex animals, humans maintain their boundaries, move, respond to environmental changes, take in and digest nutrients, carry out metabolism, dispose of wastes, reproduce themselves, and grow. We will introduce these necessary life functions here and discuss them in more detail in later chapters.

We cannot emphasize too strongly that all body cells are interdependent. This interdependence is due to the fact that humans are multicellular organisms and our vital body functions are parceled out among different organ systems. Organ systems, in turn, work cooperatively to promote the well-being of the entire body. This theme is repeated throughout the book. **Figure 1.2** identifies some of the organ systems making major contributions to necessary life functions. Also, as you read this section, check **Figure 1.3** for more detailed descriptions of the body's organ systems.

Maintaining Boundaries

Every living organism must **maintain its boundaries** so that its internal environment (its inside) remains distinct from the external environment surrounding it (its outside). In single-celled organisms, the external boundary is a limiting membrane that encloses its contents and lets in needed substances while restricting entry of potentially damaging or unnecessary substances. Similarly, all the cells of our body are surrounded by a selectively permeable membrane.

Additionally, the body as a whole is enclosed and protected by the integumentary system, or skin (Figure 1.3a). This system protects our internal organs from drying out (a fatal change), bacteria, and the damaging effects of heat, sunlight, and an unbelievable number of chemicals in the external environment.

Movement

Movement includes the activities promoted by the muscular system, such as propelling ourselves from one place to another by running or swimming, and manipulating the external environment with our nimble fingers (Figure 1.3c). The skeletal system provides the bony framework that the muscles pull on as they work (Figure 1.3b). Movement also occurs when substances such as blood, foodstuffs, and urine are propelled through internal organs of the cardiovascular, digestive, and urinary systems, respectively. On the cellular level, the muscle cell's ability to move by shortening is more precisely called **contractility**.

Responsiveness

Responsiveness, or **irritability**, is the ability to sense changes (which serve as stimuli) in the environment and then respond to them. For example, if you cut your hand on broken glass, a withdrawal reflex occurs—you involuntarily pull your hand away from the painful stimulus (the broken glass). You don't have to think about it—it just happens! Likewise, when carbon dioxide in your blood rises to dangerously high levels, chemical sensors respond by sending messages to brain centers controlling respiration, and you breathe more rapidly.

Because nerve cells are highly irritable and communicate rapidly with each other via electrical impulses, the nervous system is most involved with responsiveness (Figure 1.3d). However, all body cells are irritable to some extent.

Digestion

Digestion is the breaking down of ingested foodstuffs to simple molecules that can be absorbed into the blood. The nutrient-rich blood is then distributed to all body cells by the cardiovascular system. In a simple, one-celled organism such as an amoeba, the cell itself is the "digestion factory," but in the multicellular human body, the digestive system performs this function for the entire body (Figure 1.3i).

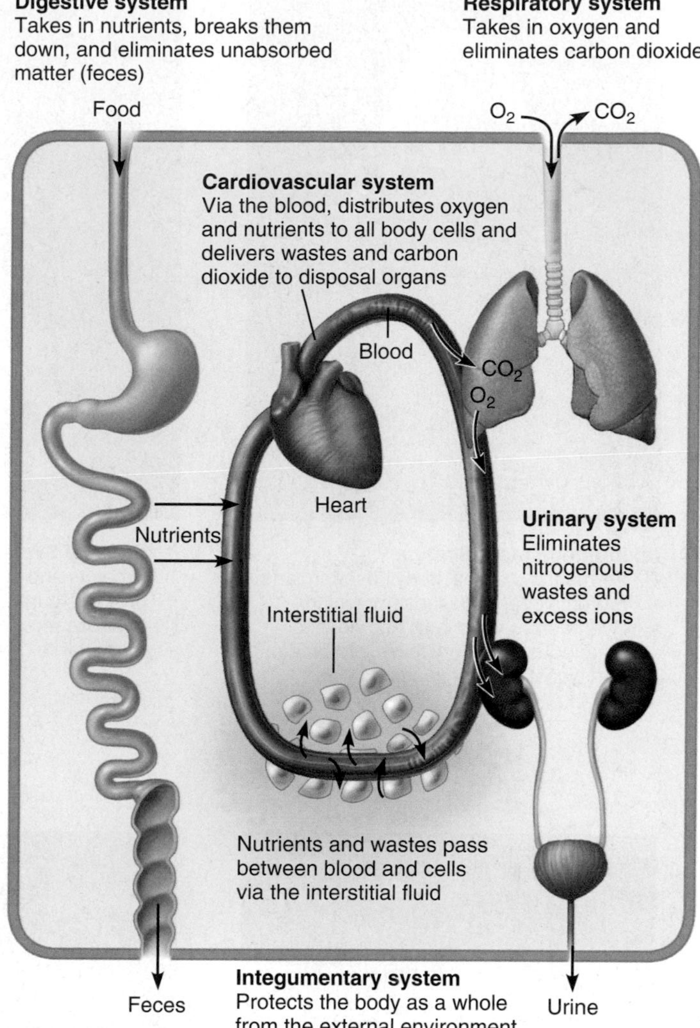
Digestive system
Takes in nutrients, breaks them down, and eliminates unabsorbed matter (feces)
Respiratory system
Takes in oxygen and eliminates carbon dioxide
Food
O_2
CO_2
Cardiovascular system
Via the blood, distributes oxygen and nutrients to all body cells and delivers wastes and carbon dioxide to disposal organs
Blood
CO_2
O_2
Heart
Nutrients
Urinary system
Eliminates nitrogenous wastes and excess ions
Interstitial fluid
Nutrients and wastes pass between blood and cells via the interstitial fluid
Feces
Integumentary system
Protects the body as a whole from the external environment
Urine

Figure 1.2 Examples of interrelationships among body organ systems.

Metabolism

Metabolism (mĕ-tab′o-lizm; "a state of change") is a broad term that includes all chemical reactions that occur within body cells. It includes breaking down substances into their simpler building blocks (more specifically, the process of *catabolism*), synthesizing more complex cellular structures from simpler substances (*anabolism*), and using nutrients and oxygen to produce (via *cellular respiration*) ATP, the energy-rich molecules that power cellular activities. Metabolism depends on the digestive and respiratory systems to make nutrients and oxygen available to the blood and on the cardiovascular system to distribute them throughout the body (Figure 1.3i, h, and f, respectively). Metabolism is regulated largely by hormones secreted by endocrine system glands (Figure 1.3e).

1

1

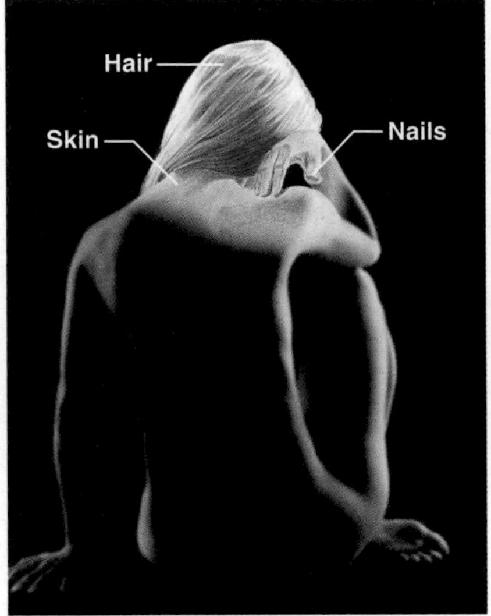

(a) Integumentary System
Forms the external body covering, and protects deeper tissues from injury. Synthesizes vitamin D, and houses cutaneous (pain, pressure, etc.) receptors and sweat and oil glands.

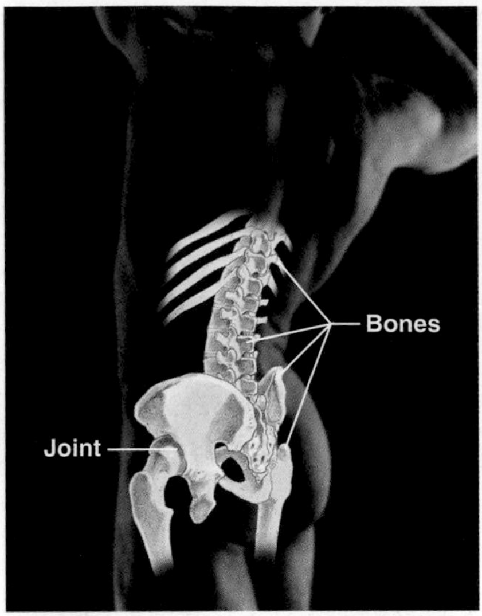

(b) Skeletal System
Protects and supports body organs, and provides a framework the muscles use to cause movement. Blood cells are formed within bones. Bones store minerals.

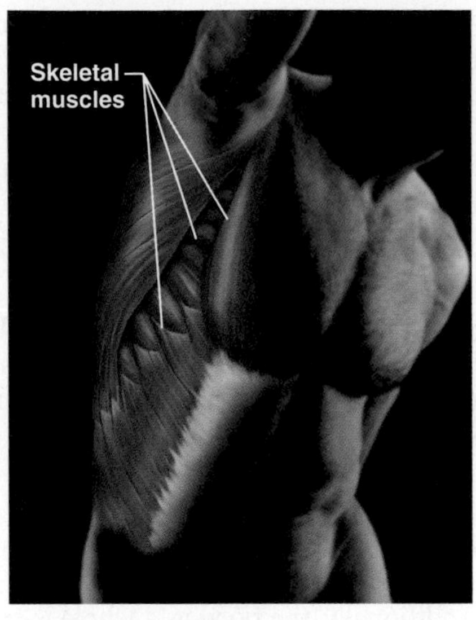

(c) Muscular System
Allows manipulation of the environment, locomotion, and facial expression. Maintains posture, and produces heat.

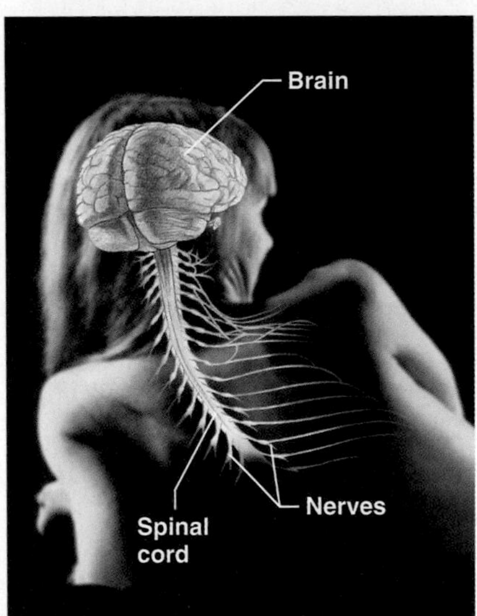

(d) Nervous System
As the fast-acting control system of the body, it responds to internal and external changes by activating appropriate muscles and glands.

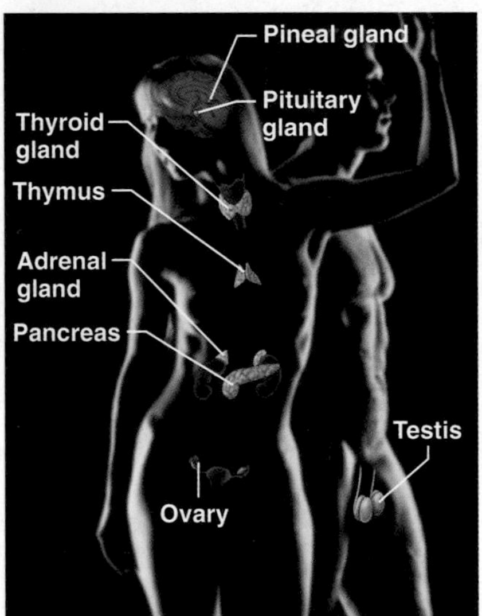

(e) Endocrine System
Glands secrete hormones that regulate processes such as growth, reproduction, and nutrient use (metabolism) by body cells.

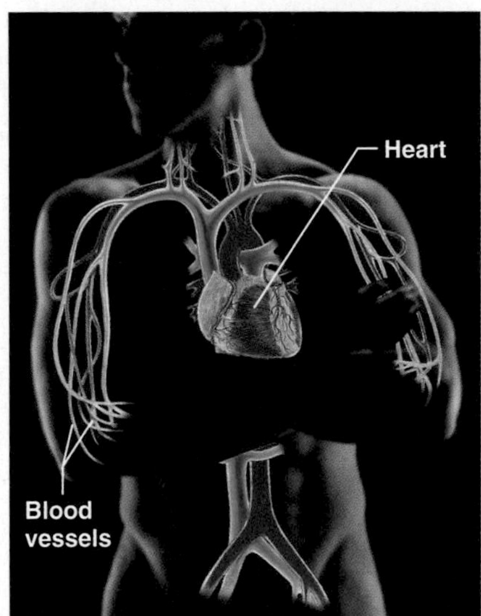

(f) Cardiovascular System
Blood vessels transport blood, which carries oxygen, carbon dioxide, nutrients, wastes, etc. The heart pumps blood.

Figure 1.3 The body's organ systems and their major functions.

Excretion

Excretion is the process of removing wastes, or *excreta* (ek-skre′tah), from the body. If the body is to operate as we expect it to, it must get rid of nonuseful substances produced during digestion and metabolism.

Several organ systems participate in excretion. For example, the digestive system rids the body of indigestible food residues in feces, and the urinary system disposes of nitrogen-containing metabolic wastes, such as urea, in urine (Figure 1.3i and j). Carbon dioxide, a by-product of cellular respiration, is carried in

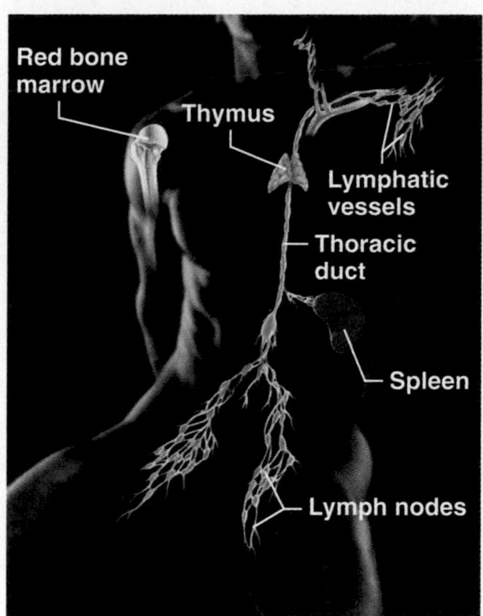

(g) Lymphatic System/Immunity
Picks up fluid leaked from blood vessels and returns it to blood. Disposes of debris in the lymphatic stream. Houses white blood cells (lymphocytes) involved in immunity. The immune response mounts the attack against foreign substances within the body.

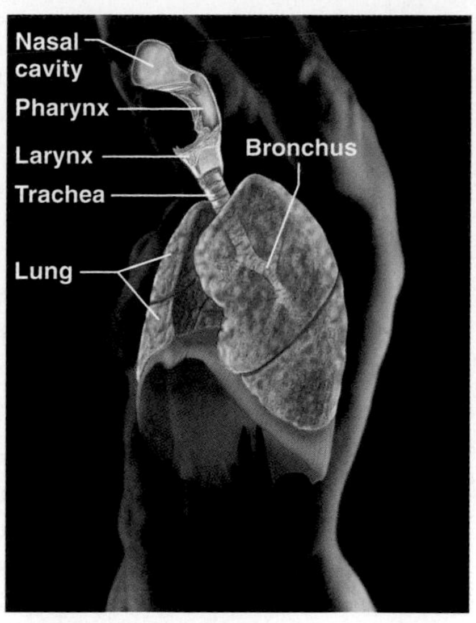

(h) Respiratory System
Keeps blood constantly supplied with oxygen and removes carbon dioxide. The gaseous exchanges occur through the walls of the air sacs of the lungs.

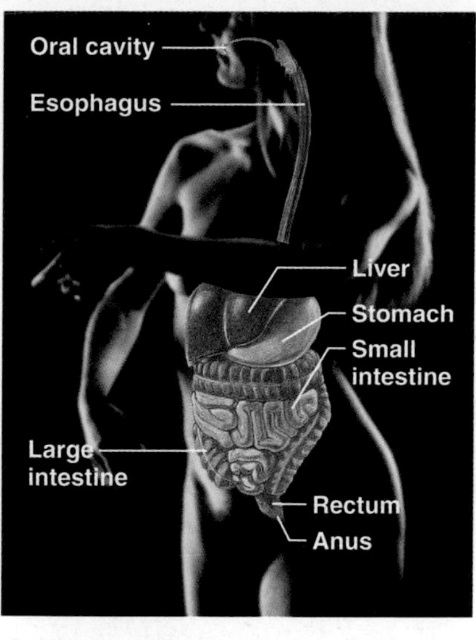

(i) Digestive System
Breaks down food into absorbable units that enter the blood for distribution to body cells. Indigestible foodstuffs are eliminated as feces.

1

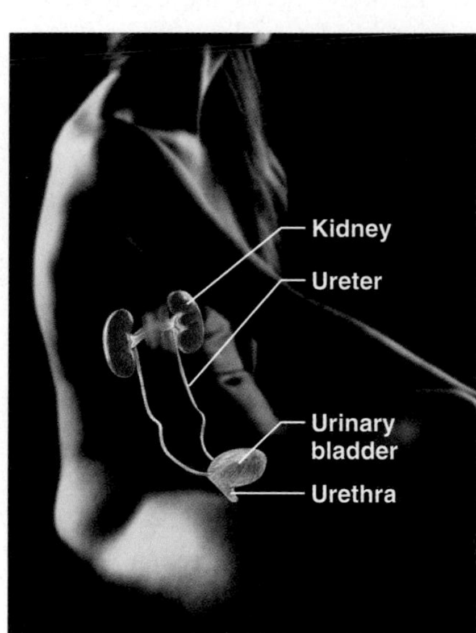

(j) Urinary System
Eliminates nitrogenous wastes from the body. Regulates water, electrolyte and acid-base balance of the blood.

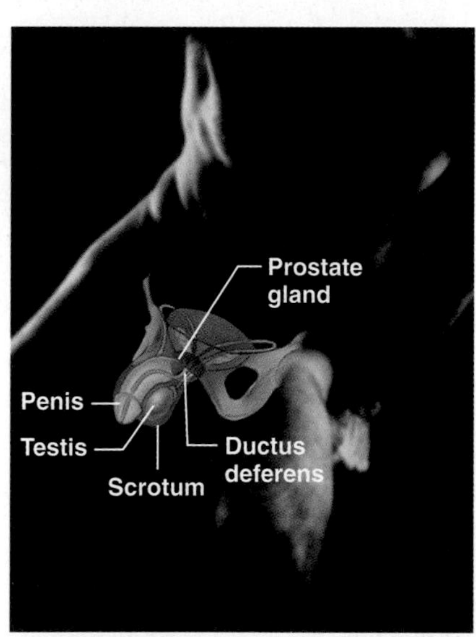

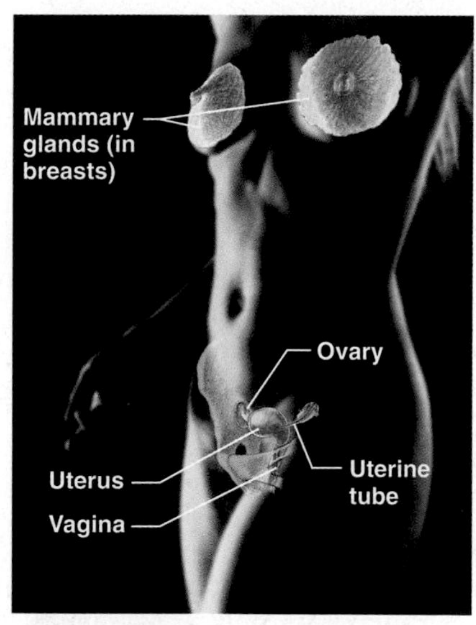

(k) Male Reproductive System **(l) Female Reproductive System**
Overall function is production of offspring. Testes produce sperm and male sex hormone, and male ducts and glands aid in delivery of sperm to the female reproductive tract. Ovaries produce eggs and female sex hormones. The remaining female structures serve as sites for fertilization and development of the fetus. Mammary glands of female breasts produce milk to nourish the newborn.

the blood to the lungs, where it leaves the body in exhaled air (Figure 1.3h).

Reproduction

Reproduction occurs at the cellular and the organismal level. In cellular reproduction, the original cell divides, producing two identical daughter cells that may then be used for body growth or repair. Reproduction of the human organism, or making a whole new person, is the major task of the reproductive system. When a sperm unites with an egg, a fertilized egg forms and develops into a baby within the mother's body. The reproductive system is directly responsible for producing offspring, but its function is exquisitely regulated by hormones of the endocrine system (Figure 1.3e).

Because males produce sperm and females produce eggs (ova), there is a division of labor in reproduction, and the reproductive organs of males and females are different (Figure 1.3k, l). Additionally, the female's reproductive structures provide the site for fertilization of eggs by sperm, and then protect and nurture the developing fetus until birth.

Growth

Growth is an increase in size of a body part or the organism. It is usually accomplished by increasing the number of cells. However, individual cells also increase in size when not dividing. For true growth to occur, constructive activities must occur at a faster rate than destructive ones.

Survival Needs

The ultimate goal of all body systems is to maintain life. However, life is extraordinarily fragile and requires several factors. These factors, which we will call *survival needs*, include nutrients (food), oxygen, water, and appropriate temperature and atmospheric pressure.

Nutrients

Nutrients, taken in via the diet, contain the chemical substances used for energy and cell building. Most plant-derived foods are rich in carbohydrates, vitamins, and minerals, whereas most animal foods are richer in proteins and fats.

Carbohydrates are the major energy fuel for body cells. Proteins, and to a lesser extent fats, are essential for building cell structures. Fats also provide a reserve of energy-rich fuel. Selected minerals and vitamins are required for the chemical reactions that go on in cells and for oxygen transport in the blood. The mineral calcium helps to make bones hard and is required for blood clotting.

Oxygen

All the nutrients in the world are useless unless **oxygen** is also available. Because the chemical reactions that release energy from foods are *oxidative* reactions that require oxygen, human cells can survive for only a few minutes without oxygen. Approximately 20% of the air we breathe is oxygen. The cooperative efforts of the respiratory and cardiovascular systems make oxygen available to the blood and body cells.

Water

Water accounts for 60–80% of our body weight and is the single most abundant chemical substance in the body. It provides the watery environment necessary for chemical reactions and the fluid base for body secretions and excretions. We obtain water chiefly from ingested foods or liquids. We lose it from the body by evaporation from the lungs and skin and in body excretions.

Normal Body Temperature

If chemical reactions are to continue at life-sustaining rates, **normal body temperature** must be maintained. As body temperature drops below 37°C (98.6°F), metabolic reactions become slower and slower, and finally stop. When body temperature is too high, chemical reactions occur at a frantic pace and body proteins lose their characteristic shape and stop functioning. At either extreme, death occurs. The activity of the muscular system generates most body heat.

Appropriate Atmospheric Pressure

Atmospheric pressure is the force that air exerts on the surface of the body. Breathing and gas exchange in the lungs depend on *appropriate* atmospheric pressure. At high altitudes, where atmospheric pressure is lower and the air is thin, gas exchange may be inadequate to support cellular metabolism.

The mere presence of these survival factors is not sufficient to sustain life. They must be present in *appropriate* amounts. Excesses and deficits may be equally harmful. For example, oxygen is essential, but excessive amounts are toxic to body cells. Similarly, the food we eat must be of high quality and in proper amounts. Otherwise, nutritional disease, obesity, or starvation is likely. Also, while the needs listed above are the most crucial, they do not even begin to encompass all of the body's needs. For example, we can live without gravity if we must, but the quality of life suffers.

CHECK YOUR UNDERSTANDING

6. What separates living beings from nonliving objects?
7. What name is given to all chemical reactions that occur within body cells?
8. Why is it necessary to be in a pressurized cabin when flying at 30,000 feet?

For answers, see Appendix G.

Homeostasis

- ▶ Define homeostasis and explain its significance.
- ▶ Describe how negative and positive feedback maintain body homeostasis.
- ▶ Describe the relationship between homeostatic imbalance and disease.

When you think about the fact that your body contains trillions of cells in nearly constant activity, and that remarkably little usually goes wrong with it, you begin to appreciate what a marvelous machine your body is. Walter Cannon, an American physiologist of the early twentieth century, spoke of the "wisdom of the body," and he coined the word **homeostasis** (ho″me-o-staʹsis) to describe its ability to maintain relatively stable internal conditions even though the outside world changes continuously.

Although the literal translation of homeostasis is "unchanging," the term does not really mean a static, or unchanging, state.

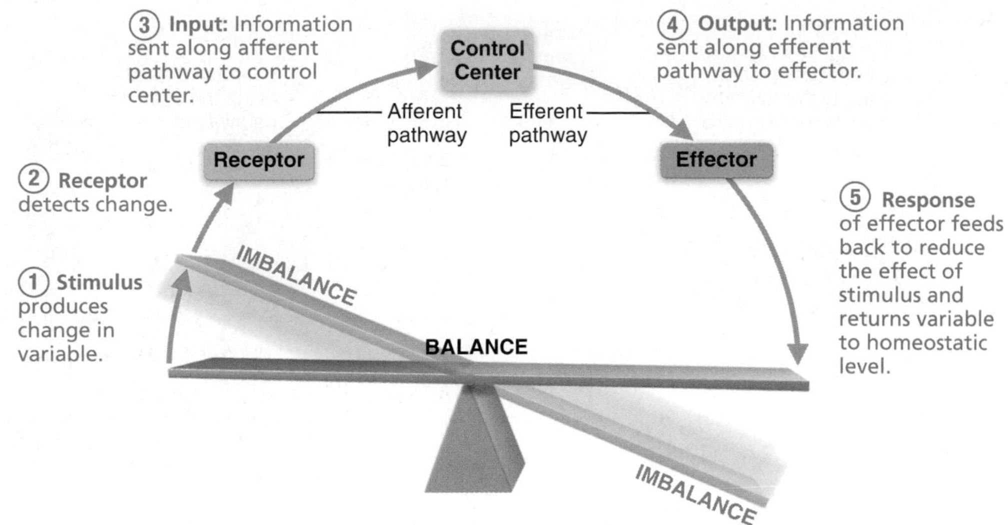
③ Input: Information sent along afferent pathway to control center.
Control Center
④ Output: Information sent along efferent pathway to effector.
Afferent pathway
Efferent pathway
Receptor
Effector
② Receptor detects change.
⑤ Response of effector feeds back to reduce the effect of stimulus and returns variable to homeostatic level.
① Stimulus produces change in variable.
IMBALANCE
BALANCE
IMBALANCE

Figure 1.4 Interaction among the elements of a homeostatic control system.

Rather, it indicates a *dynamic* state of equilibrium, or a balance, in which internal conditions vary, but always within relatively narrow limits. In general, the body is in homeostasis when its needs are adequately met and it is functioning smoothly.

Maintaining homeostasis is more complicated than it appears at first glance. Virtually every organ system plays a role in maintaining the constancy of the internal environment. Adequate blood levels of vital nutrients must be continuously present, and heart activity and blood pressure must be constantly monitored and adjusted so that the blood is propelled to all body tissues. Also, wastes must not be allowed to accumulate, and body temperature must be precisely controlled. A wide variety of chemical, thermal, and neural factors act and interact in complex ways—sometimes helping and sometimes hindering the body as it works to maintain its "steady rudder."

Homeostatic Control

Communication within the body is essential for homeostasis. Communication is accomplished chiefly by the nervous and endocrine systems, which use neural electrical impulses or bloodborne hormones, respectively, as information carriers. We cover the details of how these two great regulating systems operate in later chapters, but here we explain the basic characteristics of control systems that promote homeostasis.

Regardless of the factor or event being regulated—the **variable**—all homeostatic control mechanisms are processes involving at least three components that work together **(Figure 1.4)**. The first component, the **receptor**, is some type of sensor that monitors the environment and responds to changes, called *stimuli*, by sending information (input) to the second component, the *control center*. Input flows from the receptor to the control center along the so-called *afferent pathway*.

The **control center** determines the *set point*, which is the level or range at which a variable is to be maintained. It also analyzes the input it receives and determines the appropriate response or course of action. Information (output) then flows from the control center to the third component, the *effector*, along the *efferent pathway*. (To help you remember the difference between "afferent" and "efferent," you might note that information traveling along the afferent pathway **approaches** the control center and efferent information **exits** from the control center.)

The **effector** provides the means for the control center's response (output) to the stimulus. The results of the response then *feed back* to influence the effect of the stimulus, either reducing it (in negative feedback) so that the whole control process is shut off, or enhancing it (in positive feedback) so that the whole process continues at an even faster rate.

Negative Feedback Mechanisms

Most homeostatic control mechanisms are **negative feedback mechanisms**. In these systems, the output shuts off the original effect of the stimulus or reduces its intensity. These mechanisms cause the variable to change in a direction *opposite* to that of the initial change, returning it to its "ideal" value; thus the name "negative" feedback mechanisms.

Let's start with an example of a nonbiological negative feedback system: a home heating system connected to a temperature-sensing thermostat. The thermostat houses both the receptor (thermometer) and the control center. If the thermostat is set at 20°C (68°F), the heating system (effector) is triggered ON when the house temperature drops below that setting. As the furnace produces heat and warms the air, the temperature rises, and when it reaches 20°C or slightly higher, the thermostat triggers the furnace OFF. This process results in a cycling of "furnace-ON" and "furnace-OFF" so that the temperature in the house stays very near the desired temperature of 20°C.

1

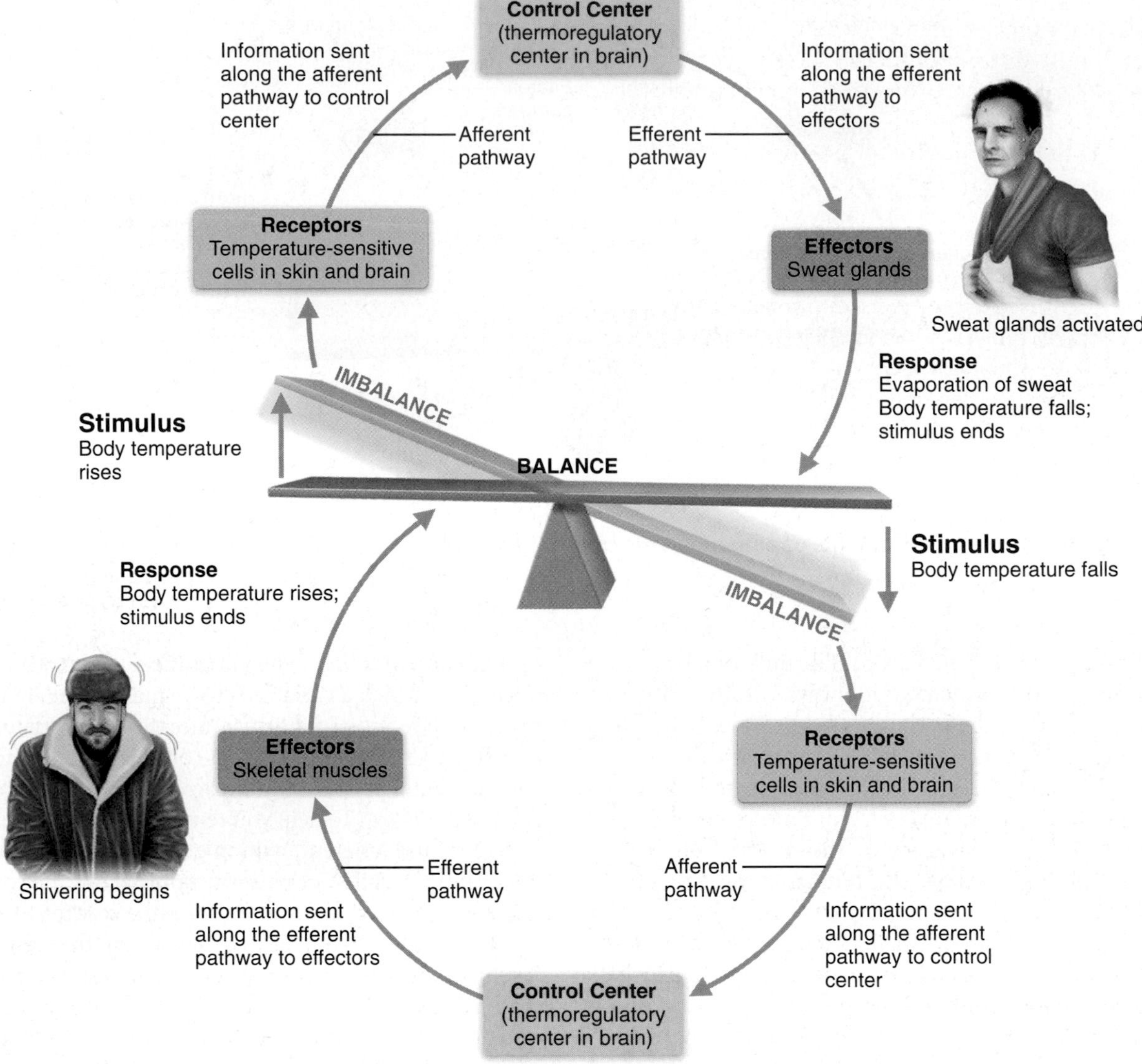

Figure 1.5 Regulation of body temperature by a negative feedback mechanism.

Your body "thermostat," located in a part of your brain called the hypothalamus, operates in a similar fashion **(Figure 1.5)**. Regulation of body temperature is only one of the many ways the nervous system maintains the constancy of the internal environment. Another type of neural control mechanism is seen in the *withdrawal reflex* mentioned earlier, in which the hand is jerked away from a painful stimulus such as broken glass.

The endocrine system is equally important in maintaining homeostasis. A good example of a hormonal negative feedback mechanism is the control of blood volume by antidiuretic hormone (ADH). As blood volume drops, receptors in the body sense this change, and the hypothalamus of the brain (the control center) stimulates the release of ADH to the blood. This change in turn prompts the kidneys to reabsorb more water and return it to the bloodstream. The rising blood volume then ends the stimulus for ADH release.

The body's ability to regulate its internal environment is fundamental. All negative feedback mechanisms have the same goal: preventing sudden severe changes within the body. Body temperature and blood volume are only two of the variables that need to be regulated. There are hundreds! Other negative feedback mechanisms regulate heart rate, blood pressure, the rate and depth of breathing, and blood levels of oxygen, carbon dioxide, and minerals. Now, let's take a look at the other type of feedback control mechanism—positive feedback.

Positive Feedback Mechanisms

In **positive feedback mechanisms**, the result or response enhances the original stimulus so that the response is accelerated. This feedback mechanism is "positive" because the change that results proceeds in the *same* direction as the initial change, causing the variable to deviate further and further from its original value or range.

In contrast to negative feedback controls, which maintain some physiological function or keep blood chemicals within narrow ranges, positive feedback mechanisms usually control infrequent events that do not require continuous adjustments. Typically, they set off a series of events that may be self-perpetuating and that, once initiated, have an amplifying or waterfall effect. Because of these characteristics, positive feedback mechanisms are often referred to as *cascades* (from the Italian word meaning "to fall").

Positive feedback mechanisms are likely to race out of control, so they are rarely used to promote the moment-to-moment well-being of the body. However, two familiar examples of their use as homeostatic mechanisms are the enhancement of labor contractions during birth and blood clotting.

Chapter 28 describes the positive feedback mechanism in which oxytocin, a hypothalamic hormone, intensifies labor contractions during the birth of a baby (see Figure 28.17). Oxytocin causes the contractions to become both more frequent and more powerful. The increased contractions cause more oxytocin to be released, which causes more contractions, and so on until the baby is finally born. The birth ends the stimulus for oxytocin release and shuts off the positive feedback mechanism.

Blood clotting is a normal response to a break in the wall of a blood vessel and is an excellent example of an important body function controlled by positive feedback. Basically, once a vessel has been damaged, blood elements called platelets immediately begin to cling to the injured site and release chemicals that attract more platelets. This rapidly growing pileup of platelets temporarily "plugs" the tear and initiates the sequence of events that finally forms a clot **(Figure 1.6)**.

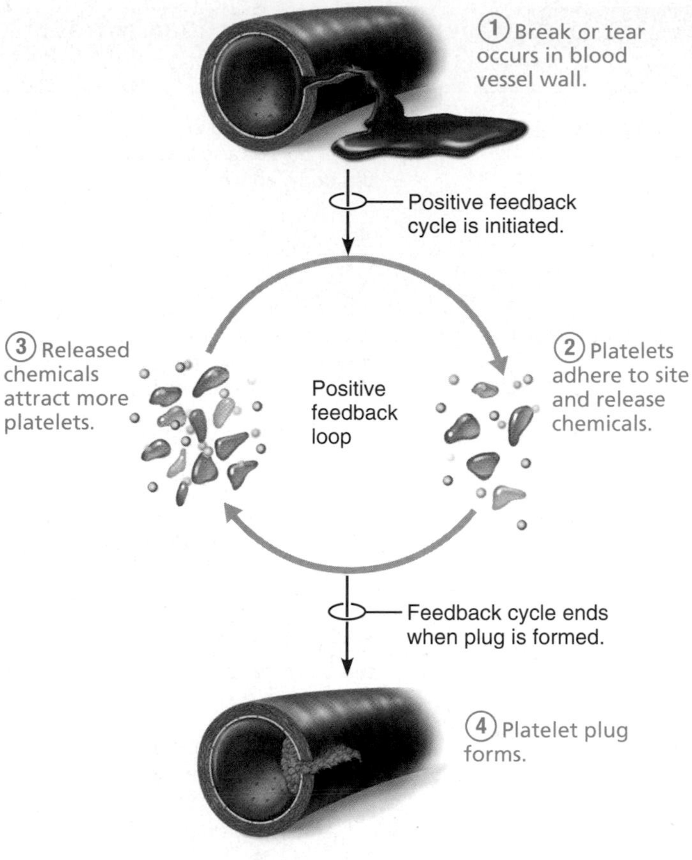

Figure 1.6 Summary of the positive feedback mechanism regulating formation of a platelet plug.

Homeostatic Imbalance

Homeostasis is so important that most disease can be regarded as a result of its disturbance, a condition called **homeostatic imbalance.** As we age, our body's control systems become less efficient, and our internal environment becomes less and less stable. These events increase our risk for illness and produce the changes we associate with aging.

Another important source of homeostatic imbalance occurs when the usual negative feedback mechanisms are overwhelmed and destructive positive feedback mechanisms take over. Some instances of heart failure reflect this phenomenon.

Examples of homeostatic imbalance appear throughout this book to enhance your understanding of normal physiological mechanisms. This symbol  introduces the homeostatic imbalance sections and alerts you to the fact that we are describing an abnormal condition.

CHECK YOUR UNDERSTANDING

9. What process allows us to adjust to either extreme heat or extreme cold?
10. When we begin to get dehydrated, we usually get thirsty, which causes us to drink fluids. Is thirst part of a negative or a positive feedback control system? Defend your choice.
11. Why is the control mechanism shown in Figure 1.6 called a positive feedback system? What event ends it?

For answers, see Appendix G.

The Language of Anatomy

- Describe the anatomical position.
- Use correct anatomical terms to describe body directions, regions, and body planes or sections.

Most of us are naturally curious about our bodies, but our interest sometimes dwindles when we are confronted with the terminology of anatomy and physiology. Let's face it—you can't just pick up an anatomy and physiology book and read it as though it were a novel.

Unfortunately, confusion is likely without precise, specialized terminology. To prevent misunderstanding, anatomists use universally accepted terms to identify body structures precisely and with a minimum of words. We present and explain the language of anatomy next.

1

TABLE 1.1 Orientation and Directional Terms

TERM	DEFINITION	EXAMPLE	
Superior (cranial)	Toward the head end or upper part of a structure or the body; above		The head is superior to the abdomen.
Inferior (caudal)	Away from the head end or toward the lower part of a structure or the body; below		The navel is inferior to the chin.
Ventral (anterior)*	Toward or at the front of the body; in front of		The breastbone is anterior to the spine.
Dorsal (posterior)*	Toward or at the back of the body; behind		The heart is posterior to the breast-bone.
Medial	Toward or at the midline of the body; on the inner side of		The heart is medial to the arm.
Lateral	Away from the midline of the body; on the outer side of		The arms are lateral to the chest.
Intermediate	Between a more medial and a more lateral structure		The collarbone is intermediate between the breastbone and shoulder.
Proximal	Closer to the origin of the body part or the point of attachment of a limb to the body trunk		The elbow is proximal to the wrist.
Distal	Farther from the origin of a body part or the point of attachment of a limb to the body trunk		The knee is distal to the thigh.
Superficial (external)	Toward or at the body surface		The skin is superficial to the skeletal muscles.
Deep (internal)	Away from the body surface; more internal		The lungs are deep to the skin.

*The terms *ventral* and *anterior* are synonymous in humans, but this is not the case in four-legged animals. *Anterior* refers to the leading portion of the body (abdominal surface in humans, head in a cat), but *ventral* specifically refers to the "belly" of a vertebrate animal, so it is the inferior surface of four-legged animals. Likewise, although the dorsal and posterior surfaces are the same in humans, the term *dorsal* specifically refers to an animal's back. Thus, the dorsal surface of four-legged animals is their superior surface.

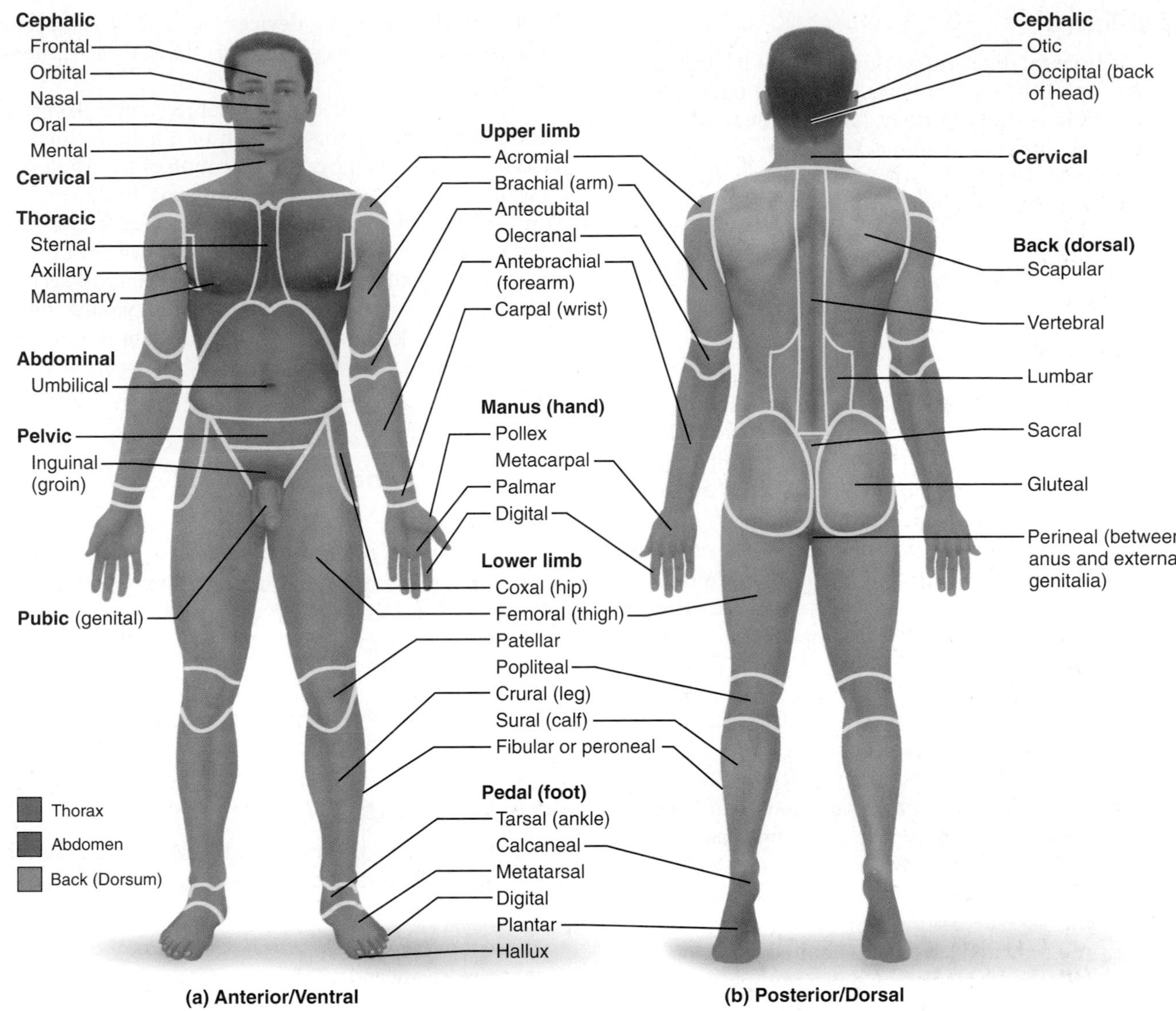

Figure 1.7 Regional terms used to designate specific body areas. (a) The anatomical position. **(b)** The heels are raised to show the plantar surface of the foot, which is actually on the inferior surface of the body.

Anatomical Position and Directional Terms

To describe body parts and position accurately, we need an initial reference point, and we must indicate direction. The anatomical reference point is a standard body position called the **anatomical position**. In the anatomical position, the body is erect with feet slightly apart. This position is easy to remember because it resembles "standing at attention," except that the palms face forward and the thumbs point away from the body. You can see the anatomical position in **Table 1.1** (top) and **Figure 1.7a**.

It is essential to understand the anatomical position because most of the directional terms used in this book refer to the body *as if it were in this position, regardless of its actual position.* Another point to remember is that the terms "right" and "left" refer to those sides of the person or the cadaver (body of a deceased person) being viewed—not those of the observer.

Directional terms allow us to explain where one body structure is in relation to another. For example, we could describe the relationship between the ears and the nose informally by stating, "The ears are located on each side of the head to the right and left of the nose." Using anatomical terminology, we can condense this to "The ears are lateral to the nose." Using anatomical terms saves words and is less ambiguous.

Commonly used orientation and directional terms are defined and illustrated in Table 1.1. Many of these terms are also used in everyday conversation, but keep in mind as you study them that their anatomical meanings are very precise.

Regional Terms

The two fundamental divisions of our body are its *axial* and *appendicular* (ap″en-dik′u-lar) parts. The **axial part**, which makes up the main *axis* of our body, includes the head, neck, and trunk. The **appendicular part** consists of the *appendages*, or *limbs*, which are attached to the body's axis. **Regional terms** used to designate specific areas within these major body divisions are indicated in Figure 1.7. The figure also gives the common term for each of these body regions
1 (in parentheses).

Anatomical Variability

Although we use common directional and regional terms to refer to all human bodies, you know from observing the faces and body shapes of people around you that humans differ in their external anatomy. The same kind of variability holds for internal organs as well. In some bodies, for example, a nerve or blood vessel may be somewhat out of place, or a small muscle may be missing. Nonetheless, well over 90% of all structures present in any human body match the textbook descriptions. We seldom see extreme anatomical variations because they are incompatible with life.

Body Planes and Sections

For anatomical studies, the body is often cut, or *sectioned*, along a flat surface called a *plane*. The most frequently used body planes are *sagittal*, *frontal*, and *transverse* planes, which lie at right angles to one another **(Figure 1.8)**. A section is named for the plane along which it is cut. Thus, a cut along a sagittal plane produces a sagittal section.

A **sagittal plane** (saj′ĭ-tal; "arrow") is a vertical plane that divides the body into right and left parts. A sagittal plane that lies exactly in the midline is the **median plane**, or **midsagittal plane** (Figure 1.8c). All other sagittal planes, offset from the midline, are **parasagittal planes** (*para* = near).

Frontal planes, like sagittal planes, lie vertically. Frontal planes, however, divide the body into anterior and posterior parts (Figure 1.8a). A frontal plane is also called a **coronal plane** (kŏ-ro′nal; "crown").

A **transverse**, or **horizontal**, **plane** runs horizontally from right to left, dividing the body into superior and inferior parts (Figure 1.8b). Of course, many different transverse planes exist, at every possible level from head to foot. A transverse section is also called a **cross section**.

Oblique sections are cuts made diagonally between the horizontal and the vertical planes. Because oblique sections are often confusing and difficult to interpret, they are seldom used.

At the bottom of Figure 1.8, you can see examples of magnetic resonance imaging (MRI) scans that correspond to the three different sections shown in the figure. In the clinical sciences, the ability to interpret sections made through the body, especially transverse sections, is important. Additionally, the new medical imaging devices (*A Closer Look*, pp. 18–19) produce sectional images rather than three-dimensional images.

It takes practice to decipher an object's overall shape from sectioned material. A cross section of a banana, for example, looks like a circle and gives no indication of the whole banana's crescent shape. Likewise, sectioning the body or an organ along different planes often results in very different views. For example, a transverse section of the body trunk at the level of the kidneys would show kidney structure in cross section very nicely. A frontal section of the body trunk would show a different view of kidney anatomy, and a midsagittal section would miss the kidneys completely. With experience, you will gradually learn to relate two-dimensional sections to three-dimensional shapes.

CHECK YOUR UNDERSTANDING

12. What is the anatomical position? Why is it important that *you* learn this position?
13. The axillary and acromial regions are both in the general area of the shoulder. Where specifically is each located?
14. What type of cut would separate the brain into anterior and posterior parts?

For answers, see Appendix G.

Body Cavities and Membranes

- ▶ Locate and name the major body cavities and their subdivisions and associated membranes, and list the major organs contained within them.
- ▶ Name the four quadrants or nine regions of the abdominopelvic cavity and list the organs they contain.

Anatomy and physiology textbooks typically describe two sets of internal body cavities called the dorsal and ventral body cavities. These cavities are closed to the outside and provide different degrees of protection to the organs contained within them. Because these two cavities differ in their mode of embryonic development, and their lining membranes, the dorsal body cavity is not recognized as such in many anatomical references. However, the idea of two sets of internal body cavities is a useful learning concept and we use it here.

Dorsal Body Cavity

The **dorsal body cavity**, which protects the fragile nervous system organs, has two subdivisions (**Figure 1.9**, gold areas). The **cranial cavity**, in the skull, encases the brain. The **vertebral**, or **spinal**, **cavity**, which runs within the bony vertebral column, encloses the delicate spinal cord. The spinal cord is essentially a continuation of the brain, and the cranial and spinal cavities are continuous with one another.

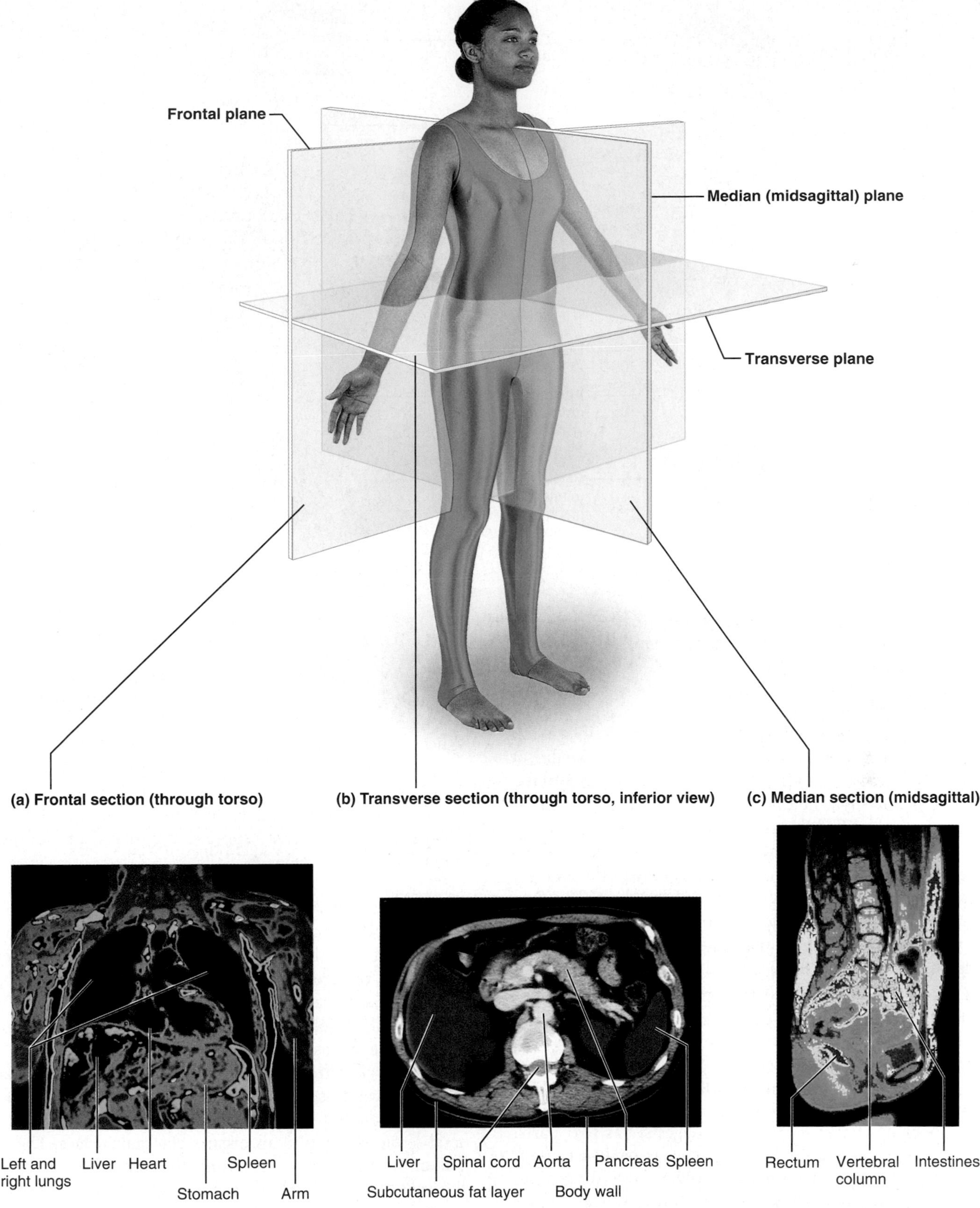

Figure 1.8 Planes of the body with corresponding magnetic resonance imaging (MRI) scans.

1

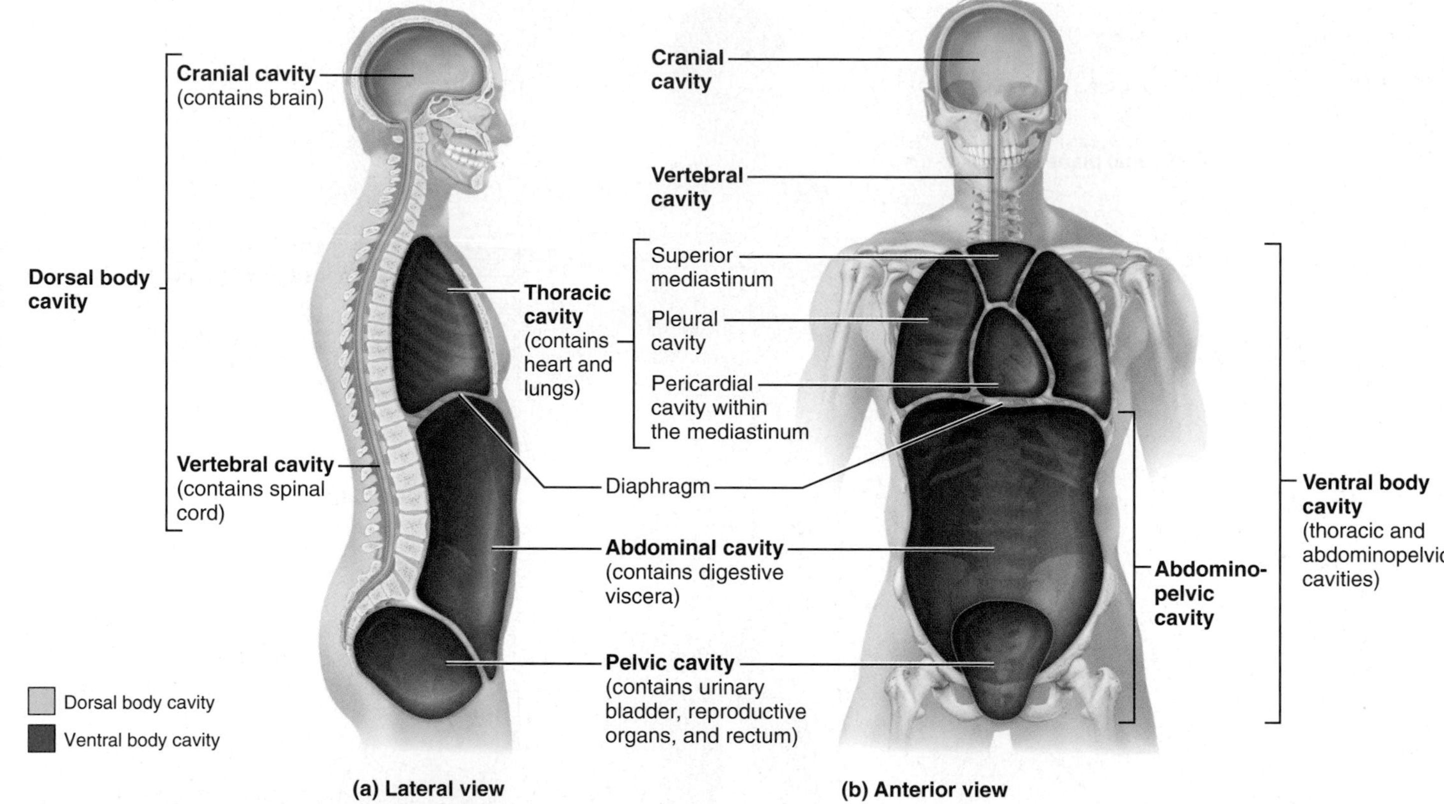

Figure 1.9 Dorsal and ventral body cavities and their subdivisions.

Ventral Body Cavity

The more anterior and larger of the closed body cavities is the **ventral body cavity** (Figure 1.9, rust-red areas). Like the dorsal cavity, it has two major subdivisions, the *thoracic cavity* and the *abdominopelvic cavity*. The ventral body cavity houses internal organs collectively called the **viscera** (vis′er-ah; *viscus* = an organ in a body cavity), or visceral organs.

The superior subdivision, the **thoracic cavity** (tho-ras′ik), is surrounded by the ribs and muscles of the chest. The thoracic cavity is further subdivided into lateral **pleural cavities** (ploo′ral), each enveloping a lung, and the medial **mediastinum** (me″de-ah-sti′num). The mediastinum contains the **pericardial cavity** (per″ĭ-kar′de-al), which encloses the heart, and it also surrounds the remaining thoracic organs (esophagus, trachea, and others).

The thoracic cavity is separated from the more inferior **abdominopelvic cavity** (ab-dom′ĭ-no-pel′-vic) by the diaphragm, a dome-shaped muscle important in breathing. The abdominopelvic cavity, as its name suggests, has two parts. However, these regions are not physically separated by a muscular or membrane wall. Its superior portion, the **abdominal cavity**, contains the stomach, intestines, spleen, liver, and other organs. The inferior part, the **pelvic cavity**, lies in the bony pelvis and contains the urinary bladder, some reproductive organs, and the rectum. The abdominal and pelvic cavities are not aligned with each other. Instead, the bowl-shaped pelvis tips away from the perpendicular.

HOMEOSTATIC IMBALANCE

When the body is subjected to physical trauma (as in an automobile accident), the abdominopelvic organs are most vulnerable. Why? This is because the walls of the abdominal cavity are formed only by trunk muscles and are not reinforced by bone. The pelvic organs receive a somewhat greater degree of protection from the bony pelvis. ■

Membranes in the Ventral Body Cavity The walls of the ventral body cavity and the outer surfaces of the organs it contains are covered by a thin, double-layered membrane, the **serosa** (se-ro′sah), or **serous membrane**. The part of the membrane lining the cavity walls is called the **parietal serosa** (pah-ri′ĕ-tal; *parie* = wall). It folds in on itself to form the **visceral serosa**, covering the organs in the cavity.

You can visualize the relationship between the serosal layers by pushing your fist into a limp balloon **(Figure 1.10a)**. The part of the balloon that clings to your fist can be compared to the visceral serosa clinging to an organ's external surface. The outer wall of the balloon then represents the parietal serosa that lines the walls of the cavity. (However, unlike the balloon, the parietal serosa is never exposed but is always fused to the cavity wall.) In the body, the serous membranes are separated not by air but by a thin layer of lubricating fluid, called **serous fluid**, which is

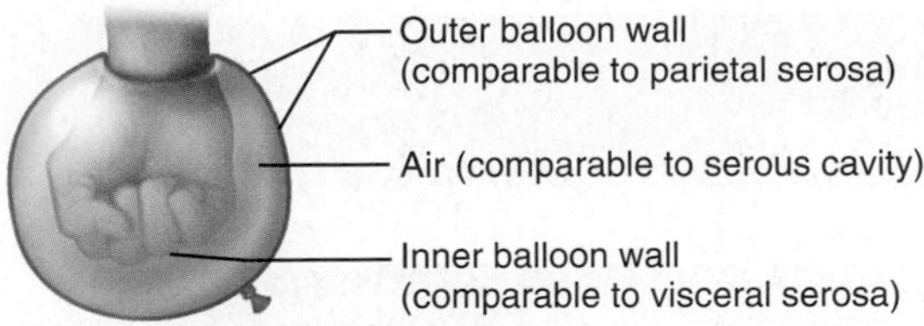

(a) A fist thrust into a flaccid balloon demonstrates the relationship between the parietal and visceral serous membrane layers.

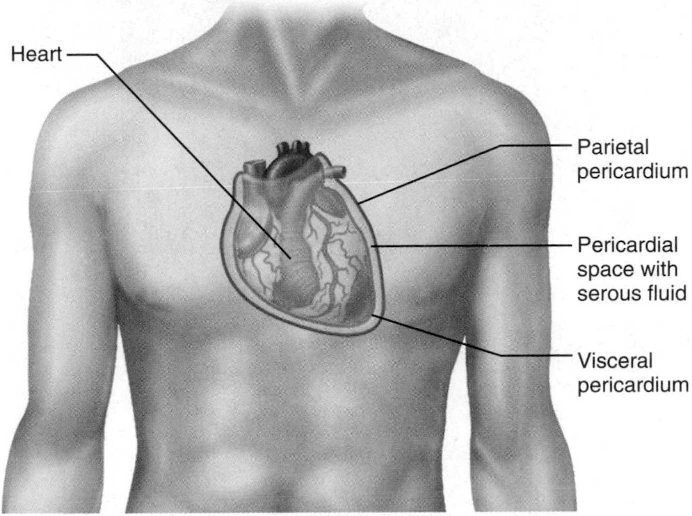

(b) The serosae associated with the heart.

Figure 1.10 **Serous membrane relationships.**

secreted by both membranes. Although there is a potential space between the two membranes, the barely present, slitlike cavity is filled with serous fluid.

The slippery serous fluid allows the organs to slide without friction across the cavity walls and one another as they carry out their routine functions. This freedom of movement is especially important for mobile organs such as the pumping heart and the churning stomach.

The serous membranes are named for the specific cavity and organs with which they are associated. For example, as shown in Figure 1.10b, the *parietal pericardium* lines the pericardial cavity and folds back as the *visceral pericardium*, which covers the heart. Likewise, the *parietal pleura* (ploo′rah) lines the walls of the thoracic cavity, and the *visceral pleurae* cover the lungs. The *parietal peritoneum* (per″ĭ-to-ne′um) is associated with the walls of the abdominopelvic cavity, while the *visceral peritoneum* covers most of the organs within that cavity. (The pleural and peritoneal serosae are illustrated in Figure 4.11c on p. 139.)

HOMEOSTATIC IMBALANCE

When serous membranes are inflamed, their normally smooth surfaces become roughened. This roughness causes the organs to stick together and drag across one another, leading to excruciating pain, as anyone who has experienced *pleurisy* (inflammation of the pleurae) or *peritonitis* (inflammation of the peritonea) knows. ■

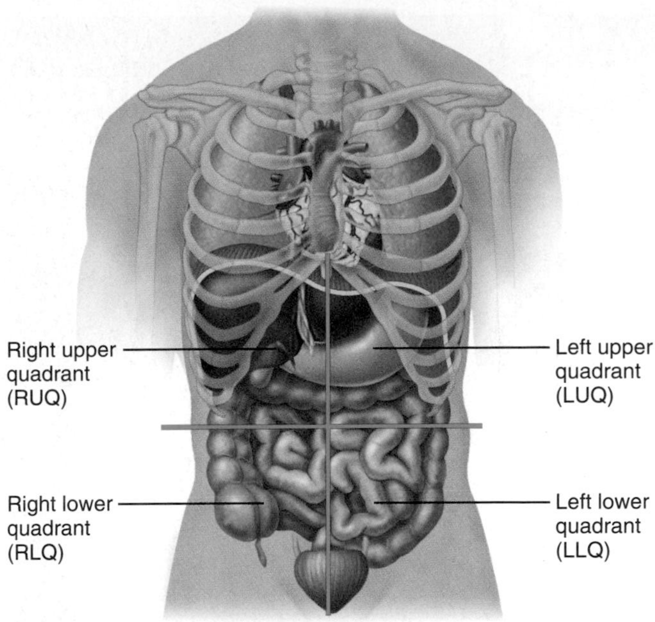

Figure 1.11 **The four abdominopelvic quadrants.** In this scheme, the abdominopelvic cavity is divided into four quadrants by two planes.

Abdominopelvic Regions and Quadrants Because the abdominopelvic cavity is large and contains several organs, it helps to divide it into smaller areas for study. Medical personnel usually use a simple scheme to locate the abdominopelvic cavity organs **(Figure 1.11)**. In this scheme, one transverse and one median midsagittal plane pass through the umbilicus at right angles. The four resulting quadrants are named according to their positions from the subject's point of view: the **right upper quadrant** (**RUQ**), **left upper quadrant** (**LUQ**), **right lower quadrant** (**RLQ**), and **left lower quadrant** (**LLQ**). (See Figure 1.12 for organs located in different areas of the abdomen.) Another division method, used primarily by anatomists, uses two transverse and two parasagittal planes. These planes, positioned like a tic-tac-toe grid on the abdomen, divide the cavity into nine regions **(Figure 1.12)**:

- The **umbilical region** is the centermost region deep to and surrounding the umbilicus (navel).
- The **epigastric region** is located superior to the umbilical region (*epi* = upon, above; *gastri* = belly).
- The **hypogastric (pubic) region** is located inferior to the umbilical region (*hypo* = below).
- The **right** and **left iliac**, or **inguinal**, **regions** (ing′gwĭ-nal) are located lateral to the hypogastric region (*iliac* = superior part of the hip bone).
- The **right** and **left lumbar regions** lie lateral to the umbilical region (*lumbus* = loin).
- The **right** and **left hypochondriac regions** lie lateral to the epigastric region (*chondro* = cartilage).

1

CLOSER LOOK

Medical Imaging: Illuminating the Body

Until 50 years ago, the magical but murky X ray was the only nonsurgical means to extract information from within a living body. Produced by directing *X rays*, electromagnetic waves of very short wavelength, at the body, an **X ray** or **radiograph** is essentially a shadowy negative image of internal structures. Dense structures absorb the X rays most and so appear as light areas. Hollow air-containing organs and fat, which absorb the X rays less, show up as dark areas. What X rays do best is visualize hard, bony structures and locate abnormally dense structures (tumors, tuberculosis nodules) in the lungs.

The 1950s saw the advent of nuclear medicine, which uses radioisotopes to scan the body, and ultrasound techniques, which use sound waves. The 1970s brought CT, PET, and MRI scans. These technologies not only reveal the structure of our "insides" but also wring out information about the hidden workings of their molecules.

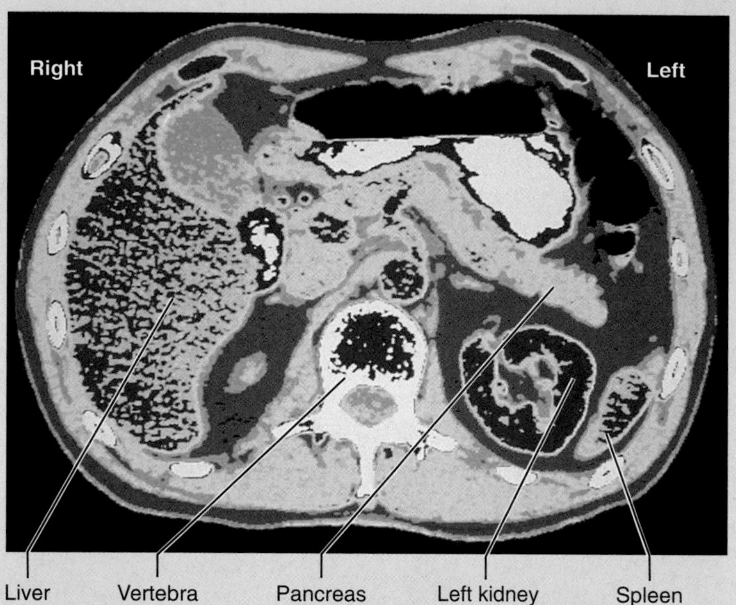

(a) A CT scan through the superior abdomen.

Computed tomography (**CT**, formerly called **computerized axial tomography, CAT**) uses a refined version of X-ray equipment. As the patient is slowly moved through the doughnut-shaped CT machine, its X-ray tube rotates around the body and sends beams from all directions to a specific level of the patient's body. Because at any moment its beam is confined to a "slice" of the body about as thick as a dime, CT ends the confusion resulting from overlapping structures seen in conventional X rays. The device's computer translates this information into a detailed, cross-sectional picture of each body region scanned. CT scans are at the forefront for evaluating most problems that affect the brain and abdomen. Their clarity, illustrated in photo (a), has all but eliminated exploratory surgery.

Xenon CT is a CT brain scan enhanced with radioactive xenon gas to quickly trace blood flow. Inhaled xenon rapidly enters the bloodstream and distributes to different body tissues in proportion to their blood flow. Absence of xenon from part of the brain indicates that a stroke is occurring there, information that aids treatment.

Dynamic spatial reconstruction (DSR) uses ultrafast CT scanners to provide three-dimensional images of body organs from any angle, and scrutinize their movements and changes in their internal volumes at normal speed, in slow motion, and at a specific moment. DSR's greatest value has been to visualize the heart beating and blood flowing through blood vessels. This information allows clinicians to evaluate heart defects, constricted or blocked blood vessels, and the status of coronary bypass grafts.

Another computer-assisted X-ray technique, **digital subtraction angiography (DSA)** (*angiography* = vessel pictures), provides an unobstructed view of small arteries. Conventional radiographs are taken before and after a contrast medium is injected into an artery. The computer subtracts the "before" image from the "after" image, eliminating all traces of body structures that obscure the vessel. DSA is often used to identify blockages in the arteries that supply the heart wall, as in photo (b), and in the brain.

Just as the X ray spawned related technologies, so too did nuclear medicine in the form of **positron emission tomography (PET)**. PET excels in observing *metabolic processes*. The patient is given an injection of radioisotopes tagged to biological molecules (such as glucose) and is then positioned in the PET scanner. As the radioisotopes are absorbed by the most active brain cells, high-energy gamma rays are produced. The computer analyzes the gamma-ray emission and produces a live-action picture of the brain's biochemical activity in vivid colors. PET's greatest value has been its ability to provide insights into brain activity in people affected by mental illness, stroke, Alzheimer's disease, and epilepsy. One of its most exciting uses has been to determine which areas of the healthy brain are most active during certain tasks (e.g., speaking, listening to music, or figuring out a mathematical problem), providing direct evidence of the functions of specific brain regions. Currently PET can reveal signs of trouble in those with undiagnosed Alzheimer's disease (AD) because regions of beta-amyloid accumulation (a defining characteristic of AD) show up in brilliant red and yellow, as in photo (c). PET scans can also help to predict who may develop AD in the future by identifying areas of decreased metabolism in crucial memory areas of the brain.

Sonography, or **ultrasound imaging**, has some distinct advantages over the approaches examined so far. The equipment is inexpensive, and the ultrasound used as its energy source seems to be safer than the ionizing forms of radiation used in nuclear medicine. The body is probed with pulses of sound waves that cause echoes when reflected and scattered by body tissues. A computer analyzes these echoes to construct somewhat blurry outlines of body organs. A single easy-to-use handheld device

A CLOSER LOOK
(continued)

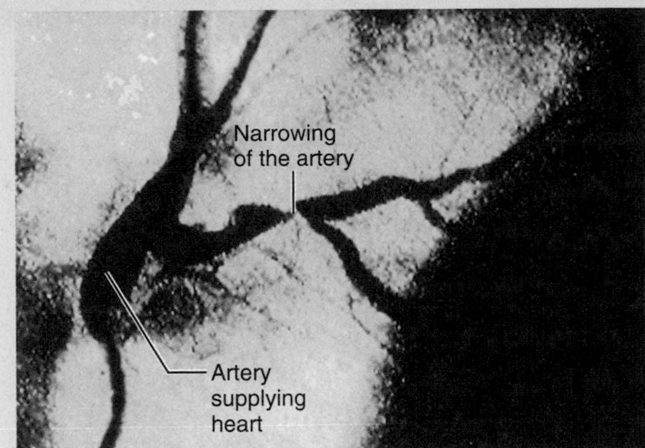

(b) A DSA image of the arteries that supply the heart.

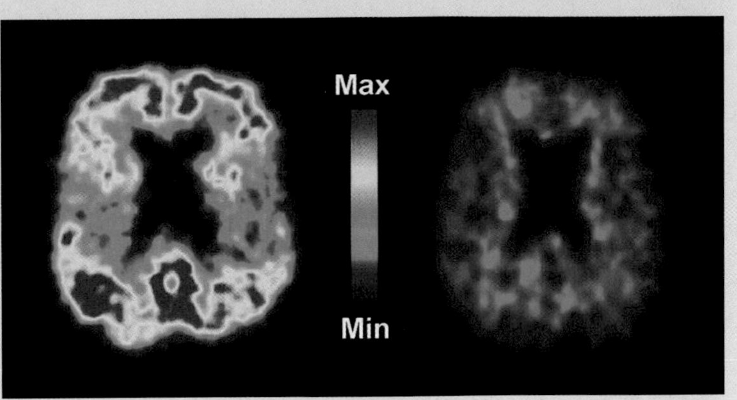

(c) In a PET scan, regions of beta-amyloid accumulation "light up" (red-yellow) in an Alzheimer's patient (*left*) but not in a healthy person (*right*).

1

emits the sound and picks up the echoes, so sections can be scanned from many different body planes.

Because of its safety, ultrasound is the imaging technique of choice in obstetrics for determining fetal age and position and locating the placenta. However, sound waves have low penetrating power and rapidly dissipate in air, so sonography is of little value for looking at air-filled structures (the lungs) or those surrounded by bone (the brain and spinal cord).

Magnetic resonance imaging (MRI) produces high-contrast images of our soft tissues, an area in which X rays and CT scans are weak. As initially developed, MRI primarily maps the body's content of hydrogen, most of which is in water. The technique subjects the body to magnetic fields up to 60,000 times stronger than that of the earth to pry information from the body's molecules. The patient lies in a chamber within a huge magnet. Hydrogen molecules act like tiny magnets, spinning like tops in the magnetic field. Their energy is further enhanced by radio waves, and when the radio waves are turned off, the energy released is translated into a visual image.

MRI distinguishes body tissues based on their water content, so it can differentiate between the fatty white matter and the more watery gray matter of the brain. Because dense structures do not show up at all in MRI, it peers easily into the skull and vertebral column, enabling the delicate nerve fibers of the spinal cord to be seen. MRI is also particularly good at detecting tumors and degenerative disease. Multiple sclerosis plaques do not show up well in CT scans, but are dazzlingly clear in MRI scans. MRI can also tune in on metabolic reactions, such as processes that generate energy-rich ATP molecules.

Until recently, trying to diagnose asthma and other lung problems has been off limits to MRI scans because the lungs have a low water content. However, an alternate tack—filling the lungs with a gas that can be magnetized (hyperpolarized helium-3 or xenon-129)—has yielded spectacular pictures of the lungs in just the few seconds it takes the patient to inhale, hold the breath briefly, and then exhale. This technique offers a distinct improvement over the hours required for conventional MRI and it has the additional advantage of using a magnetic field as little as one-tenth that of the conventional MRI.

Newer variations of MRI include **magnetic resonance spectroscopy (MRS)**, which maps the distribution of elements other than hydrogen to reveal more about how disease changes body chemistry. Other advances in computer techniques display MRI scans in three dimensions to guide laser surgery.

The **functional MRI** tracks blood flow into the brain in real time. Matching thoughts, deeds, and disease to brain activity has been the sole domain of PET. Because functional MRI does not require injections of tracers and can pinpoint much smaller brain areas than PET, it may provide a desirable alternative. Clinical studies are also using functional MRI to determine if a patient in the vegetative state has conscious thought.

Despite its advantages, the powerful magnets of the clanging, claustrophobia-inducing MRI present some thorny problems. For example, they can "suck" metal objects, such as implanted pacemakers and loose tooth fillings, through the body. Moreover, although such strong magnetic fields are currently considered safe, there is no convincing evidence that they are risk free.

Although stunning, medical images other than straight X rays are abstractions assembled within the "mind" of a computer. They are artificially enhanced for sharpness and artificially colored to increase contrast (all their colors are "phony"). The images are several steps removed from direct observation.

As you can see, medical science offers remarkable diagnostic tools. Consider the M2A Swallowable Imaging Capsule, a tiny camera that a patient swallows like a pill, and then excretes normally 8–72 hours later. As the M2A travels through the digestive tract, it photographs the small intestine and beams the color images to a Walkman-sized receiver. A study found the device to be 60% effective at detecting intestinal problems, compared to a 35% success rate with other imaging techniques. At present the M2A can provide images only of the small intestine because the battery gives out before it enters the large intestine.

New imaging technologies also make long-distance surgery possible. Visual images of a diseased organ travel via fiber-optic cable to surgeons at another location (even a different country), who manipulate delicate robotic instruments to remove the organ.

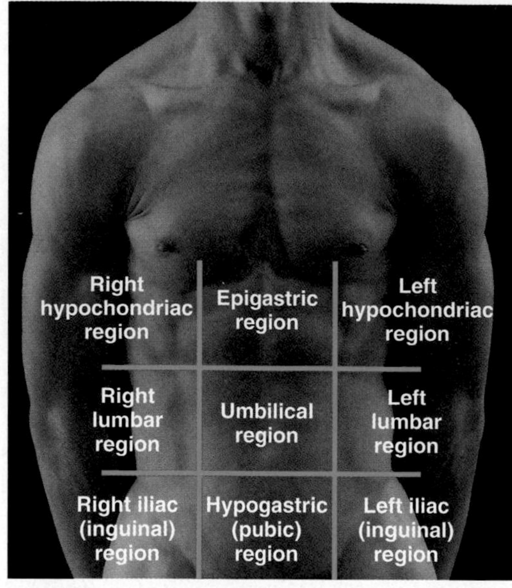

(a) Nine regions delineated by four planes

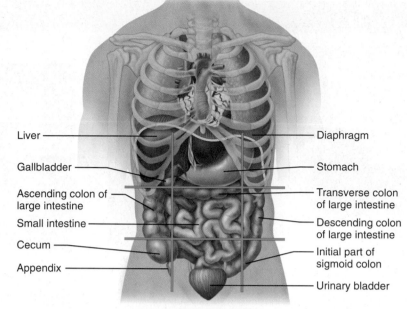

(b) Anterior view of the nine regions showing the superficial organs

Figure 1.12 The nine abdominopelvic regions. In **(a)** the superior transverse plane is just inferior to the ribs; the inferior transverse plane is just superior to the hip bones; and the parasagittal planes lie just medial to the nipples.

Other Body Cavities

In addition to the large closed body cavities, there are several smaller body cavities. Most of these are in the head and most open to the body exterior. Figure 1.7 provides the terms that will help you locate all but the last two cavities mentioned here.

1. **Oral and digestive cavities.** The oral cavity, commonly called the mouth, contains the teeth and tongue. This cavity is part of and continuous with the cavity of the digestive organs, which opens to the exterior at the anus.
2. **Nasal cavity.** Located within and posterior to the nose, the nasal cavity is part of the respiratory system passageways.
3. **Orbital cavities.** The orbital cavities (orbits) in the skull house the eyes and present them in an anterior position.
4. **Middle ear cavities.** The middle ear cavities in the skull lie just medial to the eardrums. These cavities contain tiny bones that transmit sound vibrations to the hearing receptors in the inner ears.
5. **Synovial cavities.** Synovial (sĭ-no′ve-al) cavities are joint cavities. They are enclosed within fibrous capsules that surround freely movable joints of the body (such as the elbow and knee joints). Like the serous membranes, membranes lining synovial cavities secrete a lubricating fluid that reduces friction as the bones move across one another.

CHECK YOUR UNDERSTANDING

15. Joe went to the emergency room where he complained of severe pains in the lower right quadrant of his abdomen. What might be his problem?
16. Of the uterus, small intestine, spinal cord, and heart, which is/are in the dorsal body cavity?
17. When you rub your cold hands together, the friction between them results in heat that warms your hands. Why doesn't warming friction result during movements of the heart, lungs, and digestive organs?

For answers, see Appendix G.

CHAPTER SUMMARY

An Overview of Anatomy and Physiology (pp. 2–3)

1. Anatomy is the study of body structures and their relationships. Physiology is the science of how body parts function.

Topics of Anatomy (p. 2)

2. Major subdivisions of anatomy include gross anatomy, microscopic anatomy, and developmental anatomy.

Topics of Physiology (pp. 2–3)

3. Typically, physiology concerns the functioning of specific organs or organ systems. Examples include cardiac physiology, renal physiology, and muscle physiology.
4. Physiology is explained by chemical and physical principles.

Complementarity of Structure and Function (p. 3)

5. Anatomy and physiology are inseparable: What a body can do depends on the unique architecture of its parts. This principle is called the complementarity of structure and function.

Levels of Structural Organization (pp. 3–4)

1. The levels of structural organization of the body, from simplest to most complex, are: chemical, cellular, tissue, organ, organ system, and organismal.
2. The 11 organ systems of the body are the integumentary, skeletal, muscular, nervous, endocrine, cardiovascular, lymphatic, respiratory, digestive, urinary, and reproductive systems. The immune system is a functional system closely associated with the lymphatic system. (For functions of these systems see pp. 6–7.)

Maintaining Life (pp. 4–8)

Necessary Life Functions (pp. 4–8)

1. All living organisms carry out certain vital functional activities necessary for life, including maintenance of boundaries, movement, responsiveness, digestion, metabolism, excretion, reproduction, and growth.

Survival Needs (p. 8)

2. Survival needs include nutrients, water, oxygen, and appropriate temperature and atmospheric pressure.

Homeostasis (pp. 8–11)

1. Homeostasis is a dynamic equilibrium of the internal environment. All body systems contribute to homeostasis, but the nervous and endocrine systems are most important. Homeostasis is necessary for health.

Homeostatic Control (pp. 9–11)

2. Control mechanisms of the body contain at least three elements that work together: receptor(s), control center, and effector(s).
3. Negative feedback mechanisms reduce the effect of the original stimulus, and are essential for maintaining homeostasis. Body temperature, heart rate, breathing rate and depth, and blood levels of glucose and certain ions are regulated by negative feedback mechanisms.
4. Positive feedback mechanisms intensify the initial stimulus, leading to an enhancement of the response. They rarely contribute to homeostasis, but blood clotting and labor contractions are regulated by such mechanisms.

Homeostatic Imbalance (p. 11)

5. With age, the efficiency of negative feedback mechanisms declines, and positive feedback mechanisms occur more frequently. These changes underlie certain disease conditions.

The Language of Anatomy (pp. 11–20)

Anatomical Position and Directional Terms (p. 13)

1. In the anatomical position, the body is erect, facing forward, feet slightly apart, arms at sides with palms forward.
2. Directional terms allow body parts to be located precisely. Terms that describe body directions and orientation include: superior/inferior; anterior/posterior; ventral/dorsal; medial/lateral; intermediate; proximal/distal; and superficial/deep.

Regional Terms (p. 14)

3. Regional terms are used to designate specific areas of the body (see Figure 1.7).

Anatomical Variability (p. 14)

4. People vary internally as well as externally, but extreme variations are rare.

Body Planes and Sections (p. 14)

5. The body or its organs may be cut along planes, or imaginary lines, to produce different types of sections. Frequently used planes are sagittal, frontal, and transverse.

Body Cavities and Membranes (pp. 14–20)

6. The body contains two major closed cavities. The dorsal cavity, subdivided into the cranial and spinal cavities, contains the brain and spinal cord. The ventral cavity is subdivided into the thoracic cavity, which houses the heart and lungs, and the abdominopelvic cavity, which contains the liver, digestive organs, and reproductive structures.
7. The walls of the ventral cavity and the surfaces of the organs it contains are covered with thin membranes, the parietal and visceral serosae, respectively. The serosae produce a thin fluid that decreases friction during organ functioning.
8. The abdominopelvic cavity may be divided by four planes into nine abdominopelvic regions (epigastric, umbilical, hypogastric, right and left iliac, right and left lumbar, and right and left hypochondriac), or by two planes into four quadrants. (For boundaries and organs contained, see Figures 1.11 and 1.12.)
9. There are several smaller body cavities. Most of these are in the head and open to the exterior.

REVIEW QUESTIONS

Multiple Choice/Matching

(Some questions have more than one correct answer. Select the best answer or answers from the choices given.)

1. The correct sequence of levels forming the structural hierarchy is **(a)** organ, organ system, cellular, chemical, tissue, organismal; **(b)** chemical, cellular, tissue, organismal, organ, organ system; **(c)** chemical, cellular, tissue, organ, organ system, organismal; **(d)** organismal, organ system, organ, tissue, cellular, chemical.
2. The structural and functional unit of life is **(a)** a cell, **(b)** an organ, **(c)** the organism, **(d)** a molecule.
3. Which of the following is a *major* functional characteristic of all organisms? **(a)** movement, **(b)** growth, **(c)** metabolism, **(d)** responsiveness, **(e)** all of these.
4. Two of these organ systems bear the *major* responsibility for ensuring homeostasis of the internal environment. Which two? **(a)** nervous system, **(b)** digestive system, **(c)** cardiovascular system, **(d)** endocrine system, **(e)** reproductive system.
5. In (a)–(e), a directional term [e.g., distal in (a)] is followed by terms indicating different body structures or locations (e.g., the elbow/the wrist). In each case, choose the structure or organ that matches the given directional term.
   (a) distal: the elbow/the wrist
   (b) lateral: the hip bone/the umbilicus
   (c) superior: the nose/the chin
   (d) anterior: the toes/the heel
   (e) superficial: the scalp/the skull
6. Assume that the body has been sectioned along three planes: (1) a median plane, (2) a frontal plane, and (3) a transverse plane made at the level of each of the organs listed below. Which organs would not be visible in all three cases? **(a)** urinary bladder, **(b)** brain, **(c)** lungs, **(d)** kidneys, **(e)** small intestine, **(f)** heart.
7. Relate each of the following conditions or statements to either the dorsal body cavity or the ventral body cavity.
   (a) surrounded by the bony skull and the vertebral column
   (b) includes the thoracic and abdominopelvic cavities
   (c) contains the brain and spinal cord
   (d) contains the heart, lungs, and digestive organs
8. Which of the following relationships is *incorrect*?
   (a) visceral peritoneum/outer surface of small intestine
   (b) parietal pericardium/outer surface of heart
   (c) parietal pleura/wall of thoracic cavity
9. Which ventral cavity subdivision has no bony protection? **(a)** thoracic cavity, **(b)** abdominal cavity, **(c)** pelvic cavity.
10. Terms that apply to the backside of the body in the anatomical position include:
    (a) ventral; anterior
    (b) back; rear
    (c) posterior; dorsal
    (d) medial; lateral

Short Answer Essay Questions

11. According to the principle of complementarity, how does anatomy relate to physiology?
12. Construct a table that lists the 11 systems of the body, names two organs of each system (if appropriate), and describes the overall or major function of each system.
13. List and describe briefly five external factors that must be present or provided to sustain life.
14. Define homeostasis.
15. Compare and contrast the operation of negative and positive feedback mechanisms in maintaining homeostasis. Provide two examples of variables controlled by negative feedback mechanisms and one example of a process regulated by a positive feedback mechanism.
16. Why is an understanding of the anatomical position important?
17. Define plane and section.
18. Provide the anatomical term that correctly names each of the following body regions: (a) arm, (b) thigh, (c) chest, (d) fingers and toes, (e) anterior aspect of the knee.
19. Use as many directional terms as you can to describe the relationship between the elbow's olecranal region and your palm.
20. (a) Make a diagram showing the nine abdominopelvic regions, and name each region. Name two organs (or parts of organs) that could be located in each of the named regions. (b) Make a similar sketch illustrating how the abdominopelvic cavity may be divided into quadrants, and name each quadrant.

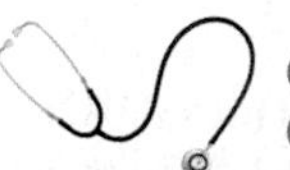

Critical Thinking and Clinical Application Questions

1. John has been suffering agonizing pain with each breath and has been informed by the physician that he has pleurisy. (a) Specifically, what membranes are involved in this condition? (b) What is their usual role in the body? (c) Explain why John's condition is so painful.
2. At the clinic, Harry was told that blood would be drawn from his antecubital region. What body part was Harry asked to hold out? Later, the nurse came in and gave Harry a shot of penicillin in the area just distal to his acromial region. Did Harry take off his shirt or drop his pants to receive the injection? Before Harry left, the nurse noticed that Harry had a nasty bruise on his gluteal region. What part of his body was black and blue?
3. A man is behaving abnormally, and his physician suspects that he has a brain tumor. Which of the following medical imaging techniques would best localize the tumor in the man's brain (and why)? Conventional X ray, DSA, PET, sonography, MRI.
4. Calcium levels in Mr. Gallariani's blood are dropping to dangerously low levels. The hormone PTH is released and soon blood calcium levels begin to rise. Shortly after, PTH release slows. Is this an example of a positive or negative feedback mechanism? What is the initial stimulus? What is the result?
5. Mr. Harvey, a computer programmer, has been complaining of numbness and pain in his right hand. The nurse practitioner diagnosed his problem as carpal tunnel syndrome and prescribed use of a splint. Where will Mr. Harvey apply the splint?

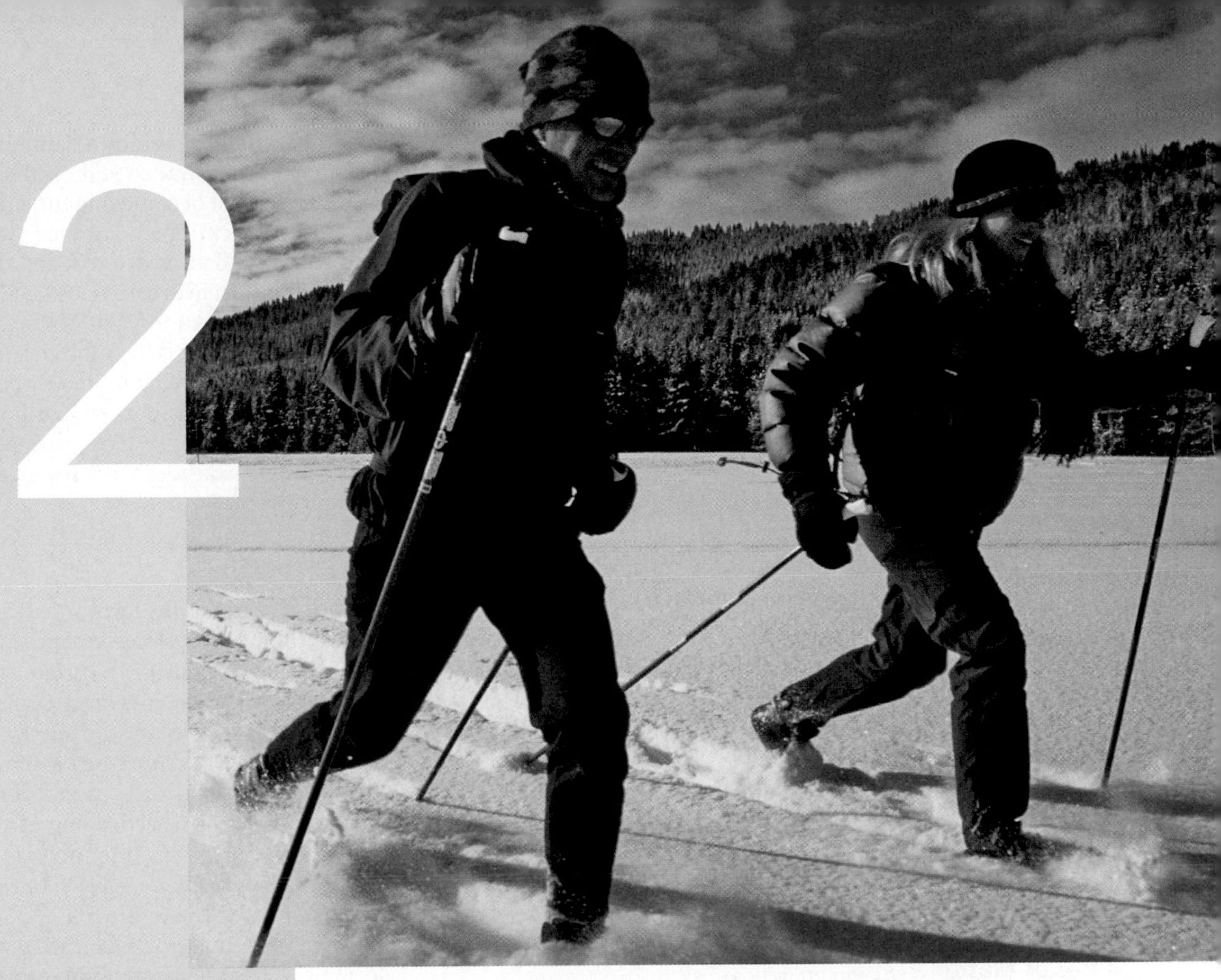

2 Chemistry Comes Alive

Why study chemistry in an anatomy and physiology course? The answer is simple. Your entire body is made up of chemicals, thousands of them, continuously interacting with one another at an incredible pace. Although it is possible to study anatomy without much reference to chemistry, chemical reactions underlie all physiological processes—movement, digestion, the pumping of your heart, and even your thoughts. This chapter presents the basic chemistry and biochemistry (the chemistry of living material) you need to understand body functions.

PART 1

BASIC CHEMISTRY

Definition of Concepts: Matter and Energy

▶ Differentiate between matter and energy and between
2 potential energy and kinetic energy.

▶ Describe the major energy forms.

Matter

Matter is the "stuff" of the universe. More precisely, **matter** is anything that occupies space and has mass. With some exceptions, it can be seen, smelled, and felt.

For all practical purposes, we can consider mass to be the same as weight. However, this usage is not quite accurate. The *mass* of an object is equal to the actual amount of matter in the object, and it remains constant wherever the object is. In contrast, weight varies with gravity. So while your mass is the same at sea level and on a mountaintop, you weigh just slightly less on that mountaintop. The science of chemistry studies the nature of matter, especially how its building blocks are put together and interact.

States of Matter

Matter exists in *solid*, *liquid*, and *gaseous states*. Examples of each state are found in the human body. Solids, like bones and teeth, have a definite shape and volume. Liquids such as blood plasma have a definite volume, but they conform to the shape of their container. Gases have neither a definite shape nor a definite volume. The air we breathe is a gas.

Energy

Compared with matter, energy is less tangible. It has no mass, does not take up space, and we can measure it only by its effects on matter. **Energy** is defined as the capacity to do work, or to put matter into motion. The greater the work done, the more energy is used doing it. A baseball player who has just hit the ball over the fence uses much more energy than a batter who bunts the ball back to the pitcher.

Kinetic Versus Potential Energy

Energy exists in two forms, or work capacities, and each can be transformed to the other. **Kinetic energy** (ki-net′ik) is energy in action. We see evidence of kinetic energy in the constant movement of the tiniest particles of matter (atoms) as well as in larger objects (a bouncing ball). Kinetic energy does work by moving objects, which in turn can do work by moving or pushing on other objects. For example, a push on a swinging door sets it into motion.

Potential energy is stored energy, that is, inactive energy that has the *potential*, or capability, to do work but is not presently doing so. The batteries in an unused toy have potential energy, as does water confined behind a dam. Your leg muscles have potential energy when you sit still on the couch. When potential energy is released, it becomes kinetic energy and so is capable of doing work. For example, dammed water becomes a rushing torrent when the dam is opened, and that rushing torrent can move a turbine at a hydroelectric plant, or charge a battery.

Actually, energy is a topic of physics, but matter and energy are inseparable. Matter is the substance, and energy is the mover of the substance. All living things are composed of matter and they all require energy to grow and function. The release and use of energy by living systems gives us the elusive quality we call life. Now let's consider the forms of energy used by the body as it does its work.

Forms of Energy

- **Chemical energy** is the form stored in the bonds of chemical substances. When chemical reactions occur that rearrange the atoms of the chemicals in a certain way, the potential energy is unleashed and becomes kinetic energy, or energy in action.

  For example, some of the energy in the foods you eat is eventually converted into the kinetic energy of your moving arm. However, food fuels cannot be used to energize body activities directly. Instead, some of the food energy is captured temporarily in the bonds of a chemical called *adenosine triphosphate (ATP)* (ah-den′o-sēn tri″fos′fāt). Later, ATP's bonds are broken and the stored energy is released as needed to do cellular work. Chemical energy in the form of ATP is the most useful form of energy in living systems because it is used to run almost all functional processes.
- **Electrical energy** results from the movement of charged particles. In your home, electrical energy is found in the flow of electrons along the household wiring. In your body, electrical currents are generated when charged particles called *ions* move along or across cell membranes. The nervous system uses electrical currents, called *nerve impulses*, to transmit messages from one part of the body to another. Electrical currents traveling across the heart stimulate it to contract (beat) and pump blood. (This is why a strong electrical shock, which interferes with such currents, can cause death.)
- **Mechanical energy** is energy *directly* involved in moving matter. When you ride a bicycle, your legs provide the mechanical energy that moves the pedals.
- **Radiant energy**, or **electromagnetic energy** (e-lek″tro-mag-net′ik), is energy that travels in waves. These waves, which vary in length, are collectively called the *electromagnetic spectrum*. They include visible light, infrared waves, radio waves, ultraviolet waves, and X rays. Light energy, which stimulates the retinas of our eyes, is important in vision. Ultraviolet waves cause sunburn, but they also stimulate our body to make vitamin D.

Energy Form Conversions

With few exceptions, energy is easily converted from one form to another. For example, the chemical energy (in gasoline) that powers the motor of a speedboat is converted into the mechanical energy of the whirling propeller that makes the boat skim across the water.

Energy conversions are quite inefficient. Some of the initial energy supply is always "lost" to the environment as heat. (It is not really lost because energy cannot be created or destroyed, but that portion given off as heat is at least partly *unusable.*) It is easy to demonstrate this principle. Electrical energy is converted into light energy in a lightbulb. But if you touch a lit bulb, you will soon discover that some of the electrical energy is producing heat instead.

Likewise, all energy conversions in the body liberate heat. This heat helps to maintain our relatively high body temperature, which influences body functioning. For example, when matter is heated, the kinetic energy of its particles increases and they begin to move more quickly. The higher the temperature, the faster the body's chemical reactions occur. We will learn more about this later.

CHECK YOUR UNDERSTANDING

1. What form of energy is found in the food we eat?
2. What form of energy is used to transmit messages from one part of the body to another?
3. What type of energy is available when we are still? When we are exercising?

For answers, see Appendix G.

Composition of Matter: Atoms and Elements

- ▶ Define chemical element and list the four elements that form the bulk of body matter.
- ▶ Define atom. List the subatomic particles, and describe their relative masses, charges, and positions in the atom.
- ▶ Define atomic number, atomic mass, atomic weight, isotope, and radioisotope.

All matter is composed of **elements**, unique substances that cannot be broken down into simpler substances by ordinary chemical methods. Among the well-known elements are oxygen, carbon, gold, silver, copper, and iron.

At present, 112 elements are known with certainty (and numbers 113, 114, 115, 116, and most recently 118 are alleged). Of these, 92 occur in nature. The rest are made artificially in particle accelerator devices.

Four elements—carbon, oxygen, hydrogen, and nitrogen—make up about 96% of body weight, and 20 others are present in the body, some in trace amounts. **Table 2.1** lists the elements contributing to body mass and gives their importance. In Appendix E, an oddly shaped checkerboard called the **periodic table** provides a more complete listing of the known elements.

Each element is composed of more or less identical particles or building blocks, called **atoms**. The smallest atoms are less than 0.1 nanometer (nm) in diameter, and the largest are only about five times as large. [1 nm = 0.0000001 (or 10^{-7}) centimeter (cm), or 40 billionths of an inch!]

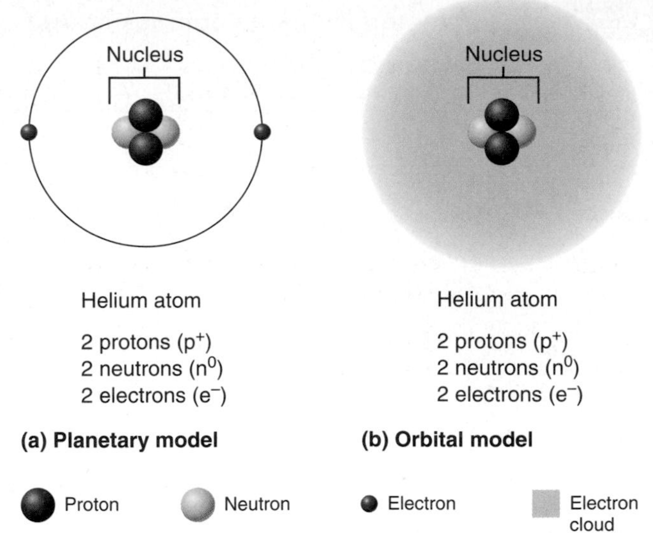

Figure 2.1 Two models of the structure of an atom.

Every element's atoms differ from those of all other elements and give the element its unique physical and chemical properties. *Physical properties* are those we can detect with our senses (such as color and texture) or measure (such as boiling point and freezing point). *Chemical properties* pertain to the way atoms interact with other atoms (bonding behavior) and account for the facts that iron rusts, animals can digest their food, and so on.

We designate each element by a one- or two-letter chemical shorthand called an **atomic symbol**, usually the first letter(s) of the element's name. For example, C stands for carbon, O for oxygen, and Ca for calcium. In a few cases, the atomic symbol is taken from the Latin name for the element. For example, sodium is indicated by Na, from the Latin word *natrium.*

Atomic Structure

The word *atom* comes from the Greek word meaning "indivisible." However, we now know that atoms are clusters of even smaller particles called protons, neutrons, and electrons and that even those subatomic particles can be subdivided with high-technology tools. Still, the old idea of atomic indivisibility is useful because an atom loses the unique properties of its element when it is split into its subatomic particles.

An atom's subatomic particles differ in mass, electrical charge, and position in the atom. An atom has a central **nucleus** containing protons and neutrons tightly bound together. The nucleus, in turn, is surrounded by orbiting electrons **(Figure 2.1)**. **Protons** (p^+) bear a positive electrical charge, and **neutrons** (n^0) are neutral, so the nucleus is positively charged overall. Protons and neutrons are heavy particles and have approximately the same mass, arbitrarily designated as 1 **atomic mass unit** (1 amu). Since all of the heavy subatomic particles are concentrated in the nucleus, the nucleus is fantastically dense. It accounts for nearly the entire mass (99.9%) of the atom.

2

TABLE 2.1 Common Elements Composing the Human Body*

ELEMENT	ATOMIC SYMBOL	APPROX. % BODY MASS†	FUNCTIONS
Major (96.1%)			
Oxygen	O	65.0	A major component of both organic (carbon-containing) and inorganic (non-carbon-containing) molecules. As a gas, it is needed for the production of cellular energy (ATP).
Carbon	C	18.5	A primary component of all organic molecules, which include carbohydrates, lipids (fats), proteins, and nucleic acids.
Hydrogen	H	9.5	A component of all organic molecules. As an ion (proton), it influences the pH of body fluids.
Nitrogen	N	3.2	A component of proteins and nucleic acids (genetic material).
Lesser (3.9%)			
Calcium	Ca	1.5	Found as a salt in bones and teeth. Its ionic (Ca^{2+}) form is required for muscle contraction, conduction of nerve impulses, and blood clotting.
Phosphorus	P	1.0	Part of calcium phosphate salts in bones and teeth. Also present in nucleic acids, and part of ATP.
Potassium	K	0.4	Its ion (K^+) is the major positive ion (cation) in cells. Necessary for conduction of nerve impulses and muscle contraction.
Sulfur	S	0.3	Component of proteins, particularly muscle proteins.
Sodium	Na	0.2	As an ion (Na^+), sodium is the major positive ion found in extracellular fluids (fluids outside of cells). Important for water balance, conduction of nerve impulses, and muscle contraction.
Chlorine	Cl	0.2	Its ion (chloride, Cl^-) is the most abundant negative ion (anion) in extracellular fluids.
Magnesium	Mg	0.1	Present in bone. Also an important cofactor in a number of metabolic reactions.
Iodine	I	0.1	Needed to make functional thyroid hormones.
Iron	Fe	0.1	Component of hemoglobin (which transports oxygen within red blood cells) and some enzymes.
Trace (less than 0.01%)			
Chromium (Cr); cobalt (Co); copper (Cu); fluorine (F); manganese (Mn); molybdenum (Mo); selenium (Se); silicon (Si); tin (Sn); vanadium (V); zinc (Zn)			
These elements are referred to as *trace elements* because they are required in very minute amounts; many are found as part of enzymes or are required for enzyme activation.			

*A listing of the elements by ascending order of atomic number appears in the periodic table, Appendix E.
†Percentage of "wet" body mass; includes water.

The tiny **electrons** (e^-) bear a negative charge equal in strength to the positive charge of the proton. However, an electron has only about 1/2000 the mass of a proton, and the mass of an electron is usually designated as 0 amu.

All atoms are electrically neutral because the number of protons in an atom is precisely balanced by its number of electrons (the + and − charges will then cancel the effect of each other). For example, hydrogen has one proton and one electron, and iron has 26 protons and 26 electrons. For any atom, the number of protons and electrons is always equal.

The **planetary model** of the atom, illustrated in Figure 2.1a, is a simplified (and now outdated) model of atomic structure. As you can see, it depicts electrons moving around the nucleus in fixed, generally circular orbits. But we can never determine the exact location of electrons at a particular time because they jump around following unknown trajectories. So, instead of speaking of specific orbits, chemists talk about **orbitals**—regions around the nucleus in which a given electron or electron pair is likely to be found most of the time. This more modern model of atomic structure, called the **orbital model**, is more useful for predicting the chemical behavior of atoms. As illustrated in Figure 2.1b, the orbital model depicts *probable* regions of greatest electron density by denser shading (this haze is called the *electron cloud*). However, the planetary model is simpler to depict, so we will use that model in most illustrations of atomic structure in this text.

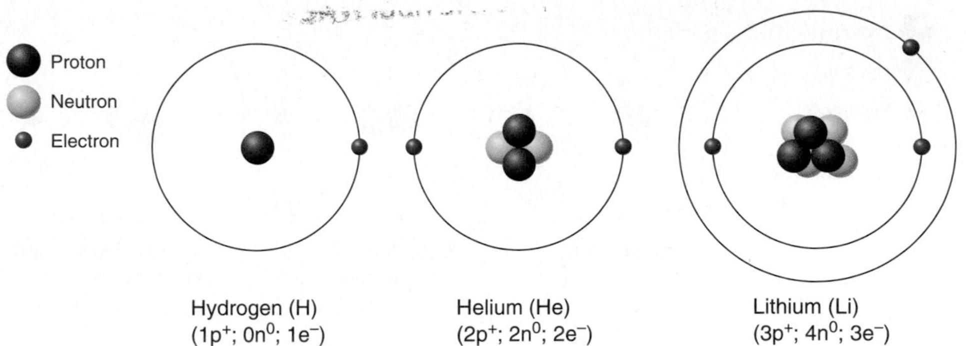

2

Figure 2.2 Atomic structure of the three smallest atoms.

Hydrogen, with just one proton and one electron, is the simplest atom. You can visualize the spatial relationships in the hydrogen atom by imagining it as a sphere enlarged until its diameter equals the length of a football field. In that case, the nucleus could be represented by a lead ball the size of a gumdrop in the exact center of the sphere. Its lone electron could be pictured as a fly buzzing about unpredictably within the sphere. Though not completely accurate, this mental image demonstrates that most of the volume of an atom is empty space, and nearly all of its mass is concentrated in the central nucleus.

Identifying Elements

All protons are alike, regardless of the atom considered. The same is true of all neutrons and all electrons. So what determines the unique properties of each element? The answer is that atoms of different elements are composed of *different numbers* of protons, neutrons, and electrons.

The simplest and smallest atom, hydrogen, has one proton, one electron, and no neutrons **(Figure 2.2)**. Next in size is the helium atom, with two protons, two neutrons, and two orbiting electrons. Lithium follows with three protons, four neutrons, and three electrons. If we continued this step-by-step progression, we would get a graded series of atoms containing from 1 to 112 protons, an equal number of electrons, and a slightly larger number of neutrons at each step.

All we really need to know to identify a particular element, however, are its atomic number, mass number, and atomic weight. Taken together, these provide a fairly complete picture of each element.

Atomic Number

The **atomic number** of any atom is equal to the number of protons in its nucleus and is written as a subscript to the left of its atomic symbol. Hydrogen, with one proton, has an atomic number of 1 ($_1H$). Helium, with two protons, has an atomic number of 2 ($_2He$), and so on. The number of protons is always equal to the number of electrons in an atom, so the atomic number *indirectly* tells us the number of electrons in the atom as well. As we will see shortly, this information is important indeed, because electrons determine the chemical behavior of atoms.

Mass Number and Isotopes

The **mass number** of an atom is the sum of the masses of its protons and neutrons. (The mass of the electrons is so small that it is ignored.) Recall that protons and neutrons have a mass of 1 amu. Hydrogen has only one proton in its nucleus, so its atomic and mass numbers are the same: 1. Helium, with two protons and two neutrons, has a mass number of 4.

The mass number is usually indicated by a superscript to the left of the atomic symbol. For example, helium is ^4_2He. This simple notation allows us to deduce the total number and kinds of subatomic particles in any atom because it indicates the number of protons (the atomic number), the number of electrons (equal to the atomic number), and the number of neutrons (mass number minus atomic number). In our example, we can do the subtraction to find that ^4_2He has two neutrons.

From what we have said so far, it may appear as if each element has one, and only one, type of atom representing it. This is not the case. Nearly all known elements have two or more structural variations called **isotopes** (i′so-tōps), which have the same number of protons (and electrons), but differ in the number of neutrons they contain. Earlier, when we said that hydrogen has a mass number of 1, we were speaking of ^1H, its most abundant isotope. Some hydrogen atoms have a mass of 2 or 3 amu (atomic mass units), which means that they have one proton and, respectively, one or two neutrons **(Figure 2.3)**.

Carbon has several isotopes. The most abundant of these are ^{12}C, ^{13}C, and ^{14}C. Each of the carbon isotopes has six protons (otherwise it would not be carbon), but ^{12}C has six neutrons, ^{13}C has seven, and ^{14}C has eight. Isotopes can also be written with the mass number following the symbol: C-14, for example.

Atomic Weight

You might think that atomic weight should be the same as atomic mass, and this would be so if atomic weight referred to the weight of a single atom. However, **atomic weight** is an average of the relative weights (mass numbers) of *all* the isotopes of an element, taking into account their relative abundance in

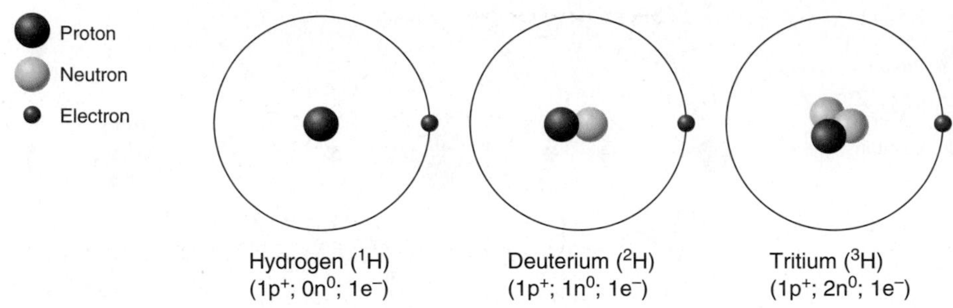

Figure 2.3 Isotopes of hydrogen.

nature. As a rule, the atomic weight of an element is approximately equal to the mass number of its most abundant isotope. For example, the atomic weight of hydrogen is 1.008, which reveals that its lightest isotope (^1H) is present in much greater amounts in our world than its ^2H or ^3H forms.

Radioisotopes

The heavier isotopes of many elements are unstable, and their atoms decompose spontaneously into more stable forms. This process of atomic decay is called *radioactivity,* and isotopes that exhibit this behavior are called **radioisotopes** (ra″de-o-i′so-tōps). The disintegration of a radioactive nucleus may be compared to a tiny explosion. It occurs when subatomic *alpha* (α) *particles* (packets of 2p + 2n), *beta* (β) *particles* (electron-like negative particles), or *gamma* (γ) *rays* (electromagnetic energy) are ejected from the atomic nucleus.

Why does this happen? The answer is complex, but for our purposes, the important point to know is that the dense nuclear particles are composed of even smaller particles called *quarks* that associate in one way to form protons and in another way to form neutrons. Apparently, the "glue" that holds these nuclear particles together is weaker in the heavier isotopes. When radioisotopes disintegrate, the element may transform to a different element.

Because we can detect radioactivity with scanners, and radioactive isotopes share the same chemistry as their more stable isotopes, radioisotopes are valuable tools for biological research and medicine. Most radioisotopes used in the clinical setting are used for diagnosis, that is, to localize and illuminate damaged or cancerous tissues. For example, iodine-131 is used to determine the size and activity of the thyroid gland and to detect thyroid cancer. The sophisticated PET scans described in *A Closer Look* in Chapter 1 use radioisotopes to probe the workings of molecules deep within our bodies. All radioisotopes, regardless of the purpose for which they are used, damage living tissue, and all radioisotopes gradually lose their radioactive behavior. The time required for a radioisotope to lose one-half of its activity is called its *half-life.* The half-lives of radioisotopes vary dramatically from hours to thousands of years.

Alpha emission has the lowest penetrating power and is least damaging to living tissue. Nonetheless, inhaled alpha particles from decaying radon are second only to smoking as a cause of lung cancer. (Radon results naturally from decay of uranium in the ground.) Gamma emission has the greatest penetrating power. Radium-226, cobalt-60, and certain other radioisotopes that decay by gamma emission are used to destroy localized cancers.

Contrary to what some believe, ionizing radiation does not damage organic molecules directly. Instead, it knocks electrons out of other atoms and sends them flying, like bowling balls smashing through pins all along their path. It is the electron energy and the unstable molecules left behind that do the damage.

CHECK YOUR UNDERSTANDING

4. What two elements besides H and N make up the bulk of living matter?
5. An element has a mass of 207 and has 125 neutrons in its nucleus. How many protons and electrons does it have and where are they located?
6. How do the terms atomic mass and atomic weight differ?

For answers, see Appendix G.

How Matter Is Combined: Molecules and Mixtures

- ▶ Define molecule, and distinguish between a compound and a mixture.
- ▶ Compare solutions, colloids, and suspensions.

Molecules and Compounds

Most atoms do not exist in the free state, but instead are chemically combined with other atoms. Such a combination of two or more atoms held together by chemical bonds is called a **molecule.**

If two or more atoms of the *same* element combine, the resulting substance is called a *molecule of that element.* When two hydrogen atoms bond, the product is a molecule of hydrogen gas and is written as H_2. Similarly, when two oxygen atoms combine, a molecule of oxygen gas (O_2) is formed. Sulfur atoms commonly combine to form sulfur molecules containing eight sulfur atoms (S_8).

When two or more *different* kinds of atoms bind, they form molecules of a **compound.** Two hydrogen atoms combine with one oxygen atom to form the compound water (H_2O). Four hydrogen atoms combine with one carbon atom to form the compound methane (CH_4). Notice again that molecules of

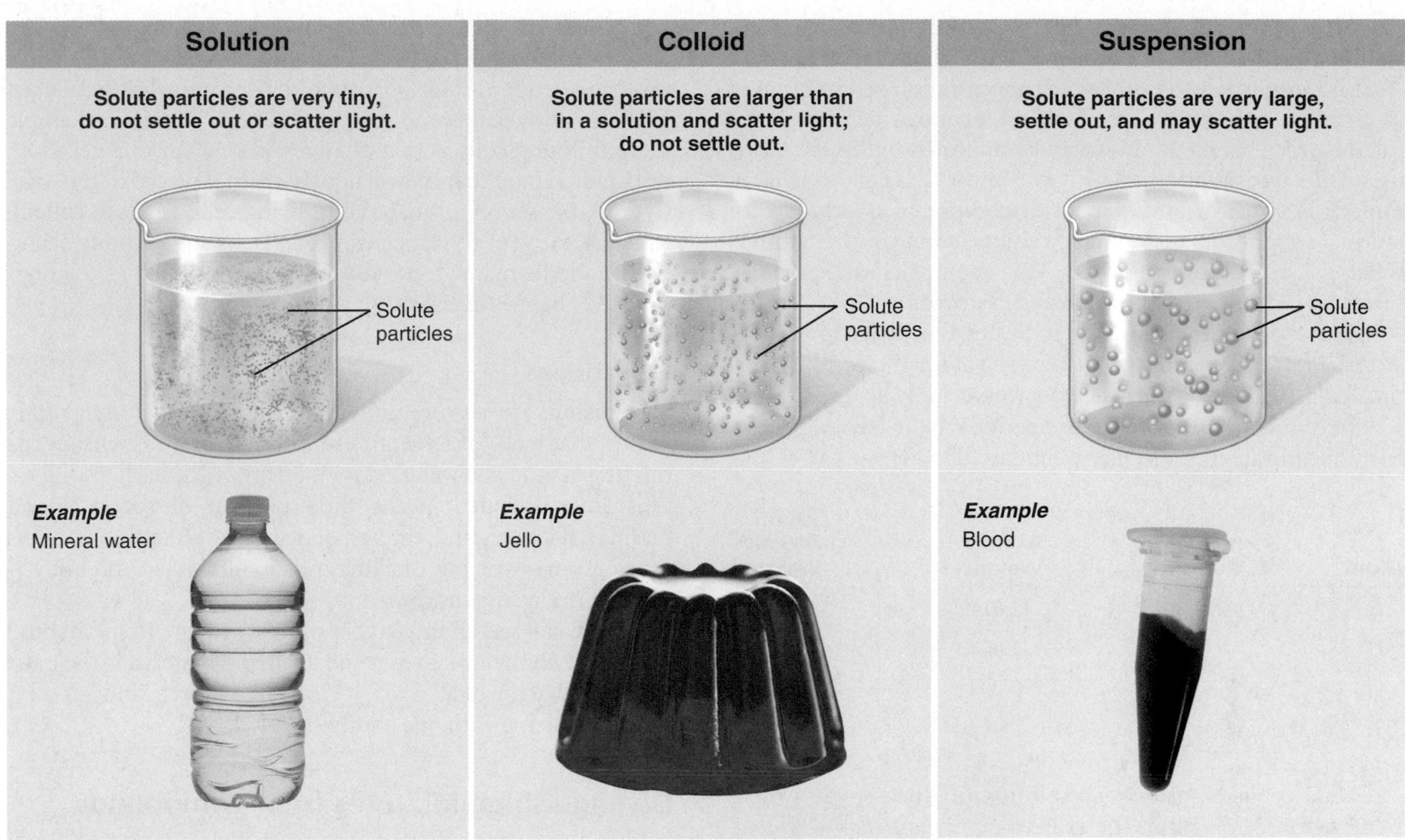

Solution
Solute particles are very tiny, do not settle out or scatter light.
Solute particles
Example
Mineral water
Colloid
Solute particles are larger than in a solution and scatter light; do not settle out.
Solute particles
Example
Jello
Suspension
Solute particles are very large, settle out, and may scatter light.
Solute particles
Example
Blood

Figure 2.4 The three basic types of mixtures.

methane and water are compounds, but molecules of hydrogen gas are not, because compounds always contain atoms of at least two different elements.

Compounds are chemically pure, and all of their molecules are identical. So, just as an atom is the smallest particle of an element that still has the properties of the element, a molecule is the smallest particle of a compound that still has the specific characteristics of the compound. This concept is important because the properties of compounds are usually very different from those of the atoms they contain. Water, for example, is very different from the elements hydrogen and oxygen. Indeed, it is next to impossible to tell what atoms are in a compound without analyzing it chemically.

Mixtures

Mixtures are substances composed of two or more components *physically intermixed.* Most matter in nature exists in the form of mixtures, but there are only three basic types: *solutions, colloids,* and *suspensions* **(Figure 2.4)**.

Solutions

Solutions are homogeneous mixtures of components that may be gases, liquids, or solids. *Homogeneous* means that the mixture has exactly the same composition or makeup throughout—a sample taken from any part of the mixture has the same composition (in terms of the atoms or molecules it contains) as a sample taken from any other part of the mixture. Examples include the air we breathe (a mixture of gases) and seawater (a mixture of salts, which are solids, and water). The substance present in the greatest amount is called the **solvent** (or dissolving medium). Solvents are usually liquids. Substances present in smaller amounts are called **solutes.**

Water is the body's chief solvent. Most solutions in the body are *true solutions* containing gases, liquids, or solids dissolved in water. True solutions are usually transparent. Examples are saline solution [table salt (NaCl) and water], a mixture of glucose and water, and mineral water. The solutes of true solutions are minute, usually in the form of individual atoms and molecules. Consequently, they are not visible to the naked eye, do not settle out, and do not scatter light. In other words, if a beam of light is passed through a true solution, you will not see the path of light.

Concentration of Solutions We describe true solutions in terms of their *concentration,* which may be indicated in various ways. Solutions used in a college laboratory or a hospital are often described in terms of the **percent** (parts per 100 parts) of the solute in the total solution. This designation always refers to the solute percentage, and unless otherwise noted, water is assumed to be the solvent.

Milligrams per deciliter (*mg/dl*) is another common concentration measurement. (A deciliter is 100 milliliters or 0.1 liter.)

Still another way to express the concentration of a solution is in terms of its **molarity** (mo-lar′ĭ-te), or moles per liter, indicated by *M*. This method is more complicated but much more useful. To understand molarity, you must understand what a mole is. A **mole** of any element or compound is equal to its atomic weight or **molecular weight** (sum of the atomic weights) weighed out in grams. This concept is easier than it seems, as illustrated by the following example.

2

Glucose is $C_6H_{12}O_6$, which indicates that it has 6 carbon atoms, 12 hydrogen atoms, and 6 oxygen atoms. To compute the molecular weight of glucose, you would look up the atomic weight of each of its atoms in the periodic table (see Appendix E) and compute its molecular weight as follows:

Atom	Number of Atoms		Atomic Weight		Total Atomic Weight
C	6	×	12.011	=	72.066
H	12	×	1.008	=	12.096
O	6	×	15.999	=	95.994
					180.156

Then, to make a *one-molar* solution of glucose, you would weigh out 180.156 grams (g), called a *gram molecular weight*, of glucose and add enough water to make 1 liter (L) of solution. In short, a one-molar solution (abbreviated 1.0 *M*) of a chemical substance is one gram molecular weight of the substance (or one gram atomic weight in the case of elemental substances) in 1 L (1000 milliliters) of solution.

The beauty of using the mole as the basis of preparing solutions is its precision. One mole of any substance always contains exactly the same number of solute particles, that is, 6.02×10^{23}. This number is called **Avogadro's number** (av″o-gad′rōz). So whether you weigh out 1 mole of glucose (180 g) or 1 mole of water (18 g) or 1 mole of methane (16 g), in each case you will have 6.02×10^{23} molecules of that substance.* This allows almost mind-boggling precision to be achieved.

Because solute concentrations in body fluids tend to be quite low, those values are usually reported in terms of millimoles (m*M*; 1/1000 mole).

Colloids

Colloids (kol′oidz), also called *emulsions*, are *heterogeneous* mixtures, which means that their composition is dissimilar in different areas of the mixture. Colloids often appear translucent or milky and although the solute particles are larger than those in true solutions, they still do not settle out. However, they do scatter light, so the path of a light beam shining through a colloidal mixture is visible.

Colloids have many unique properties, including the ability of some to undergo **sol-gel transformations**, that is, to change reversibly from a fluid (sol) state to a more solid (gel) state. Jell-O, or any gelatin product (Figure 2.4), is a familiar example of a nonliving colloid that changes from a sol to a gel when refrigerated (and that gel will liquefy again if placed in the sun). Cytosol, the semifluid material in living cells, is also a colloid, largely because of its dispersed proteins. Its sol-gel transformations underlie many important cell activities, such as cell division and changes in cell shape.

Suspensions

Suspensions are *heterogeneous* mixtures with large, often visible solutes that tend to settle out. An example of a suspension is a mixture of sand and water. So is blood, in which the living blood cells are suspended in the fluid portion of blood (blood plasma). If left to stand, the suspended cells will settle out unless some means—mixing, shaking, or circulation in the body—keeps them in suspension.

As you can see, all three types of mixtures are found in both living and nonliving systems. In fact, living material is the most complex mixture of all, since it contains all three kinds of mixtures interacting with one another.

Distinguishing Mixtures from Compounds

Now let's zero in on how to distinguish mixtures and compounds from one another. Mixtures differ from compounds in several important ways:

1. The chief difference between mixtures and compounds is that no chemical bonding occurs between the components of a mixture. The properties of atoms and molecules are not changed when they become part of a mixture. Remember they are only physically intermixed.
2. Depending on the mixture, its components can be separated by physical means—straining, filtering, evaporation, and so on. Compounds, by contrast, can be separated into their constituent atoms only by chemical means (breaking bonds).
3. Some mixtures are homogeneous, whereas others are heterogeneous. A bar of 100% pure (elemental) iron is homogeneous, as are all compounds. As already mentioned, heterogeneous substances vary in their makeup from place to place. For example, iron ore is a heterogeneous mixture that contains iron and many other elements.

CHECK YOUR UNDERSTANDING

7. What is the meaning of the term "molecule"?
8. Why is sodium chloride (NaCl) considered a compound, but oxygen gas is not?
9. Blood contains a liquid component and living cells. Would it be classified as a compound or a mixture? Why?

For answers, see Appendix G.

*The important exception to this rule concerns molecules that ionize and break up into charged particles (ions) in water, such as salts, acids, and bases (see p. 39). For example, simple table salt (sodium chloride) breaks up into two types of charged particles. Therefore, in a 1.0*M* solution of sodium chloride, *2 moles* of solute particles are actually in solution.

Chemical Bonds

- Explain the role of electrons in chemical bonding and in relation to the octet rule.
- Differentiate among ionic, covalent, and hydrogen bonds.
- Compare and contrast polar and nonpolar compounds.

As noted earlier, when atoms combine with other atoms, they are held together by **chemical bonds**. A chemical bond is not a physical structure like a pair of handcuffs linking two people together. Instead, it is an energy relationship between the electrons of the reacting atoms, and it is made or broken in less than a trillionth of a second.

The Role of Electrons in Chemical Bonding

Electrons forming the electron cloud around the nucleus of an atom occupy regions of space called **electron shells** that consecutively surround the atomic nucleus. The atoms known so far can have electrons in seven shells (numbered 1 to 7 from the nucleus outward), but the actual number of electron shells occupied in a given atom depends on the number of electrons that atom has. Each electron shell contains one or more orbitals. (Recall from our earlier discussion that *orbitals* are regions around the nucleus in which a given electron is likely to be found most of the time.)

It is important to understand that each electron shell represents a different **energy level**, because this prompts you to think of electrons as particles with a certain amount of potential energy. In general, the terms *electron shell* and *energy level* are used interchangeably.

How much potential energy does an electron have? The answer depends on the energy level that an electron occupies. The attraction between the positively charged nucleus and negatively charged electrons is greatest when electrons are closest to the nucleus and falls off with increasing distance. This statement explains why electrons farthest from the nucleus (1) have the greatest potential energy (it takes more energy for them to overcome the nuclear attraction and reach the more distant energy levels) and (2) are most likely to interact chemically with other atoms. (They are the least tightly held by their own atomic nucleus and the most easily influenced by other atoms and molecules.)

Each electron shell can hold a specific number of electrons. Shell 1, the shell immediately surrounding the nucleus, accommodates only 2 electrons. Shell 2 holds a maximum of 8, and shell 3 has room for 18. Subsequent shells hold larger and larger numbers of electrons, and the shells tend to be filled with electrons consecutively. For example, shell 1 fills completely before any electrons appear in shell 2.

Which electrons are involved in chemical bonding? When we consider bonding behavior, the only electrons that are important are those in the atom's outermost energy level. Inner electrons usually do not take part in bonding because they are more tightly held by the atomic nucleus.

When the outermost energy level of an atom is filled to capacity or contains eight electrons, the atom is stable. Such atoms are *chemically inert*, that is, unreactive. A group of elements called the *noble gases*, which include helium and neon, typify this condition **(Figure 2.5a)**. On the other hand, atoms in which the outermost energy level contains fewer than eight electrons tend to gain, lose, or share electrons with other atoms to achieve stability (Figure 2.5b).

What about atoms that have more than 20 electrons, in which the energy levels beyond shell 2 can contain *more* than eight electrons? The number of electrons that can participate in bonding is still limited to a total of eight. The term **valence shell** (va′lens) specifically indicates an atom's outermost energy level *or that portion of it* containing the electrons that are chemically reactive. Hence, the key to chemical reactivity is the **octet rule**

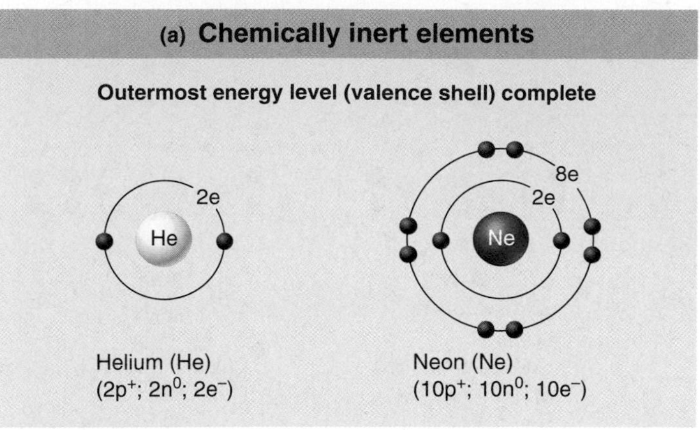

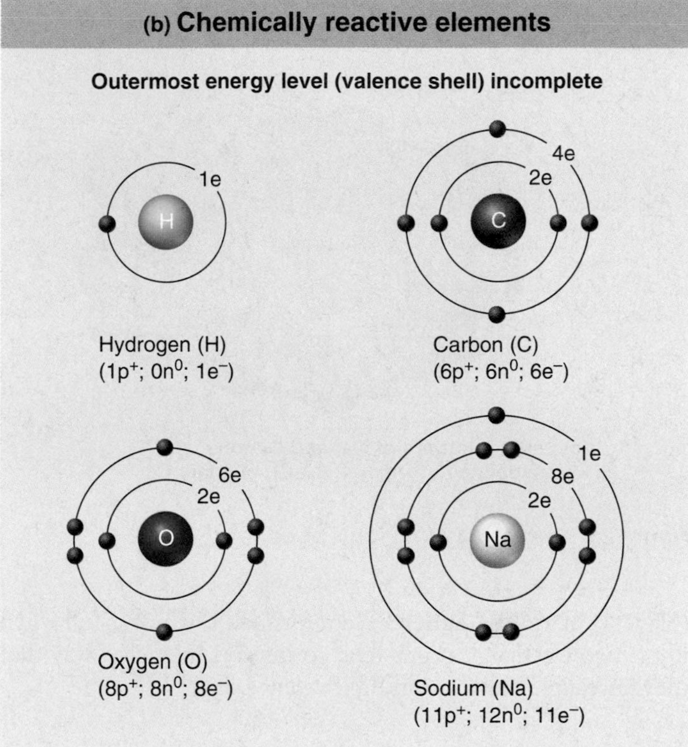

Figure 2.5 Chemically inert and reactive elements. (*Note*: For simplicity, each atomic nucleus is shown as a sphere with the atom's symbol; individual protons and neutrons are not shown.)

2

2

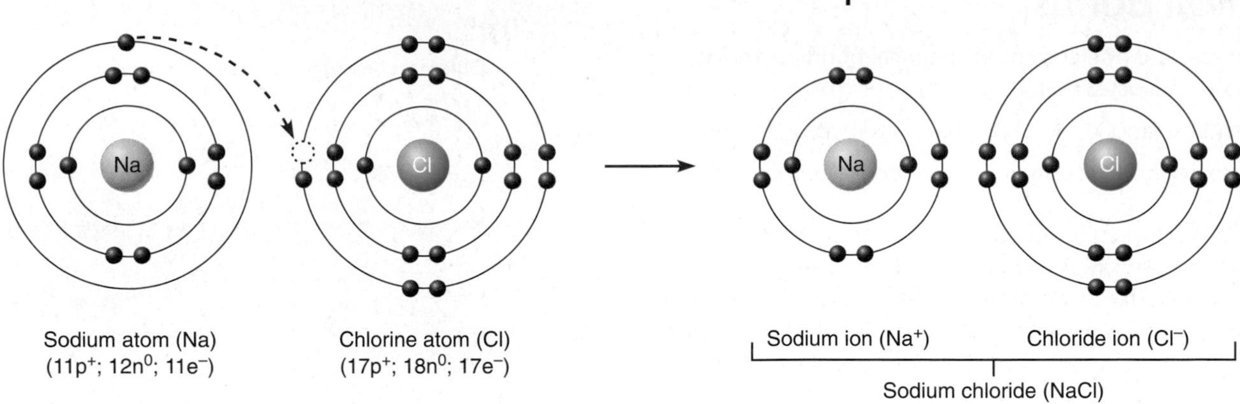

(a) Sodium gains stability by losing one electron, and chlorine becomes stable by gaining one electron.

(b) After electron transfer, the oppositely charged ions formed attract each other.

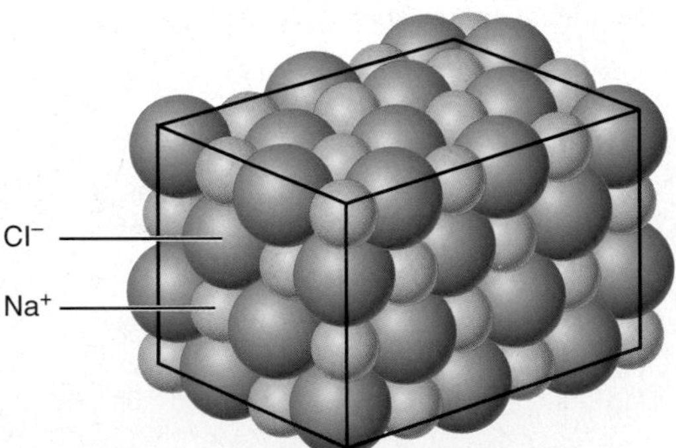

(c) Large numbers of Na^+ and Cl^- ions associate to form salt (NaCl) crystals.

Figure 2.6 **Formation of an ionic bond.**

(ok-tet′), or **rule of eights**. Except for shell 1, which is full when it has two electrons, atoms tend to interact in such a way that they have eight electrons in their valence shell.

Types of Chemical Bonds

Three major types of chemical bonds—*ionic, covalent,* and *hydrogen bonds*—result from attractive forces between atoms.

Ionic Bonds

Recall that atoms are electrically neutral. However, electrons can be transferred from one atom to another, and when this happens, the precise balance of + and − charges is lost so that charged particles called **ions** are formed. An **ionic bond** (i-on′ik) is a chemical bond between atoms formed by the transfer of one or more electrons from one atom to the other. The atom that gains one or more electrons is the *electron acceptor*. It acquires a net negative charge and is called an **anion** (an′i-on). The atom that loses electrons is the *electron donor*. It acquires a net positive charge and is called a **cation** (kat′i-on). (To remember this term, think of the "t" in "cation" as a + sign.) Both anions and cations are formed whenever electron transfer between atoms occurs. Because opposite charges attract, these ions tend to stay close together, resulting in an ionic bond.

One example of ionic bonding is the formation of table salt, or sodium chloride (NaCl), by interaction of sodium and chlorine atoms **(Figure 2.6)**. Sodium, with an atomic number of 11, has only one electron in its valence shell. It would be very difficult to attempt to fill this shell by adding seven more. However, if this single electron is lost, shell 2 with eight electrons becomes the valence shell (outermost energy level containing electrons) and is full. Thus, by losing the lone electron in its third energy level, sodium achieves stability and becomes a cation (Na^+). On the other hand, chlorine, atomic number 17, needs only one electron to fill its valence shell. By accepting an electron, chlorine achieves stability and becomes an anion.

When sodium and chlorine atoms interact, this is exactly what happens. Sodium donates an electron to chlorine (Figure 2.6a), and the oppositely charged ions created in this exchange attract each other, forming sodium chloride (Figure 2.6b). Ionic bonds are commonly formed between atoms with one or two valence shell electrons (the metallic elements, such as sodium, calcium, and potassium) and atoms with seven valence shell electrons (such as chlorine, fluorine, and iodine).

Most ionic compounds fall in the chemical category called *salts*. In the dry state, salts such as sodium chloride do not exist as individual molecules. Instead, they form **crystals**, large arrays of cations and anions held together by ionic bonds (Figure 2.6c).

Sodium chloride is an excellent example of the difference in properties between a compound and its constituent atoms. Sodium is a silvery white metal, and chlorine in its molecular state is a poisonous green gas used to make bleach. However, sodium chloride is a white crystalline solid that we sprinkle on our food.

Covalent Bonds

Electrons do not have to be completely transferred for atoms to achieve stability. Instead, they may be *shared* so that each atom is able to fill its outer electron shell at least part of the time. Electron sharing produces molecules in which the shared electrons occupy a single orbital common to both atoms, which constitutes a **covalent bond** (ko-va′lent).

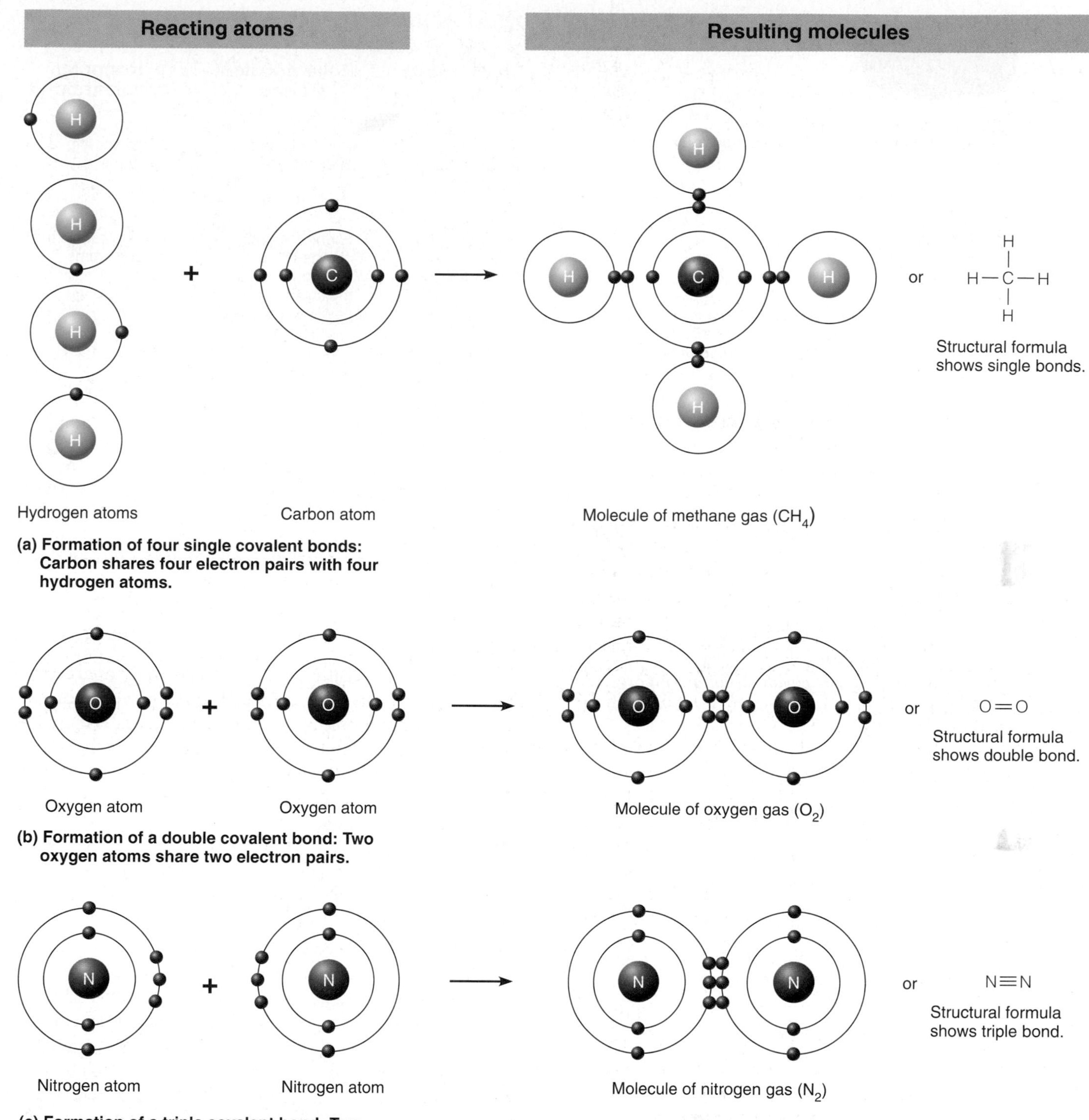
Reacting atoms
Resulting molecules
H
C
+
or
H—C—H
Structural formula shows single bonds.
Hydrogen atoms
Carbon atom
Molecule of methane gas (CH_4)
(a) Formation of four single covalent bonds: Carbon shares four electron pairs with four hydrogen atoms.
O
O=O
Structural formula shows double bond.
Oxygen atom
Oxygen atom
Molecule of oxygen gas (O_2)
(b) Formation of a double covalent bond: Two oxygen atoms share two electron pairs.
N
N≡N
Structural formula shows triple bond.
Nitrogen atom
Nitrogen atom
Molecule of nitrogen gas (N_2)
(c) Formation of a triple covalent bond: Two nitrogen atoms share three electron pairs.

Figure 2.7 Formation of covalent bonds.

Hydrogen with its single electron can fill its only shell (shell 1) by sharing a pair of electrons with another atom. When it shares with another hydrogen atom, a molecule of hydrogen gas is formed. The shared electron pair orbits around the molecule as a whole, satisfying the stability needs of each atom.

Hydrogen can also share an electron pair with different kinds of atoms to form a compound **(Figure 2.7a)**. Carbon has four electrons in its outermost shell, but needs eight to achieve stability. Hydrogen has one electron, but needs two. When a methane molecule (CH_4) is formed, carbon shares four pairs of electrons

O=C=O

(a) Carbon dioxide (CO_2) molecules are linear and symmetrical. They are nonpolar.

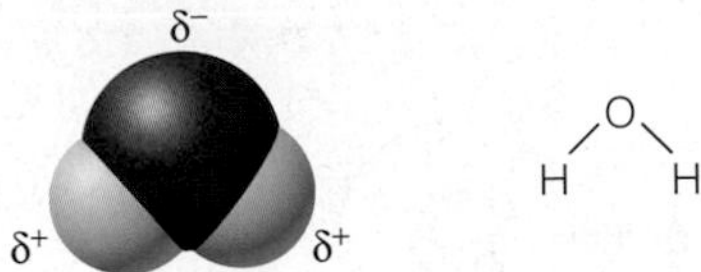

2

H–O–H

(b) V-shaped water (H_2O) molecules have two poles of charge—a slightly more negative oxygen end (δ^-) and a slightly more positive hydrogen end (δ^+).

Figure 2.8 **Carbon dioxide and water molecules have different shapes, as illustrated by molecular models.**

Ionic bond	Polar covalent bond	Nonpolar covalent bond
Complete transfer of electrons	Unequal sharing of electrons	Equal sharing of electrons
Separate ions (charged particles) form	Slight negative charge (δ^-) at one end of molecule, slight positive charge (δ^+) at other end	Charge balanced among atoms
Na^+ Cl^-	δ^- O, H δ^+, H δ^+	O=C=O
Sodium chloride	Water	Carbon dioxide

Figure 2.9 **Ionic, polar covalent, and nonpolar covalent bonds compared along a continuum.**

with four hydrogen atoms (one pair with each hydrogen). Again, the shared electrons orbit and "belong to" the whole molecule, ensuring the stability of each atom.

When two atoms share one pair of electrons, a *single covalent bond* is formed (indicated by a single line connecting the atoms, such as H—H). In some cases, atoms share two or three electron pairs, resulting in *double* or *triple covalent bonds* (Figure 2.7b and c). (These bonds are indicated by double or triple connecting lines such as O=O or N≡N.)

Polar and Nonpolar Molecules In the covalent bonds we have discussed, the shared electrons are shared equally between the atoms of the molecule for the most part. The molecules formed are electrically balanced and are called **nonpolar molecules** (because they do not have separate + and − poles of charge).

Such electrical balance is not always the case. When covalent bonds are formed, the resulting molecule always has a specific three-dimensional shape, with the bonds formed at definite angles. A molecule's shape helps determine what other molecules or atoms it can interact with. It may also result in unequal electron pair sharing, creating a **polar molecule**, especially in nonsymmetrical molecules containing atoms with different electron-attracting abilities.

In general, *small* atoms with six or seven valence shell electrons, such as oxygen, nitrogen, and chlorine, are electron-hungry and attract electrons very strongly, a capability called **electronegativity**. On the other hand, most atoms with only one or two valence shell electrons tend to be **electropositive**. In other words, their electron-attracting ability is so low that they usually lose *their* valence shell electrons to other atoms. Potassium and sodium, each with one valence shell electron, are good examples of electropositive atoms.

Carbon dioxide and water illustrate how molecular shape and the relative electron-attracting abilities of atoms determine whether a covalently bonded molecule is nonpolar or polar. In carbon dioxide (CO_2), carbon shares four electron pairs with two oxygen atoms (two pairs are shared with each oxygen). Oxygen is very electronegative and so attracts the shared electrons much more strongly than does carbon. However, because the carbon dioxide molecule is linear and symmetrical **(Figure 2.8a)**, the electron-pulling ability of one oxygen atom offsets that of the other, like a standoff between equally strong teams in a game of tug-of-war. As a result, the shared electrons orbit the entire molecule and carbon dioxide is a nonpolar compound.

In contrast, a water molecule (H_2O) is bent, or V shaped (Figure 2.8b). The two electropositive hydrogen atoms are located at the same end of the molecule, and the very electronegative oxygen is at the opposite end. This arrangement allows oxygen to pull the shared electrons toward itself and away from the two hydrogen atoms. In this case, the electron pairs are *not* shared equally, but spend more time in the vicinity of oxygen. Because electrons are negatively charged, the oxygen end of the molecule is slightly more negative (the charge is indicated with a delta and minus as δ^-) and the hydrogen end slightly more positive (indicated by δ^+). Because water has two poles of charge, it is a *polar molecule*, or **dipole** (di′pōl).

Polar molecules orient themselves toward other dipoles or toward charged particles (such as ions and some proteins), and they play essential roles in chemical reactions in body cells. The polarity of water is particularly significant, as you will see later in this chapter.

Different molecules exhibit different degrees of polarity, and we can see a gradual change from ionic to nonpolar covalent bonding as summarized in **Figure 2.9**. Ionic bonds (complete electron transfer) and nonpolar covalent bonds (equal electron sharing) are the extremes of a continuum, with various degrees of unequal electron sharing in between.

Hydrogen Bonds

Unlike the stronger ionic and covalent bonds, hydrogen bonds are more like attractions than true bonds. Hydrogen bonds form when a hydrogen atom, already covalently linked to one electronegative atom (usually nitrogen or oxygen), is attracted by another electron-hungry atom, so that a "bridge" forms between them.

Hydrogen bonding is common between dipoles such as water molecules, because the slightly negative oxygen atoms of one molecule attract the slightly positive hydrogen atoms of other molecules **(Figure 2.10a)**. Hydrogen bonding is responsible for the tendency of water molecules to cling together and form films, referred to as *surface tension.* This tendency helps explain why water beads up into spheres when it sits on a hard surface and why water striders can walk on a pond's surface (Figure 2.10b).

Although hydrogen bonds are too weak to bind atoms together to form molecules, they are important *intramolecular bonds* (literally, bonds within molecules), which hold different parts of a single large molecule in a specific three-dimensional shape. Some large biological molecules, such as proteins and DNA, have numerous hydrogen bonds that help maintain and stabilize their structures.

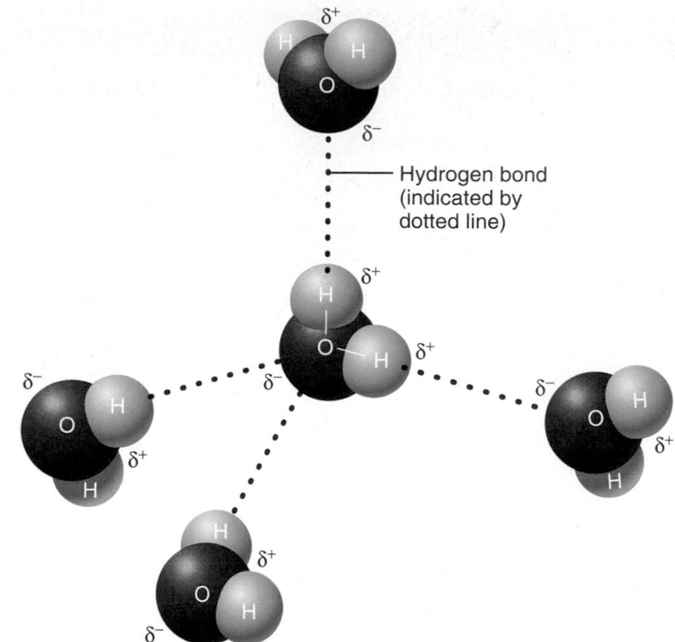

(a) The slightly positive ends (δ^+) of the water molecules become aligned with the slightly negative ends (δ^-) of other water molecules.

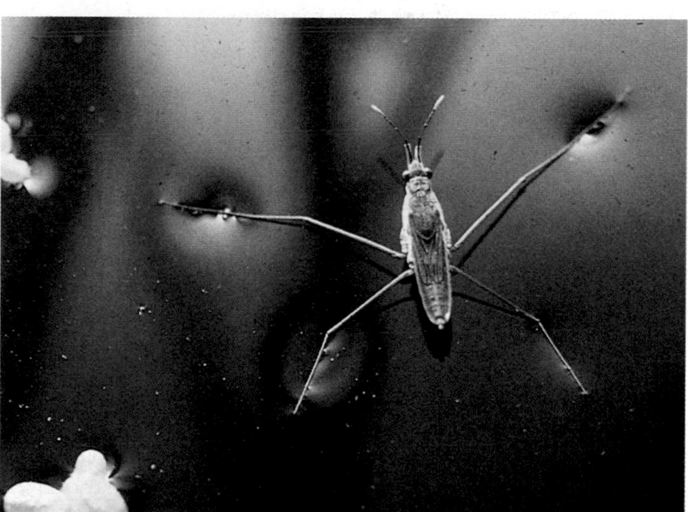

(b) A water strider can walk on a pond because of the high surface tension of water, a result of the combined strength of its hydrogen bonds.

Figure 2.10 **Hydrogen bonding between polar water molecules.**

CHECK YOUR UNDERSTANDING

10. What kinds of bonds form between water molecules?
11. Oxygen ($_8$O) and argon ($_{18}$A) are both gases. Oxygen combines readily with other elements, but argon does not. What accounts for this difference?
12. Assume imaginary compound XY has a polar covalent bond. How does its charge distribution differ from that of XX molecules?

For answers, see Appendix G.

Chemical Reactions

- Define the three major types of chemical reactions: synthesis, decomposition, and exchange. Comment on the nature of oxidation-reduction reactions and their importance.
- Explain why chemical reactions in the body are often irreversible.
- Describe factors that affect chemical reaction rates.

As we noted earlier, all particles of matter are in constant motion because of their kinetic energy. Movement of atoms or molecules in a solid is usually limited to vibration because the particles are united by fairly rigid bonds. But in liquids or gases, particles dart about randomly, sometimes colliding with one another and interacting to undergo chemical reactions. A **chemical reaction** occurs whenever chemical bonds are formed, rearranged, or broken.

Chemical Equations

We can write chemical reactions in symbolic form as **chemical equations.** For example, we indicate the joining of two hydrogen atoms to form hydrogen gas as

$$\underset{\text{reactants}}{H + H} \rightarrow \underset{\text{product}}{H_2} \text{ (hydrogen gas)}$$

2

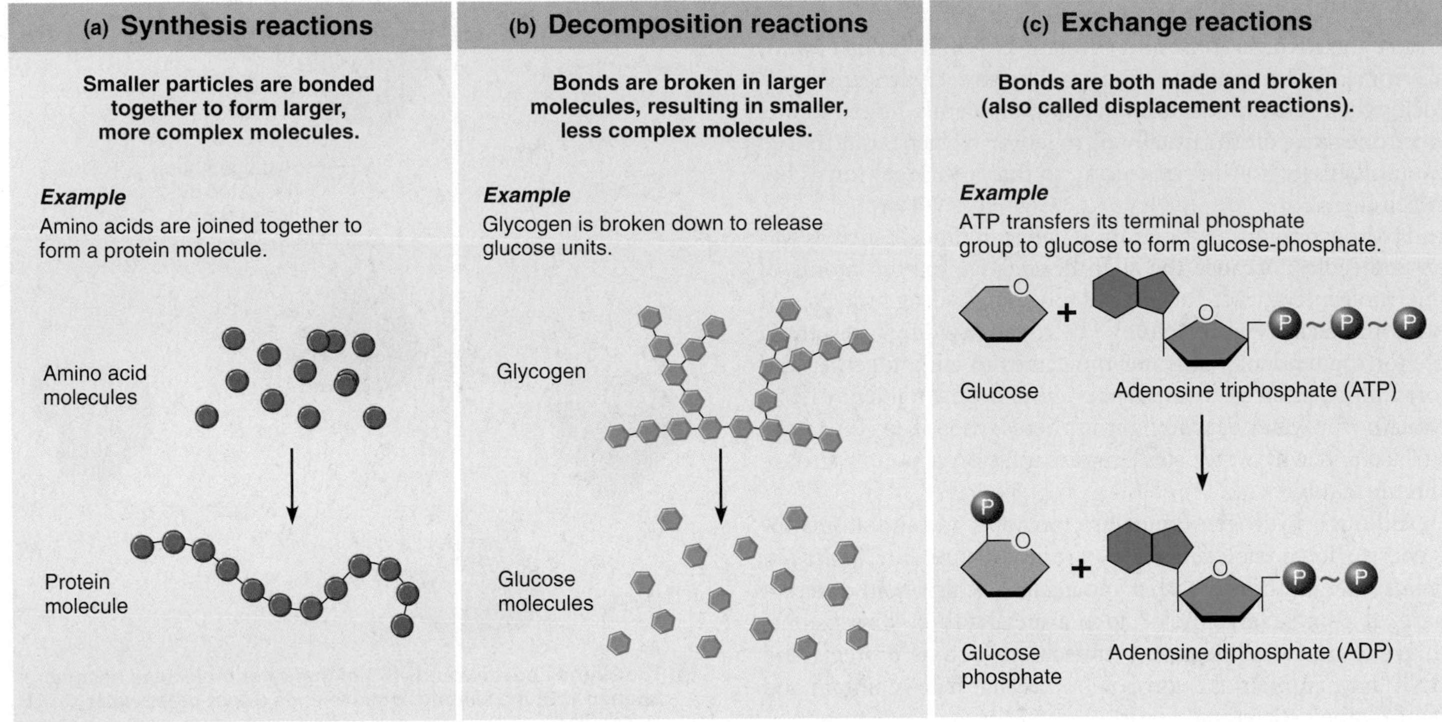

Figure 2.11 Patterns of chemical reactions.

and the combining of four hydrogen atoms and one carbon atom to form methane as

$$\underset{\text{reactants}}{4H + C} \rightarrow \underset{\text{product}}{CH_4 \text{ (methane)}}$$

Notice that in equations, a number written as a *subscript* indicates that the atoms are joined by chemical bonds. But a number written as a *prefix* denotes the number of *unjoined* atoms or molecules. For example, CH_4 reveals that four hydrogen atoms are bonded together with carbon to form the methane molecule, but 4H signifies four unjoined hydrogen atoms.

A chemical equation is like a sentence describing what happens in a reaction. It contains the following information: the number and kinds of reacting substances, or **reactants**; the chemical composition of the **product(s)**; and in balanced equations, the relative proportion of each reactant and product.

In the equations above, the reactants are atoms, as indicated by their atomic symbols (H, C). The product in each case is a molecule, as represented by its **molecular formula** (H_2, CH_4). The equation for the formation of methane may be read in terms of molecules or moles—as *either* "four hydrogen atoms plus one carbon atom yield one molecule of methane" *or* "four moles of hydrogen atoms plus one mole of carbon yield one mole of methane." Using moles is more practical because it is impossible to measure out one atom or one molecule of anything!

Patterns of Chemical Reactions

Most chemical reactions exhibit one of three recognizable patterns: They are either *synthesis, decomposition,* or *exchange reactions.*

When atoms or molecules combine to form a larger, more complex molecule, the process is a **synthesis**, or **combination, reaction**. A synthesis reaction always involves bond formation. It can be represented (using arbitrary letters) as

$$A + B \rightarrow AB$$

Synthesis reactions are the basis of constructive, or **anabolic**, activities in body cells, such as joining small molecules called amino acids into large protein molecules **(Figure 2.11a)**. Synthesis reactions are conspicuous in rapidly growing tissues.

A **decomposition reaction** occurs when a molecule is broken down into smaller molecules or its constituent atoms:

$$AB \rightarrow A + B$$

Essentially, decomposition reactions are reverse synthesis reactions: Bonds are broken. Decomposition reactions underlie all degradative, or **catabolic**, processes in body cells. For example, the bonds of glycogen molecules are broken to release simpler molecules of glucose sugar (Figure 2.11b).

Exchange, or **displacement**, **reactions** involve both synthesis and decomposition. Bonds are both made and broken. In an exchange reaction, parts of the reactant molecules change partners, so to speak, producing different product molecules:

$$AB + C \rightarrow AC + B \quad \text{and} \quad AB + CD \rightarrow AD + CB$$

An exchange reaction occurs when ATP reacts with glucose and transfers its end phosphate group (indicated by a circled P in Figure 2.11c) to glucose, forming glucose-phosphate. At the same time, the ATP becomes ADP. This important reaction occurs whenever glucose enters a body cell, and it effectively traps the glucose fuel molecule inside the cell.

Another group of important chemical reactions in living systems is **oxidation-reduction reactions**, called **redox reactions** for short. Oxidation-reduction reactions are decomposition reactions in that they are the basis of all reactions in which food fuels are broken down for energy (that is, in which ATP is produced). They are also a special type of exchange reaction because electrons are exchanged between the reactants. The reactant losing the electrons is referred to as the *electron donor* and is said to be **oxidized**. The reactant taking up the transferred electrons is called the *electron acceptor* and is said to become **reduced**.

Redox reactions also occur when ionic compounds are formed. Recall that in the formation of NaCl (see Figure 2.6), sodium loses an electron to chlorine. Consequently, sodium is oxidized and becomes a sodium ion, and chlorine is reduced and becomes a chloride ion. However, not all oxidation-reduction reactions involve *complete transfer* of electrons—some simply change the pattern of electron sharing in covalent bonds. For example, a substance is oxidized both by losing hydrogen atoms and by combining with oxygen. The common factor in these events is that electrons that formerly "belonged" to the reactant molecule are lost. The electrons are lost either entirely (as hydrogen is removed and takes its electron with it) or relatively (as the shared electrons spend more time in the vicinity of the very electronegative oxygen atom).

To understand the importance of oxidation-reduction reactions in living systems, take a look at the overall equation for *cellular respiration*, which represents the major pathway by which glucose is broken down for energy in body cells:

$$C_6H_{12}O_6 + 6O_2 \rightarrow 6CO_2 + 6H_2O + ATP$$

glucose oxygen carbon dioxide water cellular energy

As you can see, it is an oxidation-reduction reaction. Consider what happens to the hydrogen atoms (and their electrons). Glucose is oxidized to carbon dioxide as it loses hydrogen atoms, and oxygen is reduced to water as it accepts the hydrogen atoms. This reaction is described in detail in Chapter 24, along with other topics of cellular metabolism.

Energy Flow in Chemical Reactions

Because all chemical bonds represent stored chemical energy, all chemical reactions ultimately result in net absorption or release of energy. Reactions that release energy are called **exergonic reactions**. These reactions yield products with less energy than the initial reactants, along with energy that can be harvested for other uses. With a few exceptions, catabolic and oxidative reactions are exergonic.

In contrast, the products of energy-absorbing, or **endergonic**, reactions contain more potential energy in their chemical bonds than did the reactants. Anabolic reactions are typically energy-absorbing endergonic reactions. Essentially this is a case of "one hand washing the other"—the energy released when fuel molecules are broken down (oxidized) is captured in ATP molecules and then used to synthesize the complex biological molecules the body needs to sustain life.

Reversibility of Chemical Reactions

All chemical reactions are theoretically reversible. If chemical bonds can be made, they can be broken, and vice versa. Reversibility is indicated by a double arrow. When the arrows differ in length, the longer arrow indicates the major direction in which the reaction proceeds:

$$A + B \rightleftharpoons AB$$

In this example, the forward reaction (reaction going to the right) predominates. Over time, the product (AB) accumulates and the reactants (A and B) decrease in amount.

When the arrows are of equal length, as in

$$A + B \rightleftharpoons AB$$

neither the forward reaction nor the reverse reaction is dominant. In other words, for each molecule of product (AB) formed, one product molecule breaks down, releasing the reactants A and B. Such a chemical reaction is said to be in a state of **chemical equilibrium**.

Once chemical equilibrium is reached, there is no further *net change* in the amounts of reactants and products unless more of either are added to the mix. Product molecules are still formed and broken down, but the balance established when equilibrium was reached (such as greater numbers of product molecules) remains unchanged.

Chemical equilibrium is analogous to the admission scheme used by many large museums in which tickets are sold according to time of entry. If 300 tickets are issued for the 9 AM admission, 300 people will be admitted when the doors open. Thereafter, when 6 people leave, 6 are admitted, and when another 15 people leave, 15 more are allowed in. There is a continual turnover, but the museum contains about 300 art lovers throughout the day.

All chemical reactions are reversible, but many biological reactions show so little tendency to go in the reverse direction that they are irreversible for all practical purposes. Chemical reactions that release energy will not go in the opposite direction unless energy is put back into the system. For example, when our cells break down glucose via the reactions of cellular respiration to yield carbon dioxide and water, some of the energy released is trapped in the bonds of ATP. Because the cells then use ATP's energy for various functions (and more glucose will be along with the next meal), this particular reaction is never reversed in our cells. Furthermore, if a product of a reaction is continuously removed from the reaction site, it is unavailable to take part in the reverse reaction. This situation occurs when the carbon dioxide that is released during glucose breakdown leaves the cell, enters the blood, and is eventually removed from the body by the lungs.

Factors Influencing the Rate of Chemical Reactions

What influences how quickly chemical reactions go? For atoms and molecules to react chemically in the first place, they must *collide* with enough force to overcome the repulsion between their electrons. Interactions between valence shell electrons—the basis of bond making and breaking—cannot occur long distance. The force of collisions depends on how fast the particles are moving. Solid, forceful collisions between rapidly moving particles in which valence shells overlap are much more likely to cause reactions than are collisions in which the particles graze each other lightly.

Temperature Increasing the temperature of a substance increases the kinetic energy of its particles and the force of their collisions. For this reason, chemical reactions proceed more quickly at higher temperatures.

Concentration Chemical reactions progress most rapidly when the reacting particles are present in high numbers, because the chance of successful collisions is greater. As the concentration of the reactants declines, the reaction slows. Chemical equilibrium eventually occurs unless additional reactants are added or products are removed from the reaction site.

Particle Size Smaller particles move faster than larger ones (at the same temperature) and tend to collide more frequently and more forcefully. Hence, the smaller the reacting particles, the faster a chemical reaction goes at a given temperature and concentration.

Catalysts Many chemical reactions in nonliving systems can be speeded up simply by heating, but drastic increases in body temperature are life threatening because important biological molecules are destroyed. Still, at normal body temperatures, most chemical reactions would proceed far too slowly to maintain life were it not for the presence of catalysts. **Catalysts** (kat′ah-lists) are substances that increase the rate of chemical reactions without themselves becoming chemically changed or part of the product. Biological catalysts are called *enzymes* (en′zīmz). Later in this chapter we describe how enzymes work.

CHECK YOUR UNDERSTANDING

13. Which reaction type—synthesis, decomposition, or exchange—occurs when fats are digested in your small intestine?
14. Why are many reactions that occur in living systems irreversible for all intents and purposes?
15. What specific name is given to decomposition reactions in which food fuels are broken down for energy?

For answers, see Appendix G.

PART 2

BIOCHEMISTRY

Biochemistry is the study of the chemical composition and reactions of living matter. All chemicals in the body fall into one of two major classes: organic or inorganic compounds. **Organic compounds** contain carbon. All organic compounds are covalently bonded molecules, and many are large.

All other chemicals in the body are considered **inorganic compounds**. These include water, salts, and many acids and bases. Organic and inorganic compounds are equally essential for life. Trying to decide which is more valuable is like trying to decide whether the ignition system or the engine is more essential to run your car!

Inorganic Compounds

- ▶ Explain the importance of water and salts to body homeostasis.
- ▶ Define acid and base, and explain the concept of pH.

Water

Water is the most abundant and important inorganic compound in living material. It makes up 60–80% of the volume of most living cells. What makes water so vital to life? The answer lies in several properties:

1. **High heat capacity.** Water has a high heat capacity. In other words, it absorbs and releases large amounts of heat before changing appreciably in temperature itself. This property of water prevents sudden changes in temperature caused by external factors, such as sun or wind exposure, or by internal conditions that release heat rapidly, such as vigorous muscle activity. As part of blood, water redistributes heat among body tissues, ensuring temperature homeostasis.
2. **High heat of vaporization.** When water evaporates, or vaporizes, it changes from a liquid to a gas (water vapor). This transformation requires that large amounts of heat be absorbed to break the hydrogen bonds that hold water molecules together. This property is extremely beneficial when we sweat. As perspiration (mostly water) evaporates from our skin, large amounts of heat are removed from the body, providing efficient cooling.
3. **Polar solvent properties.** Water is an unparalleled solvent. Indeed, it is often called the **universal solvent**. Biochemistry is "wet chemistry." Biological molecules do not react chemically unless they are in solution, and virtually all chemical reactions occurring in the body depend on water's solvent properties.

   Because water molecules are polar, they orient themselves with their slightly negative ends toward the positive ends of the solutes, and vice versa, first attracting the solute molecules, and then surrounding them. This polarity of water explains why ionic compounds and other small reactive molecules (such as acids and bases) *dissociate* in water, their ions separating from each other and becoming evenly scattered in the water, forming true solutions **(Figure 2.12)**.

   Water also forms layers of water molecules, called **hydration layers**, around large charged molecules such as proteins, shielding them from the effects of other charged substances in the vicinity and preventing them from settling out of solution. Such protein-water mixtures are *biological colloids*. Blood plasma and cerebrospinal fluid (which surrounds the brain and spinal cord) are examples of colloids.

   Water is the body's major transport medium because it is such an excellent solvent. Nutrients, respiratory gases, and metabolic wastes carried throughout the body are dissolved in blood plasma, and many metabolic wastes are excreted from the body in urine, another watery fluid. Specialized molecules that lubricate the body (e.g., mucus) also use water as their dissolving medium.

4. **Reactivity.** Water is an important *reactant* in many chemical reactions. For example, foods are digested to their building blocks by adding a water molecule to each bond to be broken. Such decomposition reactions are more specifically called **hydrolysis reactions** (hi-drol′ĭ-sis; "water splitting"). Conversely, when large carbohydrate or protein molecules are synthesized from smaller molecules, a water molecule is removed for every bond formed, a reaction called **dehydration synthesis**.
5. **Cushioning.** By forming a resilient cushion around certain body organs, water helps protect them from physical trauma. The cerebrospinal fluid surrounding the brain exemplifies water's cushioning role.

Salts

A **salt** is an ionic compound containing cations other than H^+ and anions other than the hydroxyl ion (OH^-). As already noted, when salts are dissolved in water, they dissociate into their component ions (Figure 2.12). For example, sodium sulfate (Na_2SO_4) dissociates into two Na^+ ions and one SO_4^{2-} ion. It dissociates easily because the ions are already formed. All that remains is for water to overcome the attraction between the oppositely charged ions.

All ions are **electrolytes** (e-lek′tro-līts), substances that conduct an electrical current in solution. (Note that groups of atoms that bear an overall charge, such as sulfate, are called *polyatomic ions.*)

Salts commonly found in the body include NaCl, $CaCO_3$ (calcium carbonate), and KCl (potassium chloride). However, the most plentiful salts are the calcium phosphates that make bones and teeth hard. In their ionized form, salts play vital roles in body function. For instance, the electrolyte properties of sodium and potassium ions are essential for nerve impulse transmission and muscle contraction. Ionic iron forms part of the hemoglobin molecules that transport oxygen within red blood cells, and zinc and copper ions are important to the activity of some enzymes. Other important functions of the elements found in body salts are summarized in Table 2.1 on p. 26.

HOMEOSTATIC IMBALANCE

Maintaining proper ionic balance in our body fluids is one of the most crucial homeostatic roles of the kidneys. When this balance is severely disturbed, virtually nothing in the body works. All the physiological activities listed above and thousands of others are disrupted and grind to a stop. ■

Acids and Bases

Like salts, acids and bases are electrolytes. They ionize and dissociate in water and can then conduct an electrical current.

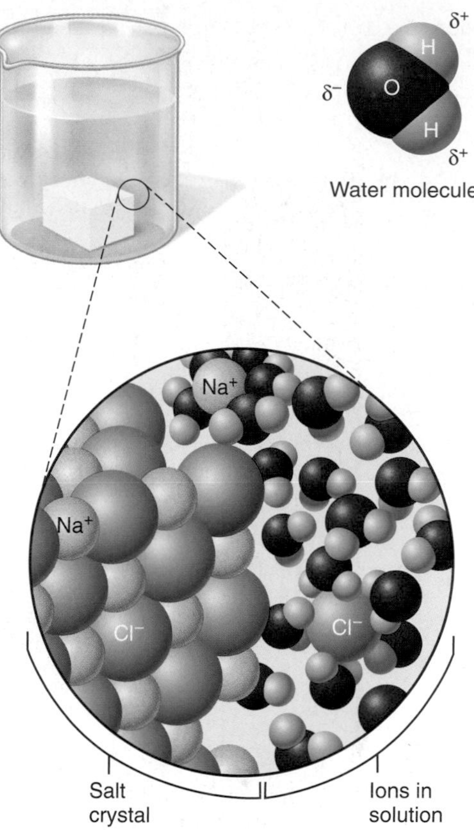

Figure 2.12 Dissociation of salt in water.

Acids

Acids have a sour taste, can react with (dissolve) many metals, and "burn" a hole in your rug. But for our purposes the most useful definition of an acid is a substance that releases **hydrogen ions** (H^+) in detectable amounts. Because a hydrogen ion is just a hydrogen nucleus, or "naked" proton, acids are also defined as **proton donors**.

When acids dissolve in water, they release hydrogen ions (protons) and anions. It is the concentration of protons that determines the acidity of a solution. The anions have little or no effect on acidity. For example, hydrochloric acid (HCl), an acid produced by stomach cells that aids digestion, dissociates into a proton and a chloride ion:

$$\underset{}{HCl} \rightarrow \underset{\text{proton}}{H^+} + \underset{\text{anion}}{Cl^-}$$

Other acids found or produced in the body include acetic acid ($HC_2H_3O_2$, commonly abbreviated as HAc), which is the acidic portion of vinegar; and carbonic acid (H_2CO_3). The molecular formula for an acid is easy to recognize because the hydrogen is written first.

Bases

Bases have a bitter taste, feel slippery, and are **proton acceptors**—that is, they take up hydrogen ions (H^+) in detectable amounts. Common inorganic bases include the *hydroxides* (hi-drok′sīds), such as magnesium hydroxide (milk of magnesia) and sodium

2

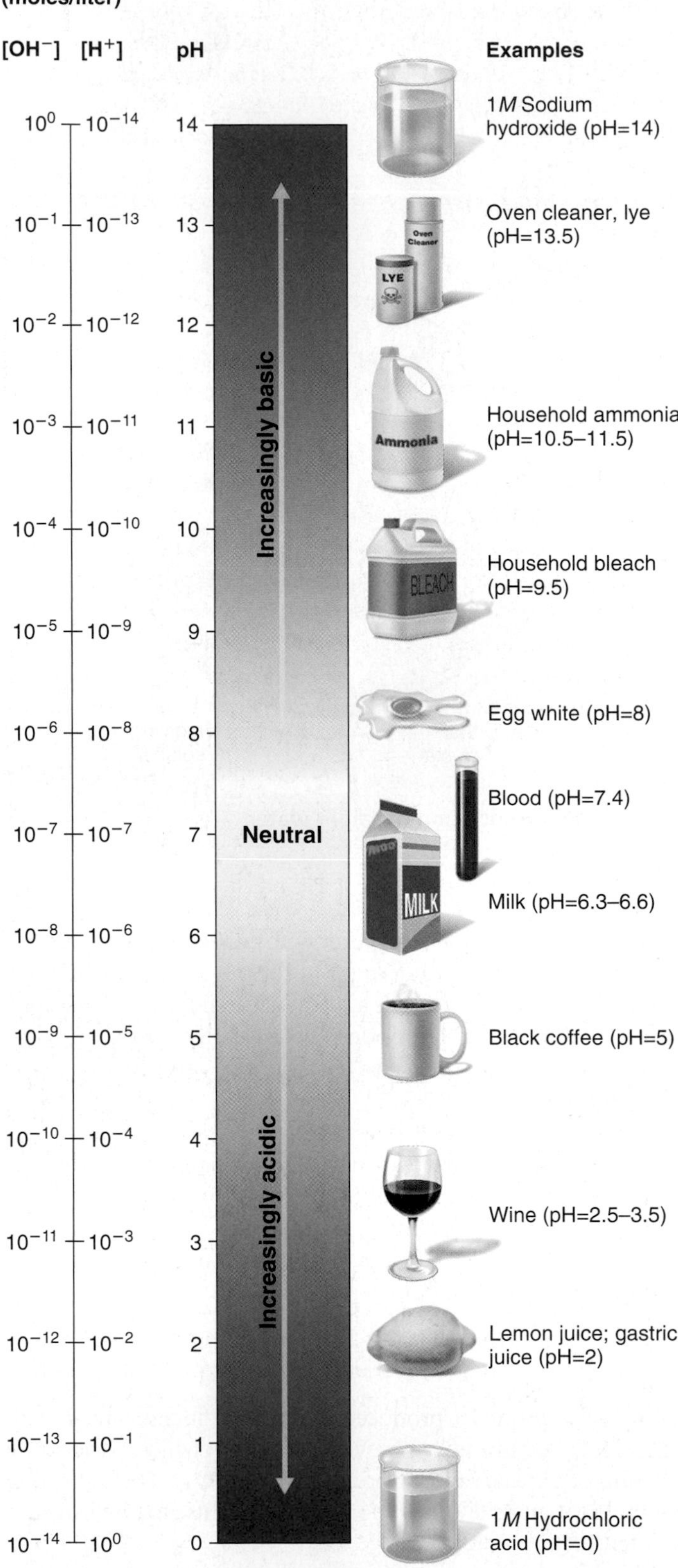

Figure 2.13 The pH scale and pH values of representative substances. The pH scale is based on the number of hydrogen ions in solution. The actual concentrations of hydrogen ions, $[H^+]$, and hydroxyl ions, $[OH^-]$, in moles per liter are indicated for each pH value noted. At a pH of 7, $[H^+] = [OH^-]$ and the solution is neutral.

hydroxide (lye). Like acids, hydroxides dissociate when dissolved in water, but in this case **hydroxyl ions (OH^-)** (hi-drok′sil) and cations are liberated. For example, ionization of sodium hydroxide (NaOH) produces a hydroxyl ion and a sodium ion, and the hydroxyl ion then binds to (accepts) a proton present in the solution. This reaction produces water and simultaneously reduces the acidity (hydrogen ion concentration) of the solution:

$$\underset{}{NaOH} \rightarrow \underset{\text{cation}}{Na^+} + \underset{\text{hydroxyl ion}}{OH^-}$$

and then

$$OH^- + H^+ \rightarrow \underset{\text{water}}{H_2O}$$

Bicarbonate ion (HCO_3^-), an important base in the body, is particularly abundant in blood. **Ammonia (NH_3)**, a common waste product of protein breakdown in the body, is also a base. It has one pair of unshared electrons that strongly attracts protons. By accepting a proton, ammonia becomes an ammonium ion:

$$NH_3 + H^+ \rightarrow \underset{\text{ammonium ion}}{NH_4^+}$$

pH: Acid-Base Concentration

The more hydrogen ions in a solution, the more acidic the solution is. Conversely, the greater the concentration of hydroxyl ions (the lower the concentration of H^+), the more basic, or *alkaline* (al′kuh-līn), the solution becomes. The relative concentration of hydrogen ions in various body fluids is measured in concentration units called **pH units** (pe-āch′).

The idea for a pH scale was devised by a Danish biochemist and part-time beer brewer named Sören Sörensen in 1909. He was searching for a convenient means of checking the acidity of his alcoholic product to prevent its spoilage by bacterial action. (Acidic conditions inhibit many bacteria.) The pH scale that resulted is based on the concentration of hydrogen ions in a solution, expressed in terms of moles per liter, or molarity. The pH scale runs from 0 to 14 and is *logarithmic.* In other words, each successive change of one pH unit represents a tenfold change in hydrogen ion concentration **(Figure 2.13)**. The pH of a solution is thus defined as the negative logarithm of the hydrogen ion concentration $[H^+]$ in moles per liter, or $-\log[H^+]$. (Note that brackets [ ] indicate concentration of a substance.)

At a pH of 7 (at which $[H^+]$ is 10^{-7} *M*), the solution is *neutral*—neither acidic nor basic. The number of hydrogen ions exactly equals the number of hydroxyl ions (pH = pOH). Absolutely pure (distilled) water has a pH of 7.

Solutions with a pH below 7 are acidic—the hydrogen ions outnumber the hydroxyl ions. The lower the pH, the more acidic the solution. A solution with a pH of 6 has ten times as many hydrogen ions as a solution with a pH of 7.

Solutions with a pH higher than 7 are alkaline, and the relative concentration of hydrogen ions decreases by a factor of 10 with each higher pH unit. Thus, solutions with pH values of 8 and 12 have, respectively, 1/10 and 1/100,000 ($1/10 \times 1/10 \times 1/10 \times 1/10 \times 1/10$) as many hydrogen ions as a solution of pH 7.

The approximate pH of several body fluids and of a number of common substances appears in Figure 2.13. Notice that as the hydrogen ion concentration decreases, the hydroxyl ion concentration rises, and vice versa.

Neutralization

What happens when acids and bases are mixed? They react with each other in displacement reactions to form water and a salt. For example, when hydrochloric acid and sodium hydroxide interact, sodium chloride (a salt) and water are formed.

$$\underset{\text{acid}}{HCl} + \underset{\text{base}}{NaOH} \rightarrow \underset{\text{salt}}{NaCl} + \underset{\text{water}}{H_2O}$$

This type of reaction is called a **neutralization reaction**, because the joining of H^+ and OH^- to form water neutralizes the solution. Although the salt produced is written in molecular form (NaCl), remember that it actually exists as dissociated sodium and chloride ions when dissolved in water.

Buffers

Living cells are extraordinarily sensitive to even slight changes in the pH of the environment. In high concentrations, acids and bases are extremely damaging to living tissue. Imagine what would happen to all those hydrogen bonds in biological molecules with large numbers of free H^+ running around. (Can't you just hear those molecules saying "Why share hydrogen when I can have my own?")

Homeostasis of acid-base balance is carefully regulated by the kidneys and lungs and by chemical systems (proteins and other types of molecules) called **buffers.** Buffers resist abrupt and large swings in the pH of body fluids by releasing hydrogen ions (acting as acids) when the pH begins to rise and by binding hydrogen ions (acting as bases) when the pH drops. Because blood comes into close contact with nearly every body cell, regulation of its pH is particularly critical. Normally, blood pH varies within a very narrow range (7.35 to 7.45). If blood pH varies from these limits by more than a few tenths of a unit, it may be fatal.

To comprehend how chemical buffer systems operate, you must thoroughly understand strong and weak acids and bases. The first important concept is that the acidity of a solution reflects *only* the free hydrogen ions, not those still bound to anions. Consequently, acids that dissociate completely and irreversibly in water are called **strong acids**, because they can dramatically change the pH of a solution. Examples are hydrochloric acid and sulfuric acid. If we could count out 100 hydrochloric acid molecules and place them in 1 milliliter (ml) of water, we could expect to end up with 100 H^+, 100 Cl^-, and no undissociated hydrochloric acid molecules in that solution.

Acids that do not dissociate completely, like carbonic acid (H_2CO_3) and acetic acid (HAc), are **weak acids.** If we were to place 100 acetic acid molecules in 1 ml of water, the reaction would be something like this:

$$100\ HAc \rightarrow 90\ HAc + 10\ H^+ + 10\ Ac^-$$

Because undissociated acids do not affect pH, the acetic acid solution is much less acidic than the HCl solution. Weak acids dissociate in a predictable way, and molecules of the intact acid are in dynamic equilibrium with the dissociated ions. Consequently, the dissociation of acetic acid may also be written as

$$HAc \rightleftharpoons H^+ + Ac^-$$

This viewpoint allows us to see that if H^+ (released by a strong acid) is added to the acetic acid solution, the equilibrium will shift to the left and some H^+ and Ac^- will recombine to form HAc. On the other hand, if a strong base is added and the pH begins to rise, the equilibrium shifts to the right and more HAc molecules dissociate to release H^+. This characteristic of weak acids allows them to play important roles in the chemical buffer systems of the body.

The concept of strong and weak bases is more easily explained. Remember that bases are proton acceptors. Thus, **strong bases** are those, like hydroxides, that dissociate easily in water and quickly tie up H^+. On the other hand, sodium bicarbonate (commonly known as baking soda) ionizes incompletely and reversibly. Because it accepts relatively few protons, its released bicarbonate ion is considered a **weak base.**

Now let's examine how one buffer system helps to maintain pH homeostasis of the blood. Although there are other chemical blood buffers, the **carbonic acid–bicarbonate system** is a major one. Carbonic acid (H_2CO_3) dissociates reversibly, releasing bicarbonate ions (HCO_3^-) and protons (H^+):

$$\underset{\substack{H^+\text{ donor}\\ \text{(weak acid)}}}{H_2CO_3} \underset{\text{Response to drop in pH}}{\overset{\text{Response to rise in pH}}{\rightleftharpoons}} \underset{\substack{H^+\text{ acceptor}\\ \text{(weak base)}}}{HCO_3^-} + \underset{\text{proton}}{H^+}$$

The chemical equilibrium between carbonic acid (a weak acid) and bicarbonate ion (a weak base) resists changes in blood pH by shifting to the right or left as H^+ ions are added to or removed from the blood. As blood pH rises (becomes more alkaline due to the addition of a strong base), the equilibrium shifts to the right, forcing more carbonic acid to dissociate. Similarly, as blood pH begins to drop (becomes more acidic due to the addition of a strong acid), the equilibrium shifts to the left as more bicarbonate ions begin to bind with protons. As you can see, strong bases are replaced by a weak base (bicarbonate ion) and protons released by strong acids are tied up in a weak one (carbonic acid). In either case, the blood pH changes much less than it would in the absence of the buffering system. We discuss acid-base balance and buffers in more detail in Chapter 26.

CHECK YOUR UNDERSTANDING

16. Water makes up 60–80% of living matter. What property makes it an excellent solvent?

17. Salts are electrolytes. What does that mean?

18. Which ion is responsible for increased acidity?

19. To minimize the sharp pH shift that occurs when a strong acid is added to a solution, is it better to add a weak base or a strong base? Why?

For answers, see Appendix G.

2

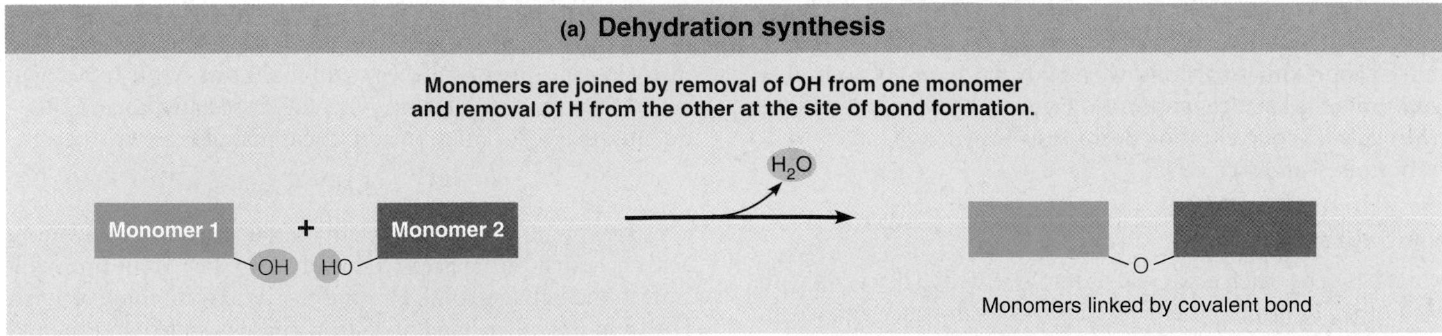
(a) Dehydration synthesis
Monomers are joined by removal of OH from one monomer and removal of H from the other at the site of bond formation.
H2O
Monomer 1
+
Monomer 2
OH
HO
O
Monomers linked by covalent bond

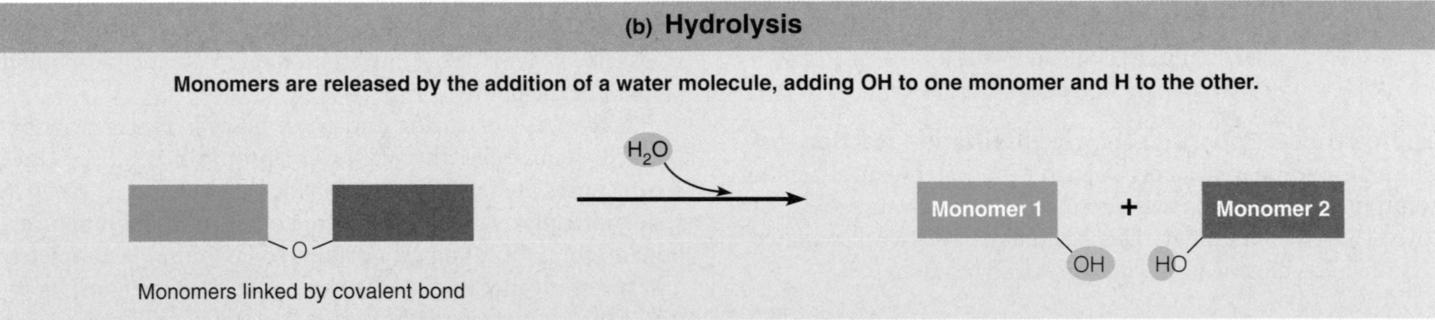
(b) Hydrolysis
Monomers are released by the addition of a water molecule, adding OH to one monomer and H to the other.
H2O
O
Monomers linked by covalent bond
Monomer 1
+
Monomer 2
OH
HO

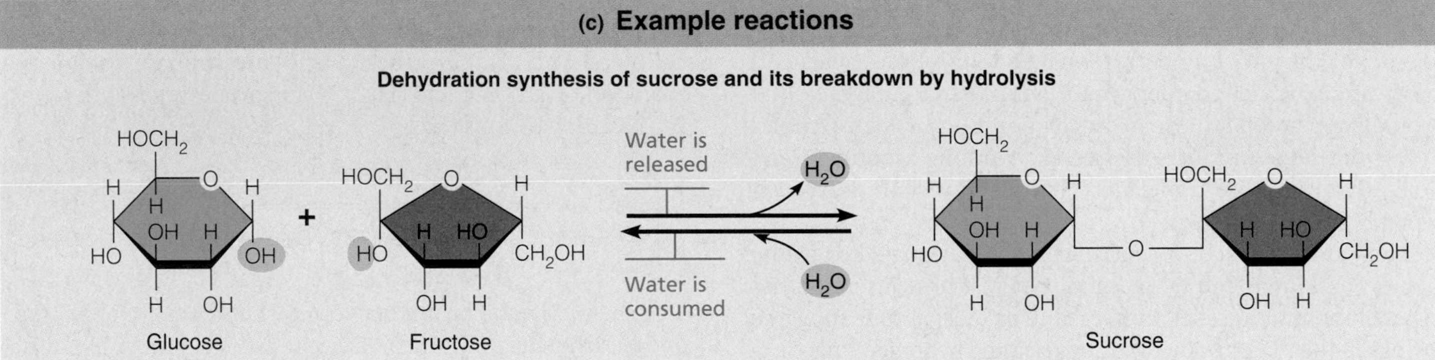
(c) Example reactions
Dehydration synthesis of sucrose and its breakdown by hydrolysis
HOCH2
Glucose
Fructose
Water is released
Water is consumed
H2O
H2O
Sucrose
CH2OH

Figure 2.14 Biological molecules are formed from their monomers or units by dehydration synthesis and broken down to the monomers by hydrolysis reactions.

Organic Compounds

- Describe and compare the building blocks, general structures, and biological functions of carbohydrates and lipids.
- Explain the role of dehydration synthesis and hydrolysis in the formation and breakdown of organic molecules.

Molecules unique to living systems—carbohydrates, lipids (fats), proteins, and nucleic acids—all contain carbon and hence are organic compounds. Organic compounds are generally distinguished by the fact that they contain carbon, and inorganic compounds are defined as compounds that lack carbon. You should be aware of a few irrational exceptions to this generalization: Carbon dioxide and carbon monoxide, for example, contain carbon but are considered inorganic compounds.

For the most part, organic molecules are very large molecules, but their interactions with other molecules typically involve only small, reactive parts of their structure called *functional groups* (acid groups, amines, and others). The most important functional groups involved in biochemical reactions are illustrated in Appendix B.

What makes carbon so special that "living" chemistry depends on its presence? To begin with, no other *small* atom is so precisely **electroneutral.** The consequence of its electroneutrality is that carbon never loses or gains electrons. Instead, it always shares them. Furthermore, with four valence shell electrons, carbon forms four covalent bonds with other elements, as well as with other carbon atoms. As a result, carbon can help form long, chainlike molecules (common in fats), ring structures (typical of carbohydrates and steroids), and many other structures that are uniquely suited for specific roles in the body.

As you will see shortly, many biological molecules (carbohydrates and proteins for example) are polymers. **Polymers** are chainlike molecules made of many similar or repeating units (**monomers**), which are joined together by dehydration synthesis **(Figure 2.14)**. During dehydration synthesis, a hydrogen atom is removed from one monomer and a hydroxyl group is removed from the monomer it is to be joined with. As a covalent bond unites the monomers, a water molecule is released. This removal of a water molecule at the bond site occurs each time a monomer is added to the growing polymer chain.

Carbohydrates

Carbohydrates, a group of molecules that includes sugars and starches, represent 1–2% of cell mass. Carbohydrates contain carbon, hydrogen, and oxygen, and generally the hydrogen and oxygen atoms occur in the same 2:1 ratio as in water. This ratio is reflected in the word *carbohydrate* ("hydrated carbon").

A carbohydrate can be classified according to size and solubility as a monosaccharide ("one sugar"), disaccharide ("two sugars"), or polysaccharide ("many sugars"). Monosaccharides are the monomers, or building blocks, of the other carbohydrates. In general, the larger the carbohydrate molecule, the less soluble it is in water.

Monosaccharides

Monosaccharides (mon″o-sak′ah-rīdz), or *simple sugars*, are single-chain or single-ring structures containing from three to seven carbon atoms **(Figure 2.15a)**. Usually the carbon, hydrogen, and oxygen atoms occur in the ratio 1:2:1, so a general formula for a monosaccharide is $(CH_2O)n$, where *n* is the number of carbons in the sugar. Glucose, for example, has six carbon atoms, and its molecular formula is $C_6H_{12}O_6$. Ribose, with five carbons, is $C_5H_{10}O_5$.

Monosaccharides are named generically according to the number of carbon atoms they contain. Most important in the body are the pentose (five-carbon) and hexose (six-carbon) sugars. For example, the pentose *deoxyribose* (de-ok″sĭ-ri′bōs) is part of DNA, and *glucose*, a hexose, is blood sugar.

Two other hexoses, *galactose* and *fructose*, are **isomers** (i′so-mers) of glucose. That is, they have the same molecular formula ($C_6H_{12}O_6$), but as you can see in Figure 2.15a, their atoms are arranged differently, giving them different chemical properties.

Disaccharides

A **disaccharide** (di-sak′ah-rīd), or *double sugar*, is formed when two monosaccharides are joined by *dehydration synthesis* (Figure 2.14a). In this synthesis reaction, a water molecule is lost as the bond is made, as illustrated by the synthesis of sucrose (soo′krōs):

$$\underset{\text{glucose + fructose}}{2C_6H_{12}O_6} \rightarrow \underset{\text{sucrose}}{C_{12}H_{22}O_{11}} + \underset{\text{water}}{H_2O}$$

Notice that the molecular formula for sucrose contains two hydrogen atoms and one oxygen atom less than the total number of hydrogen and oxygen atoms in glucose and fructose, because a water molecule is released during bond formation.

Important disaccharides in the diet are *sucrose* (glucose + fructose), which is cane or table sugar; *lactose* (glucose + galactose), found in milk; and *maltose* (glucose + glucose), also called malt sugar (Figure 2.15b). Disaccharides are too large to pass through cell membranes, so they must be digested to their simple sugar units to be absorbed from the digestive tract into the blood. This decomposition process is *hydrolysis*, essentially the reverse of dehydration synthesis (Figure 2.14a, b). A water molecule is added to each bond, breaking the bonds and releasing the simple sugar units.

Polysaccharides

Polysaccharides (pol″e-sak′ah-rīdz) are polymers of simple sugars linked together by dehydration synthesis. Because polysaccharides are large, fairly insoluble molecules, they are ideal storage products. Another consequence of their large size is that they lack the sweetness of the simple and double sugars.

Only two polysaccharides are of major importance to the body: starch and glycogen. Both are polymers of glucose. Only their degree of branching differs.

Starch is the storage carbohydrate formed by plants. The number of glucose units composing a starch molecule is high and variable. When we eat starchy foods such as grain products and potatoes, the starch must be digested for its glucose units to be absorbed. We are unable to digest *cellulose*, another polysaccharide found in all plant products. However, it is important in providing the *bulk* (one form of fiber) that helps move feces through the colon.

Glycogen (gli′ko-jen), the storage carbohydrate of animal tissues, is stored primarily in skeletal muscle and liver cells. Like starch, it is highly branched and is a very large molecule (Figure 2.15c). When blood sugar levels drop sharply, liver cells break down glycogen and release its glucose units to the blood. Since there are many branch endings from which glucose can be released simultaneously, body cells have almost instant access to glucose fuel.

Carbohydrate Functions

The major function of carbohydrates in the body is to provide a ready, easily used source of cellular fuel. Most cells can use only a few types of simple sugars, and glucose is at the top of the "cellular menu." As described in our earlier discussion of oxidation-reduction reactions (p. 37), glucose is broken down and oxidized within cells. During these chemical reactions, electrons are transferred. This relocation of electrons releases the bond energy stored in glucose, and this energy is used to synthesize ATP. When ATP supplies are sufficient, dietary carbohydrates are converted to glycogen or fat and stored. Those of us who have gained weight from eating too many carbohydrate-rich snacks have personal experience with this conversion process!

Only small amounts of carbohydrates are used for structural purposes. For example, some sugars are found in our genes. Others are attached to the external surfaces of cells where they act as "road signs" to guide cellular interactions.

Lipids

Lipids are insoluble in water but dissolve readily in other lipids and in organic solvents such as alcohol and ether. Like carbohydrates, all lipids contain carbon, hydrogen, and oxygen, but the proportion of oxygen in lipids is much lower. In addition,

2

(a) Monosaccharides

Monomers of carbohydrates

Example
Hexose sugars (the hexoses shown here are isomers)

HOCH$_2$ H O H H OH H HO OH H OH
Glucose

HOCH$_2$ O H H HO HO CH$_2$OH OH H
Fructose

HOCH$_2$ HO O OH H OH H H H H OH
Galactose

Example
Pentose sugars

HOCH$_2$ O OH H H H H OH H
Deoxyribose

HOCH$_2$ O OH H H H H OH OH
Ribose

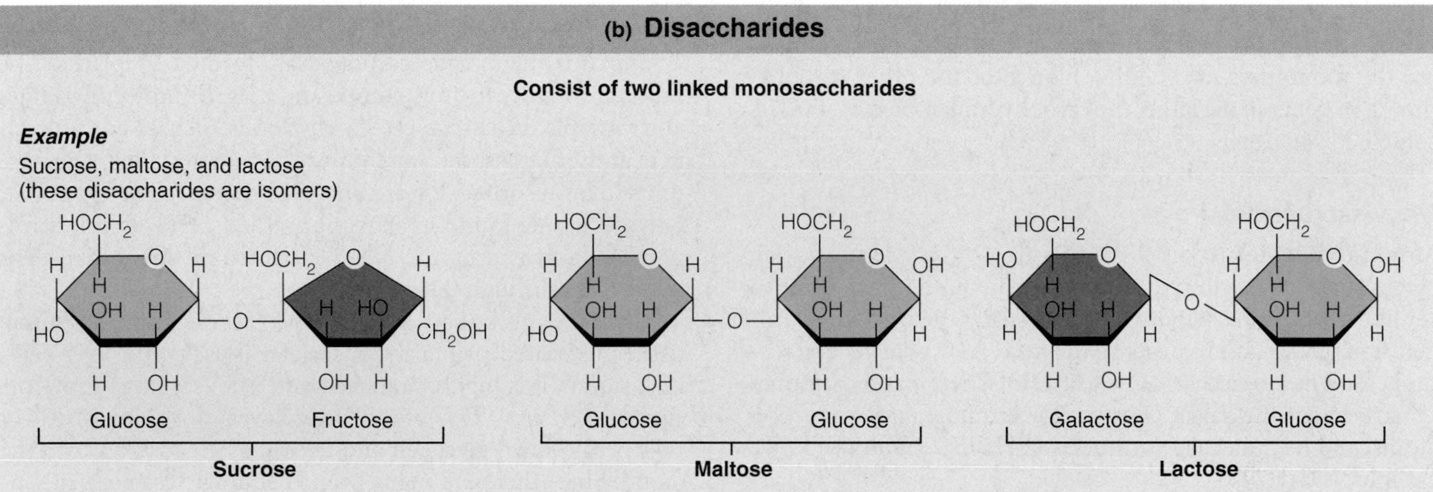

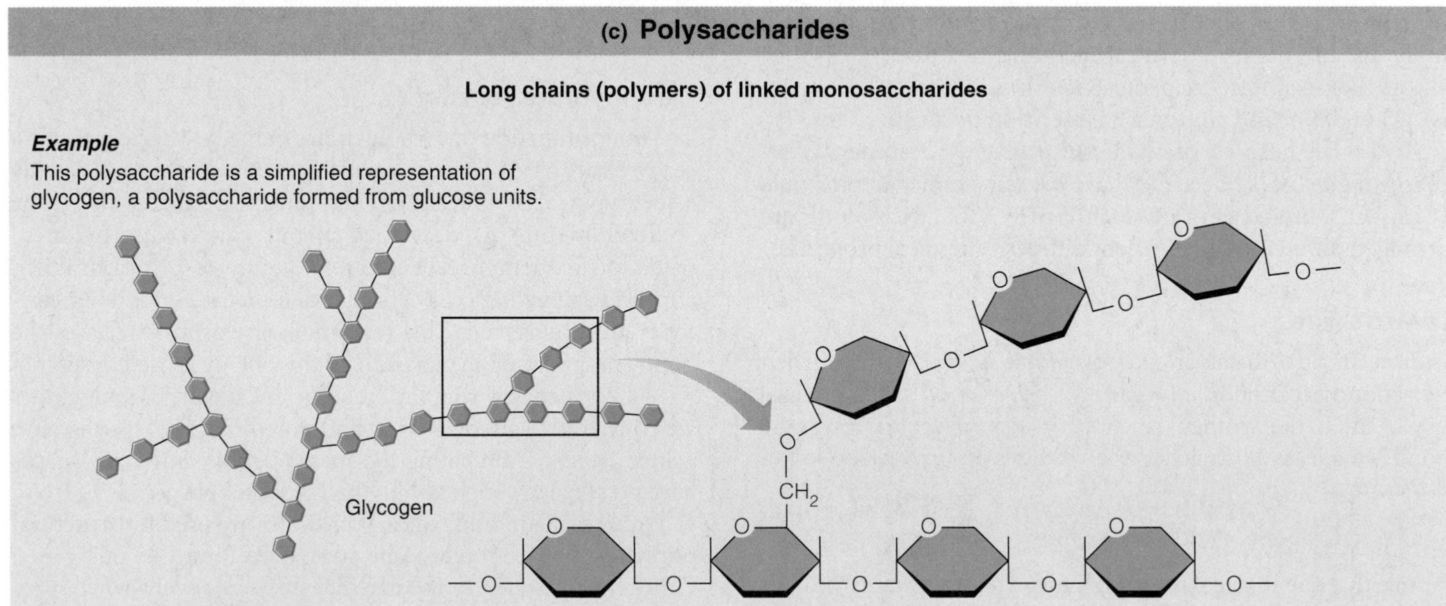

(c) Polysaccharides
Long chains (polymers) of linked monosaccharides
Example
This polysaccharide is a simplified representation of
glycogen, a polysaccharide formed from glucose units.
Glycogen
O
CH$_2$

Figure 2.15 Carbohydrate molecules important to the body.*

*Notice that in Figure 2.15 the carbon (C) atoms present at the angles of the carbohydrate ring structures are not illustrated and in Figure 2.15c only the oxygen atoms and one CH_2 group are shown. The illustrations at right give an example of this shorthand style: The full structure of glucose is on the left and the shorthand structure on the right. This style is used for nearly all organic ringlike structures illustrated in this chapter.

CH$_2$OH H C O H C OH H C HO C C OH H OH

TABLE 2.2 Representative Lipids Found in the Body

LIPID TYPE	LOCATION/FUNCTION
Triglycerides (Neutral Fats)	
	Fat deposits (in subcutaneous tissue and around organs) protect and insulate body organs, and are the major source of *stored* energy in the body.
Phospholipids (phosphatidylcholine; cephalin; others)	
	Chief components of cell membranes. Participate in the transport of lipids in plasma. Prevalent in nervous tissue.
Steroids	
Cholesterol	The structural basis for manufacture of all body steroids. A component of cell membranes.
Bile salts	These breakdown products of cholesterol are released by the liver into the digestive tract, where they aid fat digestion and absorption.
Vitamin D	A fat-soluble vitamin produced in the skin on exposure to UV radiation. Necessary for normal bone growth and function.
Sex hormones	Estrogen and progesterone (female hormones) and testosterone (a male hormone) are produced in the gonads. Necessary for normal reproductive function.
Adrenocortical hormones	Cortisol, a glucocorticoid, is a metabolic hormone necessary for maintaining normal blood glucose levels. Aldosterone helps to regulate salt and water balance of the body by targeting the kidneys.
Other Lipoid Substances	
Fat-soluble vitamins:	
A	Ingested in orange-pigmented vegetables and fruits. Converted in the retina to retinal, a part of the photoreceptor pigment involved in vision.
E	Ingested in plant products such as wheat germ and green leafy vegetables. Claims have been made (but not proved in humans) that it promotes wound healing, contributes to fertility, and may help to neutralize highly reactive particles called free radicals believed to be involved in triggering some types of cancer.
K	Made available to humans largely by the action of intestinal bacteria. Also prevalent in a wide variety of foods. Necessary for proper clotting of blood.
Eicosanoids (prostaglandins; leukotrienes; thromboxanes)	Group of molecules derived from fatty acids found in all cell membranes. The potent prostaglandins have diverse effects, including stimulation of uterine contractions, regulation of blood pressure, control of gastrointestinal tract motility, and secretory activity. Both prostaglandins and leukotrienes are involved in inflammation. Thromboxanes are powerful vasoconstrictors.
Lipoproteins	Lipoid and protein-based substances that transport fatty acids and cholesterol in the bloodstream. Major varieties are high-density lipoproteins (HDLs) and low-density lipoproteins (LDLs).

phosphorus is found in some of the more complex lipids. Lipids include *triglycerides*, *phospholipids* (fos″fo-lip′idz), *steroids* (stĕ′roidz), and a number of other lipoid substances. **Table 2.2** gives the locations and functions of some lipids found in the body.

Triglycerides (Neutral Fats)

Triglycerides (tri-glis′er-īdz), also called **neutral fats**, are commonly known as *fats* when solid or *oils* when liquid. A triglyceride is composed of two types of building blocks, **fatty acids** and **glycerol** (glis′er-ol), in a 3:1 ratio of fatty acids to glycerol **(Figure 2.16a)**. Fatty acids are linear chains of carbon and hydrogen atoms (hydrocarbon chains) with an organic acid group (—COOH) at one end. Glycerol is a modified simple sugar (a sugar alcohol).

Fat synthesis involves attaching three fatty acid chains to a single glycerol molecule by dehydration synthesis. The result is an E-shaped molecule. The glycerol backbone is the same in all triglycerides, but the fatty acid chains vary, resulting in different kinds of fats and oils.

These are large molecules, often consisting of hundreds of atoms, and ingested fats and oils must be broken down to their building blocks before they can be absorbed. Their hydrocarbon chains make the triglycerides nonpolar molecules. Because polar and nonpolar molecules do not interact, oil (or

2

(a) **Triglyceride formation**

Three fatty acid chains are bound to glycerol by dehydration synthesis.

Glycerol + 3 fatty acid chains → Triglyceride, or neutral fat + 3 water molecules

$$3\ HO{-}C(=O){-}CH_2{-}CH_2{\cdots}CH_2{-}CH_2{-}CH_3 + \text{glycerol} \rightarrow \text{triglyceride} + 3H_2O$$

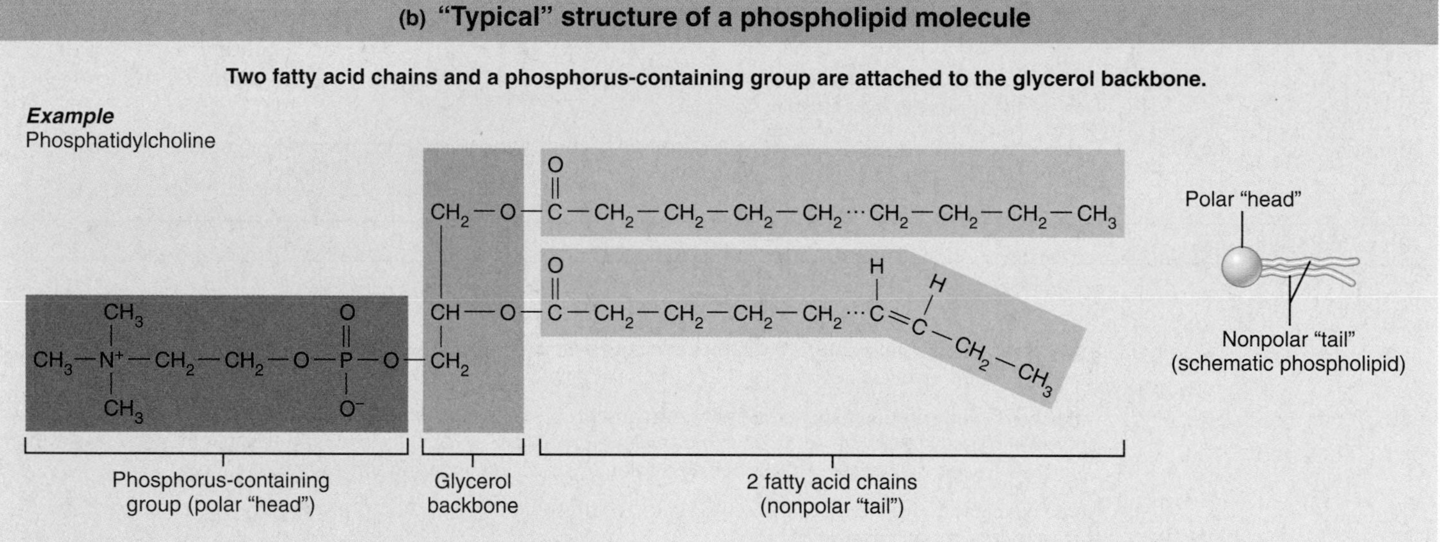

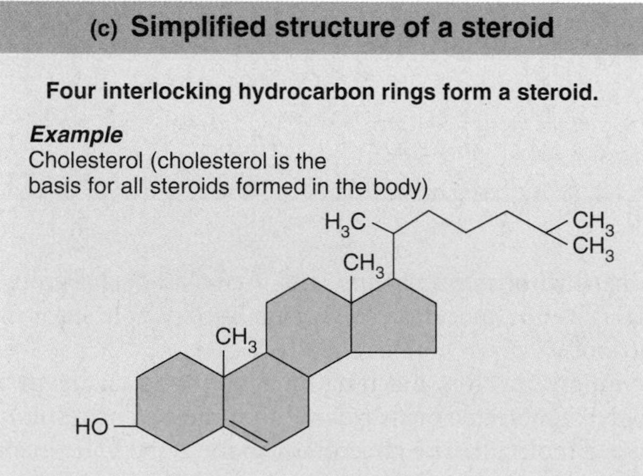

(c) Simplified structure of a steroid
Four interlocking hydrocarbon rings form a steroid.
Example
Cholesterol (cholesterol is the basis for all steroids formed in the body)

Figure 2.16 Lipids. The general structure of **(a)** triglycerides, or neutral fats, **(b)** phospholipids, and **(c)** cholesterol.

fats) and water do not mix. Consequently, triglycerides provide the body's most efficient and compact form of stored energy, and when they are oxidized, they yield large amounts of energy.

Triglycerides are found mainly beneath the skin, where they insulate the deeper body tissues from heat loss and protect them from mechanical trauma. For example, women are usually more successful English Channel swimmers than men. Their success is due partly to their thicker subcutaneous fatty layer, which helps insulate them from the bitterly cold water of the Channel.

The length of a triglyceride's fatty acid chains and their degree of *saturation* with H atoms determine how solid the molecule is at a given temperature. Fatty acid chains with only single covalent bonds between carbon atoms are referred to as **saturated.** Their fatty acid chains are straight and, at room temperature, the molecules of a saturated fat are packed closely together, forming a solid. Fatty acids that contain one or more double bonds between carbon atoms are said to be **unsaturated** (**monounsaturated** and **polyunsaturated**, respectively). The double bonds cause the fatty acid chains to kink so that they cannot be packed closely enough to solidify. Hence, triglycerides with short fatty acid chains or unsaturated fatty acids are oils (liquid at room temperature) and are typical of plant lipids. Examples include olive and peanut oils (rich in monounsaturated fats) and corn, soybean, and safflower oils, which contain a high percentage of polyunsaturated fatty acids. Longer fatty acid chains and more saturated fatty acids are common in animal fats such as butterfat and the fat of meats, which are solid at room temperature. Of the two types of fatty acids, the unsaturated variety, especially olive oil, is said to be more "heart healthy."

Trans fats, common in many margarines and baked products, are oils that have been solidified by addition of H atoms at sites of double carbon bonds. They have recently been branded as increasing the risk of heart disease even more than the solid animal fats. Conversely, the **omega-3 fatty acids**, found naturally in cold-water fish, appear to decrease the risk of heart disease and some inflammatory diseases.

Phospholipids

Phospholipids are modified triglycerides. Specifically, they are diglycerides with a phosphorus-containing group and two, rather than three, fatty acid chains (Figure 2.16b). The phosphorus-containing group gives phospholipids their distinctive chemical properties. Although the hydrocarbon portion (the "tail") of the molecule is nonpolar and interacts only with nonpolar molecules, the phosphorus-containing part (the "head") is polar and attracts other polar or charged particles, such as water or ions. This unique characteristic of phospholipids allows them to be used as the chief material for building cellular membranes. Some biologically important phospholipids and their functions are listed in Table 2.2.

Steroids

Structurally, steroids differ quite a bit from fats and oils. **Steroids** are basically flat molecules made of four interlocking hydrocarbon rings. Like triglycerides, steroids are fat soluble and contain little oxygen. The single most important molecule in our steroid chemistry is *cholesterol* (ko-les′ter-ol) (Figure 2.16c). We ingest cholesterol in animal products such as eggs, meat, and cheese, and our liver produces some.

Cholesterol has earned bad press because of its role in arteriosclerosis, but it is essential for human life. Cholesterol is found in cell membranes and is the raw material for synthesis of vitamin D, steroid hormones, and bile salts. Although steroid hormones are present in the body in only small quantities, they are vital to homeostasis. Without sex hormones, reproduction would be impossible, and a total lack of the corticosteroids produced by the adrenal glands is fatal.

Eicosanoids

The **eicosanoids** (i-ko′sah-noyds) are diverse lipids chiefly derived from a 20-carbon fatty acid (arachidonic acid) found in all cell membranes. Most important of these are the *prostaglandins* and their relatives, which play roles in various body processes including blood clotting, regulation of blood pressure, inflammation, and labor contractions (Table 2.2). Their synthesis and inflammatory actions are blocked by NSAIDs (nonsteroidal anti-inflammatory drugs) and the newer COX inhibitors.

CHECK YOUR UNDERSTANDING

20. What are the monomers of carbohydrates called? Which monomer is blood sugar?
21. What is the animal form of stored carbohydrate called?
22. How do triglycerides differ from phospholipids in body function and location?
23. What is the result of hydrolysis reactions and how are these reactions accomplished in the body?

For answers, see Appendix G.

2

Proteins

- Describe the four levels of protein structure.
- Indicate the function of molecular chaperones.
- Describe enzyme action.

The full set of proteins made by the body, called the *proteome*, and the way those proteins network in the body or change with disease, is a matter of intense biotech research.

Protein composes 10–30% of cell mass and is the basic structural material of the body. However, not all proteins are construction materials. Many play vital roles in cell function. Proteins, which include enzymes (biological catalysts), hemoglobin of the blood, and contractile proteins of muscle, have the most varied functions of any molecules in the body. All proteins contain carbon, oxygen, hydrogen, and nitrogen, and many contain sulfur and phosphorus as well.

Amino Acids and Peptide Bonds

The building blocks of proteins are molecules called **amino acids**, of which there are 20 common types (see Appendix C). All amino acids have two important functional groups: a basic group called an *amine* (ah′mēn) *group* (—NH_2), and an organic *acid group* (—COOH). An amino acid may therefore act either as a base (proton acceptor) or an acid (proton donor). All amino acids are identical except for a single group of atoms called their *R group*. Hence, it is differences in the R group that make each amino acid chemically unique, as the examples in **Figure 2.17** show.

Proteins are long chains of amino acids joined together by dehydration synthesis, with the amine end of one amino acid linked to the acid end of the next. The resulting bond produces a characteristic arrangement of linked atoms called a **peptide bond (Figure 2.18)**. Two united amino acids form a *dipeptide*, three a *tripeptide*, and ten or more a *polypeptide*. Although polypeptides containing more than 50 amino acids are called proteins, most proteins are **macromolecules**, large, complex molecules containing from 100 to over 10,000 amino acids.

Because each type of amino acid has distinct properties, the sequence in which they are bound together produces proteins that vary widely in both structure and function. We can think of

2

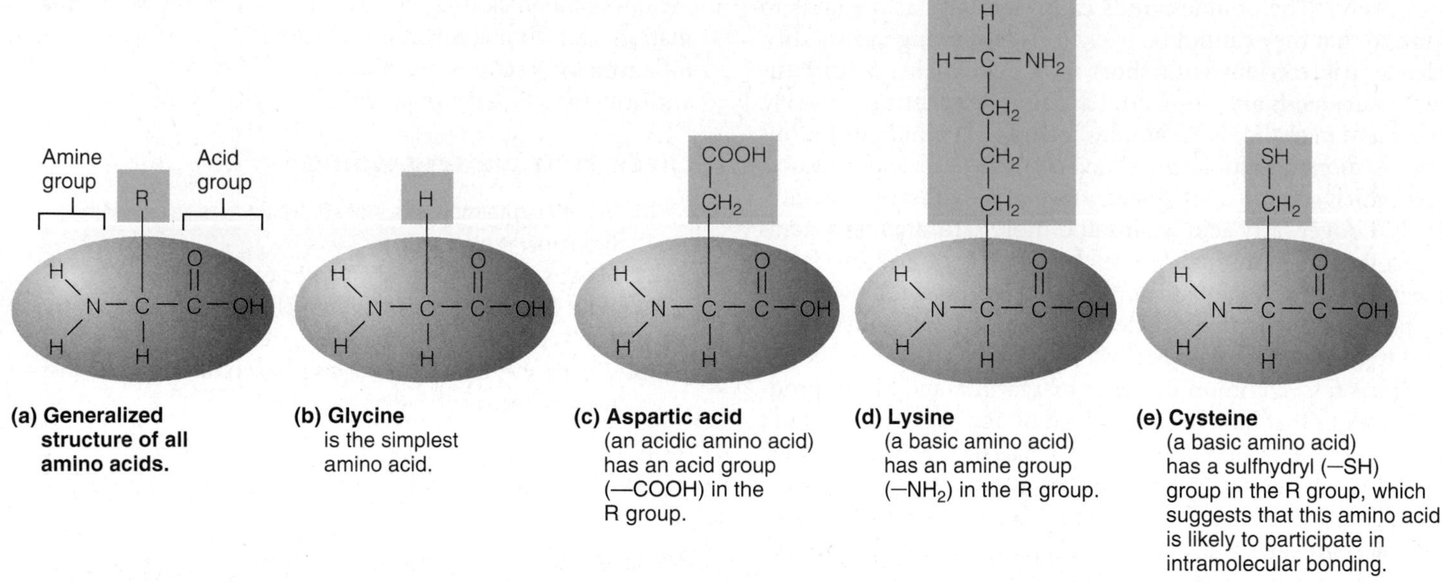

Amine group
Acid group
R
H
COOH
CH2
H—C—NH2
CH2
CH2
CH2
SH
CH2
N—C—C—OH
(a) Generalized structure of all amino acids.
(b) Glycine is the simplest amino acid.
(c) Aspartic acid (an acidic amino acid) has an acid group (—COOH) in the R group.
(d) Lysine (a basic amino acid) has an amine group (—NH2) in the R group.
(e) Cysteine (a basic amino acid) has a sulfhydryl (–SH) group in the R group, which suggests that this amino acid is likely to participate in intramolecular bonding.

Figure 2.17 Amino acid structures. All amino acids have both an amine group (—NH_2) and an acid group (—COOH). They differ only in their R groups (green). Differences in their R groups allow them to function differently in the body.

the 20 amino acids as a 20-letter "alphabet" used in specific combinations to form "words" (proteins). Just as a change in one letter can produce a word with an entirely different meaning (flour → floor) or that is nonsensical (flour → fllur), changes in the kinds or positions of amino acids can yield proteins with different functions or proteins that are nonfunctional. Nevertheless, there are thousands of different proteins in the body, each with distinct functional properties, and all constructed from different combinations of the 20 common amino acids.

Structural Levels of Proteins

Proteins can be described in terms of four structural levels. The linear sequence of amino acids composing the polypeptide chain is called the *primary structure* of a protein. This structure, which resembles a strand of amino acid "beads," is the backbone of the protein molecule **(Figure 2.19a).**

Proteins do not normally exist as simple, linear chains of amino acids. Instead, they twist or bend upon themselves to form a more complex *secondary structure.* The most common type of secondary structure is the **alpha (α)-helix**, which resembles a Slinky toy or the coils of a telephone cord (Figure 2.19b). The α-helix is formed by coiling of the primary chain and is stabilized by hydrogen bonds formed between NH and CO groups in amino acids in the primary chain which are approximately four amino acids apart. Hydrogen bonds in α-helices always link different parts of the *same* chain together.

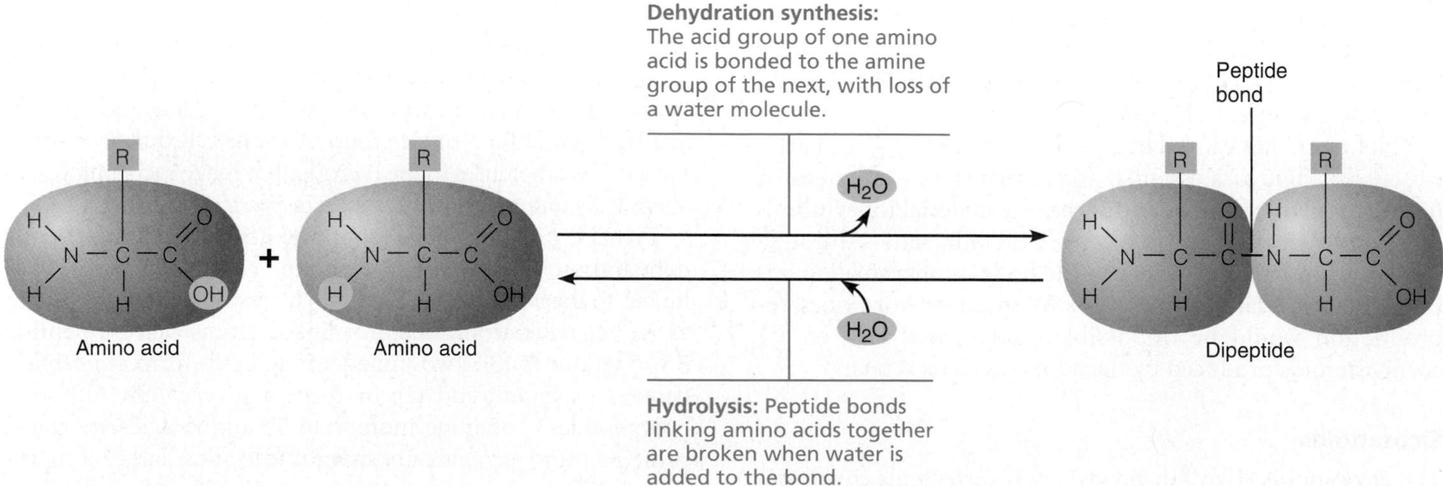

Dehydration synthesis:
The acid group of one amino acid is bonded to the amine group of the next, with loss of a water molecule.
Peptide bond
R
H2O
Amino acid
Amino acid
Dipeptide
Hydrolysis: Peptide bonds linking amino acids together are broken when water is added to the bond.

Figure 2.18 Amino acids are linked together by peptide bonds. Peptide bonds are formed by dehydration synthesis and broken by hydrolysis reactions.

(a) Primary structure:
The sequence of amino acids forms the polypeptide chain.

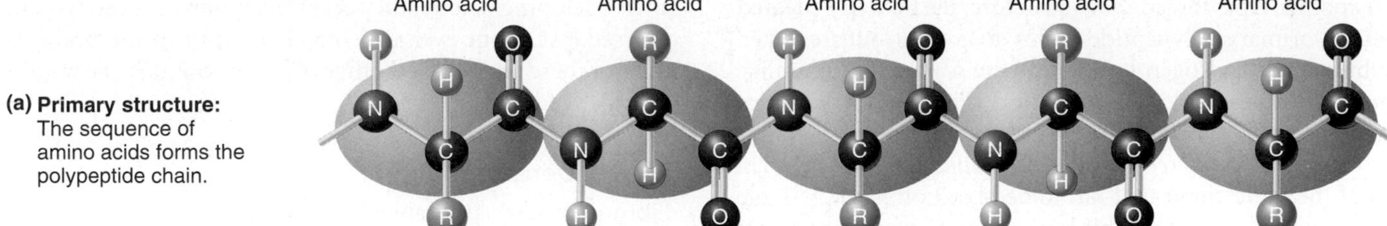

(b) Secondary structure:
The primary chain forms spirals (α-helices) and sheets (β-sheets).

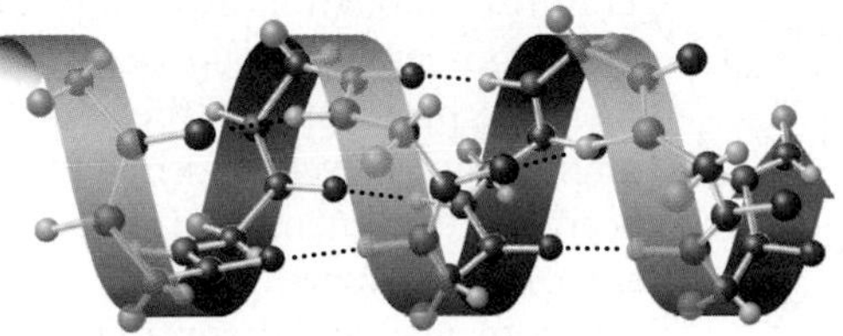

α-Helix: The primary chain is coiled to form a spiral structure, which is stabilized by hydrogen bonds.

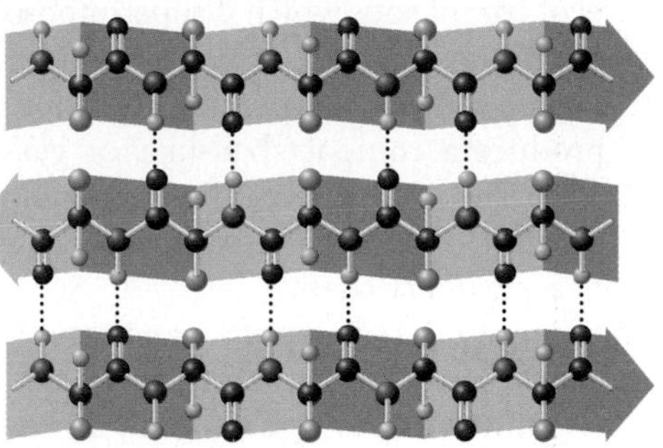

β-Sheet: The primary chain "zig-zags" back and forth forming a "pleated" sheet. Adjacent strands are held together by hydrogen bonds.

(c) Tertiary structure:
Superimposed on secondary structure. α-Helices and/or β-sheets are folded up to form a compact globular molecule held together by intramolecular bonds.

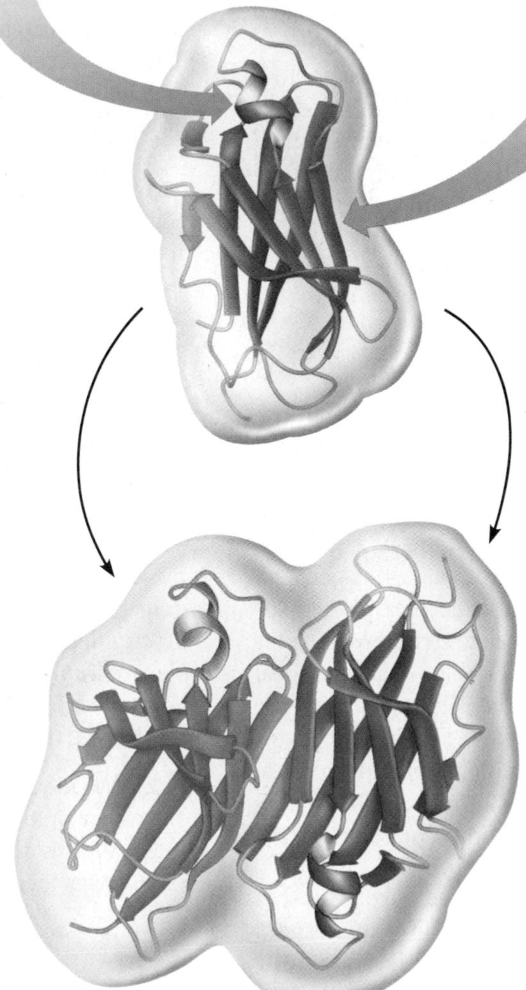

Tertiary structure of prealbumin (transthyretin), a protein that transports the thyroid hormone thyroxine in serum and cerebrospinal fluid.

(d) Quaternary structure:
Two or more polypeptide chains, each with its own tertiary structure, combine to form a functional protein.

Quaternary structure of a functional prealbumin molecule. Two identical prealbumin subunits join head to tail to form the dimer.

Figure 2.19 **Levels of protein structure.**

In another type of secondary structure, the **beta (β)-pleated sheet**, the primary polypeptide chains do not coil, but are linked side by side by hydrogen bonds to form a pleated, ribbonlike structure that resembles an accordion (Figure 2.19b). Notice that in this type of secondary structure, the hydrogen bonds may link together *different polypeptide chains* as well as *different parts* of the same chain that has folded back on itself. A single polypeptide chain may exhibit both types of secondary structure at various places along its length.

2

Many proteins have *tertiary structure* (ter′she-a″re), the next higher level of complexity, which is superimposed on secondary structure. Tertiary structure is achieved when α-helical or β-pleated regions of the polypeptide chain fold upon one another to produce a compact ball-like, or *globular*, molecule (Figure 2.19c). The unique structure is maintained by both covalent and hydrogen bonds between amino acids that are often far apart in the primary chain.

When two or more polypeptide chains aggregate in a regular manner to form a complex protein, the protein has *quaternary structure* (kwah′ter-na″re). Prealbumin, a protein that transports thyroid hormone in the blood, exhibits this structural level (Figure 2.19d).

How do these different levels of structure arise? Although a protein with tertiary or quaternary structure looks a bit like a clump of congealed pasta, the ultimate overall structure of any protein is very specific and is dictated by its primary structure. In other words, the types and relative positions of amino acids in the protein backbone determine where bonds can form to produce the complex coiled or folded structures that keep water-loving amino acids near the surface and water-fleeing amino acids buried in the protein's core. In addition, cells decorate many proteins by attaching sugars or fatty acids to them in ways that are difficult to imagine or predict.

Fibrous and Globular Proteins

The overall structure of a protein determines its biological function. In general, proteins are classified according to their overall appearance and shape as either fibrous or globular.

Fibrous proteins are extended and strandlike. Some exhibit only secondary structure, but most have tertiary or even quaternary structure as well. For example, *collagen* (kol′ah-jen) is a composite of the helical tropocollagen molecules packed together side by side to form a strong ropelike structure. Fibrous proteins are insoluble in water, and very stable—qualities ideal for providing mechanical support and tensile strength to the body's tissues. Besides collagen, which is the single most abundant protein in the body, the fibrous proteins include keratin, elastin, and certain contractile proteins of muscle **(Table 2.3)**. Because fibrous proteins are the chief building materials of the body, they are also known as **structural proteins**.

Globular proteins are compact, spherical proteins that have at least tertiary structure. Some also exhibit quaternary structure. The globular proteins are water-soluble, chemically active molecules, and they play crucial roles in virtually all biological processes. Consequently, this group is also called **functional proteins**. Some (antibodies) help to provide immunity, others (protein-based hormones) regulate growth and development, and still others (enzymes) are catalysts that oversee just about every chemical reaction in the body. The roles of these and selected other proteins found in the body are summarized in Table 2.3.

Protein Denaturation

Fibrous proteins are stable, but globular proteins are quite the opposite. The activity of a protein depends on its specific three-dimensional structure, and intramolecular bonds, particularly hydrogen bonds, are important in maintaining that structure. However, hydrogen bonds are fragile and easily broken by many chemical and physical factors, such as excessive acidity or temperature. Although individual proteins vary in their sensitivity to environmental conditions, hydrogen bonds begin to break when the pH drops or the temperature rises above normal (physiological) levels, causing proteins to unfold and lose their specific three-dimensional shape. In this condition, a protein is said to be **denatured**.

Fortunately, the disruption is reversible in most cases, and the "scrambled" protein regains its native structure when desirable conditions are restored. However, if the temperature or pH change is so extreme that protein structure is damaged beyond repair, the protein is *irreversibly denatured*. The coagulation of egg white (primarily albumin protein) that occurs when you boil or fry an egg is an example of irreversible protein denaturation. There is no way to restore the white, rubbery protein to its original translucent form.

When globular proteins are denatured, they can no longer perform their physiological roles because their function depends on the presence of specific arrangements of atoms, called **active sites**, on their surfaces. The active sites are regions that fit and interact chemically with other molecules of complementary shape and charge. Because atoms contributing to an active site may actually be far apart in the primary chain, disruption of intramolecular bonds separates them and destroys the active site. For example, hemoglobin becomes totally unable to bind and transport oxygen when blood pH is too acidic, because the structure needed for its function has been destroyed.

We will describe most types of body proteins in conjunction with the organ systems or functional processes to which they are closely related. However, two groups of proteins—*molecular chaperones* and *enzymes*—are intimately involved in the normal functioning of all cells, so we will consider these incredibly complex molecules here.

CHECK YOUR UNDERSTANDING

24. What does the name "amino acid" tell you about the structure of this molecule?

25. What is the primary structure of proteins?

26. What are the two types of secondary structure in proteins?

For answers, see Appendix G.

Molecular Chaperones

In addition to enzymes, all cells contain a class of unrelated globular proteins called **molecular chaperones** which, among

TABLE 2.3 Representative Types of Proteins in the Body

CLASSIFICATION ACCORDING TO		
OVERALL STRUCTURE	**GENERAL FUNCTION**	**EXAMPLES FROM THE BODY**
Fibrous		
	Structural framework/ mechanical support	*Collagen*, found in all connective tissues, is the single most abundant protein in the body. It is responsible for the tensile strength of bones, tendons, and ligaments. *Keratin* is the structural protein of hair and nails and a water-resistant material of skin. *Elastin* is found, along with collagen, where durability and flexibility are needed, such as in the ligaments that bind bones together. *Spectrin* internally reinforces and stabilizes the surface membrane of some cells, particularly red blood cells. *Dystrophin* reinforces and stabilizes the surface membrane of muscle cells. *Titin* helps organize the intracellular structure of muscle cells and accounts for the elasticity of skeletal muscles.
	Movement	*Actin* and *myosin*, contractile proteins, are found in substantial amounts in muscle cells, where they cause muscle cell shortening (contraction); they also function in cell division in all cell types. Actin is important in intracellular transport, particularly in nerve cells.
Globular		
	Catalysis	Protein enzymes are essential to virtually every biochemical reaction in the body; they increase the rates of chemical reactions by at least a millionfold. Examples include salivary amylase (in saliva), which catalyzes the breakdown of starch, and oxidase enzymes, which act to oxidize food fuels.
	Transport	*Hemoglobin* transports oxygen in blood, and *lipoproteins* transport lipids and cholesterol. Other transport proteins in the blood carry iron, hormones, or other substances. Some globular proteins in plasma membranes are involved in membrane transport (as carriers or channels).
	Regulation of pH	Many plasma proteins, such as *albumin*, function reversibly as acids or bases, thus acting as buffers to prevent wide swings in blood pH.
	Regulation of metabolism	*Peptide* and *protein hormones* help to regulate metabolic activity, growth, and development. For example, *growth hormone* is an anabolic hormone necessary for optimal growth; *insulin* helps regulate blood sugar levels.
	Body defense	*Antibodies* (immunoglobulins) are specialized proteins released by immune cells that recognize and inactivate foreign substances (bacteria, toxins, some viruses). *Complement proteins*, which circulate in blood, enhance both immune and inflammatory responses.
	Protein management	*Molecular chaperones* aid folding of new proteins in both healthy and damaged cells and transport of metal ions into and within the cell. They also promote breakdown of damaged proteins.

other things, help proteins to achieve their functional three-dimensional structure. Although its amino acid sequence determines the precise way a protein folds, the folding process also requires the help of molecular chaperones to ensure that the folding is quick and accurate. Apparently, molecular chaperones have numerous protein-related roles to play. For example, specific molecular chaperonens

- Prevent accidental, premature, or incorrect folding of polypeptide chains or their association with other polypeptides
- Aid the desired folding and association process
- Help to translocate proteins and certain metal ions (copper, iron, zinc) across cell membranes
- Promote the breakdown of damaged or denatured proteins
- Interact with other cells to trigger the immune response to diseased cells in the body

The first such proteins discovered were called *heat shock proteins* (*hsp*) because they seemed to protect cells from the destructive effects of heat. It was later found that these proteins are produced in response to a variety of traumatizing stimuli—for example, in the oxygen-deprived cells of a heart attack patient—and the name *stress proteins* replaced hsp for that particular group of molecular chaperones. It is now clear that these proteins are vitally important to cell function in all types of stressful circumstances.

Enzymes and Enzyme Activity

Enzymes are globular proteins that act as biological catalysts. *Catalysts* are substances that regulate and accelerate the rate of biochemical reactions but are not used up or changed in those reactions. More specifically, enzymes can be thought of as chemical

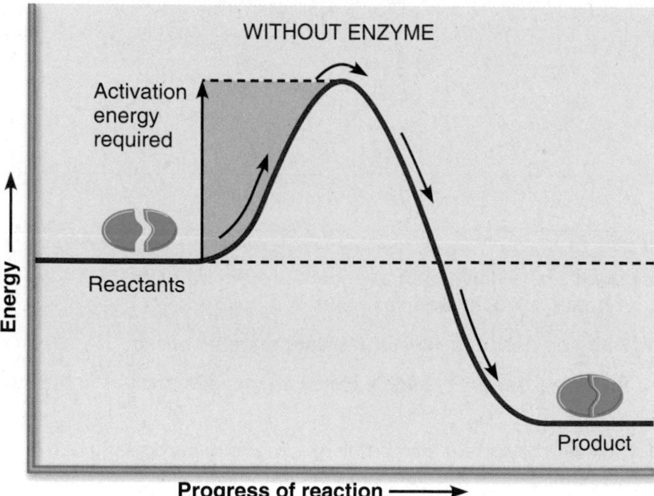

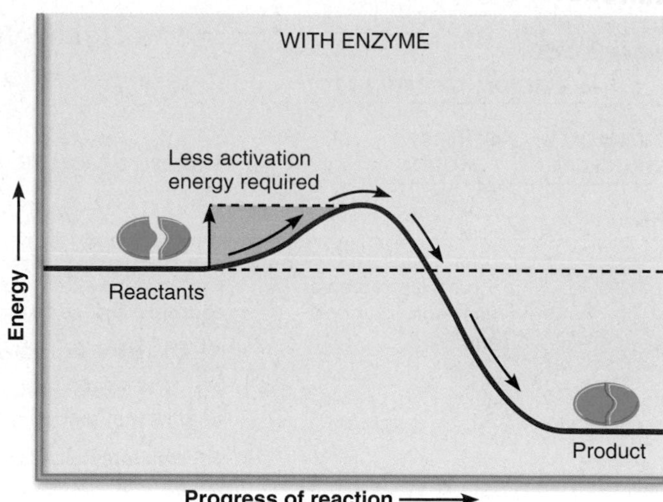

Figure 2.20 **Enzymes lower the activation energy required for a reaction to proceed rapidly.**

traffic cops that keep our metabolic pathways flowing. Enzymes cannot force chemical reactions to occur between molecules that would not otherwise react; they can only increase the speed of reaction. Without enzymes, biochemical reactions proceed so slowly that for practical purposes they do not occur at all.

Characteristics of Enzymes Some enzymes are purely protein. In other cases, the functional enzyme consists of two parts, collectively called a **holoenzyme**: an **apoenzyme** (the protein portion) and a **cofactor**. Depending on the enzyme, the cofactor may be an ion of a metal element such as copper or iron, or an organic molecule needed to assist the reaction in some particular way. Most organic cofactors are derived from vitamins (especially the B complex vitamins). This type of cofactor is more precisely called a **coenzyme**.

Each enzyme is chemically specific. Some enzymes control only a single chemical reaction. Others exhibit a broader specificity in that they can bind with similar (but not identical) molecules and thus regulate a small group of related reactions. The substance on which an enzyme acts is called a **substrate**.

The presence of specific enzymes determines not only which reactions will be speeded up, but also which reactions will occur—no enzyme, no reaction. This also means that unwanted or unnecessary chemical reactions do not occur.

Most enzymes are named for the type of reaction they catalyze. *Hydrolases* (hi′druh-lās-es) add water during hydrolysis reactions, *oxidases* (ok′sĭ-dās-es) add oxygen, and so on. You can recognize most enzyme names by the suffix *-ase*.

In many cases, enzymes are part of cellular membranes in a bucket-brigade type of arrangement. The product of one enzyme-catalyzed reaction becomes the substrate of the neighboring enzyme, and so on. Some enzymes are produced in an inactive form and must be activated in some way before they can function, often by a change in the pH of their surroundings. For example, digestive enzymes produced in the pancreas are activated in the small intestine, where they actually do their work. If they were produced in active form, the pancreas would digest itself.

Sometimes, enzymes are inactivated immediately after they have performed their catalytic function. This is true of enzymes that promote blood clot formation when the wall of a blood vessel is damaged. Once clotting is triggered, those enzymes are inactivated. Otherwise, you would have blood vessels full of solid blood instead of one protective clot. (Eek!)

Enzyme Action How do enzymes perform their catalytic role? Every chemical reaction requires that a certain amount of energy, called **activation energy**, be absorbed to prime the reaction. The activation energy is the amount of energy needed to break the bonds of the reactants so that they can rearrange themselves and become the product. It is present when kinetic energy pushes the reactants to an energy level where their random collisions are forceful enough to ensure interaction. Activation energy is needed regardless of whether the overall reaction is ultimately energy absorbing or energy releasing.

One way to increase kinetic energy is to increase the temperature, but higher temperatures denature proteins. (This is why a high fever can be a serious event.) Enzymes allow reactions to occur at normal body temperature by decreasing the amount of activation energy required **(Figure 2.20)**.

Exactly how do enzymes accomplish this remarkable feat? The answer is not fully understood. However, we know that, due to structural and electrostatic factors, they decrease the randomness of reactions by binding to the reacting molecules temporarily and presenting them to each other in the proper position for chemical interaction to occur.

Three basic steps appear to be involved in enzyme action **(Figure 2.21)**.

① **The enzyme's active site binds to the substrate(s) on which it acts, temporarily forming an enzyme-substrate complex.** Substrate binding causes the active site to change shape so that the substrate and the active site fit together

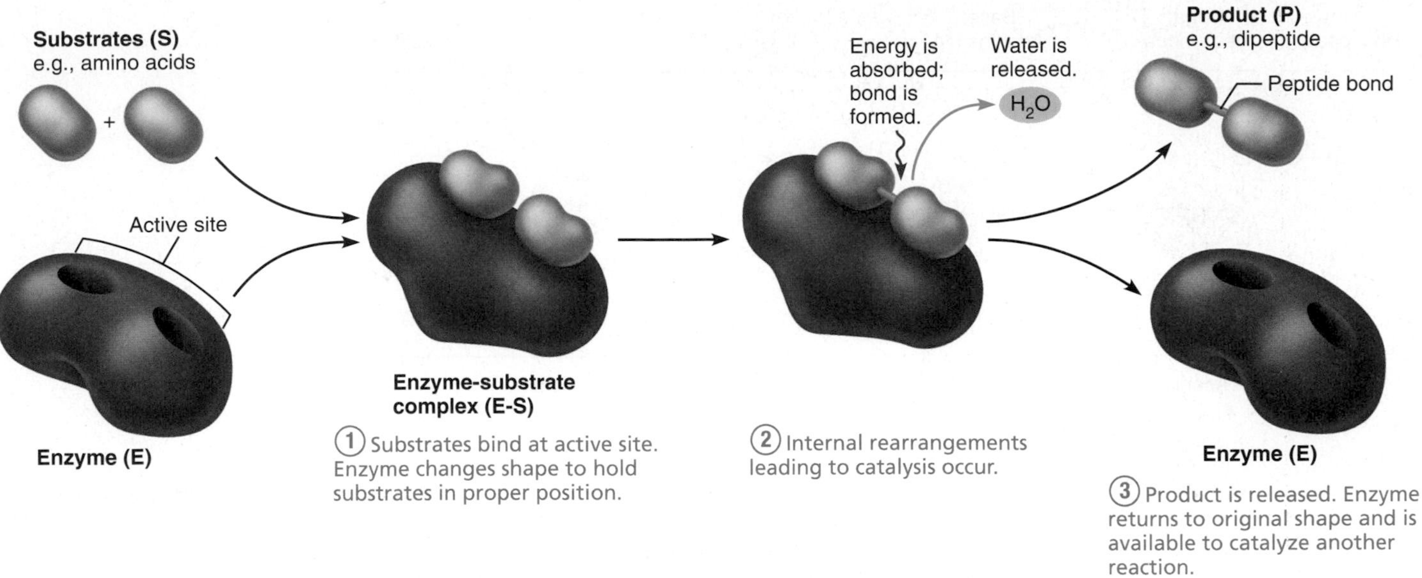
Substrates (S)
e.g., amino acids
Active site
Enzyme (E)
Enzyme-substrate complex (E-S)
① Substrates bind at active site. Enzyme changes shape to hold substrates in proper position.
Energy is absorbed; bond is formed.
Water is released.
H2O
② Internal rearrangements leading to catalysis occur.
Product (P)
e.g., dipeptide
Peptide bond
Enzyme (E)
③ Product is released. Enzyme returns to original shape and is available to catalyze another reaction.

Figure 2.21 Mechanism of enzyme action. In this example, the enzyme catalyzes the formation of a dipeptide from specific amino acids. *Summary*: E + S → E-S → P + E

precisely. Although enzymes are specific for particular substrates, other (nonsubstrate) molecules may act as *enzyme inhibitors* if their structure is similar enough to occupy or block the enzyme's active site.

② **The enzyme-substrate complex undergoes internal rearrangements that form the product(s).** This step shows the catalytic role of an enzyme.

③ **The enzyme releases the product(s) of the reaction.** The enzyme is not changed. If the enzyme became part of the product, it would be a reactant and not a catalyst.

Because enzymes are unchanged by their catalytic role and can act again and again, cells need only small amounts of each enzyme. Catalysis occurs with incredible speed. Most enzymes can catalyze millions of reactions per minute.

CHECK YOUR UNDERSTANDING

27. What is the main event that molecular chaperones prevent?
28. How do enzymes reduce the amount of activation energy needed to make a chemical reaction go?

For answers, see Appendix G.

Nucleic Acids (DNA and RNA)

▶ Compare and contrast DNA and RNA.

The **nucleic acids** (nu-kle′ic), composed of carbon, oxygen, hydrogen, nitrogen, and phosphorus, are the largest molecules in the body. The nucleic acids include two major classes of molecules, **deoxyribonucleic acid (DNA)** (de-ok″sĭ-ri″bo-nu-kle′ik) and **ribonucleic acid (RNA)**.

The structural units of nucleic acids, called **nucleotides**, are quite complex. Each nucleotide consists of three components: a nitrogen-containing base, a pentose sugar, and a phosphate group (**Figure 2.22a**). Five major varieties of nitrogen-containing bases can contribute to nucleotide structure: **adenine**, abbreviated **A** (ad′ĕ-nēn); **guanine**, **G**(gwan′ēn); **cytosine**, **C** (si′to-sēn); **thymine**, **T** (thi′mēn); and **uracil**, **U** (u′rah-sil). Adenine and guanine are large, two-ring bases (called purines), whereas cytosine, thymine, and uracil are smaller, single-ring bases (called pyrimidines).

The stepwise synthesis of a nucleotide involves the attachment of a base to the pentose sugar to form first a *nucleoside*, named for the nitrogenous base it contains. The nucleotide is formed when a phosphate group is bonded to the sugar of the nucleoside.

Although DNA and RNA are both composed of nucleotides, they differ in many respects, as summarized in **Table 2.4**. Typically, DNA is found in the nucleus (control center) of the cell, where it constitutes the *genetic material*, also called the *genes*, or more recently the *genome*. DNA has two fundamental roles: It replicates (reproduces) itself before a cell divides, ensuring that the genetic information in the descendant cells is identical, and it provides the basic instructions for building every protein in the body. Although we have said that enzymes govern all chemical reactions, remember that enzymes, too, are proteins formed at the direction of DNA.

By providing the information for protein synthesis, DNA determines what type of organism you will be—frog, human, oak tree—directs your growth and development, and accounts for your uniqueness. A technique called DNA fingerprinting can help solve forensic mysteries (for example, verify one's presence at a crime scene), identify badly burned or mangled bodies at a disaster scene, and establish or disprove paternity. DNA fingerprinting analyzes tiny samples of DNA taken from blood, semen, or other body tissues and shows the results as a "genetic barcode" that distinguishes each of us from all others.

2

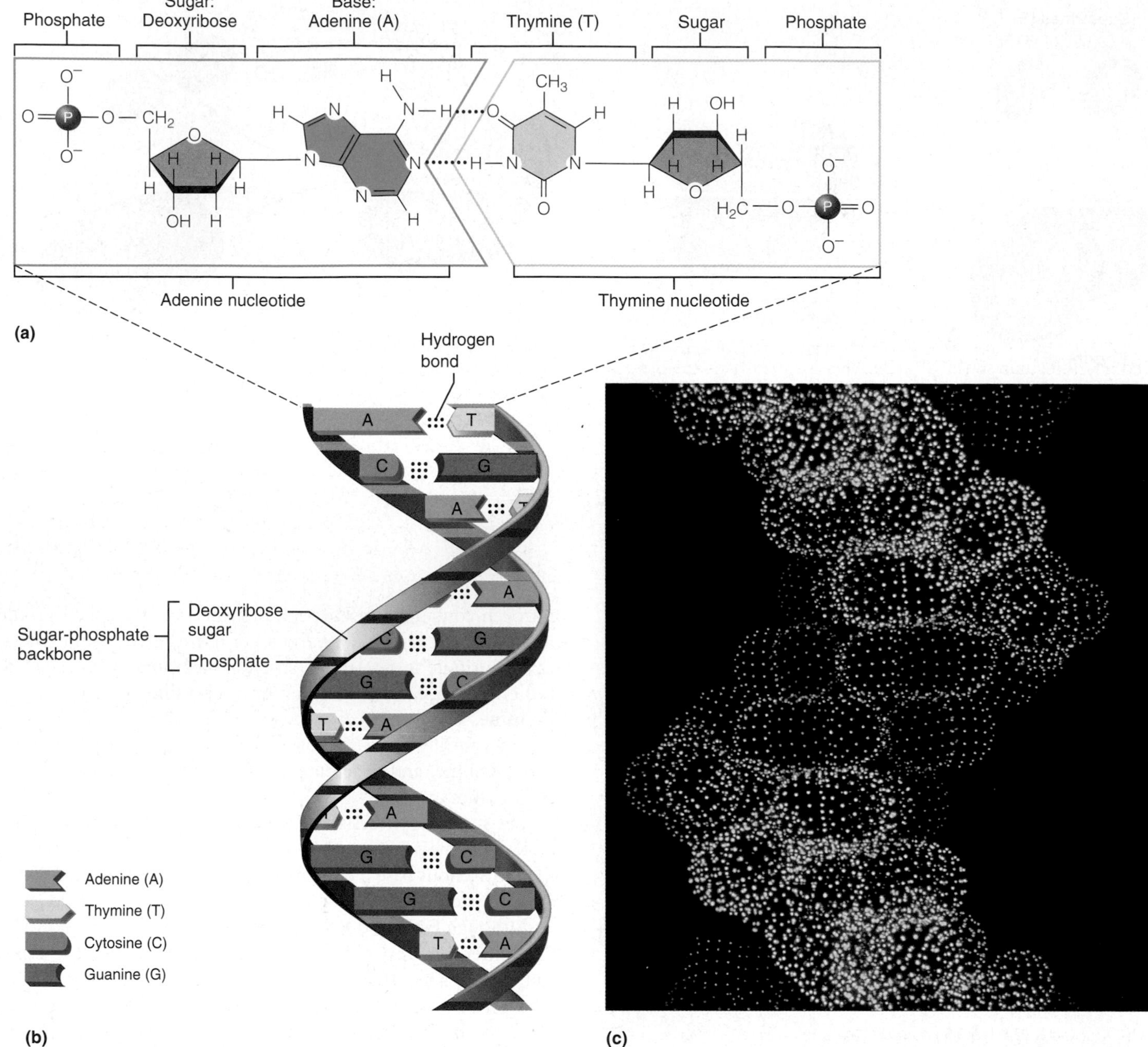

Figure 2.22 Structure of DNA. (a) The unit of DNA is the nucleotide, which is composed of a deoxyribose sugar molecule linked to a phosphate group, with a base attached to the sugar. Two nucleotides, linked by hydrogen bonds between their complementary bases, are illustrated. **(b)** DNA is a coiled double polymer of nucleotides (a double helix). The backbones of the ladderlike molecule are formed by alternating sugar and phosphate units. The rungs are formed by the binding together of complementary bases (A-T and G-C) by hydrogen bonds (shown by dotted lines). **(c)** Computer-generated image of a DNA molecule.

DNA is a long, double-stranded polymer—a double chain of nucleotides (Figure 2.22b and c). The bases in DNA are A, G, C, and T, and its pentose sugar is *deoxyribose* (as reflected in its name). Its two nucleotide chains are held together by hydrogen bonds between the bases, so that a ladderlike molecule is formed. Alternating sugar and phosphate components of each chain form the *backbones* or "uprights" of the "ladder," and the joined bases form the "rungs." The whole molecule is coiled into a spiral staircase–like structure called a **double helix**.

Bonding of the bases is very specific: A always bonds to T, and G always bonds to C. A and T are therefore called **complementary bases**, as are C and G. According to these base-pairing rules, ATGA on one DNA nucleotide strand would necessarily be bonded to TACT (a complementary base sequence) on the other strand.

TABLE 2.4 Comparison of DNA and RNA

CHARACTERISTIC	DNA	RNA
Major cellular site	Nucleus	Cytoplasm (cell area outside the nucleus)
Major functions	Is the genetic material; directs protein synthesis; replicates itself before cell division	Carries out the genetic instructions for protein synthesis
Sugar	Deoxyribose	Ribose
Bases	Adenine, guanine, cytosine, thymine	Adenine, guanine, cytosine, uracil
Structure	Double strand coiled into a double helix	Single strand, straight or folded

RNA is located chiefly outside the nucleus and can be considered a "molecular slave" of DNA. That is, RNA carries out the orders for protein synthesis issued by DNA. [Viruses in which RNA (rather than DNA) is the genetic material are an exception to this generalization.]

RNA molecules are single strands of nucleotides. RNA bases include A, G, C, and U (U replaces the T found in DNA), and its sugar is *ribose* instead of deoxyribose. The three major varieties of RNA (messenger RNA, ribosomal RNA, and transfer RNA) are distinguished by their relative size and shape, and each has a specific role to play in carrying out DNA's instructions for protein synthesis. In addition to these three RNAs, small RNA molecules called *microRNAs* (*miRNAs*) appear to control genetic expression by shutting down genes or altering their expression. We discuss DNA replication and the relative roles of DNA and RNA in protein synthesis in Chapter 3.

CHECK YOUR UNDERSTANDING

29. How do DNA and RNA differ in the bases and sugars they contain?

30. What are two important roles of DNA?

For answers, see Appendix G.

Adenosine Triphosphate (ATP)

▶ Explain the role of ATP in cell metabolism.

Glucose is the most important cellular fuel, but none of the chemical energy contained in its bonds is used directly to power cellular work. Instead, energy released during glucose catabolism is coupled to the synthesis of **adenosine triphosphate (ATP)**. In other words, some of this energy is captured and stored as small packets of energy in the bonds of ATP. ATP is the primary energy-transferring molecule in cells and it provides a form of energy that is immediately usable by all body cells.

Structurally, ATP is an adenine-containing RNA nucleotide to which two additional phosphate groups have been added (**Figure 2.23**). Chemically, the triphosphate tail of ATP can be compared to a tightly coiled spring ready to uncoil with tremendous energy when the catch is released. Actually, ATP is a very unstable energy-storing molecule because its three negatively charged phosphate groups are closely packed and repel each other. When its terminal high-energy phosphate bonds are broken (hydrolyzed), the chemical "spring" relaxes and the molecule as a whole becomes more stable.

Cells tap ATP's bond energy during coupled reactions by using enzymes to transfer the terminal phosphate groups from ATP to other compounds. These newly *phosphorylated* molecules are said to be "primed" and temporarily become more energetic and capable of performing some type of cellular work. In the process of doing their work, they lose the phosphate group. The amount of energy released and transferred during ATP hydrolysis corresponds closely to that needed to drive most biochemical reactions. As a result, cells are protected from excessive energy release that might be damaging, and energy squandering is kept to a minimum.

Cleaving the terminal phosphate bond of ATP yields a molecule with two phosphate groups—*adenosine diphosphate*

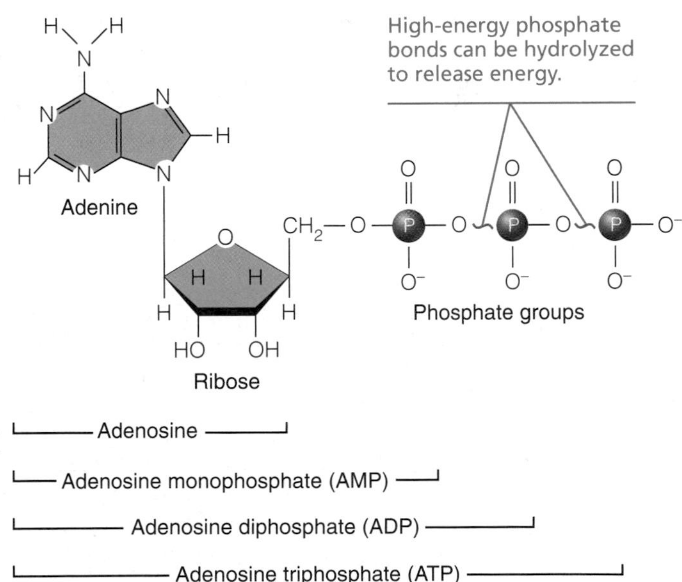

Figure 2.23 Structure of ATP (adenosine triphosphate). ATP is an adenine nucleotide to which two additional phosphate groups have been attached during breakdown of food fuels. When the terminal phosphate group is cleaved off, energy is released to do useful work and ADP (adenosine diphosphate) is formed. When the terminal phosphate group is cleaved off ADP, a similar amount of energy is released and AMP (adenosine monophosphate) is formed.

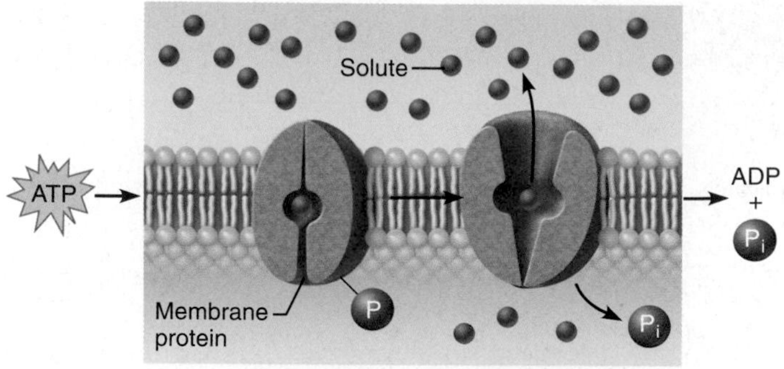

(a) Transport work: ATP phosphorylates transport proteins, activating them to transport solutes (ions, for example) across cell membranes.

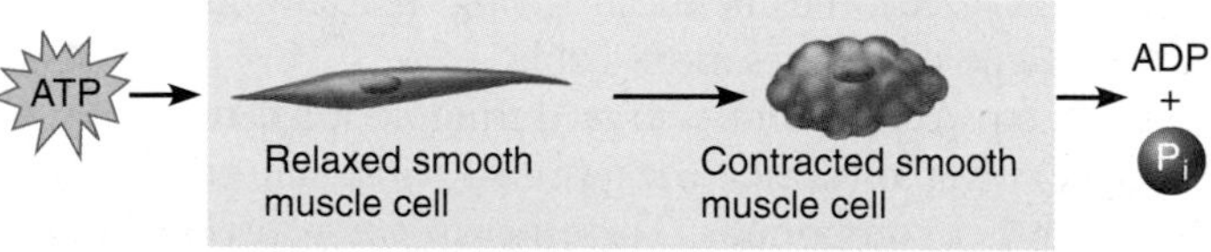

(b) Mechanical work: ATP phosphorylates contractile proteins in muscle cells so the cells can shorten.

(c) Chemical work: ATP phosphorylates key reactants, providing energy to drive energy-absorbing chemical reactions.

Figure 2.24 Three examples of cellular work driven by energy from ATP.

(*ADP*)—and an inorganic phosphate group, indicated by Pi, accompanied by a transfer of energy:

$$\text{ATP} \underset{H_2O}{\overset{H_2O}{\rightleftharpoons}} \text{ADP} + P_i + \text{energy}$$

As ATP is hydrolyzed to provide energy for cellular needs, ADP accumulates. Cleavage of the terminal phosphate bond of ADP liberates a similar amount of energy and produces adenosine monophosphate (AMP).

The cell's ATP supplies are replenished as glucose and other fuel molecules are oxidized and their bond energy is released. The same amount of energy that is liberated when ATP's terminal phosphates are cleaved off must be captured and used to reverse the reaction to reattach phosphates and re-form the energy-transferring phosphate bonds. Without ATP, molecules cannot be made or degraded, cells cannot transport substances across their membrane boundaries, muscles cannot shorten to tug on other structures, and life processes cease **(Figure 2.24).**

CHECK YOUR UNDERSTANDING

31. Glucose is an energy-rich molecule. So why do body cells need ATP?

32. What change occurs in ATP when it releases energy?

For answers, see Appendix G.

RELATED CLINICAL TERMS

Acidosis (as″ĭ-do′sis; *acid* = sour, sharp) A condition of acidity or low pH (below 7.35) of the blood; high hydrogen ion concentration.

Alkalosis (al″kah-lo′sis) A condition of basicity or high pH (above 7.45) of the blood; low hydrogen ion concentration.

Heavy metals Metals with toxic effects on the body, including arsenic, mercury, and lead. Iron, also included in this group, is toxic in high concentrations.

Ionizing radiation Radiation that causes atoms to ionize; for example, radioisotope emissions and X rays.

Ketosis (ke-to′sis) A condition resulting from excessive ketones (breakdown products of fats) in the blood; common during starvation and acute attacks of diabetes mellitus.

Radiation sickness Disease resulting from exposure of the body to radioactivity; digestive system organs are most affected.

CHAPTER SUMMARY

Media study tools that could provide you additional help in reviewing specific key topics of Chapter 2 are referenced below.

iP = ***Interactive Physiology***

PART 1: BASIC CHEMISTRY

Definition of Concepts: Matter and Energy (pp. 24–25)

Matter (p. 24)

1. Matter is anything that takes up space and has mass.

Energy (pp. 24–25)

2. Energy is the capacity to do work or put matter into motion.
3. Energy exists as potential energy (stored energy or energy of position) and kinetic energy (active or working energy).
4. Forms of energy involved in body functioning are chemical, electrical, radiant, and mechanical. Of these, chemical (bond) energy is most important.
5. Energy may be converted from one form to another, but some energy is always unusable (lost as heat) in such transformations.

Composition of Matter: Atoms and Elements (pp. 25–28)

1. Elements are unique substances that cannot be decomposed into simpler substances by ordinary chemical methods. Four elements (carbon, hydrogen, oxygen, and nitrogen) make up 96% of body weight.

Atomic Structure (pp. 25–27)

2. The building blocks of elements are atoms.
3. Atoms are composed of positively charged protons, negatively charged electrons, and uncharged neutrons. Protons and neutrons are located in the atomic nucleus, constituting essentially the atom's total mass. Electrons are outside the nucleus in the electron shells. In any atom, the number of electrons equals the number of protons.

Identifying Elements (pp. 27–28)

4. Atoms may be identified by their atomic number (p^+) and mass number ($p^+ + n^0$). The notation ^4_2He means that helium (He) has an atomic number of 2 and a mass number of 4.
5. Isotopes of an element differ in the number of neutrons they contain. The atomic weight of any element is approximately equal to the mass number of its most abundant isotope.

Radioisotopes (p. 28)

6. Many heavy isotopes are unstable (radioactive). These so-called radioisotopes decompose to more stable forms by emitting alpha or beta particles or gamma rays. Radioisotopes are useful in medical diagnosis and treatment and in biochemical research.

How Matter Is Combined: Molecules and Mixtures (pp. 28–30)

Molecules and Compounds (pp. 28–29)

1. A molecule is the smallest unit resulting from the chemical bonding of two or more atoms. If the atoms are different, they form a molecule of a compound.

Mixtures (pp. 29–30)

2. Mixtures are physical combinations of solutes in a solvent. Mixture components retain their individual properties.
3. The types of mixtures, in order of increasing solute size, are solutions, colloids, and suspensions.
4. Solution concentrations are typically designated in terms of percent or molarity.

Distinguishing Mixtures from Compounds (p. 30)

5. Compounds are homogeneous; their elements are chemically bonded. Mixtures may be homogeneous or heterogeneous; their components are physically combined and separable.

Chemical Bonds (pp. 31–35)

The Role of Electrons in Chemical Bonding (pp. 31–32)

1. Electrons of an atom occupy areas of space called electron shells or energy levels. Electrons in the shell farthest from the nucleus (valence shell) are most energetic.
2. Chemical bonds are energy relationships between valence shell electrons of the reacting atoms. Atoms with a full valence shell or eight valence shell electrons are chemically unreactive (inert). Those with an incomplete valence shell interact with other atoms to achieve stability.

Types of Chemical Bonds (pp. 32–35)

3. Ionic bonds are formed when valence shell electrons are completely transferred from one atom to another.
4. Covalent bonds are formed when atoms share electron pairs. If the electron pairs are shared equally, the molecule is nonpolar. If they are shared unequally, it is polar (a dipole).
5. Hydrogen bonds are weak bonds formed between one hydrogen atom, already covalently linked to an electronegative atom, and another electronegative atom (such as nitrogen or hydrogen and oxygen). They bind together different molecules (e.g., water molecules) or different parts of the same molecule (as in protein molecules).

Chemical Reactions (pp. 35–38)

Chemical Equations (pp. 35–36)

1. Chemical reactions involve the formation, breaking, or rearrangement of chemical bonds.

Patterns of Chemical Reactions (pp. 36–37)

2. Chemical reactions are either anabolic (constructive) or catabolic (destructive). They include synthesis, decomposition, and exchange reactions. Oxidation-reduction reactions may be considered a special type of exchange (or decomposition) reaction.

Energy Flow in Chemical Reactions (p. 37)

3. Bonds are energy relationships and there is a net loss or gain of energy in every chemical reaction.
4. In exergonic reactions, energy is liberated. In endergonic reactions, energy is absorbed.

Reversibility of Chemical Reactions (p. 37)

5. If reaction conditions remain unchanged, all chemical reactions eventually reach a state of chemical equilibrium in which the reaction proceeds in both directions at the same rate.

2

6. All chemical reactions are theoretically reversible, but many biological reactions go in only one direction because of energy requirements or the removal of reaction products.

Factors Influencing the Rate of Chemical Reactions (pp. 37–38)

7. Chemical reactions occur only when particles collide and valence shell electrons interact.
8. The smaller the reacting particles, the greater their kinetic energy and the faster the reaction rate. Higher temperature or reactant concentration, as well as the presence of catalysts, increases chemical reaction rates.

PART 2: BIOCHEMISTRY

Inorganic Compounds (pp. 38–41)

1. Most inorganic compounds do not contain carbon. Those found in the body include water, salts, and inorganic acids and bases.

Water (pp. 38–39)

2. Water is the single most abundant compound in the body. It absorbs and releases heat slowly, acts as a universal solvent, participates in chemical reactions, and cushions body organs.

Salts (p. 39)

3. Salts are ionic compounds that dissolve in water and act as electrolytes. Calcium and phosphorus salts contribute to the hardness of bones and teeth. Ions of salts are involved in many physiological processes.

Acids and Bases (pp. 39–41)

4. Acids are proton donors; in water, they ionize and dissociate, releasing hydrogen ions (which account for their properties) and anions.

iP Fluid, Electrolyte and Acid/Base Balance; Topic: Acid Base Homeostasis, pp. 1–12, 16, 17.

5. Bases are proton acceptors. The most common inorganic bases are the hydroxides; bicarbonate ion and ammonia are important bases in the body.
6. pH is a measure of hydrogen ion concentration of a solution (in moles per liter). A pH of 7 is neutral; a higher pH is alkaline, and a lower pH is acidic. Normal blood pH is 7.35–7.45. Buffers help to prevent excessive changes in the pH of body fluids.

iP Fluid, Electrolyte and Acid/Base Balance; Topic: Introduction to Body Fluids, pp. 1–8.

Organic Compounds (pp. 42–56)

1. Organic compounds contain carbon. Those found in the body include carbohydrates, lipids, proteins, and nucleic acids, all of which are synthesized by dehydration synthesis and digested by hydrolysis. All of these biological molecules contain C, H, and O. Proteins and nucleic acids also contain N.

Carbohydrates (pp. 43–44)

2. Carbohydrate building blocks are monosaccharides, the most important of which are hexoses (glucose, fructose, galactose) and pentoses (ribose, deoxyribose).
3. Disaccharides (sucrose, lactose, maltose) and polysaccharides (starch, glycogen) are composed of linked monosaccharide units.
4. Carbohydrates, particularly glucose, are the major energy fuel for forming ATP. Excess carbohydrates are stored as glycogen or converted to fat for storage.

Lipids (pp. 43–47)

5. Lipids dissolve in fats or organic solvents, but not in water.
6. Triglycerides are composed of fatty acid chains and glycerol. They are found chiefly in fatty tissue where they provide insulation and reserve body fuel. Unsaturated fatty acid chains produce oils. Saturated fatty acids produce solid fats typical of animal fats.
7. Phospholipids are modified phosphorus-containing triglycerides that have polar and nonpolar portions. They are found in all plasma membranes.
8. The steroid cholesterol is found in cell membranes and is the basis of steroid hormones, bile salts, and vitamin D.

Proteins (pp. 47–53)

9. The unit of proteins is the amino acid, and 20 common amino acids are found in the body.
10. Many amino acids joined by peptide bonds form a polypeptide. A protein (one or more polypeptides) is distinguished by the number and sequence of amino acids in its chain(s) and by the complexity of its three-dimensional structure.
11. Fibrous proteins, such as keratin and collagen, have secondary (α-helix or β-pleated sheet) and perhaps tertiary and quaternary structure. Fibrous proteins are used as structural materials.
12. Globular proteins achieve tertiary and sometimes quaternary structure and are generally spherical, soluble molecules. Globular proteins (e.g., enzymes, some hormones, antibodies, hemoglobin) perform special functional roles for the cell (e.g., catalysis, molecule transport).
13. Proteins are denatured by extremes of temperature or pH. Denatured globular proteins are unable to perform their usual function.
14. Molecular chaperones assist in folding proteins into their functional 3-D shape. They are synthesized in greater amounts when cells are stressed by environmental factors.
15. Enzymes are biological catalysts. They increase the rate of chemical reactions by decreasing the amount of activation energy needed. They do this by combining with the reactants and holding them in the proper position to interact. Many enzymes require cofactors to function.

Nucleic Acids (DNA and RNA) (pp. 53–55)

16. Nucleic acids include deoxyribonucleic acid (DNA) and ribonucleic acid (RNA). The structural unit of nucleic acids is the nucleotide, which consists of a nitrogenous base (adenine, guanine, cytosine, thymine, or uracil), a sugar (ribose or deoxyribose), and a phosphate group.
17. DNA is a double-stranded helix. It contains deoxyribose and the bases A, G, C, and T. DNA specifies protein structure and replicates itself exactly before cell division.
18. RNA is single stranded. It contains ribose and the bases A, G, C, and U. RNAs involved in carrying out DNA's instructions for protein synthesis include messenger, ribosomal, and transfer RNA.

Adenosine Triphosphate (ATP) (pp. 55–56)

19. ATP is the universal energy compound of body cells. Some of the energy liberated by the breakdown of glucose and other food fuels is captured in the bonds of ATP molecules and transferred via coupled reactions to energy-consuming reactions.

REVIEW QUESTIONS

Multiple Choice/Matching

(Some questions have more than one correct answer. Select the best answer or answers from the choices given.)

1. Which of the following forms of energy is the stimulus for vision? (a) chemical, (b) electrical, (c) mechanical, (d) radiant.
2. All of the following are examples of the four major elements contributing to body mass except (a) hydrogen, (b) carbon, (c) nitrogen, (d) sodium, (e) oxygen.
3. The mass number of an atom is (a) equal to the number of protons it contains, (b) the sum of its protons and neutrons, (c) the sum of all of its subatomic particles, (d) the average of the mass numbers of all of its isotopes.
4. A deficiency in this element can be expected to reduce the hemoglobin content of blood: (a) Fe, (b) I, (c) F, (d) Ca, (e) K.
5. Which set of terms best describes a proton? (a) negative charge, massless, in the orbital; (b) positive charge, 1 amu, in the nucleus; (c) uncharged, 1 amu, in the nucleus.
6. The subatomic particles responsible for the chemical behavior of atoms are (a) electrons, (b) ions, (c) neutrons, (d) protons.
7. In the body, carbohydrates are stored in the form of (a) glycogen, (b) starch, (c) cholesterol, (d) polypeptides.
8. Which of the following does *not* describe a mixture? (a) properties of its components are retained, (b) chemical bonds are formed, (c) components can be separated physically, (d) includes both heterogeneous and homogeneous examples.
9. In a beaker of water, the water-water bonds can properly be called (a) ionic bonds, (b) polar covalent bonds, (c) nonpolar covalent bonds, (d) hydrogen bonds.
10. When a pair of electrons is shared between two atoms, the bond formed is called (a) a single covalent bond, (b) a double covalent bond, (c) a triple covalent bond, (d) an ionic bond.
11. Molecules formed when electrons are shared unequally are (a) salts, (b) polar molecules, (c) nonpolar molecules.
12. Which of the following covalently bonded molecules are polar?

H—Cl	H—C—H (with H above and H below C)	Cl—C—Cl (with H above and Cl below C)	N≡N
(a)	**(b)**	**(c)**	**(d)**

13. Identify each reaction as one of the following: (a) a synthesis reaction (b) a decomposition reaction (c) an exchange reaction
    ______(1) $2Hg + O_2 \longrightarrow 2HgO$
    ______(2) $HCl + NaOH \longrightarrow NaCl + H_2O$
14. Factors that accelerate the rate of chemical reactions include all but (a) the presence of catalysts, (b) increasing the temperature, (c) increasing the particle size, (d) increasing the concentration of the reactants.
15. Which of the following molecules is an inorganic molecule? (a) sucrose, (b) cholesterol, (c) collagen, (d) sodium chloride.
16. Water's importance to living systems reflects (a) its polarity and solvent properties, (b) its high heat capacity, (c) its high heat of vaporization, (d) its chemical reactivity, (e) all of these.
17. Acids (a) release hydroxyl ions when dissolved in water, (b) are proton acceptors, (c) cause the pH of a solution to rise, (d) release protons when dissolved in water.
18. A chemist, during the course of an analysis, runs across a chemical composed of carbon, hydrogen, and oxygen in the proportion 1:2:1 and having a six-sided molecular shape. It is probably (a) a pentose, (b) an amino acid, (c) a fatty acid, (d) a monosaccharide, (e) a nucleic acid.
19. A triglyceride consists of (a) glycerol plus three fatty acids, (b) a sugar-phosphate backbone to which two amino groups are attached, (c) two to several hexoses, (d) amino acids that have been thoroughly saturated with hydrogen.
20. A chemical has an amine group and an organic acid group. It does not, however, have any peptide bonds. It is (a) a monosaccharide, (b) an amino acid, (c) a protein, (d) a fat.
21. The lipid(s) used as the basis of vitamin D, sex hormones, and bile salts is/are (a) triglycerides, (b) cholesterol, (c) phospholipids, (d) prostaglandin.
22. Enzymes are organic catalysts that (a) alter the direction in which a chemical reaction proceeds, (b) determine the nature of the products of a reaction, (c) increase the speed of a chemical reaction, (d) are essential raw materials for a chemical reaction that are converted into some of its products.

Short Answer Essay Questions

23. Define or describe energy, and explain the relationship between potential and kinetic energy.
24. Some energy is lost in every energy conversion. Explain the meaning of this statement. (Direct your response to answering the question: Is it really lost? If not, what then?)
25. Provide the atomic symbol for each of the following elements: (a) calcium, (b) carbon, (c) hydrogen, (d) iron, (e) nitrogen, (f) oxygen, (g) potassium, (h) sodium.
26. Consider the following information about three atoms:

    $^{12}_{6}C \quad ^{13}_{6}C \quad ^{14}_{6}C$

    (a) How are they similar to one another? (b) How do they differ from one another? (c) What are the members of such a group of atoms called? (d) Using the planetary model, draw the atomic configuration of $^{12}_{6}C$ showing the relative position and numbers of its subatomic particles.
27. How many moles of aspirin, $C_9H_8O_4$, are in a bottle containing 450 g by weight? (*Note*: The approximate atomic weights of its atoms are C = 12, H = 1, and O = 16.)
28. Given the following types of atoms, decide which type of bonding, ionic or covalent, is most likely to occur: (a) two oxygen atoms; (b) four hydrogen atoms and one carbon atom; (c) a potassium atom ($^{39}_{19}K$) and a fluorine atom ($^{19}_{9}F$).
29. What are hydrogen bonds and how are they important in the body?
30. The following equation, which represents the oxidative breakdown of glucose by body cells, is a reversible reaction.

    Glucose + oxygen ⟶ carbon dioxide + water + ATP

    (a) How can you indicate that the reaction is reversible?
    (b) How can you indicate that the reaction is in chemical equilibrium?
    (c) Define chemical equilibrium.
31. Differentiate clearly between primary, secondary, and tertiary protein structure.
32. Dehydration and hydrolysis reactions are essentially opposite reactions. How are they related to the synthesis and degradation (breakdown) of biological molecules?

2

33. Describe the mechanism of enzyme action.
34. Explain the importance of molecular chaperones.
35. Explain why, if you pour water into a glass very carefully, you can "stack" the water slightly above the rim of the glass.

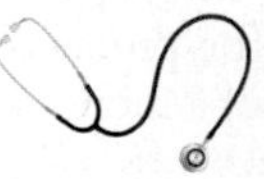

Critical Thinking and Clinical Application Questions

2

1. As Ben jumped on his bike and headed for the freshwater lake, his mother called after him, "Don't swim if we have an electrical storm—it looks threatening." This was a valid request. Why?
2. Some antibiotics act by binding to certain essential enzymes in the target bacteria. (**a**) How might these antibiotics influence the chemical reactions controlled by the enzymes? (**b**) What is the anticipated effect on the bacteria? On the person taking the antibiotic prescription?
3. Mrs. Roberts, in a diabetic coma, has just been admitted to Noble Hospital. Her blood pH indicates that she is in severe acidosis, and measures are quickly instituted to bring her blood pH back within normal limits. (**a**) Define pH and note the normal pH of blood. (**b**) Why is severe acidosis a problem?
4. Jimmy, a 12-year-old boy, was awakened suddenly by a loud crash. As he sat up in bed, straining to listen, his fright was revealed by his rapid breathing (hyperventilation), a breathing pattern effective in ridding the blood of CO_2. At this point, was his blood pH rising or falling?
5. After you eat a protein bar, which chemical reactions introduced in this chapter must occur for the amino acids in the protein bar to be converted into proteins in your body cells?

3

Cells: The Living Units

Just as bricks and timbers are the structural units of a house, **cells** are the structural units of all living things, from one-celled "generalists" like amoebas to complex multicellular organisms such as humans, dogs, and trees. The human body has 50 to 100 trillion of these tiny building blocks.

This chapter focuses on structures and functions shared by all cells. We address specialized cells and their unique functions in later chapters.

Overview of the Cellular Basis of Life

- ▶ Define cell.
- ▶ List the three major regions of a generalized cell and indicate the function of each.

The English scientist Robert Hooke first observed plant cells with a crude microscope in the late 1600s. However, it was not until the 1830s that two German scientists, Matthias Schleiden and Theodor Schwann, were bold enough to insist that all living things are composed of cells. The German pathologist Rudolf Virchow extended this idea by contending that cells arise only from other cells. Virchow's proclamation was revolutionary because it openly challenged the widely accepted *theory of spontaneous generation*, which held that organisms arise spontaneously from garbage or other nonliving matter.

Since the late 1800s, cell research has been exceptionally fruitful and provided us with four concepts collectively known as the **cell theory**:

1. A *cell* is the basic structural and functional unit of living organisms. So when you define cell properties you are in fact defining the properties of life.
2. The activity of an organism depends on both the individual and the collective activities of its cells.
3. According to the *principle of complementarity of structure and function*, the biochemical activities of cells are dictated by the relative number of their specific subcellular structures.
4. Continuity of life from one generation to another has a cellular basis.

We will expand on all of these concepts as we progress. Let us begin with the idea that the cell is the smallest living unit. Whatever its form, however it behaves, the cell is the microscopic package that contains all the parts necessary to survive in an ever-changing world. It follows then that loss of cellular homeostasis underlies virtually every disease.

The trillions of cells in the human body include over 200 different cell types that vary greatly in shape, size, and function (**Figure 3.1**). The spherical fat cells, disc-shaped red blood cells, branching nerve cells, and cubelike cells of kidney tubules are just a few examples of the shapes cells take. Depending on type, cells also vary greatly in length—ranging from 2 micrometers (1/12,000 of an inch) in the smallest cells to over a meter in the nerve cells that cause you to wiggle your toes. A cell's shape reflects its function. For example, the flat, tilelike epithelial cells that line the inside of your cheek fit closely together, forming a living barrier that protects underlying tissues from bacterial invasion.

Regardless of type, all cells are composed chiefly of carbon, hydrogen, nitrogen, oxygen, and trace amounts of several other elements. In addition, all cells have the same basic parts and some common functions. For this reason, it is possible to speak of a **generalized**, or **composite, cell** (**Figure 3.2**).

Human cells have three main parts: the plasma membrane, the cytoplasm, and the nucleus. The *plasma membrane*, a fragile barrier, is the outer boundary of the cell. Internal to this membrane is the *cytoplasm* (si′to-plazm), the intracellular fluid that is packed with organelles, small structures that perform specific cell functions. The *nucleus* (nu′kle-us) controls cellular activities and typically it lies near the cell's center. We use these three main parts of the cell to organize the summary in Table 3.3 (pp. 94–95), and we describe them in greater detail next.

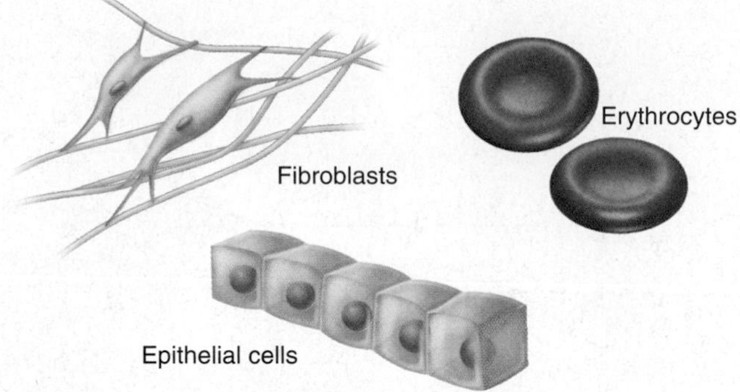

(a) Cells that connect body parts, form linings, or transport gases

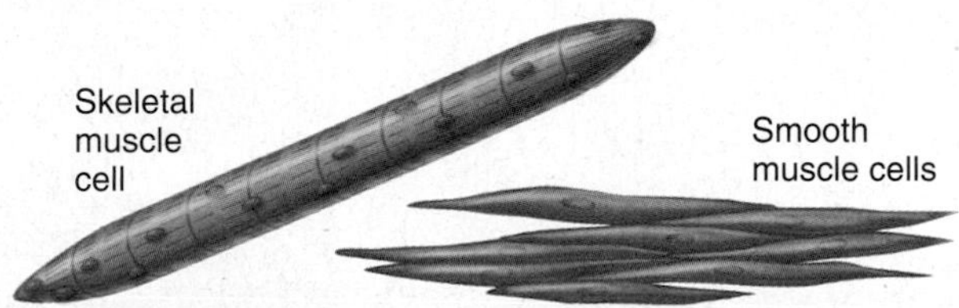

(b) Cells that move organs and body parts

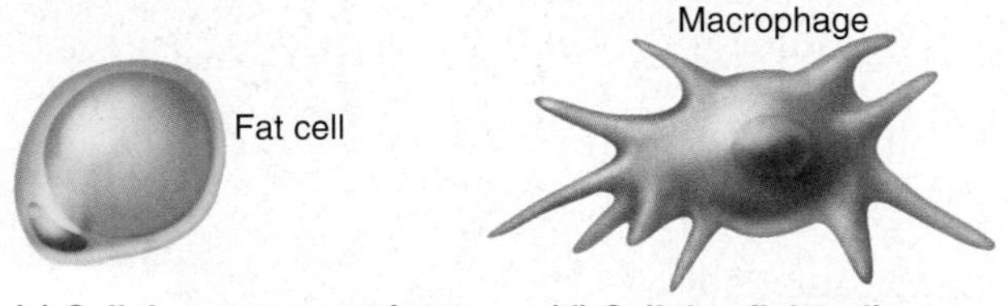

(c) Cell that stores nutrients **(d) Cell that fights disease**

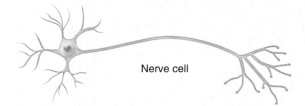

Nerve cell

(e) Cell that gathers information and controls body functions

(f) Cell of reproduction

Figure 3.1 Cell diversity. (Note that cells are not drawn to the same scale.)

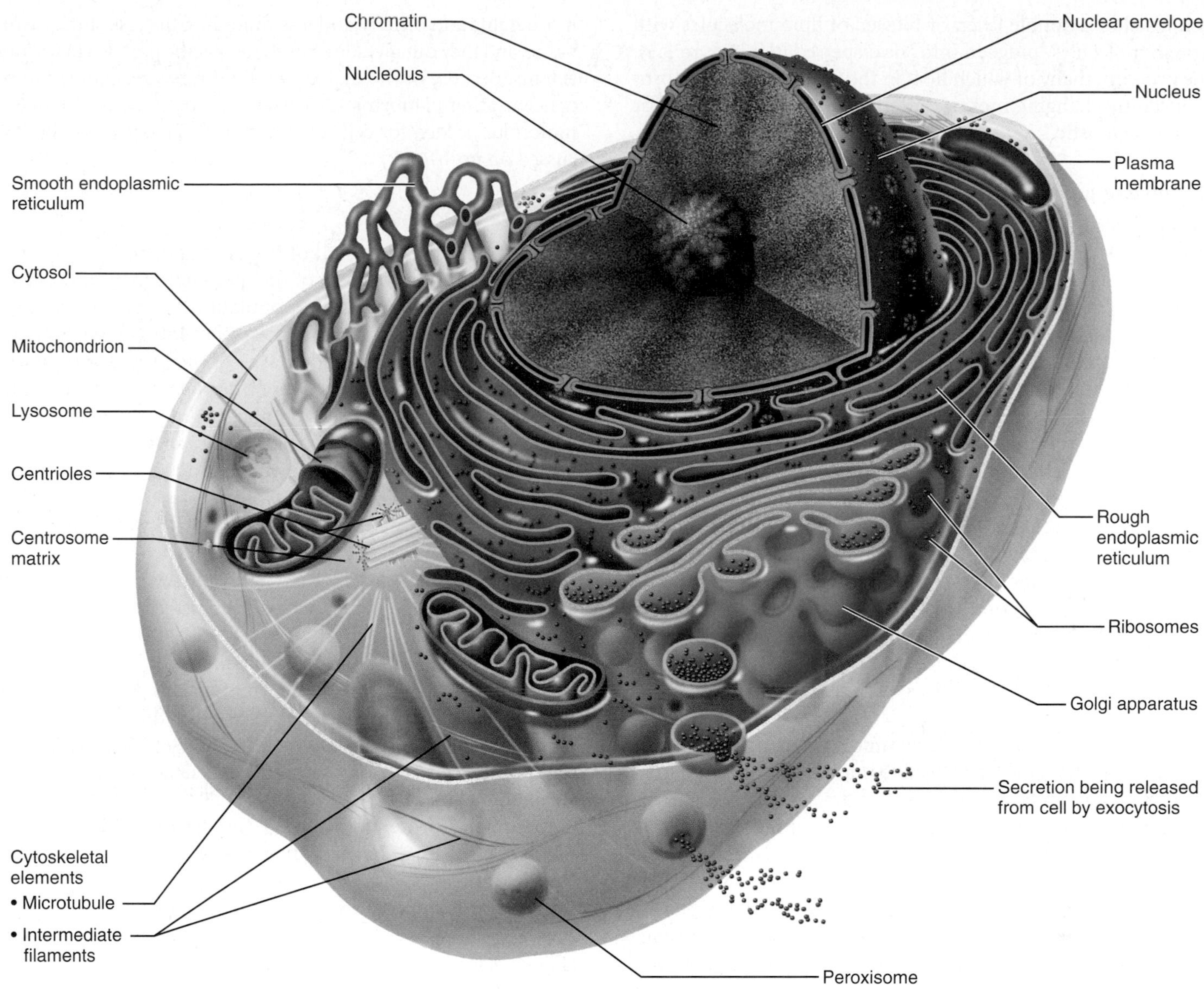

Figure 3.2 **Structure of the generalized cell.** No cell is exactly like this one, but this composite illustrates features common to many human cells. Note that not all of the organelles are drawn to the same scale in this illustration.

3

CHECK YOUR UNDERSTANDING

1. Name the three basic parts of a cell and describe the function of each.
2. How would you explain the meaning of a "generalized cell" to a classmate?

For answers, see Appendix G.

The Plasma Membrane: Structure

- Describe the chemical composition of the plasma membrane and relate it to membrane functions.
- Compare the structure and function of tight junctions, desmosomes, and gap junctions.

The flexible **plasma membrane** defines the extent of a cell, thereby separating two of the body's major fluid compartments—the *intracellular* fluid within cells and the *extracellular* fluid outside cells. The term *cell membrane* is commonly used as a synonym for plasma membrane, but because nearly all cellular organelles are enclosed in a membrane, in this book we will always refer to the cell's surface, or outer limiting membrane, as the plasma membrane. The plasma membrane is much more than a passive envelope. As you will see, its unique structure allows it to play a dynamic role in many cellular activities.

The Fluid Mosaic Model

The **fluid mosaic model** of membrane structure depicts the plasma membrane as an exceedingly thin (7–10 nm) structure composed of a double layer, or bilayer, of lipid molecules with protein molecules "plugged into" or dispersed in it **(Figure 3.3)**. The proteins, many of which float in the fluid *lipid bilayer*, form a constantly changing mosaic pattern. The model is named for this characteristic.

3

Membrane Lipids

The lipid bilayer forms the basic "fabric" of the membrane. It is constructed largely of *phospholipids*, with smaller amounts of *cholesterol* and *glycolipids*. Each lollipop-shaped phospholipid molecule has a polar "head" that is charged and is **hydrophilic** (*hydro* = water, *philic* = loving), and an uncharged, nonpolar "tail" that is made of two fatty acid chains and is **hydrophobic** (*phobia* = fear). The polar heads are attracted to water—the main constituent of both the intracellular and extracellular fluids—and so they lie on both the inner and outer surfaces of the membrane. The nonpolar tails, being hydrophobic, avoid water and line up in the center of the membrane.

The result is that all biological membranes share a common sandwich-like structure: They are composed of two parallel sheets of phospholipid molecules lying tail to tail, with their polar heads exposed to water both inside and outside the cell. This self-orienting property of phospholipids encourages biological membranes to self-assemble into closed, generally spherical, structures and to reseal themselves quickly when torn.

The plasma membrane is a dynamic fluid structure that is in constant flux. Its consistency is like that of olive oil. The lipid molecules of the bilayer move freely from side to side, parallel to the membrane surface, but their polar-nonpolar interactions prevent them from flip-flopping or moving from one phospholipid layer (half of the bilayer) to the other. The inward-facing and outward-facing surfaces of the plasma membrane differ in the kinds and amounts of lipids they contain, and these variations are important in determining local membrane structure and function. The majority of membrane phospholipids are unsaturated, a condition which kinks their tails (increasing the space between them) and increases membrane fluidity. (See the illustration of phosphatidylcholine in Figure 2.16b, p. 46.)

Glycolipids (gli″ko-lip′idz) are lipids with attached sugar groups. They are found only on the outer plasma membrane surface and account for about 5% of the total membrane lipid. Their sugar groups, like the phosphate-containing groups of phospholipids, make that end of the glycolipid molecule polar, whereas the fatty acid tails are nonpolar.

Some 20% of membrane lipid is cholesterol. Like phospholipids, cholesterol has a polar region (its hydroxyl group) and a nonpolar region (its fused ring system). It wedges its platelike hydrocarbon rings between the phospholipid tails, stabilizing the membrane, while increasing the mobility of the phospholipids and the fluidity of the membrane.

About 20% of the outer membrane surface contains **lipid rafts**, dynamic assemblies of saturated phospholipids (which pack together tightly) associated with unique lipids called sphingolipids and lots of cholesterol. These quiltlike patches are more stable and orderly and less fluid than the rest of the membrane, and they can include or exclude specific proteins to various extents. Because of these qualities, lipid rafts are assumed to be concentrating platforms for certain receptor molecules or for molecules needed for cell signaling. (Cell signaling will be discussed on pp. 81–83.)

Membrane Proteins

Proteins make up about half of the plasma membrane by mass and are responsible for most of the specialized membrane functions. There are two distinct populations of membrane proteins, integral and peripheral (Figure 3.3). **Integral proteins** are firmly inserted into the lipid bilayer. Some protrude from one membrane face only, but most are *transmembrane proteins* that span the entire width of the membrane and protrude on both sides. Whether transmembrane or not, all integral proteins have both hydrophobic and hydrophilic regions. This structural feature allows them to interact both with the nonpolar lipid tails buried in the membrane and with water inside and outside the cell.

Although some are enzymes, most transmembrane proteins are involved in transport. Some cluster together to form *channels*, or pores, through which small, water-soluble molecules or ions can move, thus bypassing the lipid part of the membrane. Others act as *carriers* that bind to a substance and then move it through the membrane. Still others are receptors for hormones or other chemical messengers and relay messages to the cell interior (a process called *signal transduction*) **(Figure 3.4a, b)**.

Peripheral proteins (Figure 3.3), in contrast, are not embedded in the lipid. Instead, they attach rather loosely only to integral proteins and are easily removed without disrupting the membrane. Peripheral proteins include a network of filaments that helps support the membrane from its cytoplasmic side (Figure 3.4c). Some peripheral proteins are enzymes. Others are motor proteins involved in mechanical functions, such as changing cell shape during cell division and muscle cell contraction. Still others link cells together.

Some of the proteins float freely. Others, particularly the peripheral proteins, are restricted in their movements because they are "tethered" to intracellular structures that make up the *cytoskeleton*. Many of the proteins that abut the extracellular fluid are glycoproteins with branching sugar groups. The term **glycocalyx** (gli″ko-kal′iks; "sugar covering") is used to describe the fuzzy, sticky, carbohydrate-rich area at the cell surface. You can think of your cells as sugar-coated. The glycocalyx that clings to each cell's surface is enriched both by glycolipids and by glycoproteins secreted by the cell.

Because every cell type has a different pattern of sugars in its glycocalyx, the glycocalyx provides highly specific biological markers by which approaching cells recognize each other (Figure 3.4f). For example, a sperm recognizes an ovum (egg cell) by the ovum's unique glycocalyx. Cells of the immune system identify a bacterium by binding to certain membrane glycoproteins in the bacterial glycocalyx.

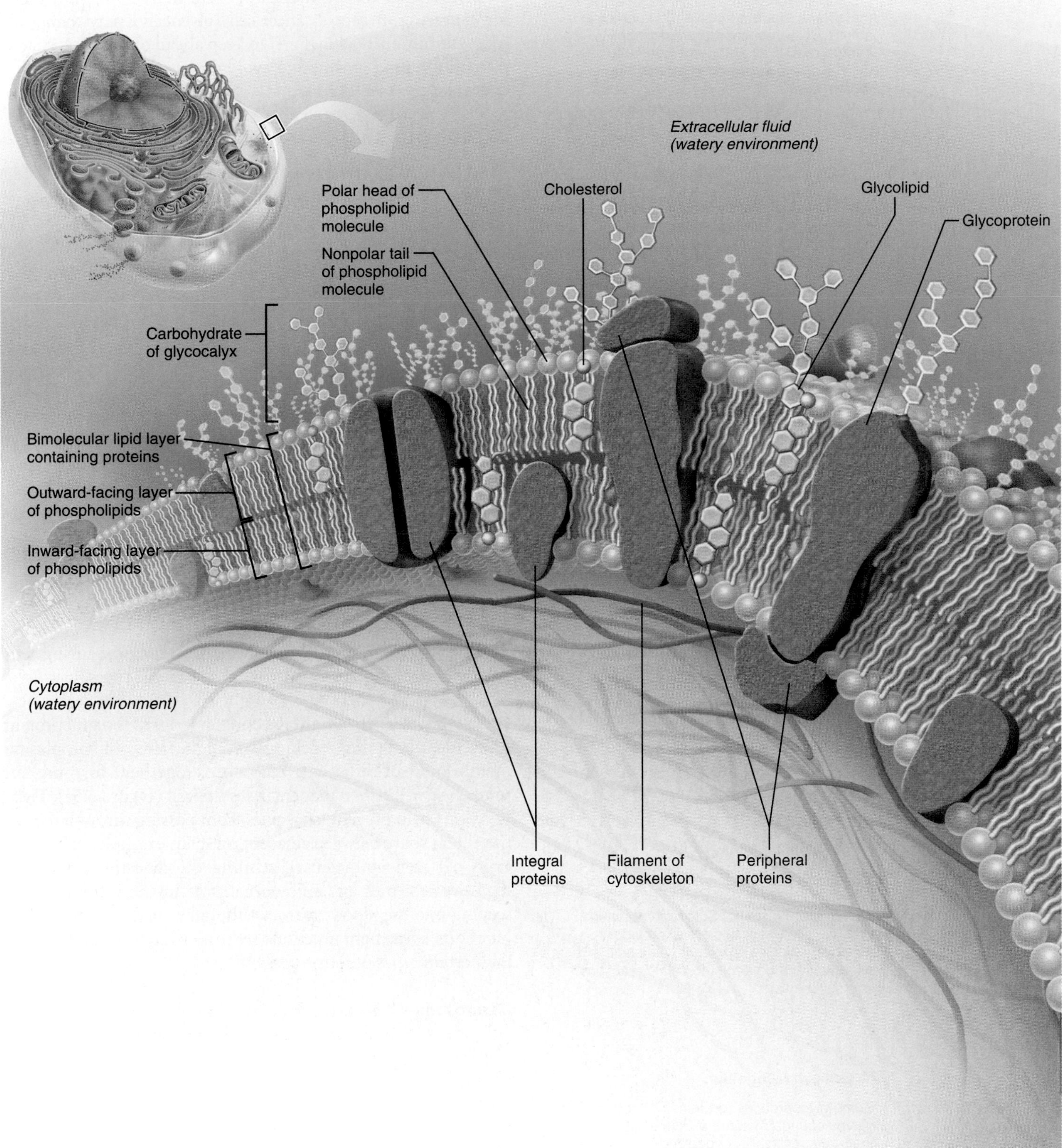

Figure 3.3 Structure of the plasma membrane according to the fluid mosaic model. The lipid bilayer forms the basic structure of the membrane. The associated proteins are involved in membrane functions such as membrane transport, catalysis, and cell-to-cell recognition.

3

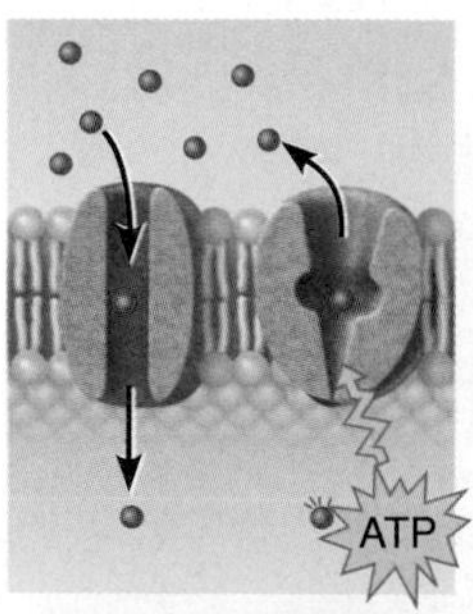

(a) Transport

A protein (left) that spans the membrane may provide a hydrophilic channel across the membrane that is selective for a particular solute. Some transport proteins (right) hydrolyze ATP as an energy source to actively pump substances across the membrane.

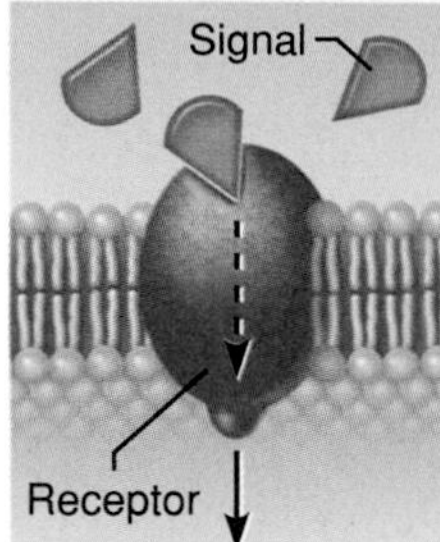

(b) Receptors for signal transduction

A membrane protein exposed to the outside of the cell may have a binding site with a specific shape that fits the shape of a chemical messenger, such as a hormone. The external signal may cause a change in shape in the protein that initiates a chain of chemical reactions in the cell.

(c) Attachment to the cytoskeleton and extracellular matrix (ECM)

Elements of the cytoskeleton (cell's internal supports) and the extracellular matrix (fibers and other substances outside the cell) may be anchored to membrane proteins, which help maintain cell shape and fix the location of certain membrane proteins. Others play a role in cell movement or bind adjacent cells together.

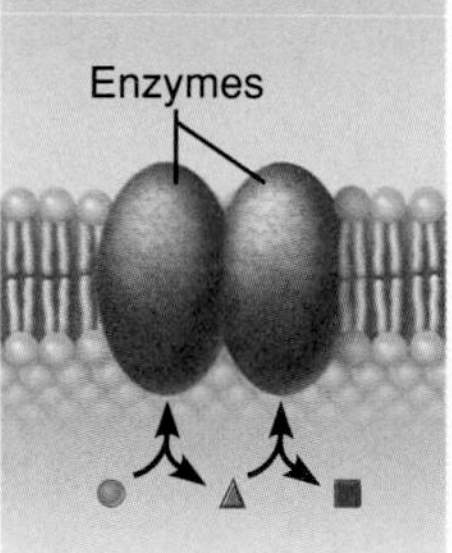

(d) Enzymatic activity

A protein built into the membrane may be an enzyme with its active site exposed to substances in the adjacent solution. In some cases, several enzymes in a membrane act as a team that catalyzes sequential steps of a metabolic pathway as indicated (left to right) here.

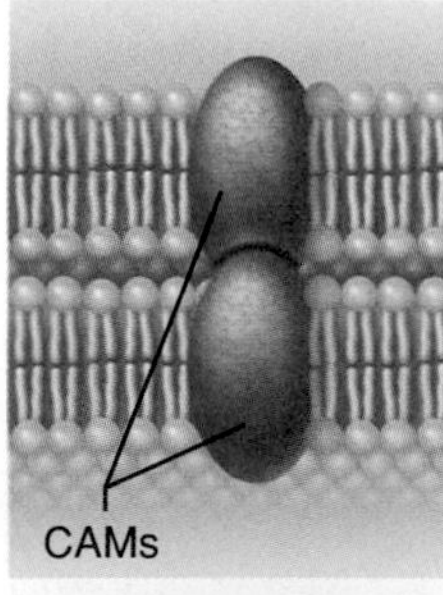

(e) Intercellular joining

Membrane proteins of adjacent cells may be hooked together in various kinds of intercellular junctions. Some membrane proteins (CAMs) of this group provide temporary binding sites that guide cell migration and other cell-to-cell interactions.

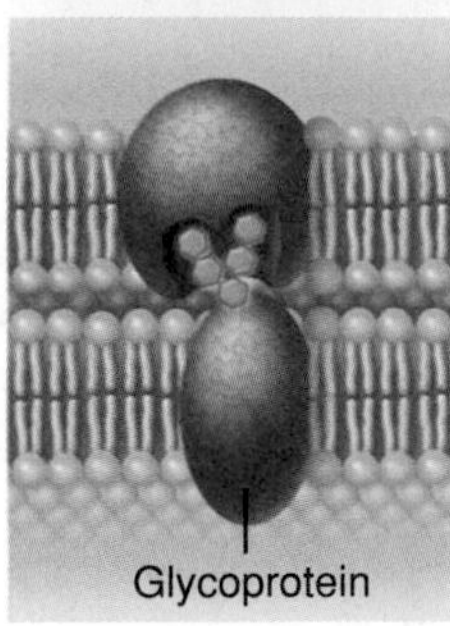

(f) Cell-cell recognition

Some glycoproteins (proteins bonded to short chains of sugars) serve as identification tags that are specifically recognized by other cells.

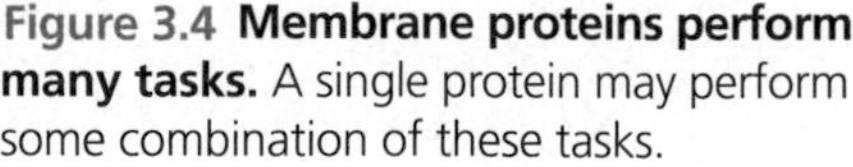

Figure 3.4 **Membrane proteins perform many tasks.** A single protein may perform some combination of these tasks.

HOMEOSTATIC IMBALANCE

Definite changes in the glycocalyx occur in a cell that is becoming cancerous. In fact, a cancer cell's glycocalyx may change almost continuously, allowing it to keep ahead of immune system recognition mechanisms and avoid destruction. (Cancer is discussed on pp. 142–143.) ■

CHECK YOUR UNDERSTANDING

3. What basic structure do all cellular membranes share?
4. Why do phospholipids, which form the greater part of cell membranes, organize into a bilayer—tail-to-tail—in a watery environment?
5. What is the importance of the glycocalyx in cell interactions?

For answers, see Appendix G.

Membrane Junctions

Although certain cell types—blood cells, sperm cells, and some phagocytic cells (which ingest and destroy bacteria and other substances)—are "footloose" in the body, many other types, particularly epithelial cells, are knit into tight communities. Typically, three factors act to bind cells together:

1. Glycoproteins in the glycocalyx act as an adhesive.
2. Wavy contours of the membranes of adjacent cells fit together in a tongue-and-groove fashion.
3. Special membrane junctions are formed **(Figure 3.5)**.

Because junctions are the most important factor securing cells together, let us look more closely at the various types.

Tight Junctions In a **tight junction**, a series of integral protein molecules (including occludins and claudins) in the plasma membranes of adjacent cells fuse together, forming an *impermeable junction* that encircles the cell (Figure 3.5a). Tight junctions help prevent molecules from passing through the extracellular space between adjacent cells. For example, tight junctions between epithelial cells lining the digestive tract keep digestive enzymes and microorganisms in the intestine from seeping into the bloodstream. (Although called "impermeable" junctions, some tight junctions are somewhat leaky and may allow certain types of ions to pass.)

Desmosomes **Desmosomes** (des′mo-sōmz; "binding bodies") are *anchoring junctions*—mechanical couplings scattered like rivets along the sides of abutting cells that prevent their separation (Figure 3.5b). On the cytoplasmic face of each plasma membrane is a buttonlike thickening called a *plaque*. Adjacent cells are held together by thin linker protein filaments (cadherins) that extend from the plaques and fit together like the teeth of a zipper in the intercellular space. Thicker keratin filaments (intermediate filaments, which form part of the cytoskeleton) extend from the cytoplasmic side of the plaque across the width of the cell to anchor to the plaque on the cell's opposite side. In this way, desmosomes not only bind neighboring cells together,

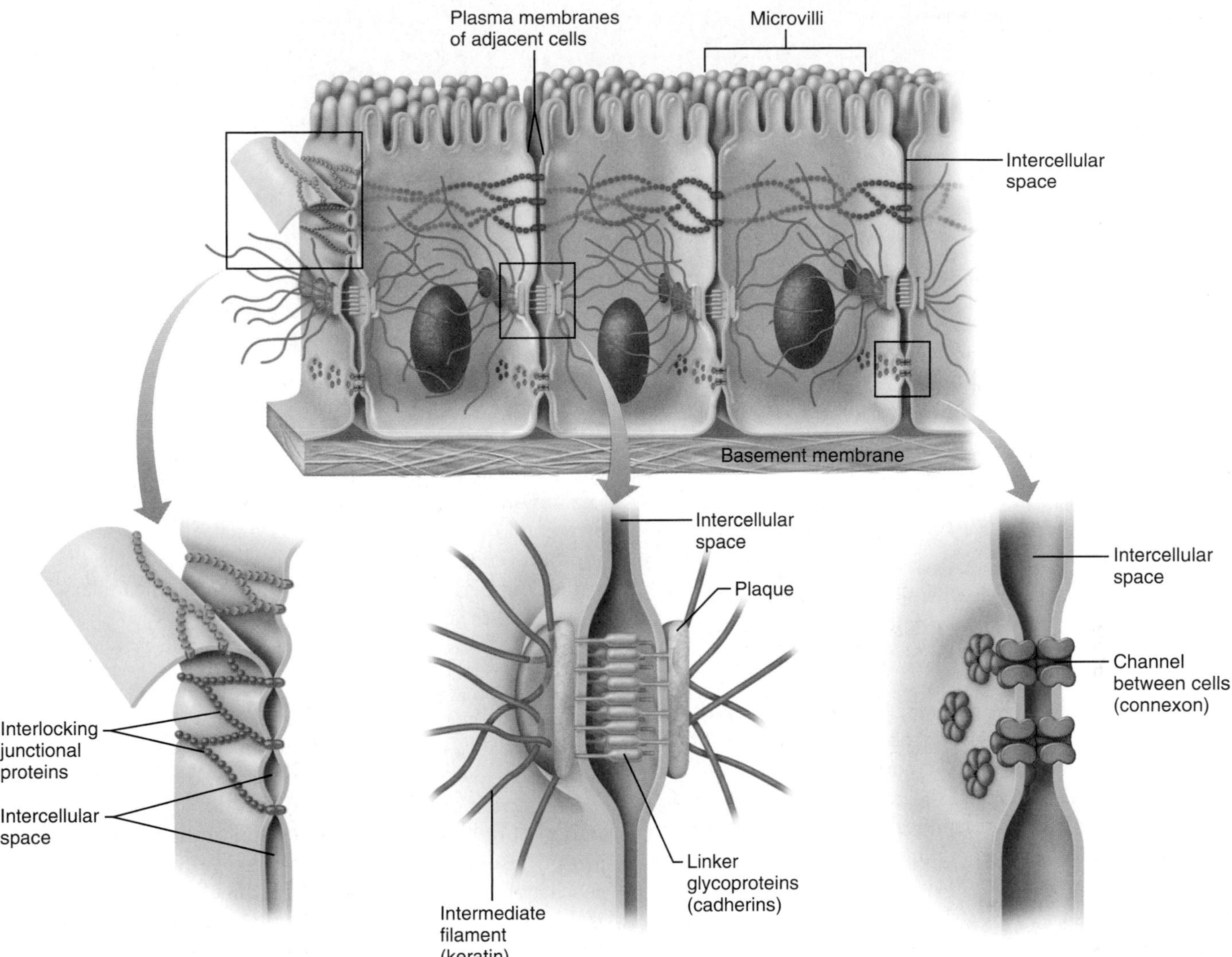

(a) Tight junctions: Impermeable junctions prevent molecules from passing through the intercellular space.

(b) Desmosomes: Anchoring junctions bind adjacent cells together and help form an internal tension-reducing network of fibers.

(c) Gap junctions: Communicating junctions allow ions and small molecules to pass from one cell to the next for intercellular communication.

Figure 3.5 **Cell junctions.** An epithelial cell is shown joined to adjacent cells by three common types of cell junctions. (Note: Except for epithelia, it is unlikely that a single cell will have all three junction types.)

but they also contribute to a continuous internal network of strong "guy-wires."

This arrangement distributes tension throughout a cellular sheet and reduces the chance of tearing when it is subjected to pulling forces. Desmosomes are abundant in tissues subjected to great mechanical stress, such as skin and heart muscle.

Gap Junctions A **gap junction**, or *nexus* (nek′sus; "bond"), is a communicating junction between adjacent cells. At gap junctions the adjacent plasma membranes are very close, and the cells are connected by hollow cylinders called *connexons* (kō-nek′sonz), composed of transmembrane proteins. The many different types of connexon proteins vary the selectivity of the gap junction channels. Ions, simple sugars, and other small molecules pass through these water-filled channels from one cell to the next (Figure 3.5c).

Gap junctions are present in electrically excitable tissues, such as the heart and smooth muscle, where ion passage from cell to cell helps synchronize their electrical activity and contraction.

CHECK YOUR UNDERSTANDING

6. What two types of membrane junctions would you expect to find between muscle cells of the heart?

For answer, see Appendix G.

3

Figure 3.6 **Diffusion.** Molecules in solution move continuously and collide constantly with other molecules, causing them to move away from areas of their highest concentration and become evenly distributed. From left to right, molecules from a dye pellet diffuse into the surrounding water down their concentration gradient.

The Plasma Membrane: Membrane Transport

- Relate plasma membrane structure to active and passive transport processes.
- Compare and contrast simple diffusion, facilitated diffusion, and osmosis relative to substances transported, direction, and mechanism.

Our cells are bathed in an extracellular fluid called **interstitial fluid** (in″ter-stish′al) that is derived from the blood. Interstitial fluid is like a rich, nutritious "soup." It contains thousands of ingredients, including amino acids, sugars, fatty acids, vitamins, regulatory substances such as hormones and neurotransmitters, salts, and waste products. To remain healthy, each cell must extract from this mix the exact amounts of the substances it needs at specific times.

Although there is continuous traffic across the plasma membrane, it is a **selectively**, or **differentially**, **permeable** barrier, meaning that it allows some substances to pass while excluding others. It allows nutrients to enter the cell, but keeps many undesirable substances out. At the same time, it keeps valuable cell proteins and other substances in the cell, but allows wastes to exit.

Substances move through the plasma membrane in essentially two ways—passively or actively. In **passive processes**, substances cross the membrane without any energy input from the cell. In **active processes**, the cell provides the metabolic energy (ATP) needed to move substances across the membrane. The various transport processes that occur in cells are summarized in Table 3.1 on p. 72 and Table 3.2 on p. 80. Let's examine each of these types of membrane transport.

HOMEOSTATIC IMBALANCE

Selective permeability is a characteristic of healthy, intact cells. When a cell (or its plasma membrane) is severely damaged, the membrane becomes permeable to virtually everything, and substances flow into and out of the cell freely. This phenomenon is evident when someone has been severely burned. Precious fluids, proteins, and ions "weep" from the dead and damaged cells. ■

Passive Processes

The two main types of passive transport are *diffusion* (di-fu′zhun) and *filtration.* Diffusion is an important means of passive membrane transport for every cell of the body. Because filtration generally occurs only across capillary walls, that topic is more properly covered in conjunction with capillary transport processes later in the book.

Diffusion

Diffusion is the tendency of molecules or ions to move from an area where they are in higher concentration to an area where they are in lower concentration, that is, down or along their **concentration gradient**. The constant random and high-speed motion of molecules and ions (a result of their intrinsic kinetic energy) results in collisions. With each collision, the particles ricochet off one another and change direction. The overall effect of this erratic movement is the scattering or dispersion of the particles throughout the environment **(Figure 3.6)**. The greater the difference in concentration of the diffusing molecules and ions between the two areas, the more collisions occur and the faster the net diffusion of the particles.

Because the driving force for diffusion is the kinetic energy of the molecules themselves, the speed of diffusion is influenced by molecular *size* (the smaller, the faster) and by *temperature* (the warmer, the faster). In a closed container, diffusion eventually produces a uniform mixture of molecules. In other words, the system reaches equilibrium, with molecules moving equally in all directions (no *net* movement).

Diffusion is occurring all around us, but obvious examples of pure diffusion are almost impossible to see. The reason is that any diffusion process that occurs over an easily observable distance takes a long time, and is often accompanied by other processes (convection, for example) that affect the movement of molecules and ions. In fact, one should suspect any readily observable "example" of diffusion. Nonetheless, diffusion is immensely important in physiological systems and it occurs rapidly because the distances molecules are moving are very short, perhaps 1/1000 (or less) the thickness of this page! Examples include the movement of ions across cell membranes and the movement of neurotransmitters between two nerve cells.

The plasma membrane is a physical barrier to free diffusion because of its hydrophobic core. However, a molecule *will* diffuse

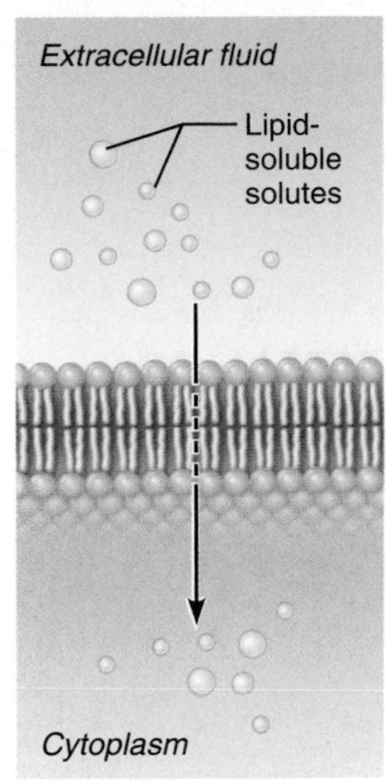

(a) **Simple diffusion** of fat-soluble molecules directly through the phospholipid bilayer

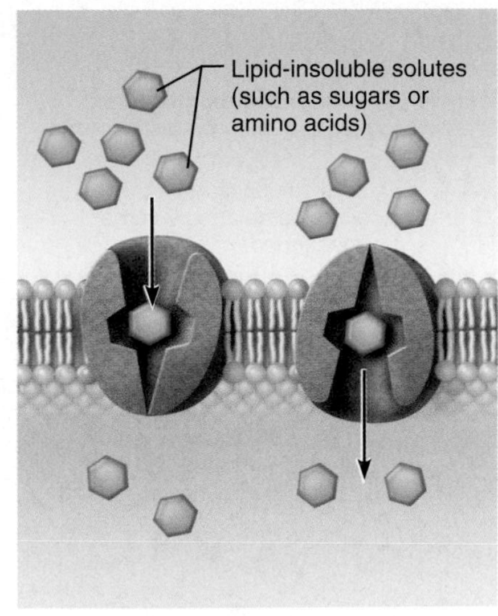

(b) **Carrier-mediated facilitated diffusion** via protein carrier specific for one chemical; binding of substrate causes shape change in transport protein

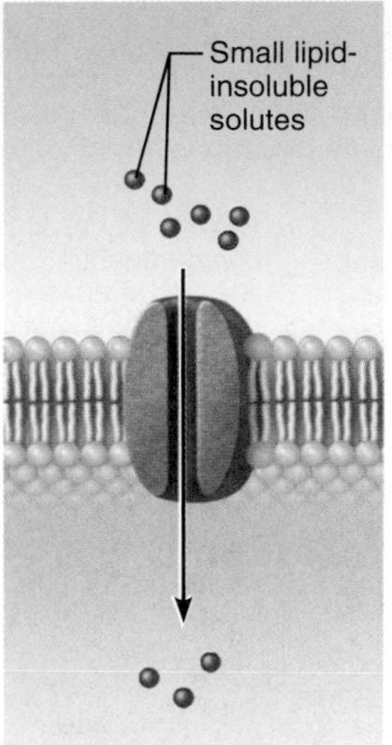

(c) **Channel-mediated facilitated diffusion** through a channel protein; mostly ions selected on basis of size and charge

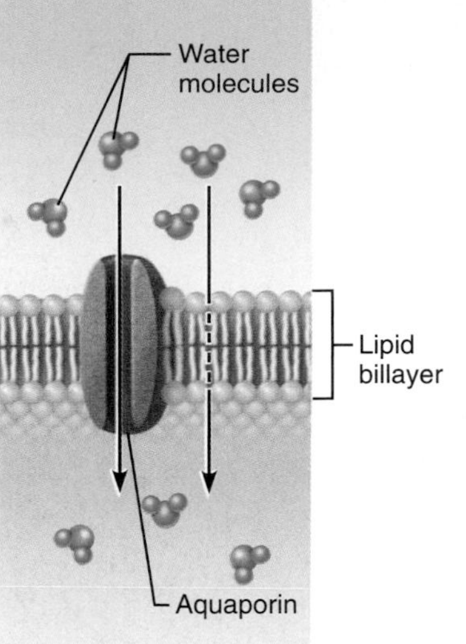

(d) **Osmosis**, diffusion of a solvent such as water through a specific channel protein (aquaporin) or through the lipid bilayer

Figure 3.7 **Diffusion through the plasma membrane.**

through the membrane if the molecule is (1) lipid soluble, (2) small enough to pass through membrane channels, or (3) assisted by a carrier molecule. The unassisted diffusion of lipid-soluble or very small particles is called *simple diffusion.* A special name, *osmosis,* is given to the unassisted diffusion of a solvent (usually water) through a membrane. Assisted diffusion is known as *facilitated diffusion.*

Simple Diffusion In **simple diffusion**, nonpolar and lipid-soluble substances diffuse directly through the lipid bilayer (Figure 3.7a). Such substances include oxygen, carbon dioxide, and fat-soluble vitamins. Because oxygen concentration is always higher in the blood than in tissue cells, oxygen continuously diffuses from the blood into the cells. Carbon dioxide, on the other hand, is in higher concentration within the cells, so it diffuses from tissue cells into the blood.

Facilitated Diffusion Certain molecules, notably glucose and other sugars, some amino acids, and ions are transported passively even though they are unable to pass through the lipid bilayer. Instead they move through the membrane by a passive transport process called **facilitated diffusion** in which the transported substance either (1) binds to protein carriers in the membrane and is ferried across or (2) moves through water-filled protein channels.

- **Carriers** are transmembrane integral proteins that show specificity for molecules of a certain polar substance or class of substances that are too large to pass through membrane channels, such as sugars and amino acids. The most popular model for the action of carriers indicates that changes in the shape of the carrier allow it to first envelop and then release the transported substance, shielding it en route from the nonpolar regions of the membrane. Essentially, the binding site is moved from one face of the membrane to the other by changes in the conformation of the carrier protein (Figure 3.7b and Table 3.1).

  Note that a substance transported by carrier-mediated facilitated diffusion, such as glucose, moves down its concentration gradient, just as in simple diffusion. Glucose is normally in higher concentrations in the blood than in the cells, where it is rapidly used for ATP synthesis. So, glucose transport within the body is *typically* unidirectional—into the cells. However, carrier-mediated transport is limited by the number of protein carriers present. For example, when all the glucose carriers are "engaged," they are said to be *saturated*, and glucose transport is occurring at its maximum rate.
- **Channels** are transmembrane proteins that serve to transport substances, usually ions or water, through aqueous channels from one side of the membrane to the other (Figure 3.7c and d). Binding or association sites exist within the channels, and the channels are selective due to pore size and the charges of the amino acids lining the channel. Some channels, the so-called *leakage channels*, are always open and simply allow ion or water fluxes according to concentration gradients. Others are gated and are controlled (opened or closed) by various chemical or electrical signals.

  Like carriers, many channels can be inhibited by certain molecules, show saturation, and tend to be specific. Substances moving through them also follow the concentration gradient (always moving down the gradient). When a substance crosses the membrane by simple diffusion, the rate of diffusion is not controllable because the lipid solubility of the membrane is not immediately changeable. By contrast, the rate of facilitated diffusion *is* controllable because the

3

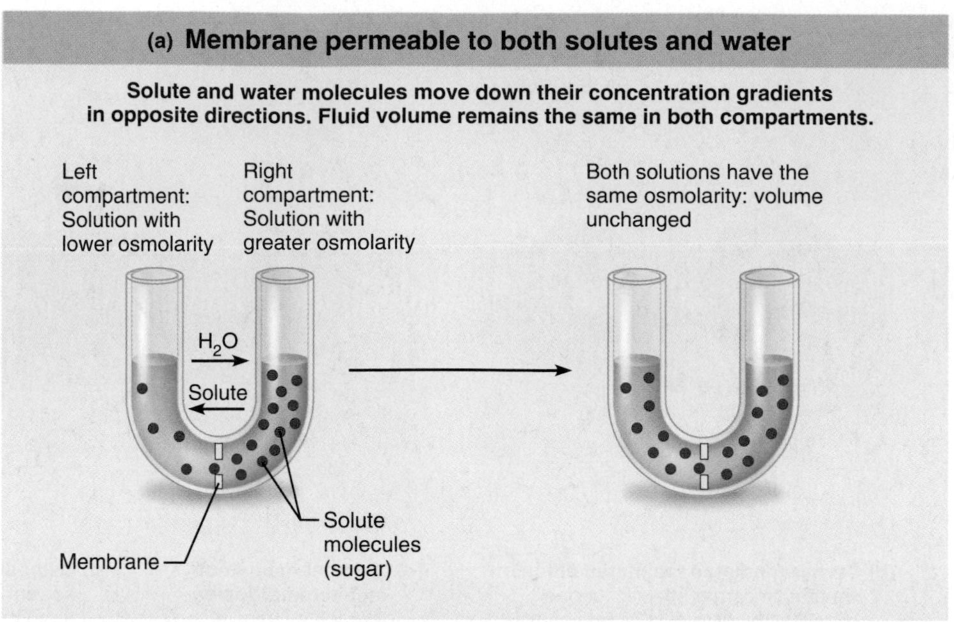

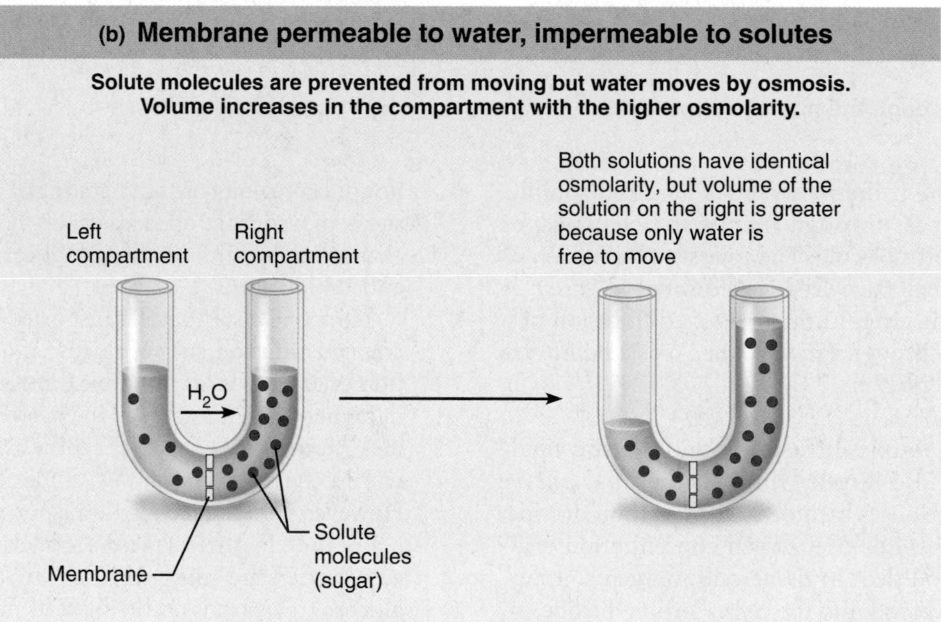

(b) Membrane permeable to water, impermeable to solutes
Solute molecules are prevented from moving but water moves by osmosis. Volume increases in the compartment with the higher osmolarity.
Left compartment
Right compartment
Both solutions have identical osmolarity, but volume of the solution on the right is greater because only water is free to move
H2O
Membrane
Solute molecules (sugar)

Figure 3.8 Influence of membrane permeability on diffusion and osmosis.

permeability of the membrane can be altered by regulating the activity or number of individual carriers or channels.

Oxygen, water, glucose, and various ions are vitally important to cellular homeostasis. Their passive transport by diffusion (either simple or facilitated) represents a tremendous saving of cellular energy. Indeed, if these substances had to be transported actively, cell expenditures of ATP would increase exponentially!

Osmosis The diffusion of a solvent, such as water, through a selectively permeable membrane is **osmosis** (oz-mo′sis; *osmos* = pushing). Even though water is highly polar, it passes via osmosis through the lipid bilayer (Figure 3.7d). This is surprising because you'd expect water to be repelled by the hydrophobic lipid tails. Although still hypothetical, one explanation is that random movements of the membrane lipids open small gaps between their wiggling tails, allowing water to slip and slide its way through the membrane by moving from gap to gap.

Water also moves freely and reversibly through water-specific channels constructed by transmembrane proteins called **aquaporins (AQPs)**. Although aquaporins are believed to be present in all cell types, they are particularly abundant in red blood cells and in cells involved in water balance such as kidney tubule cells.

Osmosis occurs whenever the water concentration differs on the two sides of a membrane. If distilled water is present on both sides of a selectively permeable membrane, no *net* osmosis occurs, even though water molecules move in both directions through the membrane. If the solute concentration on the two sides of

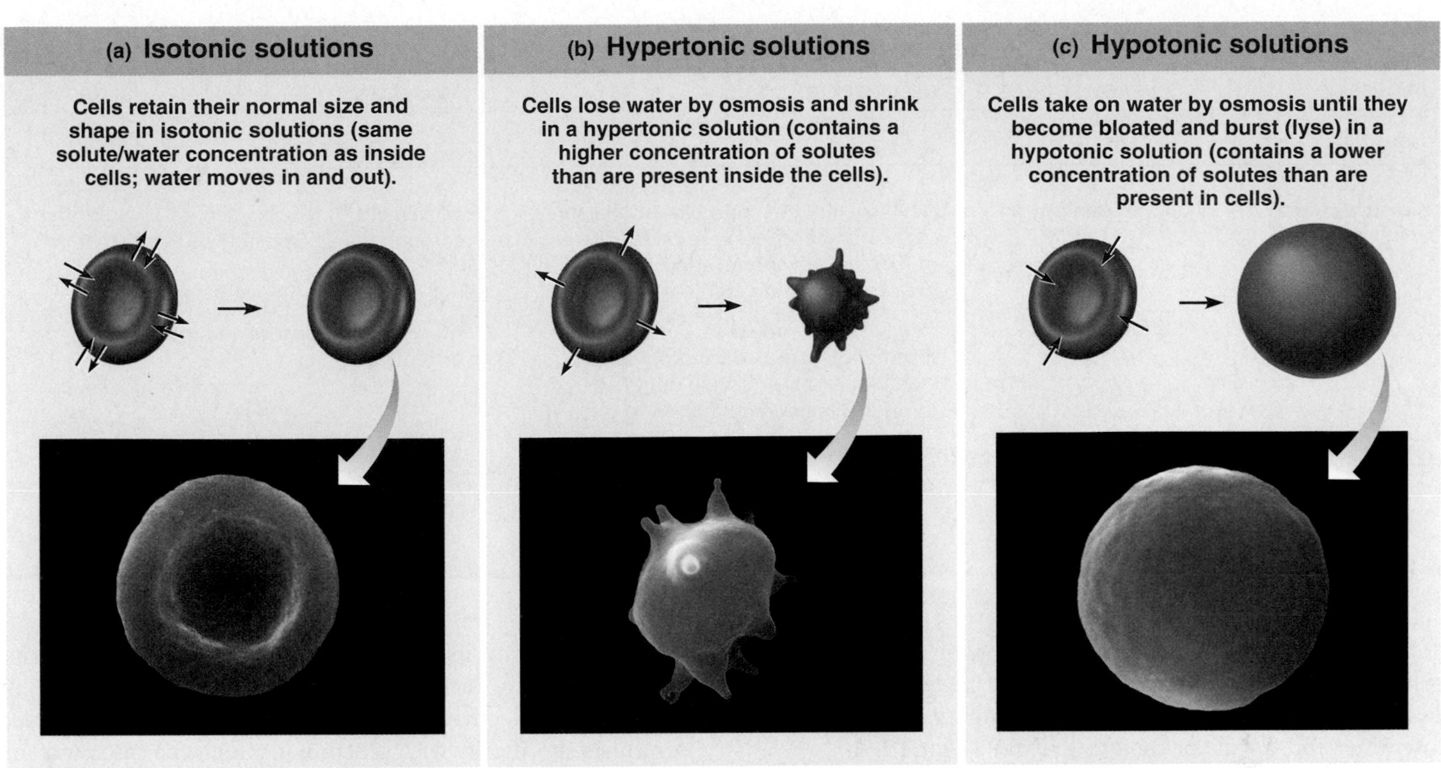

(a) Isotonic solutions
Cells retain their normal size and shape in isotonic solutions (same solute/water concentration as inside cells; water moves in and out).
(b) Hypertonic solutions
Cells lose water by osmosis and shrink in a hypertonic solution (contains a higher concentration of solutes than are present inside the cells).
(c) Hypotonic solutions
Cells take on water by osmosis until they become bloated and burst (lyse) in a hypotonic solution (contains a lower concentration of solutes than are present in cells).

3

Figure 3.9 The effect of solutions of varying tonicities on living red blood cells.

the membrane differs, water concentration differs as well (as solute concentration increases, water concentration decreases).

The extent to which water's concentration is decreased by solutes depends on the *number*, not the *type*, of solute particles, because one molecule or one ion of solute (theoretically) displaces one water molecule. The total concentration of all solute particles in a solution is referred to as the solution's **osmolarity** (oz″mo-lar′ĭ-te). When equal volumes of aqueous solutions of different osmolarity are separated by a membrane that is *permeable to all molecules* in the system, net diffusion of both solute and water occurs, each moving down its own concentration gradient. Eventually, equilibrium is reached when the water concentration on the left equals that on the right, and the solute concentration on both sides is the same **(Figure 3.8a)**.

If we consider the same system, but make the membrane *impermeable to solute molecules*, we see quite a different result (Figure 3.8b). Water quickly diffuses from the left to the right compartment and continues to do so until its concentration is the same on the two sides of the membrane. Notice that in this case equilibrium results from the movement of water alone (the solutes are prevented from moving). Notice also that the movement of water leads to dramatic changes in the volumes of the two compartments.

The last example mimics osmosis across plasma membranes of living cells, with one major difference. In our examples, the volumes of the compartments are infinitely expandable and the effect of pressure exerted by the added weight of the higher fluid column is not considered. In living plant cells, which have rigid cell walls external to their plasma membranes, this is not the case. As water diffuses into the cell, the point is finally reached where the **hydrostatic pressure** (the back pressure exerted by water against the membrane) within the cell is equal to its **osmotic pressure** (the tendency of water to move into the cell by osmosis). At this point, there is no further (net) water entry. As a rule, the higher the amount of nondiffusible, or *nonpenetrating*, solutes in a cell, the higher the osmotic pressure and the greater the hydrostatic pressure that must be developed to resist further net water entry.

However, such major changes in hydrostatic (and osmotic) pressures do not occur in living animal cells, which lack rigid cell walls. Osmotic imbalances cause animal cells to swell or shrink (due to net water gain or loss) until either the solute concentration is the same on both sides of the plasma membrane, or the membrane is stretched to its breaking point.

Such changes in animal cells lead us to the important concept of *tonicity* (to-nis′ĭ-te). As noted, many molecules, particularly intracellular proteins and selected ions, are prevented from diffusing through the plasma membrane. Consequently, any change in their concentration alters the water concentration on the two sides of the membrane and results in a net loss or gain of water by the cell.

The ability of a solution to change the shape or tone of cells by altering their internal water volume is called **tonicity** (*tono* = tension). Solutions with the same concentrations of nonpenetrating solutes as those found in cells (0.9% saline or 5% glucose) are **isotonic** ("the same tonicity"). Cells exposed to such solutions retain their normal shape, and exhibit no net loss or gain of water **(Figure 3.9a)**. As you might expect, the body's extracellular fluids and most intravenous solutions (solutions infused into the body via a vein) are isotonic.

3

TABLE 3.1 Passive Membrane Transport Processes

PROCESS	ENERGY SOURCE	DESCRIPTION	EXAMPLES
Diffusion			
Simple diffusion	Kinetic energy	Net movement of molecules from an area of their higher concentration to an area of their lower concentration, that is, along their concentration gradient	Movement of fats, oxygen, carbon dioxide through the lipid portion of the membrane
Facilitated diffusion	Kinetic energy	Same as simple diffusion, but the diffusing substance is attached to a lipid-soluble membrane carrier protein or moves through a membrane channel	Movement of glucose and some ions into cells
Osmosis	Kinetic energy	Simple diffusion of water through a selectively permeable membrane	Movement of water into and out of cells directly through the lipid phase of the membrane or via membrane channels (aquaporins)

Solutions with a higher concentration of nonpenetrating solutes than seen in the cell (for example, a strong saline solution) are **hypertonic**. Cells immersed in hypertonic solutions lose water and shrink, or *crenate* (kre′nat) (Figure 3.9b).

Solutions that are more dilute (contain a lower concentration of nonpenetrating solutes) than cells are called **hypotonic**. Cells placed in a hypotonic solution plump up rapidly as water rushes into them (Figure 3.9c). Distilled water represents the most extreme example of hypotonicity. Because it contains *no* solutes, water continues to enter cells until they finally burst or *lyse*.

Notice that osmolarity and tonicity are not the same thing. A solution's osmolarity is based solely on its total solute concentration. In contrast, its tonicity is based on how the solution affects cell volume, which depends on (1) solute concentration and (2) solute permeability of the plasma membrane. Osmolarity is expressed as osmoles per liter (osmol/L) where 1 osmol is equal to 1 mole of nonionizing molecules.* A 0.3-osmol/L solution of NaCl is isotonic because sodium ions are usually prevented from diffusing through the plasma membrane. But if the cell is immersed in a 0.3-osmol/L solution of a penetrating solute, the solute will enter the cell and water will follow. The cell will swell and burst, just as if it had been placed in pure water.

Osmosis is extremely important in determining distribution of water in the various fluid-containing compartments of the body (in cells, in blood, and so on). In general, osmosis continues until osmotic and hydrostatic pressures acting at the membrane are equal. For example, water is forced out of capillary blood by the hydrostatic pressure of the blood against the capillary wall, but the presence in blood of solutes that are too large to cross the capillary membrane draws water back into the bloodstream. As a result, very little net loss of plasma fluid occurs.

Simple diffusion and osmosis occurring directly through the plasma membrane are not selective processes. In those processes, whether a molecule can pass through the membrane depends chiefly on its size or its solubility in lipid, not on its unique structure. Facilitated diffusion, on the other hand, is often highly selective. The carrier for glucose, for example, combines specifically with glucose, in much the same way an enzyme binds to its specific substrate and ion channels allow only selected ions to pass.

*Osmolarity (Osm) is determined by multiplying molarity (moles per liter, or *M*) by the number of particles resulting from ionization. For example, since NaCl ionizes to $Na^+ + Cl^-$, a 1*M* solution of NaCl is a 2-Osm solution. For substances that do not ionize (e.g., glucose), molarity and osmolarity are the same.

HOMEOSTATIC IMBALANCE

Hypertonic solutions are sometimes infused intravenously into the bloodstream of edematous patients (those swollen because water is retained in their tissues) to draw excess water out of the extracellular space and move it into the bloodstream so that it can be eliminated by the kidneys. Hypotonic solutions may be used (with care) to rehydrate the tissues of extremely dehydrated patients. In less extreme cases of dehydration, drinking hypotonic fluids (colas, apple juice, and sports drinks) usually does the trick. ■

Table 3.1 summarizes passive membrane transport processes.

CHECK YOUR UNDERSTANDING

7. What is the energy source for all types of diffusion?
8. What determines the direction of any diffusion process?
9. What are the two types of facilitated diffusion and how do they differ?

For answers, see Appendix G.

Active Processes

- Differentiate between primary and secondary active transport.
- Compare and contrast endocytosis and exocytosis in terms of function and direction.
- Compare and contrast pinocytosis, phagocytosis, and receptor-mediated endocytosis.

Whenever a cell uses the bond energy of ATP to move solutes across the membrane, the process is referred to as *active.* Substances moved actively across the plasma membrane are usually unable to pass in the necessary direction by passive transport processes. The substance may be too large to pass through the channels, incapable of dissolving in the lipid bilayer, or unable to move down its concentration gradient. There are two major means of active membrane transport: active transport and vesicular transport.

Active Transport

Active transport, like carrier-mediated facilitated diffusion, requires carrier proteins that combine *specifically* and *reversibly* with the transported substances. However, facilitated diffusion always follows concentration gradients because its driving force is kinetic energy. In contrast, the active transporters or **solute pumps** move solutes, most importantly ions (such as Na^+, K^+, and Ca^{2+}), "uphill" *against* a concentration gradient. To do this work, cells must expend the energy of ATP.

Active transport processes are distinguished according to their source of energy. In *primary active transport,* the energy to do work comes *directly from hydrolysis of ATP.* In *secondary active transport,* transport is driven indirectly *by energy stored in ionic gradients* created by operation of primary active transport pumps. Secondary active transport systems are all *coupled systems*; that is, they move more than one substance at a time. If the two transported substances are moved in the same direction, the system is a **symport system** (*sym* = same). If the transported substances "wave to each other" as they cross the membrane in opposite directions, the system is an **antiport system** (*anti* = opposite, against). Let's examine these processes more carefully.

Primary Active Transport In **primary active transport**, hydrolysis of ATP results in the phosphorylation of the transport protein. This step causes the protein to change its shape in such a manner that it "pumps" the bound solute across the membrane.

Primary active transport systems include calcium and hydrogen pumps, but the most investigated example of a primary active transport system is the operation of the **sodium-potassium pump**, for which the carrier, or "pump," is an enzyme called **Na^+-K^+ ATPase.** In the body, the concentration of K^+ inside the cell is some 10 times higher than that outside, and the reverse is true of Na^+. These ionic concentration differences are essential for excitable cells like muscle and nerve cells to function normally and for all body cells to maintain their normal fluid volume. Because Na^+ and K^+ leak slowly but continuously through leakage channels in the plasma membrane along their concentration gradient (and cross more rapidly in stimulated muscle and nerve cells), the Na^+-K^+ pump operates more or less continuously as an antiporter. It simultaneously drives Na^+ out of the cell against a steep concentration gradient and pumps K^+ back in.

The electrochemical gradients maintained by the Na^+-K^+ pump underlie most primary and secondary active transport of nutrients and ions, and are crucial for cardiac and skeletal muscle and neuron function.

The step-by-step operation of the Na^+-K^+ pump is described in *Focus on Primary Active Transport: The Na^+-K^+ Pump* **(Figure 3.10)** on p. 74. Make sure you understand this process thoroughly before moving on to the topic of secondary active transport.

Secondary Active Transport A single ATP-powered pump, such as the Na^+-K^+ pump, can indirectly drive the **secondary active transport** of several other solutes. By moving sodium across the plasma membrane against its concentration gradient, the pump stores energy (in the ion gradient). Then, just as water pumped uphill can do work as it flows back down (to turn a turbine or water wheel), a substance pumped across a membrane can do work as it leaks back, propelled "downhill" along its concentration gradient. In this way, as sodium moves back into the cell with the help of a carrier protein (facilitated diffusion), other substances are "dragged along," or cotransported, by a common carrier protein **(Figure 3.11)**. A carrier moving two substances in the same direction is a symport system.

For example, some sugars, amino acids, and many ions are cotransported in this way into cells lining the small intestine. Both cotransported substances move passively because the energy for this type of transport is the concentration gradient of the ion (in this case Na^+). Na^+ has to be pumped back out into the lumen (cavity) of the intestine to maintain its diffusion gradient. Ion gradients can also be used to drive antiport systems such as those that help to regulate intracellular pH by using the sodium gradient to expel hydrogen ions.

Regardless of whether the energy is provided directly (primary active transport) or indirectly (secondary active transport), each membrane pump or cotransporter transports only specific substances. For this reason, active transport systems provide a way for the cell to be very selective in cases where substances cannot pass by diffusion. (No pump—no transport.)

Vesicular Transport

In **vesicular transport**, fluids containing large particles and macromolecules are transported across cellular membranes inside membranous sacs called *vesicles.* Vesicular transport processes that eject substances from the cell interior into the extracellular fluid are called **exocytosis** (ek″so-si-to′sis; "out of the cell"). Those in which the cell ingests small patches of the plasma

3

Figure 3.10 FOCUS **Primary Active Transport: The Na^+-K^+ Pump**

Primary active transport is the process in which ions are moved across cell membranes against electrochemical gradients using energy supplied directly by ATP. The action of the Na^+-K^+ pump is an important example of primary active transport.

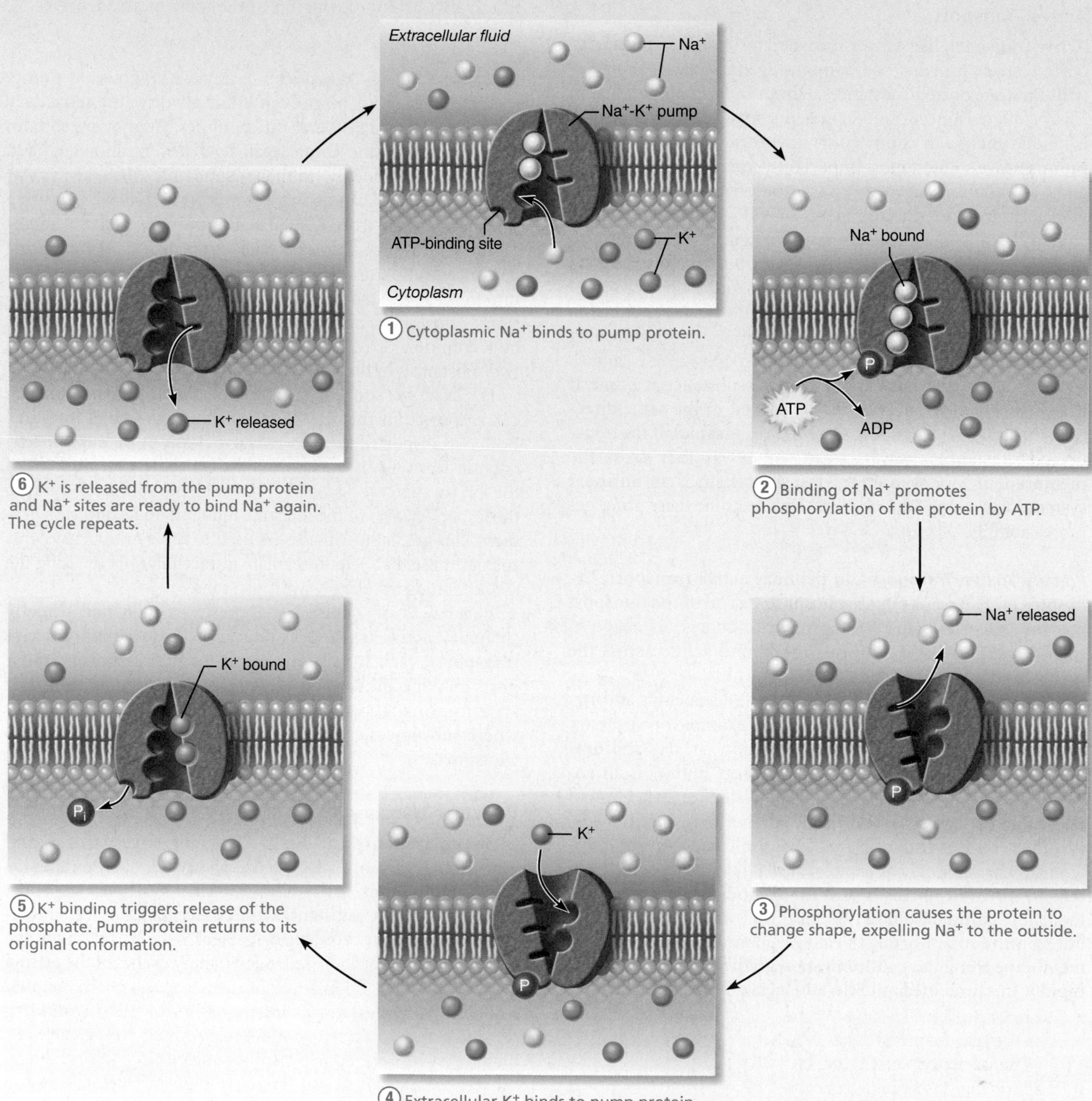

① Cytoplasmic Na^+ binds to pump protein.

② Binding of Na^+ promotes phosphorylation of the protein by ATP.

③ Phosphorylation causes the protein to change shape, expelling Na^+ to the outside.

④ Extracellular K^+ binds to pump protein.

⑤ K^+ binding triggers release of the phosphate. Pump protein returns to its original conformation.

⑥ K^+ is released from the pump protein and Na^+ sites are ready to bind Na^+ again. The cycle repeats.

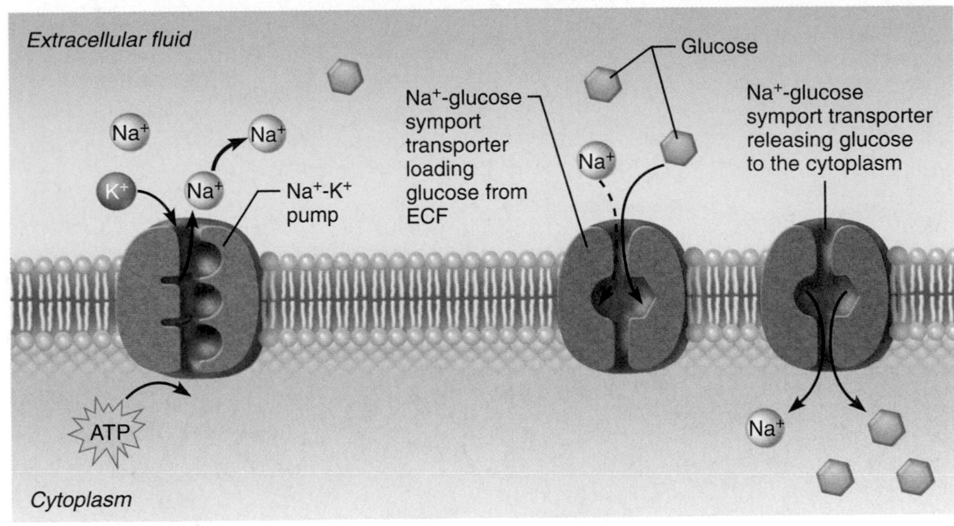

① The ATP-driven Na^+-K^+ pump stores energy by creating a steep concentration gradient for Na^+ entry into the cell.

② As Na^+ diffuses back across the membrane through a membrane cotransporter protein, it drives glucose against its concentration gradient into the cell. (ECF = extracellular fluid)

Figure 3.11 Secondary active transport.

membrane and moves substances from the cell exterior to the cell interior are called **endocytosis** (en″do-si-to′sis; "within the cell").

Vesicular transport is also used for combination processes such as *transcytosis*, moving substances into, across, and then out of the cell, and *substance*, or *vesicular*, *trafficking*, moving substances from one area (or organelle) in the cell to another.

Like solute pumping, vesicular transport processes are energized by ATP (or in some cases another energy-rich compound, *GTP*—guanosine triphosphate).

Endocytosis, Transcytosis, and Vesicular Trafficking Virtually all forms of vesicular transport involve an assortment of protein-coated vesicles of three types and, with some exceptions, all are mediated by membrane receptors. Before we get specific about each type of coated vesicular transport, let's look at the general scheme of endocytosis.

Protein-coated vesicles provide the main route for endocytosis and transcytosis of bulk solids, most macromolecules, and fluids. On occasion, these vesicles are also hijacked by pathogens seeking entry into a cell.

Figure 3.12 shows the basic steps in endocytosis and transcytosis. ① The substance to be taken into the cell by endocytosis is progressively enclosed by an infolding portion of the plasma membrane called a *coated pit*. The coating is most often the bristlelike **clathrin** (kla′thrin; "lattice clad") protein coating found on the cytoplasmic face of the pit. The clathrin coat (clathrin and some accessory proteins) acts both in cargo selection and in deforming the membrane to produce the vesicle. ② The vesicle detaches, and ③ the coat proteins are recycled back to the plasma membrane.

④ The uncoated vesicle then typically fuses with a processing and sorting vesicle called an *endosome*. ⑤ Some membrane components and receptors of the fused vesicle may be recycled back to the plasma membrane in a transport vesicle. ⑥ The remaining contents of the vesicle may (a) combine with a *lysosome* (li′sō-sōm), a specialized cell structure containing digestive enzymes, where the ingested substance is degraded or released (if iron or cholesterol), or (b) be transported completely across the cell and released by exocytosis on the opposite side (*transcytosis*). Transcytosis is common in the endothelial cells lining blood vessels because it provides a quick means to get substances from the blood to the interstitial fluid.

Based on the nature and quantity of material taken up and the means of uptake, three types of endocytosis that use clathrin-coated vesicles are recognized: phagocytosis, pinocytosis, and receptor-mediated endocytosis.

Phagocytosis (fag″o-si-to′sis; "cell eating") is the type of endocytosis in which the cell engulfs some relatively large or solid material, such as a clump of bacteria, cell debris, or inanimate particles (asbestos fibers or glass, for example) **(Figure 3.13a)**. When a particle binds to receptors on the cell's surface, cytoplasmic extensions called pseudopods (soo′do-pahdz; *pseudo* = false, *pod* = foot) form and flow around the particle and engulf it. The endocytotic vesicle formed in this way is called a **phagosome** (fag′o-sōm; "eaten body"). In most cases, the phagosome then fuses with a lysosome and its contents are digested. Any indigestible contents are ejected from the cell by exocytosis.

In the human body, only macrophages and certain white blood cells are "experts" at phagocytosis. Commonly referred to as *phagocytes*, these cells help police and protect the body by ingesting and disposing of bacteria, other foreign substances, and dead tissue cells. The disposal of dying cells is crucial, because dead cell remnants trigger inflammation in the surrounding area or may stimulate an undesirable immune response. Most phagocytes move about by **amoeboid motion** (ah-me′boyd; "changing shape"); that is, the flowing of their cytoplasm into temporary pseudopods allows them to creep along.

3

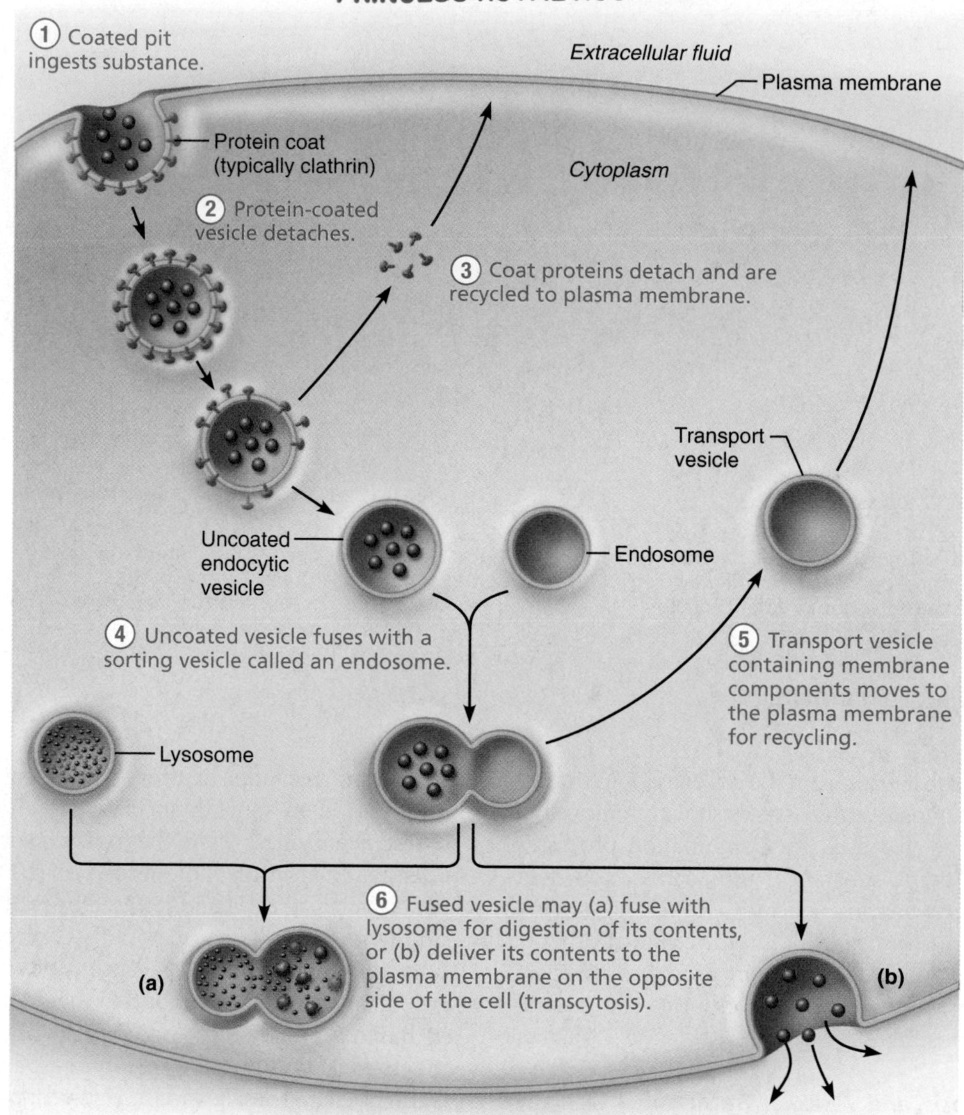

1 Coated pit ingests substance.
Extracellular fluid
Plasma membrane
Protein coat (typically clathrin)
Cytoplasm
2 Protein-coated vesicle detaches.
3 Coat proteins detach and are recycled to plasma membrane.
Transport vesicle
Uncoated endocytic vesicle
Endosome
4 Uncoated vesicle fuses with a sorting vesicle called an endosome.
5 Transport vesicle containing membrane components moves to the plasma membrane for recycling.
Lysosome
6 Fused vesicle may (a) fuse with lysosome for digestion of its contents, or (b) deliver its contents to the plasma membrane on the opposite side of the cell (transcytosis).
(a)
(b)

Figure 3.12 Events of endocytosis mediated by protein-coated pits. Note the three possible fates for a vesicle and its contents, shown in ⑤ and ⑥.

In **pinocytosis** ("cell drinking"), also called **fluid-phase endocytosis**, a bit of infolding plasma membrane (which begins as a clathrin-coated pit) surrounds a very small volume of extracellular fluid containing dissolved molecules (Figure 3.13b). This droplet enters the cell and fuses with an endosome. Unlike phagocytosis, pinocytosis is a routine activity of most cells, affording them a nonselective way of sampling the extracellular fluid. It is particularly important in cells that absorb nutrients, such as cells that line the intestines.

As mentioned, bits of the plasma membrane are removed when the membranous sacs are internalized. However, these membranes are recycled back to the plasma membrane by exocytosis as described shortly, so the surface area of the plasma membrane remains remarkably constant.

Receptor-mediated endocytosis is the main mechanism for the specific endocytosis and transcytosis of most macromolecules by body cells, and it is exquisitely selective (Figure 3.13c). It is also the mechanism that allows cells to concentrate material that is present only in very small amounts in the extracellular fluid. The receptors for this process are plasma membrane proteins that bind only certain substances. Both the receptors and attached molecules are internalized in a clathrin-coated pit and then dealt with in one of the ways discussed above. Substances taken up by receptor-mediated endocytosis include enzymes, insulin (and some other hormones), low-density lipoproteins (such as cholesterol attached to a transport protein), and iron. Unfortunately, flu viruses, diphtheria, and cholera toxins use this route to enter and attack our cells.

Other coat proteins are also used for certain types of vesicular transport. **Caveolae** (ka″ve-o′le; "little caves"), tubular or flask-shaped inpocketings of the plasma membrane seen in many cell types, are involved in a unique kind of receptor-mediated endocytosis called **potosis**. Like clathrin-coated pits, caveolae capture specific molecules (folic acid, tetanus toxin) from the extracellular fluid in coated vesicles and participate in some forms of transcytosis. However, caveolae are smaller than clathrin-coated

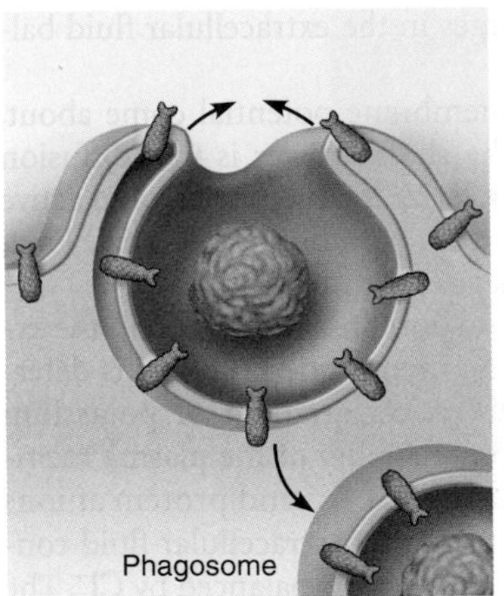

(a) Phagocytosis
The cell engulfs a large particle by forming projecting pseudopods ("false feet") around it and enclosing it within a membrane sac called a phagosome. The phagosome is combined with a lysosome. Undigested contents remain in the vesicle (now called a residual body) or are ejected by exocytosis. Vesicle may or may not be protein-coated but has receptors capable of binding to microorganisms or solid particles.

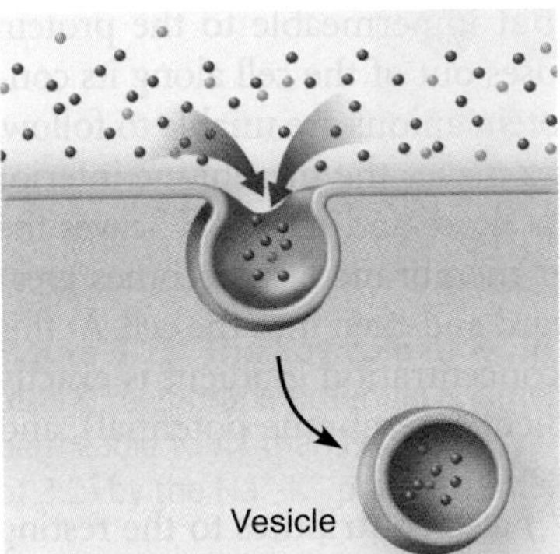

(b) Pinocytosis
The cell "gulps" drops of extracellular fluid containing solutes into tiny vesicles. No receptors are used, so the process is nonspecific. Most vesicles are protein-coated.

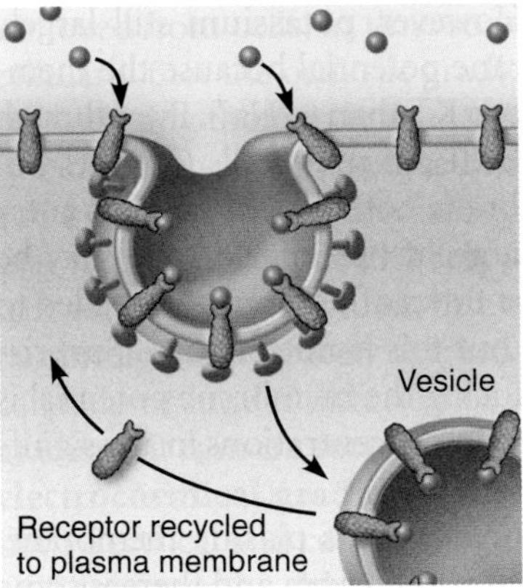

(c) Receptor-mediated endocytosis
Extracellular substances bind to specific receptor proteins in regions of coated pits, enabling the cell to ingest and concentrate specific substances (ligands) in protein-coated vesicles. Ligands may simply be released inside the cell, or combined with a lysosome to digest contents. Receptors are recycled to the plasma membrane in vesicles.

Figure 3.13 **Comparison of three types of endocytosis.**

vesicles. Additionally, their cage-like protein coat is thinner and composed of a different protein called **caveolin**.

Caveolae are closely associated with lipid rafts that are platforms for G proteins, receptors for hormones (for example, insulin), and enzymes involved in cell regulation. These vesicles appear to provide sites for cell signaling and cross talk between signaling pathways. Their precise role in the cell is still being worked out.

Vesicles coated with **coatomer (COP1** and **COP2) proteins** are used in most types of intracellular *vesicular trafficking*, in which vesicles transport substances between organelles.

Exocytosis The process of exocytosis, typically stimulated by a cell-surface signal such as binding of a hormone to a membrane receptor or a change in membrane voltage, accounts for hormone secretion, neurotransmitter release, mucus secretion, and in some cases, ejection of wastes. The substance to be removed from the cell is first enclosed in a protein-coated membranous sac called a **vesicle**. In most cases, the vesicle migrates to the plasma membrane, fuses with it, and then ruptures, spilling the sac contents out of the cell **(Figure 3.14)**.

Exocytosis, like other cases in which vesicles are targeted to their destinations, involves a "docking" process in which transmembrane proteins on the vesicles, fancifully called v-SNAREs (*v* for vesicle), recognize certain plasma membrane proteins, called t-SNAREs (*t* for target), and bind with them. This binding causes the membranes to "corkscrew" together and fuse, rearranging the lipid monolayers without mixing them (Figure 3.14a). As described, membrane material added by exocytosis is removed by endocytosis—the reverse process.

Table 3.2 summarizes active membrane transport processes.

CHECK YOUR UNDERSTANDING

10. What happens when the Na^+-K^+ pump is phosphorylated? When K^+ binds to the pump protein?
11. As a cell grows, its plasma membrane expands. Does this membrane expansion involve endocytosis or exocytosis?
12. Phagocytic cells gather in the lungs, particularly in the lungs of smokers. What is the connection?
13. What vesicular transport process allows a cell to take in cholesterol from the extracellular fluid?

For answers, see Appendix G.

The Plasma Membrane: Generation of a Resting Membrane Potential

▶ Define membrane potential and explain how the resting membrane potential is established and maintained.

As you're now aware, the selective permeability of the plasma membrane can lead to dramatic osmotic flows, but that is not its only consequence. An equally important result is the generation of a **membrane potential**, or voltage, across the membrane. A *voltage* is electrical potential energy resulting from the separation of oppositely charged particles. In cells, the oppositely charged particles are ions, and the barrier that keeps them apart is the plasma membrane.

In their resting state, all body cells exhibit a **resting membrane potential** that typically ranges from –50 to –100 millivolts (mV), depending on cell type. For this reason, all cells are said to be **polarized**. The minus sign before the voltage indicates that the *inside* of the cell is negative compared to its outside. This voltage (or charge separation) exists *only at the membrane*. If we were to add up all the negative and positive charges in the cytoplasm, we would find that the cell interior is electrically neutral. Likewise,

3

TABLE 3.2 Active Membrane Transport Processes

PROCESS	ENERGY SOURCE	DESCRIPTION	EXAMPLES
Active Transport			
Primary active transport	ATP	Transport of substances against a concentration (or electrochemical) gradient. Performed across the plasma membrane by a solute pump, directly using energy of ATP hydrolysis.	Ions (Na^+, K^+, H^+, Ca^{2+}, and others)
Secondary active transport	Ion concentration gradient maintained with ATP	Cotransport (coupled transport) of two solutes across the membrane. Energy is supplied indirectly by the ion gradient created by primary active transport. Symporters move the transported substances in the same direction; antiporters move transported substances in opposite directions across the membrane.	Movement of polar or charged solutes, e.g., amino acids (into cell by symporters); Ca^{2+}, H^+ (out of cells via antiporters)
Vesicular Transport			
Exocytosis	ATP	Secretion or ejection of substances from a cell. The substance is enclosed in a membranous vesicle, which fuses with the plasma membrane and ruptures, releasing the substance to the exterior.	Secretion of neurotransmitters, hormones, mucus, etc.; ejection of cell wastes
Endocytosis ▪ Via clathrin-coated vesicles	ATP		
Phagocytosis	ATP	"Cell eating": A large external particle (proteins, bacteria, dead cell debris) is surrounded by a "seizing foot" and becomes enclosed in a vesicle (phagosome).	In the human body, occurs primarily in protective phagocytes (some white blood cells and macrophages)
Pinocytosis (fluid-phase endocytosis)	ATP	Plasma membrane sinks beneath an external fluid droplet containing small solutes. Membrane edges fuse, forming a fluid-filled vesicle.	Occurs in most cells; important for taking in dissolved solutes by absorptive cells of the kidney and intestine
Receptor-mediated endocytosis	ATP	Selective endocytosis and transcytosis. External substance binds to membrane receptors.	Means of intake of some hormones, cholesterol, iron, and most macromolecules
▪ Via caveolin-coated vesicles (caveolae)	ATP	Selective endocytosis (and transcytosis). External substance binds to membrane receptors (often associated with lipid rafts).	Roles not fully known; proposed roles include cholesterol regulation and trafficking, and platforms for signal transduction
Intracellular vesicular trafficking			
▪ Via coatomer-coated vesicles	ATP	Vesicles pinch off from organelles and travel to other organelles to deliver their cargo.	Accounts for nearly all intracellular trafficking between certain organelles (endoplasmic reticulum and Golgi apparatus). Exceptions include vesicles budding from the trans face of the Golgi apparatus, which are clathrin-coated.

is important) and in immunity. These sticky glycoproteins (*cadherins* and *integrins*) act as

1. The molecular "Velcro" that cells use to anchor themselves to molecules in the extracellular space and to each other (see desmosome discussion on pp. 66–67)
2. The "arms" that migrating cells use to haul themselves past one another
3. SOS signals sticking out from the blood vessel lining that rally protective white blood cells to a nearby infected or injured area

4. Mechanical sensors that respond to local tension at the cell surface by stimulating synthesis or degradation of adhesive membrane junctions
5. Transmitters of intracellular signals that direct cell migration, proliferation, and specialization

Roles of Membrane Receptors

A huge and diverse group of integral proteins and glycoproteins that serve as binding sites are collectively known as **membrane receptors**. Some function in contact signaling, and others in chemical signaling. Let's take a look.

Contact Signaling *Contact signaling* is the actual coming together and touching of cells, and it is the means by which cells recognize one another. It is particularly important for normal development and immunity. Some bacteria and other infectious agents use contact signaling to identify their "preferred" target tissues or organs.

Chemical Signaling Most plasma membrane receptors are involved in *chemical signaling*, and this group will receive the bulk of our attention. Signaling chemicals that bind specifically to plasma membrane receptors are called **ligands**. Among these ligands are most *neurotransmitters* (nervous system signals), *hormones* (endocrine system signals), and *paracrines* (chemicals that act locally and are rapidly destroyed).

Different cells respond in different ways to the same ligand. Acetylcholine, for instance, stimulates skeletal muscle cells to contract, but inhibits heart muscle. Why do different cells have such different responses? The reason is that a target cell's response depends on the internal machinery that the receptor is linked to, not the specific ligand that binds to it.

Though cell responses to receptor binding vary widely, there is a fundamental similarity. When a ligand binds to a membrane receptor, the receptor's structure changes, and cell proteins are altered in some way. For example, muscle proteins change shape to generate force. Some membrane receptor proteins are *catalytic proteins* that function as enzymes. Others, such as the *chemically gated channel-linked receptors* common in muscle and nerve cells, respond to ligands by transiently opening or closing ion gates, which in turn changes the excitability of the cell.

Still other receptors are coupled to enzymes or ion channels by a regulatory molecule called a G protein. Because nearly every cell in the body displays at least some of these receptors, we will spend a bit more time with them. The lipid rafts mentioned earlier group together many receptor-mediated elements, thus facilitating cell signaling.

G protein–linked receptors exert their effect indirectly through a **G protein**, which acts as a middleman or relay to activate (or inactivate) a membrane-bound enzyme or ion channel. As a result, one or more intracellular chemical signals, commonly called **second messengers**, are generated and connect plasma membrane events to the internal metabolic machinery of the cell. Two very important second messengers are **cyclic AMP** and ionic calcium, both of which typically activate *protein kinase enzymes*, which transfer phosphate groups from ATP to other proteins. In this way, the protein kinases can activate a whole series of enzymes that bring about the desired cellular activity. Because a single enzyme can catalyze hundreds of reactions, the amplification effect of such a chain of events is tremendous. *Focus on G Proteins* **(Figure 3.16)** on p. 82 describes how a G protein signaling system works. Take a moment to study this carefully because this key signaling pathway is involved in neurotransmission, smell, vision, and hormone action (Chapters 11, 15, and 16).

One important signaling molecule must be mentioned even though it doesn't act in any of the ways we have already described. *Nitric oxide* (*NO*), one of nature's simplest molecules, is made of a single atom of nitrogen and one of oxygen. It is also an environmental pollutant and the first gas known to act as a biological messenger. Because of its tiny size, it slips into and out of cells easily. Its unpaired electron makes it highly reactive and it reacts with head-spinning speed with other key molecules to spur cells into a broad array of activities. You will be hearing more about NO later (in the neural, cardiovascular, and immune system chapters).

Role of Voltage-Sensitive Membrane Channel Proteins

Electrical Signaling In the process known as *electrical signaling*, certain plasma membrane proteins are channel proteins that respond to changes in membrane potential by opening or closing the channel. Such voltage-gated channels are common in excitable tissues like neural and muscle tissues, and are indispensable to their normal functioning.

CHECK YOUR UNDERSTANDING

16. What term is used to indicate signaling chemicals that bind to membrane receptors? Which type of membrane receptor is most important in directing intracellular events by promoting formation of second messengers?

For answer, see Appendix G.

The Cytoplasm

- ▶ Describe the composition of the cytosol. Define inclusions and list several types.
- ▶ Discuss the structure and function of mitochondria.
- ▶ Discuss the structure and function of ribosomes, the endoplasmic reticulum, and the Golgi apparatus, including functional interrelationships among these organelles.
- ▶ Compare the functions of lysosomes and peroxisomes.

Cytoplasm ("cell-forming material") is the cellular material between the plasma membrane and the nucleus. It is the site where most cellular activities are accomplished. Although early microscopists thought that the cytoplasm was a structureless gel, the electron microscope reveals that it consists of three major elements: the cytosol, organelles, and inclusions.

Figure 3.16 FOCUS G Proteins

G proteins act as middlemen or relays between extracellular first messengers and intracellular second messengers that cause responses within the cell.

② **The activated receptor binds to a G protein and activates it.** During activation, a G protein changes shape (turns "on") causing it to release GDP and bind GTP.
③ **Activated G protein activates (or inactivates) effector protein (e.g., an enzyme) by causing its shape to change.**

Effector protein (e.g., an enzyme)

④ **Activated effector enzymes catalyze reactions that produce 2nd messengers in the cell.** (Common 2nd messengers include cyclic AMP and Ca^{2+}.)

Cascade of cellular responses (metabolic and structural changes)

The sequence described here is like a molecular relay race. Instead of a baton passed from runner to runner, the message is passed from molecule to molecule as it makes its way across the cell membrane from outside to inside the cell.

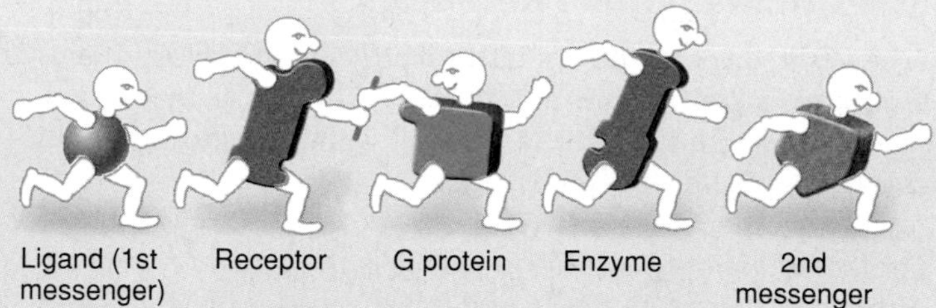

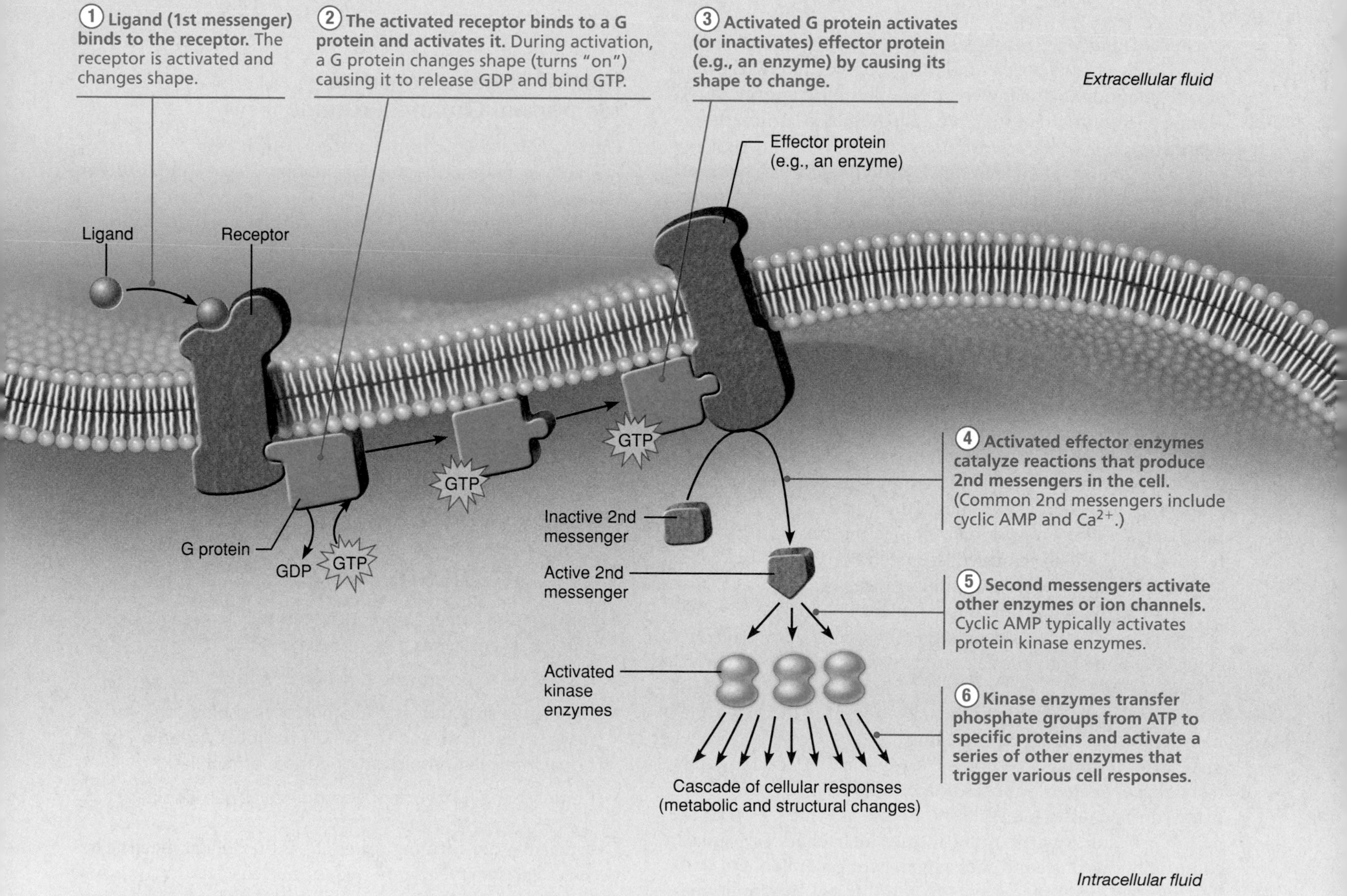

The **cytosol** (si′to-sol) is the viscous, semitransparent fluid in which the other cytoplasmic elements are suspended. It is a complex mixture with properties of both a colloid and a true solution. Dissolved in the cytosol, which is largely water, are proteins, salts, sugars, and a variety of other solutes.

The **cytoplasmic organelles** are the metabolic machinery of the cell. Each type of organelle carries out a specific function for the cell—some synthesize proteins, others package those proteins, and so on.

Inclusions are chemical substances that may or may not be present, depending on cell type. Examples include stored nutrients, such as the glycogen granules abundant in liver and muscle cells; lipid droplets common in fat cells; pigment (melanin) granules seen in certain cells of skin and hair; water-containing vacuoles; and crystals of various types.

Cytoplasmic Organelles

The cytoplasmic organelles ("little organs") are specialized cellular compartments, each performing its own job to maintain the life of the cell. Some organelles, the *nonmembranous organelles*, lack membranes. Examples are the cytoskeleton, centrioles, and ribosomes.

Most organelles, however, are bounded by a membrane similar in composition to the plasma membrane. This membrane enables such *membranous organelles* to maintain an internal environment different from that of the surrounding cytosol. This compartmentalization is crucial to cell functioning. Without it, thousands of enzymes would be randomly mixed and biochemical activity would be chaotic. The cell's membranous organelles include the mitochondria, peroxisomes, lysosomes, endoplasmic reticulum, and Golgi apparatus.

Besides providing "splendid isolation," an organelle's membrane often unites it with the rest of an interactive intracellular system called the *endomembrane system* (see p. 87) and the lipid makeup of its membrane allows it to recognize and interact with other organelles. Now, let us consider what goes on in each of the workshops of our cellular factory.

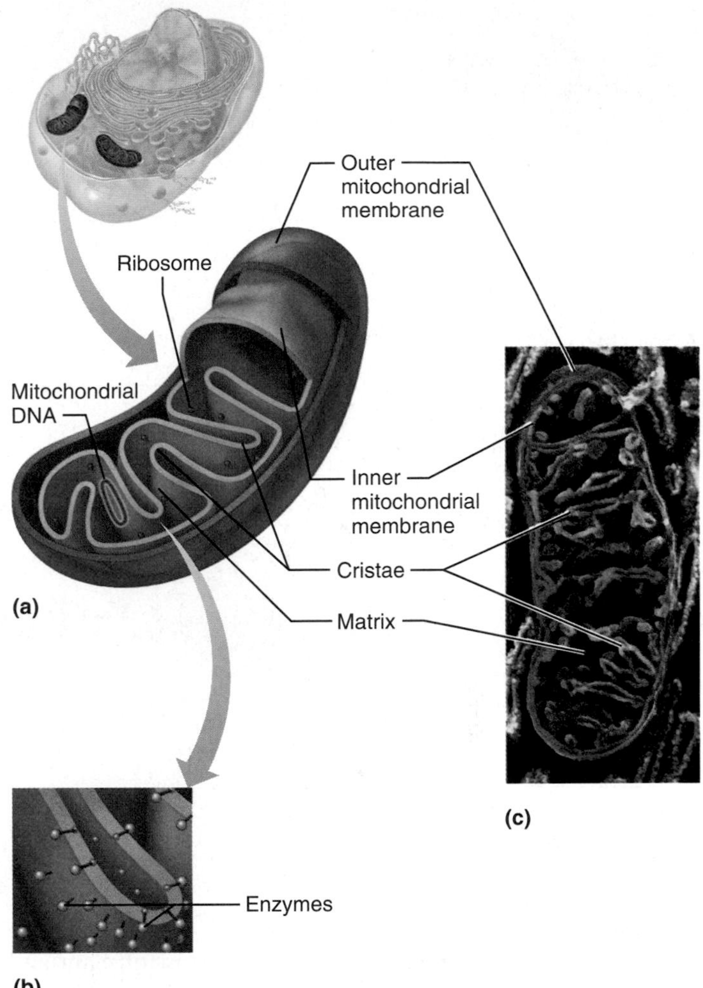

Figure 3.17 Mitochondrion. (a) Diagrammatic view of a longitudinally sectioned mitochondrion. **(b)** Close-up view of a crista showing enzymes (stalked particles). **(c)** Electron micrograph of a mitochondrion (50,000×).

Mitochondria

Mitochondria (mi″to-kon′dre-ah) are threadlike (*mitos* = thread) or lozenge-shaped membranous organelles. In living cells they squirm, elongate, and change shape almost continuously. They are the power plants of a cell, providing most of its ATP supply. The density of mitochondria in a particular cell reflects that cell's energy requirements, and mitochondria are generally clustered where the action is. Busy cells like kidney and liver cells have hundreds of mitochondria, whereas relatively inactive cells (such as unchallenged lymphocytes) have just a few.

Mitochondria are enclosed by *two* membranes, each with the general structure of the plasma membrane **(Figure 3.17)**. The *outer membrane* is smooth and featureless, but the *inner membrane* folds inward, forming shelflike **cristae** (krĭ′ste; "crests") that protrude into the *matrix*, the gel-like substance within the mitochondrion. Intermediate products of food fuels (glucose and others) are broken down to water and carbon dioxide by teams of enzymes, some dissolved in the mitochondrial matrix and others forming part of the crista membrane.

As the metabolites are broken down and oxidized, some of the energy released is captured and used to attach phosphate groups to ADP molecules to form ATP. This multistep mitochondrial process (described in Chapter 24) is generally referred to as *aerobic cellular respiration* (a-er-o′bik) because it requires oxygen.

Mitochondria are complex organelles: They contain their own DNA, RNA, and ribosomes and are able to reproduce themselves. Mitochondrial genes (some 37 of them) direct the synthesis of 1% of the proteins required for mitochondrial function, and the DNA of the cell's nucleus encodes the remaining proteins needed to carry out cellular respiration. When cellular requirements for ATP increase, the mitochondria synthesize more cristae or simply pinch in half (a process called *fission*) to increase their number, then grow to their former size.

Intriguingly, mitochondria are similar to a specific group of bacteria (the purple bacteria phylum), and mitochondrial DNA

3

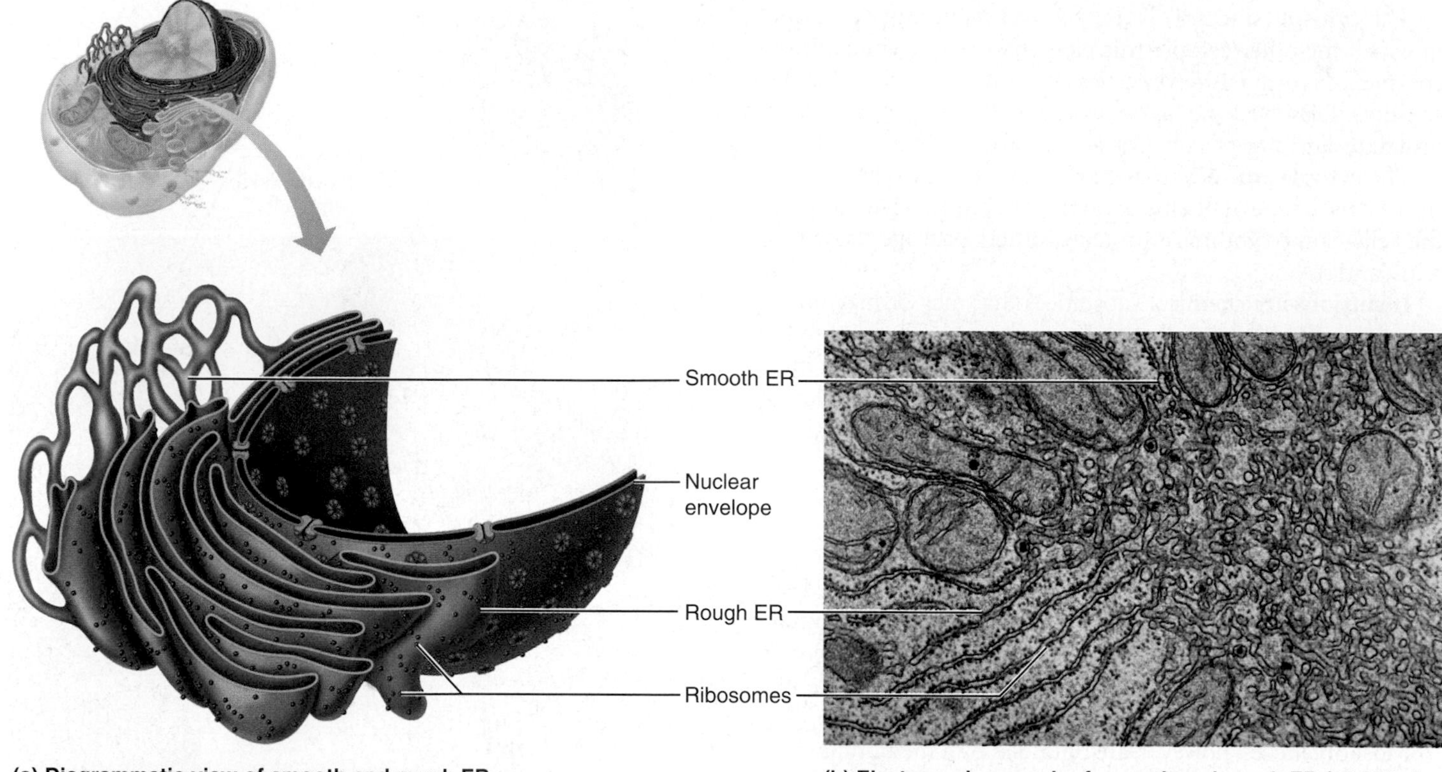

(a) Diagrammatic view of smooth and rough ER

(b) Electron micrograph of smooth and rough ER (10,000×)

Figure 3.18 The endoplasmic reticulum.

is bacteria-like. It is widely believed that mitochondria arose from bacteria that invaded the ancient ancestors of plant and animal cells, and that this unique merger gave rise to all complex cells.

Ribosomes

Ribosomes (ri′bo-sōmz) are small, dark-staining granules composed of proteins and a variety of RNAs called *ribosomal RNAs*. Each ribosome has two globular subunits that fit together like the body and cap of an acorn. Ribosomes are sites of protein synthesis, a function we discuss in detail later in this chapter.

Some ribosomes float freely in the cytoplasm. Others are attached to membranes, forming a complex called the *rough endoplasmic reticulum* **(Figure 3.18)**. These two ribosomal populations appear to divide the chore of protein synthesis. *Free ribosomes* make soluble proteins that function in the cytosol, as well as those imported into mitochondria and some other organelles. *Membrane-bound ribosomes* synthesize proteins destined either for incorporation into cell membranes or for export from the cell. Ribosomes can switch back and forth between these two functions, attaching to and detaching from the membranes of the endoplasmic reticulum, according to the type of protein they are making at a given time.

Endoplasmic Reticulum

The **endoplasmic reticulum (ER)** (en″do-plaz′mik re-tik′u-lum; "network within the cytoplasm") is an extensive system of interconnected tubes and parallel membranes enclosing fluid-filled cavities, or **cisternae** (sis-ter′ne). Coiling and twisting through the cytosol, the ER is continuous with the nuclear membrane and accounts for about half of the cell's membranes. There are two distinct varieties of ER: rough ER and smooth ER.

Rough Endoplasmic Reticulum The external surface of the **rough ER** is studded with ribosomes (Figure 3.18a, b). Proteins assembled on these ribosomes thread their way into the fluid-filled interior of the ER cisternae, where various fates await them (as described on p. 106). When complete, the newly made proteins are enclosed in coatomer-coated vesicles for their journey to the Golgi apparatus where they undergo further processing.

The rough ER has several functions. Its ribosomes manufacture all proteins secreted from cells. For this reason, the rough ER is particularly abundant and well developed in most secretory cells, antibody-producing plasma cells, and liver cells, which produce most blood proteins. It is also the cell's "membrane factory" because integral proteins and phospholipids that form part of all cellular membranes are manufactured there. The enzymes needed to catalyze lipid synthesis have their active sites on the external (cytosolic) face of the ER membrane, where the needed substrates are readily available.

Smooth Endoplasmic Reticulum The **smooth ER** (see Figures 3.2 and 3.18a, b) is continuous with the rough ER and consists of tubules arranged in a looping network. Its enzymes (all integral proteins forming part of its membranes) play no role in

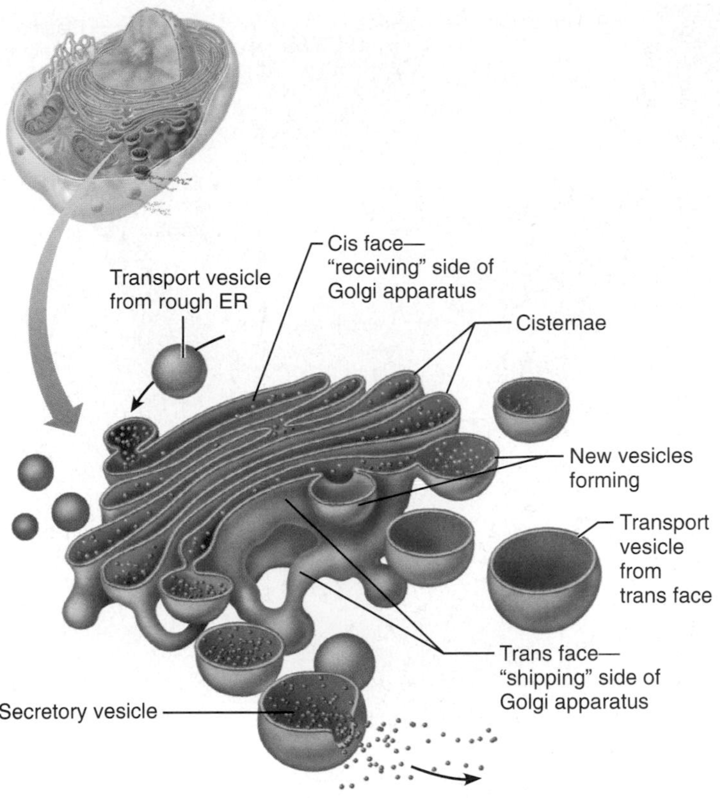

(a) Many vesicles in the process of pinching off from the membranous Golgi apparatus.

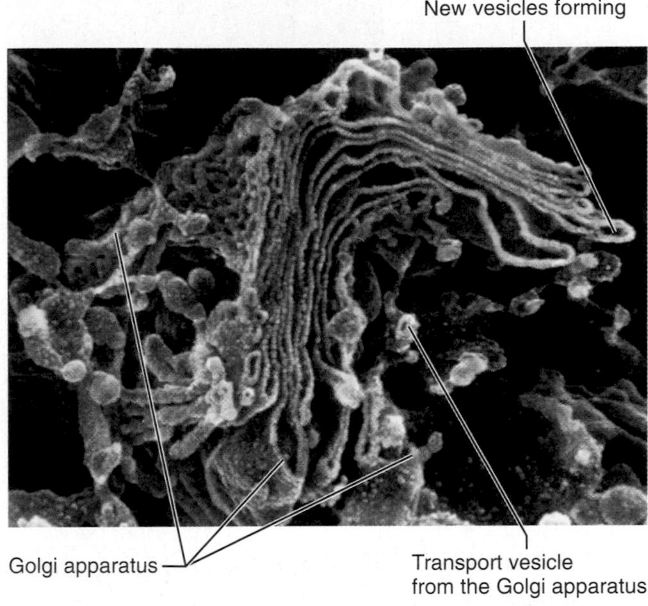

(b) Electron micrograph of the Golgi apparatus (90,000×)

Figure 3.19 Golgi apparatus. Note: In **(a)**, the various and abundant vesicles shown in the process of pinching off from the membranous Golgi apparatus would have a protein coating on their external surfaces. These proteins are omitted from the diagram for simplicity.

protein synthesis. Instead, they catalyze reactions involved with the following processes:

1. Lipid metabolism, cholesterol synthesis, and synthesis of the lipid components of lipoproteins (in liver cells)
2. Synthesis of steroid-based hormones such as sex hormones (testosterone-synthesizing cells of the testes are full of smooth ER)
3. Absorption, synthesis, and transport of fats (in intestinal cells)
4. Detoxification of drugs, certain pesticides, and carcinogens (in liver and kidneys)
5. Breakdown of stored glycogen to form free glucose (in liver cells especially)

Additionally, skeletal and cardiac muscle cells have an elaborate smooth ER (called the sarcoplasmic reticulum) that plays an important role in calcium ion storage and release during muscle contraction. Except for the examples given above, most body cells contain little, if any, true smooth ER.

Golgi Apparatus

The **Golgi apparatus** (gol′je) consists of stacked and flattened membranous sacs, shaped like hollow dinner plates, associated with swarms of tiny membranous vesicles **(Figure 3.19)**. The Golgi apparatus is the principal "traffic director" for cellular proteins. Its major function is to modify, concentrate, and package the proteins and lipids made at the rough ER.

The transport vesicles that bud off from the rough ER move to and fuse with the membranes at the convex *cis face*, the "receiving" side, of the Golgi apparatus. Inside the apparatus, the proteins are modified: Some sugar groups are trimmed while others are added, and in some cases, phosphate groups are added. The various proteins are "tagged" for delivery to a specific address, sorted, and packaged in at least three types of vesicles that bud from the concave *trans face* (the "shipping" side) of the Golgi stack.

Vesicles containing proteins destined for export pinch off from the trans face as **secretory vesicles**, or **granules**, which migrate to the plasma membrane and discharge their contents from the cell by exocytosis (pathway A, **Figure 3.20**). Specialized secretory cells, such as the enzyme-producing cells of the pancreas, have a very prominent Golgi apparatus. Besides its packaging-for-release function, the Golgi apparatus pinches off vesicles containing lipids and transmembrane proteins destined for a "home" in the plasma membrane (pathway B, Figure 3.20) or other membranous organelles. It also packages digestive

3

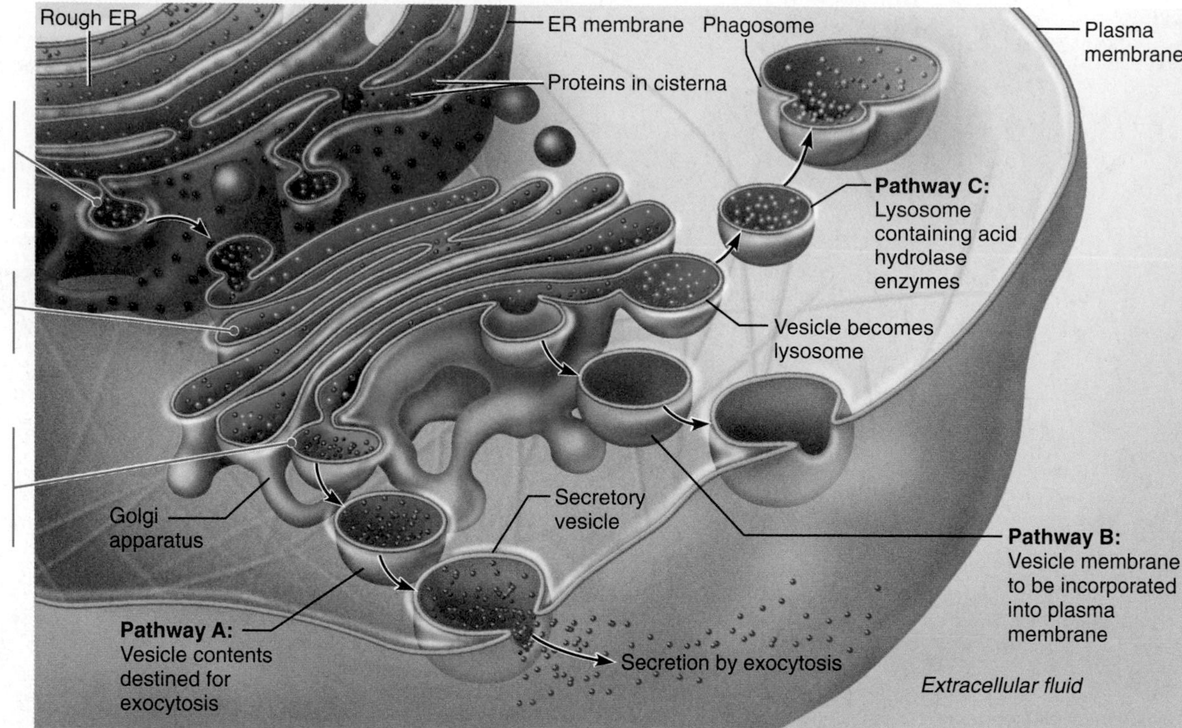
Rough ER
ER membrane
Phagosome
Plasma membrane
Proteins in cisterna
① Protein-containing vesicles pinch off rough ER and migrate to fuse with membranes of Golgi apparatus.
② Proteins are modified within the Golgi compartments.
③ Proteins are then packaged within different vesicle types, depending on their ultimate destination.
Pathway C: Lysosome containing acid hydrolase enzymes
Vesicle becomes lysosome
Golgi apparatus
Secretory vesicle
Pathway B: Vesicle membrane to be incorporated into plasma membrane
Pathway A: Vesicle contents destined for exocytosis
Secretion by exocytosis
Extracellular fluid

Figure 3.20 The sequence of events from protein synthesis on the rough ER to the final distribution of those proteins. The protein coats on the transport vesicles are not illustrated.

enzymes into membranous lysosomes that remain in the cell (pathway C in Figure 3.20, and discussed next).

Lysosomes

Born as endosomes which contain inactive enzymes, **lysosomes** ("disintegrator bodies") are spherical membranous organelles containing activated digestive enzymes **(Figure 3.21)**. As you might guess, lysosomes are large and abundant within phagocytes, the cells that dispose of invading bacteria and cell debris. Lysosomal enzymes can digest almost all kinds of biological molecules. They work best in acidic conditions and so are called *acid hydrolases.*

The lysosomal membrane is adapted to serve lysosomal functions in two important ways. First, it contains H^+ (proton) "pumps," which are ATPases that gather hydrogen ions from the surrounding cytosol to maintain the organelle's acidic pH. Second, it retains the dangerous acid hydrolases while permitting the final products of digestion to escape so that they can be used by the cell or excreted. In this way, lysosomes provide sites where digestion can proceed *safely* within a cell.

Lysosomes function as a cell's "demolition crew" by

1. Digesting particles taken in by endocytosis, particularly ingested bacteria, viruses, and toxins
2. Degrading worn-out or nonfunctional organelles
3. Performing metabolic functions, such as glycogen breakdown and release
4. Breaking down nonuseful tissues, such as the webs between the fingers and toes of a developing fetus and the uterine lining during menstruation
5. Breaking down bone to release calcium ions into the blood

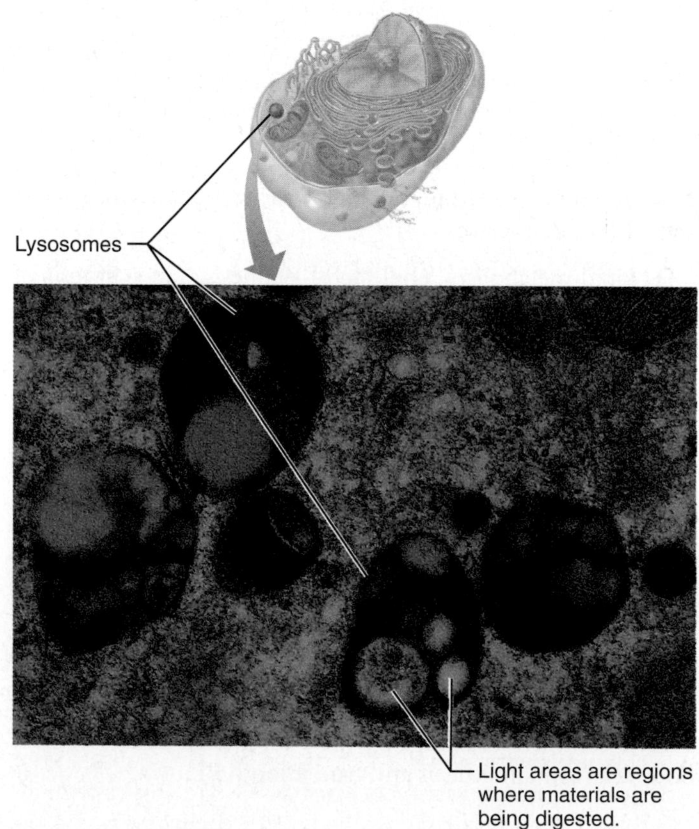
Lysosomes
Light areas are regions where materials are being digested.

Figure 3.21 Electron micrograph of a cell containing lysosomes (12,000×).

The lysosomal membrane is ordinarily quite stable, but it becomes fragile when the cell is injured or deprived of oxygen and when excessive amounts of vitamin A are present. When lysosomes rupture, the cell digests itself, a process called **autolysis** (aw″tol′ĭ-sis). Autolysis is the basis for desirable destruction of cells, as in the fourth point listed above.

HOMEOSTATIC IMBALANCE

Glycogen and certain lipids in the brain are degraded by lysosomes at a relatively constant rate. In *Tay-Sachs disease*, an inherited condition seen mostly in Jews from Central Europe, the lysosomes lack an enzyme needed to break down a glycolipid abundant in nerve cell membranes. As a result, the nerve cell lysosomes swell with the undigested lipids, which interfere with nervous system functioning. Affected infants typically have doll-like features and pink translucent skin. At 3 to 6 months of age, the first signs of disease appear (listlessness, motor weakness). These symptoms progress to mental retardation, seizures, blindness, and ultimately death within 18 months. ■

Summary of Interactions in the Endomembrane System

The **endomembrane system** is a system of organelles (most described above) that work together mainly (1) to produce, store, and export biological molecules, and (2) to degrade potentially harmful substances **(Figure 3.22)**. It includes the ER, Golgi apparatus, secretory vesicles, and lysosomes, as well as the nuclear membrane—that is, all of the membranous organelles or elements that are either structurally continuous or arise via forming or fusing transport vesicles. There are continuities between the nuclear envelope (itself an extension of the rough ER) and the rough and smooth ER (Figure 3.18). The plasma membrane, though not actually an *endo*membrane, is also functionally part of this system.

Besides these direct structural relationships, a wide variety of indirect interactions (indicated by arrows in Figure 3.22) occur among the members of the system. Some of the vesicles "born" in the ER migrate to and fuse with the Golgi apparatus or the plasma membrane, and vesicles arising from the Golgi apparatus can become part of the plasma membrane, secretory vesicles, or lysosomes.

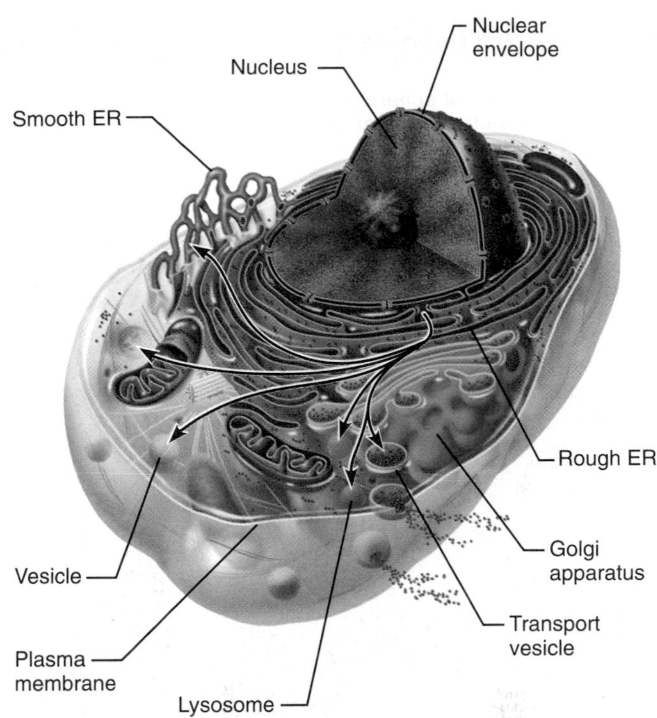

Figure 3.22 The endomembrane system.

Peroxisomes

Peroxisomes (pĕ-roks′ĭ-sōmz; "peroxide bodies") are membranous sacs containing a variety of powerful enzymes, the most important of which are oxidases and catalases. Oxidases use molecular oxygen (O_2) to detoxify harmful substances, including alcohol and formaldehyde. Their most important function is to neutralize dangerous **free radicals**, highly reactive chemicals with unpaired electrons that can scramble the structure of biological molecules. Oxidases convert free radicals to hydrogen peroxide, which is also reactive and dangerous but is quickly converted to water by catalase enzymes. Free radicals and hydrogen peroxide are normal by-products of cellular metabolism, but they have devastating effects on cells if allowed to accumulate. Peroxisomes are especially numerous in liver and kidney cells, which are very active in detoxification. A significant amount of fatty acid oxidation also occurs in peroxisomes. Hence, they also play a role in energy metabolism.

Peroxisomes look like small lysosomes (see Figure 3.2), and for many years it was thought that they were self-replicating organelles formed when existing peroxisomes simply pinch in half. Recent evidence, however, suggests that they form by budding off of the endoplasmic reticulum.

CHECK YOUR UNDERSTANDING

17. What organelle is the major site of ATP synthesis?
18. What are three organelles involved in protein synthesis and how do these organelles interact in that process?
19. How does the function of lysosomes compare to that of peroxisomes?

For answers, see Appendix G.

Cytoskeleton

▶ Name and describe the structure and function of cytoskeletal elements.

The **cytoskeleton**, literally, "cell skeleton," is an elaborate network of rods running through the cytosol. It acts as a cell's "bones," "muscles," and "ligaments" by supporting cellular structures and providing the machinery to generate various cell movements. The three types of rods in order of increasing size in the cytoskeleton are *microfilaments, intermediate filaments,* and *microtubules.* None of these is membrane covered.

3

TABLE 3.3 Parts of the Cell: Structure and Function

CELL PART	STRUCTURE	FUNCTIONS
Plasma Membrane (Figure 3.3)		
	Membrane made of a double layer of lipids (phospholipids, cholesterol, and so on) within which proteins are embedded. Proteins may extend entirely through the lipid bilayer or protrude on only one face. Externally facing proteins and some lipids have attached sugar groups.	Serves as an external cell barrier, and acts in transport of substances into or out of the cell. Maintains a resting potential that is essential for functioning of excitable cells. Externally facing proteins act as receptors (for hormones, neurotransmitters, and so on) and in cell-to-cell recognition.
Cytoplasm		
	Cellular region between the nuclear and plasma membranes. Consists of fluid **cytosol** containing dissolved solutes, **organelles** (the metabolic machinery of the cytoplasm), and **inclusions** (stored nutrients, secretory products, pigment granules).	
Cytoplasmic Organelles		
▪ Mitochondria (Figure 3.17)	Rodlike, double-membrane structures; inner membrane folded into projections called cristae.	Site of ATP synthesis; powerhouse of the cell.
▪ Ribosomes (Figures 3.18, 3.37–3.39)	Dense particles consisting of two subunits, each composed of ribosomal RNA and protein. Free or attached to rough endoplasmic reticulum.	The sites of protein synthesis.
▪ Rough endoplasmic reticulum (Figures 3.18, 3.39)	Membrane system enclosing a cavity, the cisterna, and coiling through the cytoplasm. Externally studded with ribosomes.	Sugar groups are attached to proteins within the cisternae. Proteins are bound in vesicles for transport to the Golgi apparatus and other sites. External face synthesizes phospholipids.
▪ Smooth endoplasmic reticulum (Figure 3.18)	Membranous system of sacs and tubules; free of ribosomes.	Site of lipid and steroid (cholesterol) synthesis, lipid metabolism, and drug detoxification.
▪ Golgi apparatus (Figures 3.19, 3.20)	A stack of smooth membrane sacs and associated vesicles close to the nucleus.	Packages, modifies, and segregates proteins for secretion from the cell, inclusion in lysosomes, and incorporation into the plasma membrane.
▪ Lysosomes (Figure 3.21)	Membranous sacs containing acid hydrolases.	Sites of intracellular digestion.
▪ Peroxisomes (Figure 3.2)	Membranous sacs of oxidase enzymes.	The enzymes detoxify a number of toxic substances. The most important enzyme, catalase, breaks down hydrogen peroxide.
▪ Microtubules (Figures 3.23–3.25)	Cylindrical structures made of tubulin proteins.	Support the cell and give it shape. Involved in intracellular and cellular movements. Form centrioles and cilia and flagella, if present.
▪ Microfilaments (Figures 3.23, 3.24)	Fine filaments composed of the protein actin.	Involved in muscle contraction and other types of intracellular movement, help form the cell's cytoskeleton.
▪ Intermediate filaments (Figure 3.23)	Protein fibers; composition varies.	The stable cytoskeletal elements; resist mechanical forces acting on the cell.
▪ Centrioles (Figure 3.25)	Paired cylindrical bodies, each composed of nine triplets of microtubules.	Organize a microtubule network during mitosis to form the spindle and asters. Form the bases of cilia and flagella.
Inclusions	Varied; includes stored nutrients such as lipid droplets and glycogen granules, protein crystals, pigment granules.	Storage for nutrients, wastes, and cell products.

TABLE 3.3 *(continued)*

CELL PART	STRUCTURE	FUNCTIONS
Cellular Extensions		
▪ Cilia (Figure 3.26, 3.27)	Short cell-surface projections; each cilium composed of nine pairs of microtubules surrounding a central pair.	Coordinated movement creates a unidirectional current that propels substances across cell surfaces.
▪ Flagella	Like a cilium, but longer; only example in humans is the sperm tail.	Propel the cell.
▪ Microvilli (Figure 3.28)	Tubular extensions of the plasma membrane; contain a bundle of actin filaments.	Increase surface area for absorption.
Nucleus (Figure 3.2, 3.29)		
	Largest organelle. Surrounded by the nuclear envelope; contains fluid nucleoplasm, nucleoli, and chromatin.	Control center of the cell; responsible for transmitting genetic information and providing the instructions for protein synthesis.
▪ Nuclear envelope (Figure 3.29)	Double-membrane structure pierced by pores. Outer membrane continuous with the endoplasmic reticulum.	Separates the nucleoplasm from the cytoplasm and regulates passage of substances to and from the nucleus.
▪ Nucleoli (Figure 3.29)	Dense spherical (non-membrane-bounded) bodies, composed of ribosomal RNA and proteins.	Site of ribosome subunit manufacture.
▪ Chromatin (Figure 3.30)	Granular, threadlike material composed of DNA and histone proteins.	DNA constitutes the genes.

chromosomes ("colored bodies") (Figure 3.30, bottom). Chromosome compactness prevents the delicate chromatin strands from tangling and breaking during the movements that occur during cell division. We describe the functions of DNA and the events of cell division in the next section.

Table 3.3 summarizes the parts of the cell.

CHECK YOUR UNDERSTANDING

23. If a cell ejects or loses its nucleus, what is its fate and why?
24. What is the role of nucleoli?
25. What is the importance of the histone proteins present in the nucleus?

For answers, see Appendix G.

Cell Growth and Reproduction

The Cell Life Cycle

▶ List the phases of the cell life cycle and describe the key events of each phase.

▶ Describe the process of DNA replication.

The **cell life cycle** is the series of changes a cell goes through from the time it is formed until it reproduces. The outer ring of **Figure 3.31** shows the two major periods of the cell cycle: *interphase* (in green), in which the cell grows and carries on its usual activities, and *cell division* or the *mitotic phase* (in yellow), during which it divides into two cells.

Interphase

Interphase is the period from cell formation to cell division. Early cytologists, unaware of the constant molecular activity in cells and impressed by the obvious movements of cell division, called interphase the resting phase of the cell cycle. (The term *interphase* reflects this idea of a stage *between* cell divisions.) However, this image is grossly misleading because during interphase a cell is carrying out all its routine activities and is "resting" only from dividing. Perhaps a more accurate name for this phase would be *metabolic phase* or *growth phase.*

Subphases In addition to carrying on its life-sustaining reactions, an interphase cell prepares for the next cell division. Interphase is divided into G_1, S, and G_2 subphases (the Gs stand for *gaps* before and after the S phase, and S is for *synthetic*). In all three subphases, the cell grows by producing proteins and organelles, but chromatin is reproduced only during the S subphase.

During **G_1 (gap 1)**, the cell is metabolically active, synthesizing proteins rapidly and growing vigorously (Figure 3.31, light green area). This is the most variable phase in terms of length. In cells that divide rapidly, G_1 typically lasts several minutes to hours, but in those that divide slowly, it may last for days or even

3

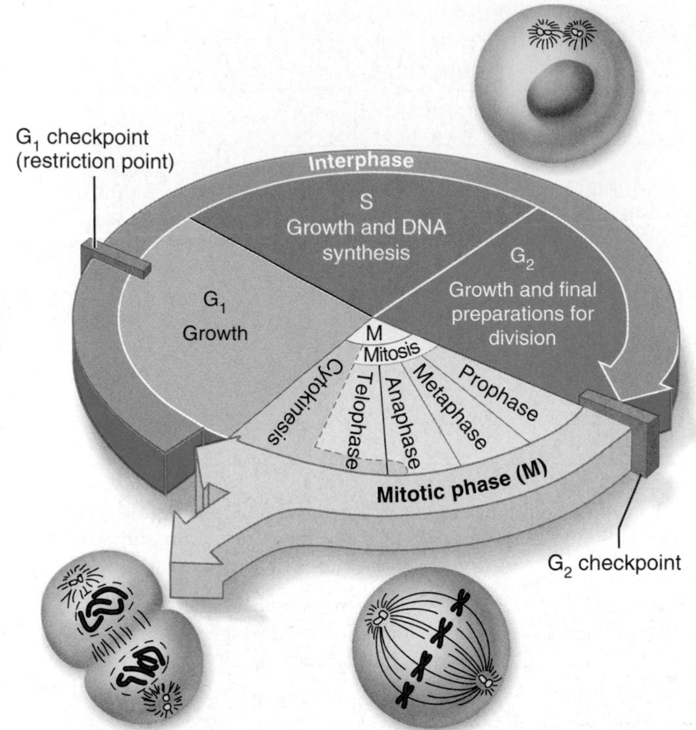

Figure 3.31 The cell cycle. During G_1, cells grow rapidly and carry out their routine functions. The S phase is the period of DNA synthesis. In G_2, materials needed for cell division are synthesized and growth continues. During the M phase (cell division), mitosis and cytokinesis occur, producing two daughter cells. Important checkpoints at which mitosis may be prevented from occurring are found throughout interphase; two are shown on the diagram.

years. Cells that permanently cease dividing are said to be in the **G_0 phase**. For most of G_1, virtually no activities directly related to cell division occur. However, as G_1 ends, the centrioles start to replicate in preparation for cell division.

During the **S phase**, DNA is replicated, ensuring that the two future cells being created will receive identical copies of the genetic material (Figure 3.31, blue area). New histones are made and assembled into chromatin. One thing is sure: Without a proper S phase, there can be no correct mitotic phase. (We will describe DNA replication next.)

The final phase of interphase, called G_2, is brief (Figure 3.31, dark green area). Enzymes and other proteins needed for division are synthesized and moved to their proper sites. By the end of G_2, centriole replication (begun in G_1) is complete. The cell is now ready to divide. Throughout S and G_2, the cell continues to grow and carries on with business as usual.

DNA Replication Before a cell can divide, its DNA must be replicated exactly, so that identical copies of the cell's genes can be passed on to each of its offspring. During the S phase, replication begins simultaneously on several chromatin threads and continues until all the DNA has been replicated.

Replication is still being studied but appears to involve the following events:

1. The DNA helices begin unwinding from the histones of the nucleosomes.
2. In an ATP-requiring process, a *helicase* enzyme untwists the double helix and gradually separates the DNA molecule into two complementary nucleotide chains, exposing the nitrogenous bases. The Y-shaped site of separation, where active DNA replication will soon begin, is called the *replication fork* **(Figure 3.32)**.
3. Each nucleotide strand then serves as a *template*, or set of instructions, for building a new complementary nucleotide strand from free DNA precursors dissolved in the nucleoplasm.
4. At sites where DNA synthesis is to occur, the needed "machinery" gradually accumulates until several different proteins (mostly enzymes) are present in a large complex called a **replisome**. Primase enzymes, which are part of the replisome, catalyze the formation of short (about ten bases long) **RNA primers**, which actually initiate DNA synthesis.
5. Once the primer is in place, the enzyme **DNA polymerase** comes into the picture. Continuing from the primer, it positions complementary nucleotides along the template strand and then covalently links them together. So, if the DNA polymerase encounters the sequence of bases GCT on the template strand, it assembles the bases CGA to bind to it. DNA polymerase works only in one direction. Consequently, one strand, the *leading strand*, is synthesized continuously (once primed by the RNA primer) following the movement of the replication fork. The other strand, called the *lagging strand*, is constructed in segments in the opposite direction and requires that a primer initiate replication of each segment. These primers are eventually replaced with DNA nucleotides by DNA polymerases.
6. The short segments of DNA are then spliced together by **DNA ligase**. The end result is that two DNA molecules are formed from the original DNA helix and are identical to it. Because each new molecule consists of one old and one new nucleotide strand, this mechanism of DNA replication is called **semiconservative replication** (Figure 3.32). Replication also involves the generation of two new *telomeres* (*tel* = end; *mer* = piece), snugly fitting nucleoprotein caps that prevent degradation of the ends of the chromatin strands.

As soon as replication ends, histones (synthesized in the cytoplasm and imported into the nucleus) associate with the DNA, completing the formation of two new chromatin strands. The chromatin strands, united by a buttonlike centromere (a stretch of repetitive DNA), remain attached, held together by the centromere and a protein complex called *cohesin*, until the cell enters the anaphase stage of mitotic cell division (see p. 99). They are then distributed to the daughter cells as described next, ensuring that each cell has identical genetic information.

The progression from DNA synthesis into the events of cell division presumes that the newly synthesized DNA is not damaged or broken in any way. If damage occurs, progress through

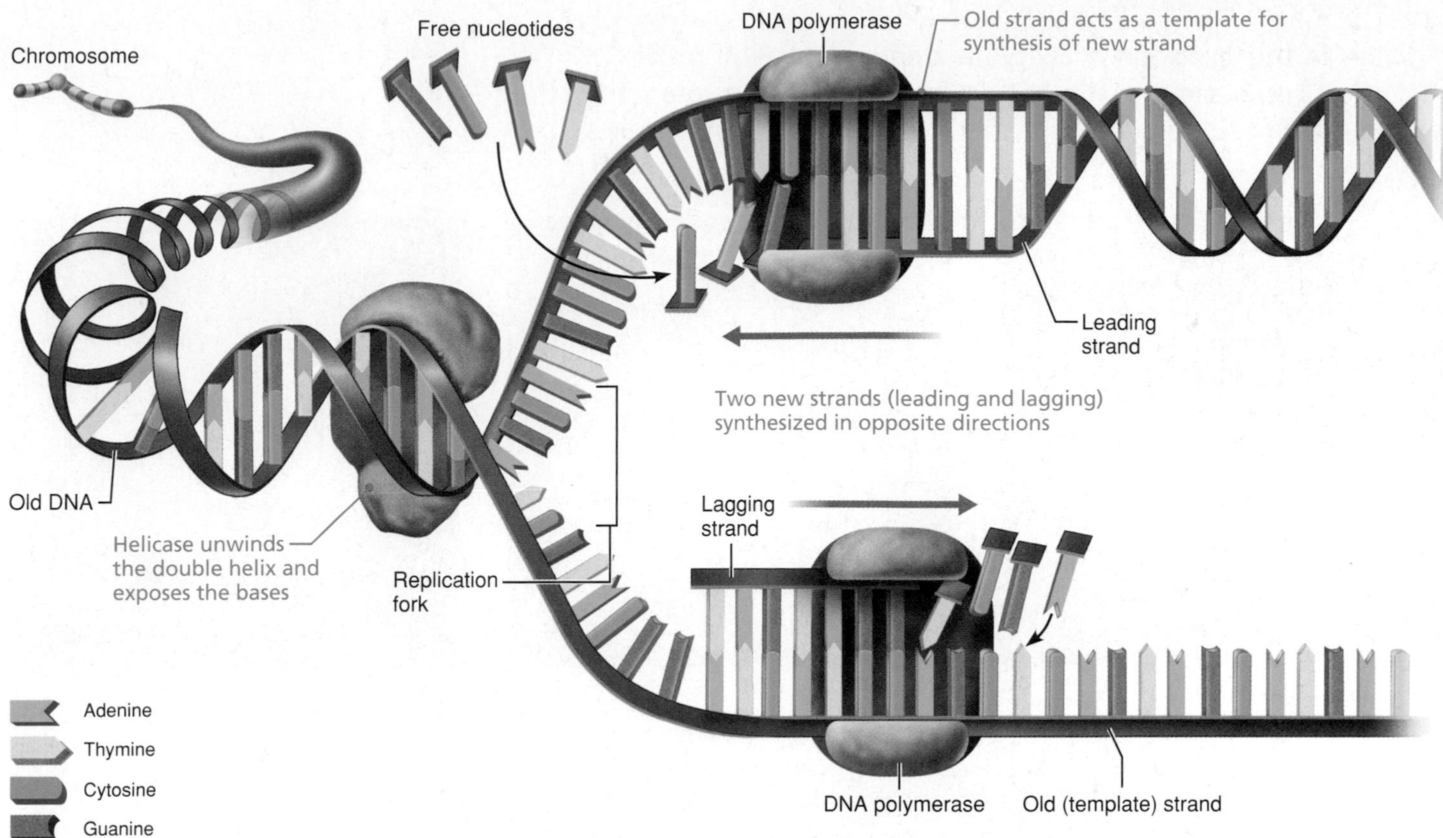

Figure 3.32 Replication of DNA. The DNA helix uncoils, and the hydrogen bonds between its base pairs are broken. Then, each nucleotide strand of the DNA acts as a template for constructing a complementary strand, as illustrated on the right-hand side of the diagram. DNA polymerases work in one direction only, so the two new strands (leading and lagging) are synthesized in opposite directions. (The DNA ligase enzymes that join the DNA fragments on the lagging strand are not illustrated.) Each DNA molecule formed consists of one old (template) strand and one newly assembled strand and constitutes a chromatid of a chromosome.

the cell cycle is arrested until the DNA repair mechanism has fixed the problem.

Cell Division

Cell division is essential for body growth and tissue repair. Cells that continually wear away, such as cells of the skin and intestinal lining, reproduce themselves almost continuously. Others, such as liver cells, divide more slowly (to maintain the particular size of the organ they compose) but retain the ability to reproduce quickly if the organ is damaged. Most cells of nervous tissue, skeletal muscle, and heart muscle lose their ability to divide when they are fully mature, and repairs are made with scar tissue (a fibrous type of connective tissue).

Events of Cell Division In most body cells, cell division, which is called the **M (mitotic) phase** of the cell life cycle, involves two distinct events: **mitosis** (mi-to′sis; *mit* = thread; *osis* = process), or division of the nucleus, and **cytokinesis** (si-to-kĭ-ne′sis; *kines* = movement), or division of the cytoplasm (see Figure 3.31, yellow area). A somewhat different process of nuclear division called *meiosis* (mi-o′sis) produces sex cells (ova and sperm) with only half the number of genes found in other body cells. We discuss the details of meiosis in Chapter 27. Here we concentrate on mitotic cell division.

Mitosis Mitosis is the series of events that parcel out the replicated DNA of the mother cell to two daughter cells. It is described in terms of four phases: **prophase**, **metaphase**, **anaphase**, and **telophase**, but it is actually a continuous process, with one phase merging smoothly into the next. Its duration varies according to cell type, but in human cells it typically lasts about an hour or less from start to finish. *Focus on Mitosis* (**Figure 3.33**), pp. 98–99, describes the phases of mitosis in detail.

Cytokinesis Cytokinesis, or the division of the cytoplasm, begins during late anaphase and is completed after mitosis ends. The plasma membrane over the center of the cell (the spindle equator) is drawn inward to form a **cleavage furrow** by the activity of a *contractile ring* (Figure 3.33) made of actin filaments. The furrow deepens until the cytoplasmic mass is pinched into two parts, so that at the end of cytokinesis there are two daughter cells. Each is smaller and has less cytoplasm than the mother

Figure 3.33 FOCUS **Mitosis**

Mitosis is the process of nuclear division in which the chromosomes are distributed to two daughter nuclei. Together with cytokinesis, it produces two identical daughter cells.

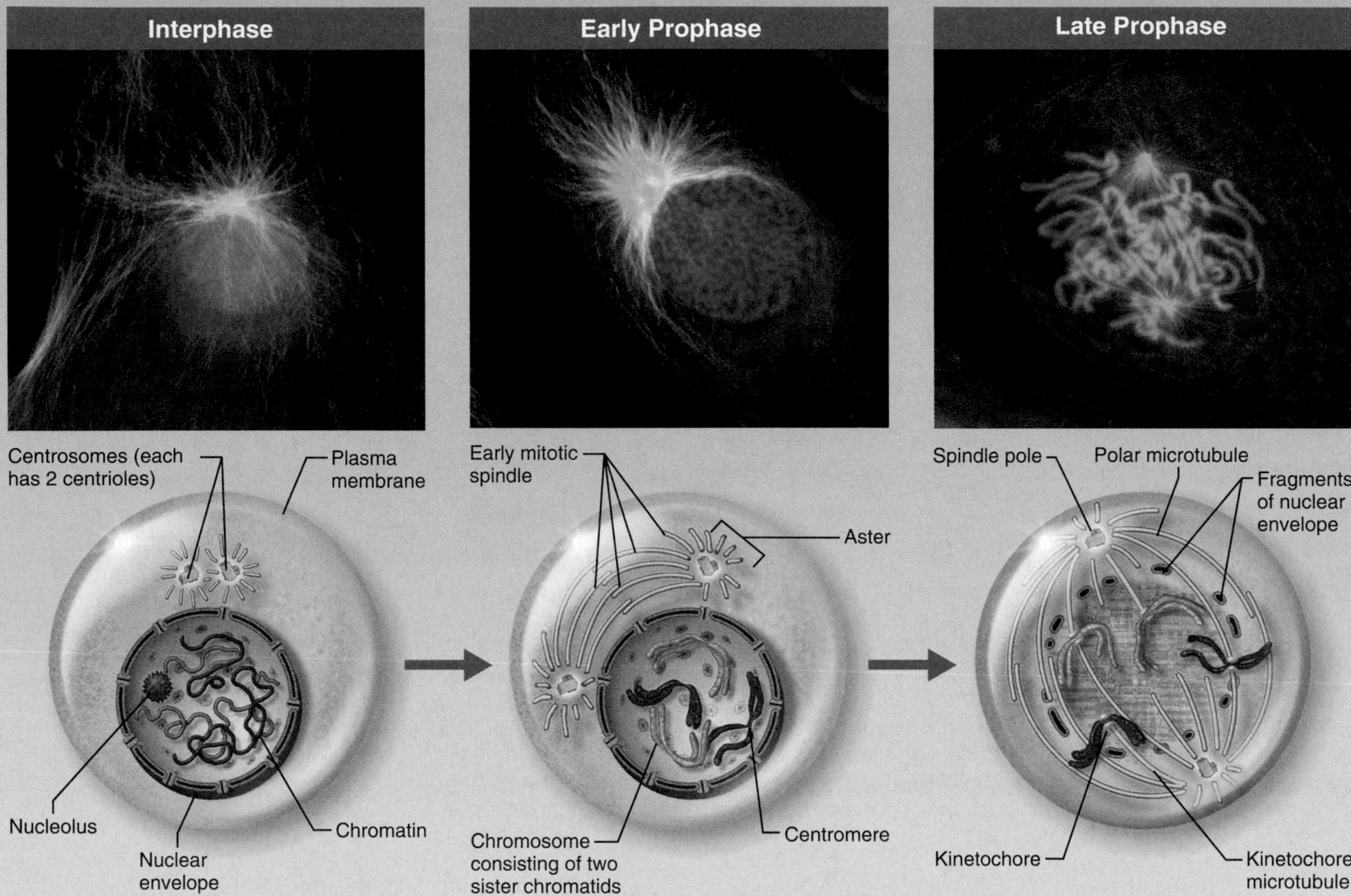

Interphase

Interphase is the period of a cell's life when it carries out its normal metabolic activities and grows.

- During interphase, the DNA-containing material is in the form of chromatin. The nuclear envelope and one or more nucleoli are intact and visible.
- There are three distinct periods of this phase:
  G_1: The centrioles begin replicating.
  S: DNA is replicated.
  G_2: Final preparations for mitosis are completed and centrioles finish replicating.

The light micrographs show dividing lung cells from a newt. The chromosomes appear blue and the microtubules green. (The red fibers are intermediate filaments.) The schematic drawings show details not visible in the micrographs. For simplicity, only four chromosomes are drawn.

Early Prophase

- The chromatin condenses, forming barlike *chromosomes* that are visible with a light microscope.
- Each duplicated chromosome appears as two identical threads, now called *sister chromatids*, held together at a small, constricted region called a *centromere*. (After the chromatids separate, each is considered a new chromosome.)
- As the chromosomes appear, the nucleoli disappear, and the two centrosomes separate from one another.
- The centrosomes act as focal points for growth of a microtubule assembly called the **mitotic spindle**. As these microtubules lengthen, they propel the centrosomes toward opposite ends (poles) of the cell.
- Microtubule arrays called *asters* ("stars") are seen extending from the matrix around the centrosomes.

Late Prophase

- While the centrosomes are still moving apart, the nuclear envelope fragments, allowing the spindle to interact with the chromosomes.
- Some of the growing spindle microtubules attach to *kinetochores* (ki-ne′-to-korz), special protein structures at each chromosome's centromere. Such microtubules are called *kinetochore microtubules*.
- The remaining spindle microtubules (not attached to any chromosomes) are called *polar microtubules*. The microtubules slide past each other, forcing the poles apart.
- The kinetochore microtubules pull on each chromosome from both poles in a kind of tug-of-war that ultimately draws the chromosomes to the exact center or equator of the cell.

A&P Flix *View this animation at myA&P*

Metaphase

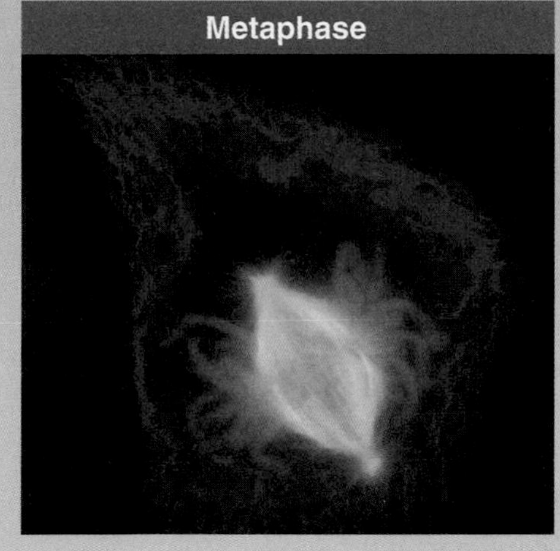

Anaphase

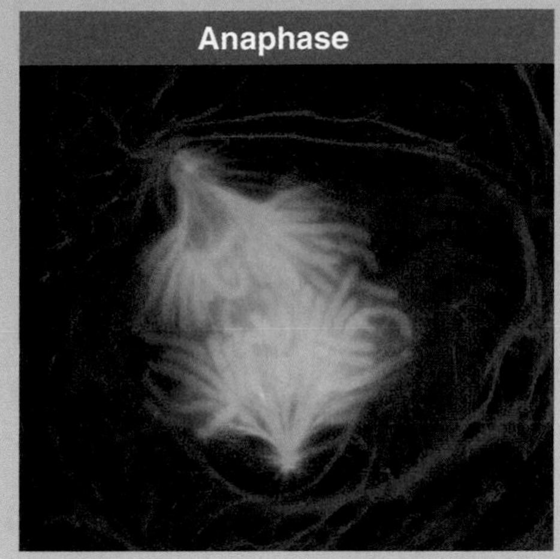

Telophase and Cytokinesis

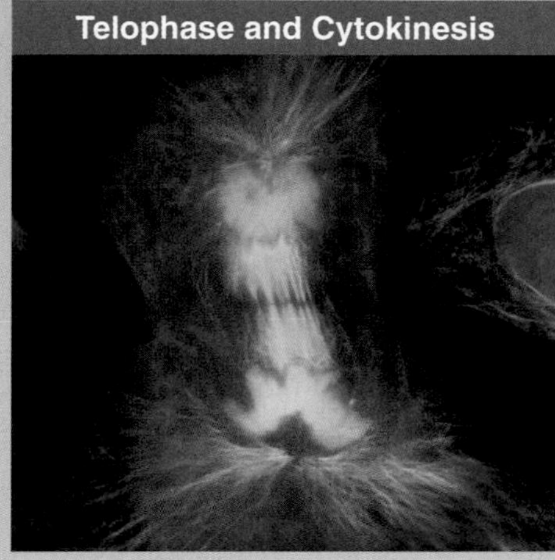

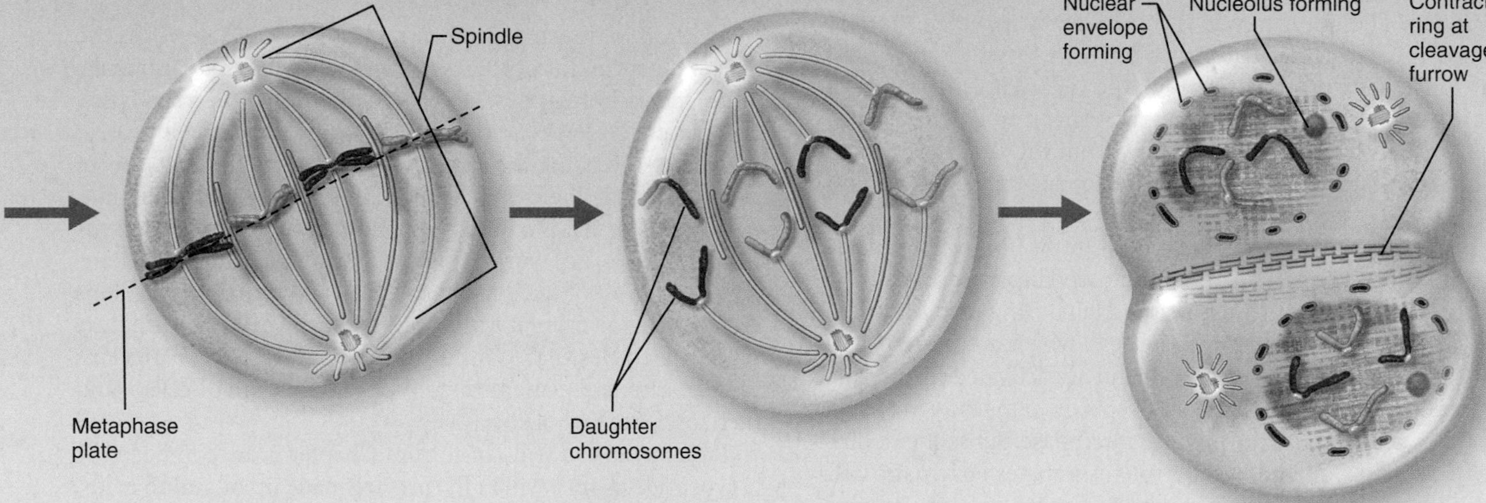

Metaphase
Metaphase is the second phase of mitosis.

- The two centrosomes are at opposite poles of the cell.
- The chromosomes cluster at the middle of the cell, with their centromeres precisely aligned at the *equator* of the spindle. This imaginary plane midway between the poles is called the *metaphase plate*.

Anaphase
Anaphase is the third and shortest phase of mitosis. Anaphase begins abruptly as the centromeres of the chromosomes split simultaneously. Each chromatid now becomes a chromosome in its own right.

- The kinetochore microtubules, moved along by motor proteins in the kinetochores, gradually pull each chromosome toward the pole it faces.
- At the same time, the polar microtubules slide past each other, lengthen, and push the two poles of the cell apart.
- Anaphase is easy to recognize because the moving chromosomes look V shaped. The centromeres lead the way, and the chromosomal "arms" dangle behind them.
- This process of moving and separating the chromosomes is helped by the fact that the chromosomes are short, compact bodies. Diffuse threads of chromatin would tangle, trail, and break, resulting in imprecise "parceling out" to the daughter cells.

Telophase
Telophase begins as soon as chromosomal movement stops. This final phase is like prophase in reverse.

- The identical sets of chromosomes at the opposite poles of the cell uncoil and resume their threadlike chromatin form.
- A new nuclear envelope forms around each chromatin mass, nucleoli reappear within the nuclei, and the spindle breaks down and disappears.
- Mitosis is now ended. The cell, for just a brief period, is binucleate (has two nuclei) and each new nucleus is identical to the original mother nucleus.

Cytokinesis
- As a rule, as mitosis draws to a close, cytokinesis completes the division of the cell into two identical daughter cells. Cytokinesis occurs as a contractile ring of actin microfilaments forms the *cleavage furrow* and pinches the cell apart. It begins during late anaphase and continues through and beyond telophase.

cell, but is genetically identical to it. The daughter cells then enter the interphase portion of the life cycle until it is their turn to divide.

Control of Cell Division The signals that prod cells to divide are incompletely understood, but we know that the ratio of cell surface area to cell volume is important. The amount of nutrients a growing cell requires is directly related to its volume. Volume increases with the cube of cell radius, whereas surface area increases more slowly with the square of the radius. For example, a 64-fold (4^3) increase in cell volume is accompanied by only a 16-fold (4^2) increase in surface area. Consequently, the surface area of the plasma membrane becomes inadequate for nutrient and waste exchange when a cell reaches a certain critical size. Cell division solves this problem because the smaller daughter cells have a favorable ratio of surface area to volume. These surface-volume relationships help explain why most cells are microscopic in size.

Two other factors that influence when cells divide are chemical signals (growth factors, hormones, and others) released by other cells and the availability of space. Normal cells stop proliferating when they begin touching, a phenomenon called *contact inhibition.* The cell life cycle is controlled by a system that has been compared to an automatic washer's timer control. Like that timer, the control system for the cell cycle is driven by a built-in clock. However, just as the washer's cycle is subject to adjustments (by regulating the flow from the faucet, say, or by an internal water-level sensor), the cell cycle is regulated by both internal and external factors.

Two groups of proteins are crucial to the ability of a cell to accomplish the S phase and enter mitosis. They are **cyclins** (regulatory proteins whose levels rise and fall during each life cycle) and **Cdks (cyclin-dependent kinases)**, which are present in a constant concentration in the cell and are activated by binding to particular cyclins. In response to specific signals, a new batch of cyclins accumulates during each interphase. Subsequent joining of specific Cdk and cyclin proteins initiates enzymatic cascades that phosphorylate histones and other proteins needed for the various stages of cell division. At the end of mitosis, the cyclins are abruptly destroyed by enzymes.

A number of "switches" and crucial checkpoints for cell division occur throughout interphase. These built-in stop signals halt the cell cycle until overridden by internal or external go-ahead signals. In many cells, a G_1 checkpoint, called the *restriction point,* seems to be most important (see Figure 3.31). If the cell is diverted from dividing at this checkpoint, it enters the nondividing state (G_0). Another important checkpoint, and the first to be understood, occurs late in G_2, when a threshold amount of a protein complex called **MPF (M-phase promoting factor)** is required to give the okay signal to pass the G_2 checkpoint and enter the M phase. Later in M phase, MPF is inactivated.

Besides these "go" signals, there are a number of so-called repressor genes that inhibit cell division. One example is the *p53* gene that initiates a series of enzymatic events that produce growth-inhibiting factors. Roughly half of all cancers have abnormal *p53* genes. These cancers are not inhibited by contact with other cells and divide wildly, making them dangerous to their host.

CHECK YOUR UNDERSTANDING

26. If one of the DNA strands being replicated "reads" CGAATG, what will be the base sequence of the corresponding DNA strand?
27. During what phase of the cell cycle is DNA synthesized?
28. What are three events occurring in prophase that are undone in telophase?

For answers, see Appendix G.

Protein Synthesis

- **Define gene and genetic code and explain the function of genes.**
- **Name the two phases of protein synthesis and describe the roles of DNA, mRNA, tRNA, and rRNA in each phase.**
- **Contrast triplets, codons, and anticodons.**

In addition to directing its own replication, DNA serves as the master blueprint for protein synthesis. Although cells also make lipids and carbohydrates, DNA does not dictate their structure. Historically, DNA is said to specify *only* the structure of protein molecules, including the enzymes that catalyze the synthesis of all classes of biological molecules. Most of the metabolic machinery of the cell is concerned in some way with protein synthesis. This is not surprising, seeing as structural proteins constitute most of the dry cell material, and functional proteins direct almost all cellular activities. Essentially, cells are miniature protein factories that synthesize the huge variety of proteins that determine the chemical and physical nature of cells—and therefore of the whole body.

Proteins, as you will recall from Chapter 2, are composed of polypeptide chains, which in turn are made up of amino acids. For purposes of this discussion, we can define a **gene** as a segment of a DNA molecule that carries instructions for creating one polypeptide chain. (Note, however, that some genes specify the structure of certain varieties of RNA as their final product.)

The four nucleotide bases (A, G, T, and C) are the "letters" used in the genetic dictionary, and the information of DNA is found in the sequence of these bases. Each sequence of three bases, called a **triplet**, can be thought of as a "word" that specifies a particular amino acid. For example, the triplet AAA calls for the amino acid phenylalanine, and CCT calls for glycine. The sequence of triplets in each gene forms a "sentence" that tells exactly how a particular polypeptide is to be made: It specifies the number, kinds, and order of amino acids needed to build a particular polypetide.

Variations in the arrangement of A, T, C, and G allow our cells to make all the different kinds of proteins needed. Even an unusually "small" gene has an estimated 210 base pairs in sequence. The ratio between DNA bases in the gene and amino acids in the polypeptide is 3:1 (because each triplet stands for one amino acid), so we would expect the polypeptide specified by such a gene to contain 70 amino acids.

Most genes of higher organisms contain **exons**, which are amino acid–specifying informational sequences. These exons are often separated by **introns**, which are noncoding, often repetitive, segments that range from 60 to 100,000 nucleotides. Once considered a type of "junk DNA," intron DNA is now believed to represent a genome scrapyard that provides a reservoir of ready-to-use DNA segments for genome evolution, as well as a rich source of a large variety of small RNA molecules. The rest of the DNA (the great majority of it) is essentially "dark matter" whose function is a mystery. It is in these regions that *pseudogenes* (false genes) are found. Pseudogenes "look like" real genes but have deficits that render them apparently functionless.

The Role of RNA

By itself, DNA is like a CD recording: The information it contains cannot be used without a decoding mechanism (a CD player). Furthermore, most polypeptides are manufactured at ribosomes in the cytoplasm, but in interphase cells, DNA never leaves the nucleus. So, DNA requires not only a decoder, but a messenger as well. The decoding and messenger functions are carried out by RNA, the second type of nucleic acid.

As you learned in Chapter 2, RNA differs from DNA in being single stranded and in having the sugar ribose instead of deoxyribose and the base uracil (U) in place of thymine (T). Three forms of RNA typically act together to carry out DNA's instructions for polypeptide synthesis:

1. **Messenger RNA (mRNA)**, relatively long nucleotide strands resembling "half-DNA" molecules (one of the two strands of a DNA molecule coding for protein structure)
2. **Ribosomal RNA (rRNA)**, part of the ribosomes
3. **Transfer RNA (tRNA)**, small, roughly L-shaped molecules

All types of RNA are formed on the DNA in the nucleus in much the same way as DNA replicates itself: The DNA helix separates and one of its strands serves as a template for synthesizing a complementary RNA strand. Once formed, the RNA molecule is released from the DNA template and migrates into the cytoplasm. Its job done, the DNA simply recoils into its helical, inactive form.

Approximately 2% of the nuclear DNA codes for the synthesis of short-lived mRNA, so named because it carries, from gene to ribosome, the "message" containing the instructions for building a polypeptide. (Kind of a genetic memo for protein structure.) DNA in the nucleolar organizer regions (mentioned previously) codes for the synthesis of rRNA, which is long-lived and stable, as is tRNA coded by other DNA sequences. Because rRNA and tRNA do not transport codes for synthesizing other molecules, they are the final products of the genes that code for them. Ribosomal RNA and tRNA act together to "translate" the message carried by mRNA.

Essentially, polypeptide synthesis involves two major steps: (1) *transcription*, in which DNA's information is encoded in mRNA, and (2) *translation*, in which the information carried by mRNA is decoded and used to assemble polypeptides. Figure 3.34 shows an overview of the information flow in these two major steps. The figure also indicates the "RNA processing" that removes introns from mRNA before this molecule leaves the nucleus and moves into the cytoplasm.

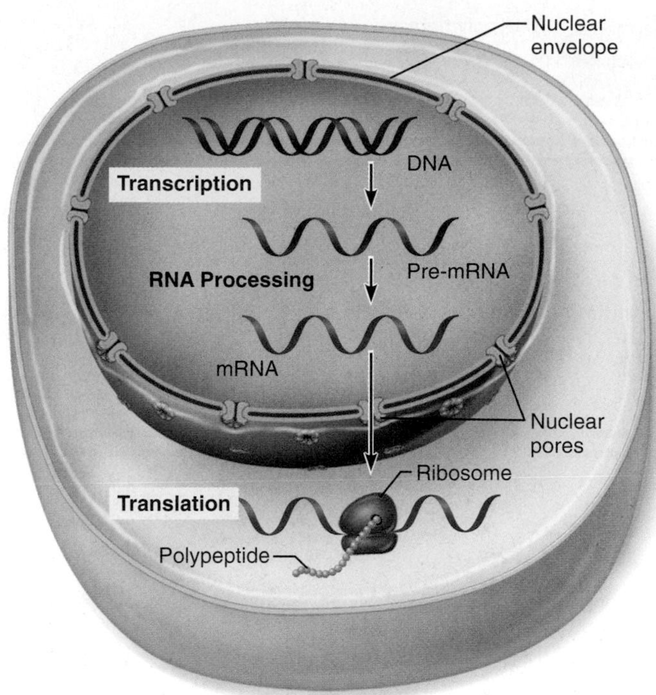

Figure 3.34 Simplified scheme of information flow from the DNA gene to mRNA to protein structure during transcription and translation. (Note that mRNA is first synthesized as pre-mRNA, which is processed by enzymes before leaving the nucleus.)

Transcription

A transcriptionist converts a message from a recording or shorthand notes into a word-processed or electronic copy. In other words, information is transformed, or transferred, from one form or format to another. In cells, **transcription** involves the transfer of information from a DNA's base sequence to the complementary base sequence of an mRNA molecule. The form is different, but the same information is being conveyed. Once the mRNA molecule is made, it detaches and leaves the nucleus via a nuclear pore.

Essentially, three basic phases are involved in transcription: (1) initiation, (2) elongation, and (3) termination. In addition to looking at the synthesis of mRNA, we will also examine the editing and further processing needed to clean up the mRNA transcript.

How does transcription get started? Transcription cannot begin until gene-activating chemicals called *transcription factors* stimulate loosening of the histones at the site-to-be of gene

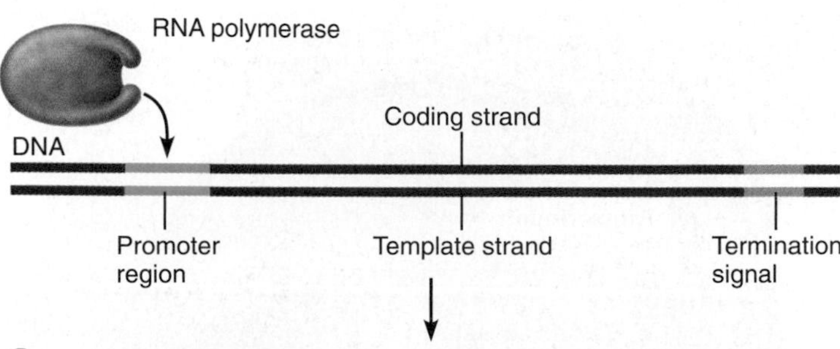

① **Initiation:** With the help of transcription factors, RNA polymerase binds to the promoter, pries apart the two DNA strands, and initiates mRNA synthesis at the start point on the template strand.

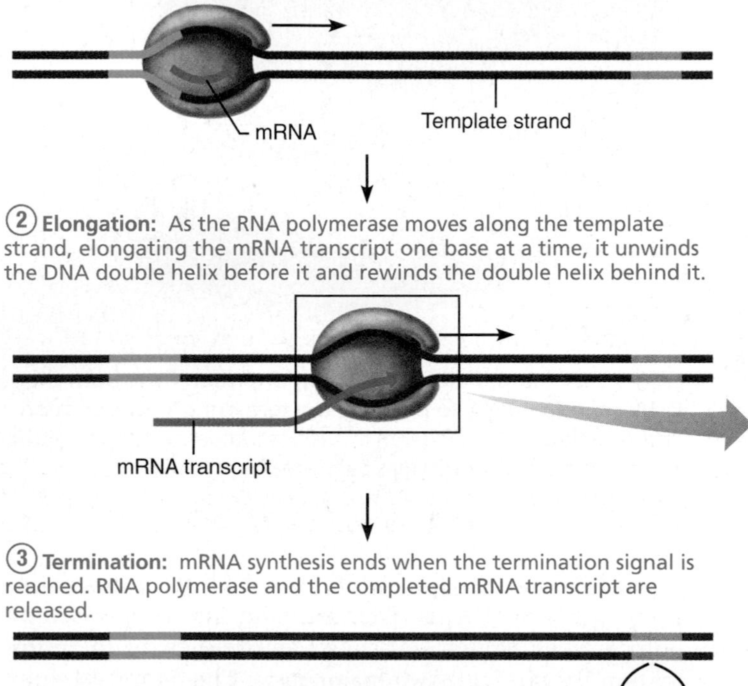

② **Elongation:** As the RNA polymerase moves along the template strand, elongating the mRNA transcript one base at a time, it unwinds the DNA double helix before it and rewinds the double helix behind it.

③ **Termination:** mRNA synthesis ends when the termination signal is reached. RNA polymerase and the completed mRNA transcript are released.

Completed mRNA transcript

RNA polymerase

Coding strand of DNA

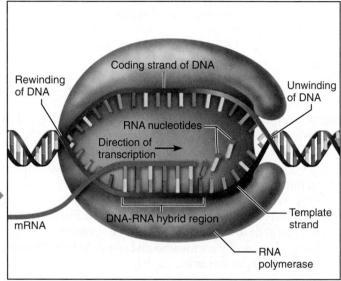

Rewinding of DNA

Unwinding of DNA

RNA nucleotides

Direction of transcription

mRNA

DNA-RNA hybrid region

Template strand

RNA polymerase

The DNA-RNA hybrid: At any given moment, 16–18 base pairs of DNA are unwound and the most recently made RNA is still bound to DNA. This small region is called the DNA-RNA hybrid.

Figure 3.35 Overview of stages of transcription.

transcription and then bind to the promoter. The **promoter** is a special DNA sequence that contains the *start point* (beginning of the structural gene to be transcribed). It specifies where mRNA synthesis starts and which DNA strand is going to serve as the *template strand* (Figure 3.35, top). The uncoiled DNA strand not used as a template is called the *coding strand* because it has the same (coded) sequence as the mRNA to be built (except for the use of U in mRNA in place of T in DNA). The transcription factors also help position **RNA polymerase**, the enzyme that oversees the synthesis of mRNA, correctly at the promoter. Once these preparations are made, RNA polymerase can initiate transcription.

Figure 3.35 illustrates the following steps in transcription:

① **Initiation**. Once bound with the help of the transcription factors, RNA polymerase pulls apart the strands of the DNA double helix so that transcription can begin at the start point in the promoter.

② **Elongation**. Using incoming RNA nucleotides as substrates, the RNA polymerase aligns them with complementary DNA bases on the template strand and then links them together. As RNA polymerase elongates the mRNA strand one base at a time, it unwinds the DNA helix in front of it, and rewinds the helix behind it. At any given moment, 16 to 18 base pairs of DNA are unwound and the most recently made mRNA is still hydrogen-bonded (H-bonded) to the template DNA. This small region—called the **DNA-RNA hybrid**—is up to 12 base pairs long.

③ **Termination**. When the polymerase reaches a special base sequence called a **termination signal**, transcription ends and the newly formed mRNA pulls off the DNA template.

Processing of mRNA Although it would appear that translation can begin as soon as the mRNA is made, that is not the case. The process is a bit more complex. Recall that mammalian DNA like ours has coding regions (exons) separated by non-protein-coding regions (introns). Because the DNA is transcribed sequentially, the mRNA initially made, called *pre-mRNA* or *primary transcript,* is littered with intron "junk" or "nonsense." Before the newly formed RNA can be used as a messenger, it must be processed, or edited—that is, sections corresponding to introns must be removed. This job is done by *spliceosomes.* These large RNA-protein complexes snip out the introns and splice together the remaining exon-coded sections in the order in which they occurred in the DNA, producing the functional mRNA that directs translation at the ribosome.

Although many introns naturally degrade, some contain active segments (such as microRNAs) that can function to control, interfere with, or silence other genes. Additionally, before the edited mRNA can function in protein synthesis, a number of specific RNA-binding proteins, called *mRNA complex proteins,* must become associated with it. These mRNA complex proteins guide its export from the nucleus, determine its localization, translation, and stability, and check it for premature termination codons.

Translation

A translator takes a message in one language and restates it in another. In the **translation** step of protein synthesis, the language of nucleic acids (base sequence) is translated into the language of proteins (amino acid sequence).

The rules by which the base sequence of a gene is translated into an amino acid sequence are called the **genetic code**. For each triplet, or three-base sequence on DNA, the corresponding three-base sequence on mRNA is called a **codon**. Since there are four kinds of RNA (or DNA) nucleotides, there are 4^3, or 64, possible codons. Three of these 64 codons are "stop signs" that call for termination of polypeptide synthesis. All the rest code for amino acids. Because there are only about 20 amino acids, some are specified by more than one codon. This redundancy in the genetic code helps protect against problems due to transcription (and translation) errors. The genetic code and a complete codon list are provided in **Figure 3.36**.

The process of translation occurs in the cytoplasm and involves the mRNAs, tRNAs, and rRNAs mentioned above, and occurs in the sequence of events described next and illustrated in **Figure 3.37**.

When it reaches the cytoplasm after processing in the nucleus, the mRNA molecule carrying instructions for a particular protein binds to a small ribosomal subunit (Figure 3.37, ①). Then tRNA comes into the picture, its job being to *transfer* amino acids, dissolved in the cytosol, to the ribosome.

Shaped like a handheld drill, tRNA is well suited to its dual function of binding to both an amino acid and an mRNA codon. The amino acid is bound to one end of tRNA, at a region called the stem. At the other end, the head, is its **anticodon** (an″ti-ko′don), a three-base sequence complementary to the mRNA codon calling for the amino acid carried by that particular tRNA. Because anticodons form hydrogen bonds with complementary codons, tRNA is the link between the language of nucleic acids and the language of proteins. For example, if the mRNA codon is AUA, which specifies isoleucine, the tRNAs carrying isoleucine will have the anticodon UAU, which can bind to the AUA codons.

There are approximately 45 different types of tRNA, each capable of binding with a specific amino acid. The attachment

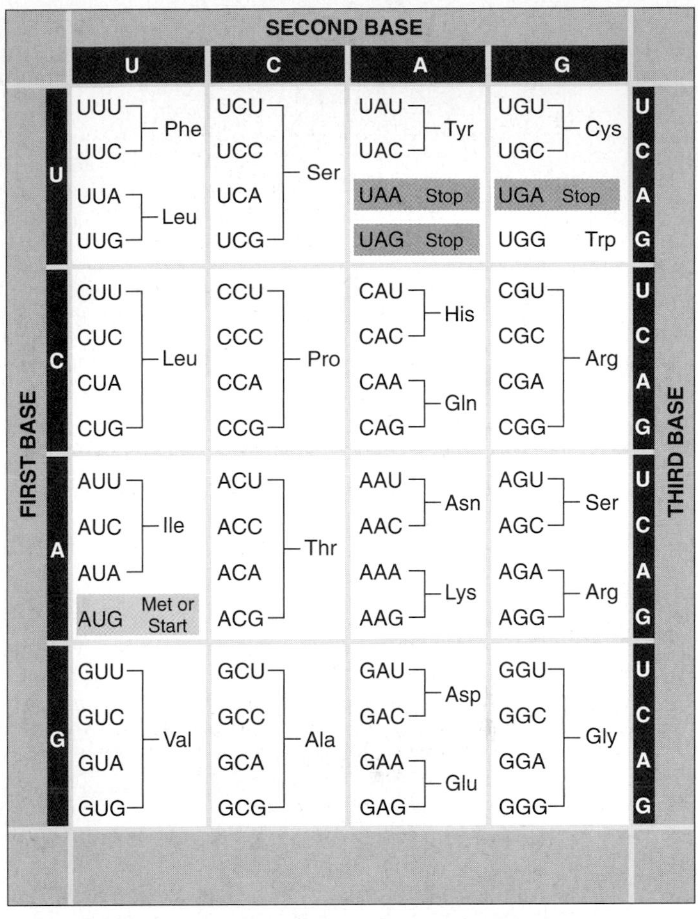

Figure 3.36 The genetic code. The three bases in an mRNA codon are designated as the first, second, and third. Each set of three specifies a particular amino acid, represented here by an abbreviation (see list below). The codon AUG (which specifies the amino acid methionine) is the usual start signal for protein synthesis. The word *stop* indicates the codons that serve as signals to terminate protein synthesis.

Abb.*	Amino acid	Abb.*	Amino acid
Ala	alanine	Leu	leucine
Arg	arginine	Lys	lysine
Asn	asparagine	Met	methionine
Asp	aspartic acid	Phe	phenylalanine
Cys	cysteine	Pro	proline
Glu	glutamic acid	Ser	serine
Gln	glutamine	Thr	threonine
Gly	glycine	Trp	tryptophan
His	histidine	Tyr	tyrosine
Ile	isoleucine	Val	valine

*Abbreviation for the amino acid

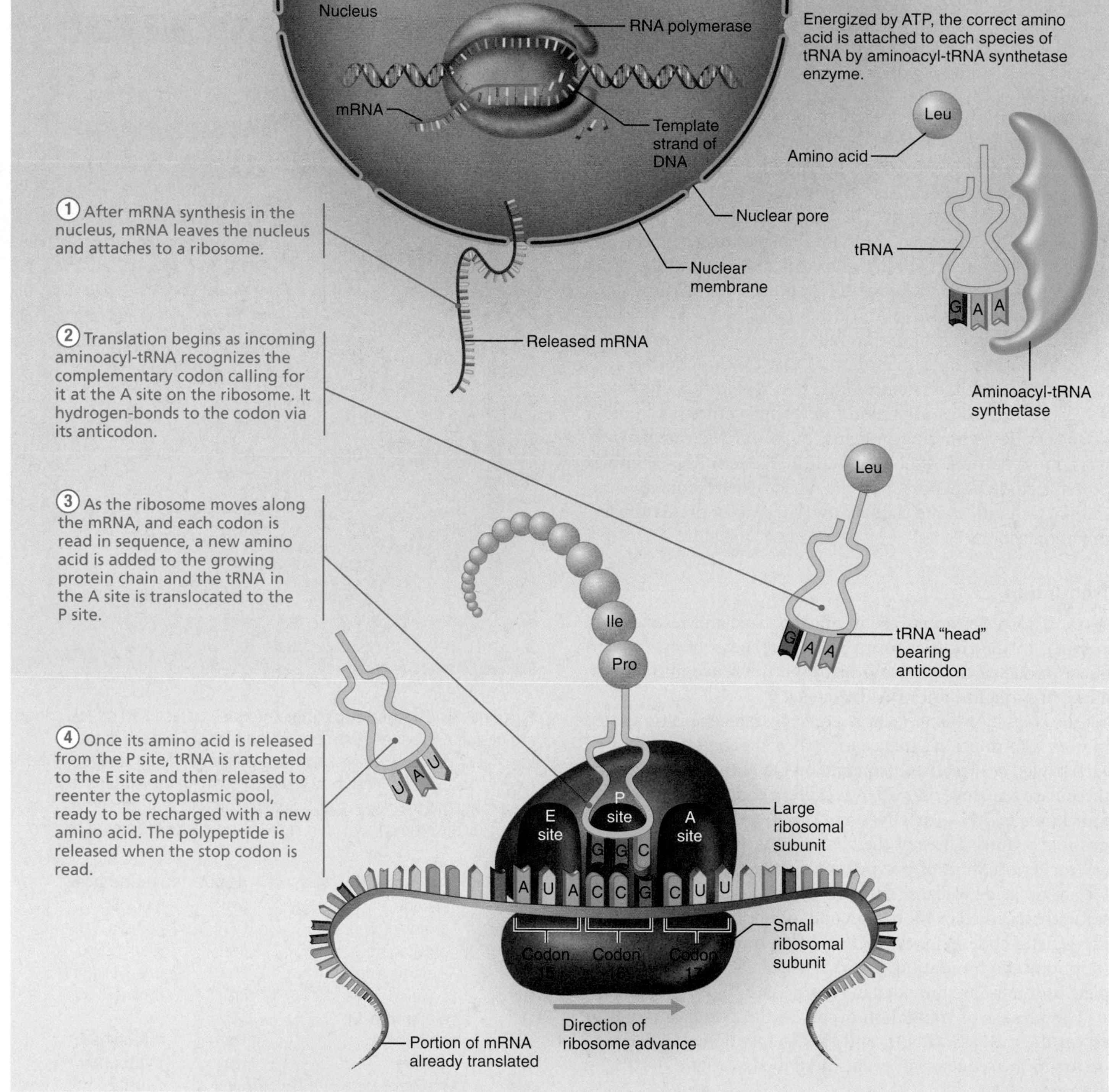

Nucleus
RNA polymerase
mRNA
Template strand of DNA
Nuclear pore
Nuclear membrane
Energized by ATP, the correct amino acid is attached to each species of tRNA by aminoacyl-tRNA synthetase enzyme.
Leu
Amino acid
tRNA
G A A
Aminoacyl-tRNA synthetase
① After mRNA synthesis in the nucleus, mRNA leaves the nucleus and attaches to a ribosome.
② Translation begins as incoming aminoacyl-tRNA recognizes the complementary codon calling for it at the A site on the ribosome. It hydrogen-bonds to the codon via its anticodon.
Released mRNA
Leu
③ As the ribosome moves along the mRNA, and each codon is read in sequence, a new amino acid is added to the growing protein chain and the tRNA in the A site is translocated to the P site.
Ile
Pro
tRNA "head" bearing anticodon
④ Once its amino acid is released from the P site, tRNA is ratcheted to the E site and then released to reenter the cytoplasmic pool, ready to be recharged with a new amino acid. The polypeptide is released when the stop codon is read.
U A U
E site
P site
A site
G G C
Large ribosomal subunit
A U A C C G C U U
Codon 15
Codon 16
Codon 17
Small ribosomal subunit
Direction of ribosome advance
Portion of mRNA already translated

Figure 3.37 The basic steps of translation. The diagram presumes that initiation was accomplished and the 17th amino acid as dictated by the mRNA codons is being brought to the ribosome.

process is controlled by an aminoacyl-tRNA synthetase enzyme and is activated by ATP (see the top right side of Figure 3.37). Once its amino acid is loaded, the tRNA (now called an *aminoacyl-tRNA* because of its amino acid cargo) migrates to the ribosome, where it maneuvers the amino acid into the proper position, as specified by the mRNA codons (Figure 3.37, ②).

The ribosome is more than just a passive attachment site for mRNA and tRNA. Like a vise, the ribosome holds the tRNA and mRNA close together to coordinate the coupling of codons and anticodons, and positions the next (incoming) amino acid for addition to the growing polypeptide chain. To do its job, the ribosome has a binding site for mRNA and three binding sites for tRNA: an A (aminoacyl) site for an incoming aminoacyl-tRNA, a P (peptidyl) site for the tRNA holding the growing polypeptide chain, and an E (exit) site for an outgoing tRNA (Figure 3.37).

How does translation get started? When a special methionine-charged **initiator tRNA** binds to the middle (P) site on the small ribosomal subunit, the translation process officially begins. The newly formed mRNA has an initial base sequence, called a *leader sequence*, that allows it to attach to its binding site on the small ribosomal subunit. With the initiator tRNA still in tow, the small ribosomal subunit scans along the mRNA until it encounters the *start codon* (the first AUG triplet it meets). When the initiator tRNA's UAC anticodon "recognizes" and binds to the start codon, a large ribosomal subunit unites with the small one, forming a functional ribosome. With the mRNA firmly positioned in the "groove" between the two ribosomal subunits, translation begins in earnest. This initiation process requires the help of a number of initiation factors and is energized by GTP.

Now the ribosome slides the mRNA strand along, bringing the next codon into position to be "read" by an aminoacyl-tRNA coming into the A site (Figure 3.37, ②). It is at this point that the ribosome does a little "proofreading" to make sure of the codon-anticodon match. That accomplished, an enzymatic component in the large ribosomal particle peptide-bonds the amino acid of the initiator tRNA in the P site to that of the tRNA at the A site. The ribosome then translocates the tRNA that is now carrying two amino acids to the P site (as in Figure 3.37, ③). At the same time, it ratchets the initiator tRNA to the E site, from which it leaves the ribosome (as in Figure 3.37, ④).

This orderly "musical chairs" process continues: the peptidyl-tRNAs transferring their polypeptide cargo to the aminoacyl-tRNAs, and then the P-site-to-E-site and A-site-to-P-site movements of the tRNAs (see again Figure 3.37, ③ and ④). As the ribosome "chugs" along the mRNA track and the mRNA is progressively read, its initial portion passes through the ribosome and may ultimately become attached successively to several other ribosomes, all reading the same message simultaneously and sequentially. Such a multiple ribosome–mRNA complex is called a *polyribosome*, and it provides an efficient system for producing many copies of the same protein **(Figure 3.38)**.

The mRNA strand continues to be read sequentially until its last codon, the *stop codon* (one of UGA, UAA, or UAG) enters the ribosomal groove. The stop codon is the "period" at the end of the mRNA sentence that tells the ribosome that translation of that mRNA is finished. The completed polypeptide chain is then released from the ribosome, and the ribosome separates into its two subunits (Figure 3.38a). The released protein folds into a complex 3-D structure and floats off, ready to work. When the message of the mRNA that directed its formation becomes outdated, it is degraded in structures called *processing (P) bodies.*

As noted earlier in the chapter, ribosomes attach to and detach from the rough ER. When a short "leader" peptide called an ER **signal sequence** is present in a protein being synthesized, the associated ribosome attaches to the membrane of the rough ER. This signal sequence, with its attached cargo of a ribosome and mRNA, is guided to appropriate receptor sites on the ER membrane by a signal-recognition particle (SRP), a chaperonin

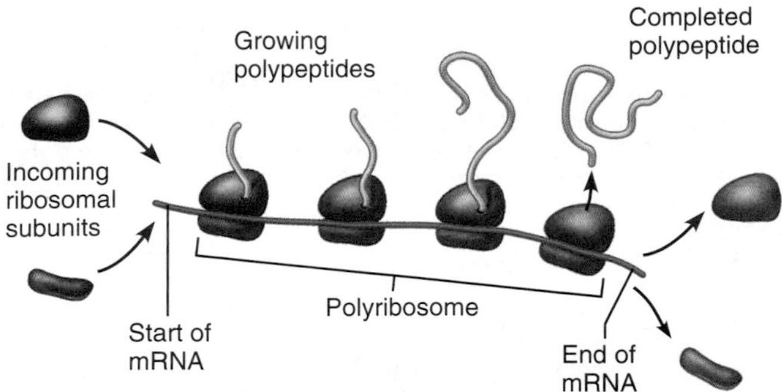

(a) Each polyribosome consists of one strand of mRNA being read by several ribosomes simultaneously. In this diagram, the mRNA is moving to the left and the "oldest" functional ribosome is farthest to the right.

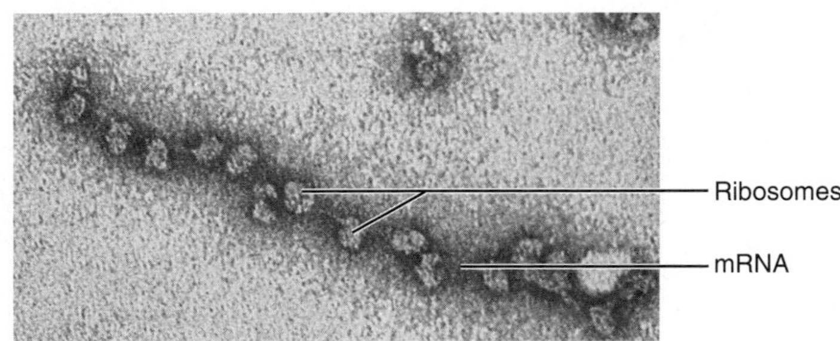

(b) This transmission electron micrograph shows a large polyribosome (400,000×).

Figure 3.38 Polyribosome arrays allow a single strand of mRNA to be translated into hundreds of the same polypeptide molecules in a short time.

that cycles between the ER and the cytosol. The subsequent events occurring at the ER are detailed in **Figure 3.39**.

In summary, the genetic information of a cell is translated into the production of proteins via a sequence of information transfer that is completely directed by complementary base pairing. The transfer of information goes from DNA base sequence (triplets) to the complementary base sequence of mRNA (codons) and then to the tRNA base sequence (anticodons), which is identical to the DNA sequence except for the substitution of uracil (U) for thymine (T) **(Figure 3.40)**.

Other Roles of DNA

The story of DNA doesn't end with the production of proteins encoded by exons. Scientists are finding that other intron or "junk" DNA actually codes for a surprising variety of active RNA species, including the following:

- **Antisense RNAs**, made on the DNA strand complementary to the template strand for mRNA, can intercept and bind to the protein-coding mRNA strand and prevent it from being translated into protein.
- **MicroRNAs** are small RNAs that can use RNA interference machinery to interfere with and suppress mRNAs made by certain exons, thus effectively silencing them.
- **Riboswitches** are folded RNAs that code, like mRNA, for a particular protein. What sets them apart from other mRNAs is

Developmental Aspects of Cells

- ▶ Discuss some theories of cell differentiation and aging.
- ▶ Indicate the value of apoptosis to the body.

We all begin life as a single cell, the fertilized egg, and all the cells of our body arise from it. Very early in development, cells begin to specialize. Some become liver cells, some nerve cells, and so on. All our cells carry the same genes, so how can one cell become so different from another? This is a fascinating question.

Apparently, cells in various regions of the embryo are exposed to different chemical signals that channel them into specific pathways of development. When the embryo consists of just a few cells (these are the embryonic *stem cells* you keep hearing about), the major signals may be nothing more than slight differences in oxygen and carbon dioxide concentrations between the more superficial and the deeper cells. But as development continues, cells release chemicals that influence development of neighboring cells by triggering processes that switch some genes "off." Some genes are active in all cells. For example, genes for rRNA and ATP synthesis are "on" in all cells, but genes for synthesizing the enzymes needed to produce thyroxine are "on" only in cells that are going to be part of the thyroid gland. Hence, the secret of cell specialization lies in the kinds of proteins made and reflects the activation of different genes in different cell types.

Cell specialization leads to *structural* variation—different organelles come to predominate in different cells. For example, muscle cells make large amounts of actin and myosin, and their cytoplasm fills with microfilaments. Liver and phagocytic cells produce more lysosomes. The development of specific and distinctive features in cells is called **cell differentiation**.

During early development, cell death and destruction are normal events. Nature takes few chances. More cells than needed are produced, and excesses are eliminated later in a type of programmed cell death called **apoptosis** (ap″o-to′sis; "falling away"). This process of controlled cellular suicide also eliminates cells that are stressed, no longer needed, injured, or aged.

How does apoptosis work? In response to damaged macromolecules within the cell or to some extracellular signal, mitochondrial membranes become permeable, allowing cytochrome c and other factors to leak into the cytosol. These chemicals, in turn, trigger apoptosis by causing activation of a series of intracellular enzymes called *caspases.* These enzymes are normally dormant, but when they are activated, they unleash a torrent of protein digestion activity within the cell, destroying the cell's DNA, cytoskeleton, and so on, producing a quick, neat death. The apoptotic cell shrinks without leaking its contents into the surrounding tissue, detaches from other cells, and rounds up. Because the dying cell releases a chemical (lysophosphatidylcholine) that attracts macrophages, and sprouts "eat me" signals, it is immediately phagocytized.

Cancer cells do not undergo apoptosis, but oxygen-starved cells do so excessively (heart-muscle and brain cells during heart attacks and strokes, for example). Apoptosis is particularly common in the developing nervous system. It is also responsible for "carving out" fingers and toes from their embryonic precursors.

Most organs are well formed and functional long before birth, but the body continues to grow and enlarge by forming new cells throughout childhood and adolescence. Once we reach adult size, cell division is important mainly to replace short-lived cells and repair wounds.

During young adulthood, cell numbers remain fairly constant. However, local changes in the rate of cell division are common. For example, when a person is anemic, his or her bone marrow undergoes **hyperplasia** (hi″per-pla′ze-ah), or accelerated growth (*hyper* = over; *plas* = grow), so that red blood cells are produced at a faster rate. If the anemia is remedied, the excessive marrow activity ceases. **Atrophy** (at′ro-fe), a decrease in size of an organ or body tissue, can result from loss of normal stimulation. Muscles that lose their nerve supply atrophy and waste away, and lack of exercise leads to thinned, brittle bones.

Cell aging occurs and it accounts for most problems associated with old age. Cell aging is a complicated process with many causes. The *wear-and-tear theory* attributes aging to little chemical insults and formation of free radicals, both of which have cumulative effects. For example, environmental toxins such as pesticides, alcohol, and bacterial toxins may damage cell membranes, poison enzyme systems, or cause "mistakes" in DNA replication. Temporary lack of oxygen, which occurs increasingly with age as our blood vessels clog with fatty materials, leads to accelerated rates of cell death throughout the body.

Most free radicals are produced in the mitochondria, because these organelles have the highest metabolic rate. This finding implies that diminished energy production by radical-damaged mitochondria, due perhaps to an increasing burden of mutations in mitochondrial DNA, weakens (and ages) cells. X rays and other types of radiation, and some chemicals, also generate huge numbers of free radicals, which can overwhelm the peroxisomal enzymes. Vitamins C and E act as antioxidants in the body and may help to prevent excessive free radical formation. With age, glucose (blood sugar) becomes party to cross-linking proteins together, a condition that severely disrupts protein function and accelerates the course of arteriosclerosis.

Another theory attributes cell aging to progressive disorders in the immune system. According to this theory, cell damage results from (1) autoimmune responses, which means the immune system turns against one's own tissues, and (2) a progressive weakening of the immune response, so that the body is less and less able to get rid of cell-damaging pathogens.

Perhaps the most widely accepted theory of cell aging is the *genetic theory,* which suggests that cessation of mitosis and cell aging are "programmed into our genes." One interesting notion here is that a *telomere clock* determines the number of times a cell can divide. **Telomeres** (*telo* = end; *mer* = piece) are strings of nucleotides that cap the ends of chromosomes, protecting them from fraying or fusing with other chromosomes, much like plastic caps preserve the ends of shoelaces. In human telomeres, the base sequence TTAGGG is repeated a thousand times or more (like a DNA stutter). Though telomeres carry no genes, they appear to be vital for chromosomal survival, because each time DNA is replicated, 50 to 100 of the end nucleotides are lost and the telomeres get a bit shorter. When telomeres reach a certain minimum length, the stop-division signal is given.

The idea that cell longevity depends on telomere integrity was supported by the discovery of *telomerase*, an enzyme that protects telomeres from degrading. Pegged as the "immortality enzyme," telomerase is found in germ line cells (cells that give rise to sperm and ova), but it is barely detectable or absent in other adult cell types.

CHECK YOUR UNDERSTANDING

34. What is apoptosis and what is its importance in the body?
35. What is the wear-and-tear theory of aging?

For answers, see Appendix G.

In this chapter we have described the structure and function of the generalized cell. One of the wonders of the cell is the disparity between its minute size and its extraordinary activity, which reflects the diversity of its organelles. The evidence for division of labor and functional specialization among organelles is inescapable. Only ribosomes synthesize proteins, while protein packaging is the bailiwick of the Golgi apparatus. Membranes compartmentalize most organelles, allowing them to work without hindering or being hindered by other cell activities, and the plasma membrane regulates molecular traffic across the cell's boundary. Now that you know what cells have in common, you are ready to explore how they differ in the various body tissues, the topic of Chapter 4.

3

RELATED CLINICAL TERMS

Anaplasia (an′ah-pla′ze-ah; *an* = without, not; *plas* = to grow) Abnormalities in cell structure and loss of differentiation; for example, cancer cells typically lose the appearance of the parent cells and come to resemble undifferentiated or embryonic cells.

Dysplasia (dis-pla′ze-ah; *dys* = abnormal) A change in cell size, shape, or arrangement due to chronic irritation or inflammation (infections, etc.).

Hypertrophy (hi-per′tro-fe) Growth of an organ or tissue due to an increase in the size of its cells. Hypertrophy is a normal response of skeletal muscle cells when they are challenged to lift excessive weight; differs from hyperplasia, which is an increase in size due to an increase in cell number.

Liposomes (lip′o-sōmz) Hollow microscopic sacs formed of phospholipids that can be filled with a variety of drugs. Serve as multipurpose vehicles for drugs, genetic material, and cosmetics.

Mutation A change in DNA base sequence that may lead to incorporation of incorrect amino acids in particular positions in the resulting protein; the affected protein may remain unimpaired or may function abnormally or not at all, leading to disease.

Necrosis (nĕ-kro′sis; *necros* = death; *osis* = process) Death of a cell or group of cells due to injury or disease. Acute injury causes the cells to swell and burst, and induces the inflammatory response. (This is *uncontrolled* cell death, in contrast to apoptosis described in the text.)

CHAPTER SUMMARY

Media study tools that could provide you additional help in reviewing specific key topics of Chapter 3 are referenced below.

iP = ***Interactive Physiology***

Overview of the Cellular Basis of Life (pp. 62–63)

1. All living organisms are composed of cells—the basic structural and functional units of life. Cells vary widely in both shape and size.
2. The principle of complementarity states that the biochemical activity of cells reflects the operation of their organelles.
3. The generalized cell is a concept that typifies all cells. The generalized cell has three major regions—the nucleus, cytoplasm, and plasma membrane.

The Plasma Membrane: Structure (pp. 63–67)

1. The plasma membrane encloses cell contents, mediates exchanges with the extracellular environment, and plays a role in cellular communication.

The Fluid Mosaic Model (pp. 63–66)

2. The fluid mosaic model depicts the plasma membrane as a fluid bilayer of lipids (phospholipids, cholesterol, and glycolipids) within which proteins are inserted.
3. The lipids have both hydrophilic and hydrophobic regions that organize their aggregation and self-repair. The lipids form the structural part of the plasma membrane.
4. Most proteins are integral transmembrane proteins that extend entirely through the membrane. Some, appended to the integral proteins, are peripheral proteins.
5. Proteins are responsible for most specialized membrane functions: Some are enzymes, some are receptors, and others mediate membrane transport functions. Externally facing glycoproteins contribute to the glycocalyx.

Membrane Junctions (p. 66–67)

6. Membrane junctions join cells together and may aid or inhibit movement of molecules between or past cells.
7. Tight junctions are impermeable junctions. Desmosomes mechanically couple cells into a functional community. Gap junctions allow joined cells to communicate.

The Plasma Membrane: Membrane Transport (pp. 68–77)

1. The plasma membrane acts as a selectively permeable barrier. Substances move across the plasma membrane by passive processes, which depend on the kinetic energy of molecules, and by active processes, which depend on the use of cellular energy (ATP).

3

Passive Processes (pp. 68–72)

2. Diffusion is the movement of molecules (driven by kinetic energy) down a concentration gradient. Fat-soluble solutes can diffuse directly through the membrane by dissolving in the lipid.
3. Facilitated diffusion is the passive movement of certain solutes across the membrane either by their binding with a membrane carrier protein or by their moving through a membrane channel. As with other diffusion processes, it is driven by kinetic energy, but the carriers and channels are selective.
4. Osmosis is the diffusion of a solvent, such as water, through a selectively permeable membrane. Water diffuses through membrane channels (aquaporins) or directly through the lipid portion of the membrane from a solution of lesser osmolarity (total concentration of all solute particles) to a solution of greater osmolarity.
5. The presence of solutes unable to permeate the plasma membrane leads to changes in cell tone that may cause the cell to swell or shrink. Net osmosis ceases when the solute concentration on both sides of the plasma membrane reaches equilibrium.
6. Solutions that cause a net loss of water from cells are hypertonic. Those causing net water gain are hypotonic. Those causing neither gain nor loss of water are isotonic.

Active Processes (pp. 72–77)

7. Active transport (solute pumping) depends on a carrier protein and ATP. Substances transported move against concentration or electrical gradients. In primary active transport, such as that provided by the Na^+-K^+ pump, ATP directly provides the energy.
8. In secondary active transport, the energy of an ion gradient (produced by a primary active transport process) is used to transport a substance passively. Many active transport systems are coupled, and cotransported substances move in either the same (symport) or opposite (antiport) directions across the membrane.
9. Vesicular transport also requires ATP. Endocytosis brings substances into the cell, typically in protein-coated vesicles. If the substance is particulate, the process is called phagocytosis. If the substance is dissolved molecules, the process is pinocytosis. Receptor-mediated endocytosis is selective: Engulfed molecules attach to receptors on the membrane before endocytosis occurs. Exocytosis, which uses SNAREs to anchor the vesicles to the plasma membrane, ejects substances (hormones, wastes, secretions) from the cell.
10. Most endocytosis (and transcytosis) is mediated by clathrin-coated vesicles. Caveolin-coated vesicles engulf some substances but appear to be more important as sites that accumulate receptors involved in cell signaling. Coatomer-coated vesicles mediate most types of substance (vesicular) trafficking within the cell.

The Plasma Membrane: Generation of a Resting Membrane Potential (pp. 77–79)

1. All cells in the resting stage exhibit a voltage across their membrane, called the resting membrane potential. Because of the membrane potential, both concentration and electrical gradients determine the ease of an ion's diffusion.
2. The resting membrane potential is generated by concentration gradients of ions and the differential permeability of the plasma membrane to ions, particularly potassium ions. Sodium is in high extracellular concentration and low intracellular concentration, and the membrane is poorly permeable to it. Potassium is in high concentration in the cell and low concentration in the extracellular fluid. The membrane is more permeable to potassium than to sodium. Protein anions in the cell are too large to cross the membrane and Cl^-, the main anion in extracellular fluid, is repelled by the negative charge on the inner membrane face.
3. Essentially, a negative membrane potential is established when the movement of K^+ out of the cell equals K^+ movement into the cell. Na^+ movements across the membrane contribute minimally to establishing the membrane potential. The greater outward diffusion of potassium (than inward diffusion of sodium) leads to a charge separation at the membrane (inside negative). This charge separation is maintained by the operation of the sodium-potassium pump.

iP Nervous System I; Topics: Ion Channels, pp. 3, 8, 9; The Membrane Potential, pp. 1–17.

The Plasma Membrane: Cell-Environment Interactions (pp. 79–81)

1. Cells interact directly and indirectly with other cells. Indirect interactions involve extracellular chemicals carried in body fluids or forming part of the extracellular matrix.
2. Molecules of the glycocalyx are intimately involved in cell-environment interactions. Most are cell adhesion molecules or membrane receptors.
3. Activated membrane receptors act as catalysts, regulate channels, or, like G protein–linked receptors, act through second messengers such as cyclic AMP and Ca^{2+}. Ligand binding results in changes in protein structure or function within the targeted cell.

The Cytoplasm (pp. 81–91)

1. The cytoplasm, the cellular region between the nuclear and plasma membranes, consists of the cytosol (fluid cytoplasmic environment), inclusions (nonliving nutrient stores, pigment granules, crystals, etc.), and cytoplasmic organelles.

Cytoplasmic Organelles (pp. 83–89)

2. The cytoplasm is the major functional area of the cell. These functions are mediated by cytoplasmic organelles.
3. Mitochondria, organelles limited by a double membrane, are sites of ATP formation. Their internal enzymes carry out the oxidative reactions of cellular respiration.
4. Ribosomes, composed of two subunits containing ribosomal RNA and proteins, are the sites of protein synthesis. They may be free or attached to membranes.
5. The rough endoplasmic reticulum is a ribosome-studded membrane system. Its cisternae act as sites for protein modification. Its external face acts in phospholipid synthesis. Vesicles pinched off from the ER transport the proteins to other cell sites.
6. The smooth endoplasmic reticulum synthesizes lipid and steroid molecules. It also acts in fat metabolism and in drug detoxification. In muscle cells, it is a calcium ion depot.
7. The Golgi apparatus is a membranous system close to the nucleus that packages protein secretions for export, packages enzymes into lysosomes for cellular use, and modifies proteins destined to become part of cellular membranes.
8. Lysosomes are membranous sacs of acid hydrolases packaged by the Golgi apparatus. Sites of intracellular digestion, they degrade worn-out organelles and tissues that are no longer useful, and they release ionic calcium from bone.
9. Peroxisomes are membranous sacs containing oxidase enzymes that protect the cell from the destructive effects of free radicals and other toxic substances by converting them first to hydrogen peroxide and then water.
10. The cytoskeleton includes microtubules, intermediate filaments, and microfilaments. Microtubules organize the cytoskeleton and

are important in intracellular transport. Microfilaments are important in cell motility or movement of cell parts. Motility functions involve motor proteins. Intermediate filaments help cells resist mechanical stress and connect other elements.

Cellular Extensions (pp. 90–91)

11. Centrioles organize the mitotic spindle and are the bases of cilia and flagella.
12. Microvilli are extensions of the plasma membrane that increase its surface area for absorption.

The Nucleus (pp. 91–95)

1. The nucleus is the control center of the cell. Most cells have a single nucleus. Without a nucleus, a cell cannot divide or synthesize more proteins, and is destined to die.
2. The nucleus is surrounded by the nuclear envelope, a double membrane penetrated by fairly large pores.
3. Nucleoli are nuclear sites of ribosome subunit synthesis.
4. Chromatin is a complex network of slender threads containing histone proteins and DNA. The chromatin units are called nucleosomes. When a cell begins to divide, the chromatin coils and condenses, forming chromosomes.

Cell Growth and Reproduction (pp. 95–107)

The Cell Life Cycle (pp. 95–100)

1. The cell life cycle is the series of changes that a cell goes through from the time it is formed until it divides.
2. Interphase is the nondividing phase of the cell life cycle. Interphase consists of G_1, S, and G_2 subphases. During G_1, the cell grows and centriole replication begins. During the S phase, DNA replicates. During G_2, the final preparations for division are made. Many checkpoints occur during interphase at which the cell gets the go-ahead signal to go through mitosis or is prevented from continuing to mitosis.
3. DNA replication occurs before cell division, ensuring that both daughter cells have identical genes. The DNA helix uncoils, and each DNA nucleotide strand acts as a template for the formation of a complementary strand. Base pairing provides the guide for the proper positioning of nucleotides.
4. The semiconservative replication of a DNA molecule produces two DNA molecules identical to the parent molecule, each formed of one "old" and one "new" strand.
5. Cell division, essential for body growth and repair, occurs during the M phase. Cell division consists of two distinct phases: mitosis (nuclear division) and cytokinesis (division of the cytoplasm).
6. Mitosis, consisting of prophase, metaphase, anaphase, and telophase, parcels out the replicated chromosomes to two daughter nuclei, each genetically identical to the mother nucleus. Cytokinesis, which begins late in mitosis, divides the cytoplasmic mass into two parts.
7. Cell division is stimulated by certain chemicals (including growth factors and some hormones) and increasing cell size. Lack of space and inhibitory chemicals deter cell division. Cell division is regulated by cyclin-Cdk complexes.

Protein Synthesis (pp. 100–105)

8. A gene is defined as a DNA segment that provides the instructions for the synthesis of one polypeptide chain. Since the major structural materials of the body are proteins, and all enzymes are proteins, this amply covers the synthesis of all biological molecules.
9. The base sequence of exon DNA provides the information for protein structure. Each three-base sequence (triplet) calls for a particular amino acid to be built into a polypeptide chain.
10. The RNA molecules acting in protein synthesis are synthesized on single strands of the DNA template. RNA nucleotides are joined according to base-pairing rules.
11. Messenger RNA carries instructions for making a polypeptide chain from the DNA to the ribosomes. Ribosomal RNA forms part of the protein synthesis sites. A transfer RNA ferries each amino acid to the ribosome and binds to a codon on the mRNA strand specifying its amino acid.
12. Protein synthesis involves (a) transcription, synthesis of a complementary mRNA, and (b) translation, "reading" of the mRNA by tRNA and peptide bonding of the amino acids into the polypeptide chain. Ribosomes coordinate translation.

Other Roles of DNA (pp. 105–106)

13. Intron and other "junk" DNA encodes many RNA species that may interfere with or promote the function of specific genes.

Cytosolic Protein Degradation (pp. 106–107)

14. Soluble proteins that are damaged or no longer needed are targeted for destruction by attachment of ubiquitin. Cytosolic enzymes or proteasomes then degrade these proteins.

Extracellular Materials (p. 107)

1. Extracellular materials are substances found outside the cells. They include body fluids, cellular secretions, and extracellular matrix. Extracellular matrix is particularly abundant in connective tissues.

Developmental Aspects of Cells (pp. 108–109)

1. The first cell of an organism is the fertilized egg. Early in development, cell specialization begins and reflects differential gene activation.
2. Apoptosis is programmed cell death. Its function is to dispose of damaged or unnecessary cells.
3. During adulthood, cell numbers remain fairly constant. Cell division occurs primarily to replace lost cells.
4. Cellular aging may reflect chemical insults, progressive disorders of immunity, or a genetically programmed decline in the rate of cell division with age.

REVIEW QUESTIONS

Multiple Choice/Matching

(Some questions have more than one correct answer. Select the best answer or answers from the choices given.)

1. The smallest unit capable of life by itself is **(a)** the organ, **(b)** the organelle, **(c)** the tissue, **(d)** the cell, **(e)** the nucleus.
2. The major types of lipid found in the plasma membranes are (choose two) **(a)** cholesterol, **(b)** triglycerides, **(c)** phospholipids, **(d)** fat-soluble vitamins.
3. Membrane junctions that allow nutrients or ions to flow from cell to cell are **(a)** desmosomes, **(b)** gap junctions, **(c)** tight junctions, **(d)** all of these.

3

4. The term used to describe the type of solution in which cells will lose water to their environment is **(a)** isotonic, **(b)** hypertonic, **(c)** hypotonic, **(d)** catatonic.
5. Osmosis always involves **(a)** a selectively permeable membrane, **(b)** a difference in solvent concentration, **(c)** diffusion, **(d)** active transport, **(e)** a, b, and c.
6. A physiologist observes that the concentration of sodium inside a cell is decidedly lower than that outside the cell. Sodium diffuses easily across the plasma membrane of such cells when they are dead, but *not* when they are alive. What cellular function that is lacking in dead cells explains the difference? **(a)** osmosis, **(b)** diffusion, **(c)** active transport (solute pumping), **(d)** dialysis.
7. The solute-pumping type of active transport is accomplished by **(a)** exocytosis, **(b)** phagocytosis, **(c)** electrical forces in the cell membrane, **(d)** changes in shape and position of carrier molecules in the plasma membrane.
8. The endocytotic process in which a sampling of particulate matter is engulfed and brought into the cell is called **(a)** phagocytosis, **(b)** fluid-phase endocytosis, **(c)** exocytosis.
9. Which is *not* true of centrioles? **(a)** they start to duplicate in G_1, **(b)** they lie in the centrosome, **(c)** they are made of microtubules, **(d)** they are membrane-walled barrels lying parallel to each other.
10. The nuclear substance composed of histone proteins and DNA is **(a)** chromatin, **(b)** the nucleolus, **(c)** nuclear sap, or nucleoplasm, **(d)** nuclear pores.
11. The information sequence that determines the nature of a protein is the **(a)** nucleotide, **(b)** gene, **(c)** triplet, **(d)** codon.
12. Mutations may be caused by **(a)** X rays, **(b)** certain chemicals, **(c)** radiation from ionizing radioisotopes, **(d)** all of these.
13. The phase of mitosis during which centrioles reach the poles and chromosomes attach to the spindle is **(a)** anaphase, **(b)** metaphase, **(c)** prophase, **(d)** telophase.
14. Final preparations for cell division are made during the life cycle subphase called **(a)** G_1, **(b)** G_2, **(c)** M, **(d)** S.
15. The RNA synthesized on one of the DNA strands is **(a)** mRNA, **(b)** tRNA, **(c)** rRNA, **(d)** all of these.
16. The RNA species that carries the coded message, specifying the sequence of amino acids in the protein to be made, from the nucleus to the cytoplasm is **(a)** mRNA, **(b)** tRNA, **(c)** rRNA, **(d)** all of these.
17. If DNA has a sequence of AAA, then a segment of mRNA synthesized on it will have a sequence of **(a)** TTT, **(b)** UUU, **(c)** GGG, **(d)** CCC.
18. A nerve cell and a lymphocyte are presumed to differ in their **(a)** specialized structure, **(b)** suppressed genes and embryonic history, **(c)** genetic information, **(d)** a and b, **(e)** a and c.
19. A pancreas cell makes proteins (enzymes) which it releases to the small intestine. Which of the following best describes the path of these proteins from synthesis to exocytosis at the pancreatic cell's plasma membrane (PM)? **(a)** Golgi → rough ER → PM, **(b)** smooth ER → Golgi → lysosome → PM, **(c)** rough ER → Golgi → PM, **(d)** nucleus → Golgi → PM.

Short Answer Essay Questions

20. Which organelle is responsible for a newborn having distinctive toes and fingers instead of webbed digits?
21. Explain why mitosis can be thought of as cellular immortality.
22. Contrast the roles of ER-bound ribosomes with those free in the cytosol.
23. Cells lining the trachea have whiplike motile extensions on their free surfaces. What are these extensions, what is their source, and what is their function?
24. Name the three phases of interphase and describe an activity unique to each phase.
25. Comment on the role of the sodium-potassium pump in maintaining a cell's resting membrane potential.
26. Differentiate clearly between primary and secondary active transport processes.
27. Cell division typically yields two daughter cells, each with one nucleus. How is the occasional binucleate condition of liver cells explained?

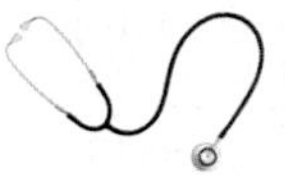

Critical Thinking and Clinical Application Questions

1. Explain why limp celery becomes crisp and the skin of your fingertips wrinkles when placed in tap water. (The principle is exactly the same.)
2. A "red-hot" bacterial infection of the intestinal tract irritates the intestinal cells and interferes with digestion. Such a condition is often accompanied by diarrhea, which causes loss of body water. On the basis of what you have learned about osmotic water flows, explain why diarrhea may occur.
3. Two examples of chemotherapeutic drugs (drugs used to treat cancer) and their cellular actions are listed below. Explain why each drug could be fatal to a cell.
   - Vincristine (Oncovin): damages the mitotic spindle
   - Doxorubicin (Adriamycin): binds to DNA and blocks mRNA synthesis
4. The normal function of one tumor suppressor gene is to prevent cells with damaged chromosomes and DNA from "progressing from G_1 to S," whereas another tumor suppressor gene prevents "passage from G_2 to M." When these tumor suppressor genes fail to work, cancer can result. Explain what the phrases in quotations mean.
5. In their anatomy lab, many students are exposed to the chemical preservatives phenol, formaldehyde, and alcohol. Our cells break down these toxins very effectively. What cellular organelle is responsible for this?
6. Dynein is missing from the cilia and flagella of individuals with a specific inherited disorder. These individuals have severe respiratory problems and, if males, are sterile. What is the structural connection between these two symptoms?
7. Explain why alcoholics are likely to have much more smooth ER than teetotalers.
8. Water is a precious natural resource in Florida and it is said that supplies are dwindling. Desalinizing (removing salt from) ocean water has been recommended as a solution to the problem. Why shouldn't we drink salt water?

4 Tissue: The Living Fabric

Amoebas and other unicellular (one-cell) organisms are rugged individualists. Each cell alone obtains and digests its food, ejects its wastes, and carries out all the other activities necessary to keep itself alive and "buzzin' around on all cylinders." But in the multicellular human body, cells do not operate independently. Instead, they form tight cell communities that live and work together.

Individual body cells are specialized, with each type performing specific functions that help maintain homeostasis and benefit the body as a whole. Cell specialization is obvious: How muscle cells look and act

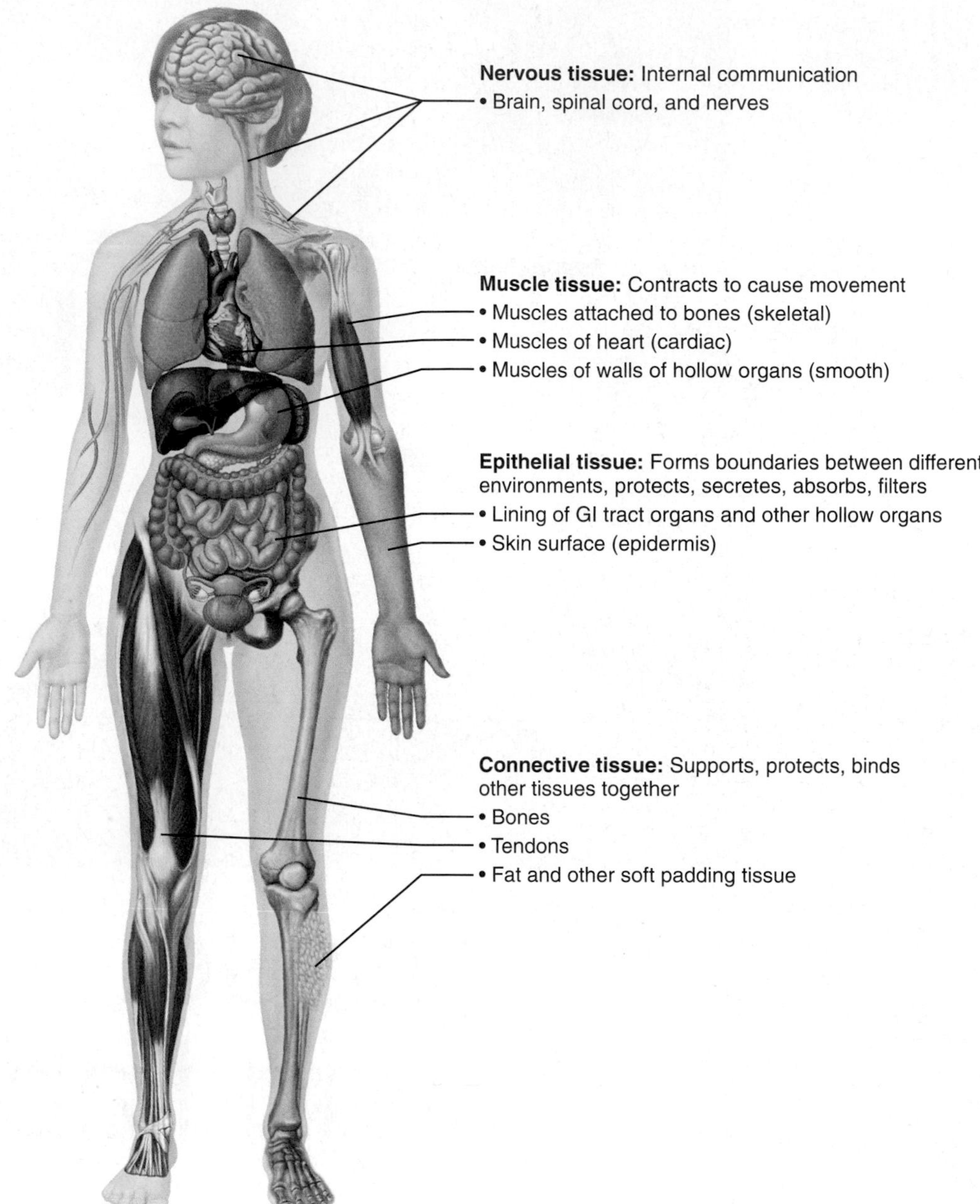
Nervous tissue: Internal communication
• Brain, spinal cord, and nerves
Muscle tissue: Contracts to cause movement
• Muscles attached to bones (skeletal)
• Muscles of heart (cardiac)
• Muscles of walls of hollow organs (smooth)
Epithelial tissue: Forms boundaries between different environments, protects, secretes, absorbs, filters
• Lining of GI tract organs and other hollow organs
• Skin surface (epidermis)
Connective tissue: Supports, protects, binds other tissues together
• Bones
• Tendons
• Fat and other soft padding tissue

Figure 4.1 Overview of four tissue types: epithelial, connective, muscle, and nervous tissues.

differs greatly from skin cells, which in turn are easy to distinguish from brain cells. Cell specialization allows the body to function in sophisticated ways, but division of labor has certain hazards. When a particular group of cells is indispensable, its loss or injury can severely disable or even destroy the body.

Groups of cells that are similar in structure and perform a common or related function are called **tissues** (*tissu* = woven). Four primary tissue types interweave to form the "fabric" of the body. These basic tissues are epithelial (ep″i-the′le-ul), connective, muscle, and nervous tissue, and each has numerous subclasses or varieties. If we had to describe the general role of each primary tissue type in a single word, the words would most likely be *covering* (epithelial), *support* (connective), *movement* (muscle), and *control* (nervous). However, these words reveal only a fraction of the functions that each tissue performs **(Figure 4.1)**.

As we explained in Chapter 1, tissues are organized into organs such as the kidneys and the heart. Most organs contain all four tissue types, and their arrangement determines the organ's structure and capabilities. The study of tissues, or **histology**, complements the study of gross anatomy. Together they provide the structural basis for understanding organ physiology.

Preparing Human Tissue for Microscopy

▶ List the steps involved in preparing animal tissue for microscopic viewing.

Elaborate steps are taken to prepare human or animal tissue for microscopic viewing. The specimen must be **fixed** (preserved) and then cut into **sections** (slices) thin enough to transmit light or electrons. Finally the specimen must be **stained** to enhance contrast.

The stains used in light microscopy are beautifully colored organic dyes, most of which were originally developed by clothing manufacturers in the mid-1800s. Many dyes consist of negatively or positively charged molecules (acidic and basic stains, respectively) that bind within the tissue to macromolecules of the opposite charge. The stains distinguish different anatomical structures because different parts of cells and tissues take up different dyes.

For transmission electron microscopy (TEM), tissue sections are "stained" with heavy metal salts. These metals deflect electrons in the beam to different extents, providing contrast in the image. Electron-microscope images are in shades of gray because color is a property of light, not of electron waves, but the image may be artificially colored to enhance contrast. Another kind of electron microscopy, scanning electron microscopy (SEM), provides three-dimensional pictures of an unsectioned tissue surface. These striking images are scattered throughout this book.

Preserved tissue we see under the microscope has been exposed to many procedures that alter its original condition and introduce minor distortions called **artifacts**. For this reason, keep in mind that most microscopic structures we view are not exactly like those in living tissue.

CHECK YOUR UNDERSTANDING

1. What is the purpose of fixing tissue for microscopic viewing?
2. What types of stains are used to stain tissues to be viewed with an electron microscope?

For answers, see Appendix G.

Epithelial Tissue

- List several structural and functional characteristics of epithelial tissue.
- Name, classify, and describe the various types of epithelia, and indicate their chief function(s) and location(s).

Epithelial tissue, or an **epithelium** (plural: epithelia), is a sheet of cells that covers a body surface or lines a body cavity (*epithe* = laid on, covering). It occurs in the body as (1) *covering and lining epithelium* and (2) *glandular epithelium*. Covering and lining epithelium forms the outer layer of the skin, dips into and lines the open cavities of the cardiovascular, digestive, and respiratory systems, and covers the walls and organs of the closed ventral body cavity. Glandular epithelium fashions the glands of the body.

Epithelia form boundaries between different environments, and nearly all substances received or given off by the body must pass through an epithelium. For example, the epidermis of the skin lies between the inside and the outside of the body. Epithelium lining the urinary bladder separates underlying cells of the bladder wall from urine.

In its role as an interface tissue, epithelium accomplishes many functions, including (1) protection, (2) absorption, (3) filtration, (4) excretion, (5) secretion, and (6) sensory reception. We describe each of these functions in detail later, but here we illustrate these functions briefly: The epithelium of the skin protects underlying tissues from mechanical and chemical injury and bacterial invasion and contains nerve endings that respond to various stimuli acting at the skin surface (pressure, heat, etc.). The epithelium lining the digestive tract is specialized to absorb substances. That found in the kidneys performs nearly the whole functional "menu"—excretion, absorption, secretion, and filtration. Secretion is the specialty of glands.

Special Characteristics of Epithelium

Epithelial tissues have many characteristics that distinguish them from other tissue types.

1. **Polarity**. All epithelia have an **apical surface**, an upper free surface exposed to the body exterior or the cavity of an internal organ, and a lower attached **basal surface**. For this reason, all epithelia exhibit *apical-basal polarity*, meaning that cell regions near the apical surface differ from those near the basal surface in both structure and function. This situation is maintained, at least in part, by the highly ordered cytoskeleton of epithelial cells.

   Although some apical surfaces are smooth and slick, most have **microvilli**, fingerlike extensions of the plasma membrane. Microvilli tremendously increase the exposed surface area. In epithelia that absorb or secrete substances (those lining the intestine or kidney tubules, for instance), the microvilli are often so dense that the cell apices have a fuzzy appearance called a *brush border*. Some epithelia, such as that lining the trachea, have motile **cilia** (tiny hairlike projections) that propel substances along their free surface.

   Lying adjacent to the basal surface of an epithelium is a thin supporting sheet called the **basal lamina** (lam′ĭ-nah; "sheet"). This noncellular, adhesive sheet consists largely of glycoproteins secreted by the epithelial cells plus some fine collagen fibers. The basal lamina acts as a selective filter that determines which molecules diffusing from the underlying connective tissue are allowed to enter the epithelium. The basal lamina also acts as a scaffolding along which epithelial cells can migrate to repair a wound.
2. **Specialized contacts**. Except for glandular epithelia (discussed on pp. 121–124), epithelial cells fit close together to form continuous sheets. Adjacent cells are bound together at many points by lateral contacts, including *tight junctions* and *desmosomes* (see Chapter 3). The tight junctions help keep proteins in the apical region of the plasma membrane from diffusing into the basal region, and thus help to maintain epithelial polarity.
3. **Supported by connective tissue**. All epithelial sheets rest upon and are supported by connective tissue. Just deep to the basal lamina is the **reticular lamina**, a layer of

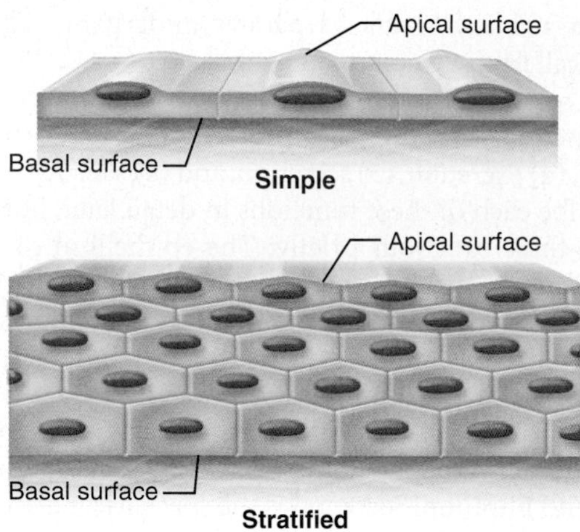

(a) Classification based on number of cell layers.

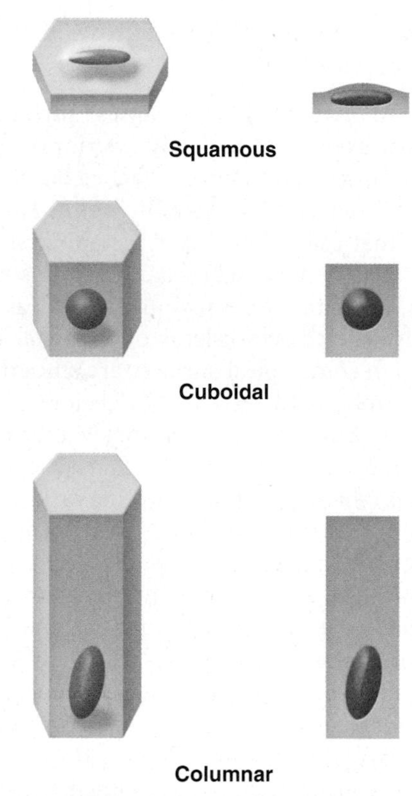

(b) Classification based on cell shape.

Figure 4.2 Classification of epithelia.

extracellular material containing a fine network of collagen protein fibers that "belongs to" the underlying connective tissue. Together the two laminae form the **basement membrane**. The basement membrane reinforces the epithelial sheet, helping it to resist stretching and tearing forces, and defines the epithelial boundary.

HOMEOSTATIC IMBALANCE

An important characteristic of cancerous epithelial cells is their failure to respect the basement membrane boundary, which they penetrate to invade the tissues beneath. ■

4. **Avascular but innervated.** Although epithelium is *innervated* (supplied by nerve fibers), it is *avascular* (contains no blood vessels). Epithelial cells are nourished by substances diffusing from blood vessels in the underlying connective tissue.
5. **Regeneration.** Epithelium has a high regenerative capacity. Some epithelia are exposed to friction and their surface cells rub off. Others are damaged by hostile substances in the external environment (bacteria, acids, smoke). If and when their apical-basal polarity and lateral contacts are destroyed, epithelial cells begin to reproduce themselves rapidly. As long as epithelial cells receive adequate nutrition, they can replace lost cells by cell division.

CHECK YOUR UNDERSTANDING

3. Epithelial tissue is the only tissue type that has polarity, that is, an apical and a basal surface. Why is this important?
4. Which of the following properties apply to epithelial tissue? Has blood vessels, can repair itself (regenerates), cells joined by lateral contacts.

For answers, see Appendix G.

Classification of Epithelia

Each epithelium is given two names. The first name indicates the number of cell layers present, and the second describes the shape of its cells. Based on the number of cell layers, there are simple and stratified epithelia **(Figure 4.2a)**. **Simple epithelia** consist of a single cell layer. They are typically found where absorption, secretion, and filtration occur and a thin epithelial barrier is desirable. **Stratified epithelia**, composed of two or more cell layers stacked one on top of the other, are common in high-abrasion areas where protection is important, such as the skin surface and the lining of the mouth.

In cross section, all epithelial cells have six (somewhat irregular) sides, and an apical surface view of an epithelial sheet looks like a honeycomb. This polyhedral shape allows the cells to be closely packed. However, epithelial cells vary in height, and on that basis, there are three common shapes of epithelial cells (Figure 4.2b). **Squamous cells** (skwa′mus) are flattened and scalelike (*squam* = scale). **Cuboidal cells** (ku-boi′dahl) are boxlike, approximately as tall as they are wide, and **columnar cells** (kŏ-lum′nar) are tall and column shaped.

In each case, the shape of the nucleus conforms to that of the cell. The nucleus of a squamous cell is a flattened disc; that of a cuboidal cell is spherical; and a columnar cell nucleus is elongated from top to bottom and usually located close to the cell base. Keep nuclear shape in mind when you attempt to identify epithelial types.

Simple epithelia are easy to classify by cell shape because all cells in the layer usually have the same shape. In stratified epithelia, however, the cell shapes usually differ among the different cell layers. To avoid ambiguity, stratified epithelia are named according to the shape of the cells in the *apical* layer. This naming system will become clearer as we explore the specific epithelial types.

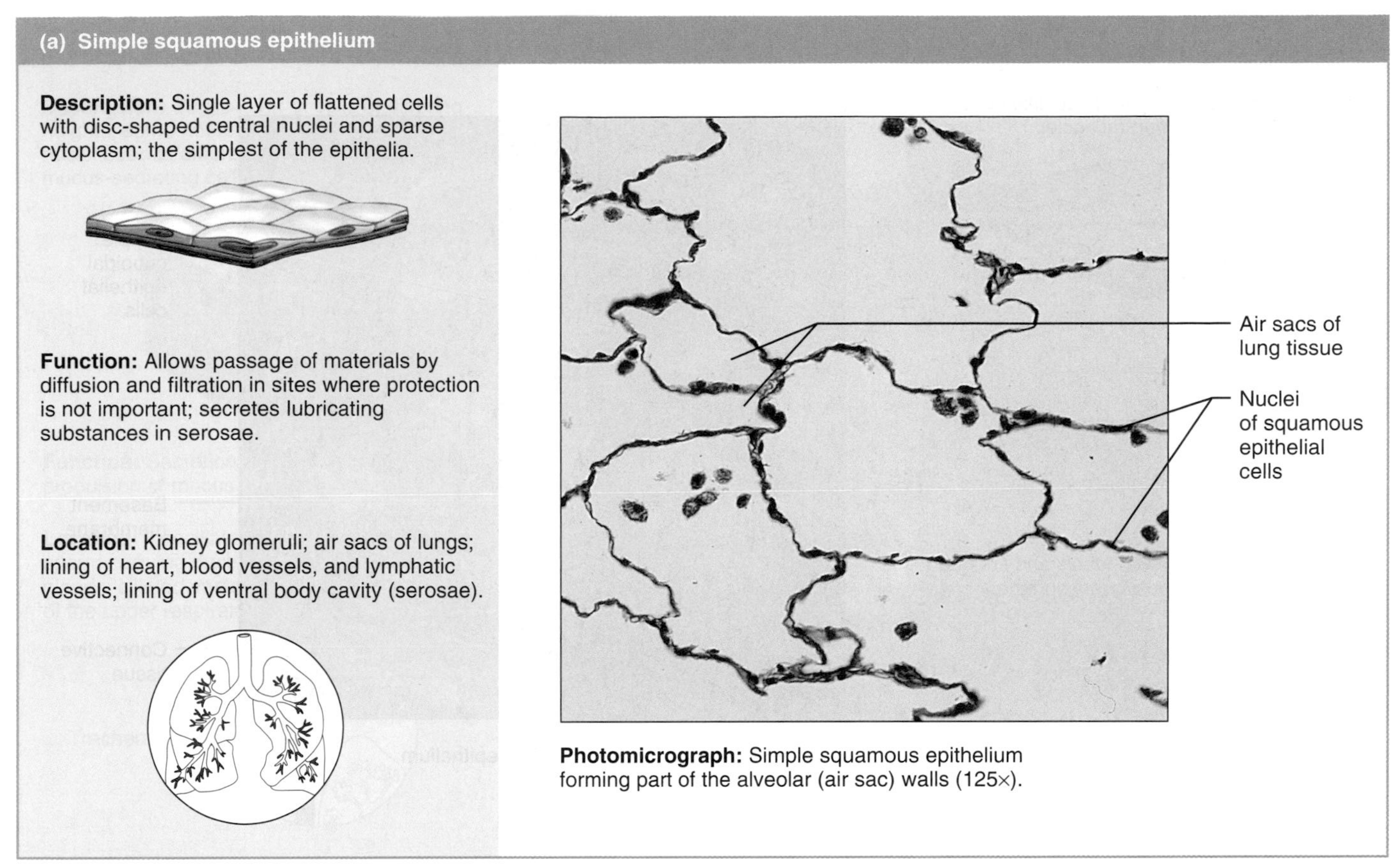

(a) Simple squamous epithelium
Description: Single layer of flattened cells with disc-shaped central nuclei and sparse cytoplasm; the simplest of the epithelia.
Function: Allows passage of materials by diffusion and filtration in sites where protection is not important; secretes lubricating substances in serosae.
Location: Kidney glomeruli; air sacs of lungs; lining of heart, blood vessels, and lymphatic vessels; lining of ventral body cavity (serosae).
Air sacs of lung tissue
Nuclei of squamous epithelial cells
Photomicrograph: Simple squamous epithelium forming part of the alveolar (air sac) walls (125×).

Figure 4.3 Epithelial tissues. (a) Simple epithelium. (See *A Brief Atlas of the Human Body*, Plates 1 and 2.)

As you read about the epithelial classes, study **Figure 4.3**. Using the photomicrographs, try to pick out the individual cells within each epithelium. This is not always easy, because the boundaries between epithelial cells often are indistinct. Furthermore, the nucleus of a particular cell may or may not be visible, depending on the precise plane of the cut made to prepare the tissue slides.

Simple Epithelia

The simple epithelia are most concerned with absorption, secretion, and filtration. Because they consist of a single cell layer and are usually very thin, protection is not one of their specialties.

Simple Squamous Epithelium The cells of a **simple squamous epithelium** are flattened laterally, and their cytoplasm is sparse (Figure 4.3a). In a surface view, the close-fitting cells resemble a tiled floor. When the cells are cut perpendicular to their free surface, they resemble fried eggs seen from the side, with their cytoplasm wisping out from the slightly bulging nucleus. Thin and often permeable, this epithelium is found where filtration or the exchange of substances by rapid diffusion is a priority. In the kidneys, simple squamous epithelium forms part of the filtration membrane. In the lungs, it forms the walls of the air sacs across which gas exchange occurs.

Two simple squamous epithelia in the body have special names that reflect their location. **Endothelium** (en″do-the′le-um; "inner covering") provides a slick, friction-reducing lining in lymphatic vessels and in all hollow organs of the cardiovascular system—blood vessels and the heart. Capillaries consist exclusively of endothelium, and its exceptional thinness encourages the efficient exchange of nutrients and wastes between the bloodstream and surrounding tissue cells. **Mesothelium** (mez″o-the′le-um; "middlecovering") is the epithelium found in serous membranes lining the ventral body cavity and covering its organs.

Simple Cuboidal Epithelium **Simple cuboidal epithelium** consists of a single layer of cells as tall as they are wide (Figure 4.3b). The spherical nuclei stain darkly, causing the cell layer to look like a string of beads when viewed microscopically. Important functions of simple cuboidal epithelium are secretion and absorption. This epithelium forms the walls of the smallest ducts of glands and of many kidney tubules.

Simple Columnar Epithelium **Simple columnar epithelium** is seen as a single layer of tall, closely packed cells, aligned like soldiers in a row (Figure 4.3c). It lines the digestive tract from the stomach through the rectum. Columnar cells are mostly associated with absorption and secretion, and the digestive tract lining has two distinct modifications that make it ideal for that

(b) Connective tissue proper: loose connective tissue, adipose

Description: Matrix as in areolar, but very sparse; closely packed adipocytes, or fat cells, have nucleus pushed to the side by large fat droplet.

Function: Provides reserve food fuel; insulates against heat loss; supports and protects organs.

Location: Under skin in the hypodermis; around kidneys and eyeballs; within abdomen; in breasts.

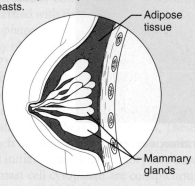

Adipose tissue

Mammary glands

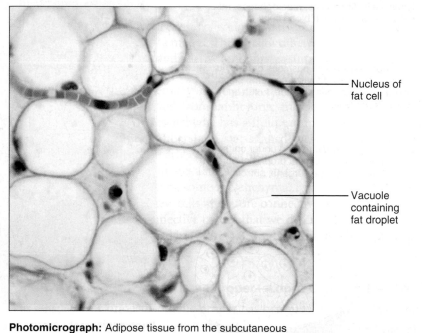

Nucleus of fat cell

Vacuole containing fat droplet

Photomicrograph: Adipose tissue from the subcutaneous layer under the skin (350×).

(c) Connective tissue proper: loose connective tissue, reticular

Description: Network of reticular fibers in a typical loose ground substance; reticular cells lie on the network.

Function: Fibers form a soft internal skeleton (stroma) that supports other cell types including white blood cells, mast cells, and macrophages.

Location: Lymphoid organs (lymph nodes, bone marrow, and spleen).

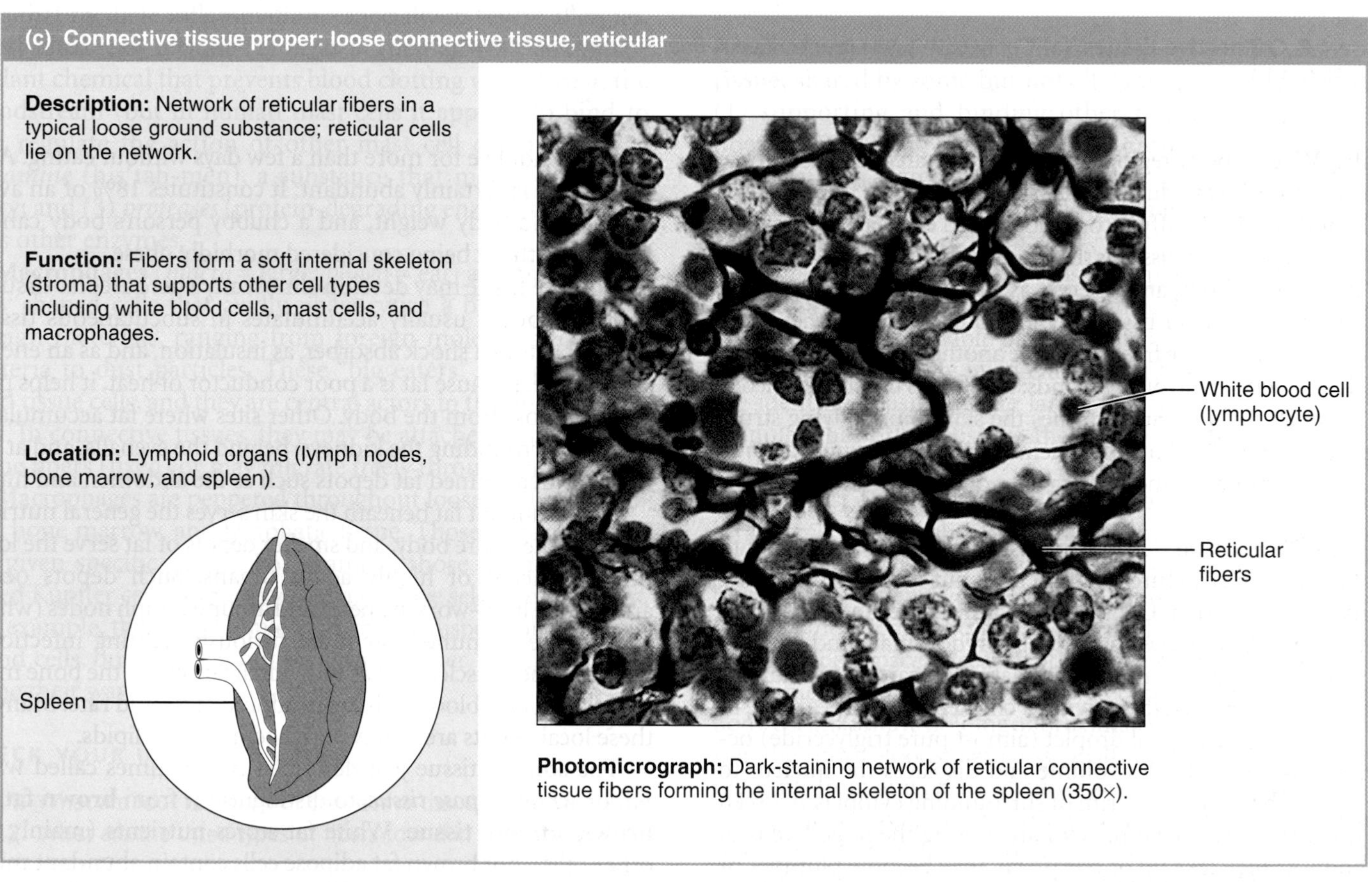

Photomicrograph: Dark-staining network of reticular connective tissue fibers forming the internal skeleton of the spleen (350×).

Figure 4.8 *(continued)* **Connective tissues. (b)** and **(c)** Connective tissue proper. (See *A Brief Atlas of the Human Body*, Plates 12 and 13.)

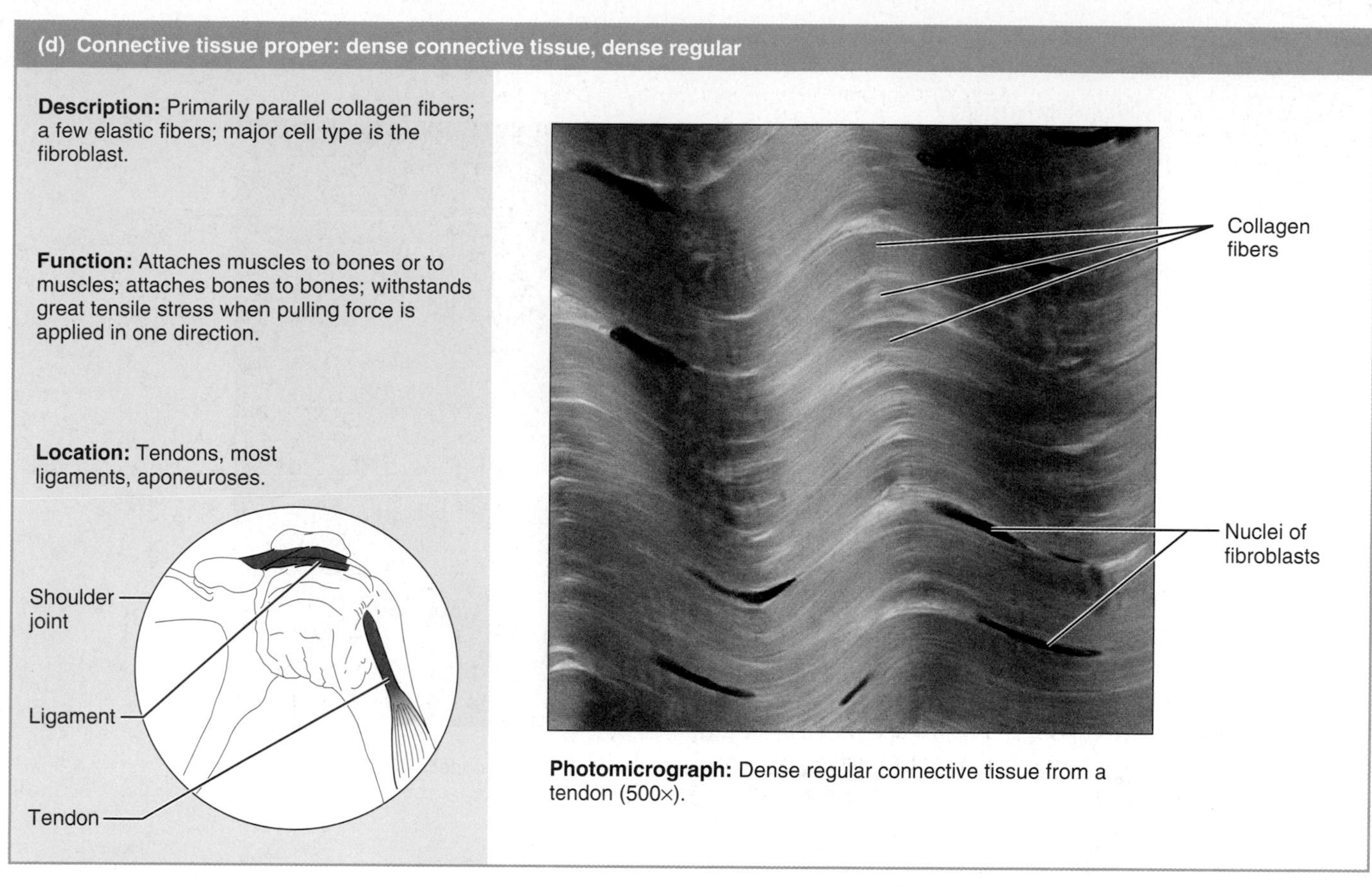

(d) Connective tissue proper: dense connective tissue, dense regular
Description: Primarily parallel collagen fibers; a few elastic fibers; major cell type is the fibroblast.
Function: Attaches muscles to bones or to muscles; attaches bones to bones; withstands great tensile stress when pulling force is applied in one direction.
Location: Tendons, most ligaments, aponeuroses.
Shoulder joint
Ligament
Tendon
Collagen fibers
Nuclei of fibroblasts
Photomicrograph: Dense regular connective tissue from a tendon (500×).

Figure 4.8 *(continued)* **(d)** Connective tissue proper. (See *A Brief Atlas of the Human Body,* Plate 15.)

the ability to produce body heat by shivering. Most such deposits are located between the shoulder blades, on the anterolateral neck, and on the anterior abdominal wall.

Reticular Connective Tissue **Reticular connective tissue** resembles areolar connective tissue, but the only fibers in its matrix are reticular fibers, which form a delicate network along which fibroblasts called **reticular cells** are scattered (Figure 4.8c). Although reticular *fibers* are widely distributed in the body, reticular *tissue* is limited to certain sites. It forms a labyrinth-like **stroma** (literally, "bed" or "mattress"), or internal framework, that can support many free blood cells (largely lymphocytes) in lymph nodes, the spleen, and bone marrow.

Connective Tissue Proper—Dense Connective Tissues

The three varieties of dense connective tissue have fibers as their prominent element. For this reason, the dense connective tissues are often referred to as **fibrous connective tissues.**

Dense Regular Connective Tissue **Dense regular connective tissue** contains closely packed bundles of collagen fibers running in the same direction, parallel to the direction of pull (Figure 4.8d). This arrangement results in white, flexible structures with great resistance to tension (pulling forces) where the tension is exerted in a single direction. Crowded between the collagen fibers are rows of fibroblasts that continuously manufacture the fibers and scant ground substance.

As seen in Figure 4.8d, collagen fibers are slightly wavy. This allows the tissue to stretch a little, but once the fibers are straightened out by a pulling force, there is no further "give" to this tissue. Unlike our model (areolar) connective tissue, this tissue has few cells other than fibroblasts and is poorly vascularized.

With its enormous tensile strength, dense regular connective tissue forms the *tendons,* which are cords that attach muscles to bones, and flat, sheetlike tendons called *aponeuroses* (ap″o-nu-ro′sēz) that attach muscles to other muscles or to bones. It also forms fascia (fash′e-ah; "a bond"), a fibrous membrane that wraps around muscles, groups of muscles, blood vessels, and nerves, binding those structures together like plastic sandwich wrap; and the ligaments that bind bones together at joints. Ligaments contain more elastic fibers than tendons and are slightly more stretchy.

Dense Irregular Connective Tissue **Dense irregular connective tissue** has the same structural elements as the regular variety. However, the bundles of collagen fibers are much thicker and they are arranged irregularly; that is, they run in more than one plane (Figure 4.8e). This type of tissue forms sheets in body areas where tension is exerted from many different directions. It is found in the skin as the leathery *dermis,* and it forms fibrous joint capsules and the fibrous coverings that surround some organs (kidneys, bones, cartilages, muscles, and nerves).

4

(e) Connective tissue proper: dense connective tissue, dense irregular

Description: Primarily irregularly arranged collagen fibers; some elastic fibers; major cell type is the fibroblast.

Function: Able to withstand tension exerted in many directions; provides structural strength.

Location: Fibrous capsules of organs and of joints; dermis of the skin; submucosa of digestive tract.

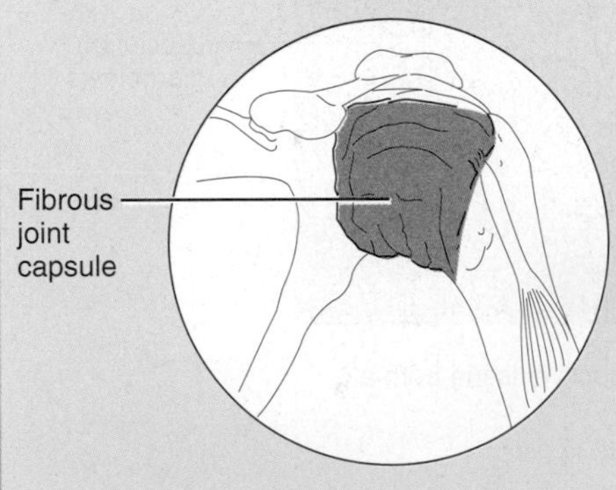

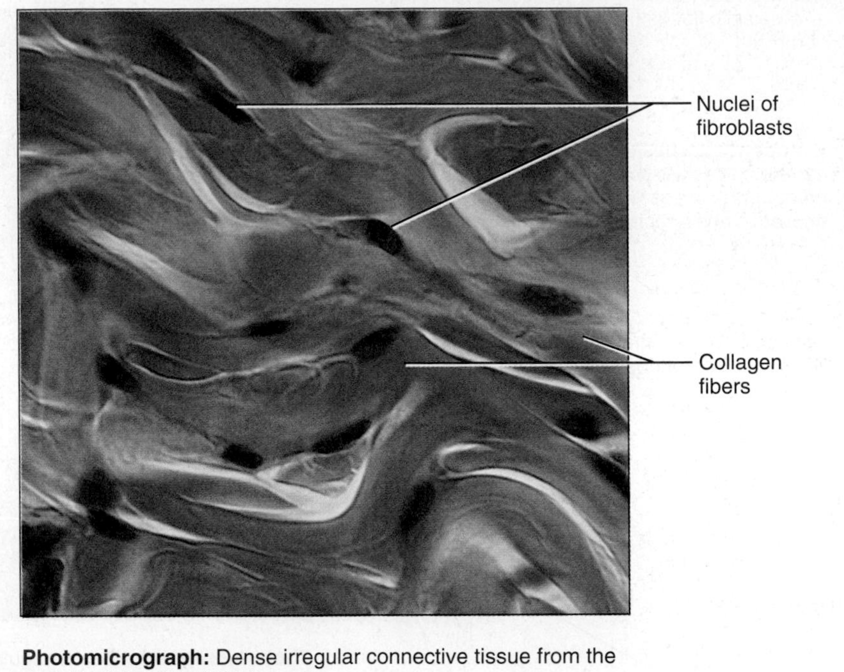

Photomicrograph: Dense irregular connective tissue from the dermis of the skin (400×).

(f) Connective tissue proper: dense connective tissue, elastic

Description: Dense regular connective tissue containing a high proportion of elastic fibers.

Function: Allows recoil of tissue following stretching; maintains pulsatile flow of blood through arteries; aids passive recoil of lungs following inspiration.

Location: Walls of large arteries; within certain ligaments associated with the vertebral column; within the walls of the bronchial tubes.

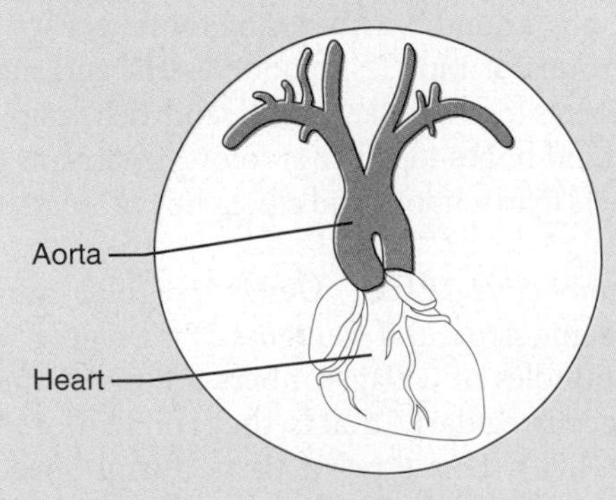

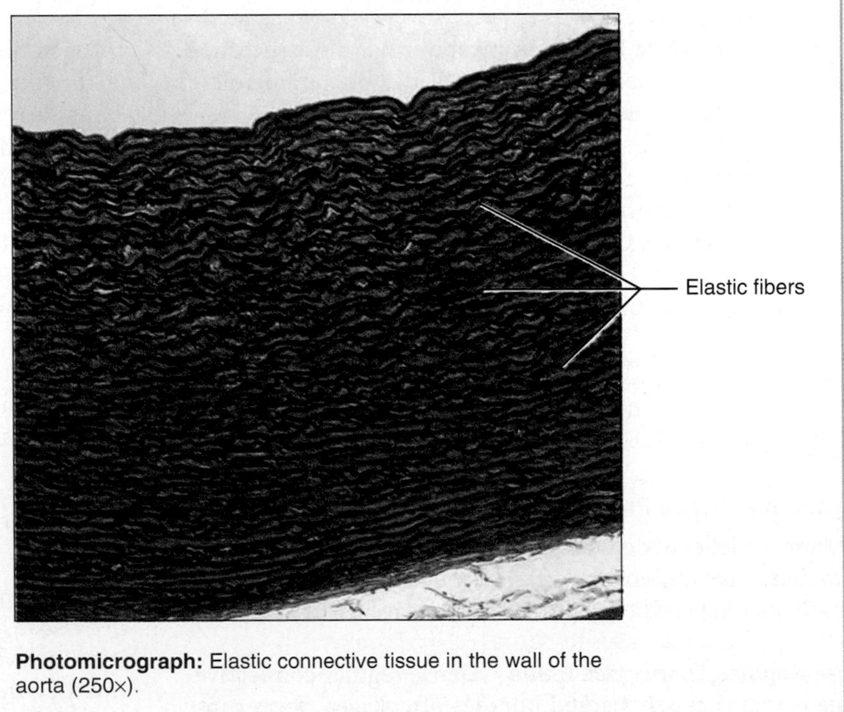

Photomicrograph: Elastic connective tissue in the wall of the aorta (250×).

Figure 4.8 *(continued)* **Connective tissues. (e)** and **(f)** Connective tissue proper. (See *A Brief Atlas of the Human Body*, Plates 14 and 16.)

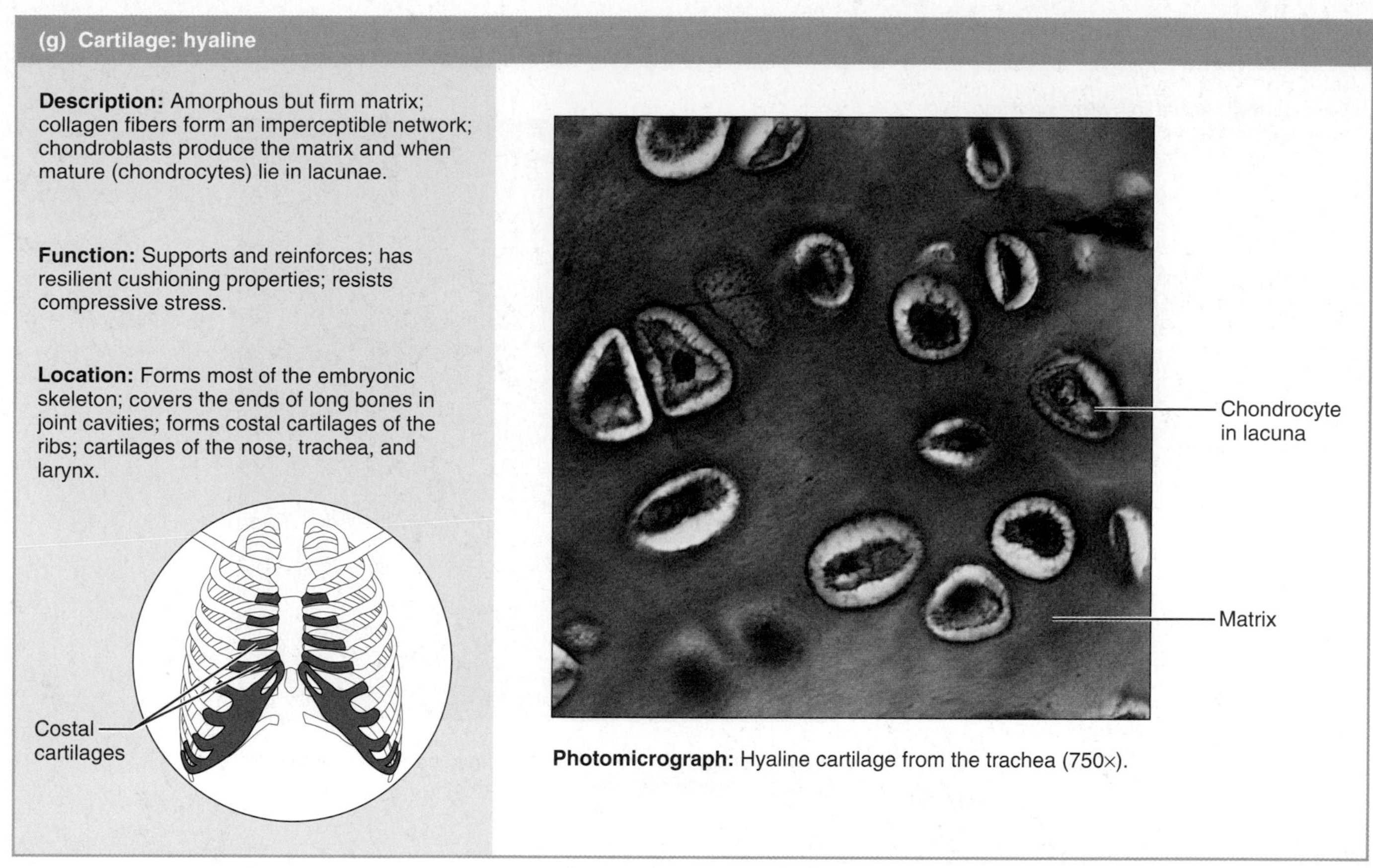

(g) Cartilage: hyaline
Description: Amorphous but firm matrix; collagen fibers form an imperceptible network; chondroblasts produce the matrix and when mature (chondrocytes) lie in lacunae.
Function: Supports and reinforces; has resilient cushioning properties; resists compressive stress.
Location: Forms most of the embryonic skeleton; covers the ends of long bones in joint cavities; forms costal cartilages of the ribs; cartilages of the nose, trachea, and larynx.
Costal cartilages
Chondrocyte in lacuna
Matrix
Photomicrograph: Hyaline cartilage from the trachea (750×).

Figure 4.8 *(continued)* **(g)** Cartilage. (See *A Brief Atlas of the Human Body*, Plate 17.)

Elastic Connective Tissue A few ligaments, such as the *ligamenta nuchae* and *flava* connecting adjacent vertebrae, are very elastic, so much so that the dense regular connective tissue in those structures is referred to more specifically as **elastic connective tissue** (Figure 4.8f).

Cartilage

Cartilage (kar′tĭ-lij), which stands up to both tension *and* compression, has qualities intermediate between dense connective tissue and bone. It is tough but flexible, providing a resilient rigidity to the structures it supports. Cartilage lacks nerve fibers and is avascular. It receives its nutrients by diffusion from blood vessels located in the connective tissue membrane (perichondrium) surrounding it. Its ground substance contains large amounts of the GAGs chrondroitin sulfate and hyaluronic acid, firmly bound collagen fibers (and in some cases elastic fibers), and is quite firm. Cartilage matrix also contains an exceptional amount of tissue fluid. In fact, cartilage is up to 80% water! The movement of tissue fluid in its matrix enables cartilage to rebound after being compressed and also helps to nourish the cartilage cells.

Chondroblasts, the predominant cell type in growing cartilage, produce new matrix until the skeleton stops growing at the end of adolescence. The firmness of the cartilage matrix prevents the cells from becoming widely separated, so **chondrocytes**, or mature cartilage cells, are typically found in small groups within cavities called *lacunae* (lah-ku′ne; "pits").

HOMEOSTATIC IMBALANCE

Because cartilage is avascular and aging cartilage cells lose their ability to divide, injured cartilages heal slowly. This phenomenon is excruciatingly familiar to those who have experienced sports injuries. During later life, cartilages tend to calcify or even ossify (become bony). In such cases, the chondrocytes are poorly nourished and die. ■

There are three varieties of cartilage: *hyaline cartilage, elastic cartilage,* and *fibrocartilage,* each dominated by a particular fiber type.

Hyaline Cartilage **Hyaline cartilage** (hi′ah-līn), or *gristle,* is the most abundant cartilage type in the body. Although it contains large numbers of collagen fibers, they are not apparent and the matrix appears amorphous and glassy (*hyalin* = glass) blue-white when viewed by the unaided eye (Figure 4.8g). Chondrocytes account for only 1–10% of the cartilage volume.

Hyaline cartilage provides firm support with some pliability. It covers the ends of long bones as *articular cartilage,* providing springy pads that absorb compression at joints. Hyaline cartilage also supports the tip of the nose, connects the ribs to the sternum, and supports most of the respiratory system passages. Most of the embryonic skeleton is formed of hyaline cartilage before bone is formed. Skeletal hyaline cartilage persists during childhood as the *epiphyseal plates* (e″pĭ-fis′e-ul), actively growing

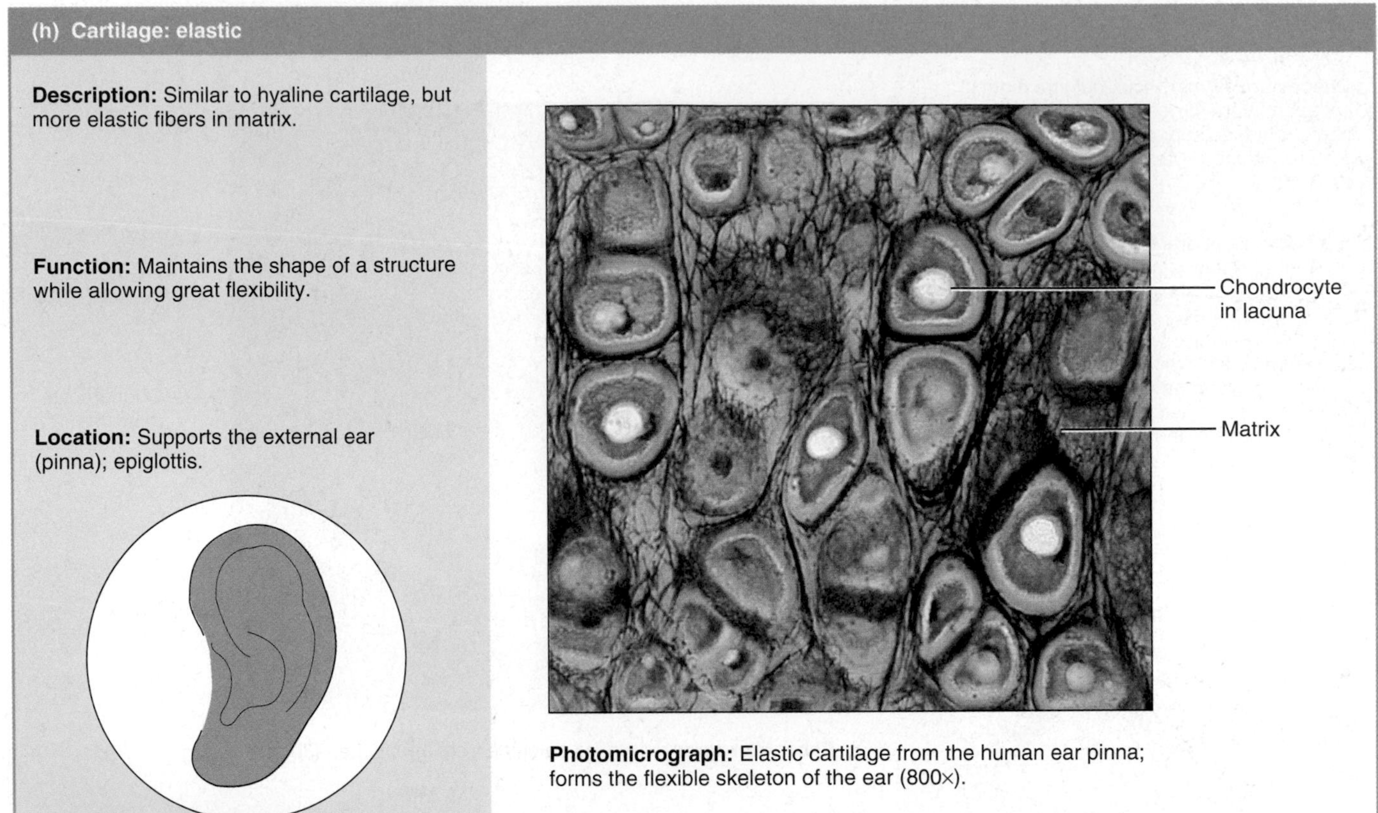

(h) Cartilage: elastic
Description: Similar to hyaline cartilage, but more elastic fibers in matrix.
Function: Maintains the shape of a structure while allowing great flexibility.
Location: Supports the external ear (pinna); epiglottis.
Chondrocyte in lacuna
Matrix
Photomicrograph: Elastic cartilage from the human ear pinna; forms the flexible skeleton of the ear (800×).

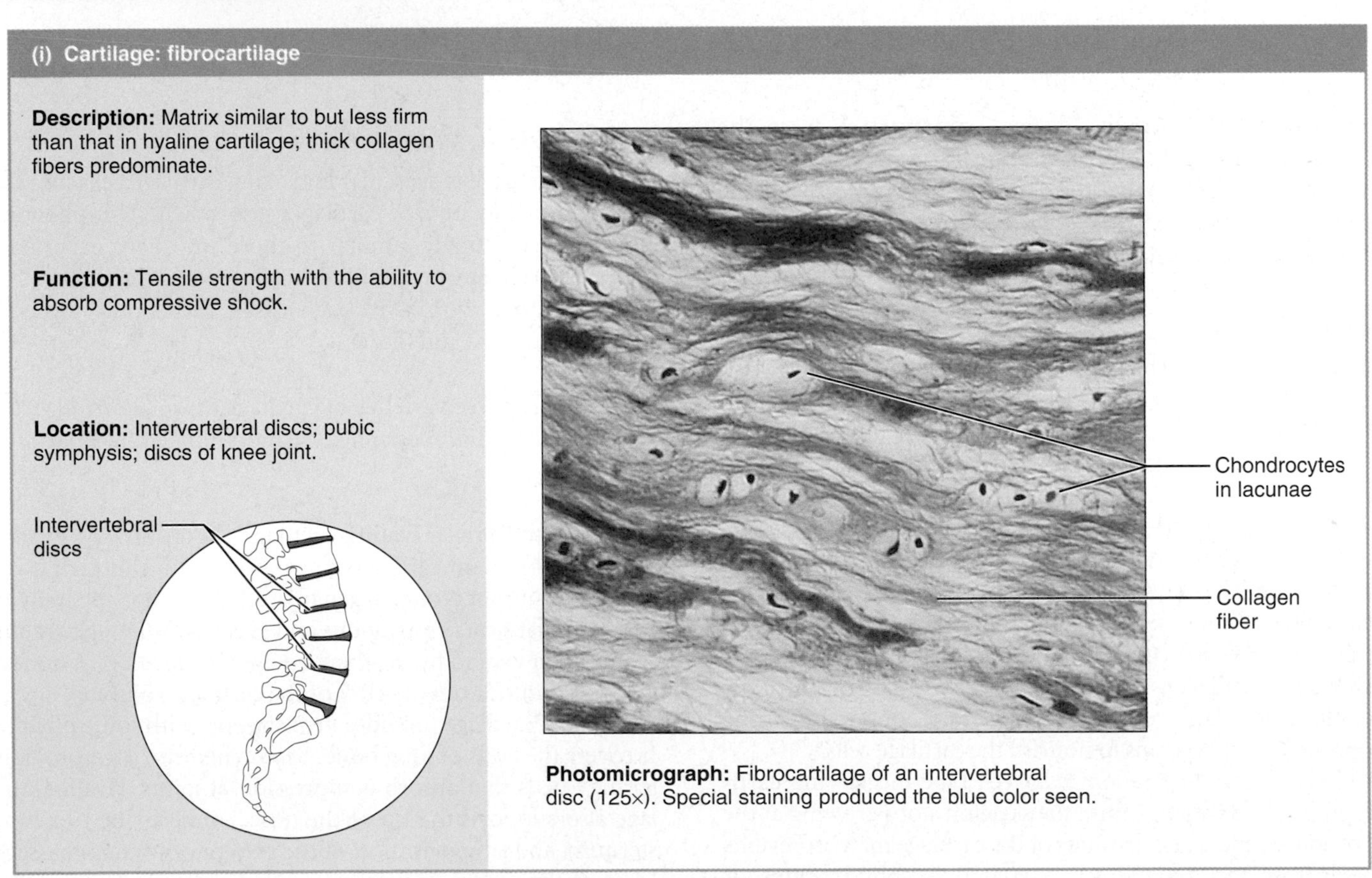

(i) Cartilage: fibrocartilage
Description: Matrix similar to but less firm than that in hyaline cartilage; thick collagen fibers predominate.
Function: Tensile strength with the ability to absorb compressive shock.
Location: Intervertebral discs; pubic symphysis; discs of knee joint.
Intervertebral discs
Chondrocytes in lacunae
Collagen fiber
Photomicrograph: Fibrocartilage of an intervertebral disc (125×). Special staining produced the blue color seen.

Figure 4.8 *(continued)* **Connective tissues. (h)** and **(i)** Cartilage. (See *A Brief Atlas of the Human Body*, Plates 18 and 19.)

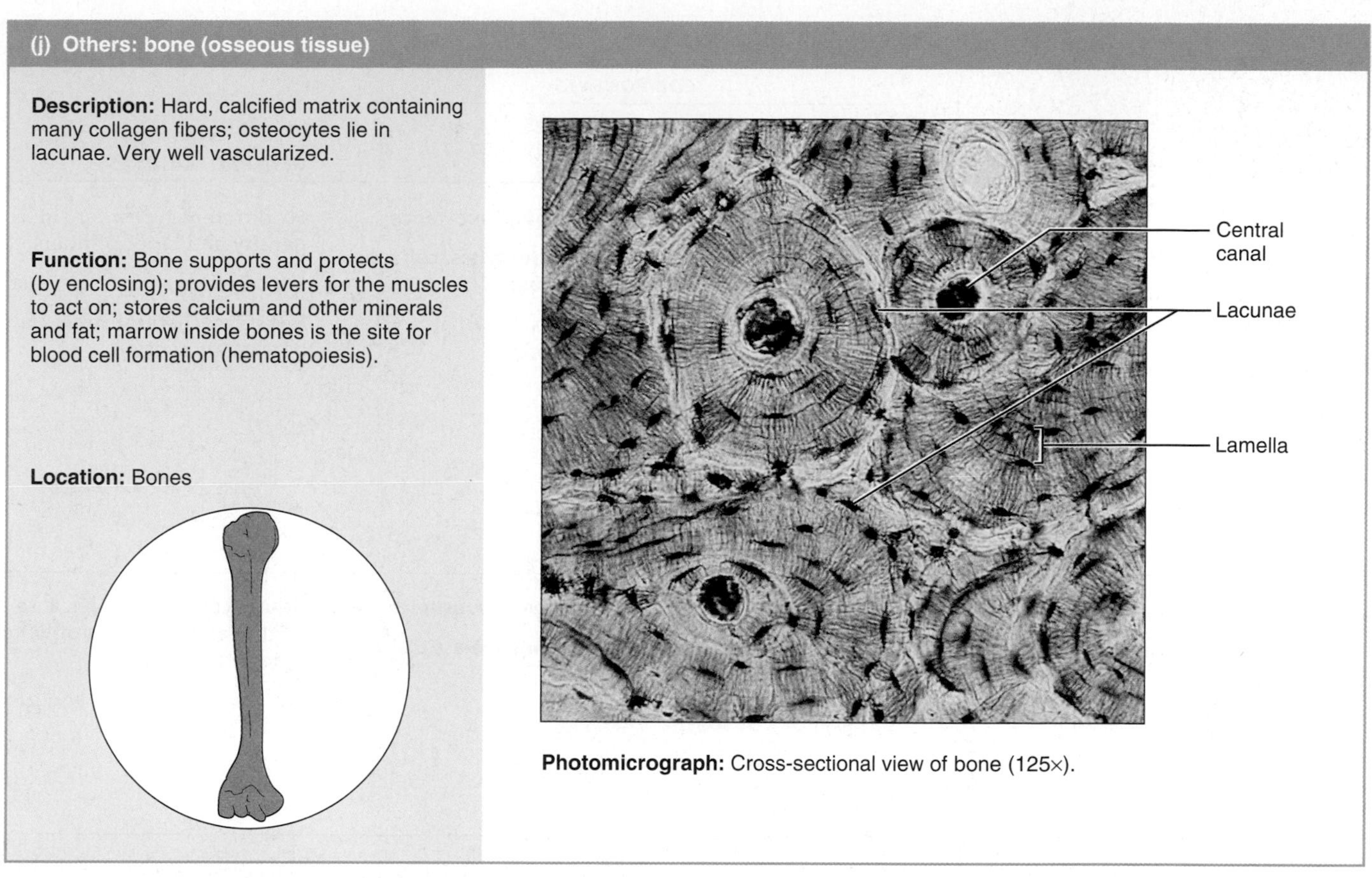

(j) Others: bone (osseous tissue)
Description: Hard, calcified matrix containing many collagen fibers; osteocytes lie in lacunae. Very well vascularized.
Function: Bone supports and protects (by enclosing); provides levers for the muscles to act on; stores calcium and other minerals and fat; marrow inside bones is the site for blood cell formation (hematopoiesis).
Location: Bones
Central canal
Lacunae
Lamella
Photomicrograph: Cross-sectional view of bone (125×).

Figure 4.8 *(continued)* **(j)** Bone. (See *A Brief Atlas of the Human Body*, Plate 20.)

regions near the ends of long bones that provide for continued growth in length.

Elastic Cartilage Histologically, **elastic cartilage** (Figure 4.8h) is nearly identical to hyaline cartilage. However, there are many more elastic fibers in elastic cartilage. Found where strength and exceptional stretchability are needed, elastic cartilage forms the "skeletons" of the external ear and the epiglottis. (The epiglottis is the flap that covers the opening to the respiratory passageway when we swallow, preventing food or fluids from entering the lungs.)

Fibrocartilage **Fibrocartilage** is a perfect structural intermediate between hyaline cartilage and dense regular connective tissues. Its rows of chondrocytes (a cartilage feature) alternate with rows of thick collagen fibers (a feature of dense regular connective tissue) (Figure 4.8i). Because it is compressible and resists tension well, fibrocartilage is found where strong support and the ability to withstand heavy pressure are required. For example, the intervertebral discs (resilient cushions between the bony vertebrae) and the spongy cartilages of the knee (menisci) are fibrocartilage structures (see Figure 6.1, p. 174).

Bone (Osseous Tissue)

Because of its rocklike hardness, **bone**, or **osseous tissue** (os′e-us), has an exceptional ability to support and protect body structures. Bones of the skeleton also provide cavities for fat storage and synthesis of blood cells. Bone matrix is similar to that of cartilage but is harder and more rigid because, in addition to its more abundant collagen fibers, bone has an added matrix element—inorganic calcium salts (bone salts).

Osteoblasts produce the organic portion of the matrix, and then bone salts are deposited on and between the fibers. Mature bone cells, or **osteocytes**, reside in the lacunae within the matrix they have made (Figure 4.8j). In cross section, bone tissue is seen as closely packed structural units called *osteons* formed of concentric rings of bony matrix (lamellae) surrounding central canals containing the blood vessels and nerves serving the bone. Unlike cartilage, the next firmest connective tissue, bone is well supplied by invading blood vessels.

Blood

Blood, the fluid within blood vessels, is the most atypical connective tissue. It does *not* connect things or give mechanical support. It is classified as a connective tissue because it develops from mesenchyme and consists of *blood cells*, surrounded by a nonliving fluid matrix called *blood plasma* (Figure 4.8k). The vast majority of blood cells are red blood cells or erythrocytes, but scattered white blood cells (neutrophils, lymphocytes, monocytes, eosinophils, basophils) are also seen. The "fibers" of blood are soluble protein molecules that precipitate, forming visible fiberlike structures during blood clotting. Blood functions as the transport vehicle for the cardiovascular system, carrying nutrients, wastes, respiratory gases, and many other substances throughout the body.

4

TABLE 4.1 Comparison of Classes of Connective Tissues

TISSUE CLASS AND EXAMPLE	SUBCLASSES	COMPONENTS: CELLS	COMPONENTS: MATRIX	GENERAL FEATURES
Connective Tissue Proper *Dense regular connective tissue*	1. Loose connective tissue ▪ Areolar ▪ Adipose ▪ Reticular 2. Dense connective tissue ▪ Regular ▪ Irregular ▪ Elastic	Fibroblasts Fibrocytes Defense cells Fat cells	Gel-like ground substance All three fiber types: collagen, reticular, elastic	Six different types; vary in density and types of fibers Functions as a binding tissue Resists mechanical stress, particulary tension
Cartilage *Hyaline cartilage*	1. Hyaline cartilage 2. Elastic cartilage 3. Fibrocartilage	Chondroblasts found in growing cartilage Chondrocytes	Gel-like ground substance Fibers: collagen, elastic fibers in some	Resists compression because of the large amounts of water held in the matrix Functions to cushion and support body structures
Bone Tissue *Compact bone*	1. Compact bone 2. Spongy bone	Osteoblasts Osteocytes	Gel-like ground substance calcified with inorganic salts Fibers: collagen	Hard tissue that resists both compression and tension Functions in support
Blood	Blood cell formation and differentiation are quite complex. Details are provided in Chapter 17.	Erythrocytes (RBC) Leukocytes (WBC) Platelets	Plasma No fibers	A fluid tissue Functions to carry O_2, CO_2, nutrients, wastes, and other substances (hormones, for example)

CHECK YOUR UNDERSTANDING

13. Which connective tissue has a soft weblike matrix capable of serving as a fluid reservoir?
14. What type of connective tissue is damaged when you lacerate your index finger tendon?
15. John wants to become a professional basketball player. Unfortunately he is short for his age and his epiphyseal plates have already fused. What type of connective tissue forms the epiphyseal plates?

For answers, see Appendix G.

Nervous Tissue

▶ Indicate the general characteristics of nervous tissue.

Nervous tissue is the main component of the nervous system—the brain, spinal cord, and nerves—which regulates and controls body functions. It contains two major cell types. **Neurons** are highly specialized nerve cells that generate and conduct nerve impulses **(Figure 4.9)**. Typically, they are branching cells with cytoplasmic extensions or processes. Their processes allow them to (1) respond to stimuli (a role of the processes called *dendrites*) and (2) to transmit electrical impulses over substantial distances within the body (the job of *axons*, which may be very long and myelinated, that is,

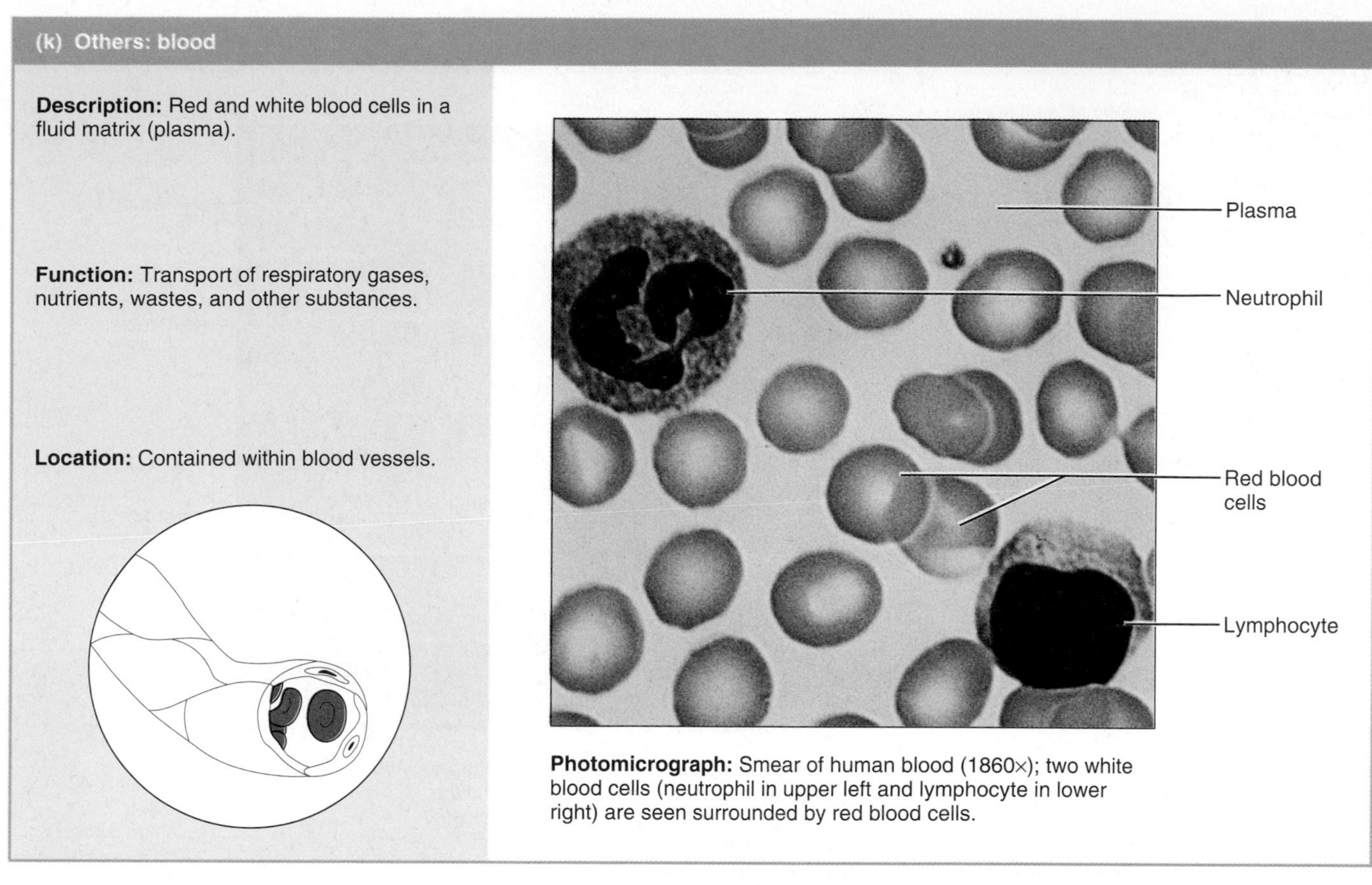

(k) Others: blood
Description: Red and white blood cells in a fluid matrix (plasma).
Function: Transport of respiratory gases, nutrients, wastes, and other substances.
Location: Contained within blood vessels.
Plasma
Neutrophil
Red blood cells
Lymphocyte
Photomicrograph: Smear of human blood (1860×); two white blood cells (neutrophil in upper left and lymphocyte in lower right) are seen surrounded by red blood cells.

Figure 4.8 *(continued)* **Connective tissues. (k)** Blood. (See *A Brief Atlas of the Human Body,* Plates 22–27.)

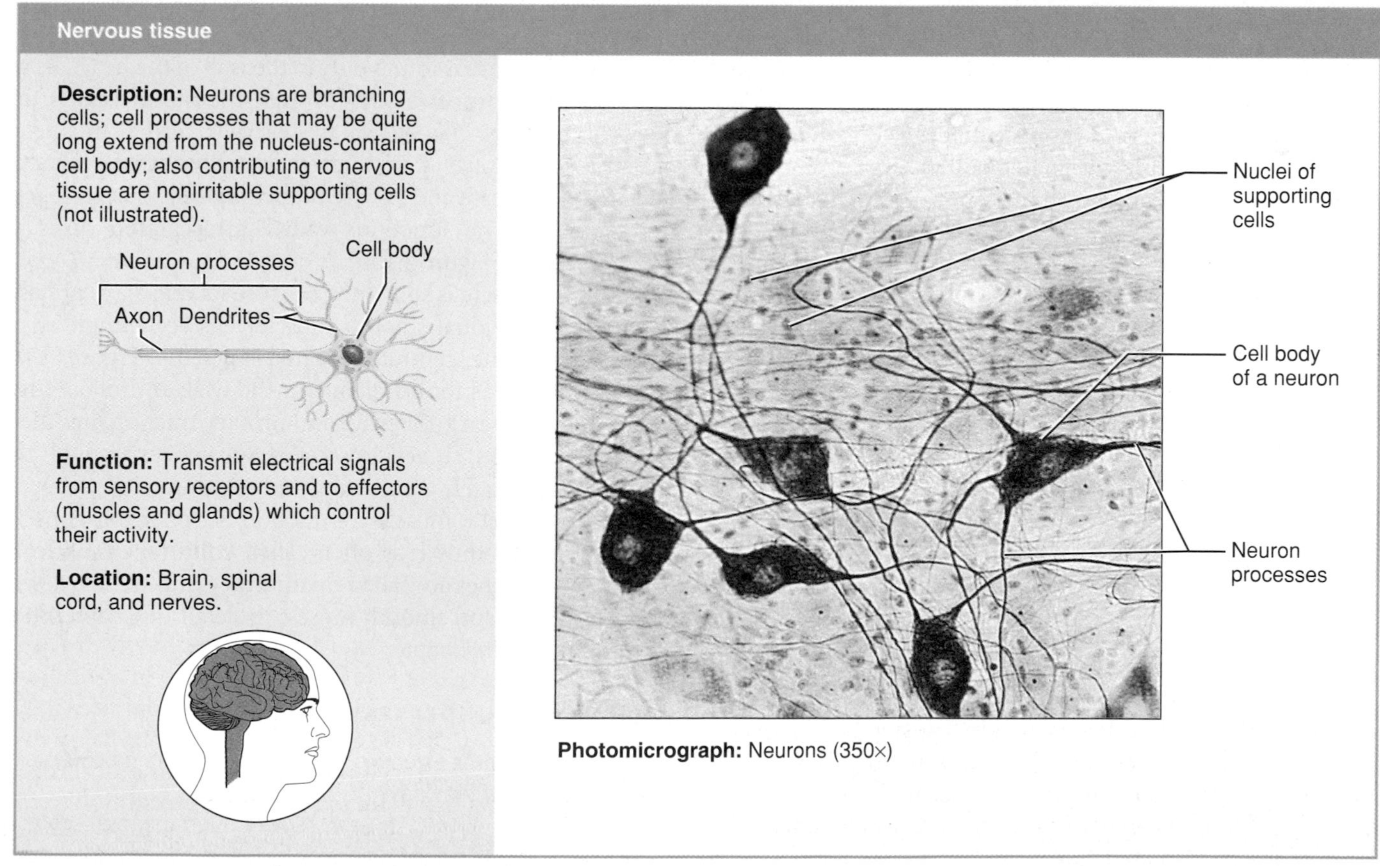

Nervous tissue
Description: Neurons are branching cells; cell processes that may be quite long extend from the nucleus-containing cell body; also contributing to nervous tissue are nonirritable supporting cells (not illustrated).
Neuron processes
Cell body
Axon
Dendrites
Function: Transmit electrical signals from sensory receptors and to effectors (muscles and glands) which control their activity.
Location: Brain, spinal cord, and nerves.
Nuclei of supporting cells
Cell body of a neuron
Neuron processes
Photomicrograph: Neurons (350×)

Figure 4.9 Nervous tissue. (See *A Brief Atlas of the Human Body,* Plate 33.)

4

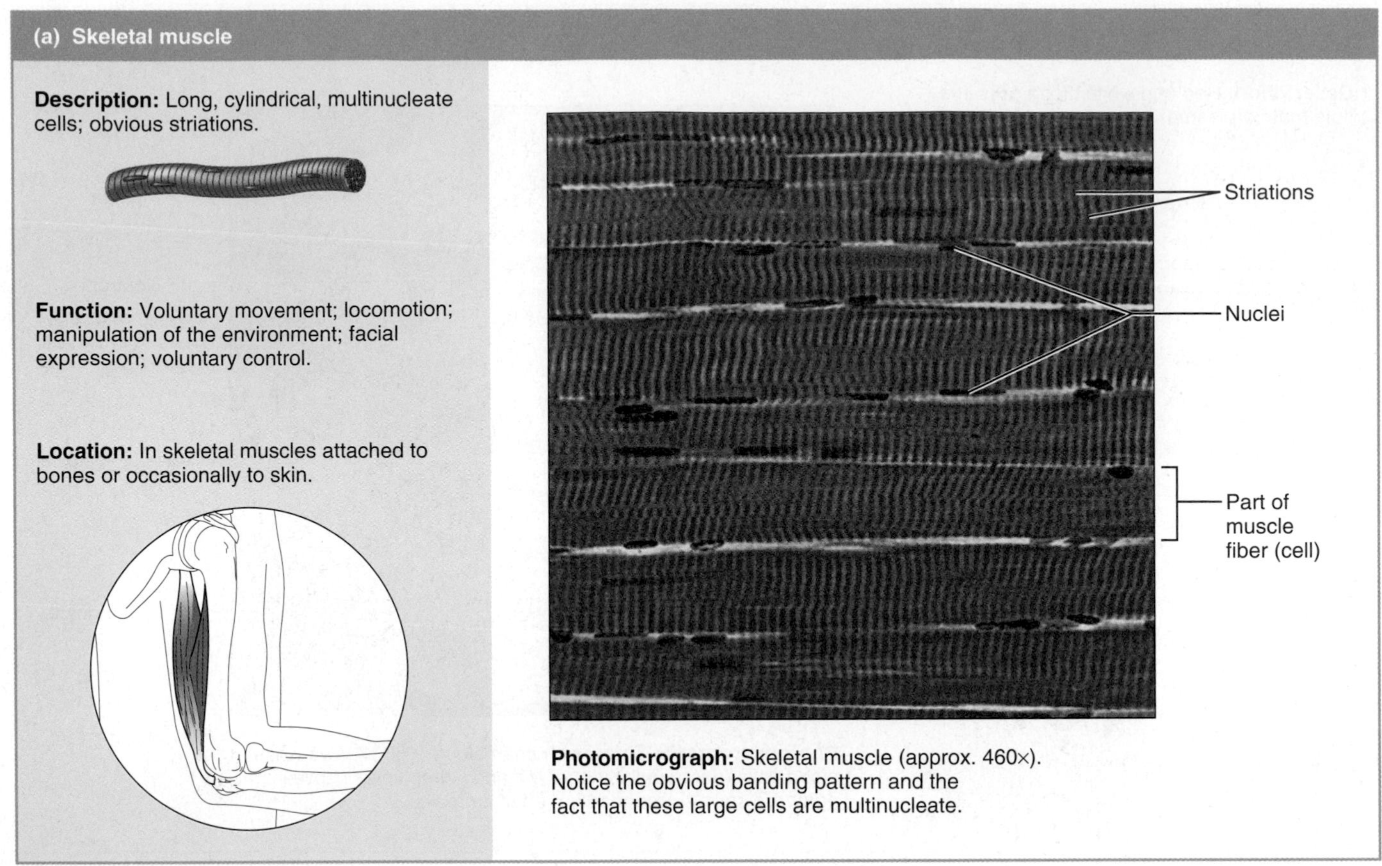

(a) Skeletal muscle
Description: Long, cylindrical, multinucleate cells; obvious striations.
Function: Voluntary movement; locomotion; manipulation of the environment; facial expression; voluntary control.
Location: In skeletal muscles attached to bones or occasionally to skin.
Striations
Nuclei
Part of muscle fiber (cell)
Photomicrograph: Skeletal muscle (approx. 460×). Notice the obvious banding pattern and the fact that these large cells are multinucleate.

Figure 4.10 Muscle tissues. (a) Skeletal muscle tissue. (See *A Brief Atlas of the Human Body,* Plate 28.)

covered with a fatty sheath that increases the speed of nerve transmission). The balance of nervous tissue consists of various types of supporting cells, nonconducting cells that support, insulate, and protect the delicate neurons. A more complete discussion of nervous tissue appears in Chapter 11.

Muscle Tissue

▶ Compare and contrast the structures and body locations of the three types of muscle tissue.

Muscle tissues are highly cellular, well-vascularized tissues that are responsible for most types of body movement. Muscle cells possess **myofilaments**, elaborate versions of the *actin* and *myosin* filaments that bring about movement or contraction in all cell types. There are three kinds of muscle tissue: skeletal, cardiac, and smooth.

Skeletal muscle tissue is packaged by connective tissue sheets into organs called *skeletal muscles* that are attached to the bones of the skeleton. These muscles form the flesh of the body, and as they contract, they pull on bones or skin, causing body movements. Skeletal muscle cells, also called **muscle fibers**, are long, cylindrical cells that contain many nuclei. Their obvious banded, or *striated,* appearance reflects the precise alignment of their myofilaments **(Figure 4.10a)**.

Cardiac muscle is found only in the wall of the heart. Its contractions help propel blood through the blood vessels to all parts of the body. Like skeletal muscle cells, cardiac muscle cells are striated. However, they differ structurally in that cardiac cells (1) are uninucleate and (2) are branching cells that fit together tightly at unique junctions called **intercalated discs** (in-ter′kah-la″ted) (Figure 4.10b).

Smooth muscle is so named because its cells have no visible striations. Individual smooth muscle cells are spindle shaped and contain one centrally located nucleus (Figure 4.10c). Smooth muscle is found mainly in the walls of hollow organs other than the heart (digestive and urinary tract organs, uterus, and blood vessels). It acts to squeeze substances through these organs by alternately contracting and relaxing.

Because skeletal muscle contraction is under our conscious control, skeletal muscle is often called **voluntary muscle**, and the other two types are called **involuntary muscle.** We describe skeletal muscle and smooth muscle in detail in Chapter 9, and cardiac muscle in Chapter 18.

CHECK YOUR UNDERSTANDING

16. How does the extended length of a neuron's processes aid its function in the body?

17. You are looking at muscle tissue through the microscope and you see striped branching cells that connect with one another. What type of muscle are you viewing?

(Text continues on p. 138.)

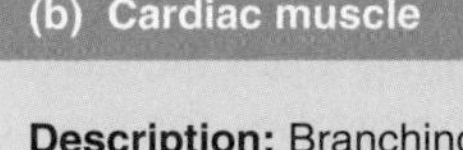

(b) Cardiac muscle

Description: Branching, striated, generally uninucleate cells that interdigitate at specialized junctions (intercalated discs).

Function: As it contracts, it propels blood into the circulation; involuntary control.

Location: The walls of the heart.

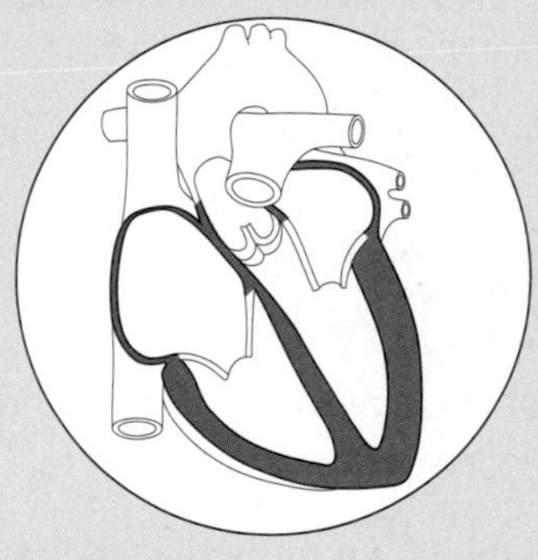

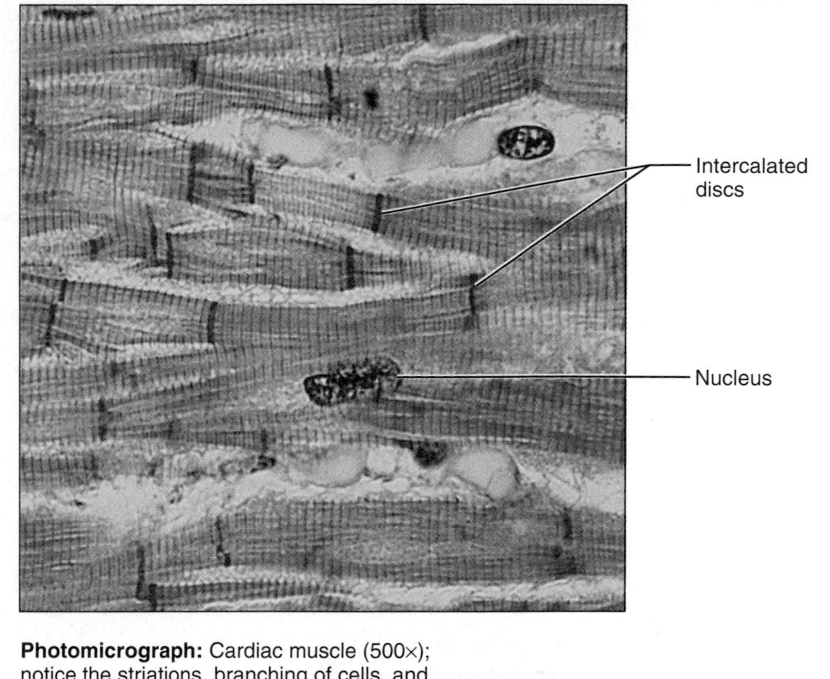

Photomicrograph: Cardiac muscle (500×); notice the striations, branching of cells, and the intercalated discs.

(c) Smooth muscle

Description: Spindle-shaped cells with central nuclei; no striations; cells arranged closely to form sheets.

Function: Propels substances or objects (foodstuffs, urine, a baby) along internal passageways; involuntary control.

Location: Mostly in the walls of hollow organs.

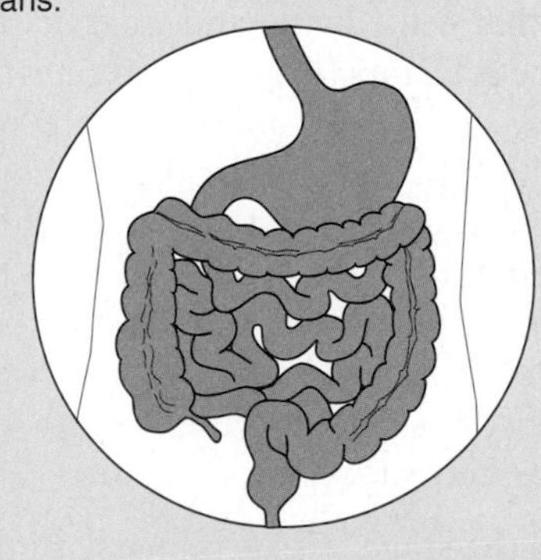

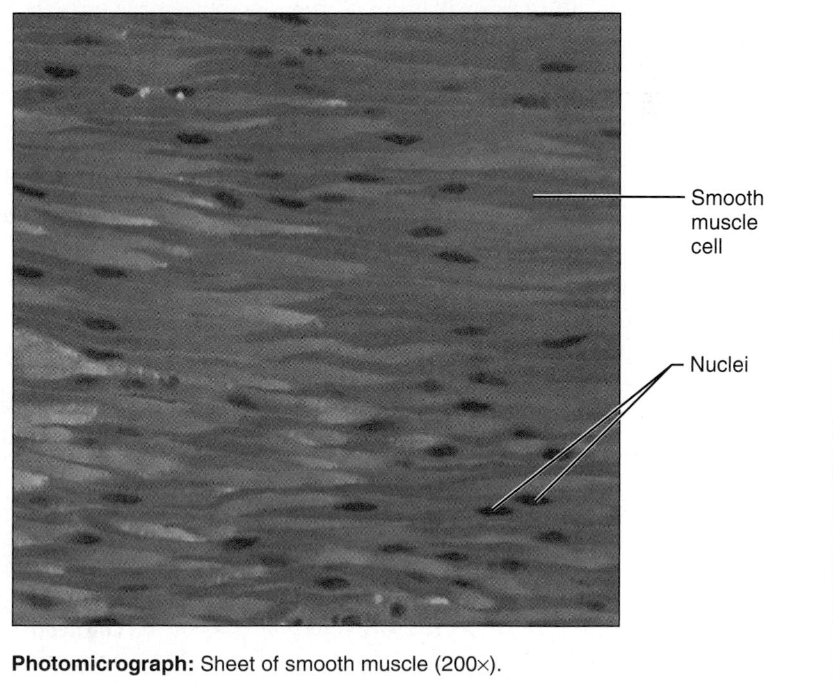

Photomicrograph: Sheet of smooth muscle (200×).

Figure 4.10 *(continued)* **(b)** Cardiac muscle tissue. (See *A Brief Atlas of the Human Body*, Plate 31.) **(c)** Smooth muscle tissue. (See *A Brief Atlas of the Human Body*, Plate 32.)

18. Which muscle type(s) is voluntary? Injured when you pull a muscle while exercising?

For answers, see Appendix G.

4

Covering and Lining Membranes

▶ Describe the structure and function of cutaneous, mucous, and serous membranes.

Now that we have described all four primary tissues, we can consider the body's membranes that incorporate more than one type of tissue. The covering and lining membranes are of three types: *cutaneous, mucous,* or *serous.* Essentially they all are continuous multicellular sheets composed of at least two primary tissue types: an epithelium bound to an underlying layer of connective tissue proper. Hence, these membranes are simple organs. We describe the *synovial membranes,* which line joint cavities and consist of connective tissue only, in Chapter 8.

Cutaneous Membrane

The **cutaneous membrane** (ku-ta′ne-us; *cutis* = skin) is your skin **(Figure 4.11a)**. It is an organ system consisting of a keratinized stratified squamous epithelium (epidermis) firmly attached to a thick layer of dense irregular connective tissue (dermis). Unlike other epithelial membranes, the cutaneous membrane is exposed to the air and is a dry membrane. Chapter 5 is devoted to this unique organ system.

Mucous Membranes

Mucous membranes, or **mucosae** (mu-ko′se), line body cavities that open to the exterior, such as those of the hollow organs of the digestive, respiratory, and urogenital tracts (Figure 4.11b). In all cases, they are "wet," or moist, membranes bathed by secretions or, in the case of the urinary mucosa, urine. Notice that the term *mucosa* refers to the location of the membrane, *not* its cell composition, which varies. However, most mucosae contain either stratified squamous or simple columnar epithelia. The epithelial sheet is directly underlain by a layer of loose connective tissue called the **lamina propria** (lam′ĭ-nah pro′pre-ah; "one's own layer"). In some mucosae, the lamina propria rests on a third (deeper) layer of smooth muscle cells.

Mucous membranes are often adapted for absorption and secretion. Although many mucosae secrete mucus, this is not a requirement. The mucosae of both the digestive and respiratory tracts secrete copious amounts of lubricating mucus, but that of the urinary tract does not.

Serous Membranes

Serous membranes, or **serosae** (se-ro′se), introduced in Chapter 1, are the moist membranes found in closed ventral body cavities (Figure 4.11c). A serous membrane consists of simple squamous epithelium (a mesothelium) resting on a thin layer of loose connective (areolar) tissue. The mesothelial cells add hyaluronic acid to the fluid that filters from the capillaries in the associated connective tissue. The result is the thin, clear *serous fluid* that lubricates the facing surfaces of the parietal and visceral layers, so that they slide across each other easily.

The serosae are named according to their site and specific organ associations. For example, the serosa lining the thoracic wall and covering the lungs is the **pleura**; that enclosing the heart is the **pericardium**; and those of the abdominopelvic cavity and viscera are the **peritoneums**.

CHECK YOUR UNDERSTANDING

19. What type of membrane consists of epithelium and connective tissue, and lines body cavities open to the exterior?
20. What type of membrane lines the thoracic walls and covers the lungs, and what is it called?

For answers, see Appendix G.

Tissue Repair

▶ Outline the process of tissue repair involved in normal healing of a superficial wound.

The body has many techniques for protecting itself from uninvited "guests" or injury. Intact mechanical barriers such as the skin and mucosae, the cilia of epithelial cells lining the respiratory tract, and the strong acid (chemical barrier) produced by stomach glands represent three defenses exerted at the body's external boundaries.

When tissue injury occurs, these barriers are penetrated. This stimulates the body's inflammatory and immune responses, which wage their battles largely in the connective tissues of the body. The *inflammatory response* is a relatively nonspecific reaction that develops quickly wherever tissues are injured, while the *immune response* is extremely specific, but takes longer to swing into action. We consider the inflammatory and immune responses in detail in Chapter 21.

Steps of Tissue Repair

Tissue repair requires that cells divide and migrate, activities that are initiated by growth factors (wound hormones) released by injured cells. Repair occurs in two major ways: by regeneration and by fibrosis. Which of these occurs depends on (1) the type of tissue damaged and (2) the severity of the injury. **Regeneration** is replacement of destroyed tissue with the same kind of tissue, whereas **fibrosis** involves proliferation of fibrous connective tissue called **scar tissue**. In skin, the tissue we will use as our example, repair involves both activities. **Figure 4.12** illustrates the following steps in tissue repair.

① **Inflammation sets the stage.** Tissue injury sets inflammatory events into motion. First, the tissue trauma causes injured tissue cells, macrophages, mast cells, and others to release inflammatory chemicals, which cause the capillar-

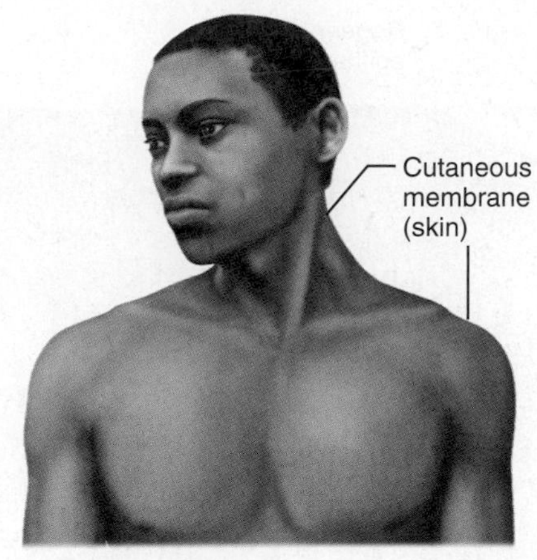

(a) Cutaneous membrane (the skin) covers the body surface.

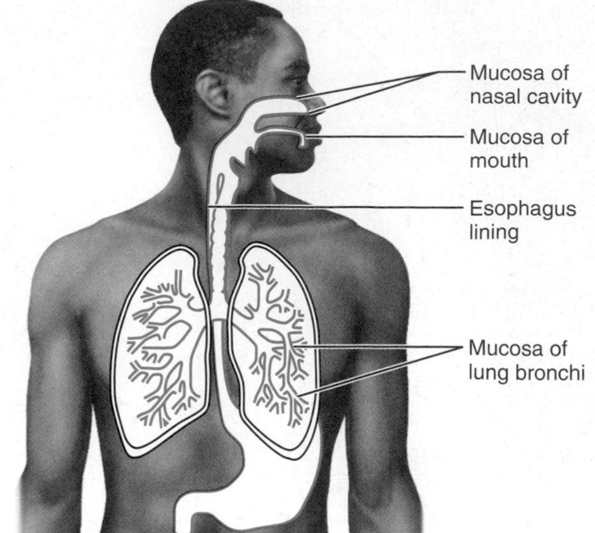

(b) Mucous membranes line body cavities open to the exterior.

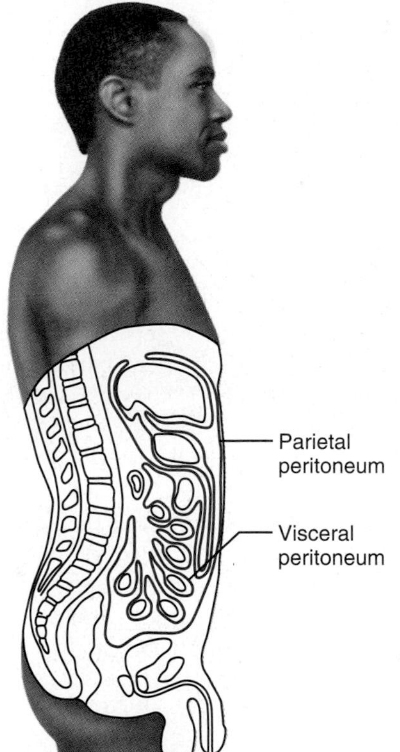

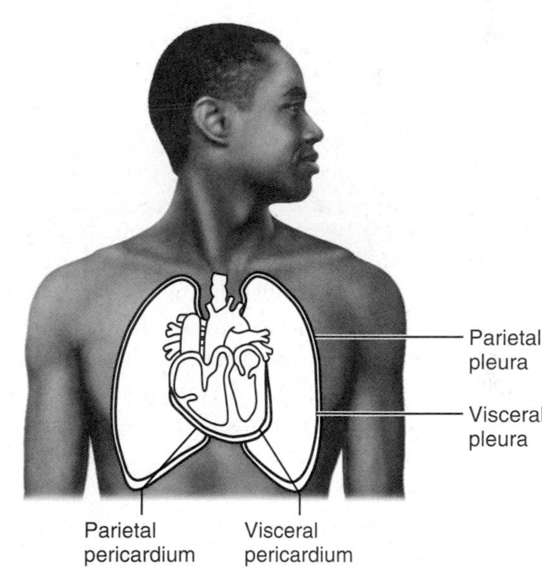

(c) Serous membranes line body cavities closed to the exterior.

Figure 4.11 Classes of membranes.

ies to dilate and become very permeable. This allows white blood cells (neutrophils, monocytes) and plasma fluid rich in clotting proteins, antibodies, and other substances to seep into the injured area. The leaked clotting proteins construct a clot, which stops the loss of blood, holds the edges of the wound together, and effectively walls in, or isolates, the injured area, preventing bacteria, toxins, or other harmful substances from spreading to surrounding tissues. The part of the clot exposed to air quickly dries and hardens, forming a *scab*. The inflammatory events leave behind excess fluid, bits of destroyed cells, and other debris, which are eventually removed via lymphatic vessels or phagocytized by macrophages.

② **Organization restores the blood supply.** Even while the inflammatory process is going on, the first phase of tissue repair, called **organization**, begins. During organization the blood clot is replaced by granulation tissue. **Granulation tissue** is a delicate pink tissue composed of several elements. It contains capillaries that grow in from nearby areas and lay down a new capillary bed. Granulation tissue is actually named for these capillaries, which protrude nublike from its surface, giving it a granular appearance. These capillaries are

4

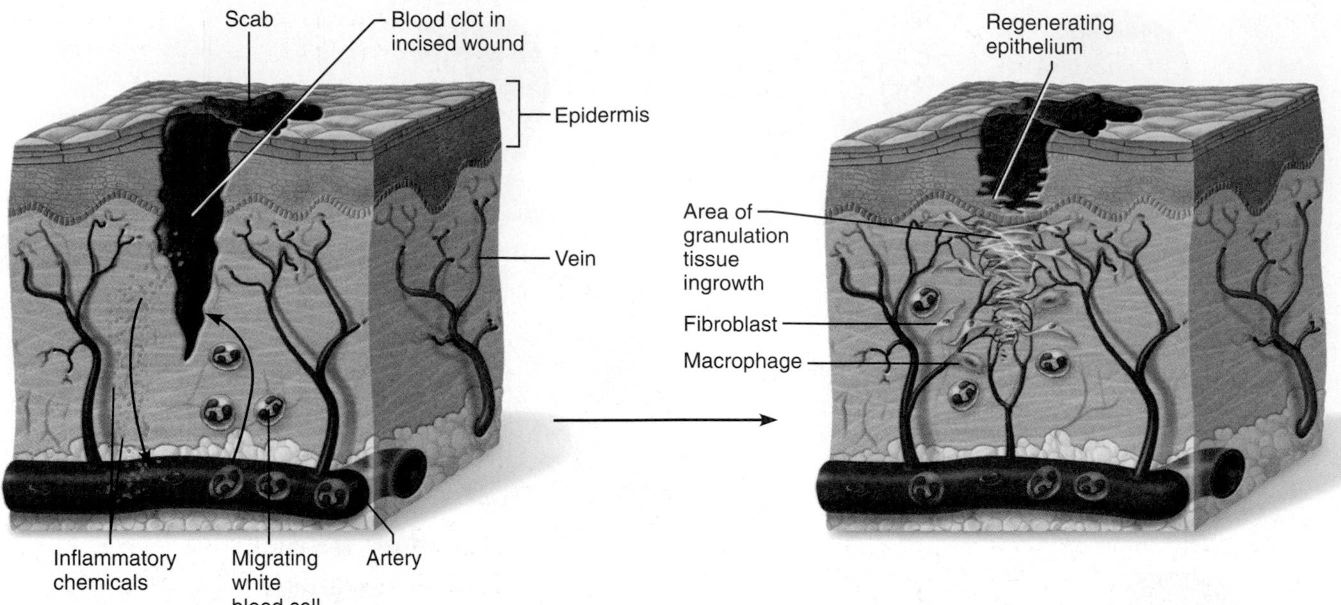

① **Inflammation sets the stage:**
- Severed blood vessels bleed and inflammatory chemicals are released.
- Local blood vessels become more permeable, allowing white blood cells, fluid, clotting proteins and other plasma proteins to seep into the injured area.
- Clotting occurs; surface dries and forms a scab.

② **Organization restores the blood supply:**
- The clot is replaced by granulation tissue, which restores the vascular supply.
- Fibroblasts produce collagen fibers that bridge the gap.
- Macrophages phagocytize cell debris.
- Surface epithelial cells multiply and migrate over the granulation tissue.

fragile and bleed freely, as we see when someone picks at a scab. Proliferating fibroblasts in granulation tissue produce growth factors as well as new collagen fibers to bridge the gap. Some of these fibroblasts have contractile properties that pull the margins of the wound together. As organization proceeds, macrophages digest the original blood clot and collagen fiber deposit continues. The granulation tissue, destined to become scar tissue (a permanent fibrous patch), is highly resistant to infection because it produces bacteria-inhibiting substances. As a rule, wound healing is a self-limited response. Once enough matrix has accumulated in the injured area, the fibroblasts either revert to the resting stage or undergo apoptosis.

③ **Regeneration and fibrosis effect permanent repair.** During organization, the surface epithelium begins to *regenerate,* growing under the scab, which soon detaches. As the fibrous tissue beneath matures and contracts, the regenerating epithelium thickens until it finally resembles that of the adjacent skin. The end result is a fully regenerated epithelium, and an underlying area of scar tissue. The scar may be invisible, or visible as a thin white line, depending on the severity of the wound.

The repair process that we have just described follows healing of a wound (cut, scrape, puncture) that breaches an epithelial barrier. In simple *infections* (a pimple or sore throat), healing is solely by regeneration. There is usually no clot formation or scarring. Only severe (destructive) infections lead to scarring.

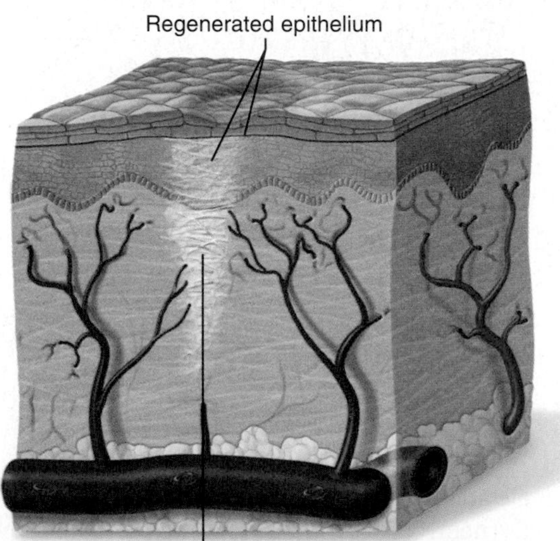

③ **Regeneration and fibrosis effect permanent repair:**
- The fibrosed area matures and contracts; the epithelium thickens.
- A fully regenerated epithelium with an underlying area of scar tissue results.

Figure 4.12 Tissue repair of a nonextensive skin wound: regeneration and fibrosis.

Regenerative Capacity of Different Tissues

The different tissues vary widely in their capacity for regeneration. Epithelial tissues, bone, areolar connective tissue, dense irregular connective tissue, and blood-forming tissue regenerate extremely well. Smooth muscle and dense regular connective tissue have a moderate capacity for regeneration, but skeletal muscle and cartilage have a weak regenerative capacity. Cardiac muscle and the nervous tissue in the brain and spinal cord have virtually no *functional* regenerative capacity, and they are routinely replaced by scar tissue. However, recent studies have shown that some unexpected (and highly selective) cellular division occurs in both these tissues after damage, and efforts are under way to coax them to regenerate better.

In nonregenerating tissues and in exceptionally severe wounds, fibrosis totally replaces the lost tissue. Over a period of months, the fibrous mass shrinks and becomes more and more compact. The resulting scar appears as a pale, often shiny area composed mostly of collagen fibers. Scar tissue is strong, but it lacks the flexibility and elasticity of most normal tissues. Also, it cannot perform the normal functions of the tissue it has replaced.

HOMEOSTATIC IMBALANCE

Scar tissue that forms in the wall of the urinary bladder, heart, or other muscular organ may severely hamper the function of that organ. The normal shrinking of the scar reduces the internal volume of an organ and may hinder or even block movement of substances through a hollow organ. Scar tissue hampers muscle's ability to contract and may interfere with its normal excitation by the nervous system. In the heart, these problems may lead to progressive heart failure. In irritated visceral organs, particularly following abdominal surgery, *adhesions* may form as the newly forming scar tissue connects adjacent organs together. Such adhesions can prevent the normal shifting about (churning) of loops of the intestine, dangerously obstructing the flow of foodstuffs through it. Adhesions can also restrict heart movements and immobilize joints. ■

CHECK YOUR UNDERSTANDING

21. What are the three main steps of tissue repair?
22. Why does a deep injury to the skin result in abundant scar tissue formation?

For answers, see Appendix G.

4

Developmental Aspects of Tissues

- Indicate the embryonic origin of each tissue class.
- Briefly describe tissue changes that occur with age.

One of the first events of embryonic development is the formation of the three **primary germ layers**, which lie one atop the next like a three-layered cellular pancake. From superficial to deep, these layers are the **ectoderm**, **mesoderm** (mez′o-derm), and **endoderm**. As shown in Figure 4.13, these primary germ layers then specialize to form the four primary tissues—epithelium, nervous tissue, muscle, and connective tissue—that make up all body organs.

By the end of the second month of development, the primary tissues have appeared, and all major organs are in place. In general, tissue cells remain mitotic and produce the rapid growth that occurs before birth. The division of nerve cells, however, stops or nearly stops during the fetal period. After birth, the cells of most other tissues continue to divide until adult body size is achieved. Cellular division then slows greatly,

(Text continues on p. 144.)

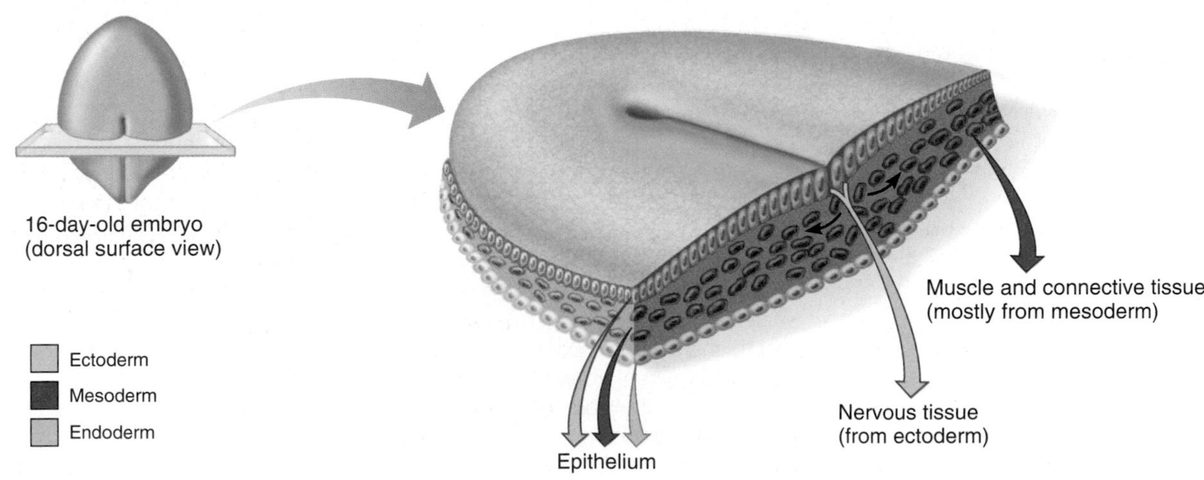

Figure 4.13 Embryonic germ layers and the primary tissue types they produce. The three embryonic layers collectively form the very early embryonic body.

4

CLOSER LOOK

Cancer—The Intimate Enemy

The word *cancer* elicits dread in everyone. Why does cancer strike some and not others?

Although once perceived as disorganized cell growth, this disease is now known to be a logical, coordinated process in which a precise sequence of tiny alterations changes a normal cell into a killer. Let's take a closer look at what cancer really is.

When cells fail to follow normal controls of cell division and multiply excessively, an abnormal mass of proliferating cells called a **neoplasm** (ne′o-plazm, "new growth") results. Neoplasms are classified as **benign** ("kindly") or **malignant** ("bad"). A benign neoplasm is strictly a local affair. Its cells remain compacted, are often encapsulated, tend to grow slowly, and seldom kill their hosts if removed before they compress vital organs.

In contrast, cancers are malignant neoplasms, nonencapsulated masses that grow relentlessly and may become killers. Their cells resemble immature cells, and they invade their surroundings rather than pushing them aside, as reflected in the name *cancer*, from the Latin word for "crab." Whereas normal cells become fatally "homesick" and die when they lose contact with the surrounding matrix, malignant cells tend to break away from the parent mass, the primary tumor, and travel via blood or lymph to other body organs, where they form *secondary cancer masses*.

This capability for traveling to other parts of the body, called **metastasis** (mĕ-tas′tah-sis), probably has a lot to do with signaling molecules and the cell-surface glycoproteins the cancer cells bear. Metastasis and invasiveness distinguish cancer cells from the cells of benign neoplasms. Cancer cells consume an exceptional amount of the body's nutrients, leading to weight loss and tissue wasting that contribute to death.

Carcinogenesis

Autopsies on individuals aged 50–70 who died of another cause have revealed that most of us have microscopic (but dormant) in situ neoplasms. So what causes a normal cell to **transform** or change into a cancerous one? Some physical factors (radiation, mechanical trauma), certain viral infections, chronic inflammations, and many chemicals (tobacco tars, saccharine, some natural food chemicals) can act as **carcinogens** (cancer-causers). What do these factors have in common? They all cause *mutations*—changes in DNA that alter the expression of certain genes. However, not all carcinogens do damage because most are eliminated by peroxisomal or lysosomal enzymes or by the immune system. Furthermore, one mutation usually isn't enough. It takes several genetic changes to transform a normal cell into a cancerous cell.

A clue to the role of genes in cancer was provided by the discovery of **oncogenes** (Greek *onco* = tumor), or cancer-causing genes, in rapidly spreading cancers. **Proto-oncogenes**, benign forms of oncogenes in normal cells, were discovered later. Proto-oncogenes code for proteins that are essential for cell division, growth, and cellular adhesion, among other things. Many have fragile sites that break when exposed to carcinogens, converting them to oncogenes. Failure to code for certain proteins may lead to loss of an enzyme that controls an important metabolic process. Oncogenes may also "switch on" dormant genes that allow cells to become invasive and metastasize. Known oncogenes now number over 100.

Oncogenes have been detected in only 15–20% of human cancers, so investigators were not too surprised by the discovery of **tumor suppressor genes**, or **anti-oncogenes**, which suppress cancer by inactivating carcinogens, aiding DNA repair, or enhancing the immune system's counterattack. In fact, over half of all cancers involve malfunction or loss of just 2 of the 15 identified tumor suppressor genes—*p53* and *p16*. This is not surprising when you learn that *p53* prompts most cells to make proteins that stop cell division in stressed cells by promoting apoptosis or cell cycle arrest. Its impairment invites uncontrolled division and cancer.

Furthermore, although each type of cancer is genetically distinct, human cancers appear to share a common master set of genes—an activated group of 67 genes—and almost all cancer cells have gained or lost entire chromosomes. Whatever genetic factors are at work, the "seeds" of cancer do appear to be in our own genes. Cancer is an intimate enemy indeed.

The illustration depicts some of the mutations involved in colorectal cancer, one of the best-understood human cancers. As with most cancers, a metastasis develops gradually. One of the first signs is a polyp, a small benign growth consisting of apparently normal mucosa cells. As cell division continues, the growth enlarges, becoming an adenoma (a term for any neoplasm of glandular epithelium). As various tumor suppressor genes are inactivated and the *K-ras* oncogene is mobilized, the mutations pile up and the adenoma becomes increasingly abnormal. The final consequence is colon carcinoma, a form of cancer that metastasizes quickly.

Cancer Prevalence

Almost half of all Americans develop cancer in their lifetime and a fifth of us will die of it. Cancer can arise from almost any cell type, but the most common cancers originate in the skin, lung, colon, breast, and prostate gland. Although stomach and colon cancer incidence is down, skin and lymphoid cancer rates are up.

Many cancers are preceded by observable lumps or other structural changes in tissue—for instance, *leukoplakia*, white patches in the mouth caused by the chronic irritation of ill-fitting dentures or heavy smoking. Although these lesions sometimes progress to cancer, in many cases they remain stable or even revert to normal if the environmental stimulus is removed.

Diagnosis and Staging

Screening procedures are vital for early detection. Examples include *mammography* (X-ray examination of the breasts), examining breasts or testicles for lumps, and checking fecal samples for blood. Unfortunately, most cancers are diagnosed only after symptoms have already appeared. In this case the diagnostic method is usually a **biopsy**: removing a tissue sample surgically and examining it microscopically for malignant cells. Increasingly, diagnosis is made by chemical or genetic analysis of the sample. Typing cancer cells by what genes are switched on or off tells clinicians which drugs to use. For example, taxol, quite successful with breast and ovarian cancer, works only against tumors with a specific genetic makeup.

Several techniques (physical and histological examinations, lab tests, and imaging techniques [MR, CT]) are used to determine the extent of the disease (size

A CLOSER LOOK
(continued)

of the neoplasm, degree of metastasis, etc.). Then, the cancer is assigned a **stage** from 1 to 4 according to the probability of cure (stage 1 has the best probability, stage 4 the worst).

Cancer Treatments

Most cancers are removed surgically if possible. To destroy metastasized cells, surgery is commonly followed by radiation therapy (X irradiation and/or treatment with radioisotopes) and chemotherapy (treatment with cytotoxic drugs). Recently, some oncologists have been using heat therapy (just a slight upward temperature change) to put the cancer cells on the "cliff's edge," so that they are sensitized and much more vulnerable to chemotherapy or radiation.

Chemotherapy is beset with the problem of resistance. Some cancer cells can eject the drugs in tiny bubbles or flattened vesicles dubbed exosomes, and these cells proliferate, forming new tumors that are invulnerable to chemotherapy. Furthermore, anticancer drugs have unpleasant side effects—nausea, vomiting, hair loss—because they kill all rapidly dividing cells, including normal tissues and cells. The anticancer drugs also can severely damage the brain, producing a phenomenon called chemobrain—a mental fuzziness and memory loss reported by many cancer patients. X rays also have side effects because, in passing through the body, they destroy healthy tissue in their path as well as cancer cells.

Promising New Therapies

Traditional cancer treatments—"cut, burn, and poison"—are widely recognized as crude and painful. Promising new therapies focus on

- *Targeted drugs that interrupt the signaling pathways that fuel the cancer's growth*. Examples include imatinib (Gleevec), which incapacitates a mutated enzyme that triggers uncontrolled division of cells in two rare blood and digestive system cancers, and trastuzumab (Herceptin), used to treat breast cancer patients. These drugs have been strikingly successful in providing a few extra weeks of life, before their protective effects wear off and the disease progresses again.
- *Delivering drugs or radiation more precisely to the cancer while sparing normal tissue*. One approach is to inject the patient with tiny drug-coated metal beads, which are guided to the tumor by a powerful magnet positioned over the body site. Or, a patient might take light-sensitive drugs that are drawn naturally into rapidly dividing cancer cells. Then, exposure to certain frequencies of laser light sets off reactions that kill the malignant cells. Another new procedure, proton therapy, delivers highly targeted killing doses of protons (radiation) that strike at cancer cells with incredible precision and with greater effectiveness than traditional X rays. Unlike X rays, which pass through the cancer and onward through the patient's body, protons can be slowed down and even directed to stop in the neoplasm.
- *Using genetically modified immune cells to target cancer cells*. One promising technique harvests a patient's most aggressive cancer-killing immune cells (T lymphocytes), inserts modified genes into them that make them even more efficient killers, multiplies the cells in the lab, and then infuses the immune cells back into the patient.
- *Using drugs that target cancer cell bioenergetics*. The fact that many cancers use glucose as their energy fuel almost exclusively has suggested a pharmaceutical approach that limits glucose use. In theory, this approach would kill cancer cells while sparing normal cells, which can also use amino acids and fats as energy fuels.

Other experimental treatments seek to starve cancer cells by cutting off their blood supply, fix defective tumor suppressor genes and oncogenes, destroy cancer cells with viruses, or signal cancer cells to commit suicide by apoptosis. Additionally, a cancer vaccine (TRICOM) contains genetically engineered viruses carrying genes for a cancer protein called carcinoembryonic antigen (CEA). When these proteins are delivered into the patient's body, they stimulate an immune response that orchestrates an attack on all CEA-bearing cancer cells.

At present, about half of all cancer cases are cured. Although average survival rates have not increased, the quality of life of cancer patients has improved in the last decade. We can offer better treatments for cancer-associated pain, and antinausea drugs and other helpful medicines can soothe the side effects of chemotherapy.

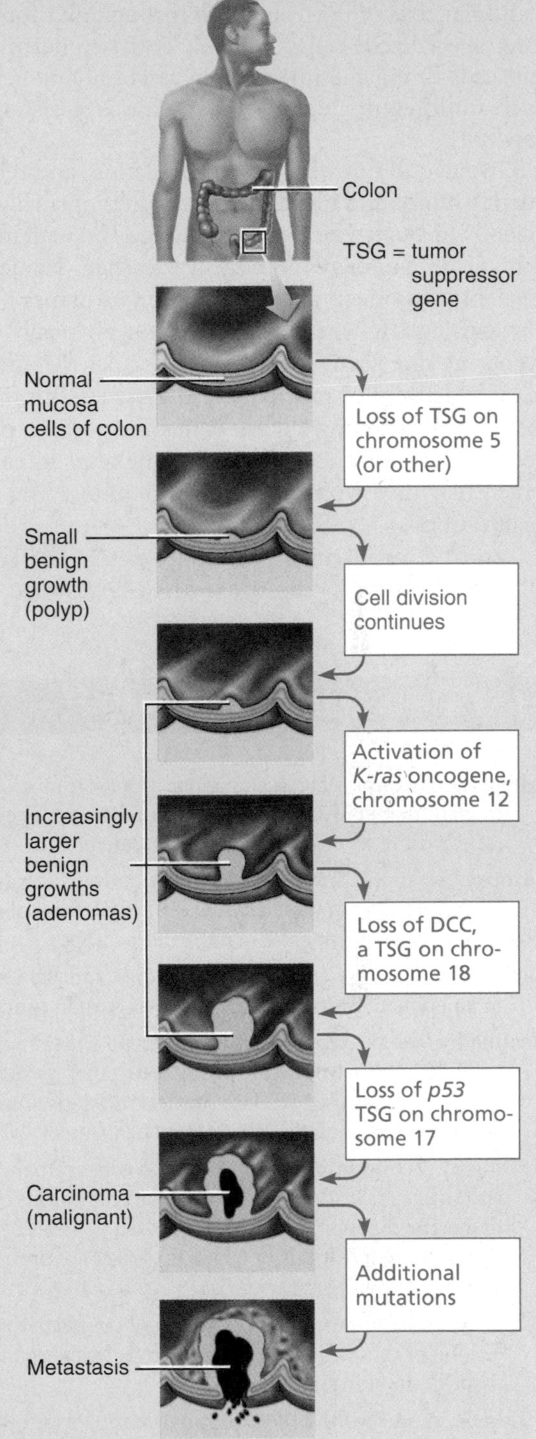

Some of the mutations involved in the development of colon cancer.

4

although many tissues retain the ability to regenerate. In adults, only epithelia and blood-forming tissues are highly mitotic. Some tissues that regenerate through life, such as the glandular cells of the liver, do so through division of their mature (specialized) cells. Others, like the epidermis of the skin and cells lining the intestine, have abundant *stem cells,* relatively undifferentiated cells that divide as necessary to produce new cells.

4

Given good nutrition, good circulation, and relatively infrequent wounds and infections, our tissues normally function efficiently through youth and middle age. But with increasing age, epithelia thin and are more easily breached. Tissue repair is less efficient, and bone, muscle, and nervous tissues begin to atrophy, particularly when a person is not physically active. These events are due partly to decreased circulatory efficiency, which reduces delivery of nutrients to the tissues, but in some cases, diet is a contributing factor. As income declines or as chewing becomes more difficult, older people tend to eat soft foods, which may be low in protein and vitamins. As a result, tissue health suffers.

Another problem of aging tissues is the likelihood of DNA mutations in the most actively mitotic cells, which increases the risk of cancer (as indicated in *A Closer Look* on p. 142).

CHECK YOUR UNDERSTANDING

23. What are the names of the three embryonic germ layers?
24. Which germ layer gives rise to the nervous system?
25. Which two tissue types remain highly mitotic throughout life?

For answers, see Appendix G.

As we have seen, body cells combine to form four discrete tissue types: epithelial, connective, muscle, and nervous tissues. The cells making up each of these tissues share certain features but are by no means identical. They "belong" together because they have basic functional similarities. The connective tissues assume many guises, but perhaps the most versatile cells are those of epithelium: They protect our outer and inner surfaces, permit us to obtain oxygen, absorb vital nutrients into the blood, and allow our kidneys to excrete wastes. The important concept to carry away with you is that tissues, despite their unique abilities, cooperate to keep the body safe, healthy, and whole.

RELATED CLINICAL TERMS

Adenoma (ad″ĕ-no′mah; *aden* = gland, *oma* = tumor) Any neoplasm of glandular epithelium, benign or malignant. The malignant type is more specifically called *adenocarcinoma.*

Autopsy (aw′top-se) Examination of the body, its organs, and its tissues after death to determine the actual cause of death; also called postmortem examination and necropsy.

Carcinoma (kar″sĭ-no′mah; *karkinos* = crab, cancer) Cancer arising in an epithelium; accounts for 90% of human cancers.

Healing by first intention The simplest type of healing; occurs when the edges of the wound are brought together by sutures, staples, or other means used to close surgical incisions. Only small amounts of granulation tissue need be formed.

Healing by second intention The wound edges remain separated, and the gap is bridged by relatively large amounts of granulation tissue; the manner in which unattended wounds heal. Healing is slower than in wounds in which the edges are brought together, and larger scars result.

Keloid (ke′loid) Abnormal proliferation of connective tissue during healing of skin wounds; results in large, unsightly mass of scar tissue at the skin surface.

Lesion (le′zhun; "wound") Any injury, wound, or infection that affects tissue over an area of a definite size (as opposed to being widely spread throughout the body).

Marfan's syndrome Genetic disease resulting in abnormalities of connective tissues due to a defect in fibrillin, a protein that is associated with elastin in elastic fibers. Clinical signs include loose-jointedness, long limbs and spiderlike fingers and toes, visual problems, and weakened blood vessels (especially the aorta) due to poor connective tissue reinforcement.

Pathology (pah-thol′o-je) Scientific study of changes in organs and tissues produced by disease.

Pus A collection of tissue fluid, bacteria, dead and dying tissue cells, white blood cells, and macrophages in an inflamed area.

Sarcoma (sar-ko′mah; *sarkos* = flesh; *oma* = tumor) Cancer arising in the mesenchyme-derived tissues, that is, in connective tissues and muscle.

Scurvy A nutritional deficiency caused by lack of adequate vitamin C needed to synthesize collagen; signs and symptoms include blood vessel disruption, delay in wound healing, weakness of scar tissue, and loosening of teeth.

VAC (vacuum-assisted closure) Innovative healing process for open-skin wounds and skin ulcers. Often induces healing when all other methods fail. Involves covering the wound with a special sponge, and then applying suction through the sponge. In response to the subsequent skin stretching, fibroblasts in the wound form more collagen tissue and new blood vessels proliferate, bringing more blood into the injured area, which also promotes healing.

CHAPTER SUMMARY

Media study tools that could provide you additional help in reviewing specific key topics of Chapter 4 are referenced below.

iP = ***Interactive Physiology***

Tissues are collections of structurally similar cells with related functions. The four primary tissues are epithelial, connective, muscle, and nervous tissues.

Preparing Human Tissue for Microscopy (pp. 114–115)

1. Preparation of tissues for microscopic examination involves cutting thin sections of the tissue and using dyes to stain the tissue. Minor distortions called artifacts can be introduced by the tissue preparation process.

Epithelial Tissue (pp. 115–124)

1. Epithelial tissue is the covering, lining, and glandular tissue of the body. Its functions include protection, absorption, excretion, filtration, secretion, and sensory reception.

Special Characteristics of Epithelium (pp. 115–116)

2. Epithelial tissues exhibit specialized contacts, polarity, avascularity, support from connective tissue, and high regenerative capacity.

Classification of Epithelia (pp. 116–121)

3. Epithelium is classified by arrangement as simple (one layer) or stratified (more than one layer) and by cell shape as squamous, cuboidal, or columnar. The terms denoting cell shape and arrangement are combined to describe the epithelium fully.
4. Simple squamous epithelium is a single layer of squamous cells. Highly adapted for filtration and exchange of substances, it forms walls of air sacs of the lungs and lines blood vessels. It contributes to serosae as mesothelium and lines all hollow circulatory system organs as endothelium.
5. Simple cuboidal epithelium, commonly active in secretion and absorption, is found in glands and in kidney tubules.
6. Simple columnar epithelium, specialized for secretion and absorption, consists of a single layer of tall columnar cells that exhibit microvilli and often mucus-producing cells. It lines most of the digestive tract.
7. Pseudostratified columnar epithelium is a simple columnar epithelium that appears stratified. Its ciliated variety, rich in mucus-secreting cells, lines most of the upper respiratory passages.
8. Stratified squamous epithelium is multilayered; cells at the free surface are squamous. It is adapted to resist abrasion. It lines the esophagus and vagina; its keratinized variety forms the skin epidermis.
9. Stratified cuboidal epithelia are rare in the body, and are found chiefly in ducts of large glands. Stratified columnar epithelium has a very limited distribution, found mainly in the male urethra and at transition areas between other epithelia types.
10. Transitional epithelium is a modified stratified squamous epithelium, adapted for responding to stretch. It lines hollow urinary system organs.

Glandular Epithelia (pp. 121–124)

11. A gland is one or more cells specialized to secrete a product.
12. On the basis of site of product release, glands are classified as exocrine or endocrine. Glands are classified structurally as multicellular or unicellular.
13. Unicellular glands, typified by goblet cells and mucous cells, are mucus-secreting single-celled glands.
14. Multicellular exocrine glands are classified according to duct structure as simple or compound, and according to the structure of their secretory parts as tubular, alveolar, or tubuloalveolar.
15. Multicellular exocrine glands of humans are classified functionally as merocrine or holocrine.

4

Connective Tissue (pp. 124–135)

1. Connective tissue is the most abundant and widely distributed tissue of the body. Its functions include support, protection, binding, insulation, and transportation (blood).

Common Characteristics of Connective Tissue (p. 124)

2. Connective tissues originate from embryonic mesenchyme and exhibit matrix. Depending on type, a connective tissue may be well vascularized (most), poorly vascularized (dense connective tissue), or avascular (cartilage).

Structural Elements of Connective Tissue (pp. 124–126)

3. The structural elements of all connective tissues are extracellular matrix and cells.
4. The extracellular matrix consists of ground substance and fibers (collagen, elastic, and reticular). It may be fluid, gel-like, or firm.
5. Each connective tissue type has a primary cell type that can exist as a mitotic, matrix-secreting cell (blast) or as a mature cell (cyte) responsible for maintaining the matrix. The undifferentiated cell type of connective tissue proper is the fibroblast; that of cartilage is the chondroblast; that of bone is the osteoblast; and that of blood-forming tissue is the hematopoietic stem cell.

Types of Connective Tissue (pp. 126–135)

6. Embryonic connective tissue is called mesenchyme.
7. Connective tissue proper consists of loose and dense varieties. The loose connective tissues are
   - Areolar: gel-like ground substance; all three fiber types loosely interwoven; a variety of cells; forms the lamina propria and soft packing around body organs; the prototype.
   - Adipose: consists largely of adipocytes; scant matrix; insulates and protects body organs; provides reserve energy fuel. Brown fat, present only in infants, is more important for generating body heat.
   - Reticular: finely woven reticular fibers in soft ground substance; the stroma of lymphoid organs and bone marrow.
8. Dense connective tissue proper includes
   - Dense regular: dense parallel bundles of collagen fibers; few cells, little ground substance; high tensile strength; forms tendons, ligaments, aponeuroses; in cases where this tissue also contains numerous elastic fibers it is called elastic connective tissue.
   - Dense irregular: like regular variety, but fibers are arranged in different planes; resists tension exerted from many different directions; forms the dermis of the skin and organ capsules.
9. Cartilage exists as
   - Hyaline: firm ground substance containing collagen fibers; resists compression well; found in fetal skeleton, at articulating surfaces of bones, and trachea; most abundant type.
   - Elastic cartilage: elastic fibers predominate; provides flexible support of the external ear and epiglottis.

- Fibrocartilage: coarse parallel collagen fibers; provides support with compressibility; forms intervertebral discs and knee cartilages.

10. Bone (osseous tissue) consists of a hard, collagen-containing matrix embedded with calcium salts; forms the bony skeleton.
11. Blood consists of blood cells in a fluid matrix (plasma).

Nervous Tissue (pp. 134–136)

1. Nervous tissue forms organs of the nervous system. It is composed of neurons and supporting cells.
2. Neurons are branching cells that receive and transmit electrical impulses. They are involved in body regulation.

iP Nervous System I; Topic: Anatomy Review, pp. 1, 3.

Muscle Tissue (pp. 136–138)

1. Muscle tissue consists of elongated cells specialized to contract and cause movement.
2. Based on structure and function, the muscle tissues are
   - Skeletal muscle: attached to and moves the bony skeleton; cells are cylindrical and striated.
   - Cardiac muscle: forms the walls of the heart; pumps blood; cells are branched and striated.
   - Smooth muscle: in the walls of hollow organs; propels substances through the organs; cells are spindle shaped and lack striations.

Covering and Lining Membranes (pp. 138–139)

1. Membranes are simple organs, consisting of an epithelium bound to an underlying connective tissue layer. They include mucosae, serosae, and the cutaneous membrane.

Tissue Repair (pp. 138–141)

1. Inflammation is the body's response to injury. Tissue repair begins during the inflammatory process. It may lead to regeneration, fibrosis, or both.
2. Tissue repair begins with organization, during which the blood clot is replaced by granulation tissue. If the wound is small and the damaged tissue is actively mitotic, the tissue will regenerate and cover the fibrous tissue. When a wound is extensive or the damaged tissue amitotic, it is repaired only by fibrous connective (scar) tissue.

Developmental Aspects of Tissues (pp. 141–144)

1. Epithelium arises from all three primary germ layers (ectoderm, mesoderm, endoderm); muscle and connective tissue from mesoderm; and nervous tissue from ectoderm.
2. The decrease in mass and viability seen in most tissues during old age often reflects circulatory deficits or poor nutrition.

REVIEW QUESTIONS

Multiple Choice/Matching

(Some questions have more than one correct answer. Select the best answer or answers from the choices given.)

1. Use the key to classify each of the following described tissue types into one of the four major tissue categories.

   Key: **(a)** connective tissue **(c)** muscle
   (b) epithelium **(d)** nervous tissue

   _____ (1) Tissue type composed largely of nonliving extracellular matrix; important in protection and support

   _____ (2) The tissue immediately responsible for body movement

   _____ (3) The tissue that enables us to be aware of the external environment and to react to it

   _____ (4) The tissue that lines body cavities and covers surfaces

2. An epithelium that has several layers, with an apical layer of flattened cells, is called (choose all that apply): **(a)** ciliated, **(b)** columnar, **(c)** stratified, **(d)** simple, **(e)** squamous.
3. Match the epithelial types named in column B with the appropriate description(s) in column A.

Column A	Column B
____ (1) Lines most of the digestive tract	**(a)** pseudostratified ciliated columnar
____ (2) Lines the esophagus	**(b)** simple columnar
____ (3) Lines much of the respiratory tract	**(c)** simple cuboidal
	(d) simple squamous
____ (4) Forms the walls of the air sacs of the lungs	**(e)** stratified columnar
	(f) stratified squamous
____ (5) Found in urinary tract organs	**(g)** transitional
____ (6) Endothelium and mesothelium	

4. The gland type that secretes products such as milk, saliva, bile, or sweat through a duct is **(a)** an endocrine gland, **(b)** an exocrine gland.
5. The membrane which lines body cavities that open to the exterior is a(n) **(a)** endothelium, **(b)** cutaneous membrane, **(c)** mucous membrane, **(d)** serous membrane.
6. Scar tissue is a variety of **(a)** epithelium, **(b)** connective tissue, **(c)** muscle tissue, **(d)** nervous tissue, **(e)** all of these.

Short Answer Essay Questions

7. Define tissue.
8. Name four important functions of epithelial tissue and provide at least one example of a tissue that exemplifies each function.
9. Describe the criteria used to classify covering and lining epithelia.
10. Explain the functional classification of multicellular exocrine glands and supply an example for each class.
11. Provide examples from the body that illustrate four of the major functions of connective tissue.
12. Name the primary cell type in connective tissue proper; in cartilage; in bone.
13. Name the two major components of matrix and, if applicable, subclasses of each component.
14. Matrix is extracellular. How does the matrix get to its characteristic position?
15. Name the specific connective tissue type found in the following body locations: (a) forming the soft packing around organs, (b) supporting the ear pinna, (c) forming "stretchy" ligaments, (d) first connective tissue in the embryo, (e) forming the intervertebral discs, (f) covering the ends of bones at joint surfaces, (g) main component of subcutaneous tissue.
16. What is the function of macrophages?

17. Differentiate clearly between the roles of neurons and the supporting cells of nervous tissue.
18. Compare and contrast skeletal, cardiac, and smooth muscle tissue relative to structure, body location, and specific function.
19. Describe the process of tissue repair, making sure you indicate factors that influence this process.
20. Indicate which primary tissue classes derive from each embryonic germ layer.
21. In what ways are adipose tissue and bone similar? How are they different?

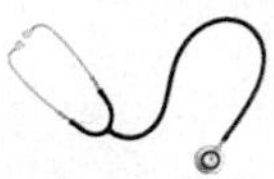

Critical Thinking and Clinical Application Questions

1. John has sustained a severe injury during football practice and is told that he has a torn knee cartilage. Can he expect a quick, uneventful recovery? Explain your response.
2. The epidermis (epithelium of the cutaneous membrane or skin) is a keratinized stratified squamous epithelium. Explain why that epithelium is much better suited for protecting the body's external surface than a mucosa consisting of a simple columnar epithelium would be.
3. Your friend is trying to convince you that if the ligaments binding the bones together at your freely movable joints (such as your knee, shoulder, and hip joints) contained more elastic fibers, you would be much more flexible. Although there is *some* truth to this statement, such a condition would present serious problems. Why?
4. In adults, over 90% of all cancers are either adenomas (adenocarcinomas) or carcinomas. (See Related Clinical Terms for this chapter.) In fact, cancers of the skin, lung, colon, breast, and prostate are all in these categories. Which one of the four basic tissue types gives rise to most cancers? Why do you think this is so?
5. Cindy, an overweight high school student, is overheard telling her friend that she's going to research how she can transform some of her white fat to brown fat. What is her rationale here (assuming it is possible)?
6. Mrs. Delancy went to the local meat market and bought a beef tenderloin (cut from the loin, the region along the steer's vertebral column) and some tripe (cow's stomach). What type of muscle was she preparing to eat in each case?

4

5 The Integumentary System

Would you be enticed by an advertisement for a coat that is waterproof, stretchable, washable, and permanent-press, that automatically repairs small cuts, rips, and burns, and that is guaranteed to last a lifetime with reasonable care? Sounds too good to be true, but you already have such a coat—your skin. The skin and its derivatives (sweat and oil glands, hairs, and nails) make up a complex set of organs that serves several functions, mostly protective. Together, these organs form the **integumentary system** (in-teg″u-men′tar-e).

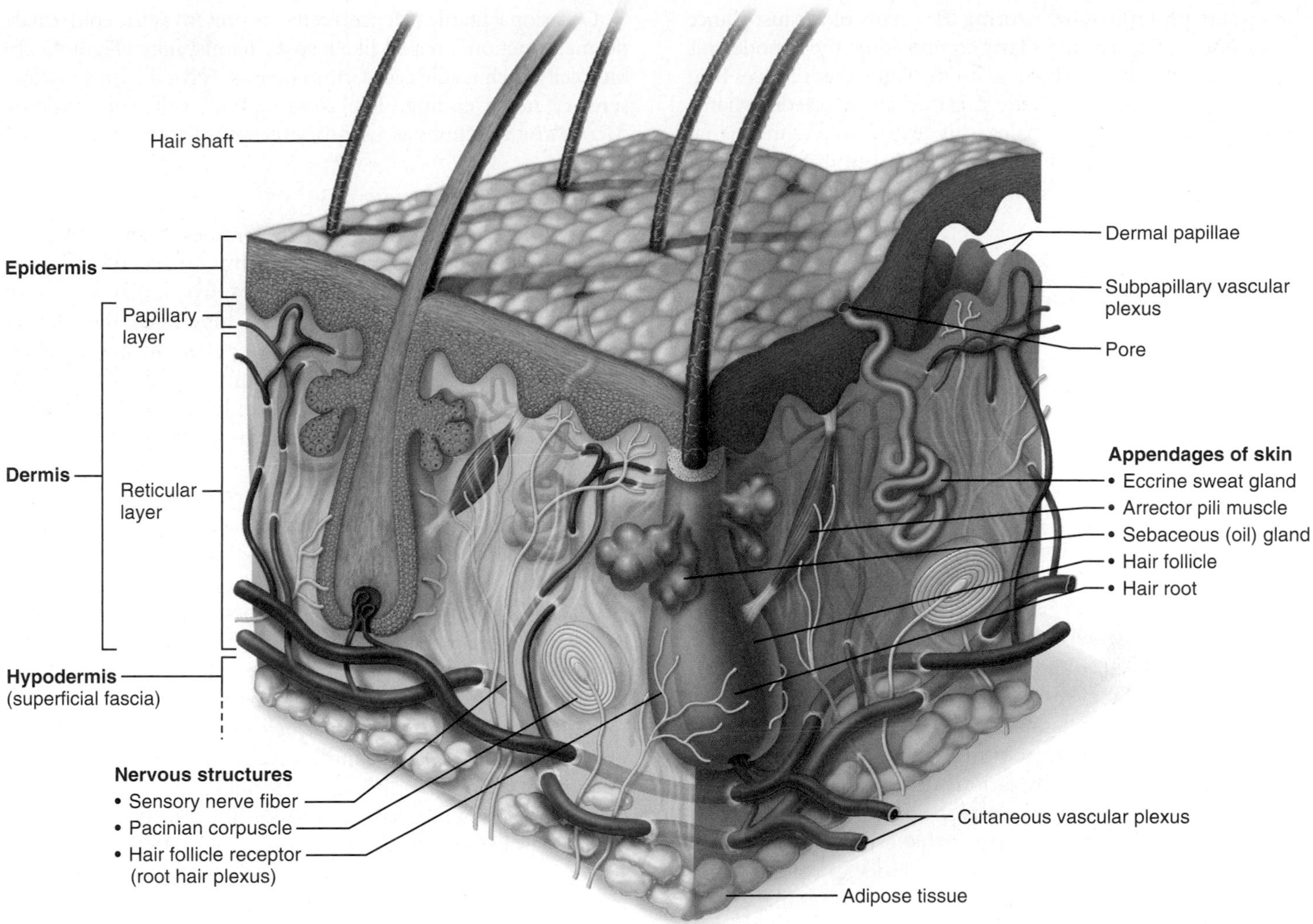

Figure 5.1 Skin structure. Three-dimensional view of the skin and underlying subcutaneous tissue. The epidermal and dermal layers have been pulled apart at the right corner to reveal the dermal papillae.

5

The Skin

- Name the tissue types composing the epidermis and dermis. List the major layers of each and describe the functions of each layer.
- Describe the factors that normally contribute to skin color. Briefly describe how changes in skin color may be used as clinical signs of certain disease states.

The skin ordinarily receives little respect from its inhabitants, but architecturally it is a marvel. It covers the entire body, has a surface area of 1.2 to 2.2 square meters, weighs 4 to 5 kilograms (4–5 kg = 9–11 lb), and accounts for about 7% of total body weight in the average adult. Also called the integument, which simply means "covering," the skin multitasks. Its functions go well beyond serving as a large, opaque bag for the body contents. It is pliable yet tough, allowing it to take constant punishment from external agents. Without our skin, we would quickly fall prey to bacteria and perish from water and heat loss.

The skin, which varies in thickness from 1.5 to 4.0 millimeters (mm) or more in different parts of the body, is composed of two distinct regions, the *epidermis* (ep″ĭ-der′mis) and the *dermis* **(Figure 5.1)**. The epidermis (*epi* = upon), composed of epithelial cells, is the outermost protective shield of the body. The underlying dermis, making up the bulk of the skin, is a tough, leathery layer composed mostly of fibrous connective tissue. Only the dermis is vascularized. Nutrients reach the epidermis by diffusing through the tissue fluid from blood vessels in the dermis.

The subcutaneous tissue just deep to the skin is known as the **hypodermis** (Figure 5.1). Strictly speaking, the hypodermis is not part of the skin, but it shares some of the skin's protective functions. The hypodermis, also called **superficial fascia** because it is superficial to the tough connective tissue wrapping (fascia) of the skeletal muscles, consists mostly of adipose tissue.

Besides storing fat, the hypodermis anchors the skin to the underlying structures (mostly to muscles), but loosely enough that the skin can slide relatively freely over those structures.

Sliding skin protects us by ensuring that many blows just glance off our bodies. Because of its fatty composition, the hypodermis also acts as a shock absorber and an insulator that reduces heat loss from the body. The hypodermis thickens markedly when a person gains weight. In females, this "extra" subcutaneous fat accumulates first in the thighs and breasts, but in males it first collects in the anterior abdomen (as a "beer belly").

Epidermis

Structurally, the **epidermis** is a keratinized stratified squamous epithelium consisting of four distinct cell types and four or five distinct layers.

Cells of the Epidermis

The cells populating the epidermis include *keratinocytes, melanocytes, epidermal dendritic cells,* and *tactile cells.* Most epidermal cells are keratinocytes, so we will consider them first.
5 The chief role of **keratinocytes** (kĕ-rat′ĭ-no-sītz″; "keratin cells") is to produce **keratin**, the fibrous protein that helps give the epidermis its protective properties (Greek *kera* = horn) (**Figure 5.2b**, orange cells).

Tightly connected to one another by desmosomes, the keratinocytes arise in the deepest part of the epidermis from a cell layer called the stratum basale. These cells undergo almost continuous mitosis in response to prompting by epidermal growth factor, a peptide produced by various cells throughout the body. As these cells are pushed upward by the production of new cells beneath them, they make the keratin that eventually dominates their cell contents. By the time the keratinocytes reach the free surface of the skin, they are dead, scalelike structures that are little more than keratin-filled plasma membranes.

Millions of these dead cells rub off every day, giving us a totally new epidermis every 25 to 45 days. In body areas regularly subjected to friction, such as the hands and feet, both cell production and keratin formation are accelerated. Persistent friction (from a poorly fitting shoe, for example) causes a thickening of the epidermis called a *callus.*

Melanocytes (mel′ah-no-sītz), the spider-shaped epithelial cells that synthesize the pigment **melanin** (mel′ah-nin; *melan* = black), are found in the deepest layer of the epidermis (Figure 5.2b, gray cells). As melanin is made, it is accumulated in membrane-bound granules called *melanosomes* that are moved along actin filaments by motor proteins to the ends of the melanocyte's processes (the "spider arms"). From there they are taken up by nearby keratinocytes. The melanin granules accumulate on the superficial, or "sunny," side of the keratinocyte nucleus, forming a pigment shield that protects the nucleus from the damaging effects of ultraviolet (UV) radiation in sunlight.

The star-shaped **epidermal dendritic cells** arise from bone marrow and migrate to the epidermis. Also called **Langerhans cells** (lahng′er-hanz) after a German anatomist, they ingest foreign substances and are key activators of our immune system, as described later in this chapter. Their slender processes extend among the surrounding keratinocytes, forming a more or less continuous network (Figure 5.2b, purple cell).

Occasional **tactile (Merkel) cells** are present at the epidermal-dermal junction. Shaped like a spiky hemisphere (Figure 5.2b, blue cell), each tactile cell is intimately associated with a disclike sensory nerve ending. The combination, called a *tactile* or *Merkel disc,* functions as a sensory receptor for touch.

Layers of the Epidermis

Variation in epidermal thickness determines if skin is *thick* or *thin.* In **thick skin**, which covers the palms, fingertips, and soles of the feet, the epidermis consists of five layers, or *strata* (stra′tah; "bed sheets"). From deep to superficial, these layers are stratum basale, stratum spinosum, stratum granulosum, stratum lucidum, and stratum corneum. In **thin skin**, which covers the rest of the body, the stratum lucidum appears to be absent and the other strata are thinner (Figure 5.2a, b).

Stratum Basale (Basal Layer) The **stratum basale** (stra′tum bah-sa′le), the deepest epidermal layer, is attached to the underlying dermis along a wavy borderline that reminds one of corrugated cardboard. For the most part, it consists of a single row of stem cells—a continually renewing cell population—representing the youngest keratinocytes. The many mitotic nuclei seen in this layer reflect the rapid division of these cells and account for its alternate name, **stratum germinativum** (jer′mĭ-nă″tiv-um; "germinating layer"). Each time one of these basal cells divides, one daughter cell is pushed into the cell layer just above to begin its specialization into a mature keratinocyte. The other daughter cell remains in the basal layer to continue the process of producing new keratinocytes.

Some 10–25% of the cells in the stratum basale are melanocytes, and their branching processes extend among the surrounding cells, reaching well into the more superficial stratum spinosum layer. Occasional tactile cells are also seen in this stratum.

Stratum Spinosum (Prickly Layer) The **stratum spinosum** (spi′no-sum; "prickly") is several cell layers thick. These cells contain a weblike system of intermediate filaments, mainly tension-resisting bundles of pre-keratin filaments, which span their cytosol to attach to desmosomes. Looking like tiny versions of the spiked iron balls used in medieval warfare, the keratinocytes in this layer appear to have spines, causing them to be called *prickle cells.* The spines do not exist in the living cells; they are artifacts that arise during tissue preparation when these cells shrink but their numerous desmosomes hold tight. Scattered among the keratinocytes are melanin granules and epidermal dendritic cells, which are most abundant in this epidermal layer.

Stratum Granulosum (Granular Layer) The thin **stratum granulosum** (gran″u-lo′sum) consists of three to five cell layers in which keratinocyte appearance changes drastically, and the process of **keratinization** (in which the cells fill with the protein keratin) begins. These cells flatten, their nuclei and organelles begin to disintegrate, and they accumulate two types of granules. The *keratohyaline granules* (ker″ah-to-hi′ah-lin) help to form keratin in the upper layers, as we will see. The *lamellated*

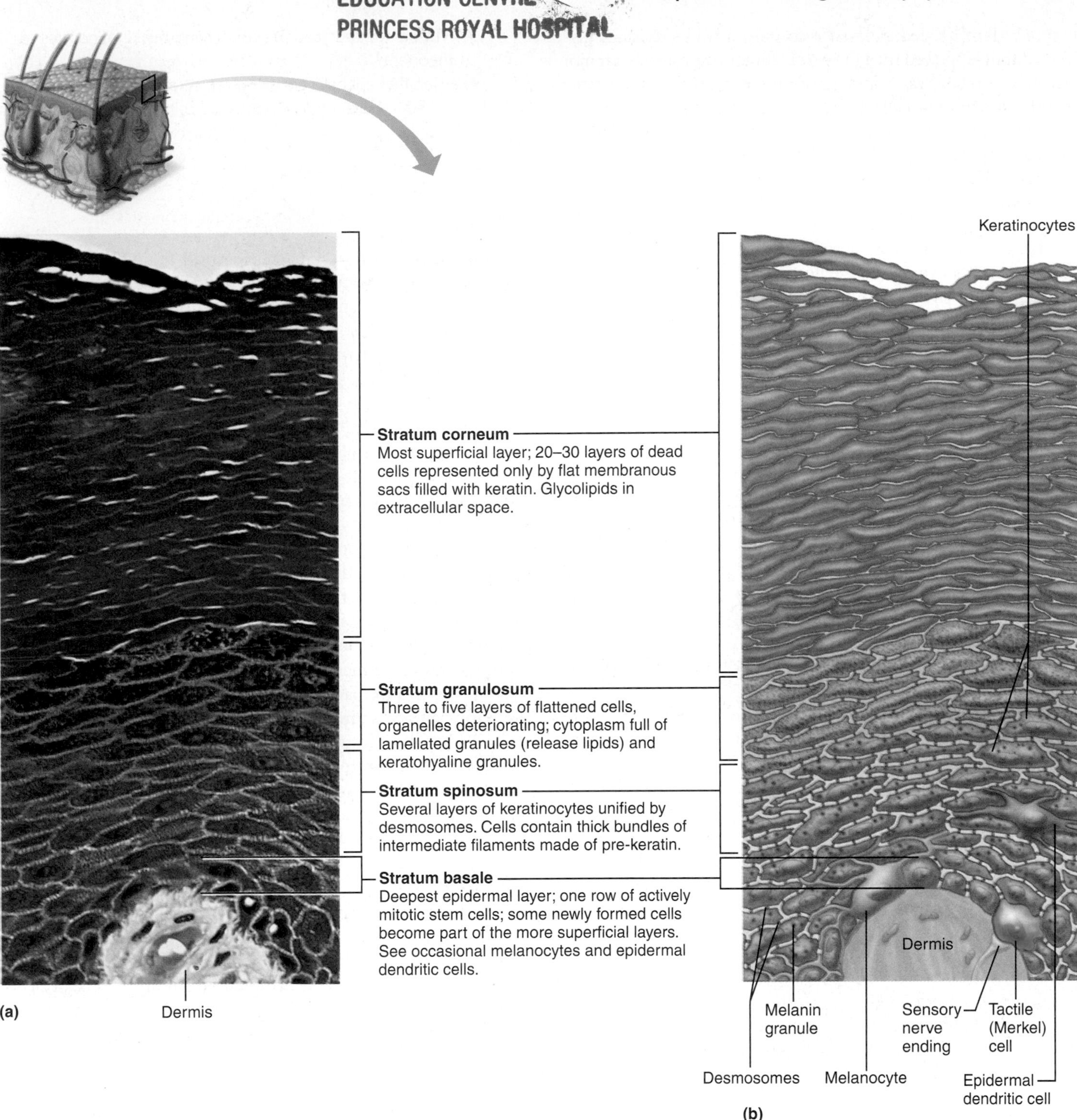

Keratinocytes
Stratum corneum
Most superficial layer; 20–30 layers of dead cells represented only by flat membranous sacs filled with keratin. Glycolipids in extracellular space.
Stratum granulosum
Three to five layers of flattened cells, organelles deteriorating; cytoplasm full of lamellated granules (release lipids) and keratohyaline granules.
Stratum spinosum
Several layers of keratinocytes unified by desmosomes. Cells contain thick bundles of intermediate filaments made of pre-keratin.
Stratum basale
Deepest epidermal layer; one row of actively mitotic stem cells; some newly formed cells become part of the more superficial layers. See occasional melanocytes and epidermal dendritic cells.
(a)
Dermis
Dermis
Melanin granule
Sensory nerve ending
Tactile (Merkel) cell
Desmosomes
Melanocyte
Epidermal dendritic cell
(b)

5

Figure 5.2 The main structural features of the skin epidermis. (a) Photomicrograph of the four major epidermal layers (200×). **(b)** Diagram showing these four layers and the distribution of different cell types. The four cell types are keratinocytes (orange), melanocytes (gray), epidermal dendritic cells (purple), and tactile cells (blue). A sensory nerve ending (yellow), extending from the dermis (pink), is shown associated with the tactile cell forming a tactile disc (touch receptor). Notice that the keratinocytes are joined by numerous desmosomes. The stratum lucidum, present in thick skin, is not illustrated here.

granules (lam′ĭ-la-ted; "plated") contain a water-resistant glycolipid that is spewed into the extracellular space and is a major factor in slowing water loss across the epidermis. The plasma membranes of these cells thicken as cytosol proteins bind to the inner membrane face and lipids released by the lamellated granules coat their external surfaces. This makes them more resistant to destruction, so you might say that the keratinocytes are "toughening up" to make the outer strata the strongest skin region.

Like all epithelia, the epidermis relies on capillaries in the underlying connective tissue (the dermis in this case) for its nutrients. Above the stratum granulosum, the epidermal cells are too far from the dermal capillaries and are cut off from nutrients by the glycolipids that coat their external surfaces, so they die. This is a completely normal sequence of events.

Stratum Lucidum (Clear Layer) Through the light microscope, the **stratum lucidum** (loo′sid-um; "light") appears as a thin translucent band just above the stratum granulosum. It consists of two or three rows of clear, flat, dead keratinocytes with indistinct boundaries. Here, or in the stratum corneum above, the gummy substance of the keratohyaline granules clings to the keratin filaments in the cells, causing them to aggregate in large, cable-like, parallel arrays. As mentioned before, the stratum lucidum is visible only in thick skin.

Stratum Corneum (Horny Layer) The outermost **stratum corneum** (kor′ne-um) is a broad zone 20 to 30 cell layers thick that accounts for up to three-quarters of the epidermal thickness. Keratin and the thickened plasma membranes of cells in this stratum protect the skin against abrasion and penetration, and the glycolipid between its cells nearly waterproofs this layer. For these reasons, the stratum corneum provides a durable "overcoat" for the body, protecting deeper cells from the hostile external environment (air) and from water loss, and rendering the body relatively insensitive to biological, chemical, and physical assaults. It is amazing that a layer of dead cells can still play so many roles.

The shingle-like cell remnants of the stratum corneum are referred to as *cornified*, or *horny, cells* (*cornu* = horn). They are familiar to everyone as the dandruff shed from the scalp and dander, the loose flakes that slough off dry skin. The average person sheds 18 kg (40 lb) of these skin flakes in a lifetime, providing a lot of fodder for the dust mites that inhabit our homes and bed linens. The common saying "Beauty is only skin deep" is especially interesting in light of the fact that nearly everything we see when we look at someone is dead!

CHECK YOUR UNDERSTANDING

1. While walking barefoot in the barn, Jeremy stepped on a rusty nail that penetrated the depth of the epidermis on the sole of his foot. Name the layers the nail pierced from the superficial skin surface to the junction with the dermis.
2. The stratum basale is also called the stratum germinativum, a name that refers to the major function of this cell layer. What is that function?
3. Why are the desmosomes connecting the keratinocytes so important?
4. Given that epithelia are avascular, what layer would be expected to have the best-nourished cells?

For answers, see Appendix G.

Dermis

The **dermis** (*derm* = skin), the second major skin region, is strong, flexible connective tissue. Its cells are typical of those found in any connective tissue proper: fibroblasts, macrophages, and occasional mast cells and white blood cells. Its semifluid matrix, embedded with fibers, binds the entire body together like a body stocking. It is your "hide" and corresponds exactly to animal hides used to make leather products.

The dermis is richly supplied with nerve fibers, blood vessels, and lymphatic vessels. The major portions of hair follicles, as well as oil and sweat glands, are derived from epidermal tissue but reside in the dermis.

The dermis has two layers, the papillary and reticular, which abut one another along an indistinct boundary **(Figure 5.3)**. The thin superficial **papillary layer** (pap′il-er-e) is areolar connective tissue in which fine interlacing collagen and elastic fibers form a loosely woven mat that is heavily invested with small blood vessels. The looseness of this connective tissue allows phagocytes and other defensive cells to wander freely as they patrol the area for bacteria that may have breached the skin. Its superior surface is thrown into peglike projections called **dermal papillae** (pah-pil′e; *papill* = nipple) that indent the overlying epidermis (see Figure 5.1). Many dermal papillae contain capillary loops (of the *subpapillary plexus*). Others house free nerve endings (pain receptors) and touch receptors called *Meissner's corpuscles* (mīs′nerz kor′pus-lz). On the palms of the hands and soles of the feet, these papillae lie atop larger mounds called *dermal ridges*, which in turn cause the overlying epidermis to form *epidermal ridges* **(Figure 5.4)**. Collectively, these skin ridges, referred to as **friction ridges**, increase friction and enhance the gripping ability of the fingers and feet. Friction ridge patterns are genetically determined and unique to each of us. Because sweat pores open along their crests, our fingertips leave identifying films of sweat called *fingerprints* on almost anything they touch.

The deeper **reticular layer**, accounting for about 80% of the thickness of the dermis, is coarse, irregularly arranged, dense fibrous connective tissue (Figure 5.3c). The network of blood vessels that nourishes this layer, the *cutaneous plexus*, lies between this layer and the hypodermis. Its extracellular matrix contains pockets of adipose cells here and there, and thick bundles of interlacing collagen fibers. The collagen fibers run in various planes, but most run parallel to the skin surface. Separations, or less dense regions, between these bundles form **cleavage**, or **tension**, **lines** in the skin. These externally invisible lines tend to run longitudinally in the skin of the head and limbs and in circular patterns around the neck and trunk (Figure 5.4b).

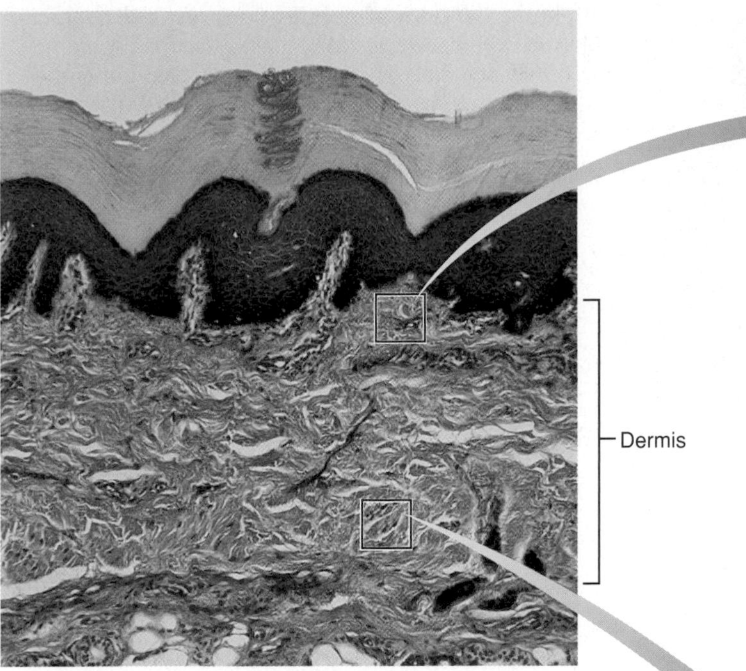

(a) Light micrograph of thick skin identifying the extent of the dermis, (50×)

(b) Papillary layer of dermis, SEM (22,700×)

(c) Reticular layer of dermis, SEM (38,500×)

Figure 5.3 The two regions of the dermis. The superficial papillary layer consists of areolar connective tissue, and the deeper reticular layer is dense irregular fibrous connective tissue.
SOURCE: Kessel and Kardon/Visuals Unlimited.

Cleavage lines are important to both surgeons and their patients. When an incision is made *parallel* to these lines, the skin gapes less and heals more readily than when the incision is made *across* cleavage lines.

The collagen fibers of the dermis give skin strength and resiliency that prevent most jabs and scrapes from penetrating the dermis. In addition, collagen binds water, helping to keep skin hydrated. Elastic fibers provide the stretch-recoil properties of skin.

In addition to the epidermal ridges and cleavage lines, a third type of skin marking, flexure lines, reflects dermal modifications. **Flexure lines** are dermal folds that occur at or near joints, where the dermis is tightly secured to deeper structures (notice the deep creases on your palms). Since the skin cannot slide easily to accommodate joint movement in such regions, the dermis folds and deep skin creases form. Flexure lines are also visible on the wrists, fingers, soles, and toes.

HOMEOSTATIC IMBALANCE

Extreme stretching of the skin, such as occurs during pregnancy, can tear the dermis. Dermal tearing is indicated by silvery white scars called *striae* (stri′e; "streaks"), commonly called "stretch marks." Short-term but acute trauma (as from a burn or wielding a hoe) can cause a *blister*, the separation of the epidermal and dermal layers by a fluid-filled pocket. ■

CHECK YOUR UNDERSTANDING

5. What layer of the dermis is responsible for producing fingerprint patterns?
6. What cell component of the hypodermis makes it a good shock absorber?
7. You have just gotten a paper cut. It is very painful, but it doesn't bleed. Has the cut penetrated into the dermis or just the epidermis?

For answers, see Appendix G.

5

(a)

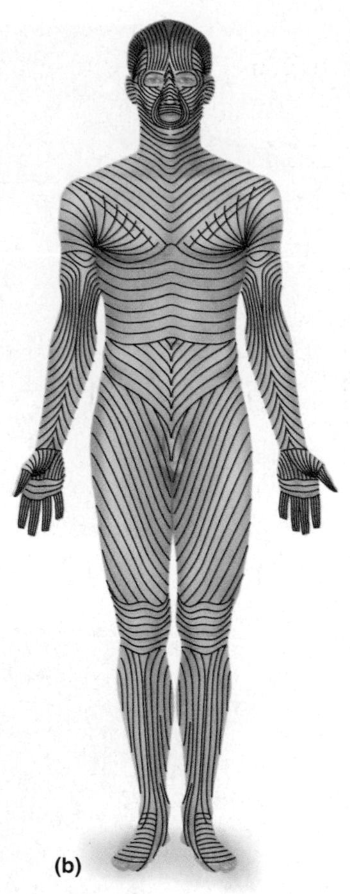

(b)

Figure 5.4 Dermal modifications result in characteristic skin markings. (a) Scanning electron micrograph of friction ridges (epidermal ridges topping the deeper dermal papillary ridges; 200×). Notice the sweat duct openings along the crests of the ridges, which are responsible for fingerprints. **(b)** Cleavage (tension) lines represent separations between underlying collagen fiber bundles in the reticular region of the dermis. They tend to run circularly around the trunk and longitudinally in the limbs.

SOURCE: Kessel and Kardon/Visuals Unlimited.

Skin Color

Three pigments contribute to skin color: melanin, carotene, and hemoglobin. Of these, only melanin is made in the skin. **Melanin** is a polymer made of tyrosine amino acids. Its two forms range in color from yellow to tan to reddish-brown to black. Its synthesis depends on an enzyme in melanocytes called tyrosinase (ti-ro′sĭ-nās) and, as noted earlier, it passes from melanocytes to the basal keratinocytes. Eventually, the melanosomes are broken down by lysosomes, so melanin pigment is found only in the deeper layers of the epidermis.

Human skin comes in different colors. However, distribution of those colors is not random—populations of darker-skinned people tend to be found nearer the equator (where greater protection from the sun is needed), and those with the lightest skin are found closer to the poles. Since all humans have the same relative number of melanocytes, individual and racial differences in skin coloring reflect the relative kind and amount of melanin made and retained. Melanocytes of black- and brown-skinned people produce many more and darker melanosomes than those of fair-skinned individuals, and their keratinocytes retain it longer. *Freckles* and *pigmented nevi* (*moles*) are local accumulations of melanin.

Melanocytes are stimulated to greater activity by chemicals secreted by the surrounding keratinocytes when we expose our skin to sunlight. Prolonged sun exposure causes a substantial melanin buildup, which helps protect the DNA of viable skin cells from UV radiation by absorbing the rays and dissipating the energy as heat. Indeed, the initial signal for speeding up melanin synthesis seems to be a faster rate of repair of photodamaged DNA. In all but the darkest people, this response causes visible darkening of the skin (a tan).

HOMEOSTATIC IMBALANCE

Despite melanin's protective effects, excessive sun exposure eventually damages the skin. It causes clumping of elastic fibers, which results in leathery skin; temporarily depresses the immune system; and can alter the DNA of skin cells and in this way lead to skin cancer. The fact that dark-skinned people get skin cancer less often than fair-skinned people and get it in areas with less pigment—the soles of the feet and nail beds—attests to melanin's effectiveness as a natural sunscreen.

Ultraviolet radiation has other consequences as well. It destroys the body's folic acid stores necessary for DNA synthesis, which can have serious consequences, particularly in pregnant women because the deficit may impair the development of the embryo's nervous system. Many chemicals induce photosensitivity; that is, they heighten the skin's sensitivity to UV radiation, setting sun worshippers up for an unsightly skin rash. Such substances include some antibiotic and antihistamine drugs, and many chemicals in perfumes and detergents. Small, itchy, blisterlike lesions erupt all over the body; then the peeling begins, in sheets! ■

Carotene (kar′o-tēn) is a yellow to orange pigment found in certain plant products such as carrots. It tends to accumulate in the stratum corneum and in fatty tissue of the hypodermis. Its color is most obvious in the palms and soles, where the stratum corneum is thickest (for example the skin of the heels), and most intense when large amounts of carotene-rich foods are eaten. However, the yellowish tinge of the skin of some Asian peoples is due to variations in melanin, as well as to carotene. In the body, carotene can be converted to vitamin A, a vitamin that is essential for normal vision, as well as for epidermal health.

The pinkish hue of fair skin reflects the crimson color of the oxygenated pigment **hemoglobin** (he′mo-glo″bin) in the red blood cells circulating through the dermal capillaries. Because Caucasian skin contains only small amounts of melanin, the epidermis is nearly transparent and allows hemoglobin's color to show through.

HOMEOSTATIC IMBALANCE

When hemoglobin is poorly oxygenated, both the blood and the skin of Caucasians appear blue, a condition called *cyanosis* (si″ah-no′sis; *cyan* = dark blue). Skin often becomes cyanotic during heart failure and severe respiratory disorders. In dark-skinned individuals, the skin does not appear cyanotic because of the masking effects of melanin, but cyanosis is apparent in their mucous membranes and nail beds (the same sites where the red cast of normally oxygenated blood is visible).

Many alterations in skin color signal certain disease states, and in many people emotional states:

- *Redness*, or *erythema* (er″ĭ-the′mah): Reddened skin may indicate embarrassment (blushing), fever, hypertension, inflammation, or allergy.
- *Pallor*, or *blanching*: During fear, anger, and certain other types of emotional stress, some people become pale. Pale skin may also signify anemia or low blood pressure.
- *Jaundice* (jawn′dis), or *yellow cast*: An abnormal yellow skin tone usually signifies a liver disorder, in which yellow bile pigments accumulate in the blood and are deposited in body tissues. [Normally, the liver cells secrete the bile pigments (bilirubin) as a component of bile.]
- *Bronzing*: A bronze, almost metallic appearance of the skin is a sign of Addison's disease, in which the adrenal cortex is producing inadequate amounts of its steroid hormones; or a sign of the presence of pituitary gland tumors that inappropriately secrete melanocyte-stimulating hormone (MSH).
- *Black-and-blue marks*, or *bruises*: Black-and-blue marks reveal where blood escaped from the circulation and clotted beneath the skin. Such clotted blood masses are called *hematomas* (he″mah-to′mah; "blood swelling"). ■

CHECK YOUR UNDERSTANDING

8. Melanin and carotene are two pigments that contribute to skin color. What is the third and where is it found?
9. What is cyanosis and what does it indicate?
10. What alteration in skin color may indicate a liver disorder?

For answers, see Appendix G.

Appendages of the Skin

- ▶ Compare the structure and locations of sweat and oil glands. Also compare the composition and functions of their secretions.
- ▶ Compare and contrast eccrine and apocrine glands.
- ▶ List the parts of a hair follicle and explain the function of each part. Also describe the functional relationship of arrector pili muscles to the hair follicles.
- ▶ Name the regions of a hair and explain the basis of hair color. Describe the distribution, growth, replacement, and changing nature of hair during the life span.
- ▶ Describe the structure of nails.

Along with the skin itself, the integumentary system includes several derivatives of the epidermis. These **skin appendages** include the nails, sweat glands, sebaceous (oil) glands, and hair follicles and hair. Each of these plays a unique role in maintaining body homeostasis.

A key step in beginning to form any of the skin's appendages is formation of an epithelial bud. This process is stimulated by a reduced production of cell adhesion factor (cadherin). Once the cell-to-cell attractions are broken, the cells can move about and rearrange themselves, allowing an epithelial bud to form.

Sweat (Sudoriferous) Glands

Sweat glands, also called **sudoriferous glands** (su″do-rif′er-us; *sudor* = sweat), are distributed over the entire skin surface except the nipples and parts of the external genitalia. Their number is staggering—up to 3 million of them per person. We have two types of sweat glands: eccrine and apocrine. Regardless of type, the secretory cells are associated with myoepithelial cells, specialized cells that contract when stimulated by the nervous system. Their contraction forces the sweat into and through the gland's duct system to the skin surface.

Eccrine sweat glands (ek′rin; "secreting"), also called **merocrine sweat glands**, are far more numerous and are particularly abundant on the palms, soles of the feet, and forehead. Each is a simple, coiled, tubular gland. The secretory part lies coiled in the dermis, and the duct extends to open in a funnel-shaped *pore* (*por* = channel) at the skin surface **(Figure 5.5b)**. (These sweat pores are different from the so-called pores of a person's complexion, which are actually the external outlets of hair follicles.)

Eccrine gland secretion, commonly called sweat, is a hypotonic filtrate of the blood that passes through the secretory cells of the sweat glands and is released by exocytosis. It is 99% water, with some salts (mostly sodium chloride), vitamin C, antibodies, a microbe-killing peptide dubbed *dermcidin*, and traces of metabolic wastes (urea, uric acid, and ammonia). The exact composition depends on heredity and diet. Small amounts of ingested drugs may also be excreted by this route. Normally, sweat is acidic with a pH between 4 and 6.

Sweating is regulated by the sympathetic division of the autonomic nervous system, over which we have little control. Its

5

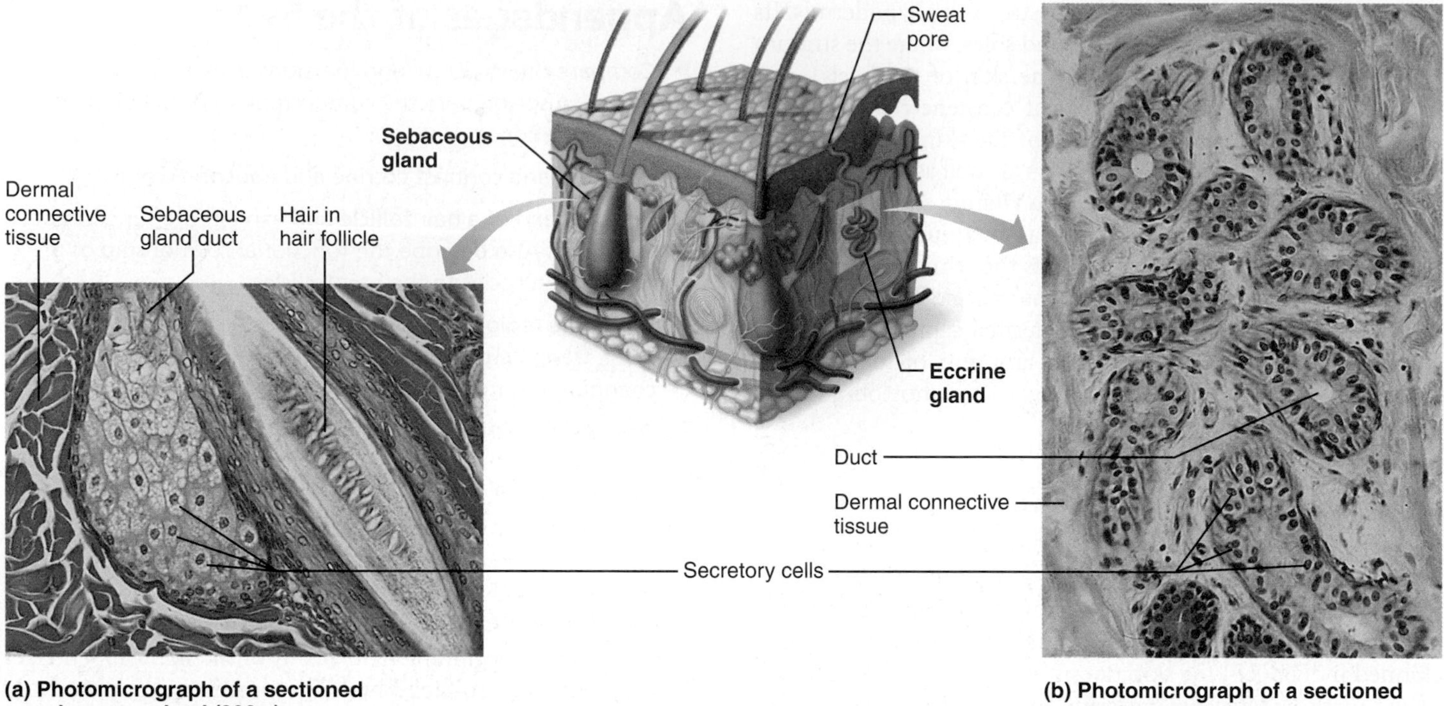

(a) Photomicrograph of a sectioned sebaceous gland (220×)

(b) Photomicrograph of a sectioned eccrine gland (220×)

Figure 5.5 **Cutaneous glands.**

major role is to prevent overheating of the body. Heat-induced sweating begins on the forehead and then spreads inferiorly over the remainder of the body. Emotionally induced sweating—the so-called "cold sweat" brought on by fright, embarrassment, or nervousness—begins on the palms, soles, and axillae (armpits) and then spreads to other body areas.

Apocrine sweat glands (ap′o-krin), approximately 2000 of them, are largely confined to the axillary and anogenital areas. In spite of their name, they are merocrine glands, which release their product by exocytosis like the eccrine sweat glands. They are larger than eccrine glands, tend to lie deeper in the dermis or even in the hypodermis, and their ducts empty into hair follicles. Apocrine secretion contains the same basic components as true sweat, plus fatty substances and proteins. Consequently, it is quite viscous and sometimes has a milky or yellowish color. The secretion is odorless, but when its organic molecules are decomposed by bacteria on the skin, it takes on a musky and generally unpleasant odor, the basis of body odor.

Apocrine glands begin functioning at puberty under the influence of androgens and have little role to play in thermoregulation. Their precise function is not yet known, but they are activated by sympathetic nerve fibers during pain and stress. Because their activity is increased by sexual foreplay, and they enlarge and recede with the phases of a woman's menstrual cycle, they may be the human equivalent of the sexual scent glands of other animals.

Ceruminous glands (sĕ-roo′mĭ-nus; *cera* = wax) are modified apocrine glands found in the lining of the external ear canal. Their secretion mixes with sebum produced by nearby sebaceous glands to form a sticky, bitter substance called *cerumen*, or earwax, that is thought to deter insects and block entry of foreign material.

Mammary glands, another variety of specialized sweat glands, secrete milk. Although they are properly part of the integumentary system, we consider the mammary glands in Chapter 27, along with female reproductive organs.

Sebaceous (Oil) Glands

The **sebaceous glands** (se-ba′shus; "greasy"), or **oil glands** (Figure 5.5a), are simple branched alveolar glands that are found all over the body except in the thick skin of the palms and soles. They are small on the body trunk and limbs, but quite large on the face, neck, and upper chest. These glands secrete an oily substance called **sebum** (se′bum). The central cells of the alveoli accumulate oily lipids until they become so engorged that they burst, so functionally these glands are *holocrine glands* (see p. 122). The accumulated lipids and cell fragments constitute sebum.

Most, but not all, sebaceous glands develop from hair follicles and sebum is secreted into a hair follicle, or occasionally to a pore on the skin surface. Sebum softens and lubricates the hair and skin, prevents hair from becoming brittle, and slows water loss from the skin when the external humidity is low. Perhaps even more important is its *bactericidal* (bacterium-killing) action.

The secretion of sebum is stimulated by hormones, especially androgens. Sebaceous glands are relatively inactive during childhood but are activated in both sexes during puberty, when androgen production begins to rise.

Additionally, and more important to humans physiologically, is the fact that arrector pili contractions force sebum out of the hair follicles to the skin surface.

HOMEOSTATIC IMBALANCE

If a sebaceous gland duct is blocked by accumulated sebum, a *whitehead* appears on the skin surface. If the material oxidizes and dries, it darkens to form a *blackhead*. *Acne* is an active inflammation of the sebaceous glands accompanied by "pimples" (pustules or cysts) on the skin. It is usually caused by bacterial infection, particularly by staphylococcus, and can be mild or extremely severe, leading to permanent scarring.

Seborrhea (seb″o-re′ah; "fast-flowing sebum"), known as "cradle cap" in infants, is caused by overactive sebaceous glands. It begins on the scalp as pink, raised lesions that gradually become yellow to brown and begin to slough off oily scales. ■

CHECK YOUR UNDERSTANDING

11. Which cutaneous glands are associated with hair follicles?
12. When Anthony returned home from a run in 85°F weather, his face was dripping with sweat. Why?
13. What is the difference between heat-induced sweating and a "cold sweat," and which variety of sweat glands is involved?
14. Sebaceous glands are not found in thick skin. Why is their absence in those body regions desirable?

For answers, see Appendix G.

Hairs and Hair Follicles

Hair is an important part of our body image—consider, for example, the spiky hair style of punk rockers and the flowing, glossy manes of some high-fashion models. Millions of hairs are distributed over our entire skin surface except our palms, soles, lips, nipples, and parts of the external genitalia (the head of the penis, for instance). Although hair helps to keep other mammals warm, our sparse body hair is far less luxuriant and useful. Its main function in humans is to sense insects on the skin before they bite or sting us. Hair on the scalp guards the head against physical trauma, heat loss, and sunlight. (Pity the bald man.) Eyelashes shield the eyes, and nose hairs filter large particles like lint and insects from the air we inhale.

Structure of a Hair

Hairs, or **pili** (pi′li), are flexible strands produced by hair follicles and consist largely of dead, keratinized cells. The *hard keratin* that dominates hairs and nails has two advantages over the *soft keratin* found in typical epidermal cells: (1) It is tougher and more durable, and (2) its individual cells do not flake off.

The chief regions of a hair are the *shaft*, the portion in which keratinization is complete, and the *root*, where keratinization is still ongoing. The shaft, which projects from the skin, extends about halfway down the portion of the hair embedded in the skin **(Figure 5.6)**. The root is the remainder of the hair deep within the follicle. If the shaft is flat and ribbonlike in cross section, the hair is kinky; if it is oval, the hair is silky and wavy; if it is perfectly round, the hair is straight and tends to be coarse.

A hair has three concentric layers of keratinized cells (Figure 5.6a, b). Its central core, the *medulla* (mĕ-dul′ah; "middle"), consists of large cells and air spaces. The medulla, which is the only part of the hair that contains soft keratin, is absent in fine hairs. The *cortex*, a bulky layer surrounding the medulla, consists of several layers of flattened cells. The outermost **cuticle** is formed from a single layer of cells that overlap one another from below like shingles on a roof. This arrangement helps to keep neighboring hairs apart so that the hair does not mat. (Hair conditioners smooth out the rough surface of the cuticle and make our hair look shiny.) The most heavily keratinized part of the hair, the cuticle provides strength and helps keep the inner layers tightly compacted. Because it is subjected to the most abrasion, the cuticle tends to wear away at the tip of the hair shaft, allowing the keratin fibrils in the cortex and medulla to frizz out, creating "split ends."

Hair pigment is made by melanocytes at the base of the hair follicle and transferred to the cortical cells. Various proportions of melanins of different colors (yellow, rust, brown, and black) combine to produce hair color from blond to pitch black. Additionally, red hair is colored by the iron-containing pigment *trichosiderin*. Gray or white hair results from decreased melanin production (mediated by delayed-action genes) and from the replacement of melanin by air bubbles in the hair shaft.

Structure of a Hair Follicle

Hair follicles (*folli* = bag) fold down from the epidermal surface into the dermis. In the scalp, they may even extend into the hypodermis. The deep end of the follicle, located about 4 mm (1/6 in.) below the skin surface, is expanded, forming a **hair bulb** (Figure 5.6c, d). A knot of sensory nerve endings called a **hair follicle receptor**, or **root hair plexus**, wraps around each hair bulb (see Figure 5.1). Bending the hair stimulates these endings. Consequently, our hairs act as sensitive touch receptors.

- Feel the tickle as you run your hand over the hairs on your forearm.

A *hair papilla*, a nipple-like bit of dermal tissue, protrudes into the hair bulb. This papilla contains a knot of capillaries that supplies nutrients to the growing hair and signals it to grow. Except for its specific location, this papilla is similar to the dermal papillae underlying other epidermal regions.

The wall of a hair follicle is composed of an outer **connective tissue root sheath**, derived from the dermis; a thickened basement membrane called the *glassy membrane*; and an inner **epithelial root sheath**, derived mainly from an invagination of the epidermis (Figure 5.6). The epithelial root sheath, which has external and internal parts, thins as it approaches the hair bulb, so that only a single layer of epithelial cells covers the papilla. However, the cells that compose the **hair matrix**, or actively dividing area of the hair bulb that produces the hair, originate in a region called the *hair bulge* located a fraction of a millimeter above the hair bulb. When chemical signals diffusing from the papilla reach the hair bulge, some of its cells migrate toward the papilla, where they divide to produce the hair cells. As new hair cells are produced by the matrix, the older part of the hair is pushed upward, and its fused cells become increasingly keratinized and die.

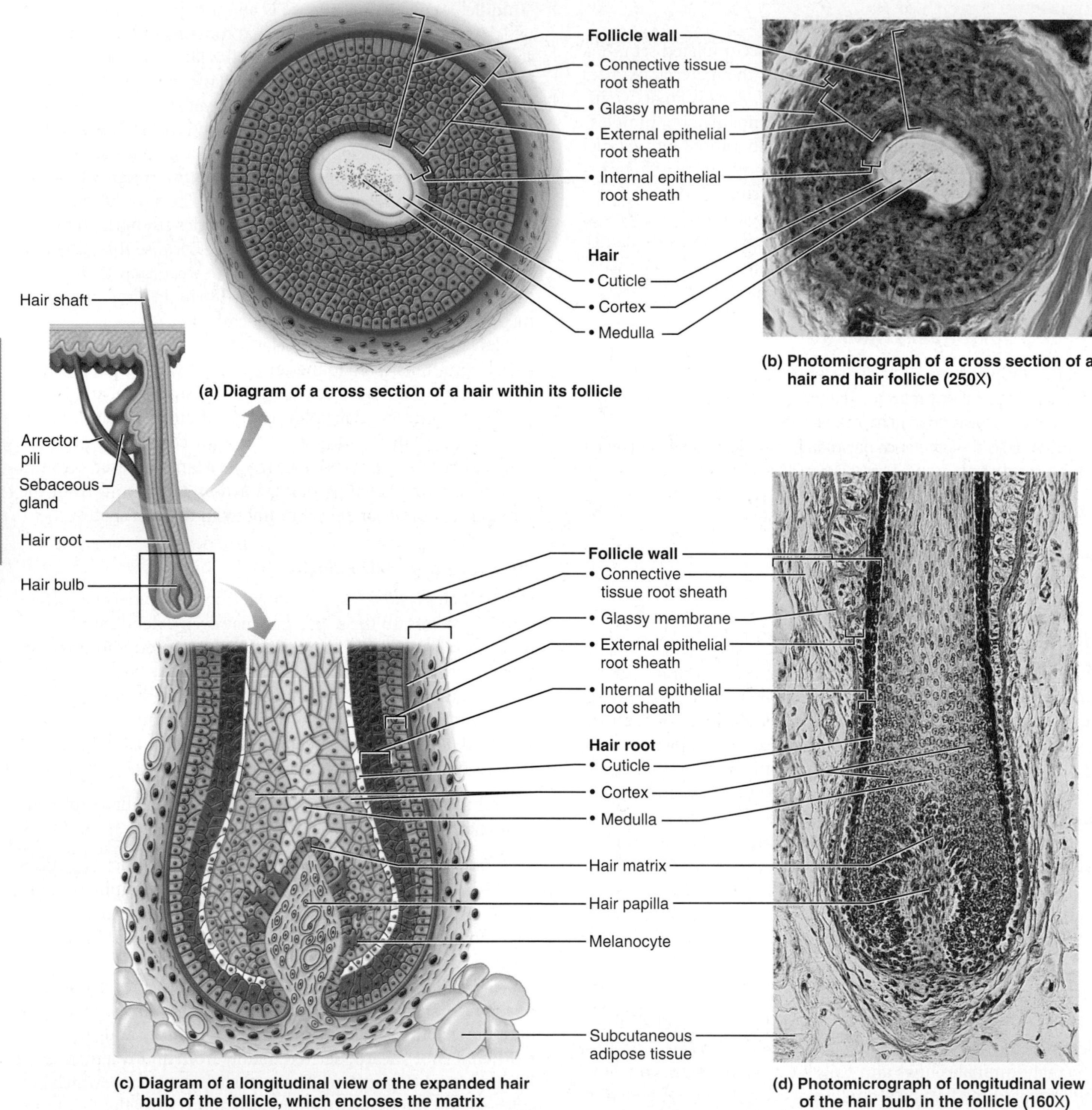

(a) Diagram of a cross section of a hair within its follicle

(b) Photomicrograph of a cross section of a hair and hair follicle (250X)

(c) Diagram of a longitudinal view of the expanded hair bulb of the follicle, which encloses the matrix

(d) Photomicrograph of longitudinal view of the hair bulb in the follicle (160X)

Figure 5.6 Structure of a hair and hair follicle.

Associated with each hair follicle is a bundle of smooth muscle cells called an **arrector pili** (ah-rek′tor pi′li; "raiser of hair") muscle. As you can see in Figure 5.1, most hair follicles approach the skin surface at a slight angle. The arrector pili muscle is attached in such a way that its contraction pulls the hair follicle into an upright position and dimples the skin surface to produce goose bumps in response to cold external temperatures or fear. This "hair-raising" response is not very useful to humans, with our short sparse hairs, but it is an important way for other animals to retain heat and protect themselves. Furry animals can stay warmer by trapping a layer of insulating air in their fur; and a scared animal with its hair on end looks larger and more formidable to its enemy.

Types and Growth of Hair

Hairs come in various sizes and shapes, but as a rule, they can be classified as vellus or terminal. The body hair of children and adult females is of the pale, fine **vellus hair** (vel′us; *vell* = wool, fleece) variety. The coarser, longer hair of the eyebrows and scalp is **terminal hair**, which may also be darker. At puberty, terminal hairs appear in the axillary and pubic regions of both sexes and on the face and chest (and typically the arms and legs) of males. These terminal hairs grow in response to the stimulating effects of male sex hormones called *androgens* (of which *testosterone* is the most important), and when male hormones are present in large amounts, terminal hair growth is luxuriant.

Hair growth and density are influenced by many factors, but most importantly by nutrition and hormones. Poor nutrition means poor hair growth, whereas conditions that increase local dermal blood flow (such as chronic physical irritation or inflammation) may enhance local hair growth. Many old-time bricklayers who carried their hod on one shoulder all the time developed one hairy shoulder. Undesirable hair growth (such as on a woman's upper lip) may be arrested by *electrolysis* or laser treatments, which use electricity or light energy, respectively, to destroy the hair roots.

HOMEOSTATIC IMBALANCE

In women, small amounts of androgens are normally produced by both the ovaries and the adrenal glands. Excessive hairiness, or **hirsutism** (her′soot-izm; *hirsut* = hairy), as well as other signs of masculinization, may result from an adrenal gland or ovarian tumor that secretes abnormally large amounts of androgens. Since few women want a beard or hairy chest, such tumors are surgically removed as soon as possible. ■

The rate of hair growth varies from one body region to another and with sex and age, but it averages 2.5 mm per week. Each follicle goes through *growth cycles.* In each cycle, an active growth phase, ranging from weeks to years, is followed by a regressive phase. During the regressive phase, the hair matrix cells die and the follicle base and hair bulb shrivel somewhat, dragging the hair papilla upward to abut the region of the follicle that does not regress. The follicle then enters a resting phase for one to three months. After the resting phase, the cycling part of the follicle regenerates and activated bulge cells migrate toward the papilla. As a result, the matrix proliferates again and forms a new hair to replace the old one that has fallen out or will be pushed out by the new hair.

The life span of hairs varies and appears to be under the control of a slew of proteins. The follicles of the scalp remain active for six to ten years before becoming inactive for a few months. Because only a small percentage of the hair is shed at any one time, we lose an average of 90 scalp hairs daily. The follicles of the eyebrow hairs remain active for only three to four months, which explains why your eyebrows are never as long as the hairs on your head.

Hair Thinning and Baldness

A follicle has only a limited number of cycles in it. Given ideal conditions, hair grows fastest from the teen years to the 40s, and then its growth slows. The fact that hairs are not replaced as fast as they are shed leads to hair thinning and some degree of baldness, or **alopecia** (al″o-pe′she-ah), in both sexes. Much less dramatic in women, the process usually begins at the anterior hairline and progresses posteriorly. Coarse terminal hairs are replaced by vellus hairs, and the hair becomes increasingly wispy.

True, or *frank, baldness* is a different story entirely. The most common type, **male pattern baldness**, is a genetically determined, sex-influenced condition. It is thought to be caused by a delayed-action gene that "switches on" in adulthood and changes the response of the hair follicles to DHT (dihydrotestosterone), a metabolite of testosterone. As a result, the follicular growth cycles become so short that many hairs never even emerge from their follicles before shedding, and those that do are fine vellus hairs that look like peach fuzz in the "bald" area.

Until recently, the only cure for male pattern baldness was drugs that inhibit testosterone production, but they also cause loss of sex drive—a trade-off few men would choose. Quite by accident, it was discovered that minoxidil, a drug used to reduce high blood pressure, has the interesting side effect in some bald men of stimulating hair regrowth. Although its results are variable, minoxidil is available over the counter in dropper bottles or spray form for application to the scalp. Finasteride, according to some the most promising cure ever developed for male pattern baldness, hit pharmacy shelves in early 1998 and has had moderate success. Available only by prescription in once-a-day pill form, it must be taken for the rest of a person's life. Once the patient stops taking it, all of the new growth falls out.

HOMEOSTATIC IMBALANCE

Hair thinning can be induced by a number of factors that upset the normal balance between hair loss and replacement. Outstanding examples are acutely high fever, surgery, severe emotional trauma, and certain drugs (excessive vitamin A, some antidepressants and blood thinners, anabolic steroids, and most chemotherapy drugs). Protein-deficient diets and lactation lead to hair thinning because new hair growth stops when protein needed for keratin synthesis is not available or is being used for milk production. In all of these cases, hair regrows if the cause of thinning is removed or corrected. In the rare condition called *alopecia areata,* the immune system attacks the follicles and the hair falls out in patches. But again, the follicles survive. Hair loss due to severe burns, excessive radiation, or other factors that eliminate the follicles is permanent. ■

CHECK YOUR UNDERSTANDING

15. What are the concentric regions of a hair shaft, from the outside in?
16. Why is having your hair cut painless?
17. What is the role of an arrector pili muscle?
18. What is the function of the hair papilla?

For answers, see Appendix G.

5

5

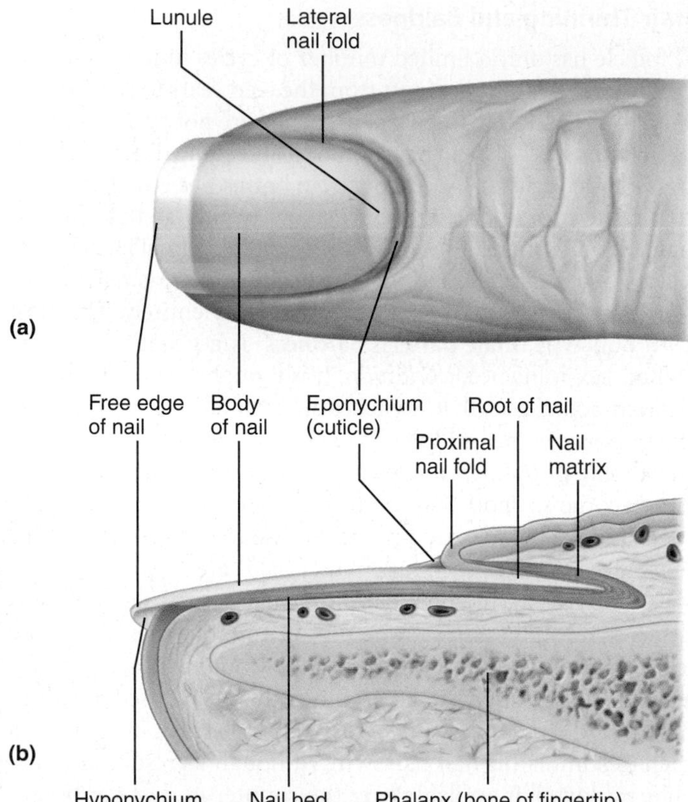

Figure 5.7 Structure of a nail. (a) Surface view of the distal part of a finger. **(b)** Sagittal section of the fingertip. The nail matrix that forms the nail lies beneath the lunule.

Nails

A **nail** is a scalelike modification of the epidermis that forms a clear protective covering on the dorsal surface of the distal part of a finger or toe **(Figure 5.7)**. Nails, which correspond to the hooves or claws of other animals, are particularly useful as "tools" to help pick up small objects and to scratch an itch. In contrast to *soft keratin* of the epidermis, nails contain *hard keratin*. Each nail has a *free edge*, a *body* (visible attached portion), and a proximal *root* (embedded in the skin). The deeper layers of the epidermis extend beneath the nail as the *nail bed*, and the nail itself corresponds to the superficial keratinized layers. The thickened proximal portion of the nail bed, called the **nail matrix**, is responsible for nail growth. As the nail cells produced by the matrix become heavily keratinized, the nail body slides distally over the nail bed.

Nails normally appear pink because of the rich bed of capillaries in the underlying dermis. However, the region that lies over the thick nail matrix appears as a white crescent called the *lunule* (lu′nool; "little moon"). The proximal and lateral borders of the nail are overlapped by skin folds, called **nail folds**. The proximal nail fold projects onto the nail body as the **cuticle** or **eponychium** (ep″o-nik′e-um; "on the nail"). The region beneath the free edge of the nail where dirt and debris tend to accumulate is the **hyponychium** ("below nail"), informally called the quick.

Changes in nail appearance may help diagnose certain conditions. For example, yellow-tinged nails may indicate a respiratory or thyroid gland disorder, and if combined with thickening of the nail, a fungus infection. An outward concavity of the nail (spoon nail) may signal an iron deficiency, and horizontal lines (Beau's lines) across the nails may hint of malnutrition.

CHECK YOUR UNDERSTANDING

19. Why is the lunule of a nail white instead of pink like the rest of the nail?

20. Why are nails so hard?

For answers, see Appendix G.

Functions of the Integumentary System

▶ Describe how the skin accomplishes at least five different functions.

The skin and its derivatives perform a variety of functions that affect body metabolism and prevent external factors from upsetting body homeostasis. Given its superficial location it is our most vulnerable organ system, exposed to bacteria, abrasion, temperature extremes, and harmful chemicals.

Protection

The skin constitutes at least three types of barriers: chemical, physical, and biological.

Chemical Barriers

The chemical barriers include skin secretions and melanin. Although the skin's surface teems with bacteria, the low pH of skin secretions—the so-called **acid mantle**—retards their multiplication. In addition, many bacteria are killed outright by dermcidin in sweat and bactericidal substances in sebum. Skin cells also secrete natural antibiotics called *defensins* that literally punch holes in bacteria, making them look like sieves. Wounded skin releases large quantities of protective peptides called *cathelicidins* that are particularly effective in preventing infection by group A streptococcus bacteria. As discussed earlier, melanin provides a chemical pigment shield to prevent UV damage to the viable skin cells.

Physical/Mechanical Barriers

Physical, or mechanical, barriers are provided by the continuity of skin and the hardness of its keratinized cells. As a physical barrier, the skin is a remarkable compromise. A thicker epidermis would be more impenetrable, but we would pay the price in loss of suppleness and agility. Epidermal continuity works hand in hand with the acid mantle and certain chemicals in skin secretions to ward off bacterial invasion. The water-resistant glycolipids of the epidermis block most diffusion of water and water-soluble substances between cells, preventing both their loss from and entry into the body through the skin. However, there is a continual small loss of water through the epidermis, and if immersed in water (other than salt water), the skin will take in some water and swell slightly.

Other substances that *do* penetrate the skin in limited amounts include (1) *lipid-soluble substances*, such as oxygen, carbon dioxide, fat-soluble vitamins (A, D, E, and K), and steroids (estrogens); (2) *oleoresins* (o″le-o-rez′inz) of certain plants, such as poison ivy and poison oak; (3) *organic solvents*, such as acetone, dry-cleaning fluid, and paint thinner, which dissolve the cell lipids; (4) *salts of heavy metals*, such as lead and mercury; (5) selected drugs (nitroglycerine, nicotine), and (6) drug agents called *penetration enhancers* that help ferry other drugs into the body. Skin permeability is dramatically enhanced by alcoholic drinks for at least 24 hours after their ingestion.

HOMEOSTATIC IMBALANCE

Organic solvents and heavy metals are devastating to the body and can be lethal. Passage of organic solvents through the skin into the blood can cause the kidneys to shut down and can also cause brain damage. Absorption of lead results in anemia and neurological defects. These substances should never be handled with bare hands. ■

Biological Barriers

Biological barriers include the dendritic cells of the epidermis, macrophages in the dermis, and DNA itself. Epidermal dendritic cells are active elements of the immune system. For the immune response to be activated, the foreign substances, or *antigens*, must be presented to specialized white blood cells called lymphocytes. In the epidermis, the dendritic cells play this role. Dermal macrophages constitute a second line of defense to dispose of viruses and bacteria that have managed to penetrate the epidermis. They, too, act as antigen "presenters."

Although melanin provides a fairly good chemical sunscreen, DNA itself is a remarkably effective biologically based sunscreen. Electrons in DNA molecules absorb UV radiation and transfer it to the atomic nuclei, which heat up and vibrate vigorously. However, since the heat dissipates to surrounding water molecules instantaneously, the DNA converts potentially destructive radiation into harmless heat.

Body Temperature Regulation

The body works best when its temperature remains within homeostatic limits. Like car engines, we need to get rid of the heat generated by our internal reactions. As long as the external temperature is lower than body temperature, the skin surface loses heat to the air and to cooler objects in its environment, just as a car radiator loses heat to the air and other nearby engine parts.

Under normal resting conditions, and as long as the environmental temperature is below 31–32°C (88–90°F), sweat glands secrete about 500 ml (0.5 L) of sweat per day. This routine and unnoticeable sweating is called *insensible perspiration*. When body temperature rises, the nervous system stimulates the dermal blood vessels to dilate and the sweat glands into vigorous secretory activity. Indeed, on a hot day, sweat becomes noticeable and can account for the loss of up to 12 L (about 3 gallons) of body water in one day. This visible output of sweat is referred to as *sensible perspiration*. Evaporation of sweat from the skin surface dissipates body heat and efficiently cools the body, preventing overheating.

When the external environment is cold, dermal blood vessels constrict. Their constriction causes the warm blood to bypass the skin temporarily and allows skin temperature to drop to that of the external environment. Once this has happened, passive heat loss from the body is slowed, conserving body heat. Body temperature regulation is discussed in Chapter 24.

Cutaneous Sensation

The skin is richly supplied with **cutaneous sensory receptors**, which are actually part of the nervous system. The cutaneous receptors are classified as *exteroceptors* (ek″ster-o-sep′torz) because they respond to stimuli arising outside the body. For example, Meissner's corpuscles (in the dermal papillae) and tactile discs allow us to become aware of a caress or the feel of our clothing against our skin, whereas pacinian corpuscles (in the deeper dermis or hypodermis) alert us to bumps or contacts involving deep pressure. Hair follicle receptors report on wind blowing through our hair and a playful tug on a pigtail. Free nerve endings that meander throughout the skin sense painful stimuli (irritating chemicals, extreme heat or cold, and others). We defer detailed discussion of these cutaneous receptors to Chapter 13. Except for Meissner's corpuscles, which are found only in skin that lacks hairs, the cutaneous receptors mentioned above are illustrated in Figure 5.1. One, a tactile disc, is shown in Figure 5.2b.

Metabolic Functions

The skin is a chemical factory, fueled in part by the sun's rays. When sunlight bombards the skin, modified cholesterol molecules circulating through dermal blood vessels are converted to a vitamin D precursor, and transported via the blood to other body areas to be ultimately converted to vitamin D, which plays various roles in calcium metabolism. For example, calcium cannot be absorbed from the digestive tract without vitamin D.

Besides synthesizing the vitamin D precursor, the epidermis has a host of other metabolic functions. It makes chemical conversions that supplement those of the liver. For example, keratinocyte enzymes can (1) "disarm" many cancer-causing chemicals that penetrate the epidermis; (2) convert some harmless chemicals into carcinogens; and (3) activate some steroid hormones—for instance, they can transform cortisone applied to irritated skin into hydrocortisone, a potent anti-inflammatory drug. Skin cells also make several biologically important proteins, including collagenase, an enzyme that aids the natural turnover of collagen (and deters wrinkles).

Blood Reservoir

The dermal vascular supply is extensive and can hold large volumes of blood (about 5% of the body's entire blood volume).

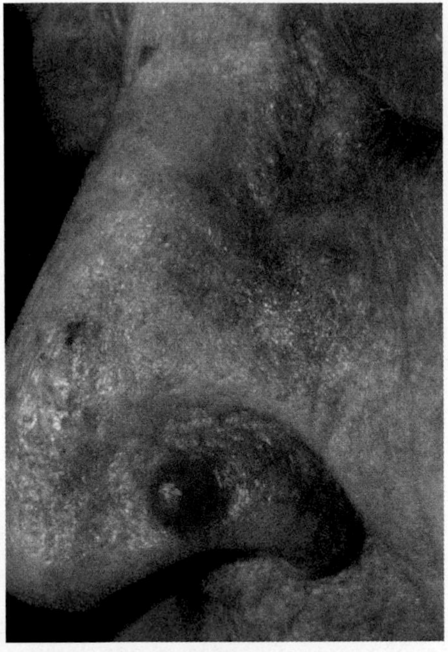

(a) Basal cell carcinoma

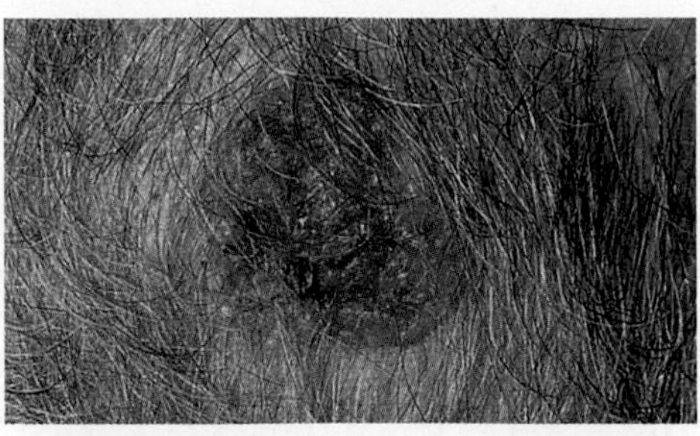

(b) Squamous cell carcinoma

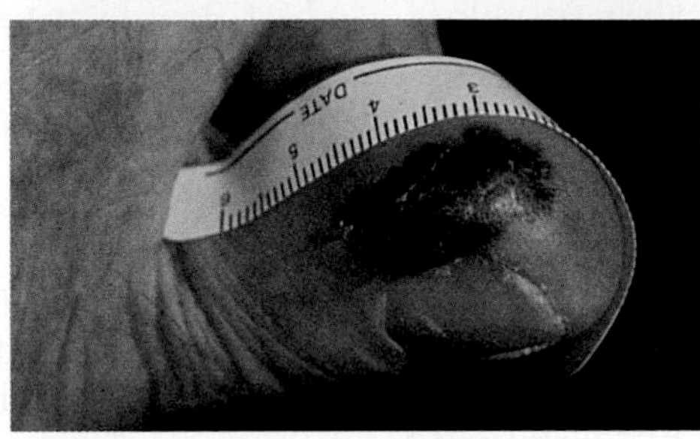

(c) Melanoma

Figure 5.8 Photographs of skin cancers.

When other body organs, such as vigorously working muscles, need a greater blood supply, the nervous system constricts the dermal blood vessels. This constriction shunts more blood into the general circulation, making it available to the muscles and other body organs.

Excretion

Limited amounts of nitrogen-containing wastes (ammonia, urea, and uric acid) are eliminated from the body in sweat, although most such wastes are excreted in urine. Profuse sweating is an important avenue for water and salt (sodium chloride) loss.

CHECK YOUR UNDERSTANDING

21. What chemicals produced in the skin help provide barriers to bacteria? List at least three and explain how the chemicals are protective.
22. What epidermal cells play a role in body immunity?
23. How is sunlight important to bone health?
24. How does the skin contribute to body metabolism?

For answers, see Appendix G.

Homeostatic Imbalances of Skin

- ▶ Summarize the characteristics of the three major types of skin cancers.
- ▶ Explain why serious burns are life threatening. Describe how to determine the extent of a burn and differentiate first-, second-, and third-degree burns.

When skin rebels, it is quite a visible rebellion. Loss of homeostasis in body cells and organs reveals itself on the skin, sometimes in startling ways. The skin can develop more than 1000 different conditions and ailments. The most common skin disorders are bacterial, viral, or yeast infections. A number of these are summarized in Related Clinical Terms on p. 168. Less common, but far more damaging to body well-being, are skin cancer and burns, considered next.

Skin Cancer

One in five Americans develops skin cancer at some point. Most tumors that arise in the skin are benign and do not spread (metastasize) to other body areas. (A wart, a neoplasm caused by a virus, is one example.) However, some skin tumors are malignant, or cancerous, and invade other body areas.

A crucial risk factor for skin cancer is overexposure to the UV radiation in sunlight, which damages DNA bases. Adjacent pyrimidine bases often respond by fusing, forming lesions called *dimers.* UV radiation also appears to disable a tumor suppressor gene [*p53* or the patched (*ptc*) gene]. In limited numbers of cases, however, frequent irritation of the skin by infections, chemicals, or physical trauma seems to be a predisposing factor.

Interestingly, sunburned skin accelerates its production of Fas, a protein that causes genetically damaged skin cells to commit suicide, reducing the risk of mutations that will cause sun-linked skin cancer. It is the death of these gene-damaged cells that causes the skin to peel after a sunburn.

There is no such thing as a "healthy tan," but the good news for sun worshippers is the newly developed skin lotions that can fix damaged DNA before the involved cells can develop into cancer cells. These lotions contain tiny oily vesicles (liposomes) filled with enzymes that initiate repair of the DNA mutations most commonly caused by sunlight. The liposomes penetrate the epidermis and enter the keratinocytes, ultimately making their way into the nuclei to bind to specific sites where two DNA bases have fused. There, by selective cutting of the DNA strands, they begin a DNA repair process that is completed by cellular enzymes.

Basal Cell Carcinoma

Basal cell carcinoma (kar″sĭ-no′mah) is the least malignant and most common skin cancer. It accounts for nearly 80% of skin cancers. Stratum basale cells proliferate, invading the dermis and hypodermis. The cancer lesions occur most often on sun-exposed areas of the face and appear as shiny, dome-shaped nodules that later develop a central ulcer with a pearly, beaded edge (**Figure 5.8a**). Basal cell carcinoma is relatively slow-growing, and metastasis seldom occurs before it is noticed. Full cure by surgical excision is the rule in 99% of cases.

Squamous Cell Carcinoma

Squamous cell carcinoma, the second most common skin cancer, arises from the keratinocytes of the stratum spinosum. The lesion appears as a scaly reddened papule (small, rounded elevation) that arises most often on the head (scalp, ears, and lower lip), and hands (Figure 5.8b). It tends to grow rapidly and metastasize if not removed. If it is caught early and removed surgically or by radiation therapy, the chance of complete cure is good.

Melanoma

Melanoma (mel″ah-no′mah), cancer of melanocytes, is the most dangerous skin cancer because it is highly metastatic and resistant to chemotherapy. It accounts for only 2–3% of skin cancers, but its incidence is increasing rapidly (by 3–8% per year in the United States). Melanoma can begin wherever there is pigment. Most such cancers appear spontaneously, and about one-third develop from preexisting moles. It usually appears as a spreading brown to black patch (Figure 5.8c) that metastasizes rapidly to surrounding lymph and blood vessels.

The key to surviving melanoma is early detection. The chance of survival is poor if the lesion is over 4 mm thick. The usual therapy for melanoma is wide surgical excision accompanied by immunotherapy (immunizing the body against its cancer cells).

The American Cancer Society suggests that sun worshippers regularly examine their skin for new moles or pigmented spots and apply the **ABCD rule** for recognizing melanoma. **A. Asymmetry**: The two sides of the pigmented spot or mole do not match. **B. Border irregularity**: The borders of the lesion exhibit indentations. **C. Color**: The pigmented spot contains several colors (blacks, browns, tans, and sometimes blues and reds). **D. Diameter**: The spot is larger than 6 mm in diameter (the size of a pencil eraser). Some experts have found that adding an E, for elevation above the skin surface, improves diagnosis, so they use the ABCD(E) rule.

Burns

Burns are a devastating threat to the body primarily because of their effects on the skin. A **burn** is tissue damage inflicted by intense heat, electricity, radiation, or certain chemicals, all of which denature cell proteins and cause cell death in the affected areas.

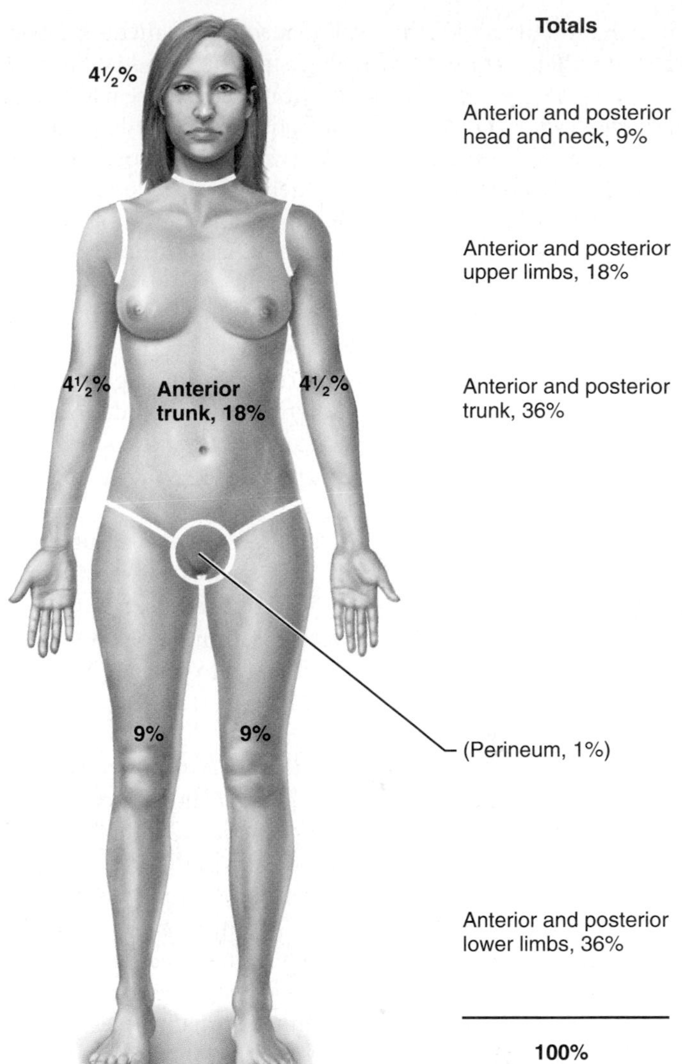

Figure 5.9 **Estimating the extent and severity of burns using the rule of nines.** Surface area values for the anterior body surface are indicated on the human figure. Total surface area (anterior and posterior body surfaces) for each body region is indicated to the right of the figure.

The immediate threat to life resulting from severe burns is a catastrophic loss of body fluids containing proteins and electrolytes, resulting in dehydration and electrolyte imbalance. These conditions, in turn, lead to renal shutdown and circulatory shock (inadequate blood circulation due to reduced blood volume). To save the patient, the lost fluids must be replaced immediately via the intravenous (IV) route.

In adults, the volume of fluid lost can be estimated by computing the percentage of body surface burned (extent of the burns) using the **rule of nines**. This method divides the body into 11 areas, each accounting for 9% of total body area, plus an additional area surrounding the genitals accounting for 1% of body surface area (**Figure 5.9**). This method is only approximate, so special tables are used when greater accuracy is desired.

Burn patients also need thousands of extra food calories daily to replace lost proteins and allow tissue repair. No one can eat

5

enough food to provide these calories, so burn patients are given supplementary nutrients through gastric tubes and IV lines. After the initial crisis has passed, infection becomes the main threat and *sepsis* (widespread bacterial infection) is the leading cause of death in burn victims. Burned skin is sterile for about 24 hours. Thereafter, bacteria, fungi, and other pathogens easily invade areas where the skin barrier is destroyed, and they multiply rapidly in the nutrient-rich environment of dead tissues. Adding to this problem is the fact that the immune system becomes deficient within one to two days after severe burn injury.

Burns are classified according to their severity (depth) as first-, second-, or third-degree burns. In **first-degree burns**, only the epidermis is damaged. Symptoms include localized redness, swelling, and pain. First-degree burns tend to heal in two to three days without special attention. Sunburn is usually a first-degree burn.

5

Second-degree burns injure the epidermis and the upper region of the dermis. Symptoms mimic those of first-degree burns, but blisters also appear. The burned area is red and painful, but skin regeneration occurs with little or no scarring within three to four weeks if care is taken to prevent infection. First- and second-degree burns are referred to as *partial-thickness burns* (**Figure 5.10a**).

Third-degree burns are *full-thickness burns*, involving the entire thickness of the skin (Figure 5.10b). The burned area appears gray-white, cherry red, or blackened, and initially there is little or no edema. Since the nerve endings in the area have been destroyed, the burned area is not painful. Although skin regeneration might eventually occur by proliferation of epithelial cells at the edges of the burn or from stem cells in hair follicles, it is usually impossible to wait that long because of fluid loss and infection. For this reason, skin grafting is usually necessary.

To prepare the burned area for grafting, the *eschar* (es′kar), or burned skin, must first be debrided (removed). To prevent infection and fluid loss, the area is then flooded with antibiotics and covered temporarily with a synthetic membrane, animal (pig) skin, cadaver skin, or "living bandage" made from the thin amniotic sac membrane that surrounds a fetus. Then healthy skin is transplanted to the burned site. Unless the graft is taken from the patient (an autograft), however, there is a good chance that it will be rejected by the patient's immune system (see p. 792 in Chapter 21). Even if the graft "takes," extensive scar tissue often forms in the burned areas.

An exciting technique is eliminating many of the traditional problems of skin grafting and rejection. Synthetic skin made of a silicone "epidermis" bound to a spongy "dermal" layer composed of collagen and ground cartilage is applied to the debrided area. In time, the patient's own dermal tissue absorbs and replaces the artificial one. Then the silicone sheet is peeled off and replaced with a network of epidermal cells cultured from the patient's own skin. The artificial skin is not rejected by the body, saves lives, and results in minimal scarring. However, it is more likely to become infected than is an autograft.

In general, burns are considered critical if any of the following conditions exists: (1) over 25% of the body has second-degree burns, (2) over 10% of the body has third-degree burns, or (3) there are third-degree burns of the face, hands, or feet. Facial burns introduce the possibility of burned respiratory pas-

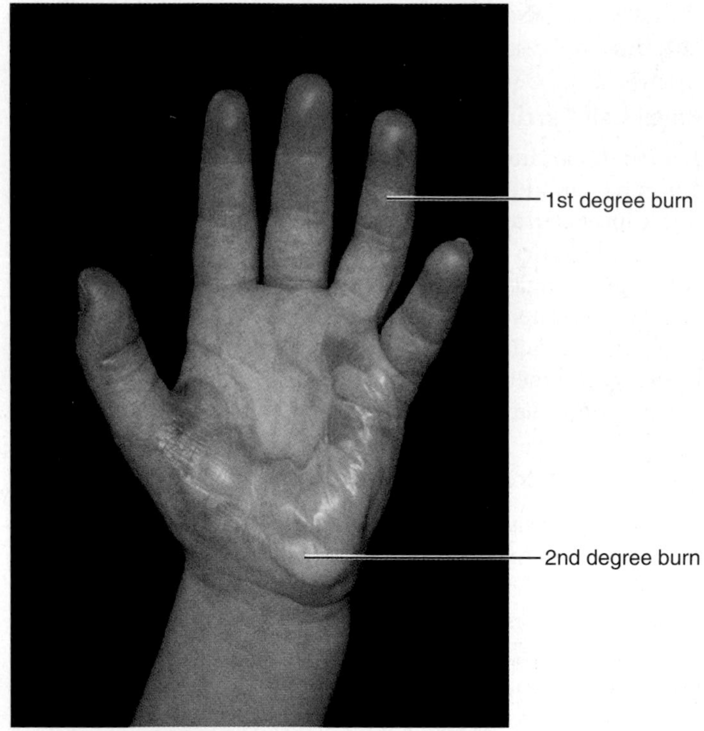

(a) Skin bearing partial thickness burn (1st and 2nd degree burns)

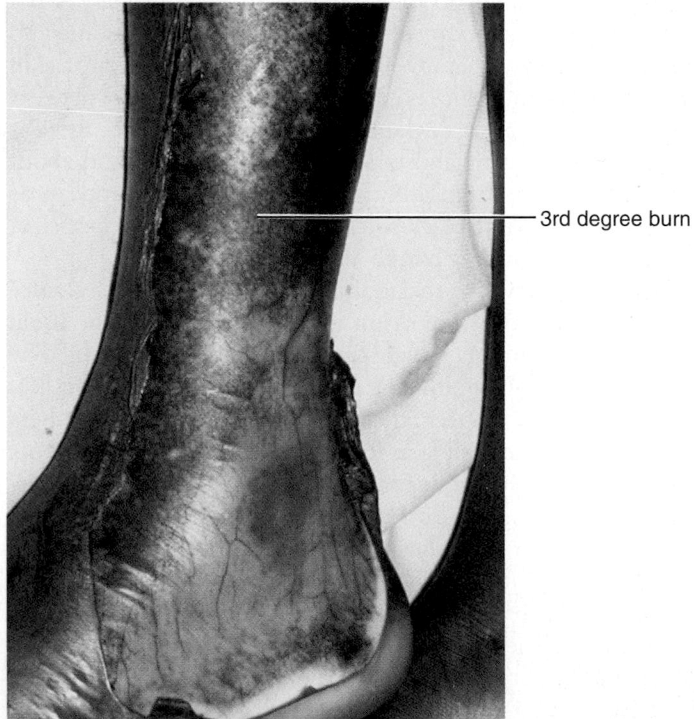

(b) Skin bearing full thickness burn (3rd degree burn)

Figure 5.10 Partial thickness and full thickness burns.

sageways, which can swell and cause suffocation. Burns at joints are also troublesome because scar tissue formation can severely limit joint mobility.

CHECK YOUR UNDERSTANDING

25. Which type of skin cancer develops from the youngest epidermal cells?
26. What name is given to the rule for recognizing the signs of melanoma?
27. The healing of burns and epidermal regeneration is usually uneventful unless the burn is a third-degree burn. What accounts for this difference?
28. Although the anterior head and face represent only a small percentage of the body surface, burns to this area are often much more serious than those to the body trunk. Why?

For answers, see Appendix G.

Developmental Aspects of the Integumentary System

▶ Describe and attempt to explain the causes of changes that occur in the skin from birth to old age.

The epidermis develops from the embryonic ectoderm, and the dermis and hypodermis develop from mesoderm. By the end of the fourth month of development, the skin is fairly well formed. The epidermis has all its strata, dermal papillae are obvious, and rudimentary epidermal derivatives formed by downward projections of cells from the basal layer are present. During the fifth and sixth months, the fetus is covered with a downy coat of delicate colorless hairs called the *lanugo coat* (lah-nu′go; "wool"). This hairy cloak is shed by the seventh month, and vellus hairs make their appearance.

When a baby is born, its skin is covered with *vernix caseosa* (ver′niks kă-se-o′sah; "varnish of cheese"), a white, cheesy-looking substance produced by the sebaceous glands that protects the fetus's skin within the water-filled amnion. The newborn's skin is very thin and often has accumulations in the sebaceous glands on the forehead and nose that appear as small white spots called *milia* (mil′e-ah). These normally disappear by the third week after birth.

During infancy and childhood, the skin thickens, and more subcutaneous fat is deposited. Although we all have approximately the same number of sweat glands, the number that begin to function increases in the first two years after birth and is determined by climate. For this reason, people who grow up in hot climates have more active sweat glands than those raised in cooler areas of the world.

During adolescence, the skin and hair become oilier as sebaceous glands are activated, and acne may appear. Acne generally subsides in early adulthood, and skin reaches its optimal appearance when we reach our 20s and 30s. Thereafter, the skin starts to show the effects of cumulative environmental assaults (abrasion, wind, sun, chemicals). Scaling and various kinds of skin inflammation, or **dermatitis** (der″mah-ti′tis), become more common.

As old age approaches, the rate of epidermal cell replacement slows, the skin thins, and its susceptibility to bruises and other types of injury increases. All of the lubricating substances produced by the skin glands that make young skin so soft start to become deficient. As a result, the skin becomes dry and itchy. However, people with naturally oily skin seem to postpone this dryness until later in life. Elastic fibers clump, and collagen fibers become fewer and stiffer. The subcutaneous fat layer diminishes, leading to the intolerance to cold so common in elderly people. Additionally, declining levels of sex hormones result in similar fat distribution in elderly men and women. The decreasing elasticity of the skin, along with the loss of subcutaneous tissue, inevitably leads to wrinkling. Decreasing numbers of melanocytes and dendritic cells enhance the risk and incidence of skin cancer in this age group. As a rule, redheads and fair-skinned individuals, who have less melanin to begin with, show age-related changes more rapidly than do those with darker skin and hair.

By the age of 50 years, the number of active hair follicles has declined by two-thirds and continues to fall, resulting in hair thinning. Hair loses its luster in old age, and the delayed-action genes responsible for graying and male pattern baldness become active.

Although there is no known way to avoid the aging of the skin, one of the best ways to slow the process is to shield your skin from the sun with protective clothing, and sunscreens or sunblocks with a sun protection factor (SPF) of 15 or higher. Remember, the same sunlight that produces that fashionable tan also causes the sagging, blotchy, wrinkled skin of old age complete with pigmented "liver spots." (It is instructive to note that aged skin that has been protected from the sun has lost some elasticity and is thinned, but it remains unwrinkled and unmarked.) Much of this havoc is due to UVA activation of enzymes called matrix metalloproteinases, which degrade collagen and other dermal components. A drug called tretinoin, related to vitamin A, inhibits these enzymes and is now being used in some skin creams to slow photo-aging. Good nutrition, plenty of fluids, and cleanliness may also delay the process.

CHECK YOUR UNDERSTANDING

29. What is the source of vernix caseosa that covers the skin of the newborn baby?
30. What change in the skin leads to cold intolerance in the elderly?
31. How does UV radiation contribute to wrinkling of the skin?

For answers, see Appendix G.

The skin is only about as thick as a paper towel—not too impressive as organ systems go. Yet, when it is severely damaged, nearly every body system reacts. Metabolism accelerates or may be impaired, immune system changes occur, bones may soften, the cardiovascular system may fail—the list goes on and on. On the other hand, when the skin is intact and performing its functions, the body as a whole benefits. Homeostatic interrelationships between the integumentary system and other organ systems are summarized in *Making Connections* on pp. 166–167.

RELATED CLINICAL TERMS

Albinism (al′bĭ-nizm; *alb* = white) Inherited condition in which melanocytes do not synthesize melanin owing to a lack of tyrosinase. An albino's skin is pink, the hair pale or white, and the irises of the eyes unpigmented or poorly so.

Boils and carbuncles (kar′bung-klz; "little glowing embers") Inflammation of hair follicles and sebaceous glands in which an infection has spread to the underlying hypodermis; common on the dorsal neck. Carbuncles are composite boils. A common cause is bacterial infection.

Cold sores (fever blisters) Small fluid-filled blisters that itch and smart; usually occur around the lips and in the mucosa of the mouth; caused by a herpes simplex infection. The virus localizes in a cutaneous nerve, where it remains dormant until activated by emotional upset, fever, or UV radiation.

Contact dermatitis Itching, redness, and swelling, progressing to blister formation; caused by exposure of the skin to chemicals (e.g., poison ivy oleoresin) that provoke an allergic response in sensitive individuals.

Decubitus ulcer (de-ku′bĭ-tus) Localized breakdown and ulceration of skin due to interference with its blood supply. Usually occurs over a bony prominence, such as the hip or heel, that is subjected to continuous pressure; also called a bedsore.

Dermatology The branch of medicine that studies and treats disorders of the skin.

Eczema (ek′ze-mah) A skin rash characterized by itching, blistering, oozing, and scaling of the skin. A common allergic reaction in children, but also occurs (typically in a more severe form) in adults. Frequent causes include allergic reactions to certain foods (fish, eggs, and others) or to inhaled dust or pollen. Treated by methods used for other allergic disorders.

Epidermolysis bullosa (EB) A group of hereditary disorders characterized by inadequate or faulty synthesis of keratin, collagen, and/or basement membrane "cement" that results in lack of cohesion between layers of the skin and mucosa; a simple touch causes layers to separate and blister. For this reason, EB victims are called "touch-me-nots." In severe cases fatal blistering occurs in major vital organs. Because the blisters rupture easily, victims suffer frequent infections; treatments are aimed at relieving the symptoms and preventing infection.

Impetigo (im″pĕ-ti′go; *impet* = an attack) Pink, fluid-filled, raised lesions (common around the mouth and nose) that develop a yellow crust and eventually rupture. Caused by staphylococcus infection, it is contagious, and common in school-age children.

Porphyria (por-fer′e-ah; "purple") An inherited condition in which certain enzymes needed to form the heme of hemoglobin of blood are lacking. Without these enzymes, metabolic intermediates of the heme pathway called porphyrins build up, spill into the circulation, and eventually cause lesions throughout the body, especially when exposed to sunlight. The skin becomes lesioned and scarred; fingers, toes, and nose are disfigured; gums degenerate and teeth become prominent; for this reason it's believed to be the basis of folklore about vampires.

Psoriasis (so-ri′ah-sis) A chronic autoimmune condition characterized by raised, reddened epidermal patches covered with silvery scales that itch or burn, crack, and sometimes bleed or become infected. When severe, it may be disfiguring and debilitating. The autoimmune attacks are often triggered by trauma, infection, hormonal changes, and stress. Cortisone-containing topicals may control mild cases. For more severe cases, self-injected drugs called biologicals and/or phototherapy with UV light in conjunction with chemotherapeutic drugs provides some relief.

Rosacea (ro-za′she-ah) A chronic skin eruption produced by dilation of small blood vessels of the face, particularly the nose and cheeks. Papules and acne-like pustules may or may not occur. More common in women, but tends to be more severe when it occurs in men. Cause is unknown, but stress, some endocrine disorders, and anything that produces flushing (hot beverages, alcohol, sunlight, etc.) can aggravate this condition.

Vitiligo (vit″ĭ-li′go; *viti* = a vine, winding) The most prevalent skin pigmentation disorder, characterized by a loss of melanocytes and uneven dispersal of melanin, so that unpigmented skin regions (light spots) are surrounded by normally pigmented areas. An autoimmune disorder.

CHAPTER SUMMARY

The Skin (pp. 149–155)

1. The skin, or integument, is composed of two discrete tissue layers, an outer epidermis and a deeper dermis, resting on subcutaneous tissue, the hypodermis.

Epidermis (pp. 150–152)

2. The epidermis is an avascular, keratinized sheet of stratified squamous epithelium. Most epidermal cells are keratinocytes. Scattered among the keratinocytes in the deepest epidermal layers are melanocytes, epidermal dendritic cells, and tactile cells.
3. From deep to superficial, the strata, or layers of the epidermis, are the basale, spinosum, granulosum, lucidum, and corneum. The stratum lucidum is absent in thin skin. The mitotically active stratum basale is the source of new cells for epidermal growth. The most superficial layers are increasingly keratinized and less viable.

Dermis (pp. 152–153)

4. The dermis, composed mainly of dense, irregular connective tissue, is well supplied with blood vessels, lymphatic vessels, and nerves. Cutaneous receptors, glands, and hair follicles reside within the dermis.
5. The more superficial papillary layer exhibits dermal papillae that protrude into the epidermis above, as well as dermal ridges. Dermal ridges and epidermal ridges together form the friction ridges that produce fingerprints.
6. In the deeper, thicker reticular layer, the connective tissue fibers are much more densely interwoven. Less dense regions between the collagen bundles produce cleavage, or tension, lines in the skin. Points of tight dermal attachment to the hypodermis produce dermal folds, or flexure lines.

Skin Color (pp. 154–155)

7. Skin color reflects the amount of pigments (melanin and carotene) in the skin and the oxygenation level of hemoglobin in blood.
8. Melanin production is stimulated by exposure to ultraviolet radiation in sunlight. Melanin, produced by melanocytes and transferred to keratinocytes, protects the keratinocyte nuclei from the damaging effects of UV radiation.
9. Skin color is affected by emotional state. Alterations in normal skin color (jaundice, bronzing, erythema, and others) may indicate certain disease states.

Appendages of the Skin (pp. 155–160)

1. Skin appendages, which derive from the epidermis, include glands (sweat and sebaceous), hairs and hair follicles, and nails.

Sweat (Sudoriferous) Glands (pp. 155–156)

2. Eccrine (merocrine) sweat glands, with a few exceptions, are distributed over the entire body surface. Their primary function is thermoregulation. They are simple coiled tubular glands that secrete a salt solution containing small amounts of other solutes. Their ducts usually empty to the skin surface via pores.
3. Apocrine sweat glands, which may function as scent glands, are found primarily in the axillary and anogenital areas. Their secretion is similar to eccrine secretion, but it also contains proteins and fatty substances on which bacteria thrive.

Sebaceous (Oil) Glands (pp. 156–157)

4. Sebaceous glands occur all over the body surface except for the palms and soles. They are simple alveolar glands; their oily holocrine secretion is called sebum. Sebaceous gland ducts usually empty into hair follicles.
5. Sebum lubricates the skin and hair, prevents water loss from the skin, and acts as a bactericidal agent. Sebaceous glands are activated (at puberty) and controlled by androgens.

Hairs and Hair Follicles (pp. 157–159)

6. A hair, produced by a hair follicle, consists of heavily keratinized cells. A typical hair has a central medulla, a cortex, and an outer cuticle and root and shaft portions. Hair color reflects the amount and kind of melanin present.
7. A hair follicle consists of an inner epithelial root sheath and an outer connective tissue root sheath derived from the dermis. The base of the hair follicle is a hair bulb with a matrix that produces the hair. A hair follicle is richly vascularized and well supplied with nerve fibers. Arrector pili muscles pull the follicles into an upright position and produce goose bumps.
8. Except for hairs of the scalp and around the eyes, hairs formed initially are fine vellus hairs; at puberty, under the influence of androgens, coarser, darker terminal hairs appear in the axillae and the genital region.
9. The rate of hair growth varies in different body regions and with sex and age. Differences in life span of hairs account for differences in length on different body regions. Hair thinning reflects factors that lengthen follicular resting periods, age-related atrophy of hair follicles, and a delayed-action gene.

Nails (p. 160)

10. A nail is a scalelike modification of the epidermis that covers the dorsum of a finger (or toe) tip. The actively growing region is the nail matrix.

Functions of the Integumentary System (pp. 160–162)

1. Protection. The skin protects by chemical barriers (the antibacterial nature of sebum, defensins, cathelicidins, the acid mantle, and the UV shield of melanin), physical barriers (the hardened keratinized and lipid-rich surface), and biological barriers (dendritic cells, macrophages, and DNA).
2. Body temperature regulation. The skin vasculature and sweat glands, regulated by the nervous system, play an important role in maintaining body temperature homeostasis.
3. Cutaneous sensation. Cutaneous sensory receptors respond to temperature, touch, pressure, and pain stimuli.
4. Metabolic functions. A vitamin D precursor is synthesized from cholesterol by epidermal cells. Skin cells also play a role in some chemical conversions.
5. Blood reservoir. The extensive vascular supply of the dermis allows the skin to act as a blood reservoir.
6. Excretion. Sweat contains small amounts of nitrogenous wastes and plays a minor role in excretion.

Homeostatic Imbalances of Skin (pp. 162–165)

1. The most common skin disorders result from infections.

Skin Cancer (pp. 162–163)

2. The most common cause of skin cancer is exposure to ultraviolet radiation.
3. Basal cell carcinoma and squamous cell carcinoma are cured if they are removed before metastasis. Melanoma, a cancer of melanocytes, is less common but more dangerous.

Burns (pp. 163–165)

4. In severe burns, the initial threat is loss of protein- and electrolyte-rich body fluids, which may lead to circulatory collapse. The second threat is overwhelming bacterial infection.
5. The extent of a burn may be evaluated by using the rule of nines. The severity of burns is indicated by the terms first degree, second degree, and third degree. Third-degree burns are full-thickness burns that require grafting for successful recovery.

Developmental Aspects of the Integumentary System (p. 165)

1. The epidermis develops from embryonic ectoderm; the dermis (and hypodermis) develops from mesoderm.
2. The fetus exhibits a downy lanugo coat. Fetal sebaceous glands produce vernix caseosa, which helps protect the fetus's skin from its watery environment.
3. A newborn's skin is thin. During childhood the skin thickens and more subcutaneous fat is deposited. At puberty, sebaceous glands are activated and terminal hairs appear in greater numbers.
4. In old age, the rate of epidermal cell replacement declines and the skin and hair thin. Skin glands become less active. Loss of collagen and elastic fibers and subcutaneous fat leads to wrinkling; delayed-action genes cause graying and balding. Photodamage is a major cause of skin aging.

6 Bones and Skeletal Tissues

All of us have heard the expressions "bone tired" and "bag of bones"—rather unflattering and inaccurate images of one of our most phenomenal tissues and our main skeletal elements. Our brains, not our bones, convey feelings of fatigue. As for "bag of bones," they are indeed more prominent in some of us, but without bones to form our internal skeleton we would all creep along the ground like slugs, lacking any definite shape or form. Along with its bones, the skeleton contains resilient cartilages, which we briefly discuss in this chapter. However, our major focus is the structure and function of bone tissue and the dynamics of its formation and remodeling throughout life.

Skeletal Cartilages

- Describe the functional properties of the three types of cartilage tissue.
- Locate the major cartilages of the adult skeleton.
- Explain how cartilage grows.

The human skeleton is initially made up of cartilages and fibrous membranes, but most of these early supports are soon replaced by bone. The few cartilages that remain in adults are found mainly in regions where flexible skeletal tissue is needed.

Basic Structure, Types, and Locations

A **skeletal cartilage** is made of some variety of *cartilage tissue*, which consists primarily of water. The high water content of cartilage accounts for its resilience, that is, its ability to spring back to its original shape after being compressed.

The cartilage, which contains no nerves or blood vessels, is surrounded by a layer of dense irregular connective tissue, the *perichondrium* (per″ĭ-kon′dre-um; "around the cartilage"). The perichondrium acts like a girdle to resist outward expansion when the cartilage is compressed. Additionally, the perichondrium contains the blood vessels from which nutrients diffuse through the matrix to reach the cartilage cells. This mode of nutrient delivery limits cartilage thickness.

As we described in Chapter 4, there are three types of cartilage tissue in the body: hyaline, elastic, and fibrocartilage. All three types have the same basic components—cells called *chondrocytes*, encased in small cavities (lacunae) within an *extracellular matrix* containing a jellylike ground substance and fibers. The skeletal cartilages contain representatives from all three types.

Hyaline cartilages, which look like frosted glass when freshly exposed, provide support with flexibility and resilience. They are the most abundant skeletal cartilages. When viewed under the microscope, their chondrocytes appear spherical (see Figure 4.8g). The only fiber type in their matrix is fine collagen fibers (which, however, are not detectable microscopically). Colored blue in **Figure 6.1**, skeletal hyaline cartilages include (1) *articular cartilages*, which cover the ends of most bones at movable joints; (2) *costal cartilages*, which connect the ribs to the sternum (breastbone); (3) *respiratory cartilages*, which form the skeleton of the *larynx* (*voicebox*) and reinforce other respiratory passageways; and (4) *nasal cartilages*, which support the external nose.

Elastic cartilages look very much like hyaline cartilages (see Figure 4.8h), but they contain more stretchy elastic fibers and so are better able to stand up to repeated bending. They are found in only two skeletal locations, shown in green in Figure 6.1—the external ear and the epiglottis (the flap that bends to cover the opening of the larynx each time we swallow).

Fibrocartilages are highly compressible and have great tensile strength. The perfect intermediate between hyaline and elastic cartilages, fibrocartilages consist of roughly parallel rows of chondrocytes alternating with thick collagen fibers (see Figure 4.8i). Fibrocartilages occur in sites that are subjected to both heavy pressure and stretch, such as the padlike cartilages (menisci) of the knee and the discs between vertebrae, colored red in Figure 6.1.

Growth of Cartilage

Unlike bone, which has a hard matrix, cartilage has a flexible matrix which can accommodate mitosis. It is the ideal tissue to use to lay down the embryonic skeleton and to provide for new skeletal growth. Cartilage grows in two ways. In **appositional growth** (ap″o-zish′un-al; "growth from outside"), cartilage-forming cells in the surrounding perichondrium secrete new matrix against the external face of the existing cartilage tissue. In **interstitial growth** (in″ter-stish′al; "growth from inside"), the lacunae-bound chondrocytes divide and secrete new matrix, expanding the cartilage from within. Typically, cartilage growth ends during adolescence when the skeleton stops growing.

Under certain conditions—during normal bone growth in youth and during old age—calcium salts may be deposited in the matrix and cause it to harden, a process called calcification. Note, however, that calcified cartilage is not bone; cartilage and bone are always distinct tissues.

CHECK YOUR UNDERSTANDING

1. Which type of cartilage is most plentiful in the adult body?
2. What two body structures contain flexible elastic cartilage?
3. Cartilage grows by interstitial growth. What does this mean?

For answers, see Appendix G.

Classification of Bones

- Name the major regions of the skeleton and describe their relative functions.
- Compare and contrast the structure of the four bone classes and provide examples of each class.

The 206 named bones of the human skeleton are divided into two groups: axial and appendicular. The **axial skeleton** forms the long axis of the body and includes the bones of the skull, vertebral column, and rib cage, shown in orange in Figure 6.1. Generally speaking these bones are most involved in protecting, supporting, or carrying other body parts.

The **appendicular skeleton** (ap″en-dik′u-lar) consists of the bones of the upper and lower limbs and the girdles (shoulder bones and hip bones) that attach the limbs to the axial skeleton. These bones are colored gold in Figure 6.1. Bones of the limbs help us to get from place to place (locomotion) and to manipulate our environment.

Bones come in many sizes and shapes. For example, the pisiform bone of the wrist is the size and shape of a pea, whereas the femur (thigh bone) is nearly 2 feet long in some people and has a large, ball-shaped head. The unique shape of each bone fulfills a particular need. The femur, for example, withstands great weight and pressure, and its hollow-cylinder design provides maximum strength with minimum weight.

6

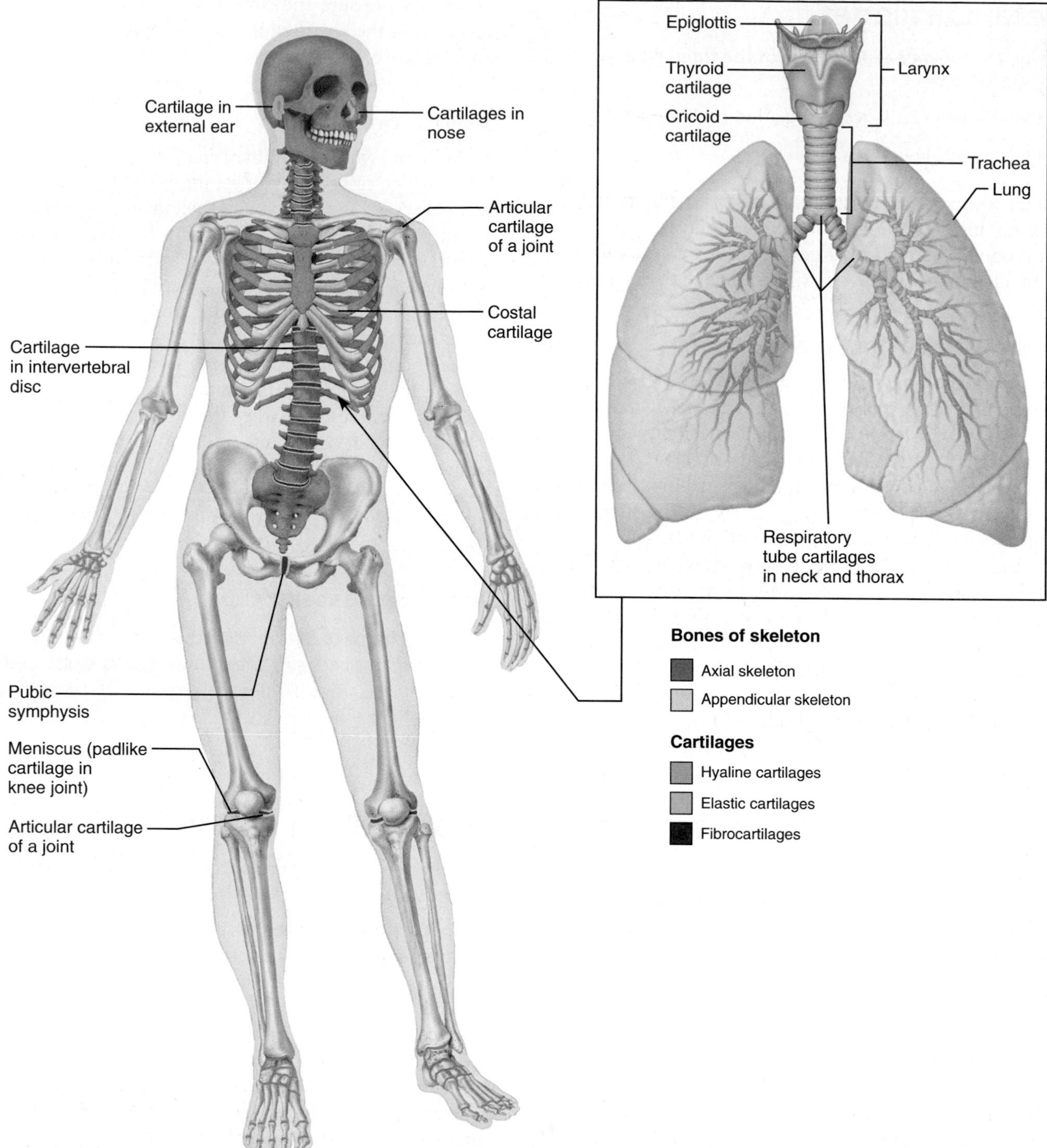

Figure 6.1 The bones and cartilages of the human skeleton. The cartilages that support the respiratory tubes and larynx are drawn separately at the right.

For the most part, bones are classified by their shape as long, short, flat, and irregular **(Figure 6.2)**.

1. **Long bones**, as their name suggests, are considerably longer than they are wide (Figure 6.2a). A long bone has a shaft plus two ends. All limb bones except the patella (kneecap) and the wrist and ankle bones are long bones. Notice that these bones are named for their elongated shape, *not* their overall size. The three bones in each of your fingers are long bones, even though they are very small.
2. **Short bones** are roughly cube shaped. The bones of the wrist and ankle are examples (Figure 6.2d).

   Sesamoid bones (ses′ah-moid; "shaped like a sesame seed") are a special type of short bone that form in a tendon (for example, the patella). They vary in size and number in different individuals. Some sesamoid bones clearly act to

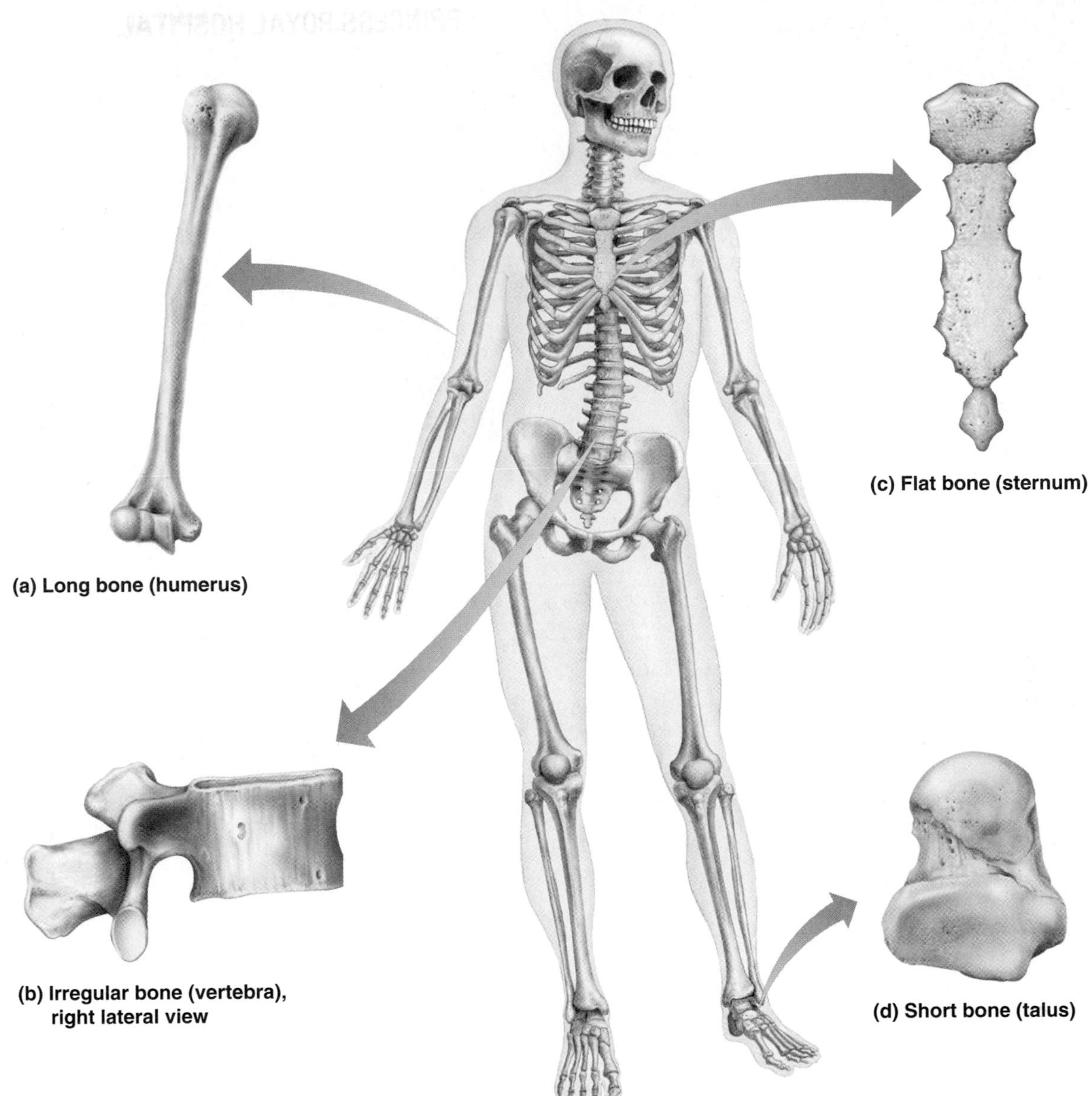

Figure 6.2 Classification of bones on the basis of shape.

alter the direction of pull of a tendon. The function of others is not known.

3. **Flat bones** are thin, flattened, and usually a bit curved. The sternum (breastbone), scapulae (shoulder blades), ribs, and most skull bones are flat bones (Figure 6.2c).
4. **Irregular bones** have complicated shapes that fit none of the preceding classes. Examples include the vertebrae and the hip bones (Figure 6.2b).

CHECK YOUR UNDERSTANDING

4. What are the components of the axial skeleton?
5. Contrast the general function of the axial skeleton to that of the appendicular skeleton.
6. What bone class do the ribs and skull bones fall into?

For answers, see Appendix G.

Functions of Bones

▶ List and describe five important functions of bones.

Besides contributing to body shape and form, our bones perform several other important functions:

1. **Support.** Bones provide a framework that supports the body and cradles its soft organs. For example, bones of lower limbs act as pillars to support the body trunk when we stand, and the rib cage supports the thoracic wall.
2. **Protection.** The fused bones of the skull protect the brain. The vertebrae surround the spinal cord, and the rib cage helps protect the vital organs of the thorax.
3. **Movement.** Skeletal muscles, which attach to bones by tendons, use bones as levers to move the body and its parts. As a result, we can walk, grasp objects, and breathe. The design of joints determines the types of movement possible.

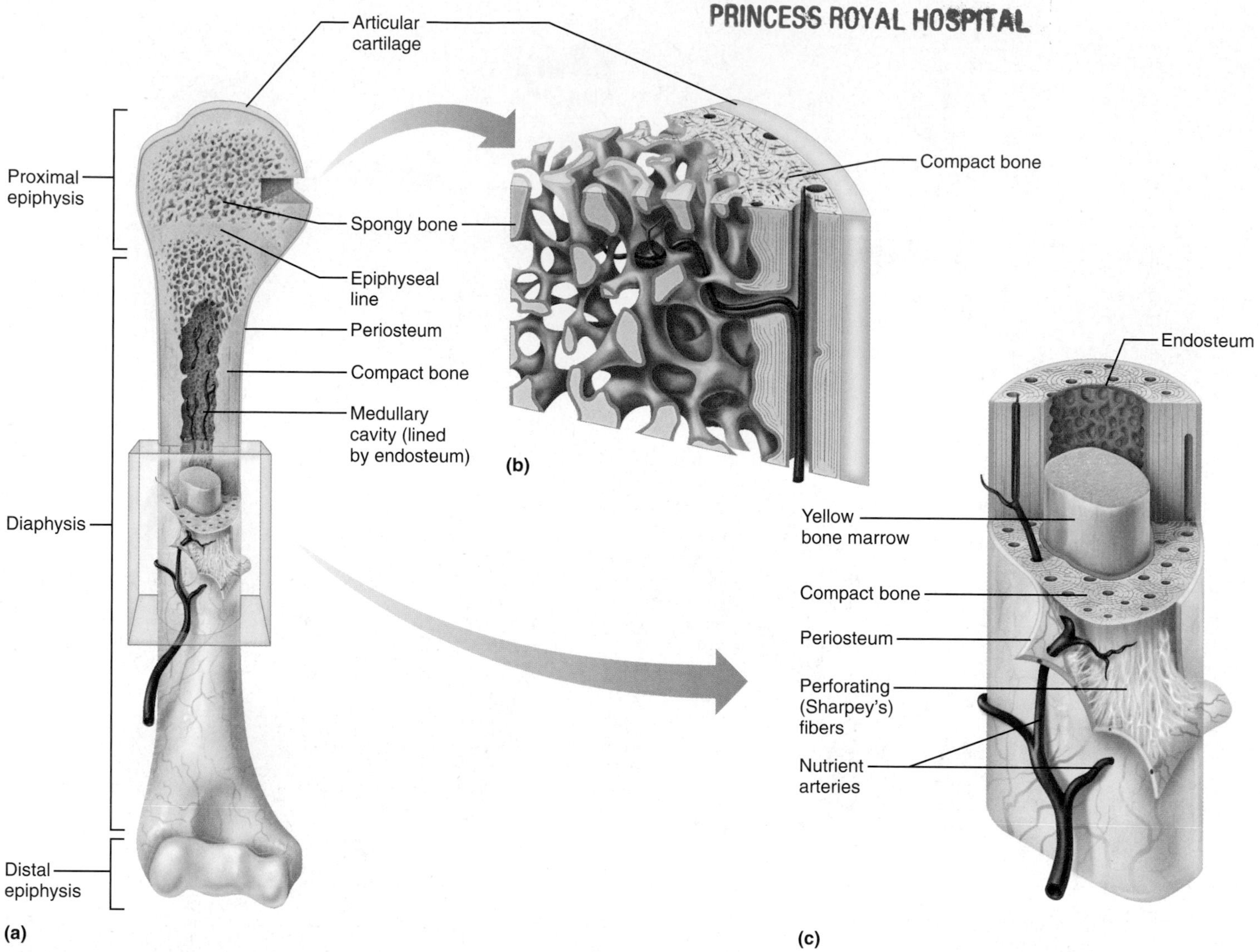

Figure 6.3 The structure of a long bone (humerus of arm). (a) Anterior view with bone sectioned frontally to show the interior at the proximal end. **(b)** Enlarged view of spongy bone and compact bone of the epiphysis of (a). (See *A Brief Atlas of the Human Body*, Plates 20 and 21.) **(c)** Enlarged cross-sectional view of the shaft (diaphysis) of (a). Note that the external surface of the diaphysis is covered by periosteum, but the articular surface of the epiphysis is covered with hyaline cartilage.

4. **Mineral and growth factor storage.** Bone is a reservoir for minerals, most importantly calcium and phosphate. The stored minerals are released into the bloodstream as needed for distribution to all parts of the body. Indeed, "deposits" and "withdrawals" of minerals to and from the bones go on almost continuously. Additionally, mineralized bone matrix stores important growth factors such as insulin-like growth factors, transforming growth factor, bone morphogenic proteins, and others.
5. **Blood cell formation.** Most blood cell formation, or **hematopoiesis** (hem″ah-to-poi-e′sis), occurs in the marrow cavities of certain bones.
6. **Triglyceride (fat) storage.** Fat is stored in bone cavities and represents a source of stored energy for the body.

CHECK YOUR UNDERSTANDING

7. What is the functional relationship between skeletal muscles and bones?
8. What two types of substances are stored in bone matrix?
9. What are two functions of a bone's marrow cavities?

For answers, see Appendix G.

Bone Structure

- ▶ Indicate the functional importance of bone markings.
- ▶ Describe the gross anatomy of a typical long bone and a flat bone. Indicate the locations and functions of red and yellow marrow, articular cartilage, periosteum, and endosteum.
- ▶ Describe the histology of compact and spongy bone.
- ▶ Discuss the chemical composition of bone and the advantages conferred by the organic and inorganic components.

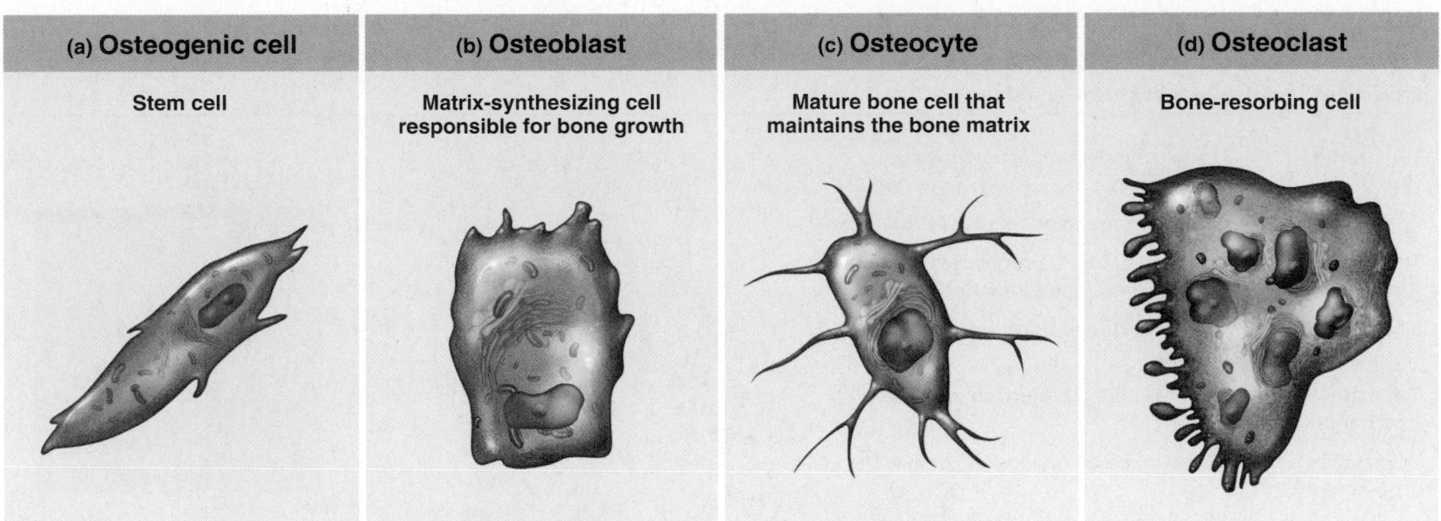

Figure 6.4 Comparison of different types of bone cells.

Because they contain various types of tissue, bones are *organs.* (Recall that an organ contains several different tissues.) Although bone (osseous) tissue dominates bones, they also contain nervous tissue in their nerves, cartilage in their articular cartilages, fibrous connective tissue lining their cavities, and muscle and epithelial tissues in their blood vessels. We will consider bone structure at three levels: gross, microscopic, and chemical.

Gross Anatomy

Bone Markings

The external surfaces of bones are rarely smooth and featureless. Instead, they display projections, depressions, and openings that serve as sites of muscle, ligament, and tendon attachment, as joint surfaces, or as conduits for blood vessels and nerves. These **bone markings** are named in different ways.

Projections (bulges) that grow outward from the bone surface include heads, trochanters, spines, and others. Each has distinguishing features and functions. In most cases, bone projections are indications of the stresses created by muscles attached to and pulling on them or are modified surfaces where bones meet and form joints.

Depressions and openings include fossae, sinuses, foramina, and grooves. They usually serve to allow passage of nerves and blood vessels. The most important types of bone markings are described in **Table 6.1**. You should familiarize yourself with these terms because you will meet them again as identifying marks of the individual bones studied in the lab.

Bone Textures: Compact and Spongy Bone

Every bone has a dense outer layer that looks smooth and solid to the naked eye. This external layer is **compact bone** (Figures 6.3 and 6.5). Internal to this is **spongy bone** (also called *cancellous bone*), a honeycomb of small needle-like or flat pieces called **trabeculae** (trah-bek′u-le; "little beams"). In living bones the open spaces between trabeculae are filled with red or yellow bone marrow.

Structure of a Typical Long Bone

With few exceptions, all long bones have the same general structure, which includes a shaft, bone ends, and membranes **(Figure 6.3)**.

Diaphysis A tubular **diaphysis** (di-af′ĭ-sis; *dia* = through, *physis* = growth), or shaft, forms the long axis of the bone. It is constructed of a relatively thick *collar* of compact bone that surrounds a central **medullary cavity** (med′u-lar-e; "middle"), or *marrow cavity.* In adults, the medullary cavity contains fat (yellow marrow) and is called the **yellow marrow cavity**.

Epiphyses The **epiphyses** (e-pif′ĭ-sēz; singular: epiphysis) are the bone ends (*epi* = upon). In many cases, they are more expanded than the diaphysis. Compact bone forms the exterior of epiphyses, and their interior contains spongy bone. The joint surface of each epiphysis is covered with a thin layer of articular (hyaline) cartilage, which cushions the opposing bone ends during joint movement and absorbs stress. Between the diaphysis and each epiphysis of an adult long bone is an **epiphyseal line**, a remnant of the **epiphyseal plate**, a disc of hyaline cartilage that grows during childhood to lengthen the bone. The region where the diaphysis and epiphysis meet, whether it is the epiphyseal plate or line, is sometimes called the *metaphysis.*

Membranes A third structural feature of long bones is membranes. The external surface of the entire bone except the joint surfaces is covered by a glistening white, double-layered membrane called the **periosteum** (per″e-os′te-um; *peri* = around, *osteo* = bone). The outer *fibrous layer* is dense irregular connective tissue. The inner *osteogenic layer,* abutting the bone surface, consists primarily of bone-forming cells, called **osteoblasts** (os′te-o-blasts; "bone germinators"), which secrete bone matrix elements, and bone-destroying cells, called **osteoclasts** ("bone breakers"). In addition, there are primitive stem cells, **osteogenic cells**, that give rise to the osteoblasts **(Figure 6.4)**.

TABLE 6.1 Bone Markings

NAME OF BONE MARKING	DESCRIPTION	ILLUSTRATIONS
Projections That Are Sites of Muscle and Ligament Attachment		
Tuberosity (too″bĕ-ros′ĭ-te)	Large rounded projection; may be roughened	
Crest	Narrow ridge of bone; usually prominent	
Trochanter (tro-kan′ter)	Very large, blunt, irregularly shaped process (the only examples are on the femur)	
Line	Narrow ridge of bone; less prominent than a crest	
Tubercle (too′ber-kl)	Small rounded projection or process	
Epicondyle (ep″ĭ-kon′dīl)	Raised area on or above a condyle	
Spine	Sharp, slender, often pointed projection	
Process	Any bony prominence	
Projections That Help to Form Joints		
Head	Bony expansion carried on a narrow neck	
Facet	Smooth, nearly flat articular surface	
Condyle (kon′dīl)	Rounded articular projection	
Ramus (ra′mus)	Armlike bar of bone	
Depressions and Openings		
For Passage of Blood Vessels and Nerves		
Groove	Furrow	
Fissure	Narrow, slitlike opening	
Foramen (fo-ra′men)	Round or oval opening through a bone	
Notch	Indentation at the edge of a structure	
Others		
Meatus (me-a′tus)	Canal-like passageway	
Sinus	Cavity within a bone, filled with air and lined with mucous membrane	
Fossa (fos′ah)	Shallow, basinlike depression in a bone, often serving as an articular surface	

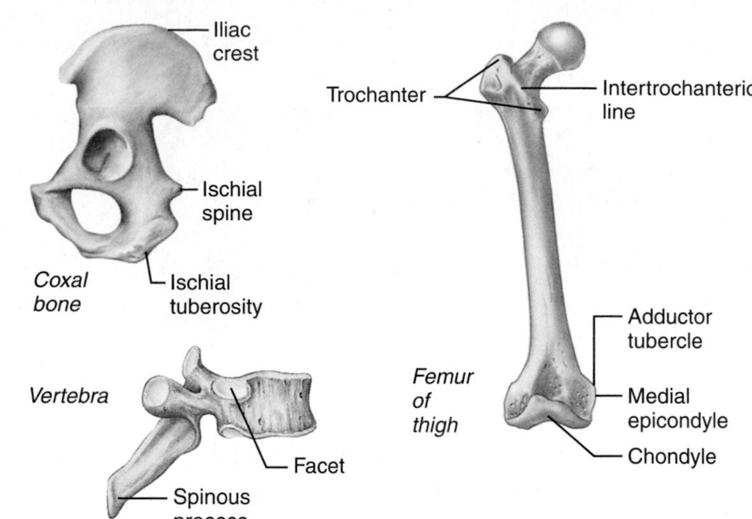

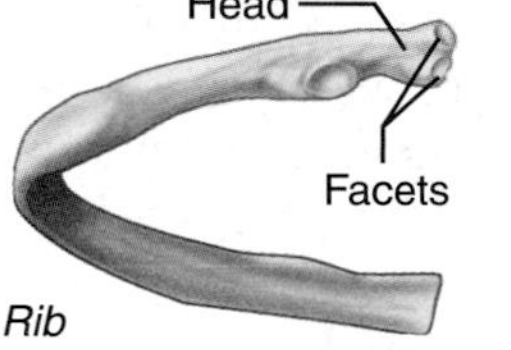

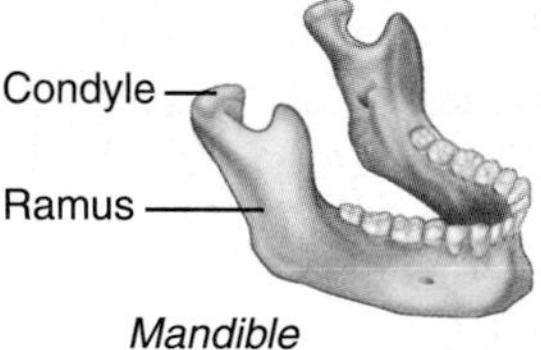

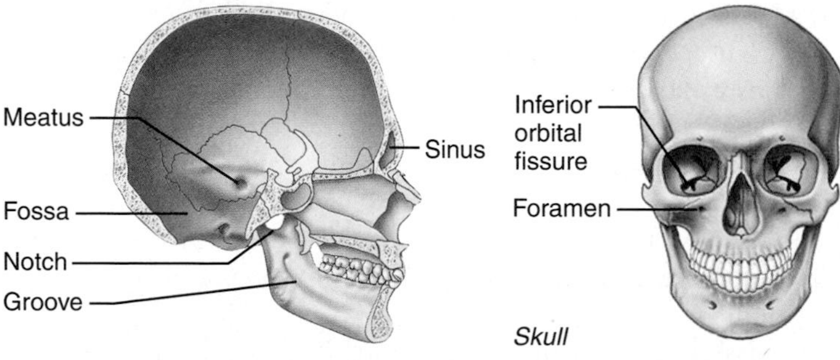

The periosteum is richly supplied with nerve fibers, lymphatic vessels, and blood vessels, which enter the diaphysis via **nutrient foramina** (fo-ra″me-nah; "opening").

The periosteum is secured to the underlying bone by *perforating* (*Sharpey's*) *fibers* (Figure 6.3), tufts of collagen fibers that extend from its fibrous layer into the bone matrix. The periosteum also provides anchoring points for tendons and ligaments. At these points the perforating fibers are exceptionally dense.

Internal bone surfaces are covered with a delicate connective tissue membrane called the **endosteum** (en-dos′te-um; "within the bone") (Figure 6.3). The endosteum covers the trabeculae of spongy bone and lines the canals that pass through the compact bone. Like the periosteum, the endosteum contains both bone-forming and bone-destroying cells.

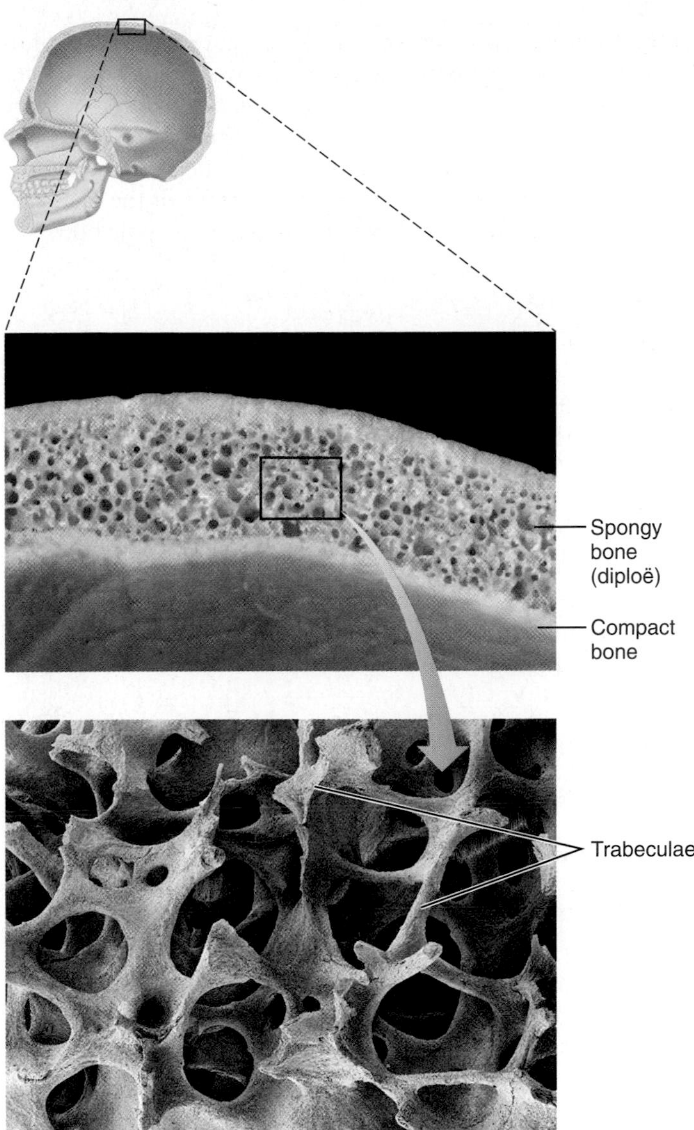

Figure 6.5 **Flat bones consist of a layer of spongy bone sandwiched between two thin layers of compact bone.** (Photomicrograph at bottom, 4×.)

Structure of Short, Irregular, and Flat Bones

Short, irregular, and flat bones share a simple design: They all consist of thin plates of periosteum-covered compact bone on the outside and endosteum-covered spongy bone within. However, these bones are not cylindrical and so they have no shaft or epiphyses. They contain bone marrow (between their trabeculae), but no significant marrow cavity is present.

Figure 6.5 shows a typical flat bone of the skull. In flat bones, the spongy bone is called the **diploë** (dip′lo-e; "folded") and the whole arrangement resembles a stiffened sandwich.

Location of Hematopoietic Tissue in Bones

Hematopoietic tissue, **red marrow**, is typically found within the trabecular cavities of spongy bone of long bones and in the diploë of flat bones. For this reason, both these cavities are often referred to as **red marrow cavities**. In newborn infants, the medullary cavity of the diaphysis and all areas of spongy bone contain red bone marrow. In most adult long bones, the fat-containing medullary cavity extends well into the epiphysis, and little red marrow is present in the spongy bone cavities. For this reason, blood cell production in adult long bones routinely occurs only in the heads of the femur and humerus (the long bone of the arm).

The red marrow found in the diploë of flat bones (such as the sternum) and in some irregular bones (such as the hip bone) is much more active in hematopoiesis, and these sites are routinely used for obtaining red marrow samples when problems with the blood-forming tissue are suspected. However, yellow marrow in the medullary cavity can revert to red marrow if a person becomes very anemic and needs enhanced red blood cell production.

Microscopic Anatomy of Bone

Essentially, four major cell types populate bone tissue: osteogenic cells, osteoblasts, osteocytes, and osteoclasts. These, like other connective tissue cells, are surrounded by an extracellular matrix of their making. The *osteogenic cells,* also called *osteoprogenitor cells,* are mitotically active stem cells found in the membranous periosteum and endosteum. Some of their progeny differentiate into **osteoblasts** (bone-forming cells) while others persist as bone stem cells to provide osteoblasts in the future. We describe the structure and function of the remaining two types of bone cells below.

Compact Bone

Although compact bone looks dense and solid, a microscope reveals that it is riddled with passageways that serve as conduits for nerves, blood vessels, and lymphatic vessels (see Figure 6.7). The structural unit of compact bone is called either the **osteon** (os′te-on) or the **Haversian system** (ha-ver′zhen). Each osteon is an elongated cylinder oriented parallel to the long axis of the bone. Functionally, osteons are tiny weight-bearing pillars.

As shown in the "exploded" view in Figure 6.6, an osteon is a group of hollow tubes of bone matrix, one placed outside the next like the growth rings of a tree trunk. Each matrix tube is a **lamella** (lah-mel′ah; "little plate"), and for this reason compact bone is often called **lamellar bone**. Although all of the collagen fibers in a particular lamella run in a single direction, the collagen fibers in adjacent lamellae always run in different directions. This alternating pattern is beautifully designed to withstand torsion stresses—the adjacent lamellae reinforce one another to resist twisting. You can think of the osteon's design as a "twister resister." Collagen fibers are not the only part of bone lamellae that are beautifully ordered. The tiny crystals of bone salts align with the collagen fibers and thus also alternate their direction in adjacent lamellae.

Running through the core of each osteon is the **central canal**, or **Haversian canal**, containing small blood vessels and nerve fibers that serve the needs of the osteon's cells. Canals of a sec-

6

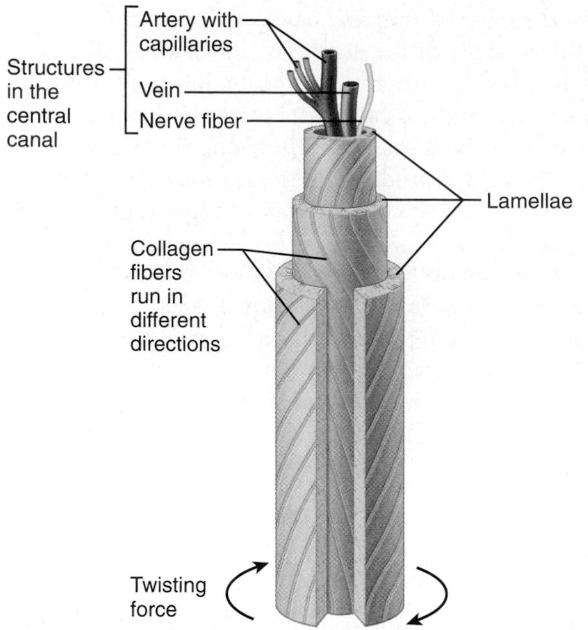

Figure 6.6 A single osteon. The osteon is drawn as if pulled out like a telescope to illustrate the individual lamellae.

ond type called **perforating canals**, or **Volkmann's canals** (folk′mahnz), lie at right angles to the long axis of the bone and connect the blood and nerve supply of the periosteum to those in the central canals and the medullary cavity **(Figure 6.7a)**. Like all other internal bone cavities, these canals are lined with endosteum.

Spider-shaped **osteocytes** (Figures 6.4c and 6.7b) occupy **lacunae** (*lac* = hollow; *una* = little) at the junctions of the lamellae. Hairlike canals called **canaliculi** (kan″ah-lik′u-li) connect the lacunae to each other and to the central canal. The manner in which canaliculi are formed is interesting. When bone is being formed, the osteoblasts secreting bone matrix surround blood vessels and maintain contact with one another by tentacle-like projections containing gap junctions. Then, as the newly secreted matrix hardens and the maturing cells become trapped within it, a system of tiny canals—the canaliculi, filled with tissue fluid and containing the osteocyte extensions—is formed. The canaliculi tie all the osteocytes in an osteon together, permitting nutrients and wastes to be relayed from one osteocyte to the next throughout the osteon. Although bone matrix is hard and impermeable to nutrients, its canaliculi and cell-to-cell relays (via gap junctions) allow bone cells to be well nourished.

One function of osteocytes is to maintain the bone matrix. If they die, the surrounding matrix is resorbed. The osteocytes also act as stress or strain "sensors" in cases of bone deformation or other damaging stimuli. They communicate this information to the cells responsible for bone remodeling (osteoblasts and osteoclasts) so that countermeasures can be taken or repairs made. We discuss the bone-destroying *osteoclast* on p. 186 in conjunction with the topic of bone remodeling.

Not all the lamellae in compact bone are part of osteons. Lying between intact osteons are incomplete lamellae called **interstitial lamellae** (in″ter-stish′al) (Figure 6.7c, right photomicrograph). They either fill the gaps between forming osteons or are remnants of osteons that have been cut through by bone remodeling (discussed later). **Circumferential lamellae,** located just deep to the periosteum and just superficial to the endosteum, extend around the entire circumference of the diaphysis (Figure 6.7a) and effectively resist twisting of the long bone.

Spongy Bone

In contrast to compact bone, spongy bone looks like a poorly organized, even haphazard, tissue (see Figure 6.5 and Figure 6.3b). However, the trabeculae in spongy bone align precisely along lines of stress and help the bone resist stress as much as possible. These tiny bone struts are as carefully positioned as the flying buttresses that help to support a Gothic cathedral.

Only a few cells thick, trabeculae contain irregularly arranged lamellae and osteocytes interconnected by canaliculi. No osteons are present. Nutrients reach the osteocytes of spongy bone by diffusing through the canaliculi from capillaries in the endosteum surrounding the trabeculae.

Chemical Composition of Bone

Bone has both organic and inorganic components. Its *organic components* include the cells (osteogenic cells, osteoblasts, osteocytes, and osteoclasts) and **osteoid** (os′te-oid), the organic part of the matrix. Osteoid, which makes up approximately one-third of the matrix, includes ground substance (composed of proteoglycans and glycoproteins) and collagen fibers, both of which are made and secreted by osteoblasts. These organic substances, particularly collagen, contribute not only to a bone's structure but also to the flexibility and great tensile strength that allow the bone to resist stretch and twisting.

Bone's exceptional toughness and tensile strength has been the subject of intense research. It now appears that this resilience comes from the presence of *sacrificial bonds* in or between collagen molecules. These bonds break easily on impact, dissipating energy to prevent the force from rising to a fracture value. In the absence of continued or additional trauma, most of the sacrificial bonds re-form.

The balance of bone tissue (65% by mass) consists of inorganic *hydroxyapatites* (hi-drok″se-ap′ah-tītz), or *mineral salts,* largely calcium phosphates present in the form of tiny, tightly packed, needle-like crystals in and around the collagen fibers in the extracellular matrix. The crystals account for the most notable characteristic of bone—its exceptional hardness, which allows it to resist compression.

The proper combination of organic and inorganic matrix elements allows bones to be exceedingly durable and strong without being brittle. Healthy bone is half as strong as steel in resisting compression and fully as strong as steel in resisting tension.

Because of the salts they contain, bones last long after death and provide an enduring "monument." In fact, skeletal remains many centuries old have revealed the shapes and sizes of ancient peoples, the kinds of work they did, and many of the ailments they suffered, such as arthritis.

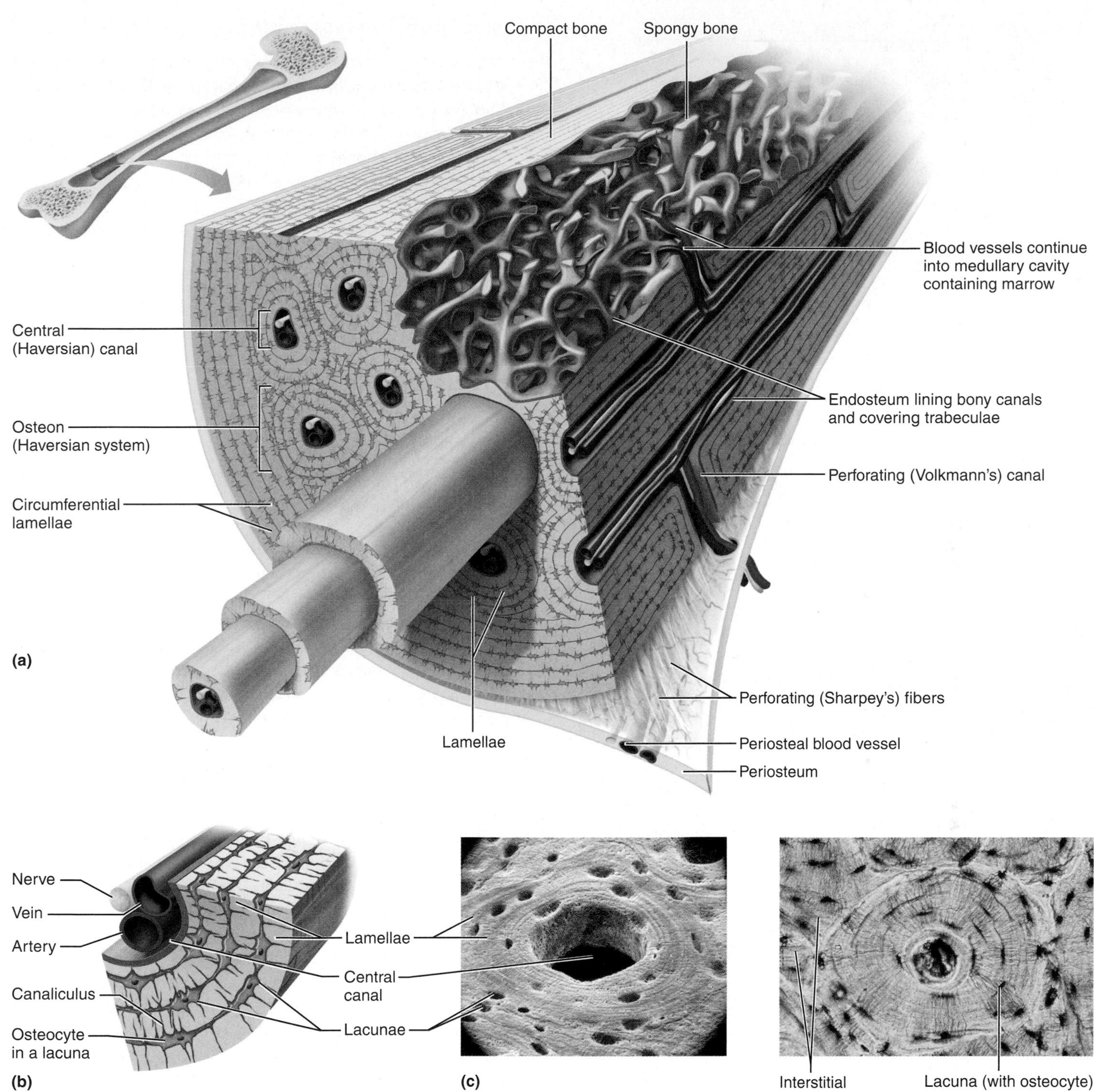

Figure 6.7 Microscopic anatomy of compact bone. (a) Diagrammatic view of a pie-shaped segment of compact bone. **(b)** Close-up of a portion of one osteon. Note the position of osteocytes in the lacunae. **(c)** SEM (left) and light photomicrographs (right) of a cross-sectional view of an osteon (180× and 160×, respectively).

SOURCE: Kessel and Kardon/Visuals Unlimited.

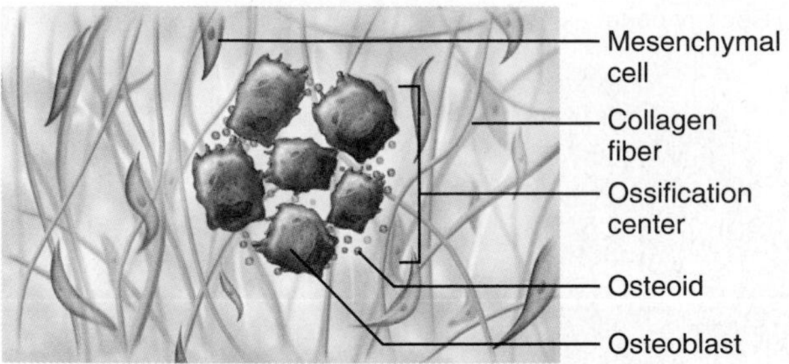

① **Ossification centers appear in the fibrous connective tissue membrane.**
- Selected centrally located mesenchymal cells cluster and differentiate into osteoblasts, forming an ossification center.

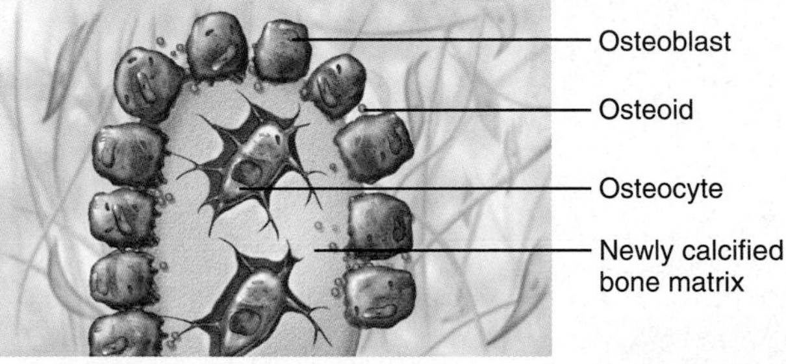

② **Bone matrix (osteoid) is secreted within the fibrous membrane and calcifies.**
- Osteoblasts begin to secrete osteoid, which is calcified within a few days.
- Trapped osteoblasts become osteocytes.

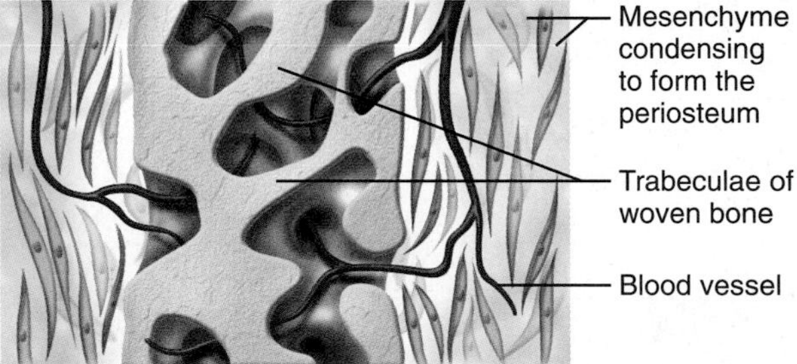

③ **Woven bone and periosteum form.**
- Accumulating osteoid is laid down between embryonic blood vessels in a random manner. The result is a network (instead of lamellae) of trabeculae called woven bone.
- Vascularized mesenchyme condenses on the external face of the woven bone and becomes the periosteum.

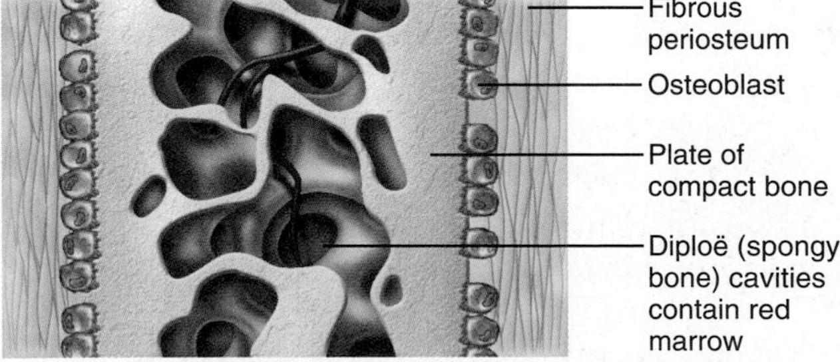

④ **Lamellar bone replaces woven bone, just deep to the periosteum. Red marrow appears.**
- Trabeculae just deep to the periosteum thicken, and are later replaced with mature lamellar bone, forming compact bone plates.
- Spongy bone (diploë), consisting of distinct trabeculae, persists internally and its vascular tissue becomes red marrow.

CHECK YOUR UNDERSTANDING

10. Are crests, tubercles, and spines bony projections or concavities?
11. How does the structure of compact bone differ from that of spongy bone when viewed with the naked eye?
12. What membrane lines the internal canals and covers the trabeculae of a bone?
13. Which component of bone—organic or inorganic—makes it hard?
14. What name is given to a cell that has a ruffled border and acts to break down bone matrix?

For answers, see Appendix G.

Bone Development

- ▶ Compare and contrast intramembranous ossification and endochondral ossification.
- ▶ Describe the process of long bone growth that occurs at the epiphyseal plates.

Ossification and **osteogenesis** (os″te-o-jen′ĕ-sis) are synonyms meaning the process of bone formation (*os* = bone, *genesis* = beginning). In embryos this process leads to the *formation of the bony skeleton.* Later another form of ossification known as *bone growth* goes on until early adulthood as the body continues to increase in size. Bones are capable of growing in thickness throughout life. However, ossification in adults serves mainly for bone *remodeling* and repair.

Formation of the Bony Skeleton

Before week 8, the skeleton of a human embryo is constructed entirely from fibrous membranes and hyaline cartilage. Bone tissue begins to develop at about this time and eventually replaces most of the existing fibrous or cartilage structures. When a bone develops from a fibrous membrane, the process is *intramembranous ossification,* and the bone is called a **membrane bone**. Bone development by replacing hyaline cartilage is called *endochondral ossification* (*endo* = within, *chondro* = cartilage), and the resulting bone is called a **cartilage,** or **endochondral, bone.** The beauty of using structures (membranes and cartilages) that are flexible and resilient to fashion the embryonic skeleton is that they can accommodate mitosis. Were the early skeleton composed of bone tissue from the outset, growth would be much more difficult.

Intramembranous Ossification

Intramembranous ossification results in the formation of cranial bones of the skull (frontal, parietal, occipital, and temporal bones) and the clavicles. Most bones formed by this process are flat bones. At about week 8 of development, ossification begins on fibrous connective tissue membranes formed by *mesenchymal*

Figure 6.8 Intramembranous ossification. Diagrams ③ and ④ represent much lower magnification than diagrams ① and ②.

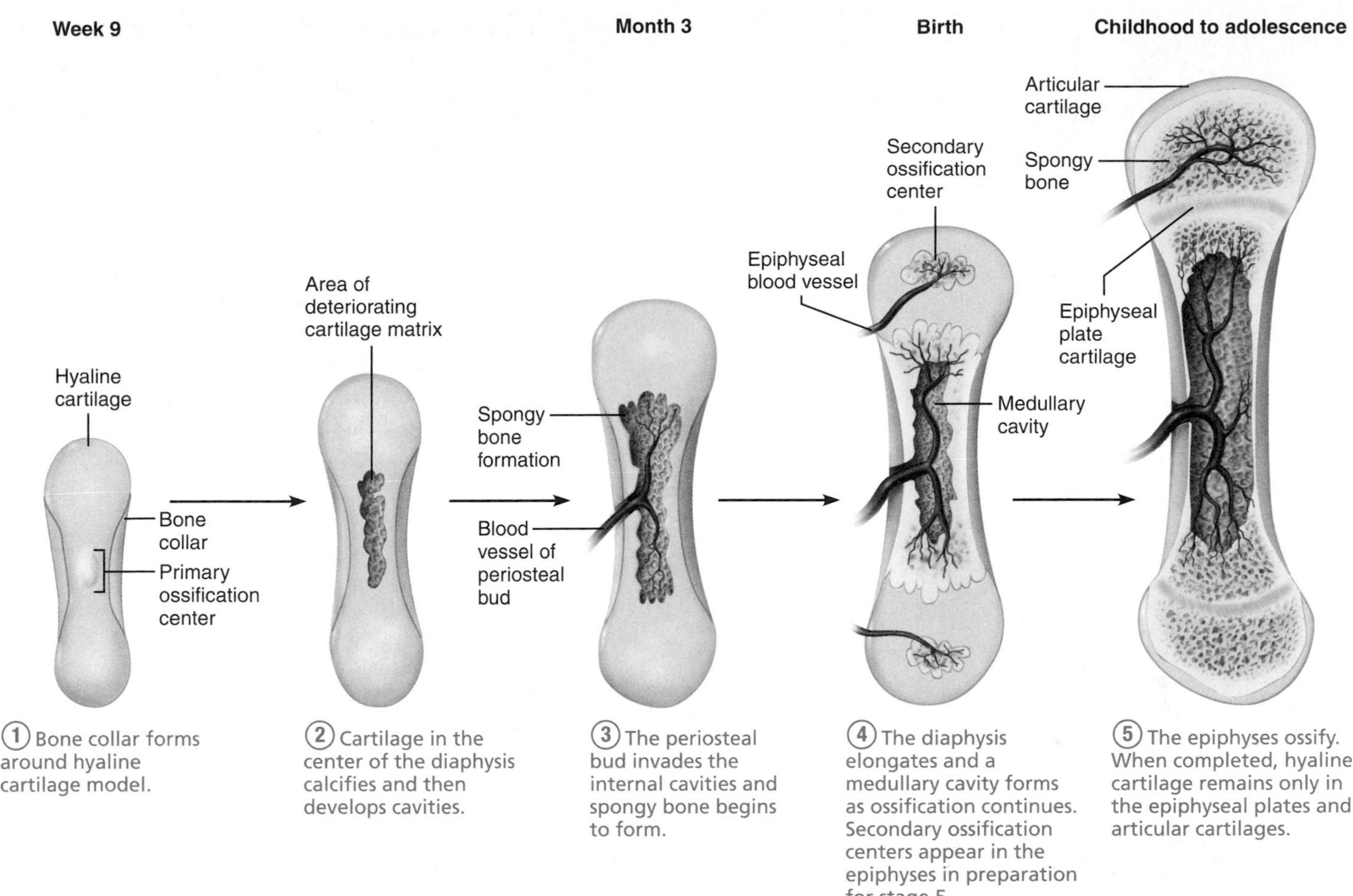

Week 9
Month 3
Birth
Childhood to adolescence
Hyaline cartilage
Bone collar
Primary ossification center
Area of deteriorating cartilage matrix
Spongy bone formation
Blood vessel of periosteal bud
Epiphyseal blood vessel
Secondary ossification center
Medullary cavity
Articular cartilage
Spongy bone
Epiphyseal plate cartilage
① Bone collar forms around hyaline cartilage model.
② Cartilage in the center of the diaphysis calcifies and then develops cavities.
③ The periosteal bud invades the internal cavities and spongy bone begins to form.
④ The diaphysis elongates and a medullary cavity forms as ossification continues. Secondary ossification centers appear in the epiphyses in preparation for stage 5.
⑤ The epiphyses ossify. When completed, hyaline cartilage remains only in the epiphyseal plates and articular cartilages.

Figure 6.9 Endochondral ossification in a long bone.

cells. Essentially, the process involves the four major steps depicted in **Figure 6.8**.

Endochondral Ossification

Except for the clavicles, essentially all bones of the skeleton below the base of the skull form by **endochondral ossification** (en″do-kon′dral). Beginning in the second month of development, this process uses hyaline cartilage "bones" formed earlier as models, or patterns, for bone construction. It is more complex than intramembranous ossification because the hyaline cartilage must be broken down as ossification proceeds. We will use a forming long bone as our example.

The formation of a long bone typically begins in the center of the hyaline cartilage shaft at a region called the **primary ossification center**. First, the perichondrium covering the hyaline cartilage "bone" is infiltrated with blood vessels, converting it to a vascularized periosteum. As a result of this change in nutrition, the underlying mesenchymal cells specialize into osteoblasts. The stage is now set for ossification to begin, as illustrated in **Figure 6.9**:

① **A bone collar is laid down around the diaphysis of the hyaline cartilage model.** Osteoblasts of the newly converted periosteum secrete osteoid against the hyaline cartilage diaphysis, encasing it in bone. This freshly formed layer of bone is called the *periosteal bone collar.*

② **Cartilage in the center of the diaphysis calcifies and then develops cavities.** As the bone collar forms, chondrocytes within the shaft hypertrophy (enlarge) and signal the surrounding cartilage matrix to calcify. Then, because calcified cartilage matrix is impermeable to diffusing nutrients, the chondrocytes die and the matrix begins to deteriorate. This deterioration opens up cavities, but the hyaline cartilage model is stabilized by the bone collar. Elsewhere, the cartilage remains healthy and continues to grow briskly, causing the cartilage model to elongate.

③ **The periosteal bud invades the internal cavities and spongy bone forms.** In month 3, the forming cavities are invaded by a collection of elements called the **periosteal bud**, which contains a nutrient artery and vein, lymphatic vessels, nerve fibers, red marrow elements, osteoblasts, and osteoclasts. The entering osteoclasts partially erode the calcified cartilage matrix, and the osteoblasts secrete osteoid around the remaining fragments of hyaline cartilage, forming bone-covered cartilage trabeculae. In this way, the earliest version of spongy bone in a developing long bone forms.

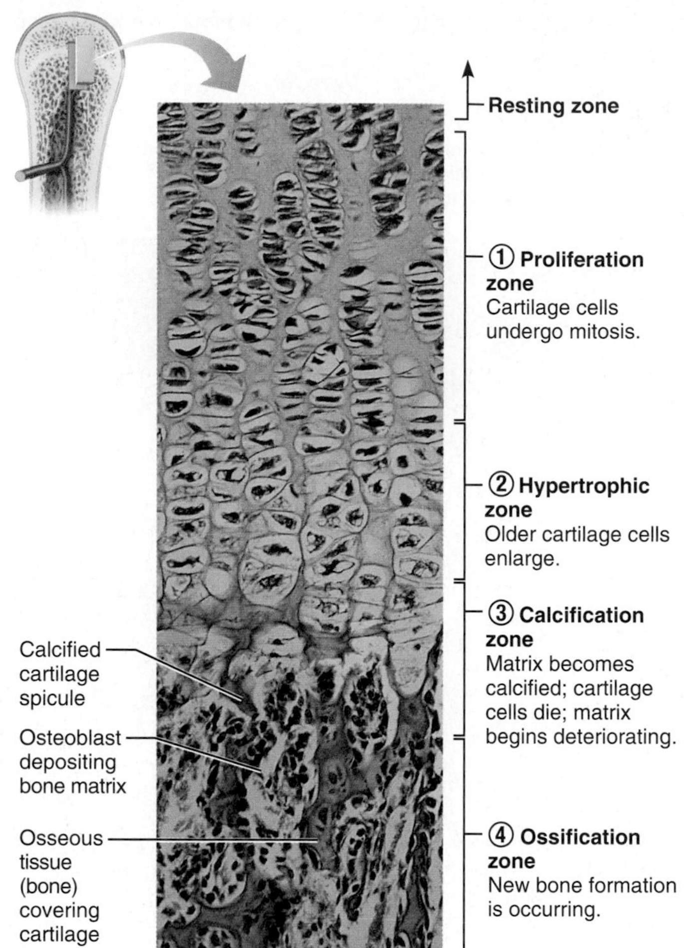

Figure 6.10 Growth in length of a long bone occurs at the epiphyseal plate. The side of the epiphyseal plate facing the epiphysis (distal face) contains resting cartilage cells. The cells of the epiphyseal plate proximal to the resting cartilage area are arranged in four zones—proliferation, hypertrophic, calcification, and ossification—from the region of the earliest stage of growth ① to the region where bone is replacing the cartilage ④ (150×).

④ **The diaphysis elongates and a medullary cavity forms.** As the primary ossification center enlarges, osteoclasts break down the newly formed spongy bone and open up a medullary cavity in the center of the diaphysis. Throughout the fetal period (week 9 until birth), the rapidly growing epiphyses consist only of cartilage, and the hyaline cartilage models continue to elongate by division of viable cartilage cells at the epiphyses. Ossification "chases" cartilage formation along the length of the shaft as cartilage calcifies, is eroded, and then is replaced by bony spicules on the epiphyseal surfaces facing the medullary cavity.

⑤ **The epiphyses ossify.** At birth, most of our long bones have a bony diaphysis surrounding remnants of spongy bone, a widening medullary cavity, and two cartilaginous epiphyses. Shortly before or after birth, **secondary ossification centers** appear in one or both epiphyses, and the epiphyses gain bony tissue. (Typically, the large long bones form secondary centers in both epiphyses, whereas the small long bones form only one secondary ossification center.) The cartilage in the center of the epiphysis calcifies and deteriorates, opening up cavities that allow a periosteal bud to enter. Then bone trabeculae appear, just as they did earlier in the primary ossification center. (In short bones, only the primary ossification center is formed. Most irregular bones develop from several distinct ossification centers.)

Secondary ossification reproduces almost exactly the events of primary ossification, except that the spongy bone in the interior is retained and no medullary cavity forms in the epiphyses. When secondary ossification is complete, hyaline cartilage remains only at two places: (1) on the epiphyseal surfaces, as the *articular cartilages*, and (2) at the junction of the diaphysis and epiphysis, where it forms the *epiphyseal plates.*

Postnatal Bone Growth

During infancy and youth, long bones lengthen entirely by interstitial growth of the epiphyseal plate cartilage and its replacement by bone, and all bones grow in thickness by appositional growth. Most bones stop growing during adolescence. However, some facial bones, such as those of the nose and lower jaw, continue to grow almost imperceptibly throughout life.

Growth in Length of Long Bones

Longitudinal bone growth mimics many of the events of endochondral ossification. The cartilage is relatively inactive on the side of the epiphyseal plate facing the epiphysis, a region called the *resting* or *quiescent zone.* But the epiphyseal plate cartilage abutting the diaphysis organizes into a pattern that allows fast, efficient growth. The cartilage cells here form tall columns, like coins in a stack. The cells at the "top" (epiphysis-facing side) of the stack abutting the resting zone comprise the *proliferation* or *growth zone* **(Figure 6.10)**. These cells divide quickly, pushing the epiphysis away from the diaphysis, causing the entire long bone to lengthen.

Meanwhile, the older chondrocytes in the stack, which are closer to the diaphysis (*hypertrophic zone* in Figure 6.10), hypertrophy, and their lacunae erode and enlarge, leaving large interconnecting spaces. Subsequently, the surrounding cartilage matrix calcifies and these chondrocytes die and deteriorate, producing the *calcification zone.* This leaves long slender spicules of calcified cartilage at the epiphysis-diaphysis junction, which look like stalactites hanging from the roof of a cave. These calcified spicules ultimately become part of the *ossification* or *osteogenic zone*, and are invaded by marrow elements from the medullary cavity. The cartilage spicules are partly eroded by osteoclasts, then quickly covered with new bone—called woven bone—by osteoblasts, and ultimately replaced by spongy bone. The spicule tips are eventually digested by osteoclasts, and in this way, the medullary cavity also grows longer as the long bone lengthens. During growth, the epiphyseal plate maintains a constant thickness because the rate of cartilage growth on its epiphysis-facing side is balanced by its replacement with bony tissue on its diaphysis-facing side.

Longitudinal growth is accompanied by almost continuous remodeling of the epiphyseal ends to maintain the proper proportions between the diaphysis and epiphyses **(Figure 6.11)**. Bone remodeling involves both new bone formation and bone resorption (destruction). It is described in more detail later in conjunction with the changes that occur in adult bones.

As adolescence ends, the chondroblasts of the epiphyseal plates divide less often and the plates become thinner and thinner until they are entirely replaced by bone tissue. Longitudinal bone growth ends when the bone of the epiphysis and diaphysis fuses. This process, called *epiphyseal plate closure*, happens at about 18 years of age in females and 21 years of age in males. However, as noted earlier, an adult bone can still increase in diameter or thickness by appositional growth if stressed by excessive muscle activity or body weight.

Growth in Width (Thickness)

Growing bones widen as they lengthen. As with cartilages, bones increase in thickness or, in the case of long bones, diameter, by appositional growth. Osteoblasts beneath the periosteum secrete bone matrix on the external bone surface as osteoclasts on the endosteal surface of the diaphysis remove bone (Figure 6.11). However, there is normally slightly less breaking down than building up. This unequal process produces a thicker, stronger bone but prevents it from becoming too heavy.

CHECK YOUR UNDERSTANDING

15. Bones don't begin as bones. What do they begin as?
16. When describing endochondral ossification, some say "bone chases cartilage." What does that mean?
17. Where is the primary ossification center located in a long bone? Where is (are) the secondary ossification center(s) located?
18. As a long bone grows in length, what is happening in the hypertrophic zone of the epiphyseal plate?

For answers, see Appendix G.

Hormonal Regulation of Bone Growth

The growth of bones that occurs until young adulthood is exquisitely controlled by a symphony of hormones. During infancy and childhood, the single most important stimulus of epiphyseal plate activity is *growth hormone* released by the anterior pituitary gland. Thyroid hormones modulate the activity of growth hormone, ensuring that the skeleton has proper proportions as it grows. At puberty, male and female sex hormones (testosterone and estrogens, respectively) are released in increasing amounts. Initially these sex hormones promote the growth spurt typical of adolescence, as well as the masculinization or feminization of specific parts of the skeleton. Later the hormones induce epiphyseal plate closure, ending longitudinal bone growth.

Excesses or deficits of any of these hormones can result in obviously abnormal skeletal growth. For example, hypersecretion of growth hormone in children results in excessive height (gigantism), and deficits of growth hormone or thyroid hormone produce characteristic types of dwarfism.

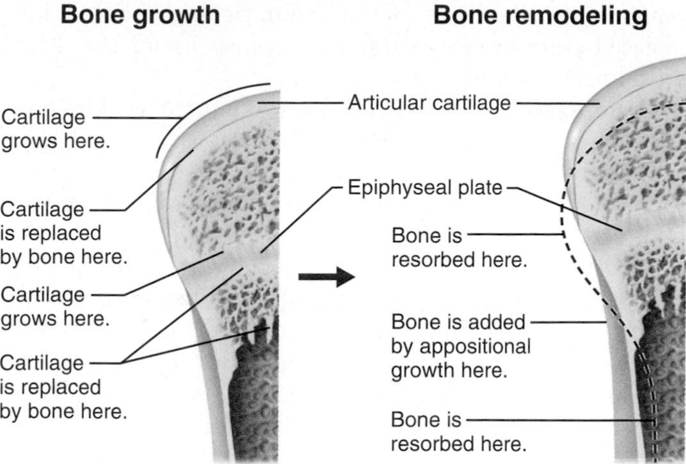

Figure 6.11 Long bone growth and remodeling during youth. The events at the left depict endochondral ossification that occurs at the articular cartilages and epiphyseal plates as the bone grows in length. Events at the right show bone remodeling during long bone growth to maintain proper bone proportions.

6

Bone Homeostasis: Remodeling and Repair

▶ Compare the locations and remodeling functions of the osteoblasts, osteocytes, and osteoclasts.

▶ Explain how hormones and physical stress regulate bone remodeling.

▶ Describe the steps of fracture repair.

Bones appear to be the most lifeless of body organs, and may even summon images of a graveyard. But as you have just learned, this appearance is deceiving. Bone is a dynamic and active tissue, and small-scale changes in bone architecture occur continually. Every week we recycle 5–7% of our bone mass, and as much as half a gram of calcium may enter or leave the adult skeleton each day! Spongy bone is replaced every three to four years; compact bone, every ten years or so. This is fortunate because when bone remains in place for long periods more of the calcium salts crystallize (see description below) and the bone becomes more brittle—ripe conditions for fracture. And when we break bones—the most common disorder of bone homeostasis—they undergo a remarkable process of self-repair.

Bone Remodeling

In the adult skeleton, bone deposit and bone resorption (removal) occur both at the surface of the periosteum and the surface of the endosteum. Together, the two processes constitute **bone remodeling**, and they are coupled and coordinated by "packets" of adjacent osteoblasts and osteoclasts called *remodeling units* (with help from the stress-sensing osteocytes). In healthy young adults, total bone mass remains constant, an indication that the rates of bone deposit and resorption are essentially equal. Remodeling does not occur uniformly, however. For example, the distal part of the femur, or thigh bone, is fully replaced every five to six months, whereas its shaft is altered much more slowly.

Bone deposit occurs wherever bone is injured or added bone strength is required. For optimal bone deposit, a healthy diet rich in proteins, vitamin C, vitamin D, vitamin A, and several minerals (calcium, phosphorus, magnesium, and manganese, to name a few) is essential.

New matrix deposits by osteocytes are marked by the presence of an *osteoid seam*, an unmineralized band of gauzy-looking bone matrix 10–12 micrometers (μm) wide. Between the osteoid seam and the older mineralized bone, there is an abrupt transition called the *calcification front*. Because the osteoid seam is always of constant width and the change from unmineralized to mineralized matrix is sudden, it seems that the osteoid must mature for about a week before it can calcify.

The precise trigger for calcification is still controversial. However, one critical factor is the product of the local concentrations of calcium and phosphate (P_i) ions (the $Ca^{2+} \cdot P_i$ product). Initially the bone salts are laid down in a noncrystalline form, but when the $Ca^{2+} \cdot P_i$ product reaches a certain level, tiny crystals of hydroxyapatite form spontaneously and then catalyze further crystallization of calcium salts in the area. Other factors involved are matrix proteins that bind and concentrate calcium, and the enzyme *alkaline phosphatase* (shed by the osteoblasts), which is essential for mineralization. Once proper conditions are present, calcium salts are deposited all at once and with great precision throughout the "matured" matrix. Normally, a small percentage of the calcified salts remain in the noncrystallized form to provide a readily available source of calcium ions when blood calcium levels decline toward nonhomeostatic values.

Bone resorption is accomplished by **osteoclasts**, giant multinucleate cells that arise from the same hematopoietic stem cells that differentiate into macrophages. Osteoclasts move along a bone surface, digging grooves as they break down the bone matrix. The part of the osteoclast that touches the bone is highly folded to form a ruffled membrane that clings tightly to the bone, sealing off the area of bone destruction. The ruffled border secretes (1) *lysosomal enzymes* that digest the organic matrix and (2) *hydrochloric acid* that converts the calcium salts into soluble forms that pass easily into solution. Osteoclasts may also phagocytize the demineralized matrix and dead osteocytes. The digested matrix end products, growth factors, and dissolved minerals are then endocytosed, transported across the osteoclast (by transcytosis), and released at the opposite side where they enter first the interstitial fluid and then the blood. There is much to learn about osteoclast activation, but proteins secreted by T cells of the immune system appear to be important.

Control of Remodeling

The remodeling that goes on continuously in the skeleton is regulated by two control loops that serve different "masters." One is a negative feedback hormonal loop that maintains Ca^{2+} homeostasis in the blood. The other involves responses to mechanical and gravitational forces acting on the skeleton.

The hormonal feedback becomes much more meaningful when you understand calcium's importance in the body. Ionic calcium is necessary for an amazing number of physiological processes, including transmission of nerve impulses, muscle contraction, blood coagulation, secretion by glands and nerve cells, and cell division. The human body contains 1200–1400 g of calcium, more than 99% present as bone minerals. Most of the remainder is in body cells. Less than 1.5 g is present in blood, and the hormonal control loop normally maintains blood Ca^{2+} within the very narrow range of 9–11 mg per dl (100 ml) of blood. Calcium is absorbed from the intestine under the control of vitamin D metabolites. The daily calcium requirement is 400–800 mg from birth until the age of 10, and 1200–1500 mg from ages 11 to 24.

Hormonal Controls The hormonal controls primarily involve **parathyroid hormone (PTH)**, produced by the parathyroid glands. To a much lesser extent **calcitonin** (kal″sĭ-to′nin), produced by parafollicular cells (C cells) of the thyroid gland, may be involved. As **Figure 6.12** illustrates, PTH is released when blood levels of ionic calcium decline. The increased PTH level stimulates osteoclasts to resorb bone, releasing calcium to the blood. Osteoclasts are no respecters of matrix age. When activated, they break down both old and new matrix. Only osteoid, which lacks calcium salts, escapes digestion. As blood concentrations of calcium rise, the stimulus for PTH release ends. The decline of PTH reverses its effects and causes blood Ca^{2+} levels to fall.

In humans, calcitonin appears to be a hormone in search of a function because its effects on calcium homeostasis are negligible. When administered at pharmacological (abnormally high) doses, it does lower blood calcium levels temporarily.

These hormonal controls act not to preserve the skeleton's strength or well-being but rather to maintain blood calcium homeostasis. In fact, if blood calcium levels are low for an extended time, the bones become so demineralized that they develop large, punched-out-looking holes. Thus, the bones serve as a storehouse from which ionic calcium is drawn as needed.

HOMEOSTATIC IMBALANCE

Minute changes from the homeostatic range for blood calcium can lead to severe neuromuscular problems ranging from hyperexcitability (when blood Ca^{2+} levels are too low) to nonresponsiveness and inability to function (with high blood Ca^{2+} levels). In addition, sustained high blood levels of Ca^{2+}, a condition known as *hypercalcemia* (hi″per-kal-se′me-ah), can lead to undesirable deposits of calcium salts in the blood vessels, kidneys, and other soft organs, which may hamper the functioning of these organs. ■

In addition to the hormones that regulate bone remodeling in response to blood calcium levels, it is now established that *leptin*, a hormone released by adipose tissue, plays a role in regulating bone density. Best known for its effects on weight and energy balance (see pp. 946–947), in animal studies leptin appears to inhibit osteoblasts through an additional pathway mediated by the hypothalamus which activates sympathetic nerves serving

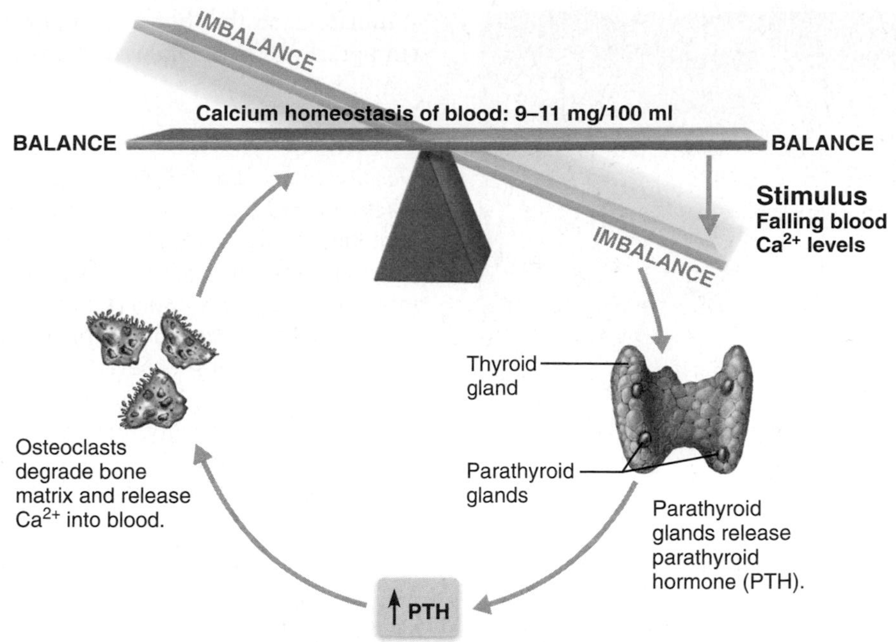

Figure 6.12 Parathyroid hormone (PTH) control of blood calcium levels.

bones. However, the full scope of leptin's bone-modifying activity in humans is still being worked out.

Response to Mechanical Stress The second set of controls regulating bone remodeling, bone's response to mechanical stress (muscle pull) and gravity, serves the needs of the skeleton by keeping the bones strong where stressors are acting. *Wolff's law* holds that a bone grows or remodels in response to the demands placed on it. The first thing to understand is that a bone's anatomy reflects the common stresses it encounters. For example, a bone is loaded (stressed) whenever weight bears down on it or muscles pull on it. This loading is usually off center, however, and tends to *bend* the bone. Bending compresses the bone on one side and subjects it to tension (stretching) on the other **(Figure 6.13)**. As a result of these mechanical stressors, long bones are thickest midway along the diaphysis, exactly where bending stresses are greatest (bend a stick and it will split near the middle). Both compression and tension are minimal toward the center of the bone (they cancel each other out), so a bone can "hollow out" for lightness (using spongy bone instead of compact bone) without jeopardy.

Other observations explained by Wolff's law include these: (1) Handedness (being right or left handed) results in the bones of one upper limb being thicker than those of the less-used limb, and vigorous exercise of the most-used limb leads to large increases in bone strength **(Figure 6.14)**. (2) Curved bones are thickest where they are most likely to buckle. (3) The trabeculae of spongy bone form trusses, or struts, along lines of compression. (4) Large, bony projections occur where heavy, active muscles attach. (The bones of weight lifters have enormous thickenings at the attachment sites of the most-used muscles.) Wolff's law also explains the featureless bones of the fetus and the atrophied bones of bedridden people—situations in which bones are not stressed.

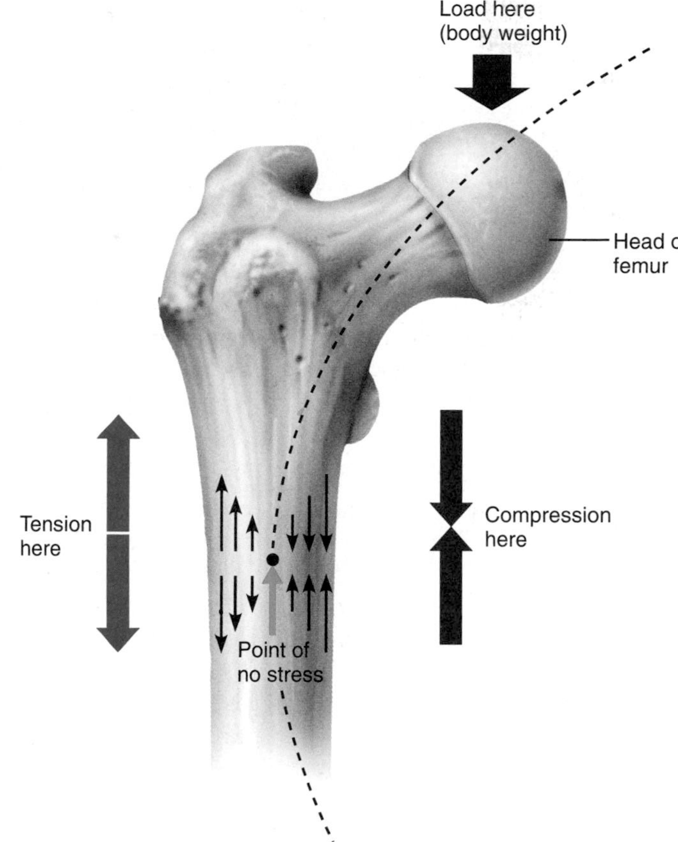

Figure 6.13 Bone anatomy and bending stress. Body weight transmitted to the head of the femur (thigh bone) threatens to bend the bone along the indicated arc, compressing it on one side (converging arrows on right) and stretching it on the other side (diverging arrows on left). Because these two forces cancel each other internally, much less bone material is needed internally than superficially.

(a)

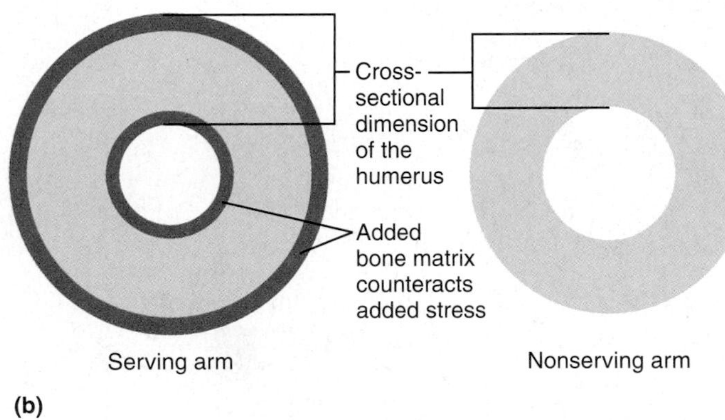

(b)

Figure 6.14 Vigorous exercise can lead to large increases in bone strength. The diagrams show the average difference in the cross-sectional dimensions of the humerus of the arm in the serving and nonserving arms of professional tennis players. The data revealed average increases in bone rigidity and strength of 62% and 45%, respectively, in the serving arms. The structural changes were more pronounced in players who began training at an early age.

SOURCE: C. B. Ruff, "Gracilization of the Modern Human Skeleton," *American Scientist 94*(6): p. 513, Nov–Dec 2006.

How do mechanical forces communicate with the cells responsible for remodeling? Although the mechanisms by which bone responds to mechanical stimuli are still uncertain, we do know that deforming a bone produces an electrical current. Because compressed and stretched regions are oppositely charged, it has been suggested that electrical signals direct remodeling. This principle underlies some of the devices currently used to speed bone repair and the healing of fractures.

The skeleton is continuously subjected to both hormonal influences and mechanical forces. At the risk of constructing too large a building on too small a foundation, we can speculate that the hormonal loop determines *whether* and *when* remodeling occurs in response to changing blood calcium levels, and mechanical stress determines *where* it occurs. For example, when bone must be broken down to increase blood calcium levels, PTH is released and targets the osteoclasts. However, mechanical forces determine *which* osteoclasts are most sensitive to PTH stimulation, so that bone in the *least* stressed areas (which is temporarily dispensable) is broken down.

Bone Repair

Despite their remarkable strength, bones are susceptible to **fractures**, or breaks. During youth, most fractures result from exceptional trauma that twists or smashes the bones (sports injuries, automobile accidents, and falls, for example). Excessive intake of vitamin A appears to increase fracture risk in some people. Elevated blood levels of the amino acid derivative homocysteine were also believed to increase fracture risk, but recent studies indicate that it is actually a marker of low bone density and bone frailty. In old age, most fractures occur as bones thin and weaken.

Fractures may be classified by

1. Position of the bone ends after fracture. In *nondisplaced fractures* the bone ends retain their normal position; in *displaced fractures* the bone ends are out of normal alignment.
2. Completeness of the break. If the bone is broken through, the fracture is a *complete fracture*; if not, it is an *incomplete fracture*.
3. Orientation of the break relative to the long axis of the bone. If the break parallels the long axis, the fracture is *linear*; if the break is perpendicular to the bone's long axis, it is *transverse*.
4. Whether the bone ends penetrate the skin. If so, the fracture is an *open (compound) fracture*; if not, it is a *closed (simple) fracture*.

In addition to these four either-or classifications, all fractures can be described in terms of the location of the fracture, the external appearance of the fracture, and/or the nature of the break. **Table 6.2** summarizes the various descriptions.

A fracture is treated by *reduction*, the realignment of the broken bone ends. In *closed (external) reduction*, the bone ends are coaxed into position by the physician's hands. In *open (internal) reduction*, the bone ends are secured together surgically with pins or wires. After the broken bone is reduced, it is immobilized either by a cast or traction to allow the healing process to begin. For a simple fracture the healing time is six to eight weeks for small or medium-sized bones in young adults, but it is much longer for large, weight-bearing bones and for bones of elderly people (because of their poorer circulation).

Repair in a simple fracture involves four major stages **(Figure 6.15)**:

① **A hematoma forms.** When a bone breaks, blood vessels in the bone and periosteum, and perhaps in surrounding tissues, are torn and hemorrhage. As a result, a **hematoma** (he″mah-to′mah), a mass of clotted blood, forms at the fracture site. Soon, bone cells deprived of nutrition die, and the tissue at the site becomes swollen, painful, and inflamed.

② **Fibrocartilaginous callus forms.** Within a few days, several events lead to the formation of soft *granulation tissue*, also called the *soft callus* (kal′us; "hard skin"). Capillaries grow into the hematoma and phagocytic cells invade the area and

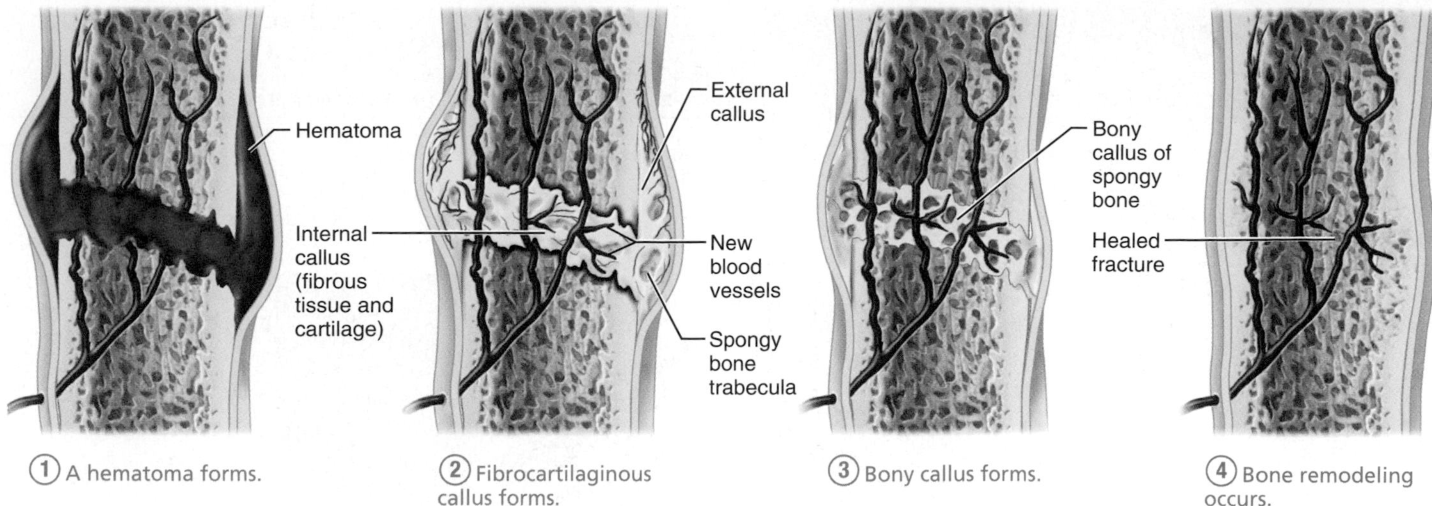

Figure 6.15 Stages in the healing of a bone fracture.

6

begin cleaning up the debris. Meanwhile, fibroblasts and osteoblasts invade the fracture site from the nearby periosteum and endosteum and begin reconstructing the bone. The fibroblasts produce collagen fibers that span the break and connect the broken bone ends, and some differentiate into chondroblasts that secrete cartilage matrix. Within this mass of repair tissue, osteoblasts begin forming spongy bone, but those farthest from the capillary supply secrete an externally bulging cartilaginous matrix that later calcifies. This entire mass of repair tissue, now called the **fibrocartilaginous callus**, splints the broken bone.

③ **Bony callus forms.** Within a week, new bone trabeculae begin to appear in the fibrocartilaginous callus and gradually convert it to a **bony (hard) callus** of spongy bone. Bony callus formation continues until a firm union is formed about two months later.

④ **Bone remodeling occurs.** Beginning during bony callus formation and continuing for several months after, the bony callus is remodeled. The excess material on the diaphysis exterior and within the medullary cavity is removed, and compact bone is laid down to reconstruct the shaft walls. The final structure of the remodeled area resembles that of the original unbroken bony region because it responds to the same set of mechanical stressors.

CHECK YOUR UNDERSTANDING

19. If osteoclasts in a long bone are more active than osteoblasts, what change in bone mass is likely?

20. Which stimulus—PTH (a hormone) or mechanical forces acting on the skeleton—is more important in maintaining homeostatic blood calcium levels?

21. How does an open fracture differ from a closed fracture?

22. How do bone growth and bone remodeling differ?

For answers, see Appendix G.

Homeostatic Imbalances of Bone

▶ Contrast the disorders of bone remodeling seen in osteoporosis, osteomalacia, and Paget's disease.

Imbalances between bone deposit and bone resorption underlie nearly every disease that affects the adult skeleton.

Osteomalacia and Rickets

Osteomalacia (os″te-o-mah-la′she-ah; "soft bones") includes a number of disorders in which the bones are inadequately mineralized. Osteoid is produced, but calcium salts are not deposited, so bones soften and weaken. The main symptom is pain when weight is put on the affected bones.

Rickets is the analogous disease in children. Because young bones are still growing rapidly, rickets is much more severe than adult osteomalacia. Bowed legs and deformities of the pelvis, skull, and rib cage are common. Because the epiphyseal plates cannot be calcified, they continue to widen, and the ends of long bones become visibly enlarged and abnormally long.

Osteomalacia and rickets are caused by insufficient calcium in the diet or by a vitamin D deficiency. For this reason, drinking vitamin D–fortified milk and exposing the skin to sunlight (which spurs the body to form vitamin D) usually cure these disorders. Although the seeming elimination of rickets in the United States has been heralded as a public health success, rickets still rears its head in isolated situations. For example, if a mother who breast-feeds her infant becomes vitamin D deficient because of dreary winter weather, the infant too will be vitamin D deficient and will develop rickets.

Osteoporosis

For most of us, the phrase "bone problems of the elderly" brings to mind the stereotype of a victim of osteoporosis—a hunched-over old woman shuffling behind her walker.

TABLE 6.2 Common Types of Fractures

FRACTURE TYPE	DESCRIPTION AND COMMENTS	FRACTURE TYPE	DESCRIPTION AND COMMENTS
Comminuted	Bone fragments into three or more pieces. Particularly common in the aged, whose bones are more brittle	**Compression**	Bone is crushed. Common in porous bones (i.e., osteoporotic bones) subjected to extreme trauma, as in a fall Crushed vertebra
Spiral	Ragged break occurs when excessive twisting forces are applied to a bone. Common sports fracture 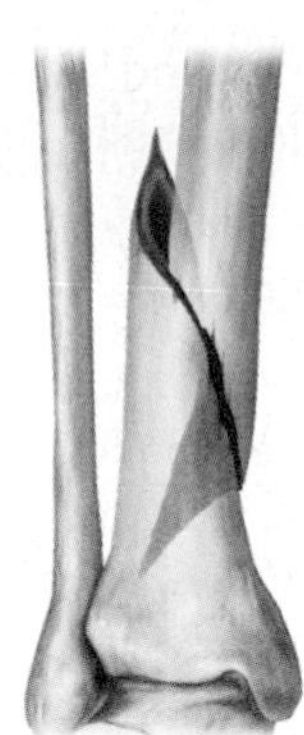 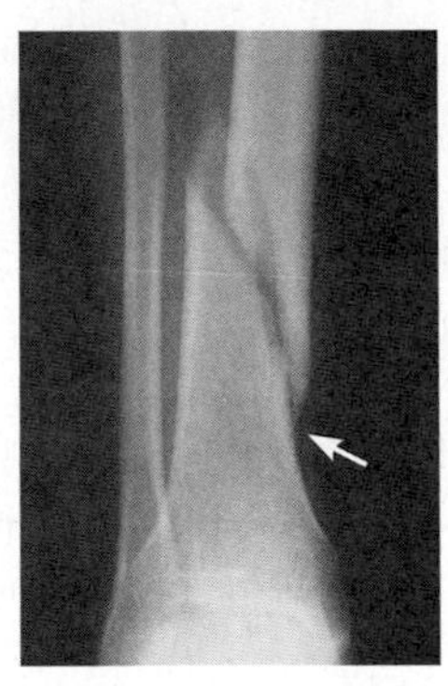	**Epiphyseal**	Epiphysis separates from the diaphysis along the epiphyseal plate. Tends to occur where cartilage cells are dying and calcification of the matrix is occurring 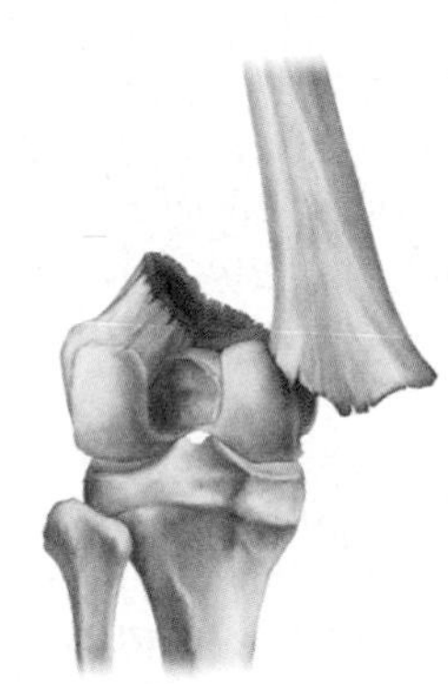 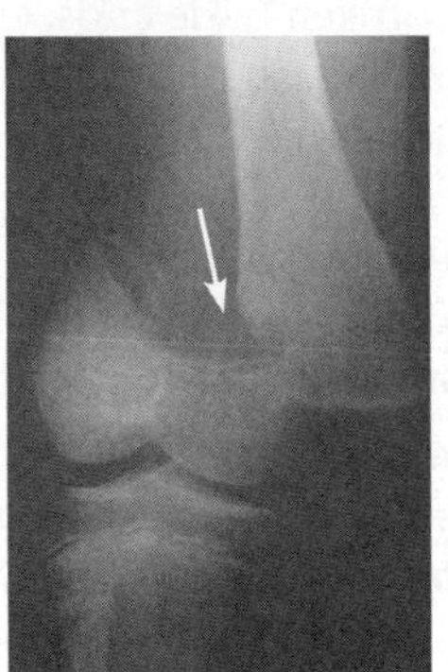
Depressed	Broken bone portion is pressed inward. Typical of skull fracture 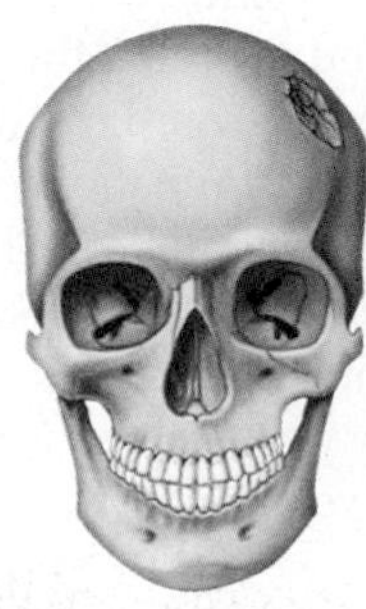 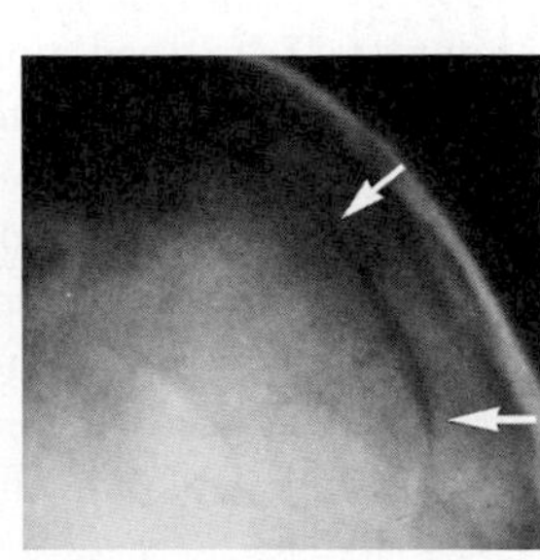	**Greenstick**	Bone breaks incompletely, much in the way a green twig breaks. Only one side of the shaft breaks; the other side bends. 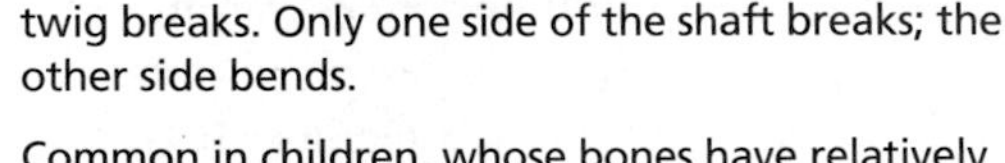 Common in children, whose bones have relatively more organic matrix and are more flexible than those of adults 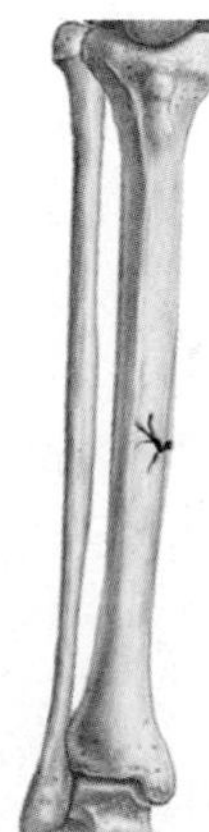 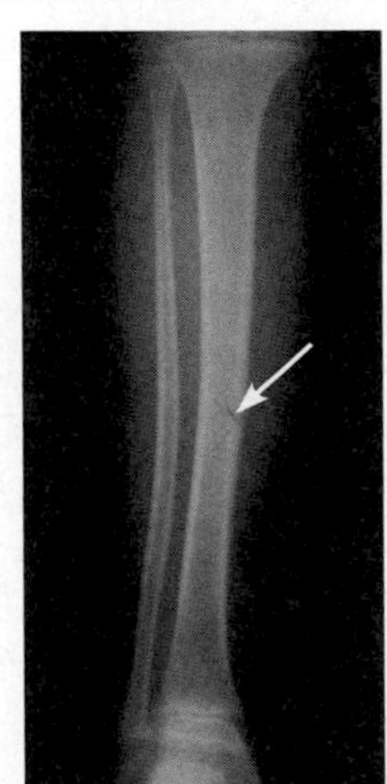

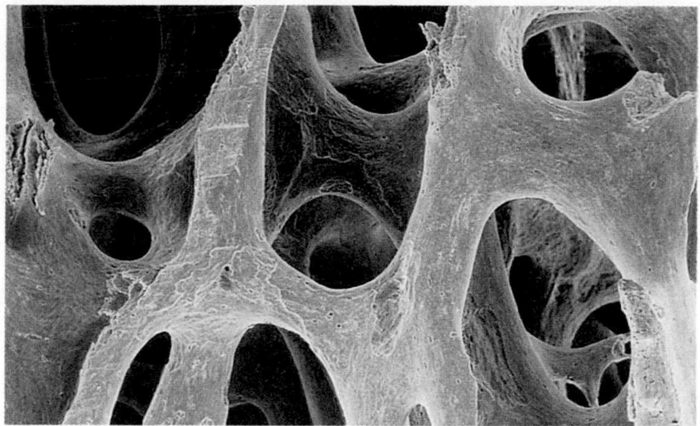

(a) Normal bone

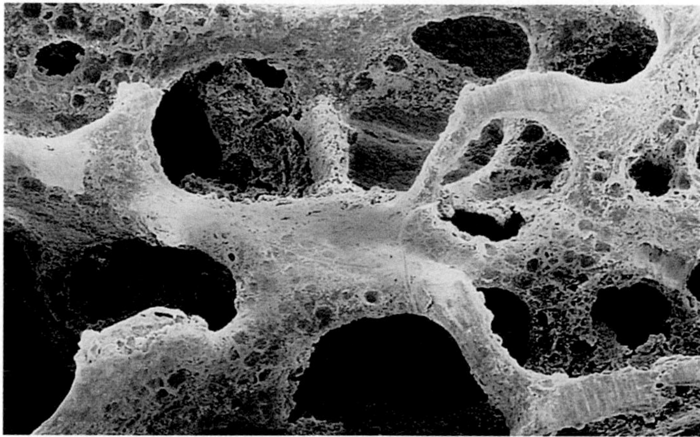

(b) Osteoporotic bone

Figure 6.16 The contrasting architecture of normal versus osteoporotic bone. Scanning electron micrographs, 10×.

Osteoporosis (os″te-o-po-ro′sis) refers to a group of diseases in which bone resorption outpaces bone deposit. The bones become so fragile that something as simple as a hearty sneeze or stepping off a curb can cause them to break. The composition of the matrix remains normal but bone mass is reduced, and the bones become porous and light **(Figure 6.16)**. Even though osteoporosis affects the entire skeleton, the spongy bone of the spine is most vulnerable, and compression fractures of the vertebrae are common. The femur, particularly its neck, is also very susceptible to fracture (called a *broken hip*) in people with osteoporosis.

Osteoporosis occurs most often in the aged, particularly in postmenopausal women. Although men develop it to a lesser degree, 30% of American women between the ages of 60 and 70 have osteoporosis, and 70% have it by age 80. Moreover, 30% of all Caucasian women (the most susceptible group) will experience a bone fracture due to osteoporosis. Sex hormones, particularly estrogen, help to maintain the health and normal density of the skeleton by restraining osteoclast activity and by promoting deposit of new bone. After menopause, however, estrogen secretion wanes, and estrogen deficiency is strongly implicated in osteoporosis in older women. Other factors that contribute to osteoporosis include a petite body form, insufficient exercise to stress the bones, a diet poor in calcium and protein, abnormal vitamin D receptors, smoking (which reduces estrogen levels), and hormone-related conditions such as hyperthyroidism, low blood levels of thyroid-stimulating hormone (better known for its role in stimulating the secretion of thyroid hormones), and diabetes mellitus. In addition, osteoporosis can occur at any age as a result of immobility.

Osteoporosis has traditionally been treated with calcium and vitamin D supplements, weight-bearing exercise, and *hormone (estrogen) replacement therapy* (*HRT*). Frustratingly, HRT only slows the loss of bone but does not reverse it. Additionally, because of the increased risk of heart attack, stroke, and breast cancer associated with estrogen replacement therapy, it is a controversial treatment these days.

Newer drugs are available. These include alendronate (Fosamax), a drug that decreases osteoclast activity and number, and shows promise in reversing osteoporosis in the spine; and selective estrogen receptor modulators (SERMs), such as raloxifene, dubbed "estrogen light" because it mimics estrogen's beneficial bone-sparing properties without targeting the uterus or breast. Additionally, *statins*, drugs used by tens of thousands of people to lower cholesterol levels, have been shown to have an unexpected side effect of increasing bone mineral density up to 8% over four years. Although not a substitute for HRT, estrogenic compounds in soy protein (principally the isoflavones daidzein and genistein) offer a good addition or adjunct for some patients.

6

How can osteoporosis be prevented (or at least delayed)? The first requirement is to get enough calcium while your bones are still increasing in density (bones reach their peak density during early adulthood). Second, drinking fluoridated water hardens bones (as well as teeth). Conversely, excessive intake of carbonated beverages leaches minerals from bone and decreases bone density. Finally, getting plenty of weight-bearing exercise (walking, jogging, tennis, etc.) throughout life will increase bone mass above normal values and provide a greater buffer against age-related bone loss.

Paget's Disease

Often discovered by accident when X rays are taken for some other reason, **Paget's disease** (paj′ets) is characterized by excessive and haphazard bone deposit and resorption. The newly formed bone, called *Pagetic bone*, is hastily made and has an abnormally high ratio of spongy bone to compact bone. This, along with reduced mineralization, causes a spotty weakening of the bones. Late in the disease, osteoclast activity wanes, but osteoblasts continue to work, often forming irregular bone thickenings or filling the marrow cavity with Pagetic bone.

Paget's disease may affect any part of the skeleton, but it is usually a localized condition. The spine, pelvis, femur, and skull are most often involved and become increasingly deformed and painful. It rarely occurs before the age of 40, and it affects about 3% of North American elderly people. Its cause is unknown, but

System Connections

Homeostatic Interrelationships Between the Skeletal System and Other Body Systems

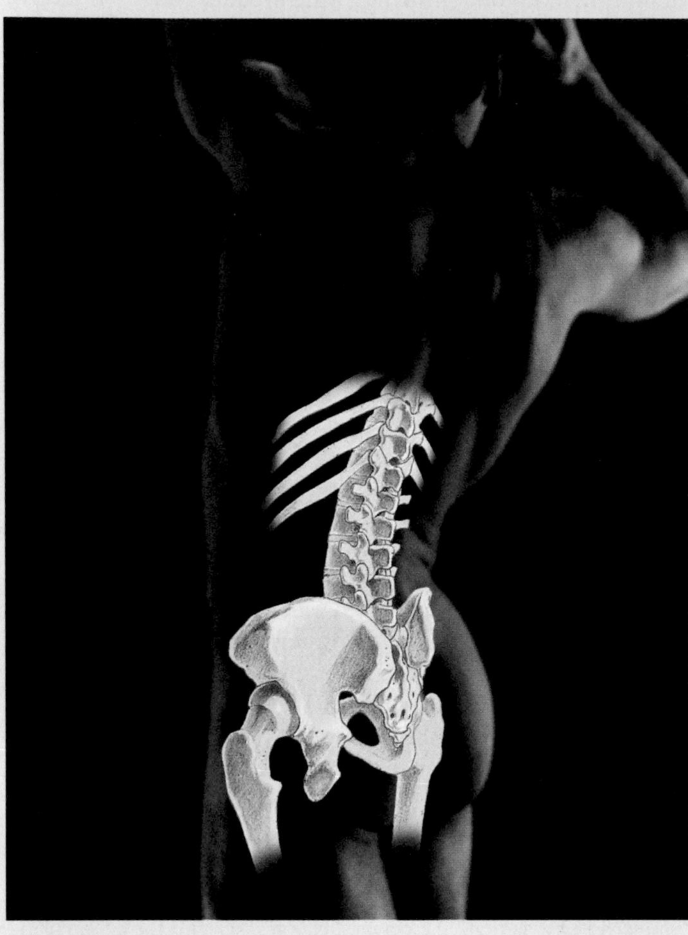

Integumentary System

- Skeletal system provides support for body organs including the skin
- Skin provides vitamin D needed for proper calcium absorption and use

Muscular System

- Skeletal system provides levers plus ionic calcium for muscle activity
- Muscle pull on bones increases bone strength and viability; helps determine bone shape

Nervous System

- Skeletal system protects brain and spinal cord; provides depot for calcium ions needed for neural function
- Nerves innervate bone and joint capsules, providing for pain and joint sense

Endocrine System

- Skeletal system provides some bony protection; stores calcium needed for second-messenger signaling mechanisms
- Hormones regulate uptake and release of calcium from bone; promote long bone growth and maturation

Cardiovascular System

- Bone marrow cavities provide site for blood cell formation; matrix stores calcium needed for cardiac muscle activity
- Cardiovascular system delivers nutrients and oxygen to bones; carries away wastes

Lymphatic System/Immunity

- Skeletal system provides some protection to lymphatic organs; bone marrow is site of origin for lymphocytes involved in immune response
- Lymphatic system drains leaked tissue fluids; immune cells protect against pathogens

Respiratory System

- Skeletal system protects lungs by enclosure (rib cage)
- Respiratory system provides oxygen; disposes of carbon dioxide

Digestive System

- Skeletal system provides some bony protection to intestines, pelvic organs, and liver
- Digestive system provides nutrients needed for bone health and growth

Urinary System

- Skeletal system protects pelvic organs (urinary bladder, etc.)
- Urinary system activates vitamin D; disposes of nitrogenous wastes

Reproductive System

- Skeletal system protects some reproductive organs by enclosure
- Gonads produce hormones that influence the form of the skeleton and epiphyseal closure

Closer Connections

The Skeletal System and Interrelationships with the Muscular, Endocrine, and Integumentary Systems

Our skeleton supports us, protects our "innards" (the protection our brain gets from the skull is indispensable), gives us stature (for some reason, tall people get more respect), contributes to our shape (women are shaped differently than men), and allows us to move. Obviously, the skeletal system has important interactions with many other body systems, not the least of which are the endocrine and integumentary systems. However, its most intimate and mutually beneficial relationship is with the muscular system, so we will consider that first.

Muscular System

The codependence of the skeletal and muscular systems is striking—as one system goes, so goes the other. If we participate in weight-bearing exercise (run, play tennis, do aerobics) regularly, our muscles become more efficient and exert more force on our bones. As a result, our bones stay healthy and strong and increase their mass to assume the added stress.

Since both spongy and compact bone reach peak density during midlife, weight-bearing exercise during youth is important, especially in females who have less bone mass than males and lose it faster.

Regular exercise also stretches the connective tissues binding bones to muscles and to other bones, and reinforcing joints. Since this increases overall flexibility, we have fewer injuries, allowing us to stay active well into old age. (Pain makes couch potatoes.)

Endocrine System

Although mechanical factors are undeniably important in shaping the skeleton and helping to keep it strong, hormones acting individually and in concert direct skeletal growth during youth, and enhance (or impair) skeletal strength in adults. Growth hormone is essential for normal skeletal growth and maintenance throughout life, whereas thyroid and sex hormones ensure that normal skeletal proportions are established during childhood and adolescence. Conversely, PTH serves not the skeleton but a different master—homeostasis of blood calcium levels. Any interference with normal hormonal functioning is soon apparent as a skeletal abnormality or malproportion.

Integumentary System

The skeletal system is absolutely dependent on the integumentary system (the skin) for the calcium that keeps the bones hard and strong. The relationship is indirect: In the presence of sunlight, a vitamin D precursor is produced in the dermal capillary blood. It is activated elsewhere, and (among its other roles) it regulates the carrier system that absorbs calcium from ingested foods into the blood. Because calcium is required for so many body functions, and bones provide the "calcium bank," the bones become increasingly soft and weak in the absence of vitamin D because no daily rations of calcium are allowed to enter the blood from the digestive tract.

Clinical Connections

Skeletal System

Case study: Remember Mrs. DeStephano? When we last heard about her she was being admitted for further studies. Relative to her skeletal system, the following notes have been added to her chart.

- Fracture of superior right tibia (shinbone of leg); skin lacerated; area cleaned and protruding bone fragments subjected to internal (open) reduction and casted
- Nutrient artery of tibia damaged
- Medial meniscus (fibrocartilage disc) of right knee joint crushed; knee joint inflamed and painful

Relative to these notes:

1. What type of fracture does Mrs. DeStephano have?
2. What problems can be predicted with such fractures and how are they treated?
3. What is internal reduction? Why was a cast applied?
4. Given an uncomplicated recovery, approximately how long should it take before Mrs. DeStephano has a good solid bony callus?
5. What complications might be predicted by the fact that the nutrient artery is damaged?
6. What new techniques might be used to enhance fracture repair if healing is delayed or impaired?
7. How likely is it that Mrs. DeStephano's knee cartilage will regenerate? Why?

(Answers in Appendix G)

6

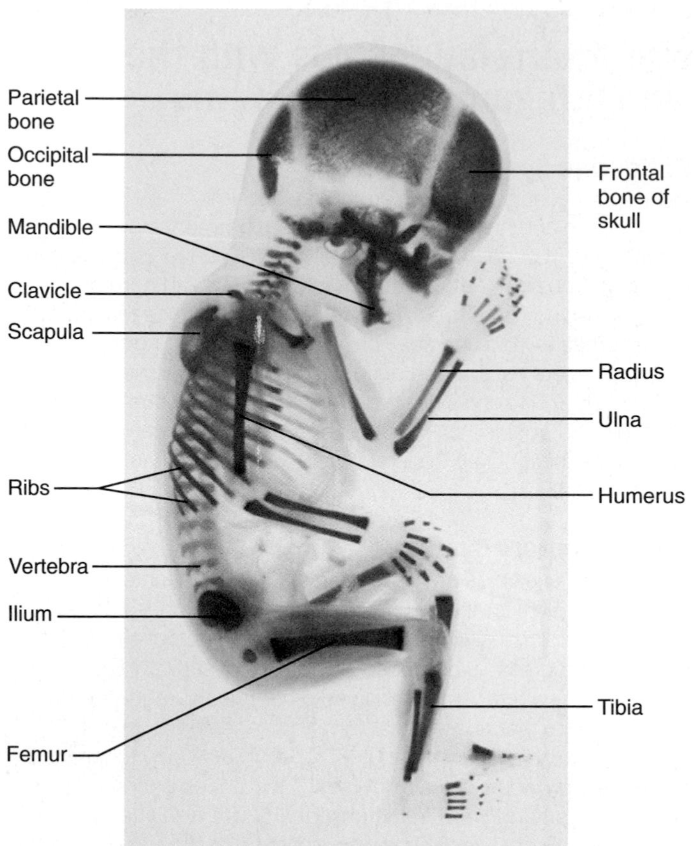

Figure 6.17 Fetal primary ossification centers at 12 weeks. The darker areas indicate primary ossification centers in the skeleton of a 12-week-old fetus.

it may be initiated by a virus. Drug therapies include calcitonin (now administered by a nasal inhaler), and the newer bisphosphonates (etidronate, alendronate, and others) which have shown success in preventing bone breakdown.

CHECK YOUR UNDERSTANDING

23. Which bone disorder is characterized by excessive deposit of weak, poorly mineralized bone?
24. What are three measures that may help to maintain healthy bone density?
25. What name is given to "adult rickets"?

For answers, see Appendix G.

Developmental Aspects of Bones: Timing of Events

▶ Describe the timing and cause of changes in bone architecture and bone mass throughout life.

Bones are on a precise schedule from the time they form until death. The mesoderm germ layer gives rise to embryonic mesenchymal cells, which in turn produce the membranes and cartilages that form the embryonic skeleton. These structures then ossify according to an amazingly predictable timetable that allows fetal age to be determined easily from either X rays or sonograms. Although each bone has its own developmental schedule, most long bones begin ossifying by 8 weeks after conception and have well-developed primary ossification centers by 12 weeks **(Figure 6.17)**.

At birth, most long bones of the skeleton are well ossified except for their epiphyses. After birth, secondary ossification centers develop in a predictable sequence. The epiphyseal plates persist and provide for long bone growth all through childhood and the sex hormone–mediated growth spurt at adolescence. By the age of 25 years, nearly all bones are completely ossified and skeletal growth ceases.

In children and adolescents, bone formation exceeds bone resorption. In young adults, these processes are in balance, and in old age, resorption predominates. Despite the environmental factors (discussed earlier) that influence bone density, genetics still plays the major role in determining how much a person's bone density will change over a lifetime. A single gene that codes for vitamin D's cellular docking site helps determine both the tendency to accumulate bone mass during early life and a person's risk of osteoporosis later in life.

Beginning in the fourth decade of life, bone mass decreases with age. The only exception appears to be in bones of the skull. Among young adults, skeletal mass is generally greater in males than in females, and greater in blacks than in whites. Age-related bone loss is faster in whites than in blacks (who have greater bone density to begin with) and faster in females than in males. Qualitative changes also occur: More osteons remain incompletely formed, mineralization is less complete, and the amount of nonviable bone increases, reflecting a diminished blood supply to the bones in old age. These age-related changes are also bad news because fractures heal more slowly in old people. Daily ultrasound treatments are helpful in hastening repair of fractures, and electrical stimulation of fracture sites dramatically increases the speed of healing. (Presumably electrical fields inhibit PTH stimulation of osteoclasts and induce formation of growth factors that stimulate osteoblasts at the fracture site.)

CHECK YOUR UNDERSTANDING

26. What is the status of bone structure at birth?
27. The decrease in bone mass that begins in the fourth decade of life affects nearly all bones. What are the exceptions?

For answers, see Appendix G.

Skeletal cartilages and bones—their architecture, composition, and dynamic nature—have been examined in this chapter. We have also discussed the role of bones in maintaining overall body homeostasis, as summarized in *Making Connections.* Now we are ready to look at the individual bones of the skeleton and how they contribute to its functions, both collectively and individually.

RELATED CLINICAL TERMS

Achondroplasia (a-kon″dro-pla′ze-ah; *a* = without; *chondro* = cartilage; *plasi* = mold, shape) A congenital condition involving defective cartilage and endochondral bone growth so that the limbs are too short but the membrane bones are of normal size; a type of dwarfism.

Bony spur Abnormal projection from a bone due to bony overgrowth; common in aging bones.

Ostealgia (os″te-al′je-ah; *algia* = pain) Pain in a bone.

Osteitis (os″te-i′tis; *itis* = inflammation) Inflammation of bony tissue.

Osteogenesis imperfecta Also called brittle bone disease, a disorder in which the bone matrix contains inadequate amounts of collagen, putting it at risk for shattering.

Osteomyelitis (os″te-o-mi″ĕ-l-li′tis) Inflammation of bone and bone marrow caused by pus-forming bacteria that enter the body via a wound (e.g., compound bone fracture), or spread from an infection near the bone. Commonly affects the long bones, causing acute pain and fever. May result in joint stiffness, bone destruction, and shortening of a limb. Treatment involves antibiotic therapy, draining of any abscesses (local collections of pus) formed, and removal of dead bone fragments (which prevent healing).

Osteosarcoma (os″te-o-sar-ko′mah) A form of bone cancer typically arising in a long bone of a limb and most often in those 10–25 years of age. Grows aggressively, painfully eroding the bone; tends to metastasize to the lungs and cause secondary lung tumors. Usual treatment is amputation of the affected bone or limb, followed by chemotherapy and surgical removal of any metastases. Survival rate is about 50% if detected early.

Pathologic fracture Fracture in a diseased bone involving slight (coughing or a quick turn) or no physical trauma. For example, a hip bone weakened by osteoporosis may break and cause the person to fall, rather than breaking because of the fall.

Traction ("pulling") Placing sustained tension on a body region to keep the parts of a fractured bone in proper alignment; also prevents spasms of skeletal muscles, which would separate the fractured bone ends or crush the spinal cord in the case of vertebral column fractures.

6

CHAPTER SUMMARY

Skeletal Cartilages (p. 173)

Basic Structure, Types, and Locations (p. 173)

1. A skeletal cartilage exhibits chondrocytes housed in lacunae (cavities) within the extracellular matrix (ground substance and fibers). It contains large amounts of water (which accounts for its resilience), lacks nerve fibers, is avascular, and is surrounded by a fibrous perichondrium that resists expansion.
2. Hyaline cartilages appear glassy; the fibers are collagenic. They provide support with flexibility and resilience and are the most abundant skeletal cartilages, accounting for the articular, costal, respiratory, and nasal cartilages.
3. Elastic cartilages contain abundant elastic fibers, in addition to collagen fibers, and are more flexible than hyaline cartilages. They support the outer ear and epiglottis.
4. Fibrocartilages, which contain thick collagen fibers, are the most compressible cartilages and are resistant to stretch. They form intervertebral discs and knee joint cartilages.

Growth of Cartilage (p. 173)

5. Cartilages grow from within (interstitial growth) and by addition of new cartilage tissue at the periphery (appositional growth).

Classification of Bones (pp. 173–175)

1. Bones are classified as long, short, flat, or irregular on the basis of their shape and their proportion of compact or spongy bone.

Functions of Bones (pp. 175–176)

1. Bones give the body shape; protect and support body organs; provide levers for muscles to pull on; store calcium and other minerals; and are the site of blood cell production.

Bone Structure (pp. 176–182)

Gross Anatomy (pp. 177–179)

1. Bone markings are important anatomical landmarks that reveal sites of muscle attachment, points of articulation, and sites of blood vessel and nerve passage.
2. A long bone is composed of a diaphysis (shaft) and epiphyses (ends). The medullary cavity of the diaphysis contains yellow marrow; the epiphyses contain spongy bone. The epiphyseal line is the remnant of the epiphyseal plate. Periosteum covers the diaphysis; endosteum lines inner bone cavities. Hyaline cartilage covers joint surfaces.
3. Flat bones consist of two thin plates of compact bone enclosing a diploë (spongy bone layer). Short and irregular bones resemble flat bones structurally.
4. In adults, hematopoietic tissue (red marrow) is found within the diploë of flat bones and occasionally within the epiphyses of long bones. In infants, red marrow is also found in the medullary cavity.

Microscopic Anatomy of Bone (pp. 179–180)

5. The structural unit of compact bone, the osteon, consists of a central canal surrounded by concentric lamellae of bone matrix. Osteocytes, embedded in lacunae, are connected to each other and the central canal by canaliculi.
6. Spongy bone has slender trabeculae containing irregular lamellae, which enclose red marrow–filled cavities.

Chemical Composition of Bone (pp. 180–182)

7. Bone is composed of living cells (osteogenic cells, osteoblasts, osteocytes, and osteoclasts) and matrix. The matrix includes osteoid, organic substances that are secreted by osteoblasts and give the bone tensile strength. Its inorganic components, the hydroxyapatites (calcium salts), make bone hard.

Bone Development (pp. 182–185)

Formation of the Bony Skeleton (pp. 182–184)

1. Intramembranous ossification forms the clavicles and most skull bones. The ground substance of the bone matrix is deposited between collagen fibers within the fibrous membrane to form woven bone. Eventually, compact bone plates enclose the diploë.
2. Most bones are formed by endochondral ossification of a hyaline cartilage model. Osteoblasts beneath the periosteum secrete bone matrix on the cartilage model, forming the bone collar. Deterioration of the cartilage model internally opens up cavities, allowing periosteal bud entry. Bone matrix is deposited around the cartilage remnants but is later broken down.

Postnatal Bone Growth (pp. 184–185)

3. Long bones increase in length by interstitial growth of the epiphyseal plate cartilage and its replacement by bone.
4. Appositional growth increases bone diameter/thickness.

Bone Homeostasis: Remodeling and Repair (pp. 185–189)

Bone Remodeling (pp. 185–188)

1. Bone is continually deposited and resorbed in response to hormonal and mechanical stimuli. Together these processes constitute bone remodeling.
2. An osteoid seam appears at areas of new bone deposit; calcium salts are deposited a few days later.
3. Osteoclasts release lysosomal enzymes and acids on bone surfaces to be resorbed. The dissolved products are transcytosed to the opposite face of the osteoclast for release to the extracellular fluid.
4. The hormonal controls of bone remodeling serve blood calcium homeostasis. When blood calcium levels decline, PTH is released and stimulates osteoclasts to digest bone matrix, releasing ionic calcium. As blood calcium levels rise, PTH secretion declines.
5. Mechanical stress and gravity acting on the skeleton help maintain skeletal strength. Bones thicken, develop heavier prominences, or rearrange their trabeculae in sites where stressed.

Bone Repair (pp. 188–189)

6. Fractures are treated by open or closed reduction. The healing process involves formation of a hematoma, a fibrocartilaginous callus, a bony callus, and bone remodeling, in succession.

Homeostatic Imbalances of Bone (pp. 189–191, 194)

1. Imbalances between bone formation and resorption underlie all skeletal disorders.
2. Osteomalacia and rickets occur when bones are inadequately mineralized. The bones become soft and deformed. The most frequent cause is inadequate vitamin D.
3. Osteoporosis is any condition in which bone breakdown outpaces bone formation, causing bones to become weak and porous. Postmenopausal women are particularly susceptible.
4. Paget's disease is characterized by excessive and abnormal bone remodeling.

Developmental Aspects of Bones: Timing of Events (p. 194)

1. Osteogenesis is predictable and precisely timed.
2. Longitudinal long bone growth continues until the end of adolescence. Skeletal mass increases dramatically during puberty and adolescence, when formation exceeds resorption.
3. Bone mass is fairly constant in young adulthood, but beginning in the 40s, bone resorption exceeds formation.

REVIEW QUESTIONS

Multiple Choice/Matching

(Some questions have more than one correct answer. Select the best answer or answers from the choices given.)

1. Which is a function of the skeletal system? **(a)** support, **(b)** hematopoietic site, **(c)** storage, **(d)** providing levers for muscle activity, **(e)** all of these.
2. A bone with approximately the same width, length, and height is most likely **(a)** a long bone, **(b)** a short bone, **(c)** a flat bone, **(d)** an irregular bone.
3. The shaft of a long bone is properly called the **(a)** epiphysis, **(b)** periosteum, **(c)** diaphysis, **(d)** compact bone.
4. Sites of hematopoiesis include all but **(a)** red marrow cavities of spongy bone, **(b)** the diploë of flat bones, **(c)** medullary cavities in bones of infants, **(d)** medullary cavities in bones of a healthy adult.
5. An osteon has **(a)** a central canal carrying blood vessels, **(b)** concentric lamellae, **(c)** osteocytes in lacunae, **(d)** canaliculi that connect lacunae to the central canal, **(e)** all of these.
6. The organic portion of matrix is important in providing all but **(a)** tensile strength, **(b)** hardness, **(c)** ability to resist stretch, **(d)** flexibility.
7. The flat bones of the skull develop from **(a)** areolar tissue, **(b)** hyaline cartilage, **(c)** fibrous connective tissue, **(d)** compact bone.
8. The remodeling of bone is a function of which cells? **(a)** chondrocytes and osteocytes, **(b)** osteoblasts and osteoclasts, **(c)** chondroblasts and osteoclasts, **(d)** osteoblasts and osteocytes.
9. Bone remodeling in adults is regulated and directed mainly by **(a)** growth hormone, **(b)** thyroid hormones, **(c)** sex hormones, **(d)** mechanical stress, **(e)** PTH.
10. Where within the epiphyseal plate are the dividing cartilage cells located? **(a)** nearest the shaft, **(b)** in the marrow cavity, **(c)** farthest from the shaft, **(d)** in the primary ossification center.
11. Wolff's law is concerned with **(a)** calcium homeostasis of the blood, **(b)** the thickness and shape of a bone being determined by mechanical and gravitational stresses placed on it, **(c)** the electrical charge on bone surfaces.
12. Formation of the bony callus in fracture repair is followed by **(a)** hematoma formation, **(b)** fibrocartilaginous callus formation, **(c)** bone remodeling to convert woven bone to compact bone, **(d)** formation of granulation tissue.
13. The fracture type in which the bone ends are incompletely separated is **(a)** greenstick, **(b)** compound, **(c)** simple, **(d)** comminuted, **(e)** compression.
14. The disorder in which bones are porous and thin but bone composition is normal is **(a)** osteomalacia, **(b)** osteoporosis, **(c)** Paget's disease.

Short Answer Essay Questions

15. Compare bone to cartilage tissue relative to its resilience, speed of regeneration, and access to nutrients.
16. Describe in proper order the events of endochondral ossification.

17. Osteocytes residing in lacunae of osteons of healthy compact bone are located quite a distance from the blood vessels in the central canals, yet they are well nourished. How can this be explained?
18. As we grow, our long bones increase in diameter, but the thickness of the compact bone of the shaft remains relatively constant. Explain this phenomenon.
19. Describe the process of new bone formation in an adult bone. Use the terms osteoid seam and calcification front in your discussion.
20. Compare and contrast controls of bone remodeling exerted by hormones and by mechanical and gravitational forces, including the actual purpose of each control system and changes in bone architecture that might occur.
21. (a) During what period of life does skeletal mass increase dramatically? Begin to decline? (b) Why are fractures most common in elderly individuals? (c) Why are greenstick fractures most common in children?
22. Yolanda is asked to review a bone slide that her professor has set up under the microscope. She sees concentric layers surrounding a central cavity. Is this bone section taken from the diaphysis or the epiphyseal plate of the specimen?

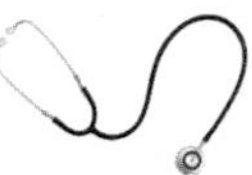

Critical Thinking and Clinical Application Questions

1. Following a motorcycle accident, a 22-year-old man was rushed to the emergency room. X rays revealed a spiral fracture of his right tibia (main bone of the leg). Two months later, X rays revealed good bony callus formation. What is bony callus?
2. Mrs. Abbruzzo brought her 4-year-old daughter to the doctor, complaining that she didn't "look right." The child's forehead was enlarged, her rib cage was knobby, and her lower limbs were bent and deformed. X rays revealed very thick epiphyseal plates. Mrs. Abbruzzo was advised to increase dietary amounts of vitamin D and milk and to "shoo" the girl outside to play in the sun. Considering the child's signs and symptoms, what disease do you think she has? Explain the doctor's instructions.
3. You overhear some anatomy students imagining out loud what their bones would look like if they had compact bone on the inside and spongy bone on the outside, instead of the other way around. You tell them that such imaginary bones would be poorly designed mechanically and would break easily. Explain your reason for saying this.
4. What would a long bone look like at the end of adolescence if bone remodeling did not occur?
5. Why do you think wheelchair-bound people with paralyzed lower limbs have thin, weak, leg and thigh bones?
6. Jay Beckenstein went to weight-lifting camp in the summer between seventh and eighth grade. He noticed that the camp trainer put tremendous pressure on him and his friends to improve their strength. After an especially vigorous workout, Jay's arm felt extremely sore and weak around the elbow. He went to the camp doctor, who took X rays and then told him that the injury was serious, for the "end of his upper arm bone was starting to twist off." What had happened? Could the same thing happen to Jay's 23-year-old sister, Trixie, who was also starting a program of weight lifting? Why or why not?
7. Old Norse stories tell of a famous Viking named Egil, who lived around 900 AD. His skull was greatly enlarged and misshapen, and the cranial bones were thickened (6 cm, more than 2 inches, thick). After he died, his skull was dug up and it withstood the blow of an ax without damage. In life, he had headaches from the pressure exerted by enlarged vertebrae on his spinal cord. So much blood was diverted to his bones to support their extensive remodeling that his fingers and toes always felt cold and his heart was damaged through overexertion. What bone disorder did Egil probably have?

6

7

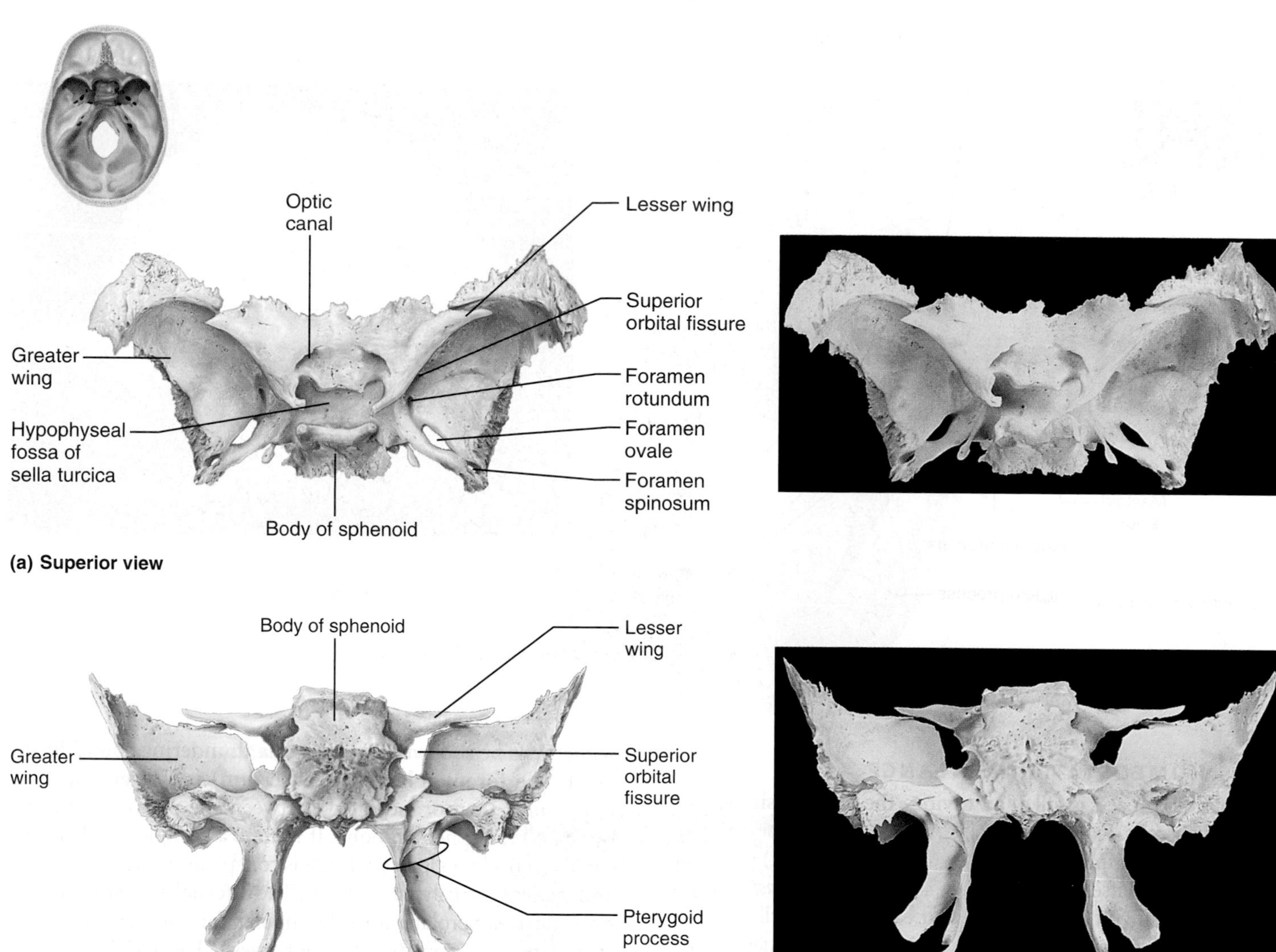

Figure 7.9 The sphenoid bone. (See *A Brief Atlas of the Human Body*, Figures 5 and 9).

junction of the body and greater wings (Figure 7.9b). They anchor the pterygoid muscles, which are important in chewing.

A number of openings in the sphenoid bone are visible in Figures 7.7 and 7.9. The **optic canals** lie anterior to the sella turcica; they allow the optic nerves (cranial nerves II) to pass to the eyes. On each side of the sphenoid body is a crescent-shaped row of four openings. The anteriormost of these, the **superior orbital fissure**, is a long slit between the greater and lesser wings. It allows cranial nerves that control eye movements (III, IV, VI) to enter the orbit. This fissure is most obvious in an anterior view of the skull (Figure 7.4. See also Figure 7.9b.). The **foramen rotundum** and **foramen ovale** (o-va′le) provide passageways for branches of cranial nerve V (the maxillary and mandibular nerves, respectively) to reach the face (Figure 7.7). The foramen rotundum is in the medial part of the greater wing and is usually oval, despite its name meaning "round opening." The foramen ovale, a large, oval foramen posterior to the foramen rotundum, is also visible in an inferior view of the skull (Figure 7.6). Posterolateral to the foramen ovale is the small **foramen spinosum** (Figure 7.7); it transmits the *middle meningeal artery*, which serves the internal faces of some cranial bones.

Ethmoid Bone

Like the temporal and sphenoid bones, the delicate **ethmoid bone** has a complex shape **(Figure 7.10)**. Lying between the sphenoid and the nasal bones of the face, it is the most deeply situated bone of the skull. It forms most of the bony area between the nasal cavity and the orbits.

The superior surface of the ethmoid is formed by the paired horizontal **cribriform plates** (krib′rĭ-form) (see also Figure 7.7), which help form the roof of the nasal cavities and the floor of the anterior cranial fossa. The cribriform plates are punctured by tiny holes (*cribr* = sieve) called *olfactory foramina* that allow the filaments of the olfactory nerves to pass from the smell receptors in the nasal cavities to the brain. Projecting superiorly between the cribriform plates is a triangular process called the **crista galli** (kris′tah gah′le; "rooster's comb"). The outermost

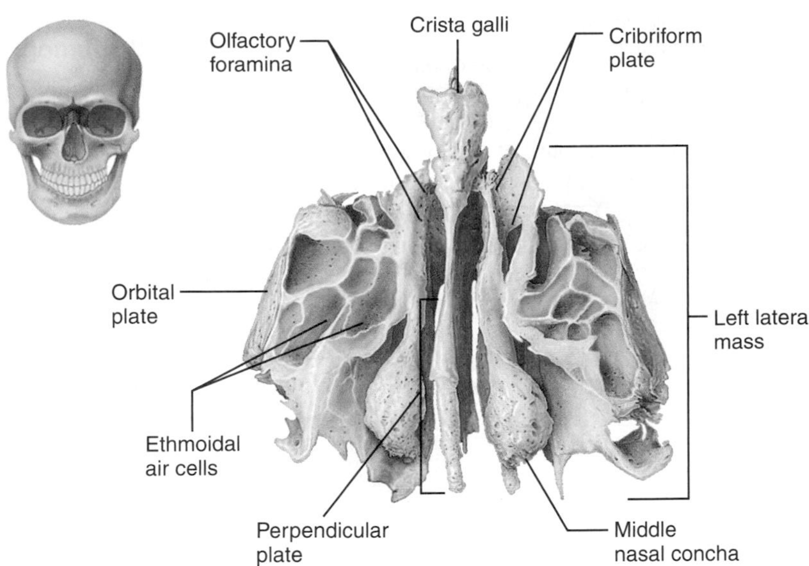

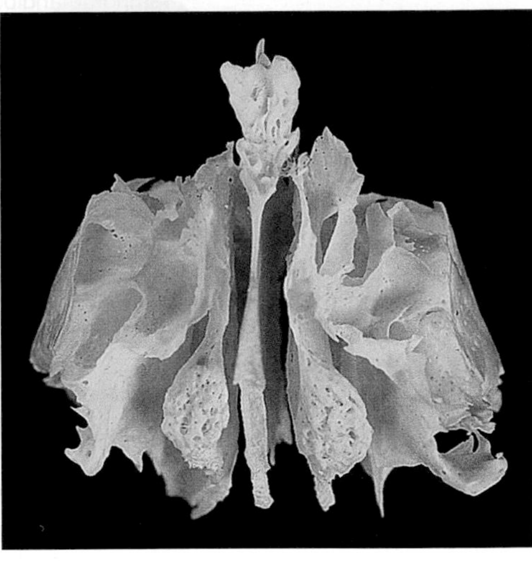

Figure 7.10 The ethmoid bone. Anterior view. (See *A Brief Atlas of the Human Body*, Figures 3 and 10.)

7

covering of the brain (the dura mater) attaches to the crista galli and helps secure the brain in the cranial cavity.

The **perpendicular plate** of the ethmoid bone projects inferiorly in the median plane and forms the superior part of the nasal septum, which divides the nasal cavity into right and left halves (Figure 7.5b and c). Flanking the perpendicular plate on each side is a **lateral mass** riddled with the **ethmoid sinuses**, also called the **ethmoidal air cells** (Figures 7.10 and 7.15), for which the bone itself is named (*ethmos* = sieve). Extending medially from the lateral masses, the delicately coiled **superior** and **middle nasal conchae** (kong′ke; *concha* = shell), named after the conch shells found on warm ocean beaches, protrude into the nasal cavity (Figures 7.10 and 7.14a). The lateral surfaces of the ethmoid's lateral masses are called **orbital plates** because they contribute to the medial walls of the orbits.

Sutural Bones

Sutural bones are tiny irregularly shaped bones or bone clusters that occur within sutures, most often in the lambdoid suture (Figure 7.4b). Structurally unimportant, their number varies, and not all skulls exhibit them. The significance of these tiny bones is unknown.

CHECK YOUR UNDERSTANDING

3. Look at Figure 7.4. Which of the skull bones illustrated in view (a) are cranial bones?
4. Which bone forms the crista galli?
5. Which skull bones house the external ear canals?
6. What bones abut one another at the sagittal suture? At the lambdoid suture?

For answers, see Appendix G.

Facial Bones

The facial skeleton is made up of 14 bones (see Figures 7.4a and 7.5a), of which only the mandible and the vomer are unpaired. The maxillae, zygomatics, nasals, lacrimals, palatines, and inferior nasal conchae are paired bones. As a rule, the facial skeleton of men is more elongated than that of women. Women's faces tend to be rounder and less angular.

Mandible

The U-shaped **mandible** (man′dĭ-bl), or lower jawbone (Figures 7.4a and 7.5, and **Figure 7.11a**), is the largest, strongest bone of the face. It has a body, which forms the chin, and two upright *rami* (*rami* = branches). Each ramus meets the body posteriorly at a **mandibular angle**. At the superior margin of each ramus are two processes separated by the **mandibular notch**. The anterior **coronoid process** (kor′o-noid; "crown-shaped") is an insertion point for the large temporalis muscle that elevates the lower jaw during chewing. The posterior **mandibular condyle** articulates with the mandibular fossa of the temporal bone, forming the *temporomandibular joint* on the same side.

The mandibular **body** anchors the lower teeth. Its superior border, called the **alveolar margin** (al-ve′o-lar), contains the sockets (*alveoli*) in which the teeth are embedded. In the midline of the mandibular body is a slight depression, the **mandibular symphysis** (sim′fih-sis), indicating where the two mandibular bones fused during infancy (Figure 7.4a).

Large **mandibular foramina**, one on the medial surface of each ramus, permit the nerves responsible for tooth sensation to pass to the teeth in the lower jaw. Dentists inject lidocaine into these foramina to prevent pain while working on the lower teeth. The **mental foramina**, openings on the lateral aspects of the mandibular body, allow blood vessels and nerves to pass to the skin of the chin (*ment* = chin) and lower lip.

7

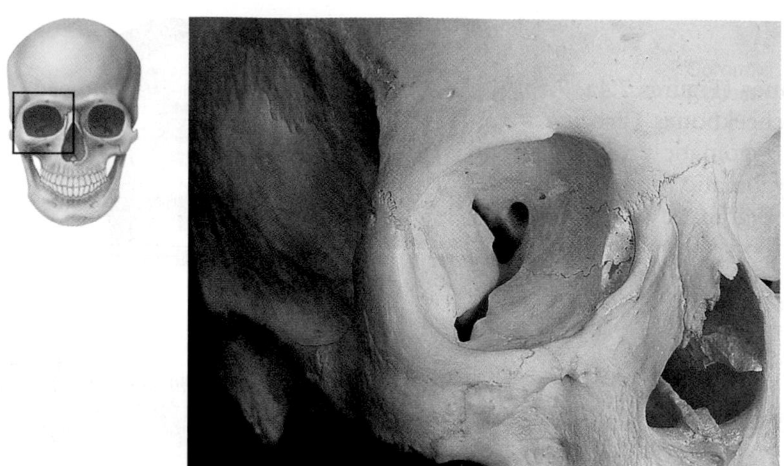

(a) Photograph, right orbit

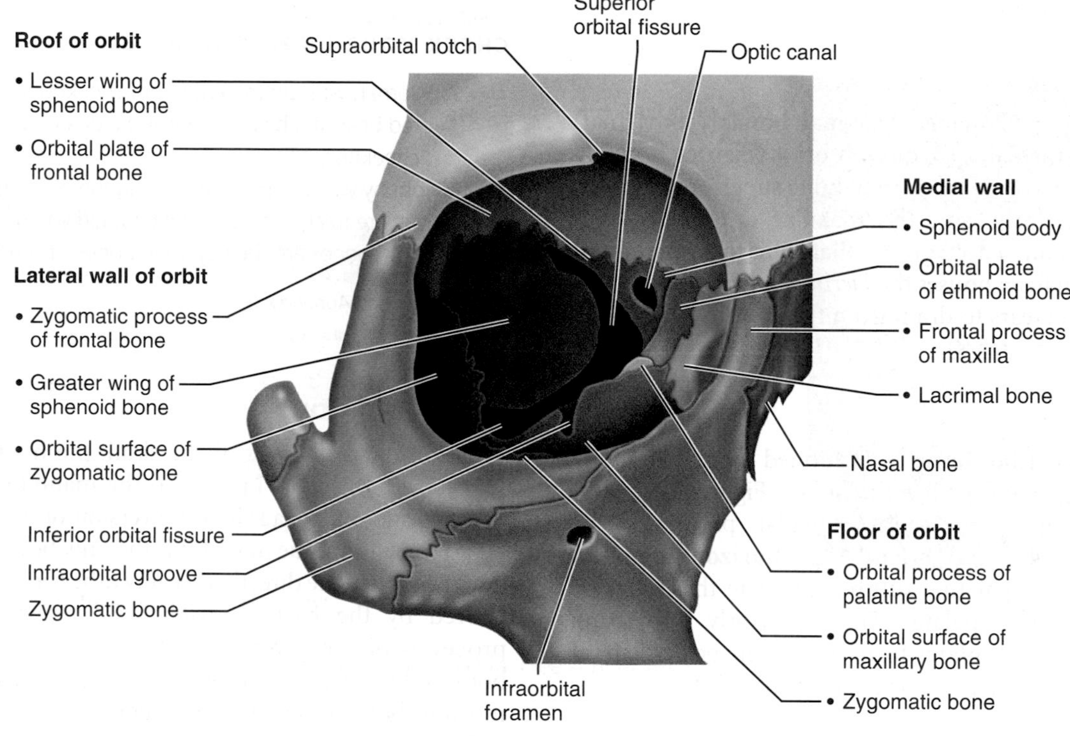

(b) Contribution of each of the seven bones forming the right orbit

Figure 7.13 Bones that form the orbits. (See *A Brief Atlas of the Human Body*, Figure 14.)

move the eyes and the tear-producing lacrimal glands are also housed in the orbits. The walls of each orbit are formed by parts of seven bones—the frontal, sphenoid, zygomatic, maxilla, palatine, lacrimal, and ethmoid bones. Their relationships are shown in **Figure 7.13**. Also seen in the orbits are the superior and inferior orbital fissures and the optic canals, described earlier.

The Nasal Cavity

The **nasal cavity** is constructed of bone and hyaline cartilage **(Figure 7.14)**. The *roof* of the nasal cavity is formed by the cribriform plates of the ethmoid. The *lateral walls* are largely shaped by the superior and middle conchae of the ethmoid bone, the perpendicular plates of the palatine bones, and the inferior nasal conchae. The depressions under cover of the conchae on the lateral walls are called *meatuses* (*meatus* = passage), so there are superior, middle, and inferior meatuses. The *floor* of the nasal cavity is formed by the palatine processes of the maxillae and the palatine bones. The nasal cavity is divided into right and left parts by the *nasal septum*. The bony portion of the septum is formed by the vomer inferiorly and the perpendicular plate of the ethmoid bone superiorly (Figure 7.14b). A sheet of cartilage called the *septal cartilage* completes the septum anteriorly.

The nasal septum and conchae are covered with a mucus-secreting mucosa that moistens and warms the entering air and helps cleanse it of debris. The scroll-shaped conchae increase

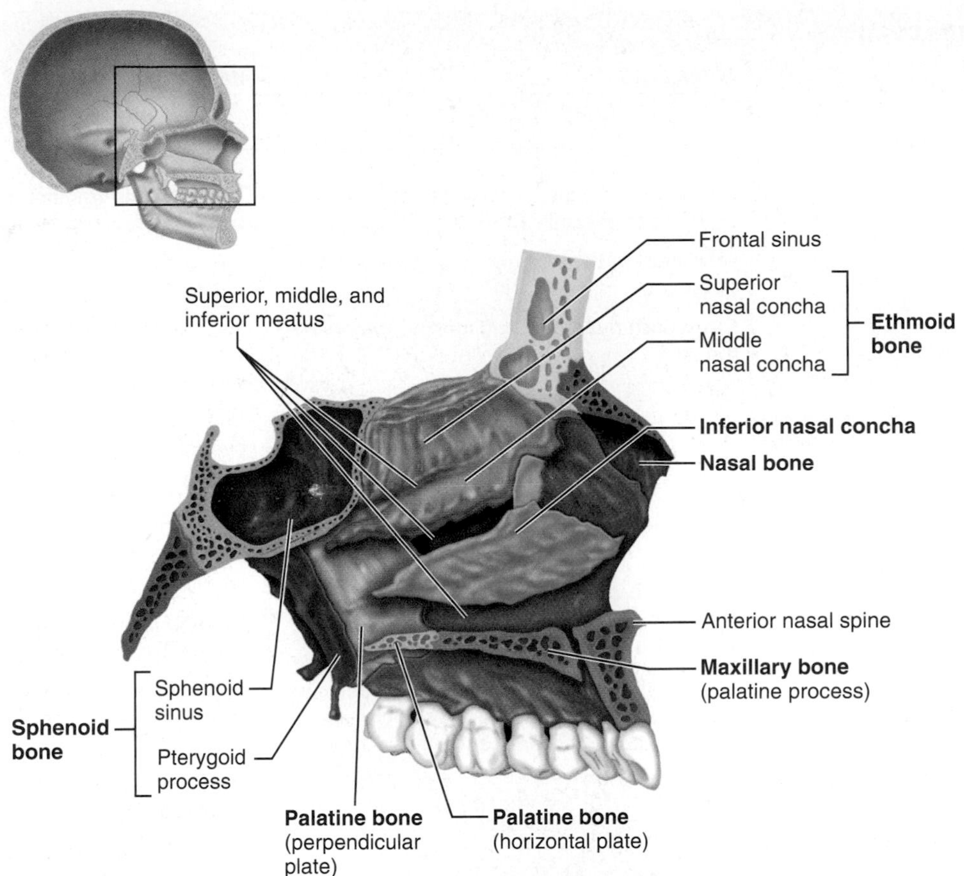

(a) Bones forming the left lateral wall of the nasal cavity (nasal septum removed)

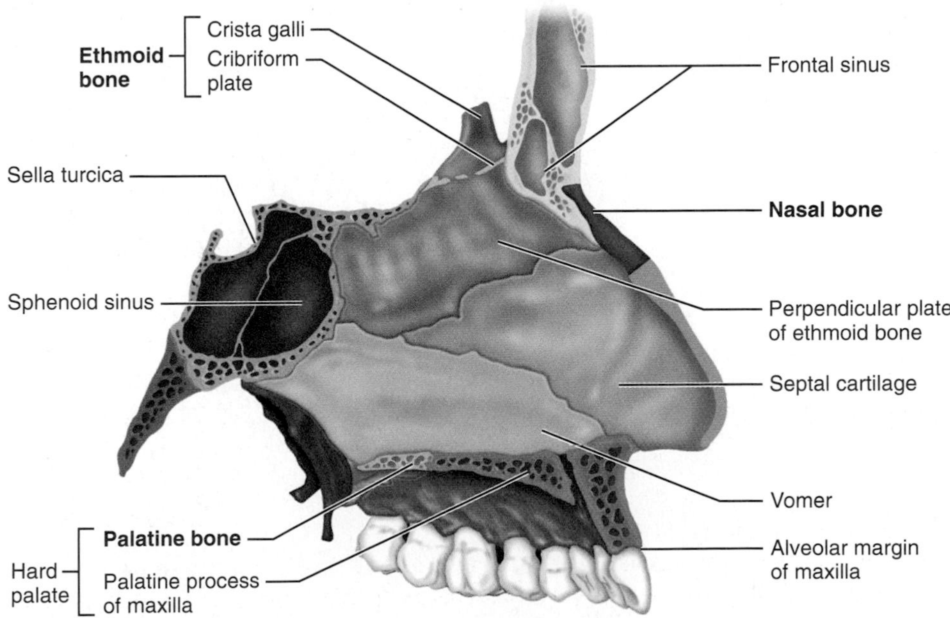

(b) Nasal cavity with septum in place showing the contributions of the ethmoid bone, the vomer, and septal cartilage

Figure 7.14 Bones of the nasal cavity. (See *A Brief Atlas of the Human Body*, Figure 15.)

7

7

TABLE 7.1 **Bones of the Skull**

BONE COLOR CODE*	COMMENTS	IMPORTANT MARKINGS
Cranial Bones		
Frontal (1) (Figures 7.4a, 7.5, and 7.7)	Forms forehead, superior part of orbits, and most of the anterior cranial fossa; contains sinuses	**Supraorbital foramina (notches):** allow the supraorbital arteries and nerves to pass
Parietal (2) (Figures 7.4 and 7.5)	Form most of the superior and lateral aspects of the skull	
Occipital (1) (Figures 7.4b, 7.5, 7.6, and 7.7)	Forms posterior aspect and most of the base of the skull	**Foramen magnum:** allows passage of the spinal cord from the brain stem to the vertebral canal **Hypoglossal canals:** allow passage of the hypoglossal nerves (cranial nerve XII) **Occipital condyles:** articulate with the atlas (first vertebra) **External occipital protuberance** and **nuchal lines:** sites of muscle attachment **External occipital crest:** attachment site of ligamentum nuchae
Temporal (2) (Figures 7.5, 7.6, 7.7, and 7.8)	Form inferolateral aspects of the skull and contributes to the middle cranial fossa; has squamous, mastoid, tympanic, and petrous regions	**Zygomatic process:** helps to form the zygomatic arch, which forms the prominence of the cheek **Mandibular fossa:** articular point of the mandibular condyle **External acoustic meatus:** canal leading from the external ear to the eardrum **Styloid process:** attachment site for several neck muscles and for a ligament to the hyoid bone **Mastoid process:** attachment site for several neck and tongue muscles **Stylomastoid foramen:** allows cranial nerve VII (facial nerve) to pass **Jugular foramen:** allows passage of the internal jugular vein and cranial nerves IX, X, and XI **Internal acoustic meatus:** allows passage of cranial nerves VII and VIII **Carotid canal:** allows passage of the internal carotid artery
Sphenoid (1) (Figures 7.4a, 7.5, 7.6, 7.7, and 7.9)	Keystone of the cranium; contributes to the middle cranial fossa and orbits; main parts are the body, greater wings, lesser wings, and pterygoid processes	**Sella turcica:** hypophyseal fossa portion is the seat of the pituitary gland **Optic canals:** allow passage of optic nerves (cranial nerves II) and the ophthalmic arteries **Superior orbital fissures:** allow passage of cranial nerves III, IV, VI, part of V (ophthalmic division), and ophthalmic vein **Foramen rotundum** (2): allows passage of the maxillary division of cranial nerve V **Foramen ovale** (2): allows passage of the mandibular division of cranial nerve V **Foramen spinosum** (2): allows passage of the middle meningeal artery

TABLE 7.1 *(continued)*

BONE COLOR CODE*	COMMENTS	IMPORTANT MARKINGS
Ethmoid (1) (Figures 7.4a, 7.5, 7.7, 7.10, and 7.14)	Helps to form the anterior cranial fossa; forms part of the nasal septum and the lateral walls and roof of the nasal cavity; contributes to the medial wall of the orbit	**Crista galli:** attachment point for the falx cerebri, a dural membrane fold **Cribriform plates:** allow passage of filaments of the olfactory nerves (cranial nerve I) **Superior** and **middle nasal conchae:** form part of lateral walls of nasal cavity; increase turbulence of air flow
Auditory ossicles (malleus, incus, and stapes) (2 each)	Found in middle ear cavity; involved in sound transmission; see Figure 15.25b, p. 575	
Facial Bones		
Mandible (1) (Figures 7.4a, 7.5, and 7.11a)	The lower jaw	**Coronoid processes:** insertion points for the temporalis muscles **Mandibular condyles:** articulate with the temporal bones in the temporomandibular joints of the jaw **Mandibular symphysis:** medial fusion point of the mandibular bones **Alveoli:** sockets for the teeth **Mandibular foramina:** permit the inferior alveolar nerves to pass **Mental foramina:** allow blood vessels and nerves to pass to the chin and lower lip
Maxilla (2) (Figure 7.4a, 7.5, 7.6, and 7.11b)	Keystone bones of the face; form the upper jaw and parts of the hard palate, orbits, and nasal cavity walls	**Alveoli:** sockets for teeth **Zygomatic processes:** help form the zygomatic arches **Palatine process:** forms the anterior hard palate; meet medially in intermaxillary suture **Frontal process:** forms part of lateral aspect of bridge of nose **Incisive fossa:** permits blood vessels and nerves to pass through anterior hard palate (fused palatine processes) **Inferior orbital fissure:** permits maxillary branch of cranial nerve V, the zygomatic nerve, and blood vessels to pass **Infraorbital foramen:** allows passage of infraorbital nerve to skin of face
Zygomatic (2) (Figures 7.4a, 7.5a, and 7.6a)	Form the cheeks and part of the orbits	
Nasal (2) (Figures 7.4a and 7.5)	Form the bridge of the nose	
Lacrimal (2) (Figures 7.4a and 7.5a)	Form part of the medial orbit walls	**Lacrimal fossa:** houses the lacrimal sac, which helps to drain tears into the nasal cavity
Palatine (2) (Figures 7.5b, 7.6a, and 7.14)	Form posterior part of the hard palate and a small part of nasal cavity and orbit walls	**Median palatine suture:** medial fusion point of the horizontal plates of the palatine bones, which form the posterior part of the hard palate
Vomer (1) (Figures 7.4a and 7.14b)	Inferior part of the nasal septum	
Inferior nasal concha (2) (Figures 7.4a and 7.14a)	Form part of the lateral walls of the nasal cavity	

*The color code beside each bone name corresponds to the bone's color in Figures 7.4 to 7.13. The number in parentheses ( ) following the bone name indicates the total number of such bones in the body.

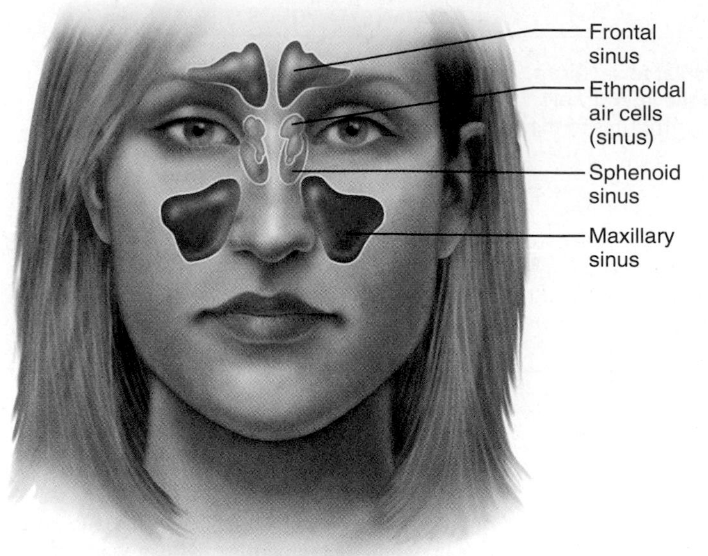

(a) Anterior aspect

7

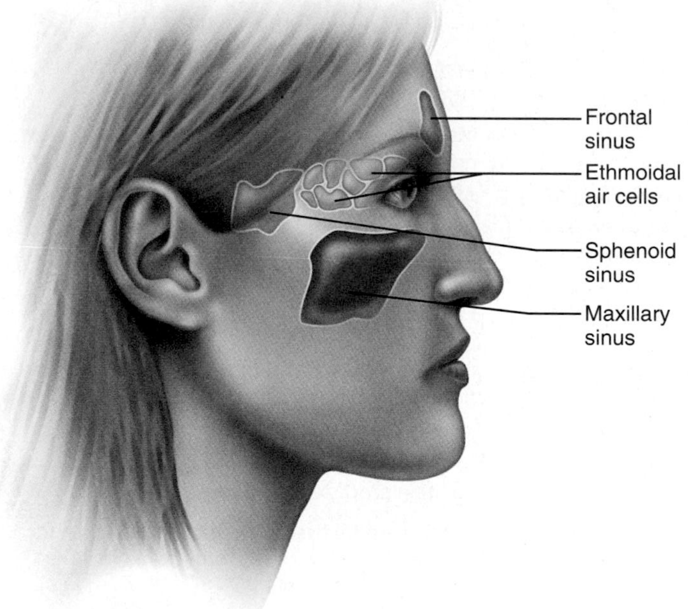

(b) Medial aspect

Figure 7.15 Paranasal sinuses.

the turbulence of air flowing through the nasal cavity. This swirling forces more of the inhaled air into contact with the warm, damp mucosa and encourages trapping of airborne particles (dust, pollen, bacteria) in the sticky mucus.

Paranasal Sinuses

Five skull bones—the frontal, sphenoid, ethmoid, and paired maxillary bones—contain mucosa-lined, air-filled sinuses that give them a rather moth-eaten appearance in an X-ray image. These particular sinuses are called **paranasal sinuses** because they cluster around the nasal cavity **(Figure 7.15)**.

Small openings connect the sinuses to the nasal cavity and act as "two-way streets": Air enters the sinuses from the nasal cavity, and mucus formed by the sinus mucosae drains into the nasal cavity. The mucosa of the sinuses also helps to warm and humidify inspired air. The paranasal sinuses lighten the skull and enhance the resonance of the voice.

CHECK YOUR UNDERSTANDING

10. What bones contain the paranasal sinuses?
11. The perpendicular plates of the palatine bones and the superior and middle conchae of the ethmoid bone form a substantial part of the nasal cavity walls. Which bone forms the roof of that cavity?
12. What bone forms the bulk of the orbit floor and what sense organ is found in the orbit of a living person?

For answers, see Appendix G.

The Vertebral Column

General Characteristics

- ▶ Describe the structure of the vertebral column, list its components, and describe its curvatures.
- ▶ Indicate a common function of the spinal curvatures and the intervertebral discs.

Some people think of the **vertebral column** as a rigid supporting rod, but this is inaccurate. Also called the **spine** or **spinal column**, the vertebral column consists of 26 irregular bones connected in such a way that a flexible, curved structure results **(Figure 7.16)**.

Serving as the axial support of the trunk, the spine extends from the skull to the pelvis, where it transmits the weight of the trunk to the lower limbs. It also surrounds and protects the delicate spinal cord and provides attachment points for the ribs and for the muscles of the back and neck.

In the fetus and infant, the vertebral column consists of 33 separate bones, or **vertebrae** (ver′tĕ-bre). Inferiorly, nine of these eventually fuse to form two composite bones, the sacrum and the tiny coccyx. The remaining 24 bones persist as individual vertebrae separated by intervertebral discs.

Regions and Curvatures

The vertebral column is about 70 cm (28 inches) long in an average adult and has five major regions (Figure 7.16). The seven vertebrae of the neck are the **cervical vertebrae** (ser′vi-kal), the next 12 are the **thoracic vertebrae** (tho-ras′ik), and the five supporting the lower back are the **lumbar vertebrae** (lum′bar). Remembering common meal times—7 AM, 12 noon, and 5 PM—will help you recall the number of bones in these three regions of the spine. The vertebrae become progressively larger from the cervical to the lumbar region, as they must support greater and greater weight.

Inferior to the lumbar vertebrae is the **sacrum** (sa′krum), which articulates with the hip bones of the pelvis. The terminus of the vertebral column is the tiny **coccyx** (kok′siks).

All of us have the same number of cervical vertebrae. Variations in numbers of vertebrae in other regions occur in about 5% of people.

When you view the vertebral column from the side, you can see the four curvatures that give it its S, or sinusoid, shape. The **cervical** and **lumbar curvatures** are concave posteriorly; the **thoracic** and **sacral curvatures** are convex posteriorly. These curvatures increase the resilience and flexibility of the spine, allowing it to function like a spring rather than a rigid rod.

HOMEOSTATIC IMBALANCE

There are several types of abnormal spinal curvatures. Some are congenital (present at birth); others result from disease, poor posture, or unequal muscle pull on the spine. *Scoliosis* (sko″le-o′sis), literally, "twisted disease," is an abnormal *lateral* curvature that occurs most often in the thoracic region. It is quite common during late childhood, particularly in girls, for some unknown reason. Other, more severe cases result from abnormal vertebral structure, lower limbs of unequal length, or muscle paralysis. If muscles on one side of the body are nonfunctional, those of the opposite side exert an unopposed pull on the spine and force it out of alignment. Scoliosis is treated (with body braces or surgically) before growth ends to prevent permanent deformity and breathing difficulties due to a compressed lung.

Kyphosis (ki-fo′sis), or hunchback, is a *dorsally* exaggerated *thoracic* curvature. It is particularly common in elderly people because of osteoporosis, but may also reflect tuberculosis of the spine, rickets, or osteomalacia.

Lordosis, or swayback, is an accentuated *lumbar* curvature. It, too, can result from spinal tuberculosis or osteomalacia. Temporary lordosis is common in those carrying a large load up front, such as men with "potbellies" and pregnant women. In an attempt to preserve their center of gravity, these individuals automatically throw back their shoulders, accentuating their lumbar curvature. ■

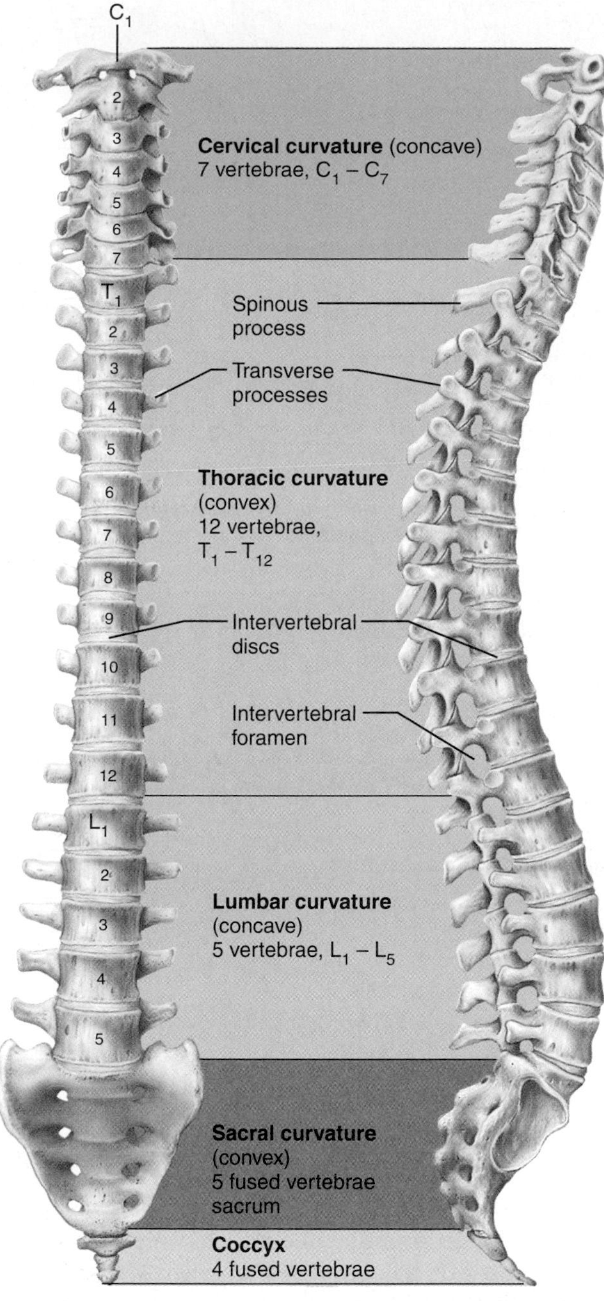

Figure 7.16 The vertebral column. Notice the curvatures in the lateral view. (The terms *convex* and *concave* refer to the curvature of the posterior aspect of the vertebral column.) (See *A Brief Atlas of the Human Body*, Figure 17.)

Ligaments

Like a tall, tremulous TV transmitting tower or cell-phone tower, the vertebral column cannot possibly stand upright by itself. It must be held in place by an elaborate system of cable-like supports. In the case of the vertebral column, straplike ligaments and the trunk muscles assume this role.

The major supporting ligaments are the **anterior** and **posterior longitudinal ligaments** (**Figure 7.17**). These run as continuous bands down the front and back surfaces of the vertebrae from the neck to the sacrum. The broad anterior ligament is strongly attached to both the bony vertebrae and the discs. Along with its supporting role, it prevents hyperextension of the spine (bending too far backward). The posterior ligament, which resists hyperflexion of the spine (bending too sharply forward), is narrow and relatively weak. It attaches only to the discs. However, the **ligamentum flavum**, which connects adjacent vertebrae, contains elastic connective tissue and is especially strong. It stretches as we bend forward and then recoils when we resume an erect posture. Short ligaments connect each vertebra to those immediately above and below.

Intervertebral Discs

Each **intervertebral disc** is a cushionlike pad composed of two parts. The inner gelatinous **nucleus pulposus** (pul-po′sus; "pulp") acts like a rubber ball, giving the disc its elasticity and

7

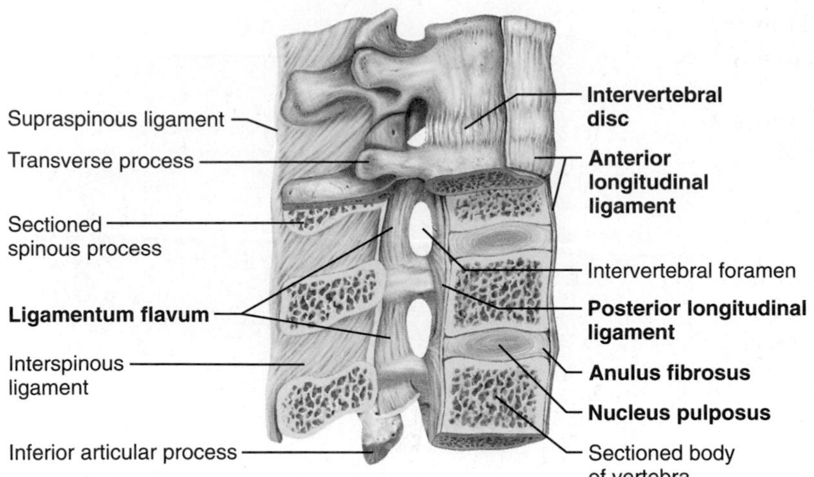

(a) Median section of three vertebrae, illustrating the composition of the discs and the ligaments

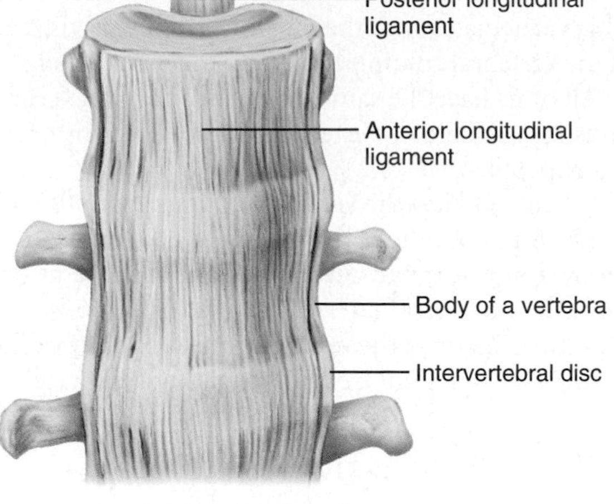

(b) Anterior view of part of the spinal column, showing the anterior longitudinal ligament

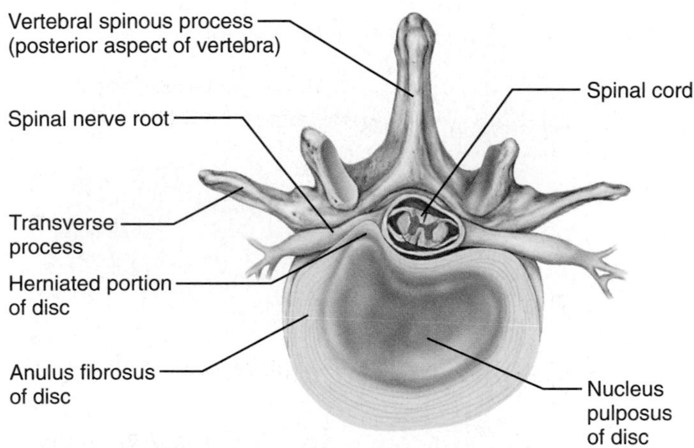

(c) Superior view of a herniated intervertebral disc

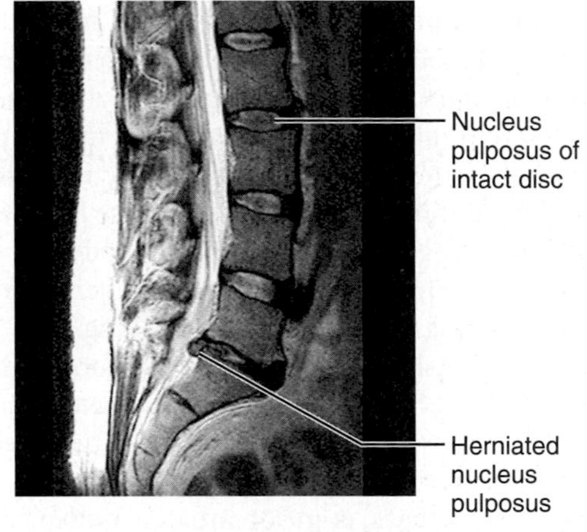

(d) MRI of lumbar region of vertebral column in sagittal section showing herniated disc

Figure 7.17 Ligaments and fibrocartilage discs uniting the vertebrae.

compressibility. Surrounding the nucleus pulposus is a strong collar composed of collagen fibers superficially and fibrocartilage internally, the **anulus fibrosus** (an′u-lus fi-bro′sus; "ring of fibers") (Figure 7.17a, c). The anulus fibrosus limits the expansion of the nucleus pulposus when the spine is compressed. It also acts like a woven strap to bind successive vertebrae together, withstands twisting forces, and resists tension in the spine.

Sandwiched between the bodies of neighboring vertebrae, the intervertebral discs act as shock absorbers during walking, jumping, and running. They allow the spine to flex and extend, and to a lesser extent to bend laterally. At points of compression, the discs flatten and bulge out a bit between the vertebrae. The discs are thickest in the lumbar and cervical regions, which enhances the flexibility of these regions.

Collectively the discs account for about 25% of the height of the vertebral column. They flatten somewhat during the course of the day, so we are always a few millimeters shorter at night than when we awake in the morning.

HOMEOSTATIC IMBALANCE

Severe or sudden physical trauma to the spine—for example, from bending forward while lifting a heavy object—may result in herniation of one or more discs. A **herniated (prolapsed) disc** (commonly called a *slipped disc*) usually involves rupture of the anulus fibrosus followed by protrusion of the spongy nucleus pulposus through the anulus (Figure 7.17c, d). If the protrusion presses on the spinal cord or on spinal nerves exiting from the cord, numbness or excruciating pain may result.

Herniated discs are generally treated with moderate exercise, massage, heat therapy, and painkillers. If this fails, the protruding disc may have to be removed surgically and a bone graft done to fuse the adjoining vertebrae. For those preferring to avoid general anesthesia, the disc can be partially vaporized with a laser in an outpatient procedure called percutaneous laser disc decompression that takes only 30 to 40 minutes. If necessary, tears in the anulus can be sealed by electrothermal means at the same time. The patient leaves with only an adhesive bandage to mark the spot. ■

CHECK YOUR UNDERSTANDING

13. What are the five major regions of the vertebral column?

14. In which two of these regions is the vertebral column concave posteriorly?

15. Besides the spinal curvatures, which skeletal elements help to make the vertebral column flexible?

For answers, see Appendix G.

General Structure of Vertebrae

▶ Discuss the structure of a typical vertebra and describe regional features of cervical, thoracic, and lumbar vertebrae.

All vertebrae have a common structural pattern (**Figure 7.18**). Each vertebra consists of a **body**, or **centrum**, anteriorly and a **vertebral arch** posteriorly. The disc-shaped body is the weight-bearing region. Together, the body and vertebral arch enclose an opening called the **vertebral foramen**. Successive vertebral foramina of the articulated vertebrae form the long **vertebral canal**, through which the spinal cord passes.

The vertebral arch is a composite structure formed by two pedicles and two laminae. The **pedicles** (ped′ĭ-kelz; "little feet"), short bony pillars projecting posteriorly from the vertebral body, form the sides of the arch. The **laminae** (lam′ĭ-ne), flattened plates that fuse in the median plane, complete the arch posteriorly. The pedicles have notches on their superior and inferior borders, providing lateral openings between adjacent vertebrae called **intervertebral foramina** (see Figure 7.16). The spinal nerves issuing from the spinal cord pass through these foramina.

Seven processes project from the vertebral arch. The **spinous process** is a median posterior projection arising at the junction of the two laminae. A **transverse process** extends laterally from each side of the vertebral arch. The spinous and transverse processes are attachment sites for muscles that move the vertebral column and for ligaments that stabilize it. The paired **superior** and **inferior articular processes** protrude superiorly and inferiorly, respectively, from the pedicle-lamina junctions. The smooth joint surfaces of the articular processes, called *facets* ("little faces"), are covered with hyaline cartilage. The inferior articular processes of each vertebra form movable joints with the superior articular processes of the vertebra immediately below. Thus, successive vertebrae join both at their bodies and at their articular processes.

Regional Vertebral Characteristics

Beyond their common structural features, vertebrae exhibit variations that allow different regions of the spine to perform slightly different functions and movements. In general, movements that can occur between vertebrae are (1) flexion and extension (anterior bending and posterior straightening of the spine), (2) lateral flexion (bending the *upper body* to the right or left), and (3) rotation (in which vertebrae rotate on one another in the longitudinal axis of the spine). The regional vertebral characteristics described in this section are illustrated and summarized in Table 7.2 on p. 222.

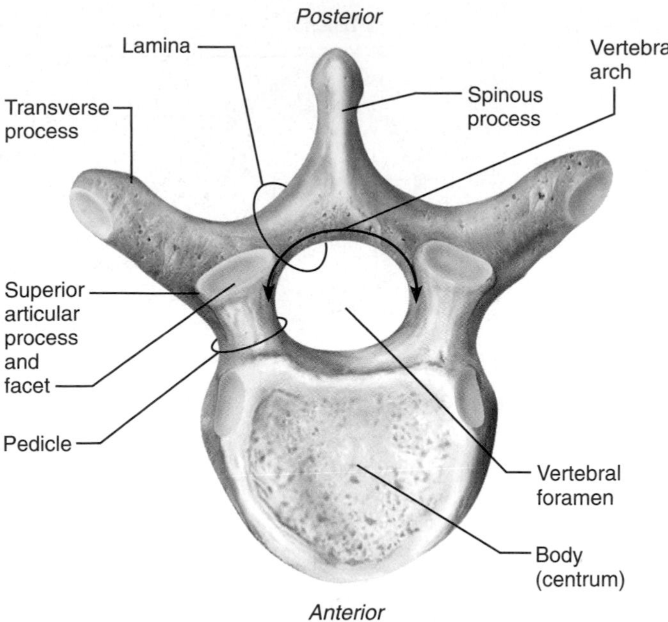

Figure 7.18 Structure of a typical vertebra. Superior view. Only bone features are illustrated in this and subsequent bone figures in this chapter. Articular cartilage is not depicted.

Cervical Vertebrae

The seven cervical vertebrae, identified as C_1–C_7, are the smallest, lightest vertebrae (see Figure 7.16). The first two (C_1 and C_2) are unusual and we will skip them for the moment. The "typical" cervical vertebrae (C_3–C_7) have the following distinguishing features (see Figure 7.20 and Table 7.2):

1. The body is oval—wider from side to side than in the anteroposterior dimension.
2. Except in C_7, the spinous process is short, projects directly back, and is *bifid* (bi′fid), or split at its tip.
3. The vertebral foramen is large and generally triangular.
4. Each transverse process contains a **transverse foramen** through which the vertebral arteries pass to service the brain.

The spinous process of C_7 is not bifid and is much larger than those of the other cervical vertebrae (see Figure 7.20a). Because its spinous process is palpable through the skin, C_7 can be used as a landmark for counting the vertebrae and is called the **vertebra prominens** ("prominent vertebra").

The first two cervical vertebrae, the atlas and the axis, are somewhat more robust than the typical cervical vertebra. They have no intervertebral disc between them, and they are highly modified, reflecting their special functions. The **atlas** (C_1) has no body and no spinous process (**Figure 7.19a and b**). Essentially, it is a ring of bone consisting of *anterior* and *posterior arches* and a *lateral mass* on each side. Each lateral mass has articular facets on both its superior and inferior surfaces. The superior articular facets receive the occipital condyles of the skull—they "carry" the skull, just as Atlas supported the heavens in Greek mythology. These joints allow you to nod "yes." The inferior articular facets form joints with the axis (C_2) below.

7

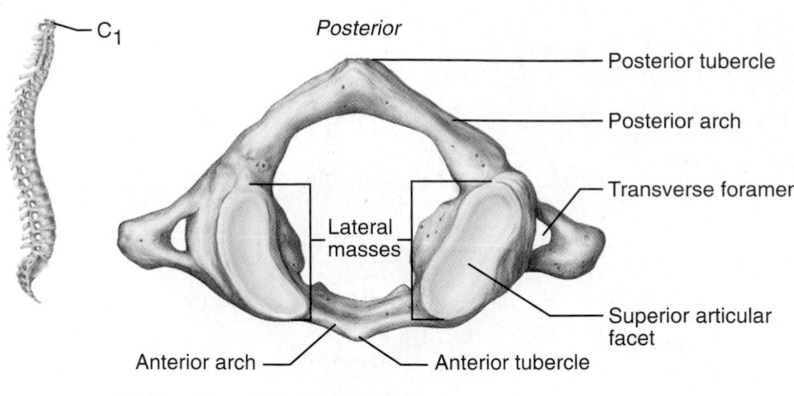

(a) Superior view of atlas (C_1)

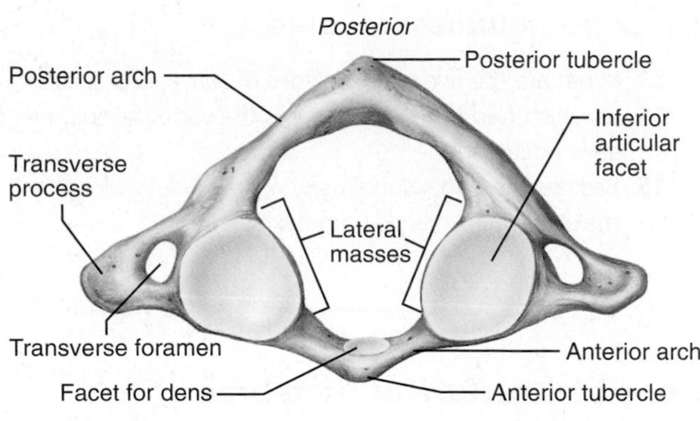

(b) Inferior view of atlas (C_1)

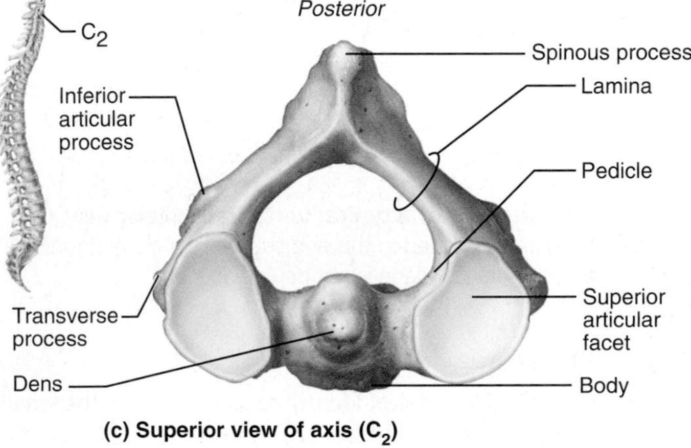

(c) Superior view of axis (C_2)

Figure 7.19 The first and second cervical vertebrae. (See *A Brief Atlas of the Human Body*, Figure 18.)

The **axis**, which has a body and the other typical vertebral processes, is not as specialized as the atlas. In fact, its only unusual feature is the knoblike **dens** (denz; "tooth") projecting superiorly from its body. The dens is actually the "missing" body of the atlas, which fuses with the axis during embryonic development. Cradled in the anterior arch of the atlas by the transverse ligaments **(Figure 7.20a)**, the dens acts as a pivot for the rotation of the atlas. Hence, this joint allows you to rotate your head from side to side to indicate "no."

Thoracic Vertebrae

The 12 thoracic vertebrae (T_1–T_{12}) all articulate with the ribs (see Table 7.2, Figure 7.16, and Figure 7.20b). The first looks much like C_7, and the last four show a progression toward lumbar vertebral structure. The thoracic vertebrae increase in size from the first to the last. Unique characteristics of these vertebrae include the following:

1. The body is roughly heart shaped. It typically bears two small *facets*, commonly called *demifacets* (half-facets), on each side, one at the superior edge (the *superior costal facet*) and the other at the inferior edge (the *inferior costal facet*). The demifacets receive the heads of the ribs. (The bodies of T_{10}–T_{12} vary from this pattern by having only a single facet to receive their respective ribs.)
2. The vertebral foramen is circular.
3. The spinous process is long and points sharply downward.
4. With the exception of T_{11} and T_{12}, the transverse processes have facets, the *transverse costal facets*, that articulate with the tubercles of the ribs.
5. The superior and inferior articular facets lie mainly in the frontal plane, a situation that prevents flexion and extension, but which allows this region of the spine to rotate. Lateral flexion, though possible, is restricted by the ribs.

Lumbar Vertebrae

The lumbar region of the vertebral column, commonly referred to as the small of the back, receives the most stress. The enhanced weight-bearing function of the five lumbar vertebrae (L_1–L_5) is reflected in their sturdier structure. Their bodies are massive and kidney shaped in a superior view (see Table 7.2, Figure 7.16, and Figure 7.20). Other characteristics typical of these vertebrae:

1. The pedicles and laminae are shorter and thicker than those of other vertebrae.
2. The spinous processes are short, flat, and hatchet shaped and are easily seen when a person bends forward. These processes are robust and project directly backward, adaptations for the attachment of the large back muscles.
3. The vertebral foramen is triangular.
4. The orientation of the facets of the articular processes of the lumbar vertebrae differs substantially from that of the other vertebra types (see Table 7.2). These modifications lock the lumbar vertebrae together and provide stability by preventing rotation of the lumbar spine. Flexion and extension are possible (as when you do sit-ups), as is lateral flexion.

Sacrum

The triangular sacrum, which shapes the posterior wall of the pelvis, is formed by five fused vertebrae (S_1–S_5) in adults (**Figure 7.21**, and see Figure 7.16). It articulates superiorly (via its **superior articular processes**) with L_5 and inferiorly with the coccyx. Laterally, the sacrum articulates, via its **auricular surfaces**, with the two hip bones to form the **sacroiliac joints** (sa″kro-il′e-ak) of the pelvis.

The **sacral promontory** (prom′on-tor″e; "high point of land projecting into the sea"), the anterosuperior margin of the first

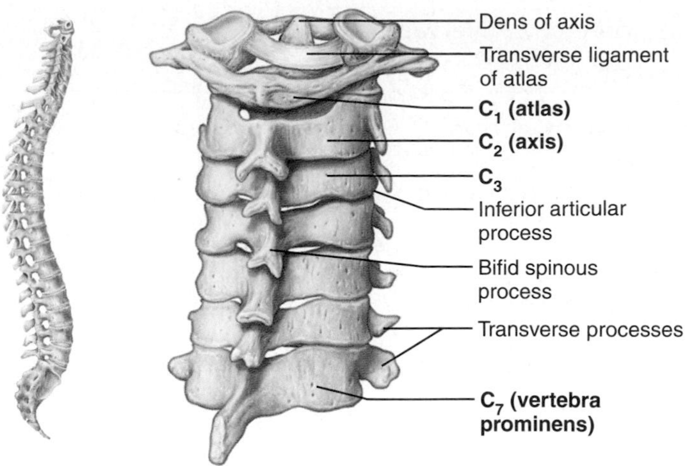

(a) Cervical vertebrae

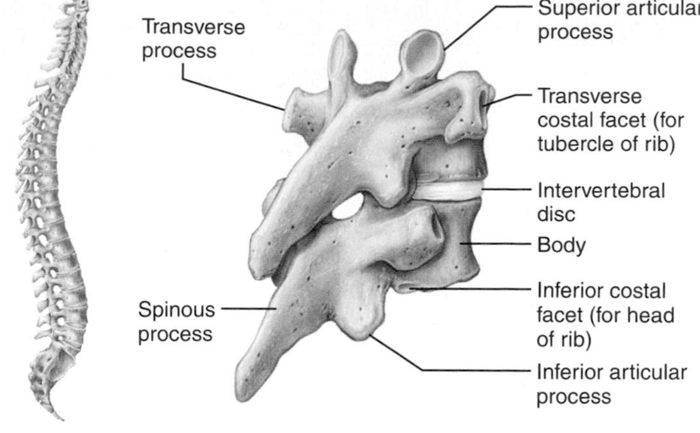

(b) Thoracic vertebrae

Superior articular process

Transverse process

Body

Intervertebral disc

Inferior articular process

Spinous process

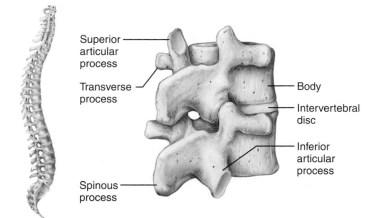

(c) Lumbar vertebrae

Figure 7.20 Posterolateral views of articulated vertebrae. Notice the bulbous tip on the spinous process of C_7, the vertebra prominens. (See *A Brief Atlas of the Human Body*, Figures 19, 20, and 21.)

Sacral promontory

Ala

Body of first sacral vertebra

Transverse ridges (sites of vertebral fusion)

Anterior sacral foramina

Apex

Coccyx

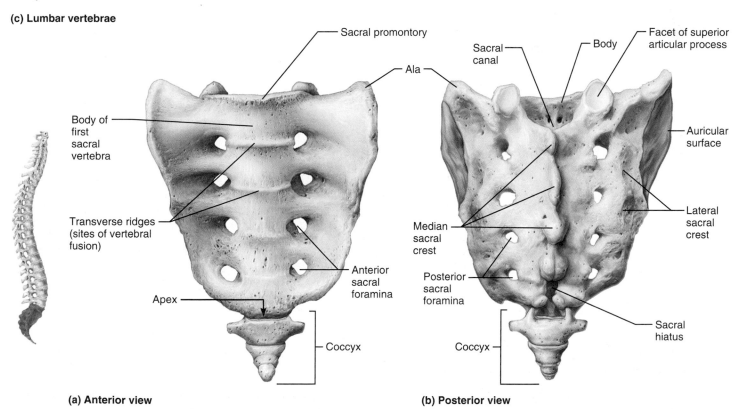

(a) Anterior view

Sacral canal

Body

Facet of superior articular process

Auricular surface

Lateral sacral crest

Median sacral crest

Posterior sacral foramina

Sacral hiatus

Coccyx

(b) Posterior view

Figure 7.21 The sacrum and coccyx. (See *A Brief Atlas of the Human Body*, Figure 22.)

7

TABLE 7.2 Regional Characteristics of Cervical, Thoracic, and Lumbar Vertebrae

CHARACTERISTIC	CERVICAL (3–7)	THORACIC	LUMBAR
Body	Small, wide side to side	Larger than cervical; heart shaped; bears two costal facets	Massive; kidney shaped
Spinous process	Short; bifid; projects directly posteriorly	Long; sharp; projects inferiorly	Short; blunt; rectangular; projects directly posteriorly
Vertebral foramen	Triangular	Circular	Triangular
Transverse processes	Contain foramina	Bear facets for ribs (except T_{11} and T_{12})	Thin and tapered
Superior and inferior articulating processes	Superior facets directed superoposteriorly	Superior facets directed posteriorly	Superior facets directed posteromedially (or medially)
	Inferior facets directed inferoanteriorly	Inferior facets directed anteriorly	Inferior facets directed anterolaterally (or laterally)
Movements allowed	Flexion and extension; lateral flexion; rotation; the spine region with the greatest range of movement	Rotation; lateral flexion possible but restricted by ribs; flexion and extension limited	Flexion and extension; some lateral flexion; rotation prevented

Superior View

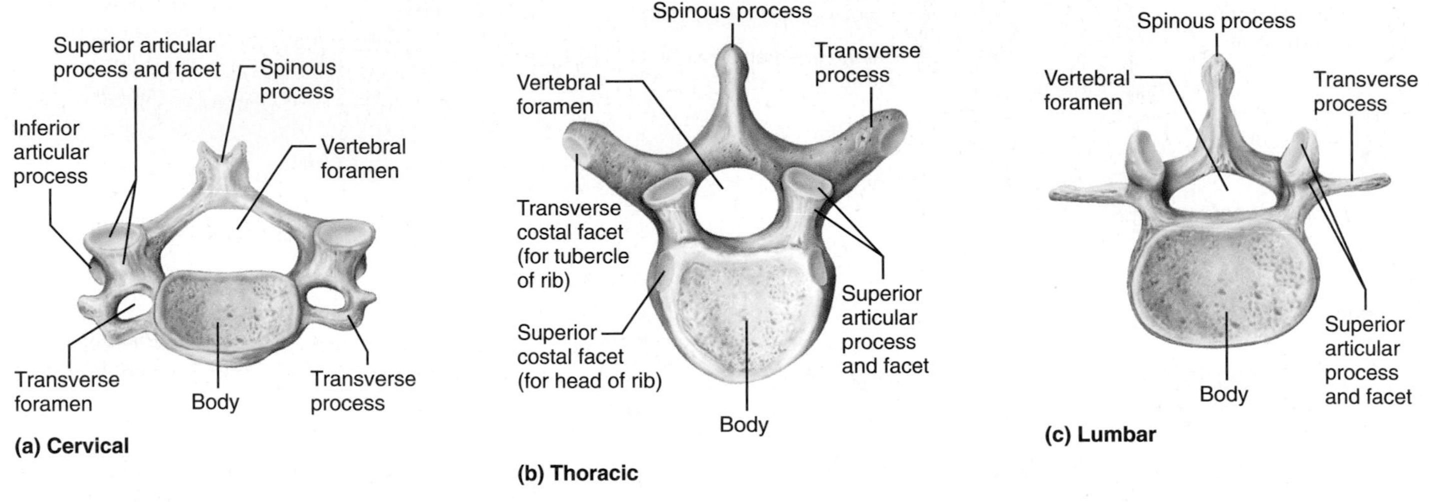

(a) Cervical (b) Thoracic (c) Lumbar

Right Lateral View

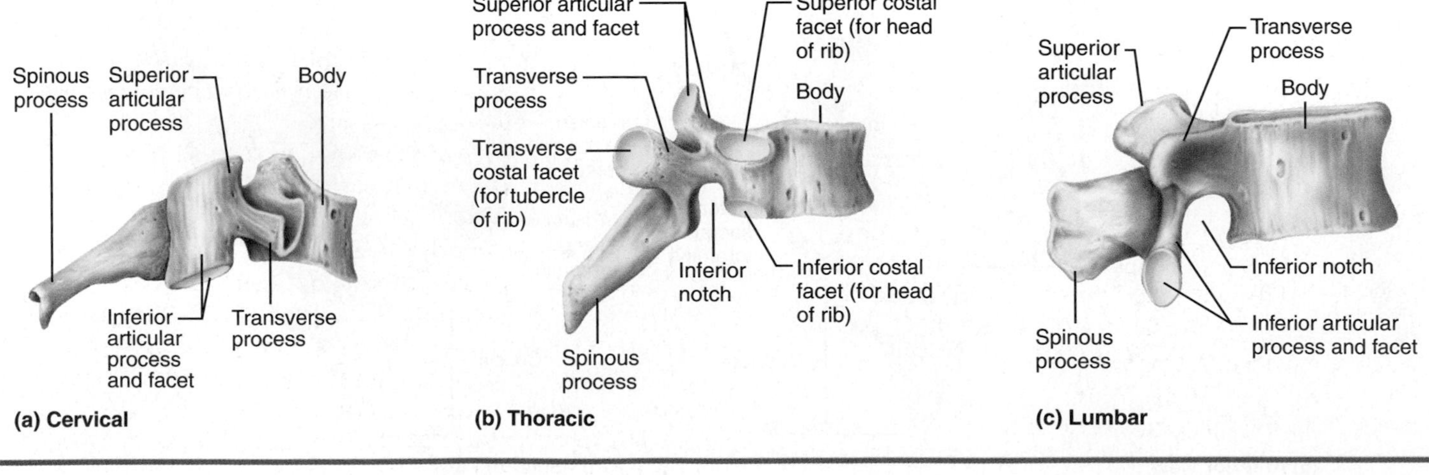

(a) Cervical (b) Thoracic (c) Lumbar

sacral vertebra, bulges anteriorly into the pelvic cavity. The body's center of gravity lies about 1 cm posterior to this landmark. Four ridges, the **transverse ridges**, cross its concave anterior aspect, marking the lines of fusion of the sacral vertebrae. The **anterior sacral foramina** lie at the lateral ends of these ridges and transmit blood vessels and anterior rami of the sacral spinal nerves. The regions lateral to these foramina expand superiorly as the winglike **alae**.

In its posterior midline the sacral surface is roughened by the **median sacral crest** (the fused spinous processes of the sacral vertebrae). This is flanked laterally by the **posterior sacral foramina**, which transmit the posterior rami of the sacral spinal nerves, and then the **lateral sacral crests** (remnants of the transverse processes of S_1–S_5).

The vertebral canal continues inside the sacrum as the **sacral canal.** Since the laminae of the fifth (and sometimes the fourth) sacral vertebrae fail to fuse medially, an enlarged external opening called the **sacral hiatus** (hi-a′tus; "gap") is obvious at the inferior end of the sacral canal.

Coccyx

The coccyx, our tailbone, is a small triangular bone (Figure 7.21, and see Figure 7.16). It consists of four (or in some cases three or five) vertebrae fused together. The coccyx articulates superiorly with the sacrum. (The name *coccyx* is from the Greek word meaning "cuckoo" and was so named because of its fancied resemblance to a bird's beak.) Except for the slight support the coccyx affords the pelvic organs, it is a nearly useless bone. Occasionally, a baby is born with an unusually long coccyx, which may need to be removed surgically.

■ ■ ■

The characteristics of the regional vertebrae are summarized in **Table 7.2.**

CHECK YOUR UNDERSTANDING

16. What is the normal number of cervical vertebrae? Of thoracic vertebrae?
17. How would a complete fracture of the dens affect the mobility of the vertebral column?
18. How can you distinguish a lumbar vertebra from a thoracic vertebra?

For answers, see Appendix G.

The Thoracic Cage

- Name and describe the bones of the thoracic cage (bony thorax).
- Differentiate true from false ribs.

Anatomically, the thorax is the chest, and its bony underpinnings are called the **thoracic cage** or **bony thorax.** Elements of the thoracic cage include the thoracic vertebrae dorsally, the ribs laterally, and the sternum and costal cartilages anteriorly. The costal cartilages secure the ribs to the sternum **(Figure 7.22a).**

Roughly cone shaped with its broad dimension positioned inferiorly, the bony thorax forms a protective cage around the vital organs of the thoracic cavity (heart, lungs, and great blood vessels), supports the shoulder girdles and upper limbs, and provides attachment points for many muscles of the neck, back, chest, and shoulders. The *intercostal spaces* between the ribs are occupied by the intercostal muscles, which lift and depress the thorax during breathing.

Sternum

The **sternum** (breastbone) lies in the anterior midline of the thorax. Vaguely resembling a dagger, it is a flat bone approximately 15 cm (6 inches) long, resulting from the fusion of three bones: the manubrium, the body, and the xiphoid process. The *manubrium* (mah-nu′bre-um; "knife handle"), is the superior portion which is shaped like the knot in a necktie. The manubrium articulates via its **clavicular notches** (klah-vik′u-lar) with the clavicles (collarbones) laterally, and just below this, it also articulates with the first two pairs of ribs. The *body*, or midportion, forms the bulk of the sternum. The sides of the body are notched where it articulates with the costal cartilages of the second to seventh ribs. The *xiphoid process* (zif′oid; "swordlike") forms the inferior end of the sternum. This small, variably shaped process is a plate of hyaline cartilage in youth, but it is usually ossified in adults over the age of 40. The xiphoid process articulates only with the sternal body and serves as an attachment point for some abdominal muscles.

HOMEOSTATIC IMBALANCE

In some people, the xiphoid process projects posteriorly. In such cases, blows to the chest can push the xiphoid into the underlying heart or liver, causing massive hemorrhage. ■

The sternum has three important anatomical landmarks: the jugular notch, the sternal angle, and the xiphisternal joint (Figure 7.22). The easily palpated **jugular** (*suprasternal*) **notch** is the central indentation in the superior border of the manubrium. If you slide your finger down the anterior surface of your neck, it will land in the jugular notch. The jugular notch is generally in line with the disc between the second and third thoracic vertebrae and the point where the left common carotid artery issues from the aorta (Figure 7.22b).

The **sternal angle** is felt as a horizontal ridge across the front of the sternum, where the manubrium joins the sternal body. This cartilaginous joint acts like a hinge, allowing the sternal body to swing anteriorly when we inhale. The sternal angle is in line with the disc between the fourth and fifth thoracic vertebrae and at the level of the second pair of ribs. It is a handy reference point for finding the second rib and thus for counting the ribs during a physical examination and for listening to sounds made by specific heart valves.

The **xiphisternal joint** (zif″ĭ-ster′nul) is the point where the sternal body and xiphoid process fuse. It lies at the level of the

7

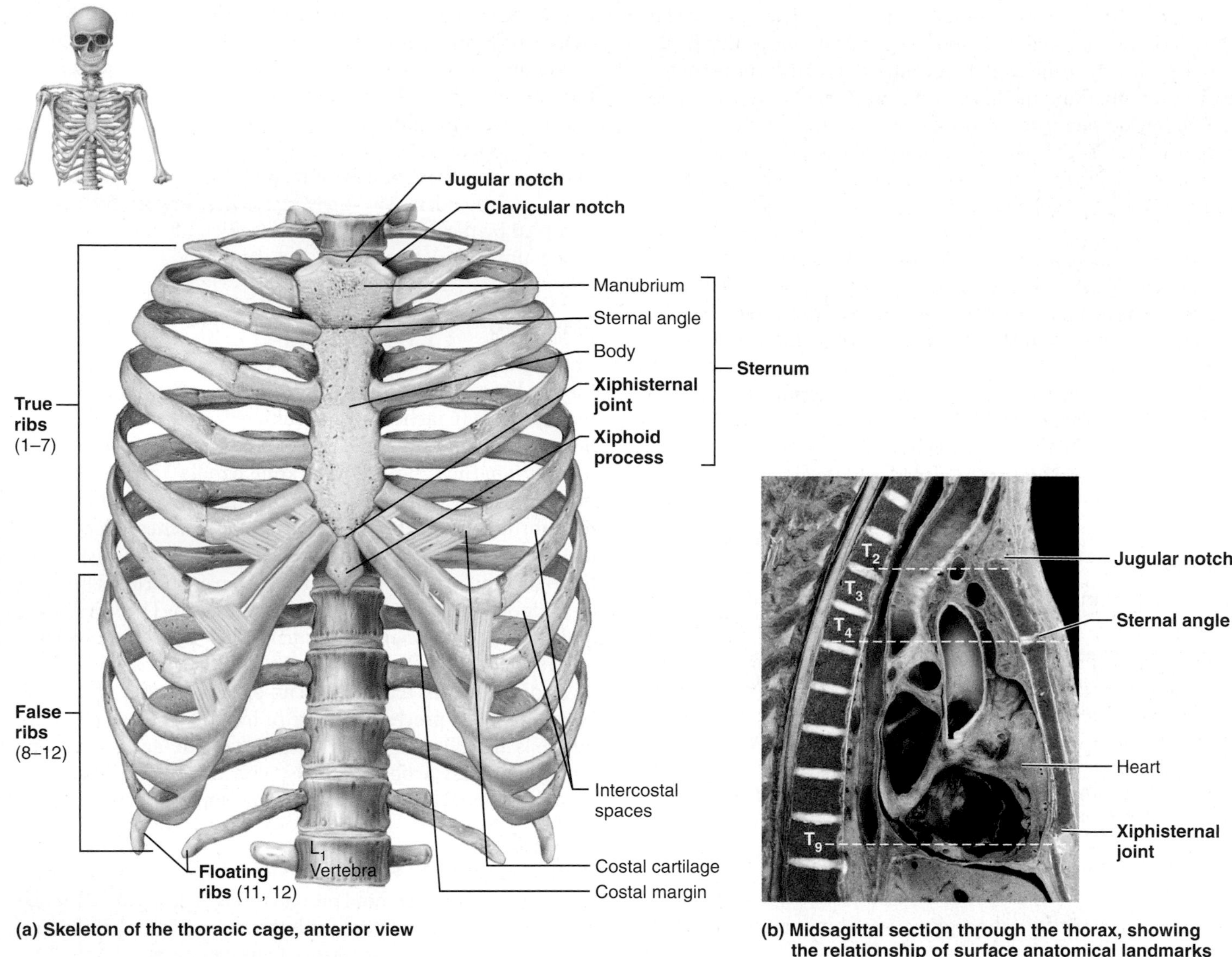

(a) Skeleton of the thoracic cage, anterior view

(b) Midsagittal section through the thorax, showing the relationship of surface anatomical landmarks of the thorax to the vertebral column

Figure 7.22 The thoracic cage. (See *A Brief Atlas of the Human Body*, Figure 23a–d.)

ninth thoracic vertebra. The heart lies on the diaphragm just deep to this joint.

Ribs

Twelve pairs of **ribs** form the flaring sides of the thoracic cage (Figure 7.22a). All ribs attach posteriorly to the thoracic vertebrae (bodies and transverse processes) and curve inferiorly toward the anterior body surface. The superior seven rib pairs attach directly to the sternum by individual costal cartilages (bars of hyaline cartilage). These are **true** or **vertebrosternal ribs** (ver″tĕ-bro-ster′nal). (Notice that the anatomical name indicates the two attachment points of a rib—the posterior attachment given first.)

The remaining five pairs of ribs are called **false ribs** because they either attach indirectly to the sternum or entirely lack a sternal attachment. Rib pairs 8–10 attach to the sternum indirectly, each joining the costal cartilage immediately above it. These ribs are also called **vertebrochondral ribs** (ver″tĕ-bro-kon′dral). The inferior margin of the rib cage, or **costal margin**, is formed by the costal cartilages of ribs 7–10. Rib pairs 11 and 12 are called **vertebral ribs** or **floating ribs** because they have no anterior attachments. Instead, their costal cartilages lie embedded in the muscles of the lateral body wall.

The ribs increase in length from pair 1 to pair 7, then decrease in length from pair 8 to pair 12. Except for the first rib, which lies deep to the clavicle, the ribs are easily felt in people of normal weight.

A typical rib is a bowed flat bone **(Figure 7.23)**. The bulk of a rib is simply called the *shaft*. Its superior border is smooth, but its inferior border is sharp and thin and has a *costal groove* on its inner face that lodges the intercostal nerves and blood vessels.

In addition to the shaft, each rib has a head, neck, and tubercle. The wedge-shaped *head*, the posterior end, articulates with the vertebral bodies by two facets: One joins the body of the same-numbered thoracic vertebra, the other articulates with the body of the vertebra immediately superior. The *neck* is the

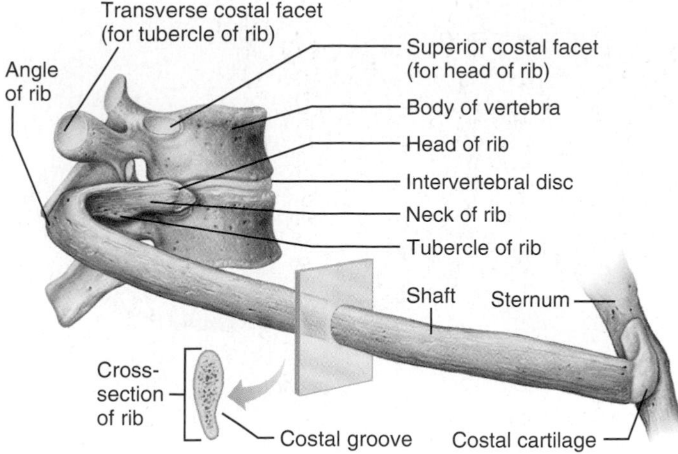

(a) Vertebral and sternal articulations of a typical true rib

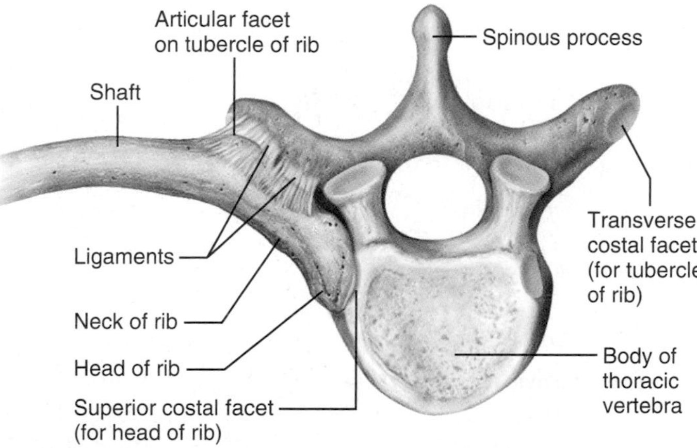

(b) Superior view of the articulation between a rib and a thoracic vertebra

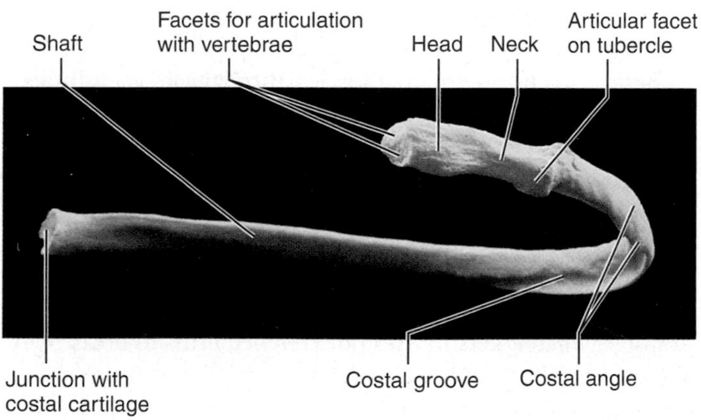

(c) A typical rib (rib 6, right), posterior view

Figure 7.23 Ribs. All ribs illustrated in this figure are right ribs. (See *A Brief Atlas of the Human Body*, Figure 23e and f.)

constricted portion of the rib just beyond the head. Lateral to this, the knoblike *tubercle* articulates with the costal facet of the transverse process of the same-numbered thoracic vertebra. Beyond the tubercle, the shaft angles sharply forward (at the angle of the rib) and then extends to attach to its costal cartilage anteriorly. The costal cartilages provide secure but flexible rib attachments to the sternum.

The first pair of ribs is quite atypical. They are flattened superiorly to inferiorly and are quite broad, forming a horizontal table that supports the subclavian blood vessels that serve the upper limbs. There are also other exceptions to the typical rib pattern. Rib 1 and ribs 10–12 articulate with only one vertebral body, and ribs 11 and 12 do not articulate with a vertebral transverse process.

CHECK YOUR UNDERSTANDING

19. How does a true rib differ from a false rib?
20. What is the sternal angle and what is its clinical importance?
21. Besides the ribs and sternum, there is a third group of bones making up the thoracic cage. What is it?

For answers, see Appendix G.

7

PART 2

THE APPENDICULAR SKELETON

Bones of the limbs and their girdles are collectively called the **appendicular skeleton** because they are *appended* to the axial skeleton that forms the longitudinal axis of the body (see Figure 7.1). The yokelike *pectoral girdles* (pek′tor-al; "chest") attach the upper limbs to the body trunk. The more sturdy *pelvic girdle* secures the lower limbs. Although the bones of the upper and lower limbs differ in their functions and mobility, they have the same fundamental plan: Each limb is composed of three major segments connected by movable joints.

The appendicular skeleton enables us to carry out the movements typical of our freewheeling and manipulative lifestyle. Each time we take a step, throw a ball, or pop a caramel into our mouth, we are making good use of our appendicular skeleton.

The Pectoral (Shoulder) Girdle

- ▶ Identify bones forming the pectoral girdle and relate their structure and arrangement to the function of this girdle.
- ▶ Identify important bone markings on the pectoral girdle.

The **pectoral**, or **shoulder**, **girdle** consists of the *clavicle* (klav′ĭ-kl) anteriorly and the *scapula* (skap′u-lah) posteriorly (**Figure 7.24** and Table 7.3 on p. 232). The paired pectoral girdles and their associated muscles form your shoulders. Although the term *girdle* usually signifies a beltlike structure encircling the body, a single pectoral girdle, or even the pair, does not quite satisfy this description. Anteriorly, the medial end of each clavicle joins the sternum; the distal ends of the clavicles meet the scapulae laterally. However, the scapulae fail to complete the ring posteriorly, because their medial borders do not join each other or the axial skeleton. Instead, the scapulae are attached to the thorax and vertebral column only by the muscles that clothe their surfaces.

The pectoral girdles attach the upper limbs to the axial skeleton and provide attachment points for many of the muscles that

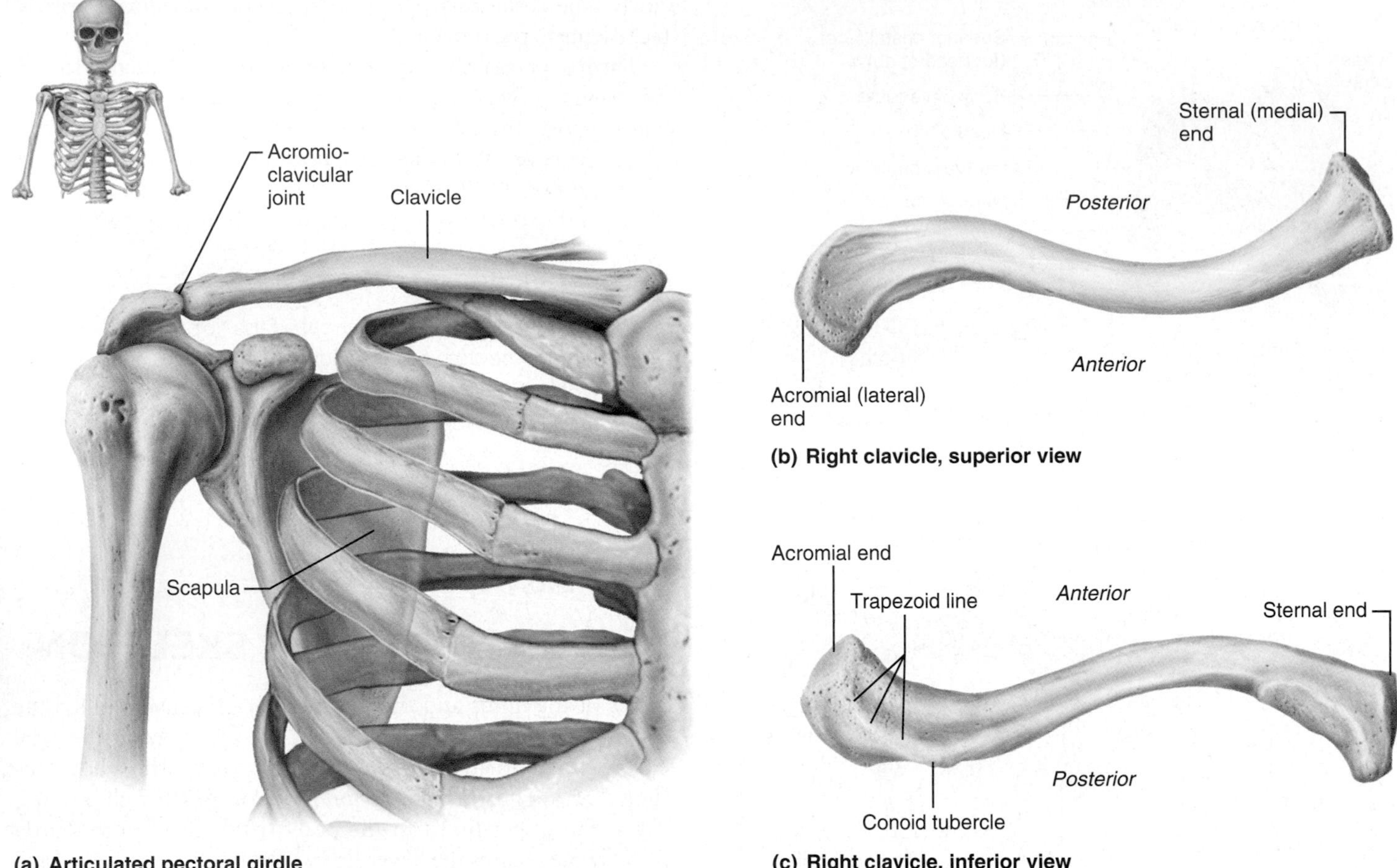

Figure 7.24 The pectoral girdle and clavicle. (See *A Brief Atlas of the Human Body*, Figure 24.)

7

move the upper limbs. These girdles are very light and allow the upper limbs a degree of mobility not seen anywhere else in the body. This mobility is due to the following factors:

1. Because only the clavicle attaches to the axial skeleton, the scapula can move quite freely across the thorax, allowing the arm to move with it.
2. The socket of the shoulder joint (the scapula's glenoid cavity) is shallow and poorly reinforced, so it does not restrict the movement of the humerus (arm bone). Although this arrangement is good for flexibility, it is bad for stability: Shoulder dislocations are fairly common.

Clavicles

The **clavicles** ("little keys"), or collarbones, are slender, doubly curved bones that can be felt along their entire course as they extend horizontally across the superior thorax (Figure 7.24). Each clavicle is cone shaped at its medial **sternal end**, which attaches to the sternal manubrium, and flattened at its lateral **acromial end** (ah-kro′me-al), which articulates with the scapula. The medial two-thirds of the clavicle is convex anteriorly; its lateral third is concave anteriorly. Its superior surface is fairly smooth, but the inferior surface is ridged and grooved by ligaments and by the action of the muscles that attach to it. The *trapezoid line* and the *conoid tubercle*, for example, are anchoring points for a ligament which runs to attach to the scapula.

Besides anchoring many muscles, the clavicles act as braces: They hold the scapulae and arms out laterally, away from the narrower superior part of the thorax. This bracing function becomes obvious when a clavicle is fractured: The entire shoulder region collapses medially. The clavicles also transmit compression forces from the upper limbs to the axial skeleton, for example, when someone pushes a car to a gas station.

The clavicles are not very strong and are likely to fracture, for example, when a person uses outstretched arms to break a fall. The curves in the clavicle ensure that it usually fractures anteriorly (outward). If it were to collapse posteriorly (inward), bone splinters would damage the subclavian artery, which passes just deep to the clavicle to serve the upper limb. The clavicles are exceptionally sensitive to muscle pull and become noticeably larger and stronger in those who perform manual labor or athletics involving the shoulder and arm muscles.

Scapulae

The **scapulae**, or *shoulder blades*, are thin, triangular flat bones (Figure 7.24a and **Figure 7.25**). Interestingly, their name derives from a word meaning "spade" or "shovel," for ancient cultures

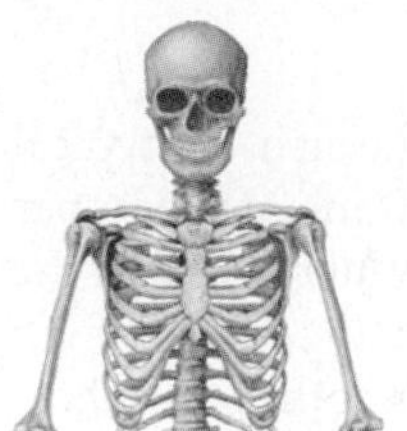

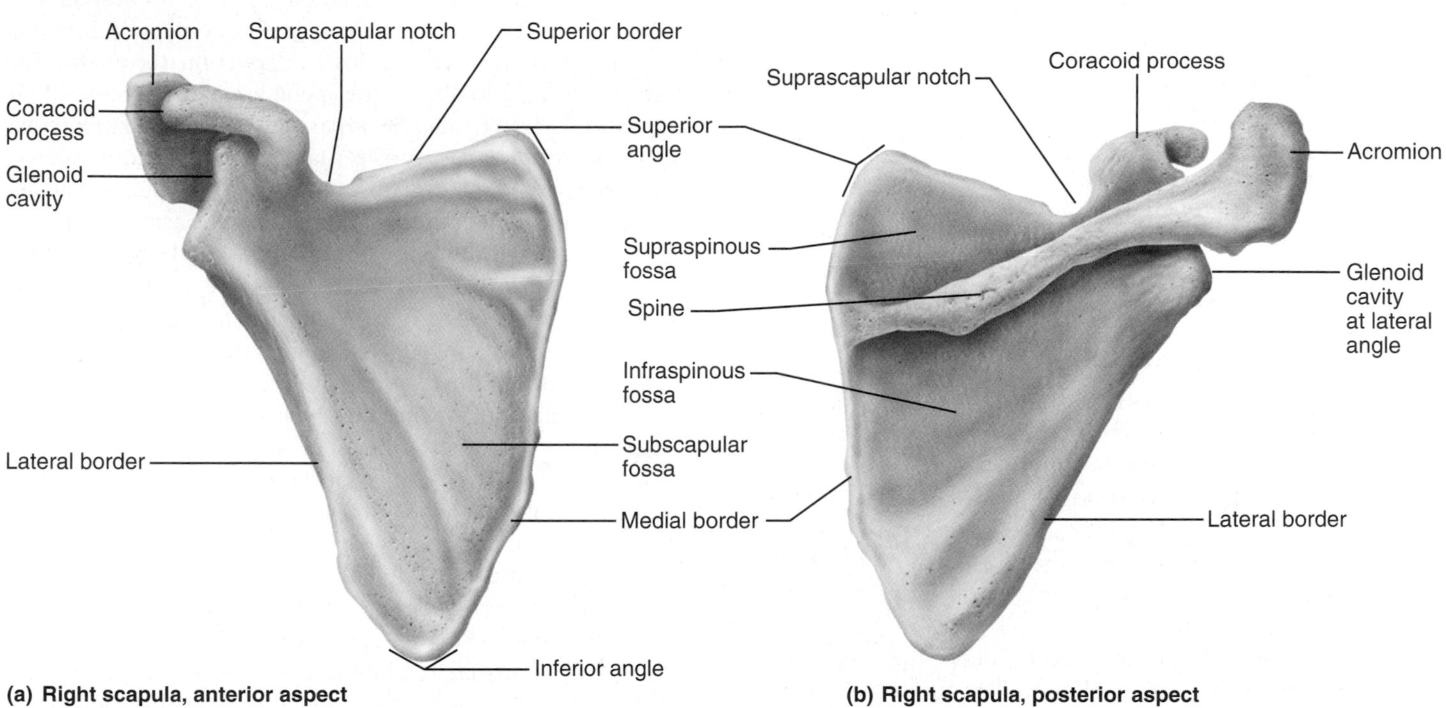

(a) Right scapula, anterior aspect

(b) Right scapula, posterior aspect

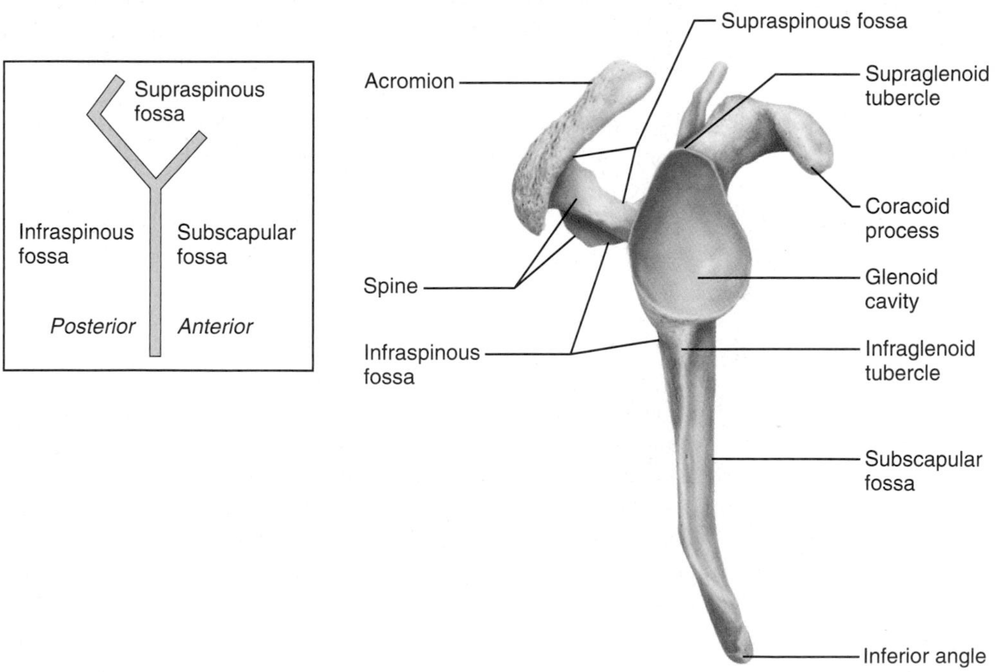

(c) Right scapula, lateral aspect

Figure 7.25 The scapula. View (c) is accompanied by a schematic representation of its orientation. (See *A Brief Atlas of the Human Body*, Figure 24.)

made spades from the shoulder blades of animals. The scapulae lie on the dorsal surface of the rib cage, between ribs 2 and 7.

Each scapula has three borders. The *superior border* is the shortest, sharpest border. The *medial*, or *vertebral, border* parallels the vertebral column. The thick *lateral*, or *axillary, border* abuts the armpit and ends superiorly in a small, shallow fossa, the **glenoid cavity** (gle′noid; "pit-shaped"). This cavity articulates with the humerus of the arm, forming the shoulder joint.

Like all triangles, the scapula has three corners or *angles*. The superior scapular border meets the medial border at the *superior angle* and the lateral border at the *lateral angle*. The medial and lateral borders join at the *inferior angle*. The inferior angle moves extensively as the arm is raised and lowered, and is an important landmark for studying scapular movements.

The anterior, or costal, surface of the scapula is concave and relatively featureless. Its posterior surface bears a prominent **spine** that is easily felt through the skin. The spine ends laterally in an enlarged, roughened triangular projection called the **acromion** (ah-kro′me-on; "point of the shoulder"). The acromion articulates with the acromial end of the clavicle, forming the **acromioclavicular joint**.

Projecting anteriorly from the superior scapular border is the **coracoid process** (kor′ah-coid); *corac* means "beaklike," but this process looks more like a bent little finger. The coracoid process helps anchor the biceps muscle of the arm. It is bounded by the **suprascapular notch** (a nerve passage) medially and by the glenoid cavity laterally.

Several large fossae appear on both sides of the scapula and are named according to location. The *infraspinous* and *supraspinous fossae* are inferior and superior, respectively, to the spine. The *subscapular fossa* is the shallow concavity formed by the entire anterior scapular surface. Lying within these fossae are muscles with similar names.

CHECK YOUR UNDERSTANDING

22. What two bones construct each pectoral girdle?
23. Where is the single point of attachment of the pectoral girdle to the axial skeleton?
24. What is the major shortcoming of the flexibility allowed by the shoulder joint?

For answers, see Appendix G.

The Upper Limb

▶ Identify or name the bones of the upper limb and their important markings.

Thirty separate bones form the bony framework of each upper limb (see Figures 7.26 to 7.28, and **Table 7.3** on p. 232). Each of these bones may be described regionally as a bone of the arm, forearm, or hand. (Keep in mind that anatomically the "arm" is only that part of the upper limb between the shoulder and elbow.)

Arm

The **humerus** (hu′mer-us), the sole bone of the arm, is a typical long bone (**Figure 7.26**). The largest, longest bone of the upper limb, it articulates with the scapula at the shoulder and with the radius and ulna (forearm bones) at the elbow.

At the proximal end of the humerus is its smooth, hemispherical **head**, which fits into the glenoid cavity of the scapula in a manner that allows the arm to hang freely at one's side. Immediately inferior to the head is a slight constriction, the **anatomical neck**. Just inferior to this are the lateral **greater tubercle** and the more medial **lesser tubercle**, separated by the **intertubercular sulcus**, or *bicipital groove* (bi-sip′ĭ-tal). These tubercles are sites of attachment of the rotator cuff muscles. The intertubercular sulcus guides a tendon of the biceps muscle of the arm to its attachment point at the rim of the glenoid cavity (the supraglenoid tubercle). Just distal to the tubercles is the **surgical neck**, so named because it is the most frequently fractured part of the humerus. About midway down the shaft on its lateral side is the V-shaped **deltoid tuberosity**, the roughened attachment site for the deltoid muscle of the shoulder. Nearby, the **radial groove** runs obliquely down the posterior aspect of the shaft, marking the course of the radial nerve, an important nerve of the upper limb.

At the distal end of the humerus are two condyles: a medial **trochlea** (trok′le-ah; "pulley"), which looks like an hourglass tipped on its side, and the lateral ball-like **capitulum** (kah-pit′u-lum). These condyles articulate with the ulna and the radius, respectively (Figure 7.26c and d). The condyle pair is flanked by the **medial** and **lateral epicondyles** (muscle attachment sites). Directly above these epicondyles are the **medial** and **lateral supracondylar ridges**. The ulnar nerve, which runs behind the medial epicondyle, is responsible for the painful, tingling sensation you experience when you hit your "funny bone."

Superior to the trochlea on the anterior surface is the **coronoid fossa**; on the posterior surface is the deeper **olecranon fossa** (o-lek′rah-non). These two depressions allow the corresponding processes of the ulna to move freely when the elbow is flexed and extended. A small **radial fossa**, lateral to the coronoid fossa, receives the head of the radius when the elbow is flexed.

Forearm

Two parallel long bones, the radius and the ulna, form the skeleton of the forearm, or *antebrachium* (an″te-bra′ke-um) (**Figure 7.27**). Unless a person's forearm muscles are very bulky, these bones are easily palpated along their entire length. Their proximal ends articulate with the humerus; their distal ends form joints with bones of the wrist. The radius and ulna articulate with each other both proximally and distally at small **radioulnar joints** (ra″de-o-ul′nar), and they are connected along their entire length by a flat, flexible ligament, the **interosseous membrane** (in″ter-os′e-us; "between the bones").

In the anatomical position, the radius lies laterally (on the thumb side) and the ulna medially. However, when you rotate your forearm so that the palm faces posteriorly (a movement called pronation), the distal end of the radius crosses over the ulna and the two bones form an X (see Figure 8.6a, p. 258).

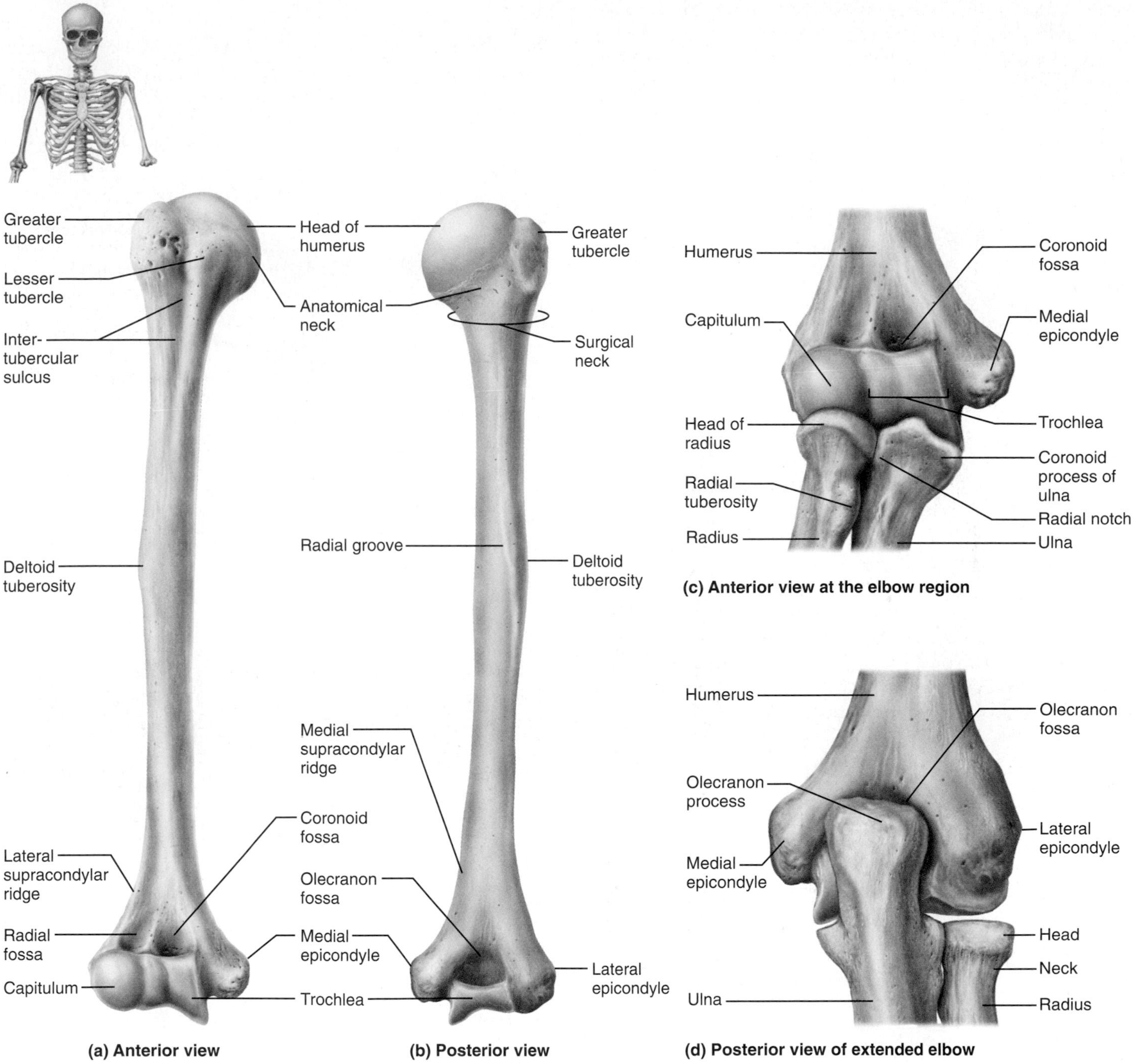

Figure 7.26 The humerus of the right arm and detailed views of articulation at the elbow. (See *A Brief Atlas of the Human Body*, Figure 25.)

Ulna

The **ulna** (ul′nah; "elbow") is slightly longer than the radius. It has the main responsibility for forming the elbow joint with the humerus. Its proximal end looks like the adjustable end of a monkey wrench: it bears two prominent processes, the **olecranon** (elbow) and **coronoid processes**, separated by a deep concavity, the **trochlear notch** (Figure 7.27c). Together, these two processes grip the trochlea of the humerus, forming a hinge joint that allows the forearm to be bent upon the arm (flexed), then straightened again (extended). When the forearm is fully extended, the olecranon process "locks" into the olecranon fossa (Figure 7.26d), keeping the forearm from hyperextending (moving posteriorly beyond the elbow joint). The posterior olecranon process forms the angle of the elbow when the forearm is flexed and is the bony part that rests on the table when you lean on your elbows. On the lateral side of the coronoid process is a small depression, the **radial notch**, where the ulna articulates with the head of the radius.

Distally the ulnar shaft narrows and ends in a knoblike **head** (Figure 7.27d). Medial to the head is a **styloid process**, from which a ligament runs to the wrist. The ulnar head is separated

7

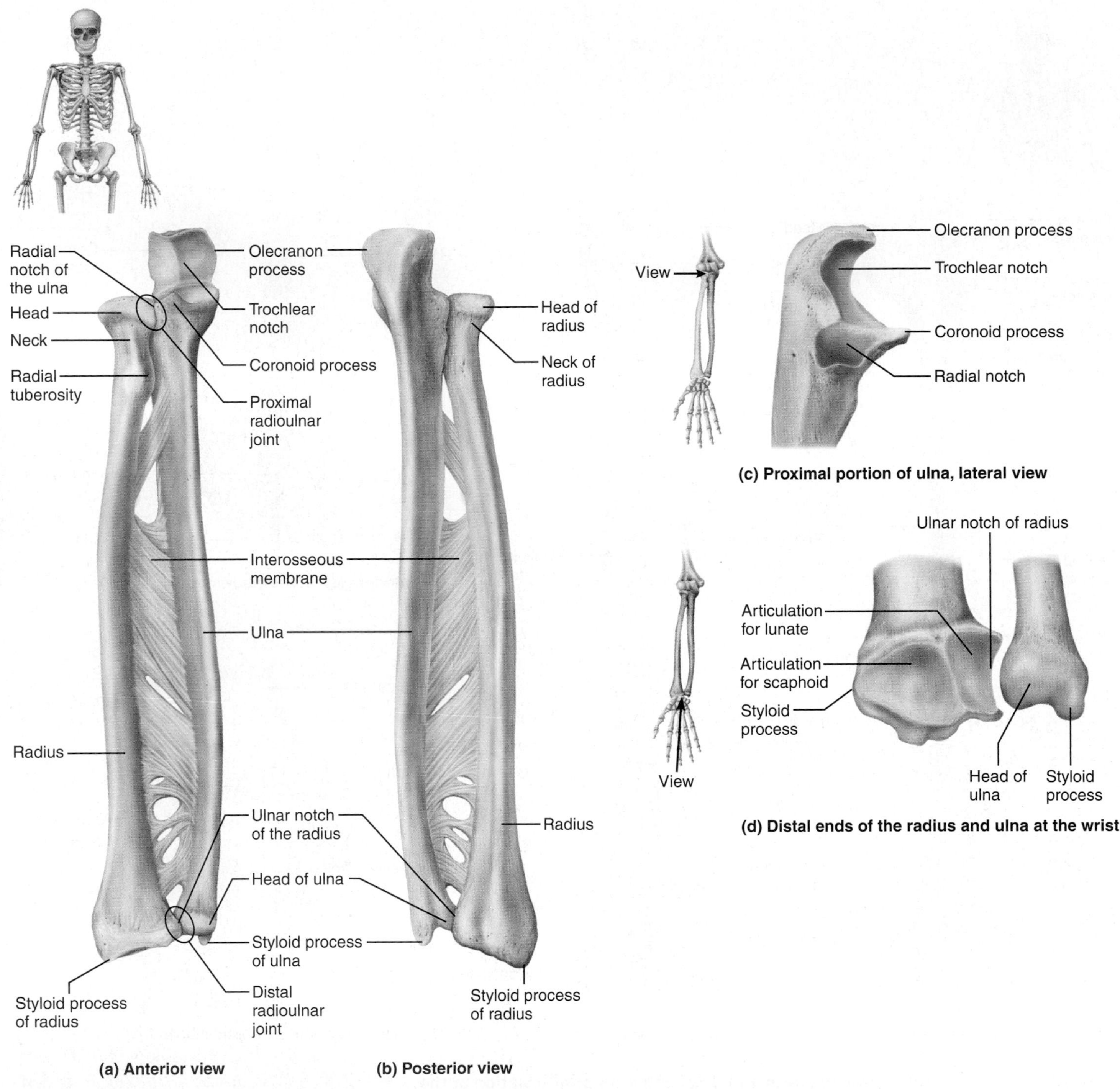

Figure 7.27 Radius and ulna of the right forearm. Note the structural details of the ulnar head and distal portion of radius and ulna. (See *A Brief Atlas of the Human Body*, Figure 26.)

from the bones of the wrist by a disc of fibrocartilage and plays little or no role in hand movements.

Radius

The **radius** ("rod") is thin at its proximal end and wide distally—the opposite of the ulna. The **head** of the radius is shaped somewhat like the head of a nail (Figure 7.27). The superior surface of this head is concave, and it articulates with the capitulum of the humerus. Medially, the head articulates with the radial notch of the ulna (Figure 7.26c). Just inferior to the head is the rough **radial tuberosity**, which anchors the biceps muscle of the arm. Distally, where the radius is expanded, it has a medial **ulnar notch** (Figure 7.27d), which articulates with the ulna, and a lateral **styloid process** (an anchoring site for ligaments that run to the wrist). Between these two markings, the radius is concave where it articulates with carpal bones of the wrist.

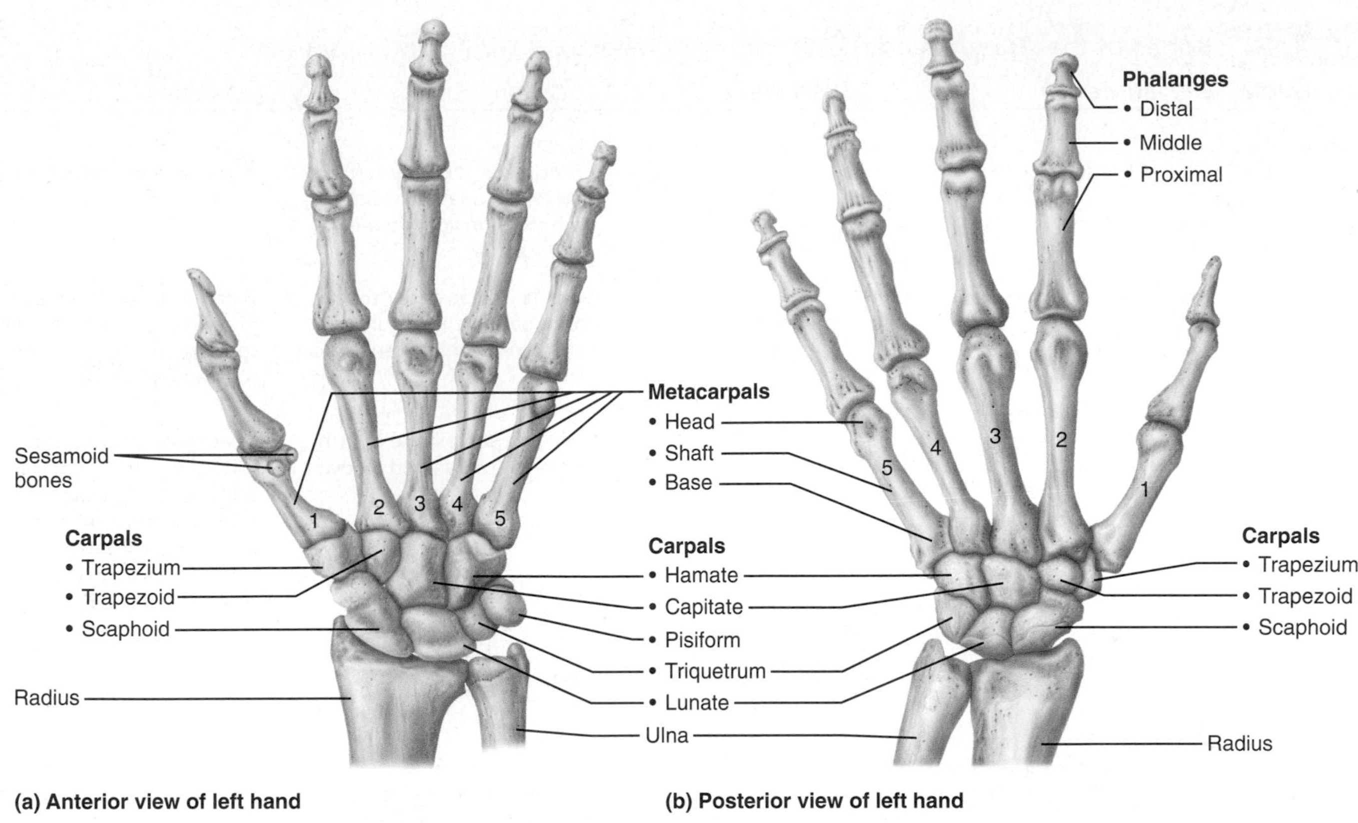

Figure 7.28 Bones of the left hand. (See *A Brief Atlas of the Human Body,* Figure 27.)

The ulna contributes more heavily to the elbow joint, and the radius is the major forearm bone contributing to the wrist joint. When the radius moves, the hand moves with it.

HOMEOSTATIC IMBALANCE

Colle's fracture is a break in the distal end of the radius. It is a common fracture when a falling person attempts to break his or her fall with outstretched hands. ■

Hand

The skeleton of the hand **(Figure 7.28)** includes the bones of the *carpus* (wrist); the bones of the *metacarpus* (palm); and the *phalanges* (bones of the fingers).

Carpus (Wrist)

A "wrist" watch is actually worn on the distal forearm (over the lower ends of the radius and ulna), not on the wrist at all. The true wrist, or carpus, is the proximal part of the structure we generally call our "hand." The carpus consists of eight marble-size short bones, or **carpals** (kar′palz), closely united by ligaments. Because gliding movements occur between these bones, the carpus as a whole is quite flexible.

The carpals are arranged in two irregular rows of four bones each (Figure 7.28). In the proximal row (lateral to medial) are the **scaphoid** (skaf′oid; "boat-shaped"), **lunate** (lu′nāt; "moon-like"), **triquetrum** (tri-kwet′rum; "triangular"), and **pisiform** (pi′sĭ-form; "pea-shaped"). Only the scaphoid and lunate articulate with the radius to form the wrist joint. The carpals of the distal row (lateral to medial) are the **trapezium** (trah-pe′ze-um; "little table"), **trapezoid** (tra′peh-zoid; "four-sided"), **capitate** ("head-shaped"), and **hamate** (ham′āt; "hooked").

There are numerous memory-jogging phrases to help you recall the carpals in the order given above. If you don't have one, try: Sally left the party to take Cindy home. As with all such memory jogs, the first letter of each word is the first letter of the term you need to remember.

HOMEOSTATIC IMBALANCE

The arrangement of its bones is such that the carpus is concave anteriorly and a ligament roofs over this concavity, forming the notorious *carpal tunnel.* Besides the median nerve (which supplies the lateral side of the hand), several long muscle tendons crowd into this tunnel. Overuse and inflammation of the tendons cause them to swell, compressing the median nerve, which causes tingling and numbness of the areas served, and movements of the thumb weaken. Pain is greatest at night. Those who repeatedly flex their wrists and fingers, such as those who work at computer keyboards all day, are particularly susceptible to this nerve impairment, called *carpal tunnel syndrome.* This condition is treated by splinting the wrist during sleep or by surgery. ■

Metacarpus (Palm)

Five **metacarpals** radiate from the wrist like spokes to form the **metacarpus** or palm of the hand (*meta* = beyond). These small long bones are not named, but instead are numbered 1 to 5

TABLE 7.3 Bones of the Appendicular Skeleton, Part 1: Pectoral Girdle and Upper Limb

BODY REGION	BONES*	ILLUSTRATION	LOCATION	MARKINGS
Pectoral girdle (Figures 7.24, 7.25)	Clavicle (2)		Clavicle is in superoanterior thorax; articulates medially with sternum and laterally with scapula	Acromial end; sternal end
	Scapula (2)		Scapula is in posterior thorax; forms part of the shoulder; articulates with humerus and clavicle	Glenoid cavity; spine; acromion; coracoid process; infraspinous, supraspinous, and subscapular fossae
Upper limb Arm (Figure 7.26)	Humerus (2)		Humerus is sole bone of arm; between scapula and elbow	Head; greater and lesser tubercles; intertubercular sulcus; radial groove; deltoid tuberosity; trochlea; capitulum; coronoid and olecranon fossae; epicondyles; radial fossa
Forearm (Figure 7.27)	Ulna (2)		Ulna is the medial bone of forearm between elbow and wrist; with the humerus forms elbow joint	Coronoid process; olecranon process; radial notch; trochlear notch; styloid process; head
Hand (Figure 7.28)	Radius (2)		Radius is the lateral bone of forearm; articulates with carpals to form part of the wrist joint	Head; radial tuberosity; styloid process; ulnar notch
	8 Carpals (16) scaphoid lunate triquetrum pisiform trapezium trapezoid capitate hamate		Carpals form a bony crescent at the wrist; arranged in two rows of four bones each	
	5 Metacarpals (10)		Metacarpals form the palm; one in line with each digit	
	14 Phalanges (28) distal middle proximal		Phalanges form the fingers; three in digits 2–5; two in digit 1 (the thumb)	

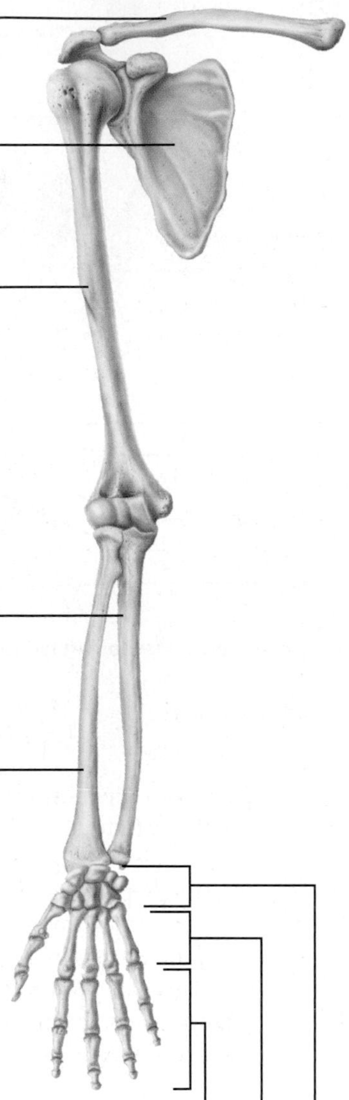

Anterior view of pectoral girdle and upper limb

*The number in parentheses ( ) following the bone name denotes the total number of such bones in the body.

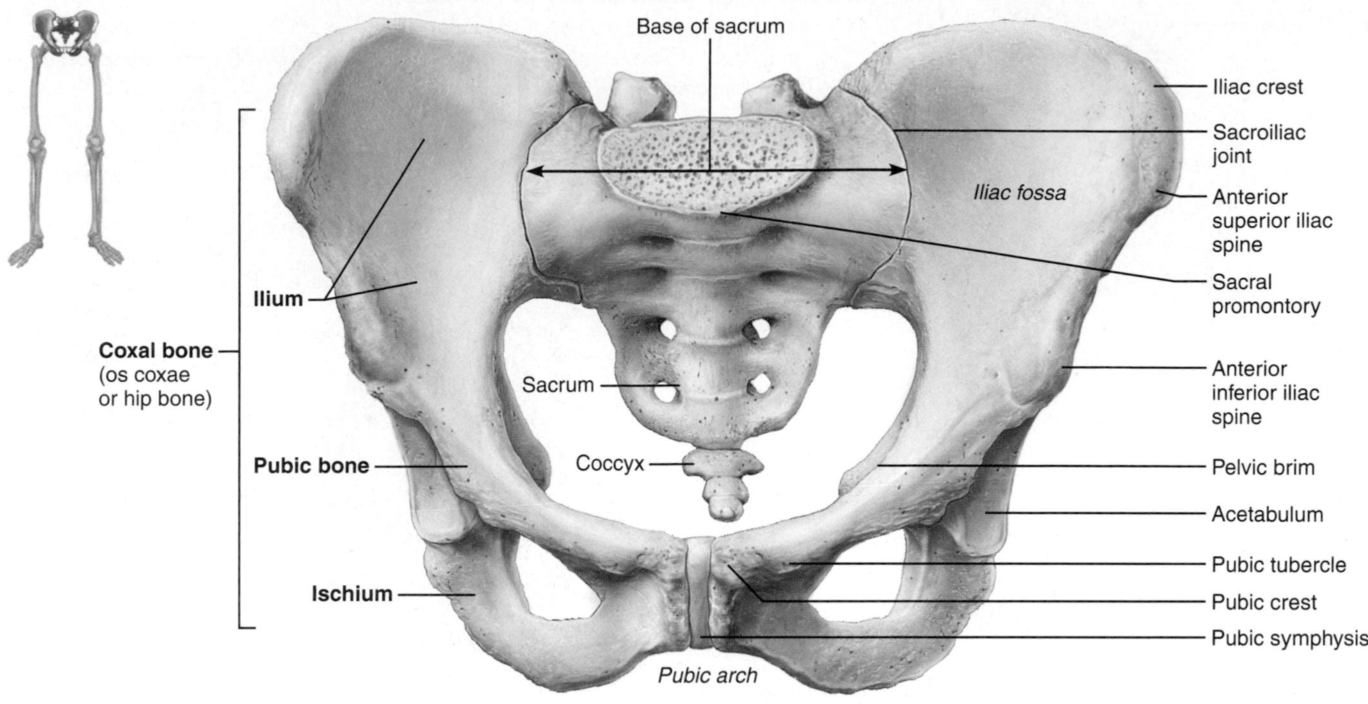

Figure 7.29 Articulated pelvis showing the two hip (coxal) bones (which together form the pelvic girdle), the sacrum, and the coccyx.

from thumb to little finger. The **bases** of the metacarpals articulate with the carpals proximally and each other medially and laterally (Figure 7.28). Their bulbous **heads** articulate with the proximal phalanges of the fingers. When you clench your fist, the heads of the metacarpals become prominent as your *knuckles.*

Metacarpal 1, associated with the thumb, is the shortest and most mobile. It occupies a more anterior position than the other metacarpals. Consequently, the joint between metacarpal 1 and the trapezium is a unique saddle joint that allows *opposition*, the action of touching your thumb to the tips of your other fingers.

Phalanges (Fingers)

The **fingers**, or **digits** of the upper limb, are numbered 1 to 5 beginning with the thumb, or **pollex** (pol′eks). In most people, the third finger is the longest. Each hand contains 14 miniature long bones called **phalanges** (fah-lan′jêz). Except for the thumb, each finger has three phalanges: *distal, middle,* and *proximal.* The thumb has no middle phalanx. [Phalanx (fa′langks; "a closely knit row of soldiers") is the singular term for phalanges.]

CHECK YOUR UNDERSTANDING

25. Which bone plays the major role in forming the elbow joint?
26. Which bones of the upper limb have a styloid process?
27. Where are carpals found and what type of bone (short, irregular, long, or flat) are they?

For answers, see Appendix G.

The Pelvic (Hip) Girdle

- Name the bones contributing to the os coxae, and relate the pelvic girdle's strength to its function.
- Describe differences in the male and female pelves and relate these to functional differences.

The **pelvic girdle**, or **hip girdle**, attaches the lower limbs to the axial skeleton, transmits the full weight of the upper body to the lower limbs, and supports the visceral organs of the pelvis (Figures 7.29 and 7.30 and Table 7.4, p. 236). Unlike the pectoral girdle, which is sparingly attached to the thoracic cage, the pelvic girdle is secured to the axial skeleton by some of the strongest ligaments in the body. And unlike the shallow glenoid cavity of the scapula, the corresponding sockets of the pelvic girdle are deep and cuplike and firmly secure the head of the femur in place. Thus, even though both the shoulder and hip joints are ball-and-socket joints, very few of us can wheel or swing our legs about with the same degree of freedom as our arms. The pelvic girdle lacks the mobility of the pectoral girdle but is far more stable.

The pelvic girdle is formed by a pair of **hip bones**, each also called an **os coxae** (ahs kok′se), or **coxal bone** (*coxa* = hip). Each hip bone unites with its partner anteriorly and with the sacrum posteriorly **(Figure 7.29)**. The deep, basinlike structure formed by the hip bones, together with the sacrum and coccyx, is called the **bony pelvis.**

Each large, irregularly shaped hip bone consists of three separate bones during childhood: the ilium, ischium, and pubis **(Figure 7.30)**. In adults, these bones are firmly fused and their

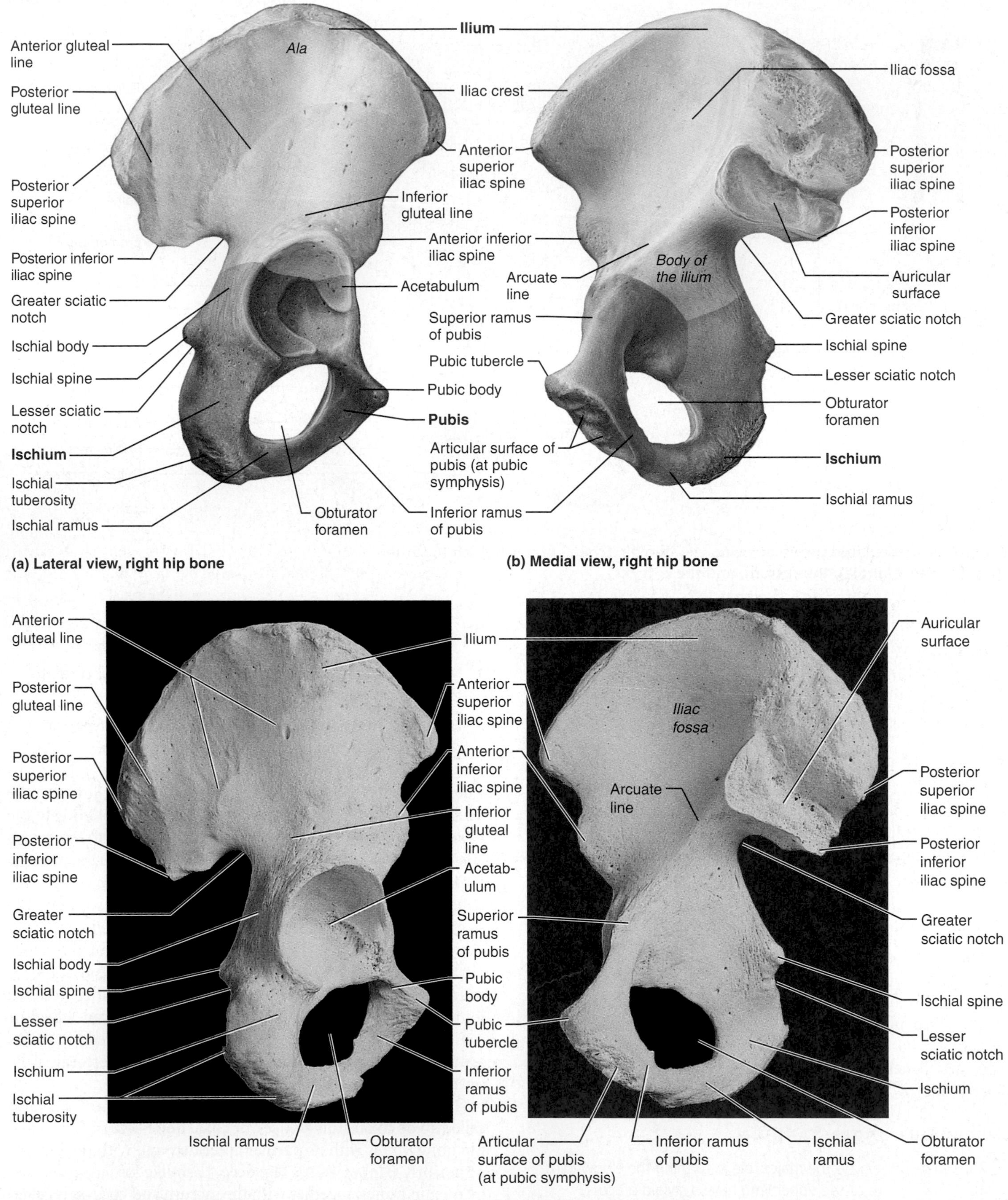

7

Figure 7.30 Bones of the bony pelvis. Lateral and medial views of the right hip bone. The point of fusion of the ilium (gold), ischium (violet), and pubic (red) bones at the acetabulum is indicated in the diagrams **(a, b)**. (See *A Brief Atlas of the Human Body*, Figure 28.)

boundaries are indistinguishable. Their names are retained, however, to refer to different regions of the composite hip bone.

At the point of fusion of the ilium, ischium, and pubis is a deep hemispherical socket called the **acetabulum** (as″ĕ-tab′u-lum; "vinegar cup") on the lateral surface of the pelvis (see Figure 7.30). The acetabulum receives the head of the femur, or thigh bone, at this *hip joint.*

Ilium

The **ilium** (il′e-um; "flank") is a large flaring bone that forms the superior region of a coxal bone. It consists of a **body** and a superior winglike portion called the **ala** (a′lah). When you rest your hands on your hips, you are resting them on the thickened superior margins of the alae, the **iliac crests**, to which many muscles attach. Each iliac crest ends anteriorly in the blunt **anterior superior iliac spine** and posteriorly in the sharp **posterior superior iliac spine**.

Located below these are the less prominent *anterior* and *posterior inferior iliac spines.* All of these spines are attachment points for the muscles of the trunk, hip, and thigh. The anterior superior iliac spine is an especially important anatomical landmark. It is easily felt through the skin and is visible in thin people. The posterior superior iliac spine is difficult to palpate, but its position is revealed by a skin dimple in the sacral region.

Just inferior to the posterior inferior iliac spine, the ilium indents deeply to form the **greater sciatic notch** (si-at′ik), through which the thick cordlike sciatic nerve passes to enter the thigh. The broad posterolateral surface of the ilium, the **gluteal surface** (gloo′te-al), is crossed by three ridges, the **posterior**, **anterior**, and **inferior gluteal lines**, to which the gluteal (buttock) muscles attach.

The medial surface of the iliac ala exhibits a concavity called the **iliac fossa**. Posterior to this, the roughened **auricular surface** (aw-rik′u-lar; "ear-shaped") articulates with the same-named surface of the sacrum, forming the *sacroiliac joint* (Figure 7.29). The weight of the body is transmitted from the spine to the pelvis through the sacroiliac joints. Running inferiorly and anteriorly from the auricular surface is a robust ridge called the **arcuate line** (ar′ku-at; "bowed"). The arcuate line helps define the **pelvic brim**, the superior margin of the *true pelvis*, which we will discuss shortly. Anteriorly, the body of the ilium joins the pubis; inferiorly it joins the ischium.

Ischium

The **ischium** (is′ke-um; "hip") forms the posteroinferior part of the hip bone (Figures 7.29 and 7.30). Roughly L- or arc-shaped, it has a thicker, superior **body** adjoining the ilium and a thinner, inferior **ramus** (*ramus* = branch). The ramus joins the pubis anteriorly. The ischium has three important markings. Its **ischial spine** projects medially into the pelvic cavity and serves as a point of attachment of the *sacrospinous ligament* running from the sacrum. Just inferior to the ischial spine is the **lesser sciatic notch**. A number of nerves and blood vessels pass through this notch to supply the anogenital area. The inferior surface of the ischial body is rough and grossly thickened as the **ischial tuberosity**. When we sit, our weight is borne entirely by the ischial tuberosities, which are the strongest parts of the hip bones.

A massive ligament runs from the sacrum to each ischial tuberosity. This *sacrotuberous ligament* (not illustrated) helps hold the pelvis together. The ischial tuberosity is also a site of attachment of the large hamstring muscles of the posterior thigh.

Pubis

The **pubis** (pu′bis; "sexually mature"), or **pubic bone**, forms the anterior portion of the hip bone (Figures 7.29 and 7.30). In the anatomical position, it lies nearly horizontally and the urinary bladder rests upon it. Essentially, the pubis is V shaped with **superior** and **inferior rami** issuing from its flattened medial **body**. The anterior border of the pubis is thickened to form the **pubic crest**. At the lateral end of the pubic crest is the **pubic tubercle**, one of the attachments for the *inguinal ligament.* As the two rami of the pubic bone run laterally to join with the body and ramus of the ischium, they define a large opening in the hip bone, the **obturator foramen** (ob″tu-ra′tor), through which a few blood vessels and nerves pass. Although the obturator foramen is large, it is nearly closed by a fibrous membrane in life (*obturator* = closed up).

The bodies of the two pubic bones are joined by a fibrocartilage disc, forming the midline **pubic symphysis** joint. Inferior to this joint, the inferior pubic rami angle laterally, forming an inverted V-shaped arch called the **pubic arch** or **subpubic angle**. The acuteness of this arch helps to differentiate the male and female pelves.

Pelvic Structure and Childbearing

The differences between the male and female pelves are striking. The female pelvis is modified for childbearing: It tends to be wider, shallower, lighter, and rounder than that of a male. The female pelvis not only accommodates a growing fetus, but it must be large enough to allow the infant's relatively large head to exit at birth. The major differences between the typical male and female pelves are summarized and illustrated in **Table 7.4**.

The pelvis is said to consist of a false (greater) pelvis and a true (lesser) pelvis separated by the **pelvic brim**, a continuous oval ridge that runs from the pubic crest through the arcuate line and sacral promontory (Figure 7.29). The **false pelvis**, that portion superior to the pelvic brim, is bounded by the alae of the ilia laterally and the lumbar vertebrae posteriorly. The false pelvis is really part of the abdomen and helps support the abdominal viscera. It does not restrict childbirth in any way.

The **true pelvis** is the region inferior to the pelvic brim that is almost entirely surrounded by bone. It forms a deep bowl containing the pelvic organs. Its dimensions, particularly those of its *inlet* and *outlet*, are critical to the uncomplicated delivery of a baby, and they are carefully measured by an obstetrician.

The **pelvic inlet** *is* the pelvic brim, and its widest dimension is from right to left along the frontal plane. As labor begins, an infant's head typically enters the inlet with its forehead facing one ilium and its occiput facing the other. A sacral promontory

7

TABLE 7.4 Comparison of the Male and Female Pelves

CHARACTERISTIC	FEMALE	MALE
General structure and functional modifications	Tilted forward; adapted for childbearing; true pelvis defines the birth canal; cavity of the true pelvis is broad, shallow, and has a greater capacity	Tilted less far forward; adapted for support of a male's heavier build and stronger muscles; cavity of the true pelvis is narrow and deep
Bone thickness	Less; bones lighter, thinner, and smoother	Greater; bones heavier and thicker, and markings are more prominent
Acetabula	Smaller; farther apart	Larger; closer
Pubic angle/arch	Broader (80° to 90°); more rounded	Angle is more acute (50° to 60°)
Anterior view	Pelvic brim Pubic arch	
Sacrum	Wider; shorter; sacral curvature is accentuated	Narrow; longer; sacral promontory more ventral
Coccyx	More movable; straighter	Less movable; curves ventrally
Greater sciatic notch	Wide and shallow	Narrow and deep
Left lateral view	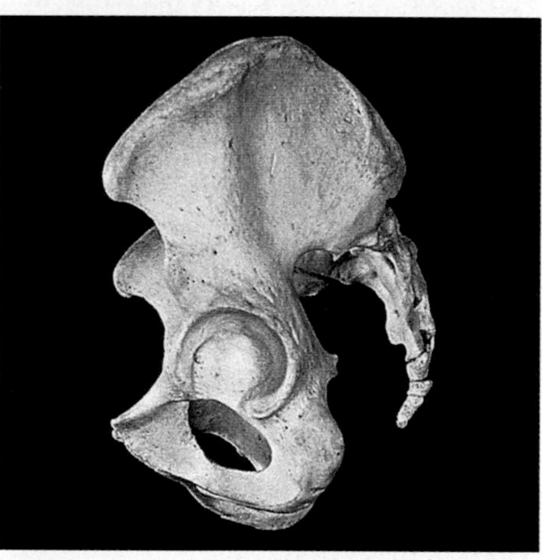	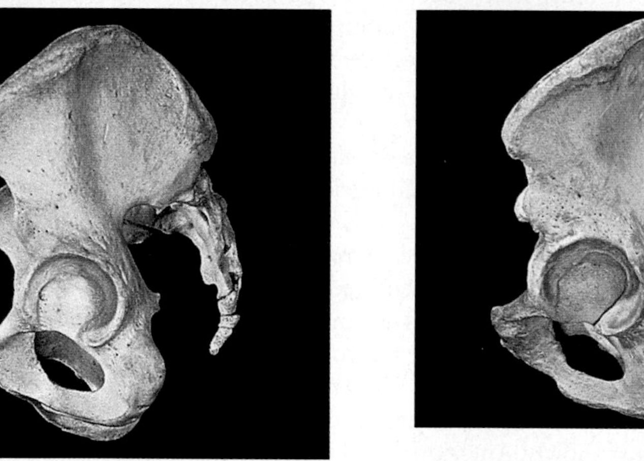
Pelvic inlet (brim)	Wider; oval from side to side	Narrow; basically heart shaped
Pelvic outlet	Wider; ischial tuberosities shorter, farther apart and everted	Narrower; ischial tuberosities longer, sharper, and point more medially
Posteroinferior view	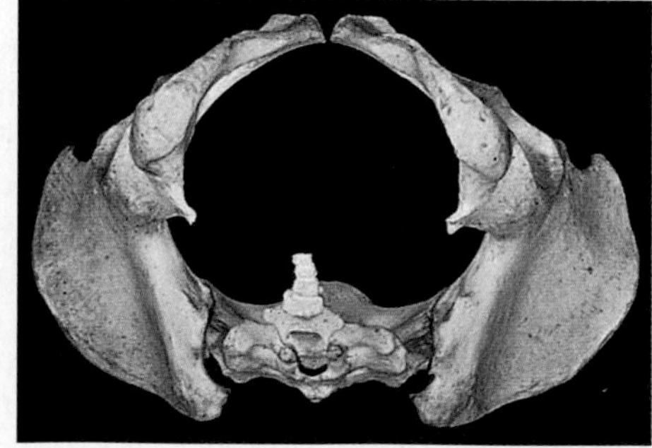	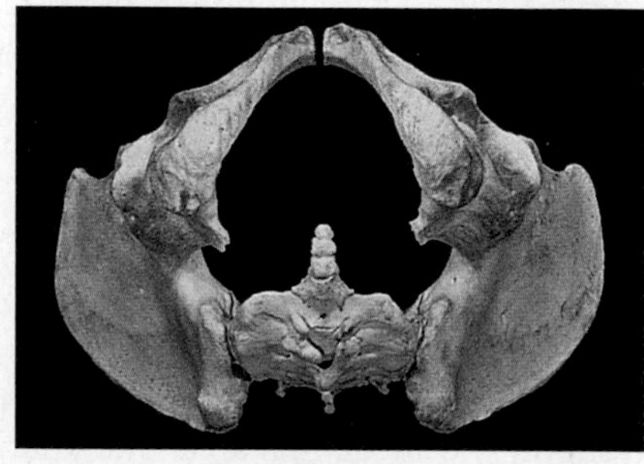

that is particularly large can impair the infant's entry into the true pelvis.

The **pelvic outlet**, illustrated in the photos at the bottom of Table 7.4, is the inferior margin of the true pelvis. It is bounded anteriorly by the pubic arch, laterally by the ischia, and posteriorly by the sacrum and coccyx. Both the coccyx and the ischial spines protrude into the outlet opening, so a sharply angled coccyx or unusually large spines can interfere with delivery. The largest dimension of the outlet is the anteroposterior diameter.

Generally, after the baby's head passes through the inlet, it rotates so that the forehead faces posteriorly and the occiput anteriorly, and this is the usual position of the baby's head as it leaves the mother's body (see Figure 28.18c). Thus, during birth, the infant's head makes a quarter turn to follow the widest dimensions of the true pelvis.

CHECK YOUR UNDERSTANDING

28. The ilium and pubis help to form the os coxae. What other bone is involved in forming the os coxae?

29. The pelvic girdle is a heavy, strong girdle. How does its structure reflect its function?

30. Which of the following terms or phrases refer to the female pelvis? Wider, shorter sacrum; cavity narrow and deep; narrow heart-shaped inlet; more movable coccyx; long ischial spines.

For answers, see Appendix G.

The Lower Limb

▶ Identify the lower limb bones and their important markings.

The lower limbs carry the entire weight of the erect body and are subjected to exceptional forces when we jump or run. Thus, it is not surprising that the bones of the lower limbs are thicker and stronger than comparable bones of the upper limbs. The three segments of each lower limb are the thigh, the leg, and the foot (see Table 7.5 on p. 242).

Thigh

The **femur** (fe′mur; "thigh"), the single bone of the thigh (**Figure 7.31**), is the largest, longest, strongest bone in the body. Its durable structure reflects the fact that the stress on the femur during vigorous jumping can reach 280 kg/cm^2 (about 2 tons per square inch)! The femur is clothed by bulky muscles that prevent us from palpating its course down the length of the thigh. Its length is roughly one-quarter of a person's height.

Proximally, the femur articulates with the hip bone and then courses medially as it descends toward the knee. This arrangement allows the knee joints to be closer to the body's center of gravity and provides for better balance. The medial course of the two femurs is more pronounced in women because of their wider pelvis, a situation that may contribute to the greater incidence of knee problems in female athletes.

The ball-like **head** of the femur has a small central pit called the **fovea capitis** (fo′ve-ah kă′pĭ-tis; "pit of the head"). The short *ligament of the head of the femur* runs from this pit to the acetabulum, where it helps secure the femur. The head is carried on a *neck* that angles *laterally* to join the shaft. This arrangement reflects the fact that the femur articulates with the lateral aspect (rather than the inferior region) of the pelvis. The neck is the weakest part of the femur and is often fractured, an injury commonly called a broken hip.

At the junction of the shaft and neck are the lateral **greater trochanter** (tro-kan′ter) and posteromedial **lesser trochanter**. These projections serve as sites of attachment for thigh and buttock muscles. The two trochanters are connected by the **intertrochanteric line** anteriorly and by the prominent **intertrochanteric crest** posteriorly.

Inferior to the intertrochanteric crest on the posterior shaft is the **gluteal tuberosity**, which blends into a long vertical ridge, the **linea aspera** (lin′e-ah as′per-ah; "rough line"), inferiorly. Distally, the linea aspera diverges, forming the **medial** and **lateral supracondylar lines**. All of these markings are sites of muscle attachment. Except for the linea aspera, the femur shaft is smooth and rounded.

Distally, the femur broadens and ends in the wheel-like **lateral** and **medial condyles**, which articulate with the tibia of the leg. The **medial** and **lateral epicondyles** (sites of muscle attachment) flank the condyles superiorly. On the superior part of the medial epicondyle is a bump, the **adductor tubercle**. The smooth **patellar surface**, between the condyles on the anterior femoral surface, articulates with the *patella* (pah-tel′ah), or kneecap (see Figure 7.31 and Table 7.5).

Between the condyles on the posterior aspect of the femur is the deep, U-shaped **intercondylar fossa**, and superior to that on the shaft is the smooth popliteal surface.

The **patella** ("small pan") is a triangular sesamoid bone enclosed in the (quadriceps) tendon that secures the anterior thigh muscles to the tibia. It protects the knee joint anteriorly and improves the leverage of the thigh muscles acting across the knee.

Leg

Two parallel bones, the tibia and fibula, form the skeleton of the leg, the region of the lower limb between the knee and the ankle (**Figure 7.32**). These two bones are connected by an *interosseous membrane* and articulate with each other both proximally and distally. Unlike the joints between the radius and ulna of the forearm, the *tibiofibular joints* (tib″e-o-fib′u-lar) of the leg allow essentially no movement. The bones of the leg thus form a less flexible but stronger and more stable limb than those of the forearm. The medial tibia articulates proximally with the femur to form the modified hinge joint of the knee and distally with the talus bone of the foot at the ankle. The fibula, by contrast, does not contribute to the knee joint and merely helps stabilize the ankle joint.

Tibia

The **tibia** (tib′e-ah; "shinbone") receives the weight of the body from the femur and transmits it to the foot. It is second only to the femur in size and strength. At its broad proximal end are the

7

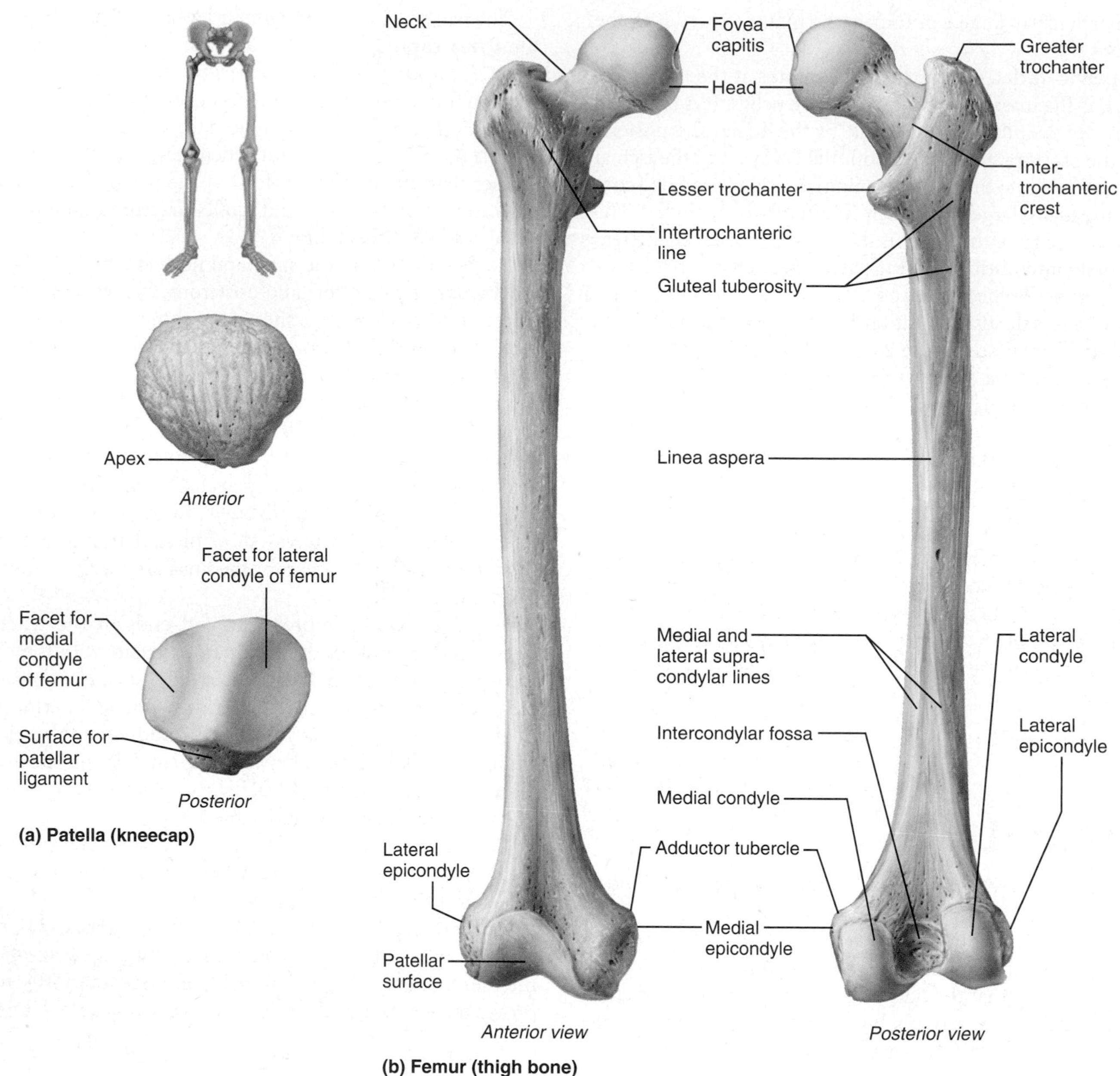

Figure 7.31 Bones of the right knee and thigh. (See *A Brief Atlas of the Human Body*, Figure 29.)

concave **medial** and **lateral condyles**, which look like two huge checkers lying side by side. These are separated by an irregular projection, the **intercondylar eminence.** The tibial condyles articulate with the corresponding condyles of the femur. The inferior region of the lateral tibial condyle bears a facet that indicates the site of the *proximal tibiofibular joint.* Just inferior to the condyles, the tibia's anterior surface displays the rough **tibial tuberosity**, to which the patellar ligament attaches.

The tibial shaft is triangular in cross section. Neither the tibia's sharp **anterior border** nor its medial surface is covered by muscles, so they can be felt just deep to the skin along their entire length. The anguish of a "bumped" shin is an experience familiar to nearly everyone. Distally the tibia is flat where it articulates with the talus bone of the foot. Medial to that joint surface is an inferior projection, the **medial malleolus** (mah-le′o-lus; "little hammer"), which forms the medial bulge of the ankle. The **fibular notch**, on the lateral surface of the tibia, participates in the *distal tibiofibular joint.*

Fibula

The **fibula** (fib′u-lah; "pin") is a sticklike bone with slightly expanded ends. It articulates proximally and distally with the lateral aspects of the tibia. Its proximal end is its **head**; its distal end is the **lateral malleolus.** The lateral malleolus forms the conspicuous lateral ankle bulge and articulates with the talus. The fibular shaft is heavily ridged and appears to have been twisted a

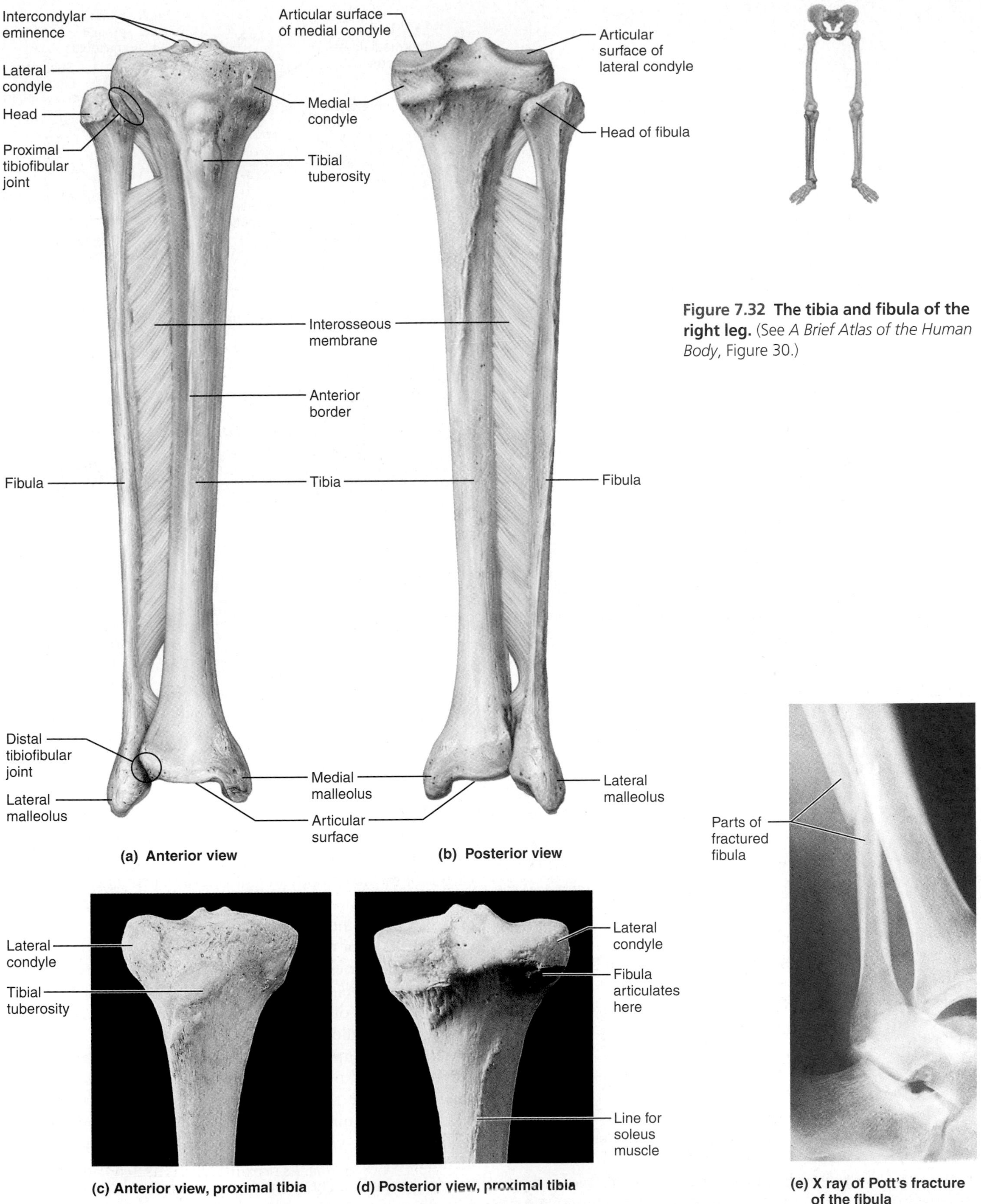

Figure 7.32 The tibia and fibula of the right leg. (See *A Brief Atlas of the Human Body*, Figure 30.)

7

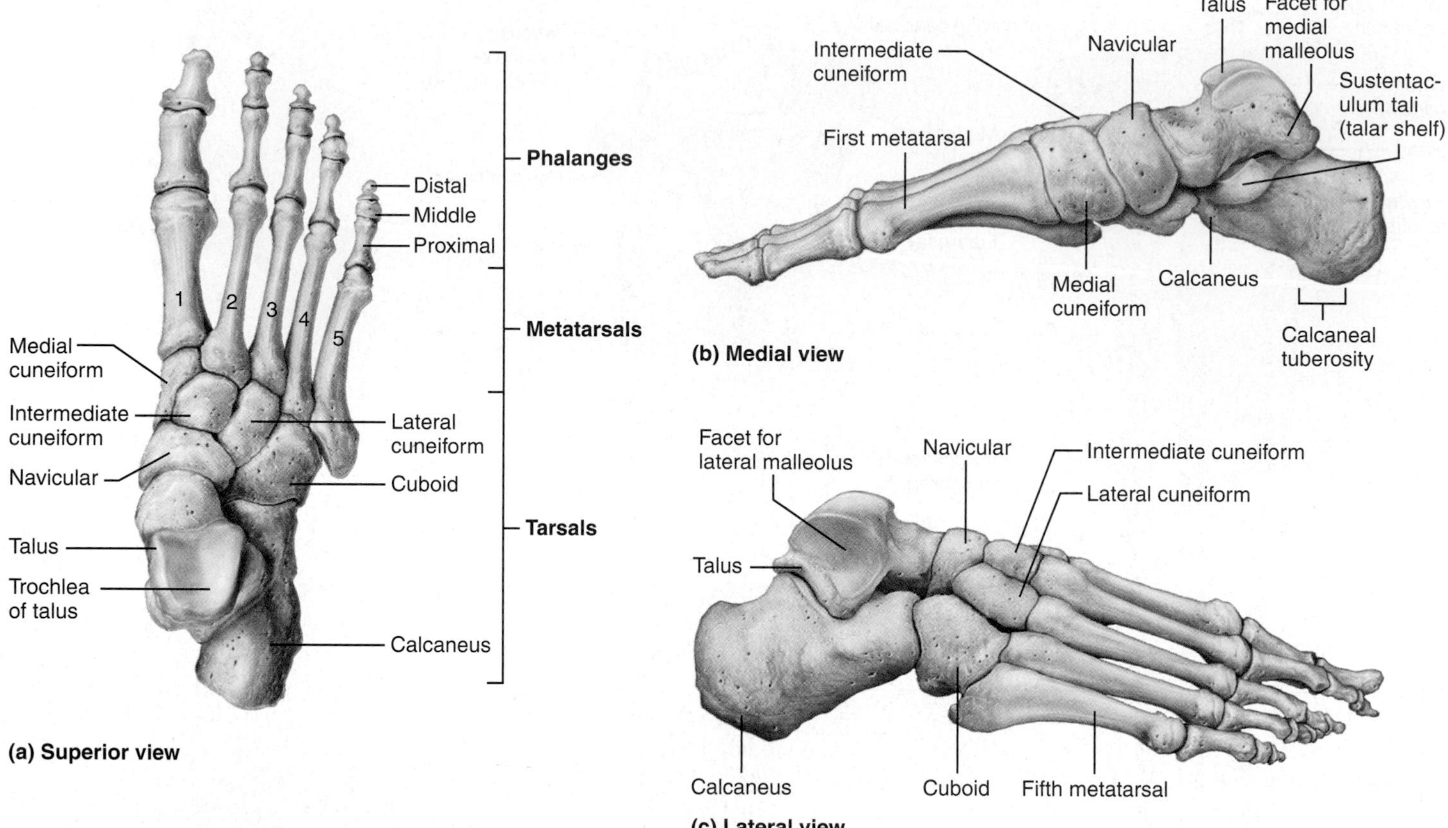

Figure 7.33 Bones of the right foot. (See *A Brief Atlas of the Human Body*, Figure 31a, c, and d.)

quarter turn. The fibula does not bear weight, but several muscles originate from it.

HOMEOSTATIC IMBALANCE

A *Pott's fracture* occurs at the distal end of the fibula, the tibia, or both. It is a common sports injury. (See Figure 7.32e.) ■

CHECK YOUR UNDERSTANDING

31. What lower limb bone is the second largest bone in the body?
32. Where is the medial malleolus located?
33. Which of the following sites is not a site of muscle attachment? Greater trochanter, lesser trochanter, gluteal tuberosity, lateral condyle.

For answers, see Appendix G.

Foot

▶ Name the arches of the foot and explain their importance.

The skeleton of the foot includes the bones of the *tarsus*, the bones of the *metatarsus*, and the *phalanges*, or toe bones **(Figure 7.33)**. The foot has two important functions: It supports our body weight, and it acts as a lever to propel the body forward when we walk and run. A single bone could serve both purposes, but it would adapt poorly to uneven ground. Segmentation makes the foot pliable, avoiding this problem.

Tarsus

The **tarsus** is made up of seven bones called **tarsals** (tar′salz) that form the posterior half of the foot. It corresponds to the carpus of the hand. Body weight is carried primarily by the two largest, most posterior tarsals: the **talus** (ta′lus; "ankle"), which articulates with the tibia and fibula superiorly, and the strong **calcaneus** (kal-ka′ne-us; "heel bone"), which forms the heel of the foot and carries the talus on its superior surface. The thick *calcaneal*, or *Achilles*, *tendon* of the calf muscles attaches to the posterior surface of the calcaneus. The part of the calcaneus that touches the ground is the **calcaneal tuberosity**, and its shelflike projection that supports part of the talus is the **sustentaculum tali** (sus″ten-tak′u-lum ta′le; "supporter of the talus") or **talar shelf**. The tibia articulates with the talus at the *trochlea* of the talus. The remaining tarsals are the lateral **cuboid**, the medial **navicular** (nah-vik′u-lar), and the anterior **medial**, **intermediate**, and **lateral cuneiform bones** (ku-ne′i-form; "wedge-shaped"). The cuboid and cuneiform bones articulate with the metatarsal bones anteriorly.

Metatarsus

The **metatarsus** consists of five small, long bones called **metatarsals.** These are numbered 1 to 5 beginning on the medial (great toe) side of the foot. The first metatarsal, which plays an important role in supporting body weight, is short and thick. The arrangement of the metatarsals is more parallel than that of the metacarpals of the hands. Distally, where the metatarsals articulate with the proximal phalanges of the toes, the enlarged head of the first metatarsal forms the "ball" of the foot.

Phalanges (Toes)

The 14 phalanges of the toes are a good deal smaller than those of the fingers and so are less nimble. But their general structure and arrangement are the same. There are three phalanges in each digit except for the great toe, the **hallux**. The hallux has only two, proximal and distal.

Arches of the Foot

A segmented structure can hold up weight only if it is arched. The foot has three arches: two *longitudinal arches* (*medial* and *lateral*) and one *transverse arch* **(Figure 7.34)**, which account for its awesome strength. These arches are maintained by the interlocking shapes of the foot bones, by strong ligaments, and by the pull of some tendons during muscle activity. The ligaments and muscle tendons provide a certain amount of springiness. In general, the arches "give," or stretch slightly, when weight is applied to the foot and spring back when the weight is removed, which makes walking and running more economical in terms of energy use than would otherwise be the case.

If you examine your wet footprints, you will see that the medial margin from the heel to the head of the first metatarsal leaves no print. This is because the **medial longitudinal arch** curves well above the ground. The talus is the keystone of this arch, which originates at the calcaneus, rises toward the talus, and then descends to the three medial metatarsals.

The **lateral longitudinal arch** is very low. It elevates the lateral part of the foot just enough to redistribute some of the weight to the calcaneus and the head of the fifth metatarsal (to the ends of the arch). The cuboid is the keystone bone of this arch.

The two longitudinal arches serve as pillars for the **transverse arch**, which runs obliquely from one side of the foot to the other, following the line of the joints between the tarsals and metatarsals. Together, the arches of the foot form a half-dome that distributes about half of a person's standing and walking weight to the heel bones and half to the heads of the metatarsals.

HOMEOSTATIC IMBALANCE

Standing immobile for extended periods places excessive strain on the tendons and ligaments of the feet (because the muscles are inactive) and can result in fallen arches, or "flat feet," particularly if a person is overweight. Running on hard surfaces can also cause arches to fall unless the runner wears shoes that give proper arch support. ■

■ ■ ■

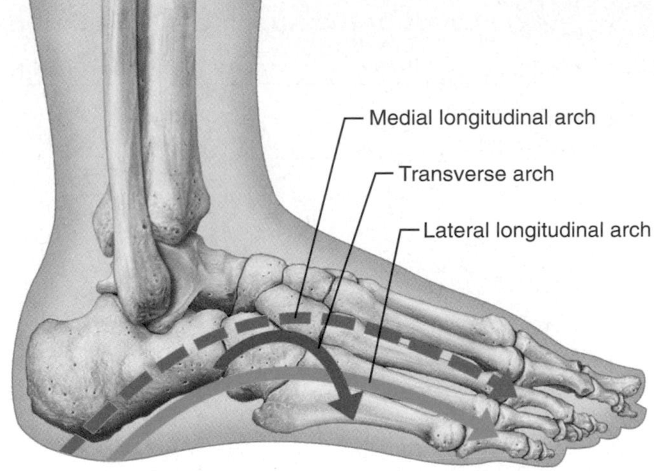

(a) Lateral aspect of right foot

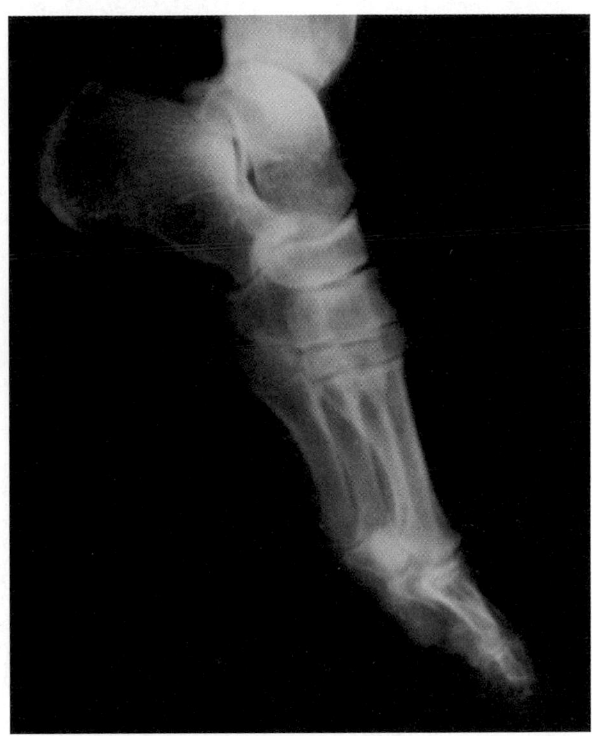

(b) X ray, medial aspect of right foot

Figure 7.34 Arches of the foot.

The bones of the thigh, leg, and foot are summarized in **Table 7.5.**

CHECK YOUR UNDERSTANDING

34. Besides supporting our weight, what is a major function of the arches of the foot?
35. What are the two largest tarsal bones in each foot, and which one forms the heel of the foot?

For answers, see Appendix G.

7

TABLE 7.5 Bones of the Appendicular Skeleton, Part 2: Pelvic Girdle and Lower Limb

BODY REGION	BONES*	ILLUSTRATION	LOCATION	MARKINGS
Pelvic girdle (Figures 7.29, 7.30)	Coxal (2) (hip)		Each coxal (hip) bone is formed by the fusion of an ilium, ischium, and pubic bone; the coxal bones articulate anteriorly at the pubic symphysis and form sacroiliac joints with the sacrum posteriorly; girdle consisting of both coxal bones is basinlike	Iliac crest; anterior and posterior iliac spines; auricular surface; greater and lesser sciatic notches; obturator foramen; ischial tuberosity and spine; acetabulum; pubic arch; pubic crest; pubic tubercle
Lower limb Thigh (Figure 7.31)	Femur (2)		Femur is the sole bone of thigh; between hip joint and knee; largest bone of the body	Head; greater and lesser trochanters; neck; lateral and medial condyles and epicondyles; gluteal tuberosity; linea aspera
Kneecap (Figure 7.31)	Patella (2)		Patella is a sesamoid bone formed within the tendon of the quadriceps (anterior thigh) muscles	
Leg (Figure 7.32)	Tibia (2)		Tibia is the larger and more medial bone of leg; between knee and foot	Medial and lateral condyles; tibial tuberosity; anterior border; medial malleolus
	Fibula (2)		Fibula is the lateral bone of leg; sticklike	Head; lateral malleolus
Foot (Figure 7.33)	7 Tarsals (14) talus calcaneus navicular cuboid lateral cuneiform intermediate cuneiform medial cuneiform		Tarsals are seven bones forming the proximal part of the foot; the talus articulates with the leg bones at the ankle joint; the calcaneus, the largest tarsal, forms the heel	
	5 Metatarsals (10)		Metatarsals are five bones numbered 1–5	
	14 Phalanges (28) distal middle proximal	*Anterior view of pelvic girdle and left lower limb*	Phalanges form the toes; three in digits 2–5, two in digit 1 (the great toe)	

*The number in parentheses ( ) following the bone name denotes the total number of such bones in the body.

Developmental Aspects of the Skeleton

- Define fontanelles and indicate their significance.
- Describe how skeletal proportions change through life.
- Discuss how age-related skeletal changes may affect health.

The membrane bones of the skull start to ossify late in the second month of development. The rapid deposit of bone matrix at the ossification centers produces cone-shaped protrusions in the developing bones. At birth, the skull bones are still incomplete and are connected by as yet unossified remnants of fibrous membranes called **fontanelles** (fon″tah-nelz′) **(Figure 7.35)**. The fontanelles allow the infant's head to be compressed slightly during birth, and they accommodate brain growth in the fetus and infant. A baby's pulse can be felt surging in these "soft spots"; hence their name (*fontanelle* = little fountain). The large, diamond-shaped *anterior fontanelle* is palpable for 1-½ to 2 years after birth. The others are replaced by bone by the end of the first year.

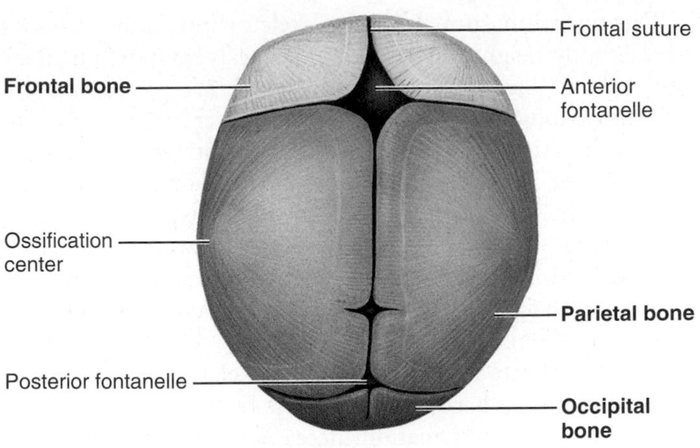

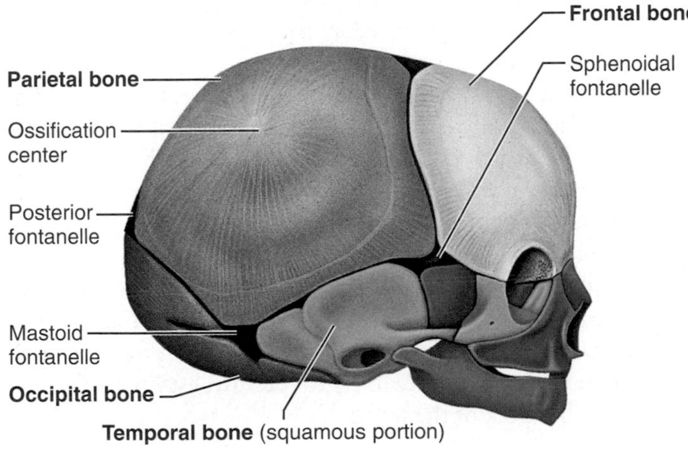

Figure 7.35 Skull of a newborn. Notice that the infant's skull has more bones than that of an adult. (See *A Brief Atlas of the Human Body*, Figure 16.)

HOMEOSTATIC IMBALANCE

Several congenital abnormalities may distort the skull. Most common is *cleft palate*, a condition in which the right and left halves of the palate fail to fuse medially **(Figure 7.36)**. The persistent opening between the oral and nasal cavities interferes with sucking and can lead to aspiration (inhalation) of food into the lungs and *aspiration pneumonia.* ■

The skeleton changes throughout life, but the changes in childhood are most dramatic. At birth, the baby's cranium is huge relative to its face, and several bones are still unfused (e.g., the mandible and frontal bones). The maxillae and mandible are foreshortened, and the contours of the face are flat (Figure 7.38). By 9 months after birth, the cranium is already half of its adult size (volume) because of the rapid growth of the brain. By 8 to 9 years, the cranium has almost reached adult proportions.

Between the ages of 6 and 13, the head appears to enlarge substantially as the face literally grows out from the skull. The jaws, cheekbones, and nose become more prominent. These facial changes are correlated with the expansion of the nose and paranasal sinuses, and development of the permanent teeth. Figure 7.38 tracks how differential bone growth alters body proportions throughout life.

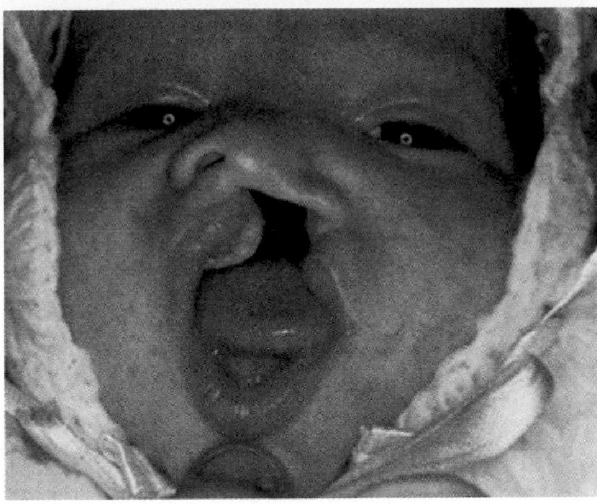

Figure 7.36 A baby born with a cleft lip and palate.

7

Only the thoracic and sacral curvatures are well developed at birth. These so-called **primary curvatures** are convex posteriorly, and an infant's spine arches, like that of a four-legged animal **(Figure 7.37)**.

The **secondary curvatures**—cervical and lumbar—are convex anteriorly and are associated with a child's development. They result from reshaping of the intervertebral discs rather than from modifications of the vertebrae. The cervical curvature is present before birth but is not pronounced until the baby starts to lift its head (at about 3 months). The lumbar curvature develops when the baby begins to walk (at about 12 months). The lumbar curvature positions the weight of the trunk over the body's center of gravity, providing optimal balance when standing.

Vertebral problems (scoliosis or lordosis) may appear during the early school years, when rapid growth of the limb bones stretches many muscles. During the preschool years, lordosis is often present, but this is usually rectified as the abdominal

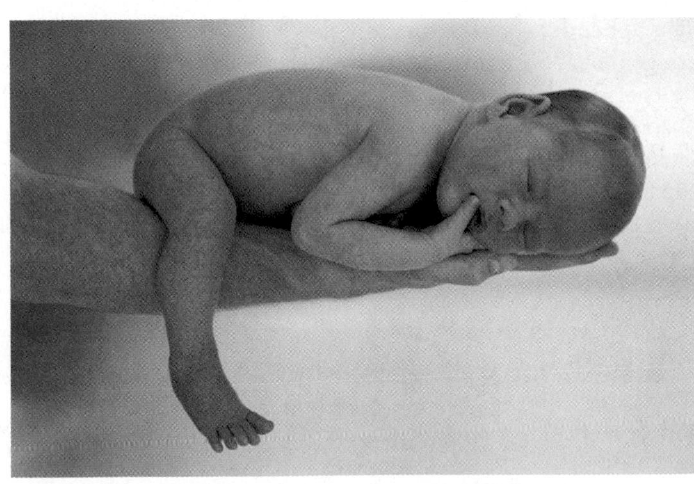

Figure 7.37 The C-shaped spine of a newborn infant.

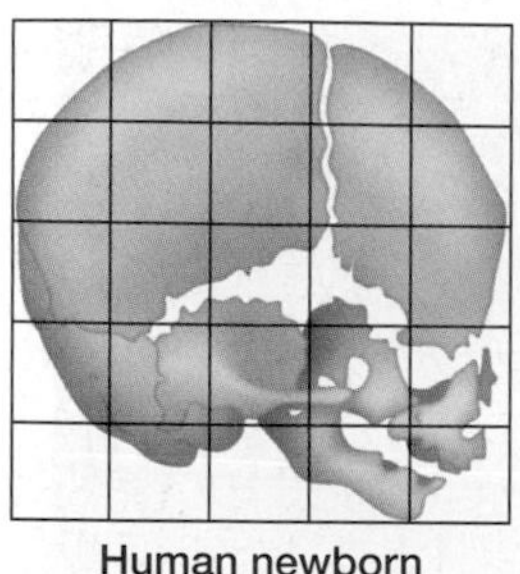

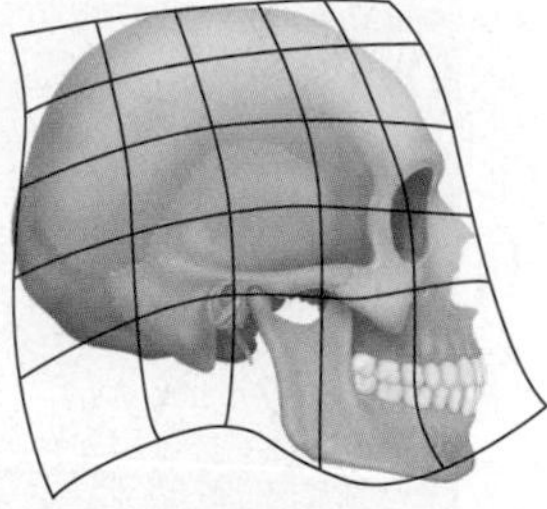

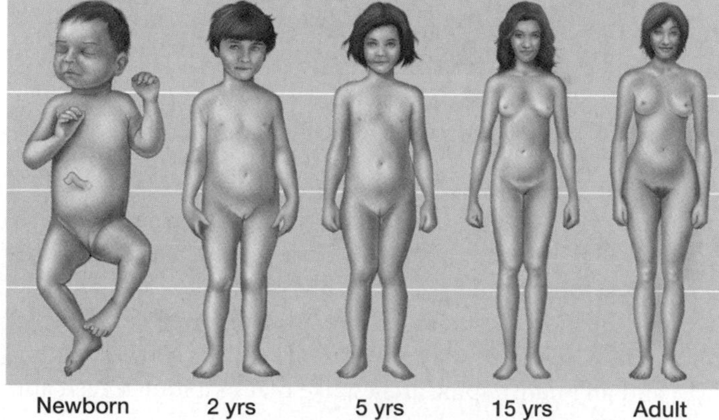

Figure 7.38 Different growth rates of body parts determine body proportions. (a) Differential growth transforms the rounded, foreshortened skull of a newborn to the sloping skull of an adult. **(b)** During growth of a human, the arms and legs grow faster than the head and trunk, as seen in this conceptualization of different-aged individuals all drawn at the same height.

muscles become stronger and the pelvis tilts forward. The thorax grows wider, but a true "military posture" (head erect, shoulders back, abdomen in, and chest out) does not develop until adolescence.

HOMEOSTATIC IMBALANCE

The appendicular skeleton can also suffer from a number of congenital abnormalities. One that occurs in just over 1% of infants and is quite severe is *dysplasia of the hip* (dis-pla′ze-ah; "bad formation"). The acetabulum forms incompletely or the ligaments of the hip joint are loose, so the head of the femur slips out of its socket. Early treatment (a splint or harness to hold the femur in place or surgery to tighten hip ligaments) is essential to prevent permanent crippling. ■

During youth, growth of the skeleton not only increases overall body height but also changes body proportions **(Figure 7.38)**. At birth, the head and trunk are approximately 1½ times as long as the lower limbs. The lower limbs grow more rapidly than the trunk from this time on, and by the age of 10, the head and trunk are approximately the same height as the lower limbs, a condition that persists thereafter. During puberty, the female pelvis broadens in preparation for childbearing, and the entire male skeleton becomes more robust. Once adult height is reached, a healthy skeleton changes very little until late middle age.

Old age affects many parts of the skeleton, especially the spine. As the discs become thinner, less hydrated, and less elastic, the risk of disc herniation increases. By 55 years, a loss of several centimeters in stature is common. Further shortening can be produced by osteoporosis of the spine or by kyphosis (called "dowager's hump" in the elderly). What was done during youth may be undone in old age as the vertebral column gradually resumes its initial arc shape.

The thorax becomes more rigid with age, largely because the costal cartilages ossify. This loss of rib cage elasticity causes shallow breathing, which leads to less efficient gas exchange.

All bones, you will recall, lose mass with age. Cranial bones lose less mass than most, but changes in facial contours with age are common. As the bony tissue of the jaws declines, the jaws look small and childlike once again. If the elderly person loses his or her teeth, this loss of bone from the jaws is accelerated, because the alveolar region bone is resorbed. As bones become more porous, they are more likely to fracture, especially the vertebrae and the neck of the femur.

CHECK YOUR UNDERSTANDING

36. What developmental events result in a dramatic enlargement of the facial skeleton between the ages of 6 and 13?
37. Under what conditions does the lumbar curvature of the spine develop?

For answers, see Appendix G.

Our skeleton is a marvelous substructure, to be sure, but it is much more than that. It is a protector and supporter of other body systems, and without it (and the joints considered in Chapter 8), our muscles would be almost useless. The homeostatic relationships between the skeletal system and other body systems are illustrated in *Making Connections* in Chapter 6 (pp. 192–193).

RELATED CLINICAL TERMS

Chiropractic (ki″ro-prak′tik) A system of treating disease by manipulating the vertebral column based on the theory that most diseases are due to pressure on nerves caused by faulty bone alignment; a specialist in this field is a chiropractor.

Clubfoot A relatively common congenital defect (1 in 700 births) in which the soles of the feet face medially and the toes point inferiorly; may be genetically induced or reflect an abnormal position of the foot during fetal development.

Laminectomy Surgical removal of a vertebral lamina; most often done to relieve the symptoms of a ruptured disc.

Orthopedist (or″tho-pe′dist) or **orthopedic surgeon** A physician who specializes in restoring lost skeletal system function or repairing damage to bones and joints.

Pelvimetry Measurement of the dimensions of the inlet and outlet of the pelvis, usually to determine whether the pelvis is of adequate size to allow normal delivery of a baby.

Spina bifida (spi′nah bĭ′fi-dah; "cleft spine") Congenital defect of the vertebral column in which one or more of the vertebral arches are incomplete; ranges in severity from inconsequential to severe conditions that impair neural functioning and encourage nervous system infections.

Spinal fusion Surgical procedure involving insertion of bone chips (or crushed bone) to immobilize and stabilize a specific region of the vertebral column, particularly in cases of vertebral fracture and herniated discs.

CHAPTER SUMMARY

1. The axial skeleton forms the longitudinal axis of the body. Its principal subdivisions are the skull, vertebral column, and thoracic cage. It provides support and protection (by enclosure).
2. The appendicular skeleton consists of the bones of the pectoral and pelvic girdles and the limbs. It allows mobility for manipulation and locomotion.

PART 1: THE AXIAL SKELETON

The Skull (pp. 200–216)

1. The skull is formed by 22 bones. The cranium forms the vault and base of the skull, which protect the brain. The facial skeleton provides openings for the respiratory and digestive passages and attachment points for facial muscles.
2. Except for the temporomandibular joints, all bones of the adult skull are joined by immovable sutures.
3. **Cranium.** The eight bones of the cranium include the paired parietal and temporal bones and the single frontal, occipital, ethmoid, and sphenoid bones (see Table 7.1, pp. 214–215).
4. **Facial bones.** The 14 bones of the face include the paired maxillae, zygomatics, nasals, lacrimals, palatines, and inferior nasal conchae and the single mandible and vomer bones (Table 7.1).
5. **Orbits and nasal cavity.** Both the orbits and the nasal cavities are complicated bony regions formed of several bones.
6. **Paranasal sinuses.** Paranasal sinuses occur in the frontal, ethmoid, sphenoid, and maxillary bones.
7. **Hyoid bone.** The hyoid bone, supported in the neck by ligaments, serves as an attachment point for tongue and neck muscles.

The Vertebral Column (pp. 216–223)

1. **General characteristics.** The vertebral column includes 24 movable vertebrae (7 cervical, 12 thoracic, and 5 lumbar) and the sacrum and coccyx.
2. The fibrocartilage intervertebral discs act as shock absorbers and provide flexibility to the vertebral column.
3. The primary curvatures of the vertebral column are the thoracic and sacral; the secondary curvatures are the cervical and lumbar. Curvatures increase spine flexibility.
4. **General structure of vertebrae.** With the exception of C_1 and C_2, all vertebrae have a body, two transverse processes, two superior and two inferior articular processes, a spinous process, and a vertebral arch.
5. **Regional vertebral characteristics.** Special features distinguish the regional vertebrae (see Table 7.2, p. 222).

The Thoracic Cage (pp. 223–225)

1. The bones of the thoracic cage include the 12 rib pairs, the sternum, and the thoracic vertebrae. The thoracic cage protects the organs of the thoracic cavity.
2. **Sternum.** The sternum consists of the fused manubrium, body, and xiphoid process.
3. **Ribs.** The first seven rib pairs are called true ribs; the rest are called false ribs. Ribs 11 and 12 are floating ribs.

PART 2: THE APPENDICULAR SKELETON

The Pectoral (Shoulder) Girdle* (pp. 225–228)

1. Each pectoral girdle consists of one clavicle and one scapula. The pectoral girdles attach the upper limbs to the axial skeleton.
2. **Clavicles.** The clavicles hold the scapulae laterally away from the thorax. The sternoclavicular joints are the only attachment points of the pectoral girdle to the axial skeleton.
3. **Scapulae.** The scapulae articulate with the clavicles and with the humerus bones of the arms.

The Upper Limb* (pp. 228–233)

1. Each upper limb consists of 30 bones and is specialized for mobility.
2. **Arm/forearm/hand.** The skeleton of the arm is composed solely of the humerus; the skeleton of the forearm is composed of the radius and ulna; and the skeleton of the hand consists of the carpals, metacarpals, and phalanges.

The Pelvic (Hip) Girdle* (pp. 233–237)

1. The pelvic girdle, a heavy structure specialized for weight bearing, is composed of two hip bones that secure the lower limbs to the axial skeleton. Together with the sacrum and coccyx, the hip bones form the basinlike bony pelvis.
2. Each hip bone consists of three fused bones: ilium, ischium, and pubis. The acetabulum occurs at the point of fusion.
3. **Ilium/ischium/pubis.** The ilium is the superior flaring portion of the hip bone. Each ilium forms a secure joint with the sacrum posteriorly. The ischium is a curved bar of bone; we sit on the ischial tuberosities. The V-shaped pubic bones articulate anteriorly at the pubic symphysis.
4. **Pelvic structure and childbearing.** The male pelvis is deep and narrow with larger, heavier bones than those of the female. The female pelvis, which forms the birth canal, is shallow and wide.

*For associated bone markings, see the pages indicated in the section heads.

7

The Lower Limb* (pp. 237–242)

1. Each lower limb consists of the thigh, leg, and foot and is specialized for weight bearing and locomotion.
2. **Thigh.** The femur is the only bone of the thigh. Its ball-shaped head articulates with the acetabulum.
3. **Leg.** The bones of the leg are the tibia, which participates in forming both the knee and ankle joints, and the fibula.
4. **Foot.** The bones of the foot include the tarsals, metatarsals, and phalanges. The most important tarsals are the calcaneus (heel bone) and the talus, which articulates with the tibia superiorly.
5. The foot is supported by three arches (lateral, medial, and transverse) that distribute body weight to the heel and ball of the foot.

Developmental Aspects of the Skeleton (pp. 242–244)

1. Fontanelles, which allow brain growth and ease birth passage, are present in the skull at birth. Growth of the cranium after birth is related to brain growth. Increase in size of the facial skeleton follows tooth development and enlargement of nose and sinus cavities.
2. The vertebral column is C shaped at birth (thoracic and sacral curvatures are present); the secondary curvatures form when the baby begins to lift its head and walk.
3. Long bones continue to grow in length until late adolescence. The head and torso, initially 1½ times the length of the lower limbs, equal their length by the age of 10.
4. Changes in the female pelvis (preparatory for childbirth) occur during puberty.
5. Once at adult height, the skeleton changes little until late middle age. With old age, the intervertebral discs thin; this, along with osteoporosis, leads to a gradual loss in height and increased risk of disc herniation. Loss of bone mass increases the risk of fractures, and thoracic cage rigidity promotes breathing difficulties.

*For associated bone markings, see the pages indicated in the section heads.

7

REVIEW QUESTIONS

Multiple Choice/Matching

(Some questions have more than one correct answer. Select the best answer or answers from the choices given.)

1. Match the bones in column B with their description in column A. (Note that some descriptions require more than a single choice.)

Column A	Column B
____ (1) connected by the coronal suture	(a) ethmoid
____ (2) keystone bone of cranium	(b) frontal
____ (3) keystone bone of the face	(c) mandible
____ (4) form the hard palate	(d) maxillary
____ (5) allows the spinal cord to pass	(e) occipital
____ (6) forms the chin	(f) palatine
____ (7) contain paranasal sinuses	(g) parietal
____ (8) contains mastoid sinuses	(h) sphenoid
	(i) temporal

2. Match the key terms with the bone descriptions that follow.

Key: (a) clavicle (b) ilium (c) ischium (d) pubis (e) sacrum (f) scapula (g) sternum

____ (1) bone of the axial skeleton to which the pectoral girdle attaches
____ (2) markings include glenoid cavity and acromion
____ (3) features include the ala, crest, and greater sciatic notch
____ (4) doubly curved; acts as a shoulder strut
____ (5) pelvic girdle bone that articulates with the axial skeleton
____ (6) the "sit-down" bone
____ (7) anteriormost bone of the pelvic girdle
____ (8) part of the vertebral column

3. Use key choices to identify the bone descriptions that follow.

Key: (a) carpals (b) femur (c) fibula (d) humerus (e) radius (f) tarsals (g) tibia (h) ulna

____ (1) articulates with the acetabulum and the tibia
____ (2) forms the lateral aspect of the ankle
____ (3) bone that "carries" the hand
____ (4) the wrist bones
____ (5) end shaped like a monkey wrench
____ (6) articulates with the capitulum of the humerus
____ (7) largest bone of this "group" is the calcaneus

Short Answer Essay Questions

4. Name the cranial and facial bones and compare and contrast the functions of the cranial and facial skeletons.
5. How do the relative proportions of the cranium and face of a fetus compare with those of an adult skull?
6. Name and diagram the normal vertebral curvatures. Which are primary and which are secondary curvatures?
7. List at least two specific anatomical characteristics each for typical cervical, thoracic, and lumbar vertebrae that would allow *anyone* to identify each type correctly.
8. What is the function of the intervertebral discs?
9. Distinguish between the anulus fibrosus and nucleus pulposus regions of a disc. Which provides durability and strength? Which provides resilience? Which part is involved in a "slipped" disc?
10. What is a true rib? A false rib?
11. The major function of the shoulder girdle is flexibility. What is the major function of the pelvic girdle? Relate these functional differences to anatomical differences seen in these girdles.
12. List three important differences between the male and female pelves.
13. Briefly describe the anatomical characteristics and impairment of function seen in cleft palate and hip dysplasia.
14. Compare a young adult skeleton to that of an extremely aged person relative to bone mass in general and the bony structure of the skull, thorax, and vertebral column in particular.
15. Peter Howell, a teaching assistant in the anatomy class, picked up a hip bone and pretended it was a telephone. He held the big hole in this bone right up to his ear and said, "Hello, obturator, obturator (operator, operator)." Name the structure he was helping the students to learn.

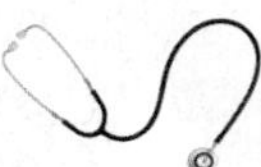

Critical Thinking and Clinical Application Questions

1. Justiniano worked in a poultry-packing plant where his job was cutting open chickens and stripping out their visceral organs. After work, he typed for long hours on his computer keyboard, writing a book about his work in the plant. Soon, his wrist and hand began to hurt whenever he flexed it, and he began to awaken at night with pain and tingling on the thumb-half of his hand. What condition did he probably have?
2. Ralph had polio as a boy and was partially paralyzed in one lower limb for over a year. Although no longer paralyzed, he now has a severe lateral curvature of the lumbar spine. Explain what has happened and identify his condition.
3. Mary's grandmother slipped on a scatter rug and fell heavily to the floor. Her left lower limb was laterally rotated and noticeably shorter than the right, and when she attempted to get up, she winced with pain. Mary surmised that her grandmother might have "fractured her hip," which later proved to be true. What bone was probably fractured and at what site? Why is a "fractured hip" a common type of fracture in the elderly?
4. Mrs. Shea came up with what she considered to be a clever idea to bypass the long lines at Disney World. She had her husband rent a wheelchair and he wheeled her around from one exhibit to another for the better part of three days. As they sat on the plane, waiting to take off for Chicago, she complained to him that she had two sore spots on her buttocks. Why? What do you suppose would happen (to her buttocks) if she was wheeled around for a few more days?

7

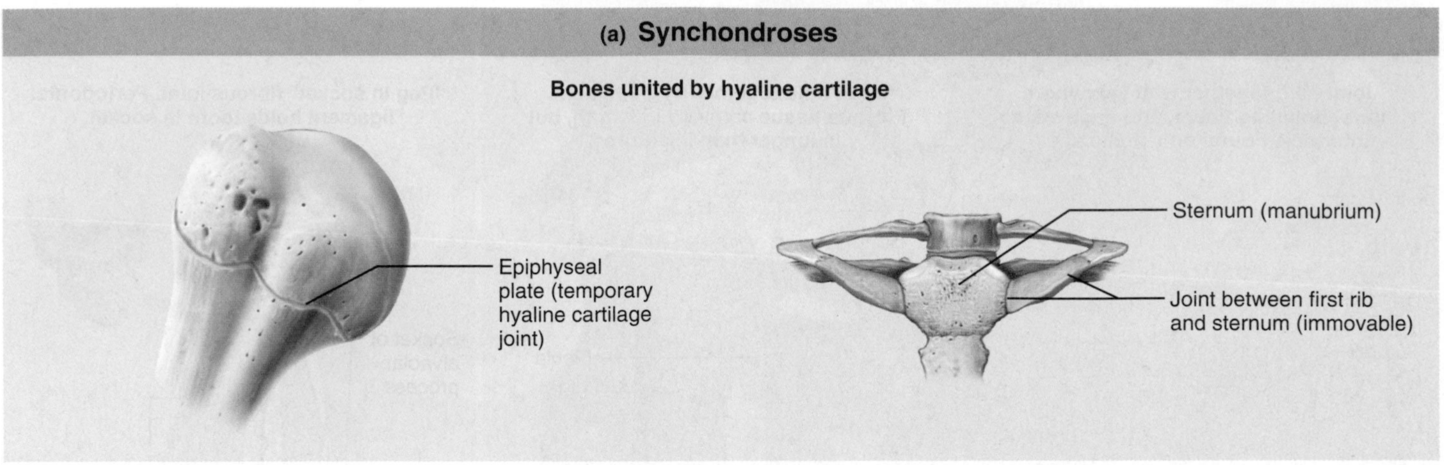

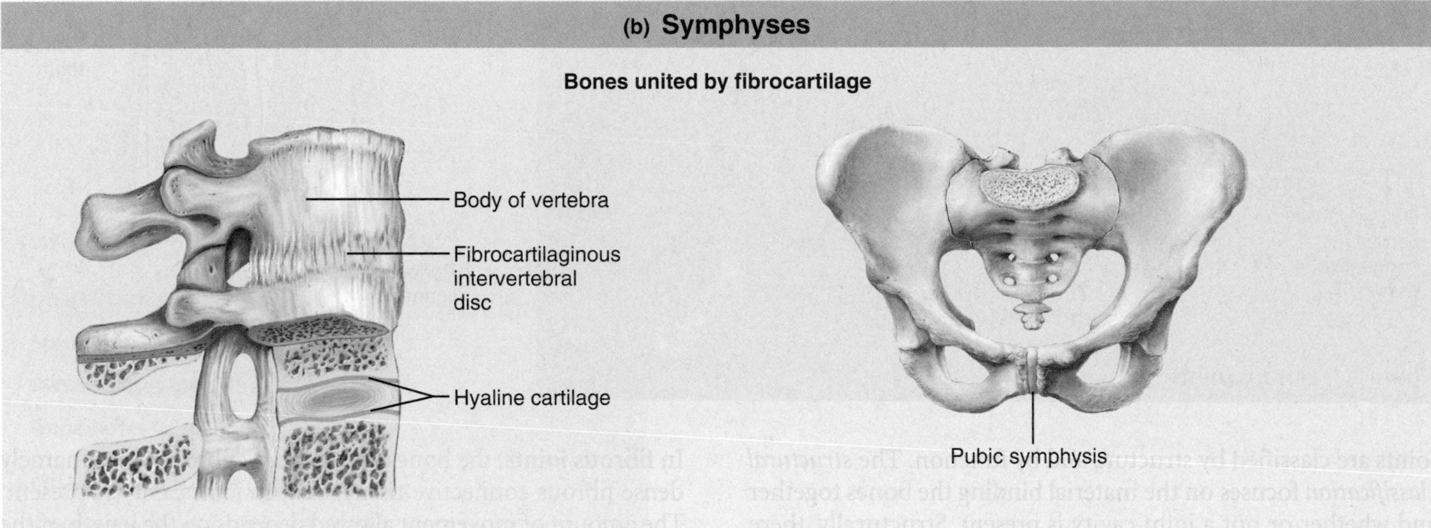

Figure 8.2 Cartilaginous joints.

The amount of movement allowed depends on the length of the connecting fibers, and slight to considerable movement is possible. For example, the ligament connecting the distal ends of the tibia and fibula is short (Figure 8.1b), and this joint allows only slightly more movement than a suture, a characteristic best described as "give." True movement is still prevented, so the joint is classed functionally as an immovable joint, or synarthrosis. (Note, however, that some authorities classify this joint as an amphiarthrosis.) On the other hand, the fibers of the ligament-like interosseous membrane connecting the radius and ulna along their length (Figure 7.27, p. 230) are long enough to permit rotation of the radius around the ulna and the joint is diarthrotic.

Gomphoses

A **gomphosis** (gom-fo′sis) is a peg-in-socket fibrous joint (Figure 8.1c). The only example is the articulation of a tooth with its bony alveolar socket. The term *gomphosis* comes from the Greek *gompho,* meaning "nail" or "bolt," and refers to the way teeth are embedded in their sockets (as if hammered in). The fibrous connection in this case is the short **periodontal ligament** (Figure 23.11, p. 863).

Cartilaginous Joints

▶ Describe the general structure of cartilaginous joints. Name and give an example of each of the two common types of cartilaginous joints.

In **cartilaginous joints** (kar″ti-laj′ĭ-nus), the articulating bones are united by cartilage. Like fibrous joints, they lack a joint cavity and are not highly movable. The two types of cartilaginous joints are *synchondroses* and *symphyses.*

Synchondroses

A bar or plate of *hyaline cartilage* unites the bones at a **synchondrosis** (sin″kon-dro′sis; "junction of cartilage"). Virtually all synchondroses are synarthrotic.

The most common examples of synchondroses are the epiphyseal plates in long bones of children **(Figure 8.2a)**. Epiphyseal plates are temporary joints and eventually become synostoses. Another example of a synchondrosis is the immovable joint between the costal cartilage of the first rib and the manubrium of the sternum (Figure 8.2a).

Symphyses

In **symphyses** (sim′fih-sēz; "growing together") the articular surfaces of the bones are covered with articular (hyaline) cartilage, which in turn is fused to an intervening pad, or plate, of *fibrocartilage*, which is the main connecting material. Since fibrocartilage is compressible and resilient, it acts as a shock absorber and permits a limited amount of movement at the joint. Symphyses are amphiarthrotic joints designed for strength with flexibility. Examples include the intervertebral joints and the pubic symphysis of the pelvis (Figure 8.2b, and see **Table 8.2** on p. 254).

CHECK YOUR UNDERSTANDING

1. What term is a synonym for "joint"?
2. What functional joint class contains the least mobile joints?
3. Of sutures, symphyses, and synchondroses, which are cartilaginous joints?
4. How are joint mobility and stability related?

For answers, see Appendix G.

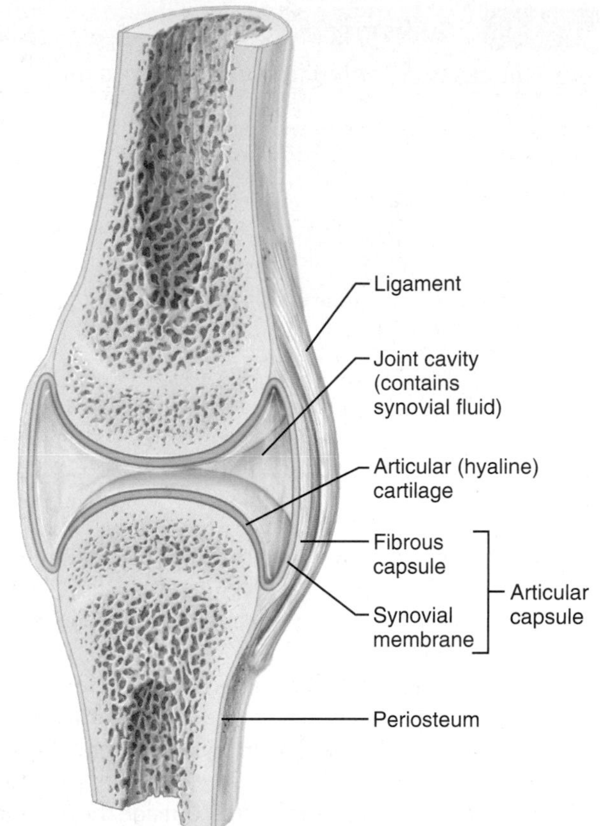

Figure 8.3 General structure of a synovial joint. The articulating bone ends are covered with articular cartilage and enclosed within an articular capsule which is typically reinforced by ligaments externally. Internally, the fibrous capsule is lined with a smooth synovial membrane that secretes synovial fluid.

8

Synovial Joints

- Describe the structural characteristics of synovial joints.
- Compare the structures and functions of bursae and tendon sheaths.
- List three natural factors that stabilize synovial joints.

Synovial joints (si-no′ve-al; "joint eggs") are those in which the articulating bones are separated by a fluid-containing joint cavity. This arrangement permits substantial freedom of movement, and all synovial joints are freely movable diarthroses. Nearly all joints of the limbs—indeed, most joints of the body—fall into this class.

General Structure

Synovial joints have six distinguishing features **(Figure 8.3)**:

1. **Articular cartilage.** Glassy-smooth hyaline cartilage covers the opposing bone surfaces as **articular cartilage**. These thin (1 mm or less) but spongy cushions absorb compression placed on the joint and thereby keep the bone ends from being crushed.
2. **Joint (synovial) cavity.** A feature unique to synovial joints, the **joint cavity** is really just a potential space that contains a small amount of synovial fluid.
3. **Articular capsule.** The joint cavity is enclosed by a two-layered **articular capsule**, or *joint capsule*. The external layer is a tough **fibrous capsule**, composed of dense irregular connective tissue, that is continuous with the periostea of the articulating bones. It strengthens the joint so that the bones are not pulled apart. The inner layer of the joint capsule is a **synovial membrane** composed of loose connective tissue. Besides lining the fibrous capsule internally, it covers all internal joint surfaces that are not hyaline cartilage.
4. **Synovial fluid.** A small amount of slippery **synovial fluid** occupies all free spaces within the joint capsule. This fluid is derived largely by filtration from blood flowing through the capillaries in the synovial membrane. Synovial fluid has a viscous, egg-white consistency (*ovum* = egg) due to hyaluronic acid secreted by cells in the synovial membrane, but it thins and becomes less viscous, as it warms during joint activity.

   Synovial fluid, which is also found *within* the articular cartilages, provides a slippery weight-bearing film that reduces friction between the cartilages. Without this lubricant, rubbing would wear away joint surfaces and excessive friction could overheat and destroy the joint tissues, essentially "cooking" them. The synovial fluid is forced from the cartilages when a joint is compressed; then as pressure on the joint is relieved, synovial fluid seeps back into the articular cartilages like water into a sponge, ready to be squeezed out again the next time the joint is loaded (put under pressure). This process, called *weeping lubrication*,

TABLE 8.1 Summary of Joint Classes

STRUCTURAL CLASS	STRUCTURAL CHARACTERISTICS	TYPES	MOBILITY
Fibrous	Bone ends/parts united by collagen fibers	Suture (short fibers)	Immobile (synarthrosis)
		Syndesmosis (longer fibers)	Slightly mobile (amphiarthrosis) and immobile
		Gomphosis (periodontal ligament)	Immobile
Cartilaginous	Bone ends/parts united by cartilage	Synchondrosis (hyaline cartilage)	Immobile
		Symphysis (fibrocartilage)	Slightly movable
Synovial	Bone ends/parts covered with articular cartilage and enclosed within an articular capsule lined with synovial membrane	(1) Plane (2) Hinge (3) Pivot (4) Condyloid (5) Saddle (6) Ball and socket	Freely movable (diarthrosis; movements depend on design of joint)

8

lubricates the free surfaces of the cartilages and nourishes their cells. (Remember, cartilage is avascular.) Synovial fluid also contains phagocytic cells that rid the joint cavity of microbes and cellular debris.

5. **Reinforcing ligaments.** Synovial joints are reinforced and strengthened by a number of bandlike **ligaments**. Most often, these are **capsular**, or **intrinsic, ligaments**, which are thickened parts of the fibrous capsule. In other cases, they remain distinct and are found outside the capsule (as **extracapsular ligaments**) or deep to it (as **intracapsular ligaments**). Since intracapsular ligaments are covered with synovial membrane, they do not actually lie *within* the joint cavity.

   People said to be double-jointed amaze the rest of us by placing both heels behind their neck. However, they have the normal number of joints. It's just that their joint capsules and ligaments are more stretchy and loose than average.

6. **Nerves and blood vessels.** Synovial joints are richly supplied with sensory nerve fibers that innervate the capsule. Some of these fibers detect pain, as anyone who has suffered joint injury is aware, but most monitor joint position and stretch, thus helping to maintain muscle tone. Stretching these structures sends nerve impulses to the central nervous system, resulting in reflexive contraction of muscles surrounding the joint. Synovial joints are also richly supplied with blood vessels, most of which supply the synovial membrane. There, extensive capillary beds produce the blood filtrate that is the basis of synovial fluid.

Besides the basic components described above, certain synovial joints have other structural features. Some, such as the hip and knee joints, have cushioning **fatty pads** between the fibrous capsule and the synovial membrane or bone. Others have discs or wedges of fibrocartilage separating the articular surfaces. Where present, these so-called **articular discs**, or **menisci** (mĕ-nis′ki; "crescents"), extend inward from the articular capsule and partially or completely divide the synovial cavity in two (see the menisci of the knee in Figure 8.8a, b, e, and f). Articular discs improve the fit between articulating bone ends, making the joint more stable and minimizing wear and tear on the joint surfaces. Besides the knees, articular discs occur in the jaw, and a few other joints (see notations in the Structural Type column in Table 8.2).

Bursae and Tendon Sheaths

Bursae and tendon sheaths are not strictly part of synovial joints, but they are often found closely associated with them **(Figure 8.4)**. Essentially bags of lubricant, they act as "ball bearings" to reduce friction between adjacent structures during joint activity. **Bursae** (ber′se; "purse") are flattened fibrous sacs lined with synovial membrane and containing a thin film of synovial fluid. They occur where ligaments, muscles, skin, tendons, or bones rub together.

A **tendon sheath** is essentially an elongated bursa that wraps completely around a tendon subjected to friction, like a bun around a hot dog. They are common where several tendons are crowded together within narrow canals (in the wrist region, for example).

Factors Influencing the Stability of Synovial Joints

Because joints are constantly stretched and compressed, they must be stabilized so that they do not dislocate (come out of alignment). The stability of a synovial joint depends chiefly on three factors: the shapes of the articular surfaces; the number and positioning of ligaments; and muscle tone.

Articular Surfaces

The shapes of articular surfaces determine what movements are possible at a joint, but surprisingly, articular surfaces play only a minor role in joint stability. Many joints have shallow sockets or noncomplementary articulating surfaces ("misfits") that actually hinder joint stability. But when articular surfaces are large and fit snugly together, or when the socket is deep, stability is vastly improved. The ball and deep socket of the hip joint provide the best example of a joint made extremely stable by the shape of its articular surfaces.

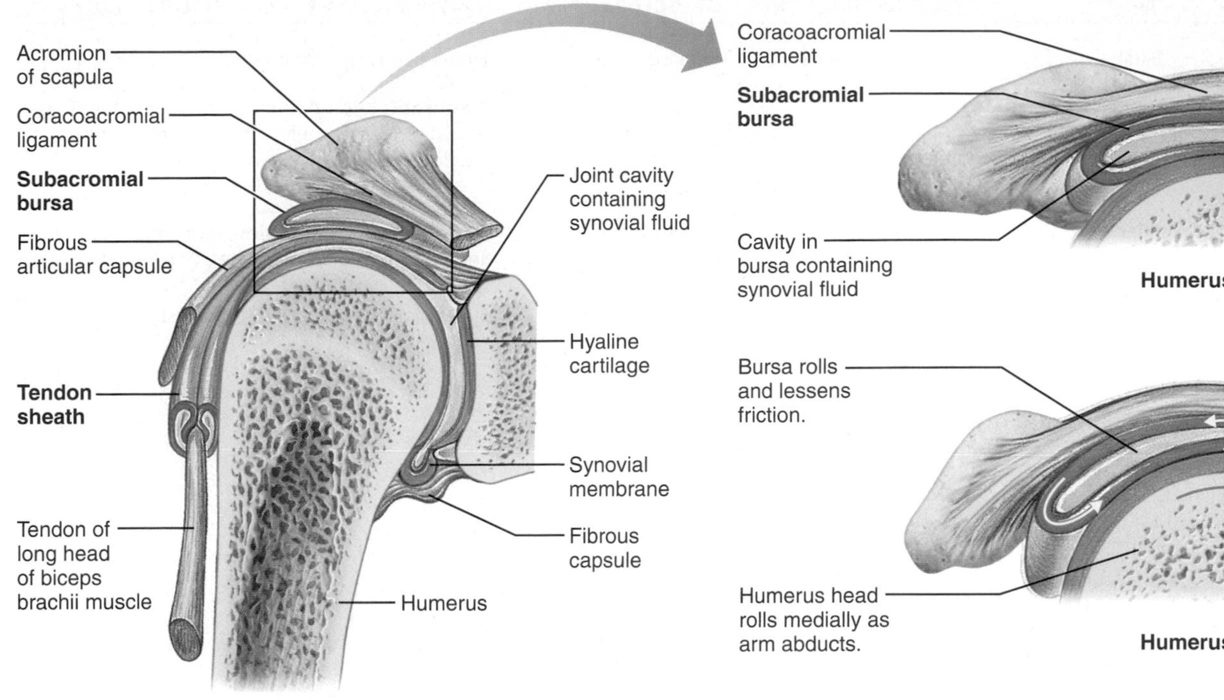

(a) Frontal section through the right shoulder joint

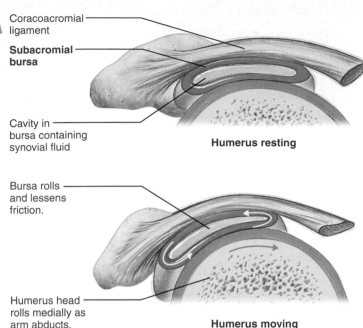

(b) Enlargement of (a), showing how a bursa eliminates friction where a ligament (or other structure) would rub against a bone

Figure 8.4 **Bursae and tendon sheaths.**

8

Ligaments

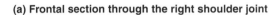

The capsules and ligaments of synovial joints unite the bones and prevent excessive or undesirable motion. As a rule, the more ligaments a joint has, the stronger it is. However, when other stabilizing factors are inadequate, undue tension is placed on the ligaments and they stretch. Stretched ligaments stay stretched, like taffy, and a ligament can stretch only about 6% of its length before it snaps. Thus, when ligaments are the major means of bracing a joint, the joint is not very stable.

Muscle Tone

For most joints, the muscle tendons that cross the joint are the most important stabilizing factor. These tendons are kept taut at all times by the tone of their muscles. (*Muscle tone* is defined as low levels of contractile activity in relaxed muscles that keep the muscles healthy and ready to react to stimulation.) Muscle tone is extremely important in reinforcing the shoulder and knee joints and the arches of the foot.

CHECK YOUR UNDERSTANDING

5. What are the two layers of the articular capsule?
6. How do bursae and tendon sheaths improve joint function?
7. Generally speaking, what factor is most important in stabilizing synovial joints?
8. What is the importance of weeping lubrication?

For answers, see Appendix G.

Movements Allowed by Synovial Joints

- ▶ Name and describe (or perform) the common body movements.
- ▶ Name and provide examples of the six types of synovial joints based on the movement(s) allowed.

Every skeletal muscle of the body is attached to bone or other connective tissue structures at no fewer than two points. The muscle's **origin** is attached to the immovable (or less movable) bone. Its other end, the **insertion**, is attached to the movable bone. Body movement occurs when muscles contract across joints and their insertion moves toward their origin. The movements can be described in directional terms relative to the lines, or *axes*, around which the body part moves and the planes of space along which the movement occurs, that is, along the transverse, frontal, or sagittal plane. (See Chapter 1 to review these planes.)

Range of motion allowed by synovial joints varies from **nonaxial movement** (slipping movements only, since there is no axis around which movement can occur) to **uniaxial movement** (movement in one plane) to **biaxial movement** (movement in two planes) to **multiaxial movement** (movement in or around all three planes of space and axes). Range of motion varies greatly in different people. In some, such as trained gymnasts or acrobats, range of joint movement may be extraordinary. The ranges of motion at the major joints are given in the far right column of Table 8.2.

There are three general types of movements: *gliding, angular movements,* and *rotation.* The most common body movements

TABLE 8.2 Structural and Functional Characteristics of Body Joints

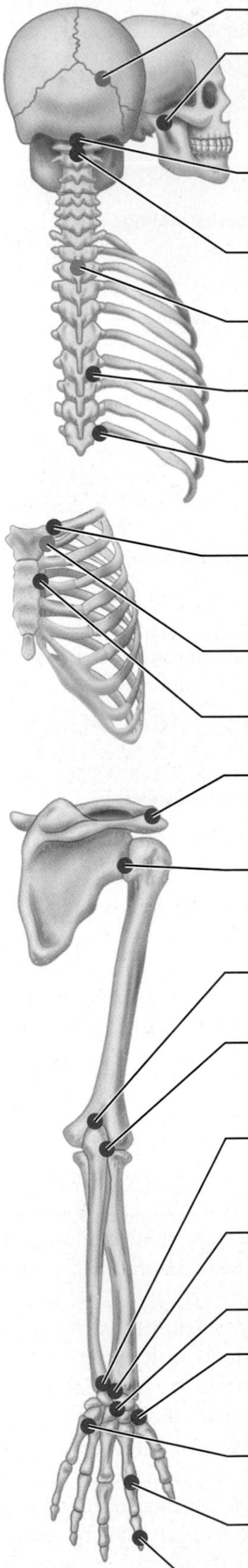

ILLUSTRATION	JOINT	ARTICULATING BONES	STRUCTURAL TYPE*	FUNCTIONAL TYPE; MOVEMENTS ALLOWED
	Skull	Cranial and facial bones	Fibrous; suture	Synarthrotic; no movement
	Temporo-mandibular	Temporal bone of skull and mandible	Synovial; modified hinge† (contains articular disc)	Diarthrotic; gliding and uniaxial rotation; slight lateral movement, elevation, depression, protraction, and retraction of mandible
	Atlanto-occipital	Occipital bone of skull and atlas	Synovial; condyloid	Diarthrotic; biaxial; flexion, extension, lateral flexion, circumduction of head on neck
	Atlantoaxial	Atlas (C_1) and axis (C_2)	Synovial; pivot	Diarthrotic; uniaxial; rotation of the head
	Intervertebral	Between adjacent vertebral bodies	Cartilaginous; symphysis	Amphiarthrotic; slight movement
	Intervertebral	Between articular processes	Synovial; plane	Diarthrotic; gliding
	Vertebrocostal	Vertebrae (transverse processes or bodies) and ribs	Synovial; plane	Diarthrotic; gliding of ribs
	Sternoclavicular	Sternum and clavicle	Synovial; shallow saddle (contains articular disc)	Diarthrotic; multiaxial (allows clavicle to move in all axes)
	Sternocostal (first)	Sternum and rib 1	Cartilaginous; synchondrosis	Synarthrotic; no movement
	Sternocostal	Sternum and ribs 2–7	Synovial; double plane	Diarthrotic; gliding
	Acromio-clavicular	Acromion of scapula and clavicle	Synovial; plane (contains articular disc)	Diarthrotic; gliding and rotation of scapula on clavicle
	Shoulder (glenohumeral)	Scapula and humerus	Synovial; ball and socket	Diarthrotic; multiaxial; flexion, extension, abduction, adduction, circumduction, rotation of humerus
	Elbow	Ulna (and radius) with humerus	Synovial; hinge	Diarthrotic; uniaxial; flexion, extension of forearm
	Radioulnar (proximal)	Radius and ulna	Synovial; pivot	Diarthrotic; uniaxial; rotation of radius around long axis of forearm to allow pronation and supination
	Radioulnar (distal)	Radius and ulna	Synovial; pivot (contains articular disc)	Diarthrotic; uniaxial; rotation (convex head of ulna rotates in ulnar notch of radius)
	Wrist (radiocarpal)	Radius and proximal carpals	Synovial; condyloid	Diarthrotic; biaxial; flexion, extension, abduction, adduction, circumduction of hand
	Intercarpal	Adjacent carpals	Synovial; plane	Diarthrotic; gliding
	Carpometacarpal of digit 1 (thumb)	Carpal (trapezium) and metacarpal 1	Synovial; saddle	Diarthrotic; biaxial; flexion, extension, abduction, adduction, circumduction, opposition of metacarpal 1
	Carpometacarpal of digits 2–5	Carpal(s) and metacarpal(s)	Synovial; plane	Diarthrotic; gliding of metacarpals
	Knuckle (metacarpo-phalangeal)	Metacarpal and proximal phalanx	Synovial; condyloid	Diarthrotic; biaxial; flexion, extension, abduction, adduction, circumduction of fingers
	Finger (interphalangeal)	Adjacent phalanges	Synovial; hinge	Diarthrotic; uniaxial; flexion, extension of fingers

TABLE 8.2 (continued)

ILLUSTRATION	JOINT	ARTICULATING BONES	STRUCTURAL TYPE*	FUNCTIONAL TYPE; MOVEMENTS ALLOWED
	Sacroiliac	Sacrum and coxal bone	Synovial; plane in childhood, increasingly fibrous in adult	Diarthrotic in child; amphiarthrotic in adult; (more movement during pregnancy)
	Pubic symphysis	Pubic bones	Cartilaginous; symphysis	Amphiarthrotic; slight movement (enhanced during pregnancy)
	Hip (coxal)	Hip bone and femur	Synovial; ball and socket	Diarthrotic; multiaxial; flexion, extension, abduction, adduction, rotation, circumduction of thigh
	Knee (tibiofemoral)	Femur and tibia	Synovial; modified hinge† (contains articular discs)	Diarthrotic; biaxial; flexion, extension of leg, some rotation allowed in flexed position
	Knee (femoropatellar)	Femur and patella	Synovial; plane	Diarthrotic; gliding of patella
	Tibiofibular (proximal)	Tibia and fibula (proximally)	Synovial; plane	Diarthrotic; gliding of fibula
	Tibiofibular (distal)	Tibia and fibula (distally)	Fibrous; syndesmosis	Synarthrotic; slight "give" during dorsiflexion
	Ankle	Tibia and fibula with talus	Synovial; hinge	Diarthrotic; uniaxial; dorsiflexion, and plantar flexion of foot
	Intertarsal	Adjacent tarsals	Synovial; plane	Diarthrotic; gliding; inversion and eversion of foot
	Tarsometatarsal	Tarsal(s) and metatarsal(s)	Synovial; plane	Diarthrotic; gliding of metatarsals
	Metatarsophalangeal	Metatarsal and proximal phalanx	Synovial; condyloid	Diarthrotic; biaxial; flexion, extension, abduction, adduction, circumduction of great toe
	Toe (interphalangeal)	Adjacent phalanges	Synovial; hinge	Diarthrotic; uniaxial; flexion; extension of toes

* **Fibrous joints** indicated by orange circles; **cartilaginous joints** by blue circles; **synovial joints** by purple circles.
† These modified hinge joints are structurally bicondylar.

allowed by synovial joints are described next and illustrated in **Figure 8.5.**

Gliding Movements

Gliding movements (Figure 8.5a) are the simplest joint movements. Gliding occurs when one flat, or nearly flat, bone surface glides or slips over another (back-and-forth and side-to-side) without appreciable angulation or rotation. Gliding movements occur at the intercarpal and intertarsal joints, and between the flat articular processes of the vertebrae (Table 8.2).

Angular Movements

Angular movements (Figure 8.5b–e) increase or decrease the angle between two bones. These movements may occur in any plane of the body and include flexion, extension, hyperextension, abduction, adduction, and circumduction.

Flexion **Flexion** (flek′shun) is a bending movement, usually along the sagittal plane, that *decreases the angle* of the joint and brings the articulating bones closer together. Examples include bending the head forward on the chest (Figure 8.5b) and bending the body trunk or the knee from a straight to an angled position (Figure 8.5c and d). As a less obvious example, the arm is flexed at the shoulder when the arm is lifted in an anterior direction (Figure 8.5d).

Extension **Extension** is the reverse of flexion and occurs at the same joints. It involves movement along the sagittal plane that *increases the angle* between the articulating bones and typically straightens a flexed limb or body part. Examples include straightening a flexed neck, body trunk, elbow, or knee (Figure 8.5b–d). Excessive extension such as extending the head or hip joint beyond anatomical position (Figure 8.5b, c) is called **hyperextension** (literally, "superextension").

Abduction **Abduction** ("moving away") is movement of a limb *away* from the midline or median plane of the body, along the frontal plane. Raising the arm or thigh laterally is an example of abduction (Figure 8.5e). For the fingers or toes, abduction means spreading them apart. In this case "midline" is the longest digit: the third finger or second toe. Notice, however, that lateral bending of the trunk away from the body midline in the frontal plane is called *lateral flexion*, not abduction.

Adduction **Adduction** ("moving toward") is the opposite of abduction, so it is the movement of a limb *toward* the body midline or, in the case of the digits, toward the midline of the hand or foot (Figure 8.5e).

Circumduction **Circumduction** (Figure 8.5e) is moving a limb so that it describes a cone in space (*circum* = around; *duco* = to draw). The distal end of the limb moves in a circle, while the point of the cone (the shoulder or hip joint) is more or less stationary. A pitcher winding up to throw a ball is actually circumducting his or her pitching arm. Because circumduction consists of flexion, abduction, extension, and adduction performed in succession, it is the quickest way to exercise the many muscles that move the hip and shoulder ball-and-socket joints.

Rotation

Rotation is the turning of a bone around its own long axis. It is the only movement allowed between the first two cervical vertebrae and is common at the hip (Figure 8.5f) and shoulder joints. Rotation may be directed toward the midline or away from it. For example, in *medial rotation* of the thigh, the femur's anterior surface moves toward the median plane of the body; *lateral rotation* is the opposite movement.

Special Movements

Certain movements do not fit into any of the above categories and occur at only a few joints. Some of these special movements are illustrated in **Figure 8.6**.

Supination and Pronation The terms **supination** (soo″pĭ-na′shun; "turning backward") and **pronation** (pro-na′shun; "turning forward") refer to the movements of the radius around the ulna (Figure 8.6a). Rotating the forearm laterally so that the palm faces anteriorly or superiorly is supination. In the anatomical position, the hand is supinated and the radius and ulna are parallel.

In pronation, the forearm rotates medially and the palm faces posteriorly or inferiorly. Pronation moves the distal end of the radius across the ulna so that the two bones form an X. This is the forearm's position when we are standing in a relaxed manner. Pronation is a much weaker movement than supination.

A trick to help you keep these terms straight: A *pro* basketball player pronates his or her forearm to dribble the ball.

Figure 8.5 Movements allowed by synovial joints.

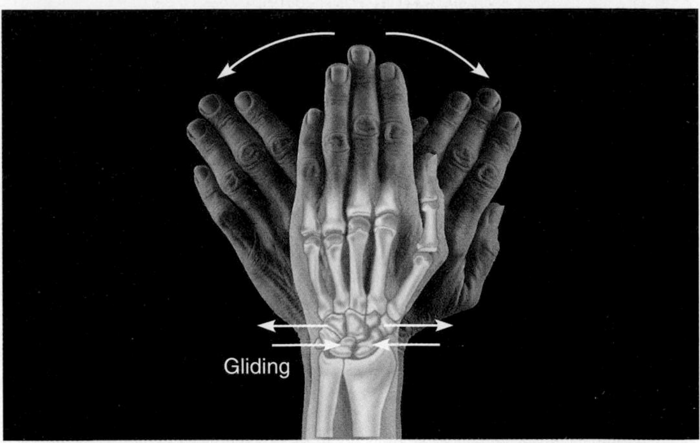

(a) Gliding movements at the wrist

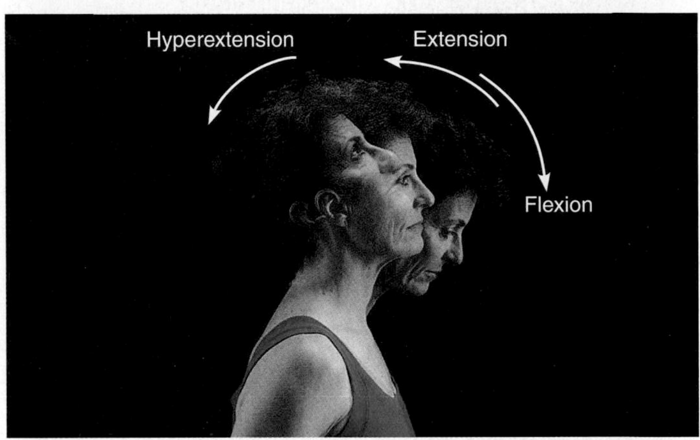

(b) Angular movements: flexion, extension, and hyperextension of the neck

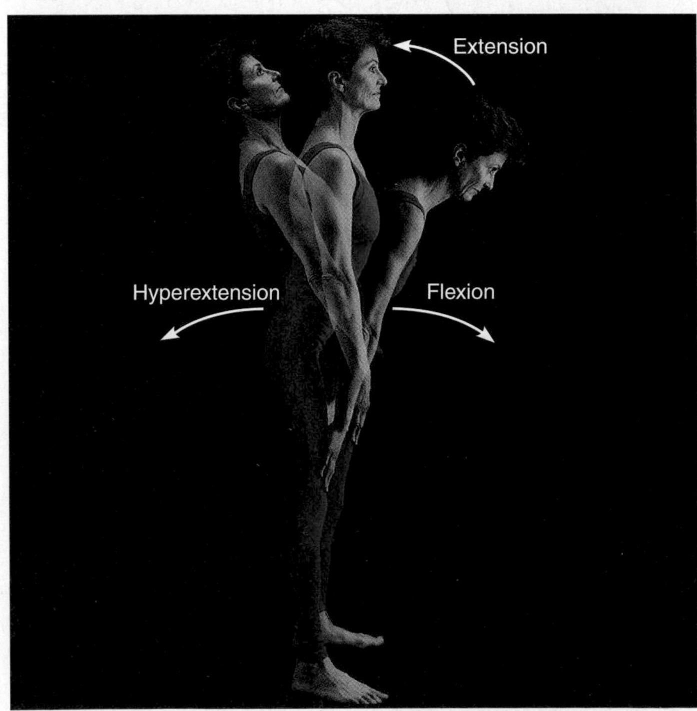

(c) Angular movements: flexion, extension, and hyperextension of the vertebral column

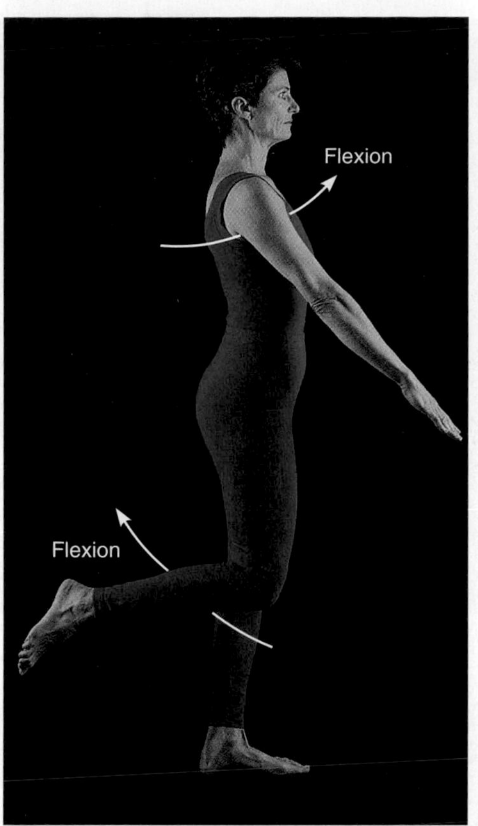

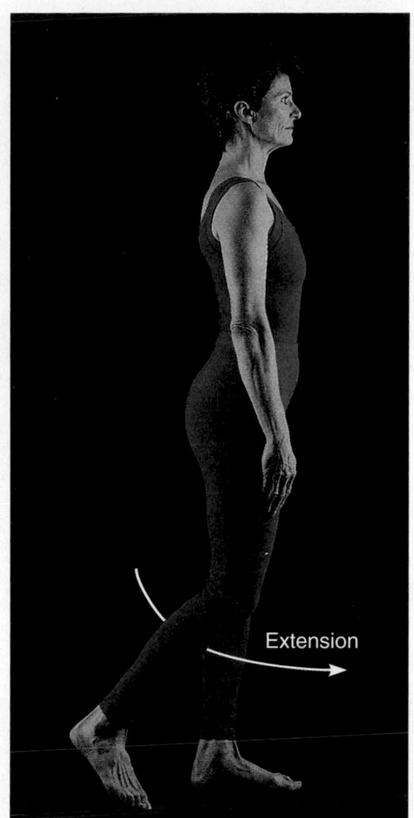

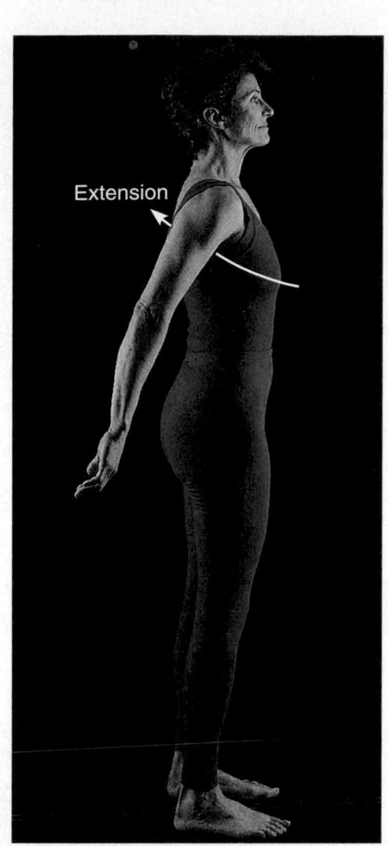

(d) Angular movements: flexion and extension at the shoulder and knee

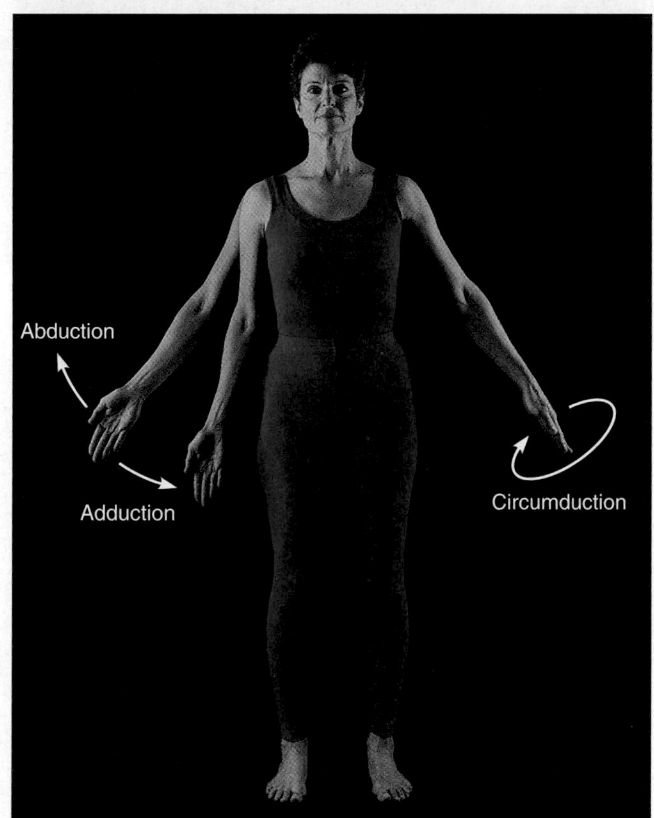

(e) Angular movements: abduction, adduction, and circumduction of the upper limb at the shoulder

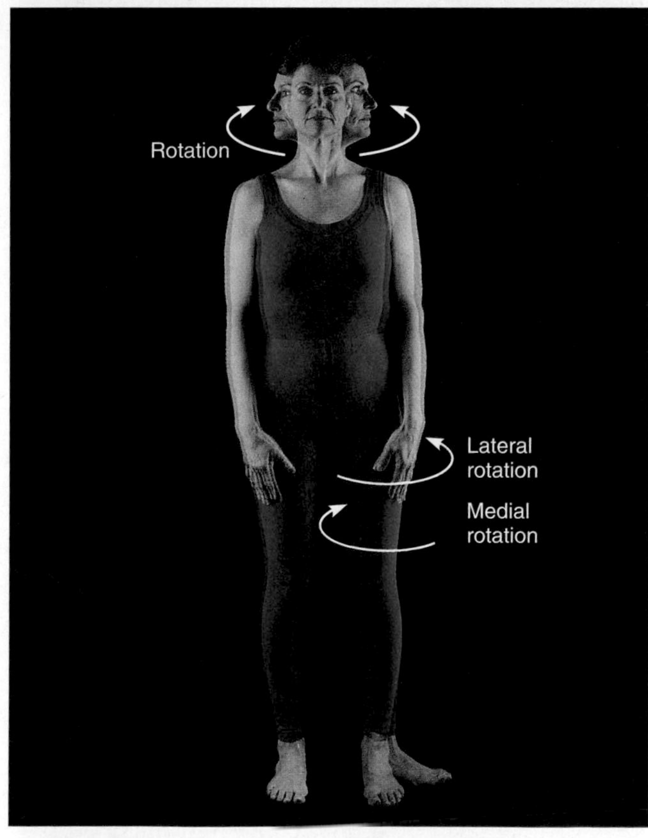

(f) Rotation of the head, neck, and lower limb

Figure 8.5 *(continued)*

8

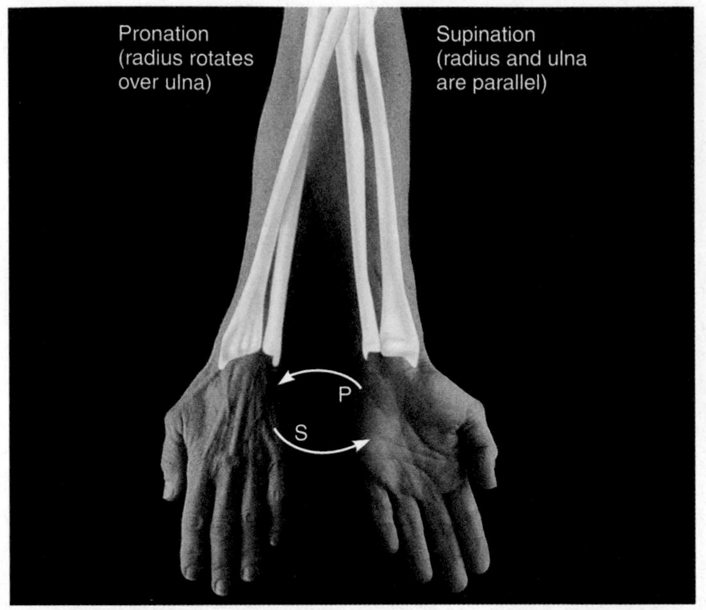

(a) Pronation (P) and supination (S)

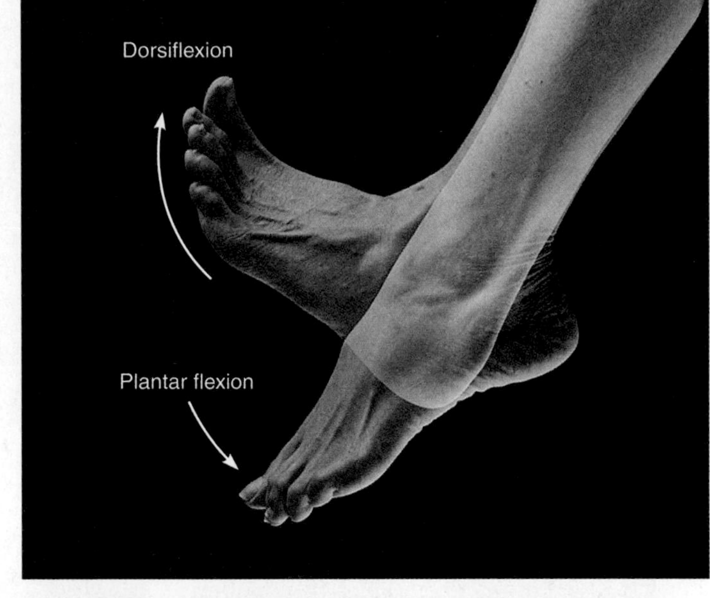

(b) Dorsiflexion and plantar flexion

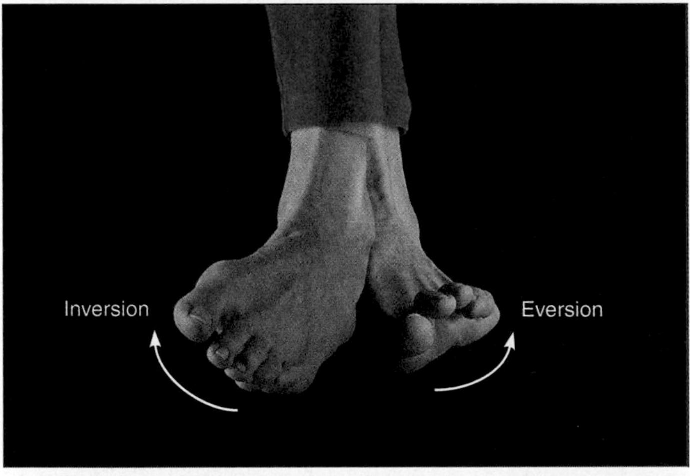

(c) Inversion and eversion

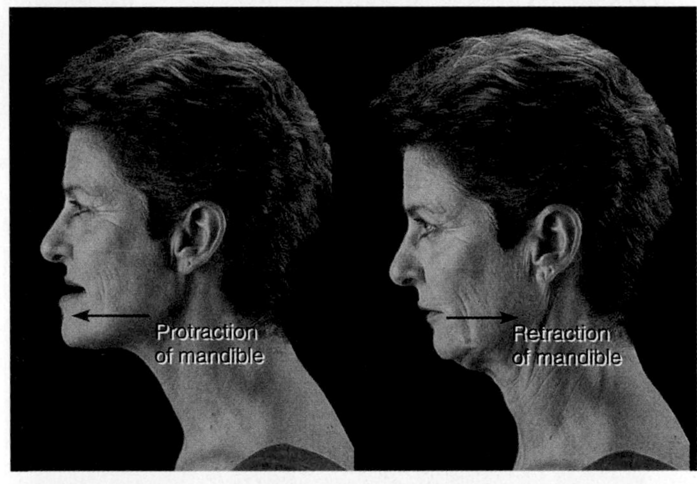

(d) Protraction and retraction

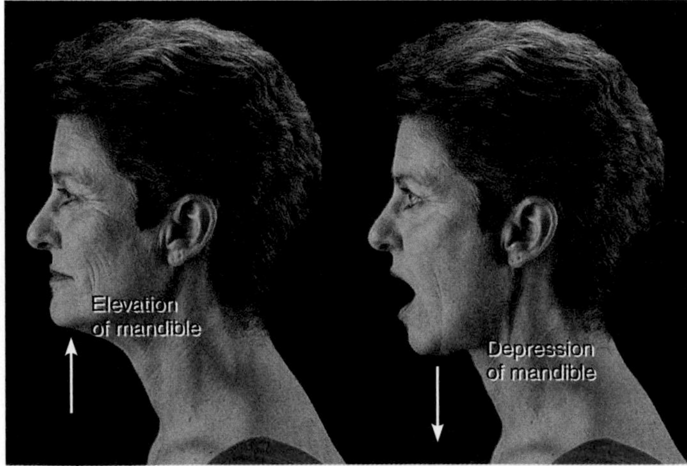

(e) Elevation and depression

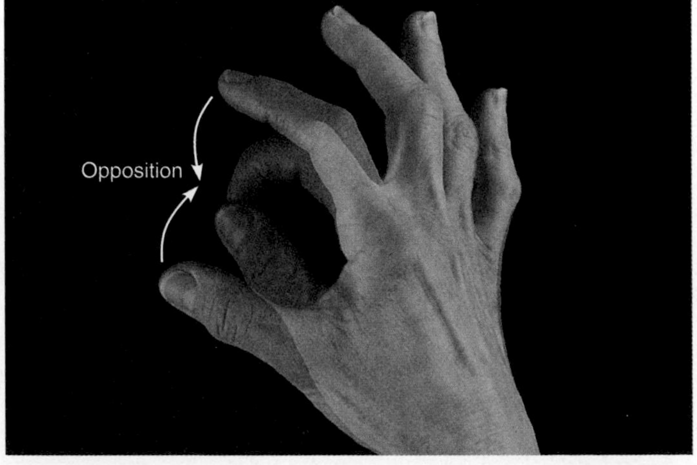

(f) Opposition

Figure 8.6 **Special body movements.**

Dorsiflexion and Plantar Flexion of the Foot The up-and-down movements of the foot at the ankle are given more specific names (Figure 8.6b). Lifting the foot so that its superior surface approaches the shin is **dorsiflexion** (corresponds to wrist extension), whereas depressing the foot (pointing the toes) is **plantar flexion** (corresponds to wrist flexion).

Inversion and Eversion **Inversion** and **eversion** are special movements of the foot (Figure 8.6c). In inversion, the sole of the foot turns medially. In eversion, the sole faces laterally.

Protraction and Retraction Nonangular anterior and posterior movements in a transverse plane are called **protraction** and **retraction**, respectively (Figure 8.6d). The mandible is protracted when you jut out your jaw and retracted when you bring it back.

Elevation and Depression **Elevation** means lifting a body part superiorly (Figure 8.6e). For example, the scapulae are elevated when you shrug your shoulders. Moving the elevated part inferiorly is **depression**. During chewing, the mandible is alternately elevated and depressed.

Opposition The saddle joint between metacarpal 1 and the trapezium allows a movement called **opposition** of the thumb (Figure 8.6f). This movement is the action taken when you touch your thumb to the tips of the other fingers on the same hand. It is opposition that makes the human hand such a fine tool for grasping and manipulating objects.

Types of Synovial Joints

Although all synovial joints have structural features in common, they do not have a common structural plan. Based on the shape of their articular surfaces, which in turn determine the movements allowed, synovial joints can be classified further into six major categories—plane, hinge, pivot, condyloid, saddle, and ball-and-socket joints.

Plane Joints

In **plane joints** **(Figure 8.7a)** the articular surfaces are essentially flat, and they allow only short nonaxial gliding movements. Examples are the gliding joints introduced earlier—the intercarpal and intertarsal joints, and the joints between vertebral articular processes. Gliding does not involve rotation around any axis, and gliding joints are the only examples of nonaxial plane joints.

Hinge Joints

In **hinge joints** (Figure 8.7b), the cylindrical end of one bone conforms to a trough-shaped surface on another. Motion is along a single plane and resembles that of a mechanical hinge. Uniaxial hinge joints permit flexion and extension only, typified by bending and straightening the elbow and interphalangeal joints.

Pivot Joints

In a **pivot joint** (Figure 8.7c), the rounded end of one bone conforms to a "sleeve" or ring composed of bone (and possibly ligaments) of another. The only movement allowed is uniaxial rotation of one bone around its own long axis. An example is the joint between the atlas and dens of the axis, which allows you to move your head from side to side to indicate "no." Another is the proximal radioulnar joint, where the head of the radius rotates within a ringlike ligament secured to the ulna.

Condyloid Joints

In **condyloid joints** (kon′dĭ-loid; "knuckle-like"), or **ellipsoidal joints**, the oval articular surface of one bone fits into a complementary depression in another (Figure 8.7d). The important characteristic is that both articulating surfaces are oval. The biaxial condyloid joints permit all *angular* motions, that is, flexion and extension, abduction and adduction, and circumduction. The radiocarpal (wrist) joints and the metacarpophalangeal (knuckle) joints are typical condyloid joints.

8

Saddle Joints

Saddle joints (Figure 8.7e) resemble condyloid joints, but they allow greater freedom of movement. Each articular surface has *both* concave and convex areas; that is, it is shaped like a saddle. The articular surfaces then fit together, concave to convex surfaces. The most clear-cut examples of saddle joints in the body are the carpometacarpal joints of the thumbs, and the movements allowed by these joints are clearly demonstrated by twiddling your thumbs.

Ball-and-Socket Joints

In **ball-and-socket joints** (Figure 8.7f), the spherical or hemispherical head of one bone articulates with the cuplike socket of another. These joints are multiaxial and the most freely moving synovial joints. Universal movement is allowed (that is, in all axes and planes, including rotation). The shoulder and hip joints are the only examples.

CHECK YOUR UNDERSTANDING

9. John bent over to pick up a dime. What movement was occurring at his hip joint, at his knees, and between his index finger and thumb?

10. On the basis of movement allowed, which of the following joints are uniaxial? Hinge, condyloid, saddle, pivot.

For answers, see Appendix G.

Selected Synovial Joints

▶ Describe the elbow, knee, hip, jaw, and shoulder joints in terms of articulating bones, anatomical characteristics of the joint, movements allowed, and joint stability.

(Text continues on p. 262.)

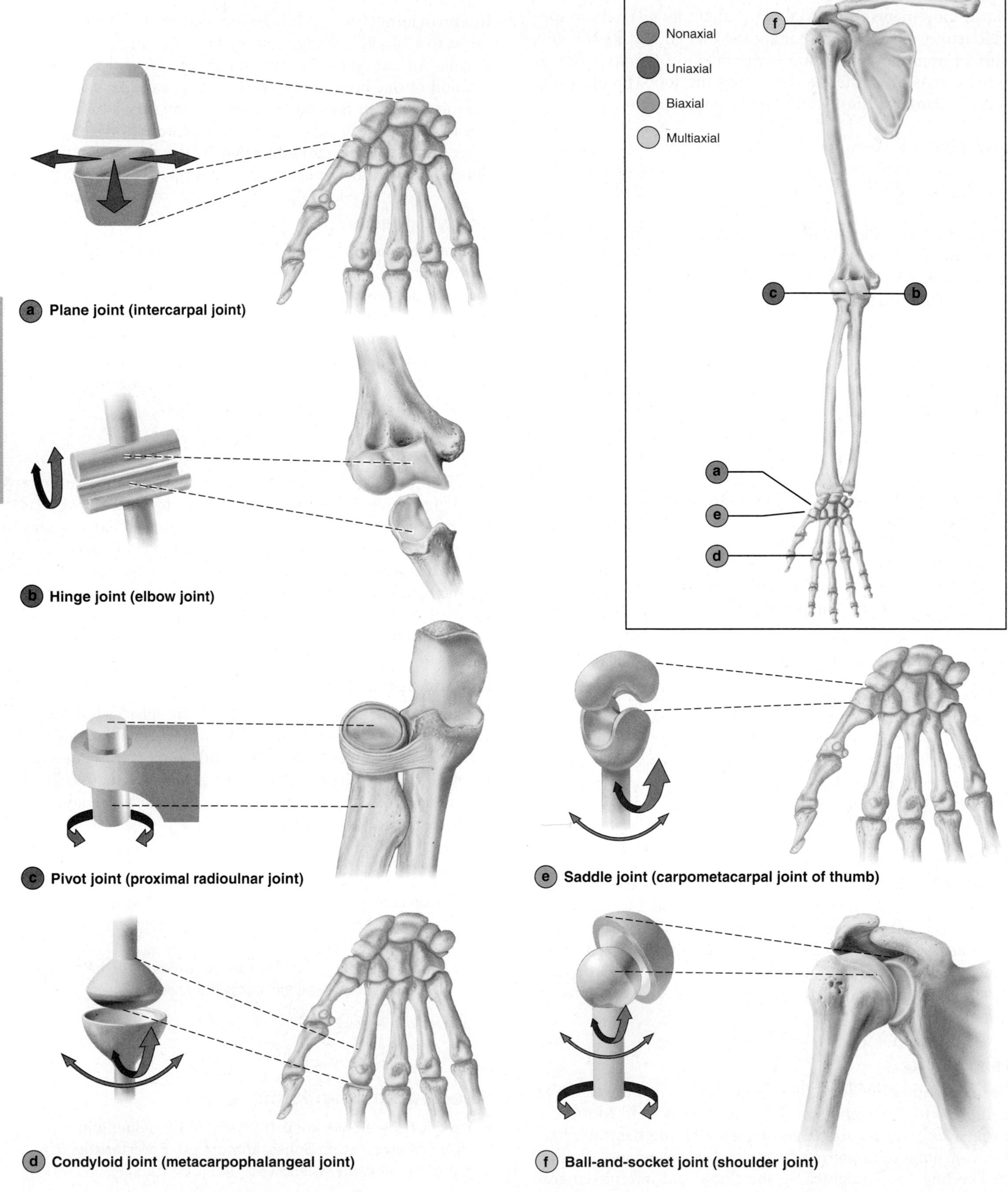

8

Figure 8.7 Types of synovial joints. Dashed lines indicate the articulating bones in each example.

CLOSER LOOK

Joints: From Knights in Shining Armor to Bionic Humans

The technology for fashioning joints in medieval suits of armor developed over centuries. The technology for creating the prostheses (artificial joints) used in medicine today developed, in relative terms, in a flash—less than 60 years. Unlike the joints in medieval armor, which was worn outside the body, today's artificial joints must function inside the body. The history of joint prostheses dates to the 1940s and 1950s, when World War II and the Korean War left large numbers of wounded who needed artificial limbs. It was predicted that by the year 2003, over a third of a million Americans would receive total joint replacements each year, mostly because of the destructive effects of osteoarthritis or rheumatoid arthritis. However, the actual number of people needing joint prostheses has burgeoned far beyond this as injured military people return from service abroad.

To produce durable, mobile joints requires a substance that is strong, nontoxic, and resistant to the corrosive effects of organic acids in blood. In 1963, Sir John Charnley, an English orthopedic surgeon, performed the first total hip replacement and revolutionized the therapy of arthritic hips. His device consisted of a metal ball on a stem and a cup-shaped polyethylene plastic socket anchored to the pelvis by methyl methacrylate cement. This cement proved to be exceptionally strong and relatively problem free. Hip prostheses were followed by knee prostheses, but not until ten years later did smoothly operating total knee joint replacements become a reality. Today, the metal parts of the prostheses are strong cobalt and titanium alloys, and the number of knee replacements equals the number of hip replacements.

Replacements are now available for many other joints, including fingers, elbows, and shoulders. Total hip and knee replacements last about 10 to 15 years in elderly patients who do not excessively stress the joint. Most such operations are done to reduce pain and restore about 80% of original joint function.

Replacement joints are not yet strong or durable enough for young, active people, but making them so is a major goal. The problem is that the prostheses work loose over time, so researchers are seeking to enhance the fit between implant and bone. One solution is to strengthen the cement that binds them (simply eliminating air bubbles from the cement increases its durability). Another solution currently being tested is a robotic surgeon, ROBODOC, to drill a better-fitting hole for the femoral prosthesis in hip surgery. In cementless prostheses, researchers are exploring better ways to get the bone to grow so that it binds more strongly to the implant. A supersmooth titanium coating seems to encourage direct bone on-growth.

Dramatic changes are also occurring in the way artificial joints are made. CAD/CAM (computer-aided design and computer-aided manufacturing) techniques have significantly reduced the time and cost of creating individualized joints. Fed the patient's X rays and medical information, the computer draws from a database of hundreds of normal joints and generates possible designs and modifications for a prosthesis. Once the best design is selected, the computer produces a program to direct the machines that shape it.

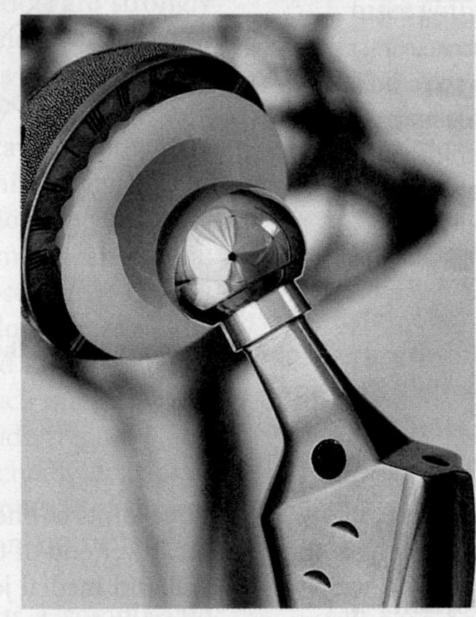

A hip prosthesis.

Joint replacement therapy is coming of age, but equally exciting are techniques that call on the ability of the patient's own tissues to regenerate.

- Osteochondral grafting: Healthy bone and cartilage are removed from one part of the body and transplanted to the injured joint.
- Autologous chondrocyte implantation: Healthy chondrocytes are removed from the body, cultivated in the lab, and implanted at the damaged joint.
- Mesenchymal stem cell regeneration: Undifferentiated mesenchymal cells are removed from bone marrow and placed in a gel, which is packed into an area of eroded cartilage.

These techniques offer hope for younger patients, since they could stave off the need for a joint prosthesis for several years.

And so, through the centuries, the focus has shifted from jointed armor to artificial joints that can be put inside the body to restore lost function. Modern technology has accomplished what the armor designers of the Middle Ages never dreamed of.

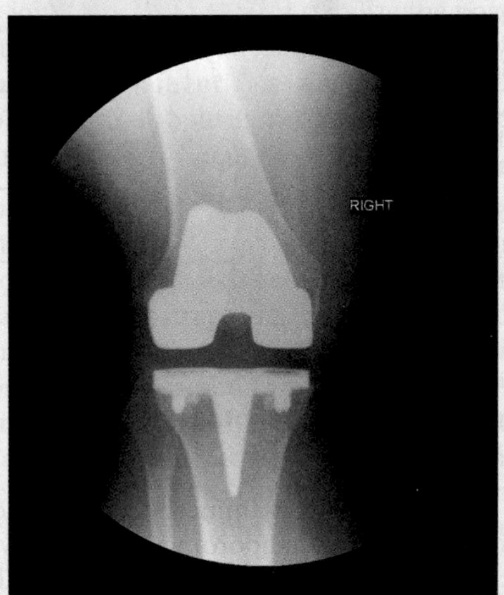

X ray of right knee showing total knee replacement prosthesis (co-designed by Kenneth Gustke, M.D., of Florida Orthopedic Institute).

Overview of Muscle Tissues

- ▶ Compare and contrast the basic types of muscle tissue.
- ▶ List four important functions of muscle tissue.

Types of Muscle Tissue

The three types of muscle tissue—*skeletal, cardiac,* and *smooth*—were introduced in Chapter 4. Now we are ready to describe the three types of muscle tissue in detail, but before we do, let's introduce some terminology. First, skeletal and smooth muscle cells (but not cardiac muscle cells) are elongated, and for this reason, are called **muscle fibers**. Second, whenever you see the prefixes **myo** or **mys** (both are word roots meaning "muscle") or **sarco** (flesh), the reference is to muscle. For example, the plasma membrane of muscle cells is called the *sarcolemma* (sar″ko-lem′ah), literally, "muscle" (sarco) "husk" (lemma), and muscle cell cytoplasm is called *sarcoplasm.* Okay, let's get to it.

Skeletal muscle tissue is packaged into the *skeletal muscles,* organs that attach to and cover the bony skeleton. Skeletal muscle fibers are the longest muscle cells and have obvious stripes called *striations* (see Figure 4.10a, p. 136). Although it is often activated by reflexes, skeletal muscle is called **voluntary muscle** because it is the only type subject to conscious control. When you think of skeletal muscle tissue, the key words to keep in mind are *skeletal, striated,* and *voluntary.*

Skeletal muscle is responsible for overall body mobility. It can contract rapidly, but it tires easily and must rest after short periods of activity. Nevertheless, it can exert tremendous power, a fact revealed by reports of people lifting cars to save their loved ones. Skeletal muscle is also remarkably adaptable. For example, your hand muscles can exert a force of a fraction of an ounce to pick up a paper clip, and the same muscles can exert a force of about 70 pounds to pick up this book!

Cardiac muscle tissue occurs only in the heart (the body's blood pump), where it constitutes the bulk of the heart walls. Like skeletal muscle cells, cardiac muscle cells are striated (see Figure 4.10b, p. 137), but cardiac muscle is not voluntary. Indeed, it can and does contract without being stimulated by the nervous system. Most of us have no conscious control over how fast our heart beats. Key words to remember for cardiac muscle are *cardiac, striated,* and *involuntary.*

Cardiac muscle usually contracts at a fairly steady rate set by the heart's pacemaker, but neural controls allow the heart to speed up for brief periods, as when you race across the tennis court to make that overhead smash.

Smooth muscle tissue is found in the walls of hollow visceral organs, such as the stomach, urinary bladder, and respiratory passages. Its role is to force fluids and other substances through internal body channels. Smooth muscle cells like skeletal muscle cells are elongated "fibers," but smooth muscle has no striations (see Figure 4.10c, p. 137). Like cardiac muscle, it is not subject to voluntary control. Contractions of smooth muscle fibers are slow and sustained. We can describe smooth muscle tissue as *visceral, nonstriated,* and *involuntary.*

Special Characteristics of Muscle Tissue

What enables muscle tissue to perform its duties? Four special characteristics or abilities are key.

Excitability, also termed **responsiveness** or **irritability,** is the ability to receive and respond to a stimulus, that is, any change in the environment either inside or outside the body. In the case of muscle, the stimulus is usually a chemical—for example, a neurotransmitter released by a nerve cell, or a local change in pH. The response (sometimes separated out as an additional characteristic called *conductivity*), is generation of an electrical impulse that passes along the plasma membrane of the muscle cell and causes the cell to contract.

Contractility is the ability to shorten forcibly when adequately stimulated. This ability sets muscle apart from all other tissue types.

Extensibility is the ability to be stretched or extended. Muscle cells shorten when contracting, but they can be stretched, even beyond their resting length, when relaxed.

Elasticity is the ability of a muscle cell to recoil and resume its resting length after being stretched.

Muscle Functions

Muscle performs at least four important functions for the body. It produces movement, maintains posture, stabilizes joints, generates heat, and more.

Producing Movement

Just about all movements of the human body and its parts result from muscle contraction. Skeletal muscles are responsible for all locomotion and manipulation. They enable you to respond quickly to changes in the external environment—for example, to jump out of the way of a car, to direct your eyeballs, and to smile or frown.

Blood courses through your body because of the rhythmically beating cardiac muscle of your heart and the smooth muscle in the walls of your blood vessels, which helps maintain blood pressure. Smooth muscle in organs of the digestive, urinary, and reproductive tracts propels, or squeezes, substances (foodstuffs, urine, a baby) through the organs and along the tract.

Maintaining Posture and Body Position

We are rarely aware of the workings of the skeletal muscles that maintain body posture. Yet these muscles function almost continuously, making one tiny adjustment after another to counteract the never-ending downward pull of gravity.

Stabilizing Joints

Even as muscles pull on bones to cause movements, they stabilize and strengthen the joints of the skeleton (Chapter 8).

Generating Heat

Muscles generate heat as they contract. This heat is vitally important in maintaining normal body temperature. Because skeletal muscle accounts for at least 40% of body mass, it is the muscle type most responsible for generating heat.

Additional Functions

What else do muscles do? Skeletal muscles protect the more fragile internal organs (the viscera) by enclosing them. Smooth muscle also forms valves to regulate the passage of substances through internal body openings, dilates and constricts the pupils of your eyes, and forms the arrector pili muscles attached to hair follicles.

■ ■ ■

In much of this chapter, we examine the structure and function of skeletal muscle. Then we consider smooth muscle more briefly, largely by comparing it with skeletal muscle. We describe cardiac muscle in detail in Chapter 18, but for easy comparison, the characteristics of all three muscle types are included in the summary in Table 9.3 on p. 309.

CHECK YOUR UNDERSTANDING

1. When describing muscle, what does "striated" mean?
2. Harry was pondering an exam question that said, "What muscle type has elongated cells and is found in the walls of the urinary bladder?" What should he have responded?

For answers, see Appendix G.

Skeletal Muscle

- ▶ Describe the gross structure of a skeletal muscle.
- ▶ Describe the microscopic structure and functional roles of the myofibrils, sarcoplasmic reticulum, and T tubules of skeletal muscle fibers.
- ▶ Describe the sliding filament model of muscle contraction.

For easy reference, **Table 9.1** summarizes the levels of skeletal muscle organization, gross to microscopic, that we describe in the following sections.

Gross Anatomy of a Skeletal Muscle

Each **skeletal muscle** is a discrete organ, made up of several kinds of tissues. Skeletal muscle fibers predominate, but blood vessels, nerve fibers, and substantial amounts of connective tissue are also present. A skeletal muscle's shape and its attachments in the body can be examined easily without the help of a microscope.

Nerve and Blood Supply

In general, each muscle is served by one nerve, an artery, and by one or more veins. These structures all enter or exit near the central part of the muscle and branch profusely through its connective tissue sheaths (described below). Unlike cells of cardiac and smooth muscle tissues, which can contract in the absence of nerve stimulation, each skeletal muscle fiber is supplied with a nerve ending that controls its activity.

Skeletal muscle has a rich blood supply. This is understandable because contracting muscle fibers use huge amounts of energy and require more or less continuous delivery of oxygen and nutrients via the arteries. Muscle cells also give off large amounts of metabolic wastes that must be removed through veins if contraction is to remain efficient. Muscle capillaries, the smallest of the body's blood vessels, are long and winding and have numerous cross-links, features that accommodate changes in muscle length. They straighten when the muscle is stretched and contort when the muscle contracts.

Connective Tissue Sheaths

In an intact muscle, the individual muscle fibers are wrapped and held together by several different connective tissue sheaths. Together these connective tissue sheaths support each cell and reinforce the muscle as a whole, preventing the bulging muscles from bursting during exceptionally strong contractions. We will consider these from external to internal (see **Figure 9.1** and the top three rows of Table 9.1).

1. **Epimysium.** The **epimysium** (ep″ĭ-mis′e-um; meaning "outside the muscle") is an "overcoat" of dense irregular connective tissue that surrounds the whole muscle. Sometimes it blends with the deep fascia that lies between neighboring muscles or the superficial fascia deep to the skin.
2. **Perimysium and fascicles.** Within each skeletal muscle, the muscle fibers are grouped into **fascicles** (fas′ĭ-klz; "bundles") that resemble bundles of sticks. Surrounding each fascicle is a layer of fibrous connective tissue called **perimysium** (per″ĭ-mis′e-um; meaning "around the muscle [fascicles]").
3. **Endomysium.** The **endomysium** (en″do-mis′e-um; meaning "within the muscle") is a whispy sheath of connective tissue that surrounds each individual muscle fiber. It consists of fine areolar connective tissue.

As shown in Figure 9.1, all of these connective tissue sheaths are continuous with one another as well as with the tendons that join muscles to bones. When muscle fibers contract, they pull on these sheaths, which in turn transmit the pulling force to the bone to be moved. The sheaths contribute somewhat to the natural elasticity of muscle tissue, and also provide entry and exit routes for the blood vessels and nerve fibers that serve the muscle.

Attachments

Recall from Chapter 8 that most skeletal muscles span joints and are attached to bones (or other structures) in at least two places, and that when a muscle contracts, the movable bone, the muscle's **insertion**, moves toward the immovable or less movable bone, the muscle's **origin**. In the muscles of the limbs, the origin typically lies proximal to the insertion.

Muscle attachments, whether origin or insertion, may be direct or indirect. In **direct**, or **fleshy**, **attachments**, the epimysium of the muscle is fused to the periosteum of a bone or

Figure 9.8 **FOCUS** **Events at the Neuromuscular Junction**

When a nerve impulse reaches a neuromuscular junction, acetylcholine (ACh) is released. Upon binding to sarcolemma receptors, ACh causes a change in sarcolemma permeability leading to a change in membrane potential.

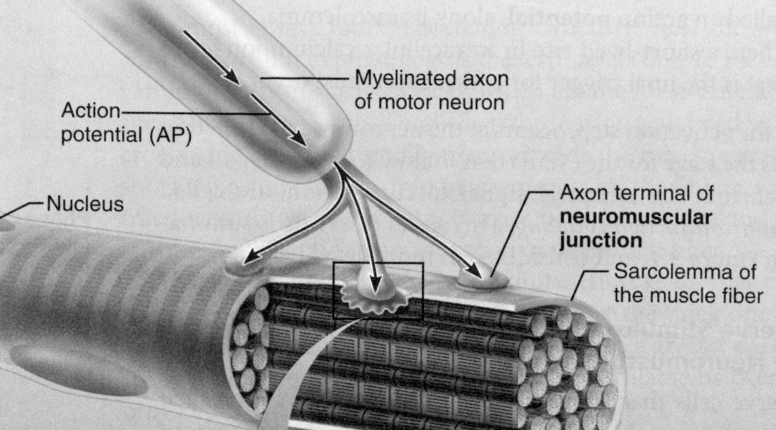

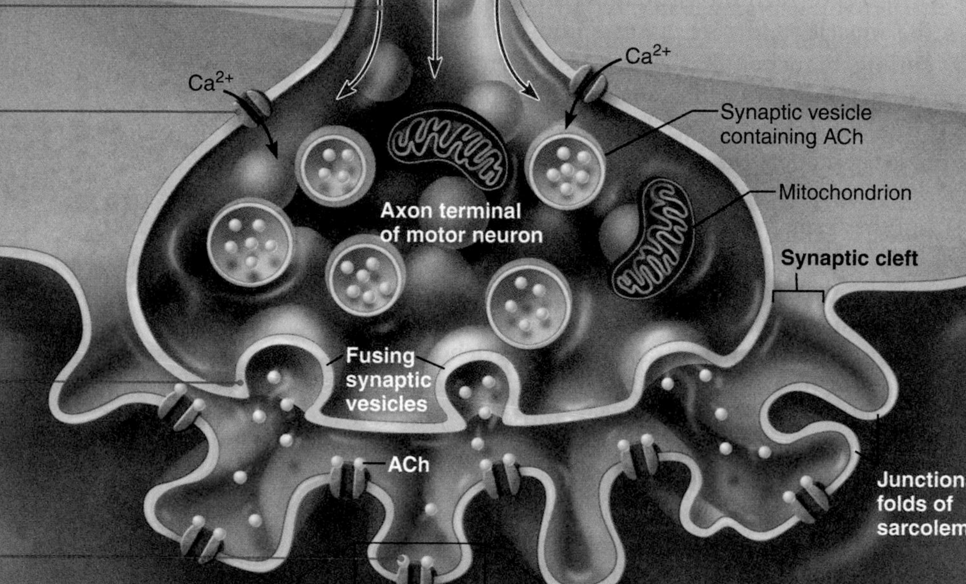

1. Action potential arrives at axon terminal of motor neuron.
2. Voltage-gated Ca^{2+} channels open and Ca^{2+} enters the axon terminal.
3. Ca^{2+} entry causes some synaptic vesicles to release their contents (acetylcholine) by exocytosis.
4. Acetylcholine, a neurotransmitter, diffuses across the synaptic cleft and binds to receptors in the sarcolemma.
5. ACh binding opens ion channels that allow simultaneous passage of Na^+ into the muscle fiber and K^+ out of the muscle fiber. More Na^+ ions enter than K^+ ions leave and this produces a local change in the membrane potential (depolarization).
6. ACh effects are terminated by its enzymatic breakdown in the synaptic cleft by acetylcholinesterase.

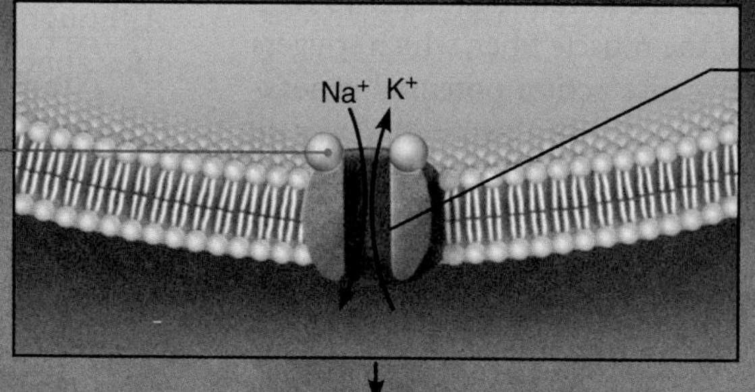

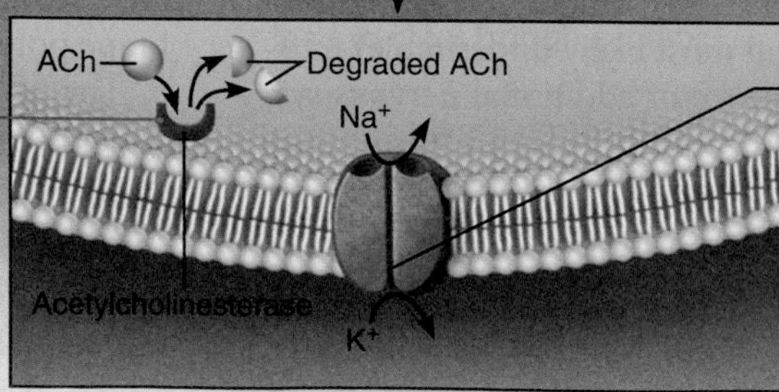

A&P Flix View this animation at myA&P

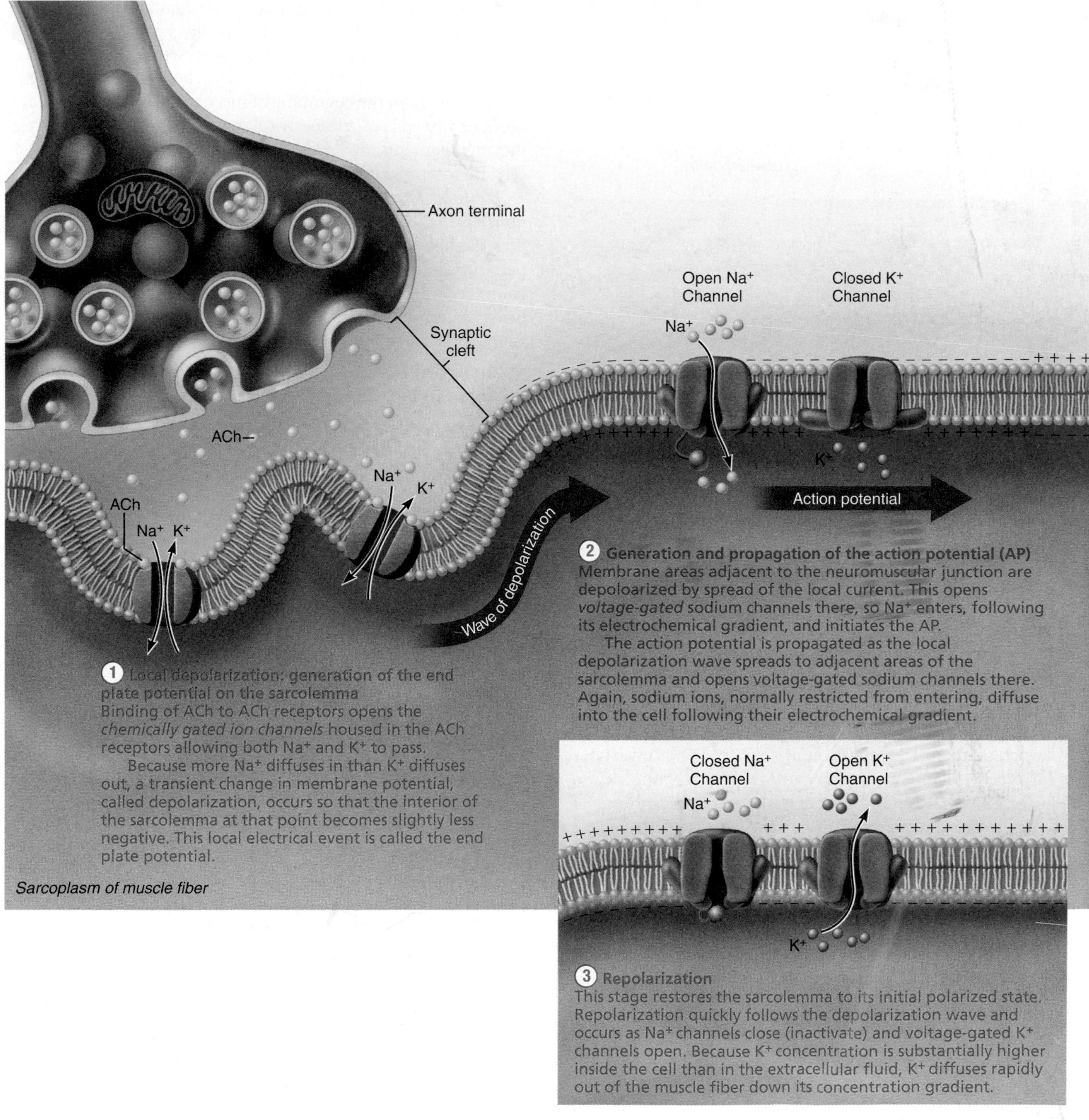

Axon terminal
Synaptic cleft
ACh
ACh
Na+ K+
Na+
K+
Wave of depolarization
Open Na+ Channel
Closed K+ Channel
Na+
K+
Action potential
① Local depolarization: generation of the end plate potential on the sarcolemma
Binding of ACh to ACh receptors opens the *chemically gated ion channels* housed in the ACh receptors allowing both Na+ and K+ to pass.
Because more Na+ diffuses in than K+ diffuses out, a transient change in membrane potential, called depolarization, occurs so that the interior of the sarcolemma at that point becomes slightly less negative. This local electrical event is called the end plate potential.
Sarcoplasm of muscle fiber
② Generation and propagation of the action potential (AP)
Membrane areas adjacent to the neuromuscular junction are depoloarized by spread of the local current. This opens *voltage-gated* sodium channels there, so Na+ enters, following its electrochemical gradient, and initiates the AP.
The action potential is propagated as the local depolarization wave spreads to adjacent areas of the sarcolemma and opens voltage-gated sodium channels there. Again, sodium ions, normally restricted from entering, diffuse into the cell following their electrochemical gradient.
Closed Na+ Channel
Open K+ Channel
Na+
K+
③ Repolarization
This stage restores the sarcolemma to its initial polarized state. Repolarization quickly follows the depolarization wave and occurs as Na+ channels close (inactivate) and voltage-gated K+ channels open. Because K+ concentration is substantially higher inside the cell than in the extracellular fluid, K+ diffuses rapidly out of the muscle fiber down its concentration gradient.

Figure 9.9 **Summary of events in the generation and propagation of an action potential in a skeletal muscle fiber.** The axon terminal plasma membrane and the sarcolemma are shown at different scales.

neuromuscular junction opens *chemically (ligand) gated ion channels* that allow Na^+ and K^+ to pass (also see Figure 9.8). Because more Na^+ diffuses in than K^+ diffuses out, a transient change in membrane potential occurs as the interior of the sarcolemma becomes slightly less negative, an event called **depolarization**.

② **Generation and propagation of the action potential.** Initially, depolarization is a local electrical event called an **end plate potential**, but it ignites the action potential that spreads in all directions from the neuromuscular junction across the sarcolemma, just as ripples move away from pebbles dropped into a stream. This local depolarization (end plate potential) then spreads to adjacent membrane areas

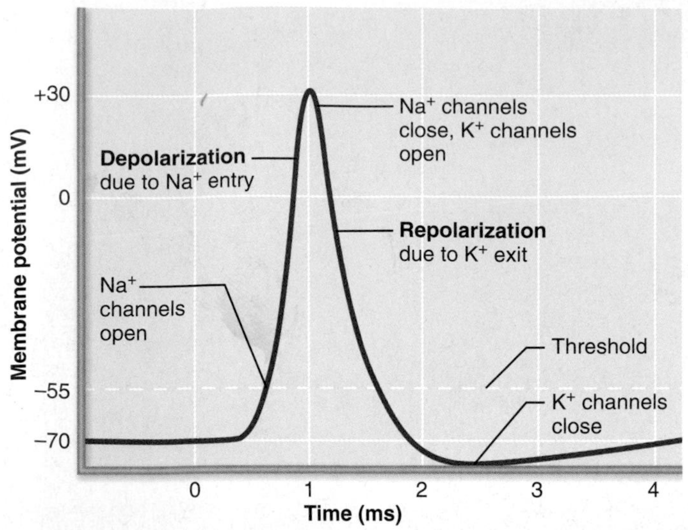

Figure 9.10 Action potential scan showing changes in Na^+ and K^+ ion channels.

and opens *voltage-gated* (p. 81) sodium channels there. Na^+ enters, following its electrochemical gradient, and once a certain membrane voltage, referred to as *threshold,* is reached, an action potential is generated (initiated).

The action potential is *propagated* (moves along the length of the sarcolemma) as the local depolarization wave spreads to adjacent areas of the sarcolemma and opens voltage-gated sodium channels there. Again, Na^+, normally restricted from entering, diffuses into the cell following its electrochemical gradient.

③ **Repolarization.** The sarcolemma is restored to its initial polarized state during **repolarization**. The repolarization wave, like the depolarization wave, is a consequence of changes in membrane permeability. In this case, Na^+ channels close and voltage-gated K^+ channels open. Since the potassium ion concentration is substantially higher inside the cell than in the extracellular fluid, K^+ diffuses rapidly out of the muscle fiber, restoring negatively charged conditions inside (also see **Figure 9.10**).

During repolarization, a muscle fiber is said to be in a **refractory period**, because the cell cannot be stimulated again until repolarization is complete. Note that repolarization restores only the *electrical conditions* of the resting (polarized) state. The ATP-dependent Na^+-K^+ pump restores the *ionic conditions* of the resting state, but hundreds of action potentials can occur before ionic imbalances interfere with contractile activity.

Once initiated, the action potential is unstoppable. It ultimately results in contraction of the muscle fiber. Although the action potential itself is very brief, only 1–2 milliseconds (ms), the contraction phase of a muscle fiber may persist for 100 ms or more and far outlasts the electrical event that triggers it.

Excitation-Contraction Coupling

Excitation-contraction (E-C) coupling is the sequence of events by which transmission of an action potential along the sarcolemma leads to the sliding of myofilaments. The action potential is brief and ends well before any signs of contraction are obvious. The events of excitation-contraction coupling occur during the *latent period* (*laten* = hidden), between action potential initiation and the beginning of mechanical activity (contraction). As you will see, the electrical signal does not act directly on the myofilaments. Instead, it causes the rise in intracellular calcium ion concentration that allows the filaments to slide.

Focus on Excitation-Contraction Coupling **(Figure 9.11)** on pp. 290–291 illustrates the steps in this process. This Focus feature also reveals how the integral proteins of the "double zippers" in the triads interact to provide the Ca^{2+} necessary for contraction to occur. Make sure you understand this material before continuing on.

Summary of Different Types of Channels Involved in Initiating Muscle Contraction

So let's summarize what has to happen from the nerve ending on to finally excite muscle cells. Essentially this process involves activation of four sets of ion channels:

1. The process is initiated when the nerve impulse reaches the neuron terminal and opens voltage-gated calcium channels in the axonal membrane. Calcium entry triggers release of ACh into the synaptic cleft.
2. Released ACh binds to ACh receptors in the sarcolemma, opening chemically gated Na^+-K^+ channels. Greater influx of Na^+ causes a local voltage change (the end plate potential).
3. Local depolarization opens voltage-gated sodium channels in the neighboring region of the sarcolemma. This allows more sodium to enter, which further depolarizes the sarcolemma, resulting in AP generation and propagation.
4. Transmission of an AP along the T tubules changes the shape of voltage-sensitive proteins in the T tubules, which in turn stimulate SR calcium release channels to release Ca^{2+} into the cytosol.

Muscle Fiber Contraction: Cross Bridge Activity

As we have noted, cross bridge formation (attachment of myosin heads to actin) requires Ca^{2+}. Let's look more closely at how calcium ions promote muscle cell contraction. When intracellular calcium levels are low, the muscle cell is relaxed, and the active (myosin-binding) sites on actin are physically blocked by tropomyosin molecules. As Ca^{2+} levels rise, the ions bind to regulatory sites on troponin. To activate its group of seven actins, a troponin must bind two calcium ions, change shape, and then roll tropomyosin into the groove of the actin helix, away from the myosin-binding sites. In short, the tropomyosin "blockade" is removed when sufficient calcium is present. Once binding sites on actin are exposed, the events of the cross bridge cycle occur in rapid succession, as depicted in *Focus on the Cross Bridge Cycle* **(Figure 9.12)** on p. 292.

Sliding of thin filaments continues as long as the calcium signal and adequate ATP are present. When nerve impulses are delivered rapidly, intracellular Ca^{2+} levels increase greatly due to successive "puffs" or rounds of Ca^{2+} released from the SR. In such cases, the muscle cells do not completely relax between successive stimuli and contraction is stronger and more sustained (within limits) until nervous stimulation ceases. As the Ca^{2+} pumps of the SR reclaim calcium ions from the cytosol and troponin again changes shape, actin's myosin-binding sites are again covered by tropomyosin. The contraction ends, and the muscle fiber relaxes.

When the cycle is back where it started, the myosin head is in its upright high-energy configuration (see step ① in Focus Figure 9.12), ready to take another "step" and attach to an actin site farther along the thin filament. This "walking" of the myosin heads along the adjacent thin filaments during muscle shortening is much like a centipede's gait. The thin filaments cannot slide backward as the cycle repeats again and again because some myosin heads ("legs") are always in contact with actin (the "ground"). Contracting muscles routinely shorten by 30–35% of their total resting length, so each myosin cross bridge attaches and detaches many times during a single contraction. It is likely that only half of the myosin heads of a thick filament are pulling at the same instant. The others are randomly seeking their next binding site.

Except for the brief period following muscle cell excitation, calcium ion concentrations in the cytosol are kept almost undetectably low. There is a reason for this: ATP is the cell's energy source, and its hydrolysis yields inorganic phosphate (P_i). P_i would combine with calcium ions to form hydroxyapatite crystals, the stony-hard salts found in bone matrix, if calcium ion concentrations were always high. Such calcified muscle cells would die.

HOMEOSTATIC IMBALANCE

Rigor mortis (death rigor) illustrates the fact that cross bridge detachment is ATP driven. Most muscles begin to stiffen 3 to 4 hours after death. Peak rigidity occurs at 12 hours and then gradually dissipates over the next 48 to 60 hours. Dying cells are unable to exclude calcium (which is in higher concentration in the extracellular fluid), and the calcium influx into muscle cells promotes formation of myosin cross bridges. Shortly after breathing stops, ATP synthesis ceases, but ATP continues to be consumed and cross bridge detachment is impossible. Actin and myosin become irreversibly cross-linked, producing the stiffness of rigor mortis, which gradually disappears as muscle proteins break down after death. ■

CHECK YOUR UNDERSTANDING

6. What are the three structural components of a neuromuscular junction?
7. What is the final trigger for contraction? What is the initial trigger?
8. What prevents the filaments from sliding back to their original position each time a myosin cross bridge detaches from actin?
9. What would happen if a muscle fiber suddenly ran out of ATP when sarcomeres had only partially contracted?

For answers, see Appendix G.

Contraction of a Skeletal Muscle

- Define motor unit and muscle twitch, and describe the events occurring during the three phases of a muscle twitch.
- Explain how smooth, graded contractions of a skeletal muscle are produced.
- Differentiate between isometric and isotonic contractions.

In its relaxed state, a muscle is soft and unimpressive, not what you would expect of a prime mover of the body. However, within a few milliseconds, it can contract to become a hard elastic structure with dynamic characteristics that intrigue not only biologists but engineers and physicists as well.

Before we consider muscle contraction on the organ level, let's note a few principles of muscle mechanics.

1. The principles governing contraction of a single muscle fiber and of a skeletal muscle consisting of a large number of fibers are pretty much the same.
2. The force exerted by a contracting muscle on an object is called **muscle tension**, and the opposing force exerted on the muscle by the weight of the object to be moved is called the **load**.
3. A contracting muscle does not always shorten and move the load. If muscle tension develops but the load is not moved, the contraction is called *isometric* ("same measure"), as when you try to lift a 2000-lb car. If the muscle tension developed overcomes the load and muscle shortening occurs, the contraction is *isotonic* ("same tension"), as when you lift a 5-lb sack of sugar. We will describe these major types of contraction in detail, but for now the important thing to remember when reading the accompanying graphs is that *increasing muscle tension* is measured for isometric contractions, whereas the *amount of muscle shortening* (distance in millimeters) is measured for isotonic contractions.
4. A skeletal muscle contracts with varying force and for different periods of time in response to stimuli of varying frequencies and intensities. To understand how this occurs, we must look at the nerve-muscle functional unit called a *motor unit.* This is our next topic.

The Motor Unit

Each muscle is served by at least one *motor nerve*, and each motor nerve contains axons (fibrous extensions) of up to hundreds of motor neurons. As an axon enters a muscle, it branches into a number of terminals, each of which forms a neuromuscular junction with a single muscle fiber. A **motor unit** consists of a motor neuron and all the muscle fibers it supplies **(Figure 9.13)**. When a motor neuron fires (transmits an action potential), all the muscle fibers it innervates contract.

The number of muscle fibers per motor unit may be as high as several hundred or as few as four. Muscles that exert fine control (such as those controlling the fingers and eyes) have small motor units. By contrast, large, weight-bearing muscles, whose movements are less precise (such as the hip muscles), have large motor units. The muscle fibers in a single motor unit are not

Excitation-contraction (E-C) coupling is the sequence of events by which transmission of an action potential along the sarcolemma leads to the sliding of myofilaments.

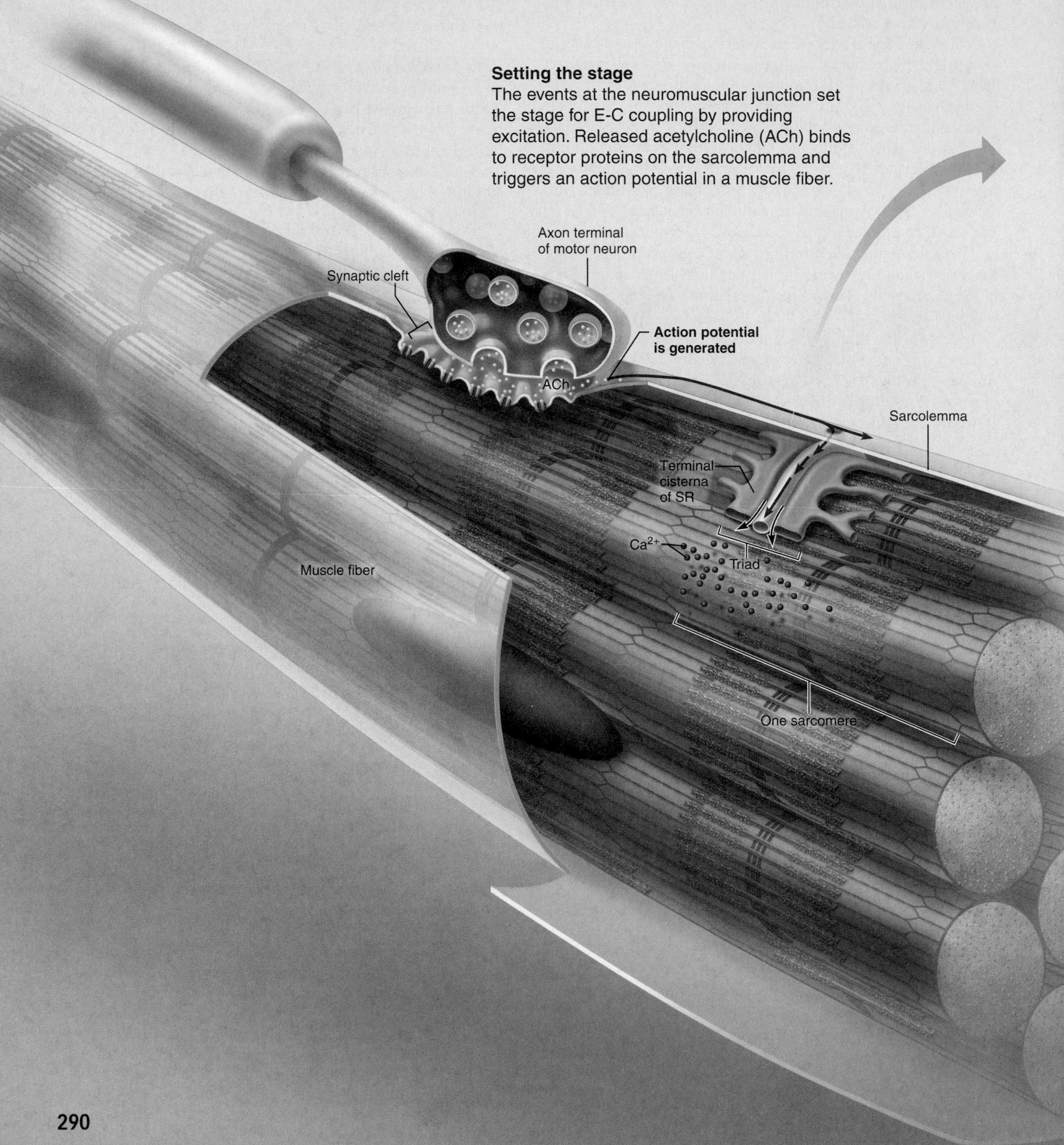

Setting the stage
The events at the neuromuscular junction set the stage for E-C coupling by providing excitation. Released acetylcholine (ACh) binds to receptor proteins on the sarcolemma and triggers an action potential in a muscle fiber.
Axon terminal of motor neuron
Synaptic cleft
Action potential is generated
ACh
Sarcolemma
Terminal cisterna of SR
Ca2+
Triad
Muscle fiber
One sarcomere

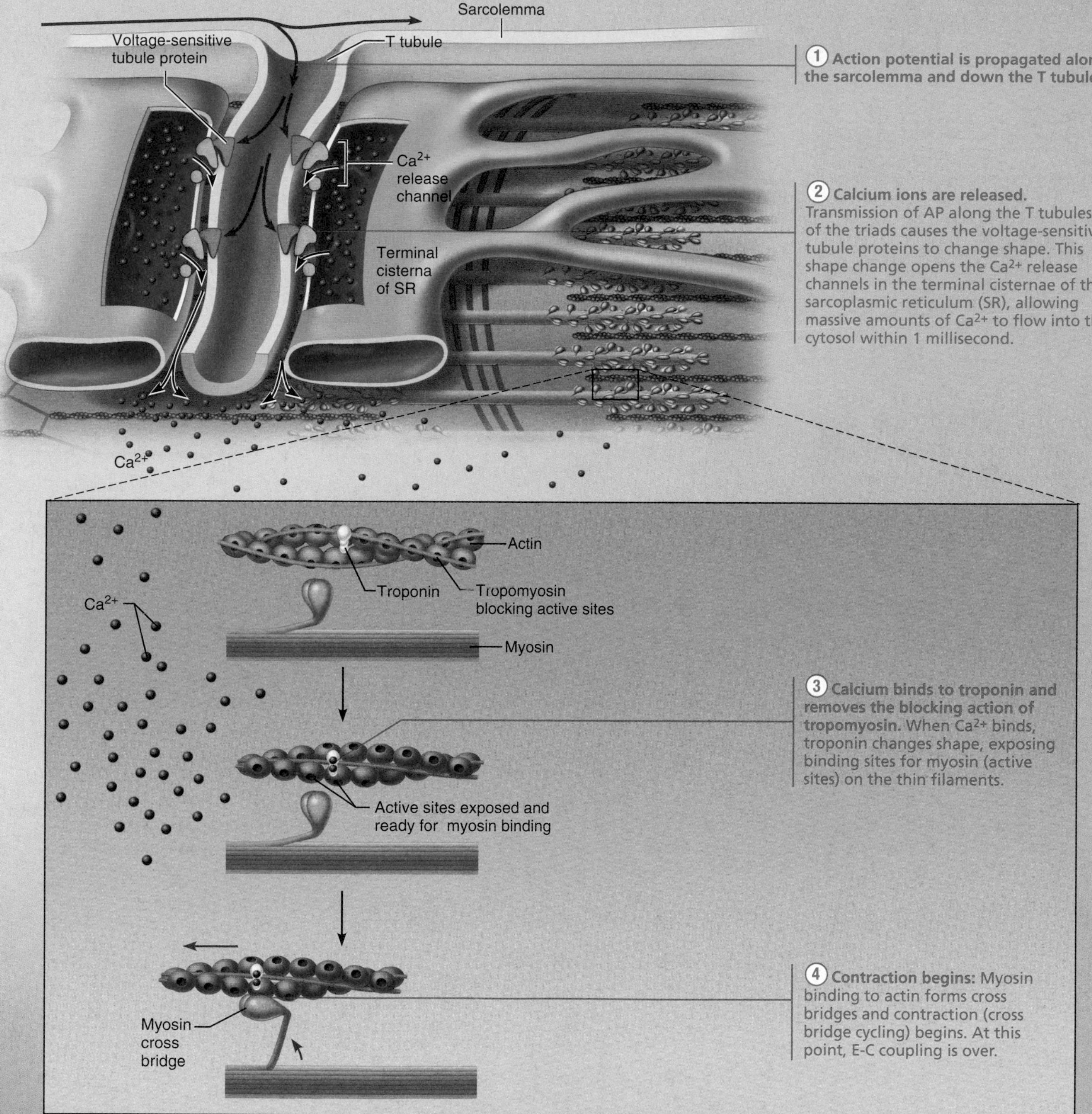
Steps in E-C Coupling:
Sarcolemma
Voltage-sensitive tubule protein
T tubule
Ca2+ release channel
Terminal cisterna of SR
Ca2+
Actin
Troponin
Tropomyosin blocking active sites
Myosin
Active sites exposed and ready for myosin binding
Myosin cross bridge
1 Action potential is propagated along the sarcolemma and down the T tubules.
2 Calcium ions are released. Transmission of AP along the T tubules of the triads causes the voltage-sensitive tubule proteins to change shape. This shape change opens the Ca2+ release channels in the terminal cisternae of the sarcoplasmic reticulum (SR), allowing massive amounts of Ca2+ to flow into the cytosol within 1 millisecond.
3 Calcium binds to troponin and removes the blocking action of tropomyosin. When Ca2+ binds, troponin changes shape, exposing binding sites for myosin (active sites) on the thin filaments.
4 Contraction begins: Myosin binding to actin forms cross bridges and contraction (cross bridge cycling) begins. At this point, E-C coupling is over.

The aftermath

The short-lived Ca^{2+} signal ends, and Ca^{2+} levels fall as Ca^{2+} is continuously pumped back into the SR by active transport. The blocking action of tropomyosin is restored, inhibiting myosin-actin interaction and relaxation occurs. The sequence of E-C coupling events followed by a drop in Ca^{2+} levels is repeated each time a nerve impulse arrives at the neuromuscular junction.

A&P Flix *View this animation at myA&P*

The cross bridge cycle is the series of events during which myosin heads pull thin filaments toward the center of the sarcomere.

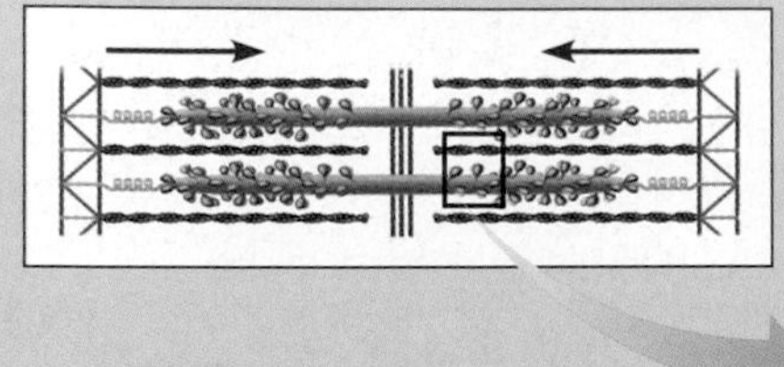

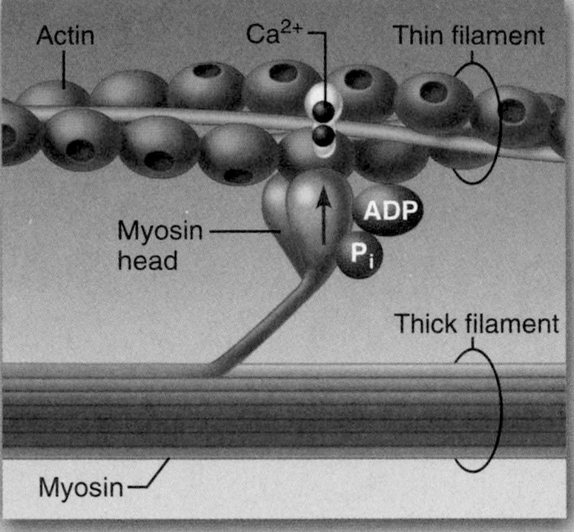

① **Cross bridge formation.** Energized myosin head attaches to actin myofilament, forming a cross bridge.

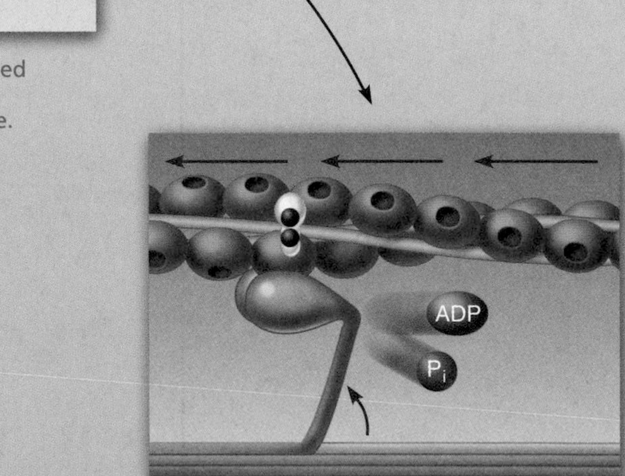

② **The power (working) stroke.** ADP and P_i are released and the myosin head pivots and bends, changing to its bent low-energy shape. As a result it pulls on the actin filament, sliding it toward the M line.

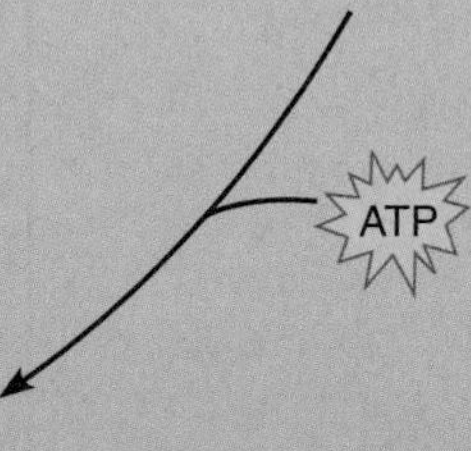

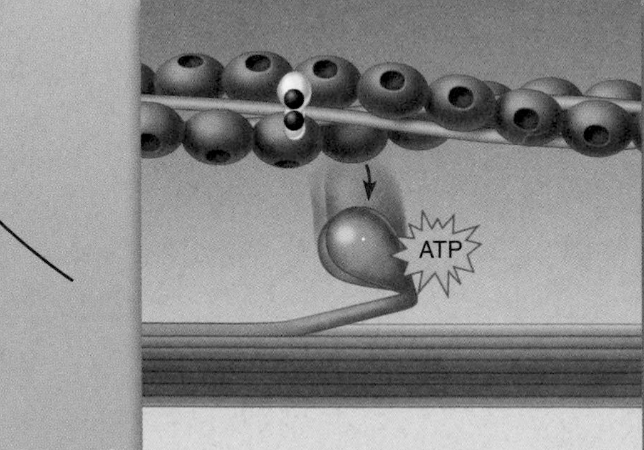

③ **Cross bridge detachment.** After ATP attaches to myosin, the link between myosin and actin weakens, and the myosin head detaches (the cross bridge "breaks").

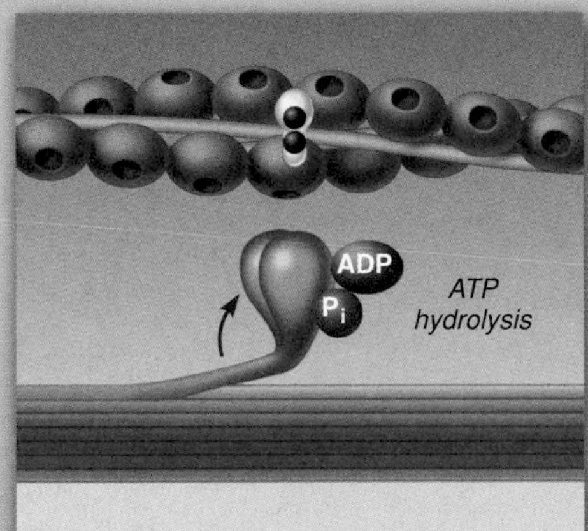

④ **Cocking of myosin head.** As ATP is hydrolyzed to ADP and P_i, the myosin head returns to its prestroke high-energy, or "cocked," position.

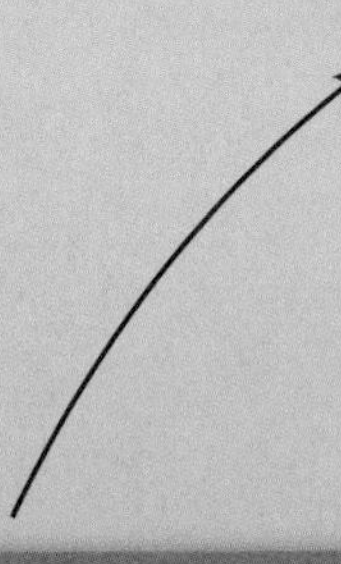

A&P Flix View this animation at myA&P

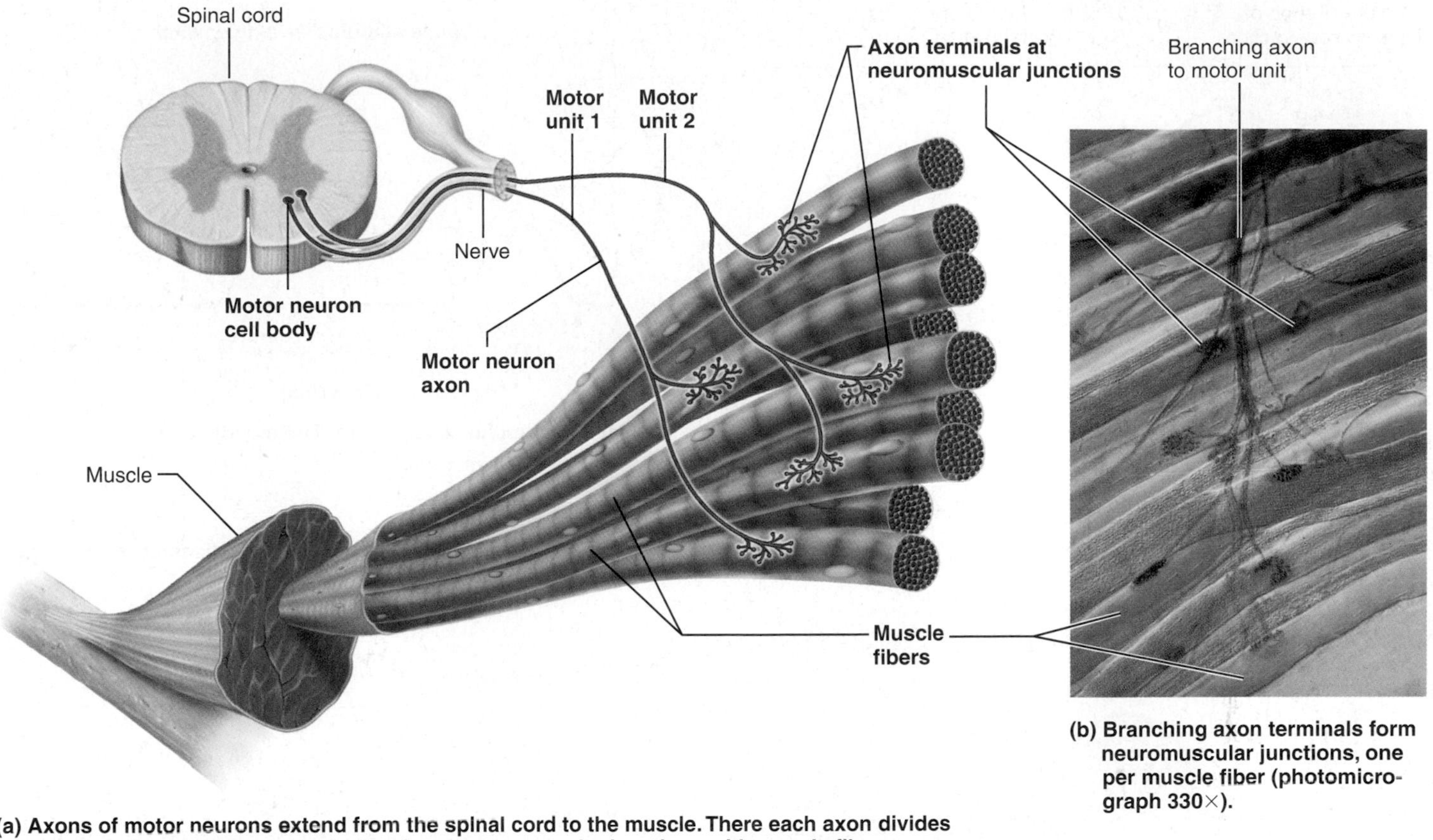

(a) Axons of motor neurons extend from the spinal cord to the muscle. There each axon divides into a number of axon terminals that form neuromuscular junctions with muscle fibers scattered throughout the muscle.

(b) Branching axon terminals form neuromuscular junctions, one per muscle fiber (photomicrograph 330×).

Figure 9.13 **A motor unit consists of a motor neuron and all the muscle fibers it innervates.** (See *A Brief Atlas of the Human Body*, Plate 30.)

clustered together but are spread throughout the muscle. As a result, stimulation of a single motor unit causes a weak contraction of the *entire* muscle.

The Muscle Twitch

Muscle contraction is easily investigated in the laboratory using an isolated muscle. The muscle is attached to an apparatus that produces a **myogram**, a graphic recording of contractile activity. The line recording the activity is called a *tracing*.

The response of a motor unit to a single action potential of its motor neuron is called a **muscle twitch**. The muscle fibers contract quickly and then relax. Every twitch myogram has three distinct phases **(Figure 9.14a)**.

1. **Latent period.** The **latent period** is the first few milliseconds following stimulation when excitation-contraction coupling is occurring. During this period, muscle tension is beginning to increase but no response is seen on the myogram.
2. **Period of contraction.** The period of contraction is when cross bridges are active, from the onset to the peak of tension development, and the myogram tracing rises to a peak. This period lasts 10–100 ms. If the tension (pull) becomes great enough to overcome the resistance of a load, the muscle shortens.
3. **Period of relaxation.** The period of contraction is followed by the period of relaxation. This final phase, lasting 10–100 ms, is initiated by reentry of Ca^{2+} into the SR. Because contractile force is declining, muscle tension decreases to zero and the tracing returns to the baseline. If the muscle shortened during contraction, it now returns to its initial length. Notice that a muscle contracts faster than it relaxes, as revealed by the asymmetric nature of the myogram tracing.

As you can see in Figure 9.14b, twitch contractions of some muscles are rapid and brief, as with the muscles controlling eye movements. In contrast, the fibers of fleshy calf muscles (gastrocnemius and soleus) contract more slowly and remain contracted for much longer periods. These differences between muscles reflect metabolic properties of the myofibrils and enzyme variations.

Graded Muscle Responses

Muscle twitches—like those single, jerky contractions provoked in a laboratory—may result from certain neuromuscular problems, but this is *not* the way our muscles normally operate. Instead, healthy muscle contractions are relatively smooth and vary in strength as different demands are placed on them. These variations, needed for proper control of skeletal movement, are

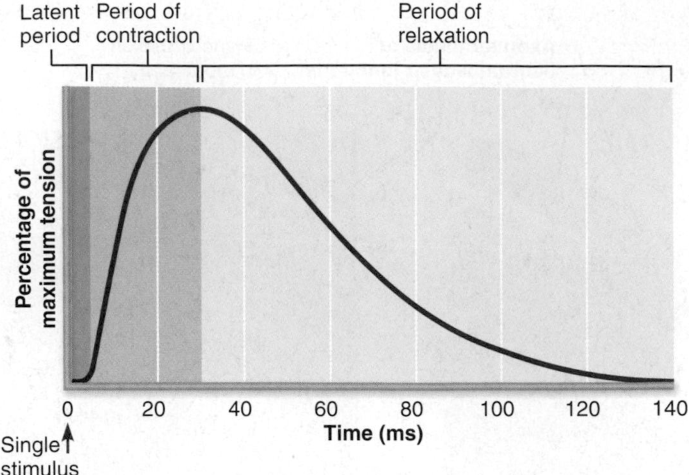

(a) Myogram showing the three phases of an isometric twitch

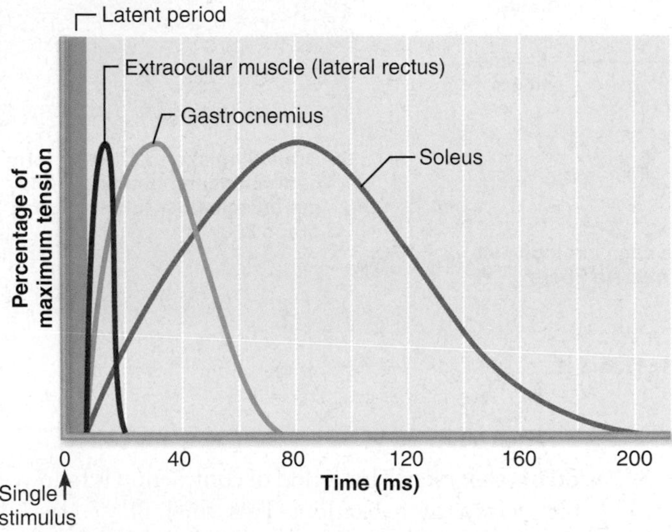

(b) Comparison of the relative duration of twitch responses of three muscles

Figure 9.14 The muscle twitch.

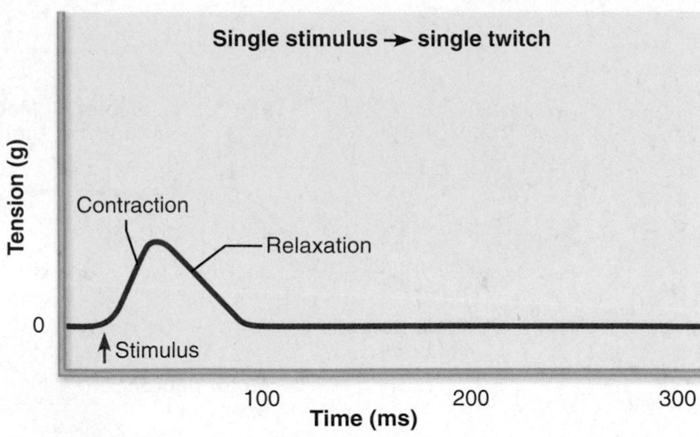

(a) A single stimulus is delivered. The muscle contracts and relaxes.

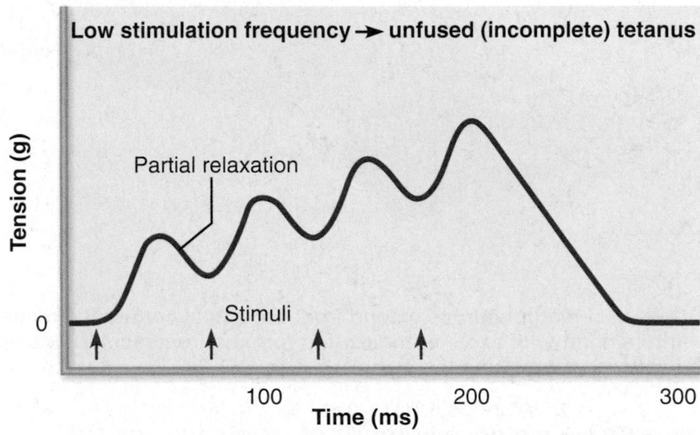

(b) If another stimulus is applied before the muscle relaxes completely, then more tension results. This is temporal (or wave) summation and results in unfused (or incomplete) tetanus.

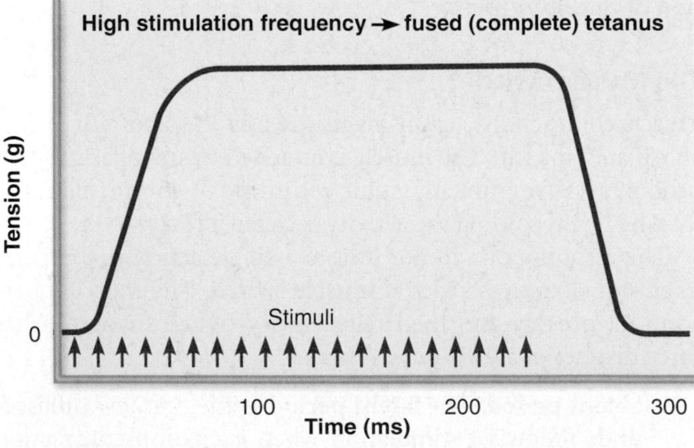

(c) At higher stimulus frequencies, there is no relaxation at all between stimuli. This is fused (complete) tetanus.

Figure 9.15 Muscle response to changes in stimulation frequency.

referred to as **graded muscle responses.** In general, muscle contraction can be graded in two ways: (1) by changing the frequency of stimulation and (2) by changing the strength of stimulation.

Muscle Response to Changes in Stimulus Frequency The nervous system achieves greater muscular force by increasing the firing rate of motor neurons. For example, if two identical stimuli (electrical shocks or nerve impulses) are delivered to a muscle in rapid succession, the second twitch will be stronger than the first. On a myogram the second twitch will appear to ride on the shoulders of the first **(Figure 9.15b)**. This phenomenon, called **temporal** or **wave summation**, occurs because the second contraction occurs before the muscle has completely relaxed. Because the muscle is already partially contracted when the next stimulus arrives and more calcium is being squirted into the cytosol to replace that being reclaimed by the SR,

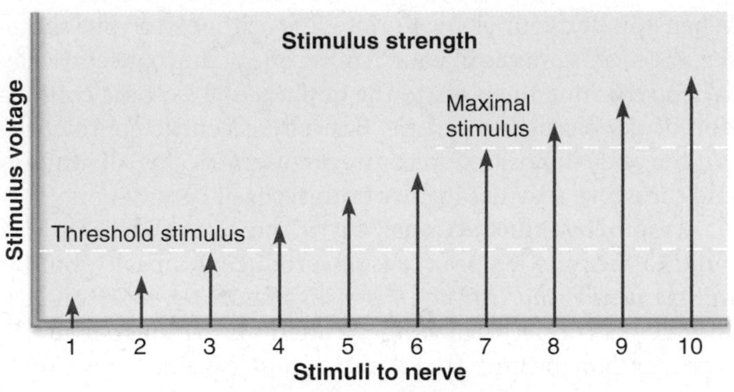

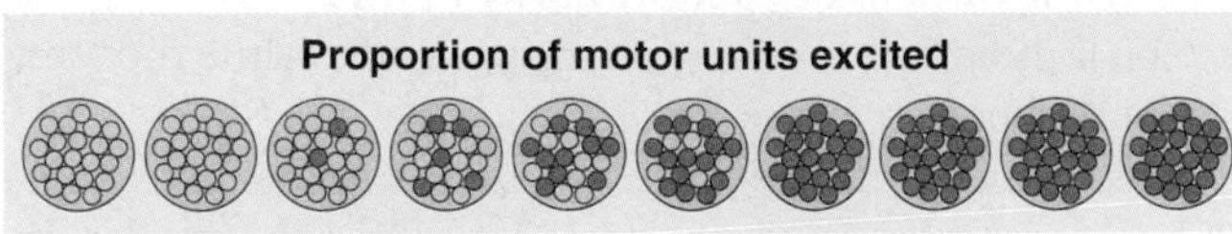

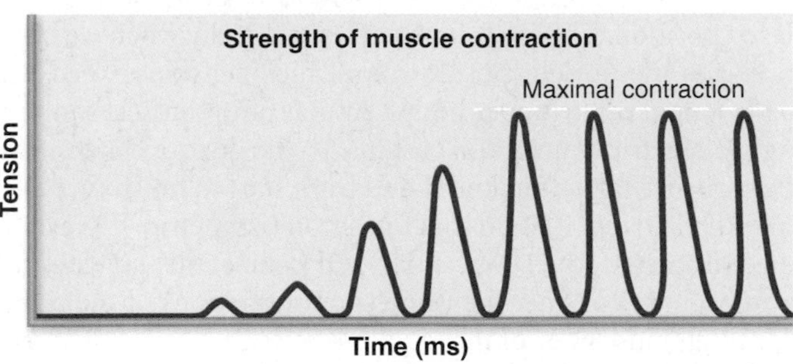

Figure 9.16 Relationship between stimulus intensity (graph at top) and muscle tension (tracing below). Below threshold voltage, no muscle response is seen on the tracing (stimuli 1 and 2). Once threshold (3) is reached, increases in voltage excite (recruit) more and more motor units until the maximal stimulus is reached (7). Further increases in stimulus voltage produce no further increase in contractile strength.

muscle tension produced during the second contraction causes more shortening than the first. In other words, the contractions are summed. (However, the refractory period is *always* honored. Thus, if a second stimulus is delivered before repolarization is complete, no wave summation occurs.)

If the stimulus strength is held constant and the muscle is stimulated at an increasingly faster rate, the relaxation time between the twitches becomes shorter and shorter, the concentration of Ca^{2+} in the cytosol higher and higher, and the degree of wave summation greater and greater, progressing to a sustained but quivering contraction referred to as **unfused** or **incomplete tetanus** (Figure 9.15b).

Finally, as the stimulation frequency continues to increase, muscle tension increases until a maximal tension is reached. At this point all evidence of muscle relaxation disappears and the contractions fuse into a smooth, sustained contraction plateau called **fused** or **complete tetanus** (tet′ah-nus; *tetan* = rigid, tense) (Figure 9.15c). (Note that this term is often confused with the bacterial disease called tetanus that causes severe involuntary contractions.) In the real world, fused tetanus happens infrequently, for example, when someone shows superhuman strength by lifting a fallen tree limb off a companion.

Vigorous muscle activity cannot continue indefinitely. Prolonged tetanus inevitably leads to *muscle fatigue*, a situation in which the muscle is unable to contract and its tension drops to zero.

Muscle Response to Changes in Stimulus Strength Wave summation contributes to contractile force, but its primary function is to produce smooth, continuous muscle contractions by rapidly stimulating a specific number of muscle cells. The force of contraction is controlled more precisely by **recruitment**, also called **multiple motor unit summation.**

In the laboratory, recruitment is achieved by delivering shocks of increasing voltage to the muscle, calling more and more muscle fibers into play. Stimuli that produce no observable contractions are called **subthreshold stimuli.** The stimulus at which the first observable contraction occurs is called the **threshold stimulus** (**Figure 9.16**). Beyond this point, the muscle contracts more and more vigorously as the stimulus strength is increased. The **maximal stimulus** is the strongest stimulus that produces increased contractile force. It represents the point at which all the muscle's motor units are recruited. Increasing the stimulus intensity beyond the maximal stimulus does not produce a stronger contraction. In the body, the same phenomenon is caused by neural activation of an increasingly large number of motor units serving the muscle.

The recruitment process is not random. Instead it is dictated by the *size principle.* In any muscle, motor units with the smallest muscle fibers are controlled by small, highly excitable motor neurons, and these motor units tend to be activated first. As motor units with larger and larger muscle fibers begin to be excited, contractile strength increases. The largest motor units, containing large, coarse muscle fibers, have as much as 50 times the contractile force of the smallest ones. They are controlled by the largest, least excitable (highest-threshold) neurons and are activated only when the most powerful contraction is necessary (**Figure 9.17**).

The size principle is important because it allows the increases in force during weak contractions (for example, those that maintain posture or slow movements) to occur in small steps, whereas gradations in muscle force are progressively greater when large amounts of force are needed for vigorous activities such as jumping or running. This principle explains how the same hand that pats your cheek can deliver a stinging slap.

Although *all* the motor units of a muscle may be recruited simultaneously to produce an exceptionally strong contraction, motor units are more commonly activated asynchronously in the body. At a given instant, some are in tetanus (usually unfused tetanus) while others are resting and recovering. This technique helps prolong a strong contraction by preventing or delaying fatigue. It also explains how weak contractions promoted by infrequent stimuli can remain smooth.

Muscle Tone

Skeletal muscles are described as voluntary, but even relaxed muscles are almost always slightly contracted, a phenomenon

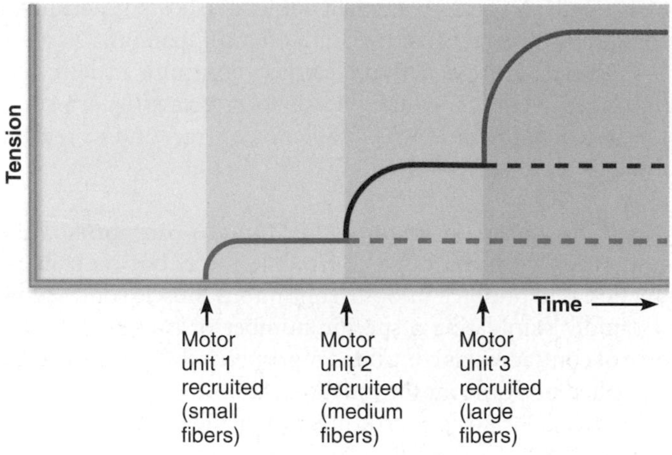

Figure 9.17 The size principle of recruitment. Recruitment of motor neurons controlling skeletal muscle fibers is orderly—small, highly excitable motor neurons are generally recruited more readily than large, less excitable ones. This phenomenon is referred to as the *size principle.* In weaker contractions, small motor units containing small-diameter muscle fibers are recruited. As contractile strength increases, larger and larger motor units containing larger numbers of increasingly larger-diameter muscle fibers are activated. Consequently, the contractions get stronger and stronger in a predictable way.

called **muscle tone.** Muscle tone is due to spinal reflexes that activate first one group of motor units and then another in response to activation of stretch receptors in the muscles. Muscle tone does not produce active movements, but it keeps the muscles firm, healthy, and ready to respond to stimulation. Skeletal muscle tone also helps stabilize joints and maintain posture.

Isotonic and Isometric Contractions

As noted earlier, there are two main categories of contractions—*isotonic* and *isometric.* In **isotonic contractions** (*iso* = same; *ton* = tension), muscle length changes and moves the load. Once sufficient tension has developed to move the load, the tension remains relatively constant through the rest of the contractile period **(Figure 9.18a).**

Isotonic contractions come in two "flavors"—*concentric* and *eccentric.* **Concentric contractions** are those in which the muscle *shortens* and does work, such as picking up a book or kicking a ball. These contractions are probably more familiar. However, **eccentric contractions,** in which the muscle generates force as it *lengthens,* are equally important for coordination and purposeful movements. Eccentric contractions occur in your calf muscle, for example, as you walk up a steep hill. Eccentric contractions are about 50% more forceful than concentric ones at the same load and more often cause delayed-onset muscle soreness. (Consider how your calf muscles *feel* the day after hiking up that hill.) Just why this is so is unclear, but it may be that the muscle stretching that occurs during such contractions causes microtears in the muscles.

Biceps curls provide a simple example of how concentric and eccentric contractions work together in our everyday activities. When you flex your elbow to raise this textbook to your shoulder, the biceps muscle in your arm is contracting concentrically. When returning the book to the desktop, the isotonic contraction of the biceps is eccentric. Basically, eccentric contractions put the body in position to contract concentrically. All jumping and throwing activities involve both types of contraction.

In **isometric contractions** (*metric* = measure), tension may build to the muscle's peak tension-producing capacity, but the muscle *neither shortens nor lengthens* (Figure 9.18b). Isometric contractions occur when a muscle attempts to move a load that is greater than the force (tension) the muscle is able to develop—think of trying to lift a piano singlehandedly. Muscles contract isometrically when they act primarily to maintain upright posture or to hold joints in stationary positions while movements occur at other joints.

Let's consider knee bends as an example. When the squat position is held for a few seconds, the quadriceps muscles of the anterior thigh contract isometrically to hold the knee in the flexed position. They also contract isometrically when we begin to rise to the upright position until their tension exceeds the load (weight of the upper body). At that point muscle shortening (concentric contraction) begins. So the quadriceps contractile sequence for a deep knee bend from start to finish is (1) flex knee (eccentric), (2) hold squat position (isometric), (3) extend knee (isometric, then concentric). Of course, this list does not even begin to consider the isometric contractions of the posterior thigh muscles or of the trunk muscles that maintain a relatively erect trunk posture during the movement.

Electrochemical and mechanical events occurring within a muscle are identical in both isotonic and isometric contractions. However, the result is different. In isotonic contractions, the thin filaments are sliding. In isometric contractions, the cross bridges are generating force but are *not* moving the thin filaments, so there is no change in the banding pattern from that of the resting state. (You could say that they are "spinning their wheels" on the same actin binding sites.)

CHECK YOUR UNDERSTANDING

10. What is a motor unit?

11. What is happening in the muscle during the latent period of a twitch contraction?

12. Jay is competing in a chin-up competition. What type of muscle contractions are occurring in his biceps muscles immediately after he grabs the bar? As his body begins to move upward toward the bar? When his body begins to approach the mat?

For answers, see Appendix G.

Muscle Metabolism

- **Describe three ways in which ATP is regenerated during skeletal muscle contraction.**
- **Define oxygen deficit and muscle fatigue. List possible causes of muscle fatigue.**

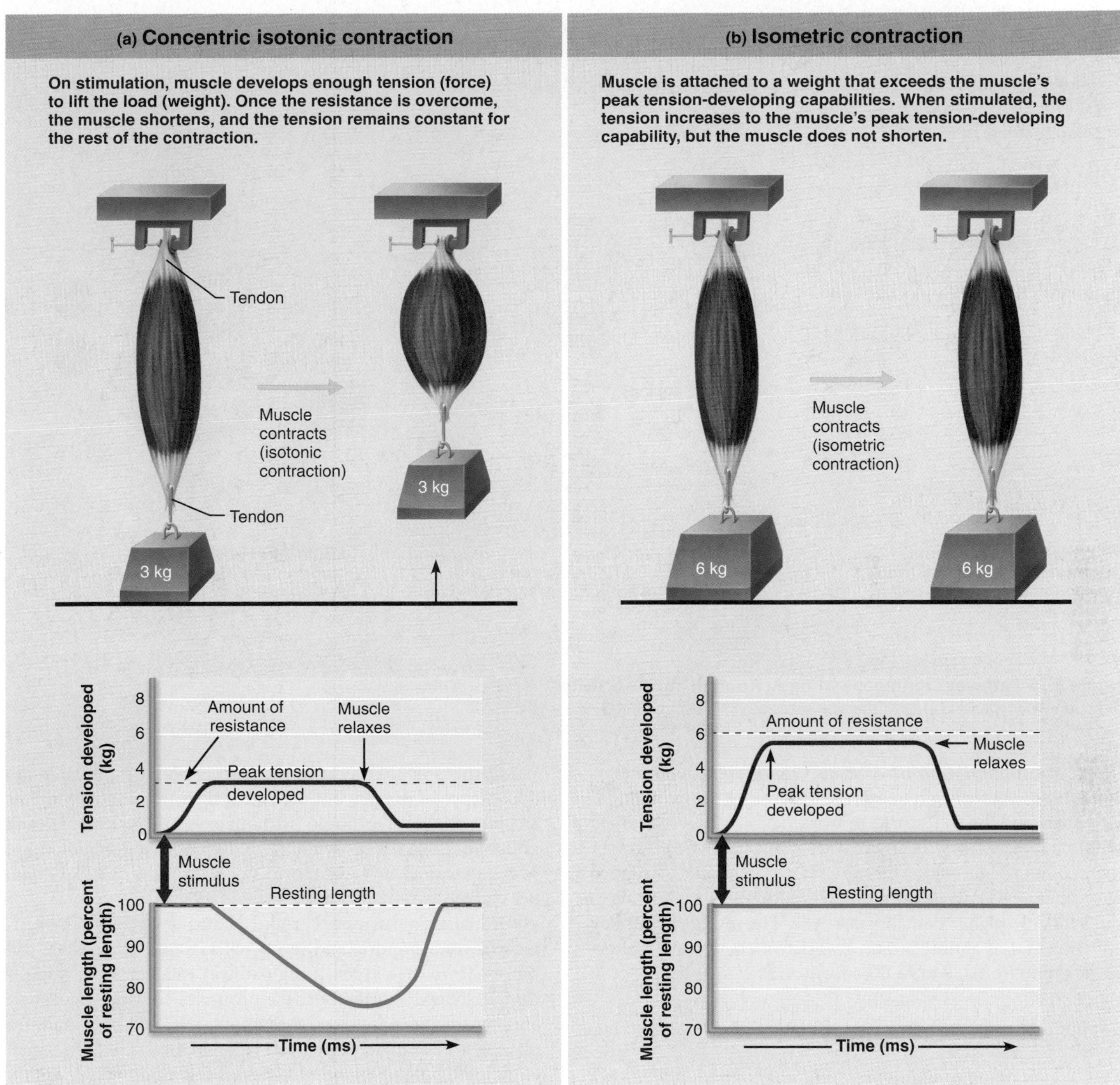

Figure 9.18 Isotonic (concentric) and isometric contractions.

Providing Energy for Contraction

How does the body provide the energy needed for contraction? As a muscle contracts, ATP supplies the energy for cross bridge movement and detachment and for operation of the calcium pump in the SR. Surprisingly, muscles store very limited reserves of ATP—4 to 6 seconds' worth at most, just enough to get you going. Because ATP is the *only* energy source used directly for contractile activities, it must be regenerated as fast as it is broken down if contraction is to continue.

Fortunately, after ATP is hydrolyzed to ADP and inorganic phosphate in muscle fibers, it is regenerated within a fraction of a second by one or more of the three pathways summarized in **Figure 9.19**: (1) direct phosphorylation of ADP by creatine phosphate, (2) the anaerobic pathway called glycolysis, which converts glucose to lactic acid, and (3) aerobic respiration. All body cells use glycolysis and aerobic respiration to produce ATP, so we touch on these metabolic pathways here but describe them in detail later, in Chapter 24.

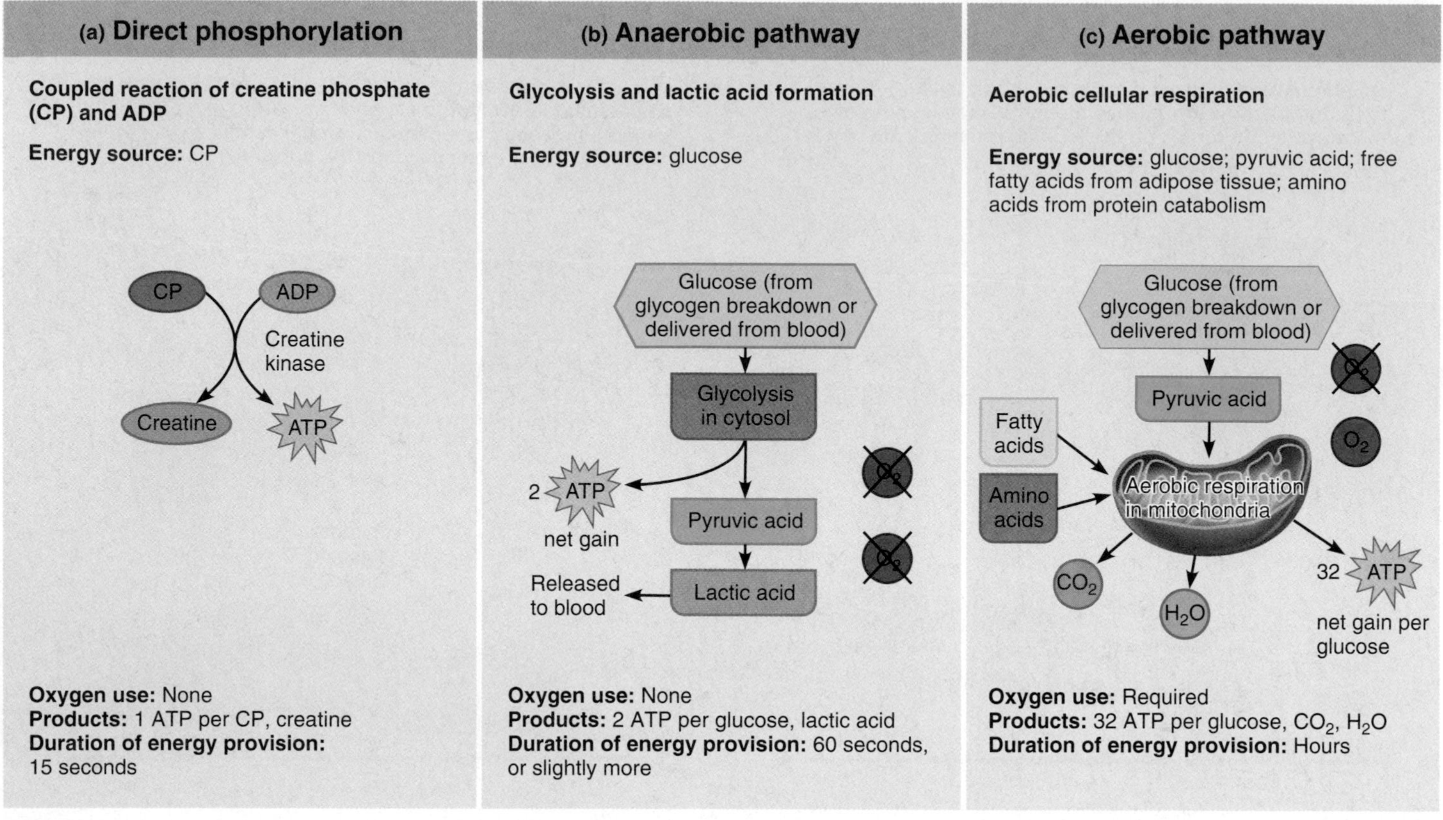

Figure 9.19 Pathways for regenerating ATP during muscle activity. The fastest pathway is direct phosphorylation **(a)**, and the slowest is aerobic respiration **(c)**.

Direct Phosphorylation of ADP by Creatine Phosphate (Figure 9.19a) As we begin to exercise vigorously, the demand for ATP soars and the ATP stored in working muscles is consumed within a few twitches. Then **creatine phosphate (CP)** (kre′ah-tin), a unique high-energy molecule stored in muscles, is tapped to regenerate ATP while the metabolic pathways are adjusting to the suddenly higher demands for ATP. The result of coupling CP with ADP is almost instant transfer of energy and a phosphate group from CP to ADP to form ATP:

$$\text{Creatine phosphate} + \text{ADP} \xrightarrow{\text{creatine kinase}} \text{creatine} + \text{ATP}$$

Muscle cells store two to three times as much CP as ATP, and the CP-ADP reaction, catalyzed by the enzyme **creatine kinase**, is so efficient that the amount of ATP in muscle cells changes very little during the initial period of contraction.

Together, stored ATP and CP provide for maximum muscle power for 14–16 seconds—long enough to energize a 100-meter dash (slightly longer if the activity is less vigorous). The coupled reaction is readily reversible, and to keep CP "on tap," CP reserves are replenished during periods of rest or inactivity.

Anaerobic Pathway: Glycolysis and Lactic Acid Formation (Figure 9.19b) As stored ATP and CP are exhausted, more ATP is generated by breakdown (catabolism) of glucose obtained from the blood or of glycogen stored in the muscle. The initial phase of glucose breakdown is **glycolysis** (gli-kol′ĭ-sis; "sugar splitting"). This pathway occurs in both the presence *and* the absence of oxygen, but because it does not *use* oxygen, it is an *anaerobic* (an-a′er-ōb-ik; "without oxygen") pathway. During glycolysis, glucose is broken down to two *pyruvic acid* molecules, releasing enough energy to form small amounts of ATP (2 ATP per glucose).

Ordinarily, pyruvic acid produced during glycolysis then enters the mitochondria and reacts with oxygen to produce still more ATP in the oxygen-using pathway called aerobic respiration, described shortly. But when muscles contract vigorously and contractile activity reaches about 70% of the maximum possible (for example, when you run 600 meters with maximal effort), the bulging muscles compress the blood vessels within them, impairing blood flow and oxygen delivery. Under these anaerobic conditions, most of the pyruvic acid produced during glycolysis is converted into **lactic acid**, and the overall process is referred to as **anaerobic glycolysis**. Thus, during oxygen deficit, lactic acid is the end product of cellular metabolism of glucose.

Most of the lactic acid diffuses out of the muscles into the bloodstream and is gone from the muscle tissue within 30 minutes after exercise stops. Subsequently, the lactic acid is picked up by liver, heart, or kidney cells, which can use it as an energy source. Additionally, liver cells can reconvert it to pyruvic acid or glucose and release it back into the bloodstream for muscle use, or convert it to glycogen for storage.

The anaerobic pathway harvests only about 5% as much ATP from each glucose molecule as the aerobic pathway, but it pro-

Short-duration exercise
Prolonged-duration exercise
6 seconds
10 seconds
30–40 seconds
End of exercise
Hours
ATP stored in muscles is used first.
ATP is formed from creatine phosphate and ADP.
Glycogen stored in muscles is broken down to glucose, which is oxidized to generate ATP.
ATP is generated by breakdown of several nutrient energy fuels by aerobic pathway. This pathway uses oxygen released from myoglobin or delivered in the blood by hemoglobin. When it ends, the oxygen deficit is paid back.

Figure 9.20 Comparison of energy sources used during short-duration exercise and prolonged-duration exercise.

duces ATP about 2½ times faster. For this reason, when large amounts of ATP are needed for moderate periods (30–40 seconds) of strenuous muscle activity, glycolysis can provide most of the ATP needed as long as the required fuels and enzymes are available. Together, stored ATP and CP and the glycolysis–lactic acid pathway can support strenuous muscle activity for nearly a minute.

Although anaerobic glycolysis readily fuels spurts of vigorous exercise, it has shortcomings. Huge amounts of glucose are used to produce relatively small harvests of ATP, and the accumulating lactic acid is partially responsible for muscle soreness during intense exercise.

Aerobic Respiration (Figure 9.19c) During rest and light to moderate exercise, even if prolonged, 95% of the ATP used for muscle activity comes from aerobic respiration. **Aerobic respiration** occurs in the mitochondria, requires oxygen, and involves a sequence of chemical reactions in which the bonds of fuel molecules are broken and the energy released is used to make ATP.

During aerobic respiration, which includes glycolysis and the reactions that take place in the mitochondria, glucose is broken down entirely, yielding water, carbon dioxide, and large amounts of ATP as the final products.

$$\text{Glucose} + \text{oxygen} \rightarrow \text{carbon dioxide} + \text{water} + \text{ATP}$$

The carbon dioxide released diffuses out of the muscle tissue into the blood and is removed from the body by the lungs.

As exercise begins, muscle glycogen provides most of the fuel. Shortly thereafter, bloodborne glucose, pyruvic acid from glycolysis, and free fatty acids are the major sources of fuels. After about 30 minutes, fatty acids become the major energy fuels. Aerobic respiration provides a high yield of ATP (about 32 ATP per glucose), but it is slow because of its many steps and it requires continuous delivery of oxygen and nutrient fuels to keep it going.

Energy Systems Used During Sports Activities Which pathways predominate during exercise? As long as it has enough oxygen, a muscle cell will form ATP by the aerobic pathway. When ATP demands are within the capacity of the aerobic pathway, light to moderate muscular activity can continue for several hours in well-conditioned individuals **(Figure 9.20)**. However, when exercise demands begin to exceed the ability of the muscle cells to carry out the necessary reactions quickly enough, glycolysis begins to contribute more and more of the total ATP generated. The length of time a muscle can continue to contract using aerobic pathways is called **aerobic endurance**, and the point at which muscle metabolism converts to anaerobic glycolysis is called **anaerobic threshold**.

Exercise physiologists have been able to estimate the relative importance of each energy-producing system to athletic performance. Activities that require a surge of power but last only a few seconds, such as weight lifting, diving, and sprinting, rely entirely on ATP and CP stores. The more on-and-off or burst-like activities of tennis, soccer, and a 100-meter swim appear to be fueled almost entirely by anaerobic glycolysis (Figure 9.20).

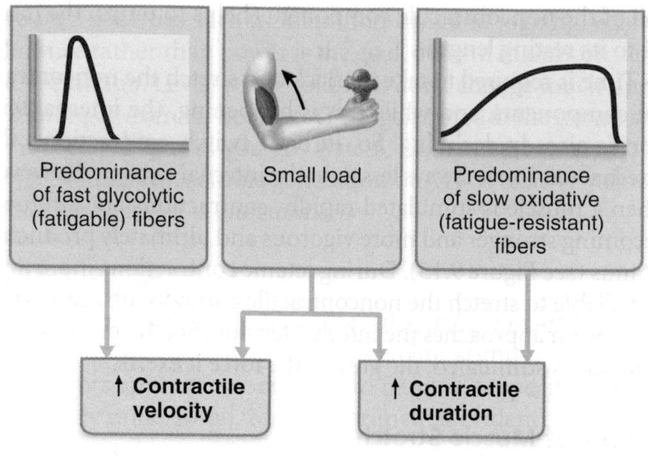

Figure 9.23 Factors influencing velocity and duration of skeletal muscle contraction.

In the body, skeletal muscles are maintained near their optimal operating length by the way they are attached to bones. The joints normally prevent bone movements that would stretch attached muscles beyond their optimal range.

Velocity and Duration of Contraction

Muscles vary in how fast they can contract and in how long they can continue to contract before they fatigue. These characteristics are influenced by muscle fiber type, load, and recruitment.

Muscle Fiber Type

There are several ways of classifying muscle fibers, but learning about these classes will be easier if you initially pay attention to just two major functional characteristics:

- **Speed of contraction.** On the basis of speed of shortening, or contraction, there are **slow fibers** and **fast fibers.** The difference in their speed reflects how fast their myosin ATPases split ATP, and on the pattern of electrical activity of their motor neurons. Duration of contraction also varies with fiber type and depends on how quickly Ca^{2+} is moved from the cytosol into the SR.
- **The major pathways for forming ATP.** The cells that rely mostly on the oxygen-using aerobic pathways for ATP generation are **oxidative fibers**, and those that rely more on anaerobic glycolysis are **glycolytic fibers.**

Using these two criteria, we can classify skeletal muscle cells as being **slow oxidative (SO) fibers, fast oxidative (FO) fibers,** or **fast glycolytic (FG) fibers.** Details about each group are given in **Table 9.2**, but a word to the wise: Do not approach this information by rote memorization—you'll just get frustrated. Instead, start with what you know for any category and see how the characteristics listed support that. For example, think about a *slow oxidative fiber* (Table 9.2, first column, and **Figure 9.23**,

TABLE 9.2 Structural and Functional Characteristics of the Three Types of Skeletal Muscle Fibers

	SLOW OXIDATIVE FIBERS	FAST OXIDATIVE FIBERS	FAST GLYCOLYTIC FIBERS
Metabolic Characteristics			
Speed of contraction	Slow	Fast	Fast
Myosin ATPase activity	Slow	Fast	Fast
Primary pathway for ATP synthesis	Aerobic	Aerobic (some anaerobic glycolysis)	Anaerobic glycolysis
Myoglobin content	High	High	Low
Glycogen stores	Low	Intermediate	High
Recruitment order	First	Second	Third
Rate of fatigue	Slow (fatigue-resistant)	Intermediate (moderately fatigue-resistant)	Fast (fatigable)
Activities Best Suited For			
	Endurance-type activities—e.g., running a marathon; maintaining posture (antigravity muscles)	Sprinting, walking	Short-term intense or powerful movements, e.g., hitting a baseball
Structural Characteristics			
Color	Red	Red to pink	White (pale)
Fiber diameter	Small	Intermediate	Large
Mitochondria	Many	Many	Few
Capillaries	Many	Many	Few

right side). We can see that it

- Contracts relatively *slowly* because its myosin ATPases are slow (a criterion)
- Depends on *oxygen* delivery and aerobic pathways (*high oxidative capacity*—a criterion)
- Is fatigue resistant and has high endurance (typical of fibers that depend on aerobic metabolism)
- Is thin (a large amount of cytoplasm impedes diffusion of O_2 and nutrients from the blood)
- Has relatively little power (a thin cell can contain only a limited number of myofibrils)
- Has many mitochondria (actual sites of oxygen use)
- Has a rich capillary supply (the better to deliver blood-borne O_2)
- Is red (its color stems from an abundant supply of myoglobin, muscle's oxygen-binding pigment that stores O_2 reserves in the cell and aids diffusion of O_2 through the cell)

Add these features together and you have muscle fibers best suited to endurance-type activities. Now think about a *fast glycolytic fiber* (Table 9.2, third column, and Figure 9.23, left side). In contrast, it

- Contracts *rapidly* due to the activity of fast myosin ATPases
- Does not use oxygen
- Depends on plentiful *glycogen* reserves for fuel rather than on blood-delivered nutrients
- Tires quickly because glycogen reserves are short-lived and lactic acid accumulates quickly, making it a fatigable fiber
- Has a large diameter, indicating the plentiful contractile myofilaments that allow it to contract powerfully before it "poops out"
- Has few mitochondria, little myoglobin and low capillary density (and so is white), and is a much thicker cell (because it doesn't depend on continuous oxygen and nutrient diffusion from the blood)

For these reasons, the fast glycolytic fibers are best suited for short-term, rapid, intense movements (moving furniture across the room, for example).

Finally, consider the less common intermediate muscle fiber types, called *fast oxidative fibers* (Table 9.2, middle column). They have many characteristics (fiber diameter and power for example) intermediate between the other two types. Like fast glycolytic fibers, they contract quickly, but like slow oxidative fibers, they are oxygen dependent and have a rich supply of myoglobin and capillaries.

Some muscles have a predominance of one fiber type, but most contain a mixture of fiber types, which gives them a range of contractile speeds and fatigue resistance **(Figure 9.24)**. For example, a calf muscle can propel us in a sprint (using its white fast glycolytic fibers) or a long-distance race (making good use of its slow and fast oxidative fibers). But, as might be expected, all muscle fibers in a particular *motor unit* are of the same type.

Although everyone's muscles contain mixtures of the three fiber types, some people have relatively more of one kind. These differences are genetically initiated, but are modified by

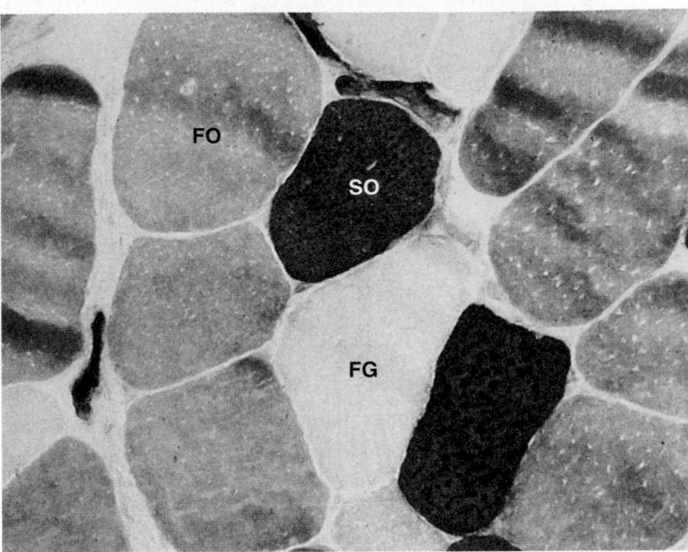

Figure 9.24 Cross section of the three types of fibers in skeletal muscle. From generally smallest to largest, they are slow oxidative fibers (SO), fast oxidative fibers (FO), and fast glycolytic fibers (FG). The staining technique used differentiates the fibers by the abundance of their mitochondria, which contain mitochondrial enzymes (600×).

exercise and no doubt determine athletic capabilities, such as endurance versus strength, to a large extent. For example, muscles of marathon runners have a high percentage of slow oxidative fibers (about 80%), while those of sprinters contain a higher percentage (about 60%) of fast oxidative and glycolytic fibers. Interconversion between the "fast" fiber types occurs as a result of specific exercise regimes, as we'll describe below.

Load Because muscles are attached to bones, they are always pitted against some resistance, or load, when they contract. As you might expect, they contract fastest when there is no added load on them. A greater load results in a longer latent period, a slower contraction, and a shorter duration of contraction **(Figure 9.25)**. If the load exceeds the muscle's maximum tension, the speed of shortening is zero and the contraction is isometric (see Figure 9.18b).

Recruitment Just as many hands on a project can get a job done more quickly and also can keep working longer, the more motor units that are contracting, the faster and more prolonged the contraction.

CHECK YOUR UNDERSTANDING

14. List two factors that influence contractile force and two that influence velocity of contraction.
15. Jim called several friends to help him move. Would he prefer to have those with more slow oxidative muscle fibers or those with more fast glycolytic fibers as his helpers? Why?

For answers, see Appendix G.

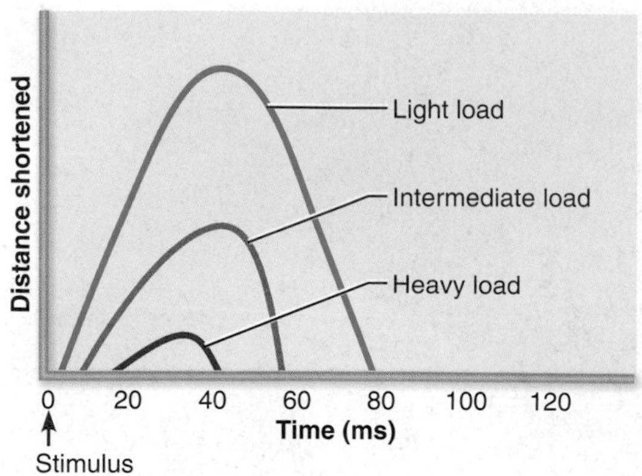

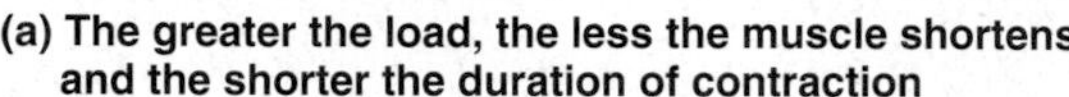

(a) The greater the load, the less the muscle shortens and the shorter the duration of contraction

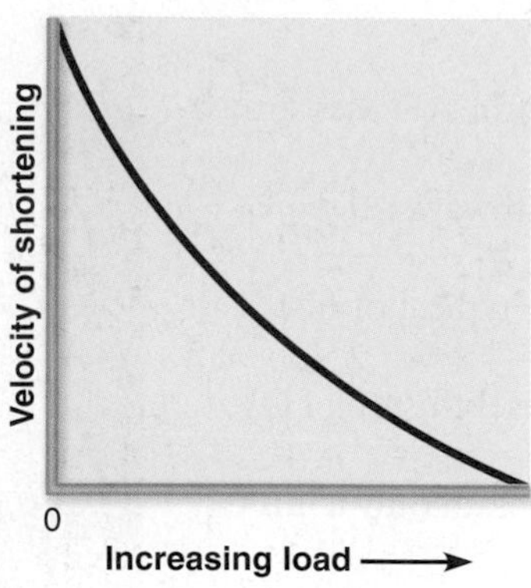

(b) The greater the load, the slower the contraction

Figure 9.25 **Influence of load on contraction duration and velocity.**

Effect of Exercise on Muscles

▶ Compare and contrast the effects of aerobic and resistance exercise on skeletal muscles and on other body systems.

The amount of work a muscle does is reflected in changes in the muscle itself. When used actively or strenuously, muscles may increase in size or strength or become more efficient and fatigue resistant. Muscle inactivity, on the other hand, *always* leads to muscle weakness and wasting.

Adaptations to Exercise

Aerobic, or **endurance**, **exercise** such as swimming, jogging, fast walking, and biking results in several recognizable changes in skeletal muscles. There is an increase in the number of capillaries surrounding the muscle fibers, and in the number of mitochondria within them, and the fibers synthesize more myoglobin. These changes occur in all fiber types, but are most dramatic in slow oxidative fibers, which depend primarily on aerobic pathways. The changes result in more efficient muscle metabolism and in greater endurance, strength, and resistance to fatigue. Additionally, regular endurance exercise may convert fast glycolytic fibers into fast oxidative fibers.

The moderately weak but sustained muscle activity required for endurance exercise does not promote significant skeletal muscle hypertrophy, even though the exercise may go on for hours. Muscle hypertrophy, illustrated by the bulging biceps and chest muscles of a professional weight lifter, results mainly from high-intensity **resistance exercise** (typically under anaerobic conditions) such as weight lifting or isometric exercise, in which the muscles are pitted against high-resistance or immovable forces. Here strength, not stamina, is important, and a few minutes every other day is sufficient to allow a proverbial weakling to put on 50% more muscle within a year.

The increased muscle bulk largely reflects increases in the size of individual muscle fibers (particularly the fast glycolytic variety) rather than an increased number of muscle fibers. [However, some of the increased muscle size may result either from longitudinal splitting or tearing of the fibers and subsequent growth of these "split" cells, or from the proliferation and fusion of satellite cells (see p. 312). The controversy is still raging.] Vigorously stressed muscle fibers contain more mitochondria, form more myofilaments and myofibrils, and store more glycogen. The amount of connective tissue between the cells also increases. Collectively these changes promote significant increases in muscle strength and size. Fast oxidative fibers can be shifted to fast glycolytic fibers in response to resistance activities. However, if the specific exercise routine is discontinued, the fibers previously converted revert to their original metabolic properties.

Resistance training can produce magnificently bulging muscles, but if done unwisely, some muscles may develop more than others. Because muscles work in antagonistic pairs or groups, opposing muscles must be equally strong to work together smoothly. When muscle training is not balanced, individuals can become *muscle-bound*, which means they lack flexibility, have a generally awkward stance, and are unable to make full use of their muscles.

Whatever the activity, exercise gains adhere to the *overload principle*. Forcing a muscle to work hard promotes increased muscle strength and endurance, and as muscles adapt to the increased demands, they must be overloaded even more to produce further gains. However, a heavy-workout day should be followed by one of rest or an easy workout to allow the muscles to recover and repair themselves. Doing too much too soon, or ignoring the warning signs of muscle or joint pain, increases the risk of **overuse injuries** that may prevent future sports activities, or even lead to lifetime disability.

Endurance and resistance exercises produce different patterns of muscular response, so it is important to know what your exercise goals are. Lifting weights will not improve your endurance for a triathlon. By the same token, jogging will do little to improve your muscle definition or to enhance your strength for moving furniture. A program that alternates aerobic activities with anaerobic ones provides the best program for optimal health.

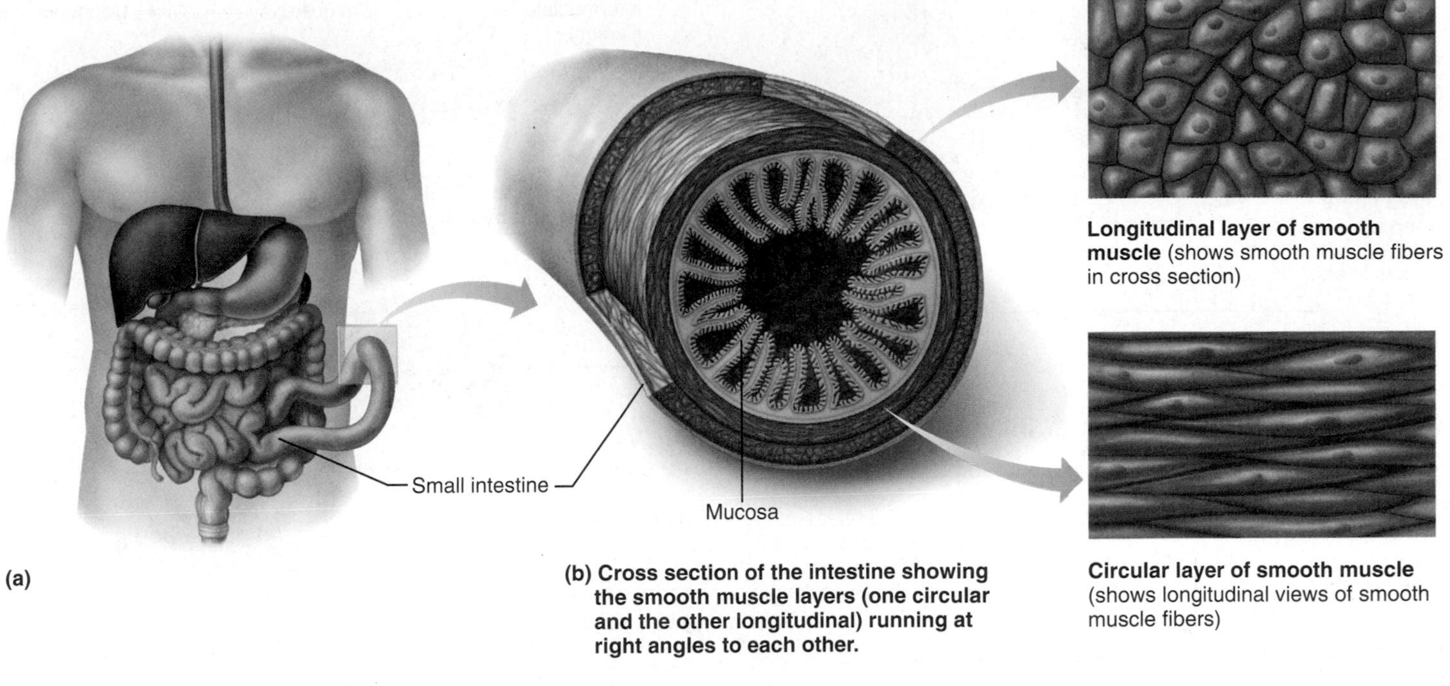

Small intestine
Mucosa
(a)
(b) Cross section of the intestine showing the smooth muscle layers (one circular and the other longitudinal) running at right angles to each other.
Longitudinal layer of smooth muscle (shows smooth muscle fibers in cross section)
Circular layer of smooth muscle (shows longitudinal views of smooth muscle fibers)

Figure 9.26 Arrangement of smooth muscle in the walls of hollow organs.

HOMEOSTATIC IMBALANCE

To remain healthy, muscles must be active. Complete immobilization due to enforced bed rest or loss of neural stimulation results in *disuse atrophy* (degeneration and loss of mass), which begins almost as soon as the muscles are immobilized. Under such conditions, muscle strength can decrease at the rate of 5% per day!

As noted earlier, even at rest, muscles receive weak intermittent stimuli from the nervous system. When totally deprived of neural stimulation, a paralyzed muscle may atrophy to one-quarter of its initial size. Lost muscle tissue is replaced by fibrous connective tissue, making muscle rehabilitation impossible. ■

CHECK YOUR UNDERSTANDING

16. Relative to their effect on muscle size and function, how do aerobic and anaerobic exercise differ?

For answers, see Appendix G.

Smooth Muscle

- ▶ Compare the gross and microscopic anatomy of smooth muscle cells to that of skeletal muscle cells.
- ▶ Compare and contrast the contractile mechanisms and the means of activation of skeletal and smooth muscles.
- ▶ Distinguish between single-unit and multiunit smooth muscle structurally and functionally.

Except for the heart, which is made of cardiac muscle, the muscle in the walls of all the body's hollow organs is almost entirely smooth muscle. The chemical and mechanical events of contraction are essentially the same in all muscle tissues, but smooth muscle is distinctive in several ways that are summarized in **Table 9.3**.

Microscopic Structure of Smooth Muscle Fibers

Smooth muscle fibers are spindle-shaped cells of variable size, each with one centrally located nucleus **(Figure 9.26b)**. Typically, they have a diameter of 5–10 μm and are 30–200 μm long. Skeletal muscle fibers are up to 10 times wider and thousands of times longer.

Smooth muscle lacks the coarse connective tissue sheaths seen in skeletal muscle. However, a small amount of fine connective tissue (endomysium), secreted by the smooth muscles themselves and containing blood vessels and nerves, is found between smooth muscle fibers.

Most smooth muscle is organized into sheets of closely apposed fibers. These sheets occur in the walls of all but the smallest blood vessels and in the walls of hollow organs of the respiratory, digestive, urinary, and reproductive tracts. In most cases, two sheets of smooth muscle are present, with their fibers oriented at right angles to each other, as in the intestine (Figure 9.26). In the *longitudinal layer*, the muscle fibers run parallel to the long axis of the organ. Consequently, when the muscle contracts, the organ dilates and shortens. In the *circular layer*, the fibers run around the circumference of the organ. Contraction of this layer constricts the *lumen* (cavity) of the organ and causes the organ to elongate.

The alternating contraction and relaxation of these opposing layers mixes substances in the lumen and squeezes them through the organ's internal pathway. This propulsive action is

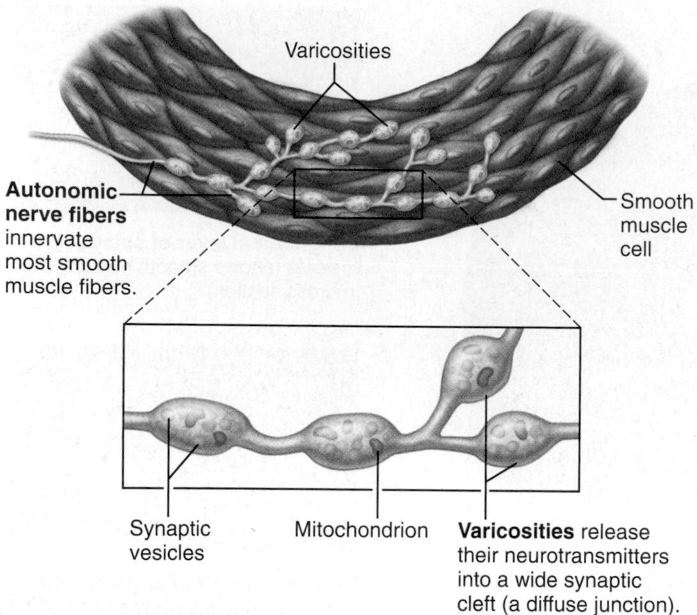

Figure 9.27 Innervation of smooth muscle.

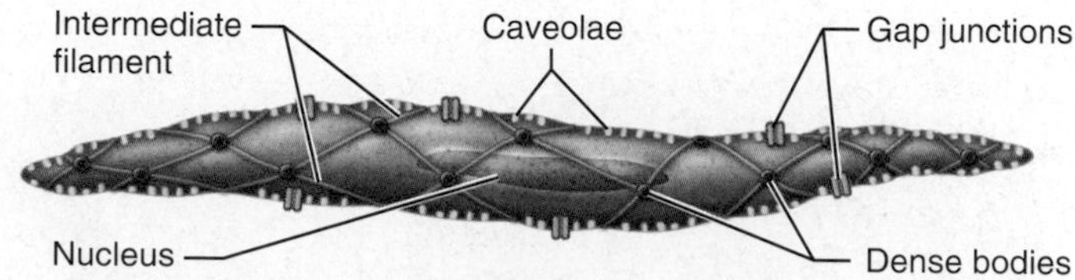

(a) Relaxed smooth muscle fiber (note that adjacent fibers are connected by gap junctions)

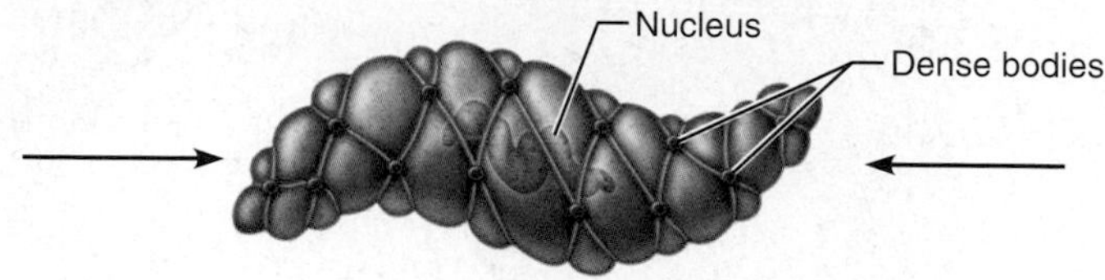

(b) Contracted smooth muscle fiber

Figure 9.28 Intermediate filaments and dense bodies of smooth muscle fibers harness the pull generated by myosin cross bridges. Intermediate filaments attach to dense bodies throughout the sarcoplasm.

called **peristalsis** (per″ĭ-stal′sis; "around contraction"). Contraction of smooth muscle in the rectum, urinary bladder, and uterus helps those organs to expel their contents. Smooth muscle contraction also accounts for the constricted breathing of asthma and for stomach cramps.

Smooth muscle lacks the highly structured, specific neuromuscular junctions of skeletal muscle. Instead, the innervating nerve fibers, which are part of the autonomic (involuntary) nervous system, have numerous bulbous swellings, called **varicosities** (**Figure 9.27**). The varicosities release neurotransmitter into a wide synaptic cleft in the general area of the smooth muscle cells. Such junctions are called **diffuse junctions**. Comparing the specificity of neural input to skeletal and smooth muscles, you could say that skeletal muscle gets priority mail while smooth muscle gets bulk mailings.

The sarcoplasmic reticulum of smooth muscle fibers is much less developed than that of skeletal muscle and lacks a specific pattern relative to the myofilaments. Some SR tubules of smooth muscle touch the sarcolemma at several sites, forming what resembles half-triads that may couple the action potential to calcium release from the SR. T tubules are notably absent, but the sarcolemma of the smooth muscle fiber has multiple **caveolae**, pouchlike infoldings that sequester bits of extracellular fluid containing a high concentration of Ca^{2+} close to the membrane (**Figure 9.28a**). Consequently, when calcium channels in the caveolae open, Ca^{2+} influx occurs rapidly. Although the SR *does* release some of the calcium that triggers contraction, most enters through calcium channels directly from the extracellular space. Contraction ends when calcium is actively transported into the SR and out of the cell. This situation is quite different from what we see in skeletal muscle, which does not depend on extracellular Ca^{2+} for excitation-contraction coupling.

There are no striations, as the name *smooth muscle* indicates, and therefore no sarcomeres. Smooth muscle fibers *do* contain interdigitating thick and thin filaments, but they are much longer than those in skeletal muscle, and the type of myosin contained differs in smooth muscle. The proportion and organization of the myofilaments are also different:

1. **Thick filaments are fewer but have myosin heads along their entire length.** The ratio of thick to thin filaments is much lower in smooth muscle than in skeletal muscle (1:13 compared to 1:2). However, thick filaments of smooth muscle contain actin-gripping myosin heads along their *entire length*, a feature that allows smooth muscle to be as powerful as a skeletal muscle of the same size.
2. **No troponin complex in thin filaments.** As opposed to skeletal muscle which has calcium-binding troponin on the thin filaments, no troponin complex is present in smooth muscle. Instead, a protein called *calmodulin* acts as the calcium-binding site.
3. **Thick and thin filaments arranged diagonally.** Bundles of contractile proteins crisscross within the smooth muscle cell so they spiral down the long axis of the cell like the stripes on a barber pole. Because of this diagonal arrangement, the smooth muscle cells contract in a twisting way so that they look like tiny corkscrews (Figure 9.28b).
4. **Intermediate filament–dense body network.** Smooth muscle fibers contain a lattice-like arrangement of noncontractile *intermediate filaments* that resist tension. They attach at regular intervals to cytoplasmic structures called **dense bodies** (Figure 9.28). The dense bodies, which are also tethered to the sarcolemma, act as anchoring points for thin filaments and therefore correspond to Z discs of skeletal muscle. The intermediate filament–dense body network forms a strong, cable-like intracellular cytoskeleton that harnesses the pull generated by the sliding of the thick

and thin filaments. During contraction, areas of the sarcolemma between the dense bodies bulge outward, giving the cell a puffy appearance, as in Figure 9.28b. Dense bodies at the sarcolemma surface also bind the muscle cell to the connective tissue fibers outside the cell (endomysium) and to adjacent cells. This arrangement transmits the pulling force to the surrounding connective tissue and partly accounts for the synchronous contraction of most smooth muscle.

Contraction of Smooth Muscle

Mechanism of Contraction

In most cases, adjacent smooth muscle fibers exhibit slow, synchronized contractions, the whole sheet responding to a stimulus in unison. This synchronization reflects electrical coupling of smooth muscle cells by *gap junctions*, specialized cell connections described in Chapter 3. Skeletal muscle fibers are electrically isolated from one another, each stimulated to contract by its own neuromuscular junction. By contrast, gap junctions allow smooth muscles to transmit action potentials from fiber to fiber.

Some smooth muscle fibers in the stomach and small intestine are *pacemaker cells* and, once excited, they act as "drummers" to set the contractile pace for the entire muscle sheet. Additionally, these pacemakers have fluctuating membrane potentials and are self-excitatory, that is, they depolarize spontaneously in the absence of external stimuli. However, both the rate and the intensity of smooth muscle contraction may be modified by neural and chemical stimuli.

Contraction in smooth muscle is like that in skeletal muscle in the following ways: (1) actin and myosin interact by the sliding filament mechanism; (2) the final trigger for contraction is a rise in the intracellular calcium ion level; and (3) the sliding process is energized by ATP.

During excitation-contraction coupling, Ca^{2+} is released by the tubules of the SR, but, as mentioned above, it also moves into the cell via membrane channels from the extracellular space. By binding to troponin, calcium ions activate myosin in all striated muscle types, but in smooth muscle, they activate myosin by interacting with a regulatory molecule called **calmodulin**, a cytoplasmic calcium-binding protein. Calmodulin, in turn, interacts with a kinase enzyme called **myosin kinase** or **myosin light chain kinase** which phosphorylates the myosin, activating it. This sequence of events is depicted in **Figure 9.29**. (Note that this is just one pathway of smooth muscle activation. There are others. For example, in some smooth muscles, regulatory proteins associated with actin appear to play a role.)

As in skeletal muscle, smooth muscle relaxes when intracellular Ca^{2+} levels drop, but getting muscle to cease its contractile activity is quite a bit more complex in smooth muscle then in skeletal muscle. Events known to be involved include calcium detachment from calmodulin, active transport of Ca^{2+} into the SR and the extracellular fluid, and dephosphorylation of myosin by a phosphorylase enzyme, which reduces the activity of the myosin ATPases.

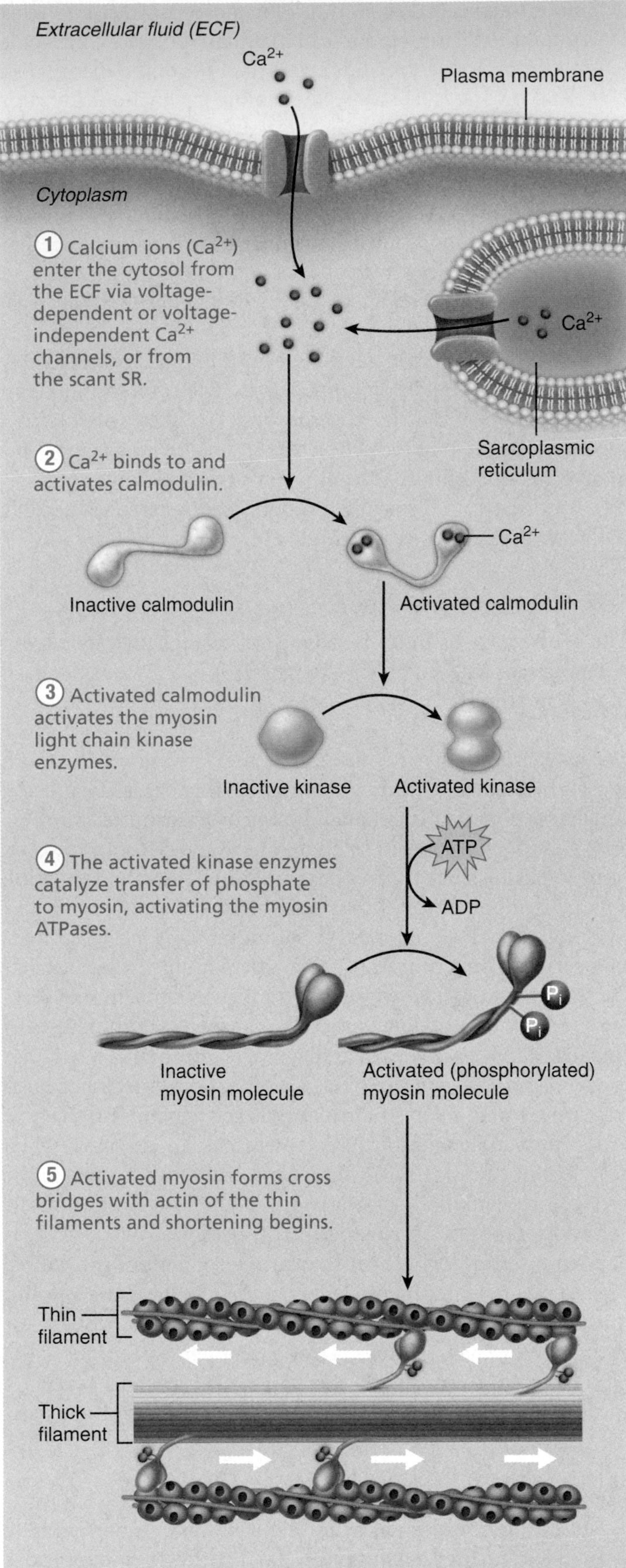

Figure 9.29 Sequence of events in excitation-contraction coupling of smooth muscle.

Smooth muscle takes 30 times longer to contract and relax than does skeletal muscle and can maintain the same contractile tension for prolonged periods at less than 1% of the energy cost. If skeletal muscle is like a speedy windup car that quickly runs down, then smooth muscle is like a steady, heavy-duty engine that lumbers along tirelessly.

Part of the striking energy economy of smooth muscle is the sluggishness of its ATPases compared to those in skeletal muscle. Moreover, smooth muscle myofilaments may latch together during prolonged contractions, saving energy in that way as well. Smooth muscle cells may maintain that *latch state* even after dephosphorylation of myosin.

The ATP-efficient contraction of smooth muscle is extremely important to overall body homeostasis. The smooth muscle in small arterioles and other visceral organs routinely maintains a moderate degree of contraction, called *smooth muscle tone*, day in and day out without fatiguing. Smooth muscle has low energy requirements. Typically, enough ATP is made via aerobic pathways to keep up with the demand.

Regulation of Contraction

The contraction of smooth muscle can be regulated by nerves, hormones, or local chemical changes. Let's briefly consider each of these methods.

Neural Regulation In some cases, the activation of smooth muscle by a neural stimulus is identical to that in skeletal muscle: An action potential is generated by neurotransmitter binding, and is coupled to a rise in calcium ions in the cytosol. However, some types of smooth muscle respond to neural stimulation with graded potentials (local electrical signals) only.

Recall that all somatic nerve endings, that is, nerve endings that serve skeletal muscle, release acetylcholine, which excites the skeletal muscle. However, different autonomic nerves serving the smooth muscle of visceral organs release different neurotransmitters, each of which may excite or inhibit a particular group of smooth muscle cells. The effect of a specific neurotransmitter on a given smooth muscle cell depends on the type of receptor molecules on its sarcolemma. For example, when acetylcholine binds to ACh receptors on smooth muscle in the bronchioles (small air passageways of the lungs), the muscle contracts strongly, narrowing the bronchioles. When norepinephrine, released by a different type of autonomic nerve fiber, binds to norepinephrine receptors on the *same* smooth muscle cells, the effect is inhibitory, so the muscle relaxes, dilating the air passageways. However, when norepinephrine binds to smooth muscle in the walls of most blood vessels, it stimulates the smooth muscle cells to contract and constrict the vessel.

Hormones and Local Chemical Factors Not all smooth muscle activation results from neural signals. Some smooth muscle layers have no nerve supply at all. Instead, they depolarize spontaneously or in response to chemical stimuli that bind to G protein–linked receptors. Others respond to both neural and chemical stimuli.

Chemical factors that cause smooth muscle contraction or relaxation without an action potential (by enhancing or inhibiting Ca^{2+} entry into the sarcoplasm) include certain hormones, lack of oxygen, histamine, excess carbon dioxide, and low pH. The direct response to these chemical stimuli alters smooth muscle activity according to local tissue needs and probably is most responsible for smooth muscle tone. For example, the hormone gastrin stimulates contraction of stomach smooth muscle so that it can churn foodstuffs more efficiently. We will consider how the smooth muscle of specific organs is activated as we discuss each organ in subsequent chapters.

Special Features of Smooth Muscle Contraction

Smooth muscle is intimately involved in the functioning of most hollow organs and has a number of unique characteristics. We have already considered some of these—smooth muscle tone; slow, prolonged contractile activity; and low energy requirements. But smooth muscle also responds differently to stretch and can shorten more than other muscle types. Let's take a look.

Response to Stretch Up to a point, when skeletal muscle is stretched, it responds with more vigorous contractions. Stretching of smooth muscle also provokes contraction, which automatically moves substances along an internal tract. However, the increased tension persists only briefly, and soon the muscle adapts to its new length and relaxes, while still retaining the ability to contract on demand. This **stress-relaxation response** allows a hollow organ to fill or expand slowly (within certain limits) to accommodate a greater volume without promoting strong contractions that would expel its contents. This is an important attribute, because organs such as the stomach and intestines must be able to store their contents temporarily to provide sufficient time for digestion and absorption of the nutrients. Likewise, your urinary bladder must be able to store the continuously made urine until it is convenient to empty your bladder, or you would spend all your time in the bathroom.

Length and Tension Changes Smooth muscle stretches much more than skeletal muscle and generates more tension than skeletal muscles stretched to a comparable extent. As we saw in Figure 9.22, precise, highly organized sarcomeres limit how far a skeletal muscle can be stretched before it is unable to generate force. In contrast, the lack of sarcomeres and the irregular, overlapping arrangement of smooth muscle filaments allow them to generate considerable force, even when they are substantially stretched. The total length change that skeletal muscles can undergo and still function efficiently is about 60% (from 30% shorter to 30% longer than resting length), but smooth muscle can contract when it is anywhere from twice to half its resting length—a total length change of 150%. This capability allows hollow organs to tolerate tremendous changes in volume without becoming flabby when they empty.

Hyperplasia Besides being able to hypertrophy (increase in cell size), which is common to all muscle cells, certain smooth

TABLE 9.3 Comparison of Skeletal, Cardiac, and Smooth Muscle

CHARACTERISTIC	SKELETAL	CARDIAC	SMOOTH
Body location	Attached to bones or (some facial muscles) to skin 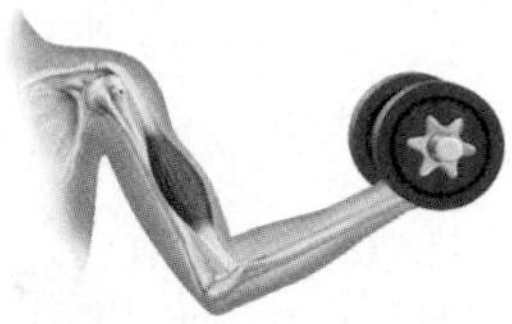	Walls of the heart 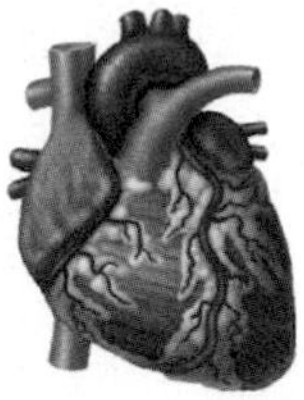	Single-unit muscle in walls of hollow visceral organs (other than the heart); multiunit muscle in intrinsic eye muscles, airways, large arteries 
Cell shape and appearance	Single, very long, cylindrical, multinucleate cells with obvious striations 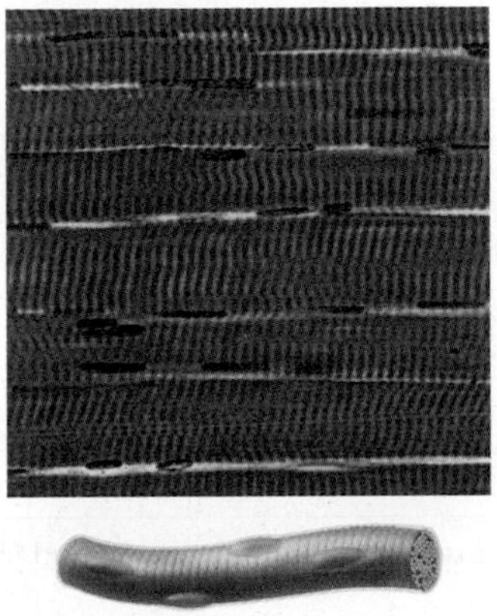	Branching chains of cells; uni- or binucleate; striations 	Single, fusiform, uninucleate; no striations 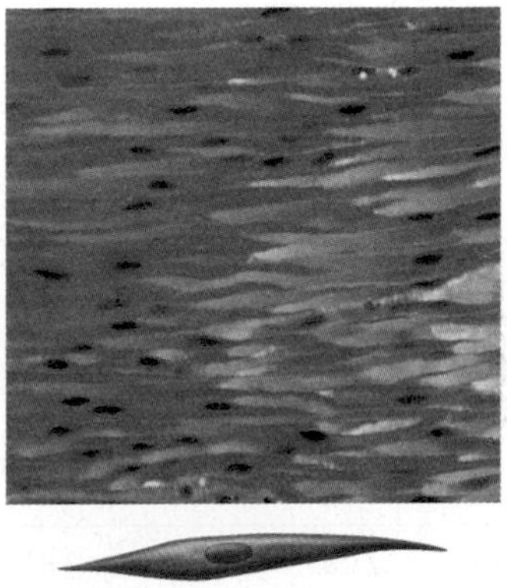
Connective tissue components	Epimysium, perimysium, and endomysium 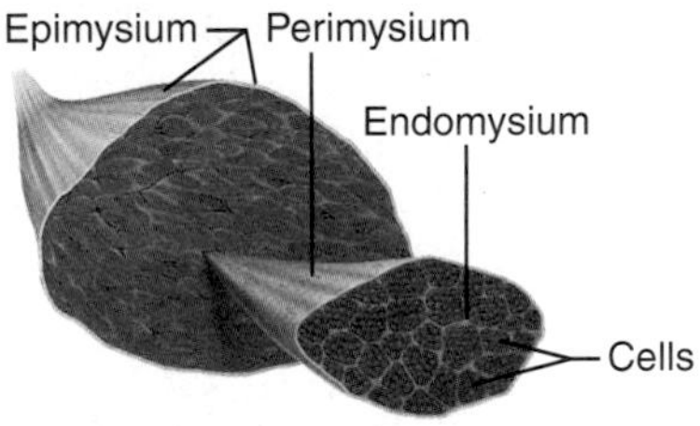	Endomysium attached to fibrous skeleton of heart 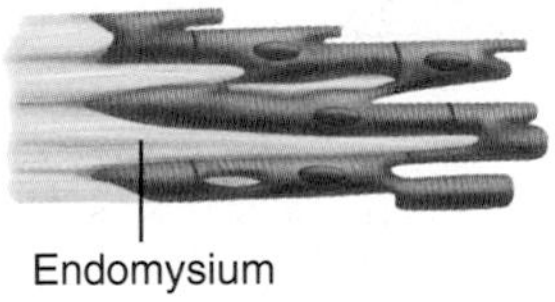	Endomysium 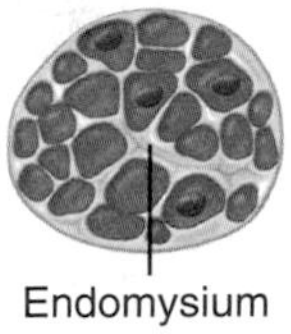
Presence of myofibrils composed of sarcomeres	Yes	Yes, but myofibrils are of irregular thickness	No, but actin and myosin filaments are present throughout; dense bodies anchor actin filaments
Presence of T tubules and site of invagination	Yes; two in each sarcomere at A-I junctions 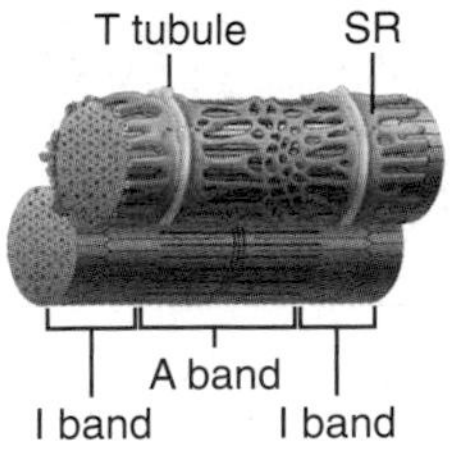	Yes; one in each sarcomere at Z disc; larger diameter than those of skeletal muscle 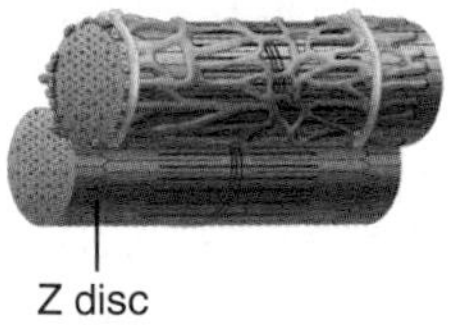	No; only caveolae

TABLE 9.3 Comparison of Skeletal, Cardiac, and Smooth Muscle *(continued)*

CHARACTERISTIC	SKELETAL	CARDIAC	SMOOTH
Elaborate sarcoplasmic reticulum	Yes	Less than skeletal muscle (1–8% of cell volume); scant terminal cisternae	Equivalent to cardiac muscle (1–8% of cell volume); some SR contacts the sarcolemma
Presence of gap junctions	No	Yes; at intercalated discs	Yes; in single-unit muscle
Cells exhibit individual neuromuscular junctions	Yes 	No	Not in single-unit muscle; yes in multiunit muscle 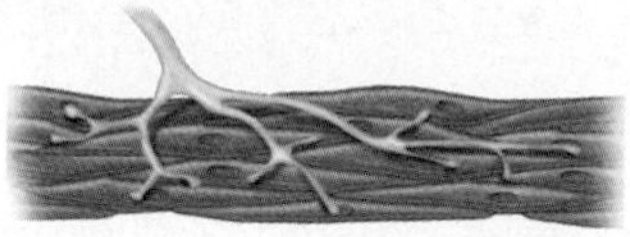
Regulation of contraction	Voluntary via axon terminals of the somatic nervous system 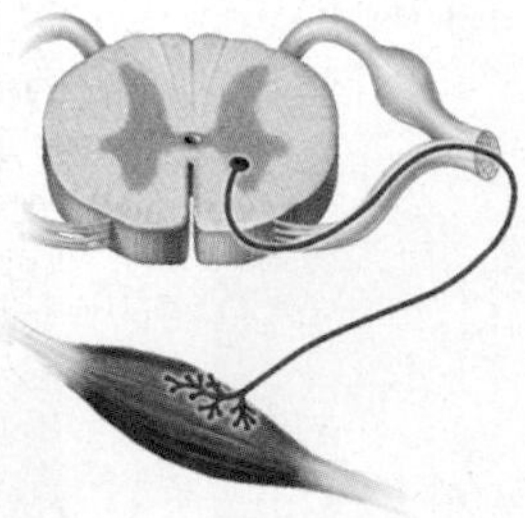	Involuntary; intrinsic system regulation; also autonomic nervous system controls; hormones; stretch 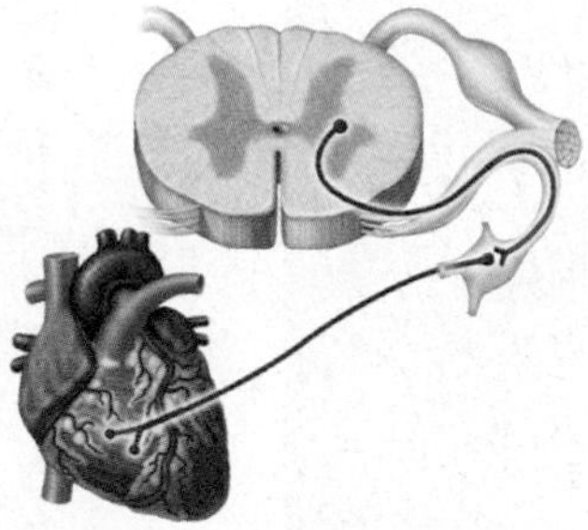	Involuntary; autonomic nerves, hormones, local chemicals; stretch 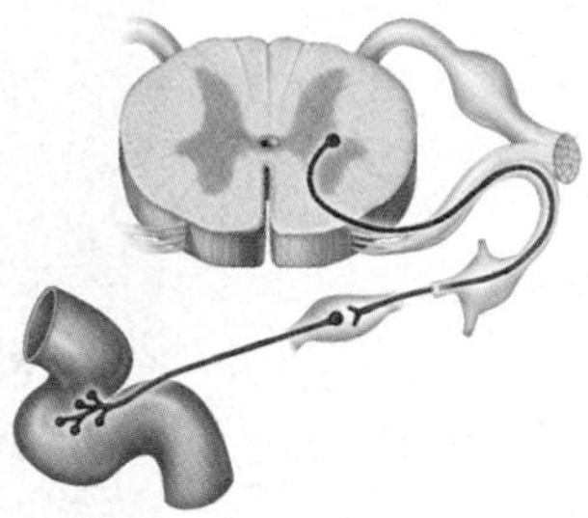
Source of Ca^{2+} for calcium pulse	Sarcoplasmic reticulum (SR)	SR and from extracellular fluid	SR and from extracellular fluid
Site of calcium regulation	Troponin on actin-containing thin filaments  Actin, Troponin	Troponin on actin-containing thin filaments  Actin, Troponin	Calmodulin in the cytosol 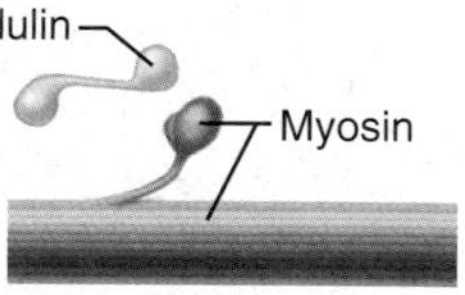 Calmodulin, Myosin
Presence of pacemaker(s)	No	Yes	Yes (in single-unit muscle only)
Effect of nervous system stimulation	Excitation	Excitation or inhibition	Excitation or inhibition
Speed of contraction	Slow to fast 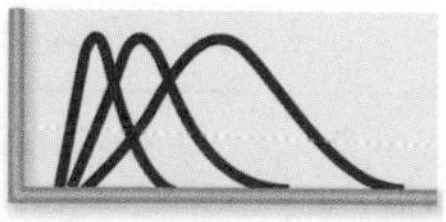	Slow 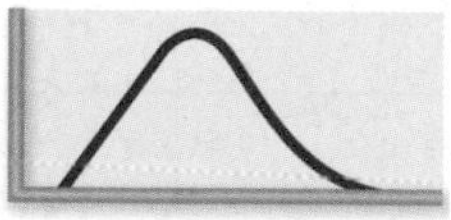	Very slow 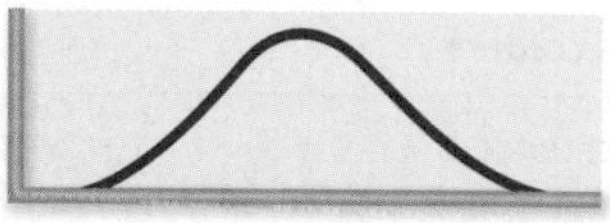
Rhythmic contraction	No	Yes	Yes in single-unit muscle
Response to stretch	Contractile strength increases with degree of stretch (to a point)	Contractile strength increases with degree of stretch	Stress-relaxation response
Respiration	Aerobic and anaerobic	Aerobic	Mainly aerobic

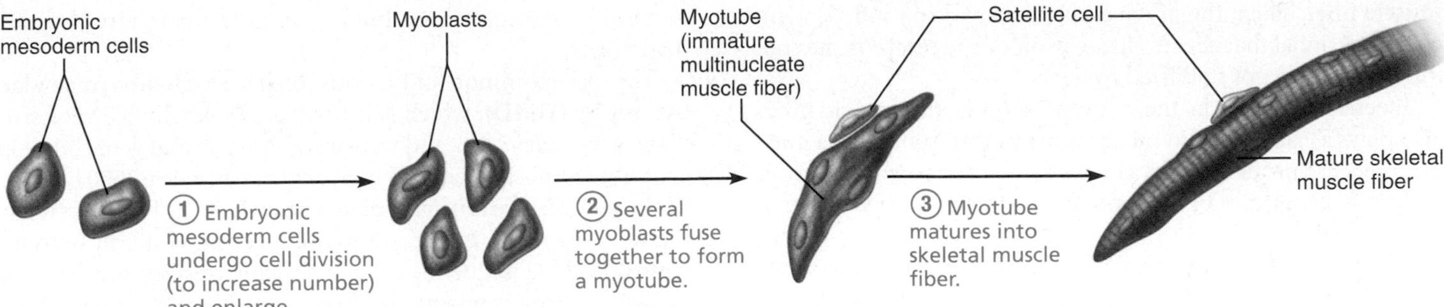

Figure 9.30 Formation of a multinucleate skeletal muscle fiber by fusion of myoblasts.

muscle fibers can divide to increase their numbers, that is, they undergo *hyperplasia.* One example is the response of the uterus to estrogen. At puberty, girls' plasma estrogen levels rise. As estrogen binds to uterine smooth muscle receptors, it stimulates the synthesis of more uterine smooth muscle, causing the uterus to grow to adult size. During pregnancy, high blood levels of estrogen stimulate uterine hyperplasia to accommodate the growing fetus.

Types of Smooth Muscle

The smooth muscle in different body organs varies substantially in its (1) fiber arrangement and organization, (2) innervation, and (3) responsiveness to various stimuli. For simplicity, however, smooth muscle is usually categorized into two major types: *single-unit* and *multiunit.*

Single-Unit Smooth Muscle

Single-unit smooth muscle, commonly called **visceral muscle** because it is in the walls of all hollow organs except the heart, is far more common. All the smooth muscle characteristics described so far pertain to single-unit smooth muscle. For example, the cells of single-unit smooth muscle

1. Are arranged in opposing (longitudinal and circular) sheets
2. Are innervated by ANS varicosities and often exhibit rhythmic spontaneous action potentials
3. Are electrically coupled by gap junctions and so contract as a unit (for this reason recruitment is not an option in smooth muscle)
4. Respond to various chemical stimuli

Multiunit Smooth Muscle

The smooth muscles in the large airways to the lungs and in large arteries, the arrector pili muscles attached to hair follicles, and the internal eye muscles that adjust pupil size and allow the eye to focus visually are all examples of **multiunit smooth muscle**.

In contrast to what we see in single-unit muscle, gap junctions are rare, and spontaneous synchronous depolarizations are infrequent. Like skeletal muscle, multiunit smooth muscle

1. Consists of muscle fibers that are structurally independent of one another
2. Is richly supplied with nerve endings, each of which forms a motor unit with a number of muscle fibers
3. Responds to neural stimulation with graded contractions that involve recruitment

However, while skeletal muscle is served by the somatic (voluntary) division of the nervous system, multiunit smooth muscle (like single-unit smooth muscle) is innervated by the autonomic (involuntary) division and is also responsive to hormonal controls.

CHECK YOUR UNDERSTANDING

17. Compare the structure of skeletal muscle fibers to that of smooth muscle fibers.
18. Calcium is the trigger for contraction of all muscle types. How does its binding site differ in skeletal and smooth muscle fibers?
19. How does the stress-relaxation response suit the role of smooth muscle in hollow organs?

For answers, see Appendix G.

Developmental Aspects of Muscles

▶ Describe embryonic development of muscle tissues and the changes that occur in skeletal muscles with age.

With rare exceptions, all three types of muscle tissue develop from embryonic mesoderm cells called **myoblasts.** In forming skeletal muscle tissue several myoblasts fuse to form multinuclear *myotubes* **(Figure 9.30).** This process is guided by the integrins (cell adhesion proteins) in the myoblast membranes. Soon functional sarcomeres are present, and skeletal muscle fibers are contracting by week 7 when the embryo is only about 2.5 cm (1 inch long).

Initially, ACh receptors "sprout" over the entire surface of the developing myoblasts. As spinal nerves invade the muscle masses, the nerve endings target individual myoblasts and release the growth factor *agrin.* This chemical activates a muscle kinase (MuSK), which stimulates clustering and maintenance of ACh receptors at the newly forming neuromuscular junction in each muscle fiber. Then, the nerve endings provide an independent chemical signal that causes elimination of the receptor sites not innervated and not stabilized by agrin.

Electrical activity in the neurons serving the muscle fibers also plays a critical role in muscle fiber maturation. As the muscle fibers are brought under the control of the somatic nervous system, the number of fast and slow contractile fiber types is determined.

Myoblasts producing cardiac and smooth muscle cells do not fuse. However, both develop gap junctions at a very early embryonic stage. Cardiac muscle is pumping blood just 3 weeks after fertilization.

Specialized skeletal and cardiac muscle cells stop dividing early on but retain the ability to lengthen and thicken in a growing child and to hypertrophy in adults. However, myoblast-like cells associated with skeletal muscle, called *satellite cells* (Figure 9.30), help repair injured fibers and allow *very limited* regeneration of dead skeletal muscle fibers. Cardiac muscle was thought to have no regenerative capability whatsoever, but recent studies suggest that cardiac cells do divide at a modest rate. Nonetheless, injured heart muscle is repaired mostly by scar tissue. Smooth muscles are able to regenerate throughout life.

At birth, a baby's movements are uncoordinated and largely reflexive. Muscular development reflects the level of neuromuscular coordination, which develops in a head-to-toe and proximal-to-distal direction. In other words, a baby can lift its head before it can walk, and gross movements precede fine ones. All through childhood, our control of our skeletal muscles becomes more and more sophisticated. By midadolescence, we reach the peak of our *natural* neural control of muscles, and can either accept that level of development or improve it by athletic or other types of training.

A frequently asked question is whether the difference in strength between women and men has a biological basis. It does. Individuals vary, but on average, women's skeletal muscles make up approximately 36% of body mass, whereas men's account for about 42%. Men's greater muscular development is due primarily to the effects of testosterone on skeletal muscle, not to the effects of exercise. Body strength per unit muscle mass, however, is the same in both sexes. Strenuous muscle exercise causes more muscle enlargement in males than in females, again because of the influence of testosterone. Some athletes take large doses of synthetic male sex hormones ("steroids") to increase their muscle mass. This illegal and physiologically dangerous practice is discussed in *A Closer Look*, opposite.

Because of its rich blood supply, skeletal muscle is amazingly resistant to infection throughout life. Given good nutrition and moderate exercise, relatively few problems afflict skeletal muscles. However, muscular dystrophy, the world's most common genetic disorder, is a serious condition that deserves more than a passing mention.

HOMEOSTATIC IMBALANCE

The term **muscular dystrophy** refers to a group of inherited muscle-destroying diseases that generally appear during childhood. The affected muscles initially enlarge due to fat and connective tissue deposit, but the muscle fibers atrophy and degenerate.

The most common and serious form is **Duchenne muscular dystrophy (DMD)**, which is inherited as a sex-linked recessive disease: Females carry and transmit the abnormal gene, but it is expressed almost exclusively in males (one in every 3500 births). This tragic disease is usually diagnosed when the boy is between 2 and 7 years old. Active, normal-appearing children become clumsy and fall frequently as their skeletal muscles weaken. The disease progresses relentlessly from the extremities upward, finally affecting the head and chest muscles and cardiac muscle of the heart. Victims rarely live beyond their early 20s, dying of respiratory failure.

Recent research has pinned down the cause of DMD: The diseased muscle fibers lack *dystrophin*, a cytoplasmic protein that links the cytoskeleton to the extracellular matrix and, like a girder, helps stabilize the sarcolemma. The fragile sarcolemma of DMD patients tears during contraction, allowing entry of excess Ca^{2+}. The deranged calcium homeostasis damages the contractile fibers which then break down, and inflammatory cells (macrophages and lymphocytes) accumulate in the surrounding connective tissue. As the regenerative capacity of the muscle is lost, and damaged cells undergo apoptosis, muscle mass drops.

There is still no cure for DMD, and thus far the only medication that has improved muscle strength and function is the steroid prednisone. One initially promising technique, *myoblast transfer therapy*, involves injecting diseased muscle with healthy myoblast cells that fuse with the unhealthy ones. The idea is that the normal gene provided would allow the fiber to produce dystrophin and so to grow normally. The therapy has shown some success in mice, but human trials have been disappointing. The large size of the dystrophin gene and of human muscles presents a huge challenge to that therapy. Two newer experimental therapies have each produced striking reversals of disease symptoms in dystrophic animal models. One is injection of adeno-associated viruses carrying pared-down microdystrophin genes. The second is infusion into the bloodstream of dystrophic mesangioblasts (stem cells harvested from blood vessels) corrected by inserting microdystrophin genes. A different approach being tested is coaxing dystrophic muscles to produce more *utrophin*, a similar protein present in low amounts in adults but at much higher levels in fetal muscles. In mice at least, utrophin can compensate for dystrophin deficiency. ■

As we age, the amount of connective tissue in our skeletal muscles increases, the number of muscle fibers decreases, and the muscles become stringier, or more sinewy. By the age of 30, even in healthy people, a gradual loss of muscle mass, called *sarcopenia* (sar-co-pe′ne-ah), begins to occur as muscle proteins start to degrade more rapidly than they can be replaced. Just why this happens is still a question, but apparently the same regulatory molecules (transcription factors, enzymes, hormones, and others) that promote muscle growth also oversee this type of muscle atrophy. Because skeletal muscles form so much of the body mass, body weight and muscle strength

(Text continues on p. 316.)

A CLOSER LOOK

Athletes Looking Good and Doing Better with Anabolic Steroids?

Society loves a winner and top athletes reap large social and monetary rewards. It is not surprising that some will grasp at anything that might increase their performance—including "juice," or anabolic steroids. These drugs are variants of the male sex hormone testosterone engineered by pharmaceutical companies. They were introduced in the 1950s to treat anemia and certain muscle-wasting diseases and to prevent muscle atrophy in patients immobilized after surgery. Testosterone is responsible for the increase in muscle and bone mass and other physical changes that occur during puberty and converts boys into men.

Convinced that megadoses of steroids could produce enhanced masculinizing effects in grown men, many athletes and bodybuilders were using them by the early 1960s. Investigations of the so-called Balco Scandal have stunned fans of major league baseball players with revelations in 2004 and since of rampant steroid use by Barry Bonds, of the San Francisco Giants, and many other elite athletes. It is still going on. In October of 2007, Marion Jones, one of the most celebrated of women athletes of all time, admitted that she was using performance-enhancing steroids when she won five gold medals in the 2000 Olympics. Furthermore, steroid use today is not confined to athletes. Indeed, it is estimated that nearly one in every 10 young men has tried them, and the practice is also spreading among young women.

It is difficult to determine the extent of anabolic steroid use because most international competitions ban the use of drugs. Users (and prescribing physicians or drug dealers) are naturally reluctant to talk about it, and users stop doping before the event, aware that evidence of drug use is hard to find a week after its use is stopped. Additionally, "underground" suppliers of performance-enhancing drugs keep producing new versions of designer steroids that evade standard antidoping tests. The Olympic Analytical Laboratory in Los Angeles rocked the sports world in November of 2003 when it revealed that a number of elite athletes tested positive for tetrahydrogestrinone (THG), a designer steroid not previously known or tested for. Nonetheless, there is little question that many professional bodybuilders and athletes competing in events that require muscle strength (e.g., shot put, discus throwing, and weight lifting) are heavy users. Sports figures such as football players have also admitted to using steroids as an adjunct to training, diet, and psychological preparation for games. These athletes claim that anabolic steroids enhance muscle mass and strength, and raise oxygen-carrying capability owing to greater red blood cell volume.

Typically, bodybuilders who use steroids combine high doses (up to 200 mg/day) via injection or transdermal skin patches with heavy resistance training. Intermittent use begins several months before an event, and commonly entails the use of many anabolic steroid supplements (a method called stacking). Doses are increased gradually as the competition nears.

Do the drugs do all that is claimed? Research studies report increased isometric strength and body weight in steroid users. While these are results weight lifters dream about, for runners and others requiring fine muscle coordination and endurance these changes may not translate into improved performance. The "jury is still out" on this question.

Do the alleged advantages of steroids outweigh their risks? Absolutely not. Physicians say they cause bloated faces (Cushingoid sign of steroid excess); acne and hair loss; shriveled testes and infertility; damage to the liver that promotes liver cancer; and changes in blood cholesterol levels that may predispose users to coronary heart disease. In addition, females can develop masculine characteristics such as smaller breasts, enlarged clitoris, excess body hair, and thinning scalp hair. The psychiatric hazards of anabolic steroid use may be equally threatening: Recent studies indicate that one-third of users suffer serious mental problems. Depression, delusions, and manic behavior—in which users undergo Jekyll-and-Hyde personality swings and become extremely violent (termed *'roid rage*)—are all common.

A more recent arrival on the scene, sold over the counter as a "nutritional performance-enhancer," is androstenedione, which is converted to testosterone in the body. Though it is taken orally (and much of it is destroyed by the liver soon after ingestion), the few milligrams that survive temporarily boost testosterone levels. Reports of its use by baseball great Mark McGwire before he retired, and of athletic wanna-bes from the fifth grade up sweeping the supplement off the drugstore shelves, are troubling, particularly since it is not regulated by the U.S. Food and Drug Administration (FDA) and its long-term effects are unpredictable and untested. A study at Massachusetts General Hospital found that males who took androstenedione developed higher levels of the female hormone estrogen as well as testosterone, raising their risk of feminizing effects such as enlarged breasts. Youths with elevated levels of estrogen or testosterone may enter puberty early, stunting bone growth and leading to shorter-than-normal adult height. Some people admit to a willingness to try almost anything to win, short of killing themselves. Are they unwittingly doing this as well?

System Connections

Homeostatic Interrelationships Between the Muscular System and Other Body Systems

Integumentary System

- Muscular exercise enhances circulation to skin and improves skin health
- Skin protects the muscles by external enclosure; helps dissipate heat generated by the muscles

Skeletal System

- Skeletal muscle activity maintains bone health and strength
- Bones provide levers for muscle activity

Nervous System

- Facial muscle activity allows emotions to be expressed
- Nervous system stimulates and regulates muscle activity

Endocrine System

- Growth hormone and androgens influence skeletal muscle strength and mass; other hormones help regulate cardiac and smooth muscle activity

Cardiovascular System

- Skeletal muscle activity increases efficiency of cardiovascular functioning; helps prevent atherosclerosis and causes cardiac hypertrophy
- Cardiovascular system delivers needed oxygen and nutrients to muscles

Lymphatic System/Immunity

- Physical exercise may enhance or depress immunity depending on its intensity
- Lymphatic vessels drain leaked tissue fluids; immune system protects muscles from disease

Respiratory System

- Muscular exercise increases respiratory capacity and efficiency of gas exchange
- Respiratory system provides oxygen and disposes of carbon dioxide

Digestive System

- Physical activity increases gastrointestinal motility and elimination when at rest
- Digestive system provides nutrients needed for muscle health; liver metabolizes lactic acid

Urinary System

- Physical activity promotes normal voiding behavior; skeletal muscle forms the voluntary sphincter of the urethra
- Urinary system disposes of nitrogenous wastes

Reproductive System

- Skeletal muscle helps support pelvic organs (e.g., uterus); assists erection of penis and clitoris
- Testicular androgen promotes increased skeletal muscle

Closer Connections

THE MUSCULAR SYSTEM and Interrelationships with the Cardiovascular, Endocrine, Lymphatic/Immunity, and Skeletal Systems

Our skeletal muscles are a marvel. In the well conditioned, they ripple with energy. In those of us who are less athletic, they enable us to perform the rather remarkable tasks of moving and getting around. No one would argue that the nervous system is indispensable for activating muscles to contract and for keeping them healthy via tone. Let's look at how skeletal muscle activity affects other body systems from the vantage point of overall health and disease prevention.

Cardiovascular System

The single most important barometer of how "well" we age is the health of our cardiovascular system. More than any other factor, regular exercise helps to maintain that health. Anything that gets you huffing and puffing on a regular basis, be it racquetball or a vigorous walk, helps to keep heart muscle healthy and strong. It also keeps blood vessels clear, delaying atherosclerosis and helping to prevent the most common type of high blood pressure and hypertensive heart disease—ailments that can lead to the deterioration of heart muscle and the kidneys. Uncluttered blood vessels also stave off painful or disabling intermittent claudication in which muscle pain due to ischemia hinders walking. Furthermore, regular exercise boosts blood levels of clot-busting enzymes, helping to ward off heart attacks and strokes, other scourges of old age.

Endocrine System

Muscle has a high rate of metabolism, and even at rest it uses much more energy than does fat. Consequently, exercise that builds moderate muscle mass helps to keep weight down and prevents obesity. Obesity is one of the risk factors for development of age-related diabetes mellitus—the metabolic disorder in which body cells are unresponsive to insulin (secreted by the pancreas) and, hence, unable to utilize glucose normally. Additionally, exercise helps to maintain normal cellular responses to insulin. On the other side of the coin, several hormones, including growth hormone, thyroid hormone, and sex hormones, are essential for normal development and maturation of the skeletal muscles.

Lymphatic System/Immunity

Physical exercise has a marked effect on immunity. Moderate or mild exercise causes a temporary rise in the number of phagocytes, T cells (a particular group of white blood cells), and antibodies, all of which populate lymphatic organs and mount the attack against infectious disease. By contrast, strenuous exercise depresses the immune system. The way in which exercise affects immunity—including these seemingly contradictory observations—is still a mystery, but the so-called stress hormones are definitely involved. Like major stressors such as surgery and serious burns, strenuous exercise increases blood levels of stress hormones such as epinephrine and glucocorticoids. These hormones depress the immune system during severe stress. This is thought to be a protective mechanism—a way of preventing large numbers of slightly damaged cells from being rejected.

Skeletal System

Last but not least, weight-bearing exercise of the skeletal muscles promotes skeletal strength and helps prevent osteoporosis. Since osteoporosis severely detracts from the quality of life by increasing risk of fractures, this is an extremely important interaction. But without their bony attachments, muscles would be ineffective in causing body movement.

Clinical Connections

Muscular System

Case study: Let's continue our tale of Mrs. DeStephano's medical problems, this time looking at the notes made detailing observations of her skeletal musculature.

- Severe lacerations of the muscles of the right leg and knee
- Damage to the blood vessels serving the right leg and knee
- Transection of the sciatic nerve (the large nerve serving most of the lower limb), just above the right knee

Her physician orders daily passive range-of-motion (ROM) exercise and electrical stimulation for her right leg and a diet high in protein, carbohydrates, and vitamin C.

1. Describe the step-by-step process of wound healing that will occur in her fleshy (muscle) wounds, and note the consequences of the specific restorative process that occurs.
2. What complications in healing can be anticipated owing to vascular (blood vessel) damage in the right leg?
3. What complications in muscle structure and function result from transection of the sciatic nerve? Why are passive ROM and electrical stimulation of her right leg muscles ordered?
4. Explain the reasoning behind the dietary recommendations.

(Answers in Appendix G)

decline in tandem. Muscle strength has usually decreased by about 50% by the age of 80 years. This "flesh wasting" condition has serious health implications for the elderly, particularly because falling becomes a common event.

But we don't have to slow up during old age. Muscle is responsive to exercise throughout life. Regular exercise helps reverse sarcopenia, and frail elders who begin to "pump iron" (lift leg and hand weights) can rebuild muscle mass and dramatically increase their strength. Performing those lifting exercises rapidly can improve a person's ability to carry out the "explosive" movements needed to rise from a chair or catch one's balance. Even moderate activity, like taking a walk daily, results in improved neuromuscular functioning and enhances independent living.

Muscles can also suffer indirectly. Aging of the cardiovascular system affects nearly every organ in the body, and muscles are no exception. As atherosclerosis takes its toll and begins to block distal arteries, a circulatory condition called *intermittent claudication* (klaw″dĭ-ka′shun; "limping") occurs in some individuals. This condition restricts blood delivery to the legs, leading to excruciating pains in the leg muscles during walking, which forces the person to stop and rest to get relief.

Smooth muscle is remarkably trouble free. The few problems that impair its functioning stem from external irritants. In the gastrointestinal tract, irritation might result from ingestion of excess alcohol, spicy foods, or bacterial infection. Under such conditions, smooth muscle motility increases in an attempt to rid the body of irritating agents, and diarrhea or vomiting occurs.

CHECK YOUR UNDERSTANDING

20. How is the multinucleate condition achieved during development of skeletal muscle fibers?
21. What does it mean when we say "muscles get stringier with age"?
22. How can we defer (or reverse) some of the effects of age on skeletal muscles?

For answers, see Appendix G.

The capacity for movement is a property of all cells but, with the exception of muscle, these movements are largely restricted to intracellular events. Skeletal muscles, the major focus of this chapter, permit us to interact with our external environment in an amazing number of ways, and they also contribute to our internal homeostasis as summarized in *Making Connections* (pp. 314–315). In this chapter we have covered muscle anatomy from the gross level to the molecular level and have considered muscle physiology in some detail. Chapter 10 continues from this point to explain how skeletal muscles interact with bones and with each other, and then describes the individual skeletal muscles that make up the muscular system of the body.

RELATED CLINICAL TERMS

Fibromyositis (*fibro* = fiber; *itis* = inflammation) Also known as **fibromyalgia**; a group of conditions involving chronic inflammation of a muscle, its connective tissue coverings and tendons, and capsules of nearby joints. Symptoms are nonspecific and involve varying degrees of tenderness associated with specific trigger points, as well as fatigue and frequent awakening from sleep.

Hernia Protrusion of an organ through its body cavity wall; may be congenital (owing to failure of muscle fusion during development), but most often is caused by heavy lifting or obesity and subsequent muscle weakening.

Myalgia (mi-al′je-ah; *algia* = pain) Muscle pain resulting from any muscle disorder.

Myofascial pain syndrome Pain caused by a tightened band of muscle fibers, which twitch when the skin over them is touched. Mostly associated with overused or strained postural muscles.

Myopathy (mi-op′ah-the; *path* = disease, suffering) Any disease of muscle.

Myotonic dystrophy A form of muscular dystrophy that is less common than DMD; in the U.S. it affects about 14 of 100,000 people. Symptoms include a gradual reduction in muscle mass and control of the skeletal muscles, abnormal heart rhythm, and diabetes mellitus. May appear at any time; not sex-linked. Underlying genetic defect is multiple repeats of a particular gene on chromosome 19. Because the number of repeats tends to increase from generation to generation, subsequent generations develop more severe symptoms. No effective treatment.

RICE Acronym for *rest, ice, compression,* and *elevation,* the standard treatment for a pulled muscle, or excessively stretched tendons or ligaments.

Spasm A sudden, involuntary twitch in smooth or skeletal muscle ranging in severity from merely irritating to very painful; may be due to chemical imbalances. In spasms of the eyelid or facial muscles, called tics, psychological factors have been implicated. Stretching and massaging the affected area may help to end the spasm. A cramp is a prolonged spasm; usually occurs at night or after exercise.

Strain Commonly called a "pulled muscle," a strain is excessive stretching and possible tearing of a muscle due to muscle overuse or abuse; the injured muscle becomes painfully inflamed (myositis), and adjacent joints are usually immobilized.

Tetanus (1) A state of sustained contraction of a muscle that is a normal aspect of skeletal muscle functioning. (2) An acute infectious disease caused by the anaerobic bacterium *Clostridium tetani* and resulting in persistent painful spasms of some of the skeletal muscles. Progresses to fixed rigidity of the jaws (lockjaw) and spasms of trunk and limb muscles; usually fatal due to respiratory failure.

CHAPTER SUMMARY

Media study tools that could provide you additional help in reviewing specific key topics of Chapter 9 are referenced below.

iP = ***Interactive Physiology***

Overview of Muscle Tissues (pp. 276–277)

Types of Muscle Tissue (p. 276)

1. Skeletal muscle is attached to the skeleton, is striated, and can be controlled voluntarily.
2. Cardiac muscle forms the heart, is striated, and is controlled involuntarily.
3. Smooth muscle, located chiefly in the walls of hollow organs, is controlled involuntarily. Its fibers are not striated.

Special Characteristics of Muscle Tissue (p. 276)

4. Special functional characteristics of muscle include excitability, contractility, extensibility, and elasticity.

Muscle Functions (pp. 276–277)

5. Muscles move internal and external body parts, maintain posture, stabilize joints, generate heat, and protect some visceral organs.

Skeletal Muscle (pp. 277–305)

Gross Anatomy of a Skeletal Muscle (pp. 277–278)

1. Skeletal muscle fibers (cells) are protected and strengthened by connective tissue coverings. Superficial to deep, these are epimysium, perimysium, and endomysium.
2. Skeletal muscle attachments (origins/insertions) may be direct or indirect via tendons or aponeuroses. Indirect attachments withstand friction better.

Microscopic Anatomy of a Skeletal Muscle Fiber (pp. 278–284)

3. Skeletal muscle fibers are long, striated, and multinucleate.
4. Myofibrils are contractile elements that occupy most of the cell volume. Their banded appearance results from a regular alternation of dark (A) and light (I) bands. Myofibrils are chains of sarcomeres; each sarcomere contains thick (myosin) and thin (actin) myofilaments arranged in a regular array. The heads of myosin molecules form cross bridges that interact with the thin filaments.
5. The sarcoplasmic reticulum (SR) is a system of membranous tubules surrounding each myofibril. Its function is to release and then sequester calcium ions.
6. T tubules are invaginations of the sarcolemma that run between the terminal cisternae of the SR. They allow an electrical stimulus to be delivered quickly deep into the cell.

Sliding Filament Model of Contraction (p. 284)

7. According to the sliding filament model, the thin filaments are pulled toward the sarcomere centers by cross bridge (myosin head) activity of the thick filaments.

Physiology of Skeletal Muscle Fibers (pp. 284–289)

8. Regulation of skeletal muscle cell contraction involves (a) generation and transmission of an action potential along the sarcolemma and (b) excitation-contraction coupling.
9. An end plate potential is set up when acetylcholine released by a nerve ending binds to ACh receptors on the sarcolemma, causing local changes in membrane permeability which allow ion flows that depolarize the membrane at that site.
10. Current flows from the locally depolarized area spread to the adjacent area of the sarcolemma, opening voltage-gated Na^+ channels which allow Na^+ influx. Then Na^+ channels close and voltage-gated K^+ channels open, repolarizing the membrane. These events generate the action potential. Once initiated, the action potential is self-propagating and unstoppable.
11. In excitation-contraction coupling, the action potential is propagated down the T tubules, causing calcium to be released from the SR into the cytosol.
12. Sliding of the filaments is triggered by a rise in intracellular calcium ion levels. Troponin binding of calcium moves tropomyosin away from myosin-binding sites on actin, allowing cross bridge binding. Myosin ATPases split ATP, which energizes the power strokes. ATP binding to the myosin head is necessary for cross bridge detachment. Cross bridge activity ends when calcium is pumped back into the SR.

iP Muscular System; Topic: Sliding Filament Theory, pp. 18–29.

Contraction of a Skeletal Muscle (pp. 289–296)

13. A motor unit is one motor neuron and all the muscle cells it innervates. The neuron's axon has several branches, each of which forms a neuromuscular junction with one muscle cell.
14. A motor unit's response to a single brief threshold stimulus is a twitch. A twitch has three phases: the latent period (preparatory events occurring), the period of contraction (the muscle tenses and may shorten), and the period of relaxation (muscle tension declines and the muscle resumes its resting length).
15. Graded responses of muscles to rapid stimuli are wave summation and unfused and fused tetanus. A graded response to increasingly strong stimuli is multiple motor unit summation, or recruitment. The type and order of motor unit recruitment follows the size principle.
16. Isotonic contractions occur when the muscle shortens (concentric contraction) or lengthens (eccentric contraction) as the load is moved. Isometric contractions occur when muscle tension produces neither shortening nor lengthening.

iP Muscular System; Topic: Contraction of Motor Units, pp. 1–11.

Muscle Metabolism (pp. 296–300)

17. The energy source for muscle contraction is ATP, obtained from a coupled reaction of creatine phosphate with ADP and from aerobic and anaerobic metabolism of glucose.
18. When ATP is produced by anaerobic pathways, lactic acid accumulates and ionic imbalances disturb the membrane potential, and an oxygen deficit occurs. To return the muscles to their resting state, ATP must be produced aerobically and used to regenerate creatine phosphate, glycogen reserves must be restored, and accumulated lactic acid must be oxidized.
19. Only about 40% of energy released during ATP hydrolysis powers contractile activity. The rest is liberated as heat.

iP Muscular System; Topic: Muscle Metabolism, pp. 1–7.

Force of Muscle Contraction (pp. 300–302)

20. The force of muscle contraction is affected by the number and size of contracting muscle cells (the more and the larger the cells, the greater the force), the frequency of stimulation, and the degree of muscle stretch.

21. In twitch contractions, the external tension exerted on the load is always less than the internal tension. When a muscle is tetanized, the external tension equals the internal tension.
22. When the thick and thin filaments are optimally overlapping, the muscle can generate maximum force. With excessive increase or decrease in muscle length, force declines.

Velocity and Duration of Contraction (pp. 302–303)

23. Factors determining the velocity and duration of muscle contraction include the load (the greater the load, the slower the contraction) and muscle fiber types.
24. The three types of muscle fibers are: (1) fast glycolytic (fatigable) fibers, (2) slow oxidative (fatigue-resistant) fibers, and (3) fast oxidative (fatigue-resistant) fibers. Most muscles contain a mixture of fiber types. The fast muscle fiber types are interconvertible with certain exercise regimens.

Effect of Exercise on Muscles (pp. 304–305)

25. Regular aerobic exercise results in increased efficiency, endurance, strength, and resistance to fatigue of skeletal muscles.
26. Resistance exercises cause skeletal muscle hypertrophy and large gains in skeletal muscle strength.
27. Immobilization of muscles leads to muscle weakness and severe atrophy.
28. Improper training and excessive exercise result in overuse injuries, which may be disabling.

Smooth Muscle (pp. 305–311)

Microscopic Structure of Smooth Muscle Fibers (pp. 305–307)

1. Smooth muscle fibers are spindle shaped and uninucleate. They display no striations.
2. Smooth muscle cells are most often arranged in sheets. They lack elaborate connective tissue coverings.
3. The SR is poorly developed and T tubules are absent. Actin and myosin filaments are present, but sarcomeres are not. Intermediate filaments and dense bodies form an intracellular network that harnesses the pull generated during cross bridge activity and transfers it to the extracellular matrix.

Contraction of Smooth Muscle (pp. 307–311)

4. Smooth muscle fibers may be electrically coupled by gap junctions, and the pace of contraction may be set by pacemaker cells.
5. Smooth muscle contraction is energized by ATP and is activated by a calcium pulse. However, calcium binds to calmodulin rather than to troponin (which is not present in smooth muscle fibers), and myosin must be phosphorylated to become active in contraction.
6. Smooth muscle contracts for extended periods at low energy cost and without fatigue.
7. Neurotransmitters of the autonomic nervous system may inhibit or stimulate smooth muscle fibers. Smooth muscle contraction may also be initiated by pacemaker cells, hormones, or local chemical factors that influence intracellular calcium levels, and by mechanical stretch.
8. Special features of smooth muscle contraction include the stress-relaxation response, the ability to generate large amounts of force when extensively stretched, and hyperplasia under certain conditions.

Types of Smooth Muscle (p. 311)

9. Single-unit smooth muscle has electrically coupled fibers that contract synchronously and often spontaneously.
10. Multiunit smooth muscle has independent, well-innervated fibers that lack gap junctions and pacemaker cells. Stimulation occurs via autonomic nerves (or hormones). Multiunit muscle contractions are rarely synchronous.

Developmental Aspects of Muscles (pp. 311–313, 316)

1. Muscle tissue develops from embryonic mesoderm cells called myoblasts. Skeletal muscle fibers are formed by the fusion of several myoblasts. Smooth and cardiac cells develop from single myoblasts and display gap junctions.
2. For the most part, specialized skeletal and cardiac muscle cells lose their ability to divide but retain the ability to hypertrophy. Smooth muscle regenerates well and undergoes hyperplasia.
3. Skeletal muscle development reflects maturation of the nervous system and occurs in cephalocaudal and proximal-to-distal directions. Natural neuromuscular control reaches its peak in midadolescence.
4. Women's muscles account for about 36% of their total body weight and men's for about 42%, a difference due chiefly to the effects of male hormones on skeletal muscle growth.
5. Skeletal muscle is richly vascularized and quite resistant to infection, but in old age, skeletal muscles become fibrous, decline in strength, and atrophy unless an appropriate exercise regimen is followed.

REVIEW QUESTIONS

Multiple Choice/Matching

(Some questions have more than one correct answer. Select the best answer or answers from the choices given.)

1. The connective tissue covering that encloses the sarcolemma of an individual muscle fiber is called the **(a)** epimysium, **(b)** perimysium, **(c)** endomysium, **(d)** periosteum.
2. A fascicle is a **(a)** muscle, **(b)** bundle of muscle fibers enclosed by a connective tissue sheath, **(c)** bundle of myofibrils, **(d)** group of myofilaments.
3. Thick and thin myofilaments have different compositions. For each descriptive phrase, indicate whether the filament is **(a)** thick or **(b)** thin.

   _____ (1) contains actin
   _____ (2) contains ATPases
   _____ (3) attaches to the Z disc
   _____ (4) contains myosin
   _____ (5) contains troponin
   _____ (6) does not lie in the I band
4. The function of the T tubules in muscle contraction is to **(a)** make and store glycogen, **(b)** release Ca^{2+} into the cell interior and then pick it up again, **(c)** transmit the action potential deep into the muscle cells, **(d)** form proteins.
5. The sites where the motor nerve impulse is transmitted from the nerve endings to the skeletal muscle cell membranes are the

(a) neuromuscular junctions, (b) sarcomeres, (c) myofilaments, (d) Z discs.

6. Contraction elicited by a single brief stimulus is called (a) a twitch, (b)wave summation, (c) multiple motor unit summation, (d) fused tetanus.
7. A smooth, sustained contraction resulting from very rapid stimulation of the muscle, in which no evidence of relaxation is seen, is called (a) a twitch, (b) wave summation, (c) multiple motor unit summation, (d) fused tetanus.
8. Characteristics of isometric contractions include all but (a) shortening, (b) increased muscle tension throughout the contraction phase, (c) absence of shortening, (d) use in resistance training.
9. During muscle contraction, ATP is provided by (a) a coupled reaction of creatine phosphate with ADP, (b) aerobic respiration of glucose, and (c) anaerobic glycolysis.
   _____ (1) Which provides ATP fastest?
   _____ (2) Which does (do) not require that oxygen be available?
   _____ (3) Which provides the highest yield of ATP per glucose molecule?
   _____ (4) Which results in the formation of lactic acid?
   _____ (5) Which has carbon dioxide and water products?
   _____ (6) Which is most important in endurance sports?
10. The neurotransmitter released by somatic motor neurons is (a) acetylcholine, (b) acetylcholinesterase, (c) norepinephrine.
11. The ions that enter the skeletal muscle cell during action potential generation are (a) calcium ions, (b) chloride ions, (c) sodium ions, (d) potassium ions.
12. Myoglobin has a special function in muscle tissue. It (a) breaks down glycogen, (b) is a contractile protein, (c) holds a reserve supply of oxygen in the muscle.
13. Aerobic exercise results in all of the following except (a) increased cardiovascular system efficiency, (b) more mitochondria in the muscle cells, (c) increased size and strength of existing muscle cells, (d) increased neuromuscular system coordination.
14. The smooth muscle type found in the walls of digestive and urinary system organs and that exhibits gap junctions and pacemaker cells is (a) multiunit, (b) single-unit.

Short Answer Essay Questions

15. Name and describe the four special functional abilities of muscle that are the basis for muscle response.
16. Distinguish between (a) direct and indirect muscle attachments and (b) a tendon and an aponeurosis.
17. (a) Describe the structure of a sarcomere and indicate the relationship of the sarcomere to myofilaments. (b) Explain the sliding filament theory of contraction using appropriately labeled diagrams of a relaxed and a contracted sarcomere.
18. What is the importance of acetylcholinesterase in muscle cell contraction?
19. Explain how a slight (but smooth) contraction differs from a vigorous contraction of the same muscle; use the concepts of multiple motor unit summation.
20. Explain what is meant by the term excitation-contraction coupling.
21. Define and draw a motor unit.
22. Describe the three distinct types of skeletal muscle fibers.
23. True or false: Most muscles contain a predominance of one skeletal muscle fiber type. Explain the reasoning behind your choice.
24. Describe some cause(s) of muscle fatigue and define this term clearly.
25. Define oxygen deficit.
26. Smooth muscle has some unique properties, such as low energy usage, and the ability to maintain contraction over long periods. Tie these properties to the function of smooth muscle in the body.

Critical Thinking and Clinical Application Questions

1. Jim Fitch decided that his physique left much to be desired, so he joined a local health club and began to "pump iron" three times weekly. After three months of training, during which he was able to lift increasingly heavier weights, he noticed that his arm and chest muscles were substantially larger. Explain the structural and functional basis of these changes.
2. When a suicide victim was found, the coroner was unable to remove the drug vial clutched in his hand. Explain the reasons for this. If the victim had been discovered three days later, would the coroner have had the same difficulty? Explain.
3. Muscle-relaxing drugs are administered to a patient during major surgery. Which of the two chemicals described next would be a good skeletal muscle relaxant and why?
   - Chemical A binds to and blocks ACh receptors of muscle cells.
   - Chemical B floods the muscle cells' cytoplasm with Ca^{2+}.
4. Michael is answering a series of questions dealing with skeletal muscle cell excitation and contraction. In response to "What protein changes shape when Ca^{2+} binds to it?" he writes "tropomyosin." What should he have responded and what is the result of that calcium ion binding?

Access everything you need to practice, review, and self-assess for both your A&P lecture and lab courses at **myA&P** (www.myaandp.com). There, you'll find powerful online resources, including chapter quizzes and tests, games, A&P Flix animations with quizzes, ***Interactive Physiology***® with quizzes, MP3 Tutor Sessions, Practice Anatomy Lab™, and more to help you get a better grade in your course.

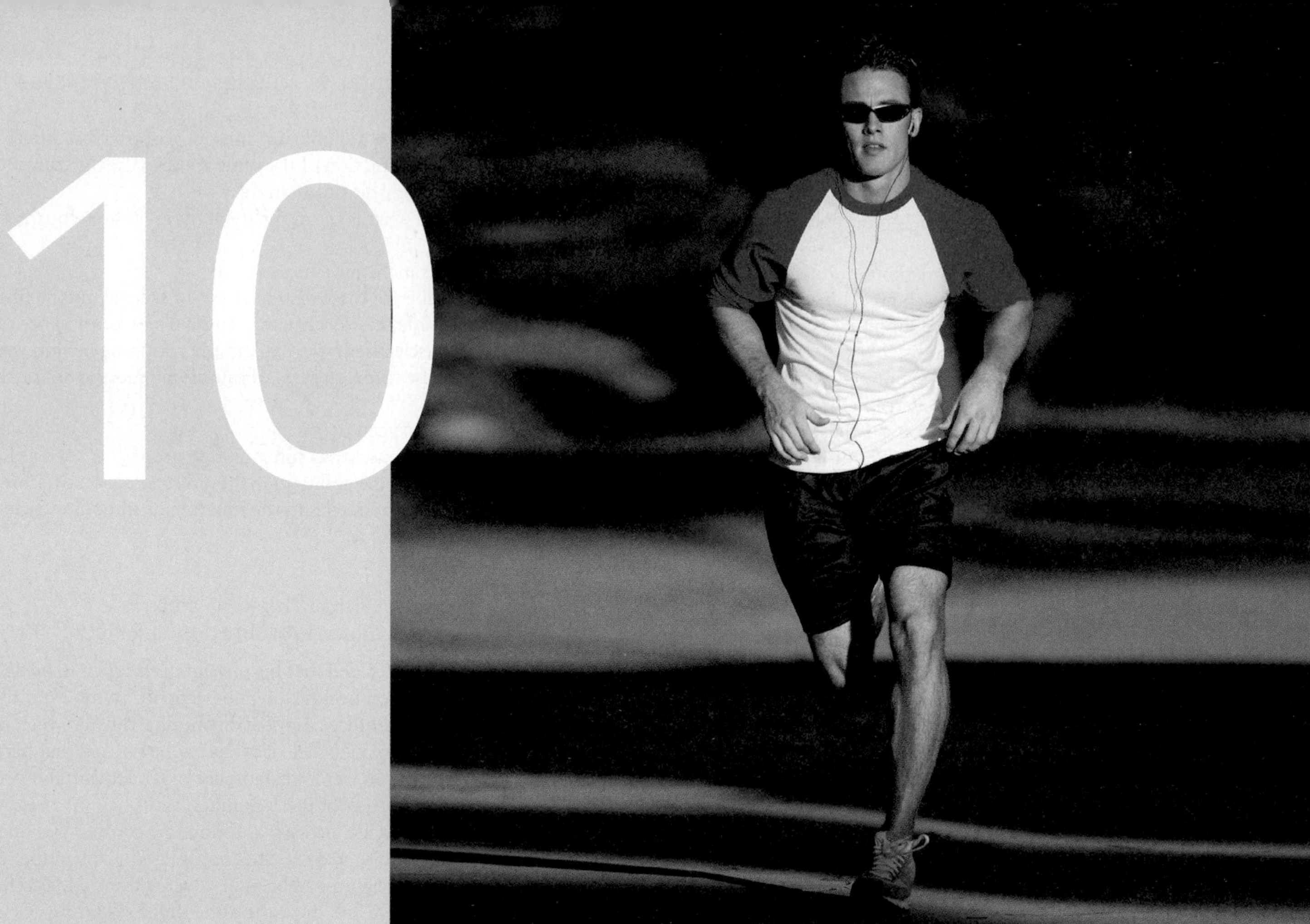

10 The Muscular System

The human body enjoys an incredibly wide range of movements. The gentle blinking of your eye, standing on tiptoe, and wielding a sledgehammer are just a small sample of the different activities promoted by your muscular system. Muscle tissue includes all contractile tissues (skeletal, cardiac, and smooth muscle), but when we study the muscular system, **skeletal muscles** take center stage. These muscular "machines" that enable us to perform so many different activities are the focus of this chapter. Before describing the individual muscles in detail, we will describe the manner in which muscles "play" with or against each other to bring about movements, consider the criteria used for naming muscles, and explain the principles of leverage.

Interactions of Skeletal Muscles in the Body

▶ Describe the function of prime movers, antagonists, synergists, and fixators.

The arrangement of body muscles permits them to work either together or in opposition to achieve a wide variety of movements. As you eat, for example, you alternately raise your fork to your lips and lower it to your plate, and both sets of actions are accomplished by your arm and hand muscles. But muscles can only *pull*; they never *push*. Generally as a muscle shortens, its *insertion* (attachment on the movable bone) moves toward its *origin* (its fixed or immovable point of attachment). Whatever one muscle or muscle group can do, another muscle or group of muscles can "undo."

Muscles can be classified into four *functional* groups: prime movers, antagonists, synergists, and fixators. A muscle that has the major responsibility for producing a specific movement is a **prime mover**, or **agonist** (ag′o-nist; "leader"), of that movement. The biceps brachii muscle, which fleshes out the anterior arm (and inserts on the radius), is a prime mover of elbow flexion.

Muscles that oppose, or reverse, a particular movement are called **antagonists** (an-tag′o-nists; "against the leaders"). When a prime mover is active, the antagonist muscles may be stretched or may remain relaxed. Usually however, antagonists help to regulate the action of a prime mover by contracting slightly to provide some resistance, thus helping to prevent overshooting the mark or to slow or stop the movement. As you might expect, a prime mover and its antagonist are located on opposite sides of the joint across which they act. Antagonists can also be prime movers in their own right. For example, flexion of the forearm by the biceps brachii muscle of the arm is antagonized by the triceps brachii, the prime mover for extending the forearm. As noted in Chapter 9, it is important that the two members of any agonist/antagonist pair be challenged and developed equally to prevent undue tension on the less developed muscle and joint inflexibility.

In addition to agonists and antagonists, most movements involve the action of one or more **synergists** (sin′er-jists; *syn* = together, *erg* = work). Synergists help prime movers by (1) adding a little extra force to the same movement or (2) reducing undesirable or unnecessary movements that might occur as the prime mover contracts. This second function deserves more explanation. When a muscle crosses two or more joints, its contraction causes movement at all of the spanned joints unless other muscles act as joint stabilizers. For example, the finger flexor muscles cross both the wrist and the interphalangeal joints, but you can make a fist without bending your wrist because synergistic muscles stabilize the wrist. Additionally, as some flexors act, they may cause several other (undesirable) movements at the same joint. Synergists can prevent this, allowing all of the prime mover's force to be exerted in the desired direction.

When synergists immobilize a bone, or a muscle's origin so that the prime mover has a stable base on which to act, they are more specifically called **fixators** (fik′sa-terz). Recall from Chapter 7 that the scapula is held to the axial skeleton only by muscles and is quite freely movable. The fixator muscles that run from the axial skeleton to the scapula can immobilize the scapula so that only the desired movements occur at the mobile shoulder joint. Additionally, muscles that help to maintain upright posture are fixators.

In summary, although prime movers seem to get all the credit for causing certain movements, antagonistic and synergistic muscles are also important in producing smooth, coordinated, and precise movements. Furthermore, a muscle may act as a prime mover in one movement, an antagonist for another movement, a synergist for a third movement, and so on.

Naming Skeletal Muscles

▶ List the criteria used in naming muscles. Provide an example to illustrate the use of each criterion.

Skeletal muscles are named according to a number of criteria, each of which describes the muscle in some way. Paying attention to these cues can simplify the task of learning muscle names and actions.

1. **Location of the muscle.** Some muscle names indicate the bone or body region with which the muscle is associated. For example, the temporalis (tem″por-ă′lis) muscle overlies the temporal bone, and intercostal (*costal* = rib) muscles run between the ribs.
2. **Shape of the muscle.** Some muscles are named for their distinctive shapes. For example, the deltoid (del′toid) muscle is roughly triangular (*deltoid* = triangle), and together the right and left trapezius (trah-pe′ze-us) muscles form a trapezoid.
3. **Relative size of the muscle.** Terms such as *maximus* (largest), *minimus* (smallest), *longus* (long), and *brevis* (short) are often used in muscle names—as in gluteus maximus and gluteus minimus (the large and small gluteus muscles, respectively).
4. **Direction of muscle fibers.** The names of some muscles reveal the direction in which their fibers (and fascicles) run in reference to some imaginary line, usually the midline of the body or the longitudinal axis of a limb bone. In muscles with the term *rectus* (straight) in their names, the fibers run parallel to that imaginary line (axis).The terms *transversus* and *oblique* indicate that the muscle fibers run respectively at right angles and obliquely to that line. Specific examples include the rectus femoris (straight muscle of the thigh, or femur) and transversus abdominis (transverse muscle of the abdomen).
5. **Number of origins.** When *biceps*, *triceps*, or *quadriceps* forms part of a muscle's name, you can assume that the muscle has two, three, or four origins, respectively. For example, the biceps brachii (bra′ke-i) muscle of the arm has two origins, or *heads*.
6. **Location of the attachments.** Some muscles are named according to their points of origin and insertion. The origin is always named first. For instance, the sternocleidomastoid

(ster″no-kli″do-mas′toid) muscle of the neck has a dual origin on the sternum (*sterno*) and clavicle (*cleido*), and it inserts on the *mastoid* process of the temporal bone.

7. **Action.** When muscles are named for the movement they produce, action words such as *flexor, extensor,* or *adductor* appear in the muscle's name. For example, the adductor longus, located on the medial thigh, brings about thigh adduction, and the supinator (soo′pĭ-na″tor) muscle supinates the forearm. (To review the terminology for various actions, see Chapter 8, Figures 8.5, 8.6, pp. 256–258.)

Often, several criteria are combined in the naming of a muscle. For instance, the name *extensor carpi radialis longus* tells us the muscle's action (extensor), what joint it acts on (*carpi* = wrist), and that it lies close to the radius of the forearm (radialis); it also hints at its size (longus) relative to other wrist extensor muscles. Unfortunately, not all muscle names are this descriptive.

CHECK YOUR UNDERSTANDING

1. The term "prime mover" is used in the business world to indicate people that get things done—the movers and shakers. What is its physiological meaning?
2. What criteria are used in naming each of the following muscles: *iliacus, adductor brevis, quadriceps femoris*?

For answers, see Appendix G.

Muscle Mechanics: Importance of Fascicle Arrangement and Leverage

- ▶ Name the common patterns of muscle fascicle arrangement and relate these to power generation.
- ▶ Define lever, and explain how a lever operating at a mechanical advantage differs from one operating at a mechanical disadvantage.
- ▶ Name the three types of lever systems and indicate the arrangement of effort, fulcrum, and load in each. Also note the advantages of each type of lever system.

We discussed most factors contributing to muscle force and speed (load, fiber type, etc.) in Chapter 9 with two important exceptions—the patterns of fascicle arrangement in muscles and lever systems. We attend to these factors next.

Arrangement of Fascicles

All skeletal muscles consist of fascicles (bundles of fibers), but fascicle arrangements vary, resulting in muscles with different shapes and functional capabilities. The most common patterns of fascicle arrangement are circular, convergent, parallel, and pennate **(Figure 10.1)**.

The fascicular pattern is **circular** when the fascicles are arranged in concentric rings (Figure 10.1a). Muscles with this arrangement surround external body openings, which they close by contracting. A general term for such muscles is *sphincters* ("squeezers"). Examples are the orbicularis muscles surrounding the eyes and the mouth.

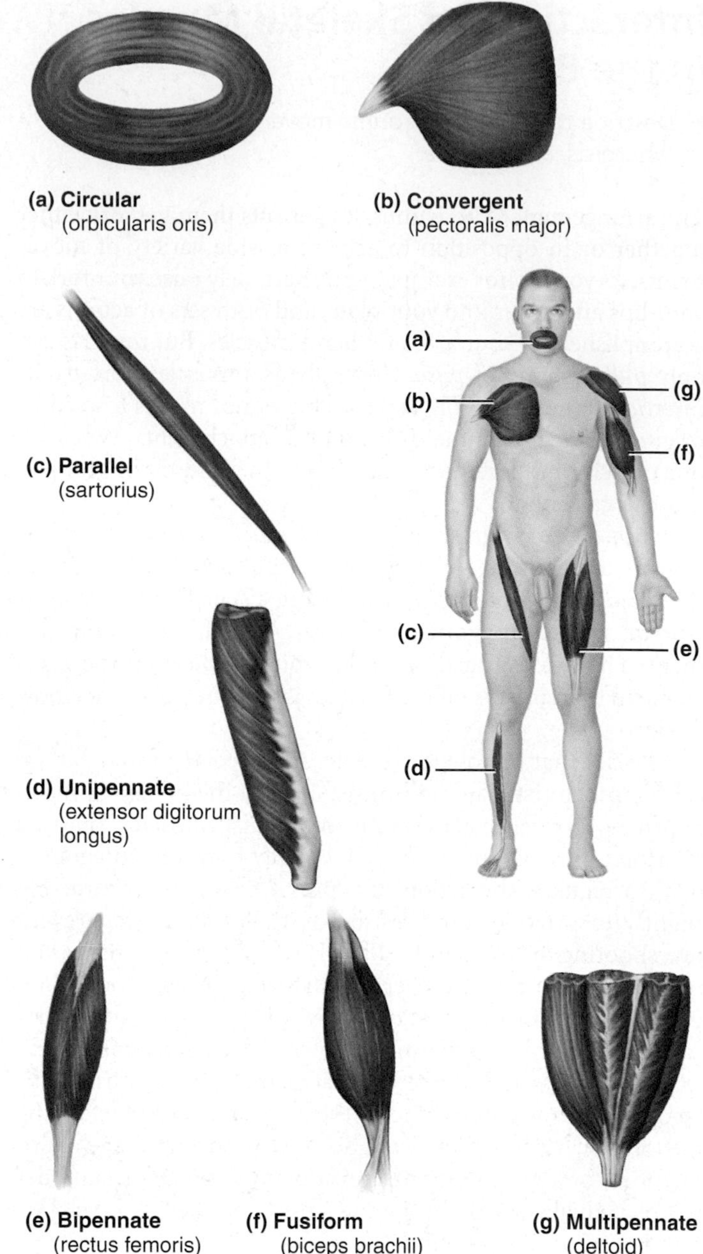

Figure 10.1 Patterns of fascicle arrangement in muscles.

A **convergent** muscle has a broad origin, and its fascicles *converge* toward a single tendon of insertion. Such a muscle is triangular or fan shaped like the pectoralis major muscle of the anterior thorax (Figure 10.1b).

In a **parallel** arrangement, the long axes of the fascicles run parallel to the long axis of the muscle. Such muscles are either *straplike* like the sartorius muscle of the thigh (Figure 10.1c), or spindle shaped with an expanded belly (midsection), like the biceps brachii muscle of the arm (Figure 10.1f). However, some authorities classify the spindle-shaped muscles into a separate class as **fusiform muscles.** This is the approach we use here.

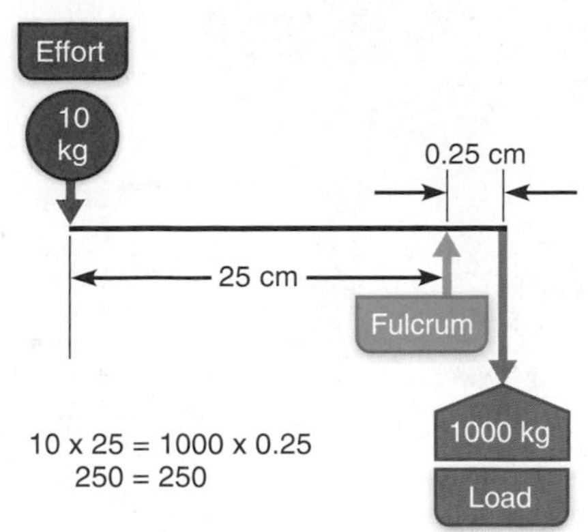

(a) Mechanical advantage with a power lever

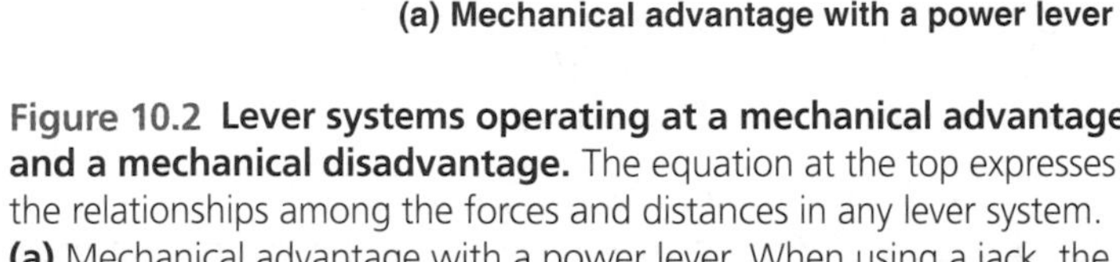
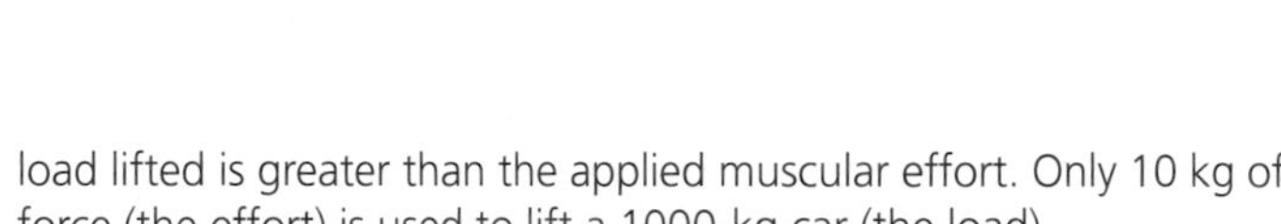

Figure 10.2 Lever systems operating at a mechanical advantage and a mechanical disadvantage. The equation at the top expresses the relationships among the forces and distances in any lever system. **(a)** Mechanical advantage with a power lever. When using a jack, the load lifted is greater than the applied muscular effort. Only 10 kg of force (the effort) is used to lift a 1000-kg car (the load).

(Continued on p. 324)

In a **pennate** (pen'āt) pattern, the fascicles (and thus the muscle fibers) are short and they attach obliquely (*penna* = feather) to a central tendon that runs the length of the muscle. If, as in the extensor digitorum longus muscle of the leg, the fascicles insert into only one side of the tendon, the muscle is *unipennate* (Figure 10.1d). If the fascicles insert into the tendon from opposite sides, so that the muscle's "grain" resembles a feather, the arrangement is *bipennate* (Figure 10.1e). The rectus femoris of the thigh is bipennate. A *multipennate* arrangement looks like many feathers situated side by side, with all their quills inserted into one large tendon. The deltoid muscle, which forms the roundness of the shoulder, is multipennate (Figure 10.1g).

The arrangement of a muscle's fascicles determines its range of motion (the amount of movement produced when a muscle shortens) and its power. Because skeletal muscle fibers may shorten to about 70% of their resting length when they contract, the longer and the more nearly parallel the muscle fibers are to a muscle's long axis, the more the muscle can shorten. Muscles with parallel fascicle arrangement shorten the most, but they are not usually very powerful. Muscle power depends more on the total number of muscle fibers in the muscle. The greater the number of muscle fibers, the greater the power. The stocky bipennate and multipennate muscles, which "pack in" the most fibers, shorten very little but are very powerful.

CHECK YOUR UNDERSTANDING

3. Of the muscles illustrated in Figure 10.1, which could shorten most? Which two would likely be most powerful? Why?

For answers, see Appendix G.

Lever Systems: Bone-Muscle Relationships

The operation of most skeletal muscles involves the use of leverage and **lever systems** (partnerships between the muscular and skeletal systems). A **lever** is a rigid bar that moves on a fixed point called the **fulcrum**, when a force is applied to it. The applied force, or **effort**, is used to move a resistance, or **load**. In your body, your joints are the fulcrums, and your bones act as levers. Muscle contraction provides the effort which is applied at the muscle's insertion point on a bone. The load is the bone itself, along with overlying tissues and anything else you are trying to move with that lever.

A lever allows a given effort to move a heavier load, or to move a load farther or faster, than it otherwise could. If, as shown in **Figure 10.2a**, the load is close to the fulcrum and the effort is applied far from the fulcrum, a small effort exerted over a relatively large distance can be used to move a large load over a small distance. Such a lever is said to operate at a **mechanical advantage** and is commonly called a *power lever*. For example, as shown to the right in Figure 10.2a, a person can lift a car with such a lever, in this case, a jack. The car moves up only a small distance with each downward "push" of the jack handle, but relatively little muscle effort is needed.

If, on the other hand, the load is far from the fulcrum and the effort is applied near the fulcrum, the force exerted by the muscle must be greater than the load to be moved or supported, as Figure 10.2b shows. This lever system operates at a **mechanical disadvantage** and is a *speed lever*. These levers are useful because they allow a load to be moved rapidly over a large distance (with a wide range of motion). Wielding a shovel is an example. As you can see, small differences in the site of a muscle's insertion (relative to the fulcrum or joint) can translate into large

10

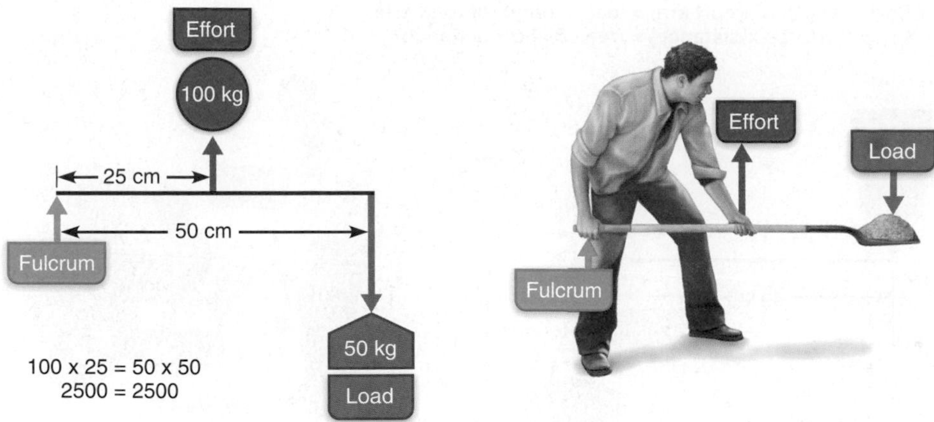

(b) Mechanical disadvantage with a speed lever

Figure 10.2 *(continued)* **Lever systems operating at a mechanical advantage and a mechanical disadvantage. (b)** Mechanical disadvantage with a speed lever. When using a shovel to lift dirt, the muscular force is greater than the load lifted. A muscular force (effort) of 100 kg is used to lift 50 kg of dirt (the load). Levers operating at a mechanical disadvantage are common in the body.

differences in the amount of force a muscle must generate to move a given load or resistance.

Regardless of type, all levers follow the same basic principle:

Effort farther than load from fulcrum	=	lever operates at a mechanical advantage
Effort nearer than load to fulcrum	=	lever operates at a mechanical disadvantage

Depending on the relative position of the three elements—effort, fulcrum, and load—a lever belongs to one of three classes. In **first-class levers**, the effort is applied at one end of the lever and the load is at the other, with the fulcrum somewhere between. Seesaws and scissors are first-class levers. First-class leverage also occurs when you lift your head off your chest (**Figure 10.3a**). Some first-class levers in the body operate at a mechanical advantage (for strength), but others, such as the action of the triceps muscle in extending the forearm against resistance, operate at a mechanical disadvantage (for speed and distance).

In a **second-class lever**, the effort is applied at one end of the lever and the fulcrum is located at the other, with the load between them. A wheelbarrow demonstrates this type of lever system. Second-class levers are uncommon in the body, but the best example is the act of standing on your toes (Figure 10.3b). All second-class levers in the body work at a mechanical advantage because the muscle insertion is always farther from the fulcrum than is the load. Second-class levers are levers of strength, but speed and range of motion are sacrificed for that strength.

In **third-class levers**, the effort is applied between the load and the fulcrum. These levers are speedy and *always* operate at a mechanical disadvantage. Tweezers or forceps provide this type of leverage. Most skeletal muscles of the body act in third-class lever systems. An example is the activity of the biceps muscle of the arm, lifting the distal forearm and anything carried in the hand (Figure 10.3c). Third-class lever systems permit a muscle to be inserted very close to the joint across which movement occurs, which allows rapid, extensive movements (as in throwing) with relatively little shortening of the muscle. Muscles involved in third-class levers tend to be thicker and more powerful.

In conclusion, differences in the positioning of the three elements modify muscle activity with respect to (1) speed of contraction, (2) range of movement, and (3) the weight of the load that can be lifted. In lever systems that operate at a mechanical disadvantage (speed levers), force is lost but speed and range of movement are gained. Systems that operate at a mechanical advantage (power levers) are slower, more stable, and used where strength is a priority.

CHECK YOUR UNDERSTANDING

4. Which of the three lever systems involved in muscle mechanics would be the fastest lever—first-, second-, or third-class?
5. What benefit is provided by a lever that operates at a mechanical advantage?

For answers, see Appendix G.

Major Skeletal Muscles of the Body

▶ **Name and identify the muscles described in Tables 10.1 to 10.17. State the origin, insertion, and action of each.**

The grand plan of the muscular system is all the more impressive because of the sheer number of skeletal muscles in the body—more than 600 of them (many more than are shown in **Figures 10.4** and **10.5** combined)! Trying to remember all the

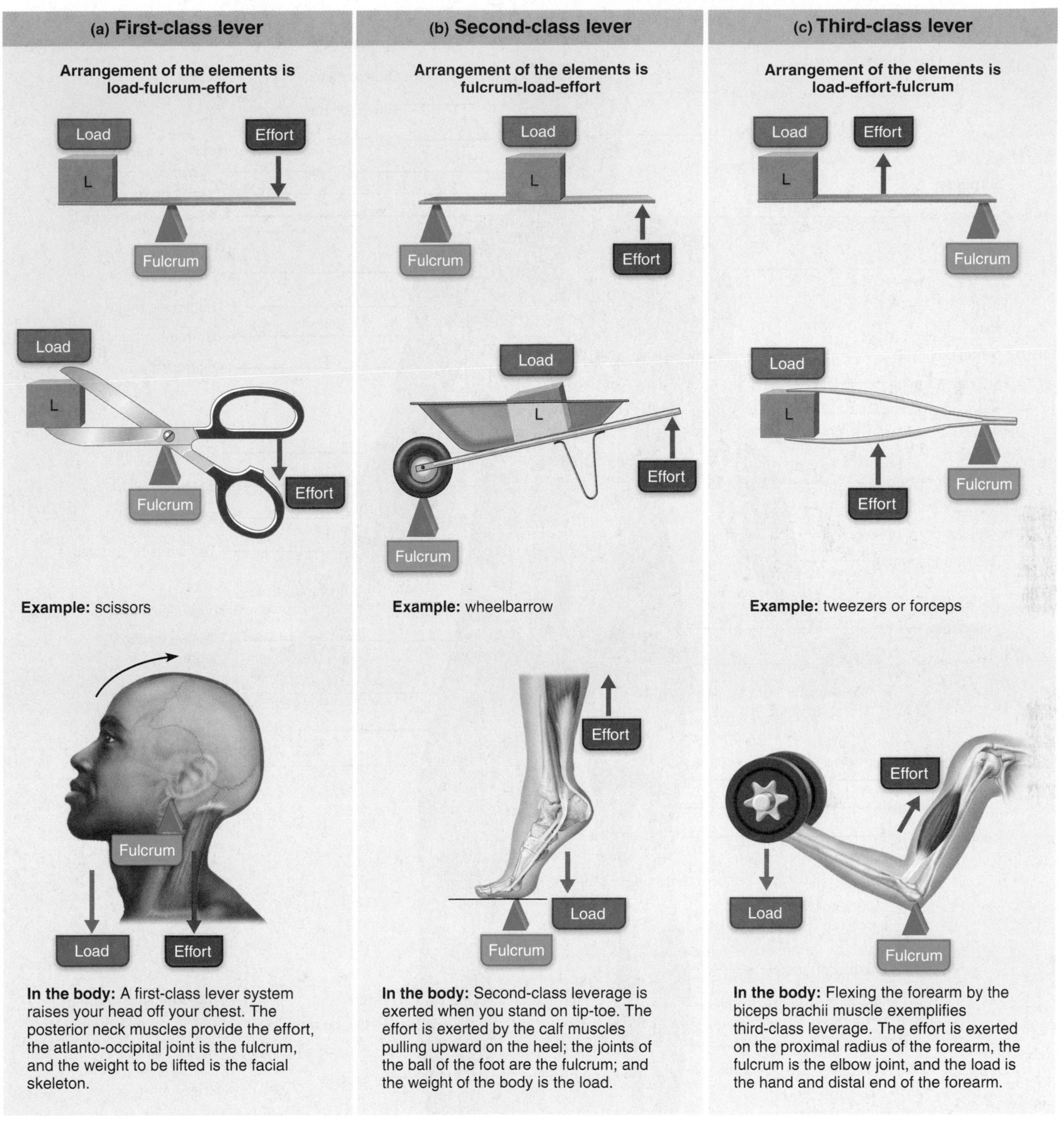

In the body: A first-class lever system raises your head off your chest. The posterior neck muscles provide the effort, the atlanto-occipital joint is the fulcrum, and the weight to be lifted is the facial skeleton.

In the body: Second-class leverage is exerted when you stand on tip-toe. The effort is exerted by the calf muscles pulling upward on the heel; the joints of the ball of the foot are the fulcrum; and the weight of the body is the load.

In the body: Flexing the forearm by the biceps brachii muscle exemplifies third-class leverage. The effort is exerted on the proximal radius of the forearm, the fulcrum is the elbow joint, and the load is the hand and distal end of the forearm.

Figure 10.3 **Lever systems.**

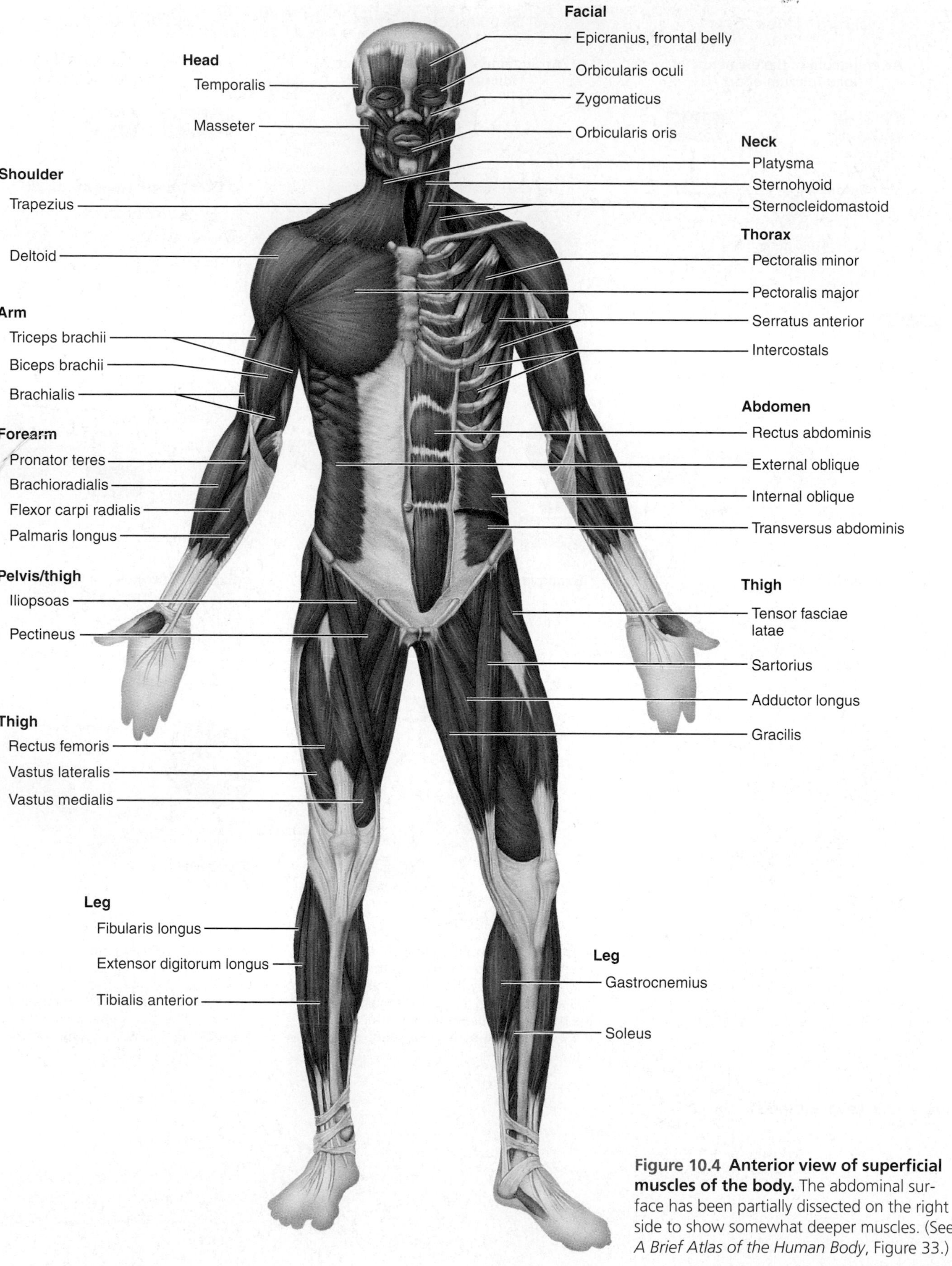

Figure 10.4 Anterior view of superficial muscles of the body. The abdominal surface has been partially dissected on the right side to show somewhat deeper muscles. (See *A Brief Atlas of the Human Body*, Figure 33.)

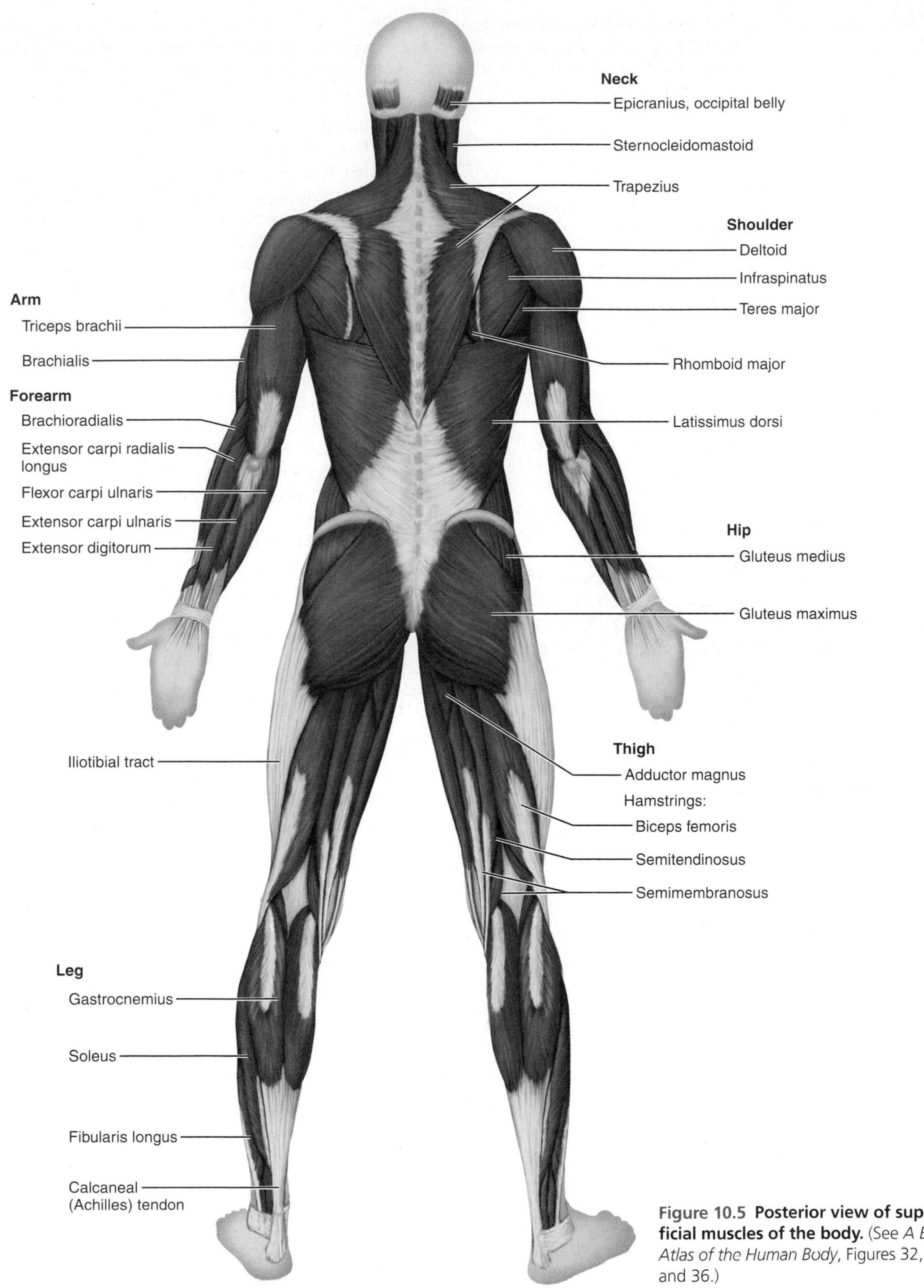

Figure 10.5 **Posterior view of superficial muscles of the body.** (See *A Brief Atlas of the Human Body*, Figures 32, 35, and 36.)

names, locations, and actions of these muscles is a monumental task. Take heart; here we consider only the principal muscles (approximately 125 pairs of them). Although this number is far fewer than 600, the job of learning about all these muscles will still require a concerted effort on your part.

Memorization will be easier if you can apply what you have learned in a practical, or clinical, way; that is, with a *functional anatomy focus.* Once you are satisfied that you have learned the name of a muscle and can identify it on a cadaver, model, or diagram, you must then flesh out your learning by asking yourself, "What does it do?" [It might be a good idea to review body movements (pp. 253–259) before you get to this point.]

In the tables that follow, the muscles of the body have been grouped by function and by location, roughly from head to foot. Each table is keyed to a particular figure or group of figures illustrating the muscles it describes. The legend at the beginning of each table provides an overview of the types of movements effected by the muscles listed and gives pointers on the way those muscles interact with one another. The table itself describes each muscle's shape, location relative to other muscles, origin and insertion, primary actions, and innervation. (Because some instructors want students to defer learning the muscle innervations until the nervous system has been studied, you might want to check on what is expected of you in this regard.)

As you consider each muscle, be alert to the information its name provides. After reading its entire description, identify the muscle on the corresponding figure and, in the case of superficial muscles, also on Figure 10.4 or Figure 10.5. Doing so will help you to link the descriptive material in the table to a visual image of the muscle's location in the body. Also try to relate a muscle's attachments and location to its actions. This will focus your attention on functional details that often escape student awareness. For example, both the elbow and knee joints are hinge joints that allow flexion and extension. However, the knee flexes to the dorsum of the body (the calf moves toward the posterior thigh), whereas elbow flexion carries the forearm toward the anterior aspect of the arm. Therefore, leg flexors are located on the posterior thigh, while forearm flexors are found on the anterior aspect of the humerus. Because many muscles have several actions, we have indicated the primary action of each muscle in blue type in the tables.

Finally, keep in mind that the *best* way to learn muscle actions is to act out their movements yourself while feeling for the muscles contracting (bulging) beneath your skin.

The sequence of the tables in this chapter is

MUSCLE GALLERY

TABLE 10.1 Muscles of the Head, Part I: Facial Expression (Figure 10.6)

The muscles that promote facial expression lie in the scalp and face just deep to the skin. They are thin and variable in shape and strength, and adjacent muscles tend to be fused. They are unusual muscles in that they insert into skin (or other muscles), not bones. In the scalp, the main muscle is the **epicranius**, which has distinct anterior and posterior parts. The lateral scalp muscles are vestigial in humans. Muscles clothing the facial bones lift the eyebrows, flare the nostrils, open and close the eyes and mouth, and provide one of the best tools for influencing others—the smile. The tremendous importance of facial muscles in nonverbal communication becomes especially clear when they are paralyzed, as in some stroke victims and in the expressionless "mask" of patients with Parkinson's disease. All muscles listed in this table are innervated by *cranial nerve VII*, the *facial nerve* (see Table 13.2). The external muscles of the eyes, which act to direct the eyeballs, and the levator palpebrae superioris muscles that raise the eyelids are described in Chapter 15.

MUSCLE	DESCRIPTION	ORIGIN (O) AND INSERTION (I)	ACTION	NERVE SUPPLY
MUSCLES OF THE SCALP				
Epicranius (occipitofrontalis) (ep″ĭ-kra′ne-us; ok-sip″ĭ-to-fron-ta′lis) (*epi* = over; *cran* = skull)	Bipartite muscle consisting of the frontal and occipital bellies connected by a cranial aponeurosis, the galea aponeurotica; the alternate actions of these two muscles pull scalp forward and backward			
▪ **Frontal belly** (fron′tal) (*front* = forehead)	Covers forehead and dome of skull; no bony attachments	O—galea aponeurotica I—skin of eyebrows and root of nose	With aponeurosis fixed, raises the eyebrows (as in surprise); wrinkles forehead skin horizontally	Facial nerve (cranial VII)
▪ **Occipital belly** (ok-sip″ĭ-tal′) (*occipito* = base of skull)	Overlies posterior occiput; by pulling on the galea, fixes origin of frontalis	O—occipital and temporal bones I—galea aponeurotica	Fixes aponeurosis and pulls scalp posteriorly	Facial nerve
MUSCLES OF THE FACE				
Corrugator supercilii (kor′ah-ga-ter soo″per-sĭ′le-i) (*corrugo* = wrinkle; *supercilium* = eyebrow)	Small muscle; activity associated with that of orbicularis oculi	O—arch of frontal bone above nasal bone I—skin of eyebrow	Draws eyebrows together and inferiorly; wrinkles skin of forehead vertically (as in frowning)	Facial nerve

➤

MUSCLE GALLERY

TABLE 10.1 Muscles of the Head, Part I: Facial Expression (Figure 10.6) *(continued)*

MUSCLE	DESCRIPTION	ORIGIN (O) AND INSERTION (I)	ACTION	NERVE SUPPLY
Orbicularis oculi (or-bik′u-lar-is ok′u-li) (*orb* = circular; *ocul* = eye)	Thin, tripartite sphincter muscle of eyelid; surrounds rim of the orbit	O—frontal and maxillary bones and ligaments around orbit I—tissue of eyelid	Closes eye; various parts can be activated individually; produces blinking, squinting, and draws eyebrows inferiorly	Facial nerve
Zygomaticus—major and minor (zi-go-mat′ĭ-kus) (*zygomatic* = cheekbone)	Muscle pair extending diagonally from cheekbone to corner of mouth	O—zygomatic bone I—skin and muscle at corner of mouth	Raises lateral corners of mouth upward (smiling muscle)	Facial nerve
Risorius (ri-zor′e-us) (*risor* = laughter)	Slender muscle inferior and lateral to zygomaticus	O—lateral fascia associated with masseter muscle I—skin at angle of mouth	Draws corner of lip laterally; tenses lips; synergist of zygomaticus	Facial nerve
Levator labii superioris (lĕ-va′tor la′be-i soo-per″e-or′is) (*leva* = raise; *labi* = lip; *superior* = above, over)	Thin muscle between orbicularis oris and inferior eye margin	O—zygomatic bone and infraorbital margin of maxilla I—skin and muscle of upper lip	Opens lips; raises and furrows the upper lip	Facial nerves
Depressor labii inferioris (de-pres′or la′be-i in-fer″e-or′is) (*depressor* = depresses; *infer* = below)	Small muscle running from mandible to lower lip	O—body of mandible lateral to its midline I—skin and muscle of lower lip	Draws lower lip inferiorly (as in a pout)	Facial nerve
Depressor anguli oris (ang′gu-li or-is) (*angul* = angle, corner; *or* = mouth)	Small muscle lateral to depressor labii inferioris	O—body of mandible below incisors I—skin and muscle at angle of mouth below insertion of zygomaticus	Draws corners of mouth downward and laterally (as in a "tragedy mask" grimace); zygomaticus antagonist	Facial nerve
Orbicularis oris	Complicated, multilayered muscle of the lips with fibers that run in many different directions; most run circularly	O—arises indirectly from maxilla and mandible; fibers blended with fibers of other facial muscles associated with the lips I—encircles mouth; inserts into muscle and skin at angles of mouth	Closes lips; purses and protrudes lips; kissing and whistling muscle	Facial nerve
Mentalis (men-ta′lis) (*ment* = chin)	One of the muscle pair forming a V-shaped muscle mass on chin	O—mandible below incisors I—skin of chin	Wrinkles chin; protrudes lower lip	Facial nerve
Buccinator (bu′sĭ-na″ter) (*bucc* = cheek or "trumpeter")	Thin, horizontal cheek muscle; principal muscle of cheek; deep to masseter (see also Figure 10.7)	O—molar region of maxilla and mandible I—orbicularis oris	Compresses cheek (as in whistling and sucking); trampoline-like action holds food between teeth during chewing; draws corner of mouth laterally; well developed in nursing infants	Facial nerve
Platysma (plah-tiz′mah) (*platy* = broad, flat)	Unpaired, thin, sheetlike superficial neck muscle; not strictly a head muscle, but plays a role in facial expression	O—fascia of chest (over pectoral muscles and deltoid) I—lower margin of mandible, and skin and muscle at corner of mouth	Tenses skin of neck (e.g., during shaving); helps depress mandible; pulls lower lip back and down, i.e., produces downward sag of mouth	Facial nerve

MUSCLE GALLERY

TABLE 10.1 *(continued)*

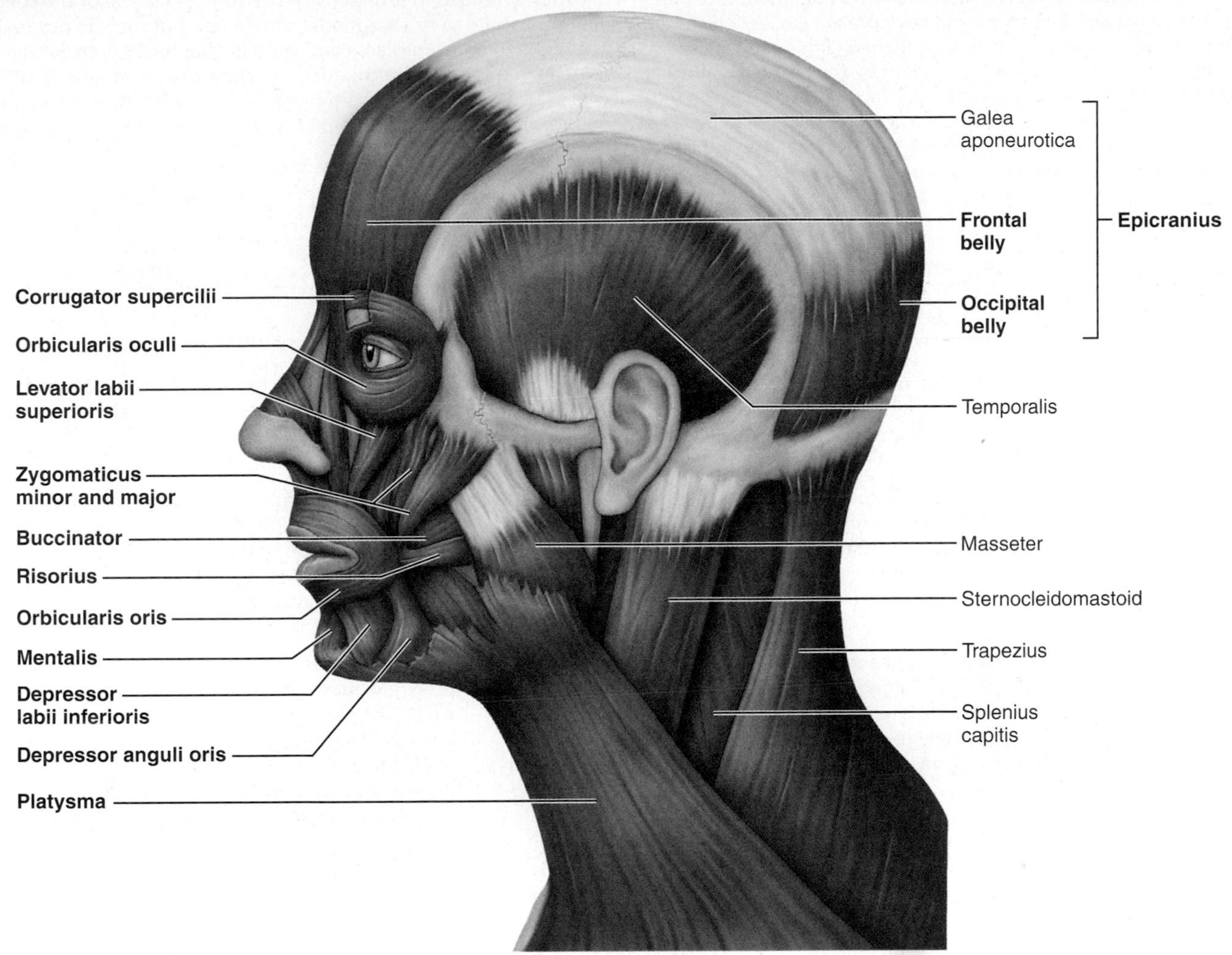

Figure 10.6 **Lateral view of muscles of the scalp, face, and neck.**

MUSCLE GALLERY

TABLE 10.2 Muscles of the Head, Part II: Mastication and Tongue Movement (Figure 10.7)

Four pairs of muscles are involved in mastication (chewing and biting) activities, and all are innervated by the *mandibular branch* of *cranial nerve V* (the *trigeminal nerve*). The prime movers of jaw closure (and biting) are the powerful **masseter** and **temporalis** muscles, which can be palpated easily when the teeth are clenched (Figure 10.7a). Side-to-side grinding movements are brought about by the **pterygoid** muscles (Figure 10.7b). The **buccinator** muscles (see Table 10.1) also play a role in chewing. Normally, gravity is sufficient to depress the mandible, but if there is resistance to jaw opening, neck muscles such as the digastric and mylohyoid muscles (see Table 10.3) are activated.

The tongue is composed of muscle fibers that curl, squeeze, and fold the tongue during speaking and chewing. These **intrinsic tongue muscles**, arranged in several planes, change the shape of the tongue and contribute to its exceptional nimbleness, but they do not really move the tongue. They are considered in Chapter 23 with the digestive system. In this table, we consider only the **extrinsic tongue muscles**, which anchor and move the tongue (Figure 10.7c). All extrinsic tongue muscles are innervated by *cranial nerve XII*, the *hypoglossal nerve* (see Table 13.2).

MUSCLE	DESCRIPTION	ORIGIN (O) AND INSERTION (I)	ACTION	NERVE SUPPLY
MUSCLES OF MASTICATION				
Masseter (mah-se'ter) (*maseter* = chewer)	Powerful muscle that covers lateral aspect of mandibular ramus	O—zygomatic arch and zygomatic bone I—angle and ramus of mandible	Prime mover of jaw closure; elevates mandible	Trigeminal nerve (cranial V)
Temporalis (tem″por-ă'lis) (*tempora* = time; pertaining to the temporal bone)	Fan-shaped muscle that covers parts of the temporal, frontal, and parietal bones	O—temporal fossa I—coronoid process of mandible via a tendon that passes deep to zygomatic arch	Closes jaw; elevates and retracts mandible; maintains position of the mandible at rest; deep anterior part may help protract mandible	Trigeminal nerve
Medial pterygoid (me'de-ul ter'ĭ-goid) (*medial* = toward median plane; *pterygoid* = winglike)	Deep two-headed muscle that runs along internal surface of mandible and is largely concealed by that bone	O—medial surface of lateral pterygoid plate of sphenoid bone, maxilla, and palatine bone I—medial surface of mandible near its angle	Acts with the lateral pterygoid muscle to protrude (protract) mandible and to promote side-to-side side (grinding) movements; synergist of temporalis and masseter muscles in elevation of the mandible	Trigeminal nerve
Lateral pterygoid (*lateral* = away from median plane)	Deep two-headed muscle; lies superior to medial pterygoid muscle	O—greater wing and lateral pterygoid plate of sphenoid bone I—condyle of mandible and capsule of temporomandibular joint	Provides forward sliding and side-to-side grinding movements of the lower teeth; protrudes mandible (pulls it anteriorly)	Trigeminal nerve
Buccinator	See Table 10.1	See Table 10.1	Compresses the cheek; helps keep food between grinding surfaces of teeth during chewing	Facial nerve (cranial VII)
MUSCLES PROMOTING TONGUE MOVEMENTS (EXTRINSIC MUSCLES)				
Genioglossus (je″ne-o-glah'sus) (*geni* = chin; *glossus* = tongue)	Fan-shaped muscle; forms bulk of inferior part of tongue; its attachment to mandible prevents tongue from falling backward and obstructing respiration	O—internal surface of mandible near symphysis I—inferior aspect of the tongue and body of hyoid bone	Protracts tongue; can depress or act in concert with other extrinsic muscles to retract tongue	Hypoglossal nerve (cranial XII)
Hyoglossus (hi'o-glos″us) (*hyo* = pertaining to hyoid bone)	Flat, quadrilateral muscle	O—body and greater horn of hyoid bone I—inferolateral tongue	Depresses tongue and draws its sides downward	Hypoglossal nerve
Styloglossus (sti-lo-glah'sus) (*stylo* = pertaining to *styloid* process)	Slender muscle running superiorly to and at right angles to hyoglossus	O—styloid process of temporal bone I—inferolateral tongue	Retracts (and elevates) tongue	Hypoglossal nerve

MUSCLE GALLERY

TABLE 10.2 *(continued)*

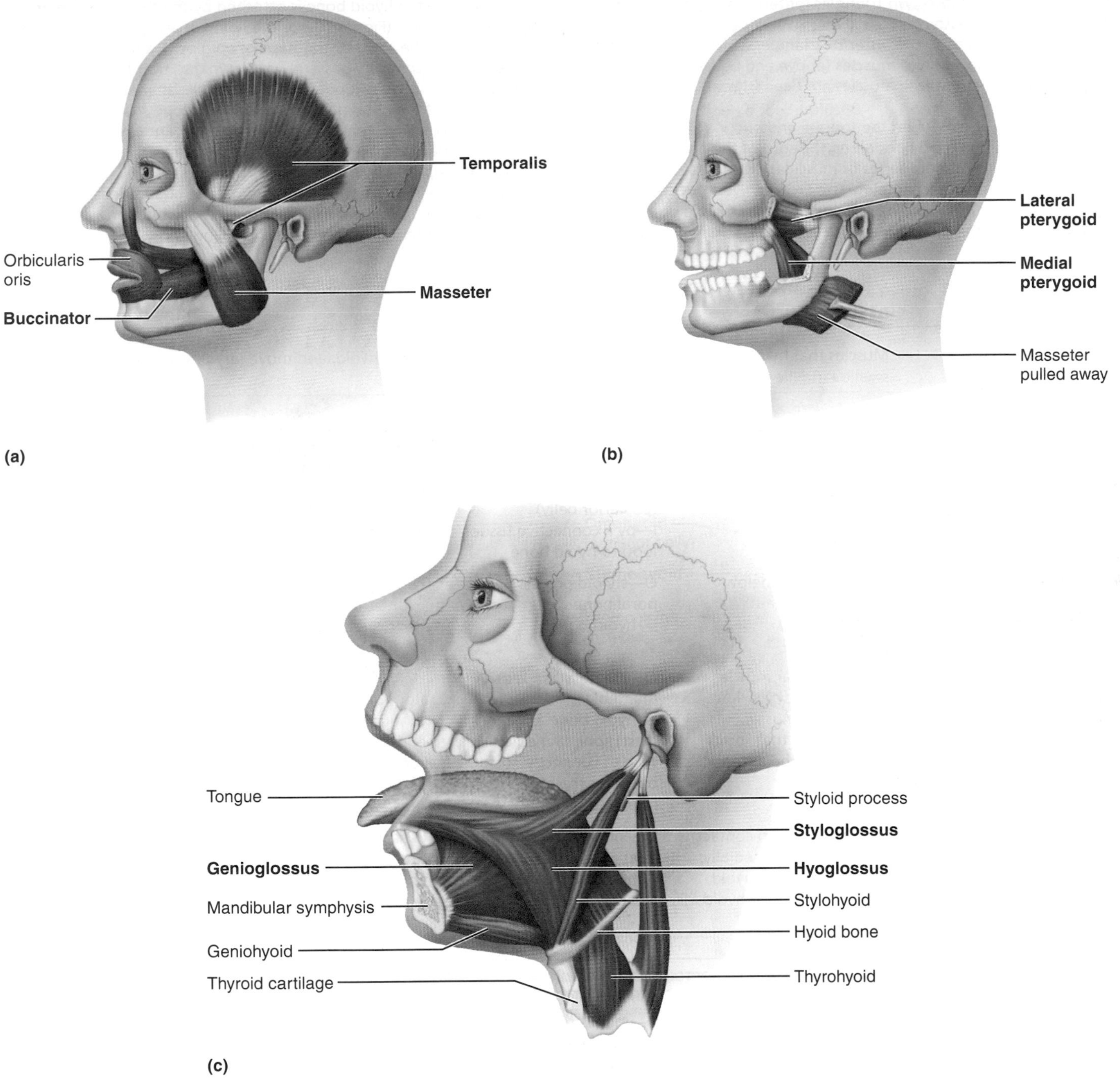

Figure 10.7 Muscles promoting mastication and tongue movements. (a) Lateral view of the temporalis, masseter, and buccinator muscles. **(b)** Lateral view of the deep chewing muscles, the medial and lateral pterygoid muscles. **(c)** Extrinsic muscles of the tongue. Some suprahyoid muscles of the throat are also illustrated.

MUSCLE GALLERY

TABLE 10.4 Muscles of the Neck and Vertebral Column: Head Movements and Trunk Extension (Figure 10.9)

Head movements

The head is moved by muscles originating from the axial skeleton. The major head flexors are the **sternocleidomastoid muscles** (Figure 10.9 a, c), with some help from the suprahyoid and infrahyoid muscles described in Table 10.3. Lateral head movements (rotating or tilting the head) result when the muscles on only one side of the neck contract. These actions are effected by the sternocleidomastoids and a number of deeper neck muscles, considered in this table. Head extension is aided by the trapezius muscles of the back, but the main extensors of the head are the **splenius** muscles deep to the trapezius muscles (Figure 10.9b).

Trunk extension

Trunk extension is effected by the *deep* or *intrinsic back muscles* associated with the bony vertebral column. These deep muscles of the back also maintain the normal curvatures of the spine, acting as postural muscles. As you consider these back muscles, keep in mind that they are deep. The superficial back muscles that cover them are concerned primarily with movements of the shoulder girdle and upper limbs (see Tables 10.8 and 10.9).

The deep muscles of the back form a broad, thick column extending from the sacrum to the skull. Many muscles of varying length contribute to this mass. It helps to regard each of these individual muscles as a string that when pulled causes one or several vertebrae to extend or to rotate on the vertebrae below. The largest of the deep back muscle groups is the **erector spinae** group (Figure 10.9d). Because the origins and insertions of the different muscle groups overlap extensively, and many of these muscles are long, large regions of the vertebral column can be moved simultaneously and smoothly. Acting in concert, the deep back muscles extend (or hyperextend) the spine, but contraction of the muscles on only one side causes lateral bending (flexion) of the spine. Lateral flexion is automatically accompanied by some degree of rotation of the vertebral column. During vertebral movements, the articular facets of the vertebrae glide on each other.

In addition to the long back muscles, a number of short muscles extend from one vertebra to the next. These small muscles act primarily as synergists in extension and rotation of the spine and as spine stabilizers. They are not described in the table but you can deduce their actions by examining their origins and insertions in Figure 10.9e.

As noted, it is the trunk *extensors* that we consider in this table. The more superficial muscles, which have other functions, are considered in subsequent tables. For example, the anterior muscles of the abdominal wall that cause trunk *flexion* are described in Table 10.6.

MUSCLE	DESCRIPTION	ORIGIN (O) AND INSERTION (I)	ACTION	NERVE SUPPLY
ANTEROLATERAL NECK MUSCLES (FIGURE 10.9a AND c)				
Sternocleidomastoid (ster″no-kli″do-mas′toid) (*sterno* = breastbone; *cleido* = clavicle; *mastoid* = mastoid process)	Two-headed muscle located deep to platysma on anterolateral surface of neck; fleshy parts on either side of neck delineate limits of anterior and posterior triangles; key muscular landmark in neck; spasms of one of these muscles may cause torticollis (wryneck)	O—manubrium of sternum and medial portion of clavicle I—mastoid process of temporal bone and superior nuchal line of occipital bone	Flexes and laterally rotates the head; simultaneous contraction of both muscles causes neck flexion, generally against resistance as when one raises head when lying on back; acting alone, each muscle rotates head toward shoulder on opposite side and tilts or laterally flexes head to its own side	Accessory nerve (cranial nerve XI) and branches of cervical spinal nerves C_2 and C_3 (ventral rami)
Scalenes (ska′lēnz)—anterior, middle, and posterior (*scalene* = uneven)	Located more laterally than anteriorly on neck; deep to platysma and sternocleidomastoid	O—transverse processes of cervical vertebrae I—anterolaterally on first two ribs	Elevate first two ribs (aid in inspiration); flex and rotate neck	Cervical spinal nerves
INTRINSIC MUSCLES OF THE BACK (FIGURE 10.9b, d, e)				
Splenius (sple′ne-us)—capitis and cervicis portions (kă′pĭ-tis; ser-vis′us) (*splenion* = bandage; *caput* = head; *cervi* = neck) (Figures 10.9b and 10.6)	Broad bipartite superficial muscle (capitis and cervicis parts) extending from upper thoracic vertebrae to skull; capitis portion known as "bandage muscle" because it covers and holds down deeper neck muscles	O—ligamentum nuchae,* spinous processes of vertebrae C_7–T_6 I—mastoid process of temporal bone and occipital bone (capitis); transverse processes of C_2–C_4 vertebrae (cervicis)	Extend or hyperextend head; when splenius muscles on one side are activated, head is rotated and bent laterally toward same side	Cervical spinal nerves (dorsal rami)

*The ligamentum nuchae (lig″ah-men′tum noo′ke) is a strong, elastic ligament extending from the occipital bone of the skull along the tips of the spinous processes of the cervical vertebrae. It binds the cervical vertebrae together and inhibits excessive head and neck flexion, thus preventing damage to the spinal cord in the vertebral canal.

MUSCLE GALLERY

TABLE 10.4 *(continued)*

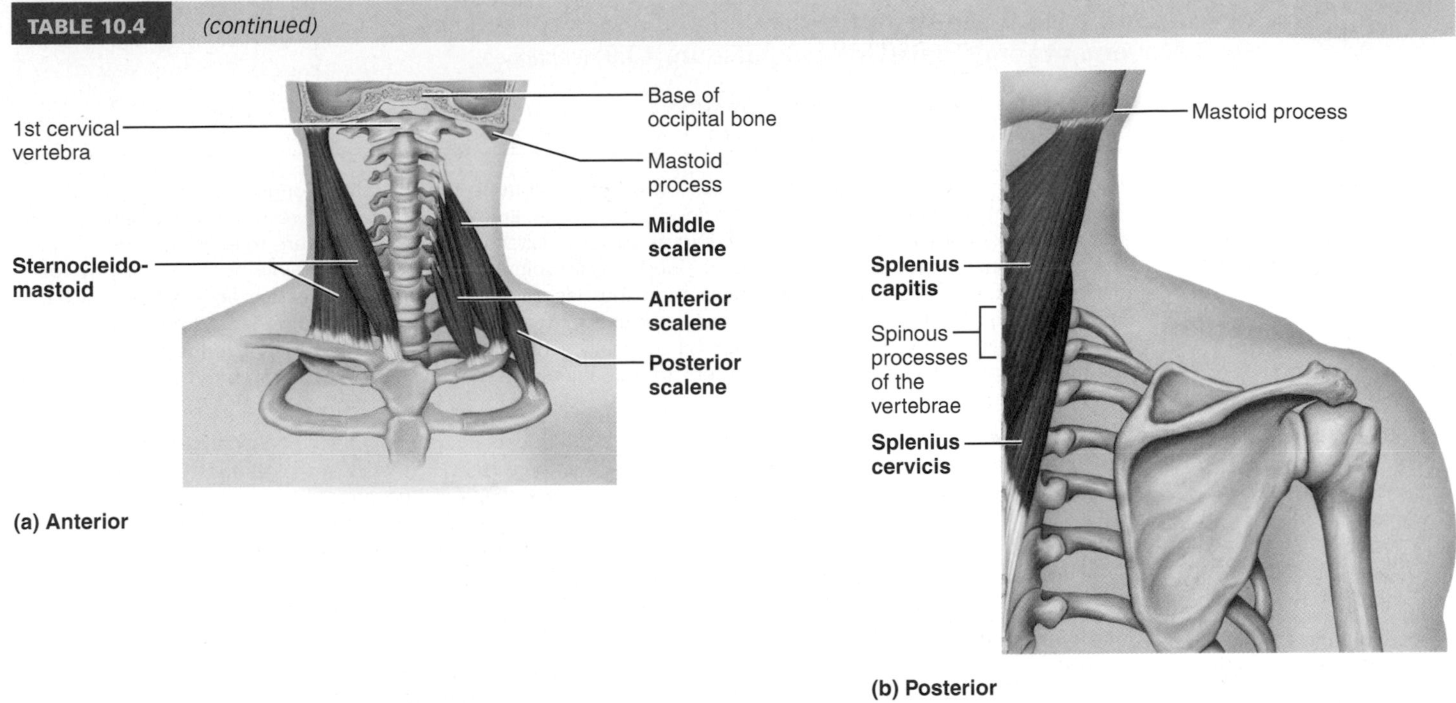

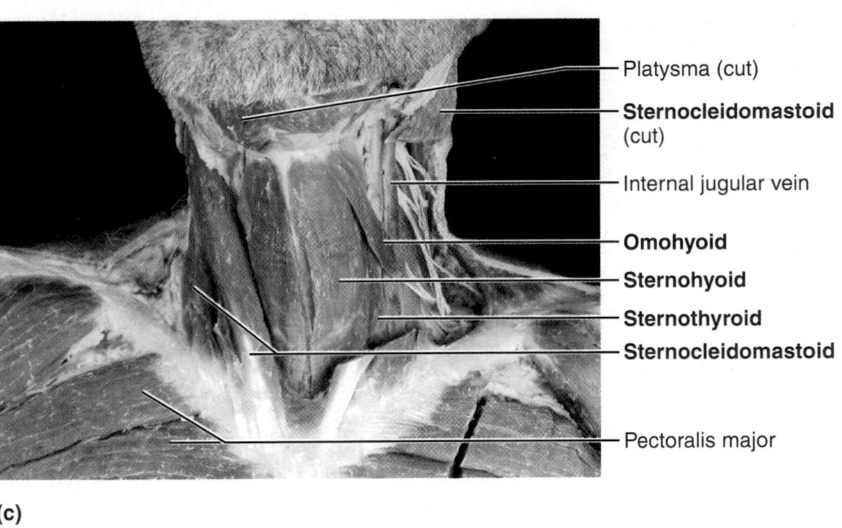

Figure 10.9 **Muscles of the neck and vertebral column causing movements of the head and trunk.** (See *A Brief Atlas of the Human Body*, Plate 44.) **(a)** Muscles of the anterolateral neck; superficial platysma muscle and the deeper neck muscles removed. **(b)** Deep muscles of the posterior neck; superficial muscles removed. **(c)** Photograph of the anterior and lateral regions of the neck.

MUSCLE GALLERY

TABLE 10.4 Muscles of the Neck and Vertebral Column: Head Movements and Trunk Extension (Figure 10.9) *(continued)*

MUSCLE	DESCRIPTION	ORIGIN (O) AND INSERTION (I)	ACTION	NERVE SUPPLY
Erector spinae (e-rek'tor spi'ne) Also called **sacrospinalis** (Figure 10.9d, left side)	Prime mover of back extension. Each side consists of three columns—the iliocostalis, longissimus, and spinalis muscles—forming intermediate layer of intrinsic back muscles. Erector spinae provide resistance that helps control action of bending forward at the waist and act as powerful extensors to promote return to erect position. During full flexion (i.e., when touching fingertips to floor), erector spinae are relaxed and strain is borne entirely by ligaments of back; on reversal of the movement, these muscles are initially inactive, and extension is initiated by hamstring muscles of thighs and gluteus maximus muscles of buttocks. As a result of this peculiarity, lifting a load or moving suddenly from a bent-over position is potentially dangerous (in terms of possible injury) to muscles and ligaments of back and intervertebral discs; erector spinae muscles readily go into painful spasms following injury to back structures.			
▪ **Iliocostalis** (il″e-o-kos-tă'lis)—lumborum, thoracis, and cervicis portions (lum'bor-um; tho-ra'sis) (*ilio* = ilium; *cost* = rib; *thorac* = thorax)	Most lateral muscle group of erector spinae muscles; extend from pelvis to neck	O—iliac crests (lumborum); inferior 6 ribs (thoracis); ribs 3 to 6 (cervicis) I—angles of ribs (lumborum and thoracis); transverse processes of cervical vertebrae C_6–C_4 (cervicis)	Extend and laterally flex the vertebral column; maintain erect posture; acting on one side, bend vertebral column to same side	Spinal nerves (dorsal rami)
▪ **Longissimus** (lon-jis'ĭ-mus)— thoracis, cervicis, and capitis parts (*longissimus* = longest)	Intermediate tripartite muscle group of erector spinae; extend by many muscle slips from lumbar region to skull; mainly pass between transverse processes of the vertebrae	O—transverse processes of lumbar through cervical vertebrae I—transverse processes of thoracic or cervical vertebrae and to ribs superior to origin as indicated by name; capitis inserts into mastoid process of temporal bone	Thoracis and cervicis act together to extend and laterally flex vertebral column; capitis extends head and turns the face toward same side	Spinal nerves (dorsal rami)
▪ **Spinalis** (spi-nă'lis)—thoracis and cervicis parts (*spin* = vertebral column, spine)	Most medial muscle column of erector spinae; cervicis usually rudimentary and poorly defined	O—spinous process of upper lumbar and lower thoracic vertebrae I—spinous process of upper thoracic and cervical vertebrae	Extends vertebral column	Spinal nerves (dorsal rami)
Semispinalis (sem'e-spĭ-nă'lis)—thoracis, cervicis, and capitis regions (*semi* = half) (Figure 10.9d, right side)	Composite muscle forming part of deep layer of intrinsic back muscles; extends from thoracic region to head	O—transverse processes of C_7–T_{12} I—occipital bone (capitis) and spinous processes of cervical (cervicis) and thoracic vertebrae T_1–T_4 (thoracis)	Extends vertebral column and head and rotates them to opposite side; acts synergistically with sternocleidomastoid muscles of opposite side	Spinal nerves (dorsal rami)
Quadratus lumborum (kwod-ra'tus lum-bor'um) (*quad* = four-sided; *lumb* = lumbar region) (See also Figure 10.19a)	Fleshy muscle forming part of posterior abdominal wall	O—iliac crest and lumbar fascia I—transverse processes of lumbar vertebrae L_1–L_4 and lower margin of 12th rib	Flexes vertebral column laterally when acting separately; when pair acts jointly, lumbar spine is extended and 12th rib is fixed; maintains upright posture; assists in forced inspiration	T_{12} and upper lumbar spinal nerves (ventral rami)

MUSCLE GALLERY

TABLE 10.4 *(continued)*

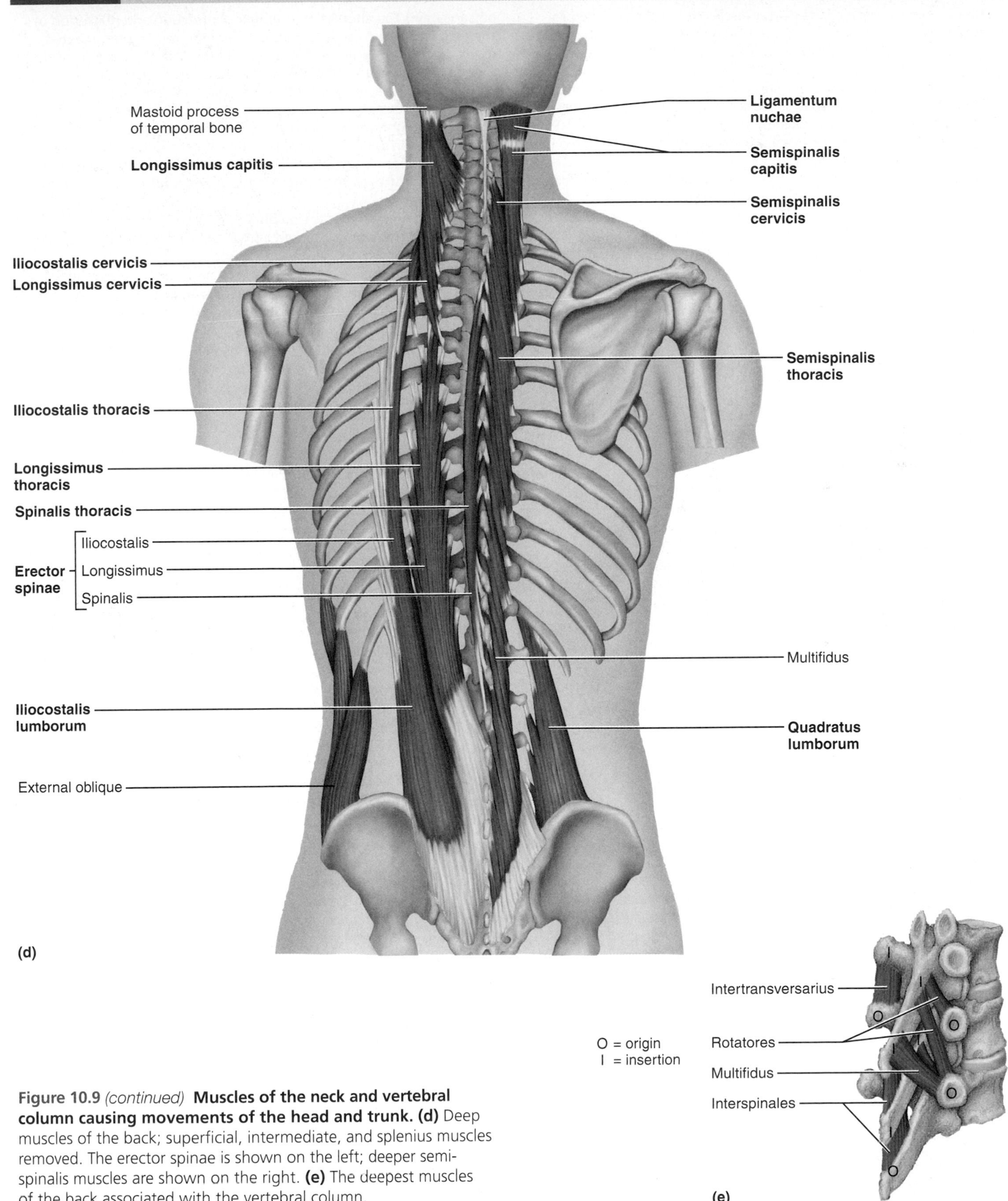

Figure 10.9 *(continued)* **Muscles of the neck and vertebral column causing movements of the head and trunk. (d)** Deep muscles of the back; superficial, intermediate, and splenius muscles removed. The erector spinae is shown on the left; deeper semispinalis muscles are shown on the right. **(e)** The deepest muscles of the back associated with the vertebral column.

MUSCLE GALLERY

TABLE 10.5 Muscles of the Thorax: Breathing (Figure 10.10)

The primary function of the deep muscles of the thorax is to promote movements necessary for breathing. Breathing consists of two phases—inspiration, or inhaling, and expiration, or exhaling—caused by cyclic changes in the volume of the thoracic cavity.

Two main layers of muscles help form the anterolateral wall of the thorax.* The thoracic muscles are very short, most extending only from one rib to the next. On contraction, they draw the somewhat flexible ribs closer together. The **external intercostal muscles**, considered inspiratory muscles, form the more superficial layer (Figure 10.10a). They lift the rib cage, an action that increases the anterior to posterior and side-to-side dimensions of the thorax. The **internal intercostal muscles** form the deeper layer and may aid active (forced) expiration by depressing the rib cage. (However, quiet expiration is largely a passive phenomenon, resulting from relaxation of the external intercostals and diaphragm and elastic recoil of the lungs.)

The **diaphragm**, the most important muscle of inspiration, forms a muscular partition between the thoracic and abdominopelvic cavities (Figure 10.10b, c). In the relaxed state, the diaphragm is dome shaped. When it contracts it moves inferiorly and flattens, increasing the volume of the thoracic cavity, which draws air into the respiratory system passageways. The alternating rhythmic contraction and relaxation of the diaphragm also causes pressure changes in the abdominopelvic cavity below that facilitate the return of blood to the heart. In addition you can contract the diaphragm voluntarily to push down on the abdominal viscera and increase the pressure in the abdominopelvic cavity to help evacuate pelvic organ contents (urine, feces, or a baby) or during weight lifting. When you take a deep breath to fix the diaphragm, the abdomen becomes a firm pillar that will not buckle under the weight being lifted. Needless to say, it is important to have good control of the urinary bladder and anal sphincters during such maneuvers.

With the exception of the diaphragm, which is served by the *phrenic nerves*, the muscles listed in this table are served by *the intercostal nerves*, which run between the ribs.

Forced breathing (as during exercise) calls into play a number of other muscles that insert into the ribs. For example, during forced inspiration the scalene and sternocleidomastoid muscles of the neck help lift the ribs. Forced expiration is aided by muscles that pull the ribs inferiorly and those that push the diaphragm superiorly by compressing the abdominal contents (abdominal wall muscles).

MUSCLE	DESCRIPTION	ORIGIN (O) AND INSERTION (I)	ACTION	NERVE SUPPLY
External intercostals (in″ter-kos′talz) (*external* = toward the outside; *inter* = between; *cost* = rib)	11 pairs lie between ribs; fibers run obliquely (down and forward) from each rib to rib below; in lower intercostal spaces, fibers are continuous with external oblique muscle, forming part of abdominal wall	O—inferior border of rib above I—superior border of rib below	With first ribs fixed by scalene muscles, pull ribs toward one another to elevate rib cage; aid in inspiration; synergists of diaphragm	Intercostal nerves
Internal intercostals (*internal* = toward the inside, deep)	11 pairs lie between ribs; fibers run deep to and at right angles to those of external intercostals (i.e., run downward and posteriorly); lower internal intercostal muscles are continuous with fibers of internal oblique muscle of abdominal wall	O—superior border of rib below I—inferior border (costal groove) of rib above	With 12th ribs fixed by quadratus lumborum, muscles of posterior abdominal wall, and oblique muscles of the abdominal wall, they draw ribs together and depress rib cage; aid in forced expiration; antagonistic to external intercostals	Intercostal nerves
Diaphragm (di′ah-fram) (*dia* = across; *phragm* = partition)	Broad muscle pierced by the aorta, inferior vena cava, and esophagus, forms floor of thoracic cavity; in relaxed state is dome shaped; fibers converge from margins of thoracic cage toward a boomerang-shaped central tendon	O—inferior, internal surface of rib cage and sternum, costal cartilages of last six ribs and lumbar vertebrae I—central tendon	Prime mover of inspiration; flattens on contraction, increasing vertical dimensions of thorax; when strongly contracted, dramatically increases intra-abdominal pressure	Phrenic nerves

*Although there is a third (deepest) muscle layer of the thoracic wall, the muscles are small and discontinuous. Additionally, their function is unclear, so they are not included in this table.

MUSCLE GALLERY

TABLE 10.5 *(continued)*

Figure 10.10 Muscles of respiration. (a) Deep muscles of the thorax. The external intercostals (inspiratory muscles) are shown on the left and the internal intercostals (expiratory muscles) are shown on the right. These two muscle layers run obliquely and at right angles to each other. **(b)** Inferior view of the diaphragm, the prime mover of inspiration. Notice that its muscle fibers converge toward a central tendon, an arrangement that causes the diaphragm to flatten and move inferiorly as it contracts. **(c)** Photograph of the diaphragm, superior view. (See *A Brief Atlas of the Human Body*, Figure 63.)

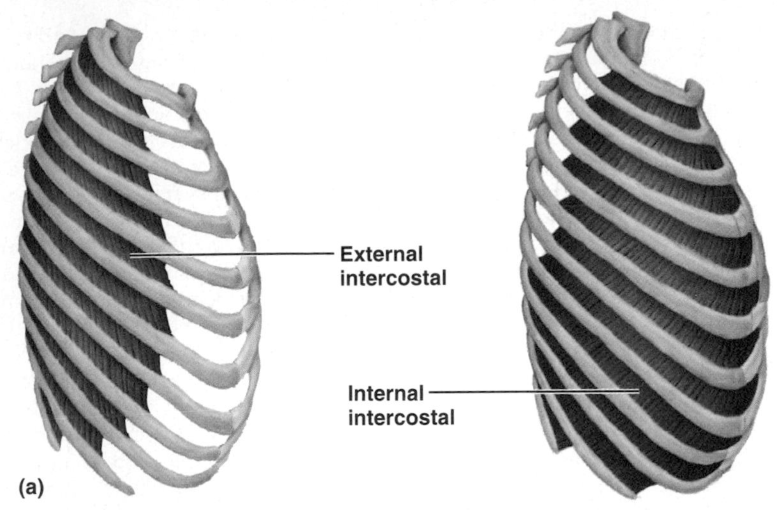

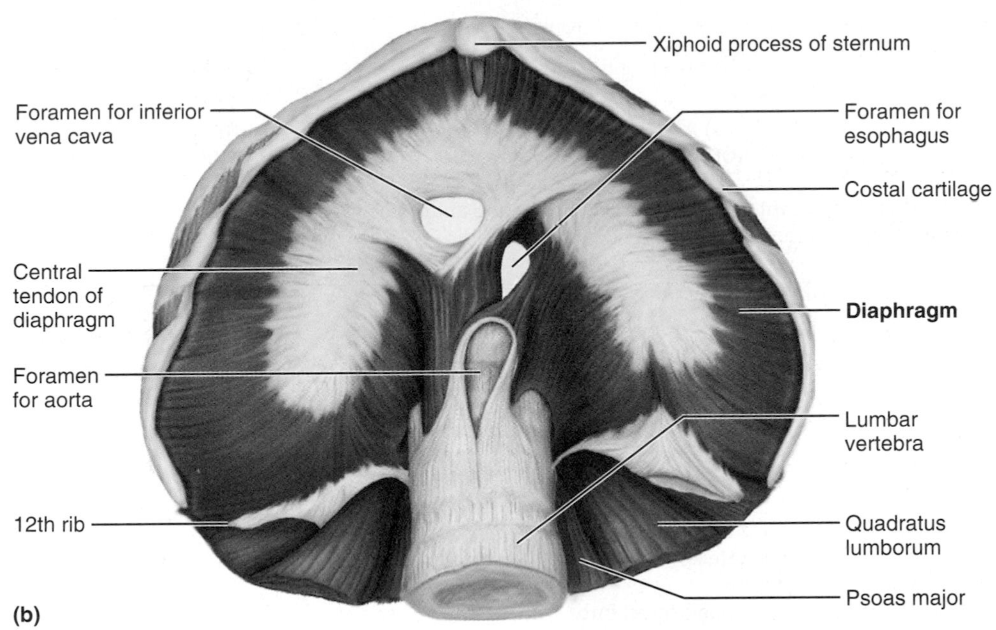

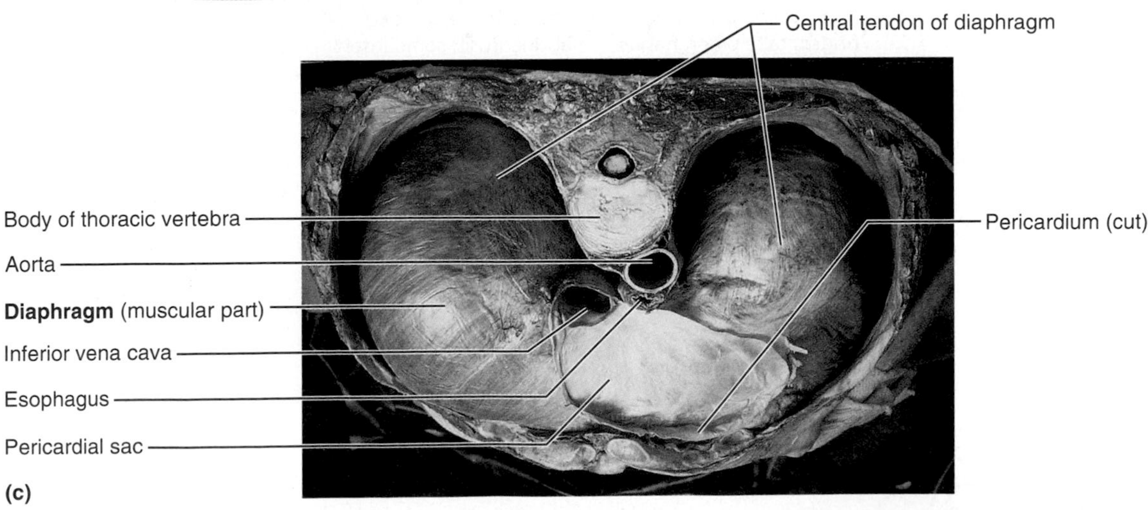

MUSCLE GALLERY

TABLE 10.6 Muscles of the Abdominal Wall: Trunk Movements and Compression of Abdominal Viscera (Figure 10.11)

Unlike the thorax, the anterior and lateral abdominal wall has no bony reinforcements (ribs). Instead, it is a composite of four paired muscles, their investing fasciae, and aponeuroses. Three broad flat muscle pairs, layered one atop the next, form the lateral abdominal walls: The fibers of the **external oblique muscle** run inferomedially and at right angles to those of the **internal oblique**, which immediately underlies it (Figure 10.11a, b). The fibers of the deep **transversus abdominis muscle** run horizontally across the abdomen at an angle to both. This alternation of fascicle directions is similar to the construction of plywood (which is made of sheets with different grains) and provides great strength. These three muscles blend into broad insertion aponeuroses anteriorly. The aponeuroses, in turn, enclose a fourth muscle pair medially, the straplike **rectus abdominis muscles**, and then fuse, forming the **linea alba** ("white line"), a tendinous raphe (seam) that runs from the sternum to the pubic symphysis (Figure 10.11a, c). The snug enclosure of the rectus abdominis muscles within the aponeuroses prevents them from "bowstringing" (protruding anteriorly) when they contract. The quadratus lumborum muscles of the *posterior* abdominal wall are covered in Table 10.4.

The abdominal muscles protect and support the viscera most effectively when they are well toned. When weak or severely stretched (as during pregnancy), they allow the abdomen to become pendulous (i.e., to form a potbelly). Additional functions include lateral flexion and rotation of the trunk and flexion of the trunk against resistance (as in sit-ups). During inspiration, the abdominal muscles relax, allowing the abdominal viscera to be pushed inferiorly by the descending diaphragm. When these abdominal muscles contract, several different activities may be effected. For example, when they contract simultaneously, the ribs are pulled inferiorly and the abdominal contents are compressed. This pushes the visceral organs upward on the diaphragm, aiding forced expiration. When the abdominal muscles contract with the diaphragm and the glottis is closed (an action called the Valsalva maneuver), the increased intra-abdominal pressure helps to promote urination, defecation, childbirth, vomiting, coughing, screaming, sneezing, burping, and nose blowing. (Next time you perform one of these activities, feel your abdominal muscles contract under your skin.) These muscles also contract during heavy lifting— sometimes so forcefully that hernias result. Contraction of the abdominal muscles as the deep back muscles contract helps prevent hyperextension of the spine and splints the entire body trunk.

MUSCLE	DESCRIPTION	ORIGIN (O) AND INSERTION (I)	ACTION	NERVE SUPPLY
MUSCLES OF THE ANTERIOR AND LATERAL ABDOMINAL WALL	Four paired flat muscles; important in supporting and protecting abdominal viscera and in promoting lateral flexion and flexion of the vertebral column			
Rectus abdominis (rek'tus ab-dom'ĭ-nis) (*rectus* = straight; *abdom* = abdomen)	Medial superficial muscle pair; extend from pubis to rib cage; ensheathed by aponeuroses of lateral muscles; segmented by three tendinous intersections	O—pubic crest and symphysis I—xiphoid process and costal cartilages of ribs 5–7	Flex and rotate lumbar region of vertebral column; fix and depress ribs, stabilize pelvis during walking, increase intra-abdominal pressure; used in sit-ups, curls	Intercostal nerves (T_6 or T_7–T_{12})
External oblique (o-blēk') (*external* = toward outside; *oblique* = running at an angle)	Largest and most superficial of the three lateral muscles; fibers run downward and medially (same direction outstretched fingers take when hands put into pants pockets); aponeurosis turns under inferiorly, forming inguinal ligament	O—by fleshy strips from outer surfaces of lower eight ribs I—most fibers insert into linea alba via a broad aponeurosis; some insert into pubic crest and tubercle and iliac crest	When pair contract simultaneously, flex vertebral column and compress abdominal wall and increase intra-abdominal pressure; acting individually, aid muscles of back in trunk rotation and lateral flexion; used in oblique curls	Intercostal nerves (T_7–T_{12})
Internal oblique (*internal* = toward the inside; deep)	Most fibers run upward and medially; however, the muscle fans so its inferior fibers run downward and medially	O—lumbar fascia, iliac crest, and inguinal ligament I—linea alba, pubic crest, last three or four ribs, and costal margin	As for external oblique	Intercostal nerves (T_7–T_{12}) and L_1
Transversus abdominis (trans-ver'sus) (*transverse* = running straight across)	Deepest (innermost) muscle of abdominal wall; fibers run horizontally	O—inguinal ligament, lumbar fascia, cartilages of last six ribs; iliac crest I—linea alba, pubic crest	Compresses abdominal contents	Intercostal nerves (T_7–T_{12}) and L_1

MUSCLE GALLERY

TABLE 10.6 *(continued)*

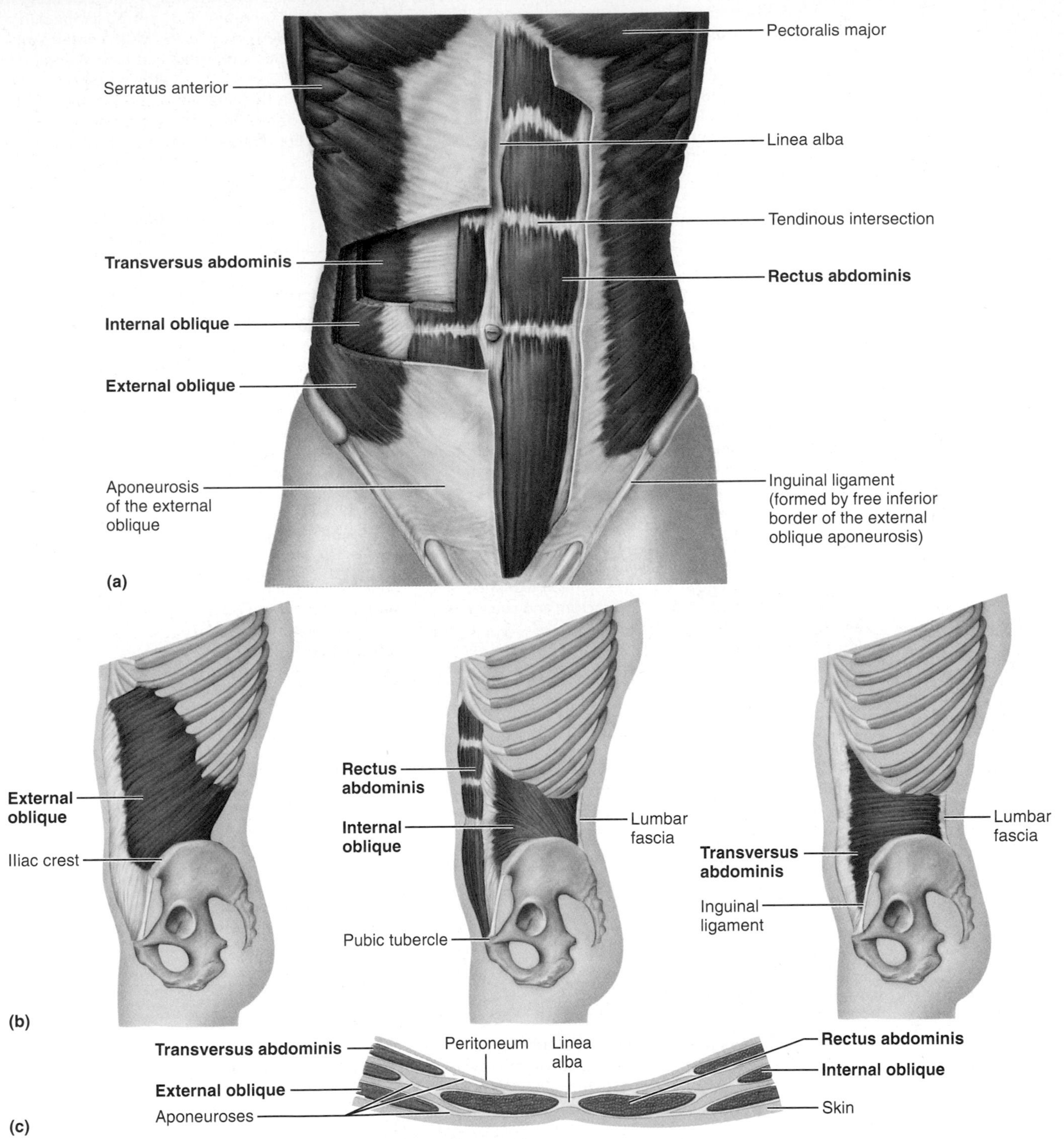

Figure 10.11 Muscles of the abdominal wall. (a) Anterior view of the muscles forming the anterolateral abdominal wall; superficial muscles partially cut away to reveal the deeper muscles. **(b)** Lateral view of the trunk, illustrating the fiber direction and attachments of the abdominal muscles. Although shown in the same view, the rectus abdominis is deep to the fascia of the internal oblique muscle. **(c)** Transverse section through the anterolateral abdominal wall (midregion), showing how the aponeuroses of the lateral abdominal muscles contribute to the rectus abdominis sheath.

TABLE 10.7 Muscles of the Pelvic Floor and Perineum: Support of Abdominopelvic Organs (Figure 10.12)

Two paired muscles, the **levator ani** and **coccygeus**, form the funnel-shaped pelvic floor, or **pelvic diaphragm** (Figure 10.12a). This diaphragm (1) seals the inferior opening of the bony pelvis, (2) supports the pelvic organs, (3) lifts the pelvic floor superiorly to release feces, and (4) resists increased intra-abdominal pressure (which would expel contents of the urinary bladder, rectum, and uterus). The pelvic diaphragm is pierced by the rectum and urethra (urinary tube), and by the vagina in females. The body region inferior to the pelvic diaphragm is the *perineum*. The relationships of the perineum are worth explaining. Inferior to the muscles of the pelvic floor, and stretching between the two sides of the pubic arch in the anterior half of the perineum, is the **urogenital diaphragm** (Figure 10.12b). This thin triangular sheet of muscle contains the **external urethral sphincter**, a sphincter muscle that surrounds the urethra and allows voluntary control of urination. Superficial to the urogenital diaphragm, and covered by the skin of the perineum, is the *superficial perineal space*, which contains muscles (**ischiocavernosus** and **bulbospongiosus**) that help maintain erection of the penis and clitoris (Figure 10.12c). In the posterior half of the perineum encircling the anus is the **external anal sphincter**, which allows voluntary control of defecation. Just anterior to this sphincter is the **central tendon of the perineum**, a strong tendon into which many of the perineal muscles insert.

MUSCLE	DESCRIPTION	ORIGIN (O) AND INSERTION (I)	ACTION	NERVE SUPPLY
MUSCLES OF THE PELVIC DIAPHRAGM (FIGURE 10.12a)				
Levator ani (lĕ-va'tor a'ne) (*levator* = raises; *ani* = anus)	Broad, thin, tripartite muscle (pubococcygeus, puborectalis, and iliococcygeus parts); its fibers extend inferomedially, forming a muscular "sling" around male prostate (or female vagina), urethra, and anorectal junction before meeting in the median plane	O—extensive linear origin inside pelvis from pubis to ischial spine I—inner surface of coccyx, levator ani of opposite side, and (in part) into the structures that penetrate it	Supports and maintains position of pelvic viscera; resists downward thrusts that accompany rises in intrapelvic pressure during coughing, vomiting, and expulsive efforts of abdominal muscles; forms sphincters at anorectal junction and vagina; lifts anal canal during defecation	S_3, S_4, and inferior rectal nerve (branch of pudendal nerve)
Coccygeus (kok-sij'e-us) (*coccy* = coccyx)	Small triangular muscle lying posterior to levator ani; forms posterior part of pelvic diaphragm	O—spine of ischium I—sacrum and coccyx	Supports pelvic viscera; supports coccyx and pulls it forward after it has been reflected posteriorly by defecation and childbirth	S_4 and S_5
MUSCLES OF THE UROGENITAL DIAPHRAGM (FIGURE 10.12b)				
Deep transverse perineal muscle (per″ĭ-ne'al) (*deep* = far from surface; *transverse* = across; *perine* = near anus)	Together the pair spans distance between ischial rami; in females, lies posterior to vagina	O—ischial rami I—midline central tendon of perineum; some fibers into vaginal wall in females	Supports pelvic organs; steadies central tendon	Pudendal nerve
External urethral sphincter (*sphin* = squeeze)	Muscle encircling urethra and vagina (female)	O—ischiopubic rami I—midline raphe	Constricts urethra; allows voluntary inhibition of urination; helps support pelvic organs	Pudendal nerve
MUSCLES OF THE SUPERFICIAL PERINEAL SPACE (FIGURE 10.12c)				
Ischiocavernosus (is'ke-o-kav'ern-o'sus) (*ischi* = hip; *caverna* = hollow chamber)	Runs from pelvis to base of penis or clitoris	O—ischial tuberosities I—crus of corpora cavernosa of male penis or female clitoris	Retards venous drainage and maintains erection of penis or clitoris	Pudendal nerve
Bulbospongiosus (bul″bo-spun″je-o'sus) (*bulbon* = bulb; *spongio* = sponge)	Encloses base of penis (bulb) in males and lies deep to labia in females	O—central tendon of perineum and midline raphe of penis I—anteriorly into corpora cavernosa of penis or clitoris	Empties male urethra; assists in erection of penis and of clitoris	Pudendal nerve
Superficial transverse perineal muscles (*superficial* = closer to surface)	Paired muscle bands posterior to urethral (and in females, vaginal) opening; variable; sometimes absent	O—ischial tuberosity I—central tendon of perineum	Stabilizes and strengthens midline tendon of perineum	Pudendal nerve

MUSCLE GALLERY

TABLE 10.7 *(continued)*

Anterior
Levator ani
Pubococcygeus
Iliococcygeus
Symphysis pubis
Urogenital diaphragm
Urethra
Vagina
Anal canal
Obturator internus
Coccyx
Piriformis
Levator ani
Coccygeus
Pelvic diaphragm
(a)
Posterior

Figure 10.12 Muscles of the pelvic floor and perineum. (a) Superior view of muscles of the pelvic diaphragm (levator ani and coccygeus) in a female pelvis. **(b)** Inferior view of muscles of the urogenital diaphragm of the perineum (external urethral sphincter and deep transverse perineal muscles). **(c)** Muscles of the superficial space of the perineum (ischiocavernosus, bulbospongiosus, and superficial transverse perineal muscle), which lie just deep to the skin of the perineum. Note that a raphe is a seam of fibrous tissue.

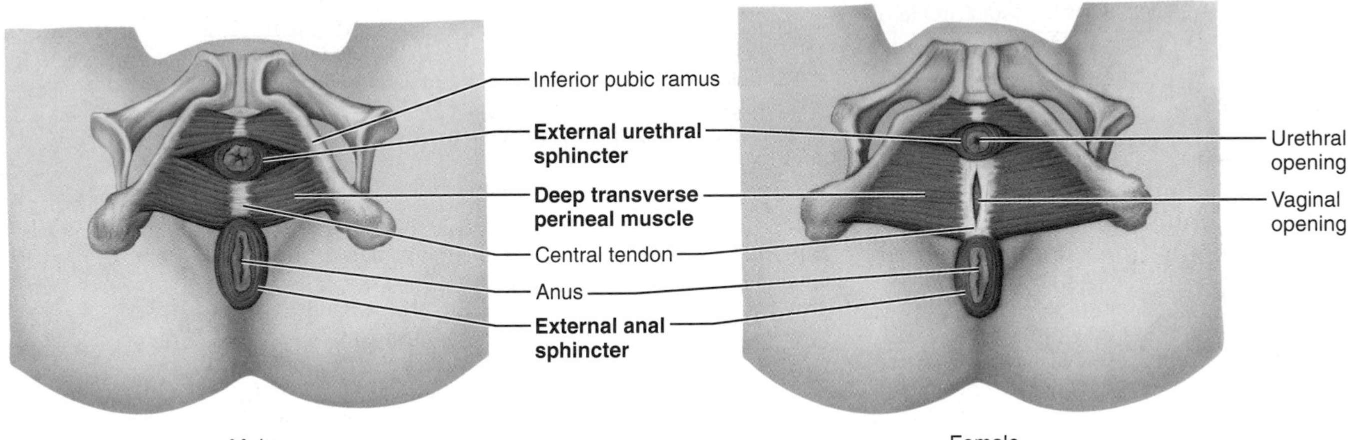

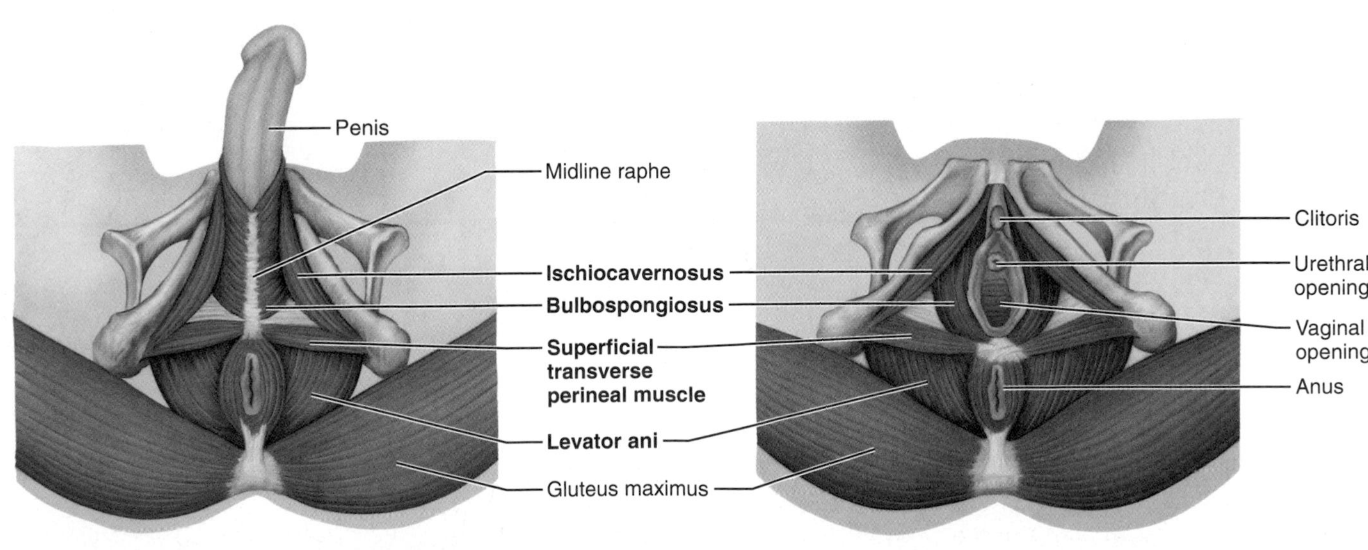

MUSCLE GALLERY

TABLE 10.8 Superficial Muscles of the Anterior and Posterior Thorax: Movements of the Scapula (Figure 10.13)

Most superficial thorax muscles are *extrinsic shoulder muscles*, which run from the ribs and vertebral column to the shoulder girdle. They both fix the scapula in place and move it to increase the range of arm movements. The anterior muscles of this group include the **pectoralis major**, **pectoralis minor**, **serratus anterior**, and **subclavius** (Figure 10.13a). Except for the pectoralis major, which inserts into the humerus, all muscles of the anterior group insert into the pectoral girdle. The posterior muscles include the **latissimus dorsi** and **trapezius muscles** superficially and the underlying **levator scapulae** and **rhomboids** (Figure 10.13c). The latissimus dorsi, like the pectoralis major muscles anteriorly, insert into the humerus and are more concerned with movements of the arm than of the scapula, so their consideration is deferred to Table 10.9 (arm-moving muscles).

The important movements of the pectoral girdle involve displacements of the scapula, i.e., its elevation and depression, rotation, lateral (forward) movements, and medial (backward) movements. The clavicles rotate around their own axes to provide both stability and precision to scapular movements.

Except for the serratus anterior, the anterior muscles stabilize and depress the shoulder girdle. Thus, most scapular movements are promoted by the serratus anterior muscles anteriorly and by posterior thoracic muscles. The arrangement of muscle attachments to the scapula is such that one muscle cannot bring about a simple (linear) movement on its own. To effect scapular movements, several muscles must act in combination.

The prime movers of shoulder (scapular) elevation are the superior trapezius fibers and the levator scapulae. When acting together to shrug the shoulder, their opposite rotational effects counterbalance each other. The scapula is depressed largely by gravity (weight of the arm), but when it is depressed against resistance, the inferior part of the trapezius, the pectoralis minor, and the serratus anterior (along with the latissimus dorsi, Table 10.9) are active. Anterolateral movements (abduction) of the scapula on the thorax wall, as in pushing or punching movements, mainly reflect serratus anterior activity. Posteromedial movement (adduction) of the scapula is effected mainly by the trapezius (midpart) and the rhomboids. Although the serratus anterior and trapezius muscles are antagonists in forward/backward movements of the scapulae, they act together to coordinate *rotational* scapular movements.

MUSCLE	DESCRIPTION	ORIGIN (O) AND INSERTION (I)	ACTION	NERVE SUPPLY
MUSCLES OF THE ANTERIOR THORAX (FIGURE 10.13a)				
Pectoralis minor (pek″to-ra′lis mi′nor) (*pectus* = chest, breast; *minor* = lesser)	Flat, thin muscle directly beneath and obscured by pectoralis major	O—anterior surfaces of ribs 3–5 (or 2–4) I—coracoid process of scapula	With ribs fixed, draws scapula forward and downward; with scapula fixed, draws rib cage superiorly	Medial and lateral pectoral nerves (C_6–C_8)
Serratus anterior (ser-a′tus) (*serratus* = saw)	Fan-shaped muscle; lies deep to scapula, deep and inferior to pectoral muscles on lateral rib cage; forms medial wall of axilla; origins have serrated, or sawtooth, appearance; paralysis results in "winging" of vertebral border of scapula away from chest wall, making arm elevation impossible	O—by a series of muscle slips from ribs 1–8 (or 9) I—entire anterior surface of vertebral border of scapula	Rotates scapula so that its inferior angle moves laterally and upward; prime mover to protract and hold scapula against chest wall; raises point of shoulder; important role in abduction and raising of arm and in horizontal arm movements (pushing, punching); called "boxer's muscle"	Long thoracic nerve (C_5–C_7)
Subclavius (sub-kla′ve-us) (*sub* = under, beneath; *clav* = clavicle)	Small cylindrical muscle extending from rib 1 to clavicle	O—costal cartilage of rib 1 I—groove on inferior surface of clavicle	Helps stabilize and depress pectoral girdle	Nerve to subclavius (C_5 and C_6)

MUSCLE GALLERY

TABLE 10.8 *(continued)*

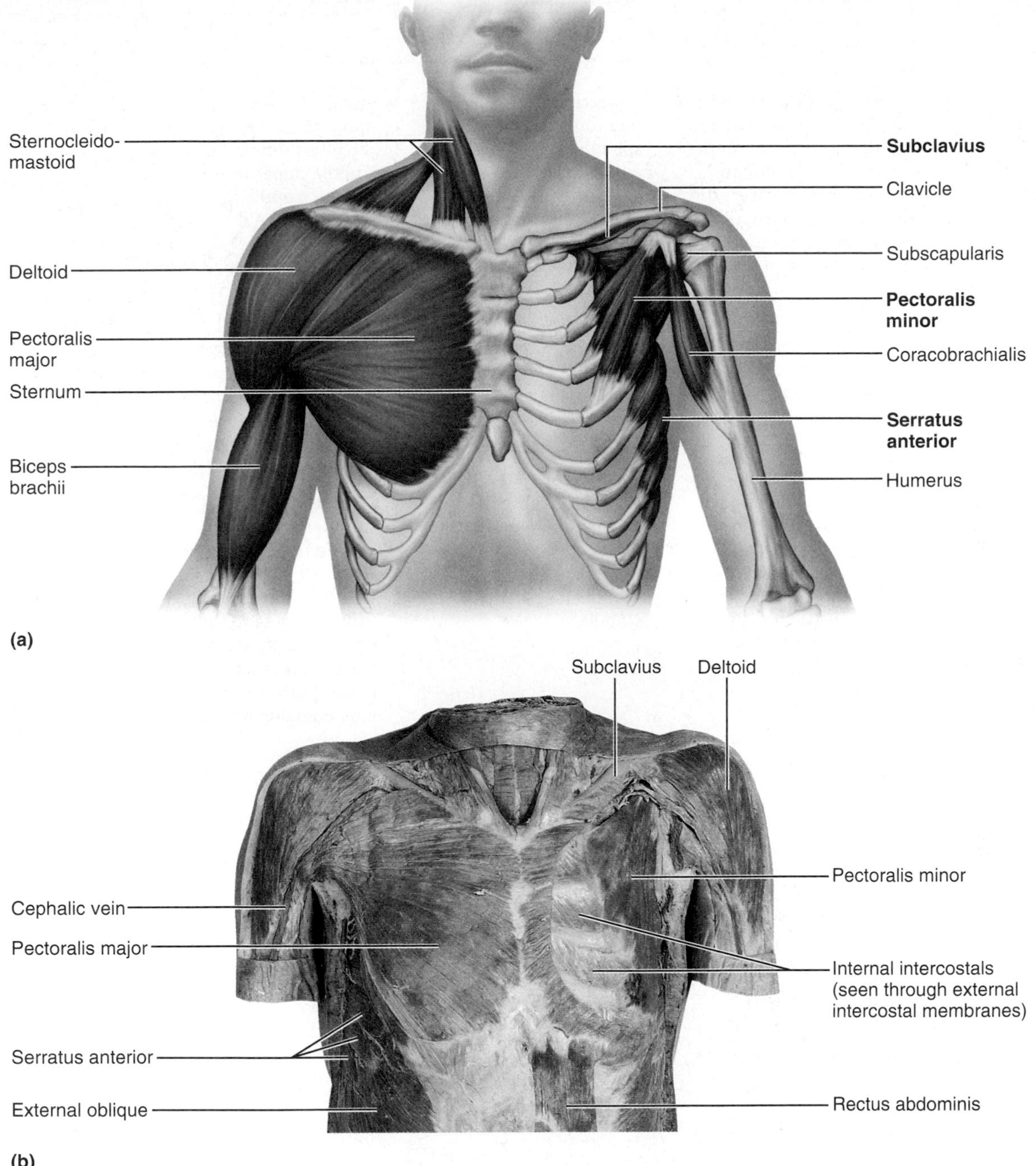

Figure 10.13 **Superficial muscles of the thorax and shoulder acting on the scapula and arm. (a)** Anterior view. The superficial muscles, which effect arm movements, are shown on the left side of the illustration. These muscles are removed on the right to show the muscles that stabilize or move the pectoral girdle. **(b)** Photo of superficial muscles of anterior thorax.

TABLE 10.8 Superficial Muscles of the Anterior and Posterior Thorax: Movements of the Scapula (Figure 10.13) *(continued)*

MUSCLE	DESCRIPTION	ORIGIN (O) AND INSERTION (I)	ACTION	NERVE SUPPLY
MUSCLES OF THE POSTERIOR THORAX (FIGURE 10.13c–e)				
Trapezius (trah-pe′ze-us) (*trapezion* = irregular four-sided figure)	Most superficial muscle of posterior thorax; flat, and triangular in shape; upper fibers run inferiorly to scapula; middle fibers run horizontally to scapula; lower fibers run superiorly to scapula	O—occipital bone, ligamentum nuchae, and spinous processes of C_7 and all thoracic vertebrae I—a continuous insertion along acromion and spine of scapula and lateral third of clavicle	Stabilizes, raises, retracts, and rotates scapula; middle fibers retract (adduct) scapula; superior fibers elevate scapula (as in shrugging the shoulders) or can help extend head with scapula fixed; inferior fibers depress scapula (and shoulder)	Accessory nerve (cranial nerve XI); C_3 and C_4
Levator scapulae (skap′u-le) (*levator* = raises)	Located at back and side of neck, deep to trapezius; thick, straplike muscle	O—transverse processes of C_1–C_4 I—medial border of the scapula, superior to the spine	Elevates/adducts scapula in concert with superior fibers of trapezius; tilts glenoid cavity downward when scapula is fixed, flexes neck to same side	Cervical spinal nerves and dorsal scapular nerve (C_3–C_5)
Rhomboids (rom′boidz)—major and minor (*rhomboid* = diamond shaped)	Two roughly diamond shaped muscles lying deep to trapezius and inferior to levator scapulae; rhomboid minor is the more superior muscle	O—spinous processes of C_7 and T_1 (minor) and spinous processes of T_2–T_5 (major) I—medial border of scapula	Stabilize scapula; act together (and with middle trapezius fibers) to retract (adduct) scapula, thus "squaring shoulders"; rotate scapula so that glenoid cavity is downward (as when arm is lowered against resistance; e.g., paddling a canoe)	Dorsal scapular nerve (C_4 and C_5)

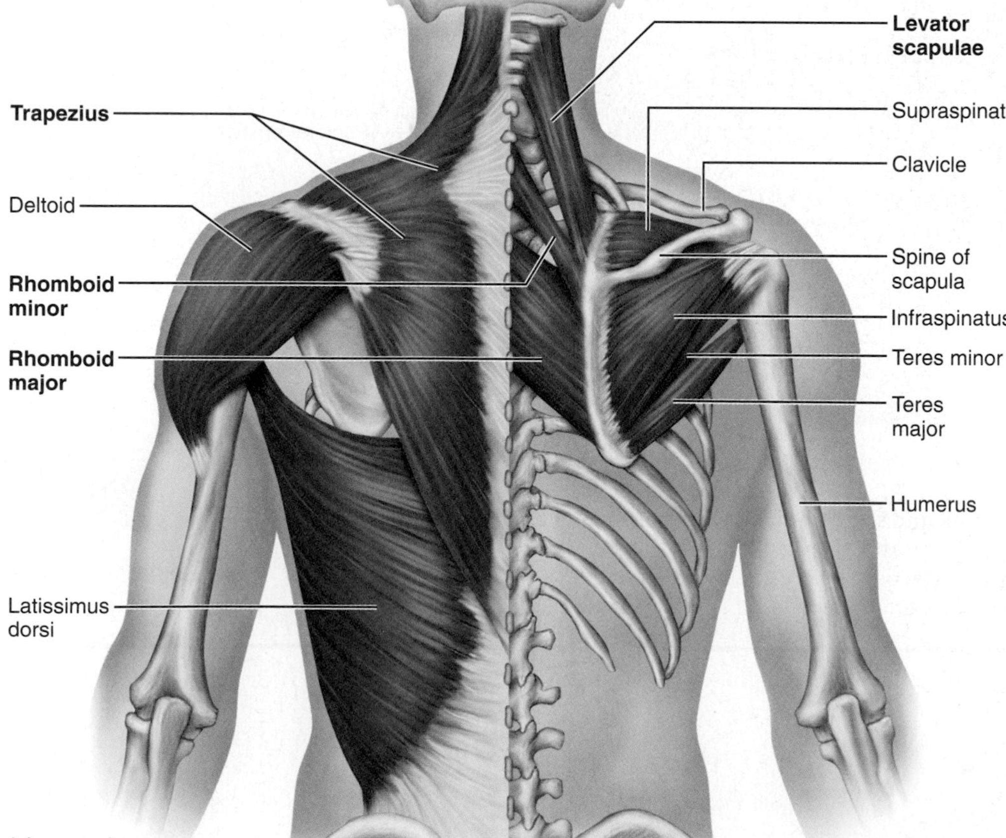

Figure 10.13 *(continued)* **Superficial muscles of the thorax and shoulder acting on the scapula and arm. (c)** Posterior view. The superficial muscles are shown on the left side of the illustration. Superficial muscles are removed on the right side to reveal the deeper muscles acting on the scapula, and the rotator cuff muscles that help stabilize the shoulder joint.

MUSCLE GALLERY

TABLE 10.8 *(continued)*

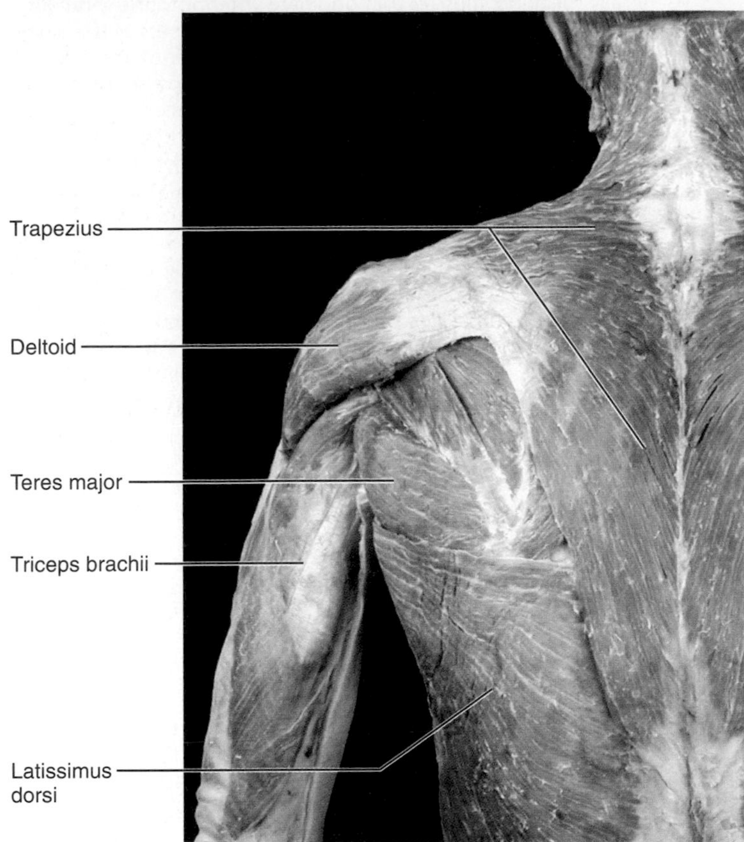

(d)

Figure 10.13 *(continued)* **(d)** Dissection showing the superficial muscles of the back, left side. **(e)** Trapezius muscle has been removed in this dissection to expose the deeper muscles of the back, right side.

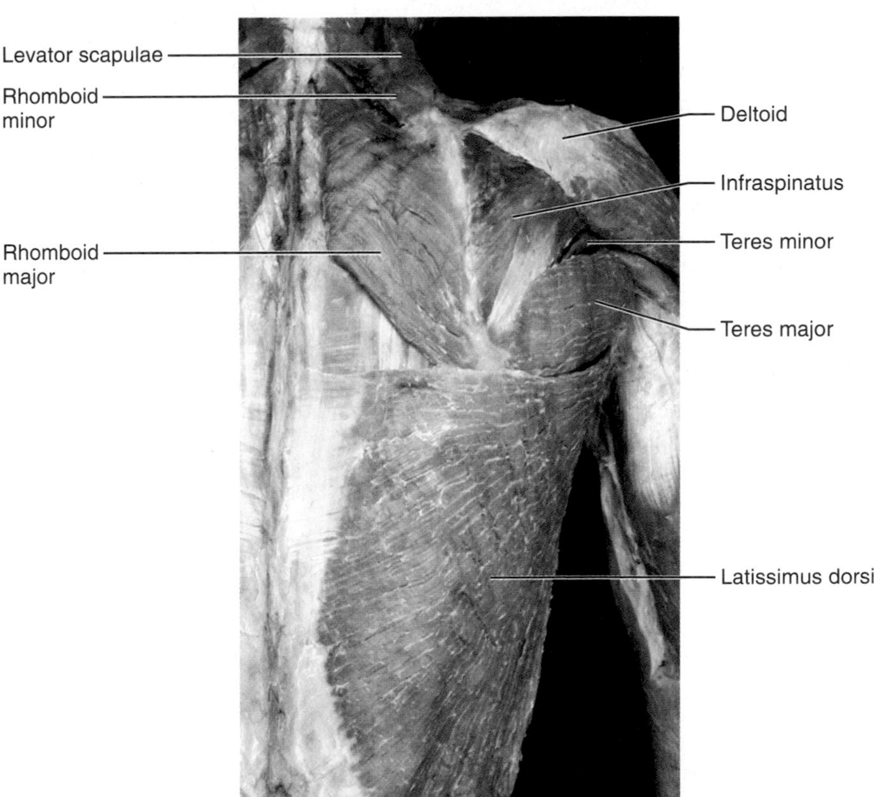

(e)

MUSCLE GALLERY

TABLE 10.9 Muscles Crossing the Shoulder Joint: Movements of the Arm (Humerus) (Figure 10.14)

Recall that the ball-and-socket shoulder joint is the most flexible joint in the body, but pays the price of instability. Several muscles cross each shoulder joint to insert on the humerus. All muscles acting on the humerus originate from the pectoral girdle; however, two of these—the latissimus dorsi and pectoralis major—primarily originate on the axial skeleton.

Of the nine muscles covered here, only the **pectoralis major, latissimus dorsi,** and **deltoid muscles** are prime movers of arm movements (Figure 10.14a, b). The remaining six are synergists and fixators. Four of these, the **supraspinatus, infraspinatus, teres minor,** and **subscapularis,** (marked with an asterisk* in Figure 10.14b, d), are *rotator cuff muscles.* They originate on the scapula, and their tendons blend with the fibrous capsule of the shoulder joint en route to the humerus. Although the rotator cuff muscles act as synergists in the angular and rotational movements of the arm, their main function is to reinforce the capsule of the shoulder joint to prevent dislocation of the humerus. The remaining two muscles, the small **teres major** and **coracobrachialis,** cross the shoulder joint but do not contribute to its reinforcement.

Generally speaking, muscles that originate *anterior* to the shoulder joint (pectoralis major, coracobrachialis, and anterior fibers of the deltoid) *flex* the arm, i.e., lift it anteriorly. The prime mover of arm flexion is the pectoralis major. The biceps brachii of the arm also assists in this action (Figure 10.14a, c; see Table 10.10). Muscles originating *posterior* to the shoulder joint extend the arm. These include the latissimus dorsi and posterior fibers of the deltoid muscles (both prime movers of arm extension) and the teres major. Note that the latissimus dorsi and pectoralis muscles are *antagonists* of one another in the flexion-extension movements of the arm.

The middle region of the fleshy deltoid muscle of the shoulder, which extends over the superolateral side of the humerus, is the prime mover of arm abduction. The main arm adductors are the pectoralis major anteriorly and latissimus dorsi posteriorly. The small muscles acting on the humerus promote lateral and medial rotation of the arm. The interactions among these nine muscles are complex and each contributes to several movements. A summary of their actions is provided in Table 10.12 (Part I).

MUSCLE	DESCRIPTION	ORIGIN (O) AND INSERTION (I)	ACTION	NERVE SUPPLY
Pectoralis major (pek″to-ra′lis ma′jer) (*pectus* = breast, chest; *major* = larger)	Large, fan-shaped muscle covering superior portion of chest; forms anterior axillary fold; divided into clavicular and sternal parts	O—sternal end of clavicle, sternum, cartilage of ribs 1–6 (or 7), and aponeurosis of external oblique muscle I—fibers converge to insert by a short tendon into intertubercular sulcus and greater tubercle of humerus	Prime mover of arm flexion; rotates arm medially; adducts arm against resistance; with scapula (and arm) fixed, pulls rib cage upward, thus can help in climbing, throwing, pushing, and in forced inspiration	Lateral and medial pectoral nerves (C_5–C_8 and T_1)
Deltoid (del′toid) (*delta* = triangular)	Thick, multipennate muscle forming rounded shoulder muscle mass; responsible for roundness of shoulder; a site commonly used for intramuscular injection, particularly in males, where it tends to be quite fleshy	O—embraces insertion of the trapezius; lateral third of clavicle; acromion and spine of scapula I—deltoid tuberosity of humerus	Prime mover of arm abduction when all its fibers contract simultaneously; antagonist of pectoralis major and latissimus dorsi, which adduct the arm; if only anterior fibers are active, can act powerfully in flexion and medial rotation of humerus, therefore synergist of pectoralis major; if only posterior fibers are active, causes extension and lateral rotation of arm; active during rhythmic arm swinging movements during walking	Axillary nerve (C_5 and C_6)

MUSCLE GALLERY

TABLE 10.9 *(continued)*

MUSCLE	DESCRIPTION	ORIGIN (O) AND INSERTION (I)	ACTION	NERVE SUPPLY
Latissimus dorsi (lah-tis'ĭ-mus dor'si) (*latissimus* = widest; *dorsi* = back)	Broad, flat, triangular muscle of lower back (lumbar region); extensive superficial origins; covered by trapezius superiorly; contributes to the posterior wall of axilla	O—indirect attachment via lumbodorsal fascia into spines of lower six thoracic vertebrae, lumbar vertebrae, lower 3 to 4 ribs, and iliac crest; also from scapula's inferior angle I—spirals around teres major to insert in floor of intertubercular sulcus of humerus	Prime mover of arm extension; powerful arm adductor; medially rotates arm at shoulder; depresses scapula; because of its power in these movements, it plays an important role in bringing the arm down in a power stroke, as in striking a blow, hammering, swimming, and rowing; with arms fixed overhead, it pulls the rest of the body upward and forward	Thoracodorsal nerve (C_6–C_8)

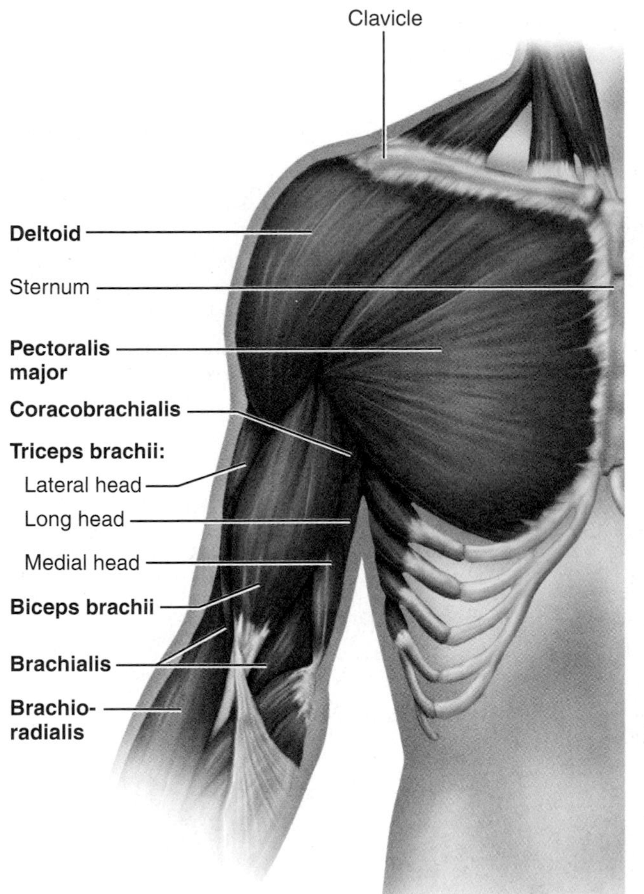

(a) Anterior view

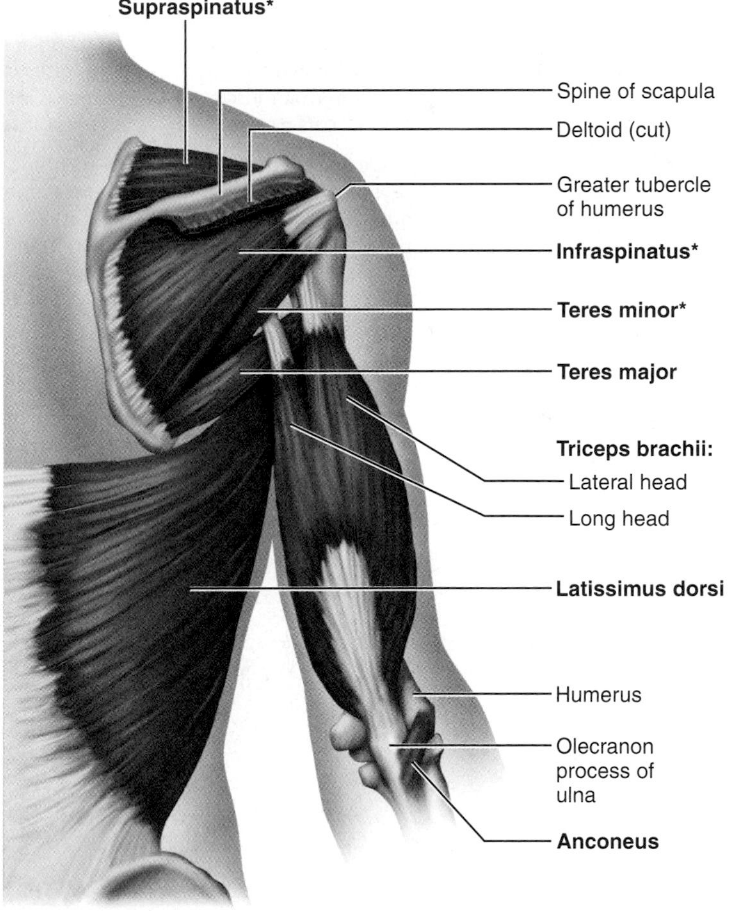

(b) Posterior view

Figure 10.14 Muscles crossing the shoulder and elbow joint, causing movements of the arm and forearm, respectively. (a) Superficial muscles of the anterior thorax, shoulder, and arm, anterior view. (See *A Brief Atlas of the Human Body,* Figures 35 and 36.) **(b)** The triceps brachii muscle of the postorior arm, shown in relation to the deep scapular muscles. The deltoid muscle of the shoulder has been removed.

*Rotator cuff muscles

MUSCLE GALLERY

TABLE 10.9 Muscles Crossing the Shoulder Joint: Movements of the Arm (Humerus) (Figure 10.14) *(continued)*

MUSCLE	DESCRIPTION	ORIGIN (O) AND INSERTION (I)	ACTION	NERVE SUPPLY
Subscapularis (sub-scap″u-lar′is) (*sub* = under; *scapular* = scapula)	Forms part of posterior wall of axilla; tendon of insertion passes in front of shoulder joint; a rotator cuff muscle	O—subscapular fossa of scapula I—lesser tubercle of humerus	Chief medial rotator of humerus; assisted by pectoralis major; helps to hold head of humerus in glenoid cavity, thereby stabilizing shoulder joint	Subscapular nerves (C_5–C_7)
Supraspinatus (soo″prah-spi-nah′tus) (*supra* = above, over; *spin* = spine)	Named for its location on posterior aspect of scapula; deep to trapezius; a rotator cuff muscle	O—supraspinous fossa of scapula I—superior part of greater tubercle of humerus	Initiates abduction; stabilizes shoulder joint; helps to prevent downward dislocation of humerus, as when carrying a heavy suitcase	Suprascapular nerve
Infraspinatus (in″frah-spi-nah′tus) (*infra* = below)	Partially covered by deltoid and trapezius; named for its scapular location; a rotator cuff muscle	O—infraspinous fossa of scapula I—greater tubercle of humerus posterior to insertion of supraspinatus	Rotates humerus laterally; helps to hold head of humerus in glenoid cavity, stabilizing the shoulder joint	Suprascapular nerve
Teres minor (te′rēz) (*teres* = round; *minor* = lesser)	Small, elongated muscle; lies inferior to infraspinatus and may be inseparable from that muscle; a rotator cuff muscle	O—lateral border of dorsal scapular surface I—greater tubercle of humerus inferior to infraspinatus insertion	Same action(s) as infraspinatus muscle	Axillary nerve
Teres major	Thick, rounded muscle; located inferior to teres minor; helps to form posterior wall of axilla (along with latissimus dorsi and subscapularis)	O—posterior surface of scapula at inferior angle I—crest of lesser tubercle on anterior humerus; insertion tendon fused with that of latissimus dorsi	Extends, medially rotates, and adducts humerus; synergist of latissimus dorsi	Lower subscapular nerve (C_6 and C_7)
Coracobrachialis (kor″ah-ko-bra″ke-al′is) (*coraco* = coracoid; *brachi* = arm)	Small, cylindrical muscle	O—coracoid process of scapula I — medial surface of humerus shaft	Flexion and adduction of the humerus; synergist of pectoralis major	Musculocutaneous nerve (C_5–C_7)

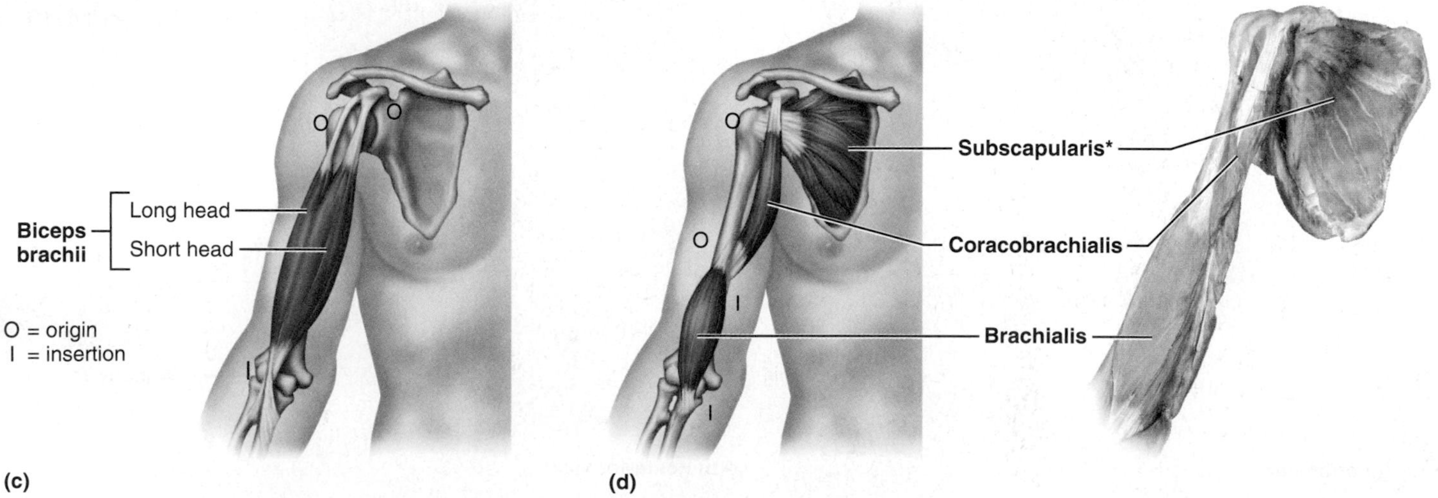

(c) (d)

Figure 10.14 *(continued)* **Muscles crossing the shoulder and elbow joint, causing movements of the arm and forearm, respectively. (c)** The isolated biceps brachii muscle of the anterior arm. **(d)** The brachialis muscle and the coracobrachialis and subscapularis muscles shown in isolation in the diagram on the left, and in a dissection on the right.

*Rotator cuff muscles

MUSCLE GALLERY

TABLE 10.10 Muscles Crossing the Elbow Joint: Flexion and Extension of the Forearm (Figure 10.14)

Muscles fleshing out the arm cross the elbow joint to insert on the forearm bones. Since the elbow is a hinge joint, movements promoted by these arm muscles are limited almost entirely to flexion and extension of the forearm. Walls of fascia divide the arm into two muscle compartments—the *posterior extensors* and *anterior flexors.* The main forearm extensor is the bulky **triceps brachii** muscle, which forms nearly the entire musculature of the posterior compartment (Figure 10.14 a, b).

All anterior arm muscles flex the forearm (elbow). In order of decreasing strength, these are the **brachialis, biceps brachii**, and **brachioradialis** (Figure 10.14a–d). The brachialis and biceps insert (respectively) into the ulna and radius and contract simultaneously during flexion; they are the chief forearm flexors. The biceps brachii, a muscle that bulges when the forearm is flexed, is familiar to almost everyone. The brachialis, which lies deep to the biceps, is less known, but is equally important in flexing the elbow. Because the brachioradialis arises from the distal humerus and inserts on the distal forearm, it resides mainly in the forearm. Its force is exerted far from the fulcrum, so the brachioradialis is a weak forearm flexor. The biceps muscle also supinates the forearm and is ineffective in flexing the elbow when the forearm *must* stay pronated. (This is why doing chin-ups with palms facing anteriorly is harder than with palms facing posteriorly.)

The actions of the muscles described here are summarized in Table 10.12 (Part II).

MUSCLE	DESCRIPTION	ORIGIN (O) AND INSERTION (I)	ACTION	NERVE SUPPLY
POSTERIOR MUSCLES				
Triceps brachii (tri'seps bra'ke-i) (*triceps* = three heads; *brachi* = arm)	Large fleshy muscle; the only muscle of posterior compartment of arm; three-headed origin; long and lateral heads lie superficial to medial head	O—long head: infraglenoid tubercle of scapula; lateral head: posterior shaft of humerus; medial head: posterior humeral shaft distal to radial groove I—by common tendon into olecranon process of ulna	Powerful forearm extensor (prime mover, particularly medial head); antagonist of forearm flexors; long and lateral heads mainly active in extension against resistance; long head tendon may help stabilize shoulder joint and assist in arm adduction	Radial nerve (C_6–C_8)
Anconeus (an-ko'ne-us) (*ancon* = elbow) (see Figure 10.16)	Short triangular muscle; partially blended with distal end of triceps on posterior humerus	O—lateral epicondyle of humerus I—lateral aspect of olecranon process of ulna	Controls ulnar abduction during forearm pronation; synergist of triceps brachii in elbow extension	Radial nerve
ANTERIOR MUSCLES				
Biceps brachii (bi'seps) (*biceps* = two heads)	Two-headed fusiform muscle; bellies unite as insertion point is approached; tendon of long head helps stabilize shoulder joint	O—short head: coracoid process; long head: supraglenoid tubercle and lip of glenoid cavity; tendon of long head runs within capsule and into intertubercular sulcus of humerus I—by common tendon into radial tuberosity	Flexes elbow joint and supinates forearm; these actions usually occur at same time (e.g., when you open a bottle of wine, it turns the corkscrew and pulls the cork); weak flexor of arm at shoulder	Musculocutaneous nerve (C_5 and C_6)
Brachialis (bra'ke-al-is)	Strong muscle that is immediately deep to biceps brachii on distal humerus	O—front of distal humerus; embraces insertion of deltoid muscle I—coronoid process of ulna and capsule of elbow joint	A major forearm flexor (lifts ulna as biceps lifts the radius)	Musculocutaneous nerve
Brachioradialis (bra"ke-o-ra"de-al'is) (*radi* = radius, ray) (also see Figure 10.15)	Superficial muscle of lateral forearm; forms lateral boundary of cubital fossa; extends from distal humerus to distal forearm	O—lateral supracondylar ridge at distal end of humerus I—base of styloid process of radius	Synergist in forearm flexion; acts to best advantage when forearm is partially flexed and semipronated; stabilizes the elbow during rapid flexion *and* extension	Radial nerve (an important exception: the radial nerve typically serves extensor muscles)

MUSCLE GALLERY

TABLE 10.11 Muscles of the Forearm: Movements of the Wrist, Hand, and Fingers (Figures 10.15 and 10.16)

The many muscles in the forearm perform several basic functions: some cause wrist movements, some move the fingers and thumb, and a few help pronate and supinate the forearms. In most cases, their fleshy portions contribute to the roundness of the proximal forearm and then they taper to long tendons distally to insert into the hand. At the wrist, these tendons are securely anchored by a bandlike thickening of deep fascia called **flexor** and **extensor retinacula** ("retainers") (Figure 10.15a). These "wrist bands" keep the tendons from jumping outward when tensed. Crowded together in the wrist and palm, the muscle tendons are surrounded by slippery tendon sheaths that minimize friction as they slide against one another.

Although many forearm muscles arise from the humerus (and thus cross both the elbow and wrist joints), their actions on the elbow are slight. Flexion and extension are the movements typically effected at both the wrist and finger joints. In addition, the wrist can be abducted and adducted by the forearm muscles.

The forearm muscles are subdivided by fascia into two main compartments (the *anterior flexors* and *posterior extensors*), each with superficial and deep muscle layers. Most flexors in the anterior compartment arise from a common tendon on the humerus and are innervated largely by the median nerve. Two anterior compartment muscles are not flexors but pronators, the **pronator teres** and **pronator quadratus** (Figure 10.15a–c). Pronation is one of the most important forearm movements.

Muscles of the posterior compartment extend the wrist and fingers. One exception is the **supinator** muscle, which assists the biceps brachii muscle of the arm in supinating the forearm (Figure 10.15b, c and 10.16b). (Also residing in the posterior compartment is the brachioradialis muscle, the weak elbow flexor considered in Table 10.10.) Most muscles of the posterior compartment arise from a common tendon on the humerus. All posterior forearm muscles are supplied by the radial nerve.

As described above, most muscles that move the hand are located in the forearm and "operate" the fingers via their long tendons, like operating a puppet by strings. This design makes the hand less bulky and enables it to perform finer movements. The hand movements promoted by the forearm muscles are assisted by the small *intrinsic* muscles of the hand, which control the most delicate and precise finger movements (see Table 10.13). The actions of the forearm muscles are summarized in Table 10.12 (Parts II and III).

MUSCLE	DESCRIPTION	ORIGIN (O) AND INSERTION (I)	ACTION	NERVE SUPPLY
PART I: ANTERIOR MUSCLES (FIGURE 10.15)	These eight muscles of the anterior fascial compartment are listed from the lateral to the medial aspect. Most arise from a common flexor tendon attached to the medial epicondyle of the humerus and have additional origins as well. Most of the tendons of insertion of these flexors are held in place at the wrist by a thickening of deep fascia called the *flexor retinaculum.*			
SUPERFICIAL MUSCLES				
Pronator teres (pro-na'tor te'rēz) (*pronation* = turning palm posteriorly, or down; *teres* = round)	Two-headed muscle; seen in superficial view between proximal margins of brachioradialis and flexor carpi radialis; forms medial boundary of cubital fossa	O—medial epicondyle of humerus; coronoid process of ulna I—by common tendon into lateral radius, midshaft	Pronates forearm; weak flexor of elbow	Median nerve
Flexor carpi radialis (flek'sor kar'pe ra"de-al'is) (*flex* = decrease angle between two bones; *carpi* = wrist; *radi* = radius)	Runs diagonally across forearm; midway, its fleshy belly is replaced by a flat tendon that becomes cordlike at wrist	O—medial epicondyle of humerus I—base of second and third metacarpals; insertion tendon easily seen and provides guide to position of radial artery (used for pulse taking) at wrist	Powerful flexor of wrist; abducts hand; weak synergist of elbow flexion	Median nerve
Palmaris longus (pahl-ma'ris lon'gus) (*palma* = palm; *longus* = long)	Small fleshy muscle with a long insertion tendon; often absent; may be used as guide to find median nerve that lies lateral to it at wrist	O—medial epicondyle of humerus I—palmar aponeurosis; (fascia of palm)	Tenses skin and fascia of palm during hand movements; weak wrist flexor; weak synergist for elbow flexion	Median nerve
Flexor carpi ulnaris (ul-na'ris) (*ulnar* = ulna)	Most medial muscle of this group; two-headed; ulnar nerve lies lateral to its tendon	O—medial epicondyle of humerus; olecranon process and posterior surface of ulna I—pisiform and hamate bones and base of fifth metacarpal	Powerful flexor of wrist; also adducts hand in concert with extensor carpi ulnaris (posterior muscle); stabilizes wrist during finger extension	Ulnar nerve (C_7 and C_8)

MUSCLE GALLERY

TABLE 10.11 *(continued)*

MUSCLE	DESCRIPTION	ORIGIN (O) AND INSERTION (I)	ACTION	NERVE SUPPLY
Flexor digitorum superficialis (dĭ″ji-tor′um soo″per-fish″e-al′is) (*digit* = finger, toe; *superficial* = close to surface)	Two-headed muscle; more deeply placed (therefore, actually forms an intermediate layer); overlain by muscles above but visible at distal end of forearm	O—medial epicondyle of humerus, coronoid process of ulna; shaft of radius I—by four tendons into middle phalanges of fingers 2–5	Flexes wrist and middle phalanges of fingers 2–5; the important finger flexor when speed and flexion against resistance are required	Median nerve (C_7, C_8, and T_1)
DEEP MUSCLES				
Flexor pollicis longus (pah′lĭ-kis) (*pollix* = thumb)	Partly covered by flexor digitorum superficialis; parallels flexor digitorum profundus laterally	O—anterior surface of radius and interosseous membrane I—distal phalanx of thumb	Flexes distal phalanx of thumb	Branch of median nerve (C_8,T_1)
Flexor digitorum profundus (pro-fun′dus) (*profund* = deep)	Extensive origin; overlain entirely by flexor digitorum superficialis	O—coronoid process, anteromedial surface of ulna, and interosseous membrane I—by four tendons into distal phalanges of fingers 2–5	Flexes distal interphalangeal joints; slow-acting flexor of any or all fingers; assists in flexing wrist	Medial half by ulnar nerve; lateral half by median nerve

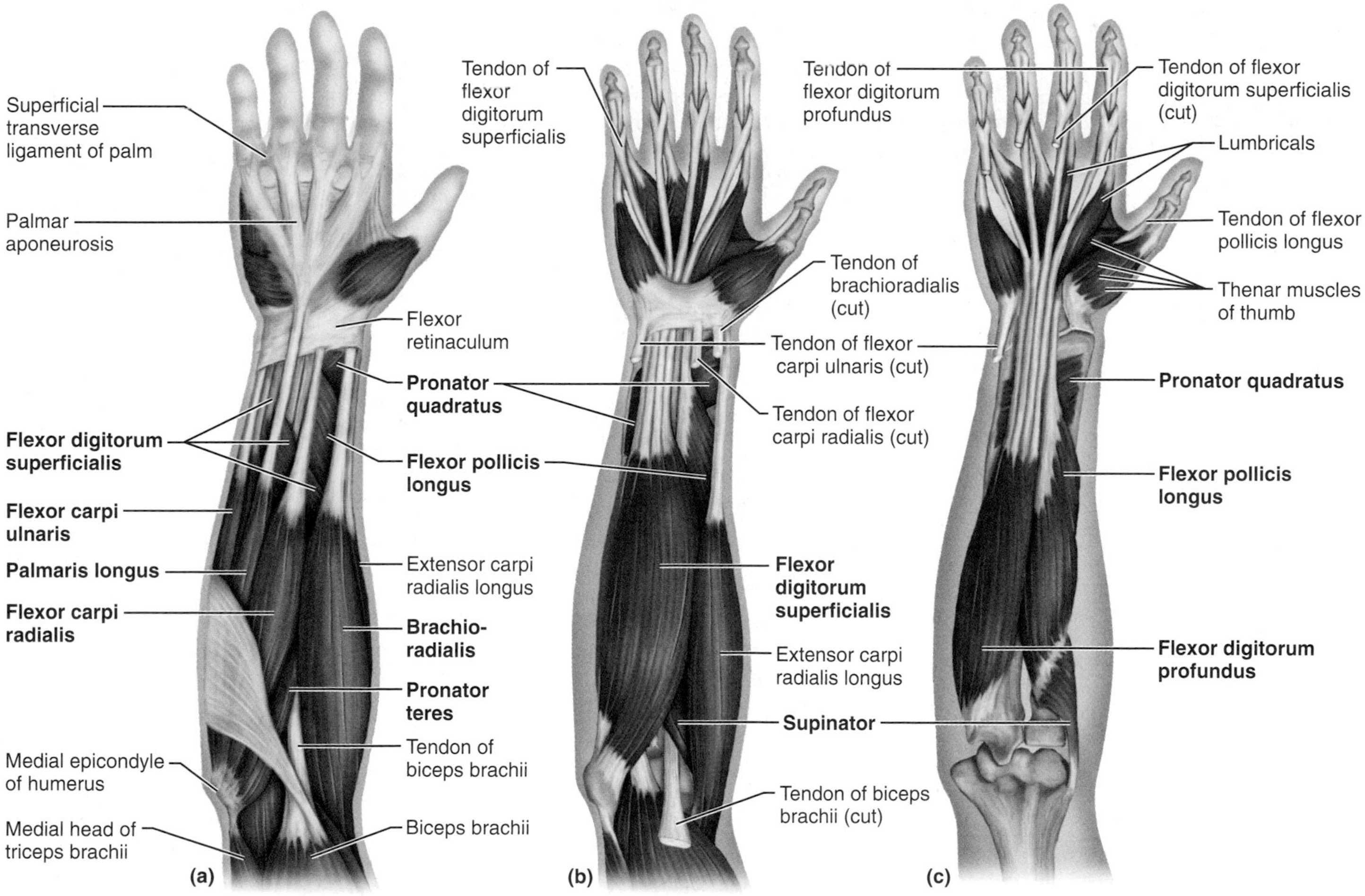

Figure 10.15 Muscles of the anterior fascial compartment of the forearm acting on the right wrist and fingers. (a) Superficial view. **(b)** The brachioradialis, flexors carpi radialis and ulnaris, and palmaris longus muscles have been removed to reveal the flexor digitorum superficialis. **(c)** Deep muscles of the anterior compartment. The lumbricals and thenar muscles (intrinsic hand muscles) are also illustrated. (See *A Brief Atlas of the Human Body*, Figure 37.)

TABLE 10.11 Muscles of the Forearm: Movements of the Wrist, Hand, and Fingers (Figures 10.15 and 10.16) *(continued)*

MUSCLE	DESCRIPTION	ORIGIN (O) AND INSERTION (I)	ACTION	NERVE SUPPLY
Pronator quadratus (kwod-ra′tus) (*quad* = square, four-sided)	Deepest muscle of distal forearm; passes downward and laterally; only muscle that arises solely from ulna and inserts solely into radius	O—distal portion of anterior ulnar shaft I—distal surface of anterior radius	Prime mover of forearm pronation; acts with pronator teres; also helps hold ulna and radius together	Median nerve (C_8 and T_1)
PART II: POSTERIOR MUSCLES (FIGURE 10.16)	These muscles of the posterior fascial compartment are listed from the lateral to the medial aspect. They are all innervated by the radial nerve or its branches. More than half of the posterior compartment muscles arise from a common extensor origin tendon attached to the posterior surface of the lateral epicondyle of the humerus and adjacent fascia. The extensor tendons are held in place at the posterior aspect of the wrist by the *extensor retinaculum,* which prevents "bowstringing" of these tendons when the wrist is hyperextended. The *extensor* muscles of the fingers end in a broad hood over the dorsal side of the digits, the extensor expansion.			
SUPERFICIAL MUSCLES				
Brachioradialis (see Table 10.10)	See Table 10.10	See Table 10.10	See Table 10.10	See Table 10.10
Extensor carpi radialis longus (ek-sten′sor) (*extend* = increase angle between two bones)	Parallels brachioradialis on lateral forearm, and may blend with it	O—lateral supracondylar ridge of humerus I—base of second metacarpal	Extends wrist in conjunction with the extensor carpi ulnaris and abducts wrist in conjunction with the flexor carpi radialis	Radial nerve (C_6 and C_7)
Extensor carpi radialis brevis (brĕ′vis) (*brevis* = short)	Somewhat shorter than extensor carpi radialis longus and lies deep to it	O—lateral epicondyle of humerus I—base of third metacarpal	Extends and abducts wrist; acts synergistically with extensor carpi radialis longus to steady wrist during finger flexion	Deep branch of radial nerve
Extensor digitorum	Lies medial to extensor carpi radialis brevis; a detached portion of this muscle, called *extensor digiti minimi,* extends little finger	O—lateral epicondyle of humerus I—by four tendons into extensor expansions and distal phalanges of fingers 2–5	Prime mover of finger extension; extends wrist; can abduct (flare) fingers	Posterior interosseous nerve, a branch of radial nerve (C_5 and C_6)
Extensor carpi ulnaris	Most medial of superficial posterior muscles; long, slender muscle	O—lateral epicondyle of humerus and posterior border of ulna I—base of fifth metacarpal	Extends wrist in conjunction with the extensor carpi radialis and adducts wrist in conjunction with flexor carpi ulnaris	Posterior interosseous nerve
DEEP MUSCLES				
Supinator (soo″pĭ-na′tor) (*supination* = turning palm anteriorly or upward)	Deep muscle at posterior aspect of elbow; largely concealed by superficial muscles	O—lateral epicondyle of humerus; proximal ulna I—proximal end of radius	Assists biceps brachii to forcibly supinate forearm; works alone in slow supination; antagonist of pronator muscles	Posterior interosseous nerve
Abductor pollicis longus (ab-duk′tor) (*abduct* = movement away from median plane)	Lateral and parallel to extensor pollicis longus; just distal to supinator	O—posterior surface of radius and ulna; interosseous membrane I—base of first metacarpal and trapezium	Abducts and extends thumb	Posterior interosseous nerve
Extensor pollicis brevis and longus	Deep muscle pair with a common origin and action; overlain by extensor carpi ulnaris	O—dorsal shaft of radius and ulna; interosseous membrane I—base of proximal (brevis) and distal (longus) phalanx of thumb	Extends thumb	Posterior interosseous nerve

MUSCLE GALLERY

TABLE 10.11 *(continued)*

MUSCLE	DESCRIPTION	ORIGIN (O) AND INSERTION (I)	ACTION	NERVE SUPPLY
Extensor indicis (in′dĭ-kis) (*indicis* = index finger)	Tiny muscle arising close to wrist	O—posterior surface of distal ulna; interosseous membrane I—extensor expansion of index finger; joins tendon of extensor digitorum	Extends index finger and assists in wrist extension	Posterior interosseous nerve

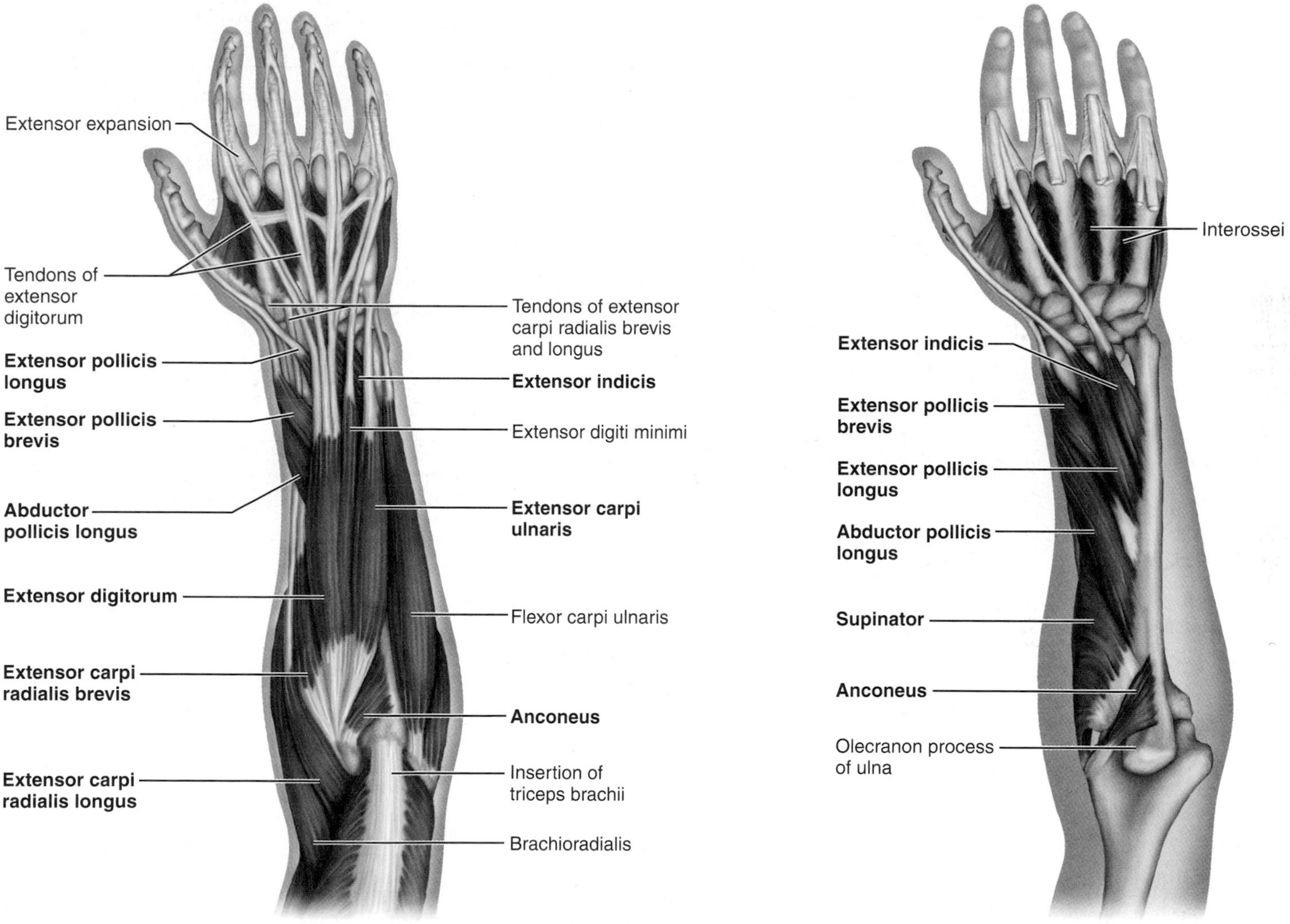

Figure 10.16 Muscles of the posterior fascial compartment of the right forearm acting on the wrist and fingers. (a) Superficial muscles, posterior view. (See *A Brief Atlas of the Human Body*, Figure 37b.) **(b)** Deep posterior muscles, superficial muscles removed. The interossei, the deepest layer of intrinsic hand muscles, are also illustrated.

MUSCLE GALLERY

TABLE 10.12 Summary of Actions of Muscles Acting on the Arm, Forearm, and Hand (Figure 10.17)

Part I: Muscles Acting on the Arm (Humerus) (PM = prime mover)	ACTIONS AT THE SHOULDER					
	Flexion	Extension	Abduction	Adduction	Medial Rotation	Lateral Rotation
Pectoralis	× (PM)			× (PM)	×	
Latissimus dorsi		× (PM)		× (PM)	×	
Teres major		×		×	×	
Deltoid	× (PM) (anterior fibers)	× (PM) (posterior fibers)	× (PM)		× (anterior fibers)	× (posterior fibers)
Subscapularis					× (PM)	
Supraspinatus			×			
Infraspinatus						× (PM)
Teres minor				× (weak)		× (PM)
Coracobrachialis	×			×		
Biceps brachii	× (weak)					
Triceps brachii				×		

Part II: Muscles Acting on the Forearm	ACTIONS ON THE FOREARM			
	Elbow Flexion	Elbow Extension	Pronation	Supination
Biceps brachii	× (PM)			×
Brachialis	× (PM)			
Triceps brachii		× (PM)		
Anconeus		×		
Pronator teres	× (weak)		×	
Pronator quadratus			× (PM)	
Supinator				×
Brachioradialis	×			

Part III: Muscles Acting on the Wrist and Fingers	ACTIONS ON THE WRIST				ACTIONS ON THE FINGERS	
	Flexion	Extension	Abduction	Adduction	Flexion	Extension
Anterior Compartment						
Flexor carpi radialis	× (PM)		×			
Palmaris longus	× (weak)					
Flexor carpi ulnaris	× (PM)			×		
Flexor digitorum superficialis	× (PM)				×	
Flexor pollicis longus					× (thumb)	
Flexor digitorum profundus	×				×	
Posterior Compartment						
Extensor carpi radialis longus and brevis		×	×			
Extensor digitorum		× (PM)				× (and abducts)
Extensor carpi ulnaris		×		×		
Abductor pollicis longus			×		(abducts thumb)	
Extensor pollicis longus and brevis						× (thumb)
Extensor indicis		× (weak)				× (index finger)

MUSCLE GALLERY

TABLE 10.12 *(continued)*

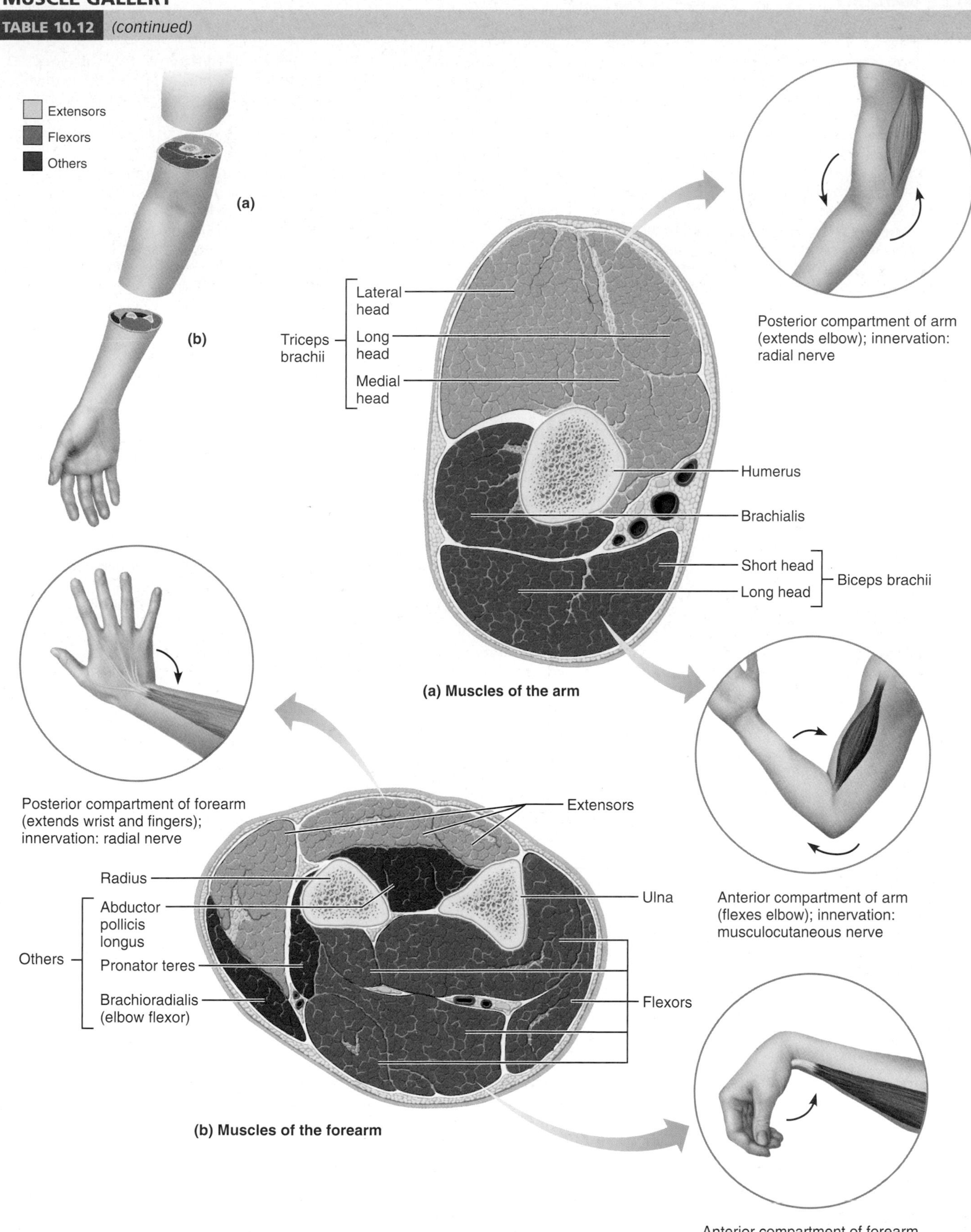

Figure 10.17 **Summary of actions of muscles of the arm and forearm.**

MUSCLE GALLERY

TABLE 10.13 Intrinsic Muscles of the Hand: Fine Movements of the Fingers (Figure 10.18)

In this table we consider the small muscles that lie entirely in the hand. All are in the palm, none on the hand's dorsal side. All move the metacarpals and fingers. Small, weak muscles, they mostly control precise movements (such as threading a needle), leaving the powerful movements of the fingers ("power grip") to the forearm muscles.

The intrinsic muscles include the main abductors and adductors of the fingers, as well as muscles that produce the movement of opposition—moving the thumb toward the little finger—that enables you to grip objects in the palm (the handle of a hammer, for example). Many palm muscles are specialized to move the thumb, and surprisingly many move the little finger. Thumb movements are defined differently from movements of other fingers because the thumb lies at a right angle to the rest of the hand. The thumb flexes by bending medially along the palm, not by bending anteriorly, as do the other fingers. (To demonstrate this difference, start with your hand in the anatomical position or this will not be clear!) The thumb extends by pointing laterally (as in hitchhiking), not posteriorly, as do the other fingers. To abduct the fingers is to splay them laterally, but to abduct the thumb is to point it anteriorly. Adduction of the thumb brings it back posteriorly.

The intrinsic muscles of the palm are divided into three groups, those in (1) the *thenar eminence* (ball of the thumb); (2) the *hypothenar eminence* (ball of the little finger); and (3) the midpalm. Thenar and hypothenar muscles are almost mirror images of each other, each containing a small flexor, an abductor, and an opponens muscle. The midpalmar muscles, called **lumbricals** and **interossei**, extend our fingers at the interphalangeal joints. The interossei are also the main finger abductors and adductors.

MUSCLE	DESCRIPTION	ORIGIN (O) AND INSERTION (I)	ACTION	NERVE SUPPLY
THENAR MUSCLES IN BALL OF THUMB (the'nar) (*thenar* = palm)				
Abductor pollicis brevis (*pollex* = thumb)	Lateral muscle of thenar group; superficial	O—flexor retinaculum and nearby carpals I—lateral base of thumb's proximal phalanx	Abducts thumb (at carpometacarpal joint)	Median nerve (C_8, T_1)
Flexor pollicis brevis	Medial and deep muscle of thenar group	O—flexor retinaculum and nearby carpals I—lateral side of base of proximal phalanx of thumb	Flexes thumb (at carpometacarpal and metacarpophalangeal joints)	Median (or occasionally ulnar) nerve (C_8, T_1)
Opponens pollicis (o-pōn'enz) (*opponens* = opposition)	Deep to abductor pollicis brevis, on metacarpal 1	O—flexor retinaculum and trapezium I—whole anterior side of metacarpal 1	Opposition: moves thumb to touch tip of little finger	Median (or occasionally ulnar) nerve
Adductor pollicis	Fan-shaped with horizontal fibers; distal to other thenar muscles; oblique and transverse heads	O—capitate bone and bases of metacarpals 2–4; front of metacarpal 3 I—medial side of base of proximal phalanx of thumb	Adducts and helps to oppose thumb	Ulnar nerve (C_8, T_1)
HYPOTHENAR MUSCLES IN BALL OF LITTLE FINGER				
Abductor digiti minimi (dĭ'jĭ-ti min'ĭ-mi) (*digiti minimi* = little finger)	Medial muscle of hypothenar group; superficial	O—pisiform bone I—medial side of proximal phalanx of little finger	Abducts little finger at metacarpophalangeal joint	Ulnar nerve
Flexor digiti minimi brevis	Lateral deep muscle of hypothenar group	O—hamate bone and flexor retinaculum I—same as abductor digiti minimi	Flexes little finger at metacarpophalangeal joint	Ulnar nerve
Opponens digiti minimi	Deep to abductor digiti minimi	O—same as flexor digiti minimi brevis I—most of length of medial side of metacarpal 5	Helps in opposition: brings metacarpal 5 toward thumb to cup the hand	Ulnar nerve

MUSCLE GALLERY

TABLE 10.13 *(continued)*

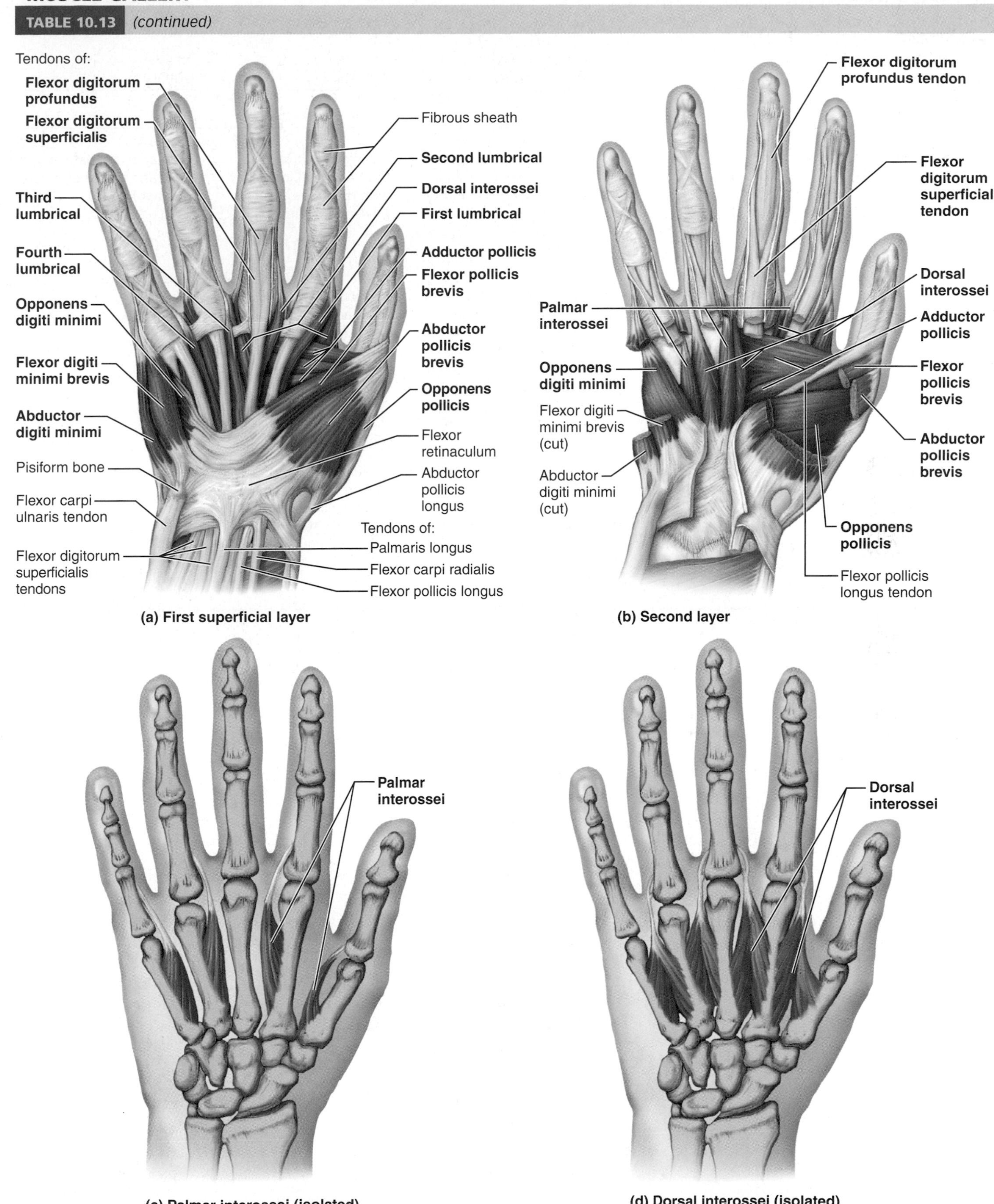

Tendons of:
Flexor digitorum profundus
Flexor digitorum superficialis
Third lumbrical
Fourth lumbrical
Opponens digiti minimi
Flexor digiti minimi brevis
Abductor digiti minimi
Pisiform bone
Flexor carpi ulnaris tendon
Flexor digitorum superficialis tendons
Fibrous sheath
Second lumbrical
Dorsal interossei
First lumbrical
Adductor pollicis
Flexor pollicis brevis
Abductor pollicis brevis
Opponens pollicis
Flexor retinaculum
Abductor pollicis longus
Tendons of:
Palmaris longus
Flexor carpi radialis
Flexor pollicis longus
(a) First superficial layer
Flexor digitorum profundus tendon
Flexor digitorum superficialis tendon
Palmar interossei
Opponens digiti minimi
Flexor digiti minimi brevis (cut)
Abductor digiti minimi (cut)
Dorsal interossei
Adductor pollicis
Flexor pollicis brevis
Abductor pollicis brevis
Opponens pollicis
Flexor pollicis longus tendon
(b) Second layer
Palmar interossei
(c) Palmar interossei (isolated)
Dorsal interossei
(d) Dorsal interossei (isolated)

Figure 10.18 Hand muscles, ventral views of right hand.

MUSCLE GALLERY

TABLE 10.13 Intrinsic Muscles of the Hand: Fine Movements of the Fingers (Figure 10.18) *(continued)*

MUSCLE	DESCRIPTION	ORIGIN (O) AND INSERTION (I)	ACTION	NERVE SUPPLY
MIDPALMAR MUSCLES				
Lumbricals (lum′brĭ-klz) (*lumbric* = earthworm)	Four worm-shaped muscles in palm, one to each finger (except thumb); unusual because they originate from the tendons of another muscle	O—lateral side of each tendon of flexor digitorum profundus in palm I—lateral edge of extensor expansion on proximal phalanx of fingers 2–5	Flex fingers at metacarpophalangeal joints but extend fingers at interphalangeal joints	Median nerve (lateral two) and ulnar nerve (medial two)
Palmar interossei (in″ter-os′e-i) (*interossei* = between bones)	Four long, cone-shaped muscles in the spaces between the metacarpals; lie ventral to the dorsal interossei	O—the side of each metacarpal that faces the midaxis of the hand (metacarpal 3) but absent from metacarpal 3 I—extensor expansion on first phalanx of each finger (except finger 3), on side facing midaxis of hand	Adductors of fingers: pull fingers in toward third digit; act with lumbricals to extend fingers at interphalangeal joints and flex them at metacarpophalangeal joints	Ulnar nerve
Dorsal interossei	Four bipennate muscles filling spaces between the metacarpals; deepest palm muscles, also visible on dorsal side of hand (Figure 10.16b)	O—sides of metacarpals I—extensor expansion over proximal phalanx of fingers 2–4 on side opposite midaxis of hand (finger 3), but on *both* sides of finger 3	Abduct (diverge) fingers; extend fingers at interphalangeal joints and flex them at metacarpophalangeal joints	Ulnar nerve

MUSCLE GALLERY

TABLE 10.14 Muscles Crossing the Hip and Knee Joints: Movements of the Thigh and Leg (Figures 10.19 and 10.20)

The muscles fleshing out the thigh are difficult to segregate into groups on the basis of action. Some thigh muscles act only at the hip joint, others only at the knee, while still others act at both joints. However, *most anterior* muscles of the hip and thigh flex the femur at the hip and extend the leg at the knee— producing the foreswing phase of walking. The *posterior* muscles of the hip and thigh, by contrast, mostly extend the thigh and flex the leg—the backswing phase of walking. A third group of muscles in this region, the *medial*, or *adductor*, muscles, all adduct the thigh; they have no effect on the leg. In the thigh, the anterior, posterior, and adductor muscles are separated by walls of fascia into *anterior*, *posterior*, and *medial compartments* (see Figure 10.25a). The deep fascia of the thigh, the *fascia lata*, surrounds and encloses all three groups of muscles like a support stocking.

Movements of the thigh (occurring at the hip joint) are accomplished largely by muscles anchored to the pelvic girdle. Like the shoulder joint, the hip joint is a ball-and-socket joint permitting flexion, extension, abduction, adduction, circumduction, and rotation. Muscles effecting these movements of the thigh are among the most powerful muscles of the body.

For the most part, the thigh *flexors* pass in front of the hip joint. The most important of these are the **iliopsoas, tensor fasciae latae,** and **rectus femoris** (Figure 10.19a). They are assisted in this action by the **adductor muscles** of the medial thigh and the straplike **sartorius**. The prime mover of thigh flexion is the iliopsoas.

Thigh *extension* is effected primarily by the massive **hamstring muscles** of the posterior thigh (Figure 10.20a), and, during forceful extension, the **gluteus maximus** of the buttock is activated. Buttock muscles that lie lateral to the hip joint (**gluteus medius** and **minimus**) *abduct* the thigh (Figure 10.20c). Thigh adduction is the role of the adductor muscles of the medial thigh. Abduction and adduction of the thighs are extremely important during walking to shift the trunk from side to side so that the body's weight is balanced over the limb that is on the ground. Many different muscles bring about medial and lateral rotation of the thigh.

At the knee joint, flexion and extension are the main movements. The sole knee *extensor* is the **quadriceps femoris** muscle of the anterior thigh, the most powerful muscle in the body (Figure 10.19a). The quadriceps is antagonized by the hamstrings of the posterior compartment, which are the prime movers of knee flexion.

The actions of these muscles are further summarized in Table 10.17 (Part I).

MUSCLE	DESCRIPTION	ORIGIN (O) AND INSERTION (I)	ACTION	NERVE SUPPLY
PART I: ANTERIOR AND MEDIAL MUSCLES (FIGURE 10.19)				
ORIGIN ON THE PELVIS OR SPINE				
Iliopsoas (il″e-o-so′us)	Iliopsoas is a composite of two closely related muscles (iliacus and psoas major) whose fibers pass under the inguinal ligament (see Figure 10.11) to insert via a common tendon on the femur.			
▪ **Iliacus** (il-e-ak′us) (*iliac* = ilium)	Large, fan-shaped, more lateral muscle	O—iliac fossa and crest, ala of sacrum I—lesser trochanter of femur via iliopsoas tendon	Iliopsoas is the prime mover for flexing thigh or for flexing trunk on thigh during a bow	Femoral nerve (L_2 and L_3)
▪ **Psoas major** (so′us) (*psoa* = loin muscle; *major* = larger)	Longer, thicker, more medial muscle of the pair (butchers refer to this muscle as the tenderloin)	O—by fleshy slips from transverse processes, bodies, and discs of lumbar vertebrae and T_{12} I—lesser trochanter of femur via iliopsoas tendon	As above; also effects lateral flexion of vertebral column; important postural muscle	Ventral rami (L_1–L_3)
Sartorius (sar-tor′e-us) (*sartor* = tailor)	Straplike superficial muscle running obliquely across anterior surface of thigh to knee; longest muscle in body; crosses both hip and knee joints	O—anterior superior iliac spine I—winds around medial aspect of knee and inserts into medial aspect of proximal tibia	Flexes, abducts, and laterally rotates thigh; flexes knee; produces the cross-legged position	Femoral nerve

TABLE 10.14 **Muscles Crossing the Hip and Knee Joints: Movements of the Thigh and Leg (Figures 10.19 and 10.20)** *(continued)*

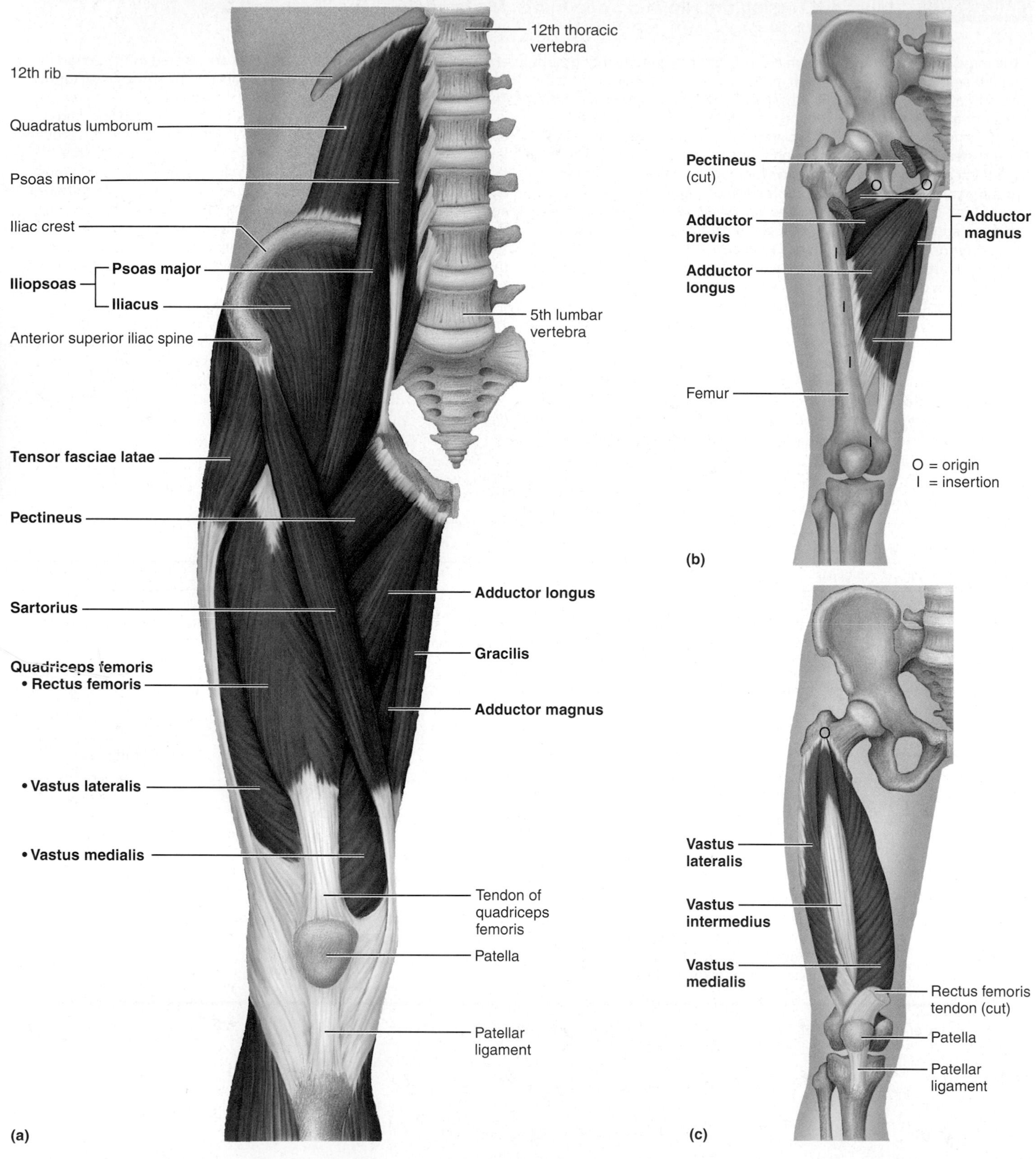

Figure 10.19 Anterior and medial muscles promoting movements of the thigh and leg. (See *A Brief Atlas of the Human Body,* Figure 40.) **(a)** Anterior view of the deep muscles of the pelvis and superficial muscles of the right thigh. **(b)** Adductor muscles of the medial compartment of the thigh, isolated. **(c)** Vastus muscles of the quadriceps group, isolated.

MUSCLE GALLERY

TABLE 10.14 *(continued)*

MUSCLES OF THE MEDIAL COMPARTMENT OF THE THIGH

Adductors (ah-duk'torz)	Large muscle mass consisting of three muscles (magnus, longus, and brevis) forming medial aspect of thigh; arise from inferior part of pelvis and insert at various levels on femur. All are used in movements that press thighs together, as when astride a horse; important in pelvic tilting movements that occur during walking and in fixing the hip when the knee is flexed and the foot is off the ground. Entire group innervated by obturator nerve. Strain or stretching of this muscle group is called a "pulled groin."			
▪ **Adductor magnus** (mag'nus) (*adduct* = move toward midline; *magnus* = large)	A triangular muscle with a broad insertion; is a composite muscle that is part adductor and part hamstring in action	O—ischial and pubic rami and ischial tuberosity I—linea aspera and adductor tubercle of femur	Anterior part adducts and medially rotates and flexes thigh; posterior part is a synergist of hamstrings in thigh extension	Obturator nerve and sciatic nerve (L_2–L_4)
▪ **Adductor longus** (*longus* = long)	Overlies middle aspect of adductor magnus; most anterior of adductor muscles	O—pubis near pubic symphysis I—linea aspera	Adducts, flexes, and medially rotates thigh	Obturator nerve (L_2–L_4)
▪ **Adductor brevis** (*brevis* = short)	In contact with obturator externus muscle; largely concealed by adductor longus and pectineus	O—body and inferior ramus of pubis I—linea aspera above adductor longus	Adducts and medially rotates thigh	Obturator nerve
Pectineus (pek-tin'e-us) (*pecten* = comb)	Short, flat muscle; overlies adductor brevis on proximal thigh; abuts adductor longus medially	O—pubis (and superior ramus) I—from lesser trochanter inferior to the linea aspera on posterior aspect of femur	Adducts, flexes, and medially rotates thigh	Femoral and sometimes obturator nerve
Gracilis (grah-sĭ'lis) (*gracilis* = slender)	Long, thin, superficial muscle of medial thigh	O—inferior ramus and body of pubis and adjacent ischial ramus I—medial surface of tibia just inferior to its medial condyle	Adducts thigh, flexes and medially rotates leg, especially during walking	Obturator nerve

MUSCLES OF THE ANTERIOR COMPARTMENT OF THE THIGH

Quadriceps femoris (kwod'rĭ-seps fem'o-ris)	Arises from four separate heads (*quadriceps* = four heads) that form the flesh of front and sides of thigh; these heads (rectus femoris, and lateral, medial, and intermediate vasti muscles) have a common insertion tendon, the *quadriceps tendon*, which inserts into the patella and then via the *patellar ligament* into tibial tuberosity. The quadriceps is a powerful knee extensor used in climbing, jumping, running, and rising from seated position. The group is innervated by femoral nerve; the tone of quadriceps plays important role in strengthening the knee joint.			
▪ **Rectus femoris** (rek'tus) (*rectus* = straight; *femoris* = femur)	Superficial muscle of anterior thigh; runs straight down thigh; longest head and only muscle of group to cross hip joint	O—anterior inferior iliac spine and superior margin of acetabulum I—patella and tibial tuberosity via patellar ligament	Extends knee and flexes thigh at hip	Femoral nerve (L_2–L_4)
▪ **Vastus lateralis** (vas'tus lat"er-a'lis) (*vastus* = large; *lateralis* = lateral)	Largest head of the group, forms lateral aspect of thigh; a common intramuscular injection site, particularly in infants (who have poorly developed buttock and arm muscles)	O—greater trochanter, intertrochanteric line, linea aspera I—as for rectus femoris	Extends and stabilizes knee	Femoral nerve
▪ **Vastus medialis** (me"de-a'lis) (*medialis* = medial)	Forms inferomedial aspect of thigh	O—linea aspera, intertrochanteric and medial supracondylar lines I—as for rectus femoris	Extends knee	Femoral nerve

10

MUSCLE GALLERY

TABLE 10.14 Muscles Crossing the Hip and Knee Joints: Movements of the Thigh and Leg (Figures 10.19 and 10.20) *(continued)*

MUSCLE	DESCRIPTION	ORIGIN (O) AND INSERTION (I)	ACTION	NERVE SUPPLY
▪ **Vastus intermedius** (in″ter-me′de-us) (*intermedius* = intermediate)	Obscured by rectus femoris; lies between vastus lateralis and vastus medialis on anterior thigh	O—anterior and lateral surfaces of proximal femur shaft I—as for rectus femoris	Extends knee	Femoral nerve
Tensor fasciae latae (ten′sor fă′she-e la′te) (*tensor* = to make tense; *fascia* = band; *lata* = wide)	Enclosed between fascia layers of anterolateral aspect of thigh; functionally associated with medial rotators and flexors of thigh	O—anterior aspect of iliac crest and anterior superior iliac spine I—iliotibial tract*	Steadies the knee and trunk on thigh by making iliotibial tract taut; flexes and abducts thigh; rotates thigh medially	Superior gluteal nerve (L_4 and L_5)
PART II: POSTERIOR MUSCLES (FIGURE 10.20)				
GLUTEAL MUSCLES—ORIGIN ON PELVIS				
Gluteus maximus (gloo′te-us mak′sĭ-mus) (*glutos* = buttock; *maximus* = largest)	Largest and most superficial of gluteus muscles; forms bulk of buttock mass; fibers are thick and coarse; important site of intramuscular injection (dorsal gluteal site); overlies large sciatic nerve; covers ischial tuberosity only when standing; when sitting, moves superiorly, leaving ischial tuberosity exposed in the subcutaneous position	O—dorsal ilium, sacrum, and coccyx I—gluteal tuberosity of femur; iliotibial tract	Major extensor of thigh; complex, powerful, and most effective when thigh is flexed and force is necessary, as in rising from a forward flexed position and in thrusting the thigh posteriorly in climbing stairs and running; generally inactive during standing and walking; laterally rotates and abducts thigh	Inferior gluteal nerve (L_5, S_1, and S_2)
Gluteus medius (me′de-us) (*medius* = middle)	Thick muscle largely covered by gluteus maximus; important site for intramuscular injections (ventral gluteal site); considered safer than dorsal gluteal site because there is less chance of injuring sciatic nerve	O—between anterior and posterior gluteal lines on lateral surface of ilium I—by short tendon into lateral aspect of greater trochanter of femur	Abducts and medially rotates thigh; steadies pelvis; its action is extremely important in walking; e.g., muscle of limb planted on ground tilts or holds pelvis in abduction so that pelvis on side of swinging limb does not sag; the foot of swinging limb can thus clear the ground	Superior gluteal nerve (L_5, S_1)
Gluteus minimus (mĭ′nĭ-mus) (minimus = smallest)	Smallest and deepest of gluteal muscles	O—between anterior and inferior gluteal lines on external surface of ilium I—anterior border of greater trochanter of femur	As for gluteus medius	Superior gluteal nerve (L_5, S_1)
LATERAL ROTATORS				
Piriformis (pir′ĭ-form-is) (*piri* = pear; *forma* = shape)	Pyramidal muscle located on posterior aspect of hip joint; inferior to gluteus minimus; issues from pelvis via greater sciatic notch	O—anterolateral surface of sacrum (opposite greater sciatic notch) I—superior border of greater trochanter of femur	Rotates extended thigh laterally; because inserted above head of femur, can also assist in abduction of thigh when hip is flexed; stabilizes hip joint	S_1 and S_2, L_5

*The iliotibial tract is a thickened lateral portion of the *fascia lata* (the fascia that ensheathes all the muscles of the thigh). It extends as a tendinous band from the iliac crest to the knee (see Figure 10.20a).

MUSCLE GALLERY

TABLE 10.14 *(continued)*

MUSCLE	DESCRIPTION	ORIGIN (O) AND INSERTION (I)	ACTION	NERVE SUPPLY

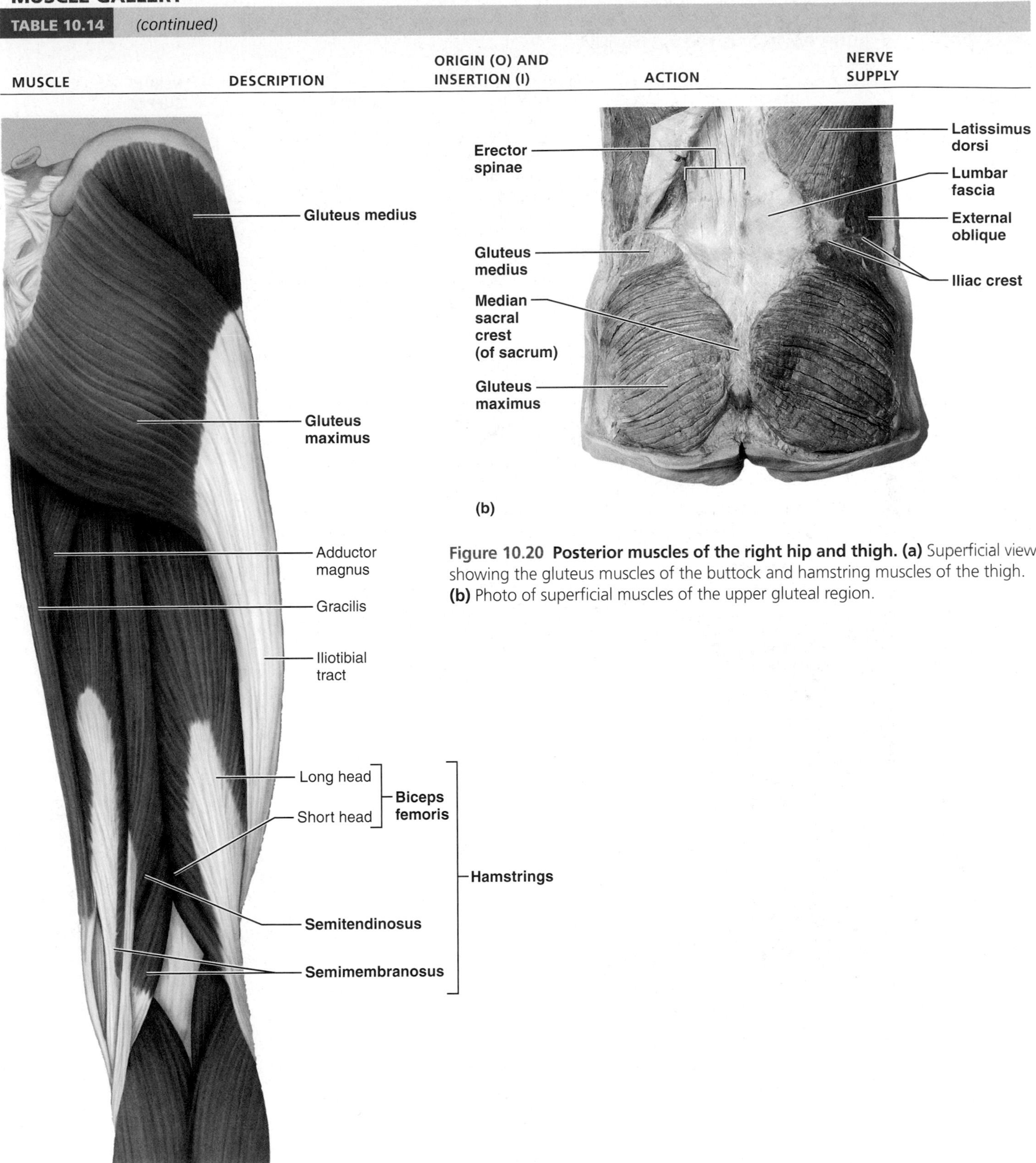

Figure 10.20 Posterior muscles of the right hip and thigh. (a) Superficial view showing the gluteus muscles of the buttock and hamstring muscles of the thigh. **(b)** Photo of superficial muscles of the upper gluteal region.

MUSCLE GALLERY

TABLE 10.14 Muscles Crossing the Hip and Knee Joints: Movements of the Thigh and Leg (Figures 10.19 and 10.20) *(continued)*

MUSCLE	DESCRIPTION	ORIGIN (O) AND INSERTION (I)	ACTION	NERVE SUPPLY
Obturator externus (ob″tu-ra′tor ek-ster′nus) (*obturator* = obturator foramen; *externus* = outside)	Flat, triangular muscle deep in superomedial aspect of thigh	O—outer surfaces of obturator membrane, pubis, and ischium, margins of obturator foramen I—by a tendon into trochanteric fossa of posterior femur	As for piriformis	Obturator nerve
Obturator internus (in-ter′nus) (*internus* = inside)	Surrounds obturator foramen within pelvis; leaves pelvis via lesser sciatic notch and turns acutely forward to insert in femur	O—inner surface of obturator membrane, greater sciatic notch, and margins of obturator foramen I—greater trochanter in front of piriformis	As for piriformis	L_5 and S_1
Gemellus (jĕ-mĕ′lis)—superior and inferior (*gemin* = twin, double; *superior* = above; *inferior* = below)	Two small muscles with common insertions and actions; considered extrapelvic portions of obturator internus	O—ischial spine (superior); ischial tuberosity (inferior) I—greater trochanter of femur	As for piriformis	L_5 and S_1
Quadratus femoris (*quad* = four-sided square)	Short, thick muscle; most inferior of lateral rotator muscles; extends laterally from pelvis	O—ischial tuberosity I—intertrochanteric crest of femur	Rotates thigh laterally and stabilizes hip joint	L_5 and S_1

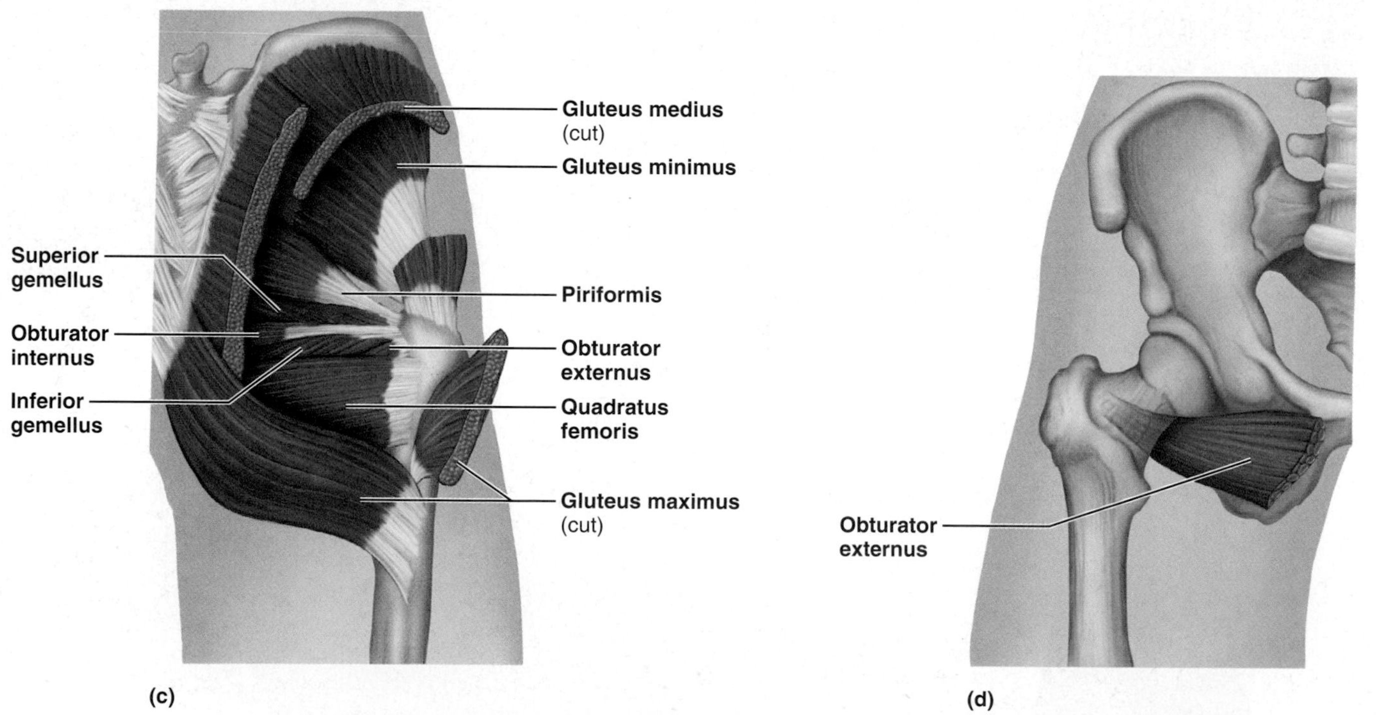

Figure 10.20 *(continued)* **Posterior muscles of the right hip and thigh. (c)** Deep muscles of the gluteal region, which act primarily to rotate the thigh laterally. **(d)** Anterior view of the isolated obturator externus muscle.

MUSCLE GALLERY

TABLE 10.14 *(continued)*

MUSCLE	DESCRIPTION	ORIGIN (O) AND INSERTION (I)	ACTION	NERVE SUPPLY
MUSCLES OF THE POSTERIOR COMPARTMENT OF THE THIGH				
Hamstrings	The hamstrings are fleshy muscles of the posterior thigh (biceps femoris, semitendinosus, and semimembranosus). They cross both the hip and knee joints and are prime movers of thigh extension and knee flexion. The group has a common origin site and is innervated by sciatic nerve (actually two nerves, the tibial and common fibular nerves wrapped in a common sheath). Ability of hamstrings to act on one of the two joints spanned depends on which joint is fixed; i.e., if knee is fixed (extended), they promote hip extension; if hip is extended, they promote knee flexion. However, when hamstrings are stretched, they tend to restrict full accomplishment of antagonistic movements; e.g., if knees are fully extended, it is difficult to flex the hip fully (and touch your toes), and when the thigh is fully flexed as in kicking a football, it is almost impossible to extend the knee fully at the same time (without considerable practice). Name of this muscle group comes from old butchers' practice of using their tendons to hang hams for smoking. "Pulled hamstrings" are common sports injuries in those who run very hard, e.g., football halfbacks.			
▪ **Biceps femoris** (*biceps* = two heads)	Most lateral muscle of the group; arises from two heads	O—ischial tuberosity (long head); linea aspera, lateral supracondylar line, and distal femur (short head) I—common tendon passes downward and laterally (forming lateral border of popliteal fossa) to insert into head of fibula and lateral condyle of tibia	Extends thigh and flexes knee; laterally rotates leg, especially when knee is flexed	Sciatic nerve—tibial nerve to long head, common fibular nerve to short head (L_5–S_2)
▪ **Semitendinosus** (sem″e-ten″dĭ-no′sus) (*semi* = half; *tendinosus* = tendon)	Lies medial to biceps femoris; although its name suggests that this muscle is largely tendinous, it is quite fleshy; its long slender tendon begins about two-thirds of the way down thigh	O—ischial tuberosity in common with long head of biceps femoris I—medial aspect of upper tibial shaft	Extends thigh and flexes knee; with semimembranosus, medially rotates leg	Sciatic nerve— tibial nerve portion (L_5–S_2)
▪ **Semimembranosus** (sem″e-mem″-brah-no′sus) (*membranosus* = membrane)	Deep to semitendinosus	O—ischial tuberosity I—medial condyle of tibia; via oblique popliteal ligament to lateral condyle of femur	Extends thigh and flexes knee; medially rotates leg	Sciatic nerve— tibial nerve portion (L_5–S_2)

MUSCLE GALLERY

TABLE 10.15 Muscles of the Leg: Movements of the Ankle and Toes (Figures 10.21 to 10.23)

The deep fascia of the leg is continuous with the fascia lata that ensheathes the thigh. Like a snug "knee sock" beneath the skin, the leg fascia binds the leg muscles tightly, helping to prevent excessive swelling of muscles during exercise and aiding venous return. Its inward extensions segregate the leg muscles into *anterior*, *lateral*, and *posterior compartments* (see Figure 10.25b), each with its own nerve and blood supply. Distally the leg fascia thickens to form the **flexor**, **extensor**, and **fibular** (or **peroneal**) **retinacula**, "ankle brackets" that hold the tendons in place where they run to the foot (Figure 10.21a, 10.22a).

The various muscles of the leg promote movements at the ankle joint (dorsiflexion and plantar flexion), at the intertarsal joints (inversion and eversion of the foot), and/or at the toes (flexion and extension). Muscles in the *anterior extensor compartment of the leg* are primarily toe extensors and ankle dorsiflexors. Although dorsiflexion is not a powerful movement, it is important in preventing the toes from dragging during walking. Lateral compartment muscles are the **fibular**, formerly *peroneal* (*peron* = fibula), **muscles** that plantar flex and evert the foot. Muscles of the *posterior flexor compartment* primarily plantar flex the foot and flex the toes (Figure 10.23b–d). Plantar flexion is the most powerful movement of the ankle (and foot) because it lifts the entire weight of our body. It is essential for standing on tiptoe and provides the forward thrust when walking and running. The **popliteus muscle**, which crosses the knee, has a unique function. It "unlocks" the extended knee in preparation for flexion (Figure 10.23b, f).

We consider the tiny intrinsic muscles of the sole of the foot (lumbricals, interossei, and others) in Table 10.16.

The actions of the muscles in this table are summarized in Table 10.17 (Part II).

MUSCLE	DESCRIPTION	ORIGIN (O) AND INSERTION (I)	ACTION	NERVE SUPPLY
PART I: MUSCLES OF THE ANTERIOR COMPARTMENT (FIGURES 10.21 AND 10.22)	All muscles of the anterior compartment are dorsiflexors of the ankle and have a common innervation, the deep fibular nerve. Paralysis of the anterior muscle group causes *foot drop*, which requires that the leg be lifted unusually high during walking to prevent tripping over one's toes. One cause of "shin splints," also called anterior compartment syndrome, is a painful inflammatory condition of the muscles of the anterior compartment.			
Tibialis anterior (tib″e-a′lis) (*tibial* = tibia; *anterior* = toward the front)	Superficial muscle of anterior leg; laterally parallels sharp anterior margin of tibia	O—lateral condyle and upper 2/3 of tibial shaft; interosseous membrane I—by tendon into inferior surface of medial cuneiform and first metatarsal bone	Prime mover of dorsiflexion; inverts foot; assists in supporting medial longitudinal arch of foot	Deep fibular nerve (L_4 and L_5)
Extensor digitorum longus (*extensor* = increases angle at a joint; *digit* = finger or toe; *longus* = long)	Unipennate muscle on anterolateral surface of leg; lateral to tibialis anterior muscle	O—lateral condyle of tibia; proximal 3/4 of fibula; interosseous membrane I—middle and distal phalanges of toes 2–5 via extensor expansion	Prime mover of toe extension (acts mainly at metatarsophalangeal joints); dorsiflexes foot	Deep fibular nerve (L_5 and S_1)
Fibularis (peroneus) tertius (fib-u-lar′ris ter′shus) (*fibular* = fibula; *tertius* = third)	Small muscle; usually continuous and fused with distal part of extensor digitorum longus; not always present	O—distal anterior surface of fibula and interosseous membrane I—tendon inserts on dorsum of fifth metatarsal	Dorsiflexes and everts foot	Deep fibular nerve (L_5 and S_1)
Extensor hallucis longus (hal′yu-kis) (*hallux* = great toe)	Deep to extensor digitorum longus and tibialis anterior; narrow origin	O—anteromedial fibula shaft and interosseous membrane I—tendon inserts on distal phalanx of great toe	Extends great toe; dorsiflexes foot	Deep fibular nerve (L_5 and S_1)

MUSCLE GALLERY

TABLE 10.15 *(continued)*

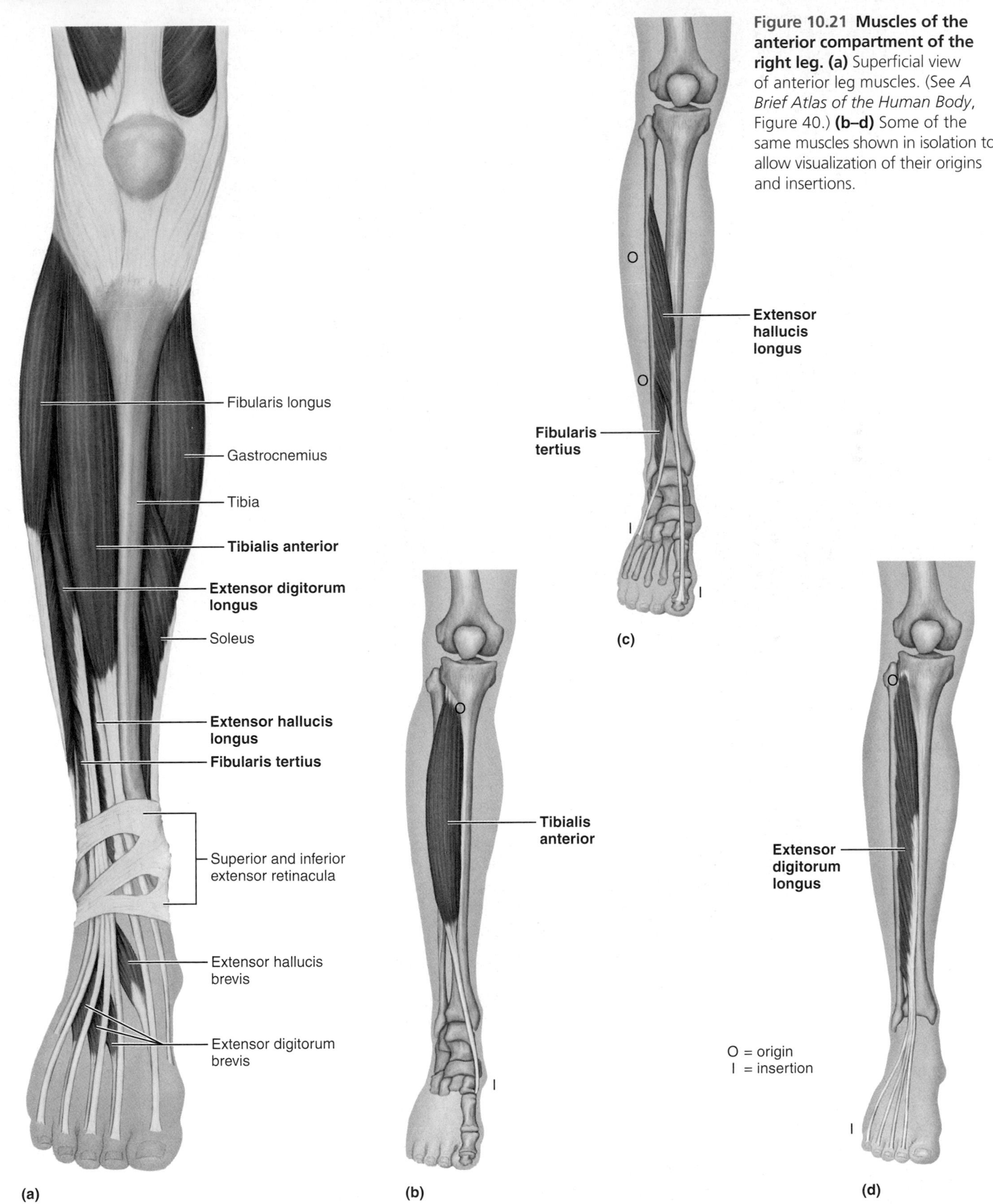

Figure 10.21 Muscles of the anterior compartment of the right leg. (a) Superficial view of anterior leg muscles. (See *A Brief Atlas of the Human Body*, Figure 40.) **(b–d)** Some of the same muscles shown in isolation to allow visualization of their origins and insertions.

MUSCLE GALLERY

TABLE 10.15 Muscles of the Leg: Movements of the Ankle and Toes (Figures 10.21 to 10.23) *(continued)*

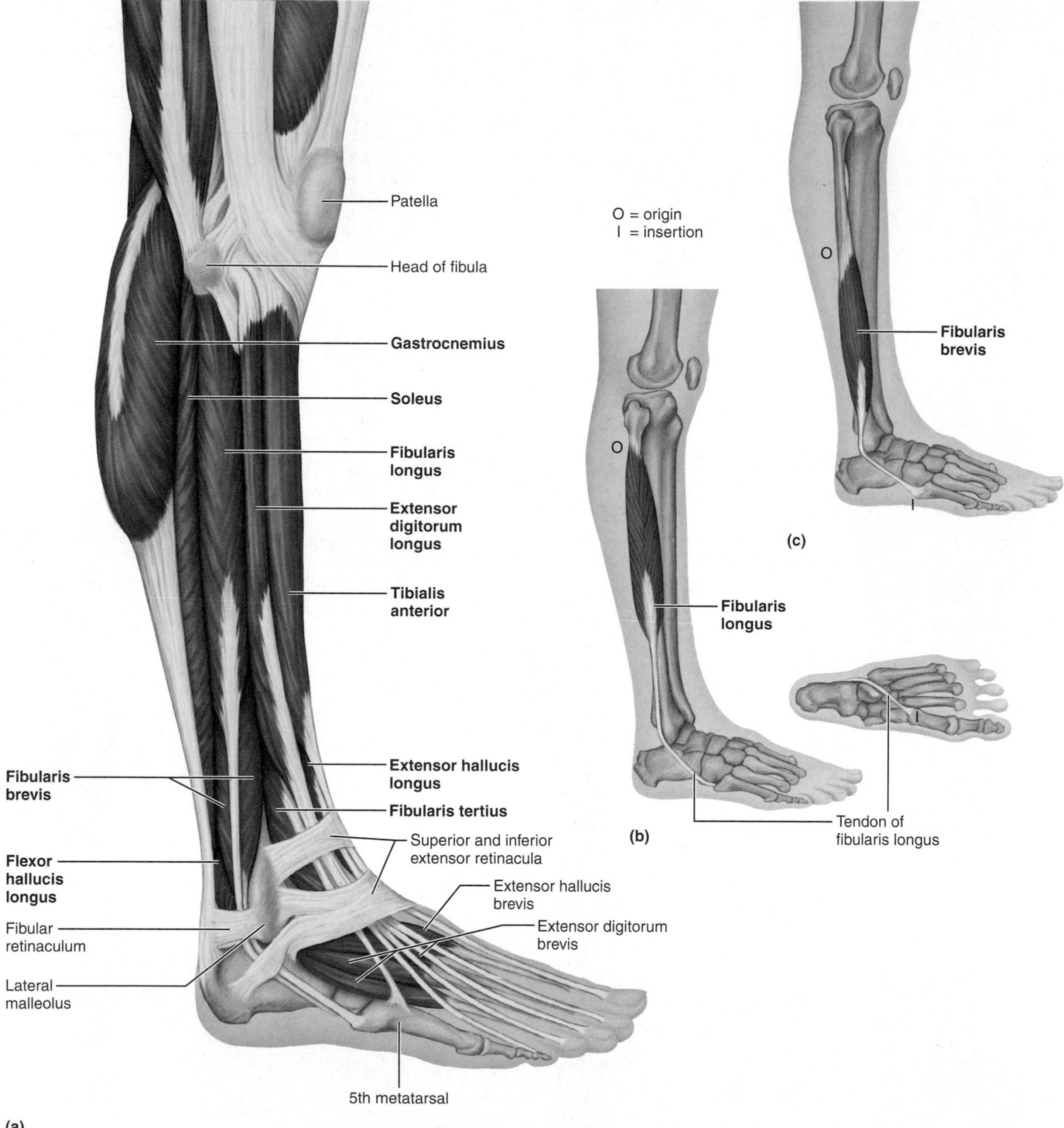

Figure 10.22 Muscles of the lateral compartment of the right leg. (a) Superficial view of lateral aspect of the leg, illustrating positions of lateral compartment muscles (fibularis longus and brevis) relative to anterior and posterior leg muscles. **(b)** Isolated view of fibularis longus; inset illustrates the insertion of the fibularis longus on the plantar surface of the foot. **(c)** Isolated view of fibularis brevis. (See *A Brief Atlas of the Human Body,* Figure 42.)

MUSCLE GALLERY

TABLE 10.15 *(continued)*

MUSCLE	DESCRIPTION	ORIGIN (O) AND INSERTION (I)	ACTION	NERVE SUPPLY
PART II: MUSCLES OF THE LATERAL COMPARTMENT (FIGURES 10.22 AND 10.23) These muscles have a common innervation, the superficial fibular nerve. Besides plantar flexion and foot eversion, these muscles stabilize the lateral ankle and lateral longitudinal arch of the foot.				
Fibularis (peroneus) longus (See also Figure 10.21)	Superficial lateral muscle; overlies fibula	O—head and upper portion of lateral fibula I—by long tendon that curves under foot to first metatarsal and medial cuneiform	Plantar flexes and everts foot; may help keep foot flat on ground	Superficial fibular nerve (L_5–S_2)
Fibularis (peroneus) brevis (*brevis* = short)	Smaller muscle; deep to fibularis longus; enclosed in a common sheath	O—distal fibula shaft I—by tendon running behind lateral malleolus to insert on proximal end of fifth metatarsal	Plantar flexes and everts foot	Superficial fibular nerve (L_5–S_2)
PART III: MUSCLES OF THE POSTERIOR COMPARTMENT (FIGURE 10.23) The muscles of the posterior compartment have a common innervation, the tibial nerve. They act in concert to plantar flex the ankle.				
SUPERFICIAL MUSCLES				
Triceps surae (tri″seps sur′e) (See also Figure 10.22)	Refers to muscle pair (gastrocnemius and soleus) that shapes the posterior calf and inserts via a common tendon into the calcaneus of the heel; this *calcaneal* or *Achilles tendon* is the largest tendon in the body. Prime movers of ankle plantar flexion.			
▪ **Gastrocnemius** (gas″truk-ne′me-us) (*gaster* = belly; *kneme* = leg)	Superficial muscle of pair; two prominent bellies that form proximal curve of calf	O—by two heads from medial and lateral condyles of femur I—posterior calcaneus via calcaneal tendon	Plantar flexes foot when knee is extended; because it also crosses knee joint, it can flex knee when foot is dorsiflexed	Tibial nerve (S_1, S_2)
▪ **Soleus** (so′le-us) (*soleus* = fish)	Broad, flat muscle, deep to gastrocnemius on posterior surface of calf	O—extensive origin from superior tibia, fibula, and interosseous membrane I—as for gastrocnemius	Plantar flexes foot; important locomotor and postural muscle during walking, running, and dancing	Tibial nerve
Plantaris (plan-tar′is) (*planta* = sole of foot)	Generally a small, feeble muscle, but varies in size and extent; may be absent	O—posterior femur above lateral condyle I—via a long, thin tendon into calcaneus or its tendon	Assists in knee flexion and plantar flexion of foot	Tibial nerve
DEEP MUSCLES (FIGURE 10.23c–f)				
Popliteus (pop-lit′e-us) (*poplit* = back of knee)	Thin, triangular muscle at posterior knee; passes inferomedially to tibial surface	O—lateral condyle of femur and lateral meniscus I—proximal tibia	Flexes and rotates leg medially to unlock extended knee when flexion begins; with tibia fixed, rotates thigh laterally	Tibial nerve (L_4–S_1)
Flexor digitorum longus (*flexor* = decreases angle at a joint)	Long, narrow muscle; runs medial to and partially overlies tibialis posterior	O—extensive origin on the posterior tibia I—tendon runs behind medial malleolus and inserts into distal phalanges of toes 2–5	Plantar flexes and inverts foot; flexes toes; helps foot "grip" ground	Tibial nerve (L_5–S_2)
Flexor hallucis longus (See also Figure 10.22)	Bipennate muscle; lies lateral to inferior aspect of tibialis posterior	O—midshaft of fibula; interosseous membrane I—tendon runs under foot to distal phalanx of great toe	Plantar flexes and inverts foot; flexes great toe at all joints; "push off" muscle during walking	Tibial nerve (L_5–S_2)
Tibialis posterior (*posterior* = toward the back)	Thick, flat muscle deep to soleus; placed between posterior flexors	O—superior tibia and fibula and interosseous membrane I—tendon passes behind medial malleolus and under arch of foot; inserts into several tarsals and metatarsals 2–4	Prime mover of foot inversion; plantar flexes foot; stabilizes medial longitudinal arch of foot (as during ice skating)	Tibial nerve (L_4 and L_5)

MUSCLE GALLERY

TABLE 10.15 Muscles of the Leg: Movements of the Ankle and Toes (Figures 10.21 to 10.23) *(continued)*

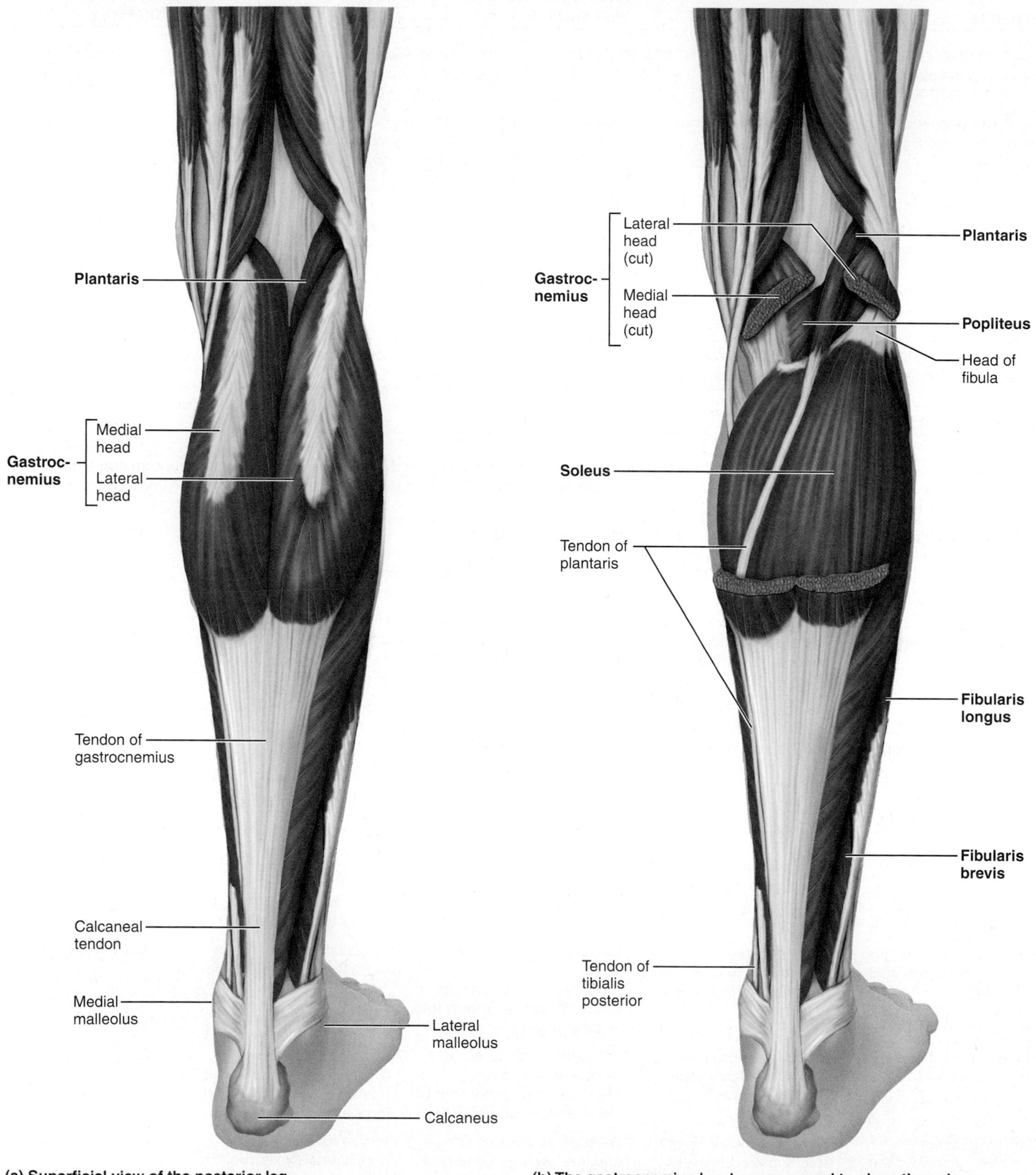

(a) Superficial view of the posterior leg.

(b) The gastrocnemius has been removed to show the soleus immediately deep to it.

Figure 10.23 **Muscles of the posterior compartment of the right leg.**

MUSCLE GALLERY

TABLE 10.15 *(continued)*

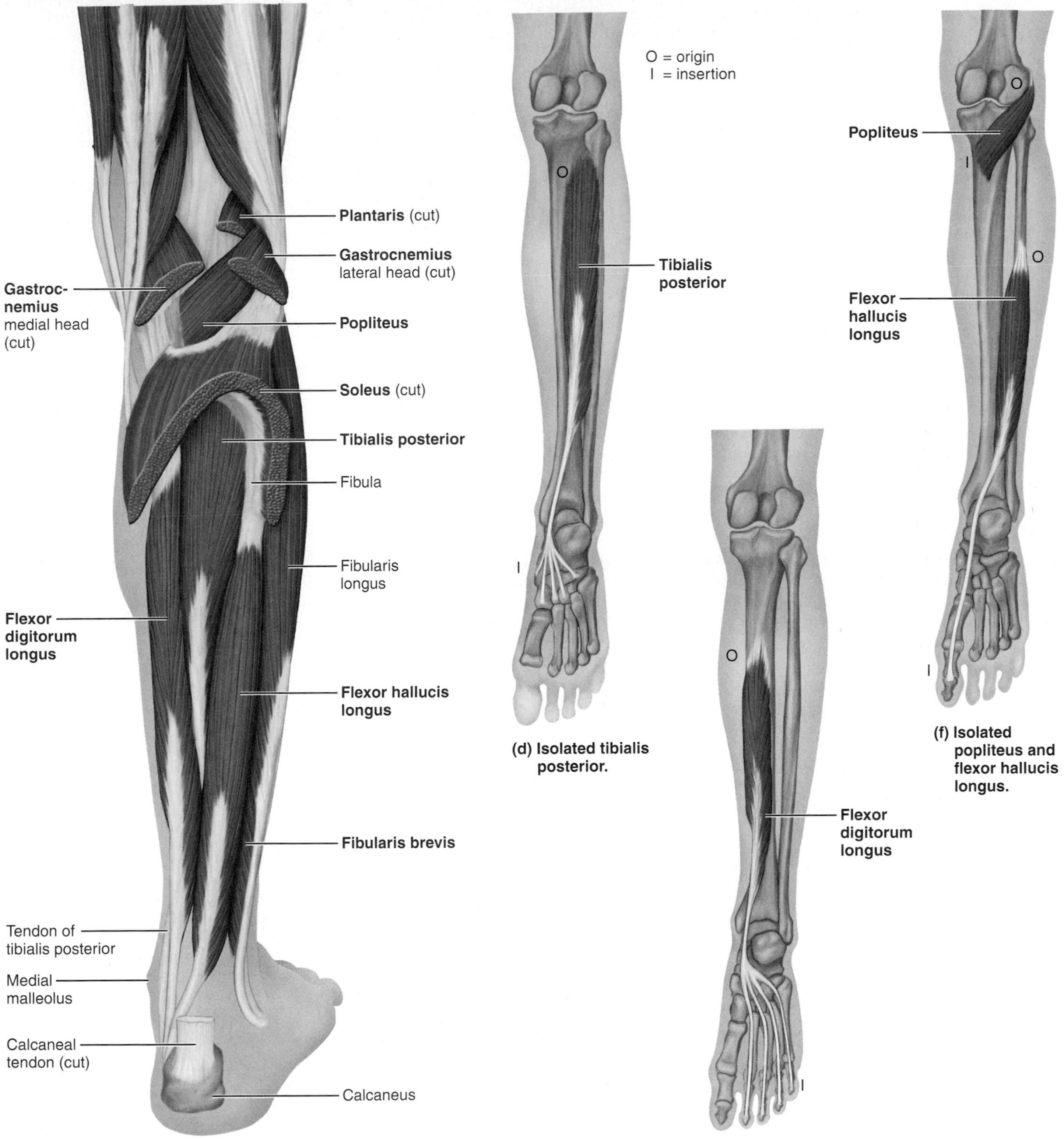

Plantaris (cut)

Gastrocnemius lateral head (cut)

Gastroc-nemius medial head (cut)

Popliteus

Soleus (cut)

Tibialis posterior

Fibula

Fibularis longus

Flexor digitorum longus

Flexor hallucis longus

Fibularis brevis

Tendon of tibialis posterior

Medial malleolus

Calcaneal tendon (cut)

Calcaneus

(c) The triceps surae has been removed to show the deep muscles of the posterior compartment.

O = origin
I = insertion

O

Tibialis posterior

I

(d) Isolated tibialis posterior.

O

Flexor digitorum longus

I

(e) Isolated flexor digitorum longus.

O

Popliteus

I

O

Flexor hallucis longus

I

(f) Isolated popliteus and flexor hallucis longus.

Figure 10.23 *(continued)*

TABLE 10.16 Intrinsic Muscles of the Foot: Toe Movement and Arch Support (Figure 10.24)

The intrinsic muscles of the foot help to flex, extend, abduct, and adduct the toes. Collectively, along with the tendons of some leg muscles that enter the sole, the foot muscles help support the arches of the foot. There is a single muscle on the foot's dorsum (superior aspect), and several muscles on the plantar aspect (the sole). The plantar muscles occur in four layers, from superficial to deep. Overall, the foot muscles are remarkably similar to those in the palm of the hand.

MUSCLE	DESCRIPTION	ORIGIN (O) AND INSERTION (I)	ACTION	NERVE SUPPLY
MUSCLES ON DORSUM OF FOOT				
Extensor digitorum brevis (Figures 10.21a and 10.22a)	Small, four-part muscle on dorsum of foot; deep to the tendons of extensor digitorum longus; corresponds to the extensor indicis and extensor pollicis muscles of forearm.	O—anterior part of calcaneus bone; extensor retinaculum I—base of proximal phalanx of great toe; extensor expansions on toes 2–5	Helps extend toes at metatarsophalangeal joints	Deep fibular nerve (L_5 and S_1)
MUSCLES ON SOLE OF FOOT—FIRST LAYER (MOST SUPERFICIAL) (FIGURE 10.24)				
Flexor digitorum brevis	Bandlike muscle in middle of sole; corresponds to flexor digitorum superficialis of forearm and inserts into digits in the same way	O—calcaneal tuberosity I—middle phalanx of toes 2–4	Helps flex toes	Medial plantar nerve (a branch of tibial nerve, S_1 and S_2)
Abductor hallucis (hal'yu-kis) (*hallux* = great toe)	Lies medial to flexor digitorum brevis (recall the similar thumb muscle, abductor pollicis brevis)	O—calcaneal tuberosity and flexor retinaculum I—proximal phalanx of great toe, medial side, in the tendon of flexor hallucis brevis (see below)	Abducts great toe	Medial plantar nerve
Abductor digiti minimi	Most lateral of the three superficial sole muscles (recall similar abductor muscle in palm)	O—calcaneal tuberosity I—lateral side of base of little toe's proximal phalanx	Abducts and flexes little toe	Lateral plantar nerve (a branch of tibial nerve, S_1, S_2, and S_3)
MUSCLES ON SOLE OF FOOT—SECOND LAYER				
Flexor accessorius (quadratus plantae)	Rectangular muscle just deep to flexor digitorum brevis in posterior half of sole; two heads (see also Figure 10.24c)	O—medial and lateral sides of calcaneus I—tendon of flexor digitorum longus in midsole	Straightens out the oblique pull of flexor digitorum longus	Lateral plantar nerve
Lumbricals	Four little "worms" (like lumbricals in hand)	O—from each tendon of flexor digitorum longus I—extensor expansion on proximal phalanx of toes 2–5, medial side	By pulling on extensor expansion, flex toes at metatarsophalangeal joints and extend toes at interphalangeal joints	Medial plantar nerve (first lumbrical) and lateral plantar nerve (second to fourth lumbrical)

MUSCLE GALLERY

TABLE 10.16 *(continued)*

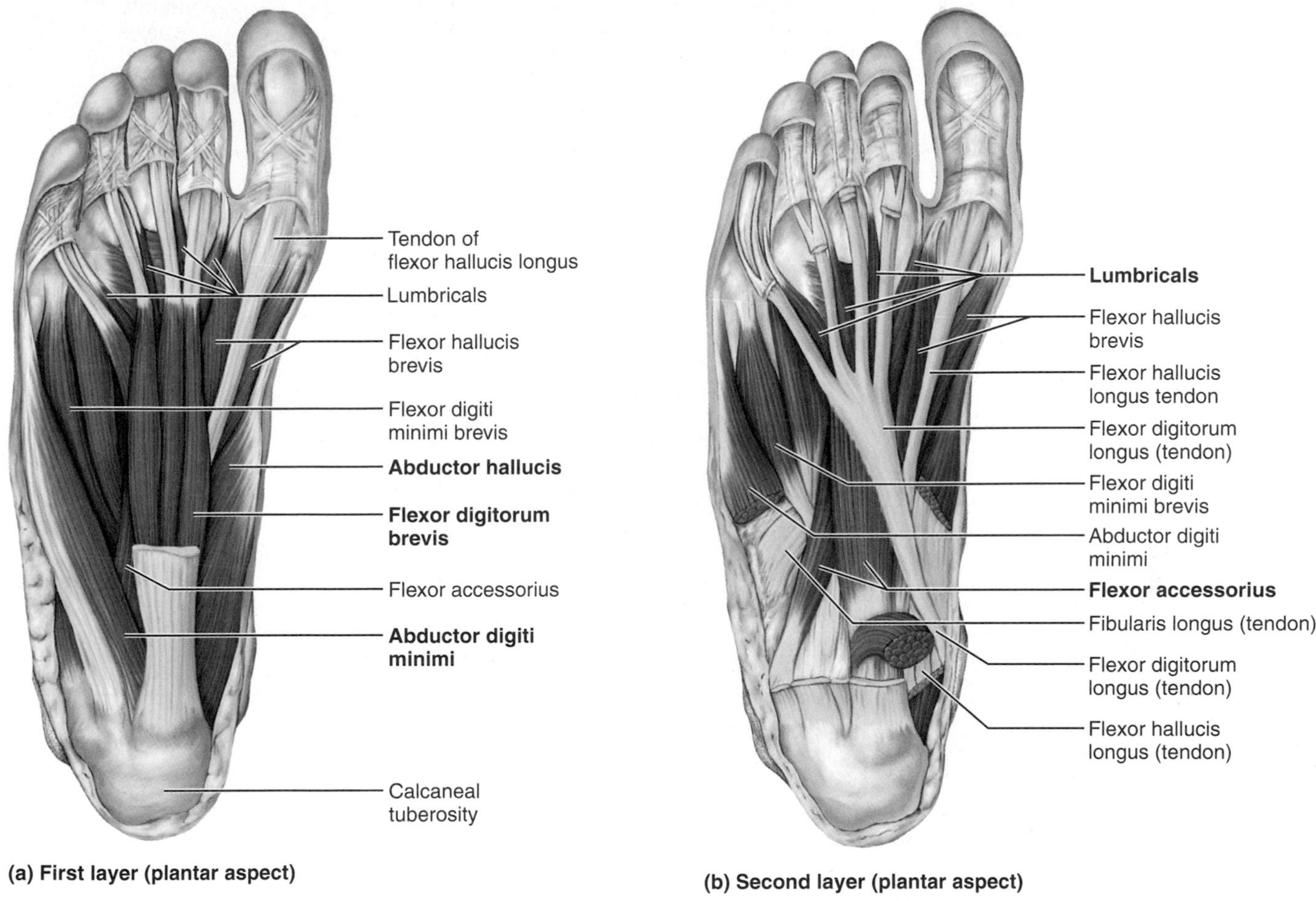

Figure 10.24 **Muscles of the right foot, plantar aspect.** (See *A Brief Atlas of the Human Body,* Figure 43.)

MUSCLE GALLERY

TABLE 10.16 Intrinsic Muscles of the Foot: Toe Movement and Arch Support (Figure 10.24) *(continued)*

MUSCLE	DESCRIPTION	ORIGIN (O) AND INSERTION (I)	ACTION	NERVE SUPPLY
MUSCLES ON SOLE OF FOOT—THIRD LAYER				
Flexor hallucis brevis	Covers metatarsal 1; splits into two bellies—recall flexor pollicis brevis of thumb	O—lateral cuneiform and cuboid bones I—via two tendons onto base of the proximal phalanx of great toe	Flexes great toe's metatarsophalangeal joint	Medial plantar nerve
Adductor hallucis	Oblique and transverse heads; deep to lumbricals (recall adductor pollicis in thumb)	O—from bases of metatarsals 2–4 and from fibularis longus tendon sheath (oblique head); from a ligament across metatarsophalangeal joints (transverse head) I—base of proximal phalanx of great toe, lateral side	Helps maintain the transverse arch of foot; weak adductor of great toe	Lateral plantar nerve (S_2 and S_3)
Flexor digiti minimi brevis	Covers metatarsal 5 (recall same muscle in hand)	O—base of metatarsal 5 and tendon sheath of fibularis longus I—base of proximal phalanx of toe 5	Flexes little toe at metatarsophalangeal joint	Lateral plantar nerve
MUSCLES ON SOLE OF FOOT—FOURTH LAYER (DEEPEST)				
Plantar (3) and dorsal interossei (4)	Similar to the palmar and dorsal interossei of hand in locations, attachments, and actions; however, the long axis of foot around which these muscles orient is the second digit, not the third	See palmar and dorsal interossei (Table 10.13)	See palmar and dorsal interossei (Table 10.13)	Lateral plantar nerve

MUSCLE GALLERY

TABLE 10.16 *(continued)*

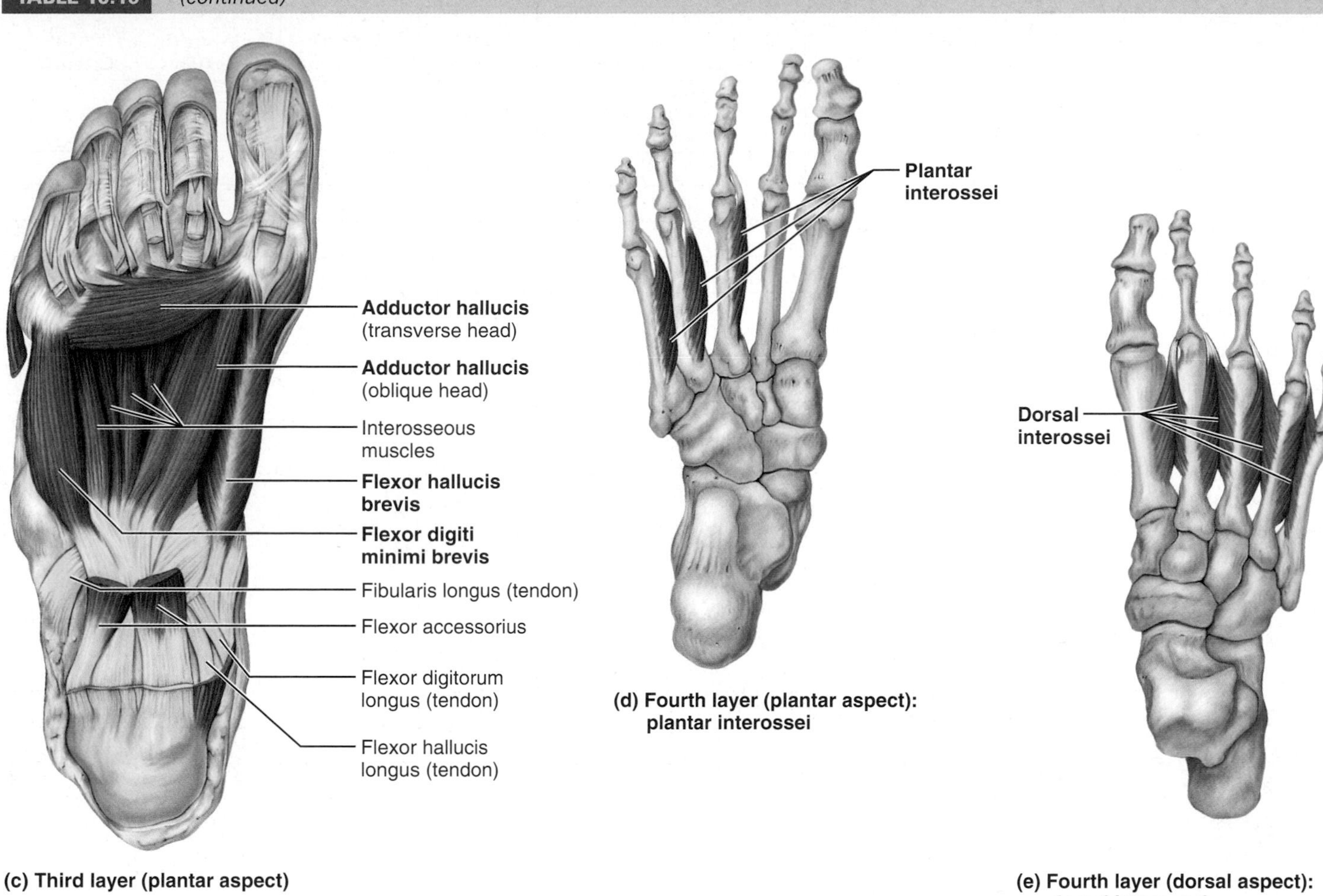

(c) Third layer (plantar aspect)

(d) Fourth layer (plantar aspect): plantar interossei

(e) Fourth layer (dorsal aspect): dorsal interossei

Figure 10.24 *(continued)* **Muscles of the right foot, plantar aspect.**

Summary of Summary of Major Actions of Muscles Acting on the Thigh, Leg, and Foot (Figure 10.25)

Part I: Muscles Acting on the Thigh and Leg (PM = prime mover)	ACTIONS AT THE HIP JOINT						ACTIONS AT THE KNEE	
	Flexion	Extension	Abduction	Adduction	Medial Rotation	Lateral Rotation	Flexion	Extension
Anterior and Medial Muscles								
Iliopsoas	× (PM)							
Sartorius	×		×			×	×	
Tensor fasciae latae	×		×		×			
Rectus femoris	×							× (PM)
Vastis muscles								× (PM)
Adductor magnus		×		×	×			
Adductor longus	×			×	×			
Adductor brevis	×			×	×			
Pectineus	×			×	×			
Gracilis				×	×		×	
Posterior Muscles								
Gluteus maximus		× (PM)	×			×		
Gluteus medius			× (PM)		×			
Gluteus minimus			×		×			
Piriformis			×			×		
Obturator internus						×		
Obturator externus						×		
Gemelli						×		
Quadratus femoris						×		
Biceps femoris		× (PM)					× (PM)	
Semitendinosus		×					× (PM)	
Semimembranosus		×					× (PM)	
Gastrocnemius							×	
Plantaris							×	
Popliteus							× (and rotates leg medially)	

Part II: Muscles Acting on the Ankle and Toes	ACTIONS AT THE ANKLE JOINT				ACTIONS AT THE TOES	
	Plantar Flexion	Dorsiflexion	Inversion	Eversion	Flexion	Extension
Anterior Compartment						
Tibialis anterior		× (PM)	×			
Extensor digitorum longus		×				× (PM)
Fibularis tertius		×		×		
Extensor hallucis longus		×	× (weak)			× (great toe)
Lateral Compartment						
Fibularis longus and brevis	×			×		
Posterior Compartment						
Gastrocnemius	× (PM)					
Soleus	× (PM)					
Plantaris	×					
Flexor digitorum longus	×		×		× (PM)	
Flexor hallucis longus	×		×		× (great toe)	
Tibialis posterior	×		× (PM)			

MUSCLE GALLERY

TABLE 10.17 (continued)

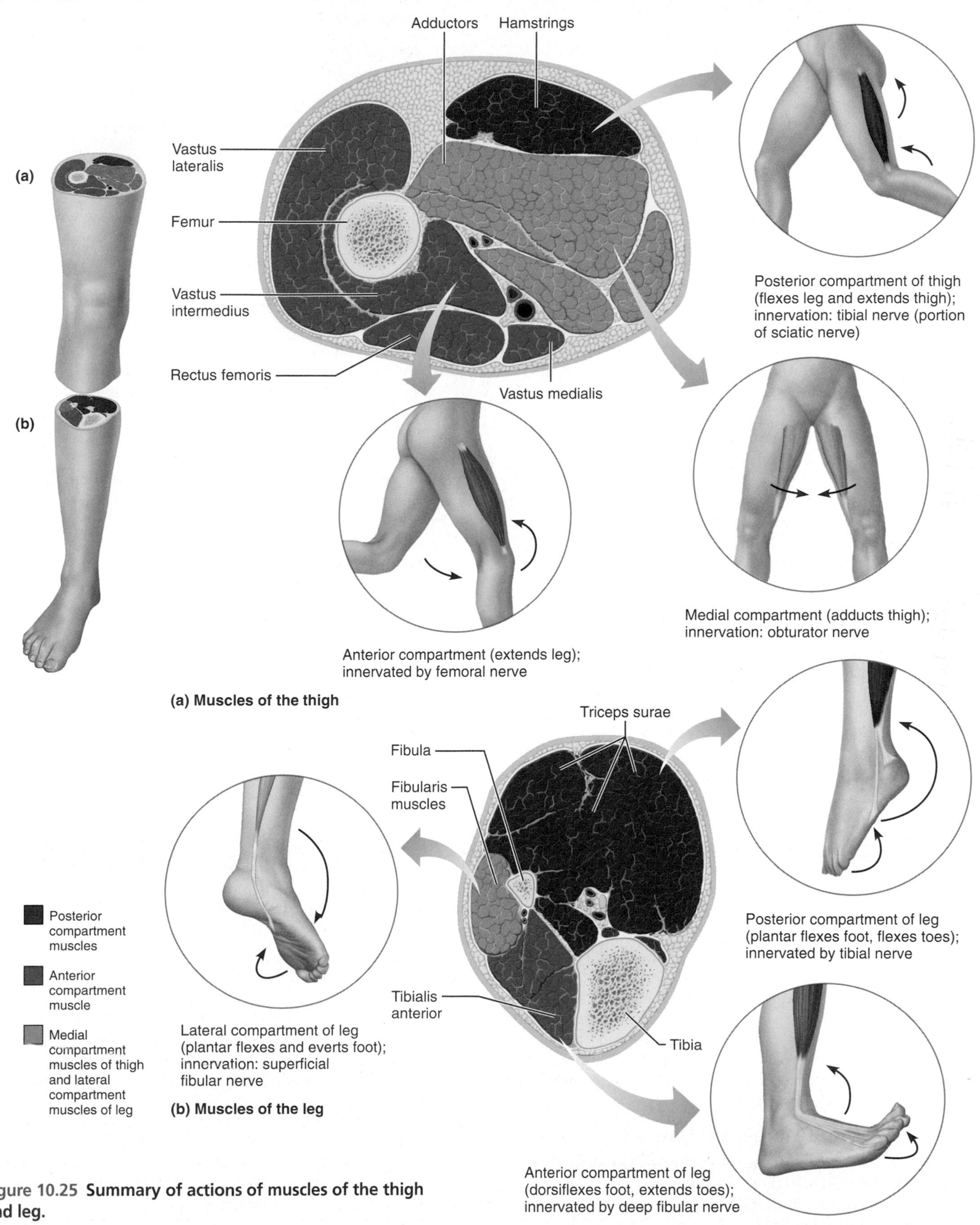

Figure 10.25 Summary of actions of muscles of the thigh and leg.

CHECK YOUR UNDERSTANDING

6. As John listened to Roger's account of how he flirted with his neighbor, he raised his eyebrows and then winked at Sarah. What facial muscles was he using?
7. What muscle would you contract to make a "sad clown's face"?
8. How can the deltoid muscles both extend and flex the arm? Aren't these antagonistic movements?
9. Which of the thenar muscles does *not* have an insertion on bones of the thumb?

For answers, see Appendix G.

RELATED CLINICAL TERMS

Charley horse A muscle contusion, i.e., tearing of muscle followed by bleeding into the tissues (hematoma) and severe, prolonged pain; a common contact sports injury; a charley horse of the quadriceps muscle of the thigh occurs frequently in football players.

Electromyography Recording and interpretation of graphic records of the electrical activity of contracting muscles. Electrodes inserted into the muscles record the impulses that pass over muscle-cell membranes to stimulate contraction. The best and most important technique for determining the functions of muscles and muscle groups.

Hernia An abnormal protrusion of abdominal contents (typically coils of the small intestine) through a weak point in the muscles of the abdominal wall. Most often caused by increased intra-abdominal pressure during lifting or straining. The hernia penetrates the muscle wall but not the skin and so appears as a visible bulge in the body surface. Common abdominal hernias include the inguinal and umbilical hernias.

Quadriceps and hamstring strains Also called quad and hamstring pulls, these conditions involve tearing of these muscles or their tendons; happen mainly in athletes who do not warm up properly and then fully extend their hip (quad pull) or knee (hamstring pull) quickly or forcefully (e.g., sprinters, tennis players); not painful at first, but pain intensifies within three to six hours (30 minutes if the tearing is severe). After a week of rest, stretching is the best therapy.

Rupture of the calcaneal tendon Although the calcaneal (Achilles) tendon is the largest, strongest tendon in the body, its rupture is surprisingly common, particularly in older people as a result of stumbling and in young sprinters when the tendon is traumatized during the takeoff. The rupture is followed by abrupt pain; a gap is seen just above the heel, and the calf bulges as the triceps surae are released from their insertion. Plantar flexion is weak or impossible, but dorsiflexion is exaggerated. Surgical repair of the tendon is usually performed.

Shin splints Common term for pain in the anterior compartment of the leg caused by irritation of the tibialis anterior muscle as might follow extreme or unusual exercise without adequate prior conditioning. Because it is tightly wrapped by fascia, the inflamed tibialis anterior cuts off its own circulation as it swells and presses painfully on its own nerves.

Tennis elbow Tenderness due to trauma or overuse of the tendon of origin of the forearm extensor muscles at the lateral epicondyle of the humerus. Caused and aggravated when these muscles contract forcefully to extend the hand at the wrist—as in executing a tennis backhand or lifting a loaded snow shovel. Despite its name, tennis elbow does not involve the elbow joint; most cases caused by work activities.

Torticollis (tor″tĭ-kol′is; *tort* = twisted) A twisting of the neck in which there is a chronic rotation and tilting of the head to one side, due to injury of the sternocleidomastoid muscle on one side; also called wryneck. Sometimes present at birth when the muscle fibers are torn during difficult delivery. Exercise that stretches the affected muscle is the usual treatment.

CHAPTER SUMMARY

Interactions of Skeletal Muscles in the Body (p. 321)

1. Skeletal muscles are arranged in opposing groups across body joints so that one group can reverse or modify the action of the other.
2. Muscles are classified as prime movers, or agonists (bear the chief responsibility for producing movement); antagonists (reverse, or oppose, the action of another muscle); synergists (aid a prime mover by effecting the same action, stabilizing joints, or preventing undesirable movements); and fixators (function to immobilize a bone or a muscle's origin).

Naming Skeletal Muscles (pp. 321–322)

1. Criteria used to name muscles include a muscle's location, shape, relative size, fiber (fascicle) direction, number of origins, attachment sites (origin/insertion), and action. Several criteria are combined to name some muscles.

Muscle Mechanics: Importance of Fascicle Arrangement and Leverage (pp. 322–323)

1. Common patterns of fascicle arrangement are parallel, fusiform, pennate, convergent, and circular. Muscles with fibers that run parallel to the long axis of the muscle shorten most; stocky pennate muscles shorten little but are the most powerful muscles.
2. A lever is a bar that moves on a fulcrum. When an effort is applied to the lever, a load is moved. In the body, bones are the levers, joints are the fulcrums, and the effort is exerted by skeletal muscles at their insertions.
3. When the effort is farther from the fulcrum than is the load, the lever operates at a mechanical advantage (is slow and strong). When the effort is exerted closer to the fulcrum than is the load, the lever operates at a mechanical disadvantage (is fast and promotes a large degree of movement).

4. First-class levers (effort-fulcrum-load) may operate at a mechanical advantage or disadvantage. Second-class levers (fulcrum-load-effort) all operate at a mechanical advantage. Third-class levers (fulcrum-effort-load) always operate at a mechanical disadvantage. Most skeletal muscles of the body act in third-class lever systems.

Major Skeletal Muscles of the Body (pp. 324–382)

1. Muscles of the head that produce facial expression tend to be small and to insert into soft tissue (skin and other muscles) rather than into bone. These muscles open and close the eyes and mouth, compress the cheeks, allow smiling and other types of facial language (see Table 10.1*).
2. Muscles of the head involved in mastication include the masseter and temporalis that elevate the mandible and two deep muscle pairs that promote grinding and sliding jaw movements (see Table 10.2*). Extrinsic muscles of the tongue anchor the tongue and control its movements.
3. Deep muscles of the anterior neck promote swallowing movements, including elevation/depression of the hyoid bone, closure of the respiratory passages, and peristalsis of the pharynx (see Table 10.3*).
4. Neck muscles and deep muscles of the vertebral column promote head and trunk movements (see Table 10.4*). The deep muscles of the posterior trunk can extend large regions of the vertebral column (and head) simultaneously. Head flexion and rotation are effected by the anteriorly located sternocleidomastoid and scalene muscles.
5. Movements of quiet breathing are promoted by the diaphragm and the external intercostal muscles of the thorax (see Table 10.5*). Downward movement of the diaphragm increases intra-abdominal pressure.
6. The four muscle pairs forming the abdominal wall are layered like plywood to form a natural muscular girdle that protects, supports, and compresses abdominal contents. These muscles also flex and laterally rotate the trunk (see Table 10.6*).
7. Muscles of the pelvic floor and perineum (see Table 10.7*) support the pelvic viscera, resist increases in intra-abdominal pressure, inhibit urination and defecation, and aid erection.
8. Except for the pectoralis major and the latissimus dorsi, the superficial muscles of the thorax act to fix or promote movements of the scapula (see Table 10.8*). Scapular movements are effected primarily by posterior thoracic muscles.
9. Nine muscles cross the shoulder joint to effect movements of the humerus (see Table 10.9*). Of these, seven originate on the scapula and two arise from the axial skeleton. Four muscles contribute to the "rotator cuff" helping to stabilize the multiaxial shoulder joint. Generally speaking, muscles located anteriorly flex, rotate, and adduct the arm. Those located posteriorly extend, rotate, and adduct the arm. The deltoid muscle of the shoulder is the prime mover of shoulder abduction.
10. Muscles causing movements of the forearm form the flesh of the arm (see Table 10.10*). Anterior arm muscles are forearm flexors; posterior muscles are forearm extensors.
11. Movements of the wrist, hand, and fingers are effected mainly by muscles originating on the forearm (see Table 10.11*). Except for the two pronator muscles, the anterior forearm muscles are wrist and/or finger flexors; those of the posterior compartment are wrist and finger extensors.
12. The intrinsic muscles of the hand aid in precise movements of the fingers (Table 10.13*) and in opposition, which helps us grip things in our palms. These small muscles are divided into thenar, hypothenar, and midpalmar groups.
13. Muscles crossing the hip and knee joints effect thigh and leg movements (see Table 10.14*). Anteromedial muscles include thigh flexors and/or adductors and knee extensors. Muscles of the posterior gluteal region extend and rotate the thigh. Posterior thigh muscles extend the hip and flex the knee.
14. Muscles in the leg act on the ankle and toes (see Table 10.15*). Anterior compartment muscles are largely ankle dorsiflexors. Lateral compartment muscles are plantar flexors and foot everters. Those of the posterior leg are plantar flexors.
15. The intrinsic muscles of the foot (Table 10.17*) support the foot arches and help move the toes. Most occur in the sole, arranged in four layers. They resemble the small muscles in the palm of the hand.

*See specific table cited for detailed description of each muscle in the group.

REVIEW QUESTIONS

Multiple Choice/Matching

(Some questions have more than one correct answer. Select the best answer or answers from the choices given.)

1. A muscle that assists an agonist by causing a like movement or by stabilizing a joint over which an agonist acts is **(a)** an antagonist, **(b)** a prime mover, **(c)** a synergist, **(d)** an agonist.
2. The arrangement of muscle fibers in which the fibers are arranged at an angle to a central longitudinal tendon is: **(a)** circular, **(b)** longitudinal, **(c)** pennate, **(d)** parallel.
3. Match the muscle names in column B to the facial muscles described in column A.

Column A	Column B
____ (1) squints the eyes	(a) corrugator supercilii
____ (2) raises the eyebrows	(b) depressor anguli oris
____ (3) smiling muscle	(c) frontal belly of epicranius
____ (4) puckers the lips	(d) occipital belly of epicranius
____ (5) pulls the scalp posteriorly	(e) orbicularis oculi
	(f) orbicularis oris
	(g) zygomaticus

4. The prime mover of inspiration is the **(a)** diaphragm, **(b)** internal intercostals, **(c)** external intercostals, **(d)** abdominal wall muscles.
5. The arm muscle that both flexes the elbow and supinates the forearm is the **(a)** brachialis, **(b)** brachioradialis, **(c)** biceps brachii, **(d)** triceps brachii.

6. The chewing muscles that protrude the mandible and produce side-to-side grinding movements are the **(a)** buccinators, **(b)** masseters, **(c)** temporalis, **(d)** pterygoids.
7. Muscles that depress the hyoid bone and larynx include all but the **(a)** sternohyoid, **(b)** omohyoid, **(c)** geniohyoid, **(d)** sternothyroid.
8. Intrinsic muscles of the back that promote extension of the spine (or head) include all but **(a)** splenius muscles, **(b)** semispinalis muscles, **(c)** scalene muscles, **(d)** erector spinae.
9. Several muscles act to move and/or stabilize the scapula. Which of the following are small rectangular muscles that square the shoulders as they act together to retract the scapula? **(a)** levator scapulae, **(b)** rhomboids, **(c)** serratus anterior, **(d)** trapezius.
10. The quadriceps include all but **(a)** vastus lateralis, **(b)** vastus intermedius, **(c)** vastus medialis, **(d)** biceps femoris, **(e)** rectus femoris.
11. A prime mover of hip flexion is the **(a)** rectus femoris, **(b)** iliopsoas, **(c)** vasti muscles, **(d)** gluteus maximus.
12. The prime mover of hip extension *against* resistance is the **(a)** gluteus maximus, **(b)** gluteus medius, **(c)** biceps femoris, **(d)** semimembranosus.
13. Muscles that cause plantar flexion include all but the **(a)** gastrocnemius, **(b)** soleus, **(c)** tibialis anterior, **(d)** tibialis posterior, **(e)** fibularis muscles.
14. In walking, which two lower limb muscles keep the forward-swinging foot from dragging on the ground? **(a)** pronator teres and popliteus, **(b)** flexor digitorum longus and popliteus, **(c)** adductor longus and abductor digiti minimi in foot, **(d)** gluteus medius and tibialis anterior.
15. What criterion (or criteria) are used in naming the gluteus medius? **(a)** relative size, **(b)** muscle location, **(c)** muscle shape, **(d)** action, **(e)** number of origins.
16. Which of the following is a large, deep muscle that protracts the scapula during punching? **(a)** serratus anterior, **(b)** rhomboids, **(c)** levator scapulae, **(d)** subscapularis.

Short Answer Essay Questions

17. Name four criteria used in naming muscles, and provide an example (other than those used in the text) that illustrates each criterion.
18. Differentiate between the arrangement of elements (load, fulcrum, and effort) in first-, second-, and third-class levers.
19. What does it mean when we say that a lever operates at a mechanical disadvantage, and what benefits does such a lever system provide?
20. What muscles act to propel a food bolus down the length of the pharynx to the esophagus?
21. Name and describe the action of muscles used to shake your head no; to nod yes.
22. (a) Name the four muscle pairs that act in unison to compress the abdominal contents. (b) How does their arrangement (fiber direction) contribute to the strength of the abdominal wall? (c) Which of these muscles can effect lateral rotation of the spine? (d) Which can act alone to flex the spine?
23. List all (six) possible movements that can occur at the shoulder joint and name the prime mover(s) of each movement. Then name their antagonists.
24. (a) Name two forearm muscles that are powerful extensors and abductors of the wrist. (b) Name the sole forearm muscle that can flex the distal interphalangeal joints.
25. Name the muscles usually grouped together as the lateral rotators of the hip.
26. Name three thigh muscles that help you keep your seat astride a horse.
27. (a) Name three muscles or muscle groups used as sites for intramuscular injections. (b) Which of these is used most often in infants, and why?
28. Name two muscles in each of the following compartments or regions: (a) thenar eminence (ball of thumb), (b) posterior compartment of forearm, (c) anterior compartment of forearm—deep muscle group, (d) anterior muscle group in the arm, (e) muscles of mastication, (f) third muscle layer of the foot, (g) posterior compartment of leg, (h) medial compartment of thigh, (i) posterior compartment of thigh.

Critical Thinking and Clinical Application Questions

1. Assume you have a 10-lb weight in your right hand. Explain why it is easier to flex the right elbow when your forearm is supinated than when it is pronated.
2. When Mrs. O'Brien returned to her doctor for a follow-up visit after childbirth, she complained that she was having problems controlling her urine flow (was incontinent) when she sneezed. The physician asked his nurse to give Mrs. O'Brien instructions on how to perform exercises to strengthen the muscles of the pelvic floor. To what muscles was he referring?
3. Mr. Ahmadi, an out-of-shape 45-year-old man, was advised by his physician to lose weight and to exercise on a regular basis. He followed his diet faithfully and began to jog daily. One day, while on his morning jog, he heard a snapping sound that was immediately followed by a severe pain in his right lower calf. When his leg was examined, a gap was seen between his swollen upper calf region and his heel, and he was unable to plantar flex that ankle. What do you think happened? Why was the upper part of his calf swollen?
4. As Peter watched Sue walk down the runway at the fashion show, he contracted his right orbicularis oculi muscle, raised his arm, and contracted his opponens pollicis. Was he pleased or displeased with her performance? How do you know?
5. What type of lever system is described by the following activities? (a) The soleus muscle plantar flexes the foot. (b) The deltoid abducts the arm. (c) The triceps brachii is strained while doing pushups.
6. While riding an unusually large horse, Chao Jung had to spread her thighs widely to span its back and she pulled the muscles in her medial thighs. Which muscles were these and what might this condition be called?

Access everything you need to practice, review, and self-assess for both your A&P lecture and lab courses at **myA&P** (www.myaandp.com). There, you'll find powerful online resources, including chapter quizzes and tests, games, A&P Flix animations with quizzes, ***Interactive Physiology***® with quizzes, MP3 Tutor Sessions, Practice Anatomy Lab™, and more to help you get a better grade in your course.

11

Fundamentals of the Nervous System and Nervous Tissue

You are driving down the freeway, and a horn blares to your right. You immediately swerve to your left. Charlie leaves a note on the kitchen table: "See you later. Have the stuff ready at 6." You know the "stuff" is chili with taco chips. You are dozing but you awaken instantly as your infant son makes a soft cry.

What do these three events have in common? They are all everyday examples of the functioning of your nervous system, which has your body cells humming with activity nearly all the time.

The **nervous system** is the master controlling and communicating system of the body. Every thought, action, and emotion reflects its activity. Its cells communicate by electrical and chemical signals, which are rapid and specific, and usually cause almost immediate responses.

We begin this chapter with a brief overview of the functions and organization of the nervous system. Then we focus on the functional anatomy of nervous tissue, especially that of nerve cells, or *neurons*, which are the key to neural communication.

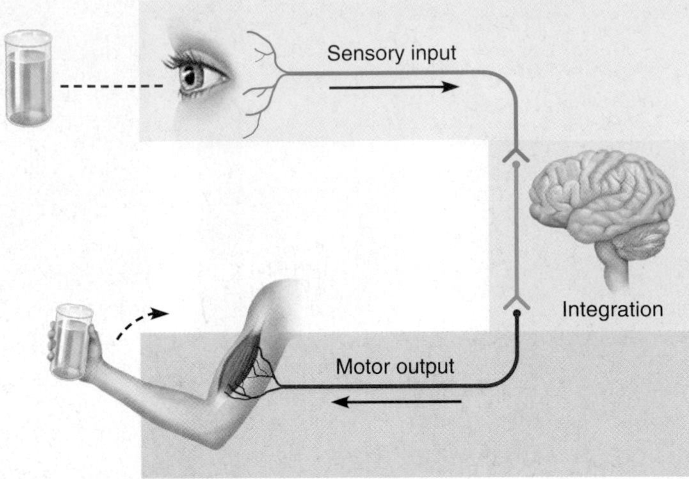

Figure 11.1 The nervous system's functions.

Functions and Divisions of the Nervous System

- ▶ List the basic functions of the nervous system.
- ▶ Explain the structural and functional divisions of the nervous system.

11

The nervous system has three overlapping functions, illustrated by the example of a thirsty person seeing and then lifting a glass of drinking water in **Figure 11.1**:

1. **Sensory input.** The nervous system uses its millions of sensory receptors to monitor changes occurring both inside and outside the body. The gathered information is called **sensory input**.
2. **Integration.** The nervous system processes and interprets sensory input and decides what should be done at each moment—a process called **integration.**
3. **Motor output.** The nervous system causes a *response*, called **motor output**, by activating *effector organs*—the muscles and glands.

In another example, when you are driving and see a red light ahead (sensory input), your nervous system integrates this information (red light means "stop"), and your foot goes for the brake (motor output).

We have only one highly integrated nervous system. For convenience, it can be divided into two principal parts. The **central nervous system (CNS)** consists of the *brain* and *spinal cord*, which occupy the dorsal body cavity. The CNS is the integrating and command center of the nervous system. It interprets sensory input and dictates motor responses based on reflexes, current conditions, and past experience **(Figure 11.2)**.

The **peripheral nervous system (PNS)** is the part of the nervous system *outside* the CNS. The PNS consists mainly of the nerves (bundles of axons) that extend from the brain and spinal cord. *Spinal nerves* carry impulses to and from the spinal cord, and *cranial nerves* carry impulses to and from the brain. These peripheral nerves serve as the communication lines that link all parts of the body to the CNS.

The PNS has two functional subdivisions, as Figure 11.2 shows. The **sensory**, or **afferent, division** (af′er-ent; "carrying toward") consists of nerve fibers (axons) that convey impulses *to* the central nervous system from sensory receptors located throughout the body (see the blue fibers in Figure 11.2). Sensory fibers conveying impulses from the skin, skeletal muscles, and joints are called *somatic afferent fibers* (*soma* = body), and those transmitting impulses from the visceral organs (organs within the ventral body cavity) are called *visceral afferent fibers.* The sensory division keeps the CNS constantly informed of events going on both inside and outside the body.

The **motor**, or **efferent, division** (ef′er-ent; "carrying away") of the PNS transmits impulses *from* the CNS to effector organs, which are the muscles and glands (see the red fibers in Figure 11.2). These impulses activate muscles to contract and glands to secrete. In other words, they *effect* (bring about) a motor response.

The motor division also has two main parts:

1. The **somatic nervous system** is composed of somatic motor nerve fibers that conduct impulses from the CNS to skeletal muscles. It is often referred to as the **voluntary nervous system** because it allows us to consciously control our skeletal muscles.
2. The **autonomic nervous system (ANS)** consists of visceral motor nerve fibers that regulate the activity of smooth muscles, cardiac muscles, and glands. *Autonomic* means "a law unto itself," and because we generally cannot control such activities as the pumping of our heart or the movement of food through our digestive tract, the ANS is also referred to as the **involuntary nervous system.** As we will describe in Chapter 14 and as Figure 11.2 shows, the ANS

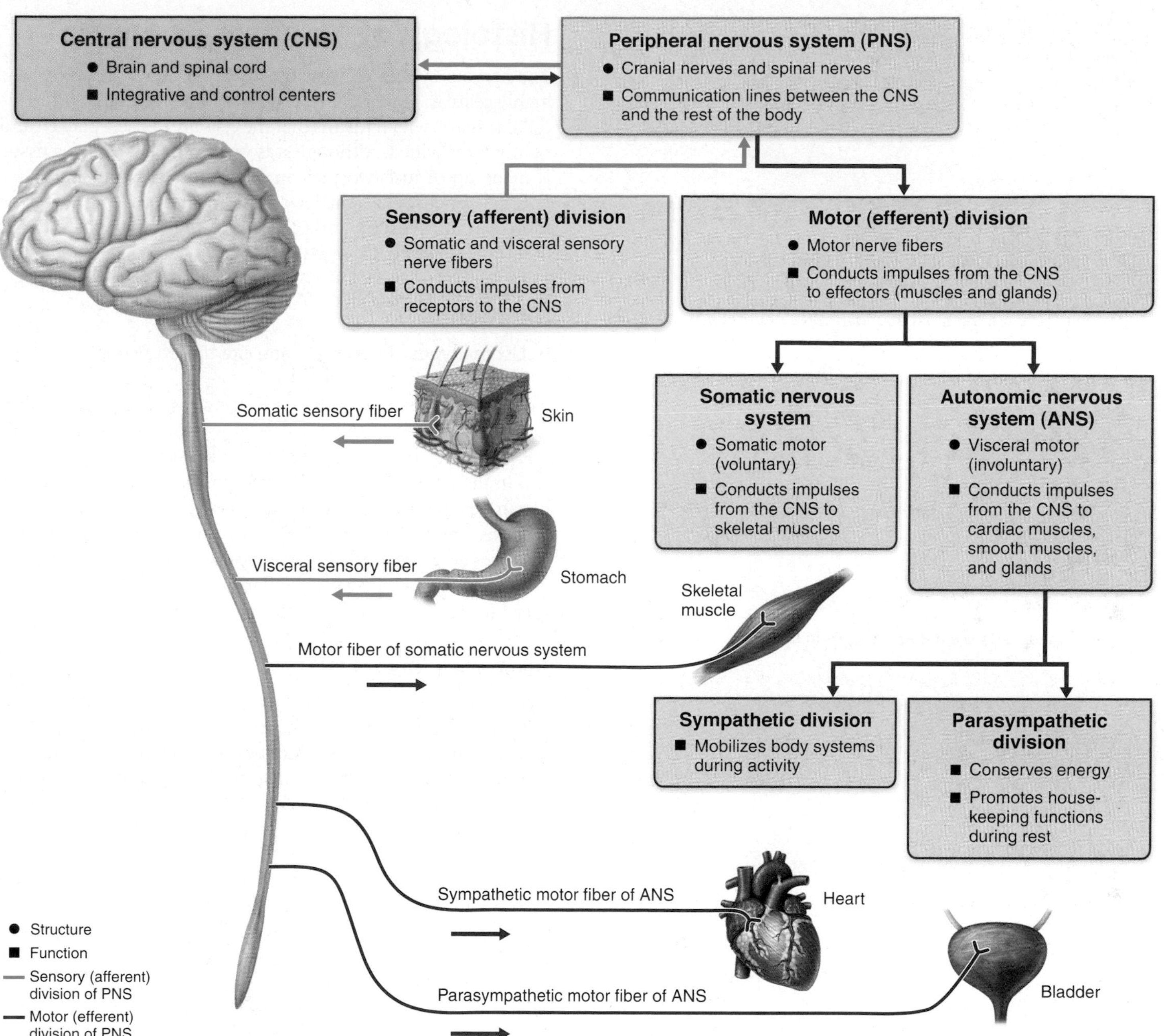

Figure 11.2 Schematic of levels of organization in the nervous system. Visceral organs (primarily located in the ventral body cavity) are served by visceral sensory fibers and by motor fibers of the autonomic nervous system. The somata (limbs and body wall) are served by motor fibers of the somatic nervous system and by somatic sensory fibers. Arrows indicate the direction of nerve impulses. (Connections to spinal cord are not anatomically accurate.)

has two functional subdivisions, the **sympathetic division** and the **parasympathetic division**, which typically work in opposition to each other—what one subdivision stimulates, the other inhibits.

CHECK YOUR UNDERSTANDING

1. What is meant by integration, and does it primarily occur in the CNS or the PNS?
2. Which subdivision of the PNS is involved in (a) relaying the feeling of a "full stomach" after a meal, (b) contracting the muscles to lift your arm, and (c) increasing your heart rate.

For answers, see Appendix G.

11

Histology of Nervous Tissue

The nervous system consists mostly of nervous tissue, which is highly cellular. For example, less than 20% of the CNS is extracellular space, which means that the cells are densely packed and tightly intertwined. Although it is very complex, nervous tissue is made up of just two principal types of cells: (1) *supporting cells* called neuroglia, smaller cells that surround and wrap the more delicate neurons, and (2) *neurons*, the excitable nerve cells that transmit electrical signals.

Neuroglia

▶ List the types of neuroglia and cite their functions.

Neurons associate closely with much smaller cells called **neuroglia** (nu-rog′le-ah; "nerve glue") or simply **glial cells** (gle′al). There are six types of neuroglia—four in the CNS and two in the PNS **(Figure 11.3)**. Each type has a unique function, but in general, these cells provide a supportive scaffolding for neurons. Some produce chemicals that guide young neurons to the proper connections, and promote neuron health and growth. Others wrap around and insulate neuronal processes to speed up action potential conduction.

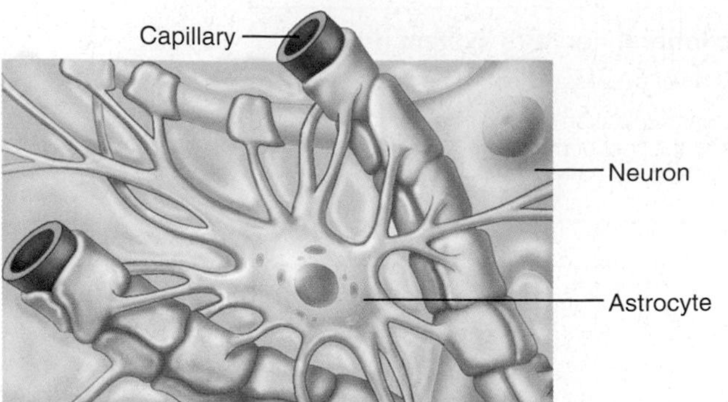

(a) Astrocytes are the most abundant CNS neuroglia.

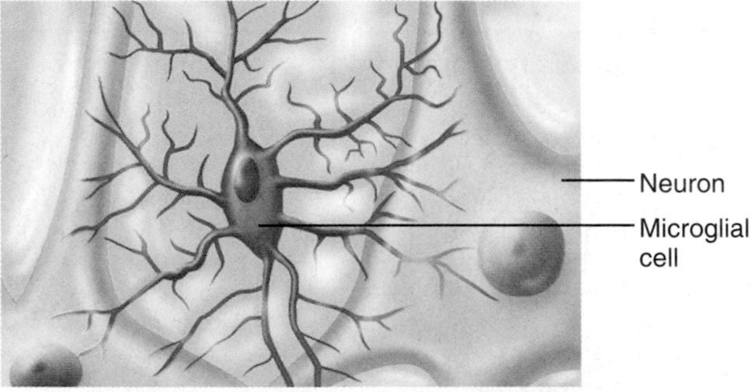

(b) Microglial cells are defensive cells in the CNS.

11

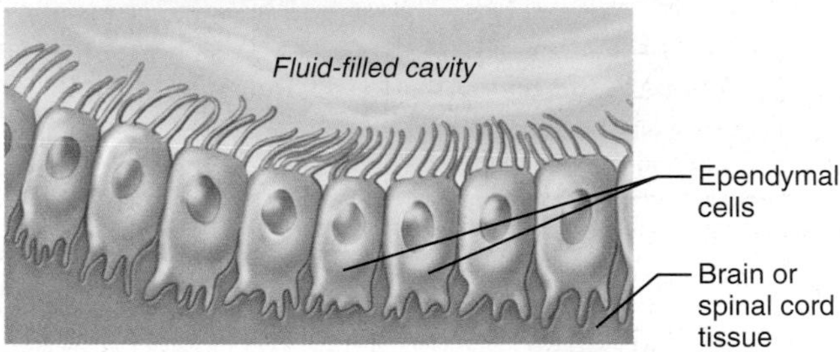

(c) Ependymal cells line cerebrospinal fluid–filled cavities.

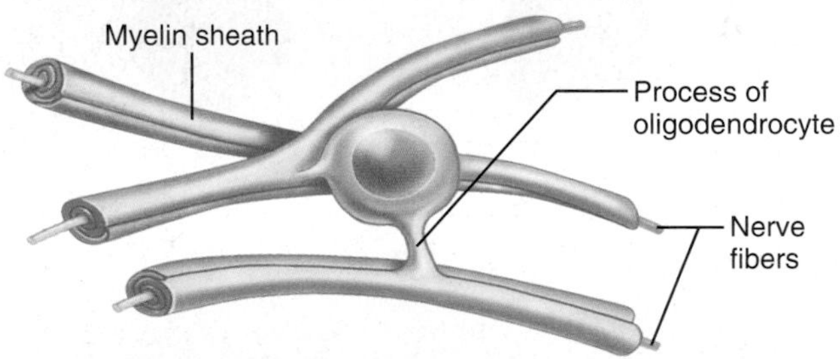

(d) Oligodendrocytes have processes that form myelin sheaths around CNS nerve fibers.

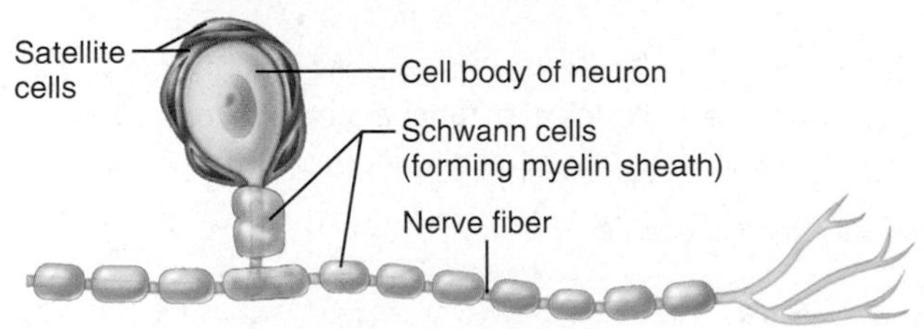

(e) Satellite cells and Schwann cells (which form myelin) surround neurons in the PNS.

Figure 11.3 **Neuroglia. (a–d)** Supporting cells of the CNS. **(e)** Supporting cells of the PNS.

Neuroglia in the CNS

Neuroglia in the CNS include *astrocytes*, *microglia*, *ependymal cells*, and *oligodendrocytes* (Figure 11.3a–d). Like neurons, most glial cells have branching processes (extensions) and a central cell body. Neuroglia can be distinguished, however, by their much smaller size and by their darker-staining nuclei. They outnumber neurons in the CNS by about 10 to 1, and make up about half the mass of the brain.

Shaped like delicate branching sea anemones, **astrocytes** (as′tro-sītz; "star cells") are the most abundant and most versatile glial cells. Their numerous radiating processes cling to neurons and their synaptic endings, and cover nearby capillaries, supporting and bracing the neurons and anchoring them to their nutrient supply lines, the blood capillaries (Figure 11.3a). Astrocytes have a role in making exchanges between capillaries and neurons, in helping to determine capillary permeability, in guiding the migration of young neurons, and in synapse formation. They also control the chemical environment around neurons, where their most important job is "mopping up" leaked potassium ions and recapturing (and recycling) released neurotransmitters. Furthermore, astrocytes have been shown to respond to nearby nerve impulses and released neurotransmitters. Astrocytes are connected together by gap junctions and signal each other both by taking in calcium, creating slow-paced intracellular calcium pulses (calcium sparks), and by releasing extracellular chemical messengers. According to recent research, astrocytes also influence neuronal functioning and therefore participate in information processing in the brain.

Microglia (mi-kro′gle-ah) are small ovoid cells with relatively long "thorny" processes (Figure 11.3b). Their processes touch nearby neurons, monitoring their health, and when they sense that certain neurons are injured or in other trouble, the microglia migrate toward them. Where invading microorganisms or dead neurons are present, the microglia transform into a special type of macrophage that phagocytizes the microorganisms or neuronal debris. This protective role of the microglia is important because cells of the immune system are denied access to the CNS.

Ependymal cells (ĕ-pen′dĭ-mul; "wrapping garment") range in shape from squamous to columnar, and many are ciliated. They line the central cavities of the brain and the spinal cord, where they form a fairly permeable barrier between the cerebrospinal fluid that fills those cavities and the tissue fluid bathing the cells of the CNS. The beating of their cilia helps to circulate the cerebrospinal fluid that cushions the brain and spinal cord (Figure 11.3c).

Though they also branch, the **oligodendrocytes** (ol″ĭ-go-den′dro-sīts) have fewer processes (*oligo* = few; *dendr* = branch) than astrocytes. Oligodendrocytes line up along the thicker neuron fibers in the CNS and wrap their processes tightly around the fibers, producing insulating coverings called *myelin sheaths* (Figure 11.3d).

Neuroglia in the PNS

The two kinds of PNS neuroglia—*satellite cells* and *Schwann cells*—differ mainly in location. **Satellite cells** surround neuron cell bodies located in the peripheral nervous system (Figure 11.3e), and are thought to have many of the same functions in the PNS as astrocytes do in the CNS. Their name comes from a fancied resemblance to the moons (satellites) around a planet.

Schwann cells (also called *neurolemmocytes*) surround and form myelin sheaths around the larger nerve fibers in the peripheral nervous system (Figures 11.3e and 11.4b). In this way, they are functionally similar to oligodendrocytes. (We describe the formation of myelin sheaths later in this chapter.) Schwann cells are vital to regeneration of damaged peripheral nerve fibers.

CHECK YOUR UNDERSTANDING

3. Which type of neuroglia controls the extracellular fluid environment around neuron cell bodies in the CNS? In the PNS?
4. Which two types of neuroglia form insulating coverings called myelin sheaths?

For answers, see Appendix G.

Neurons

- Define neuron, describe its important structural components, and relate each to a functional role.
- Differentiate between a nerve and a tract, and between a nucleus and a ganglion.
- Explain the importance of the myelin sheath and describe how it is formed in the central and peripheral nervous systems.

The billions of **neurons**, also called **nerve cells**, are the structural units of the nervous system. They are highly specialized cells that conduct messages in the form of nerve impulses from one part of the body to another. Besides their ability to conduct nerve impulses, neurons have some other special characteristics:

1. They have *extreme longevity*. Given good nutrition, neurons can function optimally for a lifetime (over 100 years).
2. They are *amitotic*. As neurons assume their roles as communicating links of the nervous system, they lose their ability to divide. We pay a high price for this neuron feature because they cannot be replaced if destroyed. There *are* exceptions to this rule. For example, olfactory epithelium and some hippocampal regions contain stem cells that can produce new neurons throughout life. (The hippocampus is a brain region involved in memory.)
3. They have an exceptionally *high metabolic rate* and require continuous and abundant supplies of oxygen and glucose. Neurons cannot survive for more than a few minutes without oxygen.

Neurons are typically large, complex cells. Although they vary in structure, they all have a *cell body* and one or more slender *processes* (**Figure 11.4**). The plasma membrane of neurons is the site of electrical signaling, and it plays a crucial role in cell-to-cell interactions that occur during development.

Cell Body

The **neuron cell body** consists of a spherical nucleus with a conspicuous nucleolus surrounded by cytoplasm. Also called the **perikaryon** (*peri* = around, *kary* = nucleus) or **soma**, the cell body ranges in diameter from 5 to 140 μm. The cell body is the major *biosynthetic center* of a neuron and it contains the usual organelles.

The neuron cell body's protein- and membrane-making machinery, consisting of clustered free ribosomes and rough endoplasmic reticulum (ER), is probably the most active and best developed in the body. This rough ER, referred to as **Nissl bodies** (nis′l) or **chromatophilic substance** (*chromatophilic* = color loving), stains darkly with basic dyes. The Golgi apparatus is also well developed and forms an arc or a complete circle around the nucleus.

Mitochondria are scattered among the other organelles. Microtubules and **neurofibrils**, which are bundles of intermediate filaments (*neurofilaments*), are important in maintaining cell shape and integrity. They form a network throughout the cell body.

The cell body of some neurons also contains pigment inclusions. For example, some contain a black melanin, a red iron-containing pigment, or a golden-brown pigment called *lipofuscin* (lip″o-fu′sin). Lipofuscin, a harmless by-product of

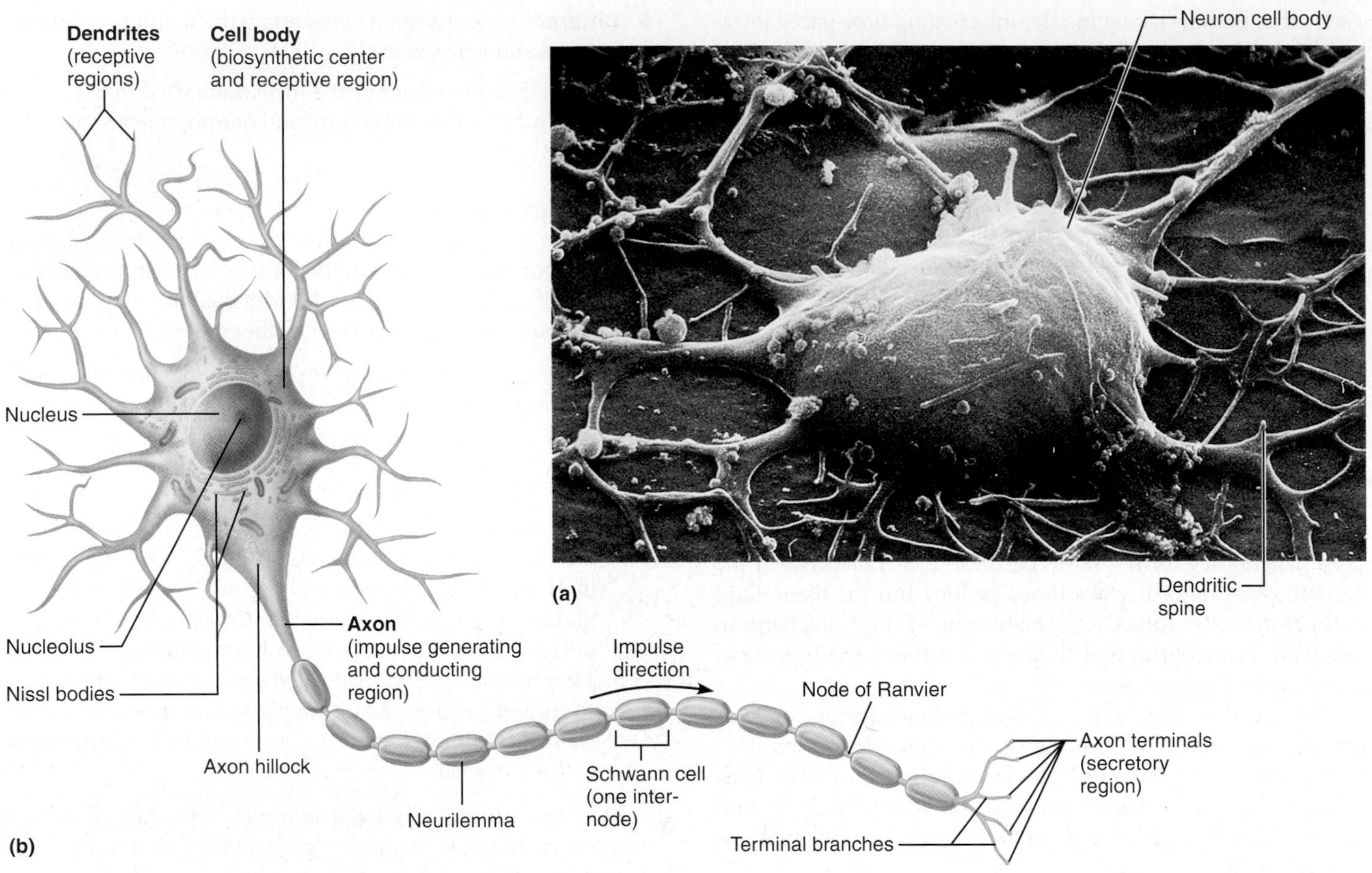

11

Figure 11.4 Structure of a motor neuron. (a) Scanning electron micrograph showing the cell body and dendrites with obvious dendritic spines (2000×). **(b)** Diagrammatic view.

lysosomal activity, is sometimes called the "aging pigment" because it accumulates in neurons of elderly individuals.

The cell body is the focal point for the outgrowth of neuron processes during embryonic development. In most neurons, the plasma membrane of the cell body also acts as *part of the receptive region* that receives information from other neurons (as shown in Table 11.1, pp. 393–394).

Most neuron cell bodies are located in the CNS, where they are protected by the bones of the skull and vertebral column. Clusters of cell bodies in the CNS are called **nuclei**, whereas those that lie along the nerves in the PNS are called **ganglia** (gang′gle-ah; *ganglion* = "knot on a string," "swelling").

Processes

Armlike **processes** extend from the cell body of all neurons. The brain and spinal cord (CNS) contain both neuron cell bodies and their processes. The PNS, for the most part, consists chiefly of neuron processes. Bundles of neuron processes are called **tracts** in the CNS and **nerves** in the PNS.

The two types of neuron processes, *dendrites* and *axons* (ak′sonz), differ from each other in the structure and function of their plasma membranes. The convention is to describe these processes using a motor neuron as an example of a typical neuron. We shall follow this practice, but keep in mind that many sensory neurons and some tiny CNS neurons differ from the "typical" pattern we present here.

Dendrites **Dendrites** of motor neurons are short, tapering, diffusely branching extensions. Typically, motor neurons have hundreds of twiglike dendrites clustering close to the cell body. Virtually all organelles present in the cell body also occur in dendrites.

Dendrites are the main **receptive** or **input regions** (Table 11.1). They provide an enormous surface area for receiving signals from other neurons. In many brain areas, the finer dendrites are highly specialized for information collection. They bristle with thorny appendages having bulbous or spiky ends called *dendritic spines*, which represent points of close contact (synapses) with other neurons (Figure 11.4a).

Dendrites convey incoming messages *toward* the cell body. These electrical signals are usually *not* action potentials (nerve impulses) but are short-distance signals called *graded potentials*, as we will describe shortly.

The Axon Each neuron has a single **axon** (*axo* = axis, axle). The initial region of the axon arises from a cone-shaped area of the cell body called the **axon hillock** ("little hill") and then narrows to form a slender process that is uniform in diameter for the rest of its length (Figure 11.4b). In some neurons, the axon is very short or absent, but in others it is long and accounts for nearly the entire length of the neuron. For example, axons of the motor neurons controlling the skeletal muscles of your great toe extend from the lumbar region of your spine to your foot, a distance of a meter or more (3–4 feet), making them among the longest cells in the body. Any long axon is called a **nerve fiber**.

Each neuron has only one axon, but axons may have occasional branches along their length. These branches, called **axon collaterals**, extend from the axon at more or less right angles. Whether an axon is undivided or has collaterals, it usually branches profusely at its end (terminus): 10,000 or more **terminal branches**, or **telodendria**, per neuron is not unusual. The knoblike distal endings of the terminal branches are variously called **axon terminals**, **synaptic knobs**, or **boutons** (boo-tonz; "buttons"). Take your pick!

Functionally, the axon is the **conducting region** of the neuron (Table 11.1). It *generates nerve impulses* and *transmits them*, typically away from the cell body, along the plasma membrane, or **axolemma** (ak″so-lem′ah). In motor neurons, the nerve impulse is generated at the junction of the axon hillock and axon (which for this reason is called the *trigger zone*) and conducted along the axon to the axon terminals, which are the **secretory region** of the neuron. When the impulse reaches the axon terminals, it causes *neurotransmitters*, signaling chemicals stored in vesicles there, to be released into the extracellular space. The neurotransmitters either excite or inhibit neurons (or effector cells) with which the axon is in close contact. Because each neuron both receives signals from and sends signals to scores of other neurons, it carries on "conversations" with many different neurons at the same time.

An axon contains the same organelles found in the dendrites and cell body with two important exceptions—it lacks Nissl bodies and a Golgi apparatus, the structures involved with protein synthesis and packaging. Consequently, an axon depends (1) on its cell body to renew the necessary proteins and membrane components, and (2) on efficient transport mechanisms to distribute them. Axons quickly decay if cut or severely damaged.

Because axons are often very long, the task of moving molecules along their length might appear difficult. However, through the cooperative effort of several types of cytoskeletal elements (microtubules, actin filaments, and so on), substances travel continuously along the axon both away from and toward the cell body. Movement toward the axon terminals is *anterograde movement*, and that in the opposite direction is *retrograde movement*.

Substances moved in the anterograde direction include mitochondria, cytoskeletal elements, membrane components used to renew the axon plasma membrane, and enzymes needed for synthesis of certain neurotransmitters. (Some neurotransmitters are synthesized in the cell body and then transported to the axon terminals.)

Substances transported through the axon in the retrograde direction are mostly organelles being returned to the cell body for degradation or recycling. Retrograde transport is also an important means of intracellular communication for "advising" the cell body of conditions at the axon terminals, and for delivering to the cell body vesicles containing signal molecules (like nerve growth factor, which activates certain nuclear genes promoting growth).

A single bidirectional transport mechanism appears to be responsible for axonal transport. It uses ATP-dependent "motor" proteins such as kinesin, dynein, and myosin. These proteins propel cellular components along the microtubules like trains along tracks at speeds up to 40 cm (15 inches) per day.

HOMEOSTATIC IMBALANCE

Certain viruses and bacterial toxins that damage neural tissues use retrograde axonal transport to reach the cell body. This transport mechanism has been demonstrated for polio, rabies, and herpes simplex viruses and for tetanus toxin. Its use as a tool to treat genetic diseases by introducing viruses containing "corrected" genes or microRNA to suppress defective genes is under investigation. ■

Myelin Sheath and Neurilemma Many nerve fibers, particularly those that are long or large in diameter, are covered with a whitish, fatty (protein-lipoid), segmented **myelin sheath** (mi′ĕ-lin). Myelin protects and electrically insulates fibers, and it increases the speed of transmission of nerve impulses. **Myelinated fibers** (axons bearing a myelin sheath) conduct nerve impulses rapidly, whereas **unmyelinated fibers** conduct impulses quite slowly. Note that myelin sheaths are associated only with axons. Dendrites are *always* unmyelinated.

Myelin sheaths in the PNS are formed by Schwann cells, which indent to receive an axon and then wrap themselves around it in a jelly roll fashion (**Figure 11.5**). Initially the wrapping is loose, but the Schwann cell cytoplasm is gradually squeezed from between the membrane layers. When the wrapping process is complete, many concentric layers of Schwann cell plasma membrane enclose the axon, much like gauze wrapped around an injured finger. This tight coil of wrapped membranes is the myelin sheath, and its thickness depends on the number of spirals.

Plasma membranes of myelinating cells contain much less protein than the plasma membranes of most body cells. Channel and carrier proteins are notably absent, a characteristic that makes myelin sheaths exceptionally good electrical insulators. Another unique characteristic of these membranes is the presence of specific protein molecules that interlock to form a sort of molecular Velcro between adjacent myelin membranes.

The nucleus and most of the cytoplasm of the Schwann cell end up as a bulge just external to the myelin sheath. This portion of the Schwann cell, which includes the exposed part of its plasma membrane, is called the **neurilemma** ("neuron husk") (Figure 11.5b). Adjacent Schwann cells along an axon do not touch one another, so there are gaps in the sheath. These gaps, called **nodes of Ranvier** (ran′vē-ā″) or **myelin sheath gaps**, occur at regular intervals (about 1 mm apart) along the myelinated axon. Axon collaterals can emerge from the axon at these nodes.

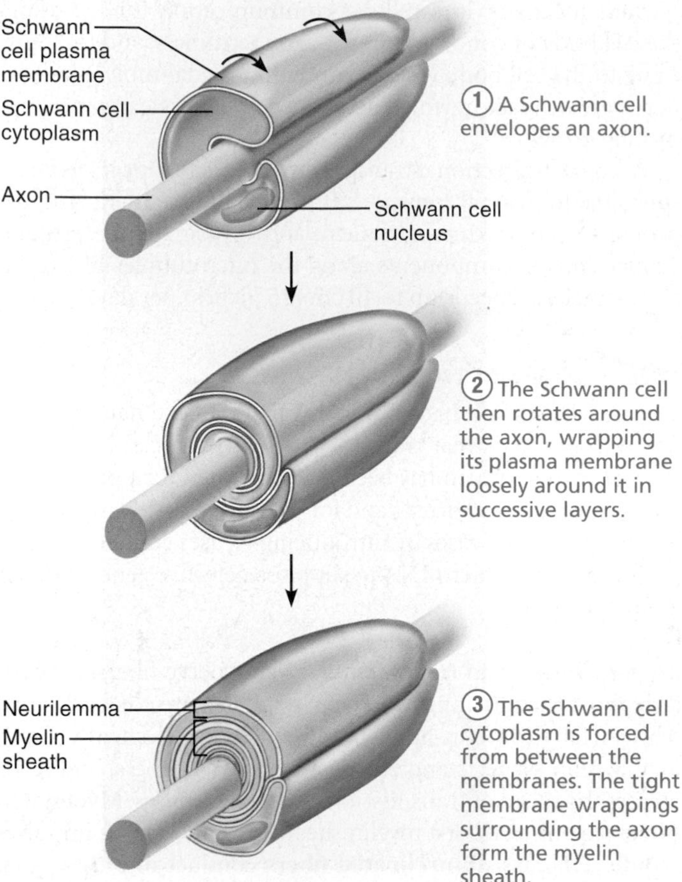

Schwann cell plasma membrane
Schwann cell cytoplasm
Axon
Schwann cell nucleus
① A Schwann cell envelopes an axon.
② The Schwann cell then rotates around the axon, wrapping its plasma membrane loosely around it in successive layers.
Neurilemma
Myelin sheath
③ The Schwann cell cytoplasm is forced from between the membranes. The tight membrane wrappings surrounding the axon form the myelin sheath.

(a) Myelination of a nerve fiber (axon)

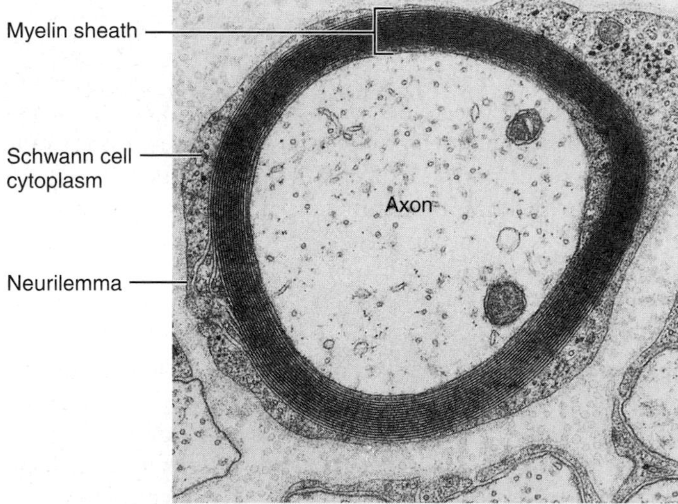

Myelin sheath
Schwann cell cytoplasm
Axon
Neurilemma

(b) Cross-sectional view of a myelinated axon (electron micrograph 24,000×)

Figure 11.5 Nerve fiber myelination by Schwann cells in the PNS.

11

Sometimes Schwann cells surround peripheral nerve fibers but the coiling process does not occur. In such instances, a single Schwann cell can partially enclose 15 or more axons, each of which occupies a separate recess in the Schwann cell surface. Nerve fibers associated with Schwann cells in this manner are said to be *unmyelinated* and are typically thin fibers.

Both myelinated and unmyelinated axons are also found in the central nervous system. However, oligodendrocytes are the cells that form CNS myelin sheaths (Figure 11.3d). In contrast to Schwann cells, each of which forms only one segment (internode) of a myelin sheath, oligodendrocytes have multiple flat processes that can coil around as many as 60 axons at the same time. As in the PNS, adjacent sections of an axon's myelin sheath are separated by nodes of Ranvier. CNS myelin sheaths lack a neurilemma because cell extensions are doing the coiling and the squeezed-out cytoplasm is forced not peripherally but back toward the centrally located nucleus. As in the PNS, the smallest-diameter axons are unmyelinated. These unmyelinated axons are covered by the long extensions of adjacent glial cells.

Regions of the brain and spinal cord containing dense collections of myelinated fibers are referred to as **white matter** and are primarily fiber tracts. **Gray matter** contains mostly nerve cell bodies and unmyelinated fibers.

CHECK YOUR UNDERSTANDING

5. Which part of the neuron is its fiber? How do nerve fibers differ from the fibers of connective tissue (see Chapter 4) and the fibers in muscle (see Chapter 9)?
6. How is a nucleus within the brain different from a nucleus within a neuron?
7. How is a myelin sheath formed in the CNS, and what is its function?

For answers, see Appendix G.

Classification of Neurons

► Classify neurons structurally and functionally.

Neurons are classified both structurally and functionally. We describe both classifications here but use the functional classification in most discussions.

Structural Classification Neurons are grouped structurally according to the number of processes extending from their cell body. Three major neuron groups make up this classification: multipolar (*polar* = end, pole), bipolar, and unipolar neurons. (**Table 11.1** is organized according to these three neuron types, and their structures are shown in the top row.)

Multipolar neurons have three or more processes—one axon and the rest dendrites. They are the most common neuron type in humans, with more than 99% of neurons belonging to this class. Multipolar neurons are the major neuron type in the CNS.

Bipolar neurons have two processes—an axon and a dendrite—that extend from opposite sides of the cell body. These rare neurons are found in some of the special sense organs. Examples include some neurons in the retina of the eye and in the olfactory mucosa.

Unipolar neurons have a single short process that emerges from the cell body and divides T-like into proximal and distal

TABLE 11.1 Comparison of Structural Classes of Neurons

NEURON TYPE		
MULTIPOLAR	**BIPOLAR**	**UNIPOLAR (PSEUDOUNIPOLAR)**
Structural Class: Neuron Type According to the Number of Processes Extending from the Cell Body		
Many processes extend from the cell body; all are dendrites except for a single axon.	Two processes extend from the cell body: One is a fused dendrite, the other is an axon.	One process extends from the cell body and forms central and peripheral processes, which together comprise an axon.
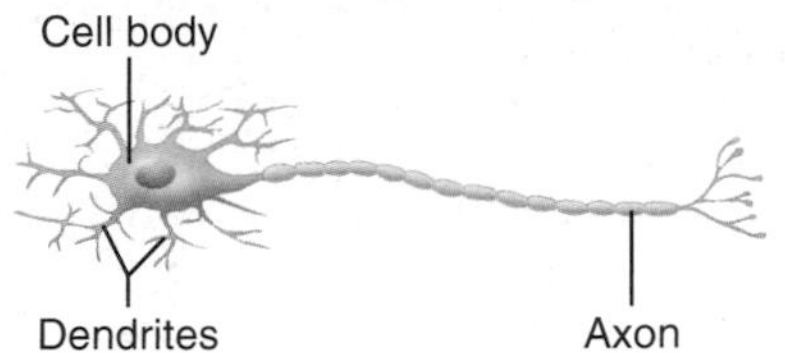	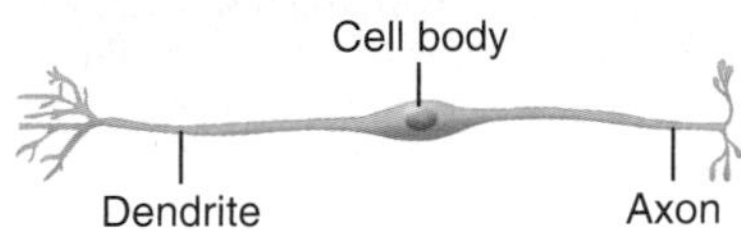	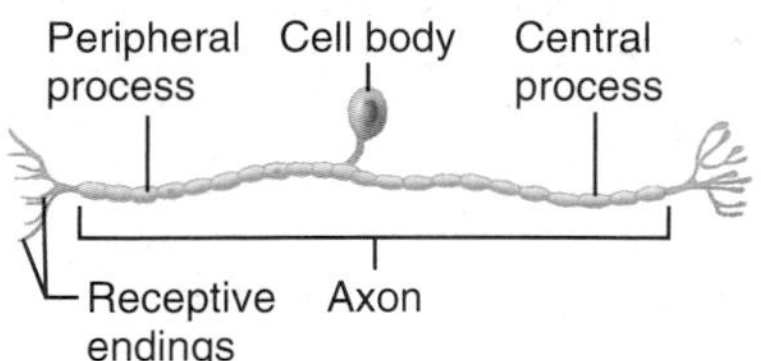
Relationship of Anatomy to the Three Functional Regions		
Receptive region (receives stimulus). Plasma membrane exhibits chemically gated ion channels.	Conducting region (generates/transmits action potential). Plasma membrane exhibits voltage-gated Na^+ and K^+ channels.	Secretory region (axon terminals release neurotransmitters). Plasma membrane exhibits voltage-gated Ca^+ channels.
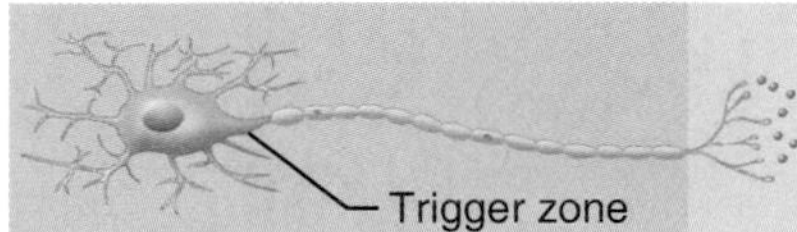	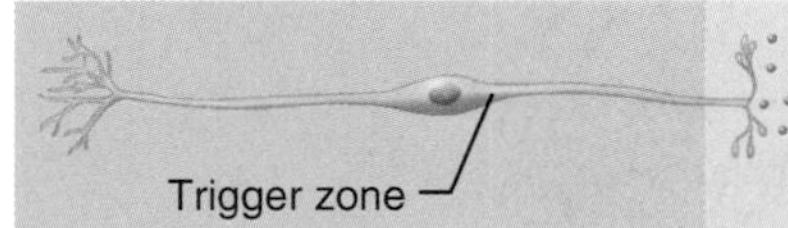 (Many bipolar neurons do not generate action potentials and, in those that do, the location of the trigger zone is not universal.)	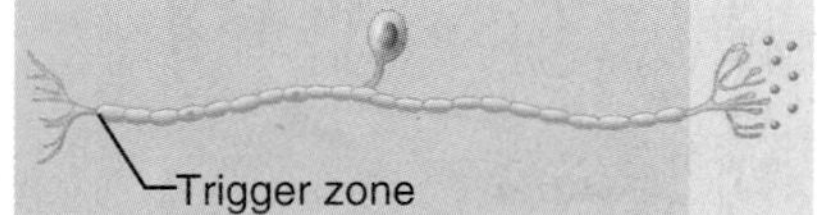
Relative Abundance and Location in Human Body		
Most abundant in body. Major neuron type in the CNS.	Rare. Found in some special sensory organs (olfactory mucosa, eye, ear).	Found mainly in the PNS. Common only in dorsal root ganglia of the spinal cord and sensory ganglia of cranial nerves.
Structural Variations		

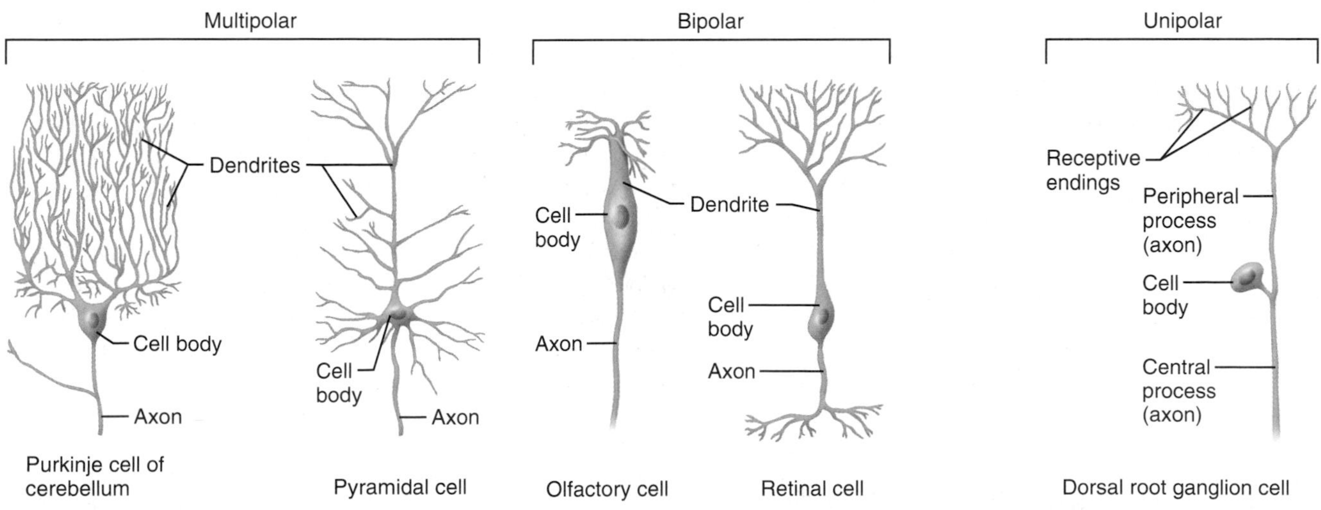

11

TABLE 11.1 **Comparison of Structural Classes of Neurons** *(continued)*

NEURON TYPE		
MULTIPOLAR	**BIPOLAR**	**UNIPOLAR (PSEUDOUNIPOLAR)**
Functional Class: Neuron Type According to Direction of Impulse Conduction		
1. Most multipolar neurons are **interneurons (association neurons)** that conduct impulses within the CNS, integrating sensory input or motor output; may be one of a chain of CNS neurons, or a single neuron connecting sensory and motor neurons. 2. Some multipolar neurons are **motor neurons** that conduct impulses along the efferent pathways from the CNS to an effector (muscle/gland).	Essentially all bipolar neurons are **sensory neurons** that are located in some special sense organs. For example, bipolar cells of the retina are involved with the transmission of visual inputs from the eye to the brain (via an intermediate chain of neurons).	Most unipolar neurons are **sensory neurons** that conduct impulses along afferent pathways to the CNS for interpretation. (These sensory neurons are called primary or first-order sensory neurons.)
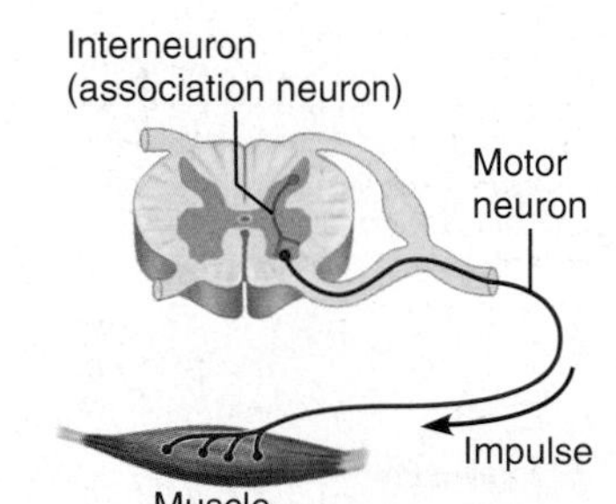	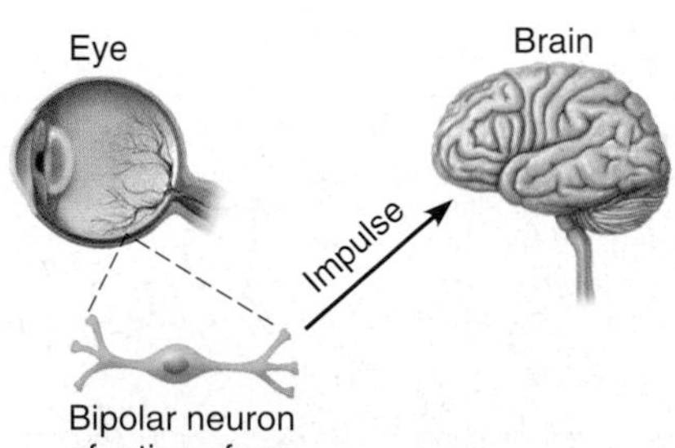	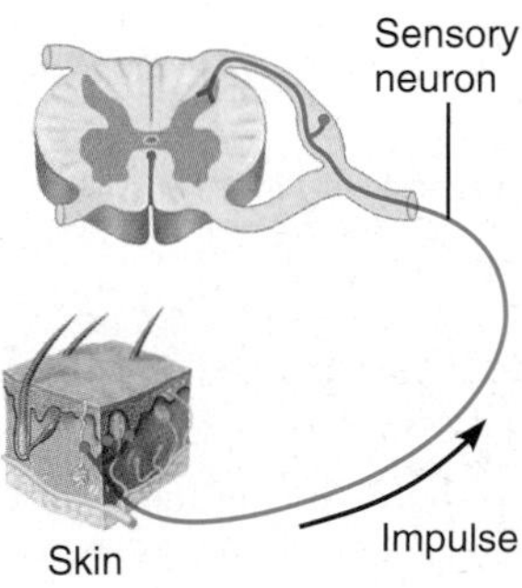

11

branches. The more distal process, which is often associated with a sensory receptor, is the **peripheral process**, whereas that entering the CNS is the **central process** (Table 11.1). Unipolar neurons are more accurately called **pseudounipolar neurons** (*pseudo* = false) because they originate as bipolar neurons. Then, during early embryonic development, the two processes converge and partially fuse to form the short single process that issues from the cell body. Unipolar neurons are found chiefly in ganglia in the PNS, where they function as sensory neurons.

The fact that the fused peripheral and central processes of unipolar neurons are continuous and function as a single fiber might make you wonder whether they are axons or dendrites. The central process is definitely an axon because it conducts impulses away from the cell body (one definition of axon). However, the peripheral process is perplexing. Three facts favor classifying it as an axon: (1) It generates and conducts an impulse (functional definition of axon); (2) when large, it is heavily myelinated; and (3) it has a uniform diameter and is indistinguishable microscopically from an axon. However, the older definition of a dendrite as a process that transmits impulses *toward* the cell body interferes with that conclusion.

So which is it? In this book, we have chosen to emphasize the newer definition of an axon as generating and transmitting an impulse. For *unipolar neurons,* we will refer to the combined length of the peripheral and central process as an axon. In place of "dendrites," unipolar neurons have *receptive endings* (sensory terminals) at the end of the peripheral process.

Functional Classification This scheme groups neurons according to the direction in which the nerve impulse travels relative to the central nervous system. Based on this criterion, there are sensory neurons, motor neurons, and interneurons (Table 11.1, last row).

Sensory, or **afferent, neurons** transmit impulses from sensory receptors in the skin or internal organs *toward* or *into* the central nervous system. Except for certain neurons found in some special sense organs, virtually all sensory neurons are unipolar, and their cell bodies are located in sensory ganglia *outside* the CNS. Only the most distal parts of these unipolar neurons act as impulse receptor sites, and the peripheral processes are often very long. For example, fibers carrying sensory impulses from the skin of your great toe travel for more than a meter before they reach their cell bodies in a ganglion close to the spinal cord.

The receptive endings of some sensory neurons are naked, in which case those terminals themselves function as sensory receptors, but many sensory neuron endings bear receptors that include other cell types. We describe the various types of general sensory receptor end organs, such as those of the skin, in Chapter 13. The special sensory receptors (of the ear, eye, etc.) are the topic of Chapter 15.

Motor, or **efferent, neurons** carry impulses *away from* the CNS to the effector organs (muscles and glands) of the body periphery. Motor neurons are multipolar. Except for some neurons of the autonomic nervous system, their cell bodies are located in the CNS.

Interneurons, or **association neurons**, lie between motor and sensory neurons in neural pathways and shuttle signals through CNS pathways where integration occurs. Most interneurons are confined within the CNS. They make up over 99% of the neurons of the body, including most of those in the CNS. Almost all interneurons are multipolar, but there is considerable diversity in both size and fiber-branching patterns. The Purkinje and pyramidal cells illustrated as structural variations in Table 11.1 are just two examples of their variety.

CHECK YOUR UNDERSTANDING

8. Which structural and functional type of neuron is activated first when you burn your finger? Which type is activated last to move your finger away from the source of heat?

For answers, see Appendix G.

Membrane Potentials

Neurons are highly *irritable* or *excitable* (responsive to stimuli). When a neuron is adequately stimulated, an electrical impulse is generated and conducted along the length of its axon. This response, called the *action potential* or *nerve impulse*, is always the same, regardless of the source or type of stimulus, and it underlies virtually all functional activities of the nervous system.

In this section, we will consider how neurons become excited or inhibited and how they communicate with other cells. First, however, we need to explore some basic principles of electricity and revisit the resting membrane potential.

Basic Principles of Electricity

The human body is electrically neutral—it has the same number of positive and negative charges. However, there are areas where one type of charge predominates, making such regions positively or negatively charged. Because opposite charges attract each other, energy must be used (work must be done) to separate them. On the other hand, the coming together of opposite charges liberates energy that can be used to do work. For this reason, situations in which there are separated electrical charges of opposite sign have potential energy.

Some Definitions: Voltage, Resistance, Current

The measure of potential energy generated by separated charge is called **voltage** and is measured in either *volts* (V) or *millivolts* (1 mV = 0.001 V). Voltage is always measured between two points and is called the **potential difference** or simply the **potential** between the points. The greater the difference in charge between two points, the higher the voltage.

The flow of electrical charge from one point to another is called a **current**, and it can be used to do work—for example, to power a flashlight. The amount of charge that moves between the two points depends on two factors: voltage and resistance. **Resistance** is the hindrance to charge flow provided by substances through which the current must pass. Substances with high electrical resistance are called *insulators*, and those with low resistance are called *conductors*.

The relationship between voltage, current, and resistance is given by **Ohm's law**:

$$\text{Current } (I) = \frac{\text{voltage } (V)}{\text{resistance}(R)}$$

which tells us that current (I) is directly proportional to voltage. This means that the greater the voltage (potential difference), the greater the current. This relationship also tells us there is no net current flow between points that have the same potential, as you can see by inserting a value of 0 V into the equation. A third thing Ohm's law tells us is that current is inversely related to resistance: The greater the resistance, the smaller the current.

In the body, electrical currents reflect the flow of ions (rather than free electrons) across cellular membranes. (Unlike the electrons flowing along your house wiring, there are no free electrons "running around" in a living system.) As we described in Chapter 3, there is a slight difference in the numbers of positive and negative ions on the two sides of cellular plasma membranes (there is a charge separation), so there is a potential across those membranes. The resistance to current flow is provided by the plasma membranes.

Role of Membrane Ion Channels

Recall that plasma membranes are peppered with a variety of membrane proteins that act as *ion channels.* Each of these channels is selective as to the type of ion (or ions) it allows to pass. For example, a potassium ion channel allows only potassium ions to pass.

Membrane channels are large proteins, often with several subunits, whose amino acid chains snake back and forth across the membrane. Some channels, **leakage** or **nongated channels**, are always open. In other channels, part of the protein forms a molecular "gate" that changes shape to open and close the channel in response to specific signals. These are called *gated* channels.

Chemically gated, or **ligand-gated, channels** open when the appropriate chemical (in this case a neurotransmitter) binds (Figure 11.6a). **Voltage-gated channels** open and close in response to changes in the membrane potential (Figure 11.6b). **Mechanically gated channels** open in response to physical deformation of the receptor (as in sensory receptors for touch and pressure).

When gated ion channels are open, ions diffuse quickly across the membrane following their electrochemical gradients, creating electrical currents and voltage changes across the membrane according to the rearranged Ohm's law equation:

$$\text{Voltage } (V) = \text{current } (I) \times \text{resistance } (R)$$

Ions move along chemical *concentration gradients* when they diffuse passively from an area of their higher concentration to an area of lower concentration, and along *electrical gradients* when they move toward an area of opposite electrical charge. Together, electrical and concentration gradients constitute the **electrochemical gradient**. It is ion flows along electrochemical gradients that underlie all electrical phenomena in neurons.

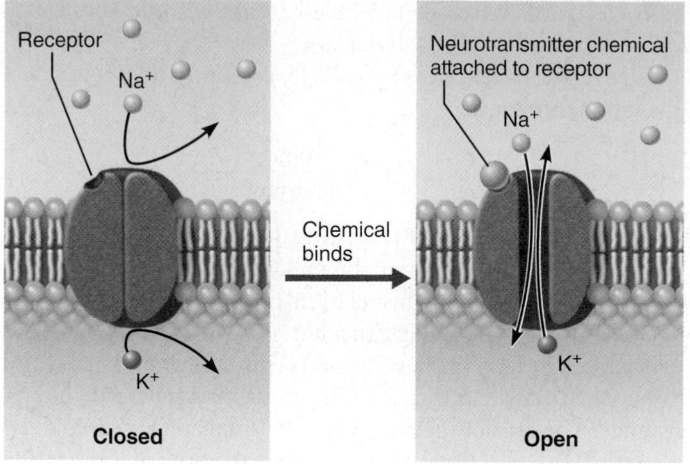

(a) Chemically (ligand) gated ion channels open when the appropriate neurotransmitter binds to the receptor, allowing (in this case) simultaneous movement of Na^+ and K^+.

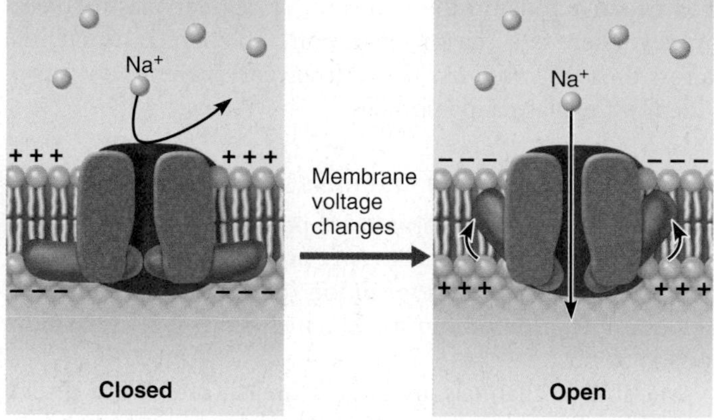

11

(b) Voltage-gated ion channels open and close in response to changes in membrane voltage.

Figure 11.6 Operation of gated channels.

The Resting Membrane Potential

▶ Define resting membrane potential and describe its electrochemical basis.

The potential difference between two points is measured with a voltmeter. When one microelectrode of the voltmeter is inserted into the neuron and the other is in the extracellular fluid, a voltage across the membrane of approximately −70 mV is recorded **(Figure 11.7)**. The minus sign indicates that the cytoplasmic side (inside) of the membrane is negatively charged relative to the outside. This potential difference in a resting neuron (V_r) is called the **resting membrane potential**, and the membrane is said to be **polarized**. The value of the resting membrane potential varies (from −40 mV to −90 mV) in different types of neurons.

The resting potential exists only across the membrane. In other words, the bulk solutions inside and outside the cell are electrically neutral. The resting membrane potential is generated by differences in the ionic makeup of the intracellular and extracellular fluids and by the differential permeability of the plasma membrane to those ions.

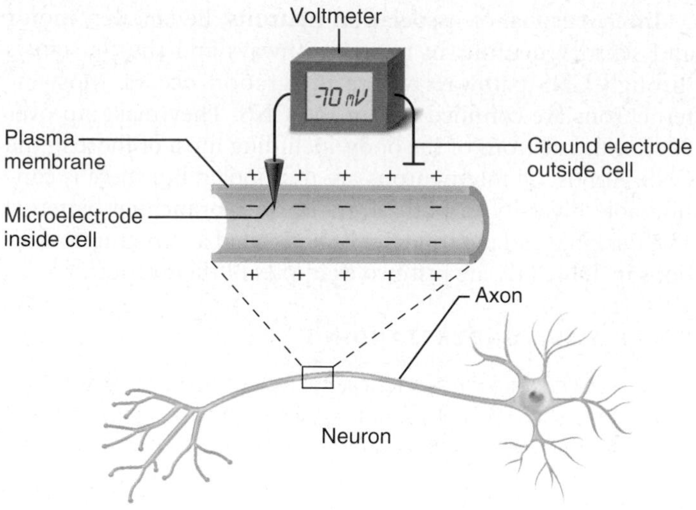

Figure 11.7 Measuring membrane potential in neurons. The potential difference between an electrode inside a neuron and the ground electrode in the extracellular fluid is approximately −70 mV (inside negative).

First, let's look at differences in ionic makeup of the intracellular and extracellular fluids, as shown in *Focus on Resting Membrane Potential* **(Figure 11.8)**. The cell cytosol contains a lower concentration of Na^+ and a higher concentration of K^+ than the extracellular fluid. Negatively charged (anionic) proteins (A^-) help to balance the positive charges of intracellular cations (primarily K^+). In the extracellular fluid, the positive charges of Na^+ and other cations are balanced chiefly by chloride ions (Cl^-). Although there are many other solutes (glucose, urea, and other ions) in both fluids, potassium (K^+) plays the most important role in generating the membrane potential.

Next, let's consider the differential permeability of the membrane to various ions (Figure 11.8, bottom). At rest the membrane is impermeable to the large anionic cytoplasmic proteins, very slightly permeable to sodium, approximately 75 times more permeable to potassium than to sodium, and quite freely permeable to chloride ions. These resting permeabilities reflect the properties of the leakage ion channels in the membrane. Potassium ions diffuse out of the cell along their *concentration gradient* much more easily than sodium ions can enter the cell along theirs. K^+ flowing out of the cell causes the cell to become more negative inside. Na^+ trickling into the cell makes the cell just slightly more positive than it would be if only K^+ flowed. Therefore, at resting membrane potential, the negative interior of the cell is due to much greater diffusion of K^+ out of the cell than Na^+ diffusion into the cell.

Because some K^+ is always leaking out of the cell and some Na^+ is always leaking in, you might think that the concentration gradients would eventually "run down," resulting in equal concentrations of Na^+ and K^+ inside and outside the cell. This does not happen because the ATP-driven **sodium-potassium pump** first ejects three Na^+ from the cell and then transports two K^+ back into the cell. In other words, the sodium-potassium pump

Resting Membrane Potential

Generating a resting membrane potential depends on (1) differences in K^+ and Na^+ concentrations inside and outside cells, and (2) differences in permeability of the plasma membrane to these ions.

The concentrations of Na^+ and K^+ on each side of the membrane are different.

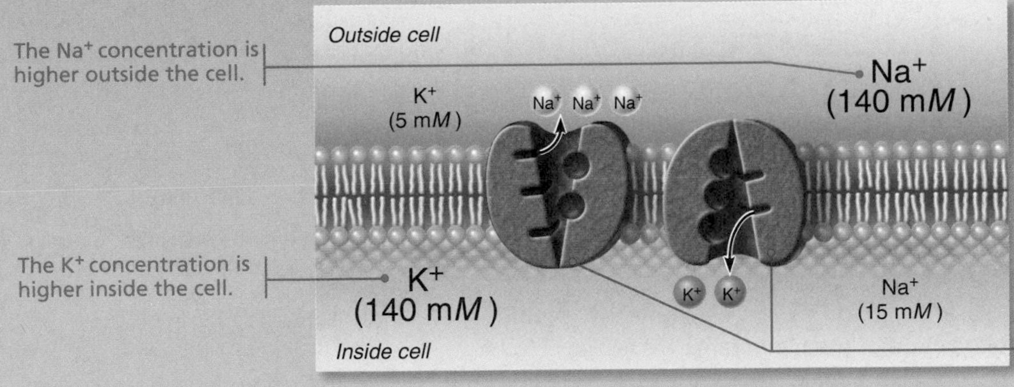

The permeabilities of Na^+ and K^+ across the membrane are different. In the next three panels, we will build the resting membrane potential step by step.

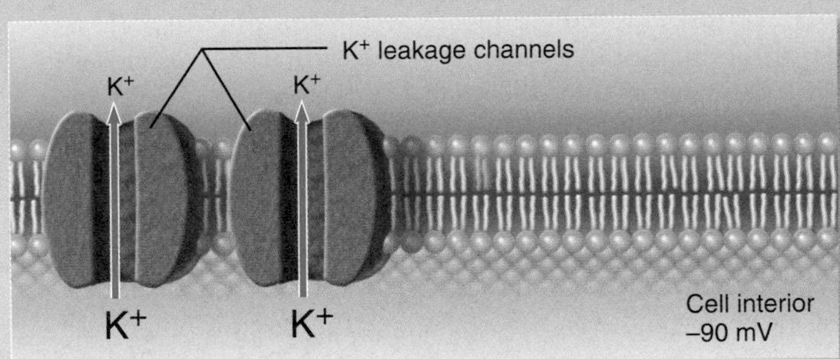

Suppose a cell has only K^+ channels...

K^+ loss through abundant leakage channels establishes a negative membrane potential. K^+ flows down its large concentration gradient because the membrane is highly permeable to K^+. As the positive K^+ ions leak out, the negative voltage that develops on the membrane interior counteracts the concentration gradient, pulling K^+ back into the cell. At –90 mV, the concentration and electrical gradients for K^+ are balanced.

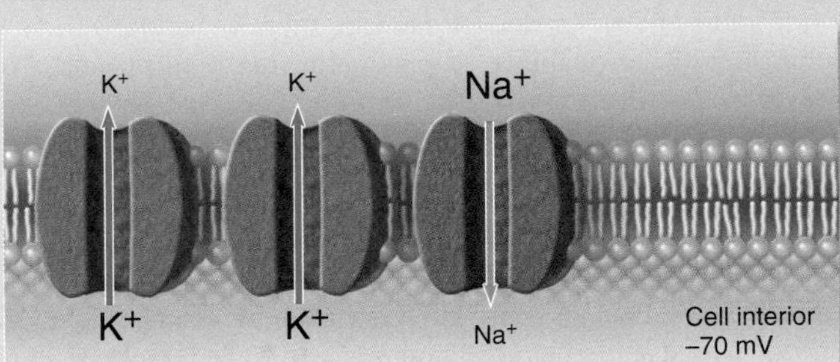

Now, let's add some Na^+ channels to our cell...

Na^+ entry through leakage channels reduces the negative membrane potential slightly. Na^+ flows down its large concentration gradient, but the membrane is only slightly permeable to Na^+. As a result, Na^+ entering the cell makes the membrane potential slightly less negative than if there were only K^+ channels.

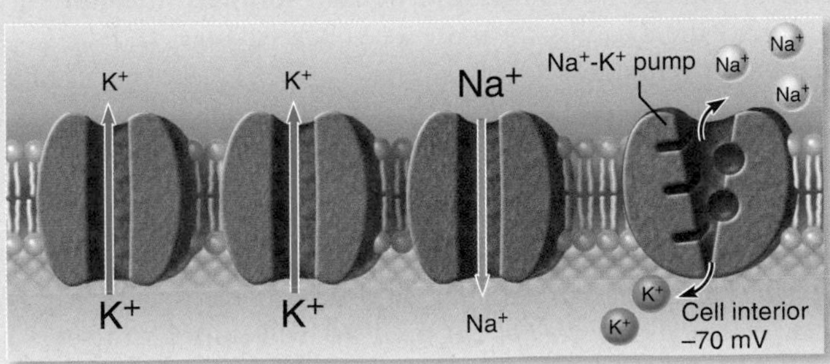

Finally, let's add a pump to compensate for leaking ions.

Na^+-K^+ ATPases (pumps) maintain the concentration gradients, resulting in the resting membrane potential. A cell at rest is like a leaky boat that is constantly leaking K^+ out and Na^+ in through open channels. The "bailing pump" for this boat is the **Na^+-K^+ ATPase** (Na^+-K^+ pump), which counteracts the leaks by transporting Na^+ out and K^+ in.

A&P Flix *View this animation at myA&P*

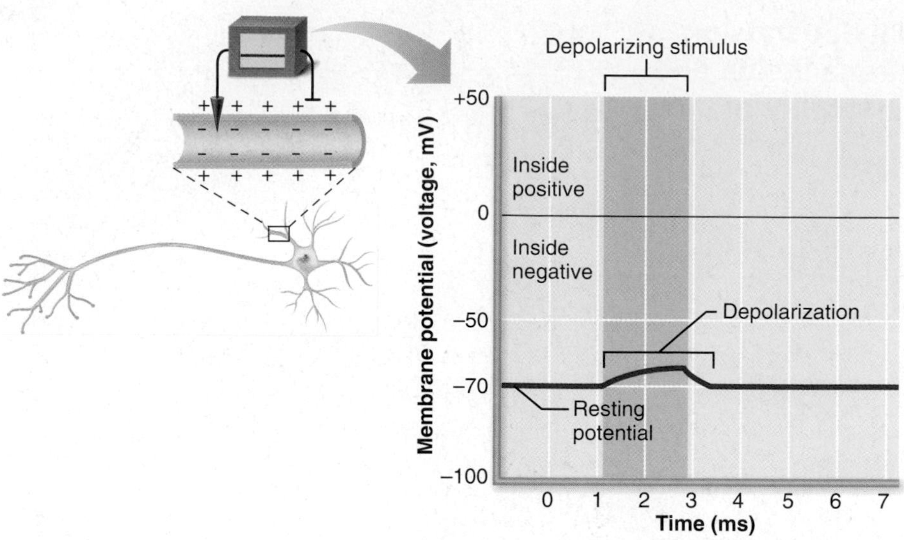

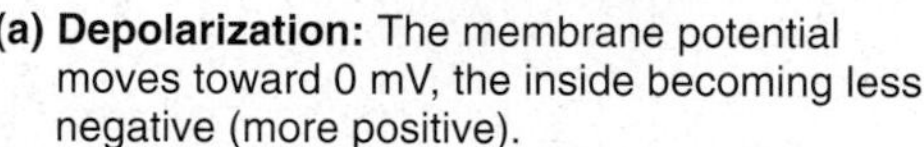
(a) Depolarization: The membrane potential moves toward 0 mV, the inside becoming less negative (more positive).

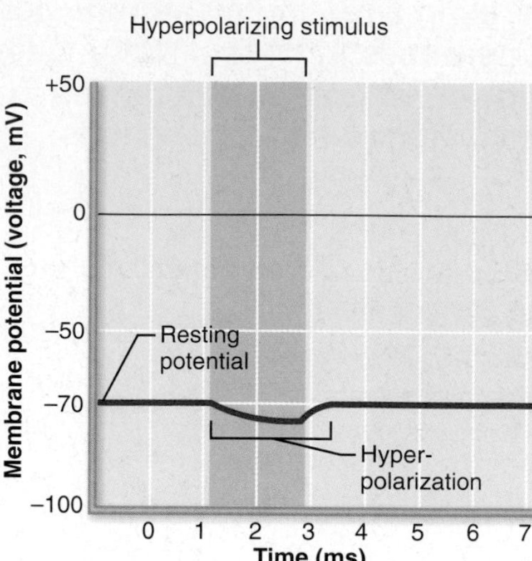

(b) Hyperpolarization: The membrane potential increases, the inside becoming more negative.

Figure 11.9 Depolarization and hyperpolarization of the membrane. The resting membrane potential is approximately −70 mV (inside negative) in neurons.

stabilizes the resting membrane potential by maintaining the concentration gradients for sodium and potassium.

CHECK YOUR UNDERSTANDING

9. For an open channel, what factors determine in which direction ions will move through that channel?
10. For which cation is there the greatest amount of leakage (through leakage channels) across the plasma membrane?

For answers, see Appendix G.

Membrane Potentials That Act as Signals

- ▶ Compare and contrast graded potentials and action potentials.
- ▶ Explain how action potentials are generated and propagated along neurons.
- ▶ Define absolute and relative refractory periods.
- ▶ Define saltatory conduction and contrast it to conduction along unmyelinated fibers.

Neurons use changes in their membrane potential as communication signals for receiving, integrating, and sending information. A change in membrane potential can be produced by (1) anything that alters ion concentrations on the two sides of the membrane or (2) anything that changes membrane permeability to any ion. However, only permeability changes are important for information transfer.

Changes in membrane potential can produce two types of signals: *graded potentials*, which are usually incoming signals operating over short distances, and *action potentials*, which are long-distance signals of axons.

The terms *depolarization* and *hyperpolarization* describe membrane potential changes *relative to resting membrane potential.* It is important to clearly understand these terms. **Depolarization** is a reduction in membrane potential: The inside of the membrane becomes *less negative* (moves closer to zero) than the resting potential. For instance, a change in resting potential from −70 mV to −65 mV is a depolarization (**Figure 11.9a**). By convention, depolarization also includes events in which the membrane potential reverses and moves above zero to become positive.

Hyperpolarization occurs when the membrane potential increases, becoming *more negative* than the resting potential. For example, a change from −70 mV to −75 mV is hyperpolarization (Figure 11.9b). As we will describe shortly, depolarization increases the probability of producing nerve impulses, whereas hyperpolarization reduces this probability.

Graded Potentials

Graded potentials are short-lived, localized changes in membrane potential that can be either depolarizations or hyperpolarizations. These changes cause current flows that decrease in magnitude with distance. Graded potentials are called "graded" because their magnitude varies directly with stimulus strength. The stronger the stimulus, the more the voltage changes and the farther the current flows.

Graded potentials are triggered by some change (a stimulus) in the neuron's environment that causes gated ion channels to open. Graded potentials are given different names, depending on where they occur and the functions they perform. When the receptor of a sensory neuron is excited by some form of energy (heat, light, or other), the resulting graded potential is called a *receptor potential* or *generator potential.* We will consider these types of graded potentials in Chapter 13. When the stimulus is a neurotransmitter released by another neuron, the graded potential is called a *postsynaptic potential*, because the neurotransmitter is released into a fluid-filled gap called a synapse and influences the neuron beyond (post) the synapse.

Fluids inside and outside cells are fairly good conductors, and current, carried by ions, flows through these fluids whenever voltage changes occur. Let us assume that a small area of a neuron's plasma membrane has been depolarized by a stimulus **(Figure 11.10a)**. Current (ions) will flow on both sides of the membrane between the depolarized (active) membrane area and the adjacent polarized (resting) areas. Positive ions migrate toward more negative areas (the direction of cation movement is the direction of current flow), and negative ions simultaneously move toward more positive areas (Figure 11.10b).

For our patch of plasma membrane, positive ions (mostly K^+) inside the cell move away from the depolarized area and accumulate on the neighboring membrane areas, where they neutralize negative ions. Meanwhile, positive ions on the outer membrane face are moving toward the region of reversed membrane polarity (the depolarized region), which is momentarily less positive. As these positive ions move, their "places" on the membrane become occupied by negative ions (such as Cl^- and HCO_3^-), sort of like ionic musical chairs. In this way, at regions abutting the depolarized region, the inside becomes less negative and the outside becomes less positive. In other words, the depolarization spreads as the neighboring membrane is, in turn, depolarized.

As just explained, the flow of current to adjacent membrane areas changes the membrane potential there as well. However, the plasma membrane is permeable like a leaky water hose, and most of the charge is quickly lost through leakage channels. Consequently, the current dies out within a few millimeters of its origin and is said to be *decremental* (Figure 11.10c).

Because the current dissipates quickly and dies out (decays) with increasing distance from the site of initial depolarization, graded potentials can act as signals only over very short distances. Nonetheless, they are essential in initiating action potentials, the long-distance signals.

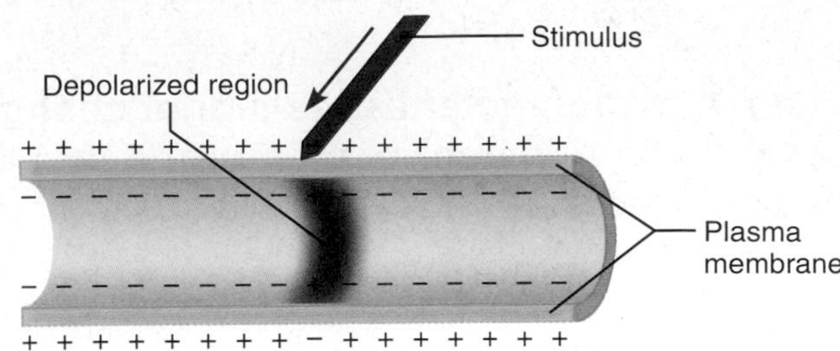

(a) Depolarization: A small patch of the membrane (red area) has become depolarized.

(b) Spread of depolarization: The local currents (black arrows) that are created depolarize adjacent membrane areas and allow the wave of depolarization to spread.

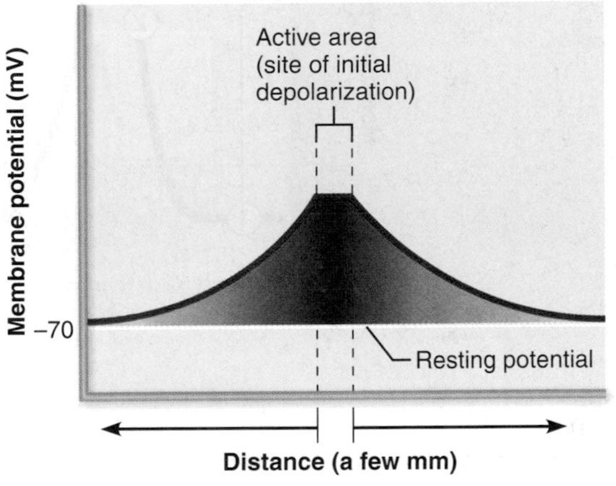

(c) Decay of membrane potential with distance: Because current is lost through the "leaky" plasma membrane, the voltage declines with distance from the stimulus (the voltage is *decremental*). Consequently, graded potentials are short-distance signals.

Figure 11.10 **The spread and decay of a graded potential.**

Action Potentials

The principal way neurons send signals over long distances is by generating and propagating action potentials (APs), and only cells with *excitable membranes*—neurons and muscle cells—can generate action potentials. An **action potential (AP)** is a brief reversal of membrane potential with a total amplitude (change in voltage) of about 100 mV (from −70 mV to +30 mV). A depolarization phase is followed by a repolarization phase and often a short period of hyperpolarization. The whole event is over in a few milliseconds. Unlike graded potentials, action potentials do not decrease in strength with distance.

The events of action potential generation and transmission are identical in skeletal muscle cells and neurons. As we have noted, in a neuron, an AP is also called a **nerve impulse**, and is typically generated *only in axons*. A neuron transmits a nerve impulse only when it is adequately stimulated. The stimulus changes the permeability of the neuron's membrane by opening specific voltage-gated channels on the axon. These channels open and close in response to changes in the membrane potential and are activated by local currents (graded potentials) that spread toward the axon along the dendritic and cell body membranes.

In many neurons, the transition from local graded potential to long-distance action potential takes place at the axon hillock. In sensory neurons, the action potential is generated by the peripheral (axonal) process just proximal to the receptor region. However, for simplicity, we will just use the term *axon* in our discussion. We'll look first at the generation of an action potential and then at its propagation.

Generation of an Action Potential *Focus on an Action Potential* **(Figure 11.11)** on pp. 400–401 describes the generation of an action potential. Generating an action potential involves three consecutive but overlapping changes in membrane permeability

resulting from the opening and closing of voltage-gated ion channels, all induced by depolarization of the axon membrane. These permeability changes are a transient increase in Na^+ permeability (Na^+ channels open), followed by restoration of Na^+ impermeability (described below), together with a short-lived increase in K^+ permeability (K^+ channels open, then close).

The first two permeability changes mark the beginning and the end of the *depolarization phase* of action potential generation, indicated by the upward-rising part of the AP curve or spike on the first graph in Figure 11.11. The third permeability change is responsible for both the *repolarization* (the downward part of the AP spike) and *hyperpolarization phases* shown in the graph. Let's examine each of these phases more carefully, starting with a neuron in the resting (polarized) state.

① **Resting state: All gated Na^+ and K^+ channels are closed.** Only the leakage channels are open, maintaining resting membrane potential.

Each Na^+ channel has two gates: a voltage-sensitive *activation gate* that is closed at rest and responds to depolarization by opening, and an *inactivation gate* that blocks the channel once it is open. Thus, *depolarization opens and then inactivates sodium channels.* Both gates must be open in order for Na^+ to enter, but the closing of *either* gate effectively closes the channel. By contrast, each active potassium channel has a single voltage-sensitive gate that is closed in the resting state and opens slowly in response to depolarization.

② **Depolarizing phase: Na^+ channels open.** As the axon membrane is depolarized by local currents, the voltage-gated sodium channels open and Na^+ rushes into the cell. This influx of positive charge depolarizes that local "patch" of membrane further, opening more Na^+ channels so that the cell interior becomes progressively less negative. When depolarization at the stimulation site reaches a certain critical level called **threshold** (often between -55 and -50 mV), depolarization becomes self-generating, urged on by positive feedback. That is, after being initiated by the stimulus, depolarization is driven by the ionic currents created by Na^+ influx. As more Na^+ enters, the membrane depolarizes further and opens still more channels until all Na^+ channels are open. At this point, Na^+ permeability is about 1000 times greater than in a resting neuron. As a result, the membrane potential becomes less and less negative and then overshoots to about $+30$ mV as Na^+ rushes in along its electrochemical gradient. This rapid depolarization and polarity reversal produces the sharply upward *spike* of the action potential.

Earlier, we stated that membrane potential depends on membrane permeability, but here we are saying that membrane permeability depends on membrane potential. Can both statements be true? Yes, because these two relationships establish a *positive feedback* cycle (increased Na^+ permeability due to increased channel openings leads to greater depolarization, which leads to increased Na^+ permeability, and so on). This explosive positive feedback cycle is responsible for the rising (depolarizing) phase of action potentials and puts the "action" in the action potential.

③ **Repolarizing phase: Na^+ channels are inactivating, and K^+ channels open.** The explosively rising phase of the action potential persists for only about 1 ms. It is self-limiting because the slow inactivation gates of the Na^+ channels begin to close at this point. As a result, the membrane permeability to Na^+ declines to resting levels, and the net influx of Na^+ stops completely. Consequently, the AP spike stops rising.

As Na^+ entry declines, the slow voltage-gated K^+ channels open and K^+ rushes out of the cell, following its electrochemical gradient. Consequently, internal negativity of the resting neuron is restored, an event called **repolarization**. Both the abrupt decline in Na^+ permeability and the increased permeability to K^+ contribute to repolarization.

④ **Hyperpolarization: Some K^+ channels remain open, and Na^+ channels reset.** The period of increased K^+ permeability typically lasts longer than needed to restore the resting state. As a result of the excessive K^+ efflux, an **after-hyperpolarization**, also called the *undershoot*, is seen on the AP curve as a slight dip following the spike (and before the potassium gates close). Also at this point, the Na^+ channels begin to reset back to their original position by changing shape to reopen their inactivation gates and close their activation gates.

Repolarization restores resting electrical conditions, but it does *not* restore resting ionic conditions. The ion redistribution is accomplished by the *sodium-potassium pump* following repolarization. While it might appear that tremendous numbers of Na^+ and K^+ ions change places during action potential generation, this is not the case. Only small amounts of sodium and potassium cross the membrane. (The Na^+ influx required to reach threshold produces only a 0.012% change in intracellular Na^+ concentration.) These small ionic changes are quickly corrected because an axon membrane has thousands of Na^+-K^+ pumps.

Propagation of an Action Potential If it is to serve as the neuron's signaling device, an AP must be **propagated** (sent or transmitted) along the axon's entire length. As we have seen, the AP is generated by the influx of Na^+ through a given area of the membrane. This influx establishes local currents that depolarize adjacent membrane areas in the forward direction (away from the origin of the nerve impulse), which opens voltage-gated channels and triggers an action potential there **(Figure 11.12)**.

Because the area where the AP originated has just generated an AP, the sodium channels in that area are inactivated and no new AP is generated there. For this reason, the AP propagates away from its point of origin. (If an *isolated* axon is stimulated by an electrode, the nerve impulse will move away from the point of stimulus in all directions along the membrane.) In the body, APs are initiated at one end of the axon and conducted away from that point toward the axon's terminals. Once initiated,

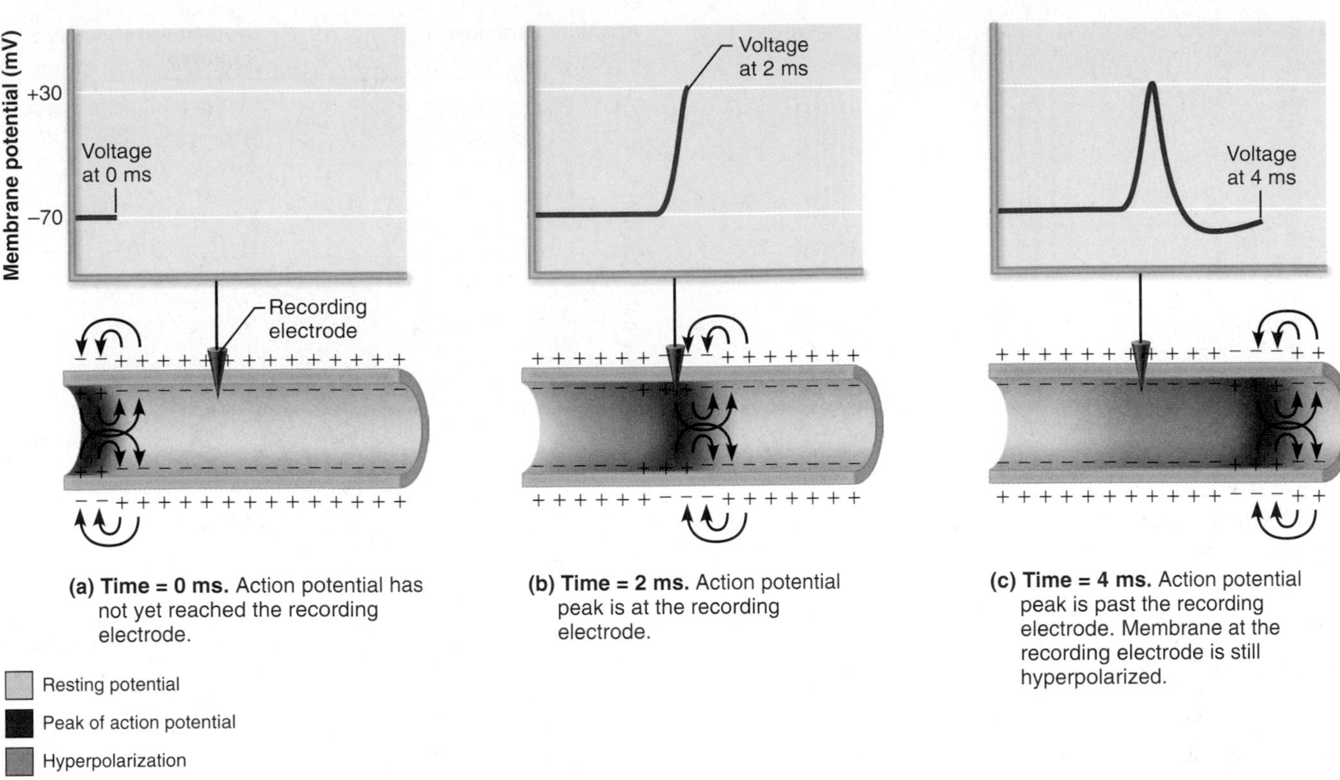

Membrane potential (mV)
+30
−70
Voltage
at 0 ms
Recording
electrode
Voltage
at 2 ms
Voltage
at 4 ms
(a) Time = 0 ms. Action potential has not yet reached the recording electrode.
(b) Time = 2 ms. Action potential peak is at the recording electrode.
(c) Time = 4 ms. Action potential peak is past the recording electrode. Membrane at the recording electrode is still hyperpolarized.
Resting potential
Peak of action potential
Hyperpolarization

Figure 11.12 Propagation of an action potential (AP). Recordings at three successive times as an AP propagates along an axon (from left to right). The arrows show the direction of local current flow generated by the movement of positive ions. This current brings the resting membrane at the leading edge of the AP to threshold, propagating the AP forward.

an AP is *self-propagating* and continues along the axon at a constant velocity—something like a domino effect.

Following depolarization, each segment of axon membrane repolarizes, which restores the resting membrane potential in that region. Because these electrical changes also set up local currents, the repolarization wave chases the depolarization wave down the length of the axon.

The propagation process we have just described occurs on unmyelinated axons. On p. 405, we will describe propagation along myelinated axons, called *saltatory conduction.*

Although the phrase *conduction of a nerve impulse* is commonly used, nerve impulses are not really conducted in the same way that an insulated wire conducts current. In fact, neurons are fairly poor conductors, and as noted earlier, local current flows decline with distance because the charges leak through the membrane. The expression *propagation of a nerve impulse* is more accurate, because the AP is *regenerated anew* at each membrane patch, and every subsequent AP is identical to the one that was generated initially.

Threshold and the All-or-None Phenomenon Not all local depolarization events produce APs. The depolarization must reach threshold values if an axon is to "fire." What determines the *threshold point*? One explanation is that threshold is the membrane potential at which the outward current created by K^+ movement is exactly equal to the inward current created by Na^+ movement. Threshold is typically reached when the membrane has been depolarized by 15 to 20 mV from the resting value. This depolarization status represents an unstable equilibrium state. If one more Na^+ enters, further depolarization occurs, opening more Na^+ channels and allowing more Na^+ entry. If, on the other hand, one more K^+ leaves, the membrane potential is driven away from threshold, Na^+ channels close, and K^+ continues to diffuse outward until the potential returns to its resting value.

Recall that local depolarizations are graded potentials and that their magnitude increases with increasing stimulus intensity. Brief weak stimuli (*subthreshold stimuli*) produce subthreshold depolarizations that are not translated into nerve impulses. On the other hand, stronger *threshold stimuli* produce depolarizing currents that push the membrane potential toward and beyond the threshold voltage. As a result, Na^+ permeability is increased to such an extent that entering sodium ions "swamp" (exceed) the outward movement of K^+, allowing the positive feedback cycle to become established and generating an AP.

The critical factor here is the total amount of current that flows through the membrane during a stimulus (electrical charge $\times$ time). Strong stimuli depolarize the membrane to threshold quickly. Weaker stimuli must be applied for longer periods to provide the crucial amount of current flow. Very weak stimuli do not trigger an AP because the local current

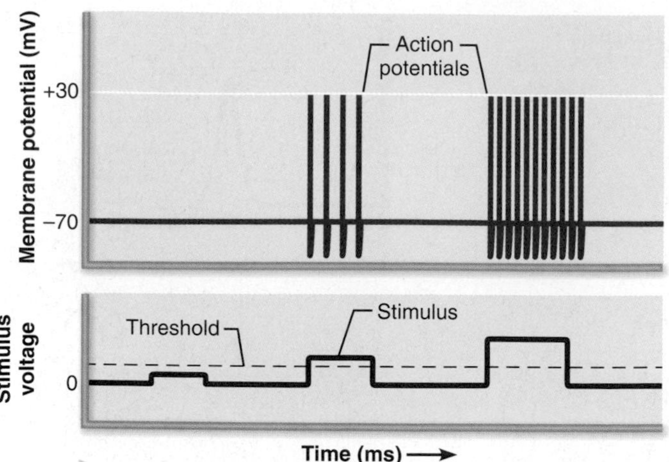

Figure 11.13 Relationship between stimulus strength and action potential frequency. APs are shown as vertical lines in the upper trace. The lower trace shows the intensity of the applied stimulus. A subthreshold stimulus does not generate an AP, but once threshold voltage is reached, the stronger the stimulus, the more frequently APs are generated.

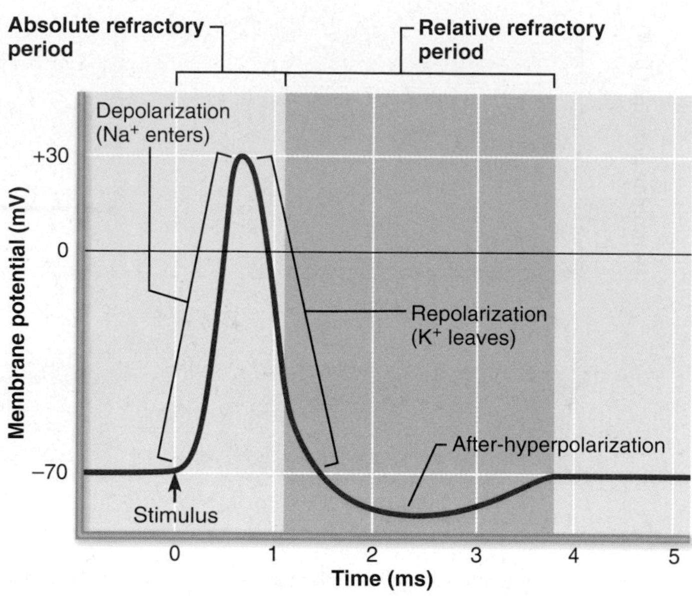

Figure 11.14 Absolute and relative refractory periods in an AP.

flows they produce are so slight that they dissipate long before threshold is reached.

11 The AP is an **all-or-none phenomenon**: It either happens completely or doesn't happen at all. Generation of an AP can be compared to lighting a match under a small dry twig. The changes occurring where the twig is heated are analogous to the change in membrane permeability that initially allows more Na^+ to enter the cell. When that part of the twig becomes hot enough (when enough Na^+ has entered the cell), the flash point (threshold) is reached and the flame consumes the entire twig, even if you blow out the match (the AP is generated and propagated whether or not the stimulus continues). But if the match is extinguished just before the twig has reached the critical temperature, ignition will not take place. Likewise, if the number of Na^+ ions entering the cell is too low to achieve threshold, no AP will occur.

Coding for Stimulus Intensity Once generated, all APs are independent of stimulus strength, and all APs are alike. So how can the CNS determine whether a particular stimulus is intense or weak—information it needs to initiate an appropriate response? The answer is really quite simple: Strong stimuli cause nerve impulses to be generated more *often* in a given time interval than do weak stimuli. In this way, stimulus intensity is coded for by the number of impulses per second—that is, by the *frequency of action potentials*—rather than by increases in the strength (amplitude) of the individual APs **(Figure 11.13)**.

Refractory Periods When a patch of neuron membrane is generating an AP and its voltage-gated sodium channels are open, the neuron cannot respond to another stimulus, no matter how strong. This period, from the opening of the Na^+ channels until the Na^+ channels begin to reset to their original resting state, is called the **absolute refractory period (Figure 11.14)**. It ensures that each AP is a separate, *all-or-none event* and enforces one-way transmission of the AP.

The **relative refractory period** is the interval following the absolute refractory period. During the relative refractory period, most Na^+ channels have returned to their resting state, some K^+ channels are still open, and repolarization is occurring. During this time, the axon's threshold for AP generation is substantially elevated. A stimulus that would normally have generated an AP is no longer sufficient, but an exceptionally strong stimulus can reopen the Na^+ channels that have already returned to their resting state and allow another AP to be generated. Strong stimuli cause more frequent generation of APs by intruding into the relative refractory period.

Conduction Velocity How fast do APs travel? Conduction velocities of neurons vary widely. Nerve fibers that transmit impulses most rapidly (100 m/s or more) are found in neural pathways where speed is essential, such as those that mediate some postural reflexes. Axons that conduct impulses more slowly typically serve internal organs (the gut, glands, blood vessels), where slower responses are not a handicap. The rate of impulse propagation depends largely on two factors:

1. **Axon diameter.** Axons vary considerably in diameter and, as a rule, the larger the axon's diameter, the faster it conducts impulses. Larger axons conduct more rapidly because they offer less resistance to the flow of local currents, and so adjacent areas of the membrane can more quickly be brought to threshold.
2. **Degree of myelination.** Action potentials propagate because they are regenerated by voltage-gated channels in the membrane **(Figure 11.15a, b)**. On unmyelinated axons, these channels are immediately adjacent to each other and conduction is relatively slow, a type of AP propagation called **continuous conduction**. The presence of a myelin

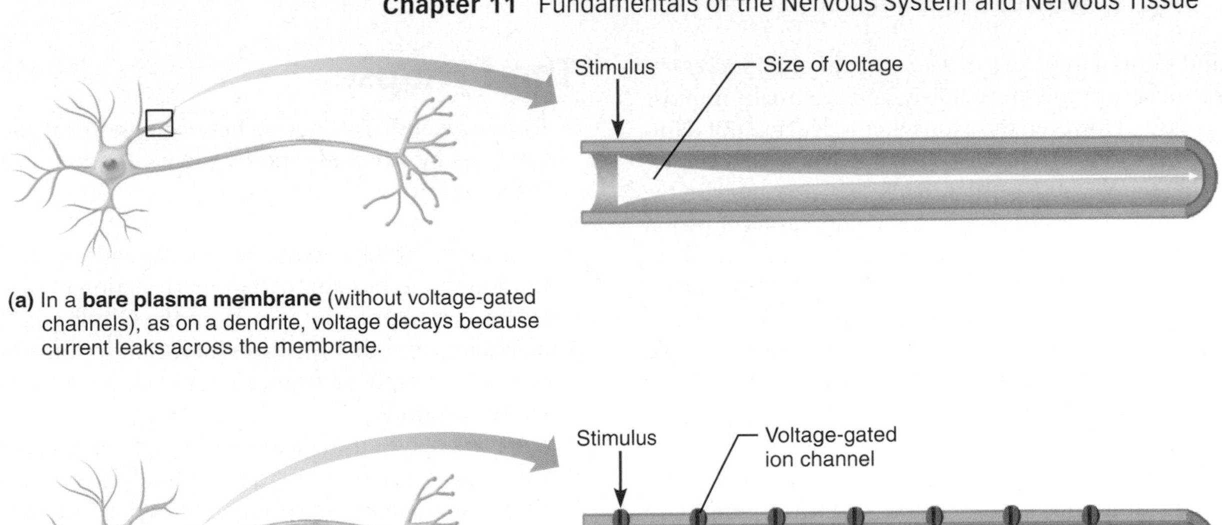

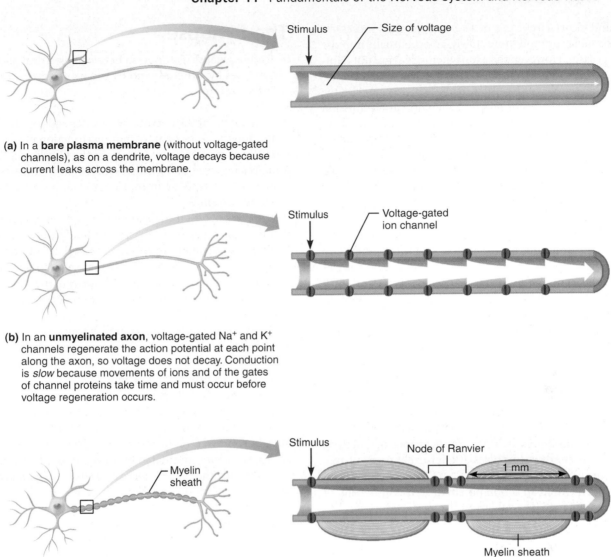

(a) In a **bare plasma membrane** (without voltage-gated channels), as on a dendrite, voltage decays because current leaks across the membrane.

(b) In an **unmyelinated axon**, voltage-gated Na^+ and K^+ channels regenerate the action potential at each point along the axon, so voltage does not decay. Conduction is *slow* because movements of ions and of the gates of channel proteins take time and must occur before voltage regeneration occurs.

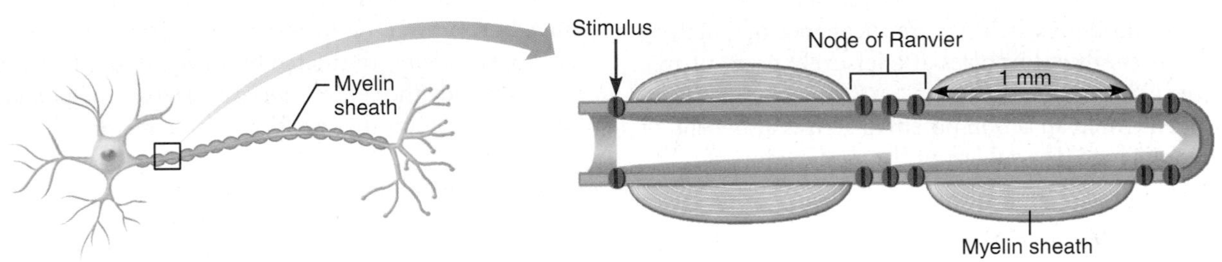

(c) In a **myelinated axon**, myelin keeps current in axons (voltage doesn't decay much). APs are generated *only* in the nodes of Ranvier and appear to jump *rapidly* from node to node.

Figure 11.15 Action potential propagation in unmyelinated and myelinated axons.

sheath dramatically increases the rate of AP propagation because myelin acts as an insulator, both preventing almost all leakage of charge from the axon and allowing the membrane voltage to change more rapidly. Current can pass through the membrane of a myelinated axon *only* at the nodes of Ranvier, where the myelin sheath is interrupted and the axon is bare, and essentially all the voltage-gated Na^+ channels are concentrated at the nodes.

When an AP is generated in a myelinated fiber, the local depolarizing current does not dissipate through the adjacent membrane regions, which are nonexcitable. Instead, the current is maintained and moves rapidly to the next node, a distance of approximately 1 mm, where it triggers another AP. Consequently, APs are triggered only at the nodes, a type of conduction called **saltatory conduction** (*saltare* = to leap) because the electrical signal jumps from node to node along the axon (Figure 11.15c). Saltatory conduction is about 30 times faster than continuous conduction.

HOMEOSTATIC IMBALANCE

The importance of myelin to nerve transmission is painfully clear to people with demyelinating diseases such as **multiple sclerosis (MS)**. This autoimmune disease affects mostly young adults. Common symptoms are visual disturbances (including blindness), problems controlling muscles (weakness, clumsiness, and ultimately paralysis), speech disturbances, and urinary incontinence. In this disease, myelin sheaths in the CNS are gradually destroyed, reduced to nonfunctional hardened lesions called *scleroses*. The loss of myelin (a result of the immune system's attack on myelin proteins) causes such substantial shunting and short-circuiting of the current that successive nodes are excited more and more slowly, and eventually impulse conduction ceases. However, the axons themselves are not damaged and growing numbers of Na^+ channels appear spontaneously in the demyelinated fibers. This may account for the remarkably variable cycles of relapse (disability) and remission (symptom-free periods) typical of this disease.

The advent of drugs that modify the immune system's activity [such as interferons and glatiramer (Copaxone)] have improved the lives of people with MS. These drugs seem to hold the symptoms at bay, reducing complications and the disability that often occurs with MS. ■

Nerve fibers may be classified according to diameter, degree of myelination, and conduction speed. **Group A fibers** are mostly somatic sensory and motor fibers serving the skin, skeletal muscles, and joints. They have the largest diameter and thick myelin sheaths, and conduct impulses at speeds ranging up to 150 m/s (over 300 miles per hour).

In the B and C fiber groups are autonomic nervous system motor fibers serving the visceral organs; visceral sensory fibers; and the smaller somatic sensory fibers transmitting afferent impulses from the skin (such as pain and small touch fibers). **Group B fibers**, lightly myelinated fibers of intermediate diameter, transmit impulses at an average rate of 15 m/s (about 30 mi/h). **Group C fibers** have the smallest diameter and are unmyelinated. Consequently, they are incapable of saltatory conduction and conduct impulses at a leisurely pace—1 m/s (2 mi/h) or less.

11

What happens when an action potential arrives at the end of a neuron's axon? That is the subject of the next section.

HOMEOSTATIC IMBALANCE

A number of chemical and physical factors impair impulse propagation. Local anesthetics like those used by your dentist act by blocking voltage-gated Na^+ channels. As we have seen, no Na^+ entry—no AP.

Cold and continuous pressure interrupt blood circulation, hindering the delivery of oxygen and nutrients to neuron processes and impairing their ability to conduct impulses. For example, your fingers get numb when you hold an ice cube for more than a few seconds, and your foot "goes to sleep" when you sit on it. When you remove the cold object or pressure, impulses are transmitted again, leading to an unpleasant prickly feeling. ■

CHECK YOUR UNDERSTANDING

11. Comparing graded potentials and action potentials, which is bigger? Which travels farthest? Which initiates the other?
12. An action potential does not get smaller as it propagates along an axon. Why not?
13. Why is conduction of action potentials faster in myelinated than in unmyelinated axons?
14. If an axon receives two stimuli close together in time, only one AP occurs. Why?

For answers, see Appendix G.

The Synapse

► Define synapse. Distinguish between electrical and chemical synapses by structure and by the way they transmit information.

The operation of the nervous system depends on the flow of information through chains of neurons functionally connected by synapses. A **synapse** (sin′aps), from the Greek *syn*, "to clasp or join," is a junction that mediates information transfer from one neuron to the next or from a neuron to an effector cell—it's where the action is.

Synapses between the axon endings of one neuron and the dendrites of other neurons are **axodendritic synapses**. Those between axon endings of one neuron and cell bodies (soma) of other neurons are **axosomatic synapses** (Figure 11.16). Less common (and far less understood) are synapses between axons (*axoaxonic*), between dendrites (*dendrodendritic*), or between dendrites and cell bodies (*dendrosomatic*).

The neuron conducting impulses toward the synapse is the **presynaptic neuron**, and the neuron transmitting the electrical signal away from the synapse is the **postsynaptic neuron**. At a given synapse, the presynaptic neuron is the information sender, and the postsynaptic neuron is the information receiver. As you might anticipate, most neurons function as both presynaptic and postsynaptic neurons. Neurons have anywhere from 1000 to 10,000 axon terminals making synapses and are stimulated by an equal number of other neurons. Outside the central nervous system, the postsynaptic cell may be either another neuron or an effector cell (a muscle cell or gland cell).

Now let's look at the two varieties of synapses: *electrical* and *chemical*.

Electrical Synapses

Electrical synapses, the less common variety, consist of gap junctions like those found between certain other body cells. They contain protein channels, called connexons, that intimately connect the cytoplasm of adjacent neurons and allow ions and small molecules to flow directly from one neuron to the next. Neurons joined in this way are said to be *electrically coupled*, and transmission across these synapses is very rapid. Depending on the nature of the synapse, communication may be unidirectional or bidirectional.

A key feature of electrical synapses between neurons is that they provide a simple means of synchronizing the activity of all interconnected neurons. In adults, electrical synapses are found in regions of the brain responsible for certain stereotyped movements, such as the normal jerky movements of the eyes, and in axoaxonic synapses in the hippocampus, a brain region intimately involved in emotions and memory. Electrical synapses are far more abundant in embryonic nervous tissue, where they permit exchange of guiding cues during early neuronal development so that neurons can connect properly with one another. As the nervous system develops, some electrical synapses are replaced by chemical synapses. Gap junctions also exist between glial cells of the CNS.

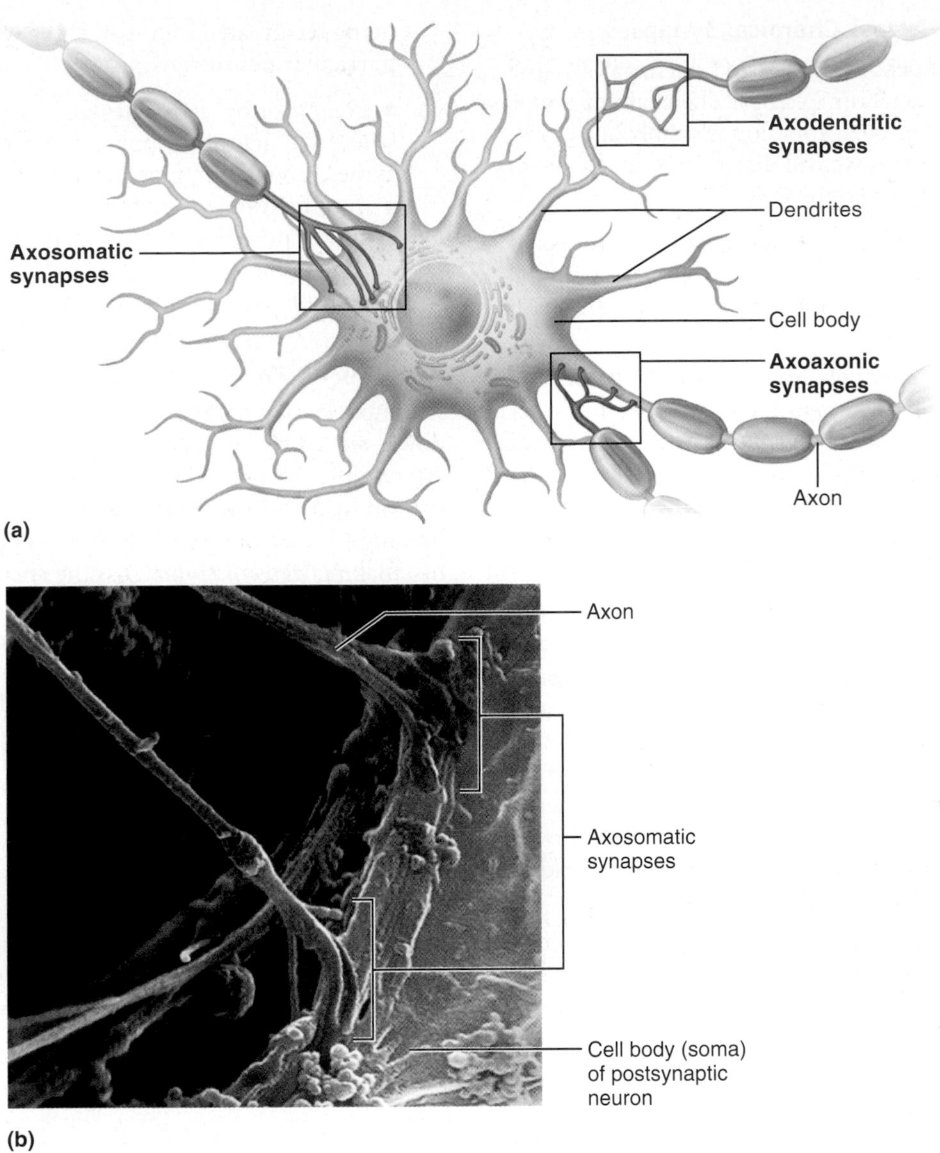

Figure 11.16 Synapses. (a) Axodendritic, axosomatic, and axoaxonic synapses. **(b)** Scanning electron micrograph of incoming fibers at axosomatic synapses (5300×).

Chemical Synapses

In contrast to electrical synapses, which are specialized to allow the flow of ions between neurons, **chemical synapses** are specialized for release and reception of chemical neurotransmitters. A typical chemical synapse is made up of two parts:

1. A knoblike *axon terminal* of the presynaptic neuron, which contains many tiny, membrane-bounded sacs called **synaptic vesicles**, each containing thousands of neurotransmitter molecules
2. A neurotransmitter *receptor region* on the membrane of a dendrite or the cell body of the postsynaptic neuron

Although close to each other, presynaptic and postsynaptic membranes are always separated by the **synaptic cleft**, a fluid-filled space approximately 30 to 50 nm (about one-millionth of an inch) wide. (If an electrical synapse is like the threshold of a doorway between neurons, a synaptic cleft is like a good-size lake intervening between them.)

Because the current from the presynaptic membrane dissipates in the fluid-filled cleft, chemical synapses effectively prevent a nerve impulse from being *directly* transmitted from one neuron to another. Instead, transmission of signals across these synapses is a *chemical event* that depends on the release, diffusion, and receptor binding of neurotransmitter molecules and results in *unidirectional communication* between neurons. In short, transmission of nerve impulses along an axon and across electrical synapses is a purely electrical event, while chemical synapses convert the electrical signals to chemical signals (neurotransmitters) that travel across the synapse to the postsynaptic cells, where they are converted back into electrical signals.

Information Transfer Across Chemical Synapses

In Chapter 9 we introduced a specialized chemical synapse called a neuromuscular junction (p. 285). The chain of events that occurs at the neuromuscular junction is simply one example of the general process that we will discuss next and show in *Focus on a Chemical Synapse* (**Figure 11.17**):

1. **Action potential arrives at axon terminal.** The process of neurotransmission at a chemical synapse begins with the arrival of an action potential at the presynaptic axon terminal.
2. **Voltage-gated Ca^{2+} channels open and Ca^{2+} enters the axon terminal.** Depolarization of the membrane by the action potential opens not only Na^{+} channels but voltage-gated Ca^{2+} channels as well. During the brief time the Ca^{2+} channels are open, Ca^{2+} floods down its electrochemical gradient into the terminal from the extracellular fluid.
3. **Ca^{2+} entry causes neurotransmitter-containing vesicles to release their contents by exocytosis.** The surge of Ca^{2+} into the axon terminal acts as an intracellular messenger. A Ca^{2+}-sensing protein (*synaptotagmin*) binds Ca^{2+} and interacts with the SNARE proteins that control membrane fusion (see Figure 3.14). As a result, synaptic vesicles fuse with the axon membrane and empty their contents by exocytosis into the synaptic cleft. Ca^{2+} is then quickly removed from the terminal—either taken up into the mitochondria or ejected from the neuron by an active Ca^{2+} pump.

   For each nerve impulse reaching the presynaptic terminal, many vesicles (perhaps 300) are emptied into the synaptic cleft. The higher the impulse frequency (that is, the more intense the stimulus), the greater the number of synaptic vesicles that fuse and spill their contents, and the greater the effect on the postsynaptic cell.
4. **Neurotransmitter diffuses across the synaptic cleft and binds to specific receptors on the postsynaptic membrane.**
5. **Binding of neurotransmitter opens ion channels, resulting in graded potentials.** When neurotransmitter binds to the receptor protein, this receptor changes its three-dimensional shape. This change in turn causes ion channels to open and creates graded potentials. Postsynaptic membranes often contain receptor proteins and ion channels packaged together as chemically gated ion channels. Depending on the receptor protein to which the neurotransmitter binds and the type of channel the receptor controls, the postsynaptic neuron may be either excited or inhibited.
6. **Neurotransmitter effects are terminated.** The binding of a neurotransmitter to its receptor is reversible. As long as it is bound to a postsynaptic receptor, a neurotransmitter continues to affect membrane permeability and to block reception of additional signals from presynaptic neurons. For this reason, some means of "wiping the postsynaptic slate clean" is necessary. The effects of neurotransmitters generally last a few milliseconds before being terminated in one of three ways, depending on the particular neurotransmitter:
   - *Reuptake* by astrocytes or the presynaptic terminal, where the neurotransmitter is stored or destroyed by enzymes, as with norepinephrine
   - *Degradation* by enzymes associated with the postsynaptic membrane or present in the synapse, as with acetylcholine
   - *Diffusion* away from the synapse

Synaptic Delay

An impulse may travel at speeds of up to 150 m/s (300 mi/h) down an axon, but neural transmission across a chemical synapse is comparatively slow. It reflects the time required for neurotransmitter release, diffusion across the synaptic cleft, and binding to receptors. Typically, this **synaptic delay** lasts 0.3–5.0 ms, making transmission across the chemical synapse the *rate-limiting* (slowest) step of neural transmission. Synaptic delay helps explain why transmission along neural pathways involving only two or three neurons occurs rapidly, but transmission along multisynaptic pathways typical of higher mental functioning occurs much more slowly. However, in practical terms these differences are not noticeable.

CHECK YOUR UNDERSTANDING

15. What is the structure that joins two neurons at an electrical synapse?
16. Events at a chemical synapse usually involve opening of both voltage-gated ion channels and chemically gated ion channels. Where are these ion channels located and what causes each to open?

For answers, see Appendix G.

Postsynaptic Potentials and Synaptic Integration

- ▶ Distinguish between excitatory and inhibitory postsynaptic potentials.
- ▶ Describe how synaptic events are integrated and modified.

Many receptors on postsynaptic membranes at chemical synapses are specialized to open ion channels, in this way converting chemical signals to electrical signals. Unlike the voltage-gated ion channels responsible for APs, however, these chemically gated channels are relatively insensitive to changes in membrane potential. Consequently, channel opening at postsynaptic membranes cannot possibly become self-amplifying or self-generating. Instead, neurotransmitter receptors mediate graded potentials—local changes in membrane potential that are *graded* (or varied in strength) according to the amount of neurotransmitter released and the time it remains in the area. APs are compared with graded potentials in **Table 11.2**.

Figure 11.17 FOCUS Chemical Synapse

Chemical synapses transmit signals from one neuron to another using neurotransmitters.

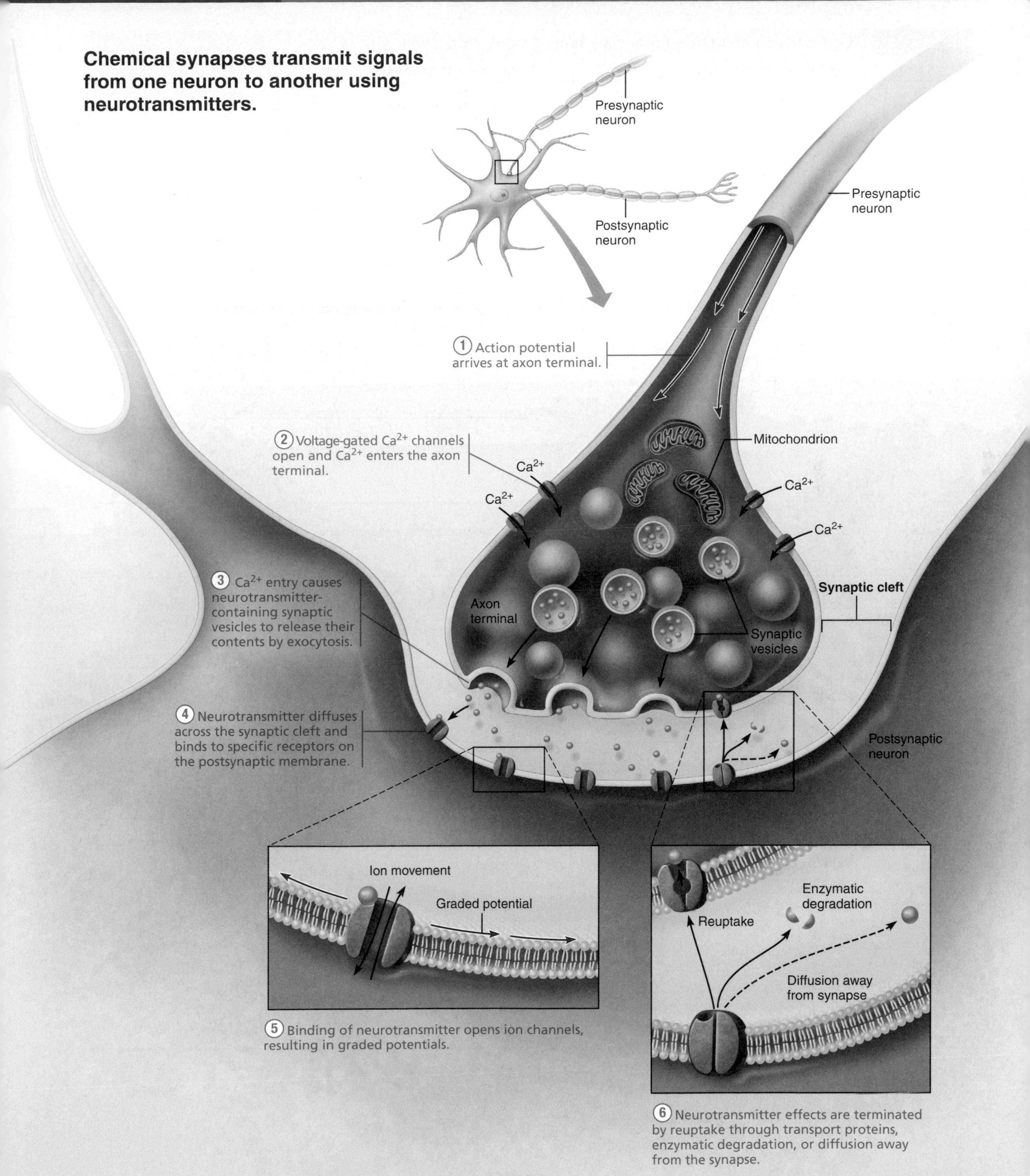

Presynaptic neuron

Postsynaptic neuron

Presynaptic neuron

(1) Action potential arrives at axon terminal.

(2) Voltage-gated Ca^{2+} channels open and Ca^{2+} enters the axon terminal.

Mitochondrion

Ca^{2+}

Ca^{2+}

Ca^{2+}

Ca^{2+}

(3) Ca^{2+} entry causes neurotransmitter-containing synaptic vesicles to release their contents by exocytosis.

Axon terminal

Synaptic cleft

Synaptic vesicles

(4) Neurotransmitter diffuses across the synaptic cleft and binds to specific receptors on the postsynaptic membrane.

Postsynaptic neuron

Ion movement

Graded potential

(5) Binding of neurotransmitter opens ion channels, resulting in graded potentials.

Reuptake

Enzymatic degradation

Diffusion away from synapse

(6) Neurotransmitter effects are terminated by reuptake through transport proteins, enzymatic degradation, or diffusion away from the synapse.

A&P Flix *View this animation at myA&P*

TABLE 11.2 Comparison of Action Potentials with Graded Potentials

<table>
<tr><th></th><th>GRADED POTENTIAL (GP)</th><th>ACTION POTENTIAL (AP)</th></tr>
<tr><td>Location of event</td><td>Cell body and dendrites, typically</td><td>Axon hillock and axon</td></tr>
<tr><td></td><td colspan="2">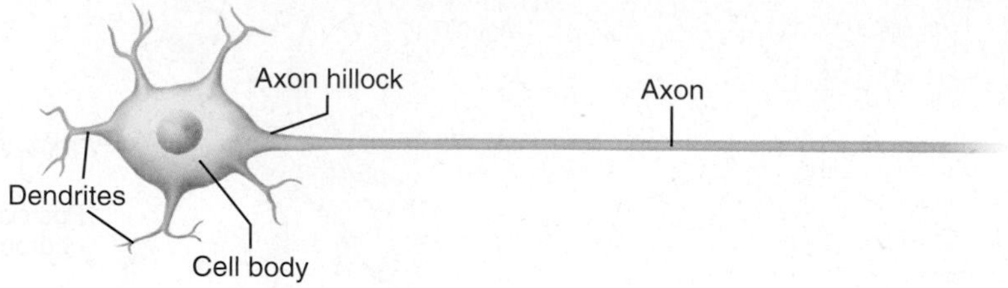
</td></tr>
<tr><td>Distance traveled</td><td>Short distance—typically within cell body to axon hillock (0.1–1.0 mm)</td><td>Long distance—from axon hillock through entire length of axon (a few mm to over a meter)</td></tr>
<tr><td></td><td colspan="2">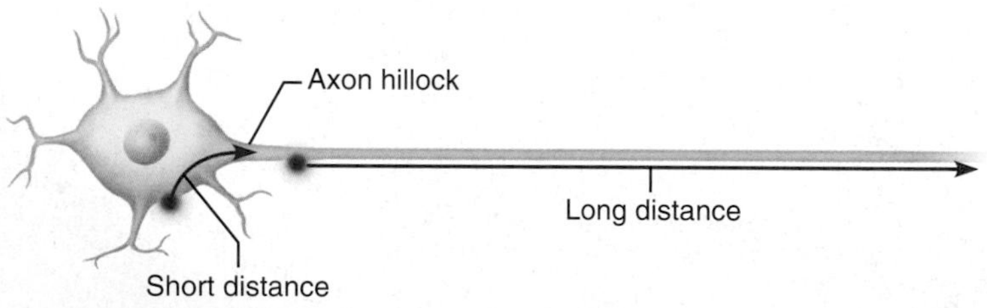
</td></tr>
<tr><td>Amplitude (size)</td><td>Various sizes (graded); declines with distance</td><td>Always the same size (all-or-none); does not decline with distance</td></tr>
<tr><td>Stimulus for opening of ion channels</td><td>Chemical (neurotransmitter) or sensory stimulus (e.g., light, pressure, temperature)</td><td>Voltage (depolarization, triggered by GP reaching threshold)</td></tr>
<tr><td>Positive feed-back cycle</td><td>Absent</td><td>Present</td></tr>
<tr><td>Repolarization</td><td>Voltage independent; occurs when stimulus is no longer present</td><td>Voltage regulated; occurs when Na^+ channels inactivate and K^+ channels open</td></tr>
<tr><td>Summation</td><td>Stimulus responses can be summed to increase amplitude of graded potential
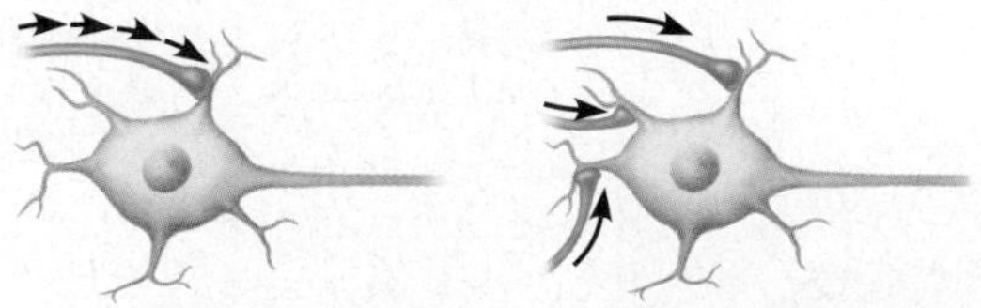
Temporal: increased frequency of stimuli
Spatial: stimuli from multiple sources</td><td>Does not occur; an all-or-none phenomenon</td></tr>
</table>

TABLE 11.2 *(continued)*

	GRADED POTENTIAL (GP)		ACTION POTENTIAL (AP)
	POSTSYNAPTIC POTENTIAL (A TYPE OF GP)		
	EXCITATORY (EPSP)	INHIBITORY (IPSP)	
Function	Short-distance signaling; depolarization that spreads to axon hillock; moves membrane potential *toward* threshold for generation of AP	Short-distance signaling; hyperpolarization that spreads to axon hillock; moves membrane potential *away from* threshold for generation of AP	Long-distance signaling; constitutes the nerve impulse
Initial effect of stimulus	Opens channels that allow simultaneous Na^+ and K^+ fluxes	Opens K^+ or Cl^- channels	First opens Na^+ channels, then K^+ channels
Peak membrane potential	Becomes depolarized; moves toward 0 mV	Becomes hyperpolarized; moves toward −90 mV	+30 to +50 mV

Chemical synapses are either excitatory or inhibitory, depending on how they affect the membrane potential of the postsynaptic neuron.

Excitatory Synapses and EPSPs

At excitatory synapses, neurotransmitter binding causes depolarization of the postsynaptic membrane. However, in contrast to what happens on axon membranes, a single type of *chemically gated* ion channel opens on postsynaptic membranes (those of dendrites and neuronal cell bodies). This channel allows Na^+ and K^+ to diffuse *simultaneously* through the membrane in opposite directions. Although this two-way cation flow may appear to be self-defeating when depolarization is the goal, remember that the electrochemical gradient for sodium is much steeper than that for potassium. As a result, Na^+ influx is greater than K^+ efflux, and *net* depolarization occurs.

If enough neurotransmitter binds, depolarization of the postsynaptic membrane can reach 0 mV, which is well above an axon's threshold (about −50 mV) for "firing off" an AP. However, *postsynaptic membranes generally do not generate APs*, unlike axons, which have voltage-gated channels that make an AP possible. The dramatic polarity reversal seen in axons never occurs in membranes containing *only* chemically gated channels because the opposite movements of K^+ and Na^+ prevent accumulation of excessive positive charge inside the cell. For this reason, instead of APs, local graded depolarization events called **excitatory postsynaptic potentials (EPSPs)** occur at excitatory postsynaptic membranes **(Figure 11.18a)**.

Each EPSP lasts a few milliseconds and then the membrane returns to its resting potential. The only function of EPSPs is to help trigger an AP distally at the axon hillock of the postsynaptic neuron. Although currents created by individual EPSPs decline with distance, they can and often do spread all the way to

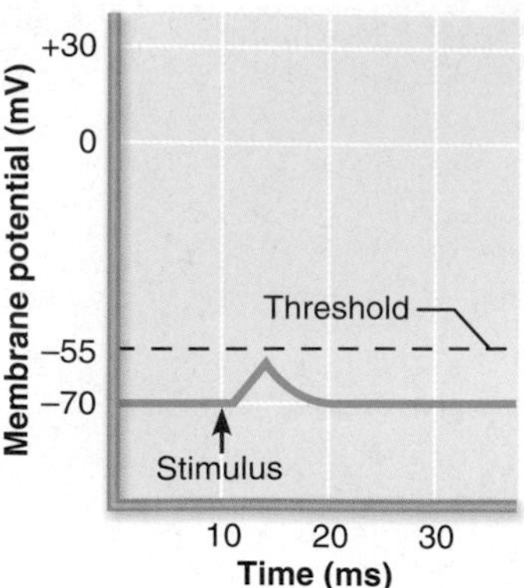

An EPSP is a local depolarization of the postsynaptic membrane that brings the neuron closer to AP threshold. Neurotransmitter binding opens chemically gated ion channels, allowing the simultaneous passage of Na^+ and K^+.

(a) Excitatory postsynaptic potential (EPSP)

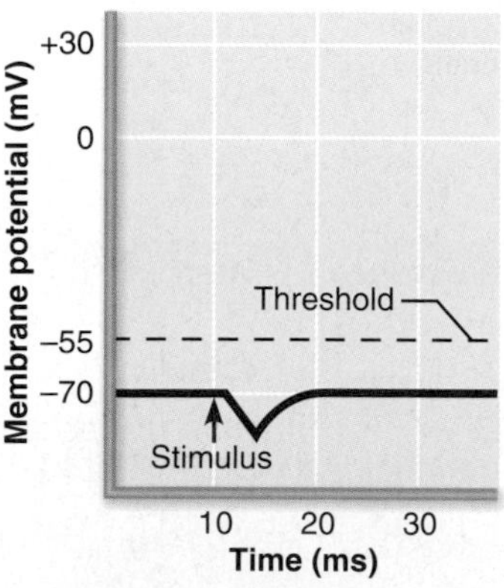

An IPSP is a local hyperpolarization of the postsynaptic membrane and drives the neuron away from AP threshold. Neurotransmitter binding opens K^+ or Cl^- channels.

(b) Inhibitory postsynaptic potential (IPSP)

Figure 11.18 **Postsynaptic potentials.**

the axon hillock. If currents reaching the hillock are strong enough to depolarize the axon to threshold, axonal voltage-gated channels open and an AP is generated.

Inhibitory Synapses and IPSPs

Binding of neurotransmitters at inhibitory synapses *reduces* a postsynaptic neuron's ability to generate an AP. Most inhibitory neurotransmitters induce hyperpolarization of the postsynaptic membrane by making the membrane more permeable to K^+ or Cl^-. Sodium ion permeability is not affected.

If K^+ channels are opened, K^+ moves out of the cell. If Cl^- channels are opened, Cl^- moves in. In either case, the charge on the inner face of the membrane becomes more negative. As the membrane potential increases and is driven farther from the axon's threshold, the postsynaptic neuron becomes *less and less likely* to "fire," and larger depolarizing currents are required to induce an AP. Such changes in potential are called **inhibitory postsynaptic potentials (IPSPs)** (Figure 11.18b).

Integration and Modification of Synaptic Events

Summation by the Postsynaptic Neuron A single EPSP cannot induce an AP in the postsynaptic neuron **(Figure 11.19a)**. But if thousands of excitatory axon terminals are firing on the same postsynaptic membrane, or if a smaller number of terminals are delivering impulses rapidly, the probability of reaching threshold depolarization increases greatly. EPSPs can add together, or **summate**, to influence the activity of a postsynaptic neuron. Nerve impulses would never be initiated if this were not so.

Two types of summation occur. **Temporal summation** (*temporal* = time) occurs when one or more presynaptic neurons transmit impulses in rapid-fire order and bursts of neurotransmitter are released in quick succession. The first impulse produces a small EPSP, and before it dissipates, successive impulses trigger more EPSPs. These summate, producing a much greater depolarization of the postsynaptic membrane than would result from a single EPSP (Figure 11.19b).

Spatial summation occurs when the postsynaptic neuron is stimulated at the same time by a large number of terminals from the same or, more commonly, different neurons. Huge numbers of its receptors bind neurotransmitter and simultaneously initiate EPSPs, which summate and dramatically enhance depolarization (Figure 11.19c).

Although we have focused on EPSPs here, IPSPs also summate, both temporally and spatially. In this case, the postsynaptic neuron is inhibited to a greater degree.

Most neurons receive both excitatory and inhibitory inputs from thousands of other neurons. Additionally, the same axon may form different types of synapses (in terms of biochemical and electrical characteristics) with different types of target neurons. How is all this conflicting information sorted out?

Each neuron's axon hillock keeps a running account of all the signals it receives. Not only do EPSPs summate and IPSPs summate, but also EPSPs summate with IPSPs. If the stimulatory effects of EPSPs dominate the membrane potential enough to reach threshold, the neuron will fire. If summation yields only subthreshold depolarization or hyperpolarization, the neuron fails to generate an AP (Figure 11.19d). However, partially depolarized neurons are **facilitated**—that is, more easily excited by successive depolarization events—because they are already near threshold. Thus, axon hillock membranes function as *neural integrators*, and their potential at any time reflects the sum of all incoming neural information.

Because EPSPs and IPSPs are graded potentials that diminish in strength the farther they spread, the most effective synapses are those closest to the axon hillock. Specifically, inhibitory synapses are most effective when located between the site of excitatory inputs and the site of action potential generation (the axon hillock). Accordingly, inhibitory synapses occur most often on the cell body and excitatory synapses occur most often on the dendrites (Figure 11.19d).

Synaptic Potentiation Repeated or continuous use of a synapse (even for short periods) enhances the presynaptic neuron's ability to excite the postsynaptic neuron, producing larger-than-expected postsynaptic potentials. This phenomenon is called **synaptic potentiation**. The presynaptic terminals at such synapses contain relatively high Ca^{2+} concentrations, a condition that (presumably) triggers the release of more neurotransmitter, which in turn produces larger EPSPs.

Furthermore, synaptic potentiation brings about Ca^{2+} influx via dendritic spines into the postsynaptic neuron as well. Brief high-frequency stimulation partially depolarizes the postsynaptic membrane. This partial depolarization causes certain

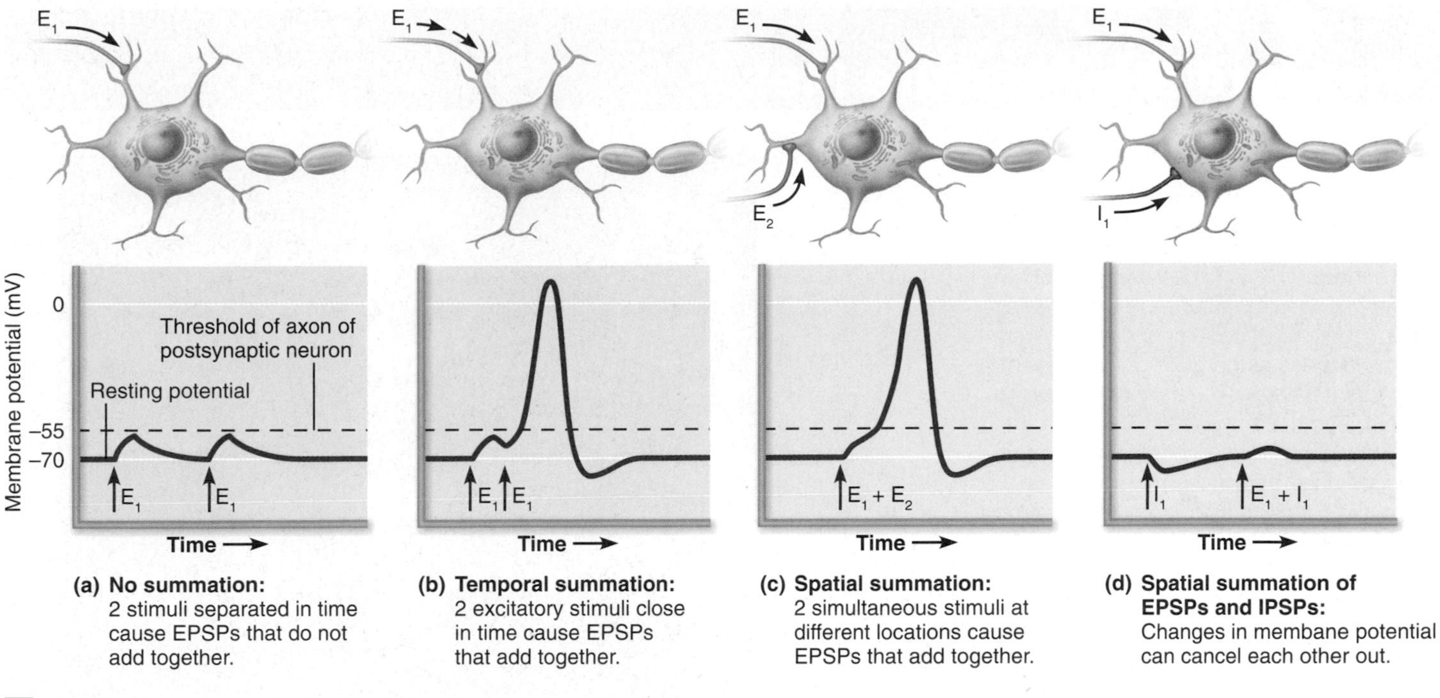

(a) No summation: 2 stimuli separated in time cause EPSPs that do not add together.

(b) Temporal summation: 2 excitatory stimuli close in time cause EPSPs that add together.

(c) Spatial summation: 2 simultaneous stimuli at different locations cause EPSPs that add together.

(d) Spatial summation of EPSPs and IPSPs: Changes in membane potential can cancel each other out.

Figure 11.19 **Neural integration of EPSPs and IPSPs.**

chemically gated channels called *NMDA* (*N-methyl-D-aspartate*) *receptors* to allow Ca^{2+} entry, something that only happens when the membrane is depolarized. As Ca^{2+} floods into the cell, it activates certain kinase enzymes that promote changes that result in more effective responses to subsequent stimuli.

In some neurons, APs generated at the axon hillock propagate back up into the dendrites. This current flow may alter the effectiveness of synapses by causing voltage-gated Ca^{2+} channels to open, again allowing Ca^{2+} into the dendrites and promoting synaptic potentiation.

Synaptic potentiation can be viewed as a learning process that increases the efficiency of neurotransmission along a particular pathway. Indeed, the hippocampus of the brain, which plays a special role in memory and learning, exhibits an important type of synaptic plasticity called *long-term potentiation* (LTP).

Presynaptic Inhibition Postsynaptic activity can also be influenced by events occurring at the presynaptic membrane. **Presynaptic inhibition** occurs when the release of excitatory neurotransmitter by one neuron is inhibited by the activity of another neuron via an axoaxonic synapse. More than one mechanism is involved, but the end result is that less neurotransmitter is released and bound, and smaller EPSPs are formed.

Notice that this is the opposite of what we see with synaptic potentiation. In contrast to postsynaptic inhibition by IPSPs, which decreases the excitability of the postsynaptic neuron, presynaptic inhibition reduces excitatory stimulation of the postsynaptic neuron. In this way, presynaptic inhibition is like a functional synaptic "pruning."

CHECK YOUR UNDERSTANDING

17. Which ions flow through chemically gated channels to produce IPSPs? EPSPs?
18. What is the difference between temporal summation and spatial summation?

For answers, see Appendix G.

Neurotransmitters and Their Receptors

▶ Define neurotransmitter and name several classes of neurotransmitters.

Neurotransmitters, along with electrical signals, are the "language" of the nervous system—the means by which each neuron communicates with others to process and send messages to the rest of the body. Sleep, thought, rage, hunger, memory, movement, and even your smile reflect the "doings" of these versatile molecules. Most factors that affect synaptic transmission do so by enhancing or inhibiting neurotransmitter release or destruction, or by blocking their binding to receptors. Just as speech defects may hinder interpersonal communication, interferences with neurotransmitter activity may short-circuit the brain's "conversations" or internal talk (see *A Closer Look* on pp. 414–415).

At present, more than 50 neurotransmitters or neurotransmitter candidates have been identified. Although some neurons

A CLOSER LOOK

Pleasure Me, Pleasure Me!

Sex! Drugs! Rock 'n' roll! Eat, drink, and be merry! Why do we find these activities so compelling? Our brains are wired to reward us with pleasure when we engage in behavior that is necessary for our own and our species' survival. This reward system consists of dopamine-releasing neurons in areas of the brain called the *ventral tegmental area* (VTA), the *nucleus accumbens*, and the *amygdala*.

Our ability to "feel good" involves brain neurotransmitters in this reward system. For example, the ecstasy of romantic love may be just a brain bath of glutamate and norepinephrine, which act on the reward system to release dopamine. Unfortunately, this powerful system can be subverted by drugs of abuse. The 1930s songwriter Cole Porter knew what he was talking about when he wrote "I get a kick out of you," because these neurotransmitters are chemical cousins of the amphetamines. People who use "crystal meth" (methamphetamine) artificially stimulate their brains to provide their highly addictive pleasure flush. However, their pleasure is short-lived, because when the brain is flooded with neurotransmitter-like chemicals from the outside, it makes less of its own (why bother?).

Cocaine, another reward system titillater, has been around since ancient times. Once a toy of the rich, its granular form is inhaled, or "snorted." The laws of supply and demand have now brought cheaper cocaine to the masses, notably "crack"—a cheaper, more potent, smokable form of cocaine. For $50 or so, a novice user can experience a rush of intense pleasure. But crack is treacherous and intensely addictive. It produces not only a higher high than the inhaled form of cocaine, but also a deeper crash that leaves the user desperate for more.

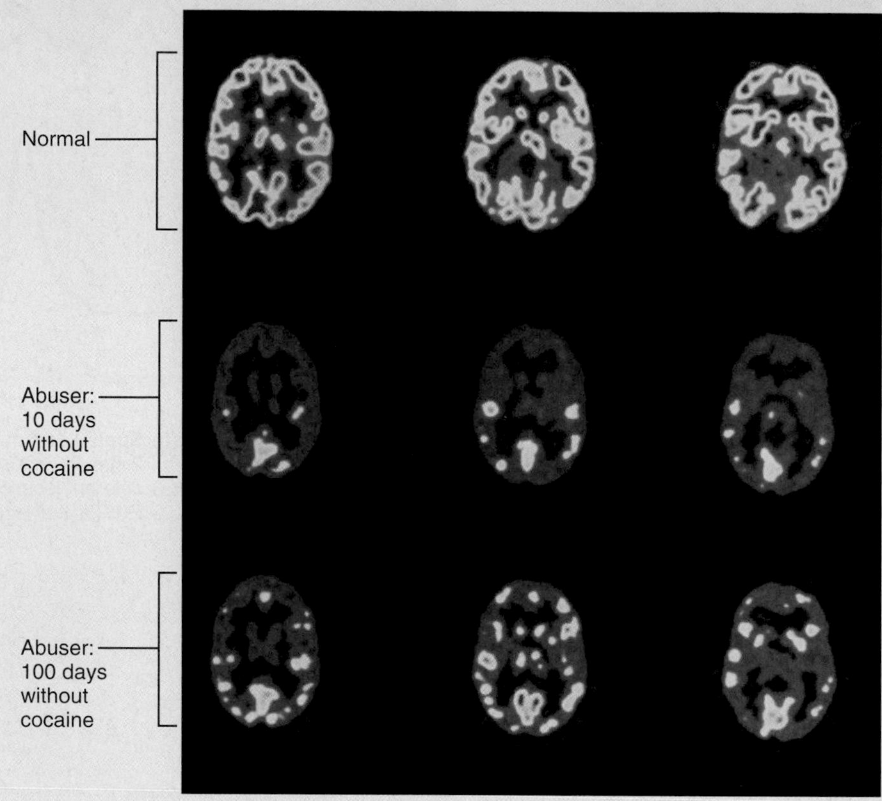

PET scans show that normal levels of brain activity (yellow and red) are depressed in cocaine users long after drug use has stopped.

How does cocaine produce its effects? Basically, the drug stimulates the reward system and then "squeezes it dry." Cocaine produces its rush by hooking up to the dopamine reuptake transporter protein, blocking the reabsorption of dopamine. The neurotransmitter remains in the synapse and stimulates the postsynaptic receptor cells again and again, allowing the body to feel its effects over a prolonged period. This sensation is accompanied by increases in heart rate, blood pressure, and sexual appetite.

As dopamine uptake continues to be blocked by repeated doses of cocaine, the system releases less and less dopamine and the reward system effectively goes dry. The cocaine user becomes anxious and, in a very real sense, unable to experience pleasure without the drug. Consequently, the postsynaptic cells become hypersensitive and sprout new receptors

produce and release only one kind of neurotransmitter, most make two or more and may release any one or all of them. It appears that in most cases, different neurotransmitters are released at different stimulation frequencies, a restriction that avoids producing a jumble of nonsense messages. However, corelease of two neurotransmitters from the same vesicles has been documented. The coexistence of more than one neurotransmitter in a single neuron makes it possible for that cell to exert several influences rather than one discrete effect.

Neurotransmitters are classified chemically and functionally. **Table 11.3** provides a fairly detailed overview of neurotransmitters, and we describe some of them here. No one expects you to memorize this table at this point, but it will be a handy reference for you to look back at when neurotransmitters are mentioned in subsequent chapters.

Classification of Neurotransmitters by Chemical Structure

Neurotransmitters fall into several chemical classes based on molecular structure.

A CLOSER LOOK
(continued)

in a desperate effort to pick up dopamine signals. A vicious cycle of addiction begins: Cocaine is needed to experience pleasure, but using it suppresses dopamine release even more.

The dopamine effect alone is not enough to establish addiction. Another neurotransmitter, glutamate, which plays an important role in learning, is required to maintain addiction. Glutamate signaling seems to cause more permanent changes in the brain (synaptic potentiation) that lead to compulsive drug-seeking behaviors elicited by external cues. Take, for example, mice genetically engineered to lack a particular glutamate receptor (GluR5). These mice are perfectly willing to try cocaine but never become addicted. (Of course, no GluR5 means they're none too bright, either.)

Current thinking, then, is that the rush of pleasure on taking cocaine is due to dopamine. Glutamate, on the other hand, is thought to be responsible for the learning that makes true the perception "once an addict, always an addict." So strong is the combined dopamine and glutamate system that, even years later, certain settings can trigger intense cravings for the drug.

These out-of-control, desperate cravings are notoriously difficult to manage. Drug abusers call it "jonesing." Traditional antiaddiction drugs take so long to reduce the cravings that users commonly drop out of treatment programs.

How can we break this cycle of addiction? One way is to prevent cocaine from ever reaching the brain. Promising results have been obtained from a vaccine that prompts the immune system to bind cocaine molecules, preventing them from entering the brain. In a clinical trial, this vaccine dampened addicts' pleasurable responses to cocaine and reduced their use of the drug. Another approach to breaking the addiction cycle is to even out the highs and lows experienced by the drug user. Clinical trials are under way with a drug (vanoxerine) that slowly binds the dopamine reuptake transporter and inhibits it in a more long-lasting manner than cocaine does. This results in a leveling-out of brain dopamine levels and keeps the user from "crashing" so badly.

A final approach to breaking the addiction cycle is to interrupt the learned reinforcement that brings on cravings. An effective ancient African folk remedy called ibogaine may do exactly this. However, ibogaine itself is too toxic for clinical use, as some unfortunate "underground" users have discovered. A close synthetic cousin, 18-methoxycoronaridine (18-MC) is much less toxic and promises to be effective against not only cocaine but also a number of other abused drugs. Future studies will show if it is truly effective.

The craving for drugs has made some who depend on them into very creative home pharmacologists, willing to experiment with practically anything, no matter how toxic or dangerous, to get the "buzz" they need. A cheap mixture of cold medications, match heads, and iodine in acetone yields crystal meth—the highly addictive and once-again popular drug that wrecks people's lives and often explodes their home laboratories. Another creative mixture, with the street name of "ill face" or "illy," involves dipping marijuana in formaldehyde, drying and then smoking it. Other remarkable combinations are the "H-bomb" (ecstasy mixed with heroin), "A-bomb" (marijuana with heroin or opium), "sextasy" [ecstasy mixed with sildenafil (Viagra)], "octane" (PCP laced with gasoline), and "ozone" (marijuana, PCP, and crack in a cigarette). Formaldehyde is a known cancer-causing agent and gasoline damages the liver, but the real damage comes from the drugs themselves.

Take, for example, ecstasy, a drug that many of its users believe to be innocuous. In reality, ecstasy (MDMA) targets serotonin-releasing neurons. The "rush" of pleasure and energy that users feel is due to release of serotonin and other neurotransmitters. However, it damages and may destroy these neurons, causing the loss of verbal and spatial memory. Depression, sleeplessness, and memory problems may be permanent consequences—a high price for a few moments of pleasure!

People who want pure, effective, and "safe" drugs of abuse don't get them on the street. They get them from doctors or "pill ladies" [female senior citizens who sell oxycodone (OxyContin), a powerful prescription opioid with effects similar to heroin]. Even people who would never dream of taking the illicit drugs can be caught in the addictive cycle of prescription drugs. Prescribed legitimately to relieve severe pain, oxycodone is meant to be swallowed whole. Abusers crush the tablets and snort the powder or dissolve it in water and inject the solution. Abuse of oxycodone and its chemical cousin hydrocodone is spreading rapidly. Medical examiners across North America report soaring rates of oxycodone-related emergency room visits and deaths.

The brain, with its complex biochemistry, always circumvents attempts to keep it in a euphoric haze. Perhaps this means that pleasure must be transient by nature, experienced only against a background of its absence.

Acetylcholine

Acetylcholine (ACh) (as″ĕ-til-ko′lēn) was the first neurotransmitter identified. It is still the best understood because it is released at neuromuscular junctions, which are much easier to study than synapses buried in the CNS. ACh is synthesized from acetic acid (as acetyl CoA) and choline by the enzyme *choline acetyltransferase*. The newly synthesized ACh is then transported into synaptic vesicles for later release. Once released by the presynaptic terminal, ACh binds briefly to the postsynaptic receptors. Then it is released and degraded to acetic acid and choline by the enzyme **acetylcholinesterase (AChE)**, located in the synaptic cleft and on postsynaptic membranes. The released choline is recaptured by the presynaptic terminals and reused to synthesize more ACh.

ACh is released by all neurons that stimulate skeletal muscles and by some neurons of the autonomic nervous system. ACh-releasing neurons are also found in the CNS.

Biogenic Amines

The **biogenic amines** (bi″o-jen′ik) include the **catecholamines** (kat″ĕ-kol′ah-mēnz), such as dopamine, norepinephrine (NE), and epinephrine, and the **indolamines**, which include serotonin

TABLE 11.3 Neurotransmitters and Neuromodulators

NEUROTRANSMITTER	FUNCTIONAL CLASSES	SITES WHERE SECRETED	COMMENTS
Acetylcholine			
▪ At *nicotinic ACh receptors* (on skeletal muscles, autonomic ganglia, and in the CNS) ▪ At *muscarinic ACh receptors* (on visceral effectors and in the CNS) $H_3C-C(=O)-O-CH_2-CH_2-\overset{+}{N}-(CH_3)_3$	Excitatory Direct action Excitatory or inhibitory depending on subtype of muscarinic receptor Indirect action via second messengers	CNS: widespread throughout cerebral cortex, hippocampus, and brain stem PNS: all neuromuscular junctions with skeletal muscle; some autonomic motor endings (all preganglionic and parasympathetic postganglionic fibers)	Effects prolonged, leading to tetanic muscle spasms, when AChE blocked by nerve gas and organophosphate insecticides (malathion). Release inhibited by botulinum toxin; binding to nicotinic ACh receptors inhibited by curare (a muscle paralytic agent) and to muscarinic ACh receptors by atropine. ACh levels decreased in certain brain areas in Alzheimer's disease; nicotinic ACh receptors destroyed in myasthenia gravis. Binding of nicotine to nicotinic receptors in the brain enhances dopamine release, which may account for the behavioral effects of nicotine in smokers.
Biogenic Amines			
Norepinephrine HO, HO (ring) $-CH(OH)-CH_2-NH_2$	Excitatory or inhibitory depending on receptor type bound Indirect action via second messengers	CNS: brain stem, particularly in the locus coeruleus of the midbrain; limbic system; some areas of cerebral cortex PNS: main neurotransmitter of ganglionic neurons in the sympathetic nervous system	A "feeling good" neurotransmitter. Release enhanced by amphetamines; removal from synapse blocked by tricyclic antidepressants [amitriptyline (Elavil) and others] and cocaine. Brain levels reduced by reserpine (an antihypertensive drug), leading to depression.
Dopamine 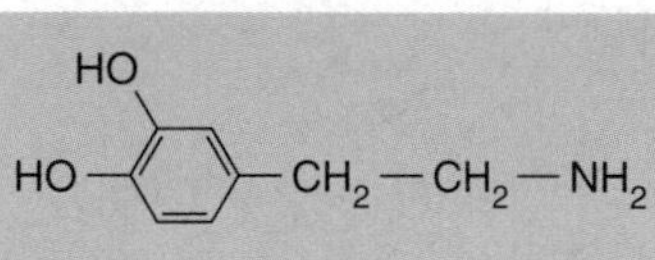	Excitatory or inhibitory depending on the receptor type bound Indirect action via second messengers	CNS: substantia nigra of midbrain; hypothalamus; is the principal neurotransmitter of extrapyramidal system PNS: some sympathetic ganglia	A "feeling good" neurotransmitter. Release enhanced by L-dopa and amphetamines; reuptake blocked by cocaine. Deficient in Parkinson's disease; dopamine neurotransmission increased in schizophrenia.
Serotonin (5-HT) HO (indole ring, N–H) $C(=CH)-CH_2-CH_2-NH_2$	Mainly inhibitory Indirect action via second messengers; direct action at 5-HT_3 receptors	CNS: brain stem, especially midbrain; hypothalamus; limbic system; cerebellum; pineal gland; spinal cord	May play a role in sleep, appetite, nausea, migraine headaches, and regulation of mood. Drugs that block its uptake [fluoxetine (Prozac)] relieve anxiety and depression. Activity blocked by LSD and enhanced by ecstasy (MDMA).
Histamine $HC=C-CH_2-CH_2-NH_2$ (ring: N, CH, NH)	Excitatory or inhibitory depending on receptor type bound Indirect action via second messengers	CNS: hypothalamus	Involved in wakefulness, appetite control, and learning and memory. Also a paracrine (local signal) released from stomach (causes acid secretion) and connective tissue mast cells (mediates inflammation and vasodilation).

TABLE 11.3 *(continued)*

NEUROTRANSMITTER	FUNCTIONAL CLASSES	SITES WHERE SECRETED	COMMENTS
Amino Acids			
GABA (γ-aminobutyric acid) $H_2N-CH_2-CH_2-CH_2-COOH$	Generally inhibitory Direct and indirect actions via second messengers	CNS: cerebral cortex, hypothalamus, Purkinje cells of cerebellum, spinal cord, granule cells of olfactory bulb, retina	Principal inhibitory neurotransmitter in the brain; important in presynaptic inhibition at axoaxonic synapses. Inhibitory effects augmented by alcohol, antianxiety drugs of the benzodiazepine class (e.g., Valium), and barbiturates, resulting in impaired motor coordination. Substances that block its synthesis, release, or action induce convulsions.
Glutamate $H_2N-CH(COOH)-CH_2-CH_2-COOH$	Generally excitatory Direct action	CNS: spinal cord; widespread in brain where it represents the major excitatory neurotransmitter	Important in learning and memory. The "stroke neurotransmitter": excessive release produces excitotoxicity—neurons literally stimulated to death; most commonly caused by ischemia (oxygen deprivation, usually due to a blocked blood vessel). When released by gliomas, aids tumor advance.
Glycine H_2N-CH_2-COOH	Generally inhibitory Direct action	CNS: spinal cord and brain stem, retina	Principal inhibitory neurotransmitter of the spinal cord. Strychnine blocks glycine receptors, resulting in uncontrolled convulsions and respiratory arrest.
Peptides			
Endorphins, e.g., dynorphin, enkephalins (illustrated) Tyr Gly Gly Phe Met	Generally inhibitory Indirect action via second messengers	CNS: widely distributed in brain; hypothalamus; limbic system; pituitary; spinal cord	Natural opiates; inhibit pain by inhibiting substance P. Effects mimicked by morphine, heroin, and methadone.
Tachykinins: Substance P (illustrated), neurokinin A (NKA) Arg Pro Lys Pro Gln Gln Phe Phe Gly Leu Met	Excitatory Indirect action via second messengers	CNS: basal nuclei, midbrain, hypothalamus, cerebral cortex PNS: certain sensory neurons of dorsal root ganglia (pain afferents), enteric neurons	Substance P mediates pain transmission in the PNS. In the CNS, tachykinins are involved in respiratory and cardiovascular controls and in mood.
Somatostatin Ala Gly Cys Lys Asn Phe Phe Trp Lys Thr Phe Thr Ser Cys	Generally inhibitory Indirect action via second messengers	CNS: hypothalamus, septum, basal nuclei, hippocampus, cerebral cortex Pancreas	Often released with GABA. A gut-brain peptide hormone. Inhibits growth hormone release.
Cholecystokinin (CCK) Asp Tyr(SO_4) Met Gly Trp Met Asp Phe	Generally excitatory Indirect action via second messengers	Throughout CNS Small intestine	Involved in anxiety, pain, memory. A gut-brain peptide hormone. Inhibits appetite.

TABLE 11.3 Neurotransmitters and Neuromodulators *(continued)*

NEUROTRANSMITTER	FUNCTIONAL CLASSES	SITES WHERE SECRETED	COMMENTS
Purines			
ATP	Excitatory or inhibitory depending on receptor type bound Direct and indirect actions via second messengers	CNS: basal nuclei, induces Ca^{2+} wave propagation in astrocytes PNS: dorsal root ganglion neurons	ATP released by sensory neurons (as well as that released by injured cells) provokes pain sensation.
Adenosine	Generally inhibitory Indirect action via second messengers	Throughout CNS	Caffeine (coffee), theophylline (tea), and theobromine (chocolate) stimulate by blocking brain adenosine receptors. May be involved in sleep-wake cycle and terminating seizures. Dilates arterioles, increasing blood flow to heart and other tissues as needed.
Gases And Lipids			
Nitric oxide (NO)	Excitatory Indirect action via second messengers	CNS: brain, spinal cord PNS: adrenal gland; nerves to penis	Its release potentiates stroke damage. Some types of male impotence treated by enhancing NO action [e.g., with sildenafil (Viagra)].
Carbon monoxide (CO)	Excitatory Indirect action via second messengers	Brain and some neuromuscular and neuroglandular synapses	
Endocannabinoids, e.g., 2-arachidonoylglycerol (illustrated), anandamide	Inhibitory Indirect action via second messengers	Throughout CNS	Involved in memory (as a retrograde messenger), appetite control, nausea and vomiting, neuronal development. Receptors also found on immune cells.

and histamine. *Dopamine* and *NE* are synthesized from the amino acid tyrosine in a common pathway consisting of several steps. The same pathway is used by the epinephrine-releasing cells of the brain and the adrenal medulla. *Serotonin* is synthesized from the amino acid tryptophan. *Histamine* is synthesized from the amino acid histidine.

Biogenic amine neurotransmitters are broadly distributed in the brain, where they play a role in emotional behavior and help regulate the biological clock. Additionally, catecholamines (particularly NE) are released by some motor neurons of the autonomic nervous system. Imbalances of these neurotransmitters are associated with mental illness. For example, overactive dopamine signaling occurs in schizophrenia. Additionally, certain psychoactive drugs (LSD and mescaline) can bind to biogenic amine receptors and induce hallucinations.

Amino Acids

It is difficult to prove a neurotransmitter role when the suspect is an amino acid, because amino acids occur in all cells of the body and are important in many biochemical reactions. The amino acids for which a neurotransmitter role is certain include **gamma (γ)-aminobutyric acid (GABA)**, **glycine**, **aspartate**, and **glutamate**, but there may be others.

Peptides

The **neuropeptides**, essentially strings of amino acids, include a broad spectrum of molecules with diverse effects. For example, a neuropeptide called **substance P** is an important mediator of pain signals. By contrast, **endorphins**, which include **beta endorphin**, **dynorphin**, and **enkephalins** (en-kef′ah-linz), act as natural opiates, reducing our perception of pain under certain stressful conditions. Enkephalin activity increases dramatically in pregnant women in labor. Endorphin release is enhanced when an athlete gets a so-called second wind and is probably responsible for the "runner's high." Additionally, some researchers claim that the placebo effect is due to endorphin release. These painkilling neurotransmitters remained undiscovered until investigators began to ask why morphine and other opiates reduce anxiety and pain, and found that these drugs attach to the same receptors that bind natural opiates, producing similar but stronger effects.

Some neuropeptides, such as somatostatin and cholecystokinin, are also produced by nonneural body tissues and are widespread in the gastrointestinal tract. Such peptides are commonly referred to as **gut-brain peptides**.

Purines

Like amino acids, another ubiquitous cellular component, **adenosine triphosphate (ATP**, the universal form of energy), is now recognized as a major neurotransmitter (perhaps the most primitive one) in both the CNS and PNS. Like glutamate and acetylcholine, it produces a fast excitatory response at certain receptors. Depending on the ATP receptor type it binds to, ATP can mediate fast excitatory responses or trigger slow, second-messenger responses. Upon binding to receptors on astrocytes, it mediates Ca^{2+} influx.

In addition to the neurotransmitter action of extracellular ATP, **adenosine**, a part of ATP, also acts outside of cells on adenosine receptors. Adenosine is a potent inhibitor in the brain. Caffeine's well-known stimulatory effects result from its block of these adenosine receptors.

Gases and Lipids

Not so long ago, it would have been scientific suicide to suggest that nitric oxide and carbon monoxide—two ubiquitous molecules—might be neurotransmitters. Nonetheless, the discovery of these unlikely messengers has opened up a whole new chapter in the story of neurotransmission.

Nitric oxide (NO), a short-lived toxic gas, defies all the official descriptions of neurotransmitters. Rather than being stored in vesicles and released by exocytosis, it is synthesized on demand and diffuses out of the cells making it. Instead of attaching to surface receptors, it zooms through the plasma membrane of nearby cells to bind with a peculiar intracellular receptor—iron in *guanylyl cyclase*, the enzyme that makes the second messenger *cyclic GMP*. NO participates in a variety of processes in the brain, including the formation of new memories by increasing the strength of certain synapses. In this process, neurotransmitter binding to the postsynaptic receptors indirectly causes the activation of *nitric oxide synthase* (*NOS*), the enzyme that makes NO. The newly synthesized NO diffuses out of the postsynaptic cell back to the presynaptic terminal, where it activates guanylyl cyclase. In this way NO is thought to act as a retrograde messenger that sends a signal to increase synaptic strength. Excessive release of NO contributes to much of the brain damage seen in stroke patients (see pp. 464–465). In the myenteric plexus of the intestine, NO causes intestinal smooth muscle to relax.

NO is the first member of a class of signaling gases that pass swiftly into cells, bind briefly to metal-containing enzymes, and then vanish. **Carbon monoxide (CO)**, another airy messenger, also stimulates synthesis of cyclic GMP. NO and CO are found in different brain regions and appear to act in different pathways, but their mode of action is similar.

Just as there are natural opiate neurotransmitters in the brain, our brains make natural neurotransmitters that act at the same receptors as the active ingredient in marijuana, tetrahydrocannabinol (THC). Surprisingly, this **endocannabinoid** (en″do-kă-nă′bĭ-noid) class of neurotransmitter has only recently been discovered. We now know that their receptors, the *cannabinoid receptors*, are the most common G protein–coupled receptors in the brain. Like NO, the endocannabinoids are lipid soluble and are synthesized on demand, rather than stored and released from vesicles. Endocannabinoids are formed by clipping the cell's own plasma membrane lipids. The newly synthesized endocannabinoids diffuse freely from the postsynaptic neuron to their receptors on presynaptic terminals where they act as a retrograde messenger to decrease neurotransmitter release. Like NO, they are thought to be involved in learning and memory. We are only beginning to understand the many other processes these neurotransmitters may be involved in, which include neuronal development, control of appetite, and suppression of nausea.

Classification of Neurotransmitters by Function

In this text we can only sample the incredible diversity of functions that neurotransmitters mediate. We limit our discussion here to two broad ways of classifying neurotransmitters according to function, adding more details in subsequent chapters.

Effects: Excitatory Versus Inhibitory

We can summarize this classification scheme by saying that some neurotransmitters are excitatory (cause depolarization), some are inhibitory (cause hyperpolarization), and others exert both effects, depending on the specific receptor types with which they interact. For example, the amino acids GABA and glycine are usually inhibitory, whereas glutamate is typically excitatory (Table 11.3). On the other hand, ACh and NE each bind to at least two receptor types that cause opposite effects. For example, acetylcholine is excitatory at neuromuscular junctions in skeletal muscle and inhibitory in cardiac muscle.

Actions: Direct Versus Indirect

Neurotransmitters that bind to and open ion channels are said to act *directly*. These neurotransmitters provoke rapid responses in postsynaptic cells by promoting changes in membrane potential. ACh and the amino acid neurotransmitters are typically direct-acting neurotransmitters.

Neurotransmitters that act *indirectly* tend to promote broader, longer-lasting effects by acting through intracellular

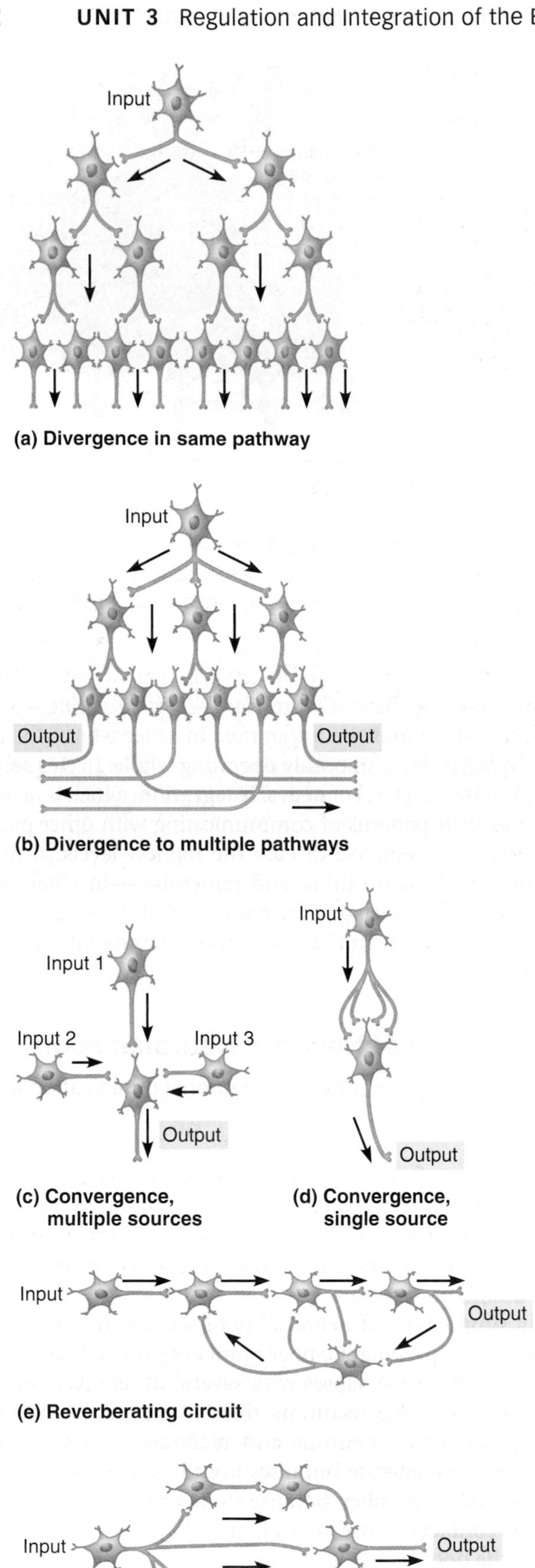

Figure 11.22 Types of circuits in neuronal pools.

11

thousands of neurons and include inhibitory as well as excitatory neurons.

Types of Circuits

Individual neurons in a neuronal pool both send and receive information, and synaptic contacts may cause either excitation or inhibition. The patterns of synaptic connections in neuronal pools are called **circuits**, and they determine the pool's functional capabilities. Four basic circuit patterns are shown in simplified form in **Figure 11.22**.

In **diverging circuits**, one incoming fiber triggers responses in ever-increasing numbers of neurons farther and farther along in the circuit. So, diverging circuits are often *amplifying* circuits. Divergence can occur along a single pathway or along several (Figure 11.22a and b). These circuits are common in both sensory and motor systems. For example, impulses traveling from a single neuron of the brain can activate a hundred or more motor neurons in the spinal cord and, consequently, thousands of skeletal muscle fibers.

The pattern of **converging circuits** is opposite that of diverging circuits, but they too are common in both sensory and motor pathways. In a converging circuit, the pool receives inputs from several presynaptic neurons, and the circuit has a funneling, or *concentrating*, effect. Incoming stimuli may converge from many different areas or from one area, resulting in strong stimulation or inhibition (Figure 11.22c and d). Convergence from different areas helps to explain how different types of sensory stimuli can have the same ultimate effect. For instance, seeing the smiling face of their infant, smelling the baby's freshly powdered skin, or hearing the baby gurgle can all trigger a flood of loving feelings in parents.

In **reverberating**, or **oscillating**, **circuits**, the incoming signal travels through a chain of neurons, each of which makes collateral synapses with neurons in a previous part of the pathway (Figure 11.22e). As a result of the positive feedback, the impulses *reverberate* (are sent through the circuit again and again), giving a continuous output signal until one neuron in the circuit fails to fire. Reverberating circuits are involved in control of rhythmic activities, such as the sleep-wake cycle, breathing, and certain motor activities (such as arm swinging when walking). Some researchers believe that such circuits underlie short-term memory. Depending on the specific circuit, reverberating circuits may continue to oscillate for seconds, hours, or (in the case of the circuit controlling the rhythm of breathing) a lifetime.

In **parallel after-discharge circuits**, the incoming fiber stimulates several neurons arranged in parallel arrays that eventually stimulate a common output cell (Figure 11.22f). Impulses reach the output cell at different times, creating a burst of impulses called an *after-discharge* that lasts 15 ms or more after the initial input has ended. This type of circuit has no positive feedback, and once all the neurons have fired, circuit activity ends. Parallel after-discharge circuits may be involved in complex, exacting types of mental processing.

Patterns of Neural Processing

▶ Distinguish between serial and parallel processing.

Rabies (*rabies* = madness) A viral infection of the nervous system transmitted by the bite of an infected mammal (such as a dog, bat, or skunk). After entry, the virus travels via axonal transport in peripheral nerve axons to the CNS, where it causes brain inflammation, delirium, and death. A vaccine- and antibody-based treatment is effective if given before symptoms appear; rabies in humans is very rare in the United States.

Shingles (herpes zoster) A viral infection of sensory neurons serving the skin. Characterized by scaly, painful blisters usually confined to a narrow strip of skin, often on one side of the body trunk. Caused by the varicella-zoster virus, which causes chicken pox (generally during childhood); during this initial infection the virus is transported from the skin lesions to the sensory cell bodies in the sensory ganglia. Typically, the virus is held in check by the immune system and remains dormant until the immune system is weakened, often by stress. Then viral particles multiply and travel back to the skin, producing the characteristic rash. Attacks last several weeks, alternating between periods of healing and relapse. Seen mostly in those over 50 years old.

CHAPTER SUMMARY

Media study tools that could provide you additional help in reviewing specific key topics of Chapter 11 are referenced below.

iP = ***Interactive Physiology***

Functions and Divisions of the Nervous System (pp. 386–388)

1. The nervous system bears a major responsibility for maintaining body homeostasis. Its chief functions are to monitor, integrate, and respond to information in the environment.
2. The nervous system is divided anatomically into the central nervous system (brain and spinal cord) and the peripheral nervous system (mainly cranial and spinal nerves).
3. The major functional divisions of the PNS are the sensory (afferent) division, which conveys impulses to the CNS, and the motor (efferent) division, which conveys impulses from the CNS.
4. The efferent division includes the somatic (voluntary) system, which serves skeletal muscles, and the autonomic (involuntary) system, which innervates smooth and cardiac muscle and glands.

Histology of Nervous Tissue (pp. 388–395)

Neuroglia (pp. 388–389)

1. Neuroglia (supporting cells) segregate and insulate neurons and assist neurons in various other ways.
2. CNS neuroglia include astrocytes, microglia, ependymal cells, and oligodendrocytes. Schwann cells and satellite cells are neuroglia found in the PNS.

Neurons (pp. 389–395)

3. Neurons have a cell body and cytoplasmic processes called axons and dendrites.
4. A bundle of nerve fibers is called a tract in the CNS and a nerve in the PNS. A collection of cell bodies is called a nucleus in the CNS and a ganglion in the PNS.
5. The cell body is the biosynthetic (and receptive) center of the neuron. Except for those found in ganglia, cell bodies are found in the CNS.
6. Most neurons have many dendrites, receptive processes that conduct signals from other neurons toward the nerve cell body. With few exceptions, all neurons have one axon, which generates and conducts nerve impulses away from the nerve cell body. Terminal endings of axons release neurotransmitter.
7. Bidirectional transport along axons uses ATP-dependent motor proteins "walking" along microtubule tracks. It moves vesicles, mitochondria, and cytosolic proteins toward the axon terminals and conducts substances destined for degradation back to the cell body.
8. Large nerve fibers (axons) are myelinated. The myelin sheath is formed in the PNS by Schwann cells and in the CNS by oligodendrocytes. The sheath has gaps called nodes of Ranvier. Unmyelinated fibers are surrounded by supporting cells, but the membrane-wrapping process does not occur.
9. Anatomically, neurons are classified according to the number of processes issuing from the cell body as multipolar, bipolar, or unipolar.
10. Functionally, neurons are classified according to the direction of nerve impulse conduction. Sensory neurons conduct impulses toward the CNS, motor neurons conduct away from the CNS, and interneurons (association neurons) lie between sensory and motor neurons in the neural pathways.

iP Nervous System I; Topic: Anatomy Review, pp. 1–12.

Membrane Potentials (pp. 395–406)

Basic Principles of Electricity (p. 395)

1. The measure of the potential energy of separated electrical charges is called voltage (V) or potential. Current (I) is the flow of electrical charge from one point to another. Resistance (R) is hindrance to current flow. The relationship among these is given by Ohm's law: $I = V/R$.
2. In the body, electrical charges are provided by ions; cellular plasma membranes provide resistance to ion flow. The membranes contain leakage channels (nongated, always open) and gated channels.

iP Nervous System I; Topic: Ion Channels, pp. 1–10.

The Resting Membrane Potential (pp. 396–398)

3. A resting neuron exhibits a resting membrane potential, which is −70 mV (inside negative). It is due both to differences in sodium and potassium ion concentrations inside and outside the cell and to differences in permeability of the membrane to these ions.
4. The ionic concentration differences result from the operation of the sodium-potassium pump, which ejects $3Na^+$ from the cell for each $2K^+$ transported in.

iP Nervous System I; Topic: The Membrane Potential, pp. 1–16.

Membrane Potentials That Act as Signals (pp. 398–406)

5. Depolarization is a reduction in membrane potential (inside becomes less negative); hyperpolarization is an increase in membrane potential (inside becomes more negative).
6. Graded potentials are small, brief, local changes in membrane potential that act as short-distance signals. The current produced dissipates with distance.
7. An action potential (AP), or nerve impulse, is a large, but brief, depolarization signal (and polarity reversal) that underlies long-distance neural communication. It is an all-or-none phenomenon.

Input processing is both *serial* and *parallel*. In serial processing, the input travels along one pathway to a specific destination. In parallel processing, the input travels along several different pathways to be integrated in different CNS regions. Each mode has unique advantages in the overall scheme of neural functioning, but as an information processor, the brain derives its power from its ability to process in parallel.

Serial Processing

In **serial processing**, the whole system works in a predictable all-or-nothing manner. One neuron stimulates the next, which stimulates the next, and so on, eventually causing a specific, anticipated response. The most clear-cut examples of serial processing are spinal reflexes, but straight-through sensory pathways from receptors to the brain are also examples. Because reflexes are the functional units of the nervous system, it is important that you understand them early on.

Reflexes are rapid, automatic responses to stimuli, in which a particular stimulus always causes the same response. Reflex activity, which produces the simplest of behaviors, is stereotyped and dependable. For example, jerking away your hand after touching a hot object is the norm, and an object approaching the eye triggers a blink. Reflexes occur over neural pathways called **reflex arcs** that have five essential components—receptor, sensory neuron, CNS integration center, motor neuron, and effector **(Figure 11.23)**.

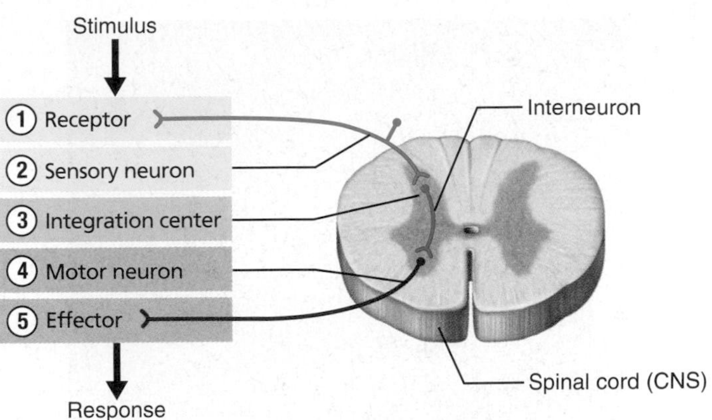

Figure 11.23 A simple reflex arc. Receptors detect changes in the internal or external environment. Effectors are muscles or glands.

Parallel Processing

In **parallel processing**, inputs are segregated into many pathways, and information delivered by each pathway is dealt with simultaneously by different parts of the neural circuitry. For example, smelling a pickle (the input) may cause you to remember picking cucumbers on a farm; or it may remind you that you don't like pickles or that you must buy some at the market; or perhaps it will call to mind *all* these thoughts. For each person, parallel processing triggers some pathways that are unique. The same stimulus—pickle smell, in our example—promotes many responses beyond simple awareness of the smell. Parallel processing is not repetitious because the circuits do different things with the information, and each pathway or "channel" is decoded in relation to all the others to produce a total picture.

Think, for example, about what happens when you step on something sharp while walking barefoot. The serially processed withdrawal reflex causes instantaneous removal of your injured foot from the sharp object (painful stimulus). At the same time, pain and pressure impulses are speeding up to the brain along parallel pathways that allow you to decide whether to simply rub the hurt spot to soothe it or to seek first aid.

Parallel processing is extremely important for higher-level mental functioning—for putting the parts together to understand the whole. For example, you can recognize a dollar bill in a split second, a task that takes a serial-based computer a fairly long time. Your recognition is quick because you use parallel processing, which allows a single neuron to send information along several pathways instead of just one, so a large amount of information is processed much more quickly.

CHECK YOUR UNDERSTANDING

21. What types of neural circuits would give a prolonged output after a single input?
22. What pattern of neural processing occurs when we blink as an object comes toward the eye? What is this response called?
23. What pattern of neural processing occurs when we smell freshly baked apple pie and remember Thanksgiving at our grandparents' house, the odor of freshly cooked turkey, and other such memories?

For answers, see Appendix G.

Developmental Aspects of Neurons

▶ Describe how neurons develop and form synapses.

We cover the nervous system in several chapters, so we limit our attention here to the development of neurons, beginning with the questions, How do nerve cells originate? and How do they mature?

The nervous system originates from a dorsal *neural tube* and the *neural crest*, formed from surface ectoderm (see Figure 12.1 ③,④, p. 430). The neural tube, whose walls begin as a layer of *neuroepithelial* cells, becomes the CNS. The neuroepithelial cells then begin a three-phase process of differentiation, which occurs largely in the second month of development. (1) They *proliferate* to produce the appropriate number of cells needed for nervous system development. (2) The potential neurons, **neuroblasts**, become amitotic and *migrate* externally into their characteristic positions. (3) The neuroblasts sprout axons to *connect with* their functional targets and in so doing become neurons.

How does a neuroblast's growing axon "know" where to go—and once it gets there, where to make the proper connection? The growth of an axon toward an appropriate target requires multiple steps and is guided by multiple signals. The growing tip of an axon, called a **growth cone**, is a prickly, fanlike structure that gives an axon the ability to interact with its

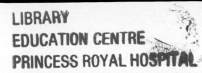

Figure 11.24 A neuronal growth cone. Fluorescent stains show the locations of cannabinoid receptors (green), tubulin (blue), and actin (pink) in this photomicrograph (1400×).

environment **(Figure 11.24)**. Extracellular and cell surface adhesion proteins such as laminin, integrin, and *nerve cell adhesion molecule* (*N-CAM*) provide anchor points for the growth cone, saying, "It's okay to grow here." *Neurotropins* are chemicals that signal to the growth cone "come this way" (netrin) or "go away" (ephrin, slit) or "stop here" (semaphorin). Throughout this growth and development, neurotrophic factors such as *nerve growth factor* (*NGF*) must be present to keep the neuroblast alive. Failure of any of these guiding signals results in catastrophic developmental problems. For example, lack of N-CAM action causes developing neural tissue to fall into a tangled, spaghetti-like mass and hopelessly impairs neural function.

The growth cone gropes along like an amoeba, with oozing processes called *filopodia* which detect the guiding signals in the surrounding environment. Receptors for these signals generate various second messengers that cause the filopodia to move by rearranging their actin protein cores. Once the axon has reached its target area, it must select the right site on the target cell to form a synapse. Special cell adhesion molecules couple the presynaptic and postsynaptic membranes together and generate intracellular signals that recruit vesicles containing preformed synaptic components. This results in the rapid formation of a synapse. In the brain and spinal cord, astrocytes seem to provide both physical support and the cholesterol essential for constructing synapses. Both dendrites and astrocytes are active partners in the process of synapse formation. In the presence of thrombospondin released by astrocytes, dendrites actually reach out and grasp migrating axons, and synapses begin sprouting.

Neurons that fail to make appropriate or functional synaptic contacts act as if they have been deprived of some essential nutrient and die. Besides cell death resulting from unsuccessful synapse formation, *apoptosis* (programmed cell death) also appears to be a normal part of the developmental process. Of the neurons formed during the embryonic period, perhaps two-thirds die before we are born. Those that remain constitute most of our neural endowment for life. The generally amitotic nature of neurons is important because their activity depends on the synapses they've formed, and if neurons were to divide, their connections might be hopelessly disrupted. This aside, there *do* appear to be some specific neuronal populations where stem cells are found and new neurons can be formed—notably olfactory neurons and some cells of the hippocampus, a brain region involved in learning and memory.

CHECK YOUR UNDERSTANDING

24. What is the name of the growing tip of an axon that "sniffs out" where to go during development? What is the general name for the chemicals that tell it where to go?

For answers, see Appendix G.

In this chapter, we have examined how the amazingly complex neurons, via electrical and chemical signals, serve the body in a variety of ways: Some serve as "lookouts," others process information for immediate use or for future reference, and still others stimulate the body's muscles and glands into activity. With this background, we are ready to study the most sophisticated mass of neural tissue in the entire body—the brain (and its continuation, the spinal cord), the focus of Chapter 12.

RELATED CLINICAL TERMS

Neuroblastoma (nu″ro-blas-to′mah; *oma* = tumor) A malignant tumor in children; arises from cells that retain a neuroblast-like structure. These tumors sometimes arise in the brain, but most occur in the peripheral nervous system.

Neurologist (nu-rol′o-jist) A medical specialist in the study of the nervous system, its functions, and its disorders.

Neuropathy (nu-rop′ah-the) Any disease of nervous tissue, but particularly degenerative disease of nerves.

Neuropharmacology (nu″ro-far″mah-kol′o-je) Scientific study of the effects of drugs on the nervous system.

Neurotoxin Substance that is poisonous or destructive to nervous tissue, e.g., botulinum and tetanus toxins.

8. An AP has three underlying phases of membrane permeability. (1) Increase in sodium permeability: Local depolarization opens voltage-gated Na^+ channels; at threshold, depolarization becomes self-generating (driven by Na^+ influx). The membrane potential is reversed to approximately +30 mV (inside positive). (2) Decrease in sodium permeability. (3) Increase in potassium permeability. Repolarization occurs during phases 2 and 3.
9. In nerve impulse propagation, each AP provides the depolarizing stimulus for triggering an AP in the next membrane patch. Regions that have just generated APs are refractory; for this reason, the nerve impulse is propagated in one direction only.
10. If threshold is reached, an AP is generated; if not, depolarization remains local.
11. APs are independent of stimulus strength: Strong stimuli cause APs to be generated more frequently but not with greater amplitude.
12. During the absolute refractory period, a neuron cannot respond to another stimulus because it is already generating an AP. During the relative refractory period, the neuron's threshold is elevated because repolarization is ongoing.
13. In unmyelinated fibers, APs are produced in a wave all along the axon, that is, by continuous conduction. In myelinated fibers, APs are generated only at nodes of Ranvier and are propagated more rapidly by saltatory conduction.

iP Nervous System I; Topic: The Action Potential, pp. 1–18.

The Synapse (pp. 406–413)

1. A synapse is a functional junction between neurons. The information-transmitting neuron is the presynaptic neuron; the information-receiving neuron is the postsynaptic neuron.

Electrical Synapses (p. 406)

2. Electrical synapses allow ions to flow directly from one neuron to another; the cells are electrically coupled.

Chemical Synapses (pp. 407–408)

3. Chemical synapses are sites of neurotransmitter release and binding. When the impulse reaches the presynaptic axon terminals, voltage-gated Ca^{2+} channels open, and Ca^{2+} enters the cell and mediates neurotransmitter release. Neurotransmitters diffuse across the synaptic cleft and attach to postsynaptic membrane receptors, opening ion channels. After binding, the neurotransmitters are removed from the synapse by enzymatic breakdown or by reuptake into the presynaptic terminal or astrocytes.

iP Nervous System II; Topics: Anatomy Review, pp. 1–9, Ion Channels, pp. 1–8, Synaptic Transmission, pp. 1–7.

Postsynaptic Potentials and Synaptic Integration (pp. 408–413)

4. Binding of neurotransmitter at excitatory chemical synapses results in local graded potentials called EPSPs, caused by the opening of channels that allow simultaneous passage of Na^+ and K^+.
5. Neurotransmitter binding at inhibitory chemical synapses results in hyperpolarizations called IPSPs, caused by the opening of K^+ or Cl^- channels. IPSPs drive the membrane potential farther from threshold.
6. EPSPs and IPSPs summate temporally and spatially. The membrane of the axon hillock acts as a neuronal integrator.
7. Synaptic potentiation, in which the postsynaptic neuron's response is enhanced, is produced by intense repeated stimulation. Ionic calcium appears to mediate such effects, which may be the basis of learning.
8. Presynaptic inhibition is mediated by axoaxonic synapses that reduce the amount of neurotransmitter released by the inhibited neuron.

iP Nervous System II; Topic: Synaptic Potentials and Cellular Integration, pp. 1–10.

Neurotransmitters and Their Receptors (pp. 413–420)

Classification of Neurotransmitters by Chemical Structure (pp. 414–419)

1. The major classes of neurotransmitters based on chemical structure are acetylcholine, biogenic amines, amino acids, peptides, purines, dissolved gases, and lipids.

Classification of Neurotransmitters by Function (pp. 419–420)

2. Functionally, neurotransmitters are classified as (1) inhibitory or excitatory (or both) and (2) direct or indirect action. Direct-acting neurotransmitters bind to and open ion channels. Indirect-acting neurotransmitters act through second messengers. Neuromodulators also act indirectly presynaptically or postsynaptically to change synaptic strength.

Neurotransmitter Receptors (p. 421)

3. Neurotransmitter receptors are either channel-linked receptors that open ion channels, leading to fast changes in membrane potential, or G protein–linked receptors that oversee slow synaptic responses mediated by G proteins and intracellular second messengers. Second messengers most often activate kinases, which in turn act on ion channels or activate other proteins.

iP Nervous System II; Topic: Synaptic Transmission, pp. 6–15.

Basic Concepts of Neural Integration (pp. 421–423)

Organization of Neurons: Neuronal Pools (pp. 421–422)

1. CNS neurons are organized into several types of neuronal pools, each with distinguishing patterns of synaptic connections called circuits.

Types of Circuits (p. 422)

2. The four basic circuit types are diverging, converging, reverberating, and parallel after-discharge.

Patterns of Neural Processing (pp. 422–423)

3. In serial processing, one neuron stimulates the next in sequence, producing specific, predictable responses, as in spinal reflexes. A reflex is a rapid, involuntary motor response to a stimulus.
4. Reflexes are mediated over neural pathways called reflex arcs. The minimum number of elements in a reflex arc is five: receptor, sensory neuron, integration center, motor neuron, and effector.
5. In parallel processing, which underlies complex mental functions, impulses are sent along several pathways to different integration centers.

Developmental Aspects of Neurons (pp. 423–424)

1. Neuron development involves proliferation, migration, and the formation of interconnections. The formation of interconnections involves axons finding their targets and forming synapses, and the synthesis of specific neurotransmitters.
2. Axon outgrowth and synapse formation are guided by other neurons, glial cells, and chemicals (such as N-CAM and nerve growth factor). Neurons that do not make appropriate synapses die, and approximately two-thirds of neurons formed in the embryo undergo programmed cell death before birth.

REVIEW QUESTIONS

Multiple Choice/Matching

(Some questions have more than one correct answer. Select the best answer or answers from the choices given.)

1. Which of the following structures is not part of the central nervous system? (a) the brain, (b) a nerve, (c) the spinal cord, (d) a tract.
2. Match the names of the supporting cells found in column B with the appropriate descriptions in column A.

Column A	Column B
____ (1) myelinates nerve fibers in the CNS	(a) astrocyte
____ (2) lines brain cavities	(b) ependymal cell
____ (3) myelinates nerve fibers in the PNS	(c) microglia
____ (4) CNS phagocytes	(d) oligodendrocyte
____ (5) helps regulate the ionic composition of CNS extracellular fluid	(e) satellite cell
	(f) Schwann cell

3. What type of current flows through the axolemma during the steep phase of repolarization? (a) chiefly a sodium current, (b) chiefly a potassium current, (c) sodium and potassium currents of approximately the same magnitude.
4. Assume that an EPSP is being generated on the dendritic membrane. Which will occur? (a) specific Na^+ channels will open, (b) specific K^+ channels will open, (c) a single type of channel will open, permitting simultaneous flow of Na^+ and K^+, (d) Na^+ channels will open first and then close as K^+ channels open.
5. The velocity of nerve impulse conduction is greatest in (a) heavily myelinated, large-diameter fibers, (b) myelinated, small-diameter fibers, (c) unmyelinated, small-diameter fibers, (d) unmyelinated, large-diameter fibers.
6. Chemical synapses are characterized by all of the following except (a) the release of neurotransmitter by the presynaptic membranes, (b) postsynaptic membranes bearing receptors that bind neurotransmitter, (c) ions flowing through protein channels from the presynaptic to the postsynaptic neuron, (d) a fluid-filled gap separating the neurons.
7. Biogenic amine neurotransmitters include all but (a) norepinephrine, (b) acetylcholine, (c) dopamine, (d) serotonin.
8. The neuropeptides that act as natural opiates are (a) substance P, (b) somatostatin, (c) cholecystokinin, (d) enkephalins.
9. Inhibition of acetylcholinesterase by poisoning blocks neurotransmission at the neuromuscular junction because (a) ACh is no longer released by the presynaptic terminal, (b) ACh synthesis in the presynaptic terminal is blocked, (c) ACh is not degraded, hence prolonged depolarization is enforced on the postsynaptic cell, (d) ACh is blocked from attaching to the postsynaptic ACh receptors.
10. The anatomical region of a multipolar neuron that has the lowest threshold for generating an AP is the (a) soma, (b) dendrites, (c) axon hillock, (d) distal axon.
11. An IPSP is inhibitory because (a) it hyperpolarizes the postsynaptic membrane, (b) it reduces the amount of neurotransmitter released by the presynaptic terminal, (c) it prevents calcium ion entry into the presynaptic terminal, (d) it changes the threshold of the neuron.
12. Identify the neuronal circuits described by choosing the correct response from the key.

Key: (a) converging (b) diverging (c) parallel after-discharge (d) reverberating

____ (1) Impulses continue around and around the circuit until one neuron stops firing.
____ (2) One or a few inputs ultimately influence large numbers of neurons.
____ (3) Many neurons influence a few neurons.
____ (4) May be involved in exacting types of mental activity.

Short Answer Essay Questions

13. Explain both the anatomical and functional divisions of the nervous system. Include the subdivisions of each.
14. (a) Describe the composition and function of the cell body. (b) How are axons and dendrites alike? In what ways (structurally and functionally) do they differ?
15. (a) What is myelin? (b) How does the myelination process differ in the CNS and PNS?
16. (a) Contrast unipolar, bipolar, and multipolar neurons structurally. (b) Indicate where each is most likely to be found.
17. What is the polarized membrane state? How is it maintained? (Note the relative roles of both passive and active mechanisms.)
18. Describe the events that must occur to generate an AP. Relate the sequence of changes in permeability to changes in the ion channels, and explain why the AP is an all-or-none phenomenon.
19. Since all APs generated by a given nerve fiber have the same magnitude, how does the CNS "know" whether a stimulus is strong or weak?
20. (a) Explain the difference between an EPSP and an IPSP. (b) What specifically determines whether an EPSP or IPSP will be generated at the postsynaptic membrane?
21. Since at any moment a neuron is likely to have thousands of neurons releasing neurotransmitters at its surface, how is neuronal activity (to fire or not to fire) determined?
22. The effects of neurotransmitter binding are very brief. Explain.
23. During a neurobiology lecture, a professor repeatedly refers to group A and group B fibers, absolute refractory period, and nodes of Ranvier. Define these terms.
24. Distinguish between serial and parallel processing.
25. Briefly describe the three stages of neuron development.
26. What factors appear to guide the outgrowth of an axon and its ability to make the "correct" synaptic contacts?

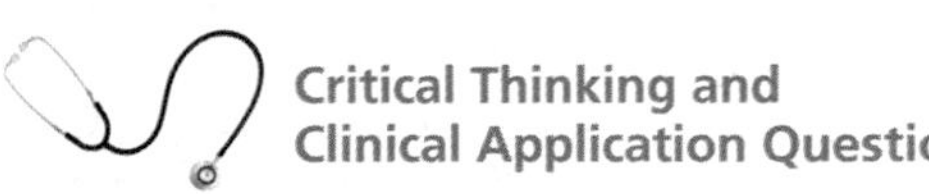

Critical Thinking and Clinical Application Questions

1. Mr. Miller is hospitalized for cardiac problems. Somehow, medical orders are mixed up and Mr. Miller is infused with a K^+-enhanced intravenous solution meant for another patient who is taking potassium-wasting diuretics (i.e., drugs that cause excessive loss of potassium from the body in urine). Mr. Miller's potassium levels are normal before the IV is administered. What do you think will happen to Mr. Miller's resting membrane potentials? To his neurons' ability to generate APs?
2. Local anesthetics block voltage-gated Na^+ channels. General anesthetics are thought to activate chemically gated Cl^- channels,

thereby rendering the nervous system quiescent while surgery is performed. What specific process do anesthetics impair, and how does this interfere with nerve impulse transmission?

3. When admitted to the emergency room, John was holding his right hand, which had a deep puncture hole in its palm. He explained that he had fallen on a nail while exploring a barn. John was given an antitetanus shot to prevent neural complications. Tetanus bacteria fester in deep, dark wounds, but how do their toxins travel in neural tissue?
4. Rochelle developed multiple sclerosis when she was 27. After eight years she had lost a good portion of her ability to control her skeletal muscles. How did this happen?
5. In the Netherlands a young man named Jan was admitted to the emergency room. He and his friends had been to a rave. His friends say he started twitching and having muscle spasms which progressed until he was "stiff as a board." On examination, staff found a marked increase in muscle tone and hyperreflexia involving facial and limb muscles. In his pocket, he had unmarked dark yellow tablets with dark flecks. Analysis of the tablets showed them to contain a mixture of ecstasy and strychnine. Ecstasy would not cause this clinical picture, but strychnine, which blocks glycine receptors, could. Explain how.

11

12

The Central Nervous System

Historically, the **central nervous system (CNS)**—brain and spinal cord—has been compared to the central switchboard of a telephone system that interconnects and directs a dizzying number of incoming and outgoing calls. Nowadays, many people compare it to a supercomputer. These analogies may explain some workings of the spinal cord, but neither does justice to the fantastic complexity of the human brain. Whether we view the brain as an evolved biological organ, an impressive computer, or simply a miracle, it is one of the most amazing things known.

During the course of animal evolution, **cephalization** (sĕ″fah-lĭ-za′shun) has occurred. That is, there has been an elaboration of the *rostral* ("toward the snout"), or anterior, portion of the CNS, along with an increase in the number of neurons in the head. This phenomenon reaches its highest level in the human brain.

In this chapter, we examine the structure of the CNS and the functions associated with its various regions. We also touch on complex integrative functions, such as sleep-wake cycles and memory.

The Brain

The unimpressive appearance of the human **brain** gives few hints of its remarkable abilities. It is about two good fistfuls of quivering pinkish gray tissue, wrinkled like a walnut, with a consistency somewhat like cold oatmeal. The average adult man's brain has a mass of about 1600 g (3.5 lb); that of a woman averages 1450 g (3.2 lb). In terms of brain mass per body mass, however, males and females have equivalent brain sizes.

Embryonic Development

- Describe the process of brain development.
- Name the major regions of the adult brain.
- Name and locate the ventricles of the brain.

12

We begin our study of the brain with brain embryology, as the terminology used for the structural divisions of the adult brain is easier to follow when you understand brain development.

The earliest phase of brain development is shown in **Figure 12.1.** Starting in the three-week-old embryo, the *ectoderm* (cell layer at the dorsal surface) thickens along the dorsal midline axis of the embryo to form the **neural plate.** The neural plate then invaginates, forming a groove flanked by **neural folds.** As this **neural groove** deepens, the superior edges of the neural folds fuse, forming the **neural tube,** which soon detaches from the surface ectoderm and sinks to a deeper position.

The neural tube, formed by the fourth week of pregnancy, differentiates rapidly into the CNS. The brain forms rostrally (anteriorly), and the spinal cord develops from the *caudal* ("toward the tail") or posterior portion of the neural tube. Small groups of neural fold cells migrate laterally from between the surface ectoderm and the neural tube, forming the **neural crest** (Figure 12.1, ③). Neural crest cells give rise (among other things) to some neurons destined to reside in ganglia.

As soon as the neural tube forms, its anterior end begins to expand and constrictions appear that mark off the three **primary brain vesicles (Figure 12.2a):** the **prosencephalon** (pros″en-sef′ah-lon), or **forebrain;** the **mesencephalon** (mes″en-sef′ah-lon), or **midbrain;** and the **rhombencephalon** (romb″en-sef′ah-lon), or **hindbrain.** (Note that *encephalo* means "brain.") The remainder of the neural tube becomes the spinal cord, which we will discuss later in the chapter.

In week 5, the primary vesicles give rise to the **secondary brain vesicles** (Figure 12.2c). The forebrain divides into the **telencephalon** ("endbrain") and **diencephalon** ("interbrain"), and the hindbrain constricts, forming the **metencephalon**

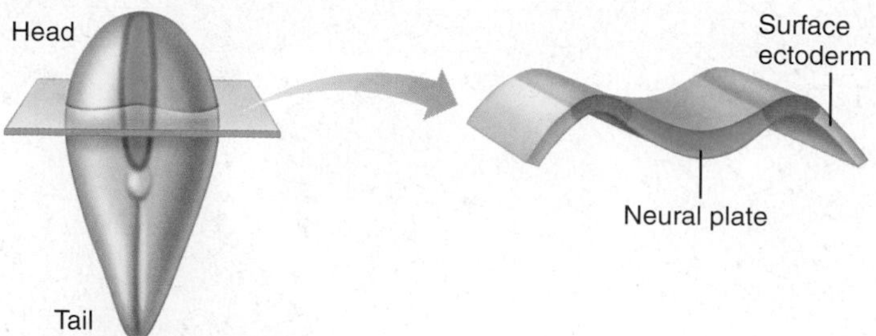

① The neural plate forms from surface ectoderm.

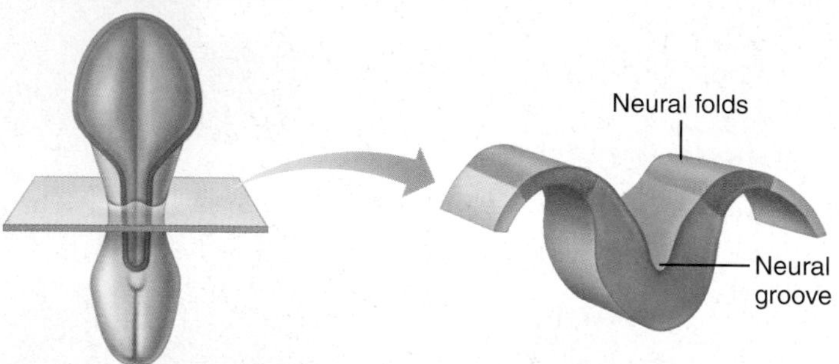

② The neural plate invaginates, forming the neural groove, flanked by neural folds.

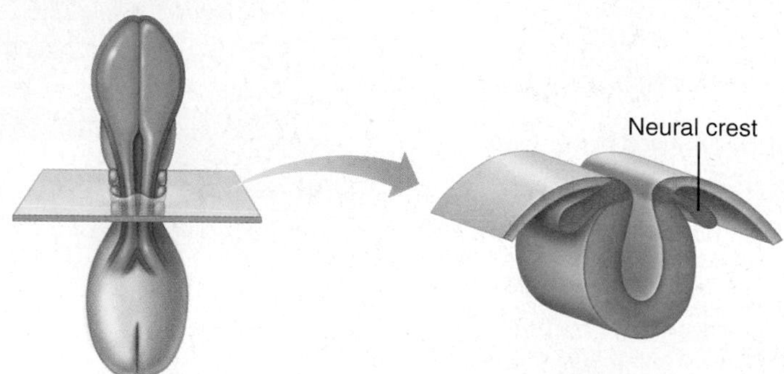

③ Neural fold cells migrate to form the neural crest, which will form much of the PNS and many other structures.

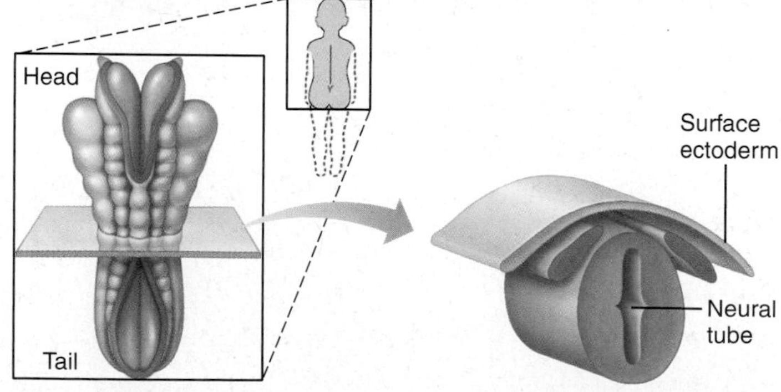

④ The neural groove becomes the neural tube, which will form CNS structures.

Figure 12.1 Development of the neural tube from embryonic ectoderm. Left: dorsal surface views of the embryo; right: transverse sections at days 17, 19, 20, and 22.

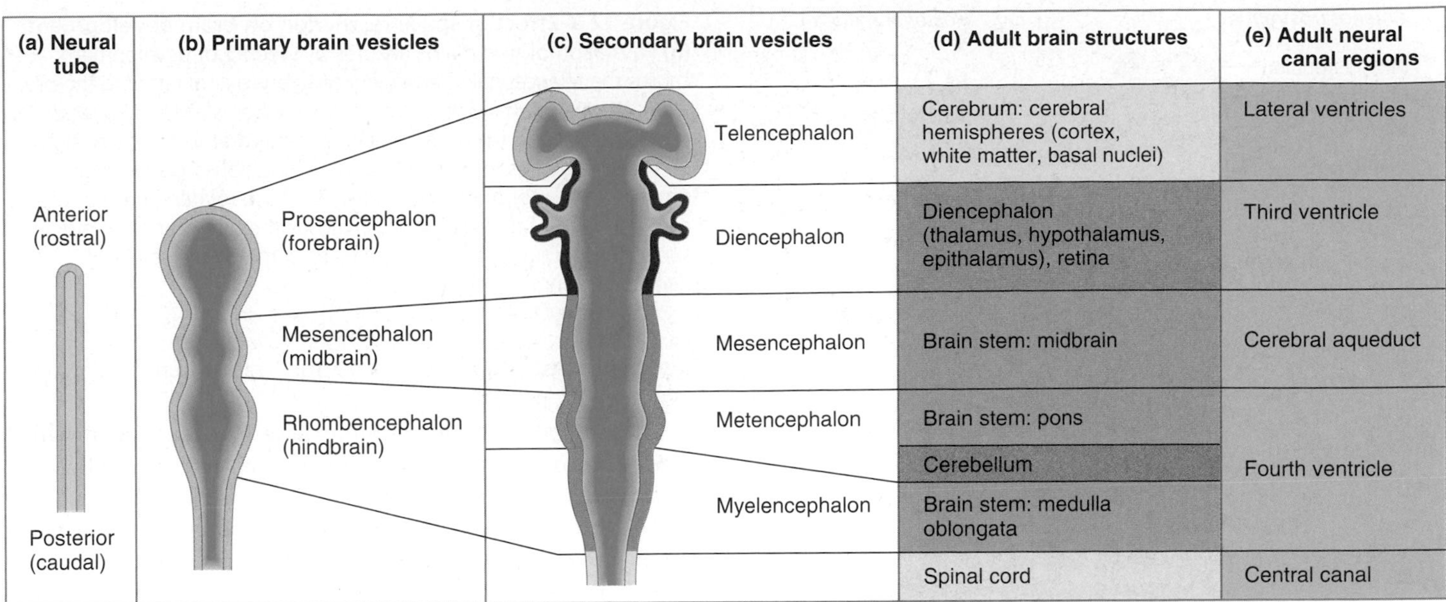

(a) Neural tube
(b) Primary brain vesicles
(c) Secondary brain vesicles
(d) Adult brain structures
(e) Adult neural canal regions
Anterior (rostral)
Posterior (caudal)
Prosencephalon (forebrain)
Mesencephalon (midbrain)
Rhombencephalon (hindbrain)
Telencephalon
Diencephalon
Mesencephalon
Metencephalon
Myelencephalon
Cerebrum: cerebral hemispheres (cortex, white matter, basal nuclei)
Diencephalon (thalamus, hypothalamus, epithalamus), retina
Brain stem: midbrain
Brain stem: pons
Cerebellum
Brain stem: medulla oblongata
Spinal cord
Lateral ventricles
Third ventricle
Cerebral aqueduct
Fourth ventricle
Central canal

Figure 12.2 Embryonic development of the human brain. (a) Formed by week 4, the neural tube quickly subdivides into **(b)** the primary brain vesicles, which subsequently form **(c)** the secondary brain vesicles by week 5. These five vesicles differentiate into **(d)** the adult brain structures. **(e)** The adult structures derived from the neural canal.

("afterbrain") and **myelencephalon** ("spinal brain"). The midbrain remains undivided.

Each of the five secondary vesicles then develops rapidly to produce the major structures of the adult brain (Figure 12.2d). The greatest change occurs in the telencephalon, which sprouts two lateral swellings that look like Mickey Mouse's ears. These become the two *cerebral hemispheres*, referred to collectively as the **cerebrum** (ser′ĕ-brum). The diencephalon part of the forebrain specializes to form the *hypothalamus* (hi″po-thal′ah-mus), *thalamus*, *epithalamus*, and *retina* of the eye. Less dramatic changes occur in the mesencephalon, metencephalon, and myelencephalon as these regions are transformed into the *midbrain*, the *pons* and *cerebellum*, and the *medulla oblongata*, respectively. All these midbrain and hindbrain structures, except the cerebellum, form the **brain stem**. The central cavity of the neural tube remains continuous and enlarges in four areas to form the fluid-filled *ventricles* (*ventr* = little belly) of the brain (Figure 12.2e). We will describe the ventricles shortly.

Because the brain grows more rapidly than the membranous skull that contains it, two major flexures develop—the *midbrain* and *cervical flexures*—which move the forebrain toward the brain stem **(Figure 12.3a)**. A second consequence of restricted space is that the cerebral hemispheres are forced to take a horseshoe-shaped course and grow posteriorly and laterally (indicated by black arrows in Figure 12.3b and c). As a result, they grow back over and almost completely envelop the diencephalon and midbrain. By week 26, the continued growth of the cerebral hemispheres causes their surfaces to crease and fold (Figure 12.3c and d), producing *convolutions* and increasing their surface area, which allows more neurons to occupy the limited space.

Regions and Organization

Some textbooks discuss brain anatomy in terms of the *embryonic scheme* (see Figure 12.2c), but in this text, we will consider the brain in terms of the medical scheme and the adult brain regions shown in Figure 12.3d: (1) cerebral hemispheres, (2) diencephalon, (3) brain stem (midbrain, pons, and medulla), and (4) cerebellum.

The basic pattern of the CNS consists of a central cavity surrounded by gray matter (mostly neuron cell bodies), external to which is white matter (myelinated fiber tracts). The brain exhibits this basic design but has additional regions of gray matter not present in the spinal cord **(Figure 12.4)**. Both the cerebral hemispheres and the cerebellum have an outer layer or "bark" of gray matter called a *cortex*. This pattern changes with descent through the brain stem—the cortex disappears, but scattered gray matter nuclei are seen within the white matter. At the caudal end of the brain stem, the basic pattern is evident.

Ventricles

As noted earlier, the brain **ventricles** arise from expansions of the lumen (cavity) of the embryonic neural tube. They are continuous with one another and with the central canal of the spinal cord **(Figure 12.5)**. The hollow ventricular chambers are filled with cerebrospinal fluid and lined by *ependymal cells*, a type of neuroglia (see Figure 11.3c on p. 388).

The paired **lateral ventricles**, one deep within each cerebral hemisphere, are large C-shaped chambers that reflect the pattern of cerebral growth. Anteriorly, the lateral ventricles lie close together, separated only by a thin median membrane called the

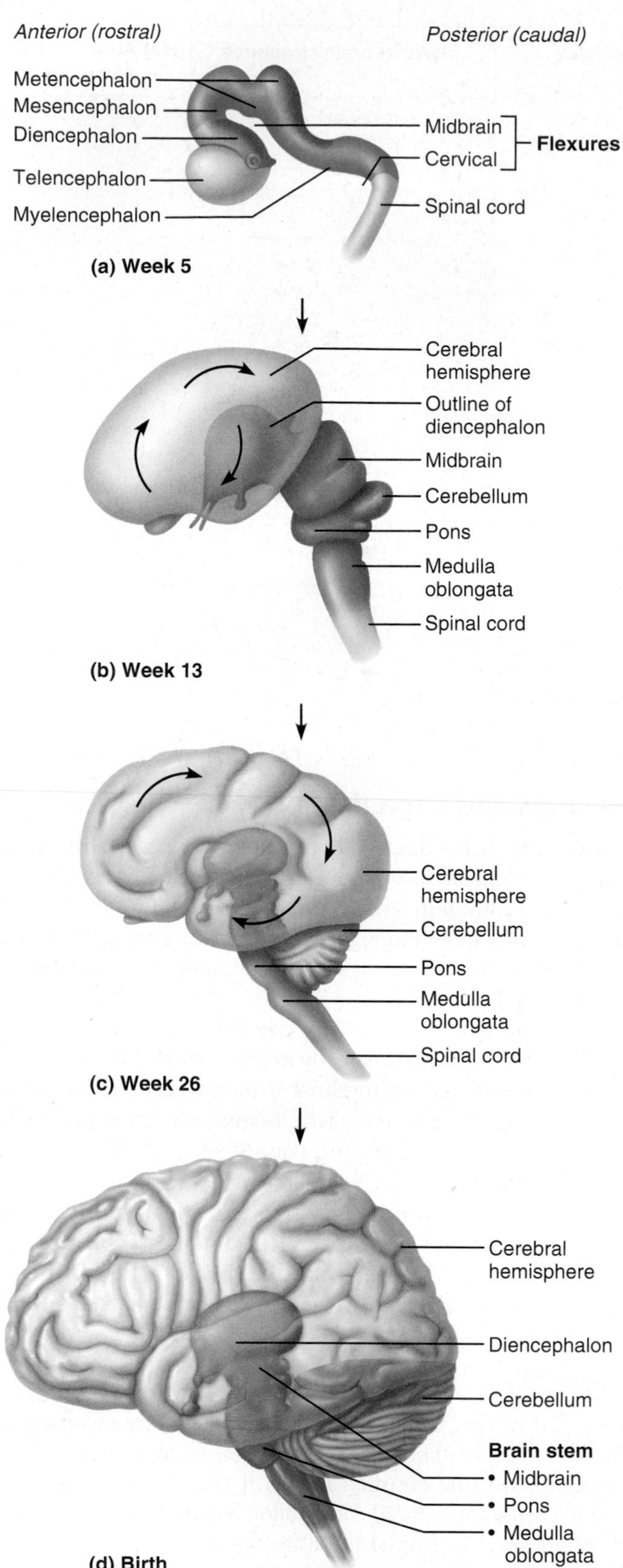

Figure 12.3 Effect of space restriction on brain development. **(a)** Formation of two major flexures by week 5 of development causes the telencephalon and diencephalon to angle toward the brain stem. Development of the cerebral hemispheres at **(b)** 13 weeks, **(c)** 26 weeks, and **(d)** birth. Initially, the cerebral surface is smooth. The folding begins in month 6, and convolutions become more obvious as development continues. The posterolateral growth of the cerebral hemispheres ultimately encloses the diencephalon and superior aspect of the brain stem (seen through the cerebral hemispheres in this see-through view).

septum pellucidum (pĕ-lu′sid-um; "transparent wall"). (See Figure 12.12, p. 443.)

Each lateral ventricle communicates with the narrow **third ventricle** in the diencephalon via a channel called an **interventricular foramen** (*foramen of Monro*).

The third ventricle is continuous with the **fourth ventricle** via the canal-like **cerebral aqueduct** that runs through the midbrain. The fourth ventricle lies in the hindbrain dorsal to the

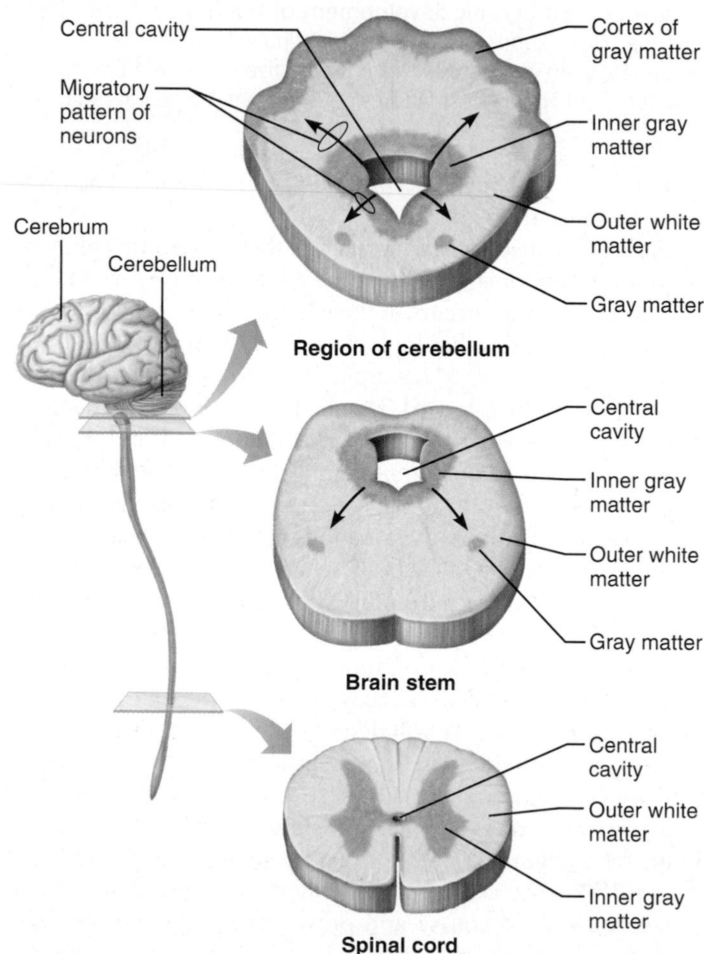

Figure 12.4 Pattern of gray and white matter in the CNS (highly simplified). In each cross section, the dorsal aspect is at the top. In general, white matter lies external to gray matter. In the developing brain, collections of gray matter migrate externally into the white matter (see black arrows). The cerebrum resembles the cerebellum in its external cortex of gray matter.

12

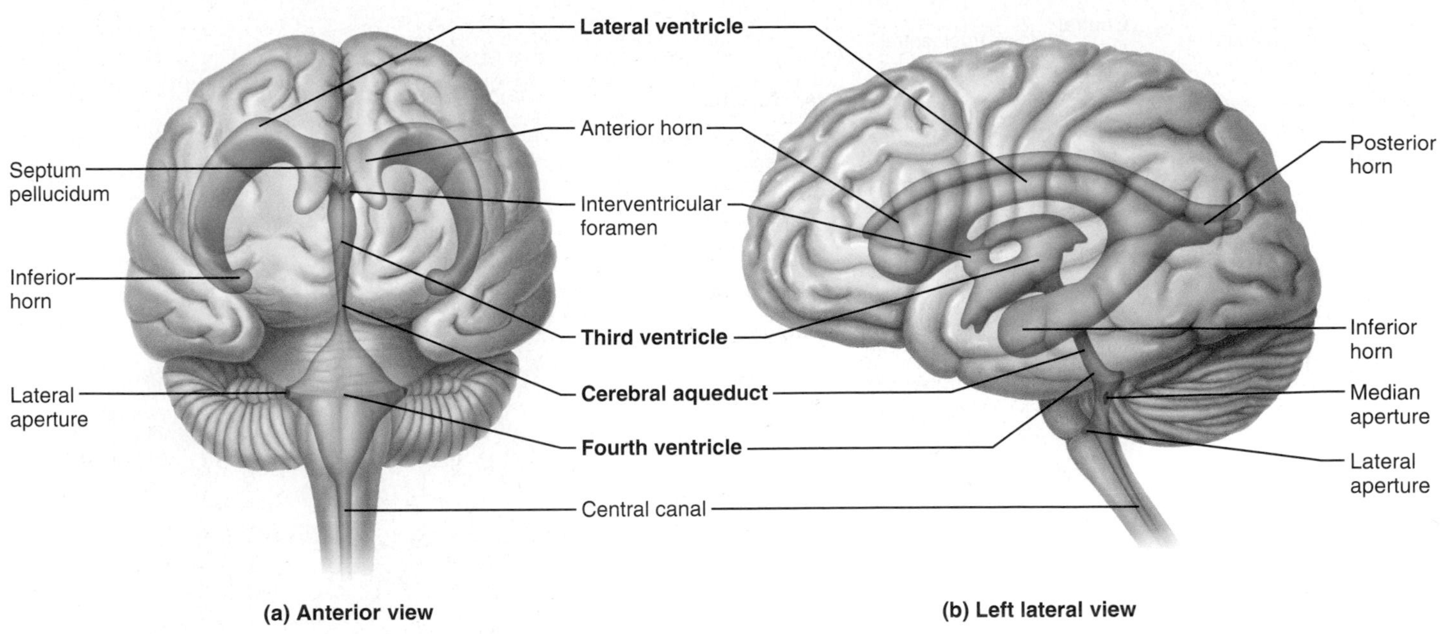

Figure 12.5 Ventricles of the brain. Different regions of the large lateral ventricles are labeled anterior horn, posterior horn, and inferior horn.

pons and superior medulla. It is continuous with the central canal of the spinal cord inferiorly. Three openings mark the walls of the fourth ventricle: the paired **lateral apertures** in its side walls and the **median aperture** in its roof. These apertures connect the ventricles to the *subarachnoid space* (sub″ah-rak′noid), a fluid-filled space surrounding the brain.

CHECK YOUR UNDERSTANDING

1. Which ventricle is surrounded by the diencephalon?
2. Which two areas of the adult brain have an outside layer of gray matter in addition to central gray matter and surrounding white matter?
3. What is the function of convolutions of the brain?

For answers, see Appendix G.

Cerebral Hemispheres

- List the major lobes, fissures, and functional areas of the cerebral cortex.
- Explain lateralization of hemisphere function.
- Differentiate between commissures, association fibers, and projection fibers.
- Describe the general function of the basal nuclei (basal ganglia).

The **cerebral hemispheres** form the superior part of the brain **(Figure 12.6)**. Together they account for about 83% of total brain mass and are the most conspicuous parts of an intact brain. Picture how a mushroom cap covers the top of its stalk, and you have a fairly good idea of how the paired cerebral hemispheres cover and obscure the diencephalon and the top of the brain stem (see Figure 12.3d).

Nearly the entire surface of the cerebral hemispheres is marked by elevated ridges of tissue called **gyri** (ji′ri; "twisters"), separated by shallow grooves called **sulci** (sul′ki; "furrows"). The singular forms of these terms are *gyrus* and *sulcus.* Deeper grooves, called **fissures**, separate large regions of the brain (Figure 12.6a).

12

The more prominent gyri and sulci are similar in all people and are important anatomical landmarks. The median **longitudinal fissure** separates the cerebral hemispheres (Figure 12.6c). Another large fissure, the **transverse cerebral fissure**, separates the cerebral hemispheres from the cerebellum below (Figure 12.6a, d).

Several sulci divide each hemisphere into five lobes—frontal, parietal, temporal, occipital, and insula (Figure 12.6a, b). All but the last are named for the cranial bones that overlie them (see Figure 7.5, pp. 203–204). The **central sulcus**, which lies in the frontal plane, separates the **frontal lobe** from the **parietal lobe**. Bordering the central sulcus are the **precentral gyrus** anteriorly and the **postcentral gyrus** posteriorly. More posteriorly, the **occipital lobe** is separated from the parietal lobe by the **parieto-occipital sulcus** (pah-ri″ĕ-to-ok-sip′ĭ-tal), located on the medial surface of the hemisphere.

The deep **lateral sulcus** outlines the flaplike **temporal lobe** and separates it from the parietal and frontal lobes. A fifth lobe of the cerebral hemisphere, the **insula** (in′su-lah; "island"), is buried deep within the lateral sulcus and forms part of its floor (Figure 12.6b). The insula is covered by portions of the temporal, parietal, and frontal lobes.

The cerebral hemispheres fit snugly in the skull. Rostrally, the frontal lobes lie in the anterior cranial fossa (see Figure 7.2b, c, p. 201). The anterior parts of the temporal lobes fill the

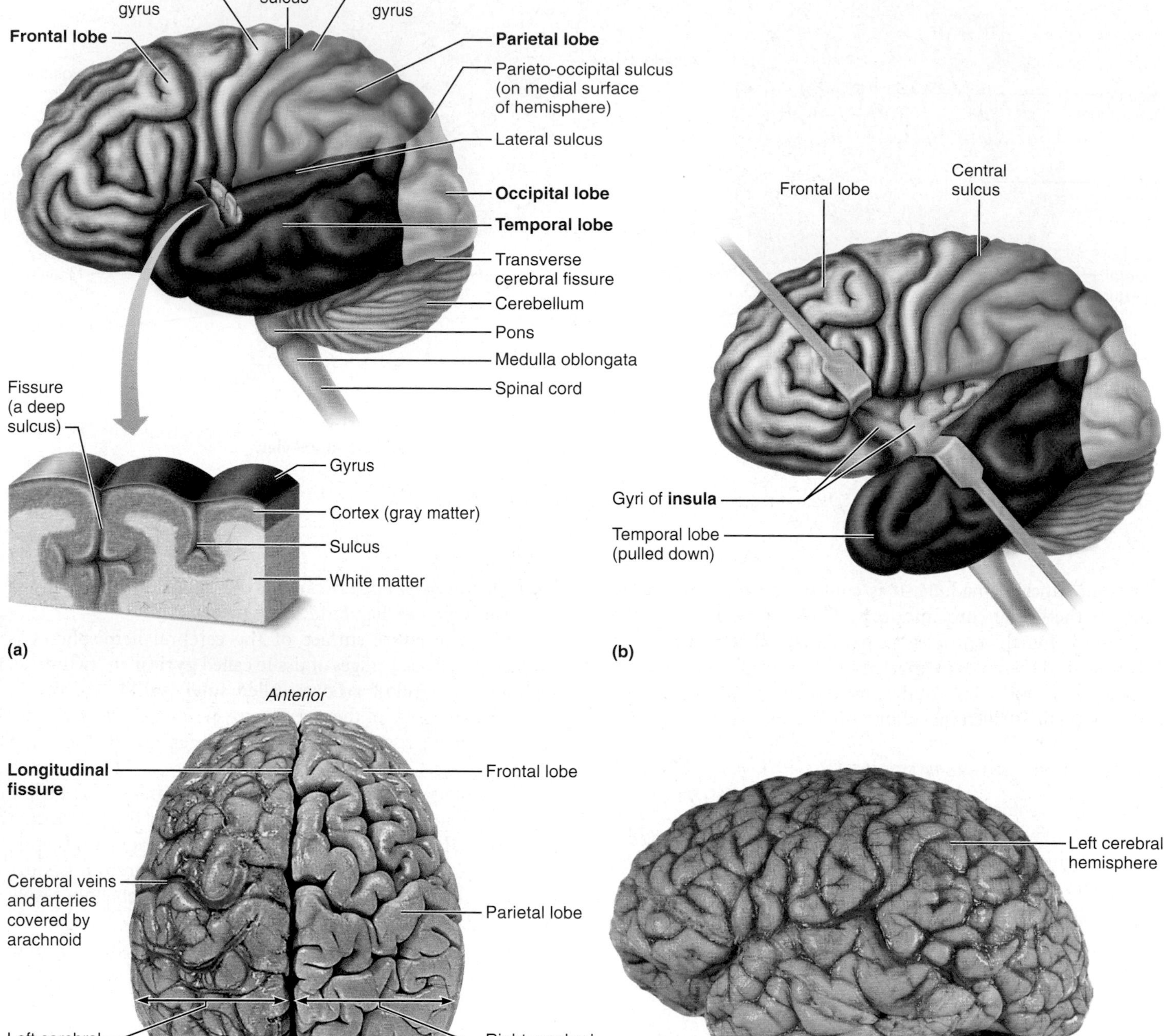

12

Figure 12.6 Lobes and fissures of the cerebral hemispheres. (a) Diagram of the lobes and major sulci and fissures of the brain. **(b)** Cortex of insula revealed by pulling back frontal and temporal lobes. **(c)** Superior surface of cerebral hemispheres; arachnoid matter has been removed from the right half. **(d)** Left lateral view of the brain.

middle cranial fossa. The posterior cranial fossa, however, houses the brain stem and cerebellum. The occipital lobes are located well superior to that cranial fossa.

Each cerebral hemisphere has three basic regions: a superficial *cortex* of gray matter, which looks gray in fresh brain tissue; an internal *white matter*; and the *basal nuclei*, islands of gray matter situated deep within the white matter. We consider these regions next.

Cerebral Cortex

The **cerebral cortex** is the "executive suite" of the nervous system, where our *conscious mind* is found. It enables us to be aware of ourselves and our sensations, to communicate, remember, and understand, and to initiate voluntary movements. The cerebral cortex is composed of gray matter: neuron cell bodies, dendrites, associated glia and blood vessels, but no fiber tracts. It contains billions of neurons arranged in six layers. Although it is only 2–4 mm (about 1/8 inch) thick, it accounts for roughly 40% of total brain mass. Its many convolutions effectively triple its surface area.

In the late 1800s, anatomists mapped subtle variations in the thickness and structure of the cerebral cortex. Most successful in these efforts was K. Brodmann, who in 1906 produced an elaborate numbered mosaic of 52 cortical areas, now called **Brodmann areas**.

With a structural map emerging, early neurologists were eager to localize *functional* regions of the cortex as well. Modern imaging techniques allow us to see the brain in action—PET scans show maximal metabolic activity in the brain, and functional MRI scans reveal blood flow **(Figure 12.7)**. They have shown that specific motor and sensory functions are localized in discrete cortical areas called *domains*. However, many higher mental functions, such as memory and language, appear to have overlapping domains and are spread over large areas of the cortex.

Before we examine the functional regions of the cerebral cortex, let's consider some generalizations about this region of the brain:

1. The cerebral cortex contains three kinds of functional areas: *motor areas*, *sensory areas*, and *association areas*. As you read about these areas, do not confuse the sensory and motor areas of the cortex with sensory and motor neurons. All neurons in the cortex are interneurons.
2. Each hemisphere is chiefly concerned with the sensory and motor functions of the opposite (contralateral) side of the body.
3. Although largely symmetrical in structure, the two hemispheres are not entirely equal in function. Instead, there is a lateralization (specialization) of cortical functions.
4. The final, and perhaps most important, generalization to keep in mind is that our approach is a gross oversimplification; no functional area of the cortex acts alone, and conscious behavior involves the entire cortex in one way or another.

Motor Areas As shown in **Figure 12.8a** (dark and light red areas), the following **motor areas** of the cortex, which control voluntary movement, lie in the posterior part of the frontal lobes: primary motor cortex, premotor cortex, Broca's area, and the frontal eye field.

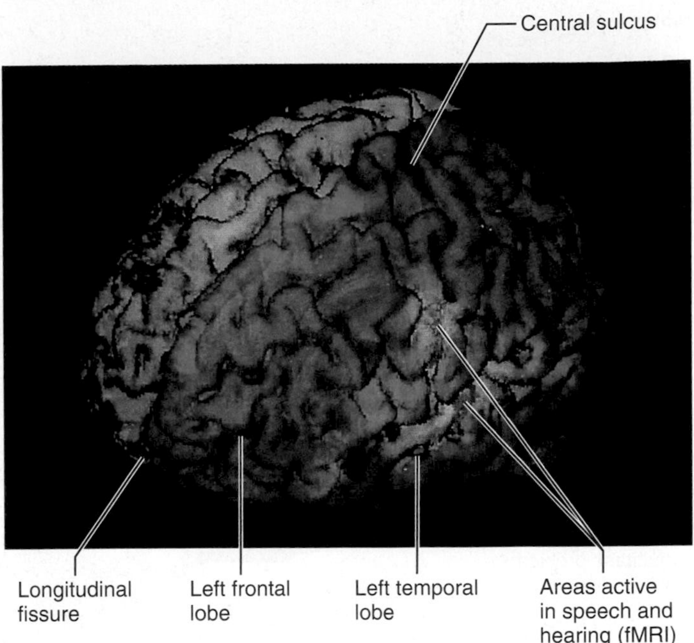

Figure 12.7 Functional neuroimaging (fMRI) of the cerebral cortex. Speaking and hearing increases activity (blood flow, yellow and orange areas) in the posterior frontal and superior temporal lobes, respectively.

1. **Primary motor cortex.** The **primary (somatic) motor cortex** is located in the precentral gyrus of the frontal lobe of each hemisphere (Figure 12.8, dark red area). Large neurons, called **pyramidal cells**, in these gyri allow us to consciously control the precise or skilled voluntary movements of our skeletal muscles. Their long axons, which project to the spinal cord, form the massive voluntary motor tracts called the *pyramidal tracts*, or *corticospinal tracts* (kor″tĭ-ko-spi′nal). All other descending motor tracts issue from brain stem nuclei and consist of chains of two or more neurons.

   The entire body is represented spatially in the primary motor cortex of each hemisphere. For example, the pyramidal cells that control foot movements are in one place and those that control hand movements are in another. Such a mapping of the body in CNS structures is called **somatotopy** (so″mah-to-to′pe).

   As illustrated in **Figure 12.9** (p. 438), the body is represented upside down—with the head at the inferolateral part of the precentral gyrus, and the toes at the superomedial end. Most of the neurons in these gyri control muscles in body areas having the most precise motor control—that is, the face, tongue, and hands. Consequently, these regions of the caricature-like **motor homunculi** (ho-mung′ku-li; singular: homunculus; "little man") drawn in Figure 12.9 are disproportionately large. The motor innervation of the

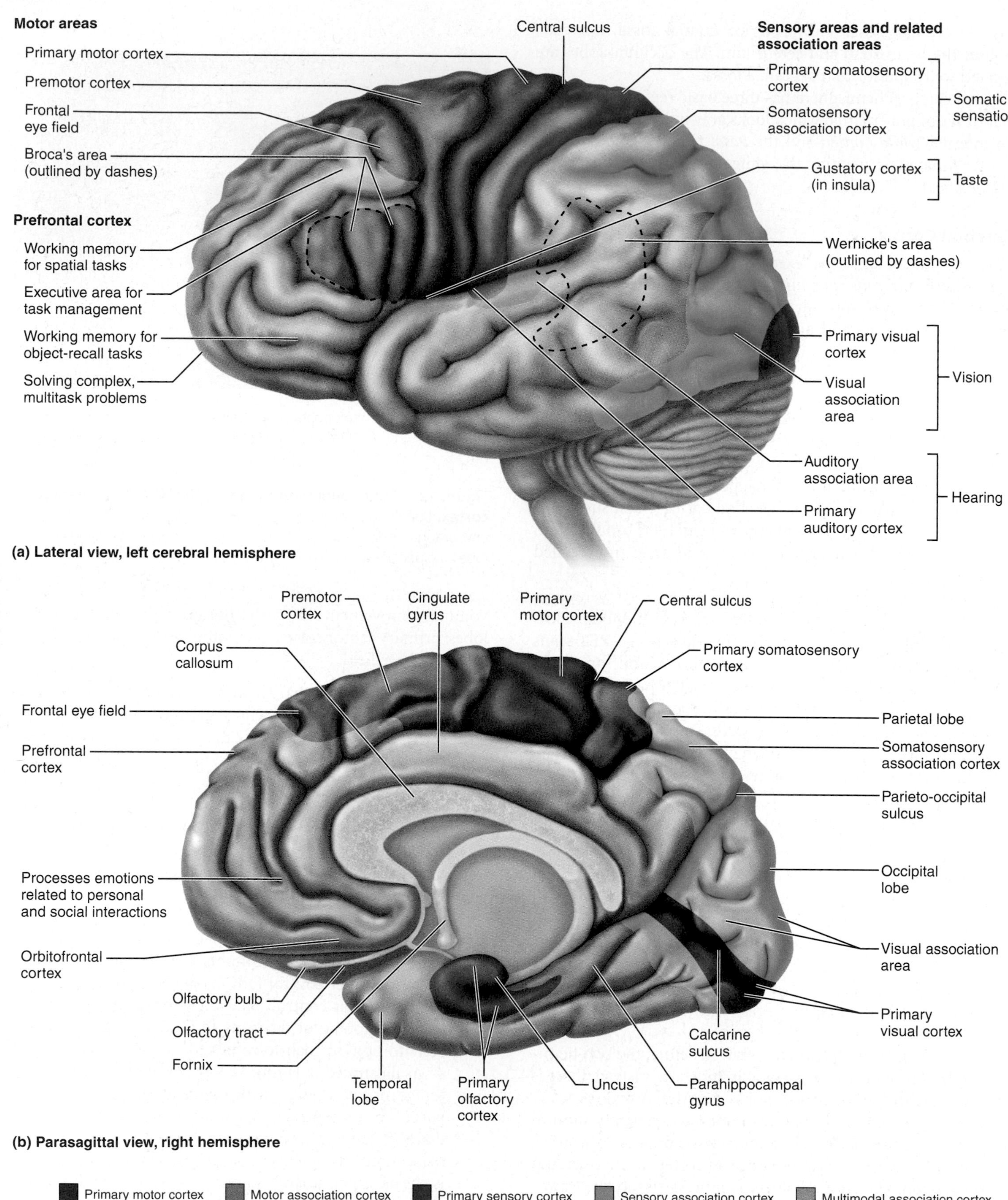
Motor areas
Primary motor cortex
Premotor cortex
Frontal eye field
Broca's area (outlined by dashes)
Prefrontal cortex
Working memory for spatial tasks
Executive area for task management
Working memory for object-recall tasks
Solving complex, multitask problems
Central sulcus
Sensory areas and related association areas
Primary somatosensory cortex
Somatosensory association cortex
Somatic sensation
Gustatory cortex (in insula)
Taste
Wernicke's area (outlined by dashes)
Primary visual cortex
Visual association area
Vision
Auditory association area
Primary auditory cortex
Hearing
(a) Lateral view, left cerebral hemisphere
Premotor cortex
Cingulate gyrus
Primary motor cortex
Central sulcus
Corpus callosum
Primary somatosensory cortex
Frontal eye field
Parietal lobe
Prefrontal cortex
Somatosensory association cortex
Parieto-occipital sulcus
Occipital lobe
Processes emotions related to personal and social interactions
Visual association area
Orbitofrontal cortex
Olfactory bulb
Primary visual cortex
Olfactory tract
Calcarine sulcus
Fornix
Temporal lobe
Primary olfactory cortex
Uncus
Parahippocampal gyrus
(b) Parasagittal view, right hemisphere
Primary motor cortex
Motor association cortex
Primary sensory cortex
Sensory association cortex
Multimodal association cortex

Figure 12.8 Functional and structural areas of the cerebral cortex.

12

body is contralateral: In other words, the left primary motor gyrus controls muscles on the right side of the body, and vice versa.

The *motor homunculus* view of the primary motor cortex, shown at the left in Figure 12.9, implies a one-to-one correspondence between certain cortical neurons and the muscles they control, but this is somewhat misleading. Current research indicates that a given muscle is controlled by multiple spots on the cortex and that individual cortical neurons actually send impulses to more than one muscle. In other words, individual pyramidal motor neurons control muscles that work together in a synergistic way to perform a given movement.

For example, reaching forward with one arm involves some muscles acting at the shoulder and some acting at the elbow. Instead of the discrete map offered by the motor homunculus, the primary motor cortex map is an orderly but fuzzy map with neurons arranged in useful ways to control and coordinate sets of muscles. Neurons controlling the arm, for instance, are intermingled and overlap with those controlling the hand and shoulder. However, neurons controlling unrelated movements, such as those controlling the arm and those controlling body trunk muscles, do not cooperate in motor activity. Thus, the motor homunculus is useful to show that broad areas of the primary cortex are devoted to the leg, arm, torso, and head, but neuron organization within those broad areas is much more diffuse than initially imagined.

2. **Premotor cortex.** Just anterior to the precentral gyrus in the frontal lobe is the **premotor cortex** (see Figure 12.8, light red area). This region controls learned motor skills of a repetitious or patterned nature, such as playing a musical instrument and typing. The premotor cortex coordinates the movement of several muscle groups either simultaneously or sequentially, mainly by sending activating impulses to the primary motor cortex. However, the premotor cortex also influences motor activity more directly by supplying about 15% of pyramidal tract fibers. Think of this region as the memory bank for skilled motor activities.

   The premotor cortex also appears to be involved in planning movements. Using highly processed sensory information received from other cortical areas, it can control voluntary actions that depend on sensory feedback, such as moving an arm through a maze to grasp a hidden object.
3. **Broca's area. Broca's area** (bro′kahz) lies anterior to the inferior region of the premotor area. It has long been considered to be (1) present in one hemisphere only (usually the left) and (2) a special *motor speech area* that directs the muscles involved in speech production. However, recent studies using PET scans to watch active areas of the brain "light up" indicate that Broca's area also becomes active as we prepare to speak and even as we think about (plan) many voluntary motor activities other than speech.
4. **Frontal eye field.** The **frontal eye field** is located partially in and anterior to the premotor cortex and superior to Broca's area. This cortical region controls voluntary movement of the eyes.

HOMEOSTATIC IMBALANCE

Damage to localized areas of the *primary motor cortex* (as from a stroke) paralyzes the body muscles controlled by those areas. If the lesion is in the right hemisphere, the left side of the body will be paralyzed. Only *voluntary* control is lost, however, as the muscles can still contract reflexively.

Destruction of the *premotor cortex*, or part of it, results in a loss of the motor skill(s) programmed in that region, but muscle strength and the ability to perform the discrete individual movements are not hindered. For example, if the premotor area controlling the flight of your fingers over a computer keyboard were damaged, you couldn't type with your usual speed, but you could still make the same movements with your fingers. Reprogramming the skill into another set of premotor neurons would require practice, just as the initial learning process did. ■

Sensory Areas Areas concerned with conscious awareness of sensation, the **sensory areas** of the cortex, occur in the parietal, insular, temporal, and occipital lobes (see Figure 12.8, dark and light blue areas).

1. **Primary somatosensory cortex.** The **primary somatosensory cortex** resides in the postcentral gyrus of the parietal lobe, just posterior to the primary motor cortex. Neurons in this gyrus receive information from the general (somatic) sensory receptors in the skin and from proprioceptors (position sense receptors) in skeletal muscles, joints, and tendons. The neurons then identify the body region being stimulated, an ability called **spatial discrimination.**

   As with the primary motor cortex, the body is represented spatially and upside-down according to the site of stimulus input, and the right hemisphere receives input from the left side of the body. The amount of sensory cortex devoted to a particular body region is related to that region's sensitivity (that is, to how many receptors it has), not to the size of the body region. In humans, the face (especially the lips) and fingertips are the most sensitive body areas. For this reason, these regions are the largest parts of the **somatosensory homunculus** shown in the right half of Figure 12.9.
2. **Somatosensory association cortex.** The **somatosensory association cortex** lies just posterior to the primary somatosensory cortex and has many connections with it. The major function of this area is to integrate sensory inputs (temperature, pressure, and so forth) relayed to it via the primary somatosensory cortex to produce an understanding of an object being felt: its size, texture, and the relationship of its parts. For example, when you reach into your pocket, your somatosensory association cortex draws upon stored memories of past sensory experiences to perceive the objects you feel as coins or keys. Someone with damage to this area could not recognize these objects without looking at them.
3. **Visual areas.** The **primary visual (striate) cortex** is seen on the extreme posterior tip of the occipital lobe, but most of it is buried deep in the *calcarine sulcus* in the medial as-

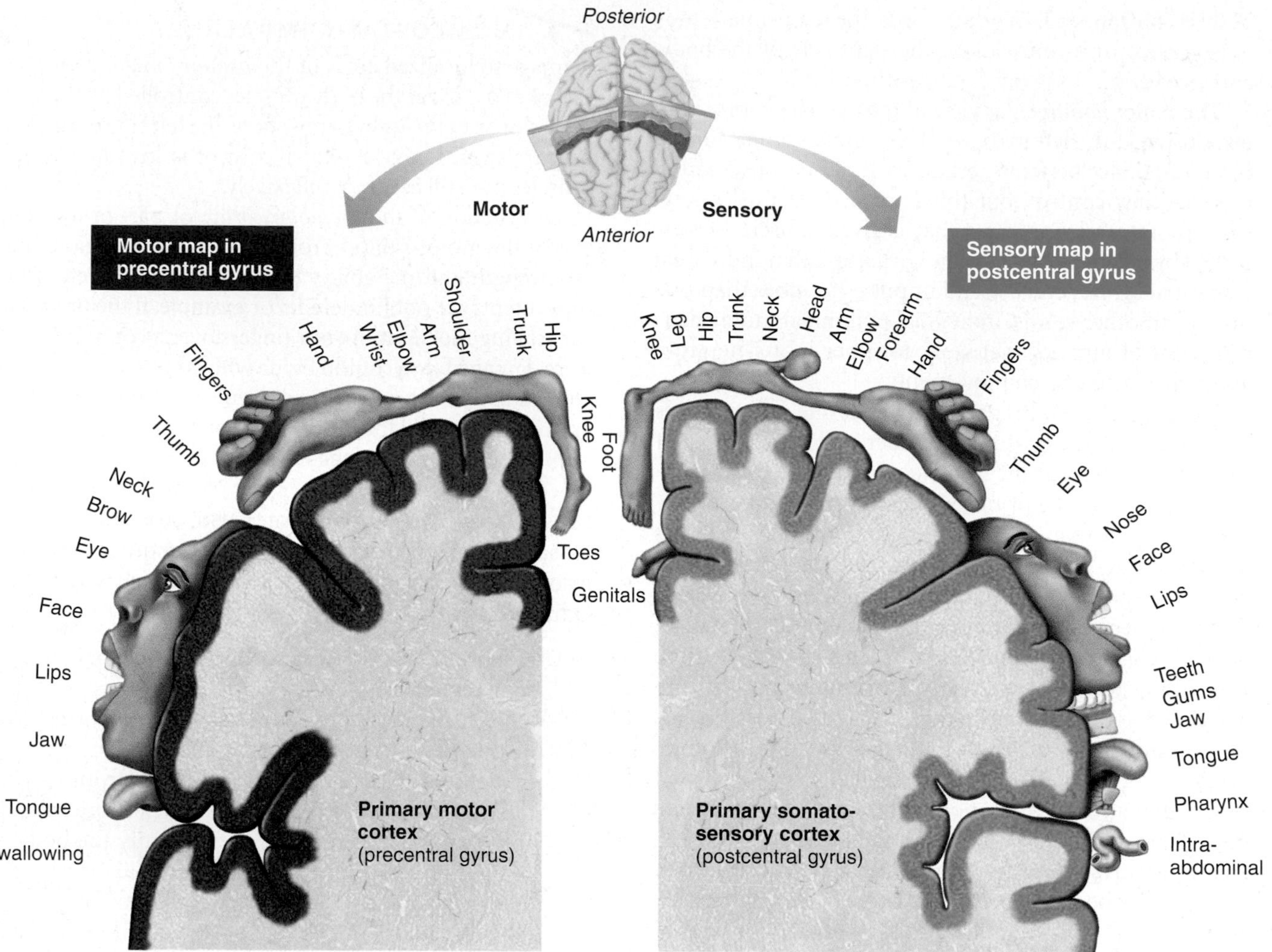

12

Figure 12.9 Body maps in the primary motor cortex and somatosensory cortex of the cerebrum. The relative amount and location of cortical tissue devoted to each function is proportional to the distorted body diagrams (homunculi).

pect of the occipital lobe (Figure 12.8b). The largest of all cortical sensory areas, the primary visual cortex receives visual information that originates on the retina of the eye. There is a contralateral map of visual space on the primary visual cortex, analogous to the body map on the somatosensory cortex.

The **visual association area** surrounds the primary visual cortex and covers much of the occipital lobe. Communicating with the primary visual cortex, the visual association area uses past visual experiences to interpret visual stimuli (color, form, and movement), enabling us to recognize a flower or a person's face and to appreciate what we are seeing. We do our "seeing" with these cortical neurons. However, experiments on monkeys and humans indicate that complex visual processing involves the entire posterior half of the cerebral hemispheres. Particularly important are two visual "streams"—one running along the top of the brain and handling spatial relationships and object location, the other taking the lower road and focusing on object identity (recognizing faces, words, and objects).

4. **Auditory areas.** Each **primary auditory cortex** is located in the superior margin of the temporal lobe abutting the lateral sulcus. Sound energy exciting the hearing receptors of the inner ear causes impulses to be transmitted to the primary auditory cortex, where they are interpreted as pitch, loudness, and location.

   The more posterior **auditory association area** then permits the perception of the sound stimulus, which we "hear" as speech, a scream, music, thunder, noise, and so on. Memories of sounds heard in the past appear to be stored here for reference. Wernicke's area, which we describe later, includes parts of the auditory cortex.
5. **Olfactory cortex.** The primary **olfactory (smell) cortex** lies on the medial aspect of the temporal lobe in a small region called the *piriform lobe* which is dominated by the hooklike *uncus* (Figure 12.8b). Afferent fibers from smell receptors in the superior nasal cavities send impulses along the olfactory tracts that are ultimately relayed to the olfactory cortices. The outcome is conscious awareness of different odors.

The olfactory cortex is part of the primitive **rhinencephalon** (ri″nen-sef′ah-lon; "nose brain"), which includes all parts of the cerebrum that receive olfactory signals—the orbitofrontal cortex, the uncus and associated regions located on or in the medial aspects of the temporal lobes, and the protruding olfactory tracts and bulbs that extend to the nose. During the course of evolution, most of the "old" rhinencephalon has taken on new functions concerned chiefly with emotions and memory. It has become part of the "newer" emotional brain, called the *limbic system*, which we will consider later in this chapter. The only portions of the human rhinencephalon still devoted to smell are the olfactory bulbs and tracts (described in Chapter 13) and the greatly reduced olfactory cortices.

6. **Gustatory cortex.** The **gustatory (taste) cortex** (gus′tah-tor-e), a region involved in the perception of taste stimuli, is located in the insula just deep to the temporal lobe (Figure 12.8a).
7. **Visceral sensory area.** The cortex of the insula just posterior to the gustatory cortex is involved in conscious perception of visceral sensations. These include upset stomach, full bladder, and the feeling that your lungs will burst when you hold your breath too long.
8. **Vestibular (equilibrium) cortex.** It has been difficult to pin down the part of the cortex responsible for conscious awareness of balance, that is, of the position of the head in space. However, imaging studies now locate this region in the posterior part of the insula and adjacent parietal cortex.

HOMEOSTATIC IMBALANCE

Damage to the *primary visual cortex* (Figure 12.8) results in functional blindness. By contrast, individuals with damage to the visual association area can see, but they do not comprehend what they are looking at. ■

Multimodal Association Areas The association areas that we have considered so far (colored light red or light blue in Figure 12.8) have all been tightly tied to one kind of primary motor or sensory cortex (colored dark red or dark blue). Most of the cortex, though, is more complexly connected, receiving inputs from multiple senses and sending outputs to multiple areas. We call these areas **multimodal association areas** (colored light violet in Figure 12.8).

In general, information flows from sensory receptors to the appropriate primary sensory cortex, then to a sensory association cortex and then on to the multimodal association cortex. Multimodal association cortex allows us to give meaning to the information that we receive, store it in memory if needed, tie it to previous experience and knowledge, and decide what action to take. Once the course of action has been decided, those decisions are relayed to the premotor cortex, which in turn communicates with the motor cortex. The multimodal association cortex seems to be where sensations, thoughts, and emotions become conscious. It is what makes us who we are.

Suppose, for example, you drop a bottle of acid in the chem lab and it splashes on you. You see the bottle shatter; hear the crash; feel your skin burning; and smell the acid fumes. These individual perceptions come together in the multimodal association areas. Along with feelings of panic, these perceptions are woven into a seamless whole, which (hopefully) recalls instructions about what to do in this situation. As a result your premotor and primary motor cortices direct your legs to propel you to the safety shower. The multimodal association areas can be broadly divided into three parts, which we describe next.

1. **Anterior association area.** The **anterior association area** in the frontal lobe, also called the **prefrontal cortex**, is the most complicated cortical region of all (Figure 12.8). It is involved with intellect, complex learning abilities (called cognition), recall, and personality. It contains working memory, which is necessary for the production of abstract ideas, judgment, reasoning, persistence, and planning. These abilities develop slowly in children, which implies that the prefrontal cortex matures slowly and depends heavily on positive and negative feedback from one's social environment.
2. **Posterior association area.** The **posterior association area** is a large region encompassing parts of the temporal, parietal, and occipital lobes. This area plays a role in recognizing patterns and faces, localizing us and our surroundings in space, and in binding different sensory inputs into a coherent whole. In the spilled acid example above, your awareness of the entire scene originates from this area. Attention to an area of space or an area of one's own body is also a function of this part of the brain. Many parts of this area (including Wernicke's area, Figure 12.8a) are also involved in understanding written and spoken language.
3. **Limbic association area.** The **limbic association area** includes the cingulate gyrus, the parahippocampal gyrus, and the hippocampus (see Figures 12.8b and 12.18). It is part of the limbic system, which we describe later. The limbic association area provides the emotional impact that makes a scene important to us. In our example above, it provides the sense of "danger" when the acid splashes on our legs. The hippocampus establishes memories that allow us to remember this incident. More on this later.

12

HOMEOSTATIC IMBALANCE

Tumors or other lesions of the *anterior association area* may cause mental and personality disorders including loss of judgment, attentiveness, and inhibitions. The affected individual may be oblivious to social restraints, perhaps becoming careless about personal appearance, or rashly attacking a 7-foot opponent rather than running.

On the other hand, individuals with lesions in the posterior parietal region of the posterior association area, which provides for awareness of self in space, may refuse to wash or dress the side of their body opposite to the lesion because "that doesn't belong to me." ■

Lateralization of Cortical Functioning We use both cerebral hemispheres for almost every activity, and the hemispheres appear nearly identical. Nonetheless, there is a division of labor,

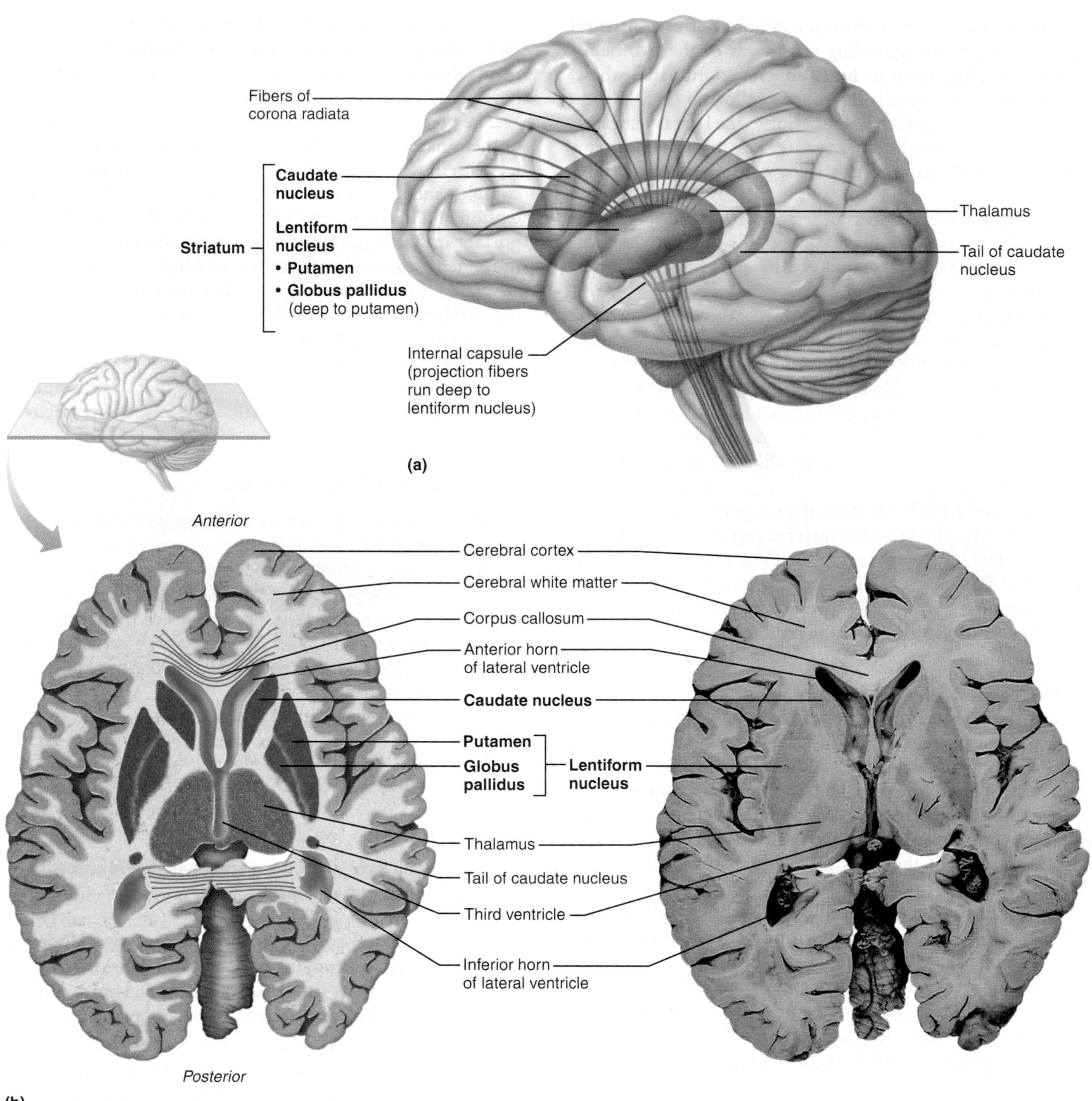

12

Figure 12.11 Basal nuclei. (a) Three-dimensional view of the basal nuclei (basal ganglia), showing their position in the cerebrum. **(b)** Transverse section of cerebrum and diencephalon showing the relationship of the basal nuclei to the thalamus and the lateral and third ventricles.

(Figure 12.13a). Each nucleus has a functional specialty, and each projects fibers to and receives fibers from a specific region of the cerebral cortex. Afferent impulses from all senses and all parts of the body converge on the thalamus and synapse with at least one of its nuclei. For example, the *ventral posterolateral nuclei* receive impulses from the general somatic sensory receptors (touch, pressure, pain, etc.), and the *lateral* and *medial geniculate bodies* (jĕ-nik′u-lāt; "knee shaped") are important visual and auditory relay centers, respectively.

Within the thalamus, information is sorted out and "edited." Impulses having to do with similar functions are relayed as a group via the internal capsule to the appropriate area of the sensory cortex as well as to specific cortical association areas. As the afferent impulses reach the thalamus, we have a crude recogni-

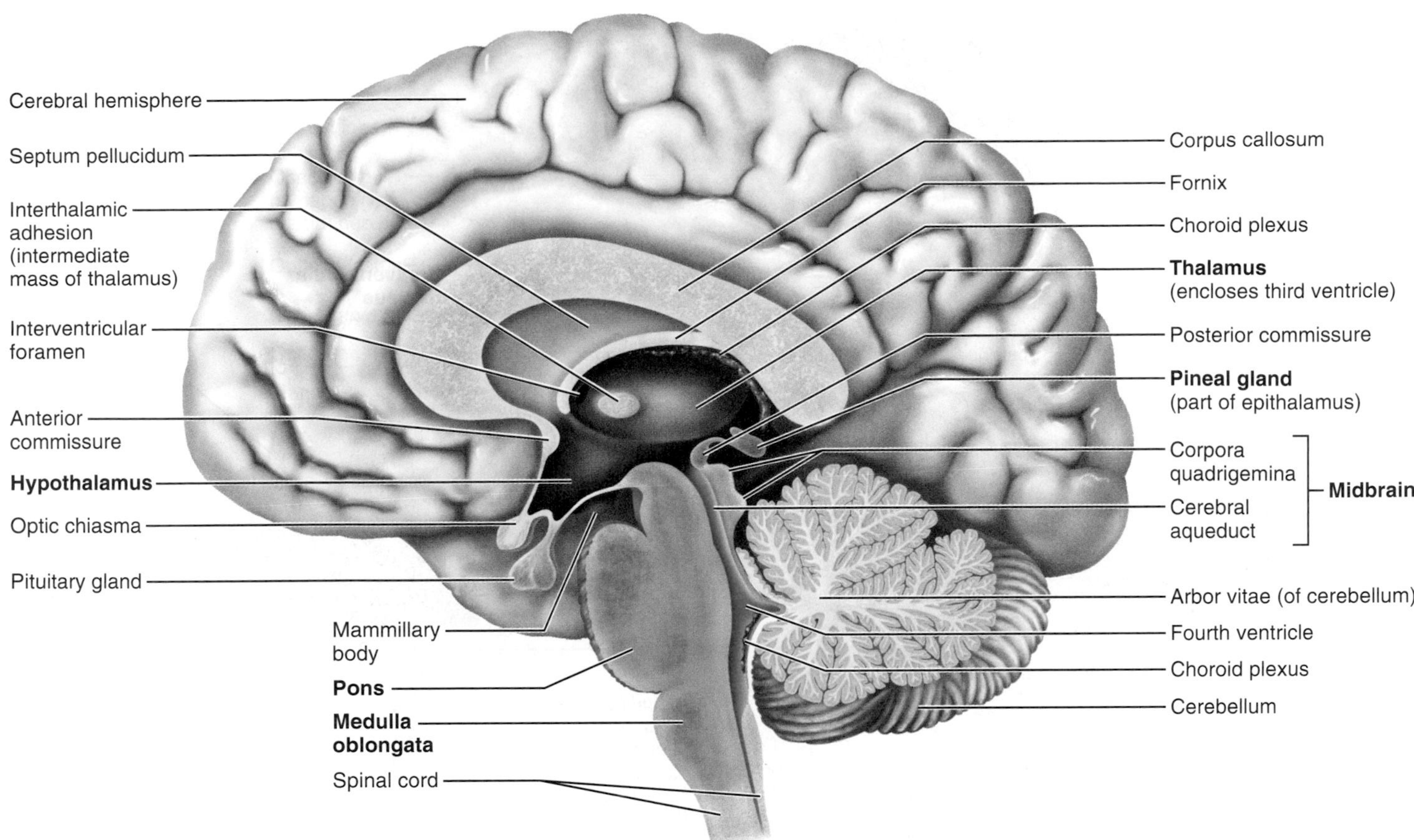

Figure 12.12 **Midsagittal section of the brain illustrating the diencephalon (purple) and brain stem (green).**

tion of the sensation as pleasant or unpleasant. However, specific stimulus localization and discrimination occur in the cerebral cortex.

In addition to sensory inputs, virtually *all* other inputs ascending to the cerebral cortex funnel through thalamic nuclei. These inputs include impulses participating in the regulation of emotion and visceral function from the hypothalamus (via the anterior nuclei), and impulses that help direct the activity of the motor cortices from the cerebellum and basal nuclei (via the ventral lateral and ventral anterior nuclei, respectively). Several thalamic nuclei (pulvinar, lateral dorsal, and lateral posterior nuclei) are involved in integration of sensory information and project to specific association cortices. In summary, the thalamus plays a key role in mediating sensation, motor activities, cortical arousal, learning, and memory. It is truly the gateway to the cerebral cortex.

Hypothalamus

Named for its position below (*hypo*) the thalamus, the **hypothalamus** caps the brain stem and forms the inferolateral walls of the third ventricle (Figure 12.12). Merging into the midbrain inferiorly, the hypothalamus extends from the optic chiasma (crossover point of the optic nerves) to the posterior margin of the mammillary bodies. The **mammillary bodies** (mam′mil-er-e; "little breast"), paired pealike nuclei that bulge anteriorly from the hypothalamus, are relay stations in the olfactory pathways. Between the optic chiasma and mammillary bodies is the **infundibulum** (in″fun-dib′u-lum), a stalk of hypothalamic tissue that connects the **pituitary gland** to the base of the hypothalamus. Like the thalamus, the hypothalamus contains many functionally important nuclei (Figure 12.13b).

Despite its small size, the hypothalamus is the main visceral control center of the body and is vitally important to overall body homeostasis. Few tissues in the body escape its influence. Its chief homeostatic roles are

1. **Autonomic control center.** As you will remember, the autonomic nervous system (ANS) is a system of peripheral nerves that regulates cardiac and smooth muscle and secretion by the glands. The hypothalamus regulates ANS activity by controlling the activity of centers in the brain stem and spinal cord. In this role, the hypothalamus influences blood pressure, rate and force of heartbeat, digestive tract motility, eye pupil size, and many other visceral activities.
2. **Center for emotional response.** The hypothalamus lies at the "heart" of the limbic system (the emotional part of the brain). Nuclei involved in the perception of pleasure, fear, and rage, as well as those involved in biological rhythms and drives (such as the sex drive), are found in the hypothalamus.

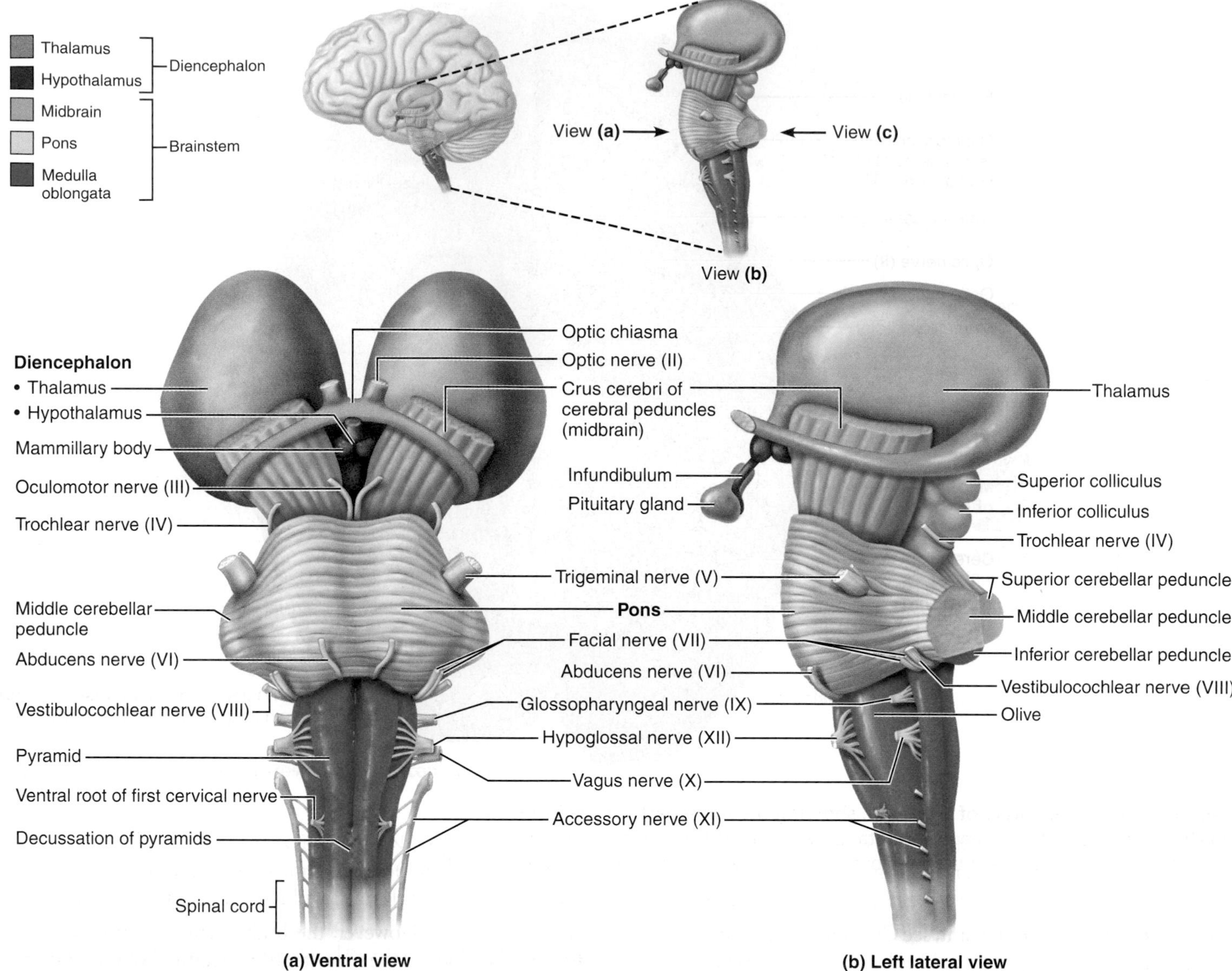

Figure 12.15 Three views of the brain stem (green) and the diencephalon (purple).

12

Running through the midbrain is the hollow *cerebral aqueduct*, which connects the third and fourth ventricles (Figures 12.12 and 12.16a). It delineates the cerebral peduncles ventrally from the *tectum*, the midbrain's roof. Surrounding the aqueduct is the *periaqueductal gray matter*, which is involved in pain suppression and serves as the link between the fear-perceiving amygdala and ANS pathways that control the "fight-or-flight" response. The periaqueductal gray matter also includes nuclei that control two cranial nerves, the *oculomotor* and the *trochlear nuclei* (trok′le-ar).

Nuclei are also scattered in the surrounding white matter of the midbrain. The largest of these are the **corpora quadrigemina** (kor′por-ah kwod″ri-jem′i-nah; "quadruplets"), which raise four domelike protrusions on the dorsal midbrain surface (Figures 12.12 and 12.15c). The superior pair, the **superior colliculi** (kŏ-lik′u-li), are visual reflex centers that coordinate head and eye movements when we visually follow a moving object, even if we are not consciously looking at the object. The **inferior colliculi** are part of the auditory relay from the hearing receptors of the ear to the sensory cortex. They also act in reflexive responses to sound, such as in the *startle reflex*, which causes you to turn your head toward an unexpected sound.

Also embedded in each side of the midbrain white matter are two pigmented nuclei, the substantia nigra and red nucleus. The bandlike **substantia nigra** (sub-stan′she-ah ni′grah) is located deep to the cerebral peduncle (Figure 12.16a). Its dark (*nigr* = black) color reflects a high content of melanin pigment, a precursor of the neurotransmitter (dopamine) released by these neurons. The substantia nigra is functionally linked to the basal nuclei (its axons project to the putamen), and is considered part of the basal nuclear complex by many authorities. Degeneration of the dopamine-releasing neurons of the substantia nigra is the ultimate cause of Parkinson's disease.

The oval **red nucleus** lies deep to the substantia nigra (Figure 12.16a). Its reddish hue is due to its rich blood supply and to

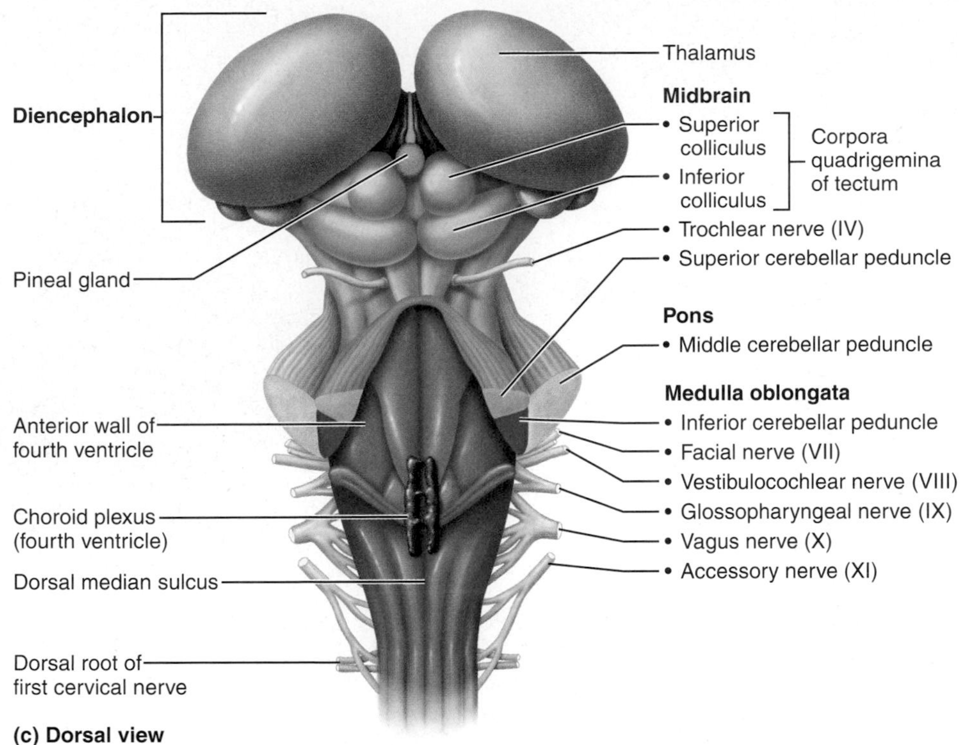

Figure 12.15 *(continued)*

the presence of iron pigment in its neurons. The red nuclei are relay nuclei in some descending motor pathways that effect limb flexion, and they are embedded in the *reticular formation,* a system of small nuclei scattered through the core of the brain stem (see pp. 452–453).

Pons

The **pons** is the bulging brain stem region wedged between the midbrain and the medulla oblongata (see Figures 12.12, 12.14, and 12.15). Dorsally, it forms part of the anterior wall of the fourth ventricle.

As its name suggests (*pons* = bridge), the pons is chiefly composed of conduction tracts. They are oriented in two different directions. The deep projection fibers run longitudinally and complete the pathway between higher brain centers and the spinal cord. The more superficial ventral fibers are oriented transversely and dorsally. They form the *middle cerebellar peduncles* and connect the pons bilaterally with the two sides of the cerebellum dorsally (Figure 12.15). These fibers issue from numerous *pontine nuclei,* which act as relays for "conversations" between the motor cortex and cerebellum.

Several cranial nerve pairs issue from pontine nuclei. They include the *trigeminal* (tri-jem′ĭ-nal), the *abducens* (ab-du′senz), and the *facial nerves* (Figures 12.15a, b and 12.16b). We discuss the cranial nerves and their functions in Chapter 13. Other important pontine nuclei are part of the reticular formation and some help the medulla maintain the normal rhythm of breathing.

Medulla Oblongata

The conical **medulla oblongata** (mĕ-dul′ah ob″long-gah′tah), or simply **medulla,** is the most inferior part of the brain stem. It blends imperceptibly into the spinal cord at the level of the foramen magnum of the skull (Figures 12.12 and 12.14; see Figure 7.6, p. 205). The central canal of the spinal cord continues upward into the medulla, where it broadens out to form the cavity of the fourth ventricle. Together, the medulla and the pons form the ventral wall of the fourth ventricle. [The dorsal ventricular wall is formed by a thin capillary-rich membrane called a choroid plexus which abuts the cerebellum dorsally (Figure 12.12).]

Flanking the midline on the medulla's ventral aspect are two longitudinal ridges called **pyramids,** formed by the large pyramidal (corticospinal) tracts descending from the motor cortex (Figure 12.16c). Just above the medulla–spinal cord junction, most of these fibers cross over to the opposite side before continuing into the spinal cord. This crossover point is called the **decussation of the pyramids** (de″kus-sa′shun; "a crossing"). As we mentioned earlier, the consequence of this crossover is that each cerebral hemisphere chiefly controls the voluntary movements of muscles on the opposite side of the body.

Also visible externally are several other structures. The *inferior cerebellar peduncles* are fiber tracts that connect the medulla to the cerebellum dorsally. Situated lateral to the pyramids, the **olives** are oval swellings (which *do* resemble olives) (Figure 12.15b). These swellings are caused mainly by the wavy folds of gray matter of the underlying **inferior olivary nuclei** (Figure 12.16c). These nuclei relay sensory information on the

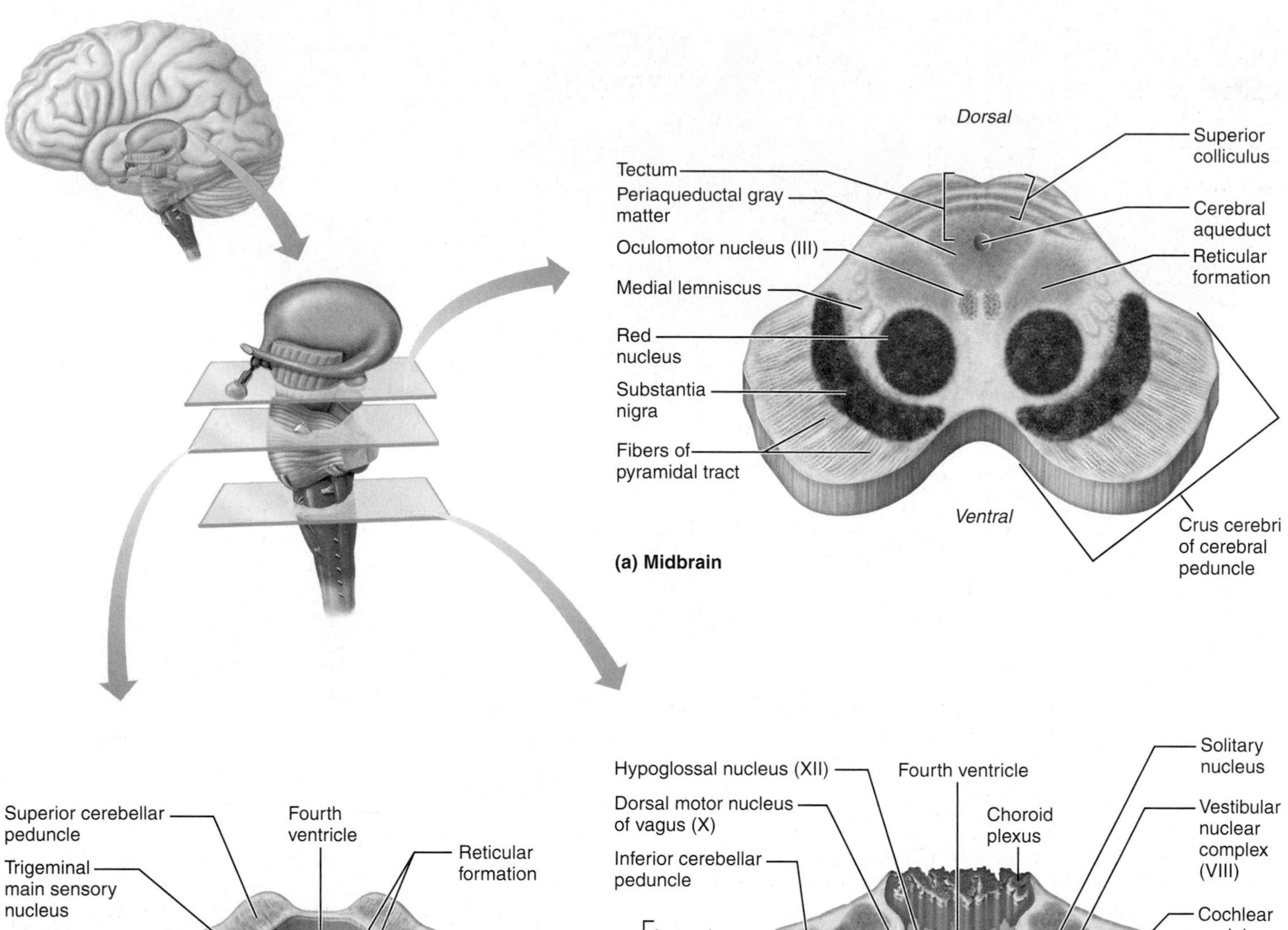

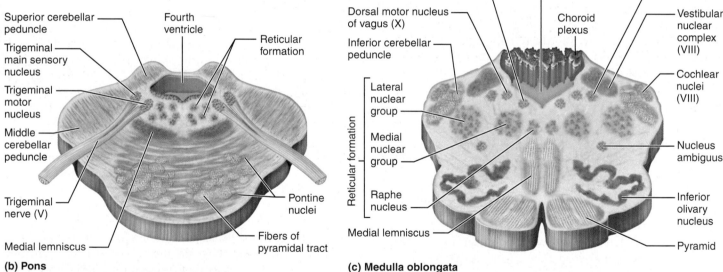

Figure 12.16 Cross sections through different regions of the brain stem.

state of stretch of muscles and joints to the cerebellum. The rootlets of the *hypoglossal nerves* emerge from the groove between the pyramid and olive on each side of the brain stem. Other cranial nerves associated with the medulla are the *glossopharyngeal nerves* and *vagus nerves*. Additionally, the fibers of the *vestibulocochlear nerves* (ves-tib″u-lo-kok′le-ar) synapse with the **cochlear nuclei** (auditory relays), and with numerous vestibular nuclei in both the pons and medulla. Collectively, the vestibular nuclei, called the **vestibular nuclear complex**, mediate responses that maintain equilibrium.

Also housed in the medulla are several nuclei associated with ascending sensory tracts. The most prominent are the dorsally located **nucleus gracilis** (grah-sĭ′lis) and **nucleus cuneatus** (ku′ne-āt-us), associated with a tract called the *medial lemniscus* (Figure 12.16). These serve as relay nuclei in a pathway by which general somatic sensory information ascends from the spinal cord to the somatosensory cortex.

The small size of the medulla belies its crucial role as an autonomic reflex center involved in maintaining body homeo-

TABLE 12.1 Functions of Major Brain Regions

REGION	FUNCTION
Cerebral Hemispheres (pp. 433–441)	
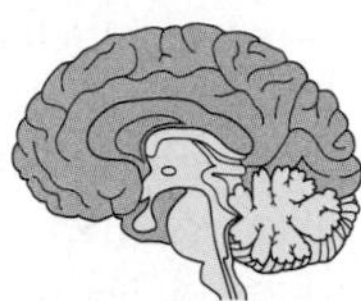	**Cortical gray matter**: Localizes and interprets sensory inputs, controls voluntary and skilled skeletal muscle activity, and functions in intellectual and emotional processing. **Basal nuclei (ganglia)**: Subcortical motor centers important in initiation of skeletal muscle movements.
Diencephalon (pp. 441–445)	
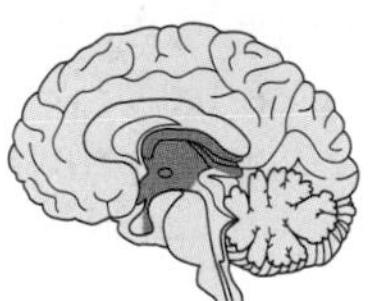	**Thalamic nuclei**: Relay stations in conduction of (1) sensory impulses to cerebral cortex for interpretation, and (2) impulses to and from cerebral motor cortex and lower (subcortical) motor centers, including cerebellum. The thalamus is also involved in memory processing. **Hypothalamus:** Chief integration center of autonomic (involuntary) nervous system; it functions in regulation of body temperature, food intake, water balance, thirst, and biological rhythms and drives. It regulates hormonal output of anterior pituitary gland and is an endocrine organ in its own right (produces ADH and oxytocin). Part of limbic system.
Limbic system (pp. 451–452)	A functional system involving cerebral and diencephalon structures that mediates emotional response; also involved in memory processing.
Brain Stem (pp. 445–449)	
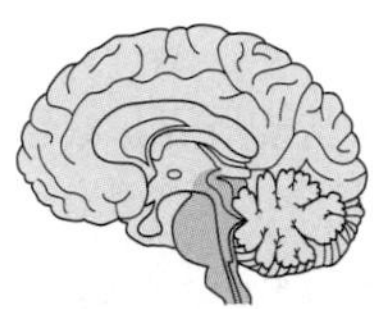	**Midbrain:** Conduction pathway between higher and lower brain centers (e.g., cerebral peduncles contain the fibers of the pyramidal tracts). Its superior and inferior colliculi are visual and auditory reflex centers; substantia nigra and red nuclei are subcortical motor centers; contains nuclei for cranial nerves III and IV. **Pons:** Conduction pathway between higher and lower brain centers; pontine nuclei relay information from the cerebrum to the cerebellum. Its respiratory nuclei cooperate with the medullary respiratory centers to control respiratory rate and depth. Houses nuclei of cranial nerves V–VII. **Medulla oblongata:** Conduction pathway between higher brain centers and spinal cord, and site of decussation of the pyramidal tracts. Houses nuclei of cranial nerves VIII–XII. Contains nuclei cuneatus and gracilis (synapse points of ascending sensory pathways transmitting sensory impulses from skin and proprioceptors), and visceral nuclei controlling heart rate, blood vessel diameter, respiratory rate, vomiting, coughing, etc. Its inferior olivary nuclei provide the sensory relay to the cerebellum.
Reticular formation (pp. 452–453)	A functional brain stem system that maintains cerebral cortical alertness (reticular activating system) and filters out repetitive stimuli. Its motor nuclei help regulate skeletal and visceral muscle activity.
Cerebellum (pp. 450–451)	
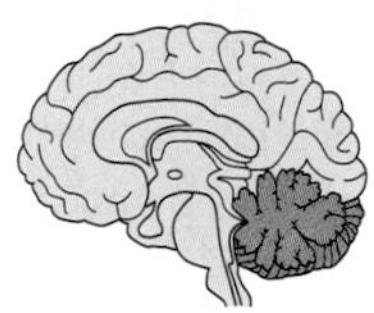	Processes information from cerebral motor cortex and from proprioceptors and visual and equilibrium pathways, and provides "instructions" to cerebral motor cortex and subcortical motor centers that result in proper balance and posture and smooth, coordinated skeletal muscle movements.

stasis, as summarized in **Table 12.1.** Important visceral motor nuclei found in the medulla include the following:

1. **Cardiovascular center.** This includes the *cardiac center*, which adjusts the force and rate of heart contraction to meet the body's needs, and the *vasomotor center*, which changes blood vessel diameter to regulate blood pressure.
2. **Respiratory centers.** These generate the respiratory rhythm and (in concert with pontine centers) control the rate and depth of breathing.
3. **Various other centers.** Additional centers regulate such activities as vomiting, hiccuping, swallowing, coughing, and sneezing.

Notice that many functions listed above are also attributed to the hypothalamus (pp. 443–444). The overlap is easily explained. The hypothalamus exerts its control over many visceral functions by relaying its instructions through medullary reticular centers, which carry them out.

CHECK YOUR UNDERSTANDING

10. What are the pyramids of the medulla? What is the result of decussation of the pyramids?
11. Which region of the brain stem is associated with the cerebral peduncles and the superior and inferior colliculi?

For answers, see Appendix G.

Cerebellum

▶ Describe the structure and function of the cerebellum.

The cauliflower-like **cerebellum** (ser″ĕ-bel′um; "small brain"), exceeded in size only by the cerebrum, accounts for about 11% of total brain mass. The cerebellum is located dorsal to the pons and medulla (and to the intervening fourth ventricle). It protrudes under the occipital lobes of the cerebral hemispheres, from which it is separated by the transverse cerebral fissure (see Figure 12.6d).

By processing inputs received from the cerebral motor cortex, various brain stem nuclei, and sensory receptors, the cerebellum provides the precise timing and appropriate patterns of skeletal muscle contraction for smooth, coordinated movements and agility needed for our daily living—driving, typing, and for some of us, playing the tuba. Cerebellar activity occurs subconsciously—we have no awareness of its functioning.

12

Anatomy

The cerebellum is bilaterally symmetrical. Its two apple-sized **cerebellar hemispheres** are connected medially by the wormlike **vermis** (Figure 12.17). Its surface is heavily convoluted, with fine, transversely oriented pleatlike gyri known as **folia** ("leaves"). Deep fissures subdivide each hemisphere into **anterior**, **posterior**, and **flocculonodular lobes** (flok″u-lo-nod′u-lar). The small propeller-shaped flocculonodular lobes, situated deep to the vermis and posterior lobe, cannot be seen in a surface view.

Like the cerebrum, the cerebellum has a thin outer cortex of gray matter, internal white matter, and small, deeply situated, paired masses of gray matter, the most familiar of which are the *dentate nuclei.* Several types of neurons populate the cerebellar cortex, including **Purkinje cells** (see Table 11.1 on p. 393). These large cells, with their extensively branched dendrites, are the only cortical neurons that send axons through the white matter to synapse with the central nuclei of the cerebellum. The distinctive pattern of white matter in the cerebellum resembles a branching tree, a pattern fancifully called the **arbor vitae** (ar′bor vi′te; "tree of life") (Figure 12.17a, b).

The anterior and posterior lobes of the cerebellum, which coordinate body movements, have three sensory maps of the entire body as indicated by the homunculi in Figure 12.17d. The part of the cerebellar cortex that receives sensory input from a body region influences motor output to that region. The medial portions influence the motor activities of the trunk and girdle muscles. The intermediate parts of each hemisphere are more concerned with the distal parts of the limbs and skilled movements. The lateralmost parts of each hemisphere integrate information from the association areas of the cerebral cortex and appear to play a role in planning rather than executing movements. The flocculonodular lobes receive inputs from the equilibrium apparatus of the inner ears, and adjust posture to maintain balance.

Cerebellar Peduncles

As noted earlier, three paired fiber tracts—the cerebellar peduncles—connect the cerebellum to the brain stem (see Figures 12.15 and 12.17b). Unlike the contralateral fiber distribution to and from the cerebral cortex, virtually all fibers entering and leaving the cerebellum are **ipsilateral** (*ipsi* = same)—from and to the same side of the body. The **superior cerebellar peduncles** connecting cerebellum and midbrain carry instructions from neurons in the deep cerebellar nuclei to the cerebral motor cortex via thalamic relays. Like the basal nuclei, the cerebellum has no *direct* connections to the cerebral cortex.

The **middle cerebellar peduncles** carry one-way communication from the pons to the cerebellum, advising the cerebellum of voluntary motor activities initiated by the motor cortex (via relays in the pontine nuclei). The **inferior cerebellar peduncles** connect medulla and cerebellum. These peduncles convey sensory information to the cerebellum from (1) muscle proprioceptors throughout the body and (2) the vestibular nuclei of the brain stem, which are concerned with equilibrium and balance.

Cerebellar Processing

The functional scheme of cerebellar processing for motor activity seems to be as follows:

1. The motor areas of the cerebral cortex, via relay nuclei in the brain stem, notify the cerebellum of their intent to initiate voluntary muscle contractions.
2. At the same time, the cerebellum receives information from proprioceptors throughout the body (regarding tension in the muscles and tendons, and joint position) and from visual and equilibrium pathways. This information enables the cerebellum to evaluate body position and momentum, that is, where the body is and where it is going.
3. The cerebellar cortex calculates the best way to coordinate the force, direction, and extent of muscle contraction to prevent overshoot, maintain posture, and ensure smooth, coordinated movements.
4. Then, via the superior peduncles, the cerebellum dispatches to the cerebral motor cortex its "blueprint" for coordinating movement. Cerebellar fibers also send information to brain stem nuclei, which in turn influence motor neurons of the spinal cord.

Just as an automatic pilot compares a plane's instrument readings with the planned course, the cerebellum continually compares the body's performance with the higher brain's intention and sends out messages to initiate the appropriate corrective measures. Cerebellar injury results in loss of muscle tone and clumsy, unsure movements.

Cognitive Function of the Cerebellum

Functional imaging studies indicate that the cerebellum plays a role in cognition. The cerebellum recognizes and predicts se-

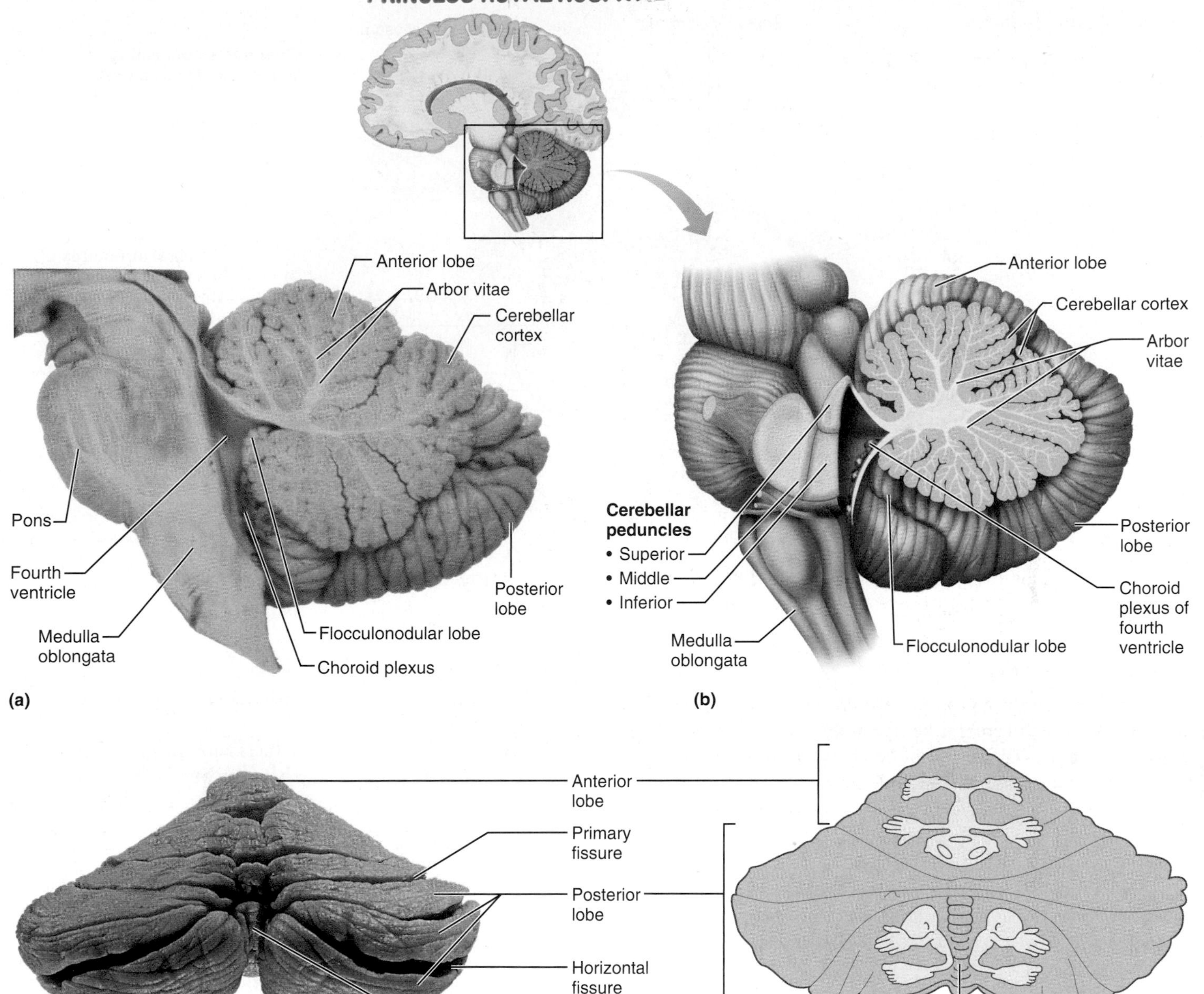

Figure 12.17 Cerebellum. (a) Photo of the midsagittal section. **(b)** Drawing of parasagittal section. **(c)** Photograph of the posterior aspect of the cerebellum. **(d)** Three body maps of the cerebellar cortex (in the form of homunculi).

quences of events so that it may adjust for the multiple forces exerted on a limb during complex movements involving several joints. Some nonmotor functions, including word association and puzzle solving, also appear to involve the cerebellum.

CHECK YOUR UNDERSTANDING

12. In what ways are the cerebellum and the cerebrum similar? In what ways are they different?

For answers, see Appendix G.

Functional Brain Systems

▶ Locate the limbic system and the reticular formation, and explain the role of each functional system.

Functional brain systems are networks of neurons that work together but span relatively large distances in the brain, so they cannot be localized to specific brain regions. The *limbic system* and the *reticular formation* are excellent examples. Table 12.1 (p. 449) summarizes their functions, as well as those of the cerebral hemispheres, diencephalon, brain stem, and cerebellum.

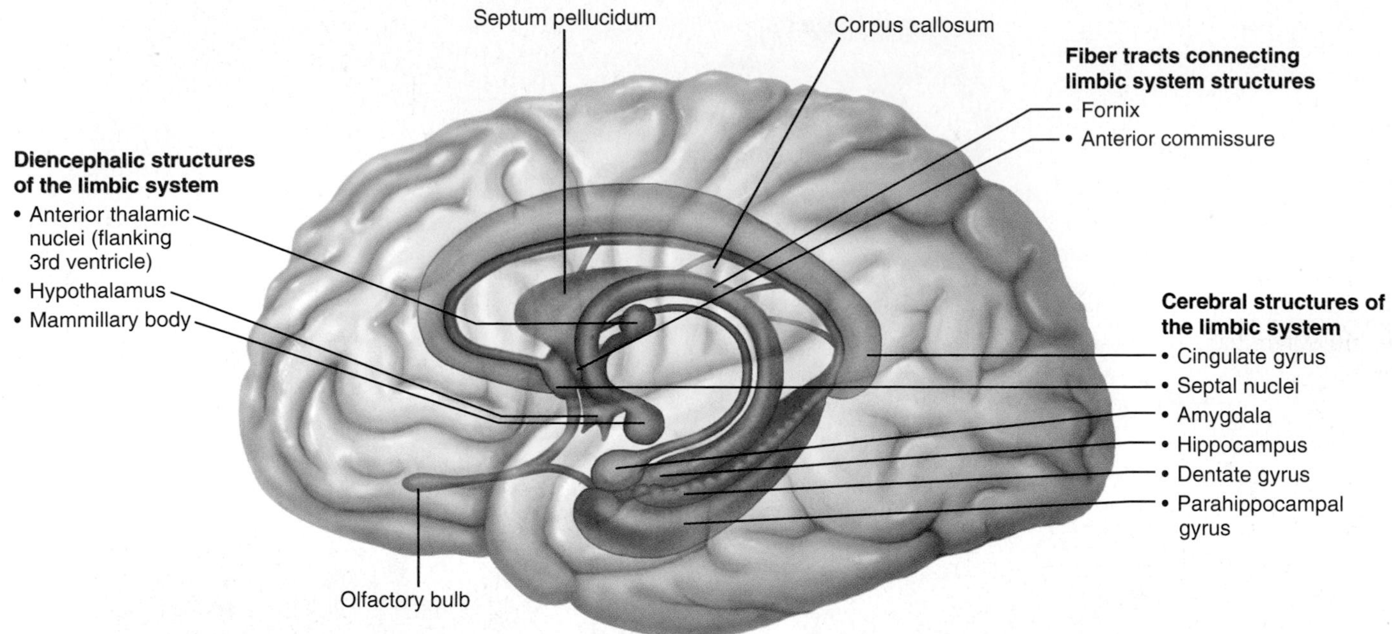

Figure 12.18 The limbic system. Lateral view of the brain, showing some of the structures of the limbic system, the emotional-visceral brain. The brain stem is not illustrated.

12

The Limbic System

The **limbic system** is a group of structures located on the medial aspect of each cerebral hemisphere and diencephalon. Its cerebral structures encircle (*limbus* = ring) the upper part of the brain stem **(Figure 12.18)**. Included are parts of the rhinencephalon (*septal nuclei, cingulate gyrus, parahippocampal gyrus, dentate gyrus*, and C-shaped *hippocampus*), and the **amygdala** (ah-mig′dah-lah), an almond-shaped nucleus that sits on the tail of the caudate nucleus. In the diencephalon, the main limbic structures are the *hypothalamus* and the *anterior thalamic nuclei*. The **fornix** ("arch") and other fiber tracts link these limbic system regions together.

The limbic system is our *emotional*, or *affective* (feelings), *brain*. Two parts seem especially important in emotions—the amygdala and the anterior part of the **cingulate gyrus**. The amygdala recognizes angry or fearful facial expressions, assesses danger, and elicits the fear response. The cingulate gyrus plays a role in expressing our emotions through gestures and in resolving mental conflicts when we are frustrated.

Odors often trigger emotional reactions and memories. These responses reflect the origin of much of the limbic system in the primitive "smell brain" (rhinencephalon). Our reactions to odors are rarely neutral (a skunk smells *bad* and repulses us), and odors often recall memories of emotion-laden events.

Extensive connections between the limbic system and lower and higher brain regions allow the system to integrate and respond to a variety of environmental stimuli. Most limbic system output is relayed through the hypothalamus. Because the hypothalamus is the neural clearinghouse for both autonomic (visceral) function and emotional response, it is not surprising that some people under acute or unrelenting emotional stress fall prey to visceral illnesses, such as high blood pressure and heartburn. Such emotion-induced illnesses are called **psychosomatic illnesses**.

Because the limbic system interacts with the prefrontal lobes, there is an intimate relationship between our feelings (mediated by the emotional brain) and our thoughts (mediated by the cognitive brain). As a result, we (1) react emotionally to things we consciously understand to be happening, and (2) are consciously aware of the emotional richness of our lives. Communication between the cerebral cortex and limbic system explains why emotions sometimes override logic and, conversely, why reason can stop us from expressing our emotions in inappropriate situations. Particular limbic system structures—the **hippocampus** and amygdala—also play a role in memory.

The Reticular Formation

The **reticular formation** extends through the central core of the medulla oblongata, pons, and midbrain **(Figure 12.19)**. It is composed of loosely clustered neurons in what is otherwise white matter. These neurons form three broad columns along the length of the brain stem (Figure 12.16c): (1) the midline **raphe nuclei** (ra′fe; *raphe* = seam or crease), which are flanked laterally by (2) the **medial (large cell) group** and then (3) the **lateral (small cell) group of nuclei**.

The outstanding feature of the reticular neurons is their far-flung axonal connections. Individual reticular neurons project to the hypothalamus, thalamus, cerebral cortex, cerebellum, and spinal cord, making reticular neurons ideal for governing the arousal of the brain as a whole. For example, certain reticular neurons, unless inhibited by other brain areas, send a continuous stream of impulses to the cerebral cortex, keeping the cortex alert and conscious and enhancing its excitability. This arm of the reticular formation is known as the **reticular activating system (RAS)**. Impulses from all the great ascending sensory tracts synapse with RAS neurons, keeping them active and enhancing their arousing effect on the cerebrum. (This may explain why many students, stimulated by a bustling environment, like to study in a crowded cafeteria.)

The RAS also acts like a filter for this flood of sensory inputs. Repetitive, familiar, or weak signals are filtered out, but unusual, significant, or strong impulses do reach consciousness. For example, you are probably unaware of your watch encircling your wrist, but would immediately notice it if the clasp broke. Between them, the RAS and the cerebral cortex disregard perhaps 99% of all sensory stimuli as unimportant. If this filtering did not occur, the sensory overload would drive us crazy. The drug LSD interferes with these sensory dampers, promoting an often overwhelming sensory overload.

- Take a moment to become aware of all the stimuli in your environment. Notice all the colors, shapes, odors, sounds, and so on. How many of these sensory stimuli are you usually aware of?

The RAS is inhibited by sleep centers located in the hypothalamus and other neural regions, and is depressed by alcohol, sleep-inducing drugs, and tranquilizers. Severe injury to this system, as might follow a knockout punch that twists the brain stem, results in permanent unconsciousness (irreversible *coma*). Although the RAS is central to wakefulness, some of its nuclei are also involved in sleep, which we will discuss later in this chapter.

The reticular formation also has a *motor* arm. Some of its motor nuclei project to motor neurons in the spinal cord via the *reticulospinal tracts*, and help control skeletal muscles during coarse limb movements. Other reticular motor nuclei, such as the vasomotor, cardiac, and respiratory centers of the medulla, are autonomic centers that regulate visceral motor functions.

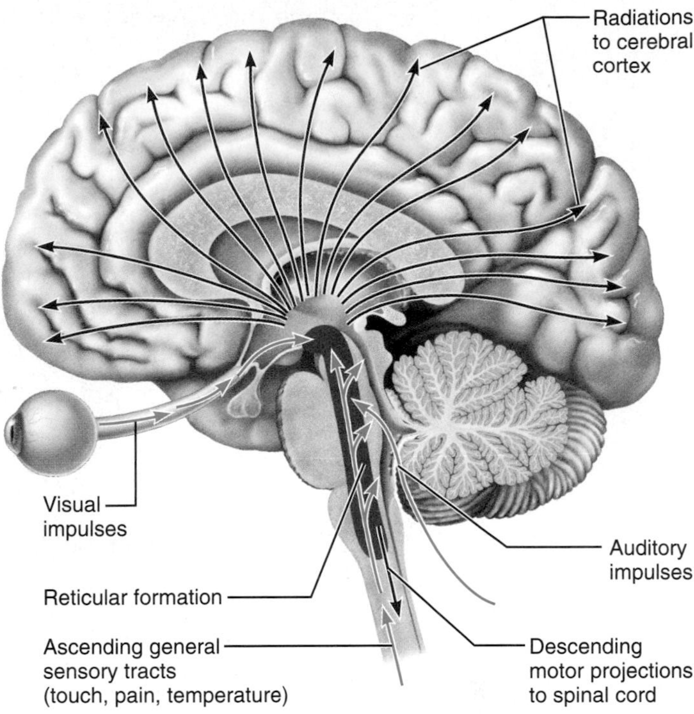

Figure 12.19 The reticular formation. This functional brain system extends the length of the brain stem. Part of this formation, the reticular activating system (RAS), maintains alert wakefulness of the cerebral cortex. Ascending blue arrows indicate input of sensory systems to the RAS. Purple arrows indicate reticular output (some via thalamic relays) to the cerebral cortex. Descending red arrow represents motor output involved in regulating muscle tone.

12

CHECK YOUR UNDERSTANDING

13. The limbic system is sometimes called the emotional-visceral brain. Which part of the limbic system is responsible for the visceral connection?
14. When Taylor begins to feel drowsy while driving, she opens her window, turns up the volume of the car stereo, and has sips of her ice-cold water. How do these actions keep her awake?

For answers, see Appendix G.

Higher Mental Functions

During the last four decades, an exciting exploration of our "inner space," or what we commonly call *the mind*, has been going on. But researchers in the field of cognition are still struggling to understand how the mind's presently incomprehensible qualities might spring from living tissue and electrical impulses. Souls and synapses are hard to reconcile!

Because brain waves reflect the electrical activity on which higher mental functions are based, we will consider them first, along with the related topics of consciousness and sleep. We will then examine language and memory, an area of ongoing research that is of particular interest to our aging population.

Brain Wave Patterns and the EEG

▶ Define EEG and distinguish between alpha, beta, theta, and delta brain waves.

Normal brain function involves continuous electrical activity of neurons. An **electroencephalogram** (e-lek″tro-en-sef′ah-lo-gram), or **EEG**, records some aspects of this activity. An EEG is made by placing electrodes on the scalp and then connecting the electrodes to an apparatus that measures electrical potential differences between various cortical areas **(Figure 12.20a)**. The patterns of neuronal electrical activity recorded, called **brain waves**, are generated by synaptic activity at the surface of the cortex, rather than by action potentials in the white matter.

Each of us has a brain wave pattern that is as unique as our fingerprints. For simplicity, however, we can group brain waves into the four frequency classes shown in Figure 12.20b. Each wave is a continuous train of peaks and troughs, and the wave

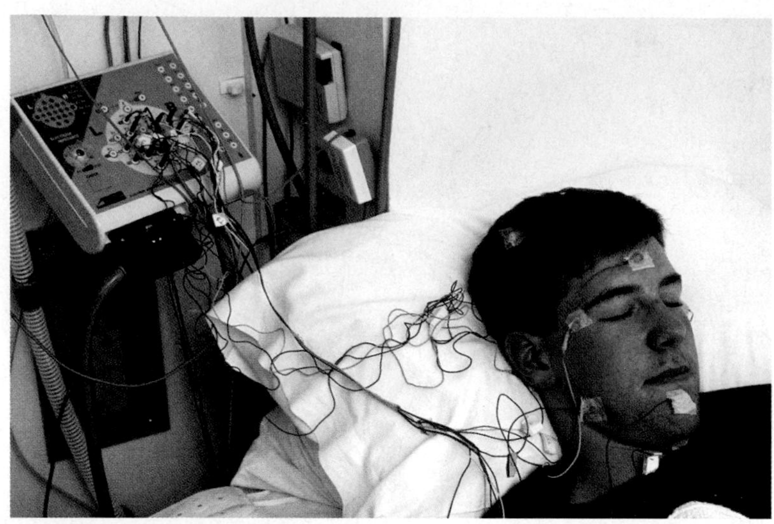

(a) Scalp electrodes are used to record brain wave activity (EEG).

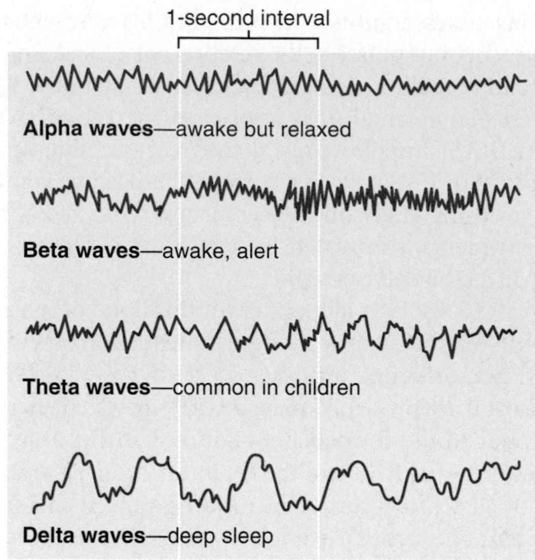

(b) Brain waves shown in EEGs fall into four general classes.

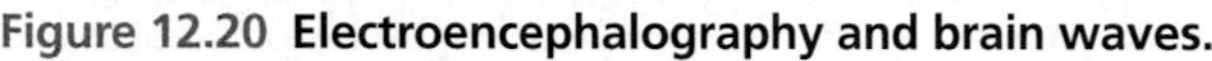

Figure 12.20 Electroencephalography and brain waves.

frequency, expressed in hertz (Hz), is the number of peaks in one second. A frequency of 1 Hz means that one peak occurs each second.

The amplitude or intensity of any wave is represented by how high the wave peaks rise and how low the troughs dip. The amplitude of brain waves reflects the synchronous activity of many neurons and not the degree of electrical activity of individual neurons. Usually, brain waves are complex and low amplitude. During some stages of sleep, neurons tend to fire synchronously, producing similar, high-amplitude brain waves.

- **Alpha waves** (8–13 Hz) are relatively regular and rhythmic, low-amplitude, synchronous waves. In most cases, they indicate a brain that is "idling"—a calm, relaxed state of wakefulness.
- **Beta waves** (14–30 Hz) are also rhythmic, but they are not as regular as alpha waves and have a higher frequency. Beta waves occur when we are mentally alert, as when concentrating on some problem or visual stimulus.
- **Theta waves** (4–7 Hz) are still more irregular. Though common in children, theta waves are uncommon in awake adults but may appear when concentrating.
- **Delta waves** (4 Hz or less) are high-amplitude waves seen during deep sleep and when the reticular activating system is damped, such as during anesthesia. In awake adults, they indicate brain damage.

Brain waves change with age, sensory stimuli, brain disease, and the chemical state of the body. EEGs are used for diagnosing epilepsy and sleep disorders, and in research on brain function. Interference with cerebral cortical functions is suggested when the frequency of brain waves is too high or too low, and unconsciousness occurs at both extremes. Because spontaneous brain waves are always present, even during unconsciousness and coma, their absence—called a "flat EEG"—is clinical evidence of brain death.

HOMEOSTATIC IMBALANCE

Almost without warning, a victim of epilepsy may lose consciousness and fall stiffly to the ground, body wracked by uncontrollable jerking. These **epileptic seizures** reflect a torrent of electrical discharges of groups of brain neurons, and while their uncontrolled activity is occurring, no other messages can get through. Epilepsy, manifested by one out of 100 of us, is not associated with, nor does it cause, intellectual impairment. Some cases of epilepsy are induced by genetic factors, but it can also result from brain injuries caused by blows to the head, stroke, infections, high fever, or tumors.

Epileptic seizures can vary tremendously in their expression and severity. *Absence seizures*, formerly known as *petit mal*, are mild forms in which the expression goes blank for a few seconds as consciousness disappears. These are typically seen in young children and usually disappear by the age of 10. In the most severe, convulsive form of epileptic seizures, *tonic-clonic* (formerly called *grand mal*), the person loses consciousness. Bones are often broken during the intense convulsions, showing the incredible strength of the muscle contractions that occur. Loss of bowel and bladder control and severe biting of the tongue are common. The seizure lasts for a few minutes, then the muscles relax and the person awakens but remains disoriented for several minutes. Many seizure sufferers experience a sensory hallucination, such as a taste, smell, or flashes of light, just before the seizure begins. This phenomenon, called an **aura**, is helpful because it gives the person time to lie down and avoid falling to the floor.

Epilepsy can usually be controlled by anticonvulsive drugs. Newer on the scene is the *vagus nerve stimulator*, which is implanted under the skin of the chest and delivers pulses via the vagus nerve to the brain at predetermined intervals to keep the electrical activity of the brain from becoming chaotic. A current line of research seeks to use electrodes implanted in the brain to detect coming seizures in time to deliver programmed electrical stimulation to avert the seizure. ■

Consciousness

▶ Describe consciousness clinically.

Consciousness encompasses conscious perception of sensations, voluntary initiation and control of movement, and capabilities associated with higher mental processing (memory, logic, judgment, perseverance, and so on). Clinically, consciousness is defined on a continuum that grades behavior in response to stimuli as (1) *alertness,* (2) *drowsiness* or *lethargy* (which proceeds to sleep), (3) *stupor,* and (4) *coma.* Alertness is the highest state of consciousness and cortical activity, and coma the most depressed.

Consciousness is difficult to define. And to be perfectly frank, reducing our response to the palette of a Key West sunset to a series of interactions between dendrites, axons, and neurotransmitters does not capture what makes that event so special. A sleeping person obviously lacks something that he or she has when awake, and we call this "something" consciousness.

The current suppositions about consciousness are as follows:

1. **It involves simultaneous activity of large areas of the cerebral cortex.**
2. **It is superimposed on other types of neural activity.** At any time, specific neurons and neuronal pools are involved both in localized activities (such as motor control) and in cognition.
3. **It is holistic and totally interconnected.** Information for "thought" can be claimed from many locations in the cerebrum simultaneously. For example, retrieval of a specific memory can be triggered by several routes—a smell, a place, a particular person, and so on.

HOMEOSTATIC IMBALANCE

Except when a person is sleeping, unconsciousness is always a signal that brain function is impaired. A brief loss of consciousness is called **fainting** or **syncope** (sing′ko-pe; "cut short"). Most often, it indicates inadequate cerebral blood flow due to low blood pressure, as might follow hemorrhage or sudden emotional stress.

Total unresponsiveness to sensory stimuli for an extended period is called **coma.** Coma is *not* deep sleep. During sleep, the brain is active and oxygen consumption resembles that of the waking state. In coma patients, in contrast, oxygen use is always below normal resting levels.

Blows to the head may induce coma by causing widespread cerebral or brain stem trauma. Tumors or infections that invade the brain stem may also produce coma. Metabolic disturbances such as hypoglycemia (abnormally low blood sugar levels), drug overdose, or liver or kidney failure interfere with overall brain function and can result in coma. Strokes rarely cause coma unless they are massive and accompanied by extreme swelling of the brain, or are located in the brain stem.

When the brain has suffered irreparable damage, irreversible coma occurs, even though life-support measures may have restored vitality to other body organs. The result is **brain death,** a dead brain in an otherwise living body. Because life support can be removed only after death, physicians must determine whether a patient in an irreversible coma is legally alive or dead. ■

Sleep and Sleep-Wake Cycles

▶ Compare and contrast the events and importance of slow-wave and REM sleep, and indicate how their patterns change through life.

Sleep is defined as a state of partial unconsciousness from which a person can be aroused by stimulation. This distinguishes sleep from coma, a state of unconsciousness from which a person *cannot* be aroused by even the most vigorous stimuli. For the most part, cortical activity is depressed during sleep, but brain stem functions, such as control of respiration, heart rate, and blood pressure, continue. Even environmental monitoring continues to some extent, as illustrated by the fact that strong stimuli ("things that go bump in the night") immediately arouse us. In fact, people who sleepwalk can avoid objects and navigate stairs while truly asleep.

Types of Sleep

The two major types of sleep, which alternate through most of the sleep cycle, are **non–rapid eye movement (NREM) sleep** and **rapid eye movement (REM) sleep,** defined in terms of their EEG patterns. During the first 30 to 45 minutes of the sleep cycle, we pass through the first two stages of NREM sleep and into NREM stages 3 and 4, also called **slow-wave sleep** (**Figure 12.21b**). As we pass through these stages and slip into deeper and deeper sleep, the frequency of the EEG waves declines, but their amplitude increases (Figure 12.21). Blood pressure and heart rate also decrease with deeper sleep.

About 90 minutes after sleep begins, after NREM stage 4 has been achieved, the EEG pattern changes abruptly. It becomes very irregular and appears to backtrack quickly through the stages until alpha waves (more typical of the awake state) appear, indicating the onset of REM sleep. This brain wave change is coupled with increases in heart rate, respiratory rate, and blood pressure and a decrease in gastrointestinal motility. Oxygen use by the brain is tremendous during REM—greater than during the awake state.

Although the eyes move rapidly under the lids during REM, most of the body's skeletal muscles are actively inhibited and go limp. This temporary paralysis prevents us from acting out our dreams. Most dreaming occurs during REM sleep, and some suggest that the flitting eye movements are following the visual imagery of our dreams. (Note, however, that most nightmares and night terrors occur during NREM stages 3 and 4.) In adolescents and adults, REM episodes are frequently associated with erection of the penis or engorgement of the clitoris.

Sleep Patterns

The alternating cycles of sleep and wakefulness reflect a natural *circadian,* or 24-hour, *rhythm.* The hypothalamus is responsible

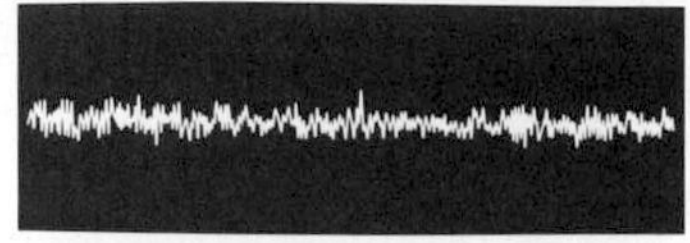

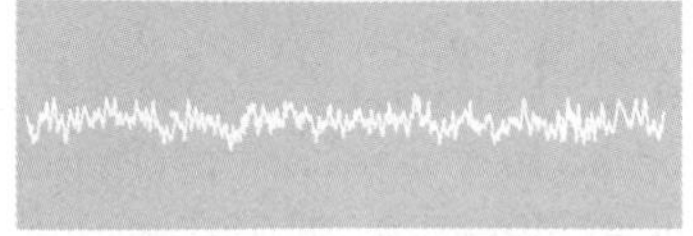

REM: Skeletal muscles (except ocular muscles and diaphragm) are actively inhibited; most dreaming occurs.

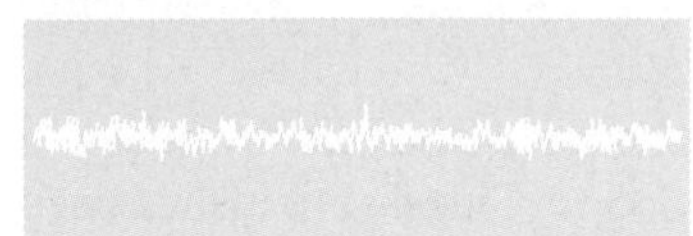

NREM stage 1: Relaxation begins; EEG shows alpha waves, arousal is easy.

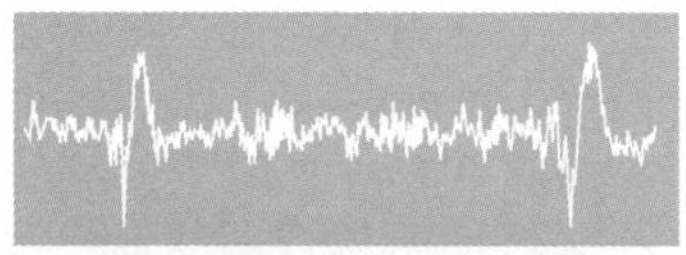

NREM stage 2: Irregular EEG with sleep spindles (short high-amplitude bursts); arousal is more difficult.

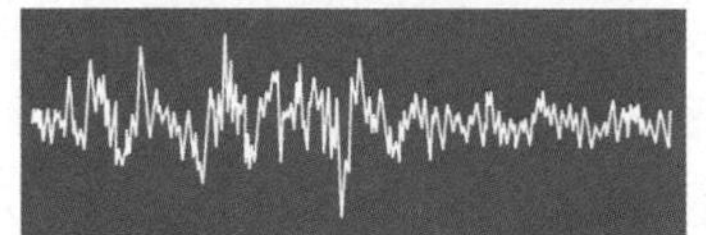

NREM stage 3: Sleep deepens; theta and delta waves appear; vital signs decline.

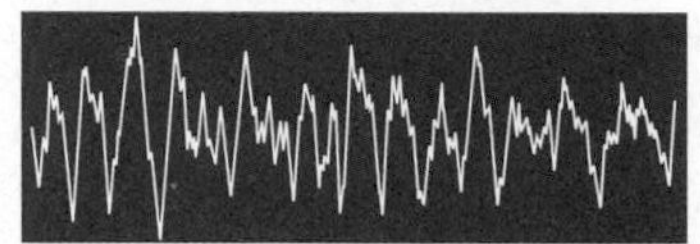

NREM stage 4: EEG is dominated by delta waves; arousal is difficult; bed-wetting, night terrors, and sleepwalking may occur.

(a) Typical EEG patterns

12

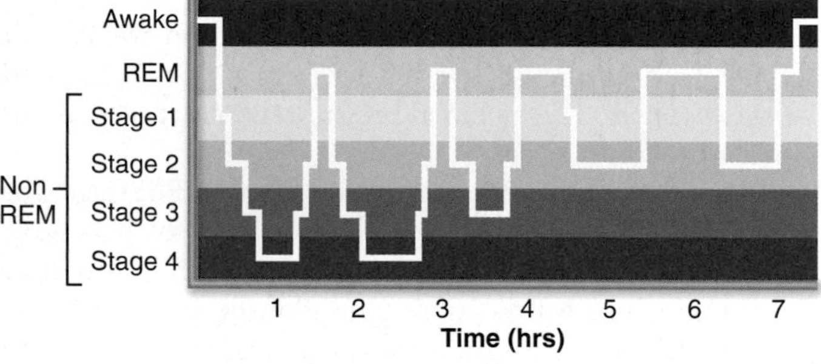

(b) Typical progression of an adult through one night's sleep stages

Figure 12.21 Types and stages of sleep. The four stages of non–rapid eye movement (NREM) sleep and rapid eye movement (REM) sleep are shown.

for the timing of the sleep cycle. Its *suprachiasmatic nucleus* (a biological clock) regulates its *preoptic nucleus* (a sleep-inducing center). By inhibiting the brain stem's reticular activating system (RAS; see Figure 12.19), the preoptic nucleus puts the cerebral cortex to sleep. However, sleep is much more than simply turning off the arousal system. RAS centers not only help maintain the awake state but also mediate some sleep stages, especially dreaming sleep.

In young and middle-aged adults, a typical night's sleep starts with the four stages of NREM sleep and then alternates between REM and NREM sleep with occasional partial arousals. Following each REM episode, the sleeper descends toward stage 4 again. REM recurs about every 90 minutes, with each REM period getting longer. The first REM of the night lasts 5–10 minutes and the final one 20 to 50 minutes (Figure 12.21b). Consequently, our longest dreams occur at the end of the sleep period. Just before we wake, hypothalamic neurons release peptides called *orexins,* which in this situation act as "wake-up" chemicals. As a result, certain neurons of the brain stem reticular formation fire at maximal rates, arousing the sleepy cortex.

The slow theta and delta waves of deep sleep are the result of synchronized firing of thalamic neurons that is normally inhibited during wakefulness by the RAS of the pons. Some pontine neurons of the reticular formation control the transition from NREM sleep to REM sleep, and others suppress motor activity, inducing paralysis. A large number of chemical substances in the body cause sleepiness, but the relative importance of these various sleep-inducing substances is not known.

Importance of Sleep

Why do we sleep? Slow-wave (NREM stages 3 and 4) and REM sleep seem to be important in different ways. Slow-wave sleep is presumed to be restorative—the time when most neural activity can wind down to basal levels. When deprived of sleep, we spend more time than usual in slow-wave sleep during the next sleep episode.

A person persistently deprived of REM sleep becomes moody and depressed, and exhibits various personality disorders. REM sleep may give the brain an opportunity to analyze the day's events and to work through emotional problems in dream imagery. Another idea is that REM sleep is reverse learning. According to this hypothesis, accidental, repetitious, and meaningless communications continually occur, and they must be eliminated from the neural networks by dreaming if the cortex is to remain a well-behaved and efficient thinking system. In other words, we dream to forget.

Alcohol and some sleep medications (barbiturates and others) suppress REM sleep but not slow-wave sleep. On the other hand, certain tranquilizers, such as diazepam (Valium) reduce slow-wave sleep much more than REM sleep.

Whatever its importance, a person's daily sleep requirement declines steadily from 16 hours or so in infants to approximately 7½ to 8½ hours in early adulthood. It then levels off before declining once again in old age. Sleep patterns also change throughout life. REM sleep occupies about half the total sleeping time in infants and then declines until the age of 10 years, when it stabilizes at about 25%. In contrast, stage 4 sleep declines steadily from birth and often disappears completely in those over 60.

HOMEOSTATIC IMBALANCE

People with **narcolepsy** lapse abruptly into REM sleep from the awake state. These sleep episodes last about 15 minutes, can occur without warning at any time, and are often triggered by a pleasurable event—a good joke, a game of poker. In most patients with narcolepsy, an emotionally intense experience can also trigger *cataplexy,* a sudden loss of voluntary muscle control similar to that seen during REM sleep. During cataplectic attacks, lasting seconds to minutes, the patient remains fully conscious but unable to move. Obviously this can be extremely hazardous when a person is driving a car or swimming! It appears that the brains of patients with narcolepsy have fewer cells in the hypothalamus that secrete peptides called orexins (hypocretins), the peptides mentioned above as a wake-up chemical. This finding may be a key to future treatments. Conversely, drugs that block the actions of orexin have been shown to promote sleep and so may help treat an entirely different sleep disorder, insomnia.

Insomnia is a chronic inability to obtain the *amount* or *quality* of sleep needed to function adequately during the day. Sleep requirements vary from four to nine hours a day in healthy people, so there is no way to determine the "right" amount. Insomniacs tend to overestimate the extent of their sleeplessness, and some come to rely on hypnotics (sleep medications), which can exacerbate the problem.

True insomnia often reflects normal age-related changes, but perhaps the most common cause is psychological disturbance. We have difficulty falling asleep when we are anxious or upset, and depression is often accompanied by early awakening.

Sleep apnea, a temporary cessation of breathing during sleep, is scary. The victim awakes abruptly due to hypoxia (lack of oxygen)—a condition that may occur repeatedly throughout the night. Obstructive sleep apnea, the most common form, occurs when the loss of muscle tone during sleep allows excess fatty tissue or other structural abnormalities to block the upper airway. It is associated with obesity and made worse by alcohol and other depressants. Aside from weight loss, effective treatments are either a mask that allows air to be blown in through the nose, keeping the airway open, or surgery to correct the problem. ■

CHECK YOUR UNDERSTANDING

15. When would you see delta waves in an EEG?
16. Which two states of consciousness are between alertness and coma?
17. During which sleep stage are most skeletal muscles actively inhibited?

For answers, see Appendix G.

Language

Language is such an important function of the brain that practically all of the association cortex on the left side is involved in one way or another. Pioneering studies of patients with *aphasias* (the loss of language abilities due to damage to specific areas of the brain) pointed to two regions that are critically important for language, Broca's area and Wernicke's area (see areas outlined by dashes in Figure 12.8a). Patients with lesions involving **Broca's area** can understand language but have difficulty speaking (and sometimes cannot write or type or use sign language). On the other hand, patients with lesions involving **Wernicke's area** are able to speak but produce a type of nonsense often referred to as a "word salad." They also have great difficulty understanding language.

Recent functional studies of the brain indicate that this picture is clinically useful, but oversimplified. In fact, Broca's and Wernicke's areas together with the basal nuclei form a single language implementation system that analyzes incoming and produces outgoing word sounds and grammatical structures. A surrounding set of cortical areas forms a bridge between this system and the regions of cortex that hold concepts and ideas, which are distributed throughout the remainder of the association cortices.

The corresponding areas in the right or non-language-dominant hemisphere are involved in "body language"—the nonverbal emotional (affective) components of language. These areas allow the lilt or tone of our voice and our gestures to express our emotions when we speak, and permit us to comprehend the emotional content of what we hear. For example, a soft, melodious response to your question conveys quite a different meaning than a sharp reply.

Memory

- Compare and contrast the stages and categories of memory.
- Describe the relative roles of the major brain structures believed to be involved in declarative and procedural memories.

Memory is the storage and retrieval of information. Memories are essential for learning and incorporating our experiences into behavior and are part and parcel of our consciousness. Stored somewhere in your 3 pounds of wrinkled brain are zip codes, the face of your grandfather, and the taste of yesterday's pizza. Your memories reflect your lifetime.

Stages of Memory

Memory storage involves two distinct stages: short-term memory and long-term memory (**Figure 12.22**). **Short-term memory (STM)**, also called *working memory*, is the preliminary step, as well as the power that lets you look up a telephone number, dial it, and then never think of it again. The capacity of STM is limited to seven or eight chunks of information, such as the digits of a telephone number or the sequence of words in an elaborate sentence.

In contrast, **long-term memory (LTM)** seems to have a limitless capacity. Although our STM cannot recall numbers much longer than a telephone number, we can remember scores of telephone numbers by committing them to LTM. However, long-term memories can be forgotten, and so our memory bank continually changes with time. Furthermore, our ability to store and to retrieve information declines with aging.

Figure 12.22 shows how information is processed for storage. We do not remember or even consciously notice much of what is going on around us. As sensory inputs flood into our cerebral cortex, they are processed (yellow box in Figure 12.22). Some 5% of this information is selected for transfer to STM (light green box in Figure 12.22). STM serves as a sort of temporary holding bin for data that we may or may not want to retain.

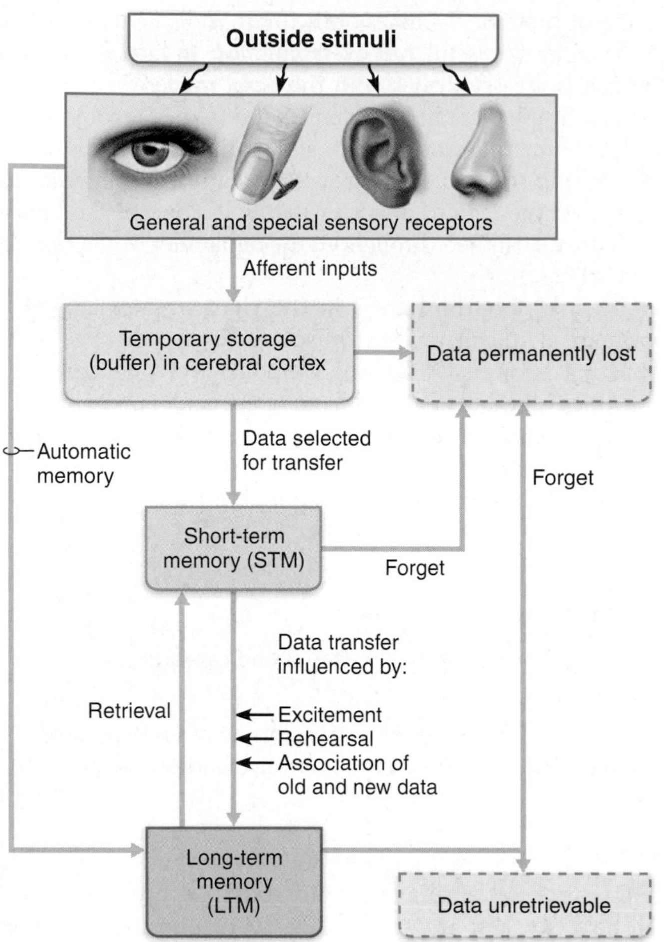

Figure 12.22 Memory processing.

12

Information is then transferred from STM to LTM (dark green box in Figure 12.22). This transfer is affected by many factors, including

1. **Emotional state.** We learn best when we are alert, motivated, surprised, and aroused. For example, when we witness shocking events, transferal is almost immediate. Norepinephrine, a neurotransmitter involved in memory processing of emotionally charged events, is released when we are excited or "stressed out," which helps to explain this phenomenon.
2. **Rehearsal.** Rehearsal or repetition of the material enhances memory.
3. **Association.** Tying "new" information to "old" information already stored in LTM appears to be important in remembering facts.
4. **Automatic memory.** Not all impressions that become part of LTM are consciously formed. A student concentrating on a lecturer's speech may record an automatic memory of the pattern of the lecturer's tie.

Memories transferred to LTM take time to become permanent. The process of **memory consolidation** apparently involves fitting new facts into the various categories of knowledge already stored in the cerebral cortex.

Categories of Memory

The brain distinguishes between factual knowledge and skills, and we process and store these different kinds of information in different ways. **Declarative (fact) memory** entails learning explicit information, such as names, faces, words, and dates. It is related to our conscious thoughts and our ability to manipulate symbols and language. When fact memories are committed to LTM, they are usually filed along with the context in which they were learned. For instance, when you think of your new acquaintance Joe, you probably picture him at the basketball game where you met him.

Nondeclarative memory is less conscious or even unconscious learning. Categories of nondeclarative memory are **procedural (skills) memory** (piano playing), **motor memory** (riding a bike), and **emotional memory** (your pounding heart when you hear a rattlesnake nearby). These kinds of memory are acquired through experience and usually repetition. They do not preserve the circumstances of learning, and in fact, they are best remembered in the doing. You do not have to think through how to tie your shoes. Once learned, nondeclarative memories are hard to unlearn.

Brain Structures Involved in Memory

Much of what scientists know about learning and memory comes from experiments with macaque monkeys, and functional imaging and studies of amnesia in humans. Such studies have revealed that different brain structures are involved in the two categories of memory.

It appears that specific pieces of each memory are stored near regions of the brain that need them so that new inputs can be quickly associated with the old. Accordingly, visual memories are stored in the occipital cortex, memories of music in the temporal cortex, and so on.

But how do we create new memories? It seems that different types of memory are created in different parts of the brain. A proposed scheme of information flow for declarative memory is shown in **Figure 12.23a**. When sensory input is processed in the association cortices, the cortical neurons dispatch impulses to the medial temporal lobe, which includes the hippocampus and surrounding temporal cortical areas. These temporal lobe areas play a major role in memory consolidation and memory access by communicating with the thalamus and the prefrontal cortex. The prefrontal cortex and medial temporal lobe receive input from acetylcholine-releasing neurons in the basal forebrain. The sprinkling of acetylcholine (ACh) onto these structures is thought to prime them to allow the formation of memories. The loss of this ACh input, for example in Alzheimer's disease, seems to disrupt both the formation of new memories and the retrieval of old ones. Memories are retrieved when the same sets of neurons that were initially involved in memory formation are stimulated.

HOMEOSTATIC IMBALANCE

Damage to the hippocampus and surrounding medial temporal lobe structures on either side results in only slight memory loss, but bilateral destruction causes widespread amnesia.

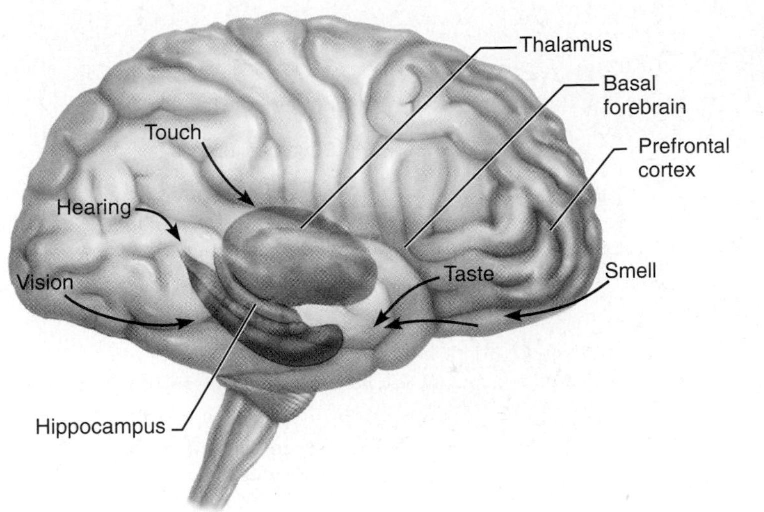

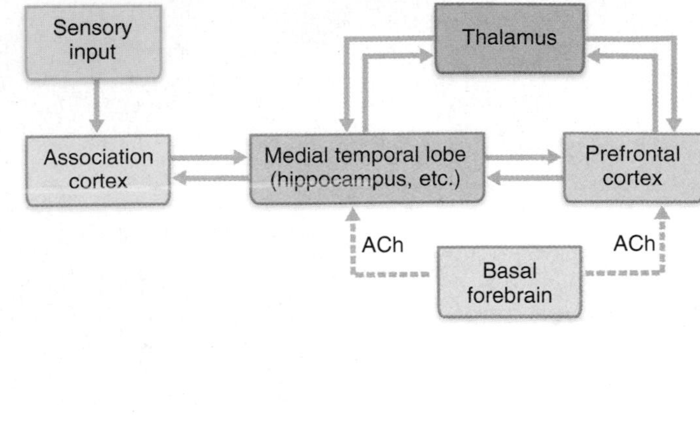

(a) Declarative memory circuits

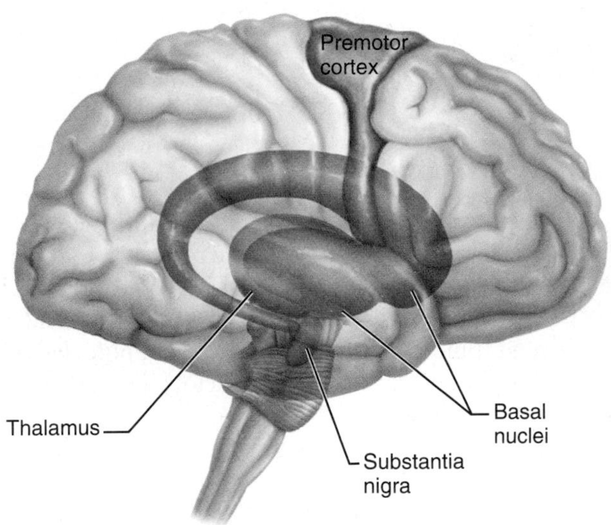

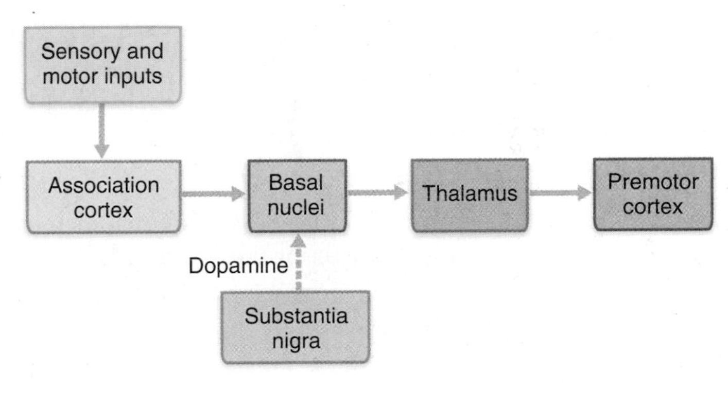

(b) Procedural (skills) memory circuits

12

Figure 12.23 **Proposed memory circuits.** **(a)** Essential structures of declarative memory formation and flowchart showing how these structures may interact. Information from association cortices flows to the medial temporal lobe (including the hippocampus), which communicates with the thalamus and prefrontal cortex. The medial temporal lobe structures then feed back to the association cortices. Acetylcholine, released from the basal forebrain, is necessary for this circuit to function. **(b)** Essential structures of procedural (skills) memory. Sensory and motor input flows through the association cortices and is relayed via the basal nuclei through the thalamus to the premotor cortex. Dopamine, released from the substantia nigra, is necessary for this circuit to function.

Consolidated memories are not lost, but new sensory inputs cannot be associated with old, and the person lives in the here and now from that point on. This condition is called *anterograde amnesia* (an′ter-o-grād″), in contrast to *retrograde amnesia*, which is the loss of memories formed in the distant past. You could carry on an animated conversation with a person with anterograde amnesia, excuse yourself, return five minutes later, and that person would not remember you. ■

Individuals suffering from anterograde amnesia can still learn skills such as drawing, which means a different learning circuit must be used for procedural memory. As Figure 12.23b shows, the basal nuclei (pink) are key players for procedural memory. Sensory and motor inputs pass through the association cortex to the basal nuclei. These inputs are then relayed via the thalamus to the premotor cortex. Note that the basal nuclei receive input from dopamine-releasing neurons in the substantia nigra of the midbrain. Just as acetylcholine is necessary for declarative memory, dopamine appears to be necessary for this procedural memory circuit to function. The loss of this dopamine input, as in Parkinson's disease, interferes with procedural memory.

The two other kinds of nondeclarative memory involve yet other brain regions. The cerebellum is involved in motor

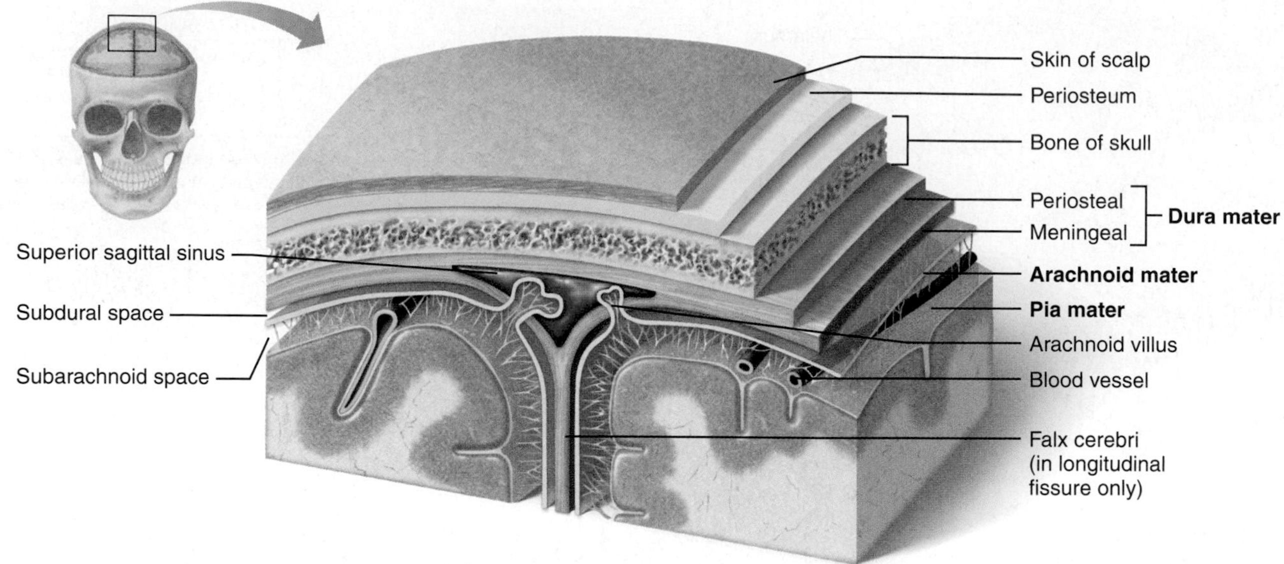

Figure 12.24 Meninges: dura mater, arachnoid mater, and pia mater. The meningeal dura forms the falx cerebri fold. A dural sinus, the superior sagittal sinus, is enclosed by the dural membranes superiorly. Arachnoid villi, which return cerebrospinal fluid to the dural sinus, are also shown. (Frontal section.)

memory, while the amygdala is crucial for emotional memory (see Figure 12.18). We will not describe these pathways here.

12

Molecular Basis of Memory

We have looked at brain structures involved in memory, but what happens at the molecular level when we form memories? Human memory is notoriously difficult to study. Animal experimental studies reveal that during learning, (1) neuronal RNA content is altered and newly synthesized mRNAs are delivered to axons and dendrites, (2) dendritic spines change shape, (3) unique extracellular proteins are deposited at synapses involved in LTM, (4) the number and size of presynaptic terminals may increase, and (5) more neurotransmitter is released by the presynaptic neurons.

Each one of these changes is an aspect of **long-term potentiation (LTP)**, a persistent increase in synaptic strength that has been shown to be crucial for memory formation. LTP was first identified in hippocampal neurons that use the amino acid glutamate as a neurotransmitter. One kind of glutamate receptor, the *NMDA receptor*, can act as a calcium channel and initiate the cellular changes that bring about LTP.

Normally, NMDA receptors are blocked, preventing calcium entry. When the postsynaptic terminal is depolarized by binding of glutamate to different receptors, as would happen upon the rapid arrival of action potentials at the synapse, this NMDA block is removed and calcium flows into the postsynaptic cell.

Calcium influx triggers activation of enzymes that carry out two main tasks. First, they modify the proteins in the postsynaptic terminal, and also in the presynaptic terminal via retrograde messengers such as nitric oxide and endocannabinoids. These changes strengthen the response to subsequent stimuli. Second, they cause the activation of genes in the postsynaptic neuron's nucleus, which leads to synthesis of synaptic proteins.

The molecular messenger that brings the news to the nucleus that more protein is needed is a molecule called CREB (cAMP response-element binding protein). A neurotrophic factor called BDNF (brain-derived neurotrophic factor) is required for the protein synthesis phase of LTP. Together, these changes create long-lasting increases in synaptic strength that are believed to underlie memory.

The events underlying memory at a cellular level suggest several approaches to enhancing memory formation. Currently several drugs to improve memory are undergoing clinical trials. Among these are drugs that enhance CREB production—the more CREB, the more protein synthesis and the stronger the synapse becomes.

CHECK YOUR UNDERSTANDING

18. Name three factors that can enhance transfer of information from STM to LTM.
19. What functional areas of the cerebrum are involved in the formation of procedural (skills) memory, but *not* involved in declarative memory formation?

For answers, see Appendix G.

Protection of the Brain

- Describe how meninges, cerebrospinal fluid, and the blood-brain barrier protect the CNS.
- Describe the formation of cerebrospinal fluid, and follow its circulatory pathway.

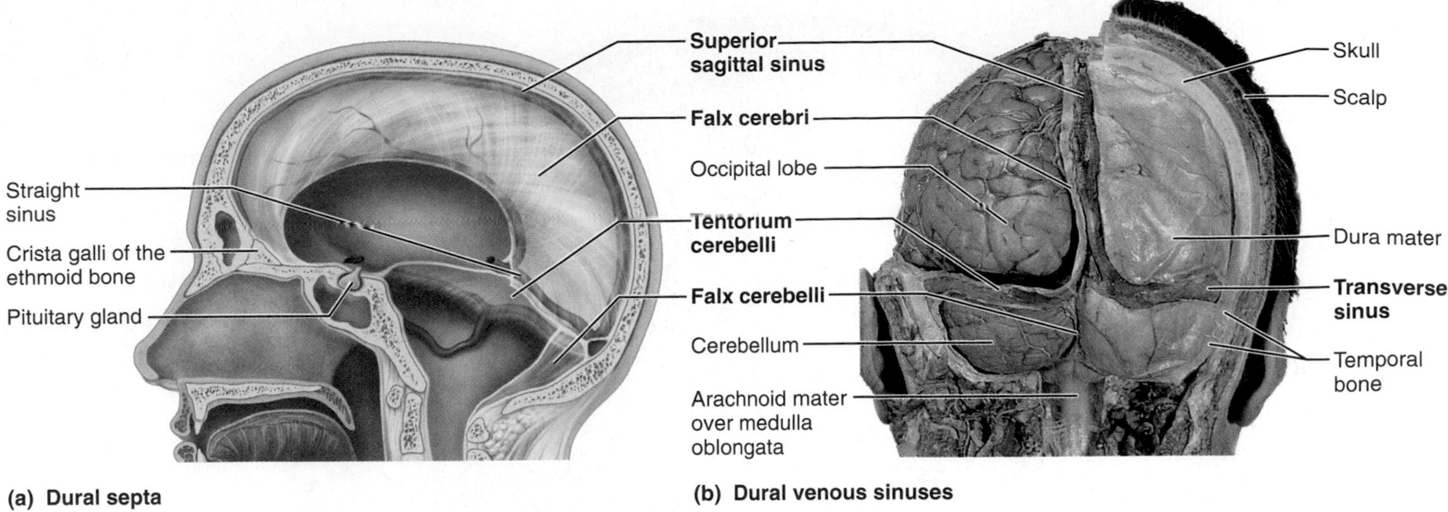

Figure 12.25 Dural septa and dural venous sinuses. (a) Dural septa are partitioning folds of dura mater in the cranial cavity. **(b)** Posterior view of the brain in place. Dural venous sinuses are spaces between the periosteal and meningeal dura containing venous blood.

▶ Describe the cause (if known) and major signs and symptoms of cerebrovascular accidents, Alzheimer's disease, Parkinson's disease, and Huntington's disease.

Nervous tissue is soft and delicate, and neurons are injured by even slight pressure. However, the brain is protected by bone (the skull), membranes (the meninges), and a watery cushion (cerebrospinal fluid). Furthermore, the brain is protected from harmful substances in the blood by the blood-brain barrier. We described the cranium, the brain's bony encasement, in Chapter 7. Here we will consider the other protective elements.

Meninges

The **meninges** (mĕ-nin′jēz; *mening* = membrane) are three connective tissue membranes that lie just external to the CNS organs. They (1) cover and protect the CNS, (2) protect blood vessels and enclose venous sinuses, (3) contain cerebrospinal fluid, and (4) form partitions in the skull. From external to internal, the meninges (singular: **meninx**) are the dura mater, arachnoid mater, and pia mater **(Figure 12.24)**.

Dura Mater

The leathery **dura mater** (du′rah ma′ter), meaning "tough mother," is the strongest meninx. Where it surrounds the brain, it is a two-layered sheet of fibrous connective tissue. The more superficial *periosteal layer* is attached to the inner surface of the skull (the periosteum). (There is no dural periosteal layer surrounding the spinal cord.) The deeper *meningeal layer* forms the true external covering of the brain and continues caudally in the vertebral canal as the spinal dura mater. The brain's two dural layers are fused together except in certain areas, where they separate to enclose **dural venous sinuses** that collect venous blood from the brain and direct it into the internal jugular veins of the neck **(Figure 12.25b)**.

In several places, the meningeal dura mater extends inward to form flat partitions that subdivide the cranial cavity. These **dural septa**, which limit excessive movement of the brain within the cranium, include the following (Figure 12.25a):

- **Falx cerebri** (falks ser′ĕ-bri). A large sickle-shaped (*falx* = sickle) fold that dips into the longitudinal fissure between the cerebral hemispheres. Anteriorly, it attaches to the crista galli of the ethmoid bone.
- **Falx cerebelli** (ser″ĕ-bel′i). Continuing inferiorly from the posterior falx cerebri, this small midline partition runs along the vermis of the cerebellum.
- **Tentorium cerebelli** (ten-to′re-um; "tent"). Resembling a tent over the cerebellum, this nearly horizontal dural fold extends into the transverse fissure between the cerebral hemispheres (which it helps to support) and the cerebellum.

12

Arachnoid Mater

The middle meninx, the **arachnoid mater**, or simply the **arachnoid** (ah-rak′noid), forms a loose brain covering, never dipping into the sulci at the cerebral surface. It is separated from the dura mater by a narrow serous cavity, the **subdural space**, which contains a film of fluid. Beneath the arachnoid membrane is the wide **subarachnoid space**. Weblike extensions span this space and secure the arachnoid mater to the underlying pia mater. (*Arachnida* means "spider," and this membrane was named for its weblike extensions.) The subarachnoid space is filled with cerebrospinal fluid and also contains the largest blood vessels serving the brain. Because the arachnoid is fine and elastic, these blood vessels are poorly protected.

Knoblike projections of the arachnoid mater called **arachnoid villi** (vil′i) protrude superiorly through the dura mater and into the superior sagittal sinus (see Figure 12.24). Cerebrospinal fluid is absorbed into the venous blood of the sinus by these valvelike villi.

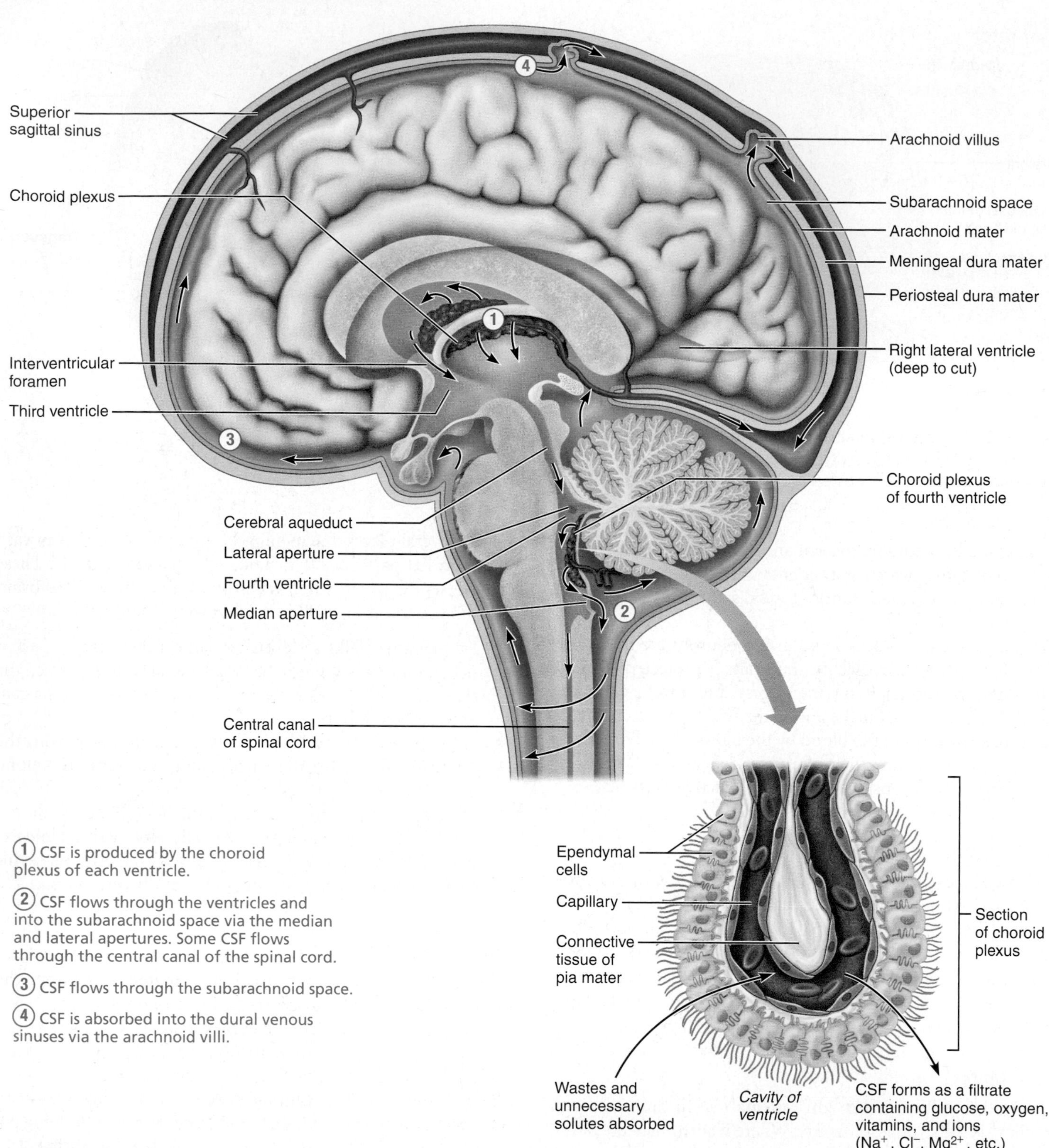

12

Figure 12.26 Formation, location, and circulation of CSF. (a) Location and circulatory pattern of cerebrospinal fluid (CSF). Arrows indicate the direction of flow. **(b)** Each choroid plexus consists of a knot of porous capillaries surrounded by a single layer of ependymal cells joined by tight junctions and bearing long cilia. Fluid leaking from porous capillaries is processed by the ependymal cells to form the CSF in the ventricles.

Pia Mater

The **pia mater** (pi′ah), meaning "gentle mother," is composed of delicate connective tissue and is richly invested with tiny blood vessels. It is the only meninx that clings tightly to the brain like cellophane wrap, following its every convolution. Small arteries entering the brain tissue carry ragged sheaths of pia mater inward with them for short distances.

HOMEOSTATIC IMBALANCE

Meningitis, inflammation of the meninges, is a serious threat to the brain because a bacterial or viral meningitis may spread to the CNS. Brain inflammation is called *encephalitis* (en-sef″ah-li′tis). Meningitis is usually diagnosed by obtaining a sample of cerebrospinal fluid via a lumbar tap (see Figure 12.30, p. 468) and examining it for microbes. ■

Cerebrospinal Fluid

Cerebrospinal fluid (CSF), found in and around the brain and spinal cord, forms a liquid cushion that gives buoyancy to the CNS structures. By floating the jellylike brain, the CSF effectively reduces brain weight by 97% and prevents the delicate brain from crushing under its own weight. CSF also protects the brain and spinal cord from blows and other trauma. Additionally, although the brain has a rich blood supply, CSF helps nourish the brain, and there is some evidence that it carries chemical signals (such as hormones and sleep- and appetite-inducing molecules) from one part of the brain to another.

CSF is a watery "broth" similar in composition to blood plasma, from which it is formed. However, it contains less protein than plasma and its ion concentrations are different. For example, CSF contains more Na^+, Cl^-, and H^+ than does blood plasma, and less Ca^{2+} and K^+.

The **choroid plexuses** that hang from the roof of each ventricle form CSF. These plexuses are frond-shaped clusters of broad, thin-walled capillaries (*plex* = interwoven) enclosed first by pia mater and then by a layer of ependymal cells lining the ventricles (**Figure 12.26b**). These capillaries are fairly permeable, and tissue fluid filters continuously from the bloodstream. However, the choroid plexus ependymal cells are joined by tight junctions, and they have ion pumps that allow them to modify this filtrate by actively transporting only certain ions across their membranes into the CSF pool. This careful regulation of CSF composition is important because CSF mixes with the extracellular fluid bathing neurons and influences the composition of this fluid. Ion pumping also sets up ionic gradients that cause water to diffuse into the ventricles.

In adults, the total CSF volume of about 150 ml (about half a cup) is replaced every 8 hours or so. About 500 ml of CSF is formed daily. The choroid plexuses also help cleanse the CSF by removing waste products and unnecessary solutes.

Once produced, CSF moves freely through the ventricles. Some CSF circulates into the central canal of the spinal cord, but most enters the subarachnoid space via the lateral and median apertures in the walls of the fourth ventricle (Figure 12.26a). The long cilia of the ependymal cells lining the ventricles help to keep the CSF in constant motion. In the subarachnoid space, CSF bathes the outer surfaces of the brain and spinal cord and then returns to the blood in the dural sinuses via the arachnoid villi.

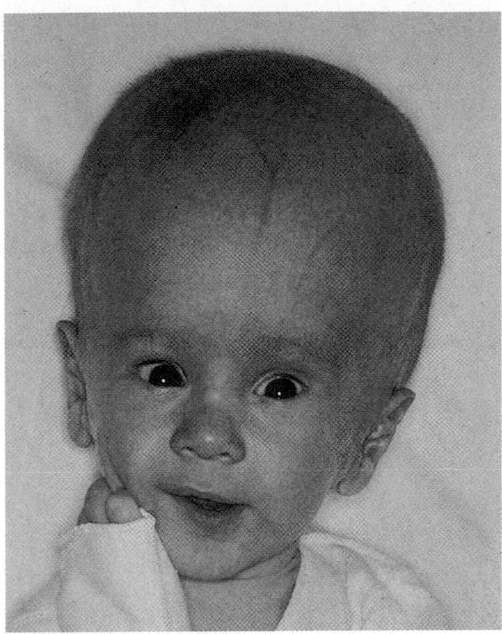

Figure 12.27 **Hydrocephalus in a newborn.**

12

HOMEOSTATIC IMBALANCE

Ordinarily, CSF is produced and drained at a constant rate. However, if something (such as a tumor) obstructs its circulation or drainage, CSF accumulates and exerts pressure on the brain. This condition is called *hydrocephalus* ("water on the brain"). Hydrocephalus in a newborn baby causes its head to enlarge (**Figure 12.27**). This enlargement is possible because the newborn's skull bones have not yet fused. In adults, however, hydrocephalus is likely to damage the brain because the skull is rigid and hard, and accumulating fluid compresses blood vessels serving the brain and crushes the soft nervous tissue. Hydrocephalus is treated by inserting a shunt into the ventricles to drain the excess fluid into the abdominal cavity. ■

Blood-Brain Barrier

The **blood-brain barrier** is a protective mechanism that helps maintain a stable environment for the brain. No other body tissue is so absolutely dependent on a constant internal environment as is the brain. In other body regions, the extracellular concentrations of hormones, amino acids, and ions are in constant flux, particularly after eating or exercise. If the brain were exposed to such chemical variations, the neurons would fire uncontrollably, because some hormones and amino acids serve as neurotransmitters and certain ions (particularly K^+) modify the threshold for neuronal firing.

Bloodborne substances in the brain's capillaries must pass through three layers before they reach the neurons: (1) the endothelium of the capillary wall, (2) a relatively thick basal lamina surrounding the external face of each capillary, and (3) the bulbous "feet" of the astrocytes clinging to the capillaries. Which of these layers constitutes the blood-brain barrier? As you might expect, the astrocyte "feet" play a role but they are not themselves the barrier. Instead, they supply required signals to the endothelial cells, causing them to make *tight junctions.* These tight junctions seamlessly join together the endothelial cells, forming the blood-brain barrier and making these the least permeable capillaries in the body.

The blood-brain barrier is selective, rather than absolute. Nutrients such as glucose, essential amino acids, and some electrolytes move passively by facilitated diffusion through the endothelial cell membranes. Bloodborne metabolic wastes, proteins, certain toxins, and most drugs are denied entry to brain tissue. Small nonessential amino acids and potassium ions not only are prevented from entering the brain, but also are actively pumped from the brain across the capillary endothelium.

The barrier is ineffective against fats, fatty acids, oxygen, carbon dioxide, and other fat-soluble molecules that diffuse easily through all plasma membranes. This explains why bloodborne alcohol, nicotine, and anesthetics can affect the brain.

The structure of the blood-brain barrier is not completely uniform. The capillaries of the choroid plexuses are very porous, but the ependymal cells surrounding them have tight junctions. In some brain areas surrounding the third and fourth ventricles, the blood-brain barrier is entirely absent and the capillary endothelium is quite permeable, allowing bloodborne molecules easy access to the neural tissue.

12

One such region is the vomiting center of the brain stem, which monitors the blood for poisonous substances. Another is in the hypothalamus, which regulates water balance, body temperature, and many other metabolic activities. Lack of a blood-brain barrier here is essential to allow the hypothalamus to sample the chemical composition of the blood. The barrier is incomplete in newborn and premature infants, and potentially toxic substances can enter the CNS and cause problems not seen in adults.

Injury to the brain, whatever the cause, may result in a localized breakdown of the blood-brain barrier. Most likely, this breakdown reflects some change in the capillary endothelial cells or their tight junctions.

Homeostatic Imbalances of the Brain

Brain dysfunctions are unbelievably varied and extensive. We have mentioned some of them already, but here we will focus on traumatic brain injuries, cerebrovascular accidents, and degenerative brain disorders.

Traumatic Brain Injury

Head injuries are a leading cause of accidental death in North America. Consider, for example, what happens if you forget to fasten your seat belt and then rear-end another car. Your head is moving and then is suddenly stopped as it hits the windshield. Brain damage is caused not only by localized injury at the site of the blow (the *coup* injury), but also by the ricocheting effect as the brain hits the opposite end of the skull (the *contrecoup* injury).

A **concussion** is an alteration in brain function, usually temporary, following a blow to the head. The victim may be dizzy or lose consciousness. Although typically mild and short-lived, even a seemingly mild concussion can be damaging, and multiple concussions over time have been shown to produce cumulative damage. More serious concussions can cause bruising of the brain and permanent neurological damage, a condition called a **contusion**. In cortical contusions, the individual may remain conscious, but severe brain stem contusions always cause coma, lasting from hours to a lifetime because of injury to the reticular activating system.

Following a head blow, death may result from **subdural** or **subarachnoid hemorrhage** (bleeding from ruptured vessels into those spaces). Individuals who are initially lucid and then begin to deteriorate neurologically are, in all probability, hemorrhaging intracranially. Blood accumulating in the skull increases intracranial pressure and compresses brain tissue. If the pressure forces the brain stem inferiorly through the foramen magnum, control of blood pressure, heart rate, and respiration is lost. Intracranial hemorrhages are treated by surgical removal of the hematoma (localized blood mass) and repair of the ruptured vessels.

Another consequence of traumatic head injury is **cerebral edema**, swelling of the brain. At best, this cerebral edema aggravates the injury—at worst, it can be fatal in and of itself.

Cerebrovascular Accidents

The single most common nervous system disorder and the third leading cause of death in North America are **cerebrovascular accidents (CVAs)** (ser″ĕ-bro-vas′ku-lar), also called *strokes.* CVAs occur when blood circulation to a brain area is blocked and brain tissue dies. [Deprivation of blood supply to any tissue is called **ischemia** (is-ke′me-ah; "to hold back blood") and results in deficient oxygen and nutrient delivery to cells.] The most common cause of CVA is blockage of a cerebral artery by a blood clot. Other causes include compression of brain tissue by hemorrhage or edema and narrowing of brain vessels by atherosclerosis.

Those who survive a CVA are typically paralyzed on one side of the body (*hemiplegia*). Many exhibit sensory deficits or have difficulty in understanding or vocalizing speech. Even so, the picture is not hopeless. Some patients recover at least part of their lost faculties, because undamaged neurons sprout new branches that spread into the injured area and take over some lost functions. Physical therapy is usually started as soon as possible to prevent muscle contractures (abnormal shortening of muscles due to differences in strength between opposing muscle groups).

Not all strokes are "completed." Temporary episodes of reversible cerebral ischemia, called **transient ischemic attacks (TIAs)**, are common. TIAs last from 5 to 50 minutes and are characterized by temporary numbness, paralysis, or impaired speech. These deficits are not permanent, but TIAs do constitute "red flags" that warn of impending, more serious CVAs.

CVAs are like undersea earthquakes. It's not the initial temblor that does most of the damage, it's the tsunami that floods the coast later. Similarly, the initial vascular blockage during a stroke is not usually disastrous because there are many blood vessels in the brain that can pick up the slack. Rather, it's the events that lead to killing of neurons not in the initial ischemic zone that wreak the most havoc.

Experimental evidence indicates that the main culprit is *glutamate*, an excitatory neurotransmitter also involved in learning and memory. Normally, glutamate binding to NMDA receptors opens NMDA channels that allow Ca^{2+} to enter the stimulated neuron. After brain injury, neurons totally deprived of oxygen begin to disintegrate, unleashing the cellular equivalent of "buckets" of glutamate. Under these conditions, glutamate acts as an *excitotoxin*, literally exciting surrounding cells to death. The initial events in excitotoxicity are identical to those in LTP— Ca^{2+} flows in through NMDA receptor channels (see p. 460). In excitotoxicity, however, the amount of Ca^{2+} swamps the cell's ability to cope and Ca^{2+} homeostasis breaks down. High levels of Ca^{2+} lead to cell death in two ways. First, Ca^{2+} damages mitochondria, causing them to produce the free radical superoxide, which can damage cells directly and can also cause programmed cell death (apoptosis). Second, Ca^{2+} turns on the synthesis of certain proteins. Some of these promote apoptosis while others are enzymes that produce the free radical NO and other powerful inflammatory agents.

Although the NMDA receptor and the enzyme that makes NO are attractive targets for drug therapies to decrease damage around the stroke epicenter, clinical trials so far have all failed. At present, the most successful treatment for stroke is tissue plasminogen activator (tPA), which dissolves blood clots in the brain. A newly approved mechanical device drills into a blood clot and pulls it from a vessel like a cork from a bottle. Other approaches currently under study focus on recovering function after a stroke. One method implants immature neurons into stroke-damaged brain regions in the hope that they will take on properties of the nearby mature neurons. Another tries to coax adult brain stem cells to replace damaged neurons. Mice with strokes can generate new tissue and recover motor function when treated with a combination of growth factors during a critical time following injury. Hopefully, this will be true for humans, too.

Degenerative Brain Disorders

Alzheimer's Disease **Alzheimer's disease (AD)** (altz′hi-merz) is a progressive degenerative disease of the brain that ultimately results in dementia (mental deterioration). Alzheimer's patients represent nearly half of the people living in nursing homes. Between 5 and 15% of people over 65 develop this condition, and for up to half of those over 85 it is a major contributing cause in their deaths.

Its victims exhibit memory loss (particularly for recent events), shortened attention span, disorientation, and eventual language loss. Over a period of several years, formerly good-natured people may become irritable, moody, and confused. Ultimately, hallucinations occur.

Examinations of brain tissue reveal senile plaques littering the brain like shrapnel between the neurons. The plaques consist of extracellular aggregations of *beta-amyloid peptide*, which has been cut from a normal membrane precursor protein (APP) by enzymes. One form of Alzheimer's disease is caused by an inherited mutation in the gene for APP, which suggests that beta-amyloid may be toxic. Some researchers believe that small clusters of these protein fragments kill cells by forming holes in their plasma membranes. Unfortunately, clinical trials of vaccines to stimulate an immune response to clear away beta-amyloid peptide have not yet been successful.

Another hallmark of Alzheimer's disease is the presence of *neurofibrillary tangles* inside neurons. These tangles involve a protein called tau, which functions like railroad ties to bind microtubule "tracks" together. In the brains of AD victims, tau abandons its microtubule-stabilizing role and grabs onto other tau molecules, forming spaghetti-like neurofibrillary tangles, which kill the neurons by disrupting their transport mechanisms.

As the brain cells die, the brain shrinks. Particularly vulnerable brain areas include the hippocampus and the basal forebrain, regions involved in thinking and memory (see Figure 12.23). Loss of neurons in the basal forebrain is associated with a shortage of the neurotransmitter acetylcholine, and drugs that inhibit breakdown of acetylcholine slightly enhance cognitive function in AD patients. Interestingly, a drug that blocks NMDA receptors, memantine, also slightly improves thinking in more advanced stages of AD, suggesting that glutamate excitotoxicity is also involved in this disease.

Parkinson's Disease Typically striking people in their 50s and 60s, **Parkinson's disease** results from a degeneration of the dopamine-releasing neurons of the substantia nigra. As those neurons deteriorate, the dopamine-deprived basal nuclei they target become overactive, causing the well-known symptoms of the disease. Afflicted individuals have a persistent tremor at rest (exhibited by head nodding and a "pill-rolling" movement of the fingers), a forward-bent walking posture and shuffling gait, and a stiff facial expression. They are slow in initiating and executing movement.

The cause of Parkinson's disease is still unknown, but the interaction of multiple factors is thought to lead to the death of dopamine-releasing neurons. Recent evidence points to abnormalities in a dopamine transport regulator and in certain mitochondrial proteins. The drug *L-dopa* often helps to alleviate some symptoms. It passes through the blood-brain barrier and is then converted into dopamine. However, it is not curative, and as more and more neurons die off, L-dopa becomes ineffective. Mixing L-dopa with drugs that inhibit the breakdown of dopamine (for example, deprenyl), can prolong the effectiveness of L-dopa. In addition, deprenyl by itself, early in the disease, slows the neurological deterioration to some extent and delays the need to administer L-dopa for up to 18 months.

Deep brain stimulation via implanted electrodes shuts down abnormal brain activity and has proved helpful in alleviating tremors. This treatment (for patients who no longer respond to drug therapy) is delicate, expensive, and risky. Another possibility is to use gene therapy to insert genes into adult brain cells, causing them to secrete the inhibitory neurotransmitter GABA. GABA then inhibits the abnormal brain activity just as the electrical stimulation does. Replacing dead or damaged cells by implanting embryonic or fetal cells also shows great promise. However, the use of embryonic or fetal tissue is controversial and riddled with ethical and legal roadblocks.

Huntington's Disease **Huntington's disease** is a fatal hereditary disorder that strikes during middle age. Mutant *huntingtin* protein accumulates in brain cells and the tissue dies, leading to massive degeneration of the basal nuclei and later of the cerebral cortex. Its initial symptoms in many are wild, jerky, almost continuous "flapping" movements called *chorea* (Greek for "dance"). Although the movements appear to be voluntary they are not. Late in the disease, marked mental deterioration occurs. Huntington's disease is progressive and usually fatal within 15 years of onset of symptoms.

The hyperkinetic manifestations of Huntington's disease are essentially the opposite of those of Parkinson's disease (overstimulation rather than inhibition of the motor drive). Huntington's is usually treated with drugs that block, rather than enhance, dopamine's effects. As with Parkinson's disease, fetal tissue implants may provide promise for its treatment in the future.

CHECK YOUR UNDERSTANDING

20. What is CSF? Where is it produced? What are its functions?

21. What is a transient ischemic attack (TIA) and how is it different from a stroke?

22. Mrs. Lee, a neurology patient, seldom smiles, has a shuffling stooped gait, and often spills her coffee. What degenerative brain disorder might she have?

For answers, see Appendix G.

The Spinal Cord

- Describe the embryonic development of the spinal cord.
- Describe the gross and microscopic structure of the spinal cord.
- List the major spinal cord tracts, and classify each as a motor or sensory tract.

Embryonic Development

The spinal cord develops from the caudal portion of the embryonic neural tube (see Figure 12.2 on p. 431). By the sixth week, each side of the developing cord has two recognizable clusters of neuroblasts that have migrated outward from the original neural tube: a dorsal **alar plate** (a′lar) and a ventral **basal plate** **(Figure 12.28)**.

Alar plate neuroblasts become interneurons. The basal plate neuroblasts develop into motor neurons and sprout axons that

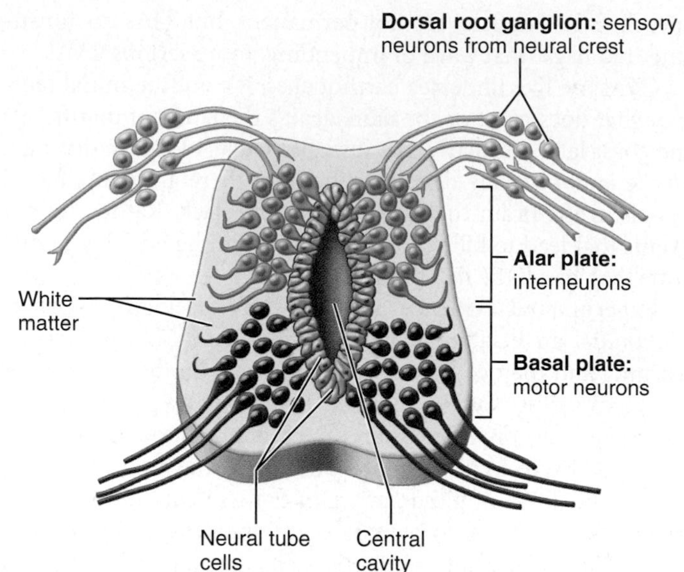

Figure 12.28 Structure of the embryonic spinal cord. At six weeks of development, aggregations of gray matter called the alar plates (future interneurons) and basal plates (future motor neurons) have formed. Dorsal root ganglia (future sensory neurons) have arisen from neural crest cells.

grow out to the effector organs. Axons that emerge from alar plate cells (and some basal plate cells) form the white matter of the cord by growing along the length of the cord. As development progresses, these plates expand dorsally and ventrally to produce the H-shaped central mass of gray matter of the adult spinal cord.

Neural crest cells that come to lie alongside the cord form the *dorsal root ganglia* containing sensory neuron cell bodies. These sensory neurons send their axons into the dorsal aspect of the cord.

Gross Anatomy and Protection

The spinal cord, enclosed in the vertebral column, extends from the foramen magnum of the skull to the level of the first or second lumbar vertebra, just inferior to the ribs **(Figure 12.29)**. About 42 cm (17 inches) long and 1.8 cm (3/4 of an inch) thick, the glistening-white **spinal cord** provides a two-way conduction pathway to and from the brain. It is a major reflex center: Spinal reflexes are initiated and completed at the spinal cord level. We discuss reflex functions and motor activity of the cord in subsequent chapters. In this section we focus on the anatomy of the cord and on the location and naming of its ascending and descending tracts.

Like the brain, the spinal cord is protected by bone, meninges, and cerebrospinal fluid. The single-layered **spinal dura mater** (Figure 12.29c) is not attached to the bony walls of the vertebral column. Between the bony vertebrae and the spinal dura mater is an **epidural space** filled with a soft padding of fat and a network of veins (see Figure 12.31a). Cerebrospinal fluid fills the subarachnoid space between the *arachnoid* and *pia mater* meninges.

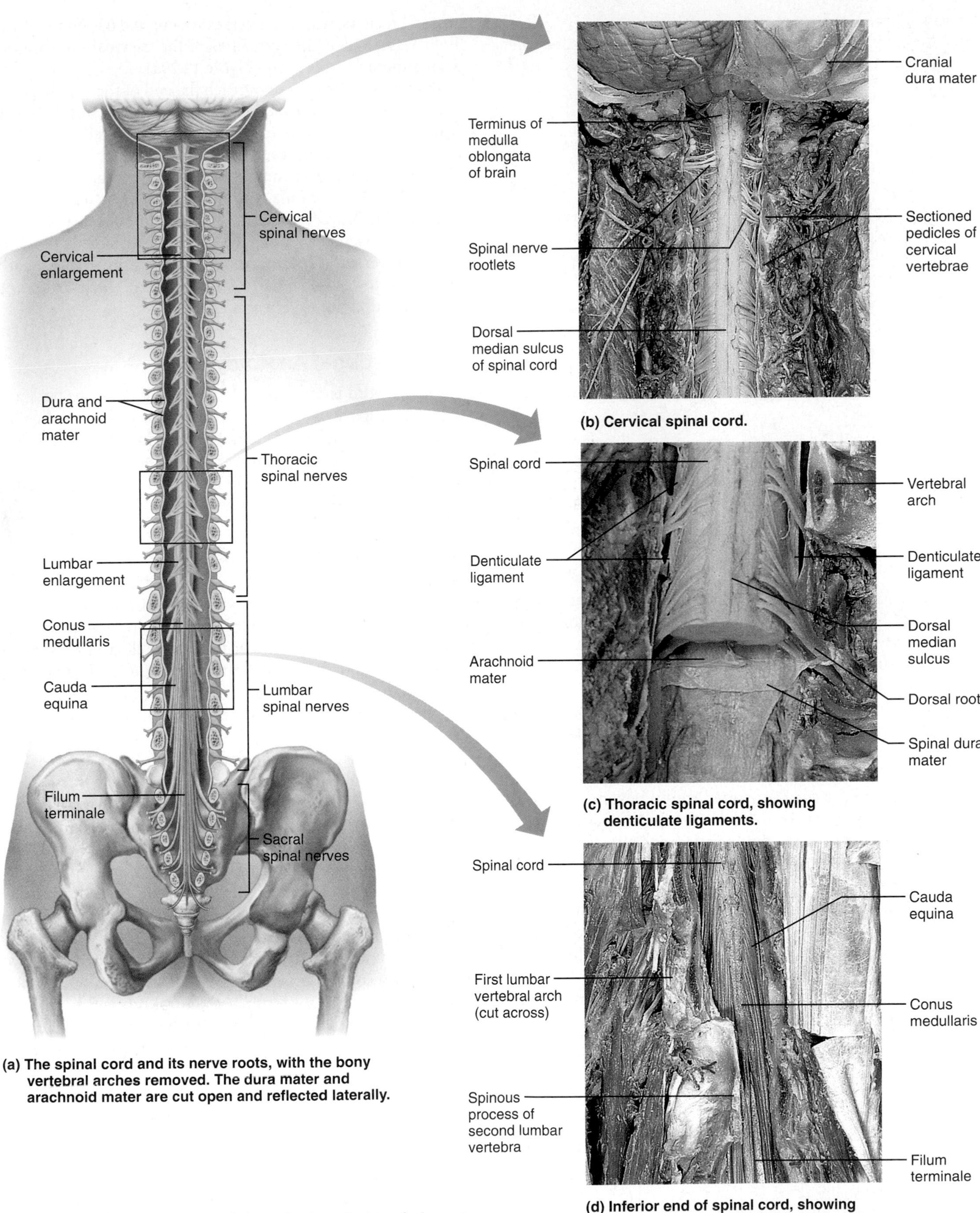

(a) The spinal cord and its nerve roots, with the bony vertebral arches removed. The dura mater and arachnoid mater are cut open and reflected laterally.

(b) Cervical spinal cord.

(c) Thoracic spinal cord, showing denticulate ligaments.

(d) Inferior end of spinal cord, showing conus medullaris, cauda equina, and filum terminale.

Figure 12.29 **Gross structure of the spinal cord, dorsal view.** (See *A Brief Atlas of the Human Body,* Figures 52, 53, and 55.)

12

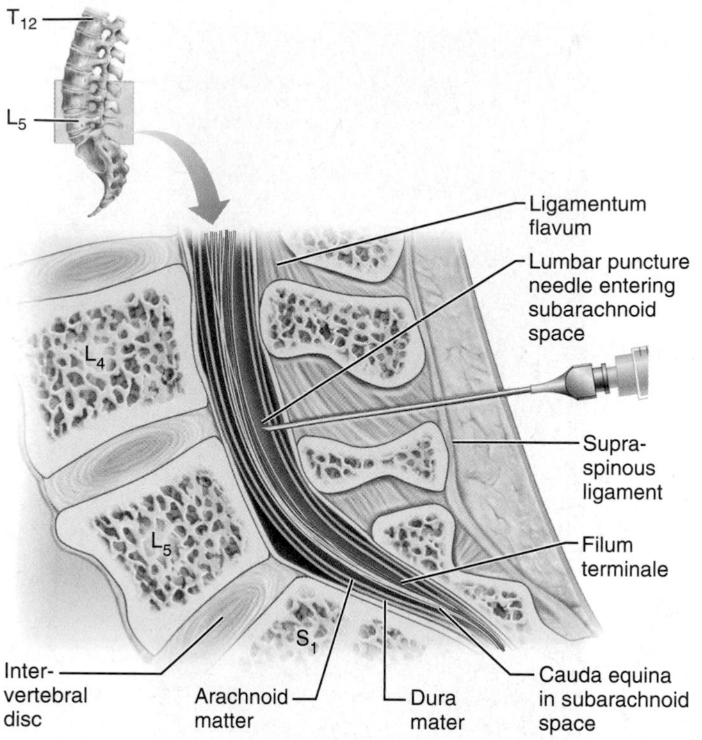

Figure 12.30 Diagrammatic view of a lumbar tap.

12

Inferiorly, the dural and arachnoid membranes extend to the level of S_2, well beyond the end of the spinal cord. The spinal cord typically ends between L_1 and L_2 (Figure 12.29a). For this reason, the subarachnoid space within the meningeal sac inferior to that point provides a nearly ideal spot for removing cerebrospinal fluid for testing, a procedure called a **lumbar puncture** or **tap** (**Figure 12.30**). Because the spinal cord is absent there and the delicate nerve roots drift away from the point of needle insertion, there is little or no danger of damaging the cord (or spinal roots) beyond L_3.

Inferiorly, the spinal cord terminates in a tapering cone-shaped structure called the **conus medullaris** (ko′nus me″dul-ar′is). The **filum terminale** (fi′lum ter″mĭ-nah′le; "terminal filament"), a fibrous extension of the conus covered by pia mater, extends inferiorly from the conus medullaris to the coccyx, where it anchors the spinal cord in place so it is not jostled by body movements (Figure 12.29a, d). Furthermore, the spinal cord is secured to the tough dura mater meninx throughout its length by saw-toothed shelves of pia mater called **denticulate ligaments** (den-tik′u-lāt; "toothed") (Figure 12.29c).

In humans, 31 pairs of *spinal nerves* attach to the cord by paired roots. Each nerve exits from the vertebral column by passing superior to its corresponding vertebra via the intervertebral foramen, and travels to the body region it serves. While each nerve pair defines a segment of the cord, the spinal cord is, in fact, continuous throughout its length and its internal structure changes gradually.

The spinal cord is about the width of a thumb for most of its length, but it has obvious enlargements in the cervical and lumbosacral regions, where the nerves serving the upper and lower limbs arise. These enlargements are the **cervical** and **lumbar enlargements**, respectively (Figure 12.29a).

Because the cord does not reach the end of the vertebral column, the lumbar and sacral spinal nerve roots angle sharply downward and travel inferiorly through the vertebral canal for some distance before reaching their intervertebral foramina. The collection of nerve roots at the inferior end of the vertebral canal is named the **cauda equina** (kaw′da e-kwi′nuh) because of its resemblance to a horse's tail (Figure 12.29a, c). This strange arrangement reflects the fact that during fetal development, the vertebral column grows faster than the spinal cord, forcing the lower spinal nerve roots to "chase" their exit points inferiorly through the vertebral canal.

Cross-Sectional Anatomy

The spinal cord is somewhat flattened from front to back and two grooves mark its surface: the **ventral (anterior) median fissure** and the shallower **dorsal (posterior) median sulcus** (**Figure 12.31b**). These grooves run the length of the cord and partially divide it into right and left halves. The gray matter of the cord is located in its core, the white matter outside.

Gray Matter and Spinal Roots

In cross section the gray matter of the cord looks like the letter H or like a butterfly (Figure 12.31b). It consists of mirror-image lateral gray masses connected by a crossbar of gray matter, the **gray commissure**, that encloses the central canal. The two dorsal projections of the gray matter are the **dorsal (posterior) horns**, and the ventral pair are the **ventral (anterior) horns**. In 3-D, these horns form columns of gray matter that run the entire length of the spinal cord. An additional pair of gray matter columns, the small **lateral horns**, is present in the thoracic and superior lumbar segments of the cord.

All neurons whose cell bodies are in the spinal cord gray matter are multipolar. The dorsal horns consist entirely of interneurons. The ventral horns have some interneurons but mainly house cell bodies of somatic motor neurons. These motor neurons send their axons out to the skeletal muscles (their effector organs) via the *ventral rootlets* that fuse together to become the **ventral roots** of the spinal cord (Figure 12.31b).

The amount of ventral gray matter present at a given level of the spinal cord reflects the amount of skeletal muscle innervated at that level. As a result, the ventral horns are largest in the limb-innervating cervical and lumbar regions of the cord and are responsible for the cord enlargements seen in those regions.

The lateral horn neurons are autonomic (sympathetic division) motor neurons that serve visceral organs. Their axons leave the cord via the ventral root along with those of the somatic motor neurons. Because the ventral roots contain both somatic and autonomic efferents, they serve both motor divisions of the peripheral nervous system (**Figure 12.32**).

Afferent fibers carrying impulses from peripheral sensory receptors form the **dorsal roots** of the spinal cord that fan out as

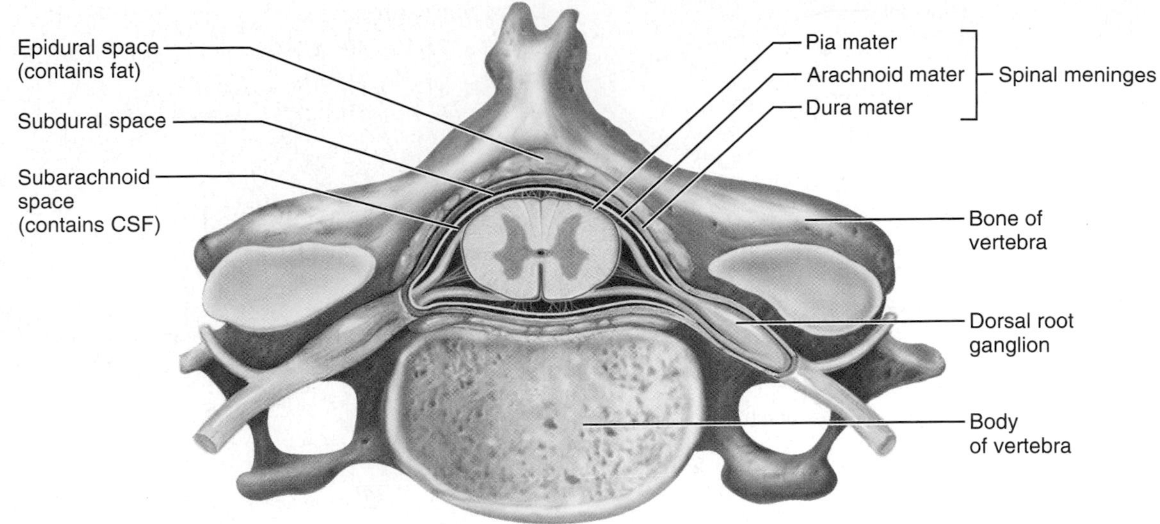

(a) Cross section of spinal cord and vertebra

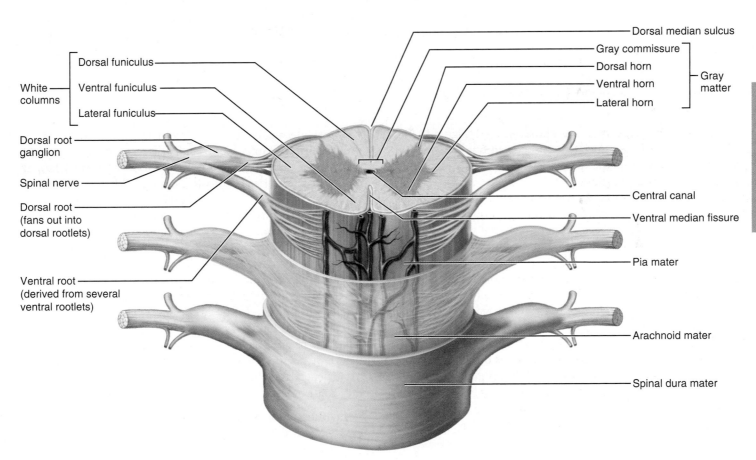

White columns
Dorsal funiculus
Ventral funiculus
Lateral funiculus
Dorsal root
ganglion
Spinal nerve
Dorsal root
(fans out into
dorsal rootlets)
Ventral root
(derived from several
ventral rootlets)
Dorsal median sulcus
Gray commissure
Dorsal horn
Ventral horn
Lateral horn
Gray
matter
Central canal
Ventral median fissure
Pia mater
Arachnoid mater
Spinal dura mater

(b) The spinal cord and its meningeal coverings

Figure 12.31 Anatomy of the spinal cord. (a) Cross section through the spinal cord illustrating its relationship to the surrounding vertebral column. **(b)** Anterior view of the spinal cord and its meningeal coverings.

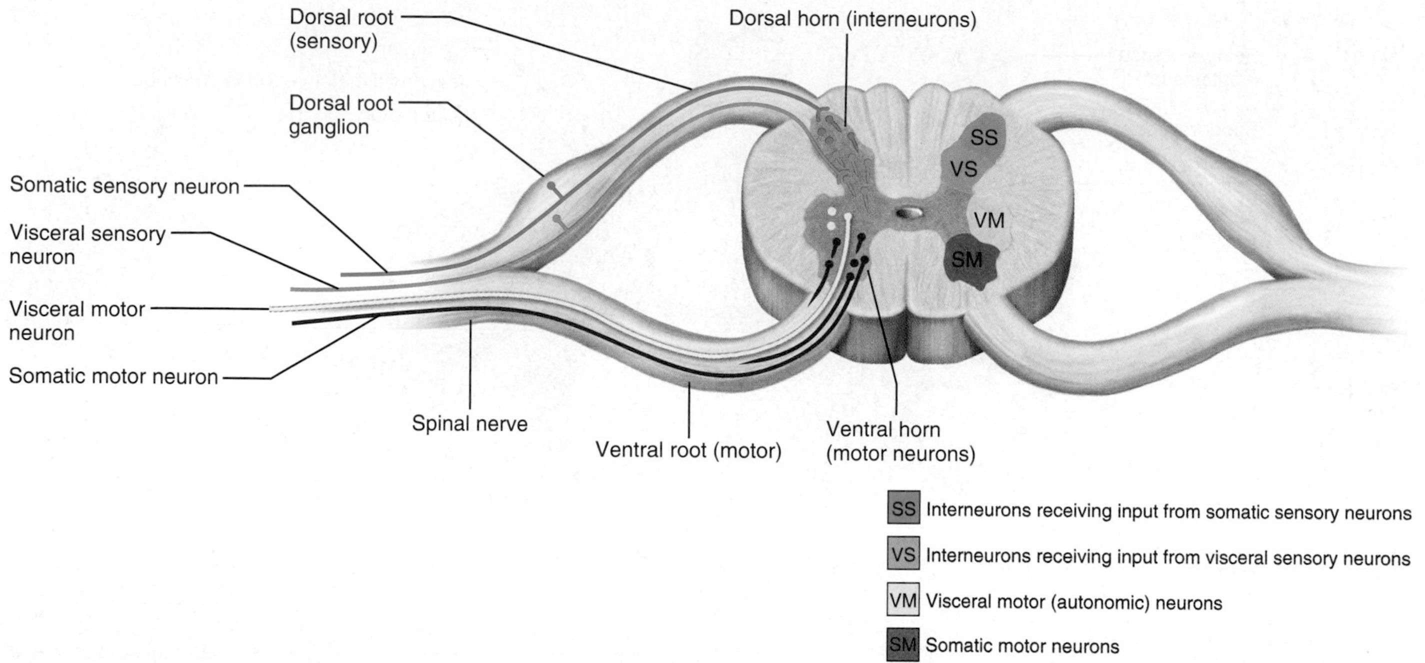

Figure 12.32 Organization of the gray matter of the spinal cord. The gray matter of the spinal cord is divided into a sensory half dorsally and a motor half ventrally. Note that the dorsal and ventral roots are part of the PNS, not of the spinal cord.

12

the *dorsal rootlets* before they enter the spinal cord (Figure 12.31). The cell bodies of the associated sensory neurons are found in an enlarged region of the dorsal root called the **dorsal root ganglion** or **spinal ganglion.** After entering the cord, the axons of these neurons may take a number of routes. Some enter the dorsal white matter of the cord directly and travel to synapse at higher cord or brain levels. Others synapse with interneurons in the dorsal horns of the spinal cord gray matter at their entry level.

The dorsal and ventral roots are very short and fuse laterally to form the **spinal nerves**. The spinal nerves, which are part of the peripheral nervous system, are considered further in Chapter 13.

The spinal gray matter can be divided further according to its neurons' relative involvement in the innervation of the somatic and visceral regions of the body. The following four zones are evident within this gray matter: **somatic sensory (SS), visceral (autonomic) sensory (VS), visceral motor (VM), somatic motor (SM)** (Figure 12.32).

White Matter

The white matter of the spinal cord is composed of myelinated and unmyelinated nerve fibers that allow communication between different parts of the spinal cord and between the cord and brain. These fibers run in three directions: (1) *ascending*—up to higher centers (sensory inputs), (2) *descending*—down to the cord from the brain or within the cord to lower levels (motor outputs), and (3) *transversely*—across from one side of the cord to the other (commissural fibers). Ascending and descending tracts make up most of the white matter.

The white matter on each side of the cord is divided into three **white columns**, or **funiculi** (fu-nik′u-li; "long ropes"), named according to their position as **dorsal (posterior), lateral,** and **ventral (anterior) funiculi** (Figure 12.31b). Each funiculus contains several fiber tracts, and each tract is made up of axons with similar destinations and functions. With a few exceptions, the names of the spinal tracts reveal both their origin and destination. The principal ascending and descending tracts of the spinal cord are illustrated schematically in the cross-sectional view of the spinal cord in **Figure 12.33**.

All major spinal tracts are *part of multineuron pathways* that connect the brain to the body periphery. These great ascending and descending pathways contain not only spinal cord neurons but also parts of peripheral neurons and neurons in the brain. Before we get specific about the individual tracts, we will make some generalizations about them and the pathways to which they contribute:

1. **Decussation.** Most pathways cross from one side of the CNS to the other (decussate) at some point along their journey.
2. **Relay.** Most pathways consist of a chain of two or three neurons (a relay) that contribute to successive tracts of the pathway.
3. **Somatotopy.** Most pathways exhibit *somatotopy*, a precise spatial relationship among the tract fibers that reflects the orderly mapping of the body. For example, in an ascending sensory tract, somatotopy refers to the fact that fibers transmitting inputs from sensory receptors in the superior parts of the body lie lateral to those conveying sensory information from inferior body regions.

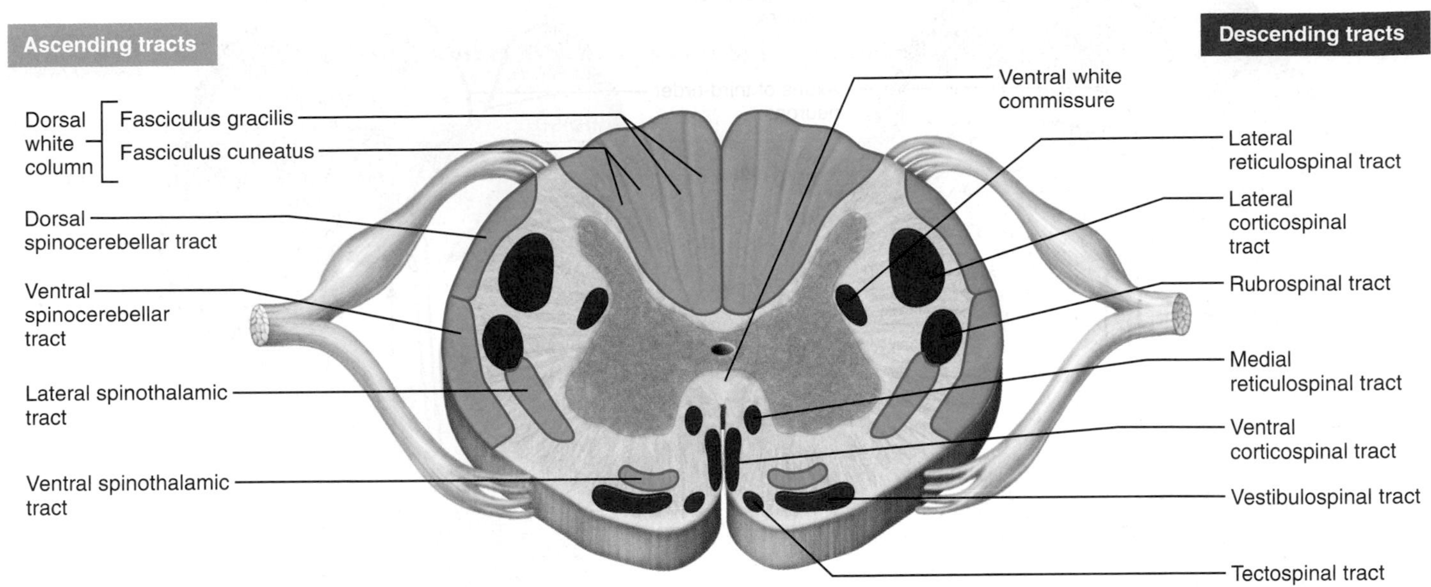

Figure 12.33 Major ascending (sensory) and descending (motor) tracts of the spinal cord, cross-sectional view.

4. **Symmetry.** All pathways and tracts are paired symmetrically (right and left), with a member of the pair present on each side of the spinal cord or brain.

Ascending Pathways to the Brain

The ascending pathways conduct sensory impulses upward, typically through chains of three successive neurons (first-, second-, and third-order neurons) to various areas of the brain. Note that both second- and third-order neurons are interneurons.

- **First-order neurons**, whose cell bodies reside in a ganglion (dorsal root or cranial), conduct impulses from the cutaneous receptors of the skin and from proprioceptors to the spinal cord or brain stem, where they synapse with second-order neurons. Impulses from the facial area are transmitted by cranial nerves, and spinal nerves conduct somatic sensory impulses from the rest of the body to the CNS. First-order neurons entering the spinal cord are shown at the bottom of Figure 12.34.
- **Second-order neurons** are shown in the middle of Figure 12.34. Their cell bodies reside in the dorsal horn of the spinal cord or in medullary nuclei, and they transmit impulses to the thalamus or to the cerebellum where they synapse.
- **Third-order neurons** have cell bodies in the thalamus (shown in the top of Figure 12.34). They relay impulses to the somatosensory cortex of the cerebrum. (There are no third-order neurons in the cerebellum.)

In general, somatosensory information is conveyed along three main pathways on each side of the spinal cord. Two of these pathways (the *dorsal column–medial lemniscal* and *spinothalamic pathways*) transmit impulses via the thalamus to the sensory cortex for conscious interpretation. Collectively the inputs of these sister tracts provide *discriminative touch* and *conscious proprioception.* Both pathways decussate—the first in the medulla and the second in the spinal cord.

The third pathway, the *spinocerebellar pathway,* terminates in the cerebellum, and does not contribute to sensory perception. Let's examine these pathways more closely.

12

1. **Dorsal column–medial lemniscal pathways.** The **dorsal column–medial lemniscal pathways** (lem-nis′kul; "ribbon") mediate precise, straight-through transmission of inputs from a single type (or a few related types) of sensory receptor that can be localized precisely on the body surface, such as discriminative touch and vibrations. These pathways are formed by the paired tracts of the **dorsal white column** of the spinal cord—**fasciculus cuneatus** and **fasciculus gracilis**—and the **medial lemniscus**. The medial lemniscus arises in the medulla and terminates in the thalamus (Figure 12.34a and **Table 12.2**). From the thalamus, impulses are forwarded to specific areas of the somatosensory cortex.
2. **Anterolateral pathways.** The **anterolateral pathways** [named for their location in the ventral (anterior) and lateral white columns of the spinal cord] receive input from many different types of sensory receptors and make multiple synapses in the brain stem. These pathways are largely formed by the **lateral** and **ventral (anterior) spinothalamic tracts** (see Figure 12.34b and Table 12.2). Their fibers cross over in the spinal cord.

   Most of the fibers in these pathways transmit pain, temperature, and coarse touch impulses, sensations that we are aware of but have difficulty localizing precisely on the body surface.
3. **Spinocerebellar tracts.** The last pair of ascending pathways, the **ventral (anterior)** and **dorsal (posterior) spinocerebellar tracts**, convey information about muscle or tendon stretch to the cerebellum, which uses this information to

TABLE 12.3 Major Descending (Motor) Pathways and Spinal Cord Tracts

SPINAL CORD TRACT	LOCATION (FUNICULUS)	ORIGIN	TERMINATION	FUNCTION
Direct (Pyramidal)				
Lateral corticospinal	Lateral	Pyramidal neurons of motor cortex of the cerebrum; decussate in pyramids of medulla	By synapse with ventral horn interneurons that influence motor neurons and occasionally with ventral horn motor neurons directly	Transmits motor impulses from cerebrum to spinal cord motor neurons (which activate skeletal muscles on opposite side of body). A voluntary motor tract.
Ventral corticospinal	Ventral	Pyramidal neurons of motor cortex; fibers cross over at the spinal cord level	Ventral horn (as above)	Same as lateral corticospinal tract.
Indirect (Extrapyramidal) Pathways				
Tectospinal	Ventral	Superior colliculus of midbrain of brain stem (fibers cross to opposite side of cord)	Ventral horn (as above)	Turns neck so eyes can follow a moving object.
Vestibulospinal	Ventral	Vestibular nuclei in medulla of brain stem (fibers descend without crossing)	Ventral horn (as above)	Transmits motor impulses that maintain muscle tone and activate ipsilateral limb and trunk extensor muscles and muscles that move head; in this way it helps maintain balance during standing and moving.
Rubrospinal	Lateral	Red nucleus of midbrain of brain stem (fibers cross to opposite side just inferior to the red nucleus)	Ventral horn (as above)	In experimental animals, transmits motor impulses concerned with muscle tone of distal limb muscles (mostly flexors) on opposite side of body. In humans, functions are largely assumed by corticospinal tracts.
Reticulospinal (ventral, medial, and lateral)	Ventral and lateral	Reticular formation of brain stem (medial nuclear group of pons and medulla); both crossed and uncrossed fibers	Ventral horn (as above)	Transmits impulses concerned with muscle tone and many visceral motor functions. May control most unskilled movements.

neurons, referred to as the upper and lower motor neurons. The pyramidal cells of the motor cortex, as well as the neurons in subcortical motor nuclei that give rise to other descending motor pathways, are called *upper motor neurons.* The ventral horn motor neurons, which actually innervate the skeletal muscles (their effectors), are called *lower motor neurons.* We briefly overview these tracts here and provide more information in **Table 12.3.**

The Direct (Pyramidal) System The direct pathways originate mainly with the pyramidal neurons located in the precentral gyri. These neurons send impulses through the brain stem via the large **pyramidal (corticospinal) tracts** (**Figure 12.35a**). The direct pathways are so called because their axons descend without synapsing from the pyramidal neurons to the spinal cord. There they synapse either with interneurons or with ventral horn motor neurons. Stimulation of the ventral horn neurons activates the skeletal muscles with which they are associated. The direct pathway primarily regulates fast and fine (or skilled) movements such as doing needlework and writing.

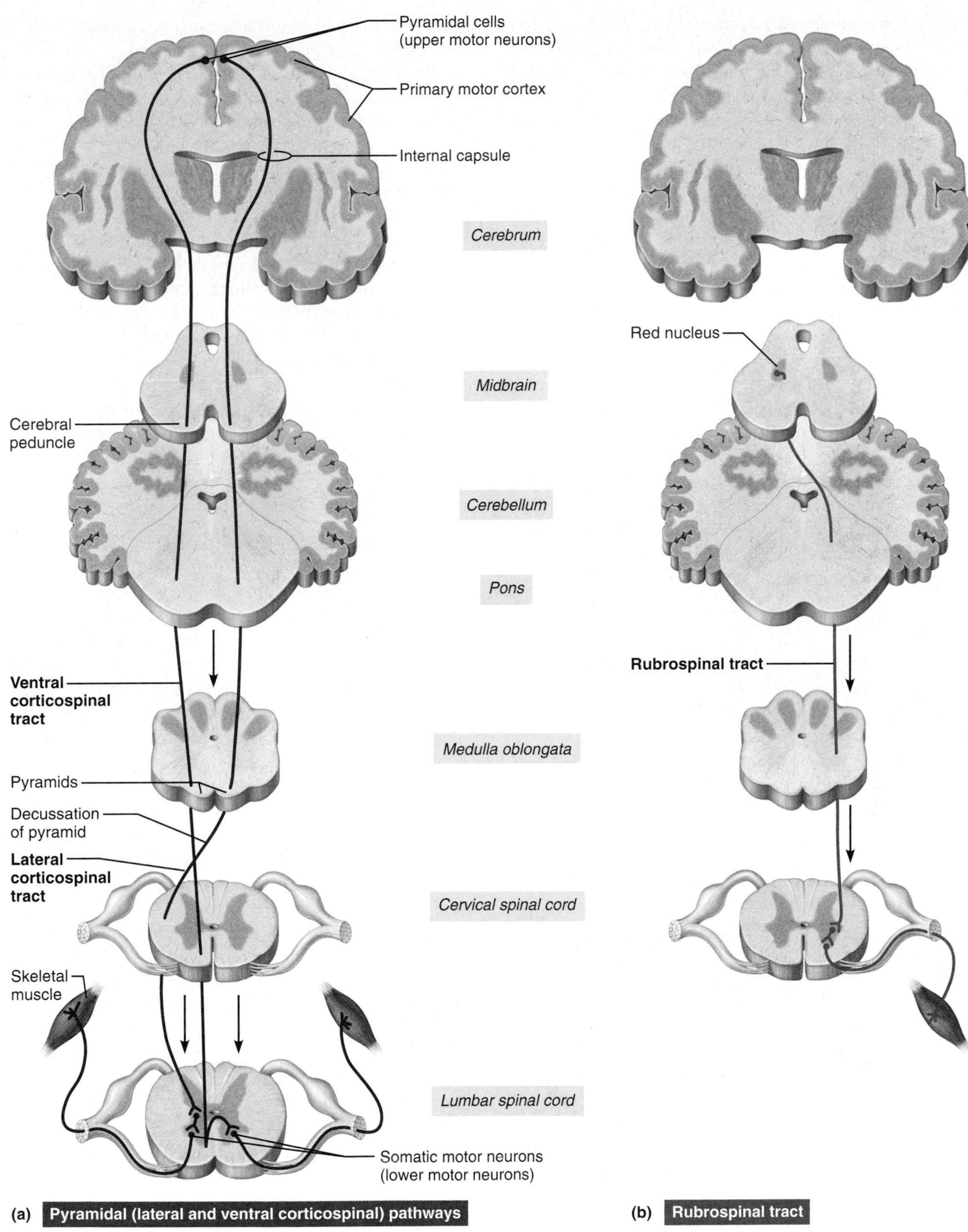

12

Figure 12.35 Three descending pathways by which the brain influences movement. Cross sections up to the cerebrum, which is shown in frontal section. **(a)** Pyramidal (lateral and ventral corticospinal pathways) are direct pathways that control skilled voluntary movements. **(b)** The rubrospinal tract, one of the indirect (or extrapyramidal) pathways, helps regulate muscle tone.

The Indirect (Extrapyramidal) System The indirect system includes brain stem motor nuclei and *all motor pathways except* the pyramidal pathways. These tracts were formerly lumped together as the **extrapyramidal system** because their nuclei of origin were presumed to be independent of ("extra to") the pyramidal tracts. This term is still widely used clinically. However, pyramidal tract neurons are now known to project to and influence the activity of most "extrapyramidal" nuclei, so modern anatomists prefer to use the term **indirect**, or **multineuronal, pathways**, or even just the names of the individual motor pathways.

These motor pathways are complex and multisynaptic. They are most involved in regulating (1) the axial muscles that maintain balance and posture, (2) the muscles controlling coarse limb movements, and (3) head, neck, and eye movements that follow objects in the visual field. Many of the activities controlled by subcortical motor nuclei depend heavily on reflex activity. One of these tracts, the rubrospinal tract, is illustrated in Figure 12.35b.

Overall, the **reticulospinal** and **vestibulospinal tracts** maintain balance by varying the tone of postural muscles (Table 12.3). The **rubrospinal tracts** control flexor muscles, whereas the **tectospinal tracts** and the *superior colliculi* mediate head movements in response to visual stimuli.

12

CHECK YOUR UNDERSTANDING

23. What functional type of neuron is derived from the alar plate? From the basal plate?
24. What is the explanation for the cervical and lumber enlargements of the spinal cord?
25. Where are the cell bodies of the first-, second-, and third-order sensory neurons in the spinothalamic pathway located?

For answers, see Appendix G.

Spinal Cord Trauma and Disorders

▶ Distinguish between flaccid and spastic paralysis, and between paralysis and paresthesia.

Spinal Cord Trauma

The spinal cord is elastic, stretching with every turn of the head or bend of the trunk, but it is exquisitely sensitive to direct pressure. Any localized damage to the spinal cord or its roots leads to some functional loss, either **paralysis** (loss of motor function) or **paresthesias** (par″es-the′ze-ahz) (sensory loss). Severe damage to ventral root or ventral horn cells results in a **flaccid paralysis** (flak′sid) of the skeletal muscles served. Nerve impulses do not reach these muscles, which consequently cannot move either voluntarily or involuntarily. Without stimulation, the muscles atrophy.

When only the upper motor neurons of the primary motor cortex are damaged, **spastic paralysis** occurs. In this case, the spinal motor neurons remain intact and the muscles continue to be stimulated irregularly by spinal reflex activity. As a result, the muscles remain healthy longer, but their movements are no longer subject to voluntary control. In many such cases, the muscles become permanently shortened.

Transection (cross sectioning) of the spinal cord at any level results in total motor and sensory loss in body regions inferior to the site of damage. If the transection occurs between T_1 and L_1, both lower limbs are affected, resulting in **paraplegia** (par″ah-ple′je-ah; *para* = beside, *plegia* = a blow). If the injury occurs in the cervical region, all four limbs are affected and the result is **quadriplegia**. *Hemiplegia*, paralysis of one side of the body, usually reflects brain injury rather than spinal cord injury.

Anyone with traumatic spinal cord injury must be watched for symptoms of **spinal shock**, a transient period of functional loss that follows the injury. Spinal shock results in immediate depression of all reflex activity caudal to the lesion site. Bowel and bladder reflexes stop, blood pressure falls, and all muscles (somatic and visceral alike) below the injury are paralyzed and insensitive. Neural function usually returns within a few hours following injury. If function does not resume within 48 hours, paralysis is permanent in most cases.

Poliomyelitis

Poliomyelitis (po″le-o-mi″ĕ-li′tis; *polio* = gray matter; *myelitis* = inflammation of the spinal cord) results from destruction of ventral horn motor neurons by the poliovirus. Early symptoms include fever, headache, muscle pain and weakness, and loss of certain somatic reflexes. Later, paralysis develops and the muscles served atrophy. The victim may die from paralysis of the respiratory muscles or from cardiac arrest if neurons in the medulla oblongata are destroyed.

In most cases, the poliovirus enters the body in feces-contaminated water (such as might occur in a public swimming pool), and the incidence of the disease has traditionally been highest in children and during the summer months. Fortunately, vaccines have nearly eliminated this disease and a global effort is ongoing to eradicate it completely from the world.

However, many of the "recovered" survivors of the great polio epidemic of the late 1940s and 1950s have begun to experience extreme lethargy, sharp burning pains in their muscles, and progressive muscle weakness and atrophy. These disturbing symptoms are referred to as **postpolio syndrome**. The cause of postpolio syndrome is not known, but a likely explanation is that its victims, like the rest of us, continue to lose neurons throughout life. While a healthy nervous system can recruit nearby neurons to compensate for the losses, polio survivors have already drawn on that "pool" and have few neurons left to take over. Ironically, those who worked hardest to overcome their disease are its newest victims.

Amyotrophic Lateral Sclerosis

Amyotrophic lateral sclerosis (ALS) (ah-mi″o-trof′ik), also called Lou Gehrig's disease, is a devastating neuromuscular condition that involves progressive destruction of ventral horn motor neurons and fibers of the pyramidal tracts. As the disease progresses, the sufferer loses the ability to speak, swallow, and breathe. Death typically occurs within five years. For 90% of cases, the cause of ALS is unknown, although ALS patients appear to have excess extracellular glutamate. Researchers now believe that motor neurons are killed as a result of glutamate excitotoxicity, attack by the immune system, or possibly a combination of the two. Riluzole, the only advance in pharmacological treatment of ALS in 50 years, prolongs life, apparently by inhibiting glutamate release.

Diagnostic Procedures for Assessing CNS Dysfunction

- List and explain several techniques used to diagnose brain disorders.

Anyone who has had a routine physical examination is familiar with the reflex tests done to assess neural function. A tap with a reflex hammer stretches your quadriceps tendon and your anterior thigh muscles contract, which results in the knee-jerk response. This response shows that the spinal cord and upper brain centers are functioning normally. Abnormal reflex test responses may indicate such serious disorders as intracranial hemorrhage, multiple sclerosis, or hydrocephalus and suggest that more sophisticated neurological tests are needed to identify the problem.

New imaging techniques have revolutionized the diagnosis of brain lesions (see *A Closer Look*, pp. 18–19). Together the various *CT* and *MRI scanning techniques* allow most tumors, intracranial lesions, multiple sclerosis plaques, and areas of dead brain tissue (infarcts) to be identified quickly. PET scans can localize brain lesions that generate seizures (epileptic tissue) and diagnose Alzheimer's disease.

Take, for example, a patient arriving in emergency with a stroke. A race against time to save the affected area of the patient's brain begins. The first step is to determine if the stroke is due to a clot or a bleed by imaging the brain, most often with CT. If the stroke is due to a clot, then the clot-busting drug tPA can be used, but only within the first hours. tPA is usually given intravenously, but a longer time window can be obtained if tPA is applied directly to the clot using a catheter guided into position. In order to visualize the location of the catheter with respect to the clot, dye is injected to make arteries stand out in an X ray, a procedure called *cerebral angiography.*

Cerebral angiography is also used for patients who have had a warning stroke or TIA. The carotid arteries of the neck feed most of the cerebral vessels and often become narrowed with age, which can lead to strokes. Cheaper and less invasive than angiography, ultrasound can be used to quickly examine the carotid arteries and even measure the flow of blood through them.

CHECK YOUR UNDERSTANDING

26. Roy was tackled while playing football. After hitting the ground, he was unable to move his lower limbs. What is this condition called? What level of his spinal cord do you think was injured (cervical, thoracic, lumbar, or sacral)? Is this a permanent injury? What diagnostic procedures might be helpful?

For answers, see Appendix G.

Developmental Aspects of the Central Nervous System

- Indicate several maternal factors that can impair development of the nervous system in an embryo.
- Explain the effects of aging on the brain.

Induced by several organizer centers and established during the first month of development, the brain and spinal cord continue to grow and mature throughout the prenatal period. During development, gender-specific areas appear in both the brain and the spinal cord. For example, certain hypothalamic nuclei concerned with regulating typical male sexual behavior and clusters of neurons in the spinal cord that serve the external genitals are much larger in males. Females have a larger corpus callosum, and gender differences are seen in the language and auditory areas of the cerebral cortex. The key to this CNS gender-specific development is whether or not testosterone is being secreted by the fetus. If it is, the male pattern develops.

Maternal exposure to radiation, various drugs (alcohol, opiates, and others), and infections can harm a developing infant's nervous system, particularly during the initial formative stages. For example, rubella (German measles) often leads to deafness and other types of CNS damage in the newborn. Smoking decreases the amount of oxygen in the blood, and lack of oxygen for even a few minutes causes death of neurons. This means that a smoking mother may be sentencing her infant to some degree of brain damage.

HOMEOSTATIC IMBALANCE

In difficult deliveries, a temporary lack of oxygen may lead to **cerebral palsy**, but any of the factors listed above may also be a cause. Cerebral palsy is a neuromuscular disability in which the voluntary muscles are poorly controlled or paralyzed as a result of brain damage. In addition to spasticity, speech difficulties, and other motor impairments, about half of cerebral palsy victims have seizures, half are mentally retarded, and about a third have some degree of deafness. Visual impairments are also common. Cerebral palsy does not get worse over time, but its deficits are irreversible. It is the largest single cause of physical disability in children, affecting three out of every 1000 births.

The CNS is plagued by a number of other congenital malformations triggered by genetic or environmental factors during early development. The most serious are congenital hydrocephalus (discussed previously), anencephaly, and spina bifida.

In **anencephaly** (an″en-sef′ah-le; "without brain") the cerebrum and part of the brain stem never develop, because the neural folds fail to fuse rostrally. The child is totally vegetative, unable to see, hear, or process sensory inputs. Muscles are flaccid, and no voluntary movement is possible. Mental life as we know it does not exist. Mercifully, death occurs soon after birth.

Spina bifida (spi′nah bif′ĭ-dah; "forked spine") results from incomplete formation of the vertebral arches and typically involves the lumbosacral region. The technical definition is that laminae and spinous processes are missing on at least one

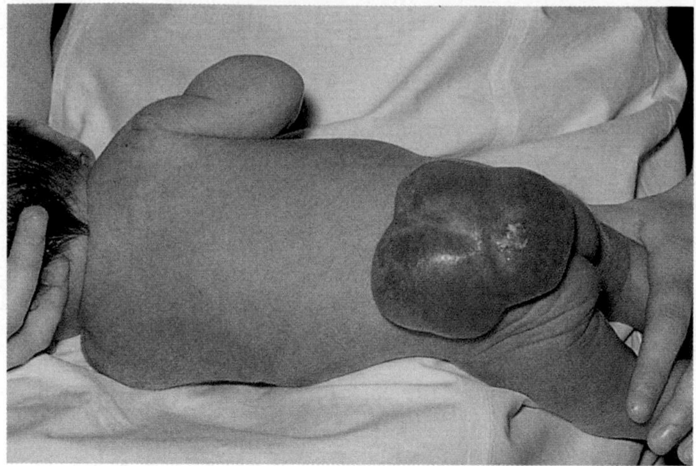

Figure 12.36 Newborn with a lumbar myelomeningocele.

vertebra. If the condition is severe, neural deficits occur as well. *Spina bifida occulta*, the least serious type, involves one or only a few vertebrae and causes no neural problems. Other than a small dimple or tuft of hair over the site of nonfusion, it has no external manifestations. In the more common and severe form, *spina bifida cystica*, a saclike cyst protrudes dorsally from the child's spine. The cyst may contain meninges and cerebrospinal fluid [a *meningocele* (mĕ-ning′go-sēl)], or even portions of the spinal cord and spinal nerve roots (a *myelomeningocele*, "spinal cord in a meningeal sac," shown in **Figure 12.36**). The larger the cyst and the more neural structures it contains, the greater the neurological impairment. In the worst case, where the inferior spinal cord is functionless, bowel incontinence, bladder muscle paralysis (which predisposes the infant to urinary tract infection and renal failure), and lower limb paralysis occur. Problems with infection are continual because the cyst wall is thin and porous and tends to rupture or leak. Spina bifida cystica is accompanied by hydrocephalus in 90% of all cases.

In the past, up to 70% of cases of spina bifida were caused by inadequate amounts of the B vitamin folic acid in the maternal diet. Recently, the number of spina bifida cases has dropped significantly in countries (such as the U.S.) where mandatory supplementation of folic acid in bread, flour, and pasta products has been introduced. ■

One of the last CNS areas to mature is the hypothalamus. Since the hypothalamus contains body temperature regulatory centers, premature babies have problems controlling their loss of body heat and must be kept in temperature-controlled environments. PET scans have revealed that the thalamus and somatosensory cortex are active in a 5-day-old baby, but that the visual cortex is not. This explains why infants of this age respond to touch but have poor vision. By 11 weeks, more of the cortex is active, and the baby can reach for a rattle. By 8 months, the cortex is very active and the child can think about what he or she sees. Growth and maturation of the nervous system continue throughout childhood and largely reflect progressive myelination. As described in Chapter 9, neuromuscular coordination progresses in a superior-to-inferior direction and in a proximal-to-distal direction, and we know that myelination also occurs in this sequence.

The brain reaches its maximum weight in the young adult. Over the next 60 years or so, neurons are damaged and die, and brain weight and volume steadily decline. However, the number of neurons lost over the decades is normally only a small percentage of the total, and the remaining neurons can change their synaptic connections, providing for continued learning throughout life.

Although age brings some cognitive declines in spatial ability, speed of perception, decision making, reaction time, and working memory, these losses are not significant in the *healthy individual* until after the seventh decade. Then the brain becomes increasingly fragile, presumably due to less efficient calcium clearance in aging neurons (as we have seen, elevated Ca^{2+} levels are neurotoxic), and a fairly rapid decline occurs in some, but not all, of these abilities. Ability to build on experience, mathematical skills, and verbal fluency do not decline with age, and many people continue to enjoy intellectual lives and to work at mentally demanding tasks their entire life. Fewer than 5% of people over 65 demonstrate true dementia. Sadly, many cases of "reversible dementia" are caused by prescription drug effects, low blood pressure, poor nutrition, hormone imbalances, depression, and/or dehydration that go undiagnosed. The best way to maintain one's mental abilities in old age may be to seek regular medical checkups throughout life.

Although eventual shrinking of the brain is normal and accelerates in old age, alcoholics and professional boxers hasten the process. Whether a boxer wins the match or not, the likelihood of brain damage and atrophy increases with every blow. Everyone recognizes that alcohol profoundly affects both the mind and the body. CT scans of alcoholics reveal a reduction in brain size and density appearing at a fairly early age. Both boxers and alcoholics exhibit signs of mental deterioration unrelated to the aging process.

CHECK YOUR UNDERSTANDING

27. Premature babies have problems controlling their body temperature. Why?

28. List several causes of reversible dementia in the elderly.

For answers, see Appendix G.

The human cerebral hemispheres—our "thinking caps"—are awesome in their complexity. But no less amazing are the brain regions that oversee our subconscious, autonomic body functions—the diencephalon and brain stem—particularly when you consider their relatively insignificant size. The spinal cord, which acts as a reflex center and a communication link between the brain and body periphery, is equally important to body homeostasis.

We have introduced a good deal of new terminology in this chapter, and much of it will come up again in one or more of the remaining nervous system chapters. Chapter 13, your next challenge, considers the structures of the peripheral nervous system that work hand in hand with the CNS to keep it informed and to deliver its orders to the effectors of the body.

RELATED CLINICAL TERMS

Autism A complex developmental neurological disorder that typically appears in the first three years of life and is characterized by difficulty in communicating, forming relationships with others, and responding appropriately to the environment. No single cause has been determined although a number of genes are thought to play a role and structural brain abnormalities are present. Occurs in about two per 1000 people, and early behavioral intervention is beneficial.

Cordotomy (kor-dot′o-me) A procedure in which a tract in the spinal cord is severed surgically; usually done to relieve unremitting pain.

Encephalopathy (en-sef″ah-lop′ah-the; *enceph* = brain; *path* = disease) Any disease or disorder of the brain.

Functional brain disorders Psychological disorders for which no structural cause can be found; include neuroses and psychoses (see entries in this list).

Hypersomnia (*hyper* = excess; *somnus* = sleep) A condition in which affected individuals sleep as much as 15 hours daily.

Microcephaly (mi″kro-sef′ah-le; *micro* = small) Congenital condition involving the formation of a small brain, as evidenced by reduced skull size; most microcephalic children are mentally retarded.

Myelitis (mi″ĕ-li′tis; *myel* = spinal cord; *itis* = inflammation) Inflammation of the spinal cord.

Myelogram (*gram* = recording) X ray of the spinal cord after injection of a contrast medium.

Myoclonus (mi″o-klo′nus; *myo* = muscle; *clon* = violent motion, tumult) Sudden contraction of a muscle or muscle part, usually involving muscles of the limbs. Myoclonal jerks can occur in normal individuals as they are falling asleep; others may be due to diseases of the reticular formation or cerebellum.

Neuroses (nu-ro′sēs) A less debilitating class of mental illness; examples include severe anxiety (panic attacks), phobias (irrational fears), and obsessive-compulsive behaviors (e.g., washing one's hands every few minutes); the affected individual, however, retains contact with reality.

Pallidectomy ("cutting the pallidus") Surgically lesioning part of the globus pallidus of the basal nuclei to relieve some symptoms of Parkinson's disease.

Psychoses (si-ko′sēs) A class of severe mental illness in which affected individuals lose touch with reality and exhibit bizarre behaviors; the *legal* word for psychotic behavior is *insanity*. Psychoses include *schizophrenia* (skit-so-fre′ne-ah), *bipolar disorder*, and some forms of *depression*.

CHAPTER SUMMARY

12

The Brain (pp. 430–453)

1. The brain provides for voluntary movements, interpretation and integration of sensation, consciousness, and cognitive function.

Embryonic Development (pp. 430–431)

2. The brain develops from the rostral portion of the embryonic neural tube.
3. Early brain development yields the three primary brain vesicles: the prosencephalon (cerebral hemispheres and diencephalon), mesencephalon (midbrain), and rhombencephalon (pons, medulla, and cerebellum).
4. Cephalization results in the envelopment of the diencephalon and superior brain stem by the cerebral hemispheres.

Regions and Organization (p. 431)

5. In a widely used system, the adult brain is divided into the cerebral hemispheres, diencephalon, brain stem, and cerebellum.
6. The cerebral hemispheres and cerebellum have gray matter nuclei surrounded by white matter and an outer cortex of gray matter. The diencephalon and brain stem lack a cortex.

Ventricles (pp. 431–433)

7. The brain contains four ventricles filled with cerebrospinal fluid. The lateral ventricles are in the cerebral hemispheres; the third ventricle is in the diencephalon; the fourth ventricle is between the brain stem and the cerebellum and connects with the central canal of the spinal cord.

Cerebral Hemispheres (pp. 433–441)

8. The two cerebral hemispheres exhibit gyri, sulci, and fissures. The longitudinal fissure partially separates the hemispheres; other fissures or sulci subdivide each hemisphere into lobes.
9. Each cerebral hemisphere consists of the cerebral cortex, the cerebral white matter, and basal nuclei (ganglia).
10. Each cerebral hemisphere receives sensory impulses from, and dispatches motor impulses to, the opposite side of the body. The body is represented in an upside-down fashion in the sensory and motor cortices.
11. Functional areas of the cerebral cortex include (1) motor areas: primary motor and premotor cortices of the frontal lobe, the frontal eye field, and Broca's area in the frontal lobe of one hemisphere (usually the left); (2) sensory areas: primary somatosensory cortex and somatosensory association cortex in the parietal lobe; visual areas in the occipital lobe; olfactory and auditory areas in the temporal lobe; gustatory, visceral, and vestibular areas in the insula; (3) association areas: anterior association area in the frontal lobe, and posterior and limbic association areas spanning several lobes.
12. The cerebral hemispheres show lateralization of cortical function. In most people, the left hemisphere is dominant (i.e., specialized for language and mathematical skills); the right hemisphere is more concerned with visual-spatial skills and creative endeavors.
13. Fiber tracts of the cerebral white matter include commissures, association fibers, and projection fibers.
14. The paired basal nuclei (also called basal ganglia) include the lentiform nucleus (globus pallidus and putamen) and caudate nucleus. The basal nuclei are subcortical nuclei that help control muscular movements. Functionally they are closely associated with the substantia nigra of the midbrain.

Diencephalon (pp. 441–445)

15. The diencephalon includes the thalamus, hypothalamus, and epithalamus and encloses the third ventricle.

16. The thalamus is the major relay station for (1) sensory impulses ascending to the sensory cortex, (2) inputs from subcortical motor nuclei and the cerebellum traveling to the cerebral motor cortex, and (3) impulses traveling to association cortices from lower centers.
17. The hypothalamus is an important autonomic nervous system control center and a pivotal part of the limbic system. It maintains water balance and regulates thirst, eating behavior, gastrointestinal activity, body temperature, and the activity of the anterior pituitary gland.
18. The epithalamus includes the pineal gland, which secretes the hormone melatonin.

Brain Stem (pp. 445–450)

19. The brain stem includes the midbrain, pons, and medulla oblongata.
20. The midbrain contains the corpora quadrigemina (visual and auditory reflex centers), the red nucleus (subcortical motor centers), and the substantia nigra. The periaqueductal gray matter is involved in pain suppression and contains the motor nuclei of cranial nerves III and IV. The cerebral peduncles on its ventral face house the pyramidal fiber tracts. The midbrain surrounds the cerebral aqueduct.
21. The pons is mainly a conduction area. Its nuclei contribute to regulation of respiration and cranial nerves V–VII.
22. The pyramids (descending corticospinal tracts) form the ventral face of the medulla oblongata; these fibers cross over (decussation of the pyramids) before entering the spinal cord. Important nuclei in the medulla regulate respiratory rhythm, heart rate, and blood pressure and serve cranial nerves VIII–XII. The olivary nuclei and cough, sneezing, swallowing, and vomiting centers are also in the medulla.

Cerebellum (pp. 450–451)

23. The cerebellum consists of two hemispheres, marked by convolutions and separated by the vermis. It is connected to the brain stem by superior, middle, and inferior peduncles.
24. The cerebellum processes and interprets impulses from the motor cortex and sensory pathways and coordinates motor activity so that smooth, well-timed movements occur. It also plays a poorly understood role in cognition.

Functional Brain Systems (pp. 451–453)

25. The limbic system consists of numerous structures that encircle the brain stem. It is the "emotional-visceral brain." It also plays a role in memory.
26. The reticular formation is a diffuse network of neurons and nuclei spanning the length of the brain stem. It maintains the alert state of the cerebral cortex (RAS), and its motor nuclei serve both somatic and visceral motor activities.

Higher Mental Functions (pp. 453–460)

Brain Wave Patterns and the EEG (pp. 453–455)

1. Patterns of electrical activity of the brain are called brain waves; a record of this activity is an electroencephalogram (EEG). Brain wave patterns, identified by their frequencies, include alpha, beta, theta, and delta waves.
2. Epilepsy results from abnormal electrical activity of brain neurons. Involuntary muscle contractions and sensory auras are typical during such seizures.

Consciousnes (p. 455)

3. Consciousness is described clinically on a continuum from alertness to drowsiness to stupor and finally to coma.
4. Human consciousness is thought to involve holistic information processing, which is (1) not localizable, (2) superimposed on other types of neural activity, and (3) totally interconnected.
5. Fainting (syncope) is a temporary loss of consciousness that usually reflects inadequate blood delivery to the brain. Coma is loss of consciousness in which the victim is unresponsive to stimuli.

Sleep and Sleep-Wake Cycles (pp. 455–457)

6. Sleep is a state of partial consciousness from which a person can be aroused by stimulation. The two major types of sleep are non–rapid eye movement (NREM) sleep and rapid eye movement (REM) sleep.
7. During stages 1–4 of NREM sleep, brain wave frequency decreases and amplitude increases until delta wave sleep (stage 4) is achieved. REM sleep is indicated by a return to a stage 1 EEG. During REM, the eyes move rapidly under the lids. NREM and REM sleep alternate throughout the night.
8. Slow-wave sleep (stages 3 and 4 of NREM) appears to be restorative. REM sleep is important for emotional stability.
9. REM occupies half of an infant's sleep time and then declines to about 25% of sleep time by the age of 10 years. Time spent in slow-wave sleep declines steadily throughout life.
10. Narcolepsy is involuntary lapses into REM sleep that occur without warning during waking periods. Insomnia is a chronic inability to obtain the amount or quality of sleep needed to function adequately. Sleep apnea is a temporary cessation of breathing during sleep, causing hypoxia.

Language (p. 457)

11. In most people language is controlled by the left hemisphere. The language implementation system, which includes Broca's and Wernicke's areas and the basal nuclei, analyzes incoming and produces outgoing language. The opposite hemisphere deals with the emotional content of language.

Memory (pp. 457–460)

12. Memory is the storage and retrieval of information. It is essential for learning and is part of consciousness.
13. Memory storage has two stages: short-term memory (STM) and long-term memory (LTM). Transfer of information from STM to LTM takes minutes to hours, but more time is required for LTM consolidation.
14. Declarative memory is the ability to learn and consciously remember information. Procedural memory is the learning of motor skills, which are then performed without conscious thought.
15. Declarative memory appears to involve the medial temporal lobe (hippocampus and surrounding temporal cortical areas), thalamus, basal forebrain, and prefrontal cortex. Skill memory (a type of procedural memory) relies on the basal nuclei.
16. The nature of memory formation at the molecular level is not fully known, but NMDA receptors (essentially calcium channels), activated by depolarization and glutamate binding, play a major role in long-term potentiation (LTP). The calcium influx that follows NMDA receptor activation mobilizes enzymes that mediate events necessary for memory formation.

Protection of the Brain (pp. 460–466)

1. The delicate brain is protected by bone, meninges, cerebrospinal fluid, and the blood-brain barrier.

Meninges (pp. 461–463)

2. The meninges from superficial to deep are the dura mater, the arachnoid mater, and the pia mater. They enclose the brain and spinal cord and their blood vessels. Inward folds of the inner layer of the dura mater secure the brain to the skull.

Cerebrospinal Fluid (p. 463)

3. Cerebrospinal fluid (CSF), formed by the choroid plexuses from blood plasma, circulates through the ventricles and into the subarachnoid space. It returns to the dural venous sinuses via the arachnoid villi. CSF supports and cushions the brain and cord and helps to nourish them.

Blood-Brain Barrier (pp. 463–464)

4. The blood-brain barrier reflects the relative impermeability of the epithelium of capillaries of the brain. It allows water, respiratory gases, essential nutrients, and fat-soluble molecules to enter the neural tissue, but prevents entry of other, water-soluble, potentially harmful substances.

Homeostatic Imbalances of the Brain (pp. 464–466)

5. Head trauma may cause brain injuries called concussions or, in severe cases, contusions (bruising). When the brain stem is affected, unconsciousness (temporary or permanent) occurs. Trauma-induced brain injuries may be aggravated by intracranial hemorrhage or cerebral edema, both of which compress brain tissue.
6. Cerebrovascular accidents (strokes) result when blood circulation to brain neurons is impaired and brain tissue dies. The result may be hemiplegia, sensory deficits, or speech impairment.
7. Alzheimer's disease is a degenerative brain disease in which beta-amyloid peptide deposits and neurofibrillary tangles appear. Marked by a deficit of ACh, it results in slow, progressive loss of memory and motor control and increasing dementia.
8. Parkinson's disease and Huntington's disease are neurodegenerative disorders of the basal nuclei. Both involve abnormalities of the neurotransmitter dopamine (too little or too much secreted) and are characterized by abnormal movements.

The Spinal Cord (pp. 466–477)

Embryonic Development (p. 466)

1. The spinal cord develops from the neural tube. Its gray matter forms from the alar and basal plates. Fiber tracts form the outer white matter. The neural crest forms the sensory (dorsal root) ganglia.

Gross Anatomy and Protection (pp. 466–468)

2. The spinal cord, a two-way impulse conduction pathway and a reflex center, resides within the vertebral column and is protected by meninges and cerebrospinal fluid. It extends from the foramen magnum to the end of the first lumbar vertebra.
3. Thirty-one pairs of spinal nerves issue from the cord. The cord is enlarged in the cervical and lumbar regions, where spinal nerves serving the limbs arise.

Cross-Sectional Anatomy (pp. 468–476)

4. The central gray matter of the cord is H shaped. Ventral horns mainly contain somatic motor neurons. Lateral horns contain visceral (autonomic) motor neurons. Dorsal horns contain interneurons.
5. Axons of neurons of the lateral and ventral horns emerge in common from the cord via the ventral roots. Axons of sensory neurons (with cell bodies located in the dorsal root ganglion) enter the dorsal aspect of the cord and form the dorsal roots. The ventral and dorsal roots combine to form the spinal nerves.
6. Each side of the white matter of the cord has dorsal, lateral, and ventral columns (funiculi), and each funiculus contains a number of ascending and descending tracts. All tracts are paired and most decussate.
7. Ascending (sensory) tracts include the fasciculi gracilis and cuneatus, spinothalamic tracts, and spinocerebellar tracts.
8. The dorsal column–medial lemniscal pathway consists of the dorsal white column (fasciculus cuneatus, fasciculus gracilis) and the medial lemniscus, which are concerned with straight-through, precise transmission of one or a few related sensory modalities. The anterolateral pathways (mostly the spinothalamic tracts) are multimodal pathways that permit brain stem processing of ascending impulses. The spinocerebellar tracts, which terminate in the cerebellum, serve muscle sense, not conscious sensory perception.
9. Descending tracts include the pyramidal tracts (ventral and lateral corticospinal tracts) and a number of motor tracts originating from subcortical motor nuclei. These descending fibers issue from the brain stem motor areas [indirect (extrapyramidal) system] and cortical motor areas [direct (pyramidal) system].

Spinal Cord Trauma and Disorders (pp. 476–477)

10. Injury to the ventral horn neurons or the ventral roots results in flaccid paralysis. (Injury to the upper motor neurons in the brain results in spastic paralysis.) If the dorsal roots or sensory tracts are damaged, paresthesias occur.
11. Poliomyelitis results from inflammation and destruction of the ventral horn neurons by the poliovirus. Paralysis and muscle atrophy ensue.
12. Amyotrophic lateral sclerosis (ALS) results from destruction of the ventral horn neurons and the pyramidal tracts. The victim eventually loses the ability to swallow, speak, and breathe. Death generally occurs within five years.

Diagnostic Procedures for Assessing CNS Dysfunction (p. 477)

1. Diagnostic procedures used to assess neurological condition and function range from routine reflex testing to sophisticated techniques such as cerebral angiography, CT scans, MRI scans, and PET scans.

Developmental Aspects of the Central Nervous System (pp. 477–478)

1. Maternal and environmental factors may impair embryonic brain development, and oxygen deprivation destroys brain cells. Severe congenital brain diseases include cerebral palsy, anencephaly, hydrocephalus, and spina bifida.
2. Premature babies have trouble regulating body temperature because the hypothalamus is one of the last brain areas to mature prenatally.
3. Development of motor control indicates progressive myelination and maturation of a child's nervous system.
4. Brain growth ends in young adulthood. Neurons die throughout life and most are not replaced; brain weight and volume decline with age.
5. Healthy elders maintain nearly optimal intellectual function. Disease—particularly cardiovascular disease—is the major cause of declining mental function with age.

12

REVIEW QUESTIONS

Multiple Choice/Matching

(Some questions have more than one correct answer. Select the best answer or answers from the choices given.)

1. The primary motor cortex, Broca's area, and the premotor cortex are located in which lobe? **(a)** frontal, **(b)** parietal, **(c)** temporal, **(d)** occipital.
2. The innermost layer of the meninges, delicate and closely apposed to the brain tissue, is the **(a)** dura mater, **(b)** corpus callosum, **(c)** arachnoid, **(d)** pia mater.
3. Cerebrospinal fluid is formed by **(a)** arachnoid villi, **(b)** the dura mater, **(c)** choroid plexuses, **(d)** all of these.
4. A patient has suffered a cerebral hemorrhage that has caused dysfunction of the precentral gyrus of his right cerebral cortex. As a result, **(a)** he cannot voluntarily move his left arm or leg, **(b)** he feels no sensation on the left side of his body, **(c)** he feels no sensation on his right side.
5. Choose the proper term from the key to respond to the statements describing various brain areas.

   Key:

(a) cerebellum	(d) striatum	(g) midbrain
(b) corpora quadrigemina	(e) hypothalamus	(h) pons
(c) corpus callosum	(f) medulla	(i) thalamus

   ______ (1) basal nuclei involved in fine control of motor activities
   ______ (2) region where there is a gross crossover of fibers of descending pyramidal tracts
   ______ (3) control of temperature, autonomic nervous system reflexes, hunger, and water balance
   ______ (4) houses the substantia nigra and cerebral aqueduct
   ______ (5) relay stations for visual and auditory stimuli input; found in midbrain
   ______ (6) houses vital centers for control of the heart, respiration, and blood pressure
   ______ (7) brain area through which all the sensory input is relayed to get to the cerebral cortex
   ______ (8) brain area most concerned with equilibrium, body posture, and coordination of motor activity
6. Which of the following tracts convey vibration and other specific sensations that can be precisely localized? **(a)** pyramidal tract, **(b)** medial lemniscus, **(c)** lateral spinothalamic tract, **(d)** reticulospinal tract.
7. Destruction of the ventral horn cells of the spinal cord results in loss of **(a)** integrating impulses, **(b)** sensory impulses, **(c)** voluntary motor impulses, **(d)** all of these.
8. Fiber tracts that allow neurons within the same cerebral hemisphere to communicate are **(a)** association tracts, **(b)** commissures, **(c)** projection tracts.
9. A number of brain structures are listed below. If an area is primarily gray matter, write **a** in the answer blank; if mostly white matter, respond with **b**.

   ______ (1) cerebral cortex
   ______ (2) corpus callosum and corona radiata
   ______ (3) red nucleus
   ______ (4) medial and lateral nuclear groups
   ______ (5) medial lemniscal tract
   ______ (6) cranial nerve nuclei
   ______ (7) spinothalamic tract
   ______ (8) fornix
   ______ (9) cingulate and precentral gyri
10. A professor unexpectedly blew a loud horn in his anatomy and physiology class. The students looked up, startled. The reflexive movements of their eyes were mediated by the **(a)** cerebral cortex, **(b)** inferior olives, **(c)** raphe nuclei, **(d)** superior colliculi, **(e)** nucleus gracilis.
11. Identify the stage of sleep described by using choices from the key. (Note that responses a–d refer to NREM sleep.)

    Key: **(a)** stage 1 **(b)** stage 2 **(c)** stage 3 **(d)** stage 4 **(e)** REM

    ______ (1) the stage when blood pressure and heart rate reach their lowest levels
    ______ (2) indicated by movement of the eyes under the lids; dreaming occurs
    ______ (3) when nightmares are likely to occur
    ______ (4) when the sleeper is very easily awakened; EEG shows alpha waves
12. All of the following descriptions refer to dorsal column–medial lemniscal ascending pathways except one: **(a)** they include the fasciculus gracilis and fasciculus cuneatus; **(b)** they include a chain of three neurons; **(c)** their connections are diffuse and polymodal; **(d)** they are concerned with precise transmission of one or a few related sensory modalities.

Short Answer Essay Questions

13. Make a diagram showing the three primary (embryonic) brain vesicles. Name each and then use clinical terminology to name the resulting adult brain regions.
14. (a) What is the advantage of having a cerebrum that is highly convoluted? (b) What term is used to indicate its grooves? Its outward folds? (c) What groove divides the cerebrum into two hemispheres? (d) What divides the parietal from the frontal lobe? The parietal from the temporal lobe?
15. (a) Make a rough drawing of the lateral aspect of the left cerebral hemisphere. (b) You may be thinking, "But I just can't draw!" So, name the hemisphere involved with most people's ability to draw. (c) On your drawing, locate the following areas and provide the major function of each: primary motor cortex, premotor cortex, somatosensory association cortex, primary somatosensory cortex, visual and auditory areas, prefrontal cortex, Wernicke's and Broca's areas.
16. (a) What does lateralization of cortical functioning mean? (b) Why is the term cerebral dominance a misnomer?
17. (a) What is the function of the basal nuclei? (b) Which basal nuclei form the lentiform nucleus? (c) Which arches over the diencephalon?
18. Explain how the cerebellum is physically connected to the brain stem.
19. Describe the role of the cerebellum in maintaining smooth, coordinated skeletal muscle activity.
20. (a) Where is the limbic system located? (b) What structures make up this system? (c) How is the limbic system important in behavior?
21. (a) Localize the reticular formation in the brain. (b) What does RAS mean, and what is its function?
22. What is an aura?
23. How do sleep patterns, extent of sleeping time, and amount of time spent in REM and NREM sleep change through life?
24. Compare and contrast short-term memory (STM) and long-term memory (LTM) relative to storage capacity and duration of the memory.

25. Define memory consolidation.
26. Compare and contrast declarative and procedural memory relative to the types of things remembered and the importance of conscious retrieval.
27. List four ways in which the CNS is protected.
28. (a) How is cerebrospinal fluid formed and drained? Describe its pathway within and around the brain. (b) What happens if CSF is not drained properly? Why is this consequence more harmful in adults?
29. What constitutes the blood-brain barrier?
30. A brain surgeon is about to make an incision. Name all the tissue layers that she cuts through from the skin to the brain.
31. (a) Define concussion and contusion. (b) Why does severe brain stem injury result in unconsciousness?
32. Describe the spinal cord, depicting its extent, its composition of gray and white matter, and its spinal roots.
33. How do the types of motor activity controlled by the direct (pyramidal) and indirect systems differ?
34. Describe the functional problems that would be experienced by a person in which these fiber tracts have been cut: (a) lateral spinothalamic, (b) ventral and dorsal spinocerebellar, (c) tectospinal.
35. Differentiate between spastic and flaccid paralysis.
36. How do the conditions paraplegia, hemiplegia, and quadriplegia differ?
37. (a) Define cerebrovascular accident or CVA. (b) Describe its possible causes and consequences.
38. (a) What factors account for brain growth after birth? (b) List some structural brain changes observed with aging.

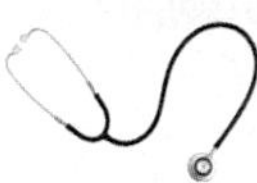

Critical Thinking and Clinical Application Questions

1. A 10-month-old infant has an enlarging head circumference and delayed overall development. She has a bulging anterior fontanelle and her CSF pressure is elevated. Based on these findings, answer the following questions: (a) What are the possible cause(s) of an enlarged head? (b) What tests might be helpful in obtaining information about this infant's problem? (c) Assuming the tests conducted showed the cerebral aqueduct to be constricted, which ventricles or CSF-containing areas would you expect to be enlarged? Which would likely not be visible? Respond to the same questions based on a finding of obstructed arachnoid villi.
2. Mrs. Jones has had a progressive decline in her mental capabilities in the last five or six years. At first her family attributed her occasional memory lapses, confusion, and agitation to grief over her husband's death six years earlier. When examined, Mrs. Jones was aware of her cognitive problems and was shown to have an IQ score approximately 30 points less than would be predicted by her work history. A CT scan showed diffuse cerebral atrophy. The physician prescribed an acetylcholinesterase inhibitor and Mrs. Jones showed slight improvement. What is Mrs. Jones's problem? Why did the acetylcholinesterase inhibitor help?
3. Robert, a brilliant computer analyst, suffered a blow to his anterior skull from a falling rock while mountain climbing. Shortly thereafter, it was obvious to his coworkers that his behavior had undergone a dramatic change. Although previously a smart dresser, he was now unkempt. One morning, he was observed defecating into the wastebasket. His supervisor ordered Robert to report to the company's doctor immediately. What region of Robert's brain was affected by the cranial blow?
4. Mrs. Adams is ready to deliver her first baby. Unfortunately, the baby appears to have a myelomeningocele. Would a vaginal or surgical (C-section) delivery be more appropriate and why?
5. The medical chart of a 68-year-old man includes the following notes: "Slight tremor of right hand at rest; stony facial expression; difficulty in initiating movements." (a) Based on your present knowledge, what is the diagnosis? (b) What brain areas are most likely involved in this man's disorder, and what is the deficiency? (c) How is this condition currently treated?
6. Cynthia, a 16-year-old girl, was rushed to the hospital after taking a bad spill off the parallel bars. After she had a complete neurological workup, her family was told that she would be permanently paralyzed from the waist down. The neurologist then outlined for Cynthia's parents the importance of preventing complications in such cases. Common complications include urinary infection, bed sores, and muscular spasms. Using your knowledge of neuroanatomy, explain the underlying reasons for these complications.
7. Mrs. Herrera suffered a stroke two weeks ago. Initially she was unable to move her right arm and the right side of her face, but her paralysis has decreased markedly. However, she still has a great deal of difficulty speaking. Mrs. Herrera's speech consists of very few words. She struggles to produce these words and they are distorted and separated by long pauses. Her comprehension of written and spoken language is unaltered. She is fully aware of what happened to her and is tearful much of the time. Which side of the brain and which particular area of the brain was affected by Mrs. Herrera's stroke?
8. Five-year-old Amy wakes her parents up at 3 AM crying and complaining of a sore neck, a severe headache, and feeling sick to her stomach. She has a temperature of 40°C (104°F) and hides her eyes, saying that the lights are too bright. The emergency physician suspects meningitis and performs a lumbar tap. Using your knowledge of neuroanatomy, explain into which space and at what level of the vertebral column the needle will be inserted to perform this test. Which fluid is being obtained and why?

12

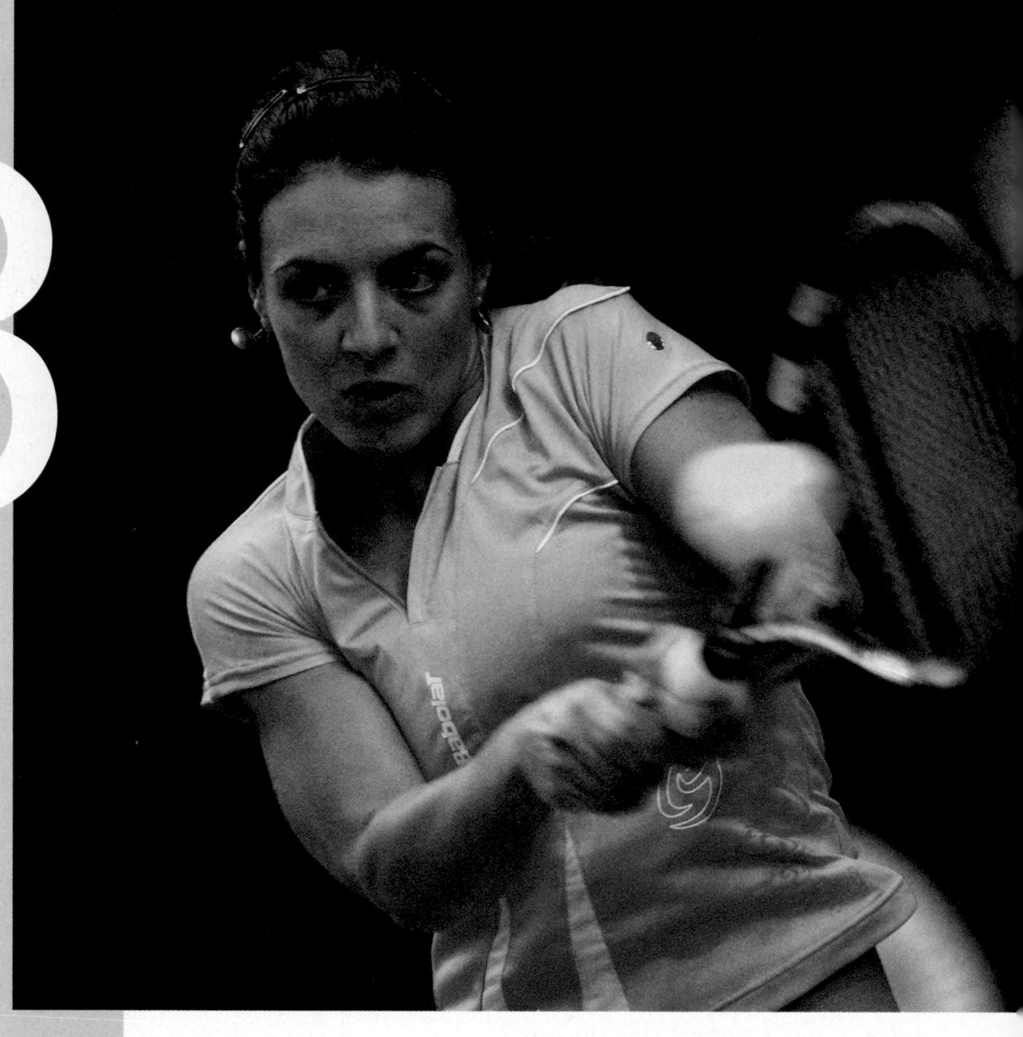

13

The Peripheral Nervous System and Reflex Activity

▶ Define peripheral nervous system and list its components.

The human brain, for all its sophistication, would be useless without its links to the outside world. In one experiment that illustrates this point, blindfolded volunteers suspended in warm water in a sensory deprivation tank (a situation that limits sensory inputs) hallucinated. One saw charging pink and purple elephants. Another heard a chorus, still others had taste hallucinations. Our very sanity depends on a continuous flow of information from the outside.

No less important to our well-being are the orders sent from the CNS to voluntary muscles and other effectors of the body, which allow us to move and to take care of our own needs. The **peripheral nervous system (PNS)** provides these links from and to the world outside our bodies. Ghostly white nerves thread through virtually every part of the body, enabling the CNS to receive information and to carry out its decisions.

The PNS includes all neural structures outside the brain and spinal cord, that is, the *sensory receptors*, peripheral *nerves* and their associated *ganglia*, and efferent *motor endings*. Its basic components are diagrammed in **Figure 13.1**. In the first portion of this chapter we deal with the functional anatomy of each PNS element. Then we consider the components of reflex arcs and some important somatic reflexes, played out almost entirely in PNS structures, that help to maintain homeostasis.

PART 1

SENSORY RECEPTORS AND SENSATION

Sensory Receptors

▶ Classify general sensory receptors by structure, stimulus detected, and body location.

Sensory receptors are specialized to respond to changes in their environment, which are called **stimuli**. Typically, activation of a sensory receptor by an adequate stimulus results in graded potentials that in turn trigger nerve impulses along the afferent PNS fibers coursing to the CNS. *Sensation* (awareness of the stimulus) and *perception* (interpretation of the meaning of the stimulus) occur in the brain. But we are getting ahead of ourselves here.

For now, let's just examine how sensory receptors are classified. Basically, there are three ways to classify sensory receptors: (1) by the type of stimulus they detect; (2) by their body location; and (3) by their structural complexity.

Classification by Stimulus Type

The receptor classes named according to the activating stimulus are easy to remember because the class name usually indicates the stimulus.

1. **Mechanoreceptors** respond to mechanical force such as touch, pressure (including blood pressure), vibration, and stretch.
2. **Thermoreceptors** are sensitive to temperature changes.
3. **Photoreceptors**, such as those of the retina of the eye, respond to light energy.
4. **Chemoreceptors** respond to chemicals in solution (molecules smelled or tasted, or changes in blood or interstitial fluid chemistry).
5. **Nociceptors** (no″se-sep′torz; *noci* = harm) respond to potentially damaging stimuli that result in pain. For example, searing heat, extreme cold, excessive pressure, and inflammatory chemicals are all interpreted as painful. These signals stimulate subtypes of thermoreceptors, mechanoreceptors, and chemoreceptors.

13

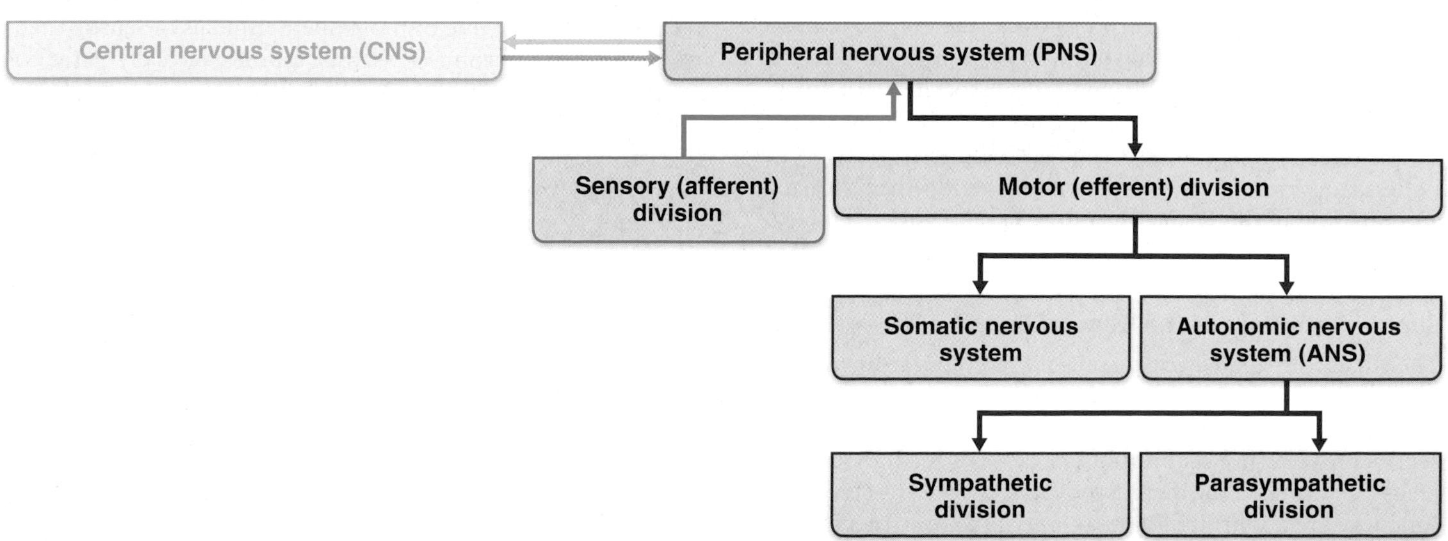

Figure 13.1 **Place of the PNS in the structural organization of the nervous system.**

Classification by Location

Three receptor classes are recognized according to either their location or the location of the activating stimulus.

1. **Exteroceptors** (ek″ster-o-sep′torz) are sensitive to stimuli arising outside the body (*extero* = outside), so most exteroceptors are near or at the body surface. They include touch, pressure, pain, and temperature receptors in the skin and most receptors of the special senses (vision, hearing, equilibrium, taste, smell).
2. **Interoceptors** (in″ter-o-sep′torz), also called *visceroceptors*, respond to stimuli within the body (*intero* = inside), such as from the internal viscera and blood vessels. They monitor a variety of stimuli, including chemical changes, tissue stretch, and temperature. Sometimes their activity causes us to feel pain, discomfort, hunger, or thirst. However, we are usually unaware of their workings.
3. **Proprioceptors** (pro″pre-o-sep′torz), like interoceptors, respond to internal stimuli; however, their location is much more restricted. Proprioceptors occur in skeletal muscles, tendons, joints, and ligaments and in connective tissue coverings of bones and muscles. (Some authorities include the equilibrium receptors of the inner ear in this class.) Proprioceptors constantly advise the brain of our body movements (*propria* = one's own) by monitoring how much the organs containing these receptors are stretched.

Classification by Structural Complexity

On the basis of overall receptor structure, there are **simple** and **complex receptors**, but the overwhelming majority are simple. The simple receptors are modified dendritic endings of sensory neurons. They are found throughout the body and monitor most types of general sensory information.

Complex receptors are actually **sense organs**, localized collections of cells (usually of many types) associated with the **special senses** (vision, hearing, equilibrium, smell, and taste). For example, the sense organ we know as the eye is composed not only of sensory neurons but also of nonneural cells that form its supporting wall, lens, and other associated structures.

Though the complex sense organs are most familiar to us, the simple sensory receptors associated with the **general senses** are no less important, and we will concentrate on their structure and function in this chapter. The special senses are the topic of Chapter 15.

Simple Receptors of the General Senses

The widely distributed general sensory receptors are involved in tactile sensation (a mix of touch, pressure, stretch, and vibration), temperature monitoring, and pain, as well as the "muscle sense" provided by proprioceptors. As you read about these receptors, notice that there is no perfect "one-receptor-one-function" relationship. Instead, one type of receptor can respond to several different kinds of stimuli. Likewise, different types of receptors can respond to similar stimuli. Anatomically, these receptors are either *unencapsulated (free) nerve endings* or *encapsulated nerve endings.* The general sensory receptors are illustrated and classified in **Table 13.1.** You may find it helpful to refer to the illustrations in this table throughout the discussion that follows.

Unencapsulated Dendritic Endings Free, or **naked, nerve endings** of sensory neurons are present nearly everywhere in the body, but they are particularly abundant in epithelia and connective tissues. Most of these sensory fibers are unmyelinated, small-diameter C fibers, and their distal endings (the sensory terminals) usually have small knoblike swellings. They respond chiefly to temperature and painful stimuli, but some respond to tissue movements caused by pressure as well. Nerve endings that respond to cold (10–40°C) are located in the superficial dermis. Those responding to heat (32–48°C) are found deeper in the dermis.

Heat or cold outside the range of thermoreceptors activates nociceptors and is perceived as painful. Nociceptors also respond to pinch and chemicals released from damaged tissue. A recently discovered receptor (*vanilloid receptor*) in the plasma membrane of nociceptive free nerve endings is an ion channel that is opened by heat, low pH, and the substance found in red peppers called capsaicin.

Another sensation mediated by free nerve endings is itch. Located in the dermis, the *itch receptor* escaped detection until 1997 because of its thin diameter. A number of chemicals—notably histamine—present at inflamed sites activate these nerve endings.

Certain free nerve endings associate with enlarged, disc-shaped epidermal cells (*tactile* or *Merkel cells*) to form **tactile (Merkel) discs**, which lie in the deeper layers of the epidermis and function as light touch receptors. **Hair follicle receptors**, free nerve endings that wrap basketlike around hair follicles, are light touch receptors that detect bending of hairs. The tickle of a mosquito landing on your skin is mediated by hair follicle receptors.

Encapsulated Dendritic Endings All **encapsulated dendritic endings** consist of one or more fiber terminals of sensory neurons enclosed in a connective tissue capsule. Virtually all the encapsulated receptors are mechanoreceptors, but they vary greatly in shape, size, and distribution in the body.

The encapsulated receptors known as **Meissner's corpuscles** (mīs′nerz kor′pus″lz) are small receptors in which a few spiraling sensory terminals are surrounded by Schwann cells and then by a thin egg-shaped connective tissue capsule. Meissner's corpuscles are found just beneath the epidermis in the dermal papillae and are especially numerous in sensitive and hairless skin areas such as the nipples, fingertips, and soles of the feet. Also called **tactile corpuscles**, they are receptors for discriminative touch, and apparently play the same role in light touch reception in hairless skin as do the hair follicle receptors in hairy skin.

Pacinian corpuscles, also called **lamellated corpuscles**, are scattered deep in the dermis, and in subcutaneous tissue underlying the skin. Although they are mechanoreceptors stimulated by deep pressure, they respond only when the pressure is first

TABLE 13.1 General Sensory Receptors Classified by Structure and Function

STRUCTURAL CLASS	ILLUSTRATION	FUNCTIONAL CLASSES ACCORDING TO LOCATION (L) AND STIMULUS TYPE (S)	BODY LOCATION
Unencapsulated			
Free nerve endings of sensory neurons	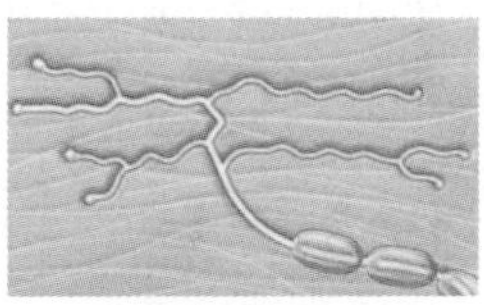	L: Exteroceptors, interoceptors, and proprioceptors S: Thermoreceptors (warm and cool), chemoreceptors (itch, pH, etc.), mechanoreceptors (pressure), nociceptors (pain, hot, cold, pinch, and chemicals)	Most body tissues; most dense in connective tissues (ligaments, tendons, dermis, joint capsules, periostea) and epithelia (epidermis, cornea, mucosae, and glands)
Modified free nerve endings: Tactile discs (Merkel discs)	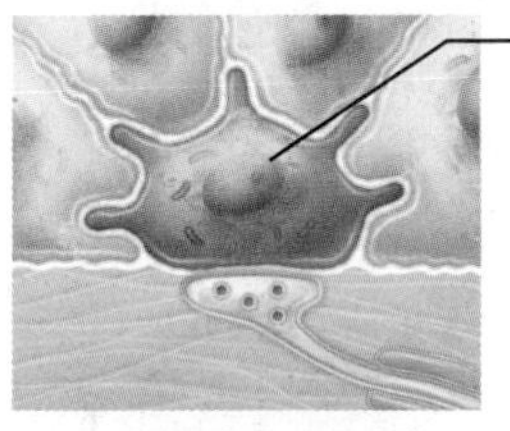	L: Exteroceptors S: Mechanoreceptors (light pressure); slowly adapting	Basal layer of epidermis
Hair follicle receptors	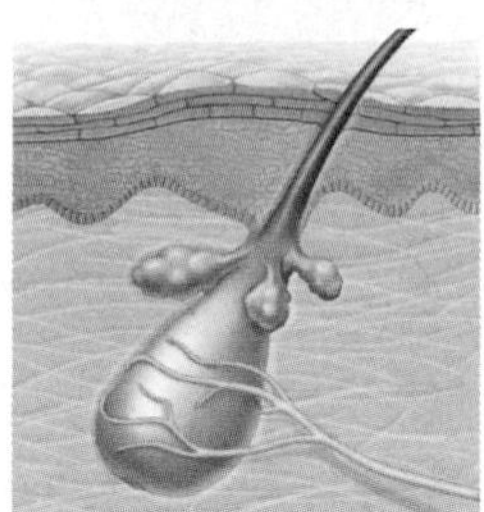	L: Exteroceptors S: Mechanoreceptors (hair deflection); rapidly adapting	In and surrounding hair follicles
Encapsulated			
Meissner's corpuscles (tactile corpuscles)	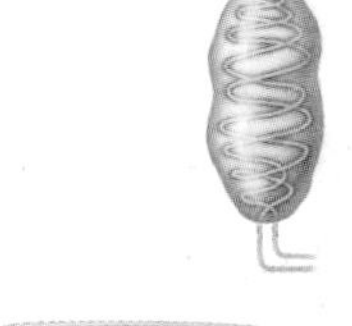	L: Exteroceptors S: Mechanoreceptors (light pressure, discriminative touch, vibration of low frequency); rapidly adapting	Dermal papillae of hairless skin, particularly nipples, external genitalia, fingertips, soles of feet, eyelids
Pacinian corpuscles (lamellated corpuscles)	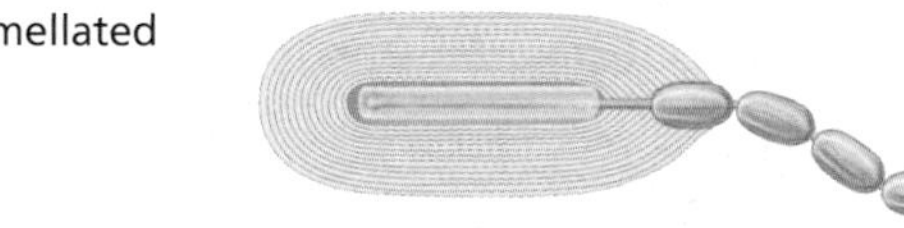	L: Exteroceptors, interoceptors, and some proprioceptors S: Mechanoreceptors (deep pressure, stretch, vibration of high frequency); rapidly adapting	Dermis and hypodermis; periostea, mesentery, tendons, ligaments, joint capsules; most abundant on fingers, soles of feet, external genitalia, nipples
Ruffini endings	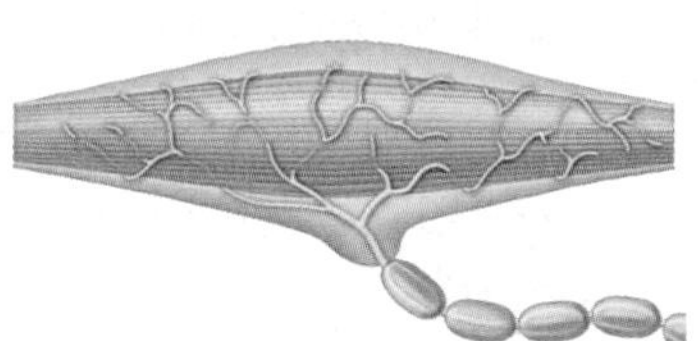	L: Exteroceptors and proprioceptors S: Mechanoreceptors (deep pressure and stretch); slowly or nonadapting	Deep in dermis, hypodermis, and joint capsules

applied, and thus are best suited to monitoring vibration (an "on/off" pressure stimulus). They are the largest corpuscular receptors. Some are over 3 mm long and half as wide and are visible to the naked eye as white, egg-shaped bodies. In section, a Pacinian corpuscle resembles a cut onion. Its single dendrite is surrounded by a capsule containing up to 60 layers of collagen fibers and flattened supporting cells.

Ruffini endings, which lie in the dermis, subcutaneous tissue, and joint capsules, contain a spray of receptor endings enclosed by a flattened capsule. They bear a striking resemblance to Golgi tendon organs (which monitor tendon stretch) and probably play a similar role in other dense connective tissues where they respond to deep and *continuous* pressure.

TABLE 13.1 General Sensory Receptors Classified by Structure and Function *(continued)*

STRUCTURAL CLASS	ILLUSTRATION	FUNCTIONAL CLASSES ACCORDING TO LOCATION (L) AND STIMULUS TYPE (S)	BODY LOCATION
Encapsulated *(continued)*			
Muscle spindles	Intrafusal fibers	L: Proprioceptors S: Mechanoreceptors (muscle stretch, length)	Skeletal muscles, particularly those of the extremities
Golgi tendon organs		L: Proprioceptors S: Mechanoreceptors (tendon stretch, tension)	Tendons
Joint kinesthetic receptors		L: Proprioceptors S: Mechanoreceptors and nociceptors	Joint capsules of synovial joints

13

Muscle spindles are fusiform (spindle-shaped) proprioceptors found throughout the perimysium of a skeletal muscle. Each muscle spindle consists of a bundle of modified skeletal muscle fibers, called *intrafusal fibers* (in″trah-fu′zal), enclosed in a connective tissue capsule, as shown in Table 13.1. They detect muscle stretch and initiate a reflex that resists the stretch. Details of muscle spindle innervation are considered later when the stretch reflex is described (see p. 515).

Golgi tendon organs are proprioceptors located in tendons, close to the skeletal muscle insertion. They consist of small bundles of tendon (collagen) fibers enclosed in a layered capsule, with sensory terminals coiling between and around the fibers. When the tendon fibers are stretched by muscle contraction, the nerve endings are activated by compression. When Golgi tendon organs are activated, the contracting muscle is inhibited, which causes it to relax.

Joint kinesthetic receptors (kin″es-thet′ik) are proprioceptors that monitor stretch in the articular capsules that enclose synovial joints. At least four receptor types (Pacinian corpuscles, Ruffini endings, free nerve endings, and receptors resembling Golgi tendon organs) contribute to this receptor category. Together these receptors provide information on joint position and motion (*kines* = movement), a sensation of which we are highly conscious.

- Close your eyes and flex and extend your fingers—you can *feel* exactly which joints are moving.

CHECK YOUR UNDERSTANDING

1. Your PNS mostly consists of nerves. What else belongs to your PNS?
2. You've cut your finger on a broken beaker in your A&P lab. Using stimulus type, location, and structural complexity, classify the sensory receptors that allow you to feel the pain.

For answers, see Appendix G.

Sensory Integration: From Sensation to Perception

- ▶ Outline the events that lead to sensation and perception.
- ▶ Describe receptor and generator potentials and sensory adaptation.
- ▶ Describe the main aspects of sensory perception.

Our survival depends not only on **sensation** (awareness of changes in the internal and external environments) but also on **perception** (conscious interpretation of those stimuli). For example, a pebble kicked up into my shoe causes the *sensation* of localized deep pressure, but my *perception* of it is an awareness of discomfort. Perception in turn determines how we will respond to stimuli. In the case of the pebble-in-my-shoe example, I'm taking off my shoe to get rid of the pesky pebble in a hurry.

General Organization of the Somatosensory System

The **somatosensory system**, or that part of the sensory system serving the body wall and limbs, receives inputs from exteroceptors, proprioceptors, and interoceptors. Consequently, it transmits information about several different sensory modalities.

As illustrated in **Figure 13.2**, three main levels of neural integration operate in the somatosensory (or any sensory) system:

① **Receptor level**: sensory receptors
② **Circuit level**: ascending pathways
③ **Perceptual level**: neuronal circuits in the cerebral cortex

Sensory input is generally relayed toward the head, but note that it is also processed along the way.

Let's examine the events that must occur at each level along the pathway.

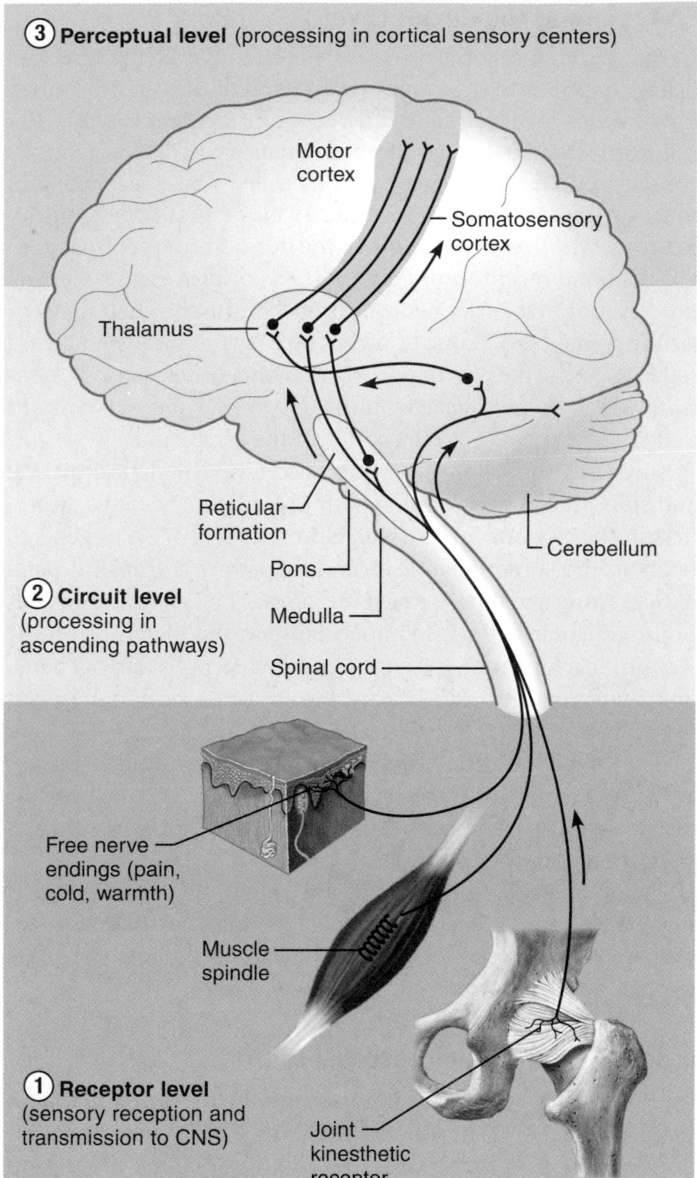

Figure 13.2 Three basic levels of neural integration in sensory systems.

Processing at the Receptor Level

For sensation to occur, a stimulus must excite a receptor and action potentials must reach the CNS (Figure 13.2, ①). For this to happen:

- The stimulus energy must match the *specificity* of the receptor. For example, a given touch receptor may be sensitive to mechanical pressure, stretch, and vibration, but not to light energy (which is the province of receptors in the eye). The more complex the sensory receptor, the greater its specificity.
- The stimulus must be applied within a sensory receptor's *receptive field*, the particular area monitored by the receptor. Typically, the smaller the receptive field, the greater the ability of the brain to accurately localize the stimulus site.
- The stimulus energy must be converted into the energy of a *graded potential* called a **receptor potential**, a process called **transduction**. This receptor potential may be a depolarizing or hyperpolarizing graded potential similar to the EPSPs or IPSPs generated at postsynaptic membranes in response to neurotransmitter binding, as described in Chapter 11 (pp. 411–412). Membrane depolarizations that summate and directly lead to generation of action potentials in an afferent fiber are called **generator potentials**.

  When the receptor region is part of a sensory neuron (as with free dendrites or the encapsulated receptors of most general sense receptors), the terms receptor potential and generator potential are synonymous. When the receptor is a separate cell, the receptor and generator potentials are completely separate events. Upon stimulation, such a depolarizing receptor cell produces a receptor potential causing neurotransmitter release. The released neurotransmitter, in turn, causes a generator potential in the associated afferent neuron.
- A generator potential in the associated sensory neuron (a first-order neuron) must reach *threshold* so that voltage-gated sodium channels on the axon (usually located just proximal to the receptor membrane, often at the first node of Ranvier) are opened and nerve impulses are generated and propagated to the CNS.

Information about the stimulus—its strength, duration, and pattern—is encoded in the frequency of nerve impulses (the greater the frequency, the stronger the stimulus). Many but not all sensory receptors exhibit **adaptation**, a change in sensitivity (and nerve impulse generation) in the presence of a constant stimulus. For example, when you step into bright sunlight from a darkened room, your eyes are initially dazzled, but your photoreceptors rapidly adapt, allowing you to see both bright areas and dark areas in the scene.

Phasic receptors are *fast adapting*, often giving bursts of impulses at the beginning and at the end of the stimulus. These act mainly to report changes in the internal or external environment. Examples of phasic receptors are Pacinian and Meissner's corpuscles.

Tonic receptors provide a sustained response with little or no adaptation. Nociceptors and most proprioceptors are tonic receptors because of the protective importance of their information.

Processing at the Circuit Level

At the second level of integration, the circuit level, the task is to deliver impulses to the appropriate region of the cerebral cortex for stimulus localization and perception (Figure 13.2, ②). Recall from Chapter 12 that ascending sensory pathways typically consist of a chain of three neurons called first-, second-, and third-order sensory neurons. The axons of first-order sensory neurons, whose cell bodies are in the dorsal root or cranial ganglia, link the receptor and circuit levels of processing. Central processes of first-order neurons branch diffusely when they enter the spinal cord. Some branches take part in local spinal cord reflexes, and others synapse with second-order sensory neurons, which then synapse with the third-order sensory neurons that take the message to the cortex of the cerebrum.

Impulses sent along the dorsal column–medial lemniscal and spinothalamic ascending pathways reach conscious awareness in the sensory cortex (see Figure 12.34, p. 472). In general, fibers in the spinothalamic ascending pathways transmit pain, temperature, and coarse touch impulses. They give off branches to the reticular formation and synapse in the thalamus on the way up, and information sent to "headquarters" is fairly general, nondiscriminatory, and heavily involved in emotional aspects of perception (pleasure, pain, etc.).

The dorsal column–medial lemniscal ascending pathways are more involved in the discriminative aspects of touch (tactile discrimination), vibration, pressure, and conscious proprioception (limb and joint position).

13

Proprioceptive impulses conducted via the *spinocerebellar tracts* end at the cerebellum, which uses this information to coordinate skeletal muscle activity. These tracts do not contribute to conscious sensation.

Processing at the Perceptual Level

Interpretation of sensory input occurs in the cerebral cortex (Figure 13.2, ③). The ability to identify and appreciate sensations depends on the specific location of the target neurons in the sensory cortex, not on the nature of the message (which is, after all, just an action potential). Each sensory fiber is analogous to a "labeled line" that tells the brain "who" is calling—a taste bud or a pressure receptor—and from "where." The brain always interprets the activity of a specific sensory receptor ("who") as a specific sensation, no matter how it is activated. For example, pressing on your eyeball activates photoreceptors, but what you "see" is light. The exact point in the cortex that is activated always refers to the same "where," regardless of how it is activated, a phenomenon called **projection**. Electrically stimulating a particular spot in the visual cortex causes you to "see" light in a particular place.

The main aspects of sensory perception include the following:

- **Perceptual detection** is the ability to detect that a stimulus has occurred. This is the simplest level of perception. As a general rule, inputs from several receptors must be summed for perceptual detection to occur.
- **Magnitude estimation** is the ability to detect how *intense* the stimulus is. Because of frequency coding, perception increases as stimulus intensity increases (see Figure 11.13).
- **Spatial discrimination** allows us to identify the site or pattern of stimulation. A common tool for studying this quality in the laboratory is the **two-point discrimination test**. The test determines how close together two points on the skin can be and still be perceived as two points rather than as one. This test provides a crude map of the density of tactile receptors in the various regions of the skin. The distance between perceived points varies from less than 5 mm on highly sensitive body areas (tip of the tongue) to more than 50 mm on less sensitive areas (back).
- **Feature abstraction** is the mechanism by which a neuron or circuit is tuned to one feature in preference to others. Sensation usually involves an interplay of several stimulus properties or features. For example, one touch tells us that velvet is warm, compressible, and smooth but not completely continuous, each a feature that contributes to our perception of "velvet." Feature abstraction enables us to identify more complex aspects of a sensation.
- **Quality discrimination** is the ability to differentiate the submodalities of a particular sensation. Each sensory modality has several **qualities**, or submodalities. For example, the submodalities of taste include sweet and bitter.
- **Pattern recognition** is the ability to take in the scene around us and recognize a familiar pattern, an unfamiliar one, or one that has special significance for us. For example, a figure made of dots may be recognized as a familiar face, and when we listen to music, we hear the melody, not just a string of notes.

Perception of Pain

Everyone has suffered pain—the smart of a bee sting, the cruel persistence of a headache, or a cut finger. Although we may not appreciate it at the time, pain is invaluable because it warns us of actual or impending tissue damage and strongly motivates us to take protective action. Managing a patient's pain can be difficult because pain is an intensely personal experience that cannot be measured objectively.

Pain receptors are activated by extremes of pressure and temperature as well as a veritable soup of chemicals released from injured tissue. Histamine, K^+, ATP, acids, and bradykinin are among the most potent pain-producing chemicals. All of these chemicals act on small-diameter fibers.

When you cut your finger, you may have noticed that you first felt a sharp pain followed some time later by burning or aching pain. Sharp pain is carried by small myelinated A delta fibers, while burning pain is carried more slowly by small unmyelinated C fibers. Both types of fibers release the neurotransmitters *glutamate* and *substance P*, which activate second-order sensory neurons. Axons from these second-order neurons ascend to the brain via the spinothalamic tract and other anterolateral pathways.

If you cut your finger while fighting off an attacker, you might not notice the cut at all. How can that be? The brain has its own pain-suppressing analgesic systems in which the endogenous opioids (*endorphins* and *enkephalins*) play a key role. Descending cortical and hypothalamic pain-suppressing signals are relayed through various nuclei in the brain stem including the periaqueductal gray matter of the midbrain. Descending fibers activate interneurons in the spinal cord, which release the opioid neurotransmitters called enkephalins. Enkephalins are inhibitory neurotransmitters that quash the pain signals generated by the nociceptive neurons.

HOMEOSTATIC IMBALANCE

Normally a steady state is maintained that correlates injury and pain. Long-lasting or very intense pain inputs, such as limb amputation, can disrupt this system, leading to **hyperalgesia** (pain amplification), chronic pain, and **phantom limb pain**. Intense or long-duration pain causes activation of *NMDA receptors*, the same receptors that strengthen neural connections during certain kinds of learning. Essentially, the spinal cord *learns* hyperalgesia. In light of this, it is crucial that pain is effectively managed early to prevent the establishment of chronic pain. *Phantom limb pain* (pain perceived in tissue that is no longer present) is a curious example of hyperalgesia. Until recently, surgical limb amputations were conducted under general anesthesia only and the spinal cord still experienced the pain of amputation. Blocking neurotransmission in the spinal cord by additionally using epidural anesthetics greatly reduces the incidence of phantom limb pain. ■

CHECK YOUR UNDERSTANDING

3. What are the three levels of sensory integration?
4. What is the key difference between tonic and phasic receptors? Why are pain receptors tonic?
5. Your cortex decodes incoming action potentials from sensory pathways. How does it tell the difference between hot and cold? Between cool and cold? Between ice on your finger and ice on your foot?

For answers, see Appendix G.

PART 2

TRANSMISSION LINES: NERVES AND THEIR STRUCTURE AND REPAIR

Nerves and Associated Ganglia

- Define ganglion and indicate the general body location of ganglia.
- Describe the general structure of a nerve.
- Follow the process of nerve regeneration.

Structure and Classification

A **nerve** is a cordlike organ that is part of the peripheral nervous system. Nerves vary in size, but every nerve consists of parallel bundles of peripheral axons (some myelinated and some not) enclosed by successive wrappings of connective tissue **(Figure 13.3)**.

Within a nerve, each axon is surrounded by **endoneurium** (en″do-nu′re-um), a delicate layer of loose connective tissue that also encloses the fiber's associated myelin sheath or neurilemma. Groups of fibers are bound into bundles or **fascicles** by a coarser connective tissue wrapping, the **perineurium**. Finally, all the fascicles are enclosed by a tough fibrous sheath, the **epineurium**, to form the nerve. Axons constitute only a small fraction of a nerve's bulk. The balance consists chiefly of myelin, the protective connective tissue wrappings, blood vessels, and lymphatic vessels.

Recall that the PNS is divided into *sensory* (afferent) and *motor* (efferent) divisions. So, too, nerves are classified according to the direction in which they transmit impulses. Nerves containing both sensory and motor fibers and transmitting impulses both to and from the central nervous system are called **mixed nerves**. Those that carry impulses only toward the CNS are **sensory (afferent) nerves**, and those carrying impulses only away from the CNS are **motor (efferent) nerves**. Most nerves are mixed. Pure sensory or motor nerves are rare.

Because mixed nerves often carry both somatic and autonomic (visceral) nervous system fibers, the fibers in them may be classified according to the region they innervate as *somatic afferent, somatic efferent, visceral afferent,* and *visceral efferent.*

For convenience, the peripheral nerves are classified as *cranial* or *spinal* depending on whether they arise from the brain or the spinal cord. Although we mention autonomic efferents of cranial nerves, in this chapter we will focus on somatic functions. We will defer the discussion of the autonomic nervous system and the visceral functions it serves to Chapter 14.

Ganglia are collections of neuron cell bodies associated with nerves in the PNS. Ganglia associated with *afferent* nerve fibers contain cell bodies of sensory neurons. (These are the *dorsal root ganglia* studied in Chapter 12.) Ganglia associated with *efferent* nerve fibers mostly contain cell bodies of autonomic motor neurons. We will describe these more complicated ganglia in Chapter 14.

Regeneration of Nerve Fibers

Damage to nervous tissue is serious because, as a rule, mature neurons do not divide. If the damage is severe or close to the cell body, the entire neuron may die, and other neurons that are normally stimulated by its axon may die as well. However, if the cell body remains intact, cut or compressed axons of peripheral nerves can regenerate successfully.

Almost immediately after a peripheral axon has been severed or crushed, the separated ends seal themselves off and then swell as substances being transported along the axon begin to accumulate in the sealed ends. Within a few hours, the axon and its myelin sheath distal to the injury site begin to disintegrate because they cannot receive nutrients from the cell body (**Figure 13.4**, ①). This process, **Wallerian degeneration**, spreads distally from the injury site, completely fragmenting the axon.

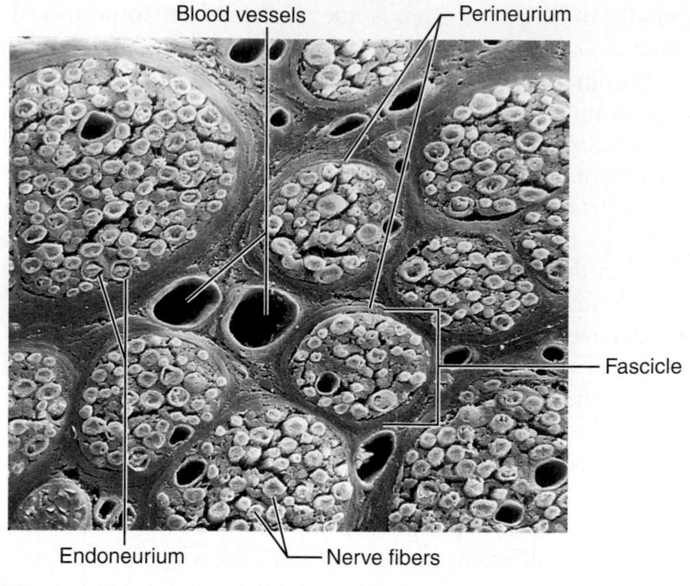

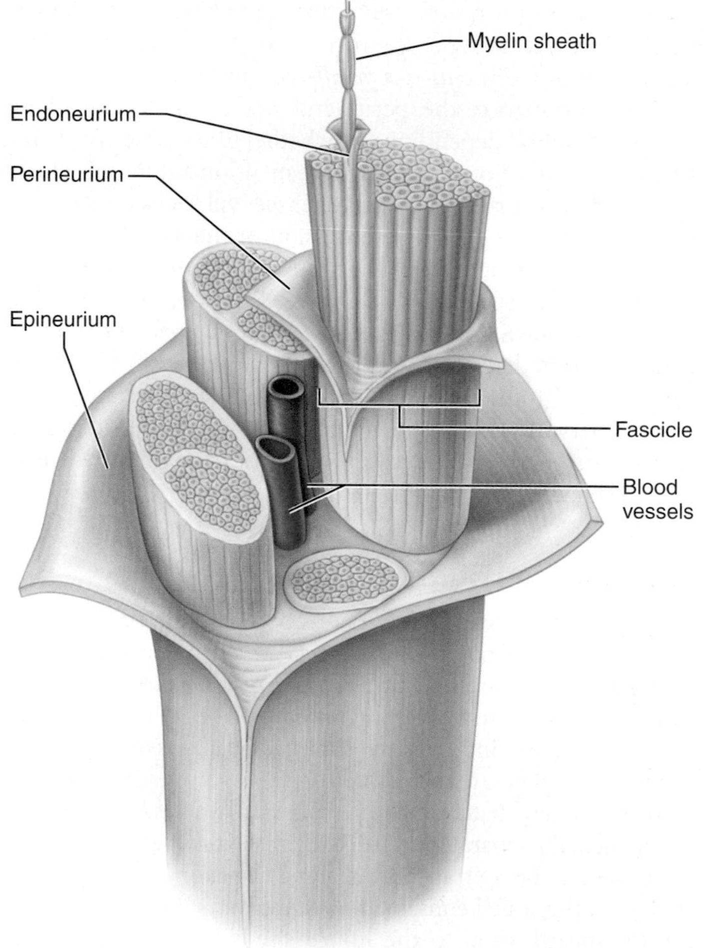

Figure 13.3 Structure of a nerve. (a) Scanning electron micrograph of a cross section of a portion of a nerve (1150×). **(b)** Three-dimensional view of a portion of a nerve, showing connective tissue wrappings.

SOURCE: Kessel and Kardon/Visuals Unlimited.

13

Generally, the entire axon distal to the injury is degraded by phagocytes within a week, but the neurilemma remains intact within the endoneurium (Figure 13.4, ②). After the debris has been disposed of, surviving Schwann cells proliferate in response to mitosis-stimulating chemicals released by the macrophages, and migrate into the injury site. Once there, they release growth factors and begin to express cell adhesion molecules (CAMs) that encourage axonal growth. Additionally, they form a *regeneration tube*, a system of cellular cords that guide the regenerating axon "sprouts" across the gap and to their original contacts (Figure 13.4, ③ and ④). The same Schwann cells protect, support, and remyelinate the regenerating axons.

Changes also occur in the neuronal cell body after the axon has been destroyed. Within two days, its chromatophilic substance breaks apart, and then the cell body swells as protein synthesis revs up to support regeneration of its axon.

Axons regenerate at the approximate rate of 1.5 mm a day. The greater the distance between the severed endings, the less the chance of recovery because adjacent tissues block growth by protruding into the gaps, and axonal sprouts fail to find the regeneration tube. Neurosurgeons align cut nerve endings surgically to enhance the chance of successful regeneration, and scaffolding devices have been successful in guiding axon growth. Whatever the measures taken, post-trauma axon regrowth never exactly matches what existed before the injury. Much of the recovery protocol involves retraining the nervous system to respond appropriately so that stimulus and response are coordinated.

Unlike peripheral nerve fibers, most CNS fibers never regenerate. Consequently, damage to the brain or spinal cord has been viewed as irreversible. This difference in regenerative capacity seems to have less to do with the neurons themselves than with the "company they keep"—oligodendrocytes, their supporting cells. Oligodendrocytes are studded with growth-inhibiting proteins (Nogo and others) that act on inhibitory neuronal receptors. Consequently, the neuronal growth cone collapses and the fiber fails to regrow. Moreover, astrocytes at the site of injury form scar tissue rich in chondroitin sulfate (see p. 124) that blocks axonal regrowth. For the clinician treating spinal cord injury, this means that multiple inhibitory processes need to be blocked simultaneously to promote axon regrowth. To date, experimentally neutralizing the myelin-bound growth inhibitors, blocking their receptors, or enzymatically destroying chondroitin sulfate have yielded promising results.

CHECK YOUR UNDERSTANDING

6. What are ganglia?
7. What is in a nerve besides axons?
8. Bill's femoral nerve was crushed while clinicians tried to control bleeding from his femoral artery. This resulted in loss of function and sensation in his leg, which gradually returned over the course of a year. Which cells were important in his recovery?

For answers, see Appendix G.

Cranial Nerves

▶ Name the 12 pairs of cranial nerves; indicate the body region and structures innervated by each.

Twelve pairs of **cranial nerves** are associated with the brain **(Figure 13.5)**. The first two pairs attach to the forebrain, and the rest are associated with the brain stem. Other than the vagus nerves, which extend into the abdomen, cranial nerves serve only head and neck structures.

In most cases, the names of the cranial nerves reveal either the structures they serve or their functions. The nerves are also generally numbered (using Roman numerals) from the most rostral to the most caudal. A mini-introduction to the cranial nerves follows.

I. Olfactory. These are the tiny sensory nerves (filaments) of smell, which run from the nasal mucosa to synapse with the olfactory bulbs. Note that the olfactory bulbs and tracts, shown in Figure 13.5a, are brain structures and not part of cranial nerve I. (See Table 13.2 art for the olfactory nerve filaments.)

II. Optic. Because this sensory nerve of vision develops as an outgrowth of the brain, it is really a brain tract.

III. Oculomotor. The name *oculomotor* means "eye mover." This nerve supplies four of the six extrinsic muscles that move the eyeball in the orbit.

IV. Trochlear. The name *trochlear* means "pulley" and it innervates an extrinsic eye muscle that loops through a pulley-shaped ligament in the orbit.

V. Trigeminal. Three (*tri*) branches spring from this, the largest of the cranial nerves. It supplies sensory fibers to the face and motor fibers to the chewing muscles.

VI. Abducens. This nerve controls the extrinsic eye muscle that *abducts* the eyeball (turns it laterally).

VII. Facial. A large nerve that innervates muscles of *facial* expression (among other things).

VIII. Vestibulocochlear. This sensory nerve for hearing and balance was formerly called the *auditory nerve.*

IX. Glossopharyngeal. The name *glossopharyngeal* means "tongue and pharynx," and reveals the structures that this nerve helps to innervate.

X. Vagus. This nerve's name means "wanderer" or "vagabond," and it is the only cranial nerve to extend beyond the head and neck to the thorax and abdomen.

XI. Accessory. Considered an *accessory* part of the vagus nerve, this nerve was formerly called the *spinal accessory nerve.*

XII. Hypoglossal. The name *hypoglossal* means under the tongue. This nerve runs inferior to the tongue and innervates the tongue muscles.

You might make up your own saying to remember the first letters of the cranial nerves in order, or use the following memory jog sent by a student: "**O**n **o**ccasion, **o**ur **t**rusty **t**ruck **a**cts **f**unny—**v**ery **g**ood **v**ehicle **a**ny**h**ow."

In the last chapter, we described how spinal nerves are formed by the fusion of ventral (motor) and dorsal (sensory)

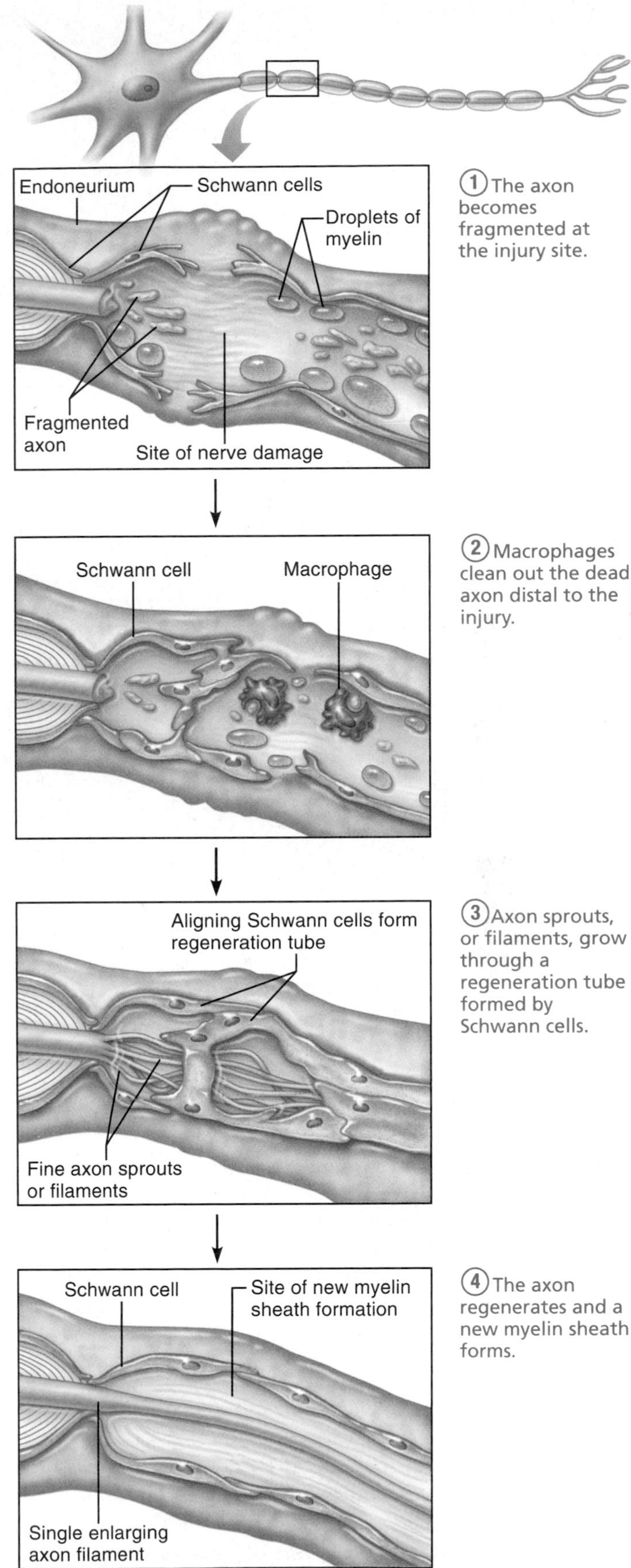
Endoneurium
Schwann cells
Droplets of myelin
Fragmented axon
Site of nerve damage
① The axon becomes fragmented at the injury site.
Schwann cell
Macrophage
② Macrophages clean out the dead axon distal to the injury.
Aligning Schwann cells form regeneration tube
Fine axon sprouts or filaments
③ Axon sprouts, or filaments, grow through a regeneration tube formed by Schwann cells.
Schwann cell
Site of new myelin sheath formation
Single enlarging axon filament
④ The axon regenerates and a new myelin sheath forms.

Figure 13.4 Regeneration of a nerve fiber in a peripheral nerve.

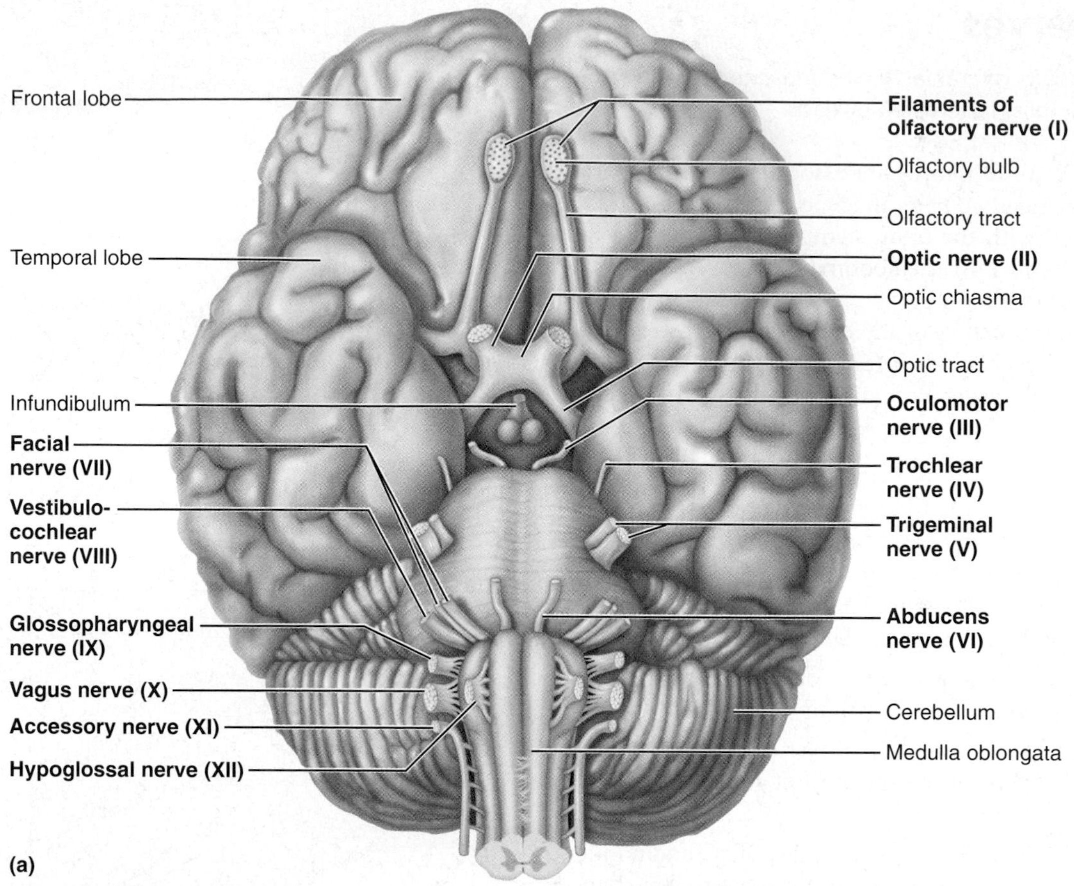

13

Cranial nerves I – VI	Sensory function	Motor function	PS* fibers
I Olfactory	Yes (smell)	No	No
II Optic	Yes (vision)	No	No
III Oculomotor	No	Yes	Yes
IV Trochlear	No	Yes	No
V Trigeminal	Yes (general sensation)	Yes	No
VI Abducens	No	Yes	No

Cranial nerves VII – XII	Sensory function	Motor function	PS* fibers
VII Facial	Yes (taste)	Yes	Yes
VIII Vestibulocochlear	Yes (hearing and balance)	Some	No
IX Glossopharyngeal	Yes (taste)	Yes	Yes
X Vagus	Yes (taste)	Yes	Yes
XI Accessory	No	Yes	No
XII Hypoglossal	No	Yes	No

(b)

*PS = parasympathetic

Figure 13.5 Location and function of cranial nerves. (a) Ventral view of the human brain, showing the cranial nerves. **(b)** Summary of cranial nerves by function. Two cranial nerves (I and II) have sensory function only, no motor function. Four nerves (III, VII, IX, and X) carry parasympathetic fibers that serve visceral muscles and glands. All cranial nerves that innervate muscles also carry afferent fibers from proprioceptors in the muscles served; only sensory functions other than proprioception are indicated.

roots. Cranial nerves, on the other hand, vary markedly in their composition. Most cranial nerves are mixed nerves, as shown in Figure 13.5b. However, two nerve pairs (the olfactory and optic) associated with special sense organs are generally considered purely sensory. The cell bodies of the sensory neurons of the olfactory and optic nerves are located *within* their respective special sense organs. In other cases of sensory neurons contributing to cranial nerves (V, VII, IX, and X), the cell bodies are located in **cranial sensory ganglia** just outside the brain. Some cranial nerves have a single sensory ganglion, others have several, and still others have none.

Several of the mixed cranial nerves contain both somatic and autonomic motor fibers and hence serve both skeletal muscles and visceral organs. Except for some autonomic motor neurons located in ganglia, the cell bodies of motor neurons contributing to the cranial nerves are located in the ventral gray matter regions (nuclei) of the brain stem.

Having read this overview, you are now ready to tackle **Table 13.2**, which provides a more detailed description of the origin, course, and function of the cranial nerves. Notice that

(Text continues on p. 501.)

TABLE 13.2 Cranial Nerves

I The Olfactory Nerves (ol-fak′to-re)

Origin and course: Olfactory nerve fibers arise from olfactory receptor cells located in olfactory epithelium of nasal cavity and pass through cribriform plate of ethmoid bone to synapse in olfactory bulb. Fibers of olfactory bulb neurons extend posteriorly as olfactory tract, which runs beneath frontal lobe to enter cerebral hemispheres and terminates in primary olfactory cortex. See also Figure 15.21.

Function: Purely sensory; carry afferent impulses for sense of smell.

Clinical testing: Person is asked to sniff aromatic substances, such as oil of cloves and vanilla, and to identify each.

Homeostatic imbalance: Fracture of ethmoid bone or lesions of olfactory fibers may result in partial or total loss of smell, a condition known as *anosmia* (an-oz′me-ah). ■

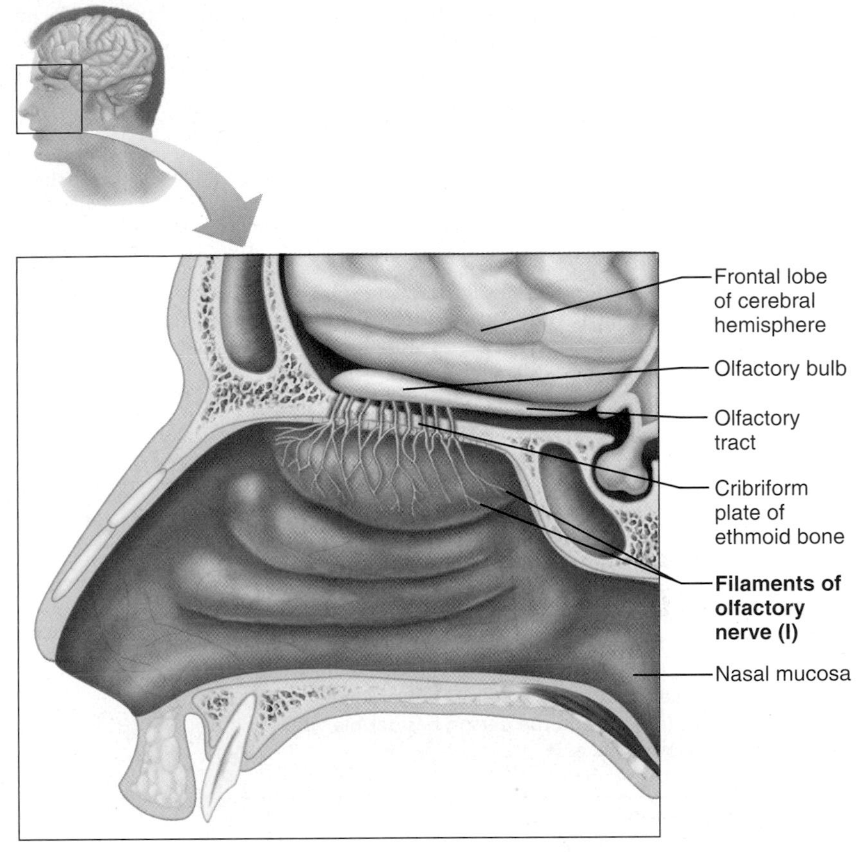

II The Optic Nerves

Origin and course: Fibers arise from retina of eye to form optic nerve, which passes through optic canal of orbit. The optic nerves converge to form the optic chiasma (ki-az′mah) where fibers partially cross over, continue on as optic tracts, enter thalamus, and synapse there. Thalamic fibers run (as the optic radiation) to occipital (visual) cortex, where visual interpretation occurs. See also Figure 15.19.

Function: Purely sensory; carry afferent impulses for vision.

Clinical testing: Vision and visual field are determined with eye chart and by testing the point at which the person first sees an object (finger) moving into the visual field. Fundus of eye viewed with ophthalmoscope to detect papilledema (swelling of optic disc, the site where the optic nerve leaves the eyeball), as well as for routine examination of the optic disc and retinal blood vessels.

Homeostatic imbalance: Damage to optic nerve results in blindness in eye served by nerve; damage to visual pathway beyond the optic chiasma results in partial visual losses; visual defects are called *anopsias* (ah-nop′se-ahz). ■

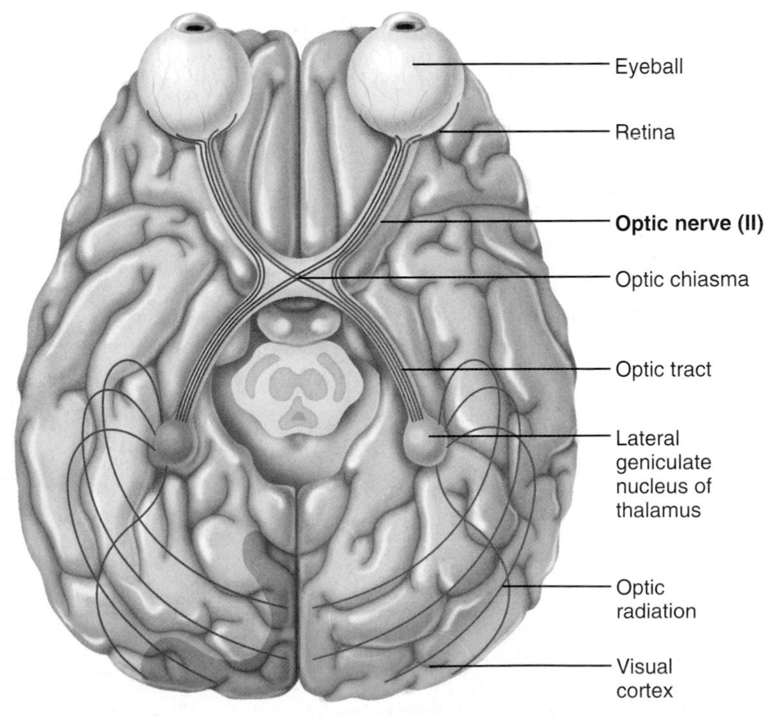

TABLE 13.2 Cranial Nerves *(continued)*

III The Oculomotor Nerves (ok″u-lo-mo′tor)

Origin and course: Fibers extend from ventral midbrain (near its junction with pons) and pass through bony orbit, via superior orbital fissure, to eye.

Function: Chiefly motor nerves (*oculomotor* = motor to the eye); contain a few proprioceptive afferents. Each nerve includes the following:

- Somatic motor fibers to four of the six extrinsic eye muscles (inferior oblique and superior, inferior, and medial rectus muscles) that help direct eyeball, and to levator palpebrae superioris muscle, which raises upper eyelid.
- Parasympathetic (autonomic) motor fibers to sphincter pupillae (circular muscles of iris), which cause pupil to constrict, and to ciliary muscle, controlling lens shape for visual focusing. Some parasympathetic cell bodies are in the ciliary ganglia.
- Sensory (proprioceptor) afferents, which run from same four extrinsic eye muscles to midbrain.

Clinical testing: Pupils are examined for size, shape, and equality. Pupillary reflex is tested with penlight (pupils should constrict when illuminated). Convergence for near vision is tested, as is subject's ability to follow objects with the eyes.

Homeostatic imbalance: In oculomotor nerve paralysis, eye cannot be moved up, down, or inward, and at rest, eye rotates laterally [*external strabismus* (strah-biz′mus)] because the actions of the two extrinsic eye muscles not served by cranial nerves III are unopposed; upper eyelid droops (*ptosis*), and the person has double vision and trouble focusing on close objects. ■

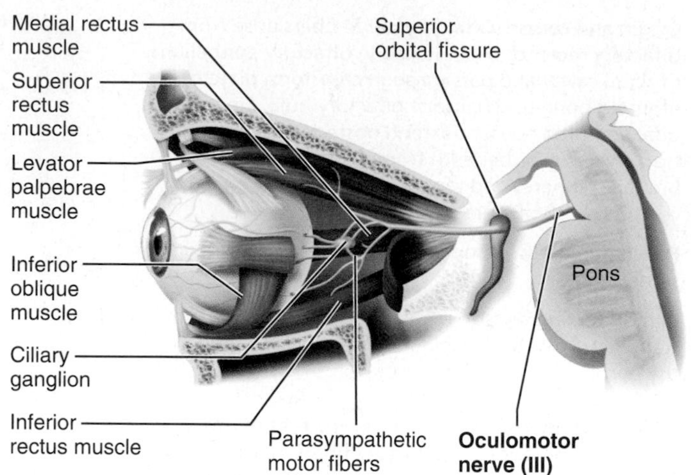

13

IV The Trochlear Nerves (trok′le-ar)

Origin and course: Fibers emerge from dorsal midbrain and course ventrally around midbrain to enter orbit through superior orbital fissure along with oculomotor nerves.

Function: Primarily motor nerves; supply somatic motor fibers to (and carry proprioceptor fibers from) one of the extrinsic eye muscles, the superior oblique muscle.

Clinical testing: Tested in common with cranial nerve III.

Homeostatic imbalance: Trauma to, or paralysis of, a trochlear nerve results in double vision and reduced ability to rotate eye inferolaterally. ■

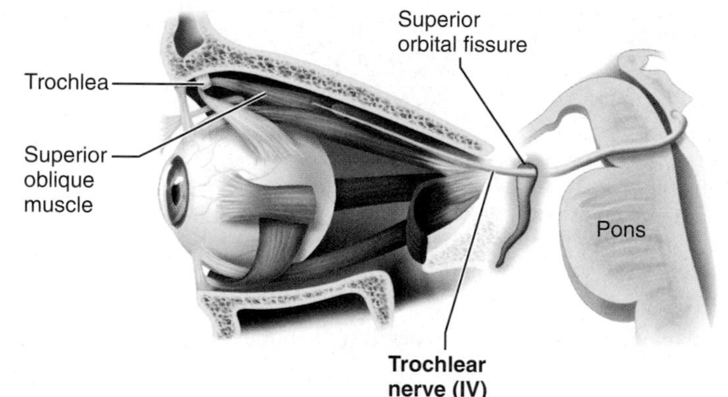

TABLE 13.2 *(continued)*

V The Trigeminal Nerves

Largest of cranial nerves; fibers extend from pons to face, and form three divisions (*trigemina* = threefold): ophthalmic, maxillary, and mandibular divisions. As major general sensory nerves of face, transmit afferent impulses from touch, temperature, and pain receptors. Cell bodies of sensory neurons of all three divisions are located in large *trigeminal ganglion.*

The mandibular division also contains motor fibers that innervate chewing muscles.

Dentists desensitize upper and lower jaws by injecting local anesthetic (such as Novocain) into alveolar branches of maxillary and mandibular divisions, respectively; since this blocks pain-transmitting fibers of teeth, the surrounding tissues become numb.

	Ophthalmic division (V_1)	Maxillary division (V_2)	Mandibular division (V_3)
Origin and course	Fibers run from face to pons via superior orbital fissure.	Fibers run from face to pons via foramen rotundum.	Fibers pass through skull via foramen ovale.
Function	Conveys sensory impulses from skin of anterior scalp, upper eyelid, and nose, and from nasal cavity mucosa, cornea, and lacrimal gland.	Conveys sensory impulses from nasal cavity mucosa, palate, upper teeth, skin of cheek, upper lip, lower eyelid.	Conveys sensory impulses from anterior tongue (except taste buds), lower teeth, skin of chin, temporal region of scalp. Supplies motor fibers to, and carries proprioceptor fibers from, muscles of mastication.
Clinical testing	Corneal reflex test: Touching cornea with wisp of cotton should elicit blinking.	Sensations of pain, touch, and temperature are tested with safety pin and hot and cold objects.	Motor branch assessed by asking person to clench his teeth, open mouth against resistance, and move jaw side to side.

Homeostatic imbalance: *Trigeminal neuralgia* (nu-ral′je-ah), or *tic douloureux* (tik doo″loo-roo′; *tic* = twitch, *douloureux* = painful), caused by inflammation of trigeminal nerve, is widely considered to produce most excruciating pain known; the stabbing pain lasts for a few seconds to a minute, but it can be relentless, occurring a hundred times a day. Usually provoked by some sensory stimulus, such as brushing teeth or even a passing breeze hitting the face. It is thought to be caused by compression of the trigeminal nerve by a loop of artery or vein close to its exit from the brain stem. Analgesics and carbamazepine (an anticonvulsant) are only partially effective. In severe cases, surgery relieves the agony—either by moving the compressing vessel or by destroying the nerve. Nerve destruction results in loss of sensation on that side of face. ■

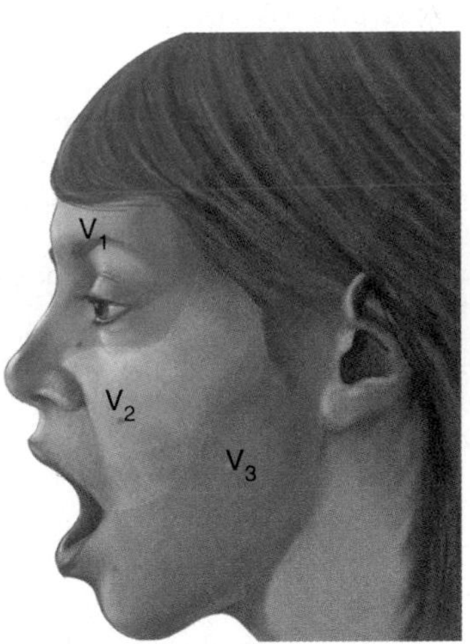

(b) Distribution of sensory fibers of each division

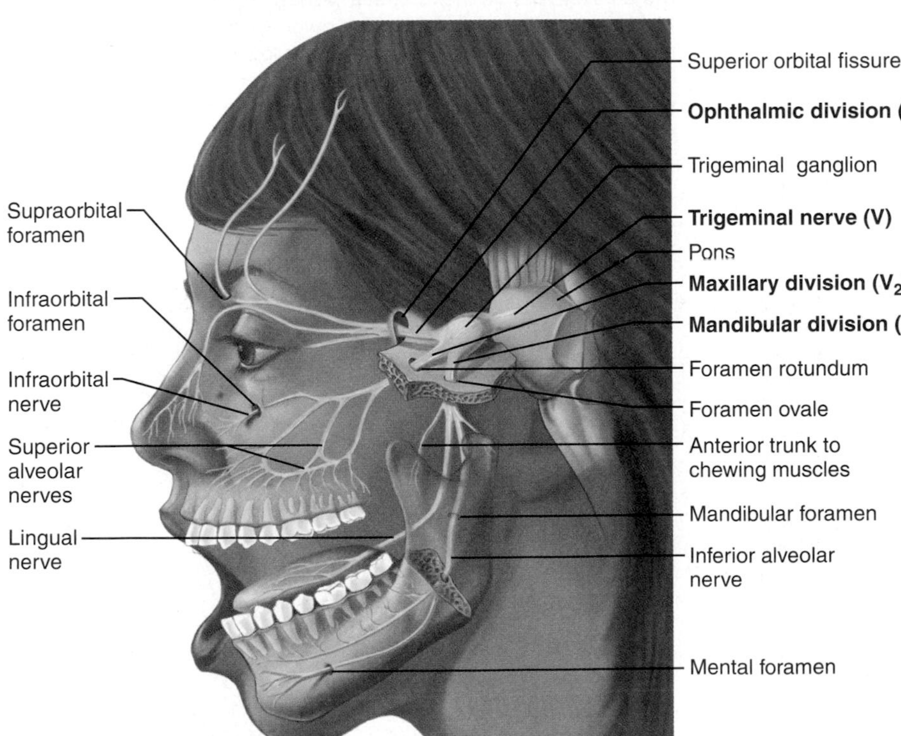

(a) Distribution of the trigeminal nerve

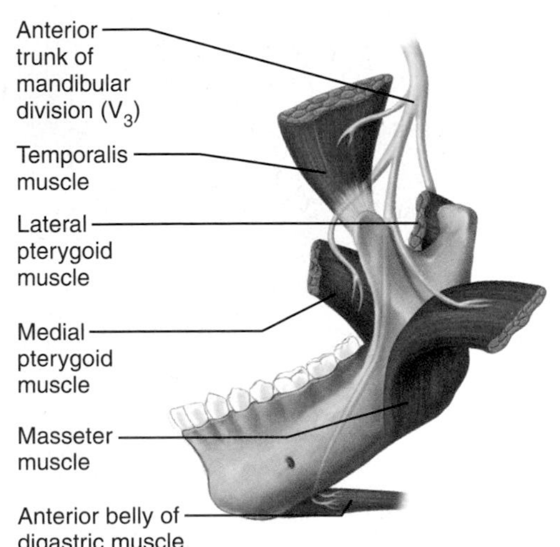

(c) Motor branches of the mandibular division (V_3)

13

TABLE 13.2 Cranial Nerves *(continued)*

VI The Abducens Nerves (ab-du'senz)

Origin and course: Fibers leave inferior pons and enter orbit via superior orbital fissure to run to eye.

Function: Primarily motor; supply somatic motor fibers to lateral rectus muscle, an extrinsic muscle of the eye; convey proprioceptor impulses from same muscle to brain.

Clinical testing: Tested in common with cranial nerve III.

Homeostatic imbalance: In abducens nerve paralysis, eye cannot be moved laterally; at rest, affected eyeball rotates medially (*internal strabismus*). ■

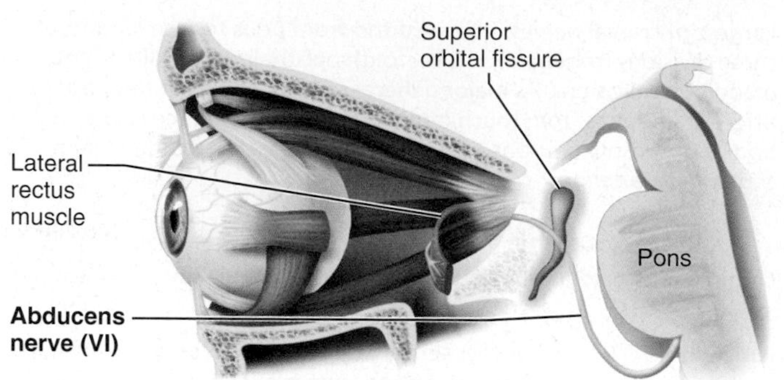

VII The Facial Nerves

Origin and course: Fibers issue from pons, just lateral to abducens nerves (see Figure 13.5), enter temporal bone via *internal acoustic meatus*, and run within bone (and through inner ear cavity) before emerging through *stylomastoid foramen*; nerve then courses to lateral aspect of face.

Function: Mixed nerves that are the chief motor nerves of face; have five major branches: temporal, zygomatic, buccal, mandibular, and cervical (see **c**).

13

- Convey motor impulses to skeletal muscles of face (muscles of facial expression), except for chewing muscles served by trigeminal nerves, and transmit proprioceptor impulses from same muscles to pons (see **b**).
- Transmit parasympathetic (autonomic) motor impulses to lacrimal (tear) glands, nasal and palatine glands, and submandibular and sublingual salivary glands. Some of the cell bodies of these parasympathetic motor neurons are in *pterygopalatine* (ter"eh-go-pal'ah-tīn) and *submandibular ganglia* on the trigeminal nerve (see **a**).
- Convey sensory impulses from taste buds of anterior two-thirds of tongue; cell bodies of these sensory neurons are in *geniculate ganglion* (see **a**).

Clinical testing: Anterior two-thirds of tongue is tested for ability to taste sweet (sugar), salty, sour (vinegar), and bitter (quinine) substances. Symmetry of face is checked. Subject is asked to close eyes, smile, whistle, and so on. Tearing is assessed with ammonia fumes.

Homeostatic imbalance: *Bell's palsy*, characterized by paralysis of facial muscles on affected side and partial loss of taste sensation, may develop rapidly (often overnight). Most often caused by herpes simplex 1 viral infection, which causes swelling and inflammation of facial nerve. Lower eyelid droops, corner of mouth sags (making it difficult to eat or speak normally), tears drip continuously from eye and eye cannot be completely closed (conversely, dry-eye syndrome may occur). Condition may disappear spontaneously without treatment. ■

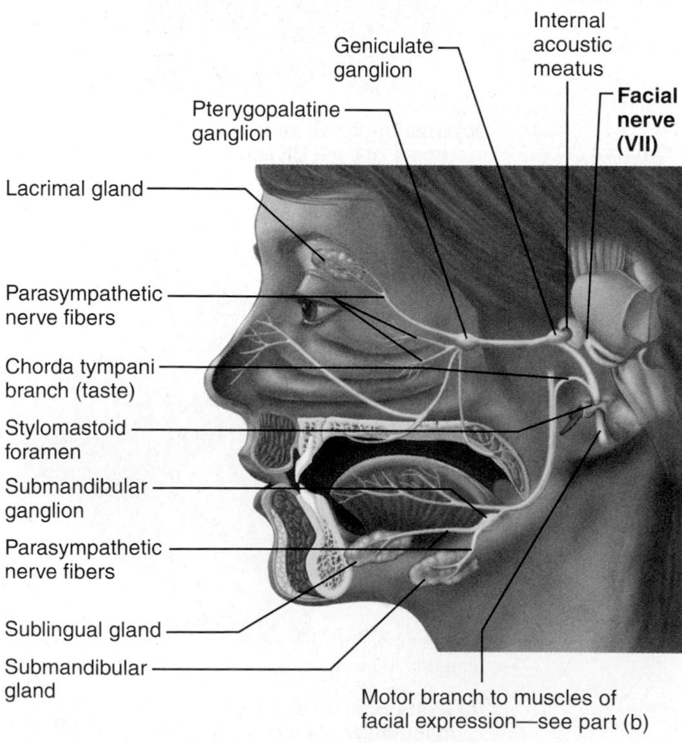

(a) Parasympathetic efferents and sensory afferents

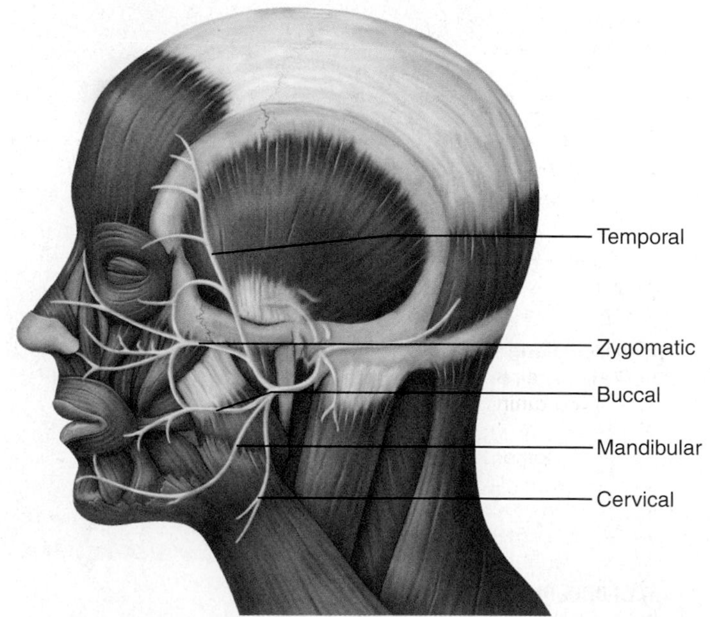

(b) Motor branches to muscles of facial expression and scalp muscles (see pp. 329–331)

TABLE 13.2 *(continued)*

VII The Facial Nerves *(continued)*

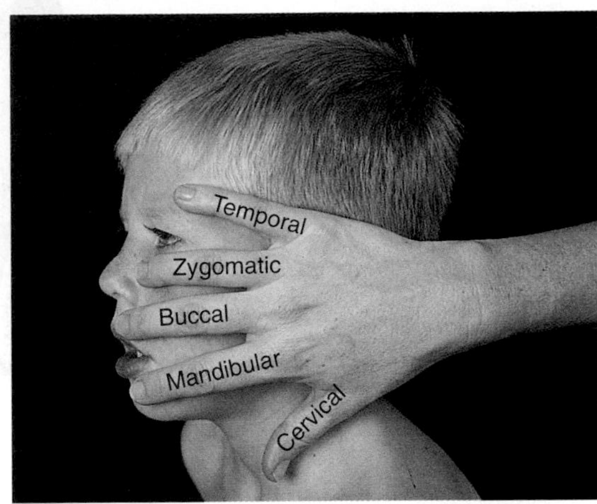

(c) A simple method of remembering the courses of the five major motor branches of the facial nerve

VIII The Vestibulocochlear Nerves (ves-tib″u-lo-kok′le-ar)

Origin and course: Fibers arise from hearing and equilibrium apparatus located within inner ear of temporal bone and pass through internal acoustic meatus to enter brain stem at pons-medulla border. Afferent fibers from hearing receptors in cochlea form the *cochlear division*; those from equilibrium receptors in semicircular canals and vestibule form the *vestibular division* (vestibular nerve); the two divisions merge to form vestibulocochlear nerve. See also Figure 15.27.

Function: Mostly sensory. Vestibular branch transmits afferent impulses for sense of equilibrium, and sensory nerve cell bodies are located in *vestibular ganglia.* Cochlear branch transmits afferent impulses for sense of hearing, and sensory nerve cell bodies are located in *spiral ganglion* within cochlea. Small motor component adjusts the sensitivity of sensory receptors. See also Figure 15.28c.

Clinical testing: Hearing is checked by air and bone conduction using tuning fork.

Homeostatic imbalance: Lesions of cochlear nerve or cochlear receptors result in *central* or *nerve deafness*, whereas damage to vestibular division produces dizziness, rapid involuntary eye movements, loss of balance, nausea, and vomiting. ■

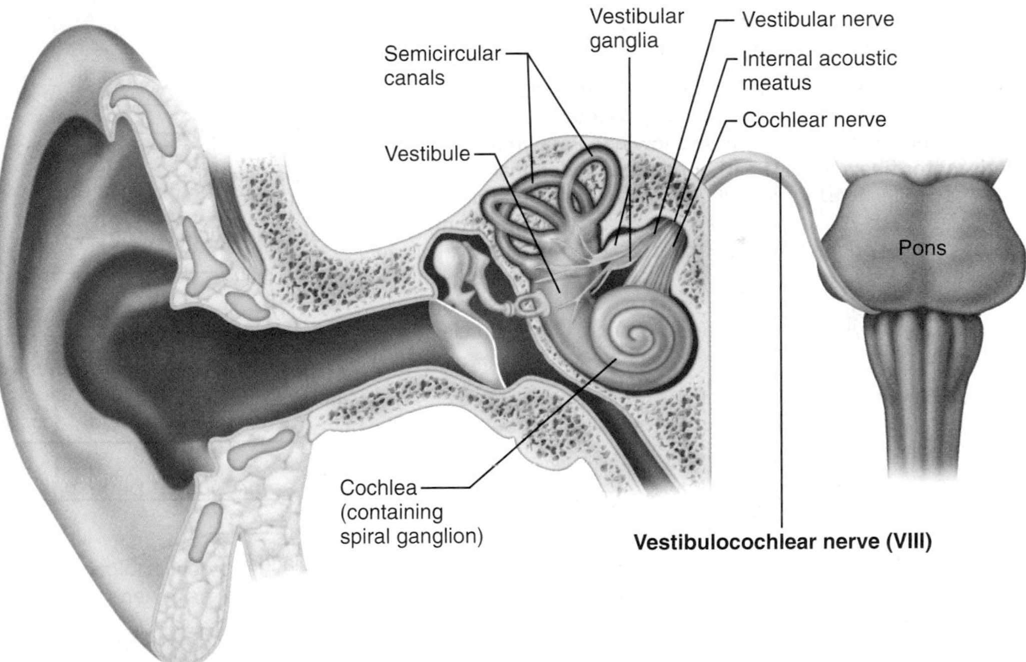

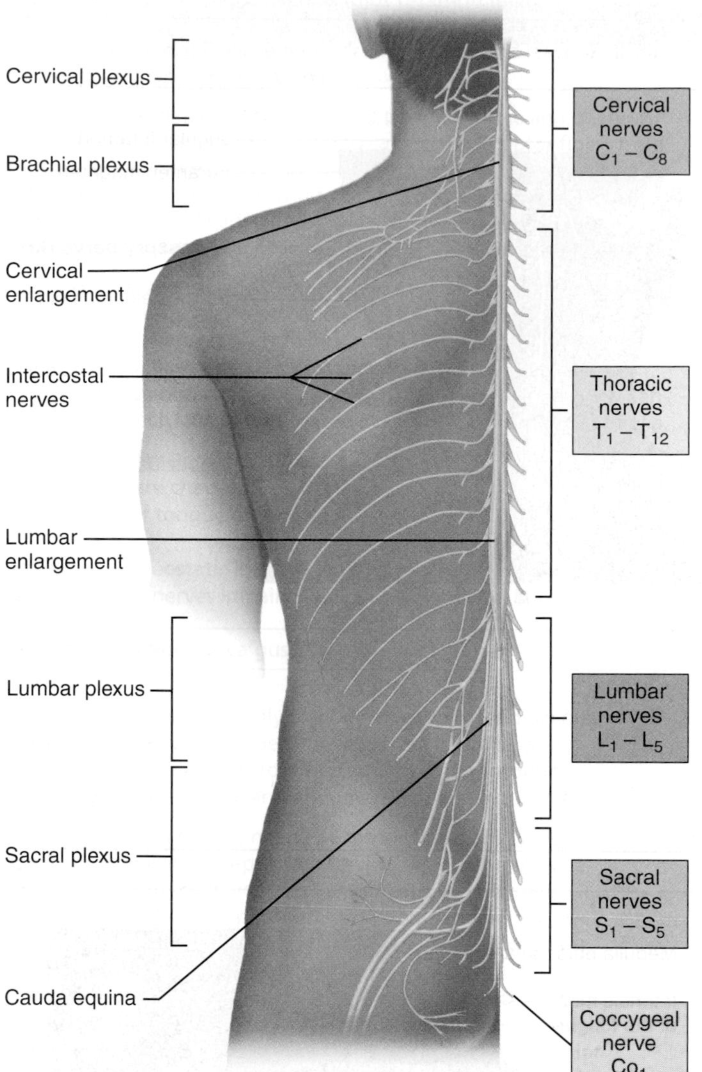

Figure 13.6 Spinal nerves. (Posterior view.) The short spinal nerves are shown at right; their ventral rami are shown at left. Most ventral rami form nerve plexuses (cervical, brachial, lumbar, and sacral).

13

Spinal Nerves

- Describe the formation of a spinal nerve and the general distribution of its rami.
- Define plexus. Name the major plexuses and describe the distribution and function of the peripheral nerves arising from each plexus.

Thirty-one pairs of **spinal nerves**, each containing thousands of nerve fibers, arise from the spinal cord and supply all parts of the body except the head and some areas of the neck. All are mixed nerves. As illustrated in **Figure 13.6**, these nerves are named according to their point of issue from the spinal cord. There are 8 pairs of cervical spinal nerves (C_1–C_8), 12 pairs of thoracic nerves (T_1–T_{12}), 5 pairs of lumbar nerves (L_1–L_5), 5 pairs of sacral nerves (S_1–S_5), and 1 pair of tiny coccygeal nerves (Co_1).

Notice that there are eight pairs of cervical nerves but only seven cervical vertebrae. This "discrepancy" is easily explained. The first seven pairs exit the vertebral canal *superior to* the vertebrae for which they are named, but C_8 emerges *inferior to* the seventh cervical vertebra (between C_7 and T_1). Below the cervical level, each spinal nerve leaves the vertebral column *inferior to* the same-numbered vertebra.

As we mentioned in Chapter 12, each spinal nerve connects to the spinal cord by a dorsal root and a ventral root **(Figure 13.7)**. Each root forms from a series of **rootlets** that attach along the length of the corresponding spinal cord segment (Figure 13.7a). The **ventral roots** contain *motor* (efferent) fibers that arise from ventral horn motor neurons and extend to and innervate the skeletal muscles. (In Chapter 14, we describe autonomic nervous system efferents that are also contained in the ventral roots.) **Dorsal roots** contain *sensory* (afferent) fibers that arise from sensory neurons in the dorsal root ganglia and conduct impulses from peripheral receptors to the spinal cord.

The spinal roots pass laterally from the cord and unite just distal to the dorsal root ganglion to form a spinal nerve before emerging from the vertebral column via their respective intervertebral foramina. Because motor and sensory fibers mingle in a spinal nerve, it contains both efferent and afferent fibers. The length of the spinal roots increases progressively from the superior to the inferior aspect of the cord. In the cervical region, the roots are short and run horizontally, but the roots of the lumbar and sacral nerves extend inferiorly for some distance through the lower vertebral canal as the *cauda equina* before exiting the vertebral column (Figure 13.6).

A spinal nerve is quite short (only 1–2 cm). Almost immediately after emerging from its foramen, it divides into a small **dorsal ramus**, a larger **ventral ramus** (ra′mus; "branch"), and a tiny **meningeal branch** (mĕ-nin′je-al) that reenters the vertebral canal to innervate the meninges and blood vessels within. Each ramus, like the spinal nerve itself, is mixed. Finally, joined to the base of the ventral rami of the thoracic spinal nerves are special rami called **rami communicantes**, which contain autonomic (visceral) nerve fibers.

Innervation of Specific Body Regions

The spinal nerve rami and their main branches supply the entire somatic region of the body (skeletal muscles and skin) from the neck down. The dorsal rami supply the posterior body trunk. The thicker ventral rami supply the rest of the trunk and the limbs.

To be clear, let's review the difference between roots and rami: Roots lie medial to and form the spinal nerves, and each root is strictly sensory or motor. Rami lie distal to and are lateral branches of the spinal nerves and, like spinal nerves, carry both sensory and motor fibers.

Before we get into the specifics of how the body is innervated, it is important for you to understand some points about the ventral rami of the spinal nerves. Except for T_2–T_{12}, all ventral rami branch and join one another lateral to the vertebral column, forming complicated interlacing nerve networks called **nerve plexuses** (Figure 13.6). Nerve plexuses occur in the cervical, brachial, lumbar, and sacral regions and primarily serve the limbs. Notice that *only ventral rami form plexuses.*

Within a plexus, fibers from the various ventral rami crisscross one another and become redistributed so that (1) each

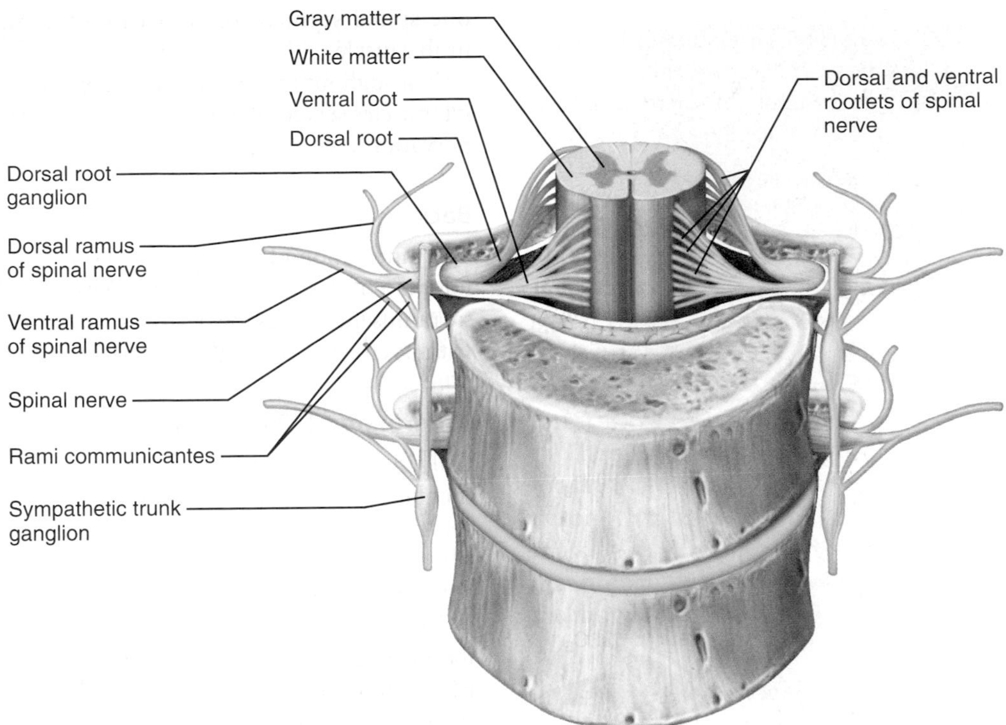

(a) Anterior view showing spinal cord, associated nerves, and vertebrae. The dorsal and ventral roots arise medially as rootlets and join laterally to form the spinal nerve.

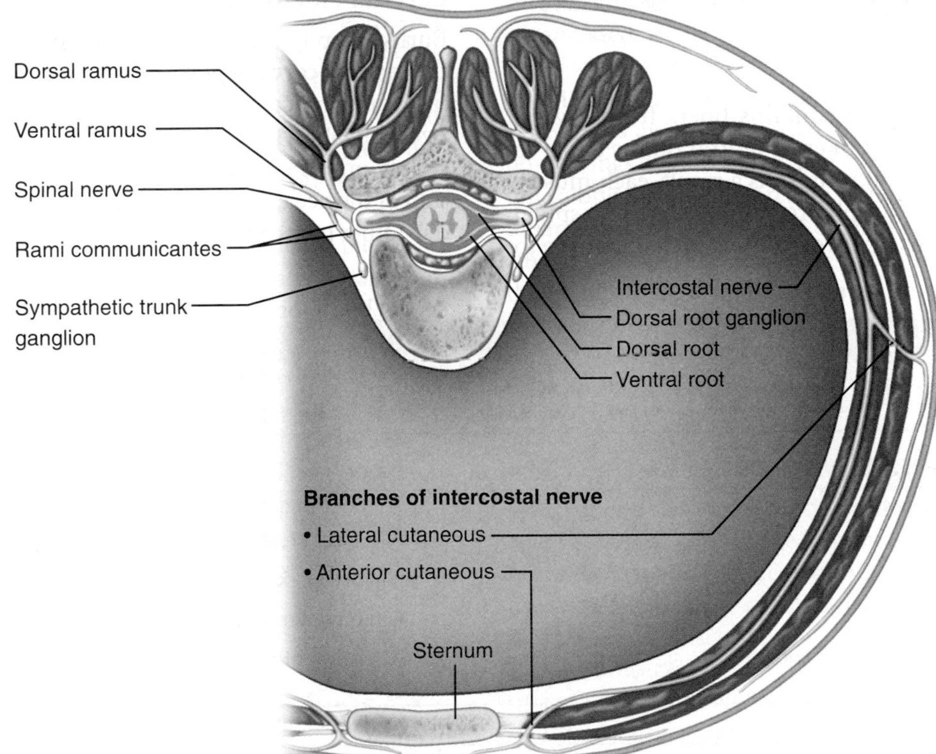

(b) Cross section of thorax showing the main roots and branches of a spinal nerve.

Figure 13.7 Formation of spinal nerves and rami distribution. Notice in (b) the dorsal and ventral roots and rami, and the rami communicantes. In the thorax, each ventral ramus continues as an intercostal nerve. (The small meningeal branch is not illustrated.)

13

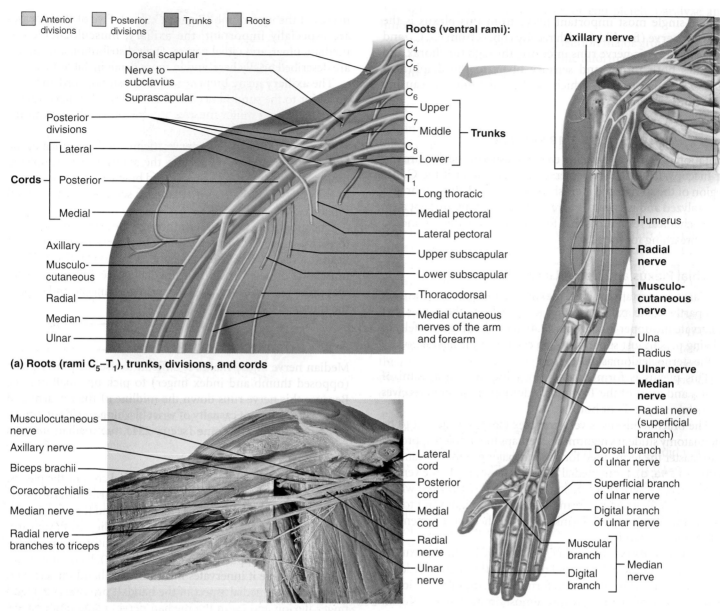

Anterior divisions
Posterior divisions
Trunks
Roots

Roots (ventral rami):
C_4
C_5
C_6
C_7
C_8
T_1
Dorsal scapular
Nerve to subclavius
Suprascapular
Posterior divisions
Cords
Lateral
Posterior
Medial
Upper
Middle
Lower
Trunks
Long thoracic
Medial pectoral
Lateral pectoral
Upper subscapular
Lower subscapular
Thoracodorsal
Medial cutaneous nerves of the arm and forearm
Axillary
Musculo-cutaneous
Radial
Median
Ulnar

(a) Roots (rami C_5–T_1), trunks, divisions, and cords

Musculocutaneous nerve
Axillary nerve
Biceps brachii
Coracobrachialis
Median nerve
Radial nerve branches to triceps
Lateral cord
Posterior cord
Medial cord
Radial nerve
Ulnar nerve

(b) Cadaver photo

Axillary nerve
Humerus
Radial nerve
Musculo-cutaneous nerve
Ulna
Radius
Ulnar nerve
Median nerve
Radial nerve (superficial branch)
Dorsal branch of ulnar nerve
Superficial branch of ulnar nerve
Digital branch of ulnar nerve
Muscular branch
Digital branch
Median nerve

(c) The major nerves of the upper limb

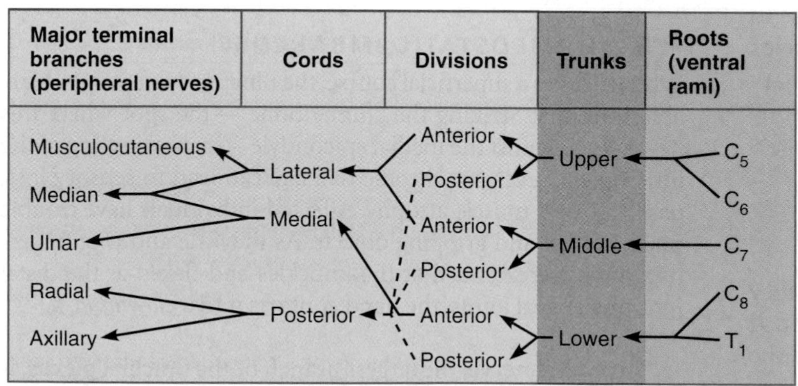

(d) Flowchart summarizing relationships within the brachial plexus

Figure 13.9 The brachial plexus. (Anterior view.)

TABLE 13.4 Branches of the Brachial Plexus (See Figure 13.9)

NERVES	CORD AND SPINAL ROOTS (VENTRAL RAMI)	STRUCTURES SERVED
Musculocutaneous	Lateral cord (C_5–C_7)	Muscular branches: flexor muscles in anterior arm (biceps brachii, brachialis, coracobrachialis) Cutaneous branches: skin on lateral forearm (extremely variable)
Median	By two branches, one from medial cord (C_8, T_1) and one from the lateral cord (C_5–C_7)	Muscular branches to flexor group of anterior forearm (palmaris longus, flexor carpi radialis, flexor digitorum superficialis, flexor pollicis longus, lateral half of flexor digitorum profundus, and pronator muscles); intrinsic muscles of lateral palm and digital branches to the fingers Cutaneous branches: skin of lateral two-thirds of hand, palm side and dorsum of fingers 2 and 3
Ulnar	Medial cord (C_8, T_1)	Muscular branches: flexor muscles in anterior forearm (flexor carpi ulnaris and medial half of flexor digitorum profundus); most intrinsic muscles of hand Cutaneous branches: skin of medial third of hand, both anterior and posterior aspects
Radial	Posterior cord (C_5–C_8, T_1)	Muscular branches: posterior muscles of arm and forearm (triceps brachii, anconeus, supinator, brachioradialis, extensors carpi radialis longus and brevis, extensor carpi ulnaris, and several muscles that extend the fingers) Cutaneous branches: skin of posterolateral surface of entire limb (except dorsum of fingers 2 and 3)
Axillary	Posterior cord (C_5, C_6)	Muscular branches: deltoid and teres minor muscles Cutaneous branches: some skin of shoulder region
Dorsal scapular	Branches of C_5 rami	Rhomboid muscles and levator scapulae
Long thoracic	Branches of C_5–C_7 rami	Serratus anterior muscle
Subscapular	Posterior cord; branches of C_5 and C_6 rami	Teres major and subscapularis muscles
Suprascapular	Upper trunk (C_5, C_6)	Shoulder joint; supraspinatus and infraspinatus muscles
Pectoral (lateral and medial)	Branches of lateral and medial cords (C_5–T_1)	Pectoralis major and minor muscles

course. Its motor branches innervate essentially all the extensor muscles of the upper limb. The radial nerve produces elbow extension, forearm supination, wrist and finger extension, and thumb abduction.

HOMEOSTATIC IMBALANCE

Trauma to the radial nerve results in *wrist drop*, an inability to extend the hand at the wrist. Improper use of a crutch or "Saturday night paralysis," in which an intoxicated person falls asleep with an arm draped over the back of a chair or sofa edge, causes radial nerve compression and ischemia (deprivation of blood supply). ■

Lumbosacral Plexus and Lower Limb

The sacral and lumbar plexuses overlap substantially. Because many fibers of the lumbar plexus contribute to the sacral plexus via the **lumbosacral trunk**, the two plexuses are often referred to as the **lumbosacral plexus**. Although the lumbosacral plexus serves mainly the lower limb, it also sends some branches to the abdomen, pelvis, and buttock.

Lumbar Plexus The **lumbar plexus** arises from the spinal nerves L_1–L_4 and lies within the psoas major muscle **(Figure 13.10)**. Its proximal branches innervate parts of the abdominal wall muscles and the psoas muscle, but its major branches descend to innervate the anterior and medial thigh.

The **femoral nerve**, the largest terminal nerve of this plexus, runs deep to the inguinal ligament to enter the thigh and then divides into several large branches. The motor branches innervate anterior thigh muscles (quadriceps), which are the principal thigh flexors and knee extensors. The cutaneous branches serve the skin of the anterior thigh and the medial surface of the leg from knee to foot.

The **obturator nerve** (ob″tu-ra′tor) enters the medial thigh via the obturator foramen and innervates the adductor muscles. These and other smaller branches of the lumbar plexus are summarized in **Table 13.5**.

HOMEOSTATIC IMBALANCE

Compression of the spinal roots of the lumbar plexus, as by a herniated disc, results in gait problems because the femoral nerve serves the prime movers of both hip flexion and knee extension. Other symptoms are pain or anesthesia of the anterior thigh and of the medial thigh if the obturator nerve is impaired. ■

13

TABLE 13.6 Branches of the Sacral Plexus (See Figure 13.11)

NERVES	SPINAL ROOTS (VENTRAL RAMI)	STRUCTURES SERVED
Sciatic nerve	L_4, L_5, S_1–S_3	Composed of two nerves (tibial and common fibular) in a common sheath; they diverge just proximal to the knee
▪ Tibial (including sural, medial and lateral plantar, and medial calcaneal branches)	L_4–S_3	Cutaneous branches: to skin of posterior surface of leg and sole of foot Motor branches: to muscles of back of thigh, leg, and foot [hamstrings (except short head of biceps femoris), posterior part of adductor magnus, triceps surae, tibialis posterior, popliteus, flexor digitorum longus, flexor hallucis longus, and intrinsic muscles of foot]
▪ Common fibular (superficial and deep branches)	L_4–S_2	Cutaneous branches: to skin of anterior and lateral surface of leg and dorsum of foot Motor branches: to short head of biceps femoris of thigh, fibular muscles of lateral compartment of leg, tibialis anterior, and extensor muscles of toes (extensor hallucis longus, extensors digitorum longus and brevis)
Superior gluteal	L_4, L_5, S_1	Motor branches: to gluteus medius and minimus and tensor fasciae latae
Inferior gluteal	L_5–S_2	Motor branches: to gluteus maximus
Posterior femoral cutaneous	S_1–S_3	Skin of buttock, posterior thigh, and popliteal region; length variable; may also innervate part of skin of calf and heel
Pudendal	S_2–S_4	Supplies most of skin and muscles of perineum (region encompassing external genitalia and anus and including clitoris, labia, and vaginal mucosa in females, and scrotum and penis in males); external anal sphincter

13

and knee flexors) and to the adductor magnus. Immediately above the knee, the two divisions of the sciatic nerve diverge.

The **tibial nerve** courses through the popliteal fossa (the region just posterior to the knee joint) and supplies the posterior compartment muscles of the leg and the skin of the posterior calf and sole of the foot. In the vicinity of the knee, the tibial nerve gives off the **sural nerve**, which serves the skin of the posterolateral leg, and at the ankle the tibial nerve divides into the **medial** and **lateral plantar nerves**, which serve most of the foot. The **common fibular nerve**, or *common peroneal nerve* (*perone* = fibula), descends from its point of origin, wraps around the neck of the fibula, and then divides into superficial and deep branches. These branches innervate the knee joint, skin of the anterior and lateral leg and dorsum of the foot, and muscles of the anterolateral leg (the extensors that dorsiflex the foot).

The next largest sacral plexus branches are the **superior** and **inferior gluteal nerves**. Together, they innervate the buttock (gluteal) and tensor fasciae latae muscles. The **pudendal nerve** (pu-den′dal; "shameful") innervates the muscles and skin of the perineum, helps stimulate erection, and is involved in voluntary control of urination (see Table 10.7). Other branches of the sacral plexus supply the thigh rotators and muscles of the pelvic floor.

HOMEOSTATIC IMBALANCE

Injury to the proximal part of the sciatic nerve, as might follow a fall, disc herniation, or improper administration of an injection into the buttock, results in a number of lower limb impairments, depending on the precise nerve roots injured. *Sciatica* (si-at′ĭ-kah), characterized by stabbing pain radiating over the course of the sciatic nerve, is common. When the nerve is transected, the leg is nearly useless. The leg cannot be flexed (because the hamstrings are paralyzed), and all foot and ankle movement is lost. The foot drops into plantar flexion (it dangles), a condition called *footdrop*. Recovery from sciatic nerve injury is usually slow and incomplete.

If the lesion occurs below the knee, thigh muscles are spared. When the tibial nerve is injured, the paralyzed calf muscles cannot plantar flex the foot and a shuffling gait develops. The common fibular nerve is susceptible to injury largely because of its superficial location at the head and neck of the fibula. Even a tight leg cast, or remaining too long in a side-lying position on a firm mattress, can compress this nerve and cause footdrop. ■

Innervation of Skin: Dermatomes

The area of skin innervated by the cutaneous branches of a single spinal nerve is called a **dermatome** (der′mah-tōm; "skin segment"). Every spinal nerve except C_1 innervates dermatomes. In patients with spinal cord injuries, you can pinpoint which nerves are damaged and locate the injured region of the spinal cord by determining which dermatomes are affected.

Adjacent dermatomes on the body trunk are fairly uniform in width, almost horizontal, and in direct line with their spinal nerves (**Figure 13.12**). The dermatome arrangement in the limbs is less obvious. (It is also more variable and different clinicians have mapped a variety of areas for the same dermatomes.) The skin of the upper limbs is supplied by ventral rami of C_5–T_1 (or T_2). The lumbar nerves supply most of the anterior surfaces of the thighs and legs, and the sacral nerves serve most of the posterior surfaces of the lower limbs. (This distribution basically reflects the areas supplied by the lumbar and sacral plexuses, respectively.)

Adjacent dermatomes are not as cleanly separated as a typical dermatome map indicates. On the trunk, neighboring dermatomes overlap considerably (about 50%). As a result, destruction of a single spinal nerve will not cause complete

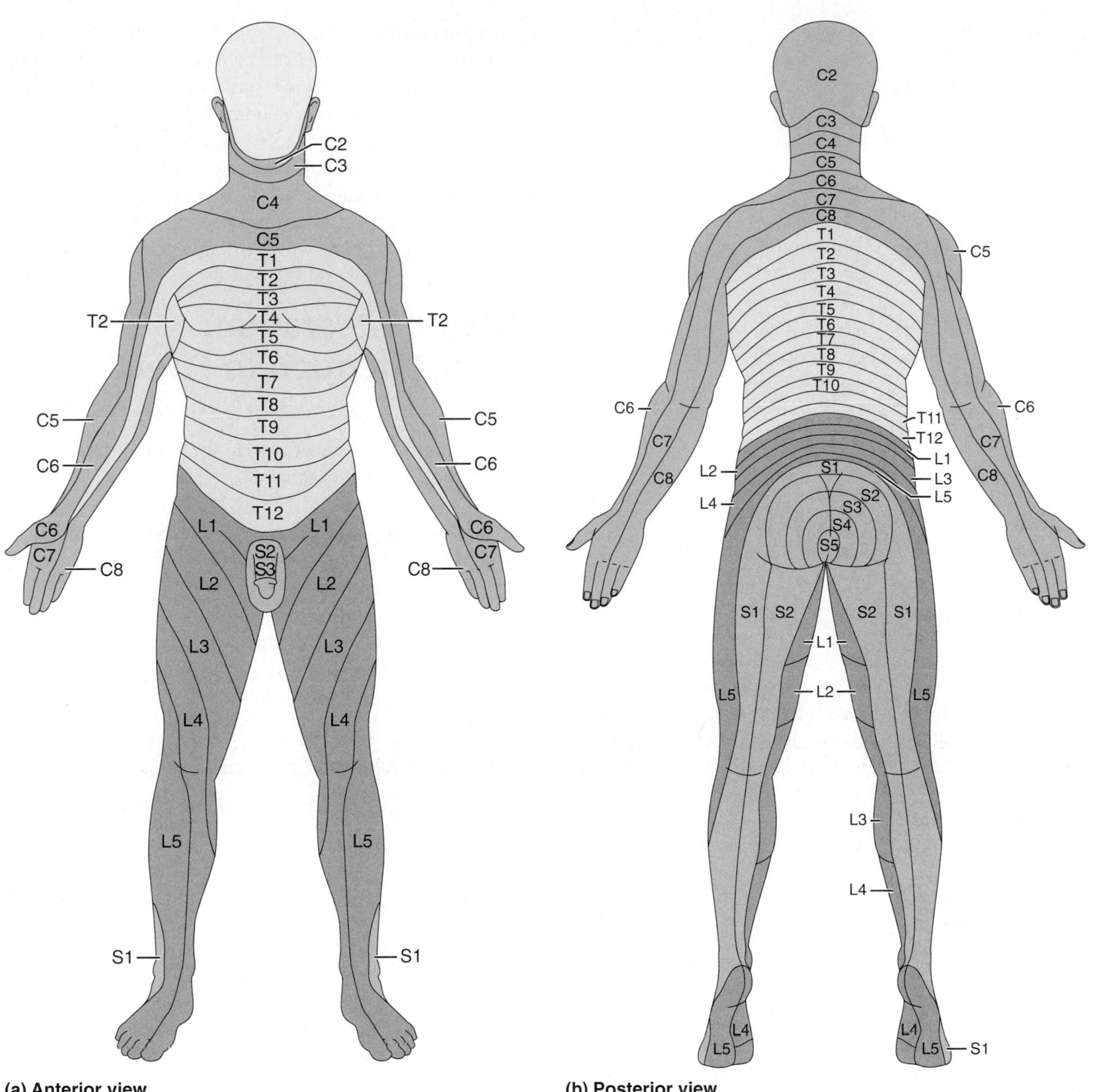

Figure 13.12 Map of dermatomes. Each dermatome is the skin segment innervated by the cutaneous sensory branches of a single spinal nerve. All spinal nerves but C_1 participate in the innervation of the dermatomes.

numbness anywhere. In the limbs, the overlap is less complete and some skin regions are innervated by just one spinal nerve.

Innervation of Joints

The easiest way to remember which nerves serve which synovial joint is to use **Hilton's law**, which says: *Any nerve serving a muscle that produces movement at a joint also innervates the joint and the skin over the joint.* Hence, once you learn which nerves serve the various major muscles and muscle groups, no new learning is necessary. For example, the knee is crossed by the quadriceps, gracilis, and hamstring muscles. The nerves to these muscles are the femoral nerve anteriorly and branches of the sciatic and obturator nerves posteriorly. Consequently, these nerves innervate the knee joint as well.

CHECK YOUR UNDERSTANDING

10. Spinal nerves have both *dorsal roots* and *dorsal rami*. How are these different from each other in location and in functional composition?
11. After his horse-riding accident, the actor Christopher Reeve was unable to breathe on his own. Which spinal nerve roots, spinal nerve, and spinal nerve plexus were involved?

For answers, see Appendix G.

PART 3

MOTOR ENDINGS AND MOTOR ACTIVITY

Peripheral Motor Endings

▶ Compare and contrast the motor endings of somatic and autonomic nerve fibers.

So far we have covered the structure of sensory receptors that detect stimuli, and that of nerves containing the afferent and efferent fibers that deliver impulses to and from the CNS. We now turn to **motor endings**, the PNS elements that activate effectors by releasing neurotransmitters. Because we discussed that topic in Chapter 9 with the innervation of body muscles, all we need to do here is recap. To balance the overview of sensory function provided earlier in this chapter, we will follow the recap with a brief overview of motor integration.

Innervation of Skeletal Muscle

As illustrated in Figure 9.8 (p. 286), the terminals of somatic motor fibers that innervate voluntary muscles form elaborate **neuromuscular junctions** with their effector cells. As each axon branch reaches its target, a single muscle fiber, the ending splits into a cluster of *axon terminals* that branch treelike over the junctional folds of the sarcolemma of the muscle fiber. The axon terminals contain mitochondria and synaptic vesicles filled with the neurotransmitter acetylcholine (ACh).

13

When a nerve impulse reaches an axon terminal, ACh is released by exocytosis, diffuses across the fluid-filled synaptic cleft (about 50 nm wide), and attaches to ACh receptors on the highly infolded sarcolemma at the junction. ACh binding results in the opening of ligand-gated channels that allow both Na^+ and K^+ to pass. Because more Na^+ enters the cell than K^+ leaves, the muscle cell interior at that point depolarizes, producing a type of graded potential called an *end plate potential*. The end plate potential spreads to adjacent areas of the membrane where it triggers the opening of voltage-gated sodium channels. This event leads to propagation of an action potential along the sarcolemma that stimulates the muscle fiber to contract. The synaptic cleft at somatic neuromuscular junctions is filled with a glycoprotein-rich basal lamina, a structure not seen at other synapses. The basal lamina contains *acetylcholinesterase*, the enzyme that breaks down ACh almost immediately after it binds.

Innervation of Visceral Muscle and Glands

The junctions between autonomic motor endings and their effectors (which are smooth and cardiac muscle and glands) are much simpler than the junctions formed between somatic fibers and skeletal muscle cells. The autonomic motor axons branch repeatedly, each branch forming *synapses en passant* with its effector cells. Instead of a cluster of bulblike terminals, an axon ending serving smooth muscle or a gland (but not cardiac muscle) has a series of **varicosities**, knoblike swellings containing mitochondria and synaptic vesicles, that make it look like a string of beads (see Figure 9.27).

The autonomic synaptic vesicles typically contain either acetylcholine or norepinephrine, both of which act indirectly on their targets via second messengers. Consequently, the visceral motor responses tend to be slower than those induced by somatic motor fibers, which directly open ion channels.

Motor Integration: From Intention to Effect

▶ Outline the three levels of the motor hierarchy.

▶ Compare the roles of the cerebellum and basal nuclei in controlling motor activity.

How does integration in the motor system compare with integration in sensory systems? In the motor system, we have motor endings serving effectors (muscle fibers) instead of sensory receptors, descending efferent circuits instead of ascending afferent circuits, and motor behavior instead of perception. However, as in sensory systems, the basic mechanisms of motor systems operate at three levels.

Levels of Motor Control

The cerebral cortex is at the highest level of our conscious motor pathways, but it is *not* the ultimate planner and coordinator of complex motor activities. The cerebellum and basal nuclei (ganglia) play this role and are therefore at the top of the motor control hierarchy. Motor control exerted by lower levels is mediated by *reflex arcs* in some cases, but complex motor behavior, such as walking and swimming, appears to depend on more complex patterns. Currently, we define three levels of motor control: the *segmental level*, the *projection level*, and the *precommand level* **(Figure 13.13)**.

The Segmental Level

The lowest level of the motor hierarchy, the **segmental level**, consists of the spinal cord circuits. A segmental circuit activates a network of ventral horn neurons in a group of cord segments, causing them to stimulate specific groups of muscles. Circuits that control locomotion and other specific and oft-repeated motor activities are called **central pattern generators (CPGs)**. CPGs consist of networks of oscillating inhibitory and excitatory neurons, which set crude rhythms and alternating patterns of movement.

The Projection Level

The spinal cord is under the direct control of the **projection level** of motor control. The projection level consists of *upper motor neurons* of the motor cortex, which initiate the *direct* (*pyramidal*) *system*, and of brain stem motor nuclei, which oversee the *indirect* (*extrapyramidal*) *system* (see Table 12.3 and pp. 473–476). Axons

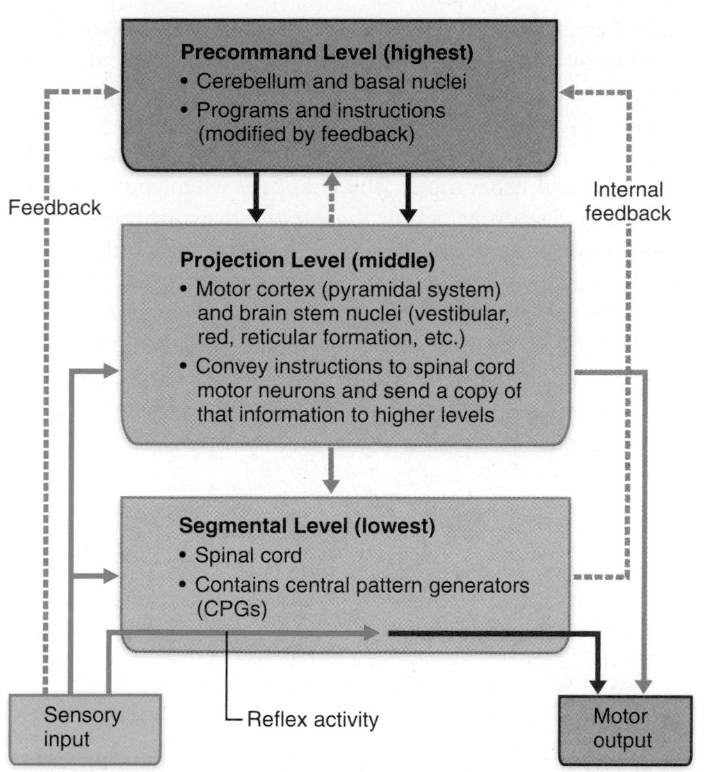

(a) Levels of motor control and their interactions

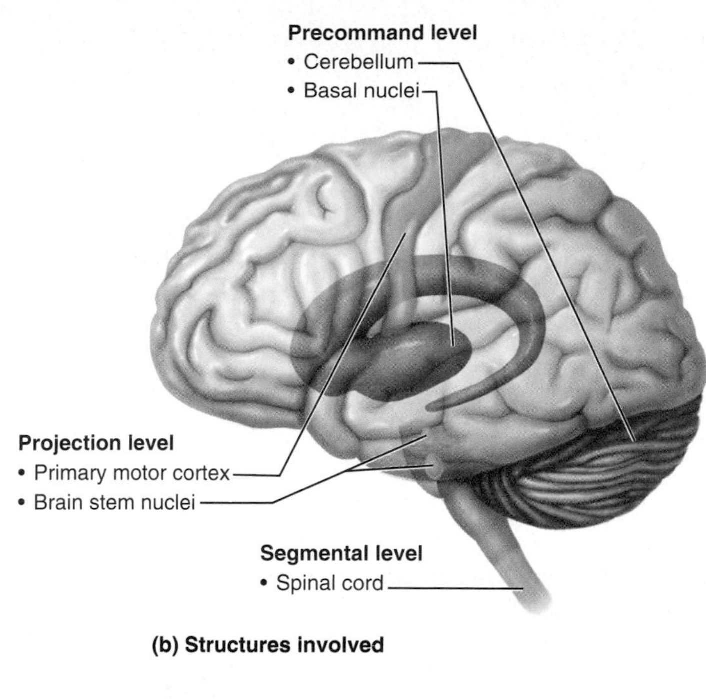

(b) Structures involved

Figure 13.13 Hierarchy of motor control.

of the direct system neurons produce discrete voluntary movements of the skeletal muscles. Axons of the indirect system help control reflex and CPG-controlled motor actions, modifying and controlling the activity of the segmental apparatus.

Projection motor pathways convey information to lower motor neurons, and send a copy of that information as *internal feedback* to higher command levels, continually informing them of what is happening. The direct and indirect systems provide separate and parallel pathways for controlling the spinal cord, but these systems are interrelated at all levels.

The Precommand Level

Two other systems of brain neurons, located in the cerebellum and basal nuclei, regulate motor activity. They precisely start or stop movements, coordinate movements with posture, block unwanted movements, and monitor muscle tone. Collectively called **precommand areas**, these systems *control the outputs* of the cortex and brain stem motor centers and stand at the highest level of the motor hierarchy.

The key center for "online" sensorimotor integration and control is the **cerebellum**. Remember the cerebellum is a target of ascending proprioceptor, tactile, equilibrium, and visual inputs—feedback that it needs for rapid correction of "errors" in motor activity. It also receives information via branches from descending pyramidal tracts, and from various brain stem nuclei. The cerebellum lacks direct connections to the spinal cord; it acts on motor pathways through the projection areas of the brain stem and on the motor cortex via the thalamus to fine-tune motor activity.

The **basal nuclei** receive inputs from *all* cortical areas and send their output back mainly to premotor and prefrontal cortical areas via the thalamus. Compared to the cerebellum, the basal nuclei appear to be involved in more complex aspects of motor control. Under resting conditions, the basal nuclei inhibit various motor centers of the brain, but when the motor centers are released from inhibition, coordinated motions can begin.

Cells in both the basal nuclei and the cerebellum are involved in this unconscious planning and discharge *in advance* of willed movements. When you actually move your fingers, both the precommand areas and the primary motor cortex are active. At the risk of oversimplifying, it appears that the cortex says, "I want to do this," and then lets the precommand areas take over to provide the proper timing and patterns to execute the movements desired. The precommand areas control the motor cortex and provide its readiness to initiate a voluntary act. The conscious cortex then chooses to act or not act, but the groundwork has already been laid.

CHECK YOUR UNDERSTANDING

12. What are varicosities and where would you find them?
13. What parts of the nervous system ultimately plan and coordinate complex motor activities?

For answers, see Appendix G.

13

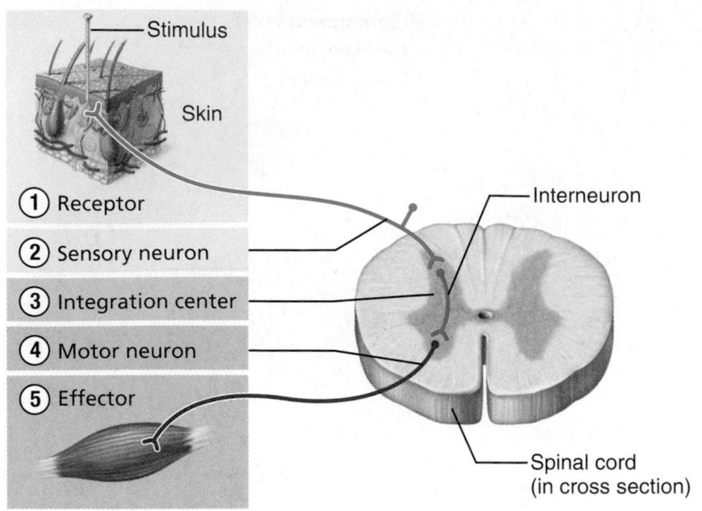

Figure 13.14 The five basic components of all reflex arcs. The reflex arc illustrated is polysynaptic.

PART 4

REFLEX ACTIVITY

The Reflex Arc

13

▶ Name the components of a reflex arc and distinguish between autonomic and somatic reflexes.

Many of the body's control systems belong to a general category known as reflexes, which can be either inborn or learned. In the most restricted sense, an *inborn*, or *intrinsic, reflex* is a rapid, predictable motor response to a stimulus. It is unlearned, unpremeditated, and involuntary, and may be considered as built into our neural anatomy. Reflexes prevent us from having to *think* about all the little details of staying upright, intact, and alive—helping us maintain posture, avoid pain, and control visceral activities.

One example of an inborn reflex is what happens when you splash a pot of boiling water on your arm; you are likely to drop the pot instantly and involuntarily even before feeling any pain. This response is triggered by a spinal reflex without any help from the brain. In many cases we are aware of the final response of a basic reflex activity (you know you've dropped the pot of boiling water). In other cases, reflex activities go on without any awareness on our part. This is typical of many visceral reflexes, which are regulated by the subconscious lower regions of the CNS, specifically the brain stem and spinal cord.

In addition to these basic, inborn types of reflexes, there are *learned*, or *acquired, reflexes* that result from practice or repetition. Take, for instance, the complex sequence of reactions that occurs when an experienced driver drives a car. The process is largely automatic, but only because substantial time and effort were expended to acquire the driving skill. In reality, the distinction between inborn and learned reflexes is not clear-cut and most inborn reflex actions are subject to modification by learning and conscious effort. For instance, if a 3-year-old child was standing by your side when you scalded your arm, you most likely would set the pot down (rather than just letting go) because of your conscious recognition of the danger to the child.

Recall the discussion in Chapter 11 about serial and parallel processing of sensory input. What happens when you scald your arm is a good example of how these two processing modes work together. You drop the pot before feeling any pain, but the pain signals picked up by the interneurons of the spinal cord are quickly transmitted to the brain, so that within the next few seconds you *do* become aware of pain, and you also know what happened to cause it. The withdrawal reflex is serial processing mediated by the spinal cord, and pain awareness reflects simultaneous parallel processing of the sensory input.

Components of a Reflex Arc

As you learned in Chapter 11, reflexes occur over highly specific neural paths called **reflex arcs**, all of which have five essential components (**Figure 13.14**):

1. **Receptor:** Site of the stimulus action.
2. **Sensory neuron:** Transmits afferent impulses to the CNS.
3. **Integration center:** In simple reflex arcs, the integration center may be a single synapse between a sensory neuron and a motor neuron (**monosynaptic reflex**). More complex reflex arcs involve multiple synapses with chains of interneurons (**polysynaptic reflex**). The integration center for the reflexes we will describe in this chapter is within the CNS.
4. **Motor neuron:** Conducts efferent impulses from the integration center to an effector organ.
5. **Effector:** Muscle fiber or gland cell that responds to the efferent impulses (by contracting or secreting).

Reflexes are classified functionally as **somatic reflexes** if they activate skeletal muscle, or as **autonomic (visceral) reflexes** if they activate visceral effectors (smooth or cardiac muscle or glands). Here we describe some common somatic reflexes mediated by the spinal cord. We will consider autonomic reflexes in later chapters along with the visceral processes they help to regulate.

Spinal Reflexes

▶ Compare and contrast stretch, flexor, crossed-extensor, and Golgi tendon reflexes.

Somatic reflexes mediated by the spinal cord are called **spinal reflexes.** Many spinal reflexes occur without the direct involvement of higher brain centers. Generally, these reflexes are even present in animals whose brains have been destroyed as long as the spinal cord is still functional. However, the brain is "advised" of most spinal reflex activity and can facilitate, inhibit, or adapt it, depending on the circumstances (as we described in the example of the hot water–filled pot). Moreover, continuous facilitating signals from the brain are required for normal spinal reflex activity. As we saw in Chapter 12, *spinal shock* occurs when the spinal cord is transected, immediately depressing all functions controlled by the cord.

Testing of somatic reflexes is important clinically to assess the condition of the nervous system. Exaggerated, distorted, or absent reflexes indicate degeneration or pathology of specific nervous system regions, often before other signs are apparent.

Stretch and Golgi Tendon Reflexes

What information does your nervous system need in order to smoothly coordinate the activity of your skeletal muscles? Two types of information about the current state of a muscle are key. First, the nervous system needs to know the length of the muscle. The *muscle spindles*, found in skeletal muscles, supply this information. Second, it needs to know the amount of tension in the muscle and its associated tendons. *Golgi tendon organs*, located in the tendons, provide this information. These two types of proprioceptors play an important role in spinal reflexes and also provide essential feedback to the cerebral cortex and cerebellum. Let's take a closer look at the functional anatomy of these proprioceptors and their roles in certain spinal reflexes.

Functional Anatomy of Muscle Spindles

Each muscle spindle consists of three to ten modified skeletal muscle fibers called **intrafusal muscle fibers** (*intra* = within; *fusal* = the spindle) enclosed in a connective tissue capsule **(Figure 13.15)**. These fibers are less than one-quarter the size of the effector fibers of the muscle, called **extrafusal muscle fibers**.

The central regions of the intrafusal fibers lack myofilaments and are noncontractile. These regions are the receptive surfaces of the spindle. They are wrapped by two types of afferent endings that send sensory inputs to the CNS. The **primary sensory endings** of large **type Ia fibers**, which innervate the spindle center, are stimulated by both the rate and degree of stretch. The **secondary sensory endings** of small **type II fibers** supply the spindle ends and are stimulated only by degree of stretch.

The intrafusal muscle fibers have contractile regions at their ends, which are the only areas containing actin and myosin myofilaments. These regions are innervated by **gamma (γ) efferent fibers** that arise from small motor neurons in the ventral horn of the spinal cord. These γ motor fibers, which maintain spindle sensitivity (as described shortly), are distinct from the **alpha (α) efferent fibers** of the large **alpha (α) motor neurons** that stimulate the extrafusal muscle fibers to contract.

The muscle spindle is stretched (and excited) in one of two ways: (1) by applying an external force that lengthens the entire muscle, such as occurs when we carry a heavy weight or when antagonistic muscles contract (external stretch); or (2) by activating the γ motor neurons that stimulate the distal ends of the intrafusal fibers to contract, thereby stretching the middle of the spindle (internal stretch). Whenever the muscle spindle is stretched, its associated sensory neurons transmit impulses at higher frequency to the spinal cord **(Figure 13.16a, b)**.

During voluntary skeletal muscle contraction, the muscle shortens. If the intrafusal muscle fibers didn't contract along

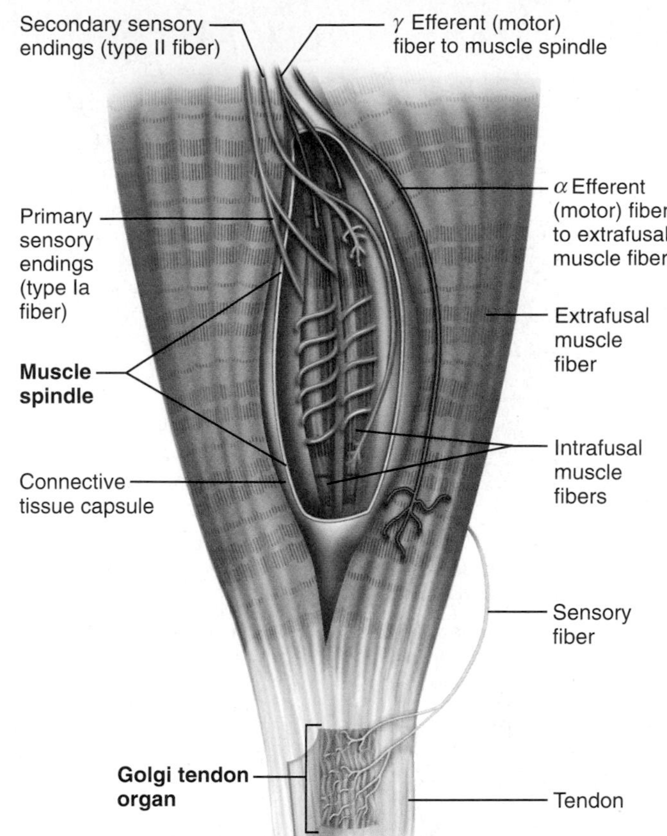

Figure 13.15 Anatomy of the muscle spindle and Golgi tendon organ. Notice the afferent fibers from and efferent fibers to the muscle spindle. Myelin has been omitted from all nerve fibers for clarity.

with the extrafusal fibers, then the muscle spindle would go slack and cease generating action potentials (Figure 13.16c). At this point it would be unable to signal further changes in muscle length, so it would be useless.

Fortunately, **α-γ coactivation** prevents this from happening. Descending fibers of motor pathways synapse with both α and γ motor neurons, and motor impulses are simultaneously sent to the large extrafusal fibers and to muscle spindle intrafusal fibers. Stimulating the intrafusal fibers maintains the spindle's tension (and sensitivity) during muscle contraction, so that the brain continues to be notified of conditions in the muscle (Figure 13.16d). Without such a system, information on changes in muscle length would cease to flow from contracting muscles.

The Stretch Reflex

By sending commands to the motor neurons, the brain essentially sets a muscle's length. The **stretch reflex** makes sure that the muscle stays at that length. For example, the **patellar** (pah-tel′ar) or **knee-jerk reflex** is a stretch reflex that helps keep your knees from buckling when you are standing upright. As your knees begin to buckle and the quadriceps lengthens, the stretch reflex causes the quadriceps to contract without your having to think about it. The stretch reflex and a specific example—the

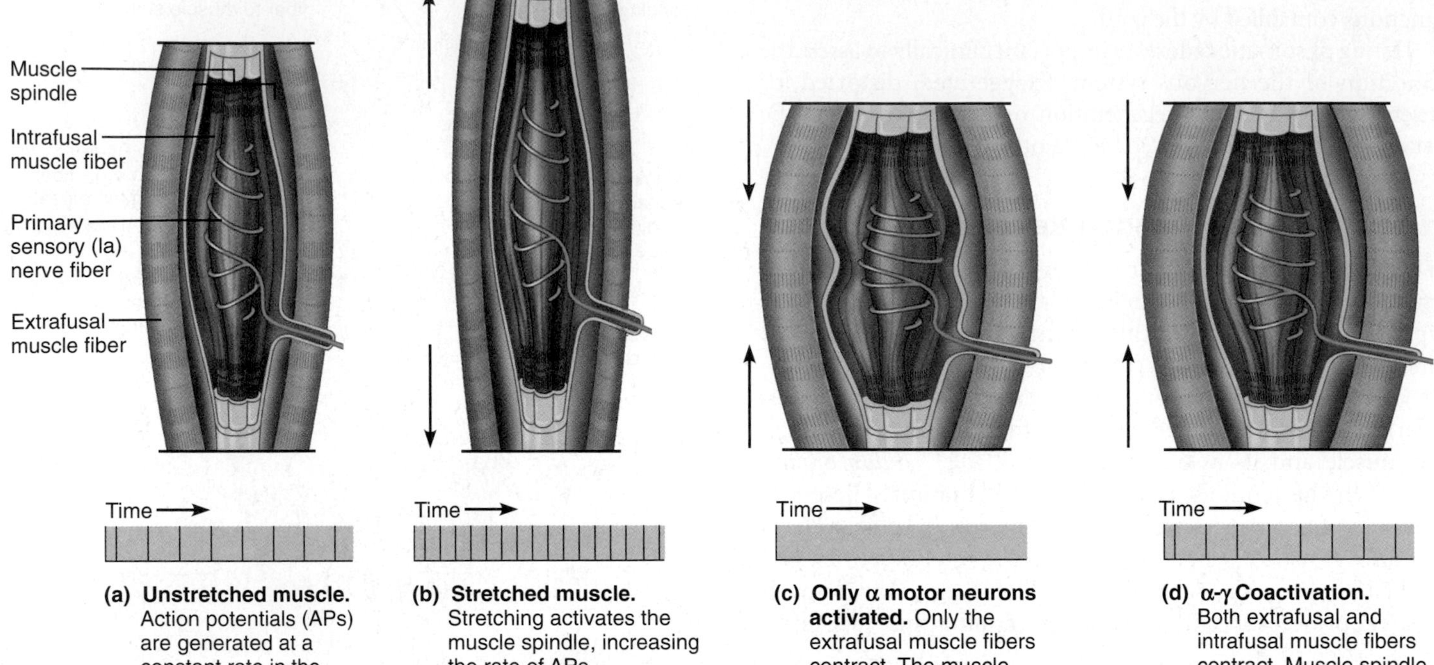

(a) Unstretched muscle. Action potentials (APs) are generated at a constant rate in the associated sensory (Ia) fiber.

(b) Stretched muscle. Stretching activates the muscle spindle, increasing the rate of APs.

(c) Only α motor neurons activated. Only the extrafusal muscle fibers contract. The muscle spindle becomes slack and no APs are fired. It is unable to signal further length changes.

(d) α-γ Coactivation. Both extrafusal and intrafusal muscle fibers contract. Muscle spindle tension is maintained and it can still signal changes in length.

Figure 13.16 Operation of the muscle spindle. The action potentials generated in the sensory (Ia) fibers are shown for each case as black lines in yellow bars.

knee-jerk reflex—are shown in *Focus on the Stretch Reflex* **(Figure 13.17)**.

The stretch reflex is important for maintaining muscle tone and adjusting it reflexively. It is most important in the large extensor muscles that sustain upright posture and in postural muscles of the trunk. For example, contractions of the postural muscles of the spine are almost continuously regulated by stretch reflexes initiated first on one side of the spine and then on the other.

Let's look at how the stretch reflex works. As we've just seen in Figure 13.16, sensory neurons of muscle spindles activated by stretch transmit impulses at a higher frequency to the spinal cord. There the sensory neurons synapse directly with α motor neurons, which rapidly excite the extrafusal muscle fibers of the stretched muscle (Figure 13.17). The reflexive muscle contraction that follows (an example of serial processing) resists further muscle stretching.

Branches of the afferent fibers also synapse with interneurons that inhibit motor neurons controlling antagonistic muscles (parallel processing), and the resulting inhibition is called **reciprocal inhibition**. Consequently, the stretch stimulus causes the antagonists to relax so that they cannot resist the shortening of the "stretched" muscle caused by the main reflex arc. While this spinal reflex is occurring, information on muscle length and the speed of muscle shortening is being relayed (mainly via the dorsal white columns) to higher brain centers (more parallel processing).

The most familiar clinical example of a stretch reflex is the knee-jerk reflex we have just described. Stretch reflexes can be elicited in any skeletal muscle by a sudden jolt to the tendon or the muscle itself. All stretch reflexes are **monosynaptic** and **ipsilateral**. In other words, they involve a single synapse and motor activity on the same side of the body. Although stretch reflexes *themselves* are monosynaptic, the reflex arcs that inhibit the motor neurons serving the antagonistic muscles are polysynaptic.

A positive knee jerk (or a positive result for any other stretch reflex test) provides two important pieces of information. First, it proves that the sensory and motor connections between that muscle and the spinal cord are intact. Second, the vigor of the response indicates the degree of excitability of the spinal cord. When the spinal motor neurons are highly facilitated by impulses descending from higher centers, just touching the muscle tendon produces a vigorous reflex response. On the other hand, when the lower motor neurons are bombarded by inhibitory signals, even pounding on the tendon may fail to cause the reflex response.

HOMEOSTATIC IMBALANCE

Stretch reflexes tend to be either hypoactive or absent in cases of peripheral nerve damage or ventral horn injury involving the tested area. These reflexes are absent in those with chronic diabetes mellitus or neurosyphilis and during coma. However, they are hyperactive when lesions of the corticospinal tract reduce the inhibitory effect of the brain on the spinal cord (as in stroke patients). ■

Figure 13.17 FOCUS The Stretch Reflex

Stretched muscle spindles initiate a stretch reflex, causing contraction of the stretched muscle and inhibition of its antagonist.

The events by which muscle stretch is damped

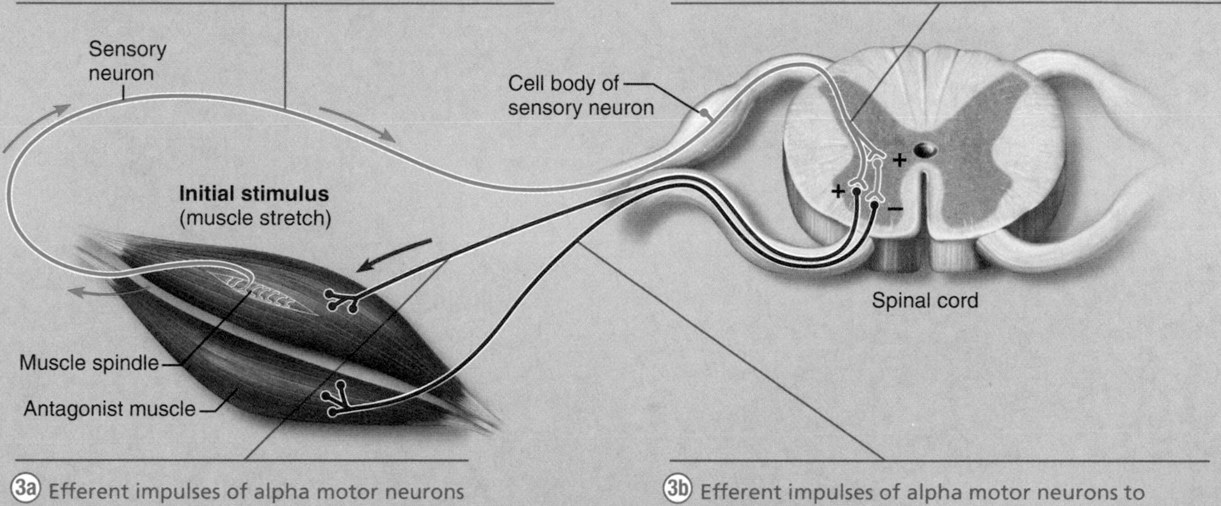

1 When muscle spindles are activated by stretch, the associated sensory neurons (blue) transmit afferent impulses at higher frequency to the spinal cord.
2 The sensory neurons synapse directly with alpha motor neurons (red), which excite extrafusal fibers of the stretched muscle. Afferent fibers also synapse with interneurons (green) that inhibit motor neurons (purple) controlling antagonistic muscles.
Sensory neuron
Cell body of sensory neuron
Initial stimulus (muscle stretch)
+
+
–
Spinal cord
Muscle spindle
Antagonist muscle
3a Efferent impulses of alpha motor neurons cause the stretched muscle to contract, which resists or reverses the stretch.
3b Efferent impulses of alpha motor neurons to antagonist muscles are reduced (reciprocal inhibition).

The patellar (knee-jerk) reflex – a specific example of a stretch reflex

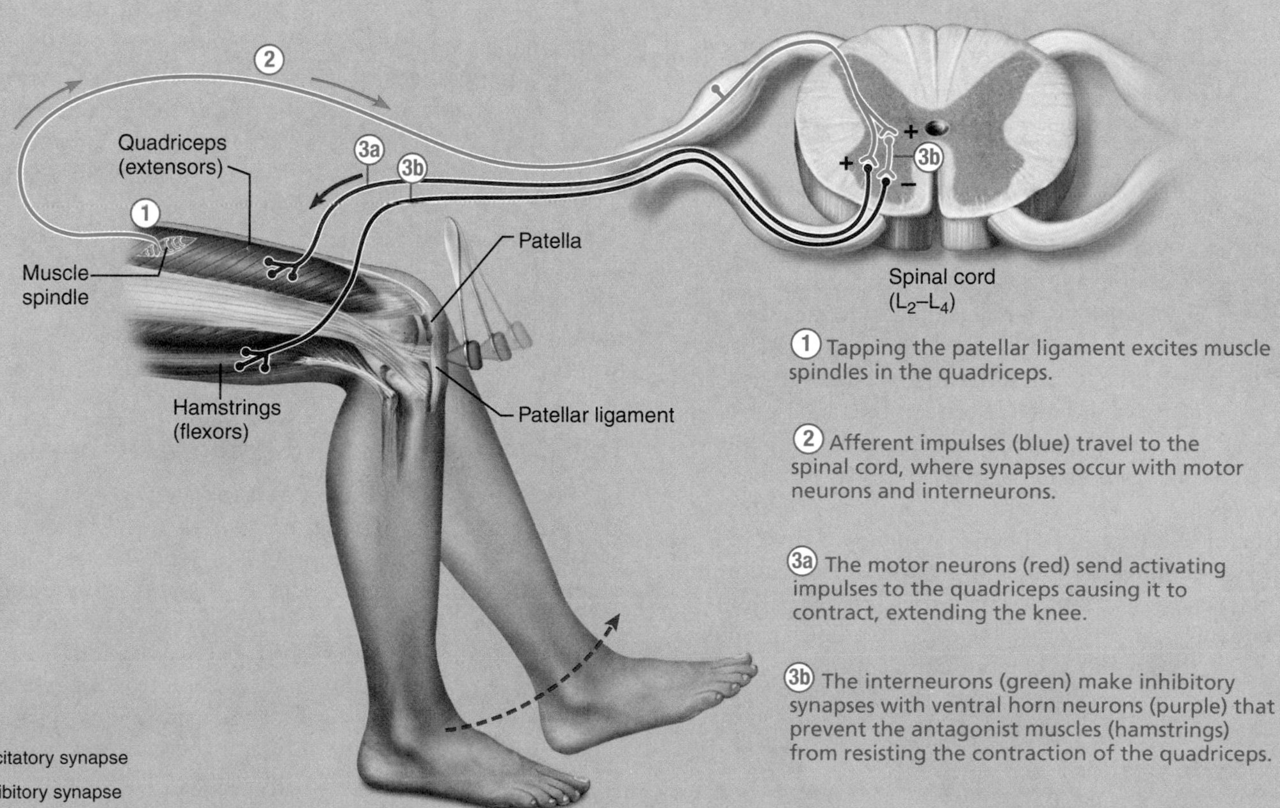

1 Tapping the patellar ligament excites muscle spindles in the quadriceps.

2 Afferent impulses (blue) travel to the spinal cord, where synapses occur with motor neurons and interneurons.

3a The motor neurons (red) send activating impulses to the quadriceps causing it to contract, extending the knee.

3b The interneurons (green) make inhibitory synapses with ventral horn neurons (purple) that prevent the antagonist muscles (hamstrings) from resisting the contraction of the quadriceps.

A&P Flix View this animation at myA&P

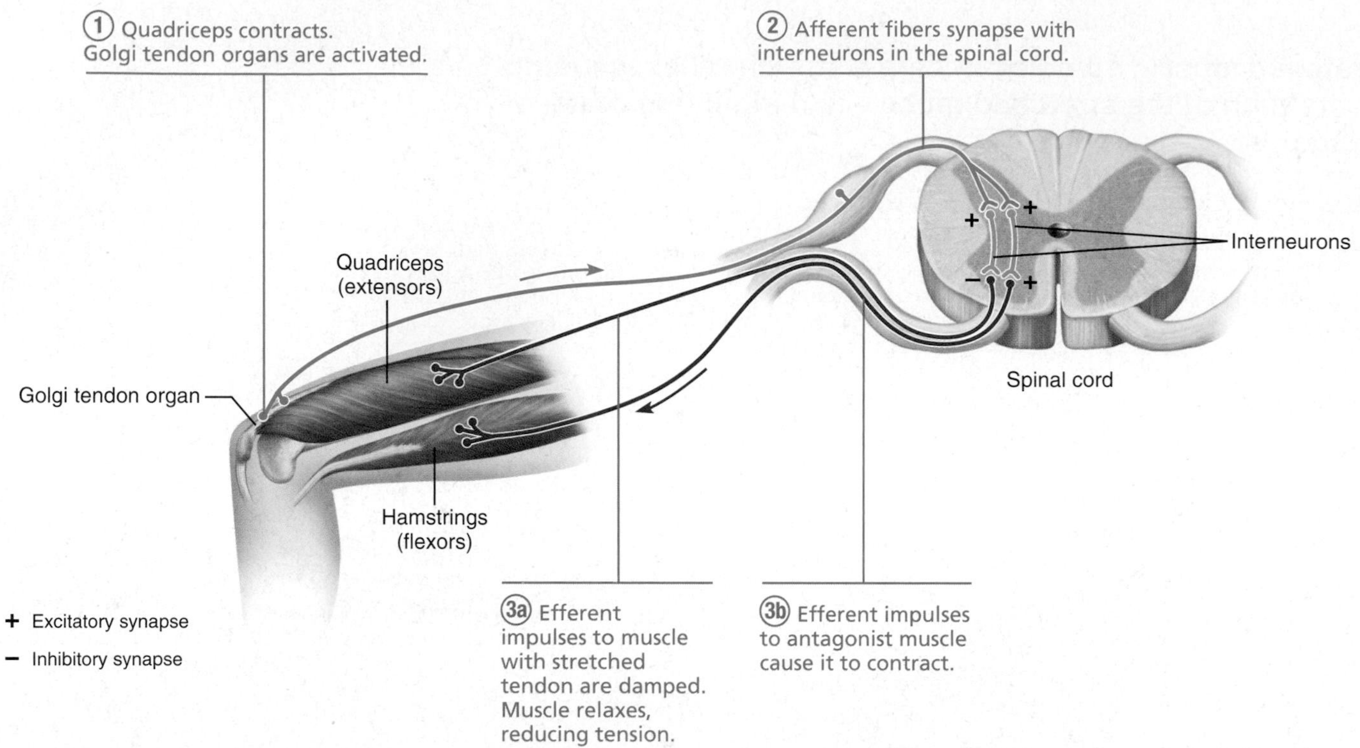
① Quadriceps contracts. Golgi tendon organs are activated.
② Afferent fibers synapse with interneurons in the spinal cord.
Interneurons
Quadriceps (extensors)
Golgi tendon organ
Spinal cord
Hamstrings (flexors)
+ Excitatory synapse
− Inhibitory synapse
3a Efferent impulses to muscle with stretched tendon are damped. Muscle relaxes, reducing tension.
3b Efferent impulses to antagonist muscle cause it to contract.

Figure 13.18 **The Golgi tendon reflex.**

13

Adjusting Muscle Spindle Sensitivity

The motor supply to the muscle spindle allows the brain to voluntarily modify the stretch reflex response and the firing rate of α motor neurons. When the γ neurons are vigorously stimulated by impulses from the brain, the spindle is stretched and highly sensitive, and muscle contraction force is maintained or increased. When the γ motor neurons are inhibited, the spindle resembles a loose rubber band and is nonresponsive, and the extrafusal muscles relax.

The ability to modify the stretch reflex is important in many situations. As the speed and difficulty of a movement increases, the brain increases γ motor output to increase the sensitivity of the muscle spindles. For example, a gymnast on a balance beam needs to have highly sensitive muscle spindles or she'll fall off. On the other hand, if you want to wind up to pitch a baseball, it is essential to suppress the stretch reflex so that your muscles can produce a large degree of motion (i.e., circumduct your pitching arm). Other athletes who require movements of maximum force learn to stretch muscles as much and as quickly as possible just before the movement. This advantage is demonstrated by the crouch that athletes assume just before jumping or running.

As we have seen, muscle tone and smooth coordination of movement depends upon having intact stretch reflex pathways. Both afferent and efferent fibers to the muscle spindle are vitally important. If either afferent or efferent fibers are cut, the muscle immediately loses its tone and becomes flaccid.

The Golgi Tendon Reflex

Stretch reflexes cause muscle contraction in response to increased muscle length (stretch). The polysynaptic **Golgi tendon reflexes**, on the other hand, produce exactly the opposite effect: muscle relaxation and lengthening in response to tension. When muscle tension increases substantially during contraction or passive stretching, high-threshold Golgi tendon organs in the tendon may be activated. Afferent impulses are transmitted to the spinal cord, and then to the cerebellum, where the information is used to adjust muscle tension. Simultaneously, motor neurons in spinal cord circuits supplying the contracting muscle are inhibited and antagonist muscles are activated, a phenomenon called **reciprocal activation**. As a result, the contracting muscle relaxes as its antagonist is activated **(Figure 13.18)**.

Golgi tendon organs help to prevent muscles and tendons from tearing when they are subjected to possibly damaging stretching force. Golgi tendon organs also function at normal muscle tensions. In the normal range, Golgi tendon organs help to ensure smooth onset and termination of muscle contraction.

The Flexor and Crossed-Extensor Reflexes

The **flexor**, or **withdrawal**, **reflex** is initiated by a painful stimulus and causes automatic withdrawal of the threatened body part from the stimulus (**Figure 13.19**, left). The response that occurs when you prick your finger is a good example. Flexor reflexes are ipsilateral and polysynaptic, the latter a necessity when several muscles must be recruited to withdraw the injured body part. Because flexor reflexes are protective reflexes important to our survival, they override the spinal pathways and prevent any other reflexes from using them at the same time. However, this reflex, like other spinal reflexes, can be overridden by descending signals from the brain. This happens when you

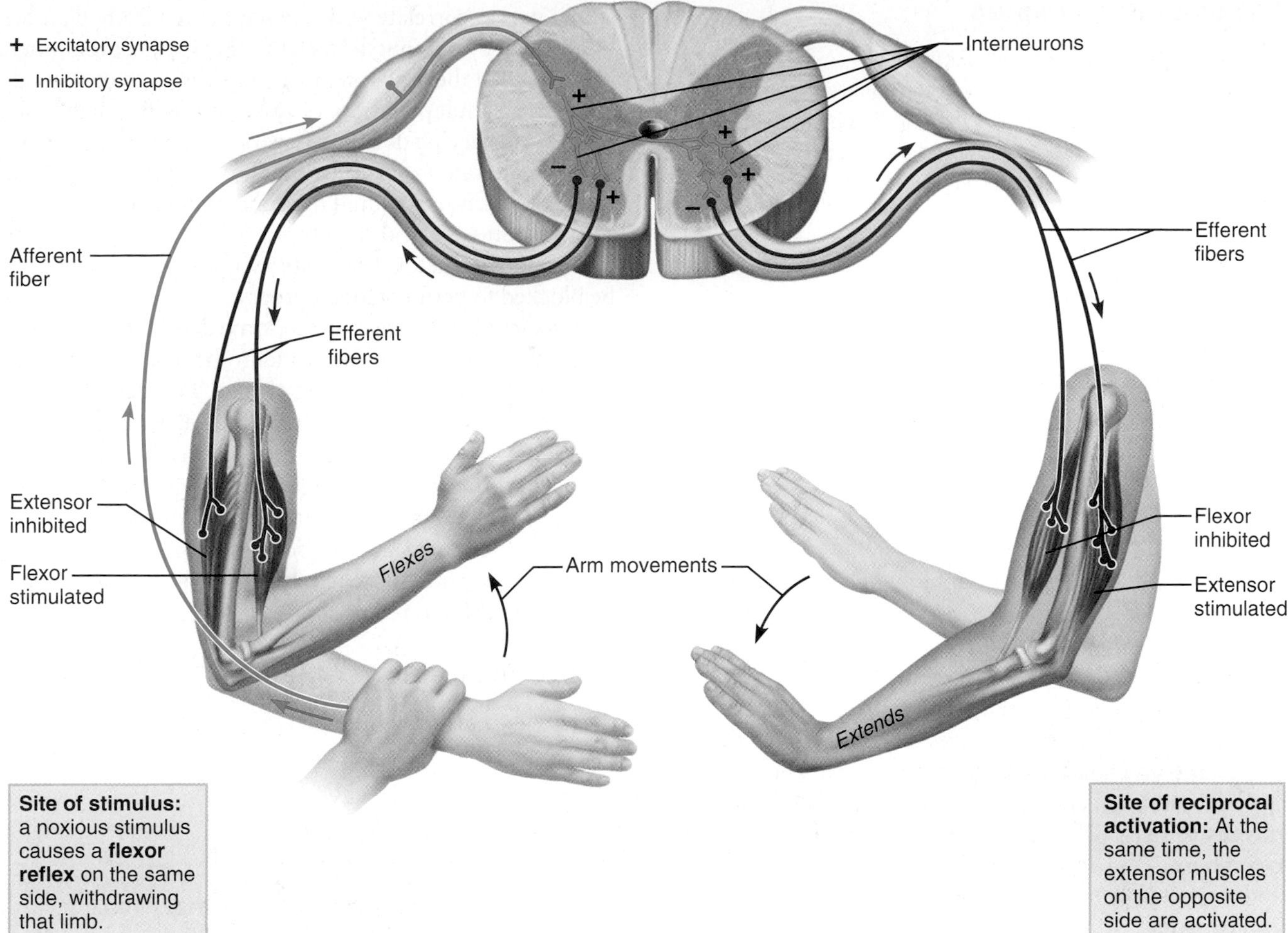
+ Excitatory synapse
– Inhibitory synapse
Interneurons
Afferent fiber
Efferent fibers
Efferent fibers
Extensor inhibited
Flexor stimulated
Flexes
Arm movements
Extends
Flexor inhibited
Extensor stimulated
Site of stimulus: a noxious stimulus causes a **flexor reflex** on the same side, withdrawing that limb.
Site of reciprocal activation: At the same time, the extensor muscles on the opposite side are activated.

Figure 13.19 The crossed-extensor reflex. In this example, a stranger suddenly grasps the right arm, which is reflexively withdrawn while the opposite (left) arm reflexively extends and pushes the stranger away.

are expecting a painful stimulus, for example a skin prick as a lab technician prepares to draw blood from a vein.

The **crossed-extensor reflex** often accompanies the flexor reflex in weight-bearing limbs and is particularly important in maintaining balance. It is a complex spinal reflex consisting of an ipsilateral withdrawal reflex and a contralateral extensor reflex. Incoming afferent fibers synapse with interneurons that control the flexor withdrawal response on the same side of the body and with other interneurons that control the extensor muscles on the opposite side.

This reflex is obvious when you step barefoot on broken glass. The ipsilateral response causes rapid lifting of the cut foot, while the contralateral response activates the extensor muscles of the opposite leg to support the weight suddenly shifted to it. The crossed-extensor reflex also occurs when someone unexpectedly grabs your arm. The grasped arm is withdrawn as the opposite arm pushes you away from the attacker (Figure 13.19).

Superficial Reflexes

Superficial reflexes are elicited by gentle cutaneous stimulation, such as that produced by stroking the skin with a tongue depressor. These clinically important reflexes depend both on functional upper motor pathways and on cord-level reflex arcs. The best known of these are the plantar and abdominal reflexes.

The **plantar reflex** tests the integrity of the spinal cord from L_4 to S_2 and indirectly determines if the corticospinal tracts are functioning properly. It is elicited by drawing a blunt object downward along the lateral aspect of the plantar surface (sole) of the foot. The normal response is a downward flexion (curling) of the toes. However, if the primary motor cortex or corticospinal tract is damaged, the plantar reflex is replaced by an abnormal reflex called **Babinski's sign**, in which the great toe dorsiflexes and the smaller toes fan laterally. Infants exhibit Babinski's sign until they are about a year old because their nervous systems are incompletely myelinated. Despite its clinical significance, the physiological mechanism of Babinski's sign is not understood.

Stroking the skin of the lateral abdomen above, to the side, or below the umbilicus induces a reflex contraction of the abdominal muscles in which the umbilicus moves toward the stimulated site. These reflexes, called **abdominal reflexes**, check the integrity of the spinal cord and ventral rami from T_8 to T_{12}. The abdominal reflexes vary in intensity from one person to another. They are absent when corticospinal tract lesions are present.

CHECK YOUR UNDERSTANDING

14. Name the five components of a reflex arc.
15. What is the role of the stretch reflex? The flexor reflex?
16. Juan injured his back in a fall. When his ER physician stroked the bottom of Juan's foot, she noted that his big toe pointed up and his other toes fanned out. What is this response called and what does it indicate?

For answers, see Appendix G.

Developmental Aspects of the Peripheral Nervous System

- **Describe the developmental relationship between the segmented arrangement of peripheral nerves, skeletal muscles, and skin dermatomes.**
- **List changes that occur in the peripheral nervous system with aging.**

Most skeletal muscles derive from paired blocks of mesoderm (somites) distributed segmentally down the posteromedial aspect of the embryo. The spinal nerves branch from the developing spinal cord and adjacent neural crest and exit between the forming vertebrae, and each nerve becomes associated with the adjacent muscle mass. The spinal nerves supply both sensory and motor fibers to the developing muscles and help direct their maturation. Cranial nerves innervate muscles of the head in a comparable manner.

13

The distribution of cutaneous nerves to the skin follows a similar pattern. Most of the scalp and facial skin is innervated by the trigeminal nerves. Spinal nerves supply cutaneous branches to specific (adjacent) dermatomes that eventually become dermal segments. As a result, the distribution and growth of the spinal nerves correlate with the segmented body plan, which is established by the fourth week of embryonic development.

Growth of the limbs and unequal growth of other body areas result in an adult pattern of dermatomes with unequal sizes and shapes and varying degrees of overlap. Because embryonic muscle cells migrate extensively, much of the early segmental pattern is lost. Understanding the general pattern of sensory nerve distribution is critical for physicians. For example, in areas of substantial dermatome overlap, two or three spinal nerves must be blocked to perform local surgery.

Sensory receptors atrophy to some degree with age, and there is some lessening of muscle tone in the face and neck. Reflexes occur a bit more slowly during old age. This deterioration seems to reflect a general loss of neurons, fewer synapses per neuron, and a slowdown in central processing rather than any major changes in the peripheral nerve fibers. In fact, the peripheral nerves remain viable and normally functional throughout life unless subjected to traumatic injury or ischemia. The most common symptom of ischemia is a tingling sensation or numbness in the affected region.

CHECK YOUR UNDERSTANDING

17. Segmentation in the embryo gives rise to segmentation in the adult. Name two examples of segmentation in the adult body.

For answers, see Appendix G.

The PNS is an essential part of any functional nervous system. Without it, the CNS would lack its rich bank of information about events of the external and internal environments. Now that we have connected the CNS to both of these environments, we are ready to consider the autonomic nervous system, the topic of Chapter 14.

RELATED CLINICAL TERMS

Analgesia (an″al-je′zeah; *an* = without; *algos* = pain) Reduced ability to feel pain, not accompanied by loss of consciousness. An analgesic is a pain-relieving drug.

Dysarthria (dis-ar′thre-ah) Difficulty in speech articulation due to motor pathway disorders that result in weakness, uncoordinated motion, or altered respiration or rhythm. For example, lesions of cranial nerves IX, X, and XII result in nasal, breathy speech, and lesions in upper motor pathways produce a hoarse, strained voice. Not to be confused with *dysphasia* or *aphasia*, which are disorders of language processing.

Dystonia (dis-to′ne-ah) Impairment of muscle tone.

Nerve conduction studies Diagnostic tests that assess nerve integrity as indicated by their conduction velocities; the nerve is stimulated at one point, activity is recorded at a second point a known distance away, and the time required for the response to reach the recording electrode is measured; used to assess suspected peripheral neuropathies.

Neuralgia (nu-ral′je-ah; *neuro* = nerve) Sharp spasmlike pain along the course of one or more nerves; may be caused by inflammation or injury to the nerve(s) (for example, trigeminal neuralgia).

Neuritis (nu-ri′tis) Inflammation of a nerve; there are many forms with different effects (e.g., increased or decreased nerve sensitivity, paralysis of structure served, and pain).

Paresthesia (par″es-the′ze-ah) An abnormal sensation (burning, numbness, tingling) in the absence of stimuli; caused by a sensory nerve disorder.

Tabes dorsalis (ta′bēz dor-sa′lis) A slowly progressive condition caused by deterioration of the dorsal tracts (gracilis and cuneatus) and associated dorsal roots; a late sign of the neurological damage caused by the syphilis bacterium. Because joint proprioceptor tracts are destroyed, affected individuals have poor muscle coordination and an unstable gait. Bacterial invasion of the sensory (dorsal) roots results in pain, which ends when dorsal root destruction is complete.

CHAPTER SUMMARY

1. The peripheral nervous system consists of sensory receptors, nerves conducting impulses to and from the CNS, their associated ganglia, and motor endings.

PART 1: SENSORY RECEPTORS AND SENSATION

Sensory Receptors (pp. 485–488)

1. Sensory receptors are specialized to respond to environmental changes (stimuli).
2. Sensory receptors include the simple (general) receptors for pain, touch, pressure, and temperature found in the skin, as well as those found in skeletal muscles and tendons and in the visceral organs. Complex receptors (sense organs), consisting of sensory receptors and other cells, serve the special senses (vision, hearing, equilibrium, smell, and taste).
3. Receptors are classified according to stimulus detected as mechanoreceptors, thermoreceptors, photoreceptors, chemoreceptors, and nociceptors, and according to location as exteroceptors, interoceptors, and proprioceptors.
4. The general sensory receptors are classified structurally as free or encapsulated nerve (receptor) endings of sensory neurons. The free endings are mainly receptors for temperature and pain, although two are for light touch (tactile discs and hair follicle receptors). The encapsulated endings, which are mechanoreceptors, include Meissner's corpuscles, Pacinian corpuscles, Ruffini endings, muscle spindles, Golgi tendon organs, and joint kinesthetic receptors.

Sensory Integration: From Sensation to Perception (pp. 488–491)

1. Sensation is awareness of internal and external stimuli; perception is conscious interpretation of those stimuli.
2. The three levels of sensory integration are the receptor, circuit, and perceptual levels. These levels are functions of the sensory receptors, the ascending pathways, and the cerebral cortex, respectively.
3. Sensory receptors transduce (convert) stimulus energy via receptor or generator potentials into action potentials. Stimulus strength is frequency coded. Adaptation (decreased response to a continuous or unchanging stimulus) is seen in all general receptors except pain and proprioceptors.
4. Some sensory fibers entering the spinal cord act in local reflex arcs. Some synapse with the dorsal horn neurons (spinothalamic ascending pathways), and others continue upward to synapse in medullary nuclei (dorsal column–medial lemniscal ascending pathways). Second-order neurons of both dorsal column–medial lemniscal and spinothalamic ascending pathways terminate in the thalamus.
5. Perception—the internal, conscious image of the stimulus that serves as the basis for response—is the result of cortical processing.
6. The main aspects of sensory perception are detection, magnitude estimation, spatial discrimination, feature abstraction, quality discrimination, and pattern recognition.

PART 2: TRANSMISSION LINES: NERVES AND THEIR STRUCTURE AND REPAIR

Nerves and Associated Ganglia (pp. 491–492)

1. A nerve is a bundle of axons in the PNS. Each fiber is enclosed by an endoneurium, fascicles of fibers are wrapped by a perineurium, and the whole nerve is bundled by the epineurium.
2. Nerves are classified according to the direction of impulse conduction as sensory, motor, or mixed; most nerves are mixed. The efferent fibers may be somatic or autonomic.
3. Ganglia are collections of neuron cell bodies associated with nerves in the PNS. Examples are the dorsal root (sensory) ganglia and autonomic (motor) ganglia.
4. Injured PNS fibers may regenerate if macrophages enter the area, phagocytize the debris, and promote Schwann cell proliferation. Schwann cells then form a channel and secrete chemicals to guide axon sprouts to their original contacts. Fibers in the CNS do not normally regenerate because the oligodendrocytes inhibit axonal sprouting and regrowth.

Cranial Nerves (pp. 493–501)

1. Twelve pairs of cranial nerves issue through the skull to innervate the head and neck. Only the vagus nerves extend into the thoracic and abdominal cavities. All but the accessory nerve originate from the brain.
2. Cranial nerves are (generally) numbered from rostral to caudal in order of emergence from the brain. Their names reflect structures served or function or both. The cranial nerves include
   - The olfactory nerves (I): purely sensory; concerned with the sense of smell.
   - The optic nerves (II): purely sensory; transmit visual impulses from the retina to the thalamus.
   - The oculomotor nerves (III): primarily motor; emerge from the midbrain and serve four extrinsic eye muscles, the levator palpebrae superioris of the eyelid, and the intrinsic ciliary muscle of the eye and constrictor fibers of the iris. Also carry proprioceptive impulses from the skeletal muscles served.
   - The trochlear nerves (IV): primarily motor; emerge from the dorsal midbrain and carry motor and proprioceptor impulses to and from superior oblique muscles of the eyeballs.
   - The trigeminal nerves (V): mixed nerves; emerge from the lateral pons as the major general sensory nerves of the face. Each has three sensory divisions: ophthalmic, maxillary, and mandibular; the mandibular branch also contains motor fibers that innervate the chewing muscles.
   - The abducens nerves (VI): primarily motor; emerge from the pons and serve the motor and proprioceptive functions of the lateral rectus muscles of the eyeballs.
   - The facial nerves (VII): mixed nerves; emerge from the pons as the major motor nerves of the face. Also carry sensory impulses from the taste buds of anterior two-thirds of the tongue.
   - The vestibulocochlear nerves (VIII): mostly sensory; transmit impulses from the hearing and equilibrium receptors of the inner ears.
   - The glossopharyngeal nerves (IX): mixed nerves; issue from the medulla. Transmit sensory impulses from the taste buds of the posterior tongue, from the pharynx, and from chemo- and

baroreceptors of the carotid bodies and sinuses. Innervate some pharyngeal muscles and parotid glands.
- The vagus nerves (X): mixed nerves; arise from the medulla. Almost all motor fibers are autonomic parasympathetic fibers; motor efferents to, and sensory fibers from, the pharynx, larynx, and visceral organs of the thoracic and abdominal cavities.
- The accessory nerves (XI): primarily motor; arise as spinal rootlets from the cervical spinal cord and enter the foramen magnum. Supply somatic efferents to the trapezius and sternocleidomastoid muscles of the neck and carry proprioceptor afferents from the same muscles.
- The hypoglossal nerves (XII): primarily motor; issue from the medulla and carry somatic motor efferents to, and proprioceptive fibers from, the tongue muscles.

Spinal Nerves (pp. 502–511)

1. The 31 pairs of spinal nerves (all mixed nerves) are numbered successively according to the region of the spinal cord from which they issue.
2. Spinal nerves are formed by the union of dorsal and ventral roots of the spinal cord and are short, confined to the intervertebral foramina.
3. Branches of each spinal nerve include dorsal and ventral rami, and a meningeal branch, and in the thoracic region, rami communicantes (ANS branches).
4. Ventral rami, except T_2–T_{12}, form plexuses that serve the limbs.
5. Dorsal rami serve the muscles and skin of the posterior body trunk. T_1–T_{12} ventral rami give rise to intercostal nerves that serve the thorax wall and abdominal surface.
6. The cervical plexus (C_1–C_4) innervates the muscles and skin of the neck and shoulder. Its phrenic nerve serves the diaphragm.
7. The brachial plexus serves the shoulder, some thorax muscles, and the upper limb. It arises primarily from C_5–T_1. Proximal to distal, the brachial plexus has roots, trunks, divisions, and cords. The main nerves arising from the cords are the axillary, musculocutaneous, median, radial, and ulnar nerves.
8. The lumbar plexus (L_1–L_4) provides the motor supply to the anterior and medial thigh muscles and the cutaneous supply to the anterior thigh and part of the leg. Its chief nerves are the femoral and obturator.
9. The sacral plexus (L_4–S_4) supplies the posterior muscles and skin of the lower limb. Its principal nerve is the large sciatic nerve composed of the tibial and common fibular nerves.
10. Joints are innervated by the same nerves that serve the muscles acting at the joint. All spinal nerves except C_1 innervate specific segments of the skin called dermatomes.

PART 3: MOTOR ENDINGS AND MOTOR ACTIVITY

Peripheral Motor Endings (p. 512)

1. Motor endings of somatic nerve fibers (axon terminals) help to form elaborate neuromuscular junctions with skeletal muscle cells. Axon terminals contain synaptic vesicles filled with acetylcholine, which (when released) signals the muscle cell to contract. An elaborate basal lamina fills the synaptic cleft.
2. Autonomic motor endings, called varicosities, are functionally similar, but structurally simpler, beaded terminals that innervate smooth muscle and glands. They do not form specialized neuromuscular junctions and the motor responses elicited are generally slower.

Motor Integration: From Intention to Effect (pp. 512–513)

1. Motor mechanisms operate at the level of the effectors (muscle fibers), descending circuits, and control levels of motor behavior.
2. The motor control hierarchy consists of the segmental level, the projection level, and the precommand level.
3. The segmental level is the spinal cord circuitry that activates ventral horn motor neurons to stimulate the muscles. It consists of reflexes and central pattern generators (CPGs), segmental circuits controlling locomotion.
4. The projection level consists of descending fibers that project to and control the segmental level. These fibers issue from the brain stem motor areas [indirect (extrapyramidal) system] and cortical motor areas [direct (pyramidal) system]. Command neurons in the brain stem appear to turn CPGs on and off, or to modulate them.
5. The cerebellum and basal nuclei constitute the precommand areas that subconsciously integrate mechanisms mediated by the projection level.

PART 4: REFLEX ACTIVITY

The Reflex Arc (p. 514)

1. A reflex is a rapid, involuntary motor response to a stimulus. The reflex arc has five elements: receptor, sensory neuron, integration center, motor neuron, and effector.

Spinal Reflexes (pp. 514–520)

1. Testing of somatic spinal reflexes provides information on the integrity of the reflex pathway and the degree of excitability of the spinal cord.
2. Somatic spinal reflexes include stretch, Golgi tendon, flexor, crossed-extensor, and superficial reflexes.
3. A stretch reflex, initiated by stretching of muscle spindles, causes contraction of the stimulated muscle and inhibits its antagonist. It is monosynaptic and ipsilateral. Stretch reflexes maintain muscle tone and body posture.
4. Golgi tendon reflexes, initiated by stimulation of Golgi tendon organs by excessive muscle tension, are polysynaptic reflexes. They cause relaxation of the stimulated muscle and contraction of its antagonist to prevent muscle and tendon damage.
5. Flexor reflexes are initiated by painful stimuli. They are polysynaptic, ipsilateral reflexes that are protective in nature.
6. Crossed-extensor reflexes consist of an ipsilateral flexor reflex and a contralateral extensor reflex.
7. Superficial reflexes (e.g., the plantar and abdominal reflexes) are elicited by cutaneous stimulation. They require functional cord reflex arcs and corticospinal pathways.

Developmental Aspects of the Peripheral Nervous System (p. 520)

1. Each spinal nerve provides the sensory and motor supply of an adjacent muscle mass (destined to become skeletal muscles) and the cutaneous supply of a dermatome (skin segment).
2. Reflexes decline in speed with age; this probably reflects neuronal loss or sluggish CNS integration circuits.

REVIEW QUESTIONS

Multiple Choice/Matching

(Some questions have more than one correct answer. Select the best answer or answers from the choices given.)

1. The large onion-shaped receptors that are found deep in the dermis and in subcutaneous tissue and that respond to deep pressure are (a) tactile discs, (b) Pacinian corpuscles, (c) free nerve endings, (d) muscle spindles.
2. Proprioceptors include all of the following except (a) muscle spindles, (b) Golgi tendon organs, (c) tactile discs, (d) joint kinesthetic receptors.
3. The aspect of sensory perception by which the cerebral cortex identifies the site or pattern of stimulation is (a) perceptual detection, (b) feature abstraction, (c) pattern recognition, (d) spatial discrimination.
4. The neural machinery of the spinal cord is at the (a) precommand level, (b) projection level, (c) segmental level.
5. Dorsal root ganglia contain (a) cell bodies of somatic motor neurons, (b) axon terminals of somatic motor neurons, (c) cell bodies of autonomic motor neurons, (d) axon terminals of sensory neurons, (e) cell bodies of sensory neurons.
6. The connective tissue sheath that surrounds a fascicle of nerve fibers is the (a) epineurium, (b) endoneurium, (c) perineurium, (d) neurilemma.
7. Match the receptor type in column B to the correct description in column A.

Column A	Column B
____ (1) pain, itch, and temperature receptors	(a) Ruffini endings
____ (2) contains intrafusal fibers and type Ia and II sensory endings	(b) Golgi tendon organ
____ (3) discriminative touch receptor in hairless skin (fingertips)	(c) muscle spindle
____ (4) contains receptor endings wrapped around thick collagen bundles	(d) free nerve endings
____ (5) rapidly adapting deep-pressure receptor	(e) Pacinian corpuscle
____ (6) slowly adapting deep-pressure receptor	(f) Meissner's corpuscle

8. Match the names of the cranial nerves in column B to the appropriate description in column A.

Column A	Column B
____ (1) causes pupillary constriction	(a) abducens
____ (2) is the major sensory nerve of the face	(b) accessory
____ (3) serves the sternocleidomastoid and trapezius muscles	(c) facial
____ (4) are purely sensory (two nerves)	(d) glossopharyngeal
____ (5) serves the tongue muscles	(e) hypoglossal
____ (6) allows you to chew your food	(f) oculomotor
____ (7) is impaired in Bell's palsy	(g) olfactory
____ (8) helps to regulate heart activity	(h) optic
____ (9) helps you to hear and to maintain your balance	(i) trigeminal
____, ____, ____, ____, (10) (contain parasympathetic motor fibers (four nerves)	(j) trochlear
	(k) vagus
	(l) vestibulocochlear

9. For each of the following muscles or body regions, identify the plexus and the peripheral nerve(s) (or branch of one) involved. Use choices from keys A and B.

____; ____ (1) the diaphragm
____; ____ (2) muscles of the posterior leg
____; ____ (3) anterior thigh muscles
____; ____ (4) medial thigh muscles
____; ____ (5) anterior arm muscles that flex the forearm
____; ____ (6) muscles that flex the wrist and digits (two nerves)
____; ____ (7) muscles that extend the wrist and digits
____; ____ (8) skin and extensor muscles of the posterior arm
____; ____ (9) fibularis muscles, tibialis anterior, and toe extensors
____; ____, ____, ____, ____ (10) elbow joint

Key A: Plexuses
(a) brachial
(b) cervical
(c) lumbar
(d) sacral

Key B: Nerves
(1) common fibular
(2) femoral
(3) median
(4) musculocutaneous
(5) obturator
(6) phrenic
(7) radial
(8) tibial
(9) ulnar

10. Characterize each receptor activity described below by choosing the appropriate letter and number(s) from keys A and B.

____, ____ (1) You are enjoying an ice cream cone.
____, ____ (2) You have just scalded yourself with hot coffee.
____, ____ (3) The retinas of your eyes are stimulated.
____, ____ (4) You bump (lightly) into someone.
____, ____ (5) You are in a completely dark room and reaching toward the light switch.
____, ____ (6) You feel uncomfortable after a large meal.

Key A:
(a) exteroceptor
(b) interoceptor
(c) proprioceptor

Key B:
(1) chemoreceptor
(2) mechanoreceptor
(3) nociceptor
(4) photoreceptor
(5) thermoreceptor

13

11. A reflex that causes reciprocal activation of the antagonist muscle is the (**a**) crossed-extensor, (**b**) flexor, (**c**) Golgi tendon, (**d**) muscle stretch.

Short Answer Essay Questions

12. What is the functional relationship of the peripheral nervous system to the central nervous system?
13. List the structural components of the peripheral nervous system, and describe the function of each component.
14. Differentiate clearly between sensation and perception.
15. Central pattern generators (CPGs) are found at the segmental level of motor control. (a) What is the job of the CPGs? (b) What controls them, and where is this control localized?
16. Make a diagram of the hierarchy of motor control. Position the CPGs, the command neurons, and the cerebellum and basal nuclei in this scheme.
17. Why are the cerebellum and basal nuclei called precommand areas?
18. Explain why damage to peripheral nerve fibers is often reversible, whereas damage to CNS fibers rarely is.
19. (a) Describe the formation and composition of a spinal nerve. (b) Name the branches of a spinal nerve (other than the rami communicantes), and indicate their distribution.
20. (a) Define plexus. (b) Indicate the spinal roots of origin of the four major nerve plexuses, and name the general body regions served by each.
21. Differentiate between ipsilateral and contralateral reflexes.
22. What is the homeostatic value of flexor reflexes?
23. Compare and contrast flexor and crossed-extensor reflexes.
24. Explain how a crossed-extensor reflex exemplifies both serial and parallel processing.
25. What clinical information can be gained by conducting somatic reflex tests?
26. What is the structural and functional relationship between spinal nerves, skeletal muscles, and dermatomes?

13

Critical Thinking and Clinical Application Questions

1. In 1962 a boy playing in a train yard fell under a train. His right arm was cut off cleanly by the train wheel. Surgeons reattached the arm, sewing nerves and vessels back together. The boy was told he should eventually regain the use of his arm but that it would never be strong enough to pitch a baseball. Explain why full recovery of strength was unlikely.
2. Jefferson, a football quarterback, suffered torn menisci in his right knee joint when tackled from the side. The same injury crushed his common fibular nerve against the head of the fibula. What locomotor problems did Jefferson have after this?
3. As Harry fell off a ladder, he grabbed a tree branch with his right hand, but unfortunately lost his grip and fell heavily to the ground. Days later, Harry complained that his upper limb was numb. What was damaged in his fall?
4. Mr. Frank, a former stroke victim who had made a remarkable recovery, suddenly began to have problems reading. He complained of seeing double and also had problems navigating steps. He was unable to move his left eye downward and laterally. What cranial nerve was the site of lesion? (Right or left?)
5. One of a group of rabbit hunters was accidentally sprayed with buckshot in both of his gluteal prominences. When his companions saw that he would survive, they laughed and joked about where he had been shot. They were horrified and ashamed a week later when it was announced that their friend would be permanently paralyzed and without sensation in both legs from the knee down, as well as on the back of his thighs. What had happened?
6. You are at a party at Mary's house. After you are blindfolded, an object (a key *or* a rabbit's foot) is placed in your hand. What spinal tracts carry the signals to the cortex that will allow you to differentiate between these objects, and what aspects of sensory perception are operating?
7. Fumiko, a 19-year-old nursing student, had had a runny nose and sore throat for several days. Upon waking, her face felt "twisted." When she examined her face in the mirror, she noticed that the right side looked "droopy" and she was unable to move the facial muscles on that side. This made it difficult to eat or speak clearly. Which cranial nerve was affected and on which side? What is a common cause of this condition?

14

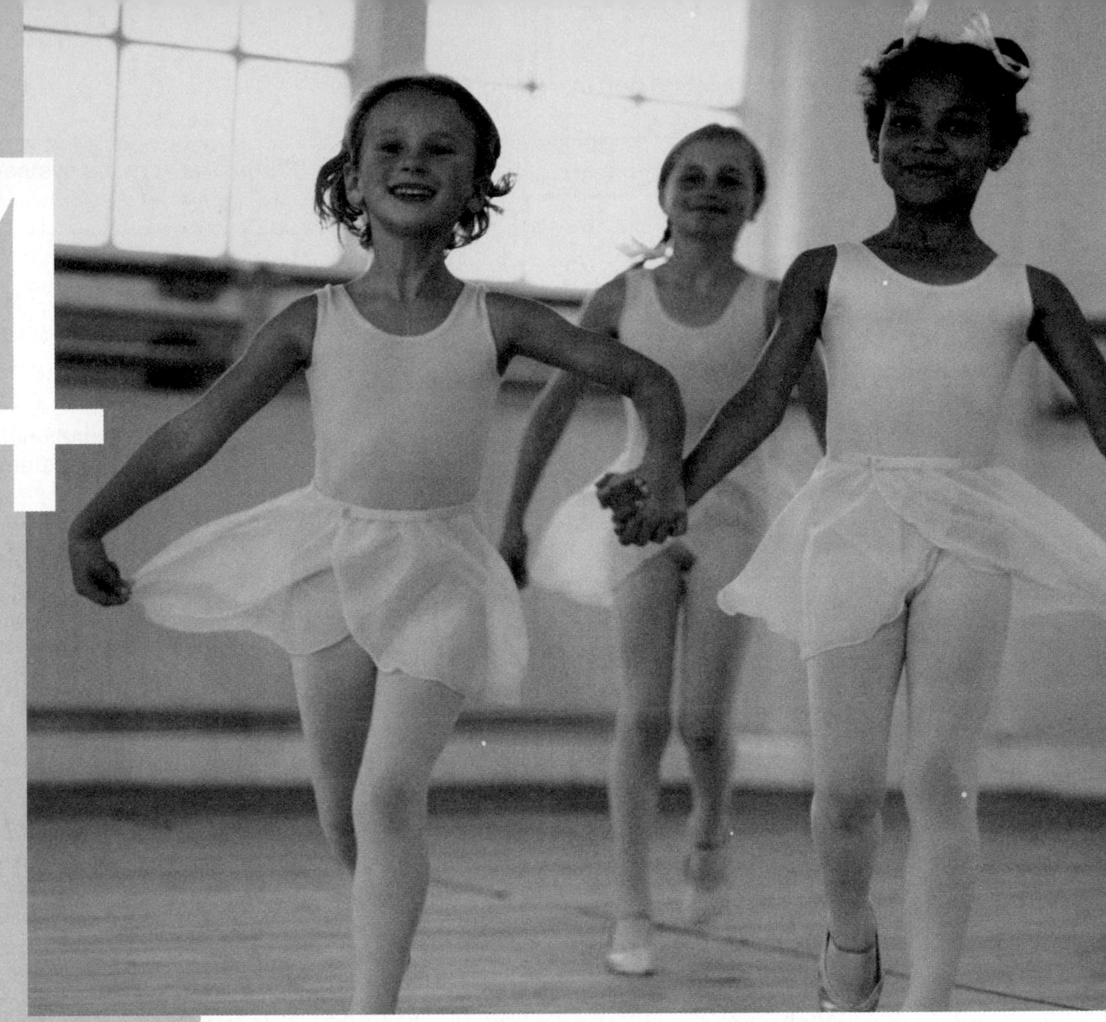

The Autonomic Nervous System

The human body is exquisitely sensitive to changes in its internal environment, and engages in a lifelong struggle to balance competing demands for resources under ever-changing conditions. Although all body systems contribute, the stability of our internal environment depends largely on the **autonomic nervous system (ANS)**, the system of motor neurons that innervates smooth and cardiac muscle and glands **(Figure 14.1)**.

At every moment, signals stream from visceral organs into the CNS, and autonomic nerves make adjustments as necessary to ensure optimal support for body activities. In response to changing conditions, the ANS shunts blood to "needy" areas, speeds or slows heart rate, adjusts

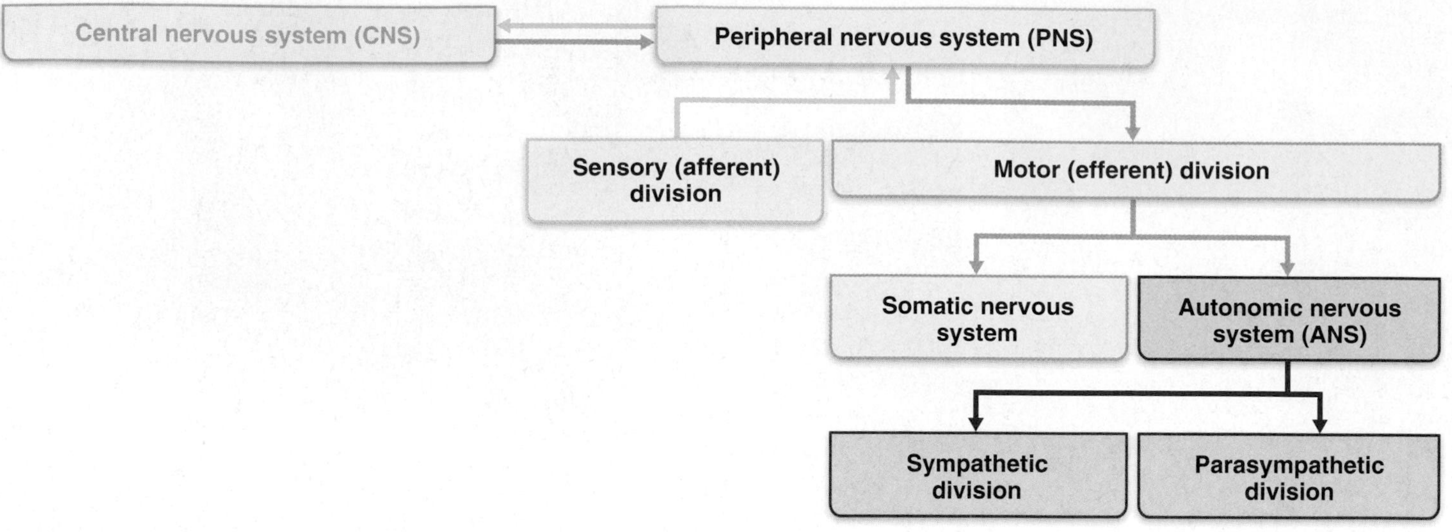

Figure 14.1 Place of the ANS in the structural organization of the nervous system.

blood pressure and body temperature, and increases or decreases stomach secretions.

Most of this fine-tuning occurs without our awareness or attention. Can you tell when your arteries are constricting or when your pupils are dilating? Probably not, but if you've ever been stuck in a checkout line, and your full bladder was contracting as if it had a mind of its own, you've been very aware of a visceral activity. These functions, both those we're aware of and those that occur without our awareness or attention, are controlled by the ANS. Indeed, as the term *autonomic* (*auto* = self; *nom* = govern) implies, this motor subdivision of the peripheral nervous system has a certain amount of functional independence. The ANS is also called the **involuntary nervous system**, which reflects its subconscious control, or the **general visceral motor system**, which indicates the location of most of its effectors.

14

Introduction

- Define autonomic nervous system and explain its relationship to the peripheral nervous system.
- Compare the somatic and autonomic nervous systems relative to effectors, efferent pathways, and neurotransmitters released.
- Compare and contrast the functions of the parasympathetic and sympathetic divisions.

Comparison of the Somatic and Autonomic Nervous Systems

In our previous discussions of motor nerves, we have focused largely on the activity of the somatic nervous system. So, before describing autonomic nervous system anatomy, we will point out the major differences between the somatic and autonomic systems as well as some areas of functional overlap.

Both systems have motor fibers, but the somatic and autonomic nervous systems differ (1) in their effectors, (2) in their efferent pathways, and (3) to some degree in target organ responses to their neurotransmitters. Consult **Figure 14.2** for a summary of the differences as we discuss them next.

Effectors

The somatic nervous system stimulates skeletal muscles, whereas the ANS innervates cardiac and smooth muscle and glands. Differences in the physiology of the effector organs account for most of the remaining differences between somatic and autonomic effects on their target organs.

Efferent Pathways and Ganglia

In the somatic nervous system, the motor neuron cell bodies are in the CNS, and their axons extend in spinal or cranial nerves all the way to the skeletal muscles they activate. Somatic motor fibers are typically thick, heavily myelinated group A fibers that conduct nerve impulses rapidly.

In contrast, the ANS uses a *two-neuron chain* to its effectors. The cell body of the first neuron, the **preganglionic neuron**, resides in the brain or spinal cord. Its axon, the **preganglionic axon**, synapses with the second motor neuron, the **ganglionic neuron**, in an **autonomic ganglion** outside the CNS. The axon of the ganglionic neuron, called the **postganglionic axon**, extends to the effector organ. If you think about the meanings of all these terms while referring to Figure 14.2, understanding the rest of the chapter will be much easier.

Preganglionic axons are lightly myelinated, thin fibers, and postganglionic axons are even thinner and are unmyelinated. Consequently, conduction through the autonomic efferent chain is slower than conduction in the somatic motor system. For most of their course, many pre- and postganglionic fibers are incorporated into spinal or cranial nerves.

Keep in mind that autonomic ganglia are *motor* ganglia, containing the cell bodies of motor neurons. Technically, they are sites of synapse and information transmission from preganglionic to ganglionic neurons. Also, remember that the somatic

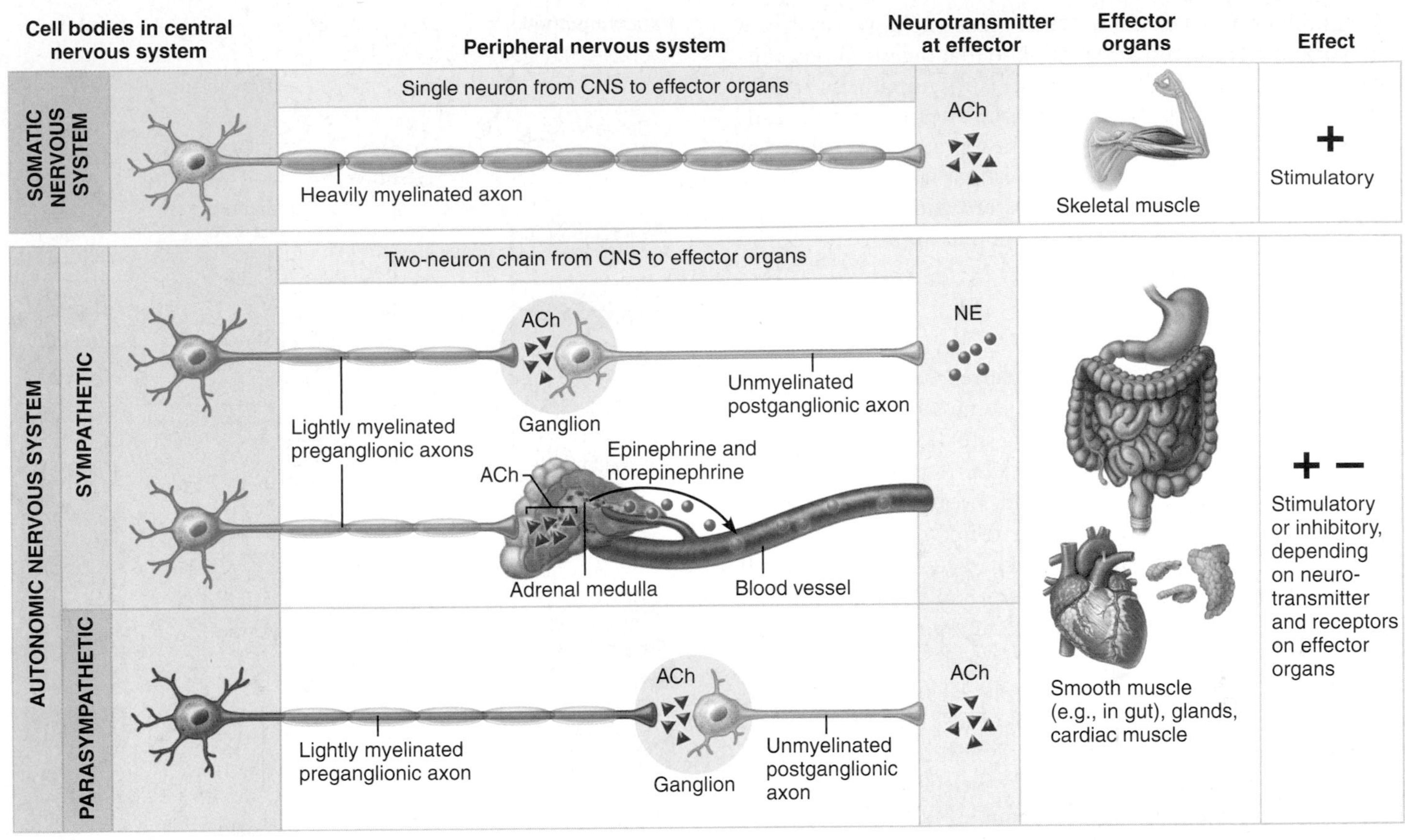

Figure 14.2 Comparison of somatic and autonomic nervous systems.

motor division *lacks* ganglia entirely. The dorsal root ganglia are part of the sensory, not the motor, division of the PNS.

Neurotransmitter Effects

All somatic motor neurons release **acetylcholine (ACh)** at their synapses with skeletal muscle fibers. The effect is always *excitatory*, and if stimulation reaches threshold, the muscle fibers contract.

Neurotransmitters released onto visceral effector organs by postganglionic autonomic fibers include **norepinephrine (NE)** secreted by most sympathetic fibers, and ACh released by parasympathetic fibers. Depending on the type of receptors present on the target organ, the organ's response may be either excitation or inhibition (Figure 14.2; see Table 14.2 on p. 536).

Overlap of Somatic and Autonomic Function

Higher brain centers regulate and coordinate both somatic and autonomic motor activities, and nearly all spinal nerves (and many cranial nerves) contain both somatic and autonomic fibers. Moreover, most of the body's adaptations to changing internal and external conditions involve both skeletal muscle activity and enhanced responses of certain visceral organs. For example, when skeletal muscles are working hard, they need more oxygen and glucose and so autonomic control mechanisms speed up heart rate and dilate airways to meet these needs and maintain homeostasis.

The ANS is only one part of our highly integrated nervous system, but according to convention we will consider it an individual entity and describe its role in isolation in the sections that follow.

ANS Divisions

The two arms of the ANS, the *parasympathetic* and *sympathetic divisions*, generally serve the same visceral organs but cause essentially opposite effects. If one division stimulates certain smooth muscles to contract or a gland to secrete, the other division inhibits that action. Through this **dual innervation**, the two divisions counterbalance each other's activities to keep body systems running smoothly. The sympathetic division mobilizes the body during activity, whereas the parasympathetic arm promotes maintenance functions and conserves body energy. Let's elaborate on these functional differences by focusing briefly on extreme situations in which each division is exerting primary control.

Role of the Parasympathetic Division

The **parasympathetic division**, sometimes called the "resting and digesting" system, keeps body energy use as low as possible, even as it directs vital "housekeeping" activities like digestion and elimination of feces and urine. (This explains why it is a good idea to relax after a heavy meal: so that digestion is not interfered with by sympathetic activity.) Parasympathetic activity is best illustrated in a person who relaxes after a meal and reads the newspaper. Blood pressure and heart rate are regulated at low normal levels, and the gastrointestinal tract is actively digesting food. In the eyes, the pupils are constricted and the lenses are accommodated for close vision to improve the clarity of the close-up image.

Role of the Sympathetic Division

The **sympathetic division** is often referred to as the "fight-or-flight" system. Its activity is evident when we are excited or find ourselves in emergency or threatening situations, such as being frightened by street toughs late at night. A rapidly pounding heart; deep breathing; dry mouth; cold, sweaty skin; and dilated eye pupils are sure signs of sympathetic nervous system mobilization. Not as obvious, but equally characteristic, are changes in brain wave patterns and in the electrical resistance of the skin (galvanic skin resistance)—events that are recorded during lie detector examinations.

During any type of vigorous physical activity, the sympathetic division also promotes a number of other adjustments. Visceral (and sometimes cutaneous) blood vessels are constricted, and blood is shunted to active skeletal muscles and the vigorously working heart. The bronchioles in the lungs dilate, increasing ventilation (and ultimately increasing oxygen delivery to body cells), and the liver releases more glucose into the blood to accommodate the increased energy needs of body cells.

14

At the same time, temporarily nonessential activities, such as gastrointestinal tract motility, are damped. If you are running from a mugger, digesting lunch can wait! It is far more important that your muscles be provided with everything they need to get you out of danger. In such active situations, the sympathetic division generates a head of steam that enables the body to cope with situations that threaten homeostasis. Its function is to provide the optimal conditions for an appropriate response to some threat, whether that response is to run, to see better, or to think more clearly.

We have just looked at two extreme situations in which one or the other branch of the ANS dominates. An easy way to remember the most important roles of the two ANS divisions is to think of the parasympathetic division as the **D** division [digestion, defecation, and diuresis (urination)], and the sympathetic division as the **E** division (exercise, excitement, emergency, embarrassment). A more detailed summary of the effects of each division on various organs is presented in Table 14.4 (p. 538).

Remember, however, that while we may find it easy to think of the two ANS divisions as working in an all-or-none fashion as described above, this is rarely the case. A dynamic antagonism exists between the divisions, and fine adjustments are made continuously by both.

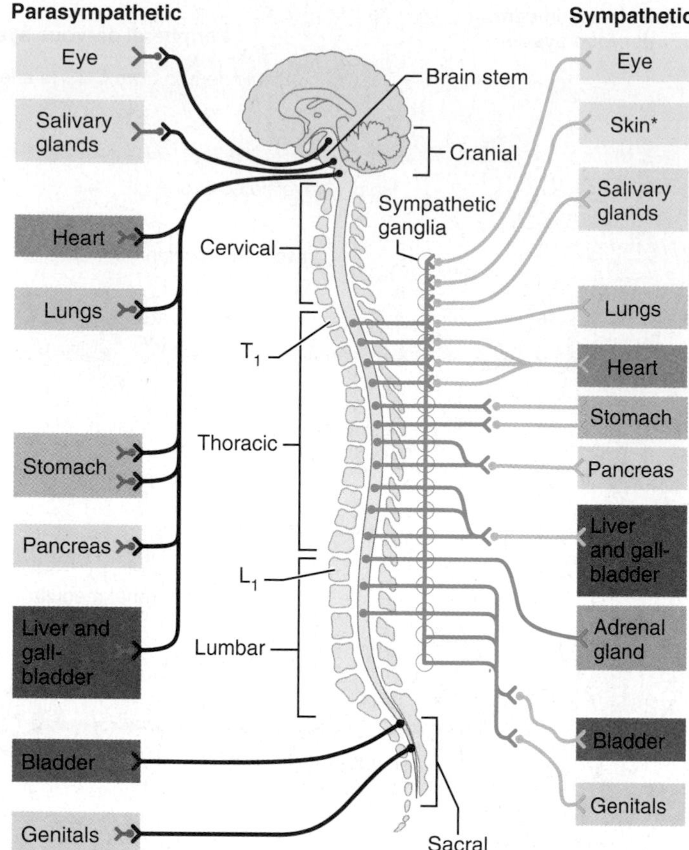

Figure 14.3 Overview of the subdivisions of the ANS. The parasympathetic and sympathetic divisions differ anatomically in (1) the sites of origin of their nerves, (2) the relative lengths of their preganglionic and postganglionic fibers, and (3) the locations of their ganglia (indicated here by synapse sites).

*Although sympathetic innervation to the skin is mapped to the cervical region here, all nerves to the periphery carry postganglionic sympathetic fibers.

CHECK YOUR UNDERSTANDING

1. Name the three types of effectors of the autonomic nervous system.
2. Which relays instructions from the CNS to muscles more quickly, the somatic nervous system or the ANS? Explain why.
3. Which branch of the ANS would predominate if you were lying on the beach enjoying the sun and the sound of the waves? Which branch would predominate if you were on a surfboard and a shark appeared within a few feet of you?

For answers, see Appendix G.

ANS Anatomy

▶ **For the parasympathetic and sympathetic divisions, describe the site of CNS origin, locations of ganglia, and general fiber pathways.**

Anatomically, the sympathetic and parasympathetic divisions differ in

1. **Their origin sites.** Parasympathetic fibers emerge from the brain and sacral spinal cord (are craniosacral). Sympa-

thetic fibers originate in the thoracolumbar region of the spinal cord.

2. **The relative lengths of their fibers.** The parasympathetic division has long preganglionic and short postganglionic fibers. The sympathetic division has the opposite condition.
3. **The location of their ganglia.** Most parasympathetic ganglia are located in the visceral effector organs. Sympathetic ganglia lie close to the spinal cord.

Note that these and other differences are illustrated in **Figure 14.3** and summarized in **Table 14.1**, p. 534.

We begin our detailed exploration of the ANS with the anatomically simpler parasympathetic division.

Parasympathetic (Craniosacral) Division

The parasympathetic division is also called the **craniosacral division** because its preganglionic fibers spring from opposite ends of the CNS—the brain stem and the sacral region of the spinal cord **(Figure 14.4)**. The preganglionic axons extend from the CNS nearly all the way to the structures to be innervated. There the axons synapse with ganglionic neurons located in **terminal ganglia** that lie very close to or within the target organs. Very short postganglionic axons issue from the terminal ganglia and synapse with effector cells in their immediate area.

Cranial Outflow

Preganglionic fibers run in the oculomotor, facial, glossopharyngeal, and vagus cranial nerves. Their cell bodies lie in associated motor cranial-nerve nuclei in the brain stem (see Figures 12.15 and 12.16). We describe the precise locations of the neurons of the cranial parasympathetics next.

1. **Oculomotor nerves (III).** The parasympathetic fibers of the oculomotor nerves innervate smooth muscles in the eyes that cause the pupils to constrict and the lenses to bulge—actions needed to focus on close objects. The preganglionic axons found in the oculomotor nerves issue from the *accessory oculomotor (Edinger-Westphal) nuclei* in the midbrain. The cell bodies of the ganglionic neurons are in the **ciliary ganglia** within the eye orbits (see Table 13.2, p. 496).
2. **Facial nerves (VII).** The parasympathetic fibers of the facial nerves stimulate many large glands in the head. Fibers that activate the nasal glands and the lacrimal glands of the eyes originate in the *lacrimal nuclei of the pons.* The preganglionic fibers synapse with ganglionic neurons in the **pterygopalatine ganglia** (ter″eh-go-pal′ah-tīn) just posterior to the maxillae. The preganglionic neurons that stimulate the submandibular and sublingual salivary glands originate in the *superior salivatory nuclei* of the pons and synapse with ganglionic neurons in the **submandibular ganglia**, deep to the mandibular angles (see Table 13.2, p. 498).
3. **Glossopharyngeal nerves (IX).** The parasympathetics in the glossopharyngeal nerves originate in the *inferior salivatory nuclei* of the medulla and synapse in the **otic ganglia**, located just inferior to the foramen ovale of the skull. The

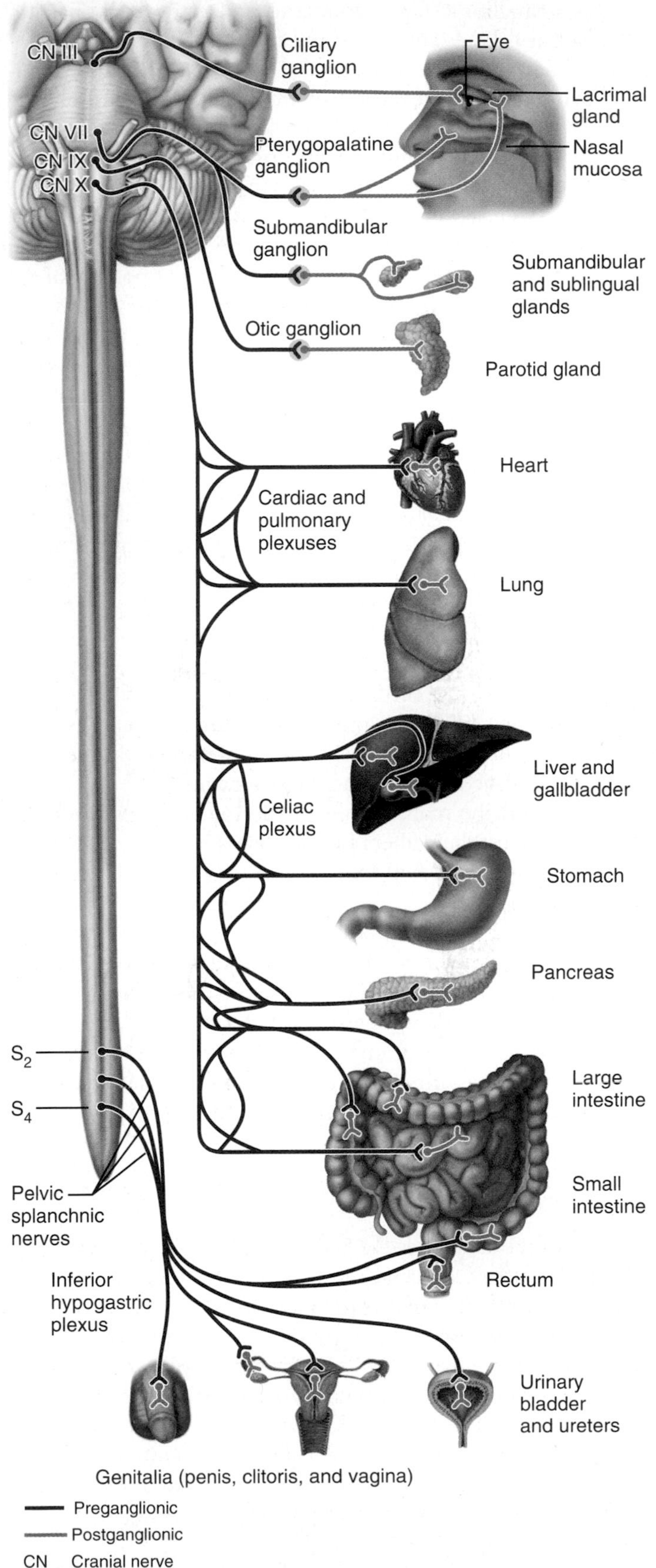

Figure 14.4 Parasympathetic division of the ANS.

14

postganglionic fibers course to and activate the parotid salivary glands anterior to the ears (see Table 13.2, p. 500).

Cranial nerves III, VII, and IX supply the entire parasympathetic innervation of the head; however, only the *preganglionic fibers* lie within these three pairs of cranial nerves—postganglionic fibers do not. Many of the postganglionic fibers "hitch a ride" with branches of the *trigeminal nerve* (*V*), taking advantage of its wide distribution, while others travel independently to their destinations.

4. **Vagus nerves (X).** The remaining and major portion of the parasympathetic cranial outflow is via the vagus (X) nerves. Between them, the two vagus nerves account for about 90% of all preganglionic parasympathetic fibers in the body. They provide fibers to the neck and to nerve plexuses (interweaving networks of nerves) that serve virtually every organ in the thoracic and abdominal cavities. The vagal nerve fibers (preganglionic axons) arise mostly from the *dorsal motor nuclei* of the medulla and synapse in terminal ganglia usually located in the walls of the target organ. Most terminal ganglia are not individually named. Instead they are collectively called *intramural ganglia*, literally, "ganglia within the walls."

   As the vagus nerves pass into the thorax, they send branches to the **cardiac plexuses** supplying fibers to the heart that slow heart rate, the **pulmonary plexuses** serving the lungs and bronchi, and the **esophageal plexuses** (ĕ-sof″ah-je′al) supplying the esophagus.

   When the main trunks of the vagus nerves reach the esophagus, their fibers intermingle, forming the **anterior** and **posterior vagal trunks**, each containing fibers from both vagus nerves. These vagal trunks then "ride" the esophagus down to the abdominal cavity. There they send fibers *through* the large **abdominal aortic plexus** [formed by a number of smaller plexuses (e.g., *celiac, superior mesenteric,* and *hypogastric*) that run along the aorta] before giving off branches to the abdominal viscera. The vagus nerves innervate the liver, gallbladder, stomach, small intestine, kidneys, pancreas, and the proximal half of the large intestine.

14

Sacral Outflow

The rest of the large intestine and the pelvic organs are served by the sacral outflow, which arises from neurons located in the lateral gray matter of spinal cord segments S_2–S_4. Axons of these neurons run in the ventral roots of the spinal nerves to the ventral rami and then branch off to form the **pelvic splanchnic nerves**, which pass through the **inferior hypogastric (pelvic) plexus** in the pelvic floor (Figure 14.4). Some preganglionic fibers synapse with ganglia in this plexus, but most synapse in intramural ganglia in the walls of the following organs: distal half of the large intestine, urinary bladder, ureters, and reproductive organs.

Sympathetic (Thoracolumbar) Division

The sympathetic division is anatomically more complex than the parasympathetic division, partly because it innervates more organs. It supplies not only the visceral organs in the internal body cavities but also all visceral structures in the superficial (somatic) part of the body. This sounds impossible, but there is an explanation—some glands and smooth muscle structures in the soma (sweat glands and the hair-raising arrector pili muscles of the skin) require autonomic innervation and are served only by sympathetic fibers. In addition, all arteries and veins (be they deep or superficial) have smooth muscle in their walls that is innervated by sympathetic fibers. But we will explain these matters later—let us get on with the anatomy of the sympathetic division.

All preganglionic fibers of the sympathetic division arise from cell bodies of preganglionic neurons in spinal cord segments T_1 through L_2 (Figure 14.3). For this reason, the sympathetic division is also referred to as the **thoracolumbar division** (tho″rah-ko-lum′bar). The presence of numerous preganglionic sympathetic neurons in the gray matter of the spinal cord produces the **lateral horns**—the so-called **visceral motor zones** (see Figures 12.31b, p. 469, and 12.32, p. 470). The lateral horns are just posterolateral to the ventral horns that house somatic motor neurons. (Parasympathetic preganglionic neurons in the sacral cord are far less abundant than the comparable sympathetic neurons in the thoracolumbar regions, and *lateral horns are absent* in the sacral region of the spinal cord. This is a major anatomical difference between the two divisions.)

After leaving the cord via the ventral root, preganglionic sympathetic fibers pass through a **white ramus communicans** [plural: **rami communicantes** (kom-mu′nĭ-kan″tēz)] to enter an adjoining **sympathetic trunk ganglion** forming part of the **sympathetic trunk** (or *sympathetic chain,* **Figure 14.5**). Looking like strands of glistening white beads, the sympathetic trunks flank each side of the vertebral column. The sympathetic trunk ganglia are also called *chain ganglia* or *paravertebral* ("near the vertebrae") *ganglia.*

Although the sympathetic *trunks* extend from neck to pelvis, sympathetic *fibers* arise only from the thoracic and lumbar cord segments, as shown in Figure 14.3. The ganglia vary in size, position, and number, but typically there are 23 in each sympathetic trunk—3 cervical, 11 thoracic, 4 lumbar, 4 sacral, and 1 coccygeal.

Once a preganglionic axon reaches a trunk ganglion, one of three things can happen to the axon, as shown by the three pathways in Figure 14.5b:

① The axon can synapse with a ganglionic neuron in the same trunk ganglion.

② The axon can ascend or descend the sympathetic trunk to synapse in another trunk ganglion. (These fibers running from one ganglion to another connect the ganglia together to form the sympathetic trunk.)

③ The axon can pass through the trunk ganglion and emerge from the sympathetic trunk without synapsing.

Preganglionic fibers following the third pathway help form several *splanchnic nerves* (splank′nik) that synapse in **collateral**, or **prevertebral**, **ganglia** located anterior to the vertebral column. Unlike sympathetic trunk ganglia, the collateral ganglia are neither paired nor segmentally arranged and occur only in the abdomen and pelvis.

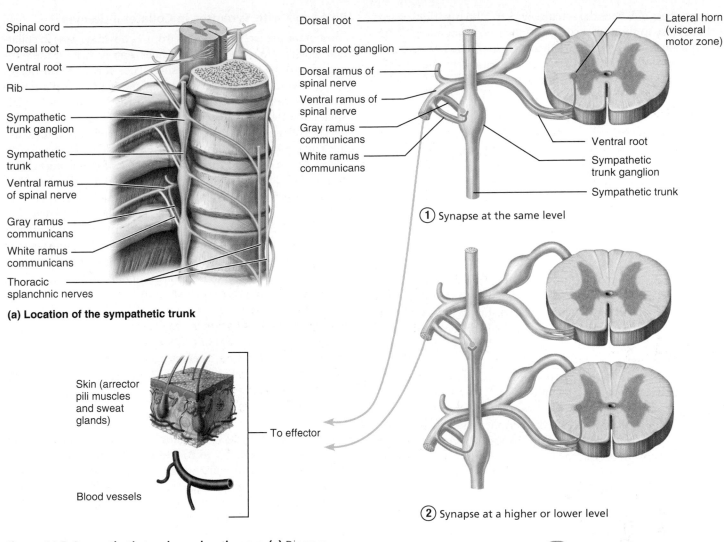

Spinal cord
Dorsal root
Ventral root
Rib
Sympathetic trunk ganglion
Sympathetic trunk
Ventral ramus of spinal nerve
Gray ramus communicans
White ramus communicans
Thoracic splanchnic nerves

(a) Location of the sympathetic trunk

Dorsal root
Dorsal root ganglion
Dorsal ramus of spinal nerve
Ventral ramus of spinal nerve
Gray ramus communicans
White ramus communicans
Lateral horn (visceral motor zone)
Ventral root
Sympathetic trunk ganglion
Sympathetic trunk

(1) Synapse at the same level

Skin (arrector pili muscles and sweat glands)
Blood vessels
To effector

(2) Synapse at a higher or lower level

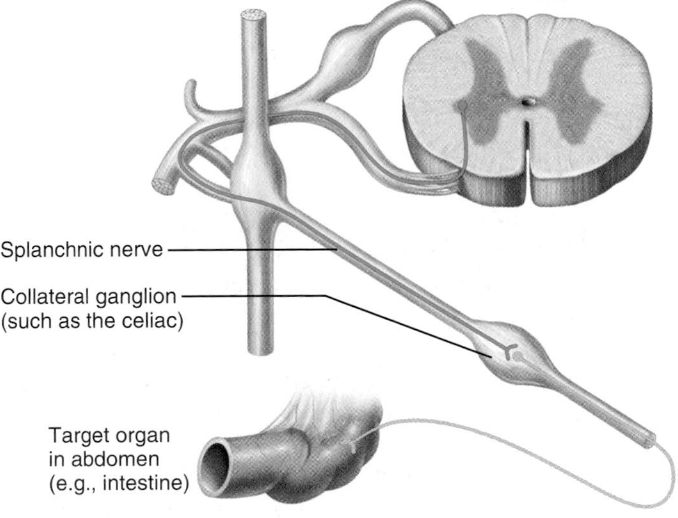

(3) Synapse in a distant collateral ganglion anterior to the vertebral column

(b) Three pathways of sympathetic innervation

Figure 14.5 Sympathetic trunks and pathways. (a) Diagram of the right sympathetic trunk in the posterior thorax, along the side of the vertebral column. **(b)** Synapses between preganglionic and ganglionic sympathetic neurons can occur at three different locations—either in a sympathetic trunk ganglion at the same or a different level, or in a collateral ganglion.

Regardless of where the synapse occurs, all sympathetic ganglia are close to the spinal cord, and their postganglionic fibers are typically much longer than their preganglionic fibers. Recall that the opposite condition exists in the parasympathetic division, an important anatomical distinction.

Pathways with Synapses in Trunk Ganglia

When synapses are made in sympathetic trunk ganglia, the postganglionic axons enter the ventral (or dorsal) ramus of the adjoining spinal nerves by way of communicating branches called **gray rami communicantes** (Figure 14.5). From there they travel via branches of the rami to their effectors, including sweat glands and arrector pili muscles of the skin. Anywhere along their path, the postganglionic axons may transfer over to nearby blood vessels and innervate the vascular smooth muscle all the way to their final branches.

Notice that the naming of the rami communicantes as *white* or *gray* reflects their appearance, revealing whether or not their fibers are myelinated (and has no relationship to the white and gray matter of the CNS). Preganglionic fibers composing the white rami are myelinated. Postganglionic axons forming the gray rami are not.

The white rami, which carry preganglionic axons to the sympathetic trunks, are found only in the T_1–L_2 cord segments, regions of sympathetic outflow. However, gray rami carrying postganglionic fibers headed for the periphery issue from every trunk ganglion from the cervical to the sacral region, allowing sympathetic output to reach all parts of the body. Note that *rami communicantes are associated only with the sympathetic division* and never carry parasympathetic fibers.

14

Pathways to the Head Sympathetic preganglionic fibers serving the head emerge from spinal cord segments T_1–T_4 and ascend the sympathetic trunk to synapse with ganglionic neurons in the **superior cervical ganglion** (Figure 14.6). This ganglion contributes sympathetic fibers that run in several cranial nerves and with the upper three or four cervical spinal nerves. Besides serving the skin and blood vessels of the head, its fibers stimulate the dilator muscles of the irises of the eyes, inhibit the nasal and salivary glands (the reason your mouth goes dry when you are scared), and innervate the smooth (tarsal) muscle that lifts the upper eyelid. The superior cervical ganglion also sends direct branches to the heart.

Pathways to the Thorax Sympathetic preganglionic fibers innervating the thoracic organs originate at T_1–T_6. From there the preganglionic fibers run to synapse in the cervical trunk ganglia. Postganglionic fibers emerging from the **middle** and **inferior cervical ganglia** enter cervical nerves C_4–C_8 (Figure 14.6). Some of these fibers innervate the heart via the cardiac plexus, and some innervate the thyroid gland, but most serve the skin. Additionally, some T_1–T_6 preganglionic fibers synapse in the nearest trunk ganglion, and the postganglionic fibers pass directly to the organ served. Fibers to the heart, aorta, lungs, and esophagus take this direct route. Along the way, they run into the plexuses associated with those organs.

Pathways with Synapses in Collateral Ganglia

Most of the preganglionic fibers from T_5 down synapse in collateral ganglia, and so most of these fibers enter and leave the sympathetic trunks without synapsing. They form several nerves called splanchnic nerves, including the thoracic **splanchnic nerves** (greater, lesser, and least) and the **lumbar** and **sacral splanchnic nerves**.

The splanchnic nerves contribute to a number of interweaving nerve plexuses known collectively as the *abdominal aortic plexus*, which clings to the surface of the abdominal aorta. This complex plexus contains several ganglia that together serve the abdominopelvic viscera (*splanchni* = viscera). From superior to inferior, the most important of these ganglia (and related subplexuses) are the **celiac**, **superior mesenteric**, and **inferior mesenteric**, named for the arteries with which they most closely associate (Figure 14.6). Postganglionic fibers issuing from these ganglia generally travel to their target organs in the company of the arteries serving these organs.

Pathways to the Abdomen Sympathetic innervation of the abdomen is via preganglionic fibers from T_5 to L_2, which travel in the thoracic splanchnic nerves to synapse mainly at the celiac and superior mesenteric ganglia. Postganglionic fibers issuing from these ganglia serve the stomach, intestines (except the distal half of the large intestine), liver, spleen, and kidneys.

Pathways to the Pelvis Preganglionic fibers innervating the pelvis originate from T_{10} to L_2 and then descend in the sympathetic trunk to the lumbar and sacral trunk ganglia. Some fibers synapse there and the postganglionic fibers run in lumbar and sacral splanchnic nerves to plexuses on the lower aorta and in the pelvis. Other preganglionic fibers pass directly to these autonomic plexuses and synapse in collateral ganglia, such as the inferior mesenteric ganglion. Postganglionic fibers proceed from these plexuses to the pelvic organs (the urinary bladder and reproductive organs) and also the distal half of the large intestine. For the most part, sympathetic fibers *inhibit* the activity of the muscles and glands in these visceral organs.

Pathways with Synapses in the Adrenal Medulla

Some fibers traveling in the thoracic splanchnic nerves pass through the celiac ganglion without synapsing and terminate by synapsing with the hormone-producing medullary cells of the adrenal gland (Figure 14.6). When stimulated by preganglionic fibers, the medullary cells secrete NE and *epinephrine* (also called *noradrenaline* and *adrenaline*, respectively) into the blood, producing the excitatory effects we have all felt as a "surge of adrenaline." Embryologically, sympathetic ganglia and the adrenal medulla arise from the same tissue. For this reason, the adrenal medulla is sometimes viewed as a "misplaced" sympathetic ganglion, and its hormone-releasing cells, although lacking nerve processes, are considered equivalent to ganglionic sympathetic neurons.

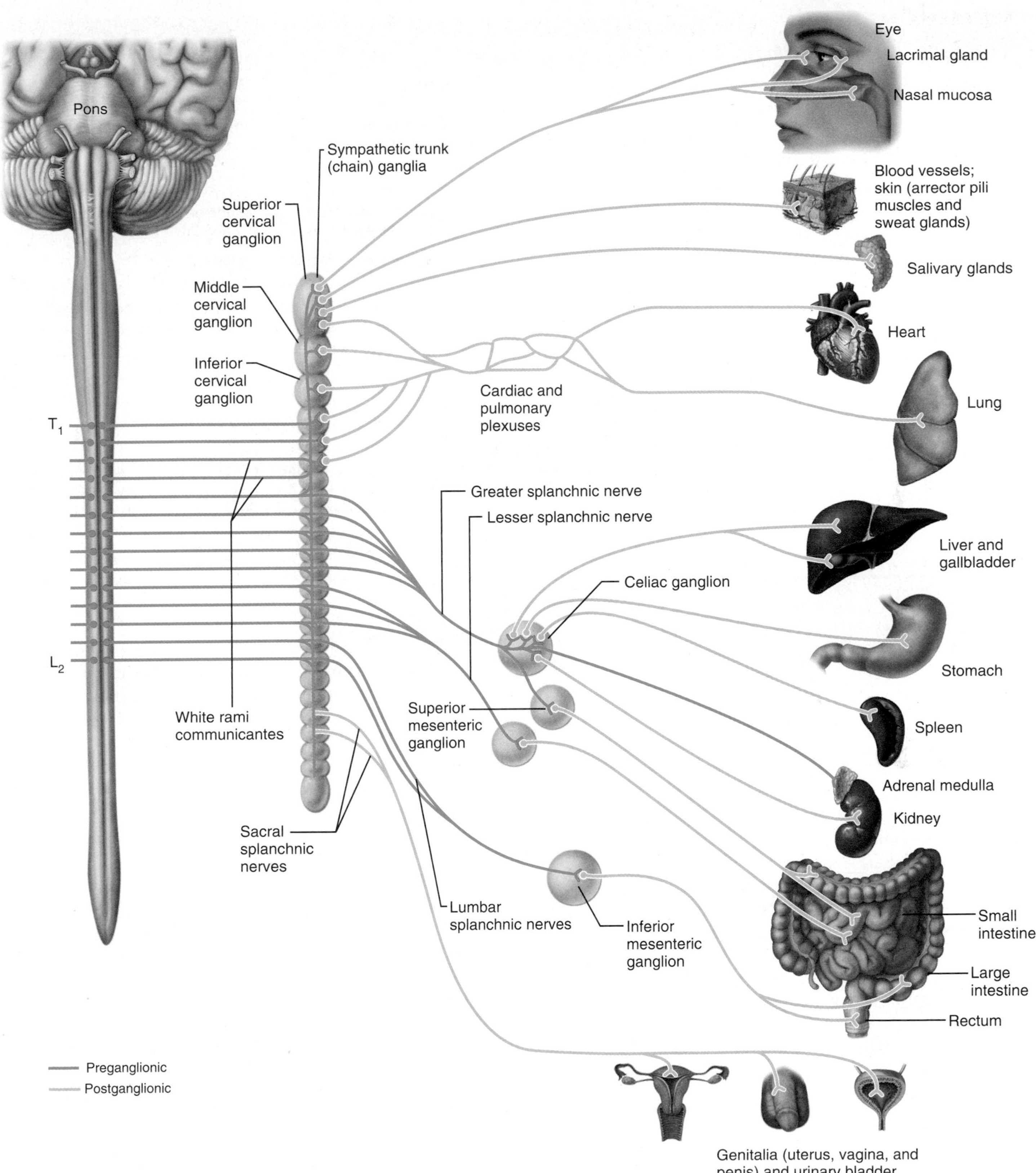

Figure 14.6 Sympathetic division of the ANS. Sympathetic innervation to peripheral structures (blood vessels, glands, and arrector pili muscles) occurs in all areas but is shown only in the cervical area.

14

TABLE 14.1 Anatomical and Physiological Differences Between the Parasympathetic and Sympathetic Divisions

CHARACTERISTIC	PARASYMPATHETIC	SYMPATHETIC
Origin	Craniosacral outflow: brain stem nuclei of cranial nerves III, VII, IX, and X; spinal cord segments S_2–S_4.	Thoracolumbar outflow: lateral horns of gray matter of spinal cord segments T_1–L_2.
Location of ganglia	Ganglia (terminal ganglia) are within the visceral organ (intramural) or close to the organ served.	Ganglia are within a few centimeters of CNS: alongside vertebral column (sympathetic trunk ganglia) and anterior to vertebral column (collateral, or prevertebral, ganglia).
Relative length of pre- and postganglionic fibers	Long preganglionic; short postganglionic.	Short preganglionic; long postganglionic.
Rami communicantes	None.	Gray and white rami communicantes. White rami contain myelinated preganglionic fibers; gray contain unmyelinated postganglionic fibers.
Degree of branching of preganglionic fibers	Minimal.	Extensive.
Functional role	Maintenance functions; conserves and stores energy; "rest and digest."	Prepares body for activity; "fight-or-flight."
Neurotransmitters	All preganglionic and postganglionic fibers release ACh (are cholinergic fibers).	All preganglionic fibers release ACh. Most postganglionic fibers release norepinephrine (are adrenergic fibers); postganglionic fibers serving sweat glands and some blood vessels of skeletal muscles release ACh. Neurotransmitter activity is augmented by release of adrenal medullary hormones (norepinephrine and epinephrine).

14

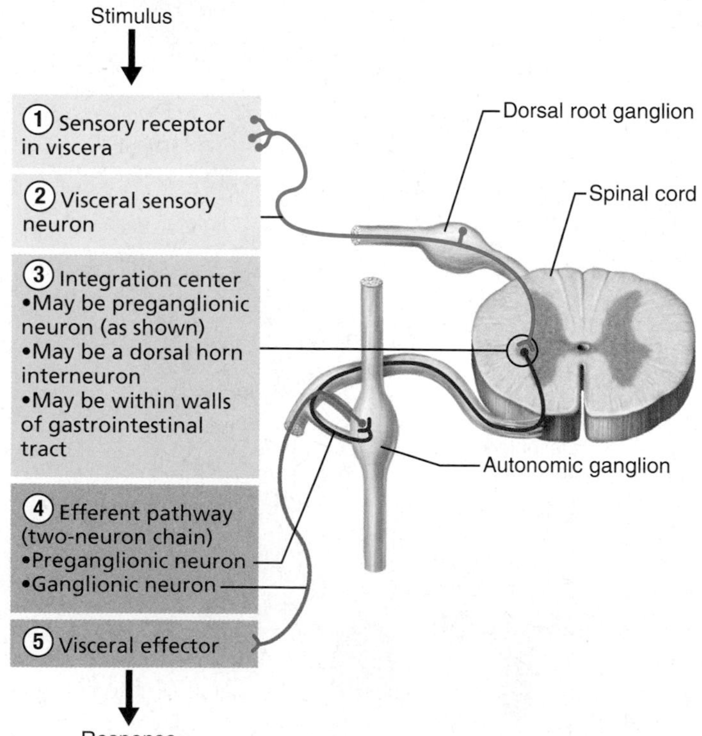

Figure 14.7 Visceral reflexes. Visceral reflex arcs have the same five elements as somatic reflex arcs. The visceral afferent (sensory) fibers are found both in spinal nerves (as depicted here) and in autonomic nerves.

Visceral Reflexes

Because most anatomists consider the ANS to be a visceral motor system, the presence of sensory fibers (mostly visceral pain afferents) is often overlooked. However, **visceral sensory neurons**, which send information concerning chemical changes, stretch, and irritation of the viscera, are the first link in autonomic reflexes. **Visceral reflex arcs** have essentially the same components as somatic reflex arcs—receptor, sensory neuron, integration center, motor neuron, effector—except that a visceral reflex arc has *two* neurons in its motor component (**Figure 14.7**; compare with Figure 13.14).

Nearly all the sympathetic and parasympathetic fibers we have described so far are accompanied by afferent fibers conducting sensory impulses from glands or muscles. This means that peripheral processes of visceral sensory neurons are found in cranial nerves VII, IX, and X, splanchnic nerves, and the sympathetic trunk, as well as in spinal nerves.

Like sensory neurons serving somatic structures (skeletal muscles and skin), the cell bodies of visceral sensory neurons are located either in the sensory ganglia of associated cranial nerves or in dorsal root ganglia of the spinal cord. Visceral sensory neurons are also found in sympathetic ganglia where preganglionic neurons synapse.

Furthermore, complete three-neuron reflex arcs (with sensory neurons, interneurons, and motor neurons) exist entirely within the walls of the gastrointestinal tract. Neurons composing these reflex arcs make up the *enteric nervous system*, which plays an important role in controlling gastrointestinal tract activity. We will discuss the enteric nervous system in more detail in Chapter 23.

The fact that visceral pain afferents travel along the same pathways as somatic pain fibers helps explain the phenomenon of **referred pain**, in which pain stimuli arising in the viscera are perceived as somatic in origin. For example, a heart attack may produce a sensation of pain that radiates to the superior thoracic wall and along the medial aspect of the left arm. Because the same spinal segments (T_1–T_5) innervate both the heart and the regions to which pain signals from heart tissue are referred, the brain interprets most such inputs as coming from the more common somatic pathway. Cutaneous areas to which visceral pain is commonly referred are shown in **Figure 14.8**.

CHECK YOUR UNDERSTANDING

4. State whether each of the following is a characteristic of the sympathetic or parasympathetic nervous system: short preganglionic fibers; origin from thoracolumbar region of spinal cord; terminal ganglia; collateral ganglia, innervates adrenal medulla.
5. How does a visceral reflex differ from a somatic reflex?

For answers, see Appendix G.

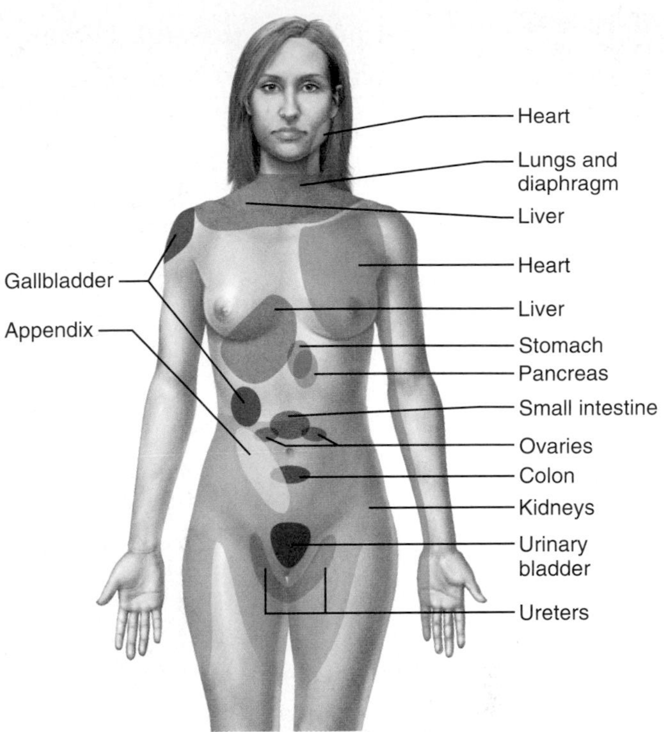

Figure 14.8 Map of referred pain. This map shows the anterior skin areas to which pain is referred from certain visceral organs.

ANS Physiology

- Define cholinergic and adrenergic fibers, and list the different types of their receptors.
- Describe the clinical importance of drugs that mimic or inhibit adrenergic or cholinergic effects.
- State the effects of the parasympathetic and sympathetic divisions on the following organs: heart, blood vessels, gastrointestinal tract, lungs, adrenal medulla, and external genitalia.
- Describe autonomic nervous system controls.

Neurotransmitters and Receptors

The major neurotransmitters released by ANS neurons are *acetylcholine* (*ACh*) and *norepinephrine* (*NE*). ACh, the same neurotransmitter secreted by somatic motor neurons, is released by (1) all ANS preganglionic axons and (2) all parasympathetic postganglionic axons at synapses with their effectors. ACh-releasing fibers are called **cholinergic fibers** (ko″lin-er′jik).

In contrast, most sympathetic postganglionic axons release NE and are classified as **adrenergic fibers** (ad″ren-er′jik). Some exceptions are sympathetic postganglionic fibers innervating sweat glands and some blood vessels in skeletal muscles. These fibers secrete ACh.

Unfortunately for memorization purposes, the effects of ACh and NE on their effectors are not consistently either excitation or inhibition. Why not? The response of visceral effectors to these neurotransmitters depends not only on the neurotransmitter but also on the receptor to which it attaches. The two or more kinds of receptors for each autonomic neurotransmitter allow it to exert different effects (activation or inhibition) at different body targets. **Table 14.2** provides a comprehensive summary of the receptor types that we introduce next.

Cholinergic Receptors

The two types of receptors that bind ACh are named for drugs that bind to them and mimic acetylcholine's effects. The first of these receptors identified were the **nicotinic receptors** (nik″o-tin′ik), which respond to nicotine. A mushroom poison, *muscarine* (mus′kah-rin), activates a different set of ACh receptors, named **muscarinic receptors**. All ACh receptors are either nicotinic or muscarinic.

Nicotinic receptors are found on (1) the sarcolemma of skeletal muscle cells at neuromuscular junctions (which, as you will recall, are somatic and not autonomic targets), (2) *all* ganglionic neurons, both sympathetic and parasympathetic, and (3) the hormone-producing cells of the adrenal medulla. The effect of ACh binding to nicotinic receptors anywhere is *always* stimulatory. Just as at the sarcolemma of skeletal muscle (examined in Chapter 9), ACh binding to any nicotinic receptor directly opens ion channels, depolarizing the postsynaptic cell.

Muscarinic receptors occur on all effector cells stimulated by postganglionic cholinergic fibers—that is, all parasympathetic target organs and a few sympathetic targets, such as eccrine sweat glands and some blood vessels of skeletal muscles. The effect of ACh binding to muscarinic receptors can be either inhibitory or stimulatory, depending on the subclass of muscarinic receptor found on the target organ. For example, binding of ACh to cardiac muscle receptors slows heart activity, whereas ACh binding to receptors on smooth muscle of the gastrointestinal tract increases its motility.

TABLE 14.2 Cholinergic and Adrenergic Receptors

NEUROTRANSMITTER	RECEPTOR TYPE	MAJOR LOCATIONS*	EFFECT OF BINDING
Acetylcholine	**Cholinergic**		
	Nicotinic	All ganglionic neurons; adrenal medullary cells (also neuromuscular junctions of skeletal muscle)	Excitation
	Muscarinic	All parasympathetic target organs	Excitation in most cases; inhibition of cardiac muscle
		Limited sympathetic targets:	
		▪ Eccrine sweat glands	Activation
		▪ Blood vessels in skeletal muscles	Vasodilation (may not occur in humans)
Norepinephrine (and epinephrine released by adrenal medulla)	**Adrenergic** β_1	Heart predominantly, but also kidneys and adipose tissue	Increases heart rate and strength; stimulates renin release by kidneys
	β_2	Lungs and most other sympathetic target organs; abundant on blood vessels serving the heart, liver and skeletal muscle	Effects mostly inhibitory; dilates blood vessels and bronchioles; relaxes smooth muscle walls of digestive and urinary visceral organs; relaxes uterus
	β_3	Adipose tissue	Stimulates lipolysis by fat cells
	α_1	Most importantly blood vessels serving the skin, mucosae, abdominal viscera, kidneys, and salivary glands; also, virtually all sympathetic target organs except heart	Constricts blood vessels and visceral organ sphincters; dilates pupils of the eyes
	α_2	Membrane of adrenergic axon terminals; pancreas; blood platelets	Inhibits NE release from adrenergic terminals; inhibits insulin secretion by pancreas; promotes blood clotting

* Note that all of these receptor subtypes are also found in the CNS.

14

Adrenergic Receptors

There are also two major classes of adrenergic (NE-binding) receptors: **alpha (α)** and **beta (β)**. These receptors are further divided into subclasses (α_1 and α_2; β_1, β_2, and β_3). Organs that respond to NE (or to epinephrine) have one or more of these receptor subtypes.

NE or epinephrine can have either excitatory or inhibitory effects on target organs depending on which subclass of receptor predominates in that organ. For example, binding of NE to the β_1 receptors of cardiac muscle prods the heart into more vigorous activity, whereas epinephrine binding to β_2 receptors in bronchiole smooth muscle causes it to relax, dilating the bronchiole.

The Effects of Drugs

Knowing the locations of the cholinergic and adrenergic receptor subtypes allows specific drugs to be prescribed to obtain the desired inhibitory or stimulatory effects on selected target organs. For example, *atropine* is an anticholinergic drug that blocks muscarinic ACh receptors. It is routinely administered before surgery to prevent salivation and to dry up respiratory system secretions. Ophthalmologists also use it to dilate the pupils for eye examination. The anticholinesterase drug *neostigmine* inhibits the enzyme acetylcholinesterase, preventing enzymatic breakdown of ACh and allowing it to accumulate in synapses. This drug is used to treat myasthenia gravis, a condition in which skeletal muscle activity is impaired for lack of ACh stimulation.

As we described in Chapter 11, NE is one of our "feeling good" neurotransmitters, and drugs that prolong the activity of NE on the postsynaptic membrane help to relieve depression. Hundreds of over-the-counter drugs used to treat colds, coughs, allergies, and nasal congestion contain sympathomimetics (phenylephrine and others), sympathetic-mimicking drugs that stimulate α-adrenergic receptors.

Much pharmaceutical research is directed toward finding drugs that affect only one subclass of receptor without upsetting the whole adrenergic or cholinergic system. An important breakthrough was finding drugs that mainly activate β_2 receptors. People with asthma use such β_2 activators to dilate their lung bronchioles without activating β_1 receptors, which would increase their heart rate. Selected drug classes that influence ANS activity are summarized in **Table 14.3**.

Interactions of the Autonomic Divisions

As we mentioned earlier, most visceral organs receive *dual innervation*. Normally, both ANS divisions are partially active, producing a dynamic antagonism that allows visceral activity to be precisely controlled. However, one division or the other usually exerts the predominant effects in given circumstances, and in a few cases, the two divisions actually cooperate with each

TABLE 14.3 Selected Drug Classes That Influence the Activity of the Autonomic Nervous System

DRUG CLASS	RECEPTOR BOUND	EFFECTS	EXAMPLE	CLINICAL USE
Nicotinic agents (little therapeutic value, but important because of presence of nicotine in tobacco)	Nicotinic ACh receptors on all ganglionic neurons and in CNS	Typically stimulation of sympathetic effects; blood pressure increases	Nicotine	Used in smoking cessation products
Parasympathomimetic agents (muscarinic agents)	Muscarinic ACh receptors	Mimic effects of ACh, enhance parasympathetic effects	Pilocarpine	Glaucoma (opens aqueous humor drainage pores)
			Bethanechol	Difficulty urinating (increases bladder contraction)
Acetylcholinesterase inhibitors	None; bind to the enzyme (AChE) that degrades ACh	Indirect effect at all ACh receptors; prolong the effect of ACh	Neostigmine	Myasthenia gravis, (increases availability of ACh)
			Sarin	Used as chemical warfare agent (similar to widely used insecticides)
Sympathomimetic agents	Adrenergic receptors	Enhance sympathetic activity by increasing NE release or binding to adrenergic receptors	Albuterol (Ventolin)	Asthma (dilates bronchioles by binding to β_2 receptors)
			Phenylephrine	Colds (nasal decongestant, binds to α_1 receptors)
Sympatholytic agents	Adrenergic receptors	Decrease sympathetic activity by blocking adrenergic receptors or inhibiting NE release	Propranolol	Hypertension (member of a class of drugs called *beta-blockers* that decrease heart rate and blood pressure)

other. **Table 14.4** contains an organ-by-organ summary of the differing effects of the two divisions.

Antagonistic Interactions

Antagonistic effects, described earlier, are most clearly seen on the activity of the heart, respiratory system, and gastrointestinal organs. In a fight-or-flight situation, the sympathetic division increases heart rate, dilates airways, and inhibits digestion and elimination. When the emergency is over, the parasympathetic division restores heart rate and airway diameter to resting levels and then attends to processes that refuel your body cells and discard wastes.

Sympathetic and Parasympathetic Tone

We have described the parasympathetic division as the "resting and digesting" division, but the sympathetic division is the major actor in controlling blood pressure, even at rest. With few exceptions, the vascular system is entirely innervated by sympathetic fibers that keep the blood vessels in a continual state of partial constriction called **sympathetic**, or **vasomotor**, **tone**. When a higher blood pressure is needed to maintain blood flow, the sympathetic fibers fire more rapidly, causing blood vessels to constrict and blood pressure to rise. When blood pressure is to be decreased, sympathetic fibers fire less rapidly and the vessels dilate. *Alpha-blockers*, drugs that interfere with the activity of these **vasomotor fibers**, are sometimes used to treat hypertension.

During circulatory shock (inadequate blood delivery to body tissues), or when more blood is needed to meet the increased needs of working skeletal muscles, blood vessels serving the skin and abdominal viscera are strongly constricted. This blood "shunting" helps maintain circulation to vital organs or enhance blood delivery to skeletal muscles.

On the other hand, parasympathetic effects normally dominate the heart and the smooth muscle of digestive and urinary tract organs. These organs exhibit **parasympathetic tone**. The parasympathetic division slows the heart and dictates the normal activity levels of the digestive and urinary tracts. However, the sympathetic division can override these parasympathetic effects during times of stress. Drugs that block parasympathetic responses increase heart rate and cause fecal and urinary retention. Except for the adrenal glands and sweat glands of the skin, most glands are activated by parasympathetic fibers.

Cooperative Effects

The best example of cooperative ANS effects is seen in controls of the external genitalia. Parasympathetic stimulation causes vasodilation of blood vessels in the external genitalia and is responsible for erection of the male penis or female clitoris during sexual excitement. (This may explain why sexual performance is sometimes impaired when people are anxious or upset and the sympathetic division is in charge.) Sympathetic stimulation

TABLE 14.4 Effects of the Parasympathetic and Sympathetic Divisions on Various Organs

TARGET ORGAN OR SYSTEM	PARASYMPATHETIC EFFECTS	SYMPATHETIC EFFECTS
Eye (iris)	Stimulates sphincter pupillae muscles; constricts pupils	Stimulates dilator pupillae muscles; dilates pupils
Eye (ciliary muscle)	Stimulates muscle, which results in bulging of the lens for close vision	Weakly inhibits muscle, which results in flattening of the lens for far vision
Glands (nasal, lacrimal, gastric, pancreas)	Stimulates secretory activity	Inhibits secretory activity; causes vasoconstriction of blood vessels supplying the glands
Salivary glands	Stimulates secretion of watery saliva	Stimulates secretion of thick, viscous saliva
Sweat glands	No effect (no innervation)	Stimulates copious sweating (cholinergic fibers)
Adrenal medulla	No effect (no innervation)	Stimulates medulla cells to secrete epinephrine and norepinephrine
Arrector pili muscles attached to hair follicles	No effect (no innervation)	Stimulates contraction (erects hairs and produces "goosebumps")
Heart (muscle)	Decreases rate; slows heart	Increases rate and force of heartbeat
Heart (coronary blood vessels)	No effect (no innervation)	Causes vasodilation*
Urinary bladder/urethra	Causes contraction of smooth muscle of bladder wall; relaxes urethral sphincter; promotes voiding	Causes relaxation of smooth muscle of bladder wall; constricts urethral sphincter; inhibits voiding
Lungs	Constricts bronchioles	Dilates bronchioles*
Digestive tract organs	Increases motility (peristalsis) and amount of secretion by digestive organs; relaxes sphincters to allow movement of foodstuffs along tract	Decreases activity of glands and muscles of digestive system and constricts sphincters (e.g., anal sphincter)
Liver	Increases glucose uptake from blood	Stimulates release of glucose to blood*
Gallbladder	Excites (gallbladder contracts to expel bile)	Inhibits (gallbladder is relaxed)
Kidney	No effect (no innervation)	Promotes renin release; causes vasoconstriction; decreases urine output
Penis	Causes erection (vasodilation)	Causes ejaculation
Vagina/clitoris	Causes erection (vasodilation) of clitoris; increases vaginal lubrication	Causes contraction of vagina
Blood vessels	Little or no effect	Constricts most vessels and increases blood pressure; constricts vessels of abdominal viscera and skin to divert blood to muscles, brain, and heart when necessary; NE constricts most vessels; epinephrine dilates vessels of the skeletal muscles during exercise*
Blood coagulation	No effect (no innervation)	Increases coagulation*
Cellular metabolism	No effect (no innervation)	Increases metabolic rate*
Adipose tissue	No effect (no innervation)	Stimulates lipolysis (fat breakdown)

*Effects are mediated by epinephrine release into the bloodstream from the adrenal medulla.

14

then causes ejaculation of semen by the penis or reflex contractions of a female's vagina.

HOMEOSTATIC IMBALANCE

Autonomic neuropathy (damage to autonomic nerves) is a common complication of diabetes mellitus. One of the earliest and most troubling symptoms is sexual dysfunction, with up to 75% of male diabetics experiencing erectile dysfunction. Women, on the other hand, often experience reduced vaginal lubrication. Other frequent manifestations of autonomic neuropathy include dizziness after standing suddenly (poor blood pressure control), urinary incontinence, sluggish eye pupil reactions, and impaired sweating. Just how the elevated blood glucose levels in diabetics damage nerves is still a mystery. ■

Unique Roles of the Sympathetic Division

The adrenal medulla, sweat glands and arrector pili muscles of the skin, the kidneys, and most blood vessels receive only sympathetic fibers. It is easy to remember that the sympathetic system innervates these structures because most of us sweat under stress, our scalp "prickles" during fear, and our blood pressure skyrockets (from widespread vasoconstriction) when we get excited. We have already described how sympathetic control of blood vessels regulates blood pressure and shunting of blood in the vascular system. We will now consider several other uniquely sympathetic functions.

Thermoregulatory Responses to Heat The sympathetic division mediates reflexes that regulate body temperature. For example, applying heat to the skin causes reflexive dilation of blood vessels in that area. When systemic body temperature is elevated, sympathetic nerves cause the skin's blood vessels to dilate, allowing the skin to become flushed with warm blood, and activate the sweat glands to help cool the body. When body temperature falls, skin blood vessels are constricted, and, as a result, blood is restricted to deeper, more vital organs.

Release of Renin from the Kidneys Sympathetic impulses stimulate the kidneys to release *renin*, a hormone that promotes an increase in blood pressure. We describe this renin-angiotensin mechanism in Chapter 25.

Metabolic Effects Through both direct neural stimulation and release of adrenal medullary hormones, the sympathetic division promotes a number of metabolic effects not reversed by parasympathetic activity. It (1) increases the metabolic rate of body cells; (2) raises blood glucose levels; and (3) mobilizes fats for use as fuels. The medullary hormones also cause skeletal muscle to contract more strongly and quickly.

As a side effect, muscle spindles are stimulated more often and, consequently, nerve impulses traveling to the muscles occur more synchronously. These neural bursts, which put muscle contractions on a "hair trigger," are great if you have to make a quick jump or run, but they can be embarrassing or even disabling to the nervous musician or surgeon.

Localized Versus Diffuse Effects

In the parasympathetic division, one preganglionic neuron synapses with one (or at most a few) ganglionic neurons. Additionally, all parasympathetic fibers release ACh, which is quickly destroyed (hydrolyzed) by acetylcholinesterase. Consequently, the parasympathetic division exerts short-lived, highly localized control over its effectors.

In contrast, preganglionic sympathetic axons branch profusely as they enter the sympathetic trunk, and they synapse with ganglionic neurons at several levels. As a result, when the sympathetic division is activated, it responds in a diffuse and highly interconnected way. Indeed, the literal translation of sympathetic (*sym* = together; *pathos* = feeling) relates to the bodywide mobilization this division provokes. Nevertheless, parts of the sympathetic nervous system can be activated individually. For example, just because your eye pupils dilate in dim light doesn't necessarily mean that your heart rate speeds up.

Effects produced by sympathetic activation are much longer-lasting than parasympathetic effects. In contrast to the parasympathetic system's ACh, NE is inactivated more slowly because it must be taken back up into the presynaptic ending

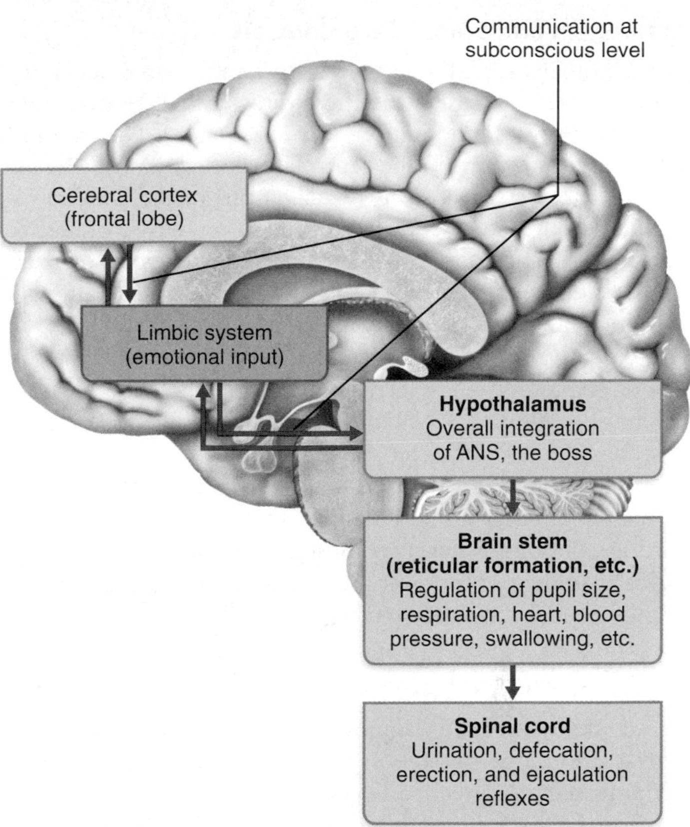

Figure 14.9 Levels of ANS control. The hypothalamus stands at the top of the control hierarchy as the integrator of ANS activity, but it is influenced by subconscious cerebral inputs via limbic system connections.

14

before being hydrolyzed or stored. More importantly, NE and epinephrine are secreted into the blood by adrenal medullary cells when the sympathetic division is mobilized. Although epinephrine is more potent at increasing heart rate and raising blood glucose levels and metabolic rate, these hormones have essentially the same effects as NE released by sympathetic neurons. In fact, circulating adrenal medullary hormones produce 25–50% of all the sympathetic effects acting on the body at a given time. These effects continue for several minutes until the hormones are destroyed by the liver.

In short, sympathetic nerve impulses act only briefly, but the hormonal effects they provoke linger. The widespread and prolonged effect of sympathetic activation helps explain why we need time to "come down" after an extremely stressful experience.

Control of Autonomic Functioning

Although the ANS is not usually considered to be under voluntary control, its activity is regulated by CNS controls in the spinal cord, brain stem, hypothalamus, and cerebral cortex (**Figure 14.9**). In general, the hypothalamus is the integrative center at the top of the ANS control hierarchy. From there, orders flow to lower and lower CNS centers for execution. Although the cerebral cortex may modify the workings of the ANS, it does so at the subconscious level and by acting through limbic system structures on hypothalamic centers.

Brain Stem and Spinal Cord Controls

The hypothalamus is the "boss," but the brain stem reticular formation appears to exert the most *direct* influence over autonomic functions (see Figure 12.16 on p. 448). For example, certain motor centers in the ventrolateral medulla (*cardiac* and *vasomotor centers*) reflexively regulate heart rate and blood vessel diameter. Other medullary regions oversee gastrointestinal activities. Most sensory impulses involved in eliciting these autonomic reflexes reach the brain stem via vagus nerve afferents. Although not considered part of the ANS, the medulla and pons also contain respiratory centers that mediate involuntary control of respiration and receive inputs from the hypothalamus. Midbrain centers (*oculomotor nuclei*) control the muscles concerned with pupil diameter and lens focus.

Defecation and micturition reflexes that promote emptying of the rectum and urinary bladder are integrated at the spinal cord level but are subject to conscious inhibition. We will describe all of these autonomic reflexes in later chapters in relation to the organ systems they serve.

Hypothalamic Controls

As we noted, the hypothalamus is the main integration center of the autonomic nervous system. In general, anterior hypothalamic regions direct parasympathetic functions, and posterior areas direct sympathetic functions. These centers exert their effects both directly and via relays through the *reticular formation*, which in turn influences the preganglionic motor neurons in the brain stem and spinal cord (Figure 14.9). The hypothalamus contains centers that coordinate heart activity, blood pressure, body temperature, water balance, and endocrine activity. It also contains centers that mediate various emotional states (rage, pleasure) and biological drives (thirst, hunger, sex).

The hypothalamus also mediates our reactions to fear via its associations with the amygdala and the periaqueductal gray matter. Emotional responses of the limbic system of the cerebrum to danger and stress signal the hypothalamus to activate the sympathetic system to fight-or-flight status. In this way, the hypothalamus serves as the keystone of the emotional and visceral brain, and through its centers emotions influence ANS functioning and behavior.

Cortical Controls

It was originally believed that the ANS is not subject to voluntary controls. However, we have all had occasions when remembering a frightening event made our heart race (sympathetic response) or just the thought of a favorite food, pecan pie for example, made our mouth water (a parasympathetic response). These inputs converge on the hypothalamus through its connections to the limbic lobe.

Additionally, studies have shown that voluntary cortical control of visceral activities is possible—a capability untapped by most people.

Influence of Biofeedback on Autonomic Function During **biofeedback training**, subjects are connected to monitoring devices that provide an awareness of what is happening in their body. This awareness is called **biofeedback**. The devices detect and amplify changes in physiological processes such as heart rate and blood pressure, and these data are "fed back" in the form of flashing lights or audible tones. Subjects are asked to try to alter or control some "involuntary" function by concentrating on calming, pleasant thoughts. The monitor allows them to identify changes in the desired direction, so they can recognize the feelings associated with these changes and learn to produce the changes at will.

Biofeedback techniques have been successful in helping individuals plagued by migraine headaches. They are also used by cardiac patients to manage stress and reduce their risk of heart attack. However, biofeedback training is time-consuming and often frustrating, and the training equipment is expensive and difficult to use.

CHECK YOUR UNDERSTANDING

6. Name the division of the ANS that does each of the following: increases digestive activity; increases blood pressure; dilates bronchioles; decreases heart rate; stimulates the adrenal medulla to release its hormones; causes ejaculation.
7. Would you find nicotinic receptors on skeletal muscle? Smooth muscle? Eccrine sweat glands? The adrenal medulla? CNS neurons?
8. Which part of the brain is the main integration center of the ANS? Which part exerts the most direct influence over autonomic functions?

For answers, see Appendix G.

Homeostatic Imbalances of the ANS

▶ Explain the relationship of some types of hypertension, Raynaud's disease, and autonomic dysreflexia to disorders of autonomic functioning.

The ANS is involved in nearly every important process that goes on in the body, so it is not surprising that abnormalities of autonomic functioning can have far-reaching effects and threaten life itself. Most autonomic disorders reflect exaggerated or deficient controls of smooth muscle activity. Of these, the most devastating involve blood vessels and include conditions such as hypertension, Raynaud's disease, and autonomic dysreflexia.

Hypertension, or high blood pressure, may result from an overactive sympathetic vasoconstrictor response promoted by continuous high levels of stress. Hypertension is always serious because it increases the workload on the heart, which may precipitate heart disease, and increases the wear and tear on artery walls. Stress-induced hypertension can be treated with adrenergic receptor–blocking drugs.

Raynaud's disease is characterized by intermittent attacks causing the skin of the fingers and toes to become pale, then cyanotic and painful. Commonly provoked by exposure to cold or emotional stress, it is an exaggerated vasoconstriction response. The severity of this disease ranges from merely uncomfortable to such severe blood vessel constriction that ischemia and gangrene (tissue death) results. Vasodilators (for example, adrenergic blockers) usually suffice, but to treat very severe cases, preganglionic sympathetic fibers serving the affected regions are severed (a procedure called *sympathectomy*). The involved vessels then dilate, reestablishing adequate blood delivery to the region.

Autonomic dysreflexia is a life-threatening condition involving uncontrolled activation of autonomic neurons. It occurs in a majority of individuals with quadriplegia and in others with spinal cord injuries above the T_6 level, usually in the first year after injury. The usual trigger is a painful stimulus to the skin or overfilling of a visceral organ, such as the urinary bladder. Arterial blood pressure skyrockets to life-threatening levels, which may rupture a blood vessel in the brain, precipitating stroke. This may be accompanied by a headache, flushed face, sweating above the level of the injury, and cold, clammy skin below. The precise mechanism of autonomic dysreflexia is not yet clear.

CHECK YOUR UNDERSTANDING

9. Jackson works long, stress-filled shifts as an air traffic controller at a busy airport. His doctor has prescribed a beta-blocker. Why might his doctor have done this? What does a beta-blocker do?

For answers, see Appendix G.

Developmental Aspects of the ANS

▶ Describe some effects of aging on the autonomic nervous system.

ANS preganglionic neurons derive from the embryonic *neural tube*, as do somatic motor neurons. ANS structures in the PNS—ganglionic neurons, the adrenal medulla, and all autonomic ganglia—derive from the **neural crest** (along with all sensory neurons) (see Figure 12.1, ③). Neural crest cells reach their ultimate destinations by migrating along growing axons. Forming ganglia receive axons from preganglionic neurons in the spinal cord or brain and send their axons to synapse with their effector cells in the body periphery. This process depends upon the presence of **nerve growth factor**, and is guided by a number of signaling chemicals similar to those acting in the CNS.

During youth, impairments of ANS function are usually due to injuries to the spinal cord or autonomic nerves. In old age the efficiency of the ANS begins to decline. At least part of the problem appears to be due to structural changes (bloating) of some preganglionic axon terminals, which become congested with neurofilaments.

Many elderly people complain of constipation (a result of reduced gastrointestinal tract motility), and of dry eyes and frequent eye infections (both a result of a diminished ability to form tears). Additionally, when they stand up they may have fainting episodes due to **orthostatic hypotension** (*ortho* = straight; *stat* = standing), a form of low blood pressure that occurs because the aging pressure receptors respond less to changes in blood pressure following changes in position, and because of slowed responses by aging sympathetic vasoconstrictor centers. These problems are distressing, but not usually life threatening, and most can be managed by lifestyle changes or artificial aids. For example, changing position slowly gives the sympathetic nervous system time to adjust the blood pressure, and eye drops (artificial tears) are available for the dry-eye problem.

CHECK YOUR UNDERSTANDING

10. Which embryonic structure gives rise to both the autonomic ganglia and the adrenal medulla?

For answers, see Appendix G.

In this chapter, we have described the structure and function of the ANS, one arm of the motor division of the peripheral nervous system. Because virtually every organ system still to be considered depends on autonomic controls, you will be hearing more about the ANS in chapters that follow. Now that we have explored most of the nervous system, this is a good time to examine how it interacts with the rest of the body as summarized in the *Making Connections* feature, pp. 542–543.

System Connections

Homeostatic Interrelationships Between the Nervous System and Other Body Systems

Endocrine System

- Sympathetic division of the ANS activates the adrenal medulla; hypothalamus helps regulate the activity of the anterior pituitary gland and produces the two posterior pituitary hormones
- Hormones influence neuronal metabolism

Cardiovascular System

- ANS helps regulate heart rate and blood pressure
- Cardiovascular system provides blood containing oxygen and nutrients to the nervous system and carries away wastes

Lymphatic System/Immunity

- Nerves innervate lymphoid organs; the brain plays a role in regulating immune function
- Lymphatic vessels carry away leaked tissue fluids from tissues surrounding nervous system structures; immune elements protect all body organs from pathogens (CNS has additional mechanisms as well)

Respiratory System

- Nervous system initiates and regulates respiratory rhythm and depth
- Respiratory system provides life-sustaining oxygen; disposes of carbon dioxide

Digestive System

- ANS (particularly the parasympathetic division) regulates digestive motility and glandular activity
- Digestive system provides nutrients needed for neuronal health

Urinary System

- ANS regulates bladder emptying and renal blood pressure
- Kidneys help to dispose of metabolic wastes and maintain proper electrolyte composition and pH of blood for neural functioning

Reproductive System

- ANS regulates sexual erection and ejaculation in males; erection of the clitoris in females
- Testosterone causes masculinization of the brain and underlies sex drive and aggressive behavior

Integumentary System

- Sympathetic division of the ANS regulates sweat glands and blood vessels of skin (therefore heat loss/retention)
- Skin serves as heat loss surface

Skeletal System

- Nerves innervate bones and joints, providing for pain and joint sense
- Bones serve as depot for calcium needed for neural function; skeletal system protects CNS structures

Muscular System

- Somatic division of nervous system activates skeletal muscles; maintains muscle health
- Skeletal muscles are the effectors of the somatic division

Closer Connections

The Nervous System and Interrelationships with the Muscular, Respiratory, and Digestive Systems

The nervous system pokes its figurative nose into the activities of virtually every organ system of the body. Hence, trying to choose the most significant interactions comes pretty close to a lesson in futility. Which is more important: digestion, elimination of wastes, or motility? Your answer probably depends on whether you are hungry or need to make a "pit stop" at the bathroom when you read this. Since neural interactions with several other systems are thoroughly plumbed in chapters to come, here we will look at the all-important interactions between the nervous system and the muscular, respiratory, and digestive systems.

Muscular System

Most simply said, the muscular system would cease to function without the nervous system. Unlike visceral or cardiac muscles, both of which have other controlling systems, somatic motor fibers are *it* for skeletal muscle activation and regulation. Somatic nerve fibers "tell" skeletal muscles not only when to contract but also how strongly. Additionally, as the nervous system makes its initial synapses with skeletal muscle fibers, it determines their fate as fast or slow fibers, which forever after affects our potential for muscle speed and endurance. The interactions of the various brain regions (basal nuclei, cerebellum, premotor cortex, etc.) and inputs of stretch receptors also determine our grace—how smooth and coordinated we are. Nonetheless, keep in mind that as long as the skeletal muscle effector cells are healthy, they help determine the viability of the neurons synapsing with them. The relationship is truly synergistic.

Respiratory System

Another system that depends entirely on the nervous system for its function is the respiratory system, which continuously refreshes the blood with oxygen and unloads the carbon dioxide waste to the sea of air that surrounds us. Neural centers in the medulla and pons both initiate and maintain the tidelike rhythm of air flushing into and out of our lungs by activating skeletal muscles that change the volume (thus the gas pressure) within the lungs. Peripheral receptors supply information about gas concentrations, lung stretch, and skeletal muscle activity to these CNS respiratory centers.

Digestive System

Although the digestive system responds to many different types of controls—for example, hormones, local pH, and irritating chemicals—the parasympathetic nervous system is crucial to its normal functioning. Without the parasympathetic inputs, sympathetic neural activity, which inhibits normal digestion, would be unopposed. So important are parasympathetic controls that some of the parasympathetic neurons are actually located in the walls of the digestive organs, in what are called intrinsic plexuses. Thus, even if all extrinsic controls are severed, the intrinsic mechanisms can still maintain this crucial body function. The role the digestive system plays for the nervous system is the same one it offers to all body systems—it sees that ingested foodstuff gets digested and loaded into the blood for cell use.

Clinical Connections

Nervous System

Case study: On arrival at Holyoke Hospital, Jimmy Chin, a 10-year-old boy, is immobilized on a rigid stretcher so that he is unable to move his head or trunk. The paramedics report that when they found him some 50 feet from the bus, he was awake and alert, but crying and complaining that he couldn't "get up to find his mom" and that he had a "wicked headache." He has severe bruises on his upper back and head, and lacerations of his back and scalp. His blood pressure is low, body temperature is below normal, lower limbs are paralyzed, and he is insensitive to painful stimuli below the nipples. Although still alert on arrival, Jimmy soon begins to drift in and out of unconsciousness.

Jimmy is immediately scheduled for a CT scan, and an operating room is reserved.

Relative to Jimmy's condition:

1. Why were his head and torso immobilized for transport to the hospital?
2. What do his worsening neurological signs (drowsiness, incoherence, etc.) probably indicate? (Relate this to the type of surgery that will be performed.)
3. Assuming that Jimmy's sensory and motor deficits are due to a spinal cord injury, at what level do you expect to find a spinal cord lesion?
4. Two days after his surgery, Jimmy is alert and his MRI scan shows no residual brain injury, but pronounced swelling and damage to the spinal cord at T_4. On physical examination, Jimmy shows no reflex activity below the level of the spinal cord injury. His blood pressure is still low. Why are there no reflexes in his lower limbs and abdomen?
5. Over the next few days, his reflexes return in his lower limbs and become exaggerated. He is incontinent. Why is Jimmy hyperreflexive and incontinent?

On one occasion, Jimmy complains of a massive headache and his blood pressure is way above normal. On examination, he is sweating intensely above the nipples but has cold, clammy skin below the nipples and his heart rate is very slow.

6. What is this condition called and what precipitates it?
7. How does Jimmy's excessively high blood pressure put him at risk?

(Answers in Appendix G)

RELATED CLINICAL TERMS

Atonic bladder (ah-ton′ik; *a* = without; *ton* = tone, tension) A condition in which the urinary bladder becomes flaccid and overfills, allowing urine to dribble through the sphincters; results from temporary loss of the micturition reflex following spinal cord injury.

Horner's syndrome A condition due to damage to the superior sympathetic trunk on one side of the body; the affected person exhibits drooping of the upper eyelid (ptosis) and constricted pupil, and does not sweat on the affected side of the head.

Vagotomy (va-got′o-me) Cutting or severing of the vagus nerve to decrease secretion of gastric juice in those with peptic ulcers that do not respond to medication.

Vasovagal syncope Vasovagal syncope, also called *neurocardiogenic syncope*, is the most common cause of fainting. Although it may be provoked by emotional stress, pain, or dehydration, it typically occurs during prolonged standing. Fainting is due to a drop in blood pressure, which decreases blood flow to the brain and results in loss of consciousness.

CHAPTER SUMMARY

Media study tools that could provide you additional help in reviewing specific key topics of Chapter 14 are referenced below.

iP = ***Interactive Physiology***

1. The autonomic nervous system (ANS) is the motor division of the PNS that controls visceral activities, with the goal of maintaining internal homeostasis.

Introduction (pp. 526–528)

Comparison of the Somatic and Autonomic Nervous Systems (pp. 526–527)

1. The somatic (voluntary) nervous system provides motor fibers to skeletal muscles. The autonomic (involuntary or visceral motor) nervous system provides motor fibers to smooth and cardiac muscles and glands.
2. In the somatic division, a single motor neuron forms the efferent pathway from the CNS to the effectors. The efferent pathway of the autonomic division consists of a two-neuron chain: the preganglionic neuron in the CNS and the ganglionic neuron in a ganglion.
3. Acetylcholine, the neurotransmitter of somatic motor neurons, is stimulatory to skeletal muscle fibers. Neurotransmitters released by autonomic motor neurons (acetylcholine and norepinephrine) may cause excitation or inhibition.

iP Nervous System II; Topic: Synaptic Transmission, pp. 8–11.

ANS Divisions (pp. 527–528)

4. The ANS consists of two divisions, the parasympathetic and sympathetic, which normally exert antagonistic effects on many of the same target organs.
5. The parasympathetic division (the resting-digesting system) conserves body energy and maintains body activities at basal levels.
6. Parasympathetic effects include pupillary constriction, glandular secretion, increased digestive tract motility, and smooth muscle activity leading to elimination of feces and urine.
7. The sympathetic division prepares the body for activity and is called the fight-or-flight system.
8. Sympathetic responses include dilated pupils, increased heart rate, increased blood pressure, dilation of the bronchioles of the lungs, increased blood glucose levels, and sweating. During exercise, sympathetic vasoconstriction shunts blood from the skin and digestive viscera to the heart, brain, and skeletal muscles.

ANS Anatomy (pp. 528–535)

Parasympathetic (Craniosacral) Division (pp. 529–530)

1. Parasympathetic preganglionic neurons arise from the brain stem and from the sacral (S_2–S_4) region of the spinal cord.
2. Preganglionic fibers synapse with ganglionic neurons in terminal ganglia located in (intramural ganglia) or close to their effector organs. Preganglionic fibers are long; postganglionic fibers are short.
3. Cranial fibers arise in the brain stem nuclei of cranial nerves III, VII, IX, and X and synapse in ganglia of the head, thorax, and abdomen. The vagus nerves serve virtually all organs of the thoracic and abdominal cavities.
4. Sacral fibers (S_2–S_4) issue from the lateral region of the cord and form pelvic splanchnic nerves that innervate the pelvic viscera. The preganglionic axons do not travel within rami communicantes.

Sympathetic (Thoracolumbar) Division (pp. 530–534)

5. Preganglionic sympathetic neurons arise from the lateral horns of the spinal cord from the level of T_1 through L_2.
6. Preganglionic axons leave the cord via white rami communicantes and enter the sympathetic trunk (chain) ganglia in the sympathetic trunk. An axon may synapse in a trunk ganglion at the same or at a different level, or it may issue from the sympathetic trunk without synapsing. Preganglionic fibers are short; postganglionic fibers are long.
7. When the synapse occurs in a trunk ganglion, the postganglionic fiber may enter the spinal nerve ramus via the gray ramus communicans to travel to the body periphery. Postganglionic fibers issuing from the cervical ganglia also serve visceral organs and blood vessels of the head, neck, and thorax.
8. When synapses do not occur in the trunk ganglia, the preganglionic fibers form splanchnic nerves (thoracic, lumbar, and sacral). Most splanchnic nerve fibers synapse in collateral ganglia, and the postganglionic fibers serve the abdominal viscera. Exceptions are that (1) some splanchnic nerve fibers synapse with cells of the adrenal medulla, and (2) some lumbar and sacral splanchnic nerve fibers *do* synapse in trunk ganglia.

Visceral Reflexes (pp. 534–535)

9. Visceral reflex arcs have the same components as somatic reflexes.
10. Cell bodies of visceral sensory neurons are located in dorsal root ganglia, sensory ganglia of cranial nerves, or autonomic ganglia. Visceral afferents are found in spinal nerves and in virtually all autonomic nerves.

ANS Physiology (pp. 535–540)

Neurotransmitters and Receptors (pp. 535–536)

1. Two major neurotransmitters, acetylcholine (ACh) and norepinephrine (NE), are released by autonomic motor neurons. On the basis of the neurotransmitter released, the fibers are classified as cholinergic or adrenergic.
2. ACh is released by all preganglionic fibers and all parasympathetic postganglionic fibers. NE is released by all sympathetic postganglionic fibers except those serving the sweat glands of the skin, and some blood vessels within skeletal muscles.
3. Neurotransmitter effects depend on the receptors to which the transmitter binds. Cholinergic (ACh) receptors are classified as nicotinic or muscarinic. Adrenergic (NE) receptors are classified as α_1 or α_2, or β_1, β_2, or β_3.

iP Nervous System II; Topic: Synaptic Transmission, pp. 8–11, 14.

The Effects of Drugs (p. 536)

4. Drugs that mimic, enhance, or inhibit the action of ANS neurotransmitters are used to treat conditions caused by excessive, inadequate, or inappropriate ANS functioning. Some drugs bind with only one receptor subtype, allowing specific ANS-mediated activities to be enhanced or blocked.

Interactions of the Autonomic Divisions (pp. 536–539)

5. Most visceral organs are innervated by both divisions; the divisions interact in various ways but usually exert a dynamic antagonism. Antagonistic interactions mainly involve the heart, respiratory system, and gastrointestinal organs. Sympathetic activity increases heart activity, dilates bronchioles, and depresses gastrointestinal activity. Parasympathetic activity reverses these effects.
6. Most blood vessels are innervated only by sympathetic fibers and exhibit vasomotor tone. Parasympathetic activity dominates the heart and muscles of the gastrointestinal tract (which normally exhibit parasympathetic tone) and glands.
7. The two ANS divisions exert cooperative effects on the external genitalia.
8. Roles unique to the sympathetic division are blood pressure regulation, shunting of blood in the vascular system, thermoregulatory responses, stimulation of renin release by the kidneys, and metabolic effects.
9. Activation of the sympathetic division can cause widespread, long-lasting mobilization of the fight-or-flight response. Parasympathetic effects are highly localized and short-lived.

Control of Autonomic Functioning (pp. 539–540)

10. Autonomic function is controlled at several levels: (1) Reflex activity is mediated by the spinal cord and brain stem (particularly medullary) centers. (2) Hypothalamic integration centers interact with both higher and lower centers to orchestrate autonomic, somatic, and endocrine responses. (3) Cortical centers influence autonomic functioning via connections with the limbic system; conscious controls of autonomic function are rare but possible, as illustrated by biofeedback training.

Homeostatic Imbalances of the ANS (pp. 540–541)

1. Most autonomic disorders reflect problems with smooth muscle control. Abnormalities in vascular control, such as occur in hypertension, Raynaud's disease, and autonomic dysreflexia, are most devastating.

Developmental Aspects of the ANS (p. 541)

1. Preganglionic neurons develop from the neural tube; ganglionic neurons develop from the embryonic neural crest.
2. The efficiency of the autonomic nervous system declines in old age, as reflected by decreased glandular secretory activity, decreased gastrointestinal motility, and slowed sympathetic vasomotor responses to changes in position.

REVIEW QUESTIONS

Multiple Choice/Matching

(Some questions have more than one correct answer. Select the best answer or answers from the choices given.)

1. All of the following characterize the ANS except **(a)** two-neuron efferent chain, **(b)** presence of nerve cell bodies in the CNS, **(c)** presence of nerve cell bodies in the ganglia, **(d)** innervation of skeletal muscles.
2. Relate each of the following terms or phrases to either the sympathetic (S) or parasympathetic (P) division of the autonomic nervous system:

_____ (1) short preganglionic, long postganglionic fibers
_____ (2) intramural ganglia
_____ (3) craniosacral outflow
_____ (4) adrenergic fibers
_____ (5) cervical ganglia
_____ (6) otic and ciliary ganglia
_____ (7) generally short-duration action
_____ (8) increases heart rate and blood pressure
_____ (9) increases gastric motility and secretion of lacrimal, salivary, and digestive juices
_____ (10) innervates blood vessels
_____ (11) most active when you are swinging in a hammock
_____ (12) active when you are running in the Boston Marathon

3. Preganglionic neurons develop from **(a)** neural crest cells, **(b)** neural tube cells, **(c)** alar plate cells, **(d)** endoderm.
4. The white rami communicantes contain what kind of fibers? **(a)** preganglionic parasympathetic, **(b)** postganglionic parasympathetic, **(c)** preganglionic sympathetic, **(d)** postganglionic sympathetic.
5. Collateral sympathetic ganglia are involved with the innervation of the **(a)** abdominal organs, **(b)** thoracic organs, **(c)** head, **(d)** arrector pili, **(e)** all of these.

Short Answer Essay Questions

6. Briefly explain why the following terms are sometimes used to refer to the autonomic nervous system: involuntary nervous system and emotional-visceral system.
7. Describe the anatomical relationship of the white and gray rami communicantes to the spinal nerve, and indicate the kind of fibers found in each ramus type.
8. Indicate the results of sympathetic activation of the following structures: sweat glands, eye pupils, adrenal medulla, heart, bronchioles of the lungs, liver, blood vessels of vigorously working skeletal muscles, blood vessels of digestive viscera, salivary glands.

9. Which of the effects listed in response to question 8 would be reversed by parasympathetic activity?
10. Which ANS fibers release acetylcholine? Which release norepinephrine?
11. Describe the meaning and importance of sympathetic tone and parasympathetic tone.
12. List the receptor subtypes for ACh and NE, and indicate the major sites where each type is found.
13. What area of the brain is most directly involved in mediating autonomic reflexes?
14. Describe the importance of the hypothalamus in controlling the autonomic nervous system.
15. Describe the basis and uses of biofeedback training.
16. What manifestations of decreased autonomic nervous system efficiency are seen in elderly individuals?
17. Ganglionic neurons are also called postganglionic neurons. Why is this a misnomer?

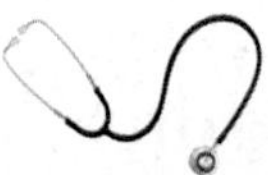

Critical Thinking and Clinical Application Questions

1. Mr. Johnson has been suffering from functional urinary retention and a hypoactive urinary bladder. Bethanechol, a drug that mimics acetylcholine's autonomic effects, is prescribed to manage his problem. First explain the rationale for prescribing bethanechol, and then predict which of the following adverse effects Mr. Johnson might experience while taking this drug (select all that apply): dizziness, low blood pressure, deficient tear formation, wheezing, increased mucus production in bronchi, deficient salivation, diarrhea, cramping, excessive sweating, undesirable erection of penis.
2. Mr. Jake was admitted to the hospital with excruciating pain in his left shoulder and arm. He was found to have suffered a heart attack. Explain the phenomenon of referred pain as exhibited by Mr. Jake.
3. A 32-year-old woman complains that she has been experiencing aching pains in the medial two fingers of both hands and that during such episodes, the fingers become blanched and then blue. Her history is taken, and it is noted that she is a heavy smoker. The physician advises her that she must stop smoking and states that he will not prescribe any medication until she has discontinued smoking for a month. What is this woman's problem, and why was she told to stop smoking?
4. Tiffany, a 21-year-old college student, had been having trouble sleeping, crying frequently, and having recurrent thoughts of suicide. She was prescribed an antidepressant. Like many such drugs, this antidepressant has anticholinergic side effects. What side effects might Tiffany experience in the first week of treatment?
5. As the aroma of freshly brewed coffee drifted by dozing Henry's nose, his mouth started to water and his stomach began to rumble. Explain his reactions in terms of ANS activity.

14

15 The Special Senses

People are responsive creatures. The aroma of freshly baked bread makes our mouths water. A sudden clap of thunder makes us jump. These stimuli and many others continually greet us and are interpreted by our nervous systems.

We are usually told that we have five senses: touch, taste, smell, sight, and hearing. Actually, touch reflects the activity of the general senses that we considered in Chapter 13. The other four traditional senses—*smell*, *taste*, *sight*, and *hearing*—are called **special senses**. Receptors for a fifth special sense, *equilibrium*, are housed in the ear, along with the organ of hearing.

In contrast to the widely distributed general receptors (most of which are modified nerve endings of sensory neurons), the **special sensory receptors** are distinct *receptor cells*. These receptor cells are

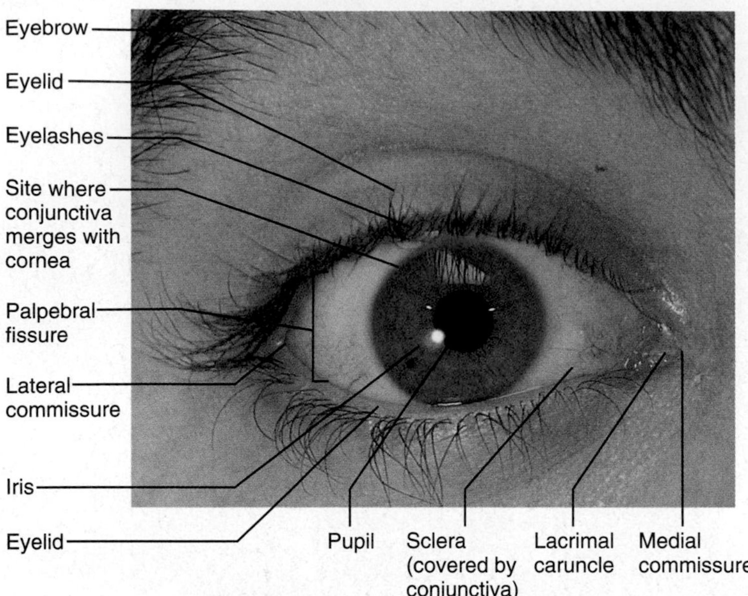

(a) Surface anatomy of the right eye

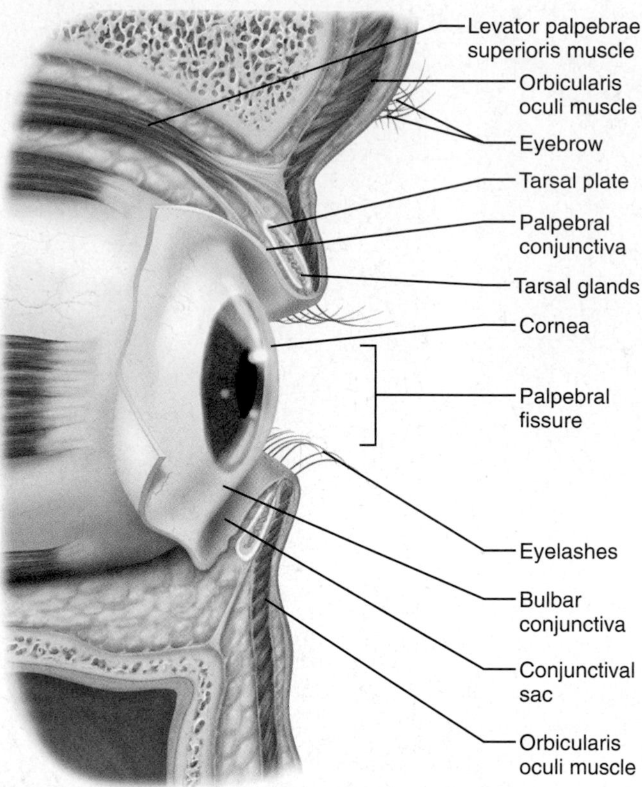

(b) Lateral view; some structures shown in sagittal section

Figure 15.1 **The eye and accessory structures.**

15

confined to the head region and are highly localized, either housed within complex sensory organs (eyes and ears) or in distinct epithelial structures (taste buds and olfactory epithelium).

In this chapter we consider the functional anatomy of each of the five special senses. But keep in mind that our perceptions of sensory inputs are overlapping. What we finally experience—our "feel" of the world—is a blending of stimulus effects.

The Eye and Vision

- Describe the structure and function of accessory eye structures, eye layers, the lens, and humors of the eye.
- Outline the causes and consequences of cataracts and glaucoma.

Vision is our dominant sense: Some 70% of all the sensory receptors in the body are in the eyes, and nearly half of the cerebral cortex is involved in some aspect of visual processing.

The adult **eye** is a sphere with a diameter of about 2.5 cm (1 inch). Only the anterior one-sixth of the eye's surface is visible (Figure 15.1a). The rest is enclosed and protected by a cushion of fat and the walls of the bony orbit. The fat pad occupies nearly all of the orbit not occupied by the eye itself. The eye is a complex structure and only a small portion of its tissues are actually involved in photoreception. Before turning our attention to the eye itself, let us consider the accessory structures that protect it or aid its functioning.

Accessory Structures of the Eye

The **accessory structures** of the eye include the eyebrows, eyelids, conjunctiva, lacrimal apparatus, and extrinsic eye muscles.

Eyebrows

The **eyebrows** are short, coarse hairs that overlie the supraorbital margins of the skull (Figure 15.1). They help shade the eyes from sunlight and prevent perspiration trickling down the forehead from reaching the eyes. Deep to the skin of the eyebrows are parts of the orbicularis oculi and corrugator muscles. Contraction of the orbicularis muscle depresses the eyebrow, whereas the corrugator moves the eyebrow medially.

Eyelids

Anteriorly, the eyes are protected by the mobile **eyelids** or **palpebrae** (pal′pĕ-bre). The eyelids are separated by the **palpebral fissure** ("eyelid slit") and meet at the medial and lateral angles of the eye—the **medial** and **lateral commissures** (*canthi*), respectively (Figure 15.1a).

The medial commissure sports a fleshy elevation called the **lacrimal caruncle** (kar′ung-kl; "a bit of flesh"). The caruncle contains sebaceous and sweat glands and produces the whitish, oily secretion (fancifully called the Sandman's eyesand) that sometimes collects at the medial commissure, especially during sleep. In most Asian peoples, a vertical fold of skin called the *epicanthic fold* commonly appears on both sides of the nose and sometimes covers the medial commissure.

The eyelids are thin, skin-covered folds supported internally by connective tissue sheets called **tarsal plates** (Figure 15.1b).

The tarsal plates also anchor the orbicularis oculi and **levator palpebrae superioris** muscles that run within the eyelid. The orbicularis muscle encircles the eye, and the eye closes when it contracts. Of the two eyelids, the larger, upper one is much more mobile, mainly because of the levator palpebrae superioris muscle, which raises that eyelid to open the eye.

The eyelid muscles are activated reflexively to cause blinking every 3–7 seconds and to protect the eye when it is threatened by foreign objects. Reflex blinking helps prevent drying of the eyes because each time we blink, accessory structure secretions (oil, mucus, and saline solution) are spread across the eyeball surface.

Projecting from the free margin of each eyelid are the **eyelashes**. The follicles of the eyelash hairs are richly innervated by nerve endings (hair follicle receptors), and anything that touches the eyelashes (even a puff of air) triggers reflex blinking.

Several types of glands are associated with the eyelids. The **tarsal glands** (*Meibomian glands*; mi-bo′me-an) are embedded in the tarsal plates (Figure 15.1b), and their ducts open at the eyelid edge just posterior to the eyelashes. These modified sebaceous glands produce an oily secretion that lubricates the eyelid and the eye and prevents the eyelids from sticking together. Associated with the eyelash follicles are a number of smaller, more typical sebaceous glands, and modified sweat glands called *ciliary glands* lie between the hair follicles (*cilium* = eyelash).

HOMEOSTATIC IMBALANCE

Infection of a tarsal gland results in an unsightly cyst called a *chalazion* (kah-la′ze-on; "swelling"). Inflammation of any of the smaller glands is called a *sty.* ■

Conjunctiva

The **conjunctiva** (kon″junk-ti′vah; "joined together") is a transparent mucous membrane. It lines the eyelids as the **palpebral conjunctiva** and reflects (folds back) over the anterior surface of the eyeball as the **bulbar conjunctiva** (Figure 15.1b). The bulbar conjunctiva covers only the white of the eye, not the cornea (the clear "window" over the iris and pupil). The bulbar conjunctiva is very thin, and blood vessels are clearly visible beneath it. (They are even more visible in irritated "bloodshot" eyes.)

When the eye is closed, a slitlike space occurs between the conjunctiva-covered eyeball and eyelids. This so-called **conjunctival sac** is where a contact lens lies, and eye medications are often administered into its inferior recess. The major function of the conjunctiva is to produce a lubricating mucus that prevents the eyes from drying out.

HOMEOSTATIC IMBALANCE

Inflammation of the conjunctiva, called *conjunctivitis*, results in reddened, irritated eyes. *Pinkeye*, a conjunctival infection caused by bacteria or viruses, is highly contagious. ■

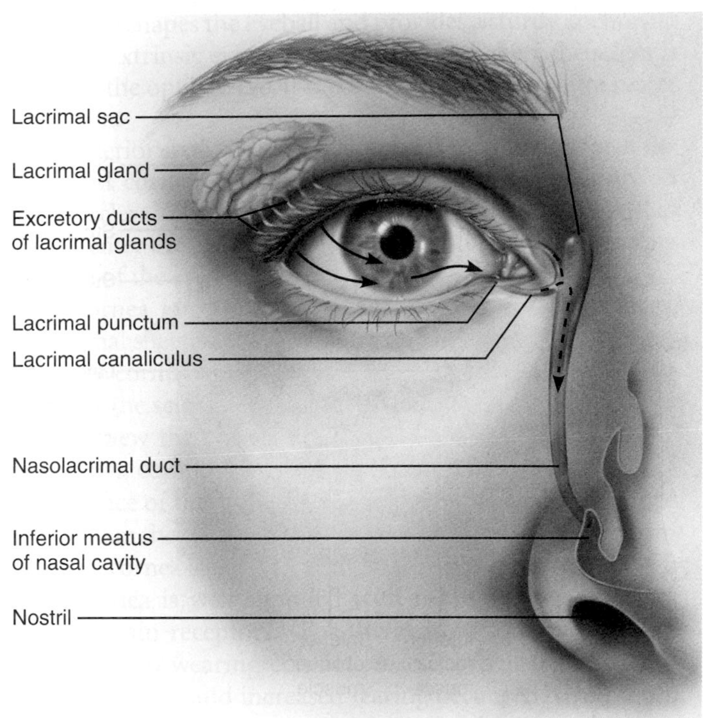

Figure 15.2 The lacrimal apparatus. Arrows indicate the flow of lacrimal fluid (tears) from the lacrimal gland to the nasal cavity.

Lacrimal Apparatus

15

The **lacrimal apparatus** (lak′rĭ-mal; "tear") consists of the lacrimal gland and the ducts that drain excess lacrimal secretions into the nasal cavity **(Figure 15.2)**. The **lacrimal gland** lies in the orbit above the lateral end of the eye and is visible through the conjunctiva when the lid is everted. It continually releases a dilute saline solution called **lacrimal secretion**—or, more commonly, **tears**—into the superior part of the conjunctival sac through several small excretory ducts.

Blinking spreads the tears downward and across the eyeball to the medial commissure, where they enter the paired **lacrimal canaliculi** via two tiny openings called **lacrimal puncta** (literally, "prick points"), visible as tiny red dots on the medial margin of each eyelid. From the lacrimal canaliculi, the tears drain into the **lacrimal sac** and then into the **nasolacrimal duct**, which empties into the nasal cavity at the inferior nasal meatus.

Lacrimal fluid contains mucus, antibodies, and **lysozyme**, an enzyme that destroys bacteria. Thus, it cleanses and protects the eye surface as it moistens and lubricates it. When lacrimal secretion increases substantially, tears spill over the eyelids and fill the nasal cavities, causing congestion and the "sniffles." This spillover (tearing) happens when the eyes are irritated and when we are emotionally upset. In the case of eye irritation, enhanced tearing serves to wash away or dilute the irritating substance. The importance of emotionally induced tears is poorly understood.

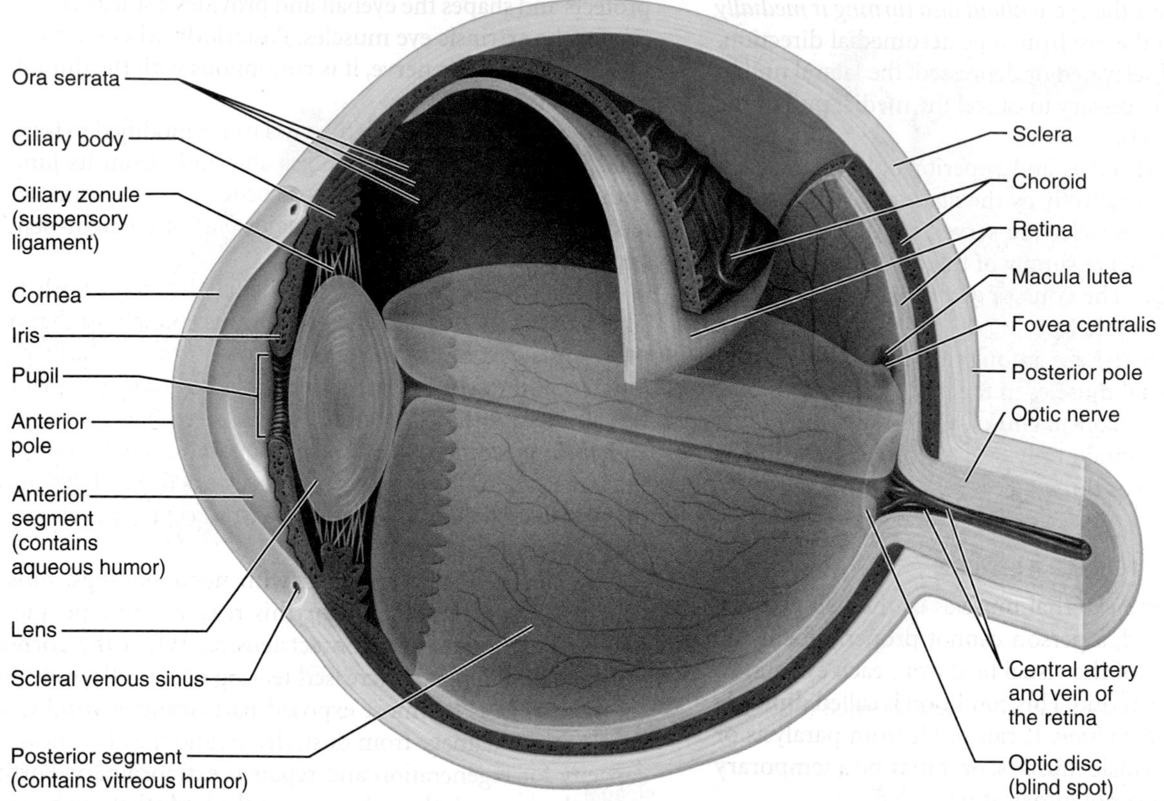

(a) Diagrammatic view. The vitreous humor is illustrated only in the bottom part of the eyeball.

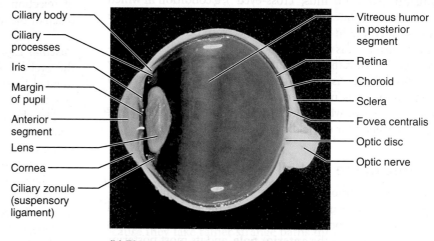

Ciliary body
Ciliary processes
Iris
Margin of pupil
Anterior segment
Lens
Cornea
Ciliary zonule (suspensory ligament)
Vitreous humor in posterior segment
Retina
Choroid
Sclera
Fovea centralis
Optic disc
Optic nerve

(b) Photograph of the human eye.

Figure 15.4 **Internal structure of the eye (sagittal section).**

congeal into a random pattern before birth. Its muscle fibers allow it to act as a reflexively activated diaphragm to vary pupil size (Figure 15.5). In close vision and bright light, the *sphincter pupillae* (circular muscles) contract and the pupil constricts. In distant vision and dim light, the *dilator pupillae* (radial muscles) contract and the pupil dilates, allowing more light to enter the eye. Pupillary dilation and constriction are controlled by sympathetic and parasympathetic fibers, respectively.

Changes in pupil size may also reflect our interests and emotional reactions to what we are seeing. Our pupils often dilate when the subject matter is appealing, in response to fear, and during problem solving. (Computing your taxes should make your pupils get bigger and bigger.) On the other hand, boredom or subject matter that is personally repulsive causes pupils to constrict.

Although irises come in different colors (*iris* = rainbow), they contain only brown pigment. When they have a lot of pigment, the eyes appear brown or black. If the amount of pigment is small and restricted to the posterior surface of the iris, the shorter wavelengths of light are scattered from the unpigmented parts, and the eyes appear blue, green, or gray. Most newborn babies' eyes are slate gray or blue because their iris pigment is not yet developed.

Inner Layer (Retina) The innermost layer of the eyeball is the delicate, two-layered **retina** (ret'ĭ-nah). Its outer **pigmented layer**, a single-cell-thick lining, abuts the choroid, and extends anteriorly to cover the ciliary body and the posterior face of the iris. These pigmented epithelial cells, like those of the choroid, absorb light and prevent it from scattering in the eye. They also act as phagocytes to remove dead or damaged photoreceptor cells, and store vitamin A needed by the photoreceptor cells. The transparent inner **neural layer** extends anteriorly to the posterior margin of the ciliary body. This junction is called the **ora serrata**, literally, the saw-toothed margin (see Figure 15.4).

Originating as an outpocketing of the brain, the retina contains millions of photoreceptors that transduce (convert) light energy, other neurons involved in the processing of light stimuli, and glia. Although the pigmented and neural layers are very close together, they are not fused. Only the neural layer of the retina plays a direct role in vision.

From posterior to anterior, the neural layer is composed of three main types of neurons: **photoreceptors**, **bipolar cells**, and **ganglion cells** (Figure 15.6). Signals are produced in response to light and spread from the photoreceptors (abutting the pigmented layer) to the bipolar cells and then to the innermost ganglion cells, where action potentials are generated. The ganglion cell axons make a right-angle turn at the inner face of the retina, then leave the posterior aspect of the eye as the thick optic nerve. The retina also contains other types of neurons—horizontal cells and amacrine cells—which play a role in visual processing. The **optic disc**, where the optic nerve exits the eye, is a weak spot in the **fundus** (posterior wall) of the eye because it is not reinforced by the sclera. The optic disc is also called the **blind spot** because it lacks photoreceptors, so light focused on it cannot be seen. Nonetheless, we do not usually notice these gaps in our vision because the brain uses a sophisticated process called *filling in* to deal with absence of input.

The quarter-billion photoreceptors found in the neural retinas are of two types: rods and cones. The more numerous **rods** are our dim-light and peripheral vision receptors. They are far more sensitive to light than cones are, but they do not provide either sharp images or color vision. This is why colors are indistinct and edges of objects appear fuzzy in dim light and at the edges of our visual field. **Cones**, by contrast, operate in bright light and provide high-acuity color vision.

Lateral to the blind spot of each eye, and located precisely at the eye's posterior pole, is an oval region called the **macula lutea** (mak'u-lah lu'te-ah; "yellow spot") with a minute (0.4 mm) pit in its center called the **fovea centralis** (see Figure 15.4). In this region, the retinal structures abutting the vitreous humor are displaced to the sides. This allows light to pass almost directly to the photoreceptors rather than through several retinal layers, greatly enhancing visual acuity. The fovea contains only cones, the macula contains mostly cones, and from the edge of the macula toward the retina periphery, cone density declines gradually. The retina periphery contains mostly rods, which continuously decrease in density from there to the macula.

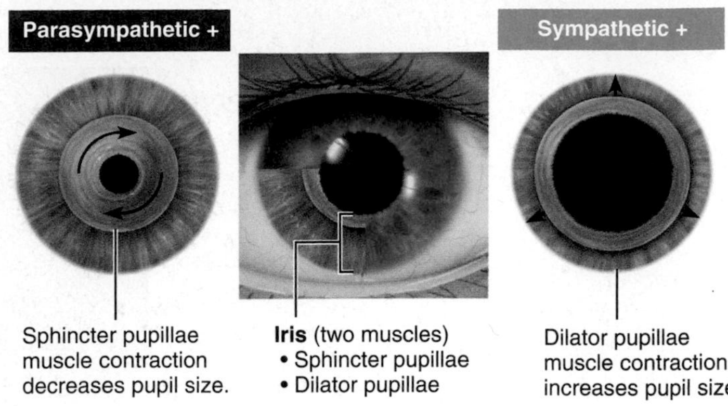

Parasympathetic +
Sympathetic +
Sphincter pupillae muscle contraction decreases pupil size.
Iris (two muscles)
• Sphincter pupillae
• Dilator pupillae
Dilator pupillae muscle contraction increases pupil size

Figure 15.5 Pupil dilation and constriction, anterior view. (+ means activation.)

Only the foveae (plural of fovea) have a sufficient cone density to provide detailed color vision, so anything we wish to view critically is focused on the foveae. Because each fovea is only about the size of the head of a pin, not more than a thousandth of the entire visual field is in *hard focus* (foveal focus) at a given moment. Consequently, for us to visually comprehend a scene that is rapidly changing (as when we drive in traffic), our eyes must flick rapidly back and forth to provide the foveae with images of different parts of the visual field.

The neural retina receives its blood supply from two sources. The outer third (containing photoreceptors) is supplied by vessels in the choroid. The inner two-thirds is served by the **central artery** and **central vein of the retina**, which enter and leave the eye through the center of the optic nerve (see Figure 15.4a). Radiating outward from the optic disc, these vessels give rise to a rich vascular network that is clearly seen when the eyeball interior is examined with an ophthalmoscope (Figure 15.7). The fundus of the eye is the only place in the body where small blood vessels can be observed directly in a living person.

15

HOMEOSTATIC IMBALANCE

The pattern of vascularization of the retina makes it susceptible to *retinal detachment*. This condition, in which the pigmented and neural layers separate (detach) and allow the jellylike vitreous humor to seep between them, can cause permanent blindness because it deprives the photoreceptors of their nutrient source. It usually happens when the retina is torn during a traumatic blow to the head or when the head stops moving suddenly and then is jerked in the opposite direction (as in bungee jumping). The symptom that victims most often describe is "a curtain being drawn across the eye," but some people see sootlike spots or light flashes. If diagnosed early, it is often possible to reattach the retina with a laser before photoreceptor damage becomes permanent. ■

Internal Chambers and Fluids

As we noted earlier, the lens and its halolike ciliary zonule divide the eye into two segments, the anterior segment in front of the lens and the larger posterior segment behind it (see Figure 15.4a).

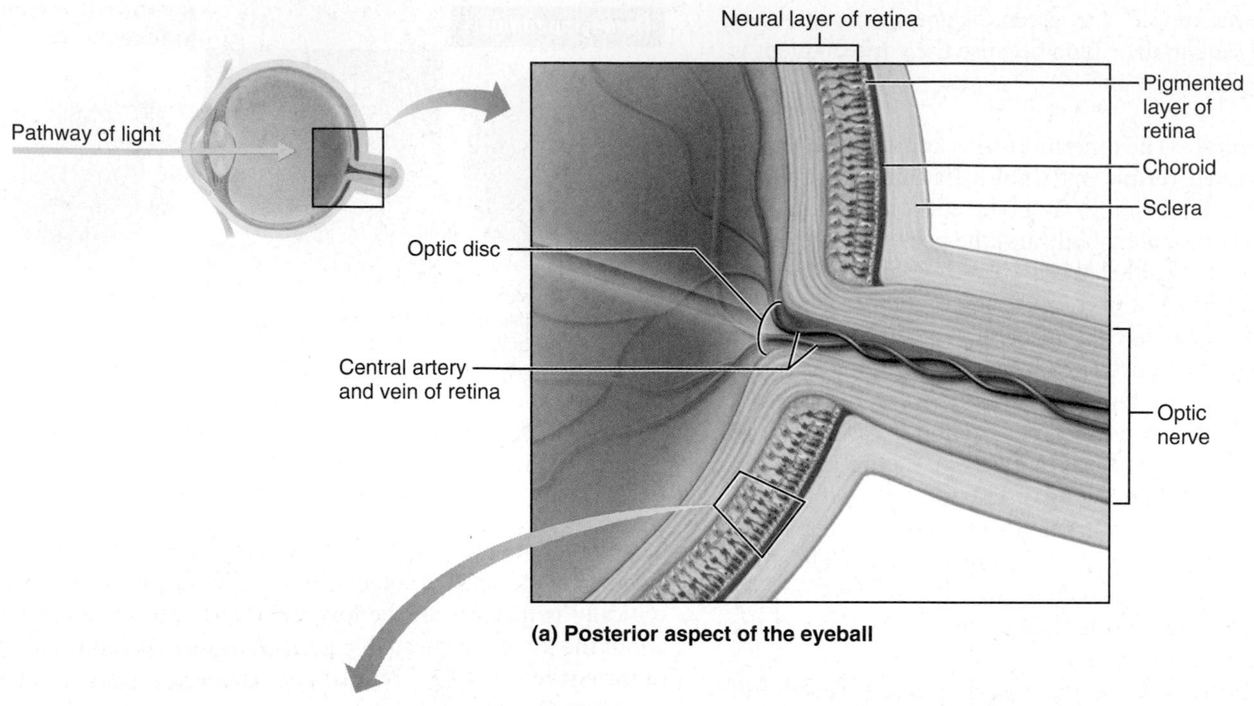

(a) Posterior aspect of the eyeball

15

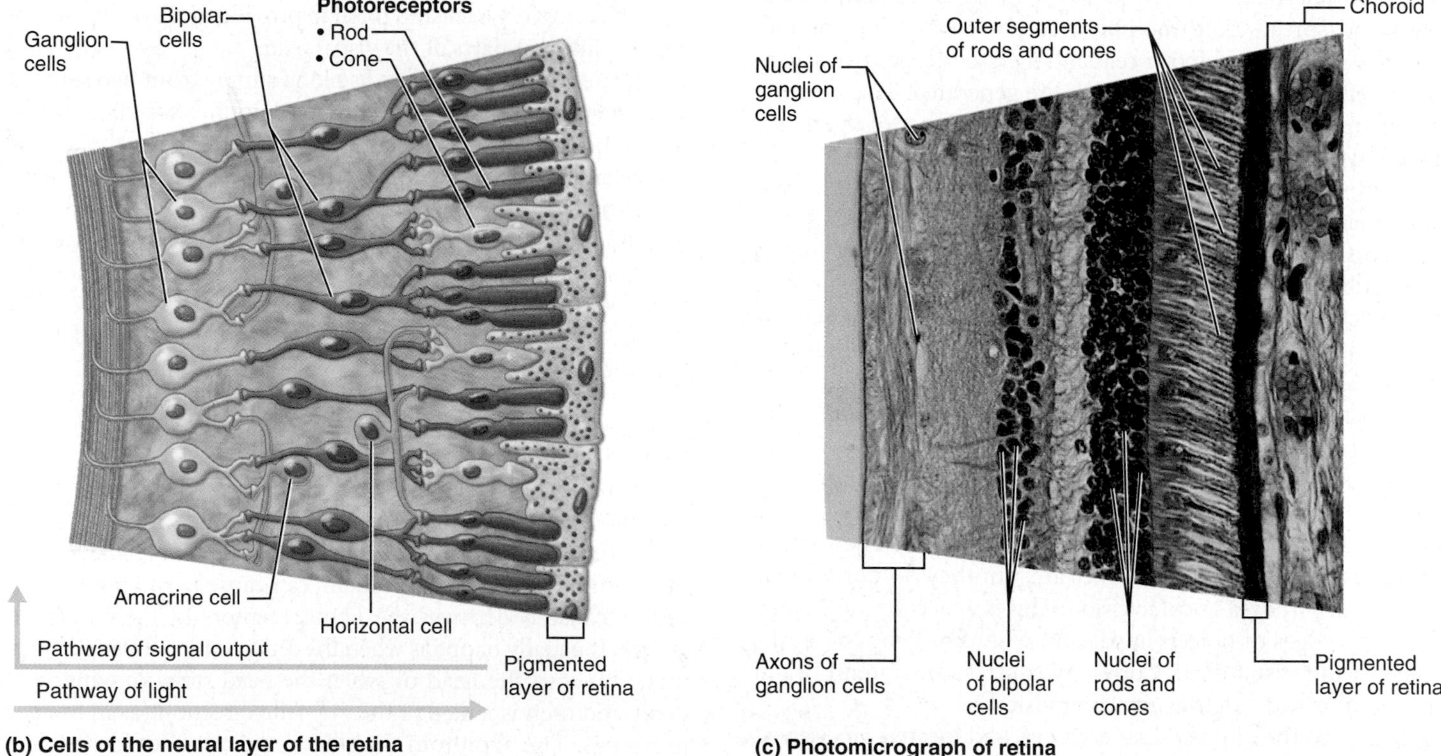

(b) Cells of the neural layer of the retina

(c) Photomicrograph of retina

Figure 15.6 Microscopic anatomy of the retina. (a) The axons of the ganglion cells form the optic nerve, which leaves the back of the eyeball at the optic disc. **(b)** Light (indicated by the yellow arrow) passes through the retina to excite the photoreceptor cells (rods and cones). Information (output signals) flows in the opposite direction via bipolar and ganglion cells. **(c)** Photomicrograph (150×).

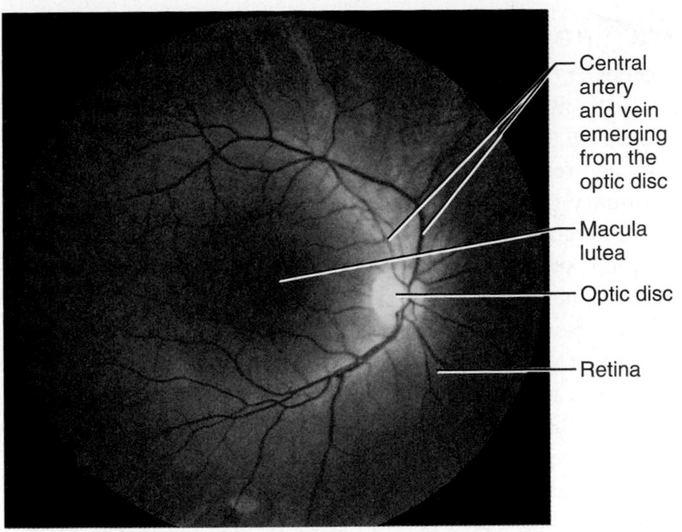

Figure 15.7 Part of the posterior wall (fundus) of the right eye as seen with an ophthalmoscope.

The **posterior segment** is filled with a clear gel called **vitreous humor** (*vitre* = glassy) that binds tremendous amounts of water. Vitreous humor (1) transmits light, (2) supports the posterior surface of the lens and holds the neural retina firmly against the pigmented layer, and (3) contributes to intraocular pressure, helping to counteract the pulling force of the extrinsic eye muscles. Vitreous humor forms in the embryo and lasts for a lifetime.

The **anterior segment** is partially subdivided by the iris into the **anterior chamber** (between the cornea and the iris) and the **posterior chamber** (between the iris and the lens) (**Figure 15.8**). The *entire* anterior segment is filled with **aqueous humor**, a clear fluid similar in composition to blood plasma. Unlike the vitreous humor, aqueous humor forms and drains continually and is in constant motion. It filters from the capillaries of the ciliary processes into the posterior chamber and a portion of it freely diffuses through the vitreous humor in the posterior segment while the remainder flows into the anterior chamber. After flowing through the pupil into the anterior chamber, it drains into the venous blood via the **scleral venous sinus** (*canal of Schlemm*), an unusual venous channel that encircles the eye in the angle at the sclera-cornea junction.

Normally, aqueous humor is produced and drained at the same rate, maintaining a constant intraocular pressure of about 16 mm Hg, which helps to support the eyeball internally. Aqueous humor supplies nutrients and oxygen to the lens and cornea and to some cells of the retina, and it carries away their metabolic wastes.

(1) Aqueous humor is formed by filtration from the capillaries in the ciliary processes.

(2) Aqueous humor flows from the posterior chamber through the pupil into the anterior chamber. Some also flows through the vitreous humor (not shown).

(3) Aqueous humor is reabsorbed into the venous blood by the scleral venous sinus.

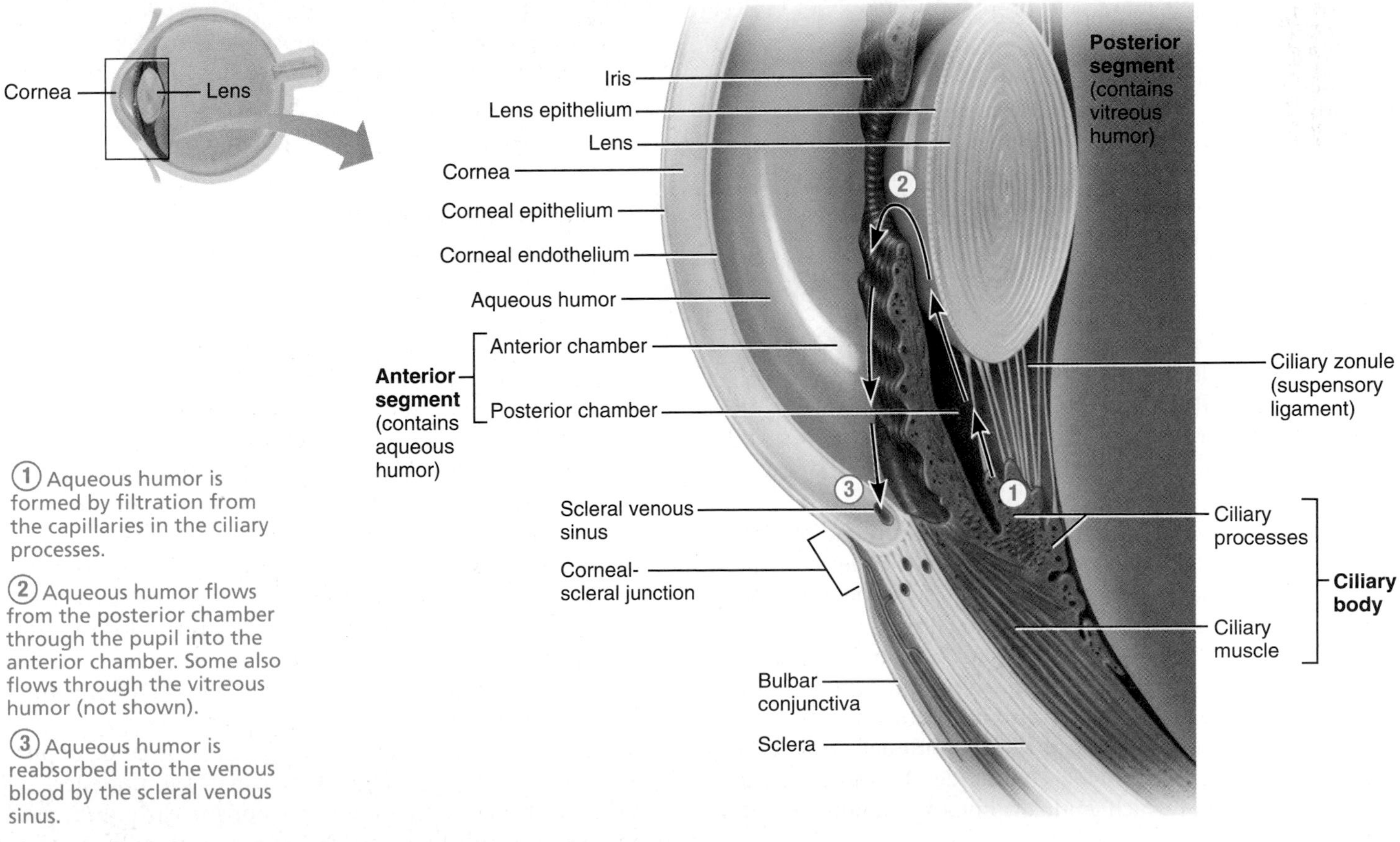

Figure 15.8 Circulation of aqueous humor. The arrows indicate the circulation pathway.

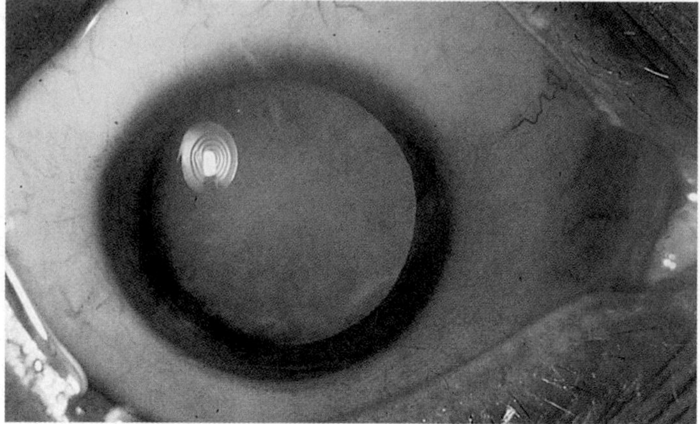

Figure 15.9 Photograph of a cataract. The lens is milky and opaque, not the cornea.

HOMEOSTATIC IMBALANCE

If the drainage of aqueous humor is blocked, fluid backs up as in a clogged sink. Pressure within the eye may increase to dangerous levels and compress the retina and optic nerve—a condition called **glaucoma** (glaw-ko′mah). The eventual result is blindness (*glaucoma* = vision growing gray) unless the condition is detected early. Unfortunately, many forms of glaucoma steal sight so slowly and painlessly that people do not realize they have a problem until the damage is done. Late signs include
15 seeing halos around lights and blurred vision.

The glaucoma examination is simple. The intraocular pressure is determined by directing a puff of air at the cornea and measuring the amount of corneal deformation it causes. This exam should be done yearly after the age of 40. The most common treatment is eye drops that increase the rate of aqueous humor drainage or decrease its production. Laser therapy or surgery can also be used. ■

Lens

The **lens** is a biconvex, transparent, flexible structure that can change shape to allow precise focusing of light on the retina. It is enclosed in a thin, elastic capsule and held in place just posterior to the iris by the ciliary zonule (Figure 15.8). Like the cornea, the lens is avascular; blood vessels interfere with transparency.

The lens has two regions: the **lens epithelium** and the lens fibers. The lens epithelium, confined to the anterior lens surface, consists of cuboidal cells that eventually differentiate into the **lens fibers** that form the bulk of the lens. The lens fibers, which are packed tightly together like the layers in an onion, contain no nuclei and few organelles. They do, however, contain transparent, precisely folded proteins called **crystallins** that form the body of the lens. Since new lens fibers are continually added, the lens enlarges throughout life, becoming denser, more convex, and less elastic, all of which gradually impair its ability to focus light properly.

HOMEOSTATIC IMBALANCE

A **cataract** ("waterfall") is a clouding of the lens that causes the world to appear distorted, as if seen through frosted glass **(Figure 15.9)**. Some cataracts are congenital, but most result from age-related hardening and thickening of the lens or are a secondary consequence of diabetes mellitus. Heavy smoking and frequent exposure to intense sunlight increase the risk for cataracts, whereas long-term dietary supplementation with vitamin C may decrease the risk.

Whatever the promoting factors, the *direct* cause of cataracts seems to be inadequate delivery of nutrients to the deeper lens fibers. The metabolic changes that result promote clumping of the crystallin proteins. Fortunately, the offending lens can be surgically removed and an artificial lens implanted to save the patient's sight. ■

CHECK YOUR UNDERSTANDING

1. What are tears and what structure secretes them?
2. What is the blind spot and why is it blind?
3. Sam's optometrist tells him that his intraocular pressure is high. What is this condition called and which fluid does it involve?

For answers, see Appendix G.

Physiology of Vision

- ▶ Trace the pathway of light through the eye to the retina, and explain how light is focused for distant and close vision.
- ▶ Outline the causes and consequences of astigmatism, myopia, hyperopia, and presbyopia.

Overview: Light and Optics

To really comprehend the function of the eye as a photoreceptor organ, we need to understand the properties of light.

Wavelength and Color **Electromagnetic radiation** includes all energy waves, from long radio waves (with wavelengths measured in meters) to very short gamma (γ) waves and X rays with wavelengths of 1 nm and less. Our eyes respond to the part of the spectrum called **visible light**, which has a wavelength range of approximately 400–700 nm **(Figure 15.10a)**. (1 nm = 10^{-9} m, or one-billionth of a meter.)

Visible light travels in the form of waves, and its wavelengths can be measured very accurately. However, light can also be envisioned as small particles or packets of energy called **photons** or **quanta**. Attempts to reconcile these two findings have led to the present concept of light as packets of energy (photons) traveling in a wavelike fashion at very high speeds (300,000 km/s or about 186,000 mi/s). We can think of light as a vibration of pure energy ("a bright wiggle") rather than a material substance.

When visible light passes through a prism, each of its component waves bends to a different degree, so that the beam of light is dispersed and a **visible spectrum**, or band of colors, is

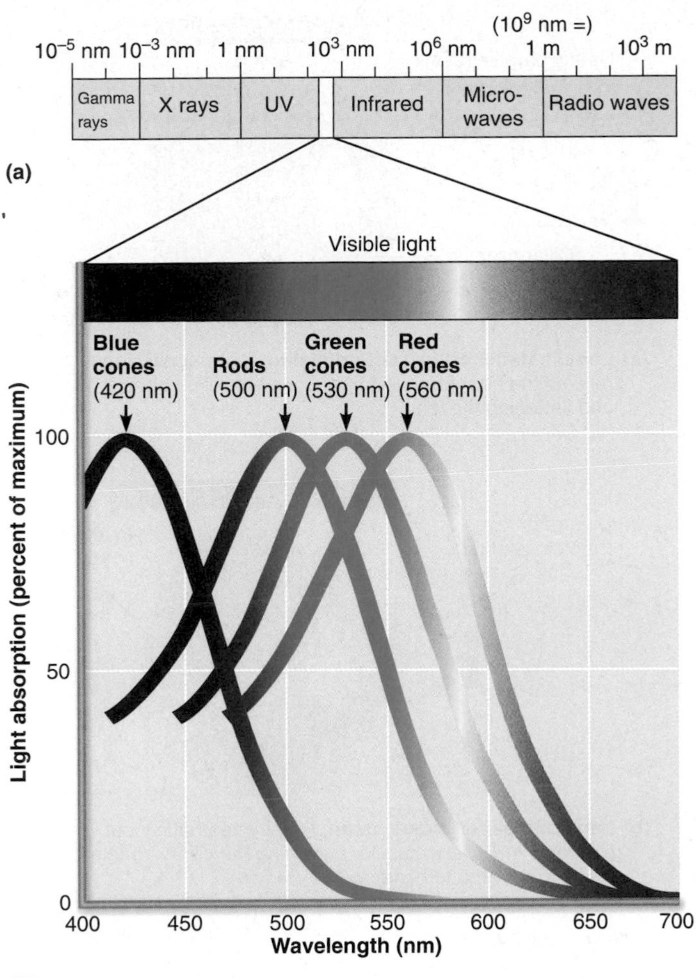

Figure 15.10 The electromagnetic spectrum and photoreceptor sensitivities. (a) The electromagnetic spectrum, of which visible light constitutes only a small portion. (nm = nanometers.) **(b)** Sensitivities of rods and the three cone types to the different wavelengths of the visible spectrum.

seen (Figure 15.10b). (In the same manner, the rainbow seen during a summer shower represents the collective prismatic effects of all the tiny water droplets suspended in air.) Red wavelengths are the longest and have the lowest energy, whereas the violet wavelengths are the shortest and most energetic. Objects have color because they absorb some wavelengths and reflect others. Things that look white reflect all wavelengths of light, whereas black objects absorb them all. A red apple reflects mostly red light, while grass reflects more of the green.

Refraction and Lenses Light travels in straight lines and is blocked by any nontransparent object. Like sound, light can reflect, or bounce, off a surface. This **reflection** of light by objects in our environment accounts for most of the light reaching our eyes.

When light travels in a given medium, its speed is constant. But when it passes from one transparent medium into another with a different density, its speed changes. Light speeds up as it passes into a less dense medium and slows as it passes into a denser medium. Because of these changes in speed, bending or

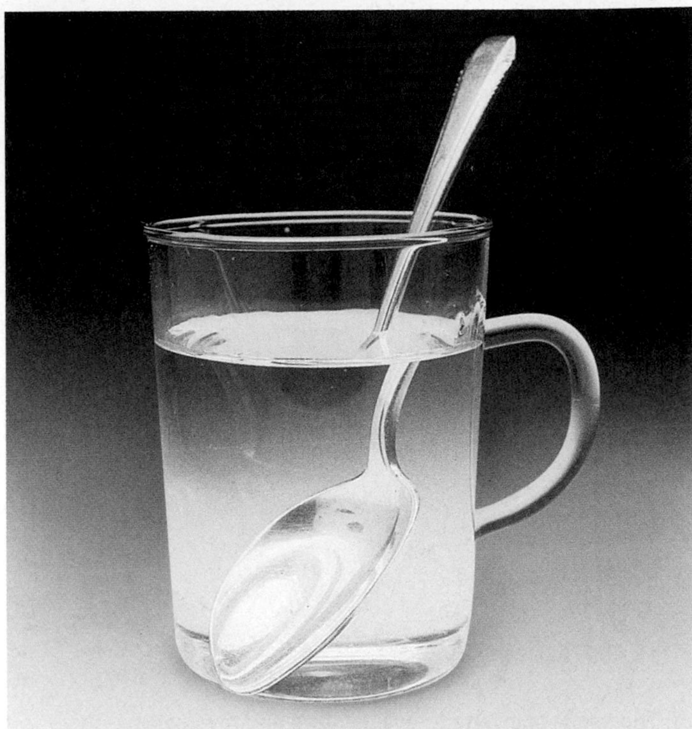

Figure 15.11 Refraction. A spoon standing in a glass of water appears to be broken at the water-air interface. This occurs because light is bent toward the perpendicular when it travels from a less dense to a more dense medium, as from air to water in this example.

15

refraction of a light ray occurs when it meets the surface of a different medium at an oblique angle rather than at a right angle (perpendicular). The greater this angle, the greater the amount of bending. **Figure 15.11** shows refraction: A spoon in a glass of water appears to break at the air-water interface.

A lens is a transparent object curved on one or both surfaces. Since light hits the curve at an angle, it is refracted. If the lens surface is convex, that is, thickest in the center like a camera lens, the light rays are bent so that they converge (come together) or intersect at a single point called the **focal point (Figure 15.12a)**. In general, the thicker (more convex) the lens, the more the light is bent and the shorter the focal distance (distance between the lens and focal point). The image formed by a convex lens, called a **real image,** is inverted—upside down and reversed from left to right (Figure 15.12b).

Concave lenses, which are thicker at the edges than at the center, diverge the light (bend it outward) so that the light rays move away from each other. Consequently, concave lenses prevent light from focusing and extend the focal distance.

Focusing of Light on the Retina

As light passes from air into the eye, it moves sequentially through the cornea, aqueous humor, lens, and vitreous humor, and then passes *through the entire thickness of the neural layer of the retina* to excite the photoreceptors that abut the pigmented layer (see Figures 15.4 and 15.6). During its passage, light is bent three times: as it enters the cornea and on entering and leaving the lens. The refractory power of the humors and cornea is

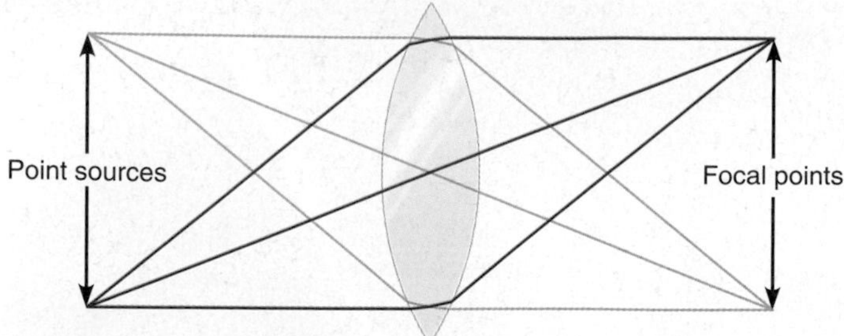

(a) Focusing of two points of light.

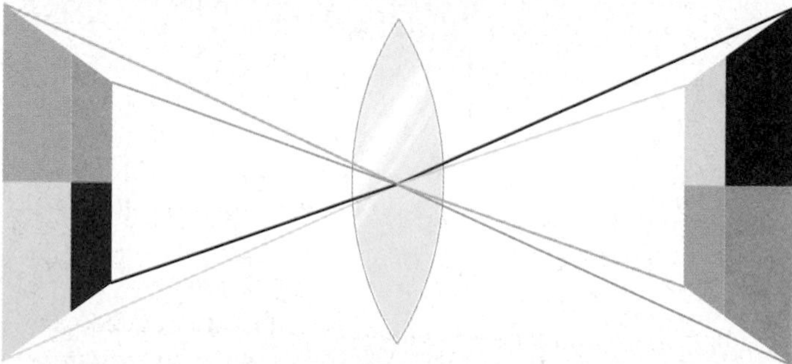

(b) The image is inverted—upside down and reversed.

Figure 15.12 **Bending of light by a convex lens.**

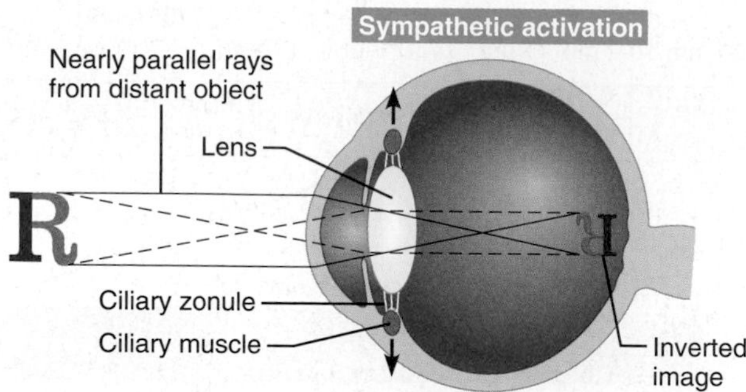

(a) Lens is flattened for distant vision. Sympathetic input relaxes the ciliary muscle, tightening the ciliary zonule, and flattening the lens.

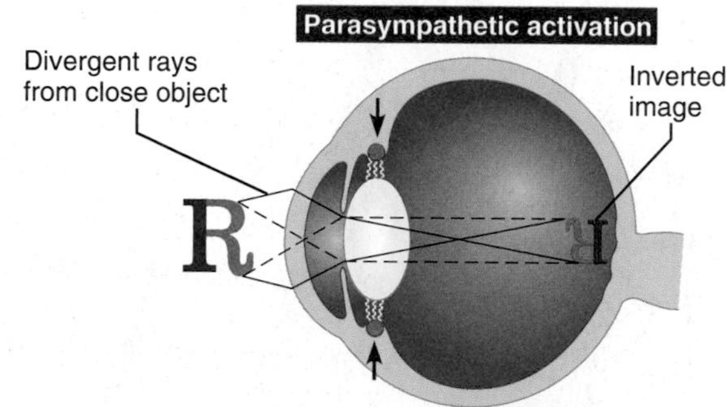

(b) Lens bulges for close vision. Parasympathetic input contracts the ciliary muscle, loosening the ciliary zonule, allowing the lens to bulge.

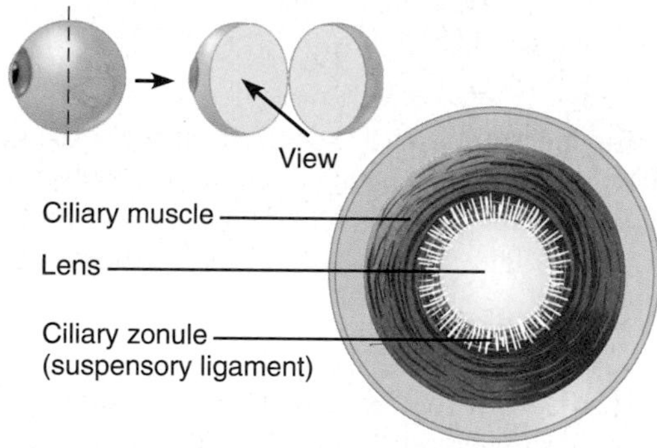

(c) The ciliary muscle and ciliary zonule are arranged sphincterlike around the lens. (Anterior segment as viewed from within the eye.)

Figure 15.13 **Focusing for distant and close vision.**

15

constant. On the other hand, the lens is highly elastic, and its curvature and light-bending power can be actively changed to allow fine focusing of the image.

Focusing for Distant Vision Our eyes are best adapted ("preset to focus") for distant vision. To look at distant objects, we need only aim our eyeballs so that they are both fixated on the same spot. The **far point of vision** is that distance beyond which no change in lens shape (accommodation) is needed for focusing. For the normal or **emmetropic** (em″ĕ-tro′pik) eye, the far point is 6 m (20 feet).

Any object being viewed can be said to consist of many small points, with light radiating outward in all directions from each point. However, because distant objects appear smaller, light from an object at or beyond the far point of vision approaches the eyes as nearly parallel rays and is focused precisely on the retina by the fixed refractory apparatus (cornea and humors) and the at-rest lens (Figure 15.13a).

During distant vision, the sphincterlike ciliary muscles are completely relaxed, and the lens is stretched flat by tension in the ciliary zonule. Consequently, the lens is as thin as it gets and is at its lowest refractory power. The ciliary muscles relax when sympathetic input to them increases and parasympathetic input decreases.

Focusing for Close Vision Light from close objects (less than 6 m away) diverges as it approaches the eyes and it comes to a focal point farther from the lens. For this reason, close vision demands that the eye make active adjustments. To restore focus, three processes must occur simultaneously: accommodation of the lenses, constriction of the pupils, and convergence of the eyeballs. The signal that induces this trio of reflex responses appears to be a blurring of the retinal image.

1. **Accommodation of the lenses. Accommodation** is the process that increases the refractory power of the lens. As the ciliary muscles contract, the ciliary body is pulled anteriorly and inward toward the pupil and releases tension in the ciliary zonule. No longer stretched, the elastic lens recoils and bulges, providing the shorter focal length needed to focus the image of a close object on the retina

(Figure 15.13b). Contraction of the ciliary muscles is controlled mainly by the parasympathetic fibers of the oculomotor nerves.

The closest point on which we can focus clearly is called the **near point of vision**, and it represents the maximum bulge the lens can achieve. In young adults with emmetropic vision, the near point is 10 cm (4 inches) from the eye. However, it is closer in children and gradually recedes with age, explaining why children can hold their books very close to their faces while many elderly people must hold the newspaper at arm's length. The gradual loss of accommodation with age reflects the lens's decreasing elasticity. In many people over the age of 50, the lens is nonaccommodating, a condition known as **presbyopia** (pres″be-o′pe-ah), literally "old person's vision."

2. **Constriction of the pupils.** The sphincter pupillae muscles of the iris enhance the effect of accommodation by reducing the size of the pupil toward 2 mm (see Figure 15.5). This **accommodation pupillary reflex**, mediated by parasympathetic fibers of the oculomotor nerves, prevents the most divergent light rays from entering the eye. Such rays would pass through the extreme edge of the lens and would not be focused properly, and so would cause blurred vision.
3. **Convergence of the eyeballs.** The visual goal is always to keep the object being viewed focused on the retinal foveae. When we look at distant objects, both eyes are directed either straight ahead or to one side to the same degree, but when we fixate on a close object, our eyes converge. **Convergence**, controlled by somatic motor fibers of the oculomotor nerves, is medial rotation of the eyeballs by the medial rectus muscles so that each is directed toward the object being viewed. The closer that object, the greater the degree of convergence required. For example, when you focus on the tip of your nose, you "go cross-eyed."

Reading or other close work requires almost continuous accommodation, pupillary constriction, and convergence. This is why prolonged periods of reading tire the eye muscles and can result in eyestrain. When you read for an extended time, it is helpful to look up and stare into the distance occasionally to relax the intrinsic eye muscles.

HOMEOSTATIC IMBALANCE

Theoretically, visual problems related to refraction could result from a hyperrefractive (overconverging) or hyporefractive (underconverging) lens or from structural abnormalities of the eyeball. In practice, 99% of refractive problems are related to eyeball shape—either too long or too short.

Myopia (mi-o′pe-ah; "short vision") occurs when distant objects are focused in front of the retina, rather than on it (**Figure 15.14**, left). Myopic people see close objects without problems because they can focus them on the retina, but distant objects are blurred. The common name for myopia is *nearsightedness.* (Notice that this terminology names the aspect of vision that is *unimpaired.*) Myopia typically results from an eyeball that is too long.

Correction has traditionally involved use of concave lenses that diverge the light before it enters the eye. Procedures to flatten the cornea slightly—laser procedures called PRK and LASIK—now offer other treatment options.

Hyperopia (hy′per-o″pe-ah; "far vision"), or *farsightedness,* occurs when the parallel light rays from distant objects are focused *behind* the retina (Figure 15.14, right). Hyperopic individuals can see distant objects perfectly well because their ciliary muscles contract almost continuously to increase the light-bending power of the lens, which moves the focal point forward onto the retina. However, diverging light rays from *nearby* objects are focused so far behind the retina that the lens cannot bring the focal point onto the retina even at its full refractory power. As a result, close objects appear blurry, and convex corrective lenses are needed to converge the light more strongly for close vision. Hyperopia usually results from an eyeball that is too short.

Unequal curvatures in different parts of the cornea or lens lead to blurry images. This refractory problem is **astigmatism** (*astigma* = not a point). Special cylindrically ground lenses, corneal implants, or laser procedures are used to correct this problem. ■

CHECK YOUR UNDERSTANDING

4. Arrange the following in the order that light passes through them to reach the photoreceptors (rods and cones) in the retina: lens, bipolar cells, vitreous humor, cornea, aqueous humor, ganglion cells. (Hint: See Figure 15.6 if you need a reminder of where ganglion cells and bipolar cells are.)
5. You have been reading this book for a while now and your eyes are beginning to tire. Which intrinsic eye muscles are relaxing as you stare thoughtfully into the distance?
6. Why does your near point of vision move farther away as you age?

For answers, see Appendix G.

15

Photoreceptors and Phototransduction

- ▶ Describe the events involved in the stimulation of photoreceptors by light, and compare and contrast the roles of rods and cones in vision.
- ▶ Compare and contrast light and dark adaptation.

Once light is focused on the retina, the photoreceptors come into play. First, we will describe the functional anatomy of the rod and cone photoreceptor cells, then the chemistry and response of their visual pigments to light, and finally, photoreceptor activation and phototransduction. **Phototransduction** is the process by which light energy is converted into a graded receptor potential.

Functional Anatomy of the Photoreceptors Photoreceptors are modified neurons, but structurally they resemble tall

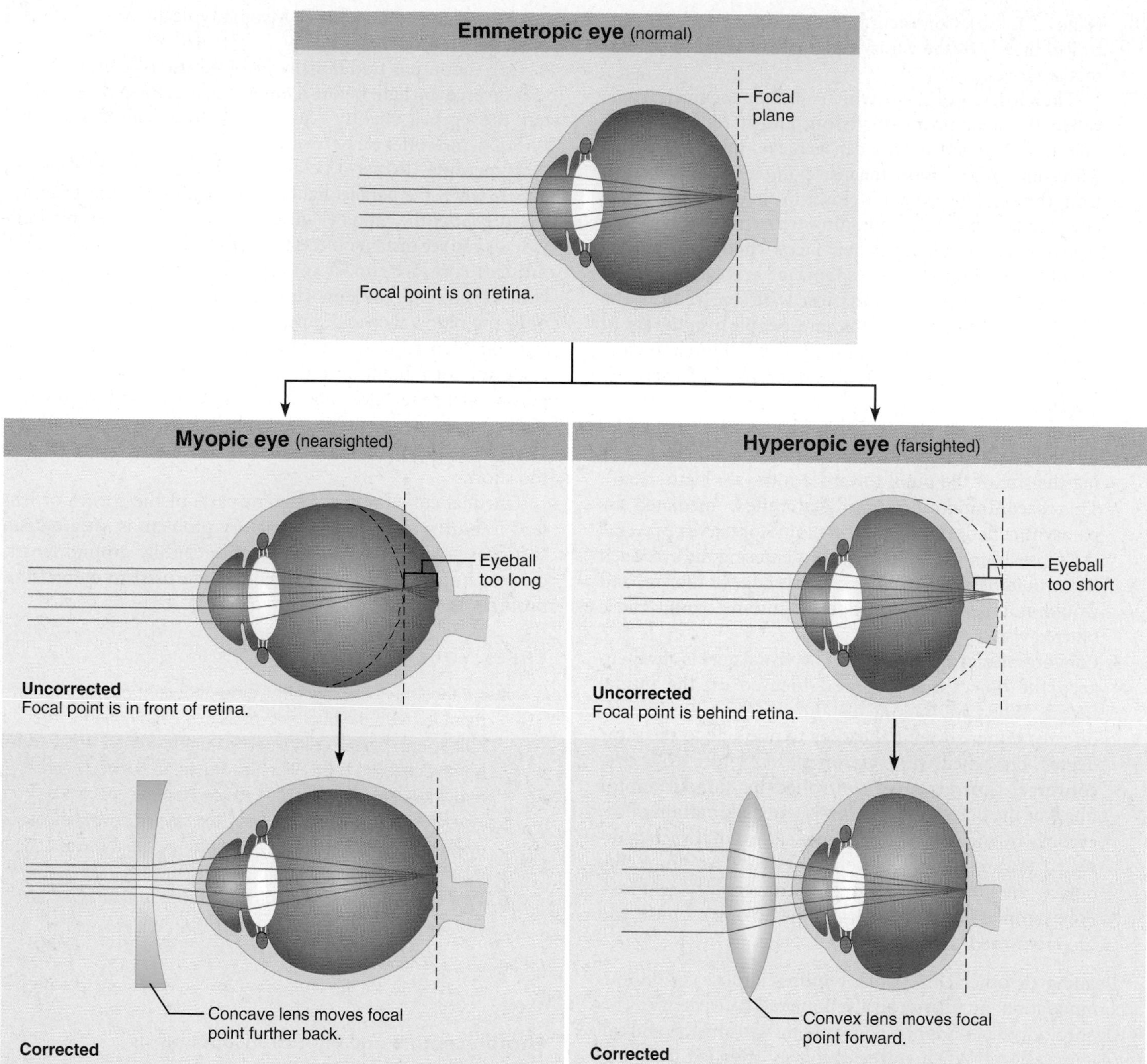

Figure 15.14 Problems of refraction. The refractive power of the cornea, which normally supplies about two-thirds of the light-bending power of the eye, is ignored here.

15

epithelial cells turned upside down with their "tips" immersed in the pigmented layer of the retina **(Figure 15.15a)**. These "tips" are the receptive regions of rods and cones and are called the **outer segments**. Moving from the pigmented layer into the neural layer, the outer segment of a rod or cone is joined to an **inner segment** by a connecting cilium. The inner segment then connects to the *cell body*, which is continuous with an *inner fiber* bearing *synaptic terminals*. In rods, the outer segment is slender and rod shaped (hence their name) and the inner segment connects to the cell body by the *outer fiber*. By contrast, the cones have a short conical outer segment and the inner segment directly joins the cell body.

The outer segments contain an elaborate array of **visual pigments (photopigments)** that change shape as they absorb light. These pigments are embedded in areas of the plasma membrane that form discs (Figure 15.15b). Folding the plasma membrane into discs increases the surface area available for trapping light. In cones, the disc membranes are continuous with the plasma membrane, so the interiors of the cone discs are continuous with the extracellular space. In rods, the discs are

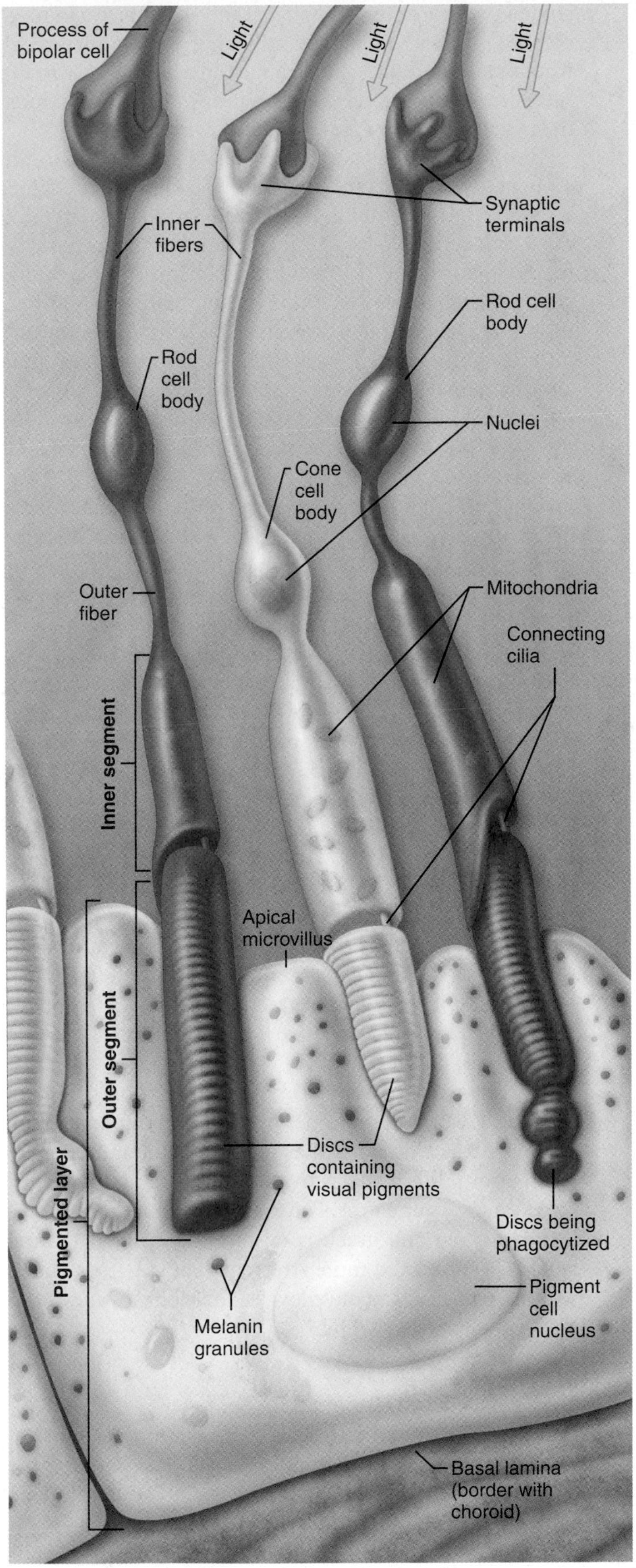

(a) The outer segments of rods and cones are embedded in the pigmented layer of the retina.

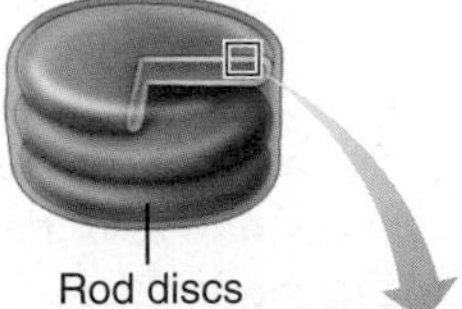

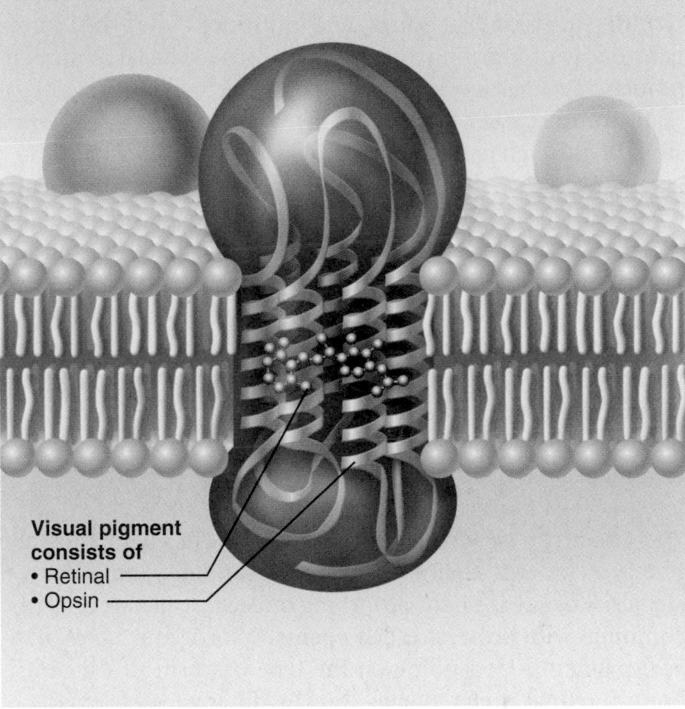

(b) Rhodopsin, the visual pigment in rods, is embedded in the membrane that forms discs in the outer segment.

Figure 15.15 **Photoreceptors of the retina.**

discontinuous—stacked inside a cylinder of plasma membrane like pennies in a coin wrapper.

The photoreceptor cells are highly vulnerable to damage and immediately begin to degenerate if the retina becomes detached. They are also destroyed by intense light, the very energy they are designed to detect. How is it, then, that we do not all gradually go blind? The answer lies in the photoreceptors' unique system for renewing their light-trapping outer segment. In rods, new discs are assembled at the proximal end of the outer segment from materials synthesized in the cell body at the end of each night. As new discs are made, they push the others peripherally. The discs at the tip of the outer segment continually fragment off and are phagocytized by cells of the pigmented layer. The tips of the outer segments of cones are also removed and renewed, but at the end of each day.

The rods and each of the three cone types contain unique visual pigments that absorb different wavelengths of light. Rods and cones also have different thresholds for activation. Rods, for example, (1) are very sensitive (respond to very dim light—a single photon), making them best suited for night vision and peripheral vision, and (2) contain a single kind of visual pigment so their inputs are perceived only in gray tones. Cones (1) need bright light for activation (have low sensitivity), but (2) have one of three different pigments that furnish a vividly colored view of the world.

Rods and cones are also "wired" differently to other retinal neurons. Rods participate in converging pathways, and as many as 100 rods may ultimately feed into each ganglion cell. As a result, rod effects are summated and considered collectively, resulting in vision that is fuzzy and indistinct. (The visual cortex has no way of knowing exactly *which* rods of the large number influencing a particular ganglion cell are actually activated.)

In contrast, each cone in the fovea (or at most a few) has a straight-through pathway via its "own personal bipolar cell" to a ganglion cell (see Figure 15.6b). Essentially, each cone has its own "labeled line" to the higher visual centers. This straight-through pathway accounts for the detailed, high-resolution views of very small areas of the visual field provided by cones.

Because rods are absent from the foveae and cones do not respond to low-intensity light, we see dimly lit objects best when we do not look at them directly, and we recognize them best when they move. If you doubt this, go out into your backyard on a moonlit evening and see how much you can actually discriminate.

15

The Chemistry of Visual Pigments How does light affect the photoreceptors so that it is ultimately translated into electrical signals? The key is a light-absorbing molecule called **retinal** that combines with proteins called **opsins** to form four types of visual pigments. Depending on the type of opsin to which it is bound, retinal preferentially absorbs different wavelengths of the visible spectrum.

Retinal is chemically related to vitamin A and is made from it. The liver stores vitamin A and releases it as needed by the photoreceptors to make their visual pigments. The cells of the pigmented layer of the retina absorb vitamin A from the blood and serve as the local vitamin A depot for the rods and cones.

Retinal can assume a variety of distinct three-dimensional forms, each form called an isomer. When bound to opsin, retinal has a bent, or kinked, shape called **11-*cis*-retinal**, as shown at the top of **Figure 15.16**. However, when the pigment is struck by light and absorbs photons, retinal twists and snaps into a new configuration, **all-*trans*-retinal**, shown at the bottom of Figure 15.16. This change, in turn, causes opsin to change shape and assume its activated form.

The capture of light by visual pigments is the *only* light-dependent stage, and this simple photochemical event initiates a whole chain of chemical and electrical reactions in rods and cones that ultimately causes electrical impulses to be transmitted along the optic nerve. Let's look more closely at these events in rods and in cones.

Stimulation of the Photoreceptors

1. **Excitation of rods.** The visual pigment of rods is a deep purple pigment called **rhodopsin** (ro-dop′sin; *rhodo* = rose, *opsis* = vision). (Do you suppose the person who coined the expression "looking at the world through rose-colored glasses" knew the meaning of "rhodopsin"?) Rhodopsin molecules are arranged in a single layer in the membranes of each of the thousands of discs in the rods' outer segments (see Figure 15.15b).

   Rhodopsin forms and accumulates in the dark in the sequence of reactions shown on the left side of Figure 15.16. As illustrated, vitamin A is oxidized (and isomerized) to the 11-*cis*-retinal form and then combined with opsin to form rhodopsin. When rhodopsin absorbs light, retinal changes shape to its all-*trans* isomer, allowing the surrounding protein to quickly relax like an uncoiling spring into its light-activated form (metarhodopsin II). Eventually, the retinal-opsin combination breaks down, allowing retinal and opsin to separate. This sequence is known as the **bleaching of the pigment**, shown on the right side of Figure 15.16.

   Once the light-struck all-*trans*-retinal detaches from opsin, it is reconverted by enzymes within the pigmented epithelium to its 11-*cis* isomer in an ATP-requiring process. Then, retinal heads "homeward" again to the photoreceptor cells' outer segments. Rhodopsin is regenerated when 11-*cis*-retinal is rejoined to opsin.
2. **Excitation of cones.** The breakdown and regeneration of visual pigments in cones is essentially the same as for rhodopsin. However, cones are about a hundred times less sensitive than rods, which means that it takes higher-intensity (brighter) light to activate cones.

   Visual pigments of the three types of cones, like those of rods, are a combination of retinal and opsins. However, the cone opsins differ both from the opsin of the rods and from one another. The naming of cones reflects the colors (that is, wavelengths) of light that each cone variety absorbs best. The blue cones respond maximally to wavelengths around 420 nm, the green cones to wavelengths of 530 nm, and the red cones to wavelengths at or close to 560 nm (see Figure 15.10b).

   How do we see other colors? The absorption spectra of the blue, green, and red cones overlap, and our perception of intermediate hues, such as orange, yellow, and purple, results from differential activation of more than one type of cone at the same time. For example, yellow light stimulates both red and green cone receptors, but if the red cones are stimulated more strongly than the green cones, we see orange instead of yellow. When all cones are stimulated equally, we see white.

HOMEOSTATIC IMBALANCE

Color blindness is due to a congenital lack of one or more of the cone types. Inherited as an X-linked condition, it is far more common in males than in females. As many as 8–10% of males have some form of color blindness.

The most common type is red-green color blindness, resulting from a deficit or absolute absence of either red or green cones. Red and green are seen as the same color—either red or green, depending on the cone type present. Many color-blind

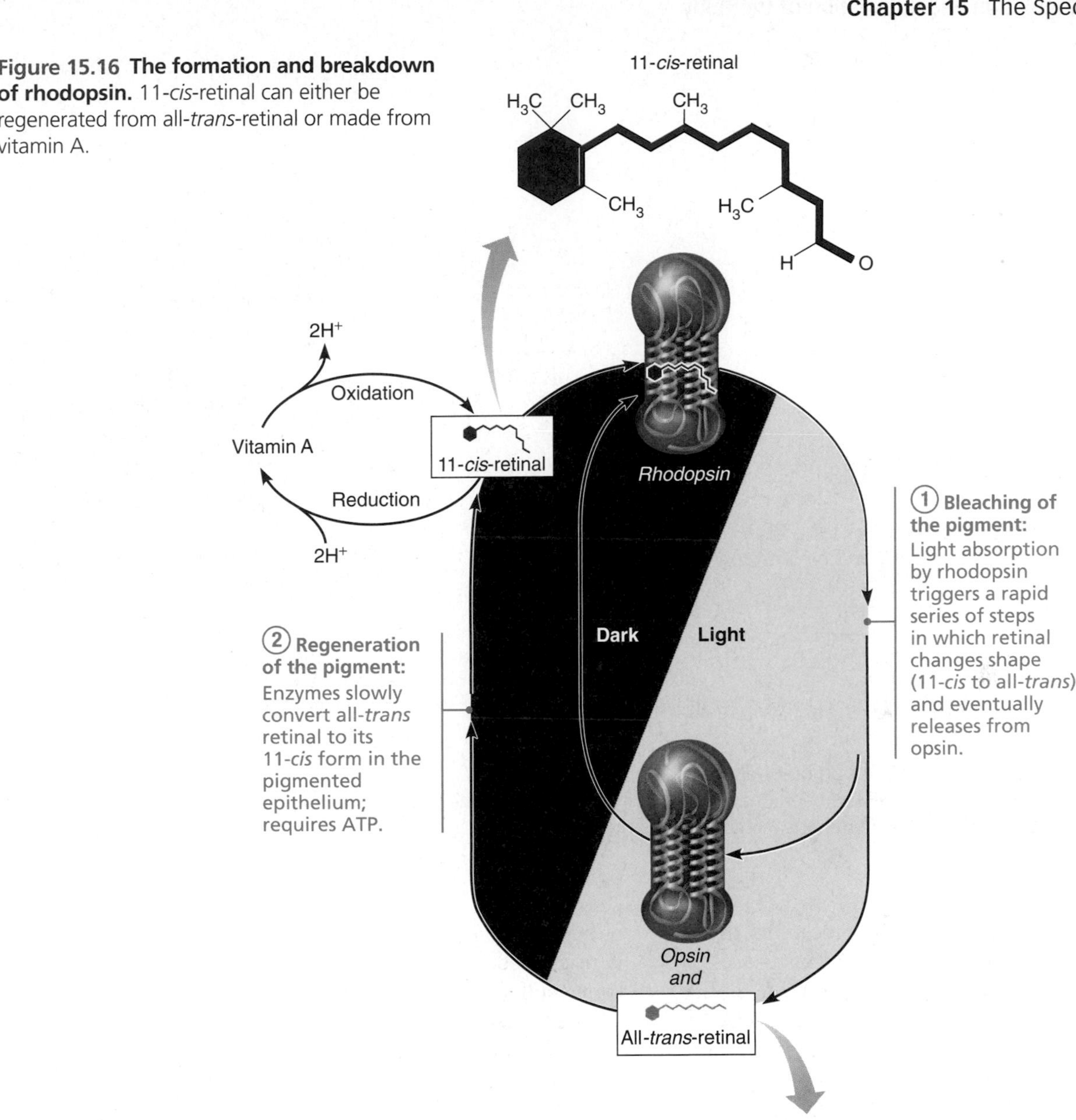
11-*cis*-retinal
H_3C CH_3 CH_3
CH_3 H_3C
H O
$2H^+$
Oxidation
Vitamin A
11-*cis*-retinal
Reduction
$2H^+$
Rhodopsin
① Bleaching of the pigment:
Light absorption by rhodopsin triggers a rapid series of steps in which retinal changes shape (11-*cis* to all-*trans*) and eventually releases from opsin.
② Regeneration of the pigment:
Enzymes slowly convert all-*trans* retinal to its 11-*cis* form in the pigmented epithelium; requires ATP.
Dark
Light
Opsin and
All-*trans*-retinal
H_3C CH_3 CH_3 CH_3 H
O
CH_3
All-*trans*-retinal

Figure 15.16 The formation and breakdown of rhodopsin. 11-*cis*-retinal can either be regenerated from all-*trans*-retinal or made from vitamin A.

15

people are unaware of their condition because they have learned to rely on other cues—such as differences in intensities of the same color—to distinguish something green from something red, such as traffic signals. ■

Light Transduction Reactions What happens when light triggers pigment breakdown? An enzymatic cascade occurs that ultimately results in closing cation channels that are normally kept open in the dark. This process is illustrated in more detail in **Figure 15.17**, but in short, light-activated rhodopsin activates a G protein called **transducin**. Transducin, in turn, activates *PDE* (*phosphodiesterase*), the enzyme that breaks down **cyclic GMP (cGMP)**. In the dark, cGMP binds to cation channels in the outer segments of photoreceptor cells, holding them open. This allows Na^+ and Ca^{2+} to enter, depolarizing the cell to its *dark potential* of about −40 mV. In the light, cGMP breaks down, the cation channels close, Na^+ and Ca^{2+} stop entering the cell, and the cell hyperpolarizes to about −70 mV.

This arrangement can seem bewildering, to say the least. Here we have receptors built to detect light that depolarize in the dark and hyperpolarize when exposed to light! However, all that is required is a signal and hyperpolarization is just as good a signal as depolarization.

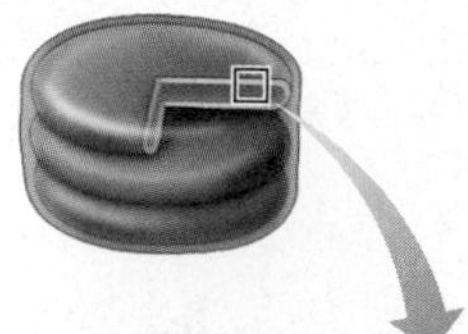

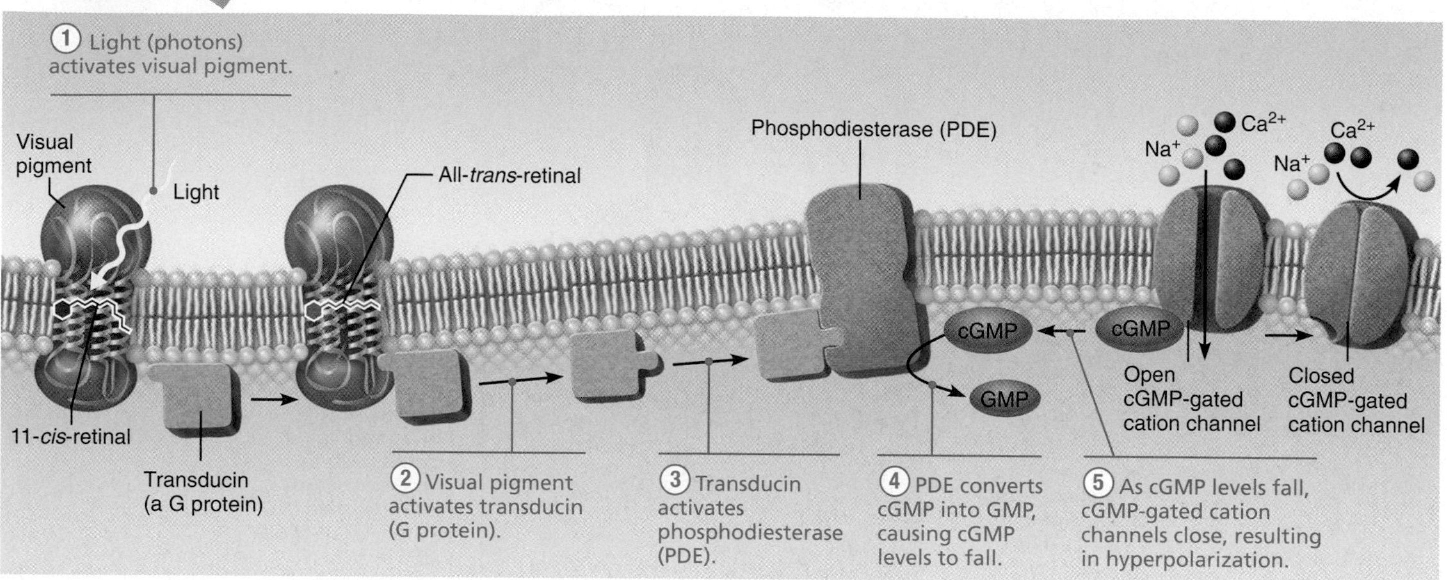

1 Light (photons) activates visual pigment.
Visual pigment
Light
All-*trans*-retinal
Phosphodiesterase (PDE)
Na+
Ca2+
Na+
Ca2+
cGMP
cGMP
GMP
Open cGMP-gated cation channel
Closed cGMP-gated cation channel
11-*cis*-retinal
Transducin (a G protein)
2 Visual pigment activates transducin (G protein).
3 Transducin activates phosphodiesterase (PDE).
4 PDE converts cGMP into GMP, causing cGMP levels to fall.
5 As cGMP levels fall, cGMP-gated cation channels close, resulting in hyperpolarization.

Figure 15.17 Events of phototransduction. A portion of photoreceptor disc membrane is shown. The G protein conversion of GTP to GDP has been omitted for clarity. For simplicity, the cGMP-gated channels are shown on the same membrane as the visual pigment instead of in the plasma membrane.

15

How is the hyperpolarization of the photoreceptors transmitted through the retina and on to the brain? This process is illustrated in **Figure 15.18**. As you study this sequence, notice that the photoreceptors do not generate action potentials (APs), and neither do the bipolar cells that lie between them and the ganglion cells. Photoreceptors and bipolar cells only generate graded potentials—excitatory postsynaptic potentials (EPSPs) and inhibitory postsynaptic potentials (IPSPs). This is not surprising if you remember that the primary function of APs is to carry information rapidly over long distances. Retinal cells are small cells that are very close together. Graded potentials can serve quite adequately as signals that directly regulate neurotransmitter release at the synapse by opening or closing voltage-gated Ca^{2+} channels. As shown in the right panel of Figure 15.18, for example, light hyperpolarizes photoreceptors, which stop releasing their inhibitory neurotransmitter (glutamate). No longer inhibited, bipolar cells depolarize and release neurotransmitter onto ganglion cells. Once the signal reaches the ganglion cells, it is converted into an AP. This AP is transmitted to the brain along the ganglion cell axons that make up the optic nerve.

Light and Dark Adaptation Rhodopsin is amazingly sensitive. Even starlight causes some of the molecules to become bleached. As long as the light is low intensity, relatively little rhodopsin is bleached and the retina continues to respond to light stimuli. However, in high-intensity light there is wholesale bleaching of the pigment, and rhodopsin is bleached as fast as it is re-formed. At this point, the rods are nonfunctional, but cones still respond. Hence, retinal sensitivity automatically adjusts to the amount of light present.

Light adaptation occurs when we move from darkness into bright light, as when leaving a movie matinee. We are momentarily dazzled—all we see is white light—because the sensitivity of the retina is still "set" for dim light. Both rods and cones are strongly stimulated, and large amounts of the visual pigments are broken down almost instantaneously, producing a flood of signals that accounts for the glare.

Under such conditions, compensations occur: The rod system turns off—all of the transducins "pack up and move" to the inner segment, uncoupling rhodopsin from the rest of the transduction cascade. Without transducin in the outer segment, light hitting rhodopsin cannot produce a signal. At the same time, the less sensitive cone system and other retinal neurons rapidly adapt, and retinal sensitivity decreases dramatically. Within about 60 seconds, the cones, initially overexcited by the bright light, are sufficiently desensitized to take over. Visual acuity and color vision continue to improve over the next 5–10 minutes. Thus, during light adaptation, retinal sensitivity (rod function) is lost, but visual acuity (the ability to resolve detail) is gained.

Dark adaptation is essentially the reverse of light adaptation. It occurs when we go from a well-lit area into a dark one. Initially, we see nothing but velvety blackness because (1) our

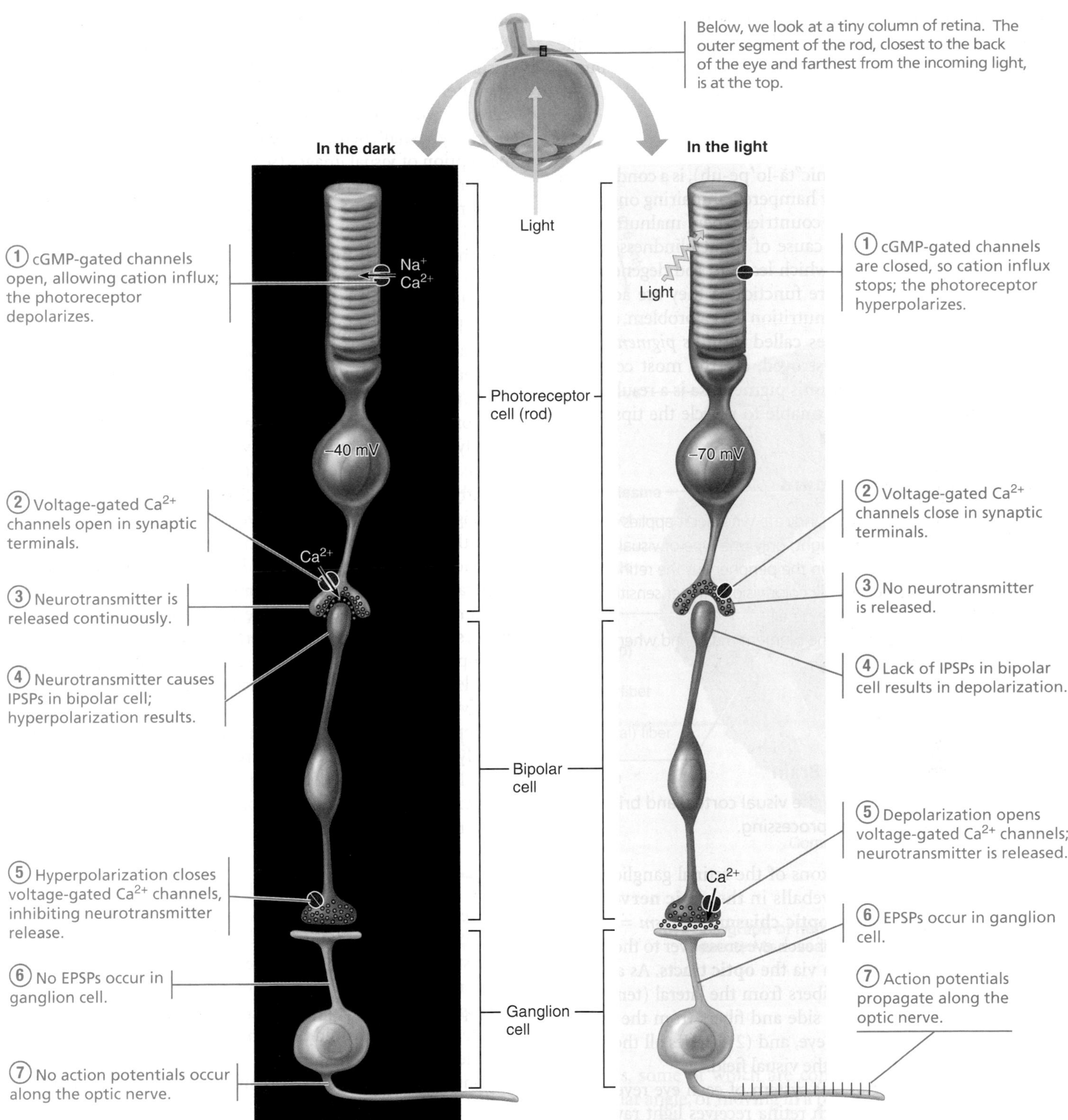

Below, we look at a tiny column of retina. The outer segment of the rod, closest to the back of the eye and farthest from the incoming light, is at the top.
In the dark
In the light
Light
① cGMP-gated channels open, allowing cation influx; the photoreceptor depolarizes.
Na+
Ca2+
Light
① cGMP-gated channels are closed, so cation influx stops; the photoreceptor hyperpolarizes.
Photoreceptor cell (rod)
–40 mV
–70 mV
② Voltage-gated Ca2+ channels open in synaptic terminals.
② Voltage-gated Ca2+ channels close in synaptic terminals.
Ca2+
③ Neurotransmitter is released continuously.
③ No neurotransmitter is released.
④ Neurotransmitter causes IPSPs in bipolar cell; hyperpolarization results.
④ Lack of IPSPs in bipolar cell results in depolarization.
Bipolar cell
⑤ Depolarization opens voltage-gated Ca2+ channels; neurotransmitter is released.
⑤ Hyperpolarization closes voltage-gated Ca2+ channels, inhibiting neurotransmitter release.
Ca2+
⑥ EPSPs occur in ganglion cell.
⑥ No EPSPs occur in ganglion cell.
Ganglion cell
⑦ Action potentials propagate along the optic nerve.
⑦ No action potentials occur along the optic nerve.

Figure 15.18 Signal transmission in the retina. EPSP = excitatory postsynaptic potential; IPSP = inhibitory postsynaptic potential.

cones stop functioning in low-intensity light, and (2) our rod pigments have been bleached out by the bright light, and the rods are still turned off. But once we are in the dark, rhodopsin accumulates, transducin returns to the outer segment, and retinal sensitivity increases. Dark adaptation is much slower than light adaptation and can go on for hours. However, there is usually enough rhodopsin within 20–30 minutes to allow adequate dim-light vision.

During these adaptations, reflexive changes also occur in pupil size. Bright light shining in one or both eyes causes both pupils to

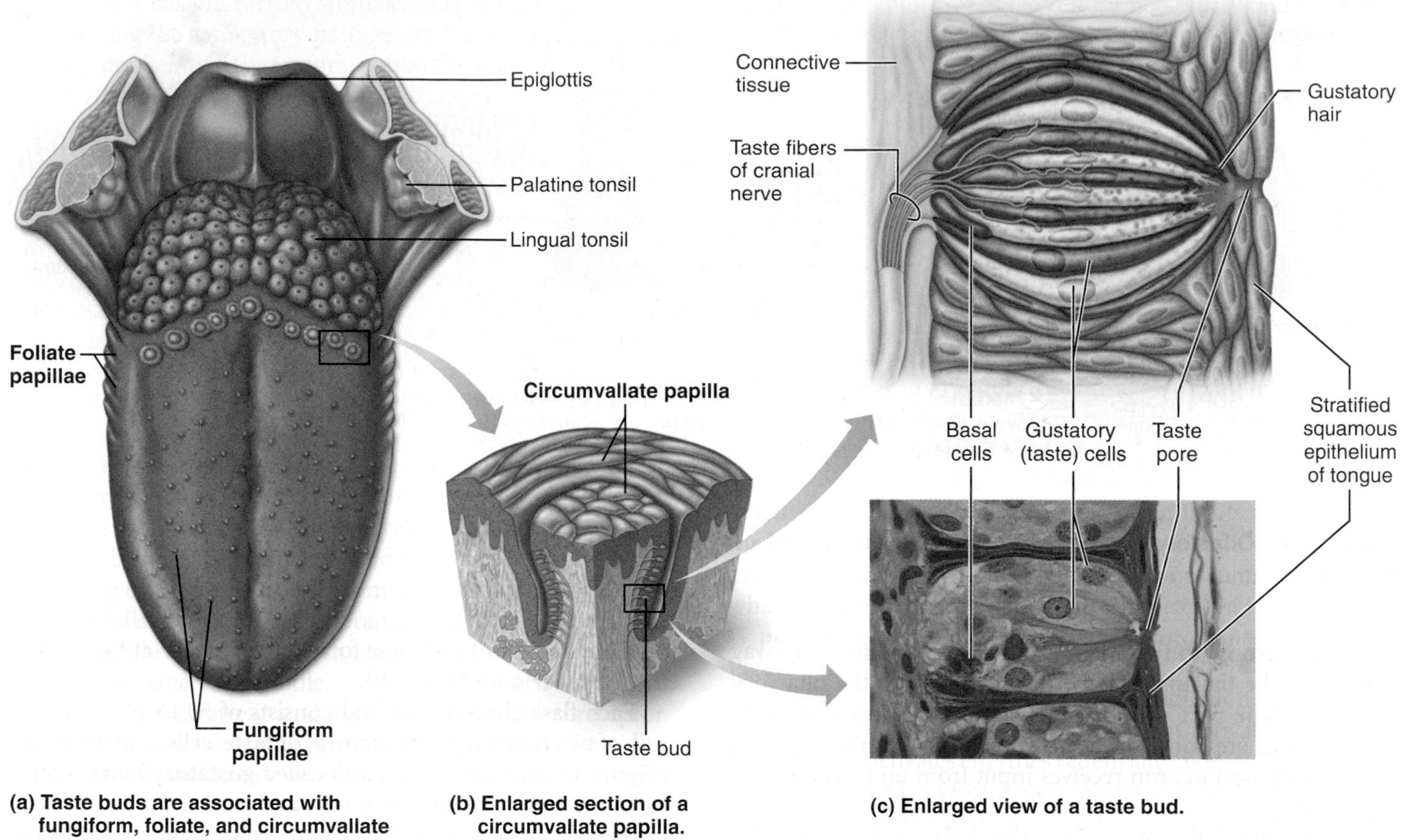

(a) Taste buds are associated with fungiform, foliate, and circumvallate (vallate) papillae.

(b) Enlarged section of a circumvallate papilla.

(c) Enlarged view of a taste bud.

15

Figure 15.23 Location and structure of taste buds on the tongue. (See *A Brief Atlas of the Human Body*, Figure 62.)

- *Bitter* taste is elicited by alkaloids (such as quinine, nicotine, caffeine, morphine, and strychnine) as well as a number of nonalkaloid substances, such as aspirin.
- *Umami* (u-mam′e; "delicious"), a new taste discovered by the Japanese, is elicited by the amino acids glutamate and aspartate, which appear to be responsible for the "beef taste" of steak, the characteristic tang of aging cheese, and the flavor of the food additive monosodium glutamate.

Keep in mind that many substances produce a mixture of the basic taste sensations, and taste buds generally respond to all five taste qualities. However, it appears that a single taste cell has receptors for only one taste quality.

Taste maps that spatially assign sweet receptors to the tip of the tongue, salty and sour receptors to the sides, bitter receptors to the back, and umami receptors to the pharynx are common in older textbooks. However, researchers have known for years that these maps are dubious. Indeed, recent molecular and functional data show that all modalities of taste can be elicited from all areas that contain taste buds.

Taste likes and dislikes have homeostatic value. Umami guides the intake of proteins, and a liking for sugar and salt helps satisfy the body's need for carbohydrates and minerals (as well as some amino acids). Many sour, naturally acidic foods (such as oranges, lemons, and tomatoes) are rich sources of vitamin C, an essential vitamin. On the other hand, intensely sour tastes warn us of spoilage. Likewise, many natural poisons and spoiled foods are bitter. Consequently, our dislike for sourness and bitterness is protective.

Physiology of Taste

For a chemical to be tasted it must dissolve in saliva, diffuse into the taste pore, and contact the gustatory hairs.

Activation of Taste Receptors The gustatory cells contain neurotransmitters, and binding of the food chemical, or *tastant,* to the receptors in the gustatory cell membrane induces a graded depolarizing potential that causes neurotransmitter release. Binding of the neurotransmitter to the associated sensory dendrites triggers generator potentials that elicit action potentials in these fibers.

The different gustatory cells have different thresholds for activation. In line with their protective nature, the bitter receptors detect substances present in minute amounts. The other receptors are less sensitive. Taste receptors adapt rapidly, with partial adaptation in 3–5 seconds and complete adaptation in 1–5 minutes.

Taste Transduction The mechanisms of taste transduction are only now beginning to become clear. Three different mechanisms underlie how we taste.

- Salty taste is due to Na^+ influx through Na^+ channels, which directly depolarizes the gustatory cell.
- Sour is mediated by H^+, which acts intracellularly to open channels that allow other cations to enter.
- Bitter, sweet, and umami responses share a common mechanism, but each occurs in a different cell. Each taste's unique receptor is coupled to a common G protein called *gustducin.* Activation leads to the release of Ca^{2+} from intracellular stores, which causes the opening of cation channels in the plasma membrane, thereby depolarizing the cell.

The Gustatory Pathway

Afferent fibers carrying taste information from the tongue are found primarily in two cranial nerve pairs. As Figure 15.24 shows, a branch of the **facial nerve** (VII), the *chorda tympani,* transmits impulses from taste receptors in the anterior two-thirds of the tongue. The lingual branch of the **glossopharyngeal nerve** (IX) services the posterior third and the pharynx just behind. Taste impulses from the few taste buds in the epiglottis and the lower pharynx are conducted primarily by the **vagus nerve** (X). These afferent fibers synapse in the **solitary nucleus** of the medulla, and from there impulses stream to the thalamus and ultimately to the *gustatory cortex* in the insula. Fibers also project to the hypothalamus and limbic system structures, regions that determine our appreciation of what we are tasting.

An important role of taste is to trigger reflexes involved in digestion. As taste impulses pass through the solitary nucleus, they initiate reflexes (via synapses with parasympathetic nuclei) that increase secretion of saliva into the mouth and of gastric juice into the stomach. Saliva contains mucus that moistens food and digestive enzymes that begin the digestion of starch. Acidic foods are particularly strong stimulants of the salivary reflex. Additionally, when we eat revolting or foul-tasting substances, gagging or even reflexive vomiting may be initiated.

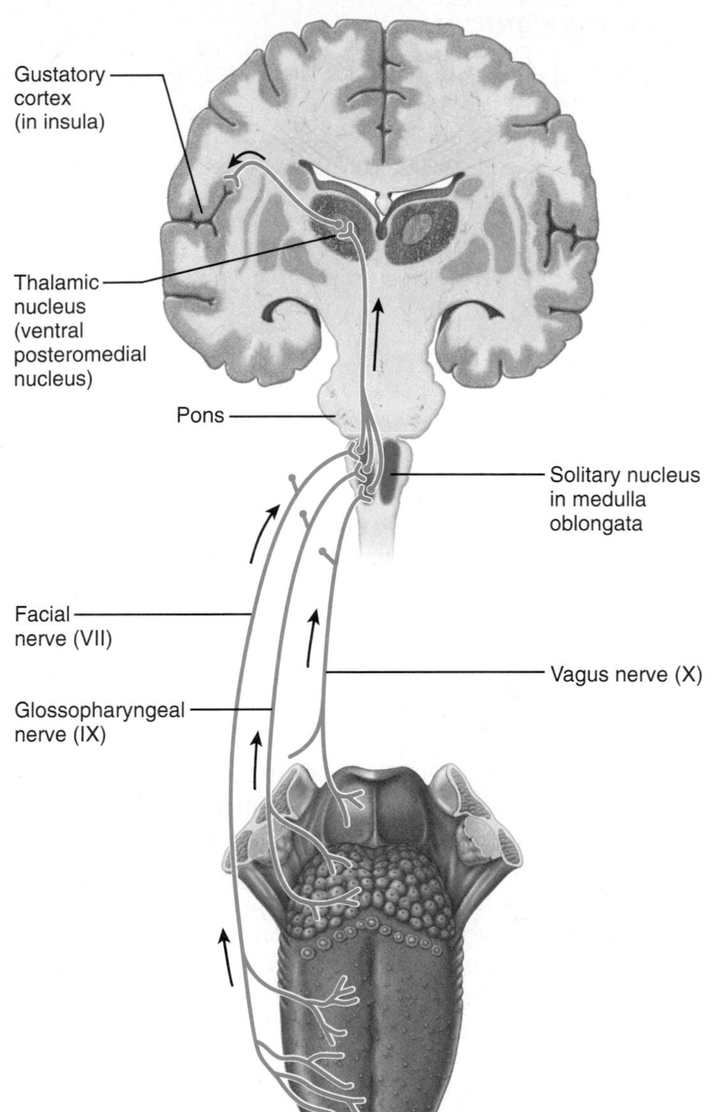

Figure 15.24 The gustatory pathway. Taste signals are relayed from the taste buds to the gustatory area of the cerebral cortex.

Influence of Other Sensations on Taste

Taste is 80% smell. When the olfactory receptors in the nasal cavity are blocked by nasal congestion (or pinching your nostrils), food is bland. Without smell, our morning coffee would lack its richness and simply taste bitter.

The mouth also contains thermoreceptors, mechanoreceptors, and nociceptors, and the temperature and texture of foods can enhance or detract from their taste. "Hot" foods such as chili peppers actually bring about their pleasurable effects by exciting pain receptors in the mouth.

Homeostatic Imbalances of the Chemical Senses

Most olfactory disorders, or *anosmias* (an-oz′me-ahz; "without smells"), result from head injuries that tear the olfactory nerves, the aftereffects of nasal cavity inflammation, and aging. Brain disorders can distort the sense of smell. Some people have *uncinate fits* (uns′ih-nāt), olfactory hallucinations during which they experience a particular (usually unpleasant) odor, such as rotting meat. Some such cases are undeniably psychological, but many result from irritation of the olfactory pathway by brain surgery or head trauma. *Olfactory auras* experienced by some epileptics just before they have a seizure are transient uncinate fits.

Taste disorders are less common than disorders of smell, in part because the taste receptors are served by three different nerves and thus are less likely to be "put out of business" completely. Causes of taste disorders include upper respiratory tract infections, head injuries, chemicals or medications, or head and neck radiation for cancer treatment. Zinc supplements have been found to help in some cases of radiation-induced taste disorders.

CHECK YOUR UNDERSTANDING

10. Name the five taste modalities. Name the three types of papillae that have taste buds.
11. Olfactory receptor cells have cilia and taste cells have hairs. How do these structures help the cells perform their functions?

For answers, see Appendix G.

The Ear: Hearing and Balance

- Describe the structure and general function of the outer, middle, and internal ears.
- Describe the sound conduction pathway to the fluids of the internal ear, and follow the auditory pathway from the spiral organ (of Corti) to the temporal cortex.
- Explain how one is able to differentiate pitch and loudness, and localize the source of sounds.
- List possible causes and symptoms of otitis media, deafness, and Ménière's syndrome.

15

At first glance, the machinery for hearing and balance appears very crude. Fluids must be stirred to stimulate the mechanoreceptors of the internal ear. Nevertheless, our hearing apparatus allows us to hear an extraordinary range of sound, and our equilibrium receptors keep the nervous system continually informed of head movements and position. Although the organs serving these two senses are structurally interconnected within the ear, their receptors respond to different stimuli and are activated independently of one another.

Structure of the Ear

The ear is divided into three major areas: external ear, middle ear, and internal ear **(Figure 15.25a)**. The external and middle ear structures are involved with hearing only and are rather simply engineered. The internal ear functions in both equilibrium and hearing and is extremely complex.

External Ear

The **external (outer) ear** consists of the auricle and the external acoustic meatus. The **auricle**, or **pinna**, is what most people call the ear—the shell-shaped projection surrounding the opening of the external acoustic meatus. The auricle is composed of elastic cartilage covered with thin skin and an occasional hair. Its rim, the **helix**, is somewhat thicker, and its fleshy, dangling **lobule** ("earlobe") lacks supporting cartilage. The function of the auricle is to direct sound waves into the external acoustic meatus.

The **external acoustic meatus** (auditory canal) is a short, curved tube (about 2.5 cm long by 0.6 cm wide) that extends from the auricle to the eardrum. Near the auricle, its framework is elastic cartilage; the remainder of the canal is carved into the temporal bone. The entire canal is lined with skin bearing hairs, sebaceous glands, and modified apocrine sweat glands called **ceruminous glands** (sĕ-roo′mĭ-nus). These glands secrete yellow-brown waxy **cerumen**, or earwax (*cere* = wax), which provides a sticky trap for foreign bodies and repels insects.

In many people, the ear is naturally cleansed as the cerumen dries and then falls out of the external acoustic meatus. Jaw movements as a person eats, talks, and so on, create an unnoticeable conveyor-belt effect that moves the wax out. In other people, cerumen builds up excessively and becomes compacted, a condition that can impair hearing.

Sound waves entering the external acoustic meatus eventually hit the **tympanic membrane**, or **eardrum** (*tympanum* = drum), the boundary between the outer and middle ears. The eardrum is a thin, translucent, connective tissue membrane, covered by skin on its external face and by a mucosa internally. It is shaped like a flattened cone, with its apex protruding medially into the middle ear. Sound waves make the eardrum vibrate. The eardrum, in turn, transfers the sound energy to the tiny bones of the middle ear and sets them into vibration.

Middle Ear

The **middle ear**, or **tympanic cavity**, is a small, air-filled, mucosa-lined cavity in the petrous portion of the temporal bone. It is flanked laterally by the eardrum and medially by a bony wall with two openings, the superior **oval (vestibular) window** and the inferior **round (cochlear) window**. Superiorly the tympanic cavity arches upward as the **epitympanic recess**, the "roof" of the middle ear cavity. The **mastoid antrum**, a canal in the posterior wall of the tympanic cavity, allows it to communicate with *mastoid air cells* housed in the mastoid process.

The anterior wall of the middle ear abuts the internal carotid artery (the main artery supplying the brain) and contains the opening of the **pharyngotympanic (auditory) tube**. This tube, formerly called the eustachian tube, runs obliquely downward to link the middle ear cavity with the nasopharynx (the superiormost part of the throat), and the mucosa of the middle ear is continuous with that lining the pharynx (throat).

Normally, the pharyngotympanic tube is flattened and closed, but swallowing or yawning opens it briefly to equalize pressure in the middle ear cavity with external air pressure. This is important because the eardrum vibrates freely only if the pressure on both of its surfaces is the same; otherwise sounds are distorted. The ear-popping sensation of the pressures equalizing is familiar to anyone who has flown in an airplane.

HOMEOSTATIC IMBALANCE

Otitis media (me′de-ah), or middle ear inflammation, is a fairly common result of a sore throat, especially in children, whose pharyngotympanic tubes are shorter and run more horizontally. Otitis media is the most frequent cause of hearing loss in children. In acute infectious forms, the eardrum bulges and becomes inflamed and red. Most cases of otitis media are treated with antibiotics. When large amounts of fluid or pus accumulate in the cavity, an emergency *myringotomy* (lancing of the eardrum) may be required to relieve the pressure, and a tiny tube implanted in the eardrum permits pus to drain into the external ear. The tube falls out by itself within the year. ■

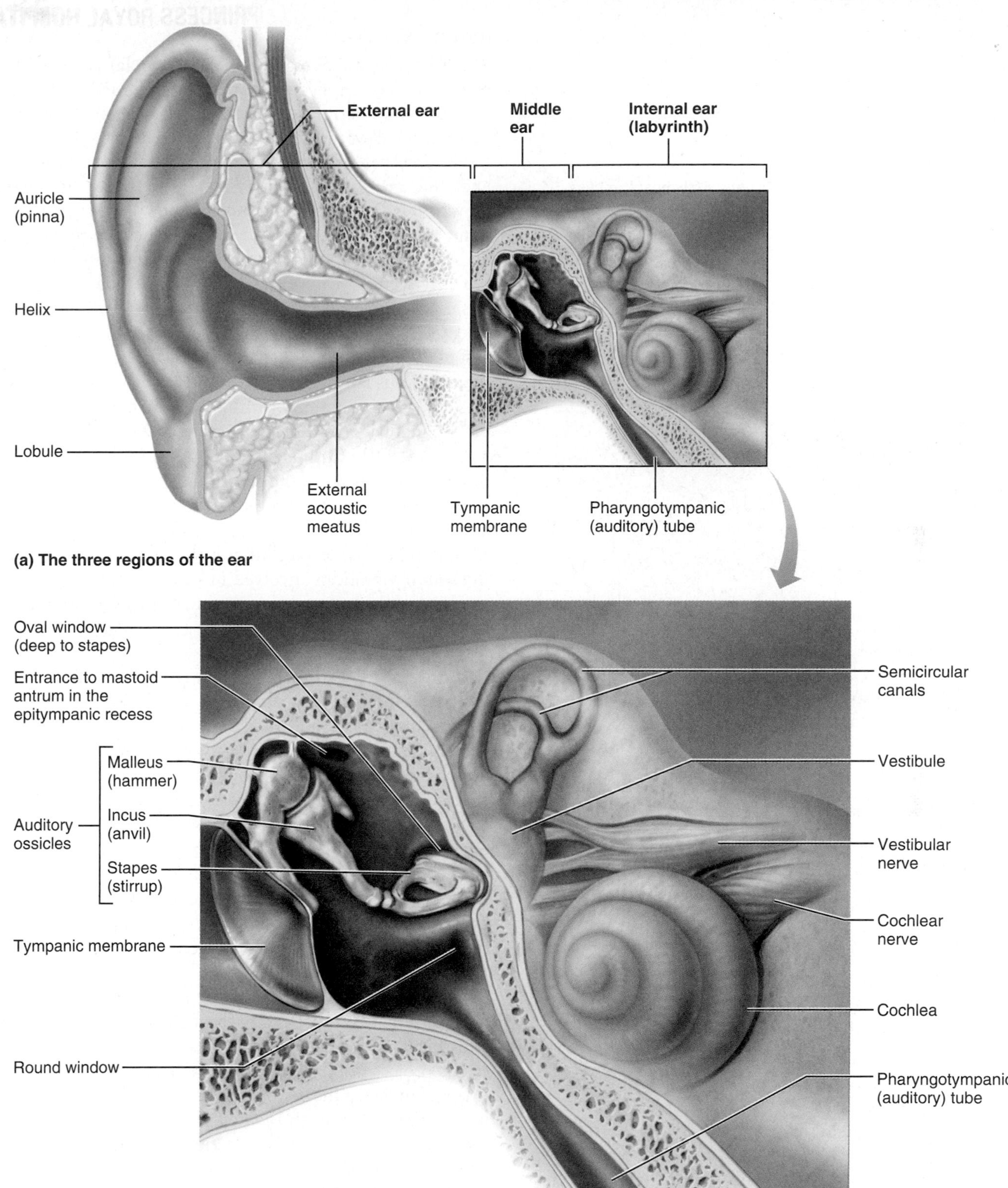

Figure 15.25 Structure of the ear. The inner ear structures in (b) appear as though they are veiled because they are cavities within the temporal bone.

15

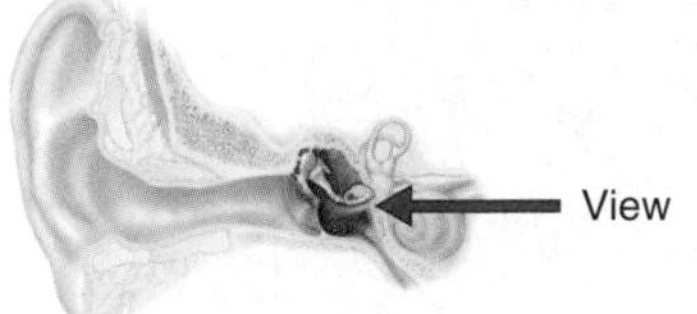

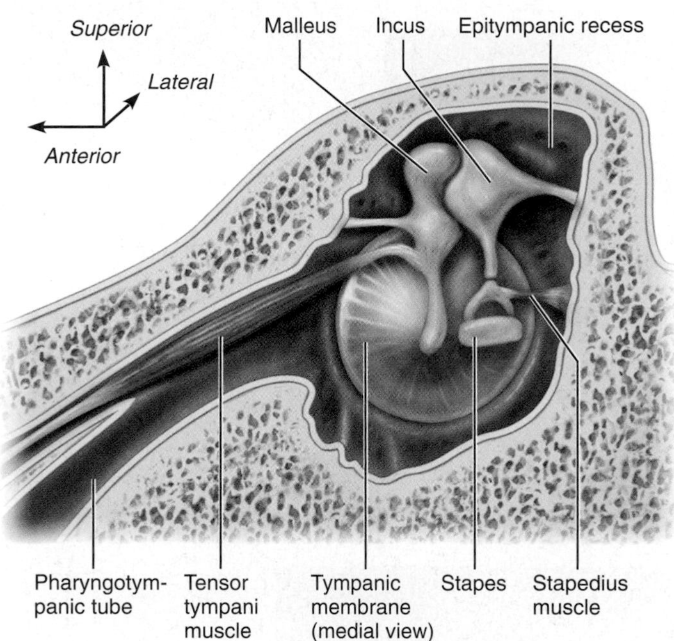

Figure 15.26 The three auditory ossicles and associated skeletal muscles. Right ear, medial view.

15

The tympanic cavity is spanned by the three smallest bones in the body: the **auditory ossicles** (Figure 15.25 and **Figure 15.26**). These bones, named for their shape, are the **malleus** (mal′e-us; "hammer"); the **incus** (ing′kus; "anvil"); and the **stapes** (sta′pēz; "stirrup"). The "handle" of the malleus is secured to the eardrum, and the base of the stapes fits into the oval window.

Tiny ligaments suspend the ossicles, and mini synovial joints link them together into a chain that spans the middle ear cavity. The incus articulates with the malleus laterally and the stapes medially. The ossicles transmit the vibratory motion of the eardrum to the oval window, which in turn sets the fluids of the internal ear into motion, eventually exciting the hearing receptors.

Two tiny skeletal muscles are associated with the ossicles (Figure 15.26). The **tensor tympani** (ten′sor tim′pah-ni) arises from the wall of the pharyngotympanic tube and inserts on the malleus. The **stapedius** (stah-pe′de-us) runs from the posterior wall of the middle ear cavity to the stapes. When the ears are assaulted by very loud sounds, these muscles contract reflexively to prevent damage to the hearing receptors. Specifically, the tensor tympani tenses the eardrum by pulling it medially; the stapedius checks vibration of the whole ossicle chain and limits the movement of the stapes in the oval window.

Internal Ear

The **internal (inner) ear** is also called the **labyrinth** ("maze") because of its complicated shape (see Figure 15.25). It lies deep in the temporal bone behind the eye socket and provides a secure site for all of the delicate receptor machinery housed there.

The internal ear has two major divisions: the bony labyrinth and the membranous labyrinth. The **bony**, or **osseous**, **labyrinth** is a system of tortuous channels worming through the bone. Its three regions are the *vestibule*, the *cochlea* (kok′le-ah), and the *semicircular canals*. The views of the bony labyrinth typically seen in textbooks, including this one, are somewhat misleading because we are talking about a *cavity* here. The representation in Figure 15.25 can be compared to a plaster of paris cast of the cavity or hollow space inside the bony labyrinth. The **membranous labyrinth** is a continuous series of membranous sacs and ducts contained within the bony labyrinth and (more or less) following its contours **(Figure 15.27)**.

The bony labyrinth is filled with **perilymph**, a fluid similar to cerebrospinal fluid and continuous with it. The membranous labyrinth is suspended in the surrounding perilymph, and its interior contains **endolymph**, which is chemically similar to K^+-rich intracellular fluid. These two fluids conduct the sound vibrations involved in hearing and respond to the mechanical forces occurring during changes in body position and acceleration.

Vestibule The **vestibule** is the central egg-shaped cavity of the bony labyrinth. It lies posterior to the cochlea, anterior to the semicircular canals, and flanks the middle ear medially. In its lateral wall is the oval window. Suspended in its perilymph and united by a small duct are two membranous labyrinth sacs, the **saccule** and **utricle** (u′trĭ-kl) (Figure 15.27). The smaller saccule is continuous with the membranous labyrinth extending anteriorly into the cochlea (the *cochlear duct*), whereas the utricle is continuous with the semicircular ducts extending into the semicircular canals posteriorly. The saccule and utricle house equilibrium receptor regions called *maculae* that respond to the pull of gravity and report on changes of head position.

Semicircular Canals The **semicircular canals** lie posterior and lateral to the vestibule, and each of these canals defines about two-thirds of a circle. The cavities of the bony semicircular canals project from the posterior aspect of the vestibule, each oriented in one of the three planes of space. Accordingly, there is an *anterior*, *posterior*, and *lateral* semicircular canal in each internal ear. The anterior and posterior canals are oriented at right angles to each other in the vertical plane, whereas the lateral canal lies horizontally (Figure 15.27).

Snaking through each semicircular canal is a corresponding membranous **semicircular duct**, which communicates with the utricle anteriorly. Each of these ducts has an enlarged swelling at one end called an **ampulla**, which houses an equilibrium receptor region called a *crista ampullaris* (literally, crest of the ampulla). These receptors respond to angular (rotational) movements of the head.

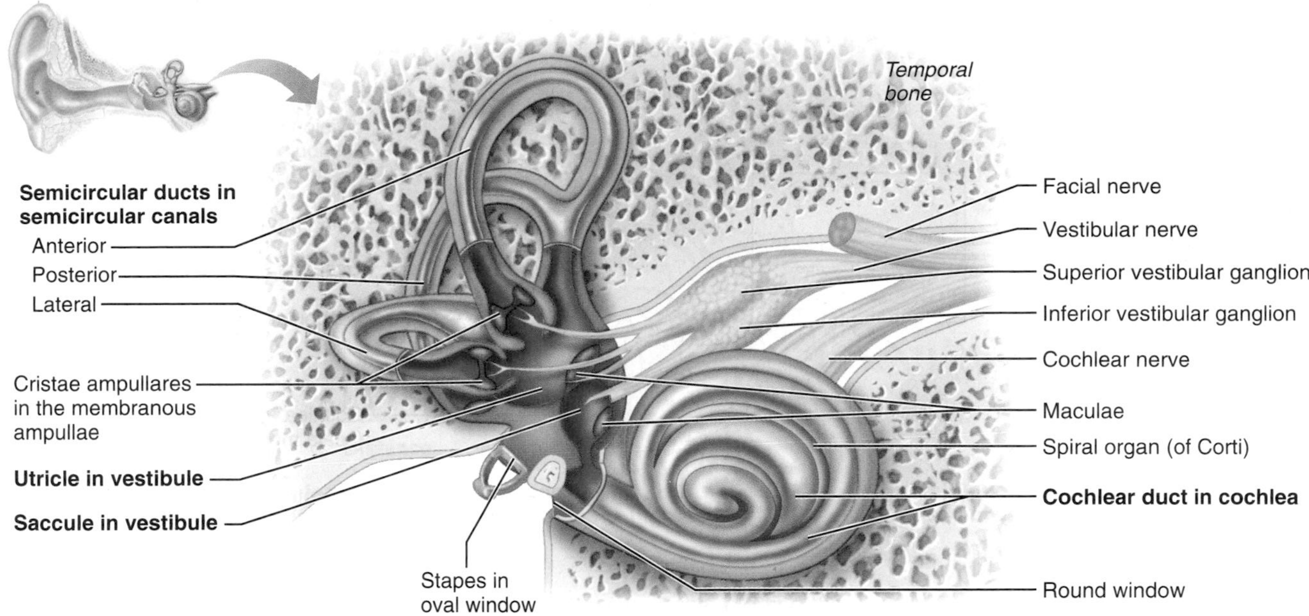

Figure 15.27 Membranous labyrinth of the internal ear. The membranous labyrinth (blue) lies within the chambers of the bony labyrinth (tan). The locations of the sensory organs for hearing [spiral organ (of Corti)] and equilibrium (maculae and cristae ampullares) are shown in purple.

15

Cochlea The **cochlea,** from the Latin "snail," is a spiral, conical, bony chamber about the size of a split pea. It extends from the anterior part of the vestibule and coils for about 2½ turns around a bony pillar called the **modiolus** (mo-di′o-lus) **(Figure 15.28a).** Running through its center like a wedge-shaped worm is the membranous **cochlear duct**, which ends blindly at the cochlear apex. The cochlear duct houses the **spiral organ (of Corti)**, the receptor organ for hearing (Figures 15.27 and 15.28b).

The cochlear duct and the **osseous spiral lamina**, a thin shelflike extension of bone that spirals up the modiolus like the thread on a screw, together divide the cavity of the bony cochlea into three separate chambers or **scalae** (*scala* = ladder). The **scala vestibuli** (ska′lah vĕs-tĭ′bu-li), which lies superior to the cochlear duct, is continuous with the vestibule and abuts the oval window. The middle **scala media** is the cochlear duct itself. The **scala tympani**, which terminates at the round window, is inferior to the cochlear duct.

Since the scala media is part of the membranous labyrinth, it is filled with endolymph. The scala vestibuli and the scala tympani, both part of the bony labyrinth, contain perilymph. The perilymph-containing chambers are continuous with each other at the cochlear apex, a region called the **helicotrema** (hel″ĭ-ko-tre′mah; "the hole in the spiral").

The "roof" of the cochlear duct, separating the scala media from the scala vestibuli, is the **vestibular membrane** (Figure 15.28b). The duct's external wall, the **stria vascularis**, is composed of an unusual richly vascularized mucosa that secretes endolymph. The "floor" of the cochlear duct is composed of the bony spiral lamina and the flexible, fibrous **basilar membrane**, which supports the spiral organ (of Corti). (We will describe the spiral organ when we discuss hearing.) The basilar membrane, which plays a critical role in sound reception, is narrow and thick near the oval window and gradually widens and thins as it approaches the cochlear apex. The *cochlear nerve*, a division of the *vestibulocochlear nerve* (VIII), runs from the spiral organ through the modiolus on its way to the brain.

Physiology of Hearing

The mechanics of human hearing can be summed up in a single sprawling sentence: Sounds set up vibrations in air that beat against the eardrum that pushes a chain of tiny bones that press fluid in the internal ear against membranes that set up shearing forces that pull on the tiny hair cells that stimulate nearby neurons that give rise to impulses that travel to the brain, which interprets them—and you hear. Before we unravel this intriguing sequence, let us describe sound, the stimulus for hearing.

Overview: Properties of Sound

Light can be transmitted through a vacuum (for instance, outer space), but sound depends on an *elastic* medium for its transmission. Sound also travels much more slowly than light. Its speed in dry air is only about 331 m/s (0.2 mi/s), as opposed to the about 300,000 km/s (186,000 mi/s) of light. A lightning flash is almost instantly visible, but the sound it creates (thunder) reaches our ears much more slowly. (For each second between the lightning bolt and the roll of thunder, the storm is 1/5 mile farther away.) The speed of sound is constant in a given medium. It is greatest in solids and lowest in gases, including air.

Sound is a pressure disturbance—alternating areas of high and low pressure—produced by a vibrating object and propagated

CHECK YOUR UNDERSTANDING

16. For each of the following statements, indicate whether it applies to a macula or a crista ampullaris: inside a semicircular canal; contains otoliths; responds to linear acceleration and deceleration; has a cupula; responds to rotational acceleration and deceleration; inside a saccule.

For answers, see Appendix G.

Developmental Aspects of the Special Senses

▶ List changes that occur in the special sense organs with aging.

Taste and Smell

All the special senses are functional, to a greater or lesser degree, at birth. Smell and taste are sharp, and infants relish food that adults consider bland or tasteless. Some researchers claim that smell is just as important as touch in guiding newborn infants to their mother's breast. However, very young children seem indifferent to odors and can play happily with their own feces.

There are few problems with the chemical senses during childhood and young adulthood. Women generally have a more acute sense of smell than men, and nonsmokers have a sharper sense of smell than smokers. Beginning in the fourth decade of life, our ability to taste and smell declines due to the gradual loss of receptors, which are replaced more slowly than in younger people. More than half of people over the age of 65 years have serious problems detecting odors, which may explain why some tend to douse themselves with large amounts of cologne, or pay little attention to formerly disagreeable odors. Additionally, their sense of taste is poor. This along with the decline in the sense of smell makes food taste bland and contributes to loss of appetite.

Vision

By the fourth week of development, the beginnings of the eyes are seen in the **optic vesicles** that protrude from the diencephalon (see Figure 12.2c on p. 431). Soon these hollow vesicles indent to form double-layered **optic cups**, and their stalks form the optic nerves and provide a pathway for blood vessels to reach the eye interior. Once an optic vesicle reaches the overlying surface ectoderm, it induces the ectoderm to thicken and then form a **lens vesicle** that pinches off into the cavity of the optic cup, where it becomes the lens. The lining (internal layer) of the optic cup becomes the neural retina, and the outer layer forms the pigmented layer of the retina. The rest of the eye tissues and the vitreous humor are formed by mesenchymal cells derived from the mesoderm that surrounds the optic cup.

In the darkness of the uterus, the fetus cannot see. Nonetheless, even before the light-sensitive portions of the photoreceptors develop, the central nervous system connections have been made and are functional. During infancy, synaptic connections are fine-tuned, and the typical cortical fields that allow binocular vision are established.

HOMEOSTATIC IMBALANCE

Congenital problems of the eyes are relatively uncommon, but their incidence is dramatically increased by certain maternal infections, particularly rubella (German measles), occurring during the critical first three months of pregnancy. Common rubella sequels include blindness and cataracts. ■

As a rule, vision is the only special sense not fully functional at birth. Because the eyeballs are foreshortened, most babies are hyperopic. The newborn infant sees only in gray tones, eye movements are uncoordinated, and often only one eye at a time is used. The lacrimal glands are not completely developed until about two weeks after birth, so babies are tearless for this period, even though they may cry lustily. By 5 months, infants can follow moving objects with their eyes, but visual acuity is still poor (20/200). By the age of 5 years, depth perception is present and color vision is well developed. Because the eyeball has grown, visual acuity has improved to about 20/30, providing a readiness to begin reading. By first grade, the initial hyperopia has usually been replaced by emmetropia, and the eye reaches its adult size at 8–9 years of age. Emmetropia usually continues until presbyopia begins to set in around age 40 owing to decreasing lens elasticity.

With age, the lens loses its crystal clarity and discolors. As a result, it begins to scatter light, causing a glare that is distressing when driving at night. The dilator pupillae muscles become less efficient, so the pupils stay partly constricted. These two changes together decrease the amount of light reaching the retina, and visual acuity is dramatically lower in people over 70. In addition, the lacrimal glands are less active and the eyes tend to be dry and more susceptible to infection. Elderly persons are also at risk for certain conditions that cause blindness, such as macular degeneration, glaucoma, cataracts, arteriosclerosis, and diabetes mellitus.

Hearing and Balance

The ear begins to develop in the three-week embryo. The internal ears develop first, from thickenings of the surface ectoderm called the **otic placodes** (o′tik), which lie lateral to the hindbrain on each side. The otic placode invaginates, forming the **otic pit** and then the **otic vesicle**, which detaches from the surface epithelium. The otic vesicle develops into the membranous labyrinth. The surrounding mesenchyme forms the bony labyrinth.

The middle ear cavity and pharyngotympanic tube of the middle ear develop from **pharyngeal pouches**, lateral outpocketings of the endoderm lining the pharynx. The auditory ossicles develop from neural crest cells.

The external acoustic meatus and external face of the tympanic membrane of the external ear differentiate from the **branchial groove** (brang′ke-al), an indentation of the surface ectoderm, and the auricle develops from swellings of the surrounding tissue.

Newborn infants can hear, but early responses to sound are mostly reflexive—for example, crying and clenching the eyelids in response to a startling noise. By the fourth month, infants will turn to the voices of family members. Critical listening begins as toddlers increase their vocabulary, and good language skills are closely tied to the ability to hear well.

HOMEOSTATIC IMBALANCE

Congenital abnormalities of the ears are fairly common. Examples include partly or completely missing pinnae and closed or absent external acoustic meatuses. Maternal rubella during the first trimester commonly results in sensorineural deafness. ■

Except for ear inflammations, mostly due to infections, few problems affect the ears during childhood and adult life. By the 60s, however, deterioration of the spiral organ becomes noticeable. We are born with approximately 40,000 hair cells, but their number decreases when they are damaged or destroyed by loud noises, disease, or drugs. The hair cells *are* replaced, but at such a slow rate that there really is no functional regeneration. It is estimated that if we were to live 140 years, we would have lost all of our hearing receptors.

The ability to hear high-pitched sounds leaves us first. This condition, called **presbycusis** (pres″bĭ-ku′sis), is a type of sensorineural deafness. Although presbycusis is considered a disability of old age, it is becoming much more common in younger people as our world grows noisier. Since noise is a stressor, one of its physiological consequences is vasoconstriction, and when delivery of blood to the ear is reduced, the ear becomes even more sensitive to the damaging effects of noise.

CHECK YOUR UNDERSTANDING

17. What age-related changes make it more difficult for the elderly to see at night?

For answers, see Appendix G.

Our abilities to see, hear, taste, and smell—and some of our responses to the effects of gravity—are largely the work of our brain. However, as we have discovered in this final nervous system chapter, the large and often elaborate sensory receptor organs that serve the special senses are works of art in and of themselves.

The concluding chapter of this unit describes how the body's functions are controlled by chemicals called hormones in a manner quite different from what we have described for neural control.

15

RELATED CLINICAL TERMS

Ageusia (ah-gu′ze-ah) Loss or impairment of the taste sense.

Age-related macular degeneration (ARMD) Progressive deterioration of the retina that affects the macula lutea and leads to loss of central vision; the main cause of vision loss in those over age 65. Early stages and mild forms involve the buildup of pigments in the macula and impaired functioning of the pigmented epithelium. Continued accumulation of pigment leads to the "dry" form of ARMD, in which many pigment epithelial cells and macular photoreceptors die. The dry form is largely untreatable, although a specific cocktail of vitamins and zinc has been shown to slow its progression. Less common is the "wet" form in which an overgrowth of new blood vessels invades the retina from the choroid. Leakage of blood and fluids from these vessels causes scarring and detachment of the retina. The cause of wet ARMD is unknown, but several treatment options can slow its progression—laser treatments that destroy some of the growing vessels and drugs that prevent blood vessel growth.

Blepharitis (blef″ah-ri′tis; *blephar* = eyelash; *itis* = inflammation) Inflammation of the margins of the eyelids.

Enucleation (e-nu″kle-a′shun) Surgical removal of an eyeball.

Exophthalmos (ek″sof-thal′mos; *exo* = out; *phthalmo* = the eye) Anteriorly bulging eyeballs; this condition is seen in some cases of hyperthyroidism.

Labyrinthitis Inflammation of the labyrinth.

Ophthalmology (of″thal-mol′o-je) The science that studies the eye and eye diseases. An ophthalmologist is a medical doctor who specializes in treating eye disorders.

Optometrist A licensed nonphysician who measures vision and prescribes corrective lenses.

Otalgia (o-tal′je-ah; *algia* = pain) Earache.

Otitis externa Inflammation and infection of the external acoustic meatus, caused by bacteria or fungi that enter the canal from outside, especially when the canal is moist (e.g., after swimming).

Papilledema (pap″il-ĕ-de′mah; *papill* = nipple; *edema* = swelling) Protrusion of the optic disc into the eyeball, which can be observed by ophthalmoscopic examination; caused by conditions that increase intracranial pressure.

Scotoma (sko-to′mah; *scoto* = darkness) A blind spot other than the normal (optic disc) blind spot; many causes, including the presence of a brain tumor pressing on fibers of the visual pathway, and stroke.

Trachoma (trah-ko′mah; *trach* = rough) A highly contagious bacterial (chlamydial) infection of the conjunctiva and cornea. Common worldwide, it blinds millions of people in poor countries of Africa and Asia; treated with eye ointments containing antibiotic drugs.

Weber's test Hearing test during which a sounding tuning fork is held to the forehead. In those with normal hearing, the tone is heard equally in both ears. The tone will be heard best in the "good" ear if sensorineural deafness is present, and in the "bad" ear if conduction deafness is present.

CHAPTER SUMMARY

The Eye and Vision (pp. 548–569)

1. The eye is enclosed in the bony orbit and cushioned by fat.

Accessory Structures of the Eye (pp. 548–551)

2. Eyebrows help to shade and protect the eyes.
3. Eyelids protect and lubricate the eyes by reflex blinking. Within the eyelids are the orbicularis oculi and levator palpebrae muscles, and modified sebaceous and sweat glands.
4. The conjunctiva is a mucosa that lines the eyelids and covers the anterior eyeball surface. Its mucus lubricates the eyeball surface.
5. The lacrimal apparatus consists of the lacrimal gland (which produces a saline solution containing mucus, lysozyme, and antibodies), the lacrimal canaliculi, the lacrimal sac, and the nasolacrimal duct.
6. The extrinsic eye muscles (superior, inferior, lateral, and medial rectus and superior and inferior oblique) move the eyeballs.

Structure of the Eyeball (pp. 551–556)

7. The wall of the eyeball is made up of three layers. The outermost fibrous layer consists of the sclera and the cornea. The sclera protects the eye and gives it shape; the cornea allows light to enter the eye.
8. The middle, pigmented vascular layer (uvea) consists of the choroid, the ciliary body, and the iris. The choroid provides nutrients to the eye and prevents light scattering within the eye. The ciliary muscles of the ciliary body control lens shape; the iris controls the size of the pupil.
9. The sensory layer, or retina, consists of an outer pigmented layer and an inner neural layer. The neural layer contains photoreceptors (rods and cones), bipolar cells, and ganglion cells. Ganglion cell axons form the optic nerve, which exits via the optic disc ("blind spot").
10. Rods respond to low-intensity light and provide night and peripheral vision. Cones are bright-light, high-discrimination receptors that provide for color vision. Anything that must be viewed precisely is focused on the cone-rich fovea centralis.
11. The posterior segment of the eyeball, behind the lens, contains vitreous humor, which helps support the eyeball and keep the retina in place. The anterior segment, anterior to the lens, is filled with aqueous humor, formed by capillaries in the ciliary processes and drained into the scleral venous sinus. Aqueous humor is a major factor in maintaining intraocular pressure.
12. The biconvex lens is suspended within the eye by the ciliary zonule attached to the ciliary body. It is the only adjustable refractory structure of the eye.

Physiology of Vision (pp. 556–569)

13. Light is made up of those wavelengths of the electromagnetic spectrum that excite the photoreceptors.
14. Light is refracted (bent) when passing from one transparent medium to another of different density. Concave lenses disperse light; convex lenses converge light and bring its rays to a focal point. The greater the lens curvature, the more light bends.
15. As light passes through the eye, it is bent by the cornea and the lens and focused on the retina. The cornea accounts for most of the refraction, but the lens allows active focusing for different distances.
16. Focusing for distance vision requires no special movements of the eye structures. Focusing for close-up vision requires accommodation (bulging of the lens), pupillary constriction, and convergence of the eyeballs. All three reflexes are controlled by cranial nerve III.
17. Refractory problems include presbyopia, myopia, hyperopia, and astigmatism.
18. The outer segments of the photoreceptors contain light-absorbing visual pigment in membrane-bounded discs.
19. The light-absorbing molecule retinal is combined with various opsins to form the visual pigments. When struck by light, retinal changes shape (11-*cis* to all-*trans*) and activates opsin. Activated opsin activates transducin (a G protein) which in turn activates PDE, an enzyme that breaks down cGMP, allowing the cation channels to close. This results in hyperpolarization of the receptor cells and inhibits their release of neurotransmitter.
20. Rod visual pigment, rhodopsin, is a combination of retinal and opsin. The light-triggered changes in retinal cause hyperpolarization of the rods. Photoreceptors and bipolar cells generate graded potentials only; action potentials are generated by the ganglion cells.
21. The three types of cones all contain retinal, but each has a different type of opsin. Each cone type responds maximally to one color of light: red, blue, or green. The chemistry of cone function is similar to that of rods.
22. During light adaptation, photopigments are bleached and rods are inactivated; then, as cones decrease their light sensitivity, high-acuity vision ensues. In dark adaptation, cones cease functioning, and visual acuity decreases; rod function begins when sufficient rhodopsin has accumulated.
23. The visual pathway to the brain begins with the optic nerve fibers (ganglion cell axons) from the retina. At the optic chiasma, fibers from the medial half of each retina cross over and continue on in the optic tracts to the thalamus. Thalamic neurons project to the visual cortex via the optic radiation. Fibers also project from the retina to the midbrain pretectal nuclei and the superior colliculi, and to the suprachiasmatic nucleus of the hypothalamus.
24. Each eye receives a slightly different view of the visual field. These views are fused by the visual cortices to provide for depth perception.
25. Retinal processing involves the selective destruction of inputs so as to emphasize bright/dark or color contrasts (edges). (The horizontal cells and amacrine cells are local integrator cells of the retina that modify and process inputs to the ganglion cells.) Thalamic processing subserves high-acuity color vision and depth perception. Cortical processing involves neurons of the striate (primary) cortex, which receive inputs from the retinal ganglion cells, and neurons of the prestriate (association) cortices, which receive inputs from striate cortical cells and mostly integrate inputs concerned with color, form, and movement. Visual processing also proceeds anteriorly in the "what" and "where" processing streams via the temporal and parietal lobes, respectively.

The Chemical Senses: Taste and Smell (pp. 569–574)

The Olfactory Epithelium and the Sense of Smell (pp. 569–571)

1. The olfactory epithelium is located in the roof of the nasal cavity. The receptor cells are ciliated bipolar neurons. Their axons are the filaments of the olfactory nerve (cranial nerve I).
2. Individual olfactory neurons show a range of responsiveness to different chemicals. Olfactory cells bearing the same odorant receptors synapse in the same glomerulus type.

3. Olfactory neurons are excited by volatile chemicals that bind to receptors in the olfactory cilia.
4. Action potentials of the olfactory nerve filaments are transmitted to the olfactory bulb where the filaments synapse with mitral cells. The mitral cells send impulses via the olfactory tract to the olfactory cortex. Fibers carrying impulses from the olfactory receptors also project to the limbic system.

Taste Buds and the Sense of Taste (pp. 571–573)

5. The taste buds are scattered in the oral cavity and pharynx but are most abundant on the tongue papillae.
6. Gustatory cells, the epithelial receptor cells of the taste buds, have gustatory hairs (microvilli) that serve as the receptor regions. The gustatory cells are excited by the binding of tastants (food chemicals) to receptors on their microvilli.
7. The five basic taste qualities are sweet, sour, salty, bitter, and umami.
8. The taste sense is served by cranial nerves VII, IX, and X, which send impulses to the solitary nucleus of the medulla. From there, impulses are sent to the thalamus and the gustatory cortex.

Homeostatic Imbalances of the Chemical Senses (pp. 573–574)

9. Most chemical sense dysfunctions are olfactory disorders (anosmias). Common causes are nasal damage or obstruction.

The Ear: Hearing and Balance (pp. 574–588)

Structure of the Ear (pp. 574–577)

1. The auricle and external acoustic meatus compose the external ear. The tympanic membrane, the boundary between the outer and middle ears, transmits sound waves to the middle ear.
2. The middle ear is a small chamber within the temporal bone, connected by the pharyngotympanic tube to the nasopharynx. The ossicles, which help to amplify sound, span the middle ear cavity and transmit sound vibrations from the tympanic membrane to the oval window.
3. The internal ear consists of the bony labyrinth, within which the membranous labyrinth is suspended. The bony labyrinth chambers contain perilymph; the membranous labyrinth ducts and sacs contain endolymph.
4. The vestibule contains the saccule and utricle. The semicircular canals extend posteriorly from the vestibule in three planes. They contain the semicircular ducts.
5. The cochlea houses the cochlear duct (scala media), containing the spiral organ (of Corti), the receptor organ for hearing. Within the cochlear duct, the hair (receptor) cells rest on the basilar membrane, and their hairs project into the gelatinous tectorial membrane.

Physiology of Hearing (pp. 577–583)

6. Sound originates from a vibrating object and travels in waves consisting of alternating areas of compression and rarefaction of the medium.
7. The distance from crest to crest on a sine wave is the sound's wavelength; the shorter the wavelength, the higher the frequency (measured in hertz). Frequency is perceived as pitch.
8. The amplitude of sound is the height of the peaks of the sine wave, which reflect the sound's intensity. Sound intensity is measured in decibels. Intensity is perceived as loudness.
9. Sound passing through the external acoustic meatus sets the tympanic membrane into vibration at the same frequency. The ossicles amplify and deliver the vibrations to the oval window.
10. Pressure waves in cochlear fluids set specific locations on the basilar membrane into resonance. At points of maximal membrane vibration, the hair cells of the spiral organ are alternately depolarized and hyperpolarized by the vibratory motion. Movements of the stereocilia toward the kinocilium depolarize the hair cells and increase the rate of impulse generation in the auditory nerve fibers. Movements away from the kinocilium have the opposite effect. High-frequency sounds stimulate hair cells near the oval window; low-frequency sounds stimulate hair cells near the apex. Most auditory inputs are sent to the brain by inner hair cells. Outer hair cells amplify responsiveness of the inner hair cells.
11. Impulses generated along the cochlear nerve travel to the cochlear nuclei of the medulla and from there through several brain stem nuclei to the medial geniculate nucleus of the thalamus and then the auditory cortex. Each auditory cortex receives impulses from both ears.
12. Auditory processing is analytic; each tone is perceived separately. Perception of pitch is related to the position of the excited hair cells along the basilar membrane. Intensity perception reflects the fact that as sound intensity increases, basilar membrane motion is increased and the frequency of impulse transmission to the cortex is enhanced. Cues for sound localization include the intensity and timing of sound arriving at each ear.

Homeostatic Imbalances of Hearing (p. 583)

13. Conduction deafness results from interference with conduction of sound vibrations to the fluids of the internal ear. Sensorineural deafness reflects damage to neural structures.
14. Tinnitus is an early sign of sensorineural deafness; it may also result from the use of certain drugs.
15. Ménière's syndrome is a disorder of the membranous labyrinth. Symptoms include tinnitus, deafness, and vertigo. Excessive endolymph accumulation is the suspected cause.

Equilibrium and Orientation (pp. 584–588)

16. The equilibrium receptor regions of the internal ear are called the vestibular apparatus.
17. The receptors for static equilibrium are the maculae of the saccule and utricle. A macula consists of hair cells with stereocilia and a kinocilium embedded in an overlying otolithic membrane. Linear movements cause the otolithic membrane to move, pulling on the hair cells and changing the rate of impulse generation in the vestibular nerve fibers.
18. The dynamic equilibrium receptor, the crista ampullaris within each semicircular duct, responds to angular or rotatory movements in one plane. It consists of a tuft of hair cells whose microvilli are embedded in the gelatinous cupula. Rotatory movements cause the endolymph to flow in the opposite direction, bending the cupula and either exciting or inhibiting the hair cells.
19. Impulses from the vestibular apparatus are sent via vestibular nerve fibers mainly to the vestibular nuclei of the brain stem and the cerebellum. These centers initiate responses that fix the eyes on objects and activate muscles to maintain balance.

Developmental Aspects of the Special Senses (pp. 588–589)

Taste and Smell (p. 588)

1. The chemical senses are sharpest at birth and gradually decline with age as replacement of the receptor cells becomes more sluggish.

Vision (p. 588)

2. Congenital eye problems are uncommon, but maternal rubella can cause blindness.
3. The eye starts as an optic vesicle, an outpocketing of the diencephalon that invaginates to form the optic cup, which becomes the retina. Overlying ectoderm folds to form the lens vesicle, which gives rise to the lens. The remaining eye tissues and the accessory structures are formed by mesenchyme.
4. The eye is foreshortened at birth and reaches adult size at the age of 8–9 years. Depth perception and color vision develop during early childhood.
5. With age, the lens loses its elasticity and clarity, there is a decline in the ability of the iris to dilate, and visual acuity decreases. The elderly are at risk for eye problems resulting from dry eyes and disease.

Hearing and Balance (pp. 588–589)

6. The membranous labyrinth develops from the otic placode, an ectodermal thickening lateral to the hindbrain. Mesenchyme forms the surrounding bony structures. Pharyngeal pouch endoderm, in conjunction with mesenchyme, forms most middle ear structures; the external ear is formed largely by ectoderm.
7. Congenital ear problems are fairly common. Maternal rubella can cause deafness.
8. Response to sound in infants is reflexive. By the fourth month, an infant can locate sound. Critical listening develops in toddlers.
9. Deterioration of the spiral organ (of Corti) occurs throughout life as noise, disease, and drugs destroy cochlear hair cells. Age-related loss of hearing (presbycusis) occurs in the 60s and 70s.

REVIEW QUESTIONS

Multiple Choice/Matching

(Some questions have more than one correct answer. Select the best answer or answers from the choices given.)

1. The accessory glands that produce an oily secretion are the (a) conjunctiva, (b) lacrimal glands, (c) tarsal glands.
2. The portion of the fibrous layer that is white and opaque is the (a) choroid, (b) cornea, (c) retina, (d) sclera.
3. Which sequence best describes a normal route for the flow of tears from the eyes into the nasal cavity? (a) lacrimal canaliculi, nasolacrimal ducts, nasal cavity; (b) lacrimal ducts, lacrimal canaliculi, nasolacrimal ducts; (c) nasolacrimal ducts, lacrimal canaliculi, lacrimal sacs.
4. Activation of the sympathetic nervous system causes (a) contraction of the sphincter pupillae muscles, (b) contraction of the dilator pupillae muscles, (c) contraction of the ciliary muscles, (d) a decrease in ciliary zonule tension.
5. Damage to the medial recti muscles would probably affect (a) accommodation, (b) refraction, (c) convergence, (d) pupil constriction.
6. The phenomenon of dark adaptation is best explained by the fact that (a) rhodopsin does not function in dim light, (b) rhodopsin breakdown occurs slowly, (c) rods exposed to intense light need time to generate rhodopsin, (d) cones are stimulated to function by bright light.
7. Blockage of the scleral venous sinus might result in (a) a sty, (b) glaucoma, (c) conjunctivitis, (d) a cataract.
8. Nearsightedness is more properly called (a) myopia, (b) hyperopia, (c) presbyopia, (d) emmetropia.
9. Of the neurons in the retina, the axons of which of these form the optic nerve? (a) bipolar cells, (b) ganglion cells, (c) cone cells, (d) horizontal cells.
10. Which sequence of reactions occurs when a person looks at a distant object? (a) pupils constrict, ciliary zonule (suspensory ligament) relaxes, lenses become less convex; (b) pupils dilate, ciliary zonule becomes taut, lenses become less convex; (c) pupils dilate, ciliary zonule becomes taut, lenses become more convex; (d) pupils constrict, ciliary zonule relaxes, lenses become more convex.
11. During embryonic development, the lens of the eye forms (a) as part of the choroid coat, (b) from the surface ectoderm overlying the optic cup, (c) as part of the sclera, (d) from mesodermal tissue.
12. The blind spot of the eye is (a) where more rods than cones are found, (b) where the macula lutea is located, (c) where only cones occur, (d) where the optic nerve leaves the eye.
13. Olfactory tract damage would probably affect your ability to (a) see, (b) hear, (c) feel pain, (d) smell.
14. Sensory impulses transmitted over the facial, glossopharyngeal, and vagus nerves are involved in the sensation of (a) taste, (b) touch, (c) equilibrium, (d) smell.
15. Taste buds are found on the (a) anterior part of the tongue, (b) posterior part of the tongue, (c) palate, (d) all of these.
16. Gustatory cells are stimulated by (a) movement of otoliths, (b) stretch, (c) substances in solution, (d) photons of light.
17. Cells in the olfactory bulb that act as local "integrators" of olfactory inputs are the (a) hair cells, (b) granule cells, (c) basal cells, (d) mitral cells, (e) supporting cells.
18. Olfactory nerve filaments are found (a) in the optic bulbs, (b) passing through the cribriform plate of the ethmoid bone, (c) in the optic tracts, (d) in the olfactory cortex.
19. Conduction of sound from the middle ear to the internal ear occurs via vibration of the (a) malleus against the tympanic membrane, (b) stapes in the oval window, (c) incus in the round window, (d) stapes against the tympanic membrane.
20. The transmission of sound vibrations through the internal ear occurs chiefly through (a) nerve fibers, (b) air, (c) fluid, (d) bone.
21. Which of the following statements does not correctly describe the spiral organ (of Corti)? (a) Sounds of high frequency stimulate hair cells at the basal end, (b) the "hairs" of the receptor cells are embedded in the tectorial membrane, (c) the basilar membrane acts as a resonator, (d) the more numerous outer hair cells are largely responsible for our perception of sound.
22. Pitch is to frequency of sound as loudness is to (a) quality, (b) intensity, (c) overtones, (d) all of these.
23. The structure that allows pressure in the middle ear to be equalized with atmospheric pressure is the (a) pinna, (b) pharyngotympanic tube, (c) tympanic membrane, (d) oval window.
24. Which of the following is important in maintaining the balance of the body? (a) visual cues, (b) semicircular canals, (c) the saccule, (d) proprioceptors, (e) all of these.
25. Static equilibrium receptors that report the position of the head in space relative to the pull of gravity are (a) spiral organs, (b) maculae, (c) cristae ampullares, (d) otoliths.

26. Which of the following is *not* a possible cause of conduction deafness? (**a**) impacted cerumen, (**b**) middle ear infection, (**c**) cochlear nerve degeneration, (**d**) otosclerosis.
27. Which of the following are intrinsic eye muscles? (**a**) superior rectus, (**b**) orbicularis oculi, (**c**) smooth muscles of the iris and ciliary body, (**d**) levator palpebrae superioris.
28. Which lies closest to the exact posterior pole of the eye? (**a**) optic nerve, (**b**) optic disc, (**c**) macula lutea, (**d**) point of entry of central artery into the eye.
29. Otoliths (ear stones) are (**a**) a cause of deafness, (**b**) a type of hearing aid, (**c**) important in equilibrium, (**d**) the rock-hard petrous temporal bones.

Short Answer Essay Questions

30. Why do you often have to blow your nose after crying?
31. How do rods and cones differ functionally?
32. Where is the fovea centralis, and why is it important?
33. Describe the response of rhodopsin to light stimuli. What is the outcome of this cascade of events?
34. Since there are only three types of cones, how can you explain the fact that we see many more colors?
35. Where are the olfactory receptors, and why is that site poorly suited for their job?
36. Each olfactory cell responds to a single odorant molecule. True or false? Explain your choice.
37. Name the five primary taste qualities and the cranial nerves that serve the sense of taste.
38. Describe the effect of aging on the special sense organs.

Critical Thinking and Clinical Application Questions

1. During an ophthalmoscopic examination, Mrs. James was found to have bilateral papilledema. Further investigation indicated that this condition resulted from a rapidly growing intracranial tumor. First, define papilledema. Then, explain its presence in terms of Mrs. James's diagnosis.
2. Sally, a 9-year-old girl, told the clinic physician that her "ear lump hurt" and she kept "getting dizzy and falling down." As she told her story, she pointed to her mastoid process. An otoscopic examination of the external acoustic meatus revealed a red, swollen eardrum, and her throat was inflamed. Her condition was described as mastoiditis with secondary labyrinthitis (inflammation of the labyrinth). Describe the most likely route of infection and the infected structures in Sally's case. Also explain the cause of her dizziness and falling.
3. Mr. Gaspe appeared at the eye clinic complaining of a chip of wood in his eye. No foreign body was found, but the conjunctiva was obviously inflamed. What name is given to this inflammatory condition, and where would you look for a foreign body that has been floating around on the eye surface for a while?
4. Mrs. Orlando has been noticing flashes of light and tiny specks in her right visual field. When she begins to see a "veil" floating before her right eye, she makes an appointment to see the eye doctor. What is your diagnosis? Is the condition serious? Explain.
5. David Norris, an engineering student, has been working in a disco to earn money to pay for his education. After about eight months, he notices that he is having problems hearing high-pitched tones. What is the cause-and-effect relationship here?
6. Assume that a tumor in the pituitary gland or hypothalamus is protruding inferiorly and compressing the optic chiasma. What could be the visual outcome?
7. Four-year-old Gary Zammer is brought to the ophthalmologist for a routine checkup on his vision. Gary is an albino. How do you think albinism affects vision?
8. Right before Jan, a senior citizen, had a large plug of earwax cleaned from her ear, she developed a constant howling sound in her ear that would not stop. It was very annoying and stressful, and she had to go to counseling to learn how to live with this awful noise. What was her condition called?
9. During an anatomy and physiology lab exercise designed to teach the use of the ophthalmoscope, you notice that your lab partner has difficulty seeing objects in the darkened room. Upon examining her retina through the ophthalmoscope, you observe streaks and patches of dark pigmentation at the back of the eye. What might this condition be?
10. Henri, a chef in a five-star French restaurant, has been diagnosed with leukemia. He is about to undergo chemotherapy, which will kill rapidly dividing cells in his body. He needs to continue working between bouts of chemotherapy. What consequences of chemotherapy would you predict that might affect his job as a chef?

CHECK YOUR UNDERSTANDING

1. For each of the following statements, indicate whether it applies more to the endocrine system or the nervous system: rapid; discrete responses; controls growth and development; long-lasting responses.
2. Which two endocrine glands are found in the neck?
3. What is the difference between a hormone and a paracrine?

For answers, see Appendix G.

Hormones

- Describe how hormones are classified chemically.
- Describe the two major mechanisms by which hormones bring about their effects on their target tissues.
- List three kinds of interaction of different hormones acting on the same target cell.
- Explain how hormone release is regulated.

The Chemistry of Hormones

16

Hormones are chemical substances, secreted by cells into the extracellular fluids, that regulate the metabolic function of other cells in the body. Although a large variety of hormones are produced, nearly all of them can be classified chemically as either amino acid based or steroids.

Most hormones are **amino acid based**. Molecular size varies widely in this group—from simple amino acid derivatives (which include amines and thyroxine constructed from the amino acid tyrosine), to peptides (short chains of amino acids), to proteins (long polymers of amino acids).

The **steroids** are synthesized from cholesterol. Of the hormones produced by the major endocrine organs, only gonadal and adrenocortical hormones are steroids.

If we also consider the **eicosanoids** (i-ko′să-noyds), which include *leukotrienes* and *prostaglandins*, we must add a third chemical class. These biologically active lipids (made from arachidonic acid) are released by nearly all cell membranes. Leukotrienes are signaling chemicals that mediate inflammation and some allergic reactions. Prostaglandins have multiple targets and effects, ranging from raising blood pressure and increasing the expulsive uterine contractions of birth to enhancing blood clotting, pain, and inflammation.

Because the effects of eicosanoids are typically highly localized, affecting only nearby cells, they generally act as paracrines and autocrines and do not fit the definition of the true *hormones*, which influence distant targets. For this reason, we will not consider this class of hormonelike chemicals here. Instead, we note their important effects in later chapters as appropriate.

Mechanisms of Hormone Action

All major hormones circulate to virtually all tissues, but a given hormone influences the activity of only certain tissue cells, referred to as its **target cells**. Hormones bring about their characteristic effects on target cells by *altering* cell activity. In other words, they increase or decrease the rates of normal cellular processes.

The precise response depends on the target cell type. For example, when the hormone epinephrine binds to certain smooth muscle cells in blood vessel walls, it stimulates them to contract. Epinephrine binding to cells other than muscle cells may have a different effect, but it does not cause those cells to contract.

A hormonal stimulus typically produces one or more of the following changes:

1. Alters plasma membrane permeability or membrane potential, or both, by opening or closing ion channels
2. Stimulates synthesis of proteins or regulatory molecules such as enzymes within the cell
3. Activates or deactivates enzymes
4. Induces secretory activity
5. Stimulates mitosis

How does a hormone communicate with its target cell? In other words, how is hormone receptor binding harnessed to the intracellular machinery needed for hormone action? The answer depends on the chemical nature of the hormone and the cellular location of the receptor, but hormones act at receptors in one of two general ways. On the one hand, *water-soluble hormones* (all amino acid–based hormones except thyroid hormone) act on *receptors in the plasma membrane.* These receptors are coupled via regulatory molecules called G proteins to one or more intracellular second messengers which mediate the target cell's response. On the other hand, *lipid-soluble hormones* (steroid and thyroid hormones) act on *intracellular receptors*, which directly activate genes.

This will be easy for you to remember if you think about *why* the hormones must bind where they do. Receptors for water-soluble hormones must be in the plasma membrane since these hormones *cannot* enter the cell, and receptors for lipid-soluble steroid and thyroid hormones are inside the cell because these hormones *can* enter the cell. Of course, things are not quite that clear-cut. Recent research has shown that steroid hormones exert some of their more immediate effects via plasma membrane receptors, and the second messengers of some water-soluble hormones can turn genes on.

Plasma Membrane Receptors and Second-Messenger Systems

With the exception of thyroid hormone, all amino acid–based hormones exert their signaling effects through intracellular **second messengers** generated when a hormone binds to a receptor on the plasma membrane. You are already familiar with one of these second messengers, **cyclic AMP**, which is used by neurotransmitters (Chapter 11) and olfactory receptors (Chapter 15).

The Cyclic AMP Signaling Mechanism As you recall, this mechanism involves the interaction of three plasma membrane components to determine intracellular levels of cyclic AMP

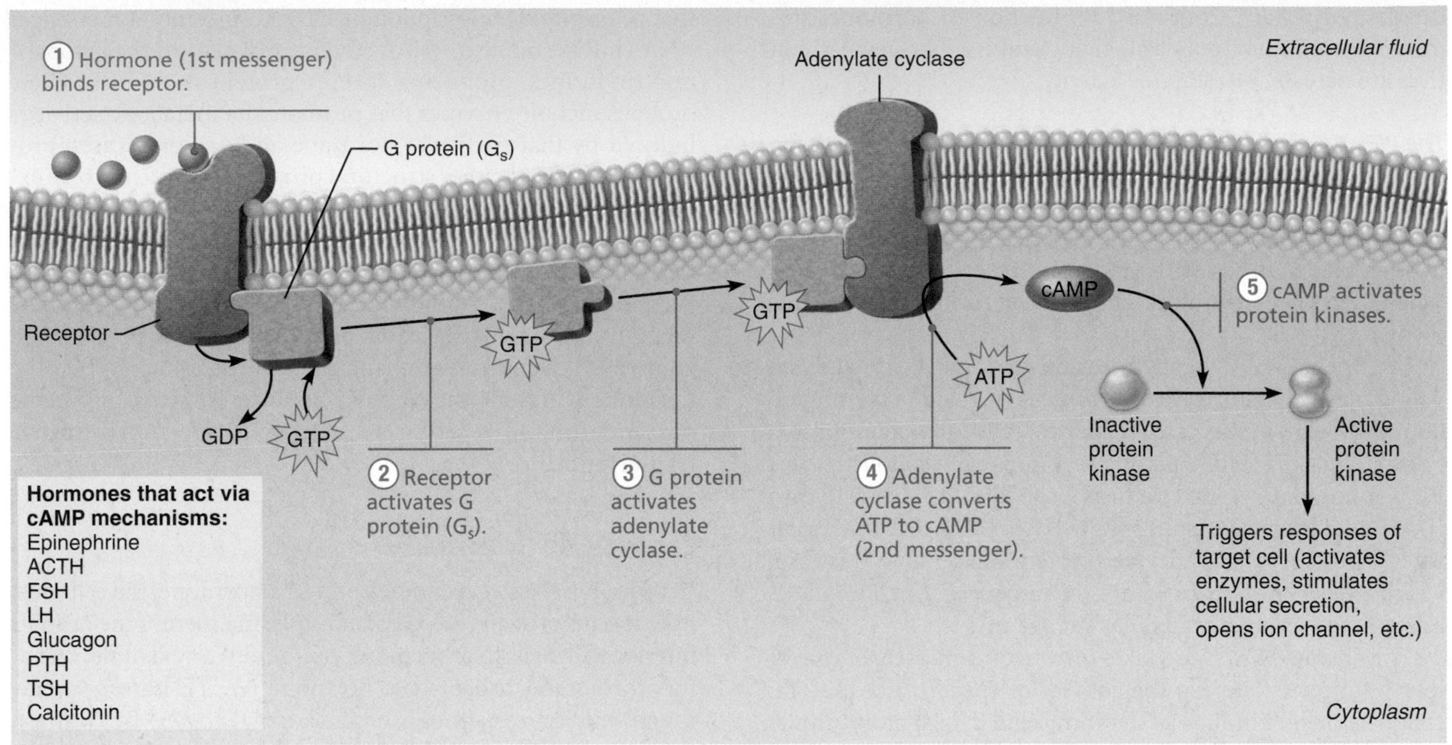

Figure 16.2 Cyclic AMP second-messenger mechanism of water-soluble hormones.

(cAMP)—a hormone receptor, a G protein, and an effector enzyme (adenylate cyclase). **Figure 16.2** illustrates these steps:

① **Hormone binds receptor.** The hormone, acting as the **first messenger**, binds to its receptor on the plasma membrane.

② **Receptor activates G protein.** Hormone binding causes the receptor to change shape, allowing it to bind a nearby inactive **G protein.** The G protein is activated as the guanosine diphosphate (GDP) bound to it is displaced by the high-energy compound *guanosine triphosphate* (*GTP*). The G protein behaves like a light switch: It is "off" when GDP is bound to it, and "on" when GTP is bound.

③ **G protein activates adenylate cyclase.** The activated G protein (moving along the membrane) binds to the effector enzyme **adenylate cyclase**. Some G proteins (G_s) *stimulate* adenylate cyclase (as shown in Figure 16.2), but others (G_i) *inhibit* adenylate cyclase. Eventually, the GTP bound to the G protein is hydrolyzed to GDP and the G protein becomes inactive once again. (The G protein cleaves the terminal phosphate group off GTP in much the same way that ATPase enzymes hydrolyze ATP.)

④ **Adenylate cyclase converts ATP to cyclic AMP.** For as long as activated G_s is bound to it, adenylate cyclase generates the *second messenger* cAMP from ATP.

⑤ **Cyclic AMP activates protein kinases.** cAMP, which is free to diffuse throughout the cell, triggers a cascade of chemical reactions by activating *protein kinase A*. **Protein kinases** are enzymes that *phosphorylate* (add a phosphate group to) various proteins, many of which are other enzymes. Because phosphorylation activates some of these proteins and inhibits others, a variety of processes may be affected in the same target cell at the same time.

This type of intracellular enzymatic cascade has a huge amplification effect. Each activated adenylate cyclase generates large numbers of cAMP molecules, and a single kinase enzyme can catalyze hundreds of reactions. As the reaction cascades through one enzyme intermediate after another, the number of product molecules increases dramatically at each step. Receptor binding of a single hormone molecule can generate millions of final product molecules!

The sequence of reactions set into motion by cAMP depends on the type of target cell, the specific protein kinases it contains, and the substrates within that cell available for phosphorylation by the protein kinase. For example, in thyroid cells, binding of thyroid-stimulating hormone promotes synthesis of the thyroid hormone thyroxine; in liver cells, binding of glucagon activates enzymes that break down glycogen, releasing glucose to the blood. Since some G proteins inhibit rather than activate adenylate cyclase, thereby reducing the cytoplasmic concentration of cAMP, even slight changes in levels of antagonistic hormones can influence a target cell's activity. Examples of hormones that act via cyclic AMP second-messenger systems are listed in Figure 16.2.

The action of cAMP persists only briefly because the molecule is rapidly degraded by the intracellular enzyme **phosphodiesterase.** While at first glance this may appear to be a problem, it is quite the opposite. Because of the amplification effect, most hormones need to be present only briefly to cause the desired results. Continued production of hormones then prompts continued cellular activity, and no extracellular controls are necessary to stop the activity.

The PIP_2-Calcium Signaling Mechanism Cyclic AMP is the activating second messenger in some tissues for at least ten amino acid–based hormones, but some of the same hormones (such as epinephrine) act through a different second-messenger system in other tissues. In one such mechanism, called the PIP_2-calcium signaling mechanism, intracellular calcium ions act as a final mediator.

Like the cAMP signaling mechanism, the PIP_2-calcium signaling mechanism involves a G protein (G_q) and a membrane-bound effector, in this case an enzyme called **phospholipase C**. Phospholipase C splits a plasma membrane phospholipid called **PIP_2** (**p**hosphatidyl **i**nositol bis**p**hosphate) into **diacylglycerol (DAG)** and **inositol trisphosphate (IP_3)**. DAG, like cAMP, activates a protein kinase enzyme (protein kinase C in this case), which triggers responses within the target cell. In addition, IP_3 releases Ca^{2+} from intracellular storage sites.

The liberated Ca^{2+} also takes on a second-messenger role, either by directly altering the activity of specific enzymes and channels or by binding to the intracellular regulatory protein **calmodulin**. Once Ca^{2+} binds to calmodulin, it activates enzymes that amplify the cellular response.

Hormones known to act on their target cells via the PIP_2 mechanism include thyrotropin-releasing hormone (TRH), antidiuretic hormone (ADH), gonadotropin-releasing hormone (GnRH), oxytocin, and epinephrine.

16

Other Signaling Mechanisms Other hormones that bind plasma membrane receptors act on their target cells through different signaling mechanisms. For example, cyclic guanosine monophosphate (cGMP) is a second messenger for selected hormones.

Insulin (and other growth factors) work without second messengers. The insulin receptor is a *tyrosine kinase* enzyme that is activated by autophosphorylation (addition of phosphate to several of its own tyrosines) when insulin binds. The activated insulin receptor provides docking sites for intracellular *relay proteins* that, in turn, initiate a series of protein phosphorylations that trigger specific cell responses.

In certain instances, any of the second messengers mentioned—and the hormone receptor itself—can cause changes in intracellular Ca^{2+} levels.

Intracellular Receptors and Direct Gene Activation

Being lipid soluble, steroid hormones (and, strangely, thyroid hormone, a small iodinated amine) diffuse into their target cells where they bind to and activate an intracellular receptor **(Figure 16.3)**. The activated receptor-hormone complex then makes its way to the nuclear chromatin, where the hormone receptor binds to a region of DNA (a *hormone response element*) specific for it. (The exception to these generalizations is that thyroid hormone receptors are always bound to DNA even in the absence of thyroid hormone.) This interaction "turns on" a gene, that is, prompts transcription of DNA to produce a messenger RNA (mRNA). The mRNA is then translated on the cytoplasmic ribosomes, producing specific protein molecules. These proteins include enzymes that promote the metabolic activities induced by that particular hormone and, in some cases, promote synthesis of either structural proteins or proteins to be exported from the target cell.

In the absence of hormone, the receptors are bound up in receptor-chaperonin complexes. These associations seem to keep the receptors from binding to DNA and perhaps protect them from proteolysis. (Check back to Chapter 2, pp. 50–51, if your recall of molecular chaperones needs a jog.) When the hormone *is* present, the complex dissociates, which allows the hormone-bound receptor to bind to DNA and influence transcription.

Target Cell Specificity

In order for a target cell to respond to a hormone, the cell must have *specific* protein receptors on its plasma membrane or in its interior to which that hormone can bind. For example, receptors for adrenocorticotropic hormone (ACTH) are normally found only on certain cells of the adrenal cortex. By contrast, thyroxine is the principal hormone stimulating cellular metabolism, and nearly all body cells have thyroxine receptors.

A hormone receptor responds to hormone binding by prompting the cell to perform, or turn on, some gene-determined "preprogrammed" function. As such, hormones are molecular triggers rather than informational molecules. Although binding of a hormone to a receptor is the crucial first step, target cell activation by hormone-receptor interaction depends equally on three factors: (1) blood levels of the hormone, (2) relative numbers of receptors for that hormone on or in the target cells, and (3) *affinity* (strength) of the binding between the hormone and the receptor. All three factors change rapidly in response to various stimuli and changes within the body. As a rule, for a given level of hormone in the blood, a large number of high-affinity receptors produces a pronounced hormonal effect, and a smaller number of low-affinity receptors results in reduced target cell response or outright endocrine dysfunction.

Receptors are dynamic structures. In some instances, target cells form more receptors in response to rising blood levels of the specific hormones to which they respond, a phenomenon called **up-regulation**. In other cases, prolonged exposure to high hormone concentrations desensitizes the target cells, so that they respond less vigorously to hormonal stimulation. This **down-regulation** involves loss of receptors and prevents the target cells from overreacting to persistently high hormone levels.

Hormones influence the number and affinity not only of their own receptors but also of receptors that respond to other hormones. For example, progesterone induces a loss of estrogen receptors in the uterus, thus antagonizing estrogen's actions. On the other hand, estrogen causes the same cells to produce more progesterone receptors, enhancing their ability to respond to progesterone.

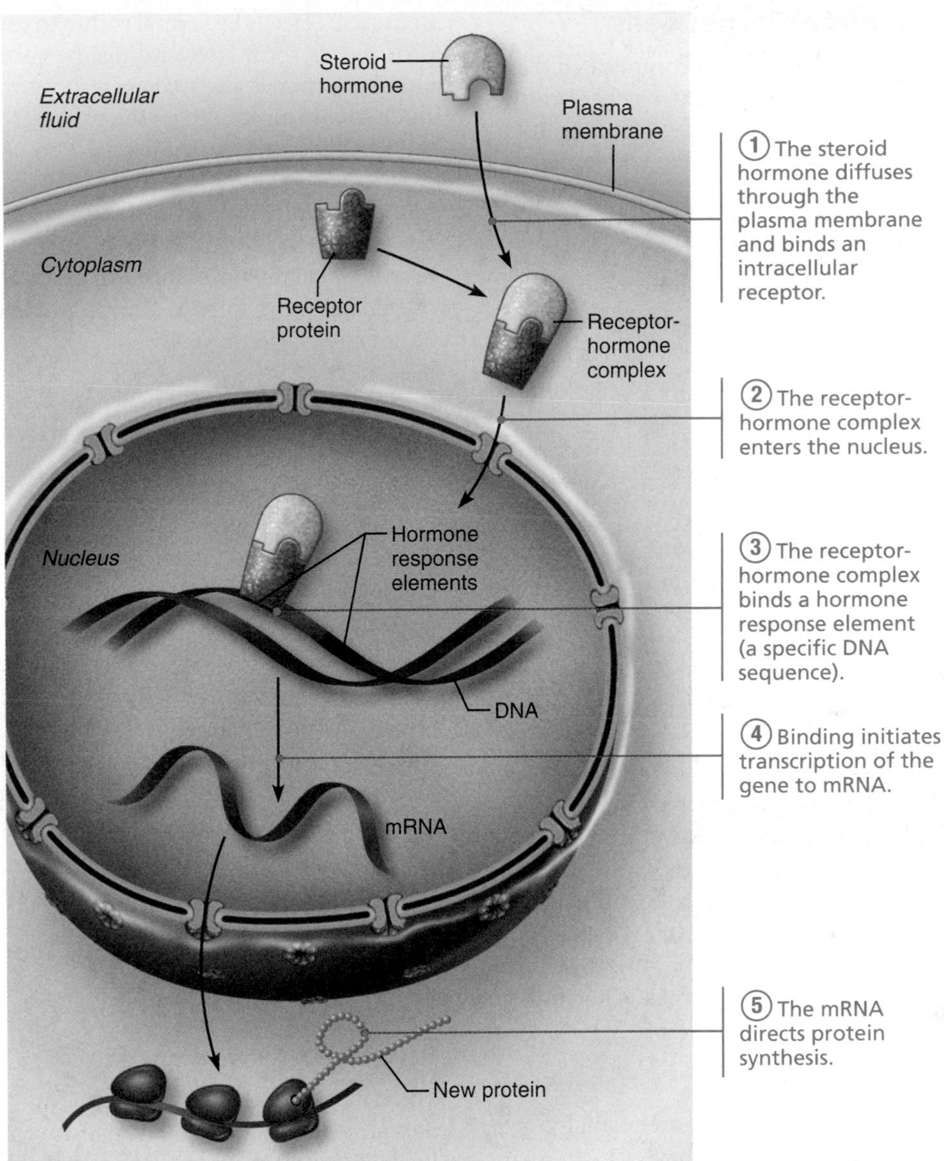
Steroid hormone
Extracellular fluid
Plasma membrane
Cytoplasm
Receptor protein
Receptor-hormone complex
Nucleus
Hormone response elements
DNA
mRNA
New protein
1 The steroid hormone diffuses through the plasma membrane and binds an intracellular receptor.
2 The receptor-hormone complex enters the nucleus.
3 The receptor-hormone complex binds a hormone response element (a specific DNA sequence).
4 Binding initiates transcription of the gene to mRNA.
5 The mRNA directs protein synthesis.

Figure 16.3 Direct gene activation mechanism of lipid-soluble hormones. Steroid hormone is illustrated. Thyroid hormone binds its receptor in the nucleus.

Half-Life, Onset, and Duration of Hormone Activity

Hormones are potent chemicals, and they exert profound effects on their target organs at very low concentrations. Hormones circulate in the blood in two forms—free or bound to a protein carrier. In general, lipid-soluble hormones (steroids and thyroid hormone) travel in the bloodstream attached to plasma proteins. Most others circulate without carriers.

The concentration of a circulating hormone in blood at any time reflects (1) its rate of release, and (2) the speed at which it is inactivated and removed from the body. Some hormones are rapidly degraded by enzymes in their target cells, but most are removed from the blood by the kidneys or liver, and their breakdown products are excreted from the body in urine or, to a lesser extent, in feces. As a result, the length of time for a hormone's blood level to decrease by half, referred to as its **half-life**, varies from a fraction of a minute to a week. The water-soluble hormones exhibit the shortest half-lives.

How long does it take for a hormone to have an effect? The time required for hormone effects to appear varies greatly. Some hormones provoke target organ responses almost immediately, while others, particularly the steroid hormones, require hours to days before their effects are seen. Additionally, some hormones are secreted in a relatively inactive form and must be activated in the target cells.

The duration of hormone action is limited, ranging from 10 seconds to several hours, depending on the hormone. Effects may disappear rapidly as blood levels drop, or they may persist for hours after very low hormone levels have been reached. Because of these many variations, hormonal blood levels must be precisely and individually controlled to meet the continuously changing needs of the body.

16

Interaction of Hormones at Target Cells

Understanding hormonal effects is a bit more complicated than you might expect because multiple hormones may act on the same target cells at the same time. In many cases the result of such an interaction is not predictable, even when you know the effects of the individual hormones. Here we will look at three types of hormone interaction—permissiveness, synergism, and antagonism.

Permissiveness is the situation when one hormone cannot exert its full effects without another hormone being present. For example, the development of the reproductive system is largely regulated by reproductive system hormones, as we might expect. However, thyroid hormone is necessary (has a permissive effect) for normal *timely* development of reproductive structures. Without thyroid hormone, reproductive system development is delayed.

Synergism of hormones occurs in situations where more than one hormone produces the same effects at the target cell and their combined effects are amplified. For example, both glucagon (produced by the pancreas) and epinephrine cause the liver to release glucose to the blood. When they act together, the amount of glucose released is about 150% of what is released when each hormone acts alone.

When one hormone opposes the action of another hormone, the interaction is called **antagonism**. For example, insulin, which lowers blood glucose levels, is antagonized by glucagon, which acts to raise blood glucose levels. How does antagonism occur? Antagonists may compete for the same receptors, act through different metabolic pathways, or even, as noted in the progesterone-estrogen interaction at the uterus, cause down-regulation of the receptors for the antagonistic hormone.

Control of Hormone Release

The synthesis and release of most hormones are regulated by some type of **negative feedback system** (see Chapter 1). In such a system, some internal or external stimulus triggers hormone secretion. As hormone levels rise, they cause target organ effects, which then inhibit further hormone release. As a result, blood levels of many hormones vary only within a narrow range.

Endocrine Gland Stimuli

Three major types of stimuli trigger endocrine glands to manufacture and release their hormones: *humoral* (hu′mer-ul), *neural*, and *hormonal* stimuli.

Humoral Stimuli Some endocrine glands secrete their hormones in direct response to changing blood levels of certain critical ions and nutrients. These stimuli are called *humoral stimuli* to distinguish them from hormonal stimuli, which are also bloodborne chemicals. The term *humoral* harks back to the ancient use of the term *humor* to refer to various body fluids (blood, bile, and others).

Humoral stimuli are the simplest endocrine controls. For example, cells of the parathyroid glands monitor the body's crucial blood Ca^{2+} levels. When they detect a decline from normal values, they secrete parathyroid hormone (PTH) **(Figure 16.4a)**. Because PTH acts by several routes to reverse that decline, blood Ca^{2+} levels soon rise, ending the initiative for PTH release. Other hormones released in response to humoral stimuli include insulin, produced by the pancreas, and aldosterone, one of the adrenal cortex hormones.

Neural Stimuli In a few cases, nerve fibers stimulate hormone release. The classic example of neural stimuli is sympathetic nervous system stimulation of the adrenal medulla to release catecholamines (norepinephrine and epinephrine) during periods of stress (Figure 16.4b).

Hormonal Stimuli Finally, many endocrine glands release their hormones in response to hormones produced by other endocrine organs, and the stimuli in these cases are called hormonal stimuli. For example, release of most anterior pituitary hormones is regulated by releasing and inhibiting hormones produced by the hypothalamus, and many anterior pituitary hormones in turn stimulate other endocrine organs to release their hormones (Figure 16.4c). As blood levels of the hormones produced by the final target glands increase, they inhibit the release of anterior pituitary hormones and thus their own release.

This hypothalamic–pituitary–target endocrine organ feedback loop lies at the very core of endocrinology, and it will come up many times in this chapter. Hormonal stimuli promote rhythmic hormone release, with hormone blood levels rising and falling in a specific pattern.

Although these three types of stimuli are typical of most systems that control hormone release, they are by no means all-inclusive or mutually exclusive, and some endocrine organs respond to multiple stimuli.

Nervous System Modulation

The nervous system can modify both "turn-on" factors (hormonal, humoral, and neural stimuli) and "turn-off" factors (feedback inhibition and others) that affect the endocrine system. Without this added safeguard, endocrine system activity would be strictly mechanical, much like a household thermostat. A thermostat can maintain the temperature at or around its set value, but it cannot sense that your grandmother visiting from Florida feels cold at that temperature and reset itself accordingly. *You* must make that adjustment. In your body, it is the nervous system that makes certain adjustments to maintain homeostasis by overriding normal endocrine controls.

For example, the action of insulin and several other hormones normally keeps blood glucose levels in the range of 90–110 mg/100 ml of blood. However, when your body is under severe stress, blood glucose levels rise because the hypothalamus and sympathetic nervous system centers are strongly activated. In this way, the nervous system ensures that body cells have sufficient fuel for the more vigorous activity required during periods of stress.

CHECK YOUR UNDERSTANDING

4. Name the two major chemical classes of hormones. Which class consists entirely of lipid-soluble hormones? Name the only hormone in the other chemical class that is lipid soluble.

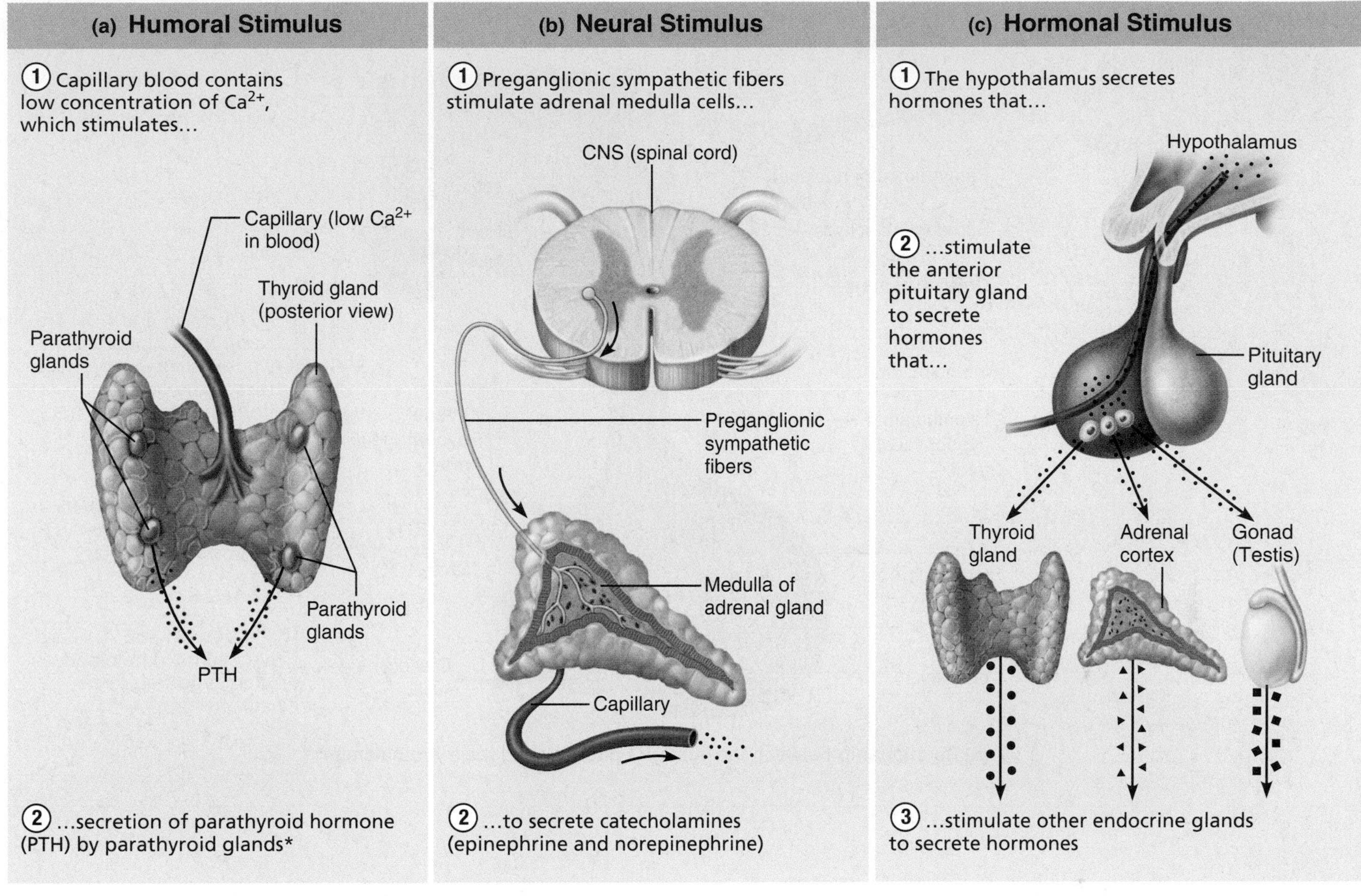
(a) Humoral Stimulus
1 Capillary blood contains low concentration of Ca^{2+}, which stimulates...
Capillary (low Ca^{2+} in blood)
Thyroid gland (posterior view)
Parathyroid glands
Parathyroid glands
PTH
2 ...secretion of parathyroid hormone (PTH) by parathyroid glands*
(b) Neural Stimulus
1 Preganglionic sympathetic fibers stimulate adrenal medulla cells...
CNS (spinal cord)
Preganglionic sympathetic fibers
Medulla of adrenal gland
Capillary
2 ...to secrete catecholamines (epinephrine and norepinephrine)
(c) Hormonal Stimulus
1 The hypothalamus secretes hormones that...
Hypothalamus
2 ...stimulate the anterior pituitary gland to secrete hormones that...
Pituitary gland
Thyroid gland
Adrenal cortex
Gonad (Testis)
3 ...stimulate other endocrine glands to secrete hormones

Figure 16.4 Three types of endocrine gland stimuli.
*PTH increases blood Ca^{2+} (see Figure 16.12).

5. Consider the signaling mechanisms of water-soluble and lipid-soluble hormones. In each case, where are the receptors found and what is the final outcome?
6. What are the three types of stimuli that control hormone release?

For answers, see Appendix G.

The Pituitary Gland and Hypothalamus

- Describe structural and functional relationships between the hypothalamus and the pituitary gland.
- List and describe the chief effects of anterior pituitary hormones.
- Discuss the structure of the posterior pituitary, and describe the effects of the two hormones it releases.

Securely seated in the sella turcica of the sphenoid bone, the tiny **pituitary gland**, or **hypophysis** (hi-pof′ĭ-sis; "to grow under"), secretes at least nine hormones. Usually said to be the size and shape of a pea, this gland is more accurately described as a pea on a stalk. Its stalk, the funnel-shaped **infundibulum**, connects the gland to the hypothalamus superiorly **(Figure 16.5)**.

In humans, the pituitary gland has two major lobes. One lobe is neural tissue and the other is glandular. The **posterior pituitary** (lobe) is composed largely of pituicytes (glia-like supporting cells) and nerve fibers (Figure 16.5). It releases **neurohormones** (hormones secreted by neurons) received ready-made from the hypothalamus. Consequently, this lobe is a hormone-storage area and not a true endocrine gland in the precise sense. The posterior lobe plus the infundibulum make up the region called the **neurohypophysis** (nu″ro-hi-pof′ĭ-sis), a term commonly used (incorrectly) to indicate the posterior lobe alone.

The **anterior pituitary** (lobe), or **adenohypophysis** (ad″ĕ-no-hi-pof′ĭ-sis), is composed of glandular tissue (*adeno* = gland). It manufactures and releases a number of hormones (**Table 16.1** on pp. 606–607).

Arterial blood is delivered to the pituitary via hypophyseal branches of the internal carotid arteries. The veins leaving the pituitary drain into the dural sinuses.

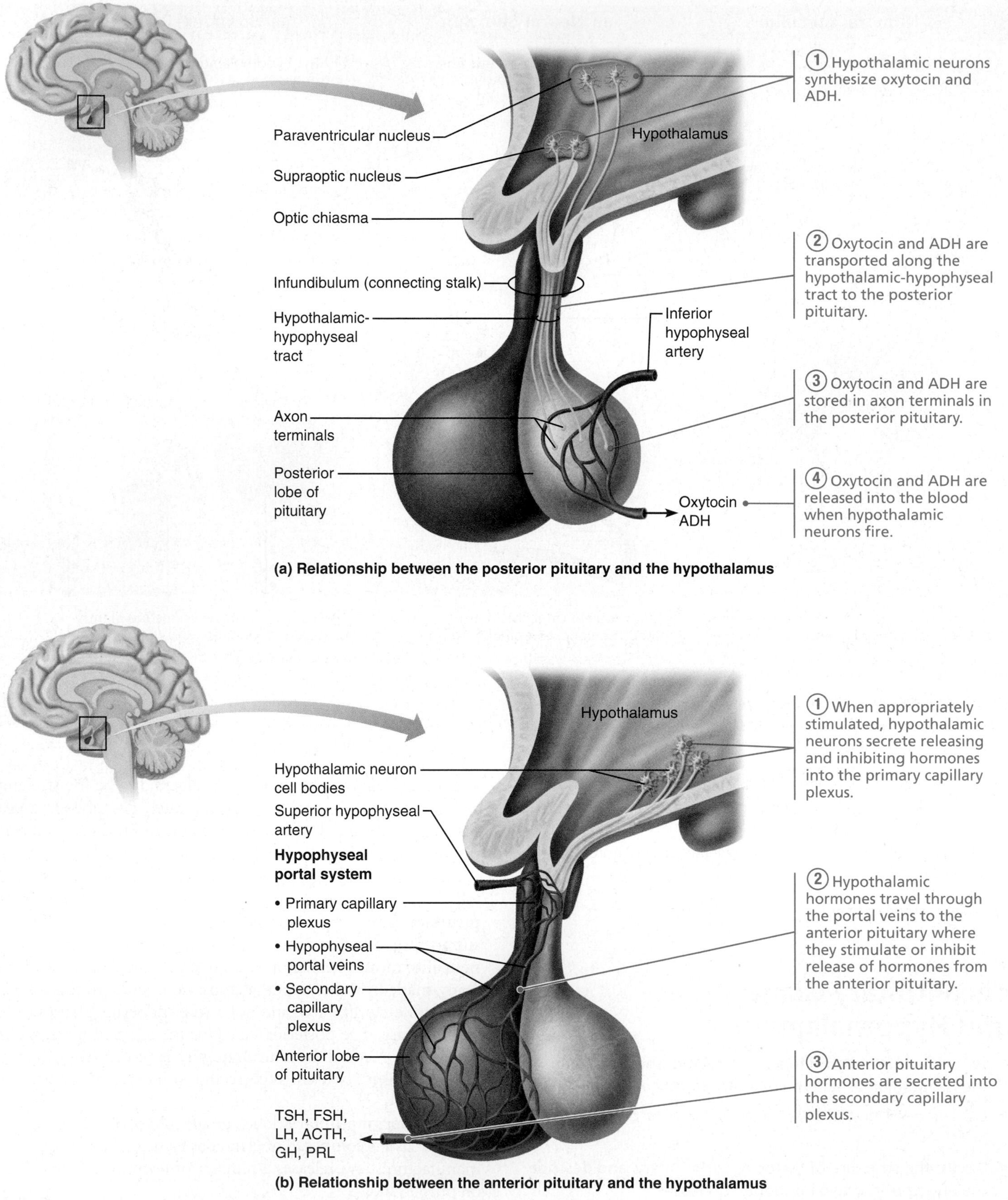

Paraventricular nucleus
Supraoptic nucleus
Optic chiasma
Infundibulum (connecting stalk)
Hypothalamic-hypophyseal tract
Axon terminals
Posterior lobe of pituitary
Hypothalamus
Inferior hypophyseal artery
Oxytocin
ADH

(1) Hypothalamic neurons synthesize oxytocin and ADH.

(2) Oxytocin and ADH are transported along the hypothalamic-hypophyseal tract to the posterior pituitary.

(3) Oxytocin and ADH are stored in axon terminals in the posterior pituitary.

(4) Oxytocin and ADH are released into the blood when hypothalamic neurons fire.

(a) Relationship between the posterior pituitary and the hypothalamus

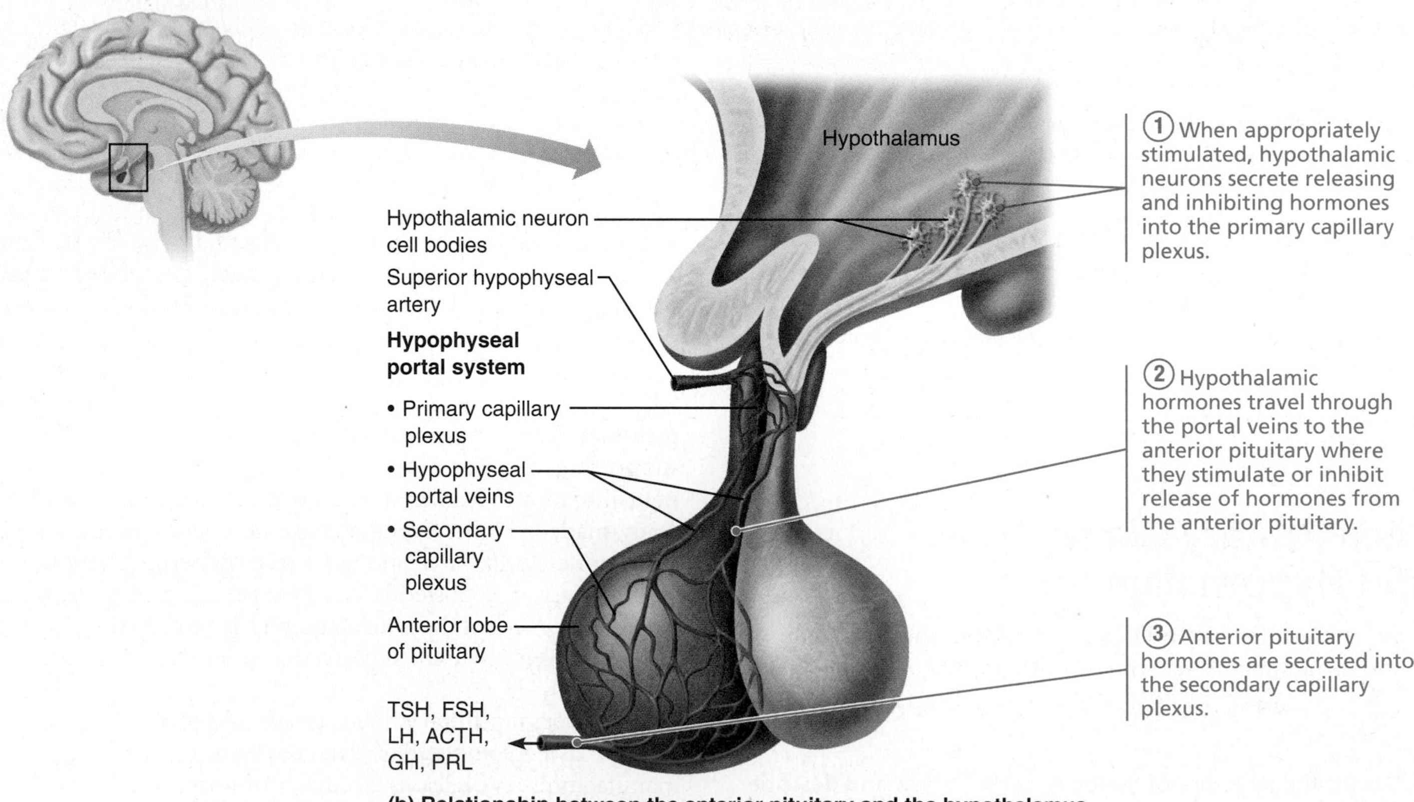

Hypothalamus
Hypothalamic neuron cell bodies
Superior hypophyseal artery
Hypophyseal portal system
• Primary capillary plexus
• Hypophyseal portal veins
• Secondary capillary plexus
Anterior lobe of pituitary
TSH, FSH, LH, ACTH, GH, PRL
(1) When appropriately stimulated, hypothalamic neurons secrete releasing and inhibiting hormones into the primary capillary plexus.
(2) Hypothalamic hormones travel through the portal veins to the anterior pituitary where they stimulate or inhibit release of hormones from the anterior pituitary.
(3) Anterior pituitary hormones are secreted into the secondary capillary plexus.

(b) Relationship between the anterior pituitary and the hypothalamus

Figure 16.5 Relationships of the pituitary gland and hypothalamus.

Pituitary-Hypothalamic Relationships

The contrasting histology of the two pituitary lobes reflects the dual origin of this tiny gland. The posterior lobe is actually part of the brain. It derives from a downgrowth of hypothalamic tissue and maintains its neural connection with the hypothalamus via a nerve bundle called the **hypothalamic-hypophyseal tract**, which runs through the infundibulum (Figure 16.5). This tract arises from neurons in the **supraoptic** and **paraventricular nuclei** of the hypothalamus. These neurosecretory cells synthesize two neurohormones and transport them along their axons to the posterior pituitary. Oxytocin (ok″sĭ-to′sin) is made primarily by the paraventricular neurons, and antidiuretic hormone (ADH) primarily by the supraoptic neurons. When these hypothalamic neurons fire, they release the stored hormones into a capillary bed in the posterior pituitary for distribution throughout the body.

The glandular anterior lobe originates from a superior outpocketing of the oral mucosa (*Rathke's pouch*) and is formed from epithelial tissue. After touching the posterior lobe, the anterior lobe loses its connection with the oral mucosa and adheres to the neurohypophysis. There is no direct neural connection between the anterior lobe and hypothalamus, but there is a vascular connection. Specifically, the **primary capillary plexus** in the infundibulum communicates inferiorly via the small **hypophyseal portal veins** with a **secondary capillary plexus** in the anterior lobe. The primary and secondary capillary plexuses and the intervening hypophyseal portal veins make up the **hypophyseal portal system** (Figure 16.5). Note that a *portal system* is an unusual arrangement of blood vessels in which a capillary bed feeds into veins, which in turn feed into another capillary bed. Via this portal system, **releasing** and **inhibiting hormones** secreted by neurons in the ventral hypothalamus circulate to the anterior pituitary, where they regulate secretion of its hormones. All these hypothalamic regulatory hormones are amino acid based, but they vary in size from a single amine to peptides to proteins.

Anterior Pituitary Hormones

The anterior pituitary has traditionally been called the "master endocrine gland" because many of the numerous hormones it produces regulate the activity of other endocrine glands. In recent years, however, it has been dethroned by the hypothalamus, which is now known to control the activity of the anterior pituitary.

Researchers have identified six distinct anterior pituitary hormones, all of them proteins (Table 16.1). In addition, a large molecule with the tongue-twisting name **pro-opiomelanocortin (POMC)** (pro″o″pe-o-mah-lan′o-kor″tin) has been isolated from the anterior pituitary. POMC is a *prohormone*, that is, a large precursor molecule that can be split enzymatically into one or more active hormones. POMC is the source of adrenocorticotropic hormone, two natural opiates (an enkephalin and a beta endorphin, described in Chapter 11), and *melanocyte-stimulating hormone* (*MSH*). In amphibians, reptiles, and other animals MSH stimulates melanocytes to increase synthesis of melanin pigment, but this is not an important function of MSH in humans. In humans and other mammals, MSH is a CNS neurotransmitter involved in the control of appetite. Although low levels of MSH are found in plasma, its role in the periphery is not yet well understood.

When the anterior pituitary receives an appropriate chemical stimulus from the hypothalamus, one or more of its hormones are released by certain of its cells. Although many different hormones pass from the hypothalamus to the anterior lobe, each target cell in the anterior lobe distinguishes the messages directed to it and responds in kind—secreting the proper hormone in response to specific releasing hormones, and shutting off hormone release in response to specific inhibiting hormones.

Four of the six anterior pituitary hormones—thyroid-stimulating hormone, adrenocorticotropic hormone, follicle-stimulating hormone, and luteinizing hormone—are **tropins** or **tropic hormones** (*tropi* = turn on, change), which are hormones that regulate the secretory action of other endocrine glands. All anterior pituitary hormones except for growth hormone affect their target cells via a cyclic AMP second-messenger system.

Growth Hormone

Growth hormone (**GH**; *somatotropin*) is produced by cells called **somatotrophs** of the anterior lobe and has both growth-promoting and metabolic actions, summarized in **Figure 16.6**. Although GH stimulates most body cells to increase in size and divide, its major targets are the bones and skeletal muscles. Stimulation of the epiphyseal plate leads to long bone growth, and stimulation of skeletal muscles promotes increased muscle mass.

Essentially an anabolic (tissue-building) hormone, GH promotes protein synthesis, and it encourages the use of fats for fuel, thus conserving glucose. Most growth-promoting effects of GH are mediated indirectly by **insulin-like growth factors (IGFs)**, a family of growth-promoting proteins produced by the liver, skeletal muscle, bone, and other tissues. IGFs produced by the liver act as hormones, while IGFs produced in other tissues act locally within those tissues (as paracrines). Specifically, IGFs stimulate actions required for growth: (1) uptake of nutrients from the blood and their incorporation into proteins and DNA, allowing growth by cell division; and (2) formation of collagen and deposition of bone matrix.

Acting directly, GH exerts metabolic effects. It mobilizes fats from fat depots for transport to cells, increasing blood levels of fatty acids. It also decreases the rate of glucose uptake and metabolism. In the liver, it encourages glycogen breakdown and release of glucose to the blood. The elevation in blood glucose levels that occurs as a result of this *glucose sparing* is called the *anti-insulin effect* of GH, because its effects are opposite to those of insulin.

Secretion of GH is regulated chiefly by two hypothalamic hormones with antagonistic effects. **Growth hormone–releasing hormone (GHRH)** stimulates GH release, while **growth hormone–inhibiting hormone (GHIH)**, also called **somatostatin** (so″mah-to-stat′in), inhibits it. GHIH release is (presumably) triggered by the feedback of GH and IGFs. Rising levels of GH also feed back to inhibit its own release. As indicated in Table 16.1, a number of secondary triggers also influence GH release. Typically, GH secretion has a daily cycle, with the highest

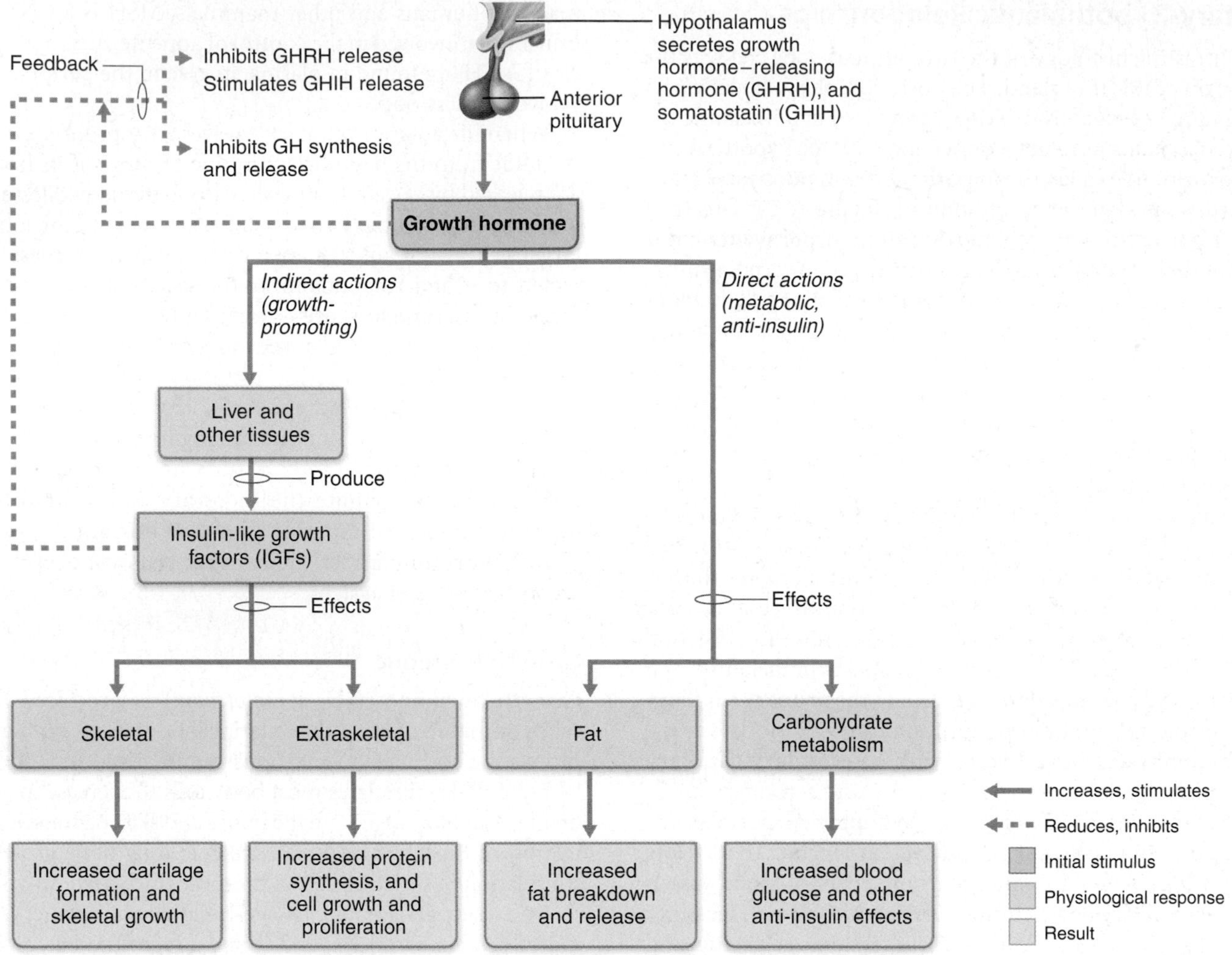

16

Figure 16.6 Growth-promoting and metabolic actions of growth hormone (GH).

levels occurring during evening sleep. The total amount secreted daily peaks during adolescence and then declines with age.

Besides inhibiting growth hormone secretion, GHIH blocks the release of thyroid-stimulating hormone. GHIH is also produced in various locations in the gut, where it inhibits the release of virtually all gastrointestinal and pancreatic secretions—both endocrine and exocrine.

HOMEOSTATIC IMBALANCE

Both hypersecretion and hyposecretion of GH may result in structural abnormalities. Hypersecretion in children results in **gigantism** because GH targets the still-active epiphyseal (growth) plates. The person becomes abnormally tall, often reaching a height of 2.4 m (8 feet), but has relatively normal body proportions. If excessive amounts of GH are secreted after the epiphyseal plates have closed, **acromegaly** (ak″ro-meg′ah-le) results. Literally translated as "enlarged extremities," this condition is characterized by overgrowth of bony areas still responsive to GH, namely bones of the hands, feet, and face. Hypersecretion usually results from an anterior pituitary tumor that churns out excessive amounts of GH. The usual treatment is surgical removal of the tumor, but this surgery does not reverse anatomical changes that have already occurred.

Hyposecretion of GH in adults usually causes no problems. GH deficiency in children results in slowed long bone growth, a condition called **pituitary dwarfism**. Such individuals attain a maximum height of 1.2 m (4 feet), but usually have fairly normal body proportions. Lack of GH is often accompanied by deficiencies of other anterior pituitary hormones, and if thyroid-stimulating hormone and gonadotropins are lacking, the individual will be malproportioned and will fail to mature sexually as well. Fortunately, human GH is produced commercially by genetic engineering techniques. When pituitary dwarfism is diagnosed before puberty, growth hormone replacement therapy can promote nearly normal somatic growth.

The availability of synthetic GH also has a downside. Athletes and the elderly have been tempted to use GH for its bodybuilding properties, and some parents have used GH in an attempt to give their children additional height. However, while muscle mass increases, there is no objective evidence for an increase in muscle strength in either athletes or the elderly, and only minimal increases in stature occur in normal children. Moreover, taking GH can lead to fluid retention, joint and muscle pain, diabetes, and may promote cancer. ■

Thyroid-Stimulating Hormone

Thyroid-stimulating hormone (TSH), or **thyrotropin**, is a tropic hormone that stimulates normal development and secretory activity of the thyroid gland. Its release follows the hypothalamic–pituitary–target endocrine organ feedback loop described earlier and shown in Figure 16.7. TSH release from cells of the anterior pituitary called **thyrotrophs** is triggered by the hypothalamic peptide **thyrotropin-releasing hormone (TRH)**. Rising blood levels of thyroid hormones act on both the pituitary and the hypothalamus to inhibit TSH secretion.

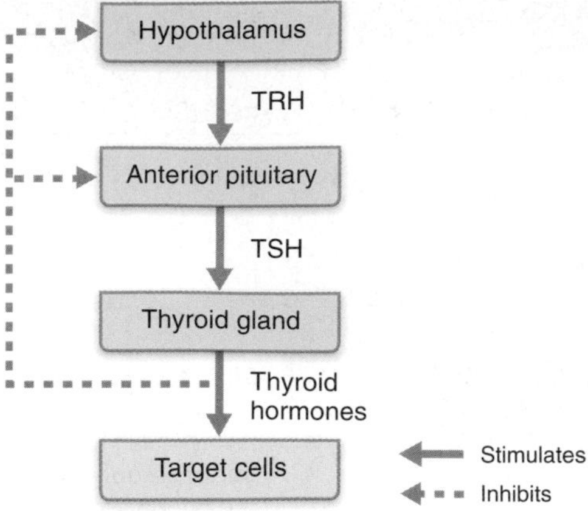

Figure 16.7 Regulation of thyroid hormone secretion. TRH = thyrotropin-releasing hormone, TSH = thyroid-stimulating hormone.

Adrenocorticotropic Hormone

Adrenocorticotropic hormone (ACTH) (ah-dre″no-kor″tĭ-ko-trōp′ik), or **corticotropin**, is secreted by the **corticotrophs** of the anterior pituitary. ACTH stimulates the adrenal cortex to release corticosteroid hormones, most importantly glucocorticoids that help the body to resist stressors. ACTH release, elicited by hypothalamic **corticotropin-releasing hormone (CRH)**, has a daily rhythm, with levels peaking in the morning, shortly before awakening. Rising levels of glucocorticoids feed back and block secretion of CRH and ACTH release. Internal and external factors that alter the normal ACTH rhythm by triggering CRH release include fever, hypoglycemia, and stressors of all types.

Gonadotropins

Follicle-stimulating hormone (FSH) and **luteinizing hormone (LH)** (lu′te-in-īz″ing) are referred to collectively as **gonadotropins**. They regulate the function of the gonads (ovaries and testes). In both sexes, FSH stimulates gamete (sperm or egg) production and LH promotes production of gonadal hormones. In females, LH works with FSH to cause an egg-containing ovarian follicle to mature. LH then independently triggers ovulation and promotes synthesis and release of ovarian hormones. In males, LH stimulates the interstitial cells of the testes to produce the male hormone testosterone.

Gonadotropins are virtually absent from the blood of prepubertal boys and girls. During puberty, the **gonadotrophs** of the anterior pituitary are activated and gonadotropin levels begin to rise, causing the gonads to mature. In both sexes, gonadotropin release by the anterior pituitary is prompted by **gonadotropin-releasing hormone (GnRH)** produced by the hypothalamus. Gonadal hormones, produced in response to the gonadotropins, feed back to suppress FSH and LH release.

Prolactin

Prolactin (PRL) is a protein hormone structurally similar to GH. Produced by the **lactotrophs**, PRL stimulates the gonads of some animals (other than humans) and is considered a gonadotropin by some researchers. Its only well-documented effect in humans is to stimulate milk production by the breasts (*pro* = for; *lact* = milk). The role of prolactin in males is not well understood.

Unlike the other anterior pituitary hormones, PRL release is controlled primarily by an inhibitory hormone, **prolactin-inhibiting hormone (PIH)**, now known to be *dopamine* (*DA*), which prevents prolactin secretion. Decreased PIH secretion leads to a surge in PRL release. There are a number of *prolactin-releasing factors*, including TRH, but their exact roles are not well understood.

In females, prolactin levels rise and fall in rhythm with estrogen blood levels. Estrogen stimulates prolactin release, both directly and indirectly. A brief rise in prolactin levels just before the menstrual period partially accounts for the breast swelling and tenderness some women experience at that time, but because this PRL stimulation is so brief, the breasts do not produce milk. In pregnant women, PRL blood levels rise dramatically toward the end of pregnancy, and milk production becomes possible. After birth, the infant's suckling stimulates release of prolactin-releasing factors in the mother, encouraging continued milk production and availability.

HOMEOSTATIC IMBALANCE

Hypersecretion of prolactin is more common than hyposecretion (which is not a problem in anyone except women who choose to nurse). In fact, hyperprolactinemia is the most frequent abnormality of anterior pituitary tumors. Clinical signs include inappropriate lactation, lack of menses, infertility in females, and impotence in males. ■

The Posterior Pituitary and Hypothalamic Hormones

The posterior pituitary, made largely of the axons of hypothalamic neurons, stores antidiuretic hormone (ADH) and oxytocin that have been synthesized and forwarded by hypothalamic neurons of the supraoptic and paraventricular nuclei. These hormones are later released "on demand" in response to nerve impulses from the same hypothalamic neurons.

ADH and oxytocin, each composed of nine amino acids, are almost identical. They differ in only two amino acids, and yet they have dramatically different physiological effects, summarized in Table 16.1. ADH influences body water balance, and

TABLE 16.1 Pituitary Hormones: Summary of Regulation and Effects

HORMONE (CHEMICAL STRUCTURE AND CELL TYPE)	REGULATION OF RELEASE	TARGET ORGAN AND EFFECTS	EFFECTS OF HYPOSECRETION ↓ AND HYPERSECRETION ↑
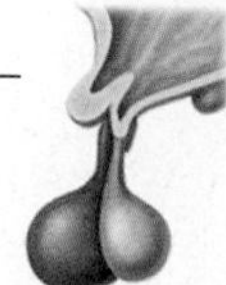 **Anterior Pituitary Hormones**			
Growth hormone (GH) (Protein, somatotroph)	**Stimulated** by GHRH* release, which is triggered by low blood levels of GH as well as by a number of secondary triggers including hypoglycemia, increases in blood levels of amino acids, low levels of fatty acids, exercise, other types of stressors, and estrogens **Inhibited** by feedback inhibition exerted by GH and IGFs, and by hyperglycemia, hyperlipidemia, obesity, and emotional deprivation via either increased GHIH* (somatostatin) or decreased GHRH* release	  Liver, muscle, bone, cartilage, and other tissues: anabolic hormone; stimulates somatic growth; mobilizes fats; spares glucose Growth-promoting effects mediated indirectly by IGFs	↓ Pituitary dwarfism in children ↑ Gigantism in children; acromegaly in adults
Thyroid-stimulating hormone (TSH) (Glycoprotein, thyrotroph)	**Stimulated** by TRH* and indirectly by pregnancy and (in infants) cold temperature **Inhibited** by feedback inhibition exerted by thyroid hormones on anterior pituitary and hypothalamus and by GHIH*	 Thyroid gland: stimulates thyroid gland to release thyroid hormones	↓ Cretinism in children; myxedema in adults ↑ Hyperthyroidism; effects similar to those of Graves' disease, in which antibodies mimic TSH
Adrenocorticotropic hormone (ACTH) (Polypeptide of 39 amino acids, corticotroph)	**Stimulated** by CRH;* stimuli that increase CRH release include fever, hypoglycemia, and other stressors **Inhibited** by feedback inhibition exerted by glucocorticoids	 Adrenal cortex: promotes release of glucocorticoids and androgens (mineralocorticoids to a lesser extent)	↓ Rare ↑ Cushing's disease
Follicle-stimulating hormone (FSH) (Glycoprotein, gonadotroph)	**Stimulated** by GnRH* **Inhibited** by feedback inhibition exerted by inhibin, and estrogen in females and testosterone in males	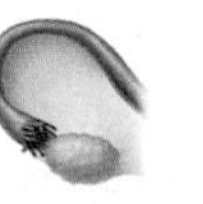 Ovaries and testes: in females, stimulates ovarian follicle maturation and estrogen production; in males, stimulates sperm production	↓ Failure of sexual maturation ↑ No important effects

oxytocin stimulates contraction of smooth muscle, particularly that of the uterus and breasts.

Oxytocin

A strong stimulant of uterine contraction, **oxytocin** is released in significantly higher amounts during childbirth (*oxy* = rapid; *tocia* = childbirth) and in nursing women. The number of oxytocin receptors in the uterus peaks near the end of pregnancy, and uterine smooth muscle becomes more and more sensitive to the hormone's stimulatory effects. Stretching of the uterus and cervix as birth nears dispatches afferent impulses to the hypothalamus, which responds by synthesizing oxytocin and triggering its release from the posterior pituitary. Oxytocin acts via the PIP_2-Ca^{2+} second-messenger system to mobilize Ca^{2+}, allowing stronger contractions. As blood levels of oxytocin rise, the expulsive contractions of labor gain momentum and finally end in birth.

TABLE 16.1 *(continued)*

HORMONE (CHEMICAL STRUCTURE AND CELL TYPE)	REGULATION OF RELEASE	TARGET ORGAN AND EFFECTS	EFFECTS OF HYPOSECRETION ↓ AND HYPERSECRETION ↑
Luteinizing hormone (LH) (Glycoprotein, gonadotroph)	**Stimulated** by GnRH* **Inhibited** by feedback inhibition exerted by estrogen and progesterone in females and testosterone in males	Ovaries and testes: in females, triggers ovulation and stimulates ovarian production of estrogen and progesterone; in males, promotes testosterone production	As for FSH
Prolactin (PRL) (Protein, lactotroph)	**Stimulated** by decreased PIH*; release enhanced by estrogens, birth control pills, breast-feeding, and dopamine-blocking drugs **Inhibited** by PIH* (dopamine)	Breast secretory tissue: promotes lactation	↓ Poor milk production in nursing women ↑ Inappropriate milk production (galactorrhea); cessation of menses in females; impotence in males
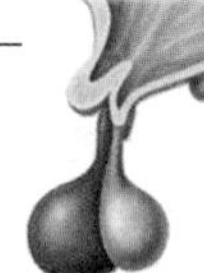 **Posterior Pituitary Hormones (Made by Hypothalamic Neurons and Stored in Posterior Pituitary)**			
Oxytocin (Peptide mostly from neurons in paraventricular nucleus of hypothalamus)	**Stimulated** by impulses from hypothalamic neurons in response to cervical/uterine stretching and suckling of infant at breast **Inhibited** by lack of appropriate neural stimuli	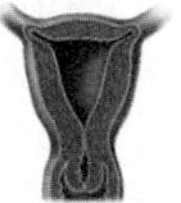  Uterus: stimulates uterine contractions; initiates labor; breast: initiates milk ejection	Unknown
Antidiuretic hormone (ADH) or **vasopressin** (Peptide, mostly from neurons in supraoptic nucleus of hypothalamus)	**Stimulated** by impulses from hypothalamic neurons in response to increased osmolality of blood or decreased blood volume; also stimulated by pain, some drugs, low blood pressure **Inhibited** by adequate hydration of the body and by alcohol	 Kidneys: stimulates kidney tubule cells to reabsorb water	↓ Diabetes insipidus ↑ Syndrome of inappropriate ADH secretion (SIADH)

*Indicates hypothalamic releasing and inhibiting hormones: GHRH = growth hormone–releasing hormone; GHIH = growth hormone–inhibiting hormone; TRH = thyrotropin-releasing hormone; CRH = corticotropin-releasing hormone; GnRH = gonadotropin-releasing hormone; PIH = prolactin-inhibiting hormone

Oxytocin also acts as the hormonal trigger for milk ejection (the "letdown" reflex) in women whose breasts are producing milk in response to prolactin. Suckling causes a reflex-initiated release of oxytocin, which targets specialized myoepithelial cells surrounding the milk-producing glands. As these cells contract, milk is forced from the breast into the infant's mouth. Both childbirth and milk ejection result from *positive feedback mechanisms* and are described in more detail in Chapter 28.

Both natural and synthetic oxytocic drugs are used to induce labor or to hasten normal labor that is progressing slowly. Less frequently, oxytocic drugs are used to stop postpartum bleeding (by compressing ruptured blood vessels at the placental site) and to stimulate the milk ejection reflex.

Until recently, oxytocin's role in males and nonpregnant, nonlactating females was unknown, but studies reveal that this potent peptide plays a role in sexual arousal and orgasm when the body is already primed for reproduction by sex hormones. Then, it may be responsible for the feeling of sexual satisfaction that results from that interaction. In nonsexual relationships, it is thought to promote nurturing and affectionate behavior, that is, it acts as a "cuddle hormone."

Antidiuretic Hormone

Diuresis (di″u-re′sis) is urine production, so an *antidiuretic* is a substance that inhibits or prevents urine formation. **Antidiuretic hormone (ADH)** prevents wide swings in water balance, helping the body avoid dehydration and water overload. Hypothalamic neurons called *osmoreceptors* continually monitor the solute concentration (and thus the water concentration) of the blood. When solutes threaten to become too concentrated (as might follow excessive perspiration or inadequate fluid intake), the osmoreceptors transmit excitatory impulses to the hypothalamic neurons, which synthesize and release ADH. Liberated into the blood by the posterior pituitary, ADH targets the kidney tubules via cAMP. The tubule cells respond by reabsorbing more water from the forming urine and returning it to the bloodstream. As a result, less urine is produced and blood solute concentration decreases. As the solute concentration of the blood declines, the osmoreceptors stop depolarizing, effectively ending ADH release. Other stimuli triggering ADH release include pain, low blood pressure, and such drugs as nicotine, morphine, and barbiturates.

16

Drinking alcoholic beverages inhibits ADH secretion and causes copious urine output. The dry mouth and intense thirst of the "hangover" reflect this dehydrating effect of alcohol. As might be expected, ADH release is also inhibited by drinking excessive amounts of water. *Diuretic drugs* antagonize the effects of ADH and cause water to be flushed from the body. These drugs are used to manage some cases of hypertension and the edema (water retention in tissues) typical of congestive heart failure.

At high blood concentrations, ADH causes vasoconstriction, primarily of the visceral blood vessels. This response targets different ADH receptors found on vascular smooth muscle. Under certain conditions, such as severe blood loss, exceptionally large amounts of ADH are released, causing a rise in blood pressure. The alternative name for this hormone, **vasopressin**, reflects this particular effect.

HOMEOSTATIC IMBALANCE

One result of ADH deficiency is **diabetes insipidus**, a syndrome marked by the output of huge amounts of urine and intense thirst. The name of this condition (*diabetes* = overflow; *insipidus* = tasteless) distinguishes it from diabetes mellitus (*mel* = honey), in which insulin deficiency causes large amounts of blood glucose to be lost in the urine. At one time, urine was tasted to determine which type of diabetes the patient was suffering from.

Diabetes insipidus can be caused by a blow to the head that damages the hypothalamus or the posterior pituitary. In either case, ADH release is deficient. Though inconvenient, the condition is not serious when the thirst center is operating properly and the person drinks enough water to prevent dehydration. However, it can be life threatening in unconscious or comatose patients, so accident victims with head trauma must be carefully monitored.

Hypersecretion of ADH occurs in children with meningitis, may follow neurosurgery or hypothalamic injury, or results from ectopic ADH secretion by cancer cells (particularly pulmonary cancers). It also may occur after general anesthesia or administration of certain drugs. The resulting condition, *syndrome of inappropriate ADH secretion* (*SIADH*), is marked by retention of fluid, headache and disorientation due to brain edema, weight gain, and decreased solute concentration in the blood. SIADH management requires fluid restriction and careful monitoring of blood sodium levels. ■

CHECK YOUR UNDERSTANDING

7. What is the key difference between the way the hypothalamus communicates with the anterior pituitary and the way it communicates with the posterior pituitary?
8. List the four anterior pituitary hormones that are tropic hormones and name their target glands.
9. Anita drank too much alcohol one night and suffered from a headache and nausea the next morning. What caused these "hangover" effects?

For answers, see Appendix G.

The Thyroid Gland

- Describe important effects of the two groups of hormones produced by the thyroid gland.
- Follow the process of thyroxine formation and release.

Location and Structure

The butterfly-shaped **thyroid gland** is located in the anterior neck, on the trachea just inferior to the larynx (Figure 16.1 and Figure 16.8a). Its two lateral *lobes* are connected by a median tissue mass called the *isthmus*. The thyroid gland is the largest pure endocrine gland in the body. Its prodigious blood supply (from the *superior* and *inferior thyroid arteries*) makes thyroid surgery a painstaking (and bloody) endeavor.

Internally, the gland is composed of hollow, spherical **follicles** (Figure 16.8b). The walls of each follicle are formed largely by cuboidal or squamous epithelial cells called *follicle cells*, which produce the glycoprotein **thyroglobulin** (thi″ro-glob′u-lin). The central cavity, or lumen, of the follicle stores **colloid**, an amber-colored, sticky material consisting of thyroglobulin molecules with attached iodine atoms. *Thyroid hormone* is derived from this iodinated thyroglobulin.

The *parafollicular cells*, another population of endocrine cells in the thyroid gland, produce *calcitonin*, an entirely different hormone. The parafollicular cells lie in the follicular epithelium but protrude into the soft connective tissue that separates and surrounds the thyroid follicles.

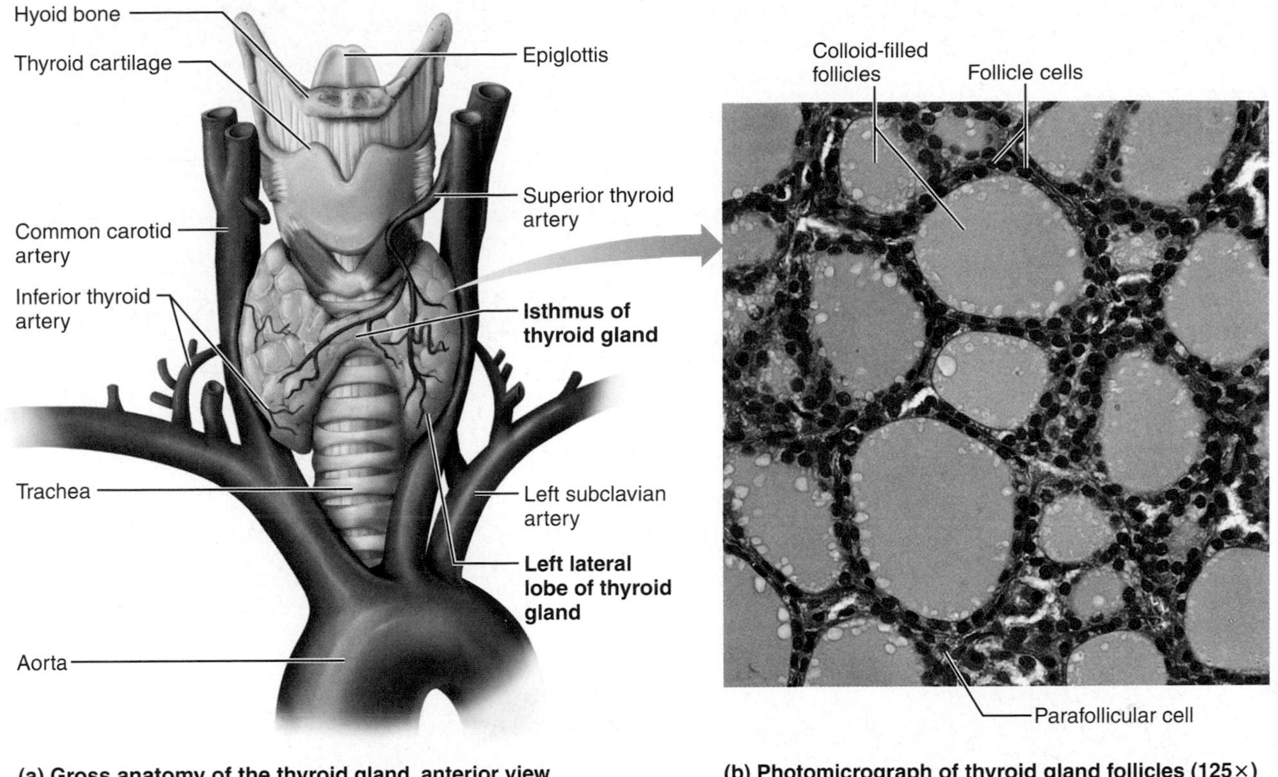

Figure 16.8 **The thyroid gland.**

16

Thyroid Hormone

Often referred to as the body's major metabolic hormone, **thyroid hormone (TH)** is actually two iodine-containing amine hormones, **thyroxine** (thi-rok′sin), or T_4, and **triiodothyronine** (tri″i-o″do-thi′ro-nēn), or T_3. T_4 is the major hormone secreted by the thyroid follicles. Most T_3 is formed at the target tissues by conversion of T_4 to T_3. Very much like one another, these hormones are constructed from two linked tyrosine amino acids. The principal difference is that T_4 has four bound iodine atoms, and T_3 has three (thus, T_4 and T_3).

TH affects virtually every cell in the body, as summarized in **Table 16.2** (p. 612). Like steroids, TH enters a target cell, binds to intracellular receptors within the cell's nucleus, and initiates transcription of mRNA for protein synthesis. By turning on transcription of genes concerned with glucose oxidation, TH increases basal metabolic rate and body heat production. This effect is known as the hormone's **calorigenic effect** (*calorigenic* = heat producing). Because TH provokes an increase in the number of adrenergic receptors in blood vessels, it plays an important role in maintaining blood pressure. Additionally, it is important in regulating tissue growth and development. It is critical for normal skeletal and nervous system development and maturation and for reproductive capabilities.

Synthesis

The thyroid gland is unique among the endocrine glands in its ability to store its hormone extracellularly and in large quantities. In the normal thyroid gland, the amount of stored colloid remains relatively constant and is sufficient to provide normal levels of hormone release for two to three months.

When TSH from the anterior pituitary binds to receptors on follicle cells, their *first* response is to secrete stored thyroid hormone. Their *second* response is to begin synthesizing more colloid to "restock" the follicle lumen. As a general rule, TSH levels are lower during the day, peak just before sleep, and remain high during the night. Consequently, thyroid hormone release and synthesis follows a similar pattern.

Let's examine how follicle cells synthesize thyroid hormone. The following numbers correspond to steps 1–7 in **Figure 16.9**:

① **Thyroglobulin is synthesized and discharged into the follicle lumen.** After being synthesized on the ribosomes of the thyroid's follicle cells, thyroglobulin is transported to the Golgi apparatus, where sugar molecules are attached and the thyroglobulin is packed into transport vesicles. These vesicles move to the apex of the follicle cell, where they discharge their contents into the follicle lumen to become part of the stored colloid.

② **Iodide is trapped.** To produce the functional iodinated hormones, the follicle cells must accumulate iodides (anions of iodine, I^-) from the blood. Iodide trapping depends on active transport. (The intracellular concentration of I^- is over 30 times higher than that in blood.) Once trapped inside the follicle cell, iodide then moves into the follicle lumen by facilitated diffusion.

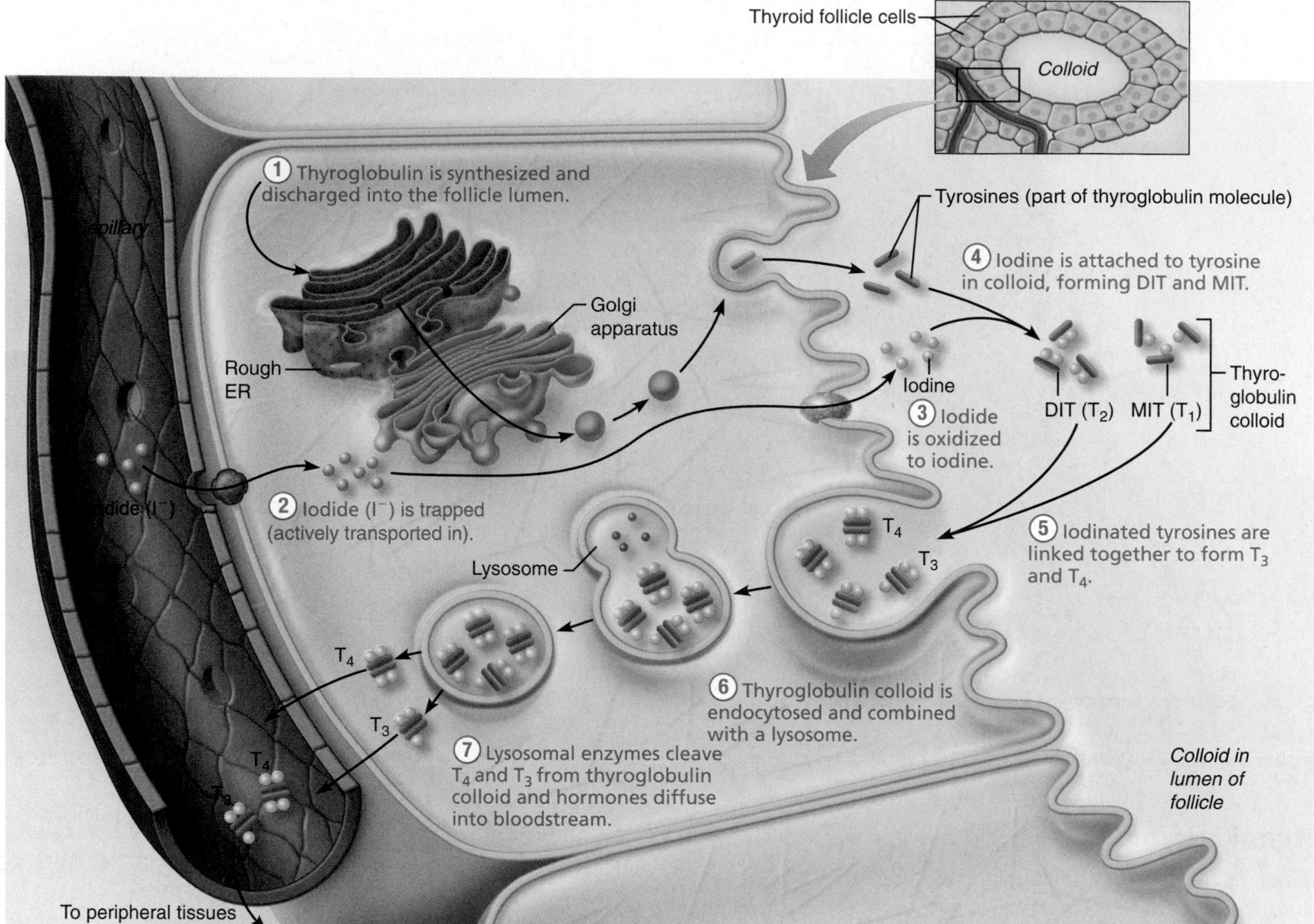

Figure 16.9 Synthesis of thyroid hormone. Only a few tyrosines of the thyroglobulins in the colloid are illustrated. The colloid is represented by the unstructured yellow substance outside of the cells.

③ **Iodide is oxidized to iodine.** At the border of the follicle cell and colloid, iodides are oxidized (by removal of electrons) and converted to iodine (I_2).

④ **Iodine is attached to tyrosine.** Once formed, iodine is attached to tyrosine amino acids that form part of the thyroglobulin colloid. This iodination reaction occurs at the junction of the follicle cell apex and the colloid and is mediated by peroxidase enzymes. Attachment of one iodine to a tyrosine produces **monoiodotyrosine (MIT or T_1)**, and attachment of two iodines produces **diiodotyrosine (DIT or T_2)**.

⑤ **Iodinated tyrosines are linked together to form T_3 and T_4.** Enzymes in the colloid link MIT and DIT together. Two linked DITs result in T_4, and coupling of MIT and DIT produces T_3. At this point, the hormones are still part of the thyroglobulin colloid.

⑥ **Thyroglobulin colloid is endocytosed.** In order to secrete the hormones, the follicle cells must reclaim iodinated thyroglobulin by endocytosis and combine the vesicles with lysosomes.

⑦ **Lysosomal enzymes cleave T_4 and T_3 from thyroglobulin and the hormones diffuse from the follicle cell into the bloodstream.** The main hormonal product secreted is T_4. Some T_4 is converted to T_3 before secretion, but most T_3 is generated in the peripheral tissues.

Transport and Regulation

Most released T_4 and T_3 immediately binds to transport proteins, most importantly *thyroxine-binding globulins* (*TBGs*) produced by the liver. Both T_4 and T_3 bind to target tissue receptors, but T_3 binds much more avidly and is about 10 times more active. Most peripheral tissues have the enzymes needed to convert T_4 to T_3 by removing one iodine atom.

The negative feedback loop that regulates blood levels of TH is shown in Figure 16.7. Falling TH blood levels trigger release of *thyroid-stimulating hormone* (*TSH*), and ultimately of more TH. Rising TH levels feed back to inhibit the hypothalamic–anterior pituitary axis, temporarily shutting off the stimulus for TSH release.

Conditions that increase body energy requirements, such as pregnancy and exposure of infants to cold, stimulate the hypothalamus to secrete *thyrotropin-releasing hormone* (*TRH*),

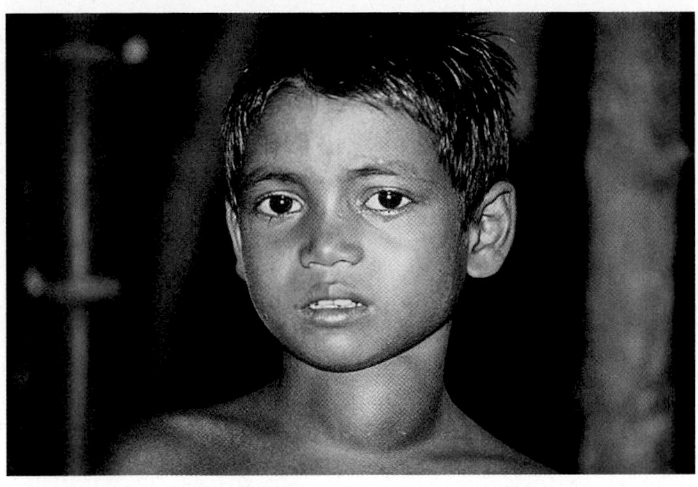
(a)

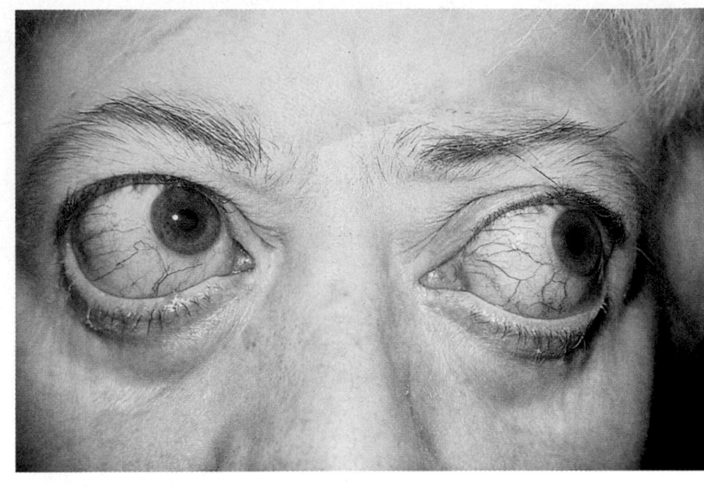
(b)

Figure 16.10 Thyroid disorders. (a) An enlarged thyroid (goiter) of a Bangladeshi boy. **(b)** Exophthalmos of Graves' disease.

which triggers TSH release, and in such instances, TRH overcomes the negative feedback controls. The thyroid gland then releases larger amounts of thyroid hormones, enhancing body metabolism and heat production.

Factors that inhibit TSH release include GHIH, dopamine, and rising levels of glucocorticoids. Excessively high blood iodide concentrations also inhibit TH release.

HOMEOSTATIC IMBALANCE

Both overactivity and underactivity of the thyroid gland can cause severe metabolic disturbances. Hypothyroid disorders may result from some thyroid gland defect or secondarily from inadequate TSH or TRH release. They also occur when the thyroid gland is removed surgically and when dietary iodine is inadequate.

In adults, the full-blown hypothyroid syndrome is called **myxedema** (mik″sĕ-de′mah; "mucous swelling"). Symptoms include a low metabolic rate; feeling chilled; constipation; thick, dry skin and puffy eyes; edema; lethargy; and mental sluggishness (but not mental retardation). If myxedema results from lack of iodine, the thyroid gland enlarges and protrudes, a condition called **endemic** (en-dem′ik), or *colloidal*, **goiter (Figure 16.10a)**. The follicle cells produce colloid but cannot iodinate it and make functional hormones. The pituitary gland secretes increasing amounts of TSH in a futile attempt to stimulate the thyroid to produce TH, but the only result is that the follicles accumulate more and more *unusable* colloid. Untreated, the thyroid cells eventually "burn out" from frantic activity, and the gland atrophies.

Before the marketing of iodized salt, parts of the midwestern United States were called the "goiter belt." Goiters were common there because these areas had iodine-poor soil and no access to iodine-rich seafood. Depending on the cause, myxedema can be reversed by iodine supplements or hormone replacement therapy.

Like many other hormones, the important effects of TH depend on a person's age and development. Severe hypothyroidism in infants is called **cretinism** (kre′tĭ-nizm). The child is mentally retarded and has a short, disproportionately sized body and a thick tongue and neck. Cretinism may reflect a genetic deficiency of the fetal thyroid gland or maternal factors, such as lack of dietary iodine. It is preventable by thyroid hormone replacement therapy if diagnosed early enough, but developmental abnormalities and mental retardation are not reversible once they appear.

The most common hyperthyroid pathology is an autoimmune disease called **Graves' disease**. In this condition, a person makes abnormal antibodies directed against thyroid follicle cells. Rather than destroying these cells, as antibodies normally do, these antibodies paradoxically mimic TSH and continuously stimulate TH release. Typical symptoms include an elevated metabolic rate; sweating; rapid, irregular heartbeat; nervousness; and weight loss despite adequate food intake. *Exophthalmos*, protrusion of the eyeballs, may occur if the tissue behind the eyes becomes edematous and then fibrous (Figure 16.10b). Treatments include surgical removal of the thyroid gland or ingestion of radioactive iodine (^{131}I), which selectively destroys the most active thyroid cells. ■

16

Calcitonin

Calcitonin is a polypeptide hormone produced by the **parafollicular**, or **C, cells** of the thyroid gland. Its effect is to lower blood Ca^{2+} levels, antagonizing the effects of parathyroid hormone produced by the parathyroid glands, which we consider in the next section. Calcitonin targets the skeleton, where it (1) inhibits osteoclast activity, inhibiting bone resorption and release of Ca^{2+} from the bony matrix, and (2) stimulates Ca^{2+} uptake and incorporation into bone matrix. As a result, calcitonin has a bone-sparing effect.

Excessive blood levels of Ca^{2+} (approximately 20% above normal) act as a humoral stimulus for calcitonin release, whereas declining blood Ca^{2+} levels inhibit C cell secretory activity. Calcitonin regulation of blood Ca^{2+} levels is short-lived but extremely rapid.

Calcitonin does not appear to play an important role in calcium homeostasis in humans. For example, calcitonin does not need to be replaced in patients whose thyroid gland has been

16

TABLE 16.2 Major Effects of Thyroid Hormone (T_4 and T_3) in the Body

PROCESS OR SYSTEM AFFECTED	NORMAL PHYSIOLOGICAL EFFECTS	EFFECTS OF HYPOSECRETION	EFFECTS OF HYPERSECRETION
Basal metabolic rate (BMR)/temperature regulation	Promotes normal oxygen use and BMR; calorigenesis; enhances effects of sympathetic nervous system	BMR below normal; decreased body temperature and cold intolerance; decreased appetite; weight gain; reduced sensitivity to catecholamines	BMR above normal; increased body temperature and heat intolerance; increased appetite; weight loss
Carbohydrate/lipid/protein metabolism	Promotes glucose catabolism; mobilizes fats; essential for protein synthesis; enhances liver's synthesis of cholesterol	Decreased glucose metabolism; elevated cholesterol/triglyceride levels in blood; decreased protein synthesis; edema	Enhanced catabolism of glucose, proteins, and fats; weight loss; loss of muscle mass
Nervous system	Promotes normal development of nervous system in fetus and infant; promotes normal adult nervous system function	In infant, slowed/deficient brain development, retardation; in adult, mental dulling, depression, paresthesias, memory impairment, hypoactive reflexes	Irritability, restlessness, insomnia, personality changes, exophthalmos (in Graves' disease)
Cardiovascular system	Promotes normal functioning of the heart	Decreased efficiency of pumping action of the heart; low heart rate and blood pressure	Increased sensitivity to catecholamines leads to rapid heart rate and possible palpitations; high blood pressure; if prolonged, heart failure
Muscular system	Promotes normal muscular development and function	Sluggish muscle action; muscle cramps; myalgia	Muscle atrophy and weakness
Skeletal system	Promotes normal growth and maturation of the skeleton	In child, growth retardation, skeletal stunting and retention of child's body proportions; in adult, joint pain	In child, excessive skeletal growth initially, followed by early epiphyseal closure and short stature; in adult, demineralization of skeleton
Gastrointestinal system	Promotes normal GI motility and tone; increases secretion of digestive juices	Depressed GI motility, tone, and secretory activity; constipation	Excessive GI motility; diarrhea; loss of appetite
Reproductive system	Promotes normal female reproductive ability and lactation	Depressed ovarian function; sterility; depressed lactation	In females, depressed ovarian function; in males, impotence
Integumentary system	Promotes normal hydration and secretory activity of skin	Skin pale, thick, and dry; facial edema; hair coarse and thick	Skin flushed, thin, and moist; hair fine and soft; nails soft and thin

removed. However, calcitonin is given therapeutically to patients to treat Paget's disease (a bone disease described in Chapter 6, p. 191).

The Parathyroid Glands

▶ Indicate general functions of parathyroid hormone.

The tiny, yellow-brown **parathyroid glands** are nearly hidden from view in the posterior aspect of the thyroid gland (**Figure 16.11a**). There are usually four of these glands, but the number varies from one individual to another—as many as eight have been reported, and some may be located in other regions of the neck or even in the thorax. The parathyroid's glandular cells are arranged in thick, branching cords containing scattered *oxyphil cells* and large numbers of smaller **chief cells** (Figure 16.11b). The chief cells secrete parathyroid hormone. The function of the oxyphil cells is unclear.

Discovery of the parathyroid glands was accidental. Years ago, surgeons were baffled by the observation that most patients recovered uneventfully after partial (or even total) thyroid gland removal, while others suffered uncontrolled muscle spasms and severe pain, and subsequently died. It was only after several such tragic deaths that the parathyroid glands were discovered and their hormonal function, quite different from that of the thyroid gland hormones, became known.

Parathyroid hormone (PTH), or *parathormone*, the protein hormone of these glands, is the single most important hormone controlling the calcium balance of the blood. Precise control of calcium levels is critical because Ca^{2+} homeostasis is essential for so many functions, including transmission of nerve impulses, muscle contraction, and blood clotting.

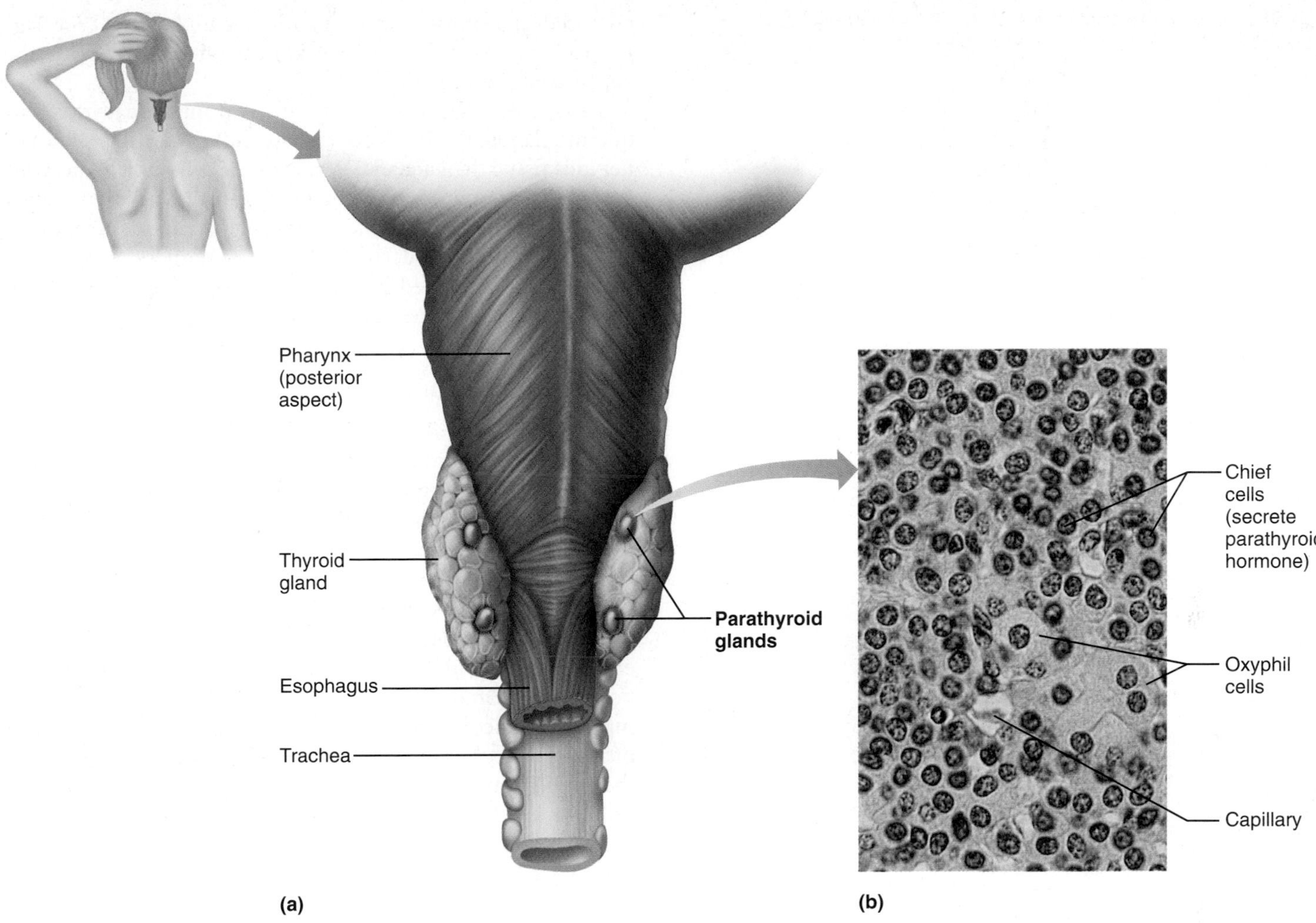

Figure 16.11 The parathyroid glands. (a) The parathyroid glands are located on the posterior aspect of the thyroid gland and may be more inconspicuous than depicted. **(b)** Photomicrograph of parathyroid gland tissue (500×).

PTH release is triggered by falling blood Ca^{2+} levels and inhibited by rising blood Ca^{2+} levels. PTH increases Ca^{2+} levels in blood by stimulating three target organs: the skeleton (which contains considerable amounts of calcium salts in its matrix), the kidneys, and the intestine **(Figure 16.12)**.

PTH release ① stimulates osteoclasts (bone-resorbing cells) to digest some of the bony matrix and release ionic calcium and phosphates to the blood; ② enhances reabsorption of Ca^{2+} [and excretion of phosphate (PO_4^{3-})] by the kidneys; and ③ promotes activation of vitamin D, thereby increasing absorption of Ca^{2+} by the intestinal mucosal cells. Vitamin D is required for absorption of Ca^{2+} from food, but the vitamin is ingested or produced by the skin in an inactive form. For vitamin D to exert its physiological effects, it must first be converted by the kidneys to its active vitamin D_3 form, *calcitriol* (1,25-dihydroxycholecalciferol). PTH stimulates this transformation.

HOMEOSTATIC IMBALANCE

Hyperparathyroidism is rare and usually results from a parathyroid gland tumor. In hyperparathyroidism, calcium is leached from the bones, and the bones soften and deform as their mineral salts are replaced by fibrous connective tissue. In *osteitis fibrosa cystica*, a severe example of this disorder, the bones have a moth-eaten appearance on X rays and tend to fracture spontaneously. The resulting abnormally elevated blood Ca^{2+} level (hypercalcemia) has many outcomes, but the two most notable are (1) depression of the nervous system, which leads to abnormal reflexes and weakness of the skeletal muscles, and (2) formation of kidney stones as excess calcium salts precipitate in the kidney tubules. Calcium deposits may also form in soft tissues throughout the body and severely impair vital organ functioning, a condition called *metastatic calcification*.

Hypoparathyroidism, or PTH deficiency, most often follows parathyroid gland trauma or removal during thyroid surgery. However, an extended deficiency of dietary magnesium (required for PTH secretion) can cause functional hypoparathyroidism. The resulting hypocalcemia (low blood Ca^{2+}) increases the excitability of neurons and accounts for the classical symptoms of *tetany* such as loss of sensation, muscle twitches, and convulsions. Untreated, the symptoms progress to respiratory paralysis and death. ■

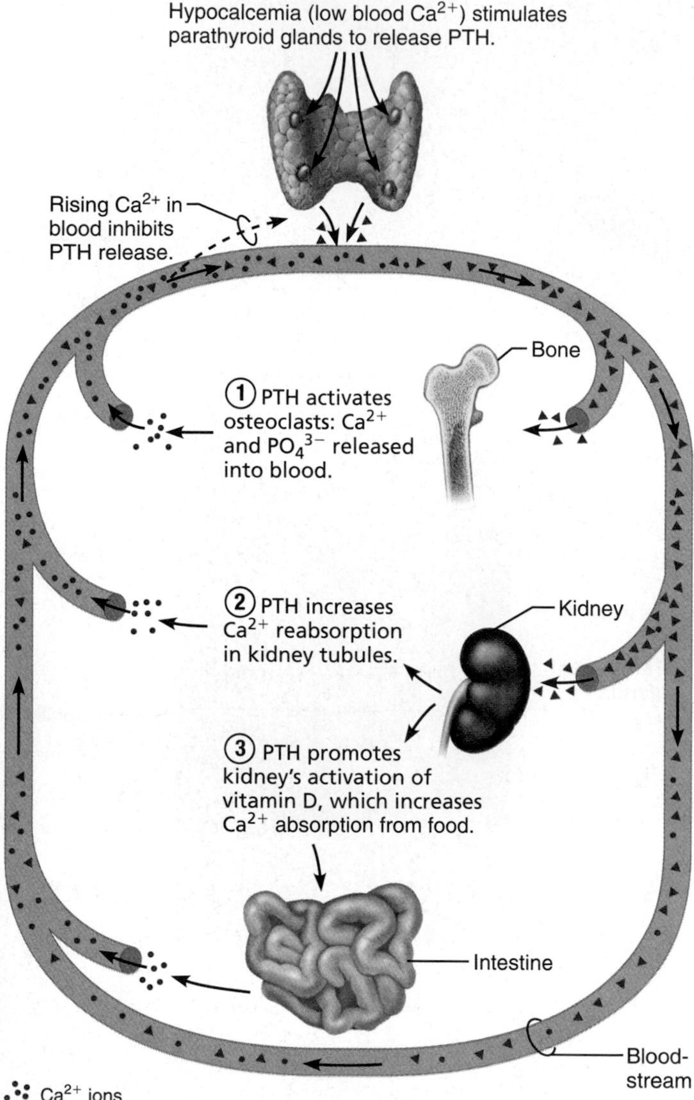
Hypocalcemia (low blood Ca2+) stimulates parathyroid glands to release PTH.
Rising Ca2+ in blood inhibits PTH release.
Bone
1 PTH activates osteoclasts: Ca2+ and PO43− released into blood.
2 PTH increases Ca2+ reabsorption in kidney tubules.
Kidney
3 PTH promotes kidney's activation of vitamin D, which increases Ca2+ absorption from food.
Intestine
Bloodstream
Ca2+ ions
PTH molecules

Figure 16.12 Effects of parathyroid hormone on bone, the kidneys, and the intestine. The effect on the intestine is indirect via activated vitamin D.

CHECK YOUR UNDERSTANDING

10. What is the major effect of thyroid hormone? Parathyroid hormone? Calcitonin?
11. Name the cells that release each of the three hormones listed above.

For answers, see Appendix G.

The Adrenal (Suprarenal) Glands

▶ List hormones produced by the adrenal gland, and cite their physiological effects.

The paired **adrenal glands** are pyramid-shaped organs perched atop the kidneys (*ad* = near; *renal* = kidney), where they are enclosed in a fibrous capsule and a cushion of fat (see Figure 16.1 and Figure 16.13). They are also often referred to as the **suprarenal glands** (*supra* = above).

Each adrenal gland is structurally and functionally two endocrine glands. The inner **adrenal medulla**, more like a knot of nervous tissue than a gland, is part of the sympathetic nervous system. The outer **adrenal cortex**, encapsulating the medulla and forming the bulk of the gland, is glandular tissue derived from embryonic mesoderm. Each region produces its own set of hormones summarized in Table 16.3 (p. 617), but all adrenal hormones help us cope with stressful situations.

The Adrenal Cortex

The adrenal cortex synthesizes well over two dozen steroid hormones, collectively called **corticosteroids.** The multistep steroid synthesis pathway begins with cholesterol, and involves varying intermediates depending on the hormone being formed. Unlike the amino acid–based hormones, steroid hormones are not stored in cells. Consequently, their rate of release in response to stimulation depends on their rate of synthesis.

The large, lipid-laden cortical cells are arranged in three layers or zones (Figure 16.13). The cell clusters forming the superficial **zona glomerulosa** (zo′nah glo-mer″u-lo′sah) produce mineralocorticoids, hormones that help control the balance of minerals and water in the blood. The principal products of the middle **zona fasciculata** (fah-sik″u-la′tah) cells, arranged in more or less linear cords, are the metabolic hormones called glucocorticoids. The cells of the innermost **zona reticularis** (rĕ-tik″u-lar′is), abutting the adrenal medulla, have a netlike arrangement. These cells mainly produce small amounts of adrenal sex hormones, or gonadocorticoids. Note, however, that the two innermost layers of the adrenal cortex share production of glucocorticoids and gonadocorticoids, although each layer predominantly produces one type.

Mineralocorticoids

The essential function of **mineralocorticoids** is to regulate the electrolyte (mineral salt) concentrations in extracellular fluids, particularly of Na^+ and K^+. The single most abundant cation in extracellular fluid is Na^+, and the amount of Na^+ in the body largely determines the volume of the extracellular fluid—where Na^+ goes, water follows. Changes in Na^+ concentration lead to changes in blood volume and blood pressure. Moreover, the regulation of a number of other ions, including K^+, H^+, HCO_3^- (bicarbonate), and Cl^- (chloride), is coupled to that of Na^+. The extracellular concentration of K^+ is also critical—it sets the resting membrane potential of all cells and determines how easily action potentials are generated in nerve and muscle. Not surprisingly, Na^+ and K^+ regulation are crucial to overall body homeostasis. Their regulation is the primary job of **aldosterone** (al-dos′ter-ōn), the most potent mineralocorticoid. Aldosterone accounts for more than 95% of the mineralocorticoids produced.

Aldosterone reduces excretion of Na^+ from the body. Its primary target is the distal parts of the kidney tubules, where it stimulates Na^+ reabsorption and water retention accompanied by elimination of K^+ and, in some instances, alterations in the

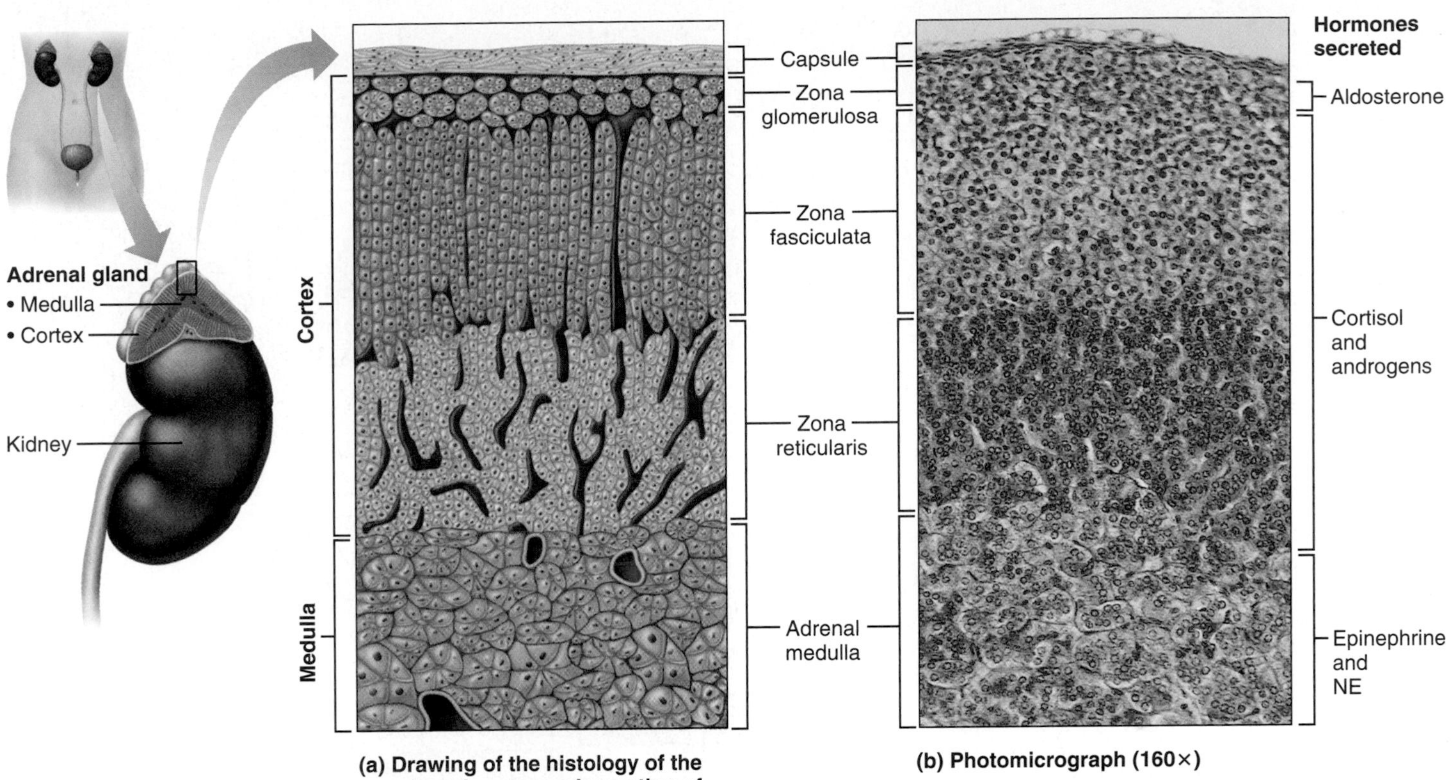

(a) Drawing of the histology of the adrenal cortex and a portion of the adrenal medulla

(b) Photomicrograph (160×)

Figure 16.13 Microscopic structure of the adrenal gland.

acid-base balance of the blood (by H^+ excretion). Aldosterone also enhances Na^+ reabsorption from perspiration, saliva, and gastric juice. Because aldosterone's regulatory effects are brief (lasting approximately 20 minutes), plasma electrolyte balance can be precisely controlled and modified continuously. The mechanism of aldosterone activity involves the synthesis and activation of proteins required for Na^+ transport such as Na^+-K^+ ATPase, the pump that exchanges Na^+ for K^+.

In this discussion we are focusing on the major roles of bloodborne aldosterone produced by the adrenal cortex, but aldosterone is also secreted by cardiovascular organs. There it is a paracrine and plays a completely different role in cardiac regulation.

Aldosterone secretion is stimulated by decreasing blood volume and blood pressure, and rising blood levels of K^+. The reverse conditions inhibit aldosterone secretion. Four mechanisms regulate aldosterone secretion, but the first two are by far the most important (**Figure 16.14** and Table 16.3):

1. **The renin-angiotensin mechanism.** The renin-angiotensin mechanism (re′nin an″je-o-ten′sin) influences both blood volume and blood pressure by regulating the release of aldosterone and therefore Na^+ and water reabsorption by the kidneys. Specialized cells of the *juxtaglomerular apparatus* in the kidneys become excited when blood pressure (or blood volume) declines. These cells respond by releasing **renin** into the blood. Renin cleaves off part of the plasma protein **angiotensinogen** (an″je-o-ten′sin-o-gen), triggering an enzymatic cascade leading to the formation of **angiotensin II**, a potent stimulator of aldosterone release by the glomerulosa cells.

   However, the renin-angiotensin mechanism does much more than trigger aldosterone release, and all of its effects are ultimately involved in raising the blood pressure. We describe these additional effects in detail in Chapters 25 and 26.
2. **Plasma concentrations of potassium.** Fluctuating blood levels of K^+ directly influence the zona glomerulosa cells in the adrenal cortex. Increased K^+ stimulates aldosterone release, whereas decreased K^+ inhibits it.
3. **ACTH.** Under normal circumstances, ACTH released by the anterior pituitary has little or no effect on aldosterone release. However, when a person is severely stressed, the hypothalamus secretes more corticotropin-releasing hormone (CRH), and the resulting rise in ACTH blood levels steps up the rate of aldosterone secretion to a small extent. The increase in blood volume and blood pressure that results helps ensure adequate delivery of nutrients and respiratory gases during the stressful period.
4. **Atrial natriuretic peptide (ANP).** Atrial natriuretic peptide, a hormone secreted by the heart when blood pressure rises, fine-tunes blood pressure and sodium-water balance of the body. One of its major effects is to inhibit the renin-angiotensin mechanism. It blocks renin and aldosterone secretion and inhibits other angiotensin-induced mechanisms that enhance water and Na^+ reabsorption. Consequently, ANP's overall influence is to decrease blood pressure by allowing Na^+ (and water) to flow out of the body in urine (*natriuretic* = producing salty urine).

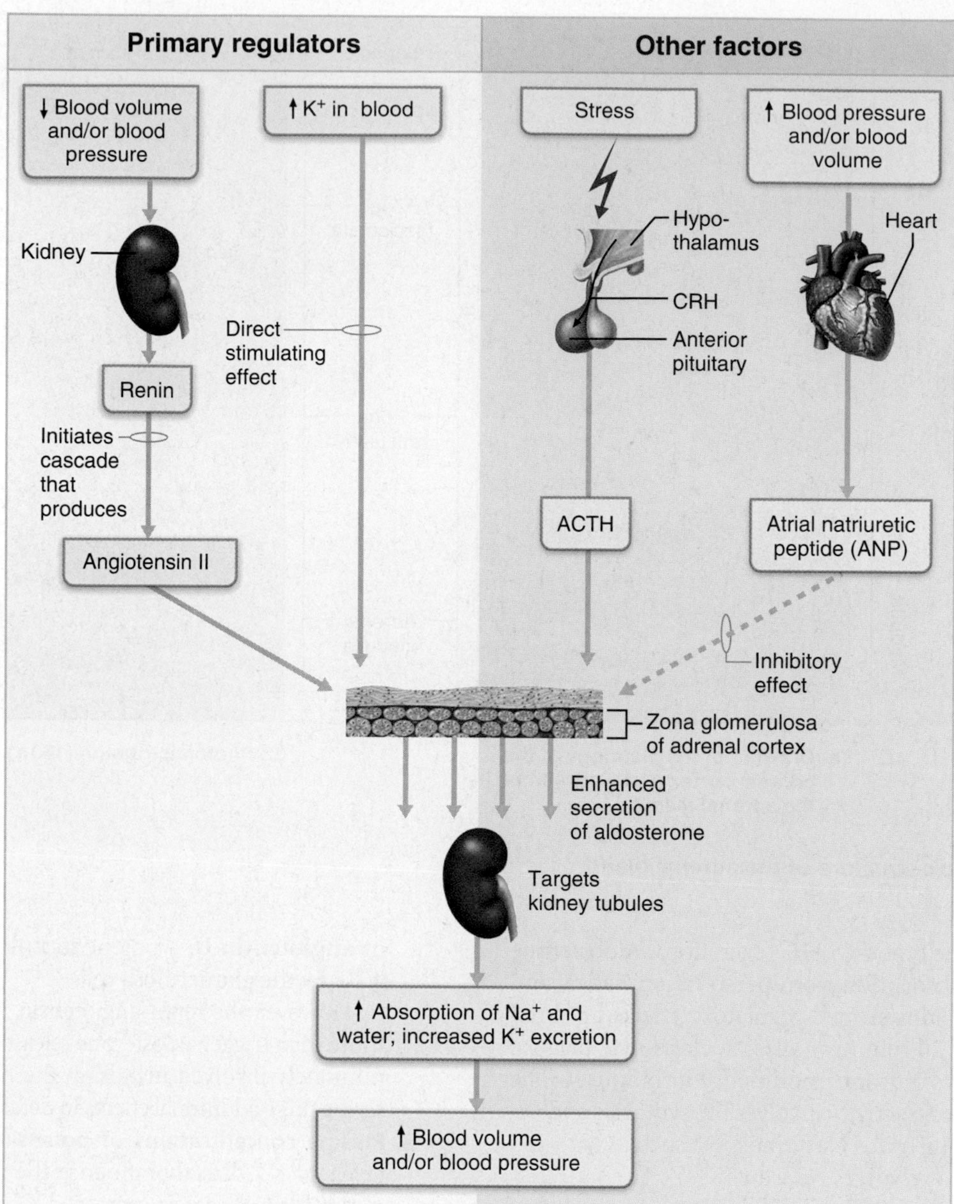

Figure 16.14 Major mechanisms controlling aldosterone release from the adrenal cortex.

16

HOMEOSTATIC IMBALANCE

Hypersecretion of aldosterone, a condition called *aldosteronism*, typically results from adrenal tumors. Two major sets of problems result: (1) hypertension and edema due to excessive Na^+ and water retention, and (2) accelerated excretion of potassium ions. If K^+ loss is extreme, neurons become nonresponsive and muscle weakness (and eventually paralysis) occurs. *Addison's disease*, a hyposecretory disease of the adrenal cortex, generally involves a deficient output of both mineralocorticoids and glucocorticoids, as we will describe shortly. ■

Glucocorticoids

Essential to life, the **glucocorticoids** influence the energy metabolism of most body cells and help us to resist stressors. Under normal circumstances, they help the body adapt to intermittent food intake by keeping blood glucose levels fairly constant, and maintain blood pressure by increasing the action of vasoconstrictors. However, severe stress due to hemorrhage, infection, or physical or emotional trauma evokes a dramatically higher output of glucocorticoids, which helps the body negotiate the crisis. Glucocorticoid hormones include **cortisol (hydrocortisone)**, **cortisone**, and **corticosterone**, but only cortisol is secreted in significant amounts in humans. As for all steroid hormones, the basic mechanism of glucocorticoid action on target cells is to modify gene activity.

Glucocorticoid secretion is regulated by negative feedback. Cortisol release is promoted by ACTH, triggered in turn by the hypothalamic releasing hormone CRH. Rising cortisol levels feed back to act on both the hypothalamus and the anterior pituitary, preventing CRH release and shutting off ACTH and cortisol secretion. Cortisol secretory bursts, driven by patterns of

TABLE 16.3 Adrenal Gland Hormones: Summary of Regulation and Effects

HORMONE	REGULATION OF RELEASE	TARGET ORGAN AND EFFECTS	EFFECTS OF HYPERSECRETION ↑ AND HYPOSECRETION ↓
Adrenocortical Hormones			
Mineralocorticoids (chiefly aldosterone)	Stimulated by renin-angiotensin mechanism (activated by decreasing blood volume or blood pressure), elevated blood K^+ levels, and ACTH (minor influence); inhibited by increased blood volume and pressure, and decreased blood K^+ levels	Kidneys: increase blood levels of Na^+ and decrease blood levels of K^+; since water reabsorption accompanies sodium retention, blood volume and blood pressure rise	↑ Aldosteronism ↓ Addison's disease
Glucocorticoids (chiefly cortisol)	Stimulated by ACTH; inhibited by feedback inhibition exerted by cortisol	Body cells: promote gluconeogenesis and hyperglycemia; mobilize fats for energy metabolism; stimulate protein catabolism; assist body to resist stressors; depress inflammatory and immune responses	↑ Cushing's syndrome ↓ Addison's disease
Gonadocorticoids (chiefly androgens, converted to testosterone or estrogens after release)	Stimulated by ACTH; mechanism of inhibition incompletely understood, but feedback inhibition not seen	Insignificant effects in males; responsible for female libido; development of pubic and axillary hair in females; source of estrogen after menopause	↑ Virilization of females (adrenogenital syndrome) ↓ No effects known
Adrenal Medullary Hormones			
Catecholamines (epinephrine and norepinephrine)	Stimulated by preganglionic fibers of the sympathetic nervous system	Sympathetic nervous system target organs: effects mimic sympathetic nervous system activation; increase heart rate and metabolic rate; increase blood pressure by promoting vasoconstriction	↑ Prolonged fight-or-flight response; hypertension ↓ Unimportant

16

eating and activity, occur in a definite pattern throughout the day and night. Cortisol blood levels peak shortly before we arise in the morning. The lowest levels occur in the evening just before and shortly after we fall asleep. The normal cortisol rhythm is interrupted by acute stress of any variety as higher CNS centers override the (usually) inhibitory effects of elevated cortisol levels and trigger CRH release. The resulting increase in ACTH blood levels causes an outpouring of cortisol from the adrenal cortex.

Stress results in a dramatic rise in blood levels of glucose, fatty acids, and amino acids, all provoked by cortisol. Cortisol's prime metabolic effect is to provoke *gluconeogenesis*, that is, the formation of glucose from fats and proteins. In order to "save" glucose for the brain, cortisol mobilizes fatty acids from adipose tissue and encourages their increased use for energy. Under cortisol's influence, stored proteins are broken down to provide building blocks for repair or for making enzymes to be used in metabolic processes. Cortisol enhances the sympathetic nervous system's vasoconstrictive effects, and the rise in blood pressure and circulatory efficiency that results helps ensure that these nutrients are quickly distributed to the cells.

Note that *ideal amounts of glucocorticoids promote normal function*, but cortisol excess is associated with significant anti-inflammatory and anti-immune effects. Excessive levels of glucocorticoids (1) depress cartilage and bone formation, (2) inhibit inflammation by decreasing the release of inflammatory chemicals, (3) depress the immune system, and (4) promote changes in cardiovascular, neural, and gastrointestinal function. Recognition of the effects of glucocorticoid hypersecretion has led to widespread use of glucocorticoid drugs to control symptoms of many chronic inflammatory disorders, such as rheumatoid arthritis or allergic responses. However, these potent drugs are a double-edged sword. Although they relieve some of the symptoms of these disorders, they also cause the undesirable effects of excessive levels of these hormones.

HOMEOSTATIC IMBALANCE

The pathology of glucocorticoid excess, **Cushing's syndrome**, may be caused by an ACTH-releasing pituitary tumor (in which case, it is called **Cushing's disease**); by an ACTH-releasing malignancy of the lungs, pancreas, or kidneys; or by a tumor of the adrenal cortex. However, it most often results from the clinical administration of pharmacological doses (doses higher than those found in the body) of glucocorticoid drugs. The syndrome is characterized by persistent elevated blood glucose levels

(a) Patient before onset.

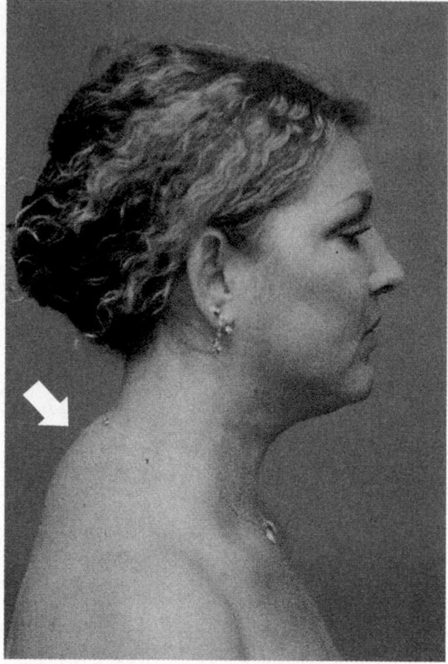

(b) Same patient with Cushing's syndrome. The white arrow shows the characteristic "buffalo hump" of fat on the upper back.

Figure 16.15 The effects of excess glucocorticoid.

(*steroid diabetes*), dramatic losses in muscle and bone protein, and water and salt retention, leading to hypertension and edema. The so-called *cushingoid signs* (Figure 16.15) include a swollen "moon" face, redistribution of fat to the abdomen and the posterior neck (causing a "buffalo hump"), a tendency to bruise, and poor wound healing. Because of enhanced anti-inflammatory effects, infections may become overwhelmingly severe before producing recognizable symptoms. Eventually, muscles weaken and spontaneous fractures force the person to become bedridden. The only treatment is removal of the cause—be it surgical removal of the offending tumor or discontinuation of the drug.

Addison's disease, the major hyposecretory disorder of the adrenal cortex, usually involves deficits in both glucocorticoids and mineralocorticoids. Its victims tend to lose weight; their plasma glucose and sodium levels drop, and potassium levels rise. Severe dehydration and hypotension are common. Corticosteroid replacement therapy at physiological doses (doses typical of those normally found in the body) is the usual treatment. ■

Gonadocorticoids (Sex Hormones)

The bulk of the **gonadocorticoids** secreted by the adrenal cortex are weak **androgens**, or male sex hormones, such as *androstenedione* and *dehydroepiandrosterone* (*DHEA*). These are converted to the more potent male hormone, *testosterone*, in the tissue cells or to estrogens (female sex homones) in females. The adrenal cortex also makes small amounts of female hormones (estradiol and other estrogens). The amount of gonadocorticoids produced by the adrenal cortex is insignificant compared with the amounts made by the gonads during late puberty and adulthood.

The exact role of the adrenal sex hormones is still in question, but because adrenal androgen levels rise continuously between the ages of 7 and 13 in both boys and girls, it is assumed that these hormones contribute to the onset of puberty and the appearance of axillary and pubic hair during that time. In adult women adrenal androgens are thought to be responsible for the sex drive, and they may account for the estrogens produced after menopause when ovarian estrogens are no longer produced. Control of gonadocorticoid secretion is not completely understood. Release seems to be stimulated by ACTH, but the gonadocorticoids do not appear to exert feedback inhibition on ACTH release.

HOMEOSTATIC IMBALANCE

Since androgens predominate, hypersecretion of gonadocorticoids causes *adrenogenital syndrome* (masculinization). In adult males, these effects of elevated gonadocorticoid levels may be obscured, since testicular testosterone has already produced virilization, but in prepubertal males and in females, the results can be dramatic. In the young man, maturation of the reproductive organs and appearance of the secondary sex characteristics occur rapidly, and the sex drive emerges with a vengeance. Females develop a beard and a masculine pattern of body hair distribution, and the clitoris grows to resemble a small penis. ■

The Adrenal Medulla

We discussed the adrenal medulla in Chapter 14 as part of the autonomic nervous system, so our coverage here is brief. The spherical **chromaffin cells** (kro′maf-in), which crowd around

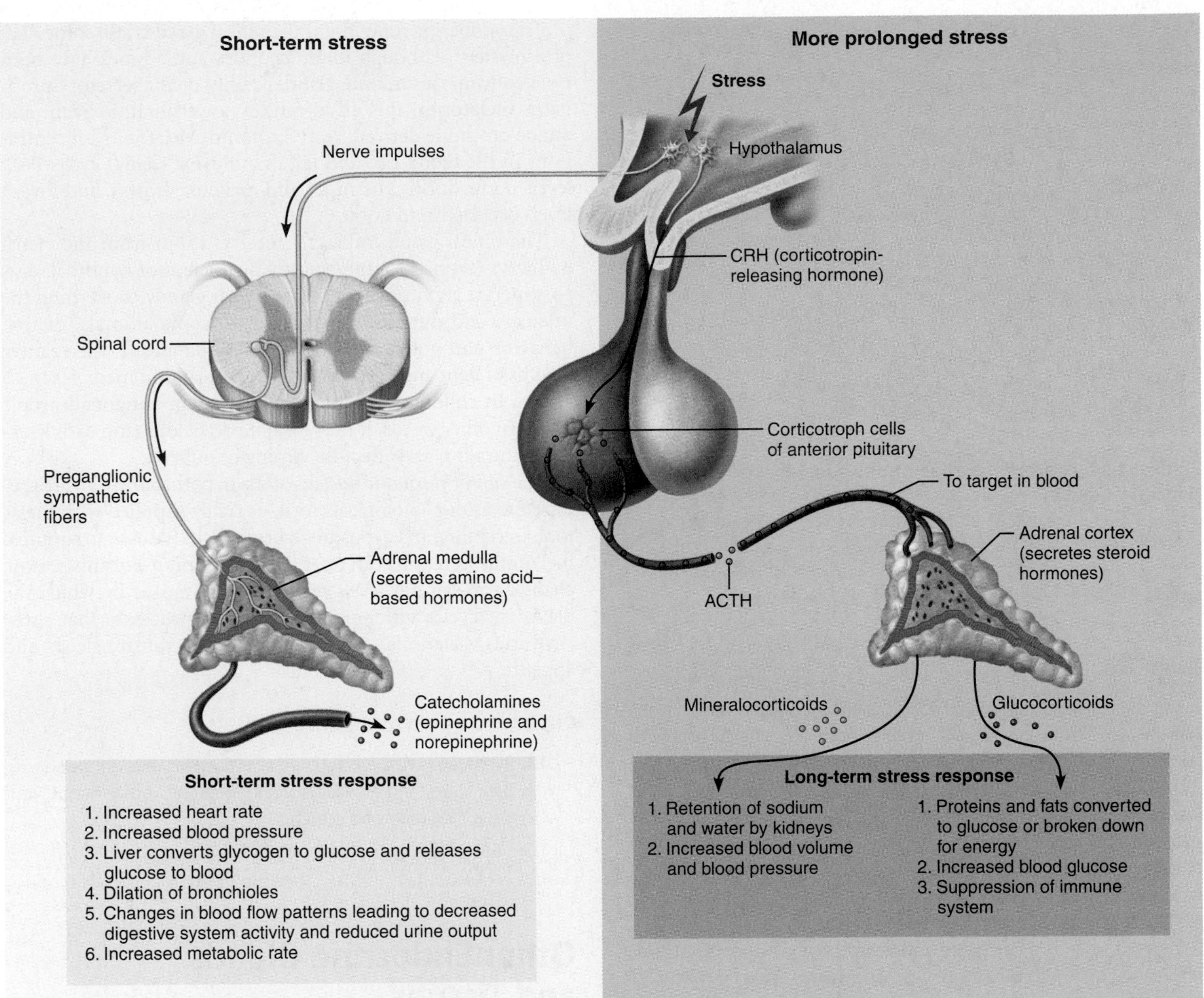

Short-term stress
More prolonged stress
Stress
Hypothalamus
Nerve impulses
CRH (corticotropin-releasing hormone)
Spinal cord
Corticotroph cells of anterior pituitary
To target in blood
Preganglionic sympathetic fibers
Adrenal cortex (secretes steroid hormones)
Adrenal medulla (secretes amino acid–based hormones)
ACTH
Catecholamines (epinephrine and norepinephrine)
Mineralocorticoids
Glucocorticoids
Short-term stress response
1. Increased heart rate
2. Increased blood pressure
3. Liver converts glycogen to glucose and releases glucose to blood
4. Dilation of bronchioles
5. Changes in blood flow patterns leading to decreased digestive system activity and reduced urine output
6. Increased metabolic rate
Long-term stress response
1. Retention of sodium and water by kidneys
2. Increased blood volume and blood pressure
1. Proteins and fats converted to glucose or broken down for energy
2. Increased blood glucose
3. Suppression of immune system

Figure 16.16 Stress and the adrenal gland. Stressful stimuli cause the hypothalamus to activate the adrenal medulla via sympathetic nerve impulses and the adrenal cortex via hormonal signals.

blood-filled capillaries and sinusoids, are modified ganglionic sympathetic neurons that synthesize the *catecholamines* **epinephrine** and **norepinephrine (NE)** via a molecular sequence from tyrosine to dopamine to NE to epinephrine.

When the body is activated to fight-or-flight status by some short-term stressor, the sympathetic nervous system is mobilized. Blood glucose levels rise, blood vessels constrict and the heart beats faster (together raising the blood pressure), blood is diverted from temporarily nonessential organs to the heart and skeletal muscles, and preganglionic sympathetic nerve endings weaving through the adrenal medulla signal for release of catecholamines, which reinforce and prolong the fight-or-flight response.

Unequal amounts of the two hormones are stored and released. Approximately 80% is epinephrine and 20% norepinephrine. With a few exceptions, the two hormones exert the same effects (see Table 14.2, p. 536). Epinephrine is the more potent stimulator of metabolic activities, bronchial dilation, and increased blood flow to skeletal muscles and the heart, but norepinephrine has the greater influence on peripheral vasoconstriction and blood pressure. Epinephrine is used clinically as a heart stimulant and to dilate the bronchioles during acute asthmatic attacks.

Unlike the adrenocortical hormones, which promote long-lasting body responses to stressors, catecholamines cause fairly brief responses. The interrelationships of the adrenal hormones and the hypothalamus, the "director" of the stress response, are depicted in **Figure 16.16**.

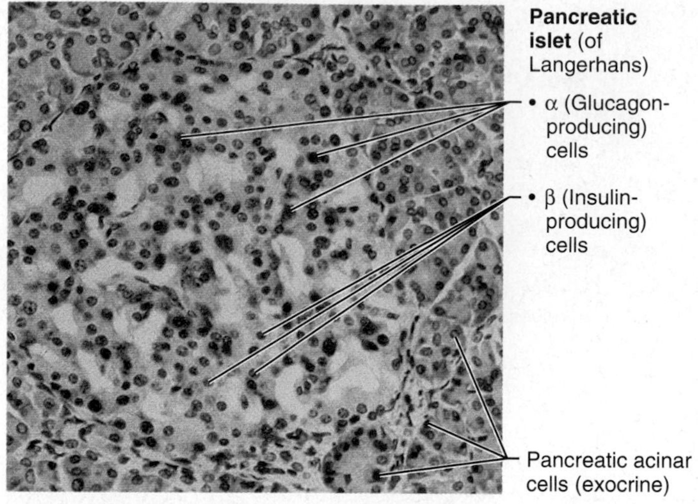

Figure 16.17 Photomicrograph of differentially stained pancreatic tissue. A pancreatic islet is surrounded by acinar cells (stained blue-gray), which produce the exocrine product (enzyme-rich pancreatic juice). The β cells of the islets that produce insulin are stained pale pink, and the α cells that produce glucagon are bright pink (230×).

HOMEOSTATIC IMBALANCE

16

A deficiency of hormones of the adrenal medulla is not a problem because these hormones merely intensify activities set into motion by the sympathetic nervous system neurons. Unlike glucocorticoids and mineralocorticoids, adrenal catecholamines are not essential for life. On the other hand, hypersecretion of catecholamines, sometimes arising from a chromaffin cell tumor called a *pheochromocytoma* (fe-o-kro″mo-si-to′mah), produces symptoms of uncontrolled sympathetic nervous system activity—**hyperglycemia** (elevated blood glucose), increased metabolic rate, rapid heartbeat and palpitations, hypertension, intense nervousness, and sweating. ■

CHECK YOUR UNDERSTANDING

12. List the three classes of hormones released from the adrenal cortex and for each briefly state its major effect(s).

For answers, see Appendix G.

The Pineal Gland

▶ Briefly describe the importance of melatonin.

The tiny, pine cone–shaped **pineal gland** hangs from the roof of the third ventricle in the diencephalon (see Figure 16.1). Its secretory cells, called **pinealocytes**, are arranged in compact cords and clusters. Lying between pinealocytes in adults are dense particles containing calcium salts ("brain-sand" or "pineal sand"). These salts are radiopaque, making the pineal gland a handy landmark for determining brain orientation in X rays.

The endocrine function of the pineal gland is still somewhat of a mystery. Although many peptides and amines have been isolated from this minute gland, its only major secretory product is **melatonin** (mel″ah-to′nin), a powerful antioxidant and amine hormone derived from serotonin. Melatonin concentrations in the blood rise and fall in a diurnal (daily) cycle. Peak levels occur during the night and make us drowsy, and lowest levels occur around noon.

The pineal gland indirectly receives input from the visual pathways (retina → suprachiasmatic nucleus of hypothalamus → superior cervical ganglion → pineal gland) concerning the intensity and duration of daylight. In some animals, mating behavior and gonadal size vary with changes in the relative lengths of light and dark periods, and melatonin mediates these effects. In children, melatonin may have an antigonadotropic effect. In other words, it may inhibit precocious (too early) sexual maturation and affect the timing of puberty.

The *suprachiasmatic nucleus* of the hypothalamus, an area referred to as our "biological clock," is richly supplied with melatonin receptors, and exposure to bright light (known to suppress melatonin secretion) can reset the clock timing. For this reason, changing melatonin levels may also be a means by which the day/night cycles influence physiological processes that show rhythmic variations, such as body temperature, sleep, and appetite.

CHECK YOUR UNDERSTANDING

13. Synthetic melatonin supplements are available, although their safety and efficacy have not been proved. What do you think they might be used for?

For answers, see Appendix G.

Other Endocrine Glands and Tissues

So far, we've examined the endocrine role of the hypothalamus and of glands dedicated solely to endocrine function. We will now consider a set of organs that contain endocrine tissue but also have other major functions. These include the pancreas, gonads, and placenta.

The Pancreas

▶ Compare and contrast the effects of the two major pancreatic hormones.

Located partially behind the stomach in the abdomen, the soft, triangular **pancreas** is a mixed gland composed of both endocrine and exocrine gland cells (see Figure 16.1). Along with the thyroid and parathyroids, it develops as an outpocketing of the epithelial lining of the gastrointestinal tract. *Acinar cells,* forming the bulk of the gland, produce an enzyme-rich juice that is carried by ducts to the small intestine during digestion.

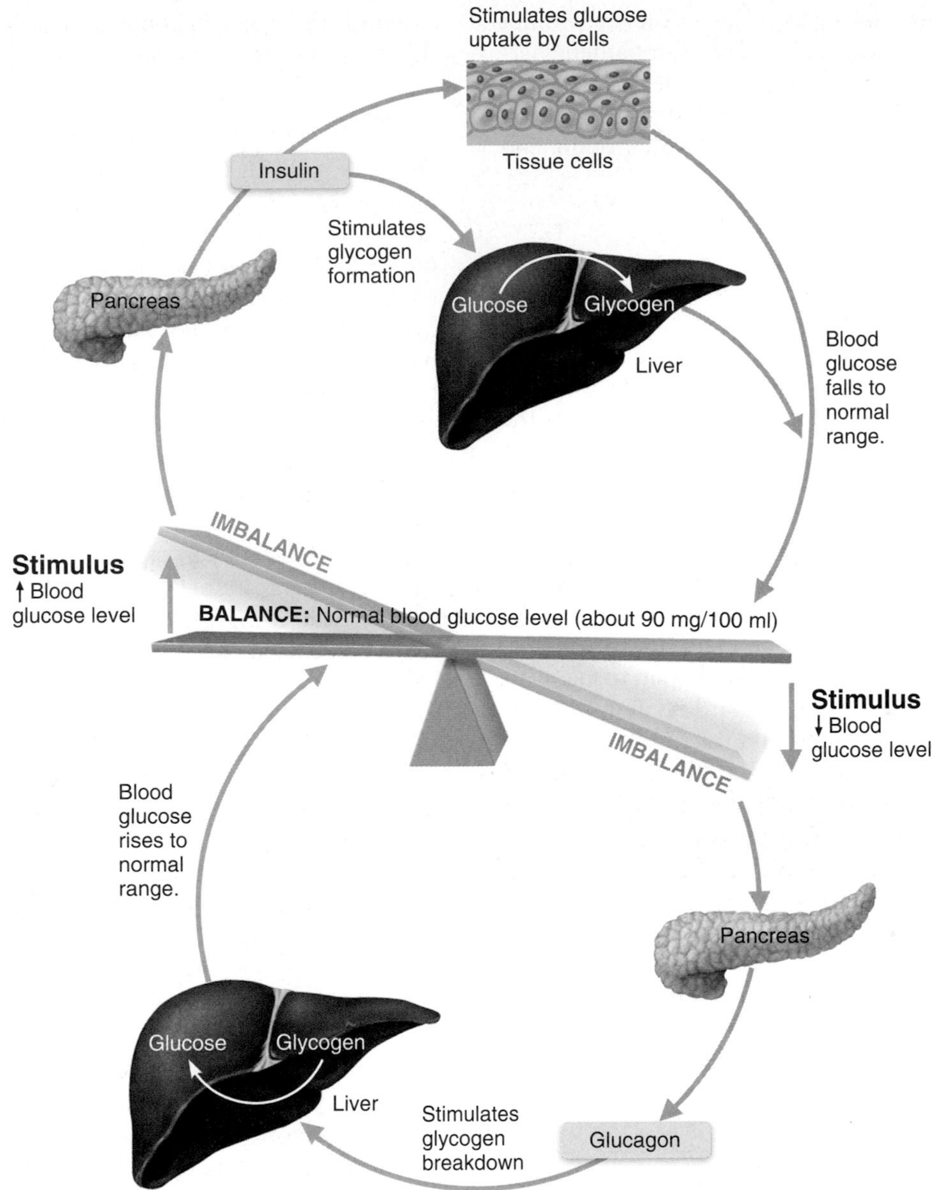

Figure 16.18 Regulation of blood glucose levels by insulin and glucagon from the pancreas.

Scattered among the acinar cells are approximately a million **pancreatic islets** (also called **islets of Langerhans**), tiny cell clusters that produce pancreatic hormones **(Figure 16.17)**. The islets contain two major populations of hormone-producing cells, the glucagon-synthesizing **alpha (α) cells** and the more numerous insulin-producing **beta (β) cells**. These cells act as tiny fuel sensors, secreting glucagon and insulin appropriately during the fasting and fed states. In this way insulin and glucagon are intimately but independently involved in the regulation of blood glucose levels. Their effects are antagonistic: Insulin is a *hypoglycemic* hormone, whereas glucagon is a *hyperglycemic* hormone **(Figure 16.18)**. Some islet cells also synthesize other peptides in small amounts. These include *somatostatin, pancreatic polypeptide* (*PP*), and others. However, we will not deal with these here.

Glucagon

Glucagon (gloo′kah-gon), a 29-amino-acid polypeptide, is an extremely potent hyperglycemic agent. One molecule of this hormone can cause the release of 100 million molecules of glucose into the blood! The major target of glucagon is the liver, where it promotes the following actions:

1. Breakdown of glycogen to glucose (*glycogenolysis*) (Figure 16.18)
2. Synthesis of glucose from lactic acid and from noncarbohydrate molecules (*gluconeogenesis*)
3. Release of glucose to the blood by liver cells, causing blood glucose levels to rise

A secondary effect is a fall in the amino acid concentration in the blood as the liver cells sequester these molecules to make new glucose molecules.

16

Humoral stimuli, mainly falling blood glucose levels, prompt the alpha cells to secrete glucagon. However, sympathetic nervous system stimulation and rising amino acid levels (as might follow a protein-rich meal) are also stimulatory. Glucagon release is suppressed by rising blood glucose levels, insulin, and somatostatin.

Insulin

Insulin is a small (51-amino-acid) protein consisting of two amino acid chains linked by disulfide (–S–S–) bonds. It is synthesized as part of a larger polypeptide chain called **proinsulin.** The middle portion of this chain is then excised by enzymes, releasing functional insulin. This "clipping" process occurs in the secretory vesicles just before insulin is released from the beta cell.

Insulin's effects are most obvious when we have just eaten. Its main effect is to lower blood glucose levels (Figure 16.18), but it also influences protein and fat metabolism. Circulating insulin lowers blood glucose levels in three main ways. (1) Insulin enhances membrane transport of glucose (and other simple sugars) into body cells, especially muscle and fat cells. (It does *not* accelerate glucose entry into liver, kidney, and brain tissue, all of which have easy access to blood glucose regardless of insulin levels. However, insulin does have important roles in the brain—it participates in neuronal development, feeding behavior, and learning and memory.) (2) Insulin inhibits the breakdown of glycogen to glucose, and (3) it inhibits the conversion of amino acids or fats to glucose. These inhibiting effects counter any metabolic activity that would increase plasma levels of glucose.

16

How does insulin act at the cellular level? Insulin activates its receptor (a tyrosine kinase enzyme), which phosphorylates specific proteins, beginning the cascade that leads to increased glucose uptake and insulin's other effects. After glucose enters a target cell, insulin binding triggers enzymatic activities that

1. Catalyze the oxidation of glucose for ATP production
2. Join glucose molecules together to form glycogen
3. Convert glucose to fat (particularly in adipose tissue)

As a rule, energy needs are met first, followed by glycogen formation. Finally, if excess glucose is still available, it is converted to fat. Insulin also stimulates amino acid uptake and protein synthesis in muscle tissue. In summary, insulin sweeps glucose out of the blood, causing it to be used for energy or converted to other forms (glycogen or fats), and it promotes protein synthesis and fat storage.

Pancreatic beta cells are stimulated to secrete insulin chiefly by elevated blood glucose levels, but also by rising plasma levels of amino acids and fatty acids, and release of acetylcholine by parasympathetic nerve fibers. As body cells take up glucose and other nutrients, and plasma levels of these substances drop, insulin secretion is suppressed.

Other hormones also influence insulin release. For example, any hyperglycemic hormone (such as glucagon, epinephrine, growth hormone, thyroxine, or glucocorticoids) called into action as blood glucose levels drop indirectly stimulates insulin release by promoting glucose entry into the bloodstream. Somatostatin and sympathetic nervous system activation depress insulin release. As you can see, blood glucose levels represent a balance of humoral, neural, and hormonal influences. Insulin is the major hypoglycemic factor that counterbalances the many hyperglycemic hormones.

HOMEOSTATIC IMBALANCE

Diabetes mellitus (DM) results from either hyposecretion or hypoactivity of insulin. When insulin is absent or deficient, blood glucose levels remain high after a meal because glucose is unable to enter most tissue cells. Ordinarily, when blood glucose levels rise, hyperglycemic hormones are not released, but when hyperglycemia becomes excessive, the person begins to feel nauseated, which precipitates the fight-or-flight response. This response results, inappropriately, in all the reactions that normally occur in the hypoglycemic (fasting) state to make glucose available—that is, glycogenolysis, lipolysis (breakdown of fat), and gluconeogenesis. Consequently, the already high blood glucose levels soar even higher, and excesses of glucose begin to be lost from the body in the urine (*glycosuria*).

When sugars cannot be used as cellular fuel, more fats are mobilized, resulting in high fatty acid levels in the blood, a condition called lipidemia or lipemia. In severe cases of diabetes mellitus, blood levels of fatty acids and their metabolites (acetoacetic acid, acetone, and others) rise dramatically. The fatty acid metabolites, collectively called **ketones** (ke′tōnz) or **ketone bodies**, are organic acids. When they accumulate in the blood, the blood pH drops, resulting in **ketoacidosis**, and ketone bodies begin to spill into the urine (*ketonuria*). Severe ketoacidosis is life threatening. The nervous system responds by initiating rapid deep breathing (hyperpnea) to blow off carbon dioxide from the blood and increase blood pH. (We will explain the physiological basis of this mechanism in Chapter 22.) If untreated, ketoacidosis disrupts heart activity and oxygen transport, and severe depression of the nervous system leads to coma and death.

The three cardinal signs of diabetes mellitus are polyuria, polydipsia, and polyphagia. The excessive glucose in the kidney filtrate acts as an osmotic diuretic (that is, it inhibits water reabsorption by the kidney tubules), resulting in **polyuria**, a huge urine output that leads to decreased blood volume and dehydration. Serious electrolyte losses also occur as the body rids itself of excess ketone bodies. Ketone bodies are negatively charged and carry positive ions out with them, and as a result, sodium and potassium ions are also lost from the body. Because of the electrolyte imbalance, the person gets abdominal pains and may vomit, and the stress reaction spirals even higher. Dehydration stimulates hypothalamic thirst centers, causing **polydipsia**, or excessive thirst. The final sign, **polyphagia**, refers to excessive hunger and food consumption, a sign that the person is "starving in the land of plenty." Although plenty of glucose is available, it cannot be used, and the body starts to utilize its fat and protein stores for energy metabolism. **Table 16.4** summarizes the consequences of insulin deficiency. DM is the focus of *A Closer Look* on pp. 626–627.

Hyperinsulinism, or excessive insulin secretion, results in low blood glucose levels, or **hypoglycemia.** This condition triggers the release of hyperglycemic hormones, which cause anxiety, nervousness, tremors, and weakness. Insufficient glucose delivery to the brain causes disorientation, progressing to convulsions, unconsciousness, and even death. In rare cases, hyperinsulinism re-

TABLE 16.4 Symptoms of Insulin Deficit (Diabetes Mellitus)

ORGANS/TISSUES INVOLVED	ORGAN/TISSUE RESPONSES TO INSULIN DEFICIENCY	RESULTING CONDITIONS IN BLOOD	RESULTING CONDITIONS IN URINE	SIGNS AND SYMPTOMS
Liver, Adipose tissue, Muscle	Decreased glucose uptake and utilization	Hyperglycemia	Glycosuria Osmotic diuresis	Polyuria (and dehydration, soft eyeballs) Polydipsia (and fatigue, weight loss) Polyphagia
Liver	Glycogenolysis			
Liver, Muscle	Protein catabolism and gluconeogenesis			
Liver, Adipose tissue	Lipolysis and ketogenesis	Lipidemia and ketoacidosis	Ketonuria Loss of Na^+, K^+; electrolyte and acid-base imbalances	Acetone breath Hyperpnea Nausea, vomiting, abdominal pain Cardiac irregularities Central nervous system depression; coma

sults from an islet cell tumor. More commonly, it is caused by an overdose of insulin and is easily treated by ingesting some sugar. ■

CHECK YOUR UNDERSTANDING

14. You've just attended a football game with your friend, Sharon, who is diabetic. While you aren't aware of Sharon having had more than one beer during the game, she is having trouble walking straight, her speech is slurred, and she is not making sense. What does it mean when we say Sharon is diabetic? What is the most likely explanation for Sharon's current behavior?

15. Diabetes mellitus and diabetes insipidus are both due to lack of a hormone. Which hormone causes which? What symptom do they have in common? What would you find in the urine of a patient with one but not the other?

For answers, see Appendix G.

The Gonads and Placenta

▶ Describe the functional roles of hormones of the testes, ovaries, and placenta.

The male and female **gonads** produce steroid sex hormones, identical to those produced by adrenal cortical cells (see Figure 16.1). The major distinction is the source and relative amounts produced. The release of gonadal hormones is regulated by gonadotropins, as described earlier.

The paired *ovaries* are small, oval organs located in the female's abdominopelvic cavity. Besides producing ova, or eggs, the ovaries produce several hormones, most importantly **estrogens** and **progesterone** (pro-jes′tĕ-rōn). Alone, the estrogens are responsible for maturation of the reproductive organs and the appearance of the secondary sex characteristics of females at puberty. Acting with progesterone, estrogens promote breast development and cyclic changes in the uterine mucosa (the menstrual cycle).

The male *testes*, located in an extra-abdominal skin pouch called the scrotum, produce sperm and male sex hormones, primarily **testosterone** (tes-tos′tĕ-rōn). During puberty, testosterone initiates the maturation of the male reproductive organs and the appearance of secondary sex characteristics and sex drive. In addition, testosterone is necessary for normal sperm production and maintains the reproductive organs in their mature functional state in adult males.

The *placenta* is a temporary endocrine organ. Besides sustaining the fetus during pregnancy, it secretes several steroid and protein hormones that influence the course of pregnancy. Placental hormones include estrogens and progesterone (hormones more often associated with the ovary), and human chorionic gonadotropin (hCG).

We will discuss the roles of the gonadal, placental, and gonadotropic hormones in detail in Chapters 27 and 28, where we consider the reproductive system and pregnancy.

CHECK YOUR UNDERSTANDING

16. Which of the two chemical classes of hormones introduced at the beginning of this chapter do the gonadal hormones belong to? Which major endocrine gland secretes hormones of this same chemical class?

For answers, see Appendix G.

Hormone Secretion by Other Organs

- Name a hormone produced by the heart.
- State the location of enteroendocrine cells.
- Briefly explain the hormonal functions of the kidney, skin, adipose tissue, bone, and thymus.

Other hormone-producing cells occur in various organs of the body, including the following, summarized in **Table 16.5**:

1. **Heart.** The atria contain some specialized cardiac muscle cells that secrete **atrial natriuretic peptide.** ANP prompts the kidneys to increase their production of salty urine and inhibits aldosterone release by the adrenal cortex. In this way, ANP decreases the amount of sodium in the extracellular fluid, thereby reducing blood volume and blood pressure (see Figure 16.14).
2. **Gastrointestinal tract.** *Enteroendocrine cells* are hormone-secreting cells sprinkled in the mucosa of the gastrointestinal (GI) tract. These scattered cells release several peptide hormones that help regulate a wide variety of digestive functions, some of which are summarized in Table 16.5. Enteroendocrine cells also release amines such as serotonin, which act as paracrines, diffusing to and influencing nearby target cells without first entering the bloodstream. Enteroendocrine cells are sometimes referred to as *paraneurons* because they are similar in certain ways to neurons and many of their hormones and paracrines are chemically identical to neurotransmitters.
3. **Kidneys.** Interstitial cells in the kidneys secrete **erythropoietin** (ĕ-rith″ro-poi′ĕ-tin; "red-maker"), a protein hormone that signals the bone marrow to increase production of red blood cells. The kidneys also release **renin**, the hormone that acts as an enzyme to initiate the renin-angiotensin mechanism of aldosterone release described earlier.
4. **Skin.** The skin produces **cholecalciferol**, an inactive form of vitamin D_3, when modified cholesterol molecules in epidermal cells are exposed to ultraviolet radiation. This compound then enters the blood via the dermal capillaries, is modified in the liver, and becomes fully activated in the kidneys. The active form of vitamin D_3, **calcitriol**, is an essential regulator of the carrier system that intestinal cells use to absorb Ca^{2+} from ingested food. Without this vitamin, the bones become soft and weak.
5. **Adipose tissue.** Adipose cells release **leptin**, which serves to tell your body how much stored energy (as fat) you have. The more fat you have, the more leptin there will be in your blood. As we describe in Chapter 24, leptin binds to CNS neurons concerned with appetite control, producing a sensation of satiety. It also appears to stimulate increased energy expenditure. Two other hormones released by adipose cells both affect the sensitivity of cells to insulin. *Resistin* is an insulin antagonist, while *adiponectin* enhances sensitivity to insulin.
6. **Skeleton.** Osteoblasts in bone secrete *osteocalcin*, a hormone that prods pancreatic beta cells to divide and secrete more insulin. It also restricts fat storage by adipocytes, and triggers the release of adiponectin. As a result, glucose handling is improved and body fat is reduced. Osteocalcin levels are low in type 2 diabetes, and increasing its level may offer a new treatment approach.
7. **Thymus.** Located deep to the sternum in the thorax is the lobulated **thymus** (see Figure 16.1). Large and conspicuous in infants and children, the thymus diminishes in size throughout adulthood. By old age, it is composed largely of adipose and fibrous connective tissues.

   Thymic epithelial cells secrete several different families of peptide hormones, including **thymulin, thymopoietins,** and **thymosins** (thi′mo-sinz). These hormones are thought to be involved in the normal development of *T lymphocytes* and the immune response, but their roles are not well understood. Although still called hormones, they mainly act locally as paracrines. We describe the thymus in Chapter 20 in our discussion of the lymphoid organs and tissues.

CHECK YOUR UNDERSTANDING

17. What hormone does the heart produce and what is its function?
18. What is the function of the hormone produced by the skin?

For answers, see Appendix G.

Developmental Aspects of the Endocrine System

- Describe the effects of aging on endocrine system functioning.

Hormone-producing glands arise from all three embryonic germ layers. Endocrine glands derived from mesoderm produce steroid hormones. All others produce amines, peptides, or protein hormones.

Though not usually considered important when describing hormone effectiveness, exposure to many environmental pollutants has been shown to disrupt endocrine function. These pollutants include many pesticides, industrial chemicals, arsenic, dioxin, and other soil and water pollutants. So far, sex hormones, thyroid hormone, and glucocorticoids have proved vulnerable to the effects of such pollutants. Interference with glucocorticoids, which turn on many genes that may suppress cancer, may help to explain the high cancer rates in certain areas of the country.

Barring exposure to environmental pollutants, and hypersecretory and hyposecretory disorders, most endocrine organs operate smoothly throughout life until old age. Aging may bring about changes in the rates of hormone secretion, breakdown, and excretion, or in the sensitivity of target cell receptors. Endocrine functioning in the elderly is difficult to research, however, because it is frequently altered by the chronic illnesses common in that age group.

TABLE 16.5 Selected Examples of Hormones Produced by Organs Other Than the Major Endocrine Organs

SOURCE	HORMONE	CHEMICAL COMPOSITION	TRIGGER	TARGET ORGAN AND EFFECTS
Adipose tissue	Leptin	Peptide	Secretion proportional to fat stores; increased by nutrient uptake	Brain: suppresses appetite; increases energy expenditure
Adipose tissue	Resistin, adiponectin	Peptides	Unknown	Fat, muscle, liver: resistin antagonizes insulin's action and adiponectin enhances it
GI tract mucosa				
▪ Stomach	Gastrin	Peptide	Secreted in response to food	Stomach: stimulates glands to release hydrochloric acid (HCl)
▪ Duodenum (of small intestine)	Intestinal gastrin	Peptide	Secreted in response to food, especially fats	Stomach: stimulates HCl secretion and gastrointestinal tract motility
▪ Duodenum	Secretin	Peptide	Secreted in response to food	Pancreas and liver: stimulates release of bicarbonate-rich juice; Stomach: inhibits secretory activity
▪ Duodenum	Cholecystokinin (CCK)	Peptide	Secreted in response to food	Pancreas: stimulates release of enzyme-rich juice; Gallbladder: stimulates expulsion of stored bile; Hepatopancreatic sphincter: causes sphincter to relax, allowing bile and pancreatic juice to enter duodenum
▪ Duodenum (and other gut regions)	Incretins [glucose-dependent insulinotropic peptide (GIP) and glucagon-like peptide 1 (GLP-1)]	Peptide	Secreted in response to glucose in intestinal lumen	Pancreas: enhances insulin release and inhibits glucagon release caused by increased blood glucose
Heart (atria)	Atrial natriuretic peptide	Peptide	Secreted in response to stretching of atria (by rising blood pressure)	Kidney: inhibits sodium ion reabsorption and renin release; adrenal cortex: inhibits secretion of aldosterone; decreases blood pressure
Kidney	Erythropoietin (EPO)	Glycoprotein	Secreted in response to hypoxia	Red bone marrow: stimulates production of red blood cells
Kidney	Renin	Peptide	Secreted in response to low blood pressure or plasma volume, or sympathetic stimulation	Acts as an enzyme to initiate renin-angiotensin mechanism of aldosterone release; returns blood pressure to normal
Skeleton	Osteocalcin	Peptide	Unknown	Increases insulin production and insulin sensitivity
Skin (epidermal cells)	Cholecalciferol (provitamin D_3)	Steroid	Activated by the kidneys to active vitamin D_3 (calcitriol) in response to parathyroid hormone	Intestine: stimulates active transport of dietary calcium across intestinal cell membranes
Thymus	Thymulin, thymopoietins, thymosins	Peptides	Unknown	Mostly act locally as paracrines; involved in T lymphocyte development and in immune responses

CLOSER LOOK

Sweet Revenge: Taming the DM Monster?

Few medical breakthroughs have been as electrifying as the discovery of insulin in 1921, an event that changed diabetes mellitus from a death sentence to a survivable disease. Nonetheless, DM is still a huge health problem: Determining blood glucose levels accurately and maintaining desirable levels sorely challenge our present biotechnology. Let's take a closer look at the characteristics and challenges of type 1 and type 2, the major forms of diabetes mellitus.

More than 1 million Americans have **type 1 diabetes mellitus**, formerly called *insulin-dependent diabetes mellitus* (*IDDM*). Symptoms appear suddenly, usually before age 15, following a long asymptomatic period during which the beta cells are destroyed by the immune system. Consequently, type 1 diabetics effectively lack insulin.

Type 1 diabetes susceptibility genes have been localized on several chromosomes, indicating that type 1 diabetes is an example of a multigene autoimmune response. However, some investigators believe that *molecular mimicry* is at least part of the problem: Some foreign substance (for example, a virus) has entered the body and is so similar to certain self (beta cell) proteins that the immune system attacks the beta cells as well as the invader.

Indeed, elegant studies on certain strains of diabetic mice, stressed by infections or other irritants, demonstrated that they produced increasing amounts of a particular stress protein (heat shock protein 60) and also churned out antibodies against that stress protein. When fragments of the stress protein called p277 were injected into other mice showing early signs of diabetes, the beta cells were spared from the autoimmune attack. Recent clinical trials in which newly diagnosed diabetics were given the fragment as a drug (called DiaPep277) have yielded conflicting results, with some studies showing less need for insulin and others showing no effect.

Type 1 diabetes is difficult to control and patients typically develop long-term vascular and neural problems. The lipidemia and high blood cholesterol levels typical of the disease can lead to severe vascular complications including atherosclerosis, strokes, heart attacks, renal shutdown, gangrene, and blindness. Nerve damage leads to loss of sensation, impaired bladder function, and impotence. Female type 1 diabetics also tend to have lumpy breasts and to undergo premature menopause, which increases their risk for cardiac problems.

Hyperglycemia is the culprit behind these complications, and the closer to normal blood glucose levels are held, the less likely complications are. Continuous glucose monitors make this much easier than relying upon finger pricks. Currently, frequent insulin injections (up to four times daily, or better yet, by a continuous infusion pump) are recommended to reduce vascular and renal complications. While some glucose sensors can talk directly to insulin pumps, problems with automatically calculating the amount of insulin to dispense mean patients must still make these decisions. In this way, such combined devices still fall short of being a true "artificial pancreas." A number of alternative insulin delivery methods exist—insulin inhalers are now approved for use and insulin patches may be on the horizon.

Pancreatic islet cell transplants have become increasingly successful in helping type 1 diabetics. Still, only 33% of patients

Structural changes in the anterior pituitary occur with age. The amount of connective tissue increases, vascularization decreases, and the number of hormone-secreting cells declines. These changes may or may not affect hormone production. In women, for example, blood levels and the release rhythm of ACTH remain constant, but levels of gonadotropins increase with age. GH levels decline in both sexes, which partially explains muscle atrophy in old age.

The adrenal glands also show structural changes with age, but normal controls of cortisol appear to persist as long as a person is healthy and not stressed. Chronic stress, on the other hand, drives up blood levels of cortisol and appears to contribute to hippocampal (and memory) deterioration. Plasma levels of aldosterone are reduced by half in old age, but this change may reflect a decline in renin release by the kidneys, which become less responsive to renin-evoking stimuli. No age-related differences have been found in the release of catecholamines by the adrenal medulla.

The gonads, particularly the ovaries, undergo significant changes with age. In late middle age, the ovaries decrease in size and weight, and they become unresponsive to gonadotropins. As female hormone production declines dramatically, the ability to bear children ends, and problems associated with estrogen deficiency, such as arteriosclerosis and osteoporosis, begin to occur. Testosterone production by the testes also wanes with age, but this effect usually is not seen until very old age.

Glucose tolerance (the ability to dispose of a glucose load effectively) begins to deteriorate as early as the fourth decade of life. Blood glucose levels rise higher and return to resting levels more slowly in the elderly than in young adults. The fact that the islet cells continue to secrete near-normal amounts of insulin leads researchers to conclude that decreasing glucose tolerance with age may reflect declining receptor sensitivity to insulin (pre–type 2 diabetes).

Thyroid hormone synthesis and release diminish somewhat with age. Typically, the follicles are loaded with colloid in the elderly, and fibrosis of the gland occurs. Basal metabolic rate declines with age. Mild hypothyroidism is only one cause of this decline. The increase in body fat relative to muscle is equally important, because muscle tissue is more active metabolically than fat.

The parathyroid glands change little with age, and PTH levels remain fairly normal throughout life. Estrogens protect women

A CLOSER LOOK
(continued)

need no injected insulin after two years. The need for long-term immunosuppression limits this treatment to only those diabetics who cannot control their blood glucose by any other means.

Over 90% of known DM cases are **type 2 diabetes mellitus**, formerly called *non-insulin-dependent diabetes mellitus (NIDDM)*, which grows increasingly common with age and with the increasing size of our waistlines. About 12 million people in the U.S. have been diagnosed with type 2 diabetes, and roughly half as many are believed to be undiagnosed victims. Type 2 diabetics are at risk for the same complications as type 1 diabetics—heart disease, amputations, kidney failure, and blindness.

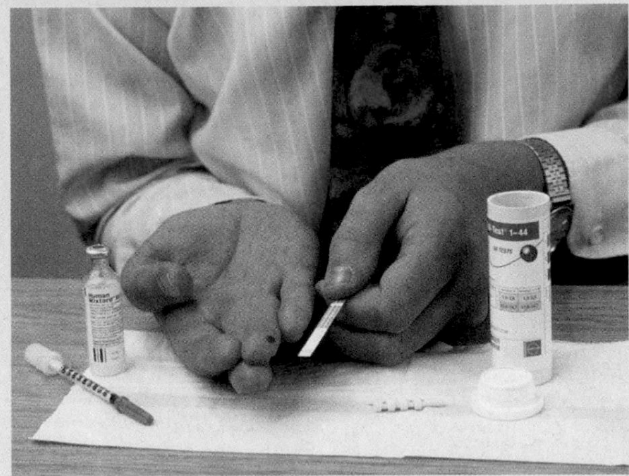

A hereditary predisposition is particularly striking in this diabetic group. About 25–30% of Americans carry a gene that predisposes them to type 2 diabetes, with nonwhites affected to a much greater extent. If an identical twin has type 2 diabetes mellitus, the probability that the other twin will have the disease is virtually 100%. Most type 2 diabetics produce insulin, but the insulin receptors are unable to respond to it, a phenomenon called **insulin resistance**. Mutations in any one of several genes could lead to insulin resistance.

Lifestyle factors also play a role: Type 2 diabetics are almost always overweight and sedentary. Adipose cells of obese people overproduce a number of signaling chemicals including *tumor necrosis factor alpha* and *resistin*, which may alter the enzymatic cascade triggered by insulin binding. The Diabetes Prevention Program, a major clinical trial, showed that weight loss and regular exercise can lower the risk of type 2 diabetes dramatically, even for people at high risk.

In many cases type 2 diabetes can be managed solely by exercise, weight loss, and a healthy diet. Some type 2 diabetics also benefit from oral medications that lower blood glucose or reduce insulin resistance. A promising new class of drugs targets the incretin hormones normally secreted by the gut, enhancing glucose-dependent insulin release from pancreatic beta cells using a normal physiological pathway. However, most type 2 diabetics must eventually inject insulin.

While we cannot yet cure diabetes, biotechnology promises to continue to improve control of blood glucose levels and thereby tame the monster that is diabetes.

16

against the demineralizing effects of PTH, but estrogen production wanes after menopause, leaving older women vulnerable to the bone-demineralizing effects of PTH and osteoporosis.

CHECK YOUR UNDERSTANDING

19. In the elderly, the decline in levels of which hormone is associated with muscle atrophy? With osteoporosis in women?

For answers, see Appendix G.

In this chapter, we have covered the general mechanisms of hormone action and have provided an overview of the major endocrine organs, their chief targets, and their most important physiological effects, as summarized in *Making Connections* on p. 628. However, every one of the hormones discussed here comes up in at least one other chapter, where its actions are described as part of the functional framework of a particular organ system. For example, we described the effects of PTH and calcitonin on bone mineralization in Chapter 6 along with the discussion of bone remodeling.

RELATED CLINICAL TERMS

Hirsutism (her′soot-izm; *hirsut* = hairy, rough) Excessive hair growth; usually refers to this phenomenon in women and reflects excessive androgen production.

Hypophysectomy (hi-pof″ĭ-sek′to-me) Surgical removal of the pituitary gland.

Prolactinoma (pro-lak″tĭ-no′mah; *oma* = tumor) The most common type (30–40% or more) of pituitary gland tumor; evidenced by hypersecretion of prolactin and menstrual disturbances in women.

Psychosocial dwarfism Dwarfism (and failure to thrive) resulting from stress and emotional disorders that suppress hypothalamic release of growth hormone–releasing hormone and thus anterior pituitary secretion of growth hormone.

Thyroid storm (thyroid crisis) A sudden and dangerous increase in all of the symptoms of hyperthyroidism due to excessive amounts of circulating TH. Symptoms of this hypermetabolic state include fever, rapid heart rate, high blood pressure, dehydration, nervousness, and tremors. Precipitating factors include severe infection, excessive intake of TH supplements, and trauma.

System Connections

Homeostatic Interrelationships Between the Endocrine System and Other Body Systems

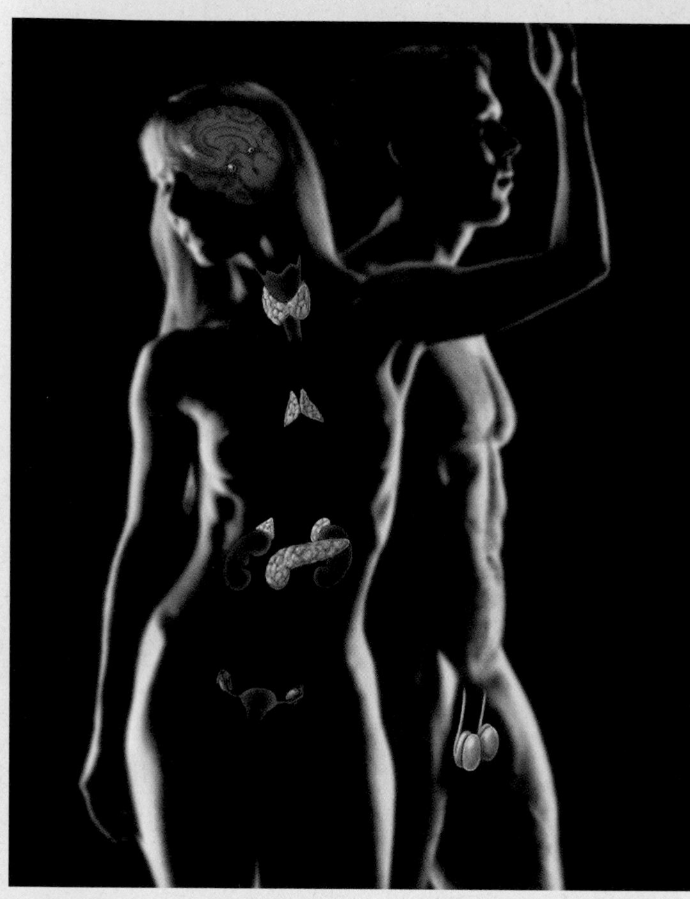

Integumentary System

- Androgens cause activation of sebaceous glands; estrogen increases skin hydration
- Skin produces cholecalciferol (provitamin D)

Skeletal System

- PTH regulates calcium blood levels; growth hormone, T_3, T_4, and sex hormones are necessary for normal skeletal development
- The skeleton provides some protection to endocrine organs, especially to those in the brain, chest, and pelvis

Muscular System

- Growth hormone is essential for normal muscular development; other hormones (thyroxine and catecholamines) influence muscle metabolism
- Muscular system mechanically protects some endocrine glands; muscular activity elicits catecholamine release

Nervous System

- Many hormones (growth hormone, thyroxine, sex hormones) influence normal maturation and function of the nervous system
- Hypothalamus controls anterior pituitary function and produces two hormones

Cardiovascular System

- Several hormones influence blood volume, blood pressure, and heart contractility; erythropoietin stimulates red blood cell production
- Blood is the main transport medium of hormones; heart produces atrial natriuretic peptide

Lymphatic System/Immunity

- Lymphocytes "programmed" by thymic hormones seed the lymph nodes; glucocorticoids depress the immune response and inflammation
- Chemical messengers of the immune system stimulate the release of cortisol and ACTH; lymph provides a route for transport of hormones

Respiratory System

- Epinephrine influences ventilation (dilates bronchioles)
- Respiratory system provides oxygen; disposes of carbon dioxide; converting enzyme in lungs converts angiotensin I to angiotensin II

Digestive System

- GI hormones and paracrines influence GI function; activated vitamin D necessary for absorption of calcium from diet; catecholamines influence digestive motility and secretory activity
- Digestive system provides nutrients to endocrine organs

Urinary System

- Aldosterone and ADH influence renal function; erythropoietin released by kidneys influences red blood cell formation
- Kidneys activate vitamin D (considered a hormone)

Reproductive System

- Hypothalamic, anterior pituitary, and gonadal hormones direct reproductive system development and function; oxytocin and prolactin involved in birth and breast-feeding
- Gonadal hormones feed back to influence endocrine system function

Closer Connections

THE ENDOCRINE SYSTEM and Interrelationships with the Nervous and Reproductive Systems

Like most body systems, the endocrine system performs many functions that benefit the body as a whole. For example, without insulin, thyroxine, and various other metabolic hormones, body cells would be unable to get or use glucose, and would die. Likewise, total body growth is beholden to the endocrine system, which coordinates the growth spurts with increases in skeletal and muscular mass so that we don't look out of proportion most of the time. But the interactions that are most noticeable and crucial are those that the endocrine system has with the nervous and the reproductive systems, and they begin before birth.

Nervous System

The influence of hormones on behavior is striking. While we are still in the wet darkness of our mother's uterus, testosterone—or lack of it—is determining the "sex" of our brain. If testosterone is produced by the exceedingly tiny male testes, then certain areas of the brain enlarge, develop large numbers of androgen receptors, and thereafter determine the so-called masculine aspects of behavior (aggressiveness, etc.). Conversely, in the absence of testosterone, the brain is feminized. At puberty, Mom and Dad's "little angels," driven by raging hormones, turn into strangers. The surge of androgens—produced first by the adrenal cortices, and then by the maturing gonads—produces an often thoughtless aggressiveness and galloping sex drive, typically long before the cognitive abilities of the brain can rein them in.

Neural involvement in hormonal affairs is no less striking. Not only is the hypothalamus an endocrine organ in its own right, but it also effectively regulates the bulk of hormonal activity via its hormonal or neural controls of the pituitary and adrenal medulla. And that's just on a normal day. The effects of trauma on the hypothalamic-pituitary axis can be far-reaching. Lack of loving care to a newborn baby results in failure to thrive; exceptionally vigorous athletic training in the pubertal female can result in bone wasting and infertility. The shadow the nervous system casts over the endocrine system is long indeed.

Reproductive System

The reproductive system is totally dependent on hormones to "order up the right organs" to match our genetic sex. Testosterone secretion by the testes of male embryos directs formation of the male reproductive tract and external genitalia. Without testosterone, female structures develop—regardless of gonadal sex. The next crucial period is puberty, when gonadal sex hormone production rises and steers maturation of the reproductive organs, bringing them to their adult structure and function. Without these hormonal signals, the reproductive organs remain childlike and the person cannot produce offspring.

Pregnancy invites more endocrine system interactions with the reproductive system. The placenta, a temporary endocrine organ, churns out estrogens and progesterone, which help maintain the pregnancy and prepare the mother's breasts for lactation, as well as a number of other hormones that influence maternal metabolism. During and after birth, oxytocin and prolactin take center stage to promote labor and delivery, and then milk production and ejection. Other than feedback inhibition exerted by its sex hormones on the hypothalamic-pituitary axis, the influence of reproductive organs on the endocrine system is negligible.

Clinical Connections

Endocrine System

Case study: We have a new patient to consider today. Mr. Gutteman, a 70-year-old male, was brought into the ER in a comatose state and has yet to come out of it. It is obvious that he suffered severe head trauma—his scalp was badly lacerated, and he has an impacted skull fracture. His initial lab tests (blood and urine) were within normal limits. His fracture was repaired and the following orders (and others) were given:

- Check qh (every hour) and record: spontaneous behavior, level of responsiveness to stimulation, movements, pupil size and reaction to light, speech, and vital signs.
- Turn patient q4h and maintain meticulous skin care and dryness.

1. Explain the rationale behind these orders.

On the second day of his hospitalization, the aide reports that Mr. Gutteman is breathing irregularly, his skin is dry and flaccid, and that she has emptied his urine reservoir several times during the day. Upon receiving this information, the physician ordered

- Blood and urine tests for presence of glucose and ketones
- Strict I&O (fluid intake and output recording)

Mr. Gutteman is found to be losing huge amounts of water in urine and the volume lost is being routinely replaced (via IV line). Mr. Gutteman's blood and urine tests are negative for glucose and ketones.

Relative to these findings:

2. What would you say Mr. Gutteman's hormonal problem is and what do you think caused it?

3. Is it life threatening? (Explain your answer.)

(Answers in Appendix G)

CHAPTER SUMMARY

Media study tools that could provide you additional help in reviewing specific key topics of Chapter 16 are referenced below.

iP = *Interactive Physiology*

1. The nervous and endocrine systems are the major controlling systems of the body. The nervous system exerts rapid controls via nerve impulses; the endocrine system's effects are mediated by hormones and are more prolonged.

The Endocrine System: An Overview (pp. 595–596)

1. Hormonally regulated processes include reproduction; growth and development; maintaining electrolyte, water, and nutrient balance; regulating cellular metabolism and energy balance; and mobilization of body defenses.
2. Endocrine organs are ductless, well-vascularized glands that release hormones directly into the blood or lymph. They are small and widely separated in the body.
3. The purely endocrine organs are the pituitary, thyroid, parathyroid, adrenal, and pineal glands. The hypothalamus is a neuroendocrine organ. The pancreas, gonads, and placenta also have endocrine tissue.
4. Local chemical messengers, not generally considered part of the endocrine system, include autocrines, which act on the cells that secrete them, and paracrines, which act on a different cell type nearby.

iP Endocrine System; Topic: Endocrine System Review, p. 3.

16

Hormones (pp. 596–601)

The Chemistry of Hormones (p. 596)

1. Most hormones are steroids or amino acid based.

iP Endocrine System; Topic: Biochemistry, Secretion, and Transport of Hormones, p. 3.

Mechanisms of Hormone Action (pp. 596–598)

2. Hormones alter cell activity by stimulating or inhibiting characteristic cellular processes of their target cells.
3. Cell responses to hormone stimulation may involve changes in membrane permeability; enzyme synthesis, activation, or inhibition; secretory activity; gene activation; and mitosis.
4. Second-messenger mechanisms employing G proteins and intracellular second messengers are a common means by which amino acid–based hormones interact with their target cells. In the cyclic AMP system, the hormone binds to a plasma membrane receptor that couples to a G protein. When the G protein is activated, it in turn couples to adenylate cyclase, which catalyzes the synthesis of cyclic AMP from ATP. Cyclic AMP initiates reactions that activate protein kinases and other enzymes, leading to cellular response. The PIP_2-calcium signaling mechanism, involving phosphatidyl inositol, is another important second-messenger system. Other second messengers are cyclic GMP and calcium.

iP Endocrine System; Topic: The Actions of Hormones on Target Cells, pp. 3–7.

5. Steroid hormones (and thyroid hormone) enter their target cells and effect responses by activating DNA, which initiates messenger RNA formation leading to protein synthesis.

Target Cell Specificity (p. 598)

6. The ability of a target cell to respond to a hormone depends on the presence of receptors, within the cell or on its plasma membrane, to which the hormone can bind.
7. Hormone receptors are dynamic structures. Changes in number and sensitivity of hormone receptors may occur in response to high or low levels of stimulating hormones.

iP Endocrine System; Topic: The Actions of Hormones on Target Cells, p. 3.

Half-Life, Onset, and Duration of Hormone Activity (p. 599)

8. Blood levels of hormones reflect a balance between secretion and degradation/excretion. The liver and kidneys are the major organs that degrade hormones; breakdown products are excreted in urine and feces.
9. Hormone half-life and duration of activity are limited and vary from hormone to hormone.

Interaction of Hormones at Target Cells (p. 600)

10. Permissiveness is the situation in which one hormone must be present in order for another hormone to exert its full effects.
11. Synergism occurs when two or more hormones produce the same effects in a target cell and their results together are amplified.
12. Antagonism occurs when a hormone opposes or reverses the effect of another hormone.

Control of Hormone Release (pp. 600–601)

13. Humoral, neural, or hormonal stimuli activate endocrine organs to release their hormones. Negative feedback is important in regulating hormone levels in the blood.
14. The nervous system, acting through hypothalamic controls, can in certain cases override or modulate hormonal effects.

iP Endocrine System; Topic: The Hypothalamic–Pituitary Axis, pp. 4 and 5.

The Pituitary Gland and Hypothalamus (pp. 601–608)

Pituitary-Hypothalamic Relationships (p. 603)

1. The pituitary gland hangs from the base of the brain by a stalk and is enclosed by bone. It consists of a hormone-producing glandular portion (anterior pituitary or adenohypophysis) and a neural portion (posterior pituitary or neurohypophysis), which is an extension of the hypothalamus.
2. The hypothalamus (a) regulates the hormonal output of the anterior pituitary via releasing and inhibiting hormones and (b) synthesizes two hormones that it exports to the posterior pituitary for storage and later release.

Anterior Pituitary Hormones (pp. 603–605)

3. Four of the six anterior pituitary hormones are tropic hormones that regulate the function of other endocrine organs. Most anterior pituitary hormones exhibit a diurnal rhythm of release, which is subject to modification by stimuli influencing the hypothalamus.
4. Growth hormone (GH) is an anabolic hormone that stimulates growth of all body tissues but especially skeletal muscle and bone. It may act directly, or indirectly, via insulin-like growth factors

(IGFs). GH mobilizes fats, stimulates protein synthesis, and inhibits glucose uptake and metabolism. Its secretion is regulated by growth hormone–releasing hormone (GHRH) and growth hormone–inhibiting hormone (GHIH), or somatostatin. Hypersecretion causes gigantism in children and acromegaly in adults; hyposecretion in children causes pituitary dwarfism.

5. Thyroid-stimulating hormone (TSH) promotes normal development and activity of the thyroid gland. Thyrotropin-releasing hormone (TRH) stimulates its release; negative feedback of thyroid hormone inhibits it.
6. Adrenocorticotropic hormone (ACTH) stimulates the adrenal cortex to release corticosteroids. ACTH release is triggered by corticotropin-releasing hormone (CRH) and inhibited by rising glucocorticoid levels.
7. The gonadotropins—follicle-stimulating hormone (FSH) and luteinizing hormone (LH)—regulate the functions of the gonads in both sexes. FSH stimulates sex cell production; LH stimulates gonadal hormone production. Gonadotropin levels rise in response to gonadotropin-releasing hormone (GnRH). Negative feedback of gonadal hormones inhibits gonadotropin release.
8. Prolactin (PRL) promotes milk production in humans. Its secretion is inhibited by prolactin-inhibiting hormone (PIH).

The Posterior Pituitary and Hypothalamic Hormones (pp. 605–608)

9. The posterior pituitary stores and releases two hypothalamic hormones, oxytocin and antidiuretic hormone (ADH).
10. Oxytocin stimulates powerful uterine contractions, which trigger labor and delivery of an infant, and milk ejection in nursing women. Its release is mediated reflexively by the hypothalamus and represents a positive feedback mechanism.
11. Antidiuretic hormone stimulates the kidney tubules to reabsorb and conserve water, resulting in small volumes of highly concentrated urine and decreased plasma osmolality. ADH is released in response to high solute concentrations in the blood and inhibited by low solute concentrations in the blood. Hyposecretion results in diabetes insipidus.

The Thyroid Gland (pp. 608–612)

1. The thyroid gland is located in the anterior neck. Thyroid follicles store colloid containing thyroglobulin, a glycoprotein from which thyroid hormone is derived.
2. Thyroid hormone (TH) includes thyroxine (T_4) and triiodothyronine (T_3), which increase the rate of cellular metabolism. Consequently, oxygen use and heat production rise.
3. Secretion of thyroid hormone, prompted by TSH, requires reuptake of the stored colloid by the follicle cells and splitting of the hormones from the colloid for release. Rising levels of thyroid hormone feed back to inhibit the anterior pituitary and hypothalamus.
4. Most T_4 is converted to T_3 (the more active form) in the target tissues. These hormones act by turning on gene transcription and protein synthesis.
5. Graves' disease is the most common cause of hyperthyroidism; hyposecretion causes cretinism in infants and myxedema in adults.
6. Calcitonin, produced by the parafollicular (C) cells of the thyroid gland in response to rising blood calcium levels, inhibits bone matrix resorption and enhances calcium deposit in bone. It is not normally important in calcium homeostasis.

iP **Endocrine System; Topic: The Hypothalamic–Pituitary Axis, p. 6.**

The Parathyroid Glands (pp. 612–614)

1. The parathyroid glands, located on the dorsal aspect of the thyroid gland, secrete parathyroid hormone (PTH), which causes an increase in blood calcium levels. It targets bone, the kidneys, and the intestine (indirectly via vitamin D activation). PTH is the key hormone for calcium homeostasis.
2. PTH release is triggered by falling blood calcium levels and is inhibited by rising blood calcium levels.
3. Hyperparathyroidism results in hypercalcemia and all its effects and in extreme bone wasting. Hypoparathyroidism leads to hypocalcemia, evidenced by tetany and respiratory paralysis.

The Adrenal (Suprarenal) Glands (pp. 614–620)

1. The paired adrenal (suprarenal) glands sit atop the kidneys. Each adrenal gland has two functional portions, the cortex and the medulla.

The Adrenal Cortex (pp. 614–618)

2. The cortex produces three groups of steroid hormones from cholesterol.
3. Mineralocorticoids (primarily aldosterone) regulate sodium ion reabsorption and potassium ion excretion by the kidneys. Sodium ion reabsorption leads to water reabsorption, and increases in blood volume and blood pressure. Release of aldosterone is stimulated by the renin-angiotensin mechanism, rising potassium ion levels in the blood, and ACTH. Atrial natriuretic peptide inhibits aldosterone release.
4. Glucocorticoids (primarily cortisol) are important metabolic hormones that help the body resist stressors by increasing blood glucose, fatty acid and amino acid levels, and blood pressure. High levels of glucocorticoids depress the immune system and the inflammatory response. ACTH is the major stimulus for glucocorticoid release.
5. Gonadocorticoids (mainly androgens) are produced in small amounts throughout life.
6. Hypoactivity of the adrenal cortex results in Addison's disease. Hypersecretion can result in aldosteronism, Cushing's syndrome, and adrenogenital syndrome.

The Adrenal Medulla (pp. 618–620)

7. The adrenal medulla produces catecholamines (epinephrine and norepinephrine) in response to sympathetic nervous system stimulation. Its catecholamines enhance and prolong the fight-or-flight response to short-term stressors. Hypersecretion leads to symptoms typical of sympathetic nervous system overactivity.

iP **Endocrine System; Topic: Response to Stress, pp. 5–8.**

The Pineal Gland (pp. 620)

1. The pineal gland is located in the diencephalon. Its primary hormone is melatonin, which influences daily rhythms and may have an antigonadotropic effect in humans.

Other Endocrine Glands and Tissues (pp. 620–624)

The Pancreas (pp. 620–623)

1. The pancreas, located in the abdomen close to the stomach, is both an exocrine and an endocrine gland. The endocrine portion (pancreatic islets) releases insulin and glucagon and smaller amounts of other hormones to the blood.
2. Glucagon, released by alpha (α) cells when blood levels of glucose are low, stimulates the liver to release glucose to the blood.

3. Insulin is released by beta (β) cells when blood levels of glucose (and amino acids) are rising. It increases the rate of glucose uptake and metabolism by most body cells. Hyposecretion or hypoactivity of insulin results in diabetes mellitus; cardinal signs are polyuria, polydipsia, and polyphagia.

iP Endocrine System; Topic: The Actions of Hormones on Target Cells, pp. 5 and 8.

The Gonads and Placenta (p. 623)

4. The ovaries of the female, located in the pelvic cavity, release two main hormones. The ovarian follicles begin to secrete estrogens at puberty under the influence of FSH. Estrogens stimulate maturation of the female reproductive system and development of the secondary sex characteristics. Progesterone is released in response to high blood levels of LH. It works with estrogens in establishing the menstrual cycle.
5. The testes of the male begin to produce testosterone at puberty in response to LH. Testosterone promotes maturation of the male reproductive organs, development of secondary sex characteristics, and production of sperm by the testes.
6. The placenta produces hormones of pregnancy—estrogens, progesterone, and others.

Hormone Secretion by Other Organs (p. 624)

7. Many body organs not normally considered endocrine organs contain cells that secrete hormones. Examples include the heart (atrial natriuretic peptide); gastrointestinal tract organs (gastrin, secretin, incretins, and others); the kidneys (erythropoietin and renin); skin (cholecalciferol); adipose tissue (leptin, resistin, and adiponectin); bone (osteocalcin); and thymus (thymic hormones).
8. The thymus, located in the upper thorax, declines in size and function with age. Its hormones, thymulin, thymosins, and thymopoietins, are important to the normal development of the immune response.

Developmental Aspects of the Endocrine System (pp. 624, 626–627)

1. Endocrine glands derive from all three germ layers. Those derived from mesoderm produce steroid hormones; the others produce the amino acid–based hormones.
2. The natural decrease in function of the female's ovaries during late middle age results in menopause.
3. The efficiency of all endocrine glands seems to decrease gradually as aging occurs. This change leads to a generalized increase in the incidence of diabetes mellitus and a lower metabolic rate.

REVIEW QUESTIONS

16

Multiple Choice/Matching

(Some questions have more than one correct answer. Select the best answer or answers from the choices given.)

1. The major stimulus for release of parathyroid hormone is **(a)** hormonal, **(b)** humoral, **(c)** neural.
2. The anterior pituitary secretes all but **(a)** antidiuretic hormone, **(b)** growth hormone, **(c)** gonadotropins, **(d)** TSH.
3. A hormone not involved in glucose metabolism is **(a)** glucagon, **(b)** cortisone, **(c)** aldosterone, **(d)** insulin.
4. Parathyroid hormone **(a)** increases bone formation and lowers blood calcium levels, **(b)** increases calcium excretion from the body, **(c)** decreases calcium absorption from the gut, **(d)** demineralizes bone and raises blood calcium levels.
5. Choose from the following key to identify the hormones described.

Key:

(a) aldosterone	(e) oxytocin
(b) antidiuretic hormone	(f) prolactin
(c) growth hormone	(g) T_4 and T_3
(d) luteinizing hormone	(h) TSH

_____ (1) important anabolic hormone; many of its effects mediated by IGFs
_____ (2) cause the kidneys to conserve water and/or salt (two choices)
_____ (3) stimulates milk production
_____ (4) tropic hormone that stimulates the gonads to secrete sex hormones
_____ (5) increases uterine contractions during birth
_____ (6) major metabolic hormone(s) of the body
_____ (7) causes reabsorption of sodium ions by the kidneys
_____ (8) tropic hormone that stimulates the thyroid gland to secrete thyroid hormone
_____ (9) secreted by the posterior pituitary (two choices)
_____ (10) the only steroid hormone in the list

6. A hypodermic injection of epinephrine would **(a)** increase heart rate, increase blood pressure, dilate the bronchi of the lungs, and increase peristalsis, **(b)** decrease heart rate, decrease blood pressure, constrict the bronchi, and increase peristalsis, **(c)** decrease heart rate, increase blood pressure, constrict the bronchi, and decrease peristalsis, **(d)** increase heart rate, increase blood pressure, dilate the bronchi, and decrease peristalsis.
7. Testosterone is to the male as what hormone is to the female? **(a)** luteinizing hormone, **(b)** progesterone, **(c)** estrogen, **(d)** prolactin.
8. If anterior pituitary secretion is deficient in a growing child, the child will **(a)** develop acromegaly, **(b)** become a dwarf but have fairly normal body proportions, **(c)** mature sexually at an earlier than normal age, **(d)** be in constant danger of becoming dehydrated.
9. If there is adequate carbohydrate intake, secretion of insulin results in **(a)** lower blood glucose levels, **(b)** increased cell utilization of glucose, **(c)** storage of glycogen, **(d)** all of these.
10. Hormones **(a)** are produced by exocrine glands, **(b)** are carried to all parts of the body in blood, **(c)** remain at constant concentration in the blood, **(d)** affect only non-hormone-producing organs.
11. Some hormones act by **(a)** increasing the synthesis of enzymes, **(b)** converting an inactive enzyme into an active enzyme, **(c)** affecting only specific target organs, **(d)** all of these.
12. Absence of thyroid hormone would result in **(a)** increased heart rate and increased force of heart contraction, **(b)** depression of the CNS and lethargy, **(c)** exophthalmos, **(d)** high metabolic rate.
13. Chromaffin cells are found in the **(a)** parathyroid gland, **(b)** anterior pituitary gland, **(c)** adrenal gland, **(d)** pineal gland.
14. Atrial natriuretic peptide secreted by the heart has exactly the opposite function of this hormone secreted by the zona glomerulosa: **(a)** antidiuretic hormone, **(b)** epinephrine, **(c)** calcitonin, **(d)** aldosterone, **(e)** androgens.

Short Answer Essay Questions

15. Define hormone.
16. Which type of hormone receptor—plasma membrane bound or intracellular—would be expected to provide the most long-lived response to hormone binding and why?
17. (a) Describe the body location of each of the following endocrine organs: anterior pituitary, pineal gland, pancreas, ovaries, testes, and adrenal glands. (b) List the hormones produced by each organ.
18. Name two endocrine glands (or regions) that are important in the stress response, and explain why they are important.
19. The anterior pituitary is often referred to as the master endocrine organ, but it, too, has a "master." What controls the release of anterior pituitary hormones?
20. The posterior pituitary is not really an endocrine gland. Why not? What is it?
21. A colloidal, or endemic, goiter is not really the result of malfunction of the thyroid gland. What does cause it?
22. How are the hyperglycemia and lipidemia of insulin deficiency linked?
23. Name a hormone secreted by a muscle cell and two hormones secreted by neurons.
24. List some problems that elderly people might have as a result of decreasing hormone production.

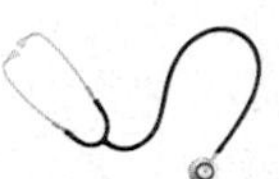

Critical Thinking and Clinical Application Questions

1. Richard Neis had symptoms of excessive secretion of PTH (high blood calcium levels), and his physicians were certain he had a parathyroid gland tumor. Yet when surgery was performed on his neck, the surgeon could not find the parathyroid glands at all. Where should the surgeon look next to find the tumorous parathyroid gland?
2. Mary Morgan has just been brought into the emergency room of City General Hospital. She is perspiring profusely and is breathing rapidly and irregularly. Her breath smells like acetone (sweet and fruity), and her blood glucose tests out at 650 mg/100 ml of blood. She is in acidosis. What hormone drug should be administered, and why?
3. Johnny, a 5-year-old boy, has been growing by leaps and bounds; his height is 100% above normal for his age. He has been complaining of headaches and vision problems. A CT scan reveals a large pituitary tumor. (a) What hormone is being secreted in excess? (b) What condition will Johnny exhibit if corrective measures are not taken? (c) What is the probable cause of his headaches and visual problems?
4. Sean, a 42-year-old single father, goes to his physician complaining of nausea and chronic fatigue. He reports having felt fatigued and listless for about half a year, but he had attributed this to stress. He has lost considerable weight and, strangely, his skin has a healthy tan, even though he spends long hours at work and rarely ventures outside. His doctor finds very low blood pressure and a rapid, weak pulse. Blood tests show that Sean does not have anemia, but his plasma glucose, cortisol, and Na^+ are low, and his plasma K^+ is high. His doctor orders an ACTH stimulation test, in which Sean's secretion of cortisol is measured after he is given a synthetic form of ACTH. (a) What would account for Sean's low plasma Na^+ and high plasma K^+? (b) What is the reason for doing an ACTH stimulation test? (c) What gland is primarily affected if ACTH *does not* cause a normal elevation of cortisol secretion? What is this abnormality called? (d) What gland is primarily affected if ACTH does cause an elevation of cortisol secretion?
5. Roger Proulx has severe arthritis and has been taking prednisone (a glucocorticoid) for two months. He isn't feeling well, complains of repeated "colds," and is extremely "puffy" (edematous). Explain the reason for these symptoms.

16

Regulation

Regulatory functions of blood include

- Maintaining appropriate body temperature by absorbing and distributing heat throughout the body and to the skin surface to encourage heat loss.
- Maintaining normal pH in body tissues. Many blood proteins and other bloodborne solutes act as buffers to prevent excessive or abrupt changes in blood pH that could jeopardize normal cell activities. Additionally, blood acts as the reservoir for the body's "alkaline reserve" of bicarbonate atoms.
- Maintaining adequate fluid volume in the circulatory system. Salts (sodium chloride and others) and blood proteins act to prevent excessive fluid loss from the bloodstream into the tissue spaces. As a result, the fluid volume in the blood vessels remains ample to support efficient blood circulation to all parts of the body.

Protection

Protective functions of blood include

- Preventing blood loss. When a blood vessel is damaged, platelets and plasma proteins initiate clot formation, halting blood loss.
- Preventing infection. Drifting along in blood are antibodies, complement proteins, and white blood cells, all of which help defend the body against foreign invaders such as bacteria and viruses.

Blood Plasma

▶ Discuss the composition and functions of plasma.

Blood **plasma** is a straw-colored, sticky fluid (Figure 17.1). Although it is mostly water (about 90%), plasma contains over 100 different dissolved solutes, including nutrients, gases, hormones, wastes and products of cell activity, ions, and proteins. **Table 17.1** summarizes the major plasma components.

Plasma proteins are the most abundant plasma solutes, accounting for about 8% by weight of plasma volume. Except for hormones and gamma globulins, most plasma proteins are produced by the liver. Plasma proteins serve a variety of functions, but they are *not* taken up by cells to be used as fuels or metabolic nutrients as are most other plasma solutes, such as glucose, fatty acids, and amino acids.

Albumin (al-bu′min) accounts for some 60% of plasma protein. It acts as a carrier to shuttle certain molecules through the circulation, is an important blood buffer, and is the major blood protein contributing to the plasma osmotic pressure (the pressure that helps to keep water in the bloodstream). (Sodium ions are the other major solute contributing to blood osmotic pressure.)

The makeup of plasma varies continuously as cells remove or add substances to the blood. However, assuming a healthy diet, plasma composition is kept relatively constant by various homeostatic mechanisms. For example, when blood protein levels drop undesirably, the liver makes more proteins. When the blood starts to become too acidic (acidosis), both the respiratory system and the kidneys are called into action to restore plasma's normal, slightly alkaline pH. Body organs make dozens of adjustments, day in and day out, to maintain the many plasma solutes at life-sustaining levels.

TABLE 17.1 Composition of Plasma

CONSTITUENT	DESCRIPTION AND IMPORTANCE
Water	90% of plasma volume; dissolving and suspending medium for solutes of blood; absorbs heat
Solutes	
Plasma proteins	8% (by weight) of plasma volume; all contribute to osmotic pressure and maintain water balance in blood and tissues; all have other functions (transport, enzymatic, etc.) as well
▪ Albumin	60% of plasma proteins; produced by liver; main contributor to osmotic pressure
▪ Globulins	36% of plasma proteins
alpha, beta	Produced by liver; most are transport proteins that bind to lipids, metal ions, and fat-soluble vitamins
gamma	Antibodies released by plasma cells during immune response
▪ Fibrinogen	4% of plasma proteins; produced by liver; forms fibrin threads of blood clot
Nonprotein nitrogenous substances	By-products of cellular metabolism, such as urea, uric acid, creatinine, and ammonium salts
Nutrients (organic)	Materials absorbed from digestive tract and transported for use throughout body; include glucose and other simple carbohydrates, amino acids (digestion products of proteins), fatty acids, glycerol and triglycerides (fat products), cholesterol, and vitamins
Electrolytes	Cations include sodium, potassium, calcium, magnesium; anions include chloride, phosphate, sulfate, and bicarbonate; help to maintain plasma osmotic pressure and normal blood pH
Respiratory gases	Oxygen and carbon dioxide; oxygen mostly bound to hemoglobin inside RBCs; carbon dioxide transported dissolved in plasma as bicarbonate ion or CO_2, or bound to hemoglobin in RBCs
Hormones	Steroid and thyroid hormones carried by plasma proteins

CHECK YOUR UNDERSTANDING

1. What is the hematocrit? What is its normal value?
2. List two protective functions of blood.
3. Are plasma proteins used as fuel for body cells? Explain your answer.

For answers, see Appendix G.

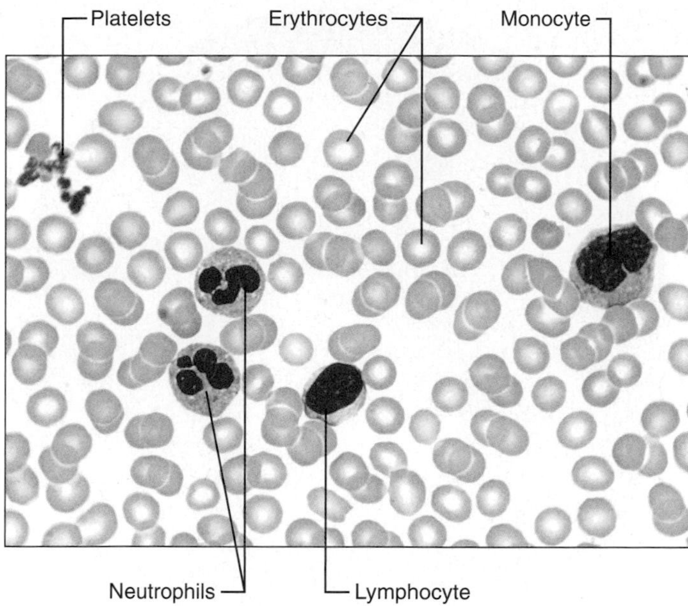

Figure 17.2 Photomicrograph of a human blood smear stained with Wright's stain. (640×)

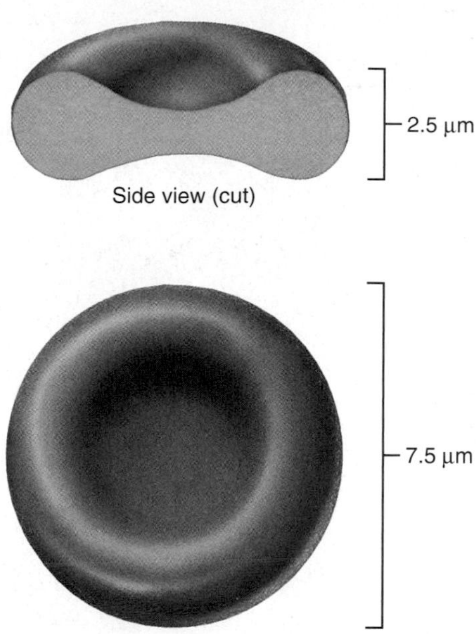

Figure 17.3 Structure of erythrocytes. Notice the distinctive biconcave shape.

Formed Elements

The **formed elements** of blood, *erythrocytes*, *leukocytes*, and *platelets*, have some unusual features. (1) Two of the three are not even true cells: Erythrocytes have no nuclei or organelles, and platelets are cell fragments. Only leukocytes are complete cells. (2) Most of the formed elements survive in the bloodstream for only a few days. (3) Most blood cells do not divide. Instead, they are continuously renewed by division of cells in red bone marrow, where they originate.

If you examine a stained smear of human blood under the light microscope, you will see disc-shaped red blood cells, a variety of gaudily stained spherical white blood cells, and some scattered platelets that look like debris **(Figure 17.2)**. Erythrocytes vastly outnumber the other types of formed elements. Table 17.2 on p. 645 summarizes the important characteristics of the formed elements.

Erythrocytes

- Describe the structure, function, and production of erythrocytes.
- Describe the chemical makeup of hemoglobin.
- Give examples of disorders caused by abnormalities of erythrocytes. Explain what goes wrong in each disorder.

Structural Characteristics

Erythrocytes or **red blood cells (RBCs)** are small cells, about 7.5 μm in diameter. Shaped like biconcave discs—flattened discs with depressed centers—they appear lighter in color at their thin centers than at their edges **(Figure 17.3)**. Consequently, erythrocytes look like miniature doughnuts when viewed with a microscope. Mature erythrocytes are bound by a plasma membrane, but lack a nucleus (are *anucleate*) and have essentially no organelles. In fact, they are little more than "bags" of *hemoglobin* (*Hb*), the RBC protein that functions in gas transport. Other proteins are present, such as antioxidant enzymes that rid the body of harmful oxygen radicals, but most function mainly to maintain the plasma membrane or promote changes in RBC shape.

For example, the biconcave shape of an erythrocyte is maintained by a network of proteins, especially one called *spectrin*, attached to the cytoplasmic face of its plasma membrane. The spectrin net is deformable, giving erythrocytes flexibility to change shape as necessary—to twist, turn, and become cup shaped as they are carried passively through capillaries with diameters smaller than themselves—and then to resume their biconcave shape.

The erythrocyte is a superb example of complementarity of structure and function. It picks up oxygen in the capillary beds of the lungs and releases it to tissue cells across other capillaries throughout the body. It also transports some 20% of the carbon dioxide released by tissue cells back to the lungs. Three structural characteristics contribute to erythrocyte gas transport functions:

1. Its small size and biconcave shape provide a huge surface area relative to volume (about 30% more surface area than comparable spherical cells). The biconcave disc shape is ideally suited for gas exchange because no point within the cytoplasm is far from the surface.
2. Discounting water content, an erythrocyte is over 97% hemoglobin, the molecule that binds to and transports respiratory gases.
3. Because erythrocytes lack mitochondria and generate ATP by anaerobic mechanisms, they do not consume any of the oxygen they are transporting, making them very efficient oxygen transporters indeed.

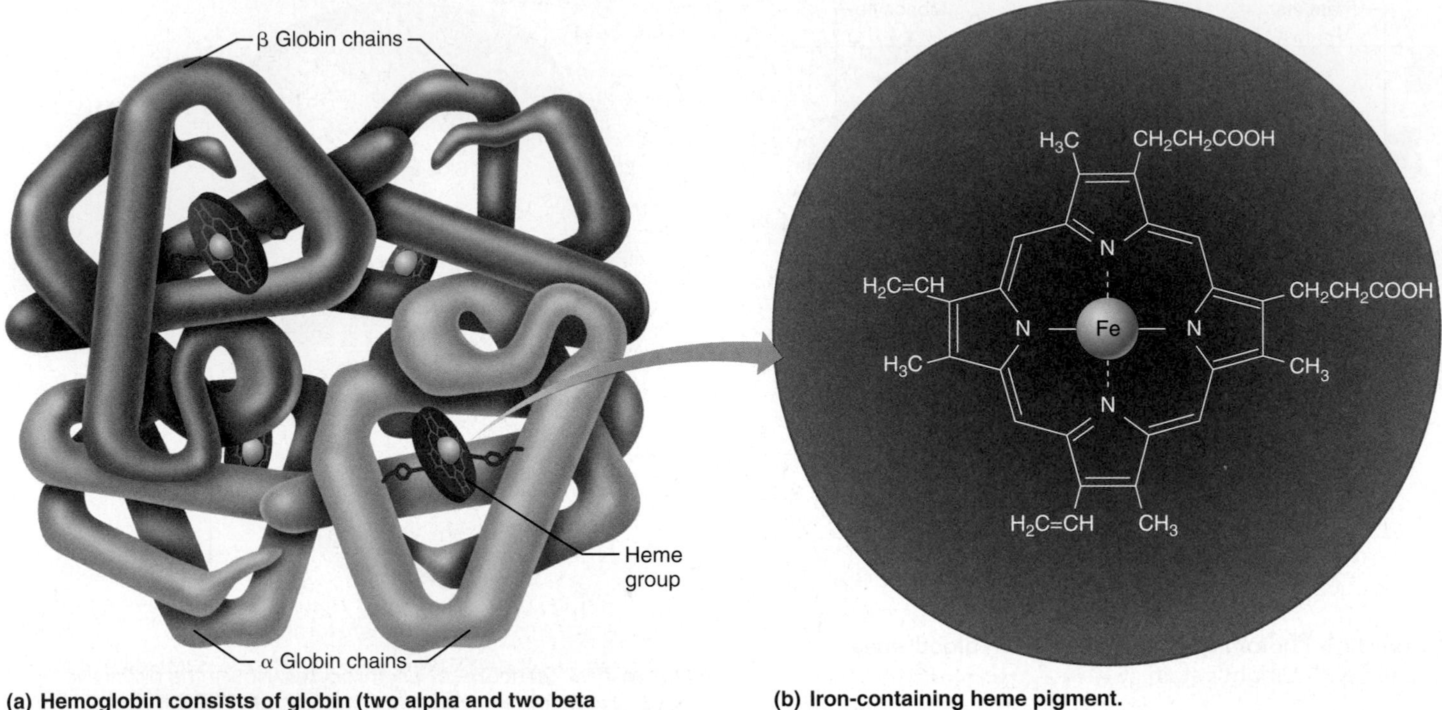

(a) Hemoglobin consists of globin (two alpha and two beta polypeptide chains) and four heme groups.

(b) Iron-containing heme pigment.

Figure 17.4 **Structure of hemoglobin.**

Erythrocytes are the major factor contributing to blood viscosity. Women typically have a lower red blood cell count than men [4.3–5.2 million cells per microliter (1 μl = 1 mm^3) of blood versus 5.1–5.8 million cells/μl respectively]. When the number of red blood cells increases beyond the normal range, blood viscosity rises and blood flows more slowly. Similarly, as the number of red blood cells drops below the lower end of the range, the blood thins and flows more rapidly.

Function

Erythrocytes are completely dedicated to their job of transporting respiratory gases (oxygen and carbon dioxide). **Hemoglobin**, the protein that makes red blood cells red, binds easily and reversibly with oxygen, and most oxygen carried in blood is bound to hemoglobin. Normal values for hemoglobin are 14–20 grams per 100 milliliters of blood (g/100 ml) in infants, 13–18 g/100 ml in adult males, and 12–16 g/100 ml in adult females.

Hemoglobin is made up of the protein **globin** bound to the red **heme** pigment. Globin consists of four polypeptide chains—two alpha (α) and two beta (β)—each binding a ringlike heme group **(Figure 17.4a)**. Each heme group bears an atom of iron set like a jewel in its center (Figure 17.4b). A hemoglobin molecule can transport four molecules of oxygen because each iron atom can combine reversibly with one molecule of oxygen. A single red blood cell contains about 250 million hemoglobin molecules, so each of these tiny cells can scoop up about 1 billion molecules of oxygen!

The fact that hemoglobin is contained in erythrocytes, rather than existing free in plasma, prevents it (1) from breaking into fragments that would leak out of the bloodstream (through the rather porous capillary membranes) and (2) from contributing to blood viscosity and osmotic pressure.

Oxygen loading occurs in the lungs, and the direction of transport is from lungs to tissue cells. As oxygen-deficient blood moves through the lungs, oxygen diffuses from the air sacs of the lungs into the blood and then into the erythrocytes, where it binds to hemoglobin. When oxygen binds to iron, the hemoglobin, now called **oxyhemoglobin**, assumes a new three-dimensional shape and becomes ruby red. In the tissues, the process is reversed. Oxygen detaches from iron, hemoglobin resumes its former shape, and the resulting **deoxyhemoglobin**, or *reduced hemoglobin*, becomes dark red. The released oxygen diffuses from the blood into the tissue fluid and then into the tissue cells.

About 20% of the carbon dioxide transported in the blood combines with hemoglobin, but it binds to globin's amino acids rather than to the heme group. This formation of **carbaminohemoglobin** (kar-bam″ ĭ-no-he″mo″-glo′bin) occurs more readily when hemoglobin is in the reduced state (dissociated from oxygen). Carbon dioxide loading occurs in the tissues, and the direction of transport is from tissues to lungs, where carbon dioxide is eliminated from the body. We describe the loading and unloading of these respiratory gases in Chapter 22.

Production of Erythrocytes

Blood cell formation is referred to as **hematopoiesis** (hem″ah-to-poi-e′sis), or **hemopoiesis** (*hemo, hemato* = blood; *poiesis* = to make). This process occurs in the **red bone marrow**, which is composed largely of a soft network of reticular connective tissue bordering on wide blood capillaries called *blood sinusoids.*

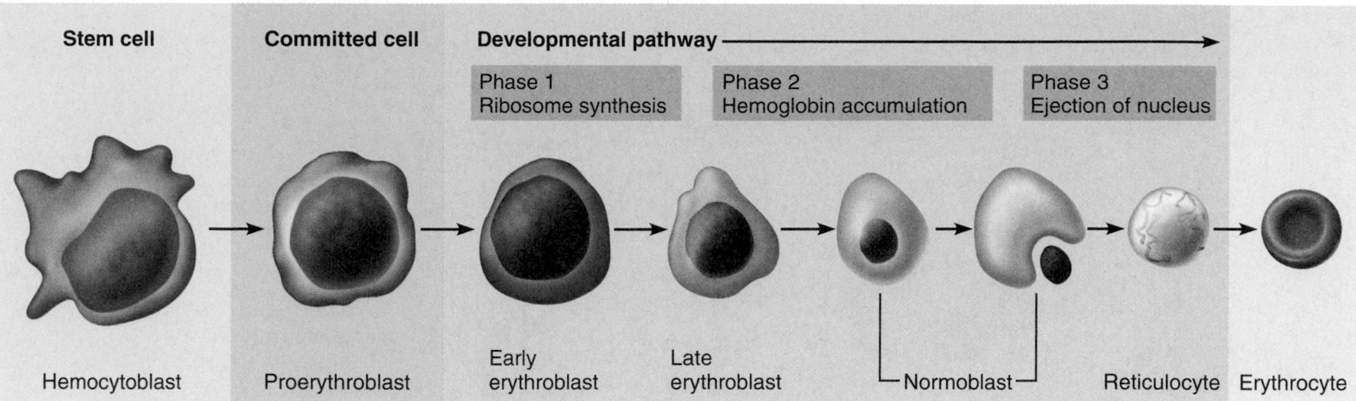

Figure 17.5 Erythropoiesis: genesis of red blood cells. Reticulocytes are released into the bloodstream. The myeloid stem cell, the phase intermediate between the hemocytoblast and the proerythroblast, is not illustrated.

Within this network are immature blood cells, macrophages, fat cells, and *reticular cells* (which secrete the fibers). In adults, red marrow is found chiefly in the bones of the axial skeleton and girdles, and in the proximal epiphyses of the humerus and femur.

Each type of blood cell is produced in different numbers in response to changing body needs and different regulatory factors. As blood cells mature, they migrate through the thin walls of the sinusoids to enter the bloodstream. On average, the marrow turns out an ounce of new blood containing some 100 billion new cells each and every day.

The various formed elements have different functions, but there are similarities in their life histories. All arise from the same type of *stem cell*, the **hemocytoblast** (*cyte* = cell, *blast* = bud), or pluripotent **hematopoietic stem cell**. These undifferentiated precursor cells reside in the red bone marrow. The maturation pathways of the various formed elements differ, however, and once a cell is *committed* to a specific blood cell pathway, it cannot change. This commitment is signaled by the appearance of membrane surface receptors that respond to specific hormones or growth factors, which in turn "push" the cell toward further specialization.

Erythrocyte production, or **erythropoiesis** (ĕ-rith″ro-poi-e′sis) begins when a hemocytoblast descendant called a **myeloid stem cell** is transformed into a **proerythroblast** (**Figure 17.5**). Proerythroblasts, in turn, give rise to the **early (basophilic) erythroblasts** that produce huge numbers of ribosomes. During these first two phases, the cells divide many times. Hemoglobin is synthesized and iron accumulates as the early erythroblast is transformed into a **late erythroblast** and then a **normoblast**. The "color" of the cell cytoplasm changes as the blue-staining ribosomes become masked by the pink color of hemoglobin. When a normoblast has accumulated almost all of its hemoglobin, it ejects most of its organelles. Additionally, its nuclear functions end and its nucleus degenerates and is pinched off, allowing the cell to collapse inward and eventually assume the biconcave shape. The result is the **reticulocyte** (essentially a young erythrocyte), so named because it still contains a scant *reticulum* (network) of clumped ribosomes.

The entire process from hemocytoblast to reticulocyte takes about 15 days. The reticulocytes, filled almost to bursting with hemoglobin, enter the bloodstream to begin their task of oxygen transport. Usually they become fully mature erythrocytes within two days of release as their ribosomes are degraded by intracellular enzymes.

Reticulocytes account for 1–2% of all erythrocytes in the blood of healthy people. **Reticulocyte counts** provide a rough index of the *rate* of RBC formation—reticulocyte counts below or above this percentage range indicate abnormal rates of erythrocyte formation.

Regulation and Requirements for Erythropoiesis

The number of circulating erythrocytes in a given individual is remarkably constant and reflects a balance between red blood cell production and destruction. This balance is important because having too few erythrocytes leads to tissue hypoxia (oxygen deprivation), whereas having too many makes the blood undesirably viscous. To ensure that the number of erythrocytes in blood remains within the homeostatic range, new cells are produced at the incredibly rapid rate of more than 2 million per second in healthy people. This process is controlled hormonally and depends on adequate supplies of iron, amino acids, and certain B vitamins.

Hormonal Controls The direct stimulus for erythrocyte formation is provided by **erythropoietin (EPO)**, a glycoprotein hormone (**Figure 17.6**). Normally, a small amount of EPO circulates in the blood at all times and sustains red blood cell production at a basal rate. The kidneys play the major role in EPO production, although the liver produces some. When certain kidney cells become *hypoxic* (i.e., have inadequate oxygen), oxygen-sensitive enzymes are unable to carry out their normal functions of degrading an intracellular signaling molecule called hypoxia-inducible factor (HIF). As HIF accumulates, it accelerates the synthesis and release of erythropoietin.

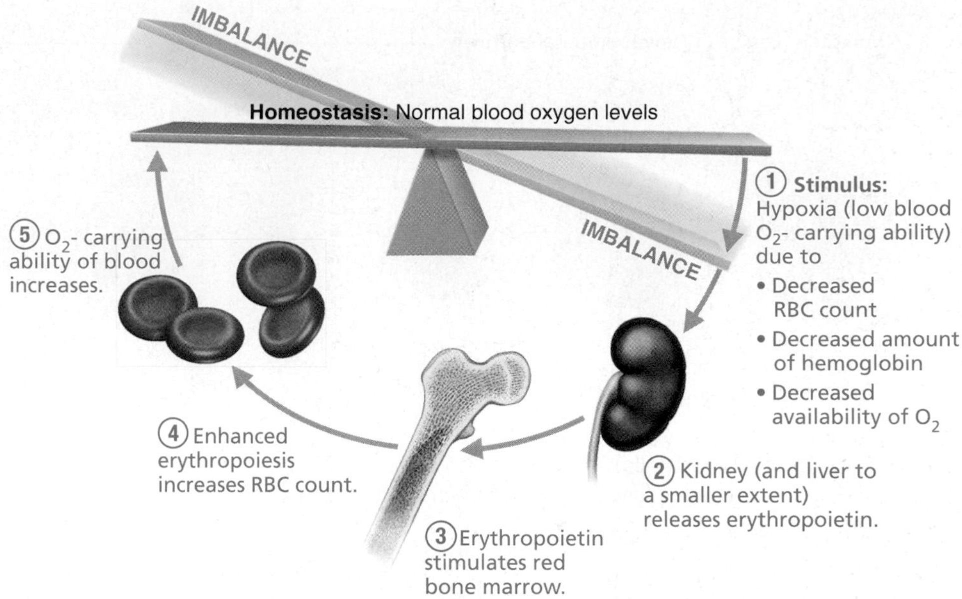

Figure 17.6 Erythropoietin mechanism for regulating erythropoiesis.

The drop in normal blood oxygen levels that triggers EPO formation can result from

1. Reduced numbers of red blood cells due to hemorrhage or excessive RBC destruction
2. Insufficient hemoglobin per RBC (as in iron deficiency)
3. Reduced availability of oxygen, as might occur at high altitudes or during pneumonia

Conversely, too many erythrocytes or excessive oxygen in the bloodstream depresses erythropoietin production. Note that it is not the number of erythrocytes in blood that controls the rate of erythropoiesis. Instead, control is based on their ability to transport enough oxygen to meet tissue demands.

Bloodborne erythropoietin stimulates red marrow cells that *are already committed* to becoming erythrocytes, causing them to mature more rapidly. One to two days after erythropoietin levels rise in the blood, a marked increase in the rate of reticulocyte release (and hence reticulocyte count) occurs. Notice that hypoxia (oxygen deficit) does not activate the bone marrow directly. Instead it stimulates the kidneys, which in turn provide the hormonal stimulus that activates the bone marrow.

HOMEOSTATIC IMBALANCE

Renal dialysis patients whose kidneys have failed produce too little EPO to support normal erythropoiesis. Consequently, they routinely have red blood cell counts less than half that of healthy individuals. Genetically engineered (recombinant) EPO has helped such patients immeasurably but has also become a substance of abuse in athletes—particularly in professional bike racers and marathon runners seeking increased stamina and performance. However, the consequences can be deadly. By injecting EPO, healthy athletes increase their normal hematocrit from 45% to as much as 65%. Then, with the dehydration that occurs in a long race, the blood concentrates even further, becoming a thick, sticky "sludge" that can cause clotting, stroke, and heart failure. ■

The male sex hormone *testosterone* also enhances EPO production by the kidneys. Because female sex hormones do not have similar stimulatory effects, testosterone may be at least partially responsible for the higher RBC counts and hemoglobin levels seen in males. Also, a wide variety of chemicals released by leukocytes, platelets, and even reticular cells stimulates bursts of RBC production.

Dietary Requirements The raw materials required for erythropoiesis include the usual nutrients and structural materials—amino acids, lipids, and carbohydrates. Iron is essential for hemoglobin synthesis. Iron is available from the diet, and its absorption into the bloodstream is precisely controlled by intestinal cells in response to changing body stores of iron.

Approximately 65% of the body's iron supply (about 4000 mg) is in hemoglobin. Most of the remainder is stored in the liver, spleen, and (to a much lesser extent) bone marrow. Free iron ions (Fe^{2+}, Fe^{3+}) are toxic, so iron is stored inside cells as protein-iron complexes such as **ferritin** (fer′ĭ-tin) and **hemosiderin** (he″mo-sid′er-in). In blood, iron is transported loosely bound to a transport protein called **transferrin**, and developing erythrocytes take up iron as needed to form hemoglobin **(Figure 17.7)**. Small amounts of iron are lost each day in feces, urine, and perspiration. The average daily loss of iron is 1.7 mg in women and 0.9 mg in men. In women, the menstrual flow accounts for the additional losses.

Two B-complex vitamins—vitamin B_{12} and folic acid—are necessary for normal DNA synthesis. Thus, even slight deficits jeopardize rapidly dividing cell populations, such as developing erythrocytes.

Fate and Destruction of Erythrocytes

Red blood cells have a useful life span of 100 to 120 days (Figure 17.7). Their anucleate condition carries with it some important limitations. Red blood cells are unable to synthesize new proteins, to grow, or to divide. Erythrocytes become "old" as they lose their flexibility and become increasingly rigid and fragile, and their contained hemoglobin begins to degenerate. They become trapped and fragment in smaller circulatory channels, particularly in those of the spleen. For this reason, the spleen is sometimes called the "red blood cell graveyard."

Dying erythrocytes are engulfed and destroyed by macrophages. The heme of their hemoglobin is split off from globin. Its core of iron is salvaged, bound to protein (as ferritin or hemosiderin), and stored for reuse. The balance of the heme group is degraded to **bilirubin** (bil″i-roo′bin), a yellow pigment that is released to the blood and binds to albumin for transport (Figure 17.7, ⑤). Bilirubin is picked up by liver cells, which in turn secrete it (in bile) into the intestine, where it is metabolized to *urobilinogen.* Most of this degraded pigment leaves the body in feces, as a brown pigment called *stercobilin.* The protein (globin) part of hemoglobin is metabolized or broken down to amino acids, which are released to the circulation.

Disposal of hemoglobin spilled from red blood cells to the blood (as occurs in sickle-cell anemia or hemolytic anemia) takes a similar but much more rapid course to avoid toxic buildup of iron in blood. Released hemoglobin is captured by the plasma protein *haptoglobin* and the complex is phagocytized by macrophages.

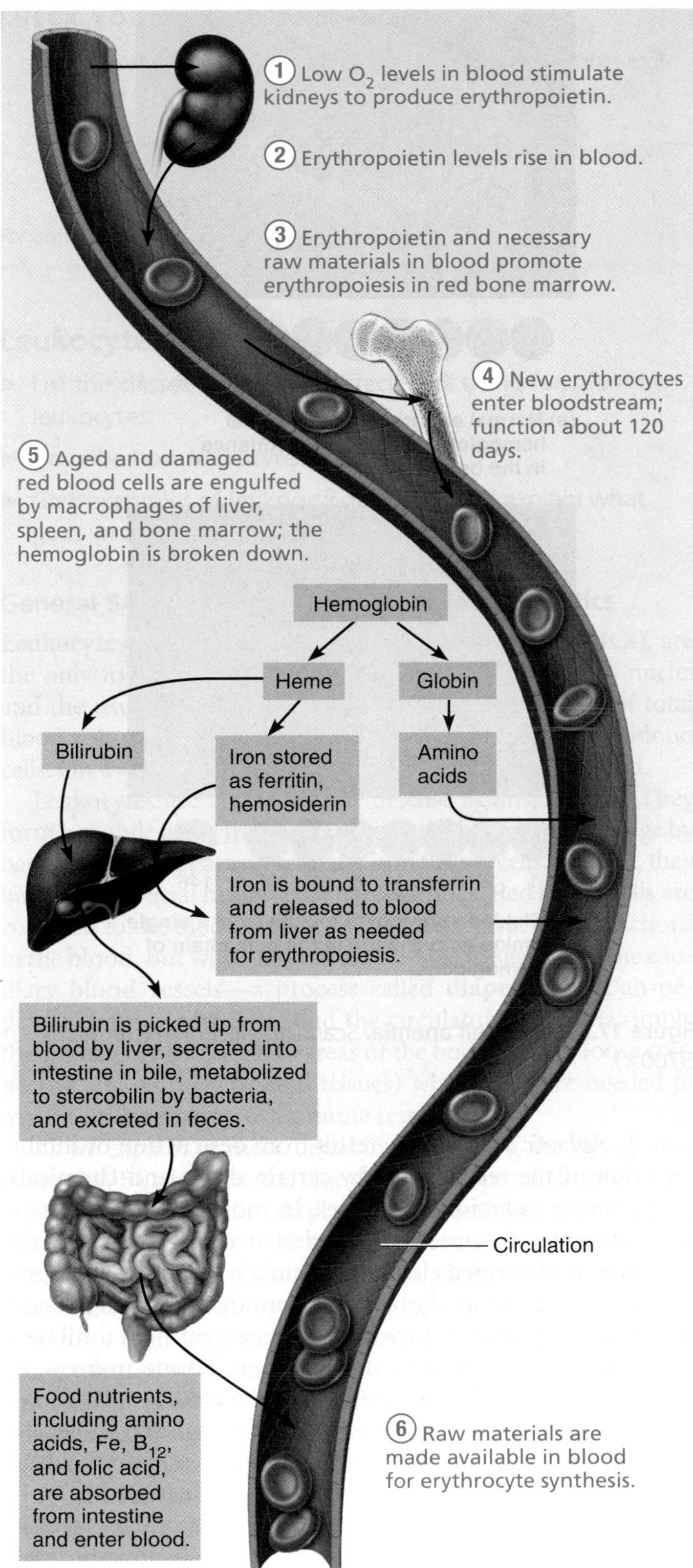

① Low O_2 levels in blood stimulate kidneys to produce erythropoietin.
② Erythropoietin levels rise in blood.
③ Erythropoietin and necessary raw materials in blood promote erythropoiesis in red bone marrow.
④ New erythrocytes enter bloodstream; function about 120 days.
⑤ Aged and damaged red blood cells are engulfed by macrophages of liver, spleen, and bone marrow; the hemoglobin is broken down.
Hemoglobin
Heme
Globin
Bilirubin
Iron stored as ferritin, hemosiderin
Amino acids
Iron is bound to transferrin and released to blood from liver as needed for erythropoiesis.
Bilirubin is picked up from blood by liver, secreted into intestine in bile, metabolized to stercobilin by bacteria, and excreted in feces.
Circulation
Food nutrients, including amino acids, Fe, B_{12}, and folic acid, are absorbed from intestine and enter blood.
⑥ Raw materials are made available in blood for erythrocyte synthesis.

Figure 17.7 Life cycle of red blood cells.

Erythrocyte Disorders

Most erythrocyte disorders can be classified as anemias or polycythemias. We describe the many varieties and causes of these conditions next.

Anemias **Anemia** (ah-ne′me-ah; "lacking blood") is a condition in which the blood has abnormally low oxygen-carrying capacity. It is a *sign* of some disorder rather than a disease in and of itself. Its hallmark is blood oxygen levels that are inadequate to support normal metabolism. Anemic individuals are fatigued, often pale, short of breath, and chilly. Common causes of anemia include the following:

1. **An insufficient number of red blood cells.** Conditions that reduce the red blood cell count include blood loss, excessive RBC destruction, and bone marrow failure.

   Hemorrhagic anemias (hem″o-raj′ik) result from blood loss. In acute hemorrhagic anemia, blood loss is rapid (as might follow a severe stab wound); it is treated by blood replacement. Slight but persistent blood loss (due to hemorrhoids or an undiagnosed bleeding ulcer, for example) causes chronic hemorrhagic anemia. Once the primary problem is resolved, normal erythropoietic mechanisms replace the deficient blood cells.

   In *hemolytic anemias* (he″mo-lit′ik), erythrocytes rupture, or lyse, prematurely. Hemoglobin abnormalities, transfusion of mismatched blood, and certain bacterial and parasitic infections are possible causes.

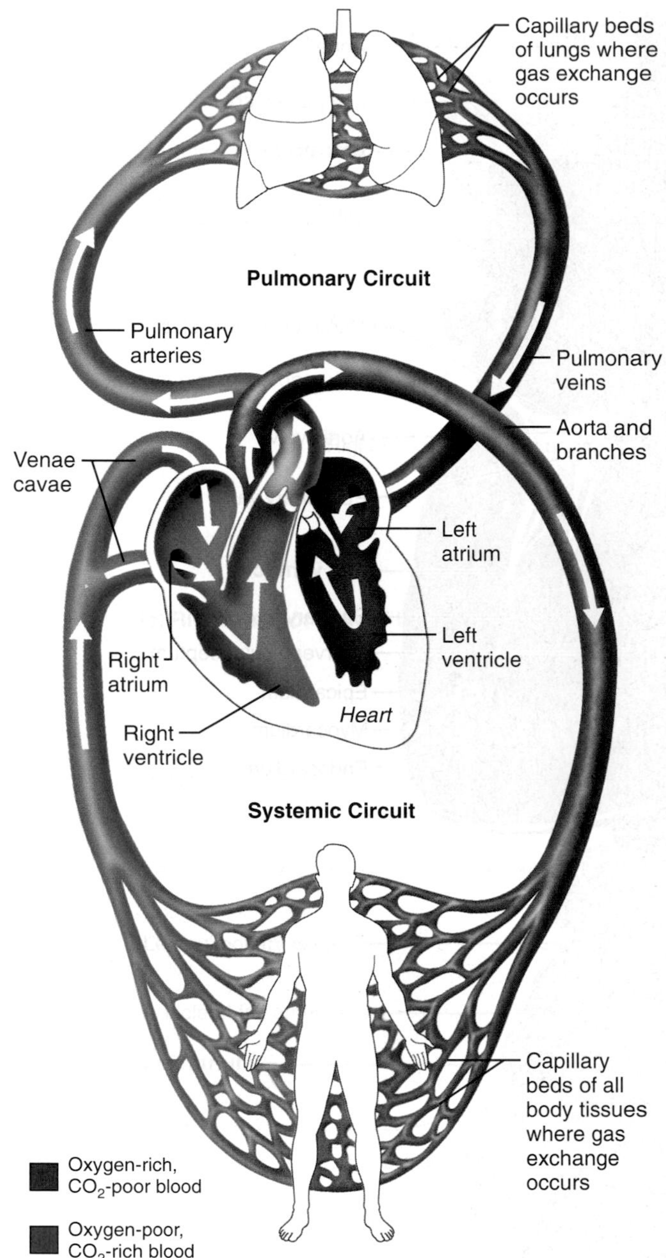

Figure 18.5 The systemic and pulmonary circuits. The right side of the heart pumps blood through the pulmonary circuit* (to the lungs and back to the left side of the heart). The left side of the heart pumps blood via the systemic circuit to all body tissues and back to the right side of the heart.

*For simplicity, the actual number of two pulmonary arteries and four pulmonary veins has been reduced to one each.

Ventricles: The Discharging Chambers

Together the ventricles (*ventr* = underside) make up most of the volume of the heart. As already mentioned, the right ventricle forms most of the heart's anterior surface and the left ventricle dominates its posteroinferior surface. Marking the internal walls of the ventricular chambers are irregular ridges of muscle called **trabeculae carneae** (trah-bek′u-le kar′ne-e; "crossbars of flesh"). Still other muscle bundles, the conelike **papillary muscles**, which play a role in valve function, project into the ventricular cavity (Figure 18.4e).

The ventricles are the discharging chambers or actual pumps of the heart. The difference in function between atria and ventricles is reflected in the much more massive ventricular walls (Figure 18.4e and f). When the ventricles contract, blood is propelled out of the heart into the circulation. The right ventricle pumps blood into the **pulmonary trunk**, which routes the blood to the lungs where gas exchange occurs. The left ventricle ejects blood into the **aorta** (a-or′tah), the largest artery in the body.

Pathway of Blood Through the Heart

▶ Trace the pathway of blood through the heart.

Until the sixteenth century, people believed that blood moved from one side of the heart to the other by seeping through pores in the septum. We now know that the heart passages open not from one side to the other but vertically. The heart is actually two side-by-side pumps, each serving a separate blood circuit **(Figure 18.5)**. The blood vessels that carry blood to and from the lungs form the **pulmonary circuit** (*pulmo* = lung), which serves gas exchange. The blood vessels that carry the functional blood supply to and from all body tissues constitute the **systemic circuit.**

The right side of the heart is the *pulmonary circuit pump.* Blood returning from the body is relatively oxygen-poor and carbon dioxide–rich. It enters the right atrium and passes into the right ventricle, which pumps it to the lungs via the pulmonary trunk (Figure 18.4e). In the lungs, the blood unloads carbon dioxide and picks up oxygen. The freshly oxygenated blood is carried by the pulmonary veins back to the left side of the heart.

Notice how unique this circulation is. Elsewhere in the body, veins carry relatively oxygen-poor blood to the heart, and arteries transport oxygen-rich blood from the heart. Exactly the opposite oxygenation conditions exist in veins and arteries of the pulmonary circuit.

The left side of the heart is the *systemic circuit pump.* Freshly oxygenated blood leaving the lungs is returned to the left atrium and passes into the left ventricle, which pumps it into the aorta. From there the blood is transported via smaller systemic arteries to the body tissues, where gases and nutrients are exchanged across the capillary walls. Then the blood, once again loaded with carbon dioxide and depleted of oxygen, returns through the systemic veins to the right side of the heart, where it enters the right atrium through the superior and inferior venae cavae. This cycle repeats itself continuously.

Equal volumes of blood are pumped to the pulmonary and systemic circuits at any moment, but the two ventricles have very unequal workloads. The pulmonary circuit, served by the right ventricle, is a short, low-pressure circulation. In contrast, the systemic circuit, associated with the left ventricle, takes a long pathway through the entire body and encounters about five times as much friction, or resistance to blood flow.

This functional difference is revealed in the anatomy of the two ventricles (Figure 18.4e and **Figure 18.6**). The walls of the

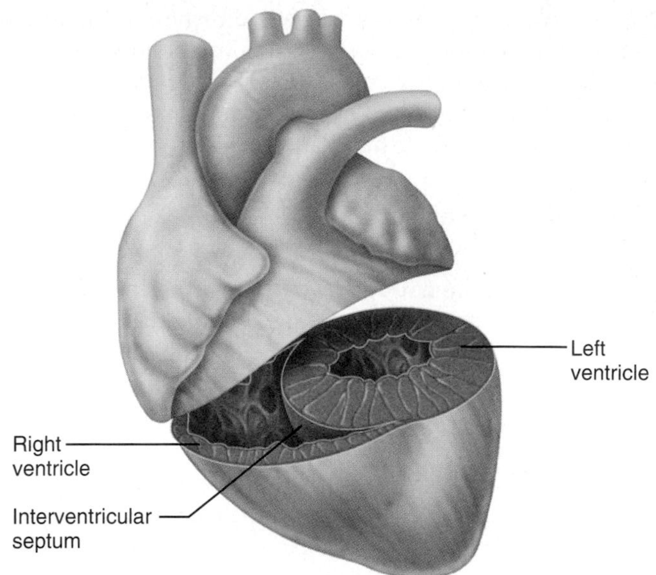

Figure 18.6 Anatomical differences between the right and left ventricles. The left ventricle has a thicker wall and its cavity is basically circular. The right ventricle cavity is crescent shaped and wraps around the left ventricle.

left ventricle are three times as thick as those of the right ventricle, and its cavity is nearly circular. The right ventricular cavity is flattened into a crescent shape that partially encloses the left ventricle, much the way a hand might loosely grasp a clenched fist. Consequently, the left ventricle can generate much more pressure than the right and is a far more powerful pump.

Coronary Circulation

▶ Name the major branches and describe the distribution of the coronary arteries.

Although the heart is more or less continuously filled with blood, this blood provides little nourishment to heart tissue. (The myocardium is too thick to make diffusion a practical means of nutrient delivery.) How, then, does the heart get nourishment?

The **coronary circulation**, the functional blood supply of the heart, is the shortest circulation in the body. The arterial supply of the coronary circulation is provided by the *right* and *left coronary arteries,* both arising from the base of the aorta and encircling the heart in the coronary sulcus **(Figure 18.7a)**. The **left coronary artery** runs toward the left side of the heart and then divides into its major branches: the **anterior interventricular artery** (also known clinically as the *left anterior descending artery*), which follows the anterior interventricular sulcus and supplies blood to the interventricular septum and anterior walls of both ventricles; and the **circumflex artery**, which supplies the left atrium and the posterior walls of the left ventricle.

The **right coronary artery** courses to the right side of the heart, where it also gives rise to two branches: the **right marginal artery**, which serves the myocardium of the lateral right side of the heart, and the **posterior interventricular artery**, which runs to the heart apex and supplies the posterior ventricular walls.

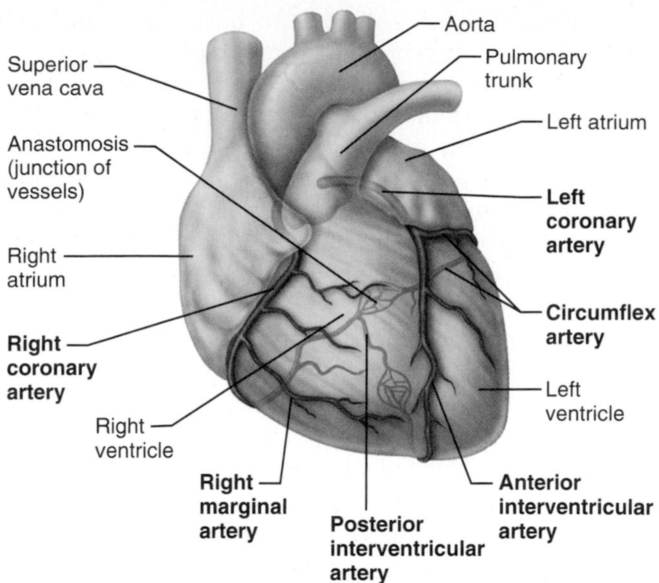

(a) The major coronary arteries

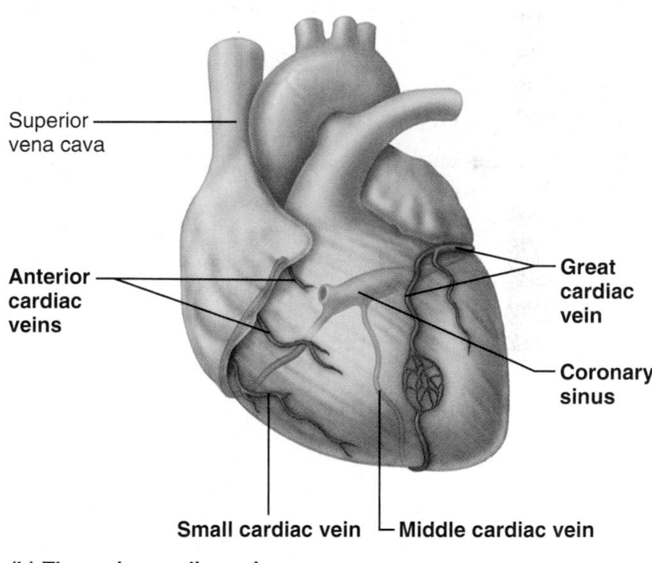

(b) The major cardiac veins

Figure 18.7 Coronary circulation. In both drawings, lighter-tinted vessels are more posterior in the heart.

Near the apex of the heart, this artery merges (anastomoses) with the anterior interventricular artery. Together the branches of the right coronary artery supply the right atrium and nearly all the right ventricle.

The arterial supply of the heart varies considerably. For example, in 15% of people, the left coronary artery gives rise to *both* the anterior and posterior interventricular arteries. In about 4% of people, a single coronary artery supplies the whole heart. Additionally, there may be both right and left marginal arteries. There are many anastomoses (junctions) among the coronary arterial branches. These fusing networks provide additional (*collateral*) routes for blood delivery to the heart muscle, but are not robust enough to supply adequate nutrition when a coronary artery is suddenly occluded (blocked). Complete blockage leads to tissue death and heart attack.

The coronary arteries provide an intermittent, pulsating blood flow to the myocardium. These vessels and their main branches lie in the epicardium and send branches inward to nourish the myocardium. They deliver blood when the heart is relaxed, but are fairly ineffective when the ventricles are contracting because they are compressed by the contracting myocardium. Although the heart represents only about 1/200 of the body's weight, it requires about 1/20 of the body's blood supply. As might be expected, the left ventricle receives the most plentiful blood supply.

After passing through the capillary beds of the myocardium, the venous blood is collected by the **cardiac veins**, whose paths roughly follow those of the coronary arteries. These veins join together to form an enlarged vessel called the **coronary sinus**, which empties the blood into the right atrium. The coronary sinus is obvious on the posterior aspect of the heart (Figure 18.7b). The sinus has three large tributaries: the **great cardiac vein** in the anterior interventricular sulcus; the **middle cardiac vein** in the posterior interventricular sulcus; and the **small cardiac vein**, running along the heart's right inferior margin. Additionally, several **anterior cardiac veins** empty directly into the right atrium anteriorly.

HOMEOSTATIC IMBALANCE

Blockage of the coronary arterial circulation can be serious and sometimes fatal. **Angina pectoris** (an-ji′nah pek′tor-is; "choked chest") is thoracic pain caused by a fleeting deficiency in blood delivery to the myocardium. It may result from stress-induced spasms of the coronary arteries or from increased physical demands on the heart. The myocardial cells are weakened by the temporary lack of oxygen but do not die. Far more serious is prolonged coronary blockage, which can lead to a **myocardial infarction (MI)**, commonly called a **heart attack**. Because adult cardiac muscle is essentially amitotic, most areas of cell death are repaired with noncontractile scar tissue. Whether or not a person survives a myocardial infarction depends on the extent and location of the damage. Damage to the left ventricle, which is the systemic pump, is most serious. ■

CHECK YOUR UNDERSTANDING

4. Which side of the heart acts as the pulmonary pump? The systemic pump?
5. Which of the following statements are true? (a) The left ventricle wall is thicker than the right ventricle wall. (b) The left ventricle pumps blood at a higher pressure than the right ventricle. (c) The left ventricle pumps more blood with each beat than the right ventricle. Explain.
6. Name the two main branches of the right coronary artery.

For answers, see Appendix G.

Heart Valves

▶ Name the heart valves and describe their location, function, and mechanism of operation.

Blood flows through the heart in one direction: from atria to ventricles and out the great arteries leaving the superior aspect of the heart. Four valves enforce this one-way traffic (Figure 18.4e and **Figure 18.8**). They open and close in response to differences in blood pressure on their two sides.

Atrioventricular Valves

The two **atrioventricular (AV) valves**, one located at each atrial-ventricular junction, prevent backflow into the atria when the ventricles are contracting. The right AV valve, the **tricuspid valve** (tri-kus′pid), has three flexible cusps (flaps of endocardium reinforced by connective tissue cores). The left AV valve, with two flaps, is called the **mitral valve** (mi′tral) because of its resemblance to the two-sided bishop's miter or hat. It is sometimes called the *bicuspid valve.* Attached to each AV valve flap are tiny white collagen cords called **chordae tendineae** (kor′de ten″dĭ′ne-e; "tendonous cords"), "heart strings" which anchor the cusps to the papillary muscles protruding from the ventricular walls (Figure 18.8c, d).

When the heart is completely relaxed, the AV valve flaps hang limply into the ventricular chambers below and blood flows into the atria and then through the open AV valves into the ventricles **(Figure 18.9a)**. When the ventricles contract, compressing the blood in their chambers, the intraventricular pressure rises, forcing the blood superiorly against the valve flaps. As a result, the flap edges meet, closing the valve (Figure 18.9b).

The chordae tendineae and the papillary muscles serve as guy-wires to anchor the valve flaps in their *closed* position. If the cusps were not anchored in this manner, they would be blown upward into the atria, in the same way an umbrella is blown inside out by a gusty wind. The papillary muscles contract with the other ventricular musculature so that they take up the slack on the chordae tendineae as the full force of ventricular contraction hurls the blood against the AV valve flaps.

Semilunar Valves

The **aortic** and **pulmonary (semilunar, SL) valves** guard the bases of the large arteries issuing from the ventricles (aorta and pulmonary trunk, respectively) and prevent backflow into the associated ventricles. Each SL valve is fashioned from three pocketlike cusps, each shaped roughly like a crescent moon (*semilunar* = half-moon).

Like the AV valves, the SL valves open and close in response to differences in pressure. In the SL case, when the ventricles are contracting and intraventricular pressure *rises above* the pressure in the aorta and pulmonary trunk, the SL valves are forced open and their cusps flatten against the arterial walls as the blood rushes past them **(Figure 18.10a)**. When the ventricles relax, and the blood (no longer propelled forward by the pressure of ventricular contraction) flows backward toward the heart, it fills the cusps and closes the valves (Figure 18.10b).

We complete the valve story by noting what seems to be an important omission—there are no valves guarding the entrances of the venae cavae and pulmonary veins into the right and left atria, respectively. Small amounts of blood *do* spurt back into these vessels during atrial contraction, but backflow is

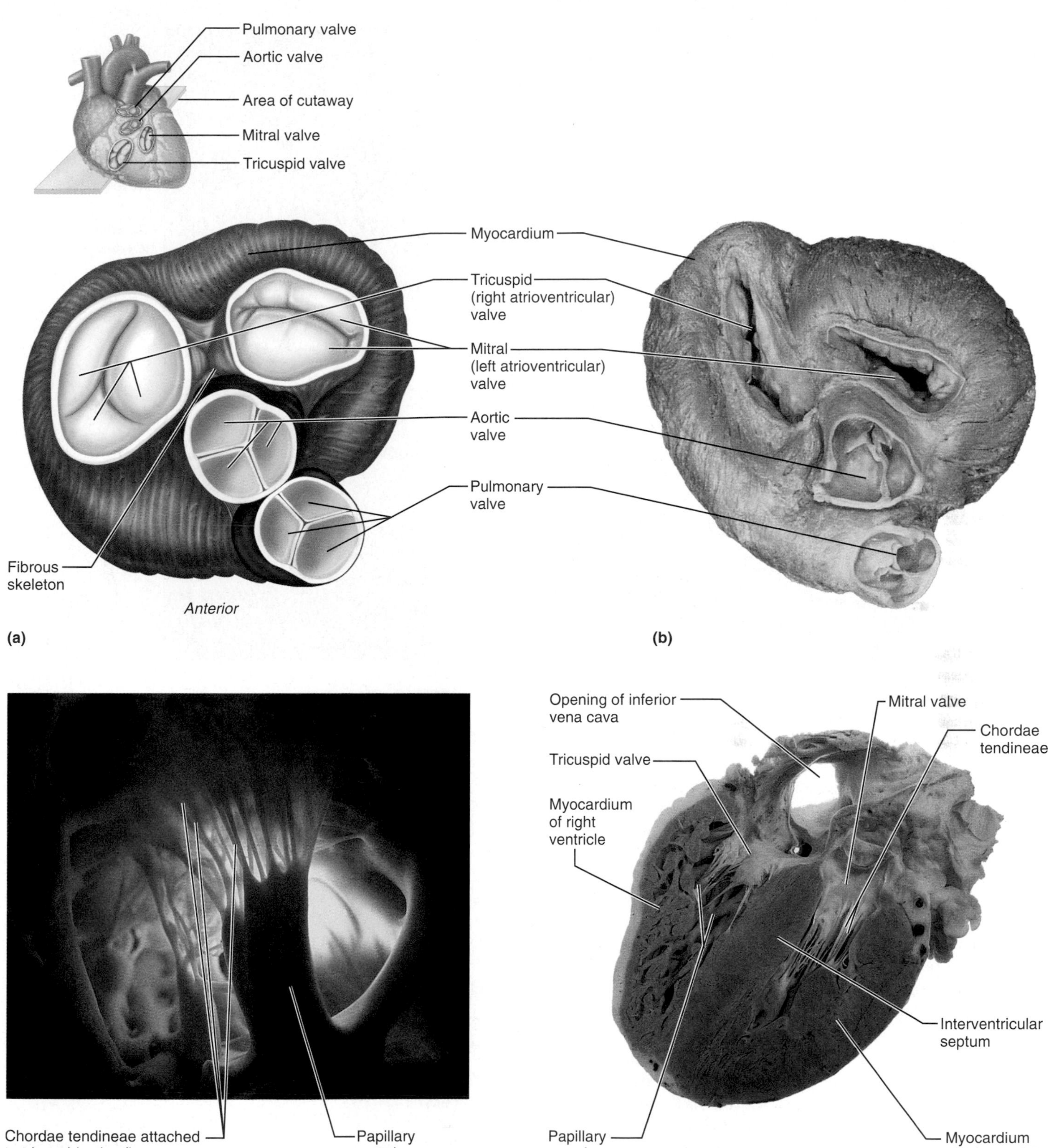

Figure 18.8 Heart valves. (a) Superior view of the two sets of heart valves (atria removed). The paired atrioventricular valves are located between atria and ventricles; the two semilunar valves are located at the junction of the ventricles and the arteries issuing from them. **(b)** Photograph of the heart valves, superior view. **(c)** Photograph of the tricuspid valve. This bottom-to-top view shows the valve as seen from the right ventricle. **(d)** Coronal section of the heart. (See also *A Brief Atlas of the Human Body*, Figures 58 and 60.)

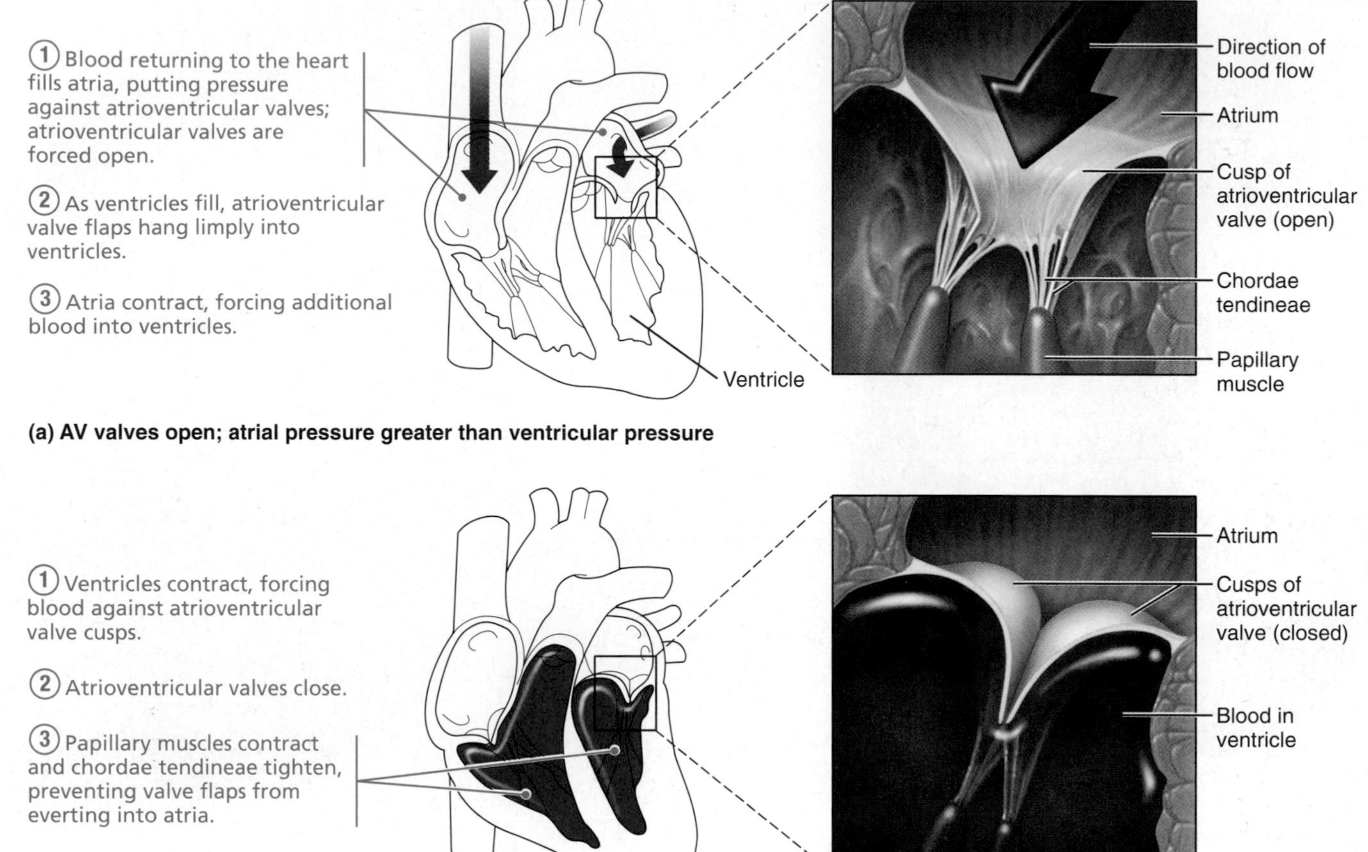

1 Blood returning to the heart fills atria, putting pressure against atrioventricular valves; atrioventricular valves are forced open.
2 As ventricles fill, atrioventricular valve flaps hang limply into ventricles.
3 Atria contract, forcing additional blood into ventricles.
Ventricle
Direction of blood flow
Atrium
Cusp of atrioventricular valve (open)
Chordae tendineae
Papillary muscle
(a) AV valves open; atrial pressure greater than ventricular pressure
1 Ventricles contract, forcing blood against atrioventricular valve cusps.
2 Atrioventricular valves close.
3 Papillary muscles contract and chordae tendineae tighten, preventing valve flaps from everting into atria.
Atrium
Cusps of atrioventricular valve (closed)
Blood in ventricle
(b) AV valves closed; atrial pressure less than ventricular pressure

Figure 18.9 The atrioventricular valves.

minimal because of the inertia of the blood and because as it contracts, the atrial myocardium compresses (and collapses) these venous entry points.

HOMEOSTATIC IMBALANCE

Heart valves are simple devices, and the heart—like any mechanical pump—can function with "leaky" valves as long as the impairment is not too great. However, severe valve deformities can seriously hamper cardiac function. An *incompetent valve* forces the heart to repump the same blood over and over because the valve does not close properly and blood backflows. In valvular *stenosis* ("narrowing"), the valve flaps become stiff (typically because of scar tissue formation following endocarditis or calcium salt deposit) and constrict the opening. This stiffness compels the heart to contract more forcibly than normal. In both instances, the heart's workload increases and, ultimately, the heart may be severely weakened.

Under such conditions, the faulty valve (most often the mitral valve) is replaced with a mechanical valve, a pig or cow heart valve chemically treated to prevent rejection, or cryopreserved valves from human cadavers. Heart valves tissue-engineered from a patient's own cells grown on a biodegradable scaffold are being developed. ■

CHECK YOUR UNDERSTANDING

7. What is the function of the papillary muscles and chordae tendineae?

For answers, see Appendix G.

Cardiac Muscle Fibers

- Describe the structural and functional properties of cardiac muscle, and explain how it differs from skeletal muscle.
- Briefly describe the events of cardiac muscle cell contraction.

Although similar to skeletal muscle, cardiac muscle displays some special anatomical features that reflect its unique blood-pumping role.

Microscopic Anatomy

Cardiac muscle, like skeletal muscle, is striated, and it contracts by the sliding filament mechanism. However, in contrast to the long, cylindrical, multinucleate skeletal muscle fibers, cardiac

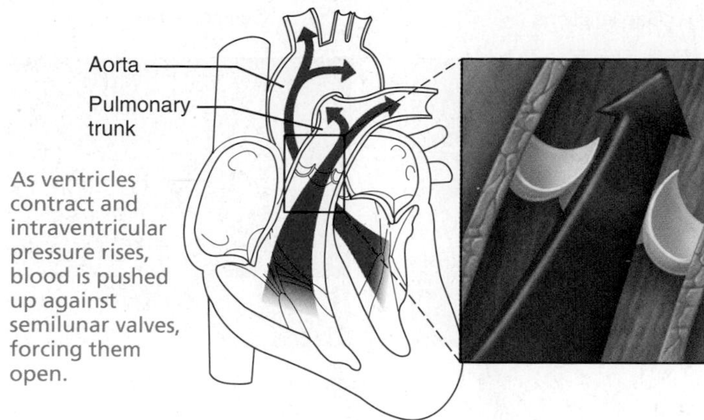

Aorta
Pulmonary trunk
As ventricles contract and intraventricular pressure rises, blood is pushed up against semilunar valves, forcing them open.

(a) Semilunar valves open

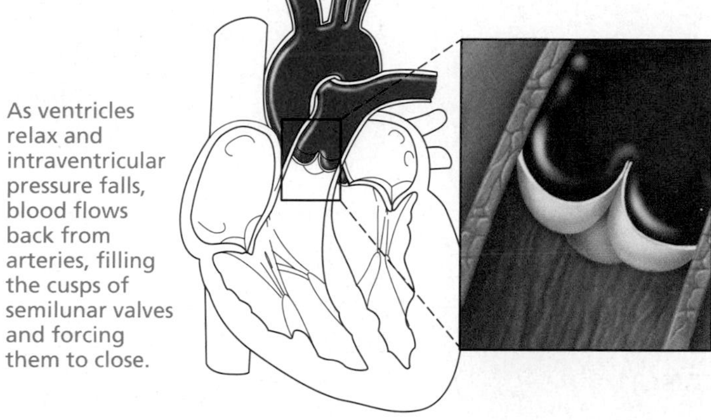

As ventricles relax and intraventricular pressure falls, blood flows back from arteries, filling the cusps of semilunar valves and forcing them to close.

(b) Semilunar valves closed

Figure 18.10 The semilunar valves.

cells are short, fat, branched, and interconnected. Each fiber contains one or at most two large, pale, *centrally* located nuclei **(Figure 18.11a)**. The intercellular spaces are filled with a loose connective tissue matrix (the *endomysium*) containing numerous capillaries. This delicate matrix is connected to the fibrous skeleton, which acts both as a tendon and as an insertion, giving the cardiac cells something to pull or exert their force against.

Skeletal muscle fibers are independent of one another both structurally and functionally. By contrast, the plasma membranes of adjacent cardiac cells interlock like the ribs of two sheets of corrugated cardboard at dark-staining junctions called **intercalated discs** (in-ter′kah-la″ted; *intercala* = insert) (Figure 18.11). These discs contain anchoring *desmosomes* and *gap junctions* (cell junctions discussed in Chapter 3). The desmosomes prevent adjacent cells from separating during contraction, and the gap junctions allow ions to pass from cell to cell, transmitting current across the entire heart. Because cardiac cells are electrically coupled by the gap junctions, the myocardium *behaves* as a single coordinated unit, or **functional syncytium.**

Large mitochondria account for 25–35% of the volume of cardiac cells (compared with only 2% in skeletal muscle) and give cardiac cells a high resistance to fatigue. Most of the remaining volume is occupied by myofibrils composed of fairly typical sarcomeres. The sarcomeres have Z discs, A bands, and I bands that reflect the arrangement of the thick (myosin) and thin (actin) filaments composing them. However, in contrast to skeletal muscle, the myofibrils of cardiac muscle cells vary greatly in diameter and branch extensively, accommodating the abundant mitochondria that lie between them. This difference produces a banding pattern less dramatic than that seen in skeletal muscle.

The system for delivering Ca^{2+} is less elaborate in cardiac muscle cells. The T tubules are wider and fewer than in skeletal muscle and they enter the cells once per sarcomere at the Z discs. (Recall that T tubules are invaginations of the sarcolemma. In skeletal muscle, the T tubules invaginate twice per sarcomere, at the A band–I band junctions.) The cardiac sarcoplasmic reticulum is simpler and lacks the large terminal cisternae seen in skeletal muscle. Consequently, no *triads* are seen in cardiac muscle fibers.

Mechanism and Events of Contraction

Although heart muscle and skeletal muscle are contractile tissues, they have some fundamental differences:

1. **Means of stimulation.** Each skeletal muscle fiber must be stimulated to contract by a nerve ending, but some cardiac muscle cells are self-excitable. These cells can initiate not only their own depolarization, but that of the rest of the heart as well, in a spontaneous and rhythmic way. We describe this property, called **automaticity**, or **autorhythmicity**, in the next section.
2. **Organ versus motor unit contraction.** In skeletal muscle, all cells of a given motor unit (but not necessarily all motor units of the muscle) are stimulated and contract at the same time. Impulses do not spread from cell to cell. In cardiac muscle, the heart either contracts as a unit or doesn't contract at all. This coordinated action occurs because gap junctions electrically tie all cardiac muscle cells together into a single contractile unit. Consequently, the depolarization wave travels across the heart from cell to cell via ion passage through the gap junctions.
3. **Length of absolute refractory period.** In cardiac muscle cells, the absolute refractory period (the inexcitable period when Na^+ channels are still open or inactivated) lasts approximately 250 ms, nearly as long as the contraction **(Figure 18.12)**. Contrast this to the short refractory periods of 1–2 ms in skeletal muscle fibers, in which contractions last 15–100 ms. The long cardiac refractory period normally prevents tetanic contractions, which would stop the heart's pumping action.

Having probed the major differences between cardiac and skeletal muscle tissues, let's now look at their similarities. As with skeletal muscle, cardiac muscle contraction is triggered by action potentials that sweep across cell membranes. About 1% of cardiac fibers are *autorhythmic* ("self-rhythm"), having the special ability to depolarize spontaneously and thus pace the heart. The bulk of heart muscle, however, is composed of

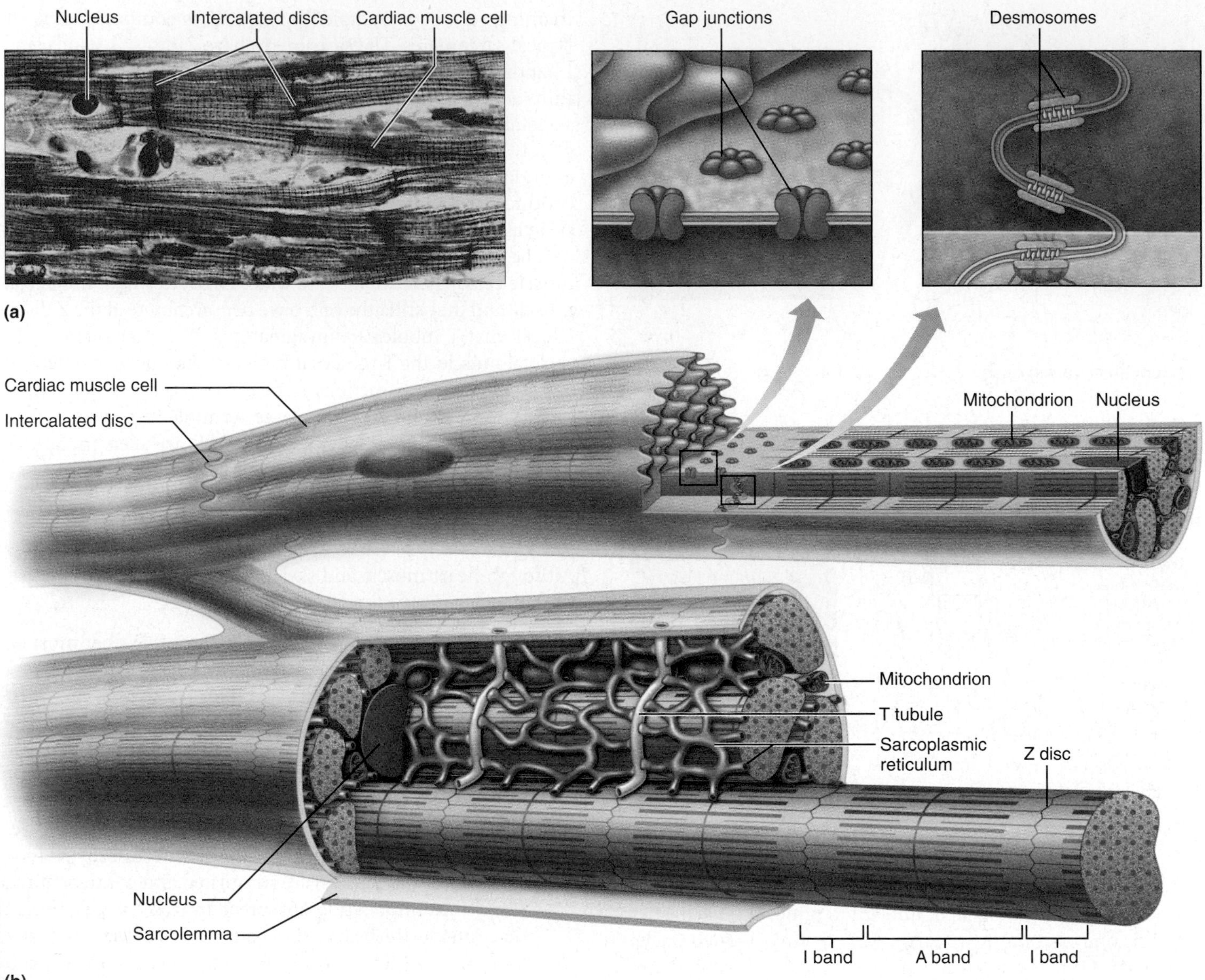

Figure 18.11 Microscopic anatomy of cardiac muscle. (a) Photomicrograph of cardiac muscle (600×). Notice that the cardiac muscle cells are short, branched, and striated. The dark-staining areas are intercalated discs, or junctions, between adjacent cells. **(b)** Components of intercalated discs and cardiac muscle fibers.

contractile muscle fibers responsible for the heart's pumping activity. In these cells, the sequence of events leading to contraction is similar to that in skeletal muscle fibers:

- Depolarization opens a few **voltage-gated fast Na^+ channels** in the sarcolemma, allowing extracellular Na^+ to enter. This influx initiates a positive feedback cycle that causes the rising phase of the action potential (and reversal of the membrane potential from −90 mV to nearly +30 mV; Figure 18.12, ①. The period of Na^+ influx is very brief, because the sodium channels quickly inactivate and the Na^+ influx stops.
- Transmission of the depolarization wave down the T tubules (ultimately) causes the sarcoplasmic reticulum (SR) to release Ca^{2+} into the sarcoplasm.
- Excitation-contraction coupling occurs as Ca^{2+} provides the signal (via troponin binding) for cross bridge activation and couples the depolarization wave to the sliding of the myofilaments.

These three steps are common to both skeletal and cardiac muscle cells (see Figure 9.11), but the two muscle types differ in how the SR is stimulated to release Ca^{2+}. Let's take a look.

Some 10–20% of the Ca^{2+} needed for the calcium pulse that triggers contraction enters the cardiac cells from the extracellular space. Once inside, it stimulates the SR to release the other 80% of the Ca^{2+} needed. Ca^{2+} is barred from entering nonstimulated cardiac fibers, but when Na^+-dependent membrane depolarization occurs, the voltage change also opens channels

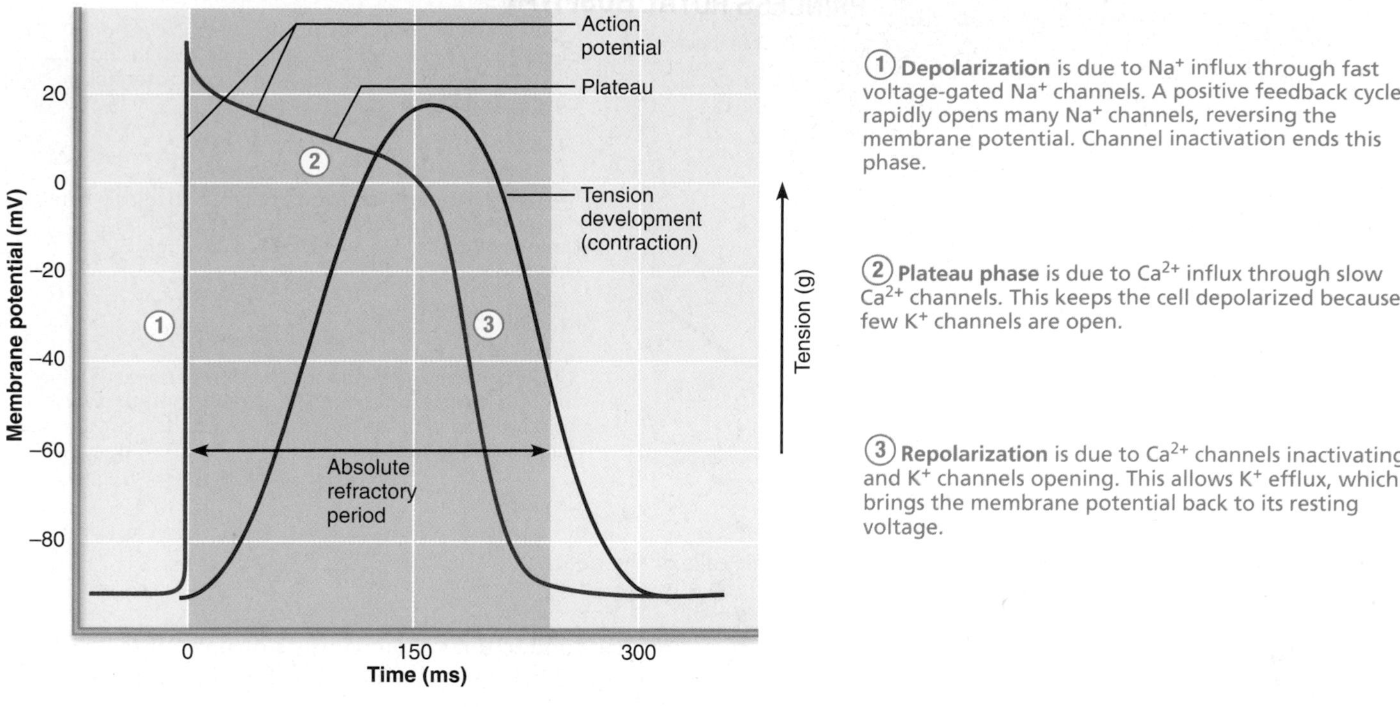

Figure 18.12 The action potential of contractile cardiac muscle cells. Relationship between the action potential, period of contraction, and absolute refractory period in a single ventricular cell.

that allow Ca^{2+} entry from the extracellular space. These channels are called **slow Ca^{2+} channels** because their opening is delayed a bit. The local influxes of Ca^{2+} through these channels trigger opening of nearby Ca^{2+}-sensitive channels in the SR tubules, which liberate bursts of Ca^{2+} ("calcium sparks") that dramatically increase the intracellular Ca^{2+} concentration.

Although Na^+ channels have inactivated and repolarization has begun by this point, the calcium surge across the sarcolemma prolongs the depolarization potential briefly, producing a **plateau** in the action potential tracing (Figure 18.12, ②). At the same time, few K^+ channels are open, which also prolongs the plateau and prevents rapid repolarization. As long as Ca^{2+} is entering, the cells continue to contract. Notice in Figure 18.12 that muscle tension develops during the plateau, and peaks just after the plateau ends.

Notice also that the duration of the action potential and contractile phase is much greater in cardiac muscle than in skeletal muscle. In skeletal muscle, the action potential typically lasts 1–2 ms and the contraction (for a single stimulus) 15–100 ms. In cardiac muscle, the action potential lasts 200 ms or more (because of the plateau), and tension development persists for 200 ms or more, providing the sustained contraction needed to eject blood from the heart.

After about 200 ms, the slope of the action potential tracing falls rapidly (Figure 18.12, ③). This repolarization results from inactivation of Ca^{2+} channels and opening of voltage-gated K^+ channels, which allows a rapid loss of potassium from the cell that restores the resting membrane potential. During repolarization, Ca^{2+} is pumped back into the SR and the extracellular space.

Energy Requirements

Cardiac muscle has more mitochondria than skeletal muscle does, reflecting its greater dependence on oxygen for its energy metabolism. Unlike skeletal muscle, which can contract for prolonged periods, even during oxygen deficits, by carrying out anaerobic respiration, the heart relies almost exclusively on aerobic respiration. As a result, cardiac muscle cannot incur much of an oxygen deficit and still operate effectively.

Both types of muscle tissue use multiple fuel molecules, including glucose and fatty acids. But cardiac muscle is much more adaptable and readily switches metabolic pathways to use whatever nutrient supply is available, including lactic acid generated by skeletal muscle activity. Consequently, the real danger of an inadequate blood supply to the myocardium is lack of oxygen, not of nutrient fuels.

HOMEOSTATIC IMBALANCE

When a region of heart muscle is deprived of blood (is ischemic), the oxygen-starved cells begin to metabolize anaerobically, producing lactic acid. The rising H^+ level that results hinders the cardiac cells' ability to produce the ATP they need to pump Ca^{2+} into the extracellular fluid. The resulting increase in intracellular H^+ and Ca^{2+} levels causes the gap junctions (which are usually open) to close, electrically isolating the damaged cells and forcing generated action potentials to find alternate routes to the cardiac cells beyond them. If the ischemic area is large, the pumping activity of the heart as a whole may be severely impaired, leading to a heart attack. ■

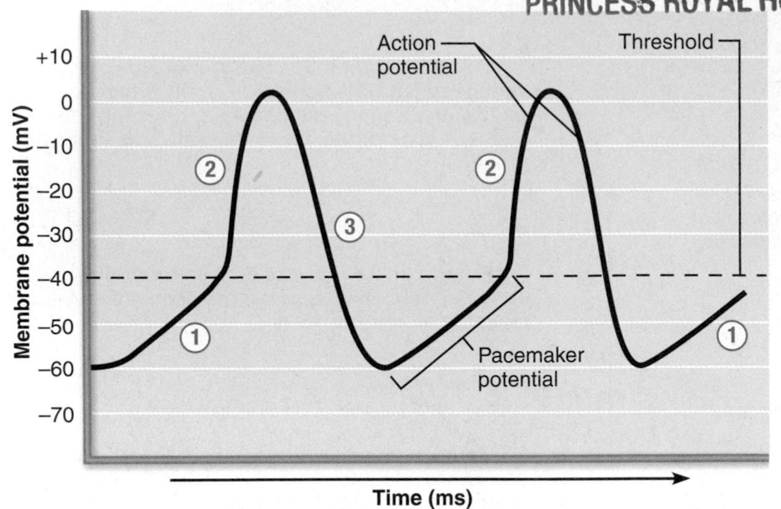

(1) **Pacemaker potential** This slow depolarization is due to both opening of Na^+ channels and closing of K^+ channels. Notice that the membrane potential is never a flat line.

(2) **Depolarization** The action potential begins when the pacemaker potential reaches threshold. Depolarization is due to Ca^{2+} influx through Ca^{2+} channels.

(3) **Repolarization** is due to Ca^{2+} channels inactivating and K^+ channels opening. This allows K^+ efflux, which brings the membrane potential back to its most negative voltage.

Figure 18.13 Pacemaker and action potentials of autorhythmic cells of the heart.

CHECK YOUR UNDERSTANDING

8. For each of the following, state whether it applies to skeletal muscle, cardiac muscle, or both: (a) refractory period is almost as long as the contraction; (b) source of Ca^{2+} for contraction is *only* SR; (c) AP exhibits a plateau phase; (d) has troponin; (e) has triads.
9. Cardiac muscle cannot go into tetany. Why?

For answers, see Appendix G.

Heart Physiology

Electrical Events

- Name the components of the conduction system of the heart, and trace the conduction pathway.
- Draw a diagram of a normal electrocardiogram tracing. Name the individual waves and intervals, and indicate what each represents.
- Name some abnormalities that can be detected on an ECG tracing.

The ability of cardiac muscle to depolarize and contract is intrinsic. In other words, it is a property of heart muscle and does not depend on the nervous system. Even if all nerve connections to the heart are severed, the heart continues to beat rhythmically, as demonstrated by transplanted hearts. Nevertheless, the healthy heart is amply supplied with autonomic nerve fibers that can alter the basic rhythm of heart activity set by intrinsic factors.

Setting the Basic Rhythm: The Intrinsic Conduction System

The independent, but coordinated, activity of the heart is a function of (1) the presence of gap junctions, and (2) the activity of the heart's "in-house" conduction system. The **intrinsic cardiac conduction system** consists of noncontractile cardiac cells specialized to initiate and distribute impulses throughout the heart, so that it depolarizes and contracts in an orderly, sequential manner. In this way, the heart beats as a coordinated unit. Let's look at how this system works.

Action Potential Initiation by Autorhythmic Cells Unlike unstimulated contractile cells of the heart (and neurons and skeletal muscle fibers), which maintain a stable resting membrane potential, the **autorhythmic cells** making up the intrinsic conduction system do *not*. Instead, they have an *unstable resting potential* that continuously depolarizes, drifting slowly toward threshold. These spontaneously changing membrane potentials, called **pacemaker potentials** or **prepotentials**, initiate the action potentials that spread throughout the heart to trigger its rhythmic contractions **(Figure 18.13)**.

The pacemaker potential is due to the special properties of the ion channels in the sarcolemma. In these cells, hyperpolarization at the end of an action potential leads to both closing of K^+ channels and opening of slow Na^+ channels. The Na^+ influx alters the balance between K^+ loss and Na^+ entry, and the membrane interior becomes less and less negative (more positive; Figure 18.13, (1)). Ultimately, at threshold (approximately -40 mV), **Ca^{2+} channels** open, allowing explosive entry of Ca^{2+} from the extracellular space. As a result, in autorhythmic cells, it is the influx of Ca^{2+} (rather than Na^+) that produces the rising phase of the action potential and reverses the membrane potential (Figure 18.13, (2)).

As in other excitable cells, the falling phase of the action potential and repolarization reflect opening of K^+ channels and K^+ efflux from the cell (Figure 18.13, (3)). Once repolarization is complete, K^+ channels close, K^+ efflux declines, and the slow depolarization to threshold begins again.

Sequence of Excitation Autorhythmic cardiac cells are found in the following areas (Figure 18.14): sinoatrial (si″no-a′tre-al) node, atrioventricular node, atrioventricular bundle, right and left bundle branches, and ventricular walls (as Purkinje fibers).

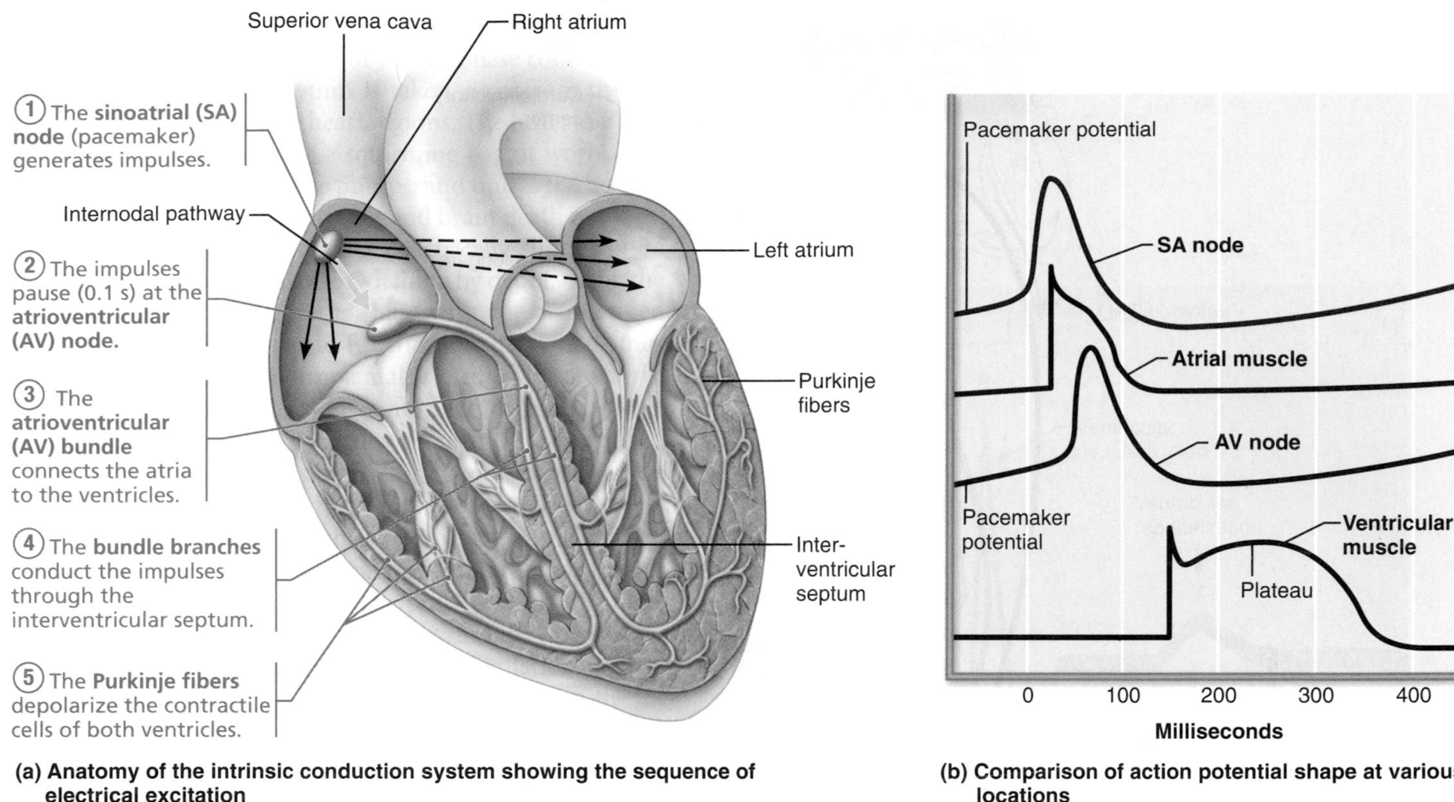

Superior vena cava
Right atrium
1 The sinoatrial (SA) node (pacemaker) generates impulses.
Internodal pathway
2 The impulses pause (0.1 s) at the atrioventricular (AV) node.
3 The atrioventricular (AV) bundle connects the atria to the ventricles.
4 The bundle branches conduct the impulses through the interventricular septum.
5 The Purkinje fibers depolarize the contractile cells of both ventricles.
Left atrium
Purkinje fibers
Inter-ventricular septum
Pacemaker potential
SA node
Atrial muscle
AV node
Pacemaker potential
Ventricular muscle
Plateau
0
100
200
300
400
Milliseconds

(a) Anatomy of the intrinsic conduction system showing the sequence of electrical excitation

(b) Comparison of action potential shape at various locations

Figure 18.14 Cardiac intrinsic conduction system and action potential succession during one heartbeat.

Impulses pass across the heart in order from ① to ⑤ following the yellow pathway in **Figure 18.14a**.

① **Sinoatrial node.** The crescent-shaped **sinoatrial (SA) node** is located in the right atrial wall, just inferior to the entrance of the superior vena cava. A minute cell mass with a mammoth job, the SA node typically generates impulses about 75 times every minute. (Its inherent rate in the absence of extrinsic neural and hormonal factors is closer to 100 times per minute.) The SA node sets the pace for the heart as a whole because no other region of the conduction system or the myocardium has a faster depolarization rate. For this reason, it is the heart's **pacemaker**, and its characteristic rhythm, called **sinus rhythm**, determines heart rate.

② **Atrioventricular node.** From the SA node, the depolarization wave spreads via gap junctions throughout the atria and via the *internodal pathway* to the **atrioventricular (AV) node**, located in the inferior portion of the interatrial septum immediately above the tricuspid valve. At the AV node, the impulse is delayed for about 0.1 s, allowing the atria to respond and complete their contraction before the ventricles contract. This delay reflects the smaller diameter of the fibers here and the fact that they have fewer gap junctions for current flow. Consequently, the AV node conducts impulses more slowly than other parts of the system, just as traffic slows when cars are forced to merge into two lanes from four. Once through the AV node, the signaling impulse passes rapidly through the rest of the system.

③ **Atrioventricular bundle.** From the AV node, the impulse sweeps to the **atrioventricular bundle** (also called the **bundle of His**) in the superior part of the interventricular septum. Although the atria and ventricles abut each other, they are *not* connected by gap junctions. The AV bundle is the *only* electrical connection between them. The balance of the AV junction is insulated by the nonconducting fibrous skeleton of the heart.

④ **Right and left bundle branches.** The AV bundle persists only briefly before splitting into two pathways—the **right** and **left bundle branches**, which course along the interventricular septum toward the heart apex.

⑤ **Purkinje fibers.** Essentially long strands of barrel-shaped cells with few myofibrils, the **Purkinje fibers** (pur-kin′je) complete the pathway through the interventricular septum, penetrate into the heart apex, and then turn superiorly into the ventricular walls. The bundle branches excite the septal cells, but the bulk of ventricular depolarization depends on the large Purkinje fibers and, ultimately, on cell-to-cell transmission of the impulse via gap junctions between the ventricular muscle cells. Because the left ventricle is much larger than the right, the Purkinje network is more elaborate in that side of the heart.

The total time between initiation of an impulse by the SA node and depolarization of the last of the ventricular muscle cells is approximately 0.22 s (220 ms) in a healthy human heart.

Mechanical Events: The Cardiac Cycle

▶ Describe the timing and events of the cardiac cycle.

The heart undergoes some dramatic writhing movements as it alternately contracts, forcing blood out of its chambers, and then relaxes, allowing its chambers to refill with blood. The terms **systole** (sis′to-le) and **diastole** (di-as′to-le) refer respectively to these *contraction* and *relaxation* periods. The **cardiac cycle** includes *all* events associated with the blood flow through the heart during one complete heartbeat—atrial systole and diastole followed by ventricular systole and diastole. These mechanical events always *follow* the electrical events seen in the ECG.

The cardiac cycle is marked by a succession of pressure and blood volume changes in the heart. Because blood circulates endlessly, we must choose an arbitrary starting point for one turn of the cardiac cycle. As shown in **Figure 18.20**, which outlines what happens in the left side of the heart, we begin our explanation with the heart in total relaxation: Atria and ventricles are quiet, and it is mid-to-late diastole.

① **Ventricular filling: mid-to-late diastole.** Pressure in the heart is low, blood returning from the circulation is flowing passively through the atria and the open AV valves into the ventricles, and the aortic and pulmonary valves are closed. Approximately 80% of ventricular filling occurs during this period, and the AV valve flaps begin to drift toward the closed position. (The remaining 20% is delivered to the ventricles when the atria contract toward the end of this phase.)

Now the stage is set for atrial systole. Following depolarization (P wave of ECG), the atria contract, compressing the blood in their chambers. This causes a sudden slight rise in atrial pressure, which propels residual blood out of the atria into the ventricles. At this point the ventricles are in the last part of their diastole and have the maximum volume of blood they will contain in the cycle, a volume called the *end diastolic volume* (*EDV*). Then the atria relax and the ventricles depolarize (QRS complex). Atrial diastole persists through the rest of the cycle.

② **Ventricular systole.** As the atria relax, the ventricles begin contracting. Their walls close in on the blood in their chambers, and ventricular pressure rises rapidly and sharply, closing the AV valves. The split-second period when the ventricles are completely closed chambers and the blood volume in the chambers remains constant as the ventricles contract is called the **isovolumetric contraction phase** (i″so-vol″u-met′rik).

Ventricular pressure continues to rise and when it finally exceeds the pressure in the large arteries issuing from the ventricles, the isovolumetric stage ends as the SL valves are forced open and blood is expelled from the ventricles into the aorta and pulmonary trunk. During this **ventricular ejection phase**, the pressure in the aorta normally reaches about 120 mm Hg.

③ **Isovolumetric relaxation: early diastole.** During this brief phase following the T wave, the ventricles relax. Because the blood remaining in their chambers, referred to as the *end systolic volume* (*ESV*), is no longer compressed, ventricular pressure drops rapidly and blood in the aorta and pulmonary trunk flows back toward the heart, closing the SL valves. Closure of the aortic valve causes a brief rise in aortic pressure as backflowing blood rebounds off the closed valve cusps, an event beginning at the **dicrotic notch** shown on the pressure graph. Once again the ventricles are totally closed chambers.

All during ventricular systole, the atria have been in diastole. They have been filling with blood and the intra-atrial pressure has been rising. When blood pressure on the atrial side of the AV valves exceeds that in the ventricles, the AV valves are forced open and ventricular filling, phase ①, begins again. Atrial pressure drops to its lowest point and ventricular pressure begins to rise, completing the cycle.

Assuming the average heart beats 75 times each minute, the length of the cardiac cycle is about 0.8 s, with atrial systole accounting for 0.1 s and ventricular systole 0.3 s. The remaining 0.4 s is a period of total heart relaxation, the **quiescent period**.

Notice two important points: (1) Blood flow through the heart is controlled entirely by pressure changes and (2) blood flows down a pressure gradient through any available opening. The pressure changes, in turn, reflect the alternating contraction and relaxation of the myocardium and cause the heart valves to open, which keeps blood flowing in the forward direction.

The situation in the right side of the heart is essentially the same as in the left side *except* for pressure. The pulmonary circulation is a low-pressure circulation as evidenced by the much thinner myocardium of its right ventricle. So, typical systolic and diastolic pressures for the pulmonary artery are 24 and 8 mm Hg as compared to systemic aortic pressures of 120 and 80 mm Hg, respectively. However, the two sides of the heart eject the same blood volume with each heartbeat.

CHECK YOUR UNDERSTANDING

12. The second heart sound is associated with the closing of which valve(s)?
13. If the mitral valve were insufficient, would you expect to hear the murmur (of blood flowing through the valve that should be closed) during ventricular systole or diastole?
14. During the cardiac cycle, there are two periods when all four valves are closed. Name these two periods.

For answers, see Appendix G.

Cardiac Output

▶ Name and explain the effects of various factors regulating stroke volume and heart rate.

▶ Explain the role of the autonomic nervous system in regulating cardiac output.

Cardiac output (CO) is the amount of blood pumped out by *each* ventricle in 1 minute. It is the product of heart rate (HR) and stroke volume (SV). **Stroke volume** is defined as the volume of blood pumped out by one ventricle with each beat. In general, stroke volume is correlated with the force of ventricular contraction.

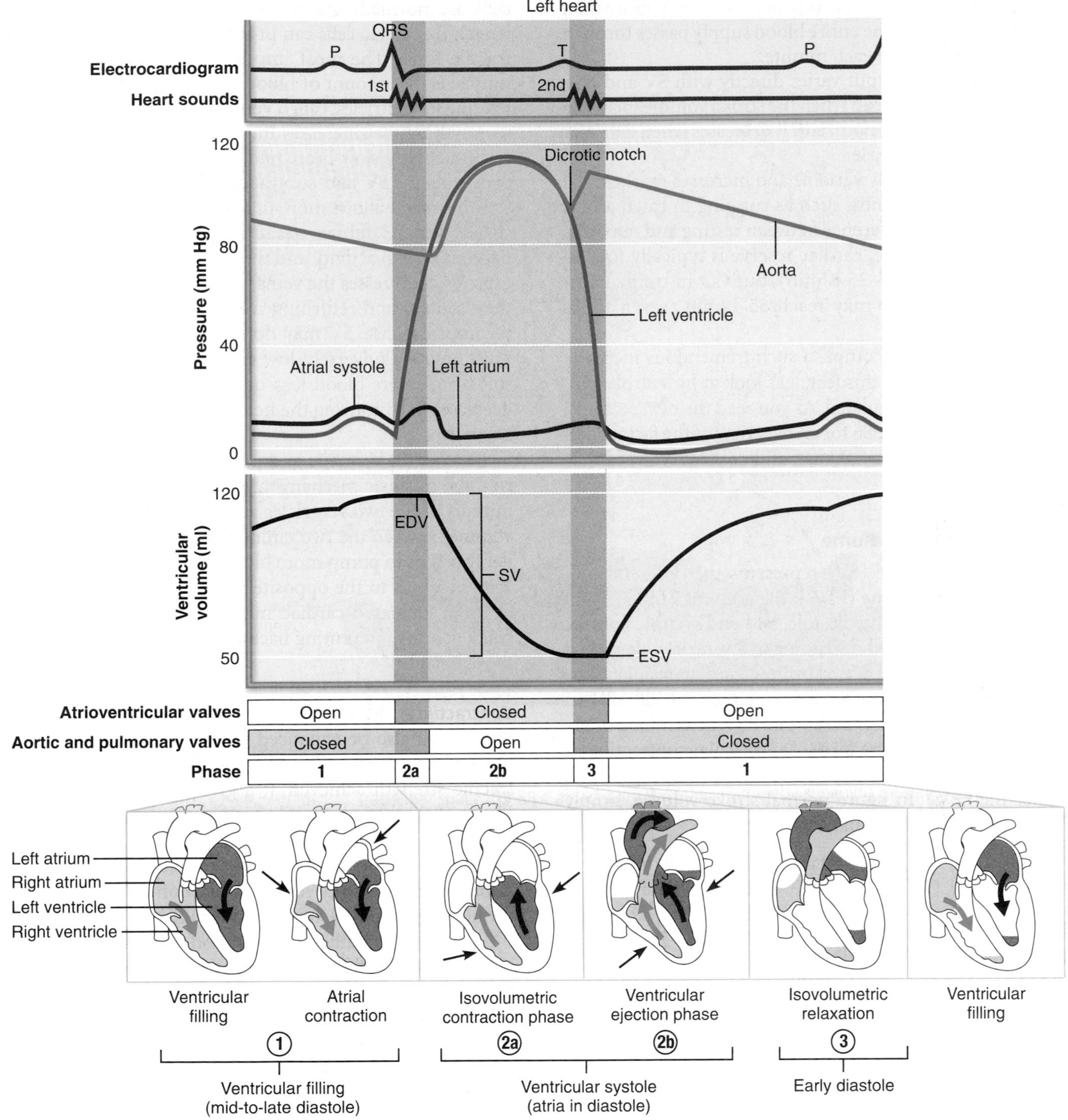

Figure 18.20 Summary of events during the cardiac cycle. An ECG tracing (*top*) correlated with graphs of pressure and volume changes (*center*) in the left side of the heart. Pressures are lower in the right side of the heart. Timing of heart sounds is also indicated. (*Bottom*) Events of phases 1 through 3 of the cardiac cycle.

Using normal resting values for heart rate (75 beats/min) and stroke volume (70 ml/beat), the average adult cardiac output can be computed:

$$CO = HR \times SV = \frac{75\ \text{beats}}{\text{min}} \times \frac{70\ \text{ml}}{\text{beat}}$$

$$= \frac{5250\ \text{ml}}{\text{min}} = \frac{5.25\ \text{L}}{\text{min}}$$

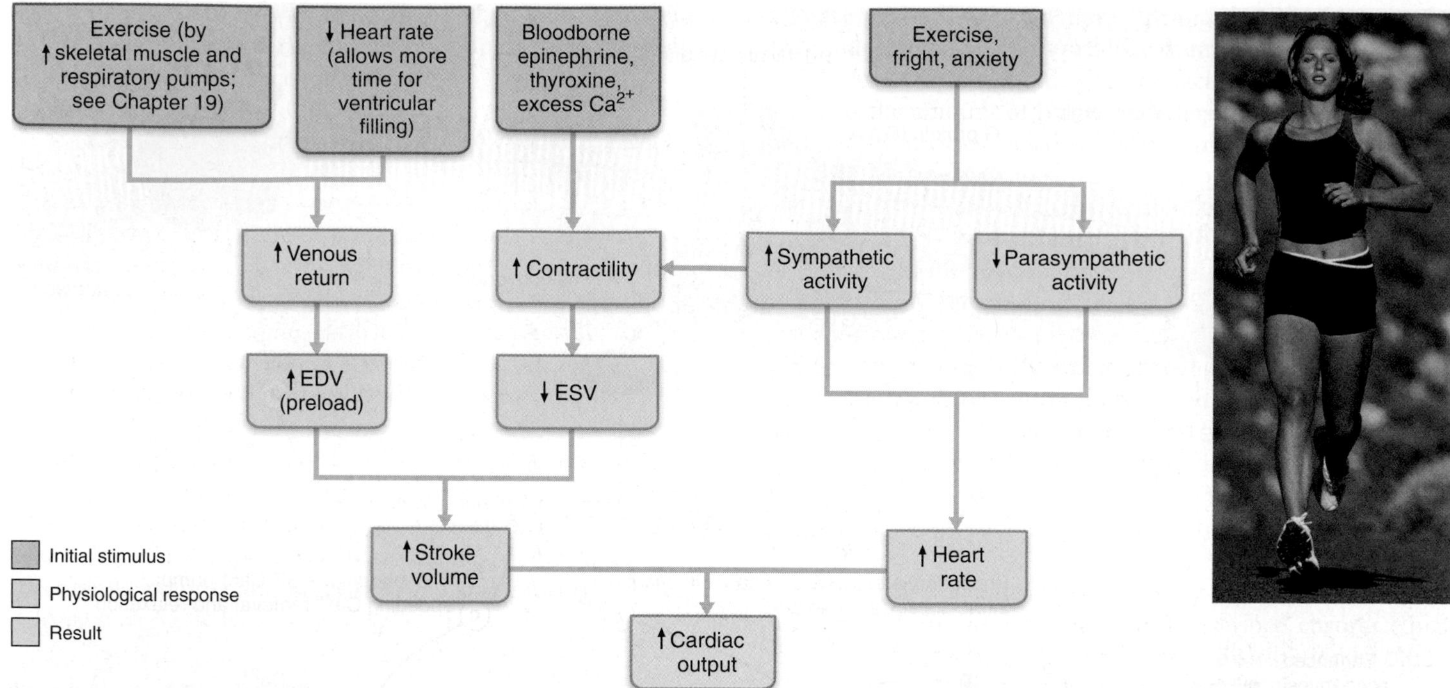

Figure 18.22 Factors involved in regulation of cardiac output.

Under resting conditions, both autonomic divisions continuously send impulses to the SA node of the heart, but the *dominant* influence is inhibitory. For this reason, the heart is said to exhibit **vagal tone**, and heart rate is generally slower than it would be if the vagal nerves were not innervating it. Cutting the vagal nerves results in an almost immediate increase in heart rate of about 25 beats/min, reflecting the inherent rate (100 beats/min) of the pacemaking SA node.

When either division of the autonomic nervous system is stimulated more strongly by sensory inputs relayed from various parts of the cardiovascular system, the other division is temporarily inhibited. Most such sensory input is generated by *baroreceptors* which respond to changes in systemic blood pressure, as we will discuss in Chapter 19. Another example, the **atrial (Bainbridge) reflex**, is a sympathetic reflex initiated by increased venous return and increased atrial filling. Stretching the atrial walls increases heart rate by stimulating both the SA node and the atrial stretch receptors, which trigger reflexive adjustments that result in increased sympathetic stimulation of the heart.

Increased or decreased CO results in corresponding changes to systemic blood pressure, so blood pressure regulation often involves reflexive controls of heart rate. In Chapter 19 we describe in more detail neural mechanisms that regulate blood pressure.

Chemical Regulation Chemicals normally present in the blood and other body fluids may influence heart rate, particularly if they become excessive or deficient.

1. **Hormones.** *Epinephrine*, liberated by the adrenal medulla during sympathetic nervous system activation, produces the same cardiac effects as norepinephrine released by the sympathetic nerves: It enhances heart rate and contractility.

   Thyroxine is a thyroid gland hormone that increases metabolic rate and body heat production. When released in large quantities, it causes a slower but more sustained increase in heart rate than that caused by epinephrine. Thyroxine also *enhances* the effects of epinephrine and norepinephrine on the heart, so chronically hyperthyroid individuals may develop a weakened heart.
2. **Ions.** Physiological relationships between intracellular and extracellular ions must be maintained for normal heart function. Plasma electrolyte imbalances pose real dangers to the heart.

HOMEOSTATIC IMBALANCE

Reduced Ca^{2+} blood levels (*hypocalcemia*) depress the heart. Conversely, above-normal levels (*hypercalcemia*) tightly couple the excitation-contraction mechanism and prolong the plateau phase of the action potential. These events dramatically increase heart irritability, which can lead to spastic heart contractions that permit the heart little rest. Many drugs used to treat heart conditions focus on calcium transport into cardiac cells.

High or low blood K^+ levels are particularly dangerous and arise in a number of clinical conditions. Excessive K^+ (*hyperkalemia*) interferes with depolarization by lowering the resting potential, and may lead to heart block and cardiac arrest. *Hypokalemia* is also life threatening, in that the heart beats feebly and arrhythmically. ■

Other Factors Age, gender, exercise, and body temperature also influence HR, although they are less important than neural factors. Resting heart rate is fastest in the fetus

(140–160 beats/min) and gradually declines throughout life. Average heart rate is faster in females (72–80 beats/min) than in males (64–72 beats/min).

Exercise raises HR by acting through the sympathetic nervous system (Figure 18.22). Exercise also increases systemic blood pressure and routes more blood to the working muscles. However, resting HR in the physically fit tends to be substantially lower than in those who are out of condition, and in trained athletes it may be as slow as 40 beats/min. We explain this apparent paradox below.

Heat increases HR by enhancing the metabolic rate of cardiac cells. This explains the rapid, pounding heartbeat you feel when you have a high fever and also accounts, in part, for the effect of exercise on HR (remember, working muscles generate heat). Cold directly decreases heart rate.

HOMEOSTATIC IMBALANCE

HR varies with changes in activity, but marked and persistent rate changes usually signal cardiovascular disease. **Tachycardia** (tak″e-kar′de-ah; "heart hurry") is an abnormally fast heart rate (more than 100 beats/min) that may result from elevated body temperature, stress, certain drugs, or heart disease. Persistent tachycardia is considered pathological because tachycardia occasionally promotes fibrillation.

Bradycardia (brad″e-kar′de-ah; *brady* = slow) is a heart rate slower than 60 beats/min. It may result from low body temperature, certain drugs, or parasympathetic nervous activation. It is a known, and desirable, consequence of endurance training. As physical and cardiovascular conditioning increases, the heart hypertrophies and SV increases, and so resting heart rate can be lower and still provide the same cardiac output. However, persistent bradycardia in poorly conditioned people may result in grossly inadequate blood circulation to body tissues, and bradycardia is often a warning of brain edema after head trauma. ■

Homeostatic Imbalance of Cardiac Output

The heart's pumping action ordinarily maintains a balance between cardiac output and venous return. Were this not so, a dangerous damming up of blood (blood congestion) would occur in the veins returning blood to the heart.

When the pumping efficiency (CO) of the heart is so low that blood circulation is inadequate to meet tissue needs, the heart is said to be in **congestive heart failure (CHF)**. This progressively worsening condition reflects weakening of the myocardium by various conditions which damage it in different ways. Let's take a look.

1. **Coronary atherosclerosis.** This condition, essentially a clogging of the coronary vessels with fatty buildup, impairs blood and oxygen delivery to the cardiac cells. The heart becomes increasingly hypoxic and begins to contract ineffectively.
2. **Persistent high blood pressure.** Normally, pressure in the aorta during diastole is 80 mm Hg, and the left ventricle exerts only slightly over that amount of force to eject blood from its chamber. When aortic diastolic blood pressure rises to 90 mm Hg or more, the myocardium must exert more force to open the aortic valve and pump out the same amount of blood. If this situation of enhanced afterload becomes chronic, the ESV rises and the myocardium hypertrophies. Eventually, the stress takes its toll and the myocardium becomes progressively weaker.
3. **Multiple myocardial infarcts.** A succession of MIs depresses pumping efficiency because the dead heart cells are replaced by noncontractile fibrous (scar) tissue.
4. **Dilated cardiomyopathy (DCM)** (kar″de-o-my-ah′path-e). The cause of this condition, in which the ventricles stretch and become flabby and the myocardium deteriorates, is often unknown. Drug toxicity (alcohol, cocaine, excess catecholamines, chemotherapeutic agents), hyperthyroidism, and inflammation of the heart following an infection are implicated in some cases. The heart's attempts to work harder result in increasing levels of Ca^{2+} in the cardiac cells. The Ca^{2+} activates calcineurin, a calcium-sensitive enzyme that initiates a cascade which switches on genes that cause heart enlargement. CO is poor because ventricular contractility is impaired, and the condition progressively worsens.

Because the heart is a double pump, each side can initially fail independently of the other. If the left side fails, **pulmonary congestion** occurs. The right side continues to propel blood to the lungs, but the left side does not adequately eject the returning blood into the systemic circulation. Blood vessels in the lungs become engorged with blood, the pressure in them increases, and fluid leaks from the circulation into the lung tissue, causing pulmonary edema. If the congestion is untreated, the person suffocates.

If the right side of the heart fails, **peripheral congestion** occurs. Blood stagnates in body organs, and pooled fluids in the tissue spaces impair the ability of body cells to obtain adequate amounts of nutrients and oxygen and to rid themselves of wastes. The resulting edema is most noticeable in the extremities (feet, ankles, and fingers).

Failure of one side of the heart puts a greater strain on the other side, and ultimately the whole heart fails. A seriously weakened, or *decompensated*, heart is irreparable. Treatment is directed primarily toward (1) removing the excess leaked fluid with *diuretics* (drugs that increase excretion of Na^+ and water by the kidneys), (2) reducing afterload with drugs that drive down blood pressure, and (3) increasing contractility with digitalis derivatives. However, heart transplants and other surgical or mechanical remedies to replace damaged heart muscle are providing additional hope for some cardiac patients.

CHECK YOUR UNDERSTANDING

15. After running to catch a bus, Josh noticed that his heart was beating faster than normally and was pounding forcefully in his chest. How did his faster HR and increased SV come about?
16. What problem of cardiac output might ensue if the heart beats far too rapidly for an extended period, that is, if tachycardia occurs? Why?

For answers, see Appendix G.

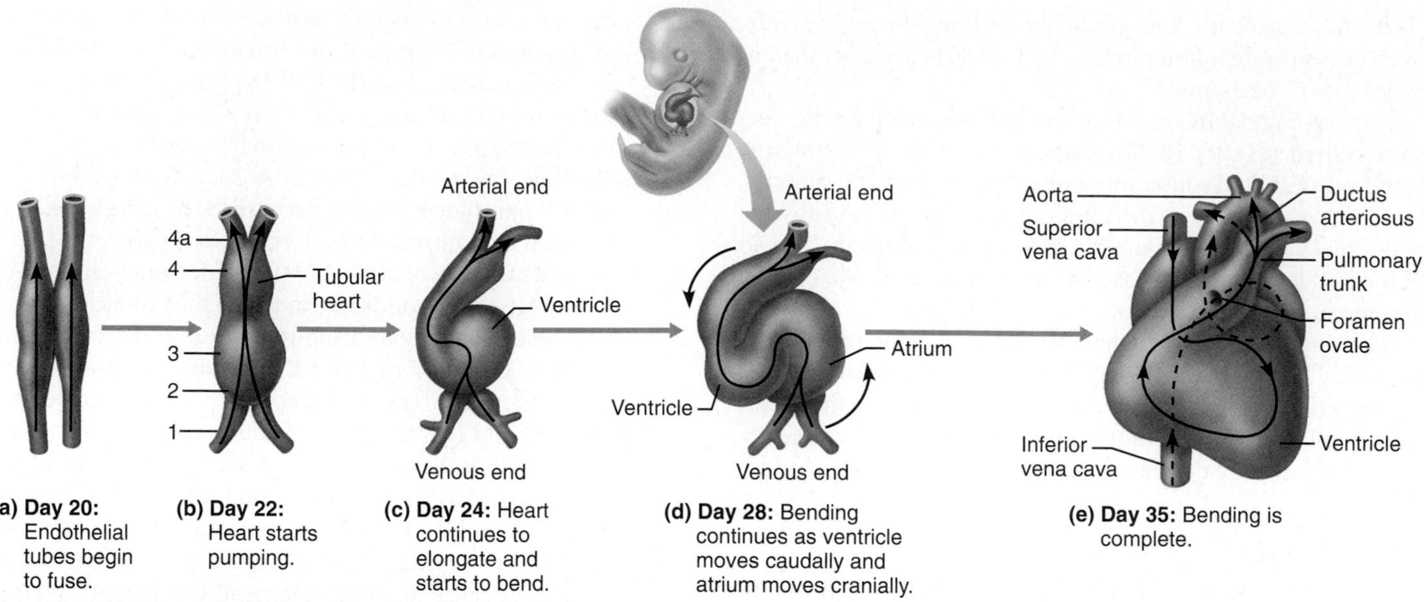

Figure 18.23 Development of the human heart. Ventral view, with the cranial direction toward the top of the figures. Arrows show the direction of blood flow. Days are approximate. In **(b)**: 1 is the sinus venosus; 2, the atrium; 3, the ventricle; 4, the bulbus cordis; and 4a, the truncus arteriosus.

Developmental Aspects of the Heart

- Describe fetal heart formation, and indicate how the fetal heart differs from the adult heart.
- Provide examples of age-related changes in heart function.

The human heart, derived from mesoderm and guided by powerful signaling molecules, begins as two simple endothelial tubes that quickly fuse to form a single chamber or heart tube that is busily pumping blood by the 22nd day of gestation **(Figure 18.23)**. The tube develops four slightly bulged areas that represent the earliest heart chambers. From tail to head, following the direction of blood flow, the four primitive chambers are the following (Figure 18.23b):

1. **Sinus venosus** (ven-o′sus). This chamber initially receives all the venous blood of the embryo. It will become the smooth-walled part of the right atrium and the coronary sinus. It also gives rise to the sinoatrial node, which "takes the baton" and sets heart rate early in embryonic development.
2. **Atrium.** This embryonic chamber eventually becomes the pectinate muscle–ridged parts of the atria.
3. **Ventricle.** The strongest pumping chamber of the early heart, the ventricle gives rise to the *left* ventricle.
4. **Bulbus cordis.** This chamber, plus its cranial extension, the *truncus arteriosus* (labeled 4a in the figure), give rise to the pulmonary trunk, the first part of the aorta, and most of the *right* ventricle.

During the next three weeks, the heart "tube" exhibits dramatic contortions as it undergoes rightward looping, and major structural changes convert it into a four-chambered organ capable of acting as a double pump—all without missing a beat! The ventricle moves caudally and the atrium cranially, assuming their adult positions. The heart divides into its four definitive chambers (via a number of stages), the midline septum forms, and the bulbus cordis splits into the pulmonary trunk and ascending aorta. After the second month, few changes other than growth occur until birth.

The interatrial septum of the fetal heart is incomplete. The **foramen ovale** (literally, "oval door") connects the two atria and allows blood entering the right heart to bypass the pulmonary circuit and the collapsed, nonfunctional fetal lungs (Figure 18.23e). Another lung bypass, the **ductus arteriosus**, exists between the pulmonary trunk and the aorta. At or shortly after birth, these shunts close, completing the separation between the right and left sides of the heart. In the adult heart, the position of the foramen ovale is revealed by the fossa ovalis, and the **ligamentum arteriosum** is the fibrous remnant of the ductus arteriosus (see Figure 18.4b). We give a more complete description of the fetal and newborn circulation in Chapter 28 (see Figure 28.14).

HOMEOSTATIC IMBALANCE

Building a perfect heart is difficult. Each year about 30,000 infants are born in the U.S. with one or more of 30 different **congenital heart defects**, making them the most common of all birth defects. Most congenital heart problems are traceable to environmental influences, such as maternal infection or drug intake during month 2 when the major events of heart formation occur.

The most prevalent abnormalities produce two basic kinds of disorders in the newborn. They either (1) lead to mixing of

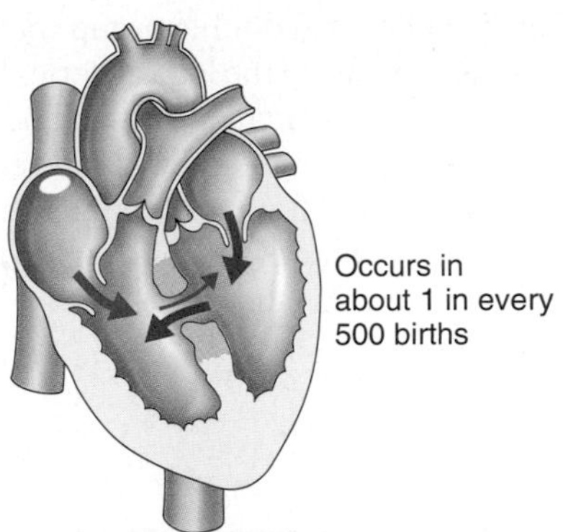

(a) Ventricular septal defect. The superior part of the interventricular septum fails to form; thus, blood mixes between the two ventricles. More blood is shunted from left to right because the left ventricle is stronger.

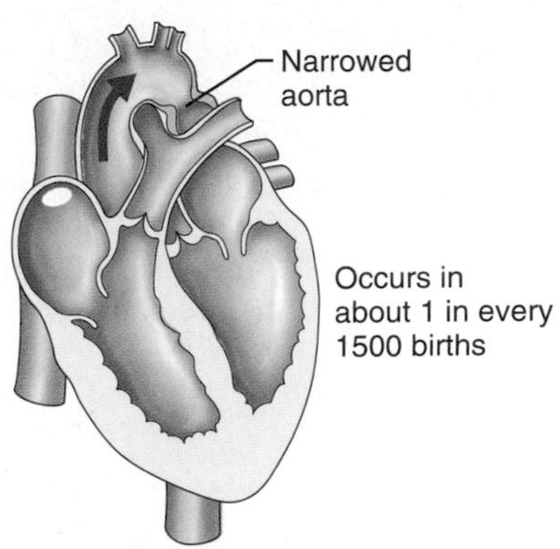

(b) Coarctation of the aorta. A part of the aorta is narrowed, increasing the workload of the left ventricle.

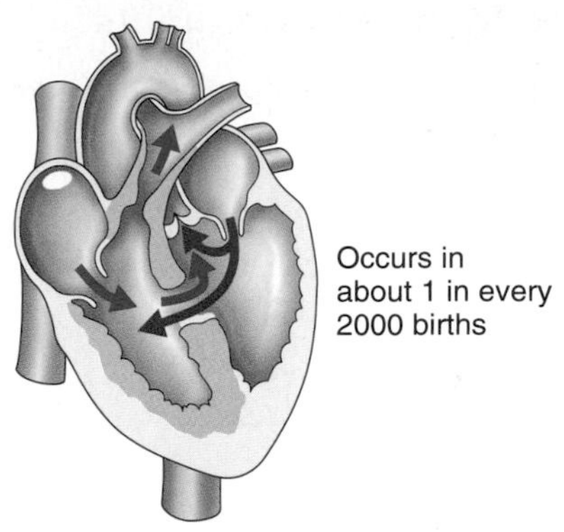

(c) Tetralogy of Fallot. Multiple defects (*tetra* = four): (1) Pulmonary trunk too narrow and pulmonary valve stenosed, resulting in (2) hypertrophied right ventricle; (3) ventricular septal defect; (4) aorta opens from both ventricles.

Figure 18.24 Three examples of congenital heart defects. Tan areas indicate the locations of the defects.

oxygen-poor systemic blood with oxygenated pulmonary blood (so that inadequately oxygenated blood reaches the body tissues) or (2) involve narrowed valves or vessels that greatly increase the workload on the heart. Examples of the first type of defect are *septal defects* **(Figure 18.24a)** and *patent ductus arteriosus*, in which the connection between the aorta and pulmonary trunk remains open. *Coarctation of the aorta* (Figure 18.24b) is an example of the second type of problem. *Tetralogy of Fallot* (te-tral′o-je ov fal-o′), a serious condition in which the baby becomes cyanotic within minutes of birth, encompasses both types of disorders (Figure 18.24c). Modern surgical techniques can correct most of these heart defects. ■

In the absence of congenital heart problems, the heart functions admirably throughout a long lifetime for most people. Homeostatic mechanisms are normally so efficient that people rarely notice when the heart is working harder. In people who exercise regularly and vigorously, the heart gradually adapts to the increased demand by increasing in size and becoming a more efficient and more powerful pump. Consequently, SV increases and resting heart rate declines. Aerobic exercise also helps to clear fatty deposits from blood vessel walls throughout the body, retarding atherosclerosis and coronary heart disease. Barring some chronic illnesses, this beneficial cardiac response to exercise persists into ripe old age.

The key word on benefiting from exercise is *regularity*. Regular exercise gradually enhances the myocardial endurance and strength. For example, 30 minutes a day of moderately vigorous exercise (brisk walking, biking, or yard work) offers significant health benefits to most adults. However, intermittent vigorous exercise, enjoyed by weekend athletes, may push an unconditioned heart beyond its ability to respond to the unexpected demands and bring on a myocardial infarct.

Because of the incredible amount of work the heart does over the course of a lifetime, certain structural changes are inevitable. These age-related changes affecting the heart include

1. **Sclerosis and thickening of valve flaps.** This change occurs particularly where the stress of blood flow is greatest (mitral valve). For this reason, heart murmurs are more common in elderly people.
2. **Decline in cardiac reserve.** Although the passing years seem to cause few changes in resting heart rate, the aged heart is less able to respond to both sudden and prolonged stressors that demand increased output. Sympathetic control of the heart becomes less efficient (in part because the amount of cAMP produced on stimulation decreases with age). HR gradually becomes more variable and there is a decline in maximum HR. These changes are less of a problem in physically active seniors.
3. **Fibrosis of cardiac muscle.** As a person ages, more and more cardiac cells die and are replaced with fibrous tissue. As a result, the heart stiffens and its filling in preparation for the next heartbeat is less efficient, resulting in reduced stroke volume. Additionally, the nodes of the heart's conduction system sometimes become fibrosed (scarred) with age, which increases the incidence of arrhythmias and other conduction problems.
4. **Atherosclerosis.** The insidious progress of atherosclerosis begins in childhood, but inactivity, smoking, and stress accelerate it. The most serious consequences to the heart are hypertensive heart disease and coronary artery occlusion, both of which increase the risk of heart attack and stroke. Although the aging process itself leads to changes in blood vessel walls that result in atherosclerosis, many investigators feel that diet, not aging, is the single most important contributor to cardiovascular disease. There is some agreement that risk is lowered if people consume less animal fat, cholesterol, and salt.

CHECK YOUR UNDERSTANDING

17. Name the two components of the fetal heart that allow blood to bypass the lungs.
18. In the past decade, many people over 70 have competed in the Ironman World Championships in Hawaii. In what way might age-related changes of the heart limit the performance of these athletes?

For answers, see Appendix G.

The heart is an exquisitely engineered double pump that operates with precision to propel blood into the large arteries leaving its chambers. However, continuous circulation of blood also depends critically on the pressure dynamics in the blood vessels. Chapter 19 considers the structure and function of these vessels and relates this information to the work of the heart to provide a complete picture of cardiovascular functioning.

RELATED CLINICAL TERMS

Asystole (a-sis′to-le) Situation in which the heart fails to contract.

Cardiac catheterization Diagnostic procedure involving passage of a fine catheter (tubing) through a blood vessel into the heart; blood samples are withdrawn, allowing oxygen content, blood flow, and pressures within the heart to be measured. Findings help to detect valve problems, heart deformities, and other heart malfunctions.

Commotio cordis ("concussion of the heart") Situation in which a relatively mild blow to the chest causes heart failure and sudden death because it occurs during a vulnerable interval (2 ms) when the heart is repolarizing. Explains those rare instances when youngsters drop dead on the playing field after being hit in the chest by a ball.

Cor pulmonale (kor pul-mun-nă-′le; *cor* = heart, *pulmo* = lung) A condition of right-sided heart failure resulting from elevated blood pressure in the pulmonary circuit (pulmonary hypertension). Acute cases may develop suddenly due to a pulmonary embolism; chronic cases are usually associated with chronic lung disorders such as emphysema.

Endocarditis (en″do-kar-di′tis) Inflammation of the endocardium, usually confined to the endocardium of the heart valves. Endocarditis often results from infection by bacteria that have entered the bloodstream but may result from fungal infection or an autoimmune response. Drug addicts may develop endocarditis by injecting themselves with contaminated needles.

Heart palpitation A heartbeat that is unusually strong, fast, or irregular so that the person becomes aware of it; may be caused by certain drugs, emotional pressures ("nervous heart"), or heart disorders.

Hypertrophic cardiomyopathy (HCM) The leading cause of sudden death in young athletes, this condition, which is usually inherited, causes the cardiac muscle cells to enlarge, thickening the heart wall. The heart pumps strongly but doesn't relax well during diastole when the heart is filling.

Mitral valve prolapse Valve disorder affecting up to 1% of the population; most often seen in young women. It appears to have a genetic basis resulting in abnormal chordae tendineae or a malfunction of the papillary muscles; one or more of the flaps of the mitral valve become incompetent and billow into the left atrium during ventricular systole, allowing blood regurgitation. Occasionally requires valve replacement surgery.

Myocarditis (mi″o-kar-di′tis; *myo* = muscle, *card* = heart, *itis* = inflammation) Inflammation of the cardiac muscle layer (myocardium) of the heart; sometimes follows an untreated streptococcal infection in children. May weaken the heart and impair its ability to act as an effective pump.

Paroxysmal atrial tachycardia (PAT) Bursts of atrial contractions with little pause between them.

Ventricular tachycardia (VT or V-tac) Rapid ventricular contractions that are not coordinated with atrial activity.

CHAPTER SUMMARY

Media study tools that could provide you additional help in reviewing specific key topics of Chapter 18 are referenced below.

iP = ***Interactive Physiology***

Heart Anatomy (pp. 662–672)

Size, Location, and Orientation (p. 663)

1. The human heart, about the size of a clenched fist, is located obliquely within the mediastinum of the thorax.

Coverings of the Heart (p. 663)

2. The heart is enclosed within a double sac made up of the outer fibrous pericardium and the inner serous pericardium (parietal and visceral layers). The pericardial cavity between the serous layers contains lubricating serous fluid.

Layers of the Heart Wall (pp. 663–664)

3. Layers of the heart wall, from the interior out, are the endocardium, the myocardium (reinforced by a fibrous skeleton), and the epicardium (visceral layer of the serous pericardium).

Chambers and Associated Great Vessels (pp. 664–668)

4. The heart has two superior atria and two inferior ventricles. Functionally, the heart is a double pump.
5. Entering the right atrium are the superior vena cava, the inferior vena cava, and the coronary sinus. Four pulmonary veins enter the left atrium.
6. The right ventricle discharges blood into the pulmonary trunk; the left ventricle pumps blood into the aorta.

Pathway of Blood Through the Heart (pp. 668–669)

7. The right heart is the pulmonary circuit pump. Oxygen-poor systemic blood enters the right atrium, passes into the right ventricle, through the pulmonary trunk to the lungs, and back to the left atrium via the pulmonary veins.
8. The left heart is the systemic circuit pump. Oxygen-laden blood entering the left atrium from the lungs flows into the left ventricle and then into the aorta, which provides the functional supply of all body organs. Systemic veins return the oxygen-depleted blood to the right atrium.

Coronary Circulation (pp. 669–670)

9. The right and left coronary arteries branch from the aorta and via their main branches (anterior and posterior interventricular, right marginal, and circumflex arteries) supply the heart itself. Venous blood, collected by the cardiac veins (great, middle, and small), is emptied into the coronary sinus.
10. Blood delivery to the myocardium occurs during heart relaxation.

Heart Valves (pp. 670–672)

11. The atrioventricular valves (tricuspid and mitral) prevent backflow into the atria when the ventricles are contracting; the pulmonary and aortic valves (semilunar valves) prevent backflow into the ventricles when the ventricles are relaxing.

iP Cardiovascular System; Topic: Anatomy Review: The Heart, pp. 1–8.

Cardiac Muscle Fibers (pp. 672–676)

Microscopic Anatomy (pp. 672–673)

1. Cardiac muscle cells are branching, striated, generally uninucleate cells. They contain myofibrils consisting of typical sarcomeres.
2. Adjacent cardiac cells are connected by intercalated discs containing desmosomes and gap junctions. The myocardium behaves as a functional syncytium because of electrical coupling provided by gap junctions.

Mechanism and Events of Contraction (pp. 673–675)

3. As in skeletal muscle, the membrane depolarization of contractile myocytes causes opening of sodium channels and sodium entry, which is responsible for the rising phase of the action potential curve. Depolarization also opens slow Ca^{2+} channels; Ca^{2+} entry prolongs the period of depolarization (creates the plateau). The action potential is coupled to sliding of the myofilaments by Ca^{2+} released by the SR and entering from the extracellular space. Compared to skeletal muscle, cardiac muscle has a prolonged refractory period that prevents tetany.

iP Cardiovascular System; Topic: Cardiac Action Potential, pp. 11–18.

Energy Requirements (pp. 675–676)

4. Cardiac muscle has abundant mitochondria and depends almost entirely on aerobic respiration to form ATP.

Heart Physiology (pp. 676–687)

Electrical Events (pp. 676–680)

1. Certain noncontractile cardiac muscle cells exhibit automaticity and rhythmicity and can independently initiate action potentials. Such cells have an unstable resting potential called a pacemaker potential that gradually depolarizes, drifting toward threshold for firing. These cells compose the intrinsic conduction system of the heart.
2. The conduction system of the heart consists of the SA and AV nodes, the AV bundle and bundle branches, and the Purkinje fibers. This system coordinates the depolarization of the heart and ensures that the heart beats as a unit. The SA node has the fastest rate of spontaneous depolarization and acts as the heart's pacemaker; it sets the sinus rhythm.
3. Defects in the intrinsic conduction system can cause arrhythmias, fibrillation, and heart block.
4. The heart is innervated by the autonomic nervous system. Cardiac centers in the medulla include the cardioacceleratory center, which projects to the T_1–T_5 region of the spinal cord, which in turn projects to the cervical and upper thoracic sympathetic trunk. Postganglionic fibers innervate the SA and AV nodes and the cardiac muscle fibers. The cardioinhibitory center exerts its influence via the parasympathetic vagus nerves (X), which project to the heart wall. Most parasympathetic fibers serve the SA and AV nodes.
5. An electrocardiogram is a graphic representation of the cardiac conduction cycle. The P wave reflects atrial depolarization. The QRS complex indicates ventricular depolarization; the T wave represents ventricular repolarization.

iP Cardiovascular System; Topic: Intrinsic Conduction System, pp. 1–7; Topic: Cardiac Action Potential, pp. 1–10.

Heart Sounds (p. 681)

6. Normal heart sounds arise chiefly from turbulent blood flow during the closing of heart valves. Abnormal heart sounds, called murmurs, usually reflect valve problems.

Mechanical Events: The Cardiac Cycle (p. 682)

7. A cardiac cycle consists of the events occurring during one heartbeat. During mid-to-late diastole, the ventricles fill and the atria contract. Ventricular systole consists of the isovolumetric contraction phase and the ventricular ejection phase. During early diastole, the ventricles are relaxed and are closed chambers until the atrial pressure exceeds the ventricular pressure, forcing the AV valves open. Then the cycle begins again. At a normal heart rate of 75 beats/min, a cardiac cycle lasts 0.8 s.
8. Pressure changes promote blood flow and valve opening and closing.

iP Cardiovascular System; Topic: Cardiac Cycle, pp. 1–19.

Cardiac Output (pp. 682–687)

9. Cardiac output, typically 5 L/min, is the amount of blood pumped out by each ventricle in 1 minute. Stroke volume is the amount of blood pumped out by a ventricle with each contraction. Cardiac output = heart rate × stroke volume.
10. Stroke volume depends to a large extent on the degree of stretch of cardiac muscle by venous return. Approximately 70 ml, it is the difference between end diastolic volume (EDV) and end systolic volume (ESV). Anything that influences heart rate or blood volume influences venous return, hence stroke volume.
11. Activation of the sympathetic nervous system increases heart rate and contractility; parasympathetic activation decreases heart rate but has little effect on contractility. Ordinarily, the heart exhibits vagal tone.
12. Chemical regulation of the heart is effected by hormones (epinephrine and thyroxine) and ions (particularly potassium and calcium). Imbalances in ions severely impair heart activity.
13. Other factors influencing heart rate are age, sex, exercise, and body temperature.

14. Congestive heart failure occurs when the pumping ability of the heart is inadequate to provide normal circulation to meet body needs. Right heart failure leads to systemic edema; left heart failure results in pulmonary edema.

iP Cardiovascular System; Topic: Cardiac Output, pp. 1–11.

Developmental Aspects of the Heart (pp. 688–690)

1. The heart begins as a simple (mesodermal) tube that is pumping blood by the fourth week of gestation. The fetal heart has two lung bypasses: the foramen ovale and the ductus arteriosus.
2. Congenital heart defects are the most common of all birth defects. The most common of these disorders lead to inadequate oxygenation of blood or increase the workload of the heart.
3. Age-related changes include sclerosis and thickening of the valve flaps, declines in cardiac reserve, fibrosis of cardiac muscle, and atherosclerosis.
4. Risk factors for cardiac disease include dietary factors, excessive stress, cigarette smoking, and lack of exercise.

REVIEW QUESTIONS

Multiple Choice/Matching

(Some questions have more than one correct answer. Select the best answer or answers from the choices given.)

1. When the semilunar valves are open, which of the following are occurring? **(a)** 2, 3, 5, 6, **(b)** 1, 2, 3, 7, **(c)** 1, 3, 5, 6, **(d)** 2, 4, 5, 7.
   (1) coronary arteries fill
   (2) AV valves are closed
   (3) ventricles are in systole
   (4) ventricles are in diastole
   (5) blood enters aorta
   (6) blood enters pulmonary arteries
   (7) atria contract
2. The portion of the intrinsic conduction system located in the interventricular septum is the **(a)** AV node, **(b)** SA node, **(c)** AV bundle, **(d)** Purkinje fibers.
3. An ECG provides information about **(a)** cardiac output, **(b)** movement of the excitation wave across the heart, **(c)** coronary circulation, **(d)** valve impairment.
4. The sequence of contraction of the heart chambers is **(a)** random, **(b)** left chambers followed by right chambers, **(c)** both atria followed by both ventricles, **(d)** right atrium, right ventricle, left atrium, left ventricle.
5. The fact that the left ventricular wall is thicker than the right reveals that it **(a)** pumps a greater volume of blood, **(b)** pumps blood against greater resistance, **(c)** expands the thoracic cage, **(d)** pumps blood through a smaller valve.
6. The chordae tendineae **(a)** close the atrioventricular valves, **(b)** prevent the AV valve flaps from everting, **(c)** contract the papillary muscles, **(d)** open the semilunar valves.
7. In the heart, which of the following apply? (1) Action potentials are conducted from cell to cell across the myocardium via gap junctions, (2) the SA node sets the pace for the heart as a whole, (3) spontaneous depolarization of cardiac cells can occur in the absence of nerve stimulation, (4) cardiac muscle can continue to contract for long periods in the absence of oxygen. **(a)** all of the above, **(b)** 1, 3, 4, **(c)** 1, 2, 3, **(d)** 2, 3.
8. The activity of the heart depends on intrinsic properties of cardiac muscle and on neural factors. Thus, **(a)** vagus nerve stimulation of the heart reduces heart rate, **(b)** sympathetic nerve stimulation of the heart decreases time available for ventricular filling, **(c)** sympathetic stimulation of the heart increases its force of contraction, **(d)** all of the above.
9. Freshly oxygenated blood is first received by the **(a)** right atrium, **(b)** left atrium, **(c)** right ventricle, **(d)** left ventricle.

Short Answer Essay Questions

10. Describe the location and position of the heart in the body.
11. Describe the pericardium and distinguish between the fibrous and the serous pericardia relative to histological structure and location.
12. Trace one drop of blood from the time it enters the right atrium until it enters the left atrium. What is this circuit called?
13. **(a)** Describe how heart contraction and relaxation influence coronary blood flow. **(b)** Name the major branches of the coronary arteries, and note the heart regions served by each.
14. The refractory period of cardiac muscle is much longer than that of skeletal muscle. Why is this a desirable functional property?
15. **(a)** Name the elements of the intrinsic conduction system of the heart in order, beginning with the pacemaker. **(b)** What is the important function of this conduction system?
16. Draw a normal ECG pattern. Label and explain the significance of its deflection waves.
17. Define cardiac cycle, and follow the events of one cycle.
18. What is cardiac output, and how is it calculated?
19. Discuss how the Frank-Starling law of the heart helps to explain the influence of venous return on stroke volume.
20. **(a)** Describe the common function of the foramen ovale and the ductus arteriosus in a fetus. **(b)** What problems result if these shunts remain patent (open) after birth?

Critical Thinking and Clinical Application Questions

1. A gang member was stabbed in the chest during a street fight. He was cyanotic and unconscious from lack of blood delivery to the brain. The diagnosis was cardiac tamponade. What is cardiac tamponade and how does it cause the observed symptoms?
2. You have been called upon to demonstrate the technique for listening to valve sounds. **(a)** Explain where you would position your stethoscope to auscultate (1) the aortic valve of a patient with severe aortic valve insufficiency and (2) a stenotic mitral valve. **(b)** During which period(s) would you hear these abnormal valve sounds most clearly? (During atrial diastole, ventricular systole, ventricular diastole, or atrial systole?) **(c)** What cues would you use to differentiate between an insufficient and a stenosed valve?
3. Florita Santos, a middle-aged woman, is admitted to the coronary care unit with a diagnosis of left ventricular failure resulting from a myocardial infarction. Her history indicated that she was aroused in the middle of the night by severe chest pain. Her skin is pale and cold, and moist sounds are heard over the lower regions of both lungs. Explain how failure of the left ventricle can cause these signs and symptoms.

4. Heather, a newborn baby, needs surgery because she was born with an aorta that arises from the right ventricle and a pulmonary trunk that issues from the left ventricle, a condition called transposition of the great vessels. What are the physiological consequences of this defect?
5. Gabriel, a heroin addict, feels tired, is weak and feverish, and has vague aches and pains. Terrified that he has AIDS, he goes to a doctor and is informed that he is suffering not from AIDS, but from a heart murmur accompanied by endocarditis. What is the most likely way that Gabriel contracted endocarditis?
6. As Cara worked at her dissection, she became frustrated that several of the structures she had to learn about had more than one common name. Provide another name for each of these structures: (a) atrioventricular groove, (b) tricuspid valve, (c) bicuspid valve (give two synonyms), and (d) atrioventricular bundle.

TABLE 19.1 Summary of Blood Vessel Anatomy

VESSEL TYPE/ ILLUSTRATION*	AVERAGE LUMEN DIAMETER (D) AND WALL THICKNESS (T)	RELATIVE TISSUE MAKEUP: Endothelium / Elastic Tissues / Smooth Muscles / Fibrous (Collagenous) Tissues
Elastic artery	D : 1.5 cm T : 1.0 mm	
Muscular artery	D : 6.0 mm T : 1.0 mm	
Arteriole	D : 37.0 µm T : 6.0 µm	
Capillary	D : 9.0 µm T : 0.5 µm	
Venule	D : 20.0 µm T : 1.0 µm	
Vein	D : 5.0 mm T : 0.5 mm	

*Size relationships are not proportional. Smaller vessels are drawn relatively larger so detail can be seen. See column 2 for actual dimensions.

Muscular (Distributing) Arteries

Distally the elastic arteries give way to the **muscular**, or **distributing, arteries**, which deliver blood to specific body organs and account for most of the named arteries studied in the anatomy laboratory. Their internal diameter ranges from that of a little finger (1 cm) to that of a pencil lead (about 0.3 mm).

Proportionately, muscular arteries have the thickest tunica media of all vessels. Their tunica media contains relatively more smooth muscle and less elastic tissue than do elastic arteries (Table 19.1). For this reason, they are more active in vasoconstriction and less distensible. In muscular arteries, however, there *is* an *elastic lamina* on each face of the tunica media.

Arterioles

The smallest of the arteries, **arterioles** have a lumen diameter ranging from 0.3 mm down to 10 µm. Larger arterioles have all three tunics, but their tunica media is chiefly smooth muscle with a few scattered elastic fibers. Smaller arterioles, which lead into the capillary beds, are little more than a single layer of smooth muscle cells spiraling around the endothelial lining.

As we will describe shortly, minute-to-minute blood flow into the capillary beds is determined by arteriole diameter, which varies in response to changing neural, hormonal, and local chemical influences. When arterioles constrict, the tissues served are largely bypassed. When arterioles dilate, blood flow into the local capillaries increases dramatically.

Capillaries

▶ Describe the structure and function of a capillary bed.

The microscopic **capillaries** are the smallest blood vessels. Their exceedingly thin walls consist of just a thin tunica intima (see Figure 19.1b). In some cases, one endothelial cell forms the entire circumference of the capillary wall. At strategic locations along the outer surface of some capillaries are spider-shaped *pericytes*, smooth muscle–like cells that stabilize the capillary wall and help control capillary permeability **(Figure 19.3a)**.

Average capillary length is 1 mm and average lumen diameter is 8–10 µm, just large enough for red blood cells to slip through in single file. Most tissues have a rich capillary supply, but there are exceptions. Tendons and ligaments are poorly vascularized. Cartilage and epithelia lack capillaries, but receive nutrients from blood vessels in nearby connective tissues, and the avascular cornea and lens of the eye receive nutrients from the aqueous humor.

If blood vessels are compared to a system of expressways and roads, the capillaries are the back alleys and driveways that provide direct access to nearly every cell in the body. Given their location and the thinness of their walls, capillaries are ideally suited for their role—exchange of materials (gases, nutrients, hormones, and so on) between the blood and the interstitial fluid (Figure 19.2 and Table 19.1). We describe these exchanges later in this chapter. Here, we focus on capillary structure.

Types of Capillaries

Structurally, there are three types of capillaries—*continuous*, *fenestrated*, and *sinusoidal*. **Continuous capillaries**, abundant in the skin and muscles, are most common (Figure 19.3a). They are continuous in the sense that their endothelial cells provide an uninterrupted lining, adjacent cells being joined laterally by *tight junctions*. However, these junctions are usually incomplete and leave gaps of unjoined membrane called **intercellular clefts**, which are just large enough to allow limited passage of fluids and small solutes. Typically, the endothelial cell cytoplasm contains numerous pinocytotic vesicles that ferry fluids across the capillary wall.

Brain capillaries, however, are unique. There the tight junctions of the continuous capillaries are complete and extend around the entire perimeter of the endothelial cells, constituting the structural basis of the *blood-brain barrier* that we described in Chapter 12.

Fenestrated capillaries (fen′es-tra-tid) are similar to the continuous variety except that some of the endothelial cells in fenestrated capillaries are riddled with oval *pores*, or *fenestrations* (*fenestra* = window) (Figure 19.3b). The fenestrations are usually covered by a delicate membrane, or diaphragm (probably condensed basal lamina material), but even so, fenestrated capillaries are much more permeable to fluids and small solutes than continuous capillaries are.

Fenestrated capillaries are found wherever active capillary absorption or filtrate formation occurs. For example, fenestrated capillaries in the small intestine receive nutrients from digested food, and those in endocrine organs allow hormones rapid entry into the blood. Fenestrated capillaries with perpetually open pores occur in the kidneys, where rapid filtration of blood plasma is essential.

Sinusoids (si′nŭ-soyds), or **sinusoidal capillaries**, are highly modified, leaky capillaries found only in the liver, bone marrow, spleen, and adrenal medulla. Sinusoids have large, irregularly shaped lumens and are usually fenestrated. Their endothelial lining has fewer tight junctions and larger intercellular clefts than ordinary capillaries (Figure 19.3c).

These structural adaptations allow large molecules and even blood cells to pass between the blood and surrounding tissues. In the liver, the endothelium of the sinusoids is *discontinuous* and its lining includes large **hepatic macrophages** (Kupffer cells), which remove and destroy any contained bacteria. In other organs, such as the spleen, phagocytes located just outside the sinusoids stick cytoplasmic extensions through the intercellular clefts into the sinusoidal lumen to get at their "prey." Blood flows sluggishly through the tortuous sinusoid channels, allowing time for it to be modified in various ways.

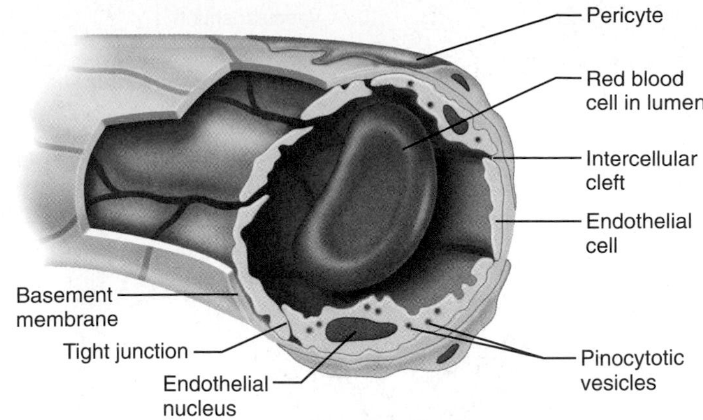

(a) Continuous capillary. Least permeable, and most common (e.g., skin, muscle).

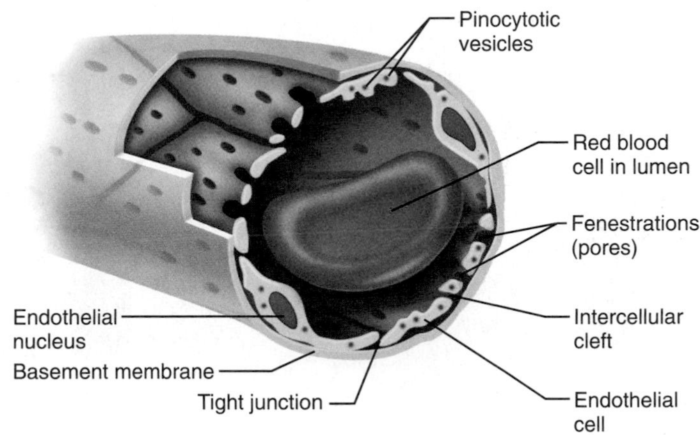

(b) Fenestrated capillary. Large fenestrations (pores) increase permeability. Occurs in special locations (e.g., kidney, small intestine).

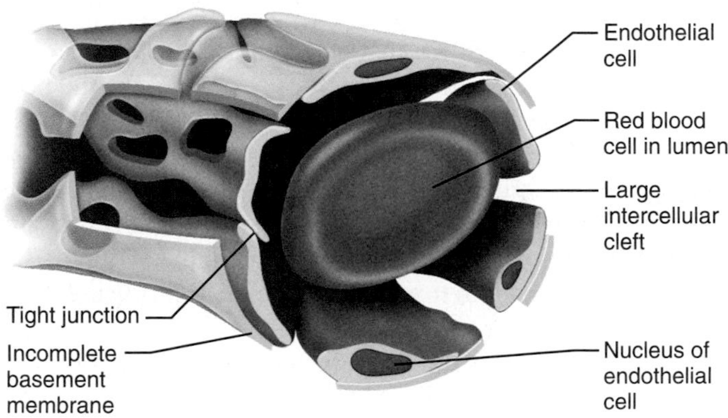

(c) Sinusoidal capillary. Most permeable. Occurs in special locations (e.g., liver, bone marrow, spleen).

Figure 19.3 Capillary structure. Note that the basement membrane is incomplete only in **(c)** and that pericytes most often occur on continuous capillaries.

Capillary Beds

Capillaries do not function independently. Instead they tend to form interweaving networks called **capillary beds**. The flow of blood from an arteriole to a venule—that is, through a capillary bed—is called the **microcirculation**. In most body regions, a capillary bed consists of two types of vessels: (1) a *vascular shunt*

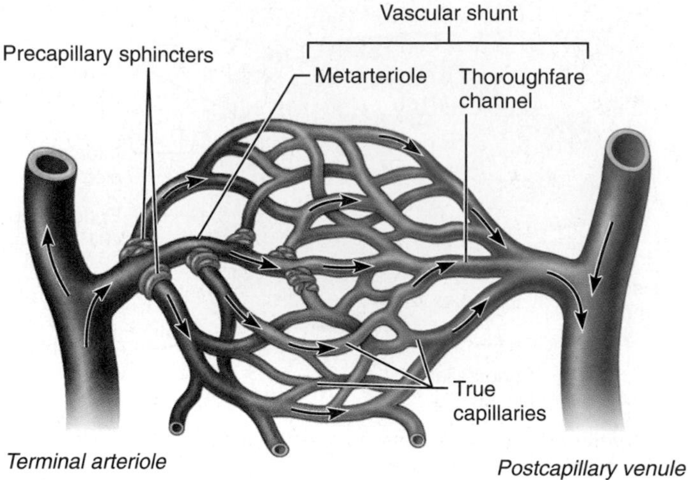

(a) Sphincters open—blood flows through true capillaries.

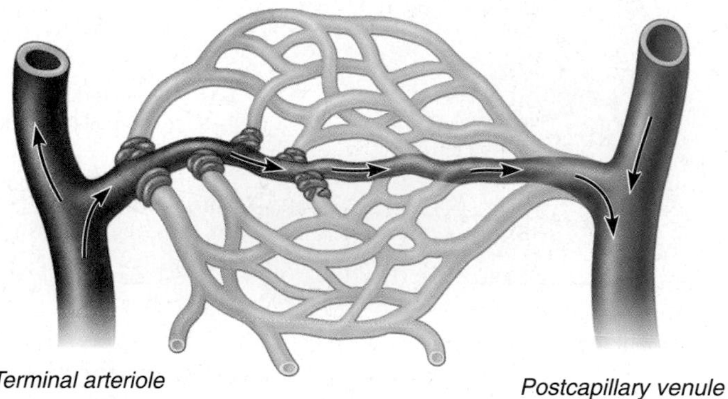

(b) Sphincters closed—blood flows through metarteriole–thoroughfare channel and bypasses true capillaries.

Figure 19.4 Anatomy of a capillary bed.

(metarteriole–thoroughfare channel), a short vessel that directly connects the arteriole and venule at opposite ends of the bed, and (2) *true capillaries*, the actual *exchange vessels* **(Figure 19.4)**.

The **terminal arteriole** feeding the bed leads into a **metarteriole** (a vessel structurally intermediate between an arteriole and a capillary), which is continuous with the **thoroughfare channel** (intermediate between a capillary and a venule). The thoroughfare channel, in turn, joins the **postcapillary venule** that drains the bed.

The **true capillaries** number 10 to 100 per capillary bed, depending on the organ or tissues served. They usually branch off the metarteriole (proximal end of the shunt) and return to the thoroughfare channel (the distal end), but occasionally they spring from the terminal arteriole and empty directly into the venule. A cuff of smooth muscle fibers, called a **precapillary sphincter**, surrounds the root of each true capillary at the metarteriole and acts as a valve to regulate blood flow into the capillary.

Blood flowing through a terminal arteriole may go either through the true capillaries or through the shunt. When the precapillary sphincters are relaxed (open), as in Figure 19.4a, blood flows through the true capillaries and takes part in exchanges with tissue cells. When the sphincters are contracted (closed), as in Figure 19.4b, blood flows through the shunts and bypasses the tissue cells.

The relative amount of blood entering a capillary bed is regulated by local chemical conditions and arteriolar vasomotor nerve fibers. A bed may be flooded with blood or almost completely bypassed, depending on conditions in the body or in that specific organ. For example, suppose you have just eaten and are sitting relaxed, listening to your favorite musical group. Food digestion is going on, and blood is circulating freely through the true capillaries of your gastrointestinal organs to receive the breakdown products of digestion. Between meals, however, most of these same capillary pathways are closed.

To take another example, when you are exercising vigorously blood is rerouted from the digestive organs (food or no food) to the capillary beds of your skeletal muscles where it is more immediately needed. This rerouting helps explain why vigorous exercise right after a meal can cause indigestion or abdominal cramps.

Venous System

▶ Describe the structure and function of veins, and explain how veins differ from arteries.

Blood is carried from the capillary beds toward the heart by veins. Along the route, the diameter of successive venous vessels increases, and their walls gradually thicken as they progress from venules to the larger and larger veins.

Venules

Ranging from 8 to 100 μm in diameter, **venules** are formed when capillaries unite. The smallest venules, the *postcapillary venules*, consist entirely of endothelium around which pericytes congregate. They are extremely porous (more like capillaries than veins in this way), and fluid and white blood cells move easily from the bloodstream through their walls. Indeed, a well-recognized sign of an inflamed area is adhesion of white blood cells to the postcapillary venule endothelium, followed by their migration through the wall into the inflamed tissue. The larger venules have one or two layers of smooth muscle cells (a scanty tunica media) and thin externa as well.

Veins

Venules join to form *veins*. Veins usually have three distinct tunics, but their walls are always thinner and their lumens larger than those of corresponding arteries (Figure 19.1 and Table 19.1). Consequently, in the view seen in routine histological preparations, veins are usually collapsed and their lumens appear slitlike.

There is relatively little smooth muscle or elastin in the tunica media, which is poorly developed and tends to be thin even in the largest veins. The tunica externa is the heaviest wall layer. Consisting of thick longitudinal bundles of collagen fibers and elastic networks, it is often several times thicker than the tunica media. In the largest veins—the venae cavae, which return

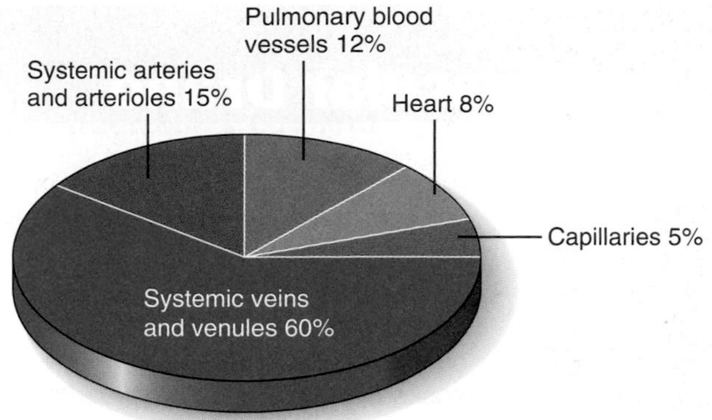

Figure 19.5 Relative proportion of blood volume throughout the cardiovascular system. The systemic veins are called capacitance vessels because they are distensible and contain a large proportion of the blood volume. Pulmonary blood vessels supply the lungs; systemic blood vessels supply the rest of the body.

blood directly to the heart—the tunica externa is further thickened by longitudinal bands of smooth muscle.

With their large lumens and thin walls, veins can accommodate a fairly large blood volume. Veins are called **capacitance vessels** and **blood reservoirs** because up to 65% of the body's blood supply is found in the veins at any time **(Figure 19.5)**. Even so, they are normally only partially filled.

The walls of veins can be much thinner than arterial walls without danger of bursting because the blood pressure in veins is low. However, the low-pressure condition demands some adaptations to ensure that blood is returned to the heart at the same rate it was pumped into the circulation. The large-diameter lumens of veins (which offer relatively little resistance to blood flow) are one structural adaptation.

Another adaptation is valves that prevent blood from flowing backward. **Venous valves** are formed from folds of the tunica intima, and resemble the semilunar valves of the heart in both structure and function (see Figure 19.1). Venous valves are most abundant in the veins of the limbs, where the upward flow of blood is opposed by gravity. They are absent in veins of the thoracic and abdominal body cavities.

A simple experiment demonstrates the effectiveness of venous valves. Hang one hand by your side until the blood vessels on its dorsal aspect distend with blood. Next place two fingertips against one of the distended veins, and pressing firmly, move the superior finger proximally along the vein and then release that finger. The vein will remain flattened and collapsed, despite the pull of gravity. Finally, remove the distal fingertip and watch the vein refill with blood.

HOMEOSTATIC IMBALANCE

Varicose veins are veins that have become tortuous and dilated because of incompetent (leaky) valves. More than 15% of adults suffer from varicose veins, usually in the lower limbs. Several factors contribute, including heredity and conditions that hinder venous return, such as prolonged standing in one position, obesity, or pregnancy. Both the "potbelly" of an overweight person and the enlarged uterus of a pregnant woman exert downward pressure on vessels of the groin, restricting return of blood to the heart. Consequently, blood pools in the lower limbs, and with time, the valves weaken and the venous walls stretch and become floppy. Superficial veins, which receive little support from surrounding tissues, are especially susceptible.

Another cause of varicose veins is elevated venous pressure. For example, straining to deliver a baby or have a bowel movement raises intra-abdominal pressure, preventing blood from draining from the veins of the anal canal. The resulting varicosities in the anal veins are called *hemorrhoids* (hem′ŏ-roidz). ■

Venous sinuses, such as the *coronary sinus* of the heart and the *dural venous sinuses* of the brain, are highly specialized, flattened veins with extremely thin walls composed only of endothelium. They are supported by the tissues that surround them, rather than by any additional tunics. The dural venous sinuses, which receive cerebrospinal fluid and blood draining from the brain, are reinforced by the tough dura mater that covers the brain surface.

Vascular Anastomoses

Where vascular channels unite, they form interconnections called **vascular anastomoses** (ah-nas″to-mo′sēz; "coming together"). Most organs receive blood from more than one arterial branch, and arteries supplying the same territory often merge, forming **arterial anastomoses**. These anastomoses provide alternate pathways, called **collateral channels**, for blood to reach a given body region. If one branch is cut or blocked by a clot, the collateral channel can often provide the area with an adequate blood supply.

Arterial anastomoses occur around joints, where active movement may hinder blood flow through one channel. They are also common in abdominal organs, the brain, and the heart. Arteries that do not anastomose or that have a poorly developed collateral circulation supply the retina, kidneys, and spleen. If their blood flow is interrupted, cells supplied by such vessels die.

The metarteriole–thoroughfare channel shunts of capillary beds that connect arterioles and venules are examples of **arteriovenous anastomoses.** Veins interconnect much more freely than arteries, and **venous anastomoses** are common. (You may be able to see venous anastomoses through the skin on the dorsum of your hand.) Because venous anastomoses are abundant, occlusion of a vein rarely blocks blood flow or leads to tissue death.

CHECK YOUR UNDERSTANDING

3. Name the type of artery that matches each description: major role in dampening the pulsatile pressure of heart contractions; vasodilation or constriction determines blood flow to individual capillary beds; have the thickest tunica media relative to their lumen size.

(Continues on p. 703.)

19

3. **Resistance** is opposition to flow and is a measure of the amount of friction blood encounters as it passes through the vessels. Because most friction is encountered in the peripheral (systemic) circulation, well away from the heart, we generally use the term **peripheral resistance**. There are three important sources of resistance: blood viscosity, vessel length, and vessel diameter.

   - *Blood viscosity.* The internal resistance to flow that exists in all fluids is *viscosity* (vis-kos'ĭ-te) and is related to the thickness or "stickiness" of a fluid. The greater the viscosity, the less easily molecules slide past one another and the more difficult it is to get and keep the fluid moving. Blood is much more viscous than water because it contains formed elements and plasma proteins. Consequently, it flows more slowly under the same conditions. Blood viscosity is fairly constant, but infrequent conditions such as polycythemia (excessive numbers of red blood cells) can increase blood viscosity and, hence, resistance. On the other hand, if the red blood cell count is low, as in some anemias, blood is less viscous and peripheral resistance declines.
   - *Total blood vessel length.* The relationship between total blood vessel length and resistance is straightforward: the longer the vessel, the greater the resistance. For example, an infant's blood vessels lengthen as he or she grows to adulthood, and so both peripheral resistance and blood pressure increase.
   - *Blood vessel diameter.* Because blood viscosity and vessel length are normally unchanging, the influence of these factors can be considered constant in healthy people. Changes in blood vessel diameter are frequent, however, and significantly alter peripheral resistance.

   How so? The answer lies in principles of fluid flow. Fluid close to the wall of a tube or channel is slowed by friction as it passes along the wall, whereas fluid in the center of the tube flows more freely and faster. You can verify this by watching the flow of water in a river. Water close to the bank hardly seems to move, while that in the middle of the river flows quite rapidly.

   In a tube of a given size, the relative speed and position of fluid in the different regions of the tube's cross section remain constant, a phenomenon called *laminar flow* or *streamlining.* The smaller the tube, the greater the friction, because relatively more of the fluid contacts the tube wall, where its movement is impeded.

   Resistance varies *inversely* with the *fourth power* of the vessel radius (one-half the diameter). This means, for example, that if the radius of a vessel is doubled, the resistance drops to one-sixteenth of its original value ($r^4 = 2 \times 2 \times 2 \times 2 = 16$ and $1/r^4 = 1/16$). For this reason, the large arteries close to the heart, which do not change dramatically in diameter, contribute little to peripheral resistance. Instead, the small-diameter arterioles, which can enlarge or constrict in response to neural and chemical controls, are the major determinants of peripheral resistance.

   When blood encounters either an abrupt change in the tube size or rough or protruding areas of the tube wall (such as the fatty plaques of atherosclerosis), the smooth laminar blood flow is replaced by *turbulent flow,* that is, irregular fluid motion where blood from the different laminae mixes. Turbulence dramatically increases resistance.

Relationship Between Flow, Pressure, and Resistance

Now that we have defined these terms, let us summarize briefly the relationships between them. Blood flow (F) is *directly* proportional to the difference in blood pressure (ΔP) between two points in the circulation, that is, the blood pressure, or hydrostatic pressure, gradient. Thus, when ΔP increases, blood flow speeds up, and when ΔP decreases, blood flow declines. Blood flow is *inversely* proportional to the peripheral resistance (R) in the systemic circulation; if R increases, blood flow decreases. These relationships are expressed by the formula

$$F = \frac{\Delta P}{R}$$

Of these two factors influencing blood flow, R is far more important than ΔP in influencing local blood flow because R can easily be changed by altering blood vessel diameter. For example, when the arterioles serving a particular tissue dilate (thus decreasing the resistance), blood flow to that tissue increases, even though the systemic pressure is unchanged or may actually be falling.

CHECK YOUR UNDERSTANDING

7. List three factors that determine resistance in a vessel. Which of these factors is physiologically most important?
8. Suppose vasoconstriction decreases the diameter of a vessel to one-third its size. What happens to the rate of flow through that vessel? (Calculate the expected size of the change.)

For answers, see Appendix G.

Systemic Blood Pressure

- ▶ List and explain the factors that influence blood pressure, and describe how blood pressure is regulated.
- ▶ Define hypertension. Describe its manifestations and consequences.

Any fluid driven by a pump through a circuit of closed channels operates under pressure, and the nearer the fluid is to the pump, the greater the pressure exerted on the fluid. The dynamics of blood flow in blood vessels is no exception, and blood flows through the blood vessels along a pressure gradient, always moving from higher- to lower-pressure areas. Fundamentally, *the pumping action of the heart generates blood flow. Pressure results when flow is opposed by resistance.*

As illustrated in **Figure 19.6**, systemic blood pressure is highest in the aorta and declines throughout the pathway to finally reach 0 mm Hg in the right atrium. The steepest drop in blood pressure occurs in the arterioles, which offer the greatest resistance to blood flow. However, as long as a pressure gradient exists, no matter how small, blood continues to flow until it completes the circuit back to the heart.

Arterial Blood Pressure

Arterial blood pressure reflects two factors: (1) how much the elastic arteries close to the heart can be stretched (their *compliance* or *distensibility*) and (2) the volume of blood forced into them at any time. If the amounts of blood entering and leaving the elastic arteries in a given period were equal, arterial pressure would be constant. Instead, as Figure 19.6 reveals, blood pressure rises and falls in a regular fashion in the elastic arteries near the heart. In other words, it is obviously *pulsatile.*

As the left ventricle contracts and expels blood into the aorta, it imparts kinetic energy to the blood, which stretches the elastic aorta as aortic pressure reaches its peak. Indeed, if the aorta were opened during this period, blood would spurt upward 5 or 6 feet! This pressure peak, called the **systolic pressure** (sis-tah′lik), averages 120 mm Hg in healthy adults. Blood moves forward into the arterial bed because the pressure in the aorta is higher than the pressure in the more distal vessels.

During diastole, the aortic valve closes, preventing blood from flowing back into the heart, and the walls of the aorta (and other elastic arteries) recoil, maintaining sufficient pressure to keep the blood flowing forward into the smaller vessels. During this time, aortic pressure drops to its lowest level (approximately 70 to 80 mm Hg in healthy adults), called the **diastolic pressure** (di-as-tah′lik). You can picture the elastic arteries as pressure reservoirs that operate as auxiliary pumps to keep blood circulating throughout the period of diastole, when the heart is relaxing. Essentially, the volume and energy of blood stored in the elastic arteries during systole are given back during diastole.

The difference between the systolic and diastolic pressures is called the **pulse pressure**. It is felt as a throbbing pulsation in an artery (a *pulse*) during systole, as the elastic arteries are expanded by the blood being forced into them by ventricular contraction. Increased stroke volume and faster blood ejection from the heart (a result of increased contractility) cause *temporary* increases in the pulse pressure. Pulse pressure is chronically increased by arteriosclerosis because the elastic arteries become less stretchy.

Because aortic pressure fluctuates up and down with each heartbeat, the important pressure figure to consider is the **mean arterial pressure (MAP)**—the pressure that propels the blood to the tissues. Diastole usually lasts longer than systole, so the MAP is not simply the value halfway between systolic and diastolic pressures. Instead, it is roughly equal to the diastolic pressure plus one-third of the pulse pressure.

$$\text{MAP} = \text{diastolic pressure} + \frac{\text{pulse pressure}}{3}$$

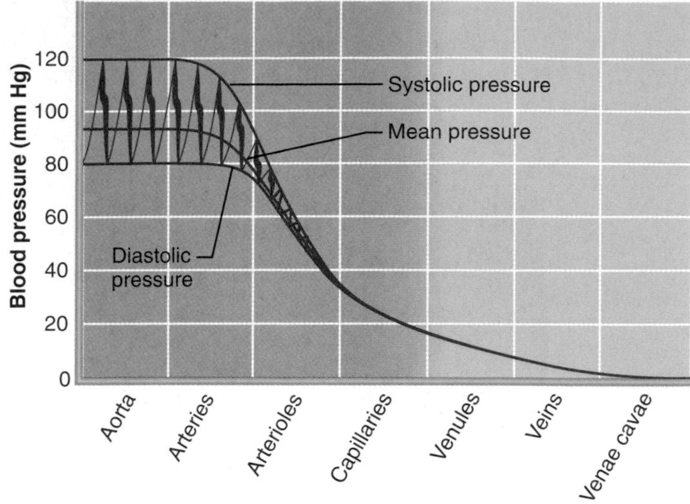

Figure 19.6 Blood pressure in various blood vessels of the systemic circulation.

For a person with a systolic blood pressure of 120 mm Hg and a diastolic pressure of 80 mm Hg:

$$\text{MAP} = 80 \text{ mm Hg} + \frac{40 \text{ mm Hg}}{3} = 93 \text{ mm Hg}$$

MAP and pulse pressure both decline with increasing distance from the heart. The MAP loses ground to the never-ending friction between the blood and the vessel walls, and the pulse pressure is gradually phased out in the less elastic muscular arteries, where elastic rebound of the vessels ceases to occur. At the end of the arterial tree, blood flow is steady and the pulse pressure has disappeared.

Capillary Blood Pressure

As Figure 19.6 shows, by the time blood reaches the capillaries, blood pressure has dropped to approximately 35 mm Hg and by the end of the capillary beds is only around 15 mm Hg. Such low capillary pressures are desirable because (1) capillaries are fragile and high pressures would rupture them, and (2) most capillaries are extremely permeable and thus even the low capillary pressure forces solute-containing fluids (filtrate) out of the bloodstream into the interstitial space. As we describe later in this chapter, these fluid flows are important for continuously refreshing the interstitial fluid.

Venous Blood Pressure

Unlike arterial pressure, which pulsates with each contraction of the left ventricle, venous blood pressure is steady and changes very little during the cardiac cycle. The pressure gradient in the veins, from venules to the termini of the venae cavae, is only about 15 mm Hg (that from the aorta to the ends of the arterioles is about 60 mm Hg). This pressure difference between arteries and veins becomes very clear when the vessels are cut. If a vein is cut, the blood flows evenly from the wound, but a lacerated

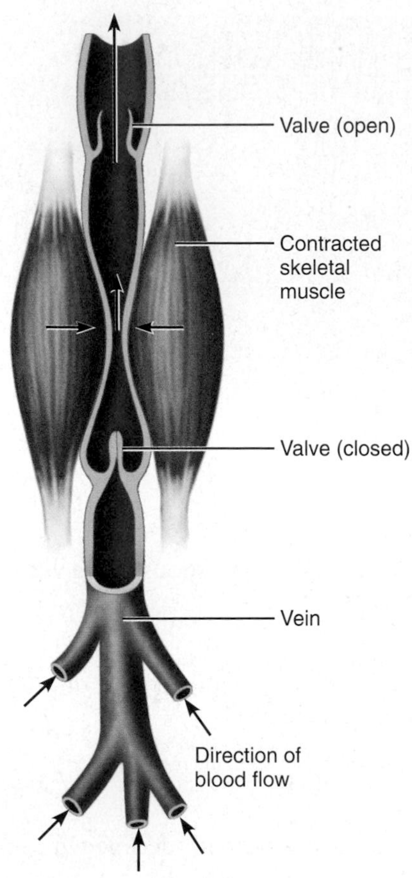

Figure 19.7 The muscular pump. When contracting skeletal muscles press against a vein, the valves proximal to the area of contraction are forced open and blood is propelled toward the heart. The valves distal to the area of contraction are closed by the backflowing blood.

artery spurts blood. The very low pressure in the venous system reflects the cumulative effects of peripheral resistance, which dissipates most of the energy of blood pressure (as heat) during each circuit.

Despite the structural modifications of veins (large lumens and valves), venous pressure is normally too low to promote adequate venous return. For this reason, three functional adaptations are critically important to venous return. The first is the **respiratory "pump."** Pressure changes occurring in the ventral body cavity during breathing create the respiratory pump that moves blood up toward the heart. As we inhale, abdominal pressure increases, squeezing the local veins and forcing blood toward the heart. At the same time, the pressure in the chest decreases, allowing thoracic veins to expand and speeding blood entry into the right atrium.

The second and most important adaptation is the **muscular "pump,"** consisting of skeletal muscle activity. As the skeletal muscles surrounding the deep veins contract and relax, they "milk" blood toward the heart, and once blood passes each successive valve, it cannot flow back **(Figure 19.7)**. People who earn their living in "standing professions," such as hairdressers and dentists, often have swollen ankles because blood pools in the feet and legs during prolonged periods of skeletal muscle inactivity.

The third adaptation is the layer of smooth muscle around the veins that constricts under sympathetic control, increasing venous return. This is yet another way in which the sympathetic nervous system increases cardiac output.

Maintaining Blood Pressure

Maintaining a steady flow of blood from the heart to the toes is vital for organs to function properly. But making sure a person jumping out of bed in the morning does not keel over from inadequate blood flow to the brain requires the finely tuned cooperation of the heart, blood vessels, and kidneys—all supervised by the brain.

Central among the homeostatic mechanisms that regulate cardiovascular dynamics are those that maintain blood pressure, principally *cardiac output, peripheral resistance,* and *blood volume.* If we rearrange the formula pertaining to blood flow presented on p. 704, we can see how cardiac output (blood flow of the entire circulation) and peripheral resistance relate to blood pressure.

$$F = \Delta P/R \quad \text{or} \quad CO = \Delta P/R \quad \text{or} \quad \Delta P = CO \times R$$

Clearly, blood pressure varies *directly* with CO and *R*. Additionally, blood pressure varies directly with blood volume because CO depends on blood volume (the heart can't pump out what doesn't enter its chambers). So in theory, a change (increase or decrease) in any of these variables would cause a corresponding change in blood pressure. However, what *really* happens in the body is that changes in one variable that threaten blood pressure homeostasis are quickly compensated for by changes in the other variables.

As we described in Chapter 18, CO is equal to *stroke volume* (ml/beat) times *heart rate* (beats/min), and normal CO is 5.0 to 5.5 L/min. **Figure 19.8** shows the main factors determining cardiac output—venous return and the neural and hormonal controls. Remember that the cardioinhibitory center in the medulla is "in charge" of heart rate most of the time and, via the parasympathetic vagus nerves, maintains the *resting heart rate.* During "resting" periods, stroke volume is controlled mainly by venous return (end diastolic volume). During stress, the cardioacceleratory center takes over, activating the sympathetic nervous system and increasing both heart rate (by acting on the SA node) and stroke volume (by enhancing cardiac muscle contractility, which decreases end systolic volume). The enhanced CO, in turn, results in an increase in MAP.

In the following discussion we focus on factors that regulate blood pressure by altering peripheral resistance and blood volume, but a flowchart showing the influence of nearly all other important factors is provided in Figure 19.11 (p. 711).

Short-Term Mechanisms: Neural Controls

The *short-term controls* of blood pressure, mediated by the nervous system and bloodborne chemicals, counteract moment-to-moment fluctuations in blood pressure by altering peripheral resistance (and CO).

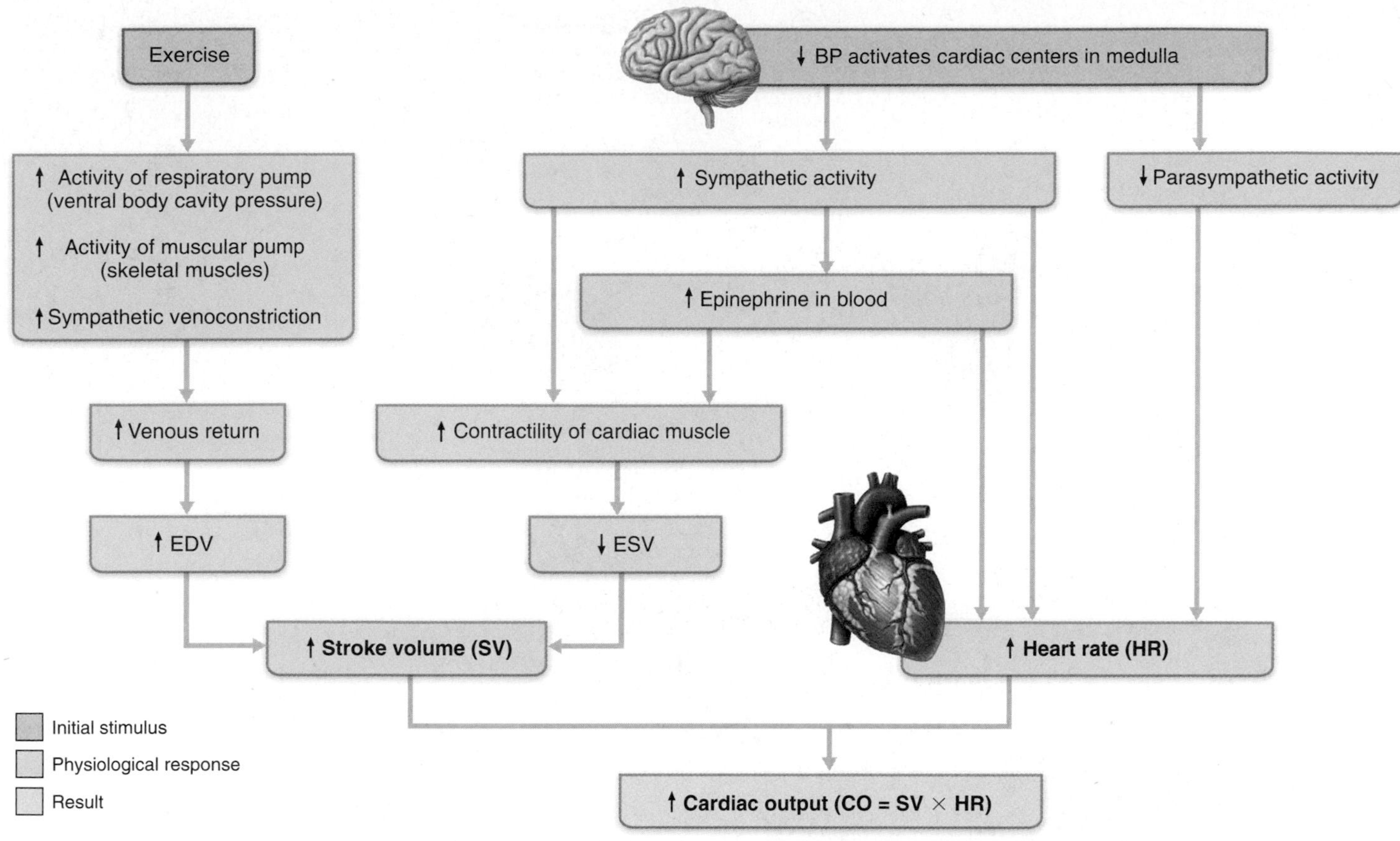

Figure 19.8 Major factors enhancing cardiac output. BP = blood pressure, EDV = end diastolic volume, ESV = end systolic volume.

Neural controls of peripheral resistance are directed at two main goals: (1) Maintaining adequate MAP by altering blood vessel diameter. (Remember, very small changes in blood vessel diameter cause substantial changes in peripheral resistance, and hence in systemic blood pressure.) Under conditions of low blood volume, all vessels except those supplying the heart and brain are constricted to allow as much blood as possible to flow to those two vital organs. (2) Altering blood distribution to respond to specific demands of various organs. For example, during exercise blood is shunted temporarily from the digestive organs to the skeletal muscles.

Most neural controls operate via reflex arcs involving *baroreceptors* and associated afferent fibers, the vasomotor center of the medulla, vasomotor fibers, and vascular smooth muscle. Occasionally, inputs from *chemoreceptors* and higher brain centers also influence the neural control mechanism.

Role of the Vasomotor Center

The neural center that oversees changes in the diameter of blood vessels is the **vasomotor center**, a cluster of neurons in the medulla. This center plus the cardiac centers described earlier make up the **cardiovascular center** that integrates blood pressure control by altering cardiac output and blood vessel diameter. The vasomotor center transmits impulses at a fairly steady rate along sympathetic efferents called **vasomotor fibers**, which exit from the T_1 through L_2 levels of the spinal cord and run to innervate the smooth muscle of blood vessels, mainly arterioles. As a result, the arterioles are almost always in a state of moderate constriction, called **vasomotor tone**.

The degree of vasomotor tone varies from organ to organ. Generally, arterioles of the skin and digestive viscera receive vasomotor impulses more frequently and tend to be more strongly constricted than those of skeletal muscles. Any increase in sympathetic activity produces generalized vasoconstriction and a rise in blood pressure. Decreased sympathetic activity allows the vascular muscle to relax somewhat and causes blood pressure to decline to basal levels.

Vasomotor activity is modified by inputs from (1) baroreceptors (pressure-sensitive mechanoreceptors that respond to changes in arterial pressure and stretch), (2) chemoreceptors (receptors that respond to changes in blood levels of carbon dioxide, H^+, and oxygen), and (3) higher brain centers. Let's take a look.

Baroreceptor-Initiated Reflexes

When arterial blood pressure rises, it stretches **baroreceptors**, neural receptors located in the *carotid sinuses* (dilations in the internal carotid arteries, which provide the major blood supply to the brain), in the *aortic arch*, and in the walls of nearly every large artery of the neck and thorax. When stretched, baroreceptors

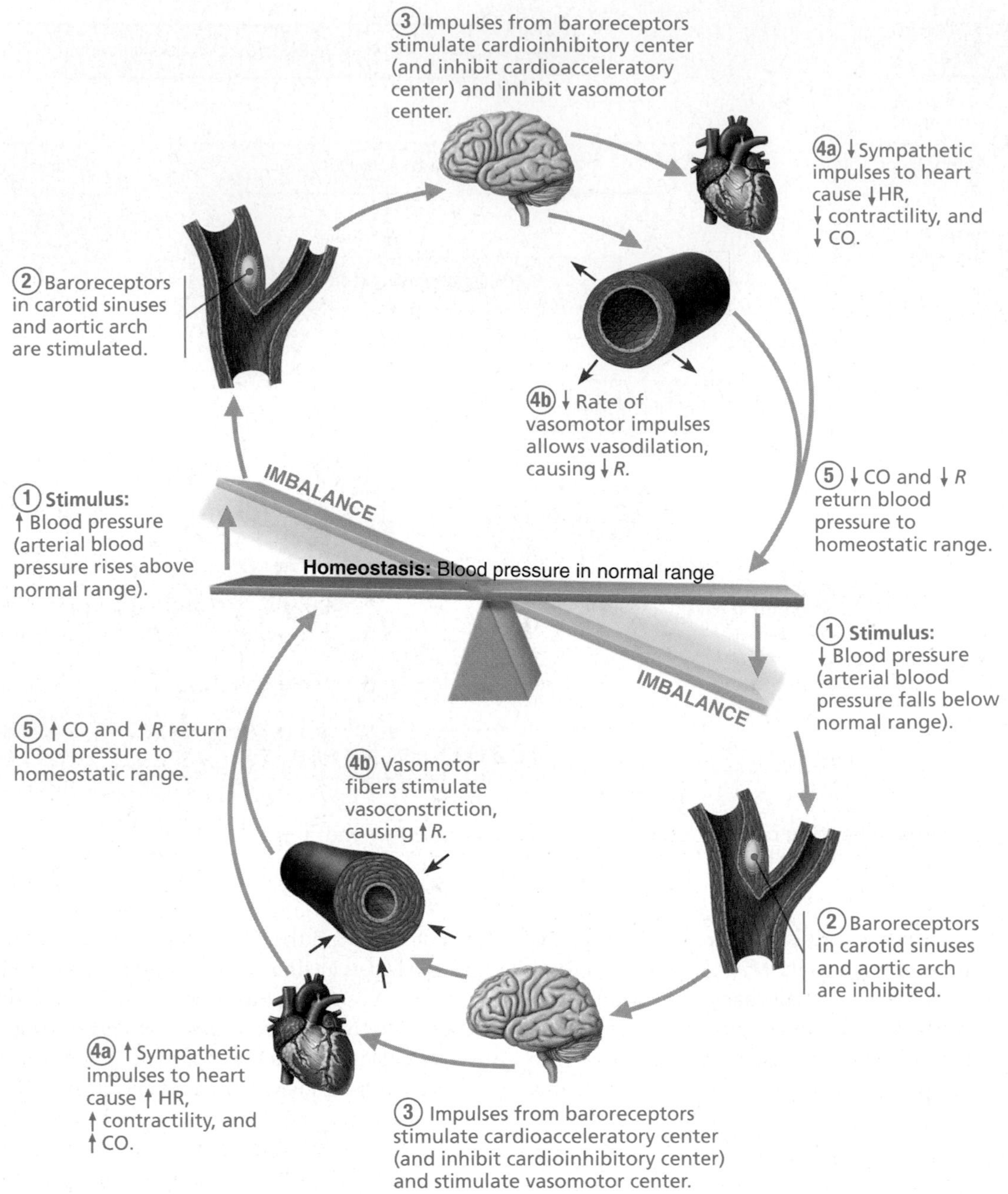
③ Impulses from baroreceptors stimulate cardioinhibitory center (and inhibit cardioacceleratory center) and inhibit vasomotor center.
④a ↓ Sympathetic impulses to heart cause ↓ HR, ↓ contractility, and ↓ CO.
② Baroreceptors in carotid sinuses and aortic arch are stimulated.
④b ↓ Rate of vasomotor impulses allows vasodilation, causing ↓ R.
IMBALANCE
① Stimulus: ↑ Blood pressure (arterial blood pressure rises above normal range).
⑤ ↓ CO and ↓ R return blood pressure to homeostatic range.
Homeostasis: Blood pressure in normal range
① Stimulus: ↓ Blood pressure (arterial blood pressure falls below normal range).
IMBALANCE
⑤ ↑ CO and ↑ R return blood pressure to homeostatic range.
④b Vasomotor fibers stimulate vasoconstriction, causing ↑ R.
② Baroreceptors in carotid sinuses and aortic arch are inhibited.
④a ↑ Sympathetic impulses to heart cause ↑ HR, ↑ contractility, and ↑ CO.
③ Impulses from baroreceptors stimulate cardioacceleratory center (and inhibit cardioinhibitory center) and stimulate vasomotor center.

Figure 19.9 Baroreceptor reflexes that help maintain blood pressure homeostasis. CO = cardiac output; *R* = peripheral resistance; HR = heart rate; BP = blood pressure.

send a rapid stream of impulses to the vasomotor center. This input inhibits the vasomotor center, resulting in vasodilation of not only arterioles but also veins, and a decline in blood pressure **(Figure 19.9)**.

While dilation of the arterioles substantially reduces peripheral resistance, venodilation shifts blood to the venous reservoirs, causing a decline in both venous return and cardiac output. Afferent impulses from the baroreceptors also reach the cardiac centers, where the impulses stimulate parasympathetic activity and inhibit the cardioacceleratory center, reducing heart rate and contractile force.

A decline in MAP initiates reflex vasoconstriction and increases cardiac output, causing blood pressure to rise. In this way, peripheral resistance and cardiac output are regulated in tandem so that changes in blood pressure are minimized.

The function of rapidly responding baroreceptors is to protect the circulation against short-term (acute) changes in blood pressure, such as those occurring when you change your posture. For example, blood pressure falls (particularly in the head) when a person stands up after reclining. Baroreceptors taking part in the **carotid sinus reflex** protect the blood supply to the brain, whereas those activated in the **aortic reflex** help maintain adequate blood pressure in the systemic circuit as a whole.

Baroreceptors are relatively *ineffective* in protecting us against sustained pressure changes, as evidenced by the fact that many people develop chronic hypertension. In such cases, the

TABLE 19.2 Influence of Selected Hormones on Variables Affecting Blood Pressure

HORMONE	EFFECT ON BP	EFFECT ON VARIABLE	SITE OF ACTION
Epinephrine and norepi-nephrine (NE)	↑	↑ CO (HR and contractility)	Heart (β_1 receptors)
		↑ Peripheral resistance (vasoconstriction)	Arterioles (α receptors)
Angiotensin II	↑	↑ Peripheral resistance (vasoconstriction)	Arterioles
Atrial natriuretic peptide (ANP)	↓	↓ Peripheral resistance (vasodilation)	Arterioles
Antidiuretic hormone (ADH)	↑	↑ Peripheral resistance (vasoconstriction)	Arterioles
		↑ Blood volume (↓ water loss)	Kidney tubule cells
Aldosterone	↑	↑ Blood volume (↓ water and salt loss)	Kidney tubule cells
Cortisol	↑	↑ Blood volume (↓ water and salt loss)	Kidney tubule cells

baroreceptors are "reprogrammed" (adapt) to monitor pressure changes at a higher set point.

Chemoreceptor-Initiated Reflexes

When the carbon dioxide levels rise, or the pH falls, or oxygen content of the blood drops sharply, **chemoreceptors** in the aortic arch and large arteries of the neck transmit impulses to the cardioacceleratory center, which then increases cardiac output, and to the vasomotor center, which causes reflex vasoconstriction. The rise in blood pressure that follows speeds the return of blood to the heart and lungs.

The most prominent chemoreceptors are the *carotid* and *aortic bodies* located close by the baroreceptors in the carotid sinuses and aortic arch. Chemoreceptors are more important in regulating respiratory rate than blood pressure, so we consider their function in Chapter 22.

Influence of Higher Brain Centers

Reflexes that regulate blood pressure are integrated in the medulla oblongata of the brain stem. Although the cerebral cortex and hypothalamus are not involved in routine controls of blood pressure, these higher brain centers can modify arterial pressure via relays to the medullary centers. For example, the fight-or-flight response mediated by the hypothalamus has profound effects on blood pressure. (Even the simple act of speaking can make your blood pressure jump if the person you are talking to makes you anxious.) The hypothalamus also mediates the redistribution of blood flow and other cardiovascular responses that occur during exercise and changes in body temperature.

Short-Term Mechanisms: Hormonal Controls

As we have seen, changing levels of carbon dioxide and oxygen help regulate blood pressure via the chemoreceptor reflexes. Hormones, too, help regulate blood pressure, both in the short term via changes in peripheral resistance and in the long term via changes in blood volume **(Table 19.2)**. Paracrines (local chemicals), on the other hand, primarily serve to match the amount of blood flow to the metabolic need of a particular tissue. In rare instances, massive release of paracrines can affect blood pressure. We will discuss these paracrines later—here we will examine the short-term effects of hormones.

- **Adrenal medulla hormones.** During periods of stress, the adrenal gland releases **norepinephrine (NE)** and **epinephrine** to the blood, and both hormones enhance the sympathetic fight-or-flight response. As noted earlier, NE has a vasoconstrictive action. Epinephrine increases cardiac output and promotes generalized vasoconstriction (except in skeletal and cardiac muscle, where it generally causes vasodilation).

  It is interesting to note that *nicotine*, an important chemical in tobacco and one of the strongest toxins known, causes intense vasoconstriction not only by directly stimulating ganglionic sympathetic neurons but also by prompting release of large amounts of epinephrine and NE.
- **Angiotensin II.** When blood pressure or blood volume are low, the kidneys release the hormone renin. Renin acts as an enzyme, ultimately generating **angiotensin II** (an″je-o-ten′sin), which stimulates intense vasoconstriction, promoting a rapid rise in systemic blood pressure. It also stimulates release of aldosterone and ADH, which act in long-term regulation of blood pressure by enhancing blood volume.
- **Atrial natriuretic peptide.** The atria of the heart produce the hormone **atrial natriuretic peptide (ANP)**, which causes blood volume and blood pressure to decline. As noted in Chapter 16, ANP antagonizes aldosterone and prods the kidneys to excrete more sodium and water from the body, causing blood volume to drop. It also causes generalized vasodilation.
- **Antidiuretic hormone.** Produced by the hypothalamus, **antidiuretic hormone (ADH,** also called **vasopressin)** stimulates the kidneys to conserve water. It is not usually important in short-term blood pressure regulation, but when blood pressure falls to dangerously low levels (as during severe hemorrhage), much more ADH is released and helps restore arterial pressure by causing intense vasoconstriction.

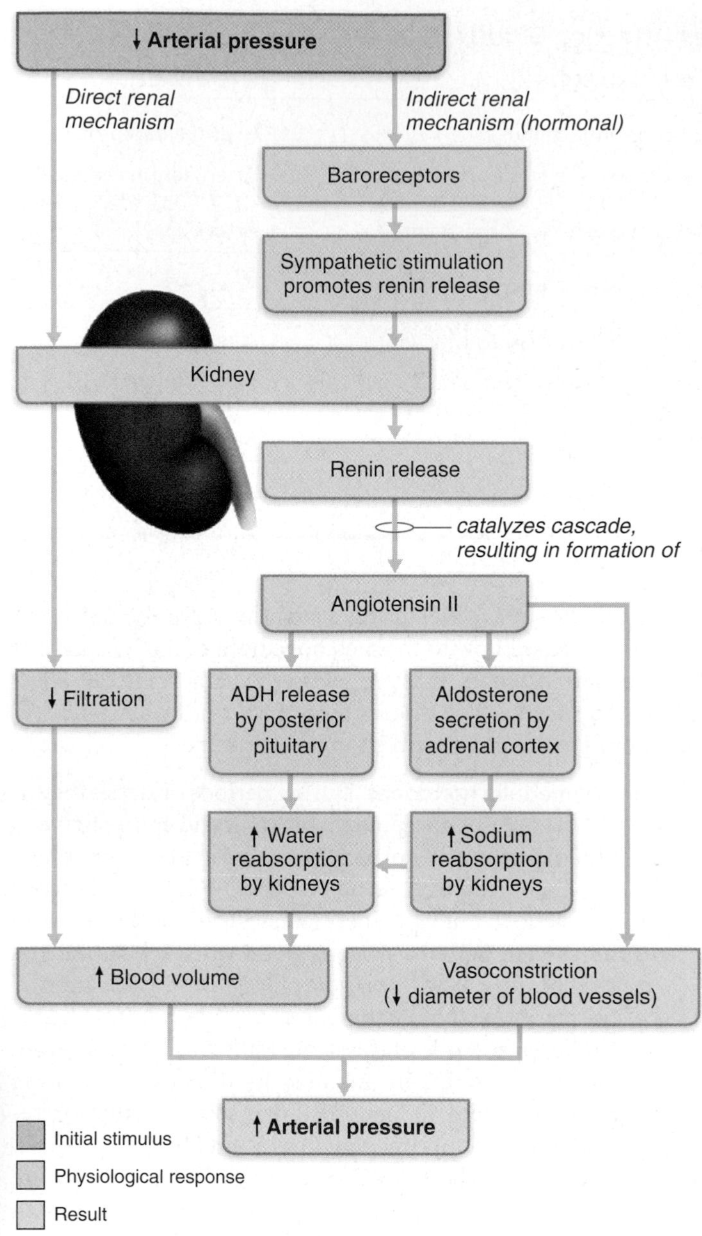

Figure 19.10 Direct and indirect (hormonal) mechanisms for renal control of blood pressure. The sympathetic nervous system acts to increase blood pressure in the long term by triggering renin release, but also acts by other mechanisms in the short term (not shown).

Long-Term Mechanisms: Renal Regulation

The long-term controls of blood pressure, mediated by renal mechanisms, counteract fluctuations in blood pressure not by altering peripheral resistance (as in short-term controls) but rather by altering blood volume.

Although baroreceptors respond to short-term changes in blood pressure, they quickly adapt to prolonged or chronic episodes of high or low pressure. This is where the kidneys step in to restore and maintain blood pressure homeostasis by regulating blood volume. Although blood volume varies with age, body size, and sex, renal mechanisms usually maintain it at close to 5 L.

As we noted earlier, blood volume is a major determinant of cardiac output (via its influence on venous return, EDV, and stroke volume). An increase in blood volume is followed by a rise in blood pressure, and anything that increases blood volume—such as excessive salt intake, which promotes water retention—raises MAP because of the greater fluid load in the vascular tree. By the same token, decreased blood volume translates to a fall in blood pressure. Blood loss and the garden-variety dehydration that occurs during vigorous exercise are common causes of reduced blood volume. A sudden drop in blood pressure often signals internal bleeding and blood volume too low to support normal circulation.

However, these assertions—increased blood volume increases BP and decreased blood volume decreases BP—do not tell the whole story because we are dealing with a dynamic system. Increases in blood volume that cause a rise in blood pressure also stimulate the kidneys to eliminate water, which reduces blood volume and consequently blood pressure. Likewise, falling blood volume triggers renal mechanisms that increase blood volume and blood pressure. As you can see, blood pressure can be stabilized or maintained within normal limits only when blood volume is stable.

The kidneys act both directly and indirectly to regulate arterial pressure and provide the major long-term mechanism of blood pressure control. The *direct renal mechanism* alters blood volume independently of hormones. When either blood volume or blood pressure rises, the rate at which fluid filters from the bloodstream into the kidney tubules is speeded up. In such situations, the kidneys cannot process the filtrate rapidly enough, and more of it leaves the body in urine. As a result, blood volume and blood pressure fall. When blood pressure or blood volume is low, water is conserved and returned to the bloodstream, and blood pressure rises **(Figures 19.10** and **19.11)**. As blood volume goes, so goes the arterial blood pressure.

The *indirect renal mechanism* is the **renin-angiotensin mechanism.** When arterial blood pressure declines, the kidneys release the enzymatic hormone *renin* into the blood. Renin triggers a series of reactions that produce angiotensin II, as mentioned previously. *Angiotensin II* increases blood pressure in three main ways (Figure 19.10). (1) It is a potent vasoconstrictor, increasing blood pressure by increasing peripheral resistance. (2) It stimulates the adrenal cortex to secrete **aldosterone**, a hormone that enhances renal reabsorption of sodium. As sodium moves into the bloodstream, water follows, and as a result, blood volume is conserved. (3) It prods the posterior pituitary to release ADH, which promotes more water reabsorption. As we will see in Chapter 26, angiotensin II also triggers the sensation of thirst, which enhances water consumption, ultimately restoring blood volume and so blood pressure.

Monitoring Circulatory Efficiency

The efficiency of a person's circulation can be assessed by taking pulse and blood pressure measurements. These measurements, along with measurements of respiratory rate and body temperature, are referred to collectively as the body's **vital signs**. Let's examine how vital signs are determined or measured.

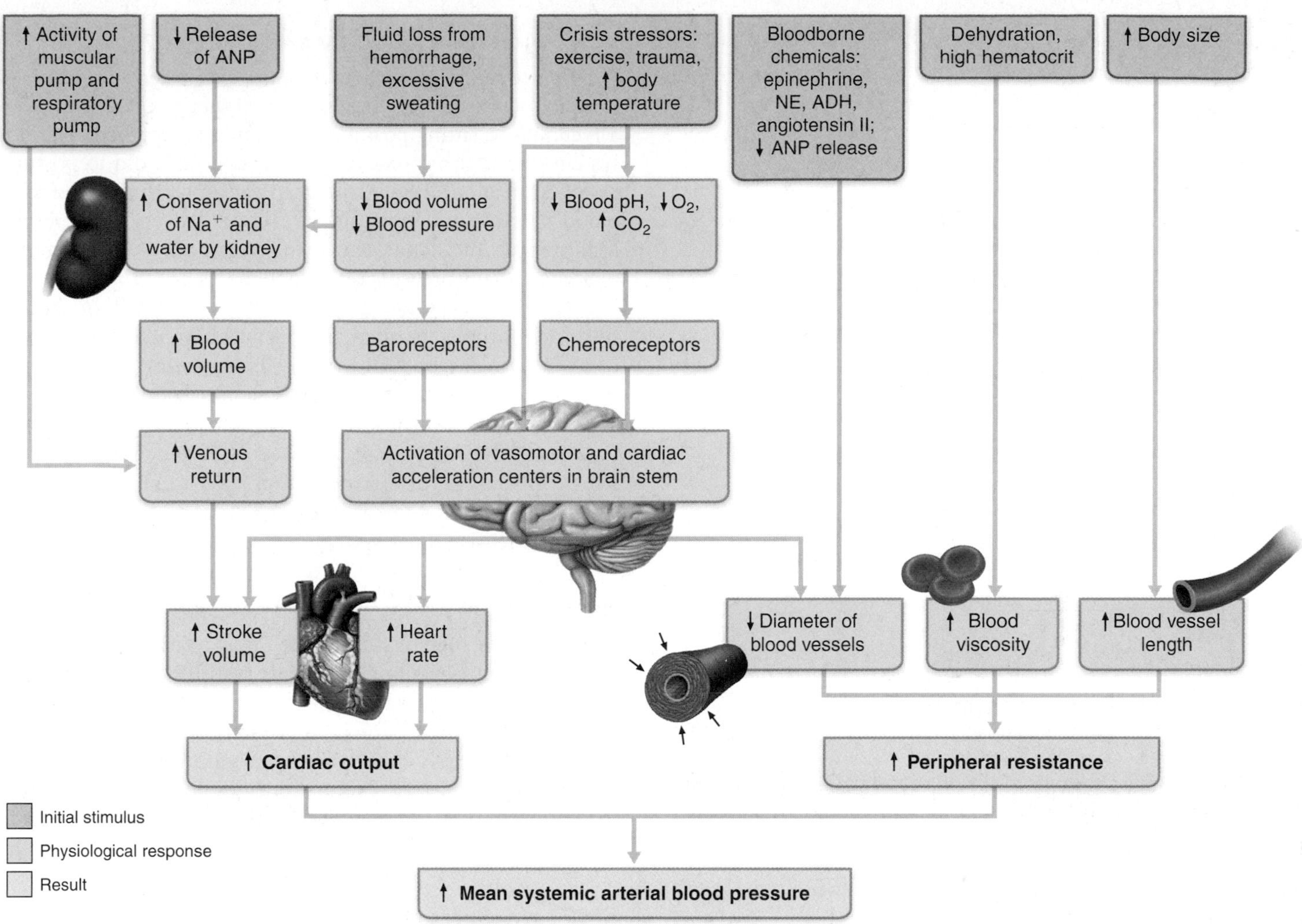

Figure 19.11 Factors causing an increase in MAP.

Taking a Pulse

The alternating expansion and recoil of arteries during each cardiac cycle allow us to feel a pressure wave—a **pulse**—that is transmitted through the arterial tree. You can feel a pulse in any artery that lies close to the body surface by compressing the artery against firm tissue, and this provides an easy way to count heart rate. Because it is so accessible, the point where the radial artery surfaces at the wrist, the *radial pulse*, is routinely used to take a pulse measurement, but there are several other clinically important arterial pulse points **(Figure 19.12)**.

These same points are also called **pressure points** because they are compressed to stop blood flow into distal tissues during hemorrhage. For example, if you seriously lacerate your hand, you can slow or stop the bleeding by compressing the radial or the brachial artery.

Monitoring pulse rate is an easy way to assess the effects of activity, postural changes, and emotions on heart rate. For example, the pulse of a healthy man may be around 66 beats per minute when he is lying down, rise to 70 when he sits up, and rise to 80 when he suddenly stands. During vigorous exercise or emotional upset, pulse rates between 140 and 180 are not unusual because of sympathetic nervous system effects on the heart.

Measuring Blood Pressure

Most often, systemic arterial blood pressure is measured indirectly in the brachial artery of the arm by the **auscultatory method** (aw-skul′tah-to″re). The *blood pressure cuff*, or *sphygmomanometer* (sfig″mo-mah-nom′ĕ-ter; *sphygmo* = pulse), is wrapped snugly around the arm just superior to the elbow and inflated until the cuff pressure exceeds systolic pressure. At this point, blood flow into the arm is stopped and a brachial pulse cannot be felt or heard. As the cuff pressure is gradually reduced, the examiner listens (auscultates) with a stethoscope for sounds in the brachial artery.

The pressure read when the first soft tapping sounds are heard (the first point at which a small amount of blood is spurting through the constricted artery) is systolic pressure. As the cuff pressure is reduced further, these sounds, called the *sounds of Korotkoff*, become louder and more distinct, but when the artery is no longer constricted and blood flows freely, the sounds

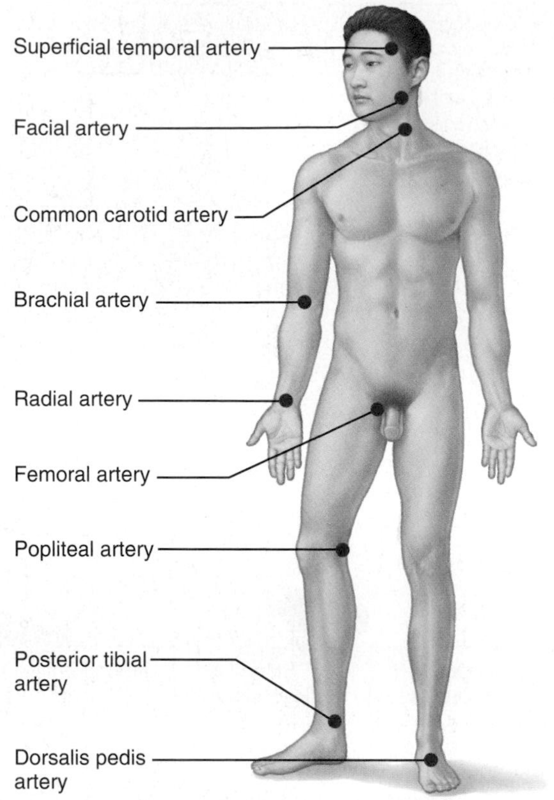

Figure 19.12 Body sites where the pulse is most easily palpated. (The specific arteries indicated are discussed on pp. 726–735.)

can no longer be heard. The pressure at which the sounds disappear is the diastolic pressure.

In normal adults at rest, systolic pressure varies between 110 and 140 mm Hg, and diastolic pressure between 70 and 80 mm Hg. Blood pressure also cycles over a 24-hour period, peaking in the morning in response to waxing and waning levels of various hormones (epinephrine, glucocorticoids, and melatonin). These hormones can either bind directly to receptors that control vasoconstriction or increase the number of receptors available.

Extrinsic factors also play a role: Blood pressure varies with age, sex, weight, race, mood, physical activity, posture, and socioeconomic status. What is "normal" for you may not be normal for your grandfather or your neighbor. Nearly all of these variations can be explained in terms of the factors affecting blood pressure discussed earlier.

Alterations in Blood Pressure

Hypotension

Hypotension, or low blood pressure, is a systolic pressure below 100 mm Hg. In many cases, hypotension simply reflects individual variations and is no cause for concern. In fact, low blood pressure is often associated with long life and an old age free of cardiovascular illness.

HOMEOSTATIC IMBALANCE

Elderly people are prone to *orthostatic hypotension*—temporary low blood pressure and dizziness when they rise suddenly from a reclining or sitting position. The aging sympathetic nervous system does not respond as quickly as it once did to postural changes, so blood pools briefly in the lower limbs, reducing blood pressure and consequently blood delivery to the brain. Making postural changes slowly to give the nervous system time to adjust usually prevents this problem.

Chronic hypotension may hint at poor nutrition. Poorly nourished people are often anemic and have inadequate levels of blood proteins, both of which reduce blood viscosity, and so their blood pressure is lower than normal. Chronic hypotension also warns of Addison's disease (inadequate adrenal cortex function), hypothyroidism, or severe tissue wasting. *Acute hypotension* is one of the most important signs of circulatory shock (p. 719) and a threat to patients undergoing surgery and those in intensive care units. ■

Hypertension

Hypertension, high blood pressure, may be transient or persistent. Transient elevations in systolic pressure occur as normal adaptations during fever, physical exertion, and emotional upset. Persistent hypertension is common in obese people for a variety of reasons that are not completely understood. It appears that hormones released from adipocytes both increase sympathetic tone and interfere with the ability of endothelial cells to induce vasodilation.

HOMEOSTATIC IMBALANCE

Chronic hypertension is a common and dangerous disease. An estimated 30% of people over the age of 50 years are hypertensive. Although this "silent killer" is usually asymptomatic for the first 10 to 20 years, it slowly but surely strains the heart and damages the arteries. Prolonged hypertension is the major cause of heart failure, vascular disease, renal failure, and stroke. Because the heart is forced to pump against greater resistance, it must work harder, and in time the myocardium enlarges. When finally strained beyond its capacity to respond, the heart weakens and its walls become flabby. Hypertension also ravages the blood vessels, accelerating the progress of atherosclerosis (see *A Closer Look* on pp. 702–703). As the vessels become increasingly blocked, blood flow to the tissues becomes inadequate and vascular complications appear in the brain, heart, kidneys, and retinas of the eyes.

Although hypertension and atherosclerosis are often linked, it is often difficult to blame hypertension on any distinct anatomical pathology. Hypertension is defined physiologically as a condition of sustained arterial pressure of 140/90 or higher, and the higher the pressure, the greater the risk for serious cardiovascular problems. Until recently, elevated diastolic pressures were considered to be more significant medically, but new studies suggest that systolic pressure is a better predictor of the likelihood of future problems in those older than 50.

About 90% of hypertensive people have **primary**, or **essential, hypertension**, in which no underlying cause has been identified. This is because primary hypertension is due to a rich interplay between hereditary predispositions and a variety of environmental factors:

1. Heredity. Hypertension runs in families. Children of hypertensive parents are twice as likely to develop hypertension as are children of normotensive parents, and more blacks than whites are hypertensive. Many of the factors listed here require a genetic predisposition, and the course of the disease varies in different population groups.
2. Diet. Dietary factors that contribute to hypertension include high intakes of salt (NaCl), saturated fat, and cholesterol and deficiencies in certain metal ions (K^+, Ca^{2+}, and Mg^{2+}).
3. Obesity.
4. Age. Clinical signs of the disease usually appear after age 40.
5. Diabetes mellitus.
6. Stress. Particularly at risk are "hot reactors," people whose blood pressure zooms upward during every stressful event.
7. Smoking. Nicotine enhances the sympathetic nervous system's vasoconstrictor effects.

Primary hypertension cannot be cured, but most cases can be controlled by restricting salt, fat, and cholesterol intake, losing weight, stopping smoking, managing stress, and taking antihypertensive drugs. Drugs commonly used are diuretics, beta-blockers, calcium channel blockers, angiotensin-converting enzyme (ACE) inhibitors, and angiotensin II receptor blockers. Inhibiting ACE or blocking receptors for angiotensin II suppresses the renin-angiotensin mechanism.

Secondary hypertension, which accounts for 10% of cases, is due to identifiable disorders, such as obstruction of the renal arteries, kidney disease, and endocrine disorders such as hyperthyroidism and Cushing's syndrome. Treatment for secondary hypertension is directed toward correcting the causative problem. ■

CHECK YOUR UNDERSTANDING

9. Describe the baroreceptor-initiated reflex changes that occur to maintain blood pressure when you rise from a lying-down to a standing position.
10. The kidneys play an important role in maintaining MAP by influencing which variable? Explain how renal artery obstruction could cause secondary hypertension.

For answers, see Appendix G.

Blood Flow Through Body Tissues: Tissue Perfusion

▶ Explain how blood flow is regulated in the body in general and in its specific organs.

Blood flow through body tissues, or **tissue perfusion**, is involved in (1) delivery of oxygen and nutrients to, and removal of

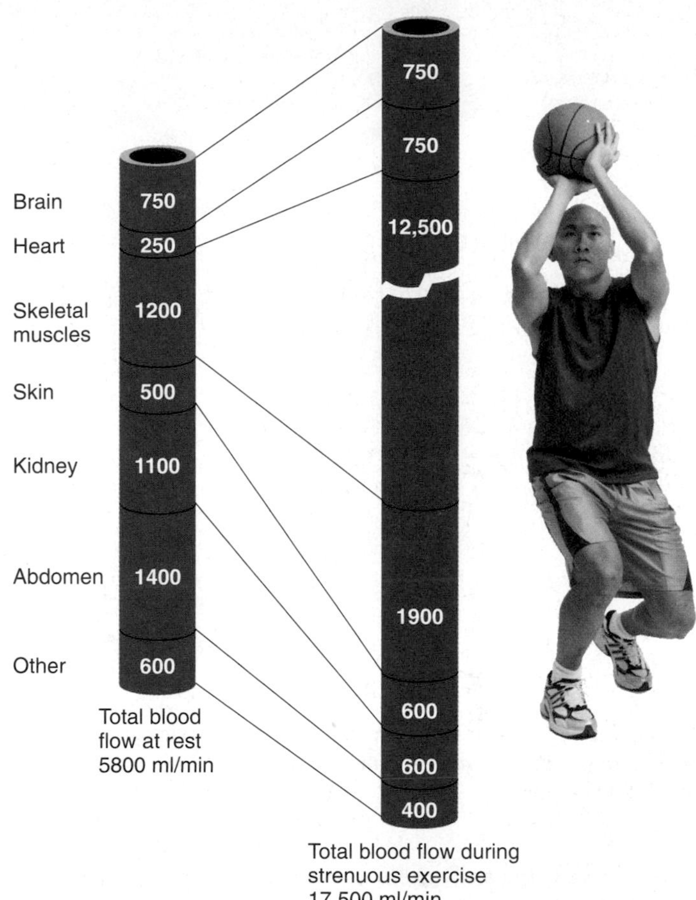

Figure 19.13 Distribution of blood flow at rest and during strenuous exercise.

wastes from, tissue cells, (2) gas exchange in the lungs, (3) absorption of nutrients from the digestive tract, and (4) urine formation in the kidneys. The rate of blood flow to each tissue and organ is almost exactly the right amount to provide for proper function—no more, no less.

When the body is at rest, the brain receives about 13% of total blood flow, the heart 4%, kidneys 20%, and abdominal organs 24%. Skeletal muscles, which make up almost half of body mass, normally receive about 20% of total blood flow. During exercise, however, nearly all of the increased cardiac output flushes into the skeletal muscles and blood flow to the kidneys and digestive organs is reduced **(Figure 19.13)**.

Velocity of Blood Flow

Have you ever watched a swift river emptying into a large lake? The water's speed decreases as it enters the lake until it becomes almost imperceptible. This is because the total cross-sectional area of the lake is much larger than that of the river. Velocity in this case is *inversely* related to cross-sectional area. The same thing happens with blood flow inside our blood vessels.

As shown in **Figure 19.14**, the speed or velocity of blood flow changes as blood travels through the systemic circulation. It is

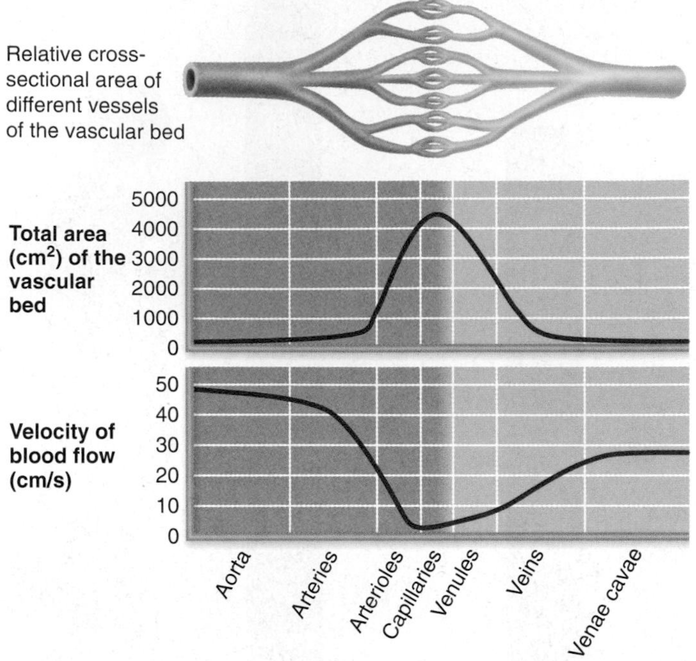

Figure 19.14 Relationship between blood flow velocity and total cross-sectional area in various blood vessels of the systemic circulation.

fastest in the aorta and other large arteries (the river), slowest in the capillaries (whose large total cross-sectional area make them analogous to the lake), and then picks up speed again in the veins (the river again).

Just as in our analogy of the river and lake, blood flows fastest where the total cross-sectional area is least. As the arterial system branches, the total cross-sectional area of the vascular bed increases, and the velocity of blood flow declines proportionately. Even though the individual branches have smaller lumens, their *combined* cross-sectional areas and thus the volume of blood they can hold are much greater than that of the aorta. For example, the cross-sectional area of the aorta is 2.5 cm^2 but that of all the capillaries is 4500 cm^2. This difference in cross-sectional area results in a fast blood flow in the aorta (40–50 cm/s) and a slow blood flow in the capillaries (about 0.03 cm/s). Slow capillary flow is beneficial because it allows adequate time for exchanges between the blood and tissue cells.

As capillaries combine to form first venules and then veins, total cross-sectional area declines and velocity increases. The cross-sectional area of the venae cavae is 8 cm^2, and the velocity of blood flow varies from 10 to 30 cm/s in those vessels, depending on the activity of the skeletal muscle pump.

Autoregulation: Local Regulation of Blood Flow

How does each organ or tissue manage to get the blood flow that it needs as our activities change throughout the day? The answer is **autoregulation**, the automatic adjustment of blood flow to each tissue in proportion to the tissue's requirements at any instant. This process is regulated by local conditions and is largely independent of systemic factors. The MAP is the same everywhere in the body and homeostatic mechanisms adjust cardiac output as needed to maintain that constant pressure. Changes in blood flow through individual organs are controlled *intrinsically* by modifying the diameter of local arterioles feeding the capillaries.

You can compare blood flow autoregulation to water use in your home. To get water in a sink, or a garden hose, you have to turn on a faucet. Whether you have several taps open or none, the pressure in the main water pipe in the street remains relatively constant, as it does in the even larger water lines closer to the pumping station. Similarly, what is happening in the arterioles feeding the capillary beds of a given organ has little effect on pressure in the muscular artery feeding the organ, or in the large elastic arteries. The pumping station is, of course, the heart. The beauty of this system is that as long as the water company (circulatory feedback mechanisms) maintains a relatively constant water pressure (MAP), local demand regulates the amount of fluid (blood) delivered to various areas.

In summary, organs regulate their own blood flows by varying the resistance of their arterioles. As we describe next, these intrinsic control mechanisms may be classed as *metabolic* (chemical) or *myogenic* (physical).

Metabolic Controls

In most tissues, declining levels of oxygen and accumulation of other substances released by metabolically active tissues serve as autoregulation stimuli. These substances include H^+ (from CO_2 and lactic acid), K^+, adenosine, and prostaglandins. The relative importance of these factors is not clear. Many of them act directly to relax vascular smooth muscle, but some may act by causing the release of nitric oxide from vascular endothelial cells.

Nitric oxide (NO) is a powerful vasodilator which acts via a cyclic GMP second-messenger system. NO is quickly destroyed and its potent vasodilator effects are very brief. However, some researchers believe that NO can be transported over longer distances by being bound to chemical buffers in the blood and that it is released where vasodilation is needed. Until recently the sympathetic nervous system was thought to rule blood vessel diameter, but we now know that its primary vascular role is to produce vasoconstriction, and that NO plays the major role in causing vasodilation. The endothelium also releases potent vasoconstrictors, including the family of peptides called **endothelins**, which are among the most potent vasoconstrictors known.

The net result of metabolically controlled autoregulation is immediate vasodilation of the arterioles serving the capillary beds of the "needy" tissues and dilation of their precapillary sphincters. This creates a temporary increase in blood flow to the area, allowing blood to surge through the true capillaries and become available to the tissue cells. Inflammatory chemicals (such as histamine, kinins, and prostaglandins) released in injury, infection, or allergic reactions also cause vasodilation.

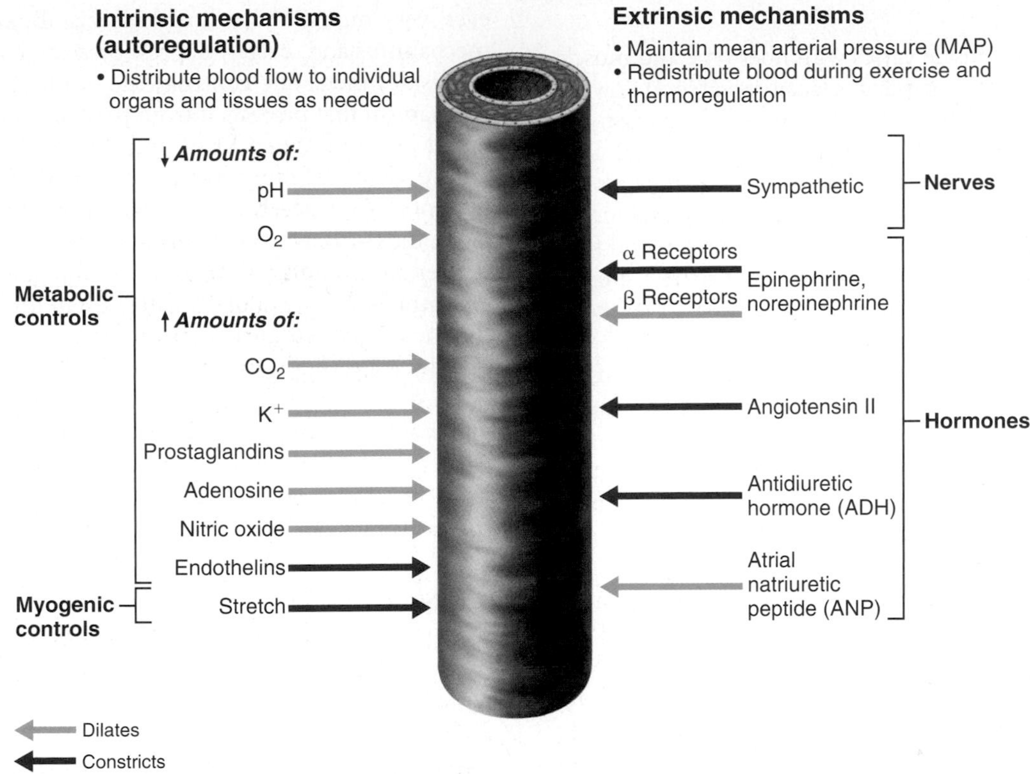

Figure 19.15 Summary of intrinsic and extrinsic control of arteriolar smooth muscle in the systemic circulation. α-Adrenergic receptors (smooth muscle contraction, arteriolar constriction) are located in arterioles supplying most tissues, whereas β-adrenergic receptors (smooth muscle relaxation, arteriolar dilation) are located primarily in arterioles supplying skeletal and heart muscle.

Inflammatory vasodilation helps the defense mechanisms clear microorganisms and toxins, and promotes healing.

Myogenic Controls

Fluctuations in systemic blood pressure would cause problems for individual organs were it not for the **myogenic responses** (*myo* = muscle; *gen* = origin) of vascular smooth muscle. Inadequate blood perfusion through an organ is quickly followed by a decline in the organ's metabolic rate and, if prolonged, organ death. Likewise, excessively high arterial pressure and tissue perfusion can be dangerous because the combination may rupture more fragile blood vessels.

Fortunately, vascular smooth muscle prevents these problems by responding directly to passive stretch (caused by increased intravascular pressure) with increased tone, which resists the stretch and causes vasoconstriction. Reduced stretch promotes vasodilation and increases blood flow into the tissue. These myogenic responses keep tissue perfusion fairly constant despite most variations in systemic pressure.

Generally, both metabolic and myogenic factors determine the final autoregulatory response of a tissue. For example, **reactive hyperemia** (hi″per-e′me-ah) refers to the dramatically increased blood flow into a tissue that occurs after the blood supply to the area has been temporarily blocked. It results both from the myogenic response and from an accumulation of metabolic wastes in the area during occlusion. The various intrinsic (local) and extrinsic controls of arteriolar diameter are summarized in **Figure 19.15**, with colored arrows indicating their effects—green for dilation and red for constriction.

Long-Term Autoregulation

If the nutrient requirements of a tissue are greater than the short-term autoregulatory mechanism can easily supply, a long-term autoregulatory mechanism may develop over a period of weeks or months to enrich the local blood flow still more. The number of blood vessels in the region increases, and existing vessels enlarge. This phenomenon, called *angiogenesis*, is particularly common in the heart when a coronary vessel is partially occluded. It occurs throughout the body in people who live in high-altitude areas, where the air contains less oxygen.

Blood Flow in Special Areas

Each organ has special requirements and functions that are revealed in its pattern of autoregulation. Autoregulation in the brain, heart, and kidneys is extraordinarily efficient. In those organs, adequate perfusion is maintained even when MAP is fluctuating.

Skeletal Muscles

Blood flow in skeletal muscle varies with fiber type and muscle activity. Generally speaking, capillary density and blood flow are greater in red (slow oxidative) fibers than in white (fast glycolytic) fibers. Resting skeletal muscles receive about 1 L of blood per minute, and only about 25% of their capillaries are open. During such periods, myogenic and general neural mechanisms predominate. When muscles become active, blood flow increases (*hyperemia*) in direct proportion to their greater *metabolic* activity, a phenomenon called **active** or **exercise hyperemia**.

This form of autoregulation occurs almost entirely in response to the decreased oxygen concentration and accumulation of metabolic factors that result from the "revved-up" metabolism of working muscles. However, systemic adjustments mediated by the vasomotor center must also occur to ensure that blood delivery to the muscles is both faster and more abundant. During exercise, sympathetic nervous system activity increases. Norepinephrine released from sympathetic nerve endings causes vasoconstriction of the vessels of blood reservoirs such as those of the digestive viscera and skin, diverting blood away from these regions temporarily and ensuring that more blood reaches the muscles. In skeletal muscles, the sympathetic nervous system and local metabolic controls have opposing effects on arteriolar diameter. During exercise, local controls override sympathetic vasoconstriction. Consequently, blood flow to skeletal muscles can increase tenfold or more during physical activity, as you saw in Figure 19.13, and virtually all capillaries in the active muscles open to accommodate the increased flow.

Epinephrine acting at beta (β) adrenergic receptors and acetylcholine acting at cholinergic receptors were once thought to contribute to arteriolar dilation during exercise. However, these appear to have little physiological importance in the control of human skeletal muscle blood flow. Without question, strenuous exercise is one of the most demanding conditions the cardiovascular system faces. Ultimately, the major factor determining how long muscles can continue to contract vigorously is the ability of the cardiovascular system to deliver adequate oxygen and nutrients and remove waste products.

19

The Brain

Blood flow to the brain averages 750 ml/min and is maintained at a relatively constant level. Constant cerebral blood flow is necessary because neurons are totally intolerant of ischemia.

Although the brain is the most metabolically active organ in the body, it is the least able to store essential nutrients. Cerebral blood flow is regulated by one of the body's most precise autoregulatory systems and is tailored to local neuronal need. For example, when you make a fist with your right hand, the neurons in the left cerebral motor cortex controlling that movement receive a more abundant blood supply than the adjoining neurons. Brain tissue is exceptionally sensitive to declining pH, and increased blood carbon dioxide levels (resulting in acidic conditions in brain tissue) cause marked vasodilation. Oxygen deficit is a much less potent stimulus for autoregulation. However, very high carbon dioxide levels abolish autoregulatory mechanisms and severely depress brain activity.

Besides metabolic controls, the brain also has a myogenic mechanism that protects it from possibly damaging changes in blood pressure. When MAP declines, cerebral vessels dilate to ensure adequate brain perfusion. When MAP rises, cerebral vessels constrict, protecting the small, more fragile vessels farther along the pathway from rupture due to excessive pressure. Under certain circumstances, such as brain ischemia caused by rising intracranial pressure (as with a brain tumor), the brain (via the medullary cardiovascular centers) regulates its own blood flow by triggering a rise in systemic blood pressure. However, when systemic pressure changes are extreme, the brain becomes vulnerable. Fainting, or *syncope* (sin′cuh-pe; "cutting short"), occurs when MAP falls below 60 mm Hg. Cerebral edema is the usual result of pressures over 160 mm Hg, which dramatically increase brain capillary permeability.

The Skin

Blood flow through the skin (1) supplies nutrients to the cells, (2) aids in body temperature regulation, and (3) provides a blood reservoir. The first function is served by autoregulation in response to the need for oxygen, but the second and third require neural intervention. The primary function of the cutaneous circulation is to help maintain body temperature, so we will concentrate on the skin's temperature regulation function here.

Below the skin surface are extensive venous plexuses. The blood flow through these plexuses can change from 50 ml/min to as much as 2500 ml/min, depending on body temperature. This capability reflects neural adjustments of blood flow through arterioles and through unique coiled arteriovenous anastomoses. These tiny A-V shunts are located mainly in the fingertips, palms of the hands, toes, soles of the feet, ears, nose, and lips. They are richly supplied with sympathetic nerve endings (a characteristic that sets them apart from the shunts of most other capillary beds), and are controlled by reflexes initiated by temperature receptors or signals from higher CNS centers. The arterioles, in addition, are responsive to metabolic autoregulatory stimuli.

When the skin surface is exposed to heat, or body temperature rises for other reasons (such as vigorous exercise), the hypothalamic "thermostat" signals for reduced vasomotor stimulation of the skin vessels. As a result, warm blood flushes into the capillary beds and heat radiates from the skin surface. Vasodilation of the arterioles is enhanced even more when we sweat, because an enzyme in perspiration acts on a protein present in tissue fluid to produce *bradykinin*, which stimulates the vessel's endothelial cells to release the potent vasodilator NO.

When the ambient temperature is cold and body temperature drops, superficial skin vessels are strongly constricted. Hence, blood almost entirely bypasses the capillaries associated with the arteriovenous anastomoses, diverting the warm blood to the deeper, more vital organs. Paradoxically, the skin may stay quite rosy because some blood gets "trapped" in the superficial capillary loops as the shunts swing into operation. The trapped blood remains red because the chilled skin cells take up less O_2.

The Lungs

Blood flow through the pulmonary circuit to and from the lungs is unusual in many ways. The pathway is relatively short, and pulmonary arteries and arterioles are structurally like veins and venules. That is, they have thin walls and large lumens. Because resistance to blood flow is low in the pulmonary arterial system, less pressure is needed to propel blood through those vessels. Consequently, arterial pressure in the pulmonary circulation is much lower than in the systemic circulation (24/8 versus 120/80 mm Hg).

Another unusual feature of the pulmonary circulation is that the autoregulatory mechanism is the *opposite* of what is seen in most tissues: Low pulmonary oxygen levels cause local vasoconstriction, and high levels promote vasodilation. While this may seem odd, it is perfectly consistent with the gas exchange role of this circulation. When the air sacs of the lungs are flooded with oxygen-rich air, the pulmonary capillaries become flushed with blood and ready to receive the oxygen load. If the air sacs are collapsed or blocked with mucus, the oxygen content in those areas is low, and blood largely bypasses those nonfunctional areas.

The Heart

Movement of blood through the smaller vessels of the coronary circulation is influenced by aortic pressure and by the pumping activity of the ventricles. When the ventricles contract and compress the coronary vessels, blood flow through the myocardium stops. As the heart relaxes, the high aortic pressure forces blood through the coronary circulation. Under normal circumstances, the myoglobin in cardiac cells stores sufficient oxygen to satisfy the cells' oxygen needs during systole. However, an abnormally rapid heartbeat seriously reduces the ability of the myocardium to receive adequate oxygen and nutrients during diastole.

Under resting conditions, blood flow through the heart is about 250 ml/min and is probably controlled by a myogenic mechanism. Consequently, blood flow remains fairly constant despite wide variations (50 to 140 mm Hg) in coronary perfusion pressure. During strenuous exercise, the coronary vessels dilate in response to local accumulation of vasodilators (particularly adenosine), and blood flow may increase three to four times (see Figure 19.13). Additionally, any event that decreases the oxygen content of the blood releases vasodilators that adjust the O_2 supply to the O_2 demand.

This enhanced blood flow during increased heart activity is important because under resting conditions, cardiac cells use as much as 65% of the oxygen carried to them in blood. (Most other tissue cells use about 25% of the delivered oxygen.) Consequently, increasing the blood flow is the only way to make sufficient additional oxygen available to a more vigorously working heart.

CHECK YOUR UNDERSTANDING

11. Suppose you are in a bicycle race. What happens to the smooth muscle in the arterioles supplying your leg muscles? What is the key mechanism in this case?
12. If many arterioles in your body dilated at once, you would expect MAP to plummet. What prevents MAP from decreasing during your bicycle race?

For answers, see Appendix G.

Blood Flow Through Capillaries and Capillary Dynamics

▶ Outline factors involved in capillary dynamics, and explain the significance of each.

Blood flow through capillary networks is slow and intermittent. This phenomenon, called **vasomotion**, reflects the on/off opening and closing of precapillary sphincters in response to local autoregulatory controls.

Capillary Exchange of Respiratory Gases and Nutrients

Oxygen, carbon dioxide, most nutrients, and metabolic wastes pass between the blood and interstitial fluid by diffusion. Recall that in **diffusion**, movement always occurs along a concentration gradient—each substance moving from an area of its higher concentration to an area of its lower concentration. Hence, oxygen and nutrients pass from the blood, where their concentration is fairly high, through the interstitial fluid to the tissue cells. Carbon dioxide and metabolic wastes leave the cells, where their content is higher, and diffuse into the capillary blood.

There are four different routes across capillaries for different types of molecules, as **Figure 19.16** shows. ① Lipid-soluble molecules, such as respiratory gases, diffuse through the lipid bilayer of the endothelial cell plasma membranes. Small water-soluble solutes, such as amino acids and sugars, pass through ② fluid-filled intercellular capillary clefts or ③ fenestrations. ④ Some larger molecules, such as proteins, are actively transported in pinocytotic vesicles or caveolae. As we mentioned earlier, capillaries differ in their "leakiness," or permeability. Liver capillaries, for instance, have large fenestrations that allow even proteins to pass freely, whereas brain capillaries are impermeable to most substances.

Fluid Movements: Bulk Flow

While nutrient and gas exchanges are occurring across the capillary walls by diffusion, bulk fluid flows are also going on. Fluid is forced out of the capillaries through the clefts at the arterial end of the bed, but most of it returns to the bloodstream at the venous end. Though relatively unimportant to capillary exchange, bulk flow is extremely important in determining the relative fluid volumes in the bloodstream and the interstitial space. (Approximately 20 L of fluid are filtered out of the capillaries each day before being returned to the blood—almost seven times the total plasma volume!) As we describe next, the

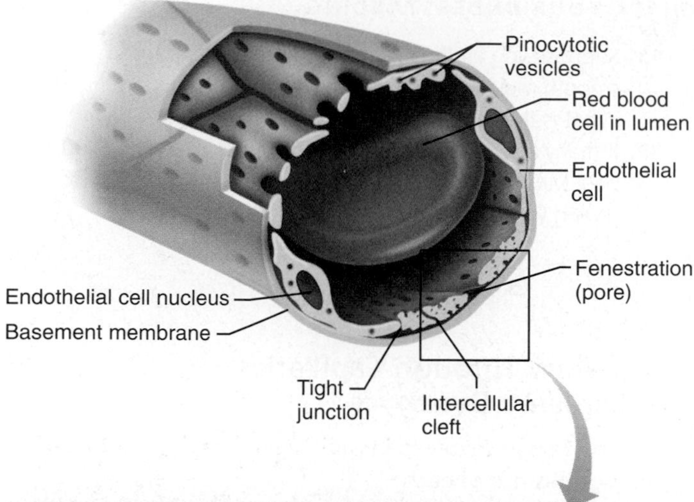

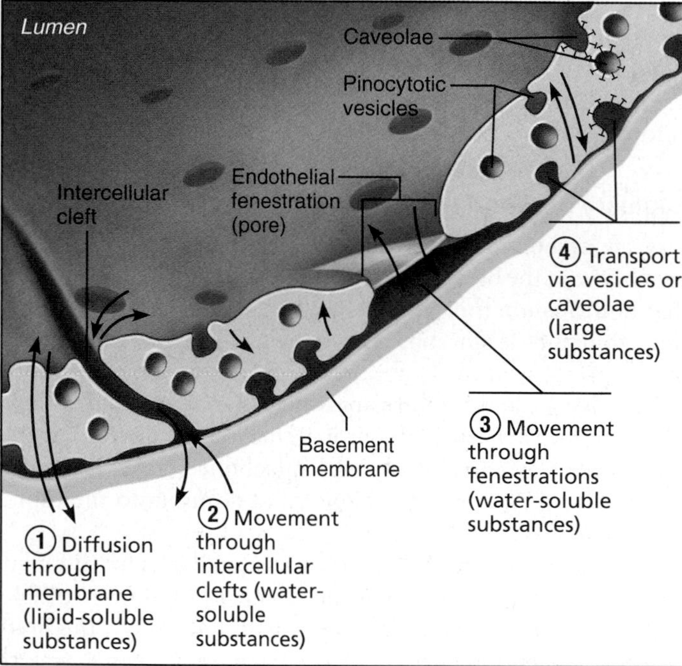

Figure 19.16 Capillary transport mechanisms. The four possible pathways or routes of transport across the endothelial cell wall of a fenestrated capillary.

direction and amount of fluid that flows across the capillary walls reflect the balance between two dynamic and opposing forces—hydrostatic and colloid osmotic pressures.

Hydrostatic Pressures **Hydrostatic pressure** is the force exerted by a fluid pressing against a wall. In capillaries, hydrostatic pressure is the same as *capillary blood pressure*—the pressure exerted by blood on the capillary walls. **Capillary hydrostatic pressure (HP_c)** tends to force fluids through the capillary walls (a process called *filtration*), leaving behind cells and most proteins. Blood pressure drops as blood flows along the length of a capillary bed, so HP_c is higher at the arterial end of the bed (35 mm Hg) than at the venous end (17 mm Hg).

In theory, blood pressure—which forces fluid out of the capillaries—is opposed by the **interstitial fluid hydrostatic pressure (HP_{if})** acting outside the capillaries and pushing fluid in. For this reason, the *net* hydrostatic pressure acting on the capillaries at any point is the difference between HP_c and HP_{if}. However, there is usually very little fluid in the interstitial space, because the lymphatic vessels constantly withdraw it. HP_{if} may vary from slightly negative to slightly positive, but traditionally it is assumed to be zero. For simplicity, that is the value we use here.

The *net effective hydrostatic pressures* at the arterial and venous ends of the capillary bed are essentially equal to HP_c (in other words, to blood pressure) at those locations. This information is summarized in **Figure 19.17**.

Colloid Osmotic Pressures **Colloid osmotic pressure**, the force opposing hydrostatic pressure, is created by the presence in a fluid of large nondiffusible molecules, such as plasma proteins, that are unable to cross the capillary wall. Such molecules draw water toward themselves. In other words, they encourage osmosis because the water concentration in their vicinity is lower than it is on the opposite side of the capillary wall. A quick and dirty way to remember this is "hydrostatic pressure pushes and osmotic pressure sucks."

The abundant plasma proteins in capillary blood (primarily albumin molecules) develop a **capillary colloid osmotic pressure (OP_c)**, also called *oncotic pressure*, of approximately 26 mm Hg. Interstitial fluid contains few proteins, so its colloid osmotic pressure (**OP_{if}**) is substantially lower—from 0.1 to 5 mm Hg. We use a value of 1 mm Hg for the OP_{if} in Figure 19.17.

Unlike HP, OP does not vary significantly from one end of the capillary bed to the other. In our example, the *net osmotic pressure* that pulls fluid back into the capillary blood is $OP_c - OP_{if} = 26\text{ mm Hg} - 1\text{ mm Hg} = 25\text{ mm Hg}$.

Hydrostatic-Osmotic Pressure Interactions To determine whether there is a net gain or net loss of fluid from the blood, we have to calculate the **net filtration pressure (NFP)**, which considers all the forces acting at the capillary bed. At any point along a capillary, fluids will leave the capillary if net HP is greater than net OP, and fluids will enter the capillary if net OP exceeds net HP. As shown in Figure 19.17, hydrostatic forces dominate at the arterial end (all values are in mm Hg):

$$\begin{aligned} NFP &= (HP_c - HP_{if}) - (OP_c - OP_{if}) \\ &= (35 - 0) - (26 - 1) \\ &= (35 - 25) = 10 \text{ mm Hg} \end{aligned}$$

As you can see in this example, a pressure of 10 mm Hg (net excess of HP) is forcing fluid out of the capillary. At the other, venous end, osmotic forces dominate:

$$\begin{aligned} NFP &= (17 - 0) - (26 - 1) \\ &= 17 - 25 = -8 \text{ mm Hg} \end{aligned}$$

The negative pressure value we see here indicates that the NFP (due to net excess of OP) is driving fluid *into* the capillary bed (a process called *reabsorption*). As a result, net fluid flow is *out* of the circulation at the arterial ends of capillary beds and *into* the circulation at the venous ends.

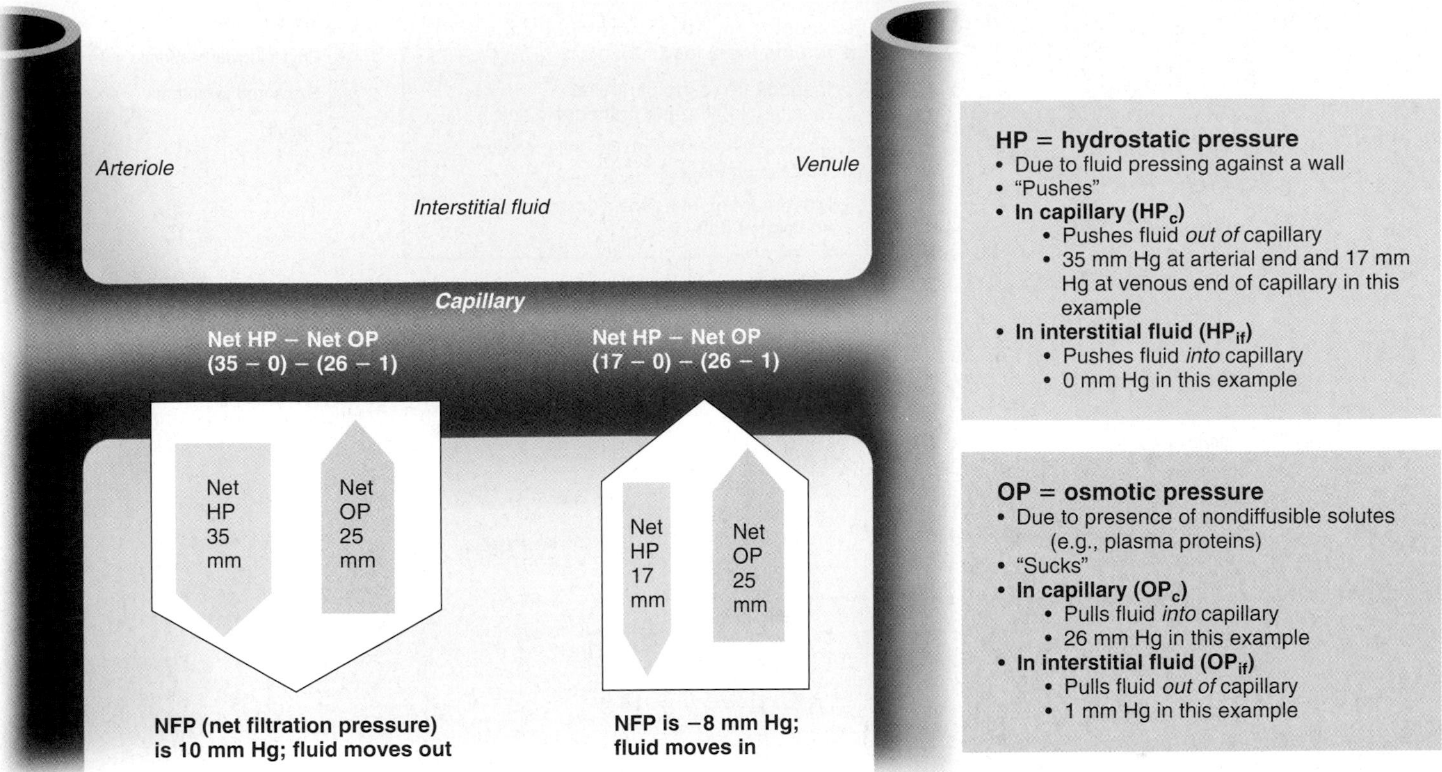

Figure 19.17 Fluid flows at capillaries.

However, more fluid enters the tissue spaces than is returned to the blood, resulting in a net loss of fluid from the circulation of about 1.5 ml/min. This fluid and any leaked proteins are picked up by the lymphatic vessels and returned to the vascular system, which accounts for the relatively low levels of both fluid and proteins in the interstitial space. Were this not so, this "insignificant" fluid loss would empty your blood vessels of plasma in about 24 hours!

CHECK YOUR UNDERSTANDING

13. Suppose that the OP of the interstitial fluid (OP_{if}) rose dramatically—say because of a severe bacterial infection in the surrounding tissue. (a) Predict how fluid flow would change in this situation. (b) Now calculate the NFP at the venous end of the capillary in Figure 19.17 if OP_{if} increased to 10 mm Hg. (c) In which direction does fluid flow at the venous end of the capillary now—in or out?

For answers, see Appendix G.

Circulatory Shock

▶ Define circulatory shock. List several possible causes.

Circulatory shock is any condition in which blood vessels are inadequately filled and blood cannot circulate normally. This results in inadequate blood flow to meet tissue needs. If the condition persists, cells die and organ damage follows.

The most common form of shock is **hypovolemic shock** (hi″po-vo-le′mik; *hypo* = low, deficient; *volemia* = blood volume), which results from large-scale loss of blood, as might follow acute hemorrhage, severe vomiting or diarrhea, or extensive burns. If blood volume drops rapidly, heart rate increases in an attempt to correct the problem. Thus, a weak, "thready" pulse is often the first sign of hypovolemic shock. Intense vasoconstriction also occurs, which shifts blood from the various blood reservoirs into the major circulatory channels and enhances venous return. Blood pressure is stable at first, but eventually drops if blood loss continues. A sharp drop in blood pressure is a serious, and late, sign of hypovolemic shock. The key to managing hypovolemic shock is to replace fluid volume as quickly as possible.

Although you have not yet explored all of the body systems that respond to hypovolemic shock, acute bleeding is such a threat to life that it seems important to have a comprehensive flowchart of its recognizable signs and symptoms and an accounting of the body's attempt to restore homeostasis. **Figure 19.18** provides such a resource. Study it in part now, and in more detail later once you have studied the remaining body systems.

In **vascular shock**, blood volume is normal, but circulation is poor as a result of an abnormal expansion of the vascular bed caused by extreme vasodilation. The huge drop in peripheral resistance that follows is revealed by rapidly falling blood pressure. A common cause of vascular shock is loss of vasomotor tone due to anaphylaxis (anaphylactic shock), a systemic allergic reaction in which bodywide vasodilation is triggered by the

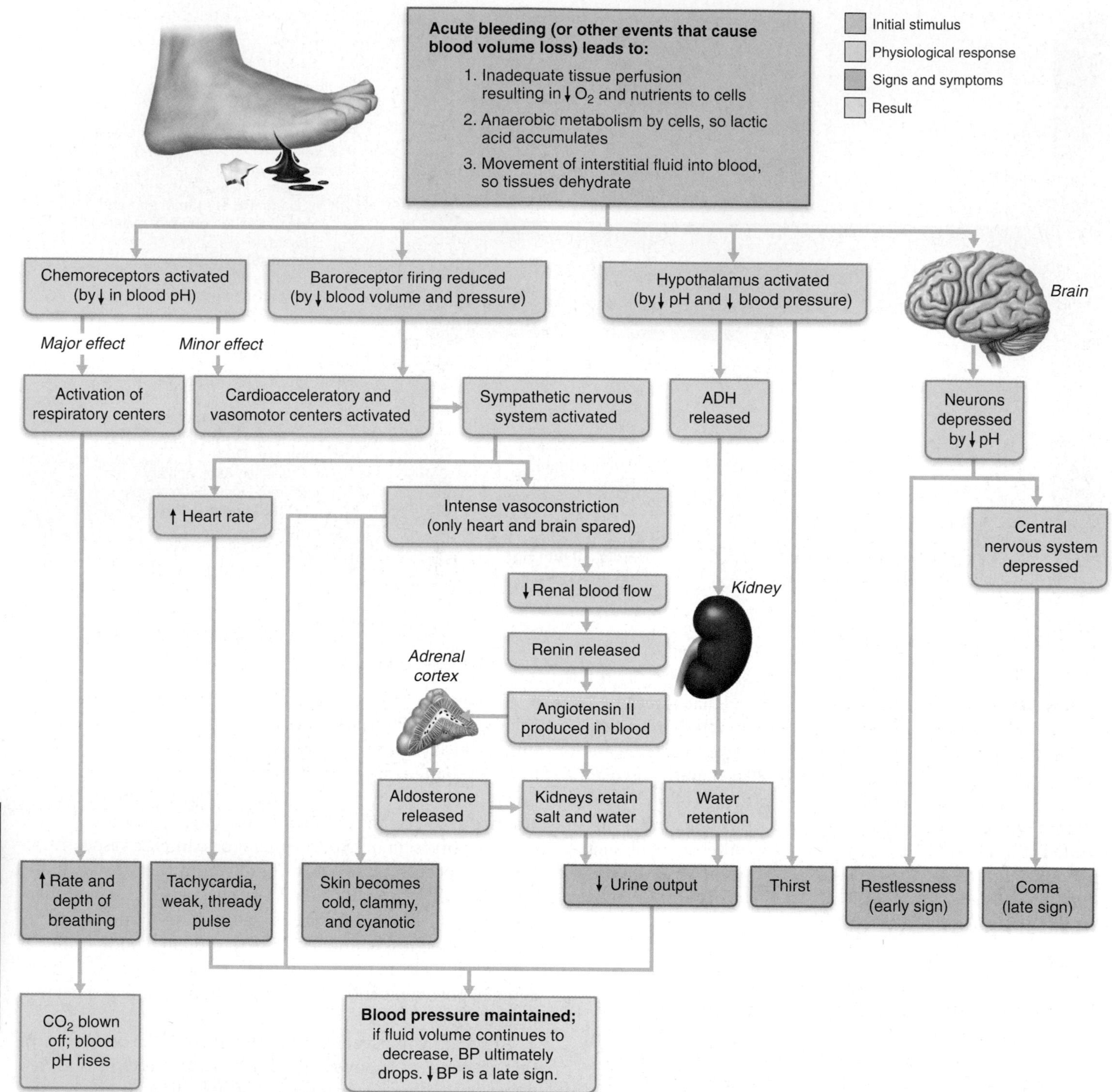

Figure 19.18 Events and signs of compensated (nonprogressive) hypovolemic shock.

massive release of histamine. Two other common causes are failure of autonomic nervous system regulation (also referred to as *neurogenic shock*), and septicemia (*septic shock*), a severe systemic bacterial infection (bacterial toxins are notorious vasodilators).

Transient vascular shock may occur when you sunbathe for a prolonged time. The heat of the sun on your skin causes cutaneous blood vessels to dilate. Then, if you stand up abruptly, blood pools briefly (because of gravity) in the dilated vessels of your lower limbs rather than returning promptly to the heart. Consequently, your blood pressure falls. The dizziness you feel at this point is a signal that your brain is not receiving enough oxygen.

Cardiogenic shock, or pump failure, occurs when the heart is so inefficient that it cannot sustain adequate circulation. Its usual cause is myocardial damage, as might follow numerous myocardial infarcts.

CHECK YOUR UNDERSTANDING

14. Your neighbor, Bob, calls you because he thinks he is having an allergic reaction to a medication. You find Bob on the verge of losing consciousness and having trouble breathing. When paramedics arrive, they note his blood pressure is 63/38 and he has a rapid, thready pulse. Explain Bob's low blood pressure and his rapid heart rate.

For answers, see Appendix G.

PART 3

CIRCULATORY PATHWAYS: BLOOD VESSELS OF THE BODY

The Two Main Circulations of the Body

- ▶ Trace the pathway of blood through the pulmonary circuit, and state the importance of this special circulation.
- ▶ Describe the general functions of the systemic circuit.

When referring to the body's complex network of blood vessels, the term **vascular system** is often used. However, the heart is actually a double pump that serves two distinct circulations, each with its own set of arteries, capillaries, and veins. The *pulmonary circulation* is the short loop that runs from the heart to the lungs and back to the heart. The *systemic circulation* routes blood through a long loop to all parts of the body before returning it to the heart. Both circuits are shown schematically in **Table 19.3** on pp. 722–723.

Systemic Arteries and Veins: Differences in Pathways and Courses

As we saw in Chapter 18, the heart pumps all of its blood into a single systemic artery—the aorta. In contrast, blood returning to the heart is delivered largely by two terminal systemic veins, the superior and inferior venae cavae. The single exception to this is the blood draining from the myocardium of the heart, which is collected by the cardiac veins and reenters the right atrium via the coronary sinus.

In addition to these differences between arteries and veins connecting to the heart, there are three important differences between systemic arteries and veins:

1. **Arteries run deep while veins are both deep and superficial.** Deep veins parallel the course of the systemic arteries and both are protected by body tissues along most of their course. With a few exceptions, the naming of these veins is identical to that of their companion arteries. Superficial veins run just beneath the skin and are readily seen, especially in the limbs, face, and neck. Because there are no superficial arteries, the names of the superficial veins do not correspond to the names of any of the arteries.
2. **Venous pathways are more interconnected.** Unlike the fairly distinct arterial pathways, venous pathways tend to have numerous interconnections, and many veins are represented by not one but two similarly named vessels. As a result, venous pathways are more difficult to follow.
3. **The brain and digestive systems have unique venous drainage systems.** Most body regions have a similar pattern for their arterial supply and venous drainage. However, the venous drainage pattern in at least two important body areas is unique. First, venous blood draining from the brain enters large *dural venous sinuses* rather than typical veins. Second, blood draining from the digestive organs enters a special subcirculation, the *hepatic portal system*, and perfuses through the liver before it reenters the general systemic circulation (see Table 19.12).

Principal Vessels of the Systemic Circulation

- ▶ Name and give the location of the major arteries and veins in the systemic circulation.
- ▶ Describe the structure and special function of the hepatic portal system.

Except for special vessels and shunts of the fetal circulation (described in Chapter 28), the principal arteries and veins of the systemic circulation are described in **Tables 19.4** through **19.13**.

Notice that by convention, oxygen-rich blood is shown red, while blood that is relatively oxygen-poor is depicted blue, regardless of vessel types. A unique convention used in the schematic flowcharts (pipe diagrams) that accompany each table is that the vessels that would be closer to the viewer are shown in brighter, more intense colors than those deeper or farther from the viewer—for example, for veins, the darker blue vessels would be closer to the viewer in the body region shown.

TABLE 19.3 Pulmonary and Systemic Circulations

Pulmonary Circulation

The pulmonary circulation **(Figure 19.19a)** functions only to bring blood into close contact with the alveoli (air sacs) of the lungs so that gases can be exchanged. It does not directly serve the metabolic needs of body tissues.

Oxygen-poor, dark red blood enters the pulmonary circulation as it is pumped from the right ventricle into the large **pulmonary trunk** (Figure 19.19b), which runs diagonally upward for about 8 cm and then divides abruptly to form the **right** and **left pulmonary arteries**. In the lungs, the pulmonary arteries subdivide into the **lobar arteries** (lo′bar) (three in the right lung and two in the left lung), each of which serves one lung lobe. The lobar arteries accompany the main bronchi into the lungs and then branch profusely, forming first arterioles and then the dense networks of **pulmonary capillaries** that surround and cling to the delicate air sacs. It is here that oxygen moves from the alveolar air to the blood and carbon dioxide moves from the blood to the alveolar air. As gases are exchanged and the oxygen content of the blood rises, the blood becomes bright red. The pulmonary capillary beds drain into venules, which join to form the two **pulmonary veins** exiting from each lung. The four pulmonary veins complete the circuit by unloading their precious cargo into the left atrium of the heart. Note that any vessel with the term *pulmonary* or *lobar* in its name is part of the pulmonary circulation. All others are part of the systemic circulation.

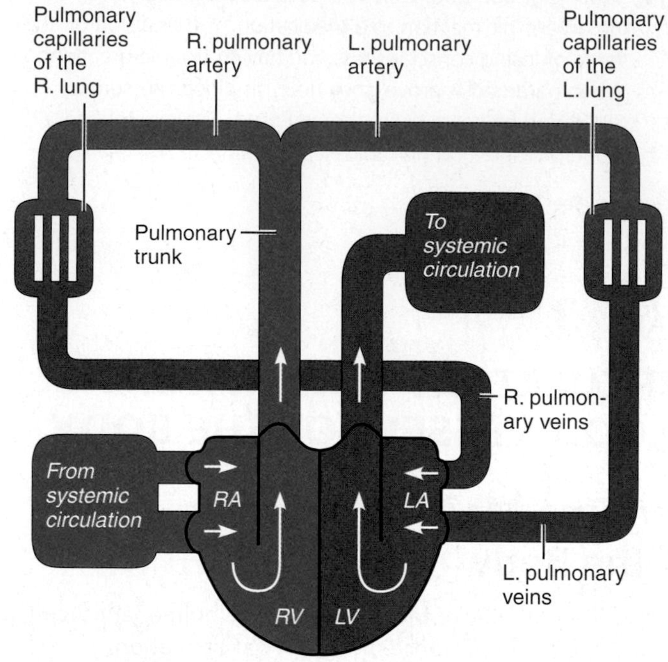

(a) Schematic flowchart.

Figure 19.19 Pulmonary circulation.

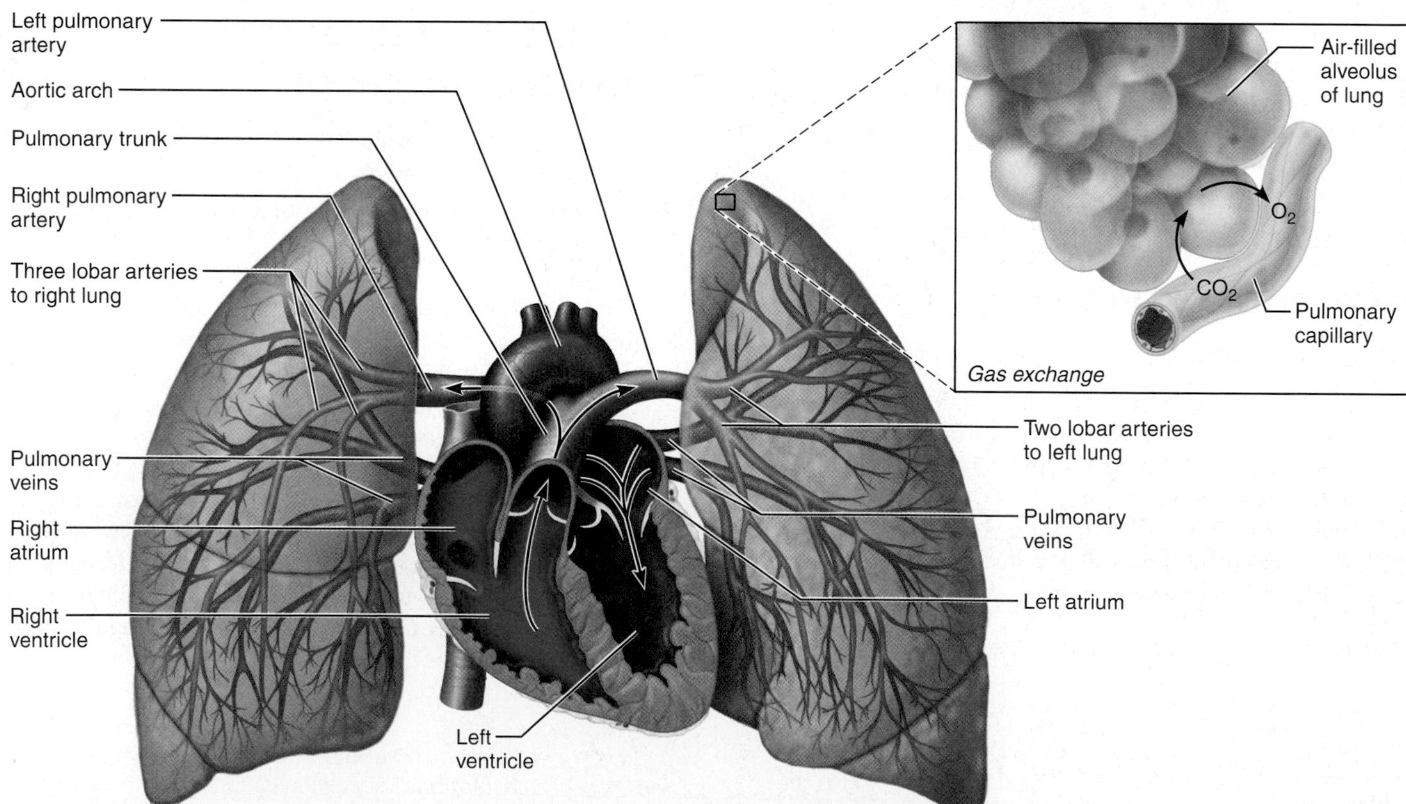

(b) Illustration. The pulmonary arterial system is shown in blue to indicate that the blood carried is oxygen-poor. The pulmonary venous drainage is shown in red to indicate that the blood transported is oxygen-rich.

TABLE 19.3 *(continued)*

Pulmonary arteries carry oxygen-poor, carbon dioxide–rich blood, and pulmonary veins carry oxygen-rich blood.* This is opposite the situation in the systemic circulation, where arteries carry oxygen-rich blood and veins carry carbon dioxide–rich, relatively oxygen-poor blood.

Systemic Circulation

The systemic circulation provides the *functional blood supply* to all body tissues; that is, it delivers oxygen, nutrients, and other needed substances while carrying away carbon dioxide and other metabolic wastes. Freshly oxygenated blood* returning from the pulmonary circuit is pumped out of the left ventricle into the aorta (**Figure 19.20**). From the aorta, blood can take various routes, because essentially all systemic arteries branch from this single great vessel. The aorta arches upward from the heart and then curves and runs downward along the body midline to its terminus in the pelvis, where it splits to form the two large arteries serving the lower extremities. The branches of the aorta continue to subdivide to produce the arterioles and, finally, the capillaries that ramify through the organs. Venous blood draining from organs inferior to the diaphragm ultimately enters the inferior vena cava.† Except for some coronary and thoracic venous drainage (which enters the azygos system of veins), body regions above the diaphragm are drained by the superior vena cava. The venae cavae empty the carbon dioxide–laden blood into the right atrium of the heart.

Two points concerning the two major circulations must be emphasized: (1) Blood passes from systemic veins to systemic arteries only after first moving through the pulmonary circuit (Figure 19.19a), and (2) although the entire cardiac output of the right ventricle passes through the pulmonary circulation, only a small fraction of the output of the left ventricle flows through any single organ (Figure 19.20). The systemic circulation can be viewed as multiple circulatory channels functioning in parallel to distribute blood to all body organs.

As you examine the tables that follow and locate the various systemic arteries and veins in the illustrations, be aware of cues that make your memorization task easier. In many cases, the name of a vessel reflects the body region traversed (axillary, brachial, femoral, etc.), the organ served (renal, hepatic, gonadal), or the bone followed (vertebral, radial, tibial). Also, notice that arteries and veins tend to run side by side and, in many places, they also run with nerves. Finally, be alert to the fact that the systemic vessels do not always match on the right and left sides of the body. Thus, while almost all vessels in the head and limbs are bilaterally symmetrical, some of the large, deep vessels of the trunk region are asymmetrical or unpaired.

*By convention, oxygen-rich blood is shown *red* and oxygen-poor blood is shown *blue*.

†Venous blood from the digestive viscera passes through the hepatic portal circulation (liver and associated veins) before entering the inferior vena cava.

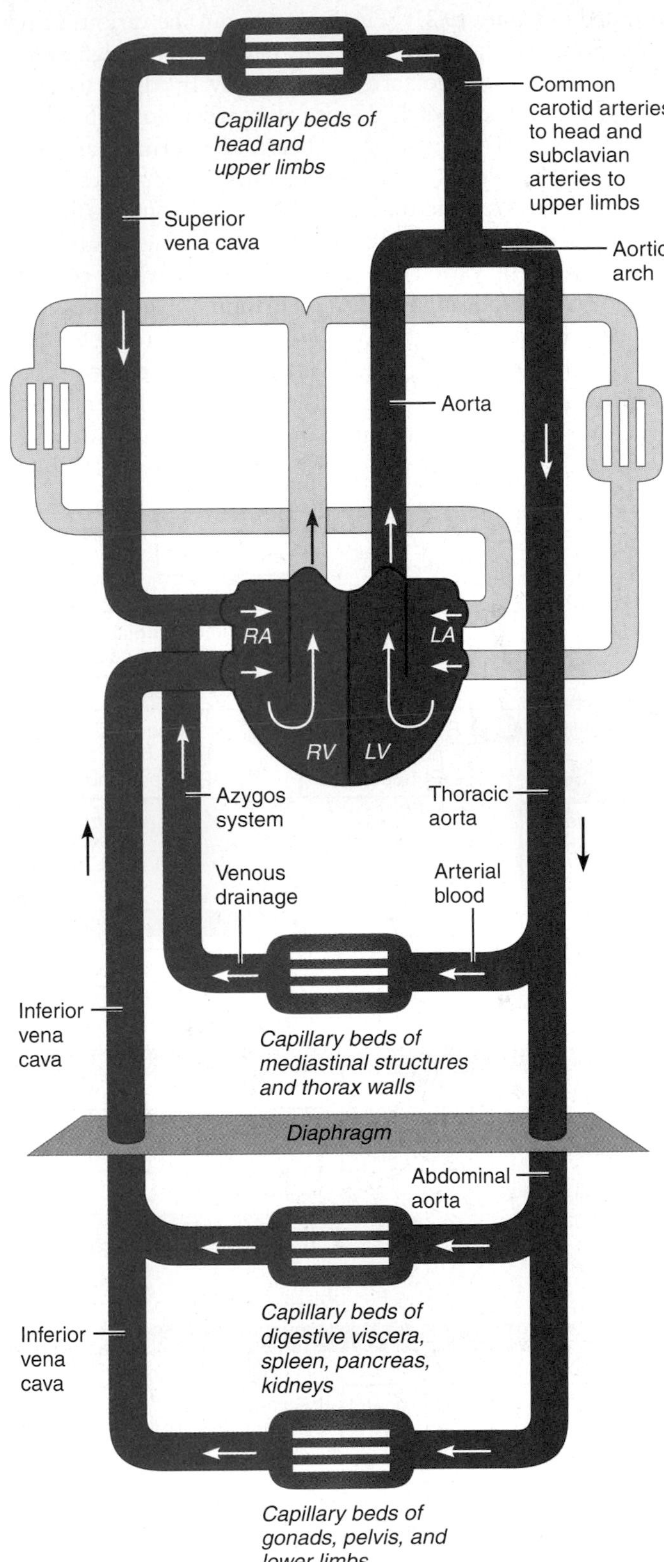

Figure 19.20 Schematic flowchart showing an overview of the systemic circulation. The pulmonary circulation is shown in gray for comparison.

TABLE 19.4 The Aorta and Major Arteries of the Systemic Circulation

The distribution of the aorta and major arteries of the systemic circulation is diagrammed in flowchart form in **Figure 19.21a** and illustrated in Figure 19.21b. Fine points about the various vessels arising from the aorta are provided in Tables 19.5 through 19.8.

The **aorta** is the largest artery in the body. In adults, the aorta (a-or′ tah) is approximately the size of a garden hose where it issues from the left ventricle of the heart. Its internal diameter is 2.5 cm, and its wall is about 2 mm thick. It decreases in size slightly as it runs to its terminus. The aortic valve guards the base of the aorta and prevents backflow of blood during diastole. Opposite each aortic valve cusp is an *aortic sinus,* which contains baroreceptors important in reflex regulation of blood pressure.

Different portions of the aorta are named according to shape or location. The first portion, the **ascending aorta**, runs posteriorly and to the right of the pulmonary trunk. It persists for only about 5 cm before curving to the left as the aortic arch. The only branches of the ascending aorta are the **right** and **left coronary arteries**, which supply the myocardium. The **aortic arch**, deep to the sternum, begins and ends at the sternal angle (T_4 level). Its three major branches (R to L) are: (1) the **brachiocephalic trunk** (bra′ke-o-sĕ-fal″ik; "armhead"), which passes superiorly under the right sternoclavicular joint and branches into the **right common carotid artery** (kah-rot′id) and the **right subclavian artery**, (2) the **left common carotid artery**, and (3) the **left subclavian artery**. These three vessels provide the arterial supply of the head, neck, upper limbs, and part of the thorax wall. The **thoracic**, or **descending, aorta** runs along the anterior spine from T_5 to T_{12}, sending off numerous small arteries to the thorax wall and viscera before piercing the diaphragm. As it enters the abdominal cavity, it becomes the **abdominal aorta**. This portion supplies the abdominal walls and viscera and ends at the L_4 level, where it splits into the **right** and **left common iliac arteries**, which supply the pelvis and lower limbs.

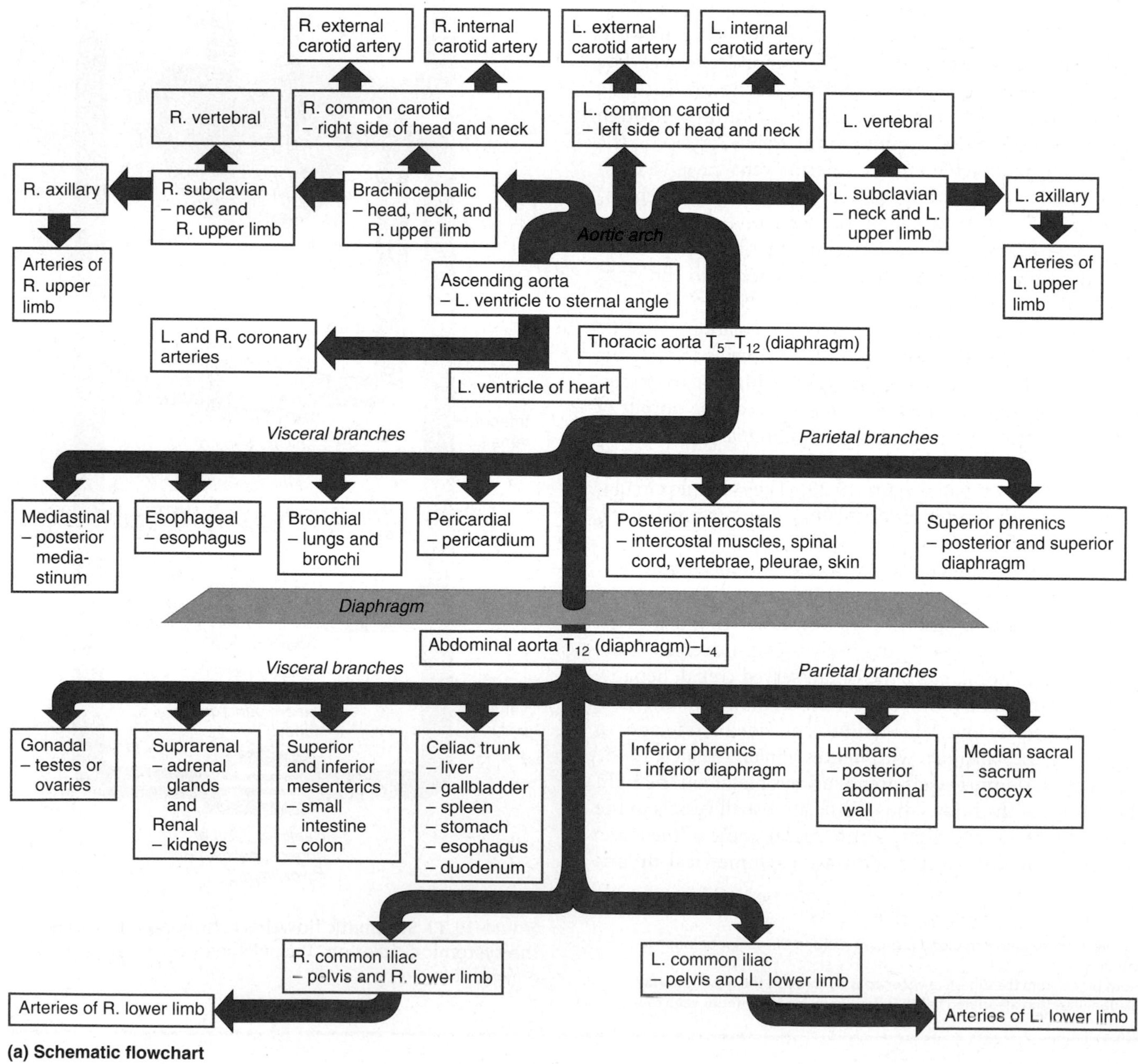

(a) Schematic flowchart

TABLE 19.4 *(continued)*

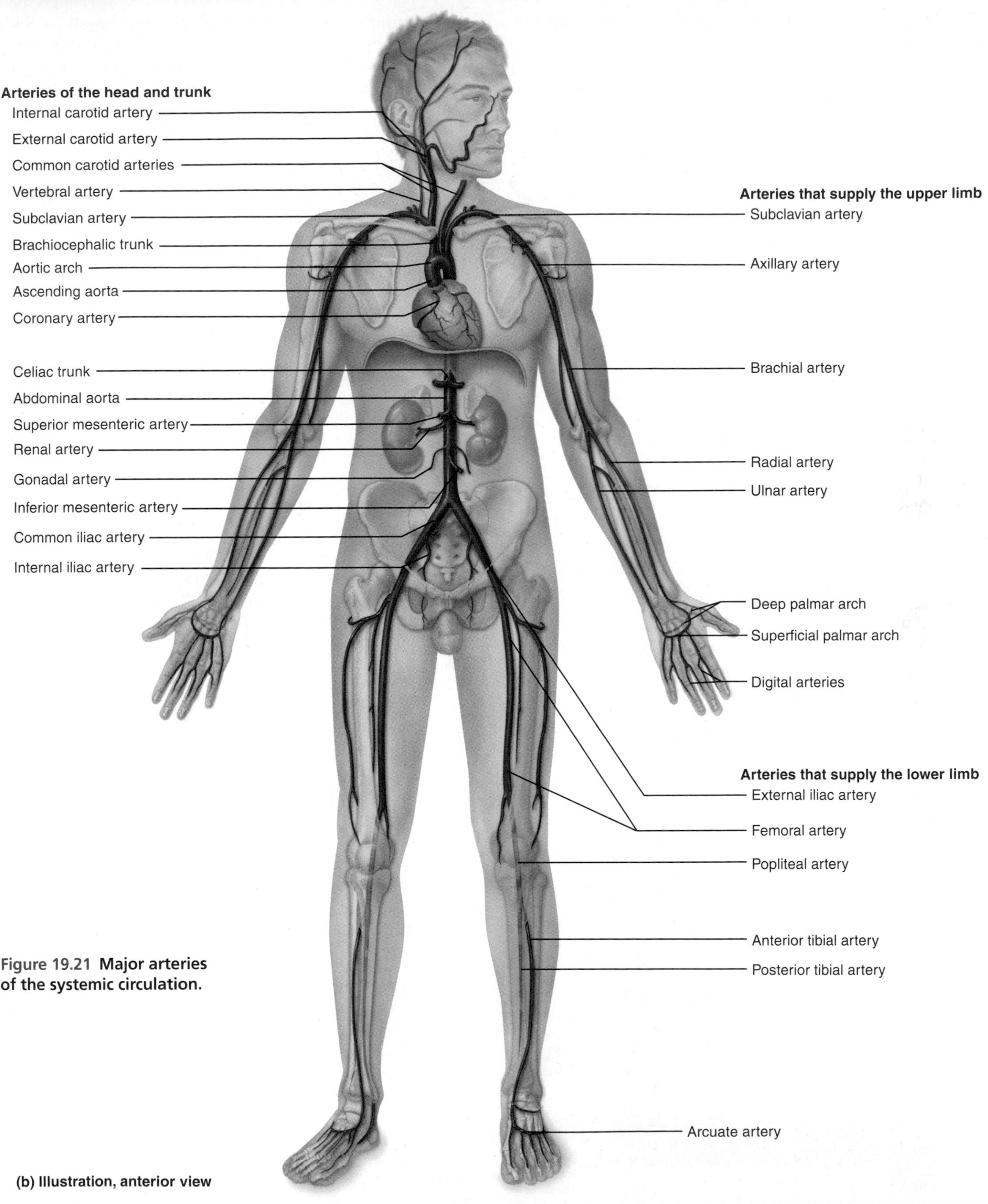
Arteries of the head and trunk
Internal carotid artery
External carotid artery
Common carotid arteries
Vertebral artery
Subclavian artery
Brachiocephalic trunk
Aortic arch
Ascending aorta
Coronary artery
Celiac trunk
Abdominal aorta
Superior mesenteric artery
Renal artery
Gonadal artery
Inferior mesenteric artery
Common iliac artery
Internal iliac artery
Arteries that supply the upper limb
Subclavian artery
Axillary artery
Brachial artery
Radial artery
Ulnar artery
Deep palmar arch
Superficial palmar arch
Digital arteries
Arteries that supply the lower limb
External iliac artery
Femoral artery
Popliteal artery
Anterior tibial artery
Posterior tibial artery
Arcuate artery
(b) Illustration, anterior view

Figure 19.21 Major arteries of the systemic circulation.

TABLE 19.5 Arteries of the Head and Neck

Four paired arteries supply the head and neck. These are the common carotid arteries, plus three branches from each subclavian artery: the vertebral arteries, the thyrocervical trunks, and the costocervical trunks **(Figure 19.22b)**. Of these, the common carotid arteries have the broadest distribution (Figure 19.22a).

Each common carotid divides into two major branches (the internal and external carotid arteries). At the division point, each internal carotid artery has a slight dilation, the **carotid sinus**, that contains baroreceptors that assist in reflex blood pressure control. The **carotid bodies**, chemoreceptors involved in the control of respiratory rate, are located close by. Pressing on the neck in the area of the carotid sinuses can cause unconsciousness (*carot* = stupor) because the pressure created mimics high blood pressure, eliciting vasodilation, which interferes with blood delivery to the brain.

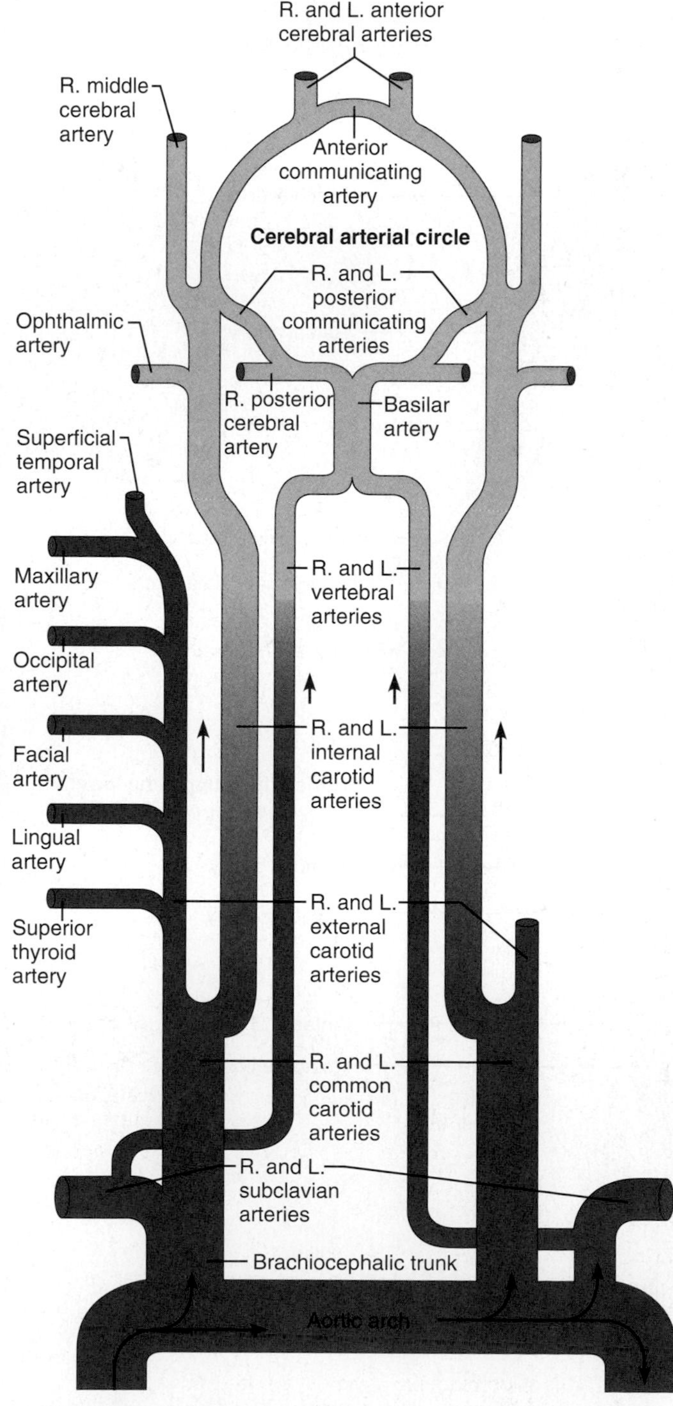

(a) Schematic flowchart

Description and Distribution

Common carotid arteries. The origins of these two arteries differ: The right common carotid artery arises from the brachiocephalic trunk; the left is the second branch of the aortic arch. The common carotid arteries ascend through the lateral neck, and at the superior border of the larynx (the level of the "Adam's apple"), each divides into its two major branches, the *external* and *internal carotid arteries.*

The **external carotid arteries** supply most tissues of the head except for the brain and orbit. As each artery runs superiorly, it sends branches to the thyroid gland and larynx (**superior thyroid artery**), the tongue (**lingual artery**), the skin and muscles of the anterior face (**facial artery**), and the posterior scalp (**occipital artery**). Each external carotid artery terminates by splitting into a **superficial temporal artery**, which supplies the parotid salivary gland and most of the scalp, and a **maxillary artery**, which supplies the upper and lower jaws and chewing muscles, the teeth, and the nasal cavity. A clinically important branch of the maxillary artery is the *middle meningeal artery* (not illustrated). It enters the skull through the foramen spinosum and supplies the inner surface of the parietal bone, squamous region of the temporal bone, and the underlying dura mater.

The larger **internal carotid arteries** supply the orbits and more than 80% of the cerebrum. They assume a deep course and enter the skull through the carotid canals of the temporal bones. Once inside the cranium, each artery gives off one main branch, the ophthalmic artery, and then divides into the anterior and middle cerebral arteries. The **ophthalmic arteries** (of-thal′mik) supply the eyes, orbits, forehead, and nose. Each **anterior cerebral artery** supplies the medial surface of the frontal and parietal lobes of the cerebral hemisphere on its side and also anastomoses with its partner on the opposite side via a short arterial shunt called the **anterior communicating artery** (Figure 19.22d). The **middle cerebral arteries** run in the lateral sulci of their respective cerebral hemispheres and supply the lateral parts of the temporal, parietal, and frontal lobes.

Vertebral arteries. These vessels spring from the subclavian arteries at the root of the neck and ascend through foramina in the transverse processes of the cervical vertebrae to enter the skull through the foramen magnum. En route, they send branches to the vertebrae and cervical spinal cord and to some deep structures of the neck. Within the cranium, the right and left vertebral arteries join to form the **basilar artery** (bas′ĭ-lar), which ascends along the anterior aspect of the brain stem, giving off branches to the cerebellum, pons, and inner ear (Figure 19.22b and d). At the pons-midbrain border, the basilar artery divides into a pair of **posterior cerebral arteries**, which supply the occipital lobes and the inferior parts of the temporal lobes.

TABLE 19.5 *(continued)*

Arterial shunts called **posterior communicating arteries** connect the posterior cerebral arteries to the middle cerebral arteries anteriorly. The two posterior and single anterior communicating arteries complete the formation of an arterial anastomosis called the **cerebral arterial circle (circle of Willis)**. This structure encircles the pituitary gland and optic chiasma and unites the brain's anterior and posterior blood supplies. It also equalizes blood pressure in the two brain areas and provides alternate routes for blood to reach the brain tissue if a carotid or vertebral artery becomes occluded.

Thyrocervical and costocervical trunks. These short vessels arise from the subclavian artery just lateral to the vertebral arteries on each side (Figures 19.22b and Figure 19.23). The thyrocervical trunk mainly supplies the thyroid gland, portions of the cervical vertebrae and spinal cord, and some scapular muscles. The costocervical trunk serves deep neck and superior intercostal muscles.

Figure 19.22 Arteries of the head, neck, and brain.

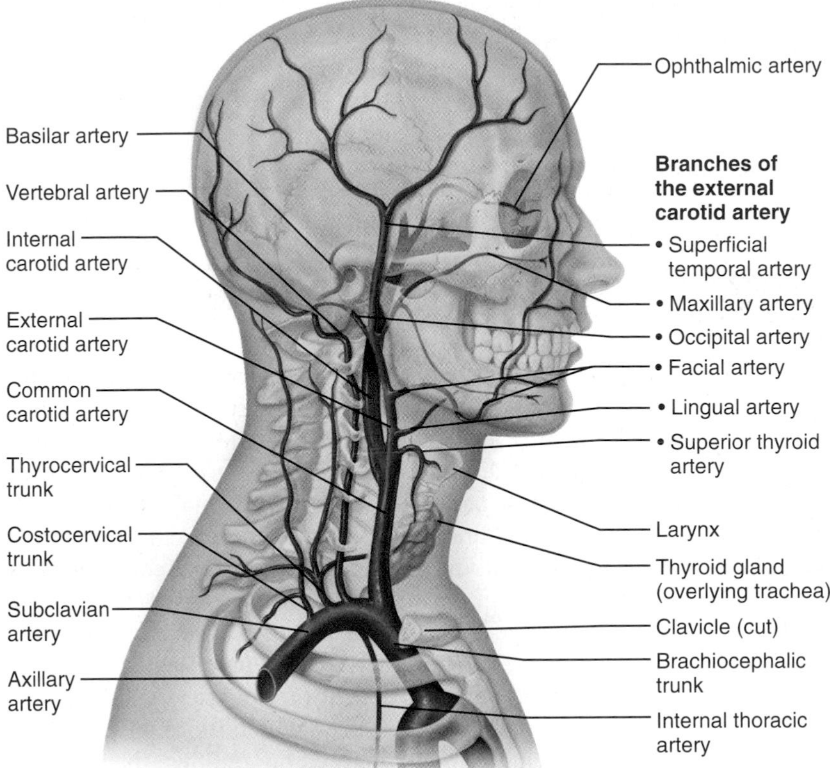

(b) Arteries of the head and neck, right aspect

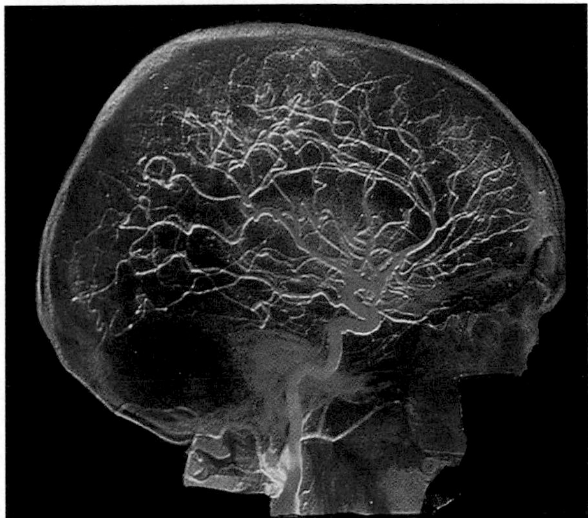

(c) Colorized arteriograph of the arterial supply of the brain

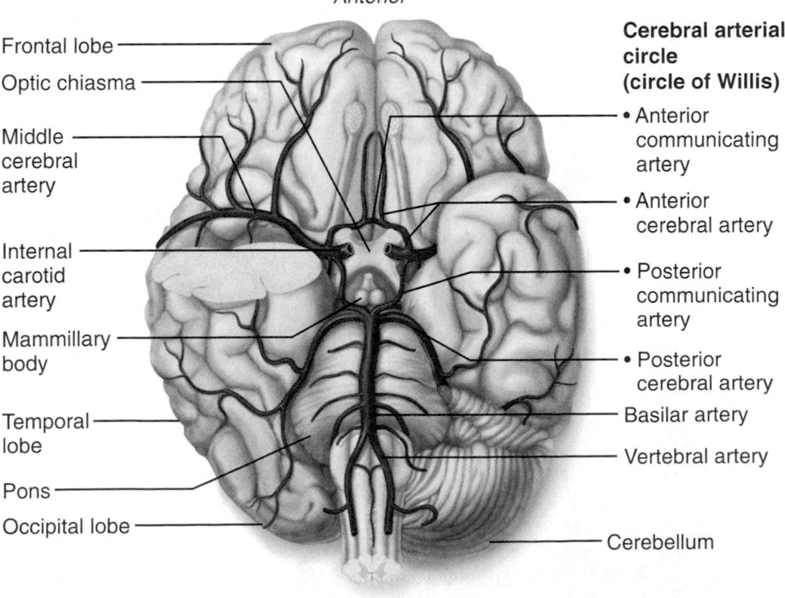

(d) Major arteries serving the brain (inferior view, right side of cerebellum and part of right temporal lobe removed)

TABLE 19.6 Arteries of the Upper Limbs and Thorax

The upper limbs are supplied entirely by arteries arising from the **subclavian arteries (Figure 19.23a)**. After giving off branches to the neck, each subclavian artery courses laterally between the clavicle and first rib to enter the axilla, where its name changes to axillary artery. The thorax wall is supplied by an array of vessels that arise either directly from the thoracic aorta or from branches of the subclavian arteries. Most visceral organs of the thorax receive their functional blood supply from small branches issuing from the thoracic aorta. Because these vessels are so small and tend to vary in number (except for the bronchial arteries), they are not illustrated in Figures 19.23a and b, but several of them are listed at the end of this table.

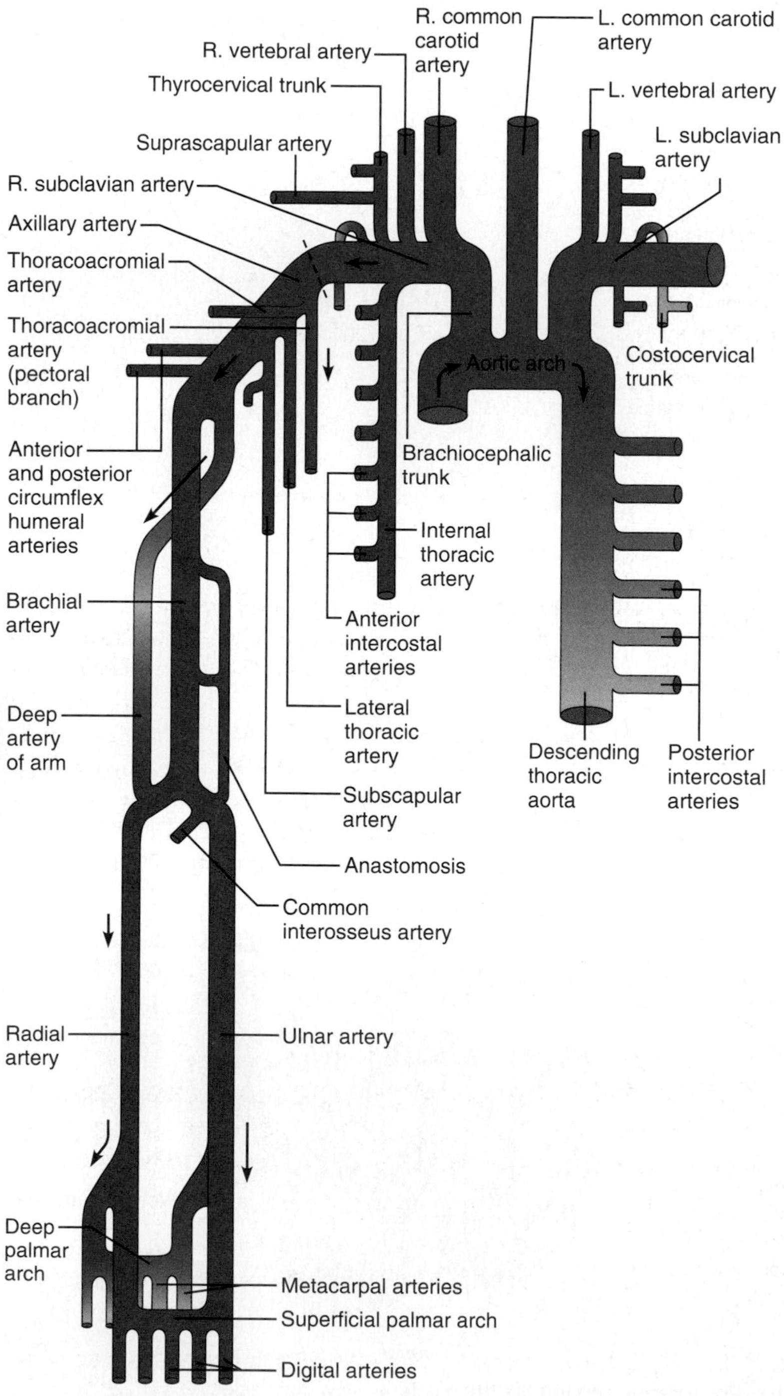

(a) Schematic flowchart

Description and Distribution

Arteries of the Upper Limb

Axillary artery. As it runs through the axilla accompanied by cords of the brachial plexus, each axillary artery gives off branches to the axilla, chest wall, and shoulder girdle. These branches include the **thoracoacromial artery** (tho″rah-ko-ah-kro′me-al), which supplies the deltoid muscle and pectoral region; the **lateral thoracic artery**, which serves the lateral chest wall and breast; the **subscapular artery** to the scapula, dorsal thorax wall, and part of the latissimus dorsi muscle; and the **anterior** and **posterior circumflex humeral arteries**, which wrap around the humeral neck and help supply the shoulder joint and the deltoid muscle. As the axillary artery emerges from the axilla, it becomes the brachial artery.

Brachial artery. The brachial artery runs down the medial aspect of the humerus and supplies the anterior flexor muscles of the arm. One major branch, the **deep artery of the arm**, serves the posterior triceps brachii muscle. As it nears the elbow, the brachial artery gives off several small branches that contribute to an anastomosis serving the elbow joint and connecting it to the arteries of the forearm. As the brachial artery crosses the anterior midline aspect of the elbow, it provides an easily palpated pulse point (brachial pulse) (see Figure 19.12). Immediately beyond the elbow, the brachial artery splits to form the radial and ulnar arteries, which more or less follow the course of similarly named bones down the length of the anterior forearm.

Radial artery. The radial artery runs from the median line of the cubital fossa to the styloid process of the radius. It supplies the lateral muscles of the forearm, the wrist, and the thumb and index finger. At the root of the thumb, the radial artery provides a convenient site for taking the radial pulse.

Ulnar artery. The ulnar artery supplies the medial aspect of the forearm, fingers 3–5, and the medial aspect of the index finger. Proximally, the ulnar artery gives off a short branch, the **common interosseous artery** (in″ter-os′e-us), which runs between the radius and ulna to serve the deep flexors and extensors of the forearm.

Palmar arches. In the palm, branches of the radial and ulnar arteries anastomose to form the **superficial** and **deep palmar arches.** The **metacarpal arteries** and the **digital arteries** that supply the fingers arise from these palmar arches.

Arteries of the Thorax Wall

Internal thoracic arteries. The internal thoracic arteries (formerly called the internal mammary arteries) arise

TABLE 19.6 *(continued)*

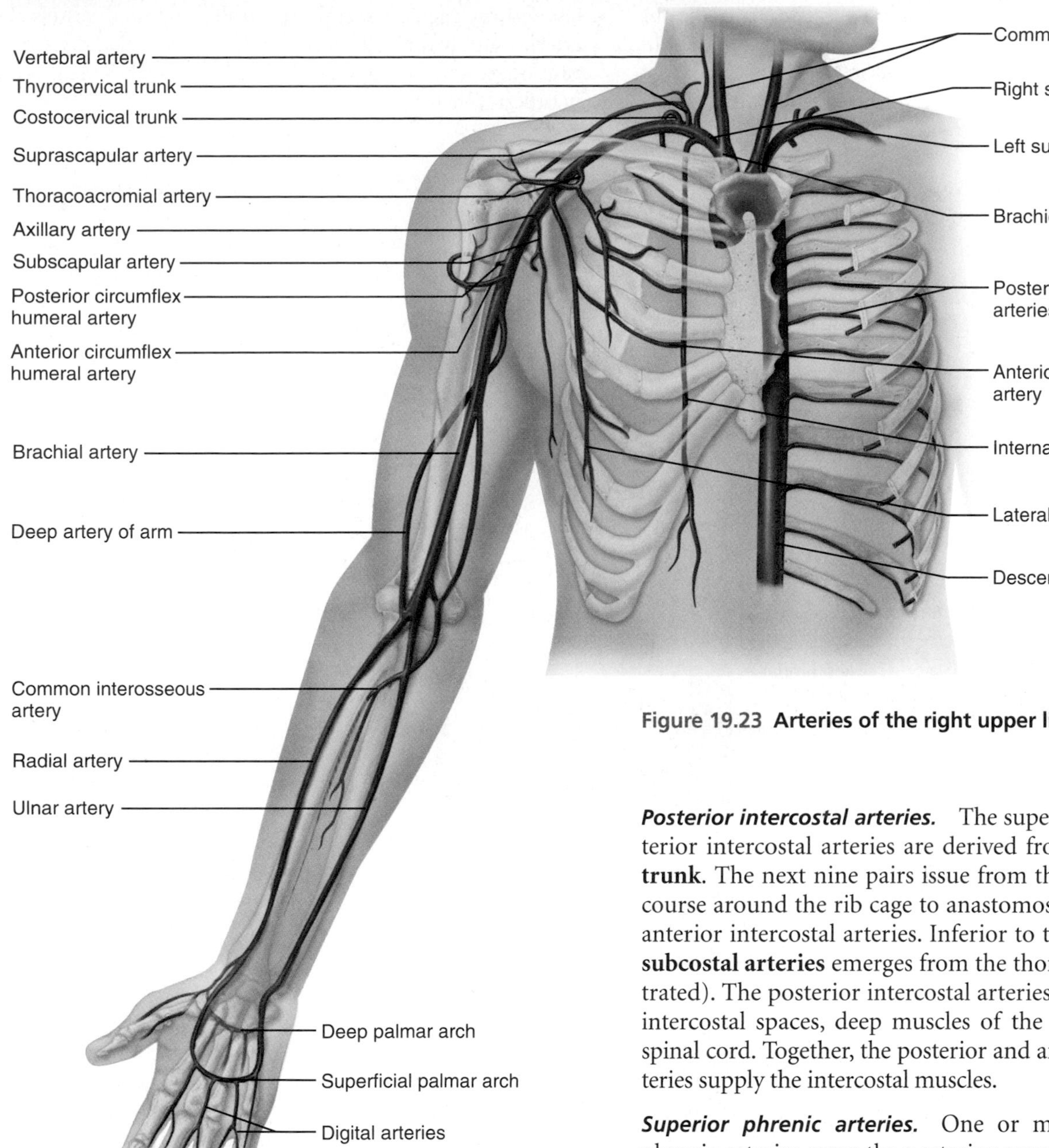

(b) Illustration, anterior view

Figure 19.23 Arteries of the right upper limb and thorax.

from the subclavian arteries and supply blood to most of the anterior thorax wall. Each of these arteries descends lateral to the sternum and gives off **anterior intercostal arteries**, which supply the intercostal spaces anteriorly. The internal thoracic artery also sends superficial branches to the skin and mammary glands and terminates in twiglike branches to the anterior abdominal wall and diaphragm.

Posterior intercostal arteries. The superior two pairs of posterior intercostal arteries are derived from the **costocervical trunk**. The next nine pairs issue from the thoracic aorta and course around the rib cage to anastomose anteriorly with the anterior intercostal arteries. Inferior to the 12th rib, a pair of **subcostal arteries** emerges from the thoracic aorta (not illustrated). The posterior intercostal arteries supply the posterior intercostal spaces, deep muscles of the back, vertebrae, and spinal cord. Together, the posterior and anterior intercostal arteries supply the intercostal muscles.

Superior phrenic arteries. One or more paired superior phrenic arteries serve the posterior superior aspect of the diaphragm surface.

Arteries of the Thoracic Viscera

Pericardial arteries. Several tiny branches supply the posterior pericardium.

Bronchial arteries. Two left and one right bronchial arteries supply systemic (oxygen-rich) blood to the lungs, bronchi, and pleurae.

Esophageal arteries. Four to five esophageal arteries supply the esophagus.

Mediastinal arteries. Many small mediastinal arteries serve the posterior mediastinum.

TABLE 19.7 Arteries of the Abdomen

The arterial supply to the abdominal organs arises from the abdominal aorta **(Figure 19.24a)**. Under resting conditions, about half of the entire arterial flow is found in these vessels. Except for the celiac trunk, the superior and inferior mesenteric arteries, and the median sacral artery, all are paired vessels. These arteries supply the abdominal wall, diaphragm, and visceral organs of the abdominopelvic cavity. The branches are given here in order of their issue.

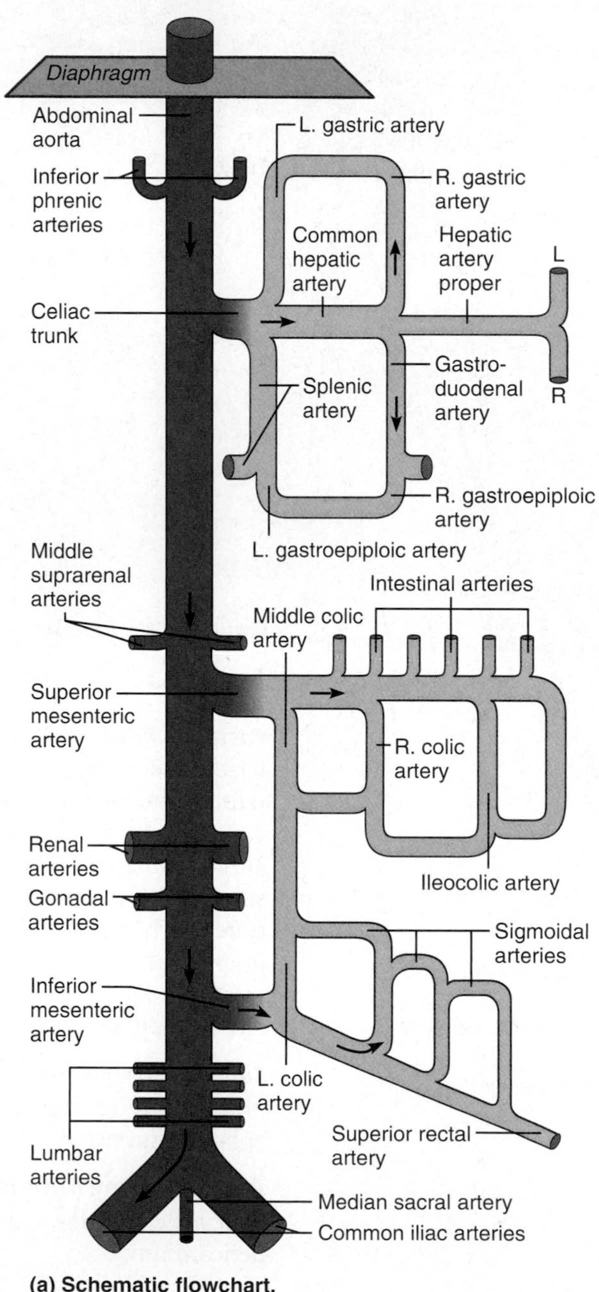

(a) Schematic flowchart.

Figure 19.24 Arteries of the abdomen.

TABLE 19.7 *(continued)*

Description and Distribution

Inferior phrenic arteries. The inferior phrenics emerge from the aorta at T_{12}, just inferior to the diaphragm. They serve the inferior diaphragm surface.

Celiac trunk. This very large unpaired branch of the abdominal aorta divides almost immediately into three branches: the common hepatic, splenic, and left gastric arteries (Figure 19.24b). The **common hepatic artery** (hĕ-pat′ik) gives off branches to the stomach, duodenum, and pancreas. Where the **gastroduodenal artery** branches off, it becomes the **hepatic artery proper**, which splits into right and left branches that serve the liver. As the **splenic artery** (splen′ik) passes deep to the stomach, it sends branches to the pancreas and stomach and terminates in branches to the spleen. The **left gastric artery** (*gaster* = stomach) supplies part of the stomach and the inferior esophagus. The **right** and **left gastroepiploic arteries** (gas″tro-ep″ĭ-plo′ik), branches of the gastroduodenal and splenic arteries, respectively, serve the greater curvature of the stomach. A **right gastric artery**, which supplies the stomach's lesser curvature, may arise from the common hepatic artery or from the hepatic artery proper.

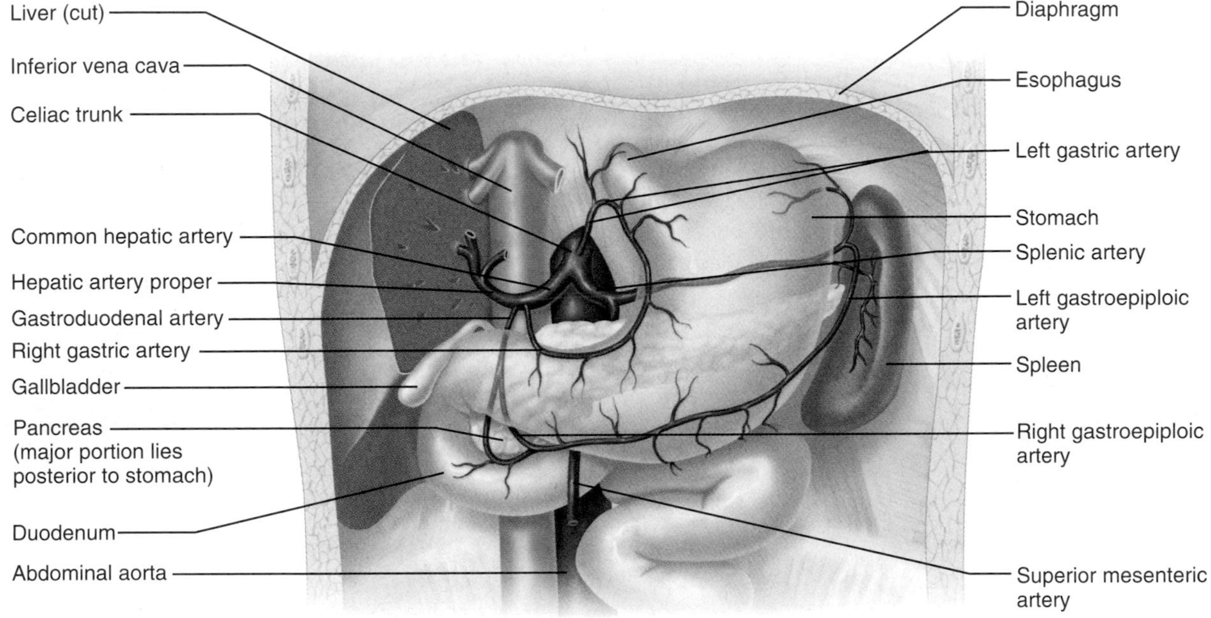

(b) The celiac trunk and its major branches. The left half of the liver has been removed.

Figure 19.24 *(continued)*

TABLE 19.7 Arteries of the Abdomen *(continued)*

Superior mesenteric artery (mes-en-ter′ik). This large, unpaired artery arises from the abdominal aorta at the L_1 level immediately below the celiac trunk (Figure 19.24d). It runs deep to the pancreas and then enters the mesentery, where its numerous anastomosing branches serve virtually all of the small intestine via the **intestinal arteries**, and most of the large intestine—the appendix, cecum, ascending colon (via the **ileocolic** and **right colic arteries**), and part of the transverse colon (via the **middle colic artery**).

Suprarenal arteries (soo″prah-re′nal). The **middle suprarenal arteries** flank the origin of the superior mesenteric artery as they emerge from the abdominal aorta (Figure 19.24c). They supply blood to the adrenal (suprarenal) glands overlying the kidneys. The adrenal glands also receive two sets of branches not illustrated: *superior suprarenal* branches from the nearby inferior phrenic arteries, and *inferior suprarenal* branches from the nearby renal arteries.

Renal arteries. The short but wide renal arteries, right and left, issue from the lateral surfaces of the aorta slightly below the superior mesenteric artery (between L_1 and L_2). Each serves the kidney on its side.

Gonadal arteries (go-nă′dul). The paired gonadal arteries are called the **testicular arteries** in males and the **ovarian arteries** in females. The ovarian arteries extend into the pelvis to serve the ovaries and part of the uterine tubes. The much longer testicular arteries descend through the pelvis and inguinal canals to enter the scrotum, where they serve the testes.

Inferior mesenteric artery. This final major branch of the abdominal aorta is unpaired and arises from the anterior aortic surface at the L_3 level. It serves the distal part of the large intestine—from the midpart of the transverse colon to the midrectum—via its **left colic**, **sigmoidal**, and **superior rectal branches** (Figure 19.24d). Looping anastomoses between the superior and inferior mesenteric arteries help ensure that blood will continue to reach the digestive viscera in cases of trauma to one of these abdominal arteries.

Lumbar arteries. Four pairs of lumbar arteries arise from the posterolateral surface of the aorta in the lumbar region. These segmental arteries supply the posterior abdominal wall.

Median sacral artery. The unpaired median sacral artery issues from the posterior surface of the abdominal aorta at its terminus. This tiny artery supplies the sacrum and coccyx.

Common iliac arteries. At the L_4 level, the aorta splits into the right and left common iliac arteries, which supply blood to the lower abdominal wall, pelvic organs, and lower limbs (Figure 19.24c).

CHECK YOUR UNDERSTANDING

15. Which paired artery supplies most of the tissues of the head except for the brain and orbits?

16. Name the arterial anastomosis at the base of the brain.

17. Name the four unpaired arteries that emerge from the abdominal aorta.

For answers, see Appendix G.

TABLE 19.7 *(continued)*

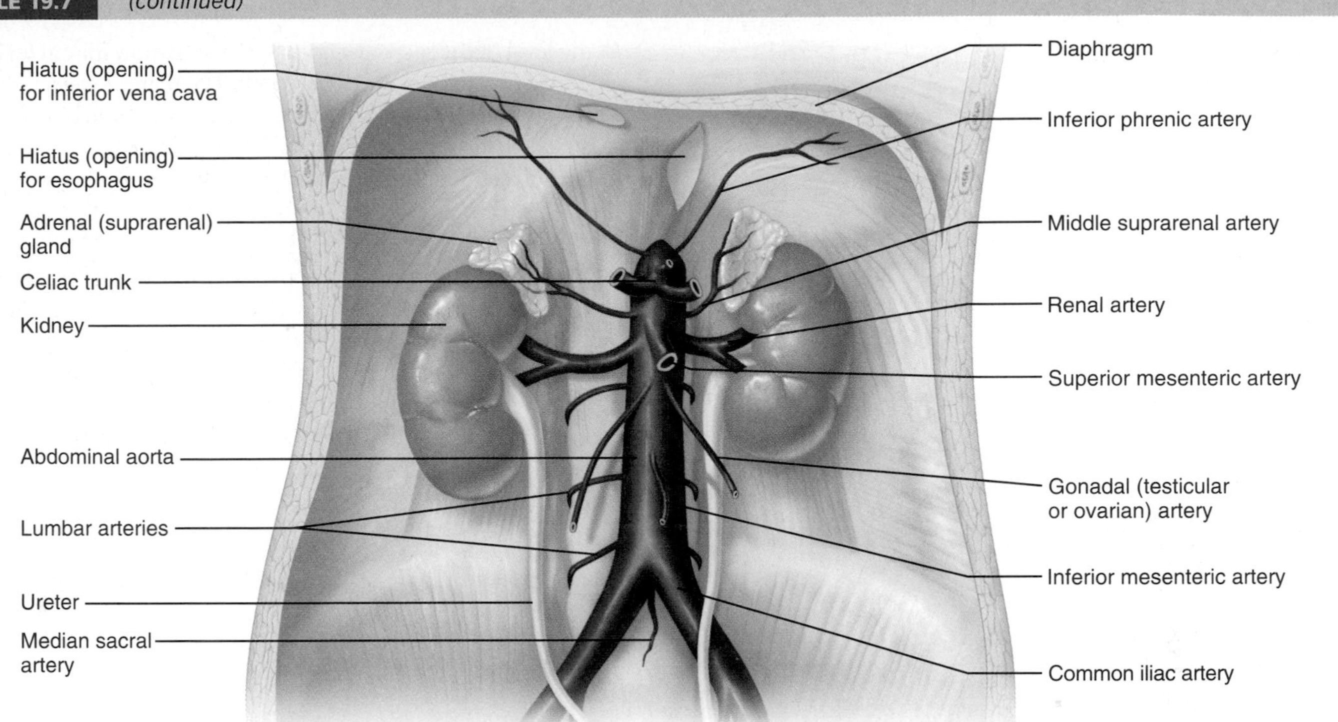

(c) Major branches of the abdominal aorta.

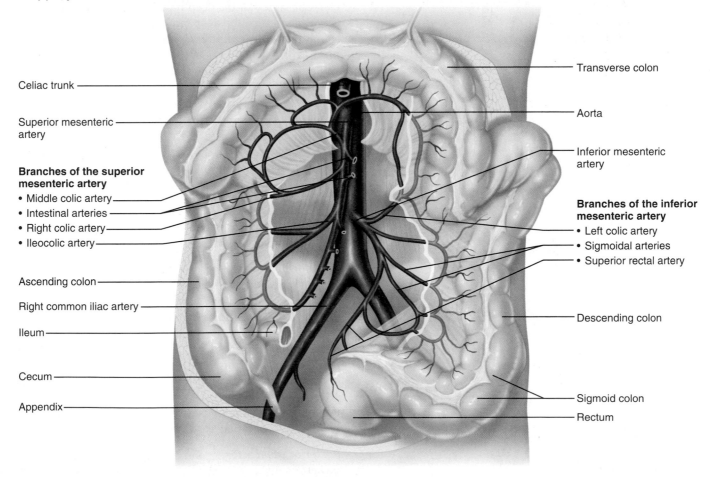

Celiac trunk
Superior mesenteric artery
Branches of the superior mesenteric artery
• Middle colic artery
• Intestinal arteries
• Right colic artery
• Ileocolic artery
Ascending colon
Right common iliac artery
Ileum
Cecum
Appendix
Transverse colon
Aorta
Inferior mesenteric artery
Branches of the inferior mesenteric artery
• Left colic artery
• Sigmoidal arteries
• Superior rectal artery
Descending colon
Sigmoid colon
Rectum

(d) Distribution of the superior and inferior mesenteric arteries. The transverse colon has been pulled superiorly.

Figure 19.24 *(continued)* **Arteries of the abdomen.**

TABLE 19.8 Arteries of the Pelvis and Lower Limbs

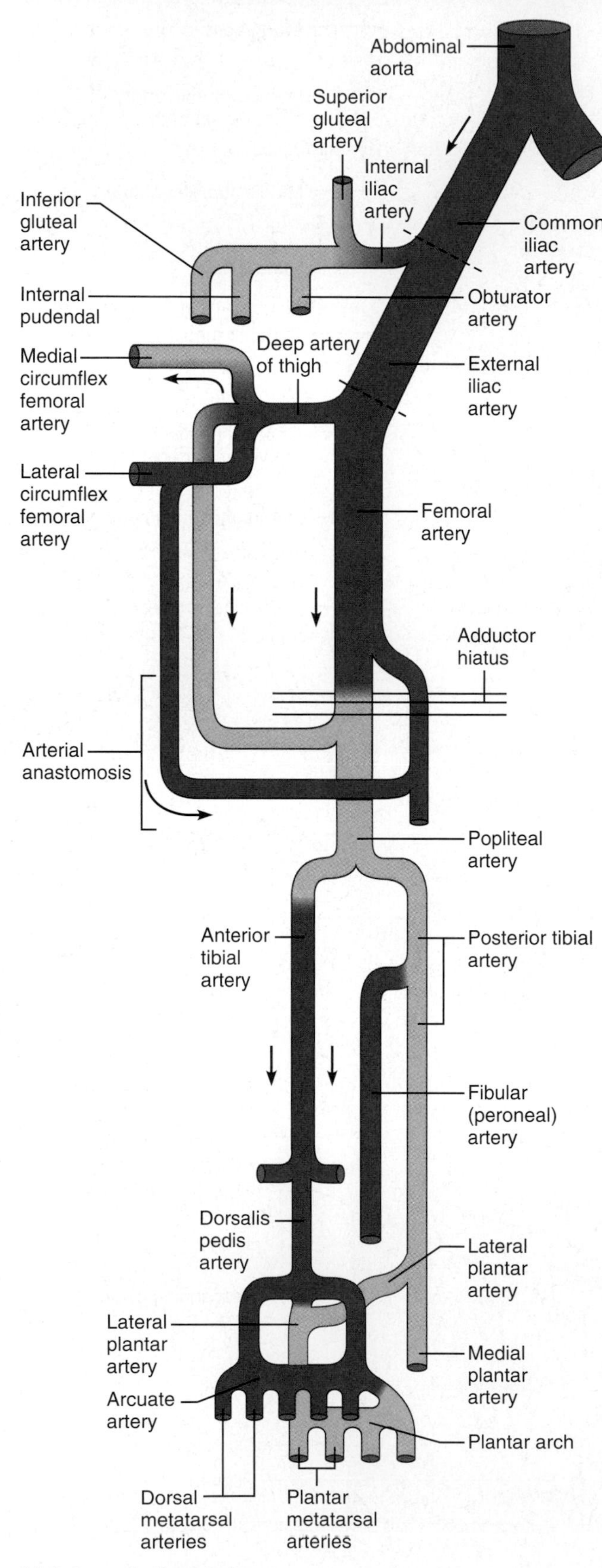

Figure 19.25 Arteries of the right pelvis and lower limb.

At the level of the sacroiliac joints, the **common iliac arteries** divide into two major branches, the internal and external iliac arteries **(Figure 19.25a)**. The internal iliacs distribute blood mainly to the pelvic region. The external iliacs primarily serve the lower limbs but also send some branches to the abdominal wall.

Description and Distribution

Internal iliac arteries. These paired arteries run into the pelvis and distribute blood to the pelvic walls and viscera (bladder, rectum, uterus, and vagina in the female and prostate and ductus deferens in the male). Additionally they serve the gluteal muscles via the **superior** and **inferior gluteal arteries**, adductor muscles of the medial thigh via the **obturator artery**, and external genitalia and perineum via the **internal pudendal artery** (not illustrated).

External iliac arteries. These arteries supply the lower limbs (Figure 19.25b). As they course through the pelvis, they give off branches to the anterior abdominal wall. After passing under the inguinal ligaments to enter the thigh, they become the femoral arteries.

Femoral arteries. As each of these arteries passes down the anteromedial thigh, it gives off several branches to the thigh muscles. The largest of the deep branches is the **deep artery of the thigh** (also called the **deep femoral artery**), which is the main supply to the thigh muscles (hamstrings, quadriceps, and adductors). Proximal branches of the deep femoral artery, the **lateral** and **medial circumflex femoral arteries**, encircle the neck of the femur. The medial circumflex artery is the major vessel to the head of the femur. If it is torn in a hip fracture, the bone tissue of the head of the femur dies. A long descending branch of the lateral circumflex artery supplies the vastus lateralis muscle. Near the knee the femoral artery passes posteriorly and through a gap in the adductor magnus muscle, the *adductor hiatus,* to enter the popliteal fossa, where its name changes to popliteal artery.

Popliteal artery. This posterior vessel contributes to an arterial anastomosis that supplies the knee region and then splits into the anterior and posterior tibial arteries of the leg.

Anterior tibial artery. The anterior tibial artery runs through the anterior compartment of the leg, supplying the extensor muscles along the way. At the ankle, it becomes the **dorsalis pedis artery**, which supplies the ankle and dorsum of the foot, and gives off a branch, the **arcuate artery**, which issues the **dorsal metatarsal arteries** to the metatarsus of the foot. The superficial dorsalis pedis ends by penetrating into the sole where it forms the medial part of the **plantar arch**. The dorsalis pedis artery provides a clinically important pulse point, the pedal pulse. If the pedal pulse is easily felt, it is fairly certain that the blood supply to the leg is good.

Posterior tibial artery. This large artery courses through the posteromedial part of the leg and supplies the flexor muscles. Proximally, it gives off a large branch, the **fibular (peroneal) artery**, which supplies the lateral fibularis muscles of the leg. On

TABLE 19.8 *(continued)*

the medial side of the foot, the posterior tibial artery divides into **lateral** and **medial plantar arteries** that serve the plantar surface of the foot. The lateral plantar artery forms the lateral end of the plantar arch. **Plantar metatarsal arteries** and **digital arteries** to the toes arise from the plantar arch.

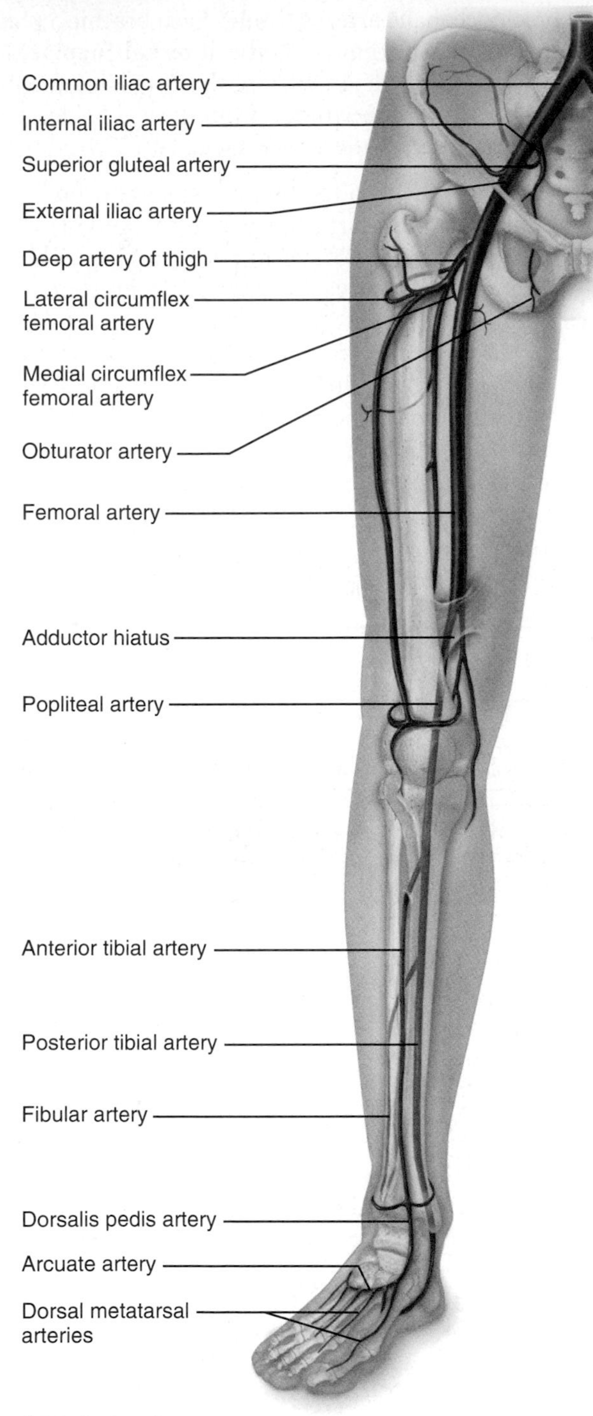

(b) Anterior view

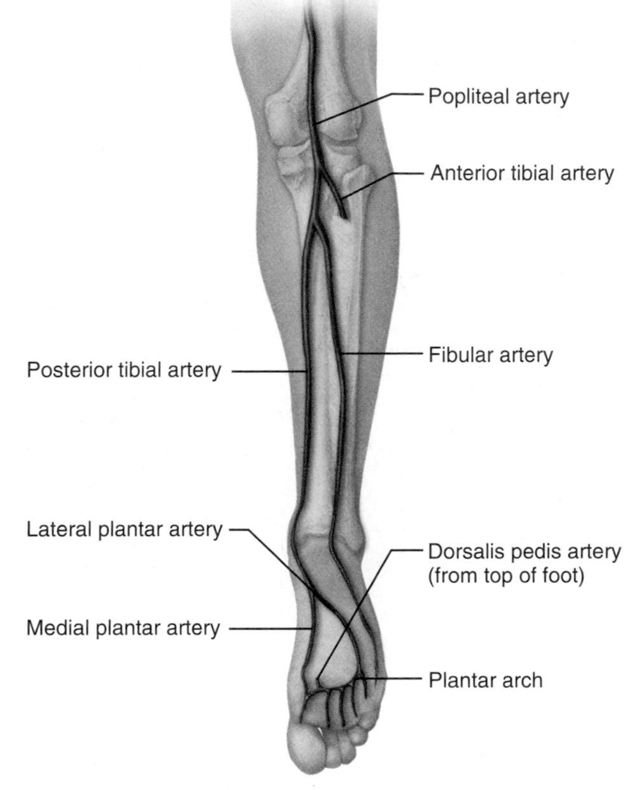

(c) Posterior view

Figure 19.25 *(continued)*

CHECK YOUR UNDERSTANDING

18. You are assessing the circulation in the leg of a diabetic patient at the clinic. Name the artery you palpate in each of these three locations: behind the knee, behind the medial malleolus of the tibia, on the dorsum of the foot.

For answers, see Appendix G.

19

TABLE 19.9 The Venae Cavae and the Major Veins of the Systemic Circulation

In our survey of the systemic veins, the major tributaries (branches) of the venae cavae are noted first in **Figure 19.26**, followed by a description in Tables 19.10 through 19.13 of the venous pattern of the various body regions. Because veins run toward the heart, the most distal veins are named first and those closest to the heart last. Deep veins generally drain the same areas served by their companion arteries, so they are not described in detail.

Description and Areas Drained

Superior vena cava. This great vein receives systemic blood draining from all areas superior to the diaphragm, except the heart wall. It is formed by the union of the **right** and **left brachiocephalic veins** and empties into the right atrium (Figure 19.26b). Notice that there are two brachiocephalic veins, but only one brachiocephalic artery (trunk). Each brachiocephalic vein is formed by the joining of the **internal jugular** and **subclavian veins** on its side. In most of the flowcharts that follow, only the vessels draining blood from the right side of the body are shown (except for the azygos circulation of the thorax).

Inferior vena cava. The widest blood vessel in the body, this vein returns blood to the heart from all body regions below the diaphragm. The abdominal aorta lies directly to its left. The distal end of the inferior vena cava is formed by the junction of the paired **common iliac veins** at L_5. From this point, it courses superiorly along the anterior aspect of the spine, receiving venous blood draining from the abdominal walls, gonads, and kidneys. Immediately above the diaphragm, the inferior vena cava ends as it enters the inferior aspect of the right atrium.

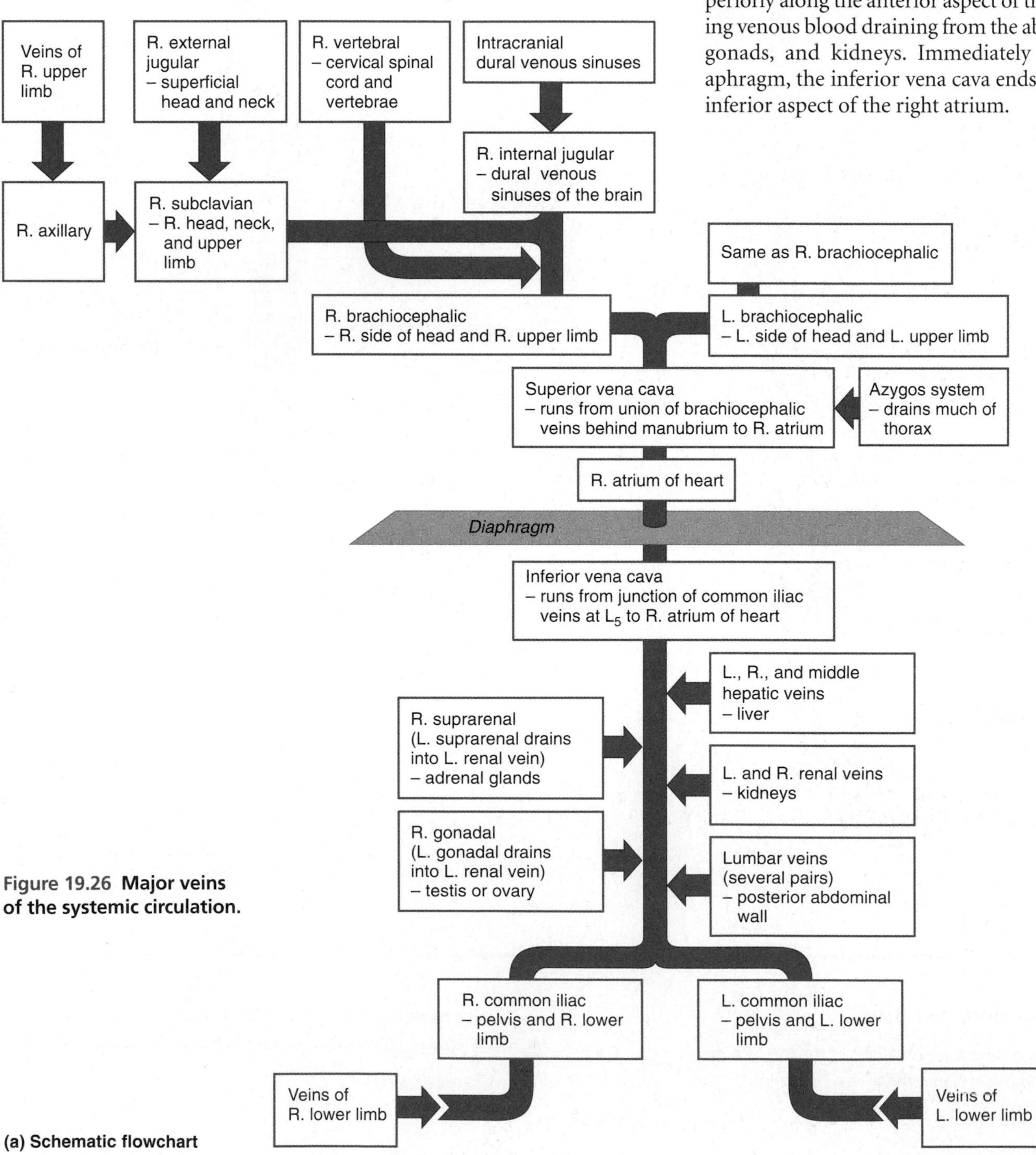

Figure 19.26 Major veins of the systemic circulation.

(a) Schematic flowchart

TABLE 19.9 *(continued)*

Figure 19.26 *(continued)*

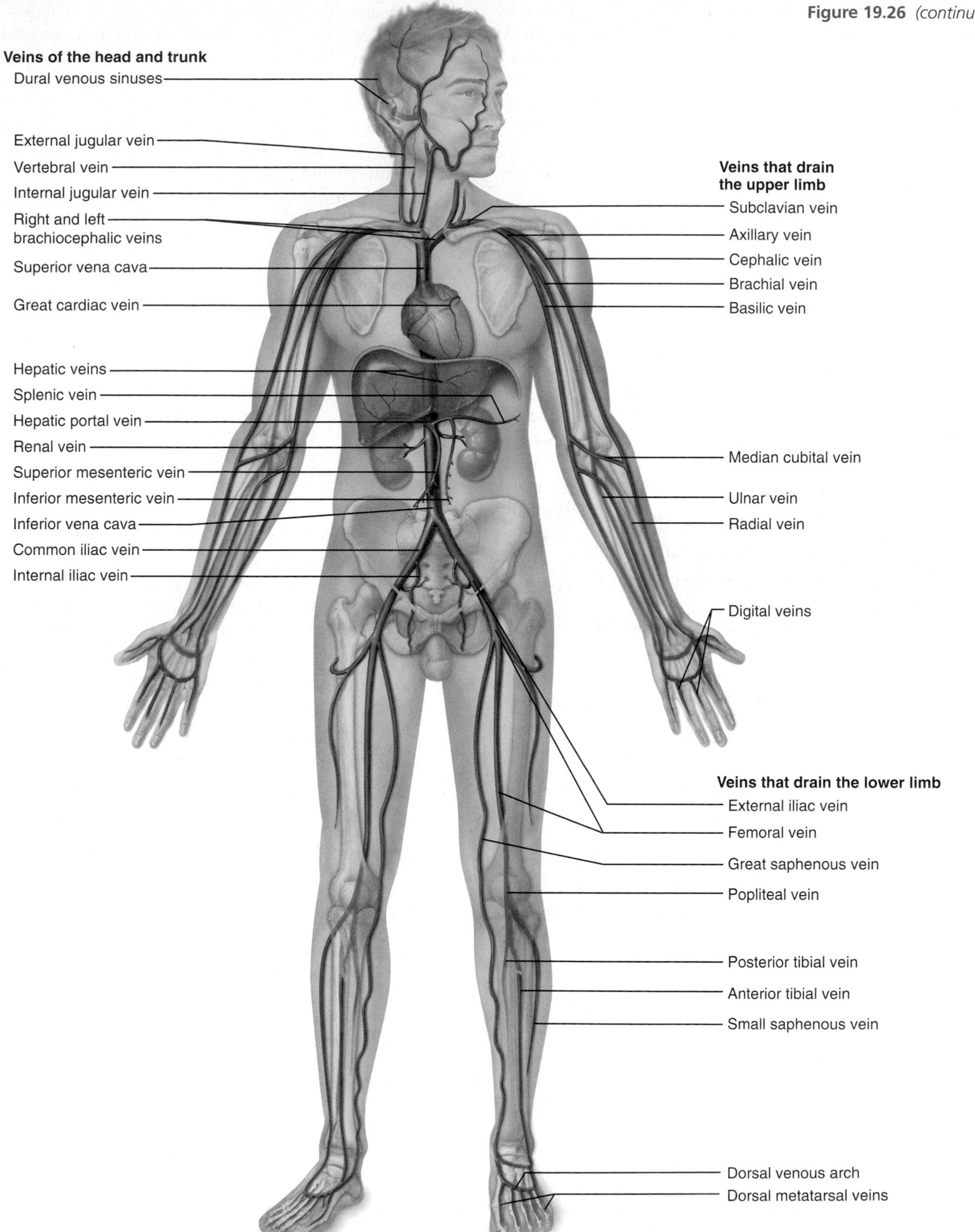

(b) Illustration, anterior view. The vessels of the pulmonary circulation are not shown.

TABLE 19.10 Veins of the Head and Neck

Most blood draining from the head and neck is collected by three pairs of veins: the external jugular veins, which empty into the subclavians; the internal jugular veins; and the vertebral veins, which drain into the brachiocephalic vein **(Figure 19.27a)**. Although most extracranial veins have the same names as the extracranial arteries, their courses and interconnections differ substantially.

Most veins of the brain drain into the **dural venous sinuses**, an interconnected series of enlarged chambers located between the dura mater layers. The **superior** and **inferior sagittal sinuses** are in the falx cerebri, which dips down between the cerebral hemispheres. The inferior sagittal sinus drains into the **straight sinus** posteriorly (Figure 19.27a and c). The superior sagittal and straight sinuses then empty into the **transverse sinuses**, which run in shallow grooves on the internal surface of the occipital bone. These drain into the S-shaped **sigmoid sinuses**, which become the *internal jugular veins* as they leave the skull through the jugular foramen. The **cavernous sinuses**, which flank the sphenoid body, receive venous blood from the **ophthalmic veins** of the orbits and the facial veins, which drain the nose and upper lip area. The internal carotid artery and cranial nerves III, IV, VI, and part of V, all run *through* the cavernous sinus on their way to the orbit and face.

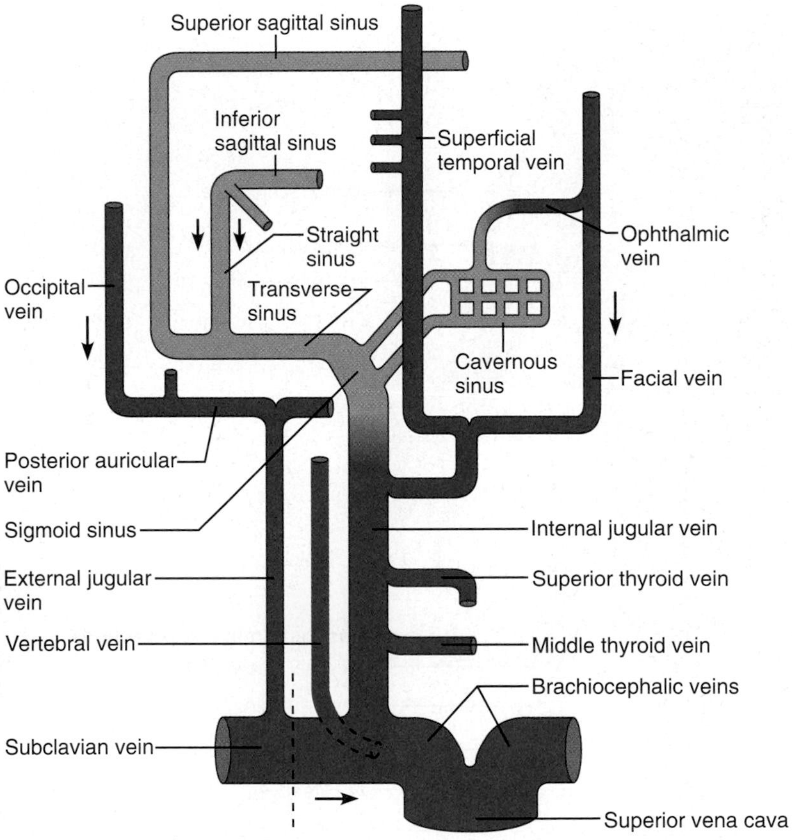

(a) Schematic flowchart

Figure 19.27 Venous drainage of the head, neck, and brain.

Description and Area Drained

External jugular veins. The right and left external jugular veins drain superficial scalp and face structures served by the external carotid arteries. However, their tributaries anastomose frequently, and some of the superficial drainage from these regions enters the internal jugular veins as well. As the external jugular veins descend through the lateral neck, they pass obliquely over the sternocleidomastoid muscles and then empty into the subclavian veins.

Vertebral veins. Unlike the vertebral arteries, the vertebral veins do not serve much of the brain. Instead they drain the cervical vertebrae, the spinal cord, and some small neck muscles. They run inferiorly through the transverse foramina of the cervical vertebrae and join the brachiocephalic veins at the root of the neck.

Internal jugular veins. The paired internal jugular veins, which receive the bulk of blood draining from the brain, are the largest of the paired veins draining the head and neck. They arise from the dural venous sinuses, exit the skull via the *jugular foramina*, and then descend through the neck alongside the internal carotid arteries. As they move inferiorly, they receive blood from some of the deep veins of the face and neck—branches of the **facial** and **superficial temporal veins** (Figure 19.27b). At the base of the neck, each internal jugular vein joins the subclavian vein on its own side to form a brachiocephalic vein. As already noted, the two brachiocephalic veins unite to form the superior vena cava.

CHECK YOUR UNDERSTANDING

19. In what important way does the area drained by the vertebral veins differ from the area served by the vertebral arteries?

20. Which veins drain the dural venous sinuses and where do these veins terminate?

For answers, see Appendix G.

TABLE 19.10 *(continued)*

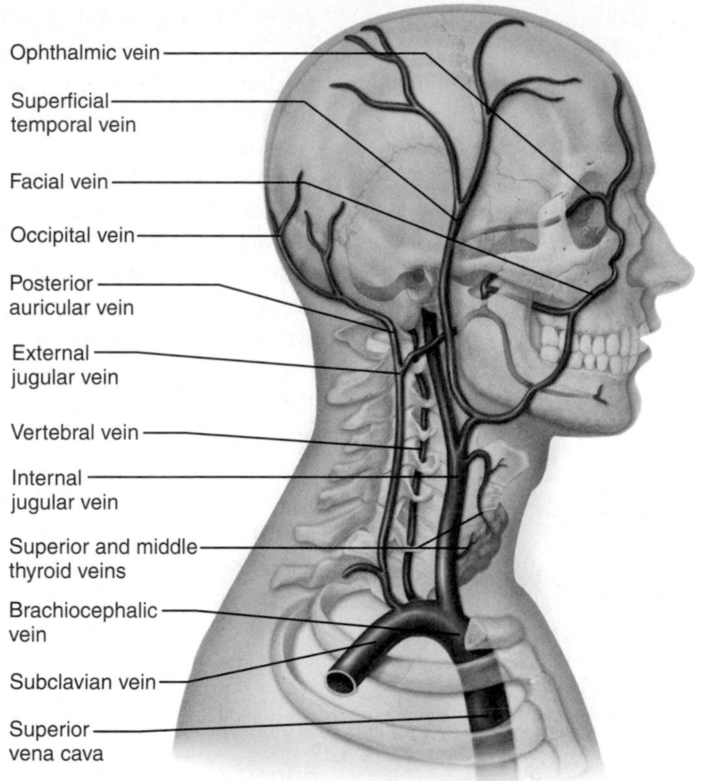

(b) Veins of the head and neck, right superficial aspect

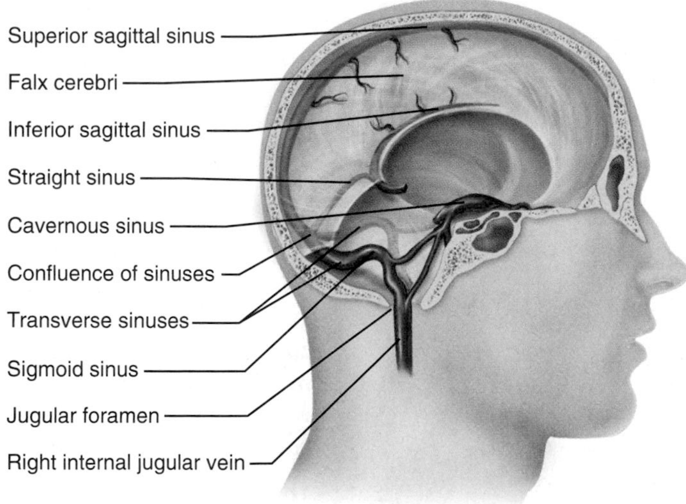

(c) Dural venous sinuses of the brain

Figure 19.27 *(continued)*

TABLE 19.11 Veins of the Upper Limbs and Thorax

The deep veins of the upper limbs follow the paths of their companion arteries and have the same names **(Figure 19.28a)**. However, except for the largest, most are paired veins that flank their artery. The superficial veins of the upper limbs are larger than the deep veins and are easily seen just beneath the skin. The median cubital vein, crossing the anterior aspect of the elbow, is commonly used to obtain blood samples or administer intravenous medications.

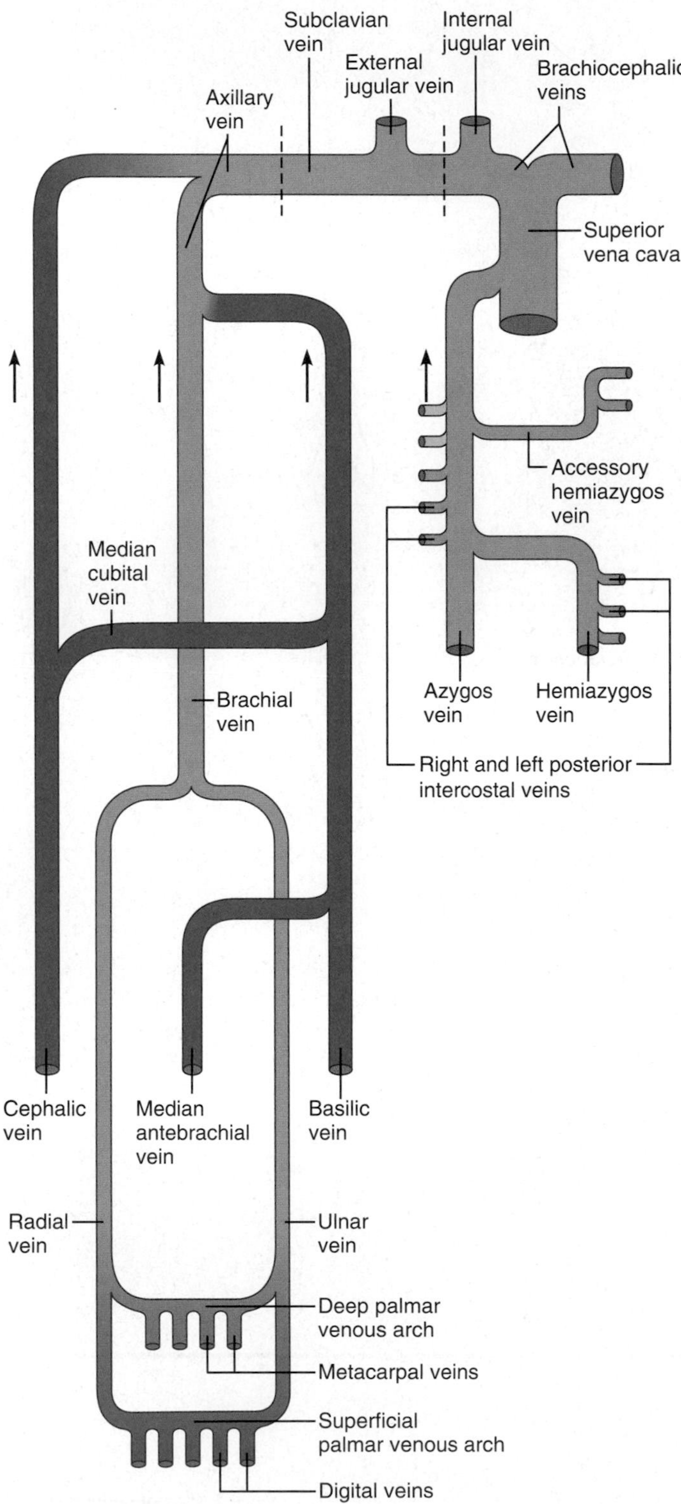

(a) Schematic flowchart

Blood draining from the mammary glands and the first two to three intercostal spaces enters the **brachiocephalic veins.** However, the vast majority of thoracic tissues and the thorax wall are drained by a complex network of veins called the **azygos system** (az′ĭ-gos). The branching nature of the azygos system provides a collateral circulation for draining the abdominal wall and other areas served by the inferior vena cava, and there are numerous anastomoses between the azygos system and the inferior vena cava.

Description and Areas Drained

Deep Veins of the Upper Limbs

The most distal deep veins of the upper limb are the radial and ulnar veins. The deep and superficial **palmar venous arches** of the hand empty into the **radial** and **ulnar veins** of the forearm, which then unite to form the **brachial vein** of the arm. As the brachial vein enters the axilla, it becomes the **axillary vein,** which becomes the **subclavian vein** at the level of the first rib.

Superficial Veins of the Upper Limbs

The superficial venous system begins with the **dorsal venous arch** (not illustrated), a plexus of superficial veins in the dorsum of the hand. In the distal forearm, this plexus drains into three major superficial veins—the cephalic and basilic veins and the median antebrachial vein—which anastomose frequently as they course upward (Figure 19.28b). The **cephalic vein** bends around the radius as it travels superiorly and then continues up the lateral superficial aspect of the arm to the shoulder, where it runs in the groove between the deltoid and pectoralis muscles to join the axillary vein. The **basilic vein** courses along the posteromedial aspect of the forearm, crosses the elbow, and then takes a deep course. In the axilla, it joins the brachial vein, forming the axillary vein. At the anterior aspect of the elbow, the **median cubital vein** connects the basilic and cephalic veins. The **median antebrachial vein** lies between the radial and ulnar veins in the forearm and terminates (variably) at the elbow by entering either the basilic or the cephalic vein.

The Azygos System

The azygos system consists of the following vessels, which flank the vertebral column laterally.

Azygos vein. Located against the right side of the vertebral column, the **azygos vein** (*azygos* = unpaired) originates in the abdomen, from the **right ascending lumbar vein** that drains most of the right abdominal cavity wall and from the **right posterior intercostal veins** (except the first) that drain the chest muscles. At the T_4 level, it arches over the great vessels that run to the right lung and empties into the superior vena cava.

TABLE 19.11 *(continued)*

Hemiazygos vein (hĕ′me-a-zi′gus; "half the azygos"). This vessel ascends on the left side of the vertebral column. Its origin, from the **left ascending lumbar vein** and the lower (9th–11th) **posterior intercostal veins,** mirrors that of the inferior portion of the azygos vein on the right. About midthorax, the hemiazygos vein passes in front of the vertebral column and joins the azygos vein.

Accessory hemiazygos vein. The accessory hemiazygos completes the venous drainage of the left (middle) thorax and can be thought of as a superior continuation of the hemiazygos vein. It receives blood from the 4th–8th posterior intercostal veins and then crosses to the right to empty into the azygos vein. Like the azygos, it receives oxygen-poor systemic blood from the bronchi of the lungs (*bronchial veins*).

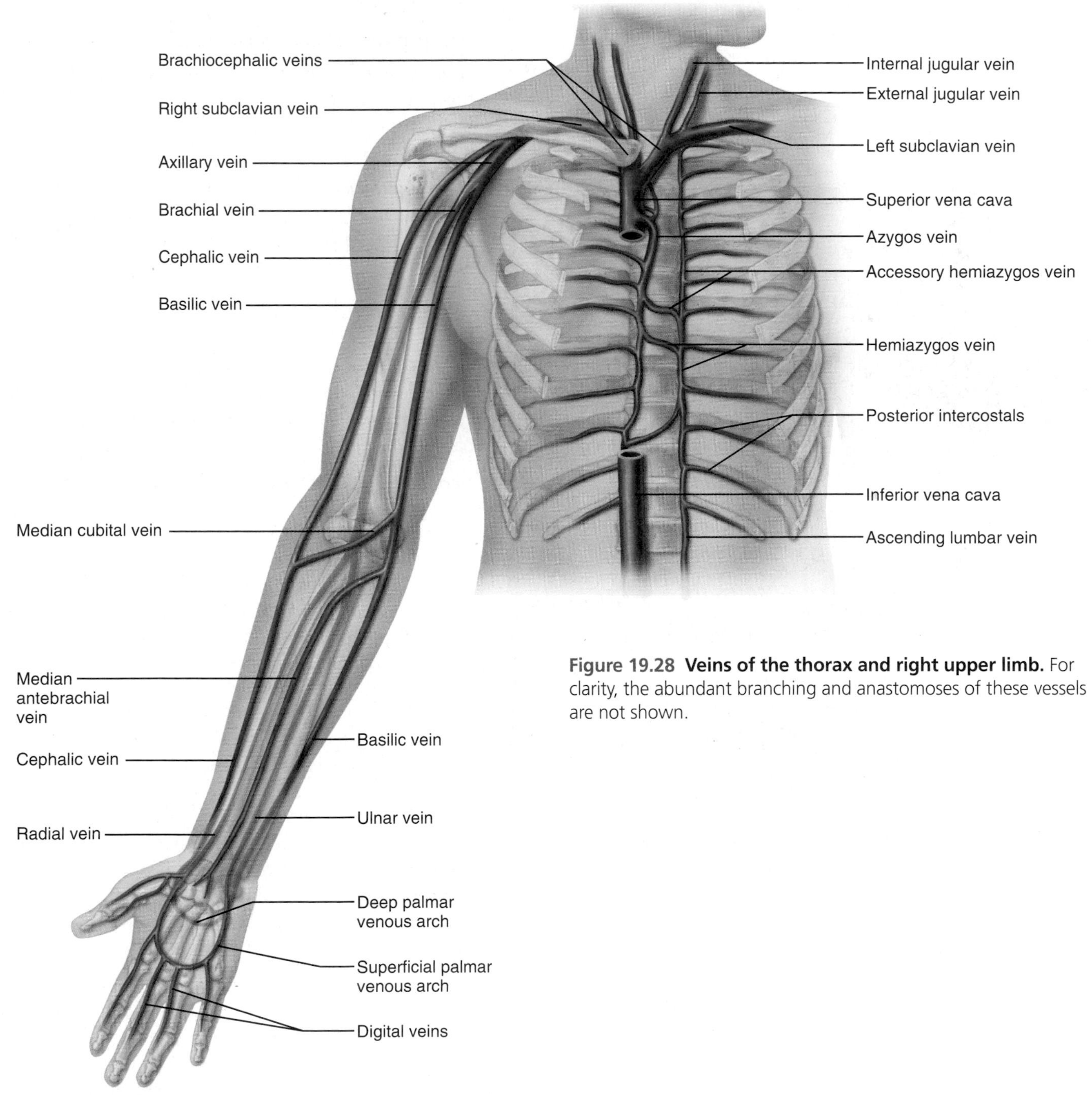

Figure 19.28 Veins of the thorax and right upper limb. For clarity, the abundant branching and anastomoses of these vessels are not shown.

TABLE 19.12 Veins of the Abdomen

Blood draining from the abdominopelvic viscera and abdominal walls is returned to the heart by the **inferior vena cava** **(Figure 19.29a)**. Most of its venous tributaries have names that correspond to the arteries serving the abdominal organs.

Veins draining the digestive viscera empty into a common vessel, the *hepatic portal vein,* which transports this venous blood into the liver before it is allowed to enter the major systemic circulation via the hepatic veins (Figure 19.29b). Such a venous system—veins connecting two capillary beds together—is called a *portal system* and always serves very specific needs. The **hepatic portal system** carries nutrient-rich blood (which may also contain toxins and microorganisms) from the digestive organs to the liver; where it can be "treated" before it reaches the rest of the body. As the blood percolates slowly through the liver sinusoidal capillaries, hepatocytes process nutrients and toxins, and phagocytic cells rid the blood of bacteria and other foreign matter. The veins of the abdomen are listed in inferior to superior order.

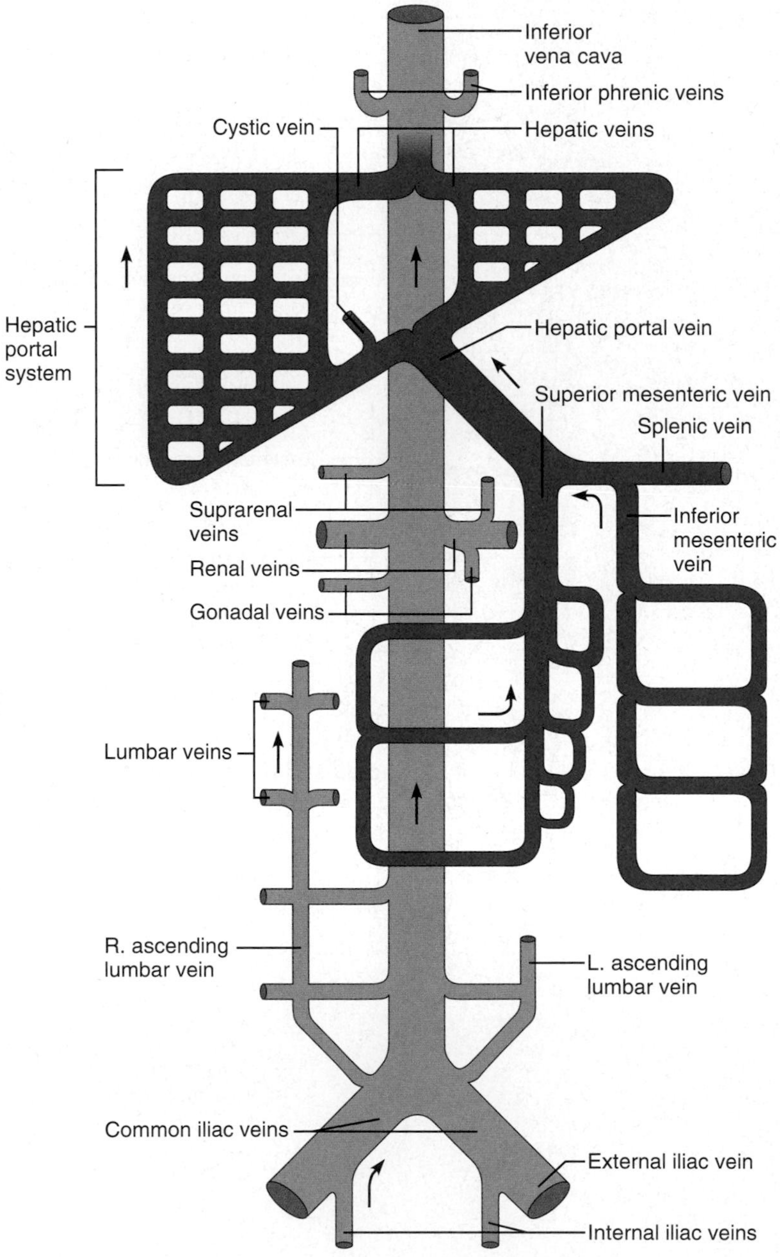

(a) Schematic flowchart.

Description and Areas Drained

Lumbar veins. Several pairs of lumbar veins drain the posterior abdominal wall. They empty both directly into the inferior vena cava and into the ascending lumbar veins of the azygos system of the thorax.

Gonadal (testicular or ovarian) veins. The right gonadal vein drains the ovary or testis on the right side of the body and empties into the inferior vena cava. The left member drains into the left renal vein superiorly.

Renal veins. The right and left renal veins drain the kidneys.

Suprarenal veins. The right suprarenal vein drains the adrenal gland on the right and empties into the inferior vena cava. The left suprarenal vein drains into the left renal vein.

Hepatic portal system. Like all portal systems, the hepatic portal system is a series of vessels in which two separate capillary beds lie between the arterial supply and the final venous drainage. In this case, the first capillary beds are in the stomach and intestines and drain into tributaries of the **hepatic portal vein**, which brings them to the second capillary bed in the liver. The short hepatic portal vein begins at the L_2 level. Numerous tributaries from the stomach and pancreas contribute to the hepatic portal system (Figure 19.29c), but the major vessels are as follows:

- **Superior mesenteric vein:** Drains the entire small intestine, part of the large intestine (ascending and transverse regions), and stomach.
- **Splenic vein:** Collects blood from the spleen, parts of the stomach and pancreas, and then joins the superior mesenteric vein to form the hepatic portal vein.
- **Inferior mesenteric vein:** Drains the distal portions of the large intestine and rectum and joins the splenic vein just before that vessel unites with the superior mesenteric vein to form the hepatic portal vein.

Hepatic veins. The right, left, and middle hepatic veins carry venous blood from the liver to the inferior vena cava.

Cystic veins. The cystic veins drain the gallbladder and join the portal veins in the liver.

Inferior phrenic veins. The inferior phrenic veins drain the inferior surface of the diaphragm.

TABLE 19.12 *(continued)*

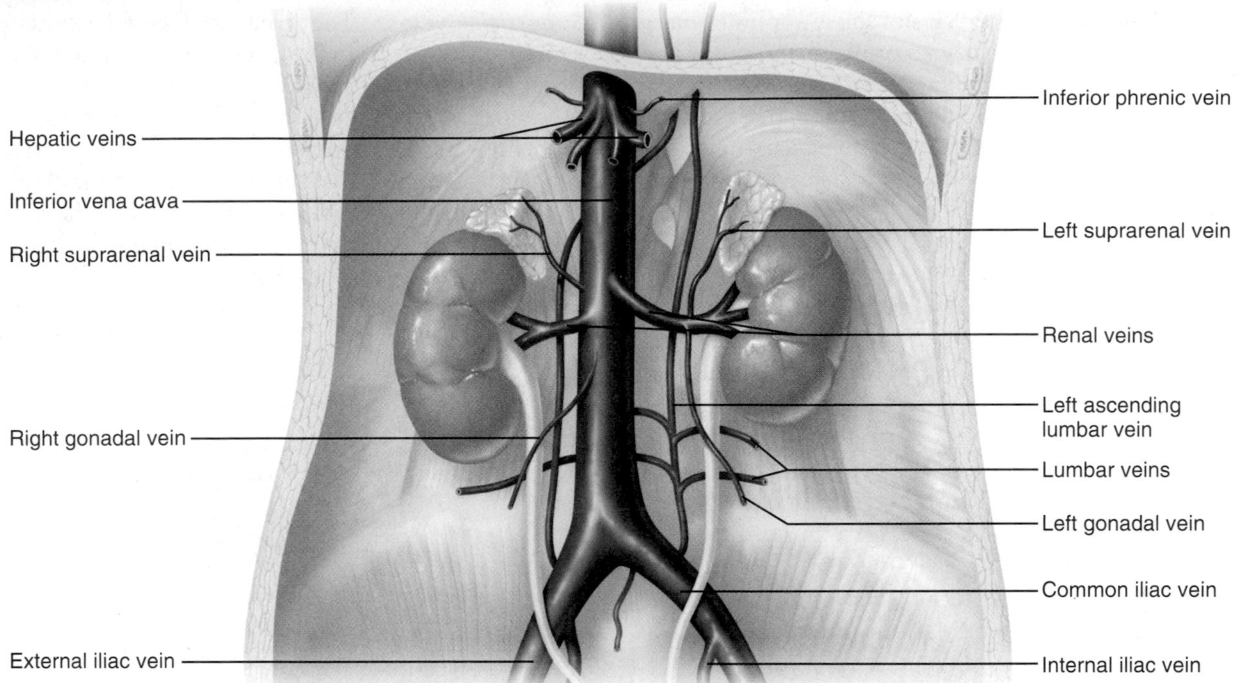

(b) Tributaries of the inferior vena cava. Venous drainage of abdominal organs not drained by the hepatic portal vein.

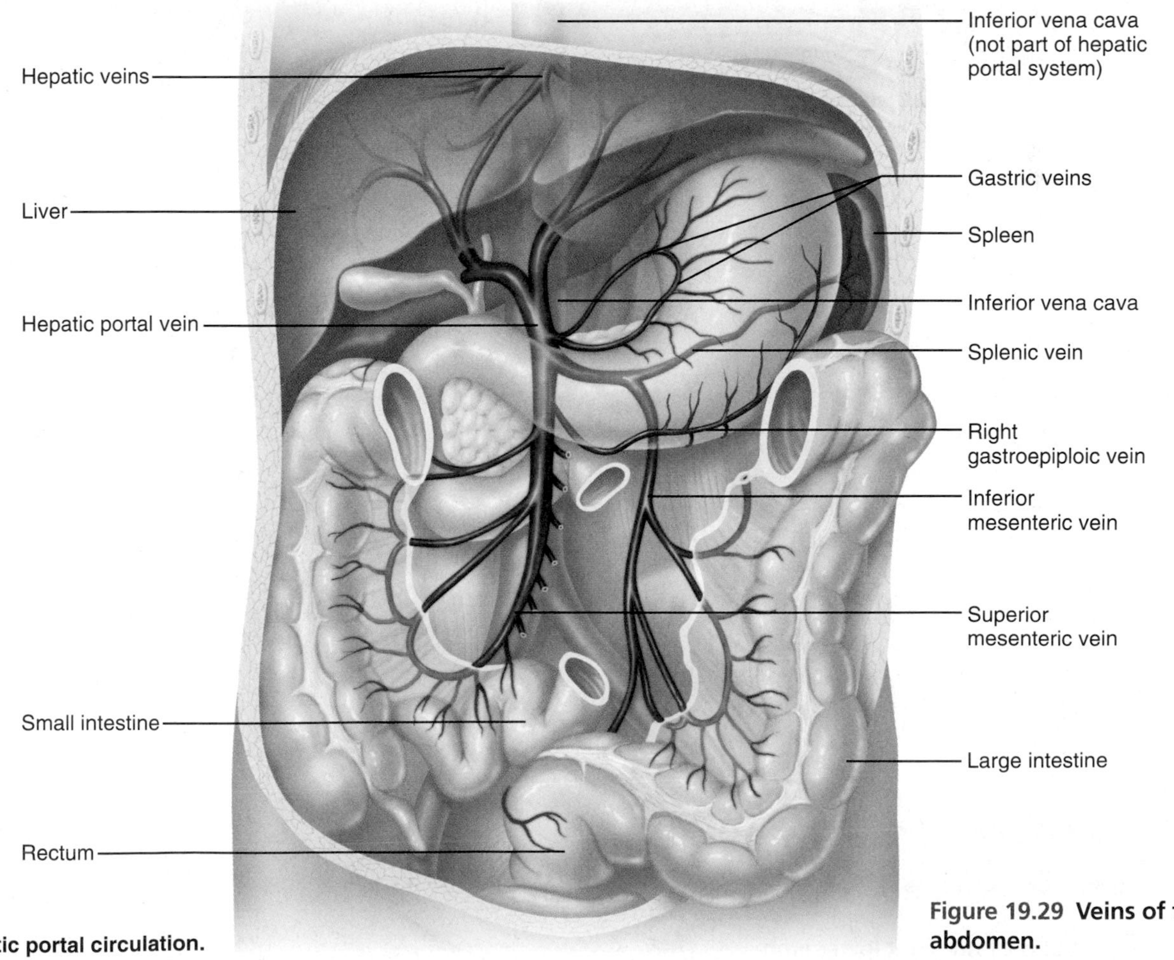

(c) The hepatic portal circulation.

Figure 19.29 Veins of the abdomen.

TABLE 19.13 Veins of the Pelvis and Lower Limbs

As in the upper limbs, most deep veins of the lower limbs have the same names as the arteries they accompany and many are double. The two superficial saphenous veins (great and small) are poorly supported by surrounding tissues, and are common sites of varicosities. The great saphenous (*saphenous* = obvious) vein is frequently excised and used as a coronary bypass vessel.

Description and Areas Drained

Deep veins. After being formed by the union of the **medial** and **lateral plantar veins**, the **posterior tibial vein** ascends deep in the calf muscle and receives the **fibular (peroneal) vein** (Figure 19.30). The **anterior tibial vein**, which is the superior continuation of the **dorsalis pedis vein** of the foot, unites at the knee with the posterior tibial vein to form the **popliteal vein**, which crosses the back of the knee. As the popliteal vein emerges from the knee, it becomes the **femoral vein**, which drains the deep structures of the thigh. The femoral vein becomes the **external iliac vein** as it enters the pelvis. In the pelvis, the external iliac vein unites with the **internal iliac vein** to form the **common iliac vein**. The distribution of the internal iliac veins parallels that of the internal iliac arteries.

Superficial veins. The **great** and **small saphenous veins** (sah-fe′nus) issue from the **dorsal venous arch** of the foot (Figure 19.30b and c). These veins anastomose frequently with each other and with the deep veins along their course. The great saphenous vein is the longest vein in the body. It travels superiorly along the medial aspect of the leg to the thigh, where it empties into the femoral vein just distal to the inguinal ligament. The small saphenous vein runs along the lateral aspect of the foot and then through the deep fascia of the calf muscles, which it drains. At the knee, it empties into the popliteal vein.

(a) Schematic flowchart of the anterior and posterior veins

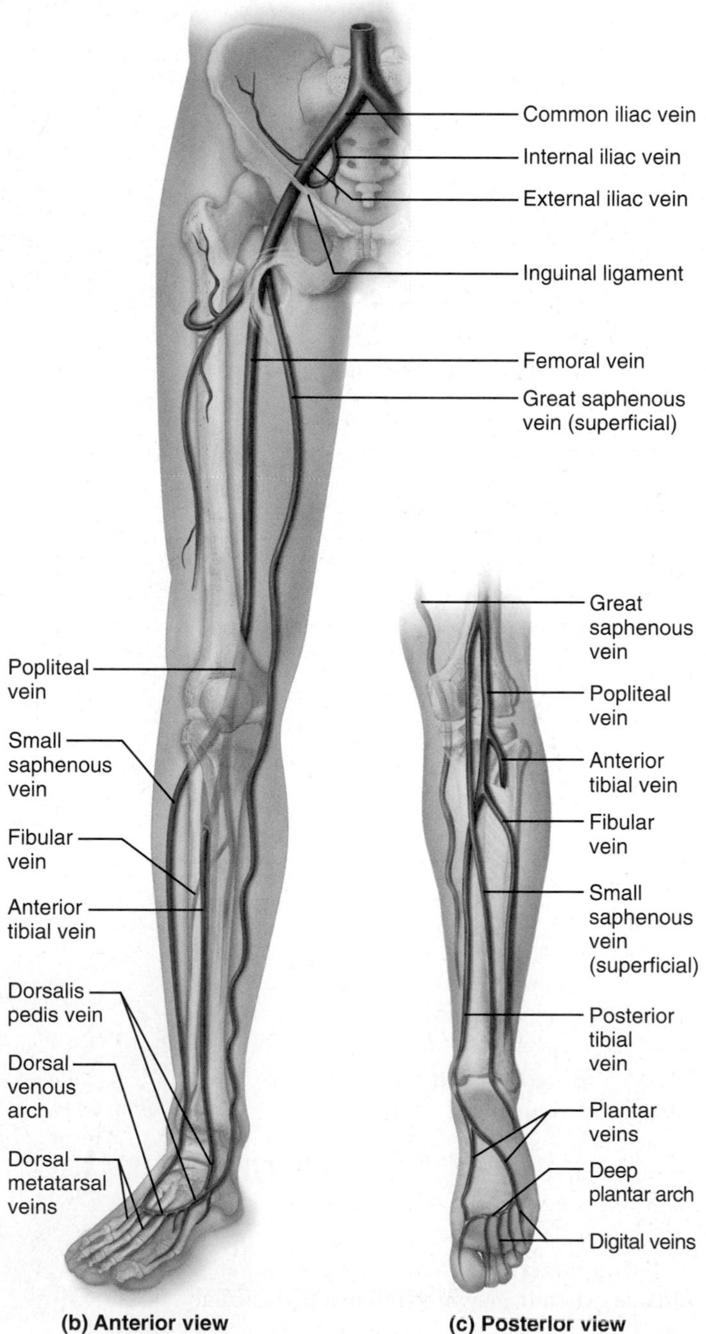

(b) Anterior view **(c) Posterior view**

Figure 19.30 Veins of the right lower limb.

CHECK YOUR UNDERSTANDING

21. What is a portal system? What is the function of the hepatic portal system?

22. Name the leg veins that often become varicosed.

For answers, see Appendix G.

Developmental Aspects of Blood Vessels

- Explain how blood vessels develop in the fetus.
- Provide examples of changes that often occur in blood vessels as a person ages.

The endothelial lining of blood vessels is formed by mesodermal cells, which collect in little masses called **blood islands** throughout the microscopic embryo. These then form fragile sprouting extensions that reach toward one another and toward the forming heart to lay down the rudimentary vascular tubes. Meanwhile, adjacent mesenchymal cells, stimulated by platelet-derived growth factor, surround the endothelial tubes, forming the stabilizing muscular and fibrous coats of the vessel walls.

How do blood vessels "know" where to grow? Many blood vessels simply follow the same guidance cues that nerves follow, which is why forming vessels often snuggle closely to nerves. Whether a vessel becomes an artery or a vein depends upon the local concentration of a differentiation factor called vascular endothelial growth factor. As noted in Chapter 18, the heart is pumping blood through the rudimentary vascular system by the fourth week of development.

In addition to the fetal shunts that bypass the nonfunctional lungs (the *foramen ovale* and *ductus arteriosus*), other vascular modifications are found in the fetus. A special vessel, the *ductus venosus*, largely bypasses the liver. Also important are the *umbilical vein* and *arteries*, large vessels that circulate blood between the fetal circulation and the placenta where gas and nutrient exchanges occur with the mother's blood (see Chapter 28). Once the fetal circulatory pattern is laid down, few vascular changes occur until birth, when the umbilical vessels and shunts are occluded.

In contrast to congenital heart diseases, congenital vascular problems are rare, and blood vessels are remarkably trouble-free during youth. Vessel formation occurs as needed to support body growth, wound healing, and to rebuild vessels lost each month during a woman's menstrual cycle. As we age, signs of vascular disease begin to appear. In some, the venous valves weaken, and purple, snakelike varicose veins appear. In others, more insidious signs of inefficient circulation appear: tingling in the fingers and toes and cramping of muscles.

Although the degenerative process of atherosclerosis begins in youth, its consequences are rarely apparent until middle to old age, when it may precipitate a myocardial infarct or stroke. Until puberty, the blood vessels of boys and girls look alike, but from puberty to about age 45, women have strikingly less atherosclerosis than men because of the protective effects of estrogen, the beneficial effects of which are well documented. By enhancing nitric oxide production, inhibiting endothelin release, and blocking voltage-gated Ca^{2+} channels, estrogen's effect on blood vessels is to reduce resistance to blood flow. Estrogen also stimulates the liver to produce enzymes that speed up catabolism of LDLs and increase the production of HDLs, thus reducing the risk of atherosclerosis (see *A Closer Look*).

Between the ages of 45 and 65, when estrogen production wanes in women, this "gap" between the sexes closes, and males and females above age 65 are equally at risk for cardiovascular disease. You might expect that giving postmenopausal women supplementary estrogen would maintain this protective effect. Surprisingly, clinical trials have shown that this is not the case.

Blood pressure changes with age. In a newborn baby, arterial pressure is about 90/55. Blood pressure rises steadily during childhood to finally reach the adult value (120/80). In old age, normal blood pressure averages 150/90, which is hypertensive in younger people. After age 40, the incidence of hypertension increases dramatically. Unlike atherosclerosis, which is increasingly important in old age, hypertension claims many youthful victims and is the single most important cause of sudden cardiovascular death in men in their early 40s and 50s.

At least some vascular disease is a product of our modern technological culture. "Blessed" with high-protein, lipid-rich diets, empty-calorie snacks, energy-saving devices, and high-stress jobs, many of us are struck down prematurely. Cardiovascular disease can be prevented somewhat by diet modifications, regular aerobic exercise, and eliminating cigarette smoking. Poor diet, lack of exercise, and smoking are probably more detrimental to your blood vessels than aging itself could ever be!

CHECK YOUR UNDERSTANDING

23. List three differences between systemic arteries and veins with respect to their general pathways and courses.

24. Name three fetal shunts that are occluded shortly after birth. Which structure is bypassed by each shunt?

25. List three common age-related vascular problems.

For answers, see Appendix G.

Now that we have described the structure and function of blood vessels, our survey of the cardiovascular system is complete. The pump, the plumbing, and the circulating fluid form a dynamic organ system that ceaselessly services every other organ system of the body, as summarized in *Making Connections* on pp. 746–747. However, our study of the circulatory system is still unfinished because we have yet to examine the lymphatic system, which acts with the cardiovascular system to ensure continuous circulation and to provide sites from which lymphocytes can police the body and provide for immunity. These are the topics of Chapter 20.

System Connections

Homeostatic Interrelationships Between the Cardiovascular System and Other Body Systems

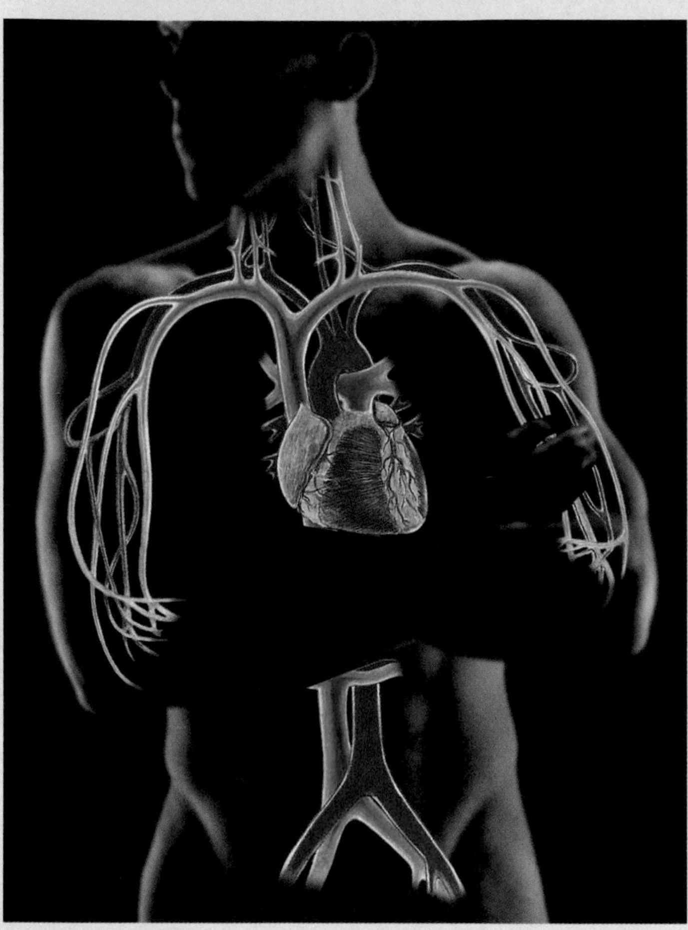

Integumentary System

- The cardiovascular system delivers oxygen and nutrients; carries away wastes
- The skin vasculature is an important blood reservoir and provides a site for heat loss from the body

Skeletal System

- The cardiovascular system delivers oxygen and nutrients; carries away wastes
- Bones are the sites of hematopoiesis; protect cardiovascular organs by enclosure; and provide a calcium depot

Muscular System

- The cardiovascular system delivers oxygen and nutrients; carries away wastes
- The muscle "pump" aids venous return; aerobic exercise enhances cardiovascular efficiency and helps prevent atherosclerosis

Nervous System

- The cardiovascular system delivers oxygen and nutrients; carries away wastes
- The ANS regulates cardiac rate and force; sympathetic division maintains blood pressure and controls blood flow to skin for thermoregulation

Endocrine System

- The cardiovascular system delivers oxygen and nutrients; carries away wastes; blood serves as a transport vehicle for hormones
- Various hormones influence blood pressure (epinephrine, ANP, angiotensin II, thyroxine, ADH); estrogen maintains vascular health in women

Lymphatic System/Immunity

- The cardiovascular system delivers oxygen and nutrients to lymphatic organs, which house immune cells; provides transport medium for lymphocytes and antibodies; carries away wastes
- The lymphatic system picks up leaked fluid and plasma proteins and returns them to the cardiovascular system; immune cells protect cardiovascular organs from specific pathogens

Respiratory System

- The cardiovascular system delivers oxygen and nutrients; carries away wastes
- The respiratory system carries out gas exchange: loads oxygen and unloads carbon dioxide from the blood; respiratory "pump" aids venous return

Digestive System

- The cardiovascular system delivers oxygen and nutrients; carries away wastes
- The digestive system provides nutrients to the blood including iron and B vitamins essential for RBC (and hemoglobin) formation

Urinary System

- The cardiovascular system delivers oxygen and nutrients; carries away wastes; blood pressure allows filtration in the kidneys
- The urinary system helps regulate blood volume and pressure by altering urine volume and releasing renin

Reproductive System

- The cardiovascular system delivers oxygen and nutrients; carries away wastes
- Estrogen maintains vascular and osseous health in women

Closer Connections

The Cardiovascular System and Interrelationships With the Muscular, Nervous, and Urinary Systems

The cardiovascular system is the "king of systems." No body system can live without the blood that surges ceaselessly through cardiovascular channels. Likewise, it is nearly impossible to find a system that does not influence the cardiovascular system in some way. The respiratory and digestive systems enrich blood with oxygen and nutrients, respectively, and in turn take a share of the various riches the blood has to offer. By returning leaked plasma fluid to the vascular system, the lymphatic system helps to keep those vessels filled with blood so that circulation is possible. However, the three cardiovascular system partnerships that we will look at more closely here are those it has with the muscular, nervous, and urinary systems.

Muscular System

Talk about one hand washing the other—that's a pretty fair analogy for the interaction between the cardiovascular and muscular systems. Muscles cramp and become nonfunctional when deprived of an adequate supply of oxygen-rich blood, and their capillary supply expands (or atrophies) along with changes in muscle mass. The arterioles of skeletal muscle even have special beta-adrenergic receptors so that they can be dilated by circulating epinephrine when most other body arterioles are responding to vasoconstrictor "orders." When muscles are active and healthy, so too is the cardiovascular system. Without exercise, the heart weakens and loses mass; but when we exercise aerobically, the heart increases in size and strength. Heart rate goes down as stroke volume rises, and a lower HR means that the heart relaxes more and beats hundreds of thousands times less in a lifetime. Aerobic exercise also enhances HDL blood levels and reduces LDL levels—helping to clear fatty deposits from the vascular walls and deferring atherosclerosis, hypertension, and heart disease. Not a bad trade!

Nervous System

The brain must have absolutely continuous oxygen and glucose delivery; thus it should come as no surprise that the brain has the body's most precise autoregulatory circulation mechanism. Acting on acutely sensitive arterioles, this mechanism protects the brain. For this we can be profoundly thankful, for when neurons die, an important part of us dies as well. The cardiovascular system is just as dependent on the nervous system. Although an intrinsic pacemaker sets sinus rhythm, it does not even begin to adapt to stimuli that mobilize the cardiovascular system to peak efficiency in maintaining blood pressure during position changes and delivering blood faster during times of stress. This is the job of the autonomic nervous system, which initiates reflexes as necessary to increase or decrease cardiac output and peripheral resistance, and to redirect blood from one organ to others to serve specific needs and protect vital organs.

Urinary System

Like the lymphatic system, the urinary system (primarily the kidneys) helps maintain blood volume and circulatory dynamics, but in a much more complicated way. The kidneys use blood pressure (courtesy of the cardiovascular system) to form the filtrate which the kidney tubule cells then process. Essentially this processing involves reclaiming needed nutrients and water while allowing metabolic waste and excess ions (including H^+) to leave the body in urine. As a result, blood is continually refreshed and blood composition and volume are carefully regulated. So dedicated are the kidneys to preserving blood volume that urine output stops entirely when blood volume is severely depressed. On the other hand, the kidneys can increase systemic BP by releasing renin when the blood pressure becomes inadequate.

Clinical Connections

Cardiovascular System

Case study: Mr. Hutchinson, another middle-aged victim of the collision on Route 91, has a tourniquet around his thigh when admitted in an unconscious state to Noble Hospital. The emergency technician who brings him in states that his right lower limb was pinned beneath the bus for at least 30 minutes. He is immediately scheduled for surgery. Admission notes include the following:

- Multiple contusions of lower limbs
- Compound fracture of the right tibia; bone ends covered with sterile gauze
- Right leg blanched and cold, no pulse
- Blood pressure 90/48; pulse 140/min and thready; patient diaphoretic (sweaty)

1. Relative to what you have learned about tissue requirements for oxygen, what is the condition of the tissues in the right lower limb?
2. Will the fracture be attended to, or will Mr. Hutchinson's other homeostatic needs take precedence? Explain your answer choice and predict his surgical treatment.
3. What do you conclude regarding Mr. Hutchinson's cardiovascular measurements (pulse and BP), and what measures do you expect will be taken to remedy the situation before commencing surgery?

(Answers in Appendix G)

RELATED CLINICAL TERMS

Aneurysm (an′u-rizm; *aneurysm* = a widening) A balloonlike outpocketing of an artery wall that places the artery at risk for rupture; most often reflects gradual weakening of the artery by chronic hypertension or arteriosclerosis. The most common sites of aneurysm formation are the abdominal aorta and arteries feeding the brain and kidneys.

Angiogram (an′je-o-gram″; *angio* = a vessel; *gram* = writing) Diagnostic technique involving the infusion of a radiopaque substance into the circulation for X-ray examination of specific blood vessels. The major technique for diagnosing coronary artery occlusion and risk of a heart attack.

Diuretic (*diure* = urinate) A chemical that promotes urine formation, thus reducing blood volume. Diuretic drugs are frequently prescribed to manage hypertension.

Phlebitis (flĕ-bi′tis; *phleb* = vein; *itis* = inflammation) Inflammation of a vein accompanied by painful throbbing and redness of the skin over the inflamed vessel. It is most often caused by bacterial infection or local physical trauma.

Phlebotomy (flĕ-bot′o-me; *tomy* = cut) A venous incision or puncture made for the purpose of withdrawing blood or bloodletting.

Sclerotherapy Procedure used for removing varicose or spider veins. Tiny needles are used to inject hardening agents into the abnormal vein. The vein scars, closes down, and is absorbed by the body.

Thrombophlebitis Condition of undesirable intravascular clotting initiated by a roughening of a venous lining; often follows severe episodes of phlebitis. An ever-present danger is that the clot may detach and form an embolus.

CHAPTER SUMMARY

Media study tools that could provide you additional help in reviewing specific key topics of Chapter 19 are referenced below.

iP = *Interactive Physiology*

PART 1: OVERVIEW OF BLOOD VESSEL STRUCTURE AND FUNCTION

1. Blood is transported throughout the body via a continuous system of blood vessels. Arteries transport blood away from the heart; veins carry blood back to the heart. Capillaries carry blood to tissue cells and are exchange sites.

Structure of Blood Vessel Walls (p. 695)

1. All blood vessels except capillaries have three layers: tunica intima, tunica media, and tunica externa. Capillary walls are composed of the tunica intima only.

Arterial System (pp. 695–698)

1. Elastic (conducting) arteries are the large arteries close to the heart that expand during systole, acting as pressure reservoirs, and then recoil during diastole to keep blood moving. Muscular (distributing) arteries carry blood to specific organs; they are less stretchy and more active in vasoconstriction. Arterioles regulate blood flow into capillary beds.
2. Arteriosclerosis is a degenerative vascular disease that decreases the elasticity of arteries. Atherosclerosis is the formation of fatty subendothelial lesions.

Capillaries (pp. 698–700)

1. Capillaries are microscopic vessels with very thin walls. Most exhibit intercellular clefts, which aid in the exchange between the blood and interstitial fluid.
2. The most permeable capillaries are sinusoids (wide, tortuous channels). Fenestrated capillaries with pores are next most permeable. Least permeable are continuous capillaries, which lack pores.
3. Vascular shunts (metarteriole–thoroughfare channels) connect the terminal arteriole and postcapillary venule at opposite ends of a capillary bed. Most true capillaries arise from and rejoin the shunt channels. The amount of blood flowing into the true capillaries is regulated by precapillary sphincters.

Venous System (pp. 700–701)

1. Veins have comparatively larger lumens than arteries, and a system of valves prevents backflow of blood.
2. Normally most veins are only partially filled; for this reason, they can serve as blood reservoirs.

Vascular Anastomoses (pp. 701–703)

1. The joining together of arteries to provide alternate channels for blood to reach the same organ is called an anastomosis. Vascular anastomoses also form between veins and between arterioles and venules.

iP Cardiovascular System; Topic: Anatomy Review: Blood Vessel Structure and Function, pp. 1–27.

PART 2: PHYSIOLOGY OF CIRCULATION

Introduction to Blood Flow, Blood Pressure, and Resistance (pp. 703–704)

1. Blood flow is the amount of blood flowing through a vessel, an organ, or the entire circulation in a given period of time. Blood pressure is the force per unit area exerted on a vessel wall by the contained blood. Resistance is opposition to blood flow; blood viscosity and blood vessel length and diameter contribute to resistance.
2. Blood flow is directly proportional to blood pressure and inversely proportional to resistance.

iP Cardiovascular System; Topic: Factors that Affect Blood Pressure, pp. 1–15.

Systemic Blood Pressure (pp. 704–706)

1. Systemic blood pressure is highest in the aorta and lowest in the venae cavae. The steepest drop in BP occurs in the arterioles, where resistance is greatest.

2. Arterial BP depends on compliance of the elastic arteries and on how much blood is forced into them. Arterial blood pressure is pulsatile, and peaks during systole; this is measured as systolic pressure. During diastole, as blood is forced distally in the circulation by the rebound of elastic arteries, arterial BP drops to its lowest value, called the diastolic pressure.
3. Pulse pressure is systolic pressure minus diastolic pressure. The mean arterial pressure (MAP) = diastolic pressure plus one-third of pulse pressure and is the pressure that keeps blood moving throughout the cardiac cycle.
4. Low capillary pressure (35 to 15 mm Hg) protects the delicate capillaries from rupture while still allowing adequate exchange across the capillary walls.
5. Venous pressure is nonpulsatile and low (declining to zero) because of the cumulative effects of resistance. Venous valves, large lumens, functional adaptations (muscular and respiratory pumps), and sympathetic nervous system activity promote venous return.

Maintaining Blood Pressure (pp. 706–713)

1. Blood pressure varies directly with CO, peripheral resistance (*R*), and blood volume. Vessel diameter is the major factor determining resistance, and small changes in the diameter of vessels (chiefly arterioles) significantly affect blood pressure.

iP Cardiovascular System; Topic: Measuring Blood Pressure, pp. 1–13.

2. BP is regulated by autonomic neural reflexes involving baroreceptors or chemoreceptors, the vasomotor center (a medullary center that regulates blood vessel diameter), and sympathetic vasomotor fibers, which act on vascular smooth muscle.
3. Activation of the receptors by falling BP (and to a lesser extent by a rise in blood CO_2, or falling blood pH or O_2 levels) stimulates the vasomotor center to increase vasoconstriction and the cardioacceleratory center to increase heart rate and contractility. Rising BP inhibits the vasomotor center (permitting vasodilation) and activates the cardioinhibitory center.
4. Higher brain centers (cerebrum and hypothalamus) may modify neural controls of BP via medullary centers.
5. Hormones that increase BP by promoting vasoconstriction include epinephrine and NE (these also increase heart rate and contractility), ADH, and angiotensin II (generated in response to renin release by kidney cells). Hormones that reduce BP by promoting vasodilation include atrial natriuretic peptide, which also causes a decline in blood volume.
6. The kidneys regulate blood pressure by regulating blood volume. Rising BP directly enhances filtrate formation and fluid losses in urine; falling BP causes the kidneys to retain more water, increasing blood volume.
7. Indirect renal regulation of blood volume involves the renin-angiotensin mechanism, a hormonal mechanism. When BP falls, the kidneys release renin, which triggers the formation of angiotensin II (a vasoconstrictor) and release of aldosterone, which causes salt and water retention.

iP Cardiovascular System; Topic: Blood Pressure Regulation, pp. 1–31.

8. Pulse and blood pressure measurements are used to assess cardiovascular efficiency.
9. The pulse is the alternating expansion and recoil of arterial walls with each heartbeat. Pulse points are also pressure points.
10. Blood pressure is routinely measured by the auscultatory method. Normal blood pressure in adults is 120/80 (systolic/diastolic).
11. Hypotension, or low blood pressure (systolic pressure below 100 mm Hg), is rarely a problem. It is a sign of health in the well conditioned. In other individuals it warns of poor nutrition, disease, or circulatory shock.
12. Chronic hypertension (high blood pressure) is persistent BP readings of 140/90 or higher. It indicates increased peripheral resistance, which strains the heart and promotes vascular complications of other organs, particularly the eyes and kidneys. It is a major cause of myocardial infarct, stroke, and renal disease. Risk factors are high-fat, high-salt diet, obesity, diabetes mellitus, advanced age, smoking, stress, and being a member of the black race or a family with a history of hypertension.

iP Cardiovascular System; Topic: Measuring Blood Pressure, pp. 11 and 12.

Blood Flow Through Body Tissues: Tissue Perfusion (pp. 713–721)

1. Blood flow is involved in delivering nutrients and wastes to and from cells, gas exchange, absorbing nutrients, and forming urine.
2. Blood flows fastest where the cross-sectional area of the vascular bed is least (aorta), and slowest where the total cross-sectional area is greatest (capillaries). The slow flow in capillaries allows time for nutrient-waste exchanges.
3. Autoregulation is the local adjustment of blood flow to individual organs based on their immediate requirements. It involves myogenic controls that maintain flow despite changes in blood pressure, and local chemical factors. Vasodilators include increased CO_2, H^+, and nitric oxide. Decreased O_2 concentrations also cause vasodilation. Other factors, including endothelins, decrease blood flow.

iP Cardiovascular; Topic: Autoregulation and Capillary Dynamics, pp. 1–13.

4. In most instances, autoregulation is controlled by oxygen deficits and accumulation of local metabolites. However, autoregulation in the brain is controlled primarily by a drop in pH and by myogenic mechanisms; and vasodilation of pulmonary circuit vessels occurs in response to high levels of oxygen.
5. Nutrients, gases, and other solutes smaller than plasma proteins cross the capillary wall by diffusion. Water-soluble substances move through the clefts or fenestrations; fat-soluble substances pass through the lipid portion of the endothelial cell membrane; larger molecules are actively transported via pinocytotic vesicles or caveolae.
6. Bulk flow of fluids at capillary beds determines the distribution of fluids between the bloodstream and the interstitial space. It reflects the relative effect of outward (net hydrostatic pressure) forces minus the effect of inward (net osmotic pressure) forces. In general, fluid flows out of the capillary bed at the arterial end and reenters the capillary blood at the venule end.

iP Cardiovascular; Topic: Autoregulation and Capillary Dynamics, pp. 14–38.

7. The small net loss of fluid and protein into the interstitial space is collected by lymphatic vessels and returned to the cardiovascular system.
8. Circulatory shock occurs when blood perfusion of body tissues is inadequate. Most cases of shock reflect low blood volume (hypovolemic shock), abnormal vasodilation (vascular shock), or pump failure (cardiogenic shock).

PART 3: CIRCULATORY PATHWAYS: BLOOD VESSELS OF THE BODY

The Two Main Circulations of the Body (p. 721)

1. The pulmonary circulation transports O_2-poor, CO_2-laden blood to the lungs for oxygenation and carbon dioxide unloading. Blood returning to the right atrium of the heart is pumped by the right ventricle to the lungs via the pulmonary trunk. Blood issuing from the lungs is returned to the left atrium by the pulmonary veins. (See Table 19.3 and Figure 19.19.)
2. The systemic circulation transports oxygenated blood from the left ventricle to all body tissues via the aorta and its branches. Venous blood returning from the systemic circuit is delivered to the right atrium via the venae cavae.

Systemic Arteries and Veins: Differences in Pathways and Courses (p. 721)

1. All arteries are deep while veins are both deep and superficial. Superficial veins tend to have numerous interconnections. Dural venous sinuses and the hepatic portal circulation are unique venous drainage patterns.

Principal Vessels of the Systemic Circulation (pp. 721–745)

1. Tables 19.3 to 19.13 and Figures 19.20 to 19.30 illustrate and describe vessels of the systemic circulation.

Developmental Aspects of Blood Vessels (p. 745)

1. The fetal vasculature develops from embryonic blood islands and mesenchyme and is functioning in blood delivery by the fourth week.
2. Fetal circulation differs from circulation after birth. The pulmonary and hepatic shunts and special umbilical vessels are normally occluded shortly after birth.
3. Blood pressure is low in infants and rises to adult values. Age-related vascular problems include varicose veins, hypertension, and atherosclerosis. Hypertension is the most important cause of sudden cardiovascular death in middle-aged men. Atherosclerosis is the most important cause of cardiovascular disease in the aged.

REVIEW QUESTIONS

Multiple Choice/Matching

(Some questions have more than one correct answer. Select the best answer or answers from the choices given.)

1. Which statement does not accurately describe veins? (a) They have less elastic tissue and smooth muscle than arteries, (b) they contain more fibrous tissue than arteries, (c) most veins in the extremities have valves, (d) they always carry deoxygenated blood.
2. Smooth muscle in the blood vessel wall (a) is found primarily in the tunica intima, (b) is mostly circularly arranged, (c) is most abundant in veins, (d) is usually innervated by the parasympathetic nervous system.
3. Peripheral resistance (a) is inversely proportional to the length of the vascular bed, (b) increases in anemia, (c) decreases in polycythemia, (d) is inversely related to the diameter of the arterioles.
4. Which of the following can lead to decreased venous return of blood to the heart? (a) an increase in blood volume, (b) an increase in venous pressure, (c) damage to the venous valves, (d) increased muscular activity.
5. Arterial blood pressure increases in response to (a) increasing stroke volume, (b) increasing heart rate, (c) arteriosclerosis, (d) rising blood volume, (e) all of these.
6. Which of the following would *not* result in the dilation of the feeder arterioles and opening of the precapillary sphincters in systemic capillary beds? (a) a decrease in local tissue O_2 content, (b) an increase in local tissue CO_2, (c) a local increase in histamine, (d) a local increase in pH.
7. The structure of a capillary wall differs from that of a vein or an artery because (a) it has two tunics instead of three, (b) there is less smooth muscle, (c) it has a single tunic—only the tunica intima, (d) none of these.
8. The baroreceptors in the carotid sinus and aortic arch are sensitive to (a) a decrease in CO_2, (b) changes in arterial pressure, (c) a decrease in O_2, (d) all of these.
9. The myocardium receives its blood supply directly from the (a) aorta, (b) coronary arteries, (c) coronary sinus, (d) pulmonary arteries.
10. Blood flow in the capillaries is steady despite the rhythmic pumping of the heart because of the (a) elasticity of the large arteries, (b) small diameter of capillaries, (c) thin walls of the veins, (d) venous valves.
11. Tracing the blood from the heart to the right hand, we find that blood leaves the heart and passes through the aorta, the right subclavian artery, the axillary and brachial arteries, and through either the radial or ulnar artery to arrive at the hand. Which artery is missing from this sequence? (a) coronary, (b) brachiocephalic, (c) cephalic, (d) right common carotid.
12. Which of the following do not drain directly into the inferior vena cava? (a) inferior phrenic veins, (b) hepatic veins, (c) inferior mesenteric vein, (d) renal veins.
13. In atherosclerosis, which layer of the vessel wall thickens most? (a) tunica media, (b) tunica intima, (c) tunica adventitia, (d) tunica externa.
14. Suppose that at a given point along a capillary, the following forces exist: capillary hydrostatic pressure (HP_c) = 30 mm Hg, interstitial fluid hydrostatic pressure (HP_{if}) = 0 mm Hg, capillary colloid osmotic pressure (OP_c) = 25 mm Hg, and interstitial fluid colloid osmotic pressure (OP_{if}) = 2 mm Hg. The net filtration pressure at this point in the capillary is (a) 3 mm Hg, (b) −3 mm Hg, (c) −7 mm Hg, (d) 7 mm Hg.
15. Using the letters from column B, match the artery descriptions in column A. (Note that some require more than a single choice.)

Column A	Column B
____ (1) unpaired branch of abdominal aorta	(a) right common carotid
	(b) superior mesenteric
____ (2) second branch of aortic arch	(c) left common carotid
	(d) external iliac
____ (3) branch of internal carotid	(e) inferior mesenteric
	(f) superficial temporal
____ (4) branch of external carotid	(g) celiac trunk
	(h) facial
____ (5) origin of femoral arteries	(i) ophthalmic
	(j) internal iliac

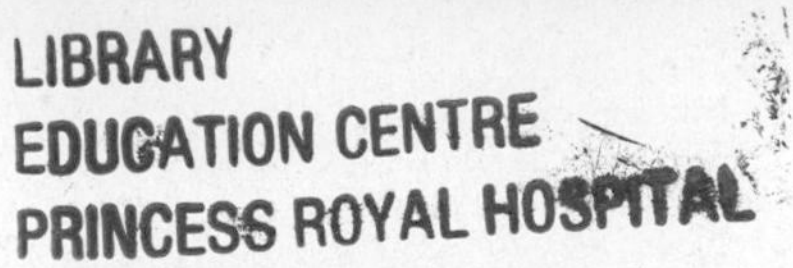

Short Answer Essay Questions

16. How is the anatomy of capillaries and capillary beds well suited to their function?
17. Distinguish between elastic arteries, muscular arteries, and arterioles relative to location, histology, and functional adaptations.
18. Write an equation showing the relationship between peripheral resistance, blood flow, and blood pressure.
19. **(a)** Define blood pressure. Differentiate between systolic and diastolic blood pressure. **(b)** What is the normal blood pressure value for a young adult?
20. Describe the neural mechanisms responsible for controlling blood pressure.
21. Explain the reasons for the observed changes in blood flow velocity in the different regions of the circulation.
22. How does the control of blood flow to the skin for the purpose of regulating body temperature differ from the control of nutrient blood flow to skin cells?
23. Describe neural and chemical (both systemic and local) effects exerted on the blood vessels when one is fleeing from a mugger. (Be careful, this is more involved than it appears at first glance.)
24. How are nutrients, wastes, and respiratory gases transported to and from the blood and tissue spaces?
25. **(a)** What blood vessels contribute to the formation of the hepatic portal circulation? **(b)** Why is a portal circulation a "strange" circulation?
26. Physiologists often consider capillaries and postcapillary venules together. **(a)** What functions do these vessels share? **(b)** Structurally, how do they differ?

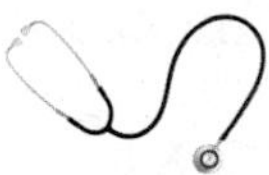

Critical Thinking and Clinical Application Questions

1. Mrs. Johnson is brought to the emergency room after being involved in an auto accident. She is hemorrhaging and has a rapid, thready pulse, but her blood pressure is still within normal limits. Describe the compensatory mechanisms that are acting to maintain her blood pressure in the face of blood loss.
2. A 60-year-old man is unable to walk more than 100 yards without experiencing severe pain in his left leg; the pain is relieved by resting for 5–10 minutes. He is told that the arteries of his leg are becoming occluded with fatty material and is advised to have the sympathetic nerves serving that body region severed. Explain how such surgery might help to relieve this man's problem.
3. Your friend Joanie, who knows little about science, is reading a magazine article about a patient who had an "aneurysm at the base of his brain that suddenly grew much larger." The surgeons' first goal was to "keep it from rupturing," and the second goal was to "relieve the pressure on the brain stem and cranial nerves." The surgeons were able to "replace the aneurysm with a section of plastic tubing," so the patient recovered. Joanie asks you what all this means. Explain. (Hint: Check this chapter's Related Clinical Terms, p. 748.)
4. The Agawam High School band is playing some lively marches while the coaches are giving pep talks to their respective football squads. Although it is September, it is unseasonably hot (88°F/31°C) and the band uniforms are wool. Suddenly, Harry the tuba player becomes light-headed and faints. Explain his fainting in terms of vascular events.
5. When one is cold or the external temperature is low, most venous blood returning from the distal part of the arm travels in the deep veins where it picks up heat (by countercurrent exchange) from the nearby brachial artery en route. However, when one is hot, and especially during exercise, venous return from the distal arm travels in the superficial veins and those veins tend to bulge superficially in a person who is working out. Explain why venous return takes a different route in the second situation.
6. Edema (swelling due to an increase in interstitial fluid) is a common clinical problem. On one of your first days of an introductory clinical experience, you encounter four patients who all have severe edema for different reasons. Your challenge is to explain the cause of the edema. In each case, try to explain the edema in terms of either an increase or a decrease in one of the four pressures that causes bulk flow at capillaries (see Figure 19.17).
   (1) First you encounter Mrs. Taylor in the medical ward awaiting a liver transplant. What is the connection between liver failure and her edema? (Hint: Think about the liver's role in producing plasma proteins.)
   (2) Next you follow a resident to the obstetric ward, where Mrs. So is experiencing premature labor. Which of the pressures that drive bulk flow might be altered here? (Hint: What might the expanded uterus be pressing on?)
   (3) Then you are called to emergency, where Mr. Herrera is in anaphylactic shock. In anaphylactic shock, the capillaries become leaky, allowing plasma proteins that are normally kept inside the blood vessels to escape into the interstitial fluid. Which of the pressures driving bulk flow is altered in this case and in what direction is the change?
   (4) Finally, you go to the oncology ward where Mrs. O'Leary is recovering from surgery for advanced breast cancer that had infiltrated her right breast and axillary lymph nodes. All of her axillary lymph nodes were removed and unfortunately, this severed most of the lymphatic vessels draining her right arm. You notice that her right arm is quite edematous. Why? Mrs. O'Leary is given a compression sleeve to wear on this arm to help relieve the edema. Which of the pressures driving bulk flow at the capillaries will be altered by the compression sleeve?

20 The Lymphatic System and Lymphoid Organs and Tissues

They can't all be superstars! When we mentally tick off the names of the body's organ systems, the lymphatic (limfat′ik) system and the lymphoid organs and tissues are probably not the first to come to mind. Yet if they failed their quiet background work, our cardiovascular system would stop working and our immune system would be hopelessly impaired.

The **lymphatic system** actually consists of three parts: (1) a meandering network of *lymphatic vessels*, (2) *lymph*, the fluid contained in those vessels, and (3) *lymph nodes* that cleanse the lymph as it passes through them. Inasmuch as the lymph nodes also form part of the *lymphoid organs and tissues*, the structures and functions of the lymphatic system overlap with those of the lymphoid organs and tissues. In addition to lymph nodes, the lymphoid organs and tissues include the spleen, thymus, tonsils, and other lymphoid tissues scattered throughout the body. The lymphoid organs house phagocytic cells and lymphocytes, which play essential roles in the body's defense mechanisms and its resistance to disease. Together, the lymphatic system and the lymphoid organs and tissues provide the structural basis of the immune system.

Lymphatic Vessels

- ▶ List the functions of the lymphatic vessels.
- ▶ Describe the structure and distribution of lymphatic vessels.
- ▶ Describe the source of lymph and mechanism(s) of lymph transport.

As blood circulates through the body, nutrients, wastes, and gases are exchanged between the blood and the interstitial fluid. As we explained in Chapter 19, the hydrostatic and colloid osmotic pressures operating at capillary beds force fluid out of the blood at the arterial ends of the beds ("upstream") and cause most of it to be reabsorbed at the venous ends ("downstream"). The fluid that remains behind in the tissue spaces, as much as 3 L daily, becomes part of the interstitial fluid.

This leaked fluid, plus any plasma proteins that escape from the bloodstream, must be carried back to the blood to ensure that the cardiovascular system has sufficient blood volume to operate properly. This problem of circulatory dynamics is resolved by the **lymphatic vessels**, or **lymphatics**, an elaborate system of drainage vessels that collect the excess protein-containing interstitial fluid and return it to the bloodstream. Once interstitial fluid enters the lymphatics, it is called **lymph** (*lymph* = clear water).

Distribution and Structure of Lymphatic Vessels

The lymphatic vessels form a one-way system in which lymph flows only toward the heart. This transport system begins in microscopic blind-ended **lymphatic capillaries** (Figure 20.1a). These capillaries weave between the tissue cells and blood capillaries in the loose connective tissues of the body. Lymphatic

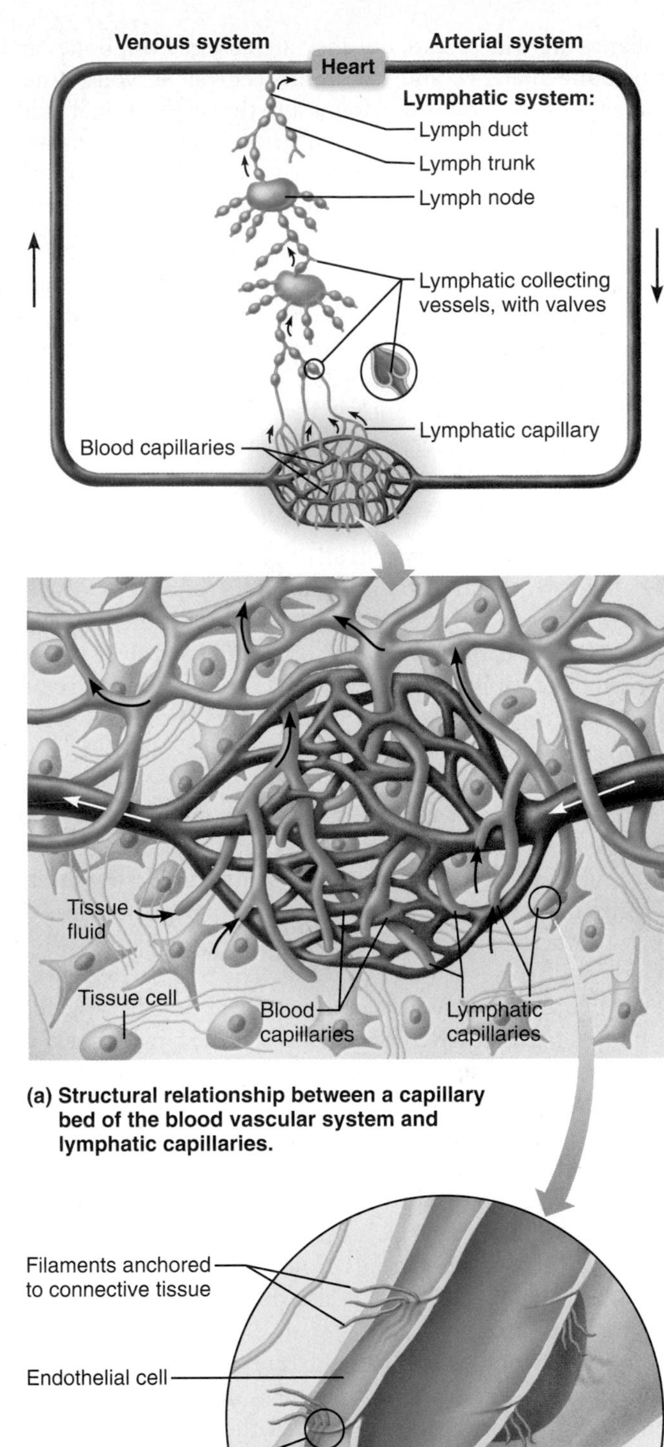

(a) Structural relationship between a capillary bed of the blood vascular system and lymphatic capillaries.

(b) Lymphatic capillaries are blind-ended tubes in which adjacent endothelial cells overlap each other, forming flaplike minivalves.

Figure 20.1 Distribution and special structural features of lymphatic capillaries. Arrows in **(a)** indicate direction of fluid movement.

20

capillaries are widespread, but they are absent from bones and teeth, bone marrow, and the entire central nervous system (where the excess tissue fluid drains into the cerebrospinal fluid).

Although similar to blood capillaries, lymphatic capillaries are so remarkably permeable that they were once thought to be open at one end like a straw. We now know that they owe their permeability to two unique structural modifications:

1. The endothelial cells forming the walls of lymphatic capillaries are not tightly joined. Instead, the edges of adjacent cells overlap each other loosely, forming easily opened, flaplike *minivalves* (Figure 20.1b).
2. Collagen filaments anchor the endothelial cells to surrounding structures so that any increase in interstitial fluid volume opens the minivalves, rather than causing the lymphatic capillaries to collapse.

So, what we have is a system analogous to one-way swinging doors in the lymphatic capillary wall. When fluid pressure in the interstitial space is greater than the pressure in the lymphatic capillary, the minivalve flaps gape open, allowing fluid to enter the lymphatic capillary. However, when the pressure is greater *inside* the lymphatic capillary, the endothelial minivalve flaps are forced closed, preventing lymph from leaking back out as the pressure moves it along the vessel.

Proteins in the interstitial space are unable to enter blood capillaries, but they enter lymphatic capillaries easily. In addition, when tissues are inflamed, lymphatic capillaries develop openings that permit uptake of even larger particles such as cell debris, pathogens (disease-causing microorganisms such as bacteria and viruses), and cancer cells. The pathogens can then use the lymphatics to travel throughout the body. This threat to the body is partly resolved by the fact that the lymph takes "detours" through the lymph nodes, where it is cleansed of debris and "examined" by cells of the immune system.

Highly specialized lymphatic capillaries called **lacteals** (lak′te-alz) are present in the fingerlike villi of the intestinal mucosa. The lymph draining from the digestive viscera is milky white
20 (*lacte* = milk) rather than clear because the lacteals play a major role in absorbing digested fats from the intestine. This fatty lymph, called **chyle** ("juice"), is also delivered to the blood via the lymphatic stream.

From the lymphatic capillaries, lymph flows through successively larger and thicker-walled channels—first collecting vessels, then trunks, and finally the largest of all, the ducts (Figure 20.1). The **lymphatic collecting vessels** have the same three tunics as veins, but the collecting vessels are thinner walled, have more internal valves, and anastomose more. In general, lymphatics in the skin travel along with superficial *veins*, while the deep lymphatic vessels of the trunk and digestive viscera travel with the deep *arteries*. The exact anatomical distribution of lymphatic vessels varies greatly between individuals, even more so than it does for veins.

The **lymphatic trunks** are formed by the union of the largest collecting vessels, and drain fairly large areas of the body. The major trunks, named mostly for the regions from which they collect lymph, are the paired **lumbar**, **bronchomediastinal**, **subclavian**, and **jugular trunks**, and the single **intestinal trunk** **(Figure 20.2b)**.

Lymph is eventually delivered to one of two large *ducts* in the thoracic region. The **right lymphatic duct** drains lymph from the right upper limb and the right side of the head and thorax (Figure 20.2a). The much larger **thoracic duct** receives lymph from the rest of the body. It arises anterior to the first two lumbar vertebrae as an enlarged sac, the **cisterna chyli** (sis-ter′nah ki′li), that collects lymph from the two large lumbar trunks that drain the lower limbs and from the intestinal trunk that drains the digestive organs. As the thoracic duct runs superiorly, it receives lymphatic drainage from the left side of the thorax, left upper limb, and the head region. Each terminal duct empties its lymph into the venous circulation at the junction of the internal jugular vein and subclavian vein on its own side of the body (Figure 20.2b).

HOMEOSTATIC IMBALANCE

Like the larger blood vessels, the larger lymphatics receive their nutrient blood supply from a branching vasa vasorum. When lymphatic vessels are severely inflamed, the related vessels of the vasa vasorum become congested with blood. As a result, the pathway of the associated superficial lymphatics becomes visible through the skin as red lines that are tender to the touch. This unpleasant condition is called *lymphangitis* (lim″fan-ji′tis; *angi* = vessel). ■

Lymph Transport

The lymphatic system lacks an organ that acts as a pump. Under normal conditions, lymphatic vessels are low-pressure conduits, and the same mechanisms that promote venous return in blood vessels act here as well—the milking action of active skeletal muscles, pressure changes in the thorax during breathing, and valves to prevent backflow. Lymphatics are usually bundled together in connective tissue sheaths along with blood vessels, and pulsations of nearby arteries also promote lymph flow. In addition to these mechanisms, smooth muscle in the walls of the lymphatic trunks and thoracic duct contracts rhythmically, helping to pump the lymph along.

Even so, lymph transport is sporadic and slow. About 3 L of lymph enters the bloodstream every 24 hours, a volume almost exactly equal to the amount of fluid lost to the tissue spaces from the bloodstream in the same time period. Movement of adjacent tissues is extremely important in propelling lymph through the lymphatics. When physical activity or passive movements increase, lymph flows much more rapidly (balancing the greater rate of fluid loss from the blood in such situations). For this reason, it is a good idea to immobilize a badly infected body part to hinder flow of inflammatory material from that region.

HOMEOSTATIC IMBALANCE

Anything that prevents the normal return of lymph to the blood, such as blockage of the lymphatics by tumors or removal of lymphatics during cancer surgery, results in short-term but

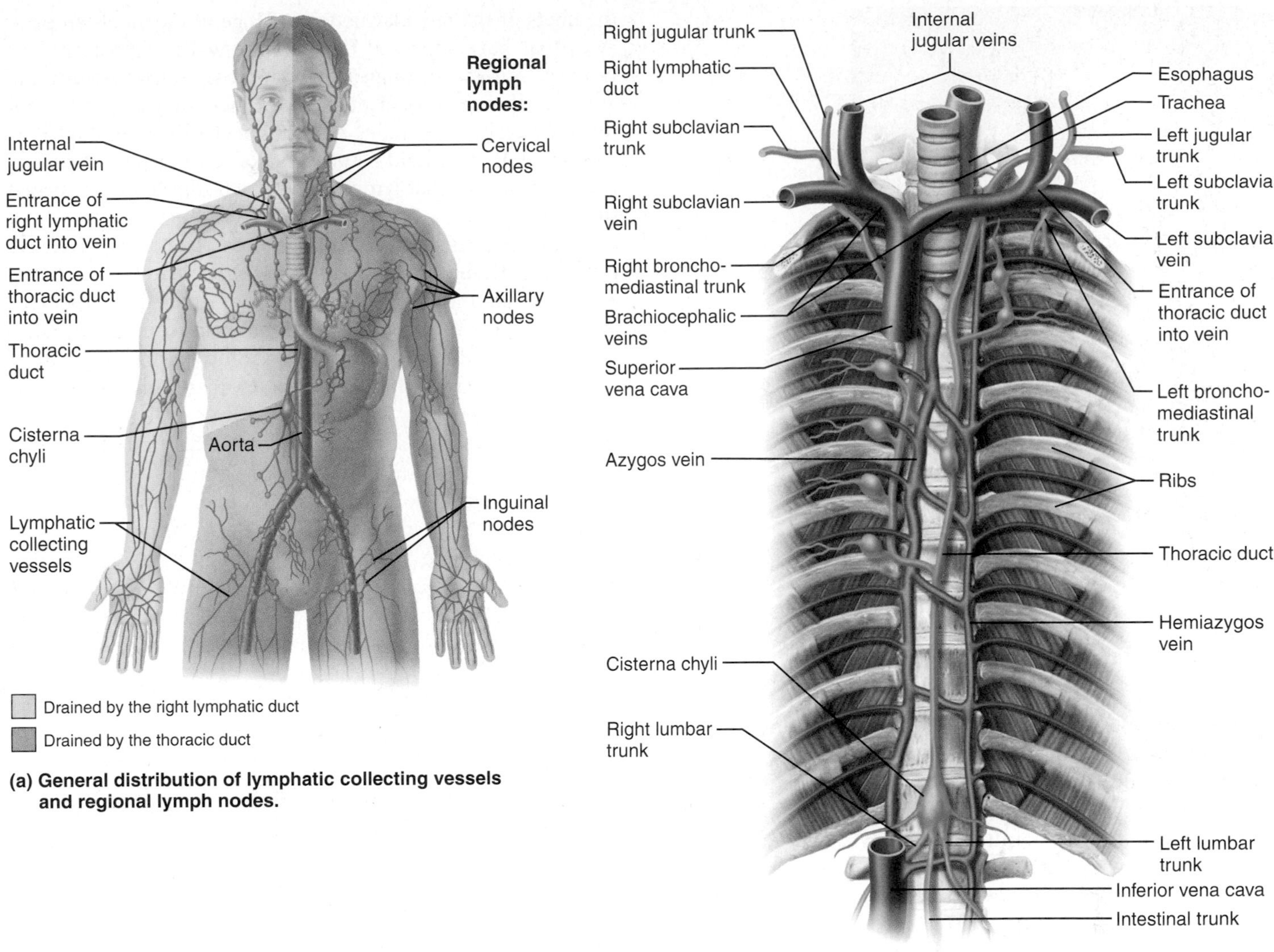

(a) General distribution of lymphatic collecting vessels and regional lymph nodes.

(b) Major lymphatic trunks and ducts in relation to veins and surrounding structures. Anterior view.

Figure 20.2 **The lymphatic system.**

severe localized edema *(lymphedema)*. Usually, however, lymphatic drainage is eventually reestablished by regrowth from the vessels remaining in the area. ■

This completes the description of the lymphatic vessels. To summarize their functions, they (1) return excess tissue fluid to the bloodstream, (2) return leaked proteins to the blood, and (3) carry absorbed fat from the intestine to the blood (through lacteals).

CHECK YOUR UNDERSTANDING

1. What is lymph? Where does it come from?
2. Name two lymphatic ducts and indicate the body regions usually drained by each.
3. What is the driving force for lymph movement?

For answers, see Appendix G.

Lymphoid Cells and Tissues

▶ Describe the basic structure and cellular population of lymphoid tissue. Differentiate between diffuse and follicular lymphoid tissues.

In order to understand the basic aspects of the role of the lymphoid organs in the body, we investigate their components—lymphoid cells and lymphoid tissues—before considering the organs themselves.

Lymphoid Cells

Infectious microorganisms that manage to penetrate the body's epithelial barriers quickly proliferate in the underlying loose connective tissues. These invaders are fought off by the inflammatory response, by phagocytes (macrophages and neutrophils), and by lymphocytes.

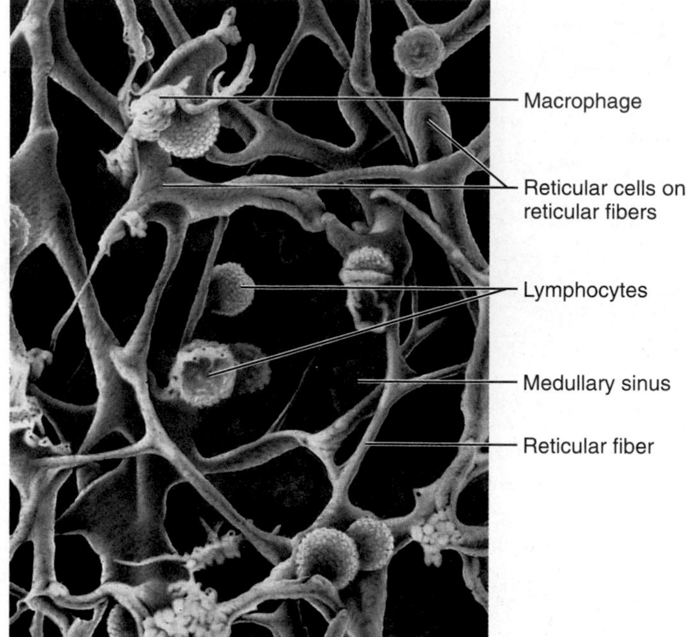

Figure 20.3 Reticular tissue in a human lymph node. Scanning electron micrograph (690×).

Lymphocytes, the main warriors of the immune system, arise in red bone marrow (along with other formed elements). They then mature into one of the two main varieties of immunocompetent cells—**T cells (T lymphocytes)** or **B cells (B lymphocytes)**—that protect the body against antigens. (*Antigens* are anything the body perceives as foreign, such as bacteria and their toxins, viruses, mismatched RBCs, or cancer cells.) Activated T cells manage the immune response, and some of them directly attack and destroy infected cells. B cells protect the body by producing **plasma cells**, daughter cells that secrete antibodies into the blood (or other body fluids). Antibodies mark antigens for destruction by phagocytes or other means. The precise roles of the lymphocytes in immunity are explored in Chapter 21.

20

Lymphoid **macrophages** play a crucial role in body protection and in the immune response by phagocytizing foreign substances and by helping to activate T cells. So, too, do the spiny-looking **dendritic cells** that capture antigens and bring them back to the lymph nodes. Last but not least are the **reticular cells**, fibroblast-like cells that produce the reticular fiber **stroma** (stro′mah), which is the network that supports the other cell types in the lymphoid organs and tissues **(Figure 20.3)**.

Lymphoid Tissue

Lymphoid (lymphatic) tissue is an important component of the immune system, mainly because it (1) houses and provides a proliferation site for lymphocytes and (2) furnishes an ideal surveillance vantage point for lymphocytes and macrophages.

Lymphoid tissue, largely composed of a type of loose connective tissue called **reticular connective tissue**, dominates all the lymphoid organs except the thymus. Macrophages live on the fibers of the reticular network. Huge numbers of lymphocytes that have squeezed through the walls of postcapillary venules coursing through this tissue temporarily occupy the spaces of this network (Figure 20.3). Then, they leave to patrol the body again. The cycling of lymphocytes between the circulatory vessels, lymphoid tissues, and loose connective tissues of the body ensures that lymphocytes reach infected or damaged sites quickly.

Lymphoid tissue comes in various "packages." **Diffuse lymphatic tissue**, consisting of a few scattered reticular tissue elements, is found in virtually every body organ, but larger collections appear in the lamina propria of mucous membranes and in lymphoid organs. **Lymphoid follicles (nodules)** represent another way lymphoid tissue is organized. Like diffuse lymphatic tissue, they lack a capsule, but follicles are solid, spherical bodies consisting of tightly packed reticular elements and cells. Follicles often have lighter-staining centers, called **germinal centers**. Proliferating B cells predominate in germinal centers, and these centers enlarge dramatically when the B cells are dividing rapidly and producing plasma cells. In many cases, the follicles form part of larger lymphoid organs, such as lymph nodes. However, isolated aggregations of lymphoid follicles occur in the intestinal wall as Peyer's patches and in the appendix.

Lymph Nodes

▶ Describe the general location, histological structure, and functions of lymph nodes.

The principal lymphoid organs in the body are the **lymph nodes**, which cluster along the lymphatic vessels of the body. There are hundreds of these small organs, but because they are usually embedded in connective tissue, they are not ordinarily seen. Large clusters of lymph nodes occur near the body surface in the inguinal, axillary, and cervical regions, places where the lymphatic collecting vessels converge to form trunks (see Figure 20.2a).

Lymph nodes have two basic functions, both concerned with body protection. (1) As lymph is transported back to the bloodstream, the lymph nodes act as lymph "filters." Macrophages in the nodes remove and destroy microorganisms and other debris that enter the lymph from the loose connective tissues, effectively preventing them from being delivered to the blood and spreading to other parts of the body. (2) They help activate the immune system. Lymph nodes and other lymphoid organs are strategically located sites where lymphocytes encounter antigens and are activated to mount an attack against them. Let's look at how the structure of a lymph node supports these defensive functions.

Structure of a Lymph Node

Lymph nodes vary in shape and size, but most are bean shaped and less than 2.5 cm (1 inch) in length. Each node is surrounded by a dense fibrous **capsule** from which connective tissue strands

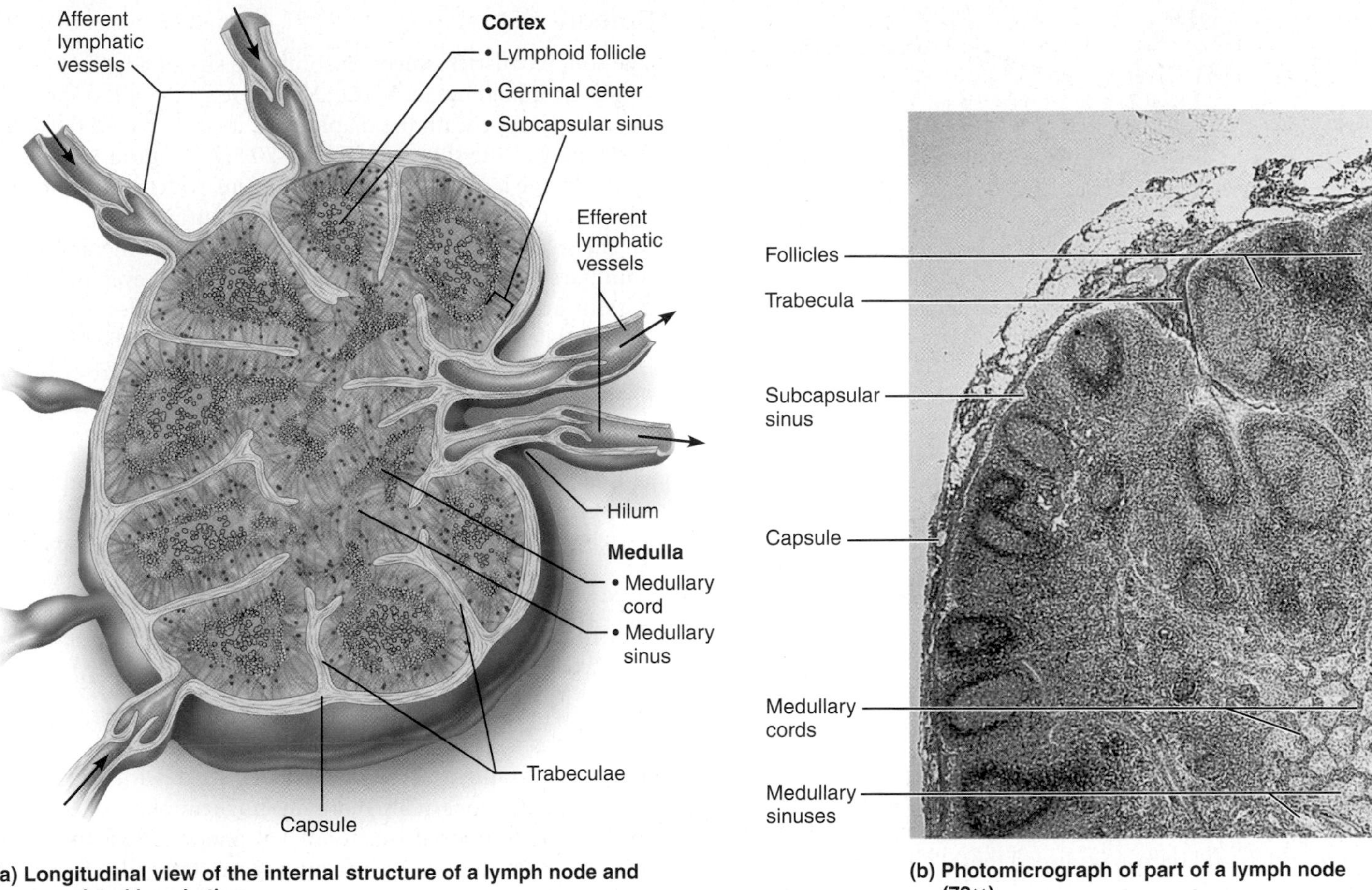

(a) Longitudinal view of the internal structure of a lymph node and associated lymphatics

(b) Photomicrograph of part of a lymph node (72×)

Figure 20.4 Lymph node. In **(a)**, notice that several afferent lymphatics converge on its convex side, whereas fewer efferent lymphatics exit at its hilum.

called **trabeculae** extend inward to divide the node into a number of compartments **(Figure 20.4)**. The node's internal framework, or stroma, of reticular fibers physically supports the ever-changing population of lymphocytes.

A lymph node has two histologically distinct regions, the **cortex** and the **medulla**. The superficial part of the cortex contains densely packed follicles, many with germinal centers heavy with dividing B cells. Dendritic cells nearly encapsulate the follicles and abut the deeper part of the cortex, which primarily houses T cells in transit. The T cells circulate continuously between the blood, lymph nodes, and lymph, performing their surveillance role.

Medullary cords are thin inward extensions from the cortical lymphoid tissue, and contain both types of lymphocytes plus plasma cells. Throughout the node are **lymph sinuses**, large lymph capillaries spanned by crisscrossing reticular fibers. Numerous macrophages reside on these reticular fibers and phagocytize foreign matter in the lymph as it flows by in the sinuses. Additionally, some of the lymph-borne antigens in the percolating lymph leak into the surrounding lymphoid tissue, where they activate lymphocytes to mount an immune attack against them.

Circulation in the Lymph Nodes

Lymph enters the convex side of a lymph node through a number of **afferent lymphatic vessels**. It then moves through a large, baglike sinus, the **subcapsular sinus**, into a number of smaller sinuses that cut through the cortex and enter the medulla. The lymph meanders through these sinuses and finally exits the node at its **hilum** (hi′lum), the indented region on the concave side, via **efferent lymphatic vessels**. There are fewer efferent vessels draining the node than afferent vessels feeding it, so the flow of lymph through the node stagnates somewhat, allowing time for the lymphocytes and macrophages to carry out their protective functions. Lymph passes through several nodes before it is completely cleansed.

HOMEOSTATIC IMBALANCE

Sometimes lymph nodes are overwhelmed by the agents they are trying to destroy. For example, when large numbers of bacteria are trapped in the nodes, the nodes become inflamed, swollen, and tender to the touch, a condition often referred to (erroneously) as swollen "glands." Such infected lymph nodes are called *buboes* (bu′bōz). (Buboes are the most obvious symptom of bubonic plague, the "Black Death" that killed much of Europe's population in the late Middle Ages.)

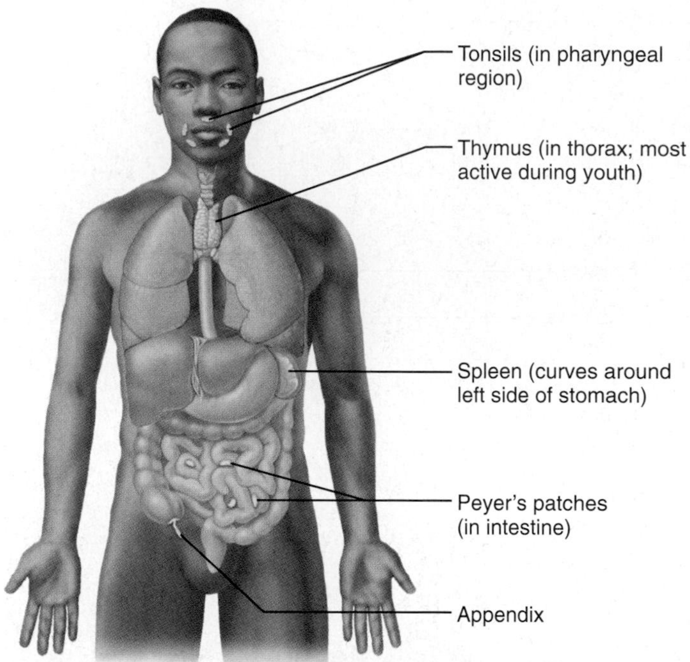

Figure 20.5 Lymphoid organs. Locations of the tonsils, spleen, thymus, Peyer's patches, and appendix.

Lymph nodes can also become secondary cancer sites, particularly in the case of metastasizing cancers that enter lymphatic vessels and become trapped there. The fact that cancer-infiltrated lymph nodes are swollen but usually not painful helps distinguish cancerous lymph nodes from those infected by microorganisms. ■

CHECK YOUR UNDERSTANDING

4. What is a lymphoid follicle? What type of lymphocyte predominates in follicles, especially in their germinal centers?
5. What is the benefit of having fewer efferent than afferent lymphatics in lymph nodes?

20

For answers, see Appendix G.

Other Lymphoid Organs

▶ Name and describe the other lymphoid organs of the body. Compare and contrast them with lymph nodes, structurally and functionally.

Lymph nodes are just one example of the many types of **lymphoid organs** or aggregates of lymphatic tissue in the body. Others are the spleen, thymus, tonsils, and Peyer's patches of the intestine **(Figure 20.5)**, as well as bits of lymphatic tissue scattered in the connective tissues. Except for the thymus, the common feature of all these organs is their tissue makeup: All are composed of *reticular connective tissue.* Although all lymphoid organs help protect the body, only the lymph nodes filter lymph. The other lymphoid organs and tissues typically have efferent lymphatics draining them, but lack afferent lymphatics.

Spleen

The soft, blood-rich **spleen** is about the size of a fist and is the largest lymphoid organ. Located in the left side of the abdominal cavity just beneath the diaphragm, it curls around the anterior aspect of the stomach (Figure 20.5 and **Figure 20.6**). It is served by the large *splenic artery* and *vein,* which enter and exit the *hilum* on its slightly concave anterior surface.

The spleen provides a site for lymphocyte proliferation and immune surveillance and response. But perhaps even more important are its blood-cleansing functions. Besides extracting aged and defective blood cells and platelets from the blood, its macrophages remove debris and foreign matter from blood flowing through its sinuses. The spleen also performs three additional, and related, functions.

1. It stores some of the breakdown products of red blood cells for later reuse (for example, it salvages iron for making hemoglobin) and releases others to the blood for processing by the liver.
2. It stores blood platelets.
3. It is thought to be a site of erythrocyte production in the fetus (a capability that normally ceases after birth).

Like lymph nodes, the spleen is surrounded by a fibrous capsule, has trabeculae that extend inward, and contains both lymphocytes and macrophages. Consistent with its blood-processing functions, it also contains huge numbers of erythrocytes. Areas composed mostly of lymphocytes suspended on reticular fibers are called **white pulp**. The white pulp clusters or forms "cuffs" around the *central arteries* (small branches of the splenic artery) in the organ and forms what appear to be islands in a sea of red pulp. **Red pulp** is essentially all remaining splenic tissue, that is, the venous sinuses (blood sinusoids) and the **splenic cords**, regions of reticular connective tissue exceptionally rich in macrophages. Red pulp is most concerned with disposing of worn-out red blood cells and bloodborne pathogens, whereas white pulp is involved with the immune functions of the spleen.

The naming of the pulp regions reflects their appearance in fresh spleen tissue rather than their staining properties. Indeed, as you can see in the photomicrograph in Figure 20.6d, the white pulp sometimes appears darker than the red pulp due to the darkly staining nuclei of the densely packed lymphocytes.

HOMEOSTATIC IMBALANCE

Because the spleen's capsule is relatively thin, a direct blow or severe infection may cause it to rupture, spilling blood into the peritoneal cavity. Once, removal of the ruptured spleen (a procedure called a *splenectomy*) was the standard treatment and thought necessary to prevent life-threatening hemorrhage and shock. However, surgeons have discovered that, if left alone, the spleen can often repair itself. At major trauma centers, the frequency of emergency splenectomies has decreased from 70% to 40%. If the spleen is removed, the liver and bone marrow take over most of its functions. In children younger than 12, the spleen will regenerate if a small part of it is left in the body. ■

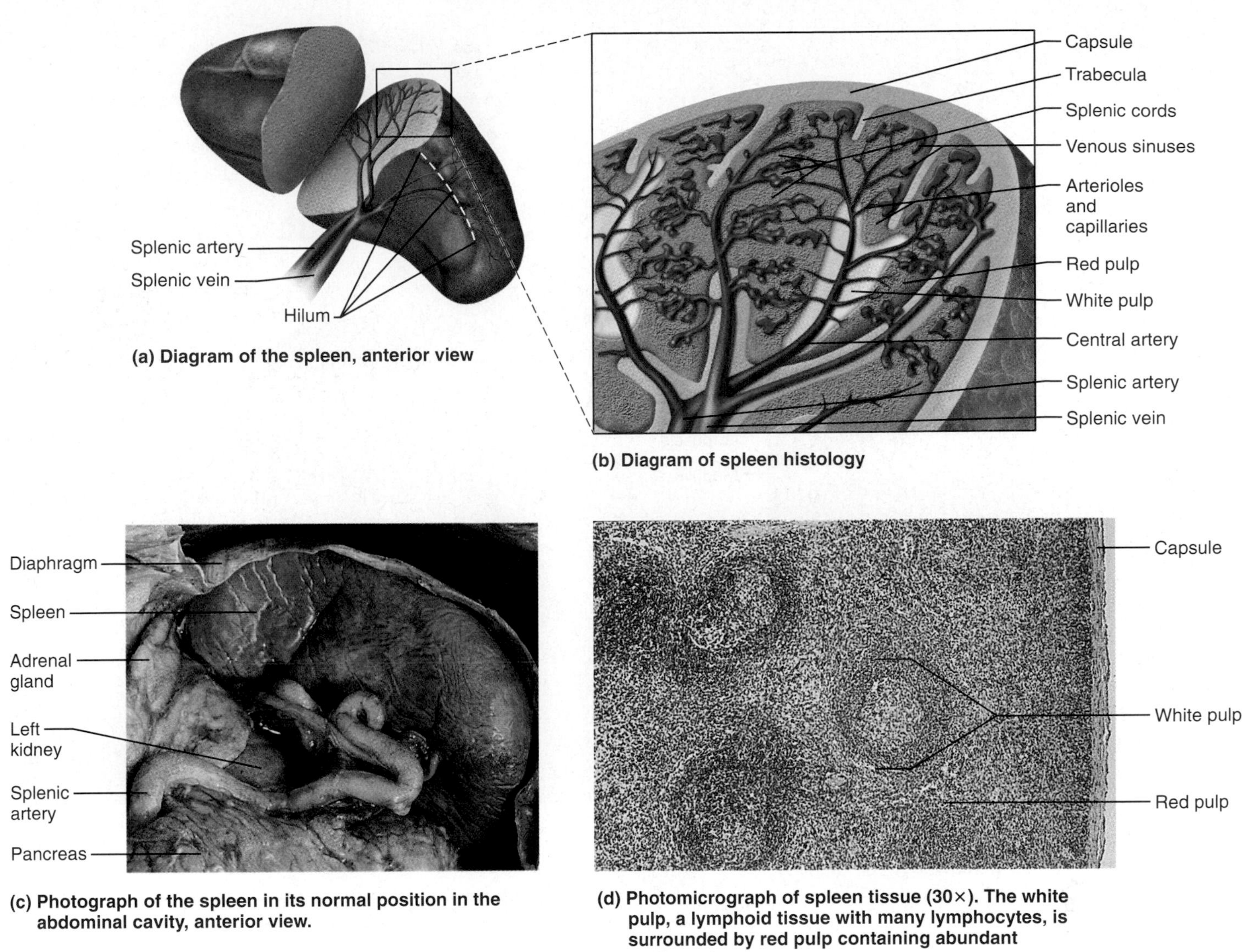

(a) Diagram of the spleen, anterior view

(b) Diagram of spleen histology

(c) Photograph of the spleen in its normal position in the abdominal cavity, anterior view.

(d) Photomicrograph of spleen tissue (30×). The white pulp, a lymphoid tissue with many lymphocytes, is surrounded by red pulp containing abundant erythrocytes.

Figure 20.6 The spleen. (See *A Brief Atlas of the Human Body*, Plate 39.)

Thymus

The bilobed **thymus** (thi′mus) has important functions primarily during the early years of life. It is found in the inferior neck and extends into the superior thorax, where it partially overlies the heart deep to the sternum (see Figure 20.5 and **Figure 20.7**). The thymus is the site where T lymphocyte precursors mature to become immunocomptetent lymphocytes. In other words, the thymus is where T lymphocytes become able to defend us against specific pathogens in the immune response.

Prominent in newborns, the thymus continues to increase in size during the first year, when it is highly active. After puberty, it starts to atrophy gradually and by old age it has been replaced almost entirely by fibrous and fatty tissue and is difficult to distinguish from surrounding connective tissue.

To understand thymic histology, it helps to compare the thymus to a cauliflower head—the flowerets represent *thymic lobules*, each containing an outer cortex and an inner medulla (Figure 20.7). Most thymic cells are lymphocytes. In the cortical regions the rapidly dividing lymphocytes are densely packed, but a few macrophages are scattered among them. The thymus has no follicles, because it lacks B cells. The lighter-staining medullary areas contain fewer lymphocytes plus some bizarre structures called **thymic (Hassall's) corpuscles**. Consisting of concentric whorls of keratinized epithelial cells, they were thought to be sites of T cell destruction. Recent evidence suggests that Hassall's corpuscles are involved in the development of a class of T lymphocytes called regulatory T cells, which are important for preventing autoimmune responses.

In addition to its lack of follicles, the thymus differs from other lymphoid organs in two other important ways. First, it functions strictly in maturation of T lymphocyte precursors and thus is the only lymphoid organ that does not *directly* fight antigens. In fact, the so-called *blood-thymus barrier* keeps bloodborne antigens from leaking into the cortical regions to prevent premature activation of the immature lymphocytes. Second, the stroma of the thymus consists of epithelial cells

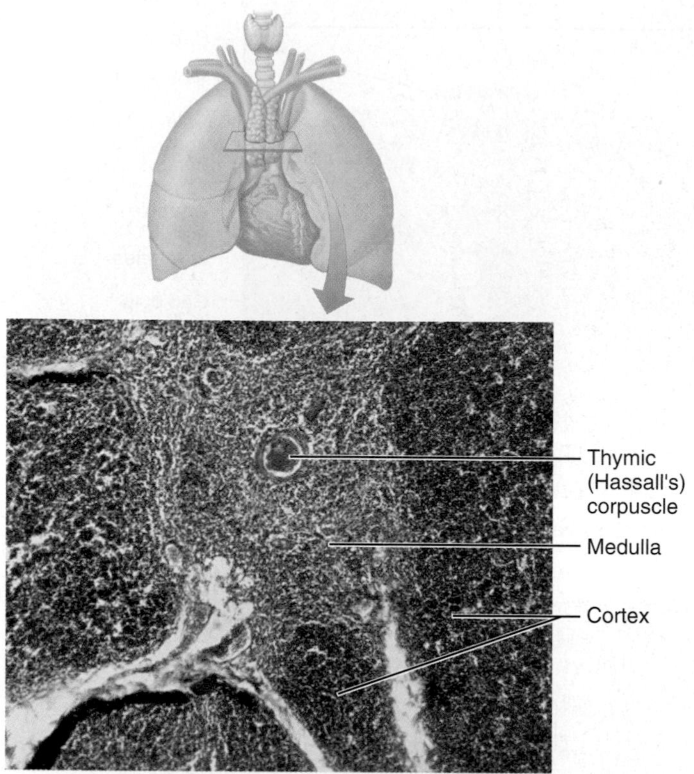

Figure 20.7 The thymus. The photomicrograph of a portion of the thymus shows part of a lobule with cortical and medullary regions (70×).

rather than reticular fibers. These epithelial cells provide the physical and chemical environment in which T lymphocytes can become immunocompetent.

Tonsils

20

The **tonsils** are the simplest lymphoid organs. They form a ring of lymphatic tissue around the entrance to the pharynx (throat), where they appear as swellings of the mucosa (**Figures 20.8** and 22.3). The tonsils are named according to location. The paired **palatine tonsils** are located on either side at the posterior end of the oral cavity. These are the largest of the tonsils and the ones most often infected. A lumpy collection of lymphoid follicles at the base of the tongue is referred to collectively as the **lingual tonsil**. The **pharyngeal tonsil** (referred to as the *adenoids* if enlarged) is in the posterior wall of the nasopharynx. The tiny **tubal tonsils** surround the openings of the auditory tubes into the pharynx. The tonsils gather and remove many of the pathogens entering the pharynx in food or in inhaled air.

The lymphoid tissue of the tonsils contains follicles with obvious germinal centers surrounded by diffusely scattered lymphocytes. The tonsils are not fully encapsulated, and the epithelium overlying them invaginates deep into their interior, forming blind-ended **tonsillar crypts** (Figure 20.8). The crypts trap bacteria and particulate matter, and the bacteria work their way through the mucosal epithelium into the lymphoid tissue, where most are destroyed. It seems a bit dangerous to "invite" infection this way, but this strategy produces a wide variety of immune cells that have a "memory" for the trapped pathogens. In other words, the body takes a calculated risk early on (during childhood) for the benefits of heightened immunity and better health later.

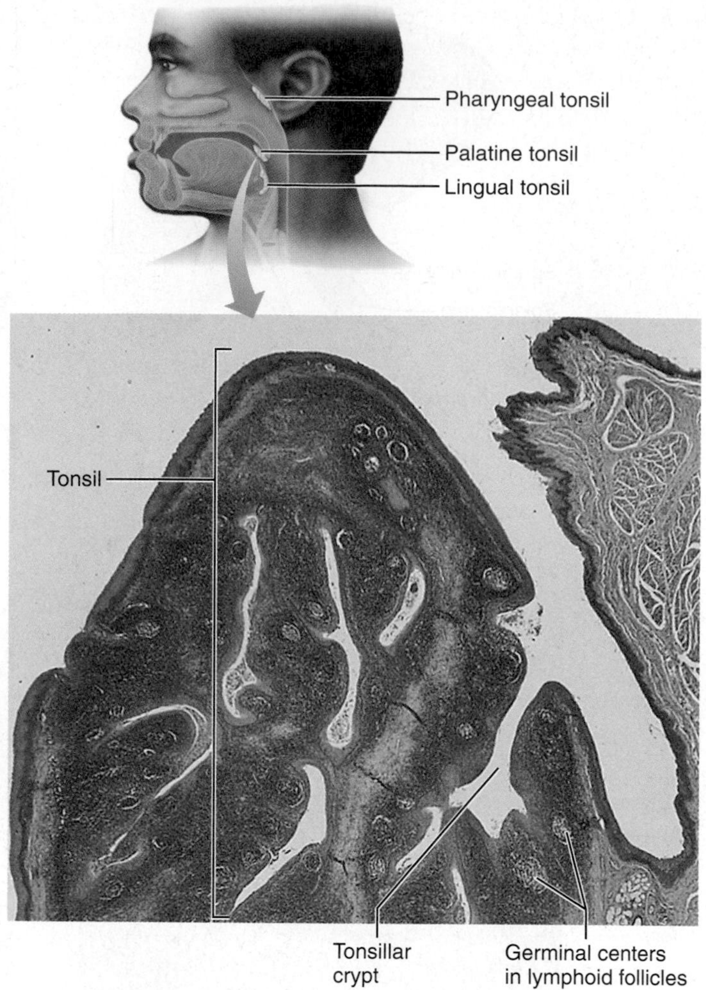

Figure 20.8 Histology of the palatine tonsil. The exterior surface of the tonsil is covered by squamous epithelium, which invaginates deeply to form tonsillar crypts (20×).

Aggregates of Lymphoid Follicles

Peyer's patches (pi′erz), or *aggregated lymphoid nodules*, are large clusters of lymphoid follicles, structurally similar to the tonsils. They are located in the wall of the distal portion of the small intestine (Figure 20.5 and **Figure 20.9**). Lymphoid follicles are also heavily concentrated in the wall of the **appendix**, a tubular offshoot of the first part of the large intestine. Peyer's patches and the appendix are in an ideal position (1) to destroy bacteria (which are present in large numbers in the intestine) before these pathogens can breach the intestinal wall, and (2) to generate many "memory" lymphocytes for long-term immunity.

Peyer's patches, the appendix, and the tonsils—all located in the digestive tract—and lymphoid follicles in the walls of the bronchi (organs of the respiratory tract) and in the mucosa of genitourinary organs are part of the collection of small lymphoid tissues referred to as **mucosa-associated lymphatic tissue (MALT)**. MALT protects passages that are open to the exterior from the never-ending onslaughts of foreign matter entering them.

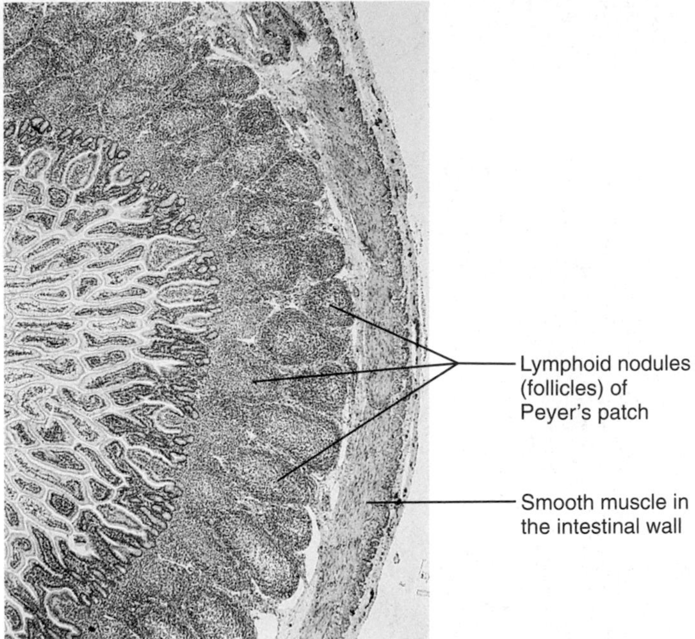

Figure 20.9 Peyer's patch. Histological structure of aggregated lymphoid nodules—a Peyer's patch—in the wall of the ileum of the small intestine (20×).

Developmental Aspects of the Lymphatic System and Lymphoid Organs and Tissues

▶ Outline the development of the lymphatic system and the lymphoid organs and tissues.

By the fifth week of embryonic development, the beginnings of the lymphatic vessels and the main clusters of lymph nodes are apparent. These arise from the budding of **lymph sacs** from developing veins. The first of these, the *jugular lymph sacs*, arise at the junctions of the internal jugular and subclavian veins and form a branching system of lymphatic vessels throughout the thorax, upper extremities, and head. The two main connections of the jugular lymph sacs to the venous system are retained and become the right lymphatic duct and, on the left, the superior part of the thoracic duct. Caudally the elaborate system of abdominal lymphatics buds largely from the primitive inferior vena cava. The lymphatics of the pelvic region and lower extremities form from sacs on the iliac veins.

Except for the thymus, which is an endodermal derivative, the lymphoid organs develop from mesodermal mesenchymal cells that migrate to particular body sites and develop into reticular tissue. The thymus, the first lymphoid organ to appear, forms as an outgrowth of the lining of the primitive pharynx. It then detaches and migrates caudally to the thorax where it becomes infiltrated with immature lymphocytes derived from hematopoietic tissues elsewhere in the embryo's body. Except for the spleen and tonsils, the lymphoid organs are poorly developed before birth. Shortly after birth, they become heavily populated by lymphocytes, and their development parallels the maturation of the immune system.

CHECK YOUR UNDERSTANDING

6. What is MALT? List several components of MALT.
7. List several functions of the spleen.
8. Which lymphoid organ develops first?

For answers, see Appendix G.

Although the functions of the lymphatic vessels and lymphoid organs overlap, each helps maintain body homeostasis in unique ways, as summarized in *Making Connections* (pp. 762–763). The lymphatic vessels help maintain blood volume. The macrophages of lymphoid organs remove and destroy foreign matter in lymph and blood. Additionally, lymphoid organs and tissues provide sites from which the immune system can be mobilized. In Chapter 21, we continue this story as we examine the inflammatory and immune responses that allow us to resist a constant barrage of pathogens.

RELATED CLINICAL TERMS

Elephantiasis (el″lĕ-fan-ti′ah-sis) Typically a tropical disease in which the lymphatics (particularly those of the lower limbs and scrotum) become clogged with parasitic roundworms, an infectious condition called filariasis. Swelling (due to edema) reaches enormous proportions.

Hodgkin's disease A malignancy of lymphoid tissue; symptoms include swollen, nonpainful lymph nodes, fatigue, and often intermittent fever and night sweats. Characterized by presence of giant malignantly transformed B cells called Reed-Sternberg cells. Infection with Epstein-Barr virus (see mononucleosis below) and genetic susceptibility appear to be predisposing factors. Treated with chemotherapy and radiation; high cure rate.

Lymphadenopathy (lim-fad″ĕ-nop′ah-the; *adeno* = a gland; *pathy* = disease) Any disease of the lymph nodes.

Lymphangiography (lim-fan″je-og′rah-fe) Diagnostic procedure in which the lymphatic vessels are injected with radiopaque dye and then visualized with X rays.

Lymphoma Any neoplasm (tumor) of the lymphoid tissue, whether benign or malignant.

System Connections

System Connections: Homeostatic Interrelationships Between the Lymphatic System/Immunity and Other Body Systems

Integumentary System

- Lymphatic vessels pick up leaked plasma fluid and proteins from the dermis; lymphocytes in lymphoid organs and tissues enhance the skin's protective role by defending against specific pathogens via the immune response
- The skin's keratinized epithelium provides a mechanical barrier to pathogens; epithelial dendritic cells and dermal macrophages act as antigen presenters in the immune response; acid pH of skin secretions inhibits growth of bacteria on the skin

Skeletal System

- Lymphatic vessels pick up leaked plasma fluid and proteins from the periostea; immune cells protect bones from pathogens
- The bones house hematopoietic tissue which produces the lymphocytes (and macrophages) that populate lymphoid organs and provide immunity

Muscular System

- Lymphatic vessels pick up leaked fluids and proteins; immune cells protect muscles from pathogens
- The skeletal muscle "pump" aids the flow of lymph; heat produced during muscle activity initiates feverlike effects

Nervous System

- Lymphatic vessels pick up leaked plasma fluid and proteins in PNS structures; immune cells protect PNS structures from specific pathogens
- The nervous system innervates larger lymphatics; opiate neuropeptides influence immune functions; the brain helps regulate immune response

Endocrine System

- Lymphatic vessels pick up leaked fluids and proteins; lymph helps distribute hormones; immune cells protect endocrine organs
- The thymus produces hormones that are thought to be involved in the maturation of T cells; stress hormones depress immune activity

Cardiovascular System

- Lymphatic vessels pick up leaked plasma and proteins; spleen removes and destroys aged RBCs and debris, and stores iron and platelets; immune cells protect cardiovascular organs from specific pathogens
- Blood is the source of lymph; lymphatics develop from veins; blood circulates immune elements

Respiratory System

- Lymphatic vessels pick up leaked fluids and proteins from respiratory organs; immune cells protect respiratory organs from specific pathogens; the tonsils and plasma cells in the respiratory mucosa (which secrete the antibody IgA) prevent pathogen invasion
- The lungs provide O_2 needed by lymphoid/immune cells and eliminate CO_2; the pharynx houses the tonsils; respiratory "pump" aids lymph flow

Digestive System

- Lymphatic vessels pick up leaked fluids and proteins from digestive organs; lymph transports some products of fat digestion to the blood; lymphoid follicles in the intestinal wall prevent invasion of pathogens
- The digestive system digests and absorbs nutrients needed by cells of lymphoid organs; gastric acidity inhibits pathogens' entry into blood

Urinary System

- Lymphatic vessels pick up leaked fluid and proteins from urinary organs; immune cells protect urinary organs from specific pathogens
- Urinary system eliminates wastes and maintains homeostatic balances of water, pH, and electrolytes in the blood for lymphoid/immune cell functioning; urine flushes some pathogens out of the urinary tract

Reproductive System

- Lymphatic vessels pick up leaked fluid and proteins; immune cells protect against pathogens
- Reproductive organs' hormones may influence immune functioning; acidity of vaginal secretions is bacteriostatic

Closer Connections

The Lymphatic System/Immunity and Interrelationships with the Cardiovascular System

All living body cells are bathed by lymph. Nonetheless, the lymphatic system with its ghostly vessels is elusive and hard to pin down. Like a ground mole that tunnels throughout your yard, you know it's there, but you never see it as it quietly goes about its business. Since hormones made by endocrine organs are released into the extracellular space, and lymphatic vessels take up the fluid containing them, there is little question that lymph is an important medium for delivering hormones throughout the body. Lymph plays a similar delivery role as it delivers fats absorbed by the digestive organs.

Our immune system, charged with protecting the body from specific pathogens, is often considered a separate and independent functional system. However, it is impossible to divorce the immune system from the lymphatic system because the lymph nodes belong to both systems. The lymphoid organs and tissues, including the lymph nodes, are the anatomical underpinnings of the immune system. The lymphoid organs are the programming sites and seedbeds for the immune cells. The lymph nodes and spleen provide crucial vantage points for monitoring blood and lymph for the presence of foreign substances. Less understood are the intimate interactions between the immune cells that populate the lymphoid tissues and the nervous and endocrine systems, but we defer this interesting topic to Chapter 21 to give you the opportunity to consider the immune system in detail.

Although the lymphatic system serves the entire body, its chief "master" is the cardiovascular system. The immune system is also intimately tied to the cardiovascular system. It is these relationships that we will explore further here.

Cardiovascular System

By now you are probably fairly comfortable with the fact that the lymphatic system picks up and returns leaked fluid and proteins to the blood vascular system. So, what's the big deal? We ingest water in virtually everything we eat and drink—lost water is replaceable in a few minutes, right? The answer, of course, is "true." But what would still be missing is the leaked proteins, and these plasma proteins (most made by the liver) take time and energy to make. Since plasma proteins play a major role in keeping fluid in blood vessels (or encouraging its return), without them our blood vessels would contain too little fluid to support blood circulation. And without blood circulation, the whole body would die for lack of oxygen and nutrients and drown in its own wastes. Lymphoid organs also help to maintain the health and purity of blood—lymph nodes filter microorganisms and other debris from lymph before it is allowed to reenter the blood, and the spleen performs the same cleansing service for blood. In addition, the spleen disposes of inefficient, deformed, or aged erythrocytes.

The benefits are not entirely one-way. Lymphatic vessels spring from veins of the cardiovascular system, and blood delivers oxygen and nutrients to all body organs, including lymphoid ones. Blood also provides a means for (1) rapid transport of lymphocytes (immune cells) that continuously patrol the body for foreign substances and (2) broad distribution of antibodies (made by descendants of B cells called plasma cells), which mark foreign substances for destruction by phagocytosis or other means. Furthermore, the endothelial cells of capillaries express surface protein "signals" (integrins/adhesion molecules) that lymphocytes can recognize when the surrounding area is injured or infected. Hence, capillaries guide immune cells to the sites where their protective services are needed, and the extremely porous walls of the postcapillary venules allow them to slip through the vessel wall to get there.

Clinical Connections

Lymphatic System/Immunity

Case study: Back to following the progress of Mr. Hutchinson, we learn that the routine complete blood count (CBC) performed on admission revealed that his leukocyte count is dangerously low and follow-up lab tests show that his lymphocytes are deficient. One day postsurgery, he complains of pain in his right ring finger (that hand had a crush injury). When examined, the affected finger and the dorsum of the right hand were edematous, and red streaks radiated superiorly on his right forearm. Higher-than-normal doses of antibiotics are prescribed, and a sling is applied to the affected arm. Nurses are instructed to wear gloves and gown when giving Mr. Hutchinson his care.

Relative to these observations:

1. What do the red streaks emanating from the bruised finger indicate? What would you conclude his problem was if there were no red streaks but the right arm was very edematous?
2. Why is it important that Mr. Hutchinson not move the affected arm excessively (i.e., why was the sling ordered)?
3. How might the low lymphocyte count, megadoses of antibiotics, and orders for additional clinical staff protection be related?
4. Do you predict that Mr. Hutchinson's recovery will be uneventful or problematic? Why?

(Answers in Appendix G)

Mononucleosis A viral disease common in adolescents and young adults; symptoms include fatigue, fever, sore throat, and swollen lymph nodes. Caused by the Epstein-Barr virus, which is transmitted in saliva ("kissing disease") and specifically attacks B lymphocytes. This attack leads to a massive activation of T lymphocytes, which in turn attack the virus-infected B cells. Large numbers of oversized T lymphocytes circulate in the bloodstream. (These lymphocytes were originally misidentified as monocytes: *mononucleosis* = a condition of monocytes.) Usually lasts four to six weeks.

Non-Hodgkin's lymphoma Includes all cancers of lymphoid tissues except Hodgkin's disease. Involves uncontrolled multiplication and metastasis of undifferentiated lymphocytes, with swelling of the lymph nodes, spleen, and Peyer's patches; other organs may eventually become involved. The fifth most common cancer. A high-grade type, which primarily affects young people, grows quickly but is responsive to chemotherapy; up to a 75% remission rate. A low-grade type, which affects the elderly, is resistant to chemotherapy and so is often fatal.

Sentinel node The first node that receives lymph drainage from a body area suspected of being cancerous. When examined for presence of cancer cells, this node gives the best indication of whether metastasis through the lymphatic vessels has occurred.

Splenomegaly (sple″no-meg′ah-le; *mega* = big) Enlargement of the spleen due to accumulation of infectious microorganisms; typically caused by septicemia, mononucleosis, malaria, or leukemia.

Tonsillitis (ton″sĭ-li′tis; *itis* = inflammation) Inflammation of the tonsils, typically due to bacterial infection; they become red, swollen, and sore.

CHAPTER SUMMARY

Media study tools that could provide you additional help in reviewing specific key topics of Chapter 20 are referenced below.

iP = *Interactive Physiology*

1. Lymphatic vessels, lymph nodes, and lymph make up the lymphatic system. Lymphatic vessels return fluids that have leaked from the blood vascular system back to the blood. Lymphoid organs and tissues protect the body by removing foreign material from the lymph and blood streams, and provide a site for immune surveillance.

Lymphatic Vessels (pp. 753–755)

Distribution and Structure of Lymphatic Vessels (pp. 753–754)

1. Lymphatic vessels form a one-way network—lymphatic capillaries, collecting vessels, trunks, and ducts—in which fluid flows only toward the heart. The right lymphatic duct drains lymph from the right arm and right side of the upper body; the thoracic duct receives lymph from the rest of the body. These ducts empty into the blood vascular system at the junction of the internal jugular and subclavian veins in the neck.

2. Lymphatic capillaries are exceptionally permeable, admitting proteins and particulate matter from the interstitial space.
3. Pathogens and cancer cells may spread through the body via the lymphatic stream.

Lymph Transport (pp. 754–755)

4. The flow of lymphatic fluid is slow; it is maintained by skeletal muscle contraction, pressure changes in the thorax, and contractions of the lymphatic vessels. Backflow is prevented by valves.

iP Immune System; Topic: Anatomy Review, pp. 7–9.

Lymphoid Cells and Tissues (pp. 755–756)

Lymphoid Cells (pp. 755–756)

1. The cells in lymphoid tissues include lymphocytes (immunocompetent cells called T cells or B cells), plasma cells (antibody-producing offspring of B cells), macrophages and dendritic cells (cells that capture antigens and initiate an immune response), and reticular cells that form the lymphoid tissue stroma.

Lymphoid Tissue (p. 756)

2. Lymphoid tissue is reticular connective tissue. It houses macrophages and a continuously changing population of lymphocytes. It is an important element of the immune system.
3. Lymphoid tissue may be diffuse or packaged into dense follicles. Follicles often display germinal centers (areas where B cells are proliferating).

iP Immune System; Topic: Anatomy Review, pp. 4–5.

Lymph Nodes (pp. 756–758)

1. Lymph nodes, the principal lymphoid organs, are discrete encapsulated structures containing both diffusely arranged and dense reticular tissue. Clustered along lymphatic vessels, lymph nodes filter lymph and help activate the immune system.

Structure of a Lymph Node (pp. 756–757)

2. Each lymph node has a fibrous capsule, a cortex, and a medulla. The cortex contains mostly lymphocytes, which act in immune responses; the medulla contains macrophages, which engulf and destroy viruses, bacteria, and other foreign debris, as well as lymphocytes and plasma cells.

Circulation in the Lymph Nodes (pp. 757–758)

3. Lymph enters the lymph nodes via afferent lymphatic vessels and exits via efferent vessels. There are fewer efferent vessels; therefore, lymph flow stagnates within the lymph node, allowing time for its cleansing.

iP Immune System; Topic: Anatomy Review, pp. 10–12.

Other Lymphoid Organs (pp. 758–763)

1. Unlike lymph nodes, the spleen, thymus, tonsils, and Peyer's patches do not filter lymph. However, most lymphoid organs contain both macrophages and lymphocytes.

Spleen (p. 758)

2. The spleen provides a site for lymphocyte proliferation and immune function, and destroys aged or defective red blood cells and bloodborne pathogens. It also stores and releases the breakdown products of hemoglobin as necessary, stores platelets, and acts as a hematopoietic site in the fetus.

Thymus (pp. 759–760)

3. The thymus is most functional during youth. The thymus provides the environment in which T lymphocytes mature and become immunocompetent.

Tonsils and Aggregates of Lymphoid Follicles (pp. 760–761)

4. Peyer's patches of the intestinal wall, lymphoid follicles of the appendix, tonsils of the pharynx and oral cavity, and follicles in the genitourinary and respiratory tract mucosae are known as MALT (mucosa-associated lymphatic tissue). They prevent pathogens in these passages from penetrating the mucous membrane lining.

iP Immune System; Topic: Anatomy Review, pp. 13–14, 16–20.

Developmental Aspects of the Lymphatic System and Lymphoid Organs and Tissues (p. 761)

1. Lymphatics develop as outpocketings of developing veins. The thymus develops from endoderm; the other lymphoid organs derive from mesenchymal cells of mesoderm.
2. The thymus is the first lymphoid organ to develop.
3. Lymphoid organs are populated by lymphocytes, which arise from hematopoietic tissue.

REVIEW QUESTIONS

Multiple Choice/Matching

(Some questions have more than one correct answer. Select the best answer or answers from the choices given.)

1. Lymphatic vessels **(a)** serve as sites for immune surveillance, **(b)** filter lymph, **(c)** transport leaked plasma proteins and fluids to the cardiovascular system, **(d)** are represented by vessels that resemble arteries, capillaries, and veins.
2. The saclike initial portion of the thoracic duct is the **(a)** lacteal, **(b)** right lymphatic duct, **(c)** cisterna chyli, **(d)** lymph sac.
3. Entry of lymph into the lymphatic capillaries is promoted by which of the following? **(a)** one-way minivalves formed by overlapping endothelial cells, **(b)** the respiratory pump, **(c)** the skeletal muscle pump, **(d)** greater fluid pressure in the interstitial space.
4. The structural framework of lymphoid organs is **(a)** areolar connective tissue, **(b)** hematopoietic tissue, **(c)** reticular tissue, **(d)** adipose tissue.
5. Lymph nodes are densely clustered in all of the following body areas *except* **(a)** the brain, **(b)** the axillae, **(c)** the groin, **(d)** the cervical region.
6. The germinal centers in lymph nodes are largely sites of **(a)** macrophages, **(b)** proliferating B lymphocytes, **(c)** T lymphocytes, **(d)** all of these.
7. The red pulp areas of the spleen are sites of **(a)** venous sinuses, macrophages, and red blood cells, **(b)** clustered lymphocytes, **(c)** connective tissue septa.
8. The lymphoid organ that functions primarily during youth and then begins to atrophy is the **(a)** spleen, **(b)** thymus, **(c)** palatine tonsils, **(d)** bone marrow.
9. Collections of lymphoid tissue (MALT) that guard mucosal surfaces include all of the following except **(a)** appendix follicles, **(b)** the tonsils, **(c)** Peyer's patches, **(d)** the thymus.

Short Answer Essay Questions

10. Compare and contrast blood, interstitial fluid, and lymph.
11. Compare the structure and functions of a lymph node to those of the spleen.
12. **(a)** What anatomical characteristic ensures that the flow of lymph through a lymph node is slow? **(b)** Why is this desirable?
13. Why doesn't the absence of lymphatic arteries represent a problem?

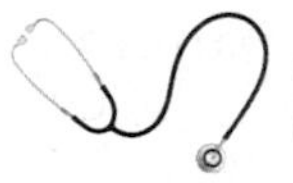

Critical Thinking and Clinical Application Questions

1. Mrs. Jackson, a 59-year-old woman, has undergone a left radical mastectomy (removal of the left breast and left axillary lymph nodes and vessels). Her left arm is severely swollen and painful, and she is unable to raise it to more than shoulder height. **(a)** Explain her signs and symptoms. **(b)** Can she expect to have relief from these symptoms in time? How so?
2. A friend tells you that she has tender, swollen "glands" along the left side of the front of her neck. You notice that she has a bandage on her left cheek that is not fully hiding a large infected cut there. Exactly what are her swollen "glands," and how did they become swollen?
3. Once almost a rite of childhood, tonsillectomy (surgical removal of the tonsils) is now rarely performed. Similarly, while ruptured spleens were once routinely removed, they are now conserved whenever possible. Why should these lymphoid organs be preserved when possible

Access everything you need to practice, review, and self-assess for both your A&P lecture and lab courses at **myA&P** (www.myaandp.com). There, you'll find powerful online resources, including chapter quizzes and tests, games, A&P Flix animations with quizzes, ***Interactive Physiology***® with quizzes, MP3 Tutor Sessions, Practice Anatomy Lab™, and more to help you get a better grade in your course.

20

21

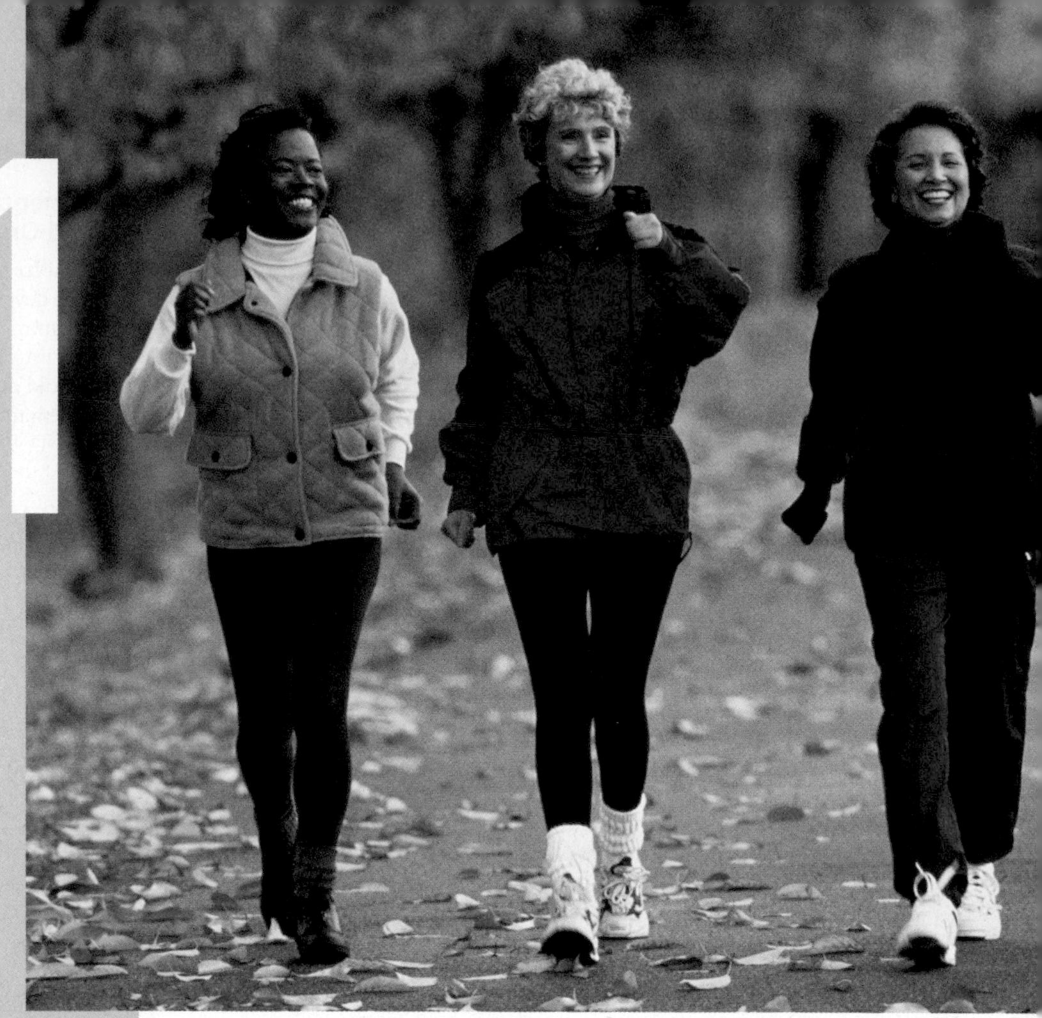

The Immune System: Innate and Adaptive Body Defenses

Every second of every day, armies of hostile bacteria, fungi, and viruses swarm on our skin and yet we stay amazingly healthy most of the time. The body seems to have evolved a single-minded approach to such foes—if you're not with us, you're against us! To implement that stance, it relies heavily on two intrinsic defense systems that act both independently and cooperatively to provide resistance to disease, or **immunity** (*immun* = free).

1. The **innate (nonspecific) defense system**, like a lowly foot soldier, is always prepared, responding within minutes to protect the body from all foreign substances. This system has two "barricades." The *first line of defense* is the external body membranes—intact skin and mucosae. The *second line of defense*, called into action whenever the first line has been penetrated, uses antimicrobial proteins, phagocytes, and other cells to inhibit the invaders' spread throughout the body. The hallmark of the second line of defense is inflammation.
2. The **adaptive** (or **specific**) **defense system** is more like an elite fighting force equipped with high-tech weapons that attacks *particular* foreign substances and provides the body's *third line of defense*. This defensive response takes considerably more time to mount than the innate response. Although we consider them separately, the adaptive and innate systems always work hand in hand. An overview of these two systems is shown in **Figure 21.1**. Small portions of this diagram will reappear in subsequent figures to let you know which part of the immune system we're dealing with.

Although certain organs of the body (notably lymphoid organs) are intimately involved in the immune response, the **immune system** is a *functional system* rather than an organ system in an anatomical sense. Its "structures" are a diverse array of molecules plus trillions of immune cells (especially lymphocytes) that inhabit lymphoid tissues and circulate in body fluids.

Once, the term *immune system* was equated with the adaptive defense system only. However, we now know that the innate and adaptive defenses are deeply intertwined. Specifically, (1) many defensive molecules are released and recognized by both the innate and adaptive arms; (2) the innate responses are not as nonspecific as once thought and have specific pathways to target certain foreign substances; and (3) proteins released during innate responses alert cells of the adaptive system to the presence of specific foreign molecules in the body.

When the immune system is operating effectively, it protects the body from most infectious microorganisms, cancer cells, and transplanted organs or grafts. It does this both directly, by cell attack, and indirectly, by releasing mobilizing chemicals and protective antibody molecules.

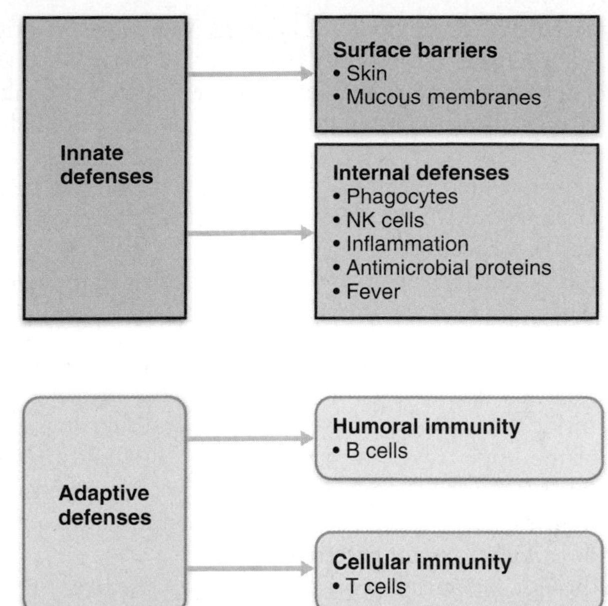

Figure 21.1 Overview of innate and adaptive defenses. Humoral immunity (primarily involving B lymphocytes) and cellular immunity (involving T lymphocytes) are distinct but overlapping areas of adaptive immunity. For simplicity, the many interactions between innate and adaptive defenses are not shown here.

PART 1

INNATE DEFENSES

Because they are part and parcel of our anatomy, you could say we come fully equipped with innate defenses. The mechanical barriers that cover body surfaces and the cells and chemicals that act on the initial internal battlefronts are in place at birth, ready to ward off invading **pathogens** (harmful or disease-causing microorganisms) and infection.

Many times, our innate defenses alone are able to destroy pathogens and ward off infection. In other cases, the adaptive immune system is called into action to reinforce and enhance the innate defenses. Either way, the innate defenses reduce the workload of the adaptive system by preventing the entry and spread of microorganisms in the body.

Surface Barriers: Skin and Mucosae

► Describe surface membrane barriers and their protective functions.

The body's first line of defense—the *skin* and the *mucous membranes*, along with the secretions these membranes produce—is highly effective. As long as the epidermis is unbroken, this heavily keratinized epithelial membrane presents a formidable physical barrier to most microorganisms that swarm on the skin. Keratin is also resistant to most weak acids and bases and to bacterial enzymes and toxins. Intact mucosae provide similar mechanical barriers within the body. Recall that mucous membranes line all body cavities that open to the exterior: the digestive, respiratory, urinary, and reproductive tracts. Besides serving as physical barriers, these epithelial membranes produce a variety of protective chemicals:

1. The acidity of skin secretions (pH 3 to 5) inhibits bacterial growth. In addition, lipids in sebum and dermcidin in

eccrine sweat are toxic to bacteria. Vaginal secretions of adult females are also very acidic.
2. The stomach mucosa secretes a concentrated hydrochloric acid solution and protein-digesting enzymes. Both kill microorganisms.
3. Saliva, which cleanses the oral cavity and teeth, and lacrimal fluid of the eye contain **lysozyme**, an enzyme that destroys bacteria.
4. Sticky mucus traps many microorganisms that enter the digestive and respiratory passageways.

The respiratory tract mucosae also have structural modifications that counteract potential invaders. Tiny mucus-coated hairs inside the nose trap inhaled particles, and cilia on the mucosa of the upper respiratory tract sweep dust- and bacteria-laden mucus toward the mouth, preventing it from entering the lower respiratory passages, where the warm, moist environment provides an ideal site for bacterial growth.

Although the surface barriers are quite effective, they are breached occasionally by small nicks and cuts resulting, for example, from brushing your teeth or shaving. When this happens and microorganisms invade deeper tissues, the *internal* innate defenses come into play.

CHECK YOUR UNDERSTANDING

1. What distinguishes the innate defense system from the adaptive defense system?
2. What is the first line of defense against disease?

For answers, see Appendix G.

Internal Defenses: Cells and Chemicals

21

▶ Explain the importance of phagocytosis and natural killer cells in innate body defense.

The body uses an enormous number of nonspecific cellular and chemical devices to protect itself, including phagocytes, natural killer cells, antimicrobial proteins, and fever. The inflammatory response enlists macrophages, mast cells, all types of white blood cells, and dozens of chemicals that kill pathogens and help repair tissue. All these protective ploys identify potentially harmful substances by recognizing surface carbohydrates unique to infectious organisms (bacteria, viruses, and fungi).

Phagocytes

Pathogens that get through the skin and mucosae into the underlying connective tissue are confronted by *phagocytes* (*phago* = eat). The chief phagocytes are **macrophages** ("big eaters"), which derive from white blood cells called **monocytes** that leave the bloodstream, enter the tissues, and develop into macrophages.

Free macrophages, like the alveolar macrophages of the lungs, wander throughout the tissue spaces in search of cellular debris or "foreign invaders." *Fixed macrophages* like *Kupffer cells* in the liver and microglia of the brain are permanent residents of particular organs. Whatever their mobility, all macrophages are similar structurally and functionally.

Neutrophils, the most abundant type of white blood cell, become phagocytic on encountering infectious material in the tissues.

Phagocytosis

A phagocyte engulfs particulate matter much the way an amoeba ingests a food particle. Flowing cytoplasmic extensions bind to the particle and then pull it inside, enclosed within a membrane-lined vesicle (**Figure 21.2a**). The resulting **phagosome** is then fused with a *lysosome* to form a **phagolysosome** (steps ①–③ in Figure 21.2b).

Phagocytic attempts are not always successful. In order for a phagocyte to accomplish ingestion, **adherence** must occur. The phagocyte must first *adhere* or cling to the pathogen, a feat made possible by recognizing the pathogen's carbohydrate "signature." Recognition is particularly difficult with microorganisms such as pneumococcus, which have an external capsule made of complex sugars. These pathogens can sometimes elude capture because phagocytes cannot bind to their capsules. Adherence is both more probable and more efficient when complement proteins or antibodies coat foreign particles, a process called **opsonization** ("to make tasty"), because the coating provides "handles" to which phagocyte receptors can bind.

Sometimes the way neutrophils and macrophages kill ingested prey is more than simple acidification and digestion by lysosomal enzymes. For example, pathogens such as the tuberculosis bacillus and certain parasites are resistant to lysosomal enzymes and can even multiply within the phagolysosome. However, when the macrophage is stimulated by chemicals released by other immune cells called helper T cells, additional enzymes are activated that produce the **respiratory burst**. This event liberates a deluge of free radicals (including nitric oxide and superoxide) that have potent cell-killing abilities. More widespread cell killing is caused by oxidizing chemicals (H_2O_2 and a substance identical to household bleach). The respiratory burst also increases the pH and osmolarity in the phagolysosome, which activates other protein-digesting enzymes that digest the invader. Neutrophils also produce antimicrobial chemicals, called *defensins*, that pierce the pathogen's membrane.

When phagocytes are unable to ingest their targets (because of size, for example), they can release their toxic chemicals into the extracellular fluid. Whether killing ingested or extracellular targets, neutrophils rapidly destroy themselves in the process, whereas macrophages are more robust and can go on to kill another day.

Natural Killer Cells

Natural killer (NK) cells, which "police" the body in blood and lymph, are a unique group of defensive cells that can lyse and kill cancer cells and virus-infected body cells before the adaptive immune system is activated. Sometimes called the "pit bulls" of the defense system, NK cells are part of a small group of *large*

Innate defenses → Internal defenses

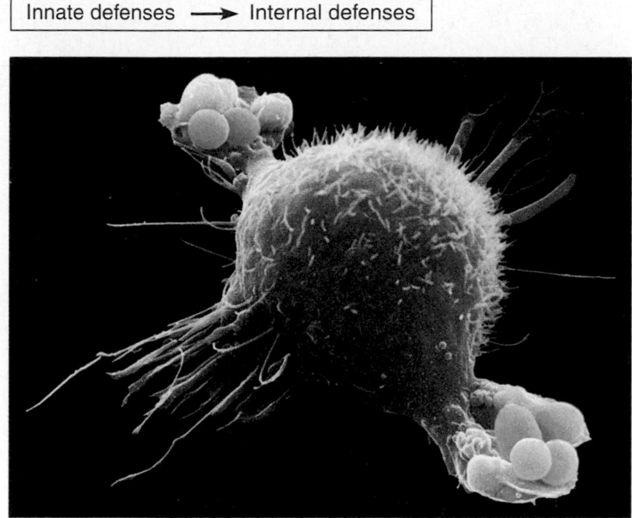

(a) **A macrophage (purple) uses its cytoplasmic extensions to pull spherical bacteria (green) toward it.** Scanning electron micrograph (1750×).

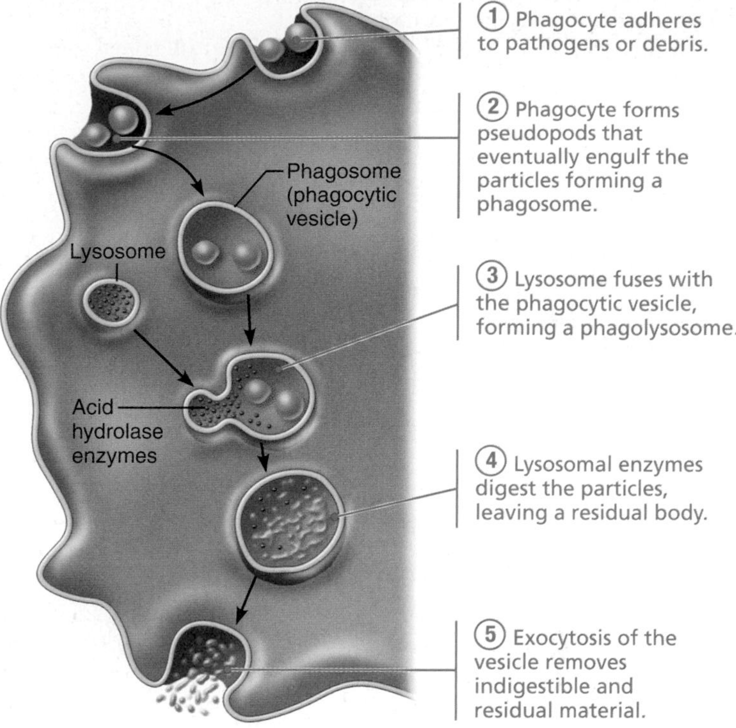

① Phagocyte adheres to pathogens or debris.
② Phagocyte forms pseudopods that eventually engulf the particles forming a phagosome.
Phagosome (phagocytic vesicle)
Lysosome
③ Lysosome fuses with the phagocytic vesicle, forming a phagolysosome.
Acid hydrolase enzymes
④ Lysosomal enzymes digest the particles, leaving a residual body.
⑤ Exocytosis of the vesicle removes indigestible and residual material.

(b) **Events of phagocytosis.**

Figure 21.2 **Phagocytosis.**

granular lymphocytes. Unlike lymphocytes of the adaptive immune system, which recognize and react only against *specific* virus-infected or tumor cells, NK cells are far less picky. They can eliminate a variety of infected or cancerous cells by detecting the lack of "self" cell-surface receptors and by recognizing certain surface sugars on the target cell. The name "natural" killer cells reflects this nonspecificity of NK cells.

NK cells are not phagocytic. Their mode of killing, which involves direct contact and induces the target cell to undergo apoptosis (programmed cell death), uses the same killing mechanisms used by cytotoxic T cells (described fully on p. 792). NK cells also secrete potent chemicals that enhance the inflammatory response.

Inflammation: Tissue Response to Injury

▶ Describe the inflammatory process. Identify several inflammatory chemicals and indicate their specific roles.

The **inflammatory response** is triggered whenever body tissues are injured by physical trauma (a blow), intense heat, irritating chemicals, or infection by viruses, fungi, or bacteria. The inflammatory response has several beneficial effects:

1. Prevents the spread of damaging agents to nearby tissues
2. Disposes of cell debris and pathogens
3. Sets the stage for repair

The four *cardinal signs* of short-term, or acute, inflammation are *redness, heat* (*inflam* = set on fire), *swelling*, and *pain.* If the inflamed area is a joint, joint movement may be hampered temporarily. This forces the injured part to rest, which aids healing. Some authorities consider *impairment of function* to be the fifth cardinal sign of acute inflammation. **Figure 21.3** presents an overview of the inflammatory process, and shows how these cardinal signs come about.

Vasodilation and Increased Vascular Permeability

The inflammatory process begins with a chemical "alarm" as a flood of inflammatory chemicals are released into the extracellular fluid. Macrophages (and cells of certain boundary tissues such as epithelial cells lining the gastrointestinal and respiratory tracts) bear surface membrane receptors, called **Toll-like receptors (TLRs)**, that play a central role in triggering immune responses. So far 11 types of human TLRs have been identified, each recognizing a specific class of attacking microbe. For example, one type responds to a glycolipid in cell walls of the tuberculosis bacterium and another to a component of gram-negative bacteria such as *Salmonella.* Once activated, a TLR triggers the release of chemicals called *cytokines* that promote inflammation and attract WBCs to the scene.

Macrophages are not the only sort of recognition "tool" in the innate system. **Mast cells**, a key component of the inflammatory response, release the potent inflammatory chemical **histamine** (his′tah-mēn). In addition, injured and stressed tissue cells, phagocytes, lymphocytes, basophils, and blood proteins are all sources of inflammatory mediators. These chemicals include not only histamine and cytokines, but **kinins** (ki′ninz), **prostaglandins** (pros″tah-glan′dinz), **leukotrienes**,

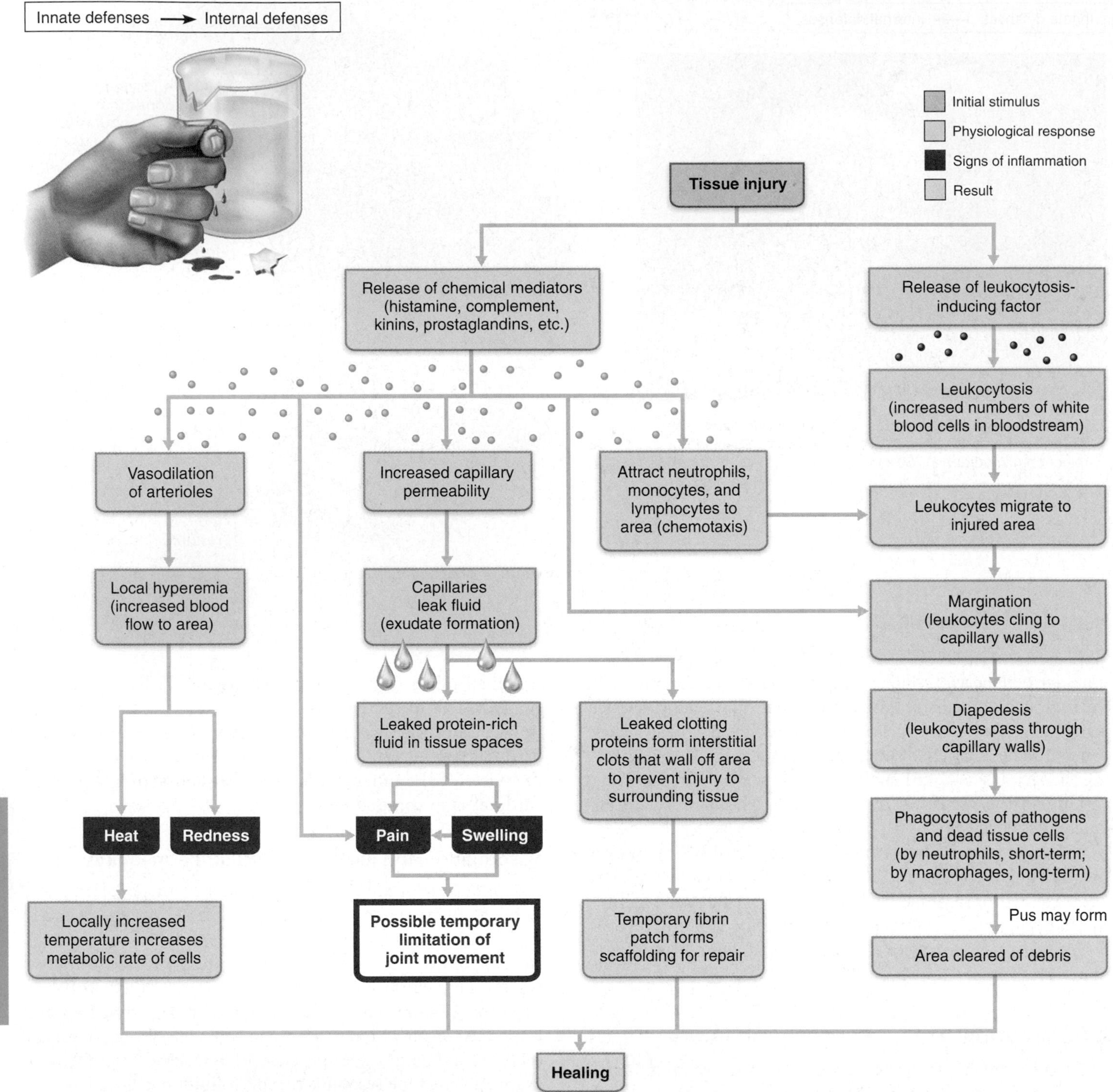

21

Figure 21.3 Inflammation: flowchart of events. The four cardinal signs of acute inflammation are shown in red boxes, as is limitation of joint movement, which in some cases constitutes a fifth cardinal sign (impairment of function).

and **complement**. They all cause arterioles in the injured area to dilate, although some of these mediators have individual inflammatory roles as well **(Table 21.1)**. As more blood flows into the area, local **hyperemia** (congestion with blood) occurs, accounting for the *redness* and *heat* of an inflamed region.

The liberated chemicals also increase the permeability of local capillaries. Consequently, **exudate**—fluid containing clotting factors and antibodies—seeps from the blood into the tissue spaces. This exudate causes the local *swelling*, also called edema, that presses on adjacent nerve endings, contributing to a sensation of *pain*. Pain also results from the release of bacterial toxins, and the sensitizing effects of released prostaglandins and kinins. Aspirin and some other anti-inflammatory drugs produce their analgesic (pain-reducing) effects by inhibiting prostaglandin synthesis.

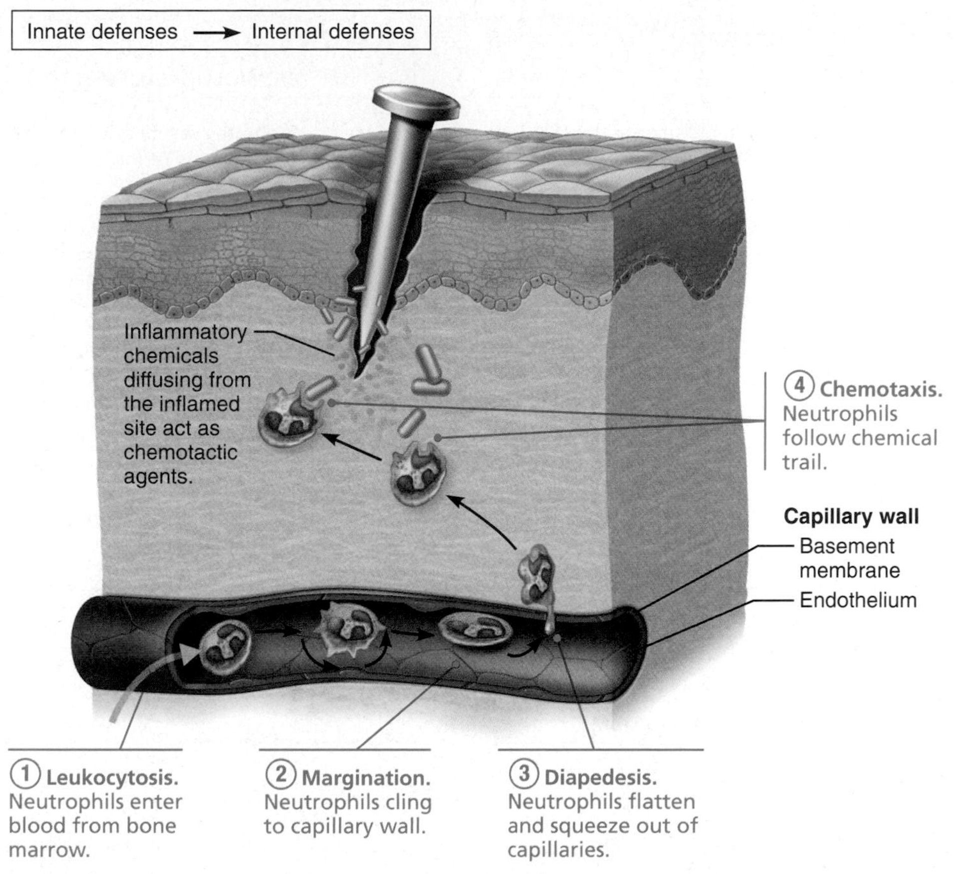

Innate defenses → Internal defenses
Inflammatory chemicals diffusing from the inflamed site act as chemotactic agents.
④ Chemotaxis. Neutrophils follow chemical trail.
Capillary wall
Basement membrane
Endothelium
① Leukocytosis. Neutrophils enter blood from bone marrow.
② Margination. Neutrophils cling to capillary wall.
③ Diapedesis. Neutrophils flatten and squeeze out of capillaries.

Figure 21.4 Phagocyte mobilization.

Although edema may seem detrimental, it isn't. The surge of protein-rich fluids into the tissue spaces sweeps any foreign material into the lymphatic vessels for processing in the lymph nodes. It also delivers important proteins such as complement and clotting factors to the interstitial fluid (Figure 21.3). The clotting proteins form a gel-like fibrin mesh that forms a scaffold for permanent repair. This isolates the injured area and prevents the spread of bacteria and other harmful agents into surrounding tissues. Walling off the injured area is such an important defense strategy that some bacteria (such as *Streptococcus*) have evolved enzymes that break down the clot, allowing the bacteria to spread rapidly into adjacent tissues.

At inflammatory sites where an epithelial barrier has been breached, additional chemicals enter the battle—β-defensins. These broad-spectrum antimicrobial chemicals are continuously present in epithelial mucosal cells in small amounts and help maintain the sterile environment of the body's internal passageways (urinary tract, respiratory bronchi, etc.). However, when the mucosal surface is abraded or penetrated and the underlying connective tissue becomes inflamed, β-defensin output increases dramatically, helping to control bacterial and fungal colonization in the exposed area.

Phagocyte Mobilization

Soon after inflammation begins, the damaged area is flooded with phagocytes. Neutrophils lead, followed by macrophages. If the inflammation was provoked by pathogens, a group of plasma proteins known as complement (to be discussed shortly) is activated and elements of adaptive immunity (lymphocytes and antibodies) also migrate to the injured site. The process by which phagocytes are mobilized to infiltrate the injured site consists of the four steps illustrated in **Figure 21.4**. 21

① **Leukocytosis.** Neutrophils enter blood from red bone marrow in response to chemicals called **leukocytosis-inducing factors** released by injured cells. Within a few hours, the number of neutrophils in blood increases four- to fivefold, resulting in **leukocytosis** (an increase in WBCs that is a characteristic of inflammation).

② **Margination.** Inflamed endothelial cells sprout cell adhesion molecules (CAMs) that signal "this is the place." As neutrophils encounter these CAMs, they slow and roll along the surface, achieving an initial foothold. When activated by inflammatory chemicals, CAMs on neutrophils bind tightly to endothelial cells. This clinging of phagocytes to the inner walls of the capillaries and postcapillary venules is called **margination**.

TABLE 21.1 Inflammatory Chemicals

CHEMICAL	SOURCE	PHYSIOLOGICAL EFFECTS
Histamine	Granules of mast cells and basophils. Released in response to mechanical injury, presence of certain microorganisms, and chemicals released by neutrophils.	Promotes vasodilation of local arterioles. Increases permeability of local capillaries, promoting exudate formation.
Kinins (bradykinin and others)	A plasma protein, kininogen, is cleaved by the enzyme kallikrein found in plasma, urine, saliva, and in lysosomes of neutrophils and other types of cells. Cleavage releases active kinin peptides.	Same as for histamine. Also induce chemotaxis of leukocytes and prompt neutrophils to release lysosomal enzymes, thereby enhancing generation of more kinins. Induce pain.
Prostaglandins	Fatty acid molecules produced from arachidonic acid found in all cell membranes; generated by enzymes of neutrophils, basophils, mast cells, and others.	Same as for histamine. Also induce neutrophil chemotaxis. Induce pain.
Platelet-derived growth factor (PDGF)	Secreted by platelets and endothelial cells.	Stimulates fibroblast activity and repair of damaged tissues.
Complement	See Table 21.2 (p. 772).	
Cytokines	See Table 21.4 (p. 790).	

TABLE 21.2 Summary of Innate Body Defenses

CATEGORY/ASSOCIATED ELEMENTS	PROTECTIVE MECHANISM
First Line of Defense: Surface Membrane Barriers	
Intact skin epidermis	Forms mechanical barrier that prevents entry of pathogens and other harmful substances into body
▪ Acid mantle	Skin secretions (sweat and sebum) make epidermal surface acidic, which inhibits bacterial growth; also contain various bactericidal chemicals
▪ Keratin	Provides resistance against acids, alkalis, and bacterial enzymes
Intact mucous membranes	Form mechanical barrier that prevents entry of pathogens
▪ Mucus	Traps microorganisms in respiratory and digestive tracts
▪ Nasal hairs	Filter and trap microorganisms in nasal passages
▪ Cilia	Propel debris-laden mucus away from nasal cavity and lower respiratory passages
▪ Gastric juice	Contains concentrated hydrochloric acid and protein-digesting enzymes that destroy pathogens in stomach
▪ Acid mantle of vagina	Inhibits growth of most bacteria and fungi in female reproductive tract
▪ Lacrimal secretion (tears); saliva	Continuously lubricate and cleanse eyes (tears) and oral cavity (saliva); contain lysozyme, an enzyme that destroys microorganisms
▪ Urine	Normally acid pH inhibits bacterial growth; cleanses the lower urinary tract as it flushes from the body
Second Line of Defense: Innate Cellular and Chemical Defenses	
Phagocytes	Engulf and destroy pathogens that breach surface membrane barriers; macrophages also contribute to adaptive immune responses
Natural killer (NK) cells	Promote apoptosis (cell suicide) by direct cell attack against virus-infected or cancerous body cells; do not require specific antigen recognition; do not exhibit a memory response
Inflammatory response	Prevents spread of injurious agents to adjacent tissues, disposes of pathogens and dead tissue cells, and promotes tissue repair; chemical mediators released attract phagocytes (and other immune cells) to the area
Antimicrobial proteins	
▪ Interferons (α, β, γ)	Proteins released by virus-infected cells and certain lymphocytes that protect uninfected tissue cells from viral takeover; mobilize immune system
▪ Complement	Lyses microorganisms, enhances phagocytosis by opsonization, and intensifies inflammatory and immune responses
Fever	Systemic response initiated by pyrogens; high body temperature inhibits microbial multiplication and enhances body repair processes

③ **Diapedesis.** Continued chemical signaling prompts the neutrophils to flatten and squeeze through the capillary walls—**diapedesis.**

④ **Chemotaxis.** Inflammatory chemicals act as homing devices, or more precisely **chemotactic agents.** Neutrophils and other WBCs migrate up the gradient of chemotactic agents to the site of injury (positive **chemotaxis**). Within an hour after the inflammatory response has begun, neutrophils have collected at the site and are devouring any foreign material present.

As the body's counterattack continues, monocytes follow neutrophils into the injured area. Monocytes are fairly poor phagocytes, but within 12 hours of leaving the blood and entering the tissues, they swell and develop large numbers of lysosomes, becoming macrophages with insatiable appetites. These late-arriving macrophages replace the neutrophils on the battlefield.

Macrophages are the central actors in the final disposal of cell debris as an acute inflammation subsides, and they predominate at sites of prolonged, or *chronic*, inflammation. The ultimate goal of an inflammatory response is to clear the injured area of pathogens, dead tissue cells, and any other debris so that tissue can be repaired. Once this is accomplished, healing usually occurs quickly.

HOMEOSTATIC IMBALANCE

In severely infected areas, the battle takes a considerable toll on both sides, and creamy yellow **pus** (a mixture of dead or dying neutrophils, broken-down tissue cells, and living and dead pathogens) may accumulate in the wound. If the inflammatory mechanism fails to clear the area of debris, the sac of pus may be walled off by collagen fibers, forming an *abscess.* Surgical drainage of abscesses is often necessary before healing can occur.

Some bacteria, such as tuberculosis bacilli, are resistant to digestion by the macrophages that engulf them. They escape the effects of prescription antibiotics by remaining snugly enclosed within their macrophage hosts. In such cases, *infectious granulomas* form. These tumorlike growths contain a central region of infected macrophages surrounded by uninfected macrophages and an outer fibrous capsule. A person may harbor pathogens walled off in granulomas for years without displaying any symptoms. However, if the person's resistance to infection is ever compromised, the bacteria may be activated and break out, leading to clinical disease symptoms. ■

Antimicrobial Proteins

▶ Name the body's antimicrobial substances and describe their function.

A variety of **antimicrobial proteins** enhance the innate defenses by attacking microorganisms directly or by hindering their ability to reproduce. The most important of these are interferons and complement proteins **(Table 21.2).**

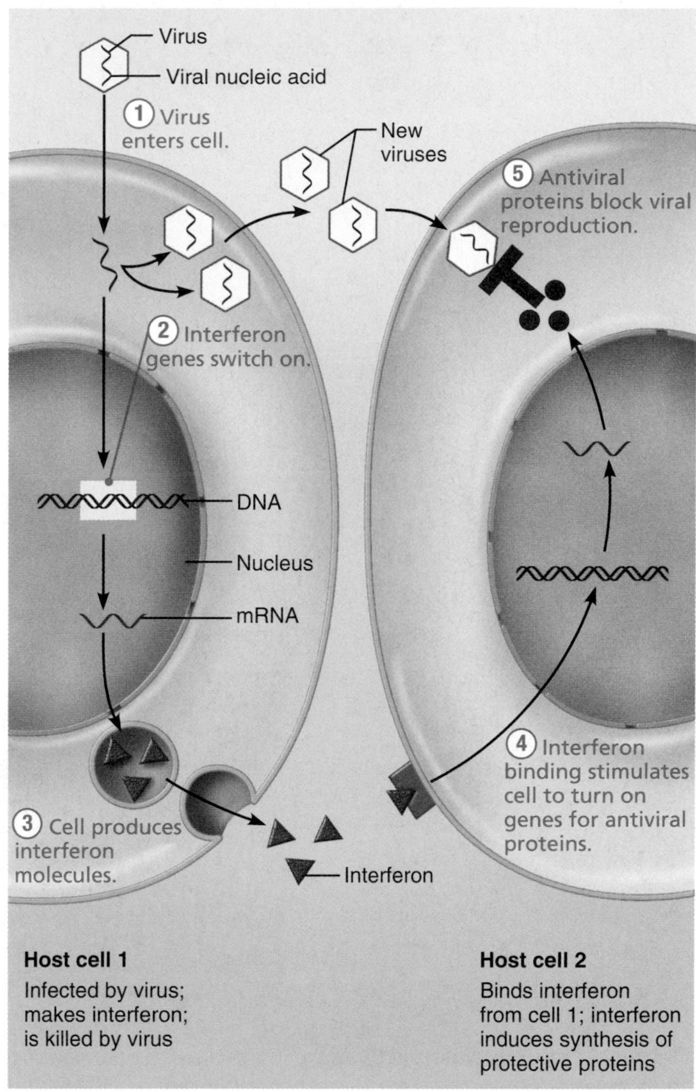

Figure 21.5 The interferon mechanism against viruses.

21

Interferons

Viruses—essentially nucleic acids surrounded by a protein coat—lack the cellular machinery to generate ATP or synthesize proteins. They do their "dirty work," or damage, in the body by invading tissue cells and taking over the cellular metabolic machinery needed to reproduce themselves. The infected cells can do little to save themselves, but some can secrete small proteins called **interferons (IFNs)** (in″ter-fēr′onz) to help protect cells that have not yet been infected. The IFNs diffuse to nearby cells, where they stimulate synthesis of proteins which then "interfere" with viral replication in the still-healthy cells by blocking protein synthesis and degrading viral RNA **(Figure 21.5).** Because IFN protection is not *virus-specific*, IFNs produced against a particular virus protect against a variety of other viruses.

The IFNs are a family of related proteins produced by a variety of body cells, each having a slightly different physiological effect. Lymphocytes secrete gamma (γ), or immune, interferon,

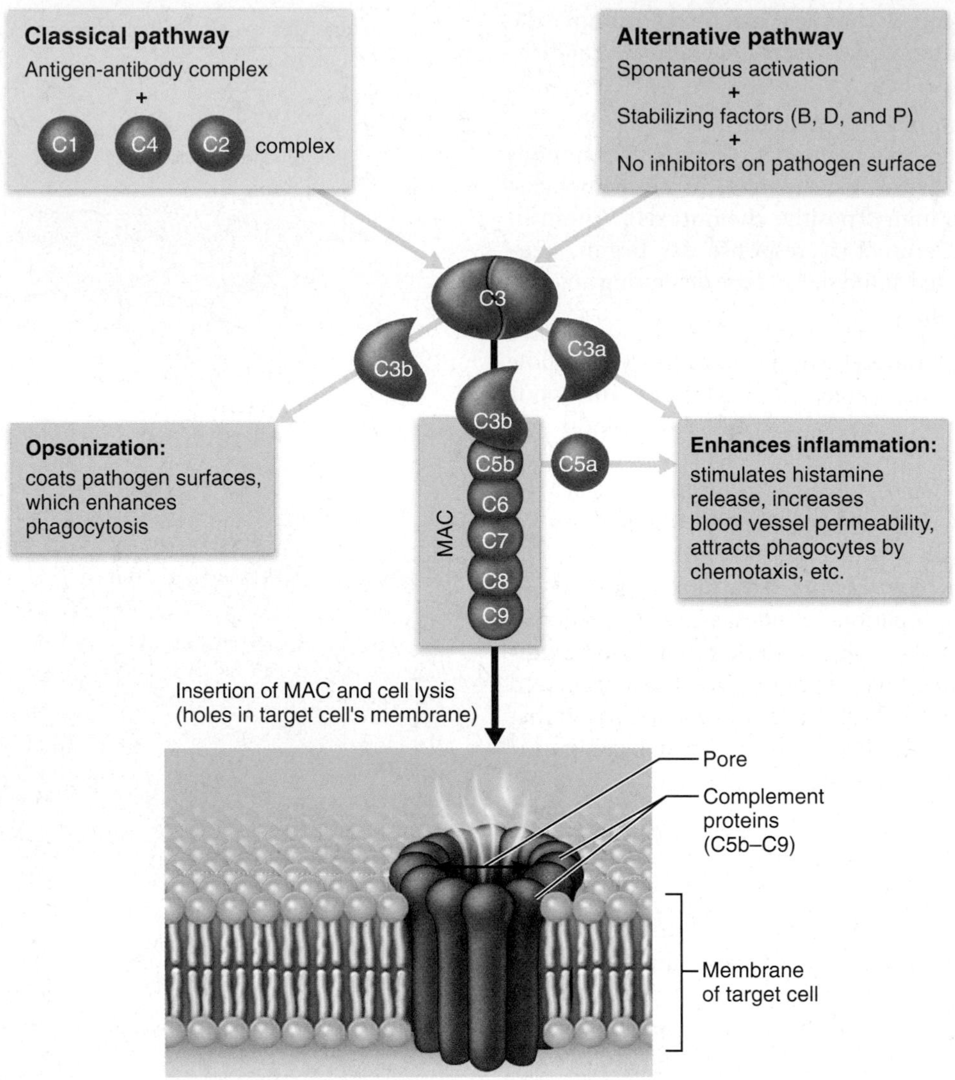

21

Figure 21.6 Complement activation. In the classical pathway of complement activation, antibodies coating the surface of a pathogen activate complement proteins C1, C4, and C2, which in turn activate C3. In the alternative pathway of complement activation, C3 spontaneously activates and attaches to pathogen membranes. (Unlike our own membranes, pathogen membranes do not have inhibitors of complement activation.) Factors B, D, and P stabilize spontaneously activated C3. The two pathways converge at C3, which splits into two active pieces: C3a and C3b. C3a promotes inflammation (with the help of C5a). C3b enhances phagocytosis by acting as an opsonin. In certain target cells (mostly bacteria), C3b also activates other complement proteins that can form a membrane attack complex (MAC). MACs form from activated complement components (C5b and C6 to C9) that insert into the target cell membrane, creating funnel-shaped pores that can lyse the target cell.

but most other leukocytes secrete alpha (α) interferon. Beta (β) interferon is secreted by fibroblasts. IFN-β and -α reduce inflammation, keeping it in control. Besides their antiviral effects, interferons activate macrophages and mobilize NK cells. Because both macrophages and NK cells can act directly against malignant cells, the interferons play some anticancer role.

Genetically engineered IFNs have found a niche as antiviral agents. For example, *IFN-α* is used to treat genital warts and hepatitis C. *IFN-β* is used to treat patients with multiple sclerosis, a devastating demyelinating disease.

Complement

The term **complement system**, or simply **complement**, refers to a group of at least 20 plasma proteins that normally circulate in the blood in an inactive state. These proteins include C1 through C9, factors B, D, and P, plus several regulatory proteins. Complement provides a major mechanism for destroying foreign substances in the body. Its activation unleashes chemical mediators that amplify virtually all aspects of the inflammatory process. Another effect of complement activation is that certain bacteria and other cell types are killed by cell lysis. (Luckily our own cells are equipped with proteins that inactivate complement.) Although complement is a nonspecific defensive mechanism, it "complements" (enhances) the effectiveness of *both* innate and adaptive defenses.

Complement can be activated by the two pathways outlined in **Figure 21.6**. The **classical pathway** involves *antibodies*, water-soluble protein molecules that the adaptive immune system produces to fight off foreign invaders. The classical pathway depends on the binding of antibodies to the invading organisms and the subsequent binding of C1 to the microorganism-antibody complexes, a step called *complement fixation*, which we will describe on p. 775. The **alternative pathway** is triggered when spontaneously activated C3 and factors B, D, and P interact on the surface of certain microorganisms.

Like the blood clotting cascade, complement activation by either pathway involves a cascade in which proteins are activated in an orderly sequence—each step catalyzing the next. The two pathways converge at C3, which cleaves into C3a and C3b. This event initiates a common terminal pathway that enhances inflammation, promotes phagocytosis, and can cause cell lysis.

The lytic events begin when C3b binds to the target cell's surface, triggering the insertion of a group of complement proteins called **MAC (membrane attack complex)** into the cell's membrane. MAC forms and stabilizes a hole in the membrane that ensures lysis of the target cell by inducing a massive influx of water.

The C3b molecules that coat the microorganism provide "handles" that receptors on macrophages and neutrophils can adhere to, allowing them to engulf the particle more rapidly. As noted earlier, this process is called *opsonization*. C3a and other cleavage products formed during complement fixation amplify the inflammatory response by stimulating mast cells and basophils to release histamine and by attracting neutrophils and other inflammatory cells to the area.

Fever

► Explain how fever helps protect the body.

Inflammation is a localized response to infection, but sometimes the body's response to the invasion of microorganisms is more widespread. **Fever**, or abnormally high body temperature, is a systemic response to invading microorganisms. As we will describe further in Chapter 24, body temperature is regulated by a cluster of neurons in the hypothalamus, commonly referred to as the body's thermostat. Normally set at approximately 37°C (98.6°F), the thermostat is reset upward in response to chemicals called **pyrogens** (*pyro* = fire), secreted by leukocytes and macrophages exposed to foreign substances in the body.

High fevers are dangerous because excess heat denatures enzymes. Mild or moderate fever, however, is an adaptive response that seems to benefit the body. In order to multiply, bacteria require large amounts of iron and zinc, but during a fever the liver and spleen sequester these nutrients, making them less available. Fever also increases the metabolic rate of tissue cells in general, speeding up repair processes.

CHECK YOUR UNDERSTANDING

3. What is opsonization and how does it help phagocytes? Give an example of a molecule that acts as an opsonin.
4. Under what circumstances are our own cells killed by NK cells?
5. What are the cardinal signs of inflammation and what causes them?

For answers, see Appendix G.

PART 2

ADAPTIVE DEFENSES

Most of us would find it wonderfully convenient if we could walk into a single clothing store and buy a complete wardrobe—hat to shoes—that fit perfectly regardless of any special figure problems. We know that such a service would be next to impossible to find. And yet, we take for granted our **adaptive immune system**, the body's built-in *specific defensive system* that stalks and eliminates with nearly equal precision almost any type of pathogen that intrudes into the body.

When it operates effectively, the adaptive immune system protects us from a wide variety of infectious agents, as well as from abnormal body cells. When it fails, or is disabled, the result is such devastating diseases as cancer and AIDS. The activity of the adaptive immune system tremendously amplifies the inflammatory response and is responsible for most complement activation.

At first glance, the adaptive system seems to have a major shortcoming. Unlike the innate system, which is always ready and able to react, the adaptive system must "meet" or be primed by an initial exposure to a specific foreign substance (antigen). Only then can it protect the body against that substance, and this priming takes precious time.

The basis of this specific immunity was revealed in the late 1800s. Researchers demonstrated that animals surviving a serious bacterial infection have in their blood protective factors (the proteins we now call *antibodies*) that defend against future attacks by the same pathogen. Furthermore, researchers found that if antibody-containing serum from the surviving animals was injected into animals that had not been exposed to the pathogen, those injected animals would also be protected. These landmark experiments were exciting because they revealed three important aspects of the adaptive immune response:

1. **It is specific.** It recognizes and is directed against *particular* pathogens or foreign substances that initiate the immune response.
2. **It is systemic.** Immunity is not restricted to the initial infection site.
3. **It has "memory."** After an initial exposure, it recognizes and mounts even stronger attacks on the previously encountered pathogens.

Antibody A
Antigen-binding sites
Antigen
Antibody B
Antibody C
Antigenic determinants

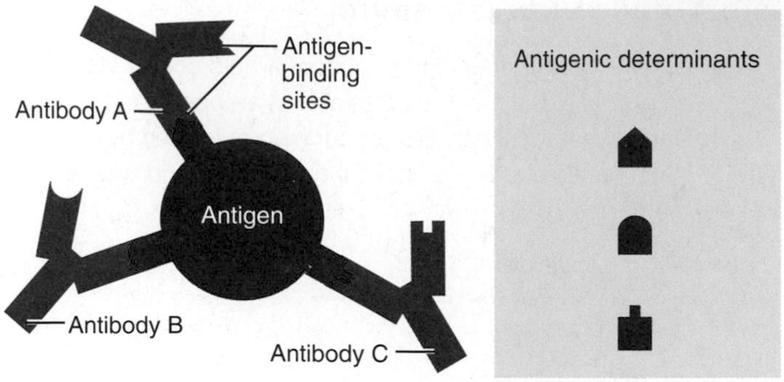

Figure 21.7 Most antigens have several different antigenic determinants. Antibodies (and related receptors on lymphocytes) bind to small areas on the antigen surface called antigenic determinants. In this example, three types of antibodies react with different antigenic determinants on the same antigen molecule.

At first antibodies were thought to be the sole artillery of the adaptive immune system, but in the mid-1900s it was discovered that injection of antibody-containing serum did *not* always protect the recipient from diseases the serum donor had survived. In such cases, however, injection of the donor's lymphocytes *did* provide immunity. As the pieces fell into place, researchers recognized two separate but overlapping arms of adaptive immunity, each using a variety of attack mechanisms that vary with the intruder.

Humoral immunity (hu′mor-ul), also called **antibody-mediated immunity**, is provided by antibodies present in the body's "humors," or fluids (blood, lymph, etc.). Though they are produced by lymphocytes (or their offspring), antibodies circulate freely in the blood and lymph, where they bind primarily to bacteria, to bacterial toxins, and to free viruses, inactivating them temporarily and marking them for destruction by phagocytes or complement.

When lymphocytes themselves rather than antibodies defend the body, the immunity is called **cellular** or **cell-mediated**
21 **immunity** because the protective factor is living cells. Cellular immunity also has cellular targets—virus-infected or parasite-infected tissue cells, cancer cells, and cells of foreign grafts. The lymphocytes act against such targets either *directly*, by killing the foreign cells, or *indirectly*, by releasing chemical mediators that enhance the inflammatory response or activate other lymphocytes or macrophages. As you can see, both arms of the adaptive immune system respond to virtually the same foreign substances, but they do it in very different ways.

Before we describe the humoral and cell-mediated responses, let's first consider the *antigens* that trigger the activity of the remarkable cells involved in these immune responses.

Antigens

- Define antigen and describe how antigens affect the adaptive defenses.
- Define complete antigen, hapten, and antigenic determinant.

Antigens (an′tĭ-jenz) are substances that can mobilize the adaptive defenses and provoke an immune response. They are the ultimate targets of all adaptive immune responses. Most antigens are large, complex molecules (both natural and synthetic) that are not normally present in the body. Consequently, as far as our immune system is concerned, they are intruders, or **nonself**.

Complete Antigens and Haptens

Antigens can be *complete* or *incomplete*. **Complete antigens** have two important functional properties:

1. **Immunogenicity**, which is the ability to stimulate proliferation of specific lymphocytes and antibodies. (*Antigen* is a contraction of "*anti*body *gen*erating," which refers to this particular antigenic property.)
2. **Reactivity**, which is the ability to react with the activated lymphocytes and the antibodies released by immunogenic reactions.

An almost limitless variety of foreign molecules can act as complete antigens, including virtually all foreign proteins, many large polysaccharides, and some lipids and nucleic acids. Of these, proteins are the strongest antigens. Pollen grains and microorganisms—such as bacteria, fungi, and virus particles—are all immunogenic because their surfaces bear many different foreign macromolecules.

As a rule, small molecules—such as peptides, nucleotides, and many hormones—are not immunogenic. But, if they link up with the body's own proteins, the adaptive immune system may recognize the *combination* as foreign and mount an attack that is harmful rather than protective. (We describe these reactions, called *hypersensitivities*, later in the chapter.) In such cases, the troublesome small molecule is called a **hapten** (hap′ten; *haptein* = grasp) or **incomplete antigen**. Unless attached to protein carriers, haptens have reactivity but not immunogenicity. Besides certain drugs (particularly penicillin), chemicals that act as haptens are found in poison ivy, animal dander, detergents, cosmetics, and a number of common household and industrial products.

Antigenic Determinants

The ability of a molecule to act as an antigen depends on both its size and its complexity. Only certain parts of an entire antigen, called **antigenic determinants**, are immunogenic. Free antibodies or lymphocyte receptors bind to these sites in much the same manner an enzyme binds to a substrate.

Most naturally occurring antigens have a variety of antigenic determinants on their surfaces, some more potent than others in provoking an immune response **(Figure 21.7)**. Different antigenic determinants are "recognized" by different lymphocytes, so a single antigen may mobilize several lymphocyte populations and may stimulate formation of many kinds of antibodies. Large proteins have hundreds of chemically different antigenic determinants, which accounts for their high immunogenicity and reactivity. However, large simple molecules such as plastics, which have many identical, regularly repeating units (and so are not chemically complex), have little or no immunogenicity. Such substances are used to make artificial implants because the substances are not seen as foreign and rejected by the body.

Self-Antigens: MHC Proteins

The external surfaces of all our cells are dotted with a huge variety of protein molecules. Assuming your immune system has been properly "programmed," your **self-antigens** are not foreign or antigenic to you, but they are strongly antigenic to other individuals. (This is the basis of transfusion reactions and graft rejection.)

Among the cell surface proteins that mark a cell as *self* is a group of glycoproteins called **MHC proteins**. Genes of the **major histocompatibility complex (MHC)** code for these proteins. Because millions of combinations of these genes are possible, it is unlikely that any two people except identical twins have the same MHC proteins.

There are two major groups of MHC proteins, distinguished by location. Class I MHC proteins are found on virtually *all* body cells, but class II MHC proteins are found *only on certain cells* that act in the immune response.

Each MHC molecule has a deep groove that typically displays a peptide. In healthy cells, peptides displayed by class I MHC come from the breakdown of cellular proteins during normal protein recycling and tend to be quite diverse, but are nevertheless all self-antigens. However, in infected cells, class I MHC also binds fragments of *foreign* antigens that come from within the infected cell. Peptides displayed by class II MHC come from outside the cell. As we will describe shortly, these displayed peptides play a crucial role in mobilizing the adaptive defenses.

CHECK YOUR UNDERSTANDING

6. Name three key characteristics of adaptive immunity.
7. What is the difference between a complete antigen and a hapten?
8. What marks a cell as "self" as opposed to "nonself"?

For answers, see Appendix G.

Cells of the Adaptive Immune System: An Overview

- Compare and contrast the origin, maturation process, and general function of B and T lymphocytes.
- Define immunocompetence and self-tolerance, and describe their development in B and T lymphocytes.
- Name several antigen-presenting cells and describe their roles in adaptive defenses.

The three crucial cell types of the adaptive immune system are two distinct populations of lymphocytes, plus *antigen-presenting cells (APCs)*. **B lymphocytes** or **B cells** oversee humoral immunity. **T lymphocytes** or **T cells** are non-antibody-producing lymphocytes that constitute the cell-mediated arm of adaptive immunity. Unlike the lymphocytes, APCs do not respond to specific antigens but instead play essential auxiliary roles.

Lymphocytes

Like all other blood cells, lymphocytes originate in red bone marrow from hematopoietic stem cells. During development, lymphocytes are "educated." The aim of this education is twofold:

1. **Immunocompetence.** Each lymphocyte must become able (competent) to recognize its one specific antigen by binding to it. This ability is called **immunocompetence**.
2. **Self-tolerance.** Each lymphocyte must be relatively unresponsive to self-antigens so that it does not attack the body's own cells. This is called **self-tolerance**.

B and T cells are educated in different parts of the body. T cells undergo this two- to three-day maturation process in the thymus. B cells become immunocompetent and self-tolerant in the bone marrow. The lymphoid organs where the lymphocytes become immunocompetent—thymus and bone marrow—are called **primary lymphoid organs**. All other lymphoid organs are referred to as **secondary lymphoid organs**.

Immunocompetent B and T cells that have not yet been exposed to antigen are called *naive*. Naive B cells and T cells are exported to the lymph nodes, spleen, and other secondary lymphoid organs, where encounters with antigens may occur. Then, when the lymphocytes bind with recognized antigens, the lymphocytes are activated to complete their differentiation into effector and memory B or T cells. **Figure 21.8** summarizes lymphocyte development.

Development of Immunocompetence and Self-Tolerance

When B or T cells become immunocompetent, they display a unique type of receptor on their surface. These receptors (some 10^5 per cell) enable the lymphocyte to recognize and bind a specific antigen. Once these receptors appear, the lymphocyte is committed to react to one distinct antigenic determinant, and one only, because *all* of its antigen receptors are the same.

21

The receptors on B cells are in fact membrane-bound antibodies, while the receptors on T cells are not antibodies but are products of the same gene superfamily. These similar structures have similar functions, and both cell types are capable of responding to the same antigens.

For T cells, the education in the thymus ensures that they meet two essential criteria for life as a successful T cell. First, the T cell must be able to bind MHC molecules, since it is on these molecules that antigens are presented to the T cell for recognition. Second, the T cell must *not* react strongly to self-antigens that are normally found in the body.

In order to cnsure that all T cells meet these criteria, their education consists of positive and negative selection **(Figure 21.9)**. **Positive selection**, which occurs in the thymic cortex, is essentially an **MHC restriction** process. It identifies T cells whose

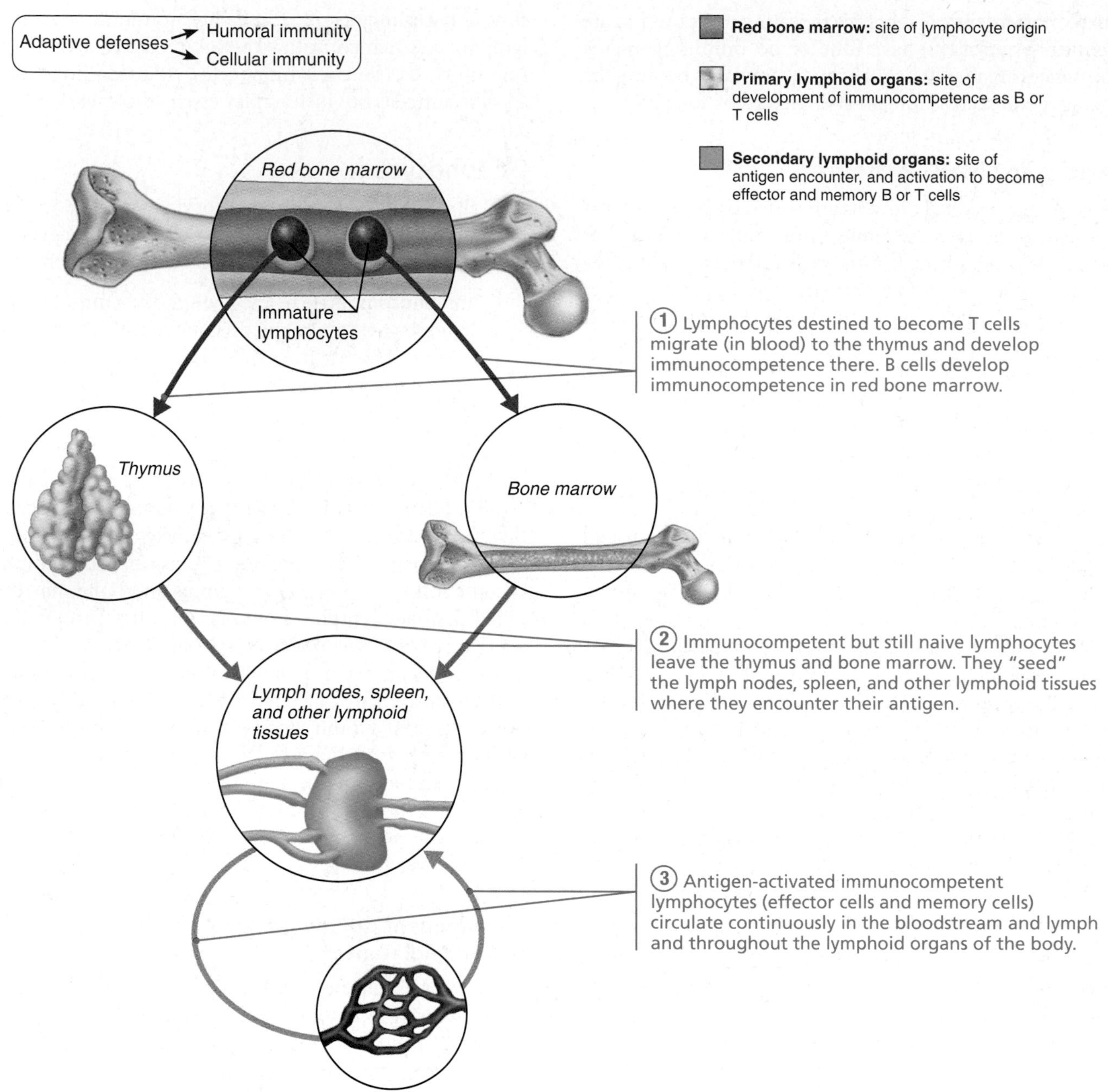
Adaptive defenses
Humoral immunity
Cellular immunity
Red bone marrow: site of lymphocyte origin
Primary lymphoid organs: site of development of immunocompetence as B or T cells
Secondary lymphoid organs: site of antigen encounter, and activation to become effector and memory B or T cells
Red bone marrow
Immature lymphocytes
① Lymphocytes destined to become T cells migrate (in blood) to the thymus and develop immunocompetence there. B cells develop immunocompetence in red bone marrow.
Thymus
Bone marrow
② Immunocompetent but still naive lymphocytes leave the thymus and bone marrow. They "seed" the lymph nodes, spleen, and other lymphoid tissues where they encounter their antigen.
Lymph nodes, spleen, and other lymphoid tissues
③ Antigen-activated immunocompetent lymphocytes (effector cells and memory cells) circulate continuously in the bloodstream and lymph and throughout the lymphoid organs of the body.

Figure 21.8 Lymphocyte traffic. Immature lymphocytes arise in red bone marrow. (Note that red marrow is not found in the medullary cavity of the diaphysis of long bones in adults.) Plasma cells (antibody-secreting effector B cells) usually do not circulate.

21

receptors are capable of recognizing (binding) self-MHC molecules and eliminates all others. In this way positive selection produces an army of self-MHC-restricted T cells.

T cells that make it through positive selection are then tested to make sure that they do not recognize (bind tightly) self-antigens displayed on self-MHC. If they do, they are eliminated by apoptosis (programmed cell death). This is **negative selection**, and it occurs in the inner edge of the thymic cortex. Negative selection ensures immunologic self-tolerance, making sure that T cells don't attack the body's own cells, which would cause autoimmune disorders. This education of T cells is expensive indeed—only about 2% of T cells survive it and continue on to become successful immunocompetent, self-tolerant T cells.

Less is known about the factors that control B cell maturation in humans. In the bone marrow, self-reactive B cells are either eliminated by apoptosis (*clonal deletion*), or are given a chance to change their self-reactive antigen receptor by *receptor editing*, in which there is another rearrangement of the antigen-binding part of the receptor. Nevertheless, some self-reacting B cells do leave the bone marrow. In the periphery, these self-reactive B cells are inactivated (a phenomenon called *anergy*).

Generation of Antigen Receptor Diversity in Lymphocytes

We know that lymphocytes become immunocompetent *before* meeting the antigens they may later attack. *It is our genes, not antigens, that determine what specific foreign substances our immune system will be able to recognize and resist.* In other words, the immune cell receptors represent our genetically acquired knowledge of the microbes that are likely to be in our environment.

An antigen simply determines which existing T or B cells will proliferate and mount the attack against it. Only some of the antigens our lymphocytes are programmed to resist will ever invade our bodies. Consequently, only some members of our army of immunocompetent cells are mobilized in our lifetime. The others are forever idle.

Our lymphocytes make up to a billion different types of antigen receptors. These receptors, like all other proteins, are specified by genes, so you might think that an individual must have billions of genes. Not so; each body cell only contains about 25,000 genes that code for all the proteins the cell must make.

How can a limited number of genes generate a seemingly limitless number of different antigen receptors? Molecular genetic studies have shown that the genes that dictate the structure of each antigen receptor are not present as such in embryonic cells. Instead of a complete set of "antigen receptor genes," embryonic cells contain a few hundred genetic bits and pieces that can be thought of as a "Lego set" for antigen receptor genes. As each lymphocyte becomes immunocompetent, these gene segments are shuffled and combined in different ways in a process called **somatic recombination**. The information of the newly assembled genes is then expressed as the surface receptors of B and T cells and as the antibodies later released by the B cell's "offspring."

Antigen-Presenting Cells

The major role of **antigen-presenting cells (APCs)** in immunity is to engulf antigens and then present fragments of them, like signal flags, on their own surfaces where they can be recognized by T cells. In other words, they *present antigens* to the cells that will deal with the antigens. The major types of cells acting as APCs are *dendritic cells* (present in connective tissues and in the epidermis, where they are also called *Langerhans cells*), *macrophages*, and *B lymphocytes*.

Notice that all these cell types are in sites that make it easy to encounter and process antigens. Dendritic cells are at the body's frontiers, best situated to act as mobile sentinels. Macrophages are widely distributed throughout the lymphoid organs and connective tissues. When they present antigens, dendritic cells and macrophages activate T cells. Activated T cells, in turn, release chemicals that prod macrophages to become *activated macrophages*, true "killers" that are insatiable phagocytes and secrete bactericidal chemicals. As you will see, interactions between various lymphocytes, and between lymphocytes and APCs, underlie virtually all phases of the immune response.

Macrophages tend to remain fixed in the lymphoid organs, as if waiting for antigens to come to them. But lymphocytes, especially the T cells (which account for 65–85% of bloodborne lymphocytes), circulate continuously throughout the body. This circulation greatly increases a lymphocyte's chance of coming into contact with antigens located in different parts of the body, as well as with huge numbers of macrophages and other lymphocytes. Although lymphocyte recirculation appears to be random, the lymphocyte emigration to the tissues where their protective services are needed is highly specific, regulated by homing signals (CAMs) displayed on vascular endothelial cells.

Immune cells in lymph nodes are in a strategic position to encounter a large variety of antigens because lymphatic capillaries pick up proteins and pathogens from nearly all body tissues. Lymphocytes and APCs in the tonsils act primarily against microorganisms that invade the oral and nasal cavities, and the spleen acts as a filter to trap bloodborne antigens.

In addition to T cell recirculation and passive delivery of antigens to lymphoid organs by lymphatics, a third delivery

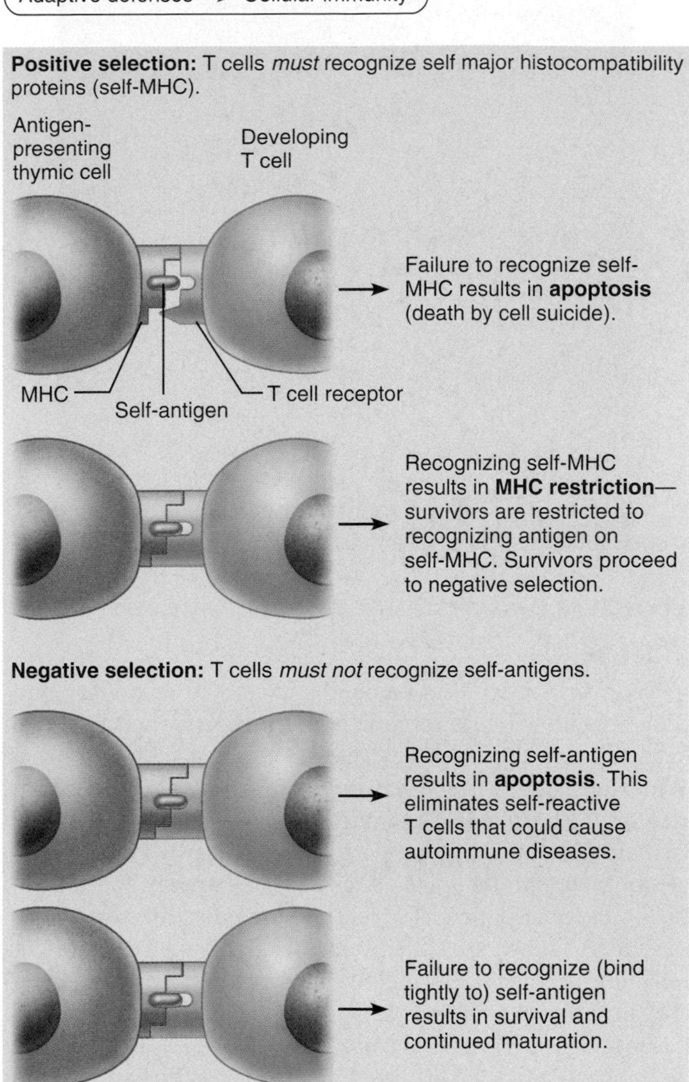

Adaptive defenses → Cellular immunity
Positive selection: T cells must recognize self major histocompatibility proteins (self-MHC).
Antigen-presenting thymic cell
Developing T cell
Failure to recognize self-MHC results in apoptosis (death by cell suicide).
MHC
Self-antigen
T cell receptor
Recognizing self-MHC results in MHC restriction—survivors are restricted to recognizing antigen on self-MHC. Survivors proceed to negative selection.
Negative selection: T cells must not recognize self-antigens.
Recognizing self-antigen results in apoptosis. This eliminates self-reactive T cells that could cause autoimmune diseases.
Failure to recognize (bind tightly to) self-antigen results in survival and continued maturation.

Figure 21.9 T cell education in the thymus.

21

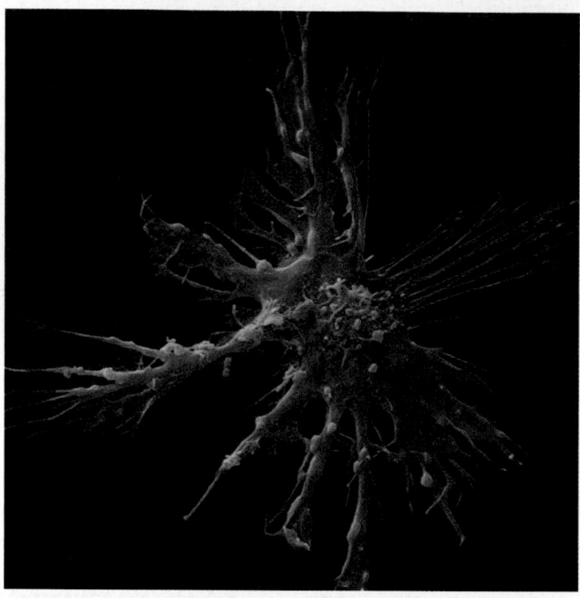

Figure 21.10 Dendritic cell. Scanning electron micrograph (1050×).

mechanism—migration of dendritic cells to secondary lymphoid organs—is now recognized as the most important way of ensuring that the immune cells encounter invading antigens. With their long, wispy extensions, dendritic cells are very efficient antigen catchers **(Figure 21.10)**. Once they have internalized antigens by phagocytosis, they enter nearby lymphatics to get to the lymphoid organ where they will present the antigens to T cells. Indeed, dendritic cells are the most effective antigen presenters known—it's their only job. Dendritic cells are a key link between innate and adaptive immunity. They initiate adaptive immune responses particularly tailored to the type of pathogen that they have encountered.

In summary, the two-fisted adaptive immune system uses lymphocytes, APCs, and specific molecules to identify and destroy all substances—both living and nonliving—that are in the body but not recognized as self. The system's response to such threats depends on the ability of its cells (1) to recognize antigens in the body by binding to them and (2) to communicate with one another so that the whole system mounts a response specific to those antigens.

21

CHECK YOUR UNDERSTANDING

9. What event (or observation) signals that a B or T cell has achieved immunocompetence?
10. Which of the following T cells would survive education in the thymus? (a) one that recognizes neither MHC nor self-antigen, (b) one that recognizes both MHC and self-antigen, (c) one that recognizes MHC but not self-antigen, (d) one that recognizes self-antigen but not MHC.
11. Name three different APCs. Which is most important for T lymphocyte activation?

For answers, see Appendix G.

Humoral Immune Response

- ▶ Define humoral immunity.
- ▶ Describe the process of clonal selection of a B cell.
- ▶ Recount the roles of plasma cells and memory cells in humoral immunity.

The **antigen challenge**, the first encounter between an immunocompetent but naive lymphocyte and an invading antigen, usually takes place in the spleen or in a lymph node, but it may happen in any secondary lymphoid organ. If the lymphocyte is a B cell, the challenging antigen provokes the *humoral immune response*, in which antibodies are produced against the challenger.

Clonal Selection and Differentiation of B Cells

An immunocompetent but naive B lymphocyte is *activated*—stimulated to complete its differentiation—when matching antigens bind to its surface receptors and cross-link adjacent receptors together. Antigen binding is quickly followed by receptor-mediated endocytosis of the cross-linked antigen-receptor complexes. This sequence of steps triggers **clonal selection** (klo′nul) because it stimulates the B cell to grow and then multiply rapidly to form an army of cells all exactly like itself and bearing the same antigen-specific receptors **(Figure 21.11)**. The resulting family of identical cells, all descended from the *same* ancestor cell, is called a **clone**. The antigen does the selecting in clonal selection by "choosing" a lymphocyte with complementary receptors. (As we will see shortly, interactions with T cells are usually required to help B cells achieve full activation.)

Most cells of the clone differentiate into **plasma cells**, the antibody-secreting *effector cells* of the humoral response. Although B cells secrete limited amounts of antibodies, plasma cells develop the elaborate internal machinery (largely rough endoplasmic reticulum) needed to secrete antibodies at the unbelievable rate of about 2000 molecules per second. Each plasma cell functions at this breakneck pace for 4 to 5 days and then dies. The secreted antibodies, each with the same antigen-binding properties as the receptor molecules on the surface of the parent B cell, circulate in the blood or lymph. There they bind to free antigens and mark them for destruction by other innate or adaptive mechanisms.

Clone cells that do not become plasma cells become long-lived **memory cells.** They can mount an almost immediate humoral response if they encounter the same antigen again at some future time (Figure 21.11, bottom).

Immunological Memory

The cellular proliferation and differentiation we have just described constitute the **primary immune response**, which occurs on first exposure to a particular antigen. The primary response typically has a lag period of 3 to 6 days after the antigen challenge. This lag period mirrors the time required for the few B cells specific for that antigen to proliferate (about 12 generations) and

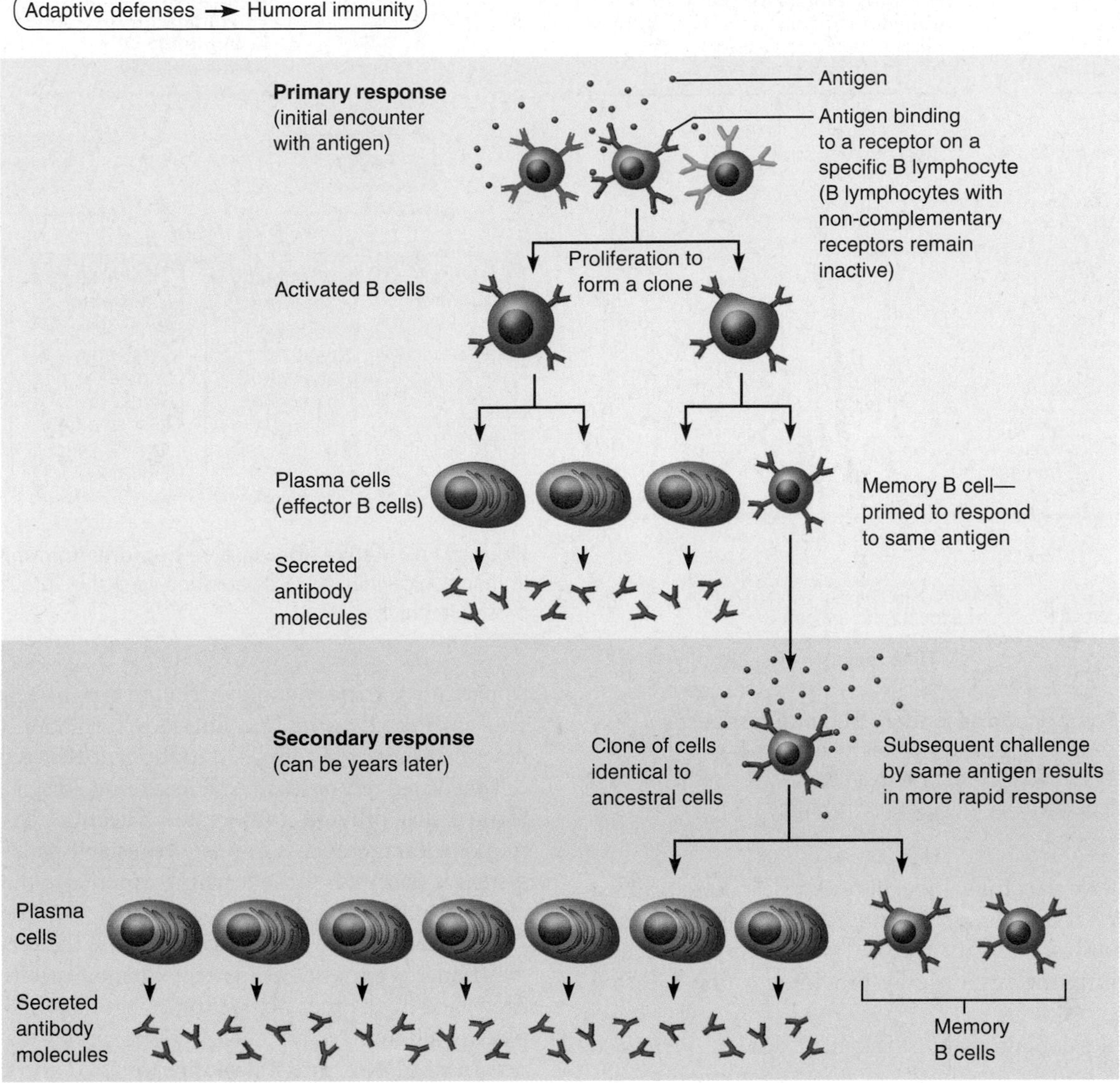

Figure 21.11 Clonal selection of a B cell.

21

for their offspring to differentiate into plasma cells. After the mobilization period, plasma antibody levels rise, reach peak levels in about 10 days, and then decline **(Figure 21.12)**.

If (and when) someone is reexposed to the same antigen, whether it's the second or the twenty-second time, a **secondary immune response** occurs. Secondary immune responses are faster, more prolonged, and more effective, because the immune system has already been primed to the antigen, and sensitized memory cells are already in place "on alert." These memory cells provide what is commonly called **immunological memory**.

Within hours after recognition of the "old enemy" antigen, a new army of plasma cells is being generated. Within 2 to 3 days the antibody concentration in the blood, called the *antibody titer*, rises steeply to reach much higher levels than were achieved in the primary response. Secondary response antibodies not only bind with greater affinity (more tightly), but their blood levels remain high for weeks to months. (When the appropriate chemical signals are present, plasma cells can keep functioning for much longer than the 4 to 5 days seen in primary responses.) Memory cells persist for long periods in humans and many retain their capacity to produce powerful secondary humoral responses for life.

The same general phenomena occur in the cellular immune response: A primary response sets up a pool of effector cells (in this case, T cells) and generates memory cells that can then mount secondary responses.

Active and Passive Humoral Immunity

▶ Compare and contrast active and passive humoral immunity.

When your B cells encounter antigens and produce antibodies against them, you are exhibiting **active humoral immunity**. Active immunity is acquired in two ways **(Figure 21.13)**. It is (1) *naturally acquired* when you get a bacterial or viral infection, during which time you may develop symptoms of the disease and suffer a little (or a lot), and (2) *artificially acquired*

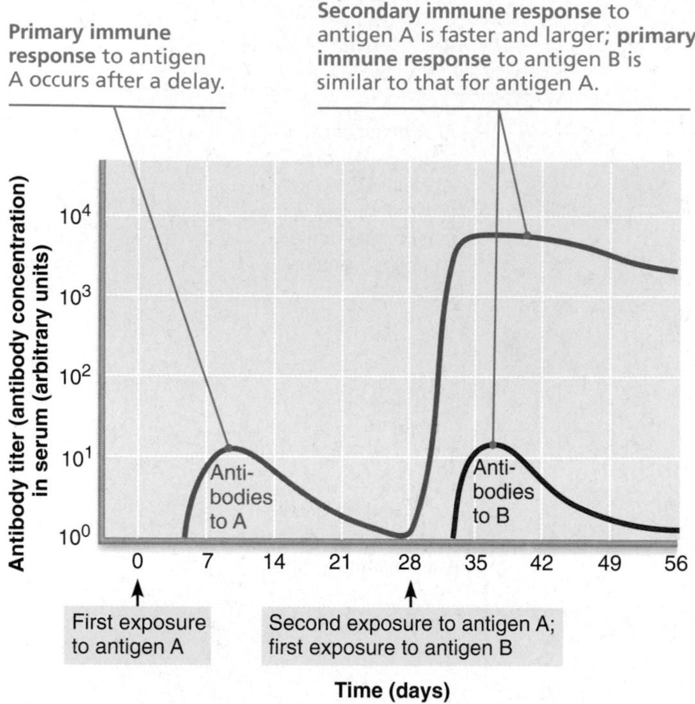

Figure 21.12 Primary and secondary humoral responses. The primary response to antigen A generates memory cells that give rise to the enhanced secondary response to antigen A. The response to antigen B is independent of the response to antigen A.

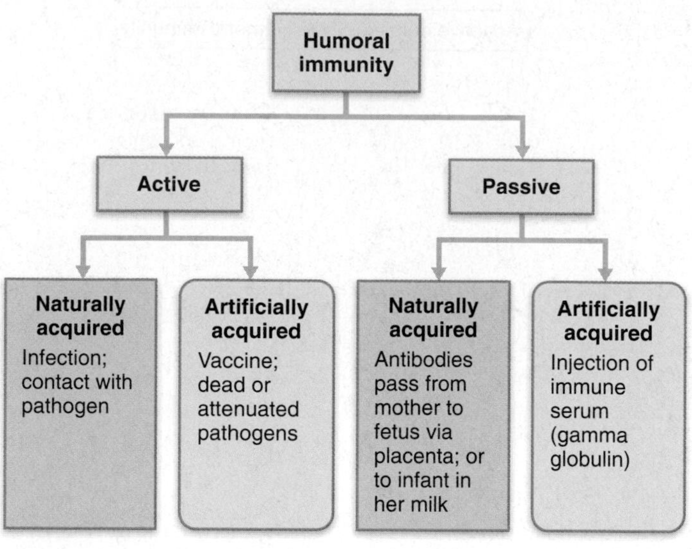

Figure 21.13 Active and passive humoral immunity. Immunological memory is established via active immunity, never via passive immunity.

when you receive **vaccines.** Indeed, once researchers realized that secondary responses are so much more vigorous than primary responses, the race was on to develop vaccines to "prime" the immune response by providing a first meeting with the antigen.

Most vaccines contain pathogens that are dead or *attenuated* (living, but extremely weakened), or their components. Vaccines provide two benefits:

21

1. They spare us most of the symptoms and discomfort of the disease that would otherwise occur during the primary response.
2. Their weakened antigens provide functional antigenic determinants that are both immunogenic and reactive.

Vaccine *booster shots,* which may intensify the immune response at later meetings with the same antigen, are also available.

Vaccines have wiped out smallpox and have substantially lessened the illness caused by such former childhood killers as whooping cough, polio, and measles. Although vaccines have dramatically reduced hepatitis B, tetanus, and pneumonia in adults, immunization of adults in the U.S. has a much lower priority than that of children and as a result more than 65,000 Americans die each year from vaccine-preventable infections (pneumonia, influenza, and hepatitis).

Conventional vaccines have shortcomings. Although it was originally believed that the immune response was about the same regardless of how an antigen got into the body (under its own power or via a vaccine), that has proved not to be the case. Apparently vaccines mainly target the type of helper T cell that revs up B cell defenses and antibody formation (the T_H2 cell, described shortly) as opposed to the type that generates strong cell-mediated responses (T_H1). As a result, lots of antibodies are formed that provide immediate protection, but cellular immunological memory is only poorly established. (The immune system is deprived of the learning experience that comes with clearing an infection via a T_H1-mediated response.)

In extremely rare cases, vaccines cause the very disease they are trying to prevent because the attenuated virus isn't weakened enough. In other cases, contaminating proteins (for example, egg albumin) cause allergic responses to the vaccine. The new "naked DNA" antiviral vaccines, blasted into the skin with a gene gun, and *edible* vaccines taken orally appear to circumvent these problems.

Passive humoral immunity differs from active immunity, both in the antibody source and in the degree of protection it provides (Figure 21.13). Instead of being made by your plasma cells, the antibodies are harvested from the serum of an immune human or animal donor. As a result, your B cells are *not* challenged by antigens, immunological memory does *not* occur, and the protection provided by the "borrowed" antibodies ends when they naturally degrade in the body.

Passive immunity is conferred *naturally* on a fetus or infant when the mother's antibodies cross the placenta or are ingested with the mother's milk. For several months after birth, the baby is protected from all the antigens to which the mother has been exposed. Passive immunity is *artificially* conferred via a serum such as *gamma globulin,* which is administered after exposure to hepatitis. Other immune sera are used to treat poisonous snake bites (antivenom), botulism, rabies, and tetanus (antitoxin) because these rapidly fatal diseases would kill a person before active immunity could be established. The donated antibodies provide immediate protection, but their effect is short-lived (two to three weeks).

Adaptive defenses → Humoral immunity

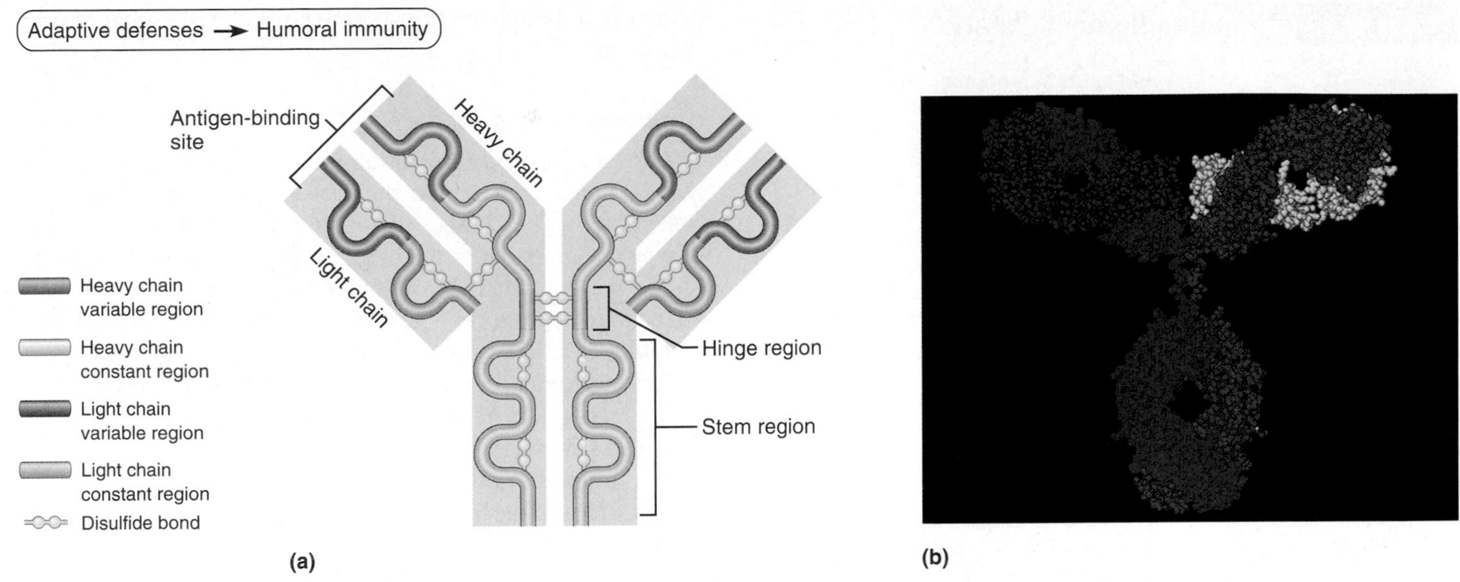

Figure 21.14 Antibody structure. (a) Schematic antibody structure (based on IgG) consists of four polypeptides—two short *light chains* and two long *heavy chains* joined together by disulfide bonds (S–S). Each chain has a V (variable) region (which differs in antibodies from different cells) and a C (constant) region (essentially identical in different antibodies of the same class). Together, the variable regions form the antigen-binding sites—two per antibody monomer. **(b)** Computer-generated image of antibody structure.

CHECK YOUR UNDERSTANDING

12. In clonal selection, "who" does the selecting? What is being selected?

13. Why is the secondary response to an antigen so much faster than the primary response?

14. How do vaccinations protect against common childhood illnesses such as chicken pox, measles, and mumps?

For answers, see Appendix G.

Antibodies

- Describe the structure of an antibody monomer, and name the five classes of antibodies.
- Explain the function(s) of antibodies and describe clinical uses of monoclonal antibodies.

Antibodies, also called **immunoglobulins (Igs)** (im″u-no-glob′u-linz), constitute the **gamma globulin** part of blood proteins. As we mentioned earlier, antibodies are proteins secreted in response to an antigen by effector B cells called plasma cells, and they are capable of binding specifically with that antigen. They are formed in response to an incredible number of different antigens.

Despite their variety, all antibodies can be grouped into one of five Ig classes, each slightly different in structure and function. Before seeing how these Ig classes differ from one another, let's take a look at how all antibodies are alike.

Basic Antibody Structure

Regardless of its class, each antibody has a basic structure consisting of four looping polypeptide chains linked together by disulfide (sulfur-to-sulfur) bonds. The four chains combined form a molecule, called an **antibody monomer** (mon′o-mer), with two identical halves. The molecule as a whole is T or Y shaped **(Figure 21.14)**.

Two of the chains, called the **heavy (H) chains**, are identical to each other and contain more than 400 amino acids each (blue chains in Figure 21.14a). The other two chains, called the **light (L) chains** (pink), are also identical to each other, but they are only about half as long as each H chain. The heavy chains have a flexible *hinge* region at their approximate "middles." The "loops" on each chain are created by disulfide bonds between amino acids that are in the same chain but about 60–70 amino acids apart. These bonds cause the intervening parts of the polypeptide chains to loop out.

Each chain forming an antibody has a **variable (V) region** at one end and a much larger **constant (C) region** at the other end. Antibodies responding to different antigens have very different V regions, but their C regions are the same (or nearly so) in all antibodies of a given class. In each arm of the monomer, the V regions of the heavy and light chains combine to form an **antigen-binding site** shaped to "fit" a specific antigenic determinant. Consequently, each antibody monomer has two such antigen-binding regions.

The C regions that form the *stem* of the antibody monomer determine the antibody class and serve common functions in

TABLE 21.3 Immunoglobulin Classes

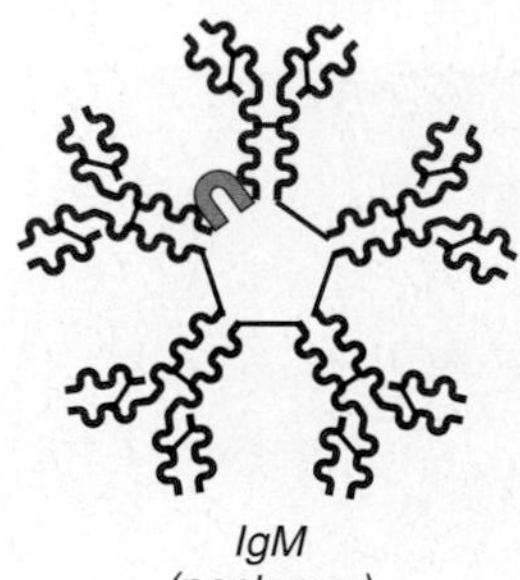 *IgM (pentamer)*	IgM exists in monomer and pentamer (five united monomers) forms. The monomer, which is attached to the B cell surface, serves as an antigen receptor. The pentamer (illustrated) circulates in blood plasma and is the first Ig class released by plasma cells during the primary response. (This fact is diagnostically useful because presence of IgM in plasma usually indicates current infection by the pathogen eliciting IgM's formation.) Its numerous antigen-binding sites make IgM a potent agglutinating agent, and it readily fixes and activates complement.
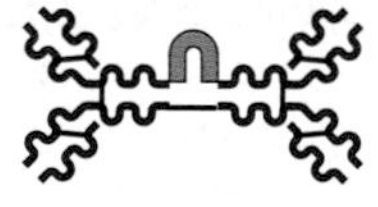 *IgA (dimer)*	IgA monomer exists in limited amounts in plasma. The dimer (illustrated), referred to as secretory IgA, is found in body secretions such as saliva, sweat, intestinal juice, and milk, and helps prevent attachment of pathogens to epithelial cell surfaces (including mucous membranes and the epidermis).
 *IgD (monomer)*	IgD is virtually always attached to the external surface of a B cell, where it functions as an antigen receptor of the B cell.
 *IgG (monomer)*	IgG is the most abundant and diverse antibody in plasma, accounting for 75–85% of circulating antibodies. It protects against bacteria, viruses, and toxins circulating in blood and lymph, readily fixes complement, and is the main antibody of both secondary and late primary responses. It crosses the placenta and confers passive immunity from the mother to the fetus.
 *IgE (monomer)*	IgE is slightly larger than the IgG antibody. It is secreted by plasma cells in skin, mucosae of the gastrointestinal and respiratory tracts, and tonsils. Its stem region becomes bound to mast cells and basophils, and when its receptor ends are triggered by an antigen, it causes the cells to release histamine and other chemicals that mediate inflammation and an allergic reaction. Typically only traces of IgE are found in plasma, but levels rise during severe allergic attacks or chronic parasitic infections of the gastrointestinal tract.

all antibodies: These are the *effector regions* of the antibody that dictate (1) the cells and chemicals of the body the antibody can bind to, and (2) how the antibody class functions in antigen elimination. For example, some antibodies can fix complement, some circulate in blood and others are found primarily in body secretions, some cross the placental barrier, and so on.

Antibody Classes

The five major immunoglobulin classes are designated IgM, IgA, IgD, IgG, and IgE, on the basis of the C regions in their heavy chains. (Remember the name MADGE to recall the five Ig types.) Compared to the other antibodies, IgM is a huge antibody. It is constructed from five Y-shaped units, or *monomers*, linked together to form a *pentamer* (*penta* = five), as illustrated in **Table 21.3**. IgA occurs in both monomer and *dimer* (two linked monomers) forms. (Only the dimer is shown in the table.) IgD, IgG, and IgE are monomers and have the same basic Y-shaped structure.

The antibodies of each class have different biological roles and locations in the body. IgM is the first antibody class *released* to the blood by plasma cells. It readily fixes complement. The IgA dimer, also called **secretory IgA**, is found primarily in mucus and other secretions that bathe body surfaces. It plays a major role in preventing pathogens from gaining entry into the body. IgD is always bound to a B cell surface, where it acts as a B cell receptor. IgG is the most abundant antibody in plasma and the only Ig class that crosses the placental barrier. For this reason, the passive immunity that a mother transfers to her fetus is courtesy of her IgG antibodies. Like IgM, IgG can fix complement, and *only* these two antibody classes can do so. IgE antibodies, found in minute quantities in blood, are the "troublemaker" antibodies involved in some allergies. These and other characteristics unique to each immunoglobulin class are summarized in Table 21.3.

Generating Antibody Diversity

Recall from p. 779 that the billions of different kinds of antibodies produced by plasma cells come about as the result of somatic recombination of a limited number of gene segments. The random mixing of gene segments that code for the antigen-binding site (variable regions) accounts for only part of the huge variability seen in antibody specificity. Certain areas of one gene segment in activated B cells contain *hypervariable regions* that are "hot spots" for somatic mutations, and that enormously increase antibody variation.

A single plasma cell can switch from making one kind of H chain to making another kind, thereby producing two or more different antibody classes having the same antigen specificity. For example, the first antibody released in the primary response is IgM, and then the plasma cell begins to secrete IgG. During secondary responses, almost all of the Ig protein is IgG.

Antibody Targets and Functions

Though antibodies themselves cannot destroy antigens, they can inactivate them and tag them for destruction **(Figure 21.15)**. The common event in all antibody-antigen interactions is formation of **antigen-antibody** (or **immune**) **complexes**. Defensive mechanisms used by antibodies include neutralization,

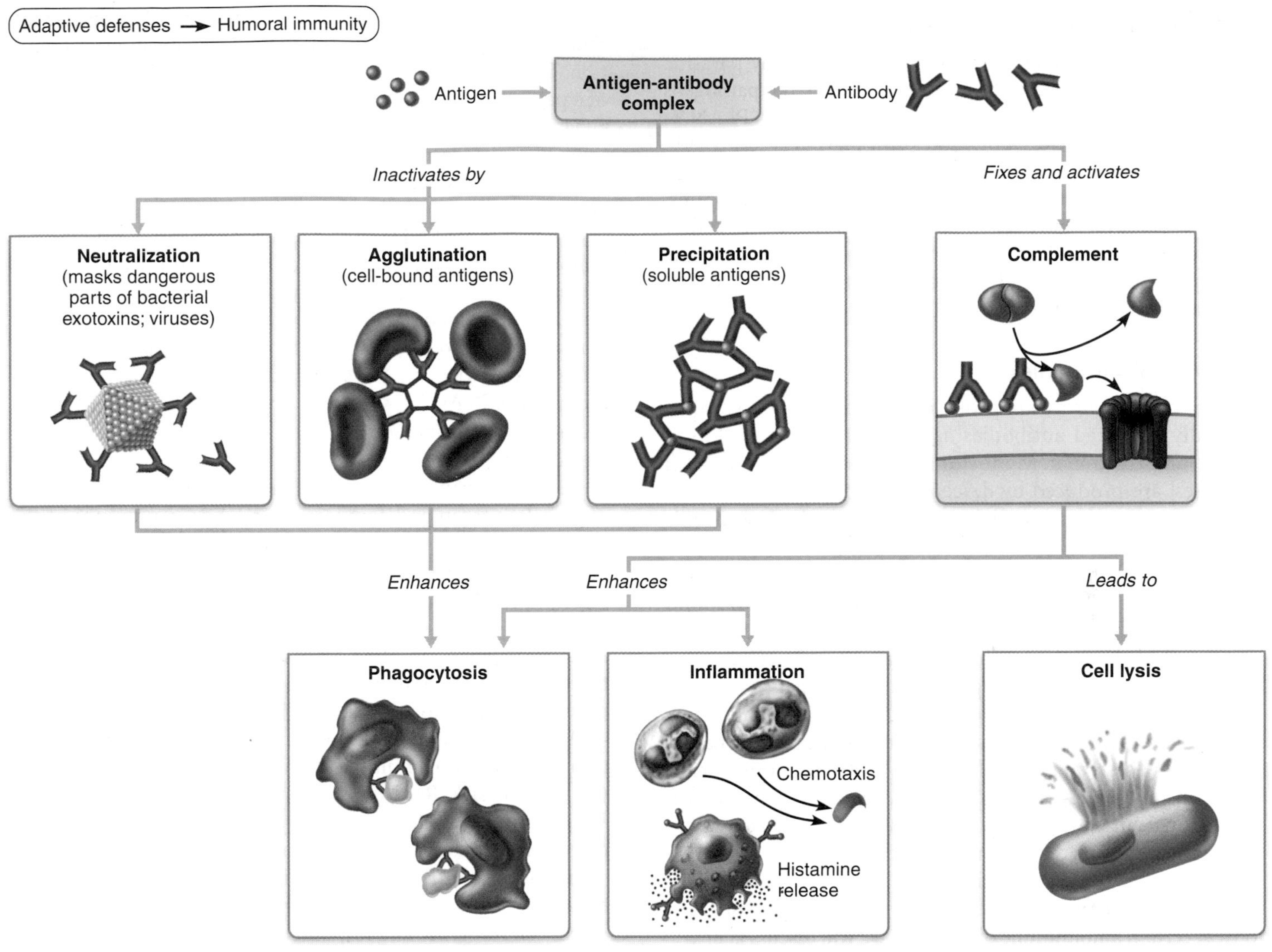

Figure 21.15 Mechanisms of antibody action. Antibodies act against free viruses, red blood cell antigens, bacterial toxins, intact bacteria, fungi, and parasitic worms.

agglutination, precipitation, and complement fixation, with the first two most important.

Neutralization, the simplest defensive mechanism, occurs when antibodies block specific sites on viruses or bacterial exotoxins (toxic chemicals secreted by bacteria). As a result, the virus or exotoxin loses its toxic effect because it cannot bind to receptors on tissue cells to cause injury. The antigen-antibody complexes are eventually destroyed by phagocytes.

Because antibodies have more than one antigen-binding site, they can bind to the same determinant on more than one antigen at a time. Consequently, antigen-antibody complexes can be cross-linked into large lattices. When cell-bound antigens are cross-linked, the process causes clumping, or **agglutination**, of the foreign cells. IgM, with 10 antigen-binding sites, is an especially potent agglutinating agent (see Table 21.3). Recall from Chapter 17 that this type of reaction occurs when mismatched blood is transfused (the foreign red blood cells are clumped) and is the basis of tests used for blood typing.

In **precipitation**, soluble molecules (instead of cells) are cross-linked into large complexes that settle out of solution. Like agglutinated bacteria, precipitated antigen molecules are much easier for phagocytes to capture and engulf than are freely moving antigens.

Complement fixation and activation is the chief antibody defense used against cellular antigens, such as bacteria or mismatched red blood cells. When several antibodies bind close together on the same cell, the complement-binding sites on their stem regions align. This triggers complement fixation into the antigenic cell's surface, followed by cell lysis. Additionally, as we described earlier, molecules released during complement activation tremendously amplify the inflammatory response and promote phagocytosis via opsonization. In this way, a positive feedback cycle that enlists more and more defensive elements is set into motion.

A quick and dirty way to remember how antibodies work is to remember they have a PLAN of action—**p**recipitation, **l**ysis (by complement), **a**gglutination, and **n**eutralization.

TABLE 21.4 Selected Cytokines

CYTOKINE	FUNCTION IN IMMUNE RESPONSE
Interferons (IFNs)	
▪ Alpha (α) and beta (β)	Secreted by many cells. Have antiviral effects; activate NK cells.
▪ Gamma (γ)	Secreted by lymphocytes. Activates macrophages; stimulates synthesis and expression of more class I and II MHC proteins; promotes differentiation of T_H cells into T_H1.
Interleukins (ILs)	
▪ IL-1	Secreted by activated macrophages. Promotes inflammation and T cell activation; causes fever (a pyrogen that resets the thermostat of the hypothalamus).
▪ IL-2	Secreted by T_H cells. Stimulates proliferation of T and B cells; activates NK cells.
▪ IL-4	Secreted by some T_H cells. Promotes differentiation to T_H2; promotes B cell activation; switches antibody production to IgE.
▪ IL-5	Secreted by some T_H cells and mast cells. Attracts and activates eosinophils; causes plasma cells to secrete IgA antibodies.
▪ IL-10	Secreted by macrophages and T_{Reg} cells. Inhibits macrophages and dendritic cells; turns down cellular and innate immune response.
▪ IL-12	Secreted by dendritic cells and macrophages. Stimulates T_C and NK cell activity; promotes T_H1 differentiation.
▪ IL-17	Secreted by T_H17 cells. Important in innate and adaptive immunity and recruitment of neutrophils. Involved in inflammation in some autoimmune diseases.
Suppressor factors	A generic term for a number of cytokines that suppress the immune system, for example TGF-β and IL-10.
Transforming growth factor beta (TGF-β)	A suppressor factor similar to IL-10.
Tumor necrosis factors (TNFs)	Produced by lymphocytes and in large amounts by macrophages. Promote inflammation; enhance phagocyte chemotaxis and nonspecific killing; slow tumor growth by selectively damaging tumor blood vessels; promote cell death by apoptosis.

After co-stimulation, cytokine chemicals such as interleukin 1 and 2 released by APCs or T cells themselves nudge the activated T cells to proliferate and differentiate. As you might guess, there are several types of cytokines, each type promoting a different response in the activated T cells.

What happens if a T cell binds to antigen without receiving the co-stimulatory signal? In this case, the T cell becomes tolerant to that antigen and is unable to divide or to secrete cytokines. This state of unresponsiveness to antigen is called **anergy**.

The two-signal sequence acts as a safeguard to prevent the immune system from destroying healthy cells. Without this safeguard, class I MHC proteins, which occur on all body cells and which display peptides from within the cell, could activate cytotoxic T cells, leading to widespread damage of healthy cells. The important thing to understand is that along with antigen binding, co-stimulation is crucial for T cell activation. To go back to our idling car analogy, the car will not go anywhere unless the car has both (1) been started and (2) put into gear.

Once activated, a T cell enlarges and proliferates to form a clone of cells that differentiate and perform functions according to their T cell class. This primary response peaks within a week of exposure to the triggering antigen. A period of apoptosis then occurs between days 7 and 30, during which time the activated T cells die off and effector activity wanes as the amount of antigen declines.

This wholesale disposal of T cells has a critical protective role because activated T cells are potential hazards. They produce huge amounts of inflammatory cytokines, which contribute to infection-driven hyperplasia, and may promote malignancies in chronically inflamed tissue. Additionally, once they've done their job, the effector T cells are unnecessary and thus disposable. Thousands of clone members become memory T cells, persisting perhaps for life, and providing a reservoir of T cells that can later mediate secondary responses to the same antigen.

Cytokines

The chemical messengers involved in cellular immunity belong to a group of molecules called **cytokines**, a general term for mediators that influence cell development, differentiation, and responses in the immune system. Cytokines include interferons and interleukins. In **Table 21.4**, you can see the large variety of these molecules and their myriad effects on target cells.

The cytokines include hormone-like or paracrine-like glycoproteins released by a variety of cells. As we mentioned previously, some cytokines act to promote T cell proliferation. For example, **interleukin 1 (IL-1)**, released by macrophages, stimulates bound T cells to liberate **interleukin 2 (IL-2)** and to synthesize more IL-2 receptors. IL-2 is a key growth factor. Acting on the cells that release it (as well as other T cells), it sets up a positive feedback cycle that encourages activated T cells to divide even more rapidly. (Therapeutically, IL-2 is used to treat melanoma and kidney cancers.)

Additionally, all activated T cells secrete one or more other cytokines that help amplify and regulate a variety of adaptive and innate immune responses. Some (such as *tumor necrosis factor*) are cell toxins. Others (e.g., *gamma interferon*) enhance the killing power of macrophages; and still others are inflammatory factors.

CHECK YOUR UNDERSTANDING

17. Class II MHC proteins display what kind of antigens? What class of T cell recognizes antigens bound to class II MHC? What types of cells display these proteins?
18. What happens when antigens are bound in the absence of co-stimulators?

For answers, see Appendix G.

Specific Effector T Cell Roles

▶ Describe T cell functions in the body.

When immunocompetent CD4 cells are activated, their offspring become effector cells (cells that carry out T cell functions) or memory cells (cells that cause faster, more prolonged responses upon a second encounter with an antigen). As we mentioned earlier, effector CD4 cells are helper or regulatory T cells, and effector CD8 cells are cytotoxic T cells. There are other minor "oddball" populations of effector T cells, but here we will focus on the three major groups.

Helper T Cells

Helper T cells play a central role in adaptive immunity, mobilizing both its cellular and its humoral arms, as the examples in **Figure 21.19** show. Once T_H cells have been primed by APC presentation of antigen, they help activate B and T cells, and induce B and T cell proliferation. In fact, without the "director" T_H cells, there is *no* adaptive immune response. They also activate macrophages to become more potent killers, and their cytokines furnish the chemical help needed to recruit other immune cells to fight off intruders.

Helper T cells interact directly with B cells displaying antigen fragments bound to class II MHC receptors (Figure 21.19a). They prod the B cells into more rapid division and then, like the boss of an assembly line, signal for antibody formation to begin. Whenever a T_H cell binds to a B cell, the T cell releases interleukin 4 and other cytokines. B cells may be activated solely by binding to certain antigens called **T cell–independent antigens**, but most antigens require T cell help to activate the B cells to which they bind. These more common antigens are called **T cell–dependent antigens**. In general, T cell–independent antigen responses are weak and short-lived. B cell division continues as long as it is stimulated by the T_H cell. In this way, helper T cells help unleash the protective potential of B cells.

Similarly, activating CD8 cells into destructive cytotoxic T cells usually requires help from T_H cells. As shown in Figure 21.19b, T_H cells cause dendritic cells to express on their surfaces the co-stimulatory molecules required for CD8 cell activation.

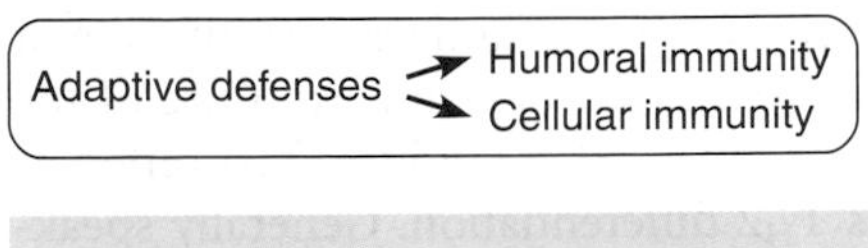
Adaptive defenses
Humoral immunity
Cellular immunity

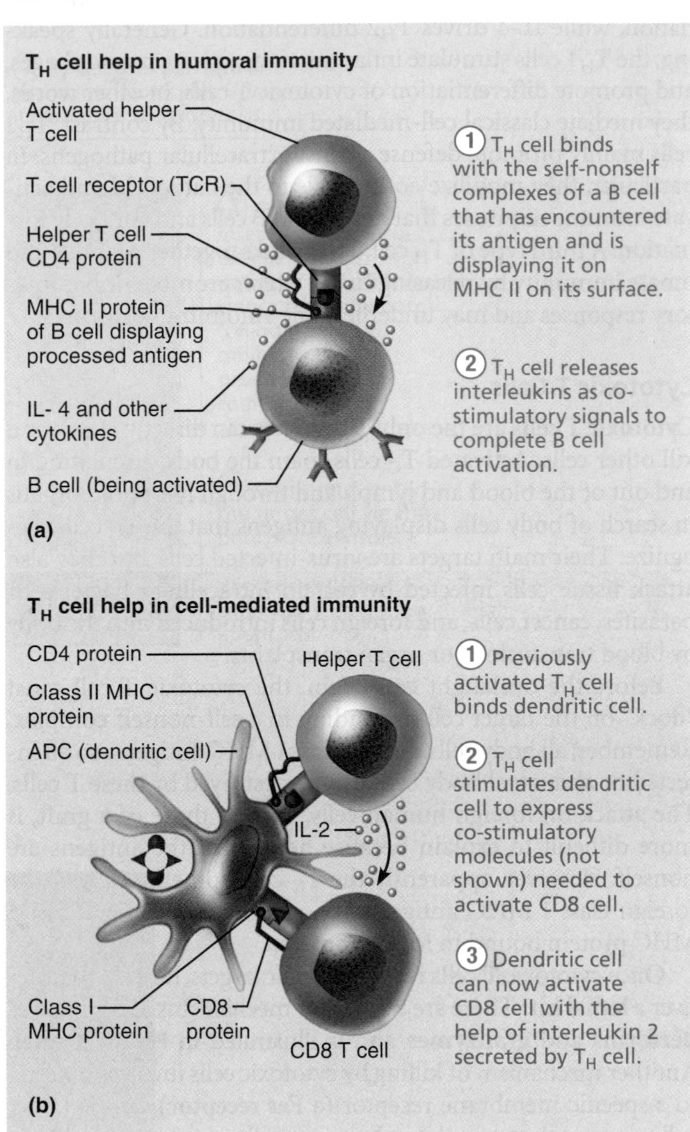
T_H cell help in humoral immunity
Activated helper T cell
T cell receptor (TCR)
Helper T cell CD4 protein
MHC II protein of B cell displaying processed antigen
IL- 4 and other cytokines
B cell (being activated)
1 T_H cell binds with the self-nonself complexes of a B cell that has encountered its antigen and is displaying it on MHC II on its surface.
2 T_H cell releases interleukins as co-stimulatory signals to complete B cell activation.
(a)
T_H cell help in cell-mediated immunity
CD4 protein
Class II MHC protein
APC (dendritic cell)
Helper T cell
IL-2
Class I MHC protein
CD8 protein
CD8 T cell
1 Previously activated T_H cell binds dendritic cell.
2 T_H cell stimulates dendritic cell to express co-stimulatory molecules (not shown) needed to activate CD8 cell.
3 Dendritic cell can now activate CD8 cell with the help of interleukin 2 secreted by T_H cell.
(b)

Figure 21.19 The central role of helper T cells in mobilizing both humoral and cellular immunity. (a) T_H and B cells usually must interact directly for full B cell activation. **(b)** Co-stimulatory molecules required for CD8 T cell activation are expressed by dendritic cells in response to T_H cell binding. (Some types of antigens induce these co-stimulatory molecules themselves, in which case T_H cell help may not be needed.) The T_H cell also produces interleukin 2, which causes the CD8 cell to proliferate and differentiate.

Cytokines released by T_H cells not only mobilize lymphocytes and macrophages but also attract other types of white blood cells into the area and tremendously amplify innate defenses. As the released chemicals summon more and more cells into the battle, the immune response gains momentum, and the antigens are overwhelmed by the sheer numbers of immune elements acting against them.

It is interesting, but not surprising, that different subsets of helper T cells exist. The subset that develops during T_H cell

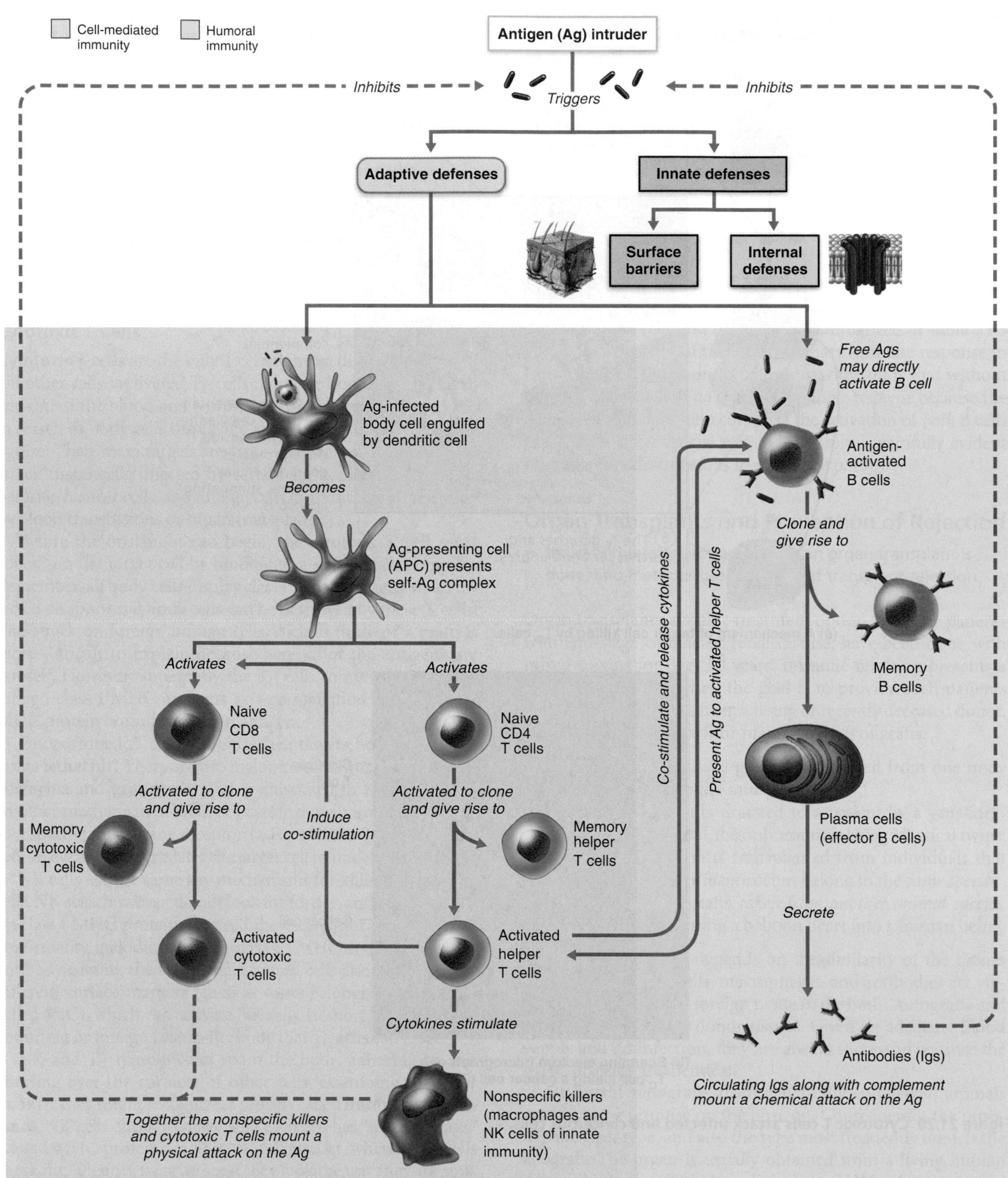

Figure 21.21 Simplified summary of the primary immune response. Co-stimulation usually requires direct cell-cell interactions; cytokines enhance these and many other events. Although complement, NK cells, and phagocytes are innate defenses, they are enlisted in the fight by cytokines. (For simplicity, only B cell receptors are illustrated.)

21

TABLE 21.5 Cells and Molecules of the Adaptive Immune Response

ELEMENT	FUNCTION IN IMMUNE RESPONSE
Cells	
B cell	Lymphocyte that matures in bone marrow. Induced to replicate by antigen binding, usually followed by helper T cell interactions in lymphoid tissues. Its progeny (clone members) form memory cells and plasma cells.
Plasma cell	Antibody-producing "machine"; produces huge numbers of antibodies (immunoglobulins) with the same antigen specificity. An effector B cell.
Helper T cell (T_H)	An effector CD4 T cell central to both humoral and cellular immunity. After binding with a specific antigen presented by an APC, it stimulates production of cytotoxic T cells and plasma cells to help fight invader, activates macrophages, and acts both directly and indirectly by releasing cytokines.
Cytotoxic T cell (T_C)	An effector CD8 cell. Activated by antigen presented by an APC, often with helper T cell involvement. Its specialty is killing virus-invaded body cells and cancer cells; also involved in rejection of foreign tissue grafts.
Regulatory T cell (T_{Reg})	Slows or stops activity of immune system. Important in controlling autoimmune diseases; several different populations probably exist.
Memory cell	Descendant of activated B cell or any class of T cell; generated during initial immune response (primary response). May exist in body for years after, enabling it to respond quickly and efficiently to subsequent infections or meetings with same antigen.
Antigen-presenting cell (APC)	Any of several cell types (dendritic cell, macrophage, B cell) that engulfs and digests antigens that it encounters, presenting parts of them on its plasma membrane (bound to an MHC protein) for recognition by T cells bearing receptors for same antigen. This function, antigen presentation, is essential for normal cell-mediated responses. Macrophages and dendritic cells also release chemicals (cytokines) that activate many other immune cells.
Molecules	
Antigen	Substance capable of provoking an immune response. Typically a large complex molecule (e.g., protein or modified protein) not normally present in the body.
Antibody (immunoglobulin)	Protein produced by B cell or by plasma cell. Antibodies produced by plasma cells are released into body fluids (blood, lymph, saliva, mucus, etc.), where they attach to antigens, causing complement fixation, neutralization, precipitation, or agglutination, which "mark" the antigens for destruction by phagocytes or complement.
Perforins, granzymes	Released by T_C cells. Perforins create large pores in the target cell's membrane, allowing entry of apoptosis-inducing granzymes.
Complement	Group of bloodborne proteins activated after binding to antibody-covered antigens or certain molecules on the surface of microorganisms; enhances inflammatory response and causes lysis of some microorganisms.
Cytokines	Small proteins that act as chemical messengers between various parts of the immune system. See Table 21.4.

bacterial and viral infection remains the most frequent cause of death in transplant patients. The key to successful graft survival is to provide enough immunosuppression to prevent graft rejection but not enough to be toxic, and to use antibiotics to keep infection under control. Even with the best conditions, by ten years after receiving a transplant, roughly 50% of patients have rejected the donor organ.

In rare cases, transplant patients have naturally achieved a state of immune tolerance and have been able to "ditch the drugs." Finding a way to induce tolerance is the goal of many research projects. One approach is to create a **chimeric immune system** (*chimer* = monster) by temporarily suppressing the recipient's bone marrow and then dousing it with bone marrow from the same donor as the new organ in the hope that this combined immune system will treat the transplanted organ as self. Another approach tries to harness the body's own tolerance-inducing cells, the regulatory T cells, to suppress only those immune reactions that lead to transplant rejection.

CHECK YOUR UNDERSTANDING

19. Which type of T cell is the most important in both cell-mediated and humoral immunity? Why?
20. Describe the killing mechanism of cytotoxic T cells that involves perforins.
21. Which proteins must be carefully matched before an organ transplant?

For answers, see Appendix G.

Homeostatic Imbalances of Immunity

- Give examples of immune deficiency diseases and of hypersensitivity states.
- Cite factors involved in autoimmune disease.

21

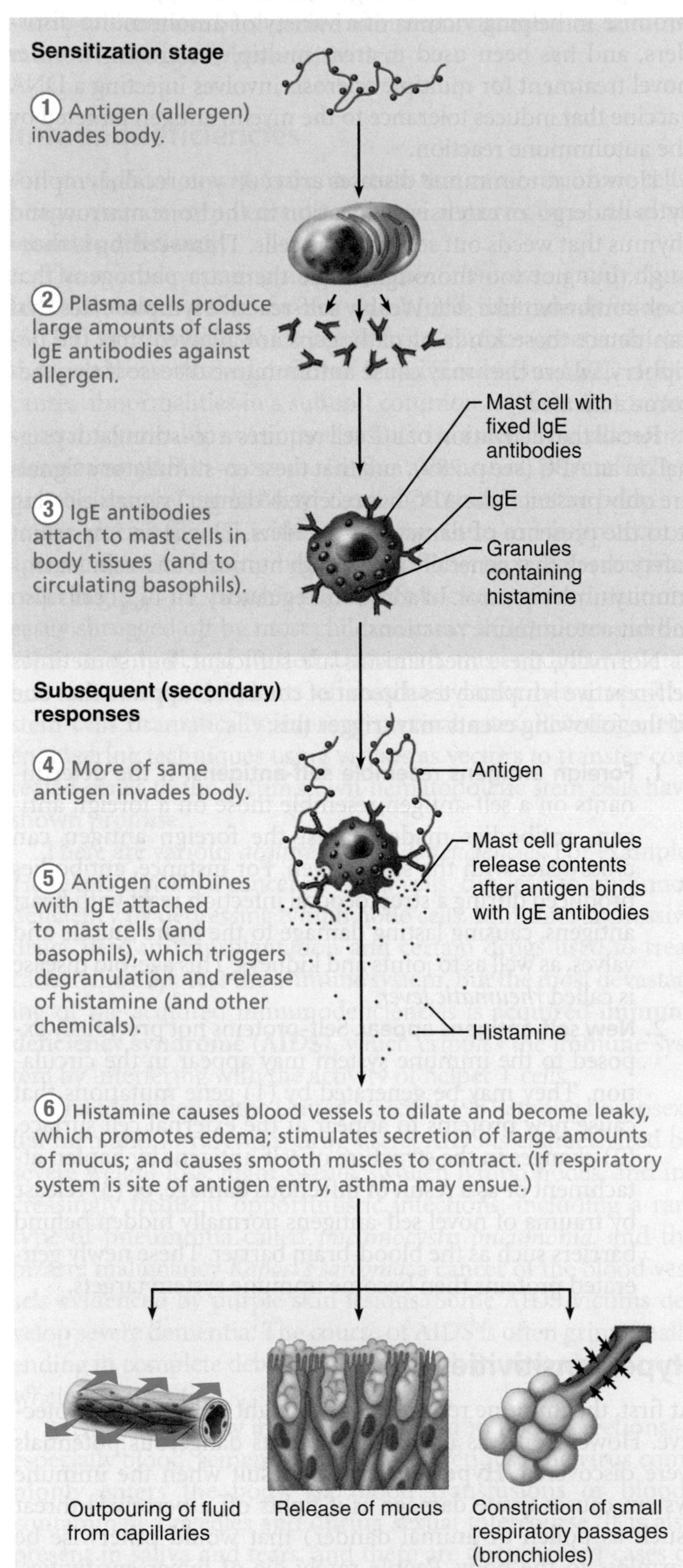

Adaptive defenses → Humoral immunity
Sensitization stage
1 Antigen (allergen) invades body.
2 Plasma cells produce large amounts of class IgE antibodies against allergen.
Mast cell with fixed IgE antibodies
IgE
Granules containing histamine
3 IgE antibodies attach to mast cells in body tissues (and to circulating basophils).
Subsequent (secondary) responses
4 More of same antigen invades body.
Antigen
Mast cell granules release contents after antigen binds with IgE antibodies
5 Antigen combines with IgE attached to mast cells (and basophils), which triggers degranulation and release of histamine (and other chemicals).
Histamine
6 Histamine causes blood vessels to dilate and become leaky, which promotes edema; stimulates secretion of large amounts of mucus; and causes smooth muscles to contract. (If respiratory system is site of antigen entry, asthma may ensue.)
Outpouring of fluid from capillaries
Release of mucus
Constriction of small respiratory passages (bronchioles)

Figure 21.22 Mechanism of an acute allergic (immediate hypersensitivity) response.

Immediate Hypersensitivities

The **immediate hypersensitivities**, also called **acute** or **type I hypersensitivities**, are simply what most of us would call **allergies** (*allo* = altered; *erg* = reaction). An **allergen** is an antigen that causes an allergic reaction. Allergic reactions begin within seconds after contact with the allergen and last about half an hour.

The initial meeting with an allergen produces no symptoms but it sensitizes the person. APCs digest the allergen and present its fragments to T_H cells as usual. In susceptible individuals, an abnormally large number of these T_H cells differentiate into IL-4-secreting T_H2 cells. IL-4 stimulates B cells to mature into IgE-secreting plasma cells, which spew out huge amounts of antibody specific for the allergen. When the IgE molecules attach to *mast cells* and *basophils*, sensitization is complete.

An allergic reaction is triggered by later encounters with the same allergen, which promptly binds and cross-links the IgE antibodies on the surfaces of the mast cells and basophils. This event induces an enzymatic cascade that causes the mast cells and basophils to degranulate, releasing a flood of *histamine* and other inflammatory chemicals that together induce the inflammatory response typical of allergy **(Figure 21.22)**.

Allergic reactions may be local or systemic. Mast cells are abundant in connective tissues of the skin and beneath the mucosa of respiratory passages and the gastrointestinal tract, and these areas are common sites of local allergic reactions. Histamine causes blood vessels to become dilated and leaky, and is largely to blame for the best recognized symptoms of allergy: runny nose, itching reddened skin (hives), and watery eyes. When the allergen is inhaled, symptoms of *asthma* appear because smooth muscle in the walls of the bronchioles contracts, constricting those small passages and restricting air flow. When the allergen is ingested in food or via drugs, gastrointestinal discomfort (cramping, vomiting, or diarrhea) occurs. Over-the-counter antiallergy drugs contain antihistamines that counteract these effects.

The bodywide or systemic response known as **anaphylactic shock** is fairly rare. It typically occurs when the allergen directly enters the blood and circulates rapidly through the body, as might happen with certain bee stings or spider bites. It may also follow injection of a foreign substance (such as penicillin or other drugs which act as haptens).

The mechanism of anaphylactic shock is essentially the same as that of local responses, but when mast cells and basophils are enlisted throughout the entire body, the outcome is life threatening. The bronchioles constrict (and the tongue may swell), making it difficult to breathe, and the sudden vasodilation and fluid loss from the bloodstream may cause circulatory collapse (hypotensive shock) and death within minutes. Epinephrine is the drug of choice to reverse these histamine-mediated effects.

Subacute Hypersensitivities

Like the immediate types, **subacute hypersensitivities** are caused by antibodies (IgG and IgM rather than IgE) and can be transferred via blood plasma or serum. However, their onset is slower (1–3 hours after antigen exposure) and the duration of the reaction is longer (10–15 hours).

21

Cytotoxic (type II) reactions occur when antibodies bind to antigens on specific body cells and subsequently stimulate phagocytosis and complement-mediated lysis of the cellular antigens. Type II hypersensitivity may occur after a patient has received a transfusion of mismatched blood and the foreign red blood cells are lysed by complement.

Immune-complex (type III) hypersensitivity results when antigens are widely distributed through the body or blood and the insoluble antigen-antibody complexes formed cannot be cleared from a particular area. (This may reflect a persistent infection or a situation in which huge amounts of antigen-antibody complexes are formed.) An intense inflammatory reaction occurs, complete with complement-mediated cell lysis and cell killing by neutrophils that severely damages local tissues. One example of type III hypersensitivity is *farmer's lung* (induced by inhaling moldy hay). Additionally, many immune-complex responses are involved in autoimmune disorders, such as glomerulonephritis, systemic lupus erythematosus, and rheumatoid arthritis.

Delayed Hypersensitivities

Delayed hypersensitivity (type IV) reactions are slower to appear (1–3 days) than antibody-mediated hypersensitivity reactions. The mechanism is basically that of a cell-mediated immune response, which depends on helper T cells. Inflammation and tissue damage comes about through the action of cytokine-activated macrophages, and sometimes cytotoxic T cells.

The most familiar examples of delayed hypersensitivity reactions are those classified as **allergic contact dermatitis** which follow skin contact with poison ivy, some metals (nickel in jewelry), and certain cosmetic and deodorant chemicals. These agents act as haptens, and after diffusing through the skin and attaching to self-proteins, they are perceived as foreign by the immune system. The *Mantoux* and *tine tests*, two skin tests for tuberculosis, depend on delayed hypersensitivity reactions. When the tubercle antigens are introduced just under the skin, a small hard lesion forms that persists for days if the person has been sensitized to the antigen.

CHECK YOUR UNDERSTANDING

22. What makes HIV particularly hard for the immune system to defeat?
23. What event triggers the release of histamine from mast cells in an allergic response?

For answers, see Appendix G.

Developmental Aspects of the Immune System

- Describe changes in immunity that occur with aging.
- Briefly describe the role of the nervous system in regulating the immune response.

Stem cells of the immune system originate in the liver and spleen during weeks 1–9 of embryonic development. Later the bone marrow becomes the predominant source of stem cells, and it persists in this role into adult life. In late fetal life and shortly after birth, the young lymphocytes develop self-tolerance and immunocompetence in their "programming organs" (thymus and bone marrow) and then populate the other lymphoid tissues. When challenged by an antigen, the naive T and B cells further differentiate into effector cells and memory cells.

The newborn's immune system depends primarily on antibodies, and hence on T_H2 lymphocytes. The T_H1 system is educated and gets stronger as a result of encounters with microbes—both harmful and harmless. If such "exercise" does not occur, immune balance is upset and the T_H2 system flourishes, causing the immune system to teeter toward allergies. Unhappily, our desire to keep our children squeaky clean with antibiotics that kill off both harmful and harmless bacteria may derail normal immune development.

The ability of the immune system to recognize foreign substances is controlled by genes. However, the nervous system also plays a role in the immune response, and studies of psychoneuroimmunology—a brain-twisting term coined to describe links between the brain and the immune system—have begun to reveal some answers. For example, the immune response is definitely impaired in individuals who are depressed or under severe stress, such as those mourning the death of a beloved family member or friend.

Our immune system normally serves us very well until late in life. Then its efficiency begins to wane, and its ability to fight infection declines. Old age is also accompanied by greater susceptibility to both immune deficiency and autoimmune diseases. The greater incidence of cancer in the elderly is assumed to be an example of the progressive failure of the immune system. We do not know why the immune system begins to fail, but we do know that the thymus begins to atrophy after puberty and the production of naive T and B cells declines with age, possibly because progenitor cells reach the limits of their ability to further divide.

■ ■ ■

The adaptive immune system provides remarkable defenses against disease. These amazingly diverse defenses are regulated by cellular interactions and a flood of chemicals. T cells and antibodies make perfect partners. Antibodies respond swiftly to toxins and molecules on the outer surfaces of foreign organisms, and T cells destroy foreign antigens hidden inside cells and our own cells that have become mutinous (cancer cells). The innate immune system exhibits a different arsenal for body defense, an arsenal that is simpler perhaps and more easily understood. The innate and adaptive defenses are tightly interlocked, each providing what the other cannot and amplifying each other's effects.

RELATED CLINICAL TERMS

Congenital thymic aplasia An immune deficiency disease in which the thymus fails to develop. Affected individuals have no T cells, hence little or no immune protection; fetal thymic and bone marrow transplants have been helpful in some cases.

Eczema (ek′zĕ-mah) A clinical term for several conditions that cause "weeping" skin lesions and intense itching. One common cause, *atopic dermatitis*, has a strong familial predisposition, has features of immediate hypersensitivity, and usually begins in the first five years of life. Recent research suggests the underlying defect may be increased leakiness of the skin.

Hashimoto's thyroiditis An autoimmune disease in which the thyroid gland is attacked by both B and T lymphocytes. It is the most common cause of hypothyroidism, affecting mostly middle-aged and elderly women. Genetic factors associated with this autoimmune disease (certain MHC variants) make individuals susceptible to environmental triggers (possibly iodine, irradiation, or trauma).

Immunization The process of rendering a subject immune (by vaccination or injection of antiserum).

Immunology The study of immunity.

Immunopathology Disease of the immune system.

Septic shock (sepsis) A dangerous condition in which the inflammatory response goes out of control. Kills 175,000 people a year in the U.S. Results from especially severe bacterial infections or more ordinary infections that grow rapidly worse in patients with weakened defenses, such as the hospitalized elderly recovering from surgery. In an inflammatory response, neutrophils and other white blood cells leave the capillaries and enter the infected connective tissue, secreting cytokines that increase capillary permeability. Secretion of cytokines is normally moderate, but in sepsis, continued cytokine release makes capillaries so leaky that the bloodstream is depleted of fluid. Blood pressure falls and the body organs shut down, causing death in 50% of all cases. Sepsis has proven difficult to control and its incidence remains high.

Systemic lupus erythematosus (SLE) (er″ĭ-the′mah-to′sus) A systemic autoimmune disorder that occurs mainly in young females. Diagnosis is helped by finding antinuclear (anti-DNA) antibodies in the patient's serum. DNA–anti-DNA complexes (typical of type III hypersensitivity) localize in the kidneys (the capillary filters, or glomeruli), in blood vessels, in the brain, and in the synovial membranes of joints, resulting in glomerulonephritis, vascular problems, loss of memory and mental sharpness, and painful arthritis. Reddened skin lesions, particularly a "butterfly rash" (the sign of the wolf, or *lupus*) on the face, are common.

CHAPTER SUMMARY

Media study tools that could provide you additional help in reviewing specific key topics of Chapter 21 are referenced below.

iP = ***Interactive Physiology***

PART 1: INNATE DEFENSES

Surface Barriers: Skin and Mucosae (pp. 767–768)

21

1. Skin and mucous membranes constitute the first line of defense. Their role is to prevent pathogens from entering the body. Protective membranes line all body cavities and organs exposed to the exterior.
2. Surface membranes provide mechanical barriers to pathogens. Some have structural modifications and produce secretions that enhance their defensive effects: The skin's acidity, lysozyme, mucus, keratin, and ciliated cells are examples.

iP IP Immune System; Topic: Innate Host Defenses, pp. 3–5.

Internal Defenses: Cells and Chemicals (pp. 768–775)

1. The innate cellular and chemical defenses provide the body's second line of defense.

Phagocytes (p. 768)

2. Phagocytes (macrophages, neutrophils, and the like) engulf and destroy pathogens that breach epithelial barriers. This process is facilitated when antibodies or complement to which the phagocyte's receptors can bind attach to the pathogen's surface. Cell killing is enhanced by the respiratory burst.

Natural Killer Cells (pp. 768–769)

3. Natural killer cells are large granular lymphocytes that act nonspecifically to kill virus-infected and malignant cells.

Inflammation: Tissue Response to Injury (pp. 769–772)

4. The inflammatory response prevents the spread of harmful agents, disposes of pathogens and dead tissue cells, and promotes healing. Exudate is formed; protective leukocytes enter the area; the area is walled off by fibrin; and tissue repair occurs.
5. The cardinal signs of inflammation are swelling, redness, heat, and pain. These result from vasodilation and increased permeability of blood vessels induced by inflammatory chemicals. If the inflamed area is a joint, movement may be impaired.

Antimicrobial Proteins (pp. 773–775)

6. Interferons are a group of related proteins synthesized by virus-infected cells and certain immune cells that prevent viruses from multiplying in other body cells.
7. Activation of complement (a group of plasma proteins) on the membrane of a foreign cell promotes phagocytosis of that cell, enhances the inflammation, and sometimes causes lysis of the target cell.

Fever (p. 775)

8. Fever enhances the body's fight against pathogens by increasing metabolism, which speeds up defensive actions and repair processes, and by prompting the liver and spleen to sequester iron and zinc needed for bacterial multiplication.

iP Immune System; Topic: Innate Host Defenses, pp. 6–23.

PART 2: ADAPTIVE DEFENSES

1. The adaptive immune system recognizes something as foreign and acts to immobilize, neutralize, or remove it. The adaptive immune response is antigen-specific, systemic, and has memory. It provides the body's third line of defense.

iP Immune System; Topic: Immune System Overview, p. 7.

Antigens (pp. 776–777)

1. Antigens are substances capable of generating an immune response.

Complete Antigens and Haptens (p. 776)

2. Complete antigens have both immunogenicity and reactivity. Incomplete antigens or haptens must combine with a body protein before becoming immunogenic.

Antigenic Determinants (pp. 776–777)

3. Antigenic determinants are the portions of antigen molecules that are recognized as foreign. Most antigens have many such sites.

Self-Antigens: MHC Proteins (p. 777)

4. Major histocompatibility complex (MHC) proteins are membrane-bound glycoproteins that mark our cells as "self." Class I MHC proteins are found on the surfaces of all body cells (except RBCs); the class II variety is found on certain cells that function in adaptive immune responses (dendritic cells, macrophages, and B cells).

iP Immune System; Topic: Immune System Overview, p. 8; Topic: Cellular Immunity, pp. 5–8.

Cells of the Adaptive Immune System: An Overview (pp. 777–780)

Lymphocytes (pp. 777–779)

1. Lymphocytes arise from hematopoietic stem cells of the bone marrow and are educated to develop immunocompetence and self-tolerance. T cells are educated in the thymus and confer cell-mediated immunity. B cells are educated in the bone marrow and provide humoral immunity. Immunocompetence is signaled by the appearance of antigen-specific receptors on the surface of the lymphocyte. Immunocompetent lymphocytes seed the secondary lymphoid organs, where the antigen challenge occurs, and circulate between the blood, lymph, and lymphoid organs.
2. In both B and T lymphocytes, antigen receptor diversity is generated by shuffling gene fragments.

Antigen-Presenting Cells (pp. 779–780)

3. Antigen-presenting cells (APCs) include dendritic cells, macrophages, and B lymphocytes. They internalize antigens and present antigenic determinants on their surfaces for recognition by T cells.

iP Immune System; Topic: Common Characteristics of B and T Lymphocytes, pp. 8–13.

Humoral Immune Response (pp. 780–786)

Clonal Selection and Differentiation of B Cells (p. 780)

1. Clonal selection and differentiation of B cells occur when antigens bind to their receptors, causing them to proliferate. Most of the clone members become effector cells called plasma cells, which secrete antibodies. This is the primary adaptive immune response.

Immunological Memory (pp. 780–781)

2. Other clone members become memory B cells, capable of mounting a rapid attack against the same antigen in subsequent encounters (secondary immune responses). The memory cells provide humoral immunological memory.

Active and Passive Humoral Immunity (pp. 781–783)

3. Active humoral immunity is acquired during an infection or via vaccination and provides immunological memory. Passive immunity is conferred when a donor's antibodies are injected into the bloodstream, or when the mother's antibodies cross the placenta. Its protection is short-lived; immunological memory is not established.

Antibodies (pp. 783–786)

4. The antibody monomer consists of four polypeptide chains, two heavy and two light, connected by disulfide bonds. Each chain has both a constant and a variable region. Constant regions determine antibody function and class. Variable regions enable the antibody to recognize its specific antigen.
5. Five classes of antibodies exist: IgM, IgA, IgD, IgG, and IgE. They differ structurally and functionally.
6. Antibody functions include complement fixation and antigen neutralization, precipitation, and agglutination.
7. Monoclonal antibodies are pure preparations of a single antibody type useful in diagnostic tests and in treatment for some types of cancer. They are prepared by fusing B cells with tumor cells to produce hybridomas.

iP Immune System; Topic: Common Characteristics of B and T Lymphocytes, pp. 14–15; Topic: Humoral Immunity, pp. 3–14.

Cell-Mediated Immune Response (pp. 786–795)

Clonal Selection and Differentiation of T Cells (pp. 786–791)

1. Immunocompetent CD4 and CD8 T cells are activated by binding to an antigen-containing MHC protein displayed on the surface of an APC. A co-stimulatory signal is also essential. Clonal selection occurs and the clone members differentiate into the appropriate effector T cells that mount the primary immune response (e.g., helper and cytotoxic T cells). Some clone members become memory T cells.

Specific Effector T Cell Roles (pp. 791–792)

2. Helper T cells are required for full activation of most B and T cells, activate macrophages, and release essential cytokines. Cytotoxic T cells directly attack and kill infected cells and cancer cells. Regulatory (T_{Reg}) T cells help to maintain tolerance in the periphery.
3. The immune response is enhanced by cytokines such as interleukin 1 released by macrophages, and interleukin 2, gamma interferon, and others released by activated T cells.

Organ Transplants and Prevention of Rejection (pp. 792–795)

4. Grafts or foreign organ transplants are rejected by cell-mediated responses unless the patient is immunosuppressed. Infections are major complications in such patients.

iP Immune System; Topic: Cellular Immunity, pp. 3–4, 9, 11–14.

Homeostatic Imbalances of Immunity (pp. 795–799)

Immunodeficiencies (pp. 796–797)

1. Immunodeficiency diseases include severe combined immunodeficiency (SCID) syndromes and acquired immune deficiency syndrome (AIDS). Overwhelming infections are fatal because the immune system is unable to combat them.

Autoimmune Diseases (p. 797)

2. Autoimmune disease occurs when the body regards its own tissues as foreign and mounts an immune attack against them. Examples include rheumatoid arthritis and multiple sclerosis.

Hypersensitivities (pp. 797–799)

3. Hypersensitivity is an abnormally intense reaction to an otherwise harmless antigen. Immediate hypersensitivities (allergies) are mounted by IgE antibodies. Subacute hypersensitivities, involving both antibodies and complement, include antibody-mediated cytotoxic and immune-complex hypersensitivities. Cell-mediated hypersensitivity is called delayed hypersensitivity.

Developmental Aspects of the Immune System (p. 799)

1. Development of the immune response occurs around the time of birth. The ability of the immune system to recognize foreign substances is genetically determined.
2. The nervous system plays an important role in regulating immune responses, possibly through common mediators. Depression impairs immune function.
3. With aging, the immune system becomes less responsive. The elderly more often suffer from immune deficiency, autoimmune diseases, and cancer.

REVIEW QUESTIONS

Multiple Choice/Matching

(Some questions have more than one correct answer. Select the best answer or answers from the choices given.)

1. All of the following are considered innate body defenses *except* (a) complement, (b) phagocytosis, (c) antibodies, (d) lysozyme, (e) inflammation.
2. The process by which neutrophils squeeze through capillary walls in response to inflammatory signals is called (a) diapedesis, (b) chemotaxis, (c) margination, (d) opsonization.
3. Antibodies released by plasma cells are involved in (a) humoral immunity, (b) immediate hypersensitivity reactions, (c) autoimmune disorders, (d) all of the above.
4. Which of the following antibodies can fix complement? (a) IgA, (b) IgD, (c) IgE, (d) IgG, (e) IgM.
5. Which antibody class is abundant in body secretions? (Use the choices from question 4.)
6. Small molecules that must combine with large proteins to become immunogenic are called (a) complete antigens, (b) kinins, (c) antigenic determinants, (d) haptens.
7. Lymphocytes that develop immunocompetence in the bone marrow are (a) T lymphocytes, (b) B lymphocytes, (c) NK cells, (d) B and T lymphocytes.

8. Cells that can directly attack target cells include all of the following *except* (a) macrophages, (b) cytotoxic T cells, (c) helper T cells, (d) natural killer cells.
9. Which of the following is involved in the activation of a B cell? (a) antigen, (b) helper T cell, (c) cytokine, (d) all of the above.
10. The cell type most often invaded by HIV is a(n) (a) eosinophil, (b) cytotoxic T cell, (c) natural killer cell, (d) helper T cell, (e) B cell.
11. Complement fixation promotes all of the following except (a) cell lysis, (b) inflammation, (c) opsonization, (d) interferon release, (e) chemotaxis of neutrophils and other cells.
12. Using the letters from column B, match the cell description in column A. (Note that all require more than a single choice.)

Column A	Column B
_____ (1) phagocyte	(a) natural killer cells
_____ (2) releases histamine	(b) neutrophil
_____ (3) releases perforins	(c) dendritic cells
_____ (4) lymphocyte	(d) mast cell
_____ (5) effector cells of adaptive immunity	(e) cytotoxic T cell
	(f) B cell
_____ (6) antigen-presenting cell	(g) macrophage
	(h) helper T cell
	(i) basophil

Short Answer Essay Questions

13. Besides acting as mechanical barriers, the skin epidermis and mucosae of the body have other attributes that contribute to their protective roles. Cite the common body locations and the importance of mucus, lysozyme, keratin, acid pH, and cilia.
14. Explain why attempts at phagocytosis are not always successful; cite factors that increase the likelihood of success.
15. What is complement? How does it cause bacterial lysis? What are some of the other roles of complement?
16. Interferons are referred to as antiviral proteins. What stimulates their production, and how do they protect uninfected cells? What cells of the body secrete interferons?
17. Differentiate between humoral and cell-mediated adaptive immunity.
18. Although the adaptive immune system has two arms, it has been said, "no T cells, no immunity." Explain.
19. Define immunocompetence and self-tolerance. How is self-tolerance achieved?
20. Differentiate between a primary and a secondary immune response. Which is more rapid and why?
21. Define antibody. Using an appropriately labeled diagram, describe the structure of an antibody monomer. Indicate and label variable and constant regions, heavy and light chains.
22. What is the role of the variable regions of an antibody? Of the constant regions?
23. Name the five antibody classes and describe where each is most likely to be found in the body.
24. How do antibodies help defend the body?
25. Do vaccines produce active or passive humoral immunity? Explain your answer. Why is passive immunity less satisfactory?
26. Describe the process of activation of a CD4 T cell.
27. Describe the specific roles of helper, regulatory, and cytotoxic T cells in normal cell-mediated immunity.
28. Name several cytokines and describe their role in the immune response.
29. Define hypersensitivity. List three types of hypersensitivity reactions. For each, note whether antibodies or T cells are involved and provide two examples.
30. What events can result in autoimmune disease?
31. What accounts for the declining efficiency of the immune system with age?

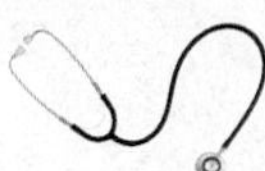

Critical Thinking and Clinical Application Questions

1. Jenny, a 6-year-old child who has been raised in a germ-free environment from birth, is a victim of one of the most severe examples of an abnormal immune system. Jenny also suffers from cancer caused by the Epstein-Barr virus. Relative to this case: **(a)** What is the usual fate of children with Jenny's condition and similar circumstances if no treatment is attempted? **(b)** Why is Jenny's brother chosen as the hematopoietic stem cell donor? **(c)** Why is her physician planning to use umbilical cord blood as a source of stem cells for transplant if her brother's stem cells fail (what are the hoped-for results)? **(d)** Attempt to explain Jenny's cancer. **(e)** What similarities and dissimilarities exist between Jenny's illness and AIDS?
2. Some people with a deficit of IgA exhibit recurrent respiratory tract infections. Explain this observation.
3. Capillary permeability increases and plasma proteins leak into the interstitial fluid as part of the inflammatory process. Why is this desirable?
4. Costanza was picking grapes in her father's arbor when she felt a short prickling pain in her finger. She ran crying to her father, who removed an insect stinger and calmed her with a glass of lemonade. Twenty minutes later Costanza's finger was red, swollen, and throbbing where she had been stung. What type of immune response was she exhibiting? What treatment would relieve her discomfort?
5. Caroline, a pregnant 29-year-old, has been HIV-positive for at least 10 years, dating back to when she was homeless and injecting heroin on a regular basis. While she currently has no symptoms of AIDS, she is taking several medications and is worried about the possibility that her baby might be infected. How do you think the HIV virus might be transferred from a mother to her offspring? Which of Caroline's cells are infected by the virus and why is the viral attack on these cells so devastating? Why is Caroline taking medication even though she has no symptoms? What types of medications might she be taking and how do they affect the virus and its replication?

21

22

The Respiratory System

Far from self-sustaining, the body depends on the external environment, both as a source of substances the body needs to survive and as a catch basin for its wastes. The trillions of cells making up the body require a continuous supply of oxygen to carry out their vital functions. We cannot do without oxygen for even a little while, as we can without food or water.

As cells use oxygen, they give off carbon dioxide, a waste product the body must get rid of. They also generate dangerous free radicals, the inescapable by-products of living in a world full of oxygen. But let's get to the topic of this chapter, the respiratory system.

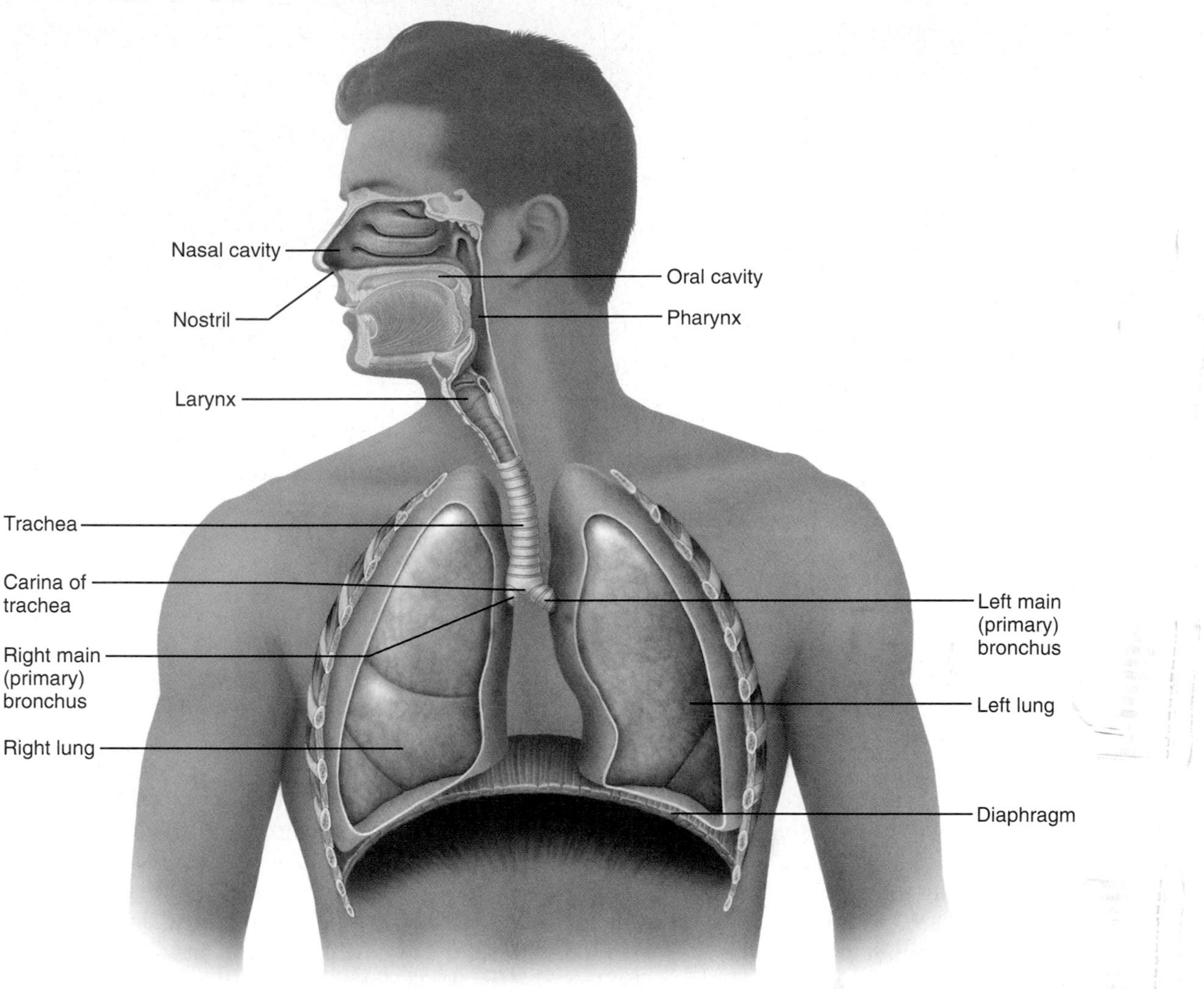

Figure 22.1 The major respiratory organs in relation to surrounding structures.

The major function of the **respiratory system** is to supply the body with oxygen and dispose of carbon dioxide. To accomplish this function, at least four processes, collectively called **respiration**, must happen:

1. **Pulmonary ventilation**: movement of air into and out of the lungs so that the gases there are continuously changed and refreshed (commonly called breathing).
2. **External respiration**: movement of oxygen from the lungs to the blood and of carbon dioxide from the blood to the lungs.
3. **Transport of respiratory gases**: transport of oxygen from the lungs to the tissue cells of the body, and of carbon dioxide from the tissue cells to the lungs. This transport is accomplished by the cardiovascular system using blood as the transporting fluid.
4. **Internal respiration**: movement of oxygen from blood to the tissue cells and of carbon dioxide from tissue cells to blood.

Only the first two processes are the special responsibility of the respiratory system **(Figure 22.1)**, but it cannot accomplish its primary goal of obtaining oxygen and eliminating carbon dioxide unless the third and fourth processes also occur. As you can see, the respiratory and circulatory systems are closely coupled, and if either system fails, the body's cells begin to die from oxygen starvation.

The actual use of oxygen and production of carbon dioxide by tissue cells, known as *cellular respiration*, is the cornerstone of all energy-producing chemical reactions in the body. We discuss cellular respiration, which is not a function of the respiratory system, in the metabolism section of Chapter 24.

Because it moves air, the respiratory system is also involved with the sense of smell and with speech.

Functional Anatomy of the Respiratory System

- Identify the organs forming the respiratory passageway(s) in descending order until the alveoli are reached.
- Describe the location, structure, and function of each of the following: nose, paranasal sinuses, pharynx, and larynx.
- List and describe several protective mechanisms of the respiratory system.

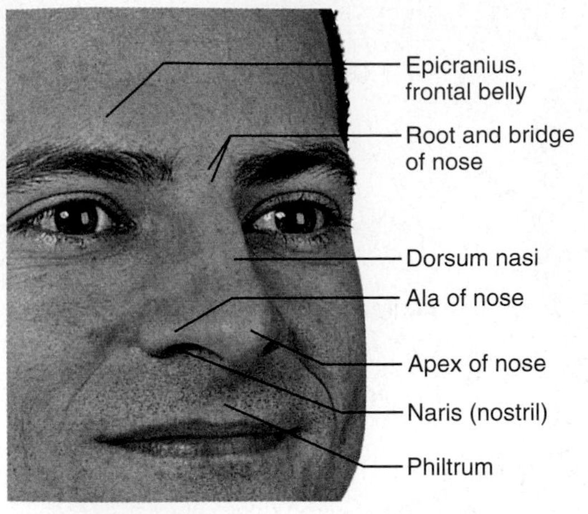

(a) Surface anatomy

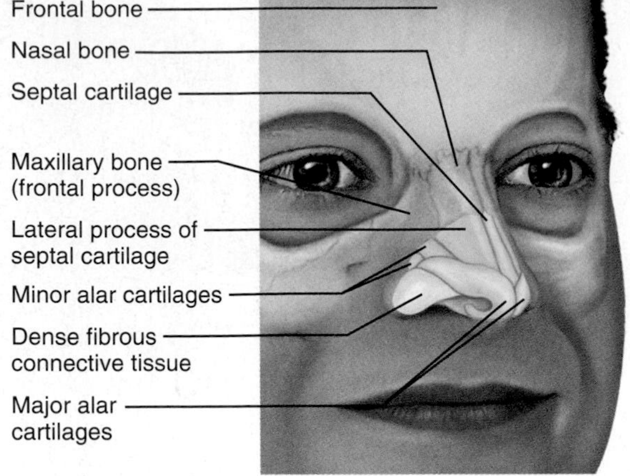

(b) External skeletal framework

Figure 22.2 The external nose.

The respiratory system includes the *nose*, *nasal cavity*, and *paranasal sinuses*; the *pharynx*; the *larynx*; the *trachea*; the *bronchi* and their smaller branches; and the *lungs*, which contain the terminal air sacs, or *alveoli* (Figure 22.1). Functionally, the system consists of two zones. The **respiratory zone**, the actual site of gas exchange, is composed of the respiratory bronchioles, alveolar ducts, and alveoli, all microscopic structures. The **conducting zone** includes all other respiratory passageways, which provide fairly rigid conduits for air to reach the gas exchange sites. The conducting zone organs also cleanse, humidify, and warm incoming air. As a result, air reaching the lungs has fewer irritants (dust, bacteria, etc.) than when it entered the system, and it is warm and damp, like the air of the tropics. The functions of the major organs of the respiratory system are summarized in **Table 22.1**.

In addition to these organs, some authorities also include the respiratory muscles (diaphragm, etc.) as part of this system. Although we will consider how these skeletal muscles bring about the volume changes that promote ventilation, we continue to classify them as part of the *muscular system*.

22

The Nose and Paranasal Sinuses

The nose is the only externally visible part of the respiratory system. Unlike the eyes and lips, facial features often referred to poetically, the nose is usually an irreverent target. We are urged to keep our nose to the grindstone and to keep it out of other people's business. Considering its important functions, however, it deserves more esteem. The nose (1) provides an airway for respiration, (2) moistens and warms entering air, (3) filters and cleans inspired air, (4) serves as a resonating chamber for speech, and (5) houses the olfactory (smell) receptors.

The structures of the nose are divided into the *external nose* and the internal *nasal cavity* for ease of consideration. The surface features of the external nose include the *root* (area between the eyebrows), *bridge*, and *dorsum nasi* (anterior margin), the latter terminating in the *apex* (tip of the nose) **(Figure 22.2a)**. Just inferior to the apex is a shallow vertical groove called the *philtrum* (fil′trum). The external openings of the nose, the *nostrils* or *nares* (na′rēz), are bounded laterally by the flared *alae*.

The skeletal framework of the external nose is fashioned by the nasal and frontal bones superiorly (forming the bridge and root, respectively), the maxillary bones laterally, and flexible plates of hyaline cartilage (the alar and septal cartilages, and the lateral processes of the septal cartilage) inferiorly (Figure 22.2b). Noses vary a great deal in size and shape, largely because of differences in the nasal cartilages. The skin covering the nose's dorsal and lateral aspects is thin and contains many sebaceous glands.

The internal **nasal cavity** lies in and posterior to the external nose. During breathing, air enters the cavity by passing through the **nostrils**, or **nares** (Figure 22.2a and **Figure 22.3c**). The nasal cavity is divided by a midline **nasal septum**, formed anteriorly by the septal cartilage and posteriorly by the vomer bone and perpendicular plate of the ethmoid bone (see Figure 7.14b, p. 213). The nasal cavity is continuous posteriorly with the nasal portion of the pharynx through the **posterior nasal apertures**, also called the *choanae* (ko-a′ne; "funnels").

The roof of the nasal cavity is formed by the ethmoid and sphenoid bones of the skull. The floor is formed by the *palate*, which separates the nasal cavity from the oral cavity below. Anteriorly, where the palate is supported by the palatine bones and processes of the maxillary bones, it is called the **hard palate**. The unsupported posterior portion is the muscular **soft palate**.

The part of the nasal cavity just superior to the nostrils, called the **nasal vestibule**, is lined with skin containing sebaceous and sweat glands and numerous hair follicles. The hairs, or **vibrissae** (vi-bris′e; *vibro* = to quiver), filter coarse particles (dust, pollen) from inspired air. The rest of the nasal cavity is lined with two types of mucous membrane. The **olfactory epithelium (mucosa)**, lining the slitlike superior region of the

TABLE 22.1 Principal Organs of the Respiratory System

STRUCTURE	DESCRIPTION, GENERAL AND DISTINCTIVE FEATURES	FUNCTION
Nose	Jutting external portion is supported by bone and cartilage. Internal nasal cavity is divided by midline nasal septum and lined with mucosa.	Produces mucus; filters, warms, and moistens incoming air; resonance chamber for speech
	Roof of nasal cavity contains olfactory epithelium.	Receptors for sense of smell
Paranasal sinuses	Mucosa-lined, air-filled cavities in cranial bones surrounding nasal cavity.	Same as for nasal cavity; also lighten skull
Pharynx	Passageway connecting nasal cavity to larynx and oral cavity to esophagus. Three subdivisions: nasopharynx, oropharynx, and laryngopharynx.	Passageway for air and food
	Houses tonsils (lymphoid tissue masses involved in protection against pathogens).	Facilitates exposure of immune system to inhaled antigens
Larynx	Connects pharynx to trachea. Has framework of cartilage and dense connective tissue. Opening (glottis) can be closed by epiglottis or vocal folds.	Air passageway; prevents food from entering lower respiratory tract
	Houses vocal folds (true vocal cords).	Voice production
Trachea	Flexible tube running from larynx and dividing inferiorly into two main bronchi. Walls contain C-shaped cartilages that are incomplete posteriorly where connected by trachealis muscle.	Air passageway; cleans, warms, and moistens incoming air
Bronchial tree	Consists of right and left main bronchi, which subdivide within the lungs to form lobar and segmental bronchi and bronchioles. Bronchiolar walls lack cartilage but contain complete layer of smooth muscle. Constriction of this muscle impedes expiration.	Air passageways connecting trachea with alveoli; cleans, warms, and moistens incoming air
Alveoli	Microscopic chambers at termini of bronchial tree. Walls of simple squamous epithelium are underlain by thin basement membrane. External surfaces are intimately associated with pulmonary capillaries.	Main sites of gas exchange
	Special alveolar cells produce surfactant.	Reduces surface tension; helps prevent lung collapse
Lungs	Paired composite organs that flank mediastinum in thorax. Composed primarily of alveoli and respiratory passageways. Stroma is fibrous elastic connective tissue, allowing lungs to recoil passively during expiration.	House respiratory passages smaller than the main bronchi
Pleurae	Serous membranes. Parietal pleura lines thoracic cavity; visceral pleura covers external lung surfaces.	Produce lubricating fluid and compartmentalize lungs

nasal cavity, contains smell receptors. The balance of the nasal cavity mucosa, the **respiratory mucosa**, is a pseudostratified ciliated columnar epithelium, containing scattered *goblet cells*, that rests on a lamina propria richly supplied with *mucous* and *serous glands*. (Mucous cells secrete mucus, and serous cells secrete a watery fluid containing enzymes.)

Each day, these glands secrete about a quart (or a liter) of mucus containing *lysozyme*, an antibacterial enzyme. The sticky mucus traps inspired dust, bacteria, and other debris, while lysozyme attacks and destroys bacteria chemically. The epithelial cells of the respiratory mucosa also secrete *defensins*, natural antibiotics that help get rid of invading microbes. Additionally, the high water content of the mucus film acts to humidify the inhaled air.

The ciliated cells of the respiratory mucosa create a gentle current that moves the sheet of contaminated mucus posteriorly toward the throat, where it is swallowed and digested by stomach juices. We are usually unaware of this important action of our nasal cilia, but when exposed to cold air they become sluggish, allowing mucus to accumulate in the nasal cavity and then dribble out the nostrils. This along with the fact that water vapor in expired air tends to condense at these lower temperatures helps explain why you might have a "runny" nose on a crisp, wintry day.

The nasal mucosa is richly supplied with sensory nerve endings, and contact with irritating particles (dust, pollen, and the like) triggers a sneeze reflex. The sneeze forces air outward in a violent burst—a somewhat crude way of expelling irritants from the nose.

Rich plexuses of capillaries and thin-walled veins underlie the nasal epithelium and warm incoming air as it flows across the mucosal surface. When the inspired air is cold, the vascular plexus becomes engorged with blood, thereby intensifying the air-heating process. Because of the abundance and superficial location of these blood vessels, nosebleeds are common and often profuse.

Protruding medially from each lateral wall of the nasal cavity are three scroll-like mucosa-covered projections, the *superior*, *middle*, and *inferior nasal conchae* (kong′ke) (Figure 22.3). The groove inferior to each concha is a *nasal meatus* (me-a′tus). The

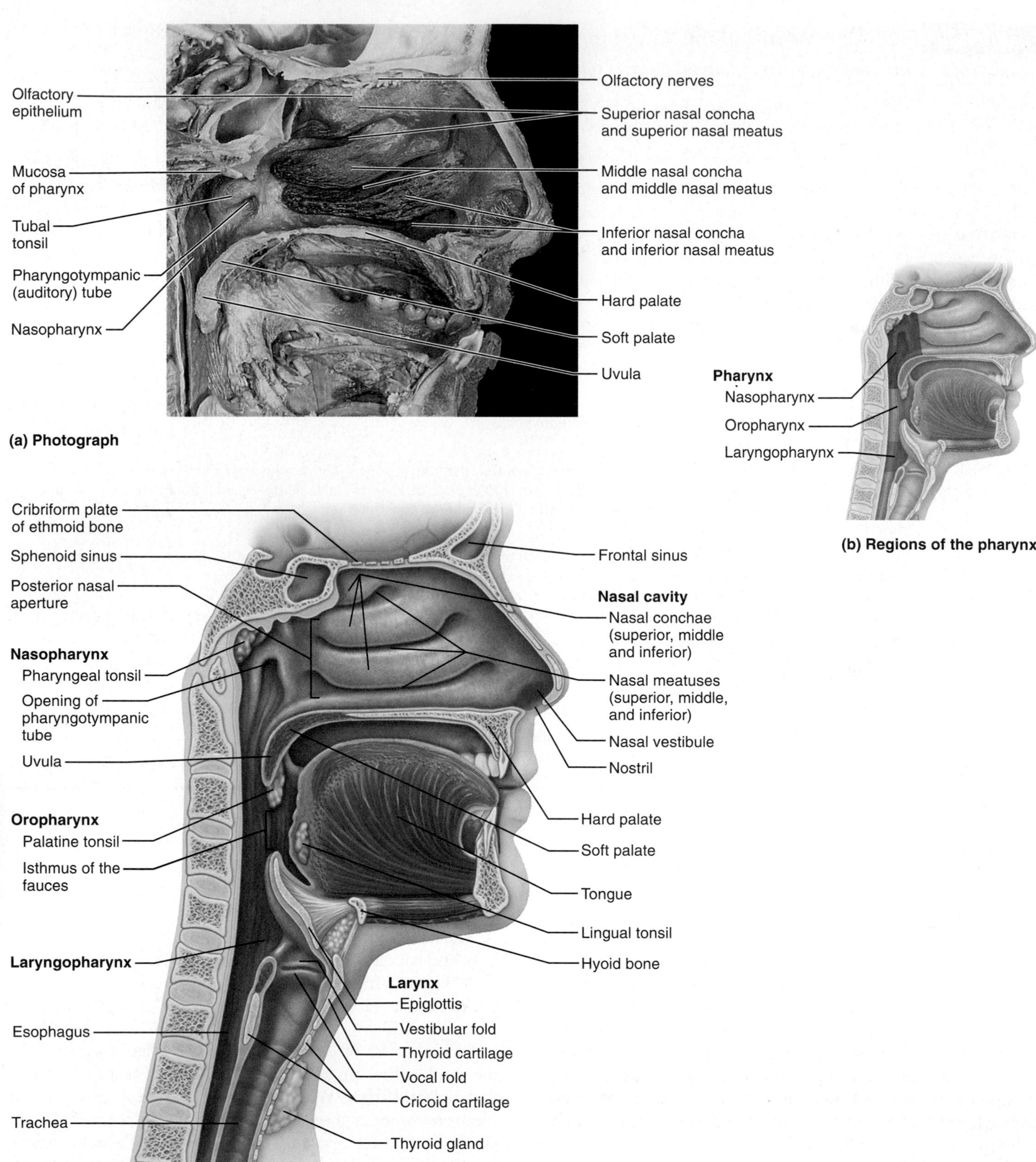

22

Figure 22.3 The upper respiratory tract. Midsagittal section of the head and neck. (See also *A Brief Atlas of the Human Body*, Figure 47.)

curved conchae greatly increase the mucosal surface area exposed to the air and enhance air turbulence in the cavity. The gases in inhaled air swirl through the twists and turns, but heavier, nongaseous particles tend to be deflected onto the mucus-coated surfaces, where they become trapped. As a result, few particles larger than 6 μm make it past the nasal cavity.

The conchae and nasal mucosa not only function during inhalation to filter, heat, and moisten the air, but also act during exhalation to reclaim this heat and moisture. In other words, the inhaled air cools the conchae, then during exhalation these cooled conchae precipitate moisture and extract heat from the humid air flowing over them. This reclamation process minimizes the amount of moisture and heat lost from the body through breathing, helping us to survive in dry and cold climates.

The nasal cavity is surrounded by a ring of **paranasal sinuses** (Figure 22.3c). They are located in the frontal, sphenoid, ethmoid, and maxillary bones (see Figure 7.15, p. 216). The sinuses lighten the skull, and together with the nasal cavity they warm and moisten the air. The mucus they produce ultimately flows into the nasal cavity, and the suctioning effect created by nose blowing helps drain the sinuses.

HOMEOSTATIC IMBALANCE

Cold viruses, streptococcal bacteria, and various allergens can cause *rhinitis* (ri-ni′tis), inflammation of the nasal mucosa accompanied by excessive mucus production, nasal congestion, and postnasal drip. The nasal mucosa is continuous with that of the rest of the respiratory tract, explaining the typical nose to throat to chest progression of colds. Because the mucosa extends tentacle-like into the nasolacrimal (tear) ducts and paranasal sinuses, nasal cavity infections often spread to those regions, causing **sinusitis** (inflamed sinuses). When the passageways connecting the sinuses to the nasal cavity are blocked with mucus or infectious material, the air in the sinus cavities is absorbed. The result is a partial vacuum and a *sinus headache* localized over the inflamed areas. ■

The Pharynx

The funnel-shaped **pharynx** (far′ingks) connects the nasal cavity and mouth superiorly to the larynx and esophagus inferiorly. Commonly called the *throat*, the pharynx vaguely resembles a short length of garden hose as it extends for about 13 cm (5 inches) from the base of the skull to the level of the sixth cervical vertebra (Figure 22.1).

From superior to inferior, the pharynx is divided into three regions—the *nasopharynx*, *oropharynx*, and *laryngopharynx* (Figure 22.3b). The muscular pharynx wall is composed of skeletal muscle throughout its length (see Table 10.3, pp. 334–335). However, the cellular composition of its mucosa varies from one pharyngeal region to another.

The Nasopharynx

The **nasopharynx** is posterior to the nasal cavity, inferior to the sphenoid bone, and superior to the level of the soft palate. Because it lies above the point where food enters the body, it serves *only* as an air passageway. During swallowing, the soft palate and its pendulous *uvula* (u′vu-lah; "little grape") move superiorly, an action that closes off the nasopharynx and prevents food from entering the nasal cavity. (When we giggle, this sealing action fails and fluids being swallowed can end up spraying out the nose.)

The nasopharynx is continuous with the nasal cavity through the posterior nasal apertures (Figure 22.3c). Its pseudostratified ciliated epithelium takes over the job of propelling mucus where the nasal mucosa leaves off. High on its posterior wall is the **pharyngeal tonsil** (far-rin′je-al) (or *adenoids*), which traps and destroys pathogens entering the nasopharynx in air.

HOMEOSTATIC IMBALANCE

Infected and swollen adenoids block air passage in the nasopharynx, making it necessary to breathe through the mouth. As a result, the air is not properly moistened, warmed, or filtered before reaching the lungs. When the adenoids are chronically enlarged, both speech and sleep may be disturbed. ■

The *pharyngotympanic (auditory) tubes*, which drain the middle ear cavities and allow middle ear pressure to equalize with atmospheric pressure, open into the lateral walls of the nasopharynx (Figure 22.3a). A ridge of pharyngeal mucosa, referred to as a *tubal tonsil*, arches over each of these openings. Because of their strategic location, the tubal tonsils help protect the middle ear against infections likely to spread from the nasopharynx. The pharyngeal tonsil, superoposterior and medial to the tubal tonsils, also plays this protective role.

The Oropharynx

The **oropharynx** lies posterior to the oral cavity and is continuous with it through an archway called the **isthmus of the fauces** (faw′sēz; "throat") (Figure 22.3c). Because the oropharynx extends inferiorly from the level of the soft palate to the epiglottis, both swallowed food and inhaled air pass through it.

As the nasopharynx blends into the oropharynx, the epithelium changes from pseudostratified columnar to a more protective stratified squamous epithelium. This structural adaptation accommodates the increased friction and greater chemical trauma accompanying food passage.

The paired **palatine tonsils** lie embedded in the oropharyngeal mucosa of the lateral walls of the fauces. The **lingual tonsil** covers the posterior surface of the tongue.

The Laryngopharynx

Like the oropharynx above it, the **laryngopharynx** (lah-ring″go-far′ingks) serves as a passageway for food and air and is lined with a stratified squamous epithelium. It lies directly posterior to the upright epiglottis and extends to the larynx, where the respiratory and digestive pathways diverge. At that point the laryngopharynx is continuous with the esophagus posteriorly. The esophagus conducts food and fluids to the stomach; air enters the larynx anteriorly. During swallowing, food has the "right of way," and air passage temporarily stops.

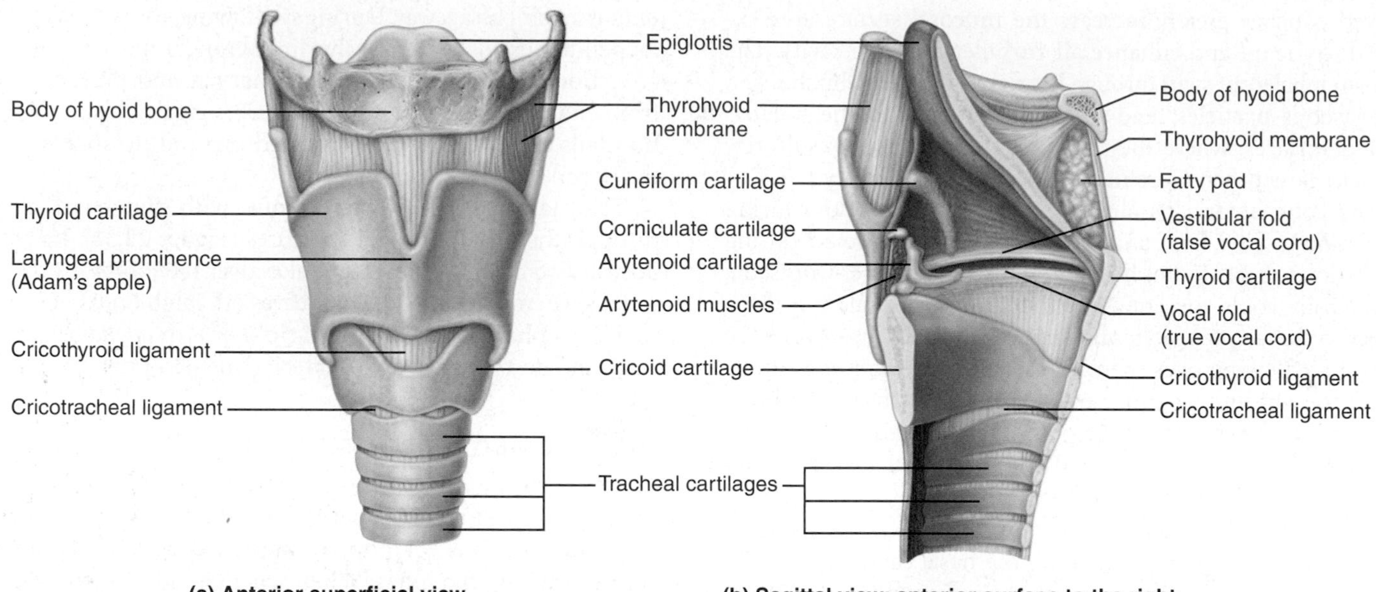

(a) Anterior superficial view

(b) Sagittal view; anterior surface to the right

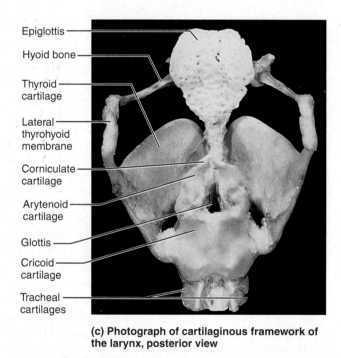

Epiglottis
Hyoid bone
Thyroid cartilage
Lateral thyrohyoid membrane
Corniculate cartilage
Arytenoid cartilage
Glottis
Cricoid cartilage
Tracheal cartilages

(c) Photograph of cartilaginous framework of the larynx, posterior view

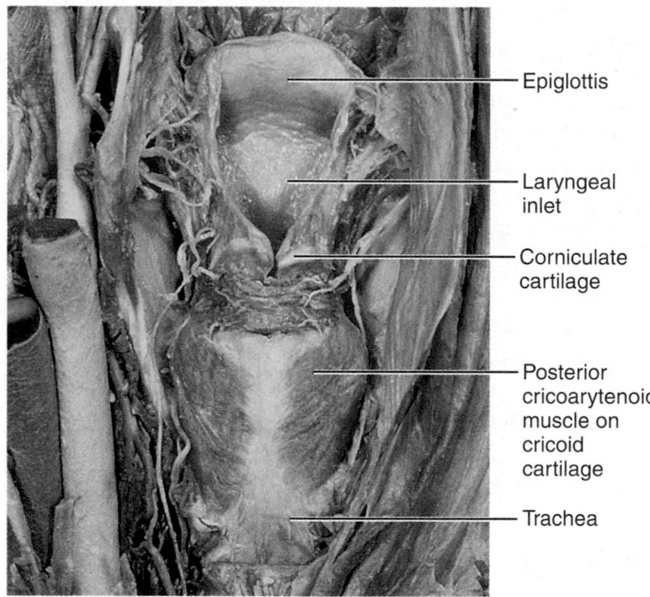

(d) Photograph of posterior aspect

Figure 22.4 The larynx.

The Larynx

Basic Anatomy

The **larynx** (lar′ingks), or voice box, extends for about 5 cm (2 inches) from the level of the third to the sixth cervical vertebra. Superiorly it attaches to the hyoid bone and opens into the laryngopharynx. Inferiorly it is continuous with the trachea (Figure 22.3c).

The larynx has three functions. Its two main tasks are to provide a *patent* (open) airway and to act as a switching mechanism to route air and food into the proper channels. Because it houses the vocal folds (vocal cords), the third function of the larynx is voice production.

The framework of the larynx is an intricate arrangement of nine cartilages connected by membranes and ligaments **(Figure 22.4)**. Except for the epiglottis, all laryngeal cartilages are hyaline cartilages. The large, shield-shaped **thyroid cartilage** is formed by the fusion of two cartilage plates. The midline **laryngeal prominence** (lah-rin′je-al), which marks the fusion point, is obvious externally as the *Adam's apple.* The thyroid cartilage is typically larger in males than in females because male sex hormones stimulate its growth during puberty. Inferior to the thyroid cartilage is the ring-shaped

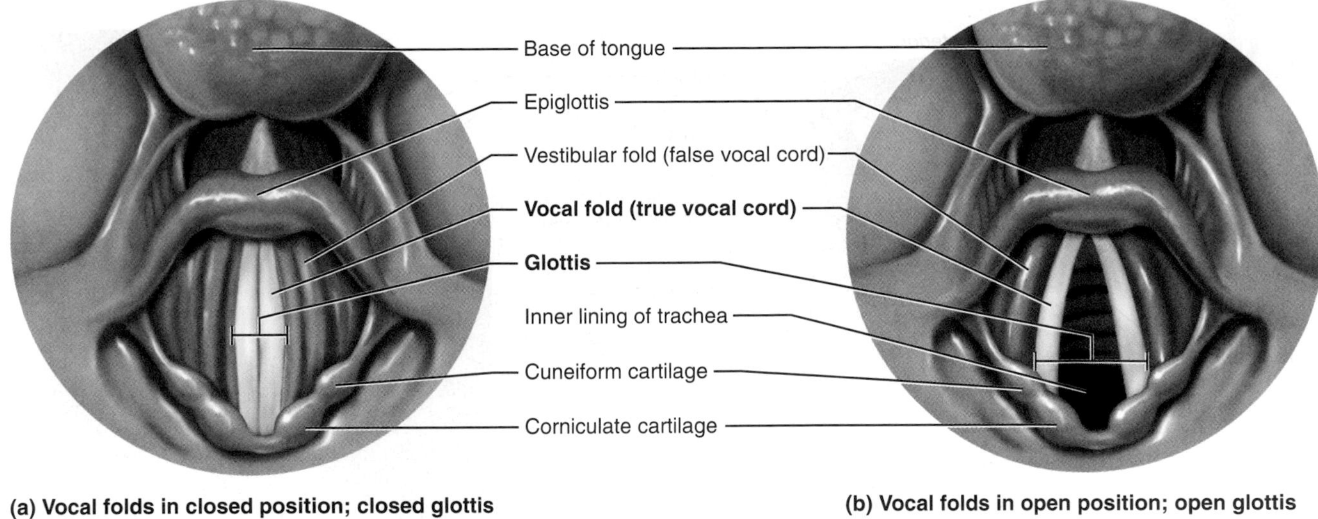

Figure 22.5 Movements of the vocal folds. Drawings of superior views of the larynx and vocal folds, as if seen through a laryngoscope.

cricoid cartilage (kri′koid), perched atop and anchored to the trachea inferiorly.

Three pairs of small cartilages, **arytenoid** (ar″ĭ-te′noid), **cuneiform** (ku-ne′ĭ-form), and **corniculate cartilages**, form part of the lateral and posterior walls of the larynx. The most important of these are the pyramid-shaped arytenoid cartilages, which anchor the vocal folds.

The ninth cartilage, the flexible, spoon-shaped **epiglottis** (ep″ĭ-glot′is; "above the glottis"), is composed of elastic cartilage and is almost entirely covered by a taste bud–containing mucosa. The epiglottis extends from the posterior aspect of the tongue to its anchoring point on the anterior rim of the thyroid cartilage (Figure 22.4b and c).

When only air is flowing into the larynx, the inlet to the larynx is open wide and the free edge of the epiglottis projects upward. During swallowing, the larynx is pulled superiorly and the epiglottis tips to cover the laryngeal inlet. Because this action keeps food out of the lower respiratory passages, the epiglottis has been called the guardian of the airways. Anything other than air entering the larynx initiates the cough reflex, which acts to expel the substance. This protective reflex does not work when we are unconscious, so it is never a good idea to administer liquids when attempting to revive an unconscious person.

Lying under the laryngeal mucosa on each side are the **vocal ligaments**, which attach the arytenoid cartilages to the thyroid cartilage. These ligaments, composed largely of elastic fibers, form the core of mucosal folds called the **vocal folds**, or **true vocal cords**, which appear pearly white because they lack blood vessels **(Figure 22.5)**.

The vocal folds vibrate, producing sounds as air rushes up from the lungs. The vocal folds and the medial opening between them through which air passes are called the **glottis.** Superior to the vocal folds is a similar pair of mucosal folds called the **vestibular folds**, or **false vocal cords**. These play no direct part in sound production but help to close the glottis when we swallow.

The superior portion of the larynx, an area subject to food contact, is lined by stratified squamous epithelium. Below the vocal folds the epithelium is a pseudostratified ciliated columnar type that acts as a dust filter. The power stroke of its cilia is directed upward toward the pharynx so that mucus is continually moved *away* from the lungs. We help to move mucus up and out of the larynx when we "clear our throat."

Voice Production

Speech involves the intermittent release of expired air and the opening and closing of the glottis. The length of the vocal folds and the size of the glottis change with the action of the intrinsic laryngeal muscles that clothe the cartilages. Most of these muscles move the arytenoid cartilages. As the length and tension of the vocal folds change, the pitch of the sound varies. Generally, the tenser the vocal folds, the faster they vibrate and the higher the pitch.

As a boy's larynx enlarges during puberty, his vocal folds become longer and thicker. Because this causes them to vibrate more slowly, his voice becomes deeper. Until the young man learns to control his newly enlarged vocal folds, his voice "cracks."

Loudness of the voice depends on the force with which the airstream rushes across the vocal folds. The greater the force, the stronger the vibration and the louder the sound. The vocal folds do not move at all when we whisper, but they vibrate vigorously when we yell. The power source for creating the airstream is the muscles of the chest, abdomen, and back.

The vocal folds actually produce buzzing sounds. The perceived quality of the voice depends on the coordinated activity of many structures above the glottis. For example, the entire length of the pharynx acts as a resonating chamber, to amplify and enhance the sound quality. The oral, nasal, and sinus cavities also contribute to vocal resonance. In addition, good enunciation depends on the "shaping" of sound into recognizable consonants and vowels by muscles in the pharynx, tongue, soft palate, and lips.

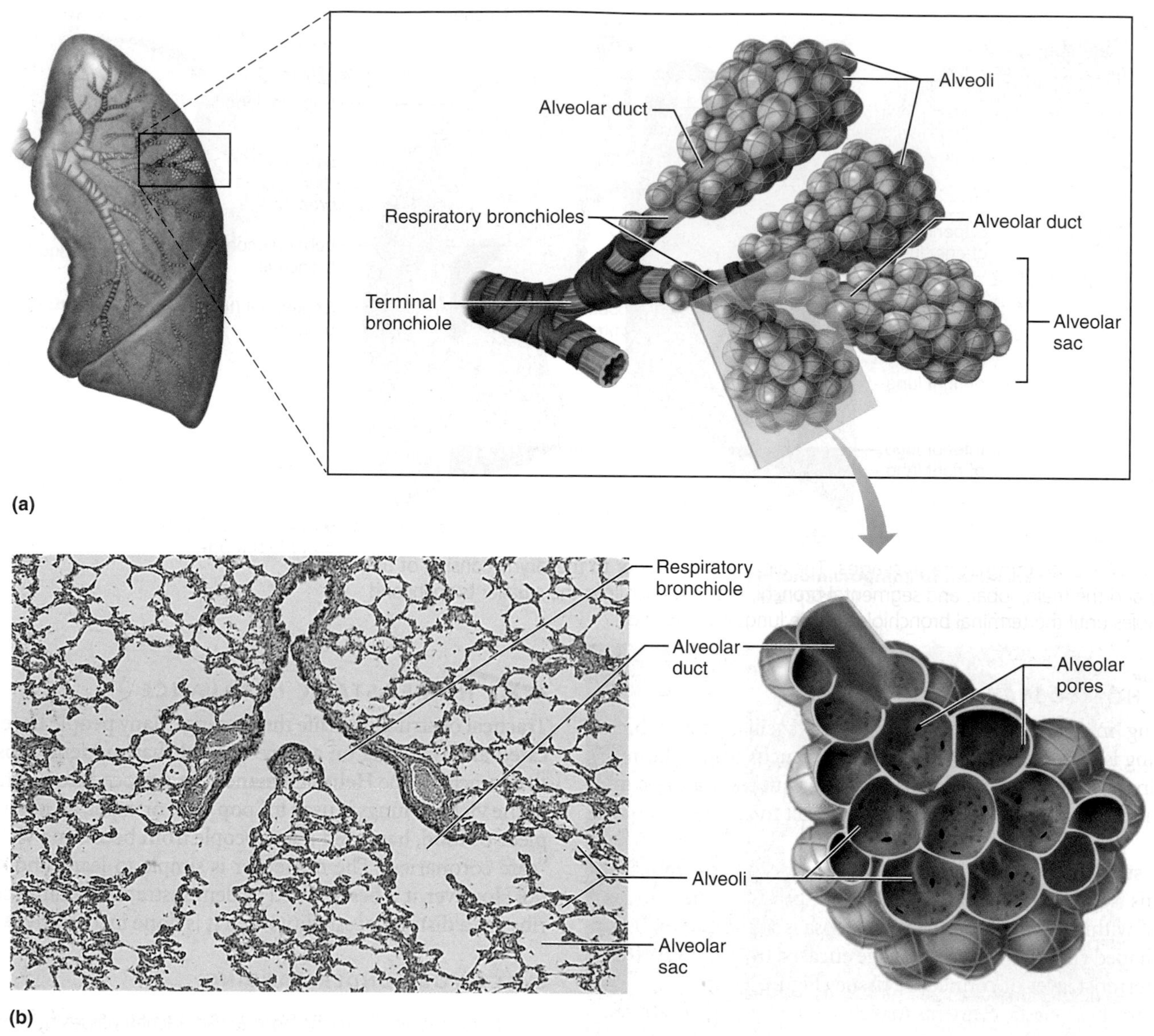

22

Figure 22.8 Respiratory zone structures. (a) Diagrammatic view of respiratory bronchioles, alveolar ducts, alveolar sacs, and alveoli. **(b)** Photomicrograph of a section of human lung, showing the respiratory structures that form the final divisions of the bronchial tree (70x). Notice the thinness of the alveolar walls.

bronchial tree is the site where conducting zone structures give way to respiratory zone structures **(Figure 22.8)**.

Conducting Zone Structures

The trachea divides to form the **right** and **left main (primary) bronchi** (brong′ki) approximately at the level of T_7 in an erect (standing) person. Each bronchus runs obliquely in the mediastinum before plunging into the medial depression (hilum) of the lung on its own side (Figure 22.7). The right main bronchus is wider, shorter, and more vertical than the left. Consequently, it is the more common site for an inhaled foreign object to become lodged.

Once inside the lungs, each main bronchus subdivides into **lobar (secondary) bronchi**—three on the right and two on the left—each supplying one lung lobe. The lobar bronchi branch into third-order **segmental (tertiary) bronchi**, which divide repeatedly into smaller and smaller bronchi (fourth-order, fifth-order, etc.). Passages smaller than 1 mm in diameter are called **bronchioles** (“little bronchi”), and the tiniest of these, the **terminal bronchioles**, are less than 0.5 mm in diameter.

The tissue composition of the walls of the main bronchi mimics that of the trachea, but as the conducting tubes become smaller, the following structural changes occur:

1. **Support structures change.** The cartilage rings are replaced by irregular *plates* of cartilage, and by the time the bronchioles are reached, supportive cartilage is no longer present in the tube walls. However, elastic fibers are found in the tube walls throughout the bronchial tree.

2. **Epithelium type changes.** The mucosal epithelium thins as it changes from pseudostratified columnar to columnar and then to cuboidal in the terminal bronchioles. Cilia are sparse, and mucus-producing cells are absent in the bronchioles. For this reason, most airborne debris found at or below the level of the bronchioles must be removed by macrophages in the alveoli.
3. **Amount of smooth muscle increases.** The relative amount of smooth muscle in the tube walls increases as the passageways become smaller. A complete layer of circular smooth muscle in the bronchioles and the lack of supporting cartilage (which would hinder constriction) allows the bronchioles to provide substantial resistance to air passage under certain conditions (as we will describe later).

Respiratory Zone Structures

Defined by the presence of thin-walled air sacs called **alveoli** (al-ve′o-li; *alveol* = small cavity), the respiratory zone begins as the terminal bronchioles feed into **respiratory bronchioles** within the lung (Figure 22.8). Protruding from these smallest bronchioles are scattered alveoli. The respiratory bronchioles lead into winding **alveolar ducts**, whose walls consist of diffusely arranged rings of smooth muscle cells, connective tissue fibers, and outpocketing alveoli. The alveolar ducts lead into terminal clusters of alveoli called **alveolar sacs.**

Many people mistakenly equate alveoli, the site of gas exchange, with alveolar sacs, but they are not the same thing. The alveolar sac is analogous to a bunch of grapes, and the alveoli are the individual grapes. The 300 million or so gas-filled alveoli in the lungs account for most of the lung volume and provide a tremendous surface area for gas exchange.

The Respiratory Membrane The walls of the alveoli are composed primarily of a single layer of squamous epithelial cells, called **type I cells**, surrounded by a flimsy basement membrane. The thinness of their walls is hard to imagine, but a sheet of tissue paper is 15 times thicker. The external surfaces of the alveoli are densely covered with a "cobweb" of pulmonary capillaries **(Figure 22.9)**. Together, the alveolar and capillary walls and their fused basement membranes form the **respiratory membrane**, a 0.5-μm-thick *air-blood barrier* that has gas on one side and blood flowing past on the other (Figure 22.9c). Gas exchanges occur readily by simple diffusion across the respiratory membrane—O_2 passes from the alveolus into the blood, and CO_2 leaves the blood to enter the gas-filled alveolus.

Scattered amid the type I squamous cells that form the major part of the alveolar walls are cuboidal **type II cells** (Figure 22.9c). The type II cells secrete a fluid containing a detergent-like substance called *surfactant* that coats the gas-exposed alveolar surfaces. (We describe surfactant's role in reducing the surface tension of the alveolar fluid later in this chapter.) Recently, type II cells have been shown to secrete a number of antimicrobial proteins that are important elements of innate immunity.

The alveoli have three other significant features: (1) They are surrounded by fine elastic fibers of the same type that surround the entire bronchial tree. (2) Open **alveolar pores** connecting adjacent alveoli allow air pressure throughout the lung to be equalized and provide alternate air routes to any alveoli whose bronchi have collapsed due to disease. (3) Remarkably efficient **alveolar macrophages** crawl freely along the internal alveolar surfaces.

Although huge numbers of infectious microorganisms are continuously carried into the alveoli, alveolar surfaces are usually sterile. Because the alveoli are "dead ends," aged and dead macrophages must be prevented from accumulating in them. Most macrophages simply get swept up by the ciliary current of superior regions and carried passively to the pharynx. In this manner, we clear and swallow over 2 million alveolar macrophages per hour!

The Lungs and Pleurae

▶ Describe the gross structure of the lungs and pleurae.

The paired **lungs** occupy all of the thoracic cavity except the mediastinum, which houses the heart, great blood vessels, bronchi, esophagus, and other organs **(Figure 22.10)**.

Gross Anatomy of the Lungs

Each cone-shaped lung is surrounded by pleurae and connected to the mediastinum by vascular and bronchial attachments, collectively called the lung **root.** The anterior, lateral, and posterior lung surfaces lie in close contact with the ribs and form the continuously curving **costal surface.** Just deep to the clavicle is the **apex**, the narrow superior tip of the lung. The concave, inferior surface that rests on the diaphragm is the **base.**

On the mediastinal surface of each lung is an indentation, the **hilum**, through which pulmonary and systemic blood vessels, bronchi, lymphatic vessels, and nerves enter and leave the lungs. Each main bronchus plunges into the hilum on its own side and begins to branch almost immediately. All conducting and respiratory passageways distal to the main bronchi are found in the lungs.

The two lungs differ slightly in shape and size because the apex of the heart is slightly to the left of the median plane. The left lung is smaller than the right, and the **cardiac notch**—a concavity in its medial aspect—is molded to and accommodates the heart (Figure 22.10a). The left lung is subdivided into superior and inferior **lobes** by the *oblique fissure*, whereas the right lung is partitioned into superior, middle, and inferior lobes by the *oblique* and *horizontal fissures*.

Each lobe contains a number of pyramid-shaped **bronchopulmonary segments** separated from one another by connective tissue septa. Each segment is served by its own artery and vein and receives air from an individual segmental (tertiary) bronchus. Initially each lung contains ten bronchopulmonary segments arranged in similar (but not identical) patterns **(Figure 22.11)**. Subsequent fusion of adjacent segmental arteries reduces the number in the left lung to eight or nine segments.

The bronchopulmonary segments are clinically important because pulmonary disease is often confined to one or a few

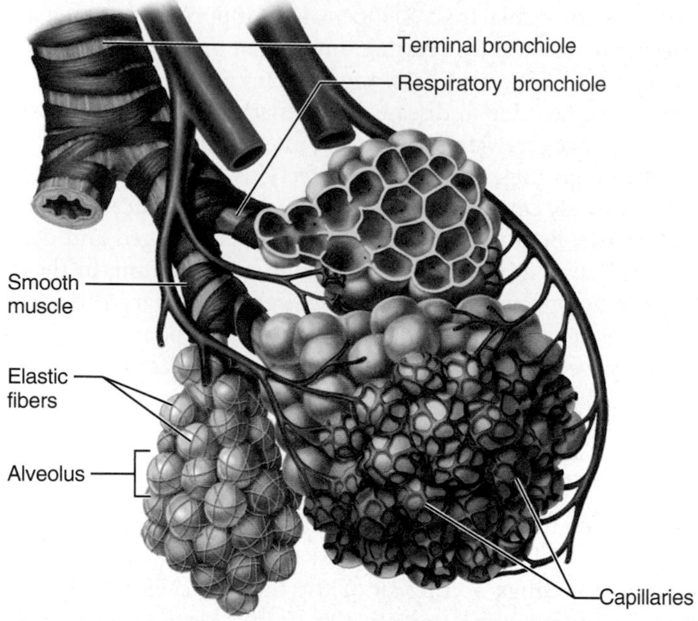

(a) Diagrammatic view of capillary-alveoli relationships

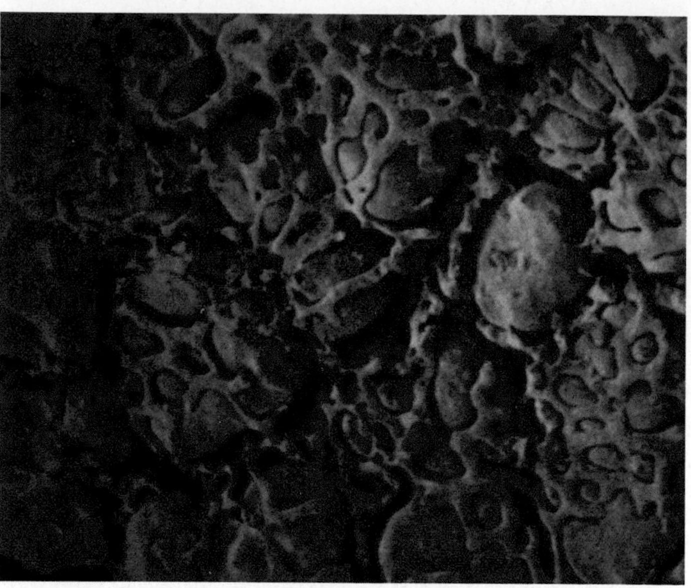

(b) Scanning electron micrograph of casts of alveoli and associated pulmonary capillaries (300x)

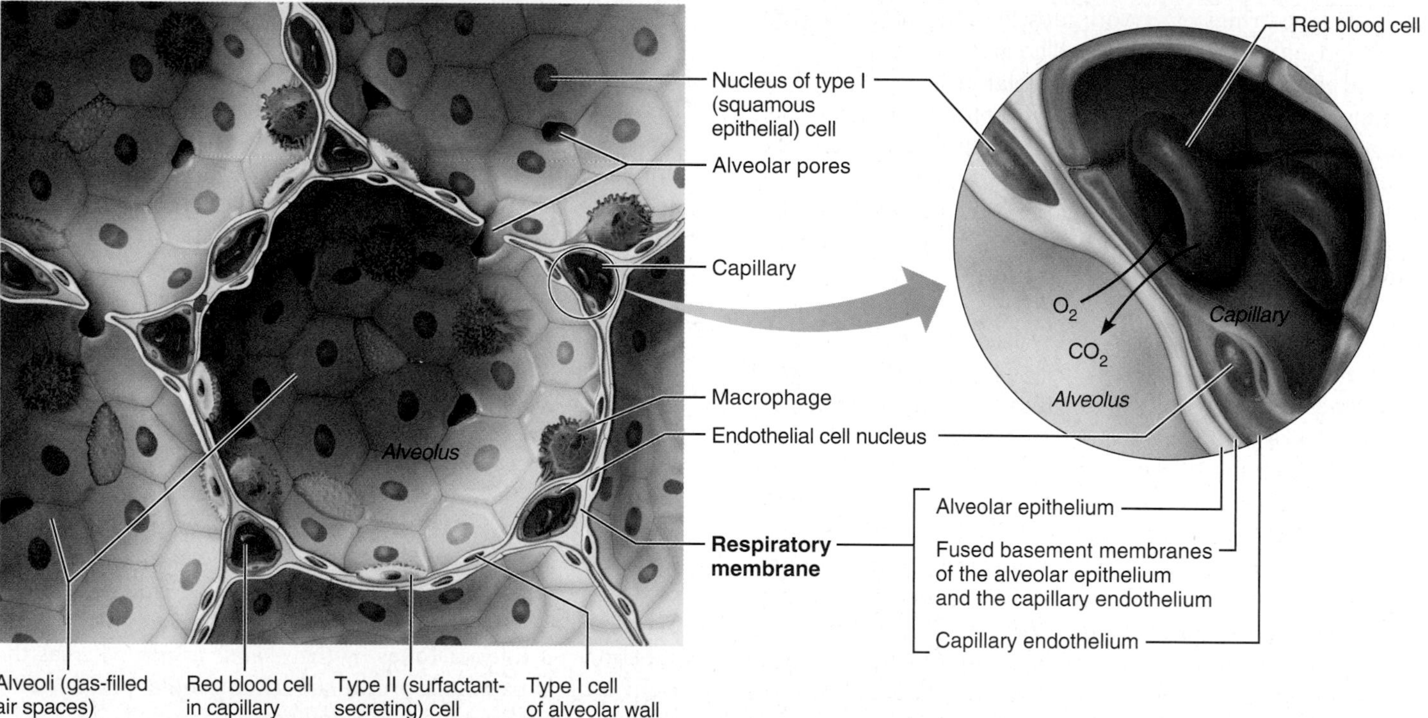

(c) Detailed anatomy of the respiratory membrane

Figure 22.9 Alveoli and the respiratory membrane. Elastic fibers and capillaries surround all alveoli, but for clarity they are shown only on some alveoli in **(a)**.

SOURCE: (b) Kessel and Kardon/Visuals Unlimited.

22

segments. Their connective tissue partitions allow diseased segments to be surgically removed without damaging neighboring healthy segments or impairing their blood supply.

The smallest subdivisions of the lung visible with the naked eye are the **lobules**, which appear at the lung surface as hexagons ranging from the size of a pencil eraser to the size of a penny (Figure 22.10b). Each lobule is served by a large bronchiole and its branches. In most city dwellers and in smokers, the connective tissue that separates the individual lobules is blackened with carbon.

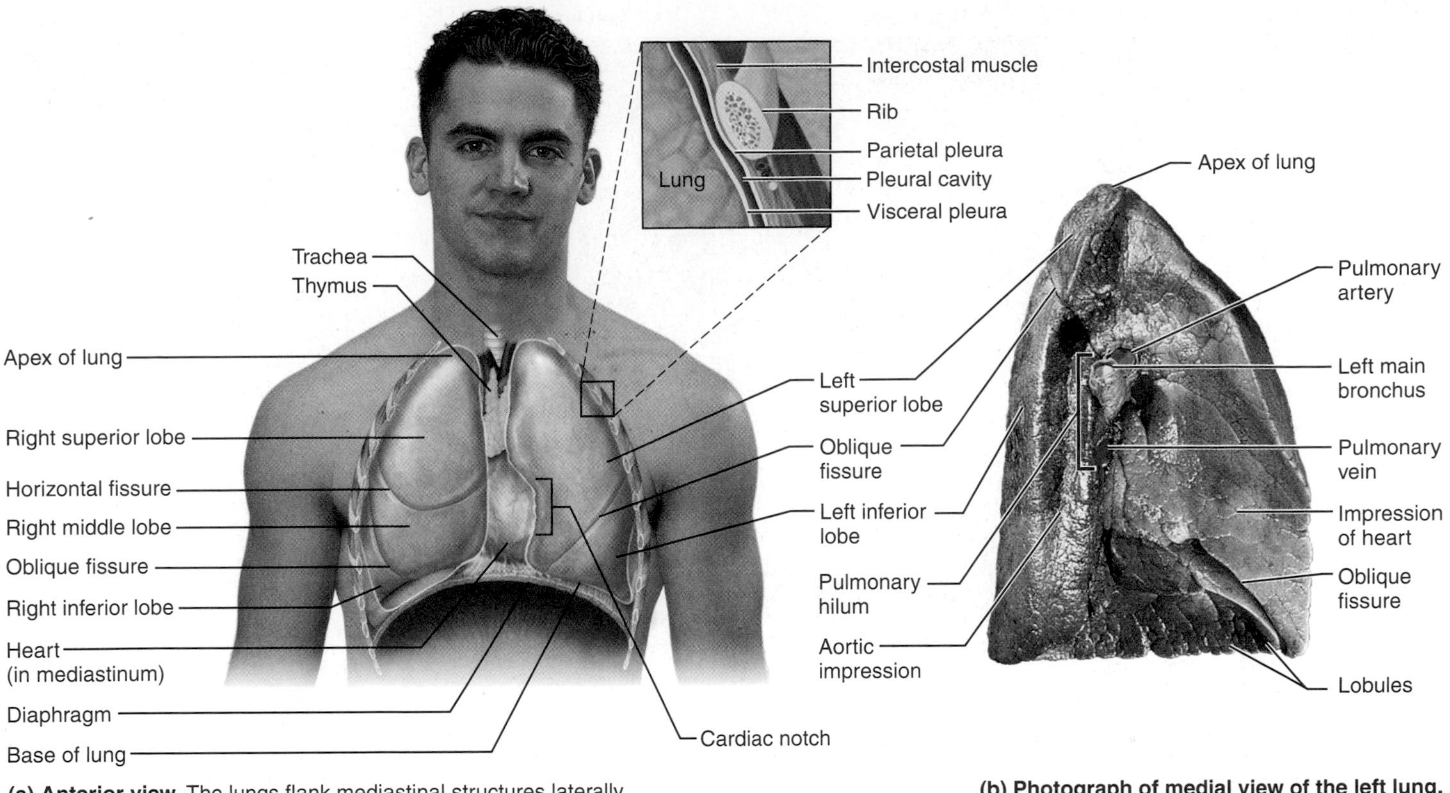

(a) Anterior view. The lungs flank mediastinal structures laterally.

(b) Photograph of medial view of the left lung.

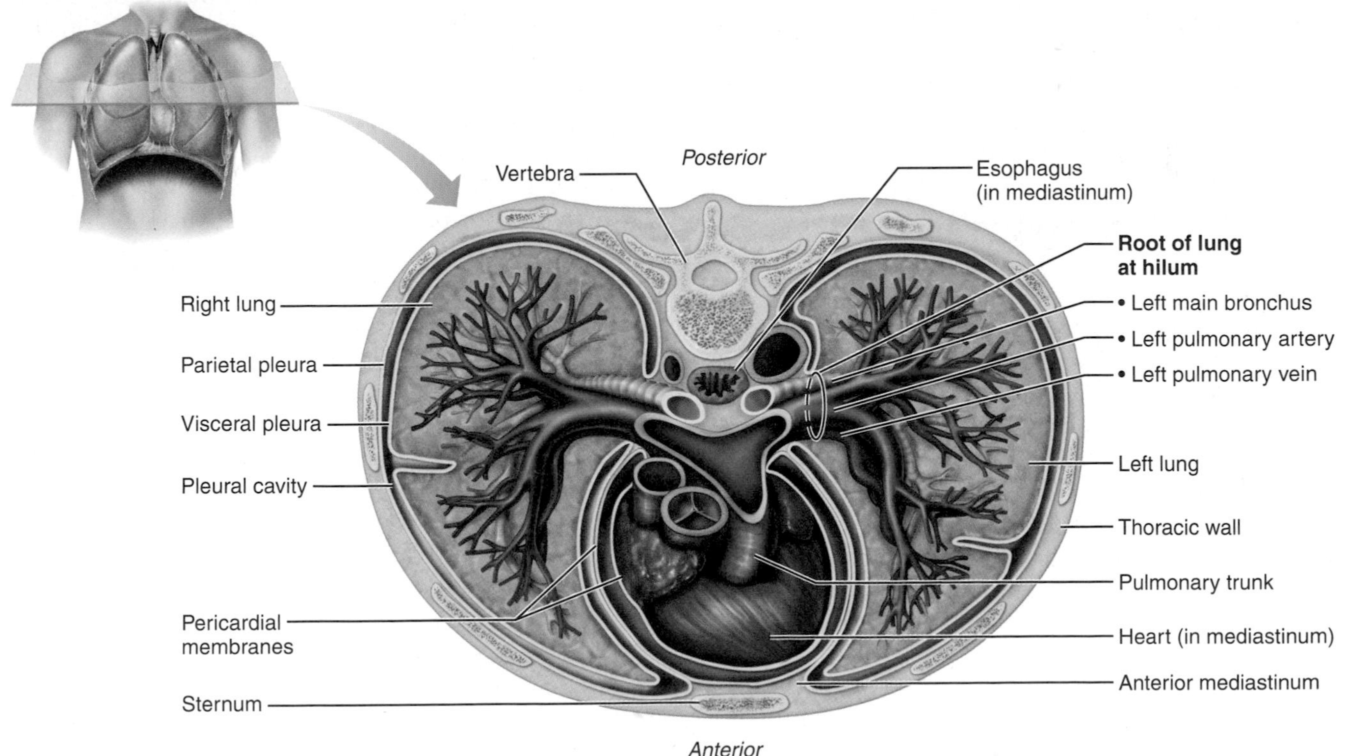

(c) Transverse section through the thorax, viewed from above. Lungs, pleural membranes, and major organs in the mediastinum are shown.

Figure 22.10 Anatomical relationships of organs in the thoracic cavity. In **(c)**, the size of the pleural cavity is exaggerated for clarity.

22

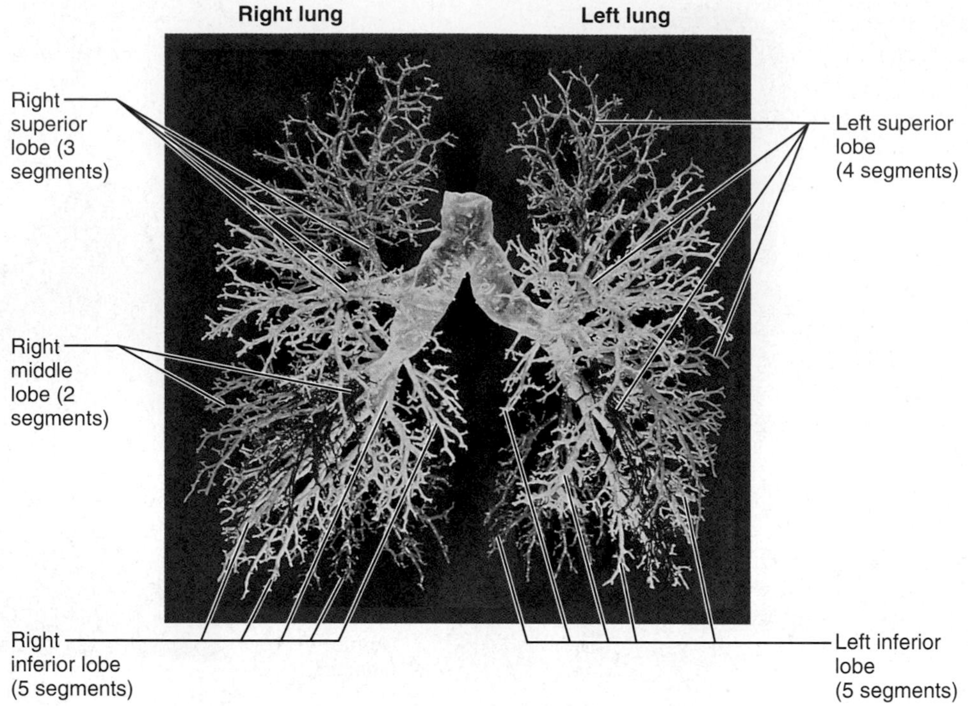

Figure 22.11 A cast of the bronchial tree. The individual bronchopulmonary segments have been painted different colors.

As we mentioned earlier, the lungs consist largely of air spaces. The balance of lung tissue, or its **stroma** (literally "mattress" or "bed"), is mostly elastic connective tissue. As a result, the lungs are soft, spongy, elastic organs that together weigh just over 1 kg (2.2 lb). The elasticity of healthy lungs helps to reduce the work of breathing, as we will describe shortly.

Blood Supply and Innervation of the Lungs

The lungs are perfused by two circulations, the pulmonary and the bronchial, which differ in size, origin, and function. Systemic venous blood that is to be oxygenated in the lungs is delivered by the **pulmonary arteries**, which lie anterior to the main bronchi (Figure 22.10c). In the lungs, the pulmonary arteries branch profusely along with the bronchi and finally feed into the **pulmonary capillary networks** surrounding the alveoli (see Figure 22.9a).

Freshly oxygenated blood is conveyed from the respiratory zones of the lungs to the heart by the **pulmonary veins**. Their tributaries course back to the hilum both with the corresponding bronchi and in the connective tissue septa separating the bronchopulmonary segments.

In contrast to the pulmonary circulation, the **bronchial arteries** provide oxygenated systemic blood to lung tissue. They arise from the aorta, enter the lungs at the hilum, and then run along the branching bronchi. They provide a high-pressure, low-volume supply of oxygenated blood to all lung tissues except the alveoli, which receive blood from the low-pressure, high-volume pulmonary circulation. Some systemic venous blood is drained from the lungs by the tiny bronchial veins, but there are multiple anastomoses between the two circulations, and most venous blood returns to the heart via the pulmonary veins.

Because *all* of the body's blood passes through the lungs about once each minute, the lung capillary endothelium is an ideal location for enzymes that act on materials in the blood. Examples include *angiotensin converting enzyme*, which activates an important blood pressure hormone, and enzymes that inactivate certain prostaglandins.

The lungs are innervated by parasympathetic and sympathetic motor fibers, and visceral sensory fibers. These nerve fibers enter each lung through the **pulmonary plexus** on the lung root and run along the bronchial tubes and blood vessels in the lungs. Parasympathetic fibers constrict the air tubes, whereas the sympathetic nervous system dilates them.

The Pleurae

The **pleurae** (ploo′re; "sides") form a thin, double-layered serosa. The layer called the **parietal pleura** covers the thoracic wall and superior face of the diaphragm (Figure 22.10a, c). It continues around the heart and between the lungs, forming the lateral walls of the mediastinal enclosure and snugly enclosing the lung root. From here, the pleura extends as the layer called the **visceral pleura** to cover the external lung surface, dipping into and lining its fissures.

The pleurae produce **pleural fluid**, which fills the slitlike **pleural cavity** between them. This lubricating secretion allows the lungs to glide easily over the thorax wall during our breathing movements. Although the pleurae slide easily across each other, their separation is strongly resisted by the surface tension of the pleural fluid. Consequently, the lungs cling tightly to the thorax wall and are forced to expand and recoil passively as the volume of the thoracic cavity alternately increases and decreases during breathing.

The pleurae also help divide the thoracic cavity into three chambers—the central mediastinum and the two lateral pleural compartments, each containing a lung. This compartmentalization helps prevent one mobile organ (for example, the lung or heart) from interfering with another. It also limits the spread of local infections.

HOMEOSTATIC IMBALANCE

Pleurisy (ploo′rĭ-se), inflammation of the pleurae, often results from pneumonia. Inflamed pleurae become rough, resulting in friction and stabbing pain with each breath. As the disease progresses, the pleurae may produce an excessive amount of fluid. This increased fluid relieves the pain caused by pleural surfaces rubbing together, but may exert pressure on the lungs and hinder breathing movements.

Other fluids that may accumulate in the pleural cavity include blood (leaked from damaged blood vessels) and blood filtrate (the watery fluid that oozes from the lung capillaries when left-sided heart failure occurs). The general term for fluid accumulation in the pleural cavity is *pleural effusion.* ■

CHECK YOUR UNDERSTANDING

4. What features of the alveoli and their respiratory membranes suit them to their function of exchanging gases by diffusion?
5. A 3-year-old boy is brought to the emergency department after aspirating (inhaling) a peanut. Bronchoscopy confirms the suspicion that the peanut is lodged in a bronchus and then it is successfully extracted. Which main bronchus was the peanut most likely to be in? Why?
6. The lungs are perfused by two different circulations. Name these circulations and indicate their roles in the lungs.

For answers, see Appendix G.

Mechanics of Breathing

Breathing, or **pulmonary ventilation**, consists of two phases: **inspiration**, the period when air flows into the lungs, and **expiration**, the period when gases exit the lungs.

Pressure Relationships in the Thoracic Cavity

▶ Explain the functional importance of the partial vacuum that exists in the intrapleural space.

Before we can begin to describe the breathing process, it is important to understand that *respiratory pressures are always described relative to* **atmospheric pressure** (**P_{atm}**), which is the pressure exerted by the air (gases) surrounding the body. At sea level, atmospheric pressure is 760 mm Hg (the pressure exerted by a column of mercury 760 mm high). This pressure can also be expressed in atmosphere units: atmospheric pressure = 760 mm Hg = 1 atm.

A negative respiratory pressure in any respiratory area, such as –4 mm Hg, indicates that the pressure in that area is lower than atmospheric pressure by 4 mm Hg (760 – 4 = 756 mm Hg).

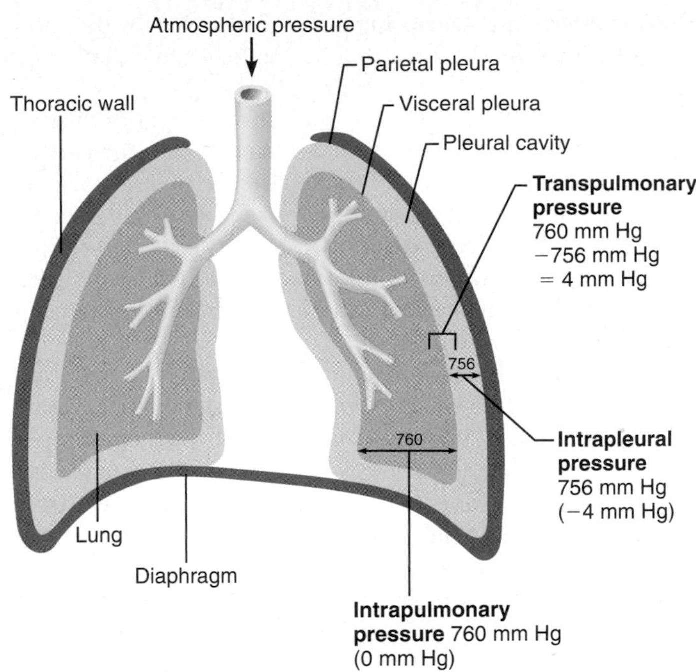

Figure 22.12 Intrapulmonary and intrapleural pressure relationships. Differences in pressure relative to atmospheric pressure (760 mm Hg) are given in parentheses. Values shown are at the end of a normal expiration. For illustration, the size of the pleural cavity has been greatly exaggerated.

A positive respiratory pressure is higher than atmospheric pressure, and zero respiratory pressure is equal to atmospheric pressure. Now we are ready to examine the pressure relationships that normally exist in the thoracic cavity.

Intrapulmonary Pressure

The **intrapulmonary** (intra-alveolar) **pressure** (**P_{pul}**) is the pressure in the alveoli. Intrapulmonary pressure rises and falls with the phases of breathing, but it *always* eventually equalizes with the atmospheric pressure (**Figure 22.12**).

Intrapleural Pressure

The pressure in the pleural cavity, the **intrapleural pressure** (**P_{ip}**), also fluctuates with breathing phases, but is always about 4 mm Hg less than P_{pul}. That is, P_{ip} is *always* negative relative to P_{pul}.

How is this negative intrapleural pressure established? In other words, what causes it? Let's examine the forces that exist in the thorax to see if we can answer these questions. First of all, we know there are opposing forces. Two forces act to pull the lungs (visceral pleura) away from the thorax wall (parietal pleura) and cause lung collapse:

1. **The lungs' natural tendency to recoil.** Because of their elasticity, lungs always assume the smallest size possible.
2. **The surface tension of the alveolar fluid.** The molecules of the fluid lining the alveoli attract each other and this produces *surface tension* that constantly acts to draw the alveoli to their smallest possible dimension.

22

However, these lung-collapsing forces are opposed by the natural elasticity of the chest wall, a force that tends to pull the thorax outward and to enlarge the lungs.

So which force wins? The answer is neither in a healthy person, because of the strong adhesive force between the parietal and visceral pleurae. Pleural fluid secures the pleurae together in the same way a drop of water holds two glass slides together. The pleurae slide from side to side easily, but they remain closely apposed, and separating them requires extreme force. The net result of the dynamic interplay between these forces is a negative P_{ip}.

The amount of pleural fluid in the pleural cavity must remain minimal in order for the negative P_{ip} to be maintained. The pleural fluid is actively pumped out of the pleural cavity into the lymphatics continuously. If it wasn't, fluid would accumulate in the intrapleural space (remember, fluids move from high to low pressure), producing a positive pressure in the pleural cavity.

We cannot overemphasize the importance of negative pressure in the intrapleural space and the tight coupling of the lungs to the thorax wall. Any condition that equalizes P_{ip} with the intrapulmonary (or atmospheric) pressure causes *immediate lung collapse.* It is the **transpulmonary pressure**—the difference between the intrapulmonary and intrapleural pressures ($P_{pul} - P_{ip}$)—that keeps the air spaces of the lungs open or, phrased another way, keeps the lungs from collapsing. Moreover, *the size of the transpulmonary pressure determines the size of the lungs* at any time—the greater the transpulmonary pressure, the larger the lungs.

HOMEOSTATIC IMBALANCE

Atelectasis (at″ĕ-lik′tah-sis), or lung collapse, occurs when a bronchiole becomes plugged (as may follow pneumonia). Its associated alveoli then absorb all of their air and collapse. Atelectasis can also occur when air enters the pleural cavity either through a chest wound, or due to rupture of the visceral pleura, which allows air to enter the pleural cavity from the respiratory tract. The presence of air in the pleural cavity is referred to as a **pneumothorax** (nu″mo-tho′raks; "air thorax"). The condition is reversed by drawing air out of the intrapleural space with chest tubes, which allows the pleurae to heal and the lung to reinflate and resume its normal function. Note that because the lungs are in separate cavities, one lung can collapse without interfering with the function of the other. ■

22

Pulmonary Ventilation

- ▶ Relate Boyle's law to events of inspiration and expiration.
- ▶ Explain the relative roles of the respiratory muscles and lung elasticity in producing the volume changes that cause air to flow into and out of the lungs.

Pulmonary ventilation, consisting of inspiration and expiration, is a mechanical process that depends on volume changes in the thoracic cavity. A rule to keep in mind throughout the following discussion is that *volume changes* lead to *pressure changes,* and pressure changes lead to the *flow of gases* to equalize the pressure.

The relationship between the pressure and volume of a gas is given by **Boyle's law**: At constant temperature, the pressure of a gas varies inversely with its volume. That is,

$$P_1V_1 = P_2V_2$$

where P is the pressure of the gas, V is its volume, and subscripts 1 and 2 represent the initial and resulting conditions respectively.

Gases always *fill* their container. Consequently, in a large container, the molecules in a given amount of gas will be far apart and the pressure will be low. But if the volume of the container is reduced, the gas molecules will be forced closer together and the pressure will rise. A good example is an inflated automobile tire. The tire is hard and strong enough to bear the weight of the car because air is compressed to about one-third of its atmospheric volume in the tire, providing the high pressure. Now let's see how this relates to inspiration and expiration.

Inspiration

Visualize the thoracic cavity as a gas-filled box with a single entrance at the top, the tubelike trachea. The volume of this box is changeable and can be increased by enlarging all of its dimensions, thereby decreasing the gas pressure inside it. This drop in pressure causes air to rush into the box from the atmosphere, because gases always flow down their pressure gradients.

The same thing happens during normal quiet inspiration, when the **inspiratory muscles**—the diaphragm and external intercostal muscles—are activated. Here's how quiet inspiration works:

1. **Action of the diaphragm.** When the dome-shaped diaphragm contracts, it moves inferiorly and flattens out (**Figure 22.13**, top). As a result, the superior-inferior dimension (height) of the thoracic cavity increases.
2. **Action of the intercostal muscles.** Contraction of the external intercostal muscles lifts the rib cage and pulls the sternum superiorly (Figure 22.13, top). Because the ribs curve downward as well as forward around the chest wall, the broadest lateral and anteroposterior dimensions of the rib cage are normally directed obliquely downward. But when the ribs are raised and drawn together, they swing outward, expanding the diameter of the thorax both laterally and in the anteroposterior plane. This is much like the action that occurs when a curved bucket handle is raised—it moves outward as it moves upward.

Although these actions expand the thoracic dimensions by only a few millimeters along each plane, this is enough to increase thoracic volume by almost 500 ml—the usual volume of air that enters the lungs during a normal quiet inspiration. Of the two types of inspiratory muscles, the diaphragm is far more important in producing the volume changes that lead to normal quiet inspiration.

As the thoracic dimensions increase during inspiration, the lungs are stretched and the intrapulmonary volume increases. As

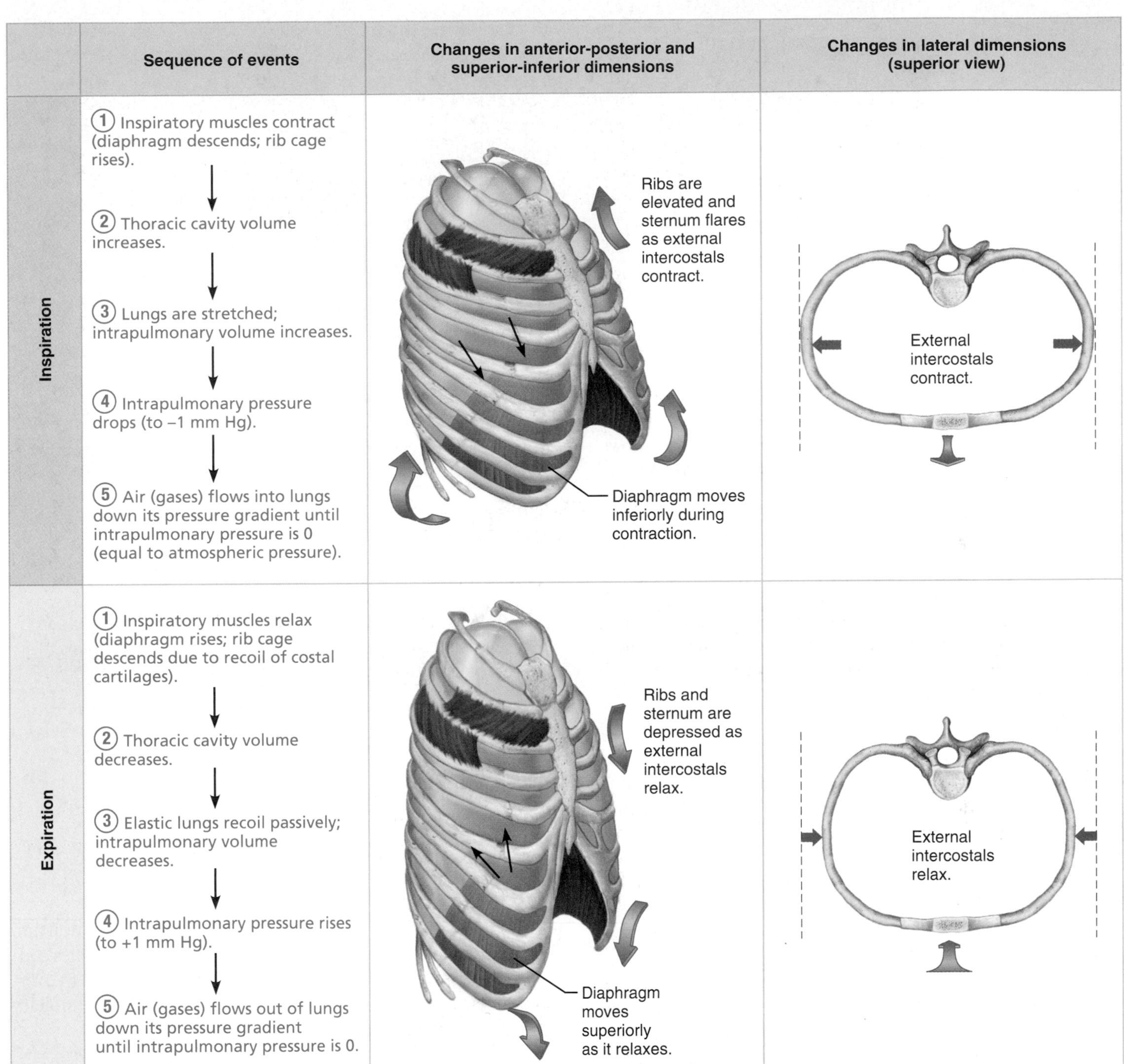

Sequence of events
Changes in anterior-posterior and superior-inferior dimensions
Changes in lateral dimensions (superior view)
Inspiration
1 Inspiratory muscles contract (diaphragm descends; rib cage rises).
2 Thoracic cavity volume increases.
3 Lungs are stretched; intrapulmonary volume increases.
4 Intrapulmonary pressure drops (to –1 mm Hg).
5 Air (gases) flows into lungs down its pressure gradient until intrapulmonary pressure is 0 (equal to atmospheric pressure).
Ribs are elevated and sternum flares as external intercostals contract.
Diaphragm moves inferiorly during contraction.
External intercostals contract.
Expiration
1 Inspiratory muscles relax (diaphragm rises; rib cage descends due to recoil of costal cartilages).
2 Thoracic cavity volume decreases.
3 Elastic lungs recoil passively; intrapulmonary volume decreases.
4 Intrapulmonary pressure rises (to +1 mm Hg).
5 Air (gases) flows out of lungs down its pressure gradient until intrapulmonary pressure is 0.
Ribs and sternum are depressed as external intercostals relax.
Diaphragm moves superiorly as it relaxes.
External intercostals relax.

Figure 22.13 Changes in thoracic volume and sequence of events during inspiration and expiration. The sequence of events in the left column includes volume changes during inspiration (top) and expiration (bottom). The lateral views in the middle column show changes in the superior-inferior dimension (as the diaphragm alternately contracts and relaxes) and in the anterior-posterior dimension (as the external intercostal muscles alternately contract and relax). The superior views of transverse thoracic sections in the right column show lateral dimension changes resulting from alternate contraction and relaxation of the external intercostal muscles.

a result, P_{pul} drops about 1 mm Hg relative to P_{atm}. Anytime the intrapulmonary pressure is less than the atmospheric pressure ($P_{pul} < P_{atm}$), air rushes into the lungs along the pressure gradient. Inspiration ends when $P_{pul} = P_{atm}$. During the same period, P_{ip} declines to about –6 mm Hg relative to P_{atm} **(Figure 22.14)**.

During the *deep* or *forced inspirations* that occur during vigorous exercise and in some chronic obstructive pulmonary diseases, the thoracic volume is further increased by activity of accessory muscles. Several muscles, including the scalenes and sternocleidomastoid muscles of the neck and the pectoralis

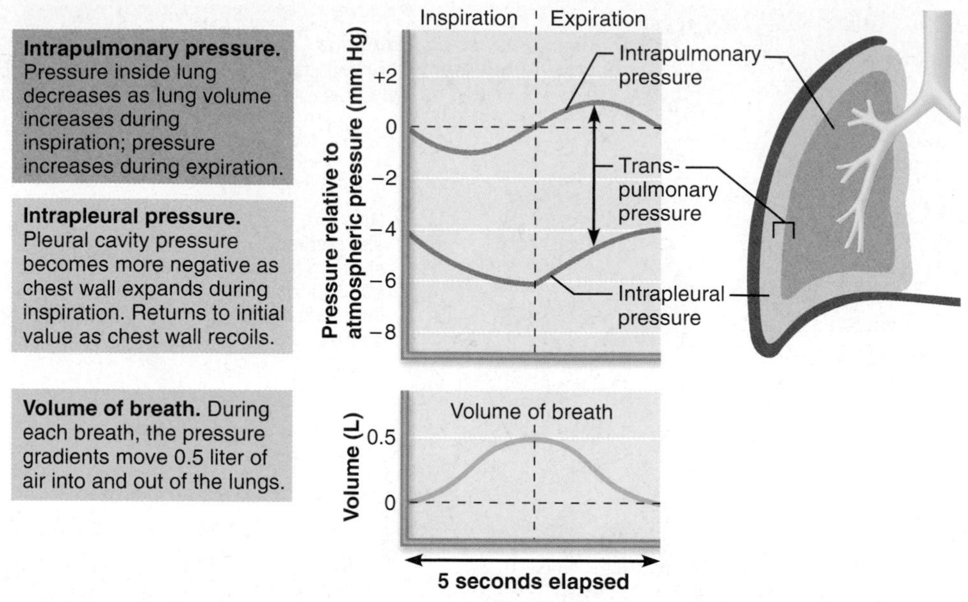
Intrapulmonary pressure. Pressure inside lung decreases as lung volume increases during inspiration; pressure increases during expiration.
Intrapleural pressure. Pleural cavity pressure becomes more negative as chest wall expands during inspiration. Returns to initial value as chest wall recoils.
Volume of breath. During each breath, the pressure gradients move 0.5 liter of air into and out of the lungs.
Inspiration
Expiration
Pressure relative to atmospheric pressure (mm Hg)
+2
0
−2
−4
−6
−8
Intrapulmonary pressure
Trans-pulmonary pressure
Intrapleural pressure
Volume of breath
Volume (L)
0.5
0
5 seconds elapsed

Figure 22.14 Changes in intrapulmonary and intrapleural pressures during inspiration and expiration. Notice that normal atmospheric pressure (760 mm Hg) is given a value of 0 on the scale.

minor of the chest, raise the ribs even more than occurs during quiet inspiration. Additionally, the back extends as the thoracic curvature is straightened by the erector spinae muscles.

Expiration

Quiet expiration in healthy individuals is a passive process that depends more on lung elasticity than on muscle contraction. As the inspiratory muscles relax and resume their resting length, the rib cage descends and the lungs recoil (Figure 22.13, bottom). As a result, both the thoracic and intrapulmonary volumes decrease. This volume decrease compresses the alveoli, and P_{pul} rises to about 1 mm Hg above atmospheric pressure (Figure 22.14). When $P_{pul} > P_{atm}$, the pressure gradient forces gases to flow out of the lungs.

22

Forced expiration is an active process produced by contraction of abdominal wall muscles, primarily the oblique and transversus muscles. These contractions (1) increase the intra-abdominal pressure, which forces the abdominal organs superiorly against the diaphragm, and (2) depress the rib cage. The internal intercostal muscles also help to depress the rib cage and decrease thoracic volume.

Control of accessory muscles of expiration is important when precise regulation of air flow from the lungs is desired. For instance, the ability of a trained vocalist to hold a musical note depends on the coordinated activity of several muscles normally used in forced expiration.

CHECK YOUR UNDERSTANDING

7. What is the driving force for ventilation?
8. What causes the intrapulmonary pressure to decrease during inspiration?
9. What causes the partial vacuum (negative pressure) inside the pleural cavity? What happens to a lung if air enters the pleural cavity? What is the clinical name for this condition?

For answers, see Appendix G.

Physical Factors Influencing Pulmonary Ventilation

▶ **List several physical factors that influence pulmonary ventilation.**

As we have seen, the lungs are stretched during inspiration and recoil passively during expiration. The inspiratory muscles consume energy to enlarge the thorax. Energy is also used to overcome various factors that hinder air passage and pulmonary ventilation. We examine these factors next.

Airway Resistance

The major *nonelastic* source of resistance to gas flow is friction, or drag, encountered in the respiratory passageways. The relationship between gas flow (F), pressure (P), and resistance (R) is given by the following equation:

$$F = \frac{\Delta P}{R}$$

Notice that the factors determining gas flow in the respiratory passages and blood flow in the cardiovascular system are equivalent. The amount of gas flowing into and out of the alveoli is directly proportional to ΔP, the *difference* in pressure, or the pressure gradient, between the external atmosphere and the alveoli. Normally, very small differences in pressure produce large changes in the volume of gas flow. The average pressure gradient during normal quiet breathing is 2 mm Hg or less, and yet it is sufficient to move 500 ml of air in and out of the lungs with each breath.

But, as the equation also indicates, gas flow changes *inversely* with resistance. In other words, gas flow decreases as resistance increases. As in the cardiovascular system, resistance in the respiratory tree is determined mostly by the diameters of the conducting tubes. However, as a rule, airway resistance is insignificant for two reasons:

1. Airway diameters in the first part of the conducting zone are huge, relative to the low viscosity of air.
2. As the airways get progressively smaller, there are progressively more branches. As a result, although individual bronchioles are tiny, there are an enormous number of them in parallel, so the total cross-sectional area is huge.

Consequently, the greatest resistance to gas flow occurs in the medium-sized bronchi (**Figure 22.15**). At the terminal bronchioles, gas flow stops and diffusion takes over as the main force driving gas movement, so resistance is no longer an issue.

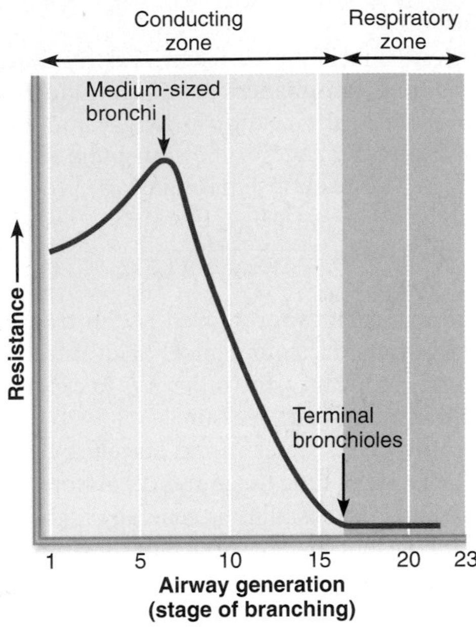

Figure 22.15 Resistance in respiratory passageways. Airway resistance peaks in the medium-sized bronchi and then declines sharply as the total cross-sectional area of the airways increases rapidly.

HOMEOSTATIC IMBALANCE

Smooth muscle of the bronchiolar walls is exquisitely sensitive to neural controls and certain chemicals. For example, inhaled irritants activate a reflex of the parasympathetic division of the nervous system that causes vigorous constriction of the bronchioles and dramatically reduces air passage. During an *acute asthma attack*, histamine and other inflammatory chemicals can cause such strong bronchoconstriction that pulmonary ventilation almost completely stops, regardless of the pressure gradient. Conversely, epinephrine released during sympathetic nervous system activation or administered as a drug dilates bronchioles and reduces airway resistance. Local accumulations of mucus, infectious material, or solid tumors in the passageways are important sources of airway resistance in those with respiratory disease.

Whenever airway resistance rises, breathing movements become more strenuous, but such compensation has its limits. When the bronchioles are severely constricted or obstructed, even the most magnificent respiratory efforts cannot restore ventilation to life-sustaining levels. ■

Alveolar Surface Tension

At any gas-liquid boundary, the molecules of the liquid are more strongly attracted to each other than to the gas molecules. This unequal attraction produces a state of tension at the liquid surface, called **surface tension**, that (1) draws the liquid molecules closer together and reduces their contact with the dissimilar gas molecules, and (2) resists any force that tends to increase the surface area of the liquid.

Water is composed of highly polar molecules and has a very high surface tension. As the major component of the liquid film that coats the alveolar walls, water is always acting to reduce the alveoli to their smallest possible size, as we noted earlier. If the film were pure water, the alveoli would collapse between breaths. But the alveolar film contains **surfactant** (ser-fak′tant), a detergent-like complex of lipids and proteins produced by the type II alveolar cells. Surfactant decreases the cohesiveness of water molecules, much the way a laundry detergent reduces the attraction of water for water, allowing water to interact with and pass through fabric. As a result, the surface tension of alveolar fluid is reduced, and less energy is needed to overcome those forces to expand the lungs and discourage alveolar collapse. Breaths that are deeper than normal stimulate type II cells to secrete more surfactant.

HOMEOSTATIC IMBALANCE

When too little surfactant is present, surface tension forces can collapse the alveoli. Once this happens, the alveoli must be completely reinflated during each inspiration, an effort that uses tremendous amounts of energy. This is the problem faced by newborns with **infant respiratory distress syndrome (IRDS)**, a condition peculiar to premature babies. Since inadequate pulmonary surfactant is produced until the last two months of fetal development, babies born prematurely often are unable to keep their alveoli inflated between breaths. IRDS is treated by spraying natural or synthetic surfactant into the newborn's respiratory passageways. In addition, devices that maintain a positive airway pressure throughout the respiratory cycle are often used to keep the alveoli open between breaths. In severe cases, mechanical ventilators are required.

Many IRDS survivors suffer from *bronchopulmonary dysplasia*, a chronic lung disease, during childhood and beyond. This condition is believed to result from inflammatory injury to respiratory zone structures caused by mechanical ventilation of the premature newborn's delicate lungs. ■

Lung Compliance

Healthy lungs are unbelievably stretchy, and this distensibility is referred to as **lung compliance**. Specifically, lung compliance (C_L) is a measure of the change in lung volume (ΔV_L) that occurs with a given change in the transpulmonary pressure [$\Delta(P_{pul} - P_{ip})$]. This relationship is stated as

$$C_L = \frac{\Delta V_L}{\Delta(P_{pul} - P_{ip})}$$

The more a lung expands for a given rise in transpulmonary pressure, the greater its compliance. Said another way, the higher the lung compliance, the easier it is to expand the lungs at any given transpulmonary pressure.

Lung compliance is determined largely by two factors: (1) distensibility of the lung tissue and (2) alveolar surface tension. Because lung distensibility is generally high and alveolar surface tension is kept low by surfactant, the lungs of healthy people tend to have high compliance, which favors efficient ventilation.

Lung compliance is diminished by a decrease in the natural resilience of the lungs. Chronic inflammation, or infections such as tuberculosis, can cause nonelastic scar tissue to replace normal lung tissue (*fibrosis*). Another factor that can decrease lung compliance is a decrease in production of surfactant. The lower the lung compliance, the more energy is needed just to breathe.

Since the lungs are contained within the thoracic cavity, we also need to consider the compliance (distensibility) of the thoracic wall. Factors that decrease the compliance of the thoracic wall hinder the expansion of the lungs. The total compliance of the respiratory system is comprised of lung compliance and thoracic wall compliance.

HOMEOSTATIC IMBALANCE

Deformities of the thorax, ossification of the costal cartilages (common during old age), and paralysis of the intercostal muscles all reduce total respiratory compliance by hindering thoracic expansion. ■

22

Respiratory Volumes and Pulmonary Function Tests

- Explain and compare the various lung volumes and capacities.
- Define dead space.
- Indicate types of information that can be gained from pulmonary function tests.

The amount of air flushed in and out of the lungs depends on the conditions of inspiration and expiration. Consequently, several respiratory volumes can be described. Specific combinations of these respiratory volumes, called *respiratory capacities*, are measured to gain information about a person's respiratory status.

Respiratory Volumes

The four **respiratory volumes** of interest are tidal, inspiratory reserve, expiratory reserve, and residual. The values recorded in **Figure 22.16a** (and used in the following text) represent normal values for a healthy 20-year-old male weighing about 70 kg (155 lb). Figure 22.16b provides average values for males and females.

During normal quiet breathing, about 500 ml of air moves into and then out of the lungs with each breath. This respiratory volume is the **tidal volume (TV)**. The amount of air that can be inspired forcibly beyond the tidal volume (2100 to 3200 ml) is called the **inspiratory reserve volume (IRV)**.

The **expiratory reserve volume (ERV)** is the amount of air—normally 1000 to 1200 ml—that can be evacuated from the lungs after a tidal expiration. Even after the most strenuous expiration, about 1200 ml of air remains in the lungs; this is the **residual volume (RV)**, which helps to keep the alveoli patent (open) and to prevent lung collapse.

Respiratory Capacities

The **respiratory capacities** include inspiratory capacity, functional residual capacity, vital capacity, and total lung capacity (Figure 22.16). As noted, the respiratory capacities always consist of two or more lung volumes.

The **inspiratory capacity (IC)** is the total amount of air that can be inspired after a tidal expiration, so it is the sum of TV and IRV. The **functional residual capacity (FRC)** represents the amount of air remaining in the lungs after a tidal expiration and is the combined RV and ERV.

Vital capacity (VC) is the total amount of exchangeable air. It is the sum of TV, IRV, and ERV. In healthy young males, VC is approximately 4800 ml. The **total lung capacity (TLC)** is the sum of all lung volumes and is normally around 6000 ml. As indicated in Figure 22.16b, lung volumes and capacities (with the possible exception of TV) tend to be smaller in women than in men because of women's smaller size.

Dead Space

Some of the inspired air fills the conducting respiratory passageways and never contributes to gas exchange in the alveoli. The volume of these conducting zone conduits, which make up the **anatomical dead space**, typically amounts to about 150 ml. (The rule of thumb is that the anatomical dead space volume in a healthy young adult is equal to 1 ml per pound of ideal body weight.) This means that if TV is 500 ml, only 350 ml of it is involved in alveolar ventilation. The remaining 150 ml of the tidal breath is in the anatomical dead space.

If some alveoli cease to act in gas exchange (due to alveolar collapse or obstruction by mucus, for example), the **alveolar dead space** is added to the anatomical dead space, and the sum of the nonuseful volumes is referred to as **total dead space**.

Pulmonary Function Tests

Because the various lung volumes and capacities are often abnormal in people with pulmonary disorders, they are routinely measured in such patients. The original clinical measuring

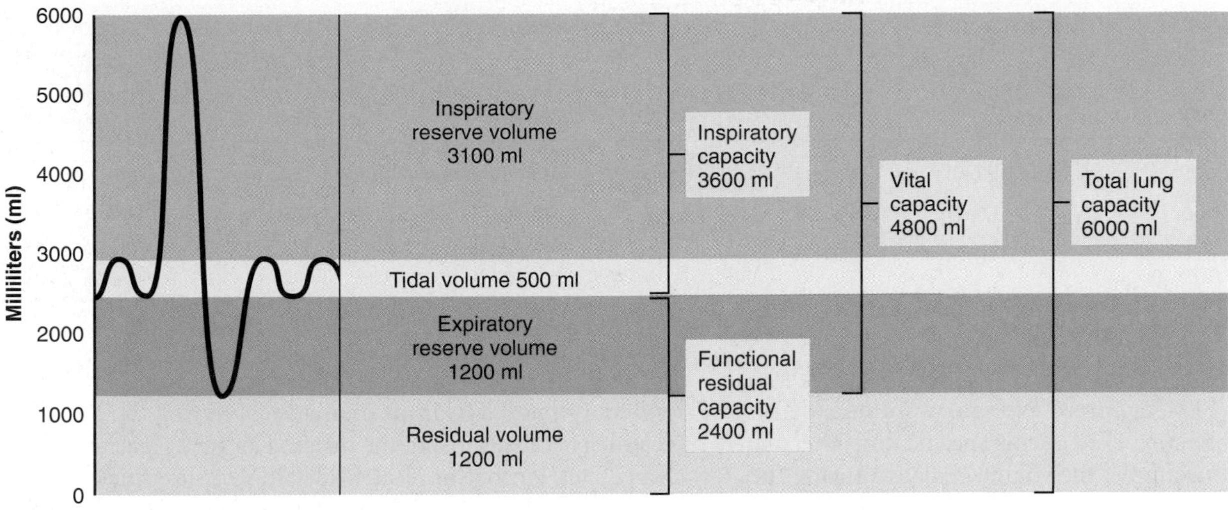

(a) Spirographic record for a male

	Measurement	Adult male average value	Adult female average value	Description
Respiratory volumes	Tidal volume (TV)	500 ml	500 ml	Amount of air inhaled or exhaled with each breath under resting conditions
	Inspiratory reserve volume (IRV)	3100 ml	1900 ml	Amount of air that can be forcefully inhaled after a normal tidal volume inhalation
	Expiratory reserve volume (ERV)	1200 ml	700 ml	Amount of air that can be forcefully exhaled after a normal tidal volume exhalation
	Residual volume (RV)	1200 ml	1100 ml	Amount of air remaining in the lungs after a forced exhalation
Respiratory capacities	Total lung capacity (TLC)	6000 ml	4200 ml	Maximum amount of air contained in lungs after a maximum inspiratory effort: TLC = TV + IRV + ERV + RV
	Vital capacity (VC)	4800 ml	3100 ml	Maximum amount of air that can be expired after a maximum inspiratory effort: VC = TV + IRV + ERV
	Inspiratory capacity (IC)	3600 ml	2400 ml	Maximum amount of air that can be inspired after a normal expiration: IC = TV + IRV
	Functional residual capacity (FRC)	2400 ml	1800 ml	Volume of air remaining in the lungs after a normal tidal volume expiration: FRC = ERV + RV

(b) Summary of respiratory volumes and capacities for males and females

Figure 22.16 Respiratory volumes and capacities. Idealized spirographic record of respiratory volumes in **(a)** is for a healthy young 70-kg adult male.

device, a **spirometer** (spi-rom′ĕ-ter), was a simple but cumbersome instrument utilizing a hollow bell inverted over water. Now patients simply blow into a small electronic measuring device.

Spirometry is most useful for evaluating losses in respiratory function and for following the course of certain respiratory diseases. It cannot provide a specific diagnosis, but it can distinguish between *obstructive pulmonary disease* involving increased airway resistance (such as chronic bronchitis) and *restrictive disorders* involving a reduction in total lung capacity resulting from structural or functional changes in the lungs. (These changes might be due to diseases such as tuberculosis, or to fibrosis due to exposure to certain environmental agents such as asbestos). Increases in TLC, FRC, and RV may occur as a result of hyperinflation of the lungs in obstructive disease, whereas VC, TLC, FRC, and RV are reduced in restrictive diseases, which limit lung expansion.

More information can be obtained about a patient's ventilation status by assessing the rate at which gas moves into and out of the lungs. The **minute ventilation** is the total amount of gas that flows into or out of the respiratory tract in 1 minute. During normal quiet breathing, the minute ventilation in healthy people is about 6 L/min (500 ml per breath multiplied by 12 breaths per minute). During vigorous exercise, the minute ventilation may reach 200 L/min.

Two other useful tests are FVC and FEV. **FVC**, or **forced vital capacity**, measures the amount of gas expelled when a subject takes a deep breath and then forcefully exhales maximally and as

TABLE 22.2 Effects of Breathing Rate and Depth on Alveolar Ventilation of Three Hypothetical Patients

BREATHING PATTERN OF HYPOTHETICAL PATIENT	DEAD SPACE VOLUME (DSV)	TIDAL VOLUME (TV)	RESPIRATORY RATE*	MINUTE VENTILATION (MVR)	ALVEOLAR VENTILATION (AVR)	% EFFECTIVE VENTILATION (AVR/MVR)
I—Normal rate and depth	150 ml	500 ml	20/min	10,000 ml/min	7000 ml/min	70%
II—Slow, deep breathing	150 ml	1000 ml	10/min	10,000 ml/min	8500 ml/min	85%
III—Rapid, shallow breathing	150 ml	250 ml	40/min	10,000 ml/min	4000 ml/min	40%

* Respiratory rate values are artificially adjusted to provide equivalent minute ventilation as a baseline for comparison of alveolar ventilation.

rapidly as possible. **FEV**, or **forced expiratory volume**, determines the amount of air expelled during specific time intervals of the FVC test. For example, the volume exhaled during the first second is FEV_1. Those with healthy lungs can exhale about 80% of the FVC within 1 second. Those with obstructive pulmonary disease exhale considerably less than 80% of the FVC within 1 second, while those with restrictive disease can exhale 80% or more of FVC in 1 second even though their FVC is reduced.

Alveolar Ventilation

Minute ventilation values provide a rough yardstick for assessing respiratory efficiency, but the **alveolar ventilation rate (AVR)** is a better index of effective ventilation. The AVR takes into account the volume of air wasted in the dead space and measures the flow of fresh gases in and out of the alveoli during a particular time interval. AVR is computed with this equation:

$$\underset{\text{(ml/min)}}{\text{AVR}} = \underset{\text{(breaths/min)}}{\text{frequency}} \times \underset{\text{(ml/breath)}}{(\text{TV} - \text{dead space})}$$

In healthy people, AVR is usually about 12 breaths per minute times the difference of 500 – 150 ml per breath, or 4200 ml/min.

Because anatomical dead space is constant in a particular individual, increasing the volume of each inspiration (breathing depth) enhances AVR and gas exchange more than raising the respiratory rate. AVR drops dramatically during rapid shallow breathing because most of the inspired air never reaches the exchange sites. Furthermore, as tidal volume approaches the dead space value, effective ventilation approaches zero, regardless of how fast a person is breathing. The effects of breathing rate and breathing depth on alveolar ventilation are summarized for three hypothetical patients in **Table 22.2**.

Nonrespiratory Air Movements

Many processes other than breathing move air into or out of the lungs, and these processes may modify the normal respiratory rhythm. Most of these **nonrespiratory air movements** result from reflex activity, but some are produced voluntarily. The most common of these movements are described in **Table 22.3**.

CHECK YOUR UNDERSTANDING

10. Resistance in the airways is typically low. Why? (Give at least two reasons.)

11. Premature infants often lack adequate surfactant. How does this affect their ability to breathe?

12. Explain why slow, deep breaths ventilate the alveoli more effectively than do rapid, shallow breaths.

For answers, see Appendix G.

22

TABLE 22.3 Nonrespiratory Air (Gas) Movements

MOVEMENT	MECHANISM AND RESULT
Cough	Taking a deep breath, closing glottis, and forcing air superiorly from lungs against glottis; glottis opens suddenly and a blast of air rushes upward. Can dislodge foreign particles or mucus from lower respiratory tract and propel such substances superiorly.
Sneeze	Similar to a cough, except that expelled air is directed through nasal cavities as well as through oral cavity; depressed uvula routes air upward through nasal cavities. Sneezes clear upper respiratory passages.
Crying	Inspiration followed by release of air in a number of short expirations. Primarily an emotionally induced mechanism.
Laughing	Essentially same as crying in terms of air movements produced. Also an emotionally induced response.
Hiccups	Sudden inspirations resulting from spasms of diaphragm; believed to be initiated by irritation of diaphragm or phrenic nerves, which serve diaphragm. Sound occurs when inspired air hits vocal folds of closing glottis.
Yawn	Very deep inspiration, taken with jaws wide open; not believed to be triggered by levels of oxygen or carbon dioxide in blood. Ventilates all alveoli (not the case in normal quiet breathing).

Gas Exchanges Between the Blood, Lungs, and Tissues

- State Dalton's law of partial pressures and Henry's law.
- Describe how atmospheric and alveolar air differ in composition, and explain these differences.
- Relate Dalton's and Henry's laws to events of external and internal respiration.

As you've discovered, during *external respiration* oxygen enters and carbon dioxide leaves the blood in the lungs by diffusion. At the body tissues, where the process is called *internal respiration*, the same gases move in opposite directions, also by diffusion. To understand these processes, let's examine some of the physical properties of gases and consider the composition of alveolar gas.

Basic Properties of Gases

Two more gas laws provide most of the information we need—*Dalton's law of partial pressures* reveals how a gas behaves when it is part of a mixture of gases, and *Henry's law* will help us understand movement of gases into and out of solution.

Dalton's Law of Partial Pressures

Dalton's law of partial pressures states that the total pressure exerted by a mixture of gases is the sum of the pressures exerted independently by each gas in the mixture. Further, the pressure exerted by each gas—its **partial pressure**—is directly proportional to the percentage of that gas in the gas mixture.

As indicated in **Table 22.4**, nitrogen makes up about 79% of air, and the partial pressure of nitrogen (P_{N_2}) is 78.6% × 760 mm Hg, or 597 mm Hg. Oxygen, which accounts for nearly 21% of the atmosphere, has a partial pressure (P_{O_2}) of 159 mm Hg (20.9 % × 760 mm Hg). Together nitrogen and oxygen contribute about 99% of the total atmospheric pressure. Air also contains 0.04% carbon dioxide, up to 0.5% water vapor, and insignificant amounts of inert gases (such as argon and helium).

At high altitudes, where the atmosphere is less influenced by gravity, partial pressures decline in direct proportion to the decrease in atmospheric pressure. For example, at 10,000 feet above sea level where the atmospheric pressure is 523 mm Hg, P_{O_2} is 110 mm Hg. Moving in the opposite direction, atmospheric pressure increases by 1 atm (760 mm Hg) for each 33 feet of descent (in water) below sea level. At 99 feet below sea level, the total pressure exerted on the body is equivalent to 4 atm, or 3040 mm Hg, and the partial pressure exerted by each component gas is also quadrupled.

Henry's Law

According to **Henry's law**, when a gas is in contact with a liquid, that gas will dissolve in the liquid in proportion to its partial pressure. Accordingly, the greater the concentration of a particular gas in the gas phase, the more and the faster that gas will go into solution in the liquid. At equilibrium, the gas partial pressures in the two phases are the same. If, however, the partial pressure of the gas later becomes greater in the liquid than in the adjacent gas phase, some of the dissolved gas molecules will reenter the gaseous phase. So the direction and amount of movement of a gas is determined by its partial pressure in the two phases. This flexible situation is exactly what occurs when gases are exchanged in the lungs and tissues.

How much of a gas will dissolve in a liquid at any given partial pressure also depends on the *solubility* of the gas in the liquid and on the *temperature* of the liquid. The gases in air have very different solubilities in water (and in plasma). Carbon dioxide is most soluble. Oxygen is only 1/20 as soluble as CO_2, and N_2 is only half as soluble as O_2. For this reason, at a given partial pressure, much more CO_2 than O_2 dissolves in water, and practically no N_2 goes into solution.

The effect of increasing the liquid's temperature is to decrease gas solubility. Think of club soda, which is produced by forcing CO_2 gas to dissolve in water under high pressure. If you take the cap off a refrigerated bottle of club soda and allow it to stand at room temperature, in just a few minutes you will have plain water—all the CO_2 gas will have escaped from solution.

22

TABLE 22.4 Comparison of Gas Partial Pressures and Approximate Percentages in the Atmosphere and in the Alveoli

	ATMOSPHERE (SEA LEVEL)		*ALVEOLI*	
GAS	APPROXIMATE PERCENTAGE	PARTIAL PRESSURE (mm Hg)	APPROXIMATE PERCENTAGE	PARTIAL PRESSURE (mm Hg)
N_2	78.6	597	74.9	569
O_2	20.9	159	13.7	104
CO_2	0.04	0.3	5.2	40
H_2O	0.46	3.7	6.2	47
	100.0%	760	100.0%	760

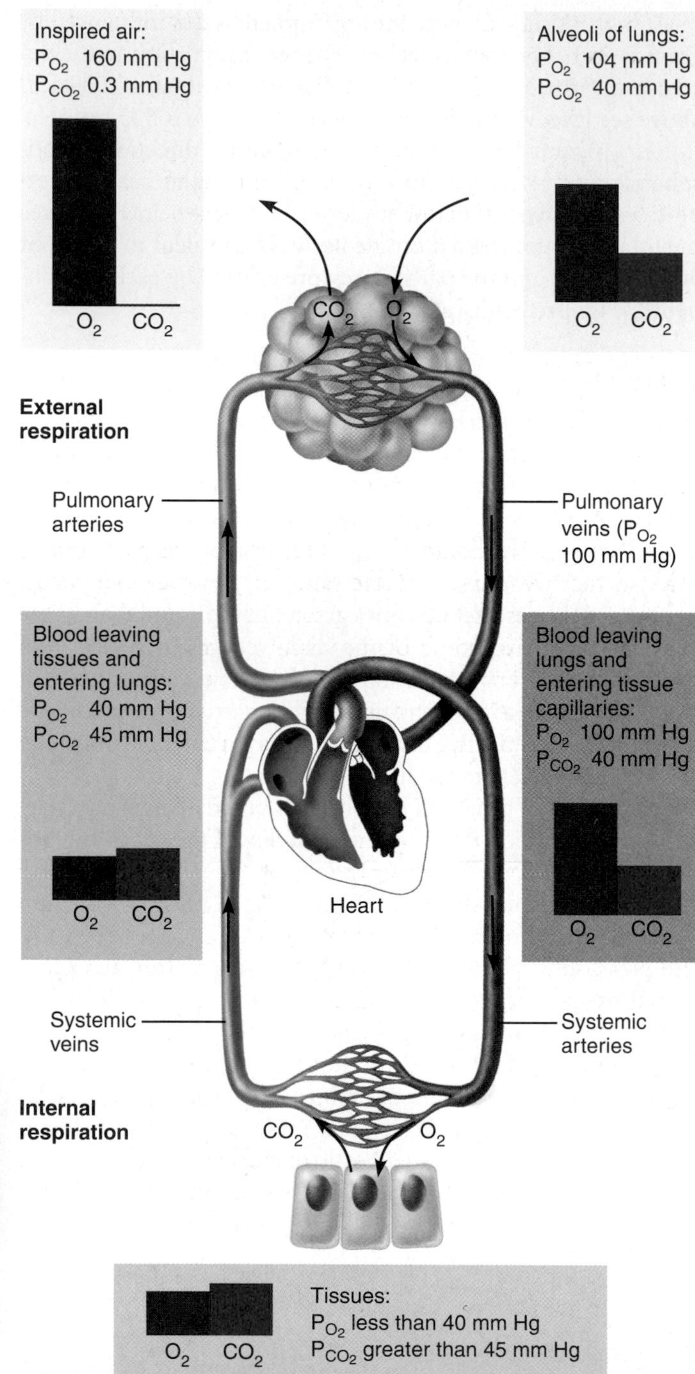

Figure 22.17 Partial pressure gradients promoting gas movements in the body. Top: Gradients promoting O_2 and CO_2 exchange across the respiratory membrane in the lungs. Bottom: Gradients promoting gas movements across systemic capillary membranes in body tissues. (Note that the small decrease in P_{O_2} in blood leaving lungs is due to partial dilution of pulmonary capillary blood with less oxygenated blood.)

Hyperbaric oxygen chambers provide clinical applications of Henry's law. These chambers contain O_2 gas at pressures higher than 1 atm and are used to force greater-than-normal amounts of O_2 into the blood of patients suffering from carbon monoxide poisoning or tissue damage following radiation therapy. Hyperbaric therapy is also used to treat individuals with gas gangrene, because the anaerobic bacteria causing this infection cannot live in the presence of high O_2 levels. Scuba diving provides another illustration of Henry's law—if divers rise rapidly from the depths, dissolved nitrogen forms bubbles in their blood, causing "the bends."

HOMEOSTATIC IMBALANCE

Although breathing O_2 gas at 2 atm is not a problem for short periods, **oxygen toxicity** develops rapidly when P_{O_2} is greater than 2.5–3 atm. Excessively high O_2 concentrations generate huge amounts of harmful free radicals, resulting in profound CNS disturbances, coma, and death. ■

Composition of Alveolar Gas

As shown in Table 22.4, the gaseous makeup of the atmosphere is quite different from that in the alveoli. The atmosphere is almost entirely O_2 and N_2; the alveoli contain more CO_2 and water vapor and much less O_2. These differences reflect the effects of (1) gas exchanges occurring in the lungs (O_2 diffuses from the alveoli into the pulmonary blood and CO_2 diffuses in the opposite direction), (2) humidification of air by conducting passages, and (3) the mixing of alveolar gas that occurs with each breath. Because only 500 ml of air is inspired with each tidal inspiration, gas in the alveoli is actually a mixture of newly inspired gases and gases remaining in the respiratory passageways between breaths.

The alveolar partial pressures of O_2 and CO_2 are easily changed by increasing breathing depth and rate. A high AVR brings more O_2 into the alveoli, increasing alveolar P_{O_2} and rapidly eliminating CO_2 from the lungs.

External Respiration

During external respiration (pulmonary gas exchange) dark red blood flowing through the pulmonary circuit is transformed into the scarlet river that is returned to the heart for distribution by systemic arteries to all body tissues. This color change is due to O_2 uptake and binding to hemoglobin in red blood cells (RBCs), but CO_2 exchange (unloading) is occurring equally fast.

The following three factors influence the movement of oxygen and carbon dioxide across the respiratory membrane:

1. Partial pressure gradients and gas solubilities
2. Matching of alveolar ventilation and pulmonary blood perfusion
3. Structural characteristics of the respiratory membrane

Let's look at these factors one by one in the sections that follow.

Partial Pressure Gradients and Gas Solubilities

Partial pressure gradients of O_2 and CO_2 drive the diffusion of these gases across the respiratory membrane. A steep oxygen partial pressure gradient exists across the respiratory membrane because the P_{O_2} of deoxygenated blood in the pulmonary arteries is only 40 mm Hg, as opposed to a P_{O_2} of approximately 104 mm Hg in the alveoli. As a result, O_2 diffuses rapidly from the alveoli into the pulmonary capillary blood **(Figure 22.17)**. Equilibrium—that is, a P_{O_2} of 104 mm Hg on both sides of the respiratory membrane—

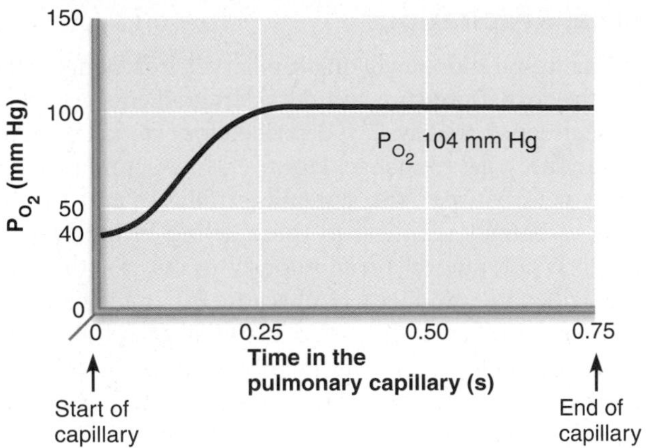

Figure 22.18 Oxygenation of blood in the pulmonary capillaries at rest. Note that the time from blood entering the pulmonary capillaries (indicated by 0) until the P_{O_2} is 104 mm Hg is approximately 0.25 second.

usually occurs in 0.25 second, which is about one-third the time a red blood cell is in a pulmonary capillary **(Figure 22.18)**. The lesson here is that the blood can flow through the pulmonary capillaries three times as quickly and still be adequately oxygenated.

Carbon dioxide diffuses in the opposite direction along a much gentler partial pressure gradient of about 5 mm Hg (45 mm Hg to 40 mm Hg) until equilibrium occurs at 40 mm Hg. Carbon dioxide is then expelled gradually from the alveoli during expiration.

Even though the O_2 pressure gradient for oxygen diffusion is much steeper than the CO_2 gradient, equal amounts of these gases are exchanged. Why? The reason is because CO_2 is 20 times more soluble in plasma and alveolar fluid than O_2.

Ventilation-Perfusion Coupling

For gas exchange to be efficient, there must be a close match, or coupling, between the amount of gas reaching the alveoli, known as *ventilation*, and the blood flow in pulmonary capillaries, known as *perfusion*. As we explained in Chapter 19, local autoregulatory mechanisms continuously respond to alveolar conditions.

Specifically, in alveoli where ventilation is inadequate, P_{O_2} is low, as **Figure 22.19a** shows. As a result, the terminal arterioles constrict, and blood is redirected to respiratory areas where P_{O_2} is high and oxygen pickup may be more efficient.

In alveoli where ventilation is maximal, pulmonary arterioles dilate, increasing blood flow into the associated pulmonary capillaries, as Figure 22.19b shows. Notice that the autoregulatory mechanism controlling pulmonary vascular muscle is the

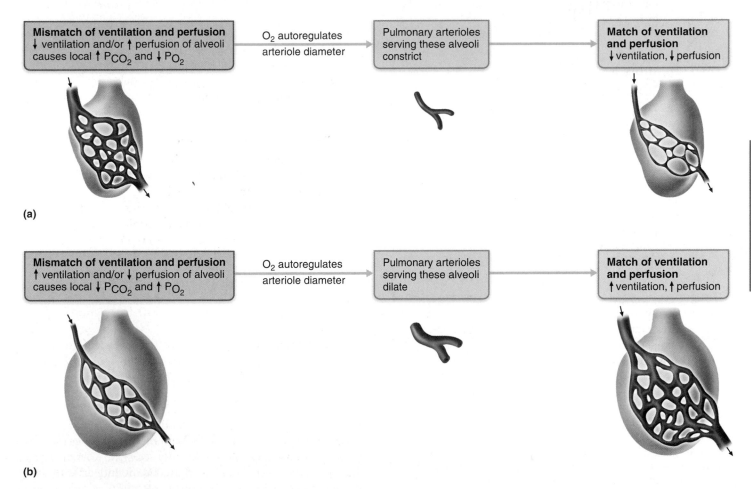

Figure 22.19 Ventilation-perfusion coupling. Autoregulatory events result in local matching of blood flow (perfusion) through the pulmonary capillaries with the amount of alveolar ventilation.

opposite of the mechanism controlling most arterioles in the systemic circulation.

While changes in alveolar P_{O_2} affect the diameter of pulmonary blood vessels (arterioles), changes in alveolar P_{CO_2} cause changes in the diameters of the *bronchioles*. Bronchioles servicing areas where alveolar CO_2 levels are high dilate, allowing CO_2 to be eliminated from the body more rapidly. Bronchioles serving areas where P_{CO_2} is low constrict.

As a result of changing the diameter of local bronchioles and arterioles, alveolar ventilation and pulmonary perfusion are synchronized. Poor alveolar ventilation results in low oxygen and high carbon dioxide levels in the alveoli. Consequently, the pulmonary arterioles constrict and the airways dilate, bringing blood flow and air flow into closer physiological match. High P_{O_2} and low P_{CO_2} in the alveoli cause bronchioles serving the alveoli to constrict, and promote flushing of blood into the pulmonary capillaries.

Although these homeostatic mechanisms provide appropriate conditions for efficient gas exchange, they never completely balance ventilation and perfusion in every alveolus due to other factors. In particular, (1) gravity causes regional variations in both blood and air flow in the lungs, and (2) the occasional alveolar duct plugged with mucus creates unventilated areas. These factors, together with the shunting of blood from bronchial veins, account for the slight drop in P_{O_2} from alveolar air (104 mm Hg) to pulmonary venous blood (100 mm Hg), as shown in Figure 22.17.

Thickness and Surface Area of the Respiratory Membrane

In healthy lungs, the respiratory membrane is only 0.5 to 1 μm thick, and gas exchange is usually very efficient.

HOMEOSTATIC IMBALANCE

The effective thickness of the respiratory membrane increases dramatically if the lungs become waterlogged and edematous, as in pneumonia or left heart failure (see p. 687). Under such conditions, even the total time (0.75 s) that red blood cells are in transit through the pulmonary capillaries may not be enough for adequate gas exchange, and body tissues begin to suffer from oxygen deprivation. ■

The greater the surface area of the respiratory membrane, the more gas can diffuse across it in a given time period. The alveolar surface area is enormous in healthy lungs. Spread flat, the total gas exchange surface of these tiny sacs in an adult male's lungs is about 90 m^2, approximately 40 times greater than the surface area of his skin!

HOMEOSTATIC IMBALANCE

In certain pulmonary diseases, the alveolar surface area actually functioning in gas exchange is drastically reduced. This reduction occurs in emphysema, when the walls of adjacent alveoli break down and the alveolar chambers become larger. It also occurs when tumors, mucus, or inflammatory material blocks gas flow into the alveoli. ■

Internal Respiration

In internal respiration, involving capillary gas exchange in body tissues, the partial pressure and diffusion gradients are reversed from the situation we have just described for external respiration and pulmonary gas exchange. However, the factors promoting gas exchanges between the systemic capillaries and the tissue cells are essentially identical to those acting in the lungs (see Figure 22.17). Tissue cells continuously use O_2 for their metabolic activities and produce CO_2. Because P_{O_2} in the tissues is always lower than that in the systemic arterial blood (40 mm Hg versus 100 mm Hg), O_2 moves rapidly from the blood into the tissues until equilibrium is reached. At the same time, CO_2 moves quickly along its pressure gradient into the blood. As a result, venous blood draining the tissue capillary beds and returning to the heart has a P_{O_2} of 40 mm Hg and a P_{CO_2} of 45 mm Hg.

In summary, the gas exchanges that occur between the blood and the alveoli and between the blood and the tissue cells take place by simple diffusion driven by the partial pressure gradients of O_2 and CO_2 that exist on the opposite sides of the exchange membranes.

CHECK YOUR UNDERSTANDING

13. You are given a sealed container of water and air. The P_{CO_2} and P_{O_2} in the air are both 100 mm Hg. What are the P_{CO_2} and P_{O_2} in the water? Which gas has more molecules dissolved in the water? Why?

14. P_{O_2} in the alveoli is about 56 mm Hg lower than in the inspired air. Explain this difference.

15. Suppose a patient is receiving oxygen by mask. Are the arterioles leading into the O_2-enriched alveoli dilated or constricted? What is the advantage of this response?

For answers, see Appendix G.

Transport of Respiratory Gases by Blood

We have considered external and internal respiration consecutively to emphasize their similarities, but keep in mind that it is the blood that transports O_2 and CO_2 between these two exchange sites.

Oxygen Transport

▶ Describe how oxygen is transported in the blood, and explain how oxygen loading and unloading is affected by temperature, pH, BPG, and P_{CO_2}.

Molecular oxygen is carried in blood in two ways: bound to hemoglobin within red blood cells and dissolved in plasma. Oxygen is poorly soluble in water, so only about 1.5% of the oxygen transported is carried in the dissolved form. Indeed, if this were the *only* means of oxygen transport, a P_{O_2} of 3 atm or a cardiac output of 15 times normal would be required to provide the oxygen levels needed by body tissues! This problem, of course, has been solved by hemoglobin, and 98.5% of the oxygen ferried from the lungs to the tissues is carried in a loose chemical combination with hemoglobin.

Association of Oxygen and Hemoglobin

As we described in Chapter 17, hemoglobin (Hb) is composed of four polypeptide chains, each bound to an iron-containing heme group (see Figure 17.4). Because the iron atoms bind oxygen, each hemoglobin molecule can combine with four molecules of O_2, and oxygen loading is rapid and reversible.

The hemoglobin-oxygen combination, called **oxyhemoglobin** (ok″sĭ-he″mo-glo′bin), is written **HbO_2.** Hemoglobin that has released oxygen is called **reduced hemoglobin**, or **deoxyhemoglobin**, and is written **HHb**. Loading and unloading of O_2 can be indicated by a single reversible equation:

$$HHb + O_2 \underset{\text{Tissues}}{\overset{\text{Lungs}}{\rightleftharpoons}} HbO_2 + H^+$$

After the first O_2 molecule binds to iron, the Hb molecule changes shape. As a result, it more readily takes up two more O_2 molecules, and uptake of the fourth is even more facilitated. When all four heme groups are bound to O_2, a hemoglobin molecule is said to be *fully saturated.* When one, two, or three oxygen molecules are bound, hemoglobin is *partially saturated.* By the same token, unloading of one oxygen molecule enhances the unloading of the next, and so on. In this way, the *affinity* (binding strength) of hemoglobin for oxygen changes with the extent of oxygen saturation, and both loading and unloading of oxygen are very efficient.

The rate at which Hb reversibly binds or releases O_2 is regulated by P_{O_2}, temperature, blood pH, P_{CO_2}, and blood concentration of an organic chemical called BPG. These factors interact to ensure adequate deliveries of O_2 to tissue cells.

Influence of P_{O_2} on Hemoglobin Saturation The relationship between the degree of hemoglobin saturation and the P_{O_2} of blood is not linear, because the affinity of hemoglobin for O_2 changes with O_2 binding, as we just described. The **oxygen-hemoglobin dissociation curve** shows this relationship **(Figure 22.20).** This S-shaped curve has a steep slope for P_{O_2} values between 10 and 50 mm Hg and then flattens out (plateaus) between 70 and 100 mm Hg.

Under normal resting conditions (P_{O_2} = 100 mm Hg), arterial blood hemoglobin is 98% saturated, and 100 ml of systemic arterial blood contains about 20 ml of O_2. This *oxygen content* of arterial blood is written as 20 vol % (volume percent). As arterial blood flows through the systemic capillaries, about 5 ml of O_2 per 100 ml of blood is released, yielding an Hb saturation of 75% and an O_2 content of 15 vol % in venous blood.

The nearly complete saturation of Hb in arterial blood explains why breathing deeply increases both the alveolar and arterial blood P_{O_2} but causes very little increase in the O_2 saturation of hemoglobin. Remember, P_{O_2} measurements indicate only the amount of O_2 dissolved in plasma, not the amount bound to hemoglobin. However, P_{O_2} values are a good index of lung function, and when arterial P_{O_2} is significantly less than alveolar P_{O_2} some degree of respiratory impairment exists.

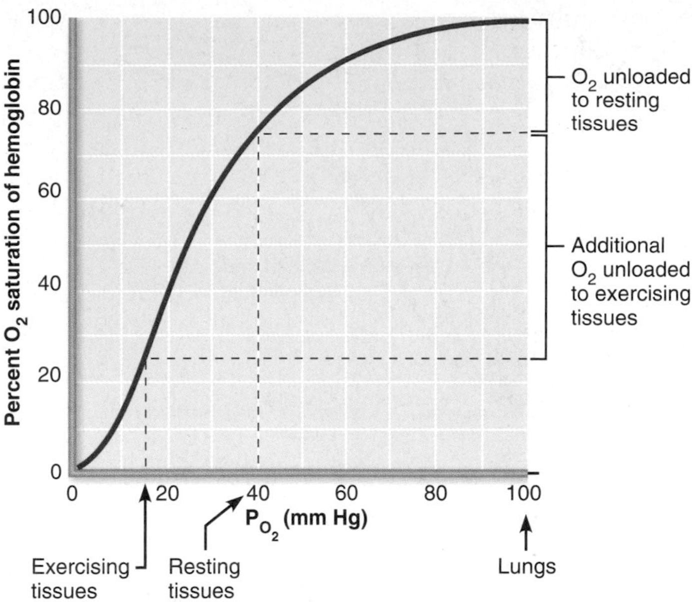

Figure 22.20 Oxygen-hemoglobin dissociation curve. Hb saturation changes as blood P_{O_2} changes. Notice that hemoglobin is almost completely saturated at a P_{O_2} of 70 mm Hg. Rapid loading and unloading of O_2 to and from hemoglobin occur at P_{O_2} values in the steep portion of the curve. In the systemic circuit, approximately 25% of Hb-bound O_2 is unloaded to resting tissues. Thus, hemoglobin of venous blood from resting tissue is still about 75% saturated with oxygen after one round through the body. During exercise, P_{O_2} can drop as low as 15 mm Hg, causing an additional 50% unloading and leaving the Hb only 25% saturated.

The oxygen-hemoglobin dissociation curve in Figure 22.20 reveals two other important facts. First, Hb is almost completely saturated at a P_{O_2} of 70 mm Hg, and further increases in P_{O_2} produce only small increases in O_2 binding. The adaptive value of this is that O_2 loading and delivery to the tissues can still be adequate when the P_{O_2} of inspired air is well below its usual levels, a situation common at higher altitudes and in people with cardiopulmonary disease. 22

Second, because most O_2 *unloading* occurs on the steep portion of the curve, a small drop in P_{O_2} will cause a large increase in unloading. Normally, only 20–25% of bound oxygen is unloaded during one systemic circuit, and substantial amounts of O_2 are still available in venous blood (the *venous reserve*). Consequently, if O_2 drops to very low levels in the tissues, as might occur during vigorous exercise, much more O_2 will dissociate from hemoglobin to be used by the tissue cells without any increase in respiratory rate or cardiac output.

Influence of Other Factors on Hemoglobin Saturation Temperature, blood pH, P_{CO_2}, and the amount of BPG in the blood all influence hemoglobin saturation at a given P_{O_2}. BPG (2,3-bisphosphoglycerate), which binds reversibly with hemoglobin, is produced by red blood cells (RBCs) as they break down glucose by the anaerobic process called glycolysis.

All of these factors influence Hb saturation by modifying hemoglobin's three-dimensional structure, and thereby changing its affinity for O_2. Generally speaking, an *increase* in

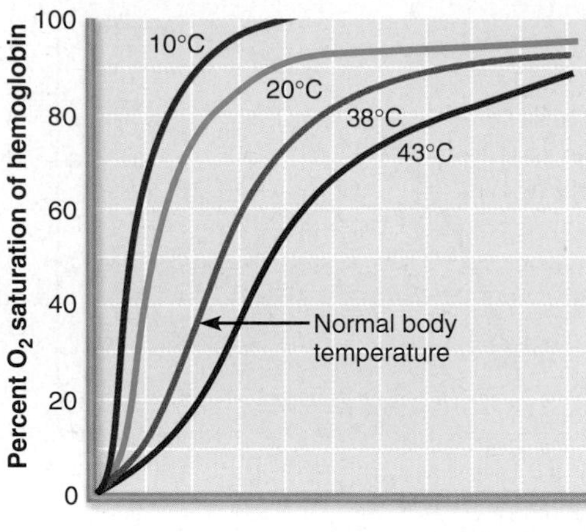

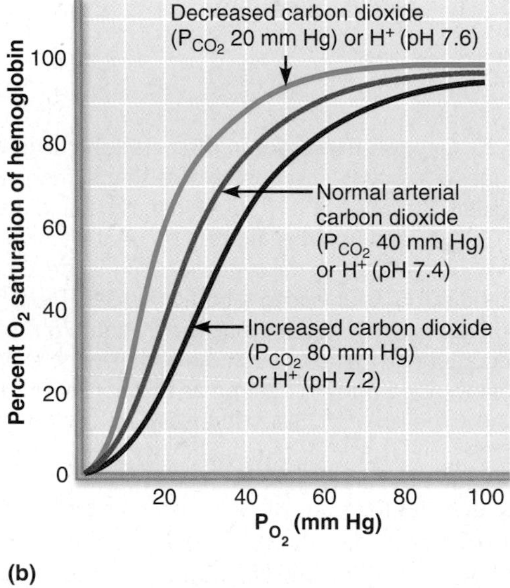

Figure 22.21 Effect of temperature, P_{CO_2}, and blood pH on the oxygen-hemoglobin dissociation curve. Oxygen unloading is enhanced at conditions of **(a)** increased temperature, **(b)** increased P_{CO_2}, and/or hydrogen ion concentration (decreased pH), causing the dissociation curve to shift to the right. This response is called the Bohr effect.

22

temperature, P_{CO_2}, H^+, or BPG levels in blood decreases Hb's affinity for O_2, enhancing oxygen unloading from the blood. This is shown by the rightward shift of the oxygen-hemoglobin dissociation curve in **Figure 22.21**. (The purple lines represent normal body conditions, and the red lines show the shift to the right.)

Conversely, a *decrease* in any of these factors increases hemoglobin's affinity for oxygen, decreasing oxygen unloading. This change shifts the dissociation curve to the left (as the blue lines show in Figure 22.21).

If you give a little thought to how these factors might be related, you'll realize that they all tend to be at their highest levels in the systemic capillaries where oxygen unloading is the goal. As cells metabolize glucose and use O_2, they release CO_2, which increases the P_{CO_2} and H^+ levels in the capillary blood. Both declining blood pH (acidosis) and increasing P_{CO_2} weaken the Hb-O_2 bond, a phenomenon called the **Bohr effect**. As a result, oxygen unloading is enhanced where it is most needed.

Additionally, heat is a by-product of metabolic activity, and active tissues are usually warmer than less active ones. A rise in temperature affects hemoglobin's affinity for O_2 both directly and indirectly (via its influence on RBC metabolism and BPG synthesis). Collectively, these factors see to it that Hb unloads much more O_2 in the vicinity of hard-working tissue cells.

HOMEOSTATIC IMBALANCE

Whatever the cause, inadequate oxygen delivery to body tissues is called **hypoxia** (hi-pok′se-ah). This condition is observed more easily in fair-skinned people because their skin and mucosae take on a bluish cast (become *cyanotic*) when Hb saturation falls below 75%. In dark-skinned individuals, this color change can be observed only in the mucosae and nail beds.

Classification of hypoxia based on cause is as follows:

- **Anemic hypoxia** reflects poor O_2 delivery resulting from too few RBCs or from RBCs that contain abnormal or too little Hb.
- **Ischemic (stagnant) hypoxia** results when blood circulation is impaired or blocked. Congestive heart failure may cause bodywide ischemic hypoxia, whereas emboli or thrombi block oxygen delivery only to tissues distal to the obstruction.
- **Histotoxic hypoxia** occurs when body cells are unable to use O_2 even though adequate amounts are delivered. This variety of hypoxia is the consequence of metabolic poisons, such as cyanide.
- **Hypoxemic hypoxia** is indicated by reduced arterial P_{O_2}. Possible causes include disordered or abnormal ventilation-perfusion coupling, pulmonary diseases that impair ventilation, and breathing air containing scant amounts of O_2.

Carbon monoxide poisoning is a unique type of hypoxemic hypoxia, and a leading cause of death from fire. Carbon monoxide (CO) is an odorless, colorless gas that competes vigorously with O_2 for heme binding sites. Moreover, because Hb's affinity for CO is more than 200 times greater than its affinity for oxygen, CO is a highly successful competitor. Even at minuscule partial pressures, carbon monoxide can displace oxygen.

CO poisoning is particularly dangerous because it does not produce the characteristic signs of hypoxia—cyanosis and respiratory distress. Instead the victim is confused and has a throbbing headache. In rare cases, fair skin becomes cherry red (the color of the Hb-CO complex), which the eye of the beholder interprets as a healthy "blush." Patients with CO poisoning are given hyperbaric therapy (if available) or 100% O_2 until the CO has been cleared from the body. ■

Carbon Dioxide Transport

▶ Describe carbon dioxide transport in the blood.

Normally active body cells produce about 200 ml of CO_2 each minute—exactly the amount excreted by the lungs. Blood

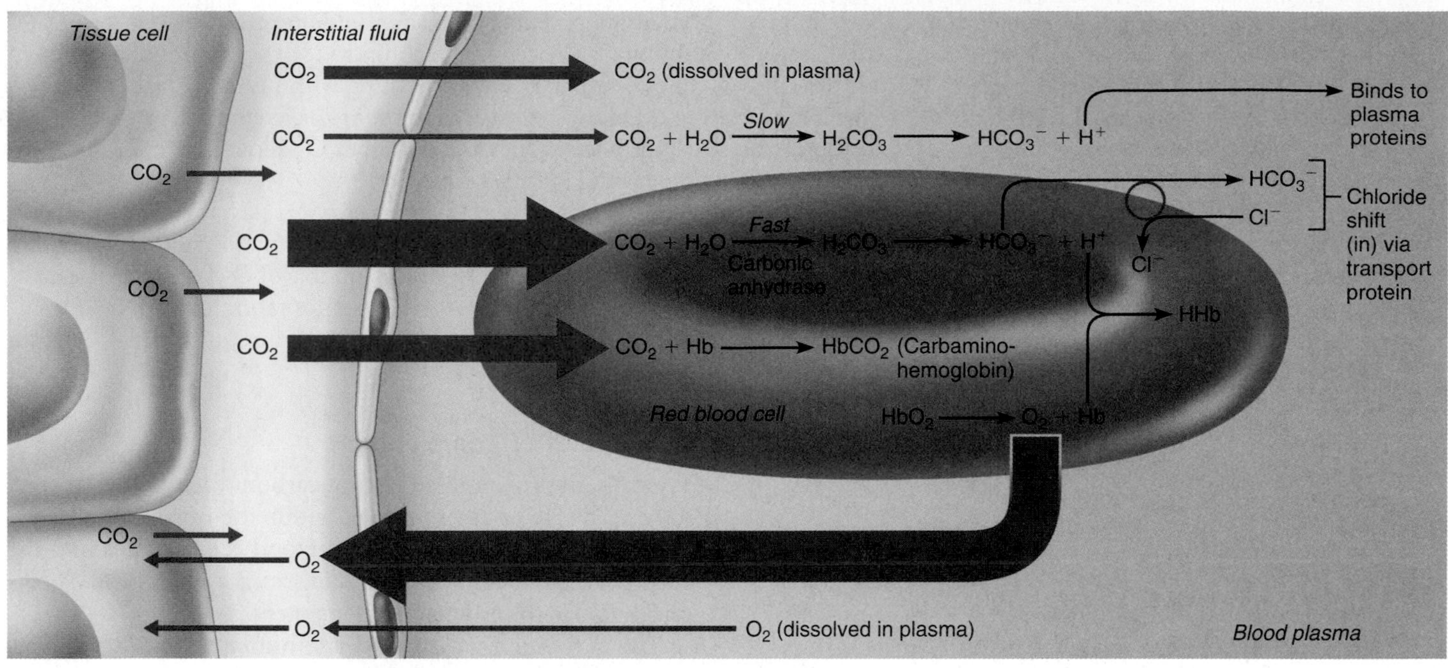

(a) Oxygen release and carbon dioxide pickup at the tissues

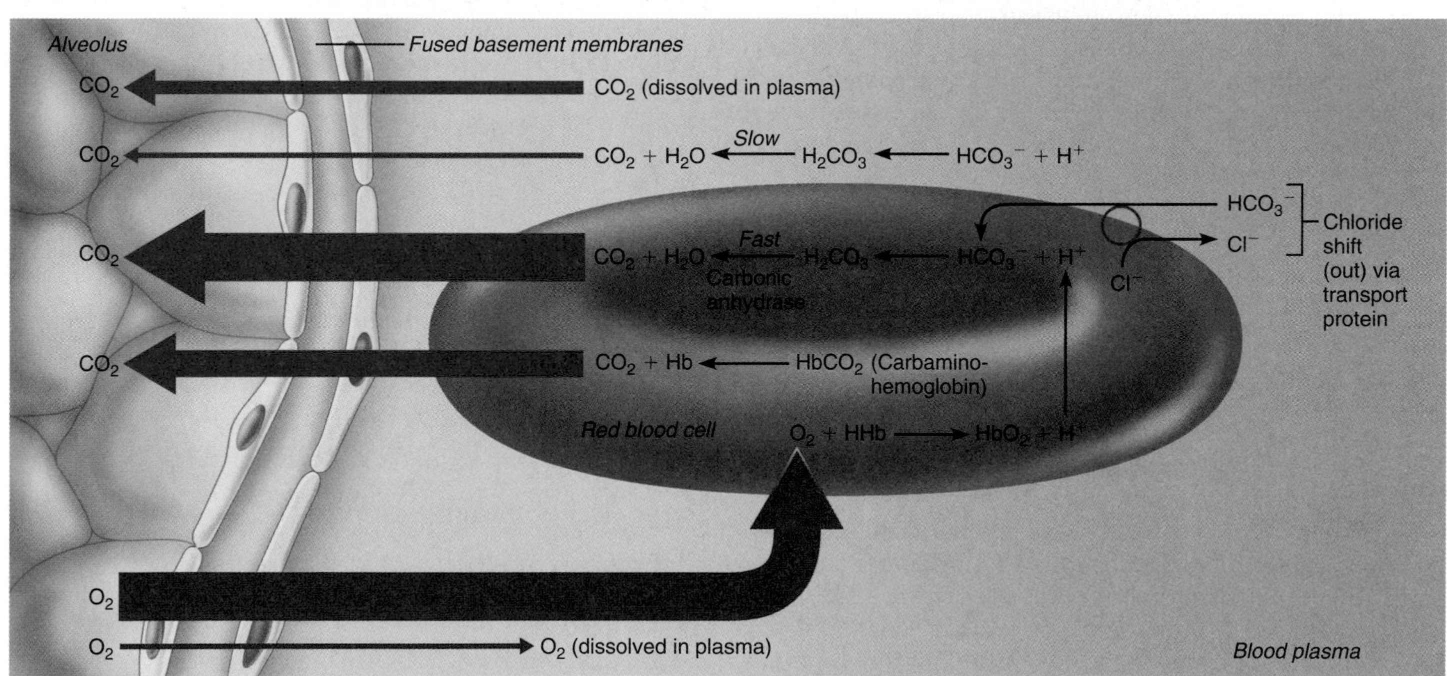

(b) Oxygen pickup and carbon dioxide release in the lungs

Figure 22.22 Transport and exchange of CO_2 and O_2. Relative sizes of the transport arrows indicate the proportionate amounts of O_2 and CO_2 moved by each method. (Reduced hemoglobin is HHb.)

transports CO_2 from the tissue cells to the lungs in three forms **(Figure 22.22)**:

1. **Dissolved in plasma** (7–10%). The smallest amount of CO_2 is transported simply dissolved in plasma.
2. **Chemically bound to hemoglobin** (just over 20%). In this form, dissolved CO_2 is bound and carried in the RBCs as **carbaminohemoglobin** (kar-bam″ĭ-no-he″mo-glo′bin):

$$CO_2 + Hb \rightleftharpoons \underset{\text{carbaminohemoglobin}}{HbCO_2}$$

This reaction is rapid and does not require a catalyst. Carbon dioxide transport in RBCs does not compete with oxyhemoglobin transport because carbon dioxide binds directly to the amino acids of *globin* (and not to the heme).

CO_2 loading and unloading to and from Hb are directly influenced by the P_{CO_2} and the degree of Hb oxygenation. Carbon dioxide rapidly dissociates from hemoglobin in the lungs, where the P_{CO_2} of alveolar air is lower than that in blood. Carbon dioxide is loaded in the tissues, where the P_{CO_2} is higher than that in the blood. Deoxygenated hemoglobin combines more readily with carbon dioxide than does oxygenated hemoglobin, as we will see in the discussion of the Haldane effect below.

3. **As bicarbonate ion in plasma** (about 70%). Most carbon dioxide molecules entering the plasma quickly enter the RBCs, where most of the reactions that prepare carbon dioxide for transport as **bicarbonate ions (HCO_3^-)** in plasma occur. As illustrated in Figure 22.22a, when dissolved CO_2 diffuses into the RBCs, it combines with water, forming carbonic acid (H_2CO_3). H_2CO_3 is unstable and dissociates into hydrogen ions and bicarbonate ions:

$$\underset{\text{carbon dioxide}}{CO_2} + \underset{\text{water}}{H_2O} \rightleftharpoons \underset{\text{carbonic acid}}{H_2CO_3} \rightleftharpoons \underset{\text{hydrogen ion}}{H^+} + \underset{\text{bicarbonate ion}}{HCO_3^-}$$

Although this reaction also occurs in plasma, it is thousands of times faster in RBCs because they (and not plasma) contain **carbonic anhydrase** (kar-bon′ik an-hi′drās), an enzyme that reversibly catalyzes the conversion of carbon dioxide and water to carbonic acid. Hydrogen ions released during the reaction (as well as CO_2 itself) bind to Hb, triggering the Bohr effect. In this way CO_2 loading enhances O_2 release. Because of the buffering effect of Hb, the liberated H^+ causes little change in pH under resting conditions. As a result, blood becomes only slightly more acidic (the pH declines from 7.4 to 7.34) as it passes through the tissues.

Once generated, HCO_3^- moves quickly from the RBCs into the plasma, where it is carried to the lungs. To counterbalance the rapid outrush of these anions from the RBCs, chloride ions (Cl^-) move from the plasma into the RBCs. This ion exchange process, called the **chloride shift**, occurs via facilitated diffusion through a RBC membrane protein.

In the lungs, the process is reversed (Figure 22.22b). As blood moves through the pulmonary capillaries, its P_{CO_2} declines from 45 mm Hg to 40 mm Hg. For this to occur, CO_2 must first be freed from its "bicarbonate housing." HCO_3^- reenters the RBCs (and Cl^- moves into the plasma) and binds with H^+ to form carbonic acid, which is then split by carbonic anhydrase to release CO_2 and water. This CO_2, along with that released from hemoglobin and from solution in plasma, then diffuses along its partial pressure gradient from the blood into the alveoli.

The Haldane Effect

The amount of carbon dioxide transported in blood is markedly affected by the degree of oxygenation of the blood. The lower the P_{O_2} and the lower the extent of Hb saturation with oxygen, the more CO_2 that can be carried in the blood. This phenomenon, called the **Haldane effect**, reflects the greater ability of reduced hemoglobin to form carbaminohemoglobin and to buffer H^+ by combining with it. As CO_2 enters the systemic bloodstream, it causes more oxygen to dissociate from Hb (Bohr effect). The dissociation of O_2 allows more CO_2 to combine with Hb (Haldane effect).

The Haldane effect encourages CO_2 exchange in both the tissues and lungs. In the pulmonary circulation, the situation that we just described is reversed—uptake of O_2 facilitates release of CO_2 (Figure 22.22b). As Hb becomes saturated with O_2, the H^+ released combines with HCO_3^-, helping to unload CO_2 from the pulmonary blood.

Influence of CO_2 on Blood pH

Typically, the H^+ released during carbonic acid dissociation is buffered by Hb or other proteins within the RBCs or in plasma. The HCO_3^- generated in the red blood cells diffuses into the plasma, where it acts as the *alkaline reserve* part of the blood's carbonic acid–bicarbonate buffer system.

The **carbonic acid–bicarbonate buffer system** is very important in resisting shifts in blood pH, as shown in the equation in point 3 concerning CO_2 transport. For example, if the hydrogen ion concentration in blood begins to rise, excess H^+ is removed by combining with HCO_3^- to form carbonic acid (a weak acid). If H^+ concentration drops below desirable levels in blood, carbonic acid dissociates, releasing hydrogen ions and lowering the pH again.

Changes in respiratory rate or depth can produce dramatic changes in blood pH by altering the amount of carbonic acid in the blood. Slow, shallow breathing allows CO_2 to accumulate in the blood. As a result, carbonic acid levels increase and blood pH drops. Conversely, rapid, deep breathing quickly flushes CO_2 out of the blood, reducing carbonic acid levels and increasing blood pH. In this way, respiratory ventilation can provide a fast-acting system to adjust blood pH (and P_{CO_2}) when it is disturbed by metabolic factors. Respiratory adjustments play a major role in the acid-base balance of the blood, as we will discuss in Chapter 26.

CHECK YOUR UNDERSTANDING

16. Rapidly metabolizing tissues generate large amounts of CO_2 and H^+. How does this affect O_2 unloading? What is this effect called?
17. List the three ways CO_2 is transported in blood and state approximate percentages of each.
18. What is the relationship between CO_2 and pH in the blood? Explain.

For answers, see Appendix G.

Control of Respiration

Although our tidelike breathing seems so beautifully simple, its control is more complex than you might think. Higher brain centers, chemoreceptors, and other reflexes all modify the basic respiratory rhythms generated in the brain stem.

Neural Mechanisms

▶ Describe the neural controls of respiration.

The control of respiration primarily involves neurons in the reticular formation of the medulla and pons. Because the medulla sets the respiratory rhythm, we will begin there.

Medullary Respiratory Centers

Clustered neurons in two areas of the medulla oblongata appear to be critically important in respiration. These are (1) the **dorsal respiratory group (DRG)**, located dorsally near the root of cranial nerve IX, and (2) the **ventral respiratory group (VRG)**, a network of neurons that extends in the ventral brain stem from the spinal cord to the pons-medulla junction **(Figure 22.23)**.

The VRG appears to be a rhythm-generating and integrative center. It contains groups of neurons that fire during inspiration and others that fire during expiration in a dance of mutual inhibition. When its inspiratory neurons fire, a burst of impulses travels along the **phrenic** and **intercostal nerves** to excite the diaphragm and external intercostal muscles, respectively (Figure 22.23). As a result, the thorax expands and air rushes into the lungs. When the VRG's expiratory neurons fire, the output stops, and expiration occurs passively as the inspiratory muscles relax and the lungs recoil.

This cyclic on/off activity of the inspiratory and expiratory neurons repeats continuously and produces a respiratory rate of 12–15 breaths per minute, with inspiratory phases lasting about 2 seconds followed by expiratory phases lasting about 3 seconds. This normal respiratory rate and rhythm is referred to as **eupnea** (ūp-ne′ah; *eu* = good, *pne* = breath). During severe hypoxia, VRG networks generate gasping (perhaps in a last-ditch effort to restore O_2 to the brain). Respiration stops completely when a certain cluster of VRG neurons is completely suppressed, as by an overdose of morphine or alcohol.

Until recently, it was thought that the DRG acts as an inspiratory center, performing many of the tasks now known to be performed by the VRG. We now know that in almost all mammals including humans, the DRG integrates input from peripheral stretch and chemoreceptors (which we will describe shortly) and communicates this information to the VRG. It may seem surprising, but many of the details of this system so essential to life are still being worked out.

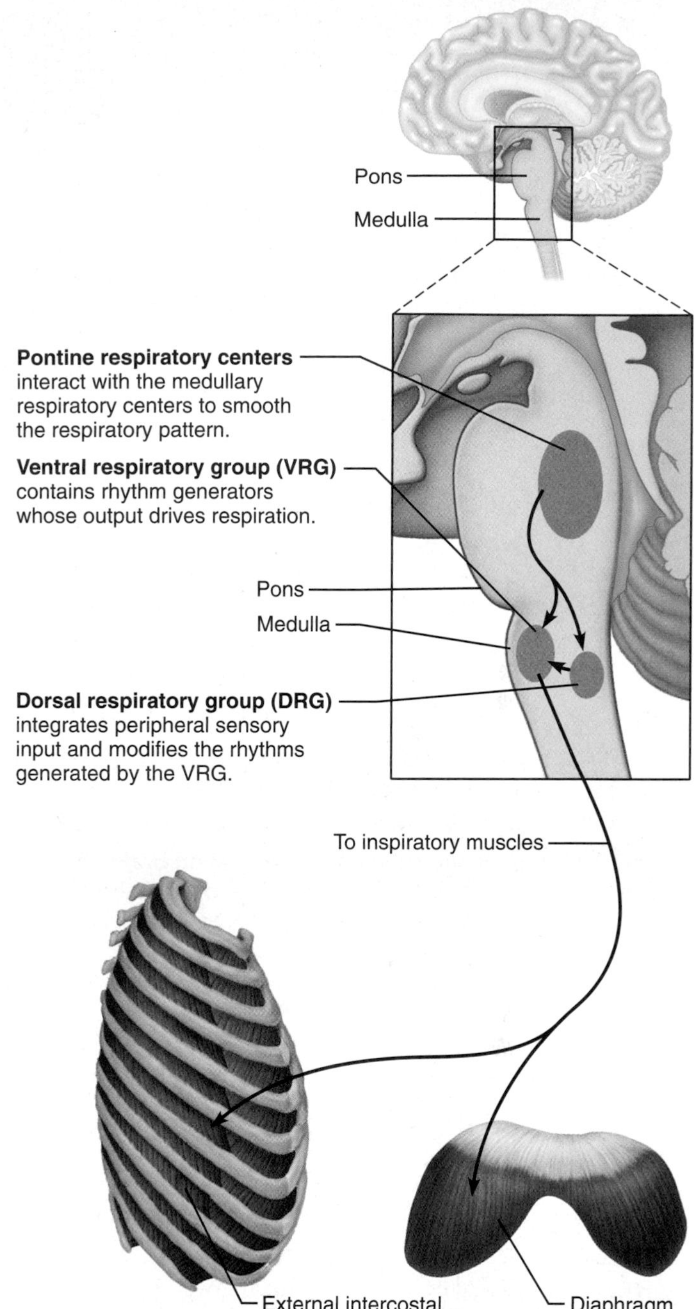
Pons
Medulla
Pontine respiratory centers interact with the medullary respiratory centers to smooth the respiratory pattern.
Ventral respiratory group (VRG) contains rhythm generators whose output drives respiration.
Pons
Medulla
Dorsal respiratory group (DRG) integrates peripheral sensory input and modifies the rhythms generated by the VRG.
To inspiratory muscles
External intercostal muscles
Diaphragm

Figure 22.23 Locations of respiratory centers and their postulated connections. The efferent pathway shown here is incomplete. Medullary neurons communicate with lower motor neurons in the spinal cord, but these are omitted for simplicity.

Pontine Respiratory Centers

Although the VRG generates the basic respiratory rhythm, the pontine respiratory centers influence and modify the activity of medullary neurons. For example, pontine centers appear to smooth out the transitions from inspiration to expiration, and vice versa. When lesions are made in its superior region, inspirations become very prolonged, a phenomenon called *apneustic breathing.*

The **pontine respiratory group**, formerly called the *pneumotaxic center* (noo″mo-tak′sik), and other pontine centers transmit impulses to the VRG of the medulla (Figure 22.23). This input modifies and fine-tunes the breathing rhythms generated by the VRG during certain activities such as vocalization, sleep, and exercise. As you would expect from these functions, the pontine respiratory centers, like the DRG, receive input from higher brain centers and from various sensory receptors in the periphery.

Genesis of the Respiratory Rhythm

There is little question that breathing is rhythmic, but we still cannot fully explain the origin of its rhythm. One hypothesis is that there are *pacemaker neurons*, which have intrinsic (automatic)

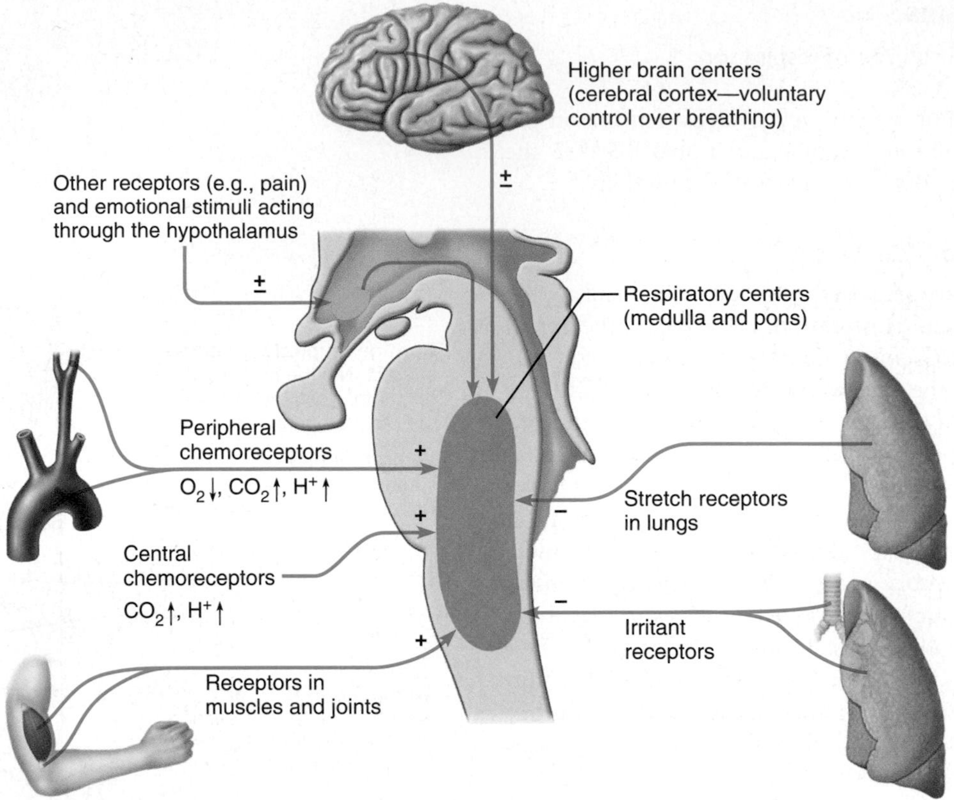

Figure 22.24 Neural and chemical influences on brain stem respiratory centers. Excitatory influences (+) increase the frequency of impulses sent to the muscles of respiration and recruit additional motor units, resulting in deeper, faster breathing. Inhibitory influences (−) have the reverse effect. In some cases, the impulses may be excitatory or inhibitory (±), depending on which receptors or brain regions are activated. The cerebral cortex also directly innervates respiratory muscle motor neurons (not shown).

22

rhythmicity like the pacemaker cells found in the heart. Pacemaker-like activity has been demonstrated in certain VRG neurons, but suppressing their activity does not abolish breathing. This leads us to the second (and more widely accepted) hypothesis: Normal respiratory rhythm is a result of reciprocal inhibition of interconnected neuronal networks in the medulla. Rather than a single set of pacemaker neurons, there are two sets that inhibit each other and so cycle their activity to generate the rhythm.

Factors Influencing Breathing Rate and Depth

- Compare and contrast the influences of arterial pH, arterial partial pressures of oxygen and carbon dioxide, lung reflexes, volition, and emotions on respiratory rate and depth.

Inspiratory depth is determined by how actively the respiratory center stimulates the motor neurons serving the respiratory muscles. The greater the stimulation, the greater the number of motor units excited and the greater the force of respiratory muscle contractions. Respiratory rate is determined by how long the inspiratory center is active or how quickly it is switched off.

Depth and rate of breathing can be modified in response to changing body demands. The respiratory centers in the medulla and pons are sensitive to both excitatory and inhibitory stimuli. We describe these stimuli next and they are summarized in **Figure 22.24**.

Chemical Factors

Among the factors that influence breathing rate and depth, the most important are changing levels of CO_2, O_2, and H^+ in arterial blood. Sensors responding to such chemical fluctuations, called **chemoreceptors**, are found in two major body locations. The **central chemoreceptors** are located throughout the brain stem, including the ventrolateral medulla. The **peripheral chemoreceptors** are found in the aortic arch and carotid arteries.

Influence of P_{CO_2} Of all the chemicals influencing respiration, CO_2 is the most potent and the most closely controlled. Normally, arterial P_{CO_2} is 40 mm Hg and is maintained within ±3 mm Hg of this level by an exquisitely sensitive homeostatic mechanism that is mediated mainly by the effect that rising CO_2 levels have on the central chemoreceptors of the brain stem **(Figure 22.25)**.

As P_{CO_2} levels rise in the blood, a condition referred to as **hypercapnia** (hi″per-kap′ne-ah), CO_2 accumulates in the brain. As CO_2 accumulates, it is hydrated to form carbonic acid. The acid dissociates, H^+ is liberated, and the pH drops. This is the same reaction that occurs when CO_2 enters RBCs (see p. 834). The increase in H^+ excites the central chemoreceptors, which make abundant synapses with the respiratory regulatory centers. As a result, the depth and rate of breathing are increased. This enhanced alveolar ventilation quickly flushes CO_2 out of the blood, increasing blood pH.

An elevation of only 5 mm Hg in arterial P_{CO_2} results in a doubling of alveolar ventilation, even when arterial O_2 levels and pH are unchanged. When P_{O_2} and pH are below normal, the response to elevated P_{CO_2} is even greater. Increased ventilation is normally self-limiting, ending when homeostatic blood P_{CO_2} levels are restored.

Notice that while rising blood CO_2 levels act as the initial stimulus, it is rising levels of H^+ generated within the brain that prod the *central* chemoreceptors into increased activity. (CO_2 readily diffuses across the blood-brain barrier between the brain and the blood, but H^+ does not.) In the final analysis, control of breathing during rest is aimed primarily at *regulating the H^+ concentration in the brain.*

HOMEOSTATIC IMBALANCE

Hyperventilation is an increase in the rate and depth of breathing that exceeds the body's need to remove CO_2. A person experiencing an anxiety attack may hyperventilate involuntarily to the point where he or she becomes dizzy or faints. This happens because low CO_2 levels in the blood (**hypocapnia**) cause cerebral blood vessels to constrict, reducing brain perfusion and producing cerebral ischemia. Earlier symptoms of hyperventilation are tingling and involuntary muscle spasms (tetany) in the hands and face caused by blood Ca^{2+} levels falling as pH rises. Such attacks may be averted by breathing into a paper bag because then the air being inspired is expired air, rich in carbon dioxide, which causes carbon dioxide to be retained in the blood. ■

When P_{CO_2} is abnormally low, respiration is inhibited and becomes slow and shallow. In fact, periods of **apnea** (breathing cessation) may occur until arterial P_{CO_2} rises and again stimulates respiration.

Sometimes swimmers voluntarily hyperventilate so that they can hold their breath longer during swim meets. This is incredibly dangerous for the following reasons. Blood O_2 content rarely drops much below 60% of normal during regular breath-holding, because as P_{O_2} drops, P_{CO_2} rises enough to make breathing unavoidable. However, strenuous hyperventilation can lower P_{CO_2} so much that a lag period occurs before it rebounds enough to stimulate respiration again. This lag may allow oxygen levels to fall well below 50 mm Hg, causing the swimmer to black out (and perhaps drown) before he or she has the urge to breathe.

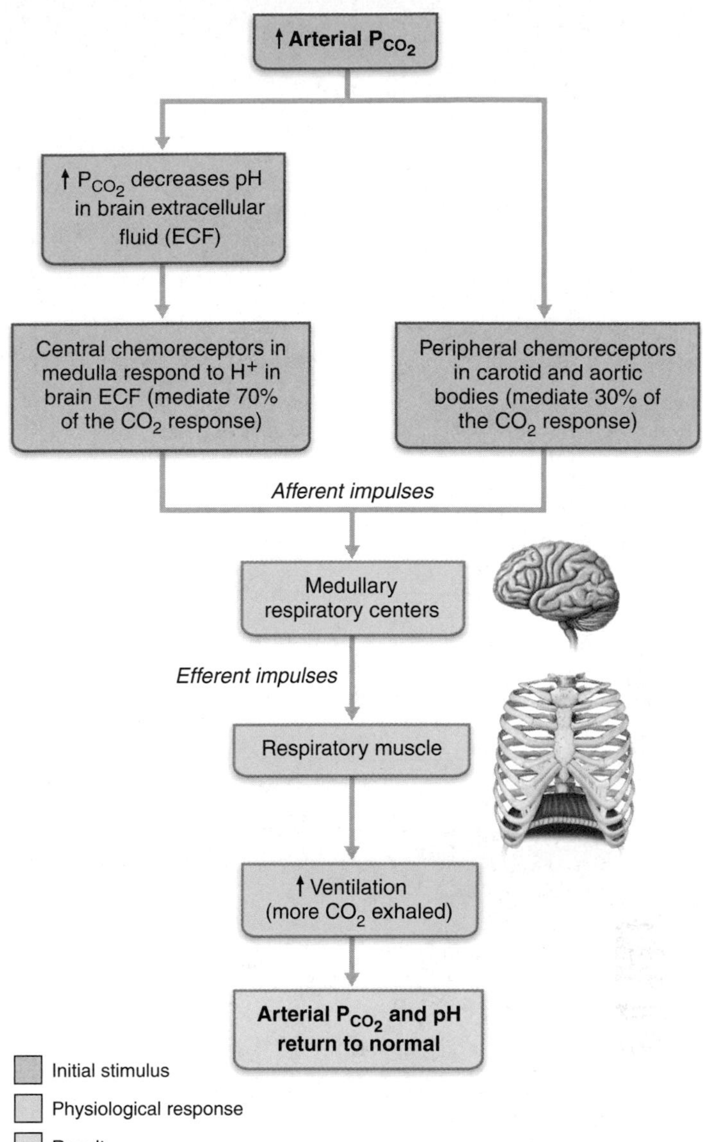

Figure 22.25 Negative feedback mechanism by which changes in P_{CO_2} and blood pH regulate ventilation.

Influence of P_{O_2} Cells sensitive to arterial O_2 levels are found in the peripheral chemoreceptors, that is, in the **aortic bodies** of the aortic arch and in the **carotid bodies** at the bifurcation of the common carotid arteries (**Figure 22.26**). Those in the carotid bodies are the main oxygen sensors.

Under normal conditions, the effect of declining P_{O_2} on ventilation is slight and mostly limited to enhancing the sensitivity of peripheral receptors to increased P_{CO_2}. Arterial P_{O_2} must drop *substantially*, to at least 60 mm Hg, before O_2 levels become a major stimulus for increased ventilation. This is not as strange as it may appear. Remember, there is a huge reservoir of O_2 bound to Hb, and Hb remains almost entirely saturated unless or until the P_{O_2} of alveolar gas and arterial blood falls below 60 mm Hg. The brain stem centers then begin to suffer from O_2 starvation, and their activity is depressed. At the same time, the peripheral chemoreceptors become excited and stimulate the respiratory centers to increase ventilation, even if P_{CO_2} is normal.

22

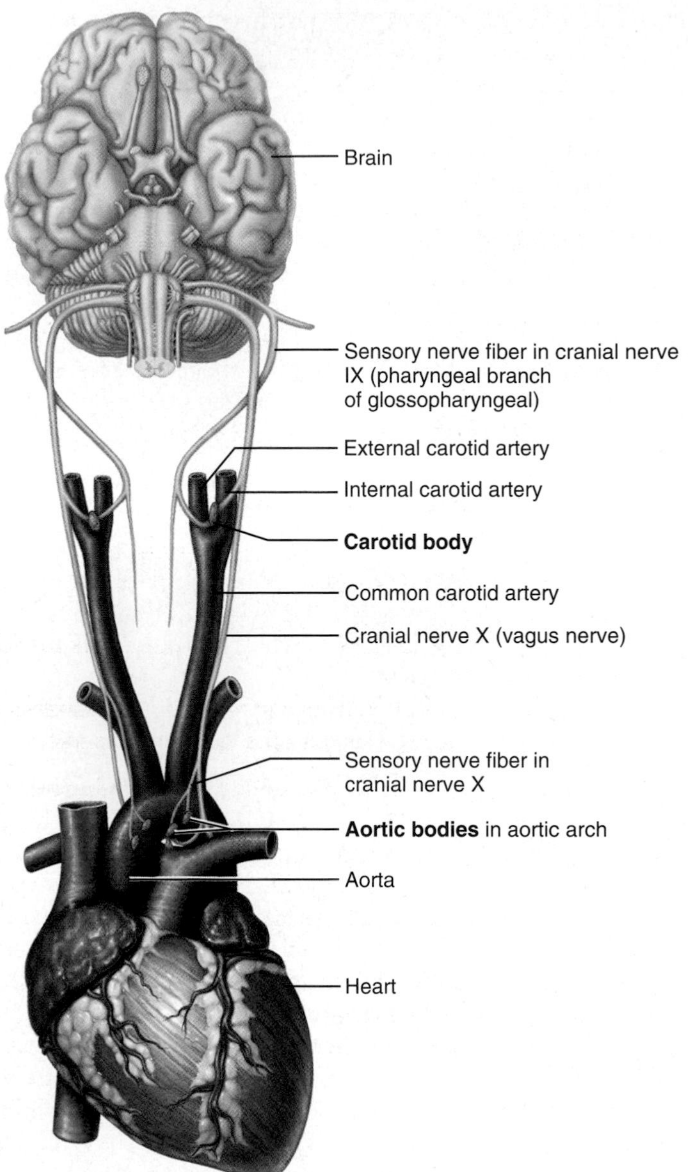

Figure 22.26 Location and innervation of the peripheral chemoreceptors in the carotid and aortic bodies.

In this way, the peripheral chemoreceptor system can maintain ventilation when alveolar O_2 levels are low even though brain stem centers are depressed by hypoxia.

Influence of Arterial pH Changes in arterial pH can modify respiratory rate and rhythm even when CO_2 and O_2 levels are normal. Because little H^+ diffuses from the blood into the brain, the direct effect of arterial H^+ concentration on central chemoreceptors is insignificant compared to the effect of H^+ generated by elevations in P_{CO_2}. The increased ventilation that occurs in response to falling arterial pH is mediated through the peripheral chemoreceptors.

Although changes in P_{CO_2} and H^+ concentration are interrelated, they are distinct stimuli. A drop in blood pH may reflect CO_2 retention, but it may also result from metabolic causes, such as accumulation of lactic acid during exercise or of fatty acid metabolites (ketone bodies) in patients with poorly controlled diabetes mellitus. Regardless of cause, as arterial pH declines, respiratory system controls attempt to compensate and raise the pH by eliminating CO_2 (and carbonic acid) from the blood by increasing respiratory rate and depth.

Summary of Interactions of P_{CO_2}, P_{O_2}, and Arterial pH Although every cell in the body must have O_2 to live, the body's need to rid itself of CO_2 is the most important stimulus for breathing in a healthy person. However, CO_2 does not act in isolation, and various chemical factors enforce or inhibit one another's effects. These interactions are summarized here:

1. *Rising CO_2 levels are the most powerful respiratory stimulant.* As CO_2 is hydrated in brain tissue, liberated H^+ acts directly on the central chemoreceptors, causing a reflexive increase in breathing rate and depth. Low P_{CO_2} levels depress respiration.
2. *Under normal conditions, blood P_{O_2} affects breathing only indirectly by influencing peripheral chemoreceptor sensitivity to changes in P_{CO_2}.* Low P_{O_2} augments P_{CO_2} effects, and high P_{O_2} levels diminish the effectiveness of CO_2 stimulation.
3. *When arterial P_{O_2} falls below 60 mm Hg, it becomes the major stimulus for respiration, and ventilation is increased via reflexes initiated by the peripheral chemoreceptors.* This may increase O_2 loading into the blood, but it also causes hypocapnia (low P_{CO_2} blood levels) and an increase in blood pH, both of which inhibit respiration.
4. *Changes in arterial pH resulting from CO_2 retention or metabolic factors act indirectly through the peripheral chemoreceptors to promote changes in ventilation, which in turn modify arterial P_{CO_2} and pH.* Arterial pH does not influence the central chemoreceptors directly.

Influence of Higher Brain Centers

Hypothalamic Controls Acting through the hypothalamus and the rest of the limbic system, strong emotions and pain send signals to the respiratory centers, modifying respiratory rate and depth. For example, have you ever touched something cold and clammy and gasped? That response was mediated through the hypothalamus. So too is the breath holding that occurs when we are angry and the increased respiratory rate that occurs when we are excited. A rise in body temperature acts to increase the respiratory rate, while a drop in body temperature produces the opposite effect. Sudden chilling of the body (a dip in the North Atlantic Ocean in late October) can cause cessation of breathing (apnea)—or at the very least, leave you gasping.

Cortical Controls Although the brain stem respiratory centers normally regulate breathing involuntarily, we can also exert conscious (volitional) control over the rate and depth of our breathing. We can choose to hold our breath or to take an extra-deep breath, for example. During voluntary control, the cerebral motor cortex sends signals to the motor neurons that stimulate the respiratory muscles, bypassing the medullary centers.

Our ability to voluntarily hold our breath is limited, however, because the brain stem respiratory centers automatically reinitiate breathing when the blood concentration of CO_2 reaches critical levels. That explains why drowning victims typically have water in their lungs.

Pulmonary Irritant Reflexes

The lungs contain receptors that respond to an enormous variety of irritants. When activated, these receptors communicate with the respiratory centers via vagal nerve afferents. Accumulated mucus, inhaled debris such as dust, or noxious fumes stimulate receptors in the bronchioles that promote reflex constriction of those air passages. The same irritants stimulate a cough when present in the trachea or bronchi, and a sneeze when present in the nasal cavity.

The Inflation Reflex

The visceral pleurae and conducting passages in the lungs contain numerous stretch receptors that are vigorously stimulated when the lungs are inflated. These receptors signal the medullary respiratory centers via afferent fibers of the vagus nerves, sending inhibitory impulses that end inspiration and allow expiration to occur. As the lungs recoil, the stretch receptors become quiet, and inspiration is initiated once again. This reflex, called the **inflation reflex**, or **Hering-Breuer reflex** (her′ing broy′er), is thought to be more a protective response (to prevent excessive stretching of the lungs) than a normal regulatory mechanism.

CHECK YOUR UNDERSTANDING

19. Which brain stem respiratory area is thought to generate the respiratory rhythm?
20. Which chemical factor in blood normally provides the most powerful stimulus to breathe? Which chemoreceptors are most important for this response?

For answers, see Appendix G.

Respiratory Adjustments

- Compare and contrast the hyperpnea of exercise with hyperventilation.
- Describe the process and effects of acclimatization to high altitude.

Exercise

Respiratory adjustments during exercise are geared both to intensity and duration of the exercise. Working muscles consume tremendous amounts of O_2 and produce large amounts of CO_2, and so ventilation can increase 10- to 20-fold during vigorous exercise. This increase in ventilation in response to metabolic needs is called **hyperpnea** (hi″perp-ne′ah).

How does it differ from hyperventilation? The respiratory changes in hyperpnea do not lead to significant changes in blood O_2 and CO_2 levels. By contrast, hyperventilation is excessive ventilation, and is characterized by low P_{CO_2} and alkalosis.

Exercise-enhanced ventilation does *not* appear to be prompted by rising P_{CO_2} and declining P_{O_2} and pH in the blood for two reasons. First, ventilation increases abruptly as exercise begins, followed by a gradual increase, and then a steady state of ventilation. When exercise stops, there is an initial small but abrupt decline in ventilation rate, followed by a gradual decrease to the pre-exercise value. Second, although venous levels change, arterial P_{CO_2} and P_{O_2} levels remain surprisingly constant during exercise. In fact, P_{CO_2} may decline to below normal and P_{O_2} may rise slightly because of the efficiency of the respiratory adjustments. Our present understanding of the mechanisms that produce these observations is sketchy, but the most accepted explanation is as follows.

The abrupt increase in ventilation that occurs as exercise begins reflects interaction of three neural factors:

1. Psychological stimuli (our conscious anticipation of exercise)
2. Simultaneous cortical motor activation of skeletal muscles and respiratory centers
3. Excitatory impulses reaching respiratory centers from proprioceptors in moving muscles, tendons, and joints

The subsequent gradual increase and then plateauing of respiration probably reflect the rate of CO_2 delivery to the lungs (the "CO_2 flow"). The small but abrupt decrease in ventilation that occurs as exercise ends reflects the shutting off of the three neural factors listed above. The subsequent gradual decline to baseline ventilation likely reflects a decline in the CO_2 flow that occurs as the oxygen deficit is being repaid. The rise in lactic acid levels that contributes to O_2 deficit is *not* a result of inadequate respiratory function, because alveolar ventilation and pulmonary perfusion are as well matched during exercise as during rest (hemoglobin remains fully saturated). Rather, it reflects cardiac output limitations or inability of the skeletal muscles to further increase their oxygen consumption.

In light of this fact, the practice of inhaling pure O_2 by mask, used by some football players to replenish their "oxygen-starved" bodies as quickly as possible, is useless. The panting athlete *does* have an O_2 deficit, but inspiring extra oxygen will not help, because the shortage is in the muscles—not the lungs.

High Altitude

Most people live between sea level and an altitude of approximately 2400 m (8000 feet). In this range, differences in atmospheric pressure are not great enough to cause healthy people any problems when they spend brief periods in the higher-altitude areas. However, when you travel quickly from sea level to elevations above 8000 ft, where atmospheric pressure and P_{O_2} are lower, your body responds with symptoms of *acute mountain sickness (AMS)*—headaches, shortness of breath, nausea, and dizziness. AMS is common in travelers to ski resorts such as Vail, Colorado (8120 ft), and Brian Head, Utah (a heart-pounding 9600 ft). In severe cases of AMS, lethal pulmonary and cerebral edema may occur.

lungs but can spread through the lymphatics to affect other organs. One-third of the world's population is infected, but most people never develop active TB because a massive inflammatory and immune response usually contains the primary infection in fibrous, or calcified, nodules (tubercles) in the lungs. However, the bacteria survive in the nodules and when the person's immunity is weakened, they may break out and cause symptomatic TB, involving fever, night sweats, weight loss, a racking cough, and coughing up blood.

Until the 1930s, TB was responsible for one-third of all deaths among 20- to 45-year-old adults in the U.S. With the advent of antibiotics in the 1940s, this killer was put into retreat and its prevalence declined dramatically. During the mid-1980s, there was an alarming increase in TB due to TB-infected HIV patients. This trend has now been brought under control and TB rates in the U.S. are again decreasing.

Of potential concern are deadly strains of drug-resistant (even multidrug-resistant) TB that develop when treatment is incomplete or inadequate. Such strains are found elsewhere in the world and have appeared in North America. Ideal breeding grounds for such strains are shelters for the homeless with their densely packed populations. The TB bacterium grows very slowly and drug therapy entails a 12-month course of antibiotics. The transient nature of shelter populations makes it difficult to track TB patients and ensure they take their medications for the full 12 months. The threat of TB epidemics is so real that health centers in some cities are detaining such patients in sanatoriums against their will for as long as it takes to complete a cure.

Lung Cancer

Lung cancer is the leading cause of cancer death for both men and women in North America, causing more deaths than breast, prostate, and colorectal cancer combined. This is tragic, because lung cancer is largely preventable—nearly 90% of lung cancers are the result of smoking. The cure rate for lung cancer is notoriously low, with most victims dying within one year of diagnosis. The overall five-year survival of those with lung cancer is about 17%. Because lung cancer is aggressive and metastasizes rapidly and widely, most cases are not diagnosed until they are well advanced.

Lung cancer appears to follow closely the oncogene-activating steps outlined in *A Closer Look* in Chapter 4. Ordinarily, nasal hairs, sticky mucus, and cilia do a fine job of protecting the lungs from chemical and biological irritants, but when a person smokes, these defenses are overwhelmed and eventually stop functioning. In particular, smoking paralyzes the cilia that clear mucus from the airways, allowing irritants and pathogens to accumulate. The "cocktail" of free radicals and other carcinogens in tobacco smoke eventually translates into lung cancer.

The three most common types of lung cancer are (1) **squamous cell carcinoma** (25–30% of cases), which arises in the epithelium of the bronchi or their larger subdivisions and tends to form masses that may cavitate (hollow out) and bleed, (2) **adenocarcinoma** (about 40%), which originates in peripheral lung areas as solitary nodules that develop from bronchial glands and alveolar cells, and (3) **small cell carcinoma** (about 20%), which contains round lymphocyte-sized cells that originate in the main bronchi and grow aggressively in small grapelike clusters within the mediastinum. Subsequent metastasis from the mediastinum is especially rapid. Some small cell carcinomas cause problems beyond their effects on the lungs because they become ectopic sites of hormone production. For example, some secrete ACTH (leading to Cushing's syndrome) or ADH (which results in the syndrome of inappropriate ADH secretion, or SIADH; see p. 608).

Complete removal of the diseased lung has the greatest potential for prolonging life and providing a cure. However, this choice is open to few lung cancer patients because the cancer has often metastasized before it is discovered. In most cases, radiation therapy and chemotherapy are the only options, but these have low success rates.

Fortunately, there are several new therapies on the horizon. These include (1) antibodies that target specific growth factors or other molecules required by the tumor or that deliver toxic agents directly to the tumor, (2) cancer vaccines to stimulate the immune system to fight the tumor, and (3) various forms of gene therapy to replace the defective genes that make tumor cells divide continuously. As clinical trials progress, we will learn which of these approaches is most effective.

Developmental Aspects of the Respiratory System

- Trace the embryonic development of the respiratory system.
- Describe normal changes that occur in the respiratory system from infancy to old age.

Because embryos develop in a cephalocaudal (head-to-tail) direction, the upper respiratory structures appear first. By the fourth week of development, two thickened plates of ectoderm, the **olfactory placodes** (plak′ōds), are present on the anterior aspect of the head (**Figure 22.28**). These quickly invaginate to form **olfactory pits** that form the nasal cavities. The olfactory pits then extend posteriorly to connect with the developing pharynx, which forms at the same time from the endodermal germ layer.

The epithelium of the lower respiratory organs develops as an outpocketing of the foregut endoderm, which becomes the pharyngeal mucosa. This protrusion, called the **laryngotracheal bud**, is present by the fifth week of development. The proximal part of the bud forms the tracheal lining, and its distal end splits and forms the mucosae of the bronchi and all their subdivisions, including (eventually) the lung alveoli. Mesoderm covers these endoderm-derived linings and forms the walls of the respiratory passageways and the stroma of the lungs.

By 28 weeks, the respiratory system has developed sufficiently to allow a baby born prematurely to breathe on its own. As we noted earlier, infants born before this time tend to exhibit infant respiratory distress syndrome resulting from inadequate surfactant production.

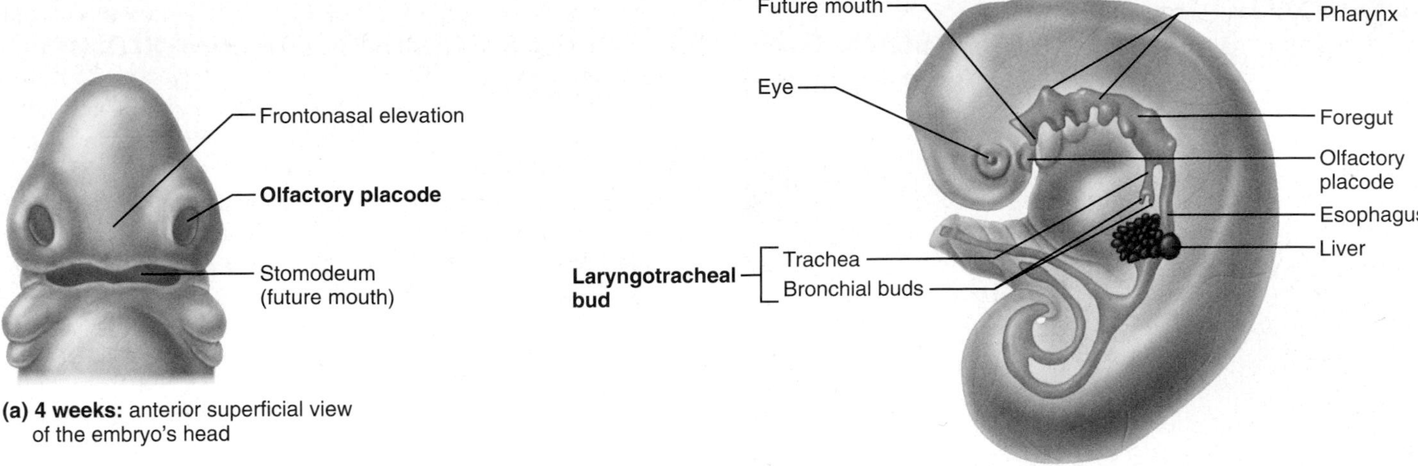

(a) 4 weeks: anterior superficial view of the embryo's head

(b) 5 weeks: left lateral view of the developing lower respiratory passageway mucosae

Figure 22.28 Embryonic development of the respiratory system.

During fetal life, the lungs are filled with fluid and all respiratory exchanges are made by the placenta. Vascular shunts cause circulating blood to largely bypass the lungs (see Chapter 28). At birth, the fluid-filled pathway empties, and the respiratory passageways fill with air. As the P_{CO_2} in the baby's blood rises, the respiratory centers are excited, causing the baby to take its first breath. The alveoli inflate and begin to function in gas exchange, but it is nearly two weeks before the lungs are fully inflated.

HOMEOSTATIC IMBALANCE

Important birth defects of the respiratory system include *cleft palate* (described in Chapter 7) and *cystic fibrosis.* **Cystic fibrosis (CF)**, the most common lethal genetic disease in North America, strikes in one out of every 2400 births. CF causes secretion of abnormally viscous mucus that clogs the respiratory passages, providing a breeding ground for airborne bacteria that predisposes the child to respiratory infections. It appears that infection of CF victims' lungs with the bacterium *Pseudomonas aeruginosa* trips a genetic switch that causes the disabled cells to churn out oceans of abnormal mucin (the primary component of mucus). The bacteria then feed on the stagnant pools of mucus and keep sending signals to the cells to make more. Toxins released by the bacteria and the local inflammatory reaction set up by the immune response both damage the lungs. Unable to reach the bacteria embedded in the mucus, the immune cells begin to attack the lung tissue, turning the air sacs into bloated cysts.

Repeated cycles of infection, chronic inflammation, and resulting tissue damage eventually result in such extensive damage that it can be treated only by a lung transplant. The disease also impairs food digestion by clogging ducts that deliver pancreatic enzymes and bile to the small intestine, and sweat glands of CF patients produce an extremely salty perspiration.

At the root of CF is a faulty gene that codes for the *CFTR* (cystic fibrosis transmembrane conductance regulator) *protein.* The normal CFTR protein works as a membrane channel to control Cl^- flow in and out of cells. In those with the mutated gene, CFTR lacks an essential amino acid and so it gets "stuck" in the endoplasmic reticulum, unable to reach the membrane and perform its normal role. Consequently, less Cl^- is secreted and less water follows, resulting in the thick mucus typical of CF.

Conventional therapy for CF has been mucus-dissolving drugs, "clapping" the chest to loosen the thick mucus, and antibiotics to prevent infection. The basic goal of CF research is to restore normal salt and water movements by (1) introducing normal CFTR genes into respiratory tract mucosa cells, (2) prodding another channel protein to take over the duties of transporting Cl^-, and (3) developing techniques to free the CFTR protein from the ER. Suppressing the inflammatory response in the lungs is another goal. Among anti-inflammatory agents being tested is a fatty acid found in fish oils, docosahexaenoic acid (DHA). A novel and surprisingly simple approach involves inhaling hypertonic saline droplets. This draws water into the mucus, making it more liquid. Alone or in combination, these therapies provide new hope to patients with CF. ■

The respiratory rate is highest in newborn infants (40–80 respirations per minute). At five years of age it is around 25 per minute, and in adults it is between 12 and 18 per minute. In old age, the rate often increases again. At birth, only about one-sixth of the final number of alveoli are present. The lungs continue to mature and more alveoli are formed until young adulthood. However, if a person begins smoking in the early teens, the lungs never completely mature, and those additional alveoli are lost forever.

In infants, the ribs take a nearly horizontal course. For this reason, infants rely almost entirely on descent of the diaphragm to increase thoracic volume for inspiration. By the second year, the ribs are positioned more obliquely, and the adult form of breathing is established.

System Connections

Homeostatic Interrelationships Between the Respiratory System and Other Body Systems

Integumentary System

- Respiratory system provides oxygen and disposes of carbon dioxide
- Skin protects respiratory system organs by forming surface barriers

Skeletal System

- Respiratory system provides oxygen and disposes of carbon dioxide
- Bones protect lungs and bronchi by enclosure

Muscular System

- Respiratory system provides oxygen needed for muscle activity and disposes of carbon dioxide
- Activity of the diaphragm and intercostal muscles essential for producing volume changes that lead to pulmonary ventilation; regular exercise increases respiratory efficiency

Nervous System

- Respiratory system provides oxygen needed for normal neuronal activity and disposes of carbon dioxide
- Medullary and pontine centers regulate respiratory rate and depth; stretch receptors in lungs and chemoreceptors provide feedback

Endocrine System

- Respiratory system provides oxygen and disposes of carbon dioxide; angiotensin converting enzyme in lungs converts angiotensin I to angiotensin II
- Epinephrine dilates the bronchioles; testosterone promotes laryngeal enlargement in pubertal males; glucocorticoids promote surfactant production

Cardiovascular System

- Respiratory system provides oxygen and disposes of carbon dioxide; carbon dioxide present in blood as HCO_3^- and H_2CO_3 contributes to blood buffering
- Blood is the transport medium for respiratory gases

Lymphatic System/Immunity

- Respiratory system provides oxygen and disposes of carbon dioxide; tonsils in pharynx house immune cells
- Lymphatic system helps to maintain blood volume required for respiratory gas transport; immune system protects respiratory organs from bacteria, bacterial toxins, viruses, protozoa, fungi, and cancer

Digestive System

- Respiratory system provides oxygen and disposes of carbon dioxide
- Digestive system provides nutrients needed by respiratory system organs

Urinary System

- Respiratory system provides oxygen and disposes of carbon dioxide to provide short-term pH homeostasis
- Kidneys dispose of metabolic wastes of respiratory system organs (other than carbon dioxide) and maintain long-term pH homeostasis

Reproductive System

- Respiratory system provides oxygen and disposes of carbon dioxide

Closer Connections

The Respiratory System and Interrelationships with the Cardiovascular, Lymphatic/Immune, and Muscular Systems

Every day we inhale and exhale nearly 10,000 L of air. This accomplishes two things—it supplies the body with the oxygen it needs to oxidize food and release energy, and it expels carbon dioxide, the major waste product of that process. As crucial as breathing is, most of us don't think very often about the importance of having fresh air on call. But the phenomenon of gasping for air is familiar to every athlete. Indeed, the respiratory rate of a competitive swimmer may jump to over 40 breaths/min and the amount of air inhaled per breath may soar from the usual 500 ml to as much as 6 to 7 L.

The respiratory system is beautifully engineered for its function. Its alveoli are flushed with new air more than 15,000 times each day and the alveolar walls are so indescribably thin that red blood cells moving single-file through the pulmonary capillaries can make the gaseous "swaps" with the air-filled alveoli within a fraction of a second. Although every cell in the body depends on this system for the oxygen it needs to live, the respiratory system interactions that we will consider here are those it has with the cardiovascular, lymphatic/immune, and muscular systems.

Cardiovascular System

The interaction between the respiratory and cardiovascular systems is so intimate that these two systems are inseparable. Respiratory system organs, as important as they are, can make only the external gas exchanges, those that occur in the lungs. Although all body cells depend on the respiratory system to provide needed oxygen, they pick up that oxygen not from the lungs but from the blood. Thus, without blood to act as the intermediary and the heart and blood vessels to act as the hardware to pump the blood around the body, all the efforts of the respiratory system would be useless.

In turn, angiotensin converting enzyme on lung capillary endothelium plays an important role in blood pressure regulation.

Lymphatic System/Immunity

Of all body systems, only the respiratory system is completely exposed to the external environment (yes, the skin is exposed but its exposed parts are all dead). Because air contains a rich mix of potentially dangerous inhabitants (bacteria, viruses, fungi, asbestos fibers, etc.), the respiratory system is continuously at risk for infection or damage from external agents. Lymphoid outposts help protect the respiratory tract and enhance the defenses (cilia, mucus) that the respiratory system itself erects. Particularly well situated to apprehend intruders at the oral-nasal-pharynx junction are the palatine, pharyngeal, lingual, and tubal tonsils. Their macrophages engulf foreign antigens and they provide sites where lymphocytes are sensitized to mount immune responses.

Muscular System

Skeletal muscle cells, like other body cells, need oxygen to live. The notable part of this interaction is that most respiratory compensations that occur service increased muscular activity. (It is difficult to think of incidents where this is not the case—the exceptions concern disease conditions.) When we are at rest, the respiratory system operates at basal levels, but any time physical activity becomes more vigorous, the respiratory rhythm picks up the beat to match supply with need and to maintain the acid-base balance of the blood.

Respiratory System

Case study: Barbara Joley was in the bus that was hit broadside. When she was freed from the wreckage, she was deeply cyanotic and her respiration had stopped. Her heart was still beating, but her pulse was fast and thready. The emergency medical technician reported that when Barbara was found, her head was cocked at a peculiar angle and it looked like she had a fracture at the level of the C_2 vertebra. The following questions refer to these observations.

1. How might the "peculiar" head position explain Barbara's cessation of breathing?
2. What procedures (do you think) should have been initiated immediately by the emergency personnel?
3. Why is Barbara cyanotic? Explain cyanosis.
4. Assuming that Barbara survives, how will her accident affect her lifestyle in the future?

Barbara survived transport to the hospital and notes recorded at admission included the following observations.

- Right thorax compressed; ribs 7 to 9 fractured
- Right lung atelectasis

Relative to these notes:

5. What is atelectasis and why is only the right lung affected?
6. How do the recorded injuries relate to the atelectasis?
7. What treatment will be done to reverse the atelectasis? What is the rationale for this treatment?

(Answers in Appendix G)

HOMEOSTATIC IMBALANCE

Most respiratory system problems are the result of external factors—for example, viral or bacterial infections or obstruction of the trachea by a piece of food. For many years, bacterial pneumonia was one of the worst killers in North America. Antibiotics have greatly decreased its lethality, but it is still a dangerous disease, particularly in the elderly. By far the most problematic diseases *at present* are those described earlier: COPD, asthma, lung cancer, and multidrug-resistant tuberculosis. ■

The maximum amount of oxygen we can use during aerobic metabolism, $\dot{V}_{O_2max}$, declines about 9% per decade in inactive people beginning in their mid-20s. In those that remain active, $\dot{V}_{O_2max}$ still declines but much less. As we age, the thoracic wall becomes more rigid and the lungs gradually lose their elasticity, resulting in a decreasing ability to ventilate the lungs. Vital capacity declines by about one-third by the age of 70. Blood O_2 levels decline slightly, and many old people tend to become hypoxic during sleep and exhibit *sleep apnea* (temporary cessation of breathing during sleep).

The number of glands in the nasal mucosa decreases as does blood flow to this mucosa. For this reason, the nose dries and produces a thick mucus that makes us want to clear our throat. Additionally, many of the respiratory system's protective mechanisms become less effective with age. Ciliary activity of the mucosa decreases, and the macrophages in the lungs become sluggish. The net result is that the elderly are more at risk for respiratory tract infections, particularly pneumonia and influenza.

CHECK YOUR UNDERSTANDING

23. What distinguishes the obstruction in asthma from that in chronic bronchitis?
24. What is the underlying defect in cystic fibrosis?
25. List two reasons for the decline in vital capacity seen with age.

For answers, see Appendix G.

Lungs, bronchial tree, heart, and connecting blood vessels—together, these organs fashion a remarkable system that ensures that blood is oxygenated and relieved of carbon dioxide and that all tissue cells have access to these services. Although the cooperation of the respiratory and cardiovascular systems is obvious, all organ systems of the body depend on the functioning of the respiratory system, as summarized in *Making Connections* on pp. 844–845.

RELATED CLINICAL TERMS

22

Adult respiratory distress syndrome (ARDS) A dangerous lung condition that can develop after severe illness or injury to the body. Neutrophils leave the body's capillaries in large numbers and then secrete chemicals that increase capillary permeability. The capillary-rich lungs are heavily affected. As the lungs fill with the fluids of edema, the patient suffocates. Even with mechanical ventilation, ARDS is hard to control and often lethal.

Adenoidectomy (adenotonsillectomy) Surgical removal of an infected pharyngeal tonsil (adenoids).

Aspiration (as″pĭ-ra′shun) (1) The act of inhaling or drawing something into the lungs or respiratory passages. Pathological aspiration in which vomit or excessive mucus is drawn into the lungs may occur when a person is unconscious or anesthetized; turning the head to one side is preventive. (2) Withdrawal of fluid by suction (use of an aspirator); done during surgery to keep an area free of blood or other body fluids; mucus is aspirated from the trachea of tracheotomy patients.

Bronchoscopy (*scopy* = viewing) Use of a viewing tube inserted through the nose or mouth to examine the internal surface of the main bronchi in the lung. Forceps attached to the tip of the tube can remove trapped objects or take samples of mucus for examination.

Cheyne-Stokes breathing (chān′stōks) Abnormal breathing pattern sometimes seen just before death (the "death rattle") and in people with combined neurological and cardiac disorders. It consists of bursts of tidal volume breaths (increasing and then decreasing in depth) alternating with periods of apnea. Trauma and hypoxia of the brain stem centers, as well as P_{CO_2} imbalances between arterial blood and brain, may be causative factors.

Deviated septum Condition in which the nasal septum takes a more lateral course than usual and may obstruct breathing; often manifests in old age or from nose trauma.

Endotracheal tube A thin plastic tube threaded into the trachea through the nose or mouth; used to deliver oxygen to patients who are breathing inadequately, in a coma, or under anesthesia.

Epistaxis (ep″ĭ-stak′sis; *epistazo* = to bleed at the nose) Nosebleed; commonly follows trauma to the nose or excessive nose blowing. Most nasal bleeding is from the highly vascularized anterior septum and can be stopped by pinching the nostrils closed or packing them with cotton.

Nasal polyps Mushroomlike benign neoplasms of the nasal mucosa; sometimes caused by infections, but most often cause is unknown; may block air flow.

Orthopnea (or″thop-ne′ah; *ortho* = straight, upright) Inability to breathe in the horizontal (lying down) position.

Otorhinolaryngology (o″to-ri″no-lar″in-gol′o-je; *oto* = ear; *rhino* = nose) Branch of medicine that deals with diagnosis and treatment of diseases of the ears, nose, and throat.

Pneumonia Infectious inflammation of the lungs, in which fluid accumulates in the alveoli; the eighth most common cause of death in the United States. Most of the more than 50 different varieties of pneumonia are viral or bacterial.

Pulmonary embolism Obstruction of the pulmonary artery or one of its branches by an embolus (most often a blood clot that has been carried from the lower limbs and through the right side of the heart into the pulmonary circulation). Symptoms are chest pain, productive bloody cough, tachycardia, and rapid, shallow breathing. Can cause sudden death unless treated quickly; usual treatment is oxygen by mask, pain relievers, and anticoagulant drugs. For severe cases, clot busters (thrombolytic drugs) are also used.

Stuttering A problem of voice production in which the first syllable of words is repeated in "machine-gun" fashion. Primarily a problem with neural control of the larynx and other voice-producing structures. Many stutterers become fluent when whispering or singing, both of which involve a change in the manner of vocalization.

Sudden infant death syndrome (SIDS) Unexpected death of an apparently healthy infant during sleep. Commonly called crib death, SIDS is one of the most frequent causes of death in infants under 1 year old. Believed to be a problem of immaturity of the respiratory control centers. Most cases occur in infants placed in a prone position (on their abdomen) to sleep—a position that may result in hypoxia and hypercapnia due to rebreathing exhaled (CO_2-rich) air. Since 1992, a campaign recommending placing infants on their back to sleep has led to a decline of 40% or more in the incidence of SIDS in the U.S.

Tracheotomy (tra″ke-ot′o-me) Surgical opening of the trachea; done to provide an alternate route for air to reach the lungs when more superior respiratory passageways are obstructed (as by food or a crushed larynx).

CHAPTER SUMMARY

Media study tools that could provide you additional help in reviewing specific key topics of Chapter 22 are referenced below.

iP = ***Interactive Physiology***

1. Respiration involves four processes: pulmonary ventilation, external respiration, transport of respiratory gases in the blood, and internal respiration. Both the respiratory system and the cardiovascular system are involved in respiration.

Functional Anatomy of the Respiratory System (pp. 805–819)

2. Respiratory system organs are divided functionally into conducting zone structures (nose to bronchioles), which filter, warm, and moisten incoming air; and respiratory zone structures (respiratory bronchioles to alveoli), where gas exchanges occur.

The Nose and Paranasal Sinuses (pp. 806–809)

3. The nose provides an airway for respiration; warms, moistens, and cleanses incoming air; and houses the olfactory receptors.
4. The external nose is shaped by bone and cartilage plates. The nasal cavity, which opens to the exterior, is divided by the nasal septum. Paranasal sinuses and nasolacrimal ducts drain into the nasal cavities.

The Pharynx (p. 809)

5. The pharynx extends from the base of the skull to the level of C_6. The nasopharynx is an air conduit; the oropharynx and laryngopharynx are common passageways for food and air. Tonsils are found in the oropharynx and nasopharynx.

The Larynx (pp. 810–812)

6. The larynx, or voice box, contains the vocal folds (cords). It also provides a patent airway and serves as a switching mechanism to route food and air into the proper channels.
7. The epiglottis prevents food or liquids from entering the respiratory channels during swallowing.

The Trachea (pp. 812–813)

8. The trachea extends from the larynx to the main bronchi. The trachea is reinforced by C-shaped cartilage rings, which keep the trachea patent, and its mucosa is ciliated.

The Bronchi and Subdivisions (pp. 813–815)

9. The right and left main bronchi run into their respective lungs, within which they continue to subdivide into smaller and smaller passageways.
10. The terminal bronchioles lead into respiratory zone structures: respiratory bronchioles, alveolar ducts, alveolar sacs, and finally alveoli. Gas exchange occurs in the alveoli, across the respiratory membrane.
11. As the respiratory conduits become smaller, cartilage is reduced in amount and finally lost; the mucosa thins, and smooth muscle in the walls increases.

iP Respiratory System; Topic: Anatomy Review: Respiratory Structures, p. 6.

The Lungs and Pleurae (pp. 815–819)

12. The lungs, the paired organs of gas exchange, flank the mediastinum in the thoracic cavity. Each is suspended in pleurae via its root and has a base, an apex, and medial and costal surfaces. The right lung has three lobes; the left has two.
13. The lungs are primarily air passageways/chambers, supported by an elastic connective tissue stroma.
14. The pulmonary arteries carry blood returned from the systemic circulation to the lungs, where gas exchange occurs. The pulmonary veins return newly oxygenated (and most bronchial venous) blood back to the heart to be distributed throughout the body. The bronchial arteries provide the nutrient blood supply of the lungs.
15. The parietal pleura lines the thoracic wall and mediastinum; the visceral pleura covers external lung surfaces. Pleural fluid reduces friction during breathing movements.

iP Respiratory System; Topic: Anatomy Review: Respiratory Structures, pp. 1–5.

Mechanics of Breathing (pp. 819–826)

Pressure Relationships in the Thoracic Cavity (pp. 819–820)

1. Intrapulmonary pressure is the pressure within the alveoli. Intrapleural pressure is the pressure within the pleural cavity; normally it is negative relative to intrapulmonary pressures.

iP Respiratory System; Topic: Pulmonary Ventilation, pp. 7–9.

Pulmonary Ventilation (pp. 820–822)

2. Gases travel from an area of higher pressure to an area of lower pressure.
3. Inspiration occurs when the diaphragm and external intercostal muscles contract, increasing the dimensions (and volume) of the thorax. As the intrapulmonary pressure drops, air rushes into the lungs until the intrapulmonary and atmospheric pressures are equalized.

4. Expiration is largely passive, occurring as the inspiratory muscles relax and the lungs recoil. When intrapulmonary pressure exceeds atmospheric pressure, gases flow from the lungs.

iP Respiratory System; Topic: Pulmonary Ventilation, pp. 3–6, 11–13.

Physical Factors Influencing Pulmonary Ventilation (pp. 822–824)

5. Friction in the air passageways causes resistance, which decreases air passage and causes breathing movements to become more strenuous. The greatest resistance to air flow occurs in the midsize bronchi.
6. Surface tension of alveolar fluid acts to reduce alveolar size and collapse the alveoli. This tendency is resisted in part by surfactant.
7. Premature infants have problems keeping their lungs inflated owing to the lack of surfactant in their alveoli, resulting in infant respiratory distress syndrome (IRDS). Surfactant formation begins late in fetal development.
8. Total respiratory compliance depends on elasticity of lung tissue and flexibility of the bony thorax. When compliance is impaired, inspiration becomes more difficult.

iP Respiratory System; Topic: Pulmonary Ventilation, pp. 14–18.

Respiratory Volumes and Pulmonary Function Tests (pp. 824–826)

9. The four respiratory volumes are tidal, inspiratory reserve, expiratory reserve, and residual. The four respiratory capacities are vital, functional residual, inspiratory, and total lung. Respiratory volumes and capacities may be measured by spirometry.
10. Anatomical dead space is the air-filled volume (about 150 ml) of the conducting passageways. If alveoli become nonfunctional in gas exchange, their volume is added to the anatomical dead space, and the sum is the total dead space.
11. Alveolar ventilation is the best index of ventilation efficiency because it accounts for anatomical dead space.

$$\text{AVR} = (\text{TV} - \text{dead space}) \times \text{respiratory rate}$$

12. The FVC and FEV tests, which determine the rate at which VC air can be expelled, are particularly valuable in distinguishing between obstructive and restrictive disease.

Nonrespiratory Air Movements (p. 826)

22

13. Nonrespiratory air movements are voluntary or reflex actions that clear the respiratory passageways or express emotions.

Gas Exchanges Between the Blood, Lungs, and Tissues (pp. 827–830)

Basic Properties of Gases (pp. 827–828)

1. Dalton's law states that each gas in a mixture of gases exerts pressure in proportion to its percentage in the total mixture.
2. Henry's law states that the amount of gas that will dissolve in a liquid is proportional to the partial pressure of the gas. Solubility of the gas in the liquid and the temperature are other important factors.

iP Respiratory System; Topic: Gas Exchange, pp. 1–6.

Composition of Alveolar Gas (p. 828)

3. Alveolar gas contains more carbon dioxide and water vapor and considerably less oxygen than atmospheric air.

External Respiration (pp. 828–830)

4. External respiration is the process of gas exchange that occurs in the lungs. Oxygen enters the pulmonary capillaries; carbon dioxide leaves the blood and enters the alveoli. Factors influencing this process include the partial pressure gradients, the thickness of the respiratory membrane, surface area available, and the matching of alveolar ventilation and pulmonary perfusion.

Internal Respiration: (p. 830)

5. Internal respiration is the gas exchange that occurs between the systemic capillaries and the tissues. Carbon dioxide enters the blood, and oxygen leaves the blood and enters the tissues.

iP Respiratory System; Topic: Gas Exchange, pp. 6–11, 15–16.

Transport of Respiratory Gases by Blood (pp. 830–834)

Oxygen Transport (pp. 830–832)

1. Molecular oxygen is carried bound to hemoglobin in the red blood cells. The amount of oxygen bound to hemoglobin depends on the P_{O_2} and P_{CO_2} of blood, blood pH, the presence of BPG, and temperature. A small amount of oxygen gas is transported dissolved in plasma.
2. Hypoxia occurs when inadequate amounts of oxygen are delivered to body tissues. When this occurs, the skin and mucosae may become cyanotic.

Carbon Dioxide Transport (pp. 832–834)

3. CO_2 is transported in the blood dissolved in plasma, chemically bound to hemoglobin, and (primarily) as bicarbonate ion in plasma. Loading and unloading of O_2 and CO_2 are mutually beneficial.
4. Accumulation of CO_2 leads to decreased pH; depletion of CO_2 from blood leads to increased blood pH.

iP Respiratory System; Topic: Gas Transport, pp. 1–15.

Control of Respiration (pp. 834–839)

Neural Mechanisms (pp. 835–836)

1. Medullary respiratory centers are the dorsal and ventral respiratory groups. The ventral respiratory group is likely responsible for the rhythmicity of breathing.
2. The pontine respiratory centers influence the activity of the medullary respiratory centers.

Factors Influencing Breathing Rate and Depth (pp. 836–839)

3. Important chemical factors modifying baseline respiratory rate and depth are arterial levels of CO_2, H^+, and O_2.
4. An increasing arterial P_{CO_2} level (hypercapnia) is the most powerful respiratory stimulant. It acts (via release of H^+ in brain tissue) on the central chemoreceptors to cause a reflexive increase in the rate and depth of breathing.
5. Hypocapnia depresses respiration and results in decreased ventilation and, possibly, apnea.
6. Arterial P_{O_2} levels below 60 mm Hg strongly stimulate peripheral chemoreceptors.
7. Decreased pH and a decline in blood P_{O_2} act on peripheral chemoreceptors and enhance the response to CO_2.
8. Emotions, pain, body temperature changes, and other stressors can alter respiration by acting through hypothalamic centers. Respiration can also be controlled voluntarily for short periods.
9. Pulmonary irritant reflexes are initiated by dust, mucus, fumes, and pollutants.
10. The inflation (Hering-Breuer) reflex is a protective reflex initiated by extreme overinflation of the lungs; it acts to terminate inspiration.

iP Respiratory System; Topic: Control of Respiration, pp. 6–14.

Respiratory Adjustments (pp. 839–840)

Exercise (p. 839)

1. As exercise begins, there is an abrupt increase in ventilation (hyperpnea) followed by a more gradual increase. When exercise stops, there is an abrupt decrease in ventilation followed by a gradual decline to baseline values.
2. P_{O_2}, P_{CO_2}, and blood pH remain quite constant during exercise and hence do not appear to account for changes in ventilation. Inputs from higher centers and proprioceptors may contribute.

High Altitude (pp. 839–840)

3. At high altitudes, there is a decrease in arterial P_{O_2} and hemoglobin saturation levels because of the decrease in atmospheric pressure compared to sea level. Increased ventilation helps restore P_{O_2} to physiological levels.
4. Long-term acclimatization involves increased erythropoiesis.

Homeostatic Imbalances of the Respiratory System (pp. 840–842)

1. Two major respiratory disorders are COPD (emphysema and chronic bronchitis) and lung cancer; a significant cause is cigarette smoking. A third major disorder is asthma. Multidrug-resistant tuberculosis may become a major health problem.

Chronic Obstructive Pulmonary Disease (pp. 840–841)

2. COPD is characterized by an irreversible decrease in the ability to force air out of the lungs. Patients may become cyanotic as a result of chronic hypoxia.
3. Emphysema is characterized by permanent enlargement and destruction of alveoli. The lungs lose their elasticity, and expiration becomes an active process.
4. Chronic bronchitis is characterized by excessive mucus production in the lower respiratory passageways, which severely impairs ventilation and gas exchange.

Asthma (p. 841)

5. Asthma is a reversible obstructive condition caused by an immune response that causes its victims to wheeze and gasp for air as their inflamed respiratory passages constrict. It is marked by exacerbations and periods of relief from symptoms.

Tuberculosis (pp. 841–842)

6. Tuberculosis (TB), an infectious disease caused by an airborne bacterium, mainly affects the lungs. Although most infected individuals remain asymptomatic by walling off the bacteria in nodules (tubercles), disease symptoms ensue when immunity is depressed. Some patients' failure to complete drug therapy have produced multidrug-resistant TB strains.

Lung Cancer (p. 842)

7. Lung cancer, promoted by free radicals and other carcinogens, is extremely aggressive and metastasizes rapidly.

Developmental Aspects of the Respiratory System (pp. 842–843, 846)

1. The superior respiratory system mucosa develops from the invagination of the ectodermal olfactory placodes; that of the inferior passageways develops from an outpocketing of the endodermal foregut lining. Mesoderm forms the walls of the respiratory conduits and the lung stroma.
2. Cystic fibrosis (CF), the most common fatal hereditary disease in North America, results from an abnormal CFTR protein that fails to form a chloride channel. The result is thick mucus which clogs respiratory passages and invites infection.
3. With age, the thorax becomes more rigid, the lungs become less elastic, and vital capacity declines. In addition, sleep apnea becomes more common, and respiratory system protective mechanisms are less effective.

REVIEW QUESTIONS

Multiple Choice/Matching

(Some questions have more than one correct answer. Select the best answer or answers from the choices given.)

1. Cutting the phrenic nerves will result in **(a)** air entering the pleural cavity, **(b)** paralysis of the diaphragm, **(c)** stimulation of the diaphragmatic reflex, **(d)** paralysis of the epiglottis.
2. Which of the following laryngeal cartilages is/are not paired? **(a)** epiglottis, **(b)** arytenoid, **(c)** cricoid, **(d)** cuneiform, **(e)** corniculate.
3. Under ordinary circumstances, the inflation reflex is initiated by **(a)** noxious chemicals, **(b)** the ventral respiratory group, **(c)** overinflation of the alveoli and bronchioles, **(d)** the pontine respiratory centers.
4. The detergent-like substance that keeps the alveoli from collapsing between breaths because it reduces the surface tension of the water film in the alveoli is called **(a)** lecithin, **(b)** bile, **(c)** surfactant, **(d)** reluctant.
5. Which of the following determines the *direction* of gas movement? **(a)** solubility in water, **(b)** partial pressure gradient, **(c)** temperature, **(d)** molecular weight and size of the gas molecule.
6. When the inspiratory muscles contract, **(a)** the size of the thoracic cavity is increased in diameter, **(b)** the size of the thoracic cavity is increased in length, **(c)** the volume of the thoracic cavity is decreased, **(d)** the size of the thoracic cavity is increased in both length and diameter.
7. The nutrient blood supply of the lungs is provided by **(a)** the pulmonary arteries, **(b)** the aorta, **(c)** the pulmonary veins, **(d)** the bronchial arteries.
8. Oxygen and carbon dioxide are exchanged in the lungs and through all cell membranes by **(a)** active transport, **(b)** diffusion, **(c)** filtration, **(d)** osmosis.
9. Which of the following would not normally be treated by 100% oxygen therapy? (Choose all that apply.) **(a)** anoxia, **(b)** carbon monoxide poisoning, **(c)** respiratory crisis in an emphysema patient, **(d)** eupnea.
10. Most oxygen carried in the blood is **(a)** in solution in the plasma, **(b)** combined with plasma proteins, **(c)** chemically combined with the heme in red blood cells, **(d)** in solution in the red blood cells.
11. Which of the following has the greatest stimulating effect on the respiratory centers in the brain? **(a)** oxygen, **(b)** carbon dioxide, **(c)** calcium, **(d)** willpower.

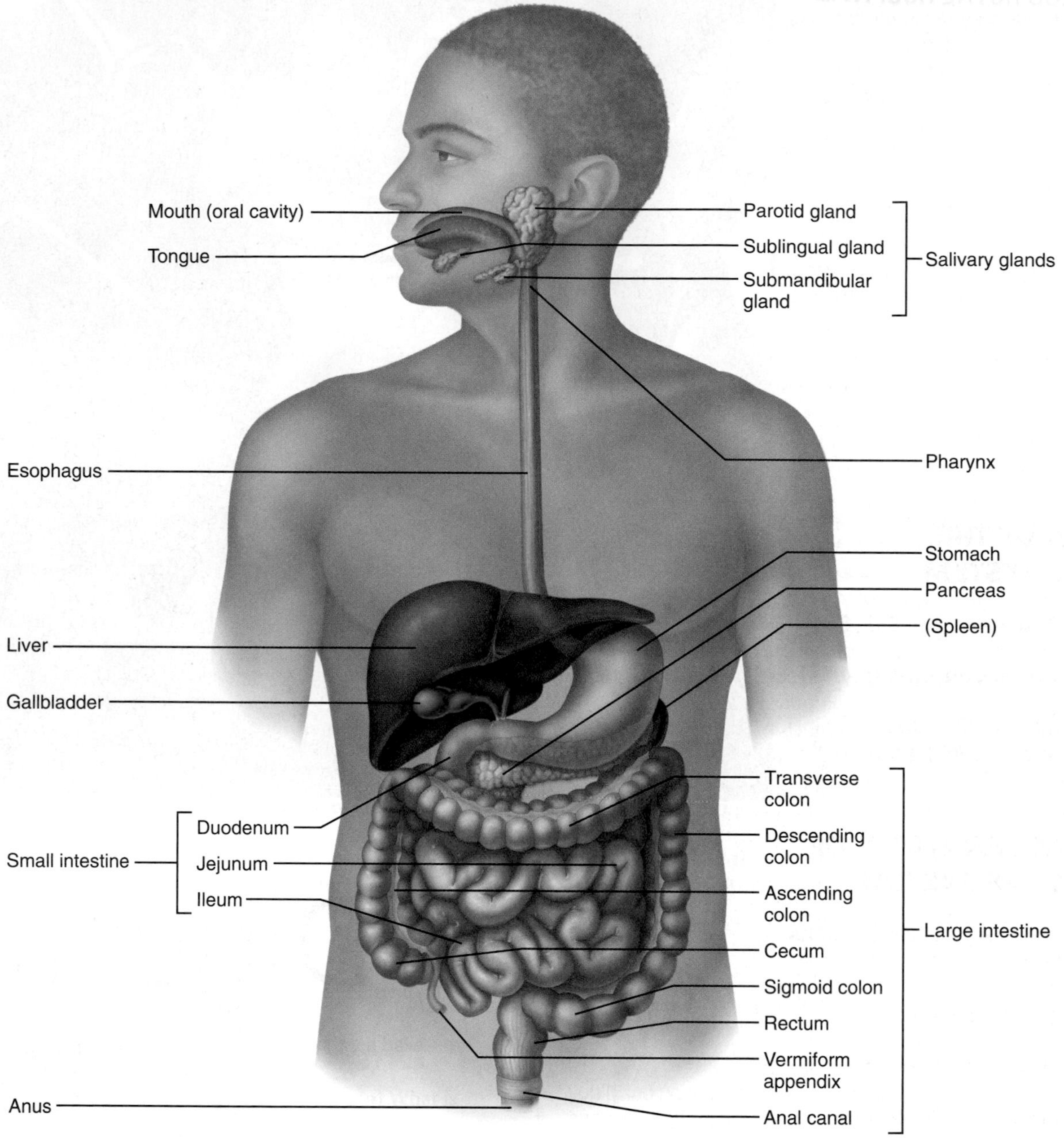

Figure 23.1 Alimentary canal and related accessory digestive organs. (See *A Brief Atlas of the Human Body*, Figure 64a.)

PART 1

OVERVIEW OF THE DIGESTIVE SYSTEM

▶ Describe the function of the digestive system, and differentiate between organs of the alimentary canal and accessory digestive organs.

The organs of the digestive system fall into two main groups: (1) those of the *alimentary canal* (al″ĭ-men′tar-e; *aliment* = nourish) and (2) *accessory digestive organs* **(Figure 23.1)**.

The **alimentary canal**, also called the **gastrointestinal (GI) tract** or **gut**, is the continuous, muscular digestive tube that winds through the body. It **digests** food—breaks it down into smaller fragments (*digest* = dissolved)—and **absorbs** the digested fragments through its lining into the blood. The organs of the alimentary canal are the *mouth, pharynx, esophagus, stomach, small intestine,* and *large intestine.* The large intestine leads to the terminal opening, or *anus.* In a cadaver, the alimentary canal is approximately 9 m (about 30 ft) long, but in a living person, it is considerably shorter because of its muscle tone. Food material in this tube is technically outside the body because the canal is open to the external environment at both ends.

The **accessory digestive organs** are the *teeth, tongue, gallbladder,* and a number of large digestive glands—the *salivary*

Ingestion
Food
Mechanical digestion
Pharynx
• Chewing (mouth)
• Churning (stomach)
• Segmentation (small intestine)
Esophagus
Propulsion
• Swallowing (oropharynx)
• Peristalsis (esophagus, stomach, small intestine, large intestine)
Chemical digestion
Stomach
Absorption
Lymph vessel
Small intestine
Large intestine
Blood vessel
Mainly H_2O
Feces
Anus
Defecation

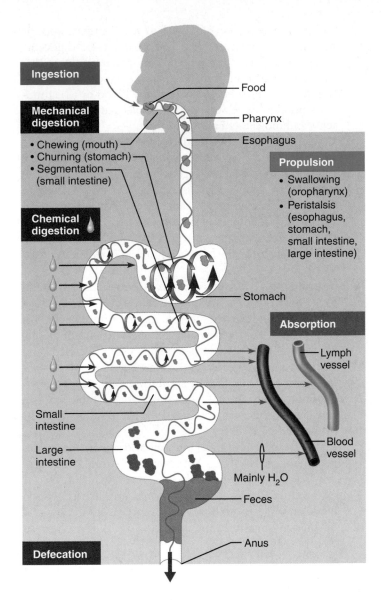

Figure 23.2 Gastrointestinal tract activities. Note that sites of chemical digestion produce enzymes or receive enzymes or other secretions made by accessory organs outside the alimentary canal.

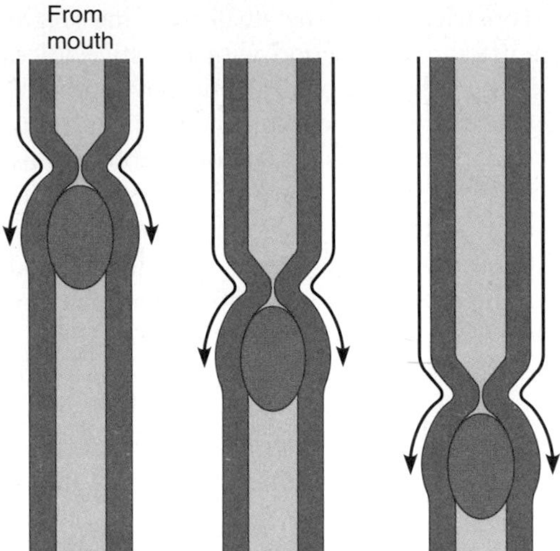

(a) Peristalsis: Adjacent segments of alimentary tract organs alternately contract and relax, which moves food along the tract distally.

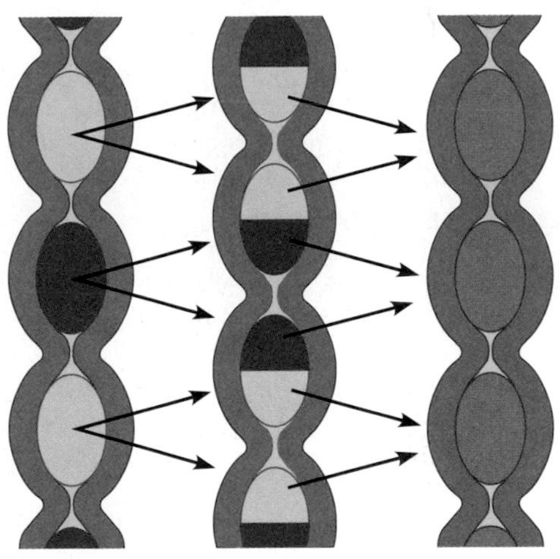

(b) Segmentation: Nonadjacent segments of alimentary tract organs alternately contract and relax, moving the food forward then backward. Food mixing and slow food propulsion occurs.

Figure 23.3 Peristalsis and segmentation.

glands, liver, and *pancreas.* The teeth and tongue are in the mouth, or oral cavity, while the digestive glands and gallbladder lie outside the GI tract and connect to it by ducts. The accessory digestive glands produce a variety of secretions that contribute to the breakdown of foodstuffs.

Digestive Processes

▶ List and define the major processes occurring during digestive system activity.

We can view the digestive tract as a "disassembly line" in which food becomes less complex at each step of processing and its nutrients become available to the body. The processing of food by the digestive system involves six essential activities: ingestion, propulsion, mechanical digestion, chemical digestion, absorption, and defecation **(Figure 23.2)**.

1. **Ingestion** is simply taking food into the digestive tract, usually via the mouth.
2. **Propulsion**, which moves food through the alimentary canal, includes *swallowing*, which is initiated voluntarily, and *peristalsis* (per″ĭ-stal′sis), an involuntary process. **Peristalsis** (*peri* = around; *stalsis* = constriction), the major means of propulsion, involves alternate waves of contraction and relaxation of muscles in the organ walls **(Figure 23.3a)**. Its main effect is to squeeze food along the tract, but some mixing occurs as well. In fact, peristaltic waves are so powerful that, once swallowed, food and fluids will reach your stomach even if you stand on your head.
3. **Mechanical digestion** physically prepares food for chemical digestion by enzymes. Mechanical processes include chewing, mixing of food with saliva by the tongue, churning food in the stomach, and **segmentation**, or rhythmic

local constrictions of the small intestine (Figure 23.3b). Segmentation mixes food with digestive juices and increases the efficiency of absorption by repeatedly moving different parts of the food mass over the intestinal wall.

4. **Chemical digestion** is a series of catabolic steps in which complex food molecules are broken down to their chemical building blocks by enzymes secreted into the lumen of the alimentary canal. Chemical digestion of foodstuffs begins in the mouth and is essentially complete in the small intestine.
5. **Absorption** is the passage of digested end products (plus vitamins, minerals, and water) from the lumen of the GI tract through the mucosal cells by active or passive transport into the blood or lymph. The small intestine is the major absorptive site.
6. **Defecation** eliminates indigestible substances from the body via the anus in the form of feces.

Some of these processes are the job of a single organ. For example, only the mouth ingests and only the large intestine defecates. But most digestive system activities require the cooperation of several organs and occur bit by bit as food moves along the tract. Later, we will consider which of these specific processes each GI tract organ performs and the neural or hormonal factors that regulate these processes.

CHECK YOUR UNDERSTANDING

1. Name one organ of the alimentary canal found in the thorax. Name three organs located in the abdominal cavity.
2. What is the usual site of ingestion?
3. Which essential digestive activity actually moves nutrients from the outside to the inside of the body?

For answers, see Appendix G.

Basic Functional Concepts

▶ Describe stimuli and controls of digestive activity.

A theme we have stressed in this book is the body's efforts to maintain the constancy of its internal environment. Most organ systems respond to changes in that environment either by attempting to restore some plasma variable to its former levels or by changing their own function. The digestive system, however, creates an optimal environment for its functioning in the lumen (cavity) of the GI tract, an area that is actually *outside* the body, and essentially all digestive tract regulatory mechanisms act to control luminal conditions so that digestion and absorption can occur there as effectively as possible.

Two facts apply to these regulatory mechanisms:

1. **Digestive activity is provoked by a range of mechanical and chemical stimuli.** Sensors (mechanoreceptors and chemoreceptors) involved in controls of GI tract activity are located in the walls of the tract organs. These sensors respond to several stimuli. The most important are stretching of the organ by food in the lumen, osmolarity (solute concentration) and pH of the contents, and the presence of substrates and end products of digestion. When stimulated, these receptors initiate reflexes that (1) activate or inhibit glands that secrete digestive juices into the lumen or hormones into the blood or (2) stimulate smooth muscle of the GI tract walls to mix lumen contents and move them along the tract.
2. **Controls of digestive activity are both intrinsic and extrinsic.** Many of the controlling systems of the digestive tract are *intrinsic*—a product of "in-house" nerve plexuses or hormone-producing cells. Between the muscle layers in the wall of the alimentary canal is the so-called **gut brain** consisting of enteric nerve plexuses which spread like chicken wire along the entire length of the GI tract and influence each other both in the same and in different digestive organs.

   As a result, two kinds of reflex activity occur, short and long. *Short reflexes* are mediated entirely by the local *enteric* or "gut" plexuses in response to stimuli arising in the GI tract. *Long reflexes* are initiated by stimuli arising inside or outside the GI tract and involve CNS centers and extrinsic autonomic nerves **(Figure 23.4)**. Generally speaking, nerve fibers that excite smooth muscle secrete acetylcholine or substance P, and those that inhibit smooth muscle release vasoactive intestinal peptide (VIP) or nitric oxide.

   The stomach and small intestine also contain hormone-producing cells. When appropriately stimulated, these cells release their products to the interstitial fluid in the extracellular space. Their hormones are distributed via blood and interstitial fluid to their target cells in the same or different digestive tract organs, which they prod to secrete or contract.

CHECK YOUR UNDERSTANDING

4. When sensors in the GI tract are stimulated, they respond via reflexes. What types of digestive activity may be put into motion via those reflexes?
5. The term "gut brain" does not really mean there is a brain in the digestive system. What does it refer to?

For answers, see Appendix G.

Digestive System Organs: Relationships

Relationship of the Digestive Organs to the Peritoneum

▶ Describe the location and function of the peritoneum.

▶ Define retroperitoneal and name the retroperitoneal organs.

Most digestive system organs reside in the abdominopelvic cavity. Recall from Chapter 1 that all ventral body cavities contain

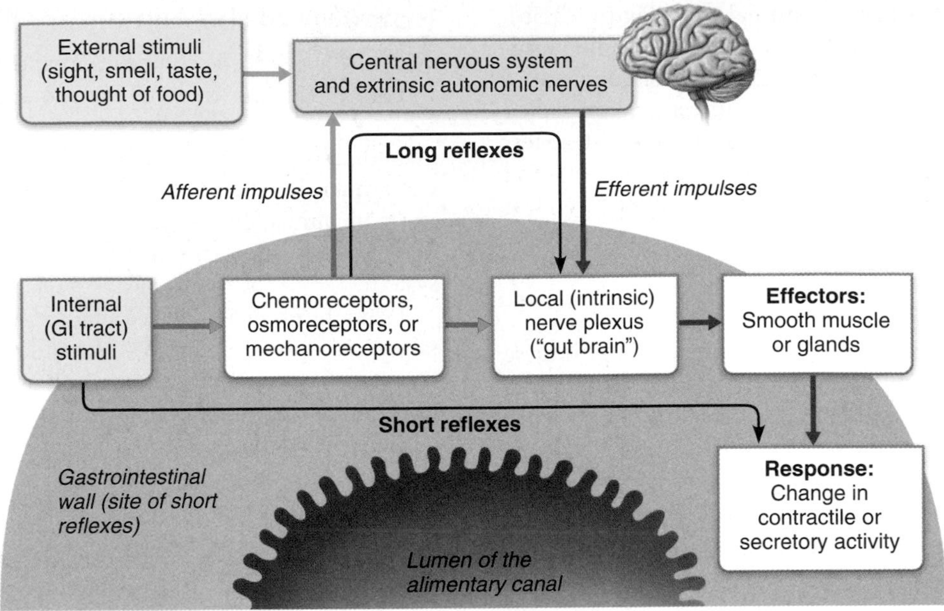

Figure 23.4 Neural reflex pathways initiated by stimuli inside or outside the gastrointestinal tract.

slippery *serous membranes.* The **peritoneum** of the abdominopelvic cavity is the most extensive of these membranes **(Figure 23.5a)**. The **visceral peritoneum** covers the external surfaces of most digestive organs and is continuous with the **parietal peritoneum** that lines the body wall. Between the two peritoneums is the **peritoneal cavity**, a slitlike potential space containing a slippery fluid secreted by the serous membranes. The serous fluid lubricates the mobile digestive organs, allowing them to glide easily across one another and along the body wall as they carry out their digestive activities.

A **mesentery** (mes′en-ter″e) is a double layer of peritoneum—a sheet of two serous membranes fused back to back—that extends to the digestive organs from the body wall. Mesenteries provide routes for blood vessels, lymphatics, and nerves to reach the digestive viscera; hold organs in place; and store fat. In most places the mesentery is *dorsal* and attaches to the posterior abdominal wall, but there are *ventral* mesenteries as well, such as the one that extends from the liver to the anterior abdominal wall (Figure 23.5a). As you read about the digestive organ mesenteries, you will see that some of them are given specific names (such as the *omenta*), or are called "ligaments" (even though these peritoneal folds are nothing like the fibrous ligaments that connect bones).

Not all alimentary canal organs are suspended by a mesentery. For example, during development, some parts of the small

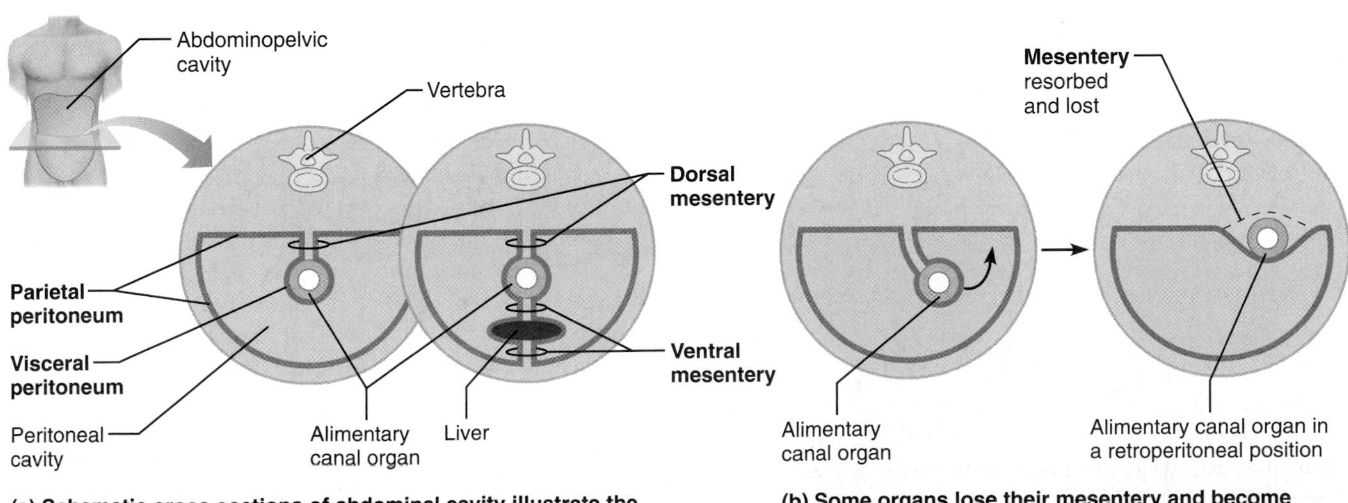

(a) Schematic cross sections of abdominal cavity illustrate the peritoneums and mesenteries.

(b) Some organs lose their mesentery and become retroperitoneal during development.

Figure 23.5 The peritoneum and the peritoneal cavity. Note that the peritoneal cavity is much smaller than depicted here.

intestine adhere to the dorsal abdominal wall (Figure 23.5b). In so doing, they lose their mesentery and come to lie posterior to the peritoneum. These organs, which include most of the pancreas and parts of the large intestine, are called **retroperitoneal organs** (*retro* = behind). By contrast, digestive organs (like the stomach) that keep their mesentery and remain in the peritoneal cavity are called **intraperitoneal** or **peritoneal organs**.

HOMEOSTATIC IMBALANCE

Peritonitis is inflammation of the peritoneum. It can arise from a piercing abdominal wound, from a perforating ulcer that leaks stomach juices into the peritoneal cavity, or from poor sterile technique during abdominal surgery, but most commonly it results from a burst appendix (that sprays bacteria-containing feces all over the peritoneum). In peritonitis, the peritoneal coverings tend to stick together around the infection site. This localizes the infection, providing time for macrophages to attack to prevent the inflammation from spreading. If peritonitis becomes widespread within the peritoneal cavity, it is dangerous and often lethal. Treatment includes removing as much infectious debris as possible from the peritoneal cavity and administering megadoses of antibiotics. ■

Blood Supply: The Splanchnic Circulation

- Define splanchnic circulation.
- Indicate the importance of the hepatic portal system.

The **splanchnic circulation** includes those arteries that branch off the abdominal aorta to serve the digestive organs and the *hepatic portal circulation.* The arterial supply—the hepatic, splenic, and left gastric branches of the celiac trunk that serve the spleen, liver, and stomach, and the mesenteric arteries (superior and inferior) that serve the small and large intestines (see pp. 877 and 890)—normally receives one-quarter of the cardiac output. This blood volume percentage increases after a meal has been eaten. The hepatic portal circulation (described on pp. 742–743) collects nutrient-rich venous blood draining from the digestive viscera and delivers it to the liver. The liver collects the absorbed nutrients for metabolic processing or for storage before releasing them back to the bloodstream for general cellular use.

CHECK YOUR UNDERSTANDING

6. How does the location of the visceral peritoneum differ from that of the parietal peritoneum?
7. Of the following organs, which is/are retroperitoneal? Stomach, pancreas, liver.
8. What name is given to the venous portion of the splanchnic circulation?

For answers, see Appendix G.

Histology of the Alimentary Canal

- Describe the tissue composition and the general function of each of the four layers of the alimentary canal.

Each digestive organ has only a share of the work of digestion. Consequently, it helps to consider structural characteristics that promote similar functions in all parts of the alimentary canal before we consider the functional anatomy of the digestive system.

From the esophagus to the anal canal, the walls of the alimentary canal have the same four basic layers, or *tunics—mucosa, submucosa, muscularis externa,* and *serosa* (**Figure 23.6**). Each layer contains a predominant tissue type that plays a specific role in food breakdown.

The Mucosa

The **mucosa**, or **mucous membrane**—the innermost layer—is a moist epithelial membrane that lines the alimentary canal lumen from mouth to anus. Its major functions are (1) to *secrete* mucus, digestive enzymes, and hormones, (2) to *absorb* the end products of digestion into the blood, and (3) to *protect* against infectious disease. The mucosa in a particular region of the GI tract may express one or all three of these capabilities.

More complex than most other mucosae in the body, the typical digestive mucosa consists of three sublayers: (1) a lining epithelium, (2) a lamina propria, and (3) a muscularis mucosae. Typically, the **epithelium** of the mucosa is a *simple columnar epithelium* rich in mucus-secreting cells. The slippery mucus it produces protects certain digestive organs from being digested themselves by enzymes working within their cavities and eases food passage along the tract. In the stomach and small intestine, the mucosa also contains both enzyme-synthesizing and hormone-secreting cells. For this reason, in such sites, the mucosa is a diffuse kind of endocrine organ as well as part of the digestive organ.

The **lamina propria** (*proprius* = one's own), which underlies the epithelium, is loose areolar connective tissue. Its capillaries nourish the epithelium and absorb digested nutrients. Its isolated lymphoid follicles, part of **MALT**, the mucosa-associated lymphatic tissue described on p. 761, help defend us against bacteria and other pathogens, which have rather free access to our digestive tract. Particularly large collections of lymphoid follicles occur within the pharynx (as the tonsils) and in the appendix.

External to the lamina propria is the **muscularis mucosae**, a scant layer of smooth muscle cells that produces local movements of the mucosa. For example, twitching of this muscle layer dislodges food particles that have adhered to the mucosa. In the small intestine, this muscle layer's tone throws the mucosa into a series of small folds that immensely increase its surface area.

The Submucosa

The **submucosa**, just external to the mucosa, is areolar connective tissue containing a rich supply of blood and lymphatic vessels, lymphoid follicles, and nerve fibers. Its abundant elastic

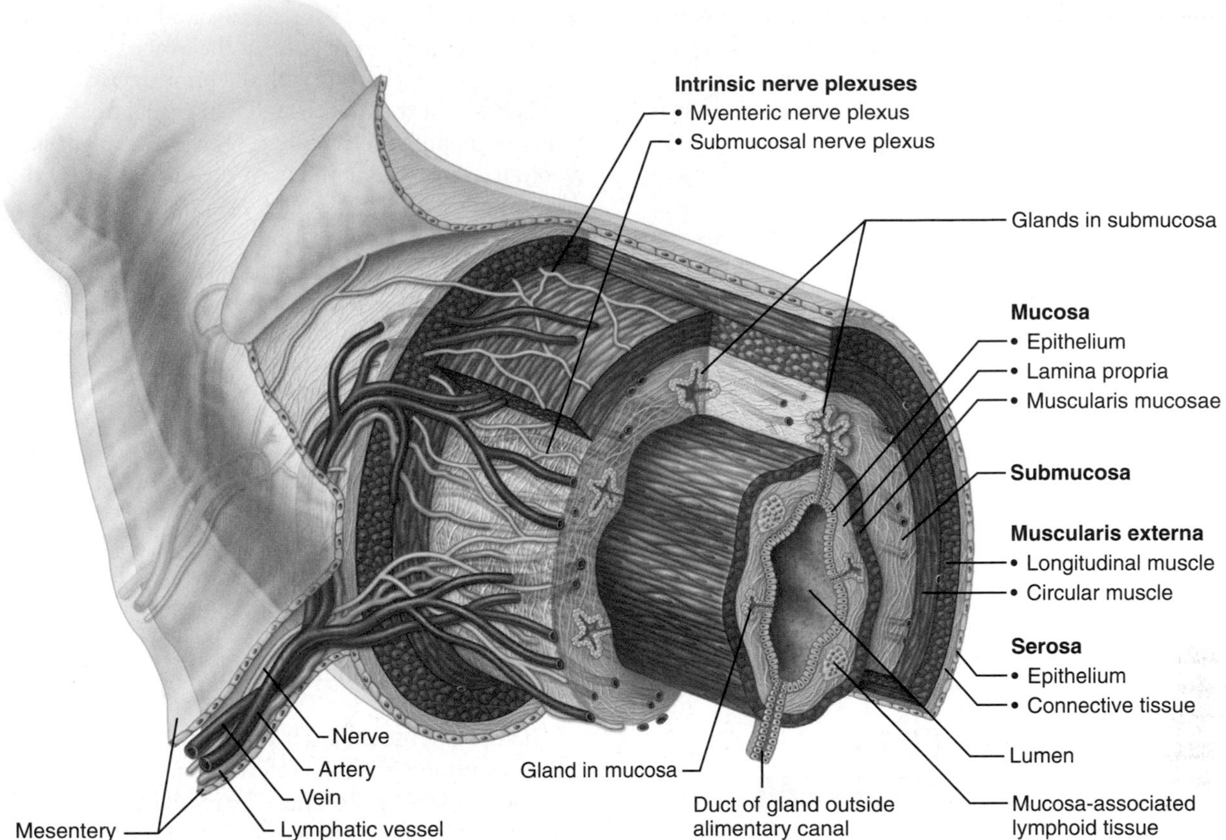

Figure 23.6 Basic structure of the alimentary canal. Its four basic layers are the mucosa, submucosa, muscularis externa, and serosa.

fibers enable the stomach to regain its normal shape after temporarily storing a large meal. Its extensive vascular network supplies surrounding tissues of the GI tract wall.

The Muscularis Externa

Surrounding the submucosa is the **muscularis externa**, also simply called the **muscularis.** This layer is responsible for segmentation and peristalsis. It typically has an inner *circular layer* and an outer *longitudinal layer* of smooth muscle cells (see Figures 4.10c and 23.6). In several places along the tract, the circular layer thickens, forming *sphincters* that act as valves to prevent backflow and control food passage from one organ to the next.

The Serosa

The **serosa,** the protective outermost layer of the intraperitoneal organs, is the *visceral peritoneum.* It is formed of areolar connective tissue covered with *mesothelium,* a single layer of squamous epithelial cells (see Figures 4.8a and 4.3a, respectively).

In the esophagus, which is located in the thoracic instead of the abdominopelvic cavity, the serosa is replaced by an **adventitia** (ad″ven-tish′e-ah). The adventitia is ordinary fibrous connective tissue that binds the esophagus to surrounding structures. Retroperitoneal organs have *both* a serosa (on the side facing the peritoneal cavity) and an adventitia (on the side abutting the dorsal body wall).

Enteric Nervous System of the Alimentary Canal

As we noted earlier, the alimentary canal has its own in-house nerve supply, staffed by the so-called **enteric neurons** (*enter* = gut), which communicate widely with one another to regulate digestive system activity. These semiautonomous enteric neurons constitute the bulk of the two major *intrinsic nerve plexuses* (ganglia interconnected by unmyelinated fiber tracts) found in the walls of the alimentary canal: the submucosal and myenteric nerve plexuses (Figure 23.6).

The **submucosal nerve plexus** occupies the submucosa. It includes sensory as well as motor neurons, and it chiefly regulates the activity of glands and smooth muscle in the mucosa.

The large **myenteric nerve plexus** (mi-en′ter-ik; "intestinal muscle") lies between the circular and longitudinal muscle layers of the muscularis externa. Enteric neurons of this plexus provide the major nerve supply to the GI tract wall and control GI tract motility. Control of the patterns of segmentation and peristalsis is largely automatic, involving pacemaker cells and local reflex arcs between enteric neurons in the same or different plexuses or (even) organs.

The enteric nervous system is also linked to the central nervous system by afferent visceral fibers and by sympathetic and parasympathetic branches (motor fibers) of the autonomic nervous system that enter the intestinal wall and synapse with neurons in the intrinsic plexuses. Hence, digestive activity is also subject to extrinsic controls exerted by autonomic fibers via long reflex arcs (see Figure 23.4). Generally speaking, parasympathetic inputs enhance secretory activity and motility, whereas sympathetic impulses inhibit digestive activities.

But the largely independent enteric ganglia are much more than just way stations for the autonomic nervous system as is the case in other organ systems. Indeed, the enteric nervous system contains over 100 million neurons, more than the entire spinal cord.

CHECK YOUR UNDERSTANDING

9. Name the layers of the alimentary canal from the inside out.
10. Jerry has been given a drug that inhibits parasympathetic stimulation of his digestive tract. Should he "eat hearty" or temporarily refrain from eating, and why?

For answers, see Appendix G.

PART 2

FUNCTIONAL ANATOMY OF THE DIGESTIVE SYSTEM

Now that we have summarized some points that unify the digestive system organs, let's consider the special structural and functional capabilities of each organ of this system. Most of the digestive organs are shown in their normal body positions in Figure 23.1, so you may find it helpful to refer back to that illustration from time to time as you read the following sections.

The Mouth and Associated Organs

- Describe the gross and microscopic anatomy and the basic functions of the mouth, pharynx, and esophagus.
- Describe the composition and functions of saliva, and explain how salivation is regulated.
- Explain the dental formula and differentiate clearly between deciduous and permanent teeth.

The mouth is the only part of the alimentary canal involved in ingestion. However, most digestive functions associated with the mouth reflect the activity of the related accessory organs, such as the teeth, salivary glands, and tongue, because in the mouth food is chewed and mixed with saliva containing enzymes that begin the process of chemical digestion. The mouth also begins the propulsive process of swallowing, which carries food through the pharynx and esophagus to the stomach.

The Mouth

The **mouth**, a mucosa-lined cavity, is also called the **oral cavity**, or *buccal cavity* (buk′al). Its boundaries are the lips anteriorly, cheeks laterally, palate superiorly, and tongue inferiorly (**Figure 23.7**). Its anterior opening is the **oral orifice**. Posteriorly, the oral cavity is continuous with the *oropharynx*.

The walls of the mouth are lined with a thick stratified squamous epithelium (see Figure 4.3e) which can withstand considerable friction. The epithelium on the gums, hard palate, and dorsum of the tongue is slightly keratinized for extra protection against abrasion during eating. Like all moist surface linings, the oral mucosa responds to injury by producing antimicrobial peptides called *defensins*, which helps to explain how the mouth, a site teeming with disease-causing microbes, remains so remarkably healthy.

The Lips and Cheeks

The **lips (labia)** and the **cheeks**, which help keep food between the teeth when we chew, are composed of a core of skeletal muscle covered externally by skin. The *orbicularis oris muscle* forms the fleshy lips; the cheeks are formed largely by the *buccinators*. The recess bounded externally by the lips and cheeks and internally by the gums and teeth is called the **vestibule** ("porch"). The area that lies within the teeth and gums is the **oral cavity proper**.

The lips are much larger than most people think. Anatomically they extend from the inferior margin of the nose to the superior boundary of the chin. The reddened area where a person applies lipstick or lands a kiss is called the **red margin**. This transitional zone, where keratinized skin meets the oral mucosa, is poorly keratinized and translucent, allowing the red color of blood in the underlying capillaries to show through. Because the red margin lacks sweat or sebaceous glands, it must be moistened with saliva periodically to prevent it from becoming dry and cracked (chapped lips). The **labial frenulum** (fren′u-lum) is a median fold that joins the internal aspect of each lip to the gum (Figure 23.7b).

The Palate

The **palate**, forming the roof of the mouth, has two distinct parts: the hard palate anteriorly and the soft palate posteriorly (Figure 23.7). The **hard palate** is underlain by the palatine bones and the palatine processes of the maxillae, and it forms a rigid surface against which the tongue forces food during chewing. The mucosa on either side of its *raphe* (ra′fe), a midline ridge, is slightly corrugated, which helps to create friction.

The **soft palate** is a mobile fold formed mostly of skeletal muscle that rises reflexively to close off the nasopharynx when we swallow.

- To demonstrate this action, try to breathe and swallow at the same time.

Laterally, the soft palate is anchored to the tongue by the **palatoglossal arches** and to the wall of the oropharynx by the more posterior **palatopharyngeal arches**. These two paired folds form the boundaries of the **fauces** (faw′sēz; *fauc* = throat), the arched area of the oropharynx that contains the palatine tonsils. Projecting downward from the free edge of the soft palate is the fingerlike **uvula** (u′vu-lah).

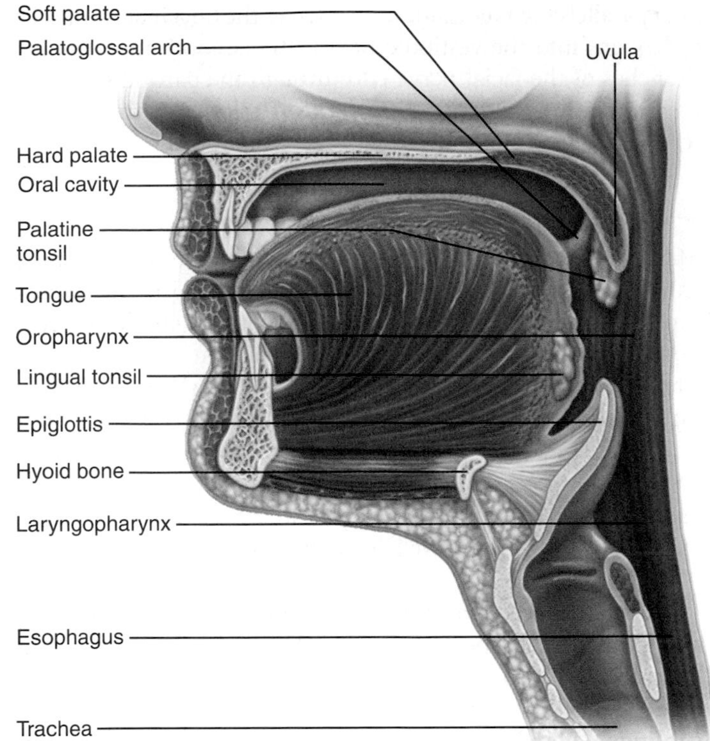

(a) Sagittal section of the oral cavity and pharynx

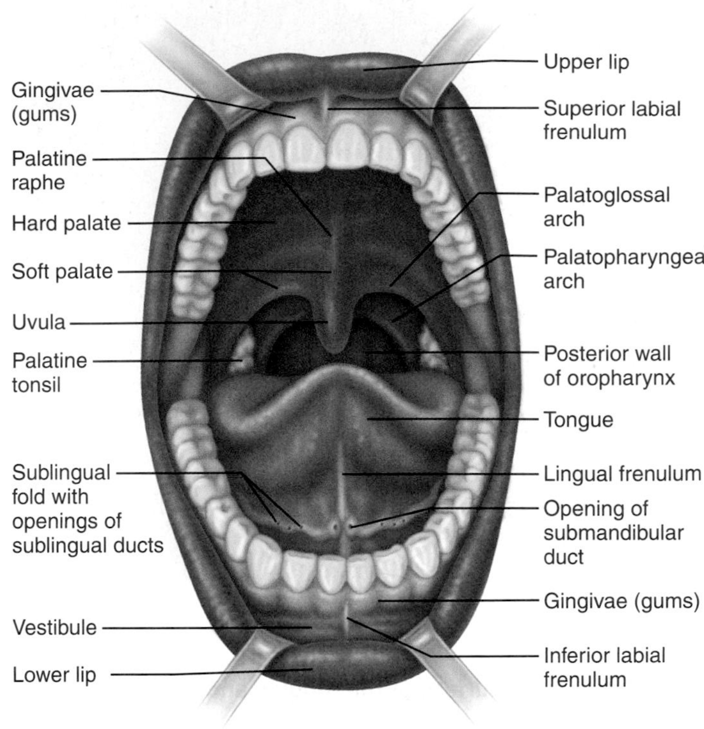

(b) Anterior view

Figure 23.7 **Anatomy of the oral cavity (mouth).**

The Tongue

The **tongue** occupies the floor of the mouth and fills most of the oral cavity when the mouth is closed (Figure 23.7). The tongue is composed of interlacing bundles of skeletal muscle fibers, and during chewing, it grips the food and constantly repositions it between the teeth. The tongue also mixes food with saliva forming it into a compact mass called a **bolus** (bo′lus; "a lump"), and then initiates swallowing by pushing the bolus posteriorly into the pharynx. The versatile tongue also helps us form consonants (k, d, t, and so on) when we speak.

The tongue has both intrinsic and extrinsic skeletal muscle fibers. The **intrinsic muscles** are confined in the tongue and are not attached to bone. Their muscle fibers, which run in several different planes, allow the tongue to change its shape (but not its position), becoming thicker, thinner, longer, or shorter as needed for speech and swallowing.

The **extrinsic muscles** extend to the tongue from their points of origin on bones of the skull or the soft palate, as described in Chapter 10 (see Table 10.2 and Figure 10.7). The extrinsic muscles alter the tongue's position. They protrude it, retract it, and move it from side to side. The tongue has a median septum of connective tissue, and each half contains identical muscle groups. A fold of mucosa called the **lingual frenulum** secures the tongue to the floor of the mouth and limits its posterior movements of the tongue.

HOMEOSTATIC IMBALANCE

Children born with an extremely short lingual frenulum are often referred to as "tongue-tied" because when tongue movement is restricted, speech is distorted. This congenital condition, called *ankyloglossia* ("fused tongue"), is corrected surgically by snipping the frenulum. ■

The superior tongue surface bears papillae, peglike projections of the underlying mucosa **(Figure 23.8)**. The conical **filiform papillae** give the tongue surface a roughness that aids in licking semisolid foods (such as ice cream) and provide friction for manipulating foods in the mouth. These papillae, the smallest and most numerous type, align in parallel rows on the tongue dorsum. They contain keratin, which stiffens them and gives the tongue its whitish appearance.

The mushroom-shaped **fungiform papillae** are scattered widely over the tongue surface. Each has a vascular core that gives it a reddish hue. Ten to twelve large **circumvallate**, or **vallate, papillae** are located in a V-shaped row at the back of the tongue. They resemble the fungiform papillae but have an additional surrounding furrow. Pleatlike **foliate papillae** are located on the lateral aspects of the posterior tongue. The fungiform, circumvallate, and foliate papillae house taste buds, but those on the foliate papillae function in taste primarily in infancy and early childhood.

Immediately posterior to the circumvallate papillae is the **terminal sulcus**, a groove that distinguishes the portion of the

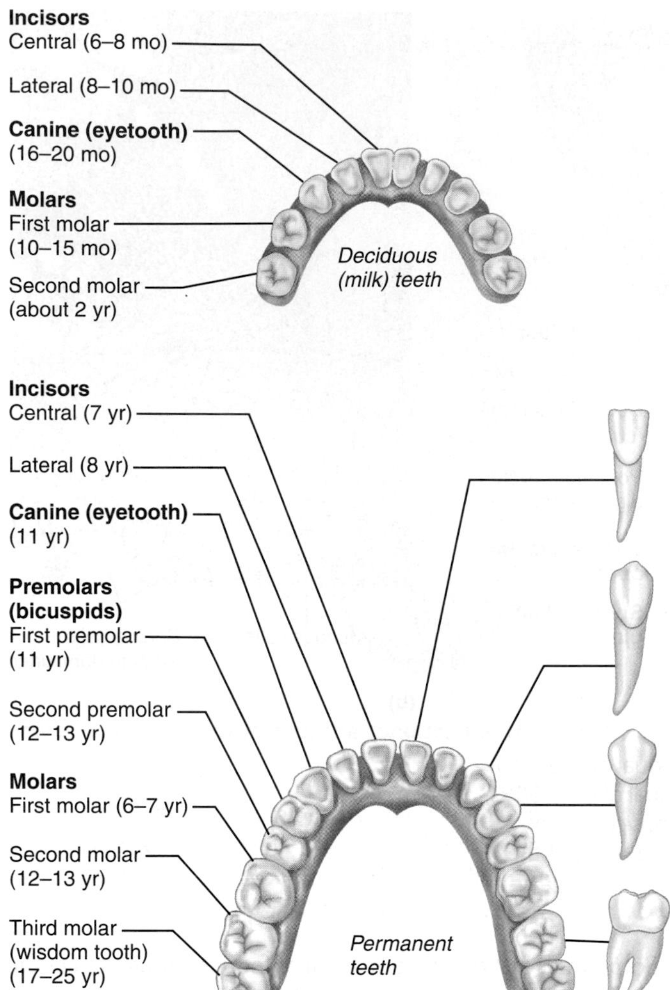

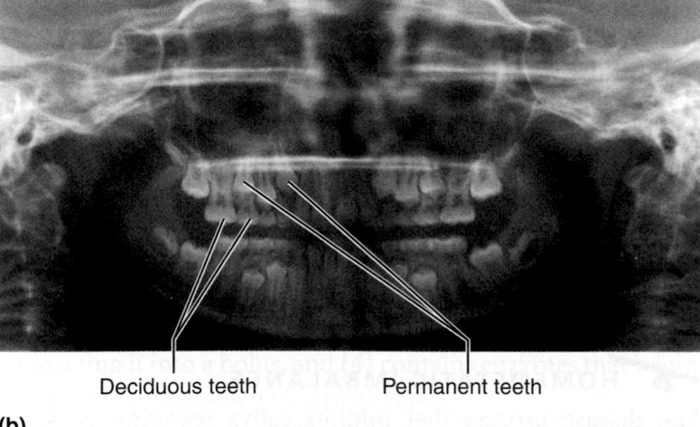

Figure 23.10 Human dentition. (a) Teeth of the lower jaw: the deciduous and permanent sets. Approximate age at which tooth erupts is shown in parentheses. The shapes of individual teeth are shown on the right. **(b)** X ray of the mouth of a child of 7 years showing permanent teeth forming deep to the deciduous teeth.

CHECK YOUR UNDERSTANDING

14. What is the importance of the serous portion of saliva?
15. Name four antimicrobial substances found in saliva.

For answers, see Appendix G.

The Teeth

The **teeth** lie in sockets (alveoli) in the gum-covered margins of the mandible and maxilla. The role of the teeth in food processing needs little introduction. We *masticate*, or chew, by opening and closing our jaws and moving them from side to side while continually using our tongue to move the food between our teeth. In the process, the teeth tear and grind the food, breaking it down into smaller fragments.

Dentition and the Dental Formula

Ordinarily by age 21, two sets of teeth, the **primary** and **permanent dentitions**, have formed **(Figure 23.10a)**. The primary dentition consists of the **deciduous teeth** (de-sid′u-us; *decid* = falling off), also called **milk** or **baby teeth.** The first teeth to appear, at about age 6 months, are the lower central incisors. Additional pairs of teeth erupt at one- to two-month intervals until about 24 months, when all 20 milk teeth have emerged.

As the deep-lying **permanent teeth** enlarge and develop, the roots of the milk teeth are resorbed from below (Figure 23.10b), causing them to loosen and fall out between the ages of 6 and 12 years. Generally, all the teeth of the permanent dentition but the third molars have erupted by the end of adolescence. The third molars, also called *wisdom teeth,* emerge between the ages of 17 and 25 years. There are usually 32 permanent teeth in a full set, but sometimes the wisdom teeth never erupt or are completely absent.

HOMEOSTATIC IMBALANCE

When a tooth remains embedded in the jawbone, it is said to be *impacted.* Impacted teeth can cause a good deal of pressure and pain and must be removed surgically. Wisdom teeth are most commonly impacted. ■

Teeth are classified according to their shape and function as incisors, canines, premolars, and molars (Figure 23.10a). The chisel-shaped **incisors** are adapted for cutting or nipping off pieces of food. The conical or fanglike **canines** (cuspids or eyeteeth) tear and pierce. The **premolars** (bicuspids) and **molars** have broad crowns with rounded cusps (tips) and are best suited for grinding or crushing. The molars (literally, "millstones"), with four or five cusps, are the best grinders. During chewing, the upper and lower molars repeatedly lock together, an action that generates tremendous crushing forces.

The **dental formula** is a shorthand way of indicating the numbers and relative positions of the different types of teeth in the mouth. This formula is written as a ratio, uppers over lowers, for *one-half* of the mouth. Since the other side is a mirror image, the total dentition is obtained by multiplying the dental formula by 2. The primary dentition consists of two incisors (I), one canine (C), and two molars (M) on each side of each jaw, and its dental formula is written as

$$\frac{\text{2I, 1C, 2M (upper jaw)}}{\text{2I, 1C, 2M (lower jaw)}} \times 2 \text{ (20 teeth)}$$

Similarly, the permanent dentition [two incisors, one canine, two premolars (PM), and three molars] is

$$\frac{\text{2I, 1C, 2PM, 3M}}{\text{2I, 1C, 2PM, 3M}} \times 2 \text{ (32 teeth)}$$

Tooth Structure

Each tooth has two major regions: the crown and the root **(Figure 23.11)**. The enamel-covered **crown** is the exposed part of the tooth above the **gingiva** (jin′jĭ-vah), or **gum**, which surrounds the tooth like a tight collar. **Enamel**, a brittle ceramic-like material thick as a dime, directly bears the force of chewing. The hardest substance in the body, it is heavily mineralized with calcium salts, and its densely packed hydroxyapatite (mineral) crystals are oriented in force-resisting columns perpendicular to the tooth's surface. The cells that produce enamel degenerate when the tooth erupts; consequently, any decayed or cracked areas of the enamel will not heal and must be artificially filled.

The portion of the tooth embedded in the jawbone is the **root**. Canine teeth, incisors, and premolars have one root, although the first upper premolars commonly have two. The first two upper molars have three roots, while the corresponding lower molars have two. The root pattern of the third molar varies, but a fused single root is most common.

The crown and root are connected by a constricted tooth region called the **neck**. The outer surface of the root is covered by **cementum**, a calcified connective tissue, which attaches the tooth to the thin **periodontal ligament** (per″e-o-don′tal; "around the tooth"). This ligament anchors the tooth in the bony alveolus of the jaw, forming a fibrous joint called a *gomphosis*. Where the gingiva borders on a tooth, it dips downward to form a shallow groove called the *gingival sulcus*.

In youth, the gingiva adheres tenaciously to the enamel covering the crown. But as the gums begin to recede with age, the gingiva adheres to the more sensitive cementum covering the superior region of the root. As a result, the teeth *appear* to get longer in old age—hence the expression "long in the tooth" sometimes applied to elderly people.

Dentin, a protein-rich bonelike material, underlies the enamel cap and forms the bulk of a tooth. More resilient than enamel, dentin acts as a shock absorber for forces acting on the enamel during biting and chewing. Dentin surrounds a central **pulp cavity** containing a number of soft tissue structures (connective tissue, blood vessels, and nerve fibers) collectively called **pulp**. Pulp supplies nutrients to the tooth tissues and provides for tooth sensation. Where the pulp cavity extends into the root, it becomes the **root canal**. At the proximal end of each root canal is an **apical foramen** that allows blood vessels, nerves, and other structures to enter the pulp cavity.

The teeth are served by the superior and inferior alveolar nerves, branches of the trigeminal nerve (see Table 13.2, p. 497).

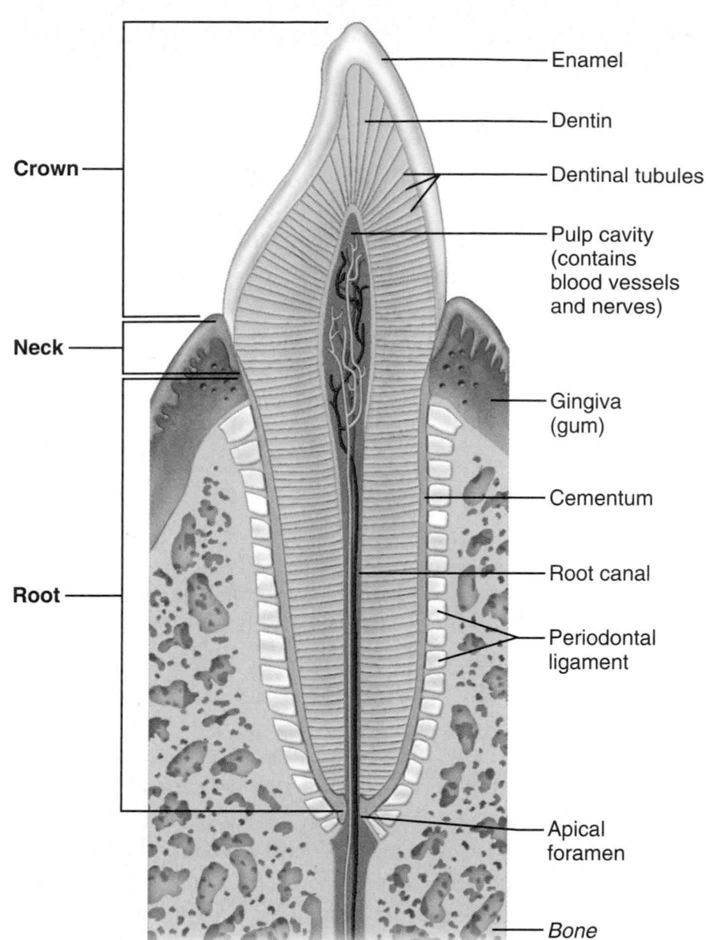

Figure 23.11 Longitudinal section of a canine tooth within its bony alveolus.

Blood is supplied by the superior and inferior alveolar arteries, branches of the maxillary artery (see Figure 19.22b, p. 727).

Dentin contains unique radial striations called *dentinal tubules* (Figure 23.11). Each tubule contains an elongated process of an **odontoblast** (o-don′to-blast; "tooth former"), the cell type that secretes and maintains the dentin. The cell bodies of odontoblasts line the pulp cavity just deep to the dentin. Dentin is formed throughout adult life and gradually encroaches on the pulp cavity. New dentin can also be laid down fairly rapidly to compensate for tooth damage or decay.

Enamel, dentin, and cementum are all calcified and resemble bone (to differing extents), but they differ from bone in that they are avascular. Enamel also differs from cementum and dentin because it lacks collagen as its main organic component and it is almost entirely mineral.

HOMEOSTATIC IMBALANCE

Death of a tooth's nerve and consequent darkening of the tooth is commonly caused by a blow to the jaw. Local swelling pinches off the blood supply to the tooth and the nerve dies. Typically the pulp becomes infected by bacteria sometime later and must be removed by *root canal therapy*. After the cavity is sterilized and filled with an inert material, the tooth is capped (covered with an artificial crown). ■

Tooth and Gum Disease

Dental caries (kār′ēz; "rottenness"), or **cavities**, result from gradual demineralization of enamel and underlying dentin by bacterial action. Decay begins when **dental plaque** (a film of sugar, bacteria, and other mouth debris) adheres to the teeth. Bacterial metabolism of the trapped sugars produces acids, which can dissolve the calcium salts of the teeth. Once the salts are leached out, the remaining organic matrix of the tooth is readily digested by protein-digesting enzymes released by the bacteria. Frequent brushing and flossing daily help prevent damage by removing forming plaque.

More serious than tooth decay is the effect of unremoved plaque on the gums. As dental plaque accumulates, it calcifies, forming **calculus** (kal′ku-lus; "stone") or tartar. These stony-hard deposits disrupt the seals between the gingivae and the teeth, deepening the sulcus and putting the gums at risk for infection by pathogenic anaerobic bacteria. In the early stages of such an infection, called **gingivitis** (jin″jĭ-vi′tis), the gums are red, sore, swollen, and may bleed.

Gingivitis is reversible if the calculus is removed, but if it is neglected the bacteria eventually form pockets of infection which become inflamed. Neutrophils and immune system cells (lymphocytes and macrophages) attack not only the intruders but also body tissues, carving deep pockets around the teeth, destroying the periodontal ligament, and activating osteoclasts which dissolve the bone away. This more serious condition, called **periodontal disease**, or **periodontitis**, affects up to 95% of all people over the age of 35 and accounts for 80–90% of tooth loss in adults. While periodontitis is widely thought to be a bacterial disease, a 14-year study of young adults indicates that those who regularly smoke marijuana are three to five times as likely as nonusers to have areas of severe gum detachment.

Regardless of its cause, tooth loss from periodontitis is not inevitable. Even advanced periodontitis can be treated by scraping the teeth, cleaning the infected pockets, then cutting the gums to shrink the pockets, and following up with anti-inflammatory and antibiotic therapy. Together, these treatments alleviate the bacterial infestations and encourage reattachment of the surrounding tissues to the teeth and bone.

Much less painful are (1) a new laser approach for destroying the diseased tissues and (2) a nonsurgical therapy now in clinical trials in which an antibiotic-impregnated film is temporarily glued to the exposed root surface. Clinical treatment is followed up by a home regimen to remove plaque by consistent frequent brushing and flossing and hydrogen peroxide rinses.

There may be more at risk than teeth in people with periodontal disease. Some contend that it increases the risk of heart disease and stroke in at least two ways: (1) the chronic inflammation promotes atherosclerotic plaque formation, and (2) bacteria entering the blood from infected gums stimulate clot formation that helps to clog coronary and cerebral arteries. Risk factors for periodontal disease include smoking, diabetes mellitus, and oral (tongue or lip) piercing.

23

CHECK YOUR UNDERSTANDING

16. Seven-year-old Tina ran to her daddy to show him her lower central incisor which she had wiggled until it "fell out." Is this a primary or secondary tooth? What name is given to teeth that (according to Tina) fall out?

17. What tooth substance is harder than bone? Which tooth part includes nervous tissue and blood vessels?

18. Which teeth are the "grinders"?

For answers, see Appendix G.

The Pharynx

From the mouth, food passes posteriorly into the **oropharynx** and then the **laryngopharynx** (see Figure 23.7a), both common passageways for food, fluids, and air. (The nasopharynx has no digestive role.)

The histology of the pharyngeal wall resembles that of the oral cavity. The mucosa contains a friction-resistant stratified squamous epithelium well supplied with mucus-producing glands. The external muscle layer consists of two *skeletal muscle* layers. The cells of the inner layer run longitudinally. Those of the outer layer, the *pharyngeal constrictor* muscles, encircle the wall like three stacked fists (see Figure 10.8b). Contractions of these muscles propel food into the esophagus below.

The Esophagus

The **esophagus** (ĕ-sof′ah-gus; "carry food"), a muscular tube about 25 cm (10 inches) long, is collapsed when not involved in food propulsion **(Figure 23.12)**. After food moves through the laryngopharynx, it is routed into the esophagus posteriorly as the epiglottis closes off the larynx to food entry.

As shown in Figure 23.1, the esophagus takes a fairly straight course through the mediastinum of the thorax. It pierces the diaphragm at the **esophageal hiatus** (hi-a′tus; "gap") to enter the abdomen. It joins the stomach at the **cardiac orifice** within the abdominal cavity. The cardiac orifice is surrounded by the **gastroesophageal** or **cardiac sphincter** (gas″tro-ĕ-sof″ah-je′al), which is a *physiological* sphincter (see Figure 23.13). That is, it acts as a valve, but the only structural evidence of this sphincter is a slight thickening of the circular smooth muscle at that point. The muscular diaphragm, which surrounds this sphincter, helps keep it closed when food is not being swallowed.

HOMEOSTATIC IMBALANCE

Heartburn, the first symptom of *gastroesophageal reflux disease (GERD)*, is the burning, radiating substernal pain that occurs when the acidic gastric juice regurgitates into the esophagus. Symptoms are so similar to those of a heart attack that many first-time sufferers of heartburn are rushed to the hospital emergency room. Heartburn is most likely to happen when a person has eaten or drunk to excess, and in conditions that force abdominal contents superiorly, such as extreme obesity,

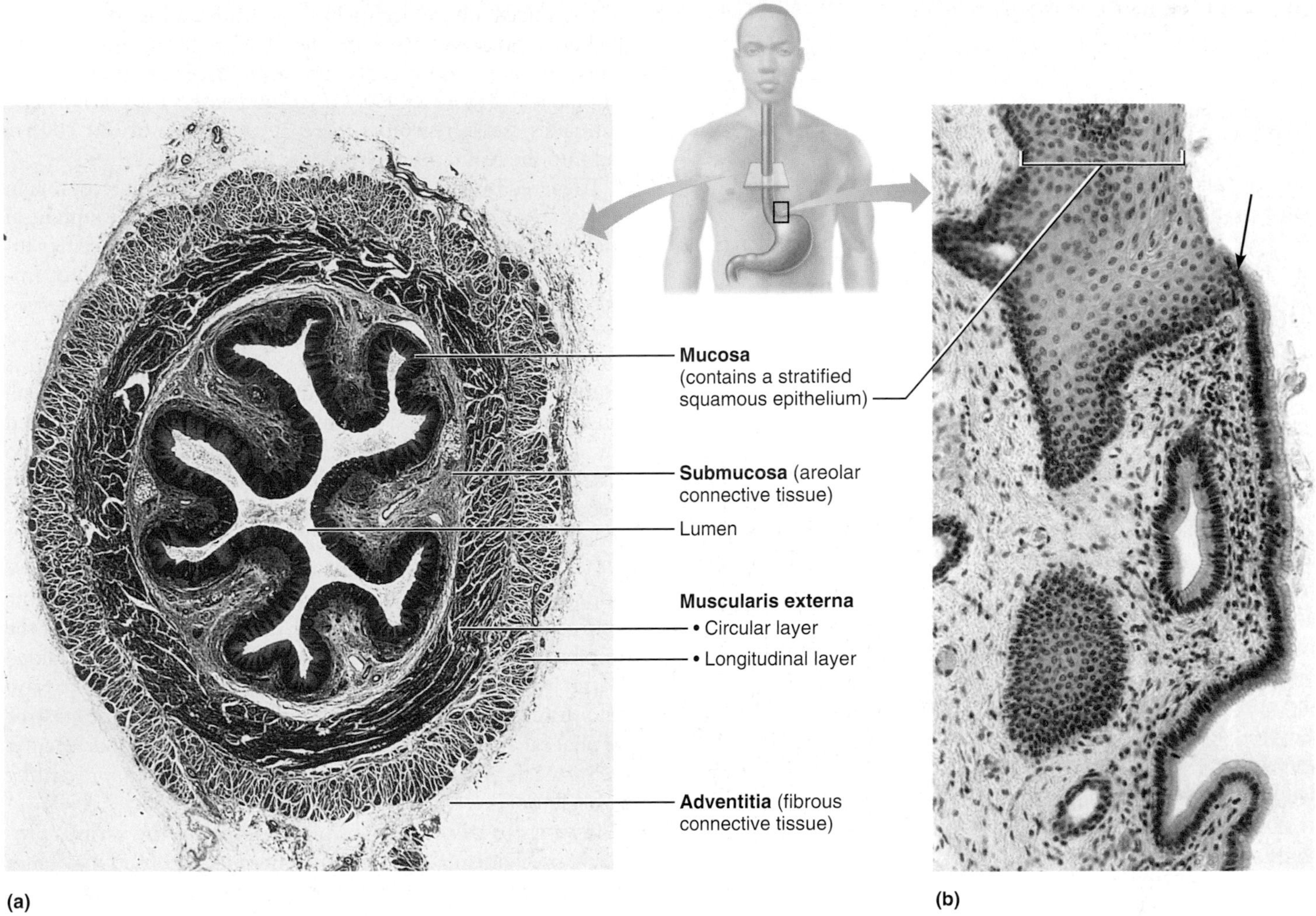

Mucosa (contains a stratified squamous epithelium)
Submucosa (areolar connective tissue)
Lumen
Muscularis externa
• Circular layer
• Longitudinal layer
Adventitia (fibrous connective tissue)

(a)

(b)

Figure 23.12 Microscopic structure of the esophagus. (a) Cross-sectional view of the esophagus taken from the region close to the stomach junction (10×). The muscularis is composed of smooth muscle. **(b)** Longitudinal section through the esophagus-stomach junction (120×). Arrow shows the point of abrupt transition from the stratified squamous epithelium of the esophagus (top) to the simple columnar epithelium of the stomach (bottom).

pregnancy, and running, which causes stomach contents to splash upward with each step (runner's reflux).

Heartburn is also common in those with a **hiatal hernia**, a structural abnormality (most often due to an abnormal relaxation or weakening of the gastroesophageal sphincter) in which the superior part of the stomach protrudes slightly above the diaphragm. Since the diaphragm no longer reinforces the sphincter, gastric juice may enter the esophagus, particularly when lying down. If the episodes are frequent and prolonged, *esophagitis* (inflammation of the esophagus) and *esophageal ulcers* may result. An even more threatening sequel is esophageal cancer. These consequences can usually be prevented or managed by avoiding late-night snacks and by using antacid preparations. ■

Unlike the mouth and pharynx, the esophagus wall has all four of the basic alimentary canal layers described earlier. Some features of interest:

1. The esophageal mucosa contains a nonkeratinized stratified squamous epithelium. At the esophagus-stomach junction, that abrasion-resistant epithelium changes abruptly to the simple columnar epithelium of the stomach, which is specialized for secretion (Figure 23.12b).
2. When the esophagus is empty, its mucosa and submucosa are thrown into longitudinal folds (Figure 23.12a). When food is in transit in the esophagus, these folds flatten out.
3. The submucosa contains mucus-secreting *esophageal glands.* As a bolus moves through the esophagus, it compresses these glands, causing them to secrete mucus that "greases" the esophageal walls and aids food passage.
4. The muscularis externa is skeletal muscle in its superior third, a mixture of skeletal and smooth muscle in its middle third, and entirely smooth muscle in its inferior third.
5. Instead of a serosa, the esophagus has a fibrous adventitia composed entirely of connective tissue, which blends with surrounding structures along its route.

CHECK YOUR UNDERSTANDING

19. To what two organ systems does the pharynx belong?
20. How is the muscularis externa of the esophagus unique in the body?
21. What is the functional significance of the epithelial change seen at the esophagus-stomach junction?

For answers, see Appendix G.

Digestive Processes: Mouth to Esophagus

▶ Describe the mechanisms of chewing and swallowing.

The mouth and its accessory digestive organs are involved in most digestive processes. The mouth (1) ingests, (2) begins mechanical digestion by chewing, (3) initiates propulsion by swallowing, and (4) starts the chemical breakdown of polysaccharides. Salivary amylase, the main enzyme in saliva, digests starch and glycogen, liberating smaller fragments of linked glucose molecules. (If you chew a piece of bread for a few minutes, it will begin to taste sweet as sugars are released.) Lingual lipase, a fat-digesting enzyme in saliva, also acts in the acidic environment of the stomach. Except for a few drugs that are absorbed through the oral mucosa (for example, nitroglycerine used to alleviate the pain of angina), essentially no absorption occurs in the mouth.

In contrast to the multifunctional mouth, the pharynx and esophagus merely serve as conduits to pass food from the mouth to the stomach. Their single digestive function is food propulsion, accomplished by the role they play in swallowing.

Since we will cover chemical digestion in a special physiology section later in the chapter, only the mechanical processes of chewing and swallowing are discussed here.

23

Mastication (Chewing)

As food enters the mouth, its mechanical breakdown begins with **mastication**, or chewing. The cheeks and closed lips hold food between the teeth, the tongue mixes food with saliva to soften it, and the teeth cut and grind solid foods into smaller morsels. Mastication is partly voluntary and partly reflexive. We voluntarily put food into our mouths and contract the muscles that close our jaws. The pattern and rhythm of continued jaw movements are controlled mainly by stretch reflexes and in response to pressure inputs from receptors in the cheeks, gums, and tongue, but they can also be voluntary if desired.

Deglutition (Swallowing)

To send food on its way from the mouth, it is first compacted by the tongue into a bolus and then swallowed. **Deglutition** (deg″loo-tish′un), or swallowing, is a complicated process that involves coordinated activity of over 22 separate muscle groups. It has two major phases, the buccal and the pharyngeal-esophageal.

The **buccal phase** occurs in the mouth and is voluntary. In the buccal phase, we place the tip of the tongue against the hard palate, and then contract the tongue to force the bolus into the oropharynx (**Figure 23.13**, ①). As food enters the pharynx and stimulates tactile receptors there, it passes out of our control and into the realm of involuntary reflex activity.

Triggered by that "bit of saliva" or food reaching receptors in the posterior pharynx, the involuntary **pharyngeal-esophageal phase** of swallowing is controlled by the swallowing center located in the brain stem (medulla and lower pons). Motor impulses from that center are transmitted via various cranial nerves, most importantly the vagus nerves, to the muscles of the pharynx and esophagus. Once food enters the pharynx, respiration is momentarily inhibited and as illustrated in Figure 23.13, ②, all routes except the desired one into the digestive tract are blocked off: The tongue blocks off the mouth. The soft palate rises to close off the nasopharynx. The larynx rises so that the epiglottis covers its opening into the respiratory passageways, and the upper esophageal sphincter relaxes.

Food is moved along through the pharynx and into the esophagus by pressure gradients created by wavelike peristaltic contractions (Figure 23.13, ③–⑤). Solid foods pass from the oropharynx to the stomach in about 8 seconds, and fluids, aided by gravity, pass in 1 to 2 seconds. Just before the peristaltic wave (and food) reaches the end of the esophagus, the gastroesophageal sphincter relaxes reflexively to allow food to enter the stomach. After food entry, that sphincter closes, preventing regurgitation.

If we try to talk or inhale while swallowing, the various protective mechanisms may be short-circuited and food may enter the respiratory passageways instead. This event typically triggers the cough reflex in an attempt to expel the food.

CHECK YOUR UNDERSTANDING

22. What role does the tongue play in swallowing?
23. How are the respiratory passages blocked during swallowing?

For answers, see Appendix G.

The Stomach

▶ Identify structural modifications of the wall of the stomach that enhance the digestive process.

▶ Name the cell types responsible for secreting the various components of gastric juice and indicate the importance of each component in stomach activity.

▶ Describe stomach structure and indicate changes in the basic alimentary canal structure that aid its digestive function.

Below the esophagus, the GI tract expands to form the **stomach** (see Figure 23.1), a temporary "storage tank" where chemical breakdown of proteins begins and food is converted to a creamy paste called **chyme** (kīm; "juice"). The stomach lies in the upper left quadrant of the peritoneal cavity, nearly hidden by the liver

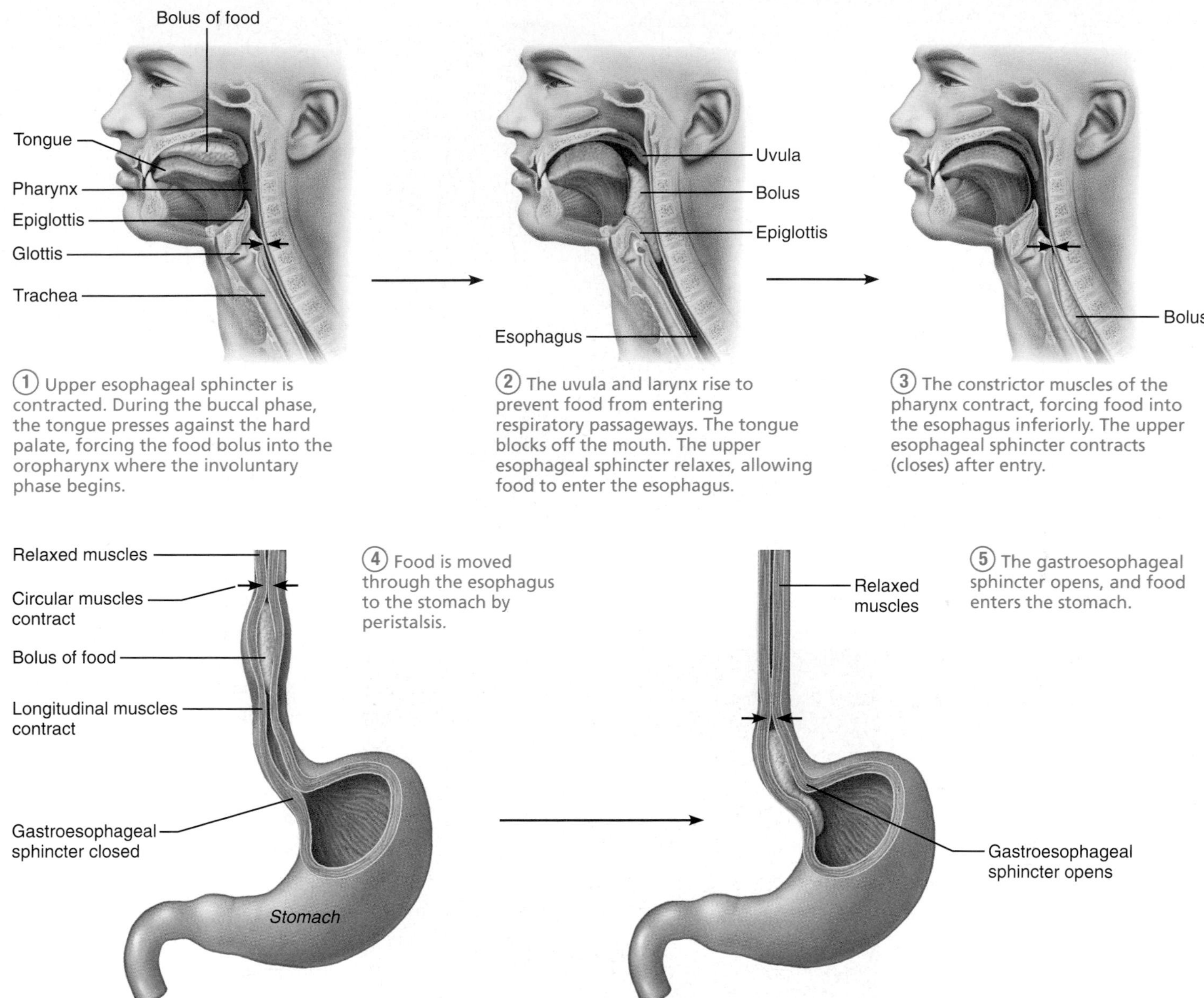
Bolus of food
Tongue
Pharynx
Epiglottis
Glottis
Trachea
Uvula
Bolus
Epiglottis
Esophagus
Bolus
① Upper esophageal sphincter is contracted. During the buccal phase, the tongue presses against the hard palate, forcing the food bolus into the oropharynx where the involuntary phase begins.
② The uvula and larynx rise to prevent food from entering respiratory passageways. The tongue blocks off the mouth. The upper esophageal sphincter relaxes, allowing food to enter the esophagus.
③ The constrictor muscles of the pharynx contract, forcing food into the esophagus inferiorly. The upper esophageal sphincter contracts (closes) after entry.
Relaxed muscles
Circular muscles contract
Bolus of food
Longitudinal muscles contract
Gastroesophageal sphincter closed
Stomach
④ Food is moved through the esophagus to the stomach by peristalsis.
Relaxed muscles
⑤ The gastroesophageal sphincter opens, and food enters the stomach.
Gastroesophageal sphincter opens

Figure 23.13 Deglutition (swallowing). The process of swallowing consists of a voluntary (buccal) phase (step ①) and an involuntary (pharyngeal-esophageal) phase (steps ②–⑤).

and diaphragm. Specifically, it lies in the left hypochondriac, epigastric, and umbilical regions of the abdomen. Though relatively fixed at both ends, the stomach is quite movable in between. It tends to lie high and run horizontally in short, stout people (a steer-horn stomach) and is often elongated vertically in tall, thin people (a J-shaped stomach).

Gross Anatomy

The adult stomach varies from 15 to 25 cm (6 to 10 inches) long, but its diameter and volume depend on how much food it contains. An empty stomach has a volume of about 50 ml and a cross-sectional diameter only slightly larger than the large intestine, but when it is really distended it can hold about 4 L (1 gallon) of food and may extend nearly all the way to the pelvis! When empty, the stomach collapses inward, throwing its mucosa (and submucosa) into large, longitudinal folds called **rugae** (roo′ge; *ruga* = wrinkle, fold).

The major regions of the stomach are shown in **Figure 23.14a**. The small **cardiac region**, or **cardia** ("near the heart"), surrounds the cardiac orifice through which food enters the stomach from the esophagus. The **fundus** is its dome-shaped part, tucked beneath the diaphragm, that bulges superolaterally to the cardia. The **body**, the midportion of the stomach, is continuous inferiorly with the funnel-shaped **pyloric region**. The wider and more superior part of the pyloric region, the **pyloric antrum** (*antrum* = cave) narrows to form the **pyloric canal**, which terminates at the **pylorus**. The pylorus is continuous with the duodenum (the first part of the small intestine) through the **pyloric valve** or **sphincter**, which controls stomach emptying (*pylorus* = gatekeeper).

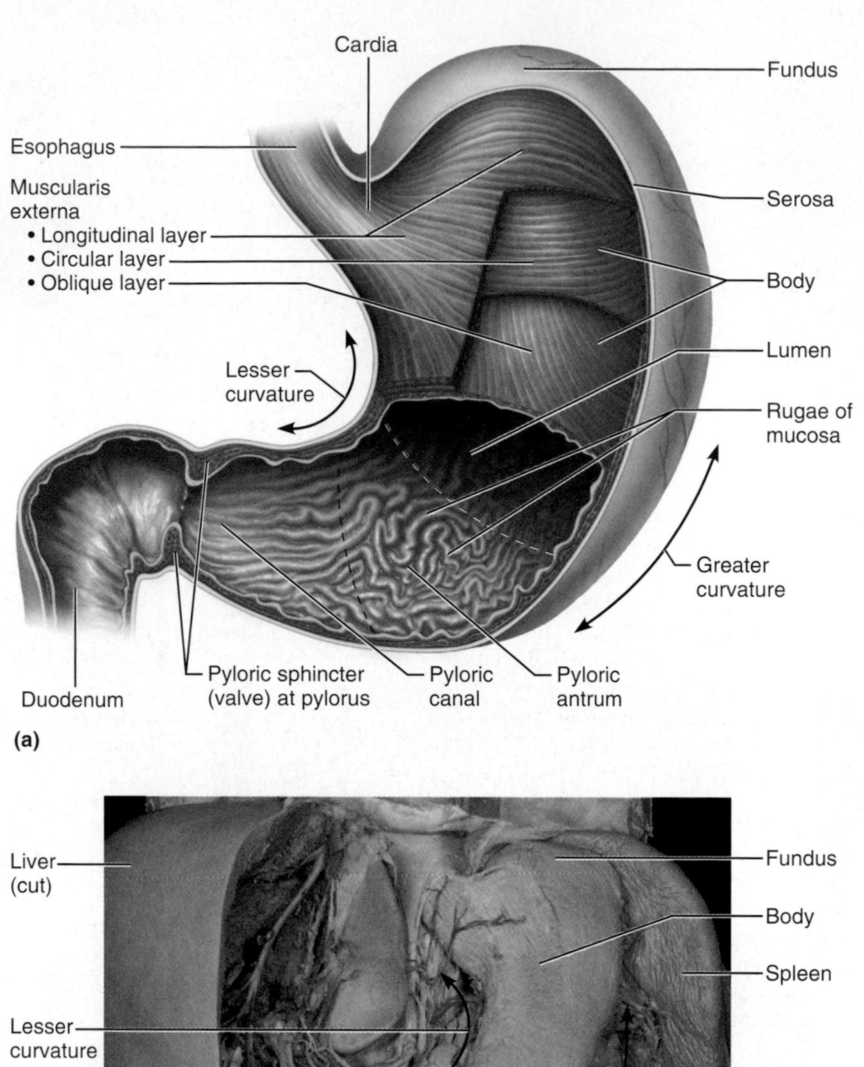

Figure 23.14 Anatomy of the stomach. (a) Gross internal anatomy (frontal section). **(b)** Photograph of external aspect of stomach. (See *A Brief Atlas of the Human Body*, Figure 69a.)

23

The convex lateral surface of the stomach is its **greater curvature**, and its concave medial surface is the **lesser curvature**. Extending from these curvatures are two mesenteries, called *omenta* (o-men′tah), that help tether the stomach to other digestive organs and the body wall (see Figure 23.30, p. 892). The **lesser omentum** runs from the liver to the lesser curvature of the stomach, where it becomes continuous with the visceral peritoneum covering the stomach. The **greater omentum** drapes inferiorly from the greater curvature of the stomach to cover the coils of the small intestine. It then runs dorsally and superiorly, wrapping the spleen and the transverse portion of the large intestine before blending with the *mesocolon*, a dorsal mesentery that secures the large intestine to the parietal peritoneum of the posterior abdominal wall. The greater omentum is riddled with fat deposits (*oment* = fatty skin) that give it the appearance of a lacy apron. It also contains large collections of lymph nodes. The immune cells and macrophages in these nodes "police" the peritoneal cavity and intraperitoneal organs.

The stomach is served by the autonomic nervous system. Sympathetic fibers from thoracic splanchnic nerves are relayed through the celiac plexus. Parasympathetic fibers are supplied

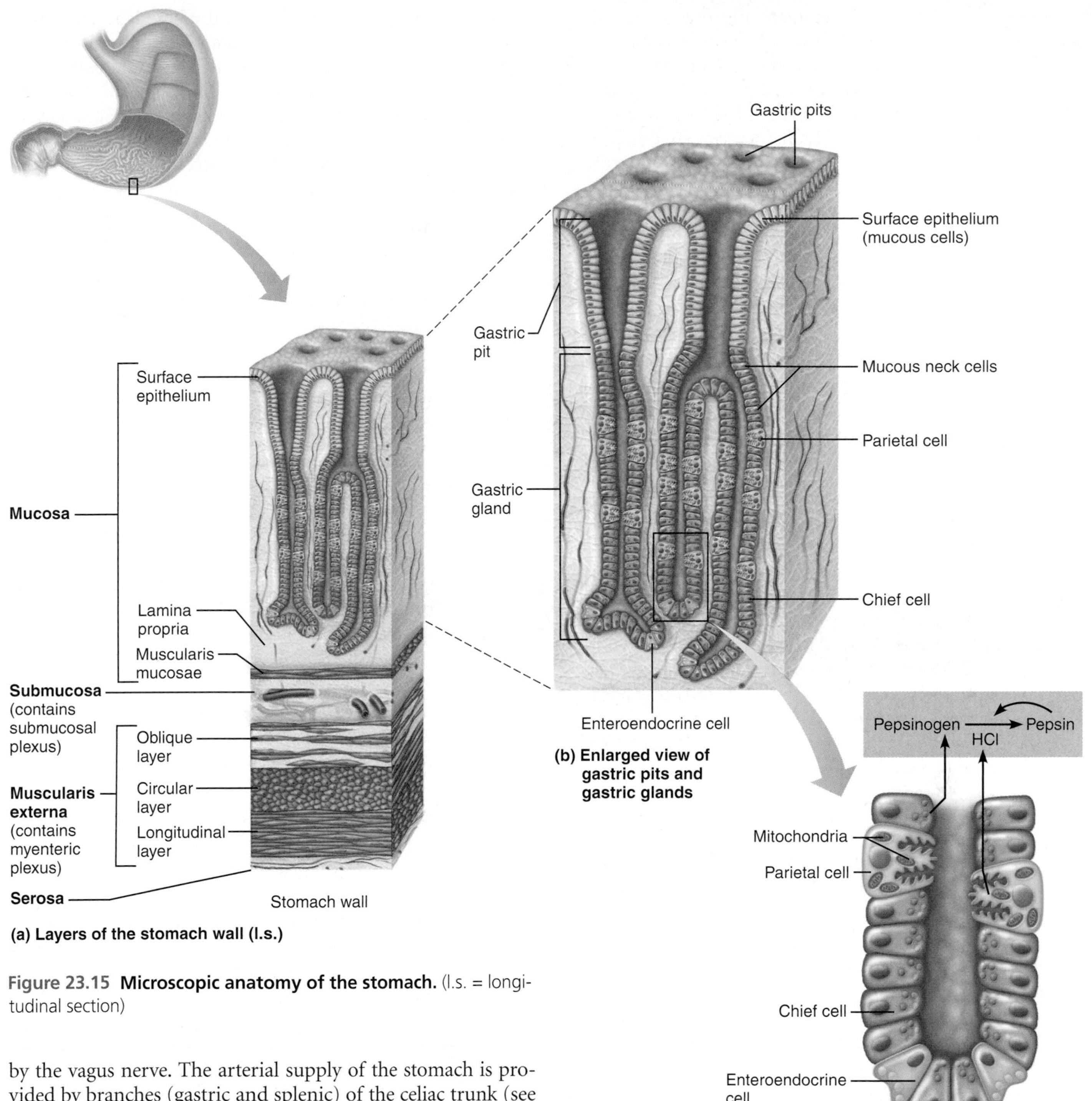

(a) Layers of the stomach wall (l.s.)

(b) Enlarged view of gastric pits and gastric glands

(c) Location of the HCl-producing parietal cells and pepsin-secreting chief cells in a gastric gland

Figure 23.15 Microscopic anatomy of the stomach. (l.s. = longitudinal section)

by the vagus nerve. The arterial supply of the stomach is provided by branches (gastric and splenic) of the celiac trunk (see Figure 19.24). The corresponding veins are part of the hepatic portal system and ultimately drain into the hepatic portal vein (see Figure 19.29c).

Microscopic Anatomy

The stomach wall contains the four tunics typical of most of the alimentary canal, but its muscularis and mucosa are modified for the special roles of the stomach. Besides the usual circular and longitudinal layers of smooth muscle, the muscularis externa has an innermost layer of smooth muscle fibers that runs *obliquely* (Figure 23.14a and **Figure 23.15a**). This arrangement allows the stomach not only to mix, churn, and move food along the tract (the job of the circular and longitudinal muscle layers), but also to pummel the food, physically breaking it down into smaller fragments, and to ram the food into the small intestine. (The oblique fibers accomplish the ramming by jackknifing the stomach into a V shape, which provides a propulsive action at the stomach terminus.)

The lining epithelium of the stomach mucosa is a simple columnar epithelium composed entirely of mucous cells. They produce a cloudy, protective two-layer coat of alkaline mucus in which the surface layer consists of viscous, insoluble mucus that traps a layer of bicarbonate-rich fluid beneath it. This otherwise smooth lining is dotted with millions of deep **gastric pits**, which lead into the tubular **gastric glands** that produce the stomach secretion called **gastric juice** (Figure 23.15).

The cells forming the walls of the gastric pits are primarily mucous cells, but those composing the gastric glands vary in different stomach regions. For example, the cells in the glands of the cardia and pylorus are primarily mucus secreting, whereas cells of the pyloric antrum produce mucus and several hormones including most of the stimulatory hormone called gastrin. Glands of the stomach fundus and body, where most chemical digestion occurs, are substantially larger and produce the majority of the stomach secretions. The glands in these regions contain a variety of secretory cells, including these four types:

1. **Mucous neck cells**, found in the upper, or "neck," regions of the glands, produce a thin, soluble mucus quite different from that secreted by the mucous cells of the surface epithelium (Figure 23.15b). It is not yet understood what special function this *acidic* mucus performs.
2. **Parietal cells**, found mainly in the middle region of the glands scattered among the chief cells (described next), simultaneously secrete *hydrochloric acid (HCl)* and *intrinsic factor* (Figure 23.15b, c). Although the parietal cells appear spherical when viewed with a light microscope, they actually have three prongs that bear dense microvilli (they look like fuzzy pitchforks!). This structure provides a huge surface area for secreting H^+ and Cl^- into the stomach lumen. HCl makes the stomach contents extremely acidic (pH 1.5–3.5), a condition necessary for activation and optimal activity of pepsin. The acidity also helps in food digestion by denaturing proteins and breaking down cell walls of plant foods, and is harsh enough to kill many of the bacteria ingested with foods. Intrinsic factor is a glycoprotein required for vitamin B_{12} absorption in the small intestine.
3. **Chief cells** occur mainly in the basal regions of the gastric glands. The chief cells produce *pepsinogen* (pep-sin′o-jen), the inactive form of the protein-digesting enzyme **pepsin**. When these cells are stimulated, the first pepsinogen molecules they release are activated by HCl encountered in the apical region of the gland (Figure 23.15c). But once pepsin is present, it also catalyzes the conversion of pepsinogen to pepsin. The activation process involves removal of a small peptide fragment from the pepsinogen molecule, causing it to change shape and expose its active site. This positive feedback process is limited only by the amount of pepsinogen present. Chief cells also secrete insignificant amounts of lipases (fat-digesting enzymes).
4. **Enteroendocrine cells** (en″ter-o-en′do-krin; "gut endocrine"), typically located deep in the gastric glands (Figure 23.15b, c), release a variety of chemical messengers directly into the interstitial fluid of the lamina propria. Some of these, for example **histamine** and **serotonin**, act locally as paracrines. Others, such as **somatostatin**, act both locally and as hormones, diffusing into the blood capillaries to influence several digestive system target organs (**Table 23.1**, p. 875). **Gastrin**, a hormone, plays essential roles in regulating stomach secretion and motility, as we will describe shortly.

The stomach mucosa is exposed to some of the harshest conditions in the entire digestive tract. Gastric juice is corrosively acidic (the H^+ concentration in the stomach can be 100,000 times that found in blood), and its protein-digesting enzymes can digest the stomach itself.

However, the stomach is not a passive victim of its formidable environment. It mounts an aggressive counterattack to protect itself, producing what is called the **mucosal barrier**. Three factors create this barrier:

1. *A thick coating of bicarbonate-rich mucus* is built up on the stomach wall.
2. *The epithelial cells of the mucosa are joined together by tight junctions* that prevent gastric juice from leaking into the underlying tissue layers.
3. *Damaged epithelial mucosal cells are shed and quickly replaced* by division of *undifferentiated stem cells* that reside where the gastric pits join the gastric glands. The stomach surface epithelium of mucous cells is completely renewed every three to six days, because these cells can survive only a few days in the stomach's harsh environment. (However, the more sheltered glandular cells deep within the gastric glands have a much longer life span.)

HOMEOSTATIC IMBALANCE

Anything that breaches the gel-like mucosal barrier causes inflammation of the stomach wall, a condition called *gastritis*. Persistent damage to the underlying tissues can promote **peptic ulcers**, specifically called **gastric ulcers** when they are erosions of the stomach wall (**Figure 23.16a**). The most distressing symptom of gastric ulcers is gnawing epigastric pain that seems to bore through to your back. The pain typically occurs 1–3 hours after eating and is often relieved by eating again. The danger posed by ulcers is perforation of the stomach wall followed by peritonitis and, perhaps, massive hemorrhage.

For years, the blame for causing ulcers was put on factors that favor high HCl production or low mucus secretion such as aspirin and nonsteroidal anti-inflammatory drugs (NSAIDs such as ibuprofen), smoking, spicy food, alcohol, coffee, and stress. Although acid conditions *are* necessary for ulcer formation, acidity in and of itself is not sufficient to cause ulcer formation. Most recurrent ulcers (90%) are the work of a certain strain of acid-resistant, corkscrew-shaped *Helicobacter pylori* bacteria (Figure 23.16b), which burrow much like a drill bit through the mucus and destroy the protective mucosal layer, leaving denuded areas.

These pathological effects occur in 10–20% of infected individuals. The antimicrobial activity of gastric mucin appears to play a major role in protecting the remaining 80–90% from *H. pylori*'s invasive attacks. The bacterial causal theory has been difficult to prove because the bacterium is also found in more than 50% of healthy people. Even more troubling are studies that link this bacterium to some stomach cancers.

These bacteria release several chemicals that help them do their "dirty work," including (1) some that curb HCl production and release ammonia (which then acts as a base to neutralize some of the stomach acid in their locale), (2) a *cytotoxin* that damages the stomach epithelium, (3) proteins that disrupt adhesion molecules and so cause the epithelial cells to detach from each other, and (4) proteins that act as chemotactic agents to attract macrophages and other defensive cells into the area, thus promoting a chronic inflammatory response.

The presence or absence of *H. pylori* is easily detected by a breath test. In ulcers colonized by it, the goal is to kill the embedded bacteria. A simple two-week-long course of antibiotics promotes healing and prevents recurrence. For active ulcers, a blocker for H_2- (histamine) receptors may also help because it inhibits HCl secretion by blocking histamine's effects. The relatively few peptic ulcers not caused by *H. pylori* generally result from the long-term use of NSAIDs. In such noninfectious cases, H_2-receptor blocker drugs such as cimetidine (Tagamet) and ranitidine (Zantac) are the therapy of choice. ■

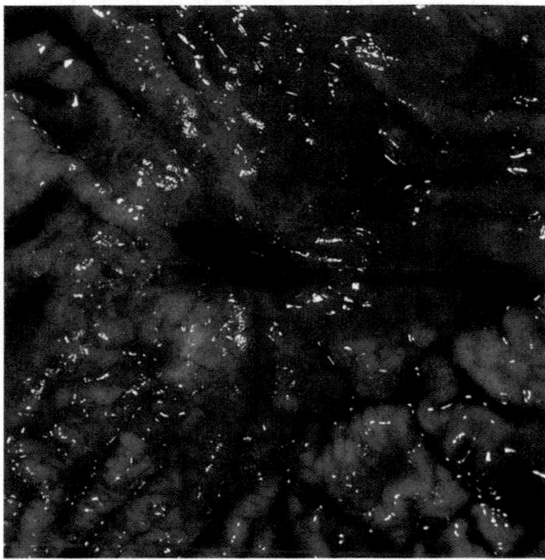

(a) A gastric ulcer lesion

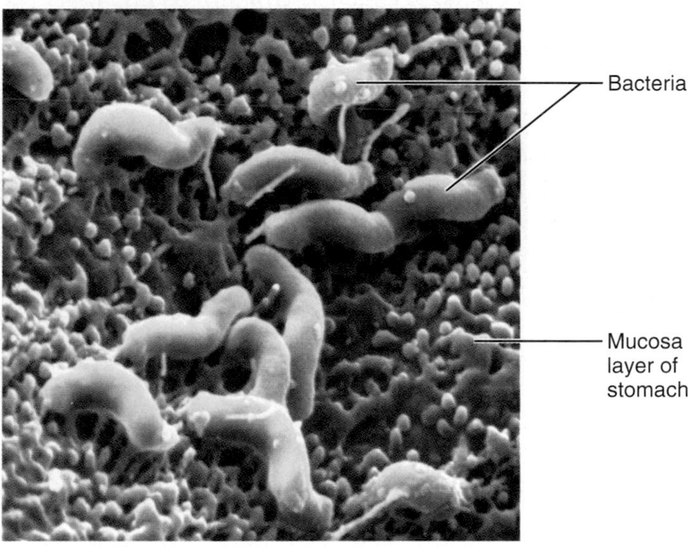

(b) *H. pylori* bacteria

Figure 23.16 **Photographs of a gastric ulcer lesion and of the bacteria that most commonly cause it.**

CHECK YOUR UNDERSTANDING

24. What structural modification of the stomach wall underlies the stomach's ability to mechanically digest food?

25. Two substances secreted by cells of the gastric glands are needed to produce the active protein-digesting enzyme pepsin. What are these substances and what cells secrete them?

26. What protective substances or activities make up the so-called mucosal barrier?

For answers, see Appendix G.

Digestive Processes Occurring in the Stomach

- Explain how gastric secretion and stomach motility are regulated.
- Define and account for the alkaline tide.

Except for ingestion and defecation, the stomach is involved in the whole "menu" of digestive activities. Besides serving as a holding area for ingested food, the stomach continues the demolition job begun in the oral cavity by further degrading food both physically and chemically. It then delivers chyme, the product of its activity, into the small intestine.

Protein digestion begins in the stomach and is the only significant type of enzymatic digestion that occurs there. Dietary proteins are denatured by HCl produced by stomach glands in preparation for enzymatic digestion. (The unfolded amino acid chain is more accessible to the enzymes.) The most important protein-digesting enzyme produced by the gastric mucosa is pepsin. In infants, however, the stomach glands also secrete **rennin**, an enzyme that acts on milk protein (casein), converting it to a curdy substance that looks like soured milk. Additionally, lingual lipase released by the intrinsic salivary glands may digest some triglycerides in the stomach.

Two common lipid-soluble substances—alcohol and aspirin—pass easily through the stomach mucosa into the blood. Alcohol and aspirin may cause gastric bleeding, so these substances should be avoided by people with gastric ulcers.

Despite the obvious benefits of preparing food to enter the intestine, the only stomach function essential to life is secretion of intrinsic factor. **Intrinsic factor** is required for intestinal absorption of vitamin B_{12}, needed to produce mature erythrocytes. In its absence, *pernicious anemia* results. However, if

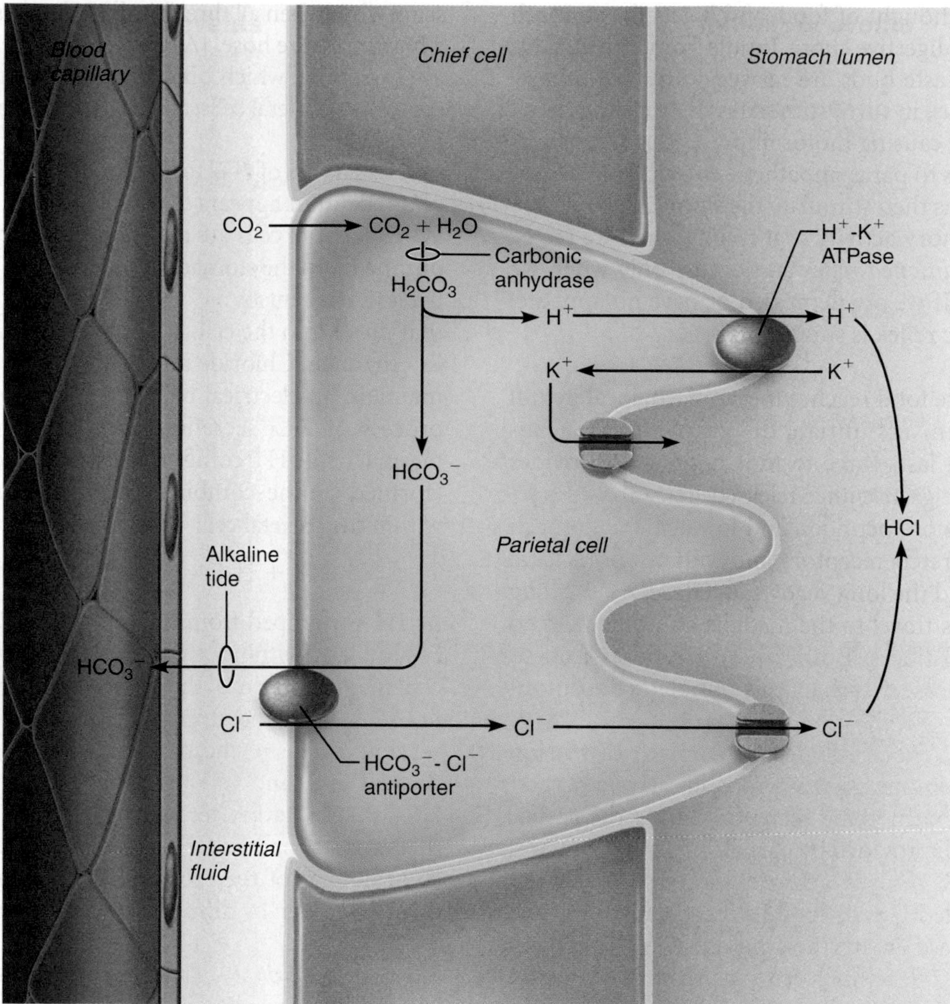

Figure 23.18 Mechanism of HCl secretion by parietal cells. H^+ and HCO_3^- (bicarbonate ions) are generated from the dissociation of carbonic acid (H_2CO_3) within the parietal cell. As H^+-K^+ ATPase pumps H^+ into the lumen, K^+ enters the cell. Meanwhile, the HCO_3^--Cl^- antiporter transports HCO_3^- into the interstitial space in exchange for chloride ions (Cl^-), establishing the alkaline tide. Cl^- and K^+ then diffuse into the lumen through membrane channels.

23

Gastric Motility and Emptying

Stomach contractions not only accommodate its filling and cause its emptying, but they also compress, knead, and continually mix the food with gastric juice to produce chyme. The processes of mechanical digestion and propulsion are inseparable in the stomach because the mixing movements are accomplished by a unique type of peristalsis. For example, in the pylorus, peristalsis is bidirectional rather than unidirectional.

Response of the Stomach to Filling The stomach stretches to accommodate incoming food, but internal stomach pressure remains constant until about 1.5 L of food has been ingested. Thereafter, the pressure rises. The relatively unchanging pressure in a filling stomach is due to (1) the reflex-mediated receptive relaxation of the stomach muscle and (2) the plasticity of visceral smooth muscle.

Receptive relaxation of smooth muscle in the stomach fundus and body occurs both in anticipation of and in response to food movement through the esophagus and into the stomach. This process is coordinated by the swallowing center of the brain stem and mediated by the vagus nerves acting on serotonin and NO-releasing enteric neurons.

Gastric accommodation, an example of smooth muscle *plasticity*, is the intrinsic ability of visceral smooth muscle to exhibit the *stress-relaxation response*, in other words, to be stretched without greatly increasing its tension and contracting expulsively. As we described in Chapter 9, this capability is very important in hollow organs, like the stomach, that must serve as temporary reservoirs.

Gastric Contractile Activity As in the esophagus, the stomach exhibits peristalsis. After a meal, peristalsis begins near the gastroesophageal sphincter, where it produces gentle rippling movements of the thin stomach wall. But as the contractions approach the pylorus, where the stomach musculature is thicker, they become much more powerful. Consequently, the

TABLE 23.1 Hormones and Paracrines That Act in Digestion*

HORMONE	SITE OF PRODUCTION	STIMULUS FOR PRODUCTION	TARGET ORGAN	ACTIVITY
Cholecystokinin (CCK)	Duodenal mucosa	Fatty chyme, in particular, but also partially digested proteins	Liver/pancreas	▪ Potentiates secretin's actions on these organs
			Pancreas	▪ Increases output of enzyme-rich pancreatic juice
			Gallbladder	▪ Stimulates organ to contract and expel stored bile
			Hepatopancreatic sphincter (of Oddi)	▪ Relaxes sphincter to allow entry of bile and pancreatic juice into duodenum
Gastric inhibitory peptide (GIP) (or glucose-dependent insulinotropic peptide)	Duodenal mucosa	Glucose, fatty acids, and amino acids in small intestine	Stomach	▪ Inhibits HCl production (minor effect)
			Pancreas (beta cells)	▪ Stimulates insulin release
Gastrin	Stomach mucosa (G cells)	Food (particularly partially digested proteins) in stomach (chemical stimulation); acetylcholine released by nerve fibers	Stomach (parietal cells)	▪ Increases HCl secretion
				▪ Stimulates gastric emptying (minor effect)
			Small intestine	▪ Stimulates contraction of intestinal muscle
			Ileocecal valve	▪ Relaxes ileocecal valve
			Large intestine	▪ Stimulates mass movements
Histamine	Stomach mucosa	Food in stomach	Stomach	▪ Activates parietal cells to release HCl
Intestinal gastrin	Duodenal mucosa	Acidic and partially digested foods in duodenum	Stomach	▪ Stimulates gastric glands and motility
Motilin	Duodenal mucosa	Fasting; periodic release every 1½–2 hours by neural stimuli	Proximal duodenum	▪ Stimulates migrating motility complex
Secretin	Duodenal mucosa	Acidic chyme (also partially digested proteins, fats, hypertonic or hypotonic fluids, or irritants in chyme)	Stomach	▪ Inhibits gastric gland secretion and gastric motility during gastric phase of secretion
			Pancreas	▪ Increases output of pancreatic juice rich in bicarbonate ions; potentiates CCK's action
			Liver	▪ Increases bile output
Serotonin	Stomach mucosa	Food in stomach	Stomach	▪ Causes contraction of stomach muscle
Somatostatin	Stomach mucosa; duodenal mucosa	Food in stomach; stimulation by sympathetic nerve fibers	Stomach	▪ Inhibits gastric secretion of all products
			Pancreas	▪ Inhibits secretion
			Small intestine	▪ Inhibits GI blood flow; thus inhibits intestinal absorption
			Gallbladder and liver	▪ Inhibits contraction and bile release
Vasoactive intestinal peptide (VIP)	Enteric neurons	Chyme containing partially digested foods	Small intestine	▪ Stimulates buffer secretion; dilates intestinal capillaries
				▪ Relaxes intestinal smooth muscle
			Pancreas	▪ Increases secretion
			Stomach	▪ Inhibits acid secretion

* Except for somatostatin, all of these polypeptides also stimulate the growth (particularly of the mucosa) of the organs they affect.

contents of the fundus and body (food storage area) remain relatively undisturbed, while foodstuffs in and around the pyloric antrum receive a truly lively pummeling and mixing.

The pyloric region of the stomach, which holds about 30 ml of chyme, acts as a "dynamic filter" that allows only liquids and small particles to pass through the barely open pyloric valve during the digestive period. Normally, each peristaltic wave reaching the pyloric muscle "spits" or squirts 3 ml or less of chyme into the small intestine. Because the contraction also *closes* the valve, which is normally partially relaxed, the rest (about 27 ml) is

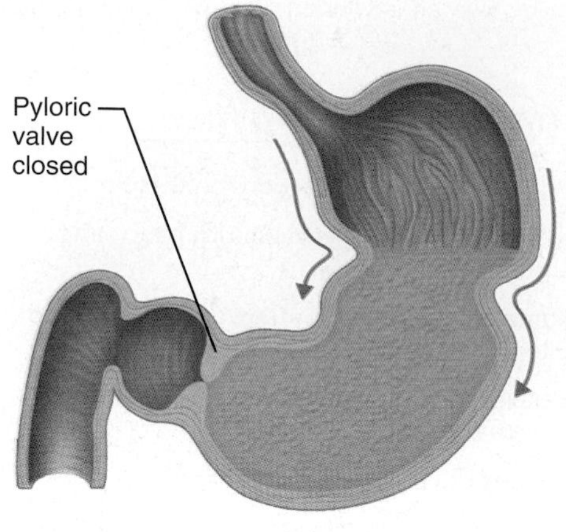

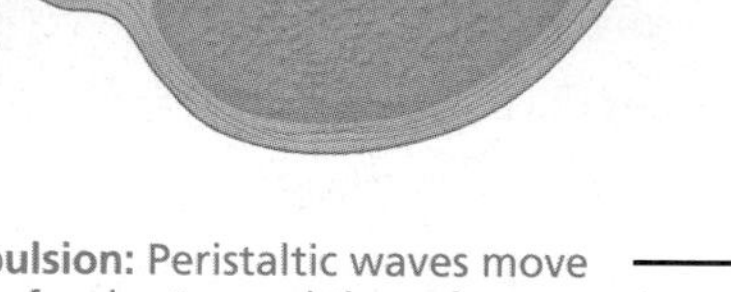

(1) **Propulsion:** Peristaltic waves move from the fundus toward the pylorus.

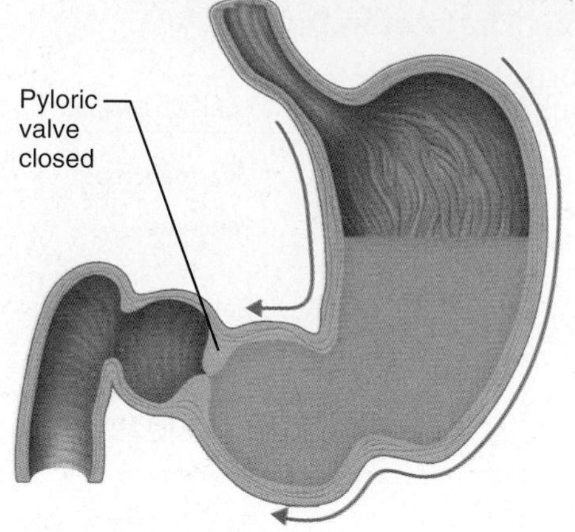

(2) **Grinding:** The most vigorous peristalsis and mixing action occur close to the pylorus.

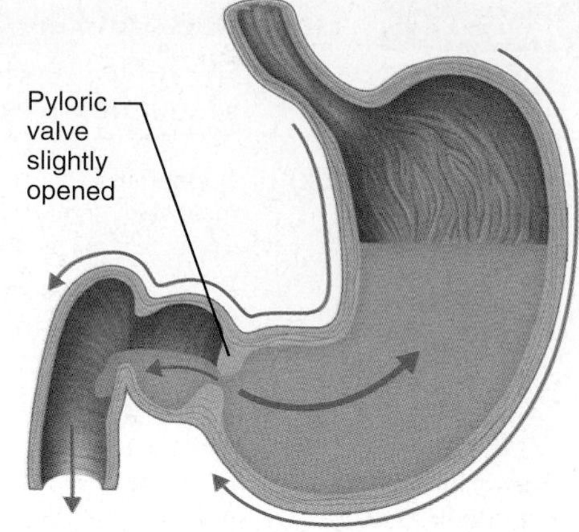

(3) **Retropulsion:** The pyloric end of the stomach acts as a pump that delivers small amounts of chyme into the duodenum, simultaneously forcing most of its contained material backward into the stomach.

Figure 23.19 Peristaltic waves in the stomach.

propelled backward into the stomach, where it is mixed further **(Figure 23.19)**. This back-and-forth pumping action (retropulsion) effectively breaks up solids in the gastric contents.

Although the intensity of the stomach's peristaltic waves can be modified, their rate is constant—always around three per minute. This contractile rhythm is set by the spontaneous activity of *pacemaker cells* located in the longitudinal smooth muscle layer. The pacemaker cells, muscle-like noncontractile cells called *interstitial cells of Cajal* (ka-hal′), depolarize and repolarize spontaneously three times each minute, establishing the so-called *cyclic slow waves* of the stomach, or its **basic electrical rhythm (BER)**. Since the pacemakers are electrically coupled to the rest of the smooth muscle sheet by gap junctions, their "beat" is transmitted efficiently and quickly to the entire muscularis.

The pacemakers set the maximum rate of contraction, but they do not initiate the contractions or regulate their force. Instead, they generate subthreshold depolarization waves, which are then "ignited" (enhanced by further depolarization and brought to threshold) by neural and hormonal factors.

Factors that increase the strength of stomach contractions are the same factors that enhance gastric secretory activity. Distension of the stomach wall by food activates stretch receptors and gastrin-secreting cells, both of which ultimately stimulate gastric smooth muscle and so increase gastric motility. For this reason, the more food there is in the stomach, the more vigorous the stomach mixing and emptying movements will be—within certain limits—as we describe next.

Regulation of Gastric Emptying The stomach usually empties completely within four hours after a meal. However, the larger the meal (the greater the stomach distension) and the more liquid its contents, the faster the stomach empties. Fluids pass quickly through the stomach. Solids linger, remaining until they are well mixed with gastric juice and converted to the liquid state.

The rate of gastric emptying depends as much—and perhaps more—on the contents of the duodenum as on what is happening in the stomach. The stomach and duodenum act in tandem like a "coupled meter" that functions at less than full capacity. As chyme enters the duodenum, receptors in its wall respond to chemical signals and to stretch, initiating the enterogastric reflex and the hormonal (enterogastrone) mechanisms that inhibit acid and pepsin secretion as we described earlier. These mechanisms inhibit gastric secretory activity and prevent further duodenal filling by reducing the force of pyloric contractions **(Figure 23.20)**.

A carbohydrate-rich meal moves through the duodenum rapidly, but fats form an oily layer at the top of the chyme and are digested more slowly by enzymes acting in the intestine. For this reason, when chyme entering the duodenum is fatty, food may remain in the stomach six hours or more.

HOMEOSTATIC IMBALANCE

Vomiting, or **emesis**, is an unpleasant experience that causes stomach emptying by a different route. Many factors signal the stomach to "launch lunch," but the most common are extreme stretching of the stomach or intestine or the presence of irritants such as bacterial toxins, excessive alcohol, spicy foods, and certain drugs in those organs. Both bloodborne molecules and sensory impulses streaming from the irritated sites to the **emetic center** (e-met′ik) of the medulla initiate a number of motor responses. The diaphragm and abdominal wall muscles contract, increasing intra-abdominal pressure, the gastroesophageal sphincter relaxes, and the soft palate rises to close off the nasal passages. As a result, the stomach (and perhaps duodenal) contents are forced upward through the esophagus and pharynx and out the mouth.

Before vomiting, an individual typically is pale, feels nauseated, and salivates. Excessive vomiting can cause dehydration and may lead to severe disturbances in the electrolyte and acid-base balance of the body. Since large amounts of HCl are lost in vomitus, the blood becomes alkaline as the stomach attempts to replace its lost acid. ■

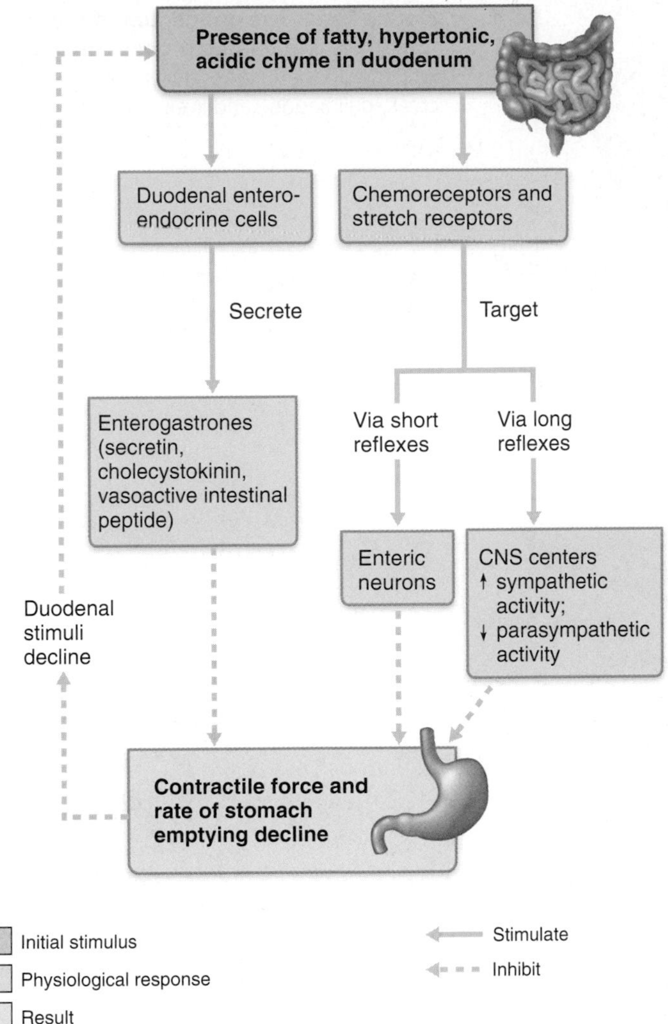

Figure 23.20 Neural and hormonal factors inhibiting gastric emptying. These controls ensure that the food will be well liquefied in the stomach and prevent the small intestine from being overwhelmed.

CHECK YOUR UNDERSTANDING

27. Name the three phases of gastric secretion.

28. How does the presence of food in the small intestine inhibit gastric secretion and motility?

29. How does the pH of venous blood leaving the stomach change during a meal?

For answers, see Appendix G.

The Small Intestine and Associated Structures

- Identify and describe structural modifications of the wall of the small intestine that enhance the digestive process.
- Differentiate between the roles of the various cell types of the intestinal mucosa.
- Describe the function of intestinal hormones and paracrines.

In the small intestine, usable food is finally prepared for its journey into the cells of the body. However, this vital function cannot be accomplished without the aid of secretions from the liver (bile) and pancreas (digestive enzymes). We will also consider these accessory organs in this section.

The Small Intestine

The **small intestine** is the body's major digestive organ. Within its twisted passageways, digestion is completed and virtually all absorption occurs.

Gross Anatomy

The small intestine is a convoluted tube extending from the pyloric sphincter in the epigastric region to the **ileocecal valve (sphincter)** (il″e-o-se′kal) in the right iliac region where it joins the large intestine. It is the longest part of the alimentary canal, but is only about half the diameter of the large intestine, ranging from 2.5 to 4 cm (1–1.6 inches). Although it is 6–7 m long (approximately 20 ft, or as tall as a two-story building) in a cadaver, the small intestine is only about 2–4 m (7–13 ft) long during life because of muscle tone.

The small intestine has three subdivisions: the duodenum, which is mostly retroperitoneal, and the jejunum and ileum, both intraperitoneal organs (see Figure 23.1). The relatively immovable **duodenum** (du″o-de′num; "twelve finger widths long"), which curves around the head of the pancreas, is about 25 cm (10 inches) long **(Figure 23.21)**. Although it is the shortest intestinal subdivision, the duodenum has the most features of interest.

The bile duct, delivering bile from the liver, and the main pancreatic duct, carrying pancreatic juice from the pancreas, unite in the wall of the duodenum in a bulblike point called the **hepatopancreatic ampulla** (hep″ah-to-pan″kre-at′ik ampul′ah; *ampulla* = flask). The ampulla opens into the duodenum via the volcano-shaped **major duodenal papilla**. The entry of bile and pancreatic juice is controlled by a smooth muscle valve called the **hepatopancreatic sphincter.**

The **jejunum** (jĕ-joo′num; "empty"), about 2.5 m (8 ft) long, extends from the duodenum to the ileum. The **ileum** (il′e-um; "twisted"), approximately 3.6 m (12 ft) in length, joins the large intestine at the ileocecal valve. The jejunum and ileum hang in sausagelike coils in the central and lower part of the abdominal cavity, suspended from the posterior abdominal wall by the fan-shaped *mesentery* (see Figure 23.30). These more distal parts of the small intestine are encircled and framed by the large intestine.

Nerve fibers serving the small intestine include parasympathetics from the vagus and sympathetics from the thoracic splanchnic nerves, both relayed through the superior mesenteric (and celiac) plexus.

The arterial supply is primarily from the superior mesenteric artery (pp. 732–733). The veins parallel the arteries and typically drain into the superior mesenteric vein. From there, the

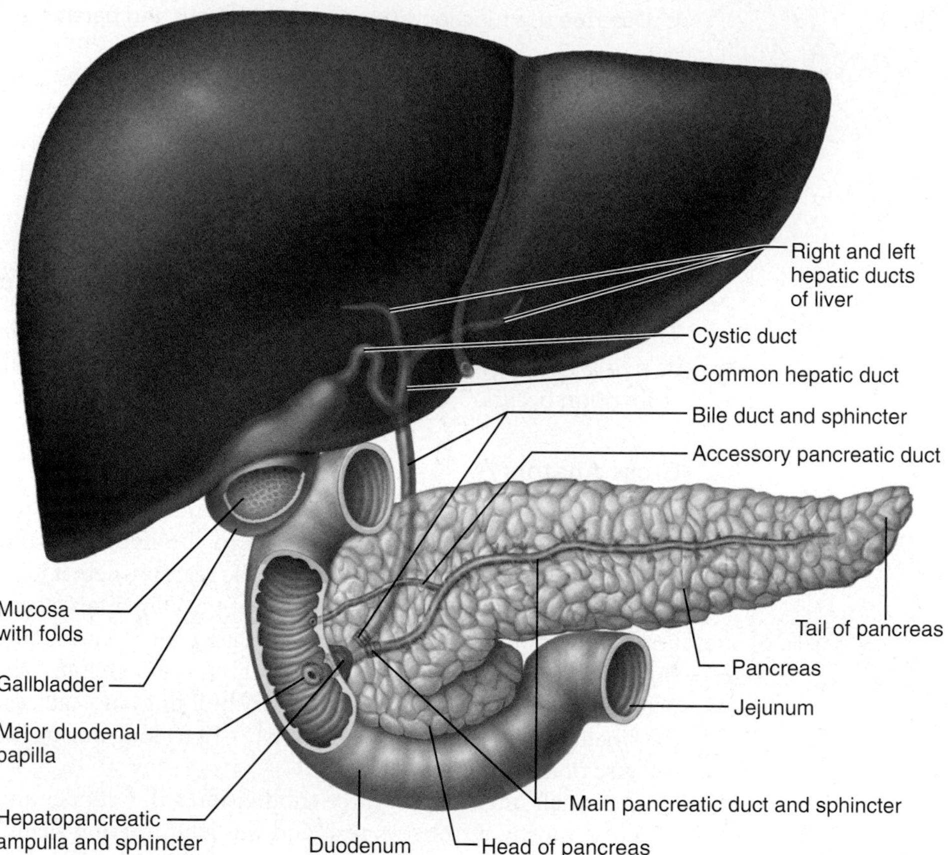

Figure 23.21 The duodenum of the small intestine, and related organs. Emptying into the duodenum are ducts from the pancreas, gallbladder, and liver.

nutrient-rich venous blood from the small intestine drains into the hepatic portal vein, which carries it to the liver.

Microscopic Anatomy

Modifications for Absorption The small intestine is highly adapted for nutrient absorption. Its length alone provides a huge surface area, and its wall has three structural modifications—circular folds, villi, and microvilli—that amplify its absorptive surface enormously (by a factor of more than 600 times). One estimate is that the intestinal surface area is about equal to 200 square meters or the surface area of a doubles tennis court. Most absorption occurs in the proximal part of the small intestine, so these specializations decrease in number toward its distal end.

The **circular folds**, or **plicae circulares** (pli′ke ser″ku-lar′ēs), are deep, permanent folds of the mucosa and submucosa **(Figure 23.22a)**. Nearly 1 cm tall, these folds force chyme to spiral through the lumen, slowing its movement and allowing time for full nutrient absorption.

Villi (vil′i; "tufts of hair") are fingerlike projections of the mucosa, over 1 mm high, that give it a velvety texture, much like the soft nap of a towel (Figures 23.22 and **Figure 23.23a**). The epithelial cells of the villi (called *enterocytes*) are chiefly absorptive columnar cells. In the core of each villus is a dense capillary bed and a wide lymph capillary called a **lacteal** (lak′te-al). Digested foodstuffs are absorbed through the epithelial cells into both the capillary blood and the lacteal.

The villi are large and leaflike in the duodenum (the intestinal site of most active absorption) and gradually narrow and shorten along the length of the small intestine. A "slip" of smooth muscle in the villus core allows it to alternately shorten and lengthen. These pulsations (1) increase the contact between the villus and the contents of the intestinal lumen, making absorption more efficient, and (2) "milk" lymph along through the lacteals.

The exceptionally long, densely packed **microvilli** of the absorptive cells of the mucosa give the mucosal surface a fuzzy appearance called the **brush border** (Figure 23.22b enlargment and Figure 23.23b). The plasma membranes of the microvilli bear enzymes referred to as **brush border enzymes**, which complete the digestion of carbohydrates and proteins in the small intestine.

Histology of the Wall Externally the subdivisions of the small intestine appear to be nearly identical, but their internal and microscopic anatomies reveal some important differences. The four tunics typical of the GI tract are also seen here, but the mucosa and submucosa are modified to reflect the intestine's functions in the digestive pathway.

The epithelium of the villus mucosa is largely simple columnar *absorptive cells* bound by tight junctions and richly endowed with microvilli. These cells bear the primary responsibility for nutrient and electrolyte absorption. The epithelium also has many mucus-secreting *goblet cells*. Between the villi, the mucosa is studded with *pits* that lead into tubular glands called

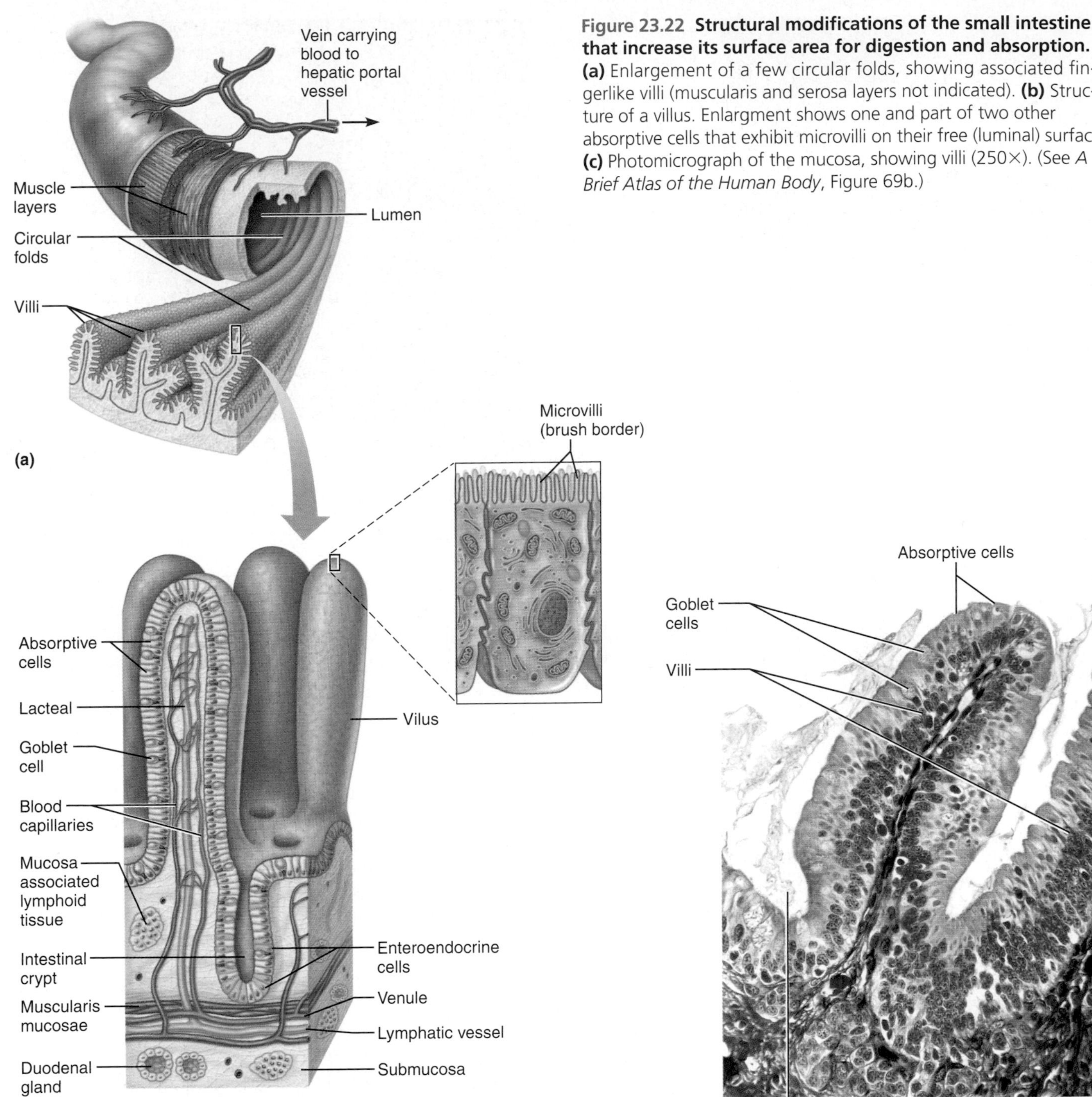

Figure 23.22 Structural modifications of the small intestine that increase its surface area for digestion and absorption. **(a)** Enlargement of a few circular folds, showing associated fingerlike villi (muscularis and serosa layers not indicated). **(b)** Structure of a villus. Enlargment shows one and part of two other absorptive cells that exhibit microvilli on their free (luminal) surface. **(c)** Photomicrograph of the mucosa, showing villi (250×). (See *A Brief Atlas of the Human Body*, Figure 69b.)

intestinal crypts, or **crypts of Lieberkühn** (le′ber-kun) (see Figure 23.22b, c). Crypt epithelial cells are primarily secretory cells that secrete *intestinal juice*, a watery mixture containing mucus that serves as a carrier fluid for absorbing nutrients from chyme. Scattered through the crypt epithelium are *enteroendocrine cells*, the source of the enterogastrones—secretin and cholecystokinin to name two—and T cells called *intraepithelial lymphocytes* (*IELs*), which represent an important immunological component. Unlike other T cells, IELs do *not* need "priming." Upon encountering antigens, they immediately release cytokines that cause killing of infected target cells.

Deep in the crypts are specialized secretory cells called *Paneth cells*, which fortify the small intestine's defenses by releasing antimicrobial agents such as defensins and *lysozyme*, an antibacterial enzyme. These secretions destroy certain bacteria and help to determine which bacteria may colonize the intestinal lumen. The crypts decrease in number along the length of the small intestine, but the goblet cells become more abundant.

The various epithelial cells arise from continuously dividing stem cells at the base of the crypts. As the daughter cells gradually migrate up the villi, they differentiate, becoming specialized cell

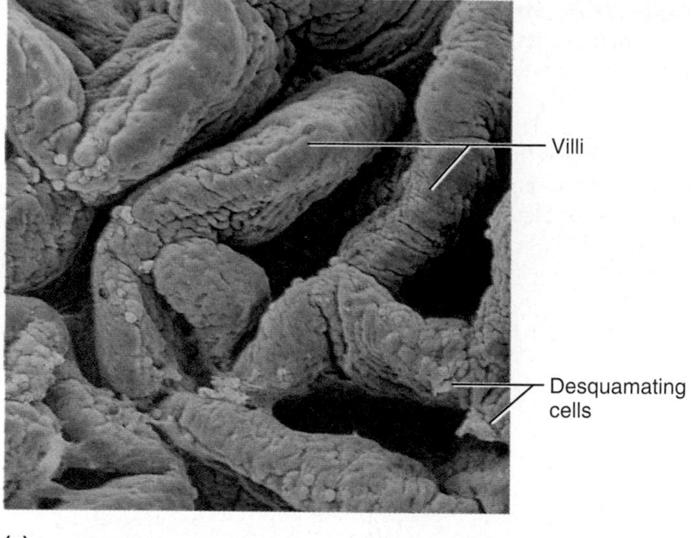

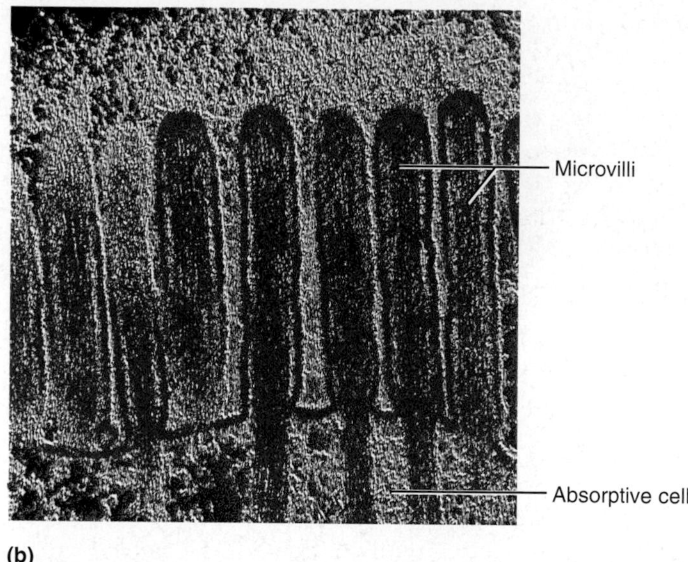

Figure 23.23 Villi and microvilli of the small intestine. False-color electron micrographs. **(a)** Villi (125×). Dead desquamating cells seen on the ridges of the villi are colored yellow and orange. **(b)** Microvilli (28,000×) of absorptive cells appear as red projections from the surface of the absorptive cell. Yellow granules are mucus granules.

types—absorptive cells, goblet cells, and enteroendocrine cells. The fourth differentiated cell type is the Paneth cells, which remain at the base of the crypts. The other three types undergo apoptosis and are shed from the villus tips, renewing the villus epithelium every two to four days.

The rapid replacement of intestinal (and gastric) epithelial cells has clinical as well as physiological significance. Treatments for cancer, such as radiation therapy and chemotherapy, preferentially target rapidly dividing cells. This kills cancer cells, but it also nearly obliterates the GI tract epithelium, causing nausea, vomiting, and diarrhea after each treatment.

The submucosa is typical areolar connective tissue. It contains both individual and *aggregated lymphoid follicles*, the latter called **Peyer's patches** (pi′erz). Peyer's patches increase in abundance toward the end of the small intestine. Their increasing abundance reflects the fact that this region of the small intestine contains huge numbers of bacteria that must be prevented from entering the bloodstream. The lymphoid tissue of the submucosa also contains proliferating lymphocytes that leave the intestine, enter the blood, and then home in on the intestinal lamina propria. There, in their new home, they release immunoglobin A (IgA), which helps provide protection against intestinal pathogens (see p. 784).

Elaborate mucus-secreting **duodenal glands** (also called *Brunner's glands*) are found in the submucosa of the duodenum only. These glands produce an alkaline (bicarbonate-rich) mucus that helps neutralize the acidic chyme moving in from the stomach. When this protective mucus barrier is inadequate, the intestinal wall erodes and *duodenal ulcers* result.

The muscularis is typical and bilayered. Except for the bulk of the duodenum, which is retroperitoneal and has an adventitia, the external intestinal surface is covered by visceral peritoneum (serosa).

Intestinal Juice: Composition and Control

The intestinal glands normally secrete 1 to 2 L of **intestinal juice** daily. The major stimulus for its production is distension or irritation of the intestinal mucosa by hypertonic or acidic chyme. Normally, intestinal juice is slightly alkaline (7.4–7.8), and isotonic with blood plasma. Intestinal juice is largely water but it also contains some mucus, which is secreted both by the duodenal glands and by goblet cells of the mucosa. Intestinal juice is enzyme-poor because intestinal enzymes are limited to the bound enzymes of the brush border.

CHECK YOUR UNDERSTANDING

30. What common advantage do circular folds, villi, and microvilli provide to the digestive process? Which of these modifications causes chyme to spiral through the lumen and slows its passage?
31. What are brush border enzymes?
32. What is a lacteal and what is its function?
33. Name three secretory products that help to protect the intestinal mucosa from bacterial damage.

For answers, see Appendix G.

The Liver and Gallbladder

- Describe the histologic anatomy of the liver.
- State the role of bile in digestion and describe how its entry into the small intestine is regulated.
- Describe the role of the gallbladder.

The *liver* and *gallbladder* are accessory organs associated with the small intestine. The liver, one of the body's most important organs, has many metabolic and regulatory roles. However, its *digestive* function is to produce bile for export to the duodenum. Bile is a fat emulsifier. In other words, it breaks up fats into tiny particles so that they are more accessible to digestive enzymes. We will describe bile and the emulsification process when we discuss the digestion and absorption of fats later in the chapter. Although the liver also processes nutrient-laden venous blood delivered to it directly from the digestive organs, this is a metabolic rather than a digestive role. (See Chapter 24.) The gallbladder is chiefly a storage organ for bile.

Gross Anatomy of the Liver

The ruddy, blood-rich **liver** is the largest gland in the body, weighing about 1.4 kg (3 lb) in the average adult. Shaped like a wedge, it occupies most of the right hypochondriac and epigastric regions, extending farther to the right of the body midline than to the left. Located under the diaphragm, the liver lies almost entirely within the rib cage, which provides some protection (see Figure 23.1 and **Figure 23.24**).

Typically, the liver is said to have four primary lobes. The largest of these, the *right lobe*, is visible on all liver surfaces and separated from the smaller *left lobe* by a deep fissure (Figure 23.24a). The posteriormost *caudate lobe* and the *quadrate lobe*, which lies inferior to the left lobe, are visible in an inferior view of the liver (Figure 23.24b).

A mesentery, the **falciform ligament**, separates the right and left lobes anteriorly and suspends the liver from the diaphragm and anterior abdominal wall. Running along the inferior edge of the falciform ligament is the **round ligament**, or **ligamentum teres** (te′rēz; "round"), a fibrous remnant of the fetal umbilical vein. Except for the superiormost liver area (the *bare area*), which touches the diaphragm, the entire liver is enclosed by the visceral peritoneum.

As we mentioned earlier, a ventral mesentery, the lesser omentum, anchors the liver to the lesser curvature of the stomach (see Figure 23.30b). The **hepatic artery** and the **hepatic portal vein**, which enter the liver at the **porta hepatis** ("gateway to the liver"), and the common hepatic duct, which runs inferiorly from the liver, all travel through the lesser omentum to reach their destinations. The gallbladder rests in a recess on the inferior surface of the right liver lobe (Figure 23.24b).

The traditional scheme of defining liver lobes (outlined above) has been criticized because it is based on superficial features of the liver. Some anatomists emphasize that the primary lobes of the liver should be defined as the territories served by the right and left hepatic ducts. These two territories are delineated by a plane drawn from the indentation (sulcus) of the inferior vena cava to the gallbladder recess (Figure 23.24b). The areas to the right of the plane are the *right lobe* and those to its left constitute the *left lobe*. According to this scheme, the small quadrate and caudate lobes are part of the left lobe.

Bile leaves the liver through several bile ducts that ultimately fuse to form the large **common hepatic duct**, which travels downward toward the duodenum. Along its course, that duct fuses with the **cystic duct** draining the gallbladder to form the **bile duct** (see Figure 23.21).

Microscopic Anatomy of the Liver

The liver is composed of sesame seed–sized structural and functional units called **liver lobules**. Each lobule is a roughly hexagonal (six-sided) structure consisting of plates of *liver cells*, or **hepatocytes** (hep′ah-to-sīts), organized like bricks in a garden wall **(Figure 23.25c)**. The hepatocyte plates radiate outward from a **central vein** (Figure 23.25c) running in the longitudinal axis of the lobule. To make a rough "model" of a liver lobule, open a thick paperback book until its two covers meet: The pages represent the plates of hepatocytes and the hollow cylinder formed by the rolled spine represents the central vein.

If you keep in mind that the liver's main function is to filter and process the nutrient-rich blood delivered to it, the description of its anatomy that follows will make a lot of sense. At each of the six corners of a lobule is a **portal triad** (*portal tract* region), so named because three basic structures are always present there: a branch of the *hepatic artery* (supplying oxygen-rich arterial blood to the liver), a branch of the *hepatic portal vein* (carrying venous blood laden with nutrients from the digestive viscera), and a *bile duct* (Figure 23.25c).

Between the hepatocyte plates are enlarged, leaky capillaries, the **liver sinusoids**. Blood from both the hepatic portal vein and the hepatic artery percolates from the triad regions through these sinusoids and empties into the central vein. From the central veins blood eventually enters the hepatic veins, which drain the liver, and empty into the inferior vena cava. Forming part of the sinusoid walls are star-shaped **hepatic macrophages**, also called **Kupffer cells** (koop′fer) (Figure 23.25c). They remove debris such as bacteria and worn-out blood cells from the blood as it flows past.

The versatile hepatocytes have large amounts of both rough and smooth ER, Golgi apparatuses, peroxisomes, and mitochondria. Equipped in this way, the hepatocytes not only can produce some 900 ml of bile daily but also can (1) process the bloodborne nutrients in various ways (e.g., they store glucose as glycogen and use amino acids to make plasma proteins); (2) store fat-soluble vitamins; and (3) play important roles in detoxification, such as ridding the blood of ammonia by converting it to urea (Chapter 24). Thanks to these various jobs, the processed blood leaving the liver contains fewer nutrients and waste materials than the blood that entered it.

In addition to providing the hepatocytes with blood for processing and nutrition, the intimate relationship between the hepatocytes and their blood supply has consequences for liver growth and repair. The regenerative capacity of the liver is exceptional. It can regenerate to its former size even after surgical removal or loss of 70% of its normal mass. During liver injury, hepatocytes secrete *vascular endothelial growth factor* (*VEGF*) which binds to specific receptors on endothelial cells lining the sinusoids. The endothelial cells proliferate and release other growth factors, such as hepatocyte growth factor (HGF) and interleukin 6, which in turn prompt the hepatocytes to multiply and replace dead and dying liver tissue.

Secreted bile flows through tiny canals, called **bile canaliculi** (kan″ah-lik′u-li; "little canals"), that run between adjacent

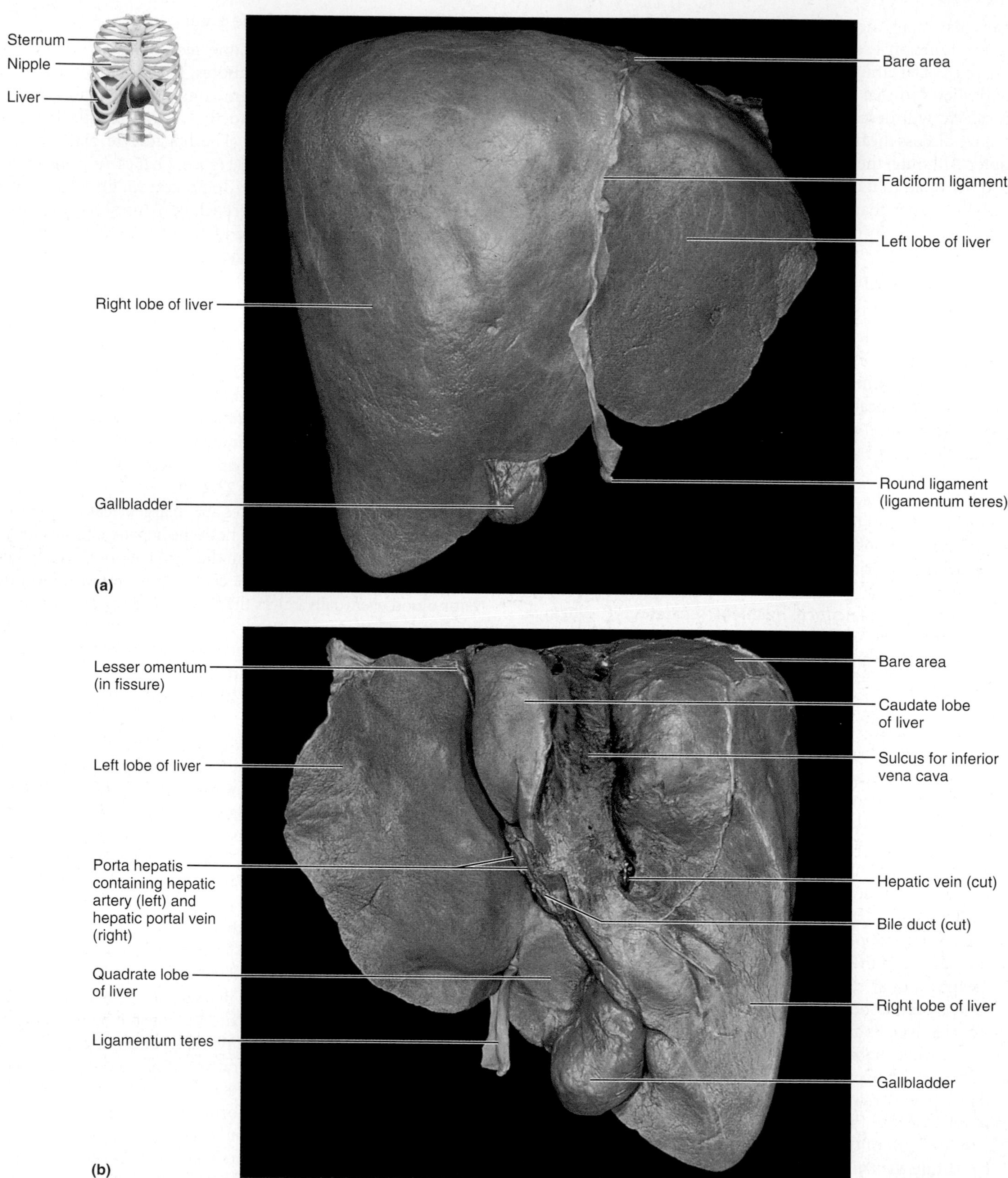

Figure 23.24 Gross anatomy of the human liver. (a) Anterior view of the liver. **(b)** Posteroinferior aspect (ventral surface) of the liver. The four liver lobes are separated by a group of fissures in this view. The porta hepatis is a deep fissure that contains the hepatic portal vein, hepatic artery, common hepatic duct, and lymphatics. (See *A Brief Atlas of the Human Body*, Figures 64 and 65.)

23

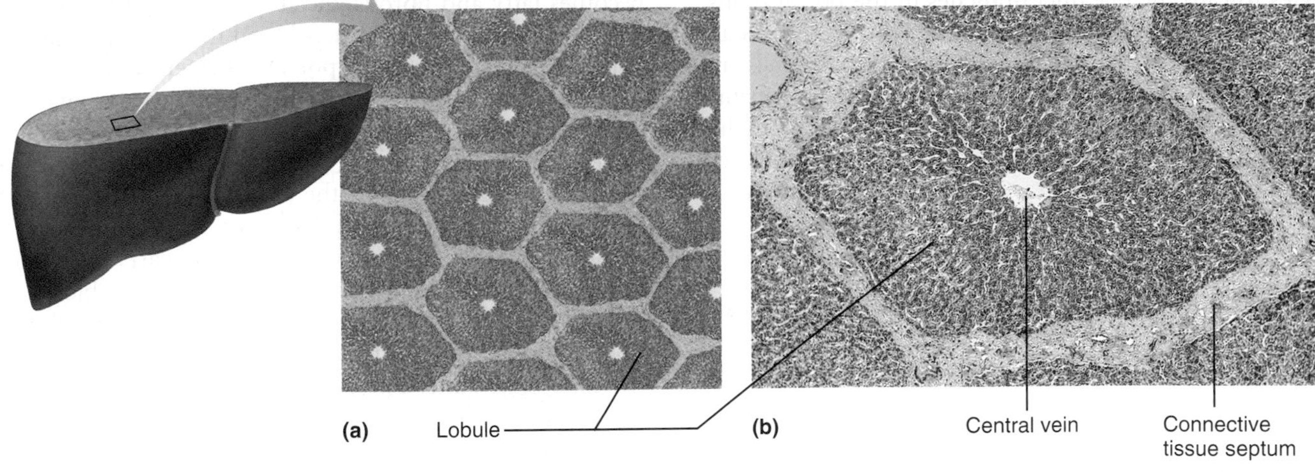

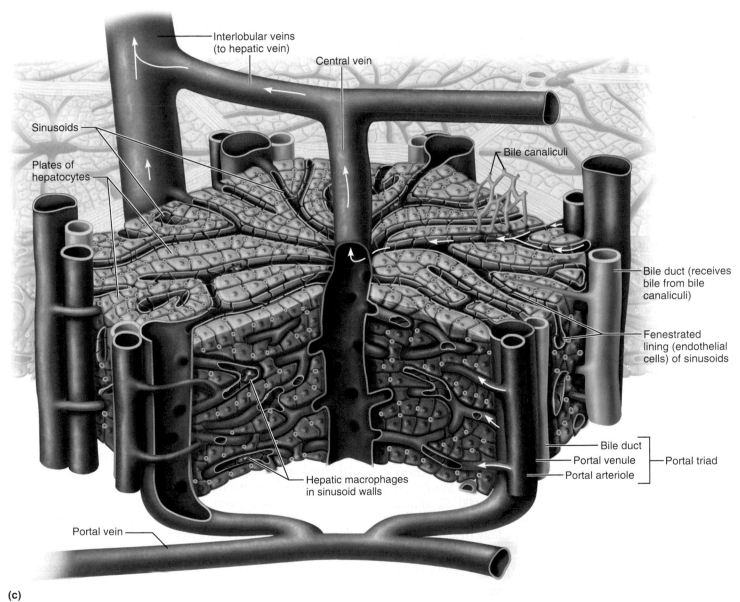

Interlobular veins (to hepatic vein)
Central vein
Sinusoids
Plates of hepatocytes
Bile canaliculi
Bile duct (receives bile from bile canaliculi)
Fenestrated lining (endothelial cells) of sinusoids
Bile duct
Portal venule
Portal arteriole
Portal triad
Hepatic macrophages in sinusoid walls
Portal vein
(c)

Figure 23.25 Microscopic anatomy of the liver. (a) Normal lobular pattern of the liver. **(b)** Enlarged view of one liver lobule. **(c)** Three-dimensional representation of a small portion of one liver lobule, showing the structure of sinusoids. Arrows indicate the direction of blood flow.

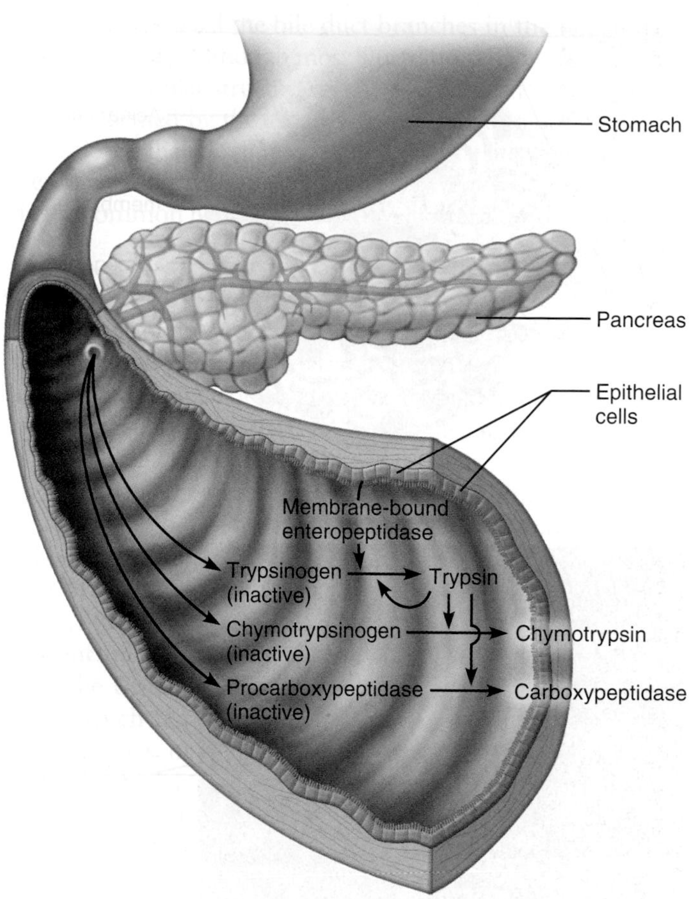

Figure 23.27 Activation of pancreatic proteases in the small intestine. Pancreatic proteases are secreted in an inactive form and are activated in the duodenum.

Scattered amid the acini are the more lightly staining *pancreatic islets* (*islets of Langerhans*). These mini-endocrine glands release insulin and glucagon, hormones that play an important role in carbohydrate metabolism, as well as several other hormones (see Chapter 16).

Composition of Pancreatic Juice

Approximately 1200 to 1500 ml of clear pancreatic juice is produced daily. It consists mainly of water, and contains enzymes and electrolytes (primarily bicarbonate ions). The acinar cells produce the enzyme-rich component of pancreatic juice. The epithelial cells lining the smallest pancreatic ducts release the bicarbonate ions that make it alkaline (about pH 8).

Normally, the amount of HCl produced in the stomach is exactly balanced by the amount of bicarbonate (HCO_3^-) secreted by the pancreas, and as HCO_3^- is secreted into the pancreatic juice, H^+ enters the blood. Consequently, the pH of venous blood returning to the heart remains relatively unchanged because alkaline blood draining from the stomach is neutralized by the acidic blood draining the pancreas.

The high pH of pancreatic fluid helps neutralize acid chyme entering the duodenum and provides the optimal environment for activity of intestinal and pancreatic enzymes. Like pepsin of the stomach, pancreatic *proteases* (protein-digesting enzymes) are produced and released in inactive forms, which are activated in the duodenum, where they do their work. This protects the pancreas from self-digestion.

For example, within the duodenum, *trypsinogen* is activated to **trypsin** by **enteropeptidase** (formerly called *enterokinase*), an intestinal brush border protease. Trypsin, in turn, activates more trypsinogen and two other pancreatic proteases (*procarboxypeptidase* and *chymotrypsinogen*) to their active forms, **carboxypeptidase** (kar-bok″se-pep′tĭ-dās) and **chymotrypsin** (ky″mo-trip′sin), respectively (**Figure 23.27**).

Other pancreatic enzymes—**amylase, lipases,** and **nucleases**—are secreted in active form, but require that ions or bile be present in the intestinal lumen for optimal activity.

CHECK YOUR UNDERSTANDING

37. What is contained in zymogen granules?

38. Maryanne has pancreatitis and her pancreas is swollen and temporarily unable to produce pancreatic juice. What type of foodstuffs will probably not be digested until she recovers?

For answers, see Appendix G.

Regulation of Bile and Pancreatic Secretion and Entry into the Small Intestine

Secretion of bile and pancreatic juice and the release of these secretions to the small intestine where they "do their business" in food digestion is regulated by the same factors—neural stimuli and, more importantly, hormones (cholecystokinin and secretin).

Bile salts themselves are the major stimulus for enhanced bile secretion (**Figure 23.28**)—when a fatty meal is eaten and the enterohepatic circulation is returning large amounts of bile salts to the liver, its output of bile rises dramatically. *Secretin*, released by intestinal cells exposed to fatty chyme, also stimulates liver cells to secrete bile.

When no digestion is occurring, the hepatopancreatic sphincter (guarding the entry of bile and pancreatic juice into the duodenum) is closed and the released bile backs up the cystic duct into the gallbladder, where it is stored until needed. Although the liver makes bile continuously, bile does not usually enter the small intestine until the gallbladder contracts. Parasympathetic impulses delivered by vagus nerve fibers is a minor stimulus for gallbladder contraction, but the major stimulus is *cholecystokinin* (*CCK*), an intestinal hormone released to the blood when acidic, fatty chyme enters the duodenum (Figure 23.28). Besides causing the gallbladder to contract, CCK (1) stimulates secretion of pancreatic juice and (2) relaxes the hepatopancreatic sphincter so that bile and pancreatic juice can enter the duodenum.

Both secretin and cholecystokinin prompt the pancreas to secrete its juice. Secretin, released in response to the presence of HCl in the intestine, mainly targets the pancreatic duct cells, resulting in a watery *bicarbonate-rich* pancreatic juice. CCK, released in response to proteins and fats in chyme, stimulates the acini to release *enzyme-rich* pancreatic juice, and potentiates

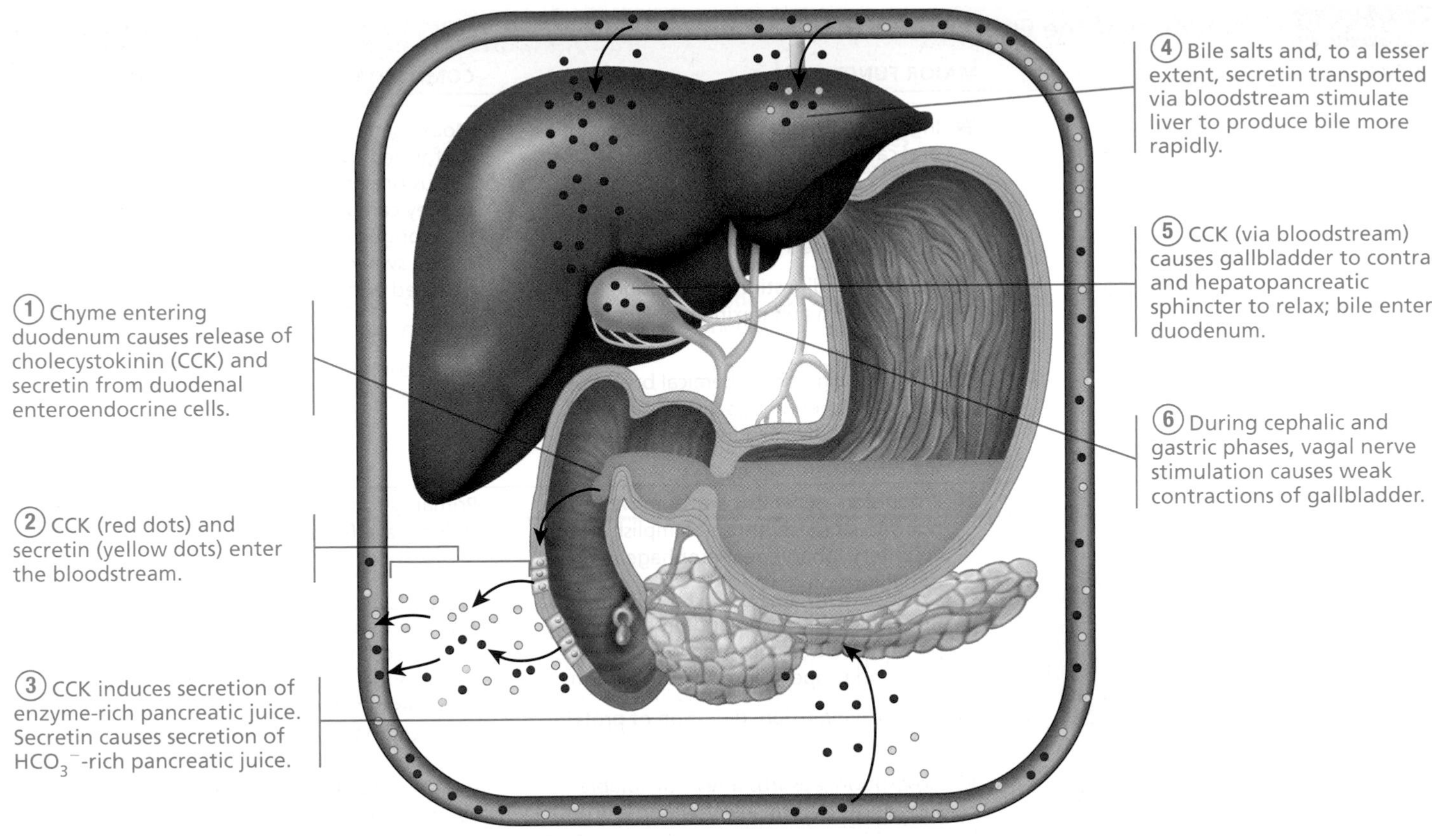

1 Chyme entering duodenum causes release of cholecystokinin (CCK) and secretin from duodenal enteroendocrine cells.
2 CCK (red dots) and secretin (yellow dots) enter the bloodstream.
3 CCK induces secretion of enzyme-rich pancreatic juice. Secretin causes secretion of HCO_3^--rich pancreatic juice.
4 Bile salts and, to a lesser extent, secretin transported via bloodstream stimulate liver to produce bile more rapidly.
5 CCK (via bloodstream) causes gallbladder to contract and hepatopancreatic sphincter to relax; bile enters duodenum.
6 During cephalic and gastric phases, vagal nerve stimulation causes weak contractions of gallbladder.

Figure 23.28 **Mechanisms promoting secretion and release of bile and pancreatic juice.** When digestion is not occurring, bile is stored and concentrated in the gallbladder. When acidic fatty chyme enters the small intestine, several mechanisms are initiated that accelerate the output of pancreatic juice and bile and cause the gallbladder to contract and the hepatopancreatic sphincter to relax. This allows bile and pancreatic juice to enter the small intestine. The single most important stimulus of bile secretion is an increased level of bile salts in the enterohepatic circulation.

the effect of secretin. CCK and the other digestive hormones are summarized in Table 23.1 (p. 875). Vagal stimulation prompts the release of pancreatic juice mainly during the cephalic and gastric phases of gastric secretion.

CHECK YOUR UNDERSTANDING

39. What is the functional difference between pancreatic acini and islets?
40. What is the makeup of the fluid in the pancreatic duct? In the cystic duct? In the bile duct?
41. What stimulates CCK release and what are its effects on the digestive process?

For answers, see Appendix G.

Digestive Processes Occurring in the Small Intestine

Food reaching the small intestine is unrecognizable, but it is far from being digested chemically. Carbohydrates and proteins are partially degraded, but virtually no fat digestion has occurred to this point. The process of food digestion is accelerated during the chyme's tortuous three- to six-hour journey through the small intestine, and it is here that absorption of most of the water and virtually all nutrients occurs. Like the stomach, the small intestine plays no part in ingestion or defecation.

Requirements for Optimal Intestinal Digestive Activity

Although a primary function of the small intestine is digestion, intestinal juice provides little of what is needed to perform this function. Most substances required for chemical digestion—bile, digestive enzymes (except for the brush border enzymes), and bicarbonate ions (to provide the proper pH for enzymatic catalysis)—are *imported* from the liver and pancreas. Hence, anything that impairs liver or pancreatic function or delivery of their juices to the small intestine severely hinders our ability to digest food and absorb nutrients. The other primary function of the small intestine—absorption—is efficiently accomplished by its absorptive cells with their rich crop of apical microvilli.

Optimal digestive activity in the small intestine also depends on a slow, measured delivery of chyme from the stomach.

As a result, as the proximal area constricts and forces chyme along the tract, the lumen of the distal part of the intestine enlarges to receive it.

Most of the time, the ileocecal sphincter is constricted and closed. However, two mechanisms—one neural and the other hormonal—cause it to relax and allow food residues to enter the cecum when ileal motility increases.

1. Enhanced activity of the stomach initiates the **gastroileal reflex** (gas″tro-il′e-ul), a long reflex that enhances the force of segmentation in the ileum.
2. Gastrin released by the stomach increases the motility of the ileum and relaxes the ileocecal sphincter. Once the chyme has passed through, it exerts backward pressure that closes the valve's flaps, preventing regurgitation into the ileum. This reflex ensures that the contents of the previous meal are swept completely out of the stomach and small intestine as the next meal is eaten.

HOMEOSTATIC IMBALANCE

Injury to the intestinal wall by severe stretching, some bacterial infections, or mechanical trauma may lead to a total shutdown of small intestine motility, a phenomenon called the intestino-intestinal reflex. ■

CHECK YOUR UNDERSTANDING

42. Distension of the stomach and duodenal walls have different effects on stomach secretory activity. What are these effects?
43. Which is more important in moving food along the small intestine—peristalsis or segmentation?
44. What is the MMC and why is it important?

For answers, see Appendix G.

The Large Intestine

- List the major functions of the large intestine.
- Describe the regulation of defecation.

The **large intestine** frames the small intestine on three sides and extends from the ileocecal valve to the anus (see Figure 23.1). Its diameter, at about 7 cm, is greater than that of the small intestine (hence, *large* intestine), but it is less than half as long (1.5 m versus 6 m). Its major digestive function is to absorb most of the remaining water from indigestible food residues (delivered to it in a fluid state), store the residues temporarily, and then eliminate them from the body as semisolid **feces** (fe′sēz).

Gross Anatomy

The large intestine exhibits three features not seen elsewhere—teniae coli, haustra, and epiploic appendages. Except for its terminal end, the longitudinal muscle layer of its muscularis is mostly reduced to three bands of smooth muscle called **teniae coli** (ten′ne-e ko′li; "ribbons of the colon"). Their tone causes the wall of the large intestine to pucker into pocketlike sacs called **haustra** (haw′strah; "to draw up"; singular *haustrum*). Another obvious feature of the large intestine is its **epiploic appendages** (ep″ĭ-plo′ik; "membrane covered"), small fat-filled pouches of visceral peritoneum that hang from its surface (**Figure 23.29a**). Their significance is not known.

Subdivisions

The large intestine has the following subdivisions: cecum, appendix, colon, rectum, and anal canal. The saclike **cecum** (se′kum; "blind pouch"), which lies below the ileocecal valve in the right iliac fossa, is the first part of the large intestine (Figure 23.29a). Attached to its posteromedial surface is the blind, wormlike **vermiform appendix**. The appendix contains masses of lymphoid tissue, and as part of MALT (see p. 761) it plays an important role in body immunity. However, it has an important structural shortcoming—its twisted structure provides an ideal location for enteric bacteria to accumulate and multiply.

HOMEOSTATIC IMBALANCE

Acute inflammation of the appendix, or **appendicitis**, results from a blockage (often by feces) that traps infectious bacteria in its lumen. Unable to empty its contents, the appendix swells, squeezing off venous drainage, which may lead to ischemia and necrosis (death and decay) of the appendix. If the appendix ruptures, feces containing bacteria spray over the abdominal contents, causing *peritonitis.*

The symptoms of appendicitis are variable, but the first is usually pain in the umbilical region. Loss of appetite, nausea and vomiting, and relocalization of pain to the lower right abdominal quadrant follow. Immediate surgical removal of the appendix (appendectomy) is the accepted treatment for suspected appendicitis. Appendicitis is most common during adolescence, when the entrance to the appendix is at its widest. ■

The **colon** has several distinct regions. As the **ascending colon**, it travels up the right side of the abdominal cavity to the level of the right kidney. Here it makes a right-angle turn—the **right colic**, or **hepatic, flexure**—and travels across the abdominal cavity as the **transverse colon**. Directly anterior to the spleen, it bends acutely at the **left colic (splenic) flexure** and descends down the left side of the posterior abdominal wall as the **descending colon**. Inferiorly, it enters the pelvis, where it becomes the S-shaped **sigmoid colon**.

The colon is retroperitoneal, except for its transverse and sigmoid parts. These parts are intraperitoneal and anchored to the posterior abdominal wall by mesentery sheets called **mesocolons** (**Figure 23.30c, d**).

In the pelvis, at the level of the third sacral vertebra, the sigmoid colon joins the **rectum**, which runs posteroinferiorly just in front of the sacrum. The position of the rectum allows a number of pelvic organs (e.g., the prostate of males) to be examined digitally (with a finger) through the anterior rectal wall. This is called a **rectal exam**.

Despite its name (*rectum* = straight), the rectum has three lateral curves or bends, represented internally as three transverse folds called **rectal valves** (Figure 23.29b). These valves

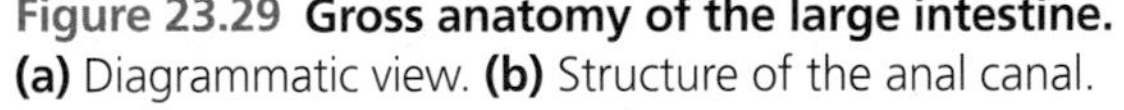

Figure 23.29 Gross anatomy of the large intestine. **(a)** Diagrammatic view. **(b)** Structure of the anal canal.

separate feces from flatus, in other words, they stop feces from being passed along with gas.

The **anal canal**, the last segment of the large intestine, lies in the perineum, entirely external to the abdominopelvic cavity. About 3 cm long, it begins where the rectum penetrates the levator ani muscle of the pelvic floor and opens to the body exterior at the **anus**. The anal canal has two sphincters, an involuntary **internal anal sphincter** composed of smooth muscle (part of the muscularis), and a voluntary **external anal sphincter** composed of skeletal muscle. The sphincters, which act rather like purse strings to open and close the anus, are ordinarily closed except during defecation.

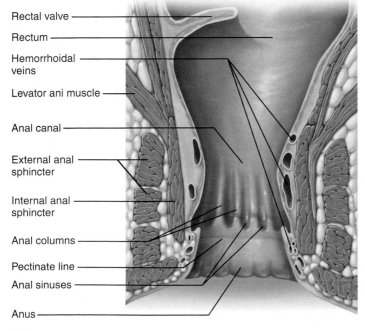

Microscopic Anatomy

The wall of the large intestine differs in several ways from that of the small intestine. The colon *mucosa* is simple columnar epithelium except in the anal canal. Because most food is absorbed before reaching the large intestine, there are no circular folds, no

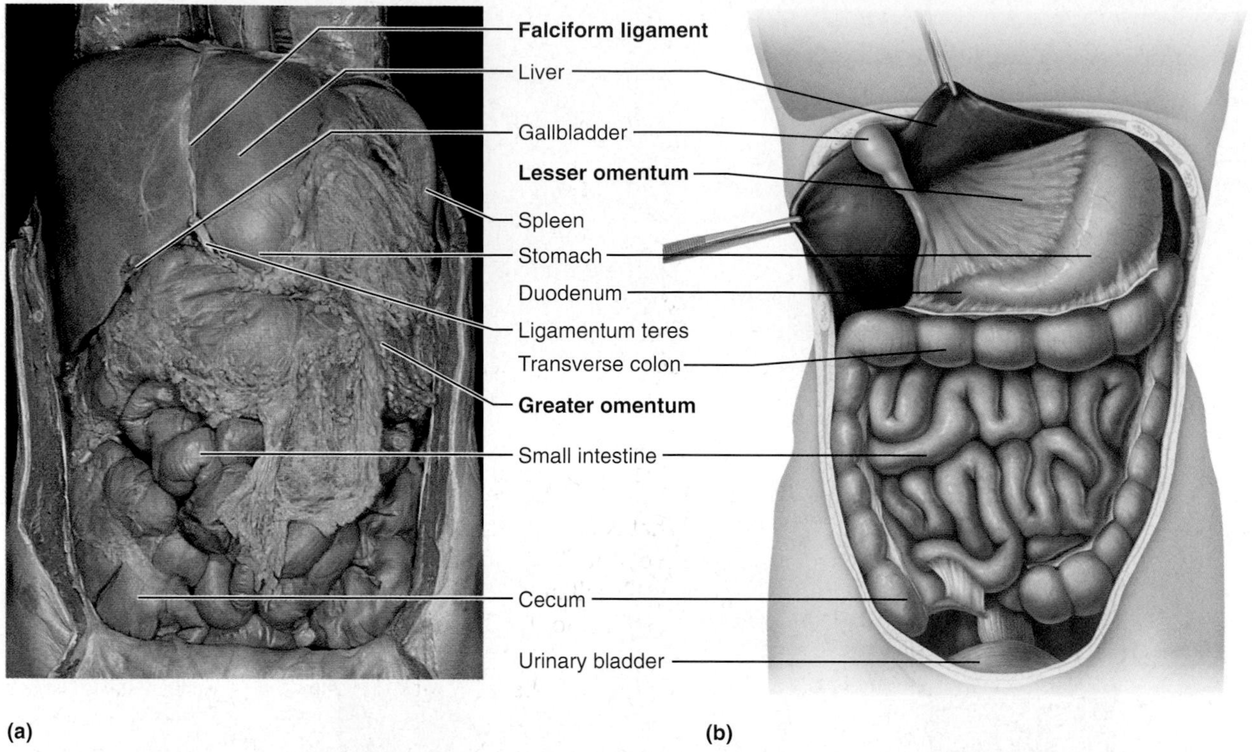

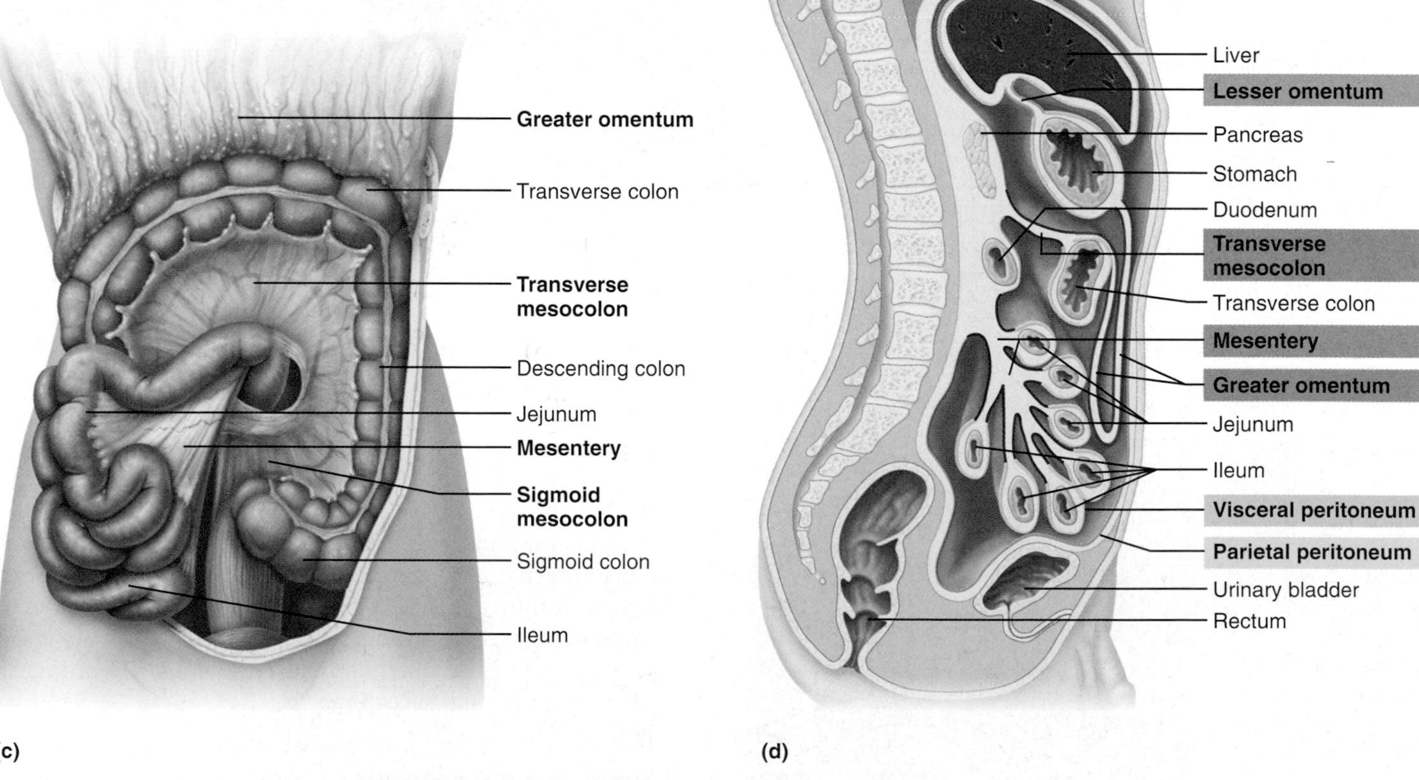

23

Figure 23.30 Mesenteries of the abdominal digestive organs. (a) The greater omentum, a dorsal mesentery, is shown in its normal position covering the abdominal viscera. **(b)** The liver and gallbladder have been reflected superiorly to reveal the lesser omentum, a ventral mesentery attaching the liver to the lesser curvature of the stomach. **(c)** The greater omentum has been reflected superiorly to reveal the mesentery attachments of the small and large intestine. **(d)** Sagittal section of the abdominopelvic cavity of a male.

villi, and virtually no cells that secrete digestive enzymes. However, its mucosa is thicker, its abundant crypts are deeper, and there are tremendous numbers of goblet cells in the crypts. Mucus produced by goblet cells eases the passage of feces and protects the intestinal wall from irritating acids and gases released by resident bacteria in the colon.

The mucosa of the anal canal, a stratified squamous epithelium, merges with the true skin surrounding the anus and is quite different from that of the rest of the colon, reflecting the greater abrasion that this region receives. Superiorly, it hangs in long ridges or folds called **anal columns**. **Anal sinuses**, recesses between the anal columns, exude mucus when compressed by feces, which aids in emptying the anal canal (Figure 23.29b).

The horizontal, tooth-shaped line that parallels the inferior margins of the anal sinuses is called the *pectinate line*. Superior to this line, the mucosa is innervated by visceral sensory fibers and is relatively insensitive to pain. The area inferior to the pectinate line is very sensitive to pain, a reflection of the somatic sensory fibers serving it.

Two superficial venous plexuses are associated with the anal canal, one with the anal columns and the other with the anus itself. If these (hemorrhoidal) veins are inflamed, itchy varicosities called *hemorrhoids* result.

Teniae coli and haustra are absent in the rectum and anal canal. However, consistent with its need to generate strong contractions to perform its expulsive role in defecation, the rectum's muscularis muscle layers are complete and well developed.

Bacterial Flora

Although most bacteria entering the cecum from the small intestine are dead (having been killed by the action of lysozyme, defensins, HCl, and protein-digesting enzymes), some are still "alive and kicking." Together with bacteria that enter the GI tract via the anus, these constitute the **bacterial flora** of the large intestine, an astonishing 10 million discrete types of them. These bacteria colonize the colon, metabolize some host-derived molecules (mucin, heparin, and hyaluronic acid) and ferment some of the indigestible carbohydrates (cellulose, xylan, and others), releasing irritating acids and a mixture of gases (including dimethyl sulfide, H_2, N_2, CH_4, and CO_2). Some of these gases (such as dimethyl sulfide) are quite odorous (smelly). About 500 ml of gas (flatus) is produced each day, much more when certain carbohydrate-rich foods (such as beans) are eaten. The bacterial flora also synthesize B complex vitamins and most of the vitamin K the liver requires to synthesize some of the clotting proteins.

Although the huge intestinal population of bacteria would seem to be enough inhabitants, the feces also have others, including viruses and protozoans. Of these, at least 20 are known pathogens.

Most enteric bacteria coexist peacefully with their host as long as they remain in the gut lumen. An elegant system keeps the bacteria from breaching the mucosal barrier. The epithelial cells of the gut mucosa respond to specific bacterial components by releasing chemicals that recruit immune cells, particularly dendritic cells, into the mucosa. The dendritic cells pry open tight junctions between the epithelial cells and send extensions into the lumen to sample the microbial antigens. They then migrate to the nearby lymphoid follicles (MALT) within the gut mucosa where they present the antigens to T cells. As a result, an IgA antibody–mediated response restricted to the gut lumen is triggered that prevents the bacteria from straying into tissues deep to the mucosa where they might elicit a much more widespread systemic response. Though beneficial in most ways, the coexistence of enteric bacteria with our immune system does sometimes fail. When that happens, the painful and debilitating conditions known as inflammatory bowel disease (see Related Clinical Terms, p. 905) may result.

Digestive Processes Occurring in the Large Intestine

What is finally delivered to the large intestine contains few nutrients, but it still has 12 to 24 hours more to spend there. Except for a small amount of digestion of that residue by the enteric bacteria, no further food breakdown occurs in the large intestine.

Although the large intestine harvests vitamins made by the bacterial flora and reclaims most of the remaining water and some of the electrolytes (particularly sodium and chloride), nutrient absorption is not the *major* function of this organ. As mentioned, the primary concerns of the large intestine are propulsive activities that force the fecal material toward the anus and then eliminate it from the body (defecation).

The large intestine is important for our comfort, but it is not essential for life. If the colon is removed, as may be necessitated by colon cancer, the terminal ileum is brought out to the abdominal wall in a procedure called an *ileostomy* (il″e-os′to-me), and food residues are eliminated from there into a sac attached to the abdominal wall. Another surgical technique, *ileoanal juncture*, links the ileum directly to the anal canal.

Motility of the Large Intestine

The large intestine musculature is inactive much of the time, but pressure in the ileum terminus opens the ileocecal sphincter and then closes it, preventing backward movement of the chyme. When presented with food residue the colon becomes motile, but its contractions are sluggish or short-lived. The movements most seen in the colon are **haustral contractions**, slow segmenting movements lasting about one minute that occur every 30 minutes or so.

These contractions, which occur mainly in the transverse and descending colon, reflect local controls of smooth muscle within the walls of the individual haustra. As a haustrum fills with food residue, the distension stimulates its muscle to contract, which propels the luminal contents into the next haustrum. These movements also mix the residue, which aids in water absorption.

Mass movements (mass peristalsis) are long, slow-moving, but powerful contractile waves that move over large areas of the colon three or four times daily and force the contents toward the rectum. Typically, they occur during or just after eating, which indicates that the presence of food in the stomach activates the

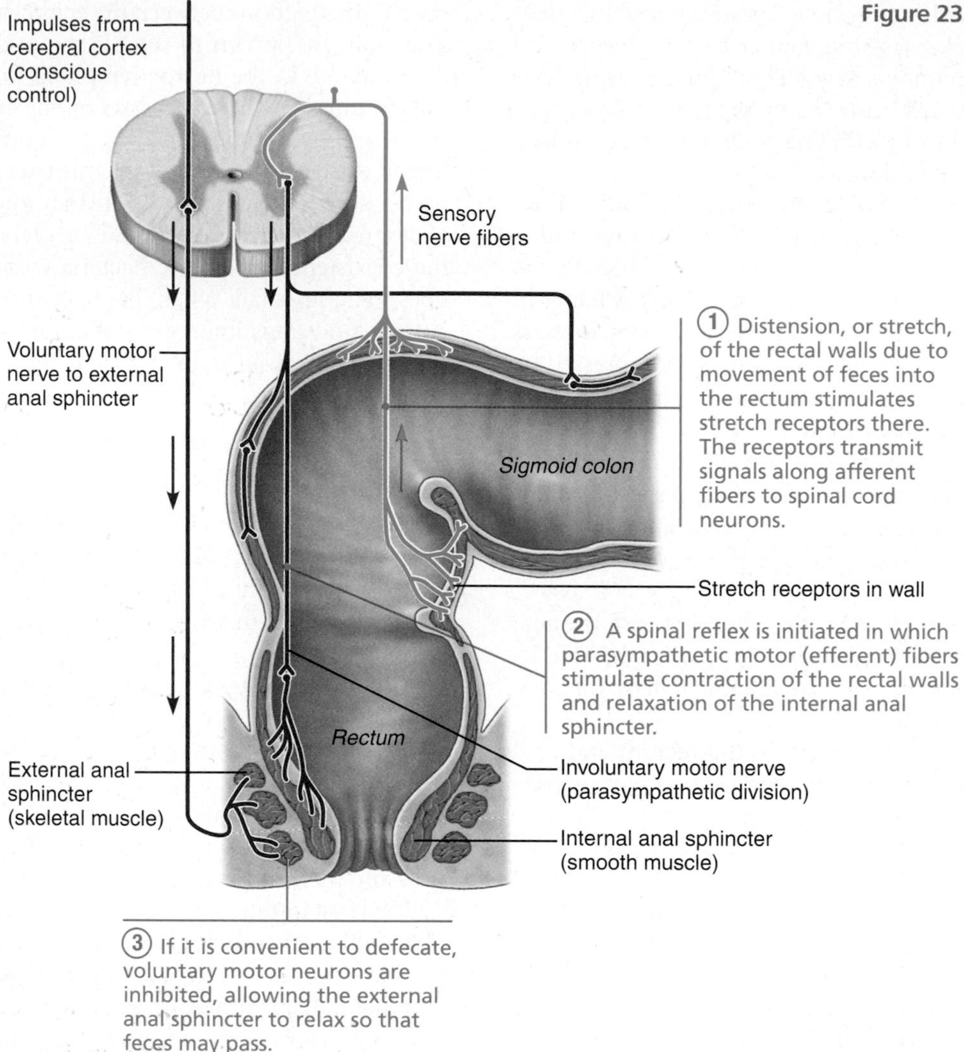
Impulses from cerebral cortex (conscious control)
Sensory nerve fibers
Voluntary motor nerve to external anal sphincter
① Distension, or stretch, of the rectal walls due to movement of feces into the rectum stimulates stretch receptors there. The receptors transmit signals along afferent fibers to spinal cord neurons.
Sigmoid colon
Stretch receptors in wall
② A spinal reflex is initiated in which parasympathetic motor (efferent) fibers stimulate contraction of the rectal walls and relaxation of the internal anal sphincter.
Rectum
External anal sphincter (skeletal muscle)
Involuntary motor nerve (parasympathetic division)
Internal anal sphincter (smooth muscle)
③ If it is convenient to defecate, voluntary motor neurons are inhibited, allowing the external anal sphincter to relax so that feces may pass.

Figure 23.31 Defecation reflex.

23

gastroileal reflex in the small intestine and the propulsive **gastrocolic reflex** in the colon. Bulk, or fiber, in the diet increases the strength of colon contractions and softens the stool, allowing the colon to act like a well-oiled machine.

HOMEOSTATIC IMBALANCE

When the diet lacks bulk and the volume of residues in the colon is small, the colon narrows and its contractions become more powerful, increasing the pressure on its walls. This promotes formation of **diverticula** (di″ver-tik′u-lah), small herniations of the mucosa through the colon walls. This condition, called **diverticulosis**, most commonly occurs in the sigmoid colon, and affects over half of people over the age of 70. In about 20% of cases, diverticulosis progresses to **diverticulitis**, in which the diverticula become inflamed and may rupture, leaking feces into the peritoneal cavity, which can be life threatening. Foods and products that increase the bulk of the stool help to prevent attacks of diverticulitis.

Irritable bowel syndrome (*IBS*) is a functional GI disorder not explained by anatomical or biochemical abnormalities. Affected individuals have recurring (or persistent) abdominal pain that is relieved by defecation, changes in the consistency (watery to stonelike) and frequency of their stools, and varying complaints of bloating, flatulence, nausea, and depression. Stress is a common precipitating factor, and stress management is an important aspect of treatment. ■

The semisolid product delivered to the rectum, called feces or the stool, contains undigested food residues, mucus, sloughed-off epithelial cells, millions of bacteria, and just enough water to allow its smooth passage. Of the 500 ml or so of food residue entering the cecum daily, approximately 150 ml becomes feces.

Defecation

The rectum is usually empty, but when mass movements force feces into it, stretching of the rectal wall initiates the **defecation reflex.** This spinal cord–mediated parasympathetic reflex causes the sigmoid colon and the rectum to contract, and the internal anal sphincter to relax (**Figure 23.31**, ① and ②). As feces are forced into the anal canal, messages reach the brain allowing us to decide whether the external (voluntary) anal sphincter should be opened or remain constricted to stop feces passage temporarily (Figure 23.31, ③). If defecation is delayed, the reflex contractions end within a few seconds, and the rectal walls relax. With the next mass movement, the defecation reflex is initiated again—and so on, until the person chooses to defecate or the urge to defecate becomes unavoidable.

During defecation, the muscles of the rectum contract to expel the feces. We aid this process voluntarily by closing the glottis and contracting our diaphragm and abdominal wall muscles to increase the intra-abdominal pressure (a procedure called *Valsalva's maneuver*). We also contract the levator ani muscle (pp. 344–345), which lifts the anal canal superiorly. This lifting action leaves the feces below the anus—and outside the body. Involuntary or automatic defecation (incontinence of feces) occurs in infants because they have not yet gained control of their external anal sphincter. It also occurs in those with spinal cord transections.

HOMEOSTATIC IMBALANCE

Watery stools, or **diarrhea**, result from any condition that rushes food residue through the large intestine before that organ has had sufficient time to absorb the remaining water. Causes include irritation of the colon by bacteria or, less commonly, prolonged physical jostling of the digestive viscera (occurs in marathon runners). Prolonged diarrhea may result in dehydration and electrolyte imbalance (acidosis and loss of potassium).

Conversely, when food remains in the colon for extended periods, too much water is absorbed and the stool becomes hard and difficult to pass. This condition, called **constipation**, may result from lack of fiber in the diet, improper bowel habits (failing to heed the "call"), lack of exercise, emotional upset, or laxative abuse. ■

CHECK YOUR UNDERSTANDING

45. What propulsive movements are unique to the large intestine?
46. What is the result of stimulation of stretch receptors in the rectal walls?
47. In what ways are enteric bacteria important to our nutrition?

For answers, see Appendix G.

PART 3

PHYSIOLOGY OF CHEMICAL DIGESTION AND ABSORPTION

- List the enzymes involved in chemical digestion; name the foodstuffs on which they act.
- List the end products of protein, fat, carbohydrate, and nucleic acid digestion.
- Describe the process of absorption of breakdown products of foodstuffs that occurs in the small intestine.

So far in this chapter, we have examined the structure and overall function of the organs that make up the digestive system. Now let's investigate the entire chemical processing (enzymatic breakdown and absorption) of each class of foodstuffs as it moves through the GI tract. As you read along, you may find it helpful to refer to the summary in **Figure 23.32**.

Chemical Digestion

After foodstuffs have spent even a short time in the stomach, they are unrecognizable. Nonetheless, they are still mostly the starchy, carbohydrate-rich foods, meat proteins, butter and other fats, and so on that we ingested. Only their appearance has been changed by mechanical digestive activities. By contrast, the products of chemical digestion are the chemical building blocks of the ingested foods and are very different molecules chemically.

Mechanism of Chemical Digestion: Enzymatic Hydrolysis

Chemical digestion is a catabolic process in which large food molecules are broken down to *monomers* (chemical building blocks), which are small enough to be absorbed by the GI tract lining. Chemical digestion is accomplished by enzymes secreted by both intrinsic and accessory glands into the lumen of the alimentary canal. The enzymatic breakdown of any type of food molecule is **hydrolysis** (hi-drol′ĭ-sis) because it involves addition of a water molecule to each molecular bond to be broken (lysed).

Chemical Digestion of Carbohydrates

Most of us ingest between 200 and 600 grams of carbohydrate foods each day. **Monosaccharides** (simple sugars), the monomers of carbohydrates, are absorbed immediately without further ado. Only three of these are common in our diet: *glucose*, *fructose*, and *galactose*. The more complex carbohydrates that our digestive system is able to break down to monosaccharides are the disaccharides *sucrose* (table sugar), *lactose* (milk sugar), and *maltose* (grain sugar) and the polysaccharides *glycogen* and *starch*.

In the average diet, most digestible carbohydrates are in the form of starch, with smaller amounts of disaccharides and monosaccharides. Humans lack enzymes capable of breaking down most other polysaccharides, such as cellulose. As a result, indigestible polysaccharides do not nourish us but they do help move the food along the GI tract by providing bulk, or fiber.

Chemical digestion of starch (and perhaps glycogen) begins in the mouth (Figure 23.32). **Salivary amylase**, present in saliva, splits starch into *oligosaccharides*, smaller fragments of two to eight linked glucose molecules. Salivary amylase works best in the slightly acid to neutral environment (pH of 6.75–7.00) maintained in the mouth by the buffering effects of bicarbonate and phosphate ions in saliva. Starch digestion continues until amylase is inactivated by stomach acid and broken apart by the stomach's protein-digesting enzymes. Generally speaking, the larger the meal, the longer amylase continues to work in the stomach because foodstuffs in its relatively immobile fundus are poorly mixed with gastric juices.

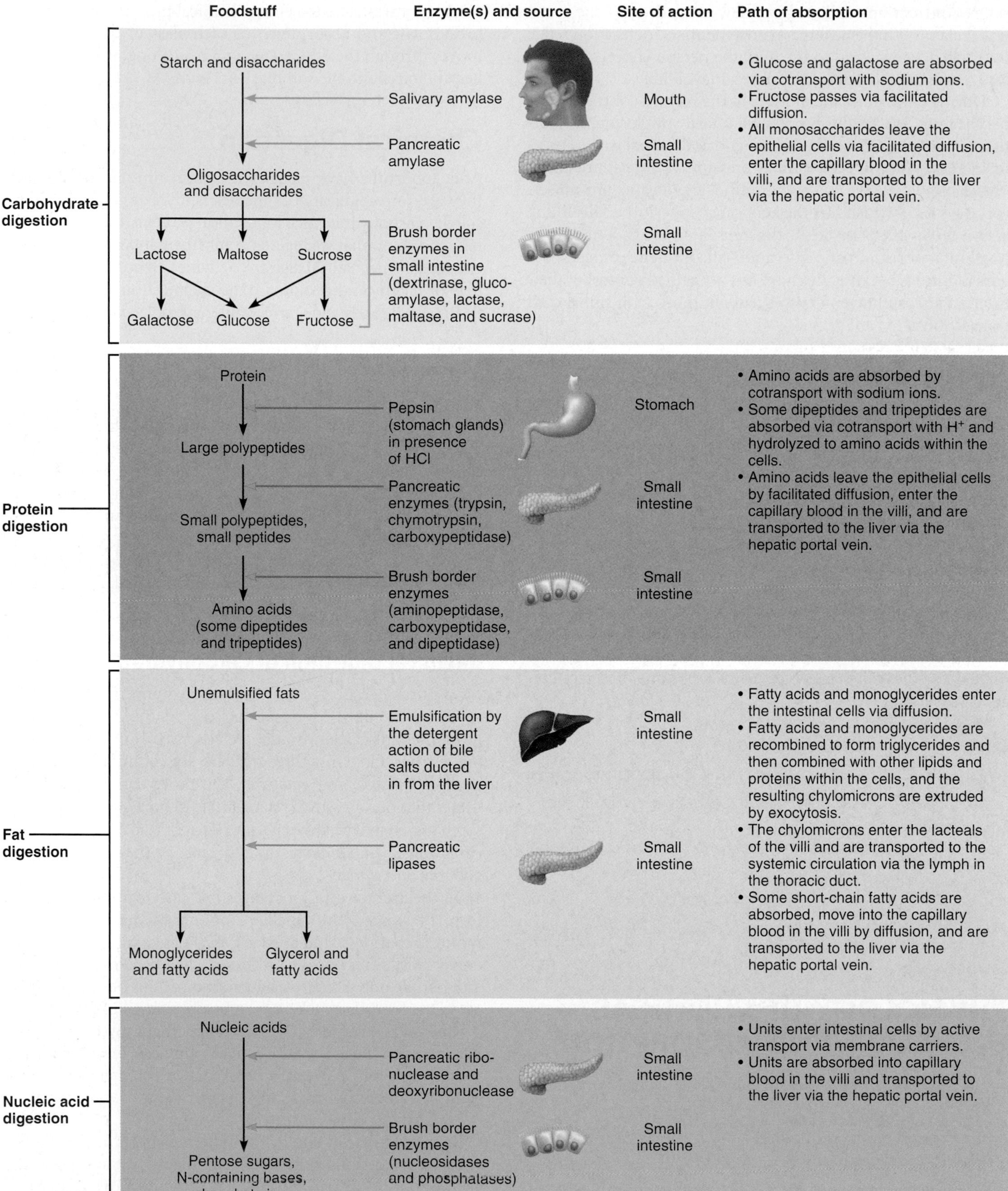
Foodstuff
Enzyme(s) and source
Site of action
Path of absorption
Carbohydrate digestion
Starch and disaccharides
Salivary amylase
Mouth
Pancreatic amylase
Small intestine
Oligosaccharides and disaccharides
Lactose
Maltose
Sucrose
Galactose
Glucose
Fructose
Brush border enzymes in small intestine (dextrinase, gluco-amylase, lactase, maltase, and sucrase)
Small intestine
• Glucose and galactose are absorbed via cotransport with sodium ions.
• Fructose passes via facilitated diffusion.
• All monosaccharides leave the epithelial cells via facilitated diffusion, enter the capillary blood in the villi, and are transported to the liver via the hepatic portal vein.
Protein digestion
Protein
Pepsin (stomach glands) in presence of HCl
Stomach
Large polypeptides
Pancreatic enzymes (trypsin, chymotrypsin, carboxypeptidase)
Small intestine
Small polypeptides, small peptides
Brush border enzymes (aminopeptidase, carboxypeptidase, and dipeptidase)
Small intestine
Amino acids (some dipeptides and tripeptides)
• Amino acids are absorbed by cotransport with sodium ions.
• Some dipeptides and tripeptides are absorbed via cotransport with H+ and hydrolyzed to amino acids within the cells.
• Amino acids leave the epithelial cells by facilitated diffusion, enter the capillary blood in the villi, and are transported to the liver via the hepatic portal vein.
Fat digestion
Unemulsified fats
Emulsification by the detergent action of bile salts ducted in from the liver
Small intestine
Pancreatic lipases
Small intestine
Monoglycerides and fatty acids
Glycerol and fatty acids
• Fatty acids and monoglycerides enter the intestinal cells via diffusion.
• Fatty acids and monoglycerides are recombined to form triglycerides and then combined with other lipids and proteins within the cells, and the resulting chylomicrons are extruded by exocytosis.
• The chylomicrons enter the lacteals of the villi and are transported to the systemic circulation via the lymph in the thoracic duct.
• Some short-chain fatty acids are absorbed, move into the capillary blood in the villi by diffusion, and are transported to the liver via the hepatic portal vein.
Nucleic acid digestion
Nucleic acids
Pancreatic ribo-nuclease and deoxyribonuclease
Small intestine
Brush border enzymes (nucleosidases and phosphatases)
Small intestine
Pentose sugars, N-containing bases, phosphate ions
• Units enter intestinal cells by active transport via membrane carriers.
• Units are absorbed into capillary blood in the villi and transported to the liver via the hepatic portal vein.
23

Figure 23.32 **Flowchart of chemical digestion and absorption of foodstuffs.**

Starchy foods and other digestible carbohydrates that escape being broken down by salivary amylase are acted on by **pancreatic amylase** in the small intestine. Within about 10 minutes of entering the small intestine, starch is entirely converted to various oligosaccharides, mostly maltose.

Intestinal brush border enzymes further digest these products to monosaccharides. The most important of the brush border enzymes are **dextrinase** and **glucoamylase**, which act on oligosaccharides composed of more than three simple sugars, and **maltase**, **sucrase**, and **lactase**, which hydrolyze maltose, sucrose, and lactose respectively into their constituent monosaccharides.

Because the colon does not secrete digestive enzymes, chemical digestion *officially ends* in the small intestine. As noted earlier, however, resident colon bacteria do break down and metabolize the residual complex carbohydrates and some proteins further, adding much to their own nutrition but essentially nothing to ours.

HOMEOSTATIC IMBALANCE

In some people, intestinal lactase is present at birth but then becomes deficient due to genetic factors. In such cases, the person becomes intolerant of milk products (the source of lactose), and undigested lactose creates osmotic gradients that not only prevent water from being absorbed in the small and large intestines but also pull water from the interstitial space into the intestines. The result is diarrhea. Bacterial metabolism of the undigested solutes produces large amounts of gas that result in bouts of bloating, flatulence, and cramping pain. For the most part, the solution to this problem is simple—add lactase enzyme "drops" to your milk or take a lactase tablet before meals containing milk products. ■

Chemical Digestion of Proteins

Proteins digested in the GI tract include not only dietary proteins (typically about 125 g per day), but also 15–25 g of enzyme proteins secreted into the GI tract lumen by its various glands and (probably) an equal amount of protein derived from sloughed and disintegrating mucosal cells. In healthy individuals, much of this protein is digested all the way to its **amino acid** monomers.

Protein digestion begins in the stomach when pepsinogen secreted by the chief cells is activated to **pepsin** (actually a group of protein-digesting enzymes) (Figure 23.32). Pepsin functions optimally in the acidic pH range found in the stomach: 1.5–2.5. It preferentially cleaves bonds involving the amino acids tyrosine and phenylalanine so that proteins are broken into polypeptides and small numbers of free amino acids. Pepsin, which hydrolyzes 10–15% of ingested protein, is inactivated by the high pH in the duodenum, so its proteolytic activity is restricted to the stomach. **Rennin** (the enzyme that coagulates milk protein) is not produced in adults.

Protein fragments entering the small intestine from the stomach are greeted by a host of proteolytic enzymes (**Figure 23.33**). **Trypsin** and **chymotrypsin** secreted by the pancreas cleave the proteins into smaller peptides, which in turn become the grist for other enzymes. The pancreatic and brush border enzyme **carboxypeptidase** splits off one amino acid at a time from the end of the polypeptide chain that bears the carboxyl group. Other brush border enzymes such as **aminopeptidase** and **dipeptidase** liberate the final amino acid products. Aminopeptidase digests a protein, one amino acid at a time, by working from the amine end. Both carboxypeptidase and aminopeptidase can independently dismantle a protein, but the teamwork between these enzymes and between trypsin and chymotrypsin, which attack the more internal parts of the protein, speeds up the process tremendously.

Chemical Digestion of Lipids

Although the American Heart Association recommends a low-fat diet, the amount of lipids (fats) ingested daily varies tremendously among American adults, ranging from 30 g to 150 g or more. Triglycerides (neutral fats or triacylglycerols) are the most abundant fats in the diet. The small intestine is essentially the sole site of lipid digestion because the pancreas is the only significant source of fat-digesting enzymes, or **lipases** (see Figure 23.32).

Because triglycerides and their breakdown products are insoluble in water, fats need special "pretreatment" with bile salts to be digested and absorbed in the watery environment of the small intestine. In aqueous solutions, triglycerides aggregate to form large fat globules, and only the triglyceride molecules at the surfaces of such fatty masses are accessible to the water-soluble lipase enzymes. However, this problem is quickly resolved because as the fat globules enter the duodenum, they are coated with detergent-like bile salts.

Bile salts have both nonpolar and polar regions. Their nonpolar (hydrophobic) parts cling to the fat molecules, and their polar (ionized hydrophilic) parts allow them to repel each other and to interact with water. As a result, fatty droplets are pulled off the large fat globules, and a stable *emulsion*—an aqueous suspension of fatty droplets, each about 1 mm in diameter—is formed (**Figure 23.34, ①**).

Emulsification does *not* break chemical bonds. It just reduces the attraction between fat molecules so that they can be more widely dispersed. This process vastly increases the number of triglyceride molecules exposed to the pancreatic lipases. Without bile, lipids would be incompletely digested in the time food is in the small intestine.

The pancreatic lipases catalyze the breakdown of fats by cleaving off two of the fatty acid chains, yielding free **fatty acids** and **monoglycerides**—glycerol with one fatty acid chain attached (Figure 23.34, ②). Fat-soluble vitamins that ride with fats require no digestion.

Chemical Digestion of Nucleic Acids

Both DNA and RNA, present in the nuclei of cells forming the ingested foods, are hydrolyzed to their **nucleotide** monomers by **pancreatic nucleases** present in pancreatic juice. The nucleotides are then broken apart by intestinal brush border enzymes (**nucleosidases** and **phosphatases**), which release their free bases, pentose sugars, and phosphate ions (see Figure 23.32).

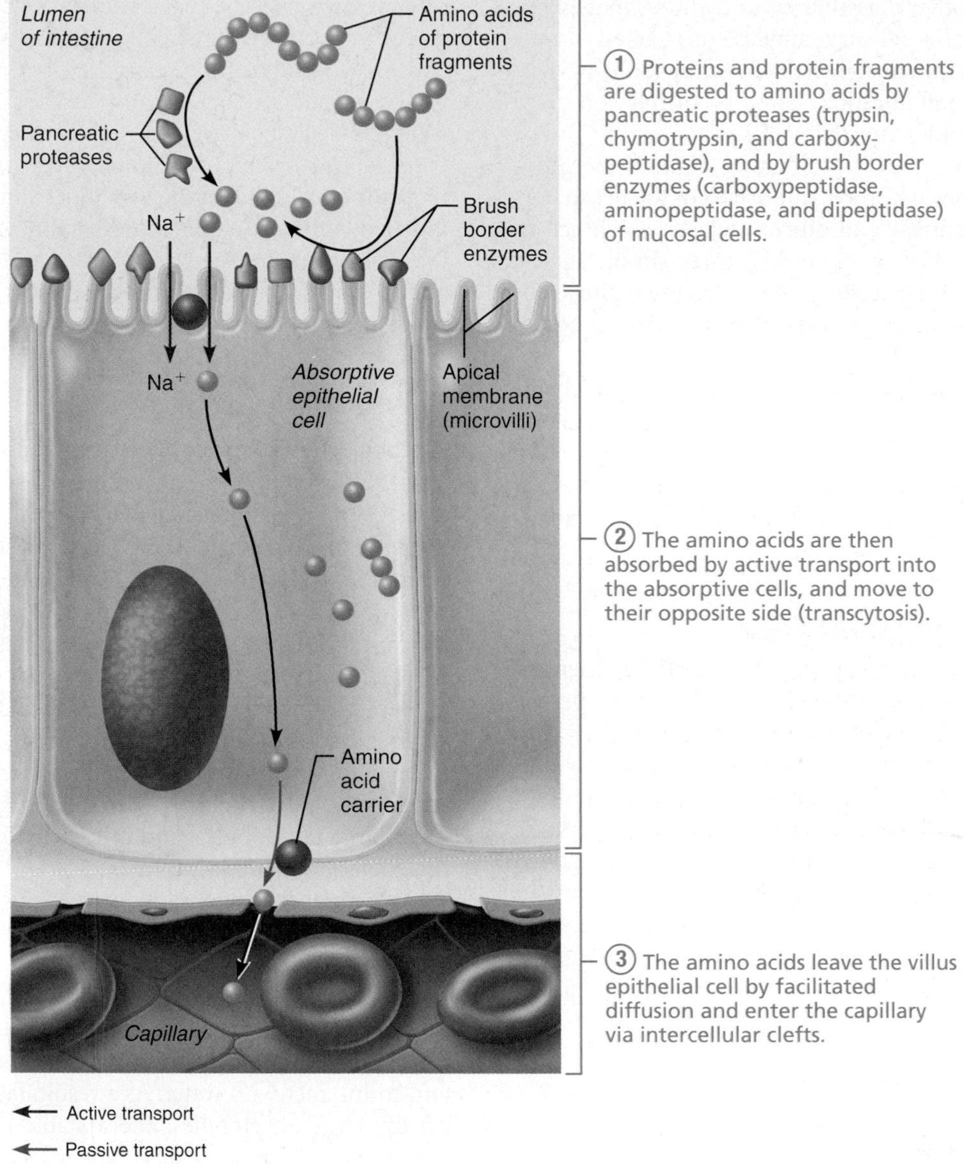

Lumen of intestine
Amino acids of protein fragments
Pancreatic proteases
Na+
Brush border enzymes
Na+
Absorptive epithelial cell
Apical membrane (microvilli)
Amino acid carrier
Capillary
① Proteins and protein fragments are digested to amino acids by pancreatic proteases (trypsin, chymotrypsin, and carboxy-peptidase), and by brush border enzymes (carboxypeptidase, aminopeptidase, and dipeptidase) of mucosal cells.
② The amino acids are then absorbed by active transport into the absorptive cells, and move to their opposite side (transcytosis).
③ The amino acids leave the villus epithelial cell by facilitated diffusion and enter the capillary via intercellular clefts.
Active transport
Passive transport

23

Figure 23.33 Protein digestion and absorption in the small intestine.

Absorption

Up to 10 L of food, drink, and GI secretions enter the alimentary canal daily, but only 1 L or less reaches the large intestine. Virtually all of the foodstuffs, 80% of the electrolytes, and most of the water (remember water follows salt) are absorbed in the small intestine. Although absorption occurs all along the length of the small intestine, most of it is completed by the time chyme reaches the ileum.

The major absorptive role of the ileum is to reclaim bile salts to be recycled back to the liver for resecretion. It is virtually impossible to exceed the absorptive capacity of the human gut, and at the end of the ileum, all that remains is some water, indigestible food materials (largely plant fibers such as cellulose), and millions of bacteria. This debris is passed on to the large intestine.

To understand absorption, remember that the fluid mosaic structure of the plasma membrane dictates that nonpolar substances, which can dissolve in the lipid core of the membrane, can be absorbed passively, and that polar substances must be absorbed by carrier mechanisms. Most nutrients are absorbed through the mucosa of the intestinal villi by *active transport* processes driven directly or indirectly (secondarily) by metabolic energy (ATP). They then enter the capillary blood in the villus to be transported in the hepatic portal vein to the liver. The exception is some lipid digestion products, which are absorbed passively by diffusion and then enter the lacteal in the villus to be carried to the blood via lymphatic fluid.

Because the epithelial cells of the intestinal mucosa are joined at their luminal surfaces by tight junctions, substances cannot move *between* the cells for the most part. Consequently, materials must pass through the epithelial cells and into the

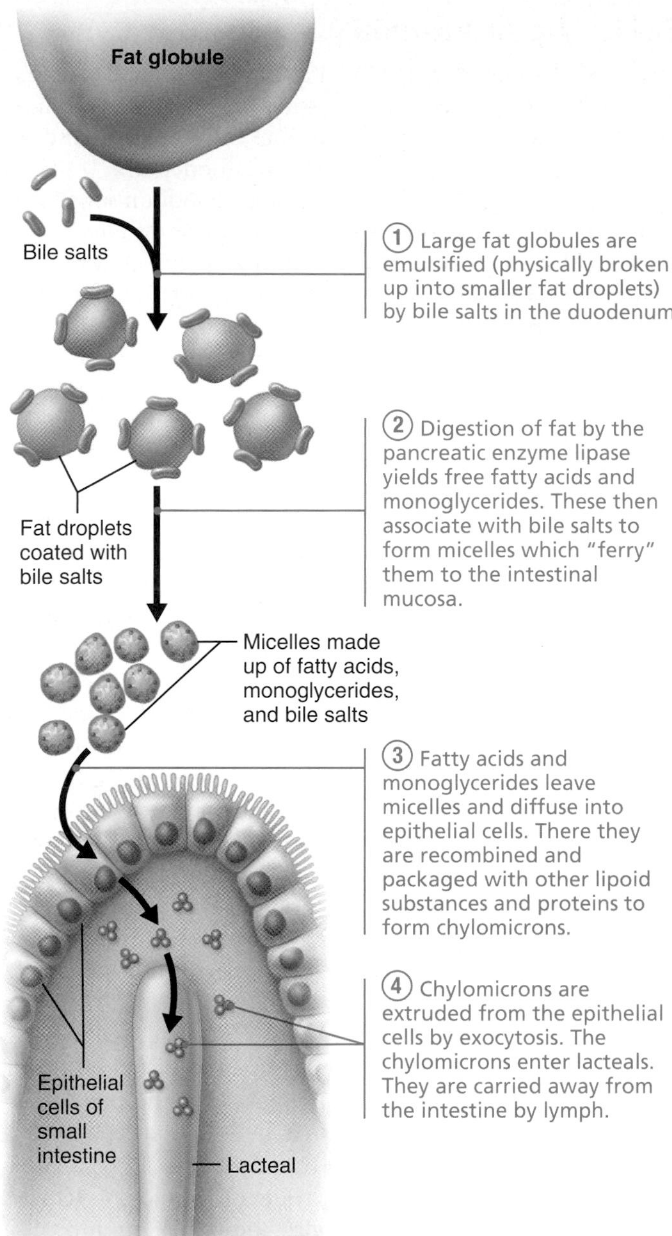

Fat globule
Bile salts
① Large fat globules are emulsified (physically broken up into smaller fat droplets) by bile salts in the duodenum.
② Digestion of fat by the pancreatic enzyme lipase yields free fatty acids and monoglycerides. These then associate with bile salts to form micelles which "ferry" them to the intestinal mucosa.
Fat droplets coated with bile salts
Micelles made up of fatty acids, monoglycerides, and bile salts
③ Fatty acids and monoglycerides leave micelles and diffuse into epithelial cells. There they are recombined and packaged with other lipoid substances and proteins to form chylomicrons.
④ Chylomicrons are extruded from the epithelial cells by exocytosis. The chylomicrons enter lacteals. They are carried away from the intestine by lymph.
Epithelial cells of small intestine
Lacteal

Figure 23.34 Emulsification, digestion, and absorption of fats.

interstitial fluid abutting their basolateral membranes (via transcytosis) if they are to enter the blood capillaries. As we describe the absorption of each nutrient class next, you may want to refer to the absorption summary on the right-hand side of Figure 23.32.

Carbohydrate Absorption

The monosaccharides glucose and galactose, liberated by the breakdown of starch and disaccharides, are shuttled by secondary active transport (cotransport with Na^+) into the epithelial cells by common protein carriers. They then move out of these cells by facilitated diffusion and pass into the capillaries via intercellular clefts. By contrast, fructose moves entirely by facilitated diffusion. The transport proteins, which are located very close to the disaccharidase enzymes on the microvilli, combine with the monosaccharides as soon as the disaccharides are broken down.

Protein Absorption

Several types of carriers transport the different amino acids resulting from protein digestion. Most of these carriers, like those for glucose and galactose, are coupled to the active transport of sodium. Short chains of two or three amino acids (dipeptides and tripeptides, respectively) are actively absorbed by H^+-dependent cotransport and are then digested to their amino acids within the epithelial cells before entering the capillary blood by diffusion (see Figure 23.33).

HOMEOSTATIC IMBALANCE

Whole proteins are not usually absorbed, but in rare cases intact proteins are taken up by endocytosis and released on the opposite side of the epithelial cell by exocytosis. This process is most common in newborn infants, reflecting the immaturity of their intestinal mucosa (gastric acid secretion does not reach normal levels until weeks after birth, and the mucosa is leakier than it is later.) Absorption of whole proteins accounts for many early food allergies. The immune system "sees" the intact proteins as antigenic and mounts an attack. These food allergies usually disappear as the mucosa matures. This mechanism may also provide a route for IgA antibodies present in breast milk to reach an infant's bloodstream. These antibodies confer some passive immunity on the infant, providing temporary protection against antigens to which the mother has been sensitized. ■

Lipid Absorption

Just as bile salts accelerate lipid digestion, they are also essential for the absorption of its end products. As the water-insoluble products of fat digestion—the monoglycerides and free fatty acids—are liberated by lipase activity, they quickly become associated with bile salts and *lecithin* (a phospholipid found in bile) to form micelles (Figure 23.34, ②). **Micelles** (mi-selz′) are collections of fatty elements clustered together with bile salts in such a way that the polar (hydrophilic) ends of the molecules face the water and the nonpolar portions form the core. Also nestled in the hydrophobic core are cholesterol molecules and fat-soluble vitamins. Although micelles are similar to emulsion droplets, they are much smaller "vehicles" and easily diffuse between microvilli to come into close contact with the luminal cell surface.

Upon reaching the epithelial cells, the various lipid substances leave the micelles and move through the lipid phase of the plasma membrane by simple diffusion (Figure 23.34, ③). Without the micelles, the lipids will simply float on the surface of the chyme (like oil on water), inaccessible to the absorptive surfaces of the epithelial cells. Generally, fat absorption is completed in the ileum, but in the absence of bile (as might occur when a gallstone blocks the cystic duct) it happens so slowly that most of the fat passes into the large intestine and is lost in feces.

Once inside the epithelial cells, the free fatty acids and monoglycerides are resynthesized into triglycerides by the smooth ER. The triglycerides are then combined with lecithin and other phospholipids and cholesterol, and coated with a "skin" of proteins to form water-soluble lipoprotein droplets called **chylomicrons** (ki″lo-mi′kronz). These are dispatched to the Golgi apparatus where they are processed for extrusion from the cell. This series of events is quite different from the absorption of amino acids and simple sugars, which pass through the epithelial cells unchanged.

The milky-white chylomicrons are too large to pass through either the plasma membranes or the basement membranes of the blood capillaries. Instead, the chylomicron-containing vesicles migrate to the basolateral membrane and are extruded by exocytosis (Figure 23.34, ④). They then enter the more permeable lacteals. Thus, most fat enters the lymphatic stream for distribution in the lymph. Eventually the chylomicrons are emptied into the venous blood in the neck region via the thoracic duct, which drains the digestive viscera.

While in the bloodstream, the triglycerides of the chylomicrons are hydrolyzed to free fatty acids and glycerol by **lipoprotein lipase**, an enzyme associated with the capillary endothelium of liver and adipose tissue. The fatty acids and glycerol can then pass through the capillary walls to be used by tissue cells for energy or stored as fats in adipose tissue. The residual chylomicron material is combined with proteins by the liver cells, and these "new" lipoproteins are used to transport cholesterol in the blood.

Passage of short-chain fatty acids is quite different from what we have just described. These fat breakdown products do not depend on the presence of bile salts or micelles, are not recombined to form triglycerides within the intestinal cells, and simply diffuse into the portal blood for distribution.

Nucleic Acid Absorption

The pentose sugars, nitrogenous bases, and phosphate ions resulting from nucleic acid digestion are transported actively across the epithelium by special carriers in the villus epithelium. They then enter the blood.

Vitamin Absorption

The small intestine absorbs dietary vitamins, and the large intestine absorbs some of the K and B vitamins made by its enteric bacterial "guests." As we already noted, fat-soluble vitamins (A, D, E, and K) dissolve in dietary fats, become incorporated into the micelles, and move across the villus epithelium passively (by diffusion). It follows that gulping pills containing fat-soluble vitamins without simultaneously eating some fat-containing food results in little or no absorption of these vitamins.

Most water-soluble vitamins (B vitamins and vitamin C) are absorbed easily by diffusion or via specific active or passive transporters. The exception is vitamin B_{12}, which is a very large, charged molecule. *Intrinsic factor*, produced by the stomach, binds to vitamin B_{12}. The vitamin B_{12}–intrinsic factor complex then binds to specific mucosal receptor sites in the terminal ileum, which trigger its active uptake by endocytosis.

Electrolyte Absorption

Absorbed electrolytes come from both ingested foods and gastrointestinal secretions. Most ions are actively absorbed along the entire length of the small intestine. However, iron and calcium absorption is largely limited to the duodenum.

As we mentioned earlier, absorption of sodium ions in the small intestine is coupled to active absorption of glucose and amino acids. For the most part, anions passively follow the electrical potential established by sodium transport. In other words, Na^+ is actively pumped out of the epithelial cells by a Na^+-K^+ pump after entering those cells. Chloride ions are also transported actively, and in the terminus of the small intestine HCO_3^- is actively secreted into the lumen in exchange for Cl^-.

Potassium ions move across the intestinal mucosa passively by facilitated diffusion (or via leaky tight junctions) in response to changing osmotic gradients. As water is absorbed from the lumen, the resulting rise in potassium levels in chyme creates a concentration gradient for its absorption. Anything that interferes with water absorption (resulting in diarrhea) not only reduces potassium absorption but also "pulls" K^+ from the interstitial space into the intestinal lumen.

For most nutrients, the amount *reaching* the intestine is the amount absorbed, regardless of the nutritional state of the body. In contrast, absorption of iron and calcium is intimately related to the body's need for them at the time.

Ionic iron, essential for hemoglobin production, is actively transported into the mucosal cells, where it binds to the protein **ferritin** (fer′ĭ-tin). This phenomenon is called the *mucosal iron barrier*. The intracellular iron-ferritin complexes then serve as local storehouses for iron. When body reserves of iron are adequate, little (only 10% to 20%) is allowed to pass into the portal blood, and most of the stored iron is lost as the epithelial cells later slough off. However, when iron reserves are depleted (as during acute or chronic hemorrhage), iron uptake from the intestine and its release to the blood are accelerated. In the blood, iron binds to **transferrin**, a plasma protein that transports it in the circulation.

Menstrual bleeding is a major route of iron loss in females, and premenopausal women require about 50% more iron in their diets. Additionally, the intestinal epithelial cells of women have about four times as many iron transport proteins as do those of men and very little iron is lost from the body other than that lost in menses.

Calcium absorption is closely related to blood levels of ionic calcium. It is locally regulated by the active form of **vitamin D**, which promotes active calcium absorption. Decreased blood levels of ionic calcium prompt *parathyroid hormone* (*PTH*) release from the parathyroid glands. Besides facilitating the release of calcium ions from bone matrix and enhancing the reabsorption of calcium by the kidneys, PTH stimulates activation of vitamin D by the kidneys, which in turn accelerates calcium ion absorption in the small intestine.

Water Absorption

Approximately 9 L of water, mostly derived from GI tract secretions, enter the small intestine daily. Water is the most abundant substance in chyme, and 95% of it is absorbed in the small

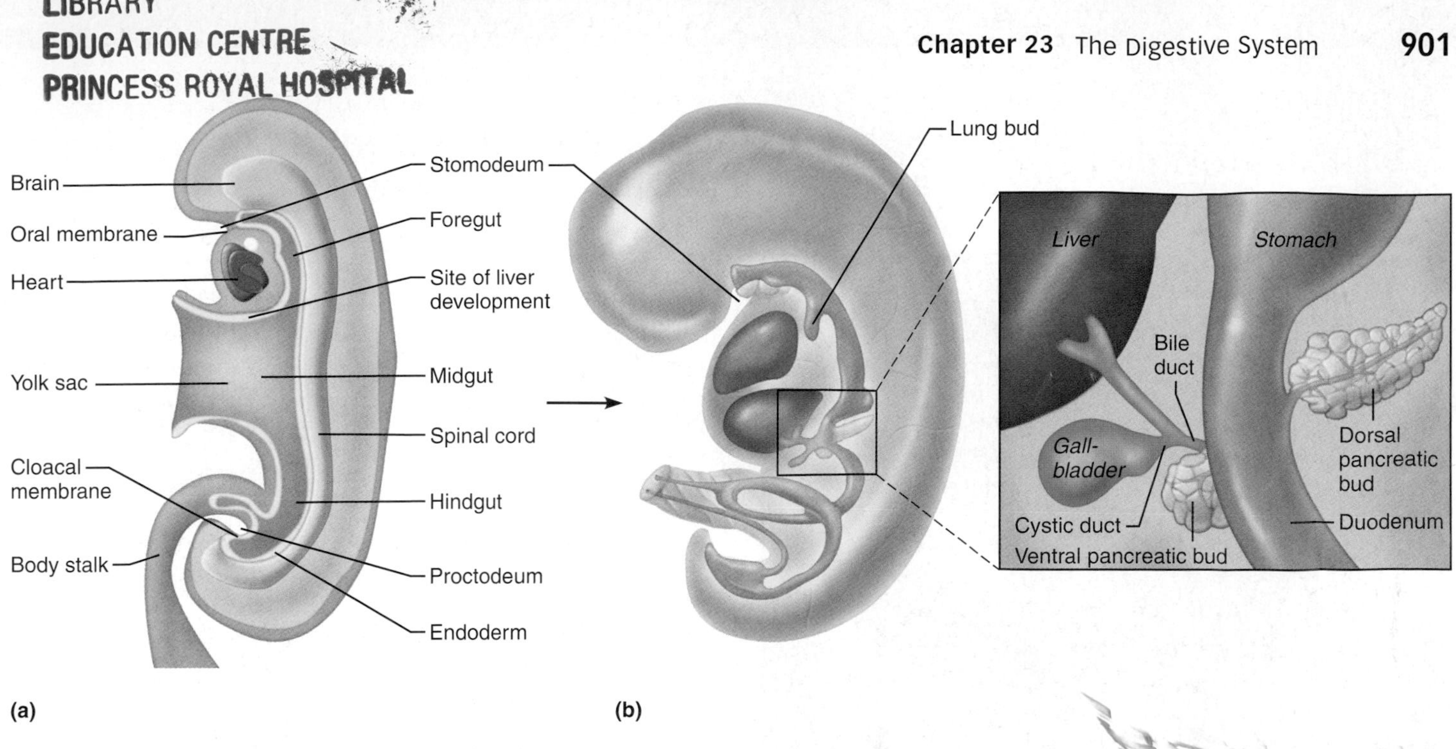

Figure 23.35 Embryonic development of the digestive system. (a) Three-week-old embryo. The endoderm has folded, and the foregut and hindgut have formed. (The midgut is still open and continuous with the yolk sac.) **(b)** By five weeks of development, the accessory organs are budding out from the endodermal layer, as shown in the enlargement.

intestine by osmosis. Most of the rest is absorbed in the large intestine, leaving only about 0.1 L to soften the feces.

The normal rate of water absorption is 300 to 400 ml per hour. Water moves freely in both directions across the intestinal mucosa, but *net osmosis* occurs whenever a concentration gradient is established by the active transport of solutes (particularly Na^+) into the mucosal cells. In this way, water uptake is effectively coupled to solute uptake and, in turn, affects the rate of absorption of substances that normally pass by diffusion. As water moves into the mucosal cells, these substances follow along their concentration gradients.

Malabsorption of Nutrients

Malabsorption, or impaired nutrient absorption, has many and varied causes. It can result from anything that interferes with the delivery of bile or pancreatic juice to the small intestine, as well as factors that damage the intestinal mucosa (severe bacterial infections and antibiotic therapy with neomycin) or reduce its absorptive surface area.

A common malabsorption syndrome is *gluten-sensitive enteropathy* or *celiac disease*, which affects one in 100 people. This chronic genetic condition is caused by an immune reaction to gluten, a protein plentiful in some grains (wheat, rye, barley). Breakdown products of gluten interact with molecules of the immune system in the GI tract, forming complexes that activate T cells, which then mount an attack on the intestinal lining. As a result, the intestinal villi are damaged and the surface area of the brush border is reduced.

The resulting bloating, diarrhea, pain, and malnutrition are usually controlled by eliminating gluten-containing grains (all grains but rice and corn) from the diet. Not many children can conceive of a world without cookies or pizza, but these are just two of the foods celiac disease victims must avoid.

CHECK YOUR UNDERSTANDING

48. What type of chemical reaction is the basis of all enzymatic food digestion?
49. Fill in the blank: Amylase is to starch as ____ is to fats.
50. What is the role of bile salts in the digestive process? In absorption?

For answers, see Appendix G.

23

Developmental Aspects of the Digestive System

- Describe embryonic development of the digestive system.
- Describe abnormalities of the gastrointestinal tract at different stages of life.

As we have described many times before, the very young embryo is flat and consists of three germ layers. From top to bottom, they are ectoderm, mesoderm, and endoderm. This flattened cell mass soon folds to form a cylindrical body, and its internal cavity becomes the cavity of the alimentary canal, which is initially closed at both ends.

The epithelial lining of the developing alimentary canal, or **primitive gut**, forms from endoderm (**Figure 23.35a**). The rest of the wall arises from mesoderm. The anteriormost endoderm

(Text continues on p. 904.)

System Connections

Homeostatic Interrelationships Between the Digestive System and Other Body Systems

Integumentary System

- Digestive system provides nutrients needed for energy, growth, and repair; supplies fats that provide insulation in the dermis and subcutaneous tissue
- The skin synthesizes vitamin D needed for calcium absorption from the intestine; protects by enclosure

Skeletal System

- Digestive system provides nutrients needed for energy, growth, and repair; absorbs calcium needed for bone salts
- Skeletal system protects some digestive organs by bone; cavities store some nutrients (e.g., calcium, fats)

Muscular System

- Digestive system provides nutrients needed for energy, growth, and repair; liver removes lactic acid, resulting from muscle activity, from the blood
- Skeletal muscle activity increases motility of GI tract

Nervous System

- Digestive system provides nutrients needed for normal neural functioning; nutrient signals influence neural regulation of satiety
- Neural controls of digestive function; in general, parasympathetic fibers accelerate and sympathetic fibers inhibit digestive activity; reflex and voluntary controls of defecation

Endocrine System

- Liver removes hormones from blood, ending their activity; digestive system provides nutrients needed for energy, growth, and repair; pancreas, stomach, and small intestine have hormone-producing cells
- Hormones and paracrines help regulate digestive function

Cardiovascular System

- Digestive system provides nutrients to heart and blood vessels; absorbs iron needed for hemoglobin synthesis; absorbs water necessary for normal plasma volume; liver secretes bilirubin resulting from hemoglobin breakdown in bile and stores iron for reuse
- Cardiovascular system transports nutrients absorbed by alimentary canal to all tissues of body; distributes hormones of the digestive tract

Lymphatic System/Immunity

- Digestive system provides nutrients for normal functioning; lysozyme of saliva and HCl of stomach provide nonspecific protection against bacteria
- Lacteals drain fatty lymph from digestive tract organs and convey it to blood; Peyer's patches and lymphoid tissue in mesentery house macrophages and immune cells that protect digestive tract organs against infection; plasma cells provide IgA in digestive tract secretions

Respiratory System

- Digestive system provides nutrients needed for energy metabolism, growth, and repair
- Respiratory system provides oxygen to and carries away carbon dioxide produced by digestive system organs

Urinary System

- Digestive system provides nutrients for energy fuel, growth, and repair; excretes some bilirubin produced by the liver
- Kidneys transform vitamin D to its active form, which is needed for calcium absorption

Reproductive System

- Digestive system provides nutrients for energy, growth, and repair and extra nutrition needed to support fetal growth

Closer Connections

The Digestive System and Interrelationships with the Cardiovascular, Lymphatic, and Endocrine Systems

As body systems go, the digestive system is very independent. Although the central nervous system can and does influence its activity, that is not a requirement. Indeed, most digestive system activities can be regulated entirely by local controls. Consider, for example, the following: (1) stomach and intestinal muscle is self-pacing, (2) the enteric neurons of the GI tract's intrinsic nerve plexuses carry out reflex activity that can regulate not only the organ they reside in but also adjacent and more distant GI tract organs, and (3) its enteroendocrine cells are probably even more important than neural influences in controlling digestive activity. Also unique is the fact that the digestive system operates in and regulates an environment that is actually outside the body. This is a difficult chore indeed—consider how quickly that environment can be changed by incoming foods (pizza and beer) or intestinal outflow (diarrhea).

Every cell in the body needs nutrients to stay healthy and grow, so the importance of the digestive system to virtually all body systems is clear-cut. But what systems are essential to the digestive system? If we ignore universal services, like gas exchange and renal regulation of body fluids, then the most crucial interactions are those with the cardiovascular, lymphatic, and endocrine systems. Let's take a look.

Cardiovascular and Lymphatic Systems

The important elements in this interaction are the pickup vessels. Indeed, all digestive activity would be for naught if the digested end products were not absorbed into the blood capillaries and lacteals, where they are accessible to all body cells. The lymphatic organs and tissues also offer protective services via the tonsils, Peyer's patches, and individual lymphoid follicles dispersed all along the length of the alimentary canal. Again, the fact that the alimentary canal is open at both ends makes it easy prey for both pathogenic and opportunistic bacteria and fungi.

Endocrine System

The digestive tract not only has major interactions with the endocrine system, it is itself the largest and most complex endocrine organ in the body. It makes hormones that regulate GI tract motility and secretory activity of the stomach, small intestine, liver, and pancreas (e.g., secretin, CCK), as well as cause local blood vessels to dilate to receive the digested food products being absorbed (VIP). Furthermore, some digestive organ hormones (CCK and glucagon, for example) help mediate hunger or satiety, and pancreatic hormones (insulin and glucagon) are involved in bodywide regulation of carbohydrate, fat, and amino acid metabolism.

Clinical Connections

Digestive System

Case study: Remember Mr. Gutteman, the gentleman who was dehydrating? It seems that his tremendous output of urine was only one of his current problems. Today, he complains of a headache, gnawing epigastric pain, and "the runs" (diarrhea). To pinpoint the problem, he is asked the following questions.

- Have you had these symptoms previously? (Response: "Yes, but never this bad.")
- Are you allergic to any foods? (Response: "Shellfish doesn't like me and milk gives me the runs.")

As a result of his responses, a lactose-free diet is ordered for Mr. Gutteman instead of the regular diet originally prescribed.

1. Why is the new diet prescribed? (What is believed to be his problem?)

Mr. Gutteman's problem continues despite the diet change. In fact, the frequency of diarrhea increases and by the end of the next day, he is complaining of severe abdominal pain. Again, he is asked some questions to probe his condition. One is whether he has traveled outside the country recently. He has not, reducing the possibility of infection with *Shigella* bacteria, which is associated with poor sanitation. Other questions:

- Do you drink alcohol and how much? (Response: "Little or none.")
- Have you recently eaten raw eggs or a salad containing mayonnaise at a gathering? (Response: "No.")
- Are there certain foods that seem to precipitate these attacks? (Response: "Yes, when I have coffee and a sandwich.")

2. On the basis of these responses, what do you think Mr. Gutteman's diarrhea stems from? How will it be diagnosed and treated?

(Answers in Appendix G)

(that of the foregut) touches a depressed area of the surface ectoderm called the **stomodeum** (sto″mo-de′um; "on the way to becoming the mouth"). The two membranes fuse, forming the **oral membrane**, which soon breaks through to form the opening of the mouth. Similarly, the end of the hindgut fuses with an ectodermal depression, called the **proctodeum** (*procto* = anus), to form the **cloacal membrane** (*cloaca* = sewer), which then breaks through to form the anus.

By week 5, the alimentary canal is a continuous "tube" extending from the mouth to the anus and is open to the external environment at each end. Shortly after, the glandular organs (salivary glands, liver with gallbladder, and pancreas) bud out from the mucosa at various points along its length (Figure 23.35b). These glands retain their connections, which become ducts leading into the digestive tract.

HOMEOSTATIC IMBALANCE

The digestive system is susceptible to many congenital defects that interfere with feeding. The most common are **cleft palate**, in which the palatine bones or palatine processes of the maxillae (or both) fail to fuse, and **cleft lip**, which often occur together. Of the two, cleft palate is far more serious because the child is unable to suck properly.

Another common defect is *tracheoesophageal fistula*, in which there is an opening between the esophagus and the trachea, and the esophagus often lacks a connection to the stomach. The baby chokes and becomes cyanotic during feedings because food enters the respiratory passageways. These defects are usually corrected surgically.

Cystic fibrosis (described in more detail in Chapter 22, p. 843) primarily affects the lungs, but it also impairs the activity of the pancreas. In this genetic disease, the mucous glands produce abnormally thick mucus, which blocks the ducts or passageways of involved organs. Blockage of the pancreatic duct prevents pancreatic juice from reaching the small intestine. As a result, digestion of chyme is less than optimal and most fats and fat-soluble vitamins are not digested or absorbed. Consequently the stools are bulky and fat laden. The pancreatic problems can be handled by administering pancreatic enzymes with meals. ■

During fetal life, the developing infant receives all of its nutrients through the placenta. Nonetheless, the fetal GI tract is "trained" in utero for future food digestion as the fetus naturally swallows some of the surrounding amniotic fluid. This fluid contains several chemicals that stimulate GI maturation, including gastrin and epidermal growth factor.

By contrast, feeding is a newborn baby's most important activity, and several reflexes enhance the infant's ability to obtain food: The *rooting reflex* helps the infant find the nipple, and the *sucking reflex* helps the baby hold onto the nipple and swallow.

Newborn babies tend to double their birth weight within six months, and their caloric intake and food processing ability are extraordinary. For example, a 6-week-old infant weighing about 4 kg (less than 9 lb) drinks about 600 ml of milk daily. A 65-kg adult (143 lb) would have to drink 10 L of milk to ingest a corresponding volume of fluid! However, the stomach of a newborn infant is very small, so feeding must be frequent (every 3–4 hours). Peristalsis is inefficient, and vomiting of the feeding is not unusual. As the teeth break through the gums, the infant progresses to solid foods and is usually eating an adult diet by the age of 2 years.

As a rule, the digestive system operates throughout childhood and adulthood with relatively few problems. However, contaminated food or extremely spicy or irritating foods sometimes cause an inflammation of the GI tract, called **gastroenteritis** (gas″tro-en″tĕ-ri′tis). Ulcers and gallbladder problems—inflammation or **cholecystitis** (ko″le-sis-ti′tis), and gallstones—are problems of middle age.

During old age, GI tract activity declines. Fewer digestive juices are produced, absorption is less efficient, and peristalsis slows. The result is less frequent bowel movements and, often, constipation. Taste and smell are less acute, and periodontal disease often develops. Many elderly people live alone or on a reduced income. These factors, along with increasing physical disability, tend to make eating less appealing, and many of our elderly citizens are poorly nourished.

Diverticulosis, fecal incontinence, and cancer of the GI tract are fairly common problems of the aged. Stomach and colon cancers rarely have early signs, and have often metastasized (making it inoperable) before a person seeks medical attention. Should metastasis occur, secondary cancer of the liver is almost guaranteed because of the "detour" the splanchnic venous blood takes through the liver via the hepatic portal circulation. However, when detected early, these cancers are treatable. Perhaps the best advice is to have regular dental and medical checkups. Most oral cancers are detected during routine dental examinations, 50% of all rectal cancers can be felt digitally, and nearly 80% of colon cancers can be seen and removed during a colonoscopy.

Presently, colon cancer is the second leading cancer killer (behind lung cancer) of males in the United States. Until now it was believed that most colorectal cancers derive from initially benign, knobby mucosal tumors called polyps. However, flat pancake-like growths which blend in with the surrounding tissue are more common in Americans than originally thought and are 10 times more likely to be cancerous than polyps. Researchers believe that the flat growths represent a separate pathway to colon cancer, instead of being polyp precursors. This information will likely have a big impact on gastroenterology and scheduling of tests for colon cancer. Currently it is recommended that an occult blood examination be done once a year and that colonoscopies be conducted at 3- to 10-year intervals depending on prior findings.

As we described in *A Closer Look* in Chapter 4, development of colon cancer is a gradual process that involves mutations in several regulatory genes. Recent studies have identified the mutant gene responsible for nearly 50% of all colon cancers. Dubbed *p53*, the gene normally acts as a watchdog to make sure that errors in DNA are not passed on before being corrected. When *p53* is damaged or inhibited, its tumor suppressor effect is lost and damaged DNA accumulates, resulting in carcinogenesis.

CHECK YOUR UNDERSTANDING

51. From what germ layer does the digestive system mucosa develop?
52. How does cystic fibrosis interfere with the digestive process?
53. Why are colon and stomach cancers so dangerous?

For answers, see Appendix G.

As summarized in *Making Connections*, the digestive system keeps the blood well supplied with the nutrients needed by all body tissues to fuel their energy needs and to synthesize new proteins for growth and maintenance of health. Now we are ready to examine how these nutrients are used by body cells, the topic of Chapter 24.

RELATED CLINICAL TERMS

Ascites (ah-si′tēz; *asci* = bag, bladder) Abnormal accumulation of fluid within the peritoneal cavity; if excessive, causes visible bloating of the abdomen. May result from portal hypertension caused by liver cirrhosis or by heart or kidney disease.

Barrett's esophagus A pathological change in the epithelium of the lower esophagus from stratified squamous to a metaplastic columnar epithelium. A possible sequel to untreated chronic gastroesophageal reflux due to hiatal hernia, it predisposes the individual to aggressive esophageal cancer (adenocarcinoma).

Bruxism (bruck′sizm) Grinding or clenching of teeth, usually at night during sleep in response to stress. Can wear down and crack the teeth.

Bulimia (bu-lim′e-ah; *bous* = ox; *limos* = hunger) Binge-purge behavior—episodes of massive overeating are followed by some method of purging (self-induced vomiting, taking laxatives or diuretics, or excessive exercise). Most common in women of high school or college age and in high school males in certain athletic activities (wrestling). Rooted in a pathological fear of being fat, and a need for control; provides a means of handling stress and depression. Consequences include erosion of tooth enamel and stomach trauma or rupture (from vomiting) and severe electrolyte disturbances which impair heart activity. Therapy includes hospitalization to control behavior, and nutritional counseling.

Dysphagia (dis-fa′je-ah; *dys* = difficult, abnormal; *phag* = eat) Difficulty swallowing; usually due to obstruction or physical trauma to the esophagus.

Endoscopy (en-dos′ko-pe; *endo* = within, inside; *scopy* = viewing) Visual examination of a ventral body cavity or the interior of a visceral organ with a flexible tubelike device called an endoscope, which contains a light source and a lens. A general term for a colonoscopy (viewing the colon), sigmoidoscopy (viewing the sigmoid colon), etc.

Enteritis (*enteron* = intestine) Inflammation of the intestine, especially the small intestine.

Hemochromatosis (he″mo-kro″mah-to′sis; *hemo* = blood; *chroma* = color; *osis* = condition of) A disorder in iron metabolism due to excessive/prolonged iron intake or a breakdown of the mucosal iron barrier; excess iron is deposited in the tissues, causing increased skin pigmentation and increased incidence of hepatic cancer and liver cirrhosis; also called *bronze diabetes* and *iron storage disease.*

Ileus (il′e-us) A condition in which all GI tract movement stops and the gut appears to be paralyzed. Can result from electrolyte imbalances and blockade of parasympathetic impulses by drugs (such as those commonly used during abdominal surgery); usually reversed when these interferences end. Restoration of motility is indicated by the reappearance of bowel (intestinal) sounds (gurgling, etc.).

Inflammatory bowel disease (IBD) A noncontagious, periodic inflammation of the intestinal wall now understood to be an abnormal immune and inflammatory response to bacterial antigens that normally occur in the intestine. Linked to a newly discovered T_H cell (T_H17), certain cytokines, and a deficit of antimicrobial substances normally secreted by the gut mucosa (lysozyme, defensins, and others). Afflicts up to two of every 1000 people. Symptoms include cramping, diarrhea, weight loss, and intestinal bleeding. Two subtypes occur: Crohn's disease, a syndrome characterized by relapsing and remitting periods, is more serious, with deep ulcers and fissures developing along the whole intestine, but mostly in the terminal ileum. Ulcerative colitis, by contrast, is characterized by a shallow inflammation of the large-intestinal mucosa, mainly in the rectum. They are treated with a special diet, reducing stress, antibiotics, and anti-inflammatory and immunosuppressant drugs. Extremely severe cases of ulcerative colitis are treated by colectomy (removal of a portion of the colon).

Laparoscopy (lap″ah-ros′ko-pe; *lapar* = the flank; *scopy* = observation) Examination of the peritoneal cavity and its organs with an endoscope inserted through the anterior abdominal wall. Often used to assess the condition of the digestive organs and the pelvic reproductive organs of females.

Orthodontics (or″tho-don′tiks; *ortho* = straight) Branch of dentistry that prevents and corrects misaligned teeth.

Pancreatitis (pan″kre-ah-ti′tis) A rare but extremely serious inflammation of the pancreas. May result from excessively high levels of fat in the blood or excessive alcohol ingestion, but most often results when a gallstone blocks the bile duct and pancreatic enzymes are activated in the pancreatic duct, causing the pancreatic tissue and duct to be digested. This painful condition can lead to nutritional deficiencies because pancreatic enzymes are essential to food digestion in the small intestine.

Peptic ulcers Term referring to gastric and duodenal ulcers.

Proctology (prok-tol′o-je; *procto* = rectum, anus; *logy* = study of) Branch of medicine dealing with treatment of diseases of the colon, rectum, and anus.

Pyloric stenosis (pi-lor′ik stě-no′sis; *stenosis* = narrowing, constriction) Congenital abnormality in which the pyloric sphincter is abnormally constricted. There is usually no problem until the baby begins to take solid food, and then projectile vomiting begins. Corrected surgically.

23

Vagotomy (va-got′o-me) Cutting or severing of the vagus nerve to decrease secretion of gastric juice in those with peptic ulcers that do not respond to medication.

Xerostomia (ze″ro-sto′me-ah; *zer* = dry, *stom* = mouth) Extreme dryness of the mouth; can be caused by blockage of salivary glands by cysts or autoimmune invasion of the salivary glands or ducts (Sjögren's syndrome).

CHAPTER SUMMARY

Media study tools that could provide you additional help in reviewing specific key topics of Chapter 23 are referenced below.

iP = ***Interactive Physiology***

PART 1: OVERVIEW OF THE DIGESTIVE SYSTEM

1. The digestive system includes organs of the alimentary canal (mouth, pharynx, esophagus, stomach, small and large intestines) and accessory digestive system organs (teeth, tongue, salivary glands, liver, gallbladder, and pancreas).

Digestive Processes (pp. 853–854)

1. Digestive system activities include six processes: ingestion, or food intake; propulsion, or movement of food through the tract; mechanical digestion, or processes that physically mix or break foods down into smaller fragments; chemical digestion, or food breakdown by enzymatic action; absorption, or transport of products of digestion through the intestinal mucosa into the blood; and defecation, or elimination of the undigested residues (feces) from the body.

Basic Functional Concepts (p. 854)

1. The digestive system controls the environment within its lumen to ensure optimal conditions for digestion and absorption of foodstuffs.
2. Receptors and hormone-secreting cells in the alimentary canal wall respond to stretch and chemical signals that result in stimulation or inhibition of GI secretory activity or motility. The alimentary canal has an intrinsic nerve supply.

iP Digestive System; Topic: Control of the Digestive System, pp. 1, 3, 4–5, 8.

23

Digestive System Organs: Relationships (pp. 854–858)

1. The parietal and visceral layers of the peritoneum are continuous with one another via several extensions (mesenteries, falciform ligament, lesser and greater omenta), and are separated by a potential space containing serous fluid, which decreases friction during organ activity.
2. The digestive viscera are served by the splanchnic circulation, consisting of arterial branches of the celiac trunk and aorta and the hepatic portal circulation.
3. All organs of the GI tract have the same basic pattern of tissue layers in their walls; all have a mucosa, a submucosa, a muscularis, and a serosa (or adventitia). Intrinsic nerve plexuses (enteric nervous system) are found within the wall.

PART 2: FUNCTIONAL ANATOMY OF THE DIGESTIVE SYSTEM

The Mouth and Associated Organs (pp. 858–864)

1. Food enters the GI tract via the mouth, which is continuous with the oropharynx posteriorly. The boundaries of the mouth are the lips and cheeks, palate, and tongue.
2. The oral mucosa is stratified squamous epithelium, an adaptation seen where abrasion occurs.
3. The tongue is mucosa-covered skeletal muscle. Its intrinsic muscles allow it to change shape; its extrinsic muscles allow it to change position.
4. Saliva is produced by many small buccal glands and three pairs of major salivary glands—parotid, submandibular, and sublingual—that secrete their product into the mouth via ducts. Largely water, saliva also contains ions, proteins, metabolic wastes, lysozyme, defensins, IgA, salivary amylase, a cyanide compound, and mucin.
5. Saliva moistens and cleanses the mouth; moistens foods, aiding their compaction; dissolves food chemicals to allow for taste; and begins chemical digestion of starch (salivary amylase). Saliva output is increased by parasympathetic reflexes initiated by activation of chemical and pressure receptors in the mouth and by conditioned reflexes. The sympathetic nervous system depresses salivation.
6. The 20 deciduous teeth begin to be shed at the age of 6 and are gradually replaced during childhood and adolescence by the 32 permanent teeth.
7. Teeth are classed as incisors, canines, premolars, and molars. Each tooth has an enamel-covered crown and a cementum-covered root. The bulk of the tooth is dentin, which surrounds the central pulp cavity. A periodontal ligament secures the tooth to the bony alveolus.

The Pharynx (p. 864)

1. Food propelled from the mouth passes through the oropharynx and laryngopharynx. The mucosa of the pharynx is stratified squamous epithelium; skeletal muscles in its wall (constrictor muscles) move food toward the esophagus.

The Esophagus (pp. 864–866)

1. The esophagus extends from the laryngopharynx and joins the stomach at the cardiac orifice, which is surrounded by the gastroesophageal sphincter.
2. The esophageal mucosa is stratified squamous epithelium. Its muscularis is skeletal muscle superiorly and changes to smooth muscle inferiorly. It has an adventitia rather than a serosa.

Digestive Processes: Mouth to Esophagus (p. 866)

1. The mouth and associated accessory organs accomplish food ingestion and mechanical digestion (chewing and mixing), initiate the chemical digestion of starch (salivary amylase), and propel food into the pharynx (buccal phase of swallowing).

2. Teeth function to masticate food. Chewing is initiated voluntarily and is then controlled reflexively.
3. The tongue mixes food with saliva, compacts it into a bolus, and initiates swallowing (the voluntary phase). The pharynx and esophagus are primarily food conduits that conduct food to the stomach by peristalsis. The swallowing center in the medulla and pons controls this phase reflexively. When the peristaltic wave approaches the gastroesophageal sphincter, the sphincter relaxes to allow food entry to the stomach.

iP Digestive System; Topic: Motility, pp. 3–5.

The Stomach (pp. 866–877)

1. The J-shaped stomach lies in the upper left quadrant of the abdomen. Its major regions are the cardia, fundus, body, and pyloric region. When empty, its internal surface exhibits rugae.
2. The stomach muscularis contains a third (oblique) layer of smooth muscle that allows it to churn and mix food.
3. The stomach mucosa is simple columnar epithelium dotted with gastric pits that lead into gastric glands. Secretory cells in the gastric glands include pepsinogen-producing chief cells; parietal cells, which secrete hydrochloric acid and intrinsic factor; mucous neck cells, which produce mucus; and enteroendocrine cells, which secrete hormones.
4. The mucosal barrier, which protects the stomach from self-digestion and HCl, reflects the fact that the mucosal cells are connected by tight junctions, secrete a thick mucus, and are quickly replaced when damaged.
5. Protein digestion is initiated in the stomach by activated pepsin and requires acidic conditions (provided by HCl). Few substances are absorbed.
6. Gastric secretory activity is controlled by both nervous and hormonal factors. The three phases of gastric secretion are the cephalic, gastric, and intestinal. Most food-related stimuli acting on the head and stomach (cephalic and gastric respectively) stimulate gastric secretion. Most stimuli acting on the small intestine trigger the enterogastric reflex and release of secretin and CCK, both of which inhibit gastric secretory activity. Sympathetic activity also inhibits gastric secretion.
7. Mechanical digestion in the stomach is triggered by stomach distension and coupled to food propulsion and stomach emptying. Food movement into the duodenum is controlled by the pylorus and feedback signals from the small intestine. Pacemaker cells in the smooth muscle sheet set the maximal rate of peristalsis.

iP Digestive System; Topic: Control of the Digestive System, pp. 3, 5–6, 8.

The Small Intestine and Associated Structures (pp. 877–890)

1. The small intestine is the major digestive and absorptive organ. It extends from the pyloric sphincter to the ileocecal valve. Its three subdivisions are the duodenum, jejunum, and ileum. The bile duct and pancreatic duct join to form the hepatopancreatic ampulla and empty their secretions into the duodenum through the hepatopancreatic sphincter.
2. Circular folds, villi, and microvilli increase the intestinal surface area for digestion and absorption.
3. The duodenal submucosa contains elaborate mucus-secreting duodenal glands (Brunner's glands); that of the ileum contains Peyer's patches (lymphoid follicles). The duodenum is covered not with a serosa but an adventitia.
4. Intestinal juice is largely water and is relatively enzyme-poor. The major stimuli for its release are stretch and chemicals.
5. The liver is a lobed organ overlying the stomach. Its digestive role is to produce bile, which it secretes into the common hepatic duct.
6. The structural and functional units of the liver are the liver lobules. Blood flowing to the liver via the hepatic artery and hepatic portal vein flows into its sinusoids, from which Kupffer cells remove debris and hepatocytes remove nutrients. Hepatocytes store glucose as glycogen, use amino acids to make plasma proteins, and detoxify metabolic wastes and drugs.
7. Bile is made continuously by the hepatocytes. Bile salts and secretin stimulate bile production.
8. The gallbladder, a muscular sac that lies beneath the right liver lobe, stores and concentrates bile.
9. Bile contains electrolytes, a variety of fatty substances, bile salts, and bile pigments in an aqueous medium. Bile salts are emulsifying agents; they disperse fats and form water-soluble micelles, which solubilize the products of fat digestion.
10. Cholecystokinin released by the small intestine stimulates the gallbladder to contract and the hepatopancreatic sphincter to relax, allowing bile (and pancreatic juice) to enter the duodenum.
11. The pancreas is retroperitoneal between the spleen and small intestine. Its exocrine product, pancreatic juice, is carried to the duodenum via the pancreatic duct.
12. Pancreatic juice is an HCO_3^--rich fluid containing enzymes that digest all categories of foods. The secretion of pancreatic juice is controlled by intestinal hormones and the vagus nerves.

iP Digestive System; Topic: Secretion, pp. 11–14.

13. Mechanical digestion and propulsion in the small intestine mix chyme with digestive juices and bile and force the residues, largely by segmentation, through the ileocecal valve. Pacemaker cells set the rate of segmentation. Ileocecal valve opening is controlled by the gastroileal reflex and gastrin.

The Large Intestine (pp. 890–895)

1. The subdivisions of the large intestine are the cecum (and appendix), colon (ascending, transverse, descending, and sigmoid portions), rectum, and anal canal. It opens to the body exterior at the anus.
2. The longitudinal muscle in the muscularis is reduced to three bands (teniae coli), which pucker its wall, producing haustra. The mucosa of most of the large intestine is simple columnar epithelium containing abundant goblet cells.
3. The major functions of the large intestine are absorption of water, some electrolytes, and vitamins made by enteric bacteria, and defecation (evacuation of food residues from the body).
4. The defecation reflex is triggered when feces enter the rectum. It involves parasympathetic reflexes leading to the contraction of the rectal walls and is aided by Valsalva's maneuver.

iP Digestive System; Topic: Anatomy Review, pp. 3–5.

PART 3: PHYSIOLOGY OF CHEMICAL DIGESTION AND ABSORPTION

Chemical Digestion (pp. 895–897)

1. Chemical digestion is accomplished by hydrolysis, catalyzed by enzymes.
2. Most chemical digestion is done in the small intestine by intestinal (brush border) enzymes and, more importantly, by pancreatic enzymes. Alkaline pancreatic juice neutralizes the acidic chyme and provides the proper environment for operation of the enzymes. Both pancreatic juice (the main source of lipases) and bile are necessary for fat breakdown.

Absorption (pp. 898–901)

1. Virtually all of the foodstuffs and most of the water and electrolytes are absorbed in the small intestine. Most nutrients are absorbed by active transport processes, except for fat digestion products, fat-soluble vitamins, and most water-soluble vitamins (which are absorbed by diffusion).
2. Fat breakdown products are solubilized by bile salts (in micelles), resynthesized to triglycerides in the intestinal mucosal cells, and combined with other lipids and protein as chylomicrons that enter the lacteals. Other absorbed substances enter the villus blood capillaries and are transported to the liver via the hepatic portal vein.

Developmental Aspects of the Digestive System (pp. 901, 904–905)

1. The glandular accessory organs (salivary glands, liver, pancreas, and gallbladder) form from outpocketings of the foregut endoderm. The mucosa of the alimentary canal develops from the endoderm, which folds to form a tube. The remaining three tunics of the alimentary canal wall are formed by mesoderm.
2. Important congenital abnormalities of the digestive tract include cleft palate/lip, tracheoesophageal fistula, and cystic fibrosis. All interfere with normal nutrition.
3. Various inflammations plague the digestive system throughout life. Appendicitis is common in adolescents, gastroenteritis and food poisoning can occur at any time (given the proper irritating factors), and ulcers and gallbladder problems increase in middle age.
4. The efficiency of all digestive system processes declines in the elderly, and periodontal disease is common. Diverticulosis, fecal incontinence, and GI tract cancers such as stomach and colon cancer appear with increasing frequency in an aging population.

REVIEW QUESTIONS

Multiple Choice/Matching

(Some questions have more than one correct answer. Select the best answer or answers from the choices given.)

1. The peritoneal cavity **(a)** is the same thing as the abdominopelvic cavity, **(b)** is filled with air, **(c)** like the pleural and pericardial cavities is a potential space containing serous fluid, **(d)** contains the pancreas and all of the duodenum.
2. Obstruction of the hepatopancreatic sphincter impairs digestion by reducing the availability of **(a)** bile and HCl, **(b)** HCl and intestinal juice, **(c)** pancreatic juice and intestinal juice, **(d)** pancreatic juice and bile.
3. The action of an enzyme is influenced by **(a)** its chemical surroundings, **(b)** its specific substrate, **(c)** the presence of needed cofactors or coenzymes, **(d)** all of these.
4. Carbohydrates are acted on by **(a)** peptidases, trypsin, and chymotrypsin, **(b)** amylase, maltase, and sucrase, **(c)** lipases, **(d)** peptidases, lipases, and galactase.
5. The parasympathetic nervous system influences digestion by **(a)** relaxing smooth muscle, **(b)** stimulating peristalsis and secretory activity, **(c)** constricting sphincters, **(d)** none of these.
6. The digestive juice product containing enzymes capable of digesting all four major foodstuff categories is **(a)** pancreatic, **(b)** gastric, **(c)** salivary, **(d)** biliary.
7. The vitamin associated with calcium absorption is **(a)** A, **(b)** K, **(c)** C, **(d)** D.
8. Someone has eaten a meal of buttered toast, cream, and eggs. Which of the following would you expect to happen? **(a)** Compared to the period shortly after the meal, gastric motility and secretion of HCl decrease when the food reaches the duodenum; **(b)** gastric motility increases even as the person is chewing the food (before swallowing); **(c)** fat will be emulsified in the duodenum by the action of bile; **(d)** all of these.
9. The site of production of VIP and cholecystokinin is **(a)** the stomach, **(b)** the small intestine, **(c)** the pancreas, **(d)** the large intestine.
10. Which of the following is not characteristic of the colon? **(a)** It is divided into ascending, transverse, and descending portions; **(b)** it contains abundant bacteria, some of which synthesize certain vitamins; **(c)** it is the main absorptive site; **(d)** it absorbs much of the water and salts remaining in the wastes.
11. The gallbladder **(a)** produces bile, **(b)** is attached to the pancreas, **(c)** stores and concentrates bile, **(d)** produces secretin.
12. The sphincter between the stomach and duodenum is **(a)** the pyloric sphincter, **(b)** the gastroesophageal sphincter, **(c)** the hepatopancreatic sphincter, **(d)** the ileocecal sphincter.

In items 13–17, trace the path of a single protein molecule that has been ingested.

13. The protein molecule will be digested by enzymes secreted by **(a)** the mouth, stomach, and colon, **(b)** the stomach, liver, and small intestine, **(c)** the small intestine, mouth, and liver, **(d)** the pancreas and stomach.
14. The protein molecule must be digested before it can be transported to and utilized by the cells because **(a)** protein is only useful directly, **(b)** protein has a low pH, **(c)** proteins in the circulating blood produce an adverse osmotic pressure, **(d)** the protein is too large to be readily absorbed.
15. The products of protein digestion enter the bloodstream largely through cells lining **(a)** the stomach, **(b)** the small intestine, **(c)** the large intestine, **(d)** the bile duct.
16. Before the blood carrying the products of protein digestion reaches the heart, it first passes through capillary networks in **(a)** the spleen, **(b)** the lungs, **(c)** the liver, **(d)** the brain.
17. Having passed through the regulatory organ selected above, the products of protein digestion are circulated throughout the body. They will enter individual body cells by **(a)** active transport, **(b)** diffusion, **(c)** osmosis, **(d)** phagocytosis.

Short Answer Essay Questions

18. Make a simple line drawing of the organs of the alimentary canal and label each organ. Then add three labels to your drawing—salivary glands, liver, and pancreas—and use arrows to show where each of these organs empties its secretion into the alimentary canal.
19. Sara was on a diet but she could not eat less and kept claiming her stomach had a mind of its own. She was joking, but indeed, there is a "gut brain" called the enteric nervous system. Is it part of the parasympathetic and sympathetic nervous system? Explain.
20. Name the layers of the alimentary canal wall. Note the tissue composition and major function of each layer.

21. What is a mesentery? Mesocolon? Greater omentum?
22. Name the six functional activities of the digestive system.
23. (**a**) Describe the boundaries of the oral cavity. (**b**) Why do you suppose its mucosa is stratified squamous epithelium rather than the more typical simple columnar epithelium?
24. (**a**) What is the normal number of permanent teeth? Of deciduous teeth? (**b**) What substance covers the tooth crown? Its root? (**c**) What substance makes up the bulk of a tooth? (**d**) What and where is pulp?
25. Describe the two phases of swallowing, noting the organs involved and the activities that occur.
26. Describe the role of these cells found in gastric glands: parietal, chief, mucous neck, and enteroendocrine.
27. Describe the regulation of the cephalic, gastric, and intestinal phases of gastric secretion.
28. (**a**) What is the relationship between the cystic, common hepatic, bile, and pancreatic ducts? (**b**) What is the point of fusion of the bile and pancreatic ducts called?
29. Explain why fatty stools result from the absence of bile or pancreatic juice.
30. Indicate the function of the Kupffer cells and the hepatocytes of the liver.
31. What are (**a**) brush border enzymes? (**b**) chylomicrons?
32. Activation of pancreatic enzymes in the small intestine is a good example of the wisdom of the body. How so?
33. Name one inflammatory condition of the digestive system particularly common to adolescents, two common in middle age, and one common in old age.
34. What are the effects of aging on digestive system activity?

Critical Thinking and Clinical Application Questions

1. You are a research assistant at a pharmaceutical company. Your group has been asked to develop an effective laxative that (1) provides bulk and (2) is nonirritating to the intestinal mucosa. Explain why these requests are important by describing what would happen if the opposite conditions were present.
2. After a heavy meal rich with fried foods, Debby Collins, an overweight 45-year-old woman, was rushed to the emergency room with severe spasmodic pains in her epigastric region that radiated to the right side of her rib cage. She indicated that the attack came on suddenly, and her abdomen was found to be tender to the touch and somewhat rigid. What do you think is this patient's problem and why is her pain discontinuous (colicky)? What are the treatment options and what might happen if the problem is not resolved?
3. A baby is admitted to the hospital with a history of diarrhea and watery feces occurring over the last three days. The baby has sunken fontanelles, indicating extreme dehydration. Tests indicate that the baby has a bacterium-induced colitis, and antibiotics are prescribed. Because of the baby's loss of intestinal juices, do you think that his blood pH would indicate acidosis or alkalosis? Explain your reasoning.
4. Gary Francis, a middle-aged salesman, complains of a burning pain in the "pit of his stomach," usually beginning about two hours after eating and abating after drinking a glass of milk. When asked to indicate the site, he points to his epigastric region. The GI tract is examined by X-ray fluoroscopy. A gastric ulcer is visualized, and drug therapy using an antibiotic-bismuth approach is recommended. (a) Why is this treatment suggested? (b) What are the possible consequences of nontreatment?
5. Dr. Dolan used an endoscope to view Mr. Habib's colon. He noted the presence of several polyps and removed them during the same procedure. What is an endoscope? Why did Dr. Dolan opt to remove the polyps immediately?
6. Mr. Holden has had severe diarrhea all day and is severely weakened. Explain why his nurse is concerned about his present condition.
7. What is the protective value of having several sets of tonsils at the oral entry to the pharynx?

24

Nutrition, Metabolism, and Body Temperature Regulation

Are you a food lover? We are too. In fact, most people can be divided into two camps according to their reactions to food—those who live to eat and those who eat to live. The saying "you are what you eat" is true in that part of the food we eat is converted to our living flesh. In other words, some nutrients are used to build cell structures, replace worn-out parts, and synthesize functional molecules. However, most nutrients we ingest are used as metabolic fuel. That is, they are oxidized and transformed to **ATP**, the chemical energy form used by cells.

The energy value of foods is measured in **kilocalories (kcal)** or "large calories" (C). One kilocalorie is the amount of heat energy needed to raise the temperature of 1 kilogram of water 1°C (1.8°F). This unit is the "calorie" that dieters count so conscientiously.

In Chapter 23, we talked about how foods are digested and absorbed, but what happens to these foods once they enter the blood? Why do we need bread, meat, and fresh vegetables? Why does everything we eat seem to turn to fat? In this chapter we will try to answer these questions as we explain both the nature of nutrients and their metabolic roles.

Diet and Nutrition

- Define nutrient, essential nutrient, and calorie.
- List the six major nutrient categories. Note important sources and main cellular uses.

A **nutrient** is a substance in food used by the body to promote normal growth, maintenance, and repair. The nutrients needed for health divide neatly into six categories. Three of these—carbohydrates, lipids, and proteins—are collectively called the **major nutrients** and make up the bulk of what we eat. The fourth and fifth categories, vitamins and minerals, though equally crucial for health, are required in minute amounts.

In a strict sense, water, which accounts for about 60% by volume of the food we eat, is also a major nutrient. However, because we described its importance in the body in Chapter 2, here we consider in detail only the five nutrient categories listed above.

Most foods offer a combination of nutrients. For example, cream of mushroom soup contains all the major nutrients plus some vitamins and minerals. A diet consisting of foods from each of the five food groups—grains, fruits, vegetables, meats and fish, and milk products—normally guarantees adequate amounts of all the needed nutrients.

There are several examples of food guide pyramids around, and two are provided in Figure 24.1. Fresh out of the oven on April 19, 2005, the U.S. Department of Agriculture's new version of the food guide pyramid—MyPyramid—is a significant departure from traditional food pyramids. MyPyramid separates the food categories vertically (rather than horizontally). Narrowing of each food group from bottom to top of the pyramid suggests moderation in selecting foods from each food group (more "picks" from food choices in the wider bands). Its weakness is that the pyramid itself lacks any notion of a hierarchical ranking of the food in a single group in terms of nutritional

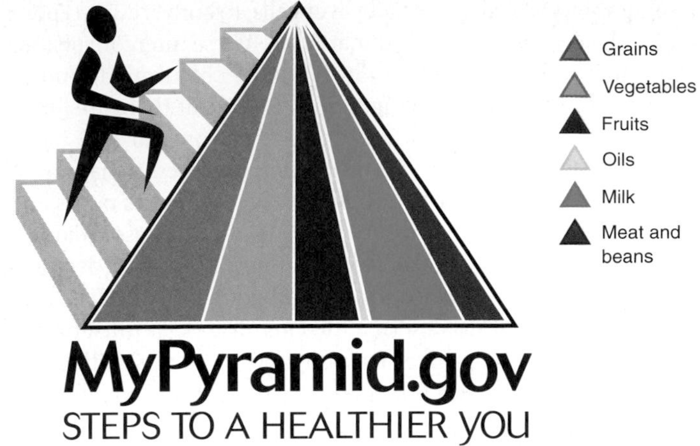

(a) USDA food guide pyramid

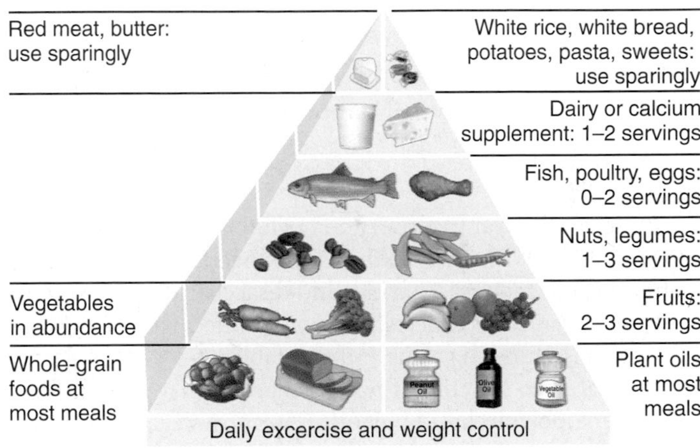

(b) Healthy eating pyramid

Figure 24.1 **Food guide pyramids.**

desirability. For example, the carbohydrate group would be whole grains, fruits and vegetables at the base of the band (best choices), pasta about halfway up, and Danish pastry at the top (eat less). Also new to this food pyramid version is the emphasis on at least 30 minutes daily of physical activity as represented by the stick person climbing the steps. A person's diet can be personalized by age, sex, and activity level by logging on to the Internet (www.mypyramid.gov), where information is available to help consumers make healthy choices and determine the size and number of servings recommended daily.

The Healthy Eating Pyramid was presented by Walter Willett of Harvard as a healthier alternative to the previous (1992) USDA food guide pyramid. It uses the traditional orientation of food groups, and emphasizes eating whole-grain foods and lots of nonstarchy fruits and vegetables. It also recommends substituting plant oils and nuts for animal fats, and restricting red meat, sweets, and starchy foods.

To be perfectly honest, nutrition advice is constantly in flux and is endlessly argued, and often mired in the self-interest of food companies. Nonetheless, basic dietary principles have not changed in years and are not in dispute—eat less; eat fruits, vegetables, and whole grains, not junk food; and get more exercise.

The ability of cells, especially liver cells, to convert one type of molecule to another is truly remarkable. These interconversions allow the body to use the wide range of chemicals found in foods and to adjust to varying food intakes. But there are limits to this ability to conjure up new molecules from old.

At least 45 and possibly 50 molecules, called **essential nutrients**, cannot be made fast enough to meet the body's needs and so must be provided by the diet. As long as all the essential nutrients are ingested, the body can synthesize the hundreds of additional molecules required for life and good health. The use of the word "essential" to describe the chemicals that must be obtained from outside sources is unfortunate and misleading, because both essential and nonessential nutrients are equally vital for normal functioning.

Carbohydrates

- ▶ Distinguish between simple and complex carbohydrate sources.
- ▶ Indicate the major uses of carbohydrates in the body.

Dietary Sources

Except for milk sugar (lactose) and negligible amounts of glycogen in meats, all the carbohydrates we ingest are derived from plants. Sugars (monosaccharides and disaccharides) come from fruits, sugar cane, sugar beets, honey, and milk. The polysaccharide starch is found in grains and vegetables.

There are two varieties of polysaccharides that provide fiber. Cellulose, a polysaccharide plentiful in most vegetables, is not digested by humans but provides roughage, or *insoluble fiber*, which increases the bulk of the stool and facilitates defecation. *Soluble fiber*, such as pectin found in apples and citrus fruits, reduces blood cholesterol levels, a desirable goal in those with cardiac disease.

Uses in the Body

24 The monosaccharide **glucose** is the carbohydrate molecule ultimately used as fuel by body cells to produce ATP. Carbohydrate digestion also yields fructose and galactose, but these monosaccharides are converted to glucose by the liver before they enter the general circulation. Many body cells use fats as energy sources, but neurons and red blood cells rely almost entirely on glucose for their energy needs. Because even a temporary shortage of blood glucose can severely depress brain function and lead to neuron death, the body carefully monitors and regulates blood glucose levels. Any glucose in excess of what is needed for ATP synthesis is converted to glycogen or fat and stored for later use.

Other uses of monosaccharides are meager. Small amounts of pentose sugars are used to synthesize nucleic acids, and a variety of sugars are attached to externally facing plasma membrane proteins and lipids.

Dietary Requirements

The low-carbohydrate diet of the Inuit (Eskimos) and the high-carbohydrate diet of peoples in the Far East indicate that humans can be healthy even with wide variations in carbohydrate intake. The minimum requirement for carbohydrates is not known, but 100 grams per day is presumed to be the smallest amount needed to maintain adequate blood glucose levels. The recommended dietary allowance (130 g/day) is based on the amount needed to fuel the brain, not the total amount need to supply all daily activities. Recommended carbohydrate intake to maintain health is 45–65% of one's total calorie intake, with the emphasis on *complex* carbohydrates (whole grains and vegetables). When less than 50 grams per day is consumed, tissue proteins and fats are used for energy fuel.

American adults typically consume about 46% of dietary food energy in the form of carbohydrates. Because starchy foods (rice, pasta, breads) are less expensive than meat and other high-protein foods, carbohydrates make up an even greater percentage of the diet in low-income groups. Vegetables, fruits, whole grains, and milk have many valuable nutrients, such as vitamins and minerals.

By contrast, highly refined carbohydrate foods such as candy and soft drinks provide concentrated energy sources only—the term "empty calories" is commonly used to describe such foods. Eating refined, sugary foods instead of more complex carbohydrates may cause nutritional deficiencies as well as obesity. Other possible consequences of excessive intake of simple carbohydrates are listed in **Table 24.1**.

Lipids

- ▶ Indicate uses of fats in the body.
- ▶ Distinguish between saturated, unsaturated, and trans fatty acid sources.

Dietary Sources

The most abundant dietary lipids are triglycerides, also called neutral fats or triacylglycerols (Chapter 2). We eat saturated fats in animal products such as meat and dairy foods, in a few tropical plant products such as coconut, and in hydrogenated oils (trans fats) such as margarine and solid shortenings used in baking. Unsaturated fats are present in seeds, nuts, olive oil, and most vegetable oils. Fats are digested to monoglycerides or all the way to fatty acids and glycerol, and then reconverted to triglycerides for transport in the lymph.

Major sources of cholesterol are egg yolk, meats and organ meats, shellfish, and milk products. However, it is estimated that the liver produces about 85% of blood cholesterol regardless of dietary intake.

The liver is adept at converting one fatty acid to another, but it cannot synthesize *linoleic acid* (lin″o-le′ik), a fatty acid component of *lecithin* (les′ĭ-thin). For this reason, linoleic acid, an omega-6 fatty acid, is an *essential fatty acid* that must be ingested. Linolenic acid, an omega-3 fatty acid, may also be essential. Fortunately, these two fatty acids are found in most vegetable oils.

Uses in the Body

Fats have fallen into disfavor, particularly among those for whom the "battle of the bulge" is constant. But fats make foods tender, flaky, or creamy, and make us feel full and satisfied. Furthermore,

TABLE 24.1 Summary of Carbohydrate, Lipid, and Protein Nutrients

FOOD SOURCES	RECOMMENDED DAILY ALLOWANCE (RDA) FOR ADULTS	PROBLEMS: EXCESSES	PROBLEMS: DEFICITS
Carbohydrates 			
Total Digestible	130 g 45–65% of total caloric intake		
▪ **Complex carbohydrates** (starches): bread, cereal, crackers, flour, pasta, nuts, rice, potatoes		Obesity; diabetes mellitus; nutritional deficits; dental caries; gastrointestinal irritation; elevated triglycerides in plasma	Tissue wasting (in extreme deprivation); metabolic acidosis resulting from accelerated fat use for energy
▪ **Simple carbohydrates (sugars):** carbonated drinks, candy, fruit, ice cream, pudding, young (immature) vegetables			
Total Fiber	25–30 g		
Lipids 			
Total	65 g 20–35% of total caloric intake	Obesity and increased risk of cardiovascular disease (particularly if excesses of saturated and trans fat)	Weight loss; fat stores and tissue proteins catabolized to provide metabolic energy; problems controlling heat loss (due to depletion of subcutaneous fat)
▪ **Animal sources:** lard, meat, poultry, eggs, milk, milk products	20g		
▪ **Plant sources:** chocolate; corn, soy, cottonseed, olive oils; coconut; corn; peanuts	11–17g		
▪ **Essential fatty acids:** fish oil, corn, cottonseed, soy oils; wheat germ; vegetable shortenings	1.1–1.6g	Excess dietary intake of omega-3 fatty acid may increase risk of stroke	Poor growth; skin lesions (eczema-like); depression
▪ **Cholesterol and trans fatty acids:** organ meats (liver, kidneys, brains), egg yolks, fish roe; smaller concentrations in milk products and meat	Not determined	Increased levels of plasma cholesterol and low-density lipoproteins, correlated with increased risk of cardiovascular disease	Increased risk of stroke (CVA) in susceptible individuals
Proteins 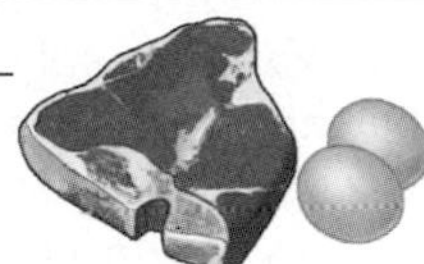			
▪ **Complete proteins:** eggs, milk, milk products, meat (fish, poultry, pork, beef, lamb)	0.8 g/kg body weight 12–20% of total caloric intake	Obesity; enhanced calcium excretion and bone loss; high cholesterol levels in blood; kidney stones	Profound weight loss and tissue wasting; growth retardation in children; anemia; edema (due to deficits of plasma proteins)
▪ **Incomplete proteins:** legumes (soybeans, lima beans, kidney beans, lentils); nuts and seeds; grains and cereals; vegetables			During pregnancy: miscarriage or premature birth

dietary fats *are* essential for several reasons. They help the body absorb fat-soluble vitamins; triglycerides are the major energy fuel of hepatocytes and skeletal muscle; and phospholipids are an integral component of myelin sheaths and cellular membranes. Fatty deposits in adipose tissue provide (1) a protective cushion around body organs, (2) an insulating layer beneath the skin, and (3) an easy-to-store concentrated source of energy fuel. Regulatory molecules called *prostaglandins* (pros″tah-glan′dinz), formed from linoleic acid via arachidonic acid (ah″rah-kĭ-don′ik), play a role in smooth muscle contraction, control of blood pressure, and inflammation.

Unlike triglycerides, cholesterol is not used for energy. It is important as a stabilizing component of plasma membranes and is the precursor from which bile salts, steroid hormones, and other essential molecules are formed.

Dietary Requirements

Fats represent over 40% of the calories in the typical American diet. There are no precise recommendations on amount or type of dietary fats, but the American Heart Association suggests that (1) fats should represent 30% or less of total caloric intake, (2) saturated fats should be limited to 10% or less of total fat intake, and (3) daily cholesterol intake should be no more than 300 mg (the amount in one egg yolk). The goal of these recommendations is to keep total blood cholesterol to less than 200 mg/dl. Because a diet high in saturated fats and cholesterol may contribute to cardiovascular disease, these are wise guidelines. Sources of the various lipid classes and consequences of their deficient or excessive intake are summarized in Table 24.1.

Fat Substitutes

24

In an attempt to reduce fat intake without losing fat's appetizing aspects, many people have turned to fat substitutes or foods prepared with them. Perhaps the oldest fat substitute is air (beaten into a product to make it fluffy). Other fat substitutes are modified starches and gums, and more recently milk whey protein. Except for those based on dietary fibers (gums), such products are metabolized and provide calories. One of the newest substitutes, Olestra, a fat-based product made from cottonseeds, is not metabolized because it is not digested or absorbed.

Most fat substitutes have two drawbacks: (1) They don't stand up to the intense heat needed to fry foods, and (2) although manufacturers claim otherwise, they don't taste nearly as good as the "real thing." The ones that are not absorbed tend to cause flatus (gas) or diarrhea, and may interfere with absorption of fat-soluble drugs, vitamins, and phytochemicals such as beta-carotene, a precursor of vitamin A.

Proteins

- ▶ Distinguish between nutritionally complete and incomplete proteins.
- ▶ Indicate uses of proteins in the body.
- ▶ Define nitrogen balance and indicate possible causes of positive and negative nitrogen balance.

Dietary Sources

Animal products contain the highest-quality proteins, in other words, those with the greatest amount and best ratios of *essential amino acids* **(Figure 24.2)**. Proteins in eggs, milk, fish, and most meats are **complete proteins** that meet all the body's amino acid requirements for tissue maintenance and growth (Table 24.1). Legumes (beans and peas), nuts, and cereals are protein-rich, but their proteins are nutritionally incomplete because they are low in one or more of the essential amino acids.

Strict vegetarians must carefully plan their diets to obtain all the essential amino acids and prevent protein malnutrition. When ingested together, cereal grains and legumes provide all the essential amino acids (Figure 24.2b). Some combination of these foods is found in the diets of all cultures (most obviously in the rice and beans seen on nearly every plate in a Mexican restaurant). For nonvegetarians, grains and legumes are useful as partial substitutes for the more expensive animal proteins.

Uses in the Body

Proteins are important structural materials of the body, including, for example, keratin in skin, collagen and elastin in connective tissues, and muscle proteins. In addition, functional proteins such as enzymes and some hormones regulate an incredible variety of body functions. Whether amino acids are used to synthesize new proteins or are burned for energy depends on a number of factors:

1. **The all-or-none rule.** All amino acids needed to make a particular protein must be present in a cell at the same time and in sufficient amounts. If one is missing, the protein cannot be made. Because essential amino acids cannot be stored, those not used immediately to build proteins are oxidized for energy or converted to carbohydrates or fats.
2. **Adequacy of caloric intake.** For optimal protein synthesis, the diet must supply sufficient carbohydrate or fat calories for ATP production. When it doesn't, dietary and tissue proteins are used for energy.
3. **Nitrogen balance.** In healthy adults the rate of protein synthesis equals the rate of protein breakdown and loss, a homeostatic state called **nitrogen balance.** The body is in nitrogen balance when the amount of nitrogen ingested in proteins equals the amount excreted in urine and feces.

   The body is in *positive nitrogen balance* when the amount of protein incorporated into tissue is greater than the amount being broken down and used for energy—the normal situation in growing children and pregnant women. A positive balance also occurs when tissues are being repaired following illness or injury.

   In *negative nitrogen balance*, protein breakdown for energy exceeds the amount of protein being incorporated into tissues. This occurs during physical and emotional stress (for example, infection, injury, or burns), or when the quality or quantity of dietary protein is poor, or during starvation.
4. **Hormonal controls.** Certain hormones, called *anabolic hormones*, accelerate protein synthesis and growth. The effects of these hormones vary continually throughout life.

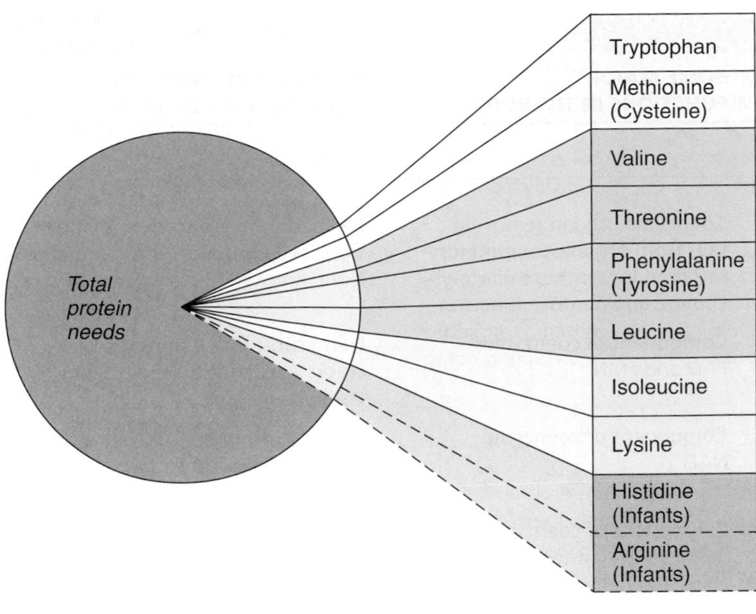

(a) Essential amino acids

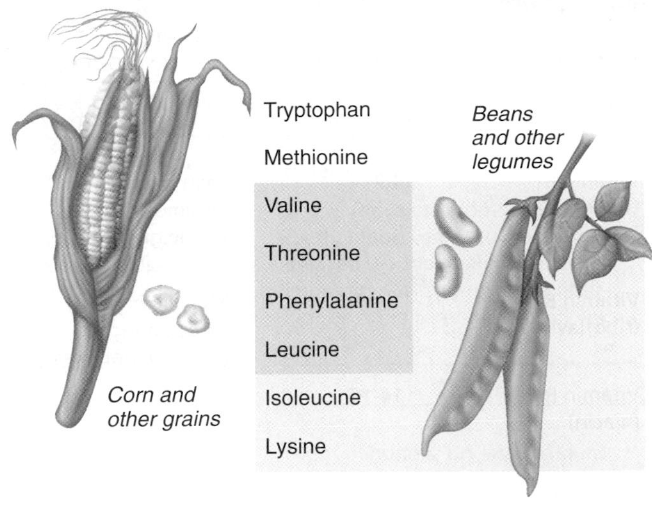

(b) Vegetarian diets providing the eight essential amino acids for humans

Figure 24.2 Essential amino acids. In order for protein synthesis to occur, 10 essential amino acids must be available simultaneously and in the correct proportions. **(a)** Relative amounts of the essential amino acids and total proteins needed by adults. Notice that the essential amino acids represent only a small percentage of the total recommended protein intake. Histidine and arginine, graphed with dashed lines, are essential in infants but not in adults. Amino acids shown in parentheses are *not* essential but can substitute in part for methionine and phenylalanine. **(b)** Vegetarian diets must be carefully constructed to provide all essential amino acids. A meal of corn and beans fills the bill: Corn provides the essential amino acids not in beans and vice versa.

For example, pituitary *growth hormone* stimulates tissue growth during childhood and conserves protein in adults, and the *sex hormones* trigger the growth spurt of adolescence. Other hormones, such as the adrenal *glucocorticoids* released during stress, enhance protein breakdown and conversion of amino acids to glucose.

Dietary Requirements

Besides supplying essential amino acids, dietary proteins furnish the raw materials for making nonessential amino acids and various nonprotein nitrogen-containing substances. The amount of protein a person needs reflects his or her age, size, metabolic rate, and current state of nitrogen balance. As a rule of thumb, nutritionists recommend a daily intake of 0.8 g per kilogram of body weight. About 30 g of protein is supplied by a small serving of fish and a glass of milk.

Most Americans eat far more protein than they need. Prolonged high protein consumption may lead to bone loss. This may occur because metabolizing sulfur-containing amino acids makes the blood more acidic and calcium is pulled from the bones to buffer these acids.

CHECK YOUR UNDERSTANDING

1. What are the six major nutrients?
2. Why is it important to include cellulose in a healthy diet even though we do not digest it?
3. How does the body use triglycerides? Cholesterol?
4. John eats nothing but baked bean sandwiches. Is he getting all the essential amino acids he needs in this restricted diet?

For answers, see Appendix G.

Vitamins

- ▶ Distinguish between fat- and water-soluble vitamins, and list the vitamins in each group.
- ▶ For each vitamin, list important sources, body functions, and important consequences of its deficit or excess.

Vitamins (*vita* = life) are potent organic compounds needed in minute amounts for growth and good health. Unlike other organic nutrients, vitamins are not used for energy and do not serve as building blocks, but they are crucial in helping the body use those nutrients that do. Without vitamins, all the carbohydrates, proteins, and fats we eat would be useless.

Most vitamins function as **coenzymes** (or parts of coenzymes), which act with an enzyme to accomplish a particular chemical task. For example, the B vitamins act as coenzymes in the oxidation of glucose for energy. We will describe the roles of some vitamins in the metabolism discussion.

Because most vitamins are not made in the body, they must be taken in via foods or vitamin supplements. The exceptions are vitamin D made in the skin, and small amounts of B vitamins and vitamin K synthesized by intestinal bacteria. In addition, the body can convert *beta-carotene* (kar′o-tēn), the orange pigment

TABLE 24.3 Minerals

MINERAL	RDA* (mg)	DIETARY SOURCES	FUNCTIONS IN THE BODY	POSSIBLE SYMPTOMS OF DEFICIENCY OR EXTREME EXCESS
Calcium (Ca)	1300	Dairy products, dark green vegetables, legumes	Bone and tooth formation, blood clotting, nerve and muscle function	Stunted growth, possibly loss of bone mass Depressed neural function, muscle weakness, calcium deposit in soft tissues
Phosphorus (P)	700	Dairy products, meats, whole grains, nuts	Bone and tooth formation, acid-base balance, nucleic acid synthesis	Weakness, loss of minerals from bone, rickets Not known
Sulfur (S)	Unknown	Sulfur-containing proteins from many sources (meats, milk, eggs)	Component of certain amino acids	Symptoms of protein deficiency Not known
Potassium (K)	4700	Meats, dairy products, many fruits and vegetables, grains	Acid-base balance, water balance, nerve function, muscle contraction	Muscular weakness, paralysis Muscular weakness, cardiac problems, alkalosis
Chlorine (Cl)	2300	Table salt, cured meats (ham)	Osmotic pressure, acid-base balance, gastric juice formation	Muscle cramps, reduced appetite Vomiting
Sodium (Na)	1500	Table salt, cured meats	Osmotic pressure, acid-base balance, water-balance, nerve function, important for pumping glucose and other nutrients	Muscle cramps, reduced appetite Hypertension, edema
Magnesium (Mg)	300–400	Whole grains, green leafy vegetables	Component of certain coenzymes in ATP formation	Nervous system disturbances, tremors, muscle weakness, hypertension, sudden cardiac death Diarrhea
Trace Minerals[†]				
Iron (Fe)	8 ♂ 16–18 ♀	Meats, liver, shellfish, eggs, legumes, whole grains, dried fruit, nuts	Component of hemoglobin and of cytochromes, (electron carriers in oxidative phosphorylation)	Iron-deficiency anemia, weakness, impaired immunity, impaired cognitive performance in children Liver damage, hemochromatosis

24

whose kidneys have a greater tendency to retain salt than do those of American whites.

Fats and sugars are practically devoid of minerals, and highly refined cereals and grains are poor sources. The most mineral-rich foods are vegetables, legumes, milk, and some meats.

CHECK YOUR UNDERSTANDING

5. Vitamins are not used for energy fuels. What are they used for?
6. Which B vitamin requires the help of a product made in the stomach in order to be absorbed? What is that gastric product?
7. What mineral is essential for thyroxine synthesis? For making bones hard? For hemoglobin synthesis?

For answers, see Appendix G.

Overview of Metabolic Reactions

- Define metabolism. Explain how catabolism and anabolism differ.
- Define oxidation and reduction and indicate the importance of these reactions in metabolism.
- Indicate the role of coenzymes used in cellular oxidation reactions.
- Explain the difference between substrate-level phosphorylation and oxidative phosphorylation.

Once inside body cells, nutrients become involved in an incredible variety of biochemical reactions known collectively as **metabolism** (*metabol* = change). During metabolism, substances are constantly being built up and torn down. Cells use energy to extract more energy from foods, and then use some of this extracted energy to drive their activities. Even at rest, the body uses energy on a grand scale.

Anabolism and Catabolism

Metabolic processes are either *anabolic* (synthetic, building up) or *catabolic* (degradative, tearing down). **Anabolism** (ah-nab′o-lizm) is the general term for all reactions in which larger mole-

TABLE 24.3 (continued)

MINERAL	RDA* (mg)	DIETARY SOURCES	FUNCTIONS IN THE BODY	POSSIBLE SYMPTOMS OF DEFICIENCY OR EXTREME EXCESS
Fluorine (F)	3–4	Fluoridated water, tea, seafood	Maintenance of tooth and (probably) bone structure	Higher frequency of tooth decay Mottling of teeth, increased risk of bone fracture, painful stiffening of joints
Zinc (Zn)	11 ♂ 8 ♀	Meats, seafood, grains, legumes	Component of several enzymes and other proteins (needed for wound healing, taste, and smell)	Growth failure, scaly skin, reproductive failure, loss of taste and smell, impaired immunity Slurred speech, tremors, difficulty in walking
Copper (Cu)	0.9–1.5	Liver, seafood, nuts, legumes, whole grains	Hemoglobin synthesis, essential for manufacture of myelin and some components of electron transport chain	Anemia, bone and cardiovascular changes (rare) Abnormal storage of copper in the body
Manganese (Mn)	2.5	Nuts, grains, vegetables, fruits, tea	Component of certain coenzymes; neural function, lactation	Abnormal bone and cartilage Appears to contribute to hallucinations and violent behavior
Iodine (I)	0.15	Cod-liver oil, seafood, dairy products, iodized salt	Component of thyroid hormones	Goiter (enlarged thyroid), cretinism, myxedema Depressed synthesis of thyroid hormones
Cobalt (Co)	Unknown	Meats, poultry, and dairy products	Component of vitamin B_{12}	None, except as B_{12} deficiency Polycythemia, heart disease
Selenium (Se)	0.05–0.07	Seafood, meats, whole grains	Component of enzymes; functions in close association with vitamin E; antioxidant	Muscle pain, maybe heart muscle deterioration Nausea, vomiting, hair loss, weight loss
Chromium (Cr)	0.035	Brewer's yeast, liver, seafood, meats, some vegetables, wine	Involved in glucose and energy metabolism, enhances effectiveness of insulin	Impaired glucose metabolism, diabetes mellitus Not known
Molybdenum (Mo)	0.045	Legumes, grains, some vegetables	Component of certain enzymes	Disorder in excretion of nitrogen-containing compounds

*RDA = recommended daily allowance
†Trace minerals together account for less than 0.005% of body weight.

cules or structures are built from smaller ones, such as the bonding together of amino acids to make proteins. **Catabolism** (kah-tab′o-lizm) refers to all processes that break down complex structures to simpler ones. One example is the hydrolysis of foods in the digestive tract. In the group of catabolic reactions collectively called **cellular respiration**, food fuels, particularly glucose, are broken down in cells and some of the energy released is captured to form ATP, the cells' energy currency. ATP then serves as the "chemical drive shaft" that links energy-releasing catabolic reactions to cellular work.

Recall from Chapter 2 that reactions driven by ATP are coupled. ATP is never hydrolyzed directly. Instead enzymes shift its high-energy phosphate groups to other molecules, which are then said to be **phosphorylated** (fos″for-ĭ-la′ted). Phosphorylation primes the molecule to change in a way that increases its activity, produces motion, or does work. For example, many regulatory enzymes that catalyze key steps in metabolic pathways are activated by phosphorylation.

Three major stages are involved in the processing of energy-containing nutrients in the body, as shown in the overview in **Figure 24.3**. (Note that the blue arrows in the figure represent catabolic reactions and the purple arrows represent anabolic reactions.)

- *Stage 1* is digestion in the gastrointestinal tract, which is described in Chapter 23. The absorbed nutrients are then transported in blood to the tissue cells.
- *Stage 2* occurs in the tissue cells. Newly delivered nutrients are either built into lipids, proteins, and glycogen by anabolic pathways or broken down by catabolic pathways to *pyruvic acid* (pi-roo′vik) and *acetyl CoA* (as′ĕ-til ko-a′) in the cell cytoplasm. Notice in Figure 24.3 that a major catabolic pathway of stage 2 is *glycolysis*, which we will discuss later in this chapter.
- *Stage 3*, which occurs in the mitochondria, is almost entirely catabolic. It requires oxygen, and completes the breakdown of foods, producing carbon dioxide and water and harvesting large amounts of ATP. As Figure 24.3 shows, the *Krebs cycle* is a key pathway in stage 3, as is *oxidative phosphorylation*. We will discuss both later.

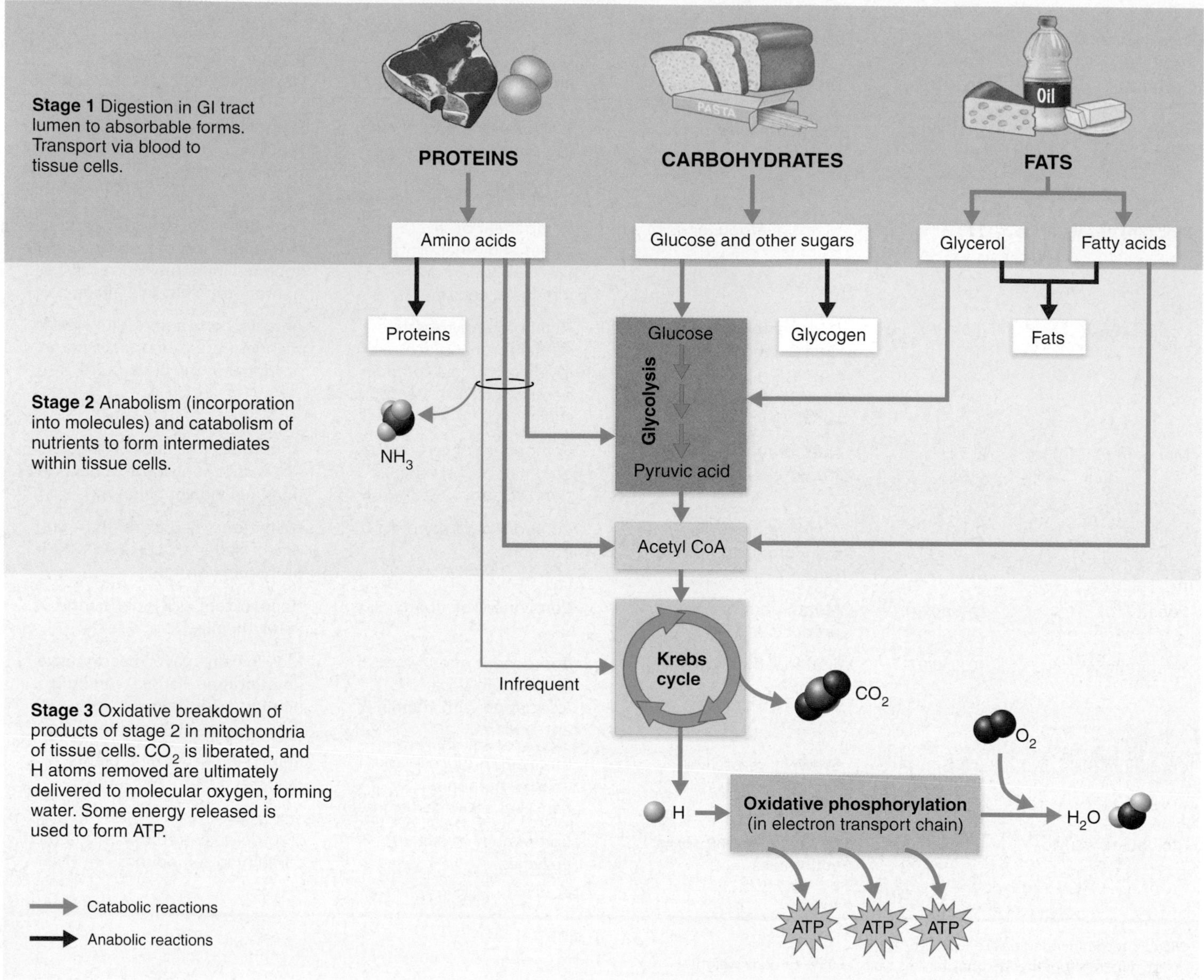
Stage 1 Digestion in GI tract lumen to absorbable forms. Transport via blood to tissue cells.
PROTEINS
CARBOHYDRATES
FATS
Oil
PASTA
Amino acids
Glucose and other sugars
Glycerol
Fatty acids
Proteins
Glucose
Glycogen
Fats
Glycolysis
Stage 2 Anabolism (incorporation into molecules) and catabolism of nutrients to form intermediates within tissue cells.
NH_3
Pyruvic acid
Acetyl CoA
Krebs cycle
Infrequent
CO_2
O_2
Stage 3 Oxidative breakdown of products of stage 2 in mitochondria of tissue cells. CO_2 is liberated, and H atoms removed are ultimately delivered to molecular oxygen, forming water. Some energy released is used to form ATP.
H
Oxidative phosphorylation (in electron transport chain)
H_2O
ATP
ATP
ATP
Catabolic reactions
Anabolic reactions

Figure 24.3 Three stages of metabolism of energy-containing nutrients.

24

The primary function of *cellular respiration*, which consists of the glycolysis of stage 2 and all events of stage 3, is to generate ATP, which traps some of the chemical energy of the original food molecules in its own high-energy bonds. The body can also store energy in fuels, such as glycogen and fats, and these stores are later mobilized to produce ATP for cellular use.

You do not need to memorize Figure 24.3, but you may want to refer to it as a cohesive summary of nutrient processing and metabolism in the body.

Oxidation-Reduction Reactions and the Role of Coenzymes

Many of the reactions that take place within cells are **oxidation reactions.** *Oxidation* was originally defined as the combination of oxygen with other elements. Examples are the rusting of iron (the slow formation of iron oxide) and the burning of wood and other fuels. In burning, oxygen combines rapidly with carbon, producing carbon dioxide, water, and an enormous amount of energy, which is liberated as heat and light. Later it was discovered that oxidation *also* occurs when hydrogen atoms are *removed* from compounds and so the definition was expanded to its current form: *Oxidation is the gain of oxygen or the loss of hydrogen.* As explained in Chapter 2, whichever way oxidation occurs, the oxidized substance always *loses* (or nearly loses) electrons as they move to (or toward) a substance that more strongly attracts them.

This loss of electrons can be explained by reviewing the consequences of different electron-attracting abilities of atoms (see p. 32). Consider a molecule made up of a hydrogen atom plus some other kinds of atoms. Hydrogen is very electropositive, so its lone electron usually spends more time orbiting the other atoms of the molecule. But when a hydrogen *atom* is removed, its electron goes with it, and the molecule as a whole loses that electron. Conversely, oxygen is very electron-hungry

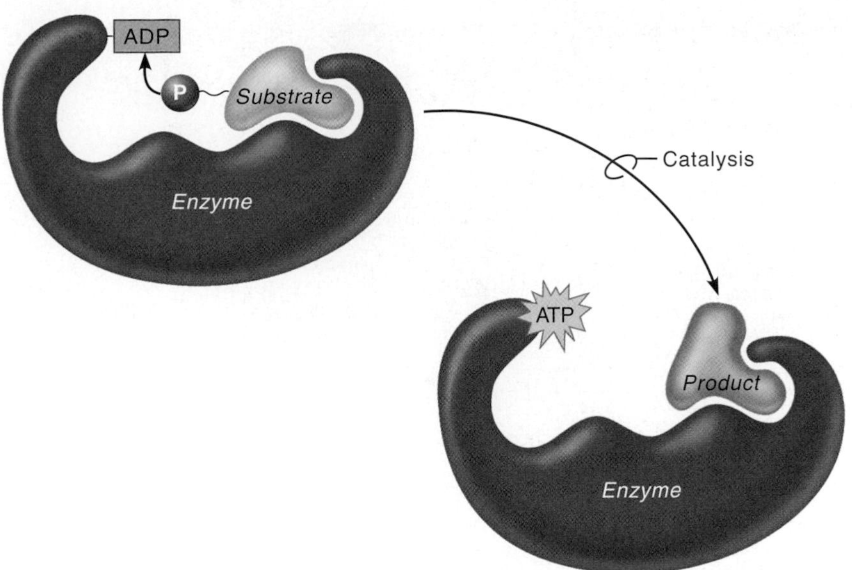

(a) Substrate-level phosphorylation

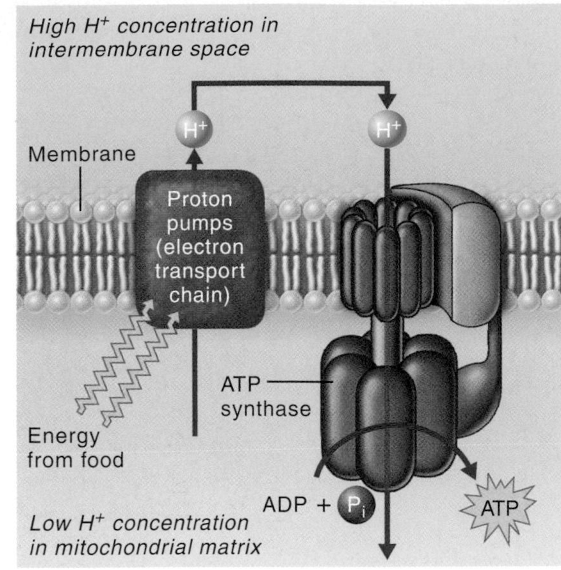

(b) Oxidative phosphorylation

Figure 24.4 Mechanisms of phosphorylation. (a) Substrate-level phosphorylation occurs when a high-energy phosphate group is transferred directly from a substrate to ADP to form ATP. This reaction occurs both in the cytosol and in the mitochondrial matrix. **(b)** Oxidative phosphorylation, which occurs in mitochondria, is carried out by electron transport proteins that act as proton "pumps" to create a proton gradient across the inner mitochondrial membranes. The source of energy for this pumping is energy released during oxidation of food molecules. As the protons flow passively back into the mitochondrial matrix through ATP synthase, some of this gradient energy is captured and used to bind phosphate groups to ADP.

(electronegative), and so when oxygen binds with other atoms the shared electrons spend more time in oxygen's vicinity. Again, the molecule as a whole loses electrons.

As you will soon see, essentially all oxidation of food fuels involves the step-by-step removal of pairs of hydrogen atoms (and also pairs of electrons) from the substrate molecules, eventually leaving only carbon dioxide (CO_2). Molecular oxygen (O_2) is the *final* electron acceptor. It combines with the removed hydrogen atoms at the very end of the process, to form water (H_2O).

Whenever one substance loses electrons (is oxidized), another substance gains them (is reduced). For this reason, oxidation and reduction are coupled reactions and we speak of **oxidation-reduction reactions**, or more commonly, **redox reactions.** The key understanding about redox reactions is that "oxidized" substances *lose* energy and "reduced" substances *gain* energy as energy-rich electrons are transferred from one substance to the next. Consequently, as food fuels are oxidized, their energy is transferred to a "bucket brigade" of other molecules and ultimately to ADP to form energy-rich ATP.

Like all other chemical reactions in the body, redox reactions are catalyzed by enzymes. Those that catalyze redox reactions in which hydrogen atoms are removed are called **dehydrogenases** (de-hi′dro-jen-ās″ez), while those catalyzing the transfer of oxygen are **oxidases.**

Most of these enzymes require the help of a specific coenzyme, typically derived from one of the B vitamins. Although the enzymes catalyze the removal of hydrogen atoms to oxidize a substance, they cannot *accept* the hydrogen (hold on or bond to it). Their *coenzymes*, however, can act as hydrogen (or electron) acceptors, becoming reduced each time a substrate is oxidized.

Two very important coenzymes of the oxidative pathways are **nicotinamide adenine dinucleotide (NAD^+)** (nik″o-tin′ah-mīd), based on *niacin*, and **flavin adenine dinucleotide (FAD)**, derived from *riboflavin.* The oxidation of succinic acid to fumaric acid and the simultaneous reduction of FAD to $FADH_2$, an example of a coupled redox reaction, is

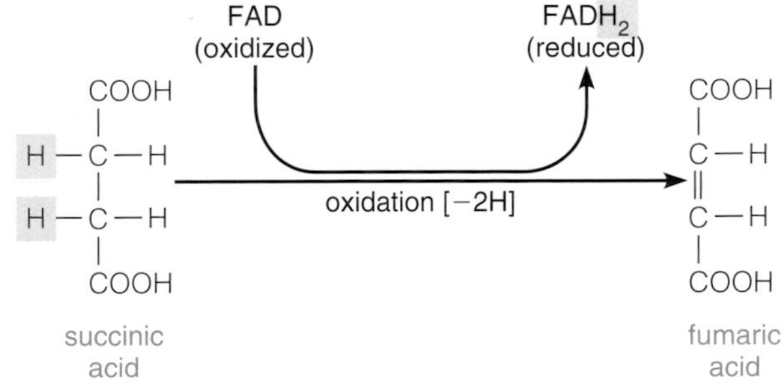

ATP Synthesis

How do our cells capture some of the energy liberated during cellular respiration to make ATP molecules? There appear to be two mechanisms—substrate-level phosphorylation and oxidative phosphorylation.

Substrate-level phosphorylation occurs when high-energy phosphate groups are transferred directly from phosphorylated substrates (metabolic intermediates such as glyceraldehyde phosphate) to ADP **(Figure 24.4a)**. Essentially, this process occurs because the high-energy bonds attaching the phosphate

24

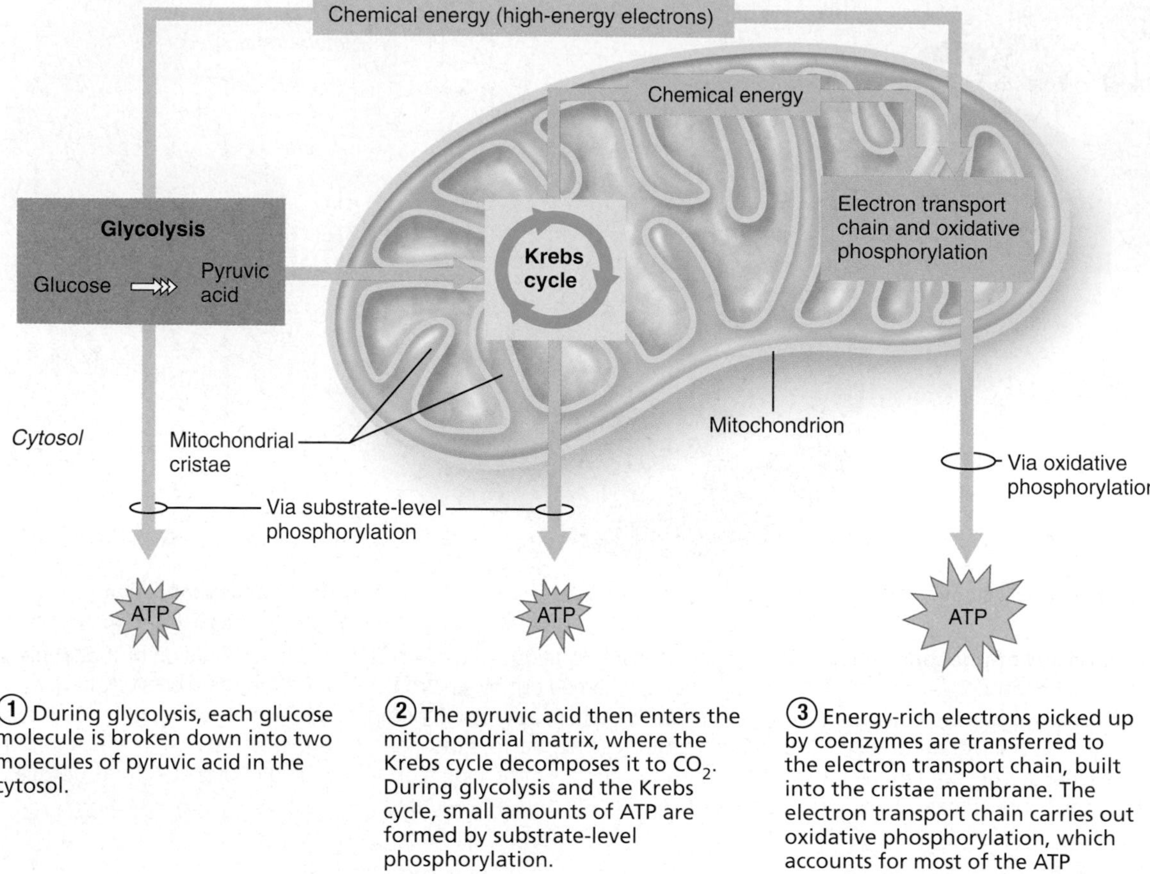
Chemical energy (high-energy electrons)
Chemical energy
Glycolysis
Glucose
Pyruvic acid
Krebs cycle
Electron transport chain and oxidative phosphorylation
Cytosol
Mitochondrial cristae
Mitochondrion
Via substrate-level phosphorylation
Via oxidative phosphorylation
ATP
ATP
ATP
① During glycolysis, each glucose molecule is broken down into two molecules of pyruvic acid in the cytosol.
② The pyruvic acid then enters the mitochondrial matrix, where the Krebs cycle decomposes it to CO2. During glycolysis and the Krebs cycle, small amounts of ATP are formed by substrate-level phosphorylation.
③ Energy-rich electrons picked up by coenzymes are transferred to the electron transport chain, built into the cristae membrane. The electron transport chain carries out oxidative phosphorylation, which accounts for most of the ATP generated by cellular respiration.

Figure 24.5 During cellular respiration, ATP is formed in the cytosol and in the mitochondria.

groups to the substrates are even more unstable than those in ATP. ATP is synthesized by this route once during glycolysis, and once during each turn of the Krebs cycle. The enzymes catalyzing substrate-level phosphorylations are located in both the cytosol (where glycolysis occurs) and in the watery matrix inside the mitochondria (where the Krebs cycle takes place) **(Figure 24.5)**.

Oxidative phosphorylation is much more complicated, but it also releases most of the energy that is eventually captured in ATP bonds during cellular respiration. This process, which is carried out by electron transport proteins forming part of the inner mitochondrial membranes, is an example of a chemiosmotic process. **Chemiosmotic processes** couple the movement of substances across membranes to chemical reactions.

In this case, some of the energy released during the oxidation of food fuels (the "chemi" part of chemiosmotic) is used to pump (*osmo* = push) protons (H^+) across the inner mitochondrial membrane into the intermembrane space (Figure 24.4b). This creates a steep concentration gradient for protons across the membrane. Then, when H^+ flows back across the membrane (through a membrane channel protein called *ATP synthase*), some of this gradient energy is captured and used to attach phosphate groups to ADP.

CHECK YOUR UNDERSTANDING

8. What is a redox reaction?
9. How are anabolism and catabolism linked by ATP?
10. What is the energy source for the proton pumps of oxidative phosphorylation?

For answers, see Appendix G.

Metabolism of Major Nutrients

Carbohydrate Metabolism

- Summarize important events and products of glycolysis, the Krebs cycle, and electron transport.
- Define glycogenesis, glycogenolysis, and gluconeogenesis.

The story of carbohydrate metabolism is really a tale of glucose metabolism because all food carbohydrates are eventually transformed to glucose. Glucose enters the tissue cells by facilitated diffusion, a process that is greatly enhanced by insulin. Immediately upon entry into the cell, glucose is phosphorylated to *glucose-6-phosphate* by transfer of a phosphate group to its sixth carbon during a coupled reaction with ATP:

$$\text{Glucose} + \text{ATP} \rightarrow \text{glucose-6-PO}_4 + \text{ADP}$$

Most body cells lack the enzymes needed to reverse this reaction, so it effectively traps glucose inside the cells. Because glucose-6-phosphate is a *different* molecule from simple glucose, the reaction also keeps intracellular glucose levels low, maintaining a concentration gradient for glucose entry. Only intestinal mucosa cells, kidney tubule cells, and liver cells have the enzymes needed to reverse this phosphorylation reaction, which reflects their central roles in glucose uptake *and* release. The catabolic and anabolic pathways for carbohydrates all begin with glucose-6-phosphate.

Oxidation of Glucose

Glucose is the pivotal fuel molecule in the oxidative (ATP-producing) pathways. Glucose is catabolized via the reaction

$$\underset{\text{glucose}}{C_6H_{12}O_6} + \underset{\text{oxygen}}{6O_2} \rightarrow \underset{\text{water}}{6H_2O} + \underset{\text{carbon dioxide}}{6CO_2} + 32\ \text{ATP} + \text{heat}$$

This equation gives few hints that glucose breakdown is complex and involves three of the pathways featured in Figures 24.3 and 24.5:

1. Glycolysis (color-coded orange throughout the chapter)
2. The Krebs cycle (color-coded green)
3. The electron transport chain and oxidative phosphorylation (color-coded lavender)

These metabolic pathways occur in a definite order, and we will consider them sequentially.

Glycolysis Also called the *glycolytic pathway*, **glycolysis** (gli-kol′ĭ-sis; "sugar splitting") occurs in the cytosol of cells. This pathway is a series of ten chemical steps by which glucose is converted to two *pyruvic acid* molecules. All steps are fully *reversible* except the first, during which glucose entering the cell is phosphorylated to glucose-6-phosphate.

Glycolysis is an *anaerobic process* (an-a′er-ōb-ik; *an* = without, *aero* = air). Although this term is sometimes mistakenly interpreted to mean the pathway occurs only in the absence of oxygen, the correct interpretation is that glycolysis *does not use oxygen and occurs whether or not oxygen is present.* **Figure 24.6** shows the three major phases of the glycolytic pathway. The complete glycolytic pathway appears in Appendix D.

Phase 1. Sugar activation. In phase 1, glucose is phosphorylated and converted to fructose-6-phosphate, which is then phosphorylated again. These three steps use two ATP molecules (which are recouped later) and yield fructose-1,6-bisphosphate. The two separate reactions of the sugar with ATP provide the *activation energy* needed to prime the later stages of the pathway, so phase 1 is sometimes called the *energy investment phase.* (Recall the importance of activation energy in preparing substances to react, as described in Chapter 2.)

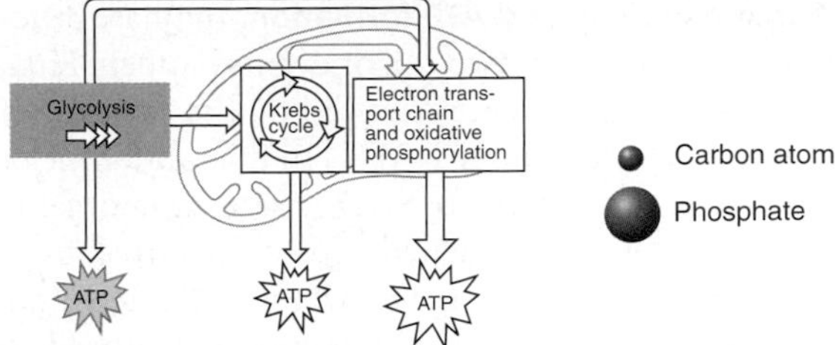

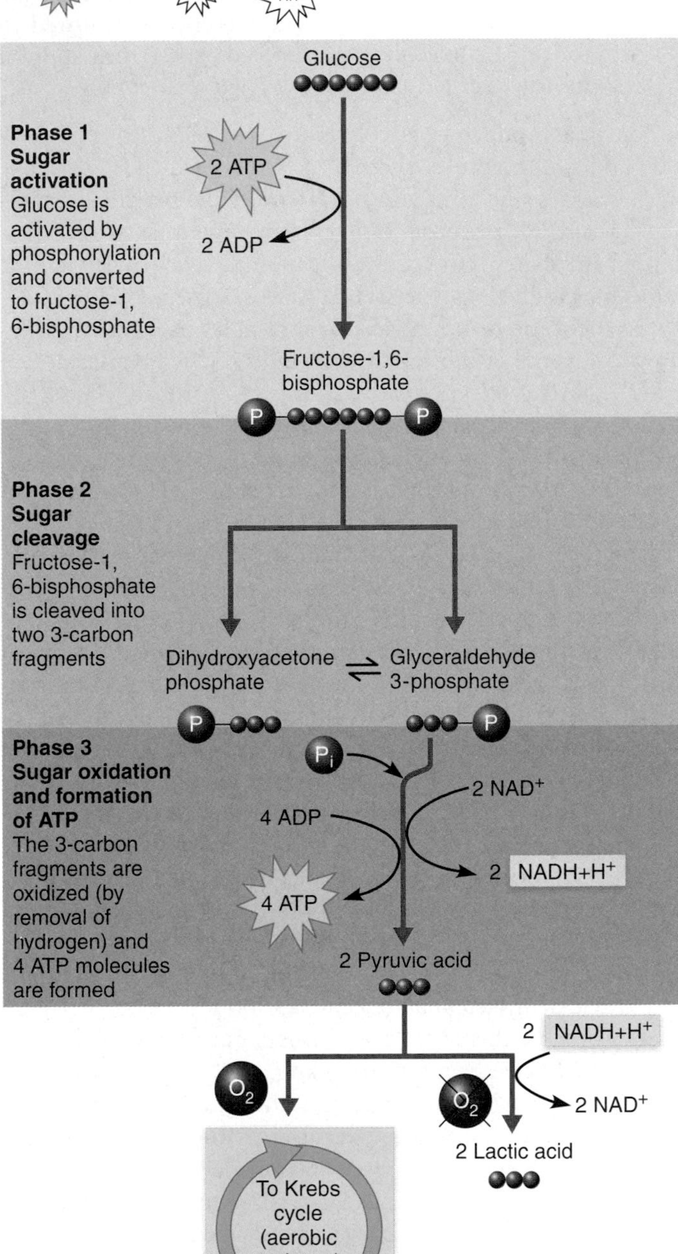

Figure 24.6 The three major phases of glycolysis. The fate of pyruvic acid depends on whether or not molecular O_2 is available.

Phase 2. Sugar cleavage. During phase 2, fructose-1,6-bisphosphate is split into two 3-carbon fragments that exist (reversibly) as one of two isomers: glyceraldehyde (glis″er-al′dĕ-hīd) 3-phosphate or dihydroxyacetone (dī″hi-drok″se-as′ĕ-tōn) phosphate.

24

Phase 3. Sugar oxidation and ATP formation. In phase 3, actually consisting of six steps, two major events happen. First, the two 3-carbon fragments are oxidized by the removal of hydrogen, which is picked up by NAD^+. In this way, some of glucose's energy is transferred to NAD^+. Second, inorganic phosphate groups (P_i) are attached to each oxidized fragment by high-energy bonds. Later, when these terminal phosphates are cleaved off, enough energy is captured to form four ATP molecules. As we noted earlier, formation of ATP this way is called *substrate-level phosphorylation.*

The final products of glycolysis are two molecules of **pyruvic acid** and two molecules of reduced NAD^+ (which is NADH + H^+). There is a net gain of two ATP molecules per glucose molecule. Four ATPs are produced, but remember that two are consumed in phase 1 to "prime the pump." Each pyruvic acid molecule has the formula $C_3H_4O_3$, and glucose is $C_6H_{12}O_6$. Between them the two pyruvic acid molecules have lost four hydrogen atoms, which are now bound to two molecules of NAD^+. NAD carries a positive charge (NAD^+), so when it accepts a hydrogen pair, NADH + H^+ is the resulting reduced product. Although a small amount of ATP has been harvested, the other two products of glucose oxidation (H_2O and CO_2) have yet to appear.

The fate of pyruvic acid, which still contains most of glucose's chemical energy, depends on the availability of oxygen at the time the pyruvic acid is produced. Because the supply of NAD^+ is limited, glycolysis can continue only if the reduced coenzymes (NADH + H^+) formed during glycolysis are relieved of their extra hydrogen. Only then can they continue to act as hydrogen acceptors.

When oxygen is readily available, this is no problem. NADH + H^+ delivers its burden of hydrogen atoms to the enzymes of the electron transport chain in the mitochondria, which deliver them to O_2, forming water. However, when oxygen is not present in sufficient amounts, as might occur during strenuous exercise, NADH + H^+ unloads its hydrogen atoms *back onto pyruvic acid,* reducing it. This addition of two hydrogen atoms to pyruvic acid yields **lactic acid** (see bottom right of Figure 24.6). Some of this lactic acid diffuses out of the cells and is transported to the liver for processing.

When oxygen is again available, lactic acid is oxidized back to pyruvic acid and enters the **aerobic pathways** (the oxygen-requiring Krebs cycle and electron transport chain within the mitochondria), and is completely oxidized to water and carbon dioxide. The liver may also convert lactic acid all the way back to glucose-6-phosphate (reverse glycolysis) and then store it as glycogen or free it of its phosphate and release it to the blood if blood sugar levels are low.

Although glycolysis generates ATP rapidly, only 2 ATP molecules are produced per glucose molecule, as compared to the 30 to 32 ATP per glucose harvested when glucose is completely oxidized. Except for red blood cells (which typically carry out *only* glycolysis), prolonged anaerobic metabolism ultimately results in acid-base problems. Consequently, *totally* anaerobic conditions resulting in lactic acid formation provide only a temporary route for rapid ATP production. It can go on without tissue damage for the longest periods in skeletal muscle, for much shorter periods in cardiac muscle, and almost not at all in the brain.

Krebs Cycle Named after its discoverer Hans Krebs, the **Krebs cycle** is the next stage of glucose oxidation. The Krebs cycle, which occurs in the mitochondrial matrix, is fueled largely by pyruvic acid produced during glycolysis and by fatty acids resulting from fat breakdown.

Because pyruvic acid is a charged molecule, it must enter the mitochondrion by active transport with the help of a transport protein. Once in the mitochondrion, the first order of business is a transitional phase that converts it to acetyl CoA. This occurs via a three-step process (**Figure 24.7**, top):

1. **Decarboxylation.** In this step, one of pyruvic acid's carbons is removed and released as carbon dioxide gas. CO_2 diffuses out of the cells into the blood to be expelled by the lungs. This is the first time that CO_2 is released during cellular respiration.
2. **Oxidation.** The remaining 2C fragment (acetic acid) is oxidized by the removal of hydrogen atoms, which are picked up by NAD^+.
3. **Formation of acetyl CoA.** Acetic acid is combined with *coenzyme A* to produce the reactive final product, **acetyl coenzyme A (acetyl CoA)**. Coenzyme A is a sulfur-containing coenzyme derived from vitamin B_5.

Acetyl CoA is now ready to enter the Krebs cycle and be broken down completely by mitochondrial enzymes. Coenzyme A shuttles the 2-carbon acetic acid to an enzyme that condenses it with a 4-carbon acid called **oxaloacetic acid** (ok″sah-lo″ah-sēt′ik) to produce the 6-carbon **citric acid**. Because citric acid is the first substrate of the cycle, biochemists prefer to call the Krebs cycle the **citric acid cycle**.

As the cycle moves through its eight successive steps, the atoms of citric acid are rearranged to produce different intermediate molecules, most called **keto acids** (Figure 24.7). The acetic acid that enters the cycle is broken apart carbon by carbon (decarboxylated) and oxidized, simultaneously generating NADH + H^+ and $FADH_2$. At the end of the cycle, acetic acid has been totally disposed of and oxaloacetic acid, the *pickup molecule,* is regenerated.

What are the products of the Krebs cycle? Because two *decarboxylations* and four *oxidations* occur, the products are two CO_2 molecules and four molecules of reduced coenzymes (3 NADH + H^+ and 1 $FADH_2$). The addition of water at certain steps accounts for some of the released hydrogen. One molecule of ATP is formed (via substrate-level phosphorylation) during each turn of the cycle. The detailed events of each of the eight steps of the Krebs cycle are described in Appendix D.

Now let's back up and account for the pyruvic acid molecules entering the mitochondria. We need to consider the products of both the transitional phase and the Krebs cycle itself. Altogether, each pyruvic acid yields three CO_2 molecules and five molecules of reduced coenzymes—1 $FADH_2$ and 4 NADH + H^+ (equal to the removal of 10 hydrogen atoms). The products of glucose oxidation in the Krebs cycle are twice that (remember 1 glucose = 2 pyruvic acids): six CO_2, ten molecules of reduced coenzymes,

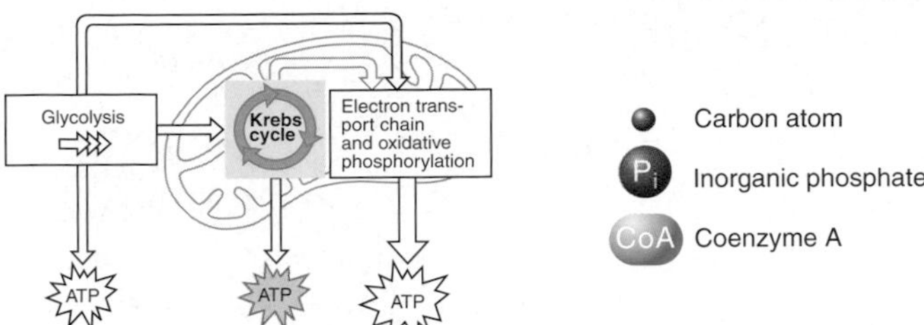

Figure 24.7 Simplified version of the Krebs (citric acid) cycle. During each turn of the cycle, two carbon atoms are removed from the substrates as CO_2 (decarboxylation reactions); four oxidations by removal of hydrogen atoms occur, producing four molecules of reduced coenzymes (3 NADH + H^+ and 1 $FADH_2$); and one ATP is synthesized by substrate-level phosphorylation. An additional decarboxylation and an oxidation reaction occur in the transitional phase that converts pyruvic acid, the product of glycolysis, to acetyl CoA, the molecule that enters the Krebs cycle pathway.

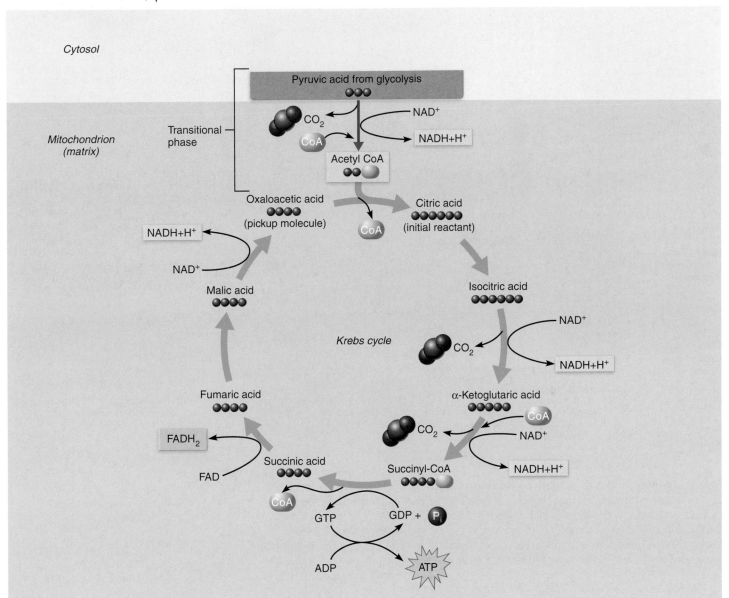

and two ATP molecules. Notice that it is these Krebs cycle reactions that produce the CO_2 evolved during glucose oxidation. The reduced coenzymes, which carry their extra electrons in high-energy linkages, must now be oxidized if the Krebs cycle and glycolysis are to continue.

Although the glycolytic pathway is exclusive to carbohydrate oxidation, breakdown products of carbohydrates, fats, and proteins can feed into the Krebs cycle to be oxidized for energy. On the other hand, some Krebs cycle intermediates can be siphoned off to make fatty acids and nonessential amino acids. Thus, the Krebs cycle is a source of building materials for anabolic reactions, in addition to serving as the final common pathway for the oxidation of food fuels.

Electron Transport Chain and Oxidative Phosphorylation Like glycolysis, none of the reactions of the Krebs cycle use oxygen directly. This is the exclusive function of the **electron transport chain**, which carries out the final catabolic reactions that occur on the mitochondrial cristae. However, because the reduced coenzymes produced in the Krebs cycle are the substrates for the

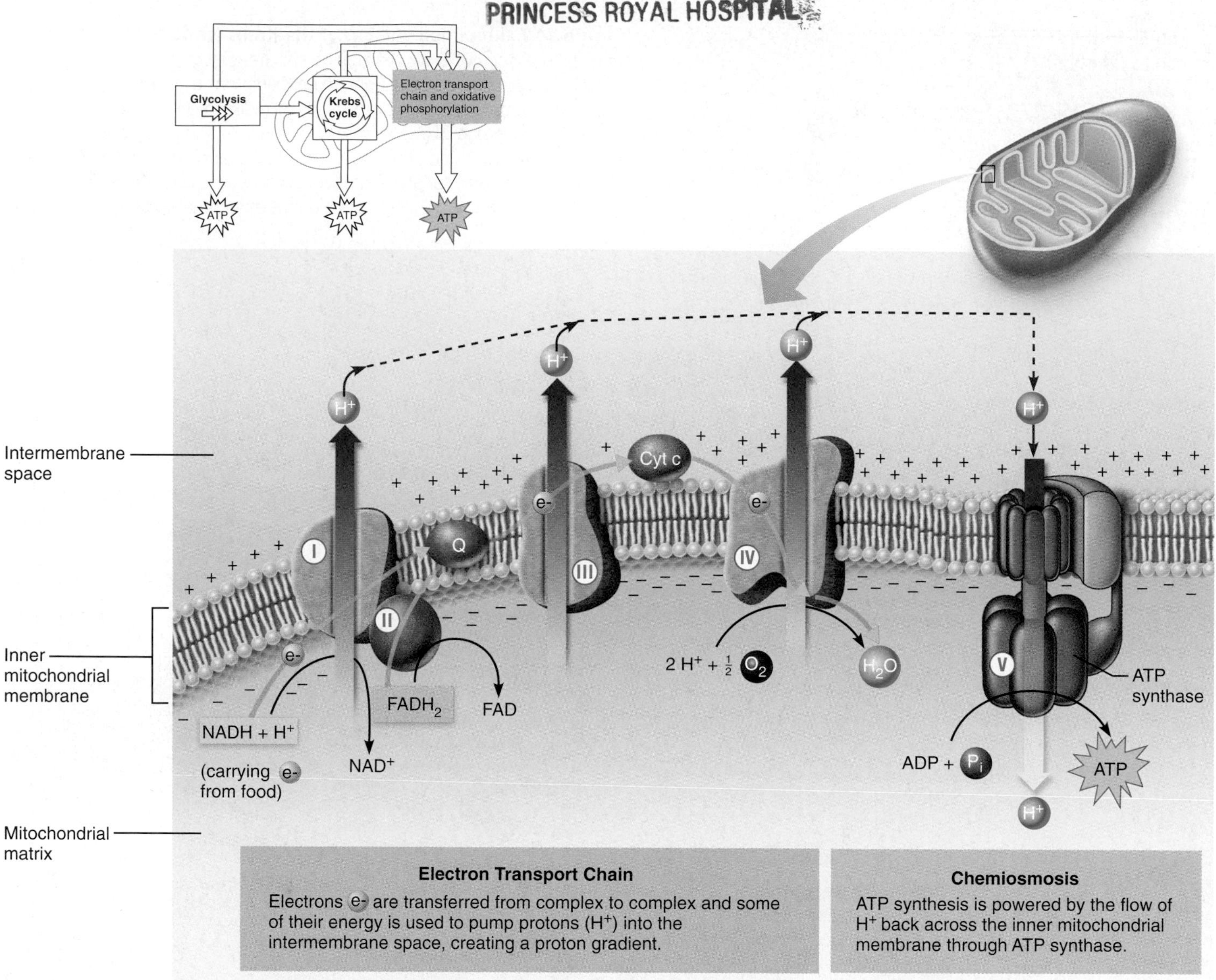

24

Figure 24.8 Mechanism of oxidative phosphorylation. Schematic diagram showing the flow of electrons through the mitochondrial respiratory enzyme complexes of the electron transport chain during the transfer of two electrons from reduced NAD^+ to oxygen. Coenzyme Q (ubiquinone) and cytochrome c are mobile and act as carriers between the complexes. Because $FADH_2$ unloads its H atoms to complex II, the small complex just beyond the first major respiratory complex, less energy is captured as a result of its oxidation.

electron transport chain, these two pathways are coupled, and both are considered to be oxygen requiring, or *aerobic.*

In the electron transport chain, the hydrogens removed during the oxidation of food fuels are combined with O_2 to form water, and the energy released during those reactions is harnessed to attach P_i groups to ADP, forming ATP. As we noted earlier, this type of phosphorylation process is called *oxidative phosphorylation.* Let's peek under the hood of a cell's power plant and see how this rather complicated process works.

Most components of the electron transport chain are proteins that are bound to metal atoms (known as *cofactors*). These proteins vary in composition and form multiprotein complexes that are firmly embedded in the inner mitochondrial membrane **(Figure 24.8)**. For example, some of the proteins, the **flavins**, contain flavin mononucleotide (FMN) derived from the vitamin riboflavin, and others contain both sulfur (S) and iron (Fe). Most, however, are brightly colored iron-containing pigments called **cytochromes** (si′to-krōmz; *cyto* = cell, *chrom* = color), including complexes III and IV depicted in Figure 24.8. Neighboring carriers are clustered together to form four **respiratory enzyme complexes** that are alternately reduced and oxidized as they pick up electrons and pass them on to the next complex in the sequence.

As Figure 24.8 shows, the first such complex accepts hydrogen atoms from NADH + H^+, oxidizing it to NAD^+. $FADH_2$ transfers its hydrogen atoms slightly farther along the chain to the small complex II. The hydrogen atoms delivered to the electron transport chain by the reduced coenzymes are quickly split into protons (H^+) plus electrons. The electrons are shuttled along the inner mitochondrial membrane from one complex to the next, losing energy with each transfer. The protons escape into the watery matrix only to be picked up and "pumped" across the inner mitochondrial membrane into the intermembrane space by one of the three major respiratory enzyme complexes (I, III, and IV).

Ultimately the electron pairs are delivered to half a molecule of O_2 (in other words, to an oxygen atom), creating oxygen ions (O^-) that strongly attract H^+ and form water as indicated by the reaction

$$2H^+ + 2e^- + \frac{1}{2}O_2 \rightarrow H_2O$$

Virtually all the water resulting from glucose oxidation is formed during oxidative phosphorylation. Because NADH + H^+ and $FADH_2$ are oxidized as they release their burden of picked-up hydrogen atoms, the net reaction for the electron transport chain is

$$\underset{\text{reduced coenzyme}}{\text{Coenzyme -2H}} + \frac{1}{2}O_2 \rightarrow \underset{\text{oxidized coenzyme}}{\text{coenzyme}} + H_2O$$

The transfer of electrons from NADH + H^+ to oxygen releases large amounts of energy. If hydrogen combined directly with molecular oxygen, the energy would be released in one big burst and most of it would be lost to the environment as heat. Instead energy is released in many small steps as the electrons stream from one electron acceptor to the next. Each successive carrier has a greater affinity for electrons than those preceding it. For this reason, the electrons cascade "downhill" from NADH + H^+ to progressively lower energy levels until they are finally delivered to oxygen, which has the greatest affinity of all for electrons. You could say that oxygen "pulls" the electrons down the chain **(Figure 24.9)**.

The electron transport chain functions as an energy converter by using the stepwise release of electronic energy to pump protons from the matrix into the intermembrane space. Because the inner mitochondrial membrane is nearly impermeable to H^+, this chemiosmotic process creates an **electrochemical proton (H^+) gradient** across that membrane that has potential energy and the capacity to do work. The proton gradient (1) creates a pH gradient, with the H^+ concentration in the matrix much lower than that in the intermembrane space; and (2) generates a voltage across the membrane that is negative on the matrix side and positive between the mitochondrial membranes. Both conditions strongly attract H^+ back into the matrix. But how can they get there?

The only areas of the membrane freely permeable to H^+ are at large enzyme-protein complexes (complex V) called **ATP synthases**. These complexes, which populate the inner mitochondrial membrane **(Figure 24.10)**, lay claim to being nature's smallest rotary motors. As the protons take this "route" they create an electrical current, and ATP synthase harnesses this electrical energy to catalyze attachment of a phosphate group to ADP to form ATP **(Figure 24.11)**. The enzyme's subunits appear to work together like gears. As the ATP synthase core rotates, ADP and inorganic phosphate are pulled in and ATP is churned out, completing the process of oxidative phosphorylation.

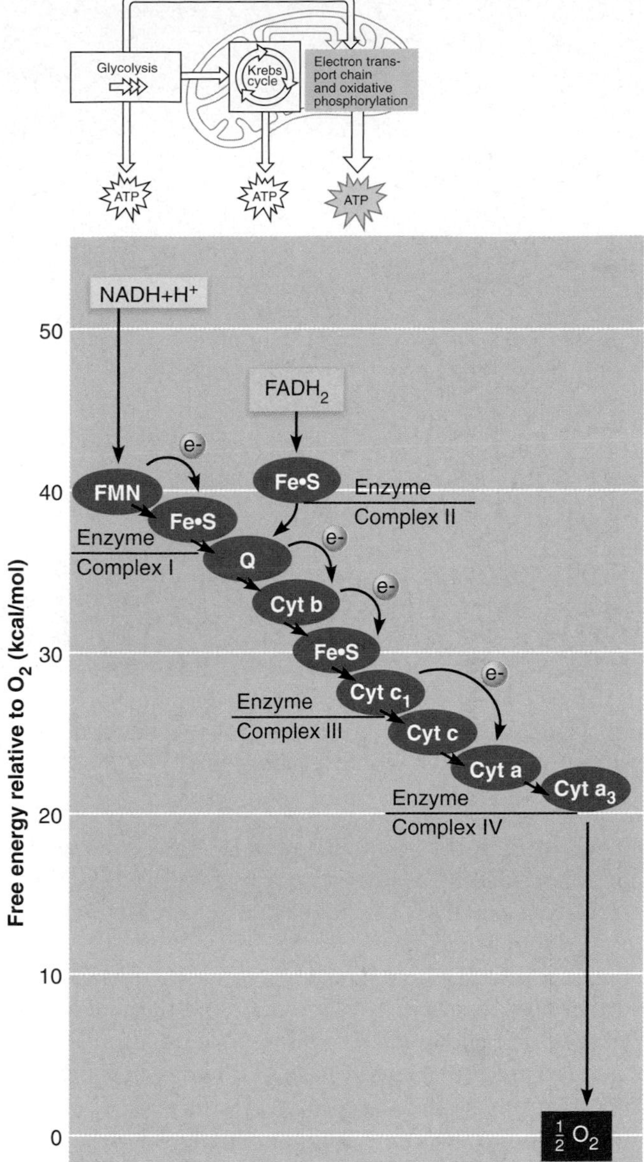

Figure 24.9 Electronic energy gradient in the electron transport chain. Each member of the chain (respiratory enzyme complex) oscillates between a reduced state and an oxidized state. A member becomes reduced by accepting electrons from its "uphill" neighbor and then reverts to its oxidized form as it passes electrons to its "downhill" neighbor. The overall energy drop is 53 kcal/mol, but this fall is broken up into smaller steps by the electron transport chain.

How exactly does ATP synthase work? Studies of its molecular structure are providing answers. The enzyme complex consists of two major linked parts, each with several protein subunits: (1) a *rotor* embedded in the inner mitochondrial membrane (Figure 24.11), and (2) a *knob* extending into the mitochondrial matrix, which is stabilized by a *stator* anchored in the membrane. A *rod* connects the rotor and the knob. The

Figure 24.10 Atomic force microscopy reveals the structure of energy-converting ATP synthase rotor rings.

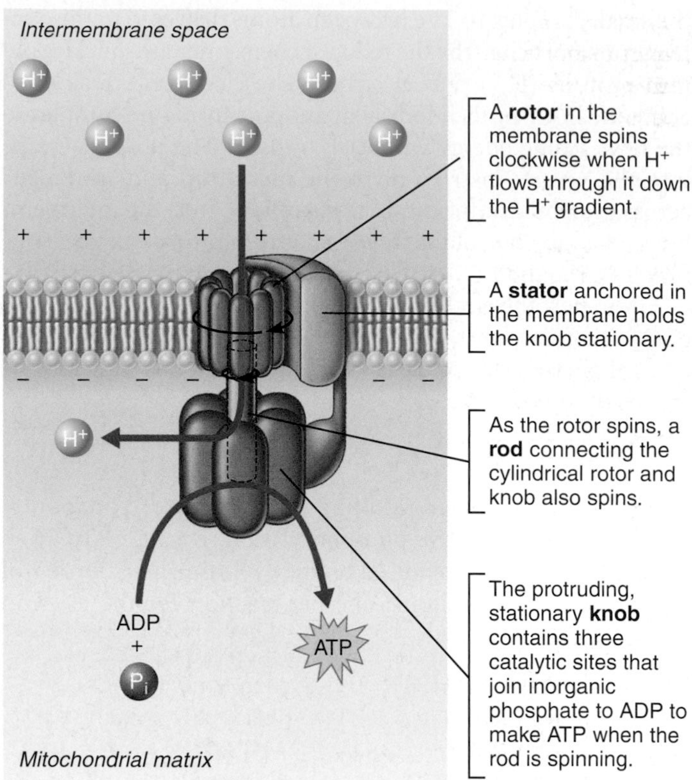

Intermembrane space
H^+
A **rotor** in the membrane spins clockwise when H^+ flows through it down the H^+ gradient.
A **stator** anchored in the membrane holds the knob stationary.
As the rotor spins, a **rod** connecting the cylindrical rotor and knob also spins.
The protruding, stationary **knob** contains three catalytic sites that join inorganic phosphate to ADP to make ATP when the rod is spinning.
ADP
+
P_i
ATP
Mitochondrial matrix

Figure 24.11 Structure of ATP synthase.

current created by the downhill flow of H^+ causes the rotor and rod to rotate, just as flowing water turns a water wheel. This rotation activates catalytic sites in the knob where ADP and P_i are combined to make ATP.

Notice something here. The ATP synthase works like an ion pump running in reverse. Recall from Chapter 3 that ion pumps use ATP as their energy source to transport ions against an electrochemical gradient. Here we have ATP synthases using the energy of a proton gradient to power ATP synthesis.

The proton gradient also supplies energy to pump needed metabolites (ADP, pyruvic acid, inorganic phosphate) and calcium ions across the relatively impermeable inner mitochondrial membrane. The outer membrane is quite freely permeable to these substances, so no "help" is needed there. The supply of energy from oxidation is not limitless however, so when more of the gradient energy is used to drive these transport processes, less is available to make ATP.

24

HOMEOSTATIC IMBALANCE

Studies of metabolic poisons support the chemiosmotic model of oxidative phosphorylation. For example, cyanide (the gas used in gas chambers) disrupts the process by binding to cytochrome oxidase and blocking electron flow from complex IV to oxygen (see Figure 24.9). Poisons commonly called "uncouplers" abolish the proton gradient by making the inner mitochondrial membrane permeable to H^+. Consequently, although the electron transport chain continues to deliver electrons to oxygen at a furious pace, and oxygen consumption rises, no ATP is made. ■

The stimulus for ATP production is entry of *ADP* into the mitochondrial matrix. As ADP is transported in, ATP is moved out in a coupled transport process.

Summary of ATP Production The average person at rest uses energy at the rate of roughly 100 kcal/hour, which is equal to 116 watts or slightly more than that of a standard lightbulb. This seems to be a minuscule amount, but from a biochemical standpoint it places a staggering power demand on our mitochondria. Luckily, they are up to the task.

When O_2 is present, cellular respiration is remarkably efficient. Of the 686 kcal of energy present in 1 mole of glucose, as much as 262 kcal can be captured in ATP bonds. (The rest is liberated as heat.) This corresponds to an energy capture of about 38%, making cells far more efficient than any human-made machines, which use only 10–30% of the energy available to them.

During cellular respiration, most energy flows in this sequence:

Glucose $\rightarrow$ NADH + H^+ $\rightarrow$ electron transport chain $\rightarrow$ proton gradient energy $\rightarrow$ ATP

Let's do a little bookkeeping to summarize the net energy gain from one glucose molecule. First, we tally the results of substrate-level phosphorylation, giving us a net gain of 4 ATP produced directly by substrate-level phosphorylations (2 during glycolysis and 2 during the Krebs cycle). Next we must calculate the much greater number of ATP molecules produced by oxidative phosphorylation **(Figure 24.12)**.

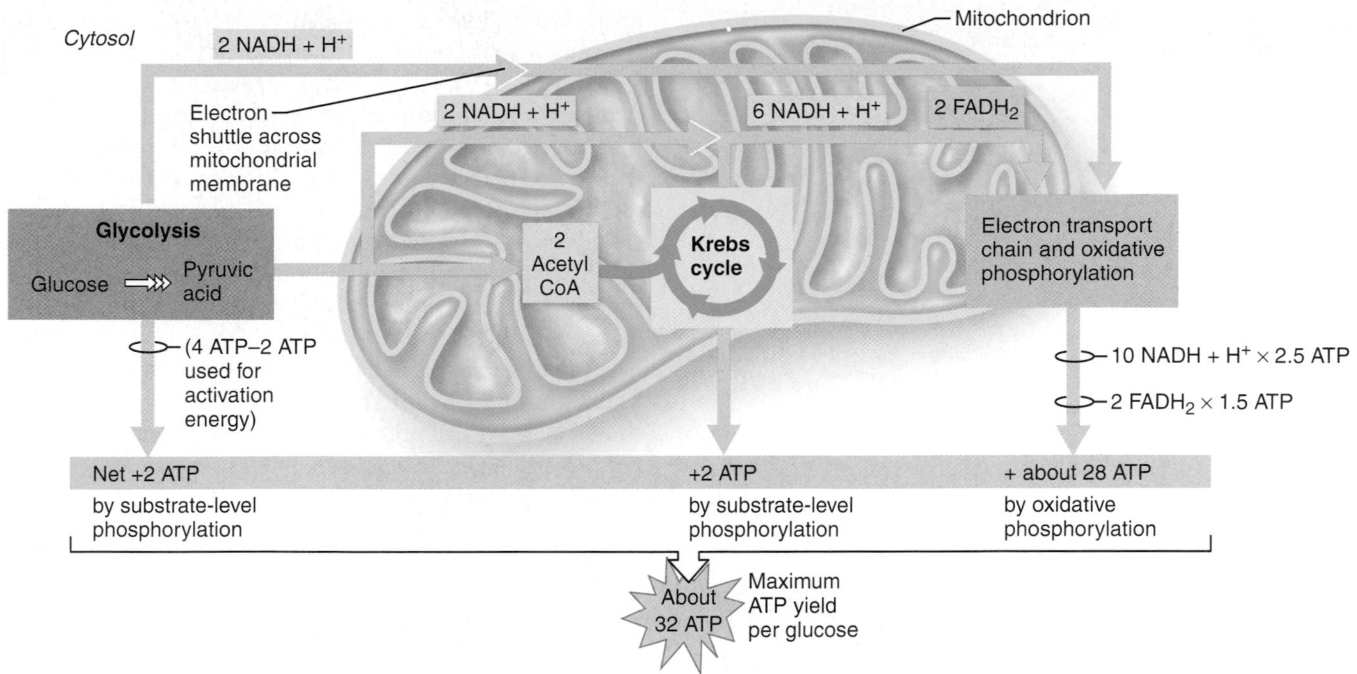

Figure 24.12 Energy yield during cellular respiration.

Each NADH + H^+ that transfers a pair of high-energy electrons to the electron transport chain contributes enough energy to the proton gradient to generate about 2½ ATP molecules. The oxidation of $FADH_2$ is less efficient because it doesn't donate electrons to the "top" of the electron transport chain as does NADH + H^+, but to a lower energy level (at complex II). So, for each 2 H delivered by $FADH_2$, just about 1½ ATP molecules are produced. Now we can tally the results of oxidative phosphorylation. The 2 NADH + H^+ generated during glycolysis yield 5 ATP molecules. The 8 NADH + H^+ and the 2 $FADH_2$ produced during the Krebs cycle are "worth" 20 and 3 ATP respectively.

Overall, complete oxidation of 1 glucose molecule to CO_2 and H_2O by both substrate-level phosphorylation and oxidative phosphorylation yields a maximum of 32 molecules of ATP (Figure 24.12). Stated another way, this adds up to some 55 kilograms of ATP every day.

However, there is an uncertainty about the energy yield of reduced NAD^+ generated *outside* the mitochondria by glycolysis. The crista membrane is not permeable to reduced NAD^+ generated in the cytosol, so NADH + H^+ formed during glycolysis uses a *shuttle molecule* to deliver its extra electron pair to the electron transport chain. It appears that cells using the malate/aspartate shuttle harvest the whole 2½ ATP from oxidation of reduced NAD^+, but in cells using a different shuttle (the glycerol phosphate shuttle for example) the shuttle has an energy cost. At present, the consensus is that the net energy yield for reoxidation of reduced NAD^+ in this case is probably the same as for $FADH_2$, that is, 1½ ATP per electron pair.

So, if we deduct 2 ATP to cover the "fare" for the shuttle, our bookkeeping comes up with a grand total of 30 ATP per glucose as the maximum possible energy yield. (Actually our figures are probably still too high because, as mentioned earlier, the proton gradient energy is also used to do other work and the electron transport chain may not work at maximum capacity all of the time.)

CHECK YOUR UNDERSTANDING

11. Briefly, how do substrate-level and oxidative phosphorylation differ?
12. What happens in the glycolytic pathway if oxygen is absent and NADH + H^+ cannot transfer its "picked-up" hydrogen to pyruvic acid?
13. What two major kinds of chemical reactions occur in the Krebs cycle, and how are these reactions indicated symbolically?

For answers, see Appendix G.

Glycogenesis, Glycogenolysis, and Gluconeogenesis

Glycogenesis and Glycogenolysis Although most glucose is used to generate ATP molecules, unlimited amounts of glucose do *not* result in unlimited ATP synthesis, because cells cannot store large amounts of ATP. When more glucose is available than can immediately be oxidized, rising intracellular ATP concentrations eventually inhibit glucose catabolism and initiate processes that store glucose as glycogen or fat. Because the body can store much more fat than glycogen, fats account for 80–85% of stored energy.

When high ATP levels begin to "turn off" glycolysis, glucose molecules are combined in long chains to form glycogen, the animal carbohydrate storage product. This process is called **glycogenesis** (*glyco* = sugar; *genesis* = origin) (**Figure 24.13**, right side). It begins as glucose entering cells is phosphorylated

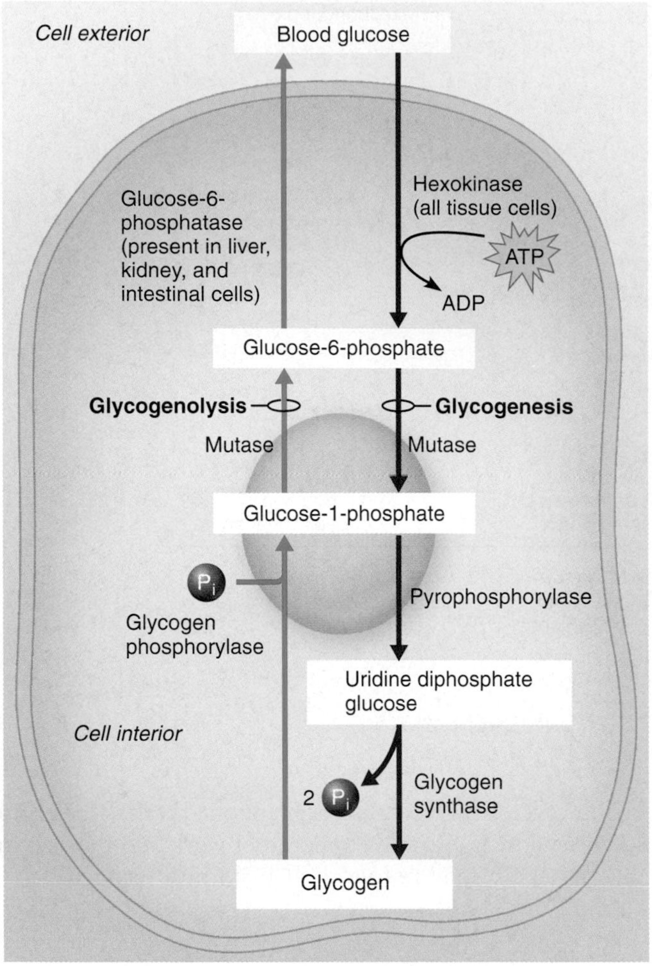

Figure 24.13 Glycogenesis and glycogenolysis. When glucose supplies exceed demands, glycogenesis (conversion of glucose to glycogen) occurs. Glycogenolysis (breakdown of glycogen to release glucose) is stimulated by falling blood glucose levels.

to glucose-6-phosphate and then converted to its isomer, *glucose-1-phosphate.* The terminal phosphate group is cleaved off as the enzyme *glycogen synthase* catalyzes the attachment of glucose to the growing glycogen chain. Liver and skeletal muscle cells are most active in glycogen synthesis and storage.

On the other hand, when blood glucose levels drop, glycogen lysis (splitting) occurs. This process is known as **glycogenolysis** (gli″ko-jĕ-nol′ĭ-sis) (Figure 24.13, left side). The enzyme *glycogen phosphorylase* oversees phosphorylation and cleavage of glycogen to release glucose-1-phosphate, which is then converted to glucose-6-phosphate, a form that can enter the glycolytic pathway to be oxidized for energy.

In muscle cells and most other cells, the glucose-6-phosphate resulting from glycogenolysis is trapped because it cannot cross the cell membrane. However, hepatocytes (and some kidney and intestinal cells) contain *glucose-6-phosphatase*, an enzyme that removes the terminal phosphate, producing free glucose. Because glucose can then readily diffuse from the cell into the blood, the liver can use its glycogen stores to provide blood sugar for the benefit of other organs when blood glucose levels drop. Liver glycogen is also an important energy source for skeletal muscles that have depleted their own glycogen reserves.

Athletes and Carbohydrates A common misconception is that athletes need to eat large amounts of protein to improve their performance and maintain their muscle mass. Actually, a diet rich in complex carbohydrates, which results in more muscle glycogen storage, is much more effective in sustaining intense muscle activity than are high-protein meals. Notice that the emphasis is on *complex* carbohydrates. Eating a candy bar before an athletic event to provide "quick" energy does more harm than good because it stimulates insulin secretion, which favors glucose use and retards fat use at a time when fat use should be maximal. Building muscle protein or avoiding its loss requires not only extra protein, but also extra (protein-sparing) complex carbohydrate calories to meet the greater energy needs of the increasingly massive muscles.

Endurance athletes, long-distance runners in particular, are well aware of the practice of glycogen loading, popularly called "carbo loading," for endurance events. Carbo loading "tricks" the muscles into storing more glycogen than they normally would. It generally involves eating a carbohydrate-rich diet (75% of energy intake) for three to four days before an endurance event while decreasing activity. This practice has been shown to increase muscle glycogen stores to as much as twice the normal amount. Carbo loading is now standard practice among marathon runners and distance cyclists, because studies have shown that it improves performance and endurance.

Gluconeogenesis When too little glucose is available to stoke the "metabolic furnace," glycerol and amino acids are converted to glucose. **Gluconeogenesis**, the process of forming new (*neo*) glucose from *noncarbohydrate* molecules, occurs in the liver. It takes place when dietary sources and glucose reserves have been depleted and blood glucose levels are beginning to drop. Gluconeogenesis protects the body, the nervous system in particular, from the damaging effects of low blood sugar (*hypoglycemia*) by ensuring that ATP synthesis can continue.

CHECK YOUR UNDERSTANDING

14. What name is given to the chemical reaction in which glycogen is broken down to its glucose subunits?
15. What does carbo loading accomplish?

For answers, see Appendix G.

Lipid Metabolism

- Describe the process by which fatty acids are oxidized for energy.
- Define ketone bodies, and indicate the stimulus for their formation.

Fats are the body's most concentrated source of energy. They contain very little water, and the energy yield from fat catabolism is approximately twice that from either glucose or protein catabolism—9 kcal per gram of fat versus 4 kcal per gram of carbohydrate or protein. Most products of fat digestion are transported in lymph in the form of fatty-protein droplets called *chylomicrons* (see Chapter 23). Eventually, the lipids in the chylomicrons are hydrolyzed by enzymes on capillary endothelium, and the resulting fatty acids and glycerol are taken up by body cells and processed in various ways.

Oxidation of Glycerol and Fatty Acids

Of the various lipids, only triglycerides are routinely oxidized for energy. Their catabolism involves the separate oxidation of their two different building blocks: glycerol and fatty acid chains **(Figure 24.14)**. Most body cells easily convert glycerol to glyceraldehyde phosphate, a glycolysis intermediate that enters the Krebs cycle. Glyceraldehyde is equal to half a glucose molecule, and ATP energy harvest from its complete oxidation is approximately half that of glucose (16 ATP/glycerol).

Beta oxidation, the initial phase of fatty acid oxidation, occurs in the mitochondria. Although oxidation and other reactions are involved, the net result is that the fatty acid chains are broken apart into two-carbon *acetic acid* fragments, and coenzymes (FAD and NAD^+) are reduced (Figure 24.14, right side). Each acetic acid molecule is fused to coenzyme A, forming acetyl CoA. The term "beta oxidation" reflects the fact that the carbon in the beta (third) position is oxidized during the process and cleavage of the fatty acid in each case occurs between the alpha and beta carbons. Acetyl CoA is then picked up by oxaloacetic acid and enters the aerobic pathways to be oxidized to CO_2 and H_2O.

Notice that unlike glycerol, which enters the glycolytic pathway, acetyl CoA resulting from fatty acid breakdown *cannot* be used for gluconeogenesis because the metabolic pathway is irreversible past pyruvic acid.

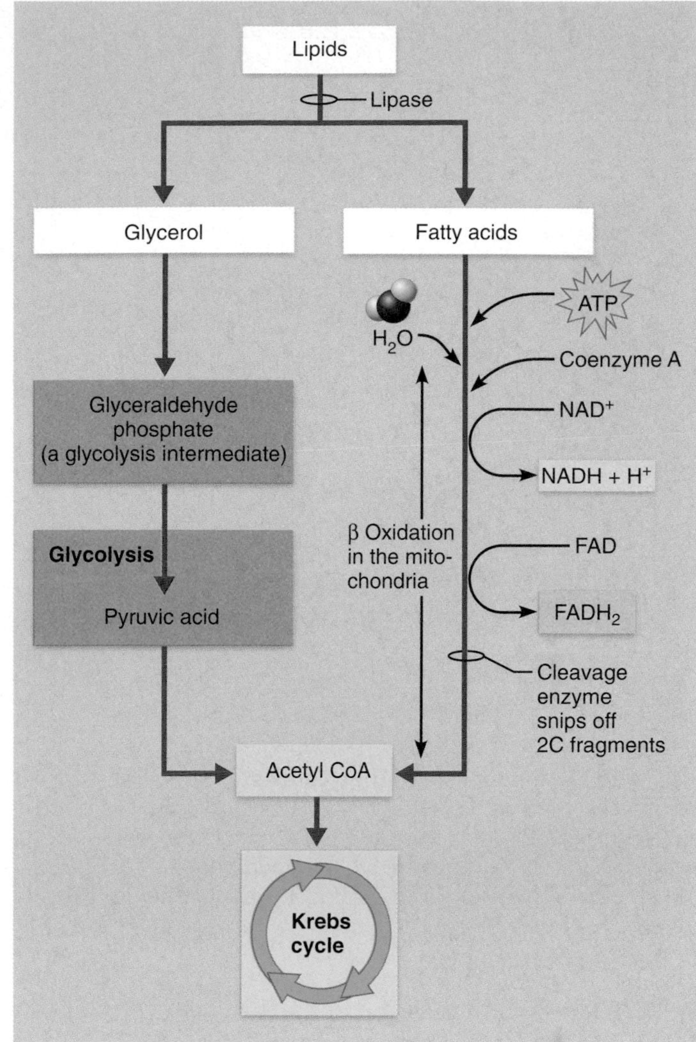

Figure 24.14 Initial phase of lipid oxidation. The glycerol portion is converted to a glycolysis intermediate, and completes the glycolytic pathway through pyruvic acid to acetyl CoA. The fatty acids undergo beta (β) oxidation in which they are first activated by a coupled reaction with ATP, combined with coenzyme A, and then oxidized twice (reducing NAD^+ and FAD). The acetyl CoA created in β oxidation is cleaved off and the process begins again.

Lipogenesis and Lipolysis

There is a continuous turnover of triglycerides in adipose tissue. New fats are "put in the larder" for later use, while stored fats are broken down and released to the blood. That bulge of fatty tissue you see today does *not* contain the same fat molecules it did a month ago.

Glycerol and fatty acids from dietary fats not immediately needed for energy are recombined into triglycerides and stored. About 50% ends up in subcutaneous tissue, and the balance is stockpiled in other fat depots of the body. Triglyceride synthesis, or **lipogenesis**, occurs when cellular ATP and glucose levels are high (**Figure 24.15**, purple arrows). Excess ATP also leads to an accumulation of acetyl CoA and glyceraldehyde-PO_4, two intermediates of glucose metabolism that would otherwise feed into the Krebs cycle. But when these two metabolites are present in excess, they are channeled into triglyceride synthesis pathways.

Acetyl CoA molecules are condensed together, forming fatty acid chains that grow two carbons at a time. (This accounts for the fact that almost all fatty acids in the body contain an even number of carbon atoms.) Because acetyl CoA, an intermediate in glucose catabolism, is also the *starting point* for fatty acid synthesis, glucose is easily converted to fat. Glyceraldehyde-PO_4 is converted to glycerol, which is condensed with fatty acids to form triglycerides. Consequently, even if the diet is fat-poor, carbohydrate intake can provide *all the raw materials* needed to form triglycerides. When blood sugar is high, lipogenesis is the major activity in adipose tissues and is also an important liver function.

Lipolysis (lĭ-pol′ĭ-sis; "fat splitting"), the breakdown of stored fats into glycerol and fatty acids, is essentially lipogenesis in reverse (Figure 24.15, blue arrows). The fatty acids and glycerol are released to the blood, helping to ensure that body organs have continuous access to fat fuels for aerobic respiration. (The liver, cardiac muscle, and resting skeletal muscles actually prefer fatty acids as an energy fuel.)

The meaning of the adage "fats burn in the flame of carbohydrates" becomes clear when carbohydrate intake is inadequate. Under such conditions, lipolysis is accelerated as the

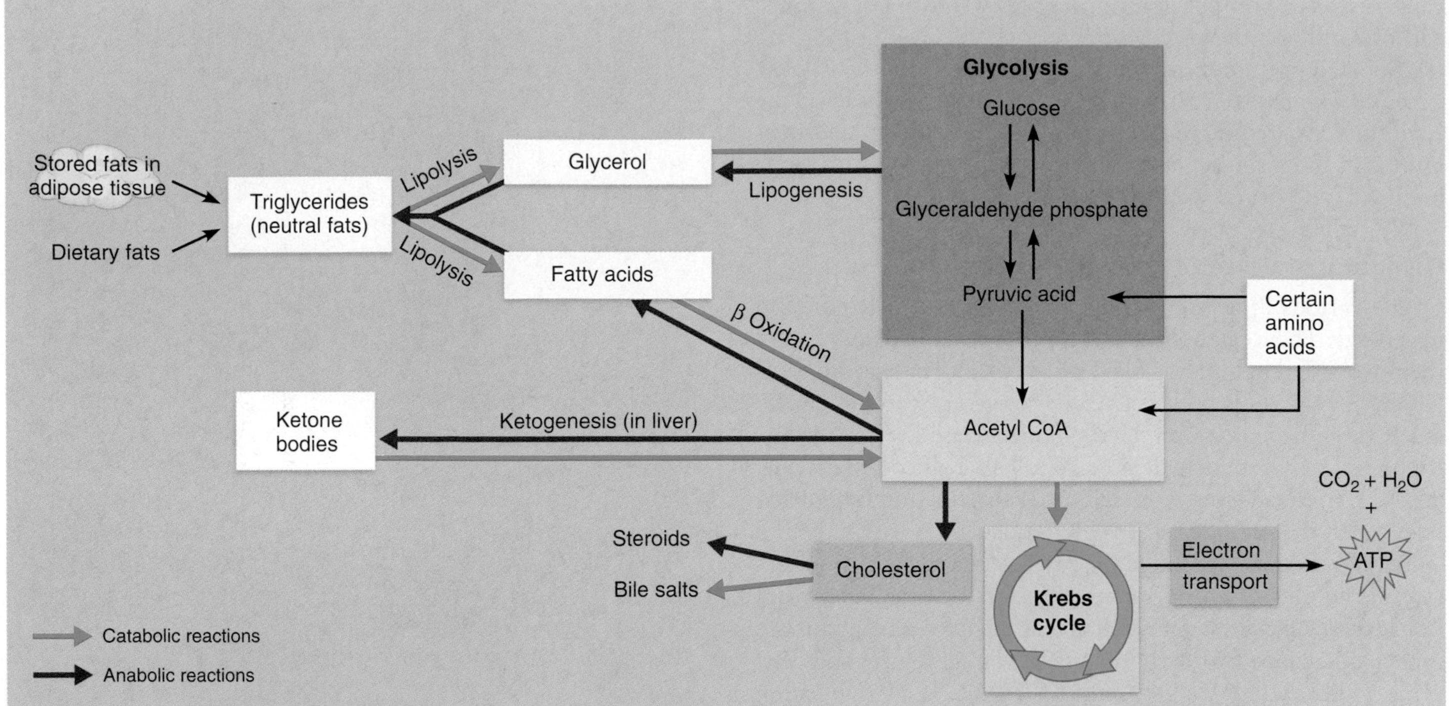

Figure 24.15 Metabolism of triglycerides. When needed for energy, fats enter catabolic pathways. Glycerol enters the glycolytic pathway and the fatty acids are broken down by beta oxidation to acetyl CoA, which enters the Krebs cycle. When fats are to be synthesized (lipogenesis) and stored, the intermediates are drawn from glycolysis and the Krebs cycle in a reversal of the processes noted above. Likewise, excess dietary fats are stored in adipose tissues. When triglycerides are in excess or are the primary energy source, the liver releases their breakdown products in the form of ketone bodies. Excessive amounts of carbohydrates and amino acids are also converted to triglycerides (lipogenesis).

body attempts to fill the fuel gap with fats. However, the ability of acetyl CoA to enter the Krebs cycle depends on the availability of oxaloacetic acid to act as the pickup molecule (see Figure 24.7). When carbohydrates are deficient, oxaloacetic acid is converted to glucose (to fuel the brain). Without oxaloacetic acid, fat oxidation is incomplete, and acetyl CoA accumulates. Via a process called **ketogenesis**, the liver converts acetyl CoA molecules to **ketone bodies**, or **ketones**, which are released into the blood. Ketone bodies include acetoacetic acid, β-hydroxybutyric acid, and acetone. (The *keto acids* cycling through the Krebs cycle and the *ketone bodies* resulting from fat metabolism are quite different and should not be confused.)

24

HOMEOSTATIC IMBALANCE

When ketone bodies accumulate in the blood, *ketosis* results and large amounts of ketone bodies are excreted in the urine. Ketosis is a common consequence of starvation, unwise dieting (in which inadequate amounts of carbohydrates are eaten), and diabetes mellitus. Because most ketone bodies are organic acids, the outcome of ketosis is *metabolic acidosis.* The body's buffer systems cannot tie up the acids (ketones) fast enough, and blood pH drops to dangerously low levels. The person's breath smells fruity as acetone vaporizes from the lungs, and breathing becomes more rapid as the respiratory system tries to reduce blood carbonic acid by blowing off CO_2 to force the blood pH up. In severe untreated cases, the person may become comatose or even die as the acid pH depresses the nervous system. ■

Synthesis of Structural Materials

All body cells use phospholipids and cholesterol to build their membranes, and phospholipids are important components of myelin sheaths of neurons. In addition, the liver (1) synthesizes lipoproteins for transport of cholesterol, fats, and other substances in the blood; (2) synthesizes cholesterol from acetyl CoA; and (3) uses cholesterol to form bile salts. The ovaries, testes, and adrenal cortex use cholesterol to synthesize their steroid hormones.

CHECK YOUR UNDERSTANDING

16. Which part of triglyceride molecules directly enters the glycolysis pathway?
17. What is the central molecule in fat metabolism?
18. What are the products of beta oxidation?

For answers, see Appendix G.

Protein Metabolism

- Describe how amino acids are metabolized for energy.
- Describe the need for protein synthesis in body cells.

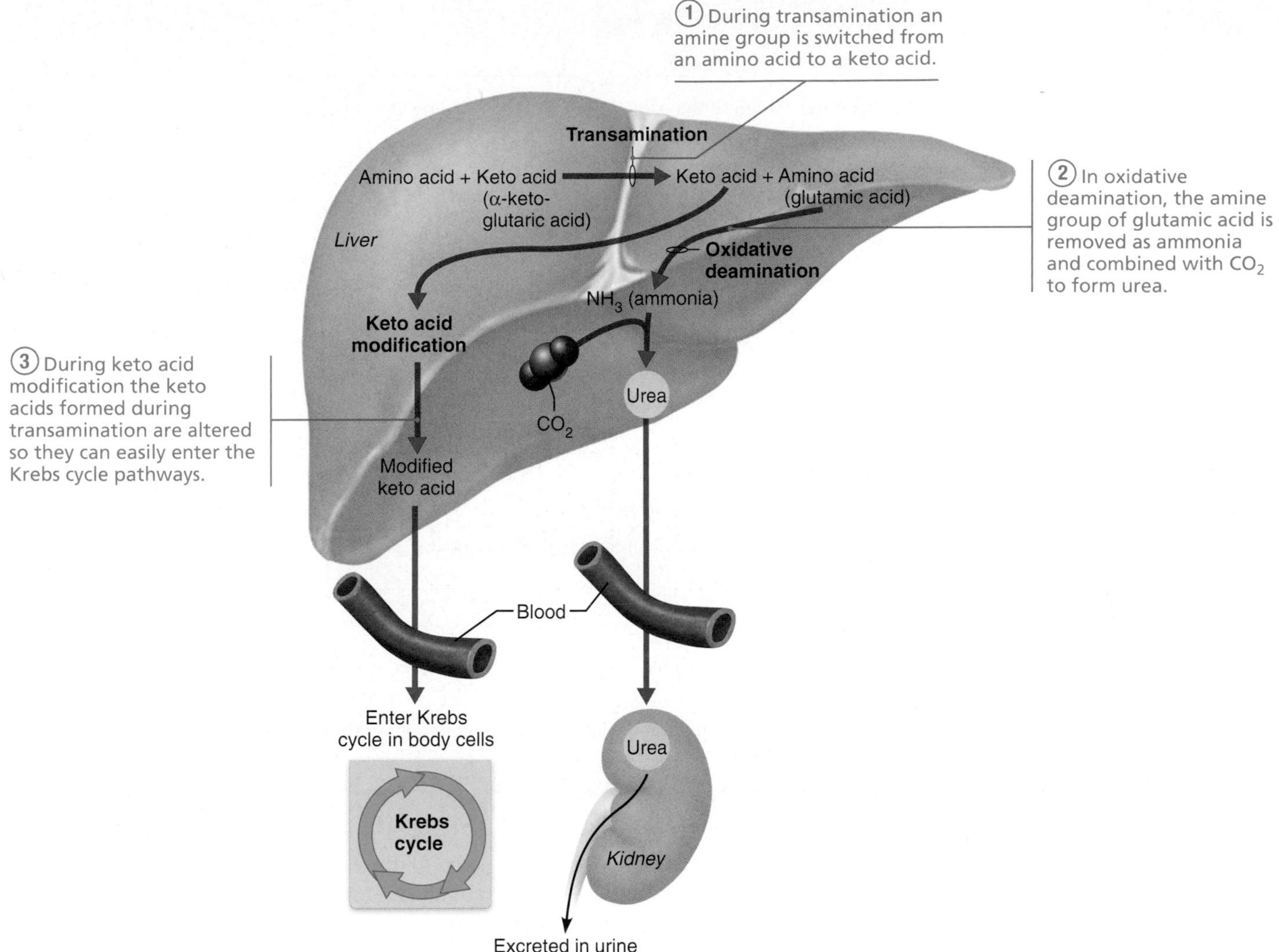

① During transamination an amine group is switched from an amino acid to a keto acid.
Transamination
Amino acid + Keto acid (α-keto-glutaric acid)
Keto acid + Amino acid (glutamic acid)
② In oxidative deamination, the amine group of glutamic acid is removed as ammonia and combined with CO_2 to form urea.
Liver
Oxidative deamination
NH_3 (ammonia)
Keto acid modification
③ During keto acid modification the keto acids formed during transamination are altered so they can easily enter the Krebs cycle pathways.
Urea
CO_2
Modified keto acid
Blood
Enter Krebs cycle in body cells
Krebs cycle
Urea
Kidney
Excreted in urine

Figure 24.16 Transamination, oxidative deamination, and keto acid modification: processes that occur when amino acids are utilized for energy.

Like all other biological molecules, proteins have a limited life span and must be broken down and replaced before they begin to deteriorate. As proteins are broken down, their amino acids are recycled and used in building new proteins or modified to form a different N-containing compound. Newly ingested amino acids transported in the blood are taken up by cells by active transport processes and used to *replace* tissue proteins at the rate of about 100 grams each day.

Although popular opinion has it that excess protein can be stored by the body, nothing is farther from the truth. When more protein is available than is needed for anabolic purposes, amino acids are oxidized for energy or converted to fat for future energy needs.

Oxidation of Amino Acids

Before amino acids can be oxidized for energy, they must be *deaminated*, that is, their amine group (NH_2) must be removed. (In sulfur-containing amino acids, such as methionine and cysteine, sulfur is released prior to deamination.) The resulting molecule is then converted to pyruvic acid or to one of the keto acid intermediates in the Krebs cycle [acetyl CoA, alpha (α) ketoglutaric acid, succinyl CoA, fumaric acid, or oxaloacetic acid]. The key molecule in these conversions is the nonessential amino acid *glutamic acid* (gloo-tam′ik). As **Figure 24.16** shows, the following events occur:

① **Transamination** (trans″am-ĭ-na′shun). A number of amino acids can transfer their amine group to α-ketoglutaric acid (a Krebs cycle keto acid), thereby transforming α-ketoglutaric acid to glutamic acid. In the process, the original amino acid becomes a keto acid (that is, it has an oxygen atom where the amine group formerly was). This reaction is fully reversible.

② **Oxidative deamination.** In the liver, the amine group of glutamic acid is removed as **ammonia (NH_3)**, and α-ketoglutaric acid is regenerated. The liberated NH_3 molecules are combined with CO_2, yielding **urea** and water. The urea is released to the blood and removed from the body in urine. Because ammonia is toxic to body cells, the ease with

TABLE 24.4 Thumbnail Summary of Metabolic Reactions

Carbohydrates	
Cellular respiration	Reactions that together complete the oxidation of glucose, yielding CO_2, H_2O, and ATP
Glycolysis	Conversion of glucose to pyruvic acid
Glycogenesis	Polymerization of glucose to form glycogen
Glycogenolysis	Hydrolysis of glycogen to glucose monomers
Gluconeogenesis	Formation of glucose from noncarbohydrate precursors
Krebs cycle	Complete breakdown of pyruvic acid to CO_2, yielding small amounts of ATP and reduced coenzymes
Electron transport chain	Energy-yielding reactions that split H removed during oxidations to H^+ and e^- and create a proton gradient used to bond ADP to P_i (forming ATP)
Lipids	
Beta oxidation	Conversion of fatty acids to acetyl CoA
Lipolysis	Breakdown of lipids to fatty acids and glycerol
Lipogenesis	Formation of lipids from acetyl CoA and glyceraldehyde phosphate
Proteins	
Transamination	Transfer of an amine group from an amino acid to α-ketoglutaric acid, thereby transforming α-ketoglutaric acid to glutamic acid
Oxidative deamination	Removal of an amine group from glutamic acid as ammonia and regeneration of α-ketoglutaric acid (NH_3 is converted to urea by the liver)

which glutamic acid funnels amine groups into the **urea cycle** is extremely important. This cycle rids the body not only of NH_3 produced during oxidative deamination, but also of bloodborne NH_3 produced by intestinal bacteria.

24

(3) **Keto acid modification.** The goal of amino acid degradation is to produce molecules that can be either oxidized in the Krebs cycle or converted to glucose. Keto acids resulting from transamination are altered as necessary to produce metabolites that can enter the Krebs cycle. The most important of these metabolites are pyruvic acid, acetyl CoA, α-ketoglutaric acid, and oxaloacetic acid (see Figure 24.7). Because the reactions of glycolysis are reversible, deaminated amino acids that are converted to pyruvic acid can be reconverted to glucose and contribute to gluconeogenesis.

Protein Synthesis

Amino acids are the most important anabolic nutrients. Not only do they form all protein structures, but they form the bulk of the body's functional molecules as well. As we described in Chapter 3, protein synthesis occurs on ribosomes, where ribosomal enzymes oversee the formation of peptide bonds linking the amino acids together into protein polymers. The amount and type of protein synthesized are precisely controlled by hormones (growth hormone, thyroxine, sex hormones, insulin-like growth factors, and others), and so protein anabolism reflects the hormonal balance at each stage of life.

During your lifetime, your cells will have synthesized 225–450 kg (about 500–1000 lb) of proteins, depending on your size. However, you do not need to consume anywhere near that amount of protein because nonessential amino acids are easily formed by siphoning keto acids from the Krebs cycle and transferring amine groups to them. Most of these transformations occur in the liver, which provides nearly all the nonessential amino acids needed to produce the relatively small amount of protein that the body synthesizes each day.

However, a complete set of amino acids must be present for protein synthesis to take place, so all essential amino acids must be provided by the diet. If some are not, the rest are oxidized for energy even though they may be needed for anabolism. In such cases, negative nitrogen balance results because body protein is broken down to supply the essential amino acids needed.

To review the various metabolic reactions described so far, consult the brief summary in **Table 24.4**.

CHECK YOUR UNDERSTANDING

19. What does the liver use as its substrates when it synthesizes nonessential amino acids?

20. What happens to the ammonia removed from amino acids when they are used for energy fuel?

For answers, see Appendix G.

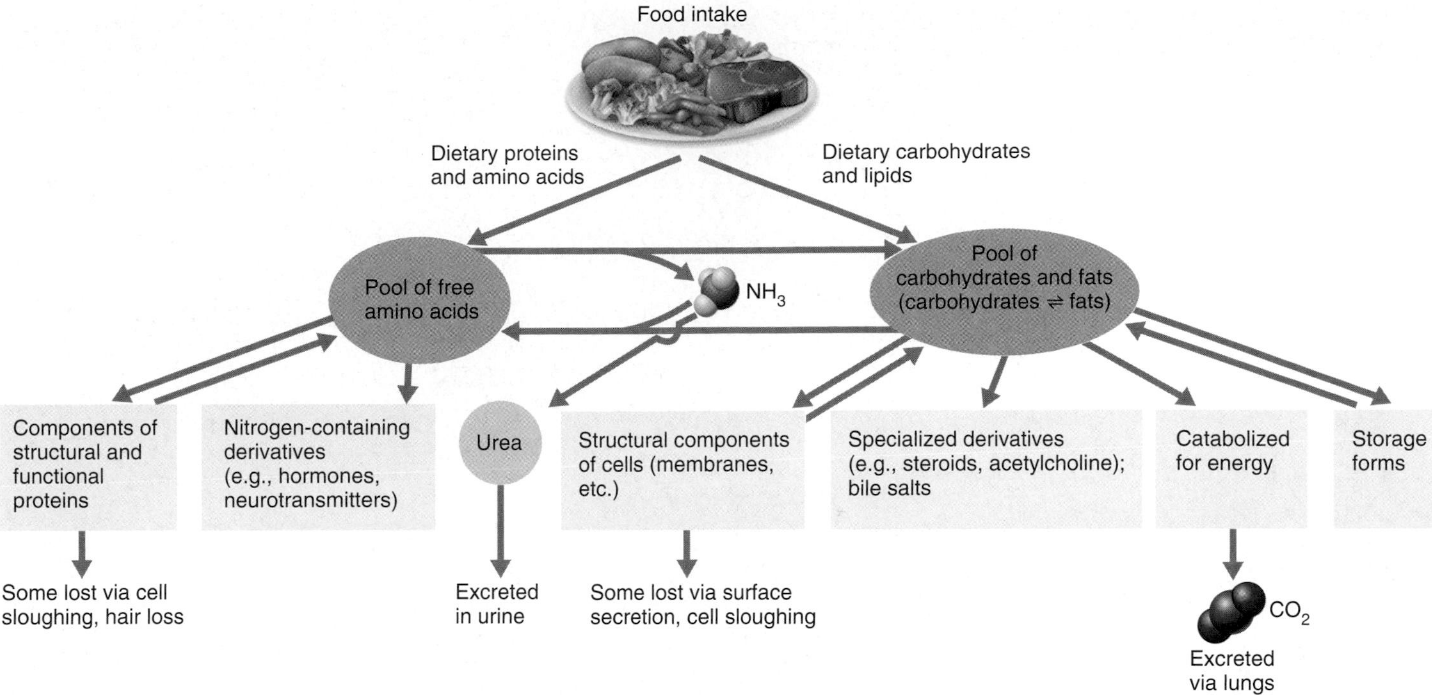

Figure 24.17 **Carbohydrate-fat and amino acid pools.**

Metabolic States of the Body

- Explain the concept of amino acid or carbohydrate-fat pools, and describe pathways by which substances in these pools can be interconverted.
- List important events of the absorptive and postabsorptive states, and explain how these events are regulated.

Catabolic-Anabolic Steady State of the Body

The body exists in a *dynamic catabolic-anabolic state* as organic molecules are continuously broken down and rebuilt—frequently at a head-spinning rate.

The blood serves as the transport pool for all body cells, and it contains many kinds of energy sources—glucose, ketone bodies, fatty acids, glycerol, and lactic acid. Some organs routinely use blood energy sources other than glucose, which saves glucose for tissues with stricter glucose requirements (**Table 24.5**).

The body can draw on its **nutrient pools**—amino acid, carbohydrate, and fat stores—to meet its varying needs **(Figure 24.17)**. These pools are interconvertible because their pathways are linked by key intermediates **(Figure 24.18)**. The liver, adipose tissue, and skeletal muscles are the primary effector organs or tissues determining the amounts and direction of the conversions shown in the figure.

The **amino acid pool** is the body's total supply of free amino acids. Small amounts of amino acids and proteins are lost daily in urine and in sloughed hairs and skin cells. Typically, these lost molecules are replaced via the diet. Otherwise, amino acids arising from tissue breakdown return to the pool. This pool is the source of amino acids used for protein synthesis and in the formation of amino acid derivatives. In addition,

24

TABLE 24.5 Profiles of the Major Body Organs in Fuel Metabolism

TISSUE	FUEL STORES	PREFERRED FUEL	FUEL SOURCES EXPORTED
Brain	None	Glucose (ketone bodies during starvation)	None
Skeletal muscle (resting)	Glycogen	Fatty acids	None
Skeletal muscle (during exertion)	None	Glucose	Lactate
Heart muscle	None	Fatty acids	None
Adipose tissue	Triglycerides	Fatty acids	Fatty acids, glycerol
Liver	Glycogen, triglycerides	Amino acids, glucose, fatty acids	Fatty acids, glucose, ketone bodies

SOURCE: Adapted from Mathews, van Holde, and Ahern, 2000, *Biochemistry* 3/e, San Francisco: Addison Wesley Longman, p. 832.

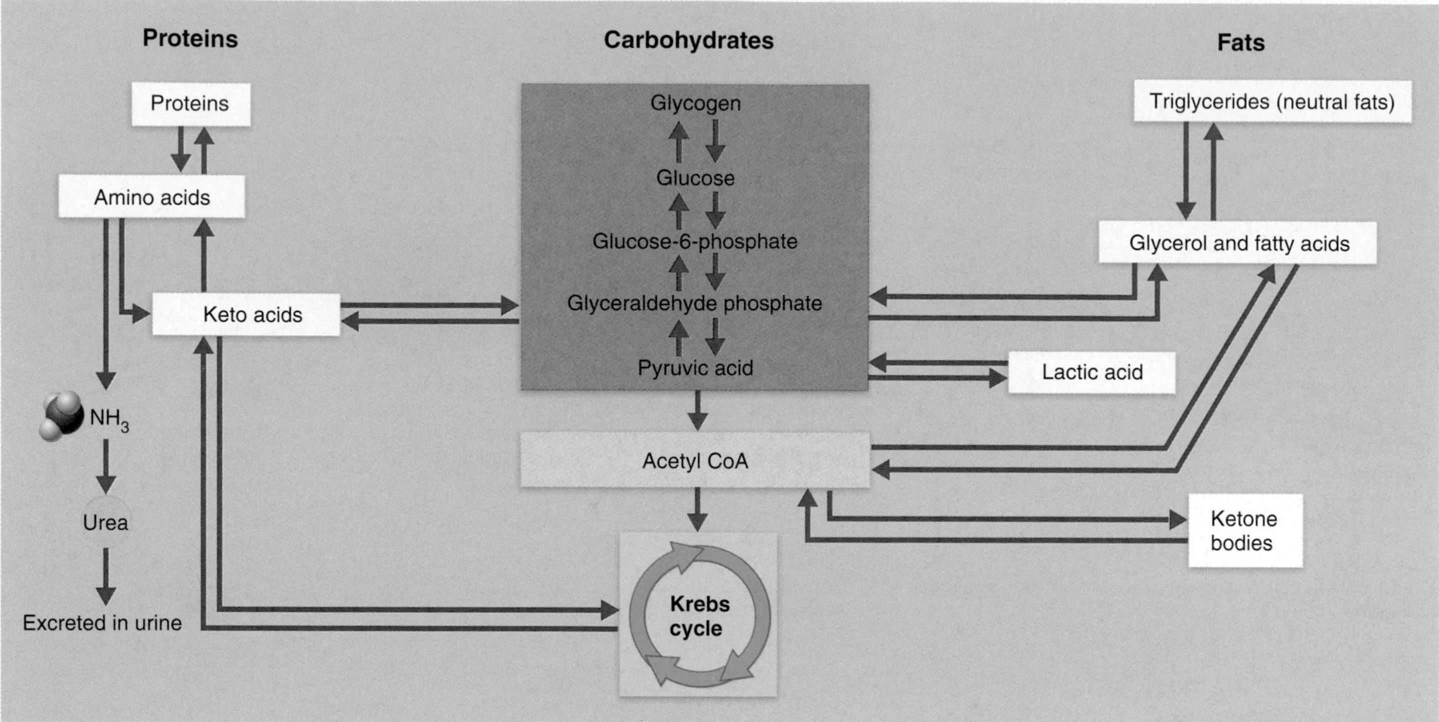

Figure 24.18 Interconversion of carbohydrates, fats, and proteins. The liver, adipose tissue, and skeletal muscles are the primary effectors determining the amounts and direction of the conversions shown.

as we described above, deaminated amino acids can participate in gluconeogenesis.

Not all events of amino acid metabolism occur in all cells. For example, *only* the liver forms urea. Nonetheless, the concept of a common amino acid pool is valid because all cells are connected by the blood.

24

Because carbohydrates are easily and frequently converted to fats, the **carbohydrate** and **fat pools** are usually considered together (Figures 24.17 and 24.18). There are two major differences between this pool and the amino acid pool: (1) Fats and carbohydrates are oxidized directly to produce cellular energy, whereas amino acids can be used to supply energy *only after being converted to a carbohydrate intermediate* (a keto acid). (2) Excess carbohydrate and fat can be stored as such, whereas excess amino acids are *not* stored as protein. Instead, they are oxidized for energy or converted to fat or glycogen for storage.

Metabolic controls act to equalize blood concentrations of energy sources between two nutritional states. Sometimes referred to as the *fed state*, the **absorptive state** is the time during and shortly after eating, when nutrients are flushing into the blood from the gastrointestinal tract. The **postabsorptive state**, or *fasting state*, is the period when the GI tract is empty and energy sources are supplied by the breakdown of body reserves.

People who eat "three squares" a day are in the absorptive state for the four hours during and after each meal and in the postabsorptive state in the late morning, late afternoon, and all night. However, postabsorptive mechanisms can sustain the body for much longer intervals if necessary—even to accommodate weeks of fasting—as long as water is taken in.

Absorptive State

During the absorptive state anabolism exceeds catabolism **(Figure 24.19)**. Glucose is the major energy fuel. Dietary amino acids and fats are used to remake degraded body protein or fat, and small amounts are oxidized to provide ATP. Excess metabolites, regardless of source, are transformed to fat if not used for anabolism. We will consider the fate and hormonal control of each nutrient group during this phase.

Carbohydrates

Absorbed monosaccharides are delivered directly to the liver, where fructose and galactose are converted to glucose. Glucose, in turn, is released to the blood or converted to glycogen and fat. Glycogen formed in the liver is stored there, but most fat synthesized there is packaged with proteins as *very low density lipoproteins* (*VLDLs*) and released to the blood to be picked up for storage by adipose tissues. Bloodborne glucose not sequestered by the liver enters body cells to be metabolized for energy, and any excess is stored in skeletal muscle cells as glycogen or in adipose cells as fat.

Triglycerides

Nearly all products of fat digestion enter the lymph in the form of chylomicrons, which are hydrolyzed to fatty acids and glycerol before they can pass through the capillary walls. *Lipoprotein lipase*, the enzyme that catalyzes fat hydrolysis, is particularly active in the capillaries of muscle and fat tissues. Adipose cells, skeletal and cardiac muscle cells, and liver cells use triglycerides

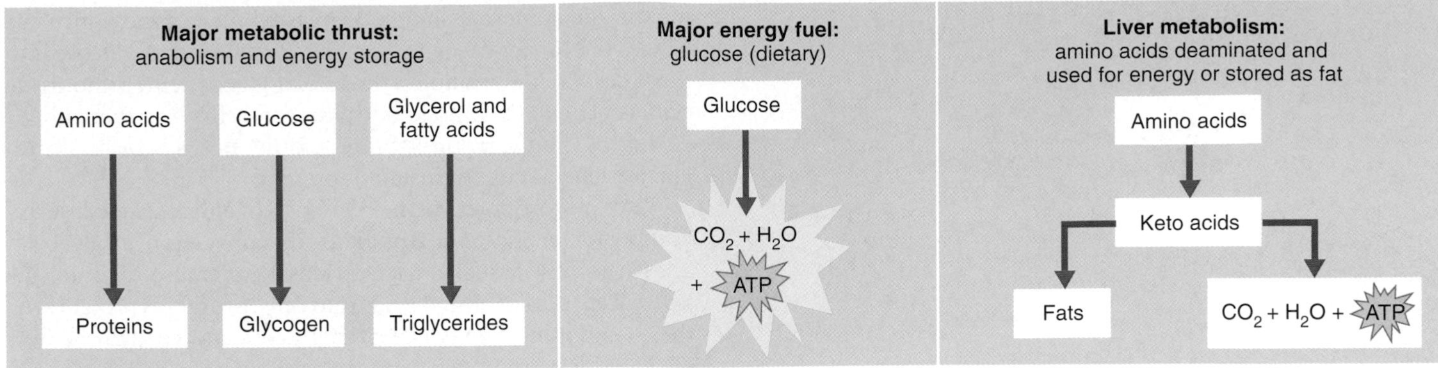

(a) Major events of the absorptive state

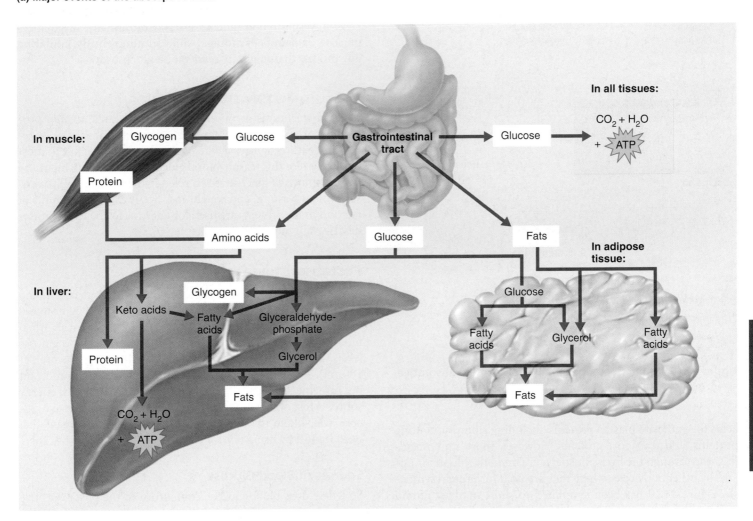

(b) Principal pathways of the absorptive state

Figure 24.19 Major events and principal metabolic pathways of the absorptive state. Although not indicated in **(b)**, amino acids are also taken up by tissue cells and used for protein synthesis, and fats (triglycerides) are the primary energy fuel of muscle, liver cells, and adipose tissue.

as their primary energy source, and when dietary carbohydrates are limited, other cells begin to oxidize more fat for energy. Although some fatty acids and glycerol are used for anabolic purposes by tissue cells, most enter adipose tissue to be reconverted to triglycerides and stored.

Amino Acids

Absorbed amino acids are delivered to the liver, which deaminates some of them to keto acids. The keto acids may flow into the Krebs cycle to be used for ATP synthesis, or they may be converted to liver fat stores. The liver also uses some of the amino

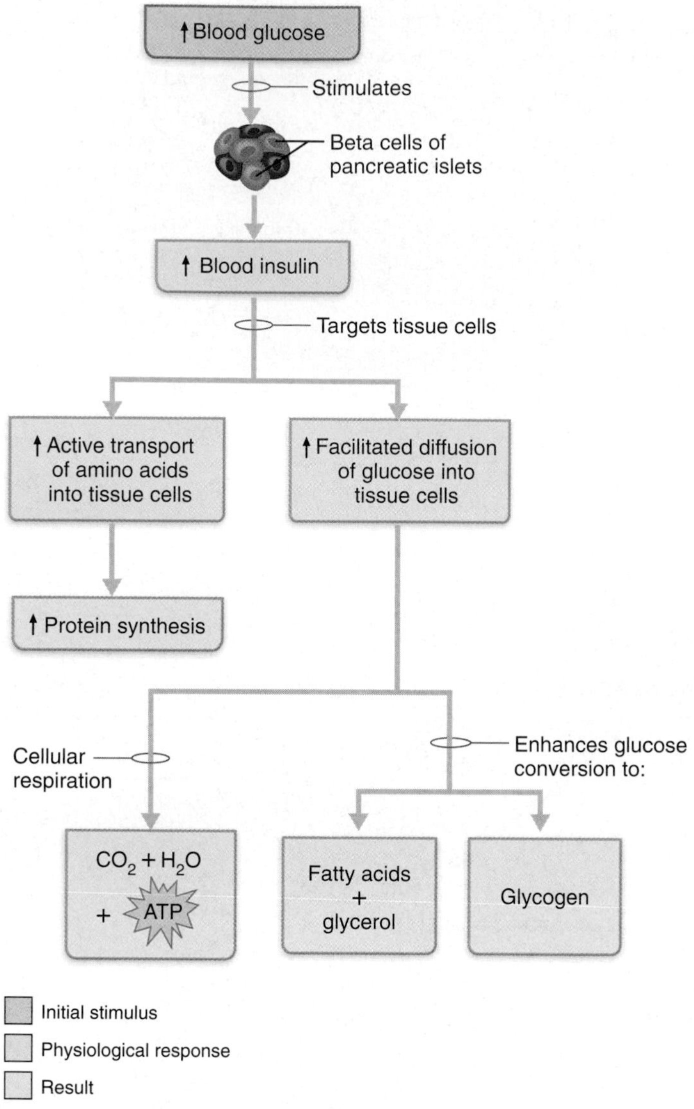

Figure 24.20 Insulin directs nearly all events of the absorptive state. (Note: Not all effects shown occur in all cells.)

acids to synthesize plasma proteins, including albumin, clotting proteins, and transport proteins. However, most amino acids flushing through the liver sinusoids remain in the blood for uptake by other body cells, where they are used for protein synthesis. Figure 24.19 has been simplified to show nonliver amino acid uptake only by muscle.

Hormonal Control

Insulin directs essentially all events of the absorptive state **(Figure 24.20)**. Rising blood glucose levels after a carbohydrate-containing meal act as a humoral stimulus that prods the beta cells of the pancreatic islets to secrete more insulin. [This glucose-induced stimulation of insulin release is enhanced by the GI tract hormone glucose-dependent insulinotropic peptide (GIP) and parasympathetic stimulation.] A second important stimulus for insulin release is elevated amino acid levels in the blood.

Insulin binding to membrane receptors of its target cells stimulates the translocation of the glucose transporter (GLUT-4 in muscle and adipose tissue) to the plasma membrane, which enhances the carrier-mediated facilitated diffusion of glucose into those cells. Within minutes, the rate of glucose entry into tissue cells (particularly muscle and adipose cells) increases about 20-fold. (The exception is brain and liver cells, which take up glucose whether or not insulin is present.)

Once glucose enters tissue cells, insulin enhances glucose oxidation for energy and stimulates its conversion to glycogen and, in adipose tissue, to triglycerides. Insulin also "revs up" the active transport of amino acids into cells, promotes protein synthesis, and inhibits liver export of glucose and virtually all liver enzymes that promote gluconeogenesis.

As you can see, insulin is a **hypoglycemic hormone** (hi″po-gli-se′mik). It sweeps glucose out of the blood into the tissue cells, lowering blood glucose levels. Additionally, it enhances glucose oxidation or storage while simultaneously inhibiting any process that might increase blood glucose levels.

HOMEOSTATIC IMBALANCE

Diabetes mellitus is a consequence of inadequate insulin production or abnormal insulin receptors. Without insulin or receptors that "recognize" it, glucose becomes unavailable to most body cells. For this reason, blood glucose levels remain high, and large amounts of glucose are excreted in urine. Metabolic acidosis, protein wasting, and weight loss occur as large amounts of fats and tissue proteins are used for energy. (Diabetes mellitus is described in more detail in Chapter 16.) ■

Postabsorptive State

In the postabsorptive state, net synthesis of fat, glycogen, and proteins ends, and catabolism of these substances begins to occur. The primary goal during this state, between meals when blood glucose levels are dropping, is to maintain blood glucose levels within the homeostatic range (70–110 mg of glucose per 100 ml). Remember that constant blood glucose is important because the brain almost always uses glucose as its energy source. Most events of the postabsorptive state either make glucose available to the blood or save glucose for the organs that need it most by using fats for energy (Figure 24.21).

Sources of Blood Glucose

So where does blood glucose come from in the postabsorptive state? Glucose can be obtained from stored glycogen, from tissue proteins, and in limited amounts from fats. Let's look at the sources, illustrated in **Figure 24.21b**, more closely.

1. **Glycogenolysis in the liver.** The liver's glycogen stores (about 100 g) are the first line of glucose reserves. They are mobilized quickly and efficiently and can maintain blood sugar levels for about four hours during the postabsorptive state.
2. **Glycogenolysis in skeletal muscle.** Glycogen stores in skeletal muscle are approximately equal to those of the liver. Before liver glycogen is exhausted, glycogenolysis begins in skeletal muscle (and to a lesser extent in other tissues). However, the glucose produced is not released to

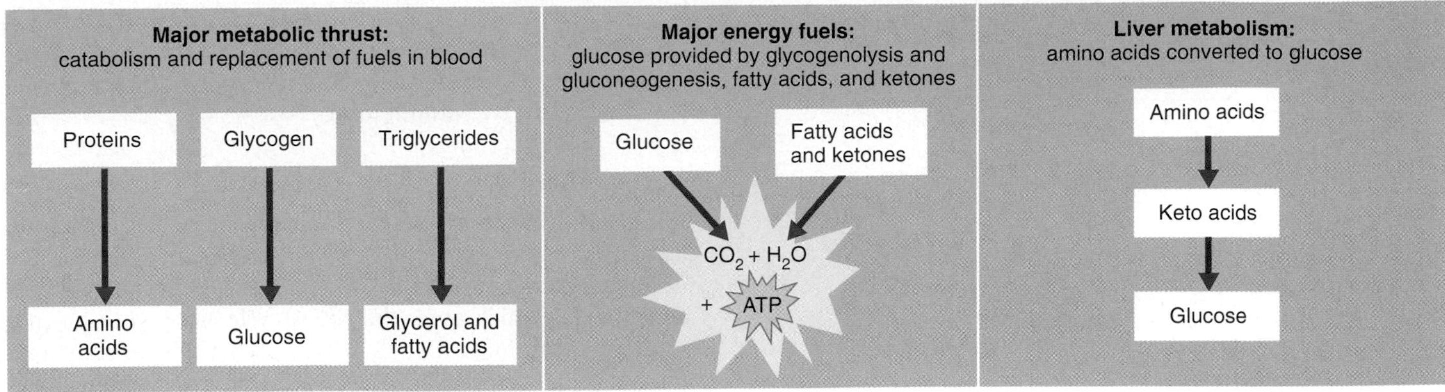

(a) Major events of the postabsorptive state

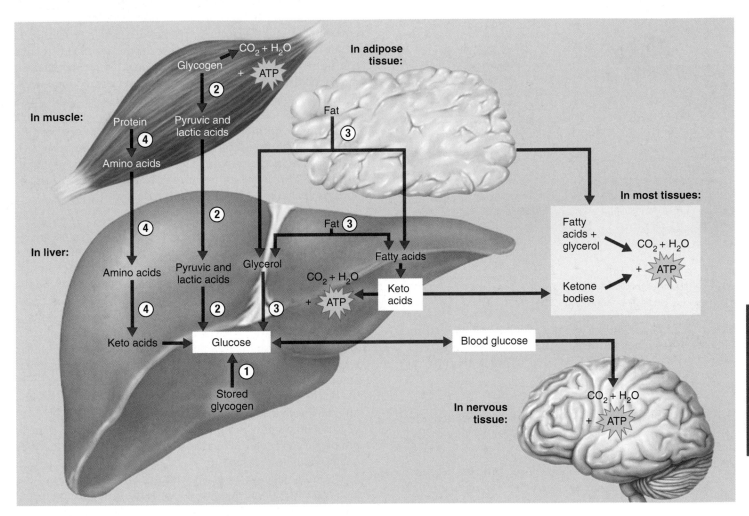

(b) Principal pathways of the postabsorptive state

Figure 24.21 Major events and principal metabolic pathways of the postabsorptive state.

the blood because, unlike the liver, skeletal muscle does not have the enzymes needed to dephosphorylate glucose. Instead, glucose is partly oxidized to pyruvic acid (or, during anaerobic conditions, lactic acid), which enters the blood, is reconverted to glucose by the liver, and is released to the blood again. Thus, skeletal muscle contributes to blood glucose homeostasis indirectly, via liver mechanisms.

③ **Lipolysis in adipose tissues and the liver.** Adipose and liver cells produce glycerol by lipolysis, and the liver converts the glycerol to glucose (gluconeogenesis) and releases it to the blood. Because acetyl CoA, a product of the beta oxidation

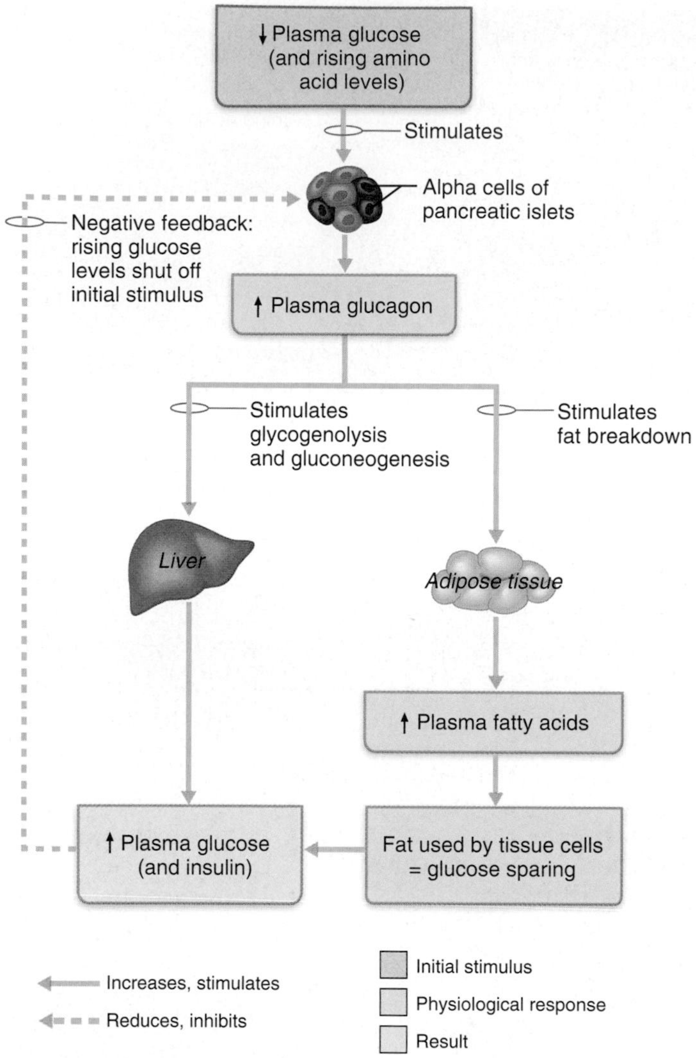

Figure 24.22 Glucagon is a hyperglycemic hormone that stimulates a rise in blood glucose levels. Negative feedback control exerted by rising plasma glucose levels on glucagon secretion is indicated by the dashed arrow.

of fatty acids, is produced beyond the *reversible* steps of glycolysis, fatty acids *cannot* be used to bolster blood glucose levels.

④ **Catabolism of cellular protein.** Tissue proteins become the major source of blood glucose when fasting is prolonged and glycogen and fat stores are nearly exhausted. Cellular amino acids (mostly from muscle) are deaminated and converted to glucose in the liver. During fasts lasting several weeks, the kidneys also carry out gluconeogenesis and contribute as much glucose to the blood as the liver.

Even during prolonged fasting or starvation, the body sets priorities. Muscle proteins are the first to go (to be catabolized). Movement is not nearly as important as maintaining wound healing and the immune response. But as long as life continues, so does the body's ability to produce the ATP needed to drive life processes.

Of course there are limits to the amount of tissue protein that can be catabolized before the body stops functioning. The heart is almost entirely muscle protein, and when it is severely catabolized, the result is death. In general, the amount of fat the body contains determines the time a person can survive without food.

Glucose Sparing

Even collectively, all of the manipulations aimed at increasing blood glucose are not enough to provide energy supplies for prolonged fasting. Luckily, the body can adapt to burn more fats and proteins, which enter the Krebs cycle along with glucose breakdown products. The increased use of noncarbohydrate fuel molecules (especially triglycerides) to conserve glucose is called **glucose sparing.**

As the body progresses from the absorptive to the postabsorptive state, the brain continues to take its share of blood glucose, but virtually every other organ switches to fatty acids as its major energy source, sparing glucose for the brain. During this transition phase, lipolysis begins in adipose tissues and released fatty acids are picked up by tissue cells and oxidized for energy. In addition, the liver oxidizes fats to ketone bodies and releases them to the blood for use by tissue cells.

If fasting continues for longer than four or five days, the brain too begins to use large quantities of ketone bodies as well as glucose as its energy fuel (Table 24.5). The brain's ability to use an alternative fuel source has survival value—much less tissue protein has to be ravaged to form glucose.

Hormonal and Neural Controls

The sympathetic nervous system and several hormones interact to control events of the postabsorptive state. Consequently, regulation of this state is much more complex than that of the absorptive state when a single hormone, insulin, holds sway.

An important trigger for initiating postabsorptive events is dampening of insulin release, which occurs as blood glucose levels drop. As insulin levels decline, all insulin-induced cellular responses are inhibited as well. Interestingly, drinking moderate amounts of beer, wine, or gin before or during a meal improves the body's use of insulin. That is, it lowers blood glucose without increasing insulin release. The advantage of this is obvious because although blood glucose naturally spikes after a meal, prolonged elevation can increase the risk of developing diabetes mellitus and heart disease.

Declining glucose levels also stimulate the alpha cells of the pancreatic islets to release the insulin antagonist **glucagon.** Like other hormones acting during the postabsorptive state, glucagon is a **hyperglycemic hormone**, in other words, it promotes a rise in blood glucose levels. Glucagon targets the liver and adipose tissue (**Figure 24.22**). The hepatocytes respond by accelerating glycogenolysis and gluconeogenesis. Adipose cells mobilize their fatty stores (lipolysis) and release fatty acids and glycerol to the blood. In this way, glucagon "refurbishes" blood energy sources by enhancing both glucose and fatty acid levels. Under certain hormonal conditions and persistent low glucose levels or prolonged fasting, most of the fat mobilized is converted to ketone bodies. Glucagon release is inhibited after the next meal or whenever blood glucose levels rise and insulin secretion begins again.

TABLE 24.6 Summary of Normal Hormonal Influences on Metabolism

HORMONE'S EFFECTS	INSULIN	GLUCAGON	EPINEPHRINE	GROWTH HORMONE	THYROXINE	CORTISOL	TESTOSTERONE
Stimulates glucose uptake by cells	✓				✓		
Stimulates amino acid uptake by cells	✓			✓			
Stimulates glucose catabolism for energy	✓				✓		
Stimulates glycogenesis	✓						
Stimulates lipogenesis and fat storage	✓						
Inhibits gluconeogenesis	✓						
Stimulates protein synthesis (anabolic)	✓			✓	✓		✓
Stimulates glycogenolysis		✓	✓				
Stimulates lipolysis and fat mobilization		✓	✓	✓	✓	✓	
Stimulates gluconeogenesis		✓	✓	✓		✓	
Stimulates protein breakdown (catabolic)						✓	

So far, the picture is pretty straightforward. Increasing blood glucose levels trigger insulin release, which "pushes" glucose out of the blood and into the cells. This drop in blood glucose stimulates secretion of glucagon, which "pulls" glucose from the cells into the blood. However, there is more than a push-pull mechanism here because *both* insulin and glucagon release are strongly stimulated by rising amino acid levels in the blood.

This effect is insignificant when we eat a balanced meal, but it has an important adaptive role when we eat a high-protein, low-carbohydrate meal. In this instance, the stimulus for insulin release is strong, and if it were not counterbalanced, the brain might be damaged by the abrupt onset of hypoglycemia as glucose is rushed out of the blood. Simultaneous release of glucagon modulates the effects of insulin and helps stabilize blood glucose levels.

The sympathetic nervous system also plays a crucial role in supplying fuel quickly when blood glucose levels drop suddenly. Adipose tissue is well supplied with sympathetic fibers, and epinephrine released by the adrenal medulla in response to sympathetic activation acts on the liver, skeletal muscle, and adipose tissues. Together, these stimuli mobilize fat and promote glycogenolysis—essentially the same effects prompted by glucagon.

Injury, anxiety, or any other stressor that mobilizes the fight-or-flight response will trigger this control pathway, as does exercise. During exercise, large amounts of fuels must be made available for muscle use, and the metabolic profile is essentially the same as that of a fasting person when glucagon and the sympathetic nervous system are in control except that the facilitated diffusion of glucose into muscle is enhanced. (The mechanism of this enhancement is not yet understood.)

In addition to glucagon and epinephrine, a number of other hormones—including growth hormone (GH), thyroxine, sex hormones, and corticosteroids—influence metabolism and nutrient flow. Growth hormone secretion is enhanced by prolonged fasting or rapid declines in blood glucose levels, and GH exerts important anti-insulin effects. For example, it reduces the ability of insulin to promote glucose uptake in fat and muscle. However, the release and activity of most of these hormones are not specifically related to absorptive or postabsorptive metabolic events. Typical metabolic effects of various hormones are summarized in **Table 24.6**.

CHECK YOUR UNDERSTANDING

21. What three organs or tissues are the primary effector organs determining the amounts and directions of interconversions in the nutrient pools?

22. Generally speaking, what kinds of reactions and events characterize the absorptive state? The postabsorptive state?

23. What hormone is glucagon's main antagonist?

24. What event increases both glucagon and insulin release?

For answers, see Appendix G.

The Metabolic Role of the Liver

- ▶ Describe several metabolic functions of the liver.
- ▶ Differentiate between LDLs and HDLs relative to their structures and major roles in the body.

TABLE 24.7 Summary of Metabolic Functions of the Liver

METABOLIC PROCESSES TARGETED	FUNCTIONS
Carbohydrate Metabolism	
Particularly important in maintaining blood glucose homeostasis	▪ Converts galactose and fructose to glucose ▪ Glucose buffer function: stores glucose as glycogen when blood glucose levels are high; in response to hormonal controls, performs glycogenolysis and releases glucose to blood ▪ Gluconeogenesis: converts amino acids and glycerol to glucose when glycogen stores are exhausted and blood glucose levels are falling ▪ Converts glucose to fats for storage
Fat Metabolism	
Although most cells are capable of some fat metabolism, liver bears the major responsibility	▪ Primary body site of beta oxidation (breakdown of fatty acids to acetyl CoA) ▪ Converts excess acetyl CoA to ketone bodies for release to tissue cells ▪ Stores fats ▪ Forms lipoproteins for transport of fatty acids, fats, and cholesterol to and from tissues ▪ Synthesizes cholesterol from acetyl CoA; catabolizes cholesterol to bile salts, which are secreted in bile
Protein Metabolism	
Other metabolic functions of the liver could be dispensed with and body could still survive; however, without liver metabolism of proteins, severe survival problems ensue: many essential clotting proteins would not be made, and ammonia would not be disposed of, for example	▪ Deaminates amino acids (required for their conversion to glucose or use for ATP synthesis); amount of deamination that occurs outside the liver is unimportant ▪ Forms urea for removal of ammonia from body; inability to perform this function (e.g., in cirrhosis or hepatitis) results in accumulation of ammonia in blood ▪ Forms most plasma proteins (exceptions are gamma globulins and some hormones and enzymes); plasma protein depletion causes rapid mitosis of the hepatocytes and actual growth of liver, which is coupled with increase in synthesis of plasma proteins until blood values are again normal ▪ Transamination: intraconversion of nonessential amino acids; amount that occurs outside liver is inconsequential
Vitamin/Mineral Storage	
	▪ Stores vitamin A (1–2 years' supply) ▪ Stores sizable amounts of vitamins D and B_{12} (1–4 months' supply) ▪ Stores iron; other than iron bound to hemoglobin, most of body's supply is stored in liver as ferritin until needed; releases iron to blood as blood levels drop
Biotransformation Functions	
	▪ Metabolizes alcohol and drugs by performing synthetic reactions yielding inactive products for excretion by the kidneys and nonsynthetic reactions that may result in products which are more active, changed in activity, or less active ▪ Processes bilirubin resulting from RBC breakdown and excretes bile pigments in bile ▪ Metabolizes bloodborne hormones to forms that can be excreted in urine

24

The liver is one of the most biochemically complex organs in the body. It processes nearly every class of nutrients and plays a major role in regulating plasma cholesterol levels. While mechanical contraptions can, in a pinch, stand in for a failed heart, lungs, or kidney, the only thing that can do the versatile liver's work is a hepatocyte.

The hepatocytes carry out some 500 or more intricate metabolic functions. A description of all of these functions is well beyond the scope of this text, but we provide a brief summary in **Table 24.7**.

Cholesterol Metabolism and Regulation of Blood Cholesterol Levels

Cholesterol, though an important dietary lipid, has received little attention in this discussion so far, primarily because it is not used as an energy source. It serves instead as the structural basis of bile salts, steroid hormones, and vitamin D and as a major component of plasma membranes. Additionally, cholesterol is part of a key signaling molecule (the *hedgehog protein*) that helps direct embryonic development.

About 15% of blood cholesterol comes from the diet. The other 85% is made from acetyl CoA by the liver, and to a lesser extent other body cells, particularly intestinal cells. Cholesterol is lost from the body when it is catabolized and secreted in bile salts, which are eventually excreted in feces.

Cholesterol Transport

Because triglycerides and cholesterol are insoluble in water, they do not circulate free in the blood. Instead, they are transported to and from tissue cells bound to small lipid-protein complexes called **lipoproteins.** These complexes solubilize the hydrophobic lipids, and the protein part of the complexes contains signals that regulate lipid entry and exit at specific target cells.

Lipoproteins vary considerably in their relative fat-protein composition, but they all contain triglycerides, phospholipids, and cholesterol in addition to protein **(Figure 24.23)**. In general, the higher the percentage of lipid in the lipoprotein, the lower its density; and the greater the proportion of protein, the higher its density. On this basis, there are **very low density lipoproteins (VLDLs)**, **low-density lipoproteins (LDLs)**, and **high-density lipoproteins (HDLs)**. *Chylomicrons*, which transport absorbed lipids from the GI tract, are a separate class and have the lowest density of all.

The liver is the primary source of VLDLs, which transport triglycerides from the liver to the peripheral tissues, mostly to *adipose tissues*. Once the triglycerides are unloaded, the VLDL residues are converted to LDLs, which are cholesterol-rich. The role of the LDLs is to transport cholesterol to *peripheral* (nonliver) *tissues*, making it available to the tissue cells for membrane or hormone synthesis and for storage for later use. The LDLs also regulate cholesterol synthesis in the tissue cells. Docking of LDL to the LDL receptor triggers receptor-mediated endocytosis of the entire particle.

The major function of HDLs, which are particularly rich in phospholipids and cholesterol, is to scoop up and transport excess cholesterol *from peripheral tissues to the liver*, where it is broken down and becomes part of bile. The liver makes the protein envelopes of the HDL particles and then ejects them into the bloodstream in collapsed form, rather like deflated beach balls. Once in the blood, these still-incomplete HDL particles fill with cholesterol picked up from the tissue cells and "pulled" from the artery walls.

HDL also provides the steroid-producing organs, like the ovaries and adrenal glands, with their raw material (cholesterol). These organs have the ability to selectively remove cholesterol from the HDL particles without engulfing them.

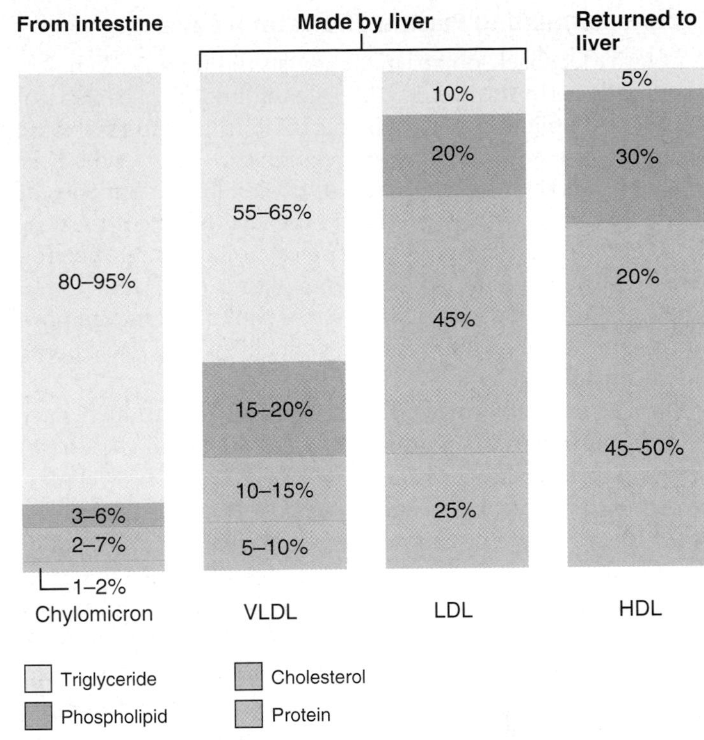

Figure 24.23 Approximate composition of lipoproteins that transport lipids in body fluids. VLDL = very low density lipoprotein, LDL = low-density lipoprotein, HDL = high-density lipoprotein.

Recommended Total Cholesterol, HDL, and LDL Levels

For adults, a total cholesterol level of 200 mg/dl of blood (or lower) is recommended. Blood cholesterol levels above 200 mg/dl have been linked to risk of atherosclerosis, which clogs the arteries and causes strokes and heart attacks. However, it is not enough to simply measure total cholesterol. The form in which cholesterol is transported in the blood is more important clinically.

As a rule, high levels of HDLs are considered *good* because the transported cholesterol is destined for degradation. In the United States, the average HDL level in males is 40–50 and in women is 50–60. Levels below 40 are considered undesirable. HDL levels above 60 are thought to protect against heart disease.

High LDL levels (130 or above) are considered *bad* because when LDLs are excessive, potentially lethal cholesterol deposits are laid down in the artery walls. The goal for LDL levels is 100 or less, but new guidelines for those at risk for cardiac disease recommend 70 mg/dl or less. A good rule of thumb is that HDL levels can't be too high and LDL levels can't be too low.

If LDLs are "bad" cholesterol, then one variety of LDL—lipoprotein (a)—is "really nasty." This lipid appears to promote plaque formation that thickens and stiffens the blood vessel walls. Additionally, lipoprotein (a) inhibits fibrinolysis, which increases the incidence of thrombus formation and can double a man's risk of a heart attack before the age of 55. It is estimated that one in five males has elevated levels of this lipoprotein in the blood.

Factors Regulating Plasma Cholesterol Levels

A negative feedback loop partially adjusts the amount of cholesterol produced by the liver according to the amount of cholesterol in the diet. A high cholesterol intake inhibits its synthesis by the liver, but it is not a one-to-one relationship because the liver produces a certain basal amount of cholesterol (about 85% of desirable values) even when dietary intake is high. For this reason, severely restricting dietary cholesterol, although helpful, does not result in a steep reduction in plasma cholesterol levels. Additionally, severely restricting dietary cholesterol may remove important nutrients found in meats, shellfish, and dairy products from a person's diet.

The relative amounts of saturated and unsaturated fatty acids in the diet have an important effect on blood cholesterol levels. Saturated fatty acids *stimulate liver synthesis* of cholesterol and *inhibit its excretion* from the body. In contrast, unsaturated (mono- and polyunsaturated) fatty acids (found in olive oil and in most vegetable oils respectively) *enhance excretion* of cholesterol and its catabolism to bile salts, thereby reducing total cholesterol levels.

The unhappy exception to this good news about unsaturated fats concerns **trans fats**, "healthy" oils that have been hardened by hydrogenation to make them more solid. Trans fats cause serum changes worse than those caused by saturated fats. The trans fatty acids spark a greater increase in LDLs and a greater reduction in HDLs, producing the unhealthiest ratio of total cholesterol to HDL.

The unsaturated omega-3 fatty acids found in especially large amounts in some cold-water fish lower the proportions of both saturated fats and cholesterol. The omega-3 fatty acids have a powerful antiarrhythmic effect on the heart and also make blood platelets less sticky, thus helping prevent spontaneous clotting that can block blood vessels. They also appear to lower blood pressure (even in nonhypertensive people). In those with moderate to high cholesterol levels, replacing half of the lipid- and cholesterol-rich animal proteins in the diet with soy protein lowers cholesterol significantly.

24

Factors other than diet also influence blood cholesterol levels. For example, cigarette smoking, coffee drinking, and stress appear to lower HDL levels, whereas regular aerobic exercise and estrogen lower LDL levels and increase HDL levels. Interestingly, body shape, that is, the distribution of body fat, provides clues to risky blood levels of cholesterol and fats. "Apples" (people with upper body and abdominal fat distribution, seen more often in men) tend to have higher levels of cholesterol in general and higher LDLs in particular than "pears" where fat is localized in the hips and thighs, a pattern more common in women.

Most cells other than liver and intestinal cells obtain the bulk of the cholesterol they need for membrane synthesis from the blood. When a cell needs cholesterol, it makes receptor proteins for LDL and inserts them in its plasma membrane. LDL binds to the receptors and is engulfed in coated pits by endocytosis. Within 15 minutes the endocytotic vesicles fuse with lysosomes, where the cholesterol is freed for use. When excessive cholesterol accumulates in a cell, it inhibits both the cell's own cholesterol synthesis and its synthesis of LDL receptors.

HOMEOSTATIC IMBALANCE

Previously, high cholesterol and LDL:HDL ratios were considered the most valid predictors of risk for atherosclerosis, cardiovascular disease, and heart attack. However, almost half of those who get heart disease have normal cholesterol levels, while others with poor lipid profiles remain free of heart disease. Presently, LDL levels and assessments of risk factors are believed to be more accurate indicators of whether treatment is needed or not, and many physicians recommend dietary changes regardless of total cholesterol or HDL levels.

Cholesterol-lowering drugs such as the *statins* (e.g., lovastatin and pravastatin) are routinely prescribed for cardiac patients with LDL levels over 130. It is estimated that more than 10 million Americans are now taking statins, and tests of a vaccine that lowers LDL while raising HDL are under way. ■

CHECK YOUR UNDERSTANDING

25. If you had your choice, would you prefer to have high blood levels of HDLs or LDLs? Explain your answer.
26. What is the maximum recommended cholesterol level for adults?
27. What are trans fats and what is their effect on LDL and HDL levels?

For answers, see Appendix G.

Energy Balance

- Explain what is meant by body energy balance.
- Describe some theories of food intake regulation.

When any fuel is burned, it consumes oxygen and liberates heat. The "burning" of food fuels by our cells is no exception. As we described in Chapter 2, energy can be neither created nor destroyed—only converted from one form to another. If we apply this principle (actually the *first law of thermodynamics*) to cell metabolism, it means that bond energy released as foods are catabolized must be precisely balanced by the total energy output of the body. For this reason, a dynamic balance exists between the body's energy intake and its energy output:

$$\text{Energy intake} = \underset{\text{(heat + work + energy storage)}}{\text{energy output}}$$

Energy intake is the energy liberated during food oxidation. (Undigested foods are not part of the equation because they contribute no energy.) **Energy output** includes energy (1) immediately lost as heat (about 60% of the total), (2) used to do work (driven by ATP), and (3) stored as fat or glycogen. Because losses of organic molecules in urine, feces, and perspiration are very small in healthy people, they are usually ignored in calculating energy output.

A close look at this situation reveals that *nearly all the energy derived from foodstuffs is eventually converted to heat.* Heat is lost during every cellular activity—when ATP bonds are formed and when they are cleaved to do work, as muscles contract, and through friction as blood flows through blood vessels. Though cells cannot use this energy to do work, the heat warms the tissues and blood and helps maintain the homeostatic body temperature that allows metabolic reactions to occur efficiently. Energy storage is an important part of the equation only during periods of growth and net fat deposit.

When energy intake and energy output are balanced, body weight remains stable. When they are not, weight is either gained or lost. Body weight in most people is surprisingly stable. This fact indicates the existence of physiological mechanisms that control food intake (and as a result, the amount of food oxidized) or heat production or both. Unhappily for many people, the body's weight-controlling systems appear to be designed more to protect us against weight loss than weight gain.

Obesity

How fat is too fat? What distinguishes a person who is obese from one that is merely overweight? Let's take a look. The bathroom scale, though helpful to determine degrees of overweight, is an inaccurate guide because body weight tells little of body composition. Dense bones and well-developed muscles can make a fit, healthy person technically overweight. Arnold Schwarzenegger, for example, has tipped the scales at a hefty 257 lb.

The official medical measure of obesity and body fatness is the **body mass index (BMI)**, an index of a person's weight relative to height. To estimate BMI, multiply weight in pounds by 705 and then divide by your height in inches squared:

$$\text{BMI} = \text{wt(lb)} \times 705/\text{ht(inches)}^2$$

Overweight is defined by a BMI between 25 and 30 and carries some health risk. Obesity is a BMI greater than 30 and has a markedly increased health risk. The most common view of obesity is that it is a condition of excessive triglyceride storage. We bewail our inability to rid ourselves of fat, but the real problem is that we keep refilling the storehoses by consuming too many calories. A body fat content of 18–20% of body weight (males and females respectively) is deemed normal for adults.

However it's defined, obesity is perplexing and poorly understood, and the economic toll of obesity-related disease is staggering. People who are obese have a higher incidence of atherosclerosis, diabetes mellitus, hypertension, heart disease, and osteoarthritis. Nonetheless, the U.S. is big and getting bigger, at least around its middle. Two out of three adults are overweight. Of that number, one out of three is obese and one in twelve has diabetes, a common sequel of weight problems. U.S. kids are getting fatter too: 20 years ago, 5% were overweight; today over 15% are and more are headed that way. Furthermore, because kids are opting for video games and nachos instead of tag or touch football and an apple, their general cardiovascular fitness is declining as well and health risks associated with excess weight start frighteningly early. For example, obesity-related heart problems, such as an enlarged heart—a risk factor for heart attack and congestive heart failure—may start developing in the early teens or even younger. Some weight-loss or weight-control methods used by people who are obese are discussed in *A Closer Look* on pp. 948–949.

Regulation of Food Intake

Control of food intake poses difficult questions to researchers. For example, what type of receptor could sense the body's total calorie content and alert one to start eating or to put down that fork? Despite heroic research efforts, no such single receptor type has been found.

It has been known for some time that the hypothalamus, particularly its *arcuate nucleus* (*ARC*) and two other areas—the *lateral hypothalamic area* (*LHA*) and the *ventromedial nucleus* (*VMN*)—release several peptides that influence feeding behavior. Most importantly, this influence ultimately reflects the activity of two distinct sets of neurons—one set that promotes hunger and the other that causes satiety. The NPY/AgRP group of ARC release neuropeptide Y (NPY) and agouti-related peptides, which collectively enhance appetite and food-seeking behavior by stimulating the second-order neurons of the LHA (**Figure 24.24**, right side). The other neuron group in ARC consists of the POMC/CART neurons, which release pro-opiomelanocortin (POMC) and cocaine- and amphetamine-regulated transcript (CART), appetite-suppressing peptides. These peptides act on the VMN, causing its neurons to release CRH (corticotropin-releasing hormone), an important appetite-suppressing peptide.

Current theories of how feeding behavior and hunger are regulated focus on several factors, most importantly neural signals from the digestive tract, bloodborne signals related to body energy stores, and hormones. To a smaller degree, body temperature and psychological factors also seem to play a role. All these factors appear to operate through feedback signals to the feeding centers of the brain. Brain receptors include thermoreceptors, chemoreceptors (for glucose, insulin, and others), and receptors that respond to a number of peptides (leptin, neuropeptide Y, and others). The hypothalamic nuclei play an essential role in regulating hunger and satiety, but brain stem areas are also involved. Sensors in peripheral locations have also been suggested, with the liver and gut itself (alimentary canal) the prime candidates. Controls of food intake come in two varieties—short term and long term.

Short-Term Regulation of Food Intake

Short-term regulation of appetite and feeding behavior involves neural signals from the GI tract, blood levels of nutrients, and GI tract hormones. For the most part, the short-term signals target hypothalamic centers via the solitary tract (and nucleus) of the brain stem (Figure 24.24).

Neural Signals from the Digestive Tract One way the brain evaluates the contents of the gut depends on vagal nerve fibers that carry on a two-way conversation between gut and brain. For example, clinical tests show that 2 kcal worth of protein produces a 30–40% larger and longer response in vagal afferents than does 2 kcal worth of glucose. Furthermore, activation of

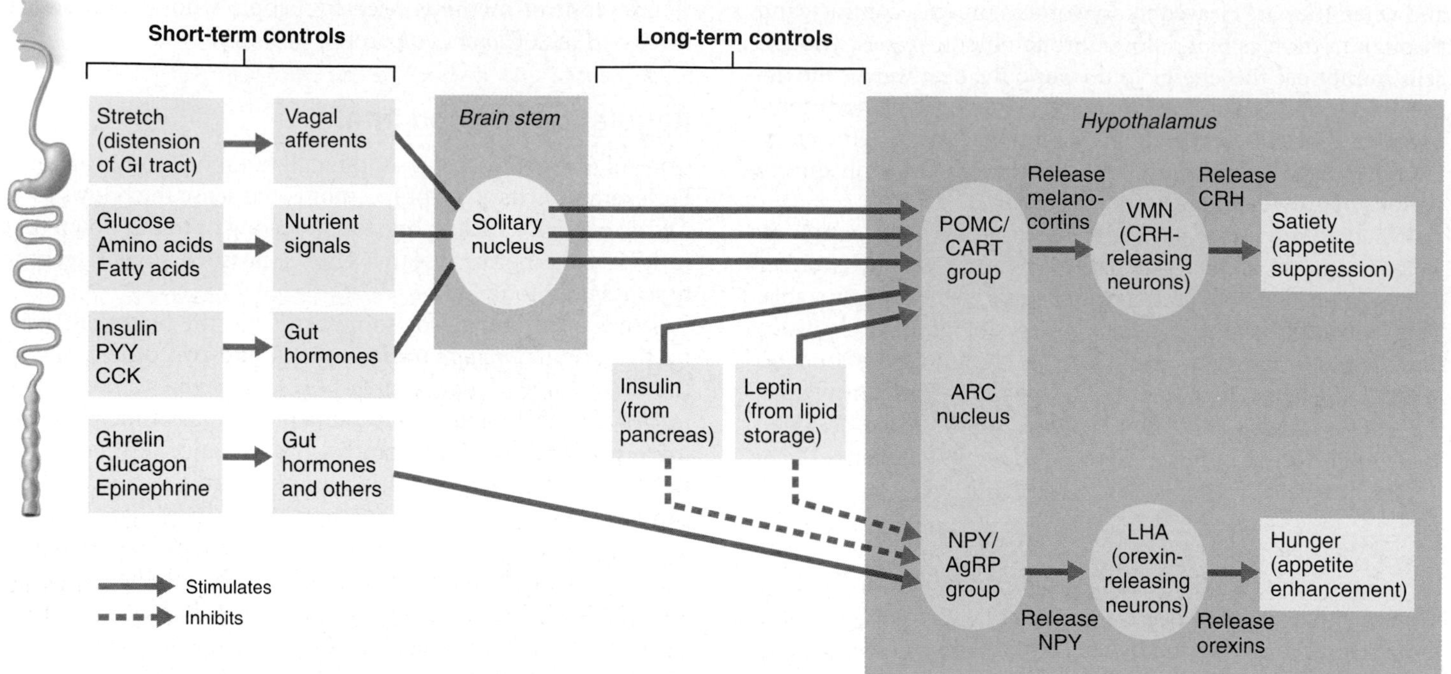

Figure 24.24 Model for hypothalamic command of appetite and food intake. The arcuate nucleus (ARC) of the brain contains two sets of neurons with opposing effects. Activation of the NPY and orexin neurons enhances appetite, whereas activation of POMC/CART neurons has the opposite effect. These centers connect with interneurons in other brain centers, and signals are transmitted through the brain stem to the body. Many appetite-regulating hormones act through the ARC, though they may also exert direct effects on other brain centers. (AgRP = agoutirelated peptide, ARC = arcuate, CART = cocaine- and amphetamine-regulated transcript, CCK = cholecystokinin, CRH = corticotropin-releasing hormone, LHA = lateral hypothalamic area, NPY = neuropeptide Y, POMC = pro-opiomelanocortin, PYY = peptide YY, VMN = ventromedial nucleus)

stretch receptors ultimately inhibits appetite, because GI tract distension sends signals along vagus nerve afferents that suppress the appetite-enhancing or hunger center. Using these signals, together with others it receives, the brain can decode what is eaten and how much.

Nutrient Signals Related to Energy Stores At any time, blood levels of glucose, amino acids, and fatty acids provide information to the brain that may help adjust energy intake to energy output. For example, nutrient signals that indicate fullness or satiety include the following:

1. *Rising blood glucose levels.* When we eat, blood glucose levels rise and subsequent activation of glucose receptors in the brain ultimately depresses eating. During fasting and hypoglycemia, this signal is absent, resulting in hunger and a "turning on" of food-seeking behavior by ultimately activating orexin-containing neurons in the LHA. Although these hypothalamic responses to the presence or absence of glucose are well known, these are not the only responses of the brain to glucose or high-caloric food. It seems that the brain's reward (pleasure) system "gives off fireworks" in the form of rising dopamine levels to a greater or lesser degree with sugar ingestion. Perhaps this response is the genesis of overeating (hedonistic) behavior.
2. *Elevated blood levels of amino acids* depress eating, but the precise mechanism mediating this effect is unknown.
3. *Blood concentrations of fatty acids* provide a mechanism for controlling hunger. The larger the amount of fatty acids in the blood, the greater the inhibition of eating behavior.

Hormones Gut hormones, including insulin and cholecystokinin (CCK) released during food absorption, act as satiety signals to depress hunger. Most important is the effect of cholecystokinin, which thwarts the appetite-inducing effect of NPY.

In contrast, glucagon and epinephrine levels rise during fasting and stimulate hunger. Ghrelin (Ghr), produced by the stomach, is a powerful appetite stimulant. In fact, ghrelin appears to be the "dinner bell" or trigger for meal initiation. Its levels peak just before mealtime, letting the brain know that it is time to eat, and then it troughs out after the meal.

Long-Term Regulation of Food Intake

A key component of the long-term controls of feeding behavior is the hormone **leptin** ("thin"). Leptin is secreted exclusively by adipose cells in response to an increase in body fat mass and it serves as an indicator of the body's total energy stores in fat tissue. (Thus adipose tissue acts as a "Fat-o-Stat" that sends chemical messages to the brain in the form of leptin.)

When leptin levels rise in the blood, it binds to receptors in ARC that specifically (1) suppress the release of NPY and

(2) stimulate the expression of CART. Neuropeptide Y is the most potent appetite stimulant known. By blocking its release, leptin prevents the release of the appetite-enhancing orexins from the LHA. This decreases appetite and subsequently food intake, eventually promoting weight loss.

When fat stores shrink, leptin blood levels drop, an event that exerts opposite effects on the two sets of ARC neurons. It activates the NPY neurons and blocks the activity of the POMC/CART neurons. Consequently, appetite and food intake increase, and (eventually) weight gain occurs.

Initially it seemed that leptin was the magic bullet that obesity researchers were looking for, but their hopes were soon dashed. Rising leptin levels do promote weight loss, but only to a certain point. Furthermore, individuals who are obese have higher-than-normal leptin blood levels, but for some unknown reason, they are resistant to its action. Now the consensus is that leptin's main role is to protect against weight loss in times of nutritional deprivation.

Although leptin has received the most attention as a long-term appetite and metabolism regulator, there are several other players. Insulin, like leptin, inhibits NPY release in non-insulin-resistant individuals, but its effect is less potent.

Additional Regulatory Factors

In actuality, there is no pat or easy answer to explain how body weight is regulated, but theories abound. Rising ambient temperature discourages food seeking, whereas cold temperature activates the hunger center. Depending on the individual, stress may increase or decrease food-seeking behavior, but chronic stress in combination with a junk food (high-fat and -sugar) diet promotes sharply increased release of NPY. Psychological factors are thought to be very important in people who are obese, but even when psychological factors contribute to obesity, individuals do not continue to gain weight endlessly. Controls still operate but at a higher weight set-point level. Certain adenovirus infections, sleep deprivation, and even the composition of gut bacteria are additional factors that may affect a person's fat mass and weight regulation, according to some clinical studies.

Whew! How does all this information fit together? So far we only have bits and pieces of the story, but the best current model based on animal studies is shown in Figure 24.24.

CHECK YOUR UNDERSTANDING

28. What three groups of stimuli influence short-term regulation of feeding behavior?

29. What is the most important long-term regulator of feeding behavior and appetite?

For answers, see Appendix G.

Metabolic Rate and Heat Production

▶ Define basal metabolic rate and total metabolic rate. Name factors that influence each.

The body's rate of energy output is called the **metabolic rate.** It is the total heat produced by all the chemical reactions and mechanical work of the body. It can be measured directly or indirectly. In the *direct method,* a person enters a chamber called a **calorimeter** and heat liberated by the body is absorbed by water circulating around the chamber. The rise in water temperature is directly related to the heat produced by the person's body. The *indirect method* uses a **respirometer** to measure oxygen consumption, which is directly proportional to heat production. For each liter of oxygen used, the body produces about 4.8 kcal of heat.

Because many factors influence metabolic rate, it is usually measured under standardized conditions. The person is in a postabsorptive state (has not eaten for at least 12 hours), is reclining, and is mentally and physically relaxed. The temperature of the room is a comfortable 20–25°C. The measurement obtained under these circumstances is called the **basal metabolic rate (BMR)** and reflects the energy the body needs to perform only its most essential activities, such as breathing and maintaining resting levels of organ function. Although named the *basal* metabolic rate, this measurement is *not* the lowest metabolic state of the body. That situation occurs during sleep, when the skeletal muscles are completely relaxed.

The BMR, often referred to as the "energy cost of living," is reported in kilocalories per square meter of body surface per hour (kcal/m^2/h). A 70-kg adult has a BMR of approximately 66 kcal/h. You can approximate your BMR by multiplying weight in kilograms (2.2 lb = 1 kg) by 1 if you are male and by 0.9 if you are female.

What factors influence BMR? There are several, including body surface area, age, gender, stress, and hormones. BMR is related to overall weight and size, but the critical factor is body surface area. As the ratio of body surface area to body volume increases, heat loss to the environment increases and the metabolic rate must be higher to replace the lost heat. For this reason, if two people weigh the same, the taller or thinner person will have the higher BMR.

In general, the younger a person, the higher the BMR. Children and adolescents require large amounts of energy for growth. In old age, BMR declines dramatically as skeletal muscles begin to atrophy. (This helps explain why those elderly who fail to reduce their caloric intake gain weight.) Gender also plays a role. Metabolic rate is disproportionately higher in males than in females, because males typically have more muscle, which is very active metabolically even during rest. Fatty tissue, present in greater relative amounts in females, is metabolically more sluggish than muscle.

Surprisingly, physical training has little effect on BMR. Although it would appear that athletes, especially those with greatly enhanced muscle mass, should have much higher BMRs than nonathletes, there is little difference between those of the same sex and surface area.

Body temperature and BMR tend to rise and fall together. Fever (hyperthermia) results in a marked increase in metabolic rate. Stress, whether physical or emotional, increases BMR by mobilizing the sympathetic nervous system. As norepinephrine

A CLOSER LOOK

Obesity: Magical Solution Wanted

Fat—unwanted, unloved, and yet often overabundant. Besides the physical toll, the social stigma and economic disadvantages of obesity are legendary. A fat person pays higher insurance premiums, is discriminated against in the job market, has fewer clothing choices, and endures frequent humiliation throughout life.

Theories of Obesity Causes

It's a fair bet that few people choose to become obese. So what causes it?

1. **Overeating during childhood promotes adult obesity.** Some believe that the "clean your plate" order sets the stage for adult obesity by increasing the number of adipose cells formed during childhood. The more adipose cells there are, the more fat can be stored.
2. **People who are obese are more fuel efficient and more effective "fat storers."** Although it is often assumed that people who are obese eat more, this is not necessarily true—many actually eat less than people of normal weight. When yo-yo dieters lose weight, their metabolic rate falls sharply. But, when they subsequently gain weight, their metabolic rate increases like a furnace being stoked. Each successive weight loss occurs more slowly, but regaining weight occurs three times as fast.

   Furthermore, fats pack more wallop per calorie (are more fattening) than proteins or carbohydrates because of the way the body processes them. If 100 excess carbohydrate calories are ingested, 23 are used in metabolic processing and 77 are stored. However, if the 100 excess calories come from fat, only 3 calories are "burned" and the rest (97) are stored.

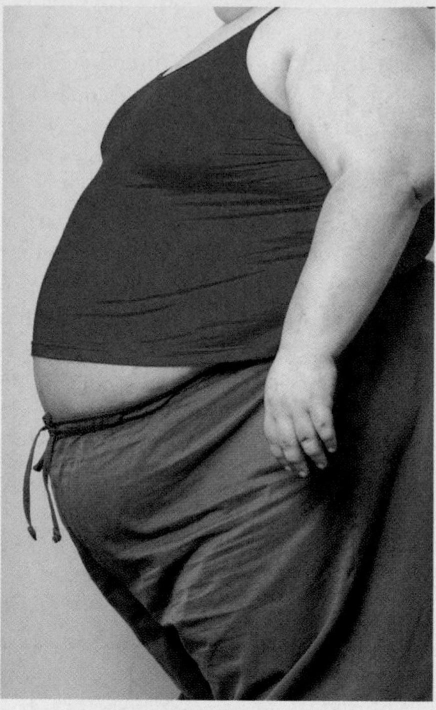

These facts apply to everyone, but when you are obese the picture is even bleaker. "Fat" fat cells of overweight people

- Sprout more alpha receptors, which favor fat accumulation.
- Send different molecular messages than "thin" fat cells. They spew out inflammatory cytokines (tumor necrosis factor and others) that can act as triggers for insulin resistance and they release less *adiponectin*, a hormone that improves the action of insulin in glucose uptake and storage.
- Have exceptionately efficient lipoprotein lipases, which unload fat from the blood (usually to fat cells).

Although it would appear that the so-called satiety chemicals (hormones and others) should prevent massive fat deposit, this seems not to be the case in people who are obese. It may be that excess weight promotes not only insulin resistance but leptin resistance as well.

3. **Genetic predisposition.** Morbid obesity is the destiny of those inheriting two obesity genes. These people, given extra calories, will always lay them down as fat, as opposed to those that lay down more muscle with some of the excess. However, a true genetic predisposition for "fatness" appears to account for only about 5% of Americans who are obese.

Treatments—The Bad and the Good

Some so-called treatments for obesity are almost more dangerous than the disease itself. Unfortunate strategies include

1. **"Water pills."** Diuretics prompt the kidneys to excrete more water and may cause a few pounds of weight loss for a few hours. They can also cause serious dehydration and electrolyte imbalance.
2. **Diets.** Diet products and books sell well, but do any of the currently popular diets work and are they safe? At the present time there is a duel between those promoting low-carbohydrate (high-protein and -fat) diets such as the Atkins and South Beach diets, and those espousing the traditional low-fat (high–complex carbohydrates) diet. Clinical studies show that people on the low-carbohydrate diets lose weight more quickly at first, but tend to plateau at 6 months. They seem to preferentially lose fat from the body trunk, a weight

and epinephrine (released by adrenal medullary cells) flood into the blood, they provoke a rise in BMR primarily by stimulating fat catabolism.

The amount of **thyroxine** produced by the thyroid gland is probably the most important hormonal factor in determining BMR. For this reason, thyroxine has been dubbed the "metabolic hormone." Its direct effect on most body cells (except brain cells) is to increase O_2 consumption and heat production, presumably by accelerating the use of ATP to operate the sodium-potassium pump. As ATP reserves decline, cellular respiration accelerates. Thus, the more thyroxine produced, the higher the BMR.

HOMEOSTATIC IMBALANCE

Hyperthyroidism causes a host of problems resulting from the excessive BMR it produces. The body catabolizes stored fats and tissue proteins, and, despite increased hunger and food intake, the person often loses weight. Bones weaken and muscles, including the heart, begin to atrophy. In contrast, *hypothyroidism* results in slowed metabolism, obesity, and diminished thought processes. ■

A CLOSER LOOK
(continued)

distribution pattern (the "apple") associated with heart disease and diabetes mellitus. (This propensity to these diseases, called "metabolic syndrome," appears to be due to the large amounts of inflammatory cytokines released by visceral fat cells.) When dieting was continued for a year, those on the low-fat diets lost just as much weight as did those on the low-carbohydrate diets. Although there was concern that the low-carbohydrate diets would promote undesirable plasma cholesterol and lipid values, for the most part this has not been the case.

Diets that have users counting the glycemic indexes of the food they eat, such as the New Glucose Revolution diet, distinguish between good carbs (whole grains, nonstarchy vegetables and fruits) and bad carbs (starches, sugary foods, refined grains) and have the approval of many physicians. The oldie-but-goodie Weight Watcher's diet, which has dieters counting points, still works and allows virtually any food choice as long as the allowed point count isn't exceeded.

Some over-the-counter liquid high-protein diets contain such poor-quality (incomplete) protein that they are actually dangerous. The worst are those that contain collagen protein instead of milk or soybean sources.

3. **Surgery.** Sometimes sheer desperation prompts surgical solutions: reducing stomach volume by banding; gastric bypass surgery, which may involve stomach stapling and intestinal bypass surgery, or the more radical biliopancreatic diversion (BPD); and liposuction (removal of fat by suctioning).

   BPD rearranges the digestive tract. Up to two-thirds of the stomach is removed and the small intestine is cut in half and one portion is sutured to the stomach opening. Pancreatic juice and bile are diverted away from the "new intestine," so far fewer nutrients and no fats are absorbed. Patients can eat anything they want without gaining weight, but exceeding the stomach capacity leads to the dumping syndrome, in which nausea and vomiting occur. Many of these procedures have been remarkably effective at promoting weight loss and restoring health. Blood pressure returns to normal in many that were originally hypertensive, and sleep apnea is reduced. Additionally, many patients with long-standing diabetes mellitus find they are suddenly *diabetes free* within weeks after the procedure. This result may prove to be the single greatest benefit of BPD.

   Liposuction reshapes the body by suctioning off fat deposits, but it is not a good choice for losing weight. It carries all the risks of surgery, and unless eating habits change, fat depots elsewhere in the body overfill.

4. **Diet drugs and weight-loss supplements.** Currently popular weight-loss drugs include phentermine, sibutramine, and orlistat. Phentermine is one-half of the former fen-phen (Redux) combination that caused heart problems, deaths, and litigation for its producer in the 1990s. It acts via increased sympathetic nervous system activity, which increases blood pressure and heart rate. Sibutramine (Meridia) has amphetamine-like effects and has come under fire because of suspected cardiovascular side effects. Orlistat (Xenical) interferes with pancreatic lipase so that part of the fat eaten is not digested or absorbed, which also interferes with the absorption of fat-soluble vitamins. It is effective as a weight-loss agent, but its side effects (diarrhea and anal leakage) are unpleasant to say the least. Currently a number of drugs are being developed that act at several different CNS sites, including neuropeptide Y inhibitors.

   Some over-the-counter weight-loss supplements that are claimed to increase metabolism and burn calories at an accelerated rate have proved to be very dangerous. For example:

   - Capsules containing usnic acid, a chemical found in some lichens, damages hepatocytes and has led to liver failure in a few cases.
   - Ephedra-containing supplements are notorious—over 100 deaths and 16,000 cases of problems including strokes, seizures, and headaches have been reported.

   When it comes to such supplements, the burden of proof is on the U.S. FDA to show that the product is unsafe. For this reason, the true extent of ills caused by weight-loss supplements is not known—while the worst cases attract attention, less serious ones go unreported or undiagnosed.

So are there any good treatments for obesity? The only realistic way to lose weight is to take in fewer calories and increase physical activity. Fidgeting helps and so does resistance exercise, which increases muscle mass. Physical exercise depresses appetite and increases metabolic rate not only during activity but also for some time after. The only way to keep the weight off is to make dietary and exercise changes lifelong habits.

The **total metabolic rate (TMR)** is the rate of kilocalorie consumption needed to fuel *all* ongoing activities—involuntary and voluntary. BMR accounts for a surprisingly large part of TMR. For example, a woman whose total energy needs per day are about 2000 kcal may spend 1400 kcal or so supporting vital body activities. Because skeletal muscles make up nearly half of body mass, skeletal muscle activity causes the most dramatic short-term changes in TMR. Even slight increases in muscular work can cause remarkable leaps in TMR and heat production. When a well-trained athlete exercises vigorously for several minutes, TMR may increase 20-fold, and it remains elevated for several hours after exercise stops.

Food ingestion also induces a rapid increase in TMR. This effect, called **food-induced thermogenesis**, is greatest when proteins are eaten. The heightened metabolic activity of the liver during such periods probably accounts for the bulk of additional energy use. In contrast, fasting or very low caloric intake depresses TMR and results in a slower breakdown of body reserves.

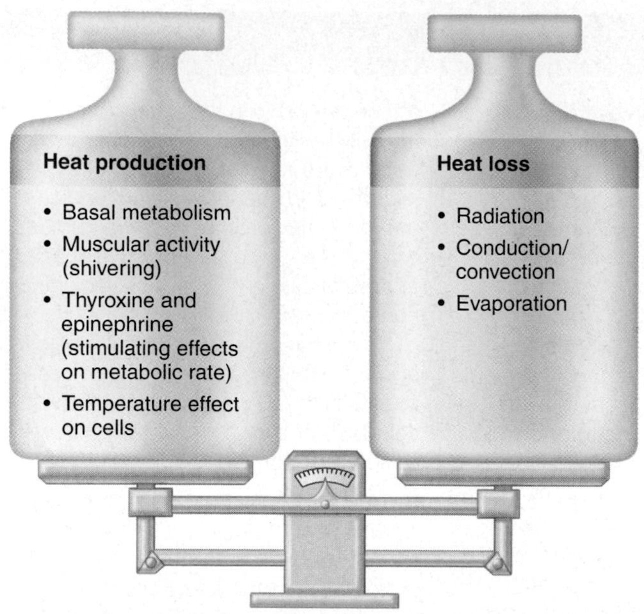

Figure 24.25 **Body temperature remains constant as long as heat production and heat loss are balanced.**

CHECK YOUR UNDERSTANDING

30. Which of the following contributes to a person's BMR? Kidney function, breathing, jogging, eating, fever.
31. Samantha is tall and slim, but athletic and well toned. Her friend, Ginger, is short and stocky, bordering on obese. Which would be expected to have a greater BMR, relatively speaking?

For answers, see Appendix G.

24

Regulation of Body Temperature

- Distinguish between core and shell body temperature.
- Describe how body temperature is regulated, and indicate the common mechanisms regulating heat production/retention and heat loss from the body.

As shown in Figure 24.25, body temperature represents the balance between heat production and heat loss. All body tissues produce heat, but those most active metabolically produce the greatest amounts. When the body is at rest, most heat is generated by the liver, heart, brain, kidneys, and endocrine organs, with the inactive skeletal muscles accounting for only 20–30%.

This situation changes dramatically with even slight alterations in muscle tone. When we are cold, shivering soon warms us up, and during vigorous exercise, heat production by the skeletal muscles can be 30 to 40 times that of the rest of the body. A change in muscle activity is one of the most important means of modifying body temperature.

Body temperature averages 37°C ± 0.5°C (98.6°F) and is usually maintained within the range 35.8–38.2°C (96–101°F), despite considerable change in external (air) temperature. A healthy individual's body temperature fluctuates approximately 1°C (1.8°F) in 24 hours, with the low occurring in early morning and the high in late afternoon or early evening.

The adaptive value of this precise temperature homeostasis becomes apparent when we consider the effect of temperature on the rate of enzymatic activity. At normal body temperature, conditions are optimal for enzymatic activity. As body temperature rises, enzymatic catalysis is accelerated: With each rise of 1°C, the rate of chemical reactions increases about 10%. As the temperature spikes above the homeostatic range, neurons are depressed and proteins begin to denature. Children below the age of 5 go into convulsions when body temperature reaches 41°C (106°F), and 43°C (about 109°F) appears to be the absolute limit for life.

In contrast, most body tissues can withstand marked reductions in temperature if other conditions are carefully controlled. This fact underlies the use of body cooling during open heart surgery when the heart must be stopped. Low body temperature reduces metabolic rate (and consequently nutrient requirements of body tissues and the heart), allowing more time for surgery without incurring tissue damage.

Core and Shell Temperatures

Different body regions have different resting temperatures. The body's **core** (organs within the skull and the thoracic and abdominal cavities) has the highest temperature and its **shell** (essentially the skin) has the lowest temperature in most circumstances. Of the two body sites used routinely to obtain body temperature clinically, the rectum typically has a temperature about 0.4°C (0.7°F) higher than the mouth and is a better indicator of core temperature.

It is core temperature that is precisely regulated. Blood serves as the major *agent of heat exchange* between the core and shell. Whenever the shell is warmer than the external environment, heat is lost from the body as warm blood is allowed to flush into the skin capillaries. On the other hand, when heat must be conserved, blood largely bypasses the skin. This reduces heat loss and allows the shell temperature to fall toward that of the environment. For this reason, the core temperature stays relatively constant, but the temperature of the shell may fluctuate substantially, for example, between 20°C (68°F) and 40°C (104°F), as it adapts to changes in body activity and external temperature. (You really *can* have cold hands and a warm heart.)

Mechanisms of Heat Exchange

Heat exchange between our skin and the external environment works in the same way as heat exchange between inanimate objects (Figure 24.26). It helps to think of an object's temperature—whether that object is the skin or a radiator—as a guide to its heat content (think "heat concentration"). Then just remember that heat always flows down its concentration gradient from a warmer region to a cooler region. The body uses four mechanisms of heat transfer—radiation, conduction, convection, and evaporation.

Radiation is the loss of heat in the form of infrared waves (thermal energy). Any object that is warmer than objects in its environment—for example, a radiator and (usually) the body—will transfer heat to those objects. Under normal conditions, close to half of your body heat loss occurs by radiation.

Because radiant energy flows from warmer to cooler, radiation explains why a cold room warms up shortly after it is filled with people (the "body heat furnace," so to speak). Your body also gains heat by radiation, as demonstrated by the warming of the skin during sunbathing.

Conduction is the transfer of heat from a warmer object to a cooler one when the two are in direct contact with each other. For example, when you step into a hot tub, some of the heat of the water is transferred to your skin, and warm buttocks transfer heat to the seat of a chair by conduction.

Convection is the process that occurs because warm air expands and rises and cool air, being denser, falls. Consequently, the warmed air enveloping the body is continually replaced by cooler air molecules. Convection substantially enhances heat transfer from the body surface to the air because the cooler air absorbs heat by conduction more rapidly than the already-warmed air. Together, conduction and convection account for 15–20% of heat loss to the environment. These processes are enhanced by anything that moves air more rapidly across the body surface, such as wind or a fan, in other words, by *forced convection.*

Evaporation is the fourth mechanism by which the body loses heat. Water evaporates because its molecules absorb heat from the environment and become energetic enough—in other words, vibrate fast enough—to escape as a gas, which we know as water vapor. The heat absorbed by water during evaporation is called **heat of vaporization**. Water absorbs a great deal of heat before vaporizing, so its evaporation from body surfaces removes large amounts of body heat. For every gram of water evaporated, about 0.58 kcal of heat is removed from the body.

There is a basal level of body heat loss due to the continuous evaporation of water from the lungs, from the oral mucosa, and through the skin. The unnoticeable water loss occurring via these routes is called **insensible water loss**, and the heat loss that accompanies it is called **insensible heat loss**. Insensible heat loss dissipates about 10% of the basal heat production of the body and is a constant not subject to body temperature controls. When necessary, however, the body's control mechanisms do initiate heat-promoting activities to counterbalance this insensible heat loss.

Evaporative heat loss becomes an active or *sensible* process when body temperature rises and sweating produces increased amounts of water for vaporization. Extreme emotional states activate the sympathetic nervous system, causing body temperature to rise one degree or so, and vigorous exercise can thrust body temperature upward as much as 2–3°C (3.6–5.4°F). During vigorous muscular activity, 1–2 L/h of perspiration can be produced and evaporated, removing 600–1200 kcal of heat from the body each hour. This is more than 30 times the amount of heat lost via insensible heat loss!

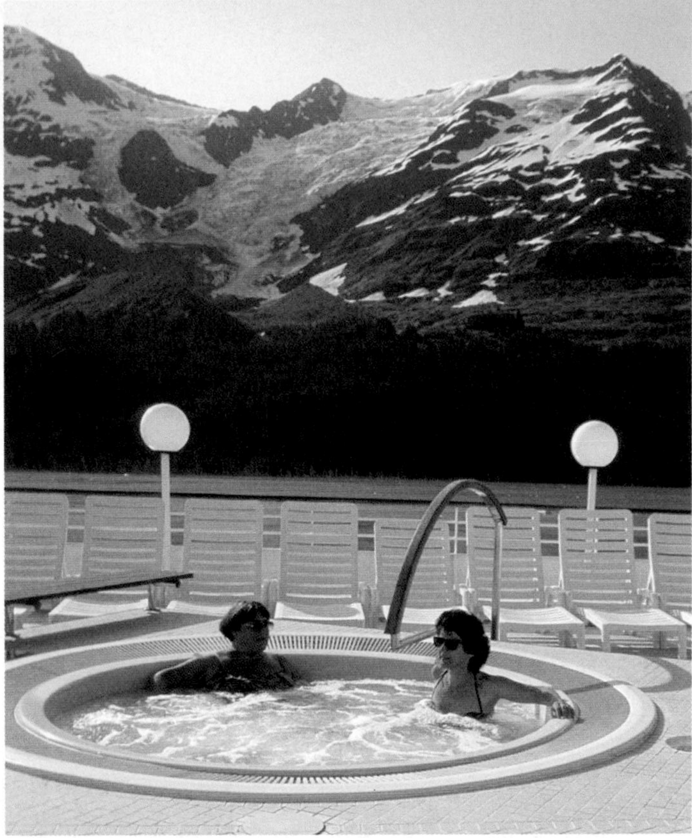

Figure 24.26 **Mechanisms of heat exchange.** Passengers enjoying a hot tub on a cruise to Alaska illustrate several mechanisms of heat exchange between the body and its environment. (1) Conduction: Heat transfers from the hot water to the skin. (2) Radiation: Heat transfers from the exposed part of the body to the cooler air. (3) Convection: Warmed air moves away from the body via breezes. There may also be (4) evaporation of perspiration: Excess body heat is carried away.

HOMEOSTATIC IMBALANCE

When sweating is heavy and prolonged, especially in untrained individuals, losses of water and NaCl may cause painful muscle spasms called *heat cramps.* This situation is easily rectified by ingesting fluids. ■

Role of the Hypothalamus

Although other brain regions contribute, the hypothalamus, particularly its *preoptic* region, is the main integrating center for thermoregulation. Together the **heat-loss center** (located more anteriorly) and the **heat-promoting center** make up the brain's **thermoregulatory centers.**

The hypothalamus receives afferent input from (1) **peripheral thermoreceptors** located in the shell (the skin), and (2) **central thermoreceptors** sensitive to blood temperature and located in the body core including the anterior portion of the hypothalamus. Much like a thermostat, the hypothalamus responds to this input by reflexively initiating appropriate heat-promoting or heat-loss activities. The central thermoreceptors

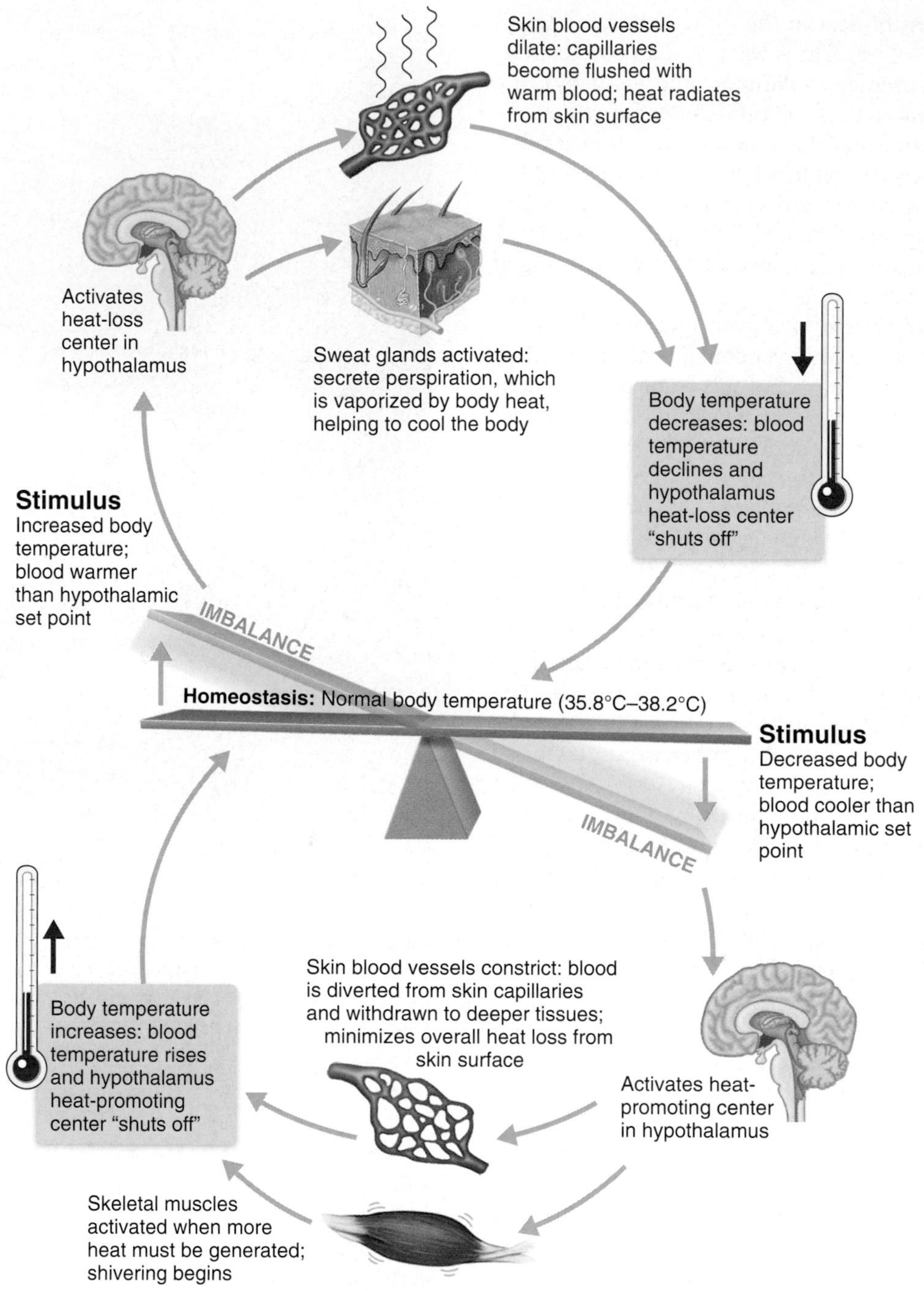

Skin blood vessels dilate: capillaries become flushed with warm blood; heat radiates from skin surface
Activates heat-loss center in hypothalamus
Sweat glands activated: secrete perspiration, which is vaporized by body heat, helping to cool the body
Body temperature decreases: blood temperature declines and hypothalamus heat-loss center "shuts off"
Stimulus
Increased body temperature; blood warmer than hypothalamic set point
IMBALANCE
Homeostasis: Normal body temperature (35.8°C–38.2°C)
Stimulus
Decreased body temperature; blood cooler than hypothalamic set point
IMBALANCE
Body temperature increases: blood temperature rises and hypothalamus heat-promoting center "shuts off"
Skin blood vessels constrict: blood is diverted from skin capillaries and withdrawn to deeper tissues; minimizes overall heat loss from skin surface
Activates heat-promoting center in hypothalamus
Skeletal muscles activated when more heat must be generated; shivering begins

Figure 24.27 Mechanisms of body temperature regulation.

24

are more critically located than the peripheral ones, but varying inputs from the shell probably alert the hypothalamus of the need to act to prevent temperature changes in the core. In other words, they allow the hypothalamus to anticipate possible changes to be made.

Heat-Promoting Mechanisms

When the external temperature is low or blood temperature falls for any reason, the heat-promoting center is activated. It triggers one or more of the following four mechanisms to maintain or increase core body temperature (**Figure 24.27**, bottom).

1. **Constriction of cutaneous blood vessels.** Activation of the sympathetic vasoconstrictor fibers serving the blood vessels of the skin causes strong vasoconstriction. As a result, blood is restricted to deep body areas and largely bypasses the skin. Because the skin is separated from deeper organs by a layer of insulating subcutaneous (fatty) tissue, heat loss from the shell is dramatically reduced and shell temperature drops toward that of the external environment.

HOMEOSTATIC IMBALANCE

Restricting blood flow to the skin is not a problem for a brief period, but if it is extended (as during prolonged exposure to very cold weather), skin cells deprived of oxygen and nutrients begin to die. This extremely serious condition is called *frostbite.* ■

2. **Shivering.** Shivering, involuntary shuddering contractions, is triggered when brain centers controlling muscle tone are activated and muscle tone reaches sufficient levels to alternately stimulate stretch receptors in antagonistic muscles. Shivering is very effective in increasing body temperature because muscle activity produces large amounts of heat.
3. **Increase in metabolic rate.** Cold stimulates the release of epinephrine and norepinephrine by the adrenal medulla in reponse to sympathetic nerve stimuli, which elevates the metabolic rate and enhances heat production. This mechanism, called **chemical**, or **nonshivering**, **thermogenesis**, is known to occur in infants but is still controversial in adults.
4. **Enhanced thyroxine release.** When environmental temperature decreases gradually, as in the transition from summer to winter, the hypothalamus of infants releases *thyrotropin-releasing hormone*. This hormone activates the anterior pituitary to release *thyroid-stimulating hormone*, which induces the thyroid to liberate larger amounts of thyroid hormone to the blood. Because thyroid hormone increases metabolic rate, body heat production increases. A similar TSH response to cold exposure does not appear to occur in adults.

Besides these involuntary adjustments, we humans make a number of *behavioral modifications* to prevent overcooling of our body core:

- Putting on more or warmer clothing to restrict heat loss (a hat, gloves, and "insulated" outer garments)
- Drinking hot fluids
- Changing posture to reduce exposed body surface area (hunching over or clasping the arms across the chest)
- Increasing physical activity to generate more heat (jumping up and down, clapping the hands)

Heat-Loss Mechanisms

Heat-loss mechanisms protect the body from excessively high temperatures. Most heat loss occurs through the skin via radiation, conduction, convection, and evaporation. How do the heat-exchange mechanisms fit into the heat-loss temperature regulation scheme? The answer is quite simple. Whenever core body temperature rises above normal, the hypothalamic heat-promoting center is inhibited. At the same time, the heat-loss center is activated and triggers one or both of the following (Figure 24.27, top):

1. **Dilation of cutaneous blood vessels.** Inhibiting the vasomotor fibers serving blood vessels of the skin allows the vessels to dilate. As the blood vessels swell with warm blood, heat is lost from the shell by radiation, conduction, and convection.
2. **Enhanced sweating.** If the body is extremely overheated or if the environment is so hot—over 33°C (about 92°F)—that heat cannot be lost by other means, evaporation becomes necessary. The sweat glands are strongly activated by sympathetic fibers and spew out large amounts of perspiration. Evaporation of perspiration is an efficient means of ridding the body of surplus heat as long as the air is dry.

However, when the relative humidity is high, evaporation occurs much more slowly. In such cases, the heat-liberating mechanisms cannot work well, and we feel miserable and irritable. Behavioral or voluntary measures commonly taken to reduce body heat in such circumstances include

- Reducing activity ("laying low")
- Seeking a cooler environment (a shady spot) or using a device to increase convection (a fan) or cooling (an air conditioner)
- Wearing light-colored, loose clothing that reflects radiant energy. (This is actually cooler than being nude because bare skin absorbs most of the radiant energy striking it.)

HOMEOSTATIC IMBALANCE

Under conditions of overexposure to a hot and humid environment, normal heat-loss processes become ineffective. The **hyperthermia** (elevated body temperature) that ensues depresses the hypothalamus. At a core temperature of around 41°C (105°F), heat-control mechanisms are suspended, creating a vicious *positive feedback cycle.* Increasing temperatures increase the metabolic rate, which increases heat production. The skin becomes hot and dry and, as the temperature continues to spiral upward, multiple organ damage becomes a distinct possibility, including brain damage. This condition, called **heat stroke**, can be fatal unless corrective measures are initiated immediately (immersing the body in cool water and administering fluids).

The terms *heat exhaustion* and *exertion-induced heat exhaustion* are often used to describe the heat-associated extreme sweating and collapse of an individual during or following vigorous physical activity. This condition, evidenced by elevated body temperature and mental confusion and/or fainting, is due to dehydration and consequent low blood pressure. In contrast to heat stroke, heat-loss mechanisms are still functional in heat exhaustion, as indicated by its symptoms. However, heat exhaustion can rapidly progress to heat stroke if the body is not cooled and rehydrated promptly.

Hypothermia (hi″po-ther′me-ah) is low body temperature resulting from prolonged uncontrolled exposure to cold. Vital signs (respiratory rate, blood pressure, and heart rate) decrease as cellular enzymes become sluggish. Drowsiness sets in and, oddly, the person becomes comfortable even though previously he or she felt extremely cold. Shivering stops at a core temperature of 30–32°C (87–90°F) when the body has exhausted its heat-generating capabilities. Uncorrected, the situation progresses to coma and finally death (by cardiac arrest), when body temperatures approach 21°C (70°F). ■

Fever

Fever is *controlled hyperthermia.* Most often, it results from infection somewhere in the body, but it may be caused by cancer, allergic reactions, or CNS injuries. Whatever the cause, macrophages and other cells release cytokines, originally called *pyrogens* (literally, "fire starters"), at least two of which are now known to be interleukins. These chemicals act on the hypothalamus, causing release of prostaglandins which reset the hypothalamic thermostat to a higher-than-normal temperature, so that heat-promoting mechanisms kick in. As a result of vasoconstriction, heat loss from the body surface declines, the skin cools, and shivering begins to generate heat. These "chills" are a sure sign body temperature is rising.

The temperature rises until it reaches the new setting, and then is maintained at that setting until natural body defenses or antibiotics reverse the disease process, and chemical messengers called *cryogens* (vasopressin and others) act to prevent fever from becoming excessive and reset the thermostat to a lower (or normal) level. Then, heat-loss mechanisms swing into action. Sweating begins and the skin becomes flushed and warm. Physicians have long recognized these signs as signals that body temperature is falling (aah, she has passed the crisis).

As we explained in Chapter 21, fever helps speed healing by increasing the metabolic rate, and it also appears to inhibit bacterial growth. The danger of fever is that if the body thermostat is set too high, proteins may be denatured and permanent brain damage may occur.

CHECK YOUR UNDERSTANDING

32. What is the body's core?

33. Cindy is flushed and her teeth are chattering even though her bedroom temperature is 72°F. Why do you think this is happening?

34. How does convection differ from conduction in causing heat loss?

For answers, see Appendix G.

24

Developmental Aspects of Nutrition and Metabolism

- Describe the effects of inadequate protein intake on the fetal nervous system.
- Describe the cause and consequences of the low metabolic rate typical of the elderly.
- List ways that medications commonly used by aged people may influence their nutrition and health.

Good nutrition is essential in utero, as well as throughout life. If the mother is ill nourished, the development of her infant is affected. Most serious is the lack of adequate calories, proteins, and vitamins needed for fetal tissue growth, especially brain growth. Additionally, inadequate nutrients during the first three years after birth will lead to mental deficits or learning disorders because brain growth continues during this time. Proteins are needed for muscle and bone growth, and calcium is required for strong bones. Although anabolic processes are less critical after growth is completed, sufficient nutrients are still essential to maintain normal tissue replacement and metabolism.

HOMEOSTATIC IMBALANCE

There are many inborn errors of metabolism (or genetic disorders), but perhaps the two most common are *cystic fibrosis* and *phenylketonuria* (*PKU*) (fen″il-ke″to-nu′re-ah). Cystic fibrosis is described in Chapter 22. In PKU, the tissue cells are unable to use the amino acid phenylalanine (fen″il-al′ah-nīn), which is present in all protein foods. The defect involves a deficiency of the enzyme that converts phenylalanine to tyrosine. Phenylalanine cannot be metabolized, so it and its deaminated products accumulate in the blood and act as neurotoxins that cause brain damage and retardation within a few months. These consequences are uncommon today because most states require a urine or blood test to identify affected newborns, and these children are put on a low-phenylalanine diet. When this special diet should be terminated is controversial. Some believe it should be continued through adolescence, but others order it discontinued when the child is 6 years old. Expectant women are sometimes encouraged to follow a controlled phenylalanine diet during their pregnancy.

There are a number of other enzyme defects, classified according to the impaired process as carbohydrate, lipid, or mineral metabolic disorders. The carbohydrate deficit *galactosemia* results from an abnormality in or lack of the liver enzymes needed to transform galactose to glucose. Galactose accumulates in the blood and leads to mental deficits.

In the carbohydrate disorder *glycogen storage disease*, glycogen synthesis is normal, but one of the enzymes needed to convert it back to glucose is missing. As excessive amounts of glycogen are stored, its storage organs (liver and skeletal muscles) become glutted with glycogen and enlarge tremendously. ■

With the exception of *insulin-dependent diabetes mellitus*, children free of genetic disorders rarely exhibit metabolic problems. However, by middle age and particularly old age, *non-insulin-dependent diabetes mellitus* becomes a major problem, particularly in people who are obese.

Metabolic rate declines throughout the life span. In old age, muscle and bone wasting and declining efficiency of the endocrine system take their toll. Because many elderly are also less active, the metabolic rate is sometimes so low that it becomes nearly impossible to obtain adequate nutrition without gaining weight. The elderly also use more medications than any other group, at a time of life when the liver has become less efficient in its detoxifying duties. Consequently, many of the agents prescribed for age-related medical problems influence nutrition. For example:

- Some diuretics prescribed for congestive heart failure or hypertension (to flush fluids out of the body) can cause severe hypokalemia by promoting excessive loss of potassium.

- Some antibiotics, for example, sulfa drugs, tetracycline, and penicillin, interfere with food digestion and absorption. They may also cause diarrhea, further decreasing nutrient absorption.
- Although its use is discouraged by physicians because it interferes with absorption of fat-soluble vitamins, mineral oil is still a popular laxative with the elderly.
- Alcohol is used by about half the elderly population in the U.S. When it is substituted for food, nutrient stores can be depleted. Excessive alcohol intake leads to absorption problems, certain vitamin and mineral deficiencies, deranged metabolism, and damage to the liver and pancreas.

In short, the elderly are at risk not only from the declining efficiency of their metabolic processes, but also from a huge variety of lifestyle and medication factors that affect their nutrition.

Although malnutrition and a waning metabolic rate present problems to some elderly, certain nutrients—notably glucose—appear to contribute to the aging process in people of all ages. Nonenzymatic reactions (the so-called *browning reactions*) between glucose and proteins, long known to discolor and toughen foods, may have the same effects on body proteins. When enzymes attach sugars to proteins, they do so at specific sites and the glycoproteins produced play well-defined roles in the body. By contrast, nonenzymatic binding of glucose to proteins (a process that increases with age) is haphazard and eventually causes cross-links between proteins. The increase in this type of binding probably contributes to lens clouding, and the general tissue stiffening and loss of elasticity so common in the aged.

CHECK YOUR UNDERSTANDING

35. List at least two reasons that the metabolic rate declines in old age.
36. List two types of drugs or over-the-counter products that can interfere with the nutrition of elderly people.

For answers, see Appendix G.

Nutrition is one of the most overlooked areas in clinical medicine. Yet, what we eat and drink influences nearly every phase of metabolism and plays a major role in our overall health. Now that we have examined the fates of nutrients in body cells, we are ready to study the urinary system, the organ system that works tirelessly to rid the body of nitrogen wastes resulting from metabolism and to maintain the purity of our internal fluids.

RELATED CLINICAL TERMS

Appetite A desire for food. A psychological phenomenon dependent on memory and associations, as opposed to hunger, which is a physiological need to eat.

Familial hypercholesterolemia (hi″per-ko-les″ter-ol-e′me-ah). An inherited condition in which the LDL receptors are absent or abnormal, the uptake of cholesterol by tissue cells is blocked, and the total concentration of cholesterol (and LDLs) in the blood is enormously elevated (e.g., 680 mg of cholesterol per 100 ml of blood). Affected individuals develop atherosclerosis at an early age, heart attacks begin in the third or fourth decade, and most die by age 60 from coronary artery disease. Treatment entails dietary modifications, exercise, and various cholesterol-reducing drugs.

Kwashiorkor (kwash″e-or′kor) Severe protein and calorie deficiency which is particularly devastating in children, resulting in mental retardation and failure to grow. A consequence of malnutrition or starvation, it is characterized by a bloated abdomen because the level of plasma proteins is inadequate to keep fluid in the bloodstream. Skin lesions and infections are likely.

Marasmus (mah-raz′mus) Protein-calorie malnutrition, accompanied by progressive wasting. Often due to ingestion of food of very poor quality. Growth for age is severely stunted.

Pica (pi′kah) Craving and eating substances not normally considered nutrients, such as clay, cornstarch, or dirt.

Skin-fold test Clinical test of body fatness. A skin fold in the back of the arm or below the scapula is measured with a caliper. A fold over 1 inch in thickness indicates excess fat. Also called the fat-fold test.

24

CHAPTER SUMMARY

Media study tools that could provide you additional help in reviewing specific key topics of Chapter 24 are referenced below.

iP = Interactive Physiology

Diet and Nutrition (pp. 911–918)

1. Nutrients include water, carbohydrates, lipids, proteins, vitamins, and minerals. The bulk of the organic nutrients is used as fuel to produce cellular energy (ATP). The energy value of foods is measured in kilocalories (kcal).
2. Essential nutrients are those that are inadequately synthesized by body cells and must be ingested in the diet.

Carbohydrates (pp. 912–913)

3. Carbohydrates are obtained primarily from plant products. Absorbed monosaccharides other than glucose are converted to glucose by the liver.
4. Monosaccharides are used primarily for cellular fuel. Small amounts are used for nucleic acid synthesis and to add sugar residues to plasma membranes.
5. Recommended carbohydrate intake for adults is 45–65% of daily caloric intake.

Lipids (pp. 912–914)

6. Most dietary lipids are triglycerides. The primary sources of saturated fats are animal products, tropical oils, and hydrogenated oils; unsaturated fats are present in plant products, nuts, and cold-water fish. The major sources of cholesterol are egg yolk, meats, and milk products.
7. Linoleic and linolenic acids are essential fatty acids.
8. Triglycerides provide reserve energy, cushion body organs, and insulate the body. Phospholipids are used to synthesize plasma membranes and myelin. Cholesterol is used in plasma membranes and is the structural basis of vitamin D, steroid hormones, and bile salts.
9. Fat intake should represent 35% or less of caloric intake, and saturated and trans fats should be replaced by mono- and polyunsaturated fats if possible.

Proteins (pp. 914–915)

10. Animal products provide high-quality complete protein containing all (10) essential amino acids. Most plant products lack one or more of the essential amino acids.
11. Amino acids are the structural building blocks of the body and of important regulatory molecules.
12. Protein synthesis can and will occur if all essential amino acids are present and sufficient carbohydrate (or fat) calories are available to produce ATP. Otherwise, amino acids will be burned for energy.
13. Nitrogen balance occurs when protein synthesis equals protein loss.
14. A dietary intake of 0.8 g of protein per kg of body weight is recommended for most healthy adults.

Vitamins (pp. 915–917)

15. Vitamins are organic compounds needed in minute amounts. Most act as coenzymes. The richest sources are whole grains, vegetables, legumes, and fruit.
16. Except for vitamin D and the K and B vitamins made by enteric bacteria, vitamins are not made in the body.
17. Water-soluble vitamins (B and C) are not stored to excess in the body. Fat-soluble vitamins include vitamins A, D, E, and K; all but vitamin K are stored and can accumulate to toxic amounts.

Minerals (pp. 917–918)

24

18. Besides calcium, phosphorus, potassium, sulfur, sodium, chloride, and magnesium, the body requires trace amounts of at least a dozen other minerals.
19. Minerals are not used for energy. Some are used to mineralize bone; others are bound to organic compounds or exist as ions in body fluids, where they play various roles in cell processes and metabolism.
20. Mineral uptake and excretion are carefully regulated to prevent mineral toxicity. The richest sources of minerals are some meats, vegetables, nuts, and legumes.

Overview of Metabolic Reactions (pp. 918–922)

1. Metabolism encompasses all chemical reactions necessary to maintain life. Metabolic processes are either anabolic or catabolic.
2. Cellular respiration refers to catabolic processes during which energy is released and some is captured in ATP bonds.
3. Energy is released when organic compounds are oxidized. Cellular oxidation is accomplished primarily by the removal of hydrogen (electrons). When molecules are oxidized, others are simultaneously reduced by accepting hydrogen (or electrons).
4. Most enzymes catalyzing oxidation-reduction reactions require coenzymes as hydrogen acceptors. Two important coenzymes in these reactions are NAD^+ and FAD.
5. In animal cells, the two mechanisms of ATP synthesis are substrate-level phosphorylation and oxidative phosphorylation.

iP Muscular System; Topic: Muscle Metabolism, pp. 3–8.

Metabolism of Major Nutrients (pp. 922–934)

Carbohydrate Metabolism (pp. 922–930)

1. Carbohydrate metabolism is essentially glucose metabolism.
2. Phosphorylation of glucose on entry into cells effectively traps it in most tissue cells.
3. Glucose is oxidized to carbon dioxide and water via three successive pathways: glycolysis, Krebs cycle, and electron transport chain. Some ATP is harvested in each pathway, but the bulk is captured in the electron transport chain.
4. Glycolysis is a reversible pathway in which glucose is converted to two pyruvic acid molecules; two molecules of reduced NAD^+ are formed, and there is a net gain of 2 ATP. Under aerobic conditions, pyruvic acid enters the Krebs cycle; under anaerobic conditions, it is reduced to lactic acid.
5. The Krebs cycle is fueled by pyruvic acid (and fatty acids). To enter the cycle, pyruvic acid is converted to acetyl CoA. The acetyl CoA is then oxidized and decarboxylated. Complete oxidation of two pyruvic acid molecules yields 6 CO_2, 8 NADH + H^+, 2 $FADH_2$, and a net gain of 2 ATP. Much of the energy originally present in the bonds of pyruvic acid is now present in the reduced coenzymes.
6. In the electron transport chain, (a) reduced coenzymes are oxidized by delivering hydrogen to a series of oxidation-reduction acceptors; (b) hydrogen is split into hydrogen ions and electrons (as electrons run downhill from acceptor to acceptor, the energy released is used to pump H^+ into the mitochondrial intermembrane space, which creates an electrochemical proton gradient); (c) the energy stored in the electrochemical proton gradient drives H^+ back through ATP synthase, which uses the energy to form ATP; (d) H^+ and electrons are combined with oxygen to form water.
7. For each glucose molecule oxidized to carbon dioxide and water, there is a net gain of 32 ATP: 4 ATP from substrate-level phosphorylation and 28 ATP from oxidative phosphorylation. The shuttle for reduced NAD^+ produced in the cytosol may use 2 ATP of that amount.
8. When cellular ATP reserves are high, glucose catabolism is inhibited and glucose is converted to glycogen (glycogenesis) or to fat (lipogenesis). Much more fat than glycogen is stored.
9. When blood glucose levels begin to fall, glycogenolysis occurs, in which glycogen stores are converted to glucose. The liver can also perform gluconeogenesis, the formation of glucose from noncarbohydrate (fat or protein) molecules.

iP Muscular System; Topic: Muscle Metabolism, pp. 10–22.

Lipid Metabolism (pp. 930–932)

10. End products of lipid digestion (and cholesterol) are transported in blood in the form of chylomicrons.
11. Glycerol is converted to glyceraldehyde-PO_4 and enters the Krebs cycle or is converted to glucose.
12. Fatty acids are oxidized by beta oxidation into acetic acid fragments. These are bound to coenzyme A and enter the Krebs cycle as acetyl CoA. Dietary fats not needed for energy or structural materials are stored in adipose tissue.
13. There is a continual turnover of fats in fat depots. Breakdown of fats to fatty acids and glycerol is called lipolysis.

14. When excessive amounts of fats are used, the liver converts acetyl CoA to ketone bodies and releases them to the blood. Excessive levels of ketone bodies (ketosis) lead to metabolic acidosis.
15. All cells use phospholipids and cholesterol to build their plasma membranes. The liver forms many functional molecules (lipoproteins, bile salts, etc.) from lipids.

Protein Metabolism (pp. 932–934)

16. To be oxidized for energy, amino acids are converted to keto acids that can enter the Krebs cycle. This involves transamination, oxidative deamination, and keto acid modification.
17. Amine groups removed during deamination (as ammonia) are combined with carbon dioxide by the liver to form urea. Urea is excreted in urine.
18. Deaminated amino acids may also be converted to fatty acids and glucose.
19. Amino acids are the body's most important building blocks. Nonessential amino acids are made in the liver by transamination.
20. In adults, most protein synthesis serves to replace tissue proteins and to maintain nitrogen balance.
21. Protein synthesis requires the presence of all 10 essential amino acids. If any are lacking, amino acids are used as energy fuels.

Metabolic States of the Body (pp. 935–941)

Catabolic-Anabolic Steady State of the Body (pp. 935–936)

1. The amino acid pool provides amino acids for synthesis of proteins and amino acid derivatives, ATP synthesis, and energy storage. To be stored, amino acids are first deaminated and then converted to fats or glycogen.
2. The carbohydrate-fat pool primarily provides fuels for ATP synthesis and other molecules that can be stored as energy reserves.
3. The nutrient pools are connected by the bloodstream; fats, carbohydrates, and proteins may be interconverted via common intermediates.

Absorptive State (pp. 936–938)

4. During the absorptive state (during and shortly after a meal), glucose is the major energy source; needed structural and functional molecules are made; excesses of carbohydrates, fats, and amino acids are stored as glycogen and fat.
5. Events of the absorptive state are controlled by insulin, which enhances the entry of glucose (and amino acids) into cells and accelerates its use for ATP synthesis or storage as glycogen or fat.

Postabsorptive State (pp. 938–941)

6. The postabsorptive state is the period when bloodborne fuels are provided by breakdown of energy reserves. Glucose is made available to the blood by glycogenolysis, lipolysis, and gluconeogenesis. Glucose sparing begins and, if fasting is prolonged (4–5 days), the brain also begins to metabolize ketone bodies.
7. Events of the postabsorptive state are controlled largely by glucagon and the sympathetic nervous system, which mobilize glycogen and fat reserves and trigger gluconeogenesis.

The Metabolic Role of the Liver (pp. 941–944)

1. The liver is the body's main metabolic organ and it plays a crucial role in processing (or storing) virtually every nutrient group. It helps maintain blood energy sources, metabolizes hormones, and detoxifies drugs and other substances.
2. The liver synthesizes cholesterol, catabolizes cholesterol and secretes it in the form of bile salts, and makes lipoproteins. The liver makes a basal amount of cholesterol (85%) even when dietary cholesterol intake is excessive.
3. LDLs transport triglycerides and cholesterol from the liver to the tissues, whereas HDLs transport cholesterol from the tissues to the liver (for catabolism and elimination).
4. Excessively high LDL levels are implicated in atherosclerosis, cardiovascular disease, and strokes.

Energy Balance (pp. 944–954)

1. Body energy intake (derived from food oxidation) is precisely balanced by energy output (heat, work, and energy storage). Eventually, all of the energy intake is converted to heat.
2. When energy balance is maintained, weight remains stable. When excess amounts of energy are stored, the result is obesity (condition of excessive fat storage resulting in a BMI above 30).

Regulation of Food Intake (pp. 945–947)

3. The hypothalamus (particularly its arcuate nucleus) and other brain centers are involved in the regulation of eating behavior.
4. Factors thought to be involved in regulating food intake include (a) neural signals from the gut to the brain, (b) nutrient signals related to total energy storage (e.g., large fatty reserves and increased plasma levels of glucose and amino acids depress hunger and eating); (c) plasma concentrations of hormones that control events of the absorptive and postabsorptive states, and hormones that provide feedback signals to brain feeding centers (leptin appears to exert the main long-term controls of appetite and energy metabolism); (d) body temperature; psychological factors; and others.

Metabolic Rate and Heat Production (pp. 947–950)

5. Energy used by the body per hour is the metabolic rate.
6. Basal metabolic rate (BMR), reported in kcal/m^2/h, is the measurement obtained under basal conditions: The person is at comfortable room temperature, supine, relaxed, and in the postabsorptive state. BMR indicates energy needed to drive only the resting body processes.
7. Factors influencing metabolic rate include age, sex, size, body surface area, thyroxine levels, specific dynamic action of foods, and muscular activity.

Regulation of Body Temperature (pp. 950–954)

8. Body temperature reflects the balance between heat production and heat loss and is normally 37°C (± 0.5°C), which is optimal for physiological activities.
9. At rest, most body heat is produced by the liver, heart, brain, kidneys, and endocrine organs. Activation of skeletal muscles causes dramatic increases in body heat production.
10. The body core (organs within the skull and the ventral body cavity) generally has the highest temperature. The shell (the skin) is the heat-exchange surface, and is usually coolest.
11. Blood serves as the major heat-exchange agent between the core and the shell. When skin capillaries are flushed with blood and the skin is warmer than the environment, heat is lost from the body. When blood is withdrawn to deep organs, heat loss from the shell is inhibited.
12. Heat-exchange mechanisms include radiation, conduction, convection, and evaporation. Evaporation, the conversion of water to water vapor, requires the absorption of heat. For each gram of water vaporized, about 0.6 kcal of heat is absorbed.
13. The hypothalamus acts as the body's thermostat. Its heat-promotion and heat-loss centers receive inputs from peripheral

and central thermoreceptors, integrate these inputs, and initiate responses leading to heat loss or heat promotion.

14. Heat-promoting mechanisms include constriction of skin vasculature, increase in metabolic rate in infants (via release of norepinephrine), and shivering. If environmental cold is prolonged, the thyroid gland may be stimulated to release more thyroxine.
15. When heat must be removed from the body, dermal blood vessels are dilated, allowing heat loss through radiation, conduction, and convection. When greater heat loss is mandated (or the environmental temperature is so high that radiation and conduction are ineffective), sweating is initiated. Evaporation of perspiration is an efficient means of heat loss as long as the humidity is low.
16. Profuse sweating can lead to heat exhaustion, indicated by a rise in temperature, a drop in blood pressure, and collapse. When the body cannot rid itself of surplus heat, body temperature rises to the point where all thermoregulatory mechanisms become ineffective—a potentially lethal condition called heat stroke.
17. Fever is controlled hyperthermia, which follows thermostat resetting to higher levels by prostaglandins and initiation of heat-promotion mechanisms, evidenced by the chills. When the disease process is reversed, heat-loss mechanisms are initiated.

Developmental Aspects of Nutrition and Metabolism (pp. 954–955)

1. Good nutrition is essential for normal fetal development and normal growth during childhood.
2. Inborn errors of metabolism include cystic fibrosis, PKU, glycogen storage disease, galactosemia, and many others. Hormonal disorders, such as lack of insulin or thyroid hormone, may also lead to metabolic abnormalities. Diabetes mellitus is the most significant metabolic disorder in middle-aged and older adults.
3. In old age, metabolic rate declines as enzyme and endocrine systems become less efficient and skeletal muscles atrophy. Reduced caloric needs make it difficult to obtain adequate nutrition without becoming overweight.
4. Elderly individuals ingest more medications than any other age group, and many of these drugs negatively affect their nutrition.

REVIEW QUESTIONS

Multiple Choice/Matching

(Some questions have more than one correct answer. Select the best answer or answers from the choices given.)

1. Which of the following reactions would liberate the most energy? **(a)** complete oxidation of a molecule of sucrose to CO_2 and water, **(b)** conversion of a molecule of ADP to ATP, **(c)** respiration of a molecule of glucose to lactic acid, **(d)** conversion of a molecule of glucose to carbon dioxide and water.
2. The formation of glucose from glycogen is **(a)** gluconeogenesis, **(b)** glycogenesis, **(c)** glycogenolysis, **(d)** glycolysis.
3. The net gain of ATP from the complete metabolism (aerobic) of glucose is closest to **(a)** 2, **(b)** 32, **(c)** 3, **(d)** 4.
4. Which of the following *best* defines cellular respiration? **(a)** intake of carbon dioxide and output of oxygen by cells, **(b)** excretion of waste products, **(c)** inhalation of oxygen and exhalation of carbon dioxide, **(d)** oxidation of substances by which energy is released in usable form to the cells.
5. What is formed during aerobic respiration when electrons are passed down the electron transport chain? **(a)** oxygen, **(b)** water, **(c)** glucose, **(d)** $NADH + H^+$.
6. Metabolic rate is relatively low in **(a)** youth, **(b)** physical exercise, **(c)** old age, **(d)** fever.
7. In a temperate climate under ordinary conditions, the greatest loss of body heat occurs through **(a)** radiation, **(b)** conduction, **(c)** evaporation, **(d)** none of the above.
8. Which of the following is *not* a function of the liver? **(a)** glycogenolysis and gluconeogenesis, **(b)** synthesis of cholesterol, **(c)** detoxification of alcohol and drugs, **(d)** synthesis of glucagon, **(e)** deamination of amino acids.
9. Amino acids are essential (and important) to the body for all the following *except* **(a)** production of some hormones, **(b)** production of antibodies, **(c)** formation of most structural materials, **(d)** as a source of quick energy.
10. A person has been on a hunger strike for seven days. Compared to normal, he has **(a)** increased release of fatty acids from adipose tissue, and ketosis, **(b)** elevated glucose concentration in the blood, **(c)** increased plasma insulin concentration, **(d)** increased glycogen synthase (enzyme) activity in the liver.
11. Transamination is a chemical process by which **(a)** protein is synthesized, **(b)** an amine group is transferred from an amino acid to a keto acid, **(c)** an amine group is cleaved from the amino acid, **(d)** amino acids are broken down for energy.
12. Three days after removal of the pancreas from an animal, the researcher finds a persistent increase in **(a)** acetoacetic acid concentration in the blood, **(b)** urine volume, **(c)** blood glucose, **(d)** all of the above.
13. Hunger, appetite, obesity, and physical activity are interrelated. Thus, **(a)** hunger sensations arise *primarily* from the stimulation of receptors in the stomach and intestinal tract in response to the absence of food in these organs; **(b)** obesity, in most cases, is a result of the abnormally high enzymatic activity of the fat-synthesizing enzymes in adipose tissue; **(c)** in all cases of obesity, the energy content of the ingested food has exceeded the energy expenditure of the body; **(d)** in a normal individual, increasing blood glucose concentration increases hunger sensations.
14. Body temperature regulation is **(a)** influenced by temperature receptors in the skin, **(b)** influenced by the temperature of the blood perfusing the heat regulation centers of the brain, **(c)** subject to both neural and hormonal control, **(d)** all of the above.
15. Which of the following yields the greatest caloric value per gram? **(a)** fats, **(b)** proteins, **(c)** carbohydrates, **(d)** all are equal in caloric value.

Short Answer Essay Questions

16. What is cellular respiration? What is the common role of FAD and NAD^+ in cellular respiration?
17. Describe the site, major events, and outcomes of glycolysis.
18. Pyruvic acid is a product of glycolysis, but it is not the substance that joins with the pickup molecule to enter the Krebs cycle. What is that substance?
19. Define glycogenesis, glycogenolysis, gluconeogenesis, and lipogenesis. Which is (are) likely to be occurring **(a)** shortly after a carbohydrate-rich meal, **(b)** just before waking up in the morning?
20. What is the harmful result when excessive amounts of fats are burned for energy? Name two conditions that might lead to this result.

21. Make a flowchart that indicates the pivotal intermediates through which glucose can be converted to fat.
22. Distinguish between the role of HDLs and that of LDLs.
23. List some factors that influence plasma cholesterol levels. Also list the sources and fates of cholesterol in the body.
24. What is meant by "body energy balance," and what happens if the balance is not precise?
25. Explain the effect of the following on metabolic rate: thyroxine levels, eating, body surface area, muscular exercise, emotional stress, starvation.
26. Explain the terms "core" and "shell" relative to body temperature balance. What serves as the heat-transfer agent from one to the other?
27. Compare and contrast mechanisms of heat loss with mechanisms of heat promotion, and explain how these mechanisms determine body temperature.

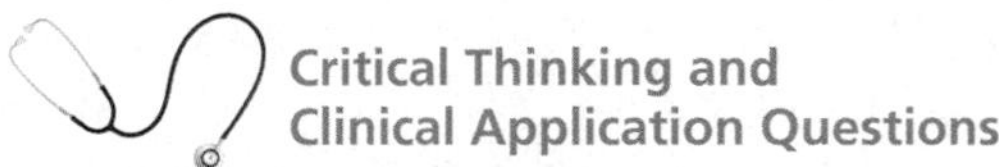

Critical Thinking and Clinical Application Questions

1. Calculate the number of ATP molecules that can be harvested during the complete oxidation of an 18-carbon fatty acid. (Take a deep breath and think about it—you *can* do it.)
2. Every year dozens of elderly people are found dead in their unheated apartments and listed as victims of hypothermia. What is hypothermia, and how does it kill? Why are the elderly more susceptible to hypothermia than the young?
3. Frank Moro has been diagnosed as having severe atherosclerosis and high blood cholesterol levels. He is told that he is at risk for a stroke or a heart attack. First, what foods would you suggest that he avoid like the plague? What foods would you suggest he add or substitute? What activities would you recommend?
4. In the 1940s, some physicians prescribed low doses of a chemical called dinitrophenol (DNP) to help patients lose weight. This drug therapy was abandoned after a few patients died. DNP uncouples the chemiosmotic machinery. Explain how this causes weight loss.
5. While attempting to sail solo from Los Angeles to Tahiti, Simon encountered a storm that marooned him on an uninhabited island. He was able, using his ingenuity and a pocket knife, to obtain plenty of fish to eat, and roots were plentiful. However, the island was barren of fruits and soon his gums began to bleed and he started to develop several infections. Analyze his problem.
6. Gregor, a large beefy man, came home from the doctor's office and complained to his wife that his blood tests "were bad." He told her that the doctor said he would have to give up some of his steaks and butter. He went on to mourn the fact that he would have to start eating more cottage cheese and olive oil instead. What kind of problem was revealed by his "bad" blood tests? How should his wife respond to his choice of food substitutes and why? What foods should she suggest?

25 The Urinary System

Every day the kidneys filter nearly 200 liters of fluid from the bloodstream, allowing toxins, metabolic wastes, and excess ions to leave the body in urine while returning needed substances to the blood. Much like a water purification plant that keeps a city's water drinkable and disposes of its wastes, the kidneys are usually unappreciated until they malfunction and body fluids become contaminated. Although the lungs and skin also participate in excretion, the kidneys are the major excretory organs.

As the kidneys perform these excretory functions, they also act as essential regulators of the volume and chemical makeup of the blood, maintaining the proper balance between water and salts and

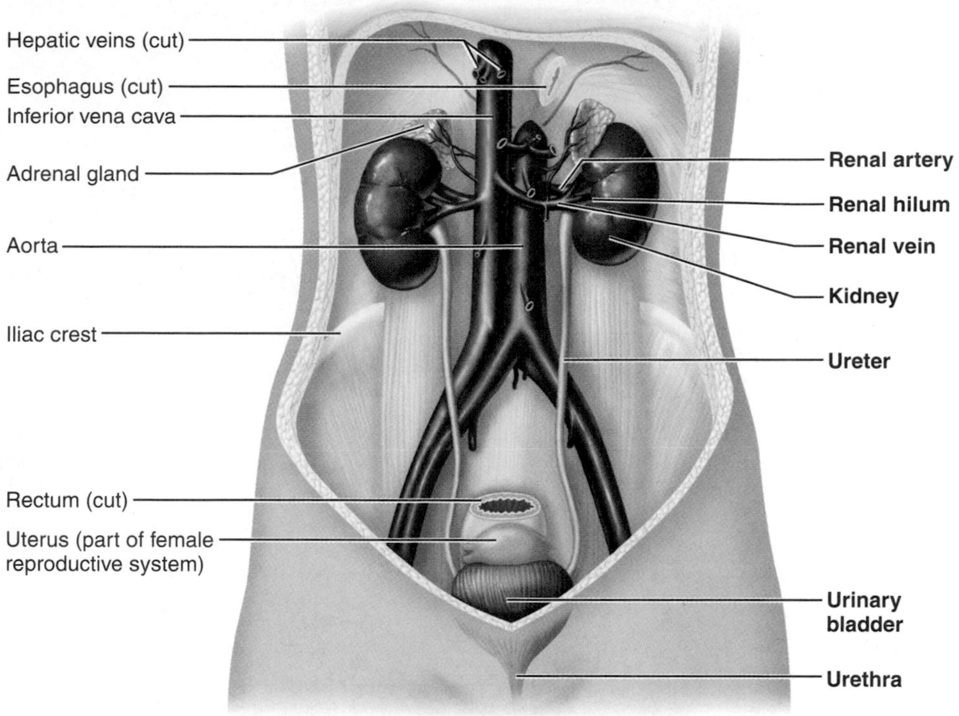

Figure 25.1 The urinary system. Anterior view of the female urinary organs. (Most unrelated abdominal organs have been omitted.)

between acids and bases. Frankly, this would be tricky work for a chemical engineer, but the kidneys do it efficiently most of the time.

Other renal functions include

- Gluconeogenesis during prolonged fasting (see p. 929).
- Producing the hormones renin and erythropoietin (see Chapter 16). *Renin* (re′nin; *ren* = kidney) acts as an enzyme to help regulate blood pressure and kidney function. *Erythropoietin* (ĕ-rith″ro-poi′ĕ-tin) stimulates red blood cell production.
- Metabolizing vitamin D to its active form (see Chapter 16).

Besides the urine-forming kidneys, the **urinary system** includes the *urinary bladder*, a temporary storage reservoir for urine, plus three tubelike organs—the paired *ureters* (u-re′terz) and the *urethra* (u-re′thrah), all three of which furnish transportation channels for urine **(Figure 25.1)**.

Kidney Anatomy

▶ Describe the gross anatomy of the kidney and its coverings.

Location and External Anatomy

The bean-shaped kidneys lie in a retroperitoneal position (between the dorsal body wall and the parietal peritoneum) in the *superior* lumbar region **(Figure 25.2)**. Extending approximately from T_{12} to L_3, the kidneys receive some protection from the lower part of the rib cage (Figure 25.2b). The right kidney is crowded by the liver and lies slightly lower than the left. An adult's kidney has a mass of about 150 g (5 ounces) and its average dimensions are 12 cm long, 6 cm wide, and 3 cm thick—about the size of a large bar of soap. The lateral surface is convex. The medial surface is concave and has a vertical cleft called the **renal hilum** that leads into an internal space within the kidney called the *renal sinus.* The ureter, renal blood vessels, lymphatics, and nerves all join each kidney at the hilum and occupy the sinus. Atop each kidney is an *adrenal* (or *suprarenal*) *gland*, an endocrine gland that is functionally unrelated to the kidney.

Three layers of supportive tissue surround each kidney (Figure 25.2a):

1. The **renal fascia**, an outer layer of dense fibrous connective tissue that anchors the kidney and the adrenal gland to surrounding structures
2. The **perirenal fat capsule**, a fatty mass that surrounds the kidney and cushions it against blows
3. The **fibrous capsule**, a transparent capsule that prevents infections in surrounding regions from spreading to the kidney

25

HOMEOSTATIC IMBALANCE

The fatty encasement of the kidneys is important in holding the kidneys in their normal body position. If the amount of fatty tissue dwindles (as with extreme emaciation or rapid weight loss), one or both kidneys may drop to a lower position, an event called *renal ptosis* (to′sis; "a fall"). Renal ptosis may cause a ureter to become kinked, which creates problems because the urine, unable to drain, backs up into the kidney and exerts

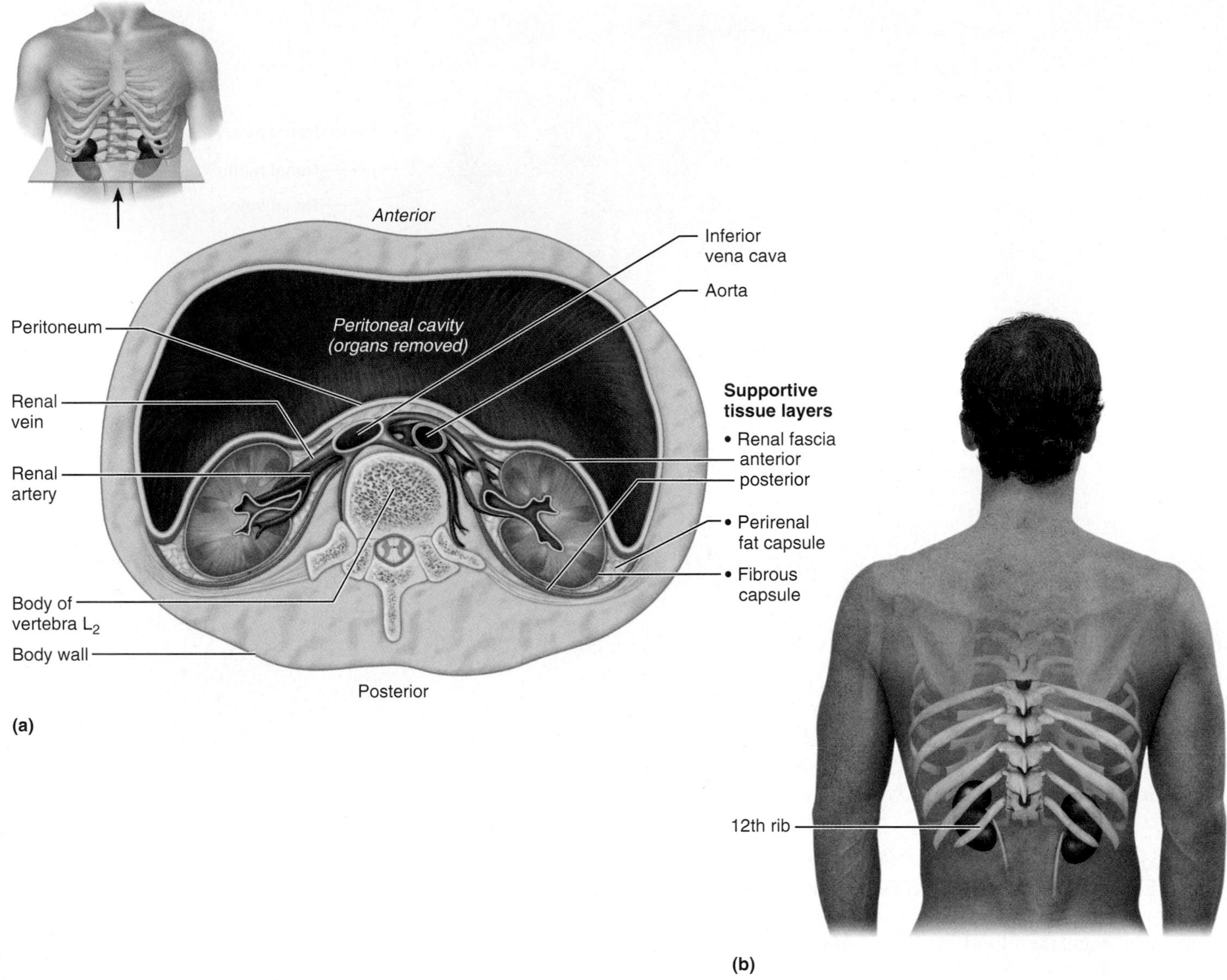

25

Figure 25.2 Position of the kidneys against the posterior body wall. (a) Cross-section viewed from inferior direction. Note the retroperitoneal position and the supportive tissue layers of the kidney. **(b)** Posterior in situ view showing relationship of the kidneys to the 12th rib pair.

pressure on its tissue. Backup of urine from ureteral obstruction or other causes is called *hydronephrosis* (hi″dro-nĕ-fro′sis; "water in the kidney"). Hydronephrosis can severely damage the kidney, leading to necrosis (tissue death) and renal failure. ■

Internal Anatomy

A frontal section through a kidney reveals three distinct regions: *cortex*, *medulla*, and *pelvis* **(Figure 25.3)**. The most superficial region, the **renal cortex**, is light in color and has a granular appearance. Deep to the cortex is the darker, reddish-brown **renal medulla**, which exhibits cone-shaped tissue masses called **medullary** or **renal pyramids**. The broad *base* of each pyramid faces toward the cortex, and its apex, or *papilla* ("nipple"), points internally. The pyramids appear striped because they are formed almost entirely of parallel bundles of microscopic urine-collecting tubules and capillaries. The **renal columns**, inward extensions of cortical tissue, separate the pyramids. Each pyramid and its surrounding cortical tissue constitutes one of approximately eight **lobes** of a kidney.

The **renal pelvis**, a funnel-shaped tube, is continuous with the ureter leaving the hilum. Branching extensions of the pelvis form two or three **major calyces** (ka′lih-sĕz; singular: calyx). Each one subdivides to form several **minor calyces**, cup-shaped areas that enclose the papillae.

The calyces collect urine, which drains continuously from the papillae, and empty it into the renal pelvis. The urine then flows through the renal pelvis and into the ureter, which moves it to the bladder to be stored. The walls of the calyces, pelvis, and ureter contain smooth muscle that contracts rhythmically to propel urine along its course by peristalsis.

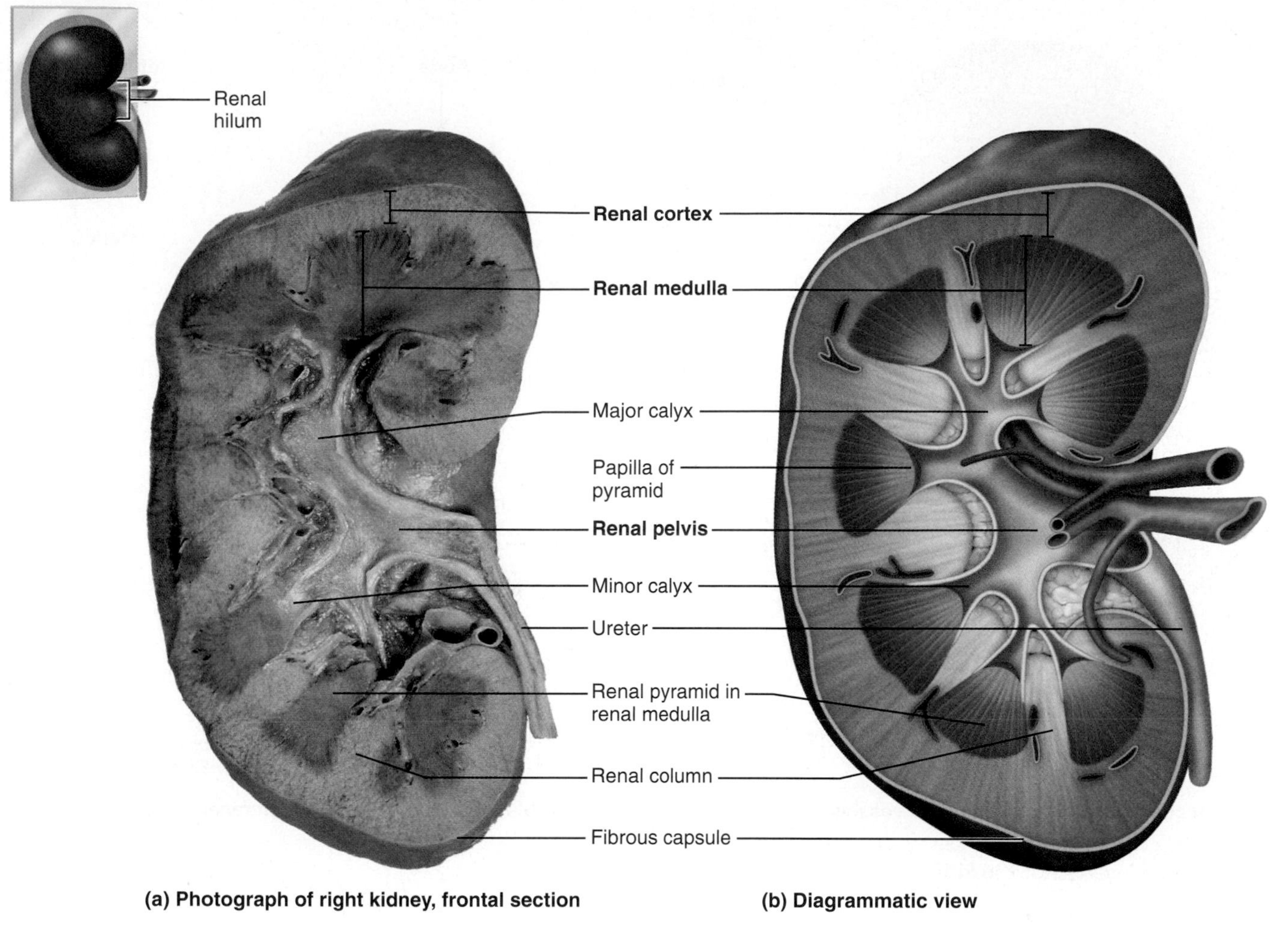

Figure 25.3 Internal anatomy of the kidney. Frontal sections. (See *A Brief Atlas of the Human Body*, Figure 71.)

HOMEOSTATIC IMBALANCE

Infection of the renal pelvis and calyces produces the condition called *pyelitis* (pi″ĕ-li′tis). Infections or inflammations that affect the entire kidney are *pyelonephritis* (pi″ĕ-lo-nĕ-fri′tis). Kidney infections in females are usually caused by fecal bacteria that spread from the anal region to the urinary tract. Less often they result from bloodborne bacteria (traveling from other infected sites) that lodge and multiply in the kidneys. In severe cases of pyelonephritis, the kidney swells, abscesses form, and the pelvis fills with pus. Untreated, the kidneys may be severely damaged, but antibiotic therapy can usually treat the infection successfully. ■

Blood and Nerve Supply

▶ Trace the blood supply through the kidney.

The kidneys continuously cleanse the blood and adjust its composition, and so it is not surprising that they have a rich blood supply. Under normal resting conditions, the large **renal arteries** deliver one-fourth of the total cardiac output (about 1200 ml) to the kidneys each minute.

The renal arteries issue at right angles from the abdominal aorta, and the right renal artery is longer than the left because the aorta lies to the left of the midline. As each renal artery approaches a kidney, it divides into five **segmental arteries** **(Figure 25.4)**. Within the renal sinus, each segmental artery branches further to form several **interlobar arteries**.

At the medulla-cortex junction, the interlobar arteries branch into the **arcuate arteries** (ar′ku-āt) that arch over the bases of the medullary pyramids. Small **cortical radiate arteries** radiate outward from the arcuate arteries to supply the cortical tissue. More than 90% of the blood entering the kidney perfuses the renal cortex.

Afferent arterioles branching from the cortical radiate arteries begin a complex arrangement of microscopic blood vessels. These vessels are key elements of kidney function, and we will examine them in detail later when we describe the nephron.

Veins pretty much trace the pathway of the arterial supply in reverse (Figure 25.4). Blood leaving the renal cortex drains

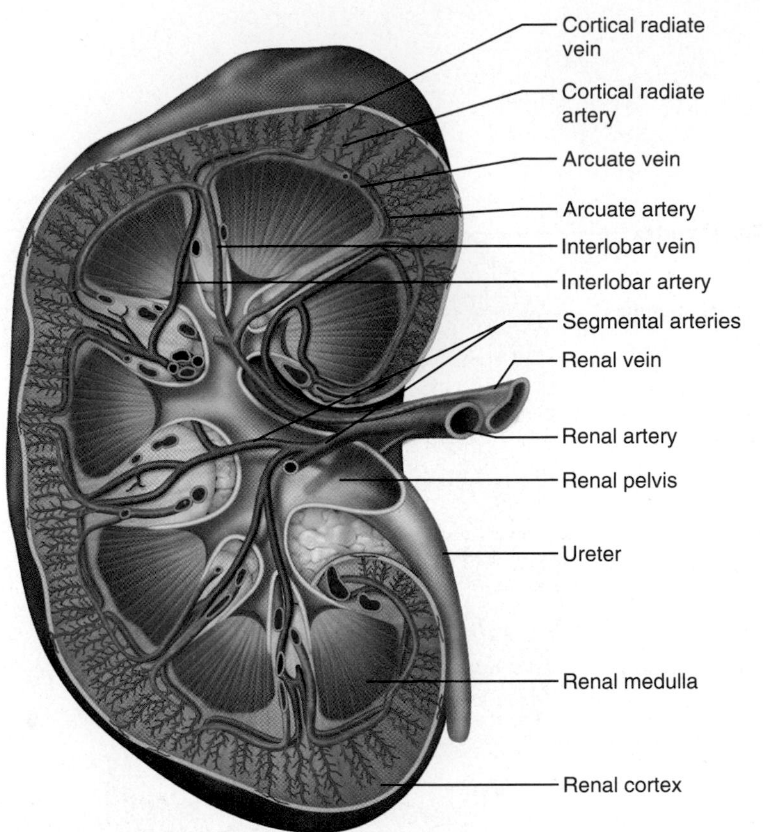

(a) Frontal section illustrating major blood vessels

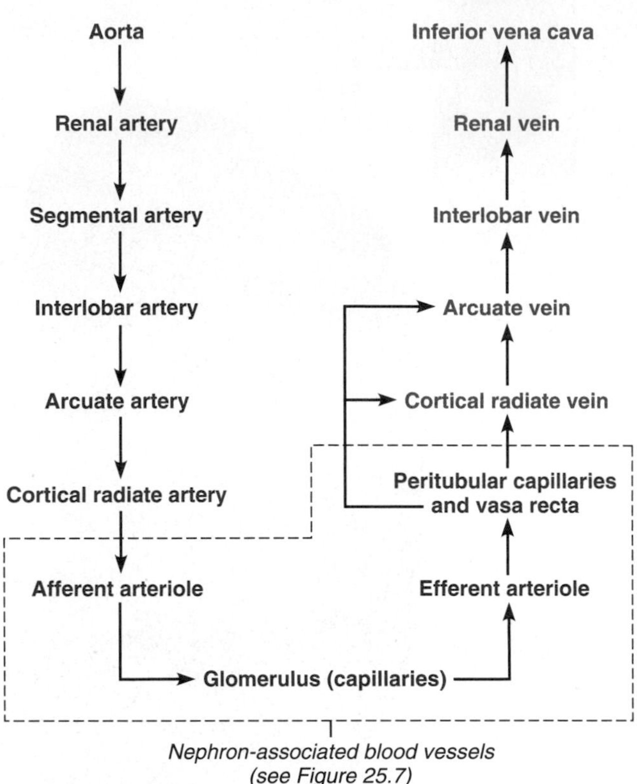

(b) Path of blood flow through renal blood vessels

Figure 25.4 **Blood vessels of the kidney.**

sequentially into the **cortical radiate**, **arcuate**, **interlobar**, and finally **renal veins**. (There are no segmental veins.) The renal veins issue from the kidneys and empty into the inferior vena cava. Because the inferior vena cava lies to the right of the vertebral column, the left renal vein is about twice as long as the right.

25

The **renal plexus**, a variable network of autonomic nerve fibers and ganglia, provides the nerve supply of the kidney and its ureter. An offshoot of the celiac plexus, the renal plexus is largely supplied by sympathetic fibers from the most inferior thoracic and first lumbar splanchnic nerves, which course along with the renal artery to reach the kidney. These sympathetic vasomotor fibers regulate renal blood flow by adjusting the diameter of renal arterioles and also influence the urine-forming role of the nephrons.

CHECK YOUR UNDERSTANDING

1. Roger is hit in the lower back by an errant baseball. What protects his kidneys from this mechanical trauma?
2. From inside to outside, list the three layers of supportive tissue that surround each kidney. Where is the parietal peritoneum in relation to these layers?
3. The lumen of the ureter is continuous with a space inside the kidney. This space has branching extensions. What are the names of this space and its extensions?

For answers, see Appendix G.

Nephrons

▶ Describe the anatomy of a nephron.

Nephrons (nef′ronz) are the structural and functional units of the kidneys. Each kidney contains over 1 million of these tiny blood-processing units, which carry out the processes that form urine **(Figure 25.5)**. In addition, there are thousands of *collecting ducts*, each of which collects fluid from several nephrons and conveys it to the renal pelvis.

Each nephron consists of a **glomerulus** (glo-mer′u-lus; *glom* = ball of yarn), which is a tuft of capillaries, and a **renal tubule**. The renal tubule has a cup-shaped end, the **glomerular capsule** (or **Bowman's capsule**), which is blind and completely surrounds the glomerulus, much as a well-worn baseball glove encloses a ball. Collectively, the glomerular capsule and the enclosed glomerulus are called the **renal corpuscle**.

The endothelium of the glomerular capillaries is *fenestrated* (penetrated by many pores), which makes them exceptionally porous. This allows large amounts of solute-rich, virtually protein-free fluid to pass from the blood into the glomerular capsule. This plasma-derived fluid or **filtrate** is the raw material that the renal tubules process to form urine.

The external *parietal layer* of the glomerular capsule is simple squamous epithelium (Figures 25.5, 25.8, and 25.9). This layer simply contributes to the capsule structure and plays no part in forming filtrate.

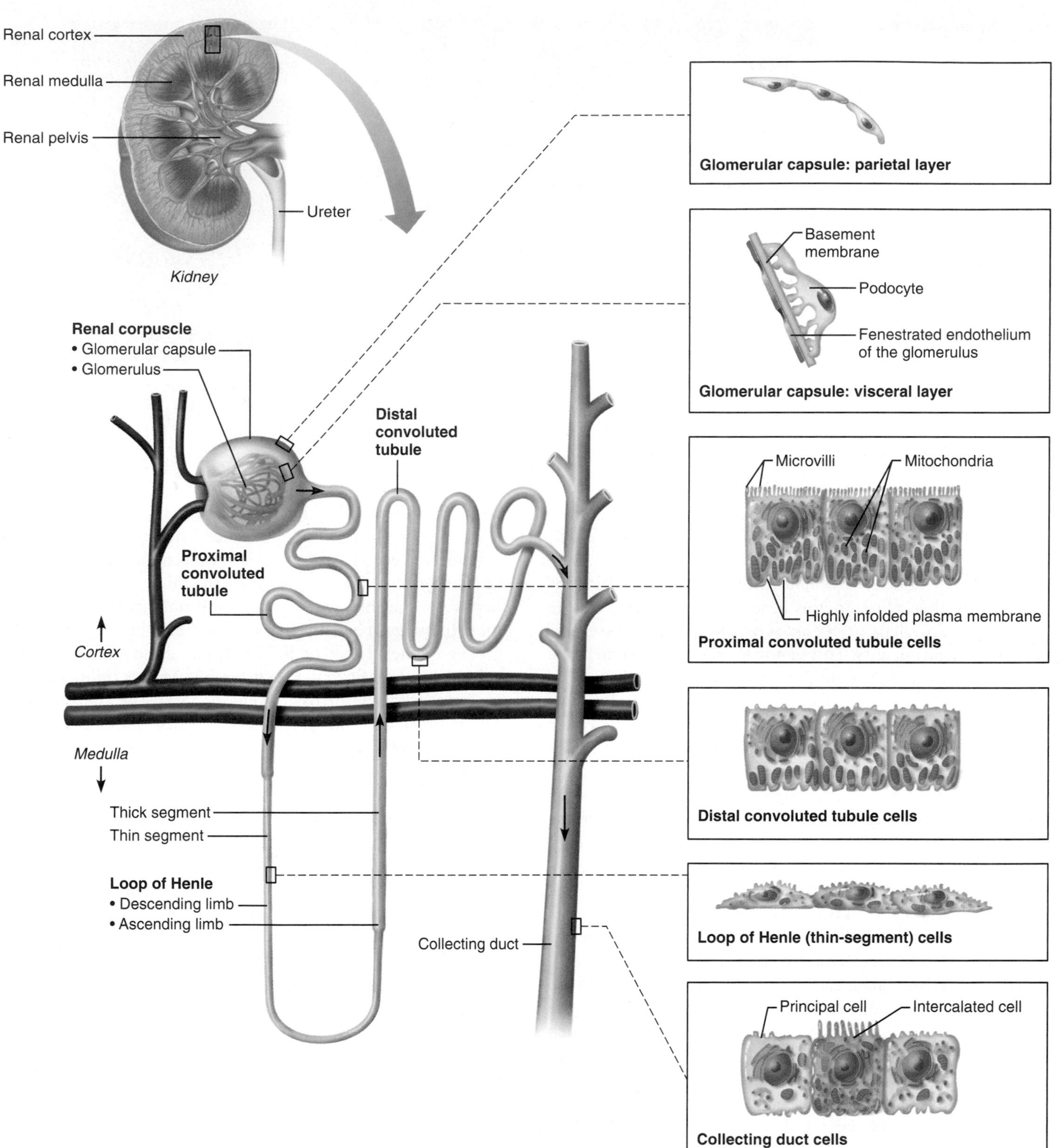

Figure 25.5 Location and structure of nephrons. Schematic view of a nephron depicting the structural characteristics of epithelial cells forming its various regions.

25

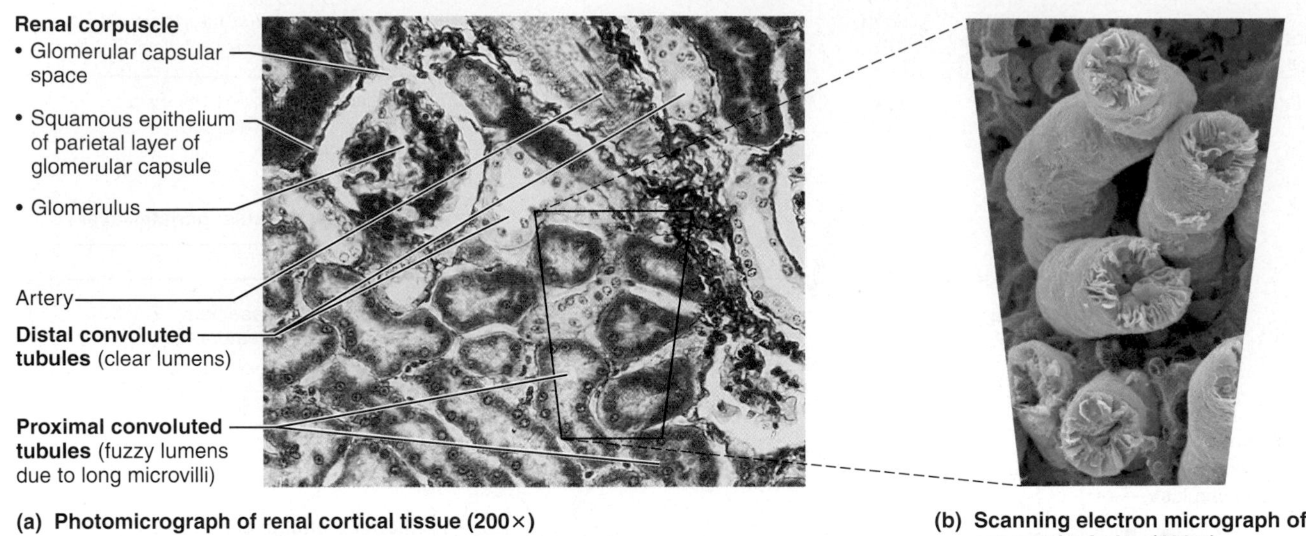

(a) Photomicrograph of renal cortical tissue (200×)

(b) Scanning electron micrograph of cut renal tubules (430×)

Figure 25.6 Renal cortical tissue and renal tubules.

The *visceral layer*, which clings to the glomerular capillaries, consists of highly modified, branching epithelial cells called **podocytes** (pod′o-sīts; "foot cells") (see Figure 25.9). The octopus-like podocytes terminate in **foot processes**, which intertwine as they cling to the basement membrane of the glomerulus. The clefts or openings between the foot processes are called **filtration slits**. Through these slits, filtrate enters the **capsular space** inside the glomerular capsule.

The remainder of the renal tubule is about 3 cm (1.2 inches) long and has three major parts. It leaves the glomerular capsule as the elaborately coiled **proximal convoluted tubule** (**PCT**), makes a hairpin loop called the **loop of Henle** (also called the **nephron loop** or **Henle's loop**), and then winds and twists again as the **distal convoluted tubule** (**DCT**) before emptying into a collecting duct. The terms *proximal* and *distal* indicate relationship of the convoluted tubules to the renal corpuscle—filtrate from the renal corpuscle passes through the PCT first and then the DCT, which is thus "further away" from the renal corpuscle. The meandering nature of the renal tubule increases its length and enhances its filtrate processing capabilities.

25

The **collecting ducts**, each of which receives filtrate from many nephrons, run through the medullary pyramids and give them their striped appearance. As the collecting ducts approach the renal pelvis, they fuse together and deliver urine into the minor calyces via papillae of the pyramids.

Throughout its length, the renal tubule consists of a single layer of polar epithelial cells on a basement membrane, but each of its regions has a unique cellular anatomy that reflects its role in processing filtrate. The walls of the PCT are formed by cuboidal epithelial cells with large mitochondria, and their luminal (exposed) surfaces bear dense microvilli (Figure 25.5 and **Figure 25.6**). Just as in the intestine, this *brush border* dramatically increases the surface area and capacity for reabsorbing water and solutes from the filtrate and secreting substances into it.

The U-shaped *loop of Henle* has *descending* and *ascending limbs*. The proximal part of the descending limb is continuous with the proximal tubule and its cells are similar. The rest of the descending limb, called the **thin segment**, is a simple squamous epithelium freely permeable to water. The epithelium becomes cuboidal or even low columnar in the ascending part of the loop of Henle, which therefore becomes the **thick segment**. In some nephrons, the thin segment is found only in the descending limb. In others, it extends into the ascending limb as well.

The epithelial cells of the DCT, like those of the PCT, are cuboidal and confined to the cortex, but they are thinner and almost entirely lack microvilli (Figure 25.5). The transition between the DCT and the collecting duct is marked by the appearance of a heterogeneous collection of cells. The two cell types seen in the collecting ducts are *intercalated* cells, cuboidal cells with abundant microvilli, and the more numerous *principal cells*, which have sparse, short microvilli. The two varieties (type A and B) of intercalated cells play a major role in maintaining the acid-base balance of the blood. The principal cells help maintain the body's water and Na^+ balance.

Nephrons are generally divided into two major groups. **Cortical nephrons** represent 85% of the nephrons in the kidneys. Except for small parts of their loops of Henle that dip into the outer medulla, they are located entirely in the cortex. The remaining **juxtamedullary nephrons** (juks″tah-mĕ′dul-ah-re) originate close to (*juxta* = near to) the cortex-medulla junction, and they play an important role in the kidneys' ability to produce concentrated urine. Their loops of Henle deeply invade the medulla, and their thin segments are much more extensive than those of cortical nephrons. **Figure 25.7a** compares the anatomy of these two types of nephrons.

Nephron Capillary Beds

The renal tubule of every nephron is closely associated with two capillary beds: the *glomerulus* and the *peritubular capillaries* (Figure 25.7). The glomerulus, in which the capillaries run in parallel, is specialized for filtration. It differs from all other capillary beds in the body in that it is both fed and drained by arterioles—the **afferent arteriole** and the **efferent arteriole**, respectively.

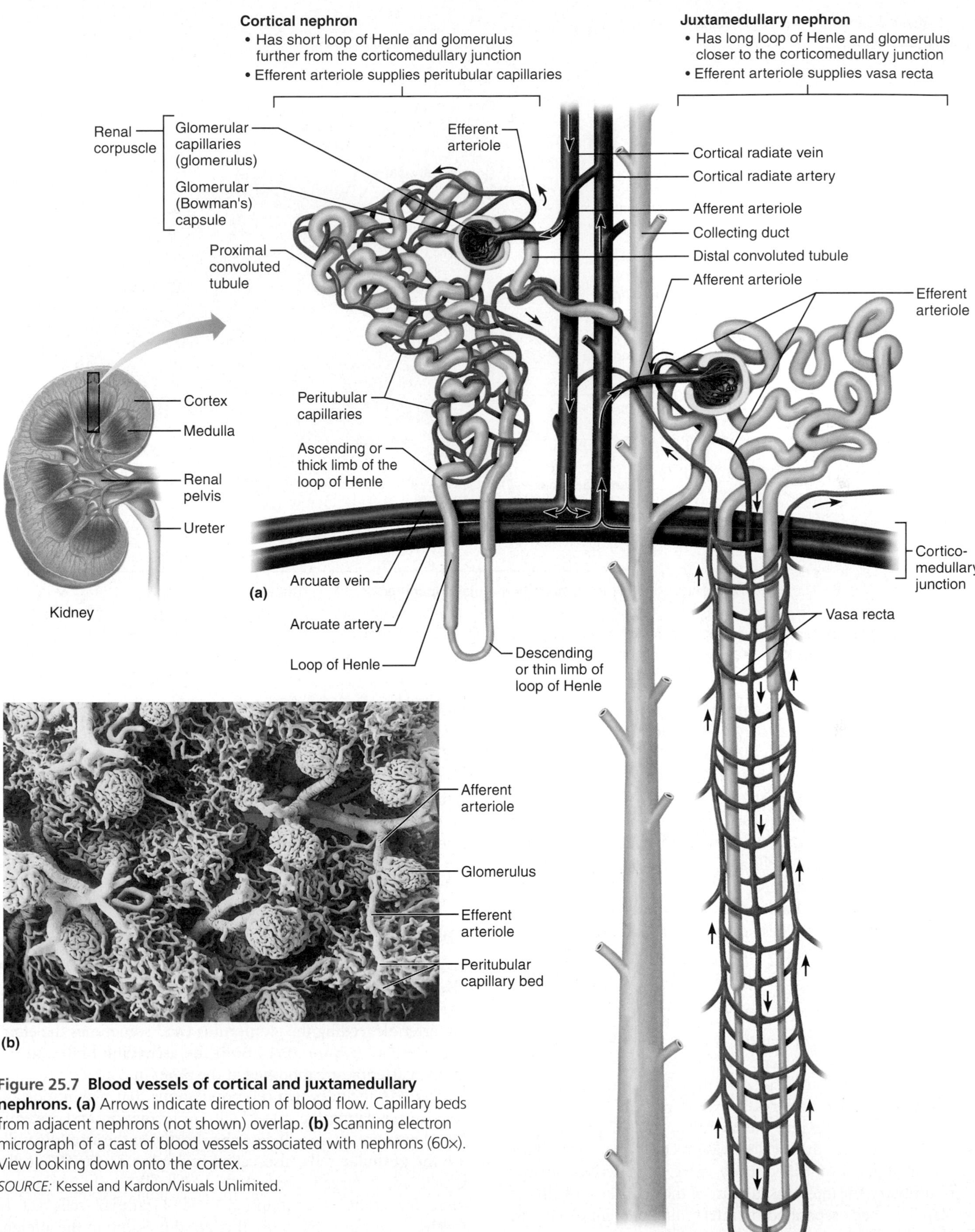

Figure 25.7 Blood vessels of cortical and juxtamedullary nephrons. (a) Arrows indicate direction of blood flow. Capillary beds from adjacent nephrons (not shown) overlap. **(b)** Scanning electron micrograph of a cast of blood vessels associated with nephrons (60×). View looking down onto the cortex.

SOURCE: Kessel and Kardon/Visuals Unlimited.

25

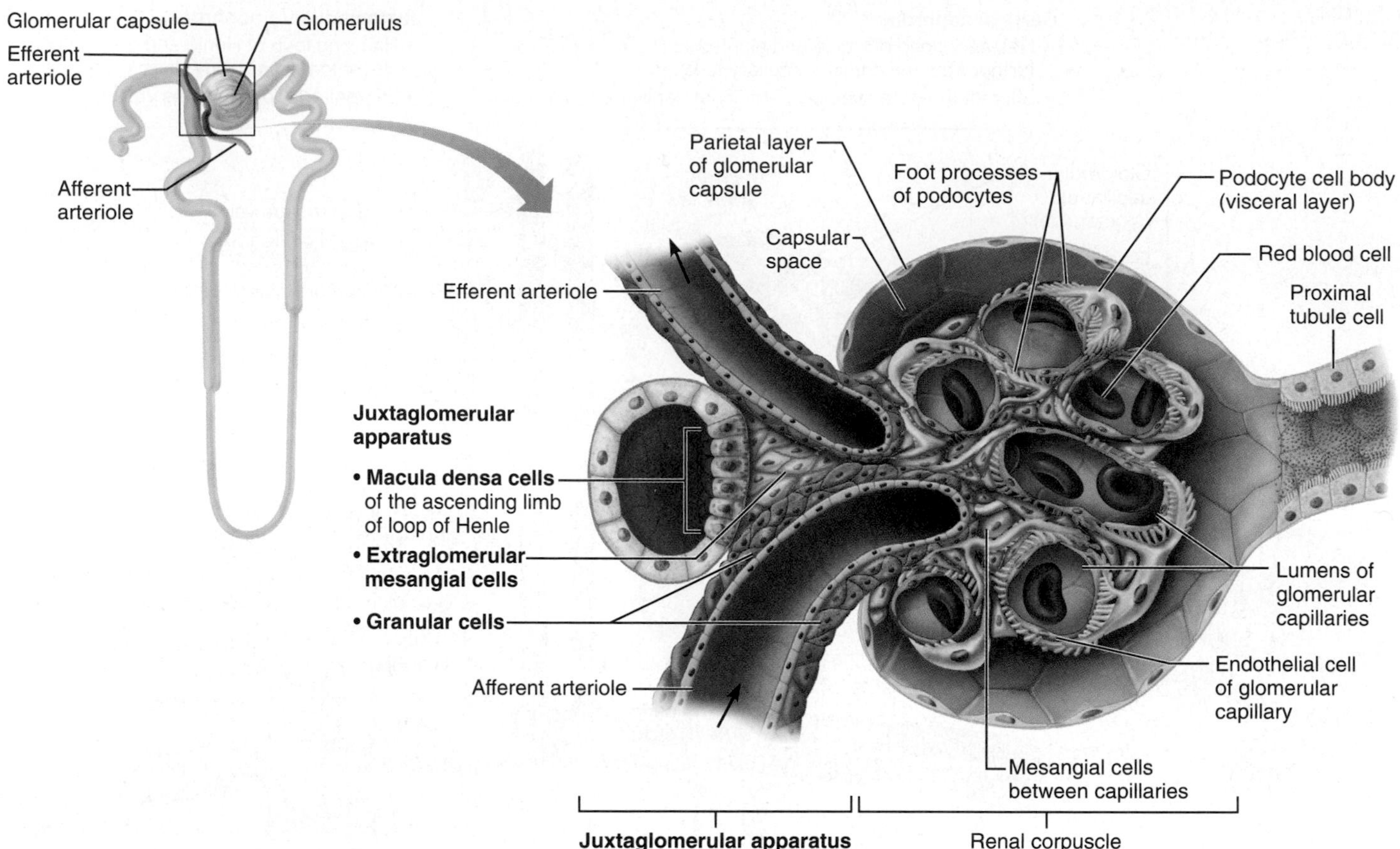

Figure 25.8 **Juxtaglomerular apparatus (JGA) of a nephron.** Mesangial cells that surround the glomerular capillaries (glomerular mesangial cells) are not part of the JGA.

The afferent arterioles arise from the *cortical radiate arteries* that run through the renal cortex. The blood pressure in the glomerulus is extraordinarily high for a capillary bed because (1) arterioles are high-resistance vessels and (2) the afferent arteriole has a larger diameter than the efferent. This high blood pressure easily forces fluid and solutes out of the blood into the
25 glomerular capsule. Most of the resulting filtrate (99%) is reabsorbed by the renal tubule cells and returned to the blood in the peritubular capillary beds.

The **peritubular capillaries** arise from the efferent arterioles draining the glomeruli. These capillaries cling closely to adjacent renal tubules and empty into nearby venules. They are low-pressure, porous capillaries that readily absorb solutes and water from the tubule cells as these substances are reclaimed from the filtrate.

Notice in Figure 25.7a that the efferent arterioles serving the juxtamedullary nephrons tend *not* to break up into meandering peritubular capillaries. Instead they form bundles of long straight vessels called **vasa recta** (va′sah rek′tah; "straight vessels") that extend deep into the medulla paralleling the longest loops of Henle. The thin-walled vasa recta play an important role in forming concentrated urine, as we will describe shortly.

In summary, the microvasculature of the nephrons consists of two capillary beds separated by intervening efferent arterioles. The first capillary bed (glomerulus) produces the filtrate. The second (peritubular capillaries) reclaims most of that filtrate.

Vascular Resistance in the Microcirculation Blood flowing through the renal circulation encounters high resistance, first in the afferent and then in the efferent arterioles. As a result, renal blood pressure declines from approximately 95 mm Hg in the renal arteries to 8 mm Hg or less in the renal veins. The resistance of the afferent arterioles protects the glomeruli from large fluctuations in systemic blood pressure. Resistance in the efferent arterioles reinforces the high glomerular pressure and reduces the hydrostatic pressure in the peritubular capillaries.

Juxtaglomerular Apparatus

Each nephron has a region called a **juxtaglomerular apparatus (JGA)** (juks″tah-glo-mer′u-lar), where the most distal portion of the ascending limb of the loop of Henle lies against the afferent arteriole feeding the glomerulus (and sometimes the efferent arteriole) **(Figure 25.8)**. Both the ascending limb and the afferent arteriole are modified at the point of contact.

The JGA includes two cell populations that play important roles in regulating the rate of filtrate formation and systemic blood pressure, as we will describe shortly. In the arteriole walls are the **granular cells**, also called **juxtaglomerular (JG) cells**, which are enlarged, smooth muscle cells with prominent secretory granules containing *renin*. Granular cells act as mechanoreceptors that sense the blood pressure in the afferent arteriole. The **macula densa** (mak′u-lah den′sah; "dense spot") is a group of tall, closely packed cells of the ascending limb of the loop of Henle that lies adjacent to the granular cells (Figure 25.8). The macula densa cells are chemoreceptors that respond to changes in the NaCl content of the filtrate. A third population of cells, the *extraglomerular mesangial cells*, is also part of the JGA. These cells are interconnected by gap junctions and may pass signals between macula densa and granular cells.

The Filtration Membrane

The **filtration membrane** lies between the blood and the interior of the glomerular capsule. It is a porous membrane that allows free passage of water and solutes smaller than plasma proteins. As **Figure 25.9c** shows, its three layers are: (1) the fenestrated endothelium of the glomerular capillaries; (2) the visceral membrane of the glomerular capsule, made of podocytes which have filtration slits between their foot processes; and between these two layers, (3) the basement membrane composed of the fused basal laminae of the two other layers.

The fenestrations (capillary pores) allow passage of all plasma components but not blood cells. The basement membrane restricts all but the smallest proteins while permitting most other solutes to pass. The structural makeup of the gel-like basement membrane also confers electrical selectivity on the filtration process. Most of the proteins in the membrane are negatively charged glycoproteins that repel other macromolecular anions and hinder their passage into the tubule. Because most plasma proteins also bear a net negative charge, this electrical repulsion reinforces the plasma protein blockage imposed by molecular size.

Almost all macromolecules that do manage to make it through the basement membrane are prevented from traveling further by thin membranes (*slit diaphragms*) that extend across the filtration slits. Macromolecules that get "hung up" in the filtration membrane are engulfed and degraded by mesangial cells within the glomerulus. These *glomerular mesangial cells* can also contract, changing the total surface area of the capillaries available for filtration.

CHECK YOUR UNDERSTANDING

4. Name the tubular components of a nephron in the order that filtrate passes through them.
5. What are the structural differences between juxtamedullary and cortical nephrons?
6. What type of capillaries are the glomerular capillaries? What is their function?

For answers, see Appendix G.

Kidney Physiology: Mechanisms of Urine Formation

Urine formation and the adjustment of blood composition involve three major processes: *glomerular filtration* by the glomeruli, and *tubular reabsorption* and *tubular secretion* in the renal tubules **(Figure 25.10)**. In addition, the collecting ducts work in concert with the nephrons to make concentrated or dilute urine.

Let's first look at the big picture to see how the kidneys use these three processes to maintain the volume and chemical makeup of the blood—in other words, how do they "clean" the blood? Conceptually, it's really very simple. The kidneys "dump" (by glomerular filtration, Figure 25.10, ❶) cell- and protein-free blood into a separate "container" (the renal tubules and collecting ducts). From this container, the kidneys reclaim (by tubular reabsorption, Figure 25.10, ❷) everything the body needs to keep. This is almost everything—all of the glucose and amino acids, and some 99% of the water, salt, and other components. Anything that is *not* reabsorbed becomes urine. In addition, some things are selectively added to the container (by tubular secretion, Figure 25.10, ❸), fine-tuning the body's chemical balance.

The volume of blood processed by the kidneys each day is enormous. Of the approximately 1200 ml of blood that passes through the glomeruli each minute, some 650 ml is plasma, and about one-fifth of this (120–125 ml) is forced into the renal tubules. This is equivalent to filtering out your entire plasma volume more than 60 times each day! Considering the magnitude of their task, it is not surprising that the kidneys (which account for only 1% of body weight) consume 20–25% of all oxygen used by the body at rest.

Filtrate and urine are quite different. Filtrate contains everything found in blood plasma except proteins. **Urine** contains mostly metabolic wastes and unneeded substances. The kidneys process about 180 L (47 gallons!) of blood-derived fluid daily. Of this amount, less than 1% (1.5 L) typically leaves the body as urine; the rest returns to the circulation.

Step 1: Glomerular Filtration

- ▶ Describe the forces (pressures) that promote or counteract glomerular filtration.
- ▶ Compare the intrinsic and extrinsic controls of the glomerular filtration rate.

Glomerular filtration is a passive process in which hydrostatic pressure forces fluids and solutes through a membrane (see Chapter 19). The glomeruli can be viewed as simple mechanical filters because filtrate formation does not consume metabolic energy.

The glomerulus is a much more efficient filter than are other capillary beds. One reason is that its *filtration membrane* has a large surface area and is thousands of times more permeable to water and solutes (Figure 25.9). Furthermore, glomerular blood pressure is much higher than that in other capillary beds (approximately 55 mm Hg as opposed to 18 mm Hg or less), resulting in a much higher *net filtration pressure*, which we will discuss shortly. As a result of these differences, the kidneys produce about 180 L of filtrate daily, in contrast to the 2 to 4 L formed daily by all other capillary beds of the body combined.

Molecules smaller than 3 nm in diameter—such as water, glucose, amino acids, and nitrogenous wastes—pass freely from the blood into the glomerular capsule. As a result, these substances

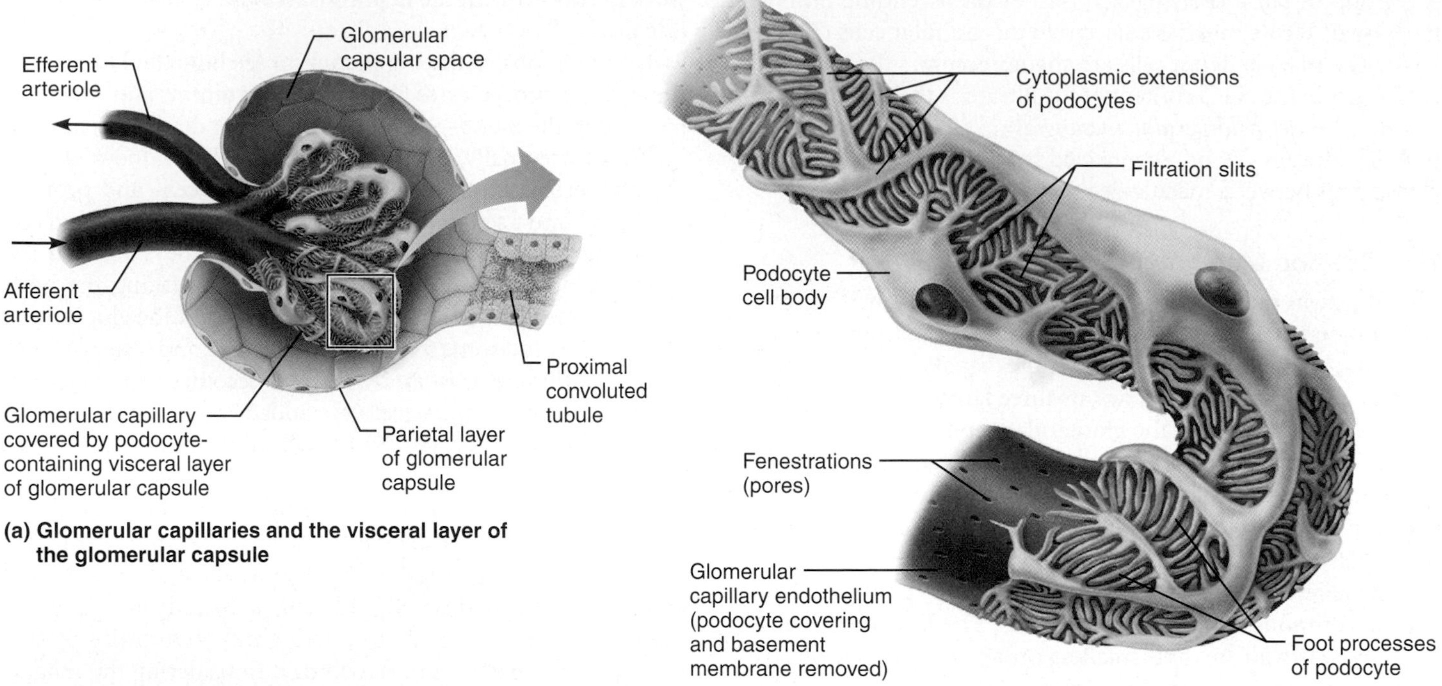

(a) Glomerular capillaries and the visceral layer of the glomerular capsule

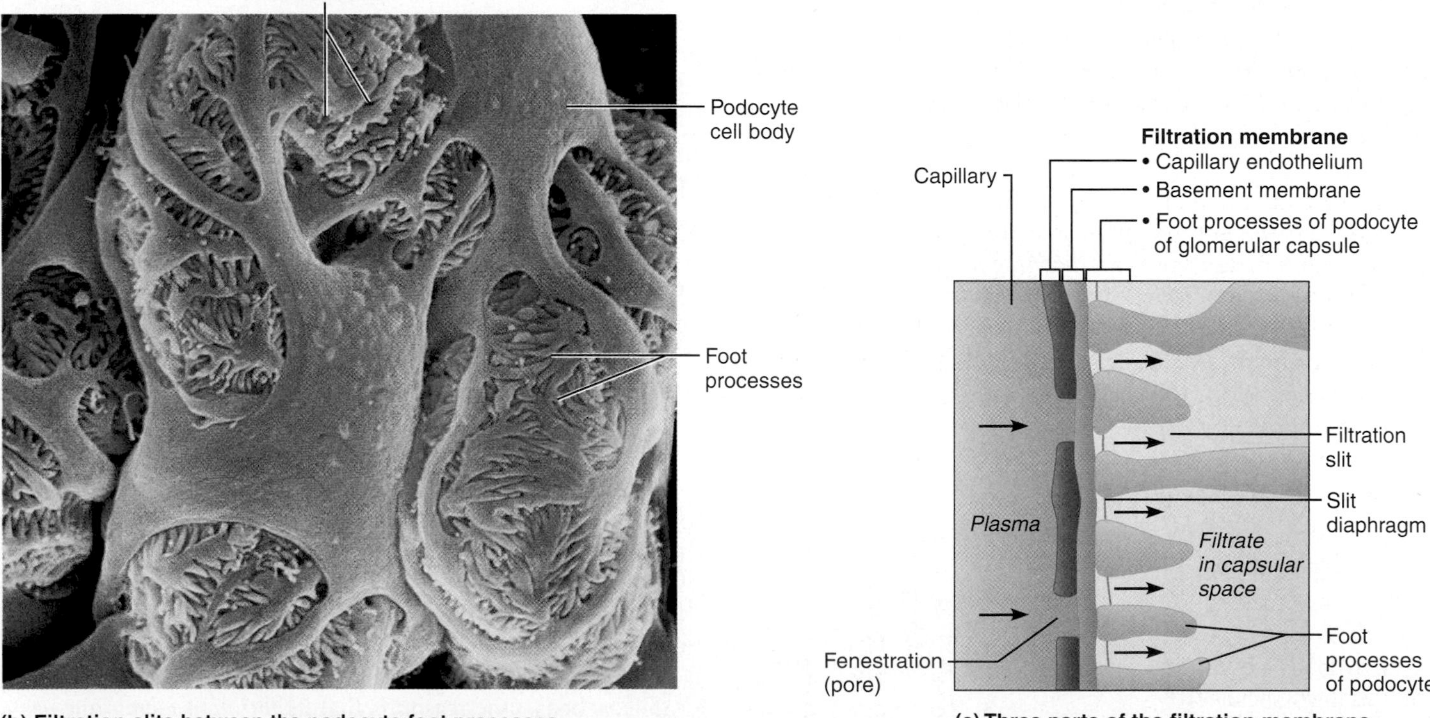

(b) Filtration slits between the podocyte foot processes

(c) Three parts of the filtration membrane

25

Figure 25.9 The filtration membrane. The filtration membrane is composed of three layers: the fenestrated endothelium of the glomerular capillaries, the podocyte-containing visceral layer of the glomerular capsule, and the intervening basement membrane. **(a)** Diagrammatic view of the relationship of the visceral layer of the glomerular capsule to the glomerular capillaries. The visceral epithelium is drawn incompletely to show the fenestrations in the underlying capillary wall. **(b)** Scanning electron micrograph of the visceral layer. Filtration slits between the podocyte foot processes are evident (6000×). **(c)** Diagrammatic view of a section through the filtration membrane showing all three structural elements.

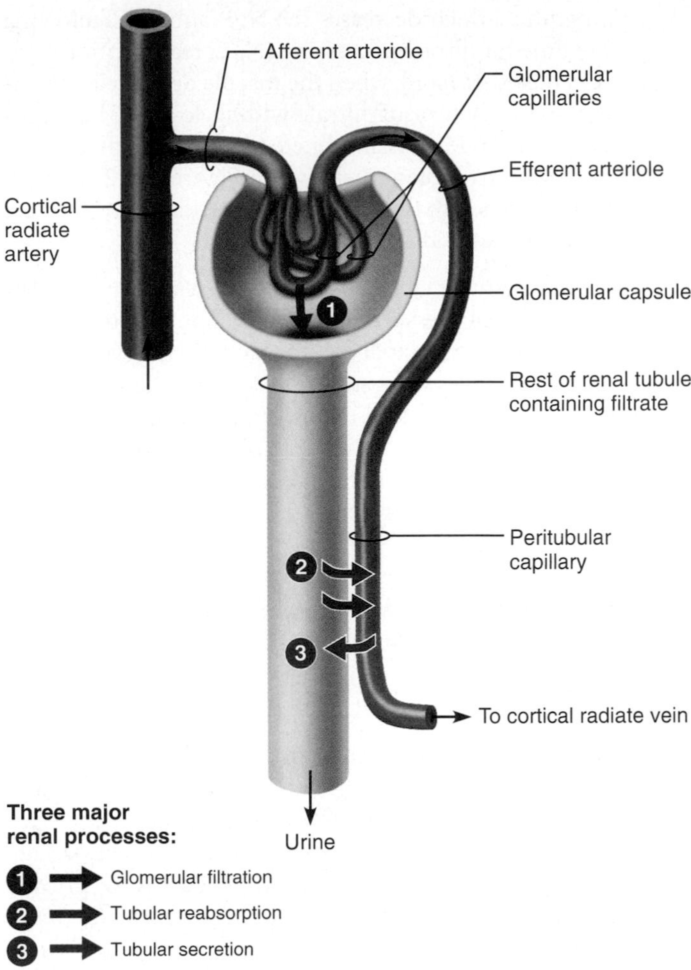

Figure 25.10 A schematic, uncoiled nephron showing the three major renal processes that adjust plasma composition. A kidney actually has more than a million nephrons acting in parallel.

usually show similar concentrations in the blood and the glomerular filtrate. Larger molecules pass with greater difficulty, and those larger than 5 nm are generally barred from entering the tubule. Keeping the plasma proteins *in* the capillaries maintains the colloid osmotic (oncotic) pressure of the glomerular blood, preventing the loss of all its water to the renal tubules. The presence of proteins or blood cells in the urine usually indicates a problem with the filtration membrane.

Net Filtration Pressure

The **net filtration pressure (NFP)**, responsible for filtrate formation, involves forces acting at the glomerular bed **(Figure 25.11)**. **Glomerular hydrostatic pressure (HPg)**, which is essentially glomerular blood pressure, is the chief force pushing water and solutes out of the blood and across the filtration membrane. Although theoretically the colloid osmotic pressure in the capsular space of the glomerular capsule "pulls" the filtrate into the tubule, this pressure is essentially zero because virtually no proteins enter the capsule.

The HPg is opposed by two forces that inhibit fluid loss from glomerular capillaries. These filtration-opposing forces are (1) **colloid osmotic (oncotic) pressure of glomerular blood (OP_g)** and (2) **capsular hydrostatic pressure (HP_c)** exerted by fluids in the glomerular capsule. Using the values shown in the key to Figure 25.11, we calculate the NFP responsible for forming renal filtrate from plasma as 10 mm Hg:

$$\begin{aligned} \text{NFP} &= \text{HP}_g - (\text{OP}_g + \text{HP}_c) \\ &= 55 \text{ mm Hg} - (30 \text{ mm Hg} + 15 \text{ mm Hg}) \\ &= 10 \text{ mm Hg} \end{aligned}$$

Glomerular Filtration Rate

The **glomerular filtration rate** or **GFR** is the volume of filtrate formed each minute by the combined activity of all 2 million glomeruli of the kidneys. Factors governing filtration rate at the capillary beds are (1) total surface area available for filtration,

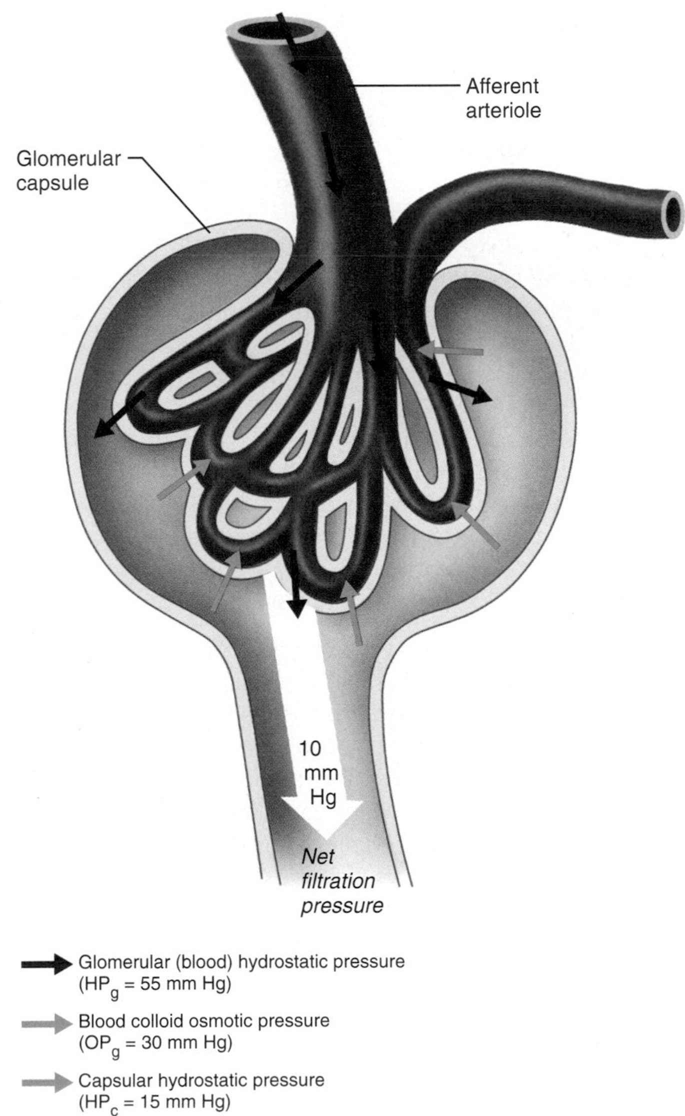

Figure 25.11 Forces determining glomerular filtration and filtration pressure. The glomerular hydrostatic (blood) pressure is the major factor forcing fluids and solutes out of the blood. This pressure is opposed by the colloid osmotic pressure of the blood and the hydrostatic pressure in the glomerular capsule. The pressure values cited in the diagram are approximate.

25

(2) filtration membrane permeability, and (3) NFP. In adults the normal GFR in both kidneys is 120–125 ml/min. Because glomerular capillaries are exceptionally permeable and have a huge surface area (collectively equal to the surface area of the skin), huge amounts of filtrate can be produced even with the usual modest NFP of 10 mm Hg. The opposite side of this "coin" is that a drop in glomerular pressure of only 18% stops filtration altogether.

The GFR is *directly proportional* to the NFP, so any change in any of the pressures acting at the filtration membrane changes both the NFP and the GFR. In the absence of regulation, an increase in arterial (and glomerular) blood pressure in the kidneys increases the GFR. As we shall see in the next section however, GFR is tightly regulated.

Regulation of Glomerular Filtration

GFR is regulated by both intrinsic and extrinsic controls. These two types of controls serve two different (and sometimes opposing) needs. The kidneys need a relatively constant GFR in order to do their job and maintain extracellular homeostasis. On the other hand, the body as a whole needs a constant blood pressure, and therefore a constant blood volume.

Intrinsic controls (*renal autoregulation*) act locally within the kidney to maintain GFR, while extrinsic controls by the nervous and endocrine systems maintain blood pressure. In extreme changes of blood pressure (mean arterial pressure less than 80 or greater than 180 mm Hg), extrinsic controls take precedence over intrinsic controls. Next we examine both types of control.

Intrinsic Controls: Renal Autoregulation By adjusting its own resistance to blood flow, a process called **renal autoregulation**, the kidney can maintain a nearly constant GFR despite fluctuations in systemic arterial blood pressure. Renal autoregulation entails two types of controls: (1) a *myogenic mechanism* and (2) a *tubuloglomerular feedback mechanism* (**Figure 25.12**, left side).

1. **Myogenic mechanism.** The **myogenic mechanism** (mi″o-jen′ik) reflects the tendency of vascular smooth muscle to contract when stretched. Increasing systemic blood pressure causes the afferent arterioles to constrict, which restricts blood flow into the glomerulus and prevents glomerular blood pressure from rising to damaging levels. Declining systemic blood pressure causes dilation of afferent arterioles and raises glomerular hydrostatic pressure. Both responses help maintain a normal GFR.
2. **Tubuloglomerular feedback mechanism.** Autoregulation by the flow-dependent **tubuloglomerular feedback mechanism** is "directed" by the *macula densa cells* of the *juxtaglomerular apparatus* (see Figure 25.8). These cells, located in the walls of the ascending limb of Henle's loop, respond to filtrate NaCl concentration (which varies directly with filtrate flow rate). When GFR increases, there is insufficient time for reabsorption and the concentration of NaCl in the filtrate remains high. This causes the macula densa cells to release a vasoconstrictor chemical (probably ATP) that causes intense constriction of the afferent arteriole. This constriction hinders blood flow into the glomerulus, which decreases the NFP and GFR, allowing more time for filtrate processing (NaCl reabsorption).

   On the other hand, when the macula densa cells are exposed to slowly flowing filtrate with its low NaCl concentration, ATP release is inhibited, causing vasodilation of the afferent arterioles, as shown in Figure 25.12. This allows more blood to flow into the glomerulus, thus increasing the NFP and GFR.

Autoregulatory mechanisms maintain a relatively constant GFR over an arterial pressure range from about 80 to 180 mm Hg. Consequently, our normal day-to-day activities (such as exercise, sleep, or changes in posture) do not cause large changes in water and solute excretion. However, the intrinsic controls cannot handle extremely low systemic blood pressure, such as might result from serious hemorrhage (*hypovolemic shock*). Once the mean arterial pressure drops below 80 mm Hg, autoregulation ceases.

Extrinsic Controls: Neural and Hormonal Mechanisms The purpose of the extrinsic controls regulating the GFR is to maintain systemic blood pressure—sometimes to the detriment of the kidneys (Figure 25.12, right side).

1. **Sympathetic nervous system controls.** Neural renal controls serve the needs of the body as a whole. When the volume of the extracellular fluid is normal and the sympathetic nervous system is at rest, the renal blood vessels are dilated and renal autoregulation mechanisms prevail. However, during extreme stress or emergency when it is necessary to shunt blood to vital organs, neural controls may overcome renal autoregulatory mechanisms.

   Norepinephrine released by sympathetic nerve fibers (and epinephrine released by the adrenal medulla) acts on alpha-adrenergic receptors on vascular smooth muscle, strongly constricting afferent arterioles, thereby inhibiting filtrate formation. This, in turn, indirectly trips the renin-angiotensin mechanism by stimulating the macula densa cells. The sympathetic nervous system also directly stimulates the granular cells to release renin, as we discuss next.
2. **Renin-angiotensin mechanism.** The **renin-angiotensin mechanism** is triggered when various stimuli cause the granular cells to release the hormone **renin**. Renin acts enzymatically on **angiotensinogen**, a plasma globulin made by the liver, converting it to **angiotensin I**. This, in turn, is converted to **angiotensin II** by **angiotensin converting enzyme (ACE)** associated with the capillary endothelium in various body tissues, particularly the lungs.

   Angiotensin II acts in five ways to stabilize systemic blood pressure and extracellular fluid volume. (1) As a potent vasoconstrictor, angiotensin II activates smooth muscle of arterioles throughout the body, raising mean arterial blood pressure. (2) Angiotensin II stimulates reabsorption of sodium, both directly by acting on renal tubules and indirectly by triggering the release of aldosterone from the adrenal cortex. Because water follows sodium osmotically, blood volume and blood pressure rise (Figure 25.12). (3) Angiotensin II stimulates the hypothalamus to release

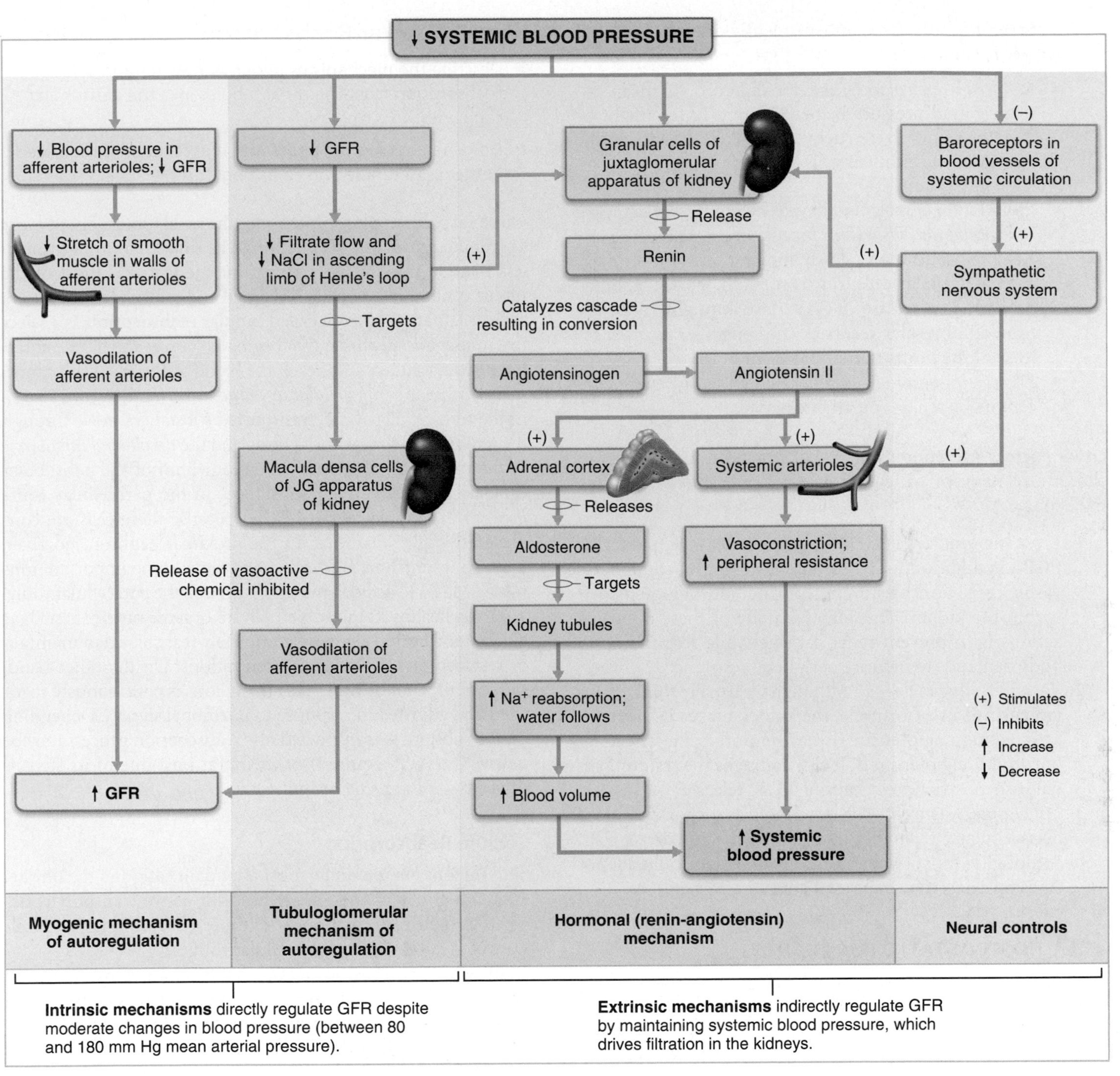

Figure 25.12 Physiological mechanisms regulating glomerular filtration rate (GFR) in the kidneys. (Note that while the extrinsic controls are aimed at maintaining blood pressure, they also maintain GFR since blood pressure drives filtration in the kidneys.)

antidiuretic hormone and activates the hypothalamic thirst center, both of which increase blood volume (see Chapter 26). (4) Angiotensin II also increases fluid reabsorption by decreasing peritubular capillary hydrostatic pressure. This pressure drop occurs because the efferent arterioles constrict, and the downstream drop in hydrostatic pressure allows more fluid to move back into the peritubular capillary bed. (5) Finally, angiotensin II targets the glomerular mesangial cells, causing them to contract and reduce the GFR by decreasing the total surface area of glomerular capillaries available for filtration.

While this seems like a daunting list at first, it will help if you remember that all of the effects of angiotensin II are aimed at restoring blood volume and blood pressure. Of angiotensin II's many effects, the first two are the most important.

25

Several factors acting independently or collectively can trigger renin release.

- *Reduced stretch of the granular cells.* A drop in mean systemic blood pressure below 80 mm Hg (as might be due to hemorrhage, dehydration, etc.) reduces the stretch of the granular cells and stimulates them to release more renin.
- *Stimulation of the granular cells by input from activated macula densa cells.* When macula densa cells sense low NaCl concentration (slowly moving filtrate), they signal the granular cells to release renin. This signal may be *decreased* release of ATP (also thought to be the tubuloglomerular feedback messenger), *increased* release of the prostaglandin PGE_2, or both.
- *Direct stimulation of granular cells* via β_1-adrenergic receptors by renal sympathetic nerves.

Other Factors Affecting GFR Renal cells produce a battery of chemicals, many of which act as paracrines (local signaling molecules).

1. *Prostaglandin E_2 (PGE_2)*: The vasodilatory paracrine PGE_2, in addition to its role discussed above, counteracts vasoconstriction by norepinephrine and angiotensin II within the kidney. The adaptive value of these opposing actions is to prevent renal damage while responding to body demands to increase peripheral resistance.
2. *Intrarenal angiotensin II*: Although we usually think of angiotensin II as a hormone, the kidney makes its own, locally acting angiotensin II that reinforces the effects of hormonal angiotensin II. It also dampens the resulting renal vasoconstriction by causing PGE_2 release.
3. *Adenosine*: Adenosine can be released as such or produced extracellularly from ATP released by macula densa cells. Although it functions as a vasodilator systemically, adenosine constricts the renal vasculature.

HOMEOSTATIC IMBALANCE

Abnormally low urinary output (less than 50 ml/day), called *anuria* (ah-nu′re-ah), may indicate that glomerular blood pressure is too low to cause filtration. However, renal failure and anuria can result from situations in which the nephrons cease to function for a variety of other reasons, including acute nephritis, transfusion reactions, and crush injuries. ■

CHECK YOUR UNDERSTANDING

7. Extrinsic and intrinsic controls of GFR serve two different purposes. What are they?
8. Calculate net filtration pressure given the following values: glomerular hydrostatic pressure = 50 mm Hg, blood colloid osmotic pressure = 25 mm Hg, capsular hydrostatic pressure = 20 Hg.
9. Describe two main ways in which angiotensin II increases blood pressure and blood volume.

For answers, see Appendix G.

Step 2: Tubular Reabsorption

- ▶ Describe the mechanisms underlying water and solute reabsorption from the renal tubules into the peritubular capillaries.
- ▶ Describe how sodium and water reabsorption is regulated in the distal tubule and collecting duct.

Our total plasma volume filters into the renal tubules about every 22 minutes, so all our plasma would be drained away as urine in less than 30 minutes were it not for the fact that most of the tubule contents are quickly reclaimed and returned to the blood. This reclamation process, called **tubular reabsorption**, is a selective *transepithelial process* that begins as soon as the filtrate enters the proximal tubules. To reach the blood, reabsorbed substances follow either the *transcellular* or *paracellular route* (**Figure 25.13**). In the transcellular route, transported substances move through the *luminal membrane*, the cytosol, and the *basolateral membrane* of the tubule cell and then the endothelium of the peritubular capillaries. Movement of substances in the paracellular route *between* the tubule cells is limited because these cells are connected by tight junctions. In the proximal nephron, however, these tight junctions are "leaky" and allow some important ions (Ca^{2+}, Mg^{2+}, K^+, and some Na^+) through the paracellular route.

Given healthy kidneys, virtually all organic nutrients such as glucose and amino acids are completely reabsorbed to maintain or restore normal plasma concentrations. On the other hand, the reabsorption of water and many ions is continuously regulated and adjusted in response to hormonal signals. Depending on the substances transported, the reabsorption process may be *passive* (no ATP required) or *active* (at least one of its steps is driven by ATP directly or indirectly).

Sodium Reabsorption

Sodium ions are the single most abundant cation in the filtrate, and about 80% of the energy used for active transport is devoted to their reabsorption. Sodium reabsorption is almost always active and via the transcellular route.

In general, two basic processes that promote active Na^+ reabsorption occur in each tubule segment. First, Na^+ is actively transported out of the tubule cell by *primary active transport*—a Na^+-K^+ ATPase pump present in the basolateral membrane (**Figure 25.14**, ①). From there, Na^+ is swept along by the bulk flow of water into adjacent peritubular capillaries. This bulk flow of water and solutes into the peritubular capillaries is rapid because the blood there has low hydrostatic pressure and high osmotic pressure (remember, most proteins remain in the blood instead of being filtered out into the tubule).

Second, active pumping of Na^+ from the tubule cells results in a strong electrochemical gradient that favors its passive entry at the luminal face via *secondary active transport* (symport or antiport) carriers (Figure 25.14, ②, ③) or via facilitated diffusion through channels. This occurs because (1) the pump maintains the intracellular Na^+ concentration at low levels, and (2) the K^+ pumped into the tubule cells almost immediately diffuses out into the interstitial fluid via leakage channels, leaving the interior of the tubule cell with a net negative charge.

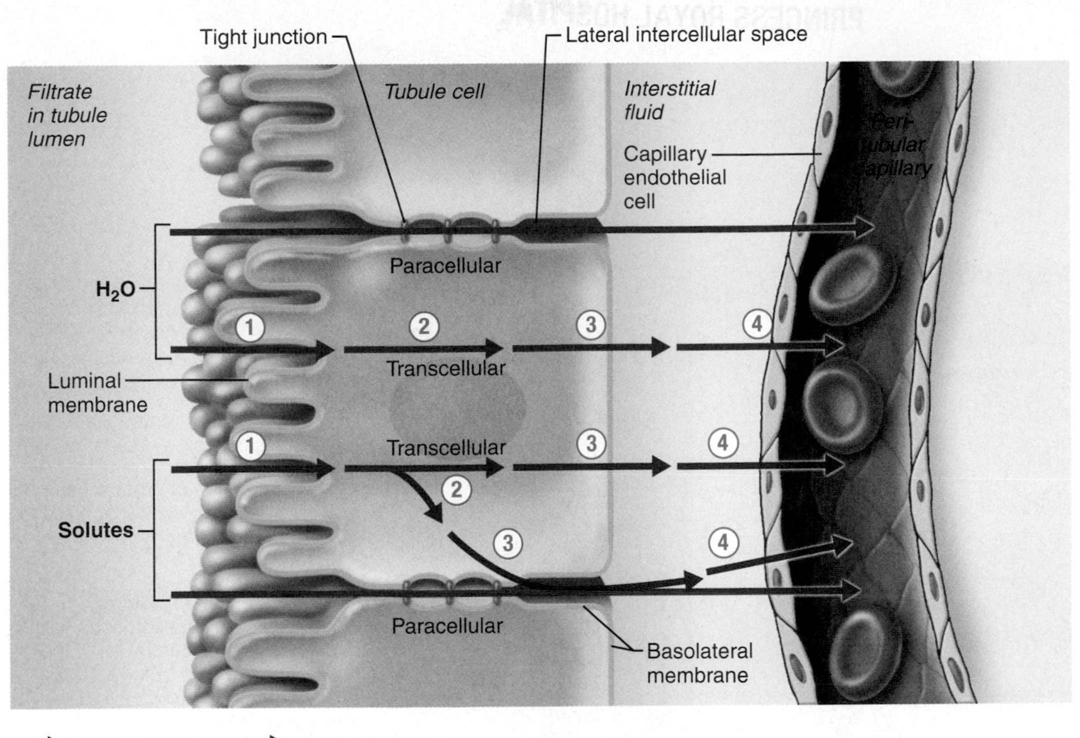

Movement via the transcellular route involves:

① Transport across the luminal membrane.

② Diffusion through the cytosol.

③ Transport across the basolateral membrane. (Often involves the lateral intercellular spaces because membrane transporters transport ions into these spaces.)

④ Movement through the interstitial fluid and into the capillary.

The paracellular route involves:

- Movement through leaky tight junctions, particularly in the PCT.

Figure 25.13 Transcellular and paracellular routes of tubular reabsorption. Generally, fluids and solutes move into the peritubular capillaries through intercellular clefts. For simplicity, transporters, ion channels, intercellular clefts, and aquaporins are not depicted.

Because each tubule segment plays a slightly different role in reabsorption, the precise mechanism by which Na^+ is reabsorbed at the luminal membrane varies.

Reabsorption of Nutrients, Water, and Ions

The reabsorption of Na^+ by primary active transport provides the energy and the means for reabsorbing almost every other substance, including water. Substances reabsorbed by **secondary active transport** (the "push" comes from the gradient created by Na^+-K^+ pumping at the basolateral membrane) include glucose, amino acids, lactate, and vitamins. In nearly all these cases, a luminal carrier moves Na^+ down its concentration gradient as it cotransports (symports) another solute (Figure 25.14, ③). Cotransported solutes diffuse (via different transport proteins) across the basolateral membrane before moving into the peritubular capillaries. Although there is some overlap of carriers, the transport systems for the various solutes are quite specific *and limited.*

There is a **transport maximum (T_m)** for nearly every substance that is reabsorbed using a transport protein in the membrane. The T_m (reported in mg/min) reflects the number of transport proteins in the renal tubules available to ferry each particular substance. In general, there are plenty of transporters and therefore high T_m values for substances such as glucose that need to be retained, and few or no transporters for substances of no use to the body.

When the transporters are saturated—that is, all bound to the substance they transport—the excess is excreted in urine. This is what happens in individuals who become hyperglycemic because of uncontrolled diabetes mellitus. As plasma levels of glucose approach and exceed 180 mg/dl, the glucose T_m is exceeded and large amounts of glucose may be lost in the urine even though the renal tubules are still functioning normally.

In **passive tubular reabsorption**, which encompasses osmosis, diffusion, and facilitated diffusion, substances move down their electrochemical gradients without the use of ATP. The movement of Na^+ and other solutes establishes a strong osmotic gradient, and water moves by osmosis into the peritubular capillaries, a process aided by transmembrane proteins called **aquaporins** that form water channels across cell membranes (Figure 25.14, ④). In continuously water-permeable regions of the renal tubules, such as the PCT, aquaporins are constant components of the tubule cell membranes. Because these channels are always present, the body is "obliged" to absorb water in the proximal nephron regardless of its state of over- or underhydration. This water flow is referred to as **obligatory water reabsorption.** Aquaporins are virtually absent in the luminal membranes of the collecting duct unless antidiuretic hormone (ADH) is present.

As water leaves the tubules, the concentration of solutes in the filtrate increases and, if able, they too begin to follow their concentration gradients into the peritubular capillaries. This phenomenon of solutes following solvent explains the passive reabsorption of a number of solutes present in the filtrate, such as lipid-soluble substances, certain ions, and some urea (Figure 25.14, ⑤, ⑥). It also explains in part why lipid-soluble drugs and environmental toxins are difficult to excrete: Since lipid-soluble compounds can generally pass through membranes,

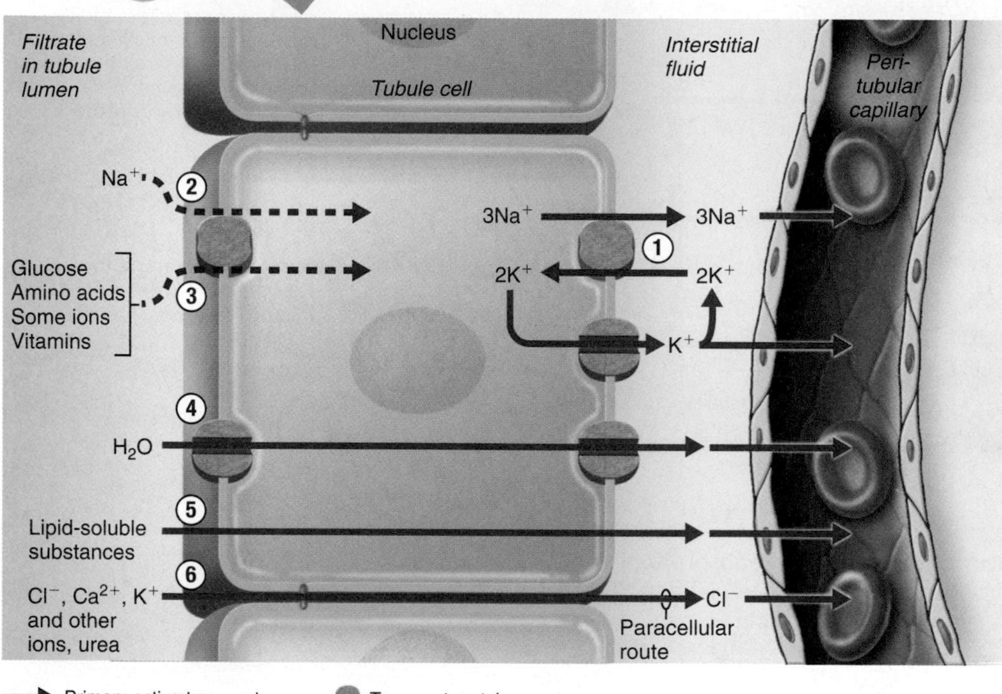

① At the basolateral membrane, Na^+ is pumped into the interstitial space by the Na^+-K^+ ATPase. Active Na^+ transport creates concentration gradients that drive:

② "Downhill" Na^+ entry at the luminal membrane.

③ Reabsorption of organic nutrients and certain ions by cotransport at the luminal membrane.

④ Reabsorption of water by osmosis. Water reabsorption increases the concentration of the solutes that are left behind. These solutes can then be reabsorbed as they move down their concentration gradients:

⑤ Lipid-soluble substances diffuse by the transcellular route.

⑥ Cl^- (and other anions), K^+, and urea diffuse by the paracellular route.

Figure 25.14 Reabsorption by PCT cells. Though not illustrated here, most organic nutrients reabsorbed in the PCT move through the basolateral membrane by facilitated diffusion. Microvilli have been omitted for simplicity.

they will follow their concentration gradients and be reabsorbed, even if this is "not desirable."

As they move through the tubule cells into the peritubular capillary blood, Na^+ ions also establish an electrical gradient that favors passive reabsorption of anions (primarily Cl^-) to restore electrical neutrality in the filtrate and plasma (Figure 25.14, ⑥).

Any plasma proteins that squeeze through the filtration membrane are removed from the filtrate in the proximal tubule by endocytosis and digested to their amino acids, which are moved into the peritubular blood.

Reabsorptive Capabilities of the Renal Tubules and Collecting Ducts

Table 25.1 compares the reabsorptive abilities of various regions of the renal tubules and collecting ducts.

Proximal Convoluted Tubule The entire renal tubule is involved in reabsorption to some degree, but the PCT cells are by far the most active "reabsorbers" and the events just described occur mainly in this tubular segment. Normally, the PCT reabsorbs *all* of the glucose, lactate, and amino acids in the filtrate and 65% of the Na^+ and water. Additionally, 80% of the filtered bicarbonate (HCO_3^-), 60% of the Cl^-, and about 55% of the K^+ are reclaimed in the PCT. The bulk of the reabsorption of electrolytes is accomplished by the time the filtrate reaches the loop of Henle. Nearly all of the uric acid and about half of the urea are reabsorbed in the proximal tubule, but both are later secreted back into the filtrate.

Loop of Henle Beyond the PCT, the permeability of the tubule epithelium changes dramatically. Here, for the first time, water reabsorption is not coupled to solute reabsorption. Water can leave the descending limb of the loop of Henle but *not* the ascending limb, where aquaporins are scarce or absent in the tubule membrane. For reasons that we will explain shortly, these permeability differences play a vital role in the kidneys' ability to form dilute and concentrated urine.

The rule for water is that it leaves the descending (but not the ascending) limb of Henle's loop, and the opposite is true for solutes. Virtually no solute reabsorption occurs in the descending

TABLE 25.1 Reabsorption Capabilities of Different Segments of the Renal Tubules and Collecting Ducts

TUBULE SEGMENT	SUBSTANCE REABSORBED	MECHANISM
Proximal Convoluted Tubule		
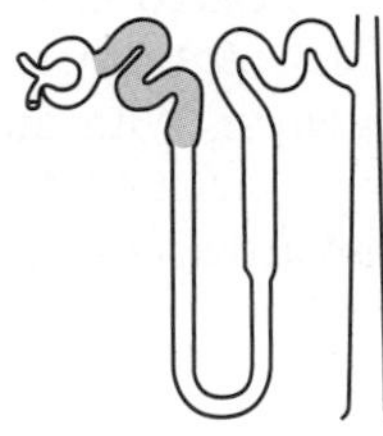	Sodium ions (Na^+)	Primary active transport via basolateral Na^+-K^+ pump; sets up electrochemical gradient for passive solute diffusion, osmosis, and secondary active transport (cotransport) with Na^+
	Virtually all nutrients (glucose, amino acids, vitamins)	Secondary active transport with Na^+
	Cations (K^+, Mg^{2+}, Ca^{2+}, and others)	Passive paracellular diffusion driven by electrochemical gradient
	Cl^-	Passive paracellular diffusion driven by electrochemical gradient
	HCO_3^-	Secondary active transport linked to H^+ secretion and Na^+ reabsorption (see Chapter 26)
	Water	Osmosis; driven by solute reabsorption (obligatory)
	Lipid-soluble solutes	Passive diffusion driven by the concentration gradient created by reabsorption of water
	Urea	Passive paracellular diffusion driven by chemical gradient; some transcellular facilitated diffusion may also occur
	Small proteins	Endocytosed by tubule cells and digested to amino acids within tubule cells
Loop of Henle		
Descending limb	Water	Osmosis
Ascending limb	Na^+, Cl^+, K^+	Secondary active transport of Cl^-, Na^+, and K^+ via Na^+-K^+-$2Cl^-$ cotransporter in thick portion; paracellular diffusion; Na^+-H^+ antiport
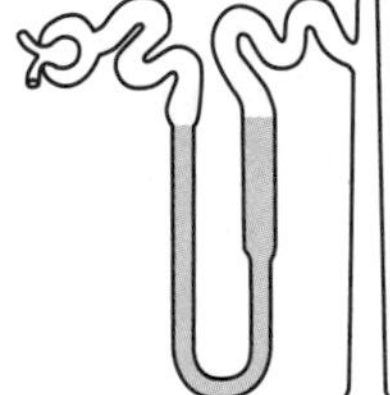	Ca^{2+}, Mg^{2+}	Passive paracellular diffusion driven by electrochemical gradient
Distal Convoluted Tubule		
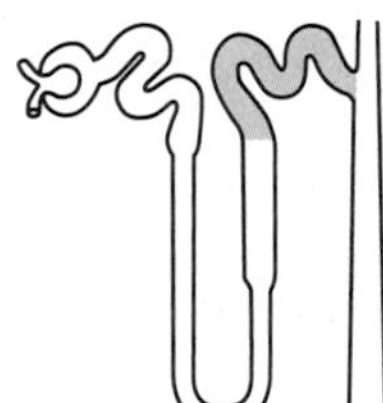	Na^+, Cl^-	Primary active Na^+ transport at basolateral membrane; secondary active transport at luminal membrane via Na^+-Cl^- symporter and channels; aldosterone-regulated at distal portion
	Ca^{2+}	Passive uptake via PTH-modulated channels in luminal membrane; primary and secondary active transport (antiport with Na^+) in basolateral membrane
Collecting Duct		
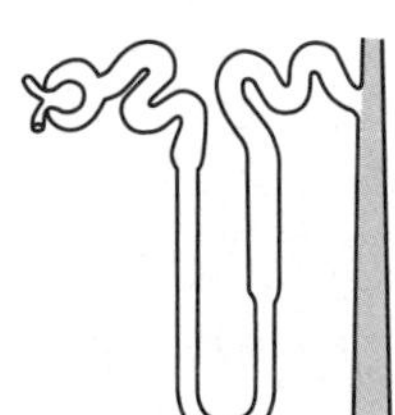	Na^+, H^+, K^+, HCO_3^-, Cl^-	Primary active transport of Na^+ (requires aldosterone); passive paracellular diffusion of some Cl^-; cotransport of H^+, Cl^-, and HCO_3^-; K^+ is both reabsorbed and secreted (aldosterone dependent), usually resulting in net K^+ secretion
	Water	Osmosis; controlled (facultative) water reabsorption; ADH required to insert aquaporins
	Urea	Facilitated diffusion in response to concentration gradient in the deep medulla region; recycles and contributes to medullary osmotic gradient

25

limb, but both active and passive reabsorption of solute occurs in the ascending limb. In the thin portion of the ascending limb, Na^+ moves passively down the concentration gradient created by water reabsorption. A Na^+-K^+-$2Cl^-$ symporter is the main means of Na^+ entry at the luminal surface in the thick portion of the ascending limb. A Na^+-K^+ ATPase operates at the basolateral membrane to create the ionic gradient that drives the symporter. The thick ascending limb also has Na^+-H^+ antiporters. In addition, some 50% of Na^+ passes via the paracellular route in this region.

Distal Convoluted Tubule and Collecting Duct By the time the DCT is reached, only about 10% of the originally filtered NaCl and 25% of the water remain in the tubule. Most reabsorption from this point on depends on the body's needs at the time and is regulated by hormones (mainly aldosterone for Na^+, ADH for water, and PTH for Ca^{2+} as we will describe in Chapter 26). If necessary, nearly all of the water and Na^+ reaching these regions can be reclaimed.

In the absence of antidiuretic hormone (ADH), the collecting ducts are relatively impermeable to water. Reabsorption of more water depends on the presence of ADH, which makes the collecting ducts more permeable to water by inserting aquaporins into the collecting duct luminal membranes.

Aldosterone "fine-tunes" reabsorption of the remaining Na^+. Decreased blood volume or blood pressure, low extracellular Na^+ concentration (hyponatremia), or high extracellular K^+ concentration (hyperkalemia) can cause the adrenal cortex to release aldosterone to the blood. Except for hyperkalemia (which *directly* stimulates the adrenal cortex to secrete aldosterone), these conditions promote the renin-angiotensin mechanism, which in turn prompts the release of aldosterone (see Figure 25.12). Aldosterone targets the principal cells of the collecting ducts and cells of the distal portion of the DCT (prodding them to synthesize and retain more luminal Na^+ and K^+ channels, and more basolateral Na^+-K^+ ATPases). As a result, little or no Na^+ leaves the body in urine. In the absence of aldosterone, much less Na^+ is reabsorbed by these segments, resulting in Na^+ losses of about 2% of Na^+ filtered daily, an amount incompatible with life.

25

Physiologically, aldosterone's role is to increase blood volume, and therefore blood pressure, by enhancing Na^+ reabsorption. In general, water follows Na^+ if it can. Aldosterone also reduces blood K^+ concentrations because aldosterone-induced reabsorption of Na^+ is coupled to K^+ secretion in the principal cells. That is, as Na^+ enters, K^+ moves into the lumen.

In contrast to aldosterone, which acts to conserve Na^+, atrial natriuretic peptide (ANP) reduces blood Na^+, thereby decreasing blood volume and blood pressure. Released by cardiac atrial cells when blood volume or blood pressure is elevated, ANP exerts several effects that lower blood Na^+ content, including direct inhibition of Na^+ reabsorption at the collecting ducts. These actions are described more fully in Chapter 26 (see Figure 26.9).

Step 3: Tubular Secretion

▶ Describe the importance of tubular secretion and list several substances that are secreted.

The failure of tubule cells to reabsorb some solutes is an important way of clearing plasma of unwanted substances. Another way is **tubular secretion**—essentially, reabsorption in reverse. Substances such as H^+, K^+, NH_4^+, creatinine, and certain organic acids either move into the filtrate from the peritubular capillaries through the tubule cells or are synthesized in the tubule cells and secreted. As a result, the urine eventually excreted contains *both filtered* and *secreted substances.* With one major exception (K^+), the PCT is the main site of secretion, but the cortical parts of the collecting ducts are also active (see Figure 25.18, p. 983).

Tubular secretion is important for

1. *Disposing of substances, such as certain drugs and metabolites, that are tightly bound to plasma proteins.* Because plasma proteins are generally not filtered, the substances they bind are not filtered and so must be secreted.
2. *Eliminating undesirable substances or end products that have been reabsorbed by passive processes.* Urea and uric acid, two nitrogenous wastes, are both handled in this way. Urea handling in the nephron is complicated and will be discussed on p. 981, but the net effect is that 40–50% of the urea in the filtrate is excreted.
3. *Ridding the body of excess K^+.* Because virtually all K^+ present in the filtrate is reabsorbed in the PCT and ascending loop of Henle, nearly all K^+ in urine is from aldosterone-driven active tubular secretion into the late DCT and collecting ducts.
4. *Controlling blood pH.* When blood pH drops toward the acidic end of its homeostatic range, the renal tubule cells actively secrete more H^+ into the filtrate and retain and generate more HCO_3^- (a base). As a result, the blood pH rises and the urine drains off the excess H^+. Conversely, when blood pH approaches the alkaline end of its range, Cl^- is reabsorbed instead of HCO_3^-, which is allowed to leave the body in urine. We will discuss the kidneys' role in pH homeostasis in more detail in Chapter 26.

CHECK YOUR UNDERSTANDING

10. In what part of the nephron does the majority of reabsorption occur?
11. How are primary and secondary active transport processes (both shown in Figure 25.14) different?
12. How does the movement of Na^+ drive the reabsorption of water and solutes?
13. List several substances that are secreted into the kidney tubules.

For answers, see Appendix G.

Regulation of Urine Concentration and Volume

▶ Describe the mechanisms responsible for the medullary osmotic gradient.

▶ Explain formation of dilute versus concentrated urine.

One crucial renal function is to keep the solute concentration of body fluids constant. We use *osmolality* to measure the amount of solutes in body fluids. A solution's **osmolality** (oz″mo-lal′ĭ-te) is the number of solute particles dissolved in 1 kg of water and reflects the solution's ability to cause osmosis. For any solution interfacing with a selectively permeable membrane, this ability, called *osmotic activity*, is determined only by the number of solute particles unable to pass through the membrane (called nonpenetrating solute particles) and is independent of their type. For example, 10 sodium ions have the same osmotic activity as 10 glucose molecules or 10 amino acids in the same volume of water.

Because 1 *osmol* (equivalent to 1 mole of particles) is a fairly large unit, the **milliosmol (mOsm)** (mil″e-oz′mōl), equal to 0.001 osmol, is generally used. In the discussion that follows, we use mOsm to mean mOsm/kg.

The kidneys keep the solute load of body fluids constant at about 300 mOsm, the osmotic concentration of blood plasma, by regulating urine concentration and volume. The kidneys accomplish this feat using **countercurrent mechanisms**. In the kidneys, the term *countercurrent* means that fluid flows in opposite directions through adjacent segments of the same tube connected by a hairpin turn (see Figure 25.16). These countercurrent mechanisms are (1) the interaction between the flow of filtrate through the ascending and descending limbs of the long loops of Henle of juxtamedullary nephrons (the *countercurrent multiplier*), and (2) the flow of blood through the ascending and descending portions of the vasa recta blood vessels (the *countercurrent exchanger*).* These countercurrent mechanisms establish and maintain an osmotic gradient extending from the cortex through the depths of the medulla **(Figure 25.15)**. This gradient allows the kidneys to vary urine concentration dramatically.

The osmolality of the filtrate entering the PCT is identical to that of plasma, about 300 mOsm. As we described earlier, because of PCT reabsorption of water *and* solutes, the filtrate is still isosmotic with plasma by the time it reaches the descending limb of the loop of Henle. However, its osmolality increases from 300 to about 1200 mOsm in the deepest part of the medulla **(Figure 25.16a)**.

How does this increase in concentration occur? The answer lies in the unique workings of the long loops of Henle of the juxtamedullary nephrons, and the vasa recta. Notice in Figure 25.16 that in each case the fluids involved—filtrate in the loops of Henle and blood in the vasa recta—first descend and then ascend through parallel limbs.

The Countercurrent Multiplier

First, we will follow filtrate processing through the loop of Henle, as portrayed in Figure 25.16a, to see how the loop functions as a **countercurrent multiplier** to establish the osmotic gradient. The countercurrent multiplier functions because of two factors:

1. **The descending limb of the loop of Henle is relatively impermeable to solutes and freely permeable to water.** Water passes osmotically out of the filtrate all along this limb because the osmolality of the medullary interstitial fluid increases all along the descending limb. (We will explain the mechanism of this increase shortly). The filtrate osmolality reaches its highest point (1200 mOsm) at the bend of the loop.
2. **The ascending limb is permeable to solutes, but not to water.** As the filtrate rounds the bend into the ascending limb, the tubule permeability changes, becoming impermeable to water and selectively permeable to salt. The Na^+ and Cl^- concentration in the filtrate entering the ascending limb is very high (and interstitial fluid concentrations of these two ions are lower). Na^+ and Cl^- reabsorption in

*The term "countercurrent" is commonly misunderstood to mean that the direction of fluid flow in the loops of Henle is opposite that of the blood in the vasa recta. In fact, there is no one-to-one relationship between individual loops of Henle and capillaries of the vasa recta as might be suggested by two-dimensional diagrams such as Figure 25.16. Instead, there are many tubules and capillaries packed together. Each tubule is surrounded by many blood vessels, whose flow is not necessarily counter to flow in that tubule (see Figure 25.7).

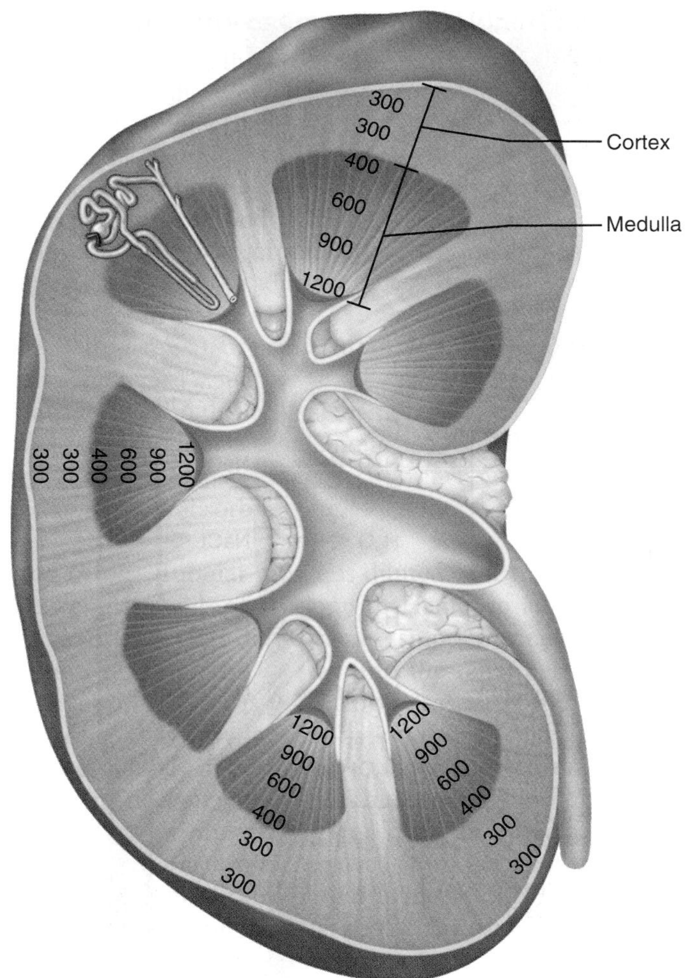

Figure 25.15 Osmotic gradient in the renal medulla. The osmolality of the interstitial fluid in the renal cortex is isotonic at 300 mOsm, but the osmolality of the interstitial fluid in the renal medulla increases progressively from 300 mOsm at the cortico-medullary junction to 1200 mOsm at the medullary-pelvis junction. One greatly enlarged nephron and its collecting duct are depicted to show their positions relative to the medullary gradient.

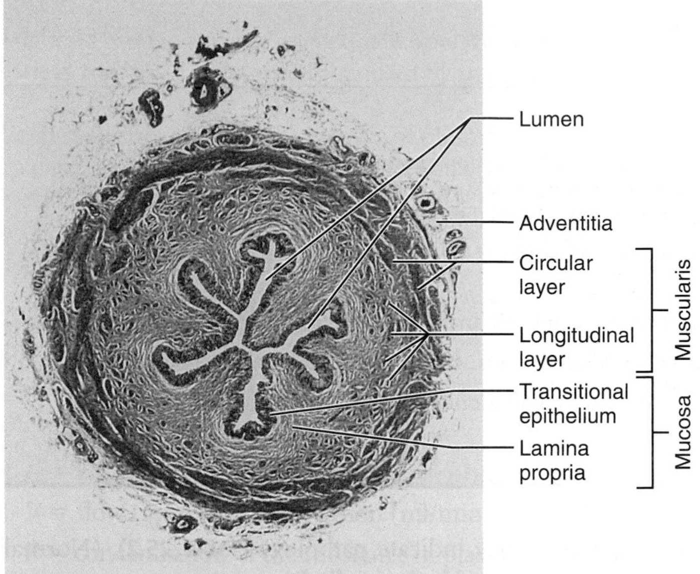

Figure 25.20 Cross-sectional view of the ureter wall (10×). The prominent mucosal folds seen in an empty ureter stretch and flatten to accommodate large pulses of urine.

From there, it descends behind the peritoneum and runs obliquely through the posterior bladder wall. This arrangement prevents backflow of urine during bladder filling because any increase in bladder pressure compresses and closes the distal ends of the ureters.

Histologically, the ureter wall has three layers. The transitional epithelium of its lining *mucosa* is continuous with that of the kidney pelvis superiorly and the bladder medially. Its middle *muscularis* is composed chiefly of two smooth muscle sheets: the internal longitudinal layer and the external circular layer. An additional smooth muscle layer, the external longitudinal layer, appears in the lower third of the ureter. The *adventitia* covering the ureter's external surface is typical fibrous connective tissue (**Figure 25.20**).

25

The ureter plays an active role in transporting urine. Incoming urine distends the ureter and stimulates its muscularis to contract, propelling urine into the bladder. (Urine does *not* reach the bladder through gravity alone.) The strength and frequency of the peristaltic waves are adjusted to the rate of urine formation. Each ureter is innervated by both sympathetic and parasympathetic fibers, but neural control of peristalsis appears to be insignificant compared to the way ureteral smooth muscle responds to stretch.

HOMEOSTATIC IMBALANCE

On occasion, calcium, magnesium, or uric acid salts in urine may crystallize and precipitate in the renal pelvis, forming **renal calculi** (kal′ku-li; *calculus* = little stone), or kidney stones. Most calculi are under 5 mm in diameter and pass through the urinary tract without causing problems. However, larger calculi can obstruct a ureter and block urine drainage. Increasing pressure in the kidney causes excruciating pain, which radiates from the flank to the anterior abdominal wall on the same side. Pain also occurs when the contracting ureter wall closes in on the sharp calculi as they are being eased through a ureter by peristalsis.

Predisposing conditions are frequent bacterial infections of the urinary tract, urine retention, high blood levels of calcium, and alkaline urine. Surgical removal of calculi has been almost entirely replaced by *shock wave lithotripsy*, a noninvasive procedure that uses ultrasonic shock waves to shatter the calculi. The pulverized, sandlike remnants of the calculi are then painlessly eliminated in the urine. People with a history of kidney stones are encouraged to acidify their urine by drinking cranberry juice and to ingest enough water to keep the urine dilute. ■

CHECK YOUR UNDERSTANDING

17. What are the three major nitrogenous wastes excreted in the urine?
18. A kidney stone blocking a ureter would interfere with urine flow to which organ? Why would the pain occur in waves?

For answers, see Appendix G.

Urinary Bladder

▶ Describe the general location, structure, and function of the urinary bladder.

The **urinary bladder** is a smooth, collapsible, muscular sac that stores urine temporarily. It is located retroperitoneally on the pelvic floor just posterior to the pubic symphysis. The prostate (part of the male reproductive system) lies inferior to the bladder neck, which empties into the urethra. In females, the bladder is anterior to the vagina and uterus.

The interior of the bladder has openings for both ureters and the urethra (**Figure 25.21**). The smooth, triangular region of the bladder base outlined by these three openings is the **trigone** (tri′gōn; *trigon* = triangle), important clinically because infections tend to persist in this region.

The bladder wall has three layers: a mucosa containing transitional epithelium, a thick muscular layer, and a fibrous adventitia (except on its superior surface, where it is covered by the peritoneum). The muscular layer, called the **detrusor muscle** (de-tru′sor; "to thrust out"), consists of intermingled smooth muscle fibers arranged in inner and outer longitudinal layers and a middle circular layer.

The bladder is very distensible and uniquely suited for its function of urine storage. When empty, the bladder collapses into its basic pyramidal shape and its walls are thick and thrown into folds (*rugae*). As urine accumulates, the bladder expands, becomes pear shaped, and rises superiorly in the abdominal cavity. The muscular wall stretches and thins, and rugae disappear. These changes allow the bladder to store more urine without a significant rise in internal pressure.

A moderately full bladder is about 12 cm (5 inches) long and holds approximately 500 ml (1 pint) of urine, but it can hold nearly double that if necessary. When tense with urine, it can be palpated well above the pubic symphysis. The maximum capacity of the bladder is 800–1000 ml and when it is

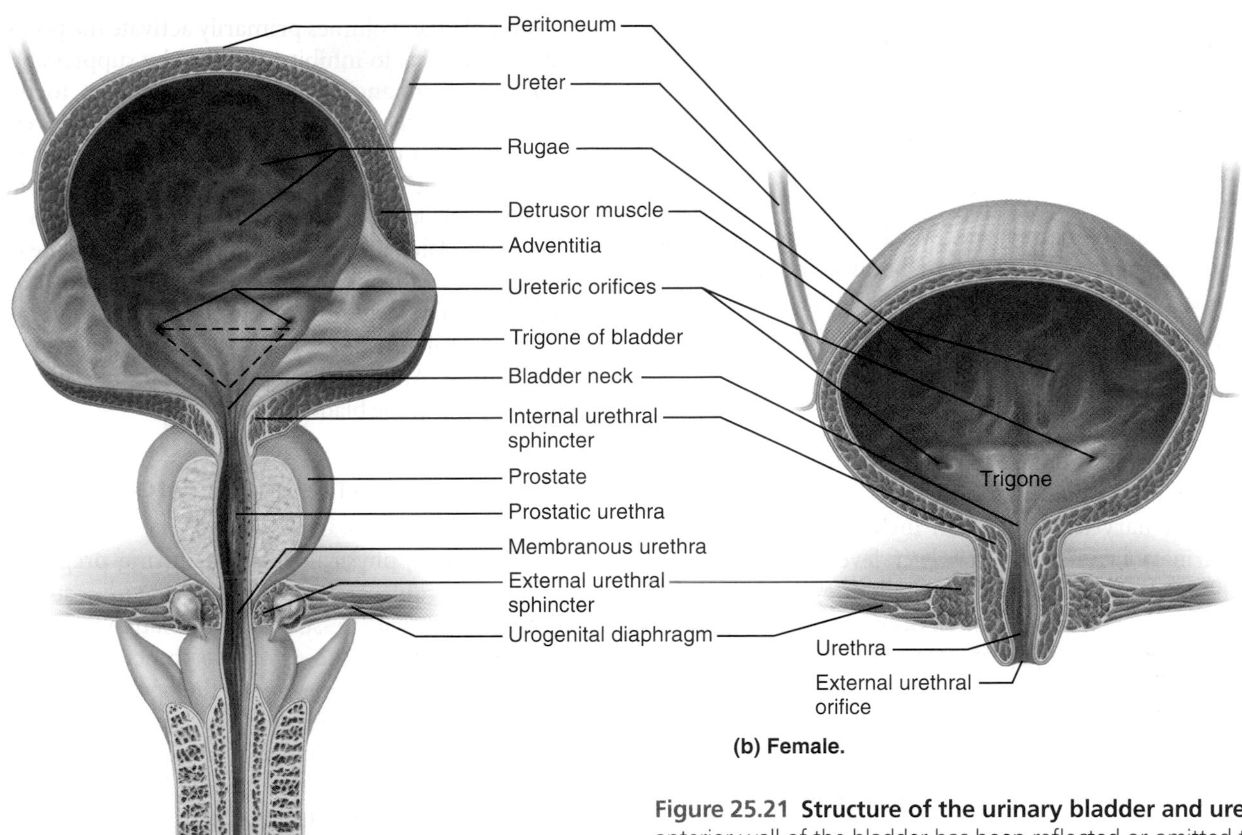

(a) Male. The long male urethra has three regions: prostatic, membranous and spongy.

(b) Female.

Figure 25.21 **Structure of the urinary bladder and urethra.** The anterior wall of the bladder has been reflected or omitted to reveal the position of the trigone.

overdistended, it may burst. Although urine is formed continuously by the kidneys, it is usually stored in the bladder until its release is convenient.

Urethra

- ▶ Describe the general location, structure, and function of the urethra.
- ▶ Compare the course, length, and functions of the male urethra with those of the female.

The **urethra** is a thin-walled muscular tube that drains urine from the bladder and conveys it out of the body. The epithelium of its mucosal lining is mostly pseudostratified columnar epithelium. However, near the bladder it becomes transitional epithelium, and near the external opening it changes to a protective stratified squamous epithelium.

At the bladder-urethra junction a thickening of the detrusor smooth muscle forms the **internal urethral sphincter** (Figure 25.21). This involuntary sphincter keeps the urethra closed when urine is not being passed and prevents leaking between voiding. This sphincter is unusual in that contraction opens it and relaxation closes it.

The **external urethral sphincter** surrounds the urethra as it passes through the *urogenital diaphragm.* This sphincter is formed of skeletal muscle and is voluntarily controlled. The *levator ani* muscle of the pelvic floor also serves as a voluntary constrictor of the urethra (see Table 10.7, pp. 344–345).

The length and functions of the urethra differ in the two sexes. In females the urethra is only 3–4 cm (1.5 inches) long and tightly bound to the anterior vaginal wall by fibrous connective tissue. Its external opening, the **external urethral orifice**, lies anterior to the vaginal opening and posterior to the clitoris.

In males the urethra is approximately 20 cm (8 inches) long and has three regions. The **prostatic urethra**, about 2.5 cm (1 inch) long, runs within the prostate. The **membranous urethra**, which runs through the urogenital diaphragm, extends about 2 cm from the prostate to the beginning of the penis. The **spongy urethra**, about 15 cm long, passes through the penis and opens at its tip via the **external urethral orifice**. The male urethra has a double function: It carries semen as well as urine out of the body. We discuss the reproductive function of the male urethra in Chapter 27.

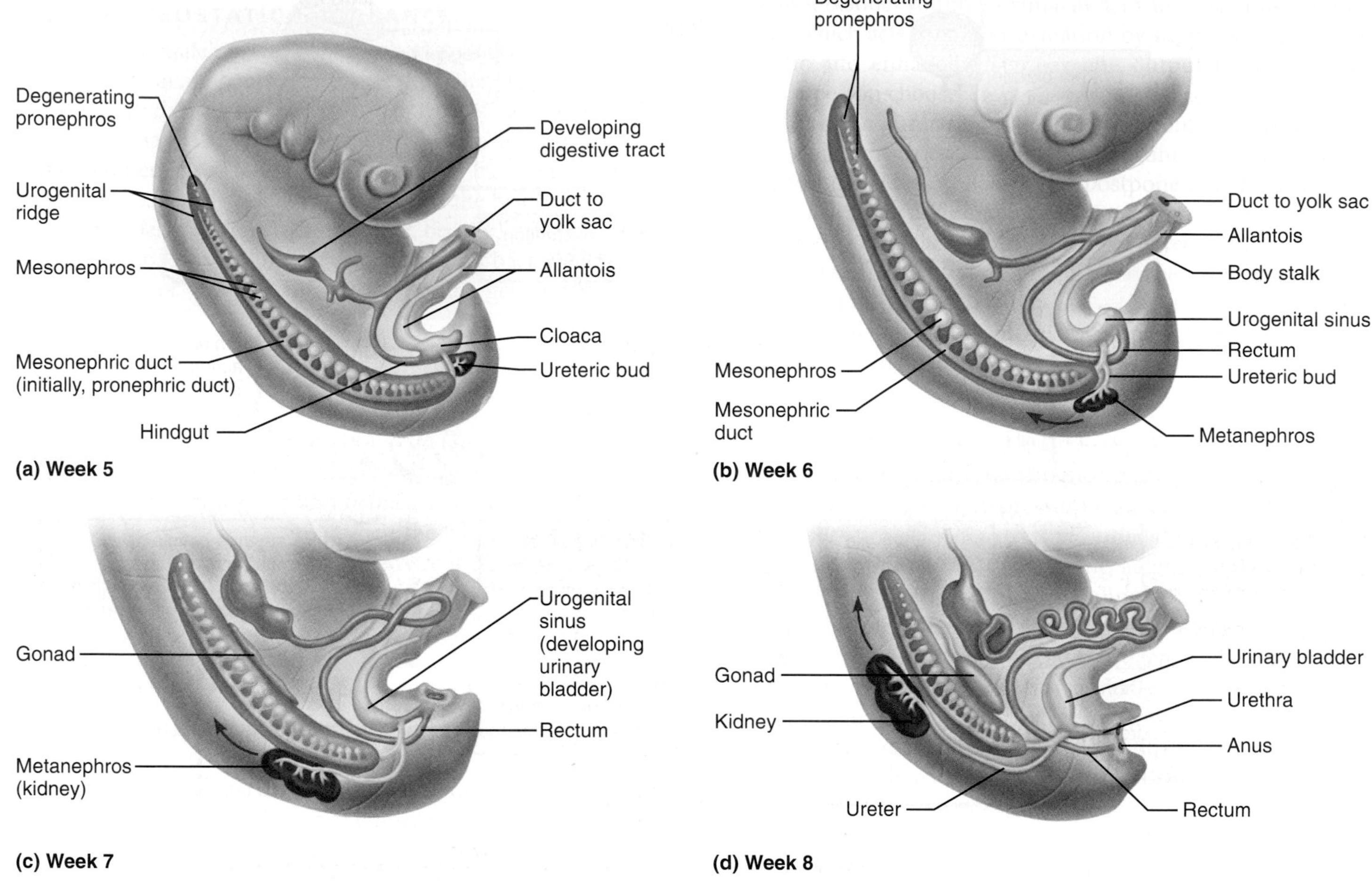

Figure 25.23 Development of the urinary system of the embryo. Direction of metanephros migration as it develops is indicated by red arrows.

25

When ascending from the pelvis the kidneys are very close together, and in 1 out of 600 people they fuse across the midline, forming a single, U-shaped *horseshoe kidney*. This condition is usually asymptomatic, but it may be associated with other kidney abnormalities, such as obstructed drainage, that place a person at risk for frequent kidney infections.

Hypospadias (hi″po-spa′de-as), found in male infants only, is the most common congenital abnormality of the urethra. It occurs when the urethral orifice is located on the ventral surface of the penis. This problem is corrected surgically when the child is around 12 months old.

Polycystic kidney disease (*PKD*) is a group of disorders characterized by the presence of many fluid-filled cysts in the kidneys, which interfere with renal function, ultimately leading to renal failure. These disorders can be grouped into two general forms. The less severe form is inherited in an autosomal dominant manner and is much more common, affecting 1 in 500 people. The cysts develop so gradually that they produce no symptoms until about 40 years of age. Then both kidneys begin to enlarge as blisterlike cysts containing fluid accumulate. The damage caused by these cysts progresses slowly, and many victims live without problems until their 60s. Ultimately, however, the kidneys become "knobby" and grossly enlarged, reaching a mass of up to 14 kg (30 lb) each.

The much less common and more severe form follows an autosomal recessive pattern of inheritance. Almost half of newborns with recessive PKD die just after birth, and survivors generally develop renal failure in early childhood. Recessive PKD results from a mutation in a single huge gene, but the dominant form of PKD (described above) is usually caused by a mutation in one of two different genes, which code for proteins involved in cell signaling. It is not yet clear how defects in these proteins lead to cyst formation. As yet, the only treatments are the usual treatments for kidney failure—renal dialysis or a kidney transplant. ■

Because its bladder is very small and its kidneys are less able to concentrate urine for the first two months, a newborn baby voids 5 to 40 times daily, depending on fluid intake. By 2 months of age, the infant is voiding approximately 400 ml/day, and the amount steadily increases until adolescence, when adult urine output (about 1500 ml/day) is achieved.

Incontinence, the inability to control micturition, is normal in infants because they have not yet learned to control the external urethral sphincter. Reflex voiding occurs each time a baby's bladder fills enough to activate the stretch receptors. Control of the voluntary urethral sphincter goes hand in hand with nervous system development. By 15 months, most toddlers know when they have voided. By 18 months, they can usually hold urine for about two hours. By 24 months, some children are ready to begin toilet training. Daytime control usually is achieved first. It is unrealistic to expect complete nighttime control before age 4.

From childhood through late middle age, most urinary system problems are infectious conditions. *Escherichia coli* (esh″ĕ-rik′e-ah ko′li) bacteria are normal residents of the digestive tract and generally cause no problems there, but these bacteria account for 80% of all urinary tract infections. *Sexually transmitted diseases* can also inflame the urinary tract and clog some of its ducts. Childhood streptococcal infections such as strep throat and scarlet fever, if not treated promptly, may cause long-term inflammatory renal damage.

Only about 3% of elderly people have histologically normal kidneys, and kidney function declines with advancing age. The kidneys shrink as the nephrons decrease in size and number, and the tubule cells become less efficient. By age 80, the GFR is only half that of young adults, possibly due to atherosclerotic narrowing of the renal arteries. Diabetics are particularly at risk for renal disease, and nearly 50% of those who have had diabetes mellitus for 20 years are in renal failure.

The bladder of an aged person is shrunken, with less than half the capacity of a young adult (250 ml versus 600 ml). Loss of bladder tone causes an annoying increase in frequency of micturition. *Nocturia* (nok-tu′re-ah), the need to get up during the night to urinate, plagues almost two-thirds of this population. Many people eventually experience incontinence, which can usually be treated with exercise, medications, or surgery.

CHECK YOUR UNDERSTANDING

22. Name the three sets of embryonic kidneys in the order that they develop.
23. List two factors that might contribute to urinary retention in elderly men.

For answers, see Appendix G.

The ureters, urinary bladder, and urethra play important roles in transporting, storing, and eliminating urine from the body, but when the term "urinary system" is used, it is the kidneys that capture center stage. As summarized in *Making Connections* in Chapter 26, other organ systems of the body contribute to the well-being of the urinary system in many ways. In turn, without continuous kidney function, the electrolyte and fluid balance of the blood is dangerously disturbed, and internal body fluids quickly become contaminated with nitrogenous wastes. No body cell can escape the harmful effects of such imbalances.

Now that we have described renal mechanisms, we are ready to integrate kidney function into the larger topic of fluid and electrolyte balance in the body—the focus of Chapter 26.

RELATED CLINICAL TERMS

Acute glomerulonephritis (GN) (glo-mer″u-lo-nef-ri′tis) Inflammation of the glomeruli, leading to increased permeability of the filtration membrane. In some cases, circulating immune complexes (antibodies bound to foreign substances, such as streptococcal bacteria) become trapped in the glomerular basement membranes. In other cases, immune responses are mounted against one's own kidney tissues, leading to glomerular damage. In either case, the inflammatory response that follows damages the filtration membrane, allowing blood proteins and even blood cells to pass into the renal tubules and into the urine. As the osmotic pressure of blood drops, fluid seeps from the bloodstream into the tissue spaces, causing bodywide edema. Renal shutdown requiring dialysis may occur temporarily, but normal renal function usually returns within a few months. If permanent glomerular damage occurs, chronic GN and ultimately renal failure result.

Bladder cancer Bladder cancer, three times more common in men than in women, accounts for about 2% of all cancer deaths. It usually involves neoplasms of the bladder's lining epithelium and may be induced by carcinogens from the environment or the workplace that end up in urine. Smoking, exposure to industrial chemicals, and arsenic in drinking water also have been linked to bladder cancer. Blood in the urine is a common warning sign.

Cystocele (sis′to-sēl; *cyst* = a sac, the bladder; *cele* = hernia, rupture) Herniation of the urinary bladder into the vagina; a common result of tearing of the pelvic floor muscles during childbirth.

Cystoscopy (sis-tos′ko-pe; *cyst* = bladder; *scopy* = observation) Procedure in which a thin viewing tube is threaded into the bladder through the urethra to examine the bladder's mucosal surface.

Diabetes insipidus (in-sĭ′pĭ-dus; *insipid* = tasteless, bland) Condition in which large amounts (up to 40 L/day) of dilute urine flush from the body; results from malfunction or deficiency of aquaporins or ADH receptors in the collecting duct (nephrogenic diabetes insipidus), or little or no ADH release due to injury to, or a tumor in, the hypothalamus or posterior pituitary. Can lead to severe dehydration and electrolyte imbalances unless the individual drinks large volumes of liquids. See Chapter 16, p. 608.

Intravenous pyelogram (IVP) (pi′ĕ-lo-gram; *pyelo* = kidney pelvis; *gram* = written) An X ray of the kidneys and ureters obtained after intravenous injection of a contrast medium (as in Figure 25.19). Allows assessment for obstructions, viewing of renal anatomy (pelvis and calyces), and determination of rate of excretion of the contrast medium.

Nephrotoxin A substance (heavy metal, organic solvent, or bacterial toxin) that is toxic to the kidney.

Nocturnal enuresis (NE) (en″u-re′sis) An inability to control urination at night during sleep; bed-wetting. In children over 6, called primary NE if control has never been achieved and secondary NE if control was achieved and then lost. Secondary NE often has psychological causes. Primary NE is more common and results from a combination of inadequate nocturnal ADH production, unusually sound sleep, or a small bladder capacity. Synthetic ADH often corrects the problem.

Renal infarct Area of dead, or necrotic, renal tissue due to blockage of the vascular supply to the kidney or hemorrhage. A common cause of localized renal infarct is obstruction of an interlobar artery. Because interlobar arteries are end arteries (do not anastomose), their obstruction leads to ischemic necrosis of the portions of the kidney they supply.

Urinalysis Analysis of urine as an aid to diagnosing health or disease. The most significant indicators of disease in urine are proteins, glucose, acetone, blood, and pus.

Urologist (u-rol′o-jist) Physician who specializes in diseases of urinary structures in both sexes and in diseases of the reproductive tract of males.

CHAPTER SUMMARY

Media study tools that could provide you additional help in reviewing specific key topics of Chapter 25 are referenced below.

iP = ***Interactive Physiology***

Kidney Anatomy (pp. 961–969)

Location and External Anatomy (pp. 961–962)

1. The paired kidneys are retroperitoneal in the superior lumbar region.
2. A fibrous capsule, a perirenal fat capsule, and renal fascia surround each kidney. The perirenal fat capsule helps hold the kidneys in position.

Internal Anatomy (pp. 962–963)

3. A kidney has a superficial cortex, a deeper medulla consisting mainly of medullary pyramids, and a medial pelvis. Extensions of the pelvis (calyces) surround and collect urine draining from the apices of the medullary pyramids.

Blood and Nerve Supply (pp. 963–964)

4. The kidneys receive 25% of the total cardiac output per minute.
5. The vascular pathway through a kidney is as follows: renal artery → segmental arteries → interlobar arteries → arcuate arteries → cortical radiate arteries → afferent arterioles → glomeruli → efferent arterioles → peritubular capillary beds → cortical radiate veins → arcuate veins → interlobar veins → renal vein.
6. The nerve supply of the kidneys is derived from the renal plexus.

Nephrons (pp. 964–969)

25

7. Nephrons are the structural and functional units of the kidneys.
8. Each nephron consists of a glomerulus (a high-pressure capillary bed) and a renal tubule. Subdivisions of the renal tubule (from the glomerulus) are the glomerular capsule, proximal convoluted tubule, loop of Henle, and distal convoluted tubule. A second capillary bed, the low-pressure peritubular capillary bed, is closely associated with the renal tubule of each nephron.
9. The more numerous cortical nephrons are located almost entirely in the cortex; only a small part of their loop of Henle penetrates into the medulla. Glomeruli of juxtamedullary nephrons are located at the cortex-medulla junction, and their loop of Henle dips deeply into the medulla. Instead of directly forming peritubular capillaries, the efferent arterioles of many of the juxtamedullary nephrons form unique bundles of straight vessels, called vasa recta, that serve tubule segments in the medulla. Juxtamedullary nephrons and the vasa recta play an important role in establishing the medullary osmotic gradient.
10. Collecting ducts receive urine from many nephrons and help concentrate urine. They form the medullary pyramids.
11. The juxtaglomerular apparatus is at the point of contact between the afferent arteriole and the most distal part of the ascending limb of the loop of Henle. It consists of the granular cells, the macula densa, and extraglomerular mesangial cells.

iP Urinary System; Topic: Anatomy Review, pp. 1–20.

12. The filtration membrane consists of the fenestrated glomerular endothelium, the intervening basement membrane, and the podocyte-containing visceral layer of the glomerular capsule. It permits free passage of substances smaller than (most) plasma proteins.

Kidney Physiology: Mechanisms of Urine Formation (pp. 969–984)

1. Functions of the nephrons include glomerular filtration, tubular reabsorption, and tubular secretion. Via these functional processes, the kidneys regulate the volume, composition, and pH of the blood, and eliminate nitrogenous metabolic wastes.

Step 1: Glomerular Filtration (pp. 969–974)

2. The glomeruli function as filters. High glomerular blood pressure (55 mm Hg) occurs because the glomeruli are fed and drained by arterioles, and the afferent arterioles are larger in diameter than the efferent arterioles.
3. About one-fifth of the plasma flowing through the kidneys is filtered from the glomeruli into the renal tubules.
4. Usually about 10 mm Hg, the net filtration pressure (NFP) is determined by the relationship between forces favoring filtration (glomerular hydrostatic pressure) and forces that oppose it (capsular hydrostatic pressure and blood colloid osmotic pressure).
5. The glomerular filtration rate (GFR) is directly proportional to the net filtration pressure and is about 125 ml/min (180 L/day).
6. Renal autoregulation, which enables the kidneys to maintain a relatively constant renal blood flow and glomerular filtration rate, involves a myogenic mechanism and a tubuloglomerular feedback mechanism mediated by the macula densa.
7. Extrinsic control of GFR, via nerves and hormones, maintains blood pressure. Strong sympathetic nervous system activation causes constriction of the afferent arterioles, which decreases filtrate formation and stimulates renin release by the granular cells.

iP Urinary System; Topic: Glomerular Filtration, pp. 1–15.

8. The renin-angiotensin mechanism mediated by the granular cells raises systemic blood pressure via generation of angiotensin II, which promotes aldosterone secretion.

Step 2: Tubular Reabsorption (pp. 974–978)

9. During tubular reabsorption, needed substances are removed from the filtrate by the tubule cells and returned to the peritubular capillary blood. The primary active transport of Na^+ by a Na^+-K^+ ATPase pump at the basolateral membrane accounts for Na^+ reabsorption and establishes the electrochemical gradient that drives the reabsorption of most other solutes and H_2O. Na^+ enters at the luminal surface of the tubule cell via facilitated diffusion through channels or as part of a cotransport mechanism.

10. Passive tubular reabsorption is driven by electrochemical gradients established by active reabsorption of sodium ions. Water, many ions, and various other substances (for example, urea) are reabsorbed passively by diffusion via transcellular or paracellular pathways.
11. Secondary active tubular reabsorption occurs by cotransport with Na^+ via transport proteins. Transport of such substances is limited by the number of carriers available. Actively reabsorbed substances include glucose, amino acids, and some ions.
12. The proximal tubule cells are most active in reabsorption. Most of the nutrients, 65% of the water and sodium ions, and the bulk of actively transported ions are reabsorbed in the proximal convoluted tubules.
13. Reabsorption of additional sodium ions and water occurs in the distal tubules and collecting ducts and is hormonally controlled. Aldosterone increases the reabsorption of sodium; antidiuretic hormone enhances water reabsorption by the collecting ducts.

Step 3: Tubular Secretion (p. 978)

14. Tubular secretion is a means of adding substances to the filtrate (from the blood or tubule cells). It is an active process that is important in eliminating drugs, certain wastes, and excess ions and in maintaining the acid-base balance of the blood.

Regulation of Urine Concentration and Volume (pp. 978–983)

15. The graduated hyperosmolality of the medullary fluids (largely due to the cycling of NaCl and urea) ensures that the filtrate reaching the distal convoluted tubule is dilute (hypo-osmolar). This allows urine with osmolalities ranging from 50 to 1200 mOsm to be formed.
    - The descending limb of the loop of Henle is permeable to water, which leaves the filtrate and enters the medullary interstitium. The filtrate and medullary fluid at the bend of the loop of Henle are hyperosmolar.
    - The ascending limb is impermeable to water. Na^+ and Cl^- move out of the filtrate into the interstitial space, passively in the thin segment and actively in the thick segment. The filtrate becomes more dilute.
    - As filtrate flows through the collecting ducts in the inner medulla, urea diffuses into the interstitial space. From here, urea reenters the ascending thin limb and is recycled.
    - The blood flow in the vasa recta is sluggish, and the contained blood equilibrates with the medullary interstitial fluid. Hence, blood exiting the medulla in the vasa recta is nearly isotonic to blood plasma and the high solute concentration of the medulla is maintained.
16. In the absence of antidiuretic hormone, dilute urine is formed because the dilute filtrate reaching the collecting duct is simply allowed to pass from the kidneys.
17. As blood levels of antidiuretic hormone rise, the collecting ducts become more permeable to water, and water moves out of the filtrate as it flows through the hyperosmotic medullary areas. Consequently, more concentrated urine is produced, and in smaller amounts.

iP Urinary System; Topics: Early Filtrate Processing, pp. 1–22; Late Filtrate Processing, pp. 1–13.

Renal Clearance (p. 984)

18. Renal clearance is the volume of plasma that is completely cleared of a particular substance per minute. Studies of renal clearance provide information about renal function or the course of renal disease.
19. Renal failure has serious consequences: The kidneys are unable to concentrate urine, acid-base and electrolyte imbalances occur, and nitrogenous wastes accumulate in the blood.

Urine (pp. 984–985)

1. Urine is typically clear, yellow, aromatic, and slightly acidic. Its specific gravity ranges from 1.001 to 1.035.
2. Urine is 95% water; solutes include nitrogenous wastes (urea, uric acid, and creatinine) and various ions (always sodium, potassium, sulfate, and phosphate).
3. Substances not normally found in urine include glucose, proteins, erythrocytes, leukocytes, hemoglobin, and bile pigments.
4. Daily urinary volume is typically 1.5–1.8 L, but this depends on the state of hydration of the body.

Ureters (pp. 985–986)

1. The ureters are slender tubes running retroperitoneally from each kidney to the bladder. They conduct urine by peristalsis from the renal pelvis to the urinary bladder.

Urinary Bladder (pp. 986–987)

1. The urinary bladder, which functions to store urine, is a distensible muscular sac that lies posterior to the pubic symphysis. It has two inlets (ureters) and one outlet (urethra) that outline the trigone. In males, the prostate surrounds the bladder outlet.
2. The bladder wall consists of a transitional epithelium–containing mucosa, a three-layered detrusor muscle, and an adventitia.

Urethra (pp. 987-988)

1. The urethra is a muscular tube that conveys urine from the bladder to the body exterior.
2. Where the urethra leaves the bladder, it is surrounded by an internal urethral sphincter, an involuntary smooth muscle sphincter. Where it passes through the urogenital diaphragm, the voluntary external urethral sphincter is formed by skeletal muscle.
3. In females the urethra is 3–4 cm long and conducts only urine. In males it is 20 cm long and conducts both urine and semen.

Micturition (p. 988)

1. Micturition is emptying of the bladder.
2. Stretching of the bladder wall by accumulating urine initiates the micturition reflex. In infants, this is a simple spinal reflex: Parasympathetic fibers are excited (and sympathetic fibers inhibited), causing the detrusor muscle to contract and the internal urethral sphincter to open. In adults, pontine storage and micturition centers can override this simple reflex.
3. Because the external sphincter is voluntarily controlled, micturition can usually be delayed temporarily.

Developmental Aspects of the Urinary System (pp. 988–991)

1. Three sets of kidneys (pronephric, mesonephric, and metanephric) develop from the intermediate mesoderm. The metanephros is excreting urine by the third month of development.
2. Common congenital abnormalities are horseshoe kidney, polycystic kidney, and hypospadias.
3. The kidneys of newborns are less able to concentrate urine; their bladder is small and voiding is frequent. Neuromuscular maturation generally allows toilet training for micturition to begin by 24 months of age.
4. The most common urinary system problems in children and young to middle-aged adults are bacterial infections.
5. With age, nephrons are lost, the filtration rate decreases, and tubule cells become less efficient at concentrating urine.
6. Bladder capacity and tone decrease with age, leading to frequent micturition and (often) incontinence. Urinary retention is a common problem of elderly men.

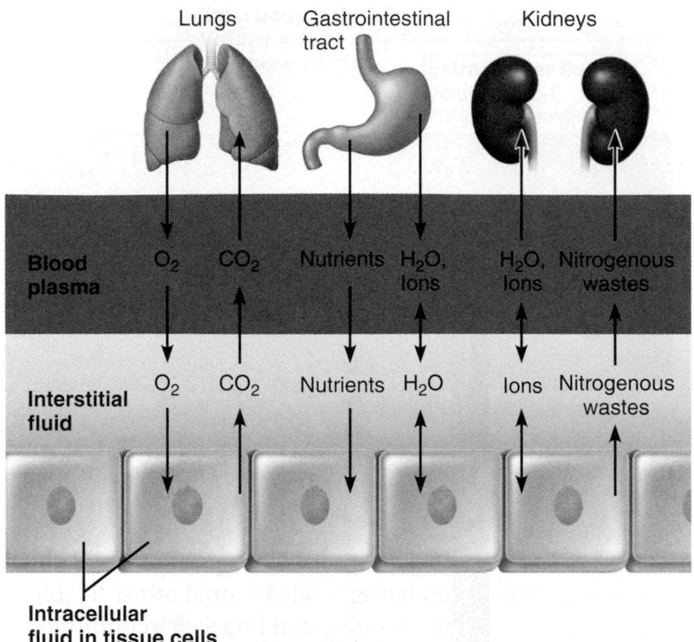

Figure 26.3 Exchange of gases, nutrients, water, and wastes between the three fluid compartments of the body.

Electrolytes are the most abundant solutes in body fluids and determine most of their chemical and physical reactions, but they do not constitute the *bulk* of dissolved solutes in these fluids. Proteins and some of the nonelectrolytes (phospholipids, cholesterol, and triglycerides) found in the ECF are large molecules. They account for about 90% of the mass of dissolved solutes in plasma, 60% in the IF, and 97% in the ICF.

Fluid Movement Among Compartments

26

The continuous exchange and mixing of body fluids are regulated by osmotic and hydrostatic pressures. Although water moves freely between the compartments along osmotic gradients, solutes are unequally distributed because of their size, electrical charge, or dependence on transport proteins. Anything that changes the solute concentration in any compartment leads to net water flows.

Figure 26.3 summarizes the exchanges of gases, solutes, and water across the body's borders and between the three fluid compartments within the body. In general, substances must pass through both the plasma and IF in order to reach the ICF. In the lungs, gastrointestinal tract, and kidneys, exchanges between the "outside world" and the plasma occur almost continuously. These exchanges alter plasma composition and volume, with the plasma serving as the "highway" for delivering substances throughout the body (see Chapter 17). Compensating adjustments between the plasma and the other two fluid compartments follow quickly so that balance is restored.

Exchanges between plasma and IF occur across capillary membranes. We described in detail the pressures driving these fluid movements in Chapter 19 on pp. 718–719. Here we will simply review the outcome of these mechanisms. Nearly protein-free plasma is forced out of the blood into the interstitial space by the hydrostatic pressure of blood. This filtered fluid is then almost completely reabsorbed into the bloodstream in response to the colloid osmotic (oncotic) pressure of plasma proteins. Under normal circumstances, the small net leakage that remains behind in the interstitial space is picked up by lymphatic vessels and returned to the blood.

Exchanges between the IF and ICF occur across plasma membranes and depend on the membranes' complex permeability properties. As a general rule, two-way osmotic flow of water is substantial. But ion fluxes are restricted and, in most cases, ions move selectively by active transport or through channels. Movements of nutrients, respiratory gases, and wastes are typically unidirectional. For example, glucose and oxygen move into the cells and metabolic wastes move out.

Many factors can change ECF and ICF volumes. Because water moves freely between compartments, however, the osmolalities of all body fluids are equal (except during the first few minutes after a change in one of the fluids occurs). Increasing the ECF solute content (mainly the NaCl concentration) can be expected to cause osmotic and volume changes in the ICF—namely, a shift of water out of the cells. Conversely, decreasing ECF osmolality causes water to move into the cells. Thus, the ICF volume is determined by the ECF solute concentration. These concepts underlie all events that control fluid balance in the body and should be understood thoroughly.

CHECK YOUR UNDERSTANDING

1. Which do you have more of, extracellular or intracellular fluid? Plasma or interstitial fluid?
2. What is the major cation in the ECF? In ICF? What are the intracellular anion counterparts of ECF's chloride ions?
3. If you eat salty pretzels without drinking, what happens to the volume of your extracellular fluid? Explain.

For answers, see Appendix G.

Water Balance and ECF Osmolality

- List the routes by which water enters and leaves the body.
- Describe feedback mechanisms that regulate water intake and hormonal controls of water output in urine.
- Explain the importance of obligatory water losses.
- Describe possible causes and consequences of dehydration, hypotonic hydration, and edema.

For the body to remain properly hydrated, water intake must equal water output. *Water intake* varies widely from person to person and is strongly influenced by habit, but it is typically about 2500 ml a day in adults **(Figure 26.4)**. Most water enters the body through ingested liquids and solid foods. Body water produced by cellular metabolism is called **metabolic water** or **water of oxidation**.

Water output occurs by several routes. Water that vaporizes out of the lungs in expired air or diffuses directly through the skin is called **insensible water loss**. Some is lost in obvious

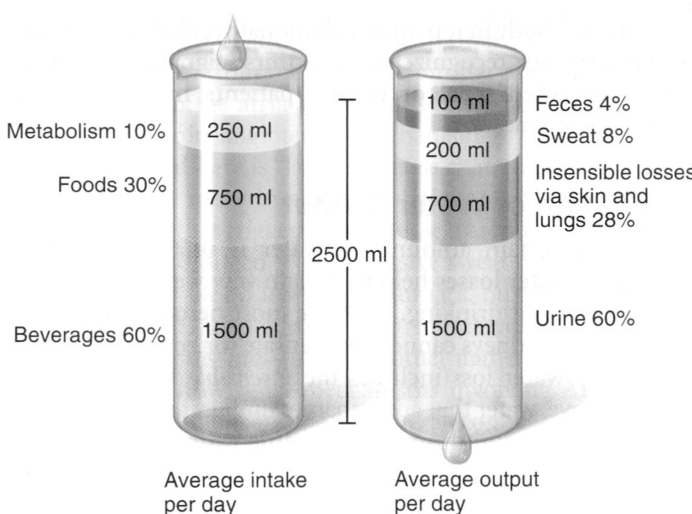

Figure 26.4 Major sources of water intake and output. When intake and output are in balance, the body is adequately hydrated.

perspiration and in feces. The balance (about 60%) is excreted by the kidneys in urine.

Healthy people have a remarkable ability to maintain the tonicity of their body fluids within very narrow limits (280–300 mOsm/kg). A rise in plasma osmolality triggers (1) thirst, which prompts us to drink water, and (2) release of antidiuretic hormone (ADH), which causes the kidneys to conserve water and excrete concentrated urine. On the other hand, a decline in osmolality inhibits both thirst and ADH release, the latter followed by output of large volumes of dilute urine.

Regulation of Water Intake

The **thirst mechanism** is the driving force for water intake. An increase in plasma osmolality of only 2–3% excites the hypothalamic *thirst center*. A dry mouth also occurs because the rise in plasma colloid osmotic pressure causes less fluid to leave the bloodstream. Because the salivary glands obtain the water they require from the blood, they produce less saliva, reinforcing the drive to drink. A decrease in blood volume (or pressure) also triggers the thirst mechanism. However, because a substantial decrease (10–15%) is required, this is the less potent stimulus.

The hypothalamic thirst center neurons are stimulated when their *osmoreceptors* lose water by osmosis to the hypertonic ECF, or are activated by angiotensin II, by baroreceptor inputs, or other stimuli. Collectively, these events cause a subjective sensation of thirst, which motivates us to get a drink **(Figure 26.5)**. This mechanism helps explain why it is that some cocktail lounges and bars provide free *salty* snacks to their patrons.

Curiously, thirst is quenched almost as soon as we begin drinking water, even though the water has yet to be absorbed into the blood. The damping of thirst begins as the mucosa of the mouth and throat is moistened and continues as stretch receptors in the stomach and intestine are activated, providing

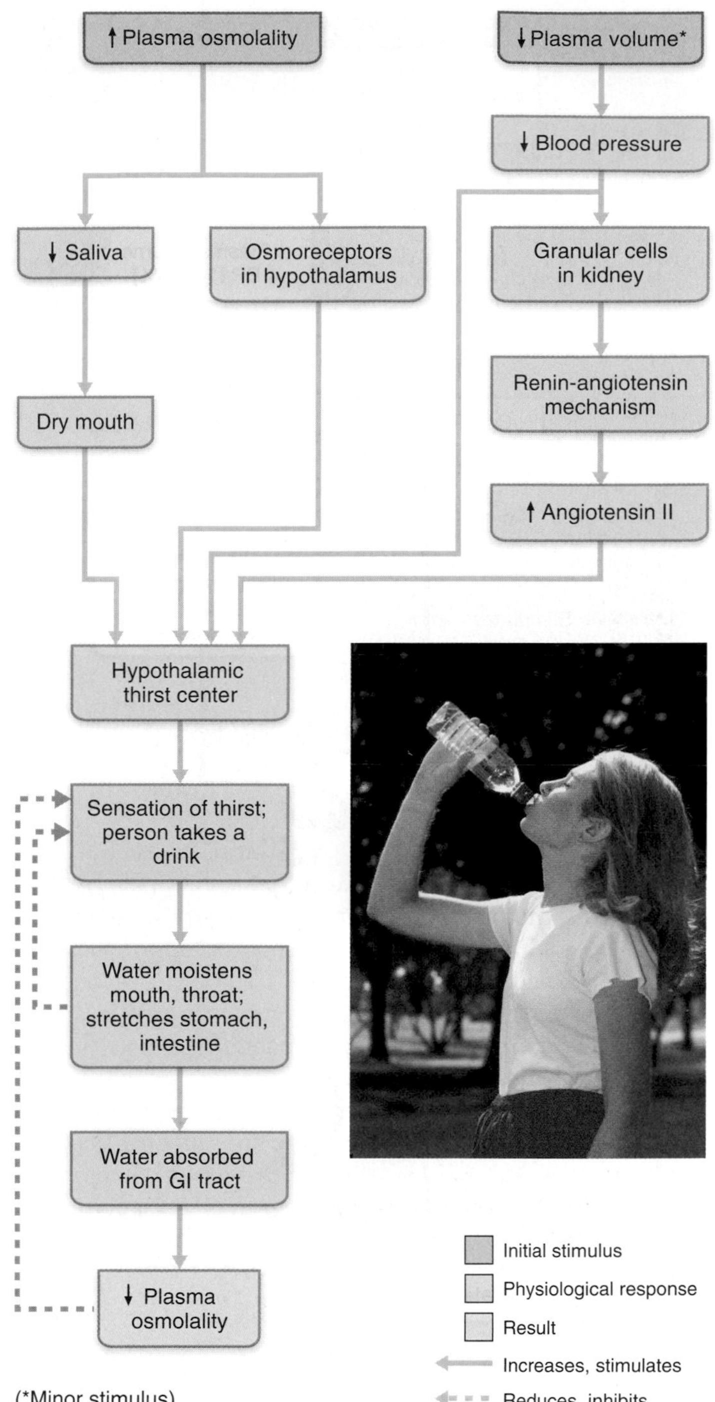

Figure 26.5 The thirst mechanism for regulating water intake. The major stimulus is increased osmolality of blood plasma. (Not all effects of angiotensin II are depicted.)

feedback signals that inhibit the thirst center. This premature quenching of thirst prevents us from drinking more than we need and overdiluting our body fluids, and allows time for the osmotic changes to come into play as regulatory factors.

As effective as thirst is, it is not always a reliable indicator of need. This is particularly true during athletic events, when thirst can be satisfied long before sufficient liquids have been drunk to

26

chemicals cause local capillaries to become very porous, allowing large amounts of exudate (containing not only clotting proteins but also other plasma proteins, nutrients, and immune elements) to form.

Edema caused by hindered fluid return to the blood usually reflects an imbalance in the colloid osmotic pressures on the two sides of the capillary membranes. For example, **hypoproteinemia** (hi″po-pro″te-ĭ-ne′me-ah), a condition of unusually low levels of plasma proteins, results in tissue edema because protein-deficient plasma has an abnormally low colloid osmotic pressure. Fluids are forced out of the capillary beds at the arterial ends by blood pressure as usual, but fail to return to the blood at the venous ends. As a result, the interstitial spaces become congested with fluid. Hypoproteinemia may result from protein malnutrition, liver disease, or *glomerulonephritis* (in which plasma proteins pass through "leaky" renal filtration membranes and are lost in urine).

Although the cause differs, the result is the same when lymphatic vessels are blocked or have been surgically removed. The small amounts of plasma proteins that seep out of the bloodstream are not returned to the blood as usual. As the leaked proteins accumulate in the IF, they exert an ever-increasing colloid osmotic pressure, which draws fluid from the blood and holds it in the interstitial space.

Edema can impair tissue function because excess fluid in the interstitial space increases the distance nutrients and oxygen must diffuse between the blood and the cells. However, the most serious problems resulting from edema affect the cardiovascular system. When fluid leaves the bloodstream and accumulates in the interstitial space, both blood volume and blood pressure decline and the efficiency of the circulation can be severely impaired.

CHECK YOUR UNDERSTANDING

4. What change in plasma is most important for triggering thirst? Where is that change sensed?
5. ADH, by itself, cannot reduce an increase in osmolality in body fluids. Why not? What other mechanism is required?
6. For each of the following, state whether it might result in dehydration, hypotonic hydration, or edema: (a) loss of plasma proteins due to liver failure; (b) copious sweating; (c) using ecstasy (MDMA), which promotes ADH secretion.

For answers, see Appendix G.

Electrolyte Balance

- Indicate routes of electrolyte entry and loss from the body.
- Describe the importance of ionic sodium in fluid and electrolyte balance of the body, and indicate its relationship to normal cardiovascular system functioning.
- Describe mechanisms involved in regulating sodium balance, blood volume, and blood pressure.
- Explain how potassium, calcium, and anion balances in plasma are regulated.

Electrolytes include salts, acids, and bases, but the term **electrolyte balance** usually refers to the salt balance in the body. Salts are important in controlling fluid movements and provide minerals essential for excitability, secretory activity, and membrane permeability. Although many electrolytes are crucial for cellular activity, here we will specifically examine the regulation of sodium, potassium, and calcium. In the next section we will consider acids and bases, which are intimately involved in determining the pH of body fluids.

Salts enter the body in foods and fluids, and small amounts are generated during metabolic activity. For example, phosphates are liberated during catabolism of nucleic acids and bone matrix. Obtaining enough electrolytes is usually not a problem. Indeed, most of us have a far greater taste than need for salt. We shake table salt (NaCl) on our food even though natural foods contain ample amounts and processed foods contain exorbitant quantities. The taste for very salty foods is learned, but some liking for salt may be innate to ensure adequate intake of these two vital ions.

Salts are lost from the body in perspiration, feces, and urine. Even though sweat is normally hypotonic, large amounts of salt can be lost on a hot day simply because more sweat is produced. Gastrointestinal disorders can also lead to large salt losses in feces or vomitus. Consequently, the flexibility of renal mechanisms that regulate the electrolyte balance of the blood is a critical asset. Some causes and consequences of electrolyte imbalances are summarized in **Table 26.1**.

HOMEOSTATIC IMBALANCE

Severe electrolyte deficiencies may prompt a craving for salty or sour foods, such as smoked meats or pickled eggs. This is common in those with *Addison's disease*, a disorder entailing deficient mineralocorticoid hormone production by the adrenal cortex. When minerals such as iron are deficient, a person may even eat substances not usually considered foods, like chalk, clay, starch, and burnt match tips. This appetite for abnormal substances is called *pica*. ■

The Central Role of Sodium in Fluid and Electrolyte Balance

Sodium holds a central position in fluid and electrolyte balance and overall body homeostasis. Indeed, regulating the balance between sodium input and output is one of the most important renal functions. The salts $NaHCO_3$ and NaCl account for 90–95% of all solutes in the ECF, and they contribute about 280 mOsm of the total ECF solute concentration (300 mOsm).

At its normal plasma concentration of about 142 mEq/L, Na^+ is the single most abundant cation in the ECF and the only one exerting *significant* osmotic pressure. Additionally, cellular plasma membranes are relatively impermeable to Na^+, but some does manage to diffuse in and must be pumped out against its electrochemical gradient. These two qualities give sodium the primary role in controlling ECF volume and water distribution in the body.

It is important to understand that while the sodium content of the body may change, *its ECF concentration normally remains*

TABLE 26.1 Causes and Consequences of Electrolyte Imbalances

ION	ABNORMALITY (SERUM VALUE)	POSSIBLE CAUSES	CONSEQUENCES
Sodium	Hypernatremia (Na^+ excess: >145 mEq/L)	Dehydration; uncommon in healthy individuals; may occur in infants or the confused aged (individuals unable to indicate thirst) or may be a result of excessive intravenous NaCl administration	Thirst. CNS dehydration leads to confusion and lethargy progressing to coma; increased neuromuscular irritability evidenced by twitching and convulsions.
	Hyponatremia (Na^+ deficit: <135 mEq/L)	Solute loss, water retention, or both (e.g., excessive Na^+ loss through vomiting, diarrhea, burned skin, tubal drainage of stomach, and as a result of excessive use of diuretics); deficiency of aldosterone (Addison's disease); renal disease; excess ADH release; excess H_2O ingestion	Most common signs are those of neurologic dysfunction due to brain swelling. If sodium amounts are actually normal but water is excessive, the symptoms are the same as those of water excess: mental confusion; giddiness; coma if development occurs slowly; muscular twitching, irritability, and convulsions if the condition develops rapidly. In hyponatremia accompanied by water loss, the main signs are decreased blood volume and blood pressure (circulatory shock).
Potassium	Hyperkalemia (K^+ excess: >5.5 mEq/L)	Renal failure; deficit of aldosterone; rapid intravenous infusion of KCl; burns or severe tissue injuries which cause K^+ to leave cells	Nausea, vomiting, diarrhea; bradycardia; cardiac arrhythmias, depression, and arrest; skeletal muscle weakness; flaccid paralysis.
	Hypokalemia (K^+ deficit: <3.5 mEq/L)	Gastrointestinal tract disturbances (vomiting, diarrhea), gastrointestinal suction; Cushing's syndrome; inadequate dietary intake (starvation); hyperaldosteronism; diuretic therapy	Cardiac arrhythmias, flattened T wave; muscular weakness; metabolic alkalosis; mental confusion; nausea; vomiting.
Phosphate	Hyperphosphatemia (HPO_4^{2-} excess: >2.9 mEq/L)	Decreased urinary loss due to renal failure; hypoparathyroidism; major tissue trauma; increased intestinal absorption	Clinical symptoms arise because of reciprocal changes in Ca^{2+} levels rather than directly from changes in plasma phosphate concentrations.
	Hypophosphatemia (HPO_4^{2-} deficit: <1.6 mEq/L)	Decreased intestinal absorption; increased urinary output; hyperparathyroidism	
Chloride	Hyperchloremia (Cl^- excess: >105 mEq/L)	Dehydration; increased retention or intake; metabolic acidosis; hyperparathyroidism	No direct clinical symptoms; symptoms generally associated with the underlying cause, which is often related to pH abnormalities.
	Hypochloremia (Cl^- deficit: <95 mEq/L)	Metabolic alkalosis (e.g., due to vomiting or excessive ingestion of alkaline substances); aldosterone deficiency	
Calcium	Hypercalcemia (Ca^{2+} excess: >5.2 mEq/L or 10.5 mg%)*	Hyperparathyroidism; excessive vitamin D; prolonged immobilization; renal disease (decreased excretion); malignancy	Decreased neuromuscular excitability leading to cardiac arrhythmias and arrest, skeletal muscle weakness, confusion, stupor, and coma; kidney stones; nausea and vomiting.
	Hypocalcemia (Ca^{2+} deficit: <4.5 mEq/L or 9 mg%)*	Burns (calcium trapped in damaged tissues); hypoparathyroidism; vitamin D deficiency; renal tubular disease; renal failure; hyperphosphatemia; diarrhea; alkalosis	Increased neuromuscular excitability leading to tingling of fingers, tremors, skeletal muscle cramps, tetany, convulsions; depressed excitability of the heart; osteomalacia; fractures.
Magnesium	Hypermagnesemia (Mg^{2+} excess: >2.2 mEq/L)	Rare; occurs in renal failure when Mg^{2+} is not excreted normally; excessive ingestion of Mg^{2+}-containing antacids	Lethargy; impaired CNS functioning, coma, respiratory depression; cardiac arrest.
	Hypomagnesemia (Mg^{2+} deficit: <1.4 mEq/L)	Alcoholism; loss of intestinal contents, severe malnutrition; diuretic therapy	Tremors, increased neuromuscular excitability, tetany, convulsions.

*1 mg% = 1 mg/100 ml

stable because of immediate movement of water into or out of the ICF and longer-term adjustments due to the ADH and thirst mechanisms. Remember, *water follows salt.* Because all body fluids are in osmotic equilibrium, a change in plasma Na^+ levels affects not only plasma volume and blood pressure, but also the ICF and IF volumes. In addition, sodium ions continuously move back and forth between the ECF and body secretions. For example, about 8 L of Na^+-containing secretions (gastric, intestinal, and pancreatic juice, saliva, bile) are spewed into the digestive tract daily, only to be almost completely reabsorbed. Finally,

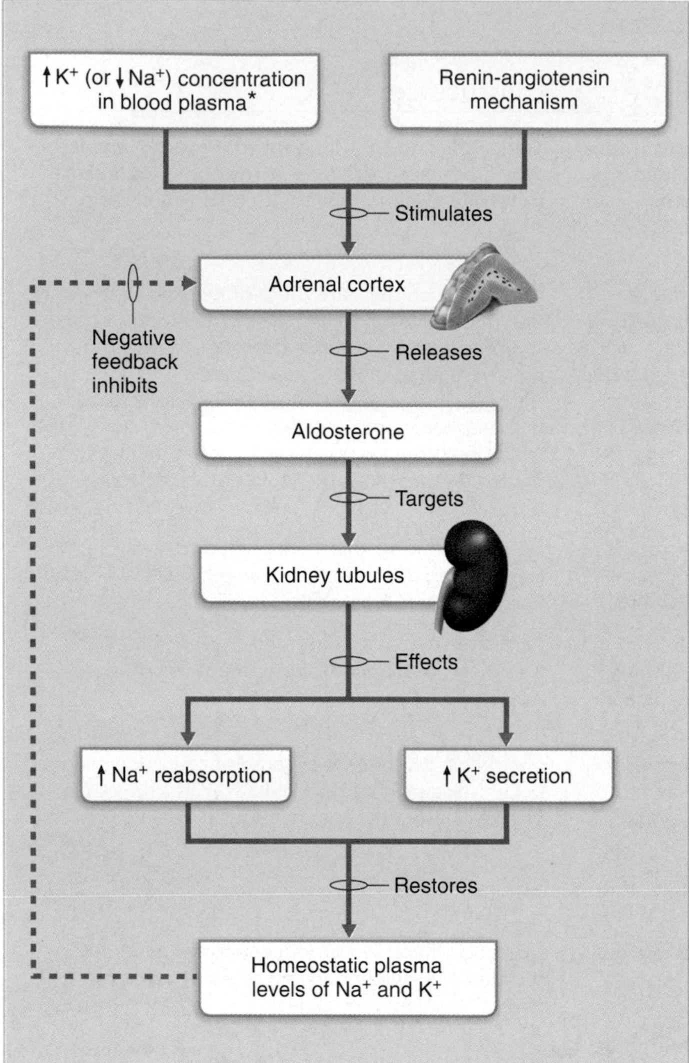

Figure 26.8 Mechanisms and consequences of aldosterone release.

*The adrenal cortex is much less sensitive to decreased plasma Na^+ than to increased plasma K^+.

26

renal acid-base control mechanisms (which we will discuss shortly) are coupled to Na^+ transport.

Regulation of Sodium Balance

Despite the crucial importance of sodium, receptors that specifically monitor Na^+ levels in body fluids have yet to be found. Regulation of the Na^+-water balance is inseparably linked to blood pressure and volume, and involves a variety of neural and hormonal controls. Reabsorption of Na^+ does *not* exhibit a transport maximum, and in healthy individuals nearly all Na^+ in the urinary filtrate can be reabsorbed. We will begin our coverage of sodium balance by reviewing the regulatory effect of aldosterone and angiotensin II. Then we will examine various feedback loops that interact to regulate sodium and water balance and blood pressure.

Influence of Aldosterone and Angiotensin II

The hormone **aldosterone** "has the most to say" about renal regulation of sodium ion concentrations in the ECF. But whether aldosterone is present or not, some 65% of the Na^+ in the renal filtrate is reabsorbed in the proximal tubules of the kidneys and another 25% is reclaimed in the loops of Henle (see Chapter 25).

When aldosterone concentrations are high, essentially all the remaining filtered Na^+ is actively reabsorbed in the distal convoluted tubules and collecting ducts. Consistent with aldosterone's central role in maintaining blood volume and blood pressure, water always follows Na^+. This water comes from either the intracellular fluid or, if ADH is present, from the filtrate in the collecting ducts. One way or another, aldosterone increases ECF volume.

When aldosterone release is inhibited, virtually no Na^+ reabsorption occurs beyond the distal tubule. Urinary excretion of large amounts of Na^+ *always* results in the excretion of large amounts of water as well, but the reverse is *not* true. Substantial amounts of nearly sodium-free urine can be eliminated as needed to achieve water balance.

The most important trigger for aldosterone release from the adrenal cortex is the renin-angiotensin mechanism mediated by the juxtaglomerular apparatus of the renal tubules (see Figures 26.8 and 26.10). When the juxtaglomerular (JG) apparatus responds to (1) sympathetic stimulation, (2) decreased filtrate NaCl concentration, or (3) decreased stretch (due to decreased blood pressure), its granular cells release renin. Renin catalyzes the initial step in the reactions that produce angiotensin II, which prompts aldosterone release. Conversely, high renal blood pressure and high filtrate NaCl concentrations depress release of renin, angiotensin II, and aldosterone. The adrenal cortical cells are also directly stimulated to release aldosterone by elevated K^+ levels in the ECF **(Figure 26.8)**.

Angiotensin II is an important intermediate in the pathway linking renin to aldosterone release. Angiotensin II prods the adrenal cortex to release aldosterone, and also directly increases Na^+ reabsorption by kidney tubules. In addition, it has a number of other actions, all aimed at raising blood volume and blood pressure. We described these actions in Chapter 25 (p. 972).

Aldosterone brings about its effects slowly, over a period of hours to days. The principal effects of aldosterone are to diminish urinary output and increase blood volume. However, before these factors can change by more than a few percent, feedback mechanisms for blood volume control come into play.

HOMEOSTATIC IMBALANCE

People with Addison's disease (hypoaldosteronism) lose tremendous amounts of NaCl and water to urine. They are perpetually teetering on the brink of hypovolemia, but as long as they ingest adequate amounts of salt and fluids, they can avoid problems with Na^+ balance. ■

Influence of Atrial Natriuretic Peptide

The influence of **atrial natriuretic peptide (ANP)** can be summarized in one sentence: It reduces blood pressure and blood

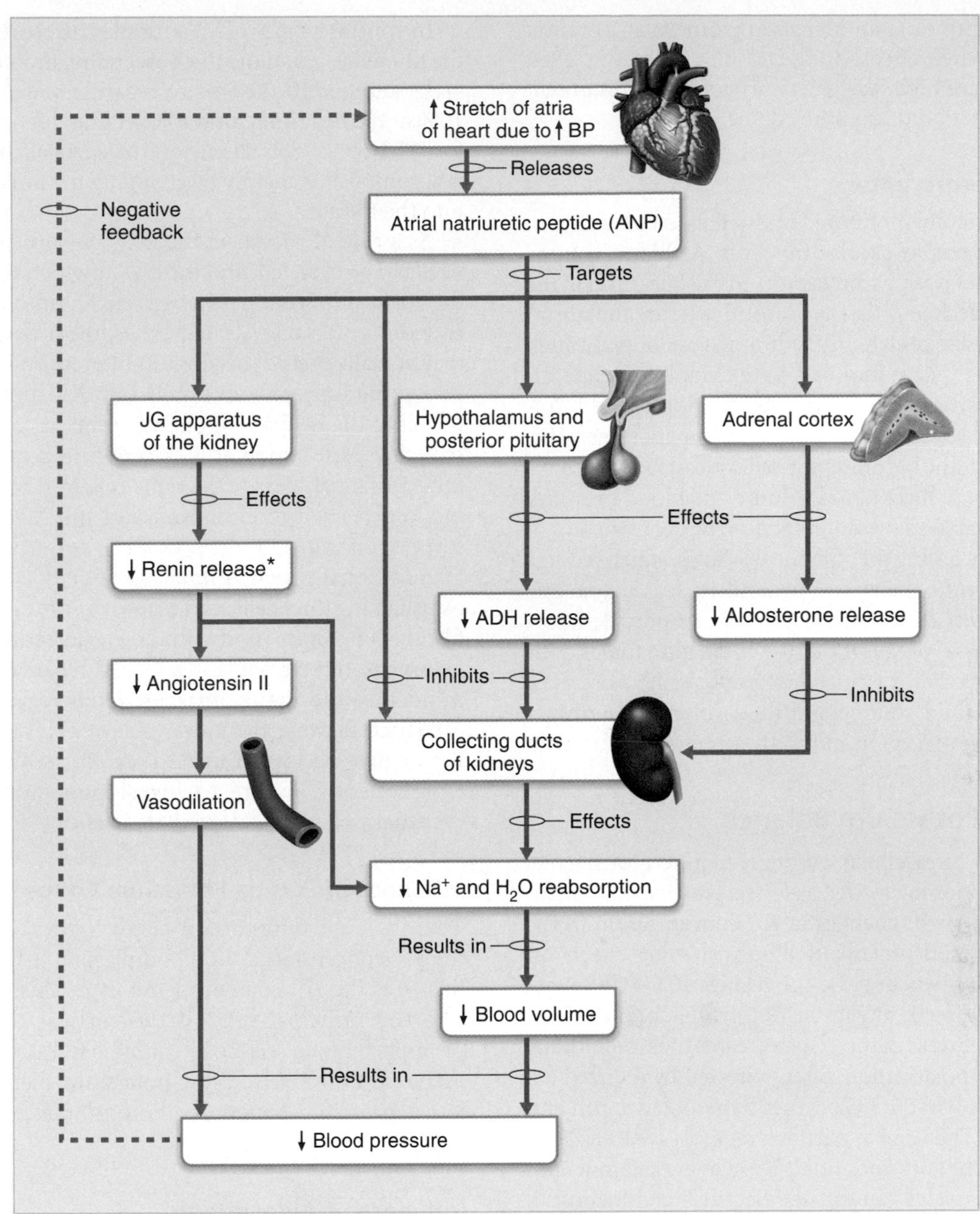

Figure 26.9 Mechanisms and consequences of ANP release.
*↓ Renin release also inhibits ADH and aldosterone release and hence the effects of those hormones.

volume by inhibiting nearly all events that promote vasoconstriction and Na^+ and water retention **(Figure 26.9)**. A hormone that is released by certain cells of the heart atria when they are stretched by the effects of elevated blood pressure, ANP has diuretic and natriuretic (salt-excreting) effects. It promotes excretion of Na^+ and water by the kidneys by inhibiting the ability of the collecting ducts to reabsorb Na^+ and by suppressing the release of ADH, renin, and aldosterone. Additionally, ANP acts both directly and indirectly (by inhibiting renin-induced generation of angiotensin II) to relax vascular smooth muscle, and in this way it causes vasodilation. Collectively, these effects reduce blood pressure.

Influence of Other Hormones

Female Sex Hormones The **estrogens** are chemically similar to aldosterone and, like aldosterone, enhance NaCl reabsorption by the renal tubules. Because water follows, many women retain fluid as their estrogen levels rise during the menstrual cycle. The edema experienced by many pregnant women is also largely due to the effect of estrogens. **Progesterone** appears to decrease Na^+ reabsorption by blocking the effect aldosterone has on the renal tubules. Thus, progesterone has a diuretic-like effect and promotes Na^+ and water loss.

Glucocorticoids The usual effect of **glucocorticoids**, such as cortisol and hydrocortisol, is to enhance tubular reabsorption of

26

Na^+, but they also promote an increased glomerular filtration rate that may mask their effects on the tubules. However, when their plasma levels are high, the glucocorticoids exhibit potent aldosterone-like effects and promote edema.

Cardiovascular Baroreceptors

Blood volume is carefully monitored and regulated to maintain blood pressure and cardiovascular function. As blood volume (and with it, pressure) rises, baroreceptors in the heart and in the large vessels of the neck and thorax (carotid arteries and aorta) alert the cardiovascular centers in the brain stem. Shortly after, sympathetic nervous system impulses to the kidneys decline, allowing the afferent arterioles to dilate. As the glomerular filtration rate rises, Na^+ output and water output increase. This phenomenon, part of the baroreceptor reflex described in Chapter 19 (pp. 707–708), reduces blood volume and blood pressure.

Drops in systemic blood pressure lead to reflex constriction of systemic arterioles including the afferent arterioles, which reduces filtrate formation and urinary output and increases systemic blood pressure **(Figure 26.10)**. The baroreceptors provide information on the "fullness" or volume of the circulation that is critical for maintaining cardiovascular homeostasis. Because Na^+ content determines fluid volume and fluid volume determines blood pressure, the baroreceptors indirectly monitor Na^+ content.

Regulation of Potassium Balance

Potassium, the chief intracellular cation, is required for normal neuromuscular functioning as well as for several essential metabolic activities. Even slight changes in K^+ concentration in the ECF have profound and potentially life-threatening effects on neurons and muscle fibers because the relative ICF-ECF potassium concentration directly affects the resting membrane potential of these cells. K^+ excess in the ECF decreases their membrane potential, causing depolarization, often followed by reduced excitability. Too little K^+ in the ECF causes hyperpolarization and nonresponsiveness. The heart is particularly sensitive to K^+ levels. Both too much and too little K^+ (hyperkalemia and hypokalemia, respectively) can disrupt electrical conduction in the heart, leading to sudden death (Table 26.1).

26

Potassium is also part of the body's buffer system, which resists changes in the pH of body fluids. Shifts of hydrogen ions (H^+) into and out of cells induce corresponding shifts of K^+ in the opposite direction to maintain cation balance. Consequently, ECF potassium levels rise with acidosis, as K^+ leaves and H^+ enters the cells, and fall with alkalosis, as K^+ moves into the cells and H^+ leaves them to enter the ECF. Although these pH-driven shifts do not change the total amount of K^+ in the body, they can seriously interfere with the activity of excitable cells.

Regulatory Site: The Cortical Collecting Duct

Like Na^+ balance, K^+ balance is maintained chiefly by renal mechanisms. However, there are important differences in the way this balance is achieved. The amount of Na^+ reabsorbed in the tubules is precisely tailored to need, and Na^+ is *never* secreted into the filtrate.

In contrast, the proximal tubules reabsorb about 60–80% of the filtered K^+, and the thick ascending limb of Henle's loop absorbs another 10–20% or so regardless of need, leaving about 10% at the beginning of the collecting ducts. The responsibility for K^+ balance falls chiefly on the cortical collecting ducts, and is accomplished mainly by changing the amount of K^+ *secreted* into the filtrate.

As a rule, K^+ levels in the ECF are sufficiently high that K^+ needs to be excreted, and the rate at which the principal cells of the cortical collecting ducts secrete K^+ into the filtrate is accelerated over basal levels. (At times, the amount of K^+ excreted may actually exceed the amount filtered.) When ECF potassium concentrations are abnormally low, K^+ moves from the tissue cells into the ECF and the renal principal cells conserve K^+ by reducing its secretion and excretion to a minimum. Note that these principal cells are the same cells that mediate aldosterone-induced reabsorption of Na^+ and the ADH-stimulated reabsorption of water.

Additionally, *type A intercalated cells,* a unique population of collecting duct cells, can reabsorb some of the K^+ left in the filtrate (in conjunction with active secretion of H^+), thereby helping to reestablish K^+ (and pH) balance. However, keep in mind that the main thrust of renal regulation of K^+ is to *excrete* it. Because the kidneys have a limited ability to retain K^+, it may be lost in urine even in the face of a deficiency. Consequently, failure to ingest potassium-rich substances eventually results in a severe deficiency.

Influence of Plasma Potassium Concentration

The single most important factor influencing K^+ secretion is the K^+ concentration in blood plasma. A high-potassium diet increases the K^+ content of the ECF. This favors entry of K^+ into the principal cells of the cortical collecting duct and prompts them to secrete K^+ into the filtrate so that more of it is excreted. Conversely, a low-potassium diet or accelerated K^+ loss depresses its secretion (and promotes its limited reabsorption) by the collecting ducts.

Influence of Aldosterone

The second factor influencing K^+ secretion into the filtrate is aldosterone. As it stimulates the principal cells to reabsorb Na^+, aldosterone simultaneously enhances K^+ secretion (see Figure 26.8). Adrenal cortical cells are *directly* sensitive to the K^+ content of the ECF bathing them. When it increases even slightly, the adrenal cortex is strongly stimulated to release aldosterone, which increases K^+ secretion by the exchange process we just described. The result is that K^+ controls its own concentrations in the ECF via feedback regulation of aldosterone release.

Aldosterone is also secreted in response to the renin-angiotensin mechanism previously described. Given the opposing effects of aldosterone on plasma Na^+ and K^+, you might expect that Na^+- and volume-driven changes in aldosterone would cause K^+ imbalances. This generally does not occur because other compensatory mechanisms in the kidneys maintain plasma K^+.

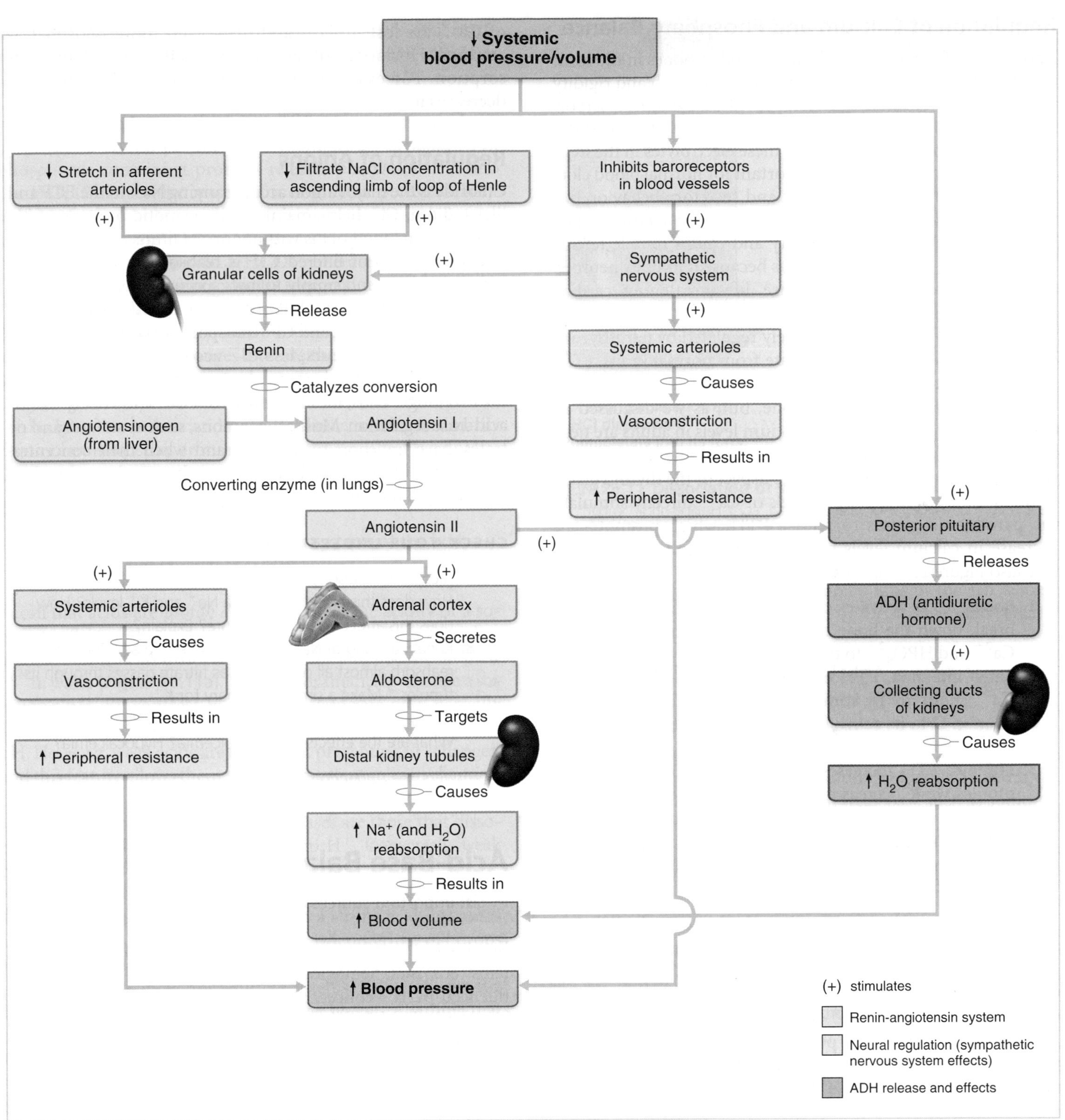

Figure 26.10 Mechanisms regulating sodium and water balance help maintain blood pressure homeostasis.

HOMEOSTATIC IMBALANCE

In an attempt to reduce NaCl intake, many people have turned to salt substitutes, which are high in potassium. However, heavy consumption of these substitutes is safe only when aldosterone release in the body is normal. In the absence of aldosterone, hyperkalemia is swift and lethal regardless of K^+ intake (Table 26.1). Conversely, when a person has an adrenocortical tumor that pumps out tremendous amounts of aldosterone, ECF potassium levels fall so low that neurons all over the body hyperpolarize and paralysis occurs. ■

26

so obtaining that member of the buffer pair is usually not a problem. The H_2CO_3 content of the blood is subject to respiratory controls.

Phosphate Buffer System

The operation of the **phosphate buffer system** is nearly identical to that of the bicarbonate buffer. The components of the phosphate system are the sodium salts of dihydrogen phosphate ($H_2PO_4^-$) and monohydrogen phosphate (HPO_4^{2-}). NaH_2PO_4 acts as a weak acid. Na_2HPO_4, with one less hydrogen atom, acts as a weak base.

Again, H^+ released by strong acids is tied up in weak acids:

$$\underset{\text{strong acid}}{HCl} + \underset{\text{weak base}}{Na_2HPO_4} \rightarrow \underset{\text{weak acid}}{NaH_2PO_4} + \underset{\text{salt}}{NaCl}$$

and strong bases are converted to weak bases:

$$\underset{\text{strong base}}{NaOH} + \underset{\text{weak acid}}{NaH_2PO_4} \rightarrow \underset{\text{weak base}}{Na_2HPO_4} + \underset{\text{water}}{H_2O}$$

Because the phosphate buffer system is present in low concentrations in the ECF (approximately one-sixth that of the bicarbonate buffer system), it is relatively unimportant for buffering blood plasma. However, it is a very effective buffer in urine and in ICF, where phosphate concentrations are usually higher.

Protein Buffer System

Proteins in plasma and in cells are the body's most plentiful and powerful source of buffers, and constitute the **protein buffer system**. In fact, at least three-quarters of all the buffering power of body fluids resides in cells, and most of this reflects the buffering activity of intracellular proteins.

As described in Chapter 2, proteins are polymers of amino acids. Some of the linked amino acids have exposed groups of atoms called *organic acid (carboxyl) groups* (—COOH), which dissociate to release H^+ when the pH begins to rise:

26

$$R\text{—}COOH \rightarrow R\text{—}COO^- + H^+$$

(Note that R indicates the rest of the organic molecule, which contains many atoms.)

Other amino acids have exposed groups that can act as bases and accept H^+. For example, an exposed $—NH_2$ group can bind with a hydrogen ion, becoming $—NH_3^+$:

$$R\text{—}NH_2 + H^+ \rightarrow R\text{—}NH_3^+$$

Because this binding removes free hydrogen ions from the solution, it prevents the solution from becoming too acidic. Consequently, a single protein molecule can function reversibly as either an acid or a base depending on the pH of its environment. Molecules with this ability are called **amphoteric molecules** (am″fo-ter′ik).

Hemoglobin of red blood cells is an excellent example of a protein that functions as an intracellular buffer. As we explained earlier, CO_2 released from the tissues forms H_2CO_3, which dissociates to liberate H^+ and HCO_3^- in the blood. Meanwhile, hemoglobin is unloading oxygen, becoming reduced hemoglobin, which carries a negative charge. Because H^+ rapidly binds to the hemoglobin anions, pH changes are minimized. In this case, carbonic acid, a weak acid, is buffered by an even weaker acid, hemoglobin.

Respiratory Regulation of H^+

► Describe the influence of the respiratory system on acid-base balance.

The respiratory and renal systems together form the *physiological buffering systems* that control pH by controlling the amount of acid or base in the body. Although such buffer systems act more slowly than the chemical buffer systems, they have many times the buffering power of all the body's chemical buffers combined.

As we described in Chapter 22, the respiratory system eliminates CO_2, an acid, from the blood while replenishing its supply of O_2. Carbon dioxide generated by cellular respiration enters erythrocytes in the circulation and is converted to bicarbonate ions for transport in the plasma:

$$CO_2 + H_2O \underset{}{\overset{\text{carbonic anhydrase}}{\rightleftharpoons}} \underset{\text{carbonic acid}}{H_2CO_3} \rightleftharpoons H^+ + \underset{\text{bicarbonate ion}}{HCO_3^-}$$

The first set of double arrows indicates a reversible equilibrium between dissolved carbon dioxide and water on the left and carbonic acid on the right. The second set indicates a reversible equilibrium between carbonic acid on the left and hydrogen and bicarbonate ions on the right. Because of these equilibria, an increase in any of these chemical species pushes the reaction in the opposite direction. Notice also that the right side of the equation is equivalent to the bicarbonate buffer system.

In healthy individuals, CO_2 is expelled from the lungs at the same rate it is formed in the tissues. During carbon dioxide unloading, the reaction shifts to the left, and H^+ generated from carbonic acid is reincorporated into water. Because of the protein buffer system, H^+ produced by CO_2 transport is not allowed to accumulate and has little or no effect on blood pH. However, when hypercapnia occurs, it activates medullary chemoreceptors (via cerebral acidosis promoted by excessive accumulation of CO_2) that respond by increasing respiratory rate and depth (see Figure 22.25, p. 837). Additionally, a rising plasma H^+ concentration resulting from any metabolic process excites the respiratory center indirectly (via peripheral chemoreceptors) to stimulate deeper, more rapid respiration. As ventilation increases, more CO_2 is removed from the blood, pushing the reaction to the left and reducing the H^+ concentration.

When blood pH rises, the respiratory center is depressed. As respiratory rate drops and respiration becomes shallower, CO_2 accumulates, and the equilibrium is pushed to the right, causing the H^+ concentration to increase. Again blood pH is restored to the normal range. These respiratory system–mediated corrections of blood pH are accomplished within a minute or so.

Changes in alveolar ventilation can produce dramatic changes in blood pH—far more than is needed. For example, a doubling or halving of alveolar ventilation can raise or lower blood pH by about 0.2 pH unit. Because normal arterial pH is 7.4, a change of 0.2 pH unit yields a blood pH of 7.6 or 7.2—both well beyond the normal limits of blood pH. Respiratory controls of blood pH have a tremendous reserve capacity because alveolar ventilation can be increased about 15-fold or reduced to zero.

Anything that impairs respiratory system functioning causes acid-base imbalances. For example, net carbon dioxide retention (hypoventilation) leads to acidosis. On the other hand, hyperventilation, which causes net elimination of CO_2, causes alkalosis. When the cause of the pH imbalance is respiratory system problems, the resulting condition is either *respiratory acidosis* or *respiratory alkalosis* (see Table 26.2 on p. 1016).

CHECK YOUR UNDERSTANDING

10. Define acidemia and alkalemia.
11. To minimize a shift in pH brought about by adding a strong acid to a solution, would it be better if the solution contained a weak base or a strong base?
12. What are the body's three major chemical buffer systems? What is the most important buffer inside cells?
13. Joanne, a diabetic patient, is at the emergency department with acidosis due to the production of ketone bodies. Would you expect her ventilation to be increased or decreased? Why?

For answers, see Appendix G.

Renal Mechanisms of Acid-Base Balance

▶ Describe how the kidneys regulate hydrogen and bicarbonate ion concentrations in the blood.

The ultimate acid-base regulatory organs are the kidneys, which act slowly but surely to compensate for acid-base imbalances resulting from variations in diet or metabolism, or from disease. Chemical buffers can tie up excess acids or bases temporarily, but they cannot eliminate them from the body. And while the lungs can dispose of the **volatile acid** carbonic acid by eliminating CO_2, only the kidneys can rid the body of other acids generated by cellular metabolism: phosphoric, uric, and lactic acids, and ketone bodies. These acids are sometimes referred to as **metabolic (fixed) acids**, but this terminology is both unfortunate and incorrect because CO_2 and carbonic acid are also products of metabolism. Additionally, only the kidneys can regulate blood levels of alkaline substances and renew chemical buffers that are used up in regulating H^+ levels in the ECF.

The most important renal mechanisms for regulating acid-base balance of the blood involve (1) conserving (reabsorbing) or generating new HCO_3^-, and (2) excreting HCO_3^-. If we look back at the equation for the operation of the carbonic acid–bicarbonate buffer system of the blood, we can see that losing a HCO_3^- from the body produces the same net effect as gaining a H^+, because it pushes the equation to the right, increasing the H^+ level. By the same token, generating or reabsorbing HCO_3^- is the same as losing H^+ because it pushes the equation to the left, decreasing the H^+ level. For this reason, to reabsorb bicarbonate, the kidney has to secrete H^+, and when it excretes excess HCO_3^-, H^+ is retained (not secreted).

Because the mechanisms for regulating acid-base balance depend on H^+ being secreted into the filtrate, we consider that process first. Secretion of H^+ occurs mainly in the PCT and in type A intercalated cells of the collecting duct. The H^+ secreted comes from the dissociation of carbonic acid, created from the combination of CO_2 and water within the tubule cells, a reaction catalyzed by *carbonic anhydrase* (**Figure 26.12**, ①, ②). For each H^+ secreted into the lumen of the PCT, one Na^+ is reabsorbed from the filtrate, maintaining the electrical balance (Figure 26.12, 3a).

The rate of H^+ secretion rises and falls with CO_2 levels in the ECF. The more CO_2 in the peritubular capillary blood, the faster the rate of H^+ secretion. Because blood CO_2 levels directly relate to blood pH, this system can respond to both rising and falling H^+ concentrations. Notice that secreted H^+ can combine with HCO_3^- in the filtrate, generating CO_2 and water (Figure 26.12, ④, ⑤). In this case, H^+ is bound in water. The rising concentration of CO_2 in the filtrate creates a steep diffusion gradient for its entry into the tubule cell, where it promotes still more H^+ secretion (Figure 26.12, ⑥).

Conserving Filtered Bicarbonate Ions: Bicarbonate Reabsorption

Bicarbonate ions (HCO_3^-) are an important part of the bicarbonate buffer system, the most important inorganic blood buffer. If this reservoir of base, or *alkaline reserve*, is to be maintained, the kidneys must do more than just eliminate enough hydrogen ions to counter rising blood H^+ levels. Depleted stores of HCO_3^- have to be replenished. This task is more complex than it seems because the tubule cells are almost completely impermeable to the HCO_3^- in the filtrate—they cannot reabsorb them.

However, the kidneys can conserve filtered HCO_3^- in a rather roundabout way, also illustrated in Figure 26.12. As you can see, dissociation of carbonic acid liberates HCO_3^- as well as H^+ (Figure 26.12, ②). Although the tubule cells cannot reclaim HCO_3^- directly from the filtrate, they can and do shunt HCO_3^- generated within them (as a result of splitting H_2CO_3) into the peritubular capillary blood (Figure 26.12, 3b). HCO_3^- leaves the tubule cell either accompanied by Na^+ or in exchange for Cl^-. For this reason, reabsorption of HCO_3^- depends on the active secretion of H^+, mostly by a Na^+-H^+ antiporter, but also by a H^+ ATPase (Figure 26.12, 3a). In the filtrate, H^+ combines with filtered HCO_3^-, as we saw earlier (Figure 26.12, ④, ⑤).

In short, for each filtered HCO_3^- that "disappears," a HCO_3^- generated within the tubule cells enters the blood—a one-for-one exchange. When large amounts of H^+ are secreted, correspondingly large amounts of HCO_3^- enter the peritubular blood. The net effect is that HCO_3^- is almost completely removed from the filtrate.

Generating New Bicarbonate Ions

Two renal mechanisms commonly carried out by cells of the PCT and collecting ducts generate *new* HCO_3^- that can be

Bicarbonate Ion Secretion

When the body is in alkalosis, another population of intercalated cells (type B) in the collecting ducts exhibit net HCO_3^- secretion (rather than net HCO_3^- reabsorption) while reclaiming H^+ to acidify the blood. Overall we can think of the type B cells as "flipped" type A cells, and we can visualize the HCO_3^- secretion process as the exact opposite of the HCO_3^- reabsorption process illustrated in Figure 26.12. However, the predominant process in the nephrons and collecting ducts is HCO_3^- reabsorption, and even during alkalosis, the amount of HCO_3^- excreted is much less than the amount conserved.

CHECK YOUR UNDERSTANDING

14. Reabsorption of HCO_3^- is always tied to the secretion of which ion?
15. What is the most important urinary buffer of H^+?
16. List the two mechanisms by which tubule and collecting duct cells generate new HCO_3^-.

For answers, see Appendix G.

Abnormalities of Acid-Base Balance

▶ Distinguish between acidosis and alkalosis resulting from respiratory and metabolic factors. Describe the importance of respiratory and renal compensations to acid-base balance.

All cases of acidosis and alkalosis can be classed according to cause as *respiratory* or *metabolic* **(Table 26.2)**. The methods for determining the cause of an acid-base disturbance and whether it is being compensated (whether the lungs or kidneys are taking steps to correct the imbalance) are described in *A Closer Look* on p. 1017.

Respiratory Acidosis and Alkalosis

Respiratory pH imbalances result from some failure of the respiratory system to perform its normal pH-balancing role. The partial pressure of carbon dioxide (P_{CO_2}) is the single most important indicator of the adequacy of respiratory function. When respiratory function is normal, the P_{CO_2} fluctuates between 35 and 45 mm Hg. Generally speaking, higher values indicate respiratory acidosis, and lower values indicate respiratory alkalosis.

Respiratory acidosis is the most common cause of acid-base imbalance. It most often occurs when a person breathes shallowly or when gas exchange is hampered by diseases such as pneumonia, cystic fibrosis, or emphysema. Under such conditions, CO_2 accumulates in the blood. Thus, respiratory acidosis is characterized by falling blood pH and rising P_{CO_2}.

Respiratory alkalosis results when carbon dioxide is eliminated from the body faster than it is produced. This is called *hyperventilation* (see p. 837), and results in the blood becoming more alkaline. While respiratory acidosis is frequently associated with respiratory system pathology, respiratory alkalosis is often due to stress or pain.

Metabolic Acidosis and Alkalosis

Metabolic pH imbalances include all abnormalities of acid-base imbalance *except* those caused by too much or too little carbon dioxide in the blood. Bicarbonate ion levels below or above the normal range of 22–26 mEq/L indicate a metabolic acid-base imbalance.

The second most common cause of acid-base imbalance, **metabolic acidosis**, is recognized by low blood pH and HCO_3^- levels. Typical causes of metabolic acidosis are ingestion of too much alcohol (which is metabolized to acetic acid) and excessive loss of HCO_3^-, as might result from persistent diarrhea. Other causes are accumulation of lactic acid during exercise or shock, the ketosis that occurs in diabetic crisis or starvation, and, infrequently, kidney failure.

Metabolic alkalosis, indicated by rising blood pH and HCO_3^- levels, is much less common than metabolic acidosis. Typical causes are vomiting of the acidic contents of the stomach (or loss of those secretions through gastrointestinal suctioning) and intake of excess base (antacids, for example).

Effects of Acidosis and Alkalosis

The absolute blood pH limits for life are a low of 7.0 and a high of 7.8. When blood pH falls below 7.0, the central nervous system is so depressed that the person goes into coma and death soon follows. When blood pH rises above 7.8, the nervous system is overexcited, leading to such characteristic signs as muscle tetany, extreme nervousness, and convulsions. Death often results from respiratory arrest.

Respiratory and Renal Compensations

When an acid-base imbalance is due to inadequate functioning of one of the physiological buffer systems (the lungs or kidneys), the other system tries to compensate. The respiratory system attempts to compensate for metabolic acid-base imbalances, and the kidneys (although much slower) work to correct imbalances caused by respiratory disease. These **respiratory** and **renal compensations** can be recognized by changes in plasma P_{CO_2} and bicarbonate ion concentrations (see *A Closer Look*). Because the compensations act to restore normal blood pH, the pH may actually be in the normal range when the patient has a significant medical problem.

Respiratory Compensations As a rule, changes in respiratory rate and depth are evident when the respiratory system is attempting to compensate for metabolic acid-base imbalances. In metabolic acidosis, the respiratory rate and depth are usually elevated—an indication that the respiratory centers are stimulated by the high H^+ levels. The blood pH is low (below 7.35) and the HCO_3^- level is below 22 mEq/L. As the respiratory system "blows off" CO_2 to rid the blood of excess acid, the P_{CO_2} falls below 35 mm Hg. By contrast, in respiratory acidosis, the respiratory rate is often depressed and *is the immediate cause of the acidosis* (with some exceptions such as pneumonia or emphysema where gas exchange is impaired).

Respiratory compensation for metabolic alkalosis involves slow, shallow breathing, which allows CO_2 to accumulate in the blood. A metabolic alkalosis being compensated by respiratory mechanisms is revealed by a pH over 7.45 (at least initially), elevated bicarbonate levels (over 26 mEq/L), and a P_{CO_2} above 45 mm Hg.

Renal Compensations When an acid-base imbalance is of respiratory origin, renal mechanisms are stepped up to compensate for the imbalance. For example, a hypoventilating individual will exhibit acidosis. When renal compensation is occurring, both the P_{CO_2} and the HCO_3^- levels are high. The high P_{CO_2} is the cause of the acidosis, and the rising HCO_3^- level indicates that the kidneys are retaining bicarbonate to offset the acidosis.

Conversely, a person with renal-compensated respiratory alkalosis will have a high blood pH and a low P_{CO_2}. Bicarbonate ion levels begin to fall as the kidneys eliminate more HCO_3^- from the body by failing to reclaim it or by actively secreting it. Note that the kidneys cannot compensate for alkalosis or acidosis if that condition reflects a *renal* problem.

CHECK YOUR UNDERSTANDING

17. What two abnormalities in plasma are key features of an uncompensated metabolic alkalosis? An uncompensated respiratory acidosis?

18. How do the kidneys compensate for respiratory acidosis?

For answers, see Appendix G.

Developmental Aspects of Fluid, Electrolyte, and Acid-Base Balance

▶ Explain why infants and the aged are at greater risk for fluid and electrolyte imbalances than are young adults.

An embryo and a very young fetus are more than 90% water, but solids accumulate as fetal development continues, and at birth an infant is "only" 70–80% water. (The average value for adults is 58%.) Infants have proportionately more ECF than adults and, consequently, a much higher NaCl content in relation to K^+, Mg^{2+}, and PO_4^{3-} salts. Distribution of body water begins to change about two months after birth and achieves the adult pattern by the time the child is 2 years old. Plasma electrolyte concentrations are similar in infants and adults, but K^+ and Ca^{2+} values are higher and Mg^{2+}, HCO_3^-, and total protein levels are lower in the first few days of life than at any other time. At puberty, sex differences in body water content become obvious as males develop relatively greater amounts of skeletal muscle.

Problems with fluid, electrolyte, and particularly acid-base balance are most common in infancy, reflecting the following conditions:

1. *The very low residual volume of infant lungs* (approximately half that of adults relative to body weight). When respiration is altered, rapid and dramatic changes in P_{CO_2} can result.
2. *The high rate of fluid intake and output in infants* (about seven times higher than in adults). Infants may exchange fully half their ECF daily. Though infants have proportionately much more body water than adults, this does not protect them from excessive fluid shifts. Even slight alterations in fluid balance can cause serious problems. Further, although adults can live without water for about ten days, infants can survive for only three to four days.
3. *The relatively high infant metabolic rate* (about twice that of adults). The higher metabolic rate yields much larger amounts of metabolic wastes and acids that need to be excreted by the kidneys. This, along with buffer systems that are not yet fully effective, results in a tendency toward acidosis.
4. *The high rate of insensible water loss in infants* because of their larger surface area relative to body volume (about three times that of adults). Infants lose substantial amounts of water through their skin.
5. *The inefficiency of infant kidneys.* At birth, the kidneys are immature, only about half as proficient as adult kidneys at concentrating urine, and notably inefficient at ridding the body of acids.

All these factors put newborns at risk for dehydration and acidosis, at least until the end of the first month when the kidneys achieve reasonable efficiency. Bouts of vomiting or diarrhea greatly amplify the risk.

In old age, total body water often decreases (the loss is largely from the intracellular compartment) because muscle mass progressively declines and body fat rises. Few changes occur in the solute concentrations of body fluids, but the speed with which homeostasis is restored after being disrupted declines with age. Elders may be unresponsive to thirst cues and thus are at risk for dehydration. Additionally, they are the most frequent prey of diseases that lead to severe fluid, electrolyte, or acid-base problems, such as congestive heart failure (and its attendant edema) and diabetes mellitus. Because most fluid, electrolyte, and acid-base imbalances occur when body water content is highest or lowest, the very young and the very old are the most frequent victims.

CHECK YOUR UNDERSTANDING

19. Infants have a higher urine output than adults relative to their body weight. In addition to their relatively higher fluid intake, what are the reasons for this?

For answers, see Appendix G.

26

TABLE 26.2 Causes and Consequences of Acid-Base Imbalances

CONDITION AND HALLMARK	POSSIBLE CAUSES; COMMENTS
Metabolic Acidosis	
If uncompensated (uncorrected): HCO_3^- <22 mEq/L; pH <7.35	**Severe diarrhea:** bicarbonate-rich intestinal (and pancreatic) secretions rushed through digestive tract before their solutes can be reabsorbed; bicarbonate ions are replaced by renal mechanisms that generate new bicarbonate ions **Renal disease:** failure of kidneys to rid body of acids formed by normal metabolic processes **Untreated diabetes mellitus:** lack of insulin or inability of tissue cells to respond to insulin, resulting in inability to use glucose; fats are used as primary energy fuel, and ketoacidosis occurs **Starvation:** lack of dietary nutrients for cellular fuels; body proteins and fat reserves are used for energy—both yield acidic metabolites as they are broken down for energy **Excess alcohol ingestion:** results in excess acids in blood **High ECF potassium concentrations:** potassium ions compete with H^+ for secretion in renal tubules; when ECF levels of K^+ are high, H^+ secretion is inhibited
Metabolic Alkalosis	
If uncompensated: HCO_3^- >26 mEq/L; pH >7.45	**Vomiting or gastric suctioning:** loss of stomach HCl requires that H^+ be withdrawn from blood to replace stomach acid; thus H^+ decreases and HCO_3^- increases proportionately **Selected diuretics:** cause K^+ depletion and H_2O loss. Low K^+ directly stimulates the tubule cells to secrete H^+. Reduced blood volume elicits the renin-angiotensin mechanism, which stimulates Na^+ reabsorption and H^+ secretion. **Ingestion of excessive sodium bicarbonate (antacid):** bicarbonate moves easily into ECF, where it enhances natural alkaline reserve **Excess aldosterone** (e.g., adrenal tumors): promotes excessive reabsorption of Na^+, which pulls increased amount of H^+ into urine. Hypovolemia promotes the same relative effect because aldosterone secretion is increased to enhance Na^+ (and H_2O) reabsorption.
Respiratory Acidosis (Hypoventilation)	
If uncompensated: P_{CO_2} >45 mm Hg; pH <7.35	**Impaired lung function** (e.g., in chronic bronchitis, cystic fibrosis, emphysema): impaired gas exchange or alveolar P_{CO_2} ventilation **Impaired ventilatory movement:** paralysis of respiratory muscles, chest injury, extreme obesity **Narcotic or barbiturate overdose or injury to brain stem:** depression of respiratory centers, resulting in hypoventilation and respiratory arrest
Respiratory Alkalosis (Hyperventilation)	
If uncompensated: P_{CO_2} <35 mm Hg; pH >7.45	**Strong emotions:** pain, anxiety, fear, panic attack **Hypoxia:** asthma, pneumonia, high altitude; represents effort to raise P_{O_2} at the expense of excessive CO_2 excretion **Brain tumor or injury:** abnormality of respiratory controls

In this chapter we have examined the chemical and physiological mechanisms that provide the optimal internal environment for survival. The kidneys are the superstars among homeostatic organs in regulating water, electrolyte, and acid-base balance, but they do not and cannot act alone. Rather, their activity is made possible by a host of hormones and enhanced both by bloodborne buffers, which give the kidneys time to react, and by the respiratory system, which shoulders a substantial responsibility for acid-base balance of the blood.

Now that we have discussed the topics relevant to renal functioning, and once you read in *Making Connections* how the urinary system interacts with other body systems, the topics in Chapters 25 and 26 should draw together in an understandable way.

A CLOSER LOOK

Sleuthing: Using Blood Values to Determine the Cause of Acidosis or Alkalosis

Students, particularly nursing students, are often given blood values and asked to determine (1) whether the patient is in acidosis or alkalosis, (2) the cause of the condition (respiratory or metabolic), and (3) whether the condition is being compensated. If such determinations are approached systematically, they are not nearly as difficult as they may appear. To analyze a person's acid-base balance, scrutinize the blood values in the following order:

1. *Note the pH*. This tells you whether the person is in acidosis (pH below 7.35) or alkalosis (pH above 7.45), but it does not tell you the cause.
2. *Check the P_{CO_2}* to see if this is the cause of the acid-base imbalance. Because the respiratory system is a fast-acting system, an excessively high or low P_{CO_2} may indicate either that the condition is respiratory system caused or that the respiratory system is compensating. For example, if the pH indicates acidosis and
   a. The P_{CO_2} is over 45 mm Hg, then the respiratory system is the cause of the problem and the condition is a respiratory acidosis
   b. The P_{CO_2} is below normal limits (below 35 mm Hg), then the respiratory system is *not the cause but is compensating*
   c. The P_{CO_2} is within normal limits, then the condition is *neither caused nor compensated* by the respiratory system
3. *Check the bicarbonate level.* If step 2 proves that the respiratory system is not responsible for the imbalance, then the condition is metabolic and should be reflected in increased or decreased bicarbonate levels. Metabolic acidosis is indicated by HCO_3^- values below 22 mEq/L, and metabolic alkalosis by values over 26 mEq/L. Notice that whereas P_{CO_2} levels vary inversely with blood pH (P_{CO_2} rises as blood pH falls), HCO_3^- levels vary directly with blood pH (increased HCO_3^- results in increased pH).

Beyond this bare-bones approach there is something else to consider when you are assessing acid-base problems. If an imbalance is fully compensated, the pH may be normal even while the patient is in trouble. Hence, when the pH is normal, carefully scrutinize the P_{CO_2} or HCO_3^- values for clues to what imbalance may be occurring.

Consider the following two examples of the three-step approach.

Problem 1

Blood values: pH 7.6; P_{CO_2} 24 mm Hg; HCO_3^- 23 mEq/L

Analysis:

1. The pH is elevated: alkalosis.
2. The P_{CO_2} is very low: the cause of the alkalosis.
3. The HCO_3^- value is within normal limits.

Conclusion: This is a respiratory alkalosis not compensated by renal mechanisms, as might occur during short-term hyperventilation.

Problem 2

Blood values: pH 7.48; P_{CO_2} 46 mm Hg; HCO_3^- 33 mEq/L

Analysis:

1. The pH is elevated: alkalosis.
2. The P_{CO_2} is elevated: the cause of acidosis, not alkalosis; thus, the respiratory system is compensating and is not the cause.
3. The HCO_3^- is elevated: the cause of the alkalosis.

Conclusion: This is a metabolic alkalosis being compensated by respiratory acidosis (retention of CO_2 to restore blood pH to the normal range).

See the accompanying simple chart to help you in your future determinations.

	Normal Range in Plasma		
Acid-Base Disturbance	**pH 7.35–7.45**	**P_{CO_2} 35–45 mm Hg**	**HCO_3^- 22–26 mEq/L**
Respiratory acidosis	↓	↑	↑ if compensating
Respiratory alkalosis	↑	↓	↓ if compensating
Metabolic acidosis	↓	↓ if compensating	↓
Metabolic alkalosis	↑	↑ if compensating	↑

RELATED CLINICAL TERMS

Antacid An agent that counteracts acidity, such as sodium bicarbonate, aluminum hydroxide gel, and magnesium trisilicate. Commonly used in the management of heartburn.

Hyperaldosteronism (Conn's disease) A condition of hypersecretion of aldosterone by adrenal cortical cells accompanied by excessive loss of potassium and generalized muscular weakness, hypernatremia, and hypertension. Usual cause is adrenal tumor; usual treatment is adrenal-suppressing drugs prior to tumor removal.

Renal tubular acidosis A metabolic acidosis resulting from impaired renal reabsorption of bicarbonate; the urine is alkaline.

Syndrome of inappropriate ADH secretion (SIADH) A group of disorders associated with excessive ADH secretion in the absence of appropriate (osmotic or nonosmotic) stimuli. Characterized by hypotonic hydration (hyponatremia, fluid retention, and weight gain) and concentrated urine. Usual causes are ectopic secretion of ADH by cancer cells (e.g., bronchogenic cancer cells) and brain disorders or trauma affecting the ADH-secreting hypothalamic neurons. Temporary management involves water restriction.

CHAPTER SUMMARY

Media study tools that could provide you additional help in reviewing specific key topics of Chapter 26 are referenced below.

iP = ***Interactive Physiology***

Body Fluids (pp. 996–998)

Body Water Content (p. 996)

1. Water accounts for 45–75% of body weight, depending on age, sex, and amount of body fat.

Fluid Compartments (p. 996)

2. About two-thirds (25 L) of body water is found within cells in the intracellular fluid (ICF) compartment; the balance (15 L) is in the extracellular fluid (ECF) compartment. The ECF includes plasma and interstitial fluid.

Composition of Body Fluids (pp. 996–998)

3. Solutes dissolved in body fluids include electrolytes and nonelectrolytes. Electrolyte concentration is expressed in mEq/L.
4. Plasma contains more proteins than does interstitial fluid; otherwise, extracellular fluids are similar. The most abundant ECF electrolytes are sodium, chloride, and bicarbonate ions.
5. Intracellular fluids contain large amounts of protein anions and potassium, phosphate, and magnesium ions.

26

iP Fluid, Electrolyte, and Acid/Base Balance; Topic: Introduction to Body Fluids, pp. 1–8.

Fluid Movement Among Compartments (p. 998)

6. Substances usually pass through the plasma and interstitial fluid to reach the intracellular fluid.
7. Fluid exchanges between compartments are regulated by osmotic and hydrostatic pressures: (a) Filtrate is forced out of the capillaries by hydrostatic pressure and pulled back in by colloid osmotic pressure. (b) Water moves freely between the ECF and the ICF by osmosis, but solute movements are restricted by size, charge, and dependence on transport proteins. (c) Water flows always follow changes in ECF osmolality.

iP Fluid, Electrolyte, and Acid/Base Balance; Topic: Introduction to Body Fluids, pp. 19–23.

Water Balance and ECF Osmolality (pp. 998–1002)

1. Sources of body water are ingested foods and fluids and metabolic water.
2. Water leaves the body primarily via the lungs, skin, gastrointestinal tract, and kidneys.

Regulation of Water Intake (pp. 999–1000)

3. Increased plasma osmolality triggers the thirst mechanism, mediated by hypothalamic osmoreceptors. Thirst, inhibited by distension of the gastrointestinal tract by ingested water and then by reduced osmolality, may be damped before body needs for water are met.

Regulation of Water Output (p. 1000)

4. Obligatory water loss is unavoidable and includes insensible water losses from the lungs, the skin, and in feces, and about 500 ml of urine output daily.
5. Beyond obligatory water loss, the volume of urinary output depends on water intake and loss via other routes and reflects the influence of antidiuretic hormone on the renal collecting ducts.

Influence of ADH (pp. 1000–1001)

6. Antidiuretic hormone causes aquaporins (water channels) to be inserted in the cell membranes of the collecting ducts, so that most filtered water is reabsorbed. ADH release is triggered if ECF osmolality is high, or if a large drop in blood volume or pressure occurs.

Disorders of Water Balance (pp. 1001–1002)

7. Dehydration occurs when water loss exceeds water intake over time. It is evidenced by thirst, dry skin, and decreased urine output. A serious consequence is hypovolemic shock.
8. Hypotonic hydration occurs when body fluids are excessively diluted and cells become swollen by water entry. The most serious consequence is cerebral edema.
9. Edema is an abnormal accumulation of fluid in the interstitial space (increased IF), which may impair blood circulation.

iP Fluid, Electrolyte, and Acid/Base Balance; Topic: Water Homeostasis, pp. 1–27.

Electrolyte Balance (pp. 1002–1008)

1. Most electrolytes (salts) are obtained from ingested foods and fluids. Salts, particularly NaCl, are often ingested in excess of need.
2. Electrolytes are lost in perspiration, feces, and urine. The kidneys are most important in regulating electrolyte balance.

The Central Role of Sodium in Fluid and Electrolyte Balance (pp. 1002–1004)

3. Sodium salts are the most abundant solutes in ECF. They exert the bulk of ECF osmotic pressure and control water volume and distribution in the body.

Regulation of Sodium Balance (pp. 1004–1006)

4. Sodium ion balance is linked to ECF volume and blood pressure regulation and involves both neural and hormonal controls.
5. Aldosterone promotes Na^+ reabsorption to maintain blood volume and blood pressure.
6. Declining blood pressure and falling filtrate NaCl concentration stimulate the granular cells to release renin. Renin, via angiotensin II, enhances systemic blood pressure, Na^+ reabsorption, and aldosterone release.
7. Cardiovascular system baroreceptors sense changing arterial blood pressure, prompting changes in sympathetic vasomotor activity. Rising arterial pressure leads to vasodilation and enhanced Na^+ and water loss in urine. Falling arterial pressure promotes vasoconstriction and conserves Na^+ and water.
8. Atrial natriuretic peptide, released by certain atrial cells in response to rising blood pressure (or blood volume), causes systemic vasodilation and inhibits renin, aldosterone, and ADH release. Hence, it enhances Na^+ and water excretion, reducing blood volume and blood pressure.
9. Estrogens and glucocorticoids increase renal retention of sodium. Progesterone promotes enhanced sodium and water excretion in urine.

Regulation of Potassium Balance (pp. 1006–1007)

10. About 90% of filtered potassium is reabsorbed by the more proximal regions of the nephrons.
11. The main thrust of renal regulation of K^+ is to excrete it. Potassium ion secretion by the principal cells of the cortical collecting ducts is enhanced by increased plasma K^+ content and aldosterone. Type A intercalated cells of the collecting duct reabsorb small amounts of K^+ during K^+ deficit.

Regulation of Calcium and Phosphate Balance (p. 1008)

12. Calcium balance is regulated primarily by parathyroid hormone (PTH), which enhances blood Ca^{2+} levels by targeting the bones, intestine, and kidneys. PTH-regulated reabsorption occurs primarily in the DCT.
13. PTH decreases renal reabsorption of phosphate ions.

Regulation of Anions (p. 1008)

14. When blood pH is normal or slightly high, chloride is the major anion accompanying sodium reabsorption. In acidosis, chloride is replaced by bicarbonate.
15. Reabsorption of most other anions is regulated by their transport maximums (T_m).

iP Fluid, Electrolyte, and Acid/Base Balance; Topic: Electrolyte Homeostasis, pp. 1–38.

Acid-Base Balance (pp. 1008–1015)

1. The homeostatic pH range of arterial blood is 7.35 to 7.45. A higher pH represents alkalosis; a lower pH reflects acidosis.
2. Some acids enter the body in foods, but most are generated by breakdown of phosphorus-containing proteins, incomplete oxidation of fats or glucose, and the loading and transport of carbon dioxide in the blood.
3. Acid-base balance is achieved by chemical buffers, respiratory regulation, and in the long term by renal regulation of bicarbonate ion (hence, hydrogen ion) concentration of body fluids.

iP Fluid, Electrolyte, and Acid/Base Balance; Topic: Acid/Base Homeostasis, pp. 1–15.

Chemical Buffer Systems (pp. 1009–1010)

4. Acids are proton (H^+) donors; bases are proton acceptors. Acids that dissociate completely in solution are strong acids; those that dissociate incompletely are weak acids. Strong bases are more effective proton acceptors than are weak bases.
5. Chemical buffers are single or paired sets (a weak acid and its salt) of molecules that act rapidly to resist excessive shifts in pH by releasing or binding H^+.
6. Chemical buffers of the body include the bicarbonate, phosphate, and protein buffer systems.

iP IP Fluid, Electrolyte, and Acid/Base Balance; Topic: Acid/Base Homeostasis, pp. 16–26.

Respiratory Regulation of H^+ (pp. 1010–1011)

7. Respiratory regulation of acid-base balance of the blood utilizes the bicarbonate buffer system and the fact that CO_2 and H_2O are in reversible equilibrium with H_2CO_3.
8. Acidosis activates the respiratory center to increase respiratory rate and depth, which eliminates more CO_2 and causes blood pH to rise. Alkalosis depresses the respiratory center, resulting in CO_2 retention and a fall in blood pH.

iP Fluid, Electrolyte, and Acid/Base Balance; Topic: Acid/Base Homeostasis, pp. 27–28.

Renal Mechanisms of Acid-Base Balance (pp. 1011–1014)

9. The kidneys provide the major long-term mechanism for controlling acid-base balance by maintaining stable HCO_3^- levels in the ECF. Metabolic acids (organic acids other than carbonic acid) can be eliminated from the body only by the kidneys.
10. Secreted hydrogen ions come from the dissociation of carbonic acid generated within the tubule cells.
11. Tubule cells are impermeable to bicarbonate in the filtrate, but they can conserve filtered bicarbonate ions indirectly by absorbing HCO_3^- generated within them (by dissociation of carbonic acid to HCO_3^- and H^+). For each HCO_3^- (and Na^+) reabsorbed, one H^+ is secreted into the filtrate, where it combines with HCO_3^-.
12. To generate and add new HCO_3^- to plasma to counteract acidosis, either of two mechanisms may be used:
    - Secreted H^+, buffered by bases other than HCO_3^-, is excreted from the body in urine (the major urine buffer is the phosphate buffer system).
    - NH_4^+ (derived from glutamine catabolism) is excreted in urine.
13. To counteract alkalosis, bicarbonate ion is secreted into the filtrate and H^+ is reabsorbed.

iP Fluid, Electrolyte, and Acid/Base Balance; Topic: Acid/Base Homeostasis, pp. 29–37.

Abnormalities of Acid-Base Balance (pp. 1014–1015)

14. Classification of acid-base imbalances as metabolic or respiratory indicates the cause of the acidosis or alkalosis.
15. Respiratory acidosis results from carbon dioxide retention; respiratory alkalosis occurs when carbon dioxide is eliminated faster than it is produced.
16. Metabolic acidosis occurs when fixed acids (lactic acid, ketone bodies, and others) accumulate in the blood or when bicarbonate

26

is lost from the body; metabolic alkalosis occurs when bicarbonate levels are excessive.
17. Extremes of pH for life are 7.0 and 7.8.
18. Compensations occur when the respiratory system or kidneys act to counteract acid-base imbalances resulting from abnormal or inadequate functioning of the alternate system. Respiratory compensations involve changes in respiratory rate and depth. Renal compensations modify blood levels of HCO_3^-.

iP Fluid, Electrolyte, and Acid/Base Balance; Topic: Acid/Base Homeostasis, pp. 38–59.

Developmental Aspects of Fluid, Electrolyte, and Acid-Base Balance (pp. 1015–1016)

1. Infants have a higher risk of dehydration and acidosis because of their low lung residual volume, high rate of fluid intake and output, high metabolic rate, relatively large body surface area, and functionally immature kidneys at birth.
2. The elderly are at risk for dehydration because of their low percentage of body water and insensitivity to thirst cues. Diseases that promote fluid and acid-base imbalances (cardiovascular disease, diabetes mellitus, and others) are most common in the aged.

REVIEW QUESTIONS

Multiple Choice/Matching

(Some questions have more than one correct answer. Select the best answer or answers from the choices given.)

1. Body water content is greatest in (a) infants, (b) young adults, (c) elderly adults.
2. Potassium, magnesium, and phosphate ions are the predominant electrolytes in (a) plasma, (b) interstitial fluid, (c) intracellular fluid.
3. Sodium balance is regulated primarily by control of amount(s) (a) ingested, (b) excreted in urine, (c) lost in perspiration, (d) lost in feces.
4. Water balance is regulated by control of amount(s) (use choices in question 3).

Answer questions 5 through 10 by choosing responses from the following:

(a) ammonium ions	(f) magnesium
(b) bicarbonate	(g) phosphate
(c) calcium	(h) potassium
(d) chloride	(i) sodium
(e) hydrogen ions	(j) water

5. Three substances regulated (at least in part) by the influence of aldosterone on the kidney tubules.
6. Two substances regulated by parathyroid hormone.
7. Two substances secreted into the proximal convoluted tubules in exchange for sodium ions.
8. Part of an important chemical buffer system in plasma.
9. Two ions produced during catabolism of glutamine.
10. Substance regulated by ADH's effects on the renal tubules.
11. Which of the following factors will enhance ADH release? (a) increase in ECF volume, (b) decrease in ECF volume, (c) decrease in ECF osmolality, (d) increase in ECF osmolality.
12. The pH of blood varies *directly* with (a) HCO_3^-, (b) P_{CO_2} (c) H^+, (d) none of the above.
13. In an individual with metabolic acidosis, a clue that the respiratory system is compensating is provided by (a) high blood bicarbonate levels, (b) low blood bicarbonate levels, (c) rapid, deep breathing, (d) slow, shallow breathing.

Short Answer Essay Questions

14. Name the body fluid compartments, noting their locations and the approximate fluid volume in each.
15. Describe the thirst mechanism, indicating how it is triggered and terminated.
16. Explain why and how ECF osmolality is maintained.
17. Explain why and how sodium balance, ECF volume, and blood pressure are jointly regulated.
18. Describe the role of the respiratory system in controlling acid-base balance.
19. Explain how the chemical buffer systems resist changes in pH.
20. Explain the relationship of the following to renal secretion and excretion of hydrogen ions: (a) plasma carbon dioxide levels, (b) phosphate, and (c) sodium bicarbonate reabsorption.
21. List several factors that place newborn babies at risk for acid-base imbalances.

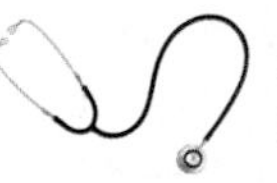

Critical Thinking and Clinical Application Questions

1. Mr. Jessup, a 55-year-old man, is operated on for a cerebral tumor. About a month later, he appears at his physician's office complaining of excessive thirst. He claims to have been drinking about 20 liters of water daily for the past week and says he has been voiding nearly continuously. A urine sample is collected and its specific gravity is reported as 1.001. What is your diagnosis of Mr. Jessup's problem? What connection might exist between his previous surgery and his present problem?
2. For each of the following sets of blood values, name the acid-base imbalance (acidosis or alkalosis), determine its cause (metabolic or respiratory), decide whether the condition is being compensated, and cite at least one possible cause of the imbalance.
   Problem 1: pH 7.63; P_{CO_2} 19 mm Hg; HCO_3^- 19.5 mEq/L
   Problem 2: pH 7.22; P_{CO_2} 30 mm Hg; HCO_3^- 12.0 mEq/L
3. Explain how emphysema and congestive heart failure can lead to acid-base imbalance.
4. Mrs. Bush, a 70-year-old woman, is admitted to the hospital. Her history states that she has been suffering from diarrhea for three weeks. On admission, she complains of severe fatigue and muscle weakness. A blood chemistry study yields the following information: Na^+ 142 mEq/L; K^+ 1.5 mEq/L; Cl^- 92 mEq/L; P_{CO_2} 32 mm Hg. Which electrolytes are within normal limits? Which are so abnormal that the patient has a medical emergency? Which of the following represents the greatest danger to Mrs. Bush? (a) a fall due to her muscular weakness, (b) edema, (c) cardiac arrhythmia and cardiac arrest.
5. During a routine medical checkup, Candace, a 26-year-old physiotherapy student, is surprised to hear that her blood pressure is 180/110. She also has a rumbling systolic and diastolic abdominal bruit (murmur) that is loudest at the mid-epigastric area. Her physician suspects renal artery stenosis (narrowing). She orders

26

an abdominal ultrasound and renal artery arteriography, which confirms that Candace has a small right kidney, and that the distal part of her right renal artery is narrowed by more than 70%. Her physician prescribes diuretics and calcium channel blockers as temporary measures, and refers Candace to a cardiovascular surgeon. Explain the connection between Candace's renal artery stenosis and her hypertension. Why is her right kidney smaller than her left? What would you expect Candace's blood levels of K^+, Na^+, aldosterone, angiotensin II, and renin to be?

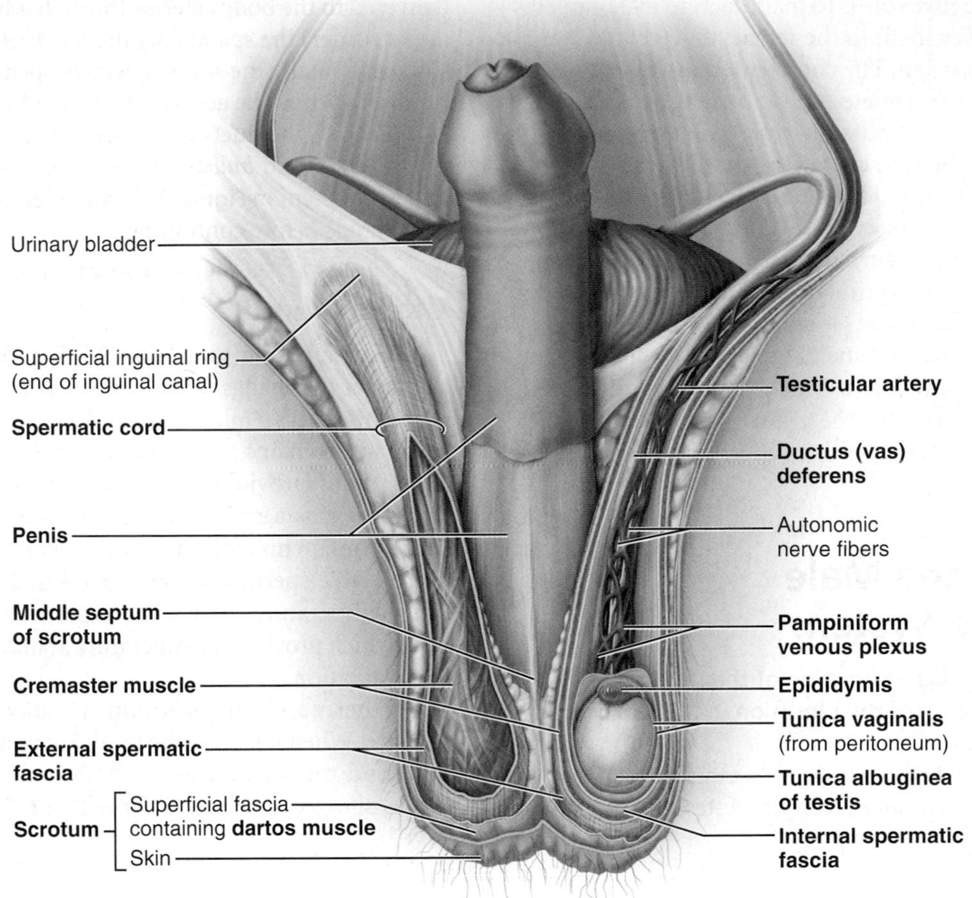

Figure 27.2 **Relationships of the testis to the scrotum and spermatic cord.** The scrotum has been opened and its anterior portion removed.

and increasing its thickness to reduce heat loss. When it is warm, the scrotal skin is flaccid and loose to increase the surface area for cooling (sweating) and the testes hang lower, away from the body trunk.

These changes in scrotal surface area help maintain a fairly constant intrascrotal temperature and reflect the activity of two sets of muscles which respond to changes in ambient temperature. The **dartos muscle** (dar′tos; "skinned"), a layer of smooth muscle in the superficial fascia, wrinkles the scrotal skin. The **cremaster muscles** (kre-mas′ter; "a suspender"), bands of skeletal muscle that arise from the internal oblique muscles of the trunk, elevate the testes.

27

The Testes

Each plum-sized testis is approximately 4 cm (1.5 inches) long and 2.5 cm (1 inch) in width and is surrounded by two tunics. The outer tunic is the two-layered **tunica vaginalis** (vaj″ĭ-nal′is), derived from an outpocketing of the peritoneum (Figure 27.2 and **Figure 27.3**). Deep to this serous layer is the **tunica albuginea** (al″bu-jin′e-ah; "white coat"), the fibrous capsule of the testis.

Septa extending from the tunica albuginea divide the testis into about 250 wedge-shaped *lobules* (Figure 27.3). Each contains one to four tightly coiled **seminiferous tubules** (sem″ĭ-nif′er-us; "sperm-carrying"), the actual "sperm factories." Surrounding each seminiferous tubule are three to five layers of smooth muscle–like **myoid cells** (Figure 27.3c). By contracting rhythmically, these cells may help to squeeze sperm and testicular fluids through the tubules and out of the testes.

The seminiferous tubules of each lobule converge to form a **straight tubule**, or **tubulus rectus**, that conveys sperm into the **rete testis** (re′te), a tubular network on the posterior side of the testis. From the rete testis, sperm leave the testis through the *efferent ductules* and enter the *epididymis* (ep″ĭ-did′ĭ-mis), which hugs the external testis surface posteriorly. The immature sperm pass through the head, the body, and then move into the tail of the epididymis, where they are stored until ejaculation.

Lying in the soft connective tissue surrounding the seminiferous tubules are the **interstitial cells**, also called *Leydig cells* (Figure 27.3c). These cells produce androgens (most importantly *testosterone*), which they secrete into the surrounding interstitial fluid. Thus, the sperm-producing and hormone-producing functions of the testis are carried out by completely different cell populations.

The long **testicular arteries**, which branch from the abdominal aorta superior to the pelvis (see *gonadal arteries* in Figure 19.24c, p. 733), supply the testes. The **testicular veins** draining

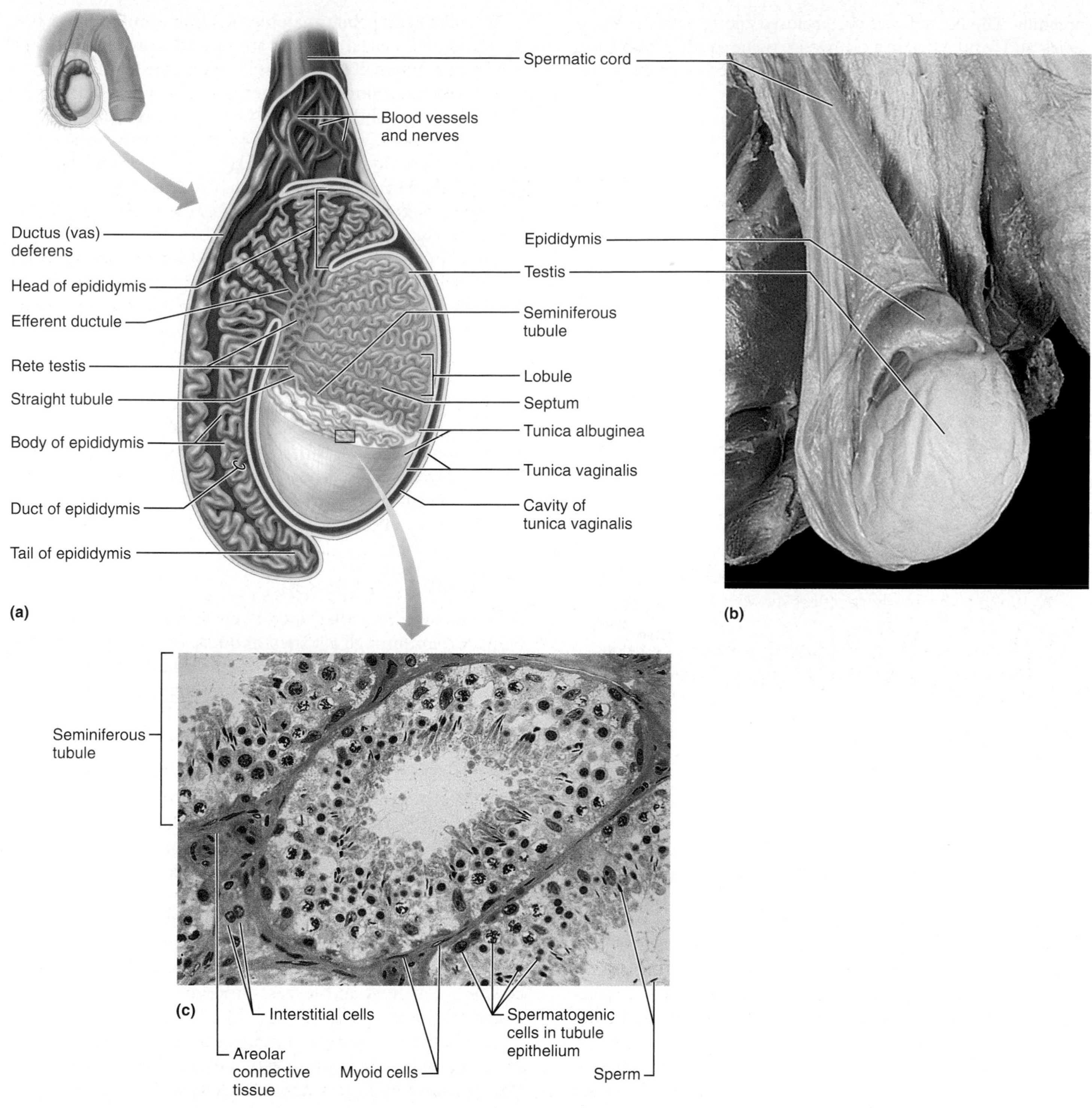

Figure 27.3 Structure of the testis. (a) Partial sagittal section through the testis and epididymis. The anterior aspect is to the right. (See *A Brief Atlas of the Human Body*, Figure 73). **(b)** External view of a testis from a cadaver; same orientation as in (a). **(c)** Seminiferous tubule in cross section (270×). Note the spermatogenic (sperm-forming) cells in the tubule epithelium and the interstitial cells in the connective tissue between the tubules.

27

the testes arise from a network called the **pampiniform venous plexus** (pam-pin′ĭ-form; "tendril-shaped") that surrounds the portion of each testicular artery within the scrotum like a climbing vine (see Figure 27.2). The cooler venous blood in each pampiniform plexus absorbs heat from the arterial blood, cooling it before it enters the testes. In this way, these plexuses help to keep the testes at their cool homeostatic temperature.

The testes are served by both divisions of the autonomic nervous system. Associated sensory nerves transmit impulses that result in agonizing pain and nausea when the testes are hit forcefully. The nerve fibers are enclosed, along with the blood vessels and lymphatics, in a connective tissue sheath. Collectively these structures make up the **spermatic cord**, which passes through the inguinal canal (see Figure 27.2).

HOMEOSTATIC IMBALANCE

Although *testicular cancer* is relatively rare (affecting one of every 50,000 males), it is the most common cancer in young men (ages 15 to 35 years). A history of mumps or orchitis (inflammation of the testis) and substantial maternal exposure to environmental toxins before birth increases the risk, but the most important risk factor for this cancer is *cryptorchidism* (nondescent of the testes, see p. 1063).

Because the most common sign of testicular cancer is a painless solid mass in the testis, self-examination of the testes should be practiced by every male. If detected early, it has an impressive cure rate. Over 90% of testicular cancers are cured by surgical removal of the cancerous testis (*orchiectomy*) either alone or in combination with radiation therapy or chemotherapy. ■

CHECK YOUR UNDERSTANDING

1. What are the two major functions of the testes?
2. Which of the tubular structures shown in Figure 27.3a are the sperm "factories"?
3. Muscle activity and the pampiniform venous plexus help to keep testicular temperature at homeostatic levels. How do they do that?

For answers, see Appendix G.

The Penis

▶ Describe the location, structure, and function of the accessory reproductive organs of the male.

The **penis** ("tail") is a copulatory organ, designed to deliver sperm into the female reproductive tract (Figure 27.1 and Figure 27.4). The penis and scrotum, which hang suspended from the perineum, make up the external reproductive structures, or **external genitalia**, of the male. The male **perineum** (per″ĭ-ne′um; "around the anus") is the diamond-shaped region located between the pubic symphysis anteriorly, the coccyx posteriorly, and the ischial tuberosities laterally. The floor of the perineum is formed by muscles described in Chapter 10 (pp. 344–345).

27

The penis consists of an attached *root* and a free *shaft* or *body* that ends in an enlarged tip, the **glans penis.** The skin covering the penis is loose, and it slides distally to form a cuff called the **prepuce** (pre′pūs), or **foreskin**, around the glans. Frequently, the foreskin is surgically removed shortly after birth, a procedure called *circumcision* ("cutting around"). Interestingly, about 60% of newborn boys in the United States are circumcised, whereas the rate in other parts of the world is around 15%.

To understand penile anatomy, it is important to know that its dorsal and ventral surfaces are named in reference to the erect penis. Internally, the penis contains the spongy urethra and three long cylindrical bodies (*corpora*) of erectile tissue, each covered by a sheath of dense fibrous connective tissue. This *erectile tissue* is a spongy network of connective tissue and smooth muscle riddled with vascular spaces. During sexual excitement, the vascular spaces fill with blood, causing the penis to enlarge and become rigid. This event, called *erection*, enables the penis to serve as a penetrating organ.

The midventral erectile body, the **corpus spongiosum** (spon″je-o′sum; "spongy body") surrounds the urethra. It expands distally to form the glans and proximally to form the part of the root called the **bulb of the penis.** The bulb is covered externally by the sheetlike bulbospongiosus muscle and is secured to the urogenital diaphragm.

The paired dorsal erectile bodies, called the **corpora cavernosa** (kor′por-ah kă-ver-no′sah; "cavernous bodies"), make up most of the penis and are bound by the fibrous tunica albuginea. Their proximal ends form the **crura of the penis** (kroo′rah; "legs"; singular: crus). Each crus is surrounded by an ischiocavernosus muscle and anchors to the pubic arch of the bony pelvis.

The Male Duct System

As we mentioned earlier, sperm travel from the testes to the body exterior through a system of ducts. In order (proximal to distal), the **accessory ducts** are the epididymis, the ductus deferens, the ejaculatory duct, and the urethra.

The Epididymis

The cup-shaped **epididymis** (*epi* = beside; *didym* = the testes) is about 3.8 cm (1.5 inches) long (Figures 27.1 and Figures 27.3a, b). Its *head*, which contains the efferent ductules, caps the superior aspect of the testis. Its *body* and *tail* are on the posterolateral area of the testis. Most of the epididymis consists of the highly coiled *duct of the epididymis* with an uncoiled length of about 6 m (20 feet). Some pseudostratified epithelial cells of the duct mucosa exhibit long, nonmotile microvilli (*stereocilia*). The huge surface area of these stereocilia allows them to absorb excess testicular fluid and to pass nutrients to the many sperm stored temporarily in the lumen.

The immature, nearly nonmotile sperm that leave the testis are moved slowly along the duct of the epididymis through fluid that contains a number of antimicrobial proteins, including several β-defensins. As they move along its tortuous course (a trip that takes about 20 days), the sperm gain the ability to swim.

Sperm are ejaculated from the epididymis, not the testes as many believe. When a male is sexually stimulated and ejaculates, the smooth muscle in the ducts of the epididymis contracts, expelling sperm into the next segment of the duct system, the *ductus deferens.* Sperm can be stored in the epididymis for several months, but if held longer, they are eventually phagocytized by epithelial cells of the epididymis. This is not a problem for the man, as sperm are generated continuously.

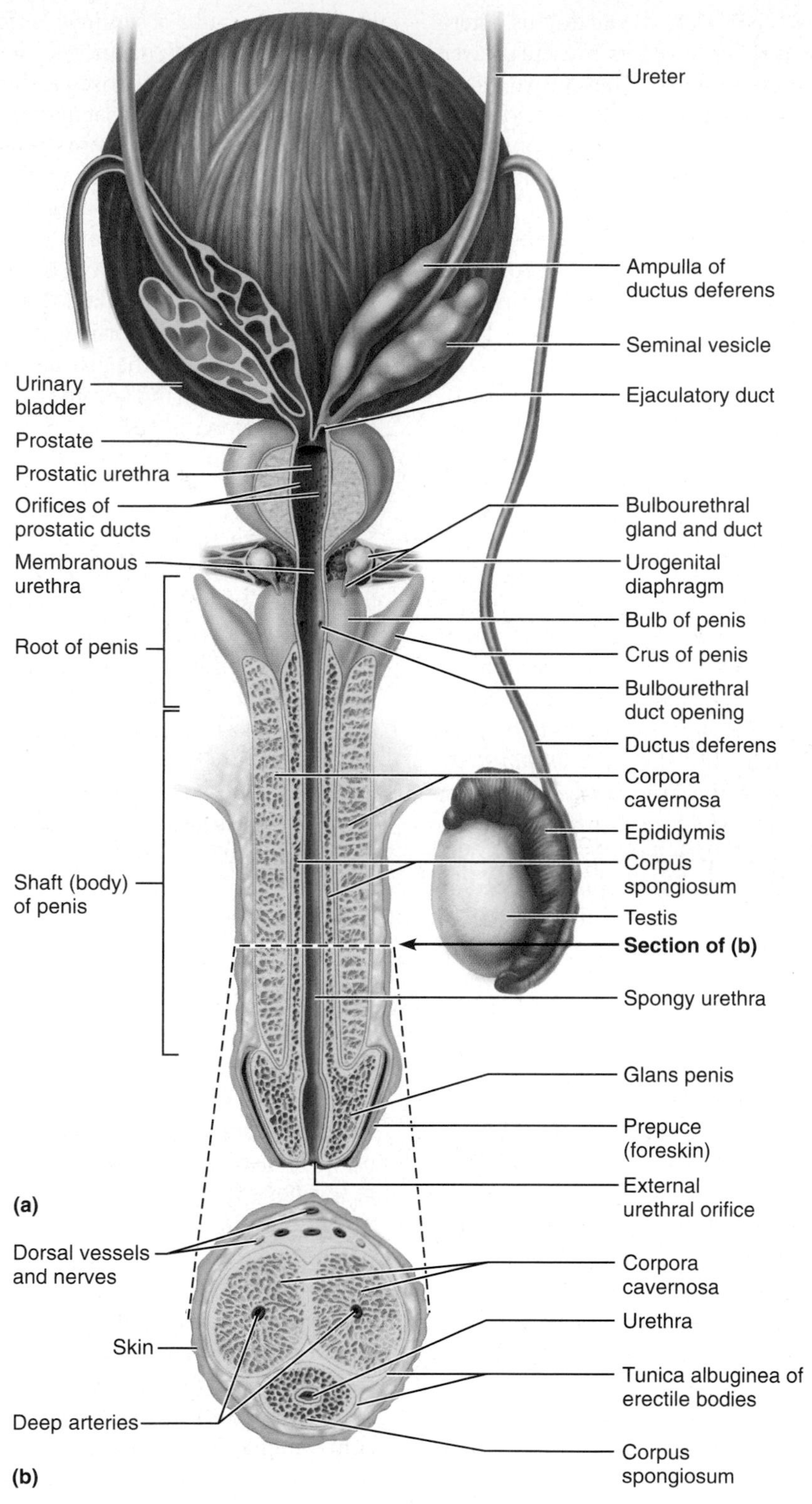

Figure 27.4 Male reproductive structures. (a) Posterior view showing longitudinal (coronal) section of the penis. **(b)** Transverse section of the penis.

The Ductus Deferens and Ejaculatory Duct

The **ductus deferens** (duk′tus def′er-ens; "carrying away"), or *vas deferens*, is about 45 cm (18 inches) long. It runs upward as part of the spermatic cord from the epididymis through the inguinal canal into the pelvic cavity (Figure 27.1). Easily palpated as it passes anterior to the pubic bone, it then loops medially over the ureter and descends along the posterior bladder wall. Its terminus expands to form the **ampulla** of the ductus deferens and then joins with the duct of the seminal vesicle (a gland) to form the short **ejaculatory duct**. Each ejaculatory duct enters the prostate, and there it empties into the urethra.

Like that of the epididymis, the mucosa of the ductus deferens is pseudostratified epithelium. However, its muscular layer is extremely thick and the duct feels like a hard wire when squeezed between the fingertips. At the moment of ejaculation, the thick layers of smooth muscle in its walls create strong peristaltic waves that rapidly squeeze the sperm forward along the tract and into the urethra.

As Figure 27.3 illustrates, part of the ductus deferens lies in the scrotal sac. Some men opt to take full responsibility for birth control by having a **vasectomy** (vah-sek′to-me; "cutting the vas"). In this relatively minor operation, the physician makes a small incision into the scrotum and then cuts through and ligates (ties off) each ductus deferens. Sperm are still produced, but they can no longer reach the body exterior. Eventually, they deteriorate and are phagocytized. Vasectomy is simple and provides highly effective birth control (close to 100%). For those wishing to reverse that procedure, the success rate is about 50%.

The Urethra

The **urethra** is the terminal portion of the male duct system (Figures 27.1 and 27.4). It conveys both urine and semen (at different times), so it serves both the urinary and reproductive systems. Its three regions are (1) the *prostatic urethra*, the portion surrounded by the prostate; (2) the *membranous* (or *intermediate part of the*) *urethra* in the urogenital diaphragm; and (3) the *spongy* (*penile*) *urethra*, which runs through the penis and opens to the outside at the *external urethral orifice*. The spongy urethra is about 15 cm (6 inches) long and accounts for 75% of urethral length. Its mucosa contains scattered *urethral glands* that secrete mucus into the lumen just before ejaculation.

CHECK YOUR UNDERSTANDING

4. What is the function of the erectile tissue of the penis?
5. Name the organs of the male duct system in order, from the epididymis to the body exterior.
6. What are two functions of the stereocilia on the epididymal epithelium?
7. Which accessory organ of the male duct system runs from the scrotum into the abdominal cavity?

For answers, see Appendix G.

27

Accessory Glands

▶ Discuss the sources and functions of semen.

The **accessory glands** include the paired seminal vesicles and bulbourethral glands and the single prostate (Figures 27.1 and 27.4). Together these glands produce the bulk of *semen* (sperm plus accessory gland secretions).

The Seminal Vesicles

The **seminal vesicles** (sem′ĭ-nul), or **seminal glands**, lie on the posterior bladder surface. Each of these fairly large, hollow glands is about the shape and length (5–7 cm) of a little finger. However, because a seminal vesicle is pouched, coiled, and folded back on itself, its uncoiled length is actually about 15 cm. Its fibrous capsule encloses a thick layer of smooth muscle which contracts during ejaculation to empty the gland.

Stored within the mucosa's honeycomb of crypts and blind alleys is a yellowish viscous alkaline fluid containing fructose sugar, ascorbic acid, a coagulating enzyme (*vesiculase*), and prostaglandins, as well as other substances that enhance sperm motility or fertilizing ability. As noted, the duct of each seminal vesicle joins that of the ductus deferens on the same side to form the ejaculatory duct. Sperm and seminal fluid mix in the ejaculatory duct and enter the prostatic urethra together during ejaculation. Seminal gland secretion accounts for some 70% of the volume of semen.

The Prostate

The **prostate** (pros′tāt) is a single doughnut-shaped gland about the size of a peach pit (Figures 27.1 and 27.4). It encircles the urethra just inferior to the bladder. Enclosed by a thick connective tissue capsule, it is made up of 20 to 30 compound tubuloalveolar glands embedded in a mass (stroma) of smooth muscle and dense connective tissue.

The prostatic gland secretion enters the prostatic urethra via several ducts when prostatic smooth muscle contracts during ejaculation. It plays a role in activating sperm and accounts for up to one-third of the semen volume. It is a milky, slightly acid fluid that contains citrate (a nutrient source), several enzymes (fibrinolysin, hyaluronidase, acid phosphatase), and prostate-specific antigen (PSA).

HOMEOSTATIC IMBALANCE

The prostate has a reputation as a health destroyer (perhaps reflected in the common mispronunciation "prostrate"). Hypertrophy of the prostate, called *benign prostatic hyperplasia* (*BPH*) which affects nearly every elderly male, distorts the urethra. Its precise cause is unknown, but it may be associated with changes in hormone levels as a result of aging. The more the man strains to urinate, the more the valvelike prostatic mass blocks the opening, enhancing the risk of bladder infections (cystitis) and kidney damage.

Traditional treatment has been surgical. Some newer options include using microwaves or drugs to shrink the prostate; inserting and inflating a small balloon to compress the prostate tissue away from the prostatic urethra; and inserting a catheter containing a tiny needle in order to incinerate excess prostate tissue with burstlike releases of radio-frequency radiation.

Finasteride, which ratchets down production of dihydrotestosterone, the hormone linked to male pattern balding and prostate enlargement, is helpful in some cases. Additionally, several drugs are available that relax the smooth muscles at the bladder outlet, facilitating bladder emptying and providing relief.

Prostatitis (pros″tah-ti′tis), inflammation of the prostate, is the most common reason for a man to consult a urologist.

Prostate cancer is the second most common cause of cancer death in men (after lung cancer) and is twice as common in blacks as in whites. Risk factors include fatty diet and genetic predisposition. Screening for prostate cancer typically involves digital examination (palpating the prostate through the anterior rectal wall) and assessment of PSA levels in the blood. Although a normal component of blood at levels below 2.5 ng/ml, PSA is a tumor marker and its rising serum level follows BPH as well as the clinical disease course of prostate cancer. Screening procedures are followed by biopsies of suspicious prostate areas, if deemed necessary, and metastases (most commonly to the pelvic lymph nodes and the skeleton) are detected by bone or MRI scans.

When possible, prostate cancer is treated surgically, alone or in conjunction with radiotherapy. Because prostate cancer is typically androgen dependent, alternative therapies for metastasized cancers involve castration or therapy with drugs that block androgen receptors (flutamide), or that inhibit gonadotropin release, such as LHRH (luteinizing hormone–releasing hormone) analogues. LHRH is another term for GnRH (gonadotropin-releasing hormone, see p. 605). Deprived of the stimulatory effects of androgens, the prostatic tissue regresses and urinary symptoms typically decline.

However, many prostate cancers are very slow growing and may never represent a threat to the patient, particularly if the patient is elderly. In such cases, frequent tests to monitor the tumor may be all that is necessary. ■

The Bulbourethral Glands

The **bulbourethral glands** (bul″bo-u-re′thral) are pea-sized glands inferior to the prostate (Figures 27.1 and 27.4). They produce a thick, clear mucus, some of which drains into the spongy urethra and lubricates the glans penis when a man becomes sexually excited. Additionally, the mucus neutralizes traces of acidic urine in the urethra just prior to ejaculation.

Semen

Semen (se′men) is a milky white, somewhat sticky mixture of sperm, testicular fluid, and accessory gland secretions. The liquid provides a transport medium and nutrients and contains chemicals that protect and activate the sperm and facilitate their movement. Mature sperm cells are streamlined cellular "missiles" containing little cytoplasm or stored nutrients. Catabolism of the fructose in seminal vesicle secretions provides nearly all the fuel needed for sperm ATP synthesis.

Prostaglandins in semen decrease the viscosity of mucus guarding the entry (cervix) of the uterus and stimulate reverse peristalsis in the uterus, facilitating sperm movement through the female reproductive tract. The presence of the hormone *relaxin* and certain enzymes in semen enhance sperm motility. The relative alkalinity of semen as a whole (pH 7.3–7.7) helps neutralize the acid environment of the male's urethra and the female's vagina, thereby protecting the delicate sperm and enhancing their motility. Sperm are very sluggish under acidic conditions particularly (below pH 6).

Semen also contains substances that suppress the immune response in the female's reproductive tract, and has antibiotic chemicals that destroy certain bacteria. Clotting factors found in semen coagulate it just after it is ejaculated. Coagulation causes the sperm to stick to the walls of the vagina and prevents their draining out of the vagina while initially immobile. Soon after, its contained fibrinolysin liquefies the sticky mass, enabling the sperm to swim out and begin their journey through the female duct system.

The amount of semen propelled out of the male duct system during ejaculation is relatively small, only 2–5 ml and only 10% sperm, but there are between 20 and 150 million sperm per milliliter.

CHECK YOUR UNDERSTANDING

8. Adolph, a 68-year-old gentleman, has trouble urinating and is scheduled for a rectal exam. What is his most probable condition and what is the purpose of the rectal exam?
9. Which glandular accessory organ produces the largest proportion of semen?
10. What is semen?

For answers, see Appendix G.

Physiology of the Male Reproductive System

Male Sexual Response

▶ Describe the phases of the male sexual response.

Although there is more to it, the chief phases of the male sexual response are (1) *erection* of the penis, which allows it to penetrate the female vagina, and (2) *ejaculation*, which expels semen into the vagina.

Erection

Erection, enlargement and stiffening of the penis, results from engorgement of the erectile bodies with blood. When a man is not sexually aroused, arterioles supplying the erectile tissue are constricted and the penis is flaccid. However, during sexual excitement a parasympathetic reflex is triggered that promotes release of nitric oxide (NO) locally. NO relaxes smooth muscle in the penile blood vessel walls, causing these arterioles to dilate, which allows the erectile bodies to fill with blood. Expansion of the corpora cavernosa of the penis compresses their drainage veins, retarding blood outflow and maintaining engorgement. The corpus spongiosum expands but not nearly as much as the cavernosa. Its main job is to keep the urethra open during ejaculation. It is important that the erect penis not kink or buckle during intercourse. This problem is prevented by the longitudinal and circular arrangement of collagen fibers surrounding the penis.

Erection of the penis is one of the rare examples of parasympathetic control of arterioles. Another parasympathetic effect is stimulation of the bulbourethral glands, the secretion of which lubricates the glans penis.

Erection is initiated by a variety of sexual stimuli, such as touching the genital skin, mechanical stimulation of the pressure receptors in the penis, and erotic sights, sounds, and smells. The CNS responds to such stimuli by activating parasympathetic neurons that innervate the internal pudendal arteries serving the penis. Sometimes erection is induced solely by emotional or higher mental activity (the thought of a sexual encounter). Emotions and thoughts can also inhibit erection, causing vasoconstriction and resumption of the flaccid penile state.

Ejaculation

Ejaculation (*ejac* = to shoot forth) is the propulsion of semen from the male duct system. Although erection is under parasympathetic control, ejaculation is under sympathetic control. When impulses provoking erection reach a certain critical level, a spinal reflex is initiated, and a massive discharge of nerve impulses occurs over the sympathetic nerves serving the genital organs (largely at the level of L_1 and L_2). As a result,

1. The bladder sphincter muscle constricts, preventing expulsion of urine or reflux of semen into the bladder.
2. The reproductive ducts and accessory glands contract, emptying their contents into the urethra.
3. Semen in the urethra triggers a spinal reflex through somatic motor neurons that cause the bulbospongiosus muscles of the penis to undergo a rapid series of contractions, propelling semen at a speed of up to 500 cm/s (200 inches/s or close to 11 mi/h) from the urethra. These rhythmic contractions are accompanied by intense pleasure and many systemic changes, such as generalized muscle contraction, rapid heartbeat, and elevated blood pressure.

The entire ejaculatory event is referred to as **climax** or **orgasm**. Orgasm is quickly followed by *resolution,* a period of muscular and psychological relaxation. Activity of sympathetic nerve fibers constricts the *internal pudendal arteries* (and penile arterioles), reducing blood flow into the penis, and activates small muscles that squeeze the cavernous bodies, forcing blood from the penis into the general circulation. Although the erection may persist a little longer, the penis soon becomes flaccid once again. After ejaculation, there is a *latent,* or *refractory, period,* ranging in time from minutes to hours, during which a man is unable to achieve another orgasm. The latent period lengthens with age.

27

CHECK YOUR UNDERSTANDING

11. What is erection and what division of the ANS regulates it?
12. What occurs during resolution and what is the result?

For answers, see Appendix G.

HOMEOSTATIC IMBALANCE

Erectile dysfunction (ED), the inability to attain an erection when desired, is due to deficient release of NO by the parasympathetic nerves serving the penis. Approximately 50% of American men over the age of 40 (some 32 million men) have some degree of erectile dysfunction. Psychological factors, alcohol, or certain drugs (antihypertensives, antidepressants, and others) can cause temporary ED. When chronic, the condition is largely the result of hormonal (diabetes mellitus), vascular (arteriosclerosis, varicose veins), or nervous system problems (stroke, penile nerve damage, MS).

Until recently, most remedies entailed using a vacuum pump to suck blood into the penis, injecting drugs into the penis that dilate penile blood vessels, or implanting a device into the penis to make it rigid. Viagra, approved in 1998, and two similar drugs (Cialis and Levitra) released in late 2003 potentiate the effect of the existing NO, a remedy that most men can live with. These drugs work within minutes to an hour to produce a sustained blood flow to the penis, are taken orally, and have essentially no significant side effects in healthy males. However, to avoid a fatal result, those with preexisting heart disease or diabetes mellitus must heed the warning that these drugs reduce systemic blood pressure. ■

Spermatogenesis

▶ Define meiosis. Compare and contrast it to mitosis.

▶ Outline events of spermatogenesis.

Spermatogenesis (sper″mah-to-jen′ĕ-sis; "sperm formation") is the sequence of events in the seminiferous tubules of the testes that produces male gametes—*sperm* or *spermatozoa.* The process begins around the age of 14 years (and often earlier) in males, and continues throughout life. Every day, a healthy adult male makes about 400 million sperm. It seems that nature has made sure that the human species will not be endangered for lack of sperm.

Before we describe the process of spermatogenesis, let's define some terms. First, having two sets of chromosomes, one from each parent, is a key factor in the human life cycle. The normal chromosome number in most body cells is referred to as the **diploid chromosomal number** (dip′loid) of the organism, symbolized as $2n$. In humans, this number is 46, and such diploid cells contain 23 pairs of similar chromosomes called **homologous chromosomes** (ho-mol′ŏ-gus) or **homologues.** One member of each pair is from the male parent (the *paternal chromosome*), and the other is from the female parent (the *maternal chromosome*).

Generally speaking, the two homologues of each chromosome pair look alike and carry genes that code for the same traits, though not necessarily for identical expression of those traits. (Consider, for example, the homologous genes controlling the expression of freckles. The paternal gene might code for an ample sprinkling of freckles and the maternal gene for no freckles.) In Chapter 29, we will consider how our mom's and dad's genes interact to produce our visible traits.

The number of chromosomes in human gametes is 23, referred to as the **haploid chromosomal number** (hap′loid), or n. Gametes contain only one member of each homologous pair. When sperm and egg fuse, they form a fertilized egg that reestablishes the typical diploid chromosomal number of human cells ($2n$).

Gamete formation in both sexes involves **meiosis** (mi-o′sis; "a lessening"), a unique kind of nuclear division that, for the most part, occurs only in the gonads. Recall that *mitosis* (the process by which most body cells divide) distributes replicated chromosomes equally between the two daughter cells. Consequently, each daughter cell receives a full set of chromosomes identical to that of the mother cell. Meiosis, on the other hand, consists of two consecutive nuclear divisions that follow one round of DNA replication. Its product is four daughter cells instead of two, each with *half* as many chromosomes as typical body cells. In other words, meiosis reduces the chromosomal number by half (from $2n$ to n) in gametes. Additionally, it introduces genetic variation because each of the haploid daughter cells have only some of the genes of each parent, as explained shortly.

Meiosis Compared to Mitosis

The two nuclear divisions of meiosis, called *meiosis I* and *meiosis II*, are divided into phases for convenience. These phases have the same names as those of mitosis (prophase, metaphase, anaphase, and telophase), but some events of meiosis I are quite different from those of mitosis. As we detail those events, you may find it helpful to refer to the comparison of mitosis and meiosis in **Figure 27.5**.

Recall that prior to mitosis all the chromosomes are replicated. Then the identical copies remain together as *sister chromatids* connected by a centromere throughout prophase and during metaphase. At anaphase, the centromeres split and the sister chromatids separate from each other so that each daughter cell inherits a copy of *every* chromosome possessed by the mother cell (Figure 27.5, left side). Let's look at how meiosis differs (Figure 27.5, right side).

Meiosis I Meiosis I is sometimes called the **reduction division of meiosis** because it reduces the chromosome number from $2n$ to n. As in mitosis, chromosomes replicate before meiosis begins and in prophase the chromosomes coil and condense, the nuclear envelope and nucleolus break down and disappear, and a spindle forms. But in prophase of meiosis I, an event never seen in mitosis (nor in meiosis II for that matter) occurs. The replicated chromosomes seek out their homologous partners and pair up with them along their entire length. This alignment takes place at discrete spots along the length of the homologues—more like buttoning together than zipping. As a result of this process, called **synapsis**, little groups of four chromatids called **tetrads** are formed (Figure 27.5 and **Figure 27.6**).

During synapsis, a second unique event, called crossover, occurs. **Crossovers**, also called **chiasmata** (singular: chiasma), are formed within each tetrad as the free ends of one maternal and one paternal chromatid wrap around each other at one or more points. Crossover allows an exchange of genetic material between the paired maternal and paternal chromosomes. Prophase I accounts for about 90% of the total period of meiosis. By its end, the tetrads are attached to the spindle and are moving toward the spindle equator.

During metaphase I, the tetrads line up *randomly* at the spindle equator, so that either the paternal or maternal chromosome can be on a given side of the equator (Figure 27.6). During anaphase I, the sister chromatids representing each homologue behave as a unit—almost as if replication had not occurred—and the *homologous chromosomes* (each still composed of two joined sister chromatids) are distributed to opposite ends of the cell.

As a result, when meiosis I is completed, the following conditions exist: Each daughter cell has (1) *two* copies of one member of each homologous pair (either the paternal or maternal) and none of the other, and (2) a *haploid* chromosomal number (because the still-united sister chromatids are considered to be a single chromosome) but twice the amount of DNA in each chromosome.

Meiosis II The second meiotic division, meiosis II, mirrors mitosis in every way, except that the chromosomes are *not* replicated before it begins. Instead, the sister chromatids in the two daughter cells of meiosis I are simply parceled out among four cells. Meiosis II is sometimes referred to as the **equational division of meiosis** because the chromatids are distributed equally to the daughter cells (as in mitosis) (Figure 27.6).

In short, meiosis accomplishes two important tasks: (1) It reduces the chromosomal number by half and (2) it introduces genetic variability. The random alignment of the homologous pairs during meiosis I provides tremendous variability in the resulting gametes by scrambling genetic characteristics of the two parents in different combinations. Variability is also increased by crossover—when, during late prophase I, the homologues break at crossover points and exchange chromosomal segments (Figure 27.6). (We will describe this process in Chapter 29.) As a result, it is likely that no two gametes are exactly alike, and all are different from the original mother cells.

Spermatogenesis: Summary of Events in the Seminiferous Tubules

A histological section of an adult testis shows that most cells making up the epithelial walls of the seminiferous tubules are in various stages of cell division (**Figure 27.7a**). These cells, collectively called **spermatogenic cells** (*spermatogenic* = sperm forming), give rise to sperm in the following series of divisions and cellular transformations (Figure 27.7b, c).

Mitosis of Spermatogonia: Forming Spermatocytes The outermost tubule cells, which are in direct contact with the epithelial basal lamina, are stem cells called **spermatogonia** (sper″mah-to-go′ne-ah; "sperm seed"). The spermatogonia divide more or less continuously by *mitosis* and, until puberty, all their daughter cells become spermatogonia.

Spermatogenesis begins during puberty, and from then on, each mitotic division of a spermatogonium results in two distinctive daughter cells—types A and B. The **type A daughter cell** remains at the basal lamina to maintain the germ cell line. The **type B cell** gets pushed toward the lumen, where it becomes a **primary spermatocyte** destined to produce four sperm. (To keep these cell types straight, remember that just as the letter A is always at the beginning of our alphabet, a type A cell is always at the tubule basal lamina ready to begin a new generation of gametes.)

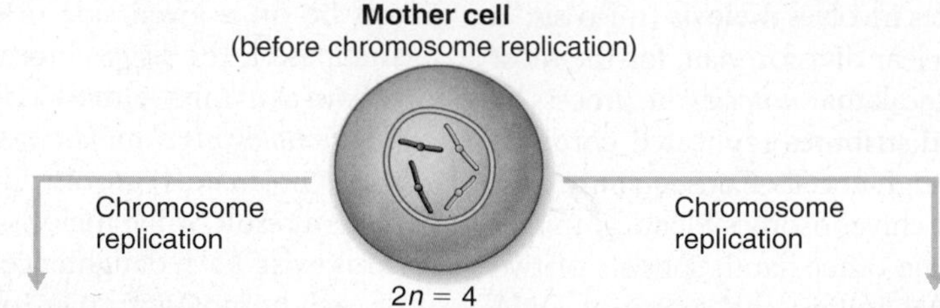

MITOSIS

Prophase

Replicated chromosome

Metaphase

Chromosomes align at the metaphase plate

Sister chromatids separate during anaphase

Daughter cells of mitosis

$2n$ $2n$

MEIOSIS

Prophase I

Tetrad formed by synapsis of replicated homologous chromosomes

Metaphase I

Tetrads align at the metaphase plate

Homologous chromosomes separate but sister chromatids remain together during anaphase I

Daughter cells of Meiosis I

Meiosis II

No further chromosomal replication; sister chromatids separate during anaphase II

n n n n

Daughter cells of meiosis II
(usually gametes)

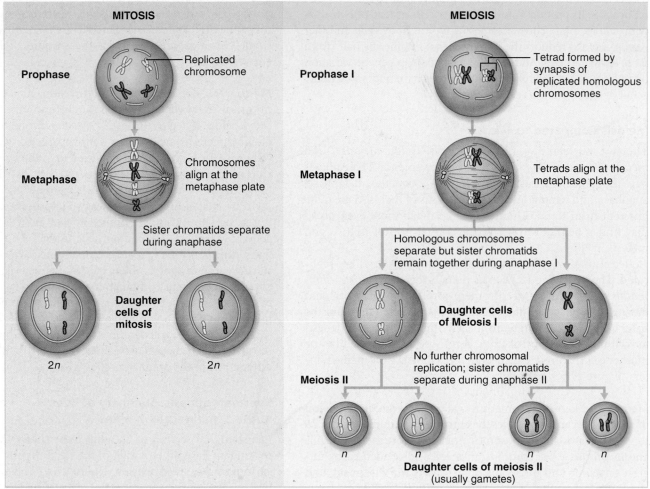

	MITOSIS	MEIOSIS
Number of divisions	One, consisting of prophase, metaphase, anaphase, and telophase.	Two, each consisting of prophase, metaphase, anaphase, and telophase. DNA replication does not occur between the two nuclear divisions.
Synapsis of homologous chromosomes	Does not occur.	Occurs during mitosis I; tetrads formed, allowing crossovers.
Daughter cell number and genetic composition	Two. Each diploid ($2n$) cell is identical to the mother cell.	Four. Each haploid (n) cell contains half as many chromosomes as the mother cell and is genetically different from the mother cell.
Roles in the body	For development of multicellular adult from zygote. Produces cells for growth and tissue repair. Ensures constancy of genetic makeup of all body cells.	Produces cells for reproduction (gametes). Introduces genetic variability in the gametes and reduces chromosomal number by half so that when fertilization occurs, the normal diploid chromosomal number is restored (in humans, $2n = 46$).

Figure 27.5 Comparison of mitosis and meiosis in a mother cell with a diploid number ($2n$) of 4. (Not all the phases of mitosis and meiosis are shown.)

Interphase cell
Nuclear envelope
Centriole pairs
Chromatin
2n = 4

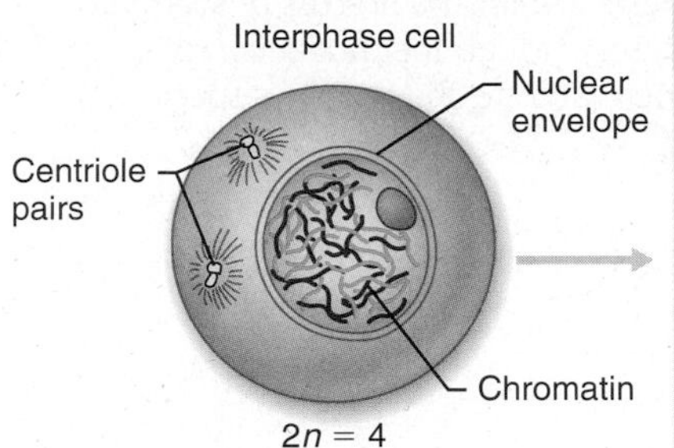

Interphase events
As in mitosis, meiosis is preceded by DNA replication and other preparations for cell division.

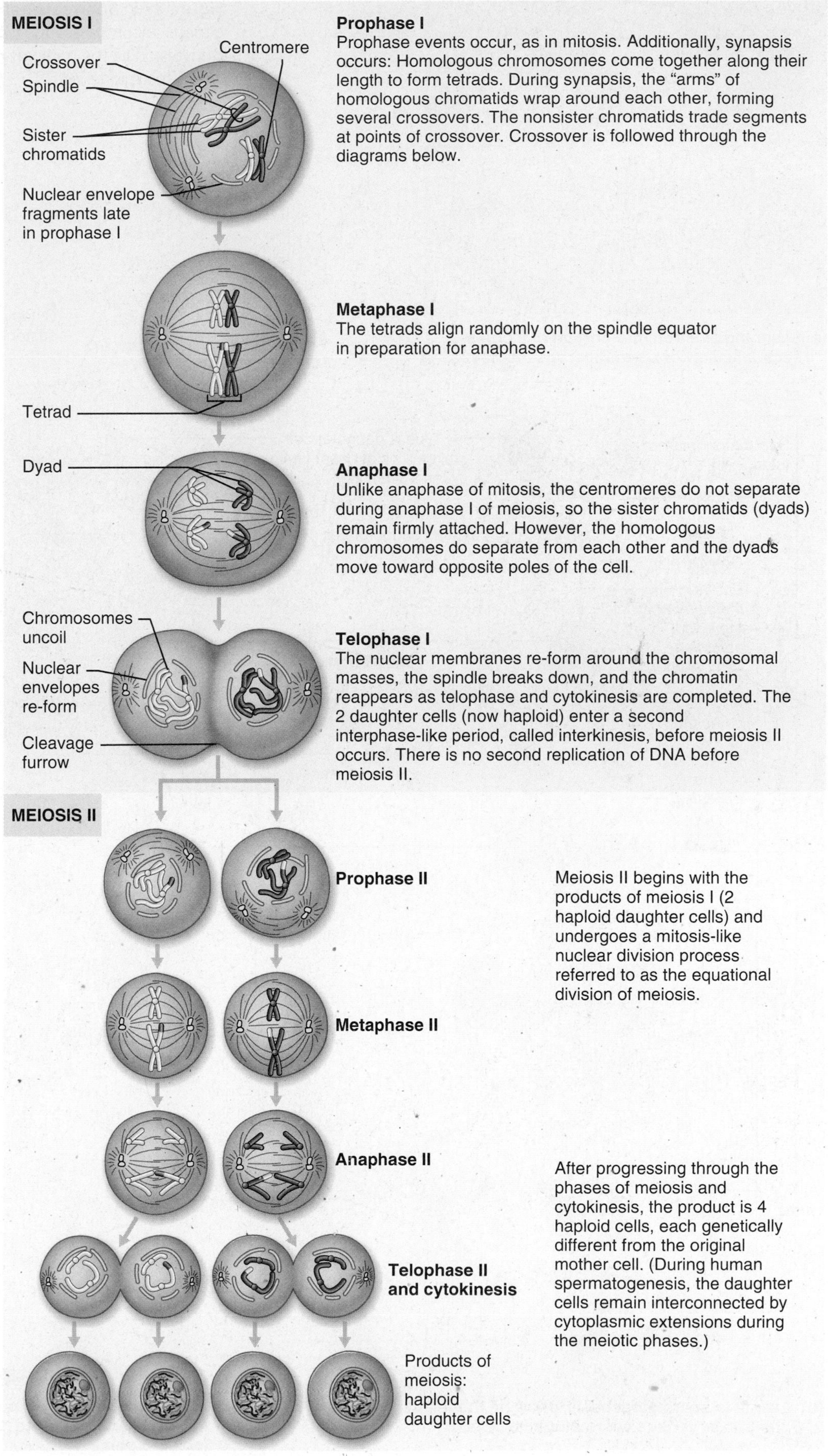

MEIOSIS I
Crossover
Spindle
Sister chromatids
Nuclear envelope fragments late in prophase I
Centromere
Prophase I
Prophase events occur, as in mitosis. Additionally, synapsis occurs: Homologous chromosomes come together along their length to form tetrads. During synapsis, the “arms” of homologous chromatids wrap around each other, forming several crossovers. The nonsister chromatids trade segments at points of crossover. Crossover is followed through the diagrams below.
Tetrad
Metaphase I
The tetrads align randomly on the spindle equator in preparation for anaphase.
Dyad
Anaphase I
Unlike anaphase of mitosis, the centromeres do not separate during anaphase I of meiosis, so the sister chromatids (dyads) remain firmly attached. However, the homologous chromosomes do separate from each other and the dyads move toward opposite poles of the cell.
Chromosomes uncoil
Nuclear envelopes re-form
Cleavage furrow
Telophase I
The nuclear membranes re-form around the chromosomal masses, the spindle breaks down, and the chromatin reappears as telophase and cytokinesis are completed. The 2 daughter cells (now haploid) enter a second interphase-like period, called interkinesis, before meiosis II occurs. There is no second replication of DNA before meiosis II.
MEIOSIS II
Prophase II
Meiosis II begins with the products of meiosis I (2 haploid daughter cells) and undergoes a mitosis-like nuclear division process referred to as the equational division of meiosis.
Metaphase II
Anaphase II
After progressing through the phases of meiosis and cytokinesis, the product is 4 haploid cells, each genetically different from the original mother cell. (During human spermatogenesis, the daughter cells remain interconnected by cytoplasmic extensions during the meiotic phases.)
Telophase II and cytokinesis
Products of meiosis: haploid daughter cells

Figure 27.6 Meiosis. The series of events in meiotic cell division for an animal cell with a diploid number (2*n*) of 4. The behavior of the chromosomes is emphasized.

27

Figure 27.7 Spermatogenesis. (Note: The process of spermatogenesis encompasses the phases through the production of sperm. Conversion of the haploid spermatids to sperm cells is specifically called spermiogenesis.)

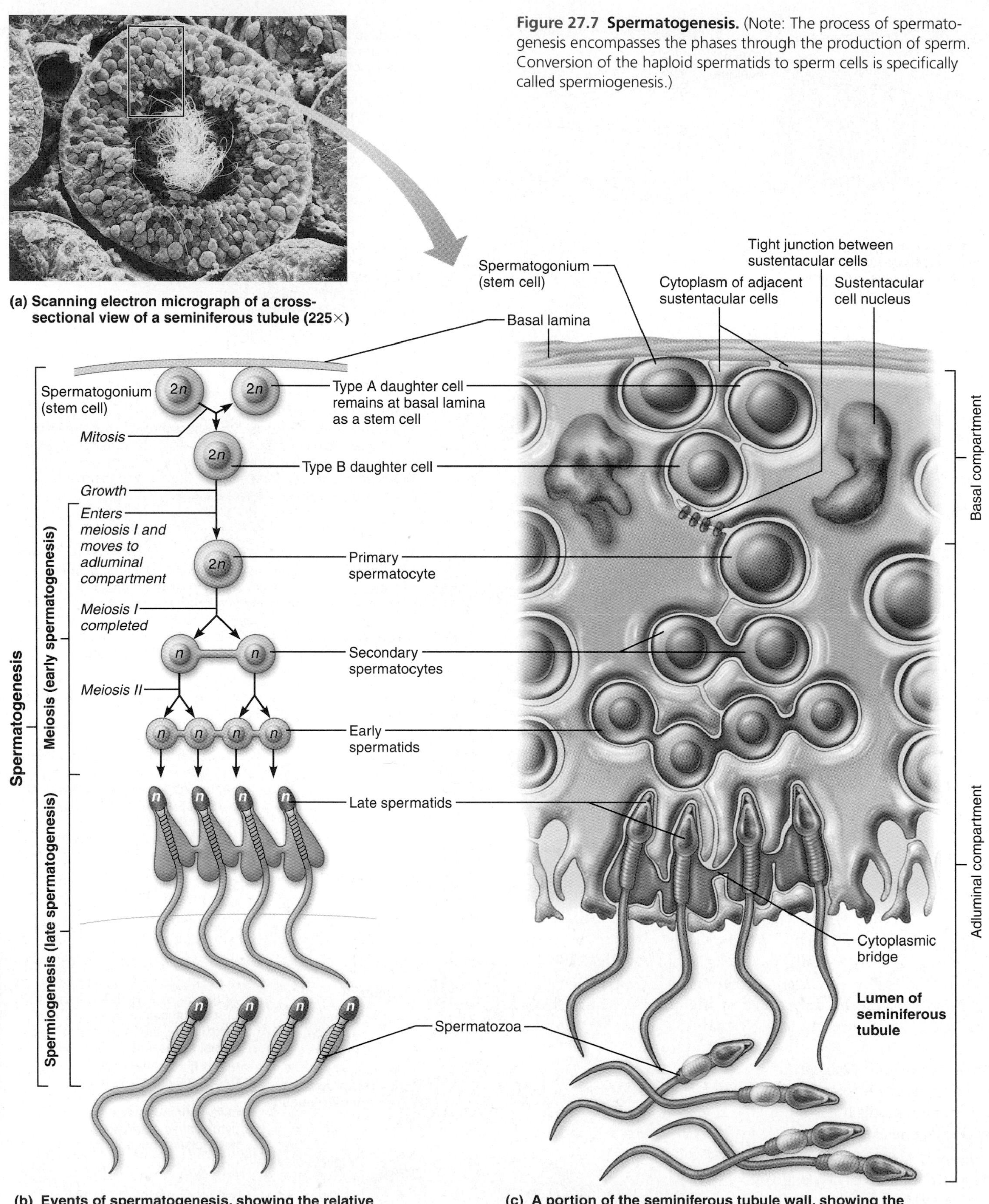

(a) Scanning electron micrograph of a cross-sectional view of a seminiferous tubule (225×)

(b) Events of spermatogenesis, showing the relative position of various spermatogenic cells

(c) A portion of the seminiferous tubule wall, showing the spermatogenic cells surrounded by sustentacular cells (colored gold)

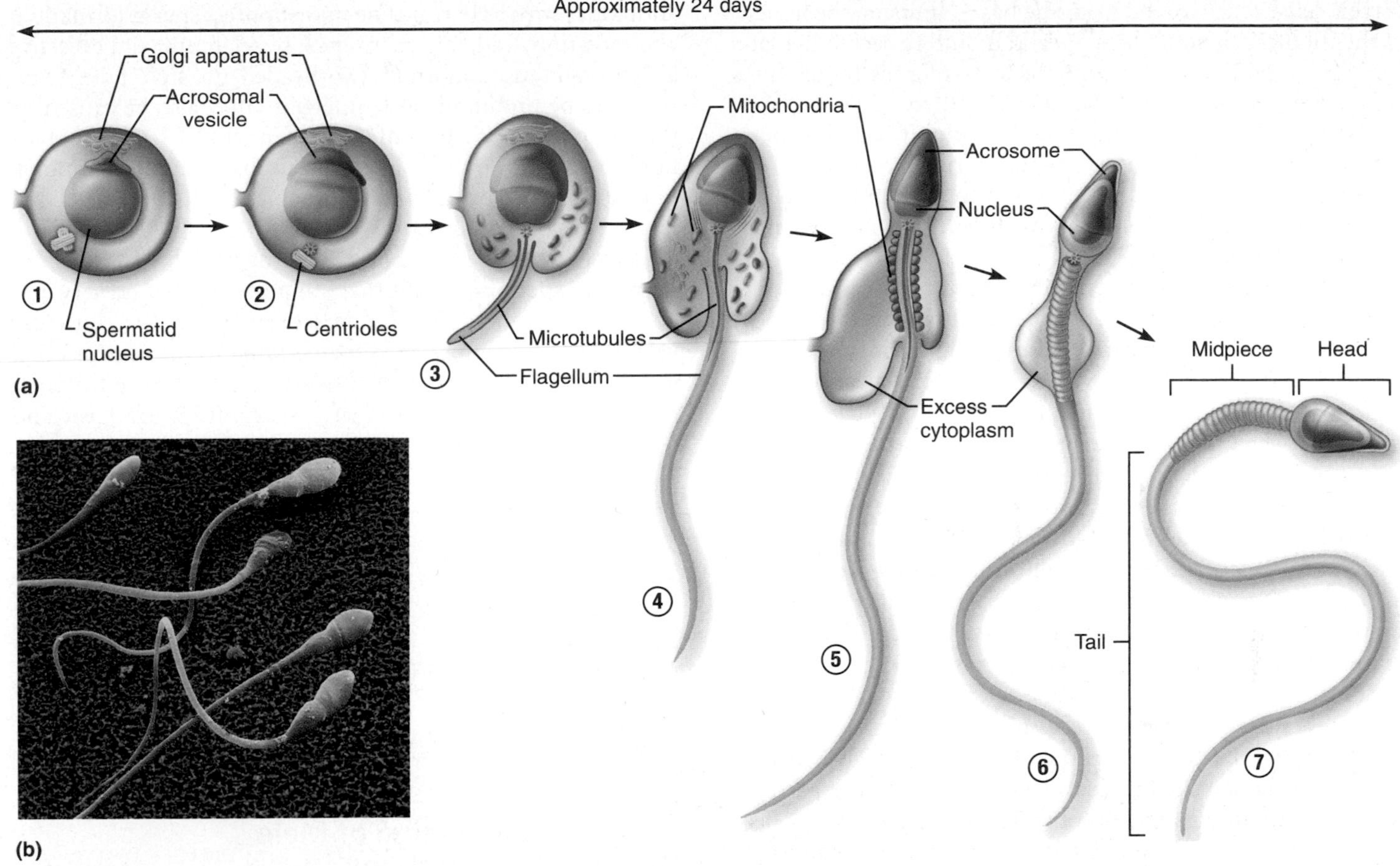

Figure 27.8 **Spermiogenesis: transformation of a spermatid into a functional sperm.** **(a)** The stepwise process of spermiogenesis consists of ① packaging of the acrosomal enzymes by the Golgi apparatus, ② forming the acrosome at the anterior end of the nucleus and positioning the centrioles at the opposite end of the nucleus, ③ elaboration of microtubules to form the flagellum, ④ mitochondrial multiplication and their positioning around the proximal portion of the flagellum, and ⑤ sloughing off excess cytoplasm. ⑥ Structure of an immature sperm that has just been released from a sustentacular cell. ⑦ Structure of a fully mature sperm. **(b)** Scanning electron micrograph of mature sperm (900×).

Meiosis: Spermatocytes to Spermatids Each primary spermatocyte generated during the first phase undergoes meiosis I, forming two smaller haploid cells called **secondary spermatocytes** (Figure 27.7b, c). The secondary spermatocytes continue on rapidly into meiosis II. Their daughter cells, called **spermatids** (sper′mah-tidz), are small round cells, with large spherical nuclei, seen closer to the lumen of the tubule. Midway through spermatogenesis, the developing sperm "turn off" nearly all their genes and compact their DNA into dense pellets.

Spermiogenesis: Spermatids to Sperm Each spermatid has the correct chromosomal number for fertilization (n), but is nonmotile. It still must undergo a streamlining process called **spermiogenesis**, during which it elongates, sheds its excess cytoplasmic baggage, and forms a tail. Follow the details of this process in **Figure 27.8a**, steps ①–⑦.

The resulting sperm, or **spermatozoon** (sper″mah-to-zo′on; "animal seed"), has a head, a midpiece, and a tail, which correspond roughly to *genetic*, *metabolic*, and *locomotor regions*, respectively (Figure 27.8a, ⑦). Sperm "pack" lightly. The **head** of a sperm consists almost entirely of its flattened nucleus, which contains the compacted DNA. Adhering to the top of the nucleus is a helmetlike **acrosome** (ak′ro-sōm; "tip piece") (Figure 27.8a, ⑤ and ⑥). The lysosome-like acrosome is produced by the Golgi apparatus and contains hydrolytic enzymes that enable the sperm to penetrate and enter an egg. The sperm **midpiece** contains mitochondria spiraled tightly around the microtubules of the tail. The long **tail** is a typical flagellum produced by one centriole (actually a basal body) near the nucleus. The mitochondria provide the metabolic energy (ATP) needed for the whiplike movements of the tail that will propel the sperm along its way in the female reproductive tract.

Role of the Sustentacular Cells Throughout spermatogenesis, descendants of the same spermatogonium remain closely attached to one another by cytoplasmic bridges (see Figure 27.7c). They are also surrounded by and connected to nonreplicating supporting cells of a special type, called **sustentacular cells** or

Sertoli cells, which extend from the basal lamina to the lumen of the tubule. The sustentacular cells, bound to each other laterally by tight junctions, divide the seminiferous tubule into two compartments. The **basal compartment** extends from the basal lamina to their tight junctions and it contains spermatogonia and the earliest primary spermatocytes. The **adluminal compartment** lies internal to the tight junctions and includes the meiotically active cells and the tubule lumen (see Figure 27.7c).

The tight junctions between the sustentacular cells form the **blood-testis barrier**. This barrier prevents the membrane antigens of differentiating sperm from escaping through the basal lamina into the bloodstream where they would activate the immune system. Because sperm are not formed until puberty, they are absent when the immune system is being programmed to recognize a person's own tissues early in life. The spermatogonia, which are recognized as "self," are outside the barrier and for this reason can be influenced by bloodborne chemical messengers that prompt spermatogenesis. Following mitosis of the spermatogonia, the tight junctions of the sustentacular cells open to allow primary spermatocytes to pass into the adluminal compartment—much as locks in a canal open to allow a boat to pass.

In the adluminal compartment, spermatocytes and spermatids are nearly enclosed in recesses in the sustentacular cells (see Figure 27.7). They are anchored there by a particular glycoprotein on the spermatogenic cell's surface. The sustentacular cells provide nutrients and essential signals to the dividing cells, even telling them to live or die. They also move the cells along to the lumen, secrete **testicular fluid** (rich in androgens and metabolic acids) that provides the transport medium for sperm in the lumen, and phagocytize faulty germ cells and the excess cytoplasm sloughed off as the spermatids transform into sperm. The sustentacular cells also produce chemical mediators (inhibin and androgen-binding protein) that help regulate spermatogenesis, as we describe below.

Given hospitable temperature conditions, spermatogenesis—from formation of a primary spermatocyte to release of immature sperm into the lumen—takes 64 to 72 days. Sperm in the lumen are unable to "swim" and are incapable of fertilizing an egg. They are pushed by the pressure of the testicular fluid through the tubular system of the testes into the epididymis, where they gain increased motility and fertilizing power.

HOMEOSTATIC IMBALANCE

Roughly one in seven American couples seek treatment for infertility, mostly because of problems with sperm quality or quantity. According to some studies, a gradual decline in male fertility has been occurring in the past 50 years.

Some believe the main cause is *xenobiotics*, alien molecules that have invaded our lives in a variety of forms including environmental toxins, PVCs, phthalates (oily solvents that make plastics flexible), certain components in pesticides and herbicides, and especially compounds with estrogenic effects, some of which are injected into beef cattle to encourage growth. These estrogen-like compounds block the action of male sex hormones as they program sexual development, and are now found in our meat supply as well as in the air. Common antibiotics such as tetracycline may suppress sperm formation, and radiation, lead, marijuana, lack of selenium, and excessive alcohol can cause abnormal (two-headed, multiple-tailed, etc.) sperm to be produced. Male infertility may also be caused by the lack of a specific type of Ca^{2+} channel (Ca^{2+} is needed for normal sperm motility), anatomical obstructions, hormonal imbalances, and oxidative stress (which contributes to DNA fragmentation in sperm).

A low sperm count accompanied by a high percentage of immature sperm may hint that a man has a *varicocele* (var′ĭ-ko-sēl″). This condition hinders drainage of the testicular vein, resulting in an elevated temperature in the scrotum that interferes with normal sperm development. Other thermal-related events that inhibit sperm maturation are fever and overuse of hot tubs. ■

CHECK YOUR UNDERSTANDING

13. How is the final product of meiosis different from that of mitosis?
14. Describe the major structural and functional regions of a sperm.
15. What is the role of sustentacular cells? Of interstitial cells?

For answers, see Appendix G.

Hormonal Regulation of Male Reproductive Function

▶ Discuss hormonal regulation of testicular function and the physiological effects of testosterone on male reproductive anatomy.

Hormonal regulation of gamete and gonadal hormone production involves interactions between the hypothalamus, anterior pituitary gland, and gonads, a relationship called the **hypothalamic-pituitary-gonadal (HPG) axis.**

The Hypothalamic-Pituitary-Gonadal Axis

The sequence of regulatory events involving the HPG axis, shown schematically in **Figure 27.9**, is as follows:

① The hypothalamus releases **gonadotropin-releasing hormone (GnRH)**, which reaches the anterior pituitary cells via the blood of the hypophyseal portal system. GnRH controls the release of the two anterior pituitary gonadotropins: **follicle-stimulating hormone (FSH)** and **luteinizing hormone (LH)**, both named for their effects on the female gonad.

② Binding of GnRH to pituitary cells (gonadotrophs) prompts them to secrete FSH and LH into the blood.

③ FSH stimulates spermatogenesis indirectly by stimulating the sustentacular cells to release **androgen-binding protein (ABP)**. ABP keeps the concentration of **testosterone** in the vicinity of the spermatogenic cells high, which in turn stimulates spermatogenesis. In this way, FSH makes these cells receptive to testosterone's stimulatory effects.

④ LH binds to the interstitial cells in the soft connective tissue surrounding the seminiferous tubules, prodding them to secrete testosterone (and a small amount of estrogen).

⑤ Locally, rising testosterone levels serve as the final trigger for spermatogenesis.

⑥ Testosterone entering the bloodstream exerts a number of effects at other body sites. It stimulates maturation of sex organs, development and maintenance of secondary sex characteristics, and libido.

⑦ Rising levels of testosterone feed back to inhibit hypothalamic release of GnRH and act directly on the anterior pituitary to inhibit gonadotropin release.

⑧ **Inhibin** (in-hib′in), a protein hormone produced by the sustentacular cells, serves as a "barometer" of the normalcy of spermatogenesis. When the sperm count is high, inhibin release increases, and it inhibits anterior pituitary release of FSH and hypothalamic release of GnRH. (The inhibitory effect of testosterone and inhibin on the hypothalamus is not illustrated in Figure 27.9.) When sperm count falls below 20 million/ml, inhibin secretion declines steeply.

As you can see, the amount of testosterone and sperm produced by the adult testes reflects a balance among the three interacting sets of hormones that make up the HPG axis: (1) GnRH, which indirectly stimulates the testes via its effect on FSH and LH release; (2) gonadotropins (FSH and LH), which directly stimulate the testes; and (3) gonadal hormones (testosterone and inhibin), which can exert negative feedback controls on the hypothalamus and anterior pituitary. Once this balance is established during puberty (a process which takes about three years), the amount of testosterone and sperm produced remains fairly stable throughout life.

Because the hypothalamus is also influenced by input from other brain areas, the whole axis is under CNS control. In the absence of GnRH and gonadotropins, the testes atrophy, and sperm and testosterone production cease.

Development of male reproductive structures (discussed later in the chapter) depends on prenatal secretion of male hormones, and for a few months after birth, a male infant has plasma gonadotropin and testosterone levels nearly equal to those of a midpubertal boy. Soon afterwards, blood levels of these hormones recede and they remain low throughout childhood. As puberty nears, much higher levels of testosterone are required to suppress hypothalamic release of GnRH. So, as more GnRH is released, more testosterone is secreted by the testes, but the threshold for hypothalamic inhibition keeps rising until the adult pattern of hormone interaction is achieved, as evidenced by the presence of mature sperm in the semen.

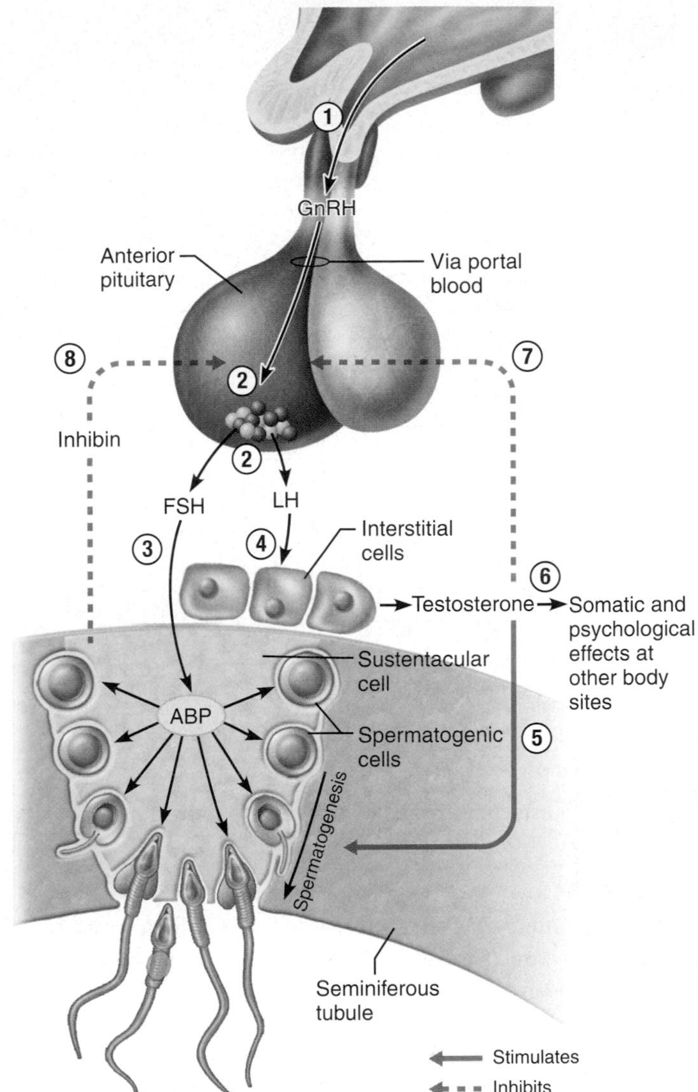

Figure 27.9 Hormonal regulation of testicular function, the hypothalamic-pituitary-gonadal axis. (Only one sustentacular cell is depicted so that its structural relationship to the spermatogenic cells it encloses is clearly seen. However, it would be flanked by sustentacular cells on each side.)

Mechanism and Effects of Testosterone Activity

Like all steroid hormones, testosterone is synthesized from cholesterol and exerts its effects by activating specific genes, which results in enhanced synthesis of certain proteins in the target cells. (See Chapter 16 for mechanisms of steroid hormone action.) In some target cells, testosterone must be transformed into another steroid to exert its effects. In the prostate, testosterone is converted to *dihydrotestosterone* (*DHT*) before it can bind in the nucleus, and in certain neurons of the brain, testosterone is converted to *estradiol* (es′trah-di-ol), a female sex hormone, to bring about its stimulatory effects. These transformations often occur in a single enzymatic step because testosterone and the other gonadal hormones are structurally very similar.

As puberty ensues, testosterone not only prompts spermatogenesis but also has multiple anabolic effects throughout the body (see Table 27.1, p. 1057). It targets accessory reproductive organs—ducts, glands, and the penis—causing them to grow and assume adult size and function. In adult males, normal plasma levels of testosterone maintain these organs. When the hormone is deficient or absent, all accessory organs atrophy, semen volume declines markedly, and erection and ejaculation are impaired. Thus, a man becomes both sterile and unable to carry out sexual intercourse. This situation is easily remedied by testosterone replacement therapy.

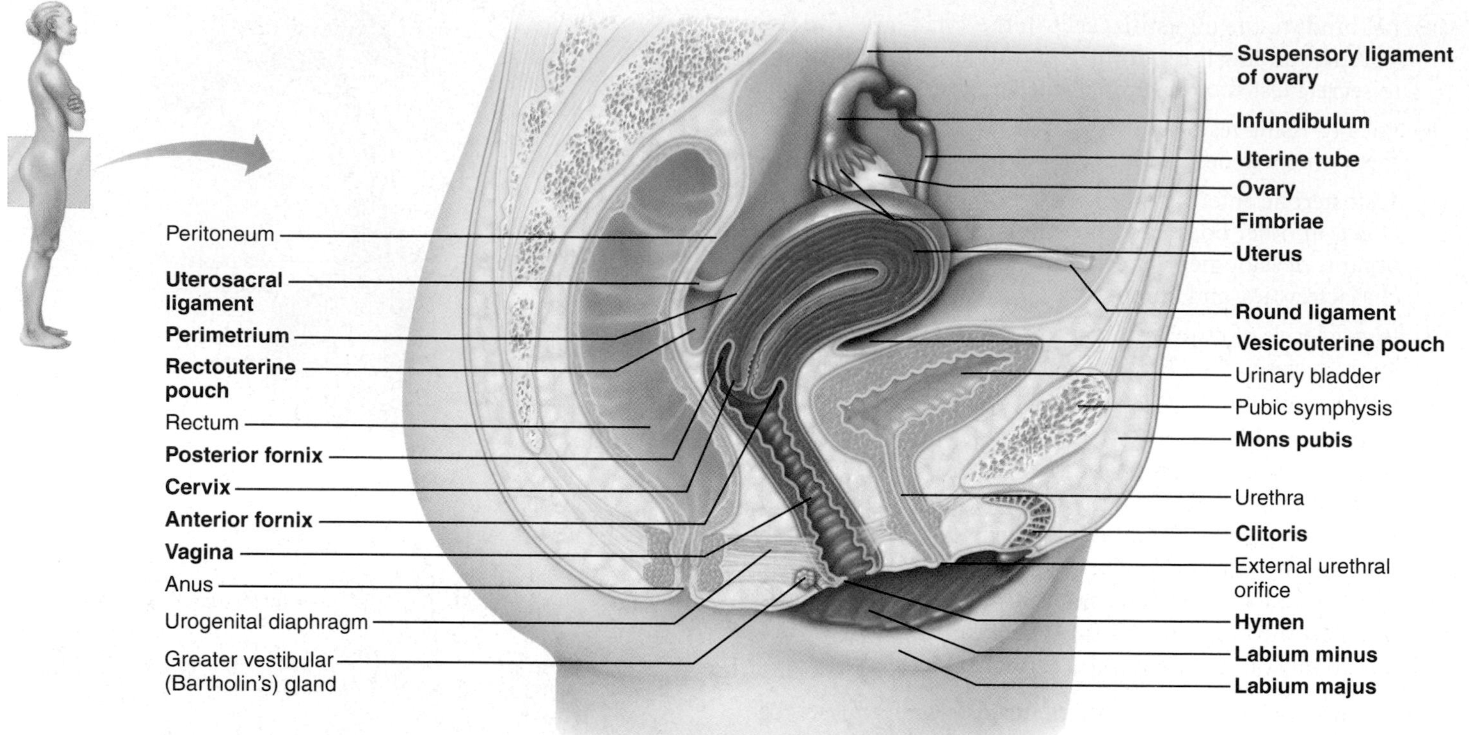

Figure 27.10 Internal organs of the female reproductive system, midsagittal section. (See *A Brief Atlas of the Human Body*, Figure 74.)

Male secondary sex characteristics—that is, features induced in the *nonreproductive* organs by the male sex hormones (mainly testosterone)—make their appearance at puberty. These include the appearance of pubic, axillary, and facial hair, enhanced hair growth on the chest or other body areas in some men, and a deepening of the voice as the larynx enlarges. The skin thickens and becomes oilier (which predisposes young men to acne), bones grow and increase in density, and skeletal muscles increase in size and mass. The last two effects are often referred to as the *somatic effects* of testosterone (*soma* = body). Epiphyseal plate closure and termination of skeletal growth in height occurs in response to rising estrogen levels late in puberty in both boys and girls. Testosterone also boosts basal metabolic rate and influences behavior. It is the basis of the sex drive (libido) in males, be they heterosexual or homosexual. As noted later (p. 1058), the adrenal androgen DHEA appears to be more important than testosterone in creating or driving the *female* libido.

27

In embryos, the presence of testosterone masculinizes the brain. Testosterone also appears to continue to shape certain regions of the male brain well into adult life, as indicated by the differences in males' and females' brain areas involved in sexual arousal (for example, the amygdala).

The testes are not the only source of androgens. The adrenal glands of both sexes also release androgens. However, the relatively small amounts of adrenal androgens are unable to support normal testosterone-mediated functions when the testes fail to produce androgens, so we can assume that it is the testosterone production by the testes that supports male reproductive function.

CHECK YOUR UNDERSTANDING

16. What is the HPG axis?
17. How does FSH indirectly stimulate spermatogenesis?
18. What are three secondary sex characteristics promoted by testosterone?

For answers, see Appendix G.

Anatomy of the Female Reproductive System

The reproductive role of the female is far more complex than that of a male. Not only must she produce gametes, but her body must prepare to nurture a developing embryo for a period of approximately nine months. **Ovaries**, the **female gonads**, are the primary reproductive organs of a female, and like the male testes, ovaries serve a dual purpose: They produce the female gametes (ova) and sex hormones, **estrogen** and **progesterone** (pro-ges′tĕ-rōn). Estrogen includes *estradiol*, *estrone*, and *estriol*, but estradiol is the most abundant and is most responsible for estrogenic effects.

As illustrated in **Figure 27.10**, the ovaries and duct system, collectively known as the **internal genitalia**, are mostly located in the pelvic cavity. The female's accessory ducts, from the vicinity of the ovary to the body exterior, are the *uterine tubes*, the *uterus*, and the *vagina*. They transport or otherwise serve the

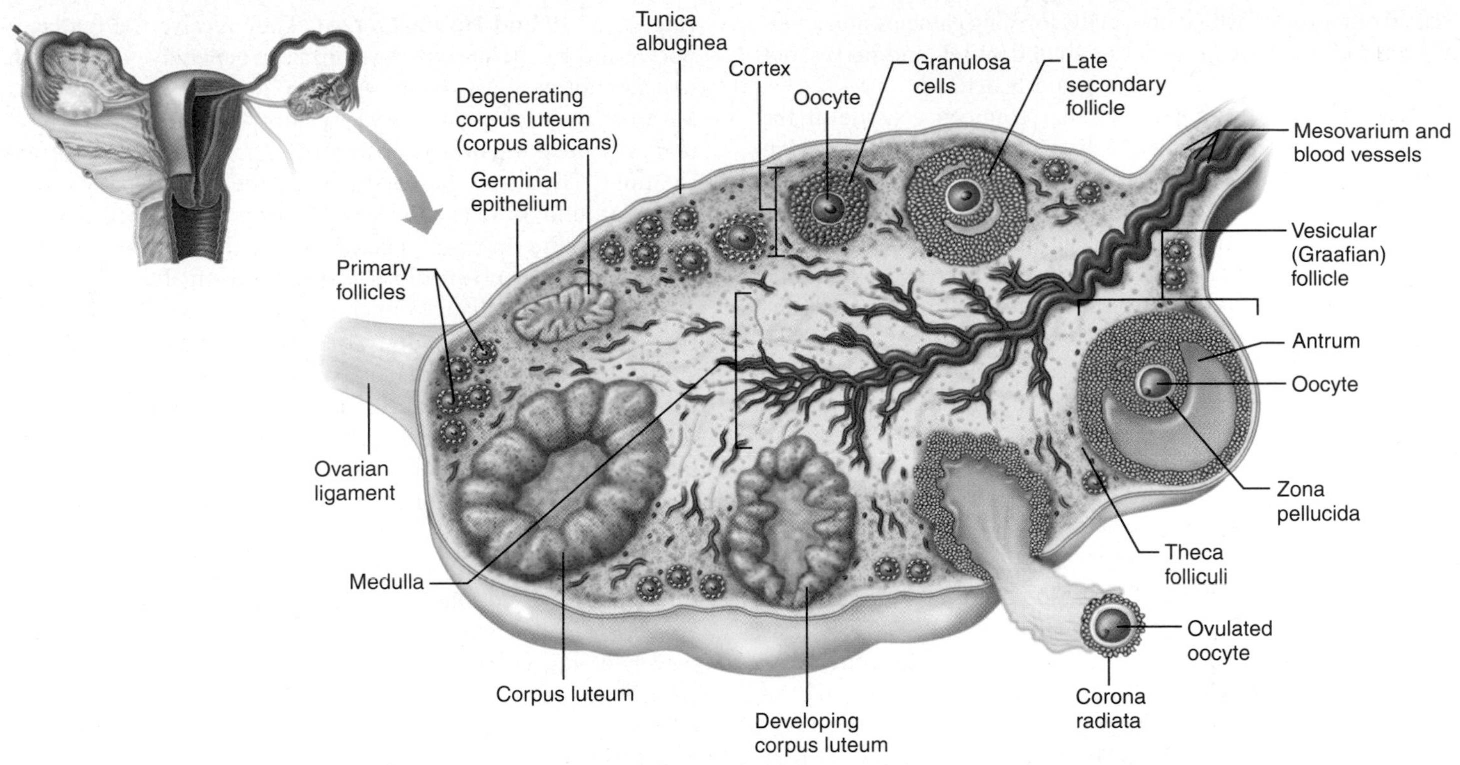

(a) Diagrammatic view of an ovary sectioned to reveal the follicles in its interior

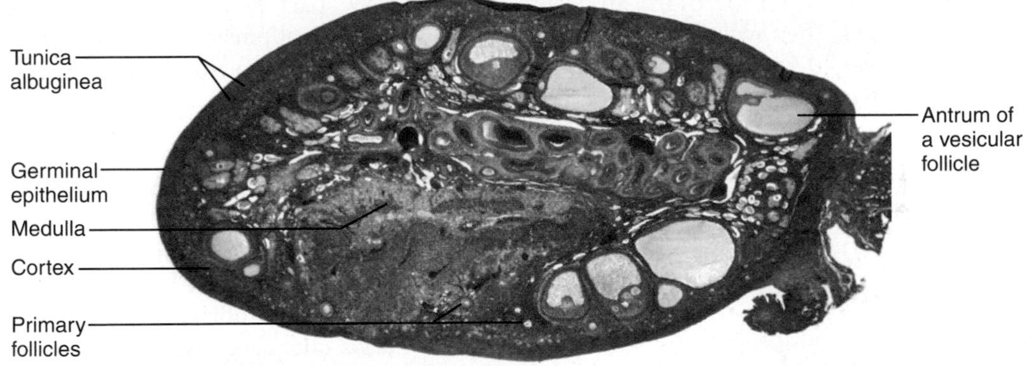

(b) Photomicrograph of a mammalian ovary showing follicles of different developmental phases

Figure 27.11 Structure of an ovary. (a) Diagrammatic view. Note that not all of these structures would appear in the ovary at the same time. **(b)** Photomicrograph (4×).

needs of the reproductive cells and a developing fetus. The external sex organs of females are referred to as the **external genitalia**.

The Ovaries

▶ Describe the location, structure, and function of the ovaries.

The paired ovaries flank the uterus on each side (Figure 27.10). Shaped like an almond and about twice as large, each ovary is held in place by several ligaments in the fork of the iliac blood vessels within the peritoneal cavity. The **ovarian ligament** anchors the ovary medially to the uterus; the **suspensory ligament** anchors it laterally to the pelvic wall; and the **mesovarium** (mez″o-va′re-um) suspends it in between (see Figures 27.10 and 27.12a). The suspensory ligament and the mesovarium are part of the **broad ligament**, a peritoneal fold that "tents" over the uterus and supports the uterine tubes, uterus, and vagina. The ovarian ligaments are enclosed by the broad ligament.

The ovaries are served by the **ovarian arteries**, branches of the abdominal aorta (see Figure 19.24c), and by the *ovarian branch of the uterine arteries.* The ovarian blood vessels reach the ovaries by traveling through the suspensory ligaments and mesovaria (see Figure 27.12a).

Like each testis, each ovary is surrounded externally by a fibrous **tunica albuginea (Figure 27.11a)**, which is in turn covered externally by a layer of cuboidal epithelial cells called the *germinal epithelium*, actually a continuation of the peritoneum. The ovary has an outer *cortex*, which houses the forming gametes, and an inner *medulla* containing the largest blood vessels and nerves, but the relative extent of each region is poorly defined.

Embedded in the highly vascular connective tissue of the ovary cortex are many tiny saclike structures called **ovarian follicles**. Each follicle consists of an immature egg, called an **oocyte** (o′o-sīt; *oo* = egg), encased by one or more layers of very different cells. The surrounding cells are called **follicle cells** if a single layer is present, and **granulosa cells** when more than one layer is present. Follicles at different stages of maturation are distinguished by their structure (Figure 27.11a). In a **primordial follicle**, one layer of squamouslike follicle cells surrounds the oocyte.

A **primary follicle** has a single layer of cuboidal or columnar-type follicle cells enclosing the oocyte.

A **secondary follicle** is formed when two or more layers of granulosa cells surround the oocyte.

A **late secondary follicle** results when small fluid-filled spaces appear between the granulosa cells.

The mature **vesicular follicle**, also called a **Graafian** (graf′e-an) or **tertiary follicle**, forms when the fluid-filled pockets coalesce to form a central fluid-filled cavity called an *antrum*. At this stage, the follicle extends from the deepest part of the ovarian cortex and bulges from the surface of the ovary (Figure 27.11) and its oocyte "sits" on a stalk of granulosa cells at one side of the antrum.

Each month in women of childbearing age, one of the ripening follicles ejects its oocyte from the ovary, an event called *ovulation* (see Figure 27.18). After ovulation, the ruptured follicle is transformed into a very different looking glandular structure called the **corpus luteum** (lu′te-um; "yellow body"; plural: corpora lutea), which eventually degenerates (Figure 27.11a). As a rule, most of these structures can be seen within the same ovary. In older women, the surfaces of the ovaries are scarred and pitted, revealing that many oocytes have been released.

CHECK YOUR UNDERSTANDING

19. Briefly, what are the internal genitalia of a woman?
20. What two roles do the ovaries assume?
21. How does a primary follicle differ from a secondary follicle? From a vesicular follicle?

For answers, see Appendix G.

27

The Female Duct System

▶ Describe the location, structure, and function of each of the organs of the female reproductive duct system.

Unlike the male duct system, which is continuous with the tubules of the testes, the female duct system has little or no actual contact with the ovaries. An ovulated oocyte is cast into the peritoneal cavity, and some oocytes are lost there.

The Uterine Tubes

The **uterine tubes** (u′ter-in), also called **fallopian tubes** or **oviducts**, form the initial part of the female duct system (Figure 27.10 and **Figure 27.12a**). They receive the ovulated oocyte and are the site where fertilization generally occurs. Each uterine tube is about 10 cm (4 inches) long and extends medially from the region of an ovary to empty into the superolateral region of the uterus via a constricted region called the **isthmus** (is′mus). The distal end of each uterine tube expands as it curves around the ovary, forming the **ampulla**, and fertilization usually occurs in this region. The ampulla ends in the **infundibulum** (in″fun-dib′u-lum), an open, funnel-shaped structure bearing ciliated, fingerlike projections called **fimbriae** (fim′bre-e; "fringe") that drape over the ovary.

Around the time of ovulation, the uterine tube performs complex movements to capture oocytes. It bends to drape over the ovary while the fimbriae stiffen and sweep the ovarian surface. The beating cilia on the fimbriae then create currents in the peritoneal fluid that tend to carry an oocyte into the uterine tube, where it begins its journey toward the uterus.

The uterine tube contains sheets of smooth muscle, and its thick, highly folded mucosa contains both ciliated and nonciliated cells. The oocyte is carried toward the uterus by a combination of muscular peristalsis and the beating of the cilia. Nonciliated cells of the mucosa have dense microvilli and produce a secretion that keeps the oocyte (and sperm, if present) moist and nourished.

Externally, the uterine tubes are covered by visceral peritoneum and supported along their length by a short mesentery (part of the broad ligament) called the **mesosalpinx** (mez″o-sal′pinks; "mesentery of the trumpet"; *salpin* = trumpet), a reference to the trumpet-shaped uterine tube it supports (Figure 27.12a).

HOMEOSTATIC IMBALANCE

The fact that the uterine tubes are not continuous with the ovaries places women at risk for *ectopic pregnancy*, in which an oocyte fertilized in the peritoneal cavity or distal portion of the uterine tube begins developing there. Because the tube lacks adequate mass and vascularization to support the full term of pregnancy, such pregnancies naturally abort, often with substantial bleeding.

Another potential problem is infection spreading into the peritoneal cavity from other parts of the reproductive tract. Gonorrhea bacteria and other sexually transmitted microorganisms sometimes infect the peritoneal cavity in this way, causing an extremely severe inflammation called **pelvic inflammatory disease (PID)**. Unless treated promptly with broad-spectrum antibiotics, PID can cause scarring of the narrow uterine tubes and of the ovaries, resulting in sterility. Indeed, scarring and closure of the uterine tubes, which have an internal diameter as small as the width of a human hair in some regions, is one of the major causes of female infertility. ■

The Uterus

The **uterus** (Latin for "womb") is located in the pelvis, anterior to the rectum and posterosuperior to the bladder (Figures 27.10 and 27.12). It is a hollow, thick-walled, muscular organ that functions to receive, retain, and nourish a fertilized ovum. In a

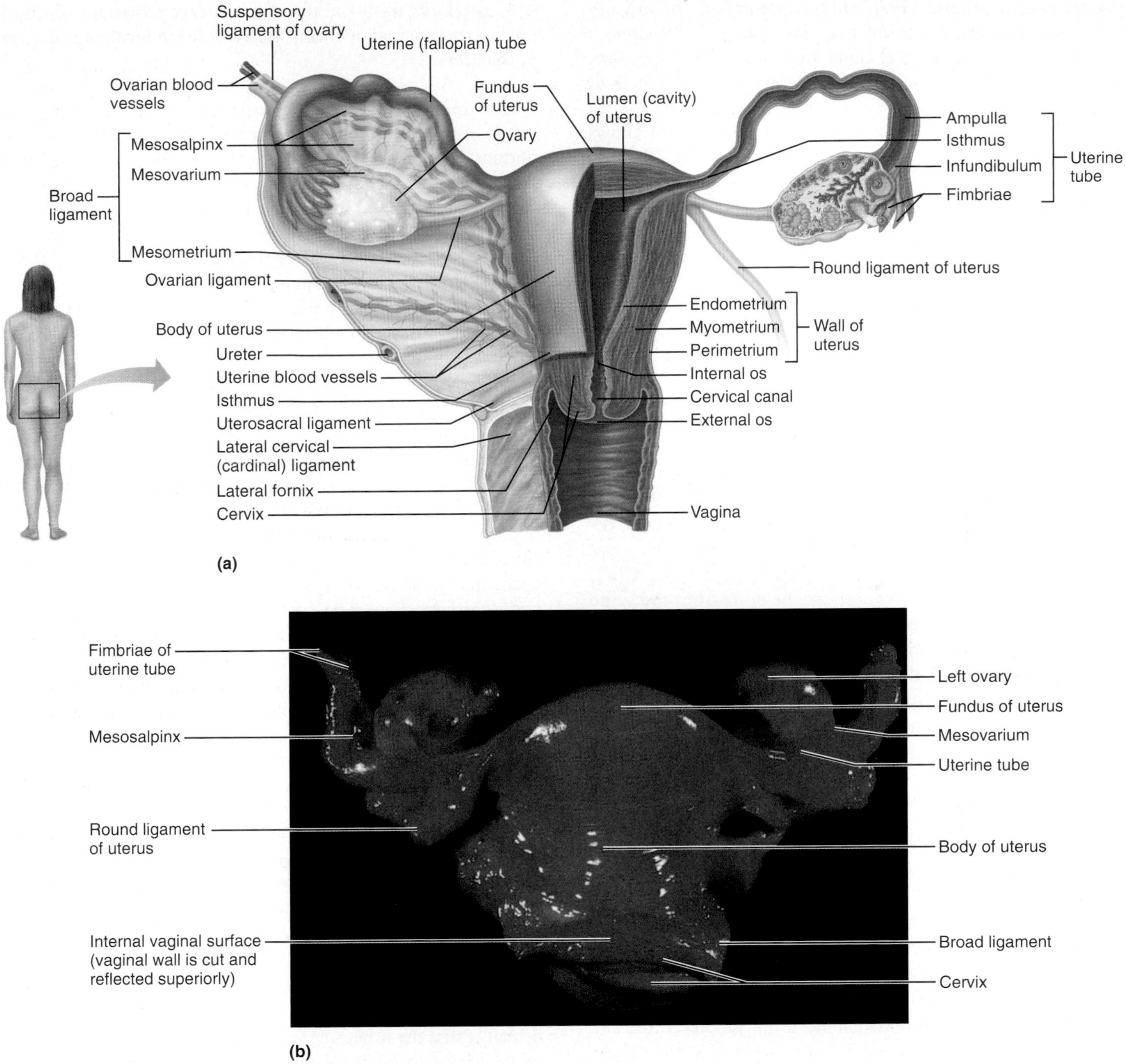

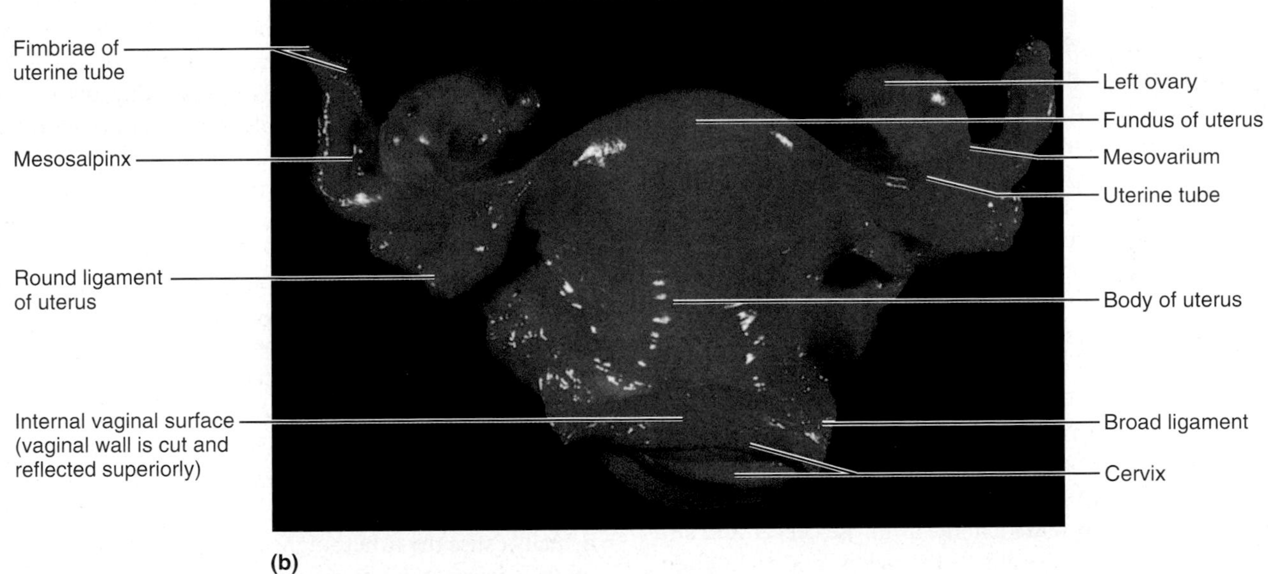

Figure 27.12 Internal reproductive organs of a female. (a) Posterior view of the female reproductive organs. The posterior walls of the vagina, uterus, and uterine tubes, and the broad ligament (a peritoneal fold) have been removed on the right side to reveal the shape of the lumen of these organs. **(b)** Anterior view of the reproductive organs of a female cadaver. (See *A Brief Atlas of the Human Body*, Figure 75.)

fertile woman who has never been pregnant, the uterus is about the size and shape of an inverted pear, but it is usually somewhat larger in women who have borne children. Normally, the uterus flexes anteriorly where it joins the vagina (see Figure 27.10), causing the uterus as a whole to be inclined forward, or *anteverted*. However, the organ is frequently turned backward, or *retroverted*, in older women.

The major portion of the uterus is referred to as the **body** (Figures 27.10 and 27.12). The rounded region superior to the entrance of the uterine tubes is the **fundus**, and the slightly narrowed region between the body and the cervix is the *isthmus*. The **cervix** of the uterus is its narrow neck, or outlet, which projects into the vagina inferiorly.

The cavity of the cervix, called the **cervical canal**, communicates with the vagina via the *external os* (*os* = mouth) and with the cavity of the uterine body via the *internal os.* The mucosa of the cervical canal contains *cervical glands* that secrete a mucus that fills the cervical canal and covers the external os, presumably to block the spread of bacteria from the vagina into the uterus. Cervical mucus also blocks the entry of sperm, except at midcycle, when it becomes less viscous and allows sperm to pass through.

HOMEOSTATIC IMBALANCE

Cancer of the cervix strikes about 450,000 women worldwide each year, killing about half. It is most common among women between the ages of 30 and 50. Risk factors include frequent cervical inflammations, sexually transmitted infections, including genital warts, and multiple pregnancies. The cancer cells arise from the epithelium covering the cervical tip.

In a *Papanicolaou (Pap) smear,* or cervical smear test, some of these cells are scraped away and then examined for abnormalities. A Pap smear is the most effective way to detect this slow-growing cancer, and American Cancer Society guidelines advise women to have one yearly until the age of 70, when this monitoring can be stopped if there have been no abnormal results in the past 10 years. When results are inconclusive, a test for the sexually transmitted human papillomavirus (HPV), the cause of most cervical cancer, can be done from the same Pap sample or from a blood sample. *Gardasil,* a three-dose vaccine that provides protection from HPV-induced cervical cancer, is the latest addition to the official childhood immunization schedule. It is recommended for all 11- and 12-year-old girls, although it may be administered to girls as young as 9 years old. In unexposed girls, the vaccine specifically blocks two cancer-causing kinds of HPV as well as two additional types which are not associated with cervical cancer. All four types of the virus are associated with genital warts and mild Pap test abnormalities. Whether or not this vaccine will become a requirement for school is presently decided on a state-to-state basis. ■

Supports of the Uterus The uterus is supported laterally by the **mesometrium** ("mesentery of the uterus") portion of the broad ligament (Figure 27.12a). More inferiorly, the **lateral cervical (cardinal) ligaments** extend from the cervix and superior vagina to the lateral walls of the pelvis, and the paired **uterosacral ligaments** secure the uterus to the sacrum posteriorly. The uterus is bound to the anterior body wall by the fibrous **round ligaments,** which run through the inguinal canals to anchor in the subcutaneous tissue of the labia
27 majora. These ligaments allow the uterus a good deal of mobility, and its position changes as the rectum and bladder fill and empty.

HOMEOSTATIC IMBALANCE

Despite the many anchoring ligaments, the principal support of the uterus is provided by the muscles of the pelvic floor, namely the muscles of the urogenital and pelvic diaphragms (Table 10.7). These muscles are stretched and sometimes torn during childbirth. Subsequently, the unsupported uterus may sink inferiorly, until the tip of the cervix protrudes through the external vaginal opening. This condition is called **prolapse of the uterus.** ■

The undulating course of the peritoneum produces several cul-de-sacs, or blind-ended peritoneal pouches. The most important of these are the *vesicouterine pouch* (ves″i-ko-u′ter-in) between the bladder and uterus, and the *rectouterine pouch* between the rectum and uterus (see Figure 27.10).

The Uterine Wall The wall of the uterus is composed of three layers (Figure 27.12a). The **perimetrium,** the incomplete outermost serous layer, is the peritoneum. The **myometrium** (mi″o-me′tre-um; "muscle of the uterus") is the bulky middle layer, composed of interlacing bundles of smooth muscle. It contracts rhythmically during childbirth to expel the baby from the mother's body. The **endometrium** is the mucosal lining of the uterine cavity **(Figure 27.13).** It is a simple columnar epithelium underlain by a thick lamina propria. If fertilization occurs, the young embryo burrows into the endometrium (implants) and resides there for the rest of its development.

The endometrium has two chief *strata* (layers). The **stratum functionalis** (fungk-shun-a′lis), or **functional layer,** undergoes cyclic changes in response to blood levels of ovarian hormones and is shed during menstruation (approximately every 28 days). The thinner, deeper **stratum basalis** (ba-să′lis), or **basal layer,** forms a new functionalis after menstruation ends. It is unresponsive to ovarian hormones. The endometrium has numerous *uterine glands* that change in length as endometrial thickness changes.

The vascular supply of the uterus is key to understanding the cyclic changes of the uterine endometrium, which we will discuss later in the chapter. The **uterine arteries** arise from the *internal iliacs* in the pelvis, ascend along the sides of the uterus, and send branches into the uterine wall (Figures 27.12a and 27.13b). These branches break up into several **arcuate arteries** (ar′ku-āt) within the myometrium. The arcuate arteries send **radial branches** into the endometrium, where they in turn give off **straight arteries** to the stratum basalis and **spiral (coiled) arteries** to the stratum functionalis. The spiral arteries repeatedly degenerate and regenerate, and it is their spasms that actually cause the functionalis layer to be shed during menstruation. Veins in the endometrium are thin-walled and form an extensive network with occasional sinusoidal enlargements.

The Vagina

The **vagina** ("sheath") is a thin-walled tube, 8–10 cm (3–4 inches) long. It lies between the bladder and the rectum and extends from the cervix to the body exterior (see Figure 27.10). The urethra parallels its course anteriorly. Often called the *birth canal,* the vagina provides a passageway for delivery of an infant and for menstrual flow. Because it receives the penis (and semen) during sexual intercourse, it is the *female organ of copulation.*

The distensible wall of the vagina consists of three coats: an outer fibroelastic *adventitia,* a smooth muscle *muscularis,* and an inner *mucosa* marked by transverse ridges or *rugae,* which stimulate the penis during intercourse. The mucosa is a stratified

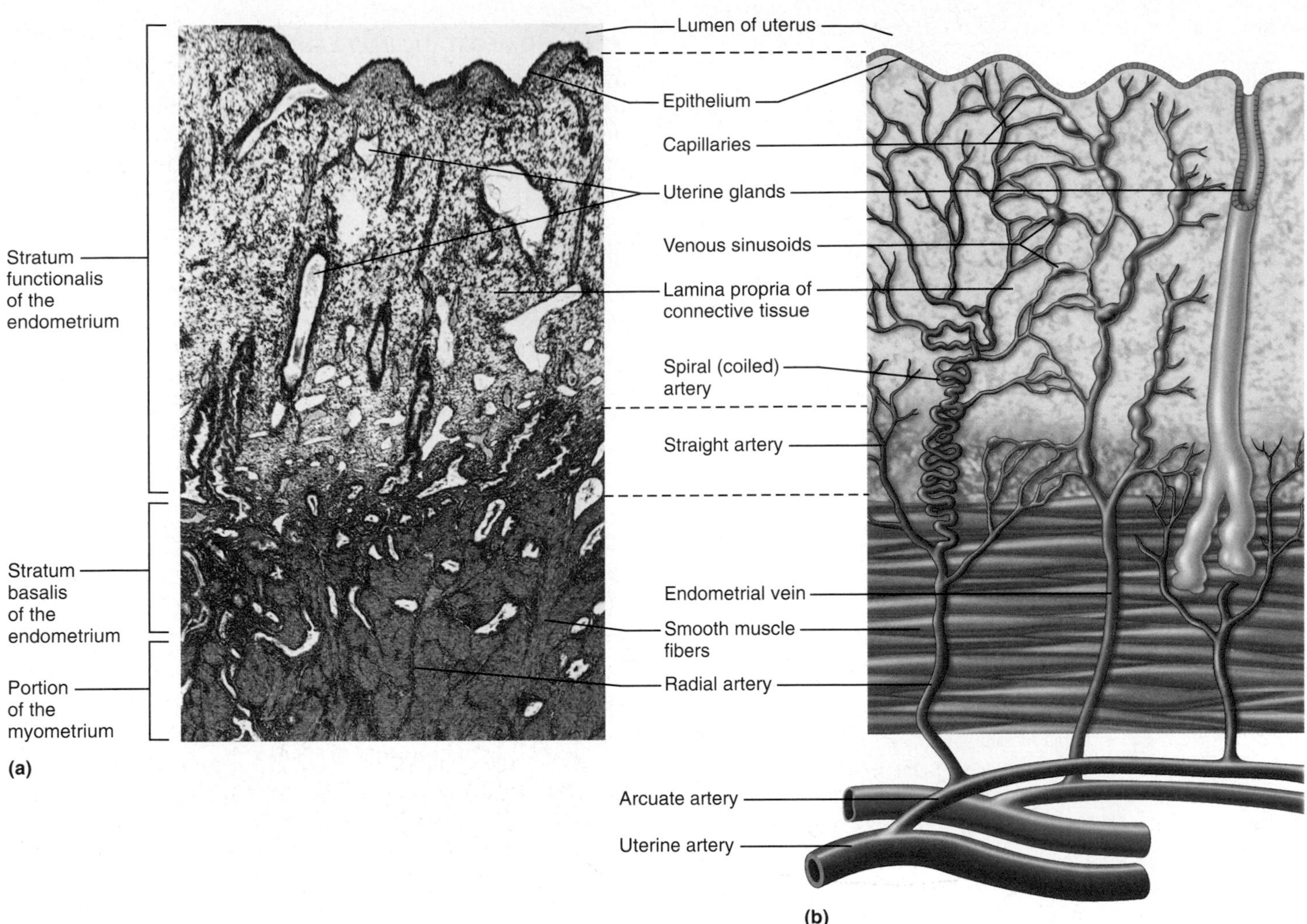

Figure 27.13 The endometrium and its blood supply. (a) Photomicrograph of the endometrium, longitudinal section, showing its functionalis and basalis regions (40×). **(b)** Diagrammatic view of the endometrium, showing the straight arteries that serve the stratum basalis and the spiral arteries that serve the stratum functionalis. The thin-walled veins and venous sinusoids are also illustrated.

squamous epithelium adapted to stand up to friction. Certain cells in the mucosa (*dendritic cells*) act as antigen-presenting cells and are thought to provide the route of HIV transmission from an infected male to the female during sexual intercourse. (AIDS, the immune deficiency disease caused by HIV, is described in Chapter 21.)

The vaginal mucosa has no glands. Instead, it is lubricated by the cervical mucous glands and the mucosal transudate that "weeps" from the vaginal walls. Its epithelial cells release large amounts of glycogen, which is anaerobically metabolized to lactic acid by resident bacteria. Consequently, the pH of a woman's vagina is normally quite acidic. This acidity helps keep the vagina healthy and free of infection, but it is also hostile to sperm. Although vaginal fluid of adult females is *acidic*, it tends to be alkaline in adolescents, predisposing sexually active teenagers to sexually transmitted infections.

In virgins (those who have never participated in sexual intercourse), the mucosa near the distal **vaginal orifice** forms an incomplete partition called the **hymen** (hi′men) **(Figure 27.14a)**. The hymen is very vascular and may bleed when it is stretched or ruptured during the first coitus (sexual intercourse). However, its durability varies. In some females, it is ruptured during a sports activity, tampon insertion, or pelvic examination. Occasionally, it is so tough that it must be breached surgically if intercourse is to occur.

The upper end of the vaginal canal loosely surrounds the cervix of the uterus, producing a vaginal recess called the **vaginal fornix**. The posterior part of this recess, the *posterior fornix*, is much deeper than the *lateral* and *anterior fornices* (see Figures 27.10 and Figures 27.12a). Generally, the lumen of the vagina is quite small and, except where it is held open by the cervix, its posterior and anterior walls are in contact with one another. The vagina stretches considerably during copulation and childbirth, but its lateral distension is limited by the ischial spines and the sacrospinous ligaments.

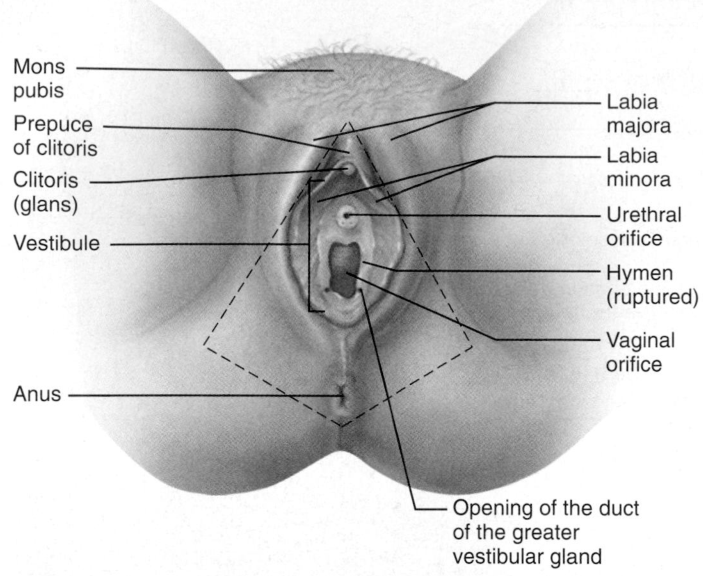

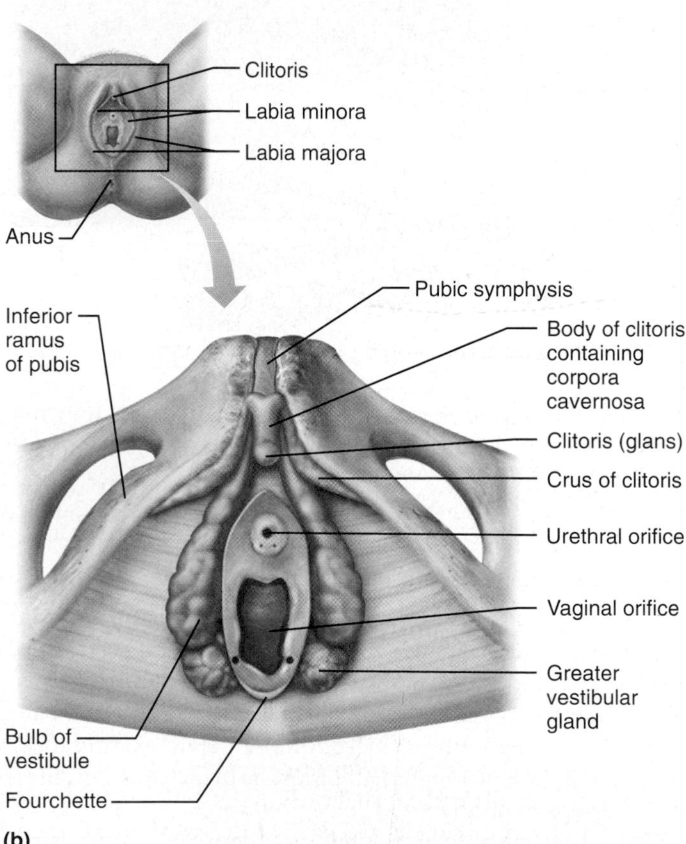

27

Figure 27.14 The external genitalia (vulva) of the female. **(a)** Superficial structures. The region enclosed by the dashed lines is the perineum. **(b)** Deep structures. The labia majora and associated skin have been removed to show the underlying erectile bodies. For the associated superficial muscles, see Figure 10.12 on p. 345.

HOMEOSTATIC IMBALANCE

The uterus tilts away from the vagina. For this reason, attempts by untrained persons to induce an abortion by entering the uterus with a surgical instrument may result in puncturing of the posterior wall of the vagina, followed by hemorrhage and—if the instrument is unsterile—peritonitis. ■

CHECK YOUR UNDERSTANDING

22. Why are women more at risk for PID than men?

23. Oocytes are ovulated into the peritoneal cavity and yet women do get pregnant. What action of the uterine tubes helps to direct the oocytes into the woman's duct system?

24. What portion of the female duct system is the usual site of fertilization? Which is the "incubator" for fetal development?

For answers, see Appendix G.

The External Genitalia and Female Perineum

▶ Describe the anatomy of the female external genitalia.

The female reproductive structures that lie external to the vagina are called the *external genitalia* (Figure 27.14). Also called the **vulva** (vul′vah; "covering") or **pudendum** ("shameful"), these structures include the mons pubis, labia, clitoris, and structures associated with the vestibule.

The **mons pubis** (mons pu′bis; "mountain on the pubis") is a fatty, rounded area overlying the pubic symphysis. After puberty, this area is covered with pubic hair. Running posteriorly from the mons pubis are two elongated, hair-covered fatty skin folds, the **labia majora** (la′be-ah mah-jor′ah; "larger lips"). These are the counterpart, or *homologue*, of the male scrotum (that is, they derive from the same embryonic tissue). The labia majora enclose the **labia minora** (mi-nor′ah; "smaller"), two thin, hair-free skin folds, homologous to the ventral penis.

The labia minora enclose a recess called the **vestibule** ("entrance hall"), which contains the external openings of the urethra and the vagina. Flanking the vaginal opening are the pea-size **greater vestibular glands**, homologous to the bulbourethral glands of males (Figure 27.14b). These glands release mucus into the vestibule and help to keep it moist and lubricated, facilitating intercourse. At the extreme posterior end of the vestibule the labia minora come together to form a ridge called the **fourchette**.

Just anterior to the vestibule is the **clitoris** (klit′o-ris; "hill"), a small, protruding structure composed largely of erectile tissue, which is homologous to the penis of the male. Its exposed portion is called the **glans clitoris** or **glans of the clitoris**. It is hooded by a skin fold called the **prepuce of the clitoris**, formed by the junction of the labia minora folds.

The clitoris is richly innervated with sensory nerve endings sensitive to touch. It becomes swollen with blood and erect during tactile stimulation, contributing to a female's sexual arousal. Like the penis, the **body of the clitoris** has dorsal erectile columns

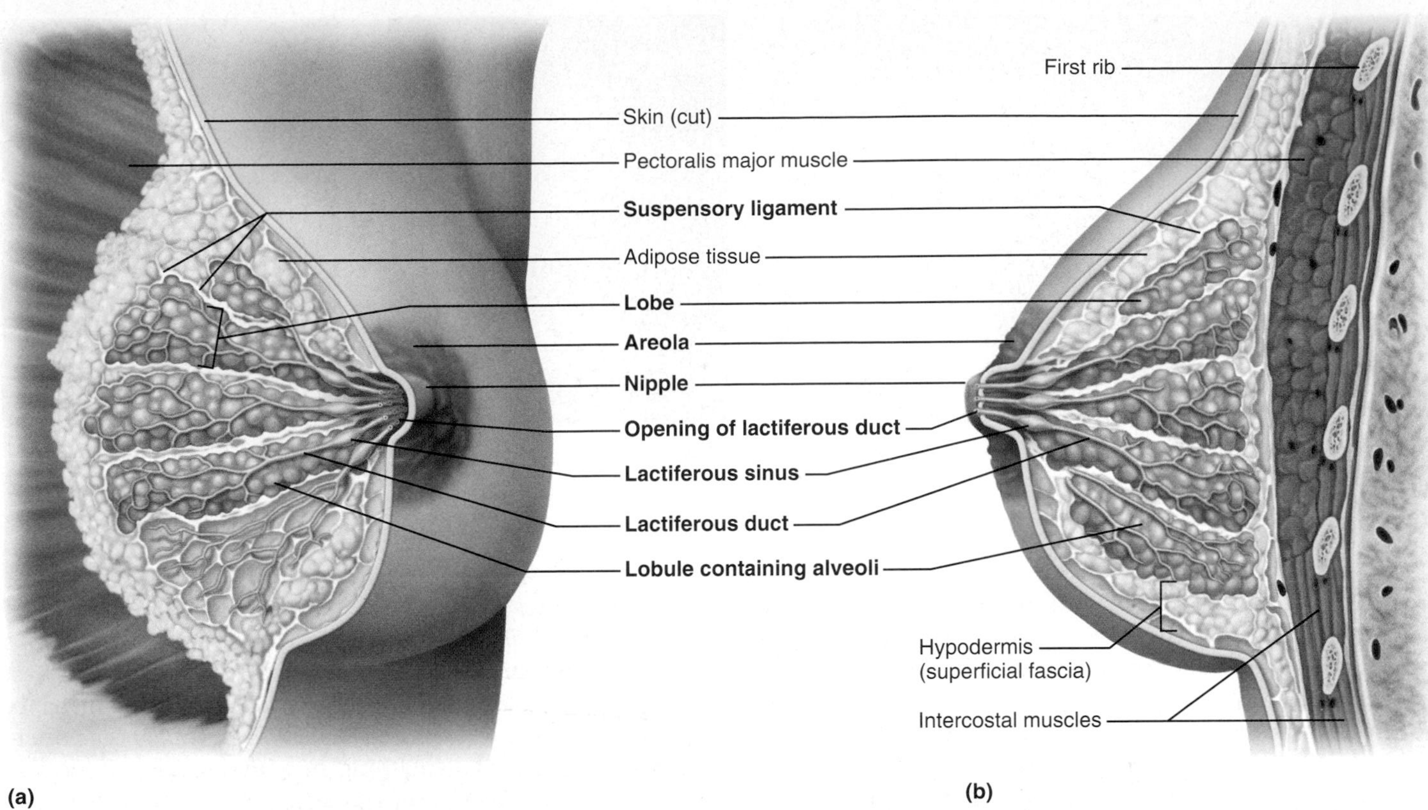

Figure 27.15 Structure of lactating mammary glands. (a) Anterior view of a partially dissected breast. **(b)** Sagittal section of a breast.

(corpora cavernosa) attached proximally by crura, but it lacks a corpus spongiosum that conveys a urethra.

In males the urethra carries both urine and semen and runs through the penis, but the female urinary and reproductive tracts are completely separate. Instead, the **bulbs of the vestibule** (Figure 27.14b), which lie along each side of the vaginal orifice and deep to the bulbospongiosus muscles, are the homologues of the single penile bulb and corpus spongiosum of the male. During sexual stimulation the bulbs of the vestibule engorge with blood. This may help grip the penis within the vagina and also squeezes the urethral orifice shut, which prevents semen (and bacteria) from traveling superiorly into the bladder during intercourse.

The female **perineum** is a diamond-shaped region located between the pubic arch anteriorly, the coccyx posteriorly, and the ischial tuberosities laterally (Figure 27.14a). The soft tissues of the perineum overlie the muscles of the pelvic outlet, and the posterior ends of the labia majora overlie the *central tendon*, into which most muscles supporting the pelvic floor insert (see Table 10.7).

CHECK YOUR UNDERSTANDING

25. What is the female homologue of the bulbourethral glands of males?
26. Cite similarities and differences between the penis and clitoris.

For answers, see Appendix G.

The Mammary Glands

▶ Discuss the structure and function of the mammary glands.

The **mammary glands** are present in both sexes, but they normally function only in females **(Figure 27.15)**. The biological role of the mammary glands is to produce milk to nourish a newborn baby, so they are important only when reproduction has already been accomplished.

Developmentally, mammary glands are modified sweat glands that are really part of the *skin*, or *integumentary system*. Each mammary gland is contained within a rounded skin-covered breast within the hypodermis (superficial fascia), anterior to the pectoral muscles of the thorax. Slightly below the center of each breast is a ring of pigmented skin, the **areola** (ah-re′o-lah), which surrounds a central protruding **nipple**. Large sebaceous glands in the areola make it slightly bumpy and produce sebum that reduces chapping and cracking of the skin of the nipple. Autonomic nervous system controls of smooth muscle fibers in the areola and nipple cause the nipple to become erect when stimulated by tactile or sexual stimuli and when exposed to cold.

Internally, each mammary gland consists of 15 to 25 **lobes** that radiate around and open at the nipple. The lobes are padded and separated from each other by fibrous connective tissue and fat. The interlobar connective tissue forms **suspensory ligaments** that attach the breast to the underlying muscle fascia and to the overlying dermis. As suggested by their name, the suspensory ligaments provide natural support for the breasts, like a built-in brassiere.

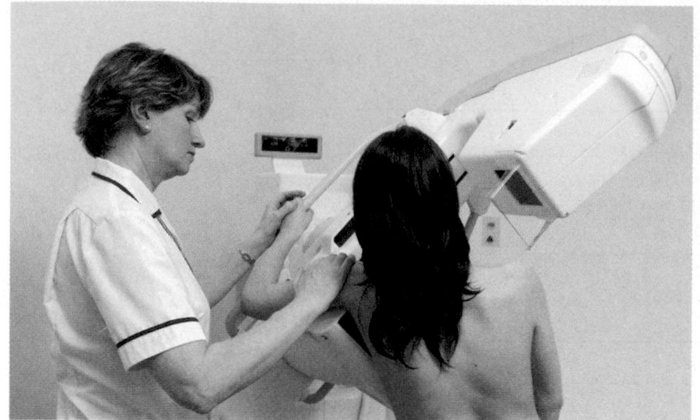

(a) Mammogram procedure

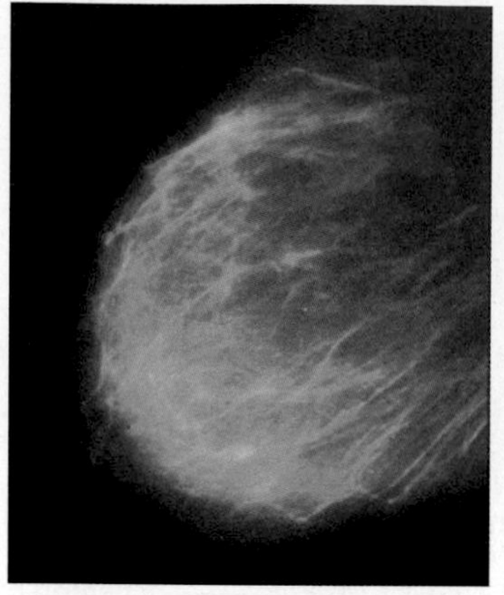

(b) Film of normal breast

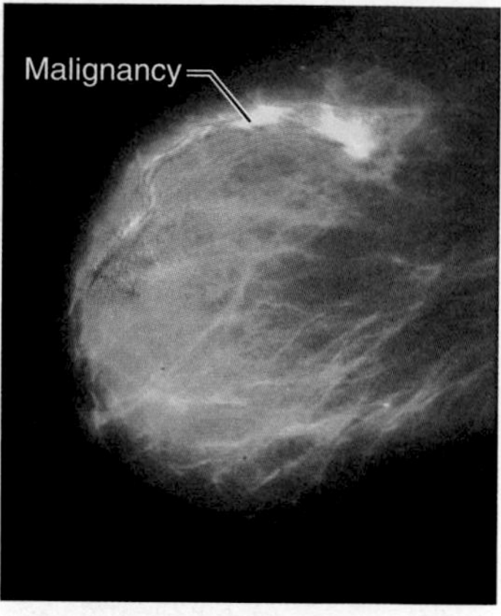

(c) Film of breast with tumor

Figure 27.16 **Mammograms.**

Within the lobes are smaller units called **lobules**, which contain glandular **alveoli** that produce milk when a woman is lactating. These compound alveolar glands pass the milk into the **lactiferous ducts** (lak-tif′er-us), which open to the outside at the nipple. Just deep to the areola, each duct has a dilated region called a **lactiferous sinus** where milk accumulates during nursing. We describe the process and regulation of lactation in Chapter 28.

The description of mammary glands that we have just given applies only to nursing women or women in the last trimester of pregnancy. In nonpregnant women, the glandular structure of the breast is largely undeveloped and the duct system is rudimentary. For this reason, breast size is largely due to the amount of fat deposits.

Breast Cancer

Invasive breast cancer is the most common malignancy and the second most common cause of cancer death of U.S. women. Thirteen percent of women in the general population (132 out of 1000 individuals) will develop this condition. Breast cancer usually arises from the epithelial cells of the smallest ducts, not from the alveoli. A small cluster of cancer cells grows into a lump in the breast from which cells eventually metastasize.

Known risk factors for developing breast cancer include (1) early onset menstruation and late menopause; (2) no pregnancies or first pregnancy later in life and no or short periods of breast feeding; and (3) family history of breast cancer (especially in a sister or mother). Some 10% of breast cancers stem from hereditary defects and half of these can be traced to dangerous mutations in a pair of genes, dubbed *BRCA1* (breast cancer 1) and *BRCA2*, which cause 80% of those who carry the altered gene to develop breast cancer. Notice that, with the possible exception of family history, these factors reflect increased lifelong exposure to estrogen. However, more than 70% of women who develop breast cancer have no known risk factors for the disease.

Breast cancer is often signaled by a change in skin texture, puckering, or leakage from the nipple. Since most breast lumps are discovered by women themselves in routine monthly breast exams, this simple examination should be a health maintenance priority in every woman's life. The American Cancer Society had recommended scheduling **mammography**—X-ray examination that detects breast cancers too small to feel (less than 1 cm)—every two years for women between 40 and 49 years old and yearly after that (**Figure 27.16**). However, some authorities suggest that yearly is too frequent. A new study suggests that diagnostic MRI scans are preferable for at-risk women who carry a mutated *BRCA* gene.

Once diagnosed, breast cancer is treated in various ways depending on specific characteristics of the lesion. Current therapies include (1) radiation therapy, (2) chemotherapy, and (3) surgery, often followed by irradiation or chemotherapy to destroy stray cancer cells. Drug therapies for estrogen-responsive cancers include trastuzumab (Herceptin), a drug containing bioengineered antibodies that jam estrogen receptors that control aggressive growth in breast cancer cells; tamoxifen, an antiestrogen compound that blocks estrogen's effects and significantly improves the outcome for premenopausal women with early- or late-stage breast cancer; and letrozole (Femara), which disables the enzyme needed to make estrogen and reduces breast cancer recurrences in women (particularly postmenopausal women) who have exhausted the usefulness of tamoxifen.

Until the 1970s, the standard treatment was **radical mastectomy** (mas-tek′to-me; "breast cutting"), removal of the entire affected breast, plus all underlying muscles, fascia, and associated lymph nodes. Most physicians now recommend **lumpectomy**, less extensive surgery in which only the cancerous part (lump) is excised, or **simple mastectomy**, removal of the breast tissue only (and perhaps some of the axillary lymph nodes). Many mastectomy patients opt for breast reconstruction to replace the excised tissue. Currently tissue "flaps," containing muscle, fat, and skin taken from the patient's abdomen or back, are used for "sculpting" a natural-looking breast.

CHECK YOUR UNDERSTANDING

27. Developmentally, mammary glands are modifications of certain skin glands. Which type?

28. From what cell types does breast cancer usually arise?

For answers, see Appendix G.

Physiology of the Female Reproductive System

Gamete production in males begins at puberty and continues throughout life, but the situation is quite different in females. It has been assumed that a female's total supply of eggs is already determined by the time she is born, and the time span during which she releases them extends only from puberty to menopause (about the age of 50). However, studies done in adult mice in 2004 and 2005 indicated that egg stem cells are alive and generating little "egglets" throughout life and there have been hopes that egg stem cells also exist in adult women. These findings might seem to overturn the assumption that the number of oocytes (the potential eggs) is limited—an idea that has been part of the bedrock of biology. As of 2008, however, these findings have not been confirmed and it has not been shown that newly formed "oocytes" restore fertility to otherwise sterile mice. Furthermore, the reproductive cycles of mice and humans are noticeably different when life span is considered. So, it is still too early to retire the "no new eggs" doctrine.

Oogenesis

▶ Describe the process of oogenesis and compare it to spermatogenesis.

Meiosis, the specialized nuclear division that occurs in the testes to produce sperm, also occurs in the ovaries. In this case, female sex cells are produced, and the process is called **oogenesis** (o″o-gen′e-sis; "the beginning of an egg"). The process of oogenesis takes years to complete, as indicated in Figure 27.17.

First, in the fetal period the **oogonia**, the diploid stem cells of the ovaries, multiply rapidly by mitosis and then enter a growth phase and lay in nutrient reserves. Gradually, *primordial follicles* begin to appear as the oogonia are transformed into **primary oocytes** and become surrounded by a single layer of flattened follicle cells. The primary oocytes begin the first meiotic division, but become "stalled" late in prophase I and do not complete it.

By birth, a female is presumed to have her lifetime supply of primary oocytes. Of the original 7 million oocytes, approximately 2 million of them escape programmed death and are already in place in the cortical region of the immature ovary. Since they remain in their state of suspended animation all through childhood, the wait is a long one—10 to 14 years at the very least!

At puberty, perhaps 250,000 oocytes remain. Beginning at this time a small number of primary oocytes are recruited (activated) each month in response to an LH surge midcycle (see Figure 27.20a). However, only one is "selected" each time to continue meiosis I, ultimately producing two haploid cells (each with 23 replicated chromosomes) that are quite dissimilar in size. The smaller cell is called the **first polar body**. The larger cell, which contains nearly all the cytoplasm of the primary oocyte, is the **secondary oocyte**.

The events of this first maturation division ensure that the polar body receives almost no cytoplasm or organelles. Notice in Figure 27.17 (left) that a spindle forms at the very edge of the oocyte. A little "nipple" also appears at that edge, and the polar body chromosomes are cast into it.

The first polar body may continue its development and undergo meiosis II, producing two even smaller polar bodies. However, in humans, the secondary oocyte arrests in metaphase II, and it is this cell (not a functional ovum) that is ovulated. If an ovulated secondary oocyte is not penetrated by a sperm, it simply deteriorates. But, if sperm penetration *does* occur, the oocyte quickly completes meiosis II, yielding one large **ovum** and a tiny **second polar body** (Figure 27.17). The union of the egg and sperm nuclei, described in Chapter 28, constitutes fertilization.

What you should realize now is that the potential end products of oogenesis are three tiny polar bodies, nearly devoid of cytoplasm, and one large ovum. All of these cells are haploid, but only the ovum is a *functional gamete.* This is quite different from spermatogenesis, where the product is four viable gametes—spermatozoa.

The unequal cytoplasmic divisions that occur during oogenesis ensure that a fertilized egg has ample nutrients for its six- to seven-day journey to the uterus. Lacking nutrient-containing cytoplasm, the polar bodies degenerate and die. Since the reproductive life of a female is at most about 40 years (from the age of 11 to approximately 51) and typically only one ovulation occurs each month, fewer than 500 oocytes out of her estimated pubertal potential of 250,000 are released during a woman's lifetime.

Perhaps the most striking difference between male and female meiosis is the error rate. As many as 20% of oocytes but only 3–4% of sperm have the wrong number of chromosomes, a situation that often results from failure of the homologues to separate during meiosis I. It appears that faced with meiotic disruption, meiosis in males grinds to a halt but in females it marches on.

CHECK YOUR UNDERSTANDING

29. How do the haploid cells arising from oogenesis differ structurally and functionally from those arising from spermatogenesis?

For answers, see Appendix G.

27

The Ovarian Cycle

▶ Describe ovarian cycle phases, and relate them to events of oogenesis.

The monthly series of events associated with the maturation of an egg is called the **ovarian cycle**. The ovarian cycle is best

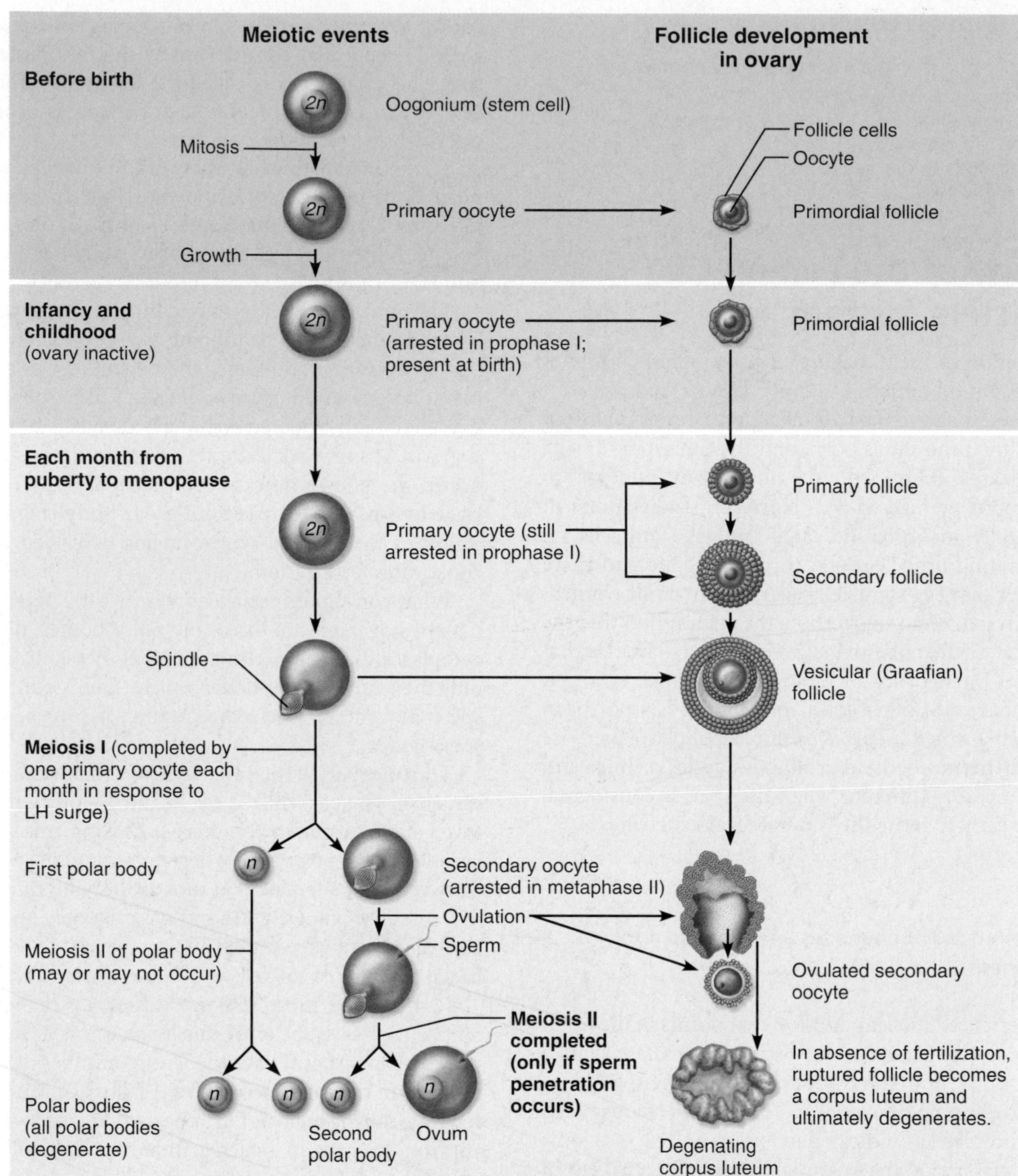

Figure 27.17 Events of oogenesis. Left, flowchart of meiotic events. Right, correlation with follicle development and ovulation in the ovary.

27

described in terms of two consecutive phases. The **follicular phase** is the period of follicle growth, typically indicated as lasting from the first to the fourteenth day of the cycle. The **luteal phase** is the period of corpus luteum activity, days 14–28. The so-called typical ovarian cycle repeats at intervals of 28 days, with *ovulation* occurring midcycle.

However, only 10–15% of women naturally have 28-day cycles, and cycles as long as 40 days or as short as 21 days are fairly common. In such cases, the length of the follicular phase and timing of ovulation vary, but the luteal phase remains constant: It is 14 days between the time of ovulation and the end of the cycle.

The Follicular Phase

Maturation of a primordial follicle occupies the first half of the cycle and involves several events as shown in **Figure 27.18**, stages ①–⑥.

A Primordial Follicle Becomes a Primary Follicle When a primordial follicle ① is activated (by a process directed by the oocyte), the squamouslike cells surrounding the primary oocyte grow, becoming cuboidal cells, and the oocyte enlarges. The follicle is now called a primary (1°) follicle ②.

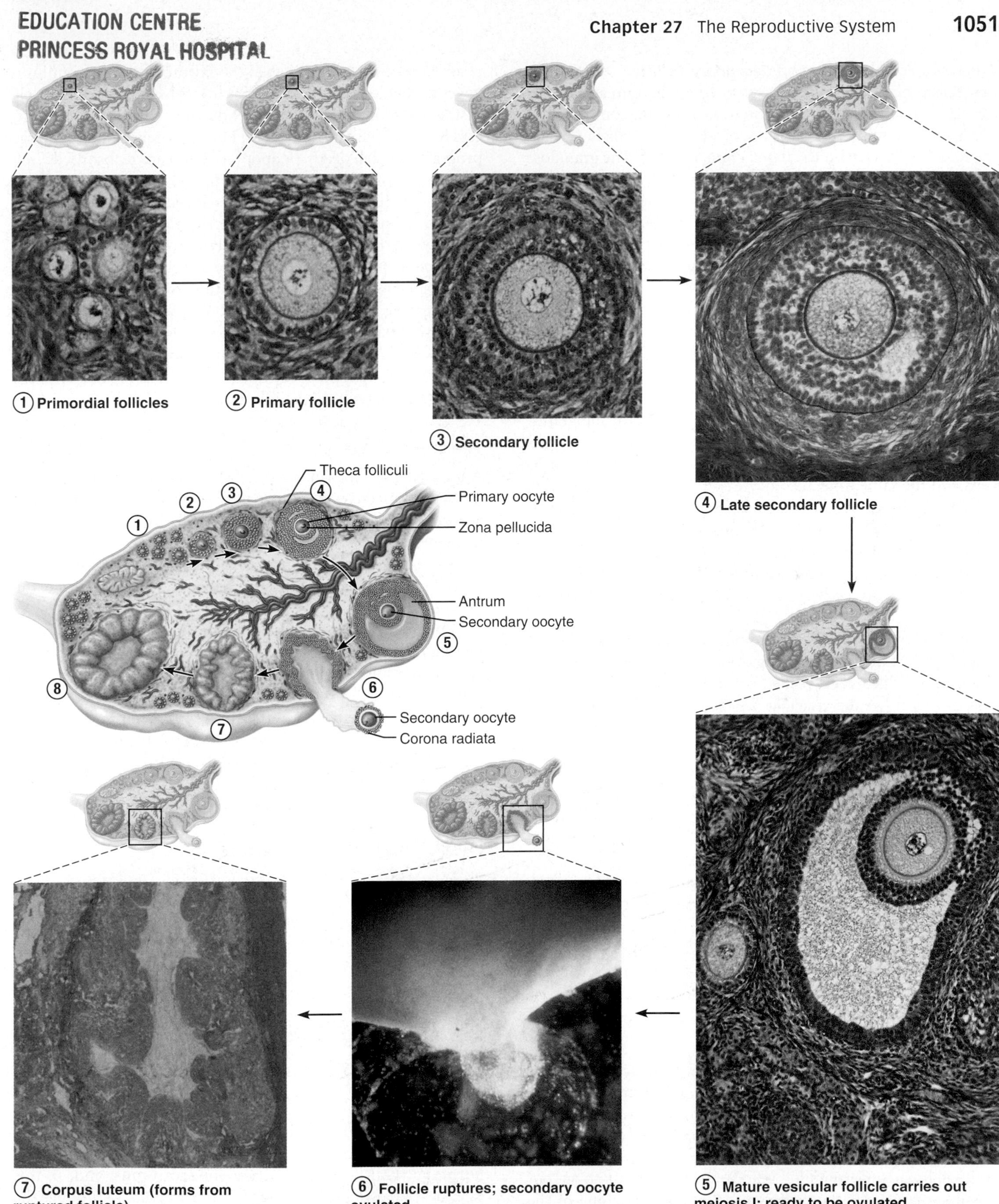

Figure 27.18 Schematic and microscopic views of the ovarian cycle: development and fate of ovarian follicles. The numbers on the diagram indicate the sequence of stages in follicle development, not the movement of a developing follicle within the ovary.

27

A Primary Follicle Becomes a Secondary Follicle Next, follicular cells proliferate, forming a stratified epithelium around the oocyte. As mentioned earlier, as soon as more than one cell layer is present, the follicle is called a secondary (2°) follicle ③ and the follicle cells take on the name *granulosa cells.* The granulosa cells are connected to the developing oocyte by gap junctions, through which ions, metabolites, and signaling molecules can pass. From this point on, bidirectional "conversations" occur between the oocyte and granulosa cells, so they guide one another's development. One of the signals passing from the granulosa cells to the oocyte "tells" the oocyte to grow. Others dictate asymmetry (polarity) in the future egg.

A Secondary Follicle Becomes a Late Secondary Follicle In stage ④, a layer of connective tissue condenses around the follicle, forming the **theca folliculi** (the′kah fah-lik′u-li; "box around the follicle"). As the follicle grows, the thecal and granulosa cells cooperate to produce estrogens. (The inner thecal cells produce androgens, which the granulosa cells convert to estrogens.) At the same time, the oocyte secretes a glycoprotein-rich substance that forms a thick transparent extracellular layer or membrane, called the **zona pellucida** (pĕ-lu′sid-ah), that encapsulates it (see Figure 27.11).

At the end of this stage, clear liquid begins to accumulate between the granulosa cells, producing the late secondary follicle.

A Late Secondary Follicle Becomes a Vesicular Follicle In stage ⑤, the fluid between the granulosa cells coalesces to form a large fluid-filled cavity called the **antrum** ("cave"), an event that distinguishes the vesicular follicle from the late secondary follicle. The antrum continues to expand with fluid until it isolates the oocyte, along with its surrounding capsule of granulosa cells called a **corona radiata** ("radiating crown"), on a stalk on one side of the follicle. When a follicle is full size (about 2.5 cm, or 1 inch, in diameter), it becomes a vesicular follicle and bulges from the external ovarian surface like an "angry boil." This usually occurs by day 14.

As one of the final events of follicle maturation, the primary oocyte completes meiosis I to form the secondary oocyte and first polar body (see Figure 27.17). Once this has occurred, the stage is set for ovulation. At this point, the granulosa cells send another important signal to the oocyte that says, in effect, "Wait, do not complete meiosis yet!"

Ovulation

Ovulation (stage ⑥) occurs when the ballooning ovary wall ruptures and expels the secondary oocyte, still surrounded by its corona radiata, into the peritoneal cavity. Some women experience a twinge of pain in the lower abdomen when ovulation occurs. The precise cause of this episode, called *mittelschmerz* (mit′el-shmärts; German for "middle pain"), is not known, but possible reasons for the pain include intense stretching of the ovarian wall during ovulation and irritation of the peritoneum by blood or fluid released from the ruptured follicle.

In the ovaries of adult females, there are always several follicles at different stages of maturation. As a rule, one follicle outstrips the others to become the *dominant follicle* and is at the peak stage of maturation when the hormonal (LH) stimulus is given for ovulation. How this follicle is selected, or selects itself, is still uncertain, but it is probably the one that attains the greatest FSH sensitivity the quickest. The others degenerate (undergo programmed cell death, or apoptosis) and are reabsorbed.

In 1–2% of all ovulations, more than one oocyte is ovulated. This phenomenon, which increases with age, can result in multiple births. Since, in such cases, different oocytes are fertilized by different sperm, the siblings are *fraternal,* or nonidentical, twins. *Identical twins* result from the fertilization of a single oocyte by a single sperm, followed by separation of the fertilized egg's daughter cells in early development. Additionally, it now appears that in some women, oocytes may be released at times unrelated to the woman's hormone levels. This timing may help to explain why a rhythm method of contraception sometimes fails and why some fraternal twins have different conception dates.

The Luteal Phase

After ovulation, the ruptured follicle collapses, and the antrum fills with clotted blood. This *corpus hemorrhagicum* is eventually absorbed. The remaining granulosa cells increase in size and along with the internal thecal cells they form a new, quite different endocrine structure, the *corpus luteum* ("yellow body") (Figure 27.18, stage ⑦). It begins to secrete progesterone and some estrogen.

If pregnancy does not occur, the corpus luteum starts degenerating in about 10 days and its hormonal output ends. In this case, all that ultimately remains is a scar called the *corpus albicans* (al′bĭ-kans; "white body"). The last two or three days of the luteal phase, when the endometrium is just beginning to erode, is sometimes called the *luteolytic* or *ischemic phase.*

On the other hand, if the oocyte is fertilized and pregnancy ensues, the corpus luteum persists until the placenta is ready to take over its hormone-producing duties in about three months.

CHECK YOUR UNDERSTANDING

30. How do identical twins differ developmentally from fraternal twins?
31. What occurs in the luteal phase of the ovarian cycle?

For answers, see Appendix G.

Hormonal Regulation of the Ovarian Cycle

▶ Describe the regulation of the ovarian and uterine cycles.

Ovarian events are much more complicated than those occurring in the testes, but the hormonal controls set into motion at puberty are similar in the two sexes. Gonadotropin-releasing hormone (GnRH), the pituitary gonadotropins, and, in this case, ovarian estrogen and progesterone interact to produce the cyclic events occurring in the ovaries.

However, in females another hormone plays an important role in stimulating the hypothalamus to release GnRH. The onset of puberty in females is linked to adiposity, and the

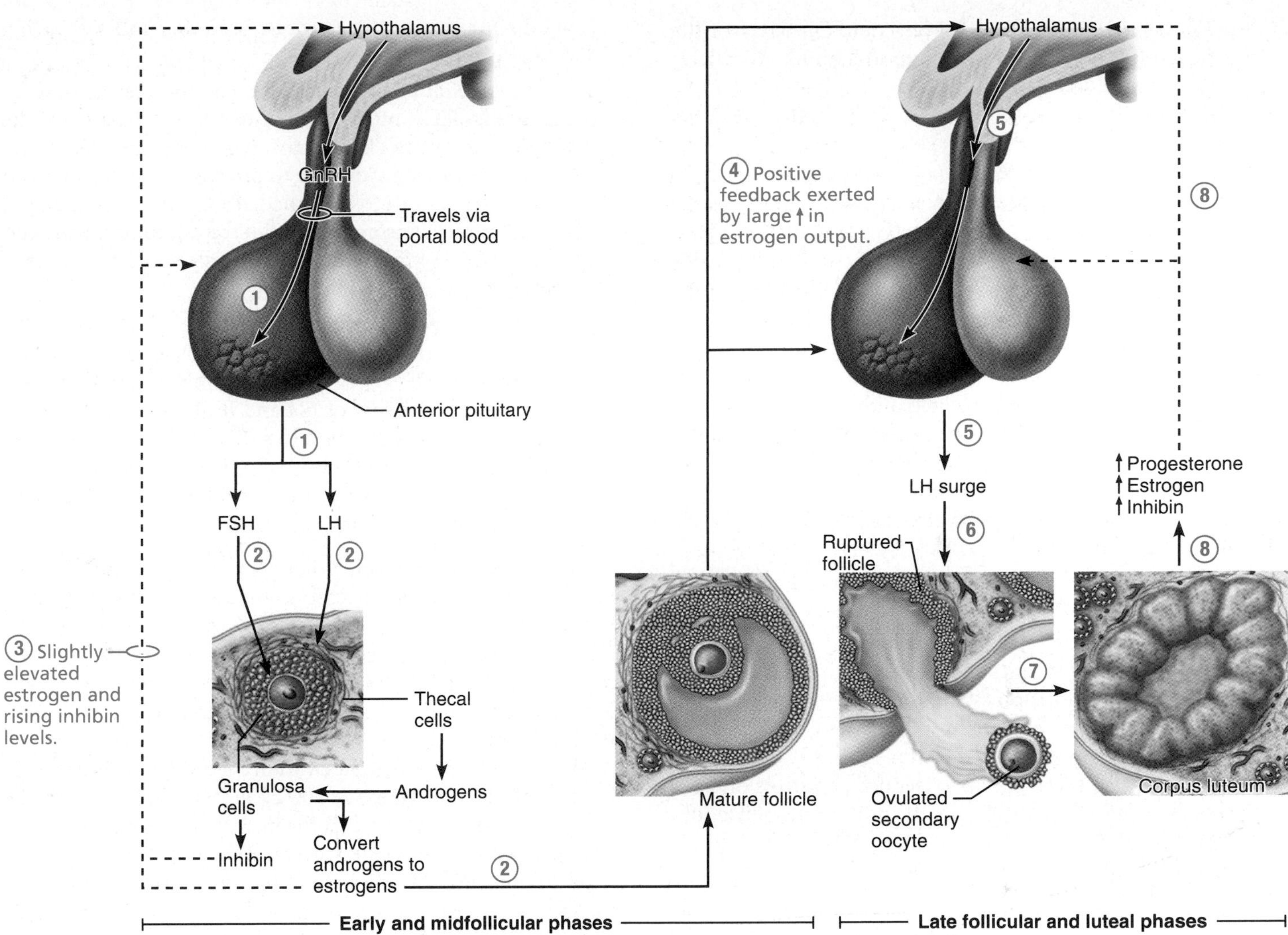

Figure 27.19 Feedback interactions in the regulation of ovarian function. Numbers refer to events listed in the text. Note that all feedback signals exerted by ovarian hormones are negative except one—that exerted by estrogen immediately before ovulation. Events that follow step ⑧ (negative feedback inhibition of the hypothalamus and anterior pituitary by progesterone and estrogens) are not depicted, but involve a gradual deterioration of the corpus luteum and, therefore, a decline in ovarian hormone production. Ovarian hormones reach their lowest blood levels around day 28.

messenger from fatty tissue to the hypothalamus is *leptin.* If blood levels of lipids and leptin (better known for its role in energy production and appetite) are low, puberty is delayed.

Establishing the Ovarian Cycle

During childhood, the ovaries grow and continuously secrete small amounts of estrogens, which inhibit hypothalamic release of GnRH. Provided that leptin levels are adequate, the hypothalamus becomes less sensitive to estrogen as puberty nears and begins to release GnRH in a rhythmic pulselike manner. GnRH, in turn, stimulates the anterior pituitary to release FSH and LH, which prompt the ovaries to secrete hormones (primarily estrogens).

Gonadotropin levels continue to increase for about four years and, during this time, pubertal girls are still not ovulating and for this reason are incapable of getting pregnant. Eventually, the adult cyclic pattern is achieved, and hormonal interactions stabilize. These events are heralded by the young woman's first menstrual period, referred to as **menarche** (mě-nar′ke; *men* = month, *arche* = first). Usually, it is not until the third year postmenarche that the cycles become regular and all are ovulatory.

Hormonal Interactions During the Ovarian Cycle

Next we describe the waxing and waning of anterior pituitary gonadotropins (FSH and LH) and ovarian hormones and the negative and positive feedback interactions that regulate ovarian function. Events ①–⑧ in the following discussion directly correspond to the same-numbered steps in **Figure 27.19**. We assume a 28-day cycle.

① *GnRH stimulates FSH and LH secretion.* On day 1 of the cycle, GnRH secreted by the hypothalamus stimulates production and release of follicle-stimulating hormone (FSH) and luteinizing hormone (LH) by the anterior pituitary.

② *FSH and LH stimulate follicle growth and maturation and estrogen secretion.* FSH exerts its main effects on the granulosa cells of the late secondary or vesicular follicles, whereas LH (at least initially) targets the thecal cells. (Why only *some* follicles respond to these hormonal stimuli is still a mystery. However, there is little doubt that enhanced responsiveness is due to formation of more gonadotropin receptors.) As the follicles enlarge, LH prods the thecal cells to produce androgens. These hormones diffuse through the basement membrane, where they are converted to estrogens by the granulosa cells. Only tiny amounts of ovarian androgens enter the blood, because they are almost completely converted to estrogens within the ovaries.

③ *Negative feedback.* The rising estrogen levels in the plasma exert *negative feedback* on the hypothalamus and anterior pituitary, inhibiting release of FSH and LH, while simultaneously prodding the pituitary to synthesize and accumulate these gonadotropins. Within the ovary, estrogen enhances estrogen output by intensifying the effect of FSH on follicle maturation. *Inhibin,* released by the granulosa cells, also exerts negative feedback controls on FSH release during this period. Only the dominant follicle survives this dip in FSH—the remaining developing follicles fail to develop further, and they deteriorate.

④ *Positive feedback.* Although the initial small rise in blood-borne estrogen inhibits the hypothalamic-pituitary axis, high estrogen levels, produced by the dominant follicle and other maturing follicles, have the opposite effect. Once estrogen reaches a critical blood concentration, it briefly exerts *positive feedback* on the brain and anterior pituitary.

⑤ *LH surge.* High estrogen levels set a cascade of events into motion. There is a sudden burstlike release of accumulated LH (and, to a lesser extent, FSH) by the anterior pituitary about midcycle (also see Figure 27.20a).

⑥ *Ovulation.* The LH surge stimulates the primary oocyte of the dominant follicle to complete the first meiotic division, forming a secondary oocyte that continues on to metaphase II. At or around day 14, LH stimulates many events that lead to ovulation: It increases local vascular permeability, stimulates release of prostaglandins, and triggers an inflammatory response that promotes the release of metalloproteinase enzymes that help to weaken the ovary wall. As a result, blood stops flowing through the protruding part of the follicle wall. Within minutes, that region of the follicle wall thins, bulges out, and then ruptures, accomplishing ovulation. The role (if any) of FSH in this process is unknown. Shortly after ovulation, estrogen levels decline. This probably reflects the damage to the dominant estrogen-secreting follicle during ovulation.

⑦ *Corpus luteum forms.* The LH surge also transforms the ruptured follicle into a corpus luteum (which gives LH the name "luteinizing" hormone). LH stimulates this newly formed endocrine structure to produce progesterone and some estrogen almost immediately after it is formed. Progesterone helps maintain the stratum functionalis and thus is essential for maintaining a pregnancy should conception occur.

⑧ *Negative feedback inhibits LH and FSH release.* Rising progesterone and estrogen blood levels exert a powerful negative feedback effect on the hypothalamus and the anterior pituitary release of LH and FSH. Release of inhibin by the corpus luteum enhances this inhibitory effect. Declining gonadotropin levels inhibit the development of new follicles and prevent additional LH surges that might cause additional oocytes to be ovulated.

In nonfertile cycles, as LH blood levels fall, the stimulus for luteal activity ends, and the corpus luteum degenerates. As goes the corpus luteum, so go the levels of ovarian hormones, and blood estrogen and progesterone levels drop sharply. The marked decline in ovarian hormones at the end of the cycle (days 26–28) ends their blockade of FSH and LH secretion, and the cycle starts anew.

We have just described the ovarian events as if we are following one follicle through the 28-day cycle, but this is not really the case. What is happening is that the increase of FSH at the beginning of each cycle activates several follicles to mature. Then, with the midcycle LH surge, one (or more) vesicular follicles undergo ovulation. However, the ovulated oocyte would actually have been activated about 110 days (some three months) before, not 14 days before.

The Uterine (Menstrual) Cycle

Although the uterus is where the young embryo implants and develops, it is receptive to implantation for only a short period each month. Not surprisingly, this brief interval is exactly the time when a developing embryo would normally begin implanting, six to seven days after ovulation. The **uterine**, or **menstrual** (men′stroo-al), **cycle** is a series of cyclic changes that the uterine endometrium goes through each month as it responds to the waxing and waning of ovarian hormones in the blood. These endometrial changes are coordinated with the phases of the ovarian cycle, which are dictated by gonadotropins released by the anterior pituitary.

The events of the uterine cycle, depicted in **Figure 27.20d**, are driven by changes in ovarian steroid hormone levels as follows:

1. **Days 1–5: Menstrual phase**. In this phase, **menstruation** (men″stroo-a′shun) or **menses**, the uterus sheds all but the deepest part of its endometrium. (Note in Figure 27.20a and c that at the beginning of this stage, ovarian hormones are at their lowest normal levels and gonadotropins are beginning to rise.) The thick, hormone-dependent functional layer of the endometrium detaches from the

27

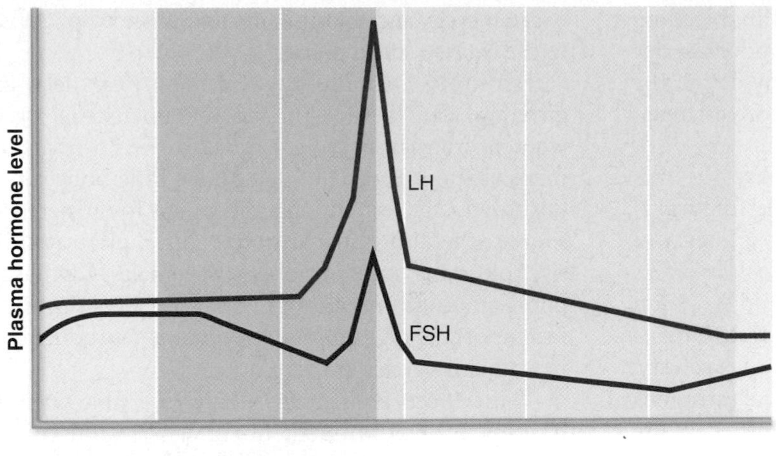

(a) Fluctuation of gonadotropin levels: Fluctuating levels of pituitary gonadotropins (follicle-stimulating hormone and luteinizing hormone) in the blood regulate the events of the ovarian cycle.

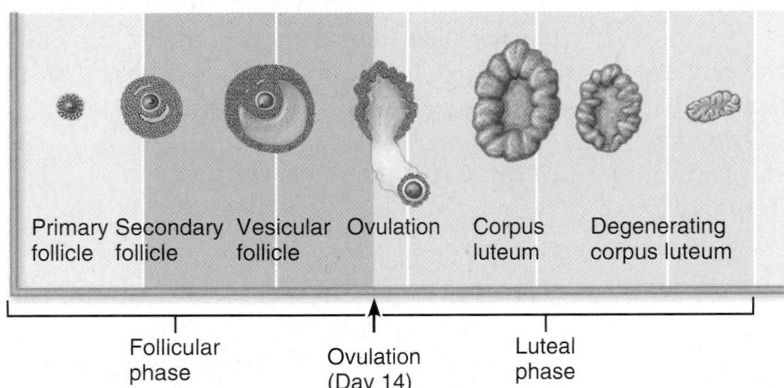

(b) Ovarian cycle: Structural changes in the ovarian follicles during the ovarian cycle are correlated with (d) changes in the endometrium of the uterus during the uterine cycle.

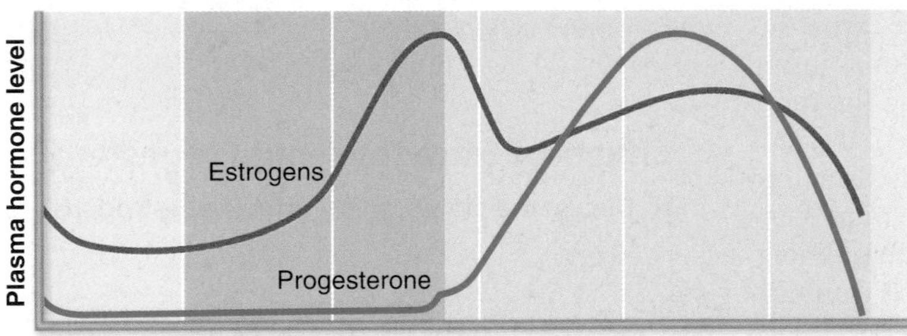

(c) Fluctuation of ovarian hormone levels: Fluctuating levels of ovarian hormones (estrogens and progesterone) cause the endometrial changes of the uterine cycle. The high estrogen levels are also responsible for the LH/FSH surge in (a).

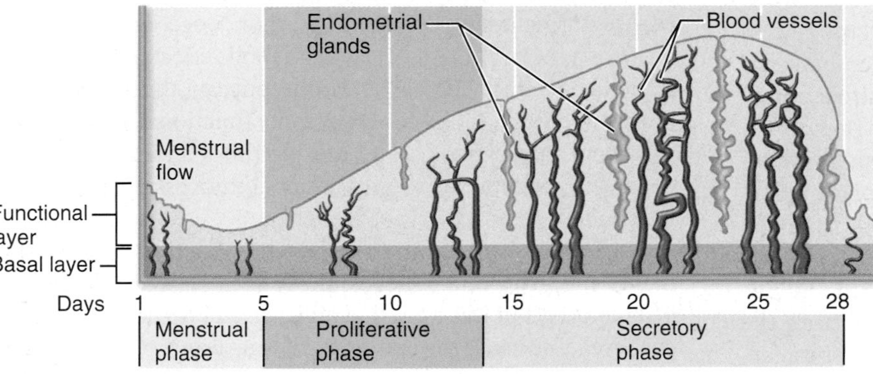

(d) The three phases of the uterine cycle:
- Menstrual: Shedding of the functional layer of the endometrium.
- Proliferative: Rebuilding of the functional layer of the endometrium.
- Secretory: Begins immediately after ovulation. Enrichment of the blood supply and glandular secretion of nutrients prepare the endometrium to receive an embryo.

Both the menstrual and proliferative phases occur before ovulation, and together they correspond to the follicular phase of the ovarian cycle. The secretory phase corresponds in time to the luteal phase of the ovarian cycle.

Figure 27.20 Correlation of anterior pituitary and ovarian hormones with structural changes of the ovary and uterus. The time bar at the bottom of the figure, reading Days 1 to 28, applies to all four parts of this figure. (See *A Brief Atlas of the Human Body*, Plate 53.)

uterine wall, a process that is accompanied by bleeding for 3–5 days. The detached tissue and blood pass out through the vagina as the menstrual flow. By day 5, the growing ovarian follicles start to produce more estrogen (Figure 27.20c).

2. **Days 6–14: Proliferative (preovulatory) phase.** In this phase, the endometrium rebuilds itself: Under the influence of rising blood levels of estrogens, the basal layer of the endometrium generates a new functional layer. As this new layer thickens, its glands enlarge and its spiral arteries increase in number (also see Figure 27.13). Consequently, the endometrium once again becomes velvety, thick, and well vascularized. During this phase, estrogens also induce synthesis of progesterone receptors in the endometrial cells, readying them for interaction with progesterone.

   Normally, cervical mucus is thick and sticky, but rising estrogen levels cause it to thin and form channels that facilitate the passage of sperm into the uterus. *Ovulation,* which takes less than five minutes, occurs in the ovary at the end of the proliferative stage (day 14) in response to the sudden release of LH from the anterior pituitary. As we saw earlier, LH also converts the ruptured follicle to a corpus luteum.

3. **Days 15–28: Secretory (postovulatory) phase.** This 14-day phase is the most constant timewise. During the secretory phase the endometrium prepares for implantation of an embryo. Rising levels of progesterone from the corpus luteum act on the estrogen-primed endometrium, causing the spiral arteries to elaborate and converting the functional layer to a secretory mucosa. The endometrial glands enlarge, coil, and begin secreting nutritious glycogen into the uterine cavity. These nutrients sustain the embryo until it has implanted in the blood-rich endometrial lining. Increasing progesterone levels also cause the cervical mucus to become viscous again, forming the *cervical plug,* which helps to block entry of sperm and pathogens or other foreign materials, and plays an important role in keeping the uterus "private" in the event an embryo has begun to implant. Rising progesterone (and estrogen) levels inhibit LH release by the anterior pituitary.

   If fertilization has not occurred, the corpus luteum degenerates toward the end of the secretory phase as LH blood levels decline. Progesterone levels fall, depriving the endometrium of hormonal support, and the spiral arteries kink and go into spasms. Denied oxygen and nutrients, the ischemic endometrial cells die, setting the stage for menstruation to begin on day 28. The spiral arteries constrict one final time and then suddenly relax and open wide. As blood gushes into the weakened capillary beds, they fragment, causing the functional layer to slough off. The menstrual cycle starts over again on this first day of menstrual flow.

Figure 27.20b and d also illustrate how the ovarian and uterine cycles fit together. Notice that the menstrual and proliferative phases overlap the follicular phase and ovulation in the ovarian cycle, and that the uterine secretory phase corresponds to the ovarian luteal phase.

Extremely strenuous physical activity can delay menarche in girls and can disrupt the normal menstrual cycle in adult women, even causing *amenorrhea* (a-men″o-re′ah), cessation of menstruation. Female athletes have little body fat, and fat deposits help convert adrenal androgens to estrogens and are the source of leptin which, as noted above, plays a critical permissive role in the onset of puberty in females. Leptin keeps the hypothalamus informed about whether energy stores are sufficient to support the high energy demands of reproduction. If not, the reproductive cycles are shut down.

These effects are usually totally reversible when the athletic training is discontinued, but a worrisome consequence of amenorrhea in young, healthy adult women is that they suffer dramatic losses in bone mass normally seen only in osteoporosis of old age. Once estrogen levels drop and the menstrual cycle stops (regardless of cause), bone loss begins.

CHECK YOUR UNDERSTANDING

32. What hormone plays an important role in "letting the brain know" that puberty may occur in girls?
33. What hormone(s) prompt follicle growth? What hormone prompts ovulation?
34. What gonadal hormone exerts positive feedback on the anterior pituitary that results in a burstlike release of LH?

For answers, see Appendix G.

Effects of Estrogens and Progesterone

▶ Discuss the physiological effects of estrogens and progesterone.

With a name meaning "generators of sexual activity," estrogens are analogous to testosterone, the male steroid. As estrogen levels rise during puberty, they (1) promote oogenesis and follicle growth in the ovary and (2) exert anabolic effects on the female reproductive tract **(Table 27.1)**. Consequently, the uterine tubes, uterus, and vagina enlarge and become functional—more ready to support a pregnancy. The uterine tubes and uterus exhibit enhanced motility; the vaginal mucosa thickens; and the external genitalia mature.

Estrogens also support the growth spurt at puberty that makes girls grow much more quickly than boys during the ages of 11 and 12. But this growth is short-lived because rising estrogen levels also cause the epiphyses of long bones to close sooner, and females reach their full height between the ages of 13 and 15 years. In contrast, the aggressive growth of males continues until the age of 15 to 19 years, at which point rising estrogen levels cause epiphyseal closure.

The estrogen-induced secondary sex characteristics of females include (1) growth of the breasts; (2) increased deposit of subcutaneous fat, especially in the hips and breasts; and (3) widening and lightening of the pelvis (adaptations for childbirth).

TABLE 27.1 Summary of Hormonal Effects of Gonadal Estrogens, Progesterone, and Testosterone

SOURCE, STIMULUS, EFFECTS	ESTROGENS	PROGESTERONE	TESTOSTERONE
Major source	Ovary: developing follicles and corpus luteum.	Ovary: mainly the corpus luteum.	Testes: interstitial cells.
Stimulus for release	FSH (and LH).	LH.	LH and declining levels of inhibin produced by the sustentacular cells.
Feedback effects exerted	Both negative and positive feedback exerted on anterior pituitary release of gonadotropins.	Negative feedback exerted on anterior pituitary release of gonadotropins.	Negative feedback suppresses release of LH by the anterior pituitary and release of GnRH by the hypothalamus.
Effects on reproductive organs	Stimulate growth and maturation of reproductive organs and breasts at puberty and maintain their adult size and function. Promote the proliferative phase of the uterine cycle. Stimulate production of watery cervical mucus and activity of fimbriae and uterine tube cilia.	Cooperates with estrogen in stimulating growth of breasts. Promotes the secretory phase of the uterine cycle. Stimulates production of viscous cervical mucus.	Stimulates formation of male reproductive ducts, glands, and external genitalia. Promotes descent of the testes. Stimulates growth and maturation of the internal and external genitalia at puberty; maintains their adult size and function.
	Promote oogenesis and ovulation by stimulating formation of FSH and LH receptors on follicle cells. Stimulate capacitation of sperm in the female reproductive tract.	During pregnancy, quiets the myometrium and acts with estrogen to cause mammary glands to achieve their mature milk-producing state.	Required for normal spermatogenesis via effects promoted by ABP, which keeps its concentration high near spermatogenic cells. Suppresses mammary gland development.
	During pregnancy stimulate growth of the uterus and enlargement of the external genitalia and mammary glands.		
Promotion of secondary sex characteristics and somatic effects	Promote long bone growth and feminization of the skeleton (particularly the pelvis); inhibit bone reabsorption and then stimulate epiphyseal closure. Promote hydration of the skin and female pattern of fat deposit.		Stimulates the growth spurt at puberty; promotes increased skeletal and muscle mass during adolescence. Promotes growth of the larynx and vocal cords and deepening of the voice. Enhances sebum secretion and hair growth, especially on the face, axillae, genital region, and chest.
	During pregnancy act with relaxin (placental hormone) to induce softening and relaxation of the pelvic ligaments and pubic symphysis.		
Metabolic effects	Generally anabolic. Stimulate Na^+ reabsorption by the renal tubules, hence inhibit diuresis. Enhance HDL (and reduce LDL) blood levels (cardiovascular sparing effect).	Promotes diuresis (antiestrogenic effect). Increases body temperature.	Generally anabolic. Stimulates hematopoiesis. Enhances the basal metabolic rate.
Neural effects	Along with DHEA (an androgen produced by the adrenal cortex) are partially responsible for female libido (sex drive).		Responsible for libido in males; promotes aggressiveness.

Estrogens have several metabolic effects, including maintaining low total blood cholesterol levels (and high HDL levels) and facilitating calcium uptake, which helps sustain the density of the skeleton. These metabolic effects are initiated under estrogen's influence during puberty, but they are not true secondary sex characteristics.

Progesterone works with estrogen to establish and then help regulate the uterine cycle and promotes changes in cervical mucus (see Table 27.1). Its other effects are exhibited largely during pregnancy, when it inhibits uterine motility and takes up where estrogen leaves off in preparing the breasts for lactation. Indeed, progesterone is named for these important roles (*pro* = for, *gestation* = pregnancy). However, the source of progesterone and estrogen during most of pregnancy is the placenta, not the ovaries.

Female Sexual Response

▶ Describe the phases of the female sexual response.

The **female sexual response** is similar to that of males in most respects. During sexual excitement, the clitoris, vaginal mucosa, bulbs of the vestibule, and breasts engorge with blood; the nipples erect; and increased activity of the vestibular glands and "sweating" of the vaginal walls lubricates the vestibule and facilitates entry of the penis. These events, though more widespread, are analogous to the *erection* phase in men. Touch and psychological stimuli promote sexual excitement, which is mediated along the same autonomic nerve pathways as in males.

The final phase of the female sexual response, *orgasm*, is not accompanied by ejaculation, but muscle tension increases throughout the body, pulse rate and blood pressure rise, and the uterus begins to contract rhythmically. As in males, orgasm is accompanied by a sensation of intense pleasure and followed by relaxation. Orgasm in females is not followed by a refractory period, so females may experience multiple orgasms during a single sexual experience. A man must achieve orgasm and ejaculate if fertilization is to occur, but female orgasm is not required for conception. Indeed, some women never experience orgasm, yet are perfectly able to conceive.

Although the female libido was formerly believed to be prompted by testosterone, new studies indicate that dehydroepiandrosterone (DHEA), an androgen produced by the adrenal cortex, is in fact the male sex hormone associated with desire or lack of it in females.

CHECK YOUR UNDERSTANDING

35. Which gonadal hormone causes the secondary sex characteristics to appear in a young woman?
36. What gonadal hormone promotes epiphyseal closure in both males and females?

For answers, see Appendix G.

Sexually Transmitted Infections

▶ Indicate the infectious agents and modes of transmission of gonorrhea, syphilis, chlamydia, trichomoniasis, genital warts, and genital herpes.

Sexually transmitted infections (STIs) also called *sexually transmitted diseases* (*STDs*) or *venereal diseases* (*VDs*), are infectious diseases spread through sexual contact. The United States has the highest rates of infection among developed countries. Over 12 million people in the United States, a quarter of them adolescents, get STIs each year.

As a group, STIs are the single most important cause of reproductive disorders. Until recently, the bacterial infections gonorrhea and syphilis were the most common STIs, but now viral diseases have taken center stage. AIDS, the most notorious STI, is caused by HIV, the virus that cripples the immune system. We describe AIDS in Chapter 21 and focus on the other important bacterial and viral STIs here. Condoms are effective in helping to prevent the spread of STIs, and their use is strongly urged, particularly since the advent of AIDS.

Gonorrhea

The causative agent of **gonorrhea** (gon″o-re′ah) is *Neisseria gonorrhoeae*, which invades the mucosae of the reproductive and urinary tracts. These bacteria are spread by contact with genital, anal, and pharyngeal mucosal surfaces. Commonly called "the clap," gonorrhea occurs most frequently in adolescents and young adults.

The most common symptom of gonorrhea in males is *urethritis*, accompanied by painful urination and discharge of pus from the penis (penile "drip"). Symptoms vary in women, ranging from none (about 20% of cases) to abdominal discomfort, vaginal discharge, abnormal uterine bleeding, and occasionally, urethral symptoms similar to those seen in males.

Untreated gonorrhea can cause urethral constriction and inflammation of the entire male duct system. In women, it causes pelvic inflammatory disease and sterility. These consequences declined with the advent in the 1950s of penicillin, tetracycline, and certain other antibiotics. However, strains resistant to those antibiotics are becoming increasingly prevalent. Currently, ceftriaxone is the antibiotic used most often to treat gonorrhea.

Syphilis

Syphilis (sif′ĭ-lis), caused by *Treponema pallidum*, a corkscrew-shaped bacterium, is usually transmitted sexually, but it can be contracted congenitally from an infected mother. Fetuses infected with syphilis are usually stillborn or die shortly after birth.

The bacterium easily penetrates intact mucosae and abraded skin. Within a few hours of exposure, an asymptomatic body-wide infection is in progress. After an incubation period of two to three weeks, a red, painless primary lesion called a *chancre* (shang′ker) appears at the site of bacterial invasion. In males, this is typically the penis, but in females the lesion often goes undetected within the vagina or on the cervix. The chancre ulcerates and becomes crusty, and then it heals spontaneously and disappears within a few weeks.

If syphilis is untreated, its secondary signs appear several weeks later. A pink skin rash all over the body is one of the first symptoms. Fever and joint pain are common. These signs and symptoms disappear spontaneously in three to twelve weeks. Then the disease enters the *latent period* and is detectable only by a blood test. The latent stage may last a person's lifetime (or the bacteria may be killed by the immune system), or it may be followed by the signs of *tertiary syphilis*. Tertiary syphilis is characterized by *gummas* (gum′ahs), destructive lesions of the CNS, blood vessels, bones, and skin. Penicillin is still the treatment of choice for all stages of syphilis.

Chlamydia

Chlamydia (klah-mid′e-ah; *chlamys* = cloak) is a largely undiagnosed, silent epidemic that is currently on the rise in college-age people. It infects perhaps 4–5 million people yearly, making it the most common bacterial sexually transmitted infection in the United States. Chlamydia is responsible for 25–50% of all diagnosed cases of pelvic inflammatory disease (and, consequently, at least a one-in-four chance of ectopic pregnancy). Each year more than 150,000 infants are born to infected mothers. About 20% of men and 30% of women infected with gonorrhea are also infected by *Chlamydia trachomatis*, the causative agent of chlamydia.

Chlamydia is a bacterium with a viruslike dependence on host cells. Its incubation period within the body cells is about one week. Symptoms include urethritis (involving painful, frequent urination and a thick penile discharge); vaginal discharge; abdominal, rectal, or testicular pain; painful intercourse; and irregular menses. In men, it can cause arthritis as well as widespread urogenital tract infection. In women, 80% of whom suffer *no* symptoms from the infection, it is a major cause of sterility. Newborns infected in the birth canal tend to develop conjunctivitis (specifically, *trachoma*, a painful eye infection that leads to corneal scarring if untreated, often the case in third-world countries) and respiratory tract inflammations including pneumonia. The disease can be diagnosed by cell culture techniques and is easily treated with tetracycline.

Trichomoniasis

Trichomoniasis is the most common *curable* STI in sexually active young women in the United States Accounting for about 7.4 million new cases of STI per year, this parasitic infection is easily and inexpensively treated once diagnosed. Trichomoniasis is indicated by a yellow-green vaginal discharge with a strong odor. However, many of its victims exhibit no symptoms.

Genital Warts

Genital warts due to the *human papillomavirus* (*HPV*)—actually a group of about 60 viruses—is the second most common STI in the United States. About 6.2 million new cases of genital warts develop in Americans each year, and it appears that HPV infection increases the risk for cancers in infected body regions. Indeed, the virus is linked to 80% of all cases of invasive cervical cancer. Importantly, most of the strains that cause genital warts do not cause cervical cancer.

Treatment is difficult and controversial, and the warts tend to disappear. Some clinicians prefer to leave the warts untreated unless they become widespread, whereas others recommend their removal by cryosurgery or laser therapy, and/or treatment with alpha interferon.

Genital Herpes

The cause of **genital herpes** is the *human herpes virus type 2*, and these viruses are among the most difficult human pathogens to control. They remain silent for weeks or years and then suddenly flare up, causing a burst of blisterlike lesions. Transmission of the virus is via infectious secretions or via direct skin-to-skin contact when the virus is shedding. The painful lesions that appear on the reproductive organs of infected adults are usually more of a nuisance than a threat. However, congenital herpes infections can cause severe malformations of a fetus.

Most people who have genital herpes do not know it, and it has been estimated that one-quarter to one-half of all adult Americans harbor the type 2 herpes simplex virus. Only about 15% of that population displays signs of infection. The antiviral *acyclovir*, which speeds healing of the lesions and reduces the frequency of flare-ups, is the drug of choice for treatment. Once contracted, genital herpes never leaves. It just goes into periodic remissions.

CHECK YOUR UNDERSTANDING

37. Which pathogen is most associated with cervical cancer?
38. What is the most common bacterial STI in the United States?

For answers, see Appendix G.

Developmental Aspects of the Reproductive System

So far, we have described the reproductive organs as they exist and operate in adults. Now we are ready to look at events that cause us to become reproductive individuals. These events begin long before birth and end, at least in women, in late middle age.

Embryological and Fetal Events

▶ Discuss the determination of genetic sex and prenatal development of male and female structures.

Determination of Genetic Sex

Aristotle believed that the "heat" of lovemaking determined maleness. Not so! Genetic sex is determined at the instant the genes of a sperm combine with those of an ovum, and the determining factor is the **sex chromosomes** each gamete contains. Of the 46 chromosomes in the fertilized egg, two (one pair) are sex chromosomes. The other 44 are called **autosomes**.

Two types of sex chromosomes, quite different in size, exist in humans: the large **X chromosome** and the much smaller **Y chromosome**. The body cells of females have two X chromosomes and are designated XX, and the ovum resulting from normal meiosis in a female always contains an X chromosome. Males have one X chromosome and one Y in each body cell (XY). Approximately half of the sperm produced by normal meiosis in males contain an X and the other half a Y.

If the fertilizing sperm delivers an X chromosome, the fertilized egg and its daughter cells will contain the female (XX) composition, and the embryo will develop ovaries. If the sperm bears the Y, the offspring will be male (XY) and will develop testes. A single gene on the Y chromosome—the *SRY* (for **s**ex-determining **r**egion of the **Y** chromosome) gene—is the master switch that initiates testes development and hence maleness. Thus, the father's gamete determines the genetic sex of the offspring. All subsequent events of sexual differentiation depend on which gonads are formed during embryonic life. However, genes present in autosomes are necessary for a male to become a functional male. Mutation of the *SRY* gene can result in XY females with the male chromosome but female sex characteristics.

HOMEOSTATIC IMBALANCE

When meiosis distributes the sex chromosomes to the gametes improperly, an event called **nondisjunction**, abnormal combinations of sex chromosomes occur in the zygote and cause striking abnormalities in sexual and reproductive system development. For example, females with a single X chromosome (XO), a condition called *Turner's syndrome*, never develop ovaries. Males with no X chromosome (YO) die during embryonic development. Most XXX females have learning disabilities and lower-than-average IQs. Females carrying four or more X chromosomes as a rule are mentally retarded and have underdeveloped ovaries and limited fertility.

Klinefelter's syndrome, which affects one out of 500 live male births, is the most common sex chromosome abnormality. Affected individuals usually have a single Y chromosome, two or more X chromosomes, and are sterile males. Although XXY males are normal (or only slightly below normal) intellectually, the incidence of mental retardation increases as the number of X chromosomes rises. ■

Sexual Differentiation of the Reproductive System

The gonads of both males and females begin their development during week 5 of gestation as masses of mesoderm called the **gonadal ridges** (**Figure 27.21**). The gonadal ridges bulge from the dorsal abdominal wall just medial to the mesonephros (a transient kidney system, see p. 989). The **paramesonephric**, or **Müllerian, ducts** (future female ducts) develop lateral to the **mesonephric (Wolffian) ducts** (future male ducts), and both sets of ducts empty into a common chamber called the *cloaca*. At this stage of development, the embryo is said to be in the **sexually indifferent stage**, because the gonadal ridge tissue can develop into either male or female gonads and both duct systems are present.

Shortly after the gonadal ridges appear, **primordial germ cells** migrate to them from a different region of the embryo, presumably guided by a gradient of chemical signals (chemokines), and seed the developing gonads with stem cells destined to become spermatogonia or oogonia. Once these cells are in residence, the gonadal ridges form testes or ovaries, depending on the genetic makeup of the embryo.

The process begins in week 7 in male embryos. Seminiferous tubules form in the internal part of the gonadal ridges and join the mesonephric duct via the efferent ductules. Further development of the mesonephric duct produces the duct system of the male. The paramesonephric ducts play no part in male development and degenerate unless the tiny testes fail to secrete a hormone variously called *Müllerian inhibitory factor* (*MIF*) or *Müllerian inhibitory substance* (*MIS*). This hormone causes the breakdown of the paramesonephric (Müllerian) ducts, which give rise to the female duct system (oviducts and uterus).

In female embryos the process begins about a week later. The outer, or cortical, part of each immature ovary forms follicles, and the paramesonephric ducts differentiate into the structures of the female duct system. The mesonephric ducts degenerate.

Like the gonads, the external genitalia arise from the same structures in both sexes (**Figure 27.22**). During the indifferent stage, all embryos exhibit a small projection called the **genital tubercle** on their external body surface. The urogenital sinus, which develops from subdivision of the cloaca (future urethra and bladder), lies deep to the tubercle. The **urethral groove**, the external opening of the urogenital sinus, is on the tubercle's inferior surface. The urethral groove is flanked laterally by the **urethral folds** and then the **labioscrotal swellings**.

During week 8, the external genitalia begin to develop rapidly. In males, the genital tubercle enlarges, forming the penis. The urethral folds fuse in the midline, forming the spongy urethra in the penis. Only the tips of the folds remain unfused to form the *urethral orifice* at the tip of the penis. The labioscrotal swellings also fuse in the body midline to form the scrotum.

In females, the genital tubercle gives rise to the clitoris and the urethral groove persists as the vestibule. The unfused urethral folds become the labia minora, and the unfused labioscrotal folds become the labia majora.

Differentiation of accessory structures and the external genitalia as male or female structures depends on the presence or absence of testosterone. When testes are formed, they quickly begin to release testosterone, which continues until four to five days after birth and which causes the development of male accessory ducts and external genitalia. In the absence of testosterone, the female ducts and external genitalia develop.

HOMEOSTATIC IMBALANCE

Any interference with the normal pattern of sex hormone production in the embryo results in abnormalities. For example, if the embryonic testes do not produce testosterone, a genetic male develops the female accessory structures and external genitalia. If the testes fail to produce MIF, both the female and male duct systems form, but the external genitalia are those of the male. On the other hand, if a genetic female is exposed to testosterone (as might happen if the mother has an androgen-producing tumor of her adrenal gland), the embryo has ovaries but develops the male ducts and glands, as well as a penis and an empty scrotum. It appears that the female pattern of reproductive structures has an intrinsic ability to develop (is the default condition), and in the absence of testosterone it proceeds to do so, regardless of the embryo's genetic makeup.

Individuals with external genitalia that do not "match" their gonads are called *pseudohermaphrodites* (soo-do-her-maf′ro-dīts).

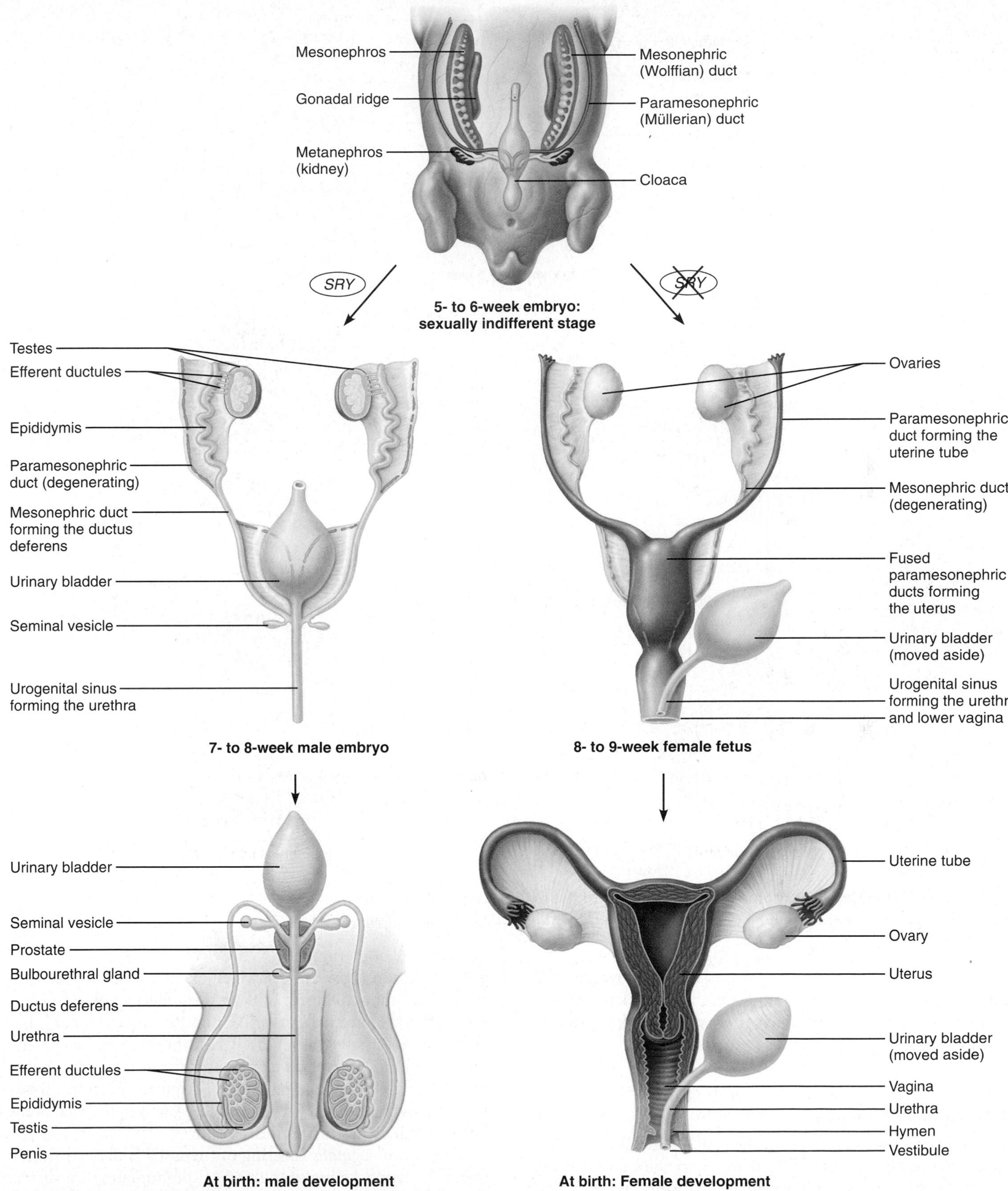
Mesonephros
Gonadal ridge
Metanephros (kidney)
Mesonephric (Wolffian) duct
Paramesonephric (Müllerian) duct
Cloaca
SRY
SRY
5- to 6-week embryo: sexually indifferent stage
Testes
Efferent ductules
Epididymis
Paramesonephric duct (degenerating)
Mesonephric duct forming the ductus deferens
Urinary bladder
Seminal vesicle
Urogenital sinus forming the urethra
7- to 8-week male embryo
Ovaries
Paramesonephric duct forming the uterine tube
Mesonephric duct (degenerating)
Fused paramesonephric ducts forming the uterus
Urinary bladder (moved aside)
Urogenital sinus forming the urethra and lower vagina
8- to 9-week female fetus
Urinary bladder
Seminal vesicle
Prostate
Bulbourethral gland
Ductus deferens
Urethra
Efferent ductules
Epididymis
Testis
Penis
At birth: male development
Uterine tube
Ovary
Uterus
Urinary bladder (moved aside)
Vagina
Urethra
Hymen
Vestibule
At birth: Female development

Figure 27.21 **Development of the internal reproductive organs.**

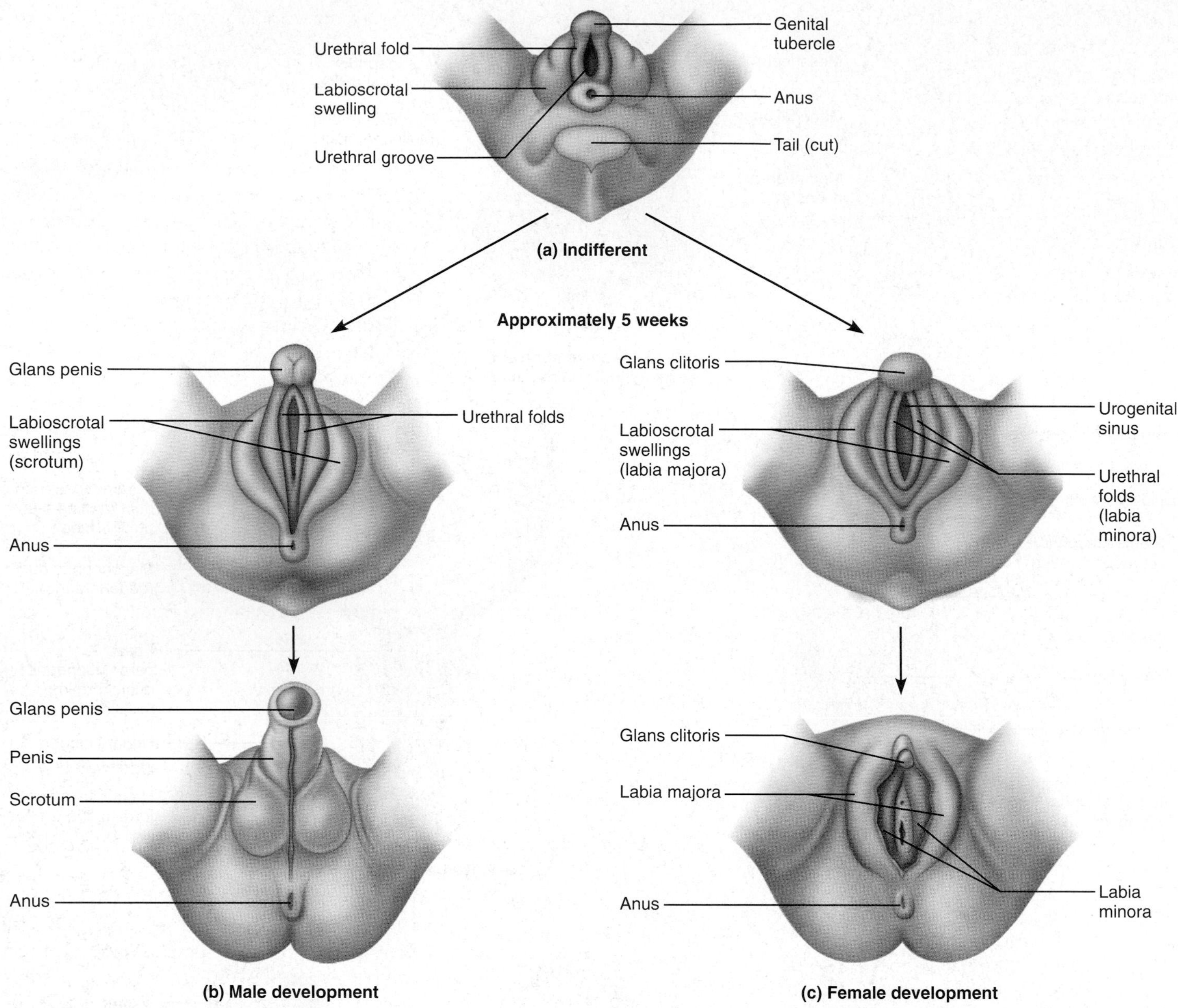

Figure 27.22 Development of homologous structures of the external genitalia in both sexes. The two pictures at the bottom of the figure show the fully developed perineal region.

27

(True *hermaphrodites* are rare and possess both ovarian and testicular tissue.) Many pseudohermaphrodites have sought sex-change operations to match their outer selves (external genitalia) with their inner selves (gonads). ■

Descent of the Gonads

About two months before birth, the testes begin their descent toward the scrotum, dragging their supplying blood vessels and nerves along behind them. They finally exit from the pelvic cavity via the inguinal canals and enter the scrotum. This migration is stimulated by testosterone made by the male fetus's testes, and is guided mechanically by a strong fibrous cord called the **gubernaculum** ("governor"), which extends from the testis to the floor of the scrotal sac. Initially the gubernaculum is a column of soft connective tissue rich in hyaluronic acid, but it becomes increasingly fibrous as it continues to grow. By the seventh month of fetal development, its growth ceases and its inferior part fills the inguinal canal. The gubernaculum's cessation of growth, coupled with the rapid growth of the fetal body, helps to pull the testes into the scrotum.

The *tunica vaginalis* covering of the testis is derived from a fingerlike outpocketing of the parietal peritoneum, the *vaginal process.* The accompanying blood vessels, nerves, and fascial layers form part of the *spermatic cord,* which helps suspend the testis within the scrotum.

Like the testes, the ovaries descend during fetal development, but in this case only to the pelvic brim, where their progress is stopped by the tentlike broad ligament. Each ovary is guided in its descent by a gubernaculum (anchored in the labium majus) that later divides, becoming the ovarian and round ligaments that help support the internal genitalia in the pelvis.

HOMEOSTATIC IMBALANCE

Failure of the testes to make their normal descent leads to *cryptorchidism* (*crypt* = hidden, concealed; *orchi* = testicle). Because this condition causes sterility and increases the risk of testicular cancer, surgery is usually performed during early childhood to rectify this problem. ■

CHECK YOUR UNDERSTANDING

39. If the fertilized egg contains X and Y sex chromosomes, a baby girl will develop, right?

40. What is the sexually indifferent stage of development?

41. What structure guides the descent of the testis into the scrotum?

For answers, see Appendix G.

Puberty

▶ Describe the significant events of puberty and menopause.

FSH and LH levels, elevated at birth, fall to low levels within a few months and remain low throughout the prepubertal years. Between the ages of 10 and 15 years, a host of interacting hormones bring on the changes of puberty. **Puberty** is the period of life when the reproductive organs grow to adult size and become functional. As puberty nears, these changes occur in response to rising levels of gonadal hormones (testosterone in males and estrogen in females). Secondary sex characteristics and regulatory events of puberty were described earlier. But it is important to remember that puberty represents the earliest time that reproduction is possible.

The events of puberty occur in the same sequence in all individuals, but the age at which they occur varies widely. In males, secretion of adrenal androgens, particularly dehydroepiandrosterone (DHEA), begins to rise several years before the testosterone surge of puberty and initiates facial, pubic, and axillary hair growth and other pubertal events. The major event that signals puberty's onset in males is enlargement of the testes and scrotum between the ages of 8 and 14. Growth of the penis goes on over the next two years, and sexual maturation is evidenced by the presence of mature sperm in the semen. In the meantime, the young man has unexpected erections and occasional nocturnal emissions ("wet dreams") as his hormones surge and the hormonal control axis struggles to achieve a normal balance.

The first sign of puberty in females is budding breasts, apparent between the ages of 8 and 13 years, followed by the appearance of axillary and pubic hair. Menarche usually occurs about two years later. Dependable ovulation and fertility await the maturation of the hormonal controls, which takes nearly two more years.

Menopause

Most women reach the peak of their reproductive abilities in their late 20s. After that, ovarian function declines gradually, presumably because the ovaries become less and less responsive to gonadotropin signals. At the age of 30, there are still some 100,000 oocytes in the ovaries but the quality (hence fertility) has begun to decline. By the age of 50, there are probably 3 eggs left (the pantry is nearly bare).

As estrogen production declines, many ovarian cycles become anovulatory, while in others 2 to 4 oocytes per month are ovulated, a sign of declining control. These increasingly frequent multiple ovulations explain why twins and triplets are more common in women who have deferred childbearing until their late 30s. In the perimenopausal period, menstrual periods become erratic and increasingly shorter. Eventually, ovulation and menstruation cease entirely. This normally occurs between the ages of 46 and 54 years, an event called *menopause.* **Menopause** is considered to have occurred when a whole year has passed without menstruation.

Although ovarian estrogen production continues for a while after menopause, the ovaries finally stop functioning as endocrine organs. Without sufficient estrogen the reproductive organs and breasts begin to atrophy, the vagina becomes dry, and vaginal infections become increasingly common. Other sequels due to lack of estrogen include irritability and depression (in some); intense vasodilation of the skin's blood vessels, which causes uncomfortable, sweat-drenching "hot flashes"; gradual thinning of the skin; and loss of bone mass.

Slowly rising total blood cholesterol levels and falling HDL levels place postmenopausal women at risk for cardiovascular disorders. At one time physicians prescribed low-dose estrogen-progesterone preparations to help women through this often difficult period and to prevent the skeletal and cardiovascular complications. These seemed like great bonuses and until July 2002, some 14 million American women were taking some form of estrogen-containing hormone replacement therapy (HRT). Then, on July 9, the Women's Health Initiative (WHI) abruptly ended a clinical trial of 16,000 postmenopausal women, reporting that in those taking a popular progesterone-estrogen hormone combination there was an increase of 51% in heart disease, 24% in invasive breast cancer, 31% in stroke, and a doubling of the risk of dementia compared to those taking placebos. The backlash of this information (aired in the popular press) is still spreading through physicians' offices and research labs and is restricting funding and the ability to find new volunteers for future studies, and has dampened enthusiasm for HRT in both the medical community and postmenopausal women. However, newer data suggest that the smallest dose of HRT for the shortest time is OK to reduce symptoms in women that do not have existing breast cancer or mutated *BRCA* gene(s).

There is no equivalent of menopause in males, and healthy men are able to father offspring well into their eighth decade

(Text continues on p. 1066.)

of life thanks to their small but enduring population of stem cells (spermatogonia). However, aging men do exhibit a steady decline in testosterone secretion and a longer latent period after orgasm, a condition sometimes called *andropause*, and testosterone replacement therapy is currently being prescribed for more older men. Additionally, there is a noticeable difference in sperm motility with aging. Sperm of a young man can make it up the uterine tubes in 20–50 minutes, whereas those of a 75-year-old take 2½ days for the same trip.

CHECK YOUR UNDERSTANDING

42. What are the early signs of puberty's onset in boys?
43. What is the definition of menopause?

For answers, see Appendix G.

The reproductive system is unique among organ systems in at least two ways: (1) It is nonfunctional during the first 10–15 years of life, and (2) it is capable of interacting with the complementary system of another person—indeed, it *must* do so to carry out its biological function of pregnancy and birth. To be sure, having a baby is not always what the interacting partners have in mind, and we humans have devised a variety of techniques for preventing this outcome (see *A Closer Look* on pp. 1094–1095).

The major goal of the reproductive system is ensuring the healthy function of its own organs so that conditions are optimal for producing offspring. However, as illustrated in *Making Connections*, gonadal hormones do influence other body organs, and the reproductive system depends on other body systems for oxygen and nutrients and to carry away and dispose of its wastes.

Now that we know how the reproductive system functions to prepare itself for childbearing, we are ready to consider the events of pregnancy and prenatal development of a new living being, the topics of Chapter 28.

RELATED CLINICAL TERMS

Dysmenorrhea (dis″men-ŏ-re′ah; *dys* = bad, *meno* = menses, a month) Painful menstruation; may reflect abnormally high prostaglandin activity during menses.

Endometrial cancer (en″do-me′tre-al; *endo* = within, *metrio* = the uterus, womb) Cancer which arises from the uterine endometrium (usually from uterine glands); the fourth most common cancer of women. Most important sign is vaginal bleeding, which allows early detection. Risk factors include obesity and HRT.

Endometriosis (en″do-me″tre-o′sis) An inflammatory condition in which endometrial tissue occurs and grows atypically in the pelvic cavity. Characterized by abnormal uterine or rectal bleeding, dysmenorrhea, and pelvic pain. May cause sterility.

Gynecology (gi″nĕ-kol′o-je; *gyneco* = woman, *ology* = study of) Specialized branch of medicine that deals with the diagnosis and treatment of female reproductive system disorders.

Gynecomastia (gi″nĕ-ko-mas′te-ah) Development of breast tissue in the male; a consequence of adrenal cortex hypersecretion of estrogens, certain drugs (cimetidine, spironolactone, and some chemotherapeutic agents), and marijuana use.

Hysterectomy (his″tĕ-rek′to-me; *hyster* = uterus, *ectomy* = cut out) Surgical removal of the uterus.

Inguinal hernia Protrusion of part of the intestine into the scrotum or through a separation in the abdominal muscles in the groin region. Since the inguinal canals represent weak points in the abdominal wall, inguinal hernia may be caused by heavy lifting or other activities that increase intra-abdominal pressure.

Laparoscopy (lap″ah-ros′ko-pe; *lapar* = the flank, *scopy* = observation) Examination of the abdominopelvic cavity with a laparoscope, a viewing device at the end of a thin tube inserted through the anterior abdominal wall. Laparoscopy is often used to assess the condition of the pelvic reproductive organs of females.

Oophorectomy (o″of-o-rek′to-me; *oophor* = ovary) Surgical removal of the ovary.

Orchitis (or-ki′tis; *orcho* = testis) Inflammation of the testes, sometimes caused by the mumps virus.

Ovarian cancer Malignancy that typically arises from the cells in the germinal epithelial covering of the ovary. The fifth most common reproductive system cancer, its incidence increases with age. Called the "disease that whispers" because early symptoms are nondescript and easily mistaken for other disorders (back pain, abdominal discomfort, nausea, bloating, and flatulence). Diagnosis may involve palpating a mass during a physical exam, visualizing it with an ultrasound probe, or conducting blood tests for a protein marker for ovarian cancer (CA-125). However, medical assessment is often delayed until after metastasis has occurred; five-year survival rate is 90% if the condition is diagnosed before metastasis.

Ovarian cysts The most common disorders of the ovary; some are tumors. Types include (1) *simple follicle retention cysts* in which single or clustered follicles become enlarged with a clear fluid; (2) *dermoid cysts*, which are filled with a thick yellow fluid and contain partially developed hair, teeth, bone, etc.; and (3) *chocolate cysts* filled with dark gelatinous material, which are the result of endometriosis of the ovary. None of these is malignant, but the latter two may become so.

Polycystic ovary syndrome (PCOS) The most common endocrinopathy in women and the most common cause of anovulatory infertility. Affects 5–10% of women; characterized by signs of androgen excess, increased cardiovascular risk (evidenced by high blood pressure, decreased HDL cholesterol levels, and high triglycerides); and linked to extreme obesity and some degree of insulin resistance. Treated with insulin-sensitizing drugs.

Salpingitis (sal″pin-ji′tis; *salpingo* = uterine tube) Inflammation of the uterine tubes.

CHAPTER SUMMARY

1. The function of the reproductive system is to produce offspring. The gonads produce gametes (sperm or ova) and sex hormones. All other reproductive organs are accessory organs.

Anatomy of the Male Reproductive System (pp. 1025–1031)

The Scrotum (pp. 1025–1026)

1. The scrotum contains the testes. It provides a temperature slightly lower than that of the body, as required for viable sperm production.

The Testes (pp. 1026–1028)

2. Each testis is covered externally by a tunica albuginea that extends internally to divide the testis into many lobules. Each lobule contains sperm-producing seminiferous tubules and interstitial cells that produce androgens.

The Penis (p. 1028)

3. The penis, the male copulatory organ, is largely erectile tissue (corpus spongiosum and corpora cavernosa). Engorgement of the erectile tissue with blood causes the penis to become rigid, an event called erection.
4. The male perineum is the region encompassed by the pubic symphysis, ischial tuberosities, and coccyx.

The Male Duct System (pp. 1028–1030)

5. The epididymis hugs the external surface of the testis and serves as a site for sperm maturation and storage.
6. The ductus (vas) deferens, extending from the epididymis to the ejaculatory duct, propels sperm into the urethra by peristalsis during ejaculation. Its terminus fuses with the duct of the seminal vesicle, forming the ejaculatory duct, which empties into the urethra within the prostate.
7. The urethra extends from the urinary bladder to the tip of the penis. It conducts semen and urine to the body exterior.

Accessory Glands (pp. 1030–1031)

8. The accessory glands produce the bulk of the semen, which contains fructose from the seminal vesicles, an activating fluid from the prostate, and mucus from the bulbourethral glands.

Semen (p. 1031)

9. Semen is an alkaline fluid that dilutes and transports sperm. Important chemicals in semen are nutrients, prostaglandins, and antibiotic chemicals. An ejaculation contains 2–5 ml of semen, with 20–150 million sperm/ml in normal adult males.

Physiology of the Male Reproductive System (pp. 1031–1040)

Male Sexual Response (pp. 1031–1032)

1. Erection is controlled by parasympathetic reflexes.
2. Ejaculation is expulsion of semen from the male duct system, promoted by the sympathetic nervous system. Ejaculation is part of male orgasm, which also includes pleasurable sensations and increased pulse and blood pressure.

Spermatogenesis (pp. 1032–1038)

3. Spermatogenesis, the production of male gametes in the seminiferous tubules, begins at puberty.
4. Meiosis, the basis of gamete production, consists of two consecutive nuclear divisions without DNA replication in between. Meiosis reduces the chromosomal number by half and introduces genetic variability. Events unique to meiosis include synapsis and crossover of homologous chromosomes.
5. Spermatogonia divide by mitosis to maintain the germ cell line. Some of their progeny become primary spermatocytes, which undergo meiosis I to produce secondary spermatocytes. Secondary spermatocytes undergo meiosis II, each producing a total of four haploid (n) spermatids.
6. Spermatids are converted to functional sperm by spermiogenesis, during which superfluous cytoplasm is stripped away and an acrosome and a flagellum (tail) are produced.
7. Sustentacular cells form the blood-testis barrier, nourish spermatogenic cells, move them toward the lumen of the tubules, and secrete fluid for sperm transport.

Hormonal Regulation of Male Reproductive Function (pp. 1038–1040)

8. GnRH, produced by the hypothalamus, stimulates the anterior pituitary gland to release FSH and LH. FSH causes sustentacular cells to produce androgen-binding protein (ABP). LH stimulates interstitial cells to release testosterone, which binds to ABP, stimulating spermatogenesis. Testosterone and inhibin (produced by sustentacular cells) feed back to inhibit the hypothalamus and anterior pituitary.
9. Maturation of hormonal controls occurs during puberty and takes about three years.
10. Testosterone stimulates maturation of the male reproductive organs and triggers the development of the secondary sex characteristics of the male. It exerts anabolic effects on the skeleton and skeletal muscles, stimulates spermatogenesis, and is responsible for sex drive.

Anatomy of the Female Reproductive System (pp. 1040–1049)

1. The female reproductive system produces gametes and sex hormones and houses a developing infant until birth.

The Ovaries (pp. 1041–1042)

2. The ovaries flank the uterus laterally and are held in position by the ovarian and suspensory ligaments and mesovaria.
3. Within each ovary are oocyte-containing follicles at different stages of development and corpora lutea.

The Female Duct System (pp. 1042–1046)

4. The uterine tube, supported by the mesosalpinx, extends from near the ovary to the uterus. Its fimbriae and ciliated distal end creates currents that help move an ovulated oocyte into the uterine tube. Cilia of the uterine tube mucosa help and peristalsis propel the oocyte toward the uterus.
5. The uterus has fundus, body, and cervical regions. It is supported by the broad, lateral cervical, uterosacral, and round ligaments.
6. The uterine wall is composed of the outer perimetrium, the myometrium, and the inner endometrium. The endometrium consists of a functional layer (stratum functionalis), which sloughs off periodically unless an embryo has implanted, and an underlying basal layer (stratum basalis), which rebuilds the functional layer.

7. The vagina extends from the uterus to the exterior. It is the copulatory organ and allows passage of the menstrual flow and a baby.

The External Genitalia and Female Perineum (pp. 1046–1047)

8. The female external genitalia (vulva) include the mons pubis, labia majora and minora, clitoris, and the urethral and vaginal orifices. The labia majora house the mucus-secreting greater vestibular glands.

The Mammary Glands (pp. 1047–1049)

9. The mammary glands lie over the pectoral muscles of the chest and are surrounded by adipose and fibrous connective tissue. Each mammary gland consists of many lobules, which contain milk-producing alveoli.

Physiology of the Female Reproductive System (pp. 1049–1058)

Oogenesis (p. 1049)

1. Oogenesis, the production of eggs, begins in the fetus. Oogonia, the diploid stem cells of female gametes, are converted to primary oocytes before birth. The infant female's ovaries contain about 2 million primary oocytes arrested in prophase of meiosis I.
2. At puberty, meiosis resumes. Each month, one primary oocyte completes meiosis I, producing a large secondary oocyte and a tiny first polar body. Meiosis II of the secondary oocyte produces a functional ovum and a second polar body, but does not occur in humans unless the secondary oocyte is penetrated by a sperm.
3. The ovum contains most of the primary oocyte's cytoplasm. The polar bodies are nonfunctional and degenerate.

The Ovarian Cycle (pp. 1049–1052)

4. During the follicular phase (days 1–14), several primary follicles begin to mature. The follicle cells proliferate and a connective tissue capsule (theca) is formed around the maturing follicle. The theca produces androgens that the follicle cells convert to estrogens. Generally, only one follicle per month completes the maturation process and protrudes from the ovarian surface. Late in this phase, the oocyte in the dominant follicle completes meiosis I. Ovulation occurs about day 14, releasing the secondary oocyte into the peritoneal cavity. Other developing follicles deteriorate.
5. In the luteal phase (days 15–28), the ruptured follicle is converted to a corpus luteum, which produces progesterone and estrogen for the remainder of the cycle. If fertilization does not occur, the corpus luteum degenerates after about 10 days.

Hormonal Regulation of the Ovarian Cycle (pp. 1052–1054)

6. Beginning at puberty, the hormones of the hypothalamus, anterior pituitary, and ovaries interact to establish and regulate the ovarian cycle. Establishment of the mature cyclic pattern, indicated by menarche, takes about four years. Leptin serves a permissive role in puberty's onset, stimulating the hypothalamus when adipose tissue is sufficient for the energy requirements of reproduction.
7. The hormonal events of each ovarian cycle are as follows: (1) GnRH stimulates the anterior pituitary to release FSH and LH, which stimulate follicle maturation and estrogen production. (2) When blood estrogen reaches a certain level, positive feedback exerted on the hypothalamic-pituitary axis causes a sudden release of LH that stimulates the primary oocyte to continue meiosis and triggers ovulation. LH then causes conversion of the ruptured follicle to a corpus luteum and stimulates its secretory activity. (3) Rising levels of progesterone and estrogen inhibit the hypothalamic-pituitary-gonadal axis, the corpus luteum deteriorates, ovarian hormones drop to their lowest levels, and the cycle begins anew.

The Uterine (Menstrual) Cycle (pp. 1054–1056)

8. Varying levels of ovarian hormones in the blood trigger events of the uterine cycle.
9. During the menstrual phase of the uterine cycle (days 1–5), the functional layer sloughs off in menses. During the proliferative phase (days 6–14), rising estrogen levels stimulate its regeneration, making the uterus receptive to implantation about one week after ovulation. During the secretory phase (days 15–28), the uterine glands secrete glycogen, and endometrial vascularity increases further.
10. Falling levels of ovarian hormones during the last few days of the ovarian cycle cause the spiral arteries to become spastic and cut off the blood supply of the functional layer, and the uterine cycle begins again with menstruation.

Effects of Estrogens and Progesterone (pp. 1056–1057)

11. Estrogen promotes oogenesis. At puberty, it stimulates the growth of the reproductive organs and the growth spurt and promotes the appearance of the secondary sex characteristics.
12. Progesterone cooperates with estrogen in breast maturation and regulation of the uterine cycle.

Female Sexual Response (p. 1058)

13. The female sexual response is similar to that of males. Orgasm in females is not accompanied by ejaculation and is not necessary for conception.

Sexually Transmitted Infections (pp. 1058–1059)

1. Sexually transmitted infections (STIs) are infectious diseases spread via sexual contact. Gonorrhea, syphilis, and chlamydia, are bacterial diseases; if untreated, they can cause sterility. Syphilis has broader consequences than most other sexually transmitted bacterial diseases since it can infect organs throughout the body. Trichomoniasis is a parasitic infection. Genital herpes and genital warts are viral infections; genital warts are implicated in cervical cancer. AIDS, a condition of immune suppression, can also be transmitted sexually.

Developmental Aspects of the Reproductive System (pp. 1059–1063, 1066)

Embryological and Fetal Events (pp. 1059–1063)

1. Genetic sex is determined by the sex chromosomes: an X from the mother, an X or a Y from the father. If the fertilized egg contains XX, it is a female and develops ovaries; if it contains XY, it is a male and develops testes.
2. Gonads of both sexes arise from the mesodermal gonadal ridges. The mesonephric ducts produce the male accessory ducts and glands. The paramesonephric ducts produce the female duct system.
3. The external genitalia arise from the genital tubercle and associated structures. The development of male accessory structures and external genitalia depends on the presence of testosterone produced by the embryonic testes. In its absence, female structures develop.
4. The testes form in the abdominal cavity and descend into the scrotum.

Puberty (p. 1063)

5. Puberty is the interval when reproductive organs mature and become functional. It begins in males with penile and scrotal growth and in females with breast development.

Menopause (pp. 1063, 1066)

6. During menopause, ovarian function declines, and ovulation and menstruation cease. Hot flashes and mood changes may occur. Postmenopausal events include atrophy of the reproductive organs, bone mass loss, and increasing risk for cardiovascular disease. The milder signs and symptoms of gonadal steroid hormone deficit in men are called andropause.

REVIEW QUESTIONS

Multiple Choice/Matching

(Some questions have more than one correct answer. Select the best answer or answers from the choices given.)

1. The structures that draw an ovulated oocyte into the female duct system are (a) cilia, (b) fimbriae, (c) microvilli, (d) stereocilia.
2. The usual site of embryo implantation is (a) the uterine tube, (b) the peritoneal cavity, (c) the vagina, (d) the uterus.
3. The male homologue of the female clitoris is (a) the penis, (b) the scrotum, (c) the penile urethra, (d) the testis.
4. Which of the following is correct relative to female anatomy? (a) The vaginal orifice is the most dorsal of the three openings in the perineum, (b) the urethra is between the vaginal orifice and the anus, (c) the anus is between the vaginal orifice and the urethra, (d) the urethra is the more ventral of the two orifices in the vulva.
5. Secondary sex characteristics are (a) present in the embryo, (b) a result of male or female sex hormones increasing in amount at puberty, (c) the testis in the male and the ovary in the female, (d) not subject to withdrawal once established.
6. Which of the following produces the male sex hormones? (a) seminal vesicles, (b) corpus luteum, (c) developing follicles of the testes, (d) interstitial cells.
7. Which will occur as a result of nondescent of the testes? (a) Male sex hormones will not be circulated in the body, (b) sperm will have no means of exit from the body, (c) inadequate blood supply will retard the development of the testes, (d) viable sperm will not be produced.
8. The normal diploid number of human chromosomes is (a) 48, (b) 47, (c) 46, (d) 23, (e) 24.
9. Relative to differences between mitosis and meiosis, choose the statements that apply *only* to events of meiosis. (a) tetrads present, (b) produces two daughter cells, (c) produces four daughter cells, (d) occurs throughout life, (e) reduces the chromosomal number by half, (f) synapsis and crossover of homologues occur.
10. Match the key choices with the descriptive phrases below.

Key:

(a) androgen-binding protein	(b) estrogens
(c) FSH	(d) GnRH
(e) inhibin	(f) LH
(g) progesterone	(h) testosterone

_____ (1) Hormones that directly regulate the ovarian cycle

_____ (2) Chemicals in males that inhibit the pituitary-testicular axis

_____ (3) Hormone that makes the cervical mucus viscous

_____ (4) Potentiates the activity of testosterone on spermatogenic cells

_____ (5) In females, exerts feedback inhibition on the hypothalamus and anterior pituitary

_____ (6) Stimulates the secretion of testosterone

11. The menstrual cycle can be divided into three continuous phases. Starting from the first day of the cycle, their consecutive order is (a) menstrual, proliferative, secretory, (b) menstrual, secretory, proliferative, (c) secretory, menstrual, proliferative, (d) proliferative, menstrual, secretory, (e) secretory, proliferative, menstrual.
12. Spermatozoa are to seminiferous tubules as oocytes are to (a) fimbriae, (b) corpus albicans, (c) ovarian follicles, (d) corpora lutea.
13. Which of the following does not add a secretion that makes a major contribution to semen? (a) prostate, (b) bulbourethral glands, (c) testes, (d) vas deferens.
14. The corpus luteum is formed at the site of (a) fertilization, (b) ovulation, (c) menstruation, (d) implantation.
15. The sex of a child is determined by (a) the sex chromosome contained in the sperm, (b) the sex chromosome contained in the oocyte, (c) the number of sperm fertilizing the oocyte, (d) the position of the fetus in the uterus.
16. FSH is to estrogen as estrogen is to (a) progesterone, (b) LH, (c) FSH, (d) testosterone.
17. A drug that "reminds the pituitary" to produce gonadotropins might be useful as (a) a contraceptive, (b) a diuretic, (c) a fertility drug, (d) an abortion stimulant.

Short Answer Essay Questions

18. Why is the term *urogenital system* more applicable to males than to females?
19. Describe the major structural (and functional) regions of a sperm.
20. Oogenesis in the female results in one functional gamete—the egg, or ovum. What other cells are produced? What is the significance of this rather wasteful type of gamete production—that is, production of a single functional gamete instead of four, as seen in males?
21. Describe the events and possible consequences of menopause.
22. Define menarche. What does it indicate?
23. Trace the pathway of a sperm from the male testes to the uterine tube of a female.
24. In menstruation, the stratum functionalis is shed from the endometrium. Explain the hormonal and physical factors responsible for this shedding. (Hint: See Figure 27.20.)
25. Both the epithelium of the vagina and the cervical glands of the uterus help prevent the invasion and spread of vaginal pathogens. Explain how each of these mechanisms works.
26. Some anatomy students were saying that the bulbourethral glands of males act like city workers who come around and clear parked cars from the street before a parade. What did they mean by this analogy?
27. A man swam in a cold lake for an hour and then noticed that his scrotum was shrunken and wrinkled. His first thought was that he had lost his testicles. What had really happened?

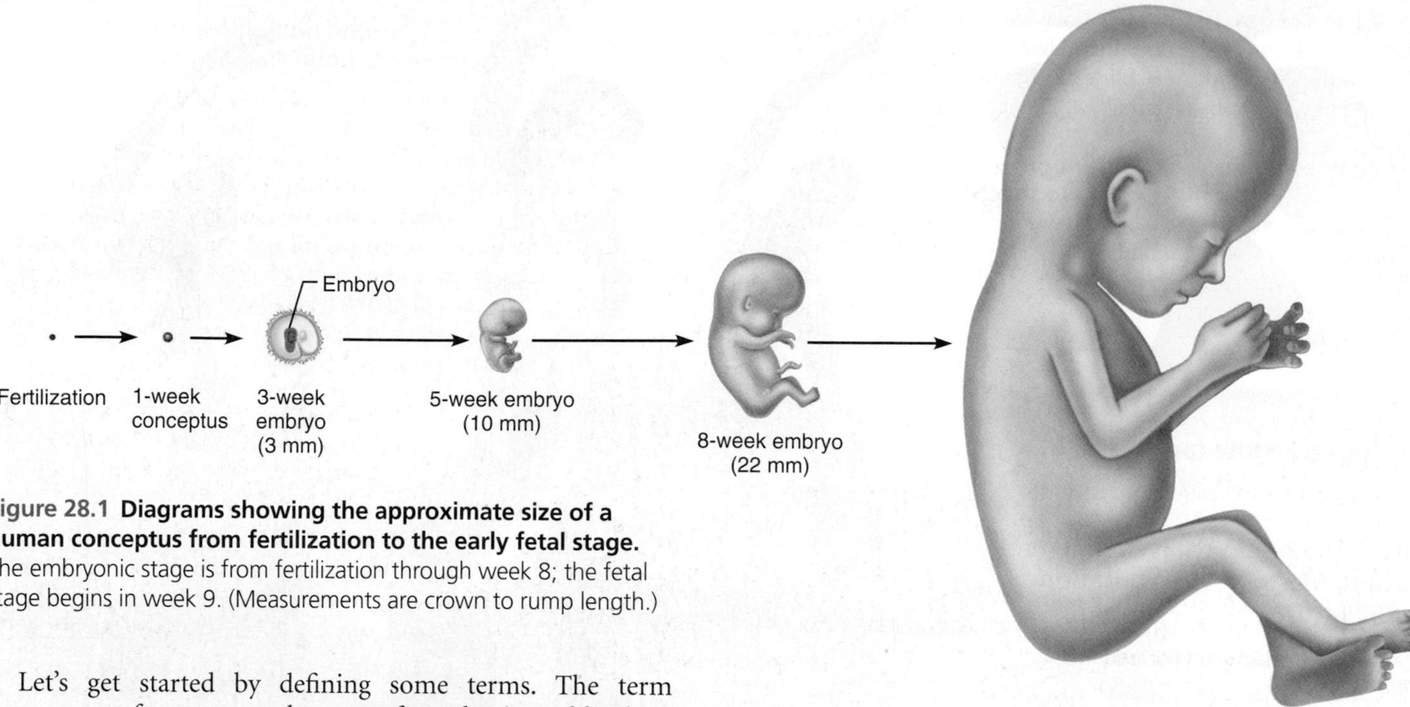

Figure 28.1 Diagrams showing the approximate size of a human conceptus from fertilization to the early fetal stage. The embryonic stage is from fertilization through week 8; the fetal stage begins in week 9. (Measurements are crown to rump length.)

Let's get started by defining some terms. The term **pregnancy** refers to events that occur from the time of fertilization (conception) until the infant is born. The pregnant woman's developing offspring is called the **conceptus** (kon-sep′tus; "that which is conceived"). Development occurs during the **gestation period** (*gestare* = to carry), which extends by convention from the last menstrual period (a date the woman is likely to remember) until birth, approximately 280 days. So, at the moment of fertilization, the mother is officially (but illogically) two weeks pregnant!

From fertilization through week 8, the *embryonic period*, the conceptus is called an **embryo**, and from week 9 through birth, the *fetal period*, the conceptus is called a **fetus** ("the young in the womb"). At birth, it is an infant. Figure 28.1 shows the changing size and shape of the conceptus as it progresses from fertilization to the early fetal stage.

From Egg to Zygote

- Describe the importance of sperm capacitation.
- Explain the mechanism of the slow block to polyspermy.
- Define fertilization.

28

Before fertilization can occur, sperm must reach the ovulated secondary oocyte. The oocyte is viable for 12 to 24 hours after it is cast out of the ovary. The chance of pregnancy drops to almost zero the next day. Most sperm retain their fertilizing power for 24 to 48 hours after ejaculation. Consequently, for successful fertilization to occur, coitus must occur no more than two days before ovulation and no later than 24 hours after. At this point the oocyte is approximately one-third of the way down the length of the uterine tube.

Accomplishing Fertilization

Fertilization occurs when a sperm's chromosomes combine with those of an egg (actually a secondary oocyte) to form a fertilized egg, or **zygote** (zi′gōt; "yoked together"), the first cell of the new individual. Let's look at the events leading to fertilization.

Sperm Transport and Capacitation

During copulation, a man expels millions of sperm with considerable force into his partner's vaginal canal. Despite this "head start," most sperm don't reach the oocyte, even though it is only about 12 cm (5 inches) away. Millions of sperm leak from the vagina almost immediately after being deposited there. Of those remaining, millions more are destroyed by the vagina's acidic environment. Millions more fail to make it through the cervix, unless the thick "curtain" of cervical mucus has been made fluid by estrogens.

Sperm that do reach the uterus, propelled by their whiplike tail movements, are then subjected to forceful uterine contractions that act in a washing machine–like manner to disperse them throughout the uterine cavity, where thousands more are destroyed by resident phagocytes. Only a few thousand (and sometimes fewer than 100) sperm, out of the millions in the male ejaculate, are conducted by reverse peristalsis into the uterine tube, where the oocyte may be moving leisurely toward the uterus.

These difficulties aside, there is still another hurdle to overcome. Sperm freshly deposited in the vagina are incapable of penetrating an oocyte. They must first be **capacitated** over the next 8 to 10 hours. Specifically, their motility must be enhanced and their membranes must become fragile so that the hydrolytic enzymes in their acrosomes can be released. As sperm swim through the cervical mucus, uterus, and uterine tubes, secretions of the female tract cause some of their membrane proteins to be removed, and the cholesterol that keeps their acrosomal membranes "tough" and stable is depleted. Even though the sperm may reach the oocyte within a few minutes, they must "wait around" (so to speak) for capacitation to occur.

This elaborate mechanism prevents the spilling of acrosomal enzymes. But consider the alternative. Fragile acrosomal membranes could rupture prematurely in the male reproductive tract, causing some degree of autolysis (self-digestion) of the male reproductive organs.

How do sperm navigate to find a released oocyte in the uterine tube? This question is an area of active research. It now appears that they "sniff" their way to the oocyte. Sperm bear proteins called *olfactory receptors* that respond to chemical stimuli. It is presumed that the oocyte or its surrounding cells release signaling molecules that direct the sperm.

Acrosomal Reaction and Sperm Penetration

The ovulated oocyte is encapsulated by the corona radiata and by the deeper zona pellucida, a transparent layer of glycoprotein-rich extracellular matrix secreted by the oocyte. Both must be breached before the oocyte can be penetrated. Once a sperm gets to the immediate vicinity of the oocyte, it weaves its way through the cells of the corona radiata. This journey is assisted by a cell-surface hyaluronidase on the sperm that digests the intercellular cement between the granulosa cells in the immediate area, causing them to fall away from the oocyte (**Figure 28.2**, ①).

After breaching the corona, the sperm head binds to the ZP3 glycoprotein of the zona pellucida, which functions as a sperm receptor. This binding leads to a rise in Ca^{2+} inside the sperm that triggers the acrosomal (ak′ro-sōm-al) reaction (Figure 28.2, ②). The **acrosomal reaction** involves the breakdown of the plasma membrane and the acrosomal membrane, and release of acrosomal enzymes (hyaluronidase, acrosin, proteases, and others) that digest holes through the zona pellucida (Figure 28.2, ③). Hundreds of acrosomes must undergo exocytosis to digest holes in the zona pellucida. This is one case that does not bear out the adage, "The early bird catches the worm." A sperm that comes along later, after hundreds of sperm have undergone acrosomal reactions to expose the oocyte membrane, is in the best position to be *the* fertilizing sperm.

Once a path is cleared, the sperm's whiplike tail gyrates, forcing the sperm head toward the oocyte membrane. At the same time, actin filaments in the sperm head form an *acrosomal process* that quickly finds and binds to the oocyte's sperm-binding membrane receptors (Figure 28.2, ④). This binding event has two consequences. (1) It causes the oocyte and sperm membranes to fuse, and then (2) the contents of the sperm enter the oocyte cytoplasm (Figure 28.2, ⑤). The gametes fuse together with such perfect contact that the contents of both cells are combined within a single membrane—all without spilling a drop.

Blocks to Polyspermy

Polyspermy (entry of several sperm into an egg) occurs in some animals, but in humans only one sperm is allowed to penetrate the oocyte, ensuring **monospermy**, the one-sperm-per-oocyte condition. Once the sperm head has entered the oocyte, waves of Ca^{2+} are released by the oocyte's endoplasmic reticulum into its cytoplasm, which activates the oocyte to prepare for cell division. These calcium surges also cause the **cortical reaction** (Figure 28.2, ⑥), in which granules located just inside the plasma membrane spill their enzymes into the extracellular space beneath the zona pellucida. These enzymes, called *zonal inhibiting proteins* (*ZIPs*), destroy the sperm receptors, preventing further sperm entry.

Additionally, the spilled material binds water, and as it swells and hardens, it detaches all sperm still bound to receptors on the oocyte membrane, accomplishing the so-called *slow block to polyspermy*. In the rare cases of polyspermy that do occur, the embryos contain too much genetic material and die.

Completion of Meiosis II and Fertilization

As the sperm's cytoplasmic contents enter the oocyte, it loses its plasma membrane. The centrosome from its midpiece elaborates microtubules which the sperm uses to locomote its DNA-rich nucleus toward the oocyte nucleus. On the way, its nucleus swells to about five times its normal size to form the **male pronucleus** (pro-nu′kle-us; *pro* = before). Meanwhile the secondary oocyte, stimulated into activity by the calcium surges, completes meiosis II, forming the ovum nucleus and the second polar body (**Figure 28.3**, ① and ②).

This accomplished, the ovum nucleus swells, becoming the **female pronucleus**, and the two pronuclei approach each other. As a mitotic spindle develops between them (Figure 28.3, ③), the pronuclei membranes rupture, releasing their chromosomes together into the immediate vicinity of the newly formed spindle.

The true moment of fertilization occurs as the maternal and paternal chromosomes combine and produce the diploid *zygote*, or fertilized egg (Figure 28.3, ④). Some sources define the term fertilization simply as the act of oocyte penetration by the sperm. However, unless the chromosomes in the male and female pronuclei are actually joined, the zygote is never formed in humans. Almost as soon as the male and female pronuclei come together, their chromosomes replicate. The zygote, the first cell of a new individual, is now ready to undergo the first mitotic division of the conceptus. 28

CHECK YOUR UNDERSTANDING

1. What has to happen before ejaculated sperm can penetrate an oocyte?
2. What is the cortical reaction and what does it accomplish?

For answers, see Appendix G.

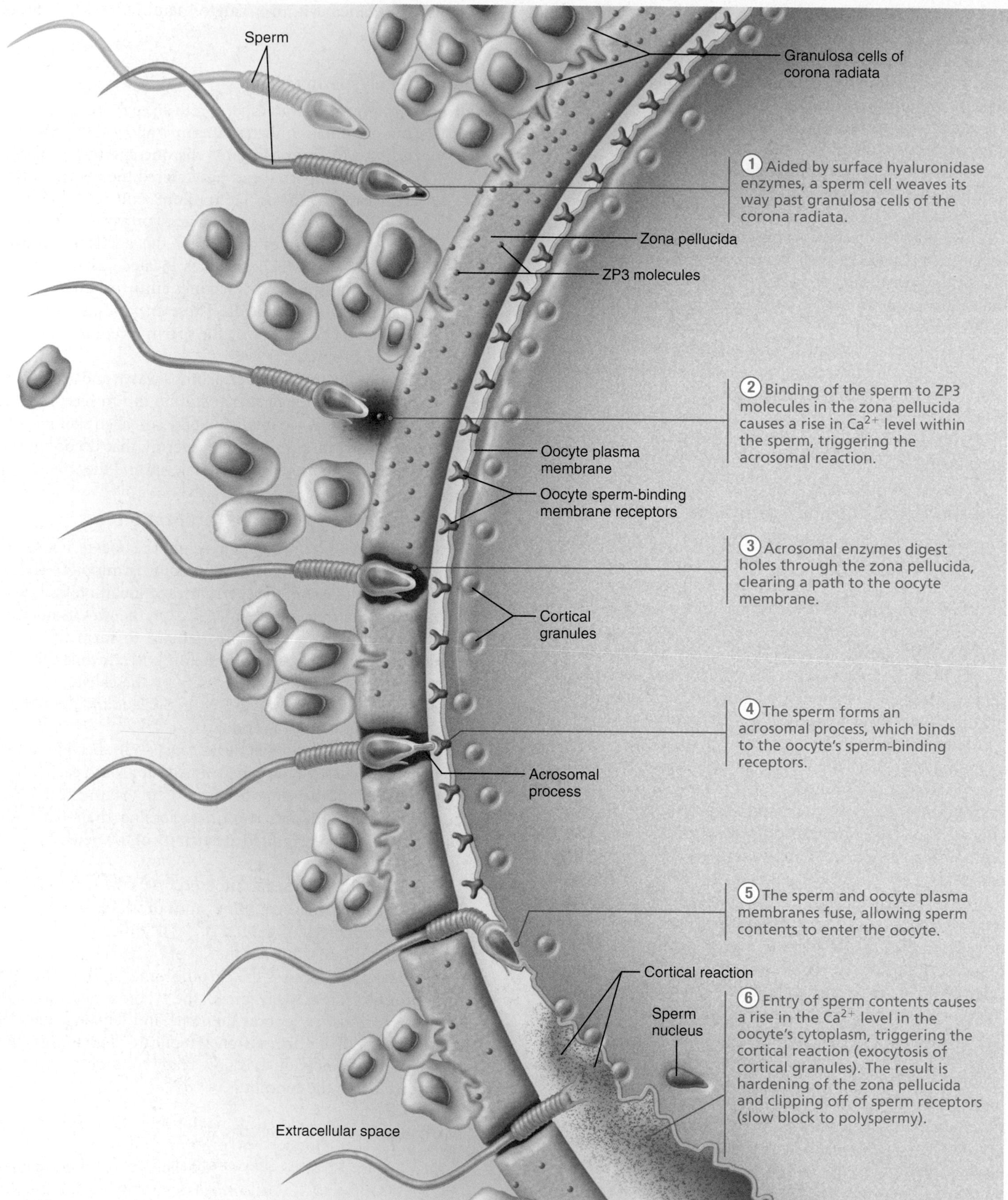

Sperm
Granulosa cells of corona radiata
① Aided by surface hyaluronidase enzymes, a sperm cell weaves its way past granulosa cells of the corona radiata.
Zona pellucida
ZP3 molecules
② Binding of the sperm to ZP3 molecules in the zona pellucida causes a rise in Ca^{2+} level within the sperm, triggering the acrosomal reaction.
Oocyte plasma membrane
Oocyte sperm-binding membrane receptors
③ Acrosomal enzymes digest holes through the zona pellucida, clearing a path to the oocyte membrane.
Cortical granules
④ The sperm forms an acrosomal process, which binds to the oocyte's sperm-binding receptors.
Acrosomal process
⑤ The sperm and oocyte plasma membranes fuse, allowing sperm contents to enter the oocyte.
Cortical reaction
Sperm nucleus
⑥ Entry of sperm contents causes a rise in the Ca^{2+} level in the oocyte's cytoplasm, triggering the cortical reaction (exocytosis of cortical granules). The result is hardening of the zona pellucida and clipping off of sperm receptors (slow block to polyspermy).
Extracellular space

Figure 28.2 Sperm penetration and the cortical reaction (slow block to polyspermy). The sequential steps of oocyte penetration by a sperm are depicted from top to bottom.

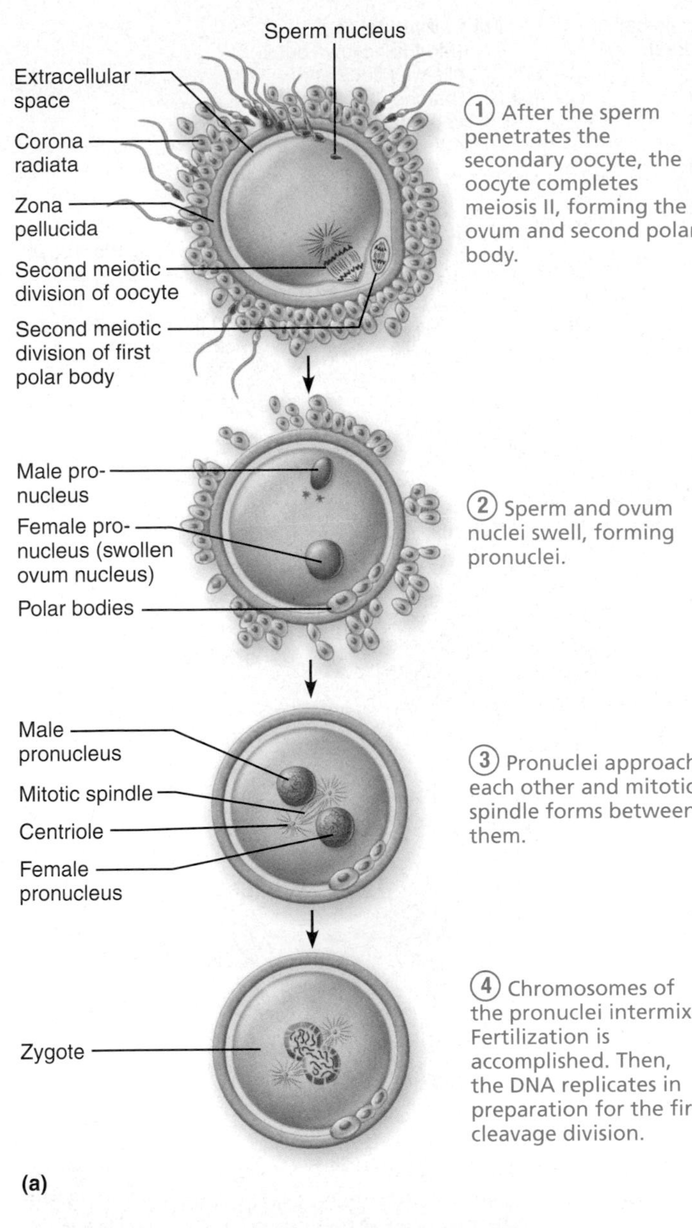
Sperm nucleus
Extracellular space
Corona radiata
Zona pellucida
Second meiotic division of oocyte
Second meiotic division of first polar body
① After the sperm penetrates the secondary oocyte, the oocyte completes meiosis II, forming the ovum and second polar body.
Male pro-nucleus
Female pro-nucleus (swollen ovum nucleus)
Polar bodies
② Sperm and ovum nuclei swell, forming pronuclei.
Male pronucleus
Mitotic spindle
Centriole
Female pronucleus
③ Pronuclei approach each other and mitotic spindle forms between them.
Zygote
④ Chromosomes of the pronuclei intermix. Fertilization is accomplished. Then, the DNA replicates in preparation for the first cleavage division.
(a)

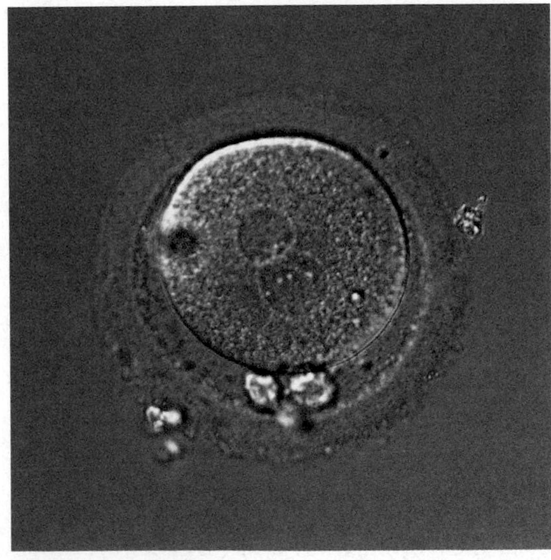
(b)

Figure 28.3 Events of fertilization. (a) Events from sperm penetration to zygote formation. **(b)** Micrograph of an oocyte in which the male and female pronuclei are beginning to fuse to accomplish fertilization and form the zygote. Occurs in time between steps ③ and ④ of **(a)**. Notice the polar bodies protruding from the bottom.

Events of Embryonic Development: Zygote to Blastocyst Implantation

Early embryonic development begins with fertilization and continues as the embryo travels through the uterine tube, floats free in the cavity of the uterus, and finally implants in the uterine wall. Significant events of this early embryonic period are *cleavage*, which produces a structure called a blastocyst, and *implantation* of the blastocyst.

Cleavage and Blastocyst Formation

▶ Explain the process and product of cleavage.

Cleavage is a period of fairly rapid mitotic divisions of the zygote without intervening growth **(Figure 28.4)**. Cleavage produces small cells with a high surface-to-volume ratio, which enhances their uptake of nutrients and oxygen and the disposal of wastes. It also provides a large number of cells to serve as building blocks for constructing the embryo. Consider, for a moment, the difficulty of trying to construct a building from one huge block of granite. If you now consider how much easier it would be if instead you could use hundreds of bricks, you will quickly grasp the importance of cleavage.

Some 36 hours after fertilization, the first cleavage division has produced two identical cells called *blastomeres*. These divide to produce four cells (Figure 28.4b), then eight, and so on. By 72 hours after fertilization, a loose collection of cells that form a berry-shaped cluster of 16 or more cells called the **morula** (mor′u-lah; "little mulberry") has been formed (Figure 28.4c). All the while, transport of the embryo toward the uterus continues.

By day 3 or 4 after fertilization, the embryo consists of about 100 cells and floats free in the uterus (Figure 28.4d). By this time, it has tightened its connections between neighboring cells and begins accumulating fluid within an internal cavity. The zona pellucida now starts to break down and the inner structure, now called a blastocyst, "hatches" from it. The **blastocyst** (blas′to-sist) is a fluid-filled hollow sphere composed of a single layer of large, flattened cells called **trophoblast cells** (trof′o-blast) and a small cluster of 20 to 30 rounded cells, called the **inner cell mass**, located at one side (Figure 28.4e).

Soon after the blastocyst forms, trophoblast cells begin to display L-selectin adhesion molecules on their surface. They take part in placenta formation, as suggested by the literal translation of "trophoblast" (nourishment generator). They also secrete and display several factors with immunosuppressive effects that protect the trophoblast (and the developing embryo) from attack by the mother's cells.

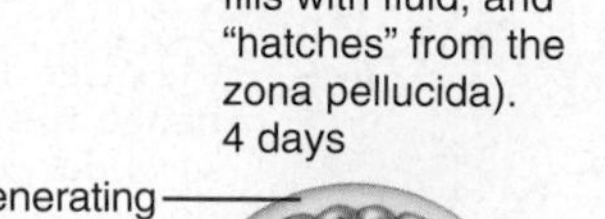
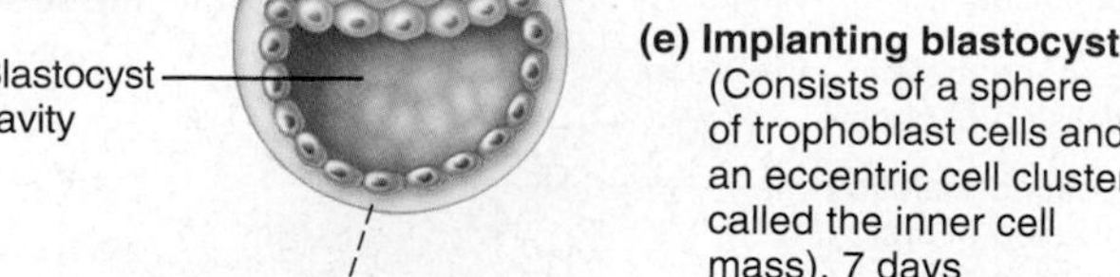
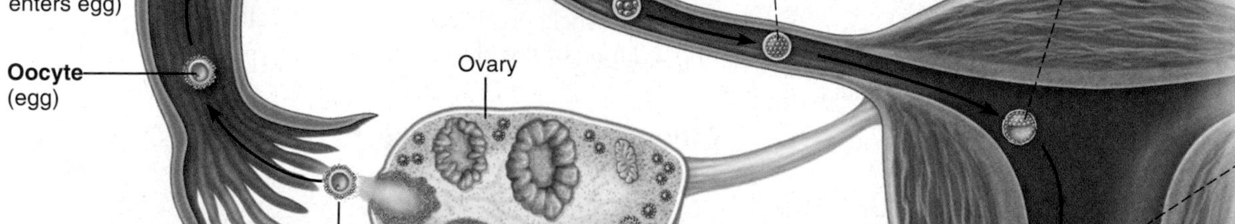

Figure 28.4 Cleavage from zygote to blastocyst. The zygote begins to divide about 24 hours after fertilization, and continues the rapid mitotic divisions of cleavage as it travels down the uterine tube. Three to four days after ovulation, the embryo reaches the uterus and floats freely for two to three days, nourished by secretions of the endometrial glands. Because there is very little time for growth between successive cleavage divisions, the resulting blastocyst is only slightly larger than the zygote. At the late blastocyst stage, the embryo implants into the endometrium; this begins at about day 7 after ovulation.

The inner cell mass becomes the *embryonic disc*, which forms the embryo proper, and three of the four extraembryonic membranes. (The fourth membrane, the chorion, is a trophoblast derivative.)

CHECK YOUR UNDERSTANDING

3. Why is the multicellular blastocyst only slightly larger than the single-cell zygote?
4. What is the function of the trophoblast cells?

For answers, see Appendix G.

Implantation

▶ Describe implantation.

While the blastocyst floats in the uterine cavity for two to three days, it is nourished by the glycogen-rich uterine secretions. Then, some six to seven days after ovulation, given a properly prepared endometrium, **implantation** begins. The receptivity of the endometrium to implantation—the so-called *window of implantation*—is opened by the surging levels of ovarian hormones (estrogens and progesterone) in the blood. If the mucosa is properly prepared, integrin and selectin proteins on the trophoblast cells bind respectively to the extracellular matrix components (collagen, fibronectin, laminin, and others) of the endometrial cells and to selectin-binding carbohydrates on the inner uterine wall, and the blastocyst implants high in the uterus. If the endometrium is not yet optimally mature, the blastocyst detaches and floats to a lower level, implanting when it finds a site with the proper receptors and chemical signals.

The trophoblast cells overlying the inner cell mass adhere to the endometrium **(Figure 28.5a, b)** and secrete digestive enzymes and growth factors against the endometrial surface. The endometrium quickly thickens at the point of contact and takes on characteristics of an acute inflammatory response—the uterine blood vessels become more permeable and leaky, and inflammatory cells including lymphocytes, natural killer cells, and macrophages invade the area.

The trophoblast then proliferates and forms two distinct layers (Figure 28.5c). The cells in the inner layer, collectively

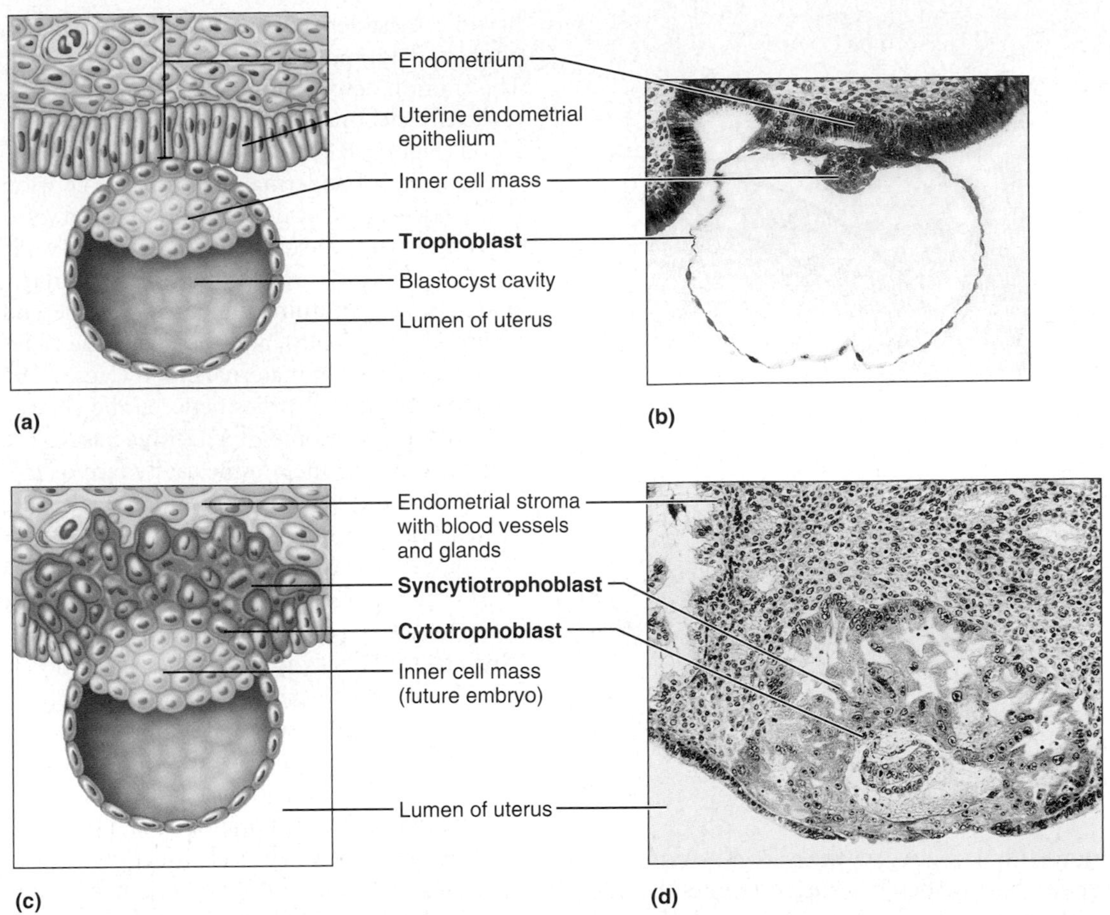

Figure 28.5 Implantation of the blastocyst. (a) Diagrammatic view of a blastocyst that has just adhered to the uterine endometrium and **(b)** microscopic view. **(c)** Slightly later stage of an implanting embryo (approximately seven days after ovulation), depicting the cytotrophoblast and syncytiotrophoblast of the eroding trophoblast. **(d)** Light micrograph of an implanted blastocyst (approximately 12 days after ovulation).

SOURCE: R. O'Rahilly and R. Muller, *Human Embryology and Teratology*, Wiley–Liss, 3rd Edition, 2001. This material is reproduced with permission of Wiley–Liss, Inc., a subsidiary of John Wiley & Sons, Inc.

called the **cytotrophoblast** (si″to-trof′o-blast) or **cellular trophoblast**, retain their cell boundaries. The cells in the outer layer lose their plasma membranes and form a multinuclear cytoplasmic mass called the **syncytiotrophoblast** (sin-sit″e-o-trof′o-blast; *syn* = together, *cyt* = cell) or **syncytial trophoblast**, which invades the endometrium and rapidly digests the uterine cells it contacts. As the endometrium is eroded, the blastocyst burrows into this thick, velvety lining and is surrounded by a pool of blood leaked from degraded endometrial blood vessels. Shortly, the implanted blastocyst is covered over and sealed off from the uterine cavity by proliferation of the endometrial cells (Figure 28.5d).

In cases where implantation fails to occur, a receptive uterus becomes nonreceptive once again. It is estimated that a minimum of two-thirds of all zygotes formed fail to implant by the end of the first week or spontaneously abort. Moreover, an estimated 30% of implanted embryos later miscarry due to genetic defects of the embryo, uterine malformation, or other, often unknown, problems.

When successful, implantation takes about five days and is usually completed by the 12th day after ovulation—just before the endometrium normally begins to slough off. Menstruation would flush away the embryo as well and must be prevented if the pregnancy is to continue. Viability of the corpus luteum is maintained by an LH-like hormone called **human chorionic gonadotropin (hCG)** (ko″re-on′ik go-nad″o-trōp′in) secreted by the trophoblast cells. hCG bypasses hypothalamic-pituitary-ovarian controls at this critical time and prompts the corpus luteum to continue secreting progesterone and estrogen. The *chorion*, the extraembryonic membrane that develops from the trophoblast after implantation, continues this hormonal stimulus. In this way, the developing conceptus takes over the hormonal control of the uterus during this early phase of development.

Usually detectable in the mother's blood one week after fertilization, blood levels of hCG continue to rise until the end of the second month. Then blood levels decline sharply to reach a low value by 4 months, a situation that persists for the remainder

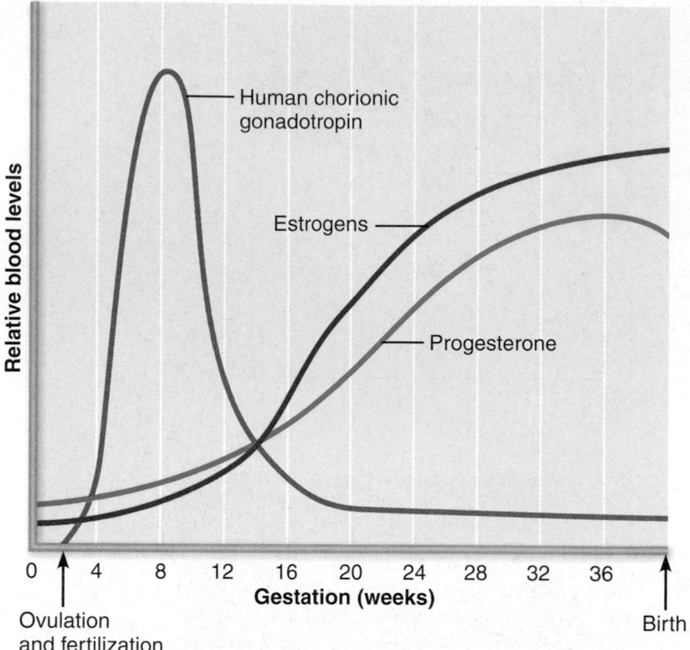

Figure 28.6 Hormonal changes during pregnancy. The relative changes in maternal blood levels of three hormones that maintain pregnancy are depicted, rather than actual blood concentrations.

of gestation **(Figure 28.6)**. Between the second and third month, the placenta (which we describe next) assumes the role of progesterone and estrogen production for the remainder of the pregnancy. The corpus luteum then degenerates and the ovaries remain inactive until after birth. All pregnancy tests used today are antibody tests that detect hCG in a woman's blood or urine.

Initially, the implanted embryo obtains its nutrition by digesting the endometrial cells, but by the second month, the placenta is providing nutrients and oxygen to the embryo and carrying away embryonic metabolic wastes. Since placenta formation is a continuation of the events of implantation, we will consider it next, although we will be getting a little ahead of ourselves as far as embryonic development is concerned.

CHECK YOUR UNDERSTANDING

5. Which portion of the trophoblast accomplishes implantation?
6. Marie, wondering if she is pregnant, buys an over-the-counter pregnancy test to assess this possibility. How will the blastocyst, if present, "make itself known?"

For answers, see Appendix G.

28

Placentation

▶ Describe placenta formation, and list placental functions.

Placentation (plas″en-ta′shun) refers to the formation of a **placenta** ("flat cake"), a temporary organ that originates from both embryonic and maternal (endometrial) tissues. Cells from the original inner cell mass give rise to a layer of extraembryonic mesoderm that lines the inner surface of the trophoblast **(Figure 28.7b)**. Together these become the **chorion**. The chorion develops fingerlike **chorionic villi**, which become especially elaborate where they are in contact with maternal blood (Figure 28.7c).

Soon the mesodermal cores of the chorionic villi become richly vascularized by newly forming blood vessels, which extend to the embryo as the umbilical arteries and vein. The continuing erosion produces large, blood-filled **lacunae**, or **intervillous spaces**, in the stratum functionalis of the endometrium (see Figure 27.13, p. 1045), and the villi come to lie in these spaces totally immersed in maternal blood (Figure 28.7d). The part of the endometrium that lies between the chorionic villi and the stratum basalis becomes the **decidua basalis** (de-sid′u-ah), and that surrounding the uterine cavity face of the implanted embryo forms the **decidua capsularis** (Figure 28.7d and e). Together, the chorionic villi and the decidua basalis form the disc-shaped placenta.

The placenta detaches and sloughs off after the infant is born, so the name of the maternal portion—*decidua* ("that which falls off")—is appropriate. During development, the decidua capsularis expands to accommodate the fetus, which eventually fills and stretches the uterine cavity. As the developing fetus grows, the villi in the decidua capsularis are compressed and degenerate, and the villi in the decidua basalis increase in number and branch even more profusely.

The placenta is usually fully functional as a nutritive, respiratory, excretory, and endocrine organ by the end of the third month of pregnancy. However, well before this time, oxygen and nutrients are diffusing from maternal to embryonic blood, and embryonic metabolic wastes are passing in the opposite direction. The barriers to free passage of substances between the two blood supplies are embryonic barriers—the membranes of the chorionic villi and the endothelium of embryonic capillaries. Although the maternal and embryonic blood supplies are very close, they normally do not intermix **(Figure 28.8)**.

The placenta secretes hCG from the beginning, but the ability of its syncytiotrophoblast cells (the "hormone manufacturers") to produce the estrogens and progesterone of pregnancy matures much more slowly. If, for some reason, placental hormones are inadequate when hCG levels wane, the endometrium degenerates and the pregnancy is aborted. Throughout pregnancy, blood levels of estrogens and progesterone continue to increase (see Figure 28.6). They encourage growth and further differentiation of the mammary glands and ready them for lactation. The placenta also produces other hormones, such as *human placental lactogen, human chorionic thyrotropin*, and *relaxin*. Shortly we will describe the effects of these hormones on the mother.

CHECK YOUR UNDERSTANDING

7. What is the composition of the chorion?
8. What endometrial decidua cooperates with the chorionic villi to form the placenta?
9. Generally speaking, when does the placenta become fully functional?

For answers, see Appendix G.

Endometrium
Maternal blood vessels
Proliferating syncytiotrophoblast
Cytotrophoblast
Amniotic cavity
Bilayered embryonic disc
• Epiblast
• Hypoblast
Endometrial epithelium
Chorion being formed
Lacuna (intervillous space) containing maternal blood
Chorionic villus
Chorion
Amnion
Yolk sac
Extraembryonic mesoderm
Lumen of uterus
Amniotic cavity
Primary germ layers
• Ectoderm
• Mesoderm
• Endoderm
Forming body stalk
Allantois
Extraembryonic coelom

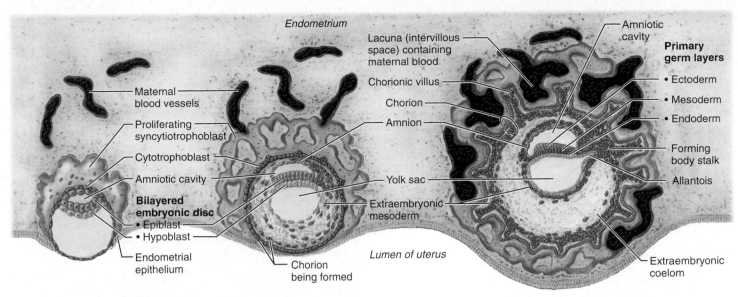

(a) Implanting 7½-day blastocyst. The syncytiotrophoblast is eroding the endometrium. Cells of the embryonic disc are now separated from the amnion by a fluid-filled space.

(b) 12-day blastocyst. Implantation is complete. Extraembryonic mesoderm is forming a discrete layer beneath the cytotrophoblast.

(c) 16-day embryo. Cytotrophoblast and associated mesoderm have become the chorion, and chorionic villi are elaborating. The embryo exhibits all three germ layers, a yolk sac and an allantois, which forms the basis of the umbilical cord.

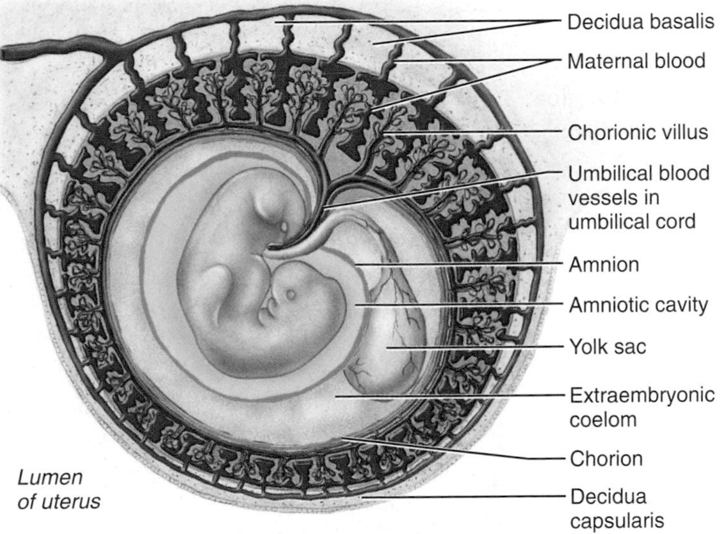

(d) 4½-week embryo. The decidua capsularis, decidua basalis, amnion, and yolk sac are well formed. The chorionic villi lie in blood-filled intervillous spaces within the endometrium. The embryo is now receiving its nutrition via the umbilical vessels that connect it (through the umbilical cord) to the placenta.

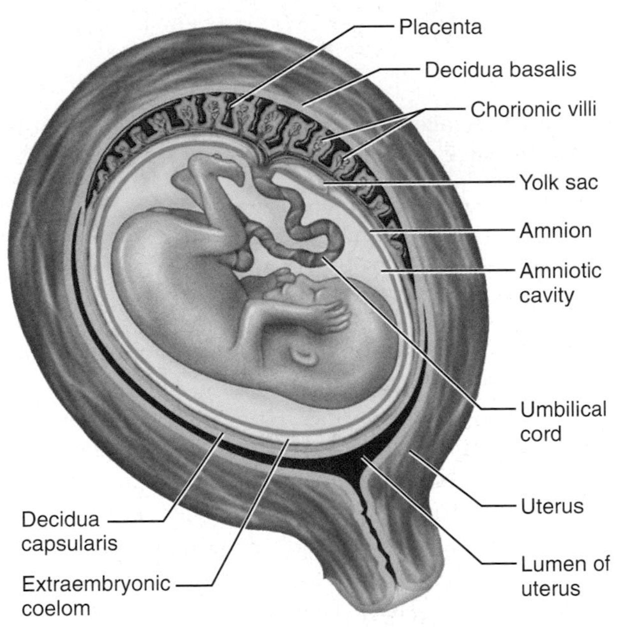

(e) 13-week fetus.

Figure 28.7 **Events of placentation, early embryonic development, and extraembryonic membrane formation.**

28

Events of Embryonic Development: Gastrula to Fetus

Having followed placental development into the fetal stage, we will now backtrack and consider development of the embryo during and after implantation, referring again to Figure 28.7. Even while implantation is occurring, the blastocyst is being converted to a **gastrula** (gas′troo-lah), in which the three primary germ layers form, and the extraembryonic membranes develop. Before becoming three-layered, the inner cell mass first subdivides into two layers—the upper *epiblast* and the lower *hypoblast* (Figure 28.7a, b). The subdivided inner cell mass is now called the **embryonic disc**.

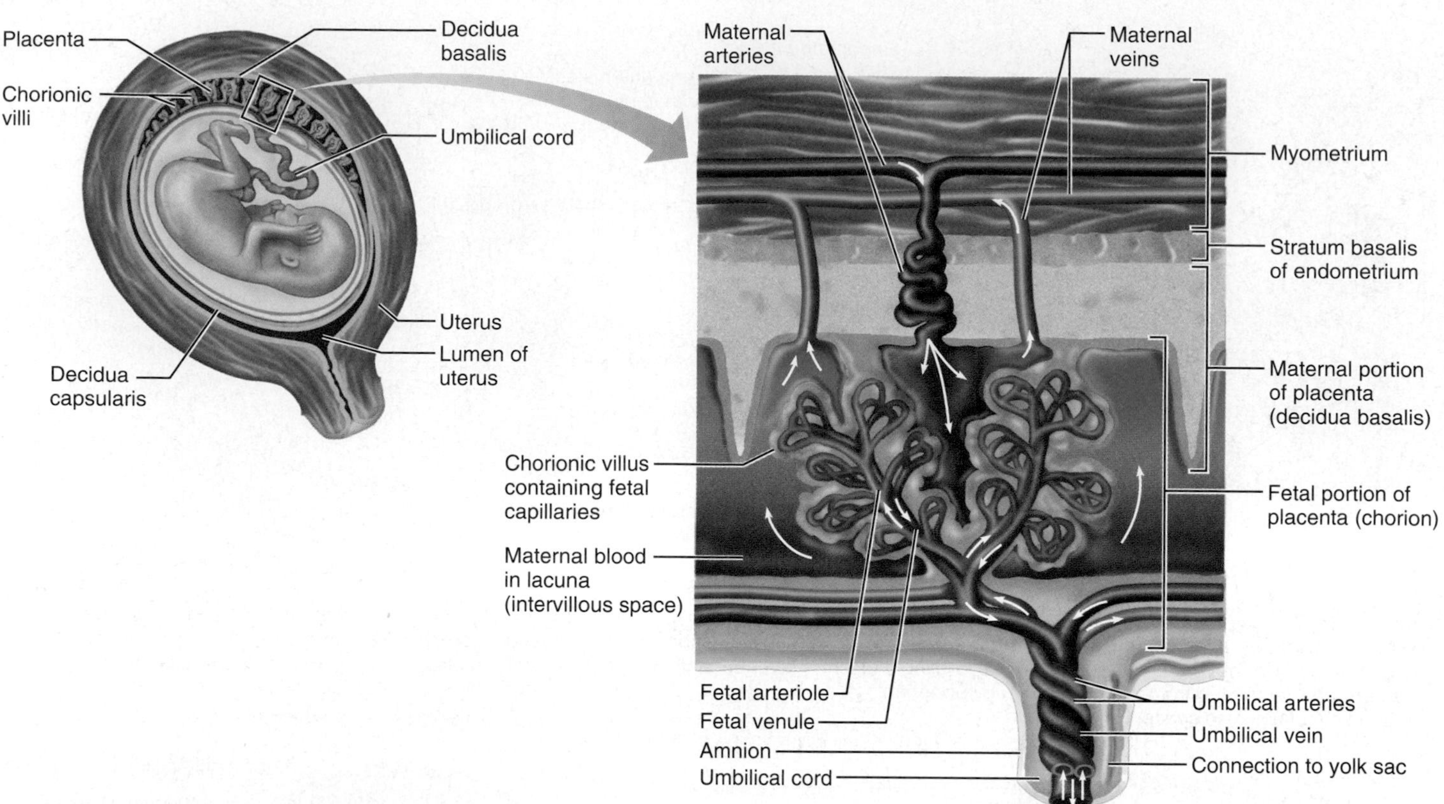

Figure 28.8 Detailed anatomy of the vascular relationships in the mature decidua basalis. This state of development has been accomplished by the end of the third month of development.

Formation and Roles of the Extraembryonic Membranes

- Name and describe the formation, location, and function of the extraembryonic membranes.

The **extraembryonic membranes** that form during the first two to three weeks of development include the amnion, yolk sac, allantois, and chorion (Figure 28.7c). The **amnion** (am′ne-on) develops when cells of the epiblast fashion themselves into a transparent membranous sac. This sac, the amnion, becomes filled with **amniotic fluid.** Later, as the embryonic disc curves to form the tubular body, the amnion curves with it. Eventually, the sac extends all the way around the embryo, broken only by the umbilical cord (Figure 28.7d).

28

Sometimes called the "bag of waters," the amnion provides a buoyant environment that protects the developing embryo against physical trauma, and helps maintain a constant homeostatic temperature. The fluid also prevents the rapidly growing embryonic parts from adhering and fusing together and allows the embryo considerable freedom of movement. Initially, the fluid is derived from the maternal blood, but as the fetal kidneys become functional later in development, fetal urine contributes to amniotic fluid volume.

The **yolk sac** forms from cells of the primitive gut (see below), which arrange themselves into a sac that hangs from the ventral surface of the embryo (Figure 28.7b–d). The amnion and yolk sac resemble two balloons touching one another with the embryonic disc at the point of contact. In many species, the yolk sac is the main source of nutrition for the embryo, but human eggs contain very little yolk and nutritive functions have been taken over by the placenta. Nevertheless, the yolk sac is important in humans because it (1) forms part of the gut (digestive tube), and (2) is the source of the earliest blood cells and blood vessels.

The **allantois** (ah-lan′to-is) forms as a small outpocketing of embryonic tissue at the caudal end of the yolk sac (Figure 28.7c). In animals that develop in shelled eggs, the allantois is a disposal site for solid metabolic wastes (excreta). In humans, the allantois is the structural base for the **umbilical cord** that links the embryo to the placenta, and ultimately it becomes part of the urinary bladder. The fully formed umbilical cord contains a core of embryonic connective tissue (Wharton's jelly), the umbilical arteries and vein, and is covered externally by amniotic membrane.

We have already described the *chorion,* which helps to form the placenta (Figure 28.7c). As the outermost membrane, the chorion encloses the embryonic body and all other membranes.

CHECK YOUR UNDERSTANDING

10. What is the function of the amnion?
11. Which extraembryonic membrane provides a path for the embryonic blood vessels to reach the placenta?

For answers, see Appendix G.

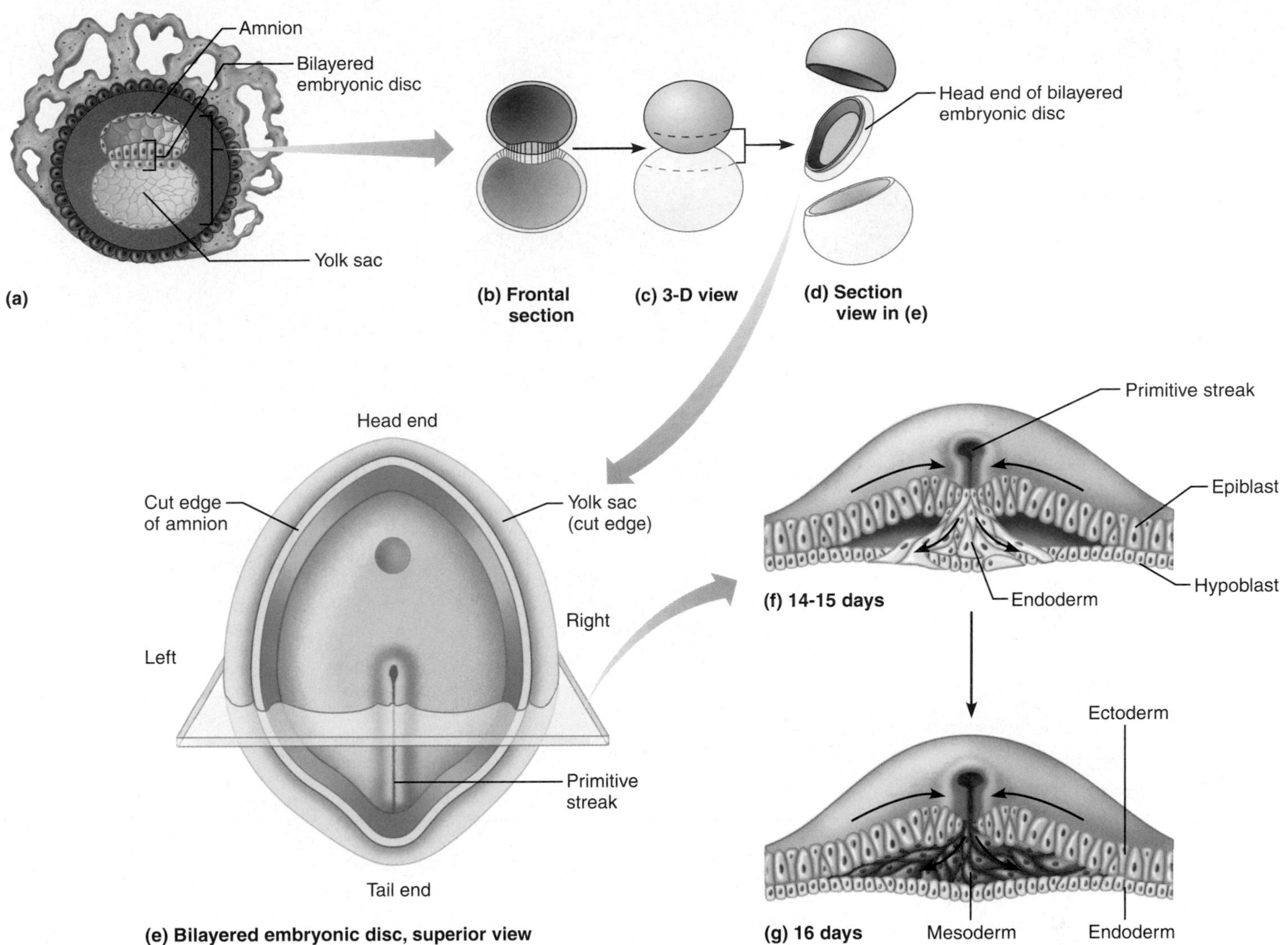

Figure 28.9 Formation of the three primary germ layers. (a–d) Orienting diagrams. **(e)** Surface view of an embryonic disc, amnion and yolk sac removed. **(f, g)** Cross sections of the embryonic disc, showing the germ layers resulting from cell migration. The first epiblast cells that migrate medially into the primitive streak **(f)** become endoderm. Those that follow **(g)** become mesoderm. The epiblast surface is now called ectoderm.

Gastrulation: Germ Layer Formation

▶ Describe gastrulation and its consequence.

During week 3, the two-layered embryonic disc transforms into a three-layered *embryo* in which the **primary germ layers**—*ectoderm, mesoderm,* and *endoderm*—are present (Figure 28.7c). This process, called **gastrulation** (gas″troo-la′shun), involves cellular rearrangements and migrations.

Figure 28.9 focuses on the changes that take place in the days between Figure 28.7b (12-day embryo) and Figure 28.7c (16-day embryo). Gastrulation begins when a groove with raised edges called the **primitive streak** appears on the dorsal surface of the embryonic disc and establishes the longitudinal axis of the embryo (Figure 28.9e). Surface (epiblast) cells of the embryonic disc then migrate medially across other cells and enter the primitive streak. The first cells to enter the groove displace the hypoblast cells of the yolk sac and form the most inferior germ layer, the **endoderm** (Figure 28.9f). Those that follow push laterally between the cells at the upper and lower surfaces, forming the **mesoderm** (Figure 28.9g). As soon as the mesoderm is formed, the mesodermal cells immediately beneath the early primitive streak aggregate, forming a rod of mesodermal cells called the **notochord** (no′to-kord), the first axial support of the embryo **(Figure 28.10a)**.The cells that remain on the embryo's dorsal surface are the **ectoderm**. At this point, the embryo is about 2 mm long.

The three primary germ layers serve as the *primitive tissues* from which all body organs derive. Ectoderm ("outer skin") fashions structures of the nervous system and the skin epidermis. Endoderm ("inner skin") forms the epithelial linings of the digestive, respiratory, and urogenital systems, and associated glands. Mesoderm ("middle skin") forms virtually everything else.

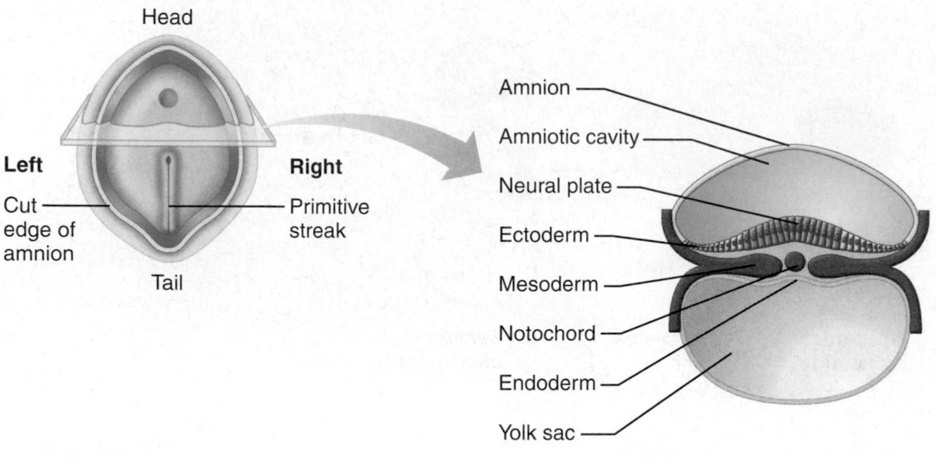

(a) 17 days. The flat three-layered embryo has completed gastrulation. Notochord and neural plate are present.

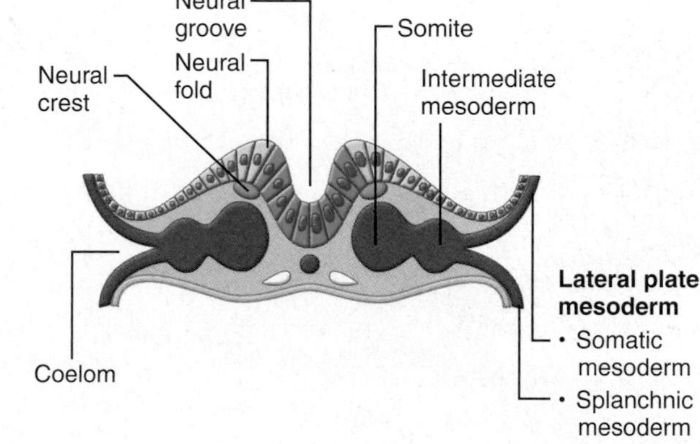

(b) 20 days. The neural folds form by folding of the neural plate, which then deepens, producing the neural groove. Three mesodermal aggregates form on each side of the notochord (somite, intermediate mesoderm, and lateral plate mesoderm).

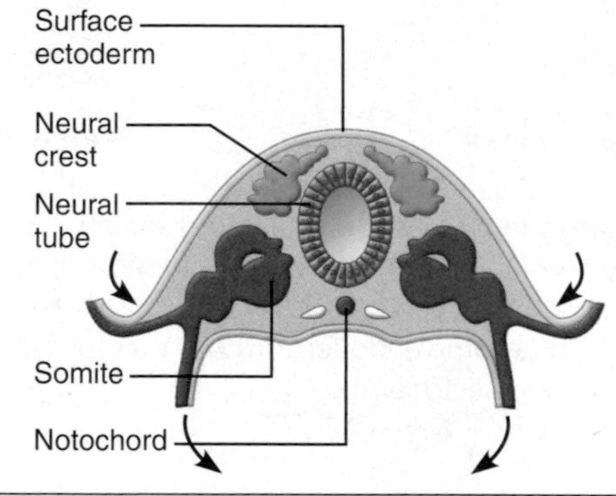

(c) 22 days. The neural folds have closed, forming the neural tube which has detached from the surface ectoderm and lies between the surface ectoderm and the notochord. Embryonic body is beginning to undercut.

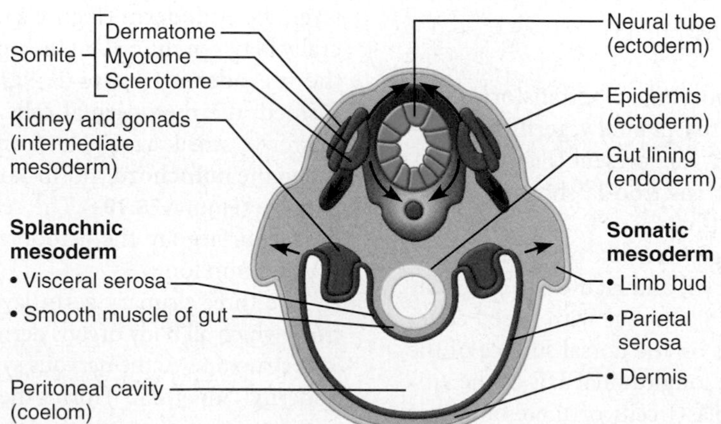

(d) End of week 4. Embryo undercutting is complete. Somites have subdivided into sclerotome, myotome, and dermatome, which form the vertebrae, skeletal muscles, and dermis respectively. Body coelom present.

Figure 28.10 Neurulation and early mesodermal differentiation.

Both ectoderm and endoderm consist mostly of cells that are securely joined to each other and are *epithelia*. Mesoderm, by contrast, is a *mesenchyme* [literally, "poured into the middle (of the embryo)"], an embryonic tissue with star-shaped cells that are free to migrate widely within the embryo. Figure 28.13 (p. 1085) lists the germ layer derivatives. Some of the details of the differentiation processes are described next.

Organogenesis: Differentiation of the Germ Layers

- Define organogenesis and indicate the important roles of the three primary germ layers in this process.
- Describe unique features of the fetal circulation.

Gastrulation lays down the basic structural framework of the embryo and sets the stage for the rearrangements that occur during **organogenesis** (or″gah-no-jen′ĕ-sis), formation of body organs and organ systems. By the end of the embryonic period at 8 weeks, when the embryo is about 22 mm (slightly less than 1 inch) long from head to buttocks (referred to as the *crown-rump measurement*), all the adult organ systems are recognizable. It is truly amazing how much organogenesis occurs in such a short time in such a small amount of living matter.

Specialization of the Ectoderm

The first major event in organogenesis is **neurulation**, the differentiation of ectoderm that produces the brain and spinal cord (Figure 28.10). This process is *induced* (stimulated to happen) by chemical signals from the *notochord*, the rod of mesoderm that defines the body axis, mentioned earlier. The ectoderm overlying the notochord thickens, forming the **neural plate** (Figure 28.10a). Then the ectoderm starts to fold inward as a **neural groove**. As the neural groove deepens it forms prominent **neural folds** (Figure 28.10b). By day 22, the superior margins of the neural folds fuse, forming a **neural tube**, which soon pinches off from the ectodermal layer and becomes covered by surface ectoderm (Figure 28.10c).

As we described in Chapter 12, the anterior end of the neural tube becomes the brain and the rest becomes the spinal cord. The associated **neural crest cells** (Figure 28.10c) migrate widely and give rise to the cranial, spinal, and sympathetic ganglia (and associated nerves), to the medulla of the adrenal gland, and to pigment cells, and contribute to some connective tissues.

By the end of the first month of development, the three primary brain vesicles (fore-, mid-, and hindbrain) are apparent. By the end of the second month, all brain flexures are evident, the cerebral hemispheres cover the top of the brain stem (see Figure 12.3), and brain waves can be recorded. Most of the remaining ectoderm forming the surface layer of the embryonic body differentiates into the epidermis of the skin. Other ectodermal derivatives are indicated in Figure 28.13.

Specialization of the Endoderm

As **Figure 28.11** shows, the embryo starts off as a flat plate, but as it grows, it folds to achieve a cylindrical body shape which lifts

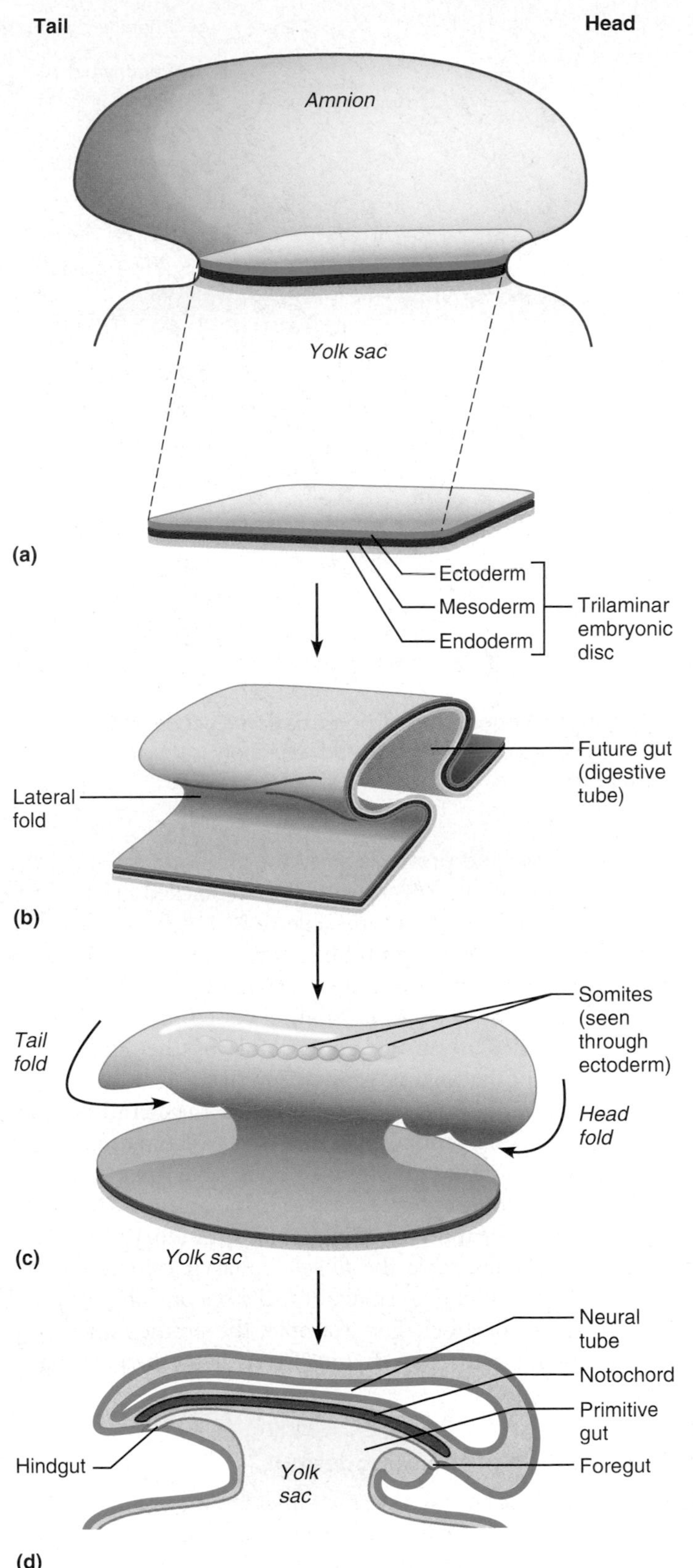

Figure 28.11 Folding of the embryonic body, lateral views. **(a)** Model of the flat three-layered embryo as three sheets of paper. **(b, c)** Folding begins with lateral folds, then head and tail folds appear. **(d)** A 24-day embryo in sagittal section. Notice the primitive gut, which derives from the yolk sac, and the notochord and neural tube dorsally.

28

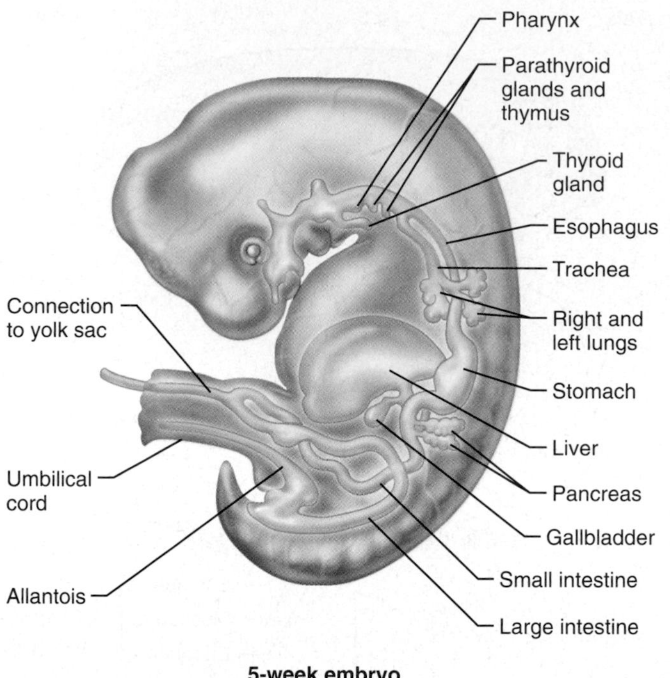

Figure 28.12 Endodermal differentiation. Endoderm forms the epithelial linings of the digestive and respiratory tracts and associated glands.

off the yolk sac and protrudes into the amniotic cavity. In the simplest sense, this process resembles three stacked sheets of paper folding laterally into a tube (Figure 28.11a, b). At the same time, the folding occurs from both ends (the head and tail regions) and progresses toward the central part of the embryonic body, where the yolk sac and umbilical vessels protrude. As the endoderm undercuts and its edges come together and fuse, it encloses part of the yolk sac (Figure 28.11d).

The tube of endoderm formed, called the **primitive gut**, forms the epithelial lining (mucosa) of the gastrointestinal tract **(Figure 28.12)**. The organs of the GI tract (pharynx, esophagus, etc.) quickly become apparent, and then the oral and anal openings perforate. The mucosal lining of the respiratory tract forms as an outpocketing from the *foregut* (pharyngeal endoderm), and glands arise as endodermal outpocketings at various points further along the tract. For example, the epithelium of the thyroid, parathyroids, and thymus forms from the pharyngeal endoderm.

28

Specialization of the Mesoderm

The first evidence of mesodermal differentiation is the appearance of the notochord in the embryonic disc (see Figure 28.10a). The notochord is eventually replaced by the vertebral column, but its remnants persist in the springy *nucleus pulposus* of the intervertebral discs. Shortly thereafter, three mesodermal aggregates appear on either side of the notochord (Figure 28.10b, c). The largest of these, the *somites* (so′mīts), are paired mesodermal blocks that hug the notochord on either side. All 40 pairs of somites are present by the end of week 4. Flanking the somites laterally are small clusters of segmented mesoderm called *intermediate mesoderm* and then double sheets of *lateral plate mesoderm.*

Each **somite** has three functional parts—*sclerotome, dermatome,* and *myotome* (Figure 28.10d). Cells of the **sclerotome** (skle′ro-tōm; "hard piece") migrate medially, gather around the notochord and neural tube, and produce the vertebra and rib at the associated level. **Dermatome** ("skin piece") cells help form the dermis of the skin in the dorsal part of the body. The **myotome** (mi′o-tōm; "muscle piece") cells develop in conjunction with the vertebrae. They form the skeletal muscles of the neck, body trunk, and, via their **limb buds**, the muscles of the limbs.

Cells of the **intermediate mesoderm** form the gonads and kidneys. The **lateral plate mesoderm** consists of paired mesodermal plates: the *somatic mesoderm* and the *splanchnic mesoderm* (Figure 28.10d). Cells of the **somatic mesoderm** (1) help to form the dermis of the skin in the ventral body region; (2) form the parietal serosa that lines the ventral body cavity; and (3) migrate into the forming limbs and produce the bones, ligaments, and dermis of the limbs (see Figure 28.10d). **Splanchnic mesoderm** provides the mesenchymal cells that form the heart and blood vessels and most connective tissues of the body. Splanchnic mesodermal cells also form the smooth muscle, connective tissues, and serosal coverings (in other words, nearly the entire wall) of the digestive and respiratory organs. Thus, the lateral mesodermal layers cooperate to form the serosae of the **coelom** (se′lom), or ventral body cavity. The mesodermal derivatives are summarized in **Figure 28.13**.

By the end of the embryonic period, the bones have begun to ossify and the skeletal muscles are well formed and contracting spontaneously. Metanephric kidneys are developing, gonads are formed, and the lungs and digestive organs are attaining their final shape and body position. Blood delivery to and from the placenta via the umbilical vessels is constant and efficient. The heart and the liver are competing for space and form a conspicuous bulge on the ventral surface of the embryo's body. All this by the end of eight weeks in an embryo about 2.5 cm (1 inch) long from crown to rump!

Development of the Fetal Circulation Embryonic development of the cardiovascular system lays the groundwork for the fetal circulatory pattern, which is converted to the adult pattern at birth. The first blood cells arise in the yolk sac. Before week 3 of development, tiny spaces appear in the splanchnic mesoderm. These are quickly lined by endothelial cells, covered with mesenchyme, and linked together into rapidly spreading vascular networks, destined to form the heart, blood vessels, and lymphatics. By the end of week 3, the embryo has a system of paired blood vessels, and the two vessels forming the heart have fused and bent into an S shape. By 3½ weeks, the miniature heart is pumping blood for an embryo less than a quarter inch long.

Unique cardiovascular modifications seen only during prenatal development include the **umbilical arteries** and **vein** and three *vascular shunts* **(Figure 28.14)**. All of these structures are occluded at birth. As you read about these vessels, keep in mind that the blood is flowing from and to the fetal heart. The large umbilical vein carries freshly oxygenated blood returning from the placenta into the embryonic body, where it is conveyed to

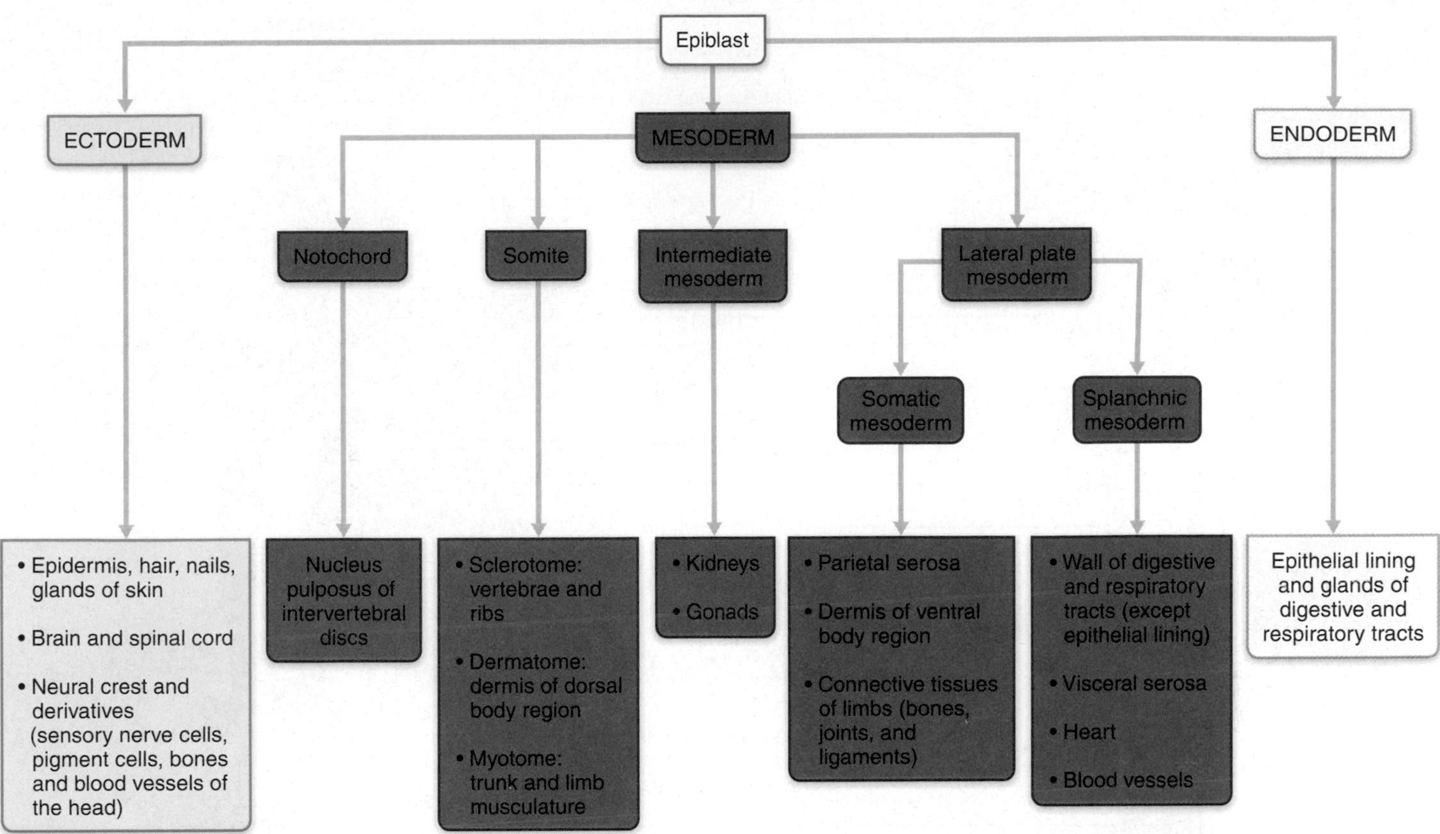

Figure 28.13 Flowchart showing major derivatives of the embryonic germ layers.

the liver. There, some of the returning blood percolates through the liver sinusoids and out the hepatic veins. Most of the blood coursing through the umbilical vein, however, enters the **ductus venosus** (duk′tus ve-no′sus), a venous shunt that bypasses the liver sinusoids. Both the hepatic veins and the ductus venosus empty into the inferior vena cava where the placental blood mixes with deoxygenated blood returning from the lower parts of the fetus's body. The vena cava in turn conveys this "mixed load" of blood directly to the right atrium of the heart.

After birth, the liver plays an important role in nutrient processing, but during embryonic life the mother's liver performs these functions. Consequently, blood flow through the fetal liver during development is important only to ensure that the liver cells remain healthy.

Blood entering and leaving the heart encounters two more shunt systems, each serving to bypass the nonfunctional lungs. Some of the blood entering the right atrium flows directly into the left atrium via the **foramen ovale** ("oval hole"), an opening in the interatrial septum loosely closed by a flap of tissue. Blood that enters the right ventricle is pumped out into the pulmonary trunk. However, the second shunt, the **ductus arteriosus**, transfers most of that blood directly into the aorta, again bypassing the pulmonary circuit. (The lungs *do* receive adequate blood to maintain their growth.) Blood enters the two pulmonary bypass shunts because the heart chamber or vessel on the other side of each shunt is a lower-pressure area, owing to the low volume of venous return from the lungs. Blood flowing distally through the aorta eventually reaches the umbilical arteries, which are branches of the internal iliac arteries serving the pelvis. From here the largely deoxygenated blood, laden with metabolic wastes, is delivered back to the capillaries in the chorionic villi of the placenta. The changes in the circulatory plan that occur at birth are illustrated in Figure 28.14b.

HOMEOSTATIC IMBALANCE

Because many potentially harmful substances can cross placental barriers and enter the fetal blood, a pregnant woman should be aware of what she is taking into her body, particularly during the embryonic period when the body's foundations are laid down. **Teratogens** (ter′ah-to-jenz; *terato* = monster), factors that may cause severe congenital abnormalities or even fetal death, include alcohol, nicotine, many drugs (anticoagulants, sedatives, antihypertensives, and some antibiotics), and maternal infections, particularly German measles. For example, when a woman drinks alcohol, her fetus becomes inebriated as well. However, the fetal consequences may be much more lasting and result in the *fetal alcohol syndrome* (*FAS*) typified by microcephaly (small head), mental retardation, and abnormal growth. Nicotine hinders oxygen delivery to the fetus, impairing normal growth and development. The sedative **thalidomide** (thah-lid′o-mīd) was used by thousands of pregnant women in the 1960s to alleviate morning sickness. When taken during the period of limb bud differentiation (days 26–56), it sometimes resulted in tragically deformed infants with short flipperlike legs and arms. ■

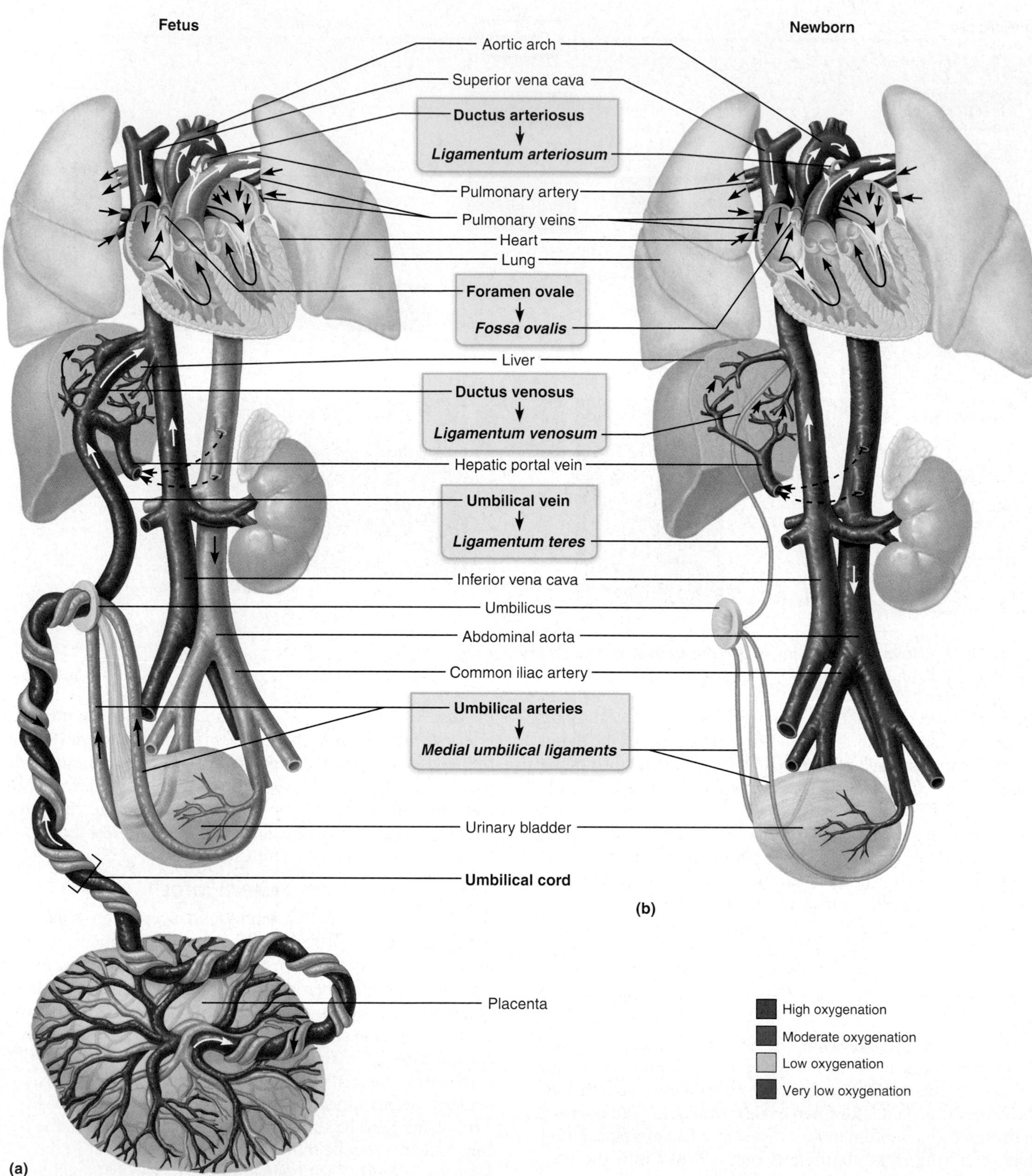

Figure 28.14 Circulation in fetus and newborn. Arrows on blood vessels indicate direction of blood flow. Arrows in color screens go from the fetal structure to what it becomes after birth. **(a)** Special adaptations for embryonic and fetal life. The umbilical vein carries oxygen- and nutrient-rich blood from the placenta to the fetus. The umbilical arteries carry waste-laden blood from the fetus to the placenta. The ductus arteriosus and foramen ovale bypass the nonfunctional lungs. The ductus venosus allows blood to partially bypass the liver. **(b)** Changes in the cardiovascular system at birth. The umbilical vessels as well as the liver and lung bypasses are occluded.

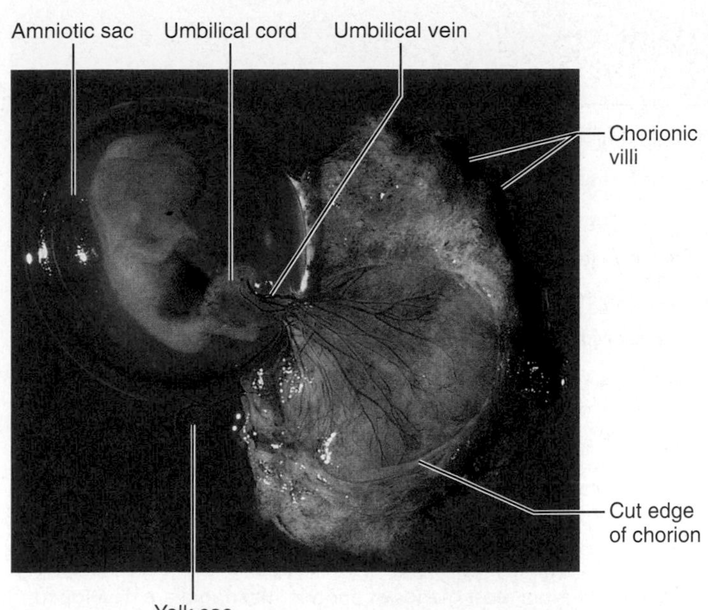

(a) Embryo at week 7, about 17 mm long.

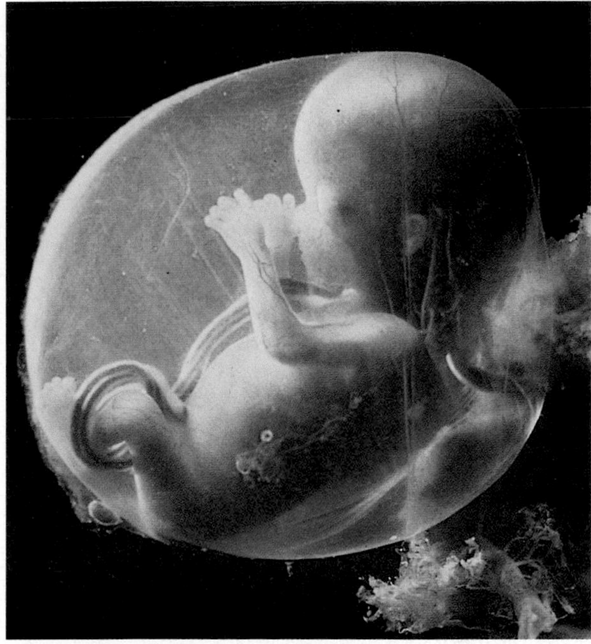

(b) Fetus in month 3, about 6 cm long.

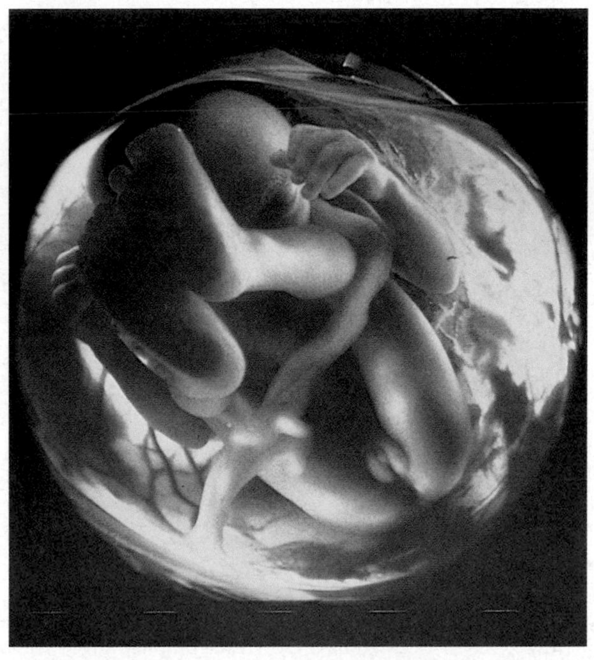

(c) Fetus late in month 5, about 19 cm long.

Figure 28.15 Photographs of a developing fetus. By birth, the fetus is typically 36 cm long from crown to rump.

CHECK YOUR UNDERSTANDING

12. The early embryo is flat like a three-layered pancake. What event must occur before organogenesis can get going in earnest?
13. What germ layer gives rise to essentially all body tissues except nervous tissue, the epidermis, and mucosae?

For answers, see Appendix G.

Events of Fetal Development

▶ Indicate the duration of the fetal period, and note the major events of fetal development.

The main events of the fetal period—weeks 9 through 38—are listed chronologically in **Table 28.1**. The fetal period is a time of rapid growth of the body structures that were established in the embryo. During the first half of this period, cells are still differentiating into specific cell types to form the body's distinctive tissues and are completing the fine details of body structure. During the fetal period, the developing fetus grows from a crown-to-rump length of about 22 mm (slightly less than 1 inch) and a weight of approximately 2 g (0.06 ounce) to about 360 mm (14 inches) and 3.2 kg (7 lb) or more. (Total body length at birth is about 550 mm, or 22 inches.) As you might expect with such tremendous growth, the changes in fetal appearance are quite dramatic **(Figure 28.15)**. Nevertheless, the greatest

28

TABLE 28.1 Developmental Events of the Fetal Period

TIME		CHANGES AND ACCOMPLISHMENTS
8 weeks (end of embryonic period)	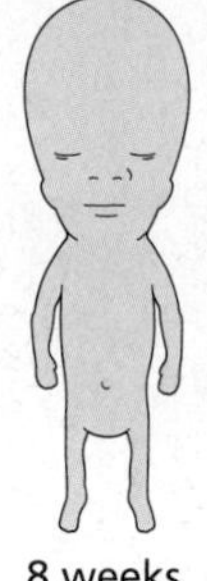 8 weeks	Head nearly as large as body; all major brain regions present; first brain waves in brain stem Liver disproportionately large and begins to form blood cells Limbs present; digits are initially webbed, but fingers and toes are free by the end of this interval Ossification just begun; weak, spontaneous muscle contractions occur Cardiovascular system fully functional (heart has been pumping blood since the fourth week) All body systems present in at least rudimentary form Approximate crown-to-rump length: 22 mm (0.9 inch); weight: 2 grams (0.06 ounce)
9–12 weeks (month 3)	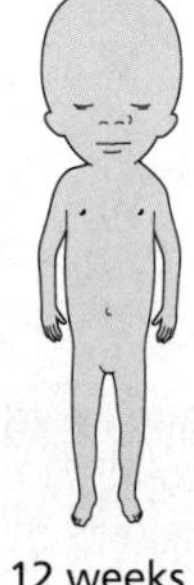 12 weeks	Head still dominant, but body elongating; brain continues to enlarge, shows its general structural features; cervical and lumbar enlargements apparent in spinal cord; retina of eye is present Skin epidermis and dermis obvious; facial features present in crude form Liver prominent and bile being secreted; palate is fusing; most glands of endodermal origin are developed; walls of hollow visceral organs gaining smooth muscle Blood cell formation begins in bone marrow Notochord degenerating and ossification accelerating; limbs well molded Sex readily detected from the genitals Approximate crown-to-rump length at end of interval: 90 mm
13–16 weeks (month 4)	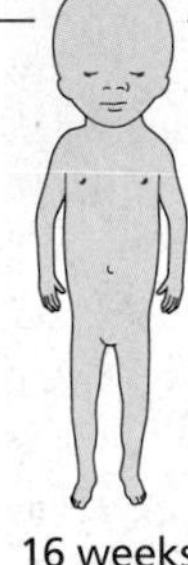 16 weeks	Cerebellum becoming prominent; general sensory organs differentiated; eyes and ears assume characteristic position and shape; blinking of eyes and sucking motions of lips occur Face looks human and growth of the body beginning to outpace that of the head Glands developed in GI tract; meconium is collecting Kidneys attain typical structure Most bones are now distinct and joint cavities are apparent Approximate crown-to-rump length at end of interval: 140 mm
17–20 weeks (month 5)		Vernix caseosa (fatty secretions of sebaceous glands) covers body; lanugo (silklike hair) covers skin Fetal position (body flexed anteriorly) assumed because of space restrictions Limbs reach near-final proportions Quickening occurs (mother feels spontaneous muscular activity of fetus) Approximate crown-to-rump length at end of interval: 190 mm
21–30 weeks (months 6 and 7)	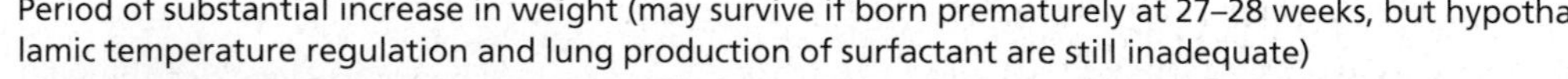	Period of substantial increase in weight (may survive if born prematurely at 27–28 weeks, but hypothalamic temperature regulation and lung production of surfactant are still inadequate) Myelination of spinal cord begins; eyes are open Distal limb bones are beginning to ossify Skin is wrinkled and red; fingernails and toenails are present; tooth enamel is forming on deciduous teeth Body is lean and well proportioned Bone marrow becomes sole site of blood cell formation Testes reach scrotum in seventh month (in males) Approximate crown-to-rump length at end of interval: 280 mm
30–40 weeks (term) (months 8 and 9)	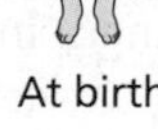 At birth	Skin whitish pink; fat laid down in subcutaneous tissue (hypodermis) Approximate crown-to-rump length at end of interval: 360 mm (14 inches); weight: 3.2 kg (7 lb)

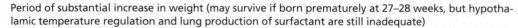

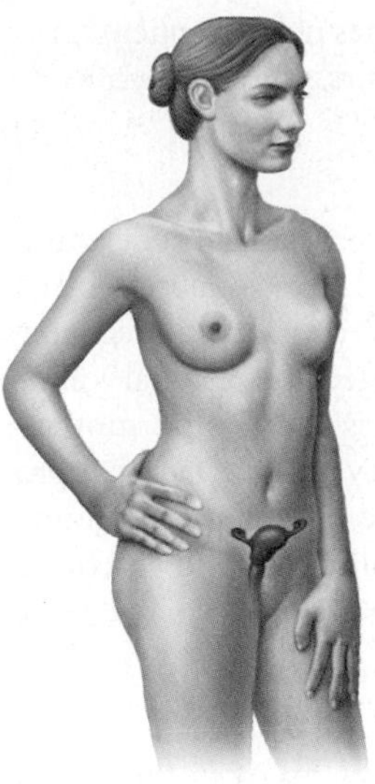

(a) **Before conception**
(Uterus the size of a fist and resides in the pelvis.)

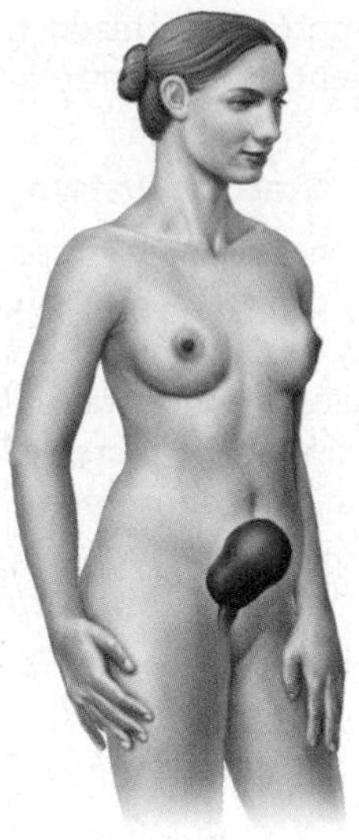

(b) **4 months**
(Fundus of the uterus is between the pubic symphysis and the umbilicus.)

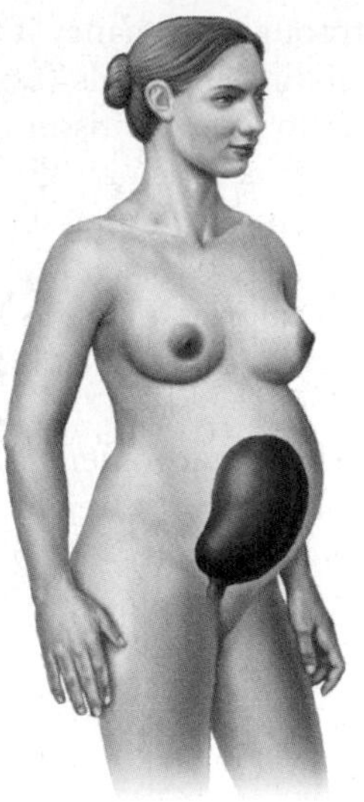

(c) **7 months**
(Fundus is well above the umbilicus.)

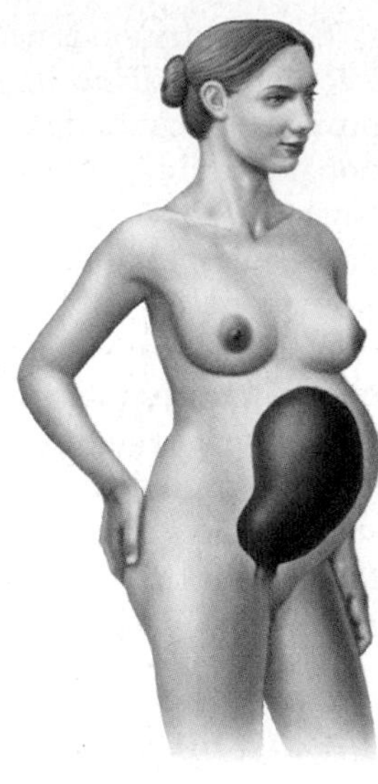

(d) **9 months**
(Fundus reaches the xiphoid process.)

Figure 28.16 **Relative size of the uterus before conception and during pregnancy.**

amount of growth occurs in the first 8 weeks of life, when the embryo grows from one cell to a fetus of 1 inch.

CHECK YOUR UNDERSTANDING

14. When does the fetal period begin?

For answers, see Appendix G.

Effects of Pregnancy on the Mother

- Describe functional changes in maternal reproductive organs and in the cardiovascular, respiratory, and urinary systems during pregnancy.
- Indicate the effects of pregnancy on maternal metabolism and posture.

Pregnancy can be a difficult time for the mother. Not only are there anatomical changes, but striking changes in her metabolism and physiology occur to support the pregnancy and prepare her body for delivery and lactation.

Anatomical Changes

As pregnancy progresses, the female reproductive organs become increasingly vascular and engorged with blood, and the vagina develops a purplish hue (*Chadwick's sign*). The enhanced vascularity increases vaginal sensitivity and sexual intensity, and some women achieve orgasm for the first time when they are pregnant. The breasts, too, engorge with blood and, prodded by rising levels of estrogen and progesterone, they enlarge and their areolae darken. Some women develop increased pigmentation of facial skin of the nose and cheeks, a condition called *chloasma* (klo-az′mah; "to be green") or the "mask of pregnancy."

The degree of uterine enlargement during pregnancy is remarkable. Starting as a fist-sized organ, the uterus fills most of the pelvic cavity by 16 weeks **(Figure 28.16a, b)**. Though the fetus is only about 140 mm long (crown-to-rump) at this time, the placenta is fully formed, uterine muscle is hypertrophied, and amniotic fluid volume is increasing.

As pregnancy continues, the uterus pushes higher into the abdominal cavity, exerting pressure on both abdominal and pelvic organs (Figure 28.16c). As birth nears, the uterus reaches the level of the xiphoid process and occupies most of the abdominal cavity (Figure 28.16d). The crowded abdominal organs press superiorly against the diaphragm, which intrudes on the thoracic cavity. As a result, the ribs flare, causing the thorax to widen.

The increasing bulkiness of the anterior abdomen changes the woman's center of gravity, and many women develop *lordosis* (accentuated lumbar curvature) and backaches during the last few months of pregnancy. Placental production of the hormone **relaxin** causes pelvic ligaments and the pubic symphysis to relax, widen, and become more flexible. This increased flexibility eases birth passage, but it may result in a waddling gait in the meantime. Considerable weight gain occurs during a normal pregnancy. Because some women are over- or underweight before pregnancy begins, it is almost impossible to state the ideal or desirable weight gain. However, summing up the weight increases resulting from fetal and placental growth, increased size of the maternal reproductive organs and breasts, and greater blood volume during pregnancy, a weight gain of approximately 13 kg (about 28 lb) usually occurs.

Good nutrition is necessary all through pregnancy if the developing fetus is to have all the building materials (especially proteins, calcium, and iron) needed to form its tissues. Additionally, multivitamins containing folic acid reduce the risk of having a baby with neurological problems, including such birth defects as spina bifida and anencephaly. However, a pregnant woman needs only 300 additional calories daily to sustain proper fetal growth. The emphasis should be on eating high-quality food, not just more food.

Not surprisingly, effects of the fetal environment may not show up until decades later. Below-normal birth weight, for instance, places females at risk for type 2 diabetes and increases the general risk of cardiovascular disease later in life for both men and women.

Metabolic Changes

As the placenta enlarges, it secretes increasing amounts of **human placental lactogen (hPL)**, also called **human chorionic somatomammotropin (hCS)**. hPL works cooperatively with estrogens and progesterone to stimulate maturation of the breasts for lactation, promotes growth of the fetus, and exerts a glucose-sparing effect in the mother. Consequently, maternal cells metabolize more fatty acids and less glucose than usual, sparing glucose for use by the fetus. Gestational diabetes mellitus occurs in about 10% of pregnancies, but over half of those women go on to develop type 2 diabetes later in life.

The placenta also releases **human chorionic thyrotropin (hCT)**, a glycoprotein hormone similar to thyroid-stimulating hormone of the anterior pituitary. hCT increases the rate of maternal metabolism throughout the pregnancy, causing hypermetabolism. Plasma levels of parathyroid hormone and activated vitamin D rise, so that pregnant women tend to be in positive calcium balance throughout pregnancy. This state ensures that the developing fetus will have adequate calcium to mineralize its bones.

Physiological Changes

Physiological changes take place in many systems during pregnancy. A few of these changes are described next.

Gastrointestinal System

28

Until their system adjusts to the elevated levels of progesterone and estrogens, many women suffer nausea, commonly called *morning sickness*, during the first few months of pregnancy. (Nausea is also a side effect of many birth control pills.) *Heartburn*, due to reflux of stomach acid into the esophagus, is common because the esophagus is displaced and the stomach is crowded by the growing uterus. *Constipation* occurs because motility of the digestive tract declines during pregnancy.

Urinary System

The kidneys produce more urine during pregnancy because of the mother's increased metabolic rate and the additional burden of disposing of fetal metabolic wastes. As the growing uterus compresses the bladder, urination becomes more frequent, more urgent, and sometimes uncontrollable (*stress incontinence*).

Respiratory System

The nasal mucosa responds to estrogens by becoming edematous and congested. Thus, nasal stuffiness and occasional nosebleeds may occur. Tidal volume increases markedly during pregnancy, while respiratory rate is relatively unchanged and residual volume declines. The increase in tidal volume is due to the mother's greater need for oxygen during pregnancy and the fact that progesterone enhances the sensitivity of the medullary respiratory center to CO_2. Many women exhibit *dyspnea* (disp-ne′ah), or difficult breathing, during the later stages of pregnancy.

Cardiovascular System

The most dramatic physiological changes occur in the cardiovascular system. Total body water rises, and blood volume increases 25–40% by the 32nd week to accommodate the additional needs of the fetus. The rise in blood volume also safeguards against blood loss during birth. Blood pressure and pulse typically rise and increase cardiac output by 20–40% at various stages of pregnancy. This helps propel the greater circulatory volume around the body. The uterus presses on the pelvic blood vessels, which may impair venous return from the lower limbs, resulting in *varicose veins* and leg edema.

HOMEOSTATIC IMBALANCE

A dangerous complication of pregnancy called **preeclampsia** results in an insufficient placental blood supply, which can starve a fetus of oxygen. The pregnant woman becomes edematous and hypertensive, and proteinuria occurs. This condition, which affects one in 10 pregnancies, is believed to be due to immunological abnormalities in some cases, because its occurrence seems to be positively correlated with the number of fetal cells that enter the maternal circulation. ■

CHECK YOUR UNDERSTANDING

15. What causes the difficult breathing that some women experience during pregnancy? What causes the waddling gait seen in some?

16. What is the cause of morning sickness?

17. What is the role of the hormone hCT?

For answers, see Appendix G.

Parturition (Birth)

▶ Explain how labor is initiated, and describe the three stages of labor.

Parturition (par″tu-rish′un; "bringing forth young") is the culmination of pregnancy—giving birth to the baby. It usually occurs within 15 days of the calculated due date (280 days from the last menstrual period). The series of events that expel the infant from the uterus are collectively called **labor.**

Initiation of Labor

Several events and hormones interlock to trigger labor. During the last few weeks of pregnancy, estrogens reach their highest levels in the mother's blood. Studies indicate that the fetus determines its own birth date. Rising levels of fetal adrenocortical hormones (especially cortisol) late in pregnancy are a major stimulus for the placenta to release such large amounts of estrogens. In addition, increased production of *surfactant protein A* (*SP-A*) by the fetal lungs in the weeks before delivery appears to trigger an inflammatory response in the cervix that stimulates its softening in preparation for labor.

The rise in estrogens has two important consequences. It stimulates the myometrial cells of the uterus to form abundant oxytocin receptors **(Figure 28.17)**. It also antagonizes progesterone's quieting influence on uterine muscle. As a result, the myometrium becomes increasingly irritable, and weak, irregular uterine contractions begin to occur. These contractions, called *Braxton Hicks contractions*, have caused many women to go to the hospital, only to be told that they were in **false labor** and sent home.

As birth nears, two more chemical signals cooperate to convert these false labor pains into the real thing. Certain fetal cells begin to produce **oxytocin** (ok″sĭ-to′sin), which causes the placenta to release **prostaglandins** (pros″tah-glan′dinz) (Figure 28.17). Both hormones are powerful uterine muscle stimulants, and since the myometrium is now highly sensitive to oxytocin, contractions become more frequent and more vigorous. While elevated levels of oxytocin and prostaglandins sustain labor once it begins, many studies indicate that it is the prostaglandins (acting as paracrines) that actually trigger the rhythmic expulsive contractions of true labor. At this point, the increasing emotional and physical stresses (pain and uterine distension respectively) activate the mother's hypothalamus, which signals for oxytocin release by the posterior pituitary.

Once the hypothalamus is involved, a *positive feedback mechanism* is propelled into action—greater distension causes the release of more oxytocin, which causes greater contractile force, and so on (Figure 28.17). These expulsive contractions are aided by the fact that *fetal fibronectin*, a natural "stickum" (adhesive protein) that binds the fetal and maternal tissues of the placenta together throughout pregnancy, changes to a lubricant just before true labor begins.

As mentioned earlier, prostaglandins are essential for initiating labor in humans, and interfering with their production will hinder onset of labor. For example, antiprostaglandin drugs such as ibuprofen can inhibit the early stages of labor and such drugs are used occasionally to prevent preterm births.

Stages of Labor

Labor includes the dilation, expulsion, and placental stages illustrated in **Figure 28.18**.

Stage 1: Dilation Stage

The **dilation stage** is the time from labor's onset until the cervix is fully dilated by the baby's head (about 10 cm in diameter) (Figure 28.18a). As labor starts, weak but regular contractions begin in the upper part of the uterus and move toward the vagina. At first, only the superior uterine muscle is active, the contractions are 15–30 minutes apart, and they last for 10–30 seconds. As labor progresses, the contractions become more vigorous and rapid, and the lower part of the uterus gets involved. As the infant's head is forced against the cervix with each contraction, the cervix softens and thins (*effaces*), and dilates. Eventually the amnion ruptures, releasing the amniotic fluid, an event commonly called "breaking the water."

The dilation stage is the longest part of labor, lasting 6–12 hours or more. Several events happen during this phase. *Engagement* occurs when the infant's head enters the true pelvis. As descent continues through the birth canal, the baby's head rotates so that its greatest dimension is in the anteroposterior line, which allows it to navigate the narrow dimensions of the pelvic outlet (Figure 28.18b).

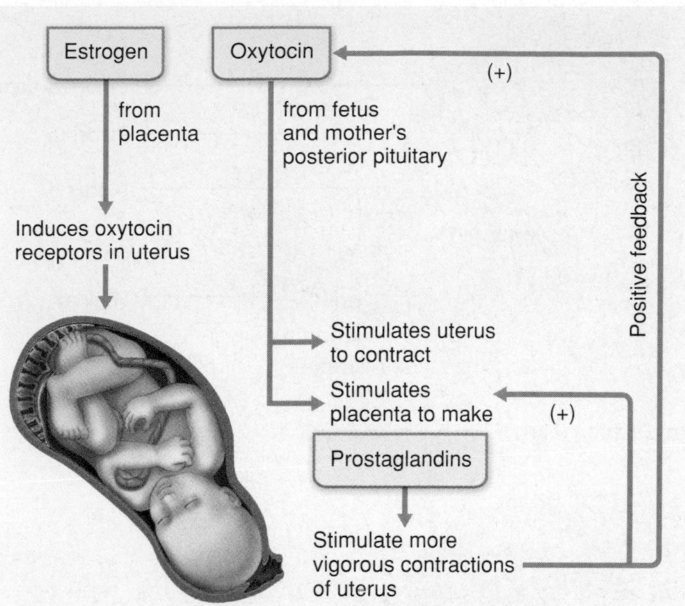

Figure 28.17 Hormonal induction of labor.

Stage 2: Expulsion Stage

The **expulsion stage** lasts from full dilation to delivery of the infant, or actual childbirth (Figure 28.18c). By the time the cervix is fully dilated, strong contractions occur every 2–3 minutes and last about 1 minute. In this stage, a mother undergoing labor without local anesthesia has an increasing urge to push or bear down with the abdominal muscles. Although this phase may last 2 hours, it is typically 50 minutes in a first birth and around 20 minutes in subsequent births.

Crowning occurs when the largest dimension of the baby's head distends the vulva. At this point, an *episiotomy* (e-piz″e-ot′o-me) may be done to reduce tissue tearing. An episiotomy is an incision made to widen the vaginal orifice. The baby's neck extends as the head exits from the perineum, and once the head has been delivered, the rest of the baby's body is delivered much more easily. After birth, the umbilical cord is clamped and cut.

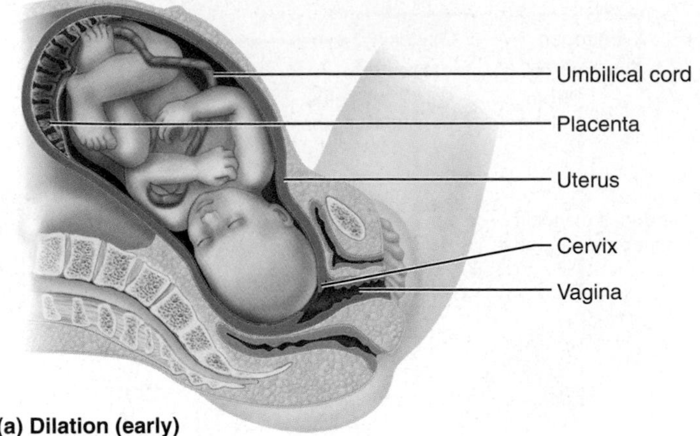

(a) Dilation (early)

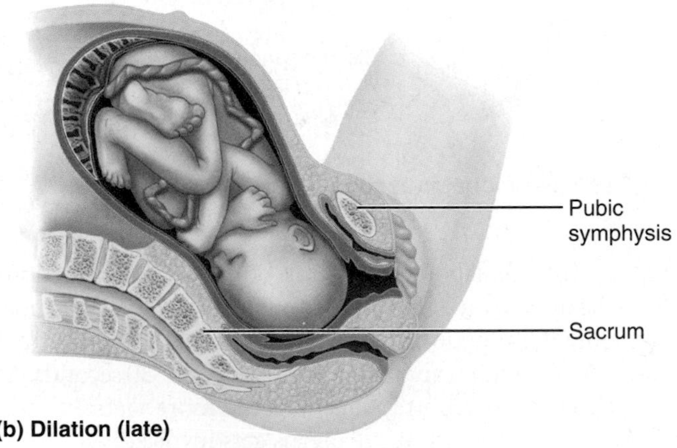

(b) Dilation (late)

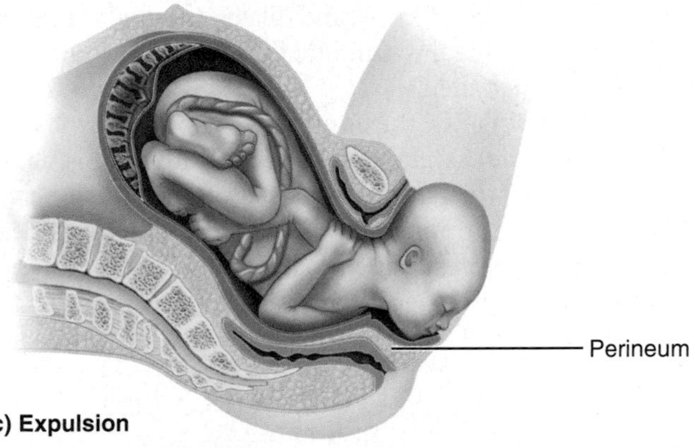

(c) Expulsion

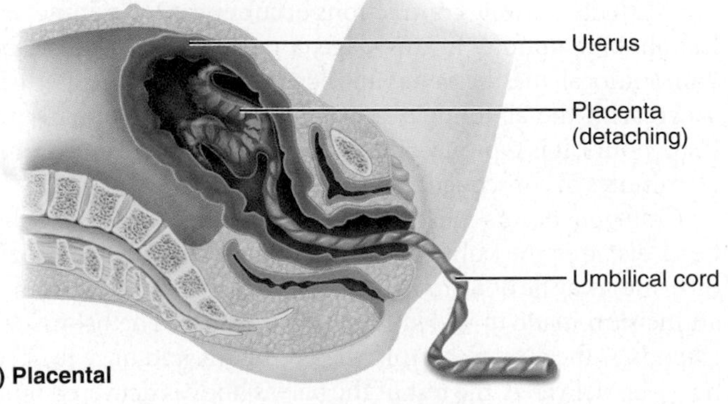

(d) Placental

28

Figure 28.18 Parturition. (a) Dilation stage (early). The baby's head is engaged in the true pelvis. The widest head dimension is along the left-right axis. **(b)** Late dilation. The baby's head rotates so that its greatest dimension is in the anteroposterior axis as it moves through the pelvic outlet. Dilation of the cervix is nearly complete. **(c)** Expulsion stage. The baby's head extends as it reaches the perineum and is delivered. **(d)** Placental stage. After the baby is delivered, the placenta is detached by the continuing uterine contractions and removed.

When the infant is in the usual *vertex*, or head-first, *presentation*, the skull (its largest diameter) acts as a wedge to dilate the cervix. The head-first presentation also allows the baby to be suctioned free of mucus and to breathe even before it has completely exited from the birth canal (Figure 28.18c). In *breech* (buttock-first) and other nonvertex presentations, these advantages are lost and delivery is much more difficult, often requiring the use of forceps, or a C-section (see below).

HOMEOSTATIC IMBALANCE

If a woman has a deformed or malelike pelvis, labor may be prolonged and difficult. This condition is called *dystocia* (dis-to′se-ah; *dys* = difficult; *toc* = birth). Besides extreme maternal fatigue, another possible consequence of dystocia is fetal brain damage, resulting in cerebral palsy or epilepsy. To prevent these outcomes, a *cesarean* (*C*-) *section* (se-sa′re-an) is performed in many such cases. A C-section is delivery of the infant through an incision made through the abdominal and uterine walls. ■

Stage 3: Placental Stage

The **placental stage**, or the delivery of the placenta and its attached fetal membranes, which are collectively called the **afterbirth**, is usually accomplished within 30 minutes after birth of the infant (Figure 28.18d). The strong uterine contractions that continue after birth compress uterine blood vessels, limit bleeding, and shear the placenta off the uterine wall (cause placental detachment). It is very important that all placental fragments be removed to prevent continued uterine bleeding after birth (*postpartum bleeding*).

CHECK YOUR UNDERSTANDING

18. What is a breech presentation?
19. What chemical is most responsible for triggering true labor?
20. Why does a baby turn as it travels through the birth canal?

For answers, see Appendix G.

Adjustments of the Infant to Extrauterine Life

- Outline the events leading to the first breath of a newborn.
- Describe changes that occur in the fetal circulation after birth.

The **neonatal period** is the four-week period immediately after birth. Here we will be concerned with the events of just the first few hours after birth in a normal infant. As you might suspect, birth represents quite a shock to the infant. Exposed to physical trauma during the birth process, it is suddenly cast out of its watery, warm environment and its placental life supports are severed. Now it must do for itself all that the mother had been doing for it—respire, obtain nutrients, excrete, and maintain its body temperature.

At 1 and 5 minutes after birth, the infant's physical status is assessed based on five signs: heart rate, respiration, color, muscle tone, and reflexes (tested by response to catheter in nostril). Each observation is given a score of 0 to 2, and the total is called the **Apgar score**. An Apgar score of 8 to 10 indicates a healthy baby. Lower scores reveal problems in one or more of the physiological functions assessed.

Taking the First Breath and Transition

The crucial first requirement is to breathe. Vasoconstriction of the umbilical arteries, initiated when they are stretched during birth, leads to loss of placental support. Once carbon dioxide is no longer removed by the placenta, it accumulates in the baby's blood, causing central acidosis. This excites respiratory control centers in the baby's brain and triggers the first inspiration. The first breath requires a tremendous effort—the airways are tiny, and the lungs are collapsed. However, once the lungs have been inflated in full-term babies, surfactant in alveolar fluid reduces surface tension in the alveoli, and breathing is easier. The rate of respiration is rapid (about 45 respirations/min) during the first two weeks and then gradually declines.

Keeping the lungs inflated is much more difficult for premature infants (those weighing less than 2500 g, or about 5.5 lb, at birth) because surfactant production occurs during the last months of prenatal life. Consequently, preemies are usually put on respiratory assistance (a ventilator) until their lungs are mature enough to function on their own.

For 6–8 hours after birth, infants pass through an unstable **transitional period** marked by alternating periods of increased activity and sleep, and during which they adjust to extrauterine life. During the activity periods, vital signs are irregular and the baby gags frequently as it regurgitates mucus and debris. After this, the infant stabilizes, with waking periods (dictated by hunger) occurring every 3–4 hours.

Occlusion of Special Fetal Blood Vessels and Vascular Shunts

After birth the special umbilical blood vessels and fetal shunts are no longer necessary (see Figure 28.14b). The umbilical arteries and vein constrict and become fibrosed. The proximal parts of the umbilical arteries persist as the *superior vesical arteries* that supply the urinary bladder, and their distal parts become the **medial umbilical ligaments**. The remnant of the umbilical vein becomes the **round ligament of the liver**, or **ligamentum teres**, that attaches the umbilicus to the liver. The ductus venosus collapses as blood stops flowing through the umbilical vein and is eventually converted to the **ligamentum venosum** on the liver's undersurface.

As the pulmonary circulation becomes functional, pressure in the left side of the heart increases and that in the right side of the heart decreases, causing the pulmonary shunts to close. The flap of the foramen ovale is pushed to the shut position, and its edges fuse to the septal wall. Ultimately, only a slight depression, the **fossa ovalis**, marks its position. The ductus arteriosus constricts and is converted to the cordlike **ligamentum arteriosum**, connecting the aorta and pulmonary trunk.

Except for the foramen ovale, all of the special circulatory adaptations of the fetus are functionally occluded within 30 minutes after birth. Closure of the foramen ovale is usually complete within the year. As we described in Chapter 18, failure of the ductus arteriosus or foramen ovale to close leads to congenital heart defects.

CHECK YOUR UNDERSTANDING

21. What two modifications of the fetal circulation allow most blood to bypass parts of the heart?
22. What happens to the special fetal circulatory modifications after birth?

For answers, see Appendix G.

Lactation

▶ Explain how the breasts are prepared for lactation.

Lactation is production of milk by the hormone-prepared mammary glands. Rising levels of (placental) estrogens, progesterone, and human placental lactogen toward the end of pregnancy stimulate the hypothalamus to release prolactin-releasing factors (PRFs). The anterior pituitary gland responds by secreting **prolactin**. (This mechanism is described below.) After a delay of two to three days following birth, true milk production begins.

During the initial delay (and during late gestation), the mammary glands secrete a yellowish fluid called **colostrum** (ko-los′trum). It has less lactose than milk and almost no fat, but it contains more protein, vitamin A, and minerals than true milk. Like milk, colostrum is rich in IgA antibodies. Since these antibodies are resistant to digestion in the stomach, they may help to protect the infant's digestive tract against bacterial infection. Additionally, these IgA antibodies are absorbed by endocytosis and subsequently enter the bloodstream to provide even broader immunity.

After birth, prolactin release gradually wanes, and continual milk production depends on mechanical stimulation of the nipples, normally provided by the suckling infant. Mechanoreceptors in the nipple send afferent nerve impulses to the hypothalamus, stimulating secretion of PRF. This results in a burstlike release of prolactin, which stimulates milk production for the next feeding.

The same afferent impulses also prompt hypothalamic release of oxytocin from the posterior pituitary via a *positive feedback mechanism*. Oxytocin causes the **let-down reflex**, the actual ejection of milk from the alveoli of the mammary glands

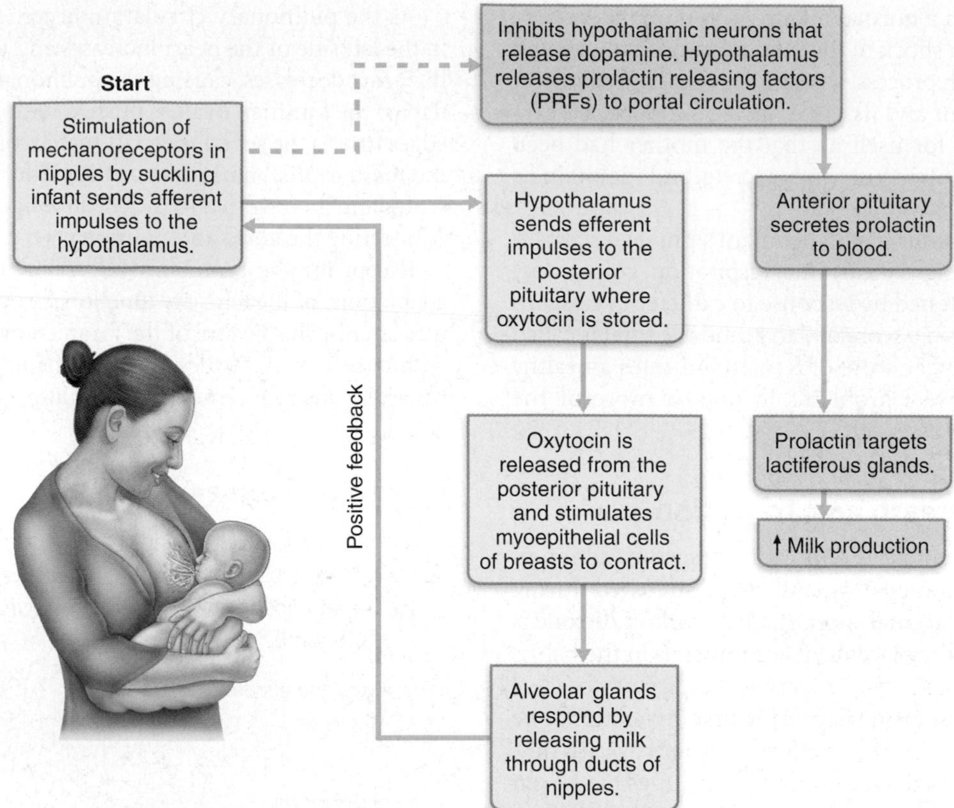

Start
Stimulation of mechanoreceptors in nipples by suckling infant sends afferent impulses to the hypothalamus.
Inhibits hypothalamic neurons that release dopamine. Hypothalamus releases prolactin releasing factors (PRFs) to portal circulation.
Hypothalamus sends efferent impulses to the posterior pituitary where oxytocin is stored.
Anterior pituitary secretes prolactin to blood.
Positive feedback
Oxytocin is released from the posterior pituitary and stimulates myoepithelial cells of breasts to contract.
Prolactin targets lactiferous glands.
↑ Milk production
Alveolar glands respond by releasing milk through ducts of nipples.

Figure 28.19 Milk production and the positive feedback mechanism of the milk let-down reflex.

(Figure 28.19). Let-down occurs when oxytocin binds to myoepithelial cells surrounding the glands, after which milk is ejected from *both* breasts, not just the suckled one. During nursing oxytocin also stimulates the recently emptied uterus to contract, helping it to return to (nearly) its prepregnant size.

Breast milk has advantages for the infant:

1. *Its fats and iron are better absorbed and its amino acids are metabolized more efficiently* than those of cow's milk.
2. *It has a host of beneficial chemicals,* including IgA, complement, lysozyme, interferon, and lactoperoxidase, that protect infants from life-threatening infections. Mother's milk also contains interleukins and prostaglandins that prevent overzealous inflammatory responses, and a glycoprotein that deters the ulcer-causing bacterium (*H. pylori*) from attaching to the stomach mucosa.
3. *Its natural laxative effect helps to cleanse the bowels of* **meconium** (mĕ-ko′ne-um), a tarry green-black paste containing sloughed-off epithelial cells, bile, and other substances. Since meconium, and later feces, provides the route for eliminating bilirubin from the body, clearing meconium as quickly as possible helps to prevent *physiological jaundice* (see the Related Clinical Terms section). It also encourages bacteria (the source of vitamin K and some B vitamins) to colonize the large intestine.

28

When nursing is discontinued, the stimulus for prolactin release and milk production ends, and the mammary glands stop producing milk. Women who nurse their infants for six months or more lose a significant amount of calcium from their bones, but those on sound diets usually replace lost bone calcium after weaning the infant.

While prolactin levels are high, the normal hypothalamic-pituitary controls of the ovarian cycle are damped, probably because stimulation of the hypothalamus by suckling causes it to release beta endorphin, a peptide hormone that inhibits hypothalamic release of GnRH and, for this reason, the release of gonadotropins by the pituitary. Because of this inhibition of ovarian function, nursing has been called natural birth control. Nonetheless, there is a good deal of "slippage" in these controls, and most women begin to ovulate even while continuing to nurse their infants.

CHECK YOUR UNDERSTANDING

23. What hormone causes the let-down reflex?

For answers, see Appendix G.

Assisted Reproductive Technology and Reproductive Cloning

▶ Describe some techniques of ART including IVF, ZIFT, and GIFT.

So far we have been describing how babies are made. But if a couple lacks that capability for some reason, what recourse do

A CLOSER LOOK

Contraception: To Be or Not To Be

In a society such as ours, where many women opt for professional careers or work for economic reasons, **contraception** (*contra* = against, *cept* = taking), commonly called birth control, is often seen as a necessity. An effective antisperm vaccine exists, but so far the only approved male contraceptive methods are vasectomy and condom use. Consequently, the burden for birth control still falls on women's shoulders, and most birth control products are female directed.

The key to birth control is dependability. As the red arrows in the accompanying flowchart show, the birth control techniques currently available have many sites of action for blocking reproduction. Let's examine a few of them more closely.

Coitus interruptus, or withdrawal of the penis just before ejaculation, is unreliable because control of ejaculation is never ensured. Additionally, sperm may be present in the preejaculatory fluid secreted by the bulbourethral glands.

Rhythm or *temporary abstinence methods* involve avoiding intercourse during periods of ovulation or fertility. This may be accomplished by (1) recording daily basal body temperatures (body temperature drops slightly immediately prior to ovulation and then rises slightly after ovulation) or (2) more simply by buying the over-the-counter *Ovulite Microscope*, which is the size of a lipstick. On awakening, a drop of saliva is placed on the microscope slide and examined for a particular pattern of crystals that indicate the optimal days for fertilization. These techniques require accurate record keeping for several cycles before they can be used with confidence, but have a high success rate for those willing to take the time necessary. With a failure rate of 10–20%, it is obvious that some people are willing and some are not.

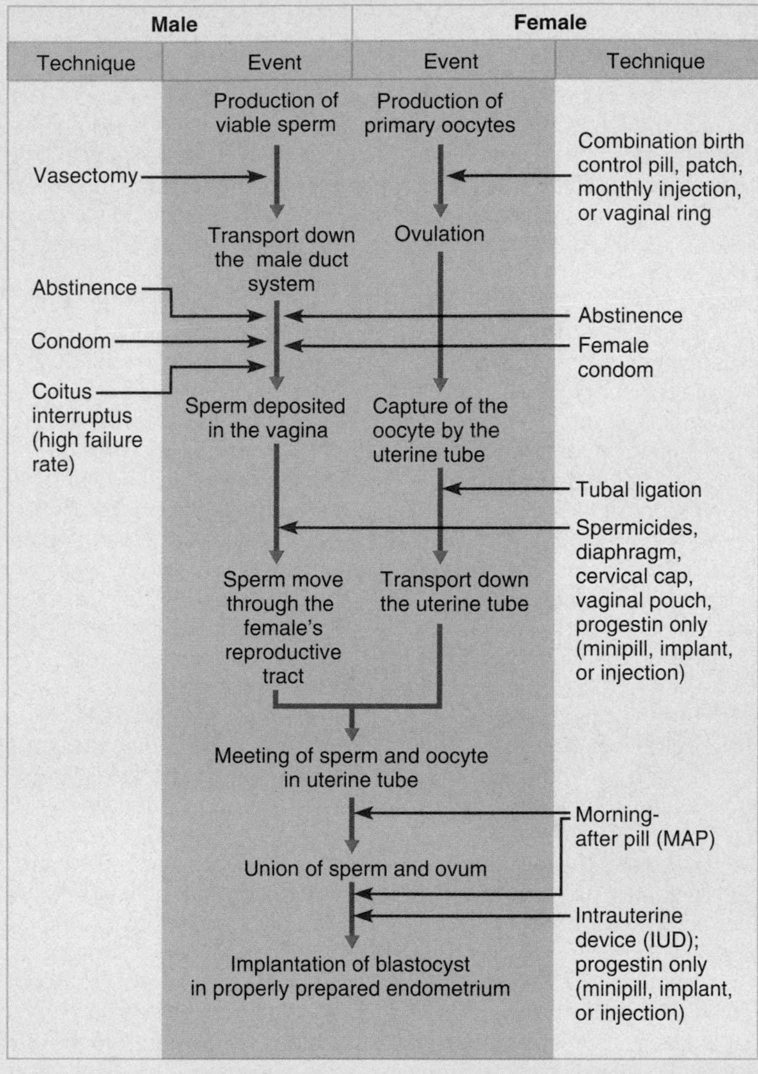

Mechanisms of contraception. Techniques or products that interfere with events from production of gametes to implantation are indicated by red arrows at the site of interference and act to prevent the next step from occurring.

Barrier methods, such as diaphragms, cervical caps, male and female condoms, spermicidal foams, gels, and sponges, are

28

they have? Hormone therapy may increase sperm or egg production in cases where that is the problem, and surgery can open blocked uterine tubes. Beyond that are *assisted reproductive technology* (*ART*), procedures that entail surgically removing oocytes from a woman's ovaries following hormone stimulation, fertilizing the oocytes, and then returning them to the woman's body. These procedures, now performed worldwide in major medical centers, have produced thousands of infants, but they are expensive, emotionally draining, and painful for the oocyte donor. Unused oocytes, sperm, and embryos can be frozen for later attempts at accomplishing pregnancy.

In the most common ART process, *in vitro fertilization* (*IVF*), harvested oocytes are incubated with sperm in culture dishes (*in vitro*) for several days to allow fertilization to occur. In cases where the quality or number of sperm is low, the oocytes are injected with sperm. Embryos reaching the two-cell or blastocyst stage are then carefully transferred into the woman's uterus in the hope that implantation will occur.

A CLOSER LOOK
(continued)

quite effective, especially when used in combination—for example, condoms and spermicides. But many avoid them because they can reduce the spontaneity of sexual encounters, and sperm may be present in male secretions that precede ejaculation.

For several years, the second most used contraceptive method was the *intrauterine device* (*IUD*), and it is still one of the most commonly used methods in the world, largely because of its economy. Developed in 1909, this plastic or metal device is inserted into the uterus and prevents the young embryo from implanting in the endometrial lining. Although IUDs' failure rate was nearly as low as that of the pill, they were taken off the U.S. market because of occasional contraceptive failure, uterine perforation, or pelvic inflammatory disease (PID). New IUD products that provide sustained local delivery of synthetic progesterone to the endometrium are particularly recommended for women who have given birth and are in monogamous relationships (i.e., who have a lower risk of developing PID).

The most-used contraceptive product in the United States is the *birth control pill*, or simply "the pill," first marketed in 1960. Supplied in 28-tablet packets, the first 20 or 21 tablets contain minute amounts of estrogens and progestins (progesterone-like hormones) taken daily; the last seven tablets are hormone free. The pill tricks the hypothalamic-pituitary axis and "lulls it to sleep," because the relatively constant blood levels of ovarian hormones make it appear that the woman is pregnant (both estrogen and progesterone are produced throughout pregnancy). Ovarian follicles do not develop, ovulation ceases, and menstrual flow is much reduced. However, since hormonal balance is one of the most precisely controlled body functions, some women cannot tolerate these changes—they become nauseated and/or hypertensive. The pill has adverse cardiovascular effects in a small number of users and there is still debate about whether it increases the risk of uterine, ovarian, or (particularly) breast cancer. Presently, well over 50 million women use the pill, and its failure rate is less than 1%.

Other delivery methods using the combination hormone approach include two slow-release products approved in 2001—a flexible ring that is inserted into the vagina, and a transdermal (skin) patch. Failure rates and side effects of the vaginal ring and skin patch are comparable to those of the pill, although in clinical trials the patch was less effective in women weighing more than 198 pounds.

Combination birth control pills with substantially higher hormone concentrations used for *postcoital contraception* have a 75% effectiveness. Taken within three days of unprotected intercourse, these *morning-after pills* (*MAPs*), or *emergency contraceptive pills* (*ECPs*) as they are also called, "mess up" normal hormonal signals enough to prevent a fertilized egg from implanting or prevent fertilization altogether.

Other hormonal approaches to contraception use progestin-only products which thicken the cervical mucus enough to block sperm entry into the uterus, decrease the frequency of ovulation, and make the endometrium inhospitable to implantion. These include a tablet form (the *minipill*), match-size silicone rods implanted just under the skin that release progestin over a five-year period (Norplant), and an injectable form that lasts for three months (Depo Provera). The failure rates of progestin treatments are even less than that of the "pill."

Abortion is the termination of a pregnancy that is in progress. Spontaneous abortion, also called miscarriage, is common and frequently occurs before a woman is aware that she has conceived. Additionally, over a million American women choose to undergo abortions performed by physicians. *Mifepristone* (RU-486), the so-called abortion pill developed in France, enables a woman to end a pregnancy during its first 7 weeks and has a 96–98% success rate with few side effects. RU-486 is an antihormone that, when taken along with a tiny amount of prostaglandin to stimulate uterine contractions, induces miscarriage by blocking progesterone's quieting effect on the uterus.

Sterilization techniques permanently prevent gamete release. *Tubal ligation* or vasectomy (cutting or cauterizing the uterine tubes or ductus deferens, respectively) are nearly foolproof and are the choice of approximately 33% of couples of childbearing age in the United States. Both procedures can be done in the physician's office. However, these techniques are usually permanent, making them unpopular with individuals who still plan to have children but want to select the time.

This summary doesn't even begin to touch on the experimental birth control drugs now awaiting clinical trials, and other methods are sure to be developed in the near future. In the final analysis, however, the only 100% effective means of birth control is the age-old one—total abstinence.

28

In *zygote intrafallopian transfer* (*ZIFT*), oocytes fertilized in vitro are immediately transferred to the woman's uterine (fallopian) tubes. The goal is to have development to the blastocyst stage occur followed by normal implantation in the uterus.

In *gamete intrafallopian transfer* (*GIFT*), no in vitro procedures are used. Instead, sperm and harvested oocytes are transferred together into the woman's uterine tubes in the hope that fertilization will take place there.

Although there is a great deal of hubbub in the scientific community about using cloning as another possible avenue to produce offspring, humans have proved notoriously difficult to create and sustain past very early (blastocyst) development. Cloning entails insertion of a somatic cell nucleus into an oocyte from which the nucleus is removed and then an incubation period to dedifferentiate the inserted nucleus. This technique has proved more successful in creating stem cells for therapeutic use in treating selected diseases than for reproductive cloning to produce whole and healthy human offspring. Furthermore, human reproductive cloning is currently fraught with legal, moral, ethical, and political roadblocks.

■ ■ ■

In this chapter, we have focused on changes that occur during human development in utero. But having a baby is not always what the interacting partners have in mind, and we humans have devised a variety of techniques for preventing this outcome (see *A Closer Look* on p. 1095).

We must admit that the description of embryonic development here has fallen short because we have barely touched upon the phenomenon of differentiation. How does an unspecialized cell that can become *anything* in the body develop into a specific *something* (a heart cell, for example)? And what paces the developmental sequence, so that if a particular process fails to occur at a precise time, it never occurs at all? Scientists are beginning to believe that there are master switches in the genes. In Chapter 29, the final chapter of this book, we describe a small part of the "how" as we examine the interaction of genes and other components that determine who we finally become.

RELATED CLINICAL TERMS

Abortion (*abort* = born prematurely) Premature removal of the embryo or fetus from the uterus; may be spontaneous or induced.

Ectopic pregnancy (ek-top′ik; *ecto* = outside) A pregnancy in which the embryo implants in any site other than the uterus; most often the site is a uterine tube (tubal pregnancy). Since the uterine tube (as well as most other ectopic sites) is unable to establish a placenta or accommodate growth, the uterine tube ruptures unless the condition is diagnosed early, or the pregnancy spontaneously aborts.

Hydatid (hydatidiform) mole (hi′dah-tid; *hydat* = watery) Developmental abnormality of the placenta; the conceptus degenerates and the chorionic villi convert into a mass of vesicles that resemble tapioca. Signs include vaginal bleeding, which contains some of the grapelike vesicles.

Physiological jaundice (jawn′dis) Jaundice sometimes occurring in normal newborns within three to four days after birth. Fetal erythrocytes are short-lived, and they break down rapidly after birth; the infant's liver may be unable to process the bilirubin (breakdown product of hemoglobin pigment) fast enough to prevent its accumulation in blood and subsequent deposit in body tissues.

Placenta abruptio (ah-brup′she-o; *abrupt* = broken away from) Premature separation of the placenta from the uterine wall; if this occurs before labor, it can result in fetal death due to anoxia.

Placenta previa (pre′ve-ah) Placental formation adjacent to or across the internal os of the uterus. Represents a problem because as the uterus and cervix stretch, tearing of the placenta may occur. Additionally, the placenta precedes the infant during labor.

Ultrasonography (ul″trah-son-og′rah-fe) Noninvasive technique that uses sound waves to visualize the position and size of the fetus and placenta (see *A Closer Look*, Chapter 1).

CHAPTER SUMMARY

1. The gestation period of approximately 280 days extends from the woman's last menstrual period to birth. The conceptus undergoes embryonic development for 8 weeks after fertilization, and fetal development from week 9 to birth.

From Egg to Zygote (pp. 1072–1075)

Accomplishing Fertilization (pp. 1072–1075)

1. An oocyte is fertilizable for up to 24 hours; most sperm are viable within the female reproductive tract for one to two days.
2. Sperm must survive the hostile environment of the vagina and become capacitated (capable of reaching and fertilizing the oocyte).
3. Hundreds of sperm must release their acrosomal enzymes to break down the egg's corona radiata and zona pellucida.
4. When one sperm binds to receptors on the egg, it triggers the slow block to polyspermy (release of cortical granules).
5. Following sperm penetration, the secondary oocyte completes meiosis II. Then the ovum and sperm pronuclei fuse (fertilization), forming a zygote.

Events of Embryonic Development: Zygote to Blastocyst Implantation (pp. 1075–1078)

Cleavage and Blastocyst Formation (pp. 1075–1076)

1. Cleavage, a rapid series of mitotic divisions without intervening growth, begins with the zygote and ends with a blastocyst. The blastocyst consists of the trophoblast and an inner cell mass. Cleavage produces a large number of cells with a favorable surface-to-volume ratio.

Implantation (pp. 1076–1078)

2. The trophoblast adheres to, digests, and implants in the endometrium. Implantation is completed when the blastocyst is entirely surrounded by endometrial tissue, about 12 days after ovulation.
3. hCG released by the blastocyst maintains hormone production by the corpus luteum, preventing menses. hCG levels decline after four months.

Placentation (p. 1078)

4. The placenta acts as the respiratory, nutritive, and excretory organ of the fetus and produces the hormones of pregnancy. It is formed

from embryonic (chorionic villi) and maternal (endometrial decidua) tissues. The chorion develops when the trophoblast becomes associated with extraembryonic mesoderm. Typically, the placenta is functional as an endocrine organ by the third month.

Events of Embryonic Development: Gastrula to Fetus (pp. 1079–1087)

Formation and Roles of the Extraembryonic Membranes (p. 1080)

1. The fluid-filled amnion forms from cells of the superior surface (epiblast) of the embryonic disc. It protects the embryo from physical trauma and adhesion formation, provides a constant temperature, and allows fetal movements.
2. The yolk sac forms from the hypoblast of the embryonic disc; it is the source of early blood cells.
3. The allantois, a caudal outpocketing adjacent to the yolk sac, forms the structural basis of the umbilical cord.
4. The chorion is the outermost membrane and takes part in placentation.

Gastrulation: Germ Layer Formation (pp. 1081–1083)

5. Gastrulation involves cellular migrations that ultimately transform the inner cell mass into a three-layered embryo (gastrula) containing ectoderm, mesoderm, and endoderm. Cells that move through the midline primitive streak become endoderm if they form the most inferior layer of the embryonic disc and mesoderm if they ultimately occupy the middle layer. Cells remaining on the superior surface become ectoderm.

Organogenesis: Differentiation of the Germ Layers (pp. 1083–1087)

6. Ectoderm forms the nervous system and the epidermis of the skin and its derivatives. The first event of organogenesis is neurulation, which produces the brain and spinal cord. By the eighth week, all major brain regions are formed.
7. Endoderm forms the mucosa of the digestive and respiratory systems, and the epithelial cells of all associated glands (thyroid, parathyroids, thymus, liver, pancreas). It becomes a continuous tube when the embryonic body undercuts and fuses ventrally.
8. Mesoderm forms all other organ systems and tissues. It segregates early into (1) a dorsal superior notochord, (2) paired somites that form the vertebrae, skeletal trunk muscles, and part of the dermis, and (3) paired masses of intermediate and lateral plate mesoderm. The intermediate mesoderm forms the kidneys and gonads. The somatic layer of the lateral plate mesoderm forms the dermis of skin, parietal serosa, and bones and muscles of the limbs; the splanchnic layer forms the cardiovascular system and the visceral serosae.
9. The fetal cardiovascular system is formed in the embryonic period. The umbilical vein delivers nutrient- and oxygen-rich blood to the embryo; the paired umbilical arteries return oxygen-poor, waste-laden blood to the placenta. The ductus venosus allows most of the blood to bypass the liver; the foramen ovale and ductus arteriosus are pulmonary shunts.

Events of Fetal Development (pp. 1087–1089)

1. All organ systems are laid down during the embryonic period; growth and tissue/organ specialization are the major events of the fetal period.
2. During the fetal period, fetal length increases from about 22 mm to 360 mm, and weight increases from less than an ounce to 7 pounds or more.

Effects of Pregnancy on the Mother (pp. 1089–1090)

Anatomical Changes (pp. 1089–1090)

1. Maternal reproductive organs and breasts become increasingly vascularized during pregnancy, and the breasts enlarge.
2. The uterus eventually occupies nearly the entire abdominopelvic cavity. Abdominal organs are pushed superiorly and encroach on the thoracic cavity, causing the ribs to flare.
3. The increased abdominal mass changes the woman's center of gravity; lordosis and backache are common. A waddling gait occurs as pelvic ligaments and joints are loosened by placental relaxin.
4. A typical weight gain during pregnancy in a woman of normal weight is 28 pounds.

Metabolic Changes (p. 1090)

5. Human placental lactogen has anabolic effects and promotes glucose sparing in the mother. Human chorionic thyrotropin results in maternal hypermetabolism.

Physiological Changes (p. 1090)

6. Many women suffer morning sickness, heartburn, and constipation during pregnancy.
7. The kidneys produce more urine, and pressure on the bladder may cause frequency, urgency, and stress incontinence.
8. Tidal volume increases, but residual volume decreases. The respiratory rate remains relatively unchanged. Dyspnea is common.
9. Total body water and blood volume increase dramatically. Heart rate and blood pressure rise, resulting in enhancement of cardiac output in the mother.

Parturition (Birth) (pp. 1090–1092)

1. Parturition encompasses a series of events called labor.

Initiation of Labor (p. 1091)

2. When estrogen levels are sufficiently high, they induce oxytocin receptors on the myometrial cells and inhibit progesterone's quieting effect on uterine muscle. Weak, irregular contractions begin.
3. Fetal cells produce oxytocin, which stimulates prostaglandin production by the placenta. Both hormones stimulate contraction of uterine muscle. Increasing distension of the uterus activates the hypothalamus, causing oxytocin release from the mother's posterior pituitary; this sets up a positive feedback loop resulting in true labor.

Stages of Labor (pp. 1091–1092)

4. The dilation stage is from the onset of rhythmic, strong contractions until the cervix is fully dilated. The head of the fetus rotates as it descends through the pelvic outlet.
5. The expulsion stage extends from full cervical dilation until birth of the infant.
6. The placental stage is the delivery of the afterbirth (the placenta and attached fetal membranes).

Adjustments of the Infant to Extrauterine Life (pp. 1092–1093)

1. The infant's Apgar score is recorded immediately after birth.

Taking the First Breath and Transition (p. 1093)

2. After the umbilical cord is clamped, carbon dioxide accumulates in the infant's blood, causing respiratory centers in the brain to trigger the first inspiration.
3. Once the lungs are inflated, breathing is eased by the presence of surfactant, which decreases the surface tension of the alveolar fluid.

4. During transition, the first 8 hours after birth, the infant is physiologically unstable and adjusting. After stabilizing, the infant wakes approximately every 3–4 hours in response to hunger.

Occlusion of Special Fetal Blood Vessels and Vascular Shunts (p. 1093)

5. Inflation of the lungs causes pressure changes in the circulation; as a result, the umbilical arteries and vein, ductus venosus, and ductus arteriosus collapse, and the foramen ovale closes. The occluded blood vessels are converted to fibrous cords; the site of the foramen ovale becomes the fossa ovalis.

Lactation (pp. 1093–1094)

1. The breasts are prepared for lactation during pregnancy by high blood levels of estrogen, progesterone, and human placental lactogen.
2. Colostrum, a premilk fluid, is a fat-poor fluid that contains more protein, vitamin A, and minerals than true milk. It is produced toward the end of pregnancy and for the first two to three days after birth.
3. True milk is produced around day 3 in response to suckling, which stimulates the hypothalamus to prompt anterior pituitary release of prolactin and posterior pituitary release of oxytocin. Prolactin stimulates milk production. Oxytocin triggers milk letdown. Continued breast-feeding is required for continued milk production.
4. At first, ovulation and menses are absent or irregular during nursing, but in most women the ovarian cycle is eventually reestablished while still nursing.

Assisted Reproductive Technology and Reproductive Cloning (pp. 1094–1097)

1. Assisted reproductive technology (ART) procedures assist infertile couples to bear children. Among the most-used techniques are IVF, ZIFT, and GIFT. IVF and ZIFT attempt to fertilize harvested oocytes in vitro and then return the embryo or zygote to the woman's body. GIFT utilizes in vivo methods—sperm and oocytes are transferred together to the woman's uterine tubes.
2. Reproductive cloning in humans has proved difficult to achieve and the practice has met several roadblocks.

REVIEW QUESTIONS

Multiple Choice/Matching

(Some questions have more than one correct answer. Select the best answer or answers from the choices given.)

1. Indicate whether each of the following statements is describing (a) cleavage or (b) gastrulation.
   _____ (1) period during which a morula forms
   _____ (2) period when vast amounts of cell migration occur
   _____ (3) period when the three embryonic germ layers appear
   _____ (4) period during which the blastocyst is formed
2. Most systems are operational in the fetus by four to six months. Which system is the exception to this generalization, affecting premature infants? (a) the circulatory system, (b) the respiratory system, (c) the urinary system, (d) the digestive system.
3. The zygote contains chromosomes from (a) the mother only, (b) the father only, (c) both the mother and father, but half from each, (d) each parent and synthesizes others.
4. The outer layer of the blastocyst, which later attaches to the uterus, is the (a) decidua, (b) trophoblast, (c) amnion, (d) inner cell mass.
5. The fetal membrane that forms the basis of the umbilical cord is the (a) allantois, (b) amnion, (c) chorion, (d) yolk sac.
6. In the fetus, the ductus arteriosus carries blood from (a) the pulmonary artery to the pulmonary vein, (b) the liver to the inferior vena cava, (c) the right ventricle to the left ventricle, (d) the pulmonary trunk to the aorta.
7. Which of the following changes occur in the baby's cardiovascular system after birth? (a) Blood clots in the umbilical vein, (b) the pulmonary vessels dilate as the lungs expand, (c) the ductus venosus becomes obliterated, as does the ductus arteriosus, (d) all of these.
8. Following delivery of the infant, the delivery of the afterbirth includes the (a) placenta only, (b) placenta and decidua, (c) placenta and attached (torn) fetal membranes, (d) chorionic villi.
9. The umbilical vein carries (a) waste products to the placenta, (b) oxygen and food to the fetus, (c) oxygen and food to the placenta, (d) oxygen and waste products to the fetus.
10. The germ layer from which the skeletal muscles, heart, and skeleton are derived is the (a) ectoderm, (b) endoderm, (c) mesoderm.
11. Which of the following cannot pass through placental barriers? (a) blood cells, (b) glucose, (c) amino acids, (d) gases, (e) antibodies.
12. The most important hormone in initiating and maintaining lactation after birth is (a) estrogen, (b) FSH, (c) prolactin, (d) oxytocin.
13. The initial stage of labor, during which the neck of the uterus is stretched, is the (a) dilation stage, (b) expulsion stage, (c) placental stage.
14. Match each adult structure in column B with the embryonic structure it derives from in column A.

Column A	Column B
_____ (1) notochord	(a) kidney
_____ (2) ectoderm (not neural tube)	(b) peritoneal cavity
	(c) pancreas, liver
_____ (3) intermediate mesoderm	(d) parietal serosa, dermis
_____ (4) splanchnic mesoderm	(e) nucleus pulposus
_____ (5) sclerotome	(f) visceral serosa
_____ (6) coelom	(g) hair and epidermis
_____ (7) neural tube	(h) brain
_____ (8) somatic mesoderm	(i) ribs
_____ (9) endoderm	

Short Answer Essay Questions

15. Why does testing for hCG work as a pregnancy test during early pregnancy but not late pregnancy?
16. Fertilization involves much more than a mere restoration of the diploid chromosome number. (a) What does the process of fertilization entail on the part of both the egg and sperm? (b) What are the effects of fertilization?
17. Cleavage is an embryonic event that mainly involves mitotic divisions. How does cleavage differ from mitosis occurring during life after birth, and what are its important functions?
18. The life span of the ovarian corpus luteum is extended for nearly three months after implantation, but otherwise it deteriorates. (a) Explain why this is so. (b) Explain why it is important that the corpus luteum remain functional following implantation.

19. The placenta is a marvelous, but temporary, organ. Starting with a description of its formation, show how it is an intimate part of both fetal and maternal anatomy and physiology during the gestation period.
20. Why is it that only one sperm out of the hundreds (or thousands) available enters the oocyte?
21. What is the function of gastrulation?
22. Cite two problems with a breech presentation.
23. What factors are believed to bring about uterine contractions at the termination of pregnancy?
24. Explain how the flat embryonic disc takes on the cylindrical shape of a tadpole.

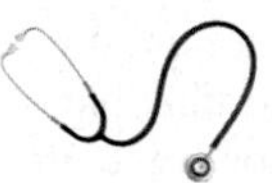

Critical Thinking and Clinical Application Questions

1. Jennie, a freshman in your dormitory, tells you she just discovered that she is three months pregnant. She recently bragged that since she came to college she has been drinking alcohol heavily and experimenting with every kind of recreational drug she could find. From the following, select the advice you would give her, and explain why it is the best choice. (a) She must stop taking drugs, but they could not have affected her fetus during these first few months of her pregnancy. (b) Harmful substances usually cannot pass from mother to embryo, so she can keep using drugs. (c) There could be defects in the fetus, so she should stop using drugs and visit a doctor as soon as possible. (d) If she has not taken any drugs in the last week, she is okay.
2. During Mrs. Jones's labor, the obstetrician decided that it was necessary to perform an episiotomy. What is an episiotomy, and why is it done?
3. A woman in substantial pain called her doctor and explained (between sobs) that she was about to have her baby "right here." The doctor calmed her and asked how she had come to that conclusion. She said that her water had broken and that her husband could see the baby's head. (a) Was she right? If so, what stage of labor was she in? (b) Do you think that she had time to make it to the hospital 60 miles away? Why or why not?
4. Mary is a heavy smoker and has ignored a friend's advice to stop smoking during her pregnancy. On the basis of what you know about the effect of smoking on physiology, describe how Mary's smoking might affect her fetus.
5. While Mortimer was cramming for his anatomy test, he read that some parts of the mesoderm become segmented. He suddenly realized that he could not remember what segmentation is. Define segmentation, and give two examples of segmented structures in the embryo.
6. Assume a sperm has penetrated a polar body and their nuclei fuse. Why would it be unlikely for the resulting cell to develop into a healthy embryo?

Access everything you need to practice, review, and self-assess for both your A&P lecture and lab courses at **myA&P** (www.myaandp.com). There, you'll find powerful online resources, including chapter quizzes and tests, games, A&P Flix animations with quizzes, ***Interactive Physiology***® with quizzes, MP3 Tutor Sessions, Practice Anatomy Lab™, and more to help you get a better grade in your course.

28

29

Heredity

The wondrous growth and development of a new individual is guided by the gene-bearing chromosomes it receives from its parents in egg and sperm. As we described in Chapter 3, individual *genes*, or DNA segments, contain the "recipes" or genetic blueprints for proteins. Many of these proteins are enzymes that dictate the synthesis of virtually all the body's molecules. Consequently, genes are ultimately expressed in your hair color, sex, blood type, and so on.

Genes, however, do not act as free agents. As you will see, a gene's ability to prompt the development of a specific trait is enhanced or inhibited by interactions with other genes, as well as by environmental factors.

The science of genetics, which studies the mechanism of heredity (*genes* = give birth to), is relatively young, but our understanding of how human genes act and interact has advanced considerably since

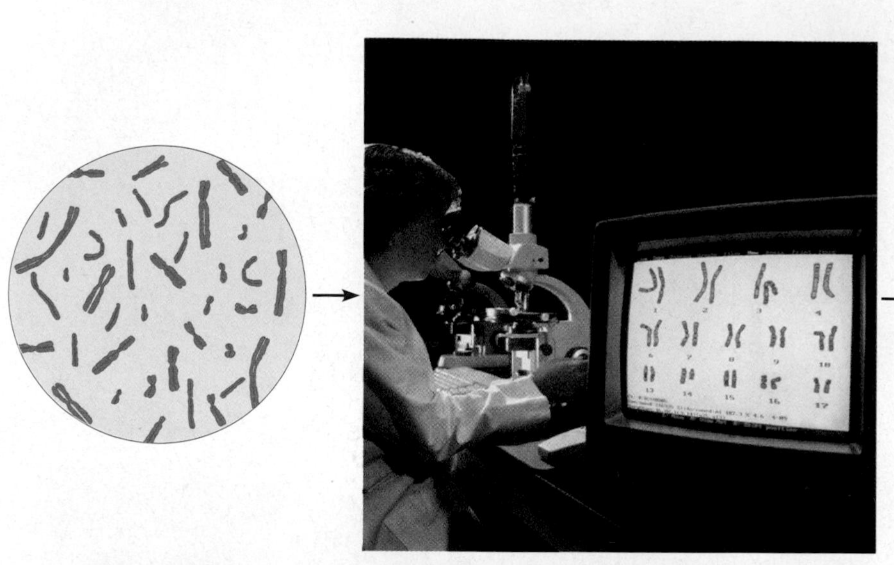

(a) The slide is viewed with a microscope, and the chromosomes are photographed.

(b) The photograph is entered into a computer, and the chromosomes are electronically rearranged into homologous pairs according to size and structure.

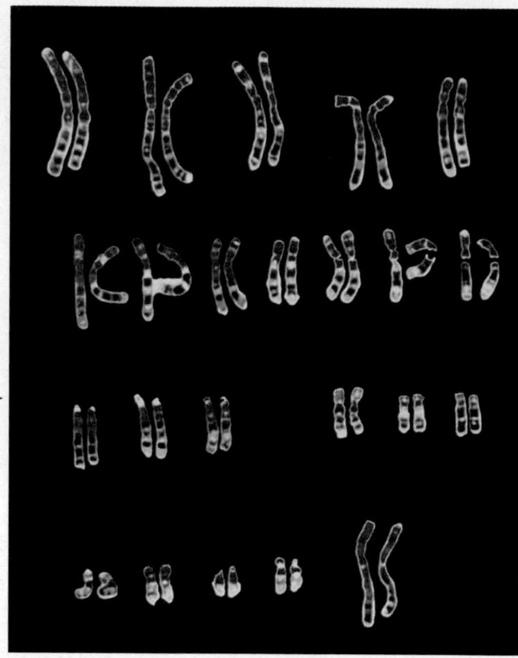

(c) The resulting display is the karyotype, which is examined for chromosome number and structure.

Figure 29.1 Preparing a karyotype. After lymphocytes are stimulated to divide and grow in culture for several days, they are treated with a drug that arrests mitosis in metaphase, a stage when the chromosomes are easily identified. The cells are harvested, treated with a solution that makes their chromosomes spread out, transferred to a microscope for photography, and then subjected to computer analysis and arrangement of the chromosomes into homologous pairs. The patterns of stained bands in the karyotype help to identify specific chromosomes and parts of chromosomes.

Gregor Mendel first proposed the basic principles of heredity in the mid-1800s. Mendel studied characteristics that vary in an either-or fashion, which is easier to understand than the more-or-less fashion in which many human traits vary.

However, the urge to understand human inheritance is powerful. Recent advances provided by the *Human Genome Project*, which scanned and determined the human DNA sequence, are enabling geneticists to manipulate and engineer human genes to examine their expression. This research has tremendous promise for more sophisticated genetic screening, and for drug development to treat or cure disease. Although we touch upon some of these newer topics, we will concentrate on the principles of heredity discovered by Mendel more than a century ago.

29

The Vocabulary of Genetics

- ▶ Define allele.
- ▶ Differentiate between genotype and phenotype.

The nuclei of all human cells except gametes contain the diploid number of chromosomes (46), consisting of 23 pairs of homologous chromosomes. Recall that *homologous chromosomes* are pairs of chromosomes—one from the father (sperm) and one from the mother (egg)—that look similar and carry genes for the same traits, but do not necessarily bring about the same expressions of those traits. Two of the 46 chromosomes are **sex chromosomes** (X and Y), which determine genetic sex (male = XY; female = XX). The other 44 are the 22 pairs of **autosomes** that guide the expression of most other traits.

The complete human **karyotype** (kar′e-o-tīp), or diploid chromosomal complement displayed in homologous pairs, is illustrated in **Figure 29.1c**. The diploid **genome** (je′nōm), or genetic (DNA) makeup, represents two sets of genetic instructions—one from the egg and the other from the sperm.

Gene Pairs (Alleles)

Because chromosomes are paired, it follows that the genes in them are paired as well. Consequently, each of us receives *two* genes, one from each parent (for the most part), that interact to dictate each trait. Matched genes, which are at the same *locus* (location) on homologous chromosomes, are called **alleles** (ah-lēlz′) of each other. Alleles may code for the same or for alternative forms of a given trait. For example, of the alleles that dictate whether or not you have loose thumb ligaments, one allele might code for tight ligaments and the other for loose ligaments (the double-jointed thumb condition). When the two

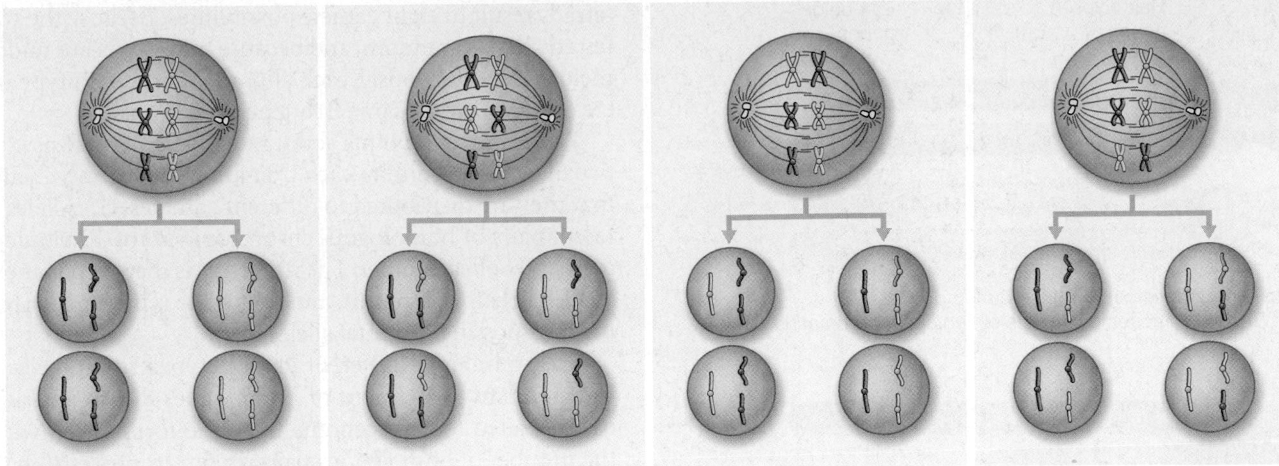

Figure 29.2 Gamete variability resulting from independent assortment. During metaphase of meiosis I, the tetrads of homologous chromosomes align independently of other tetrads. The large circles depict the possible alignments in a mother cell having a diploid number of 6 (paternal homologues purple; maternal, green). The small circles show the gametes arising from each alignment. Some gametes contain all maternal or all paternal chromosomes; others have maternal and paternal chromosomes in various combinations.

alleles controlling a trait are the same, a person is said to be **homozygous** (homo-zi′gus) for that trait. When the two alleles are different, the individual is **heterozygous** (het″er-o-zi′gus) for the trait.

Sometimes, one allele masks or suppresses the expression of its partner. Such an allele is said to be **dominant**, whereas the allele that is masked is said to be **recessive.** By convention, a dominant allele is represented by a capital letter (for example, *J*), and a recessive allele by the lowercase form of the same letter (*j*). Dominant alleles are expressed, or make themselves "known," when they are present in either single or double dose. For recessive alleles to be expressed, they must be present in double dose, that is, the homozygous condition. Returning to our thumb example, a person whose genetic makeup includes either the gene pair *JJ* (the homozygous dominant condition) or the gene pair *Jj* (the heterozygous condition) will have double-jointed thumbs. The combination *jj* (the homozygous recessive condition) is needed to produce tight thumb ligaments.

Genotype and Phenotype

A person's genetic makeup is referred to as his or her **genotype** (jen′o-tīp). The way that genotype is expressed in the body is called one's **phenotype** (fe′no-tīp). For example, the double-jointed condition is the phenotype produced by a genotype of *JJ* or *Jj*.

CHECK YOUR UNDERSTANDING

1. When a geneticist orders a karyotype, why are cells in metaphase studied instead of those in interphase?
2. What term is given to chromosomes *other than* our sex chromosomes?
3. Is an allele represented by a capital letter presumed to be dominant or recessive?
4. Harold is homozygous for the dominant alleles *HH*, *CC*, and *LL*, and heterozygous for *Bb* and *Kk*. He is blond and blue-eyed and has a very hairy chest. Which of these descriptions refer to his phenotype?

For answers, see Appendix G.

Sexual Sources of Genetic Variation

▶ Describe events that lead to genetic variability of gametes.

Before we examine how genes interact, let us consider why (with the possible exception of identical siblings) each of us is one of a kind, with a unique genotype and phenotype. This variability reflects three events that occur before we are even a twinkle in our parents' eyes: independent assortment of chromosomes, crossover of homologues, and random fertilization of eggs by sperm.

Chromosome Segregation and Independent Assortment

As we described in Chapter 27, each pair of homologous chromosomes synapses during meiosis I, forming a tetrad. This happens during both spermatogenesis and oogenesis. Because chance determines how the tetrads align (line up) on the meiosis I metaphase spindle, maternal and paternal chromosomes are randomly distributed to daughter nuclei.

As illustrated in **Figure 29.2**, this simple event leads to an amazing amount of variation in gametes. The cell in our example has a diploid number of 6, and so three tetrads form. As you can see, the possible combinations of alignments of the three

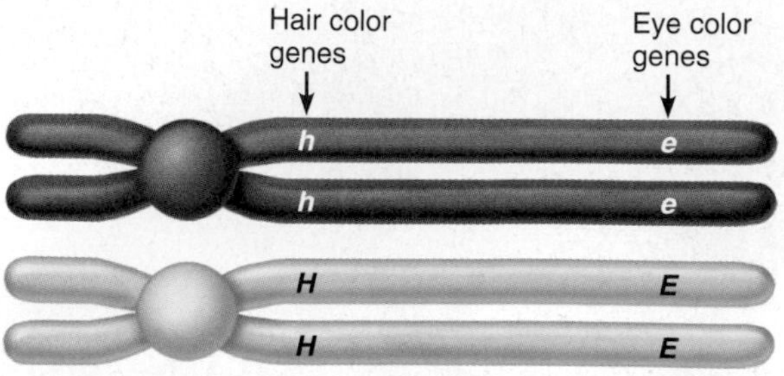

Homologous chromosomes synapse during prophase of meiosis I. Each chromosome consists of two sister chromatids.

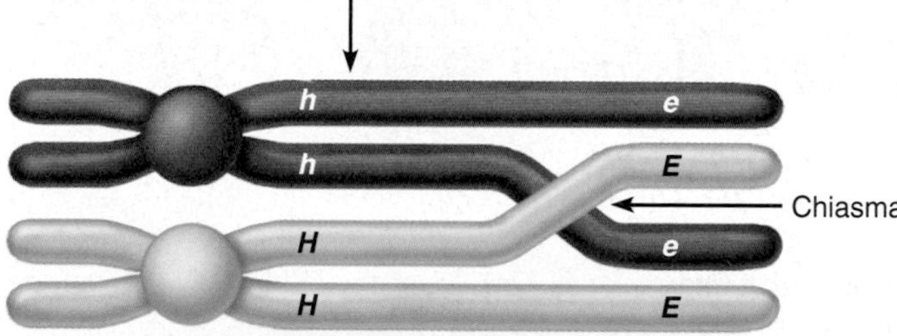

One chromatid segment exchanges positions with a homologous chromatid segment—in other words, crossing over occurs, forming a chiasma.

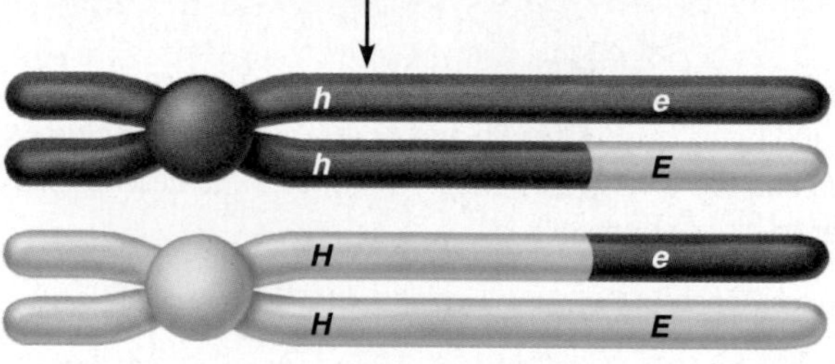

The chromatids forming the chiasma break, and the broken-off ends join their corresponding homologues.

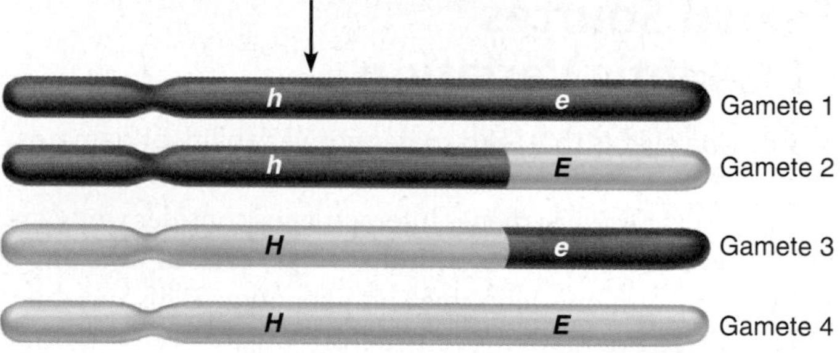

At the conclusion of meiosis, each haploid gamete has one of the four chromosomes shown. Two of the chromosomes are recombinant (they carry new combinations of genes).

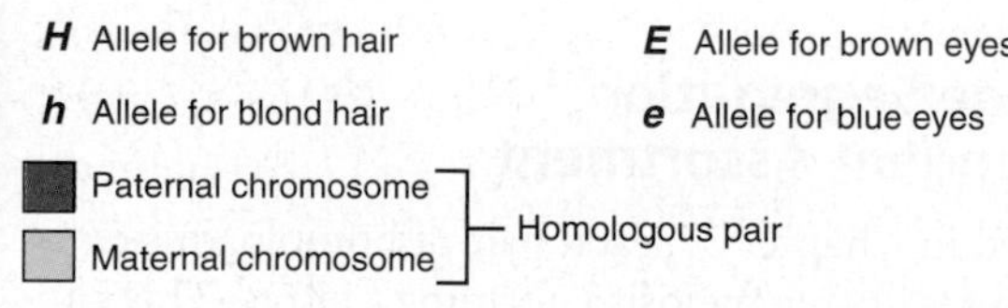

Figure 29.3 Crossover and genetic recombination. These meiosis I events increase genetic variability in the gametes formed. For simplicity, only two chromatids are shown taking part in crossover. Multiple crossovers result in more complex patterns.

tetrads result in eight gamete possibilities. Because the way each tetrad aligns is random, and because many cells are undergoing meiosis simultaneously, each alignment and each type of gamete occurs with the same frequency as all others.

Two important points about metaphase of meiosis I: (1) The two alleles determining each trait are **segregated**, which means that they are distributed to different gametes. (2) Alleles on different pairs of homologous chromosomes are distributed independently of each other. The net result is that each gamete has a single allele for each trait, and that allele represents only one of the four possible parental alleles.

The number of different gamete types resulting from this **independent assortment** of homologues during meiosis I can be calculated for any genome from the formula 2^n, where n is the number of homologous pairs. In our example, $2^n = 2^3$ ($2 \times 2 \times 2$), for a total of 8 different gamete types.

Note that the number of gamete types increases dramatically as the chromosome number increases. A cell with six pairs of homologues would produce 2^6, or 64, kinds of gametes. In a man's testes, the number of gamete types that can be produced on the basis of independent assortment alone is 2^{23}, or about 8.5 million—an incredible variety. The number of different gamete types produced simultaneously in a woman's ovaries is significantly less because her ovaries complete at most 500 reduction divisions in her lifetime. Still, each ovulated oocyte will most likely be novel genetically because of independent assortment.

Crossover of Homologues and Gene Recombination

Additional gamete variation results from the crossing over and exchange of chromosomal parts during meiosis I. Genes are arranged linearly along a chromosome's length, and genes on the same chromosome are said to be **linked** because they are transmitted as a unit to daughter cells during mitosis. However, as we described in Chapter 27, chromosomes can break and precisely exchange gene segments with their homologous counterparts during meiosis. This exchange gives rise to **recombinant chromosomes** that have mixed contributions from each parent.

In the hypothetical example shown in **Figure 29.3**, the genes for hair and eye color are linked. The paternal chromosome contains alleles coding for blond hair and blue eyes, and the maternal alleles code for brown hair and brown eyes. In the **crossover**, or **chiasma**, shown, the break occurs between these linked genes, resulting in one gamete with alleles for blond hair and brown eyes and another with alleles for brown hair and blue eyes. As a result of the crossover, two of the four chromatids present in the tetrad end up with a mixed set of alleles—some maternal and some paternal. This means that when the chromatids segregate, each gamete will receive a unique combination of parental genes.

Because humans have 23 tetrads, with crossovers going on in most of them during meiosis I, the variability resulting from this factor alone is tremendous.

Random Fertilization

At any point in time, gametogenesis is turning out gametes with all variations possible from independent assortment and random crossovers. Fertilization compounds this variety because a single human egg will be fertilized by a single sperm on a totally haphazard, or random, basis. If we consider variation resulting only from independent assortment and random fertilization, any offspring represents one out of the close to 72 trillion (8.5 million × 8.5 million) zygotes possible. The additional variation introduced by crossovers increases this number exponentially. Perhaps now you can understand why brothers and sisters are so different, and marvel at how they can also be so alike in many ways.

CHECK YOUR UNDERSTANDING

5. We said that genetic variability is introduced by independent assortment. Just what is assorting independently?
6. How does crossover increase genetic variability?

For answers, see Appendix G.

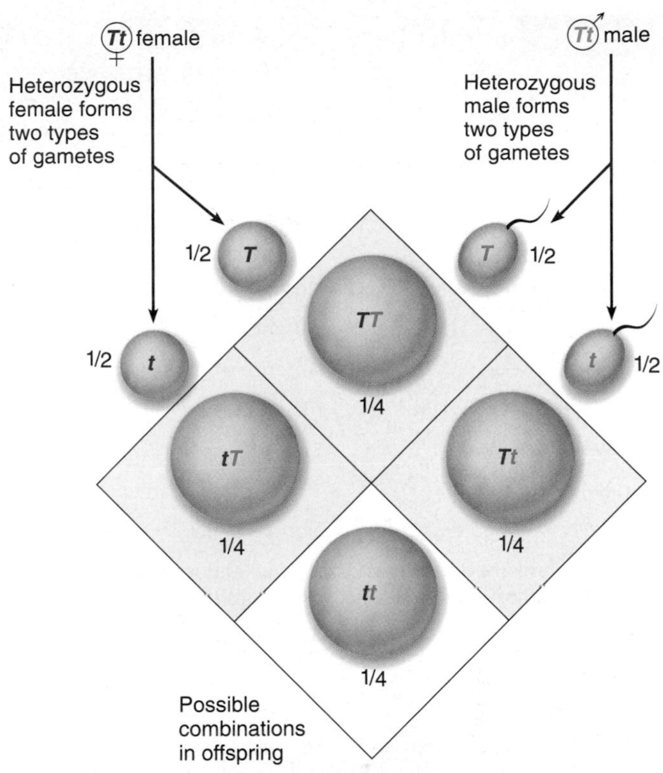

Figure 29.4 Genotype and phenotype probabilities resulting from a mating of two heterozygous parents. The Punnett square shows all possible combinations of a set of alleles in the zygote. In this example, the *T* allele is dominant and determines tongue-rolling ability; the *t* allele is recessive.

Types of Inheritance

A few human phenotypes can be traced to a single gene pair (as we will describe shortly), but most such traits are very limited in nature, or reflect variation in a single enzyme. Most human traits are determined by multiple alleles or by the interaction of several gene pairs.

Dominant-Recessive Inheritance

▶ Compare and contrast dominant-recessive inheritance with incomplete dominance and codominance.

Dominant-recessive inheritance reflects the interaction of dominant and recessive alleles. A simple diagram, called the **Punnett square**, is used to figure out, for a single trait, the possible gene combinations that would result from the mating of parents of known genotypes (**Figure 29.4**). In the example shown, both parents can roll their tongue into a U because both are heterozygous for the dominant allele (*T*) that confers this ability. In other words, each parent has the genotype *Tt*. The alleles of one parent are written along one side of the Punnett square, and the alleles for the other parent are shown along an adjacent side. The alleles are then combined down and across to determine the possible gene combinations (genotypes) and their expected frequency in the offspring of these two parents.

As the completed Punnett square shows, the probability of these parents producing a homozygous dominant child (*TT*) is 25% (1 out of 4); of producing a heterozygous child (*Tt*), 50% (2 out of 4); and of producing a homozygous recessive child, 25% (1 out of 4). The *TT* and *Tt* offspring will be tongue rollers. Only the *tt* offspring will not be able to roll their tongues.

The Punnett square predicts only the *probability* of a particular genotype (and phenotype). The larger the number of offspring, the greater the likelihood that the ratios will conform to the predicted values—just as the chances of getting heads half the time and tails half the time increase with the number of tosses of a coin. If we toss only twice, we may well get heads both times. Likewise, if the couple in our example had only two children, it would not be surprising if both children had the genotype *Tt*.

What are the chances of having two children of the same genotype? To determine the probability of two events happening in succession, we must multiply the probabilities of the separate events happening. The probability of getting heads in one coin toss is 1/2, so the probability of getting two heads in a row is $1/2 \times 1/2 = 1/4$. Now consider the chances of our couple's having two children who are both non–tongue rollers (*tt*). The probability that one child will be *tt* is 1/4, so the probability that both children will be *tt* is $1/4 \times 1/4 = 1/16$, only slightly better than 6%.

Remember that the production of each child, like each coin toss in a series, is an *independent event* that does not influence the production of any other child by the same couple. If you get heads on the first toss, the chance of getting heads the second time is still 1/2. Likewise, if our couple's first child is a *tt*, they still have a 1/4 chance of getting a *tt* the next time.

Dominant Traits

Human traits dictated by dominant alleles include widow's peaks, dimples, and freckles.

TABLE 29.1 Traits Determined by Simple Dominant-Recessive Inheritance

PHENOTYPE DUE TO EXPRESSION OF:	
DOMINANT GENES (ZZ OR Zz)	RECESSIVE GENES (zz)
Tongue roller	Inability to roll tongue into a U shape
Astigmatism	Normal vision
Freckles	Absence of freckles
Dimples in cheeks	Absence of dimples
PTC* taster	PTC nontaster
Widow's peak	Straight hairline
Double-jointed thumb	Tight thumb ligaments
Syndactyly (webbed digits)	Normal digits
Achondroplasia (heterozygous: dwarfism; homozygous: lethal)	Normal endochondral ossification
Huntington's disease	Absence of Huntington's disease
Normal skin pigmentation	Albinism
Absence of Tay-Sachs disease	Tay-Sachs disease
Absence of cystic fibrosis	Cystic fibrosis

*PTC is phenylthiocarbamide, a harmless bitter chemical.

Disorders caused by dominant genes are uncommon because *lethal dominant genes* are almost always expressed and result in the death of the embryo, fetus, or child. The deadly genes are not usually passed along to successive generations. However, in some dominant disorders the person is less impaired or at least survives long enough to reproduce. One example is *Huntington's disease*, a fatal nervous system disease involving degeneration of the basal nuclei. It involves a *delayed-action gene* that is expressed when the affected individual is about 40. Offspring of a parent with Huntington's disease have a 50% chance of inheriting the lethal gene. (The parent is heterozygous, because the dominant homozygous condition is lethal to the fetus.) Many informed offspring of such parents are opting not to become parents themselves.

These and other dominant gene–determined traits are listed in **Table 29.1**.

29

Recessive Traits

Some examples of recessive inheritance are the more desirable genetic condition. For example, normal vision is dictated by recessive alleles, whereas astigmatism is prescribed by dominant alleles. However, many if not most genetic disorders are inherited as simple recessive traits. These include conditions as different as *albinism* (lack of skin pigmentation); *cystic fibrosis*, a condition of excessive mucus production that impairs lung and pancreatic functioning; and *Tay-Sachs disease*, a disorder of brain lipid metabolism, caused by an enzyme deficit that shows itself a few months after birth.

Recessive genetic disorders are more frequent than those caused by dominant alleles because those who carry a *single* recessive allele for a recessive genetic disorder do not themselves express the disease. However, they can pass the gene on to offspring and so are called *carriers* of the disorder. Conversely, as we explained above, few individuals with dominant genetic disorders get a chance to pass on the detrimental gene.

Incomplete Dominance

In dominant-recessive inheritance, one allele variant completely masks the other. Some traits, however, exhibit **incomplete dominance**. In such instances, the heterozygote has a phenotype intermediate between those of homozygous dominant and homozygous recessive individuals. Incomplete dominance is uncommon in humans.

Perhaps the best human example is inheritance of the *sickling gene* (*s*), which causes a substitution of one amino acid in the beta chain of hemoglobin. Hemoglobin molecules containing the abnormal beta chains crystallize when blood oxygen levels are low, causing the erythrocytes to assume a sickle shape (see Figure 17.8b, p. 642). Individuals with a double dose of the sickling allele (*ss*) have **sickle-cell anemia**, and any condition that lowers their blood oxygen level, such as respiratory difficulty or excessive exercise, can precipitate a *sickle-cell crisis*. The deformed erythrocytes jam up and fragment in small capillaries, causing intense pain.

Individuals heterozygous for the sickling gene (*Ss*) have **sickle-cell trait.** They make both normal and sickling hemoglobin, and as a rule, these individuals are healthy. However, they can suffer a crisis if there is prolonged reduction in blood oxygen levels, as might happen when traveling in high-altitude areas, and they can transmit the sickling gene to their offspring.

Multiple-Allele Inheritance

Although we inherit only two alleles for each gene, some genes exhibit more than two allele forms, leading to a phenomenon called **multiple-allele inheritance**. For example, three alleles determine the ABO blood types in humans: I^A, I^B and *i*. Each of us receives two of these. The I^A and I^B alleles are *codominant*, and both are expressed when present, resulting in the AB blood type. The *i* allele is recessive to the other two. Genotypes determining the four possible ABO blood groups are shown in **Table 29.2**.

TABLE 29.2 ABO Blood Groups

BLOOD GROUP (PHENOTYPE)	FREQUENCY (% OF U.S. POPULATION)			
	GENOTYPE	WHITE	BLACK	ASIAN
O	*ii*	45	49	40
A	I^AI^A or I^Ai	40	27	28
B	I^BI^B or I^Bi	11	20	27
AB	I^AI^B	4	4	5

CHECK YOUR UNDERSTANDING

7. Why are there so few genetic disorders caused by dominant genes?
8. How does incomplete dominance differ from codominance?

For answers, see Appendix G.

Sex-Linked Inheritance

▶ Describe the mechanism of sex-linked inheritance.

Inherited traits determined by genes on the sex chromosomes are said to be **sex-linked**. The X and Y sex chromosomes are not homologous in the true sense. The Y, which contains the gene (or genes) that determines maleness, is much smaller than the X chromosome **(Figure 29.5)**. The X bears over 2500 genes, and a disproportionately large number of them code for proteins important to brain function. Because the Y carries only 78 genes as of the last count, it lacks many of the genes present on the X. For example, genes coding for certain clotting factors, cone pigments, and even testosterone receptors are present on X but not on Y. A gene found only on the X chromosome is said to be **X-linked**.

Only relatively short regions at either end of the Y chromosome (corresponding to about 5% of the Y's DNA) code for nonsexual characteristics corresponding to those on the X. Those regions are the only areas that can participate in crossovers with the X.

When a male inherits an X-linked recessive allele—for example, for hemophilia or for red-green color blindness—its expression is never masked or damped, because there is no corresponding allele on his Y chromosome. Consequently, the recessive gene is always expressed, even when present only in single dose. In contrast, females must have two X-linked recessive alleles to express such a disorder, and as a result, very few females exhibit any X-linked conditions.

X-linked traits are typically passed from mother to son, never from father to son, because males receive no X chromosome from their father. Of course, the mother can also pass the recessive allele to her daughter, but unless the daughter receives another such allele from her father, she will not express the trait.

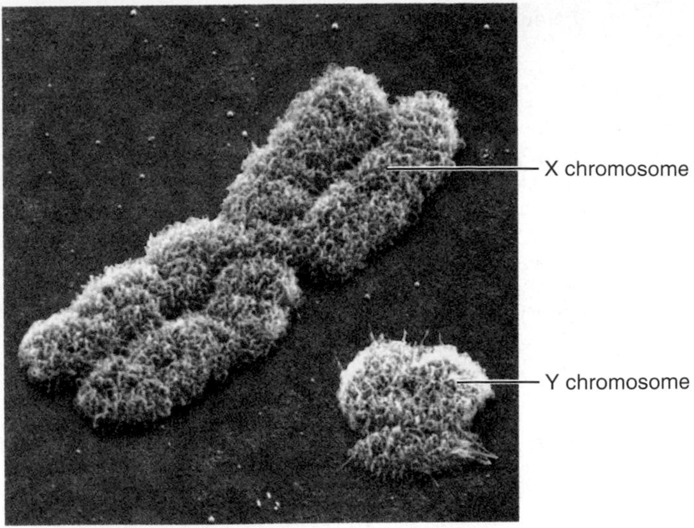

Figure 29.5 Photograph of human sex chromosomes.

Polygene Inheritance

▶ Explain how polygene inheritance differs from that resulting from the action of a single pair of alleles.

So far, we have considered only traits inherited by mechanisms of classical Mendelian genetics, which are fairly easy to understand, and such traits typically have two, or perhaps three, alternate forms. However, many phenotypes depend on several gene pairs at different locations acting in tandem. Such **polygene inheritance** results in *continuous*, or *quantitative*, phenotypic variation between two extremes and explains many human characteristics. Examples of polygene traits in humans include skin color, height, metabolic rate, and intelligence.

Skin color, for instance, is controlled by three separately inherited genes, each existing in two allelic forms: *A*, *a*; *B*, *b*; *C*, *c*. The *A*, *B*, and *C* alleles confer dark skin pigment, and their effects are additive. The *a*, *b*, and *c* alleles confer pale skin tone. An individual with an *AABBCC* genotype would be about as dark-skinned as a human can get, while an *aabbcc* person would be very fair. However, when individuals heterozygous for at least one of these gene pairs mate, a broad range of pigmentation is possible in their offspring. Such polygene inheritance results in a distribution of genotypes and phenotypes that, when plotted, yields a bell-shaped curve **(Figure 29.6)**.

CHECK YOUR UNDERSTANDING

9. Why does a male *always* express an X-linked recessive allele?
10. How can you explain why parents of average height can produce very tall or very short offspring?

For answers, see Appendix G.

Environmental Factors in Gene Expression

▶ Provide examples illustrating how gene expression may be modified by environmental factors.

In many situations environmental factors override or at least influence gene expression. Our genotype (discounting mutations) is as unchanging as the Rock of Gibraltar, but our phenotype is more like clay. If this were not the case, we would never get a tan, women bodybuilders would never be able to develop bulging muscles, and there would be no hope for treating genetic disorders.

Sometimes, such maternal factors as drugs or pathogens alter normal gene expression during embryonic development.

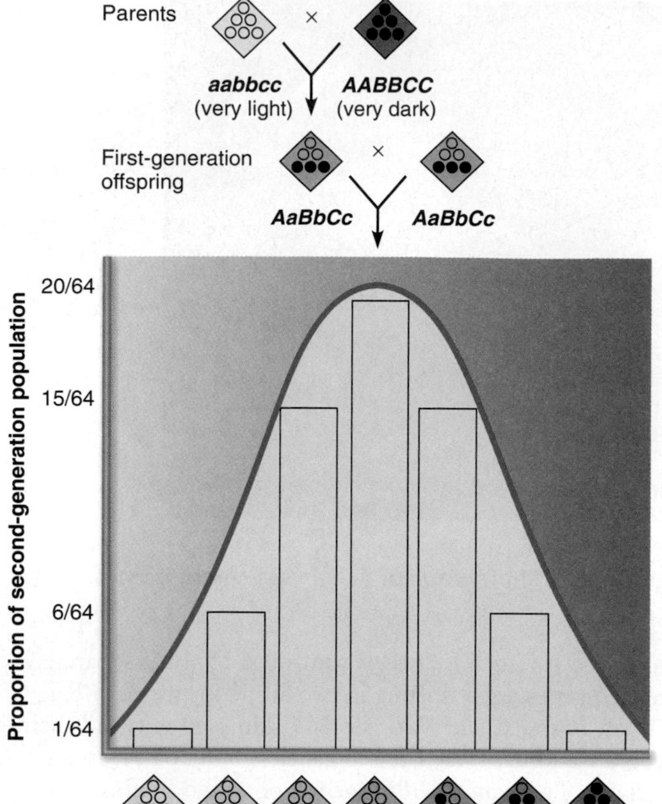

Figure 29.6 Simplified model for polygene inheritance of skin color based on three gene pairs. Alleles for dark skin are incompletely dominant over those for light skin. Each dominant gene (*A*, *B*, *C*) contributes 1 "unit" of darkness (indicated by a dark dot) to the phenotypes. Here the parents are homozygotes at opposite ends of the phenotype range.

Each child inherits 3 "units" of darkness (is a heterozygote) and has intermediate pigmentation. When they grow up and mate with those with the same alleles, their offspring (second generation) may have a wide variation in pigmentation, as shown by the plotted distribution of skin colors.

Take, for example, the case of the "thalidomide babies" (discussed on p. 1085). As a result of their mothers taking that sedative, the embryos developed phenotypes (flipperlike appendages) not directed by their genes. Such environmentally produced phenotypes that mimic conditions that may be caused by genetic mutations (permanent transmissible changes in the DNA) are called **phenocopies.**

29

Equally significant are environmental factors that may influence genetic expression after birth, such as the effect of poor infant nutrition on brain growth, general body development, and height. In this way, a person with "tall genes" can be stunted by insufficient nutrition. Furthermore, part of a gene's environment consists of the influence of other genes. For example, hormonal deficits during childhood can lead to abnormal skeletal growth and proportions, as in cretinism, a type of dwarfism resulting from hypothyroidism.

CHECK YOUR UNDERSTANDING

11. Which of the following factors may alter gene expression? Other genes, measles in a pregnant woman, lack of key nutrients in the diet.

For answers, see Appendix G.

Nontraditional Inheritance

Mendel's writings underlie mainstream thinking about heredity, but some genetic outcomes do not fit his rules. Among these nontraditional types of inheritance are influences due to RNA-only genes, to chemical groups attached to DNA or histone proteins, and to *extranuclear inheritance* conferred by mitochondrial DNA.

Beyond DNA: Regulation of Gene Expression

▶ Describe how RNA-only genes and epigenetic marks affect gene expression.

Our genome is a biochemical system of awesome complexity, with three basic levels of controls. The protein-coding genes that we have been describing to this point only make up the first level and account for less than 2% of the DNA of a human cell. This is the part of the genome traditionally considered to be a "blueprint" for protein structure.

Since the 1960s, scientists have been finding second and third layers of information important in directing development elsewhere—in the noncoding DNA and even totally outside the DNA sequences. So what are these other regulatory systems?

Small RNAs

The second layer appears to be the product of the abundant "RNA-only genes" (see p. 105), formerly believed to be "junk," that are found in the non-protein-coding DNA. They form a parallel regulatory system that generates single stranded microRNAs (miRNAs) and short interfering RNAs (siRNAs). These small RNA molecules can act directly on DNA, other RNAs, or proteins. They also can tame or inactivate aggressive (jumping) genes, called *transposons*, that tend to replicate themselves and then insert the copies into distant DNA sites, disabling or hyperactivating those genes.

Small RNAs control timing of programmed cell death during development and can also prevent translation of another gene. Mutations in these RNA-only areas have already been linked to several conditions including prostate and lung cancers and schizophrenia.

As a result of the Human Genome Project, nucleotide sequences of these RNA-specifying DNA areas are now worked out and biochemical companies are investing heavily in gene therapy research. They are especially hot on synthesizing RNA-interfering drugs to silence or shut down particular genes to treat age-related macular degeneration, Parkinson's disease, cancer, and a host of other disorders.

Epigenetic Marks

Epigenetic marks form the third layer of gene controls. This continually changing information is stored in the proteins and chemical groups that bind to the DNA and in the manner in which chromatin is packaged in the cell. Within cells, chemical tags such as methyl and acetyl groups bound to DNA segments and to histones determine whether the DNA is available for transcription (acetylation) or silenced (methylation). Epigenetic marks also account for the inactivation (by methylation) of one of the female's X chromosomes in the early embryo.

Epigenetic marks or lack of them may predispose a cell for transformation from normal to cancerous, and even slight deviations in the epigenetic marks on specific chromosomes can result in devastating human illness.

Epigenetic marks also underlie the phenomenon called genomic imprinting. For most genes, the maternal and paternal genes turn on or off at the same time. This balance is upset during gametogenesis when certain genes in both sperm and eggs are modified by the addition of a methyl (—CH_3) group, a process called **genomic imprinting**. Genomic imprinting is essential for normal development and somehow tags the genes as paternal or maternal and confers important functional differences in the resulting embryo. The developing embryo "reads" these tags, and then either expresses the mother's gene while the father's version remains idle or vice versa. The methylation process is reversible. In each generation, the old imprints are "erased" when new gametes are produced and all the chromosomes are newly imprinted according to the sex of the parents.

Mutations of imprinted genes may lead to pathology. For example, victims of Prader-Willi syndrome are mildly to moderately retarded, short, and grossly obese. Children with Angelman syndrome are severely retarded, unable to speak coherently, laugh uncontrollably, and exhibit jerky, lurching movements as if tied to a puppeteer's strings. The symptoms of these two disorders are very different, but the genetic cause is the same—deletion of a particular region of chromosome 15. If the defective chromosome comes from the father, the result is Prader-Willi syndrome; the mother's defective chromosome confers Angelman syndrome. Thus, it seems that the same allele can have different effects depending on which parent it comes from.

In short, protein-coding genes are not the only instructions to which cells refer. RNA matters, and so do the tiny chemical tags that attach to the chromatin.

Extranuclear (Mitochondrial) Inheritance

▶ Describe the basis of extranuclear (mitochondria-based) genetic disorders.

Although we have focused on the chromosomal basis of inheritance, remember that not all genes are in the cell's nucleus. Some 37 genes (referred to as mtDNA) are in mitochondria. Mitochondrial genes are transmitted to the offspring almost exclusively by the mother because the ovum donates essentially all the cytoplasm in the fertilized egg.

A growing list of disorders, all rare, is now being linked to errors (mutations) in mitochondrial genes. Most involve problems with oxidative phosphorylation within the mitochondria, but a few lead to unusual degenerative muscle disorders or neurological problems. Some researchers suggest that Alzheimer's and Parkinson's disease may be among them.

CHECK YOUR UNDERSTANDING

12. What process labels genes as paternal or maternal?
13. What is the source of the genes that confer extranuclear inheritance?

For answers, see Appendix G.

Genetic Screening, Counseling, and Therapy

▶ List and explain several techniques used to determine or predict genetic diseases.

▶ Describe briefly some approaches of gene therapy.

Genetic screening and *genetic counseling* provide information and options for prospective parents not even dreamed of 100 years ago. Newborn infants are routinely screened for a number of anatomical disorders (congenital hip dysplasia, imperforate anus, and others), and testing for phenylketonuria (PKU) and other metabolic diseases is mandated by law in many states. These tests alert the new parents that treatment is necessary to ensure the well-being of their infant. Anatomical defects are usually treated surgically, and PKU is managed by strict dietary measures that exclude most phenylalanine-containing foods.

Adult children of parents with Huntington's disease are obvious candidates for these services, but many other genetic conditions also place babies at risk. For example, a woman pregnant at age 35 may wish to know if her baby has trisomy-21 or Down syndrome (see Related Clinical Terms, p. 1111), a chromosome abnormality with a high incidence in children of older mothers.

Depending on the condition being investigated, screening can occur before conception, by carrier recognition, or during fetal testing.

Carrier Recognition

There are two major avenues for identifying carriers of detrimental genes: pedigrees and blood tests. A **pedigree** traces a genetic trait through several generations and helps predict the future. For prospective parents, a genetic counselor collects phenotype information on as many family members as possible and uses it to construct the pedigree.

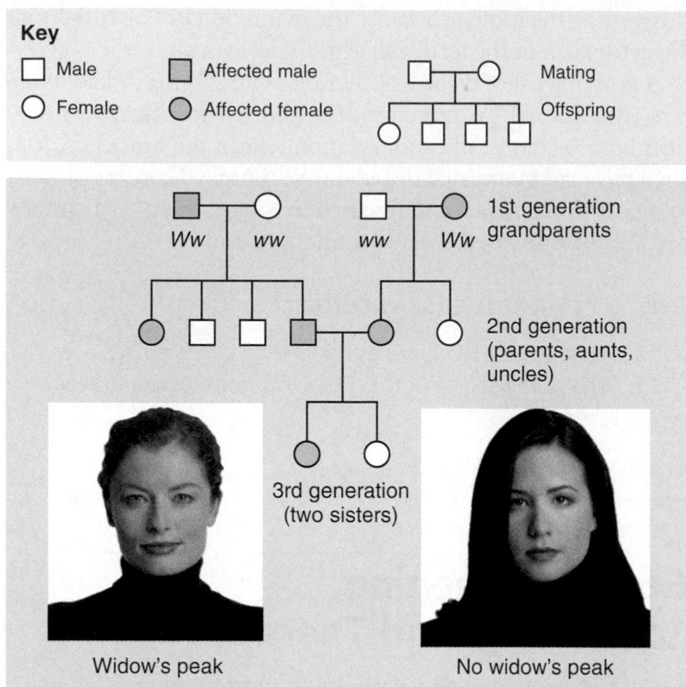

Figure 29.7 Pedigree tracing the trait called widow's peak through three generations. Notice in the third generation that the second-born daughter lacks a widow's peak, although both of her parents had the trait.

Figure 29.7 uses a normal trait, a widow's peak (pointed contour of the hair on the forehead) to illustrate how a pedigree is constructed and read. Widow's peak results from the presence of a dominant allele (*W*). Individuals with the widow's peak phenotype (blue symbols) have at least one dominant gene (*WW* or *Ww*), while those lacking a widow's peak must be homozygous recessive (*ww*).

By working backward from the bottom row and applying the rules of dominant-recessive inheritance, a counselor can deduce the genotypes of the parents (middle row). That one of their two offspring lacks a widow's peak (*ww*) tells the counselor that each parent must have at least one recessive gene (*Ww*), even though both parents exhibit a widow's peak. You should be able to take it from here and figure out the genotypes of the other individuals in their parents' generation. (Try it!)

Simple blood tests are used to screen for the sickling gene in heterozygotes, and sophisticated *blood chemistry tests* and *DNA probes* can detect the presence of other unexpressed recessive genes. At present, carriers of the Tay-Sachs and cystic fibrosis genes can be identified with such tests.

29

Fetal Testing

Fetal testing is used when there is a known risk of a genetic disorder. The most common type of fetal testing is **amniocentesis** (am″ne-o-sen-te′sis), in which a wide-bore needle is inserted into the amniotic sac through the mother's abdominal wall, and about 10 ml of fluid is withdrawn **(Figure 29.8a)**. Because there is a chance of injuring the fetus before ample amniotic fluid is present, this procedure is not normally done before the 14th week of pregnancy. Using ultrasound to visualize the position of the fetus and the amniotic sac has dramatically reduced the risk of this procedure.

The fluid is checked for enzymes and other chemicals that serve as markers for specific diseases, but most tests are done on the sloughed-off fetal cells in the fluid. These cells are isolated and cultured in laboratory dishes over a period of several weeks. Then the cells are examined for DNA markers of genetic disease and karyotyped to check for chromosomal abnormalities (see Figure 29.1).

A type of cell that floats freely in the amniotic fluid has proven to be very interesting to stem cell investigators. Armed with many of the same traits as embryonic stem (ES) cells, including pluripotence (an ability to grow into brain, vascular, muscle, and bone tissues), it may prove useful to serve as a "repair kit" for birth defects. Only time will tell.

Chorionic villus sampling (CVS) suctions off bits of the chorionic villi from the placenta for examination (Figure 29.8b). A small tube is inserted through the vagina and cervical canal and guided by ultrasound to an area where a piece of placental tissue can be removed. CVS allows testing at 8 weeks (but waiting until after the 10th week is usually recommended). Karyotyping can be done almost immediately on the rapidly dividing chorionic cells, much earlier than in amniocentesis.

Both of these procedures are invasive, and they carry with them an inherent risk to both fetus and mother. (For example, increased fetal risk of finger and toe defects is linked to CVS.) These tests are routinely ordered for pregnant women over 35 (because of the enhanced risk of Down syndrome), but they are performed on younger women when the probability of finding a severe fetal disorder is greater than the probability of doing harm during the procedure. If a serious genetic or congenital defect is detected in the developing fetus, the parents must decide whether or not to continue the pregnancy.

Human Gene Therapy

As indicated earlier, the Human Genome Project has opened an exciting world to scientists investigating gene expression. Advances in diagnosing human genetic diseases have been followed quickly by new applications of "gene therapy" to alleviate or even cure the disorders, particularly in cases traced to a single defective gene or protein. Indeed, genetic engineering has the potential to replace a defective gene with a normal version.

One approach (used successfully in treatment of severe combined immunodeficiency disorders) is to "infect" the defective cells with a virus into which a functional gene has been inserted. Another is to inject the "corrected" DNA directly into the patient's cells. Such therapies have had mixed results in treating cystic fibrosis and muscular dystrophy. However, genetic engineering processes are prohibitively expensive and raise some thorny ethical, religious, and societal questions: Who will pay? Who determines who will be treated with the new therapies? Are we playing God?

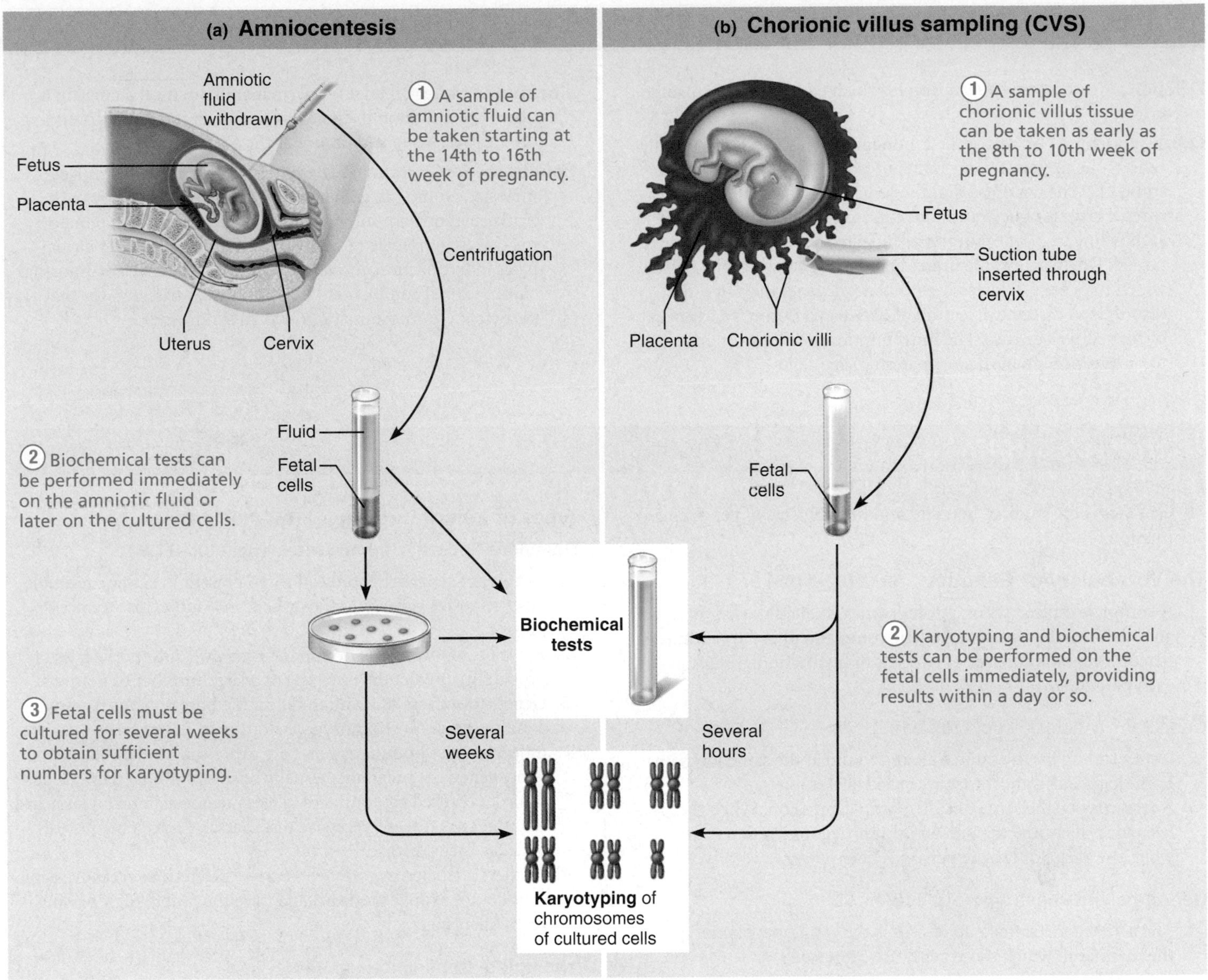

Figure 29.8 Fetal testing—amniocentesis and chorionic villus sampling.

CHECK YOUR UNDERSTANDING

14. Which fetal testing procedure depends on analyzing amniotic fluid?
15. What noninvasive imaging procedure is used to determine some aspects of fetal development? (You many want to check *A Closer Look* in Chapter 1 for this.)

For answers, see Appendix G.

In this chapter, we have explored some of the basic principles of genetics, the manner in which genes are expressed, and the means by which gene expression can be modified. Considering the precision required to make perfect copies of genes and chromosomes, and the incredible mechanical events of meiotic division, it is amazing that genetic defects are as rare as they are. Perhaps, after reading this chapter, you have a greater sense of wonder that you turned out as well as you did.

RELATED CLINICAL TERMS

Deletion Chromosomal aberration in which part of a chromosome is lost.

Down syndrome Formerly called mongolism, this condition usually reflects the presence of an extra autosome (trisomy of chromosome 21). The child has slightly slanted eyes, flattened facial features, a large tongue, and a tendency toward short stature and stubby fingers. Some, but not all, affected people are mentally retarded. Down syndrome individuals have mitochondrial defects, which have been associated with neurodegeneration in other disorders. A distinctive feature of Down syndrome is early onset of Alzheimer's disease. The most important risk factor appears to be advanced maternal (or paternal) age.

Mutation (*mutare* = change) A permanent structural change in a gene. The mutation may or may not affect function, depending on the precise site and nature of the alteration.

Nondisjunction Abnormal segregation of chromosomes during meiosis, resulting in gametes receiving two or no copies of a particular parental chromosome. It is more common in female meiosis than in male meiosis. If the abnormal gamete participates in fertilization, the resulting zygote will have an abnormal chromosomal complement (monosomy or trisomy) for that particular chromosome (as in Down syndrome).

CHAPTER SUMMARY

1. Genetics is the study of heredity and mechanisms of gene transmission.

The Vocabulary of Genetics (pp. 1102–1103)

1. A complete diploid set of chromosomes is called the karyotype of an organism; the complete genetic complement is the genome. A person's genome consists of two sets of instructions, one from each parent.

Gene Pairs (Alleles) (pp. 1102–1103)

2. Genes coding for the same trait and found at the same locus on homologous chromosomes are called alleles.
3. Alleles may be the same or different in expression. When the allele pair is identical, the person is homozygous for that trait; when the alleles differ, the person is heterozygous.

Genotype and Phenotype (p. 1103)

4. The actual genetic makeup of cells is the genotype; phenotype is the manner in which those genes are expressed.

Sexual Sources of Genetic Variation (pp. 1103–1105)

Chromosome Segregation and Independent Assortment (pp. 1103–1104)

1. During meiosis I of gametogenesis, tetrads align randomly on the metaphase plate, and chromatids are randomly distributed to the daughter cells. This is called independent assortment of the homologues. Each gamete receives only one allele of each gene pair.
2. Each different metaphase I alignment produces a different assortment of parental chromosomes in the gametes, and all combinations of maternal and paternal chromosomes are equally possible.

29

Crossover of Homologues and Gene Recombination (p. 1104)

3. During meiosis I, all four chromatids of each tetrad may cross over at one or more points and exchange corresponding gene segments. The recombinant chromosomes contain new gene combinations, adding to the variability arising from independent assortment.

Random Fertilization (p. 1105)

4. The third source of genetic variation is random fertilization of eggs by sperm.

Types of Inheritance (pp. 1105–1107)

Dominant-Recessive Inheritance (pp. 1105–1106)

1. Dominant genes are expressed when present in single or double dose; recessive genes must be present in double dose to be expressed.
2. For traits following the dominant-recessive pattern, the laws of probability predict the outcome of a large number of matings.
3. Genetic disorders more often reflect the homozygous recessive condition than the homozygous dominant or heterozygous condition because dominant genes are expressed and, if they are lethal genes, the pregnancy usually results in miscarriage. Genetic disorders caused by dominant alleles include achondroplasia and Huntington's disease; recessive disorders include cystic fibrosis and Tay-Sachs disease.
4. Carriers are heterozygotes who carry a deleterious recessive gene (but do not express the trait) and have the potential of passing it on to offspring.

Incomplete Dominance (p. 1106)

5. In incomplete dominance, the heterozygote exhibits a phenotype intermediate between those of the homozygous dominant and recessive individuals. Inheritance of sickle-cell trait is an example of incomplete dominance.

Multiple-Allele Inheritance (pp. 1106–1107)

6. Multiple-allele inheritance involves genes that exist in more than two allelic forms in a population. Only two of the alleles are inherited, but on a random basis. Inheritance of ABO blood types is an example of multiple-allele inheritance in which the I^A and I^B alleles are codominant.

Sex-Linked Inheritance (p. 1107)

7. Traits determined by genes on the X and Y chromosomes are said to be sex-linked. The small Y chromosome lacks most genes present on the X chromosome. Recessive genes located only on the X chromosome are expressed in single dose in males. Examples of such X-linked conditions, passed from mother to son, include hemophilia and red-green color blindness.

Polygene Inheritance (p. 1107)

8. Polygene inheritance occurs when several gene pairs interact to produce phenotypes that vary quantitatively over a broad range. Height, intelligence, and skin pigmentation are examples.

Environmental Factors in Gene Expression (pp. 1107–1108)

1. Environmental factors may influence the expression of genotype.
2. Maternal factors that cross the placenta may alter expression of fetal genes. Environmentally provoked phenotypes that mimic genetically determined ones are called phenocopies. Nutritional deficits or hormonal imbalances may alter anticipated growth and development during childhood.

Nontraditional Inheritance (pp. 1108–1109)

Beyond DNA: Regulation of Gene Expression (pp. 1108–1109)

1. Control of gene expression occurs at three levels: protein-coding genes, RNA-only genes, and epigenetic marks. The latter two account for many cases of inherited disease that fail to follow traditional genetics.
2. The products of the RNA-only genes (siRNA and miRNA) may silence genes or prevent their expression and appear to play a role in directing apoptosis during development.
3. Epigenetic marks involve attachment of small chemical groups (methyl or acetyl) to DNA or histone proteins. In general, methylation prevents access to the DNA whereas acetylation provides access to the DNA. Genomic imprinting, which involves methylation of certain genes during gametogenesis, confers different effects and phenotypes on maternal and paternal genes. It is reversible, and occurs anew each generation.

Extranuclear (Mitochondrial) Inheritance (p. 1109)

4. Cytoplasmic (mitochondrial) genes pass to offspring via the ovum and help to determine certain characteristics. Deletions or mutations in mitochondrial genes are responsible for some problems with oxidative phosphorylation and some rare genetic diseases.

Genetic Screening, Counseling, and Therapy (pp. 1109–1111)

Carrier Recognition (pp. 1109–1110)

1. The likelihood of an individual carrying a deleterious recessive gene may be assessed by constructing a pedigree. Some of these genes can be deduced by various blood tests and DNA probes.

Fetal Testing (p. 1110)

2. Amniocentesis is fetal testing based on aspirated samples of amniotic fluid. Fetal cells in the fluid are cultured for several weeks, then examined for chromosomal defects (karyotyped) or for DNA markers of genetic disease. Amniocentesis cannot be performed until the 14th week of pregnancy.
3. Chorionic villus sampling is fetal testing based on a sample of the chorion. This tissue is rapidly mitotic, so karyotyping can be done almost immediately. Samples may be obtained by the 8th week.

Human Gene Therapy (pp. 1110–1111)

4. The Human Genome Project has allowed research on diagnosis of genetic disease and on its treatment to surge forward. Thus far, gene therapy has been particularly useful for correcting single-gene disorders. The most common approach involves transferring a corrected gene via a virus to the affected cells to restore normal function.

REVIEW QUESTIONS

Multiple Choice/Matching

(Some questions have more than one correct answer. Select the best answer or answers from the choices given.)

1. Match one of the following terms (a–i) with each of the descriptions below:

   Key: (a) alleles (b) autosomes (c) dominant allele (d) genotype (e) heterozygote (f) homozygote (g) phenotype (h) recessive allele (i) sex chromosomes

   _____ (1) genetic makeup
   _____ (2) how genetic makeup is expressed
   _____ (3) chromosomes that dictate most body characteristics
   _____ (4) alternate forms of the same gene
   _____ (5) an individual bearing two alleles that are the same for a particular trait
   _____ (6) an allele that is expressed whether in single or double dose
   _____ (7) an individual bearing two alleles that differ for a particular trait
   _____ (8) an allele that must be present in double dose to be expressed

2. Match the following types of inheritance (a–f) with the descriptions below:

   Key: (a) dominant-recessive (b) incomplete dominance (c) multiple-allele (d) polygene (e) sex-linked (f) extranuclear

   _____ (1) only sons show the trait
   _____ (2) homozygotes and heterozygotes have the same phenotype
   _____ (3) heterozygotes exhibit a phenotype intermediate between those of the homozygotes
   _____ (4) phenotypes of offspring may be more varied than those of the parents
   _____ (5) inheritance of ABO blood types
   _____ (6) inheritance of stature
   _____ (7) reflects activity of mitochondrial DNA

Short Answer Essay Questions

3. Describe the important mechanisms that lead to genetic variations in gametes.
4. The ability to taste PTC (phenylthiocarbamide) depends on the presence of a dominant gene *T*; nontasters are homozygous for the recessive gene *t*. This is a situation of classical dominant-recessive inheritance. (a) Consider a mating between heterozygous

29

parents producing three offspring. What proportion of the offspring will be tasters? What is the chance that all three offspring will be tasters? Nontasters? What is the chance that two will be tasters and one will be a nontaster? (b) Consider a mating between *Tt* and *tt* parents. What is the anticipated percentage of tasters? Nontasters? What proportion can be expected to be homozygous recessive? Heterozygous? Homozygous dominant?

5. Most albino children are born to normally pigmented parents. Albinos are homozygous for the recessive gene (*aa*). What can you conclude about the genotypes of the nonalbino parents?
6. A woman with blood type A has two children. One has type O blood and the other has type B blood. What is the genotype of the mother? What are the genotype and phenotype of the father? What is the genotype of each child?
7. In skin color inheritance, what will be the relative range of pigmentation in offspring arising from the following parental matches? (a) *AABBCC* × *aabbcc*, (b) *AABBCC* × *AaBbCc*, (c) *Aabbcc* × *aabbcc*.
8. Compare and contrast amniocentesis and chorionic villus sampling as to the time at which they can be performed and the techniques used to obtain information on the fetus's genetic status.

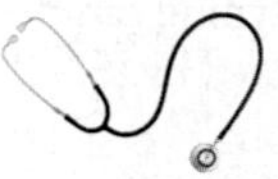

Critical Thinking and Clinical Application Questions

1. A color-blind man marries a woman with normal vision. The woman's father was also color-blind. (a) What is the chance that their first child will be a color-blind son? A color-blind daughter? (b) If they have four children, what is the chance that two will be color-blind sons? (Be careful on this one.)
2. Brian is a college student. His genetics assignment is to do a family pedigree for dimples in the cheeks. Absence of dimples is recessive; presence of dimples reflects a dominant allele. Brian has dimples, as do his three brothers. His mother and maternal grandmother are dimple free, but his father and all other grandparents have dimples. Construct a pedigree spanning three generations for Brian's family. Show the phenotype and genotype for each person.
3. Mr. and Mrs. Lehman have sought genetic counseling. Mrs. Lehman is concerned because she is unexpectedly pregnant and her husband's brother died of Tay-Sachs disease. She can recall no incidence of Tay-Sachs disease in her own family. Do you think biochemical testing should be recommended to detect the deleterious gene in Mrs. Lehman? Explain your answer.
4. The Browns are both carriers of the recessive allele that causes the metabolic disorder called phenylketonuria. What is the probability of each of the following occurring? (a) All three children will have the disorder. (b) None of their three children will have the disorder. (c) One or more of their children will have the disorder. (d) At least one of their children will be phenotypically normal.

APPENDIX A The Metric System

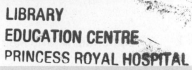

MEASUREMENT	UNIT AND ABBREVIATION	METRIC EQUIVALENT	METRIC TO ENGLISH CONVERSION FACTOR	ENGLISH TO METRIC CONVERSION FACTOR
Length	1 kilometer (km)	= 1000 (10^3) meters	1 km = 0.62 mile	1 mile = 1.61 km
	1 meter (m)	= 100 (10^2) centimeters = 1000 millimeters	1 m = 1.09 yards 1 m = 3.28 feet 1 m = 39.37 inches	1 yard = 0.914 m 1 foot = 0.305 m
	1 centimeter (cm)	= 0.01 (10^{-2}) meter	1 cm = 0.394 inch	1 foot = 30.5 cm 1 inch = 2.54 cm
	1 millimeter (mm)	= 0.001 (10^{-3}) meter	1 mm = 0.039 inch	
	1 micrometer (μm) [formerly micron (μ)]	= 0.000001 (10^{-6}) meter		
	1 nanometer (nm)	= 0.000000001 (10^{-9}) meter		
	1 angstrom (Å)	= 0.0000000001 (10^{-10}) meter		
Area	1 square meter (m^2)	= 10,000 square centimeters	1 m^2 = 1.1960 square yards 1 m^2 = 10.764 square feet	1 square yard = 0.8361 m^2 1 square foot = 0.0929 m^2
	1 square centimeter (cm^2)	= 100 square millimeters	1 cm^2 = 0.155 square inch	1 square inch = 6.4516 cm^2
Mass	1 metric ton (t)	= 1000 kilograms	1 t = 1.103 ton	1 ton = 0.907 t
	1 kilogram (kg)	= 1000 grams	1 kg = 2.205 pounds	1 pound = 0.4536 kg
	1 gram (g)	= 1000 milligrams	1 g = 0.0353 ounce 1 g = 15.432 grains	1 ounce = 28.35 g
	1 milligram (mg)	= 0.001 gram	1 mg = approx. 0.015 grain	
	1 microgram (μg)	= 0.000001 gram		
Volume (solids)	1 cubic meter (m^3)	= 1,000,000 cubic centimeters	1 m^3 = 1.3080 cubic yards 1 m^3 = 35.315 cubic feet	1 cubic yard = 0.7646 m^3 1 cubic foot = 0.0283 m^3
	1 cubic centimeter (cm^3 or cc)	= 0.000001 cubic meter = 1 milliliter	1 cm^3 = 0.0610 cubic inch	1 cubic inch = 16.387 cm^3
	1 cubic millimeter (mm^3)	= 0.000000001 cubic meter		
Volume (liquids and gases)	1 kiloliter (kl or kL)	= 1000 liters	1 kL = 264.17 gallons	1 gallon = 3.785 L
	1 liter (l or L)	= 1000 milliliters	1 L = 0.264 gallon 1 L = 1.057 quarts	1 quart = 0.946 L
	1 milliliter (ml or mL)	= 0.001 liter = 1 cubic centimeter	1 ml = 0.034 fluid ounce 1 ml = approx. $\frac{1}{5}$ teaspoon 1 ml = approx. 15–16 drops (gtt)	1 quart = 946 ml 1 pint = 473 ml 1 fluid ounce = 29.57 ml 1 teaspoon = approx. 5 ml
	1 microliter (μl or μL)	= 0.000001 liter		
Time	1 second (s)	= $\frac{1}{60}$ minute		
	1 millisecond (ms)	= 0.001 second		
Temperature	Degrees Celsius (°C)		°F = $\frac{9}{5}$(°C) + 32	°C = $\frac{5}{9}$(°F − 32)

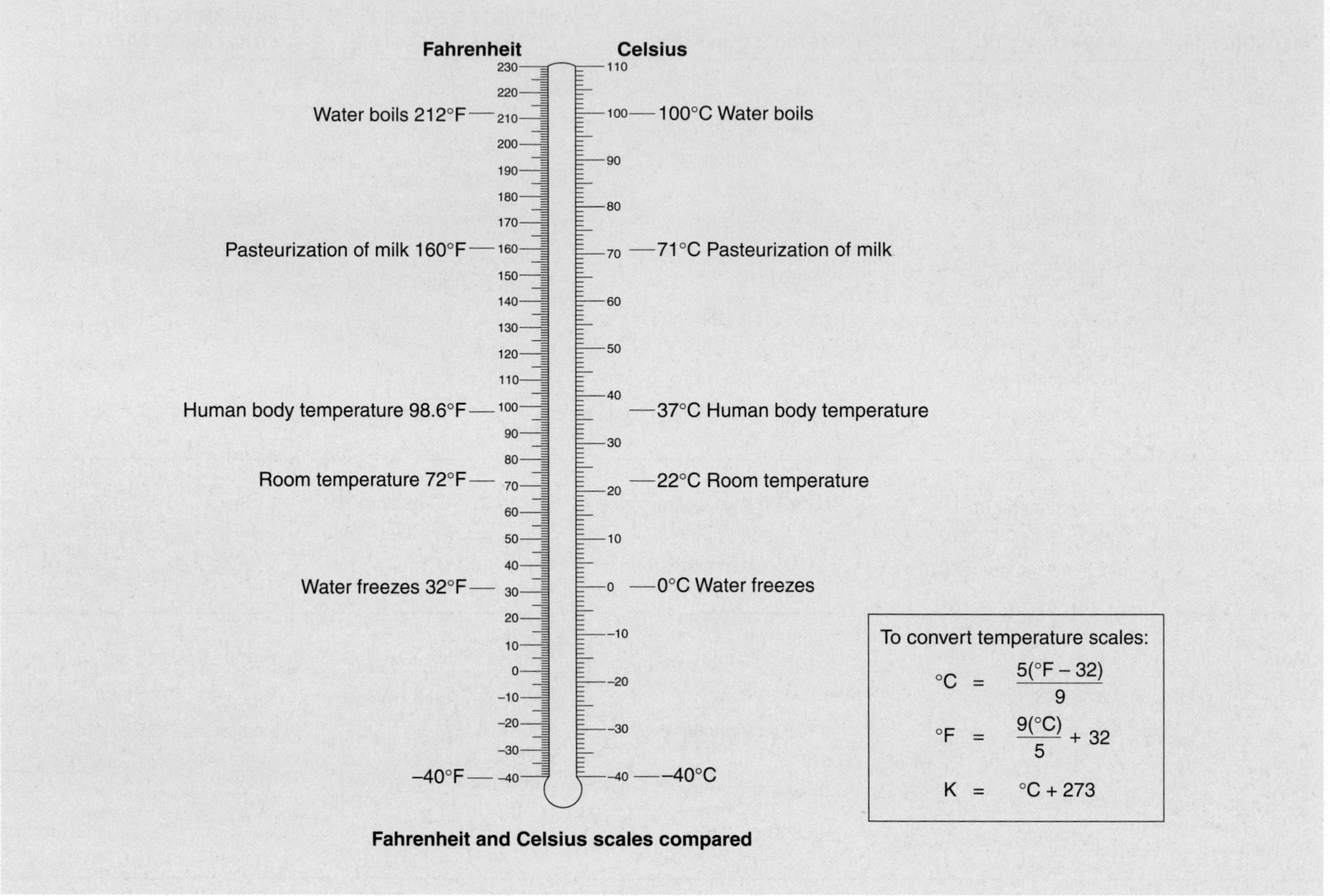

Fahrenheit and Celsius scales compared

APPENDIX B Functional Groups in Organic Molecules

Functional group	General formula	Name of compounds	Example	Where else found
Hydroxyl —OH (or HO—)	—O—H	Alcohols	$H_3C{-}CH_2{-}OH$ Ethanol	Sugars; water-soluble vitamins
Carbonyl $\rangle$CO	—C(=O)H	Aldehydes	$H_3C{-}CH_2{-}CHO$ Propanal	Some sugars; formaldehyde (a preservative)
	—C(=O)—	Ketones	$H_3C{-}C(=O){-}CH_3$ Acetone	Some sugars; "ketone bodies" in urine (from fat breakdown)
Carboxyl —COOH	—C(=O)OH	Carboxylic acids	$H_3C{-}C(=O)OH$ Acetic acid	Amino acids; proteins; some vitamins; fatty acids
Amino $—NH_2$ (or $H_2N—$)	$—NH_2$	Amines	$H_3C{-}NH_2$ Methylamine	Amino acids; proteins; urea in urine (from protein breakdown)

APPENDIX C The Amino Acids

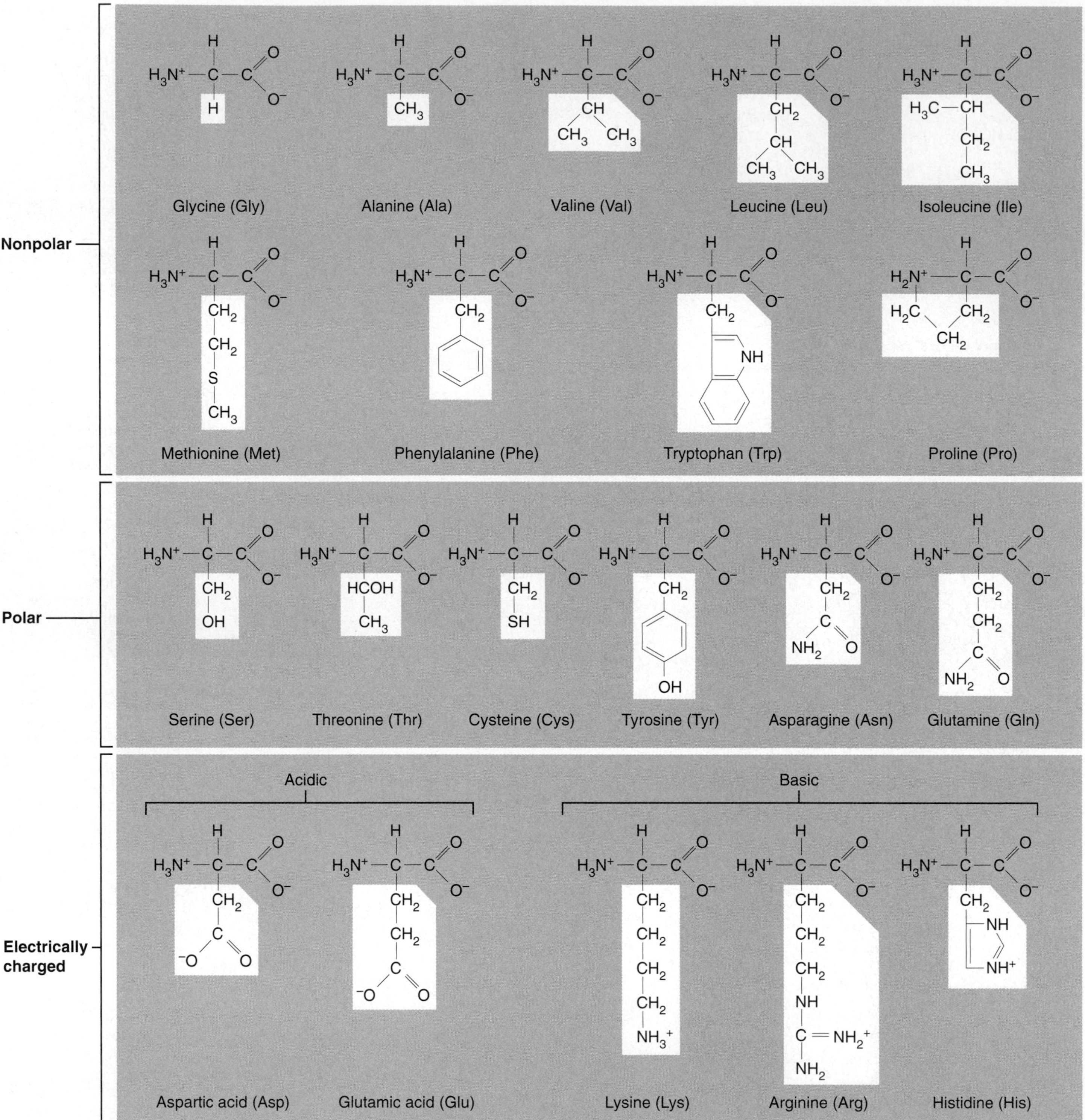

APPENDIX D Two Important Metabolic Pathways

① Glucose enters the cell and is phosphorylated by the enzyme hexokinase, which catalyzes the transfer of a phosphate group, indicated as Ⓟ, from ATP to the number six carbon of the sugar, producing glucose-6-phosphate. The electrical charge of the phosphate group traps the sugar in the cell because the plasma membrane is impermeable to ions. Phosphorylation of glucose also makes the molecule more chemically reactive. Although glycolysis is supposed to *produce* ATP, ATP is actually consumed in step 1—an energy investment that will be repaid with dividends later in glycolysis.

② Glucose-6-phosphate is rearranged and converted to its isomer, fructose-6-phosphate. Isomers, remember, have the same number and types of atoms but in different structural arrangements.

③ In this step, still another molecule of ATP is used to add a second phosphate group to the sugar, producing fructose-1,6-bisphosphate. So far, the ATP ledger shows a debit of –2. With phosphate groups on its opposite ends, the sugar is now ready to be split in half.

④ This is the reaction from which glycolysis gets its name. An enzyme cleaves the sugar molecule into two different 3-carbon sugars: glyceraldehyde 3-phosphate and dihydroxyacetone phosphate. These two sugars are isomers of one another.

⑤ An isomerase enzyme interconverts the 3-carbon sugars, and if left alone in a test tube, the reaction reaches equilibrium. This does not happen in the cell, however, because the next enzyme in glycolysis uses only glyceraldehyde phosphate as its substrate and not dihydroxyacetone phosphate. This pulls the equilibrium between the two 3-carbon sugars in the direction of glyceraldehyde phosphate, which is removed as fast as it forms. Thus, the net result of steps 4 and 5 is cleavage of a 6-carbon sugar into two molecules of glyceraldehyde phosphate; each will progress through the remaining steps of glycolysis.

⑥ An enzyme now catalyzes two sequential reactions while it holds glyceraldehyde phosphate in its active site. First, the sugar is oxidized by the transfer of H from the number one carbon of the sugar to NAD, forming NADH + H^+. Here we see in metabolic context the oxidation-reduction reaction described in Chapter 24. This reaction releases substantial amounts of energy, and the enzyme capitalizes on this by coupling the reaction to the creation of a high-energy phosphate bond at the number one carbon of the oxidized substrate. The source of the phosphate is inorganic phosphate (P_i) always present in the cytosol. The enzyme releases NADH + H^+ and 1,3-bisphosphoglyceric acid as products. Notice in the figure that the new phosphate bond is symbolized with a squiggle (~), which indicates that the bond is at least as energetic as the high-energy phosphate bonds of ATP.

THE TEN STEPS OF GLYCOLYSIS Each of the ten steps of glycolysis is catalyzed by a specific enzyme found dissolved in the cytoplasm. All steps are reversible. An abbreviated version of the three major phases of glycolysis appears in the lower right-hand corner of the next page.

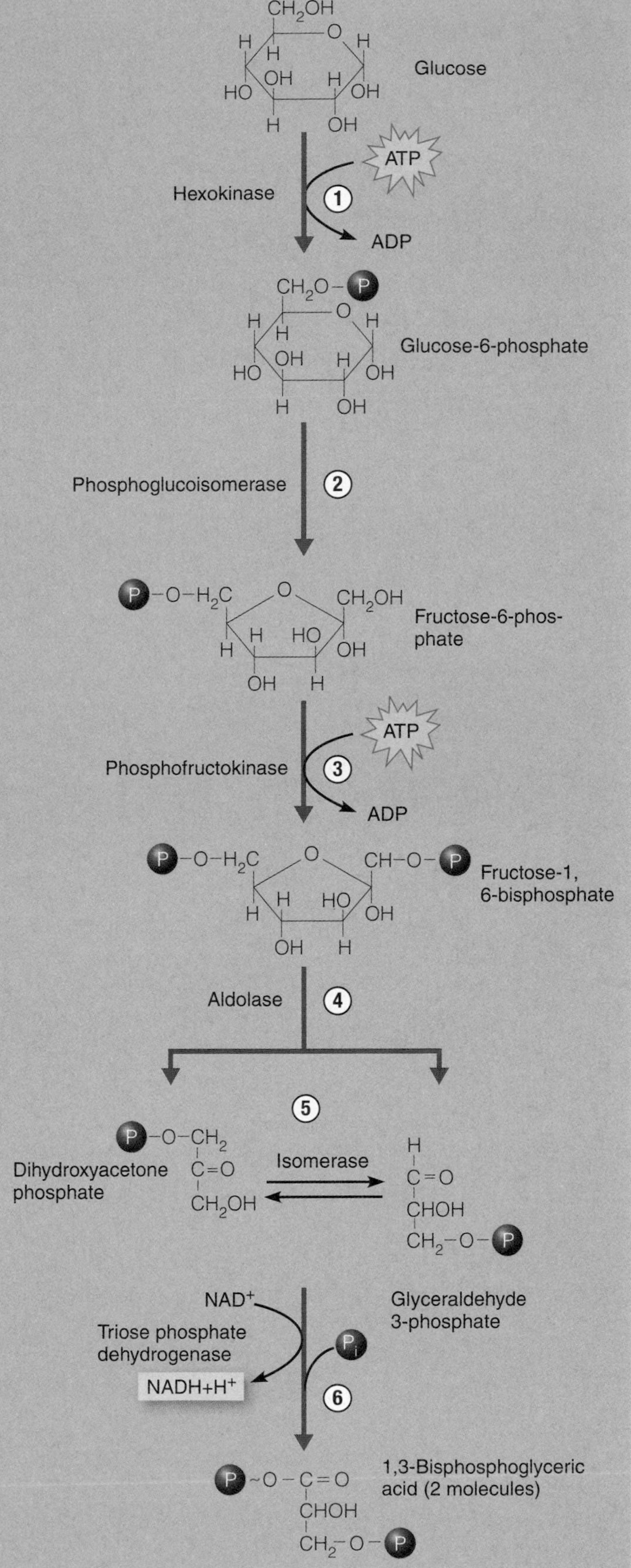

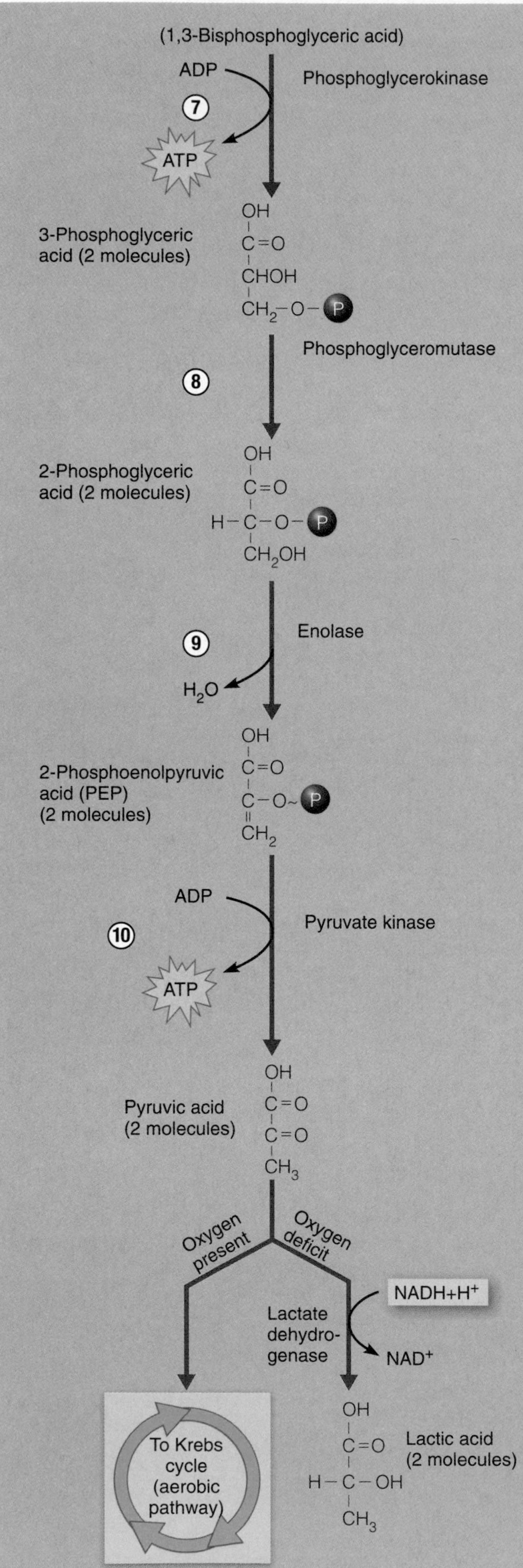

⑦ Finally, glycolysis produces ATP. The phosphate group, with its high-energy bond, is transferred from 1,3-bisphosphoglyceric acid to ADP. For each glucose molecule that began glycolysis, step 7 produces two molecules of ATP, because every product after the sugar-splitting step (step 4) is doubled. Of course, two ATPs were invested to get sugar ready for splitting. The ATP ledger now stands at zero. By the end of step 7, glucose has been converted to two molecules of 3-phosphoglyceric acid. This compound is not a sugar. The sugar was oxidized to an organic acid back in step 6, and now the energy made available by that oxidation has been used to make ATP.

⑧ Next, an enzyme relocates the remaining phosphate group of 3-phosphoglyceric acid to form 2-phosphoglyceric acid. This prepares the substrate for the next reaction.

⑨ An enzyme forms a double bond in the substrate by removing a water molecule from 2-phosphoglyceric acid to form phosphoenolpyruvic acid, or PEP. This results in the electrons of the substrate being rearranged in such a way that the remaining phosphate bond becomes very unstable; it has been upgraded to high-energy status.

⑩ The last reaction of glycolysis produces another molecule of ATP by transferring the phosphate group from PEP to ADP. Because this step occurs twice for each glucose molecule, the ATP ledger now shows a net gain of two ATPs. Steps 7 and 10 each produce two ATPs for a total credit of four, but a debt of two ATPs was incurred from steps 1 and 3. Glycolysis has repaid the ATP investment with 100% interest. In the meantime, glucose has been broken down and oxidized to two molecules of pyruvic acid, the compound produced from PEP in step 10.

Summary

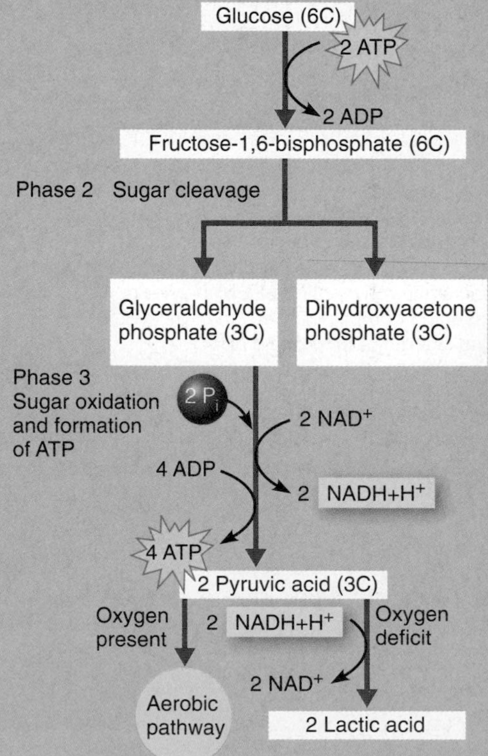

① Two-carbon acetyl CoA is combined with oxaloacetic acid, a 4-carbon compound. The unstable bond between the acetyl group and CoA is broken as oxaloacetic acid binds and CoA is freed to prime another 2-carbon fragment derived from pyruvic acid. The product is the 6-carbon citric acid, for which the cycle is named.

② A molecule of water is removed, and another is added back. The net result is the conversion of citric acid to its isomer, isocitric acid.

③ The substrate loses a CO_2 molecule, and the remaining 5-carbon compound is oxidized, forming an α-ketoglutaric acid and reducing NAD^+.

④ This step is catalyzed by a multienzyme complex very similar to the one that converts pyruvic acid to acetyl CoA. CO_2 is lost; the remaining 4-carbon compound is oxidized by the transfer of electrons to NAD^+ to form $NADH+H^+$ and is then attached to CoA by an unstable bond. The product is succinyl CoA.

⑤ Substrate-level phosphorylation occurs in this step. CoA is displaced by a phosphate group, which is then transferred to GDP to form guanosine triphosphate (GTP). GTP is similar to ATP, which is formed when GTP donates a phosphate group to ADP. The products of this step are succinic acid and ATP.

⑥ In another oxidative step, two hydrogens are removed from succinic acid (forming fumaric acid) and transferred to FAD to form $FADH_2$. The function of this coenzyme is similar to that of $NADH+H^+$, but $FADH_2$ stores less energy. The enzyme that catalyzes this oxidation-reduction reaction is the only enzyme of the cycle that is embedded in the mitochondrial membrane. All other enzymes of the citric acid cycle are dissolved in the mitochondrial matrix.

⑦ Bonds in the substrate are rearranged in this step by the addition of a water molecule. The product is malic acid.

⑧ The last oxidative step reduces another NAD^+ and regenerates oxaloacetic acid, which accepts a 2-carbon fragment from acetyl CoA for another turn of the cycle.

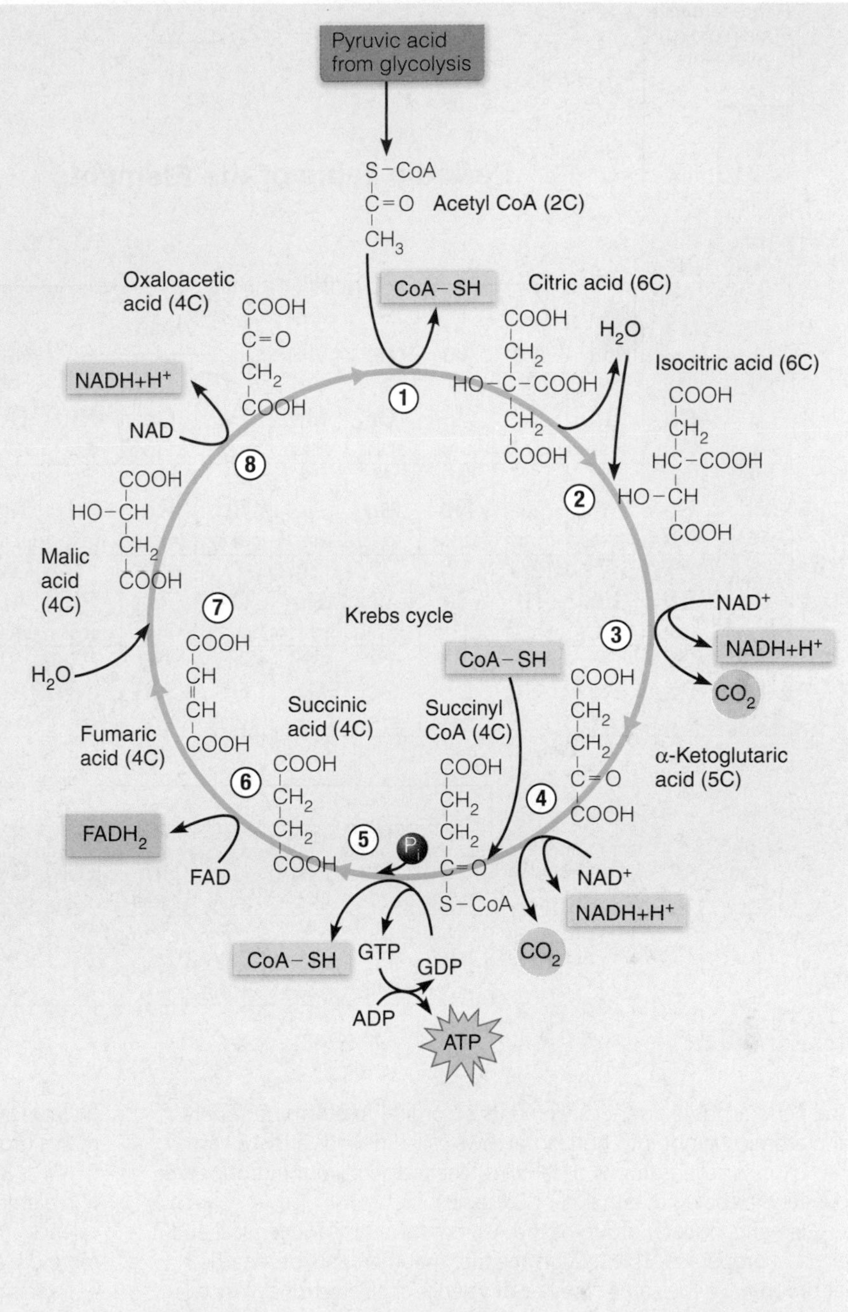

Krebs Cycle (Citric Acid Cycle) All but one of the steps (step 6) occur in the mitochondrial matrix. The preparation of pyruvic acid (by oxidation, decarboxylation, and reaction with coenzyme A) to enter the cycle as acetyl CoA is shown above the cycle. Acetyl CoA is picked up by oxaloacetic acid to form citric acid; and as it passes through the cycle, it is oxidized four more times [forming three molecules of reduced NAD ($NADH + H^+$) and one of reduced FAD ($FADH_2$)] and decarboxylated twice (releasing 2 CO_2). Energy is captured in the bonds of GTP, which then acts in a coupled reaction with ADP to generate one molecule of ATP by substrate-level phosphorylation.

APPENDIX E Periodic Table of the Elements*

Periodic Table of the Elements

Representative (main group) elements — Transition metals — Representative (main group) elements

Period	IA	IIA	IIIB	IVB	VB	VIB	VIIB	VIIIB	VIIIB	VIIIB	IB	IIB	IIIA	IVA	VA	VIA	VIIA	VIIIA
1	1 H 1.0079																	2 He 4.003
2	3 Li 6.941	4 Be 9.012											5 B 10.811	6 C 12.011	7 N 14.007	8 O 15.999	9 F 18.998	10 Ne 20.180
3	11 Na 22.990	12 Mg 24.305											13 Al 26.982	14 Si 28.086	15 P 30.974	16 S 32.065	17 Cl 35.453	18 Ar 39.948
4	19 K 39.098	20 Ca 40.078	21 Sc 44.956	22 Ti 47.867	23 V 50.942	24 Cr 51.996	25 Mn 54.938	26 Fe 55.845	27 Co 58.933	28 Ni 58.69	29 Cu 63.546	30 Zn 65.38	31 Ga 69.723	32 Ge 72.64	33 As 74.922	34 Se 78.96	35 Br 79.904	36 Kr 83.8
5	37 Rb 85.468	38 Sr 87.62	39 Y 88.906	40 Zr 91.224	41 Nb 92.906	42 Mo 95.96	43 Tc 98	44 Ru 101.07	45 Rh 102.906	46 Pd 106.42	47 Ag 107.868	48 Cd 112.411	49 In 114.82	50 Sn 118.71	51 Sb 121.76	52 Te 127.60	53 I 126.905	54 Xe 131.29
6	55 Cs 132.905	56 Ba 137.327	57 La 138.906	72 Hf 178.49	73 Ta 180.948	74 W 183.84	75 Re 186.207	76 Os 190.23	77 Ir 192.22	78 Pt 195.08	79 Au 196.967	80 Hg 200.59	81 Tl 204.383	82 Pb 207.2	83 Bi 208.980	84 Po 209	85 At 210	86 Rn 222
7	87 Fr 223	88 Ra 226	89 Ac 227	104 Rf 267	105 Db 268	106 Sg 271	107 Bh 272	108 Hs 270	109 Mt 276	110 Ds 281	111 Rg 280	112 Uub 285	113 Uut 284	114 Uuq 289	115 Uup 288	116 Uuh 293		118 Uuo 294

Rare earth elements

Lanthanides	58 Ce 140.116	59 Pr 140.908	60 Nd 144.24	61 Pm 145	62 Sm 150.36	63 Eu 151.964	64 Gd 157.25	65 Tb 158.925	66 Dy 162.5	67 Ho 164.93	68 Er 167.26	69 Tm 168.934	70 Yb 173.054	71 Lu 174.967
Actinides	90 Th 232.038	91 Pa 231.036	92 U 238.029	93 Np 237.048	94 Pu 244	95 Am 243	96 Cm 247	97 Bk 247	98 Cf 251	99 Es 252	100 Fm 257	101 Md 258	102 No 259	103 Lr 262

The periodic table arranges elements according to atomic number and atomic weight into horizontal rows called periods and 18 vertical columns called groups or families. The elements in the groups are classified as being in either A or B classes.

Elements of each group of the A series have similar chemical and physical properties. This reflects the fact that members of a particular group have the same number of valence shell electrons, which is indicated by the roman numeral of the group. For example, group IA elements have one valence shell electron, group IIA elements have two, and group VA elements have five. In contrast, as you progress across a period from left to right, the properties of the elements change in discrete steps, varying gradually from the very metallic properties of groups IA and IIA elements to the nonmetallic properties seen in group VIIA (chlorine and others), and finally to the inert elements (noble gases) in group VIIIA. This change reflects the continual increase in the number of valence shell electrons seen in elements (from left to right) within a period.

Class B elements are referred to as transition elements. All transition elements are metals, and in most cases they have one or two valence shell electrons. (In these elements, some electrons occupy more distant electron shells before the deeper shells are filled.)

In this periodic table, the colors are used to convey information about the phase (solid, liquid, or gas) in which a pure element exists under standard conditions (25 degrees centigrade and 1 atmosphere of pressure). If the element's symbol is solid black, then the element exists as a solid. If its symbol is red, then it exists as a gas. If its symbol is dark blue, then it is a liquid. If the element's symbol is green, the element does not exist in nature and must be created by some type of nuclear reaction.

*Atomic weights of the elements per IUPAC Commission on Isotopic Abundances and Atomic Weights, 2007.

APPENDIX F Reference Values for Selected Blood and Urine Studies

The reference values listed for the selected blood and urine studies are common ranges for adults, but specific "normals" are established by the laboratory performing the analysis. The values may be affected by a wide range of circumstances, including testing methods and equipment used, client age, body mass, sex, diet, activity level, medications, and extent of disease processes.

Reference values are identified in both standard or conventional units and in the system of international (SI) units. SI units (given in parentheses) are measurements of amount per volume and are used in most countries and scientific journals. SI units are often given as moles or millimoles per liter. Most clinical laboratories and textbooks in the United States use conventional or standard units, which measure mass per volume. These values are given as grams, milligrams, or milliequivalents per deciliter or liter. It is anticipated that the United States will eventually use SI units exclusively.

For enzymes, 1 international unit (IU) represents an arbitrary but defined amount of activity, whereas 1 katal (kat) is the amount of enzyme required to consume 1 mol of substrate per second.

Sample types in column 1 are serum (S), plasma (P), arterial whole blood (A), and whole blood (WB).

TEST (SAMPLE)	REFERENCE VALUES: CONVENTIONAL (SI)	PHYSIOLOGICAL INDICATION AND CLINICAL IMPLICATIONS
Blood Chemistry Studies		
Ammonia (P)	15–120 μg/dl (9–70 μmol/L)	Liver and renal function. Increased values in liver disease, renal failure, newborn hemolytic disease, heart failure, cor pulmonale. Decreased values in hypertension.
Amylase (S)	56–190 IU/L (0.4–2.1 μkat/L)	Pancreatic function. Increased values in pancreatitis, mumps, obstruction of pancreatic duct, ketoacidosis. Decreased values in kidney disease, pancreatic damage or cancer, toxemia of pregnancy.
Aspartate aminotransferase (AST, or SGOT) (S)	≤40 U/ml (≤0.7 μkat/L)	Cellular damage. Increased after myocardial infarction, acute liver disease, drug toxicity, muscle trauma. Decreased in pyridoxine (vitamin B_6) deficiency.
Bilirubin (S)	Total: 0.1–1.0 mg/dl (1.7–17.1 μmol/L) Direct: <0.4 mg/dl (<6.8 μmol/L) Indirect: 0.1–1.0 mg/dl (1.7–17.1 μmol/L) Newborn: <13.0 mg/dl (<222 μmol/L)	Liver function and red cell breakdown. Increased levels of direct in liver disease and biliary obstruction. Increased levels of indirect in hemolysis of red blood cells.
Blood urea nitrogen (S)	7–26 mg/dl (2.5–9.3 mmol/L)	Kidney function. Increased values in renal disease, dehydration, urinary obstruction, congestive heart failure, myocardial infarction, burns. Decreased values in liver failure, overhydration, impaired protein absorption, pregnancy.
Cholesterol	<200 mg/dl (<5.2 mmol/L)	Metabolism—fat utilization. Increased values in diabetes mellitus, pregnancy, use of oral contraceptives or anabolic steroids.
High-density lipoprotein (HDL) cholesterol (S)	20–30% of total >40 mg/dl (>1.0 mmol/L)	Increased levels in liver disease, aerobic exercise. Decreased levels in atherosclerotic heart disease, malnutrition.
Low-density lipoprotein (LDL) cholesterol (S)	60–70% of total <130 mg/dl (<3.4 mmol/L)	Increased values in hyperlipidemia, atherosclerotic heart disease. Decreased values in fat malabsorption and malnutrition.
Very low density lipoprotein cholesterol (VLDL) (S)	10–15% of total	Same as LDL.
Creatine kinase (CK) (S)	Female: ≤190 U/L (≤3.2 μkat/L) Male: ≤235 U/L (≤3.9 μkat/L)	Cellular damage. Increased values in myocardial infarction, muscular dystrophy, hypothyroidism, pulmonary infarction, cerebrovascular accident (CVA), shock, tissue damage, and trauma.
Creatinine (S)	0.5–1.2 mg/dl (44–106 μmol/L)	Renal function. Increased values in renal disease and acromegaly. Decreased in muscular dystrophy.

TEST (SAMPLE)	REFERENCE VALUES: CONVENTIONAL (SI)	PHYSIOLOGICAL INDICATION AND CLINICAL IMPLICATIONS
Blood Chemistry Studies *(continued)*		
Gases (A)		
Bicarbonate	22–26 mEq/L (22–26 mmol/L)	Acid-base balance. Increased values in metabolic alkalosis and respiratory acidosis. Decreased values in metabolic acidosis and respiratory alkalosis.
Carbon dioxide content	Arterial: 19–24 mEq/L (19–24 mmol/L) Venous: 22–30 mEq/L (22–30 mmol/L)	
Carbon dioxide partial pressure (P_{CO_2})	Arterial: 35–45 mm Hg Venous: 45 mm Hg	
Oxygen (O_2) saturation	95–98% (same)	Values increased slightly in hyperventilation. Decreased values (hypoxia) in pulmonary disease, hypoventilation, high altitude.
Oxygen partial pressure (P_{O_2})	80–105 mm Hg	
pH	7.35–7.45 (same)	Increased values in metabolic and respiratory alkalosis. Decreased values in metabolic and respiratory acidosis.
Glucose (S)	70–120 mg/dl (3.9–6.7 mmol/L)	Metabolic function. Increased values in diabetes mellitus, Cushing's syndrome, liver disease, acute stress, and acromegaly. Decreased levels in Addison's disease, insulinomas.
Immunoglobulins (S)		
IgG	560–1800 mg/dl (5.6–18 g/L)	Immune response. Increased levels in chronic infections, rheumatic fever, liver disease, rheumatoid arthritis. Decreased levels in amyloidosis, leukemia, and preeclampsia.
IgE	<43.2 μg/dl (<432 μg/L)	Allergic responses. Increased values in allergic responses. Decreased values in agammaglobulinemia.
IgA	85–563 mg/dl (0.85–5.6 g/L)	Immune integrity. Increased values in liver disease, rheumatic fever, chronic infection, inflammatory bowel disease. Decreased values in immunodeficiency disorders and immunosuppression.
IgM	55–375 mg/dl (0.5–3.8 g/L)	Immune integrity. Increased in autoimmune disease (e.g., rheumatoid arthritis), acute infections. Decreased in amyloidosis and leukemia.
IgD	0.5–14 mg/dl (5–140 mg/L)	Immune integrity. Increased values in myelomas.
Ketone bodies (S or P)	Negative Toxic level >20 mg/dl (0.2 g/L)	Fatty acid catabolism. Increased values (ketosis, ketoacidosis) in starvation, low-carbohydrate diet, uncontrolled diabetes mellitus, aspirin overdose.
Lactate dehydrogenase (LDH) (S)	105–333 U/L (1.7–5.6 μkat/L)	Tissue damage of organs or striated muscle. Increased in myocardial infarction, pulmonary infarction, liver disease, cerebrovascular accident, infectious mononucleosis, muscular dystrophy, fractures.
Lactic acid (lactate) (P)	9–16 mg/dl (1.0–1.8 mmol/L)	Anaerobic tissue metabolism. Increased values in congestive heart failure, shock, hemorrhage, strenuous exercise.
Osmolality (S)	280–300 mOsm/kg H_2O (280–300 mmol/kg H_2O)	Fluid and electrolyte balance. Increased levels in hypernatremia, dehydration, kidney disease, alcohol ingestion. Decreased levels in hyponatremia, overhydration, and syndrome of inappropriate ADH secretion (SIADH).
Phosphate (S) (phosphorus)	2.5–4.5 mg/dl (0.8–1.5 mmol/L)	Parathyroid function; bone disease. Increased levels in hypoparathyroidism, renal failure, bone metastasis, hypocalcemia. Decreased values in hyperparathyroidism, hypercalcemia, alcoholism, vitamin D deficiency, ketoacidosis, osteomalacia.
Potassium (S)	3.5–5.5 mEq/L (3.5–5.5 mmol/L)	Fluid and electrolyte balance. Increased levels in renal disease, Addison's disease, ketoacidosis, burns, and crush injuries. Decreased levels in vomiting, diarrhea, Cushing's syndrome, alkalosis, diuretics.

TEST (SAMPLE)	REFERENCE VALUES: CONVENTIONAL (SI)	PHYSIOLOGICAL INDICATION AND CLINICAL IMPLICATIONS
Blood Chemistry Studies *(continued)*		
Protein (S) Total	6.0–8.5 g/dl (60–85 g/L)	Osmotic pressure; immune system integrity. Increased values in multiple myeloma, dehydration, myxedema. Decreased values in protein malnutrition, burns, diarrhea, renal failure, liver failure.
Albumin	3.2–5.0 g/dl (32–50 g/L)	Osmotic pressure. Increased levels in dehydration. Decreased levels in liver disease, malnutrition, Crohn's disease, nephrotic syndrome, systemic lupus erythematosus.
Sodium (S)	135–145 mEq/L (135–145 mmol/L)	Fluid and electrolyte balance. Increased values in dehydration, diabetes insipidus, Cushing's syndrome. Decreased values in vomiting, diarrhea, burns, Addison's disease, myxedema, congestive heart failure, overhydration, syndrome of inappropriate ADH secretion (SIADH).
Triglycerides	10–150 mg/dl (0.1–1.5 g/L)	Increased values in diabetes mellitus, liver disease, nephrotic syndrome, pregnancy.
Uric acid (S)	Female: 2.0–7.3 mg/dl (119–434 μmol/L) Male: 2.1–8.5 mg/dl (125–506 μmol/L)	Renal function. Increased in lead poisoning, impaired renal function, gout, alcoholism, hematologic cancers. Decreased in Wilson's disease.
Hematology Studies		
Hemoglobin (S)	Female: 12–16 g/dl (120–160 g/L) Male: 13–18 g/dl (130–180 g/L)	Oxygenation status. Increased values in dehydration, polycythemia, congestive heart failure, chronic obstructive pulmonary disease, high altitudes. Decreased levels in anemia, hemorrhage, bone marrow cancer, renal disease, systemic lupus erythematosus, nutritional deficiency.
Hematocrit (WB)	Female: 37–47% (same) Male: 42–52% (same)	Oxygenation status. Increased levels in polycythemia, dehydration, congestive heart failure, shock, surgery. Decreased levels in anemia, hemorrhage, bone marrow disease, malnutrition, cirrhosis, rheumatoid arthritis.
Partial thromboplastin time (activated) (PTT or aPTT)	20–36 s (same)	Clotting mechanisms. Increased values in clotting factor deficiencies, cirrhosis, vitamin K deficiency, disseminated intravascular coagulation (DIC). Decreased values in early DIC, extensive cancer.
Platelet count (WB)	150,000–400,000/μl (150–400 × 10^9/L)	Clotting mechanisms. Increased values in polycythemia, cancers, rheumatoid arthritis, trauma. Decreased values in liver disease, hemolytic uremic syndrome, disseminated intravascular coagulation (DIC), idiopathic thrombocytopenic purpura (ITP), systemic lupus erythematosus (SLE).
Prothrombin time (PT) (WB)	11–12.5 s (same) 1.5–2 × control (evaluating anticoagulant treatment)	Clotting mechanisms. Increased values in liver disease, vitamin K deficiency, salicylate intoxication. Decreased values in disseminated intravascular coagulation (DIC).
Red blood cell count (RBC) (WB)	Female: 4.2–5.4 million/μl (4.2–5.4 × 10^{12}/L) Male: 4.7–6.1 million/μl (4.7–6.1 × 10^{12}/L)	Oxygenation status. Increased values in high altitudes, polycythemia, hemoconcentration, cor pulmonale. Decreased values in hemorrhage, hemolysis, anemias, chronic illness, nutritional deficiencies, leukemia, overhydration.
Reticulocyte count (WB)	0.5–2.0% (same)	Bone marrow function. Increased values in hemolytic anemia, sickle-cell anemia, leukemia, pregnancy. Decreased values in pernicious anemia, folic acid deficiency, cirrhosis, chronic infection, bone marrow depression or failure.
White blood cell count (WBC) (WB)		
Total (males)	4800–10,800/μl (4.8–10.8 × 10^9/L)	Immune system integrity. Increased values in infection, trauma, stress, tissue necrosis, inflammation. Decreased values in bone marrow depression or failure, drug toxicity, overwhelming infection, malnutrition.

TEST (SAMPLE)	REFERENCE VALUES: CONVENTIONAL (SI)	PHYSIOLOGICAL INDICATION AND CLINICAL IMPLICATIONS
Hematology Studies *(continued)*		
White blood cell count, differential (WB)		
Neutrophils	50–70% (same)	Immune system integrity. Increased values in acute bacterial infections, stress, Cushing's syndrome, inflammatory disorders, ketoacidosis, gout. Decreased levels in aplastic anemia, bone marrow suppression, and overwhelming bacterial infections.
Lymphocytes	25–40% (same)	Immune system integrity. Increased values in viral infections (e.g., mumps, rubella, infectious mononucleosis, hepatitis), lymphocytic leukemia, certain bacterial infections. Decreased levels in leukemia, immunodeficiency, lupus erythematosus, bone marrow depressive drugs.
Eosinophils	1–4% (same)	Immune system integrity. Increased levels in allergic reactions, parasitic infections, leukemia. Decreased levels in excess adrenosteroid production.
Monocytes	3–8% (same)	Immune system integrity. Increased levels in certain infections (e.g., tuberculosis, malaria), inflammatory disorders.
Basophils	0.5–1.0% (same)	Immune system integrity. Increased levels in myeloproliferative disorders, leukemia. Decreased levels in allergic reactions, hyperthyroidism, stress.
Urine Tests		
Amylase (24 h)	<6000 Somogyi units/24 h 0–500 U/24 h (0–8.3 μkat/24h)	Pancreatic function. Increased values in pancreatic disease or obstruction, inflammation of the salivary glands, and cholecystitis.
Bilirubin (random)	Negative (same)	Liver function. Increased values in liver disease, extrahepatic obstruction (gallstones, tumor, inflammation).
Blood (hemoglobin) (random)	Negative (same)	Urinary system function. Increased values in cystitis, renal disease, hemolytic anemia, transfusion reaction, prostatitis, burns.
Osmolality (random or fasting)	Random: 50–1200 mOsm/kg H_2O (50–1200 mmol/kg H_2O) Fluid restriction: ≥850 mOsm/kg H_2O (≥850 mmol/kg H_2O)	Fluid and electrolyte balance, renal function, and endocrine function. Increased levels in hypernatremia, syndrome of inappropriate ADH secretion (SIADH), congestive heart failure, metabolic acidosis. Decreased levels in diabetes insipidus, water intoxication, pyelonephritis, renal tubular necrosis, aldosteronism.
Phosphate (24 h)	0.9–1.3g/24 hr (same)	Parathyroid function. Increased levels in hyperparathyroidism, osteomalacia, certain renal diseases, vitamin D deficiency. Decreased levels in hypoparathyroidism.
Potassium (24 h)	25–120 mEq/24 h (25–120 mmol/24 h)	Fluid and electrolyte balance. Increased values in renal tubular necrosis, metabolic acidosis, dehydration, aldosteronism, Cushing's syndrome. Decreased values in Addison's disease, malabsorption, acute renal failure.
Protein (random)	<8 mg/dl (<80 mg/L)	Renal function. Increased levels in nephrotic syndrome, renal trauma, hyperthyroidism, diabetic nephropathy, lupus erythematosus.
Sodium (24 h)	40–220 mEq/24 h (40–220 mmol/24 h)	Fluid and electrolyte balance. Increased values in dehydration, ketoacidosis, syndrome of inappropriate ADH secretion (SIADH), adrenocortical insufficiency. Decreased levels in congestive heart failure, renal failure, diarrhea, aldosteronism.
Uric acid (24 h)	250–750 mg/24 h (1.5–4.5 mmol/24 h)	Renal function and metabolism. Increased in gout, leukemia, liver disease, ulcerative colitis. Decreased in renal disease, alcoholism, lead toxicity, folic acid deficiency.

TEST (SAMPLE)	REFERENCE VALUES: CONVENTIONAL (SI)	PHYSIOLOGICAL INDICATION AND CLINICAL IMPLICATIONS
Urine Tests *(continued)*		
Urinalysis (random)		
Color	Straw, yellow, amber	Fluid balance and renal function. Darker in dehydration. Lighter in overhydration, diabetes insipidus. Color varies with disease states, diet, and medications.
Odor	Aromatic	Metabolic function, infection. Abnormal odors in infection, ketonuria, rectal fistula, hepatic failure, phenylketonuria.
Specific gravity	1.001–1.035	An indirect measurement of urine concentration (osmolality). Same physiological indications and clinical implications as osmolality.
pH	4.5–8.0	A crude indicator of acid-base balance. Decreased by acidic diet (proteins). Increased by a vegetarian diet, prolonged vomiting, and bacterial infection of the urinary tract.
Urobilinogen (24 h)	0.2–1.0 mg/dl (2–10 mg/L)	Liver function. Increased in hemolytic anemias, hepatitis, cirrhosis, biliary disease. Decreased in common bile duct obstruction.
Volume (24 h)	800–2000 ml/24 h (0.8–2.0 L/24 h)	Fluid and electrolyte balance, renal function. Increased values in diabetes insipidus, diabetes mellitus, renal disease. Decreased values in dehydration, syndrome of inappropriate ADH secretion (SIADH), renal disease.

APPENDIX G Answers to Check Your Understanding, Clinical Connections, Multiple Choice, and Matching Questions

Chapter 1

Check Your Understanding 1. The operation or function of a structure is dictated (promoted or prevented) by its anatomy. For example, oxygen and carbon dioxide are exchanged across the very thin membranes of the lungs but not across the skin. **2.** Muscle shortening is a topic of physiology. The body location of the lungs is an anatomy topic. **3.** Cytologists study the cellular level of organization. **4.** The order in the structural hierarchy is cell, tissue, organ, and organism. **5.** Bones and cartilages are part of the skeletal system. The nasal cavity, lungs, and trachea are organs of the respiratory system. **6.** Living organisms are able to maintain their boundaries, move, respond to environmental changes, digest nutrients, carry out metabolism, dispose of wastes, reproduce, and grow. While inanimate objects may exhibit some of these properties, they do not exhibit all of them. **7.** Metabolism is the term that encompasses all the chemical reactions that occur in body cells. **8.** In flight, the cabin must be pressurized because the atmosphere is thinner at high altitudes and the amount of oxygen entering the blood under such conditions may be insufficient to maintain life. **9.** Negative feedback mechanisms allow us to adjust to conditions outside the normal temperature range by causing heat to be lost from the body (in hot conditions) and retained or generated by the body (in cold conditions). **10.** Thirst is part of a negative feedback control system because it prods us to drink, which ends the thirst stimulus and returns body fluid volume to the normal range. **11.** This is a positive feedback mechanism because it enhances the change (formation of a platelet plug) set into motion by the stimulus (damage to the blood vessel). The response ends when the platelet plug has plugged the hole in the blood vessel. **12.** The position in which a person is standing erect with feet slightly separated and palms facing anteriorly. Knowing the anatomical position is important because directional terms refer to the body as if it is in this position. **13.** Axillary region is the armpit. Acromial area is the tip of the shoulder. **14.** A frontal (coronal) section would separate the brain into anterior and posterior parts. **15.** He may have appendicitis if the pain is in the lower right quadrant of his abdomen. **16.** Of these organs, only the spinal cord is in the dorsal body cavity. **17.** As mobile organs (heart, lungs, digestive organs) work, friction is greatly reduced by the presence of serous fluid. Serous fluid allows the surrounding serous membranes to glide easily over one another.

Review Questions 1. c; **2.** a; **3.** e; **4.** a, d; **5.** (a) wrist (b) hip bone (c) nose (d) toes (e) scalp; **6.** neither c nor d would be visible in the median section; **7.** (a) dorsal (b) ventral (c) dorsal (d) ventral; **8.** b; **9.** b; **10.** c

Chapter 2

Check Your Understanding 1. Foods contain chemical energy. **2.** Electrical energy is the energy used by nerve cells to transmit messages in the body. **3.** Potential energy (PE) is available when we are still. PE is converted to kinetic (working) energy when we exercise. **4.** Besides hydrogen and nitrogen, carbon and oxygen help to make up the bulk of living matter. **5.** This element has 82 protons in its nucleus and 82 electrons in its orbitals (electron cloud). **6.** Atomic mass indicates the sum of the protons and neutrons in a given atom's nucleus. Atomic weight indicates the average mass of all the isotopes of a given element. **7.** A molecule is 2 or more atoms held together by chemical bonds. **8.** A compound is formed when two or more different kinds of atoms chemically bond together, as in NaCl. Oxygen gas is 2 oxygen atoms (the same kind of atom) bonded together. **9.** Blood is a mixture because its components are not changed by their combination and they can be separated by physical means. **10.** Hydrogen bonds (linking H of one water molecule to O of another) form between water molecules. **11.** Argon's valence shell is full:)2e)8e)8e. Hence it is nonreactive. **12.** Electrons would spend more time in the vicinity of the more electronegative atom in XY, whereas electrons in XX would orbit both X atoms to an equal extent. **13.** Fats are digested in the small intestine by decomposition reactions. **14.** Biochemical reactions in the body tend to be irreversible because (a) one or more of the products is removed from the reaction site or (b) the product is needed more than the reactants, so the cell would not provide energy to reverse the reaction. **15.** Decomposition reactions in which foods are broken down for energy are oxidation-reduction (O-R) reactions. **16.** Water is an excellent solvent because of its polarity. As a dipole, it can orient itself to the end of other molecules, causing them to dissociate or go into solution. **17.** Electrolytes are substances like salts that will conduct an electrical current in aqueous solution. **18.** H^+ is responsible for acidity. **19.** It is better to add a weak base, which will act to buffer the strong acid. **20.** Monomers of carbohydrates are called monosaccharides or simple sugars. Glucose is blood sugar. **21.** The animal form of stored carbohydrate is glycogen. **22.** Triglycerides, the major source of stored energy in the body, are composed of three fatty acid chains and a glycerol molecule and are found in fat tissue. Phospholipids consist of two fatty acid chains and a charged P-containing group. They are found in all cell membranes where they form the basis of those membranes. **23.** Hydrolysis reactions break down polymers or macromolecules to their monomers by adding water to each bond joining monomers. **24.** An "amino acid" has an amine group (NH_2) and a COOH group that has acidic properties. **25.** The primary structure of proteins is the stringlike chain of amino acids. **26.** The secondary structures of proteins are the alpha helix and the beta-pleated sheet. **27.** Molecular chaperones prevent inaccurate or inappropriate folding in the 3-D structure of a protein. **28.** Enzymes hold the substrate(s) in a desirable position to interact. **29.** DNA contains deoxyribose sugar and the bases A, T, G, C. RNA contains ribose sugar and the bases A, U, G, C. **30.** DNA dictates protein structure by its base sequence and reproduces itself before a cell divides to ensure that the genetic information in the daughter cells is identical. **31.** ATP stores energy in smaller packets that are more readily released and transferred (during ATP hydrolysis) than the energy stored in glucose. Hence the use of ATP as an energy source keeps energy waste to a minimum. **32.** When ATP releases energy, it loses a phosphate group and becomes ADP (also energy rich).

Review Questions 1. d; **2.** d; **3.** b; **4.** a; **5.** b; **6.** a; **7.** a; **8.** b; **9.** d; **10.** a; **11.** b; **12.** a, c; **13.** (1)a, (2)c; **14.** c; **15.** d; **16.** e; **17.** d; **18.** d; **19.** a; **20.** b; **21.** b; **22.** c

Chapter 3

Check Your Understanding 1. The three basic parts of a cell are the plasma membrane (the outer boundary of the cell), the nucleus (control center of the cell), and the cytoplasm (the fluid material between the nucleus and plasma membrane), which contains the organelles. **2.** It is the cell concept that includes structures and functions common to all cells. **3.** All celluar membranes consist of a double layer of phospholipids in which proteins are embedded. **4.** Hydrophobic regions (tails of phospholipid molecules) orient toward each other while the hydrophilic regions (phospholipid heads) orient to the aqueous fluid inside and outside the cell. **5.** The sugar residues of the glycocalyx provide recognizable biological markers for cells to recognize each other. **6.** The heart has desmosomes (anchoring junctions) that secure cardiac cells together as the heart works and gap junctions (communicating junctions) that allow ions to flow from cardiac cell to cardiac cell. **7.** Diffusion is driven by kinetic energy of the molecules. **8.** The relative concentration of the substance in different areas determines the direction of diffusion. Diffusion occurs from regions of high concentration to regions of low concentration. **9.** In channel-mediated diffusion, the diffusing substance moves through a membrane channel. In carrier-mediated diffusion, the diffusing substance attaches to a membrane (protein) carrier which moves it across the membrane. **10.** Phosphorylation of the Na^+-K^+ pump causes the pump protein to change shape so that it "pumps" Na^+ across the membrane. K^+ binding to the pump protein triggers the release of phosphate and the pump protein returns to its original shape. **11.** The plasma membrane expands as a result of exocytosis. **12.** Phagocytic cells engulf debris and a smoker's lungs would be laden with carbon particles and other debris from smoke inhalation. **13.** Cholesterol is taken in by receptor-mediated endocytosis. **14.** Diffusion of ions, mainly the diffusion of K^+ from the cell through leakage channels, establishes the resting membrane potential. **15.** In a polarized membrane, the inside is negative relative to its outside. **16.** Signaling chemicals that bind to membrane receptors are called ligands. G protein–linked receptors direct intracellular events by promoting formation of second messengers. **17.** Mitochondria are the major sites of ATP synthesis. **18.** Ribosomes are the sites of protein synthesis. The rough ER provides a site for ribosome attachment and its cisternae package in vesicles the proteins made on the ribosomes for transport to the Golgi apparatus. The Golgi apparatus modifies and packages the proteins it receives for various destinations within or outside the cell. **19.** The lysosomal enzymes digest foreign substances engulfed by the cell, nonuseful or deteriorating organelles, or even the cell itself to prevent the buildup of cellular debris. The enzymes in peroxisomes detoxify harmful chemicals and neutralize free radicals. **20.** Both microfilaments and microtubules are involved in organelle movements within the cell and/or movements of the cell as a whole. **21.** Intermediate filaments are the most important cytoskeletal elements in maintaining cell shape. **22.** The major function of microvilli is to increase the cell's surface area for absorption or filtration of substances. **23.** If a cell loses its nucleus, it is doomed to die because it will be unable to make proteins, which include the enzymes needed for all metabolic reactions. **24.** Nucleoli are the site of synthesis of ribosomal subunits. **25.** Histone proteins provide the means to pack DNA in a compact, orderly way and play a role in gene regulation. **26.** The base sequence of the corresponding strand will be GCTTAC. **27.** DNA is synthesized during the S phase. **28.** Nuclear envelope breaks up, spindle forms, nucleoli disappear, and the chromosomes coil and condense. **29.** Codons are three base sequences in mRNA, each of which specifies an amino acid. Anticodons are three-base sequences in tRNA that are complementary to the codons specifying the amino acid they transport to the ribosome during protein synthesis. **30.** A site = entry site for tRNA at the ribosome. P site = site where peptide bonds form between delivered amino acids. E site = the tRNA exit site from the ribosome. **31.** DNA provides the coded instructions (is the template) for protein synthesis via the mRNA synthesized on it. **32.** Ubiquitin attaches to misfolded, damaged, or unneeded proteins, tagging them for destruction by proteasomes. **33.** Blood plasma is extracellular fluid that transports nutrients, gases, hormones, and other substances throughout the body. Interstitial fluid is an important transport and dissolving medium. **34.** Apoptosis is a process of programmed cell death which rids the body of cells that are stressed, damaged, old, or no longer needed. **35.** The wear-and-tear theory of aging attributes aging to little chemical insults and free radicals which have cumulative detrimental effects.

Review Questions 1. d; **2.** a, c; **3.** b; **4.** b; **5.** e; **6.** c; **7.** d; **8.** a; **9.** d; **10.** a; **11.** b; **12.** d; **13.** c; **14.** b; **15.** d; **16.** a; **17.** b; **18.** d; **19.** c

Chapter 4

Check Your Understanding 1. Fixing tissue preserves it and prevents it from deteriorating. **2.** Organic dyes are used to stain tissues viewed by a light microscope. Heavy metal salts are used to stain tissues viewed by electron microscopes. **3.** Epithelial tissue lines body cavities and covers the body's external surface; thus polarity with one free (apical) surface is a requirement. **4.** Epithelial tissue can regenerate and its cells are joined by lateral contacts. **5.** Simple epithelia are "built" to provide for efficient absorption and filtration across their thin epithelial barriers. **6.** Pseudostratified epithelia appear to be stratified because their cells' nuclei lie at different distances from the basement membrane. However, all cells rest on the basement membrane. **7.** Transitional epithelium is found in the urinary bladder and other hollow urinary organs. The ability of this epithelium to thin allows the urinary organs to handle (store or transport) a larger urine volume when necessary. **8.** All unicellular exocrine glands secrete the protein mucin (which becomes mucus on mixing with water). **9.** These exocrine glands are classified by duct and secretory unit types. **10.** Holocrine glands have the highest rate of cell division. The secretory cells fragment and are lost in the secretion; thus the secretory cells must be continuously replaced. **11.** Connective tissue functions to bind, support, protect, and insulate body organs. In addition, blood acts to transport substances throughout the body. **12.** Reticular, collagen, and elastic fibers are found in the various connective tissues. **13.** Areolar connective tissue, because of its loose weblike nature, is capable of serving as a fluid reservoir. **14.** Dense regular connective tissue is damaged when you lacerate a tendon. **15.** Hyaline cartilage forms the growth plates. **16.** With extended processes, a neuron can conduct electrical signals a greater distance within the body. **17.** Cardiac muscle cells have striations and are branching cells. **18.** Skeletal muscle tissue is voluntary and is the muscle tissue injured when you "pull a muscle" while exercising. **19.** A mucous membrane consists of both connective tissue and epithelium. It lines body cavities open to the exterior. **20.** The serous membranes called pleurae line the thorax walls and cover the lungs. **21.** The three main steps of tissue repair are inflammation, organization, and regeneration and fibrosis (which is a permanent repair). **22.** More severe injuries damage and destroy more tissue, requiring greater replacement with scar tissue. **23.** The three embryonic germ layers are the ectoderm, mesoderm, and endoderm. **24.** Ectoderm gives rise to the nervous system. **25.** Epithelium and some connective tissues (areolar, dense irregular, and blood-forming tissue) remain highly mitotic all through life.

Review Questions 1. a, c, d, b; **2.** c, e; **3.** b, f, a, d, g, d; **4.** b; **5.** c; **6.** b

Chapter 5

Check Your Understanding 1. Since the sole of the foot has thick skin, the layers from most superficial to deepest would be the stratum corneum, stratum lucidum, stratum granulosum, stratum spinosum, and stratum basale. **2.** The stratum basale undergoes almost continuous mitosis to replace cells lost by abrasion. **3.** The skin is subjected to a lot of abrasion and physical trauma. The desmosomes, which are connecting junctions, help to hold the cells together during such stress. **4.** The stratum basale, which abuts the dermis where the blood supply is, has the best nourished cells. **5.** The papillary layer of the dermis gives rise to fingerprint patterns. **6.** Fatty tissue in the hypodermis gives it insulating and shock-absorbing properties. **7.** Because there is no bleeding, the cut has penetrated into the avascular epidermis only. **8.** The third pigment contributing to skin color is hemoglobin, the pigment found in red blood cells found in blood vessels of the dermis. **9.** Cyanosis is a bluish cast of the skin that indicates that hemoglobin in the red blood cells in the dermal capillaries is poorly oxygenated. **10.** Jaundice or a yellow cast to the skin due to the deposit of yellow bile pigments in body tissues may indicate a liver disorder. **11.** Sebaceous (oil) glands and apocrine glands are associated with the hair follicles. **12.** His sympathetic nervous system activated his eccrine sweat glands and caused heat-induced sweating in order to cool the body. **13.** Heat-induced sweating occurs all over the body when we are overheated. A cold sweat is emotionally induced sweating that begins on the palms, soles, and armpits and then spreads to other body areas. Both types of sweating are produced by the eccrine sweat glands, but activity of apocrine sweat glands is also likely during a cold sweat. **14.** The palms of the hands and soles of the feet are thick skin areas. It would be dangerous to have oily soles, and oily palms would decrease the ability of the hands to hang onto things. **15.** The regions of a hair from outside in are the cuticle, cortex, and medulla. **16.** There are no nerves in a hair, so cutting hair is painless. **17.** The arrector pili muscles pull the hair (normally slanted) to the upright position (when cold or scared). **18.** The hair papilla contains a knot of capillaries that supplies nutrients to cells of the hair bulb. **19.** The lunule of the nail is white because the thick nail matrix that underlies it blocks the rosy color of the dermal blood supply from showing through. **20.** Nails are hard because the keratin they contain is the hard keratin variety. **21.** The low pH of skin secretions (acid mantle) inhibits division of bacteria, and many bacteria are killed by bactericidal substances in sebum or by natural antibiotics called defensins. Damaged skin secretes cathelicidins that are effective against a certain strain of strep bacteria. **22.** The epidermal dendritic cells play a role in body immunity. **23.** Sunlight causes the skin to produce a precursor of vitamin D from cholesterol. **24.** The skin carries out chemical conversions that supplement some of the protective conversions carried out by the liver, convert some chemicals into carcinogens, activate some steroid hormones, and synthesize the vitamin D precursor. **25.** Basal cell carcinoma develops from the youngest epidermal cells. **26.** The ABCD rule helps one to recognize signs of melanoma. **27.** First- and second-degree burns can heal uneventfully by regeneration of epidermal cells as long as infection does not occur. Third-degree burns destroy the entire epidermis and regeneration is not possible. Infection and loss of body fluid and proteins are problematic. **28.** Burns to the face are serious because damage to the lungs can occur in such burns. **29.** Vernix caseosa is a product of the sebaceous glands. **30.** Loss of subcutaneous fat, common in the elderly, leads to cold intolerance seen in this group of people. **31.** UV radiation degrades collagen and leads to loss of skin elasticity and water-holding capacity.

Clinical Connections 1. The skin separates and protects the internal environment of the body from potentially dangerous elements in the external environment. Mrs. DeStephano's chart indicates epidermal abrasions, which represent the loss of this barrier. Epidermal loss will also cost Mrs. DeStephano the acid mantle of her skin, protection against UV radiation, and epidermal dendritic (Langerhans) cells, which protect against invasion by microorganisms. **2.** Macrophages found in the dermis can act as a backup system against bacterial and viral invasion when the epidermis is damaged. **3.** Suturing brings the edges of wounds close together and promotes faster healing because smaller amounts of granulation tissue need to be formed. This is termed *healing by first intention.* **4.** Cyanosis signals a decrease in the amount of oxygen carried by hemoglobin in the blood. Respiratory system and/or cardiovascular system impairments can lead to cyanosis.

Review Questions 1. a; **2.** c; **3.** d; **4.** d; **5.** b; **6.** b; **7.** c; **8.** c; **9.** b; **10.** a; **11.** d; **12.** b

Chapter 6

Check Your Understanding 1. Hyaline cartilage is the most plentiful in the adult body. **2.** The epiglottis and external ear cartilages are flexible elastic cartilage. **3.** Interstitial growth is growth from within. **4.** The components of the axial skeleton are the skull, the vertebral column, and the thoracic cage. **5.** The major function of the axial skeleton is to establish the long axis of the body and to protect structures that it encloses. The general function of the appendicular skeleton is to allow us mobility for propulsion and manipulation of our environment. **6.** The ribs and skull bones are flat bones. **7.** Skeletal muscles use bones as levers to cause movement of the body and its parts. **8.** Bone matrix stores minerals and growth tors. **9.** Bone marrow cavities serve as sites for blood cell formation t storage. **10.** Crests, tubercles, and spines are bony projections. **11.** Compact bone looks fairly solid and homogeneous whereas spongy bones have an open network of bone spicules. **12.** Endosteum lines the internal canals and covers the trabeculae. **13.** Bone's inorganic component (bone salts) makes it hard. **14.** The osteoclast fits this description. **15.** Bones begin as fibrous membranes or hyaline cartilages. **16.** The cartilage model grows, then breaks down and is replaced by bone. **17.** The primary ossification center in a long bone is in the center of the shaft. The secondary ossification centers are in the epiphyses (bone ends). **18.** The chondrocytes are enlarging and their lacunae are breaking down and leaving holes in the cartilage matrix. **19.** If bone-destroying cells (osteoclasts) are more active than bone-forming cells (osteoblasts), bone mass will decrease. **20.** The hormonal stimulus maintains homeostatic blood calcium levels. **21.** In an open fracture, the bone ends are exposed to the external environment. In a closed fracture, the bone ends do not penetrate the external boundary of the skin. **22.** Bone growth increases bone mass, as during childhood or when exceptional stress is placed on the bones. Bone remodeling follows bone growth to maintain the proper proportions of the bone considering stresses placed upon it. **23.** Paget's disease is characterized by excessive deposit of weak, poorly mineralized bone. **24.** Sufficient vitamin D, calcium, and weight-bearing exercise all help to maintain healthy bone density. **25.** Adult rickets is called osteomalacia. **26.** At birth, most bones are well formed and ossified. Two areas of hyaline cartilage remain—in the epiphyseal plates and covering the bone ends (articular cartilage). **27.** The skull bones do not appear to lose mass with aging.

Clinical Connections 1. Mrs. DeStephano's broken leg has a transverse fracture of the open variety because the broken ends of the bone are protruding through the skin. **2.** The laceration of the skin caused by the broken end of the bone creates a breach in the protective barrier created by the skin, providing an entry point for bacteria and other microorganisms. In addition, the protruding ends of the bone have now been exposed to the nonsterile external environment. This could result in the development of osteomyelitis, a bacterial infection, which can be treated with antibiotics. **3.** Reduction of a fracture is the clinical term for "setting the bone." Mrs. DeStephano's physician chose internal reduction, in which surgery is performed and the broken ends of the bone are secured together by pins or wires. A cast was applied to keep the aligned ends of the bone immobile until healing of the fracture has occurred. **4.** Healing of Mrs. DeStephano's fracture will begin as bony callus formation fills the break in the bone with bony tissue. This process begins 3–4 weeks after the break occurs and is completed within 2–3 months. **5.** Nutrient arteries supply blood to the bone tissue. In order for Mrs. DeStephano's break to heal normally, the bony tissue must be supplied with oxygen (to generate ATP for energy) and nutrients in order to rebuild the bone. Damage to a nutrient artery will decrease the delivery of these building materials and could slow the process of healing. **6.** For a fracture that is slow to heal, new techniques that promote healing include electrical stimulation, which promotes the deposition of new bone tissue; ultrasound treatments, known to speed healing; and possibly the addition of bone substitutes to the fractured area. **7.** At age 45, Mrs. DeStephano will most likely not regenerate her knee cartilage. (Cartilage growth typically ends during adolescence.) Cartilage damage that occurs during adulthood is slow to heal, due to the avascular nature of cartilage, and is usually irreparable. Surgical removal of cartilage fragments to allow improved movement of the joint is the usual treatment for this type of damage.

Review Questions 1. e; **2.** b; **3.** c; **4.** d; **5.** e; **6.** b; **7.** c; **8.** b; **9.** d, e; **10.** c; **11.** b; **12.** c; **13.** a; **14.** b

Chapter 7

Check Your Understanding 1. The three main parts of the axial skeleton are the skull, vertebral column, and thoracic cage. **2.** The axial skeleton is more important in protecting internal organs. **3.** The frontal, parietal, temporal, sphenoid, and ethmoid bones are all cranial bones. **4.** The ethmoid bone forms the crista galli. **5.** The temporal bones house the external ear canals. **6.** The parietal bones abut each other at the sagittal suture. The occipital bones abuts the parietal bones at the lambdoid suture. **7.** The zygomatic bones are the cheekbones. **8.** Eating or talking, because the only freely movable joints of the skull are the temporomandibular joints of the jaw. **9.** The maxillae are the keystone bones of the facial skeleton. **10.** The sphenoid, ethmoid, frontal, and maxillary bones contain paranasal sinuses. **11.** The cribriform plates of the ethmoid bone form the roof of the nasal cavity. **12.** The maxillae form the bulk of the orbit floor. The eye is housed in the orbit. **13.** The five major regions of the vertebral column are the cervical, thoracic, lumbar, sacral, and coccygeal regions. **14.** The cervical and lumbar regions are concave posteriorly. **15.** The fibrocartilage discs contribute to the flexibility of the vertebral column. **16.** There are 7 cervical and 12 thoracic vertebrae. **17.** The dens is the axis on which the atlas rotates. If it's broken, movements of the atlas would be less controllable. **18.** A lumbar vertebra is heavier and its massive body is kidney shaped. Its spinous processes are short and project directly back. A thoracic vertebral body is generally heart shaped, its spinous process is long, sharp, and points downward, and its transverse processes have facets for articulating with the ribs. **19.** A true rib connects to the sternum by its own costal cartilage. A false rib connects to the sternum via costal cartilages of other ribs or not at all. **20.** The sternal angle is a ridge across the front of the sternum where the manubrium joins the sternal body. It acts as a hinge allowing the sternum to swing anteriorly when we inhale. Because it is aligned with the second rib, it is a handy cue for finding that rib and then counting the ribs during a physical exam. **21.** The thoracic vertebrae also contribute to the thoracic cage. **22.** Each pectoral girdle is formed by a scapula and a clavicle. **23.** The pectoral girdle attaches to the sternal manubrium of the axial skeleton via the medial end of its clavicle. **24.** A consequence of its flexibility is that it is easily dislocated. **25.** Together the ulna and humerus form the elbow joint. **26.** The ulna and the radius each have a styloid process distally. **27.** Carpals are found in the proximal region of the palm. They are short bones. **28.** The third bone of the os coxae is the ischium. **29.** The pelvic girdle (along with the sacrum) receives the weight of the upper body (trunk, head, and upper limbs) and transmits that weight to the lower limbs. **30.** The female pelvis is wider and has a shorter sacrum and a more movable coccyx. **31.** The tibia is the second largest bone in the body. **32.** The medial malleolus is located on the distalmost medial aspect of the tibia. **33.** The lateral condyles are not sites of muscle attachment, they are articular surfaces. **34.** Because of their springiness, the foot arches save energy during locomotion. **35.** The two largest tarsals are the talus, and the calcaneus which forms the heel. **36.** The enlargement of the facial skeleton between ages 6 and 13 is due to the enlargement of the nose and paranasal sinuses and the development of the permanent teeth. **37.** The lumbar curvature develops when the baby begins to walk.

Review Questions 1. (1)b, g; (2)h; (3)d; (4)d, f; (5)e; (6)c; (7)a, b, d, h; (8)i; **2.** (1)g, (2)f, (3)b, (4)a, (5)b, (6)c, (7)d, (8)e; **3.** (1)b, (2)c, (3)e, (4)a, (5)h, (6)e, (7)f

Chapter 8

Check Your Understanding 1. Joint and articulation are synonyms. **2.** The synarthroses are the least mobile of the joint types. **3.** Symphyses and synchondroses are cartilaginous joints. **4.** In general, the more stable a joint, the less mobile it is. **5.** The fibrous capsule and the synovial membrane make up the wall of the articular capsule. **6.** Bursae and tendon sheaths help to reduce friction during joint movement. **7.** The muscle tendons that cross the joint are typically the most important factor in stabilizing synovial joints. **8.** Weeping lubrication helps keep the joint cartilages nourished and "lubricates" the joint surfaces. **9.** John's hip joint was flexed and his knees extended and his thumb was in opposition (to his index finger). **10.** The hinge and pivot joints are uniaxial joints. **11.** The knee and temporomandibular joints have menisci. The elbow and knee act mainly as a uniaxial hinge. The shoulder depends largely on muscle tendons for stability. **12.** Arthritis means inflammation of the joint. **13.** RA typically produces pain, swelling, and joint deformations that tend to be bilateral and crippling. OA patients tend to have pain, particularly on arising, which is relieved by gentle exercise, and enlarged bone ends (due to spurs) in affected joints. Affected joints may exhibit crepitus. **14.** Lyme disease is caused by spirochete bacteria and transmitted by a tick bite. **15.** Regular exercise strengthens joints and keeps them well nourished.

Review Questions 1. (1)c, (2)a, (3)a, (4)b, (5)c, (6)b, (7)b, (8)a, (9)c; **2.** b; **3.** d; **4.** d; **5.** b; **6.** d; **7.** d

Chapter 9

Check Your Understanding 1. Striated means "with stripes." **2.** He should have responded "smooth muscle," which fits the description. **3.** "Epimysium" literally translates to "outside the muscle" and this connective tissue sheath is the outermost muscle sheath which encloses the entire muscle. **4.** The thin myofilaments have binding sites for calcium on the troponin molecules forming part of those filaments. **5.** In a resting muscle fiber, the SR would have the highest concentration of calcium ions. The mitochondrion provides the ATP needed for muscle activity. **6.** The components of the neuromuscular junction are the axon terminal, the synaptic cleft and the junctional folds of the sarcolemma. **7.** The final trigger for contraction is a certain concentration of calcium ions in the cytosol. The initial trigger is depolarization of the sarcolemma. **8.** There are always some myosin cross bridges bound to the actin myofilament during the contraction phase. Thus, backward sliding of the actin filaments is prevented. **9.** Without ATP, rigor would occur because the myosin heads could not detach. **10.** A motor unit is an axon of a motor neuron and all the muscle fibers it innervates. **11.** During the latent period, events of excitation-contraction coupling are occurring. **12.** Immediately after Jay grabs the bar, his biceps muscles are contracting isometrically. As his body moves upward toward the bar, they are contracting isotonically and concentrically. As he lowers his body, the biceps are contracting isotonically and eccentrically. **13.** Eric was breathing heavily because it takes some time for his heart rate and overall metabolism to return to the resting state after exercise. Moreover, he had likely incurred an oxygen deficit. Although jogging is primarily an aerobic exercise, there is always some anaerobic respiration that occurs as well—the amount depends on exercise intensity. As fatigue occurs, potassium ions accumulate in the T tubules, and lactic acid and phosphate ions accumulate in the muscle cells. **14.** Factors that influence muscle contractile force include muscle fiber size, the number of muscle fibers stimulated, the frequency of stimulation, and the degree of muscle stretch. Factors that influence velocity of contraction include muscle fiber type, load, and the number of motor units contracting. **15.** Fast glycolytic fibers would provide for short periods of intense strength needed to lift and move furniture. **16.** To increase muscle size and strength, anaerobic exercise is best. Muscle endurance is enhanced by aerobic exercise. **17.** Both skeletal and smooth muscle fibers are elongated cells, but unlike smooth muscle cells, which are spindle shaped, uninucleate, and nonstriated, skeletal muscle cells are very large cigar-shaped, multinucleate, striated cells. **18.** Calcium binds to troponin on the actin filaments in skeletal muscle cells. In smooth muscle cells, it binds to a cytoplasmic protein called calmodulin. **19.** Hollow organs that have smooth muscle cells helping to form their walls often must temporarily store the organ's contents (urine, food residues, etc.), an ability ensured by the stress-relaxation response. **20.** During development of skeletal muscle fibers, the myoblast cells join together, forming multinucleate myotubes. **21.** The connective tissue in muscles increases with age, causing the muscles to get stringier. **22.** Regular exercise and strength training help to defer the loss in strength and muscle wasting that tends to occur with age, and improve neuromuscular function.

Clinical Connections 1. The first reaction to tissue injury is the initiation of the inflammatory response. The inflammatory chemicals increase the permeability of the capillaries in the injured area, allowing white blood cells, fluid, and other substances to reach the injured area. The next step in healing involves the formation of granulation tissue, in which the vascular supply for the injured area is regenerated and collagen fibers to knit the torn edges of the tissue together are formed. Skeletal muscle does not regenerate well, so the damaged areas of Mrs. DeStephano's muscle tissue will probably be repaired primarily by the formation of fibrous tissue, creating scar tissue. **2.** Healing is aided by good circulation of blood within the injured area. Vascular damage compromises healing because the supply of oxygen and nutrients to the tissue is reduced. **3.** Under normal circumstances, skeletal muscles receive electrical signals from the nervous system continuously. These signals help to maintain muscle tone and readiness. Severing of the sciatic nerve removes this continuous nervous input to the muscles and will lead to muscle atrophy. Immobility of muscles will lead to a replacement of contractile muscle tissue with noncontractile fibrous connective tissue. Distal to the point of transection, the muscle will begin to decrease in size within 3–7 days of becoming immobile. This process can be delayed by electrically stimulating the tissues. Passive range-of-motion exercises also help prevent loss of muscle tone and joint range, and improve circulation in the injured areas. **4.** Mrs. DeStephano's physician wants to supply her damaged tissues with the necessary building materials to encourage healing. A high-protein diet will provide plenty of amino acids to rebuild or replace damaged proteins, carbohydrates will provide the fuel molecules needed to generate the required ATP, and vitamin C is important for the regeneration of connective tissue.

Review Questions 1. c; **2.** b; **3.** (1)b, (2)a, (3)b, (4)a, (5)b, (6)a; **4.** c; **5.** a; **6.** a; **7.** d; **8.** a; **9.** (1)a, (2)a, c, (3)b, (4)c, (5)b, (6)b; **10.** a; **11.** c; **12.** c; **13.** c; **14.** b

Chapter 10

Check Your Understanding 1. The term "prime mover" refers to the muscle that bears the most responsibility for causing a particular movement. **2.** The iliacus overlies the iliac bone; the adductor brevis is a small (size) muscle that adducts (movement caused) the thigh; and the quadriceps (4 heads) femoris muscle follows the course of the femur. **3.** Of the muscles illustrated in Figure 10.1, the one with the parallel arrangement (sartorius) could shorten to the greatest degree. The stocky bipennate (rectus femoris) and multipennate (deltoid) muscles would be most powerful because they pack in the most fibers. **4.** Third-class levers are the fastest levers. **5.** A lever that operates at a mechanical advantage allows the muscle to exert less force than the load being moved. **6.** John was using the frontal belly of his epicranius to raise his eyebrows and the orbicularis oculi muscles to wink at Sarah. **7.** To make a sad clown's face you would contract your platysma, depressor anguli oris, and depressor labii inferioris muscles. **8.** The deltoid has a broad origin. When only its anterior fibers contract, it flexes and medially rotates the humerus. When only its posterior fibers contract, it extends and laterally rotates the humerus. **9.** The opponens pollicis does not have an insertion on the bones of the thumb.

Review Questions 1. c; **2.** c; **3.** (1)e, (2)c, (3)g, (4)f, (5)d; **4.** a; **5.** c; **6.** d; **7.** c; **8.** c; **9.** b; **10.** d; **11.** b; **12.** a; **13.** c; **14.** d; **15.** a, b; **16.** a

Chapter 11

Check Your Understanding 1. Integration involves processing and interpreting sensory information, and making a decision about motor output. Integration occurs primarily in the CNS. **2.** (a) This "full stomach" feeling would be relayed by the sensory (afferent) division of the PNS (via its visceral afferent fibers). (b) The somatic nervous system, which is part of the motor (efferent) division of the PNS, controls movement of skeletal muscle. (c) The autonomic nervous system, which is part of the motor (efferent) division of the PNS, controls the heart rate. **3.** Astrocytes control the extracellular environment around neuron cell bodies in the CNS, whereas satellite cells perform this function in the PNS. **4.** Oligodendrocytes and Schwann cells form myelin sheaths in the CNS and PNS, respectively. **5.** A nerve fiber is a long *axon*, an extension of the cell. In connective tissue, fibers are extracellular *proteins* that provide support. In muscle tissue, a muscle fiber is a muscle *cell.* **6.** A nucleus within the brain is a cluster of cell bodies, whereas the nucleus within each neuron is a large organelle that acts as the control center of the cell. **7.** In the CNS, a myelin sheath is formed by oligodendrocytes that wrap their plasma membranes around the axon. The myelin sheath protects and electrically insulates fibers and increases the speed of transmission of nerve impulses. **8.** Burning a finger will first activate unipolar (pseudounipolar) neurons that are sensory (afferent) neurons. The impulse to move your finger away from the heat will be carried by multipolar neurons that are motor (efferent) neurons. **9.** The concentration gradient and the electrical gradient—together called the electrochemical gradient—determine the direction in which ions flow through an open membrane channel. **10.** There is more leakage of K^+ than of any other cations. **11.** Action potentials are larger than graded potentials and travel further. Graded potentials generally initiate action potentials. **12.** An action potential is regenerated anew at each membrane patch. **13.** Conduction of action potentials is faster in myelinated axons because myelin allows the axon membrane between nodes to change its voltage rapidly, and allows current to flow only at the widely spaced nodes. **14.** If a second stimulus occurs before the end of the absolute refractory period, no AP can occur because sodium channels are still inactivated. **15.** At an electrical synapse, neurons are joined by gap junctions. **16.** Voltage-gated ion (Ca^{2+}) channels are found in the presynaptic axon terminal and open when an action potential reaches the axon terminal. Chemically gated ion channels are found in the postsynaptic membrane and open when neurotransmitter binds to the receptor protein. **17.** IPSPs result from the flow of either K^+ or Cl^- through chemically gated channels. EPSPs result from the flow of both Na^+ and K^+ through chemically gated channels. **18.** Temporal summation is summation in *time* of graded potentials occurring in quick succession at the postsynaptic membrane. It can result from EPSPs arising from just one synapse. Spatial summation is summation in *space*—a postsynaptic neuron is stimulated by a large number of terminals at the same time. **19.** ACh interacts with more than one specific receptor type, and this explains how it can excite at some synapses and inhibit at others. **20.** cAMP is called a second messenger because it relays the message between the *first messenger* (the original chemical messenger) outside of the cell and effector molecules that will ultimately bring about the desired response within the cell. **21.** Reverberating circuits and parallel after-discharge circuits both result in prolonged output. **22.** The pattern of neural processing is serial processing. The response is a reflex arc. **23.** The pattern of neural processing is parallel processing. **24.** The growing tip of an axon is called a growth cone. Neurotropins are chemicals that signal the growth cone.

Review Questions 1. b; **2.** (1)d, (2)b, (3)f, (4)c, (5)a; **3.** b; **4.** c; **5.** a; **6.** c; **7.** b; **8.** d; **9.** c; **10.** c; **11.** a; **12.** (1)d, (2)b, (3)a, (4)c

Chapter 12

Check Your Understanding 1. The third ventricle is surrounded by the diencephalon. **2.** The cerebral hemispheres and the cerebellum have an outside layer of gray matter in addition to central gray matter and its surrounding white matter. **3.** Convolutions increase surface area of the cortex, which allows more neurons to occupy the limited space within the skull. **4.** The central sulcus separates primary motor areas from somatosensory areas. **5.** Motor functions on the left side of the body are controlled by the right hemisphere of the brain because motor tracts from the right hemisphere cross over (in the medulla oblongata) to the left side of the spinal cord to go to the left side of the body. **6.** Commissural fibers (which form commissures) allow the cerebral hemispheres to "talk to each other." **7.** The caudate nucleus, putamen, and globus pallidus together form the basal nuclei. **8.** Virtually all inputs ascending to the cerebral cortex synapse in the thalamus en route. **9.** The hypothalamus oversees the autonomic nervous system. **10.** The pyramids of the medulla are the corticospinal (pyramidal) tracts, the large voluntary motor tracts descending from the motor cortex. The result of decussation (crossing over) is that each side of the motor cortex controls the opposite side of the body. **11.** The cerebral peduncles and the colliculi are associated with the midbrain. **12.** There are many possible answers to this question—here are a few: Structurally, the cerebellum and cerebrum are similar in that they both have a thin outer cortex of gray matter, internal white matter, and deep gray matter nuclei. Also, both have body maps (homunculi) and large fiber tracts connecting them to the brain stem. Both receive sensory input and influence motor output. A major difference is that the cerebellum is almost entirely concerned with motor output, whereas the cerebrum has much broader responsibilities. Also, while a cerebral hemisphere controls the opposite side of the body, a cerebellar hemisphere controls the same side of the body. **13.** The hypothalamus is part of the limbic system and also an autonomic (visceral) control center. **14.** Taylor is increasing the amount of sensory stimuli she receives, which will be relayed to the reticular activating system, which, in turn, will increase activation of the cerebral cortex. **15.** Delta waves are typically seen in deep sleep in normal adults. **16.** Drowsiness (or lethargy) and stupor are stages of consciousness between alertness and coma. **17.** Most skeletal muscles are actively inhibited during REM sleep. **18.** Transfer of memory from STM to LTM is enhanced by (1) rehearsal, (2) association (tying "new" information to "old" information), and (3) a heightened emotional state (for example, alert, motivated, surprised, or aroused). **19.** The basal nuclei and premotor cortex are involved in procedural (skills) memory, but not in declarative memory. **20.** CSF, formed by the choroid plexuses as a filtrate of blood plasma, is a watery "broth" similar in composition to plasma. It protects the brain and spinal cord from blows and other trauma, helps nourish the brain, and carries chemical signals from one part of the brain to another. **21.** A TIA is a temporary loss of blood supply to brain tissue, and it differs from a stroke in that the resultant impairment is fully reversible. **22.** Mrs. Lee might have Parkinson's disease. **23.** Alar plate neuroblasts become interneurons, whereas basal plate neuroblasts become motor (efferent) neurons. **24.** The nerves serving the limbs arise in the cervical and lumbar enlargements of the spinal cord. **25.** In the spinothalamic pathway, the cell bodies of first-order sensory neurons are outside the spinal cord in a ganglion, cell bodies of second-order sensory neurons are in the dorsal horn of the spinal cord, and cell bodies of third-order sensory neurons are in the thalamus. (See also Figure 12.34b.) **26.** Roy had lower limb paralysis. This could be caused by a spinal cord injury in the thoracic region (between T_1 and L_1). If the spinal cord is transected, the result is paraplegia. If the cord is only bruised, he may regain function in the limbs. An MRI scan (or CT scan) of the spinal cord would be helpful. **27.** Premature babies have trouble regulating body temperature because the hypothalamus is immature. **28.** Reversible causes of dementia include prescription drug effects, low blood pressure, poor nutrition, hormone imbalances, depression, and dehydration.

Review Questions 1. a; **2.** d; **3.** c; **4.** a; **5.** (1)d, (2)f, (3)e, (4)g, (5)b, (6)f, (7)i, (8)a; **6.** b; **7.** c; **8.** a; **9.** (1)a, (2)b, (3)a, (4)a, (5)b, (6)a, (7)b, (8)b, (9)a; **10.** d; **11.** (1)d, (2)e, (3)c and d, (4)a; **12.** (c)

Chapter 13

Check Your Understanding 1. In addition to nerves, the PNS also consists of sensory receptors, motor endings, and ganglia. **2.** Nociceptors respond to painful stimuli. They are exteroceptors that are simple and unencapsulated (free nerve endings). **3.** The three levels of sensory integration are receptor level, circuit level, and perceptual level. **4.** Phasic receptors adapt, whereas tonic receptors exhibit little or no adaptation. Pain receptors are tonic so that we are reminded to protect the injured body part. **5.** Hot and cold are conveyed by different sensory receptors that are parts of separate "labeled lines." Cool and cold are two different intensities of the same stimulus, detected by frequency coding—the frequency of APs would be higher for a cold than a cool stimulus. Action potentials arising in the fingers and foot arrive at different locations in the somatosensory cortex via their own "labeled lines" and in this way the cortex can determine their origin. **6.** Ganglia are collections of neuron cell bodies in the PNS. **7.** Nerves also contain connective tissue, blood vessels, lymphatic vessels, and the myelin surrounding the axons. **8.** Schwann cells, macrophages, and the neurons themselves were all important in healing the nerve. **9.** The oculomotor (III), trochlear (IV), and abducens (VI) nerves control eye movements. Sticking out your tongue involves the hypoglossal nerve (XII). The vagus nerve (X) influences heart rate and digestive activity. The accessory nerve (XI) innervates the trapezius muscle, which is involved in shoulder shrugging. **10.** Roots lie medial to spinal nerves, whereas rami lie lateral to spinal nerves. Dorsal roots are purely sensory, whereas dorsal rami carry both motor and sensory fibers. **11.** The spinal nerve roots were C_3–C_5, the spinal nerve was the phrenic nerve, the plexus was the cervical plexus. The phrenic nerve is the sole motor nerve supply to the diaphragm, the primary muscle for respiration. **12.** Varicosities are the series of knoblike swellings that are the axon endings of autonomic motor neurons. You would find them on axon endings serving smooth muscle or glands. **13.** The cerebellum and basal nuclei, which form the precommand level of motor control, plan and coordinate complex motor activities. **14.** The five components of a reflex arc are receptor, sensory neuron, integration center, motor neuron, and effector. **15.** The stretch reflex is important for maintaining muscle tone and adjusting it reflexively by causing muscle contraction in response to increased muscle length (stretch). It maintains posture. The flexor or withdrawal reflex is initiated by a painful stimulus and causes automatic withdrawal of the painful body part from the stimulus. It is protective. **16.** This response is called Babinski's sign and it indicates damage to the corticospinal tract or primary motor cortex. **17.** The vertebral column, spinal nerves, and dermatomes are all examples of segmentation in the adult.

Review Questions 1. b; **2.** c; **3.** d; **4.** c; **5.** e; **6.** c; **7.** (1)d, (2)c, (3)f, (4)b, (5)e, (6)a; **8.** (1)f, (2)i, (3)b, (4)g, h, (5)e, (6)i, (7)c, (8)k, (9)l, (10)c, d, f, k; **9.** (1)b 6; (2)d 8; (3)c 2; (4)c 5; (5)a 4; (6)a 3, 9; (7)a 7; (8)a 7; (9)d 1; (10)a 3, 4, 7, 9; **10.** (1)a, 1 and 5; (2)a, 3 and 5; (3)a, 4; (4)a, 2; (5)c, 2; (6)b, 2; **11.** c

Chapter 14

Check Your Understanding 1. The effectors of the autonomic nervous system are cardiac muscle, smooth muscle, and glands. **2.** The somatic motor system relays instructions to muscles more quickly because it involves only one motor neuron, whereas the ANS uses a two-neuron chain. Moreover, axons of somatic motor neurons are typically heavily myelinated, whereas preganglionic autonomic axons are lightly myelinated and postganglionic axons are unmyelinated. **3.** While you relax in the sun on the beach, the parasympathetic branch of the ANS would probably predominate. When you perceive danger (as in a shark), the sympathetic branch of the ANS predominates. **4.** "Short preganglionic fibers," "origin from thoracolumbar region of spinal cord," "collateral ganglia," and "innervates adrenal medulla" are all characteristic of the sympathetic nervous system. Terminal ganglia are found in the parasympathetic nervous system. **5.** The major differences are (1) the ANS has visceral afferents rather than somatic afferents, (2) the ANS has a two-neuron efferent chain, whereas the somatic nervous system (SNS) has one, and (3) the effectors of the ANS are smooth muscles, cardiac muscles, and glands, whereas the effectors of the SNS are skeletal muscles. **6.** The parasympathetic nervous system increases digestive activity and decreases heart rate. The sympathetic nervous system increases blood pressure, dilates bronchioles, stimulates the adrenal medulla to release its hormones, and causes ejaculation. **7.** You would find nicotinic receptors on skeletal muscle and the hormone-producing cells of the adrenal medulla, but not on smooth muscle or glands. Virtually all types of receptors (including nicotinic receptors) are also found in the CNS (see Table 11.3 on p. 416). **8.** The main integration center of the ANS is the hypothalamus, although the most direct influence is through the brain stem reticular formation and the reflex centers in the pons and medulla oblongata. **9.** Jackson's doctor may have prescribed a beta-blocker because Jackson has hypertension. (Chronic stress is a factor in causing hypertension.) The beta-blocker will decrease blood pressure by blocking beta-adrenergic receptors in the heart and blood vessels, thereby decreasing heart rate and dilating blood vessels. **10.** The neural crest gives rise to both autonomic ganglia and the adrenal medulla.

Clinical Connections 1. The location of Jimmy's lacerations and bruises and his inability to rise led the paramedics to suspect a head, neck, or back injury. They immobilized his head and torso to prevent any further damage to the brain and spinal cord. **2.** The worsening neurological signs indicate a probable intracranial hemorrhage. The blood escaping from the ruptured blood vessel(s) will begin to compress Jimmy's brain and increase his intracranial pressure. Jimmy's surgery will involve repair of the damaged vessel(s) and removal of the mass of clotted blood pressing on his brain. **3.** Loss of motor and sensory function below the level of the nipples indicates a lesion at T_4. See Figure 13.12. **4.** Jimmy is suffering from spinal shock, which occurs as a result of injury to the spinal cord. Spinal shock is a temporary condition in which all reflex and motor activities caudal to the level of spinal cord injury are lost, so Jimmy's muscles are paralyzed. His blood pressure is low due to the loss of sympathetic tone in his vasculature. **5.** Jimmy's exaggerated reflexes are caused by damaged upper motor neuron axons in the spinal cord. These upper motor neurons normally inhibit spinal reflexes. He is incontinent because there are no longer pathways to support voluntary control of bowel and bladder emptying. **6.** This condition is called *autonomic dysreflexia* (or *autonomic hyperreflexia*). This is a condition in which a normal stimulus triggers a massive activation of autonomic neurons. **7.** Extremely high arterial blood pressure can cause a rupture of the cerebral blood vessels (as well as other blood vessels in the body) and put Jimmy's life at risk.

Review Questions 1. d; **2.** (1)S, (2)P, (3)P, (4)S, (5)S, (6)P, (7)P, (8)S, (9)P, (10)S, (11)P, (12)S; **3.** b; **4.** c; **5.** a

Chapter 15

Check Your Understanding 1. Tears (lacrimal fluid) are a dilute saline secretion that contains mucus, antibodies, and lysozyme. They are secreted by the lacrimal glands. **2.** The blind spot of the eye is the optic disc. It is the part of the retina where the optic nerve exits the eye and it is "blind" because it is a region of the retina that lacks photoreceptors. **3.** An increase in intraocular pressure is called glaucoma and is due to an accumulation of aqueous humor, usually because of impaired drainage of the fluid. **4.** Light passes through the cornea, aqueous humor, lens, vitreous humor, ganglion cells and bipolar cells before it reaches the photoreceptors. **5.** The ciliary muscles and sphincter pupillae relax for distant vision. (If you said the medial rectus muscles also relax, this is true, but remember that the rectus muscles are *extrinsic* eye muscles, not intrinsic.) **6.** The near point moves farther away as you age because the lens becomes less flexible (presbyopia), so that it is unable to assume the more rounded shape required for near vision. **7.** The following are characteristics of cones: "vision in bright light," "color vision," and "higher acuity." The following are characteristics of rods: "only one type of visual pigment," "most abundant in the periphery of the retina," "many feed into one ganglion cell," and "higher sensitivity." **8.** Breakdown of the retinal-opsin combination is called bleaching of the pigment. It occurs with exposure to light. **9.** A tumor in the right visual cortex would affect the left visual field. A tumor compressing the right optic nerve would affect both the left and right visual fields from the right eye only. **10.** The five taste modalities are sweet, sour, bitter, salty, and umami. The fungiform, circumvallate, and foliate papillae contain taste buds. **11.** The cilia and hairs of these receptor cells greatly increase the surface area for sensory receptors. **12.** The tympanic membrane separates the external from the middle ear. The oval and round windows separate the middle from the inner ear. **13.** The basilar membrane allows us to differentiate sounds of different pitch. **14.** You would not be able to locate the origin of a sound if the brain stem did not receive input from both ears. **15.** The "fullness" in Mohammed's ears is likely due to an accumulation of fluid in the middle ear as a result of his upper respiratory infection ("cold") spreading into his ear. He has a form of conduction deafness. **16.** The following apply to a macula: "contains otholiths," "responds to linear acceleration or deceleration," and "inside a saccule." The following apply to a crista ampullaris: "inside a semicircular canal," "has a cupula," "responds to rotational acceleration and deceleration." **17.** With age, the lens discolors and loses its clarity and the dilator pupillae muscles become less efficient, decreasing the amount of light that reaches the retinas at night.

Review Questions 1. c; **2.** d; **3.** a; **4.** b; **5.** c; **6.** c; **7.** b; **8.** a; **9.** b; **10.** b; **11.** b; **12.** d; **13.** d; **14.** a; **15.** d; **16.** c; **17.** d; **18.** b; **19.** b; **20.** c; **21.** d; **22.** b; **23.** b; **24.** e; **25.** b; **26.** c; **27.** c; **28.** c; **29.** c

Chapter 16

Check Your Understanding 1. The endocrine system is more closely associated with growth and development, and its responses tend to be long-lasting, whereas nervous system responses tend to be rapid and discrete. **2.** The thyroid and parathyroid glands are found in the neck. **3.** Hormones are released into the blood and transported throughout the body, whereas paracrines act locally, generally within the same tissue. **4.** The two major chemical classes of hormones are amino acid–based hormones and steroids. Steroids are all lipid soluble. Thyroid hormones are the only amino acid–based hormones that are lipid soluble. **5.** Water-soluble hormones act on receptors in the plasma membrane coupled most often via regulatory molecules called G proteins to intracellular second messengers. Lipid-soluble hormones act on intracellular receptors, directly activating genes and stimulating synthesis of specific proteins. **6.** Hormone release can be triggered by humoral, neural, or hormonal stimuli. **7.** The hypothalamus communicates with the anterior pituitary via *hormones* released into a special portal system of blood vessels. In contrast, it communicates with the posterior pituitary via *action potentials traveling down axons* that connect the hypothalamus to the posterior pituitary. **8.** LH and FSH are tropic hormones that act on the gonads, TSH is a tropic hormone that acts on the thyroid, and ACTH is a tropic hormone that acts on the adrenal cortex. (If you said growth hormone, that's also a good answer, as GH causing the liver to release IGFs might also be considered a tropic effect.) **9.** Drinking alcoholic beverages inhibits ADH secretion from the posterior pituitary and causes copious urine output and dehydration. The dehydration causes the hangover effects. **10.** Thyroid hormone increases basal metabolic rate (and heat production) in the body. Parathyroid hormone increases blood Ca^{2+} levels in a variety of ways.

Calcitonin at high (pharmacological) levels has a Ca^{2+}-lowering, bone-sparing effect. (At normal blood levels its effects in humans are negligible.) **11.** Thyroid follicular cells release thyroid hormone, chief cells in the parathyroid gland release parathyroid hormone, and parafollicular (C) cells in the thyroid gland release calcitonin. **12.** Glucocorticoids are stress hormones that, among many effects, increase blood glucose. Mineralocorticoids increase blood Na^+ (and blood pressure) and decrease blood K^+. Gonadocorticoids are male and female sex hormones that are thought to have a variety of effects (for example, onset of puberty, sex drive in women, pubic and axillary hair development in women). **13.** Melatonin is used by some individuals as a sleep aid, particularly to counter jet lag. **14.** When we say Sharon is diabetic, this means that she has insufficient insulin action in her body. The most likely explanation for Sharon's behavior is that she has taken too much insulin and is experiencing hypoglycemia. Ingesting some sugar will likely help her. **15.** Diabetes mellitus is due to a lack of insulin production or action, whereas diabetes insipidus is due to a lack of ADH. Both conditions are characterized by production of copious amounts of urine. You would find glucose in the urine of a patient with diabetes mellitus, but not in the urine of a patient with diabetes insipidus. **16.** The gonadal hormones are steroid hormones. A major endocrine gland that also secretes steroid hormones is the adrenal cortex. **17.** The heart produces atrial natriuretic peptide (ANP). ANP decreases blood volume and blood pressure by increasing the kidneys' production of salty urine. **18.** Vitamin D_3, produced in inactive form by the skin, increases intestinal absorption of calcium. **19.** The decline in growth hormone with age contributes to muscle atrophy. The decline in estrogen contributes to osteoporosis in women.

Clinical Connections 1. Rationale for orders: As Mr. Gutteman is unconscious, the level of damage to his brain is unclear. Monitoring his responses and vital signs every hour will provide information for his care providers about the extent of his injuries. Turning him every 4 hours and providing careful skin care will prevent decubitus ulcers (bedsores) as well as stimulating his proprioceptive pathways. **2.** Mr. Gutteman's condition is termed *diabetes insipidus*, a condition in which insufficient quantities of antidiuretic hormone (ADH) are produced or released. Diabetes insipidus patients excrete large volumes of urine but do not have glucose or ketones present in the urine. The head trauma could have damaged Mr. Gutteman's hypothalamus, which produces the hormone, or injured his posterior pituitary gland, which releases ADH into the bloodstream. **3.** Diabetes insipidus is not life threatening for most individuals with normal thirst mechanisms, as they will be thirsty and drink to replenish the lost fluid. However, Mr. Gutteman is comatose, so his fluid output must be monitored closely so that the volume lost can be replaced by IV line. His subsequent recovery may be complicated if he has suffered damage to his hypothalamus, which houses the thirst center neurons.

Review Questions 1. b; **2.** a; **3.** c; **4.** d; **5.** (1)c, (2)a and b, (3)f, (4)d, (5)e, (6)g, (7)a, (8)h, (9)b and e, (10)a; **6.** d; **7.** c; **8.** b; **9.** d; **10.** b; **11.** d; **12.** b; **13.** c; **14.** d

Chapter 17

Check Your Understanding 1. The hematocrit is the percentage of blood that is occupied by erythrocytes. It is normally about 45%. **2.** Blood can prevent blood loss when a blood vessel is damaged by forming clots. Blood can prevent infection because it contains antimicrobial proteins and white blood cells. **3.** Plasma proteins are not used as fuel for body cells because their presence in blood is required to perform many key functions. **4.** Each hemoglobin molecule can transport four O_2. The heme portion of the hemoglobin binds the O_2. **5.** The kidneys' synthesis of erythropoietin is compromised in advanced kidney disease, so RBC production decreases, causing anemia. **6.** Monocytes become macrophages in tissues. Neutrophils are also voracious phagocytes. **7.** "Thrombopoietin" is derived from the same word as "thrombocyte." It is the hormone that promotes platelet formation. **8.** Amos's red bone marrow is spewing out many abnormal white blood cells, which are crowding out the production of normal bone marrow elements. The lack of normal white blood cells allows the infections, the lack of platelets fails to stop bleeding, and the lack of erythrocytes is anemia. **9.** The three steps of hemostasis are vascular spasm, platelet plug formation, and coagulation. **10.** Fibrinogen is water soluble, whereas fibrin is not. Prothrombin is an inactive precursor, whereas thrombin acts as an enzyme. Most factors are inactive in blood before activation and become enzymes upon activation. (There are exceptions, such as fibrinogen and calcium.) **11.** Thrombocytopenia (platelet deficiency) results in failure to plug the countless small tears in blood vessels, and so manifests as small purple spots. Hemophilia A results from the absence of clotting factor VIII. **12.** Nigel has anti-A antibodies in his blood and type B agglutinogens on his RBCs. He can donate blood to an AB recipient, but he should not receive blood from an AB donor because his anti-A antibodies will cause a transfusion reaction. **13.** If Emily has a bacterial meningitis, a differential WBC count would likely reveal an increase in neutrophils because neutrophils are a major body defense against bacteria. **14.** Hemoglobin F has a higher affinity (binding strength) for oxygen than adult hemoglobin does.

Review Questions 1. c; **2.** c; **3.** d; **4.** b; **5.** d; **6.** a; **7.** a; **8.** b; **9.** c; **10.** d

Chapter 18

Check Your Understanding 1. The mediastinum is the medial cavity of the thorax within which the heart, great vessels, and trachea are found. **2.** The layers of the heart wall are the endocardium, the myocardium, and the epicardium. The epicardium is also called the visceral layer of the serous pericardium. This is surrounded by the parietal layer of the serous pericardium and the fibrous pericardium. **3.** The serous fluid decreases friction caused by movement of the layers against one another. **4.** The right side of the heart acts as the pulmonary pump, whereas the left acts as the systemic pump. **5.** (a) True. The left ventricle wall is thicker than the right. (b) True. The left ventricle pumps blood at much higher pressure than the right ventricle because the left ventricle supplies the whole body, whereas the right ventricle supplies only the lungs. (c) False. Each ventricle pumps the *same amount* of blood with each beat. If this were not true, blood would back up in either the systemic or pulmonary circulation (because the two ventricles are in series). **6.** The branches of the right coronary artery are the right marginal artery and the posterior interventricular artery. **7.** The papillary muscles and chordae tendineae keep the AV valve flaps from everting into the atria as the ventricles contract. **8.** (a) The refractory period is almost as long as the contraction in cardiac muscle. (b) The source of Ca^{2+} for the contraction is *only* SR in skeletal muscle. (c) The AP exhibits a plateau phase in cardiac muscle. (d) Both skeletal muscle and cardiac muscle have troponin. (e) Only skeletal muscle has triads. **9.** Cardiac muscle cannot go into tetany because the absolute refractory period is almost as long as the contraction. **10.** The Purkinje fibers excite ventricular muscle fibers. The depolarization wave travels upward from the apex toward the atria, and from endocardium to epicardium. **11.** (a) The QRS wave occurs during ventricular depolarization. (b) The T wave of the ECG occurs during ventricular repolarization. (c) The P-Q interval of the ECG occurs during atrial depolarization and the conduction of the action potential through the rest of the intrinsic conduction system. **12.** The second heart sound is associated with the closing of the semilunar valves. **13.** The murmur of mitral insufficiency occurs during ventricular systole (because this is when the valve should be closed, and the murmur is due to blood leaking through the incompletely closed valve into the atrium). **14.** The periods when all four valves are closed are the isovolumetric contraction phase and the isovolumetric relaxation phase. **15.** Exercise activates the sympathetic nervous system.

Sympathetic nervous system activity increases heart rate. It also directly increases ventricular contractility, thereby increasing Josh's stroke volume. **16.** If the heart is beating very rapidly, the amount of time for ventricular filling between contractions is decreased. This decreases the end diastolic volume, decreases the stroke volume, and therefore decreases the cardiac output. **17.** The foramen ovale and the ductus arteriosus both allow blood to bypass the fetal lungs. **18.** Elderly athletes may be hampered by sclerosis and thickening of heart valve flaps, a decline in cardiac reserve, fibrosis of cardiac muscle, and atherosclerosis.

Review Questions 1. a; **2.** c; **3.** b; **4.** c; **5.** b; **6.** b; **7.** c; **8.** d; **9.** b

Chapter 19

Check Your Understanding 1. The sympathetic nervous system innervates blood vessels. The sympathetic nerves innervate the tunica media. The effector cells in the tunica media are smooth muscle cells. **2.** When vascular smooth muscle contracts, the diameter of the blood vessel becomes smaller. This is called vasoconstriction. **3.** *Elastic arteries* play a major role in dampening the pulsatile pressure of heart contractions. Dilation or constriction of *arterioles* determines blood flow to individual capillary beds. *Muscular arteries* have the thickest tunica media relative to their lumen size. **4.** If you were doing calf raises, your capillary bed would be in the condition depicted in part (a). The true capillaries would be flushed with blood to ensure that the working calf muscles could receive the needed nutrients and dispose of their metabolic wastes. **5.** Valves prevent blood from flowing backwards in veins. They are formed from folds of the tunica intima. **6.** In the systemic circuit, veins contain more blood than arteries (see Figure 19.5). **7.** The three factors that determine resistance are blood viscosity, vessel length, and vessel diameter. Vessel diameter is physiologically most important. **8.** The rate of flow will decrease 81-fold from its original flow ($3 \times 3 \times 3 \times 3 = 81$). **9.** When you first stand up, mean arterial pressure (MAP) temporarily decreases and this is sensed by aortic and carotid baroreceptors. Medullary cardiac and vasomotor center reflexes increase sympathetic and decrease parasympathetic outflow to the heart. Heart rate and contractility increase, increasing cardiac output, and therefore MAP. Further, sympathetic constriction of arterioles increases peripheral resistance, also increasing MAP. (In addition, increased constriction of veins increases venous return, which increases end diastolic volume, increasing stroke volume, and therefore cardiac output and MAP.) See also Figure 19.9 (bottom). **10.** The kidneys help maintain MAP by influencing blood volume. In renal artery obstruction, the blood pressure in the kidney is lower than in the rest of the body (because it is downstream of the obstruction). Low renal blood pressure triggers both direct and indirect renal mechanisms to increase blood pressure by increasing blood volume. This can cause hypertension (called "secondary hypertension" because it is secondary to a defined cause—in this case the renal artery obstruction). **11.** In a bicycle race, autoregulation by intrinsic metabolic controls causes arteriolar smooth muscle in your legs to relax, dilating the vessels and supplying more O_2 and nutrients to the exercising muscles. **12.** Extrinsic mechanisms, primarily the sympathetic nervous system, prevent blood pressure from plummeting by constricting arterioles elsewhere (such as gut, kidneys). In addition, cardiac output increases, which also helps maintain MAP. **13.** (a) An increase in interstitial fluid osmotic pressure (OP_{if}) would tend to pull more fluid out of capillaries (causing localized swelling, or edema). (b) An increase of OP_{if} to 10 mm Hg would decrease the net osmotic pressure on both the arteriole and venule ends of the capillary to 16 mm Hg (26 mm Hg − 10 mm Hg). The NFP at the venous end would become 1 mm Hg (17 mm Hg − 16 mm Hg). (c) Fluid would flow *out of* the venous end of the capillary rather than in. **14.** Bob is in vascular shock due to anaphylaxis, a systemic allergic reaction to his medication. His blood pressure is low because of widespread vasodilation triggered by the massive release of histamine. Bob's rapid heart rate is a result of the baroreceptor reflex triggered by his low blood pressure. This activates the sympathetic nervous system, increasing heart rate, in an attempt to restore blood pressure. **15.** The external carotid arteries supply most of the tissues of the head except for the brain and orbits. **16.** The cerebral arterial circle (circle of Willis) is the arterial anastomosis at the base of the brain. **17.** The four unpaired arteries that emerge from the abdominal aorta are the celiac trunk, the superior and inferior mesenteric arteries, and the median sacral artery. **18.** You would palpate the popliteal artery behind the knee, the posterior tibial artery behind the medial malleolus of the tibia, and the dorsalis pedis artery on the foot. (See also Figure 19.12.) **19.** The vertebral arteries help supply the brain, but the vertebral veins do *not* drain much blood from the brain. **20.** The internal jugular veins drain the dural venous sinuses. Each internal jugular vein joins a subclavian vein to form a brachiocephalic vein. **21.** A portal system is a system where two capillary beds occur in series. In other words, in a portal system, a capillary bed is drained by a vein that leads into a second capillary bed. The function of the hepatic portal system is to transport venous blood from the digestive organs to the liver for processing before it enters the rest of the systemic circulation. This plays an important role in defense against absorbed toxins or microorganisms and also allows direct delivery of absorbed nutrients to the liver for processing. **22.** The leg veins that often become varicosed are the great and small saphenous veins. **23.** Three differences between arteries and veins are (1) arteries run deep while veins are both deep and superficial, (2) venous pathways are more interconnected than arterial pathways, and (3) the brain and digestive systems have unique venous drainage systems, whereas their arterial supply patterns are not substantially different from those of other organs. **24.** The foramen ovale and the ductus arteriosus both bypass the fetal lungs. The ductus venosus bypasses the fetal liver. **25.** Varicose veins, atherosclerosis, and hypertension are associated with aging.

Clinical Connections 1. The tissues in Mr. Hutchinson's right leg were deprived of oxygen and nutrients for at least one-half hour. When tissues are deprived of oxygen, tissue metabolism decreases and eventually ceases, so these tissues may have died due to anoxia. **2.** Mr. Hutchinson's vital signs (low BP; rapid, thready pulse) indicate that he is facing a life-threatening problem that must be stabilized before other, less vital problems can be addressed. As for surgery, he may be scheduled for open reduction of his crushed bone, depending upon the condition of the tissues in his crushed right leg. If tissue death has occurred in his leg, he may undergo amputation of that limb. **3.** Mr. Hutchinson's rapid, thready pulse and falling blood pressure are indications of hypovolemic shock, a type of shock resulting from decreased blood volume. Because his blood volume is low, his heart rate is elevated to increase cardiac output in an effort to maintain the blood supply to his vital organs. Mr. Hutchinson's blood volume must be increased as quickly as possible with blood transfusions or intravenous saline. This will stabilize his condition and allow his physicians to continue with his surgery.

Review Questions 1. d; **2.** b; **3.** d; **4.** c; **5.** e; **6.** d; **7.** c; **8.** b; **9.** b; **10.** a; **11.** b; **12.** c; **13.** b; **14.** d; **15.** (1)b, e, g; (2)c; (3)i; (4) f, h; (5) d

Chapter 20

Check Your Understanding 1. Lymph is the fluid inside lymphatic vessels. It enters lymphatic vessels from interstitial fluid. Interstitial fluid, in turn, is a filtrate of blood plasma. **2.** The right lymphatic duct receives lymph from the right upper arm and the right side of the head and thorax. The thoracic duct drains lymph from the rest of the body. **3.** Lymph movement is driven by the contraction of adjacent skeletal muscles, pressure changes in the thorax during breathing, and the pulsations of nearby arteries. (Valves in lymphatic vessels prevent backflow of lymph.) **4.** Lymphoid follicles are solid, spherical bodies consisting of tightly packed reticular elements and cells, often with a lighter-staining central region. They are regions where B cells predominate. **5.** Having fewer efferents causes lymph to accumulate in lymph nodes, allowing more time for its cleansing. **6.** MALT (mucosa-associated lymphatic tissue) is lymphoid tissue found in the mucosa of the digestive, respiratory, and genitourinary tracts. **7.** The spleen cleanses the blood, stores breakdown products of RBCs, stores platelets, and is thought to be a site of erythrocyte production in the fetus. **8.** The thymus develops first.

Clinical Connections 1. The red streaks radiating from Mr. Hutchinson's finger indicate that his lymphatic vessels are inflamed. This inflammation may be caused by a bacterial infection. If Mr. Hutchinson's arm had exhibited edema without any accompanying red streaks, the problem would likely have been impaired lymph transport from his arm back to his trunk, due to injury or blockage of his lymphatic vessels. **2.** Mr. Hutchinson's arm was placed in a sling to immobilize it, slowing the drainage of lymph from the infected area in an attempt to limit the spread of the infection. **3.** Mr. Hutchinson's low lymphocyte count indicates that his body's ability to fight infection by bacteria or viruses is impaired. The antibiotics and additional staff protection will protect Mr. Hutchinson until his lymphocyte count increases again. Gloving and gowning also protect the staff caring for Mr. Hutchinson from any body infection he might have. **4.** Mr. Hutchinson's recovery may be problematic, as he probably already has an ongoing bacterial infection and his ability to raise a defense to this infection is impaired.

Review Questions 1. c; **2.** c; **3.** a, d; **4.** c; **5.** a; **6.** b; **7.** a; **8.** b; **9.** d

Chapter 21

Check Your Understanding 1. The innate defense system is always ready to respond immediately, whereas it takes considerable time to mount the adaptive defense system. The innate defenses consist of surface barriers and internal defenses, whereas the adaptive defenses consist of humoral and cellular immunity, which rely on B and T lymphocytes. **2.** Surface barriers (the skin and mucous membranes) constitute the first line of defense. **3.** Opsonization is the process of making pathogens more susceptible to phagocytosis by decorating their surface with molecules that phagocytes can bind. Antibodies and complement proteins are examples of molecules that act as opsonins. **4.** Our own cells are killed by NK cells when they have been infected by viruses or when they have become cancerous. **5.** Redness, heat, swelling, and pain are the cardinal signs of inflammation. Redness and local heat are both caused by vasodilation of arterioles, which increases the flow of blood (warmed by the body core) to the affected area. The swelling (edema) is due to the release of histamine and other chemical mediators of inflammation, which increase capillary permeability. This increased permeability allows proteins to leak into the interstitial fluid (IF), increasing the IF osmotic pressure and drawing more fluid out of blood vessels and into the tissues, thereby causing swelling. The pain is due to two things: (1) the actions of certain chemical mediators (kinins and prostaglandins) on nerve endings, and (2) the swelling, which can compress free nerve endings. **6.** Three key characteristics of adaptive immunity are that it is specific, it is systemic, and it has memory. **7.** A complete antigen has both immunogenicity and reactivity, whereas a hapten has reactivity but not immunogenicity. **8.** Self-antigens, particularly MHC proteins, mark a cell as self. **9.** Development of immunocompetence of a B or T cell is signaled by the appearance on its surface of specific and unique receptors for an antigen. In the case of a B cell, this receptor is a membrane-bound antibody. (In T cells, it is simply called the T cell receptor.) **10.** The T cell that would survive is (c), one that recognizes MHC but not self-antigen. **11.** Dendritic cells, macrophages, and B cells can all act as APCs. Dendritic cells are most important for T cell activation. **12.** In clonal selection, the antigen does the selecting. What is being selected is a particular clone of B or T cells that has antigen receptors corresponding to that antigen. **13.** The secondary response to an antigen is faster than the primary response because the immune system has already been "primed" and has memory cells that are specific for that particular antigen. **14.** Vaccinations protect by providing the initial encounter to an antigen—the primary response to that antigen. As a result, when the pathogen for that illness is encountered again, the pathogen elicits the much faster, more powerful secondary response, which is generally effective enough to prevent clinical illness. **15.** IgG antibody is most abundant in blood. IgM is secreted first in a primary immune response. IgA is most abundant in secretions. **16.** Antibodies can bring about destruction of pathogen via "PLAN"—**p**hagocytosis, **l**ysis (via complement), **a**gglutination, or **n**eutralization. **17.** Class II MHC proteins display exogenous antigens. Class II MHC proteins are recognized by CD4 T cells (which usually become helper T cells). APCs display class II MHC proteins. **18.** When antigens are bound in the absence of co-stimulators this causes the lymphocyte to develop *anergy*—a state of permanent unresponsiveness to that antigen. **19.** Helper T cells are central to both humoral and cell-mediated immunity because they are required for activation of both cytotoxic T cells and most B cells. **20.** The cytotoxic T cell releases perforins and granzymes onto the identified target cell. Perforins form a pore in the target cell membrane, and granzymes enter through this pore, triggering apoptosis (cell suicide). **21.** MHC proteins and blood type antigens (ABO, etc.) are carefully matched before an organ transplant. **22.** HIV is particularly hard for the immune system to defeat because (1) it destroys helper T cells, which are key players in adaptive immunity and (2) it has a high mutation rate and so it rapidly becomes resistant to drugs. **23.** Binding of an allergen onto specific IgE antibodies attached to mast cells triggers the mast cells to release histamine.

Review Questions 1. c; **2.** a; **3.** d; **4.** d, e; **5.** a; **6.** d; **7.** b; **8.** c; **9.** d; **10.** d; **11.** d; **12.** (1)b, g; (2)d, i; (3)a, e; (4)a, e, f, h; (5)e, h; (6)c, f, g

Chapter 22

Check Your Understanding 1. The structures that air passes by are the nasal cavity (nares, nasal vestibule, nasal conchae), nasopharynx (with pharyngeal tonsil), oropharynx (with palatine tonsil), laryngopharynx, and larynx (with epiglottis, vestibular fold, and vocal fold). **2.** The epiglottis seals the larynx when we swallow. **3.** The incomplete, C-shaped cartilage rings of the trachea allow it to expand and contract and yet keep it from collapsing. **4.** The many tiny alveoli together have a large surface area. This and the thinness of their respiratory membranes make them ideal for gas exchange. **5.** The peanut was most likely in the right main bronchus because it is wider and more vertical than the left. **6.** The two circulations of the lungs are the pulmonary circulation, which delivers deoxygenated blood to the lungs for oxygenation and returns oxygenated blood to the heart, and the bronchial circulation, which provides systemic (oxygenated) blood to lung tissue. **7.** The driving force for pulmonary ventilation is a pressure gradient created by changes in the thoracic volume. **8.** The intrapulmonary pressure decreases during inspiration because of the increase in thoracic cavity volume brought about by the muscles of inspiration. **9.** The partial vacuum (negative pressure) inside the pleural cavity is caused by the opposing forces acting on the visceral and parietal pleurae. The visceral pleurae are pulled inward by the lungs' natural tendency to recoil and the surface tension of the alveolar fluid. The parietal pleurae are pulled outward by the elasticity of the chest wall. If air enters the pleural cavity, the lung on that side will collapse. This condition is called pneumothorax. **10.** Airway resistance is low because (1) the diameters of most airways are relatively large, (2) for smaller airways there are many in parallel, making their combined diameter large, and (3) air has a low viscosity. **11.** A lack of surfactant increases surface tension in the alveoli and causes them to collapse between breaths. (In other words, it markedly decreases lung compliance.) **12.** Slow, deep breaths ventilate the alveoli more effectively because a smaller fraction of the tidal volume of each breath is spent moving air into and out of the dead space. **13.** In a sealed container, the air and water would be at equilibrium. Therefore, the partial pressures of CO_2 and O_2 (P_{CO_2} and P_{O_2}) will be the same in the water as in the air: 100 mm Hg each. More CO_2 than O_2 molecules will be dissolved in the water (even though they are at the same partial pressure) because CO_2 is much more soluble than O_2 in water. **14.** The difference in P_{O_2} between inspired air and alveolar air can be explained by (1) the gas exchange occurring in the lungs (O_2 continuously diffuses out of the alveoli into the blood), (2) the humidification of inspired air (which adds water molecules that dilute the O_2 molecules), and (3) the mixing of newly inspired air with gases already present in the alveoli. **15.** The arterioles leading into the O_2-enriched alveoli would be dilated. This response allows matching of blood flow to availability of oxygen. **16.** Both CO_2 and H^+ increase O_2 unloading by binding to Hb. This is called the Bohr effect. **17.** About 70% of CO_2 is transported as bicarbonate ion (HCO_3^-) in plasma. Just over 20% is transported bound to hemoglobin in the RBCs, and 7–10% is dissolved in plasma. **18.** As blood CO_2 increases, blood pH decreases. This is because CO_2 combines with water to form carbonic acid. (However, the change in pH in blood for a given increase in CO_2 is minimized by other buffer systems.) **19.** The ventral respiratory group of the medulla (VRG) is thought to be the rhythm-generating area. **20.** CO_2 in blood normally provides the most powerful stimulus to breathe. Central chemoreceptors are most important in this response (see Figure 22.25). **21.** The injured soccer player's P_{CO_2} is low. (Recall that normal P_{CO_2} = 40 mm Hg.) The low P_{CO_2} reveals that this is hyperventilation and not hyperpnea (which is not accompanied by changes in blood CO_2 levels). **22.** Long-term adjustments to altitude include an increase in erythropoiesis, resulting in a higher hematocrit; an increase in BPG, which decreases Hb affinity for oxygen; and an increase in minute respiratory volume. **23.** The obstruction in asthma is *reversible*, and acute exacerbations are typically followed by symptom-free periods. In contrast, the obstruction in chronic bronchitis is generally not reversible. **24.** The underlying defect in cystic fibrosis is an abnormality in a protein (CFTR protein) that acts as a membrane channel for chloride ions. **25.** Vital capacity declines with age because the thoracic wall becomes more rigid and the lungs lose their elasticity.

Clinical Connections 1. Spinal cord injury from a fracture at the level of the C_2 vertebra would interrupt the normal transmission of signals from the brain stem down the phrenic nerve to the diaphragm, and Barbara would be unable to breathe due to paralysis of the diaphragm. **2.** Barbara's head, neck, and torso should have been immobilized to prevent further damage to the spinal cord. In addition, she required assistance to breathe, so her airway was probably intubated to permit ventilation of her lungs. **3.** Cyanosis is a decrease in the degree of oxygen saturation of hemoglobin. As Barbara's respiratory efforts cease, her alveolar P_{O_2} will fall, so there is less oxygen to load onto hemoglobin. In her peripheral tissues, what little oxygen hemoglobin carries will be consumed leaving these tissues with a bluish tinge. **4.** Injury to the spinal cord at the level of the C_2 vertebra will cause quadriplegia (paralysis of all four limbs). **5.** Atelectasis is the collapse of a lung. Because it is the right thorax that is compressed, only her right lung is affected. Because the lungs are in separate pleural cavities, only the right lung collapsed. **6.** Barbara's fractured ribs probably punctured her lung tissue and allowed air within the lung to enter the pleural cavity. **7.** The atelectasis will be reversed by inserting a chest tube and removing the air from the pleural cavity. This will allow her lung to heal and reinflate.

Review Questions 1. b; **2.** a and c; **3.** c; **4.** c; **5.** b; **6.** d; **7.** d; **8.** b; **9.** c, d; **10.** c; **11.** b; **12.** b; **13.** b; **14.** c; **15.** b; **16.** b

Chapter 23

Check Your Understanding 1. The esophagus is found in the thorax. Three alimentary canal organs found in the abdominal cavity include the stomach, small intestine, and large intestine. **2.** The usual site of ingestion in a healthy person is the mouth. **3.** The process of absorption moves nutrients into the body. **4.** Reflexes associated with the GI tract promote muscle contraction and secretion of digestive juices or hormones. **5.** The term "gut brain" refers to the enteric nervous system or web of neurons closely associated with the digestive organs. **6.** The visceral peritoneum is the outermost layer of the digestive organ; the parietal peritoneum is the serous membrane covering the wall of the abdominal cavity. **7.** The pancreas is retroperitoneal. **8.** The hepatic portal circulation is the venous portion of the splanchnic circulation. **9.** From deep to superficial the layers of the alimentary canal are the mucosa, submucosa, muscularis externa, and serosa. **10.** He should temporarily refrain from eating because the parasympathetic nervous system oversees digestive activities. **11.** The vestibule is the region between the cheek and the teeth. The oral cavity proper is the area enclosed by the teeth. **12.** The palate forms the roof of the mouth. The hard palate supported by bone is anterior to the soft palate (no bony support). **13.** The tongue is important for speech, particularly for uttering consonants, and taste. **14.** The serous portion of saliva is rich in salivary amylase, an enzyme that chemically breaks down starch. Additionally, the serous portion of saliva helps to hydrate the foodstuffs and helps provide protection against microorganisms.
15. Antimicrobial substances found in saliva include lysozyme, defensins, a cyanide compound, and IgA antibodies. **16.** Tina's "show and tell" tooth is a primary tooth, also called a deciduous tooth. **17.** Enamel is harder than bone. Pulp consists of nervous tissue and blood vessels. **18.** The molars are grinding teeth. **19.** The pharynx is part of the digestive and respiratory systems. **20.** The esophageal mucularis externa undergoes a transformation along its length from skeletal muscle superiorly to smooth muscle near the stomach. **21.** The esophagus is merely a chute for food passage and is subjected to a good deal of abrasion, which a stratified squamous epithelium can withstand. The stomach mucosa is a secretory mucosa served well by a simple columnar epithelium. **22.** The tongue mixes the chewed food with saliva and compacts the food into a bolus. **23.** During swallowing the larynx rises and the epiglottis covers its lumen so that foodstuffs are diverted into the esophagus posteriorly. **24.** The stomach has three layers of smooth muscle—longitudinal, circular, and oblique. Addition of the oblique layer allows the stomach to pummel food in addition to its peristaltic movements. **25.** The chief cells produce pepsinogen, which is the inactive pepsin enzyme, and the parietal cells secrete HCl needed to activate pepsinogen. **26.** The mucosal barrier consists of the thick alkaline mucus secreted by the mucous cells, the fact that the epithelial cells of the mucosa are joined by tight junctions, and the quick replacement of dead or dying cells by stem cells. **27.** The three phases of gastric secretion are the cephalic, gastric, and intestinal phases. **28.** The presence of food in the duodenum inhibits gastric activity by triggering the enterogastric reflex and the secretion of certain enterogastrones (hormones). **29.** Venous blood leaving the stomach during a meal becomes more alkaline due to the alkaline tide occurring during HCl secretion. **30.** All of these modifications increase the surface area of the small intestine. The circular folds force the chyme to spiral through the lumen. **31.** Brush border enzymes are enzymes associated with the microvilli of the small intestine mucosal cells. **32.** A lacteal is a blind-ended lymphatic capillary that picks up lymph (fluid and proteins leaked from the bloodstream), which is then returned to the blood. **33.** IgA, HCl, defensins, and lysozyme protect the intestinal cells from bacterial damage. **34.** A portal triad is a region at the corner of a hepatic lobule that contains a branch of the hepatic portal vein, a branch of the hepatic artery, and a bile duct. **35.** The enterohepatic circulation is an important recycling mechanism

for retaining bile salts needed for fat absorption. **36.** Kupffer cells are macrophages that rid the blood of bacteria and dead cells. **37.** Zymogen granules contain digestive enzymes. **38.** The pancreas is the only important source of lipases, so fats will probably not be digested or absorbed during her illness. **39.** Pancreatic acini produce the exocrine products of the pancreas (digestive enzymes and bicarbonate-rich juice). The islets produce pancreatic hormones, most importantly insulin and glucagon. **40.** Fluid in the pancreatic duct is bicarbonate-rich, enzyme-rich pancreatic juice. Fluid in the cystic and bile ducts is bile. **41.** CCK is secreted in response to the entry of chyme rich in protein and fat into the duodenum. It causes the pancreatic acini to secrete digestive enzymes and stimulates the gallbladder to contract. **42.** Distension of stomach walls enhances stomach secretory activity. Distension of the walls of the small intestine reduces stomach secretory activity (to give the small intestine time to carry out its digestive and absorptive activities). **43.** Segmentation is more important for moving food along the small intestine. **44.** MMC is the migrating motility complex, a pattern of peristalsis seen in the small intestine that moves the last remnants of a meal plus bacteria and other debris into the large intestine. MMC is important to prevent the overgrowth of bacteria in the small intestine. **45.** Mass movements are unique to the large intestine. They are long, slow, powerful contractions that move over large areas of the colon three or four times a day, forcing the contents toward the rectum. **46.** Activation of stretch receptors in the rectal wall initiates the defecation reflex. **47.** Enteric bacteria synthesize B vitamins and most of the vitamin K the liver needs to synthesize clotting proteins. **48.** All food digestion depends on hydrolysis reactions. **49.** Amylase is to starch as lipase is to fats. **50.** Bile salts emulsify fats so that they can be acted on efficiently by lipase enzymes, and form micelles that aid fat absorption. **51.** The digestive system mucosa develops from the endoderm. **52.** The thick viscous mucus produced in cystic fibrosis patients clogs pancreatic ducts and prevents the delivery of pancreatic fluid to the duodenum, thus inhibiting fat digestion and absorption. **53.** Colon and stomach cancers are dangerous because they have few early signs and symptoms.

Clinical Connections 1. Mr. Gutteman's statement about the effects of milk on his digestive tract suggests that he may be deficient in lactase, a brush border enzyme that breaks down lactose (milk sugar). **2.** His responses to the questions reduced the possibility that he has gastric ulcers. Mr. Gutteman's diarrhea may be due to gluten enteropathy. To verify this diagnosis, Mr. Gutteman should be screened for specific IgA antibodies in his blood. If these tests are positive, a biopsy of the intestinal mucosa would be performed to verify or negate the diagnosis. A positive diagnosis of gluten enteropathy would lead to dietary restriction of all grains but rice and corn. In the meantime, grains should not be restricted.

Review Questions 1. c; **2.** d; **3.** d; **4.** b; **5.** b; **6.** a; **7.** d; **8.** d; **9.** b; **10.** c; **11.** c; **12.** a; **13.** d; **14.** d; **15.** b; **16.** c; **17.** a

Chapter 24

Check Your Understanding 1. The six major nutrients are carbohydrates, proteins, fats, water, minerals, and vitamins. **2.** Cellulose provides fiber, which helps in elimination. **3.** Triglycerides are used for ATP synthesis, body insulation and protective padding, and to help the body absorb fat-soluble vitamins. Cholesterol is the basis of our steroid hormones and bile salts, and stabilizes cellular membranes. **4.** Beans (legumes) and grains (rice) are good sources of protein, but neither is a complete one. Together they should provide all the essential amino acids. Like beans, the bread supplies carbohydrates. **5.** Vitamins serve as the basis for coenzymes, which work with enzymes to accomplish metabolic reactions. **6.** Vitamin B_{12} needs intrinsic factor to be absorbed by the intestine. **7.** Iodine is essential for thyroxine synthesis. Calcium in the form of bone salts is needed to make bones hard. Iron is needed to make functional hemoglobin. **8.** A redox reaction is a combination of an oxidation and a reduction reaction. As one substance is oxidized, another is reduced. **9.** Some of the energy released during catabolism is captured in the bonds of ATP, which provides the energy needed to carry out the constructive activities of anabolism. **10.** The energy released during the oxidation of food fuels is used to pump protons across the inner mitochondrial membrane. **11.** In substrate-level phosphorylation, high-energy phosphate groups are transferred directly from phosphorylated intermediates to ADP to form ATP. In oxidative phosphorylation, electron transport proteins forming part of the mitochondrial cristae use energy released during oxidation of glucose to create a steep gradient for protons across the crista membrane. Then as protons flow back through the membrane, gradient energy is captured to attach phosphate to ADP. **12.** If oxygen and pyruvic acid are not available, glycolysis will stop because the supply of NAD^+ is limited and glycolysis can continue only if the reduced coenzymes ($NADH + H^+$) formed during glycolysis are relieved of their extra hydrogen. **13.** Oxidation (via removal of H) is common in the Krebs cycle; it is indicated by the reduction of a coenzyme (either NAD^+ or FAD). Decarboxylations are also common, and are indicated by the removal of CO_2 from the cycle. **14.** Glycogenolysis is the reaction in which glycogen is broken down to its glucose units. **15.** Carbo loading forces the skeletal muscles to store more glycogen than they ordinarily would. **16.** Glycerol, a breakdown product of fat metabolism, directly enters the glycolysis pathway. **17.** Acetyl CoA is the central molecule of fat metabolism. **18.** The products of beta oxidation are acetyl CoA (acetic acid + coenzyme A), $NADH^+ + H^+$, and $FADH_2$. **19.** The liver uses keto acids drained off the Krebs cycle and amino groups (from other nonessential amino acids) as substrates to make the nonessential amino acids that the body needs. **20.** The ammonia removed from amino acids is combined with carbon dioxide to form urea, which is then eliminated by the kidneys. **21.** The three organs or tissues that regulate the directions of interconversions in the nutrient pools are the liver, skeletal muscles, and adipose tissues. **22.** Anabolic reactions and energy storage typify the absorptive state. Catabolic reactions (to increase blood sugar levels) such as lipolysis and glycogenolysis, and glucose sparing occur in the postabsorptive state. **23.** The main antagonist of glucagon is insulin. **24.** A rise in amino acid levels in blood increases both insulin and glucagon release. **25.** High HDLs would be preferable because the cholesterol these particles transport are destined for the liver and elimination from the body. **26.** A total cholesterol level of 200 mg/dl of blood or lower is recommended. **27.** Trans fats are oils that have been hydrogenated (with H atoms). They are unhealthy because they cause LDLs to increase and HDLs to decrease, exactly the opposite of what is desirable. **28.** Among short-term stimuli influencing feeding behavior are neural signals from the digestive tract, nutrient signals related to energy stores, and GI tract hormones (CCK, insulin, glucagon, and ghrelin). **29.** Leptin is the most important long-term regulator of feeding behavior. **30.** Of the factors listed, breathing and kidney function contribute to BMR. **31.** Samantha has a larger relative body surface area and well-toned muscles, both of which promote a higher BMR. **32.** The bodys core is the organs within the skull, and the thoracic and abdominal cavities. **33.** Cindy's body temperature is rising as heat-promoting mechanisms (shivering, chills) are activated. Something (an infection?) has caused the hypothalamic thermostat to be set to a higher level (fever) temporarily. **34.** In conduction, heat is transferred directly from one object to another (a hot surface to your palm). In convection, air warmed by body heat is continually removed (warm air rises) and replaced by cooler air (cool air falls), which in turn will absorb heat radiating from the body. **35.** Metabolic rate falls in old age because muscle mass declines and physical activity tends to be reduced. **36.** Elderly nutrition can be influenced and impaired by alcoholism, certain antibiotics that interfere with food absorption, other drugs that may cause electrolyte imbalances, and use of mineral oil.

Review Questions 1. a; **2.** c; **3.** b; **4.** d; **5.** b; **6.** c; **7.** a; **8.** d; **9.** d; **10.** a; **11.** b; **12.** d; **13.** c; **14.** d; **15.** a

Chapter 25

Check Your Understanding 1. The lower part of his rib cage and the perirenal fat capsule protect his kidneys from blows. **2.** The layers of supportive tissue around each kidney are the fibrous capsule, the perirenal fat capsule, and the renal fascia. The parietal peritoneum overlies the anterior renal fascia. **3.** The renal pelvis, which has extensions called calyces, is continuous with the ureter. **4.** Filtrate is formed in the glomerular capsule and then passes through the proximal convoluted tubule (PCT), the descending and ascending limbs of the loop of Henle, and the distal convoluted tubule (DCT). **5.** The structural differences are (1) juxtamedullary nephrons have long loops of Henle (with long thin segments) and renal corpuscles that are near the cortex-medulla junction, whereas cortical nephrons have short loops of Henle and renal corpuscles that lie more superficially in the cortex; (2) efferent arterioles of juxtamedullary nephrons supply vasa recta, while efferent arterioles of cortical nephrons supply peritubular capillaries. **6.** The glomerular capillaries are fenestrated capillaries. (See Figure 19.3 on p. 699 to refresh your memory of capillary types.) Their function is to filter large amounts of plasma into the glomerular capsule. **7.** Intrinsic controls serve to maintain a nearly constant GFR in spite of changes in systemic blood pressure. Extrinsic controls serve to maintain systemic blood pressure. **8.** Net filtration pressure is 5 mm Hg [50 mm Hg − (25 mm Hg + 20 mm Hg)]. **9.** The two main ways in which angiotensin II increases blood pressure and blood volume are by causing vasoconstriction and by causing aldosterone release. (In addition, angiotensin II causes ADH release, activates the thirst mechanism, increases tubular reabsorption, and contracts mesangial cells, which decreases the GFR.) **10.** The majority of reabsorption occurs in the proximal convoluted tubule. **11.** In primary active transport, the energy for the process is provided directly by the cleavage of ATP. In secondary active transport, the energy for the process is provided by the Na^+ concentration gradient. As Na^+ moves down its own concentration gradient established by the active pumping of Na^+ occurring elsewhere in the cell, it drives the movement of another substance (e.g., glucose) against its concentration gradient. **12.** The reabsorption of Na^+ by primary active transport, in turn, drives reabsorption of amino acids and glucose by secondary active transport. It also drives passive reabsorption of chloride, and reabsorption of water by osmosis. The reabsorption of water leaves behind other solutes, which become more concentrated and can therefore be reabsorbed by diffusion. **13.** H^+, K^+, NH_4^+, creatinine, urea, and uric acid are all substances that are secreted into the kidney tubules. **14.** The descending limb of the loop of Henle is permeable to water and impermeable to NaCl. The ascending limb is impermeable to water and permeable to NaCl. **15.** ADH is released from the posterior pituitary in response to hyperosmotic extracellular fluid (as sensed by hypothalamic osmoreceptors). ADH causes insertion of aquaporins into the luminal membrane of the principal cells of the collecting ducts. **16.** The normal renal clearance value for amino acids is zero. You would expect this because amino acids are valuable as nutrients and as the building blocks for protein synthesis, so it would not be good to lose them in the urine. **17.** The three major nitrogenous wastes excreted in urine are urea, creatinine, and uric acid. **18.** A kidney stone blocking the ureter would interfere with urine flow to the bladder. The pain would occur in waves that coincide with the peristaltic contractions of the smooth muscle of the ureter. **19.** The trigone is a smooth triangular region at the base of the bladder. Its borders are defined by the openings for the ureters and the urethra. **20.** The prostatic urethra, membranous urethra, and spongy urethra are the three regions of the male urethra. **21.** The detrusor muscle contracts in response to increased firing of parasympathetic nerves. Contraction of the detrusor muscle opens the internal urethral sphincter. **22.** The three sets of embryonic kidneys in the order that they develop are the pronephros, mesonephros, and metanephros. **23.** Both loss of bladder tone with age and an increase in size of the prostate can contribute to urinary retention in elderly men.

Review Questions 1. d; **2.** b; **3.** c; **4.** d; **5.** c; **6.** b; **7.** a; **8.** c; **9.** a; **10.** b

Chapter 26

Check Your Understanding 1. You have more intracellular than extracellular fluid and more interstitial fluid than plasma. **2.** Na^+ is the major cation in the ECF and K^+ is the major cation in the ICF. The intracellular counterparts to extracellular Cl^- are HPO_4^{2-} and protein anions. **3.** If you eat salty pretzels, your extracellular fluid volume will expand even if you don't ingest fluids. This is because water will flow by osmosis from the intracellular fluid to the extracellular fluid. **4.** An increases in osmolality of the plasma is most important for triggering thirst. This change is sensed by osmoreceptors in the hypothalamus. **5.** ADH cannot add water—it can only conserve what is already there. In order to reduce an increase in osmolality of body fluids, the thirst mechanism is required. **6.** (a) A loss of plasma proteins causes edema. (b) Copious sweating causes dehydration. (c) Using ecstasy (together with drinking lots of fluids) could cause hypotonic hydration because it promotes ADH secretion, which interferes with the body's ability to get rid of extra water. **7.** Insufficient aldosterone would cause Jacob's plasma Na^+ to be decreased and his plasma K^+ to be elevated. The decrease in plasma Na^+ would cause a decrease in blood pressure, because plasma Na^+ is directly related to blood volume, which is a major determinant of blood pressure. **8.** The kidneys' handling of K^+ can be summed up as "The kidneys reabsorb most of the filtered K^+ in the proximal parts of the kidney's tubules and then secrete just the right amount in the distal parts (cortical collecting duct)." **9.** The major regulator of calcium in the blood is parathyroid hormone. Hypercalcemia decreases excitability of neurons and muscle cells and may cause life-threatening cardiac arrhythmias. Hypocalcemia increases excitability and causes muscle tetany. **10.** Acidemia is an arterial pH below 7.35 and alkalemia is a pH above 7.45. **11.** A weak base would be better at minimizing the shift in pH caused by adding a strong acid to a solution because its ability to loosely tie up H^+ allows it to act as a buffer. **12.** The three major chemical buffer systems of the body are the bicarbonate buffer system, the phosphate buffer system, and the protein buffer system. The most important intracellular buffer is the protein buffer system. **13.** Joanne's ventilation would be increased. The acidosis caused by the accumulated ketone bodies will stimulate the peripheral chemoreceptors, and this will cause more CO_2 to be "blown off" in an attempt to restore pH to normal. **14.** Reabsorption of HCO_3^- is always linked with secretion of H^+. **15.** The most important urine buffer of H^+ is the phosphate buffer system. **16.** The tubule and collecting duct cells generate new HCO_3^- either by excreting ammonium ion (NH_4^+) or by excreting buffered H^+ ions. **17.** Key features of an uncompensated metabolic alkalosis are an increase in blood pH and an increase in blood HCO_3^-. Key features of an uncompensated respiratory acidosis are a decrease in blood pH and an increase in blood P_{CO_2}. **18.** The kidneys compensate for respiratory acidosis by excreting more H^+ and generating new HCO_3^- to buffer the acidosis. **19.** Infants' immature kidneys are not as proficient at concentrating urine. In addition, they have a high metabolic rate, so they produce more metabolic water and also have larger amounts of metabolic wastes and acids that need to be excreted with water.

Clinical Connections 1. Mr. Heyden's vital signs suggest that he is in hypovolemic shock, which is probably due to an internal hemorrhage. **2.** The spleen is a highly vascular organ due to its role as a blood-filtering organ. The macrophages in Mr. Heyden's liver and bone marrow will help compensate for the loss of his spleen. **3.** Elevation of renin, aldosterone, and antidiuretic hormone indicate that Mr. Heyden's body is trying to com-

APPENDIX G ANSWERS

pensate for his falling blood pressure and blood loss. *Renin*: released when renal blood flow is diminished and blood pressure falls. The angiotensin-aldosterone response is initiated by renin. The formation of angiotensin II leads to vasoconstriction, which will increase blood pressure, and to the release of aldosterone. *Aldosterone*: increases Na^+ reabsorption by the kidney. The movement of this reabsorbed Na^+ into the bloodstream will promote the movement of water from the interstitial fluid, resulting in an increase in blood volume. *Antidiuretic hormone (ADH)*: released when the hypothalamic osmoreceptors sense an increase in osmolality. ADH has two consequences: It is a potent vasoconstrictor and will increase blood pressure, and it promotes water retention by the kidney, increasing blood volume. **4.** Mr. Heyden's urine production may be decreased for several reasons. The severe drop in his blood pressure would reduce renal blood flow, thus reducing glomerular blood pressure and decreasing his glomerular filtration rate. The elevation in his ADH levels can reduce urine output due to increased water reabsorption by the kidney. He may also have damage to the kidney due to his crush injury in the left lumbar region. The presence of casts and a brownish-red color in his urine are probably damaged cells and blood. If he has suffered kidney damage due to being crushed, he may have nephron damage that would include disruption of the filtration membrane, allowing red blood cells to pass into the filtrate and therefore the urine. He may also have damaged renal tubules and peritubular capillaries, allowing entry of blood and damaged renal tubule epithelial cells into the filtrate.

Review Questions 1. a; **2.** c; **3.** b; **4.** a and b; **5.** h, i, j; **6.** c, g; **7.** a, e; **8.** b; **9.** a, b; **10.** j; **11.** b, d; **12.** a; **13.** c

Chapter 27

Check Your Understanding 1. The testes produce the male gametes (sperm) and testosterone. **2.** The sperm factories are the seminiferous tubules. **3.** When the ambient temperature is cold, the muscles contract, bringing the testes close to the warm body wall. When body temperature is high, the muscles relax, allowing the testes to hang away from the body wall. The pampiniform venous plexus absorbs heat from the arterial blood before it enters the testes. **4.** The erectile tissue of the penis allows the penis to become stiff so that it may more efficiently enter the female vagina to deliver sperm. **5.** The organs of the male duct system in order from the epididymis to the body exterior are the ductus deferens, ejaculatory duct, prostatic urethra, membranous urethra, and spongy urethra. **6.** These stereocilia pass nutrients to the sperm and absorb excess testicular fluid. **7.** The ductus deferens runs from the scrotum into the abdominal cavity. **8.** Adolph probably has a hypertrophied prostate, a condition which can be felt through the anterior wall of the rectum. **9.** The seminal vesicles produce the bulk of seminal fluid. **10.** Semen is sperm plus the secretions of the male accessory glands. **11.** Erection is the stiffening of the penis that occurs when blood in the cavernous tissue is prevented from leaving the penis. It is caused by the parasympathetic division of the autonomic nervous system. **12.** Resolution is a period of muscular and psychological relaxation that follows orgasm. It results as the sympathetic nervous system constricts the internal pudendal arteries, reducing blood flow to the penis, and activates small muscles that force blood out of the penis. **13.** Meiosis reduces the chromosomal count from $2n$ to n and introduces variability. **14.** The sperm head is the compacted DNA-containing nucleus. The acrosome that caps the head is a lysosome-like sac of enzymes. The midpiece contains the energy-producing mitochondria. The tail fashioned by a centriole is the propulsive structure. **15.** Sustentacular cells provide nutrients and essential development signals to the developing sperm and form the blood-testis barrier that prevents sperm antigens from escaping into the blood. Interstitial cells secrete testosterone. **16.** The HPG axis is the hormonal interrelationship between the hypothalamus, anterior pituitary, and gonads that regulates the production of gametes and sex hormones (e.g., sperm production and testosterone in the male). **17.** Follicle-stimulating hormone indirectly stimulates spermatogenesis by prompting the sustentacular cells to secrete androgen-binding protein. Androgen-binding protein keeps the concentration of testosterone high in the vicinity of the spermatogenic cells, which directly stimulates spermatogenesis. **18.** Secondary sex characteristics of males include appearance of pubic, axillary, and facial hair, deepening of the voice, increased oiliness of the skin, and increased size (length and mass) of the bones and skeletal muscles. **19.** The female's internal genitalia include the ovaries and duct system (uterine tubes, uterus, and vagina). **20.** The ovaries produce the female gametes and secrete female sex hormones (estrogens and progesterone). **21.** A primary follicle has one layer of cuboidal follicle cells surrounding the oocyte; a secondary follicle has more than one layer of follicle cells, and small fluid-filled spaces form between the follicle cells. A vesicular follicle has several layers of follicle cells surrounding a fluid-filled cavity (antrum) that pushes the oocyte to one side. **22.** Women are more at risk for PID than men because the duct system of women is incomplete—there is no physical connection between the ovary and the uterine tubes, which are open to the pelvic cavity. In men, the duct system is continuous from the testes to the body exterior. **23.** The waving action of the fimbriae helps to direct the ovulated oocytes into the uterine tube. **24.** The usual site of fertilization is the uterine tube. The uterus serves as the incubator for fetal development. **25.** The greater vestibular glands are the female homologue of the male bulbourethral glands. **26.** Both the penis and clitoris are hooded by a skin fold and are largely erectile tissue. However, the clitoris lacks a corpus spongiosum containing a urethra, so the urinary and reproductive systems are completely separate in females. **27.** Developmentally, the mammary glands are modified sweat glands. **28.** Breast cancer usually arises from the epithelial cells of the small ducts. **29.** The products of meiosis in females are 3 polar bodies (tiny haploid cells with essentially no cytoplasm) and 1 haploid ovum (functional gamete). Meiosis in males yields 4 functional gametes, the haploid sperm. **30.** Identical twins develop from separation of a very young embryo (the result of fertilization by a single sperm) into two parts. Fraternal twins develop when different oocytes are fertilized by different sperm. **31.** In the luteal phase, the ovulated follicle develops into a corpus luteum, which then secretes progesterone (and some estrogen). **32.** Leptin is important in advising the brain of the girl's readiness (relative to energy stores) for puberty. **33.** FSH prompts follicle growth and LH prompts ovulation. **34.** Estrogen exerts positive feedback on the anterior pituitary that leads to a burstlike release of LH. **35.** Estrogen is responsible for the secondary sex characteristics of females. **36.** Estrogen promotes epiphyseal closure in both males and females. **37.** The human papillomavirus (HPV) is most associated with cervical cancer. **38.** Chlamydia is the most common bacterial sexually transmitted infection in the U.S. **39.** No. If the sex chromosomes are XY, a male baby will develop. **40.** The sexually indifferent stage of development is the early period when the presumptive reproductive structures can produce either male or female organs. **41.** The gubernaculum guides the descent of the testis into the scrotum. **42.** Early signs of puberty's onset in boys are enlargement of the testes and scrotum. **43.** Menopause has happened when ovulation and menstruation have not occurred for a year.

Clinical Connections 1. *Carcinoma* is the term for cancer originating from epithelial tissue. The primary source of Mr. Heyden's cancer is likely to be the prostate. **2.** Elevation of serum acid phosphatase levels is diagnostic for carcinoma of the prostate. (In addition, Mr. Heyden's age places him in a group that is at relatively higher risk for this type of cancer.) **3.** Digital examination of Mr. Heyden's prostate should detect the presence of carcinoma in this tissue. In addition, serum levels of prostate-specific antigen (PSA) should be checked, as an increase in this antigen is often indicative of prostate cancer. **4.** Mr. Heyden's carcinoma has advanced to the point of metastasis. He will probably undergo a treatment that reduces the levels of androgens in his body, as androgens promote growth of the prostate-derived tissue. These treatments could include castration or administration of drugs that block the production and/or effects of androgens.

Review Questions 1. a and b; **2.** d; **3.** a; **4.** d; **5.** b; **6.** d; **7.** d; **8.** c; **9.** a, c, e, f; **10.** (1)c, f; (2)e, h; (3)g; (4)a; (5)b, g, and e; (6)f; **11.** a; **12.** c; **13.** d; **14.** b; **15.** a; **16.** b; **17.** c

Chapter 28

Check Your Understanding 1. Sperm must be capacitated before they can penetrate an oocyte. **2.** The cortical reaction involves the release of enzymes from cortical granules to the oocyte exterior, which accomplishes the slow block to polyspermy. **3.** The blastocyst is only slightly larger than the zygote because, although cell division has been going on (cleavage divisions), there is essentially no time for growth between divisions, so the resulting cells get smaller and smaller. **4.** The trophoblast cells adhere to and secrete digestive enzymes onto the uterine endometrium, accomplishing implantation, and they contribute to the formation of the placenta. **5.** The syncytiotrophoblast actually accomplishes implantation. **6.** The blastocyst secretes the hormone human chorionic gonadotropin, which is detectable in the urine. **7.** The chorion develops from the trophoblast and a layer of extraembryonic mesoderm. **8.** The decidua basalis cooperates with the chorionic villi to form the placenta. **9.** The placenta is usually fully functional by the end of the third month of pregnancy. **10.** The amnion helps to maintain a constant temperature for the developing fetus and protects it from physical trauma. **11.** The allantois provides the basis of the umbilical cord, which provides a pathway for the embryonic blood vessels to reach the placenta. **12.** Before organogenesis can occur in earnest, the embryonic body must fold and undercut to form a tubular embryo. **13.** The mesoderm gives rise to essentially all body tissues except neural and epidermal tissue and mucosae. **14.** The fetal period begins at the end of 8 weeks. **15.** Difficult breathing during late pregnancy is due to the fact that the uterus is pressing against and crowding the diaphragm (and hence the lungs). The waddling gait is due to the relaxing effect of relaxin on the pelvic ligaments and the pubic symphysis. **16.** While the exact cause of morning sickness is unknown, it is thought to be due to the rising levels of female sex hormones in the mother's blood, which sometimes takes awhile to get used to. **17.** The hormone hCT increases the pregnant woman's metabolic rate. **18.** Breech presentation is a buttock-first presentation of the baby during labor. **19.** Prostaglandins are most responsible for triggering true labor. **20.** The descent of the baby's head (the largest part of its body) follows the widest dimensions of the bony pelvis. **21.** The foramen ovale and ductus arteriosus allow most of the blood to bypass the heart. **22.** For the most part, the special fetal circulatory modifications are occluded at birth or shortly thereafter. **23.** Oxytocin causes the let-down reflex.

Review Questions 1. (1)a, (2)b, (3)b, (4)a; **2.** b; **3.** c; **4.** b; **5.** a; **6.** d; **7.** d; **8.** c; **9.** b; **10.** c; **11.** a; **12.** c; **13.** a; **14.** (1)e, (2)g, (3)a, (4)f, (5)i, (6)b, (7)h, (8)d, (9)c

Chapter 29

Check Your Understanding 1. Chromosomes are not visible during interphase. The DNA-containing material is in the form of dispersed strands of chromatin. As mitosis begins, the chromatin coils and condenses, becoming visible. Chromosomes continue to condense throughout prophase and are most visible during metaphase. **2.** Chromosomes other than sex chromosomes are called autosomes. **3.** An allele represented by a capital letter is presumed to be a dominant allele. **4.** Descriptions of his phenotype are blond, blue-eyed, and hairy chest. **5.** The alleles pertaining to the same trait are segregated independently of each other to different gametes. **6.** It causes separations of linked genes on the same chromosome, producing gametes with variable genomes. **7.** Dominant genes are expressed. Thus if the dominant gene is detrimental (lethal), the carrier will probably not live very long or will die during development. **8.** In incomplete dominance, the heterozygote has a phenotype intermediate between that of the dominant and recessive alleles [for example, for the sickling gene, the dominant homozygote (*SS*) has no evidence of sickling; the heterozygote (*Ss*) has sickle-cell trait; and the homozygote for the recessive gene (*ss*) has sickle-cell anemia]. In codominance, both dominant alleles are expressed (as in ABO blood types). **9.** A male always expresses an X-linked recessive allele because, unlike a female, he does not have a second X containing homologous alleles to blunt or counteract the effect. The Y chromosome lacks most of the genes carried on the X chromosome. **10.** Height is an example of a trait conferred by polygene inheritance in which several genes on different chromosomes contribute to the trait. Such traits show a distribution of phenotypes that yields a bell-shaped curve. **11.** Other genes, measles in a pregnant woman, and lack of key dietary nutrients all may alter gene expression. **12.** Genomic imprinting labels genes as maternal or paternal. **13.** Maternal mitochondrial DNA confers extranuclear inheritance. **14.** Amniocentesis analyzes chemicals and cells in amniotic fluid. **15.** Ultrasound imaging is used to determine some aspects of fetal development (fetal age for example) and is noninvasive.

Review Questions 1. (1)d, (2)g, (3)b, (4)a, (5)f, (6)c, (7)e, (8)h; **2.** (1)e, (2)a, (3)b, (4)d, (5)c, (6)d, (7)f

Glossary

Pronunciation Key

′ = Primary accent

″ = Secondary accent

Pronounce:

a, fa, āt	as in	fate	o, no, ōt	as in	note
ă, hă, at		hat	ŏ, frŏ, og		frog
ah		father	oo		soon
ar		tar	or		for
e, stre, ēt		street	ow		plow
ĕ, hĕ, en		hen	oy		boy
er		her	sh		she
ew		new	u, mu, ūt		mute
g		go	ŭ, sŭ, un		sun
i, bi, īt		bite	z		zebra
ĭ, hĭ, im		him	zh		measure
ng		ring			

Abduct (ab-dukt′) To move away from the midline of the body.

Absolute refractory period Period following stimulation during which no additional action potential can be evoked.

Absorption Process by which the products of digestion pass through the alimentary tube mucosa into the blood or lymph.

Accessory digestive organs Organs that contribute to the digestive process but are not part of the alimentary canal; include the tongue, teeth, salivary glands, pancreas, liver.

Accommodation The process of increasing the refractive power of the lens of the eye; focusing.

Acetabulum (as″ĕ-tab′u-lum) Cuplike cavity on lateral surface of the hip bone that receives the femur.

Acetylcholine (ACh) (as″ĕ-til-ko′lēn) Chemical transmitter substance released by some nerve endings.

Acetylcholinesterase (AChE) (as″ĕ-til-ko″lin-es′ter-ās) Enzyme present at the neuromuscular junction and synapses that degrades acetylcholine and terminates its action.

Achilles tendon *See* Calcaneal tendon.

Acid A substance that releases hydrogen ions when in solution (compare with Base); a proton donor.

Acid-base balance Situation in which the pH of the blood is maintained between 7.35 and 7.45.

Acidosis (as″ĭ-do′sis) State of abnormally high hydrogen ion concentration in the extracellular fluid.

Actin (ak′tin) A contractile protein of muscle.

Action potential A large transient depolarization event, including polarity reversal, that is conducted along the membrane of a muscle cell or a nerve fiber.

Activation energy The amount of energy required to push a reactant to the level necessary for action.

Active immunity Immunity produced by an encounter with an antigen; provides immunological memory.

Active site Region on the surface of a functional (globular) protein where it binds and interacts chemically with other molecules of complementary shape and charge.

Active (transport) processes (1) Membrane transport processes for which ATP is required, e.g., solute pumping and endocytosis. (2) "Active transport" also refers specifically to solute pumping.

Adaptation (1) Any change in structure or response to suit a new environment; (2) decline in the transmission of a sensory nerve when a receptor is stimulated continuously and without change in stimulus strength.

Adduct (a-dukt′) To move toward the midline of the body.

Adenine (A) (ad′ĕ-nēn) One of the two major purines found in both RNA and DNA; also found in various free nucleotides of importance to the body, such as ATP.

Adenohypophysis (ad″ĕ-no-hi-pof′ĭ-sis) Anterior pituitary; the glandular part of the pituitary gland.

Adenoids (ad′en-noids) Pharyngeal tonsil.

Adenosine triphosphate (ATP) (ah-den′o-sēn tri″fos′fāt) Organic molecule that stores and releases chemical energy for use in body cells.

Adipocyte (ad′ĭ-po-sīt) An adipose, or fat, cell.

Adipose tissue (ad′ĭ-pōs) Areolar connective tissue modified to store nutrients; a connective tissue consisting chiefly of fat cells.

Adrenal glands (uh-drē′nul) Hormone-producing glands located superior to the kidneys; each consists of medulla and cortex areas.

Adrenergic fibers (ad″ren-er′jik) Nerve fibers that release norepinephrine.

Adrenocorticotropic hormone (ACTH) (ah-dre′no-kor″tĭ-ko-trō′pik) Anterior pituitary hormone that influences the activity of the adrenal cortex.

Adventitia (ad″ven-tish′e-ah) Outermost layer or covering of some organs.

Aerobic (a′er-ōb″ik) Oxygen-requiring.

Aerobic endurance The length of time a muscle can continue to contract using aerobic pathways.

Aerobic respiration Respiration in which oxygen is consumed and glucose is broken down entirely; water, carbon dioxide, and large amounts of ATP are the final products.

Afferent (af′er-ent) Carrying to or toward a center.

Afferent (sensory) nerve Nerve that contains processes of sensory neurons and carries nerve impulses to the central nervous system.

Agglutination (ah-gloo″tĭ-na′shun) Clumping of (foreign) cells; induced by cross-linking of antigen-antibody complexes.

Agonist (ag′o-nist) Muscle that bears the major responsibility for effecting a particular movement; a prime mover.

AIDS Acquired immune deficiency syndrome; caused by human immunodeficiency virus (HIV); symptoms include severe weight loss, night sweats, swollen lymph nodes, opportunistic infections.

Albumin (al-bu′min) The most abundant plasma protein.

Aldosterone (al-dos′ter-ōn) Hormone produced by the adrenal cortex that regulates Na^+ reabsorption and K^+ secretion by the kidneys.

Alimentary canal (al″ĭ-men′tar-e) The continuous hollow tube extending from the mouth to the anus; its walls are constructed by the oral cavity, pharynx, esophagus, stomach, and small and large intestines.

Alkalosis (al″kah-lo′sis) State of abnormally low hydrogen ion concentration in the extracellular fluid.

Allantois (ah″lan′to-is) Embryonic membrane; its blood vessels develop into blood vessels of the umbilical cord.

Alleles Genes coding for the same trait and found at the same locus on homologous chromosomes.

Allergy A type of hypersensitivity (overzealous immune response to an otherwise harmless antigen) that involves IgE antibodies and histamine release.

Alopecia (al″o-pe′she-ah) Baldness.

Alpha (α)-helix The most common type of secondary structure of the amino acid chain in proteins; resembles the coils of a telephone cord.

Alveolar (acinar) gland (al-ve′o-lar) A gland whose secretory cells form small, flasklike sacs.

Alveolar ventilation rate (AVR) An index of respiratory efficiency; measures volume of fresh air that flows in and out of alveoli.

Alveolus (al-ve′o-lus) (1) One of the microscopic air sacs of the lungs; (2) tiny milk-producing glandular sac in the breast; (3) tooth socket.

Alzheimer's disease (AD) (altz′hi-merz) Degenerative brain disease resulting in progressive loss of memory and motor control, and increasing dementia.

Amino acid (ah-me′no) Organic compound containing nitrogen, carbon, hydrogen, and oxygen; building block of protein.

Ammonia (NH_3) Common waste product of protein breakdown in the body; a colorless volatile gas, very soluble in water and capable of forming a weak base; a proton acceptor.

Amniocentesis A common form of fetal testing in which a small sample of fluid is removed from the amniotic cavity.

Amnion (am′ne-on) Fetal membrane that forms a fluid-filled sac around the embryo.

Amoeboid motion (ah-me′boyd) The flowing movement of the cytoplasm of a phagocyte.

Amphiarthrosis (am″fe-ar-thro′sis) A slightly movable joint.

Ampulla (am-pul′lah) A localized dilation of a canal or duct.

Amylase Digestive system enzyme that breaks down starchy foods.

Anabolism (ah-nab′o-lizm) Energy-requiring building phase of metabolism in which simpler substances are combined to form more complex substances.

Anaerobic (an-a′er-ōb-ik) Not requiring oxygen.

Anaerobic glycolysis (gli-kol′ĭ-sis) Energy-yielding conversion of glucose to lactic acid in various tissues, notably muscle, when sufficient oxygen is not available.

Anaerobic threshold The point at which muscle metabolism converts to anaerobic glycolysis.

Anaphase Third stage of mitosis, meiosis I, and meiosis II in which daughter chromosomes move toward each pole of a cell.

Anastomosis (ah-nas″to-mo′sis) A union or joining of nerves, blood vessels, or lymphatics.

Anatomy Study of the structure of living organisms.

Androgen (an′dro-jen) A hormone such as testosterone that controls male secondary sex characteristics.

Anemia (ah-ne′me-ah) Reduced oxygen-carrying ability of blood resulting from too few erythrocytes or abnormal hemoglobin.

Aneurysm (an′u-rizm) Blood-filled sac in an artery wall caused by dilation or weakening of the wall.

Angina pectoris (an′jĭ-nah pek′tor-is) Severe suffocating chest pain caused by brief lack of oxygen supply to heart muscle.

Angiotensin II (an″je-o-ten′sin) A potent vasoconstrictor activated by renin; also triggers release of aldosterone.

Anion (an′i-on) An ion carrying one or more negative charges and therefore attracted to a positive pole.

Anoxia (ah-nŏk′se-ah) Deficiency of oxygen.

Antagonist (an-tag′o-nist) (1) Muscle that reverses, or opposes, the action of another muscle. (2) Hormone that opposes the action of another hormone.

Anterior pituitary *See* Adenohypophysis.

Antibody A protein molecule that is released by a plasma cell (a daughter cell of an activated B lymphocyte) and that binds specifically to an antigen; an immunoglobulin.

Anticodon (an″ti-ko′don) The three-base sequence complementary to the messenger RNA (mRNA) codon.

Antidiuretic hormone (ADH, also called **vasopressin)** (an″ti-di″yer-eh′tik) Hormone produced by the hypothalamus and released by the posterior pituitary; stimulates the kidneys to reabsorb more water, reducing urine volume.

Antigen (Ag) (an′tĭ-jen) A substance or part of a substance (living or nonliving) that is recognized as foreign by the immune system, activates the immune system, and reacts with immune cells or their products.

Antigen-presenting cell (APC) A specialized cell (dendritic cell, macrophage, or B cell) that captures, processes, and presents antigens on its surface to T lymphocytes.

Anucleate cell (a-nu′kle-āt) A cell without a nucleus.

Anus (a′nus) Distal end of digestive tract; outlet of rectum.

Aorta (a-or′tah) Major systemic artery; arises from the left ventricle of the heart.

Aortic body Receptor in the aortic arch sensitive to changing oxygen, carbon dioxide, and pH levels of the blood.

Apgar score Evaluation of an infant's physical status at 1 and 5 minutes after birth by assessing five criteria: heart rate, respiration, color, muscle tone, and reflexes.

Apnea Breathing cessation.

Apocrine sweat gland (ap′o-krin) The less numerous type of sweat gland; produces a secretion containing water, salts, proteins, and fatty acids.

Apoenzyme (ap′ō-en-zīm) The protein portion of an enzyme.

Aponeurosis (ap″o-nu-ro′sis) Fibrous or membranous sheet connecting a muscle and the part it moves.

Apoptosis A process of controlled cellular suicide; eliminates cells that are unneeded, stressed, or aged.

Appendicitis (ă-pen′dĭ-sī′tis) Inflammation of the appendix (wormlike sac attached to the cecum of the large intestine).

Appendicular Relating to the limbs; one of the two major divisions of the body.

Appositional growth Growth accomplished by the addition of new layers onto those previously formed.

Aquaporins (ă″kwă-por′ins) Transmembrane proteins that form water channels.

Aqueous humor (a′kwe-us) Watery fluid in the anterior segment of the eye.

Arachnoid (ah-rak′noid) Weblike; specifically, the weblike arachnoid mater, the middle layer of the three meninges.

Areola (ah-re′o-lah) Circular, pigmented area surrounding the nipple; any small space in a tissue.

Areolar connective tissue A type of loose connective tissue.

Arrector pili (ah-rek′tor pi′li) Tiny, smooth muscles attached to hair follicles; contraction causes the hair to stand upright.

Arrhythmia (a-rith′me-ah) Irregular heart rhythm, often caused by defects in the intrinsic conduction system.

Arteries Blood vessels that conduct blood away from the heart and into the circulation.

Arteriole (ar-tēr′e-ōl) A minute artery.

Arteriosclerosis (ar-tēr′e-o-skler-o′sis) Any of a number of proliferative and degenerative changes in the arteries leading to their decreased elasticity.

Arthritis Inflammation of the joints.

Arthroscopic surgery (ar-thro-skop′ik) Procedure enabling a surgeon to repair the interior of a joint through a small incision.

Articular capsule Double-layered capsule composed of an outer fibrous capsule lined by synovial membrane; encloses the joint cavity of a synovial joint.

Articular cartilage Hyaline cartilage covering bone ends at movable joints.

Articulation (joint) The junction of two or more bones.

Association areas Functional areas of the cerebral cortex that act mainly to integrate diverse information for purposeful action.

Astigmatism (ah-stig′mah-tizm) A condition in which unequal curvatures in different parts of the cornea or lens of the eye lead to blurred vision.

Astrocyte (as′tro-sīt) A type of CNS supporting cell; assists in exchanges between blood capillaries and neurons.

Atelectasis (at″ĕ-lik′tah-sis) Lung collapse.

Atherosclerosis (a″ther-o″skler-o′sis) Changes in the walls of large arteries consisting of lipid deposits on the artery walls; one form of arteriosclerosis.

Atmospheric pressure Force that air exerts on the surface of the body (760 mm Hg at sea level).

Atom Smallest particle of an elemental substance that exhibits the properties of that element; composed of protons, neutrons, and electrons.

Atomic number The number of protons in an atom.

Atomic symbol The one- or two-letter symbol used to indicate an element; usually the first letter(s) of the element's name.

Atomic weight The average of the mass numbers of all the isotopes of an element.

ATP (adenosine triphosphate) (ah-den′o-sēn tri″fos′fāt) Organic molecule that stores and releases chemical energy for use in body cells.

Atria (a′tre-ah) The two superior receiving chambers of the heart.

Atrial natriuretic peptide (ANP) (a′tre-al na″tre-u-ret′ik) A hormone released by certain cells of the heart atria that reduces blood pressure and blood volume by inhibiting nearly all events that promote vasoconstriction and Na^+ and water retention.

Atrioventricular (AV) bundle (a″tre-o-ven-trĭ′kyoo-ler) Bundle of specialized fibers that conduct impulses from the AV node to the right and left ventricles; also called bundle of His.

Atrioventricular (AV) node Specialized mass of conducting cells located at the atrioventricular junction in the heart.

Atrioventricular (AV) valve Valve that prevents backflow into the atrium when the connected ventricle is contracting.

Atrophy (at′ro-fe) Reduction in size or wasting away of an organ or cell resulting from disease or lack of use.

Auditory ossicles (ah′sih-kulz) The three tiny bones serving as transmitters of vibrations and located within the middle ear: the malleus, incus, and stapes.

Auditory tube *See* Pharyngotympanic tube.

Autoimmunity Production of antibodies or effector T cells that attack a person's own tissue.

Autolysis (aw″tol′ĭ-sis) Process of autodigestion (self-digestion) of cells, especially dead or degenerate cells.

Autonomic ganglion Collection of sympathetic or parasympathetic ganglionic neuronal cell bodies.

Autonomic nervous system (ANS) Efferent division of the peripheral nervous system that innervates cardiac and smooth muscles and glands; also called the involuntary or visceral motor system.

Autonomic (visceral) reflexes Reflexes that activate smooth or cardiac muscle and/or glands.

Autoregulation The automatic local adjustment of blood flow to a particular body area in response to its current requirements.

Autosomes Chromosomes number 1 to 22; do not include the sex chromosomes.

Avogadro's number (av″o-gad′rōz) The number of molecules in one mole of any substance, 6.02×10^{23}.

Axial Relating to the head, neck, and trunk; one of the two major divisions of the body.

Axolemma (ak″so-lem′ah) The plasma membrane of an axon.

Axon Neuron process that carries impulses away from the nerve cell body; efferent process; the conducting portion of a nerve cell.

Axon terminals (synaptic knobs, boutons) The bulbous distal endings of the terminal branches of an axon.

B cells Also called B lymphocytes; oversee humoral immunity; their descendants differentiate into antibody-producing plasma cells.

Baroreceptor (bayr″o-re-sep′tor) A sensory nerve ending in the wall of the carotid sinus or aortic arch sensitive to vessel stretching.

Basal body (ba′sal) An organelle structurally identical to a centriole and forming the base of a cilium or flagellum.

Basal ganglia *See* Basal nuclei.

Basal lamina (lam′ĭ-nah) Noncellular, adhesive supporting sheet consisting largely of glycoproteins secreted by epithelial cells.

Basal metabolic rate (BMR) Rate at which energy is expended (heat produced) by the body per unit time under controlled (basal) conditions: 12 hours after a meal, at rest.

Basal nuclei (basal ganglia) Specific gray matter areas located deep within the white matter of the cerebral hemispheres.

Basal surface The surface near the base or interior of a structure; nearest the lower side or bottom of a structure.

Base A substance capable of binding with hydrogen ions; a proton acceptor.

Basement membrane Extracellular material consisting of a basal lamina secreted by epithelial cells and a reticular lamina secreted by underlying connective tissue cells.

Basophil (ba′zo-fil) White blood cell whose granules stain purplish-black and nucleus purple with basic dye.

Benign (be-nīn′) Not malignant.

Bile Greenish-yellow or brownish fluid produced in and secreted by the liver, stored in the gallbladder, and released into the small intestine.

Bilirubin (bil″i-roo′bin) Yellow pigment of bile.

Bipolar neuron Neuron with axon and dendrite that extend from opposite sides of the cell body.

Blastocyst (blas′to-sist) Stage of early embryonic development; the product of cleavage.

Blood pressure (BP) Force exerted by blood against a unit area of the blood vessel walls; differences in blood pressure between different areas of the circulation provide the driving force for blood circulation.

Blood-brain barrier Mechanism that inhibits passage of materials from the blood into brain tissues; reflects relative impermeability of brain capillaries.

Bolus (bo′lus) A rounded mass of food prepared by the mouth for swallowing; any soft round mass.

Bone marrow Fat- or blood-forming tissue found within bone cavities; called yellow and red bone marrow, respectively.

Bone (osseous tissue) (os′e-us) A connective tissue that forms the bony skeleton.

Bone remodeling Process involving bone formation and destruction in response to hormonal and mechanical factors.

Bone resorption The removal of osseous tissue; part of the continuous bone remodeling process.

Bowman's capsule (bo-manz) *See* Glomerular capsule.

Boyle's law States that when the temperature is constant, the pressure of a gas varies inversely with its volume.

Bradycardia (brad″e-kar′de-ah) A heart rate below 60 beats per minute.

Brain death State of irreversible coma, even though life-support measures may have restored other body organs.

Brain stem Collectively the midbrain, pons, and medulla of the brain.

Brain ventricle Fluid-filled cavity of the brain.

Branchial groove (brang′ke-al) An indentation of the surface ectoderm in the embryo; the external acoustic meatus develops from it.

Bronchioles Smaller (<1 mm in diameter) branching air passageways inside the lungs.

Bronchus (brong′kus) One of the two large branches of the trachea that leads to the lungs.

Buffer Chemical substance or system that minimizes changes in pH by releasing or binding hydrogen ions.

Burn Tissue damage inflicted by intense heat, electricity, radiation, or certain chemicals, all of which denature cell proteins and cause cell death in the affected areas.

Bursa (ber′sa) A fibrous sac lined with synovial membrane and containing synovial fluid; occurs between bones and muscle tendons (or other structures), where it acts to decrease friction during movement.

Bursitis Inflammation of a bursa.

Calcaneal tendon (kal-ka′ne-al) Tendon that attaches the calf muscles to the heelbone (calcaneus); also called the Achilles tendon.

Calcitonin (kal″sih-to′nin) Hormone released by the thyroid. Lowers blood calcium levels only when present at high (therapeutic). levels.

Calculus (kal′ku-lus) A stone formed within various body parts.

Callus (kal′lus) (1) Localized thickening of skin epidermis resulting from physical trauma; (2) repair tissue (fibrous or bony) formed at a fracture site.

Calorie (cal) Amount of energy needed to raise the temperature of 1 gram of water 1° Celsius. Energy exchanges associated with biochemical reactions are usually reported in kilocalories (1 kcal = 1000 cal), also called large calories (Cal).

Calyx (ka′liks) A cuplike extension of the pelvis of the kidney.

Canaliculus (kan″ah-lik′u-lus) Extremely small tubular passage or channel.

Cancer A malignant, invasive cellular neoplasm that has the capability of spreading throughout the body or body parts.

Capillaries (kap′il-layr″ēs) The smallest of the blood vessels and the sites of exchange between the blood and tissue cells.

Carbohydrate (kar″bo-hi′drāt) Organic compound composed of carbon, hydrogen, and oxygen; includes starches, sugars, cellulose.

Carbonic acid–bicarbonate buffer system Chemical system that helps maintain pH homeostasis of the blood.

Carbonic anhydrase (kar-bon′ik an-hi′drās) Enzyme that facilitates the combination of carbon dioxide with water to form carbonic acid.

Carcinogen (kar″sĭ′no-jin) Cancer-causing agent.

Cardiac cycle Sequence of events encompassing one complete contraction and relaxation of the atria and ventricles of the heart.

Cardiac muscle Specialized muscle of the heart.

Cardiac output (CO) Amount of blood pumped out of a ventricle in one minute.

Cardiac reserve The difference between resting and maximal cardiac output.

Cardiogenic shock Pump failure; the heart is so inefficient that it cannot sustain adequate circulation.

Cardiovascular system Organ system that distributes the blood to deliver nutrients and remove wastes.

Carotene (kar′o-tēn) Yellow to orange pigment that accumulates in the stratum corneum epidermal layer and in fatty tissue of the hypodermis.

Carotid body (kar-rot′id) A receptor in the common carotid artery sensitive to changing oxygen, carbon dioxide, and pH levels of the blood.

Carotid sinus (si′nus) A dilation of a common carotid artery; involved in regulation of systemic blood pressure.

Carrier A transmembrane protein that changes shape to envelop and transport a polar substance across the cell membrane.

Cartilage (kar′tĭ-lij) White, semiopaque connective tissue.

Cartilage bone (endochondral bone) Bone formed by using hyaline cartilage structures as models for ossification.

Cartilaginous joints (kar″ti-laj′ĭ-nus) Bones united by cartilage; no joint cavity is present.

Catabolism (kat-tab′o-lizm) Process in which living cells break down substances into simpler substances.

Catalyst (kat′ah-list) Substance that increases the rate of a chemical reaction without itself becoming chemically changed or part of the product.

Cataract Clouding of the eye's lens; often congenital or age-related.

Catecholamines (kat″ĕ-kol′ah-mēnz) Epinephrine, norepinephrine, and dopamine; a class of amines that act as chemical transmitters.

Cation (kat′i-on) An ion with a positive charge.

Caudal (kaw′dul) Literally, toward the tail; in humans, the inferior portion of the anatomy.

Cecum (se′kum) The blind-end pouch at the beginning of the large intestine.

Cell Structural unit of all living things.

Cell differentiation The development of specific and distinctive features in cells, from a single cell (the fertilized egg) to all the specialized cells of adulthood.

Cell life cycle Series of changes a cell goes through from the time it is formed until it reproduces itself.

Cell-mediated immunity Immunity conferred by activated T cells, which directly kill infected or cancerous body cells or cells of foreign grafts and release chemicals that regulate the immune response.

Cell membrane *See* Plasma membrane.

Cellular respiration Metabolic processes in which ATP is produced.

Cellulose (sel′u-lōs) A fibrous carbohydrate that is the main structural component of plant tissues.

Central (Haversian) canal (hah-ver′zhan) The canal in the center of each osteon that contains minute blood vessels and nerve fibers that serve the needs of the osteocytes.

Central nervous system (CNS) Brain and spinal cord.

Centriole (sen′tre-ol) Minute body found near the nucleus of the cell; active in cell division.

Centrosome (cell center) A region near the nucleus which contains paired organelles called centrioles.

Cerebellum (ser″ĕ-bel′um) Brain region most involved in producing smooth, coordinated skeletal muscle activity.

Cerebral aqueduct (ser′ĕ-bral, sĕ-re′bral) The slender cavity of the midbrain that connects the third and fourth ventricles.

Cerebral arterial circle (circle of Willis) An arterial anastomosis at the base of the brain.

Cerebral cortex The outer gray matter region of the cerebral hemispheres.

Cerebral dominance Designates the hemisphere that is dominant for language.

Cerebral palsy Neuromuscular disability in which voluntary muscles are poorly controlled or paralyzed as a result of brain damage.

Cerebral white matter Consists largely of myelinated fibers bundled into large tracts; provides for communication between cerebral areas and lower CNS centers.

Cerebrospinal fluid (CSF) (ser″ĕ-bro-spi′nal) Plasmalike fluid that fills the cavities of the CNS and surrounds the CNS externally; protects the brain and spinal cord.

Cerebrovascular accident (CVA) (ser″ĕ-bro-vas′ku-lar) Condition in which brain tissue is deprived of a blood supply, as in blockage of a cerebral blood vessel; a stroke.

Cerebrum (ser′ĕ-brum) The cerebral hemispheres and the structures of the diencephalon.

Cervical vertebrae The seven vertebrae of the vertebral column located in the neck.

Cervix Lower outlet of the uterus extending into the vagina.

Channel A transmembrane protein that forms an aqueous pore, allowing substances to move from one side of the membrane to the other.

Chemical bond An energy relationship holding atoms together; involves the interaction of electrons.

Chemical digestion A series of catabolic steps in which complex food molecules are broken down to their building blocks by enzymes.

Chemical energy Energy stored in the bonds of chemical substances.

Chemical equilibrium A state of apparent repose created by two reactions proceeding in opposite directions at equal speed.

Chemical reaction Process in which molecules are formed, changed, or broken down.

Chemoreceptor (ke″mo-re-sep′ter) Receptor sensitive to various chemicals in solution.

Chemotaxis (ke″mo-tak′sis) Movement of a cell, organism, or part of an organism toward or away from a chemical substance.

Cholecystokinin (CCK) (ko″le-sis″to-ki′nin) An intestinal hormone that stimulates gallbladder contraction and pancreatic juice release.

Cholesterol (ko-les′ter-ol″) Steroid found in animal fats as well as in most body tissues; made by the liver.

Cholinergic fibers (ko″lin-er′jik) Nerve endings that, upon stimulation, release acetylcholine.

Chondroblast (kon′dro-blast) Actively mitotic cell of cartilage.

Chondrocyte (kon′dro-sīt) Mature cell of cartilage.

Chorion (kor′e-on) Outermost fetal membrane; helps form the placenta.

Chorionic villi sampling (ko″re-on′ik vil′i) Fetal testing procedure in which bits of the chorionic villi from the placenta are snipped off and the cells karyotyped. This procedure can be done as early as 8 weeks into the pregnancy.

Choroid (ko′roid) The vascular middle layer of the eye.

Choroid plexus (ko′roid plex′sus) A capillary knot that protrudes into a brain ventricle; involved in forming cerebrospinal fluid.

Chromatin (kro′mah-tin) Structures in the nucleus that carry the hereditary factors (genes).

Chromosomes (kro′mo-somz) Barlike bodies of tightly coiled chromatin; visible during cell division.

Chronic obstructive pulmonary disease (COPD) Collective term for progressive, obstructive respiratory disorders; includes emphysema, chronic bronchitis.

Chyme (kīm) Semifluid, creamy mass consisting of partially digested food and gastric juice.

Cilia (sil′e-ah) Tiny, hairlike projections on cell surfaces that move in a wavelike manner.

Circumduction (ser″kum-duk′shun) Movement of a body part so that it outlines a cone in space.

Cirrhosis (sĭ-ro′sis) Chronic disease of the liver, characterized by an overgrowth of connective tissue or fibrosis.

Cisterna chyli (sis-ter′nah ki′li) An enlarged sac at the base of the thoracic duct; the origin of the thoracic duct.

Cisternae (sis-ter′ne) Any cavity or enclosed space serving as a reservoir.

Cleavage An early embryonic phase consisting of rapid mitotic cell divisions without intervening growth periods; product is a blastocyst.

Clonal selection (klo′nul) Process during which a B cell or T cell becomes activated by binding contact with an antigen.

Clone Descendants of a single cell.

Coagulation Process in which blood is transformed from a liquid to a gel; blood clotting.

Cochlea (kok′le-ah) Snail-shaped chamber of the bony labyrinth that houses the receptor for hearing [the spiral organ (of Corti)].

Codon (ko′don) The three-base sequence on a messenger RNA molecule that provides the genetic information used in protein synthesis; code for a given amino acid.

Coenzyme (ko-en′zīm) Nonprotein substance associated with and activating an enzyme, typically a vitamin.

Cofactor Metal ion or organic molecule that is required for enzyme activity.

Collagen fiber The most abundant of the three fibers found in the matrix of connective tissue.

Colloid (kol′oid) (1) A mixture in which the solute particles (usually proteins) do not settle out readily. (2) Substance in the thyroid gland containing thyroglobulin protein.

Colloid osmotic pressure (kol′oid ahz-mah′tik) Pressure created in a fluid by large nondiffusible molecules, such as plasma proteins that are prevented from moving through a (capillary) membrane. Such substances tend to draw water to them.

Colon Regions of the large intestine; includes ascending, transverse, descending, and sigmoid portions.

Combination (synthesis) reaction Chemical reaction in which larger, more complex atoms or molecules are formed from simpler ones.

Complement A group of bloodborne proteins, which, when activated, enhance the inflammatory and immune responses and may lead to cell lysis.

Complementarity of structure and function The relationship between a structure and its function; i.e., structure determines function.

Complementary base Refers to how a given nitrogenous base of DNA or RNA bonds to another nitrogenous base. For example, adenine (A) is the complementary base of thymine (T). The result is base pairing.

Complete blood count (CBC) Clinical test that includes counts of all formed elements, a hematocrit, and measurements of erythrocyte size and hemoglobin content.

Compound Substance composed of two or more different elements, the atoms of which are chemically united.

Concentration gradient The difference in the concentration of a particular substance between two different areas.

Conducting zone Includes all respiratory passageways that provide conduits for air to reach the sites of gas exchange (the respiratory zone).

Conductivity Ability to transmit an electrical impulse.

Cones One of the two types of photoreceptor cells in the retina of the eye; provide for color vision.

Congenital (kun-jeh′nih-tul) Existing at birth.

Congestive heart failure (CHF) Condition in which the pumping efficiency of the heart is depressed so that circulation is inadequate to meet tissue needs.

Conjunctiva (kon″junk-ti′vah) Thin, protective mucous membrane lining the eyelids and covering the anterior surface of the eye itself.

Connective tissue A primary tissue; form and function vary extensively. Functions include support, storage, and protection.

Consciousness The ability to perceive, communicate, remember, understand, appreciate, and initiate voluntary movements.

Contraception The prevention of conception; birth control.

Contractility Muscle cell's ability to move by shortening.

Contraction To shorten or develop tension, an ability highly developed in muscle cells.

Contralateral Relating to the opposite side.

Cornea (kor′ne-ah) Transparent anterior portion of the eyeball; part of the fibrous layer.

Corona radiata (kor-o′nah ra-de-ah′tah) (1) Arrangement of elongated follicle cells around a mature ovum; (2) crownlike arrangement of nerve fibers radiating from the internal capsule of the brain to every part of the cerebral cortex.

Coronary circulation The functional blood supply of the heart; shortest circulation in the body.

Cortex (kor′teks) Outer surface layer of an organ.

Corticosteroids (kor″tĭ-ko-stĕ′roidz) Steroid hormones released by the adrenal cortex.

Cortisol (hydrocortisone) (kor′tih-sol) Glucocorticoid produced by the adrenal cortex.

Covalent bond (ko-va′lent) Chemical bond created by electron sharing between atoms.

Cranial nerves The 12 nerve pairs that arise from the brain.

Craniosacral division Another name for the parasympathetic division of the autonomic nervous system.

Cranium (cranial bones) (kra′ne-um) Bony protective encasement of the brain and organs of hearing and equilibrium; also called the skull.

Creatine kinase (kre′ah-tin) Enzyme that catalyzes the transfer of phosphate from creatine phosphate to ADP, forming creatine and ATP; important in muscle contraction.

Creatine phosphate (CP) (fos-fāt) Compound that serves as an alternative energy source for muscle tissue.

Creatinine (kre-at′ĭ-nin) A nitrogenous waste molecule which is not reabsorbed by the kidney; this characteristic makes it useful for measurement of the GFR and glomerular function.

Crista ampullaris Sensory receptor organ within the ampulla of each semicircular canal of the inner ear; dynamic equilibrium receptor.

Cross section A cut running horizontally from right to left, dividing the body or an organ into superior and inferior parts.

Cutaneous (ku-ta′ne-us) Pertaining to the skin.

Cutaneous sensory receptors Receptors located throughout the skin that respond to stimuli arising outside the body; part of the nervous system.

Cyclic AMP Intracellular second messenger that mediates the effects of the first (extracellular) messenger (hormone or neurotransmitter); formed from ATP by a plasma membrane enzyme (adenylate cyclase).

Cystic fibrosis (CF) Genetic disorder in which secretion of overly viscous mucus clogs the respiratory passages, predisposes to fatal respiratory infections.

Cytochromes (si′to-krōmz) Brightly colored iron-containing proteins that form part of the inner mitochondrial membrane and function as electron carriers in oxidative phosphorylation.

Cytokines Small proteins that act as chemical messengers between various parts of the immune system.

Cytokinesis (si″to-kĭ-ne′sis) The division of cytoplasm that occurs after the cell nucleus has divided.

Cytoplasm (si′to-plazm) The cellular material surrounding the nucleus and enclosed by the plasma membrane.

Cytosine (C) (si′to-sēn) Nitrogen-containing base that is part of a nucleotide structure.

Cytoskeleton Literally, cell skeleton. An elaborate series of rods running through the cystol, supporting cellular structures and providing the machinery to generate various cell movements.

Cytosol Viscous, semitransparent fluid substance of cytoplasm in which other elements are suspended.

Cytotoxic T cell (T_C cell) Effector T cell that directly kills foreign cells, cancer cells, or virus-infected body cells by inducing apoptosis (cell suicide).

Deamination (de′am″ih-na′shun) Removal of an amine group from an organic compound.

Decomposition reaction Chemical reaction in which a molecule is broken down into smaller molecules or its constituent atoms.

Defecation (def″ih-ka′shun) Elimination of the contents of the bowels (feces).

Deglutition (deg″loo-tish′un) Swallowing.

Dehydration (de″hi-dra′shun) Condition of excessive water loss.

Dehydration synthesis Process by which a large molecule is synthesized by removing water and covalently bonding smaller molecules together.

Dendrite (den′drīt) Branching neuron process that serves as a receptive, or input, region; transmits an electrical signal toward the cell body.

Dendritic cells Protective cells that phagocytize antigens, migrate to lymph nodes, and present the antigen to T cells, causing them to activate and mount an immune response; called epidermal dendritic cells or Langerhans cells in the skin.

Depolarization (de-po″ler-ah-za′shun) Loss of a state of polarity; loss or reduction of negative membrane potential.

Dermatome (der′mah-tōm) Portion of somite mesoderm that forms the dermis of the skin; also the area of skin innervated by the cutaneous branches of a single spinal nerve.

Dermis Layer of skin deep to the epidermis; composed mostly of dense irregular connective tissue.

Desmosome (dez′muh-sōm) Cell junction composed of thickened plasma membranes joined by filaments.

Diabetes insipidus (di″ah-be′tēz in-sih′pih-dus) Disease characterized by passage of a large quantity of dilute urine plus intense thirst and dehydration caused by inadequate release of antidiuretic hormone (ADH).

Diabetes mellitus (DM) (meh-li′tus) Disease caused by deficient insulin release or by insulin resistance, leading to inability of the body cells to use carbohydrates.

Dialysis (di-al′ah-sis) Diffusion of solute(s) through a semipermeable membrane.

Diapedesis (di″ah-pĕ-de′sis) Passage of white blood cells through intact vessel walls into tissue.

Diaphragm (di′ah-fram) (1) Any partition or wall separating one area from another; (2) a muscle that separates the thoracic cavity from the lower abdominopelvic cavity.

Diaphysis (di-af′ĭ-sis) Elongated shaft of a long bone.

Diarthrosis (di″ar-thro′sis) Freely movable joint.

Diastole (di-as′to-le) Period of the cardiac cycle when either the ventricles or the atria are relaxing.

Diastolic pressure (di-as-tah′lik) Arterial blood pressure reached during or as a result of diastole; lowest level of any given cardiac cycle.

Diencephalon (interbrain) (di″en-seh′fuh-lon) That part of the forebrain between the cerebral hemispheres and the midbrain including the thalamus, the epithalamus, and the hypothalamus.

Differential white blood cell count Diagnostic test to determine relative proportion of individual leukocyte types.

Diffusion (dĭ-fu′zhun) The spreading of particles in a gas or solution with a movement toward uniform distribution of particles; driven by kinetic energy.

Digestion Chemical or mechanical process of breaking down foodstuffs to substances that can be absorbed.

Digestive system System that processes food into absorbable units and eliminates indigestible wastes.

Dipeptide A combination of two amino acids united by means of a peptide bond.

Diploë (dip′lo-e) The internal layer of spongy bone in flat bones.

Diploid chromosomal number The chromosomal number characteristic of an organism, symbolized as $2n$; twice the chromosomal number (n) of the gamete; in humans, $2n = 46$.

Diplopia (dĭ-plo′pe-ah) Double vision.

Dipole (polar molecule) Nonsymmetrical molecules that contain electrically unbalanced atoms.

Disaccharide (di-sak′ah-rīd″, di-sak′ah-rid) Literally, double sugar; e.g., sucrose, lactose.

Dislocation (luxation) Occurs when bones are forced out of their normal alignment at a joint.

Displacement (exchange) reaction Chemical reaction in which bonds are both made and broken; atoms become combined with different atoms.

Distal (dis′tul) Away from the attached end of a limb or the origin of a structure.

Diuretics (di″u-ret′iks) Chemicals that enhance urinary output.

Diverticulum (di″ver-tik′u-lum) A pouch or sac in the walls of a hollow organ or structure.

DNA (deoxyribonucleic acid) (de-ok″sĭ-ri″bo-nu-kla′ik) A nucleic acid found in all living cells; it carries the organism's hereditary information.

DNA replication Process that occurs before cell division; ensures that all daughter cells have identical genes.

Dominant traits Occurs when one allele masks or suppresses the expression of its partner.

Dominant-recessive inheritance Reflects the interaction of dominant and recessive alleles.

Dorsal (dor′sul) Pertaining to the back; posterior.

Dorsal root ganglion Peripheral collection of cell bodies of first-order afferent neurons whose central axons enter the spinal cord.

Double helix The secondary structure assumed by two strands of DNA, held together throughout their length by hydrogen bonds between bases on opposite strands.

Duct (dukt) A canal or passageway; a tubular structure that provides an exit for the secretions of a gland, or for conducting any fluid.

Ductus (vas) deferens Extends from the epididymis to the urethra; propels sperm into the urethra by peristalsis during ejaculation.

Duodenum (du″o-de′num) First part of the small intestine.

Dura mater (du′rah ma′ter) Outermost and toughest of the three membranes (meninges) covering the brain and spinal cord.

Dynamic equilibrium Sense that reports on angular (rotatory) acceleration or deceleration of the head in space.

Dyskinesia (dis-kĭ-ne′ze-ah) Disorders of muscle tone, posture, or involuntary movements.

Dyspnea (disp-ne′ah) Difficult or labored breathing; air hunger.

Eccrine glands (ek′rin) Sweat glands abundant on the palms, soles of feet, and the forehead.

Ectoderm (ek′to-derm) Embryonic germ layer; forms the epidermis of the skin and its derivatives, and nervous tissues.

Edema (ĕ-de′mah) Abnormal increase in the amount of interstitial fluid; causes swelling.

Effector (ef-ek′ter) Organ, gland, or muscle capable of being activated by nerve endings.

Efferent (ef′er-ent) Carrying away or away from, especially a nerve fiber that carries impulses away from the central nervous system.

Elastic cartilage Cartilage with abundant elastic fibers; more flexible than hyaline cartilage.

Elastic fiber Fiber formed from the protein elastin, which gives a rubbery and resilient quality to the matrix of connective tissue.

Electrical energy Energy formed by the movement of charged particles across cell membranes.

Electrocardiogram (ECG or **EKG)** (e-lek″tro-car′de-o-gram″) Graphic record of the electrical activity of the heart.

Electrochemical gradient The combined difference in concentration and charge; influences the distribution and direction of diffusion of ions.

Electroencephalogram (EEG) (e-lek″tro-en-sef′ah-lo-gram″) Graphic record of the electrical activity of nerve cells in the brain.

Electrolyte (e-lek′tro-līt) Chemical substances, such as salts, acids, and bases, that ionize and dissociate in water and are capable of conducting an electrical current.

Electrolyte balance Refers to the balance between input and output of salts (sodium, potassium, calcium, magnesium) in the body.

Electromagnetic radiation Emitted photons (wave packets) of energy, e.g., light, X ray, infrared.

Electron Negatively charged subatomic particle; orbits the atom's nucleus.

Electron shells (energy levels) Regions of space that consecutively surround the nucleus of an atom.

Element One of a limited number of unique varieties of matter that composes substances of all kinds; e.g., carbon, hydrogen, oxygen.

Embolism (em′bo-lizm) Obstruction of a blood vessel by an embolus (blood clot, fatty mass, bubble of air, or other debris) floating in the blood.

Embryo (em′bre-o) Developmental stage extending from fertilization to the end of the eighth week.

Emesis Reflexive emptying of the stomach through the esophagus and pharynx; also known as vomiting.

Encephalitis (en″seh-fuh-lī′tis) Inflammation of the brain.

Endergonic reaction Chemical reaction that absorbs energy, e.g., an anabolic reaction.

Endocardium (en″do-kar′de-um) Endothelial membrane that lines the interior of the heart.

Endochondral ossification (en″do-kon′dral) Embryonic formation of bone by the replacement of calcified cartilage; most skeletal bones are formed by this process.

Endocrine glands (en′do-krin) Ductless glands that empty their hormonal products directly into the blood.

Endocrine system Body system that includes internal organs that secrete hormones.

Endocytosis (en″do-si-to′sis) Means by which fairly large extracellular molecules or particles enter cells, e.g., phagocytosis, pinocytosis, receptor-mediated endocytosis.

Endoderm (en′do-derm) Embryonic germ layer; forms the lining of the digestive tube and its associated structures.

Endogenous (en-doj′ĕ-nŭs) Originating or produced within the organism or one of its parts.

Endometrium (en″do-me′tre-um) Mucous membrane lining of the uterus.

Endomysium (en″do-mis′e-um) Thin connective tissue surrounding each muscle cell.

Endoplasmic reticulum (ER) (en″do-plaz′mik rĕ-tik′u-lum) Membranous network of tubular or saclike channels in the cytoplasm of a cell.

Endosteum (en-dos′te-um) Connective tissue membrane covering internal bone surfaces.

Endothelium (en″do-the′le-um) Single layer of simple squamous cells that line the walls of the heart, blood vessels, and lymphatic vessels.

Energy The capacity to do work; may be stored (potential energy) or in action (kinetic energy).

Energy intake Energy liberated during food oxidation.

Energy output Sum of energy lost as heat, as work, and as fat or glycogen storage.

Enzyme (en′zīm) A protein that acts as a biological catalyst to speed up a chemical reaction.

Eosinophil (e″o-sin′o-fil) Granular white blood cell whose granules readily take up an acid stain called eosin.

Ependymal cell (ĕ-pen′dĭ-mul) A type of CNS supporting cell; lines the central cavities of the brain and spinal cord.

Epidermis (ep″ĭ-der′mis) Superficial layer of the skin; composed of keratinized stratified squamous epithelium.

Epididymis (ep″ĭ-dĭ′dĭ-mis) That portion of the male duct system in which sperm mature. Empties into the ductus (or vas) deferens.

Epidural space Area between the bony vertebrae and the dura mater of the spinal cord.

Epiglottis (eh″puh-glah′tis) Elastic cartilage at the back of the throat; covers the opening of the larynx during swallowing.

Epileptic seizures Abnormal electrical discharges of groups of brain neurons, during which no other messages can get through.

Epimysium (ep″ĭ-mis′e-um) Sheath of fibrous connective tissue surrounding a muscle.

Epinephrine (ep″ĭ-nef′rin) Chief hormone produced by the adrenal medulla. Also called adrenaline.

Epiphyseal plate (e″pĭ-fis′e-ul) Plate of hyaline cartilage at the junction of the diaphysis and epiphysis that provides for growth in length of a long bone.

Epiphysis (e-pif′ĭ-sis) The end of a long bone, attached to the shaft.

Epithalamus Most dorsal portion of the diencephalon; forms the roof of the third ventricle with the pineal gland extending from its posterior border.

Epithelium (epithelial tissue) (ep″ĭ-the′le-ul) Pertaining to a primary tissue that covers the body surface, lines its internal cavities, and forms glands.

Erythrocytes (e-rith′ro-sīts) Red blood cells.

Erythropoiesis (ĕ-rith″ro-poi-e′sis) Process of erythrocyte formation.

Erythropoietin (EPO) (ĕ-rith″ro-poi′ĕ-tin) Hormone that stimulates production of red blood cells.

Esophagus (ĕ-sof′ah-gus) Muscular tube extending from the laryngopharynx through the diaphragm to join the stomach; collapses when not involved in food propulsion.

Estrogens (es′tro-jenz) Hormones that stimulate female secondary sex characteristics; female sex hormones.

Eupnea (ūp-ne′ah) Normal respiratory rate and rhythm.

Exchange (displacement) reaction Chemical reaction in which bonds are both made and broken; atoms become combined with different atoms.

Excitability (irritability) Ability to respond to stimuli.

Excitation-contraction (E-C) coupling Sequence of events by which transmission of an action potential along the sarcolemma leads to the sliding of myofilaments.

Excitatory postsynaptic potential (EPSP) Depolarizing graded potential in a postsynaptic neuron.

Excretion (ek-skre′shun) Elimination of waste products from the body.

Exergonic reaction Chemical reaction that releases energy, e.g., a catabolic or oxidative reaction.

Exocrine glands (ek′so-krin) Glands that have ducts through which their secretions are carried to a particular site.

Exocytosis (ek″so-si-to′sis) Mechanism by which substances are moved from the cell interior to the extracellular space as a secretory vesicle fuses with the plasma membrane.

Exons Amino acid–specifying informational sequences (separated by introns) in the genes of higher organisms.

Extension Movement that increases the angle of a joint, e.g., straightening a flexed knee.

Exteroceptor (ek″ster-o-sep′tor) Sensory end organ that responds to stimuli from the external world.

Extracellular fluid (ECF) Internal fluid located outside cells; includes interstitial fluid, blood plasma, and cerebrospinal fluid.

Extracellular matrix Nonliving material in connective tissue consisting of ground substance and fibers that separates the living cells.

Extrasystole (ek″strah-sis′to-le) Premature heart contraction.

Extrinsic (ek-strin′sik) Of external origin.

Extrinsic eye muscles The six skeletal muscles which attach to and move each eye.

Facilitated diffusion Passive transport process used by certain molecules, e.g., glucose and other simple sugars too large to pass through plasma membrane pores. Involves movement through channels or movement facilitated by a membrane carrier.

Fallopian tube (fah-lō′pē-un) *See* Uterine tube.

Fascia (fash′e-ah) Layers of fibrous tissue covering and separating muscle.

Fascicle (fas′ĭ-kl) Bundle of nerve or muscle fibers bound together by connective tissue.

Fatty acids Linear chains of carbon and hydrogen atoms (hydrocarbon chains) with an organic acid group at one end. A constituent of fat.

Feces (fe′sēz) Material discharged from the bowel; composed of food residue, secretions, bacteria.

Fenestrated (fen′es-tra-tid) Pierced with one or more small openings.

Fertilization Fusion of the sperm and egg nuclei.

Fetus Developmental stage extending from the ninth week of development to birth.

Fiber A slender threadlike structure or filament. *See also* Nerve fiber, Muscle fiber.

Fibrillation Condition of rapid and irregular or out-of-phase heart contractions.

Fibrin (fi′brin) Fibrous insoluble protein formed during blood clotting.

Fibrinogen (fi-brin′o-jin) A soluble blood protein that is converted to insoluble fibrin during blood clotting.

Fibrinolysis Process that removes unneeded blood clots when healing has occurred.

Fibroblast (fi′bro-blast) Young, actively mitotic cell that forms the fibers of connective tissue.

Fibrocartilage The most compressible type of cartilage; resistant to stretch. Forms vertebral discs and knee joint cartilages.

Fibrocyte (fi′bro-sīt) Mature fibroblast; maintains the matrix of fibrous types of connective tissue.

Fibrosis Proliferation of fibrous connective tissue called scar tissue.

Fibrous joints Bones joined by fibrous tissue; no joint cavity is present.

Filtrate A plasma-derived fluid that is processed by the renal tubules to form urine.

Filtration Passage of a solvent and dissolved substances through a membrane or filter.

First-degree burn A burn in which only the epidermis is damaged.

Fissure (fih′sher) (1) A groove or cleft; (2) the deepest depressions or inward folds on the brain.

Fixator (fix′a-ter) Muscle that immobilizes one or more bones, allowing other muscles to act from a stable base.

Flagellum (flah-jel′lum) Long, whiplike cellular extension containing microtubules; propels sperm and some single-celled eukaryotes.

Flexion (flek′shun) Movement that decreases the angle of the joint, e.g., bending the knee from a straight to an angled position.

Flexor (withdrawal) reflex Reflex initiated by a painful stimulus (actual or perceived); causes automatic withdrawal of the threatened body part from the stimulus.

Fluid mosaic model A depiction of the structure of the membranes of a cell as phospholipid bilayers in which proteins are dispersed.

Follicle (fah′lih-kul) (1) Ovarian structure consisting of a developing egg surrounded by one or more layers of follicle cells; (2) colloid-containing structure of the thyroid gland; (3) B cell–rich region in lymphoid tissue.

Follicle-stimulating hormone (FSH) Hormone produced by the anterior pituitary that stimulates ovarian follicle production in females and sperm production in males.

Fontanelles (fon″tah-nelz′) Fibrous membranes at the angles of cranial bones that accommodate brain growth in the fetus and infant.

Foramen (fo-ra′men) Hole or opening in a bone or between body cavities.

Forebrain (prosencephalon) Anterior portion of the brain consisting of the telencephalon and the diencephalon.

Formed elements Cellular portion of blood.

Fossa (fos′ah) A depression, often an articular surface.

Fovea (fo′ve-ah) A pit.

Fracture A break in a bone.

Free radicals Highly reactive chemicals with unpaired electrons that can scramble the structure of proteins, lipids, and nucleic acids.

Frontal (coronal) plane Longitudinal (vertical) plane that divides the body or an organ into anterior and posterior parts.

Fulcrum The fixed point on which a lever moves when a force is applied.

Fundus (fun′dus) Base of an organ; part farthest from the opening of the organ. For example, the posterior wall of the eye.

G protein Protein that relays signals between extracellular first messengers (hormones or neurotransmitters) and intracellular second messengers (such as cyclic AMP) via an effector enzyme.

Gallbladder Sac beneath the right lobe of the liver used for bile storage.

Gallstones (biliary calculi) Crystallized cholesterol that obstructs the flow of bile from the gallbladder.

Gamete (gam′ēt) Sex or germ cell.

Gametogenesis (gam″eh-to-jen′eh-sis) Formation of gametes.

Ganglion (gang′gle-on) Collection of nerve cell bodies outside the CNS.

Ganglionic neuron (gan″gle-ah′nik) Autonomic motor neuron that has its cell body in a peripheral ganglion and projects its (postganglionic) axon to an effector.

Gap junction A passageway between two adjacent cells; formed by transmembrane proteins called connexons.

Gastrin Hormone secreted in the stomach; regulates gastric juice secretion by stimulating HCl production.

Gastroenteritis Inflammation of the gastrointestinal tract.

Gastrulation (gas″troo-la′shun) Developmental process that produces the three primary germ layers (ectoderm, mesoderm, and endoderm).

Gene One of the biological units of heredity located in chromatin; transmits hereditary information.

Genetic code Refers to the rules by which the base sequence of a DNA gene is translated into protein structures (amino acid sequences).

Genitalia (jen″i-ta′le-ă) The internal and external reproductive organs.

Genome The complete set of chromosomes derived from one parent (the haploid genome); or the two sets of chromosomes, i.e., one set from the egg, the other from the sperm (the diploid genome).

Genotype (jen′o-tīp) One's genetic makeup or genes.

Germ layers Three cellular layers (ectoderm, mesoderm, and endoderm) that represent the initial specialization of cells in the embryonic body and from which all body tissues arise.

Gestation period (jes-ta′shun) The period of pregnancy; about 280 days for humans.

Gland Organ specialized to secrete or excrete substances for further use in the body or for elimination.

Glaucoma (glaw-ko′mah) Condition in which intraocular pressure increases to levels that cause compression of the retina and optic nerve; results in blindness unless detected early.

Glial cells (gle′al) *See* Neuroglia.

Glomerular capsule (glo-mer′yoo-ler) Double-walled cup at end of a renal tubule; encloses a glomerulus. Also called Bowman's capsule.

Glomerular filtration rate (GFR) Rate of filtrate formation by the kidneys.

Glomerulus (glo-mer′u-lus) (1) Cluster of capillaries forming part of the nephron; forms filtrate; (2) odor-specific processing unit in olfactory bulb.

Glottis (glah′tis) Opening between the vocal cords in the larynx.

Glucagon (gloo′kah-gon) Hormone formed by alpha cells of pancreatic islets; raises the glucose level of blood.

Glucocorticoids (gloo″ko-kor′tĭ-koidz) Adrenal cortex hormones that increase blood glucose levels and aid the body in resisting long-term stressors.

Gluconeogenesis (gloo″ko-ne″o-jen′ĕ-sis) Formation of glucose from noncarbohydrate molecules.

Glucose (gloo′kōs) Principal blood sugar; a hexose.

Glycerol (glis′er-ol) A modified simple sugar (a sugar alcohol); a building block of fats.

Glycocalyx (cell coat) (gli″ko-kal′iks) A layer of externally facing glycoproteins on a cell's plasma membrane that determines blood type; involved in the cellular interactions of fertilization, embryonic development, and immunity, and acts as an adhesive between cells.

Glycogen (gli′ko-jin) Main carbohydrate stored in animal cells; a polysaccharide.

Glycogenesis (gli″ko-jen′ĕ-sis) Formation of glycogen from glucose.

Glycogenolysis (gli″ko-jĕ-nol′ĭ-sis) Breakdown of glycogen to glucose.

Glycolipid (gli″ko-lip′id) A lipid with one or more covalently attached sugars.

Glycolysis (gli-kol′ĭ-sis) Breakdown of glucose to pyruvic acid—an anaerobic process.

Goblet cells Individual cells (unicellular glands) that produce mucus.

Golgi apparatus (gol′je) Membranous system close to the cell nucleus that packages protein secretions for export, packages enzymes into lysosomes for cellular use, and modifies proteins destined to become part of cellular membranes.

Golgi tendon organs Proprioceptors located in tendons; monitor muscle tension to prevent tearing and help smooth onset and termination of muscle contraction.

Gonad (go′nad) Primary reproductive organ; i.e., the testis of the male or the ovary of the female.

Gonadocorticoids (gon″ah-do-kor′tĭ-koidz) Sex hormones, primarily androgens, secreted by the adrenal cortex.

Gonadotropins (gon″ah-do-trōp′inz) Gonad-stimulating hormones produced by the anterior pituitary.

Graafian follicle (graf′e-an) *See* Vesicular follicle.

Graded muscle responses Variations in the degree of muscle contraction by changing either the frequency or strength of the stimulus.

Graded potential A local change in membrane potential that varies directly with the strength of the stimulus, declines with distance.

Graves' disease Disorder resulting from hyperactive thyroid gland.

Gray matter Gray area of the central nervous system; contains neuronal cell bodies and their dendrites.

Growth hormone (GH) Hormone that stimulates growth in general; produced in the anterior pituitary; also called somatotropin (STH).

Guanine (G) (gwan′ēn) One of two major purines occurring in all nucleic acids.

Gustation (gus-ta′shun) Taste.

Gyrus (ji′rus) An outward fold of the surface of the cerebral cortex.

Hair follicle Structure with outer and inner root sheaths extending from the epidermal surface into the dermis and from which new hair develops.

Hapten (hap′ten) An incomplete antigen; has reactivity but not immunogenicity.

Haversian system (hah-ver′zhen) *See* Osteon.

Heart attack (coronary) *See* Myocardial infarction.

Heart block Impaired transmission of impulses from atrium to ventricle resulting in abnormally slow heart rhythms.

Heart murmur Abnormal heart sound (usually resulting from valve problems).

Heimlich maneuver Procedure in which the air in a person's own lungs is used to expel an obstructing piece of food.

Helper T cell Type of T lymphocyte that orchestrates cellular immunity by direct contact with other immune cells and by releasing chemicals called cytokines; also helps to mediate the humoral response by interacting with B cells.

Hematocrit (he-mat′o-krit) The percentage of total blood volume occupied by erythrocytes.

Hematoma (he″mah-to′mah) Mass of clotted blood that forms at an injured site.

Hematopoiesis (hem″ah-to-poi-e′sis) Blood cell formation; hemopoiesis.

Heme (hēm) Iron-containing pigment that is essential to oxygen transport by hemoglobin.

Hemocytoblast (he″mo-si′to-blast) Bone marrow cell that gives rise to all the formed elements of blood; hematopoietic stem cell.

Hemoglobin (he′muh-glo-bin) Oxygen-transporting protein of erythrocytes.

Hemolysis (he-mah′lĕ-sis) Rupture of erythrocytes.

Hemophilia (he″mo-fil′e-ah) A term loosely applied to several different hereditary bleeding disorders that exhibit similar signs and symptoms.

Hemopoiesis (he″mo-poi-e′sis) *See* Hematopoiesis.

Hemorrhage (hem′or-ij) Loss of blood from the vessels by flow through ruptured walls; bleeding.

Hemostasis (he″mo-sta′sis) Stoppage of bleeding.

Heparin Natural anticoagulant secreted into blood plasma.

Hepatic portal system (hĕ-pat′ik) Circulation in which the hepatic portal vein carries dissolved nutrients to the liver tissues for processing.

Hepatitis (hep″ah-ti′tis) Inflammation of the liver.

Hernia (her′ne-ah) Abnormal protrusion of an organ or a body part through the containing wall of its cavity.

Heterozygous (het″er-o-zi′gus) Having different allelic genes at a given locus or (by extension) many loci.

Hilton's law Any nerve serving a muscle that produces movement at a joint also innervates the joint and the skin over the joint.

Hilum (hi′lum) The indented region of an organ from which blood and/or lymphatic vessels and nerves enter and exit.

Hippocampus Limbic system structure that plays a role in converting new information into long-term memories.

Histamine (his′tuh-mēn) A chemical messenger (neurotransmitter or paracrine); causes vasodilation and increased capillary permeability; in stomach causes acid secretion.

Histology (his-tol′o-je) Branch of anatomy dealing with the microscopic structure of tissues.

HIV (human immunodeficiency virus) Virus that destroys helper T cells, thus depressing adaptive immunity; symptomatic AIDS gradually appears when lymph nodes can no longer contain the virus.

Holocrine glands (hol′o-krin) Glands that accumulate their secretions within their cells; secretions are discharged only upon rupture and death of the cell.

Homeostasis (ho″me-o-sta′sis) A state of body equilibrium or stable internal environment of the body.

Homologous (ho-mol′ŏ-gus) Parts or organs corresponding in structure but not necessarily in function.

Homozygous (ho-mo-zi′gus) Having identical genes at one or more loci.

Hormones Steroidal or amino acid–based molecules released to the blood that act as chemical messengers to regulate specific body functions.

Humoral immunity (hu′mer-ul) Immunity conferred by antibodies present in blood plasma and other body fluids.

Huntington's disease Hereditary disorder leading to degeneration of the basal nuclei and the cerebral cortex.

Hyaline cartilage (hi′ah-līn) The most abundant cartilage type in the body; provides firm support with some pliability.

Hydrochloric acid (HCl) (hi″dro-klor′ik) Acid that aids protein digestion in the stomach; produced by parietal cells.

Hydrogen bond Weak bond in which a hydrogen atom forms a bridge between two electron-hungry atoms. An important intramolecular bond.

Hydrogen ion (H^+) A hydrogen atom minus its electron and therefore carrying a positive charge (i.e., a proton).

Hydrolysis (hi″drah′lă-sis) Process in which water is used to split a substance into smaller particles.

Hydrophilic (hi″dro-fil′ik) Refers to molecules, or portions of molecules, that interact with water and charged particles.

Hydrophobic (hi″dro-fo′bik) Refers to molecules, or portions of molecules, that interact only with nonpolar molecules.

Hydrostatic pressure (hi″dro-stă′tic) Pressure of fluid in a system.

Hydroxyl ion (OH^-) (hi-drok′sil) An ion liberated when a hydroxide (a common inorganic base) is dissolved in water.

Hyperalgesia Pain amplification.

Hypercapnia (hi″per-kap′ne-ah) High carbon dioxide levels in the blood.

Hyperemia An increase in blood flow into a tissue or organ; congested with blood.

Hyperglycemic (hi″per-gli-se′mik) Term used to describe hormones such as glucagon that elevate blood glucose level.

Hyperopia (hi″per-o′pe-ah) A condition in which visual images are routinely focused behind rather than on the retina; commonly known as farsightedness.

Hyperplasia (hi″per-pla′ze-ah) Accelerated growth, e.g., in anemia, the bone marrow produces red blood cells at a faster rate.

Hyperpnea (hi″perp-ne′ah) An increase in ventilation in response to metabolic need (e.g., during exercise).

Hyperpolarization An increase in membrane potential in which the membrane becomes more negative than resting membrane potential.

Hypersensitivity Overzealous immune response to an otherwise harmless antigen.

Hypertension (hi″per-ten′shun) High blood pressure.

Hypertonic (hi″per-ton′ik) Excessive, above normal, tone or tension.

Hypertonic solution A solution that has a higher concentration of nonpenetrating solutes than the reference cell; having greater osmotic pressure than the reference solution (blood plasma or interstitial fluid).

Hypertrophy (hi-per′trah-fe) Increase in size of a tissue or organ independent of the body's general growth.

Hyperventilation An increase in the depth and rate of breathing that is in excess of the body's need for removal of carbon dioxide.

Hypocapnia Low carbon dioxide levels in the blood.

Hypodermis (superficial fascia) Subcutaneous tissue just deep to the skin; consists of adipose plus some areolar connective tissue.

Hypoglycemic (hi″po-gli-se′mik) Term used to describe hormones such as insulin that decrease blood glucose level.

Hyponatremia Abnormally low concentrations of sodium ions in extracellular fluid.

Hypoproteinemia (hi″po-pro″te-ĭ-ne′me-ah) A condition of unusually low levels of plasma proteins causing a reduction in colloid osmotic pressure; results in tissue edema.

Hypotension Low blood pressure.

Hypothalamic-hypophyseal tract (hi″po-thah-lam′ik–hi″po-fiz′-e-al) Nerve bundles that run through the infundibulum and connect the neurohypophysis and the hypothalamus.

Hypothalamus (hi″po-thal′ah-mus) Region of the diencephalon forming the floor of the third ventricle of the brain.

Hypotonic (hi″po-ton′ik) Below normal tone or tension.

Hypotonic solution A solution that is more dilute (containing fewer nonpenetrating solutes) than the reference cell. Cells placed in hypotonic solutions plump up rapidly as water rushes into them.

Hypoventilation A decrease in the depth and rate of breathing; characterized by an increase in blood carbon dioxide.

Hypovolemic shock (hi″po-vo-le′mik) Most common form of shock; results from extreme blood loss.

Hypoxia (hi-pok′se-ah) Condition in which inadequate oxygen is available to tissues.

Ileocecal valve (il″e-o-se′kal) Site where the small intestine joins the large intestine.

Ileum (il′e-um) Terminal part of the small intestine; between the jejunum and the cecum of the large intestine.

Immune system A functional system whose components attack foreign substances or prevent their entry into the body.

Immunity (im″ūn′ĭ-te) Ability of the body to resist many agents (both living and nonliving) that can cause disease; resistance to disease.

Immunocompetence Ability of the body's immune cells to recognize (by binding) specific antigens; reflects the presence of plasma membrane–bound receptors.

Immunodeficiency Any congenital or acquired condition causing a deficiency in the production or function of immune cells or certain molecules (complement, antibodies, etc.) required for normal immunity.

In vitro (in ve′tro) In a test tube, glass, or artificial environment.

In vivo (in ve′vo) In the living body.

Incompetent valve Valve which does not close properly.

Incontinence Inability to control micturition or defecation voluntarily.

Infarct (in′farkt) Region of dead, deteriorating tissue resulting from a lack of blood supply.

Infectious mononucleosis Highly contagious viral disease; marked by excessive agranulocytes.

Inferior (caudal) Pertaining to a position toward the lower or tail end of the long axis of the body.

Inferior vena cava Vein that returns blood from body areas below the diaphragm.

Inflammation (in″flah-ma′shun) An innate (nonspecific) defensive response of the body to tissue injury; includes dilation of blood vessels and an increase in vessel permeability; indicated by redness, heat, swelling, and pain.

Infundibulum (in″fun-dib′u-lum) (1) A stalk of tissue that connects the pituitary gland to the hypothalamus; (2) the distal end of the uterine (fallopian) tube.

Inguinal (ing′wĭ-nal) Pertaining to the groin region.

Inhibitory postsynaptic potential (IPSP) A graded potential in a postsynaptic neuron that inhibits action potential generation; usually hyperpolarizing.

Inner cell mass Accumulation of cells in the blastocyst from which the embryo develops.

Innervation (in″er-va′shun) Supply of nerves to a body part.

Inorganic compound Chemical substances that do not contain carbon, including water, salts, and many acids and bases.

Insertion Movable attachment of a muscle.

Insula Lobe of the cerebral cortex that is buried in the lateral sulcus beneath portions of the parietal, frontal, and temporal lobes.

Insulin A hormone that enhances the carrier-mediated diffusion of glucose into tissue cells, thus lowering blood glucose levels.

Insulin resistance State in which a greater than normal amount of insulin is required to maintain normal glucose blood levels.

Integration The process by which the nervous system processes and interprets sensory input and makes decisions about what should be done at each moment.

Integumentary system (in-teg″u-men′tar-e) Skin and its derivatives; provides the external protective covering of the body.

Intercalated discs (in-ter′kah-la″ted) Specialized connections between myocardial cells containing gap junctions and desmosomes.

Interferons (IFNs) (in-ter-fēr′ons) Proteins released from virus-infected (and other) cells that protect uninfected cells from viral takeover. Also inhibit some cancers.

Internal capsule Band of projection fibers that runs between the basal nuclei and the thalamus.

Internal respiration Exchange of gases between blood and tissue fluid and between tissue fluid and cells.

Interneuron (association neuron) Nerve cell located between motor and sensory neurons that shuttles signals through CNS pathways where integration occurs.

Interoceptor (in″ter-o-sep′tor) Sensory receptor nerve ending in the viscera, which is sensitive to changes and stimuli within the body's internal environment; also called visceroceptor.

Interphase One of two major periods in the cell life cycle; includes the period from cell formation to cell division.

Interstitial cells Cells located in the loose connective tissue surrounding the seminiferous tubules; they produce androgens (most importantly testosterone), which are secreted into the surrounding interstitial fluid.

Interstitial fluid (IF) (in″ter-stish′al) Fluid between the cells.

Interstitial lamellae Incomplete lamellae that lie between intact osteons, filling the gaps between forming osteons, or representing the remnants of an osteon that has been cut through by bone remodeling.

Intervertebral discs (in″ter-ver′teh-brul) Discs of fibrocartilage between vertebrae.

Intracapsular ligament Ligament located within and separate from the articular capsule of a synovial joint.

Intracellular fluid (ICF) (in″trah-sel′u-ler) Fluid within a cell.

Intrinsic factor Substance produced by the stomach that is required for vitamin B_{12} absorption.

Intron Noncoding segment or portion of DNA that ranges from 60 to 100,000 nucleotides long.

Involuntary muscle Muscle that cannot ordinarily be controlled voluntarily (e.g., smooth and cardiac muscle).

Involuntary nervous system The autonomic nervous system.

Ion (ī′on) Atom with a positive or negative electric charge.

Ionic bond (ī-ah′nik) Chemical bond formed by electron transfer between atoms.

Ipsilateral (ip″sih-lă′ter-ul) Situated on the same side.

Ischemia (is-ke′me-ah) Local decrease in blood supply.

Isograft Tissue graft donated by an identical twin.

Isomer (i′so-mer) One of two or more substances that has the same molecular formula but with its atoms arranged differently.

Isometric contraction (i″so-mĕ′trik) Contraction in which the muscle does not shorten (the load is too heavy) but its internal tension increases.

Isotonic contraction (i″so-tah′nik) Contraction in which muscle tension remains constant at a given joint angle and load, and the muscle shortens.

Isotonic solution A solution with a concentration of nonpenetrating solutes equal to that found in the reference cell.

Isotopes (i′so-tōps) Different atomic forms of the same element, vary only in the number of neutrons they contain; the heavier species tend to be radioactive.

Jejunum (jĕ-joo′num) The part of the small intestine between the duodenum and the ileum.

Joint (articulation) The junction of two or more bones.

Joint kinesthetic receptor (kin″es-thet′ik) Receptor that provides information on joint position and motion.

Juxtaglomerular apparatus (JGA) (juks″tah-glo-mer′u-lar) Cells of the distal part of the ascending limb of the loop of Henle and afferent arteriole located close to the glomerulus; involved in blood pressure regulation (via release of the hormone renin) and autoregulation of GFR.

Karyotype (kar′e-o-tīp) The diploid chromosomal complement, typically shown as homologous chromosome pairs arranged from longest to shortest (X and Y are arranged by size rather than paired).

Keratin (ker′ah-tin) Fibrous protein found in the epidermis, hair, and nails that makes those structures hard and water resistant; precursor is keratohyaline.

Ketones (ketone bodies) (ke′tōnz) Fatty acid metabolites; strong organic acids.

Ketosis (kē-tō′sis) Excess levels of ketone bodies in blood. Called ketoacidosis if blood pH is low.

Killer T cell *See* Cytotoxic T cell.

Kilocalories (kcal) *See* Calorie.

Kinetic energy (ki-net′ik) The energy of motion or movement, e.g., the constant movement of atoms, or the push given to a swinging door that sets it into motion.

Krebs cycle Aerobic metabolic pathway occurring within mitochondria, in which food metabolites are oxidized and CO_2 is liberated, and coenzymes are reduced. Also called the citric acid cycle.

Labia (la′be-ah) Lips; singular: labium.

Labor Collective term for the series of events that expel the infant from the uterus.

Labyrinth (lab′ĭ-rinth″) Bony cavities and membranes of the inner ear.

Lacrimal (lak′ri-mal) Pertaining to tears.

Lactation (lak-ta′shun) Production and secretion of milk.

Lacteal (lak′te-al) Special lymphatic capillaries of the small intestine that take up lipids.

Lactic acid (lak′tik) Product of anaerobic metabolism, especially in muscle.

Lacuna (lah-ku′nah) A small space, cavity, or depression; lacunae in bone or cartilage are occupied by cells.

Lamella (lah-mel′ah) A layer, such as of bone matrix in an osteon of compact bone.

Lamina (lam′ĭ-nah) (1) A thin layer or flat plate; (2) the portion of a vertebra between the transverse process and the spinous process.

Large intestine Portion of the digestive tract extending from the ileocecal valve to the anus; includes the cecum, appendix, colon, rectum, and anal canal.

Larynx (lar′ingks) Cartilaginous organ located between the trachea and the pharynx; voice box.

Latent period Period of time between stimulation and the onset of muscle contraction.

Lateral Away from the midline of the body.

Leptin Hormone released by fat cells that signals satiety.

Leukemia Refers to a group of cancerous conditions of white blood cells.

Leukocytes (loo′ko-sīts) White blood cells; formed elements involved in body protection that take part in inflammatory and immune responses.

Leukocytosis An increase in the number of leukocytes (white blood cells); usually the result of a microbiological attack on the body.

Leukopenia (loo″ko-pe′ne-ah) Abnormally low white blood cell count.

Leukopoiesis The production of white blood cells.

Lever system Consists of a lever (bone), effort (muscle action), resistance (weight of object to be moved), and fulcrum (joint).

Ligament (lig′ah-ment) Band of regular fibrous tissue that connects bones.

Ligands Signaling chemicals that bind specifically to membrane receptors.

Limbic system (lim′bik) Functional brain system involved in emotional response and memory formation.

Lipid (lih′pid) Organic compound formed of carbon, hydrogen, and oxygen; examples are fats and cholesterol.

Lipolysis (lĭ-pol′ĭ-sis) The breakdown of stored fats into glycerol and fatty acids.

Liver Lobed accessory organ that overlies the stomach; produces bile to help digest fat, and serves other metabolic and regulatory functions.

Lumbar (lum′bar) Portion of the back between the thorax and the pelvis.

Lumbar vertebrae The five vertebrae of the lumbar region of the vertebral column, commonly called the small of the back.

Lumen (loo′min) Cavity inside a tube, blood vessel, or hollow organ.

Luteinizing hormone (LH) (lu′te-in-īz″ing) Anterior pituitary hormone that aids maturation of cells in the ovary and triggers ovulation in females. In males, causes the interstitial cells of the testis to produce testosterone.

Lymph (limf) Protein-containing fluid transported by lymphatic vessels.

Lymph node Small lymphoid organ that filters lymph; contains macrophages and lymphocytes.

Lymphatic system (lim-fat′ik) System consisting of lymphatic vessels, lymph nodes, and lymph; drains excess tissue fluid from the extracellular space. The nodes provide sites for immune surveillance.

Lymphatics General term used to designate the lymphatic vessels that collect and transport lymph.

Lymphocyte Agranular white blood cell that arises from bone marrow and becomes functionally mature in the lymphoid organs of the body.

Lysosomes (li′so-sōmz) Organelles that originate from the Golgi apparatus and contain strong digestive enzymes.

Lysozyme (li′so-zīm) Enzyme in sweat, saliva, and tears that is capable of destroying certain kinds of bacteria.

Macromolecules Large, complex molecules containing from 100 to over 10,000 subunits.

Macrophage (mak′ro-fāj″) Protective cell type common in connective tissue, lymphoid tissue, and many body organs; phagocytizes tissue cells, bacteria, and other foreign debris; presents antigens to T cells in the immune response.

Macula (mak′u-lah) (1) Static equilibrium receptor within the vestibule of the inner ear; (2) a colored area or spot.

Malignant (muh-lig′nent) Life threatening; pertains to neoplasms that spread and lead to death, such as cancer.

Malignant melanoma (mel″ah-no′mah) Cancer of the melanocytes; can begin wherever there is pigment.

Mammary glands (mam′mer-e) Milk-producing glands of the breast.

Mandible (man′dĭ-bl) Lower jawbone; U shaped, largest bone of the face.

Mass number Sum of the number of protons and neutrons in the nucleus of an atom.

Mast cells Immune cells that function to detect foreign substances in the tissue spaces and initiate local inflammatory responses against them; typically found clustered deep to an epithelium or along blood vessels.

Mastication (mas″tĭ-ka′shun) Chewing.

Meatus (me-a′tus) External opening of a canal.

Mechanical advantage (power lever) Condition that occurs when the load is close to the fulcrum and the effort is applied far from the fulcrum; allows a small effort exerted over a relatively large distance to move a large load over a small distance.

Mechanical disadvantage (speed lever) Condition that occurs when the load is far from the fulcrum and the effort is applied near the fulcrum; the effort applied must be greater than the load to be moved.

Mechanical energy The energy directly involved in moving matter; e.g., in bicycle riding, the legs provide the mechanical energy that moves the pedals.

Mechanoreceptor (meh″kĕ-no-re-sep′tor) Receptor sensitive to mechanical pressure such as touch, sound, or exerted by muscle contraction.

Medial (me′de-ahl) Toward the midline of the body.

Median (midsagittal) plane Specific sagittal plane that lies exactly in the midline.

Mediastinum (me″de-ah-sti′num) The medial cavity of the thorax containing the heart, great vessels, and trachea.

Medulla (mĕ-dul′ah) Central portion of certain organs.

Medulla oblongata (mĕ-dul′ah ob″long-gah′tah) Inferiormost part of the brain stem.

Medullary cavity Central cavity of a long bone. Contains yellow or red (bone) marrow.

Meiosis (mi-o′sis) Nuclear division process that reduces the chromosomal number by half and results in the formation of four haploid (n) cells; occurs only in certain reproductive organs.

Melanin (mel′ah-nin) Dark pigment formed by cells called melanocytes; imparts color to skin and hair.

Melatonin (mel″ah-to′nin) A hormone secreted by the pineal gland; secretion peaks at night and helps set sleep-wake cycles; also a powerful antioxidant.

Membrane potential Voltage across the plasma membrane.

Membrane receptors A large, diverse group of integral proteins and glycoproteins that serve as binding sites for signaling molecules.

Memory cells Members of T cell and B cell clones that provide for immunological memory.

Menarche (mĕ-nar′ke) Establishment of menstrual function; the first menstrual period.

Meninges (mĕ-nin′jēz) Protective coverings of the central nervous system; from the most external to the most internal, the dura mater, arachnoid mater, and pia mater.

Meningitis (mĕ-nin-ji′tis) Inflammation of the meninges.

Menopause Period of life when, prompted by hormonal changes, ovulation and menstruation cease.

Menstruation (men″stroo-a′shun) The periodic, cyclic discharge of blood, secretions, tissue, and mucus from the mature female uterus in the absence of pregnancy.

Merocrine glands (mer′o-krin) Glands that produce secretions intermittently; secretions do not accumulate in the gland.

Mesencephalon (mes″en-sef′ah-lon) One of the three primary vesicles of the developing brain; becomes the midbrain.

Mesenchyme (meh′zin-kīm) Common embryonic tissue from which all connective tissues arise.

Mesenteries (mes″en-ter′ēz) Double-layered extensions of the peritoneum that support most organs in the abdominal cavity.

Mesoderm (mez′o-derm) Primary germ layer that forms the skeleton and muscles of the body.

Mesothelium (mez″o-the′le-um) The epithelium found in serous membranes lining the ventral body cavity and covering its organs.

Messenger RNA (mRNA) Long nucleotide strands that reflect the exact nucleotide sequences of the genetically active DNA and carry the message of the latter.

Metabolic (fixed) acid Acid generated by cellular metabolism that must be eliminated by the kidneys.

Metabolic rate (mĕt″ah-bol′ik) Energy expended by the body per unit time.

Metabolic water (water of oxidation) Water produced from cellular metabolism (about 10% of our body's water).

Metabolism (mĕ-tab′o-lizm) Sum total of the chemical reactions occurring in the body cells.

Metaphase Second stage of mitosis.

Metastasis (mĕ-tas′tah-sis) The spread of cancer from one body part or organ into another not directly connected to it.

Metencephalon (afterbrain) A secondary brain vesicle; anterior portion of the rhombencephalon of the developing brain; becomes the pons and the cerebellum.

MHC (major histocompatibility complex) proteins Molecules on the outer plasma membrane of all cells; help the immune system distinguish self from nonself. T cells recognize antigens only when combined with these proteins.

Microfilaments (mi″kro-fil′ah-ments) Thin strands of the contractile protein actin.

Microglia (mi-kro′gle-ah) A type of CNS supporting cell; can transform into phagocytes in areas of neural damage or inflammation.

Microtubules (mi″kro-tu′būlz) One of three types of rods in the cytoskeleton of a cell; hollow tubes made of spherical protein that determine the cell shape as well as the distribution of cellular organelles.

Microvilli (mi″kro-vil′i) Tiny projections on the free surfaces of some epithelial cells; increase surface area for absorption.

Micturition (mik″tu-rish′un) Urination, or voiding; emptying the bladder.

Midbrain (mesencephalon) Region of the brain stem between the diencephalon and the pons.

Midsagittal (median) plane Specific sagittal plane that lies exactly in the midline.

Milliequivalents per liter (mEq/L) The units used to measure electrolyte concentrations of body fluids; a measure of the number of electrical charges in 1 liter of solution.

Mineralocorticoid (min″er-al″ō-kor′tih-koyd) Steroid hormone of the adrenal cortex that regulates Na^+ and K^+ metabolism and fluid balance.

Minerals Inorganic chemical compounds found in nature; salts.

Mitochondria (mi″to-kon′dre-ah) Cytoplasmic organelles responsible for ATP generation for cellular activities.

Mitosis Process during which the chromosomes are redistributed to two daughter nuclei; nuclear division. Consists of prophase, metaphase, anaphase, and telophase.

Mitotic (M) phase One of two major periods in the cell life cycle; involves the division of the nucleus (mitosis) and the division of the cytoplasm (cytokinesis).

Mitral (bicuspid) valve (mi′tral) The left atrioventricular valve.

Mixed nerves Nerves containing the processes of motor and sensory neurons; their impulses travel to and from the central nervous system.

Molar (mo′lar) (1) A solution concentration determined by mass of solute—1 liter of solution contains an amount of solute equal to its molecular weight in grams. (2) Broad back teeth that grind and crush.

Molarity (mo-lar′ĭ-te) A way to express the concentration of a solution; moles per liter of solution.

Mole (mōl) A mole of any element or compound is equal to its atomic weight or its molecular weight (sum of atomic weights) measured in grams.

Molecule Particle consisting of two or more atoms joined together by chemical bonds.

Monoclonal antibodies (mon″o-klo′nal) Pure preparations of identical antibodies that exhibit specificity for a single antigen.

Monocyte (mon′o-sīt) Large single-nucleus white blood cell; agranular leukocyte.

Monosaccharide (mon″o-sak′ah-rīd) Literally, one sugar; building block of carbohydrates; e.g., glucose.

Morula (mor′u-lah) The mulberry-like solid mass of blastomeres resulting from cleavage in the early conceptus.

Motor areas Functional areas in the cerebral cortex that control voluntary motor functions.

Motor (efferent) nerves Nerves that carry impulses leaving the brain and spinal cord, and destined for effectors.

Motor unit A motor neuron and all the muscle cells it stimulates.

Mucous membranes (mucosae) Membranes that form the linings of body cavities open to the exterior (digestive, respiratory, urinary, and reproductive tracts).

Mucus (myoo′kus) A sticky, thick fluid secreted by mucous glands and mucous membranes; keeps the free surface of membranes moist.

Multinucleate cell (mul″tĭ-nu′kle-āt) Cell with more than one nucleus, e.g., skeletal muscle cells, osteoclasts.

Multiple sclerosis (MS) Demyelinating disorder of the CNS; causes hardened patches (sclerosis) in the brain and spinal cord.

Multipolar neurons Neurons with three or more processes; most common neuron type in the CNS.

Muscarinic receptors (mus″kah-rin′ik) Acetylcholine-binding receptors of the autonomic nervous system's target organs; named for activation by the mushroom poison muscarine.

Muscle fiber A muscle cell.

Muscle spindle (neuromuscular spindle) Encapsulated receptor found in skeletal muscle that is sensitive to stretch.

Muscle tension The force exerted by a contracting muscle on some object.

Muscle tone Low levels of contractile activity in relaxed muscle; keeps the muscle healthy and ready to act.

Muscle twitch The response of a muscle to a single brief threshold stimulus.

Muscular dystrophy A group of inherited muscle-destroying diseases.

Muscular system The organ system consisting of the skeletal muscles of the body and their connective tissue attachments.

Myelencephalon (spinal brain) A secondary brain vesicle; lower part of the developing hindbrain, especially the medulla oblongata.

Myelin sheath (mi′ĕ-lin) Fatty insulating sheath that surrounds all but the smallest nerve fibers.

Myoblasts Embryonic mesoderm cells from which all muscle fibers develop.

Myocardial infarction (MI) (mi″o-kar′de-al in-fark′shun) Condition characterized by dead tissue areas in the myocardium; caused by interruption of blood supply to the area. Commonly called heart attack.

Myocardium (mi″o-kar′de-um) Layer of the heart wall composed of cardiac muscle.

Myofibril (mi″o-fi′bril) Rodlike bundle of contractile filaments (myofilaments) found in muscle fibers (cells).

Myofilament (mi″o-fil′ah-ment) Filament that constitutes myofibrils. Of two types: actin and myosin.

Myoglobin (mi″o-glo′bin) Oxygen-binding pigment in muscle.

Myogram A graphic recording of mechanical contractile activity produced by an apparatus that measures muscle contraction.

Myometrium (mi″o-me′tre-um) Thick uterine musculature.

Myopia (mi-o′pe-ah) A condition in which visual images are focused in front of rather than on the retina; nearsightedness.

Myosin (mi′o-sin) One of the principal contractile proteins found in muscle.

Myxedema (mik″sĕ-de′mah) Condition resulting from underactive thyroid gland.

Nares (na′rez) Nostrils.

Natural killer (NK) cell Defensive cell (a type of lymphocyte) that can kill cancer cells and virus-infected body cells before the adaptive immune system is activated.

Necrosis (nĕ-kro′sis) Death or disintegration of a cell or tissues caused by disease or injury.

Negative feedback mechanisms The most common homeostatic control mechanism. The net effect is that the output of the system shuts off the original stimulus or reduces its intensity.

Neonatal period The four-week period immediately after birth.

Neoplasm (ne′o-plazm) An abnormal mass of proliferating cells. Benign neoplasms remain localized; malignant neoplasms are cancers, which can spread to other organs.

Nephron (nef′ron) Structural and functional unit of the kidney; consists of the glomerulus and renal tubule.

Nerve A bundle of axons in the peripheral nervous system.

Nerve fiber Axon of a neuron.

Nerve growth factor (NGF) Protein that promotes survival and development of neurons; secreted by their target cells and many other cell types.

Nerve impulse A self-propagating wave of depolarization; also called an action potential.

Nerve plexuses Interlacing nerve networks that occur in the cervical, brachial, lumbar, and sacral regions and primarily serve the limbs.

Nervous system Fast-acting control system that triggers muscle contraction or gland secretion.

Neural tube Fetal structure which gives rise to the brain, spinal cord, and associated neural structures; formed from ectoderm by day 23 of embryonic development.

Neuroglia (nu-rog′le-ah) Nonexcitable cells of neural tissue that support, protect, and insulate the neurons; glial cells.

Neurohypophysis (nu″ro-hi-pof′ĭ-sis) Posterior pituitary plus infundibulum; portion of the pituitary gland derived from the brain.

Neuromuscular junction Region where a motor neuron comes into close contact with a skeletal muscle cell.

Neuron (nerve cell) (nu′ron) Cell of the nervous system specialized to generate and transmit electrical signals (action potentials and graded potentials).

Neuron cell body The biosynthetic center of a neuron; also called the perikaryon, or soma.

Neuronal pools Functional groups of neurons that process and integrate information.

Neuropeptides (nu″ro-pep′tīds) A class of neurotransmitters including beta endorphins and enkephalins (which act as euphorics and reduce perception of pain) and gut-brain peptides.

Neurotransmitter Chemical messenger released by neurons that may, upon binding to receptors of neurons or effector cells, stimulate or inhibit those neurons or effector cells.

Neutral fats Consist of fatty acid chains and glycerol; also called triglycerides or triacylglycerols. Commonly known as oils when liquid.

Neutralization reaction Displacement reaction in which mixing an acid and a base forms water and a salt.

Neutron (nu′tron) Uncharged subatomic particle; found in the atomic nucleus.

Neutrophil (nu′tro-fil) Most abundant type of white blood cell.

Nicotinic receptors (nik″o-tin′ik) Acetylcholine-binding receptors of all autonomic ganglionic neurons and skeletal muscle neuromuscular junctions; named for activation by nicotine.

Nitric oxide (NO) A gaseous chemical messenger; diverse functions include participation in memory formation in the brain, and causing vasodilation throughout the body.

Nociceptor (no″se-sep′tor) Receptor sensitive to potentially damaging stimuli that result in pain.

Nondisjunction Failure of sister chromatids to separate during mitosis or failure of homologous pairs to separate during meiosis; results in abnormal numbers of chromosomes in the resulting daughter cells.

Nonpolar molecules Electrically balanced molecules.

Norepinephrine (NE) (nor″ep-ĭ-nef′rin) A catecholamine neurotransmitter and adrenal medullary hormone, associated with sympathetic nervous system activation.

Nuclear envelope The double membrane barrier of a cell nucleus.

Nucleic acid (nu-kle′ik) Class of organic molecules that includes DNA and RNA.

Nucleoli (nu-kle′o-li) Dense spherical bodies in the cell nucleus involved with ribosomal RNA (rRNA) synthesis and ribosomal subunit assembly.

Nucleosome (nu′kle-o-sōm) Fundamental unit of chromatin; consists of a strand of DNA wound around a cluster of eight histone proteins.

Nucleotide (nu′kle-o-tīd) Building block of nucleic acids; consists of a sugar, a nitrogen-containing base, and a phosphate group.

Nucleus (nu′kle-is) (1) Control center of a cell; contains genetic material; (2) clusters of nerve cell bodies in the CNS.

Nutrients Chemical substances taken in via the diet that are used for energy and cell building.

Oblique section A cut made diagonally between the horizontal and vertical plane of the body or an organ.

Occlusion (ah-kloo′zhun) Closure or obstruction.

Octet rule (rule of eights) (ok-tet′) The tendency of atoms to interact in such a way that they have eight electrons in their valence shell.

Olfaction (ol-fak′shun) Smell.

Oligodendrocyte (ol″ĭ-go-den′dro-sīt) A type of CNS supporting cell that composes myelin sheaths.

Oocyte (o′o-sīt) Immature female gamete.

Oogenesis (o″o-jen′ĕ-sis) Process of ovum (female gamete) formation.

Ophthalmic (of-thal′mik) Pertaining to the eye.

Optic (op′tik) Pertaining to the eye or vision.

Optic chiasma (op′tik ki-az′muh) The partial crossover of fibers of the optic nerves.

Organ A part of the body formed of two or more tissues and adapted to carry out a specific function; e.g., the stomach.

Organ system A group of organs that work together to perform a vital body function; e.g., the nervous system.

Organelles (or″gah-nelz′) Small cellular structures (ribosomes, mitochondria, and others) that perform specific metabolic functions for the cell as a whole.

Organic compound Any compound composed of atoms (some of which are carbon) held together by covalent (shared electron) bonds.

Organic Pertaining to carbon-containing molecules, such as proteins, fats, and carbohydrates.

Organism The living animal (or plant), which represents the sum total of all its organ systems working together to maintain life.

Origin Attachment of a muscle that remains relatively fixed during muscular contraction.

Osmolality The number of solute particles dissolved in 1 kilogram (1000 g) of water; reflects the solution's ability to cause osmosis.

Osmolarity (oz″mo-lar′ĭ-te) The number of solute particles present in 1 liter of a solution.

Osmoreceptor (oz″mo-re-sep′tor) Structure sensitive to osmotic pressure or concentration of a solution.

Osmosis (oz-mo′sis) Diffusion of a solvent through a membrane from a dilute solution into a more concentrated one.

Osmotic pressure A measure of the tendency of water to move into a more concentrated solution.

Ossicles *See* Auditory ossicles.

Ossification (os″ĭ-fi-ka′shun) *See* Osteogenesis.

Osteoblasts (os′te-o-blasts) Bone-forming cells.

Osteoclasts (os′te-o-klasts) Large cells that resorb or break down bone matrix.

Osteocyte (os′te-o-sīt) Mature bone cell.

Osteogenesis (os″te-o-jen′e-sis) The process of bone formation; also called ossification.

Osteoid (os′te-oid) Unmineralized bone matrix.

Osteomalacia (os″te-o-mah-la′she-ah) Disorder in which bones are inadequately mineralized; soft bones.

Osteon (os′te-on) System of interconnecting canals in the microscopic structure of adult compact bone; unit of bone; also called Haversian system.

Osteoporosis (os″te-o-po-ro′sis) Decreased density and strength of bone resulting from a gradual decrease in rate of bone formation.

Ovarian cycle (o-vayr′e-an) Monthly cycle of follicle development, ovulation, and corpus luteum formation in an ovary.

Ovary (o′var-e) Female reproductive organ in which ova (eggs) are produced; female gonad.

Ovulation (ov″u-la′shun) Ejection of an immature egg (oocyte) from the ovary.

Ovum (o′vum) Female gamete; egg.

Oxidases Enzymes that catalyze the transfer of oxygen in oxidation-reduction reactions.

Oxidation (oks′ĭ-da″shun) Process of substances combining with oxygen or the removal of hydrogen.

Oxidation-reduction (redox) reaction A reaction that couples the oxidation (loss of electrons) of one substance with the reduction (gain of electrons) of another substance.

Oxidative phosphorylation (ok′sĭ-da″tiv fos″for-ĭ-la′shun) Process of ATP synthesis during which an inorganic phosphate group is attached to ADP; occurs via the electron transport chain within the mitochondria.

Oxygen deficit The volume of oxygen required after exercise to replenish stores of O_2, ATP, creatine phosphate, and glycogen and oxidize the lactic acid formed during exercise.

Oxyhemoglobin (ok″sĭ-he″mo-glo′bin) Oxygen-bound form of hemoglobin.

Oxytocin (ok″sĭ-to′sin) Hormone synthesized in the hypothalamus and secreted by the posterior pituitary; stimulates contraction of the uterus during childbirth and the ejection of milk during nursing.

Paget's disease (paj′ets) Disorder characterized by excessive bone breakdown and abnormal bone formation.

Palate (pal′at) Roof of the mouth.

Pancreas (pan′kre-us) Gland located behind the stomach, between the spleen and the duodenum; produces both endocrine and exocrine secretions.

Pancreatic juice (pan″kre-at′ik) Bicarbonate-rich secretion of the pancreas containing enzymes for digestion of all food categories.

Papilla (pah-pil′ah) Small, nipple-like projection; e.g., dermal papillae are projections of dermal tissue into the epidermis.

Paracrine (par′ah-krin) A chemical messenger that acts locally within the same tissue and is rapidly destroyed. Examples are prostaglandins and nitric oxide.

Parasagittal planes All sagittal planes offset from the midline.

Parasympathetic division The division of the autonomic nervous system that oversees digestion, elimination, and glandular function; the resting and digesting subdivision.

Parasympathetic tone Normal (background) level of parasympathetic output; sustains normal gastrointestinal and urinary tract activity, lowers heart rate.

Parathyroid glands (par″ah-thi′roid) Small endocrine glands located on the posterior aspect of the thyroid gland.

Parathyroid hormone (PTH) Hormone released by the parathyroid glands that regulates blood calcium level.

Parietal (pah-ri′ĕ-tal) Pertaining to the walls of a cavity.

Parietal serosa The part of the double-layered membrane that lines the walls of the ventral body cavity.

Parkinson's disease Neurodegenerative disorder of the basal nuclei due to insufficient secretion of the neurotransmitter dopamine; symptoms include tremor and rigid movement.

Partial pressure The pressure exerted by a single component of a mixture of gases.

Parturition (par″tu-rish′un) Culmination of pregnancy; giving birth.

Passive immunity Short-lived immunity resulting from the introduction of "borrowed antibodies" obtained from an immune animal or human donor; immunological memory is not established.

Passive (transport) processes Membrane transport processes that do not require cellular energy (ATP), e.g., diffusion, which is driven by kinetic energy.

Pathogen (path′o-jen) Disease-causing organism.

Pectoral (pek′tor-al) Pertaining to the chest.

Pectoral (shoulder) girdle Bones that attach the upper limbs to the axial skeleton; includes the clavicle and scapula.

Pedigree Traces a particular genetic trait through several generations and helps predict the genotype of future offspring.

Pelvic girdle (hip girdle) Consists of the paired coxal bones that attach the lower limbs to the axial skeleton.

Pelvis (pel′vis) (1) Basin-shaped bony structure composed of the pelvic girdle, sacrum, and coccyx; (2) funnel-shaped tube within the kidney continuous with the ureter.

Penis (pe′nis) Male organ of copulation and urination.

Pepsin Enzyme capable of digesting proteins in an acid pH.

Peptide bond (pep′tīd) Bond joining the amine group of one amino acid to the acid carboxyl group of a second amino acid with the loss of a water molecule.

Perforating canals Canals that run at right angles to the long axis of the bone, connecting the vascular and nerve supplies of the periosteum to those of the central canals and medullary cavity; also called Volkmann's canals.

Pericardium (per″ĭ-kar′de-um) Double-layered sac enclosing the heart and forming its superficial layer; has fibrous and serous layers.

Perichondrium (per″i-kon′dre-um) Fibrous, connective-tissue membrane covering the external surface of cartilaginous structures.

Perimysium (per″ĭ-mis′e-um) Connective tissue enveloping bundles of muscle fibers.

Perineum (per″ĭ-ne′um) That region of the body spanning the region between the ischial tuberosities and extending from the pubic arch to the coccyx.

Periosteum (per″e-os′te-um) Double-layered connective tissue that covers and nourishes the bone.

Peripheral congestion Condition caused by failure of the right side of the heart; results in edema in the extremities.

Peripheral nervous system (PNS) Portion of the nervous system consisting of nerves and ganglia that lie outside of the brain and spinal cord.

Peripheral resistance A measure of the amount of friction encountered by blood as it flows through the blood vessels.

Peristalsis (per″i-stal′sis) Progressive, wavelike contractions that move foodstuffs through the alimentary tube organs (or that move other substances through other hollow body organs).

Peritoneum (per″ĭ-to-ne′um) Serous membrane lining the interior of the abdominal cavity and covering the surfaces of abdominal organs.

Peritonitis (per″ĭ-to-ni′tis) Inflammation of the peritoneum.

Permeability That property of membranes that permits passage of molecules and ions.

Peroxisomes (pĕ-roks′ĭ-sōmz) Membranous sacs in cytoplasm containing powerful oxidase enzymes that use molecular oxygen to detoxify harmful or toxic substances, such as free radicals.

Peyer's patches (pi′erz) Lymphoid organs located in the small intestine; also called aggregated lymphoid nodules.

pH unit (pe-āch) The measure of the relative acidity or alkalinity of a solution.

Phagocytosis (fag″o-si-to′sis) Engulfing of foreign solids by (phagocytic) cells.

Phagosome (fag′o-sōm) Vesicle formed as a result of phagocytosis.

Pharmacological dose A drug dose that is dramatically higher than normal levels of that substance (e.g., hormone) in the body.

Pharyngotympanic tube Tube that connects the middle ear and the pharynx. Also called auditory tube, eustachian tube.

Pharynx (fayr′inks) Muscular tube extending from the region posterior to the nasal cavities to the esophagus.

Phenotype (fe′no-tīp) Observable expression of the genotype.

Phospholipid (fos″fo-lip′id) Modified lipid, contains phosphorus.

Phosphorylation A chemical reaction in which a phosphate molecule is added to a molecule; for example, phosphorylation of ADP yields ATP.

Photoreceptor (fo″to-re-sep′tor) Specialized receptor cells that respond to light energy; rods and cones.

Physiological acidosis (as″ĭ-do′sis) Arterial pH lower than 7.35 resulting from any cause.

Physiological dose A drug dose that replicates normal levels of that substance (e.g., hormone) in the body.

Physiology (fiz″e-ol′o-je) Study of the function of living organisms.

Pineal gland (body) (pin′e-al) A hormone-secreting part of the diencephalon of the brain thought to be involved in setting the biological clock and influencing reproductive function.

Pinocytosis (pe″no-si-to′sis) Engulfing of extracellular fluid by cells.

Pituitary gland (pĭ-tu′ih-tayr″e) Neuroendocrine gland located beneath the brain that serves a variety of functions including regulation of gonads, thyroid, adrenal cortex, lactation, and water balance.

Placenta (plah-sen′tah) Temporary organ formed from both fetal and maternal tissues that provides nutrients and oxygen to the developing fetus, carries away fetal metabolic wastes, and produces the hormones of pregnancy.

Plasma (plaz′mah) The nonliving fluid component of blood within which formed elements and various solutes are suspended and circulated.

Plasma cells Members of a B cell clone; specialized to produce and release antibodies.

Plasma membrane Membrane, composed of phospholipids, cholesterol, and proteins, that encloses cell contents; outer limiting cell membrane.

Platelet (plāt′let) Cell fragment found in blood; involved in clotting.

Pleurae (ploo′re) Two layers of serous membrane that line the thoracic cavity and cover the external surface of the lung.

Pleural cavity (ploo′ral) A potential space between the two layers of pleura; contains a thin film of serous fluid.

Plexus (plek′sus) A network of converging and diverging nerve fibers, blood vessels, or lymphatics.

Polar molecules Nonsymmetrical molecules that contain electrically unbalanced atoms.

Polarized State of a plasma membrane of an unstimulated neuron or muscle cell in which the inside of the cell is relatively negative in comparison to the outside; the resting state.

Polycythemia (pol″e-si-the′me-ah) An abnormally high number of erythrocytes.

Polymer A substance of high molecular weight with long, chainlike molecules consisting of many similar (repeated) units.

Polypeptide (pol″e-pep′tīd) A chain of amino acids.

Polyps Benign mucosal tumors.

Polysaccharide (pol″e-sak′ah-rīd) Literally, many sugars, a polymer of linked monosaccharides; e.g., starch, glycogen.

Pons (1) Any bridgelike structure or part; (2) the part of the brain stem connecting the medulla with the midbrain, providing linkage between upper and lower levels of the central nervous system.

Pore The surface opening of the duct of a sweat gland.

Positive feedback mechanisms Feedback that tends to cause the level of a variable to change in the same direction as an initial change.

Posterior pituitary *See* Neurohypophysis.

Postganglionic axon (fiber) (post″gang-gle-ah′nik) Axon of a ganglionic neuron, an autonomic motor neuron that has its cell body in a peripheral ganglion; the axon projects to an effector.

Potential energy Stored or inactive energy.

Preganglionic neuron Autonomic motor neuron that has its cell body in the central nervous system and projects its axon to a peripheral ganglion.

Presbyopia (pres″be-o′pe-ah) A condition that results in the loss of near focusing ability; typical onset is around age 40.

Pressure gradient Difference in pressure (hydrostatic or osmotic) that drives movement of fluid.

Primary active transport A type of active transport in which the energy needed to drive the transport process is provided directly by hydrolysis of ATP.

Prime mover Muscle that bears the major responsibility for effecting a particular movement; an agonist.

Process (1) Prominence or projection; (2) series of actions for a specific purpose.

Progesterone (pro-jes′ter-ōn) Hormone partly responsible for preparing the uterus for the fertilized ovum.

Prolactin (PRL) (pro-lak′tin) Adenohypophyseal hormone that stimulates the breasts to produce milk.

Pronation (pro-na′shun) Inward rotation of the forearm causing the radius to cross diagonally over the ulna—palms face posteriorly.

Prophase The first stage of mitosis, consisting of coiling of the chromosomes accompanied by migration of the two daughter centrioles toward the poles of the cell, and nuclear membrane breakdown.

Proprioceptor (pro″pre-o-sep′tor) Receptor located in a joint, muscle, or tendon; concerned with locomotion, posture, and muscle tone.

Prostaglandin (PG) (pros″tah-glan′din) A lipid-based chemical messenger synthesized by most tissue cells that acts locally as a paracrine.

Prostate Accessory reproductive gland; produces one-third of semen volume, including fluids that activate sperm.

Protein (pro′tēn) Complex substance containing carbon, oxygen, hydrogen, and nitrogen; composes 10–30% of cell mass.

Prothrombin time Diagnostic test to determine status of hemostasis system.

Proton (pro′ton) Subatomic particle that bears a positive charge; located in the atomic nucleus.

Proton acceptor A substance that takes up hydrogen ions in detectable amounts. Commonly referred to as a base.

Proton donor A substance that releases hydrogen ions in detectable amounts; an acid.

Proximal (prok′si-mul) Toward the attached end of a limb or the origin of a structure.

Pseudounipolar neuron (soo″do-u″nĭ-po′lar) Another term for unipolar neuron.

Puberty Period of life when reproductive maturity is achieved.

Pulmonary (pul′muh-nayr-e) Pertaining to the lungs.

Pulmonary arteries Vessels that deliver blood to the lungs to be oxygenated.

Pulmonary circuit System of blood vessels that serves gas exchange in the lungs; i.e., pulmonary arteries, capillaries, and veins.

Pulmonary edema (ĕ-de′muh) Leakage of fluid into the air sacs and tissue of the lungs.

Pulmonary veins Vessels that deliver freshly oxygenated blood from the respiratory zones of the lungs to the heart.

Pulmonary ventilation Breathing; consists of inspiration and expiration.

Pulse Rhythmic expansion and recoil of arteries resulting from heart contraction; can be felt from outside the body.

Pupil Opening in the center of the iris through which light enters the eye.

Purkinje fibers (pur-kin′je) Modified ventricular muscle fibers of the conduction system of the heart.

Pus Fluid product of inflammation composed of white blood cells, the debris of dead cells, and a thin fluid.

Pyloric sphincter (pi-lor′ik sfink′ter) Valve of the distal end of the stomach that controls food entry into the duodenum.

Pyramidal (corticospinal) tracts Major motor pathways concerned with voluntary movement; descend from pyramidal cells in the frontal lobes of each cerebral hemisphere.

Pyruvic acid An intermediate compound in the metabolism of carbohydrates.

Radioactivity The process of spontaneous decay seen in some of the heavier isotopes, during which particles or energy is emitted from the atomic nucleus; results in the atom becoming more stable.

Radioisotope (ra″de-o-i′so-tōp) Isotope that exhibits radioactive behavior.

Ramus (ra′mus) Branch of a nerve, artery, vein, or bone.

Rapid eye movement (REM) sleep Stage of sleep in which rapid eye movements, an alert EEG pattern, and dreaming occur.

Reactant A substance taking part in a chemical reaction.

Receptor (re-sep′tor) (1) A cell or nerve ending of a sensory neuron specialized to respond to particular types of stimuli; (2) protein that binds specifically with other molecules, e.g., neurotransmitters, hormones, paracrines, antigens.

Receptor-mediated endocytosis One of three types of endocytosis in which engulfed particles attach to receptors before endocytosis occurs.

Receptor potential A graded potential that occurs at a sensory receptor membrane.

Recessive traits A trait due to a particular allele that does not manifest itself in the presence of other alleles that generate traits dominant to it; must be present in double dose to be expressed.

Reduction Chemical reaction in which electrons and energy are gained by a molecule (often accompanied by gain of hydrogen ions) or oxygen is lost.

Referred pain Pain felt at a site other than the area of origin.

Reflex Automatic reaction to stimuli.

Refraction The bending of a light ray when it meets a different surface at an oblique rather than right angle.

Regeneration Replacement of destroyed tissue with the same kind of tissue.

Regulatory T cells (T_{Reg} cells) Population of T cells (usually expressing CD4) that suppress the immune response.

Relative refractory period Follows the absolute refractory period; interval when a threshold for action potential stimulation is markedly elevated.

Renal (re′nal) Pertaining to the kidney.

Renal autoregulation Process the kidney uses to maintain a nearly constant glomerular filtration rate despite fluctuations in systemic blood pressure.

Renal clearance The volume of plasma from which a particular substance is completely removed in a given time, usually 1 minute; provides information about renal function.

Renin (re′nin) Hormone released by the kidneys that is involved with raising blood pressure.

Rennin Stomach-secreted enzyme that acts on milk protein; not produced in adults.

Repolarization Movement of the membrane potential to the initial resting (polarized) state.

Reproductive system Organ system that functions to produce offspring.

Resistance exercise High-intensity exercise in which the muscles are pitted against high resistance or immovable forces and, as a result, muscle cells increase in size.

Respiration The processes involved in supplying the body with oxygen and disposing of carbon dioxide.

Respiratory system Organ system that carries out gas exchange; includes the nose, pharynx, larynx, trachea, bronchi, lungs.

Resting membrane potential The voltage that exists across the plasma membrane during the resting state of an excitable cell; ranges from −90 to −20 millivolts depending on cell type.

Reticular activating system (RAS) (re-tik′u-lar) Diffuse brain stem neural network that receives a wide variety of sensory input and maintains wakefulness of the cerebral cortex.

Reticular connective tissue Connective tissue with a fine network of reticular fibers that form the internal supporting framework of lymphoid organs.

Reticular formation Functional system that spans the brain stem; involved in regulating sensory input to the cerebral cortex, cortical arousal, and control of motor behavior.

Reticular lamina A layer of extracellular material containing a fine network of collagen protein fibers; together with the basal lamina it is a major component of the basement membrane.

Reticulocyte (rĕ-tik′u-lo-sīt) Immature erythrocyte.

Retina (ret′ĭ-nah) Neural layer of the eyeball; contains photoreceptors (rods, cones).

Rhombencephalon (hindbrain) (romb″en-sef′ah-lon) Caudal portion of the developing brain; constricts to form the metencephalon and myelencephalon; includes the pons, cerebellum, and medulla oblongata.

Ribosomal RNA (rRNA) A constituent of ribosome; exists within the ribosomes of cytoplasm and assists in protein synthesis.

Ribosomes (ri′bo-sōmz) Cytoplasmic organelles at which proteins are synthesized.

RNA (ribonucleic acid) (ri′bo-nu-kle′ik) Nucleic acid that contains ribose and the bases A, G, C, and U. Carries out DNA's instructions for protein synthesis.

Rods One of the two types of photosensitive cells in the retina.

Rotation The turning of a bone around its own long axis.

Rugae (ru′ge) Elevations or ridges, as in stomach mucosa.

Rule of nines Method of computing the extent of burns by dividing the body into a number of areas, each accounting for 9% (or a multiple thereof) of the total body area.

S (synthetic) phase The part of the interphase period of the cell life cycle in which DNA replicates itself, ensuring that the two future cells will receive identical copies of genetic material.

Sagittal plane (saj′ĭ-tal) A longitudinal (vertical) plane that divides the body or any of its parts into right and left portions.

Saliva Secretion of the salivary glands; cleanses and moistens the mouth and begins chemical digestion of starchy foods.

Saltatory conduction Transmission of an action potential along a myelinated fiber in which the nerve impulse appears to leap from node to node.

Sarcolemma The plasma membrane surface of a muscle fiber.

Sarcomere (sar′ko-mēr) The smallest contractile unit of muscle; extends from one Z disc to the next.

Sarcoplasm The nonfibrillar cytoplasm of a muscle fiber.

Sarcoplasmic reticulum (SR) (sar″ko-plaz′mik rĕ-tik′u-lum) Specialized endoplasmic reticulum of muscle cells.

Schwann cell A type of supporting cell in the PNS; forms myelin sheaths and is vital to peripheral nerve fiber regeneration.

Sclera (skle′rah) White opaque portion of the fibrous layer of the eyeball.

Scrotum (skro′tum) External sac enclosing the testes.

Sebaceous glands (oil glands) (se-ba′shus) Epidermal glands that produce an oily secretion called sebum.

Sebum (se′bum) Oily secretion of sebaceous glands.

Second-degree burn A burn in which the epidermis and the upper region of the dermis are damaged.

Second messenger Intracellular molecule generated by the binding of a chemical (hormone or neurotransmitter) to a plasma membrane receptor; mediates intracellular responses to the chemical messenger.

Secondary sex characteristics Anatomic features, not directly involved in the reproductive process, that develop under the influence of sex hormones, e.g., male or female pattern of muscle development, bone growth, body hair distribution.

Secretion (se-kre′shun) (1) The passage of material formed by a cell to its exterior; (2) cell product that is transported to the exterior of a cell.

Secretory vesicles (granules) Vesicles that migrate to the plasma membrane of a cell and discharge their contents from the cell by exocytosis.

Section A cut through the body (or an organ) that is made along a particular plane; a thin slice of tissue prepared for microscopic study.

Segregation During meiosis, the distribution of the members of the allele pair to different gametes.

Selectively permeable membrane A membrane that allows certain substances to pass while restricting the movement of others; also called differentially permeable membrane.

Semen (se′men) Fluid mixture containing sperm and secretions of the male accessory reproductive glands.

Semilunar valves (sĕ″me-loo′ner) Valves that prevent blood return to the ventricles after contraction; aortic and pulmonary valves.

Seminiferous tubules (sem″ĭ-nif′er-us) Highly convoluted tubes within the testes; form sperm.

Sense organs Localized collections of many types of cells working together to accomplish a specific receptive process.

Sensory (afferent) nerves Nerves that contain processes of sensory neurons and carry impulses to the central nervous system.

Sensory areas Functional areas of the cerebral cortex that provide for conscious awareness of sensation.

Sensory receptor Dendritic end organs, or parts of other cell types, specialized to respond to a stimulus.

Serosa (serous membrane) (se-ro′sah) The moist membrane found in closed ventral body cavities.

Serous fluid (sēr′us) Clear, watery fluid secreted by cells of a serous membrane.

Serum (sēr′um) Amber-colored fluid that exudes from clotted blood as the clot shrinks; plasma without clotting factors.

Sesamoid bones (ses′ah-moid) Short bones embedded in tendons, variable in size and number, many of which influence the action of muscles; largest is the patella (kneecap).

Severe combined immunodeficiency syndromes (SCIDs) Congenital conditions resulting in little or no protection against disease-causing organisms of any type.

Sex chromosomes The chromosomes, X and Y, that determine genetic sex (XX = female; XY = male); the 23rd pair of chromosomes.

Sex-linked inheritance Inherited traits determined by genes on the sex chromosomes, e.g., X-linked genes are passed from mother to son, Y-linked genes are passed from father to son.

Sexually transmitted infection (STI) Infectious disease spread through sexual contact.

Signal sequence A short peptide segment present in a protein being synthesized that causes the associated ribosome to attach to the membrane of rough ER.

Simple diffusion The unassisted transport across a plasma membrane of a lipid-soluble or very small particle.

Sinoatrial (SA) node (si″no-a′tre-al) Specialized myocardial cells in the wall of the right atrium; pacemaker of the heart.

Sinus (si′nus) (1) Mucous-membrane-lined, air-filled cavity in certain cranial bones; (2) dilated channel for the passage of blood or lymph.

Skeletal muscle Muscle composed of cylindrical multinucleate cells with obvious striations; the muscle(s) attached to the body's skeleton; voluntary muscle.

Skeletal system System of protection and support composed primarily of bone and cartilage.

Skull Bony protective encasement of the brain and the organs of hearing and equilibrium; includes the facial bones. Also called the cranium.

Small intestine Convoluted tube extending from the pyloric sphincter to the ileocecal valve where it joins the large intestine; the site where digestion is completed and virtually all absorption occurs.

Smooth muscle Spindle-shaped cells with one centrally located nucleus and no externally visible striations (bands). Found mainly in the walls of hollow organs.

Sodium-potassium (Na^+-K^+) pump A primary active transport system that simultaneously drives Na^+ out of the cell against a steep gradient and pumps K^+ back in.

Sol-gel transformation Reversible change of a colloid from a fluid (sol) to a more solid (gel) state.

Solute (sol′yoot) The substance that is dissolved in a solution.

Solute pump Enzyme-like protein carrier that mediates active transport of solutes such as amino acids and ions uphill against their concentration gradients.

Somatic nervous system (so-mă′tik) Division of the peripheral nervous system that provides the motor innervation of skeletal muscles; also called the voluntary nervous system.

Somatic reflexes Reflexes that activate skeletal muscle.

Somatosensory system That part of the sensory system dealing with reception in the body wall and limbs; receives inputs from exteroceptors, proprioceptors, and interoceptors.

Somite (so′mīt) A mesodermal segment of the body of an embryo that contributes to the formation of skeletal muscles, vertebrae, and dermis of skin.

Spatial discrimination The ability of neurons to identify the site or pattern of stimulation.

Special senses The senses of taste, smell, vision, hearing, and equilibrium.

Specific gravity Term used to compare the weight of a substance to the weight of an equal volume of distilled water.

Sperm (spermatozoon) Male gamete.

Spermatogenesis (sper″mah-to-jen′ĕ-sis) The process of sperm (male gamete) formation; involves meiosis.

Sphincter (sfink′ter) A circular muscle surrounding an opening; acts as a valve.

Spinal cord The bundle of nervous tissue that runs from the brain to the first to third lumbar vertebrae and provides a conduction pathway to and from the brain.

Spinal nerves The 31 nerve pairs that arise from the spinal cord.

Splanchnic circulation (splangk′nik) The blood vessels serving the digestive system.

Spleen Largest lymphoid organ; provides for lymphocyte proliferation, immune surveillance and response, and blood-cleansing functions.

Spongy bone Internal layer of skeletal bone. Also called cancellous bone.

Sprain Ligaments reinforcing a joint are stretched or torn.

Static equilibrium Sense of head position in space with respect to gravity.

Stenosis (stĕ-no′sis) Abnormal constriction or narrowing.

Steroids (stĕ′roidz) Group of chemical substances including certain hormones and cholesterol; they are fat soluble and contain little oxygen.

Stimulus (stim′u-lus) An excitant or irritant; a change in the environment that evokes a response.

Stomach Temporary reservoir in the gastrointestinal tract where chemical breakdown of proteins begins and food is converted into chyme.

Stressor Any stimulus that directly or indirectly causes the hypothalamus to initiate stress-reducing responses, such as the fight-or-flight response.

Stroke *See* Cerebrovascular accident.

Stroke volume (SV) Amount of blood pumped out of a ventricle during one contraction.

Stroma (stro′mah) The basic internal structural framework of an organ.

Structural (fibrous) proteins Consist of extended, strandlike polypeptide chains forming a strong, ropelike structure that is linear, insoluble in water, and very stable; e.g., collagen.

Subcutaneous (sub″kyu-ta′ne-us) Beneath the skin.

Substrate A reactant on which an enzyme acts to cause a chemical action to proceed.

Sudoriferous gland (su″do-rif′er-us) Epidermal gland that produces sweat.

Sulcus (sul′kus) A furrow on the brain, less deep than a fissure.

Summation Accumulation of effects, especially those of muscular, sensory, or mental stimuli.

Superficial Located close to or on the body surface.

Superior Toward the head or upper body regions.

Superior vena cava Vein that returns blood from body regions superior to the diaphragm.

Supination (soo″pĭ-na′shun) The outward rotation of the forearm causing palms to face anteriorly.

Surfactant (ser-fak′tant) Secretion produced by certain cells of the alveoli that reduces the surface tension of water molecules, thus preventing the collapse of the alveoli after each expiration.

Suspension Heterogeneous mixtures with large, often visible solutes that tend to settle out.

Suture (soo′cher) An immovable fibrous joint; with one exception, all bones of the skull are united by sutures.

Sweat gland *See* Sudoriferous gland.

Sympathetic division The division of the autonomic nervous system that prepares the body for activity or to cope with some stressor (danger, excitement, etc.); the fight, fright, and flight subdivision.

Sympathetic (vasomotor) tone State of partial vasoconstriction of the blood vessels maintained by sympathetic fibers.

Symphysis (sim′fih-sis) A joint in which the bones are connected by fibrocartilage.

Synapse (sin′aps) Functional junction or point of close contact between two neurons or between a neuron and an effector cell.

Synapsis (sĭ-nap′sis) Pairing of homologous chromosomes during the first meiotic division.

Synaptic cleft (si-nap′tik) Fluid-filled space at a synapse.

Synaptic delay Time required for an impulse to cross a synapse between two neurons.

Synaptic vesicles Small membranous sacs containing neurotransmitter.

Synarthrosis (sin″ar-thro′sis) Immovable joint.

Synchondrosis (sin″kon-dro′sis) A joint in which the bones are united by hyaline cartilage.

Syndesmosis (sin″des-mo′sis) A joint in which the bones are united by a ligament or a sheet of fibrous tissue.

Synergist (sin′er-jist) (1) Muscle that aids the action of a prime mover by effecting the same movement or by stabilizing joints across which the prime mover acts, preventing undesirable movements. (2) Hormone that amplifies the effect of another hormone at a target cell.

Synostosis (sin″os-to′sis) A completely ossified joint; a fused joint.

Synovial fluid Fluid secreted by the synovial membrane; lubricates joint surfaces and nourishes articular cartilages.

Synovial joint Freely movable joint exhibiting a joint cavity; also called a diarthrosis.

Synthesis (combination) reaction A chemical reaction in which larger, more complex atoms or molecules are formed from simpler ones.

Systemic (sis-tem′ik) Pertaining to the whole body.

Systemic circuit System of blood vessels that serves gas exchange in the body tissues.

Systole (sis′to-le) Period when either the ventricles or the atria are contracting.

Systolic pressure (sis-tah′lik) Pressure exerted by blood on the blood vessel walls during ventricular contractions.

T cells Lymphocytes that mediate cellular immunity; include helper, cytotoxic, regulatory, and memory cells. Also called T lymphocytes.

T tubule (transverse tubule) Extension of the muscle cell plasma membrane (sarcolemma) that protrudes deeply into the muscle cell.

Tachycardia (tak″e-kar′de-ah) A heart rate over 100 beats per minute.

Taste buds Sensory receptor organs that house gustatory cells, which respond to dissolved food chemicals.

Telencephalon (endbrain) (tel″en-seh′fuh-lon) Anterior subdivision of the primary forebrain that develops into olfactory lobes, cerebral cortex, and basal nuclei.

Telophase The final phase of mitosis; begins when migration of chromosomes to the poles of the cell has been completed and ends with the formation of two daughter nuclei.

Tendon (ten′dun) Cord of dense fibrous tissue attaching muscle to bone.

Tendonitis Inflammation of tendon sheaths, typically caused by overuse.

Terminal branches Branching ends of an axon that allow it to form many axon terminals; telodendria.

Testis (tes′tis) Male primary reproductive organ that produces sperm; male gonad.

Testosterone (tes-tos′tĕ-rōn) Male sex hormone produced by the testes; during puberty promotes virilization, and is necessary for normal sperm production.

Tetanus (tet′ah-nus) (1) A smooth, sustained muscle contraction resulting from high-frequency stimulation; (2) an infectious disease caused by an anaerobic bacterium.

Thalamus (thal′ah-mus) A mass of gray matter in the diencephalon of the brain.

Thermogenesis (ther″mo-jen′ĕ-sis) Heat production.

Thermoreceptor (ther″mo-re-sep′ter) Receptor sensitive to temperature changes.

Third-degree burn A burn that involves the entire thickness of the skin; also called a full-thickness burn. Usually requires skin grafting.

Thoracic cage (bony thorax) Bones that form the framework of the thorax; includes sternum, ribs, and thoracic vertebrae.

Thoracic duct Large duct that receives lymph drained from the entire lower body, the left upper extremity, and the left side of the head and thorax.

Thorax (tho′raks) That portion of the body trunk above the diaphragm and below the neck.

Threshold stimulus Weakest stimulus capable of producing a response in an irritable tissue.

Thrombin (throm′bin) Enzyme that induces clotting by converting fibrinogen to fibrin.

Thrombocyte (throm′bo-sīt) Platelet; cell fragment that participates in blood coagulation.

Thrombocytopenia (throm″bo-si″to-pe′ne-ah) A reduction in the number of platelets circulating in the blood.

Thrombus (throm′bus) A clot that develops and persists in an unbroken blood vessel.

Thymine (T) (thi′mēn) Single-ring base (a pyrimidine) in DNA.

Thymus (thi′mus) Lymphoid organ and endocrine gland active in immune response; site of maturation of T lymphocytes.

Thyroid gland (thi′roid) One of the largest of the body's endocrine glands; straddles the anterior trachea.

Thyroid hormone (TH) The major hormone secreted by thyroid follicles; stimulates enzymes concerned with glucose oxidation.

Thyroid-stimulating hormone (TSH) Adenohypophyseal hormone that regulates secretion of thyroid hormones.

Thyroxine (T_4) (thi-rok′sin) Iodine-containing hormone secreted by the thyroid gland; accelerates cellular metabolic rate in most body tissues.

Tight junction Area where plasma membranes of adjacent cells are fused.

Tissue A group of similar cells and their intercellular substance specialized to perform a specific function; primary tissue types of the body are epithelial, connective, muscle, and nervous tissue.

Tissue perfusion Blood flow through body tissues or organs.

Tonicity (to-nis′ĭ-te) A measure of the ability of a solution to cause a change in cell shape or tone by promoting osmotic flows of water.

Tonsils A ring of lymphoid tissue around the entrance to the pharynx. *See also* Adenoids.

Trabecula (trah-bek′u-lah) (1) Any of the fibrous bands extending from the capsule into the interior of an organ; (2) strut or thin plate of bone in spongy bone.

Trachea (tra′ke-ah) Windpipe; cartilage-reinforced tube extending from larynx to bronchi.

Tract A collection of axons in the central nervous system having the same origin, termination, and function.

Transcription One of the two major steps in the transfer of genetic code information from a DNA base sequence to the complementary base sequence of an mRNA molecule.

Transduction (trans-duk′shun) The conversion of the energy of a stimulus into an electrical event.

Transepithelial transport (trans-ep″ĭ-the′le-al) Movement of substances through, rather than between, adjacent epithelial cells connected by tight junctions, such as absorption of nutrients in the small intestine.

Transfer RNA (tRNA) Short-chain RNA molecules that transfer amino acids to the ribosome.

Transfusion reaction Agglutination and destruction of red blood cells following transfusion of incompatible blood.

Translation One of the two major steps in the transfer of genetic code information, in which the information carried by mRNA is decoded and used to assemble polypeptides.

Transverse (horizontal) plane A plane running from right to left, dividing the body or an organ into superior and inferior parts.

Tricuspid valve (tri-kus′pid) The right atrioventricular valve.

Triglycerides (tri-glis′er-īdz) Fats and oils composed of fatty acids and glycerol; are the body's most concentrated source of energy fuel; also known as neutral fats.

Triiodothyronine (T_3) (tri″i-o″do-thi′ro-nēn) Thyroid hormone; secretion and function similar to those of thyroxine (T_4).

Trophoblast (tro′fo-blast) Outer sphere of cells of the blastocyst.

Tropic hormone (trōp′ik) A hormone that regulates the secretory action of another endocrine organ.

Trypsin Proteolytic enzyme secreted by the pancreas.

Tubular reabsorption The movement of filtrate components from the renal tubules into the blood.

Tubular secretion The movement of substances (such as drugs, urea, excess ions) from blood into filtrate.

Tumor An abnormal growth of cells; a swelling; may be cancerous.

Tunica (too′nĭ-kah) A covering or tissue coat; membrane layer.

Tympanic membrane (tim-pan′ik) Eardrum.

Ulcer (ul′ser) Lesion or erosion of the mucous membrane, such as gastric ulcer of stomach.

Umbilical cord (um-bĭ′lĭ-kul) Structure bearing arteries and veins connecting the placenta and the fetus.

Umbilicus (um-bĭ′lĭ-kus) Navel; marks site where umbilical cord was attached in fetal stage.

Unipolar neuron Neuron in which embryological fusion of the two processes leaves only one process extending from the cell body.

Unmyelinated fibers (un-mi′ĕ-lĭ-nāt″ed) Axons lacking a myelin sheath and therefore conducting impulses quite slowly.

Uracil (U) (u′rah-sil) A smaller, single-ring base (a pyrimidine) found in RNA.

Urea (u-re′ah) Main nitrogen-containing waste excreted in urine.

Ureter (u-re′ter) Tube that carries urine from kidney to bladder.

Urethra (u-re′thrah) Canal through which urine passes from the bladder to outside the body.

Uric acid The nitrogenous waste product of nucleic acid metabolism; component of urine.

Urinary bladder A smooth, collapsible, muscular sac that stores urine temporarily.

Urinary system System primarily responsible for water, electrolyte, and acid-base balance and removal of nitrogenous wastes.

Uterine tube (u′ter-in) Tube through which the ovum is transported to the uterus. Also called fallopian tube.

Uterus (u′ter-us) Hollow, thick-walled organ that receives, retains, and nourishes fertilized egg; site where embryo/fetus develops.

Uvula (u′vu-lah) Tissue tag hanging from soft palate.

Vaccine Preparation that provides artificially acquired active immunity.

Vagina Thin-walled tube extending from the cervix to the body exterior; often called the birth canal.

Valence shell (va′lens) Outermost electron shell (energy level) of an atom that contains electrons.

Varicosities Knoblike swellings of certain autonomic axons containing mitochondria and synaptic vesicles.

Vas (vaz′) A duct; vessel.

Vasa recta (va′sah rek′tah) Capillary branches that supply loops of Henle in the medulla region of the kidney.

Vascular Pertaining to blood vessels or richly supplied with blood vessels.

Vascular spasm Immediate response to blood vessel injury; results in constriction.

Vasoconstriction (vas″o-kon-strik′shun) Narrowing of blood vessels.

Vasodilation (vas″o-di-la′shun) Relaxation of the smooth muscles of the blood vessels, producing dilation.

Vasomotion (vas″o-mo′shun) Intermittent contraction or relaxation of the precapillary sphincters, resulting in a staggered blood flow when tissue needs are not extreme.

Vasomotor center (vas″o-mo′ter) Brain area concerned with regulation of blood vessel resistance.

Vasomotor fibers Sympathetic nerve fibers that cause the contraction of smooth muscle in the walls of blood vessels, thereby regulating blood vessel diameter.

Veins (vānz″) Blood vessels that return blood toward the heart from the circulation.

Ventral Pertaining to the front; anterior.

Ventricles (1) Paired, inferiorly located heart chambers that function as the major blood pumps; (2) cavities in the brain.

Venule (ven′ūl) A small vein.

Vertebral column (spine) (ver′tĕ-brul) Formed of a number of individual bones called vertebrae and two composite bones (sacrum and coccyx).

Vesicle (vĕ′sĭ-kul) A small liquid-filled sac or bladder.

Vesicular (bulk) transport The movement of large particles and macromolecules across a plasma membrane.

Vesicular follicle Mature ovarian follicle.

Vestibule An enlarged area at the beginning of a canal, i.e., inner ear, nose, larynx.

Villus (vil′us) Fingerlike projections of the small intestinal mucosa that tremendously increase its surface area for absorption.

Visceral (vis′er-al) Pertaining to an internal organ of the body or the inner part of a structure.

Visceral muscle Type of smooth muscle; its cells contract as a unit and rhythmically, are electrically coupled by gap junctions, and often exhibit spontaneous action potentials.

Visceral organs (viscera) A group of internal organs housed in the ventral body cavity.

Visceral serosa (se-ro′sah) The part of the double-layered membrane that lines the outer surfaces of organs within the ventral body cavity.

Viscosity (vis′kos′ĭ-te) State of being sticky or thick.

Visual field The field of view seen when the head is still.

Vital capacity (VC) The volume of air that can be expelled from the lungs by forcible expiration after the deepest inspiration; total exchangeable air.

Vital signs Includes pulse, blood pressure, respiratory rate, and body temperature measurements.

Vitamins Organic compounds required by the body in minute amounts.

Vocal folds Mucosal folds that function in voice production (speech); also called the true vocal cords.

Volatile acid An acid that can be eliminated by the lungs; carbonic acid is converted to CO_2, which diffuses into the alveoli.

Volkmann's canals *See* Perforating canals.

Voluntary muscle Muscle under strict nervous control; skeletal muscle.

Voluntary nervous system The somatic nervous system.

Vulva (vul′vuh) Female external genitalia.

Wallerian degeneration (wal-er′ē-an) A process of disintegration of an axon that occurs when it is crushed or severed and cannot receive nutrients from the cell body.

White matter White substance of the central nervous system; myelinated nerve fibers.

Xenograft Tissue graft taken from another animal species.

Yolk sac (yōk) One of the extraembryonic membranes; involved in early blood cell formation.

Zygote (zi′gōt) Fertilized egg.

Photo and Illustration Credits

Photo Credits

Chapter 1

Chapter Opener: Jorge Silva/Corbis.
1.1: GoGo Images/Jupiter Images.
1.3: Pearson Science.
1.8.1: Jenny Thomas/Addison Wesley Longman.
1.8.2: Howard Sochurek.
1.8.3: James Cavallini/Photo Researchers.
1.8.4: CNRI/Science Photo Library/Photo Researchers.
1.12a: Custom Medical Stock Photo, Inc.
A Closer Look: a: Clinique Ste Catherine/CNRI/Science Photo Library/Photo Researchers.
b: Willaim Klunk, Chet Mathis, University of Pittsburgh, PET Amyloid Imaging Group.
c: Custom Medical Stock Photography.

Chapter 2

Chapter Opener: Chuck Haney Outdoor Photography/Danita Delimont.
2.4.1: Stockbyte/Getty Images.
2.4.2: Marc O. Finley, Stock Food Creative/Getty Images.
2.4.3: Cristina Pedrazzini/Photo Researchers.
2.10b: Bernard Photo Productions/Animals Animals/Earth Sciences.
2.22c: Computer Graphics Laboratory, University of California, San Francisco.

Chapter 3

Chapter Opener: Paul Burns/Getty Images.
3.9a–c: David M. Philips/Photo Researchers.
3.14b: Birgit H. Satir, Dept. of Anatomy and Structural Biology, Albert Einstein College of Medicine.
3.17c: Professors P. Motta & T. Naguro/SPL/Photo Researchers.
3.18b: R. Bolender and Donald Fawcett/Visuals Unlimited.
3.19b: P. Motta & T. Naguro/SPL/Photo Researchers.
3.21: K. G. Murti/Visuals Unlimited.
3.23a: Mary Osborn, Max Planck Institute.
3.23b: Frank Solomon and J. Dinsmore, Massachusetts Institute of Technology.
3.23c: Mark S. Ladinsky and J. Richard McIntosh, University of Colorado.
3.25b: David M. Philips/Visuals Unlimited.
3.26.1: OMIKRON/Science Source/Photo Researchers.
3.26.2 and 3: W. L. Dentler/Biological Photo Service.
3.28: Don W. Fawcett/Photo Researchers.
3.29b.1: From L. Orci and A. Perrelet, *Freeze-Etch Histology.* (Heidelberg: Springer-Verlag, 1975) © 1975 Springer-Verlag.
3.29b.2: From A. C. Faberge, *Cell Tiss. Res.* 151 (1974):403. © 1974 Springer-Verlag.
3.29b.3: U. Aebi et al. *Nature* 323 (1996):560–564, Figure 1a. Used by permission.
3.30.1: Dr. Victoria E. Foe.
3.30.2: GF Bahr, Armed Forces Institute of Pathology.
3.33.1–6: Conly Rieder.
3.38b: Barbara Hamkalo.

Chapter 4

Chapter Opener: Mark Karrass/Corbis.
4.3a: G. W. Willis/Visuals Unlimited.
4.3b, f: Allen Bell, University of New England; Pearson Science.
4.3c: Cabisco/Visuals Unlimited.
4.3d: Ed Reschke.
4.3e: Nina Zanetti, Pearson Science.
4.4a: SPL/Photo Researchers.
4.8a, e, g, k: Ed Reschke.
4.8b, f, h: Nina Zanetti, Pearson Science.
4.8c, i, j: Allen Bell, University of New England; Pearson Science.
4.8d: Ed Reschke/Peter Arnold.
4.9: Biophoto Associates/Photo Researchers.
4.10a: Eric Graves/Photo Researchers.
4.10b: Ed Reschke.
4.10c: SPL/Photo Researchers.

Chapter 5

Chapter Opener: Creasource/Corbis.
5.2a: Ed Reschke/Peter Arnold.
5.3a: Ed Reschke/Peter Arnold.
5.3b, c: Kessel and Kardon/Visuals Unlimited.
5.4a: Kessel and Kardon/Visuals Unlimited.
5.5a: Cabisco/Visuals Unlimited.
5.5b: John D. Cunningham/Visuals Unlimited.
5.6b: Carolina Biological Supply/Phototake.
5.6d: Manfred Kage/Peter Arnold.
5.8a: Bart's Medical Library/Phototake.
5.8b: P. Marazzi/SPL/Photo Researchers.
5.8c: Zeva Oelbaum/Peter Arnold.
5.10a: Scott Camazine/Photo Researchers.
5.10b: Dr. M.A. Ansary/Photo Researchers.

Chapter 6

Chapter Opener: Luciana Ellington/age fotostock.
6.5.1: seelevel.com.
6.5.2: Steve Gschmeissner/Photo Researchers.
6.7c.1: Kessel and Kardon/Visuals Unlimited.
6.7c.2: Ed Reschke/Peter Arnold.
6.10: Ed Reschke.
6.14a: Martin Rose/Bongarts/Getty Images.
6.16a, b: P. Motta, Department of Anatomy, University "La Sapienza," Rome/Science Photo Library/Photo Researchers.
6.17: Carolina Biological Supply.
Table 6.2.1: Lester Bergman/Corbis.
Table 6.2.2: ISM/Phototake.
Table 6.2.3: SIU/Peter Arnold.
Table 6.2.4: SIU/Visuals Unimited.
Table 6.2.5: William T. C. Yuh.
Table 6.2.6: Charles Stewart MD.

Chapter 7

Chapter Opener: George Shelley/Masterfile.
7.5c: Ralph T. Hutchings.
7.6b: Ralph T. Hutchings.
7.7b: Ralph T. Hutchings.
7.8: Ralph T. Hutchings.
7.9a, b: Ralph T. Hutchings.
7.10: Ralph T. Hutchings.
7.11c: Ralph T. Hutchings/Pearson Science.
7.13a: Ralph T. Hutchings.
7.17d: Neil Borden/Photo Researchers.
7.22b: Dissection by Shawn Miller, photography by Mark Nielsen and Alexa Doig.
7.23c: Pearson Science.
7.30c, d: Ralph T. Hutchings.
7.32c, d: Ralph T. Hutchings.
7.32.e: Biophoto Associates/Science Source/Photo Researchers.
7.34b: Elaine N. Marieb.
7.36: Center For Cranialfacial Anomalies, University of California, San Francisco.
7.37: Reik/age fotostock.
Table 7.4.2–5: From *A Stereoscopic Atlas of Human Anatomy* by David L. Bassett.

Chapter 8

Chapter Opener: ANP/AFP/Getty Images.
8.5a–f: John Wilson White, Pearson Science.
8.6a–f: John Wilson White, Pearson Science.
8.8f: L. Bassett/Visuals Unlimited.
8.10b: Mark Neilsen, University of Utah; Pearson Science.
8.10e: VideoSurgery/Photo Researchers.
8.12b: From *A Stereoscopic Atlas of Human Anatomy* by David L. Bassett.
8.14: Elaine Marieb.
8.15: CNRI/Science Photo Library/Photo Researchers.
A Closer Look: 1: Lawrence Livermore National Laboratory/Photo Researchers. 2: Elaine Marieb.

Chapter 9

Chapter Opener: Robert Michael/Corbis.
9.1b: John D. Cunningham/Visuals Unlimited.
9.2a: Marian Rice.
9.4: John Heuser.
9.6.1, 3: James Dennis/Phototake.
9.13b: Eric Graves/Photo Researchers.
9.24: National Library of Medicine.
A Closer Look: Moritz Steiger/Riser/Getty Images.
Table 9.3.4: Eric Graves/Photo Researchers.
Table 9.3.5: Marian Rice.
Table 9.3.6: SPL/Photo Researchers.

Chapter 10

Chapter Opener: Goodshoot/Jupiter Images.
10.8b: Creative Digital Visions, Pearson Science.
10.9c: Dissection by Shawn Miller, photography by Mark Nielsen and Alexa Doig.
10.10c: From *A Stereoscopic Atlas of Human Anatomy* by David L. Bassett.
10.13b: Ralph T. Hutchings.
10.13d, e: Dissection by Shawn Miller, photography by Mark Nielsen and Alexa Doig.
10.14d: Dissection by Shawn Miller, photography by Mark Nielsen and Alexa Doig.
10.20b: Ralph T. Hutchings.

Chapter 11

Chapter Opener: Hulton Archive Photos/Getty Images.
11.4a: Manfred Kage/Peter Arnold.
11.5d: Don W. Fawcett/Photo Researchers.
11.17b: Oliver Meckes/Ottawa/Photo Researchers.
11.24: Tibor Harkany, Science Magagzine.
A Closer Look: Brookhaven National Laboratory/PictureQuest.

Chapter 12

Chapter Opener: Getty Images.
12.6c: Robert A. Chase.
12.6d: A. Glauberman/Photo Researchers.
12.7: Volker Steger/Peter Arnold.
12.10b: Ralph T. Hutchings.
12.11b: Pat Lynch/Photo Researchers.
12.14: Ralph T. Hutchings/Visuals Unlimited.
12.17a: Ralph T. Hutchings.
12.17c: L. Bassett/Visuals Unlimited.
12.20a: Hank Morgan/Photo Researchers.
12.25b: From *A Stereoscopic Atlas of Human Anatomy* by David L. Bassett.
12.27: Carroll H. Weiss.
12.29b–d: L. Bassett/Visuals Unlimited.
12.36: Biophoto Associates/Science Source/Photo Researchers.

Chapter 13

Chapter Opener: AP Photo/Andy Wong.
13.3a: Kessel and Kardon/Visuals Unlimited.
13.9b: Ralph T. Hutchings.
13.11c: Ralph T. Hutchings.
Table 13.2.8: William Thompson, Pearson Science.

Chapter 14

Chapter Opener: Dorling Kindersley.

Chapter 15

Chapter Opener: LWA/Dann Tardif/Jupiter Images.
15.1a: Richard Tauber, Pearson Science.
15.4b: From *A Stereoscopic Atlas of Human Anatomy* by David L Bassett.
15.6c: Ed Reschke/Peter Arnold.
15.7: A. L. Blum/Visuals Unlimited.
15.9: NMSB/Custom Medical Stock Photography.
15.11: Charles D. Wintrer/Photo Researchers.
15.19b: Stephen Spector/Benjamin Cummings.
15.23d: Carolina Biological Supply Company/Phototake, NYC.
15.28d: P. Motta/Department of Anatomy/University "La Sapienza," Rome/Science Photo Library/Photo Researchers.
15.32: *Science* Vol.279 March 1998, p. 1870.
15.36b: I. M. Hunter-Duvar, Department of Otolaryngology, The Hospital for Sick Children, Toronto.

Chapter 16

Chapter Opener: Creasource/Corbis.
16.8b: Ed Reschke.
16.10a: John Paul Kay/Peter Arnold.
16.10b: Photo Researchers.
16.11b: From *Color Atlas of Histology* by Leslie P. Garner and James L. Hiatt, © Williams and Wilkins, 1990.
16.13b: Ed Reschke.
16.15a, b: Charles B. Wilson, Neurological Surgery, University of California Medical Center, San Francisco.
16.17: Carolina Biological Supply/Phototake NYC.
A Closer Look: Saturn Stills/Science Photo Library/Photo Researchers.

Chapter 17

Chapter Opener: Mark Scoggins/Jupiter Images.
17.2: Ed Reshke/Peter Arnold.
17.8a, b: Stan Flegler/Visuals Unlimited.
17.10a–e: Nina Zanetti, Pearson Science.
17.15: Meckes/Ottawa/Photo Researchers.
17.16.1–4: Jack Scanlon, Holyoke Community College.

Chapter 18

Chapter Opener: Brian Bahr/Getty Images.
18.4a: A. & F. Michler/Peter Arnold.
18.4c: L. Bassett/Visuals Unlimited.
18.4f: *Color Atlas of Anatomy: A Photographic Study of the Human Body*. Germany: © Schattauer Publishing.
18.8b: From *A Stereoscopic Atlas of Human Anatomy* by David L. Bassett.
18.8c: Lennart Nilsson, *The Body Victorious*, New York: Dell, © Boehringer Ingelheim International GmbH.
18.8d: L. Bassett/Visuals Unlimited.
18.11a: Manfred Kage/Peter Arnold.
18.19: Phil Jude/Science Photo Library/Photo Researchers.
18.22: Chris Cole/Getty Images.

Chapter 19

Chapter Opener: Jim Cummins/Getty Images.
19.1a: Gladden Willis/Visuals Unlimited.
19.13: Asia Images Group/Getty Images.
19.22c: Science Photo Library/Photo Researchers.
A Closer Look: 1: Sheila Terry/Science Photo Library/Photo Researchers. 2: ISM/Phototake.

Chapter 20

Chapter Opener: Stockbyte/Getty Images.
20.3: Francis Leroy, Biocosmos/Science Photo Library/Photo Researchers.
20.4b: Biophoto Associates/Photo Researchers.
20.6c: Mark Neilsen/Benjamin Cummings.
20.6d: LUMEN Histology, Loyola University Medical Education Network, http://www.lumen.luc.edu/lumen/MedEd/Histo/frames/histo_frames.html.
20.7: Astrid and Hanss-Freider Michler/Science Photo Library/Photo Researchers.
20.8: John Cunningham/Visuals Unlimited.
20.9: Biophoto Associates/Science Source/Photo Researchers.

Chapter 21

Chapter Opener: Jim Cummins/Getty Images.
21.2a: Dr. David M. Phillips/Visuals Unlimited/Getty Images.
21.10: David Scharf/Photo Researchers, Inc.
21.14b: Eduardo A. Padlan, National Institutes of Health.
21.20b: Andrejs Liepins/Science Photo Library/Photo Researchers.

Chapter 22

Chapter Opener: AP World Wide Photos.
22.2a: Jenny Thomas/Benjamin Cummings.
22.3a: From *A Stereoscopic Atlas of Human Anatomy* by David L. Bassett.
22.4c, d: From *A Stereoscopic Atlas of Human Anatomy* by David L. Bassett.
22.6b: Nina Zanetti, Pearson Science.
22.6c: Science Photo Library/Photo Researchers.
22.8b: Carolina Biological Supply/Phototake.
22.9b: Kessel and Kardon/Visuals Unlimited.
22.10a: Richard Tauber/Benjamin Cummings.
22.10b: From *A Stereoscopic Atlas of Human Anatomy* by David L. Bassett.

Chapter 23

Chapter Opener: Al Bello/Allsport/Getty Images.
23.9b: Science Photo Library/Photo Researchers.
23.10.b: Elaine N. Marieb.
23.12a: Biophoto Associates/Photo Researchers.
23.12b: From *Color Atlas of Histology* by Leslie P. Garner and James L. Hiatt, © Williams and Wilkins, 1990.
23.14.b: From *A Stereoscopic Atlas of Human Anatomy* by David L. Bassett.
23.16.a: Javer Domingo/Phototake NYC.
23.16.b: Oliver Meckes/Ottawa/Photo Researchers.
23.22d: Steve Gschmeissner/Photo Researchers, Inc.
23.23a: P. Motta/Department of Anatomy/University "La Sapienza," Rome/Science Photo Library/Photo Researchers.
23.23b: Secchi-Lecaque-Roussel-UCLAF/CNRI/Science Photo Library/Photo Researchers.
23.24a, b: From *A Stereoscopic Atlas of Human Anatomy* by David L. Bassett.
23.25b: From *A Stereoscopic Atlas of Human Anatomy* by David L. Bassett.
23.26b: Victor Eroschenko, University of Idaho; Benjamin Cummings.
23.30a: From *A Stereoscopic Atlas of Human Anatomy* by David L. Bassett.

Chapter 24

Chapter Opener: Stewart Cohen/Getty Images.
24.10: Andreas Engel and Daniel J. Müller.
24.CL: Bartomeu Amengual/age footstock.
24.26: Jeff Greenberg/Omni-Photo.

Chapter 25

Chapter Opener: Zefa RF/age fotostock.
25.2b: Richard Tauber, Benjamin Cummings.
25.3a: Ralph T. Hutchings.
25.6a: Biophoto Associates/Photo Researchers.
25.6b: Dennis Kunkel/PhotoTake.
25.7b: Kessel and Kardon/Visuals Unlimited.
25.9b: P. Motta and M. Castellucci/Science Photo Library/Photo Researchers.
25.19: National Institute of Health.
25.20: Biophoto Associates/Photo Researchers.

Chapter 26

Chapter Opener: Jim Cummins/Getty Images.
26.5: Spencer Grant/Stock Boston.

Chapter 27

Chapter Opener: Lisette Le Bon/SuperStock.
27.3b: From *A Stereoscopic Atlas of Human Anatomy* by David L. Bassett.
27.3c: Ed Reschke.
27.8a: Kessel and Kardon/Visuals Unlimited.
27.8b: Juergen Berger/Photo Researchers, Inc.
27.9b: Manfred Kage/Peter Arnold.
27.11b: Biophoto Associates/Photo Researchers, Inc.
27.12b: From *A Stereoscopic Atlas of Human Anatomy* by David L. Bassett.
27.13a: Carolina Biological Supply Company/Phototake.
27.16a: Mark Thomas/Photo Researchers, Inc.
27.16b, c: Leonard Lessin/Peter Arnold.
27.18.1–3: Science Pictures Ltd./Photo Researchers, Inc.
27.18.4: ISM/Phototake.
27.18.5: Ed Reschke/Peter Arnold.
27.18.6: C. Edelman/La Vilette/Photo Researchers.
27.18.7: Lester V. Bergman/Corbis.

Chapter 28
Chapter Opener: John Lund/Drew Kelly/Blend Images/Jupiter Images.
28.2b: Lennart Nilsson, *A Child Is Born*, New York: Dell, © Boehringer Ingelheim International GmbH.
28.3b: CC Studio/Science Photo Library/Photo Researchers, Inc.
28.5b: R. O'Rahilly and R. Muller, *Human Embryology and Teratology*. This material is reproduced with permission of Wiley-Liss, Inc., a subsidiary of John Wiley & Sons, Inc.
28.5d: Allen C. Enders, University of California, Davis/Carnegie Collection.
28.15a: From *A Stereoscopic Atlas of Human Anatomy* by David L. Bassett.
28.15b, c: From Lennart Nilsson/Albert Bonniers Forlag.

Chapter 29
Chapter Opener: Ariel Skelley/Corbis.
29.1b: Lester Lefkowitz/Corbis.
29.1c: L.Williatt, East Anglian Regional Genetics/SPL/Photo Researchers.
29.5: Andrew Syred/Photo Researchers.
29.7.1, 2: PhotoDisc/Getty Images.

Illustration Credits
All illustrations by Imagineering STA Media Services unless otherwise noted.

Chapter 1
1.3: Vincent Perez/Wendy Hiller Gee.
1.7: Imagineering STA Media Services/Precision Graphics
1.10a: Adapted from Seeley, Stephens, and Tate, *Anatomy & Physiology*, 4e, F1.15a, New York: WCB/McGraw-Hill, © McGraw-Hill, 1998.

Chapter 3
3.2: Imagineering STA Media Services/Precision Graphics
3.3: Imagineering STA Media Services/Precision Graphics
3.19: Tomo Narashima.
3.20: Imagineering STA Media Services/Precision Graphics
3.24: Adapted from Campbell, *Biology*, 4e, F7.21, © Benjamin Cummings, 1996.
3.27: Imagineering STA Media Services/Precision Graphics
3.29: Tomo Narashima.

Chapter 4
4.6: From Mathews, Van Holde, and Ahern, *Biochemistry*, 3e, F9.24, © Benjamin Cummings, 2000.

Chapter 5
5.1, 5.2, 5.5: Electronic Publishing Services, Inc.
Making Connections: Vincent Perez/Wendy Hiller Gee.

Chapter 6
6.9: Imagineering STA Media Services/Precision Graphics
Making Connections: Vincent Perez/Wendy Hiller Gee.

Chapter 7
7.3, 7.5, 7.6, 7.7, 7.8: Nadine Sokol.

Chapter 9
9.1: Imagineering STA Media Services/Precision Graphics
9.2: Imagineering STA Media Services/Precision Graphics
9.8, 9.9, 9.11, 9.12, 9.13, 9.26, 9.27, 9.28: Electronic Publishing Services, Inc.
Making Connections: Vincent Perez/Imagineering STA Media Services.

Chapter 11
11.2, 11.17, 11.23: Electronic Publishing Services, Inc.
11.3: Imagineering STA Media Services/Precision Graphics
11.5: Imagineering STA Media Services/Precision Graphics

Chapter 12
12.1–12.6, 12.8–12.11, 12.13, 12.15–12.19, 12.23–12.26: Electronic Publishing Services, Inc.
12.12: Electronic Publishing Services, Inc./Precision Graphics
12.31: Imagineering STA Media Services/Electronic Publishing Services, Inc.
12.32–12.35: Electronic Publishing Services, Inc.

Chapter 13
13.1, 13.3–13.5, 13.13, 13.14, 13.17, 13.18, 13.19: Electronic Publishing Services, Inc.
13.T02.06: Imagineering STA Media Services/Precision Graphics
13.15: Imagineering STA Media Services/Precision Graphics

Chapter 14
14.1–14.4, 14.6, 14.7, 14.9: Electronic Publishing Services, Inc.
Making Connections: Vincent Perez/Wendy Hiller Gee.

Chapter 15
15.1–15.6, 15.8, 15.15, 15.16, 15.18, 15.20, 15.21, 15.23–15.24, 15.26, 15.27, 15.28, 15.31, 15.33–15.36 : Electronic Publishing Services, Inc.
15.25: Electronic Publishing Services, Inc./Precision Graphics

Chapter 16
16.1: Electronic Publishing Services, Inc.
Making Connections: Vincent Perez/Wendy Hiller Gee.
16.8: Imagineering STA Media Services/Precision Graphics

Chapter 18
18.1: Electronic Publishing Services, Inc./Precision Graphics
18.4, 18.8, 18.11, 18.14: Electronic Publishing Services, Inc.
18.9: Imagineering STA Media Services/Precision Graphics

Chapter 19
19.1, 19.4, 19.21–19.30: Electronic Publishing Services, Inc.
Making Connections: Vincent Perez/Wendy Hiller Gee.

Chapter 20
Making Connections: Vincent Perez/Wendy Hiller Gee.

Chapter 21
21.16: Adapted from Johnson, *Human Biology*, 2e, F9.13, © Benjamin Cummings, 2003.

Chapter 22
22.1: Electronic Publishing Services, Inc./Precision Graphics
22.3, 22.4, 22.7–22.10, 22.26: Electronic Publishing Services, Inc.
Making Connections: Vincent Perez/Wendy Hiller Gee.

Chapter 23
23.1, 23.6–23.9, 23.14, 23.15, 23.21, 23.22, 23.29–23.31: Electronic Publishing Services, Inc.
Making Connections: Vincent Perez/Wendy Hiller Gee.
23.25: Electronic Publishing Services, Inc./Precision Graphics
23.28: Imagineering STA Media Services/Precision Graphics

Chapter 24
24.1b: Willett, W.C., Eat, Drink, and Be Healthy. Nutrition Source, Simon & Schuster, Harvard School of Public Health, Food Pyramids, 2001. www.hsph.harvard.edu/nutritionsource/pyramids.html.

Chapter 25
25.1, 25.5, 25.7–25.9, 25.21: Electronic Publishing Services, Inc.
25.12: Imagineering STA Media Services/Precision Graphics

Chapter 26
Making Connections: Vincent Perez/Wendy Hiller Gee.
26.09: Imagineering STA Media Services/Precision Graphics

Chapter 27
27.1–27.4, 27.10–27.12: Electronic Publishing Services, Inc.
Making Connections: Vincent Perez/Wendy Hiller Gee.

Chapter 28
28.4, 28.13: Electronic Publishing Services, Inc.

Index

NOTE: Page numbers in **boldface** indicate a definition. A *t* following a page number indicates tabular material, an *f* indicates an illustration, and a *b* indicates boxed material.

Word Roots, Prefixes, Suffixes, and Combining Forms

Prefixes and Combining Forms

a-, an- *absence or lack*; acardia, lack of a heart, anaerobic, in the absence of oxygen
ab- *departing from, away from*; abnormal, departing from normal
acou- *hearing*; acoustics, the science of sound
ac-, acro- *extreme or extremity, peak*; acrodermatitis, inflammation of the skin of the extremities
ad- *to or toward*; adorbital, toward the orbit
aden-, adeno- *gland*; adeniform, resembling a gland in shape
adren- *toward the kidney*; adrenal gland, adjacent to the kidney
aero- *air*; aerobic respiration, oxygen-requiring metabolism
af- *toward*; afferent neurons, which carry impulses to the central nervous system
agon- *contest*; agonistic and antagonistic muscles, which oppose each other
alb- *white*; corpus albicans of the ovary, a white scar tissue
aliment- *nourish*; alimentary canal, or digestive tract
allel- *of one another*; alleles, alternative expressions of a gene
amphi- *on both sides, of both kinds*; amphibian, an organism capable of living in water and on land
ana- *apart, up, again*; anaphase of mitosis, when the chromosomes separate
anastomos- *come together*; arteriovenous anastomosis, a connection between an artery and a vein
aneurysm *a widening*; aortic aneurysm, a weak spot that causes enlargement of the blood vessel
angi- *vessel*; angiitis, inflammation of a lymph vessel or blood vessel
angin- *choked*; angina pectoris, a choked feeling in the chest due to dysfunction of the heart
ant-, anti- *opposed to, preventing or inhibiting*; anticoagulant, a substance that prevents blood coagulation
ante- *preceding, before*; antecubital, in front of the elbow
aort- *great artery*; aorta
ap-, api- *tip, extremity*; apex of the heart
append- *hang to*; appendicular skeleton
aqua-, aque- *water*; aqueous solutions
arbor *tree*; arbor vitae of the cerebellum, the treelike pattern of white matter
areola- *open space*; areolar connective tissue, a loose connective tissue
arrect- *upright*; arrector pili muscles of the skin, which make the hairs stand erect
arthr-, arthro- *joint*; arthropathy, any joint disease
artic- *joint*; articular surfaces of bones, the points of connection
atri- *vestibule*; atria, upper chambers of the heart
auscult- *listen*; auscultatory method for measuring blood pressure
aut-, auto- *self*; autogenous, self-generated
ax-, axi-, axo- *axis, axle*; axial skeleton, axis of vertebral column
azyg- *unpaired*; azygous vein, an unpaired vessel
baro- *pressure*; baroreceptors for monitoring blood pressure
basal *base*; basal lamina of epithelial basement membrane
bi- *two*; bicuspid, having two cusps
bili- *bile*; bilirubin, a bile pigment
bio- *life*; biology, the study of life and living organisms
blast- *bud or germ*; blastocyte, undifferentiated embryonic cell
brachi- *arm*; brachial plexus of peripheral nervous system supplies the arm
brady- *slow*; bradycardia, abnormally slow heart rate
brev- *short*; peroneus brevis, a short leg muscle
broncho- *bronchus*; bronchospasm, spasmodic contraction of bronchial muscle
bucco- *cheek*; buccolabial, pertaining to the cheek and lip
calor- *heat*; calories, a measure of energy
capill- *hair*; blood and lymph capillaries
caput- *head*; decapitate, remove the head
carcin- *cancer*; carcinogen, a cancer-causing agent
cardi-, cardio- *heart*; cardiotoxic, harmful to the heart
carneo- *flesh*; trabeculae carneae, ridges of muscle in the ventricles of the heart
carot- *(1) carrot, (2) stupor*; (1) carotene, an orange pigment, (2) carotid arteries in the neck, blockage causes fainting
cata- *down*; catabolism, chemical breakdown
caud- *tail*; caudal (directional term)
cec- *blind*; cecum of large intestine, a blind-ended pouch
cele- *abdominal*; celiac artery, in the abdomen
cephal- *head*; cephalometer, an instrument for measuring the head
cerebro- *brain, especially the cerebrum*; cerebrospinal, pertaining to the brain and spinal cord
cervic-, cervix *neck*; cervix of the uterus
chiasm- *crossing*; optic chiasma, where optic nerves cross
chole- *bile*; cholesterol, cholecystokinin, a bile-secreting hormone
chondr- *cartilage*; chondrogenic, giving rise to cartilage
chrom- *colored*; chromosome, so named because they stain darkly
cili- *small hair*; ciliated epithelium
circum- *around*; circumnuclear, surrounding the nucleus
clavic- *key*; clavicle, a "skeleton key"
co-, con- *together*; concentric, common center, together in the center
coccy- *cuckoo*; coccyx, which is beak-shaped
cochlea *snail shell*; the cochlea of the inner ear, which is coiled like a snail shell
coel- *hollow*; coelom, the ventral body cavity
commis- *united*; gray commissure of the spinal cord connects the two columns of gray matter
concha *shell*; nasal conchae, coiled shelves of bone in the nasal cavity
contra- *against*; contraceptive, agent preventing conception

corn-, cornu- *horn*; stratum corneum, outer layer of the skin composed of (horny) cells
corona *crown*; coronal suture of the skull
corp- *body*; corpse, corpus luteum, hormone-secreting body in the ovary
cort- *bark*; cortex, the outer layer of the brain, kidney, adrenal glands, and lymph nodes
cost- *rib*; intercostal, between the ribs
crani- *skull*; craniotomy, a skull operation
crypt- *hidden*; cryptomenorrhea, a condition in which menstrual symptoms are experienced but no external loss of blood occurs
cusp- *pointed*; bicuspid, tricuspid valves of the heart
cutic- *skin*; cuticle of the nail
cyan- *blue*; cyanosis, blue color of the skin due to lack of oxygen
cyst- *sac, bladder*; cystitis, inflammation of the urinary bladder
cyt- *cell*; cytology, the study of cells
de- *undoing, reversal, loss, removal*; deactivation, becoming inactive
decid- *falling off*; deciduous (milk) teeth
delta *triangular*; deltoid muscle, roughly triangular in shape
den-, dent- *tooth*; dentin of the tooth
dendr- *tree, branch*; dendrites, telodendria, both branches of a neuron
derm- *skin*; dermis, deep layer of the skin
desm- *bond*; desmosome, which binds adjacent epithelial cells
di- *twice, double*; dimorphism, having two forms
dia- *through, between*; diaphragm, the wall through or between two areas
dialys- *separate, break apart*; kidney dialysis, in which waste products are removed from the blood
diastol- *stand apart*; cardiac diastole, between successive contractions of the heart
diure- *urinate*; diuretic, a drug that increases urine output
dors- *the back*; dorsal, dorsum, dorsiflexion
duc-, duct *lead, draw*; ductus deferens which carries sperm from the epididymis into the urethra during ejaculation
dura *hard*; dura mater, tough outer meninx
dys- *difficult, faulty, painful*; dyspepsia, disturbed digestion
ec-, ex-, ecto- *out, outside, away from*; excrete, to remove materials from the body
ectop- *displaced*; ectopic pregnancy, ectopic focus for initiation of heart contraction
edem- *swelling*; edema, accumulation of water in body tissues
ef- *away*; efferent nerve fibers, which carry impulses away from the central nervous system
ejac- *to shoot forth*; ejaculation of semen
embol- *wedge*; embolus, an obstructive object traveling in the bloodstream
en-, em- *in, inside*; encysted, enclosed in a cyst or capsule
enceph- *brain*; encephalitis, inflammation of the brain
endo- *within, inner*; endocytosis, taking particles into a cell
entero- *intestine*; enterologist, one who specializes in the study of intestinal disorders
epi- *over, above*; epidermis, outer layer of skin
erythr- *red*; erythema, redness of the skin, erythrocyte, red blood cell
eso- *within*; esophagus
eu- *well*; euesthesia, a normal state of the senses
excret- *separate*; excretory system
exo- *outside, outer layer*; exophthalmos, an abnormal protrusion of the eye from the orbit
extra- *outside, beyond*; extracellular, outside the body cells of an organism
extrins- *from the outside*; extrinsic regulation of the heart
fasci-, fascia- *bundle, band*; superficial and deep fascia
fenestr- *window*; fenestrated capillaries
ferr- *iron*; transferrin, ferritin, both iron-storage proteins
flagell- *whip*; flagellum, the tail of a sperm cell
flat- *blow, blown*; flatulence
folli- *bag, bellows*; hair follicle
fontan- *fountain*; fontanelles of the fetal skull
foram- *opening*; foramen magnum of the skull
foss- *ditch*; fossa ovalis of the heart, mandibular fossa of the skull
gam-, gamet- *married, spouse*; gametes, the sex cells
gangli- *swelling, or knot*; dorsal root ganglia of the spinal nerves
gastr- *stomach*; gastrin, a hormone that influences gastric acid secretion
gene *beginning, origin*; genetics
germin- *grow*; germinal epithelium of the gonads
gero-, geront- *old man*; gerontology, the study of aging
gest- *carried*; gestation, the period from conception to birth
glauc- *gray*; glaucoma, which causes gradual blindness
glom- *ball*; glomeruli, clusters of capillaries in the kidneys
glosso- *tongue*; glossopathy, any disease of the tongue
gluco-, glyco- gluconeogenesis, the production of glucose from noncarbohydrate molecules
glute- *buttock*; gluteus maximus, largest muscle of the buttock
gnost- *knowing*; the gnostic sense, a sense of awareness of self
gompho- *nail*; gomphosis, the term applied to the joint between tooth and jaw
gon-, gono- *seed, offspring*; gonads, the sex organs
gust- *taste*; gustatory sense, the sense of taste
hapt- *fasten, grasp*; hapten, a partial antigen
hema-, hemato-, hemo- *blood*; hematocyst, a cyst containing blood
hemi- *half*; hemiglossal, pertaining to one-half of the tongue
hepat- *liver*; hepatitis, inflammation of the liver
hetero- *different or other*; heterosexuality, sexual desire for a person of the opposite sex
hiat- *gap*; the hiatus of the diaphragm, the opening through which the esophagus passes
hippo- *horse*; hippocampus of the brain, shaped like a seahorse
hirsut- *hairy*; hirsutism, excessive body hair
hist- *tissue*; histology, the study of tissues
holo- *whole*; holocrine glands, whose secretions are whole cells
hom-, homo- *same*; homeoplasia, formation of tissue similar to normal tissue; homocentric, having the same center
hormon- *to excite*; hormones
humor- *a fluid*; humoral immunity, which involves antibodies circulating in the blood
hyal- *clear*; hyaline cartilage, which has no visible fibers
hydr-, hydro- *water*; dehydration, loss of body water
hyper- *excess*; hypertension, excessive tension
hypno- *sleep*; hypnosis, a sleeplike state
hypo- *below, deficient*; hypodermic, beneath the skin, hypokalemia, deficiency of potassium
hyster-, hystero- *uterus or womb*; hysterectomy, removal of the uterus, hysterodynia, pain in the womb
ile- *intestine*; ileum, the last portion of the small intestine
im- *not*; impermeable, not permitting passage, not permeable
inter- *between*; intercellular, between the cells
intercal- *insert*; intercalated discs, the end membranes between adjacent cardiac muscle cells
intra- *within, inside*; intracellular, inside the cell
iso- *equal, same*; isothermal, equal, or same, temperature
jugul- *throat*; jugular veins, prominent vessels in the neck
juxta- *near, close to*; juxtaglomerular apparatus, a cell cluster next to a glomerulus in the kidneys
karyo- *kernel, nucleus*; karyotype, the assemblage of the nuclear chromosomes
kera- *horn*; keratin, the water-repellent protein of the skin
kilo- *thousand*; kilocalories, equal to 1000 calories
kin-, kines- *move*; kinetic energy, the energy of motion

labi-, labri- *lip*; labial frenulum, the membrane which joins the lip to the gum
lact- *milk*; lactose, milk sugar
lacun- *space, cavity, lake*; lacunae, the spaces occupied by cells of cartilage and bone tissue
lamell- *small plate*; concentric lamellae, rings of bone matrix in compact bone
lamina *layer, sheet*; basal lamina, part of the epithelial basement membrane
lat- *wide*; latissimus dorsi, a broad muscle of the back
laten- *hidden*; latent period of a muscle twitch
later- *side*; lateral (directional term)
leuko- *white*; leukocyte, white blood cell
leva- *raise, elevate*; levator labii superioris, muscle that elevates upper lip
lingua- *tongue*; lingual tonsil, adjacent to the tongue
lip-, lipo- *fat, lipid*; lipophage, a cell that has taken up fat in its cytoplasm
lith- *stone*; cholelithiasis, gallstones
luci- *clear*; stratum lucidum, clear layer of the epidermis
lumen *light*; lumen, center of a hollow structure
lut- *yellow*; corpus luteum, a yellow, hormone-secreting structure in the ovary
lymph *water*; lymphatic circulation, return of clear fluid to the bloodstream
macro- *large*; macromolecule, large molecule
macula *spot*; macula lutea, yellow spot on the retina
magn- *large*; foramen magnum, largest opening of the skull
mal- *bad, abnormal*; malfunction, abnormal functioning of an organ
mamm- *breast*; mammary gland, breast
mast- *breast*; mastectomy, removal of a mammary gland
mater *mother*; dura mater, pia mater, membranes that envelop the brain
meat- *passage*; external acoustic meatus, the ear canal
medi- *middle*; medial (directional term)
medull- *marrow*; medulla, the middle portion of the kidney, adrenal gland, and lymph node
mega- *large*; megakaryocyte, large precursor cell of platelets
meio- *less*; meiosis, nuclear division that halves the chromosome number
melan- *black*; melanocytes, which secrete the black pigment melanin
men-, menstru- *month*; menses, the cyclic menstrual flow
meningo- *membrane*; meningitis, inflammation of the membranes of the brain
mer-, mero-, *a part*; merocrine glands, the secretions of which do not include the cell
meso- *middle*; mesoderm, middle germ layer
meta- *beyond, between, transition*; metatarsus, the part of the foot between the tarsus and the phalanges
metro- *uterus*; endometrium, the lining of the uterus
micro- *small*; microscope, an instrument used to make small objects appear larger
mictur- *urinate*; micturition, the act of voiding the bladder
mito- *thread, filament*; mitochondria, small, filament-like structures located in cells
mnem- *memory*; amnesia
mono- *single*; monospasm, spasm of a single limb
morpho- *form*; morphology, the study of form and structure or organisms
multi- *many*; multinuclear, having several nuclei
mur- *wall*; intramural ganglion, a nerve junction within an organ
muta- *change*; mutation, change in the base sequence of DNA
myelo- *spinal cord, marrow*; myeloblasts, cells of the bone marrow
myo- *muscle*; myocardium, heart muscle
nano- *dwarf*; nanometer, one-billionth of a meter
narco- *numbness*; narcotic, a drug producing stupor or numbed sensations
natri- *sodium*; atrial natriuretic peptide, a sodium-regulating hormone
necro- *death*; necrosis, tissue death
neo- *new*; neoplasm, an abnormal growth
nephro- *kidney*; nephritis, inflammation of the kidney
neuro- *nerve*; neurophysiology, the physiology of the nervous system
noci- *harmful*; nociceptors, receptors for pain
nom- *name*; innominate artery, innominate bone
noto- *back*; notochord, the embryonic structure that precedes the vertebral column
nucle- *pit, kernel, little nut*; nucleus
nutri- *feed, nourish*; nutrition
ob- *before, against*; obstruction, impeding or blocking up
oculo- *eye*; monocular, pertaining to one eye
odonto- *teeth*; orthodontist, one who specializes in proper positioning of the teeth in relation to each other
olfact- *smell*; olfactory nerves
oligo- *few*; oligodendrocytes, neuroglial cells with few branches
onco- *a mass*; oncology, study of cancer
oo- *egg*; oocyte, precursor of female gamete
ophthalmo- *eye*; ophthalmology, the study of the eyes and related disease
orb- *circular*; orbicularis oculi, muscle that encircles the eye
orchi- *testis*; cryptorchidism, failure of the testes to descend into the scrotum
org- *living*; organism
ortho- *straight, direct*; orthopedic, correction of deformities of the musculoskeletal system
osm- *smell*; anosmia, loss of sense of smell
osmo- *pushing*; osmosis
osteo- *bone*; osteodermia, bony formations in the skin
oto- *ear*; otoscope, a device for examining the ear
ov-, ovi- *egg*; ovum, oviduct
oxy- *oxygen*; oxygenation, the saturation of a substance with oxygen
pan- *all, universal*; panacea, a cure-all
papill- *nipple*; dermal papillae, projections of the dermis into the epidermal area
para- *beside, near*; paranuclear, beside the nucleus
pect-, pectus *breast*; pectoralis major, a large chest muscle
pelv- *a basin*; pelvic girdle, which cradles the pelvic organs
peni- *a tail*; penis; penile urethra
penna- *feather*; unipennate, bipennate muscles, whose fascicles have a feathered appearance
pent- *five*; pentose, a 5-carbon sugar
pep-, peps-, pept- *digest*; pepsin, a digestive enzyme of the stomach; peptic ulcer
per-, permea- *through*; permeate, permeable
peri- *around*; perianal, situated around the anus
phago- *eat*; phagocyte, a cell that engulfs and digests particles or cells
pheno- *show, appear*; phenotype, the physical appearance of an individual
phleb- *vein*; phlebitis, inflammation of the veins
pia *tender*; pia mater, delicate inner membrane around the brain and spinal cord
pili *hair*; arrector pili muscles of the skin, which make the hairs stand erect
pin-, pino- *drink*; pinocytosis, the engulfing of small particles by a cell
platy- *flat, broad*; platysma, broad, flat muscle of the neck
pleur- *side, rib*; pleural serosa, the membrane that lines the thoracic cavity and covers the lungs
plex-, plexus *net, network*; brachial plexus, the network of nerves that supplies the arm
pneumo- *air, wind*; pneumothorax, air in the thoracic cavity
pod- *foot*; podiatry, the treatment of foot disorders
poly- *multiple*; polymorphism, multiple forms
post- *after, behind*; posterior, places behind (a specific) part
pre-, pro- *before, ahead of*; prenatal, before birth
procto- *rectum, anus*; proctoscope, an instrument for examining the rectum
pron- *bent forward*; prone, pronate

propri- *one's own*; proprioception, awareness of body parts and movement
pseudo- *false*; pseudotumor, a false tumor
psycho- *mind, psyche*; psychogram, a chart of personality traits
ptos- *fall*; renal ptosis, a condition in which the kidneys drift below their normal position
pub- *of the pubis*; puberty
pulmo- *lung*; pulmonary artery, which brings blood to the lungs
pyo- *pus*; pyocyst, a cyst that contains pus
pyro- *fire*; pyrogen, a substance that induces fever
quad-, quadr- *four-sided*; quadratus lumborum, a muscle with a square shape
re- *back, again*; reinfect
rect- *straight*; rectus abdominis, rectum
ren- *kidney*; renal, renin, an enzyme secreted by the kidney
retin, retic- *net, network*; endoplasmic reticulum, a network of membranous sacs within a cell
retro- *backward, behind*; retrogression, to move backward in development
rheum- *watery flow, change, or flux*; rheumatoid arthritis, rheumatic fever
rhin-, rhino- *nose*; rhinitis, inflammation of the nose
ruga- *fold, wrinkle*; rugae, the folds of the stomach, gallbladder, and urinary bladder
sagitt- *arrow*; sagittal (directional term)
salta- *leap*; saltatory conduction, the rapid conduction of impulses along myelinated neurons
sanguin- *blood*; consanguineous, indicative of a genetic relationship between individuals
sarco- *flesh*; sarcomere, unit of contraction in skeletal muscle
saphen- *visible, clear*; great saphenous vein, superficial vein of the thigh and leg
sclero- *hard*; sclerodermatitis, inflammatory thickening and hardening of the skin
seb- *grease*; sebum, the oil of the skin
semen *seed, sperm*; semen, the discharge of the male reproductive system
semi- *half*; semicircular, having the form of half a circle
sens- *feeling*; sensation, sensory
septi- *rotten*; sepsis, infection, antiseptic
septum *fence*; nasal septum
sero- *serum*; serological tests, which assess blood conditions
serrat- *saw*; serratus anterior, a muscle of the chest wall that has a jagged edge
sin-, sino- *a hollow*; sinuses of the skull
soma- *body*; somatic nervous system
somn- *sleep*; insomnia, inability to sleep
sphin- *squeeze*; sphincter
splanchn- *organ*; splanchnic nerve, autonomic supply to abdominal viscera
spondyl- *vertebra*; ankylosing spondylitis, rheumatoid arthritis affecting the spine
squam- *scale, flat*; squamous epithelium, squamous suture of the skull
steno- *narrow*; stenocoriasis, narrowing of the pupil
strat- *layer*; strata of the epidermis, stratified epithelium
stria- *furrow, streak*; striations of skeletal and cardiac muscle tissue
stroma *spread out*; stroma, the connective tissue framework of some organs
sub- *beneath, under*; sublingual, beneath the tongue
sucr- *sweet*; sucrose, table sugar
sudor- *sweat*; sudoriferous glands, the sweat glands
super- *above, upon*; superior, quality or state of being above others or a part
supra- *above, upon*; supracondylar, above a condyle
sym-, syn- *together, with*; synapse, the region of communication between two neurons
synerg- *work together*; synergism
systol- *contraction*; systole, contraction of the heart
tachy- *rapid*; tachycardia, abnormally rapid heartbeat
tact- *touch*; tactile sense
telo- *the end*; telophase, the end of mitosis
templ-, tempo- *time*; temporal summation of nerve impulses
tens- *stretched*; muscle tension
terti- *third*; fibularis tertius, one of three fibularis muscles
tetan- *rigid, tense*; tetanus of muscles
therm- *heat*; thermometer, an instrument used to measure heat
thromb- *clot*; thrombocyte, thrombus
thyro- *a shield*; thyroid gland
tissu- *woven*; tissue
tono- *tension*; tonicity, hypertonic
tox- *poison*; toxicology, study of poisons
trab- *beam, timber*; trabeculae, spicules of bone in spongy bone tissue
trans- *across, through*; transpleural, through the pleura
trapez- *table*; trapezius, the four-sided muscle of the upper back
tri- *three*; trifurcation, division into three branches
trop- *turn, change*; tropic hormones, whose targets are endocrine glands
troph- *nourish*; trophoblast, from which develops the fetal portion of the placenta
tuber- *swelling*; tuberosity, a bump on a bone
tunic- *covering*; tunica albuginea, the covering of the testis
tympan- *drum*; tympanic membrane, the eardrum
ultra- *beyond*; ultraviolet radiation, beyond the band of visible light
vacc- *cow*; vaccine
vagin- *a sheath*; vagina
vagus *wanderer*; the vagus nerve, which starts at the brain and travels into the abdominopelvic cavity
valen- *strength*; valence shells of atoms
venter, ventr- *abdomen, belly*; ventral (directional term), ventricle
vent- *the wind*; pulmonary ventilation
vert- *turn*; vertebral column
vestibul- *a porch*; vestibule, the anterior entryway to the mouth and nose
vibr- *shake, quiver*; vibrissae, hairs of the nasal vestibule
villus *shaggy hair*; microvilli, which have the appearance of hair in light microscopy
viscero- *organ, viscera*; visceroinhibitory, inhibiting the movements of the viscera
viscos- *sticky*; viscosity, resistance to flow
vita- *life*; vitamin
vitre- *glass*; vitreous humor, the clear jelly of the eye
viv- *live*; in vivo
vulv- *a covering*; vulva, the female external genitalia
zyg- *a yoke, twin*; zygote

Suffixes

-able *able to, capable of*; viable, ability to live or exist
-ac *referring to*; cardiac, referring to the heart
-algia *pain in a certain part*; neuralgia, pain along the course of a nerve
-apsi *juncture*; synapse, where two neurons communicate
-ary *associated with, relating to*; coronary, associated with the heart
-asthen *weakness*; myasthenia gravis, a disease involving paralysis
-bryo *swollen*; embryo
-cide *destroy or kill*; germicide, an agent that kills germs
-cipit *head*; occipital
-clast *break*; osteoclast, a cell that dissolves bone matrix
-crine *separate*; endocrine organs, which secrete hormones into the blood
-dips *thirst, dry*; polydipsia, excessive thirst associated with diabetes
-ectomy *cutting out, surgical removal*; appendectomy, cutting out of the appendix
-ell, -elle *small*; organelle
-emia *condition of the blood*; anemia, deficiency of red blood cells
-esthesi *sensation*; anesthesia, lack of sensation